IF	Initiation factor
K_M	Michaelis constant
LDL	Low-density lipoprotein
Mb	Myoglobin
NAD^+	Nicotinamide adenine dinucleotide (oxidized form)
NADH	Nicotinamide adenine dinucleotide (reduced form)
$NADP^+$	Nicotinamide adenine dinucleotide phosphate (oxidized form)
NADPH	Nicotinamide adenine dinucleotide phosphate (reduced form)
P_i	Phosphate ion
PAGE	Polyacrylamide gel electrophoresis
PCR	Polymerase chain reaction
PEP	Phosphoenolpyruvate
PIP_2	Phosphatidylinositol 4,5-*bis*phosphate
PKU	Phenylketonuria
Pol	DNA polymerase
PP_i	Pyrophosphate ion
PRPP	Phosphoribosylpyrophosphate
PS	Photosystem
RF	Release factor
RFLPs	Restriction-fragment-length polymorphisms
RNA	Ribonucleic acid
RNase	Ribonuclease
mRNA	Messenger RNA
rRNA	Ribosomal RNA
tRNA	Transfer RNA
snRNP	Small nuclear ribounuclear protein
S	Svedberg unit
SCID	Severe combined immune deficiency
SSB	Single-strand binding protein
SV40	Simian virus 40
T	Thymine
TDP	Thymidine diphosphate
TMP	Thymidine monophosphate
TTP	Thymidine triphosphate
U	Uracil
UDP	Uridine diphosphate
UMP	Uridine monophosphate
UTP	Uridine triphosphate
V_{max}	Maximal velocity

Biochemistry

Biochemistry

10TH EDITION

Mary K. Campbell
Mount Holyoke College

Saul R. Treviño
Houston Christian University

Cengage

Australia • Brazil • Canada • Mexico • Singapore • United Kingdom • United States

Cengage

***Biochemistry,* Tenth Edition**
Mary Campbell and Saul Treviño

EVP, General Manager, Cengage Academic: Nhaim Khoury

VP, Portfolio Product Management: Mark Santee

Portfolio Product Director: Maureen McLaughlin

Senior Portfolio Product Manager: Roxanne Wang

Senior Portfolio Product Assistant: Ellie Purgavie

Portfolio Product Assistant: Sophie Yook

Senior Learning Designer: Courtney Wolstoncroft

Learning Designer: Malini Rao

Subject Matter Expert: Dakin Sharum

Senior Content Manager: Lynh Pham

Content Project Manager: Anandan Bommen

Developmental Editor: John S. Murdzek

Senior Manager, Product Marketing: Neena Bali

Product Marketing Manager: Justin Hoffman

Technical Content Program Manager: Ivan Corriher

Content Acquisition Analyst: Christy Frame

Production Service: Straive

Senior Designer: Chris Doughman

PiB & CiC Editor: Rebecca Heider

PiB & CiC Editor: Micah Newman

Cover Image Sources: Todd Easterbrook/Ascent Xmedia/Photodisc/Getty Images; LAGUNA DESIGN/Science Photo Library/Getty Images; Andrey Prokhorov/Getty Images; Aldona Griskeviciene/Shutterstock.com

Library of Congress Control Number: 2026932517

Student Edition ISBN: 979-8-214-19360-1
Loose-leaf Edition ISBN: 979-8-214-19361-8

Cengage
10650 Toebben Drive, Suite A
Independence, KY 41051
USA

Printed Number: 1 Print Year: 2026
Printed in Mexico

This book is dedicated to the memory of Mary Campbell, who was passionately involved in its creation. Her avid interest in writing and devotion to student engagement led to the publication of the first eight highly successful editions of this textbook.

To all the teachers I've had and continue to have in my life, including past, present, and future students, and especially my parents René and Marie, my wife Polly, and our sons Benjamin and Aaron.

—Saul R. Treviño

About the Authors

Mary K. Campbell

Mary K. Campbell was a professor emeritus of chemistry at Mount Holyoke College, where she taught for 36 years. Mary received her PhD from Indiana University and did postdoctoral work in biophysical chemistry at Johns Hopkins University. Her area of interest included researching the physical chemistry of biomolecules, specifically, spectroscopic studies of protein–nucleic acid interactions.

Saul R. Treviño

Saul R. Treviño is a professor of chemistry and biochemistry at Houston Christian University. He earned a biochemistry degree at Texas A&M University and a PhD at the Texas A&M University System Health Science Center in the biochemistry laboratories of C. Nick Pace and J. Martin Scholtz. Saul's love of teaching and learning led him to a chemistry faculty position at Houston Christian University in 2008. Over his years teaching general chemistry and biochemistry, Saul has also enjoyed reading and discussing works related to the scholarship of teaching and learning, especially in the areas of metacognition. Saul lives in Houston, Texas, with wife Polly, sons Benjamin and Aaron, and dog Snowy, where he enjoys playing tennis, pickleball, disc golf, and reading of all kinds. He and his family also enjoy riding roller coasters at amusement/theme parks all over the United States.

Brief Contents

Contents

Preface

This text is intended for students in any field of science or engineering who want a one-semester introduction to biochemistry but who do not intend to be biochemistry majors. Our main goal in writing this book is to make biochemistry as clear and applied as possible and to familiarize science students with the major aspects of biochemistry. For students of biology, chemistry, physics, geology, nutrition, sports physiology, and agriculture, biochemistry impacts greatly on the content of their fields, especially in the areas of medicine and biotechnology. For engineers, studying biochemistry is especially important for those who hope to enter a career in biomedical engineering or some form of biotechnology.

Students who will use this text are at an intermediate level in their studies. A beginning biology course, general chemistry, and at least one semester of organic chemistry are assumed as preparation.

What's New

All textbooks evolve to meet the interests and needs of students and instructors and to include the most current information. Numerous learner-centered updates mark this edition.

Learning-enhancement Features The 10th edition includes major changes to help students learn exquisite details of the chemistry of life, while showing them how biochemistry principles connect to their everyday lives. In each chapter, significant improvements have been made to the organization of content, the accuracy of content, and the consistency of content between the text and figures, making this edition an easier and more engaging read for students and instructors. Here are a few highlights:

- A major new feature in this edition is a reimagined amino acid structure chart that tells a story that connects groups of amino acids while revised accompanying text reviews and utilizes prerequisite functional group knowledge from organic chemistry. These features will help students deeply learn amino acid structures for quick recall as they build new knowledge of biochemistry topics. Prerequisite functional group knowledge from organic chemistry is also reviewed and utilized in learning structures of other important biochemical compounds.
- Major figure updates: Figure art, figure captions, and figure-related text have been revised and updated to help learners make connections that aid learning and understanding. Screens that highlight functional groups and chemical changes have also been added to many of the figures. Enzyme mechanisms have been updated and enhanced for easier understanding.
- Subsection letters (e.g., Section 1-2A) have been added, and numerous cross references to previous sections and subsections inserted to help students quickly review and connect biochemical patterns that appear across biochemical topics.
- The addition of bulleted text in certain areas breaks up large paragraphs of text into digestible chunks.

Hot Topic

Aging—Looking for the Biochemical Fountain of Youth

Since the beginning of humankind, people have been obsessed with finding ways to turn back the clock, or at least to stop it from advancing. In this book, we provide many practical applications of biochemistry that can lead to a higher quality of life and perhaps even a longer one. Most of these are intuitive, such as maintaining a healthy lifestyle through diet, exercise, and the avoidance of negative factors like smoking, but others can be revealed through a deeper understanding of biochemical processes. Although the exact causes of aging are still not clear, it is believed to be caused by the gradual wearing out over time of the body's natural ability to maintain itself and repair damage.

When We Think and Study Aging, What Are We Really Concerned About?

There are three major issues—namely, maximum lifespan, average life expectancy, and quality of life. Maximum lifespan is the maximum time that any member of a species has lived. For humans, this is about 120 years. Average life expectancy has been increasing dramatically for humans for the last hundred years, although it is slowing considerably now. Average life expectancy has gone up over 30 years since the early 1900s, mainly due to modern medicine and public health initiatives. Diseases that killed people early in life and complications in childbirth led to many deaths at a young age, which brought down the average life expectancy. So the first is how long you would live if nothing killed you. The second is how long you really are likely to live given all environmental factors. The third consideration is the quality of life. Whether maximum life expectancy goes up or not, people today are experiencing a much higher quality of life as they age. When people say, "70 is the new 40," they mean that we now see septuagenarians being as active today as 40-year-olds were decades ago. The goal of gerontology is to improve health near the end of life, rather than to make people live 300 years. Doctors seek to increase "health span," the number of years free of chronic illnesses and other age-related issues.

Exercise and Aging

Leading a healthy lifestyle can extend a person's life, as well as make the available years more pleasant and productive. Fitness and diet can lead to a person avoiding many of the diseases that the elderly often succumb to, such as heart disease, stroke, and some forms of cancer. Being physically fit can slow the general decline we experience as we age. The earlier the fitness begins, the better. Figure 4.37 shows how different types of fitness levels are related to the deterioration with aging.

Figure 4.37 shows that the earlier physically active lifestyles begin, the greater the person's ability to do physical work and the slower the decline is. The people who were active from age 10 years had a much higher overall work capacity. Equally important, at 85, they had the same work capacity as a 25-year-old sedentary person, and they made it past 90 before the most serious decline into death began. For many people, that is the most important statistic. When people wonder whether they would really want to live for a hundred years, their answer would undoubtedly depend on what those years looked like. As a comedian once said, "Do I really want another 20 years of wearing adult diapers?" Most people, if given the choice, would like to live a long time and would probably choose the green (top) curve in the figure, where they are relatively healthy well into their old age, and then a quick decline rather than a slow and lingering one.

Figure 4.37 The effects of exercise on lifespan and quality of life. (From Hoeger/Hoeger, Fitness and Wellness, 10E. © 2009 Cengage Learning.)

Can Longevity Be Increased with Chemistry?

What if we could have increased longevity *and* quality of life? We discovered more than 70 years ago that calorie restriction (CR) is associated with increased longevity in life forms as varied as yeast and rodents. Recently, primates were added to that list. In some species, restricting caloric intake by 30% compared to normal levels was shown to increase life span by 30% or more. In addition to the life span extension, CR leads to a higher quality of life and forestalls many diseases, such as cancer, diabetes, inflammation, and even neurodegenerative diseases.

Many mechanisms for this longevity increase have been suggested, including general health benefits of weight reduction and specific improvements in DNA stability due to lower levels of oxidative compounds that are created as by-products of metabolism. However, about 15 years ago, researchers began to pinpoint a family of genes in the yeast *Saccharomyces*

HT-112

Biochemistry Hot Topics Revised and updated "Hot Topic" articles appear at the end of select chapters to describe the latest advances in biochemistry and to show students how biochemistry is relevant and important to our lives. The Hot Topics are great to inspire in-class discussions or can be assigned for student presentations.

Biochemical Connections Revised and updated Biochemical Connections, integrated into the text, demonstrate how biochemistry topics relate to the real world in high-interest areas such as allied health, sports medicine, and biotechnology. These relevant content examples are meant to be read with the narrative, but they are specifically written in a magazine-style voice that enlivens the narrative to keep student interest at a high level and to show students how biochemistry is a part of them and their daily lives.

Pioneers in Biochemistry and Careers in Chemistry These textbook features are designed to broaden students' understanding of people who have had a significant influence in the field of chemistry. Pioneers in Biochemistry (PiB) profiles highlight influential individuals throughout history whose research, discoveries, and real-world impact have advanced the discipline in meaningful ways. These profiles celebrate trailblazers from a wide range of backgrounds, offering students a richer and more inclusive view of the scientists who have shaped biochemistry. Careers in Chemistry (CiC) demonstrates that chemistry is not a one-size-fits-all field by introducing students to the diverse professional paths available to those who study biochemistry in college. By showcasing people from different communities, identities, and lived experiences, this initiative aims to help every student see themselves reflected in the discipline—and to reinforce that if they are curious about biochemistry, there is space for them here.

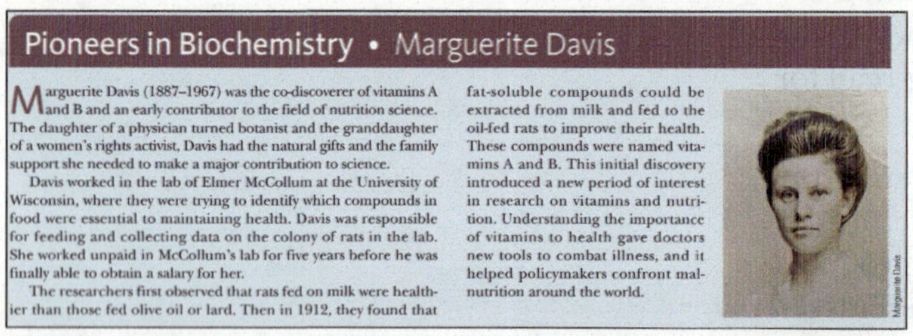

Pioneers in Biochemistry • Marguerite Davis

Marguerite Davis (1887–1967) was the co-discoverer of vitamins A and B and an early contributor to the field of nutrition science. The daughter of a physician turned botanist and the granddaughter of a women's rights activist, Davis had the natural gifts and the family support she needed to make a major contribution to science.

Davis worked in the lab of Elmer McCollum at the University of Wisconsin, where they were trying to identify which compounds in food were essential to maintaining health. Davis was responsible for feeding and collecting data on the colony of rats in the lab. She worked unpaid in McCollum's lab for five years before he was finally able to obtain a salary for her.

The researchers first observed that rats fed on milk were healthier than those fed olive oil or lard. Then in 1912, they found that fat-soluble compounds could be extracted from milk and fed to the oil-fed rats to improve their health. These compounds were named vitamins A and B. This initial discovery introduced a new period of interest in research on vitamins and nutrition. Understanding the importance of vitamins to health gave doctors new tools to combat illness, and it helped policymakers confront malnutrition around the world.

Marguerite Davis

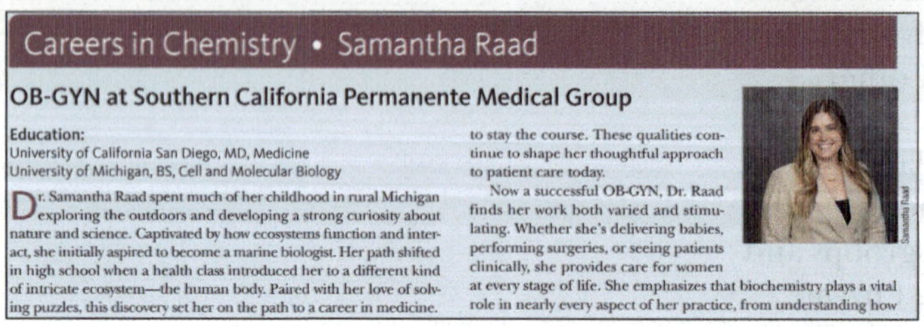

Careers in Chemistry • Samantha Raad

OB-GYN at Southern California Permanente Medical Group

Education:
University of California San Diego, MD, Medicine
University of Michigan, BS, Cell and Molecular Biology

Dr. Samantha Raad spent much of her childhood in rural Michigan exploring the outdoors and developing a strong curiosity about nature and science. Captivated by how ecosystems function and interact, she initially aspired to become a marine biologist. Her path shifted in high school when a health class introduced her to a different kind of intricate ecosystem—the human body. Paired with her love of solving puzzles, this discovery set her on the path to a career in medicine. to stay the course. These qualities continue to shape her thoughtful approach to patient care today.

Now a successful OB-GYN, Dr. Raad finds her work both varied and stimulating. Whether she's delivering babies, performing surgeries, or seeing patients clinically, she provides care for women at every stage of life. She emphasizes that biochemistry plays a vital role in nearly every aspect of her practice, from understanding how

Samantha Raad

WebAssign WebAssign is an easy-to-use platform with a wide variety of flexible settings and features. WebAssign's capabilities exist to help you and your students get the most out of the teaching and learning experience. We provide extensive resources to assist you with everything from basic course setup to customizing our more advanced features.

New Features Added to the WebAssign Course

- New Targeted Feedback Select problems in WebAssign will now have targeted feedback on common errors that students make. The targeted feedback explains why the student's answer is incorrect and offers insight into how to modify their approach toward a correct solution.

- New Problem Sets: Over 100 new digital biochemistry questions that correspond with the textbook have been added.
- Content Audit: All online exercises and randomizations have been reviewed for accuracy by the author to ensure they align with the textbook so that students are tested on content that they have learned.
- Customization: Create your own questions in WebAssign with ease. Use our built-in templates or start from scratch with detailed documentation and step-by-step video tutorials.
- Mastery Activities: Mastery activities are a flexible tool to help students develop conceptual understanding. Students are asked to answer a set number of questions about a learning objective correctly to receive credit.
- WebAssign Help: Resources have been streamlined and improved to make it easier for instructors and students to master WebAssign.

Art and Design Program The book's vibrant illustrations have been revised to even better help students visualize key processes and understand important topics. The visual style and format of the art, a hallmark of the text, has been updated to better appeal to students and is designed to improve student learning outcomes. Connections between information in the art and text have also been improved for stronger learning.

Updated Coverage Each chapter in the text has been updated with current developments and scientific findings in the biochemistry field.

Table of Notable Learner-Centered Changes by Chapter

Chapter 3 A reimagined amino acid structure chart that tells a story that connects groups of amino acids together and revised accompanying text that reviews and utilizes prerequisite functional group knowledge from organic chemistry.

Chapter 4 An updated graphical α-helix representation.

Chapter 5 Significant updates to chromatography-related text and figures.

Chapter 6 An updated representation of chymotrypsin-catalyzed reaction on a peptide that more accurately represents the peptide substrate structure and reminds learners of peptide/protein structure compared to individual amino acid structure. Significant changes to enzyme inhibition text for consistency in comparing the different types of enzyme inhibition.

Chapter 7 Significant updates to the figure of the sequential model for allosteric enzymes. See also significant updates to the chymotrypsin reaction mechanism figure that more accurately represents the protein substrate structure and the chymotrypsin active site structure. These changes remind learners of protein structure compared to individual amino acid structure.

Chapter 8 An updated phosphoacyglycerol figure that provides more complete representations of various phosphoacylglycerol structures. A separation of the discussion on waxes and sphingolipids into separate subsections and figures. An expanded discussion comparing the structures of various steroids that utilizes prerequisite functional group knowledge from organic chemistry. Updated figure showing the role of vitamin K in the modification of prothrombin.

Chapter 9 New aids for learning the pyrimidine and purine bases that utilizes prerequisite functional group knowledge from organic chemistry.

Chapter 10 An updated base excision repair figure.

Chapter 11 Significant updates to text related to the *lac* operon as well as text related to *lac* operon regulation and transcription attenuation. Significant updates to text and figures related to structural motifs in DNA-binding proteins.

Chapter 12 Significant updates to text and figure related to aminoacyl-tRNA formation. A new figure representing the peptidyl transferase mechanism. A new subsection on the rare 21st and 22nd amino acids in the genetic code.

Chapter 13 Updated figures related to recombinant DNA that more consistently illustrate sticky DNA ends. A new subsection on genome editing and the CRISPR-Cas9 system. Updated text related to PCR and qPCR.

Chapter 14 Significant updates to the lytic virus life cycle, retrovirus life cycle, and B cell development figures.

Chapter 16 Significant updates to the text and figures that discuss monosaccharide stereochemistry and cyclization. Significant updates to the figures for cellulose, starch, and peptidoglycan structures.

Chapter 17 Updates to the figures related to the glyceraldehyde-3-phosphate dehydrogenase active-site schematic and catalytic mechanism. Enzyme names added to the glycolysis control points figure for a more complete illustration of glycolysis regulation.

Chapter 18 A step-by-step update to the glycogenolysis text. Significant updates to the figures related to biotin enzyme attachment and the pyruvate carboxylase mechanism. Enzyme names added to the pentose phosphate pathway figure. Species labels added to the figure of glutathione and its reactions.

Chapter 19 Significant updates to the text and figure that discuss the pyruvate dehydrogenase mechanism. A new bulleted summary of the last three steps of the citric acid cycle and a comparison to the same three-step pattern seen in the β-oxidation pathway. Updated figure showing the unsymmetrical attachment of citrate to aconitase. Enzyme names added to the citric acid cycle control points figure for a more complete illustration of citric acid cycle regulation. Significant updates to the discussion of the control of the pyruvate dehydrogenase reaction and of the glyoxylate cycle.

Chapter 20 Significant updates to the text related to the conformational coupling mechanism in oxidative phosphorylation, the glycerol-3-phosphate shuttle, and the malate–aspartate shuttle.

Chapter 21 Significant updates to the text and figures related to the hormone-activated release of fatty acids from triacylglycerols as well as the carnitine shuttle pathway. A step-by-step update to the β-oxidation pathway text. Enzyme names added to the odd-numbered fatty acid oxidation figure. Significant updates to the figure related to the acetyl-CoA carboxylase mechanism. An overhaul of the fatty acid biosynthesis text and figure that helps students better understand the beginning reactions as well as the last reaction that completes one cycle. Enzyme names added to the figures for the biosynthesis of the following: phosphatidylethanolamine, a ceramide, and bile acids.

Chapter 22 Significant updates to the text and figures related to the Z-scheme of photosynthesis, the photophosphorylation of ADP to ATP, and cyclic photophosphorylation in photosystem I.

Chapter 23 Important updates to the text and figures related to the flow of nitrogen between the atmosphere and the biosphere as well as transamination reactions. Enzyme names added to most of the figures for nitrogen metabolism including the following: the biosynthesis of glutamate, glutamine, serine, and cysteine; the conversion of serine to glycine; the urea cycle; and the reactions of purine and pyrimidine catabolism. A new methionine to *S*-adenosylmethionine reaction figure. A step-by-step update to the urea cycle text.

Chapter 24 Significant updates to the text and figures related to second-messenger generation and effects.

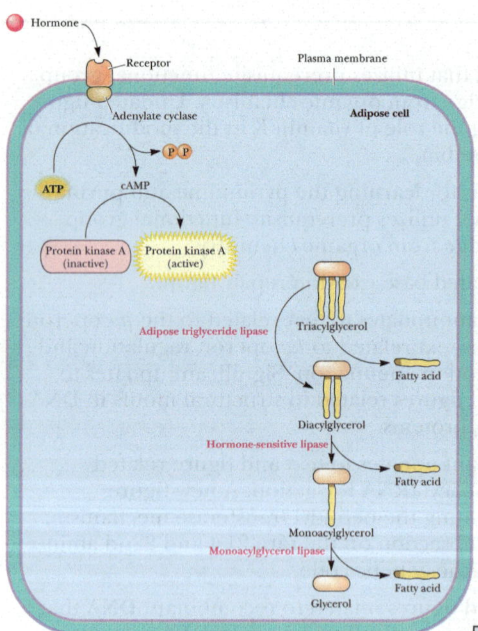

Proven Features

Visual Impact Ideal for visual learners, this book's state-of-the-art approach helps students visualize key processes and understand important topics.

Biochemical Connections The Biochemical Connections highlight special topics of particular interest to students. Topics frequently have clinical implications, such as cancer, AIDS, and nutrition. These essays help students make the connection between biochemistry and the real world. They are flowed in with the narrative and are placed exactly where they need to be read in each chapter. And although they have a different presentation than the rest of the narrative, they are meant to be read with the narrative and should not be skipped. They are like crescendos in classical music—the change in tempo from the usual narrative to the unique visual presentation and voice of the Biochemical Connections prevents the student's level of interest from dipping—students are always engaged. See a full listing of Biochemical Connections boxes in the Table of Contents.

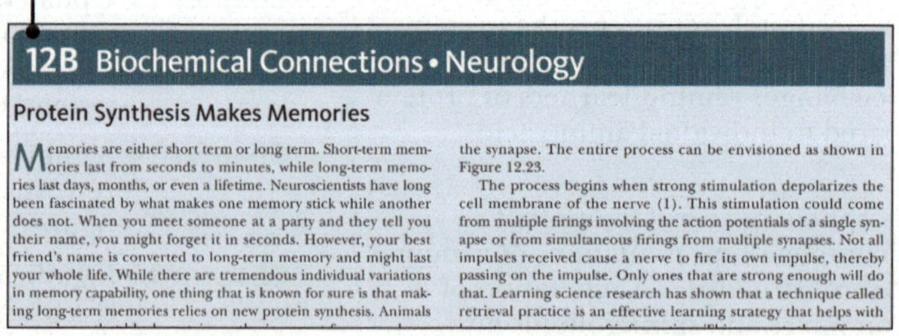
12B Biochemical Connections • Neurology

Protein Synthesis Makes Memories

Memories are either short term or long term. Short-term memories last from seconds to minutes, while long-term memories last days, months, or even a lifetime. Neuroscientists have long been fascinated by what makes one memory stick while another does not. When you meet someone at a party and they tell you their name, you might forget it in seconds. However, your best friend's name is converted to long-term memory and might last your whole life. While there are tremendous individual variations in memory capability, one thing that is known for sure is that making long-term memories relies on new protein synthesis. Animals

the synapse. The entire process can be envisioned as shown in Figure 12.23.

The process begins when strong stimulation depolarizes the cell membrane of the nerve (1). This stimulation could come from multiple firings involving the action potentials of a single synapse or from simultaneous firings from multiple synapses. Not all impulses received cause a nerve to fire its own impulse, thereby passing on the impulse. Only ones that are strong enough will do that. Learning science research has shown that a technique called retrieval practice is an effective learning strategy that helps with

Apply Your Knowledge The Apply Your Knowledge boxes are interspersed within chapters and are designed to provide students with problem-solving experience. The topics chosen are areas of study where students usually have the most difficulty. Solutions and problem-solving strategies are included, giving examples of the problem-solving approach for specific material.

> **2.1 Apply Your Knowledge**
>
> pH Calculations
>
> Calculate the pH of the following aqueous solutions:
>
> **a.** 1×10^{-3} *M* HCl
> **b.** 1×10^{-4} *M* NaOH

Marginal Glossary No flipping back and forth to read full definitions of key terms. Terms are defined in the margins.

> **titration** an experiment in which a measured amount of base is added to an acid

Summaries and Questions Each chapter closes with a concise summary, a broad selection of questions, and an annotated (and updated) bibliography that suggests sources for further reading. The Review Exercises fall into four categories: **Recall**, **Reflect and Apply**, **Biochemical Connections**, and **Mathematical**. The **Recall** questions are designed for students to quickly assess their mastery of the material, and the **Reflect and Apply** questions are for students to work through more thought-provoking questions. **Biochemical Connections** questions test students on the **Biochemical Connections** essays in that chapter. The **Mathematical** questions are quantitative in nature and focus on calculations.

> **Summary**
>
> **Biochemistry and Life**
> - Biochemistry is a multidisciplinary field that asks and tries to answer questions about the molecular nature of life processes.
> - Many chemical reactions take place simultaneously in living cells.
>
> **Origin of Life on Earth**
> - The fundamental biochemical similarities observed in all living organisms have engendered speculation about the origins of life.
>
> **Chemical Foundations of Biochemistry**
> - The four basic molecules discussed in this text are amino acids, carbohydrates, nucleic acids, and lipids.
>
> **Functional Groups Important in Biochemistry**
> - Both organic chemistry and biochemistry base their approaches on the behavior of functional groups, but their emphases differ because some functional groups important to organic chemistry do not play a role in biochemistry, and vice versa.
> - Important functional groups in biochemistry include carbonyl groups, hydroxyl groups, carboxyl groups, amines, amides, and esters; derivatives of phosphoric acid such as esters and anhydrides are also important.
>
> **Earth and Its Age**
> - Earth, along with the rest of the solar system, was formed about 4 to 5 billion years ago from elements produced by first-generation stars.
>
> **Biomolecules**
> - Important biomolecules can be produced under abiotic (nonliving) conditions from simple compounds postulated to have been present in the atmosphere of the early Earth.
> - These simple biomolecules can polymerize, also under abiotic conditions, to give rise to compounds resembling proteins and others having a less marked resemblance to nucleic acids.
>
> **Molecules to Cells**
> - All cellular activity depends on the presence of catalysts, which increase the rates of chemical reactions, and

Alternative Teaching Options

The order in which individual chapters are covered can be changed to suit the needs of specific groups of students. For example, all of the molecular biology chapters (Chapters 10 through 14) can precede metabolism or can follow it, depending on the instructor's choice. The order in which the material on molecular biology is treated can be varied according to the preference of the instructor.

Alternate Editions

Loose-Leaf Edition for *Biochemistry* 10e
ISBN: 979-8-214-19361-8
A loose-leaf (unbound, three-hole-punched) version of *Biochemistry* 10e, which can be inserted in a binder, is also available.

Supporting Materials

Please visit www.cengage.com for information about student and instructor resources for this text.

Acknowledgments

We would like to acknowledge HCU biochemistry students between Fall 2023 and Fall 2025 who contributed ideas from the learner perspective as well as colleagues who contributed their ideas and critiques of the manuscript. Some reviewers responded to specific queries regarding the text itself. We thank them for their efforts and their helpful suggestions.

Reviewers Acknowledgments

Tenth Edition Reviewers

John Alexander, *Rust College*
Deb Audino, *Great Bay Community College*
Alfredo Castro, *Felician University*
Dipak K. Ghosh, *North Carolina A & T State University*
Anureet Kaur Shah, *California State University Los Angeles*
Jim Killarney, *Colby-Sawyer College*
Ho-Jin Lee, *LeMoyne-Owen College*
Scott Lefler, *Arizona State University*
David J. Merkler, *University of South Florida*
Allyn Ontko, *Arkansas State University*
Jennifer Sniegowski, *Arizona State University*
Mikaela Stewart, *Texas Christian University*
Alexander Zestos, *American University*

Ninth Edition Reviewers

Paul Adams, *University of Kansas*
Kenneth Balazovich, *University of Michigan*
Tory Hagen, *Oregon State University*
Marcy Henrick, *Appalachian College of Pharmacy*
Deborah Heyl-Clegg, *Eastern Michigan University*

Eighth Edition Reviewers

Kenneth Balazovich, PhD, *University of Michigan*
Laurent Dejean, *California State University at Fresno*
Marcy Hernick, *Virginia Tech*
Holly Huffman, *Arizona State University*
Mark Kearley, *Florida State University*
James Knopp, *North Carolina State University*
Paul Larsen, *University of California–Riverside*
Gerry Prody, *Western Washington University*
Sandra Turchi, *Millersville University*

Seventh Edition Reviewers

Paul D. Adams, *University of Kansas*
Dan Davis, *University of Arkansas*
Nick Flynn, *Angelo State University*
Denise Greathouse, *University of Arkansas*
James R. Paulson, *University of Wisconsin–Oshkosh*
Kerry Smith, *Clemson University*
Alexandre G. Volkov, *Oakwood University*

We would also like to thank everyone at Cengage Learning who were extremely supportive and essential to the development of this book, especially the following: Roxanne Wang, Senior Portfolio Product Manager; Ellie Purgavie, Senior Portfolio Product Assistant; Sophie Yook, Portfolio Product Assistant; Courtney Wolstoncroft, Senior Learning Designer; Malini Rao, Learning Designer; Dakin Sharum, Subject Matter Expert; Lynh Pham, Senior Content Manager; Anandan Bommen, Content Project Manager; and John Murdzek, Developmental Editor.

A Final Note

We hope this tenth edition of our textbook assists and inspires professors and students of biochemistry, fostering curiosity, critical thinking, and a lasting appreciation for the wonders and mysteries of the chemistry of life.

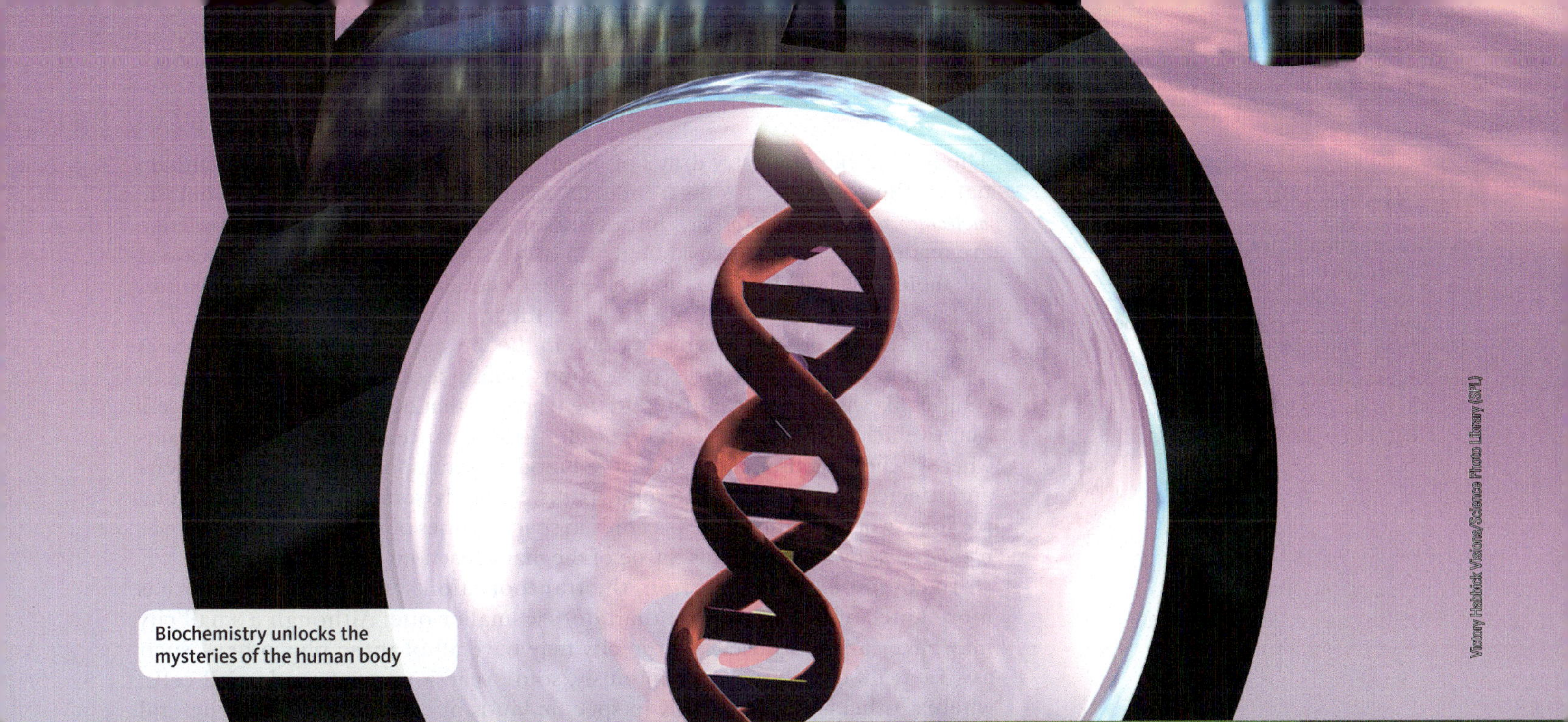

Biochemistry unlocks the mysteries of the human body

Biochemistry and the Organization of Cells 1

Chapter Outline

1-1 Basic Themes

Learning Objectives

After reading this section, students will be able to:

LO 1.1.1 List the six simple molecules that may have given rise to early molecules and cells.

LO 1.1.2 List and briefly describe the eight levels of structural organization in the human body.

1-1A Biochemistry and Life

▶ How does biochemistry describe life processes?

Living organisms, such as humans, and even the individual cells of which they are composed, are enormously complex and diverse. Nevertheless, certain unifying features are common to all living things, from the simplest bacterium to the human being. They all use the same types of *biomolecules*, and they all use energy. As a result, organisms can be studied via the methods of chemistry and physics.

Biochemistry can be defined in many ways. As the name suggests, it is the chemistry of life. It combines biology and chemistry, and any given instructor may have more of a biology focus, a chemistry focus, or anything in between.

Disciplines that appear to be unrelated to biochemistry can provide answers to important biochemical questions. For example, the magnetic resonance imaging tests that play an important role in the health sciences originated with physicists became a vital tool for chemists and currently play a large role in biochemical research.

Chapter Outline (continued)

The field of biochemistry draws on many disciplines, and its multidisciplinary nature allows it to use results from many sciences to answer questions about the *molecular nature of life processes.* Important applications of this kind of knowledge are made in medically related fields; an understanding of health and disease at the molecular level leads to more effective treatment of illnesses of many kinds.

The activities within a cell are similar to the transportation system of a city. The cars and other modes of transportation in the city correspond to the molecules involved in reactions (or series of reactions) within a cell. The routes traveled by vehicles likewise can be compared to the reactions that occur in the life of the cell. Many vehicles travel more than one route—for instance, cars can go almost anywhere—whereas other, more specialized modes of transportation, such as subways and streetcars, are confined to single paths. Similarly, some molecules play multiple roles, whereas others take part only in specific series of reactions. Also, *the routes operate simultaneously,* and this is true of the many reactions within a cell as well.

To continue the comparison, the transportation system of a large city has more kinds of transportation than does a smaller one. Although a small city may have cars and buses, a large city may have all of these plus others, such as streetcars or subways. Analogously, some reactions are found in all cells, whereas others are found only in specific kinds of cells. Also, more structural features are found in the larger, more complex cells of larger organisms than in the simpler cells of organisms such as bacteria.

An inevitable consequence of this complexity is the large number of terms needed to describe it; learning considerable new vocabulary is an essential part of the study of biochemistry. There are many cross-references in this book, too, which reflect the many connections among the processes that take place in the cell.

1-1B Origin of Life on Earth

The fundamental similarity of cells of all types makes speculating on the origins of life a worthwhile question. How did the components of our bodies come to be, and how do they do the things they do? What are the molecules of life? Even comparatively small biomolecules consist of several parts. Large biomolecules, such as proteins and nucleic acids, have complex structures, and living cells are enormously more complex (Figure 1.1). Even so, *both molecules and cells must have arisen ultimately from very simple molecules,* such as water, methane, carbon dioxide, ammonia, nitrogen, and hydrogen. In turn, these simple molecules must have arisen from atoms.

▶ How did living things originate?

The way in which the Universe itself, and the atoms of which it is composed, came to be is a topic of great interest to astrophysicists as well as other scientists. Simple molecules were formed by combining atoms, and reactions of simple molecules led in turn to more complex molecules. The molecules that play a role in living cells today are the same molecules as those encountered in organic chemistry; they simply operate in a different context.

1-2 Chemical Foundations of Biochemistry

Learning Objectives

After reading this section, students will be able to:

LO 1.2.1 Describe the characteristics of the four basic classes of small molecules found in cells.

LO 1.2.2 Write general structures and names of functional groups for the following classes of compounds of biochemical importance: alkenes, alcohols, ethers, amines, thiols, aldehydes, ketones, carboxylic acids, esters, amides, phosphoric acid esters, and phosphoric acid anhydrides.

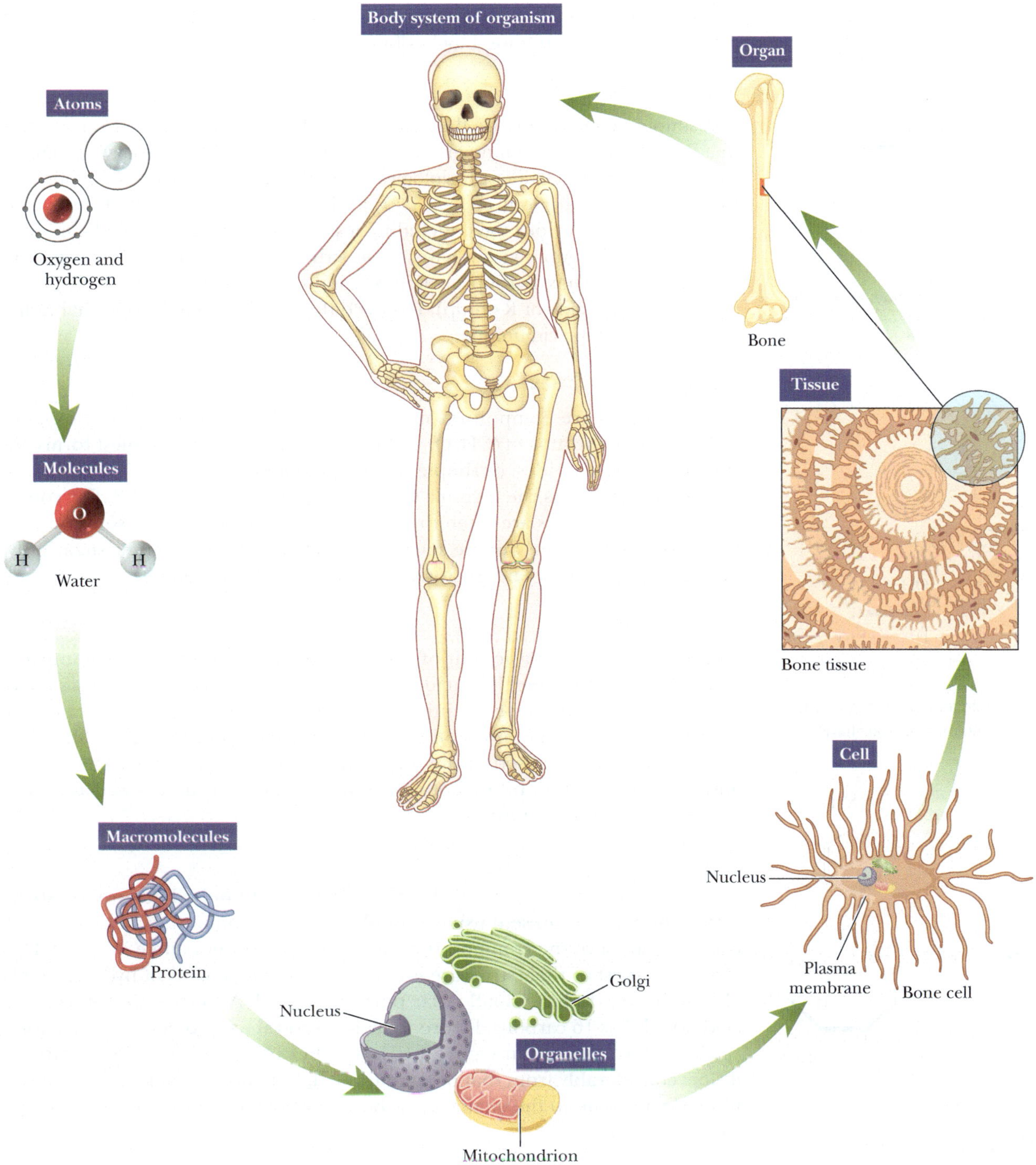

Figure 1.1 **Levels of structural organization in the human body.** The hierarchy ranges from simple to complex.

Organic chemistry is the study of compounds of carbon. Because the cellular apparatus of living organisms is made up of carbon compounds, biomolecules are part of the subject matter of organic chemistry. Additionally, many carbon compounds are not found in any organism, and many topics of importance to organic chemistry have little connection with living things. We are going to concentrate on the aspects of organic chemistry that we need in order to understand what goes on in living cells.

organic chemistry the study of compounds of carbon

The small molecules found in the cell can usually be lumped into four basic classes: amino acids, carbohydrates, nucleotides, and lipids. These four kinds

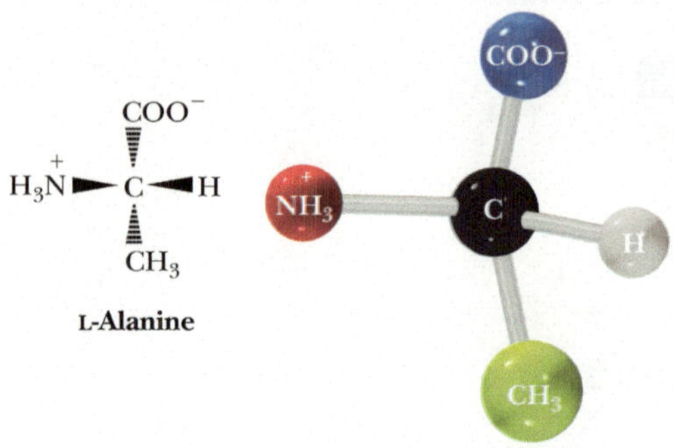

Figure 1.2 Basic structure of alanine, an amino acid.

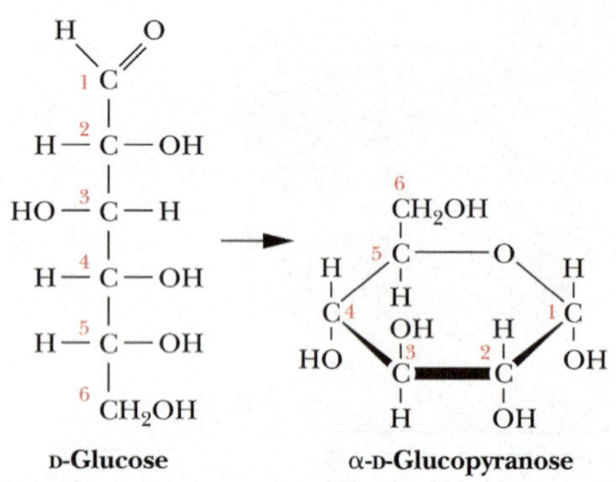

Figure 1.3 Straight chain and cyclic depictions of glucose, the most common monosaccharide.

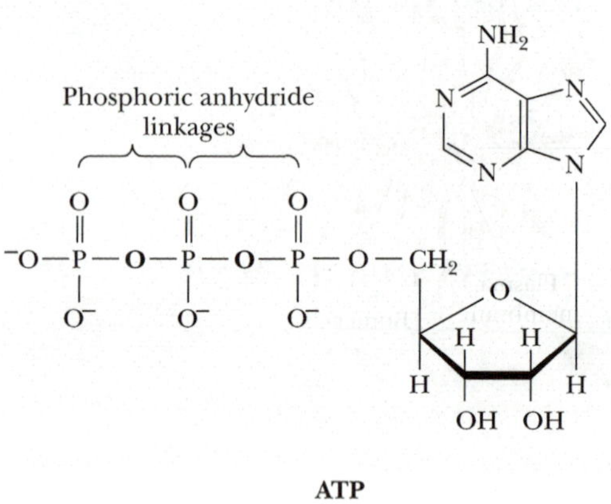

Figure 1.4 The structure of adenosine triphosphate (ATP), an important nucleotide in energy production.

of compounds come up over and over again during our study of biochemistry. They are the basic building blocks of life.

1-2A Amino Acids

The simplest compounds are the amino acids. They get their name from the fact that they all contain an amino group and a carboxylic acid (or carboxyl) group, as shown in Figure 1.2. Under physiological conditions, both the carboxyl group and the amino group are ionized ($—COO^-$ and $—NH_3^+$), respectively. Amino acids can be shown in various ways, including a structural formula or a ball-and-stick molecular model. Amino acids have a basic structure where a central carbon atom is bonded to a carboxyl group, an amino group, a hydrogen, and a variable group, called the side chain (or R group). It is the difference between the side chains that makes each amino acid unique.

1-2B Carbohydrates

Carbohydrates are compounds made up of carbon, hydrogen, and oxygen, with a general formula of $(CH_2O)_n$, where n is at least 3. The simplest forms are called monosaccharides, or simple sugars. The most common monosaccharide is glucose, which has the formula $C_6H_{12}O_6$, as shown in Figure 1.3. For convenience, simple sugars are often drawn as a straight chain, but in solution they form cyclic structures. Simple sugars often make up much larger sugar polymers and are used for energy storage and structural components.

1-2C Nucleotides

Nucleotides are the basic unit of the hereditary materials deoxyribonucleic acid (DNA) and ribonucleic acid (RNA). A nucleotide is composed of a five-carbon sugar, a nitrogen-containing ring, and one or more phosphate groups. A particularly important nucleotide is adenosine triphosphate (ATP), shown in Figure 1.4. It is composed of the nitrogenous base adenine, the sugar ribose, and three phosphates. It functions as a building block of RNA as well as molecular energy currency in the cell.

1-2D Lipids

The fourth major group of biochemicals consists of lipids. They are too structurally diverse to represent using a simple structure common to all lipids. However, they all have the common trait that they are poorly soluble in water. This is because most of their structure is composed of long chains of hydrocarbons, although they do have small polar portions as well. A simple lipid is palmitic acid, which has 16 carbons. Figure 1.5 shows several ways to depict such a lipid.

Cholesterol, shown in Figure 1.6, is another important lipid. Its structure differs considerably from that of palmitic acid, but cholesterol is still very insoluble in water due to the chains of carbon and the fact that it has only a single oxygen atom in it.

▶ Can a chemist make the molecules of life in a laboratory?

Until the early part of the nineteenth century, there was a widely held belief in "vital forces," forces presumably unique to living things. This belief included the idea that the compounds found in living organisms could not be produced in the laboratory. German chemist Friedrich Wöhler performed the critical experiment that disproved this belief in 1828. Wöhler synthesized urea, a well-known waste product of animal metabolism, from ammonium cyanate, a compound obtained from mineral (i.e., nonliving) sources.

$$\underset{\text{Ammonium cyanate}}{NH_4OCN} \rightarrow \underset{\text{Urea}}{H_2NCONH_2}$$

It has subsequently been shown that any compound that occurs in a living organism can be synthesized in the laboratory, although in many cases, the synthesis represents a considerable challenge to even the most skilled organic chemist.

1-2E Functional Groups Important in Biochemistry

The reactions of biomolecules can be described by the methods of organic chemistry, which requires the classification of compounds according to their **functional groups**. *The reactions of molecules are based on the reactions of their respective functional groups.*

Table 1.1 lists some biologically important functional groups. Most of these functional groups contain oxygen and nitrogen, which are among the most electronegative elements. As a result, many of these functional groups are polar, and their polar nature plays a crucial role in their reactivity. Some groups that are vitally important to organic chemists are missing from the table because molecules containing these groups, such as alkyl halides and acyl chlorides, do not have any particular applicability in biochemistry. Conversely, carbon-containing derivatives of phosphoric acid are mentioned infrequently in beginning courses on organic chemistry, but esters and anhydrides of phosphoric acid (Figure 1.7) are of vital importance in biochemistry. ATP, a molecule that is the energy currency of the cell, contains both ester and anhydride linkages involving phosphoric acid.

Important classes of biomolecules have characteristic functional groups that determine their reactions. We discuss the reactions of the functional groups when we consider the compounds in which they occur.

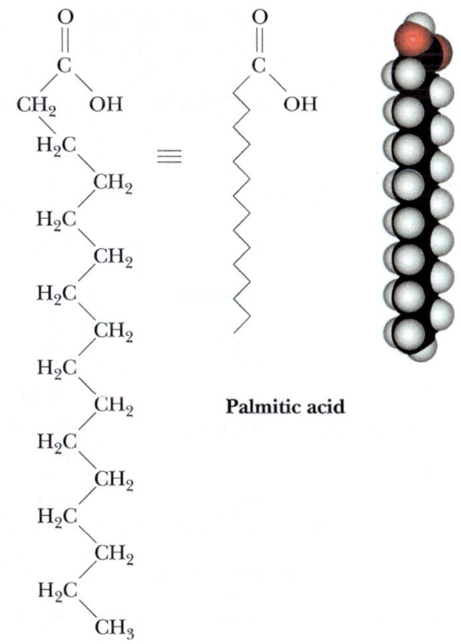

Palmitic acid

Figure 1.5 The simple lipid palmitic acid, shown with a structural formula, an abbreviated formula, and a space-filling model.

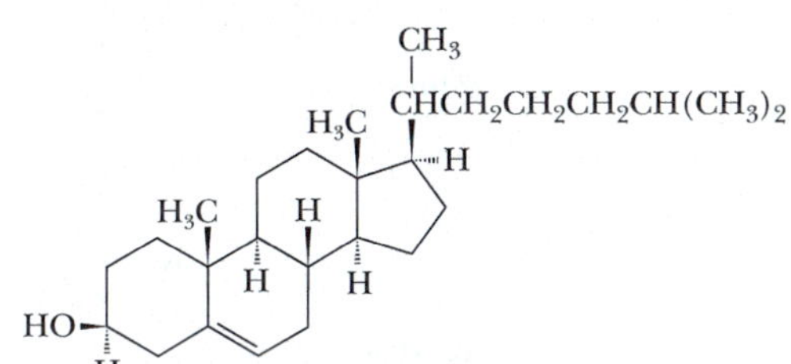

Cholesterol

Figure 1.6 The structure of cholesterol, an important lipid in biological membranes.

functional groups groups of atoms that give rise to the characteristic reactions of organic compounds

1-3 The Beginnings of Biology

Learning Objectives

After reading this section, students will be able to:

LO 1.3.1 Distinguish between *monomers* and *polymers*.

LO 1.3.2 Explain the directionality that is found in proteins, polysaccharides, and nucleic acids.

1-3A Earth and Its Age

Although humans in general and science fiction writers in particular are fascinated by the idea of life on other planets, to date, we are aware of only one planet that unequivocally supports life as we know it: our own. Earth and its waters are universally understood to be the source and mainstay of life. How, then, did Earth, along with the Universe of which it is a part, come to be?

▶ How and when did Earth come to be?

Currently, the most widely accepted cosmological theory for the origin of the Universe is a cataclysmic explosion called the *Big Bang*. According to Big Bang cosmology, all the matter in the Universe was originally confined to a comparatively small volume of space. As a result of a tremendous explosion, this "primordial fireball" started to expand with great force. Immediately after the Big Bang, the Universe was extremely hot, on the order of 15 billion (15×10^9) K. The average temperature of the Universe has been decreasing ever since as a result of expansion, and the lower temperatures have permitted the formation of stars and planets.

In its earliest stages, the Universe had a fairly simple composition. Hydrogen, helium, and some lithium (the three smallest and simplest elements in the periodic table) were present, having been formed in the original Big Bang explosion. The rest of the chemical elements are thought to have been formed in three ways: (1) by thermonuclear reactions that normally take place in stars, (2) in explosions of stars, and (3) by the action of cosmic rays outside the stars since the formation of the galaxy.

The process by which the elements are formed in stars is a topic of interest to chemists as well as to astrophysicists. For our purposes, the most abundant

Table 1.1 Functional Groups of Biochemical Importance

Class of Compound	General Structure	Characteristic Functional Group	Name of Functional Group	Example
Alkenes	$RCH{=}CH_2$ $RCH{=}CHR$ $R_2C{=}CHR$ $R_2C{=}CR_2$	$C{=}C$	**Double bond**	$CH_2{=}CH_2$
Alcohols	ROH	—OH	**Hydroxyl group**	CH_3CH_2OH
Ethers	ROR	—O—	**Ether group**	CH_3OCH_3
Amines	RNH_2 R_2NH R_3N	$—N<$	**Amino group**	CH_3NH_2
Thiols	RSH	—SH	**Sulfhydryl group**	CH_3SH
Aldehydes	R—C(=O)—H	—C(=O)—	**Carbonyl group**	$CH_3C(=O)H$
Ketones	R—C(=O)—R	—C(=O)—	**Carbonyl group**	$CH_3C(=O)CH_3$
Carboxylic acids	R—C(=O)—OH	—C(=O)—OH	**Carboxyl group**	$CH_3C(=O)OH$
Esters	R—C(=O)—OR	—C(=O)—OR	**Ester group**	$CH_3C(=O)OCH_3$
Amides	R—C(=O)—NR_2 R—C(=O)—NHR R—C(=O)—NH_2	$—C(=O)—N<$	**Amide group**	$CH_3C(=O)N(CH_3)_2$
Phosphoric acid esters	R—O—P(=O)(OH)—OH	—O—P(=O)(OH)—OH	**Phosphoric ester group**	CH_3—O—P(=O)(OH)—OH
Phosphoric acid anhydrides	R—O—P(=O)(OH)—O—P(=O)(OH)—OH	—P(=O)(OH)—O—P(=O)(OH)—	**Phosphoric anhydride group**	HO—P(=O)(OH)—O—P(=O)(OH)—OH

isotopes of biologically important elements such as carbon, oxygen, nitrogen, phosphorus, and sulfur have *particularly stable nuclei.* These elements were produced by nuclear reactions in first-generation stars, the original stars produced after the beginning of the Universe (Table 1.2). Many first-generation stars were destroyed by explosions called *supernovas,* and their stellar material was recycled to produce second-generation stars, such as our own Sun, along with our solar system. Radioactive dating, which measures the decay of unstable nuclei, indicates that the age of Earth (and the rest of the solar system) is 4 billion to 5 billion (4×10^9 to 5×10^9) years.

The atmosphere of the early Earth was very different from the one we live in, and it probably went through several stages before reaching its current composition.

1 Reaction of phosphoric acid with a hydroxyl group to form an ester, which contains a P-O-R linkage. Phosphoric acid is shown in its nonionized form in this figure. Space-filling models of phosphoric acid and its methyl ester are shown. The red spheres represent oxygen; the white, hydrogen; the green sphere represents a group containing carbon and hydrogen atoms; and the orange, phosphorus.

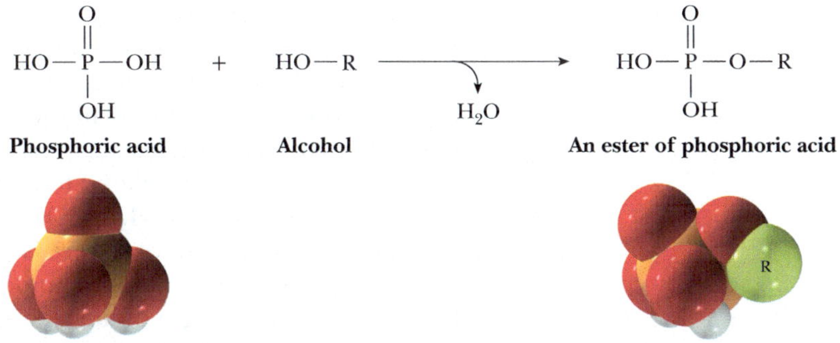

2 Reaction of two molecules of phosphoric acid to form an anhydride, which contains a P-O-P linkage. A space-filling model of the anhydride of phosphoric acid is shown.

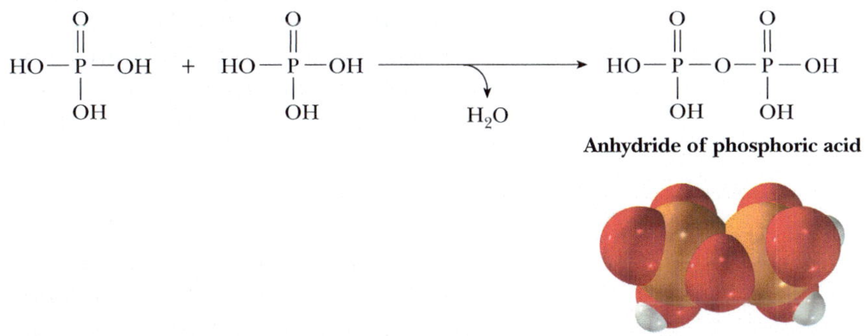

3 The structure of ATP (adenosine triphosphate), showing two anhydride linkages and one ester.

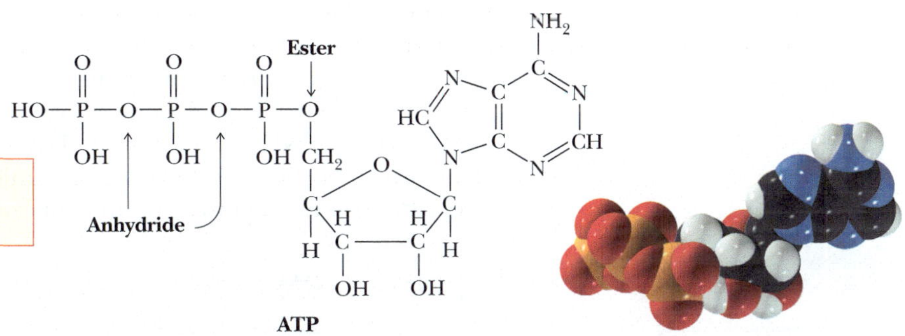

Figure 1.7 Adenosine triphosphate and the reactions for its formation.

Table 1.2 Abundance of Important Elements Relative to Carbon*

Element	Abundance in Organisms	Abundance in the Universe
Hydrogen	80–250	10,000,000
Carbon	1,000	1,000
Nitrogen	60–300	1,600
Oxygen	500–800	5,000
Sodium	10–20	12
Magnesium	2–8	200
Phosphorus	8–50	3
Sulfur	4–20	80
Potassium	6–40	0.6
Calcium	25–50	10
Manganese	0.25–0.8	1.6
Iron	0.25–0.8	100
Zinc	0.1–0.4	0.12

*Each abundance is given as the number of atoms relative to a thousand atoms of carbon.

The most important difference is that, according to most theories of the origins of Earth, very little or no free oxygen (O_2) existed in the early stages (Figure 1.8). The early Earth was constantly irradiated with ultraviolet light from the Sun because there was no ozone (O_3) layer in the atmosphere to block it. Under these conditions, the chemical reactions that produced simple biomolecules took place.

The gases usually postulated to have been present in the atmosphere of the early Earth include NH_3, H_2S, CO, CO_2, CH_4, N_2, H_2, and H_2O (in both liquid and vapor forms). However, there is less agreement on the relative amounts of these components, from which biomolecules ultimately arose. Many of the earlier theories of the origin of life postulated CH_4 as the carbon source, but more recent studies have shown that appreciable amounts of CO_2 must have existed in the atmosphere at least 3.8 billion (3.8×10^9) years ago.

This conclusion is based on geological evidence: The earliest known rocks are 3.8 billion years old, and they are carbonates, which arise from CO_2. Any NH_3 originally present must have dissolved in the oceans, leaving N_2 in the atmosphere as the nitrogen source required for the formation of proteins and nucleic acids.

1-3B Biomolecules

▶ How were biomolecules likely to have formed on the early Earth?

Experiments have been performed in which the simple compounds of the early atmosphere were allowed to react under the varied sets of conditions that might have been present on Earth at that time. The results of such experiments indicate that these simple compounds react *abiotically* or, as the word indicates, in the absence of life (*a*, "not," and *bios*, "life"), to give rise to biologically important compounds such as the components of proteins and nucleic acids. Of historic interest is the Miller–Urey experiment. In each trial of this experiment, an electric discharge, simulating lightning, is passed through a closed system that contains H_2, CH_4, and NH_3, in addition to H_2O. Simple organic molecules, such as formaldehyde (HCHO) and hydrogen cyanide (HCN), are typical products of such reactions, as are amino acids, the building blocks of proteins.

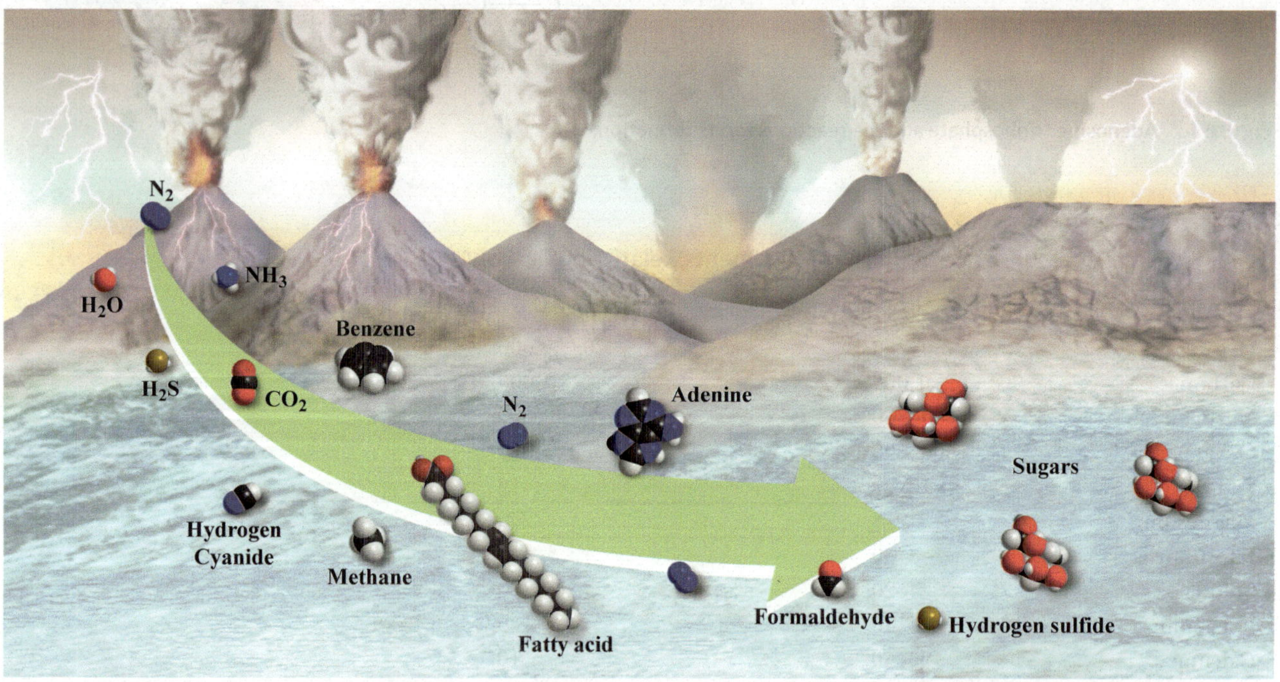

Figure 1.8 Formation of biomolecules on the early Earth. Conditions on early Earth would have been inhospitable for most of today's life forms. Very little or no oxygen (O_2) existed. Volcanoes erupted, spewing gases, and violent thunderstorms produced torrential rainfall that covered Earth. Space-filling models of biomolecules are shown. The blue spheres represent nitrogen; the red, oxygen; the white, hydrogen; the black, carbon; and the yellow, sulfur. The green arrow indicates the formation of biomolecules from simple precursors.

Careers in Chemistry • Samantha Raad

OB-GYN at Southern California Permanente Medical Group

Samantha Raad

Education:
University of California San Diego, MD, Medicine
University of Michigan, BS, Cell and Molecular Biology

Dr. Samantha Raad spent much of her childhood in rural Michigan exploring the outdoors and developing a strong curiosity about nature and science. Captivated by how ecosystems function and interact, she initially aspired to become a marine biologist. Her path shifted in high school when a health class introduced her to a different kind of intricate ecosystem—the human body. Paired with her love of solving puzzles, this discovery set her on the path to a career in medicine.

Although her journey through medical school and residency was demanding; filled with 24-hour shifts, code blues, and emergency C-sections, Dr. Raad's commitment never wavered. Fueled by a deep-rooted desire to provide care and contribute to her community, she relied on her resilience and early passion for science to stay the course. These qualities continue to shape her thoughtful approach to patient care today.

Now a successful OB-GYN, Dr. Raad finds her work both varied and stimulating. Whether she's delivering babies, performing surgeries, or seeing patients clinically, she provides care for women at every stage of life. She emphasizes that biochemistry plays a vital role in nearly every aspect of her practice, from understanding how medications alter the cellular structure of the cervix to explaining to gestational diabetic patients how a high-protein bedtime snack can help prevent rebound hyperglycemia. She notes, "each day, I am grateful to use my biochemistry knowledge to care for patients and to learn from those advancing the field of medicine."

According to one theory, reactions such as these took place in the oceans of early Earth; other researchers postulate that such reactions occurred on the surfaces of clay particles that were present on the early Earth. It is certainly true that mineral substances similar to clay can serve as catalysts in many types of reactions. Recent theories of the origin of life focus on RNA, not proteins, as the first genetic molecules. Proteins are thought to have developed later in the evolution of the earliest cells. This point does not diminish the importance of this first experiment on the abiotic synthesis of biomolecules.

Living cells today are assemblages that include very large molecules, such as proteins, nucleic acids, and polysaccharides. These molecules are larger by many powers of 10 than the smaller molecules from which they are built. Hundreds or thousands of these smaller molecules, called **monomers**, can be linked to produce macromolecules, which are also called **polymers**. The versatility of carbon is important here. Carbon is tetravalent and able to form bonds with itself and with many other elements, giving rise to different kinds of monomers, such as amino acids, nucleotides, and monosaccharides (sugar monomers).

In present-day cells, amino acid monomers combine by polymerization to form **proteins**, nucleotide monomers combine to form **nucleic acids**, and sugar monomers polymerize to form **polysaccharides**. The monomers in each case, once they become part of their respective polymers, are called residues. Polymerization experiments with amino acids carried out under early-Earth conditions

monomers small molecules that may bond to many others to form a polymer

polymers macromolecules formed by the bonding of smaller units

proteins macromolecules formed by the polymerization of amino acids

nucleic acids macromolecules formed by the polymerization of nucleotides

polysaccharides macromolecules formed by the polymerization of sugar monomers

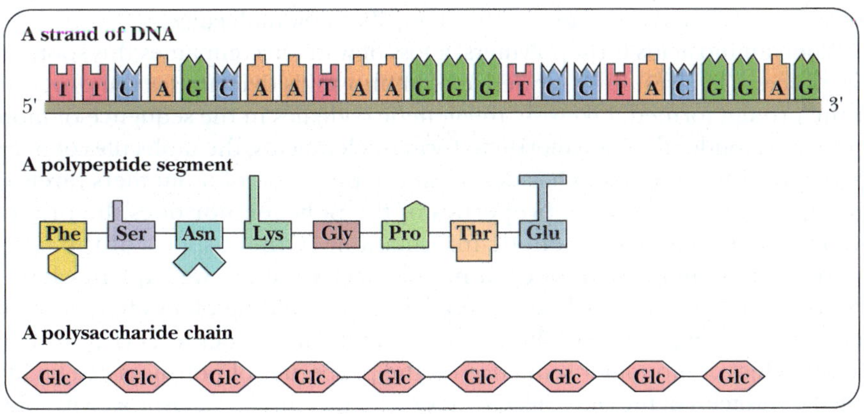

Figure 1.9 Biological macromolecules can be informational. The sequence of monomeric units in a biological polymer has the potential to contain information if the order of units is not overly repetitive. Nucleic acids and proteins are informational macromolecules; polysaccharides are not.

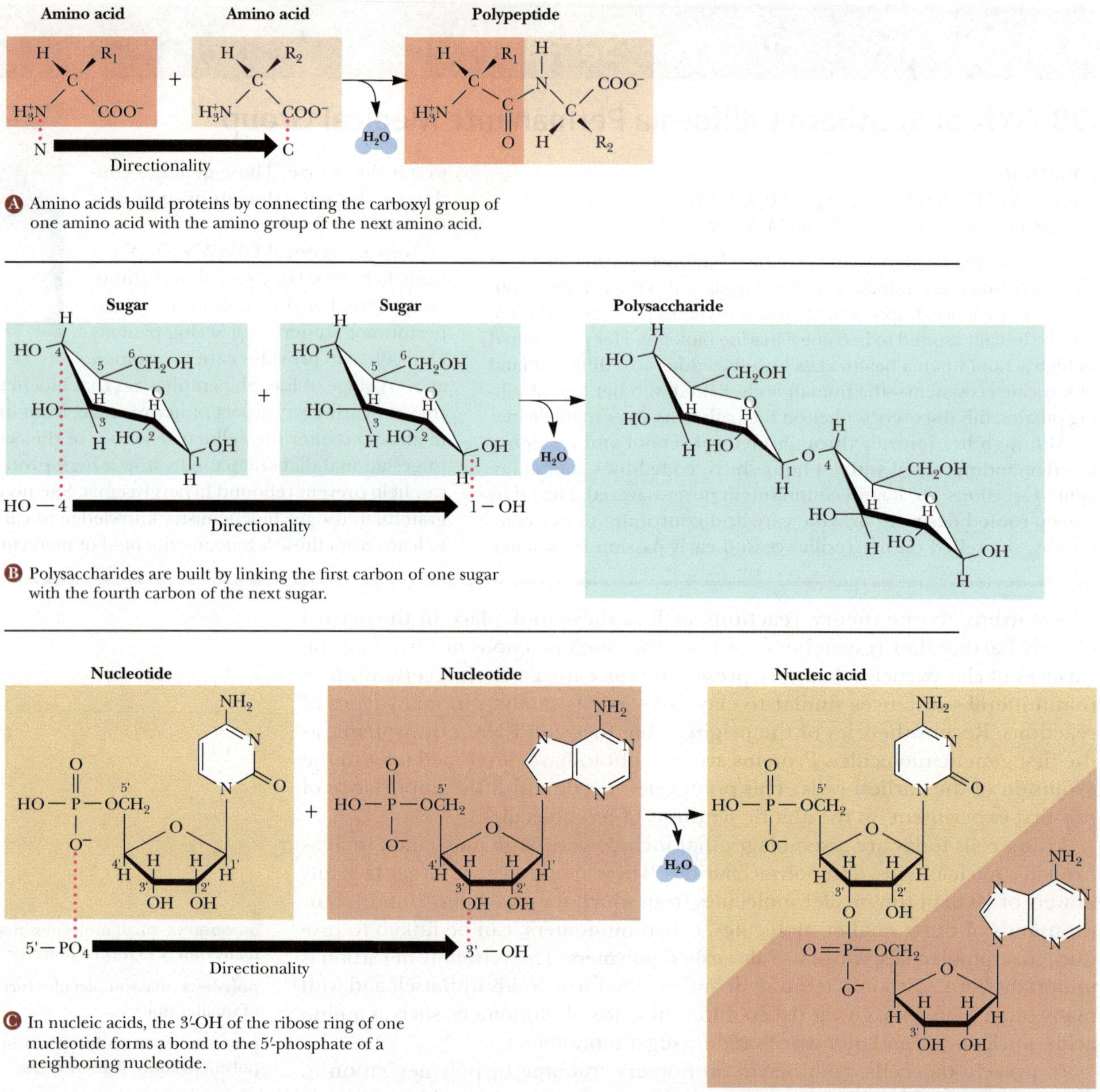

A Amino acids build proteins by connecting the carboxyl group of one amino acid with the amino group of the next amino acid.

B Polysaccharides are built by linking the first carbon of one sugar with the fourth carbon of the next sugar.

C In nucleic acids, the 3′-OH of the ribose ring of one nucleotide forms a bond to the 5′-phosphate of a neighboring nucleotide.

Figure 1.10 Directionality in macromolecules. Biological macromolecules and their building blocks have a "sense" or directionality.

have produced protein-like polymers. Similar experiments have been done on the abiotic polymerization of nucleotides and sugars, which tends to happen less readily than the polymerization of amino acids. Much of this discussion is speculative, but it is a useful way to start thinking about biomolecules.

When amino acids form polymers, a loss of water accompanies this spontaneous process. The resulting sequence of amino acids determines the properties of the protein formed. Likewise, the genetic code lies in the sequence of monomeric nucleotides that polymerize to form nucleic acids, the molecules of heredity (Figure 1.9). In polysaccharides, however, the order of monomers rarely has an important effect on the properties of the polymer, nor does the order of the monomers carry any genetic information. (Other aspects of the *linkage* between monomers are important in polysaccharides, as we explain when we discuss carbohydrates in Chapter 16.) All of the building blocks have a "head" and a "tail," giving a sense of direction even at the monomer level (Figure 1.10).

The effect of monomer sequence on the properties of polymers can be illustrated by proteins of the class called *enzymes.* Enzymes display **catalytic activity**, which

catalytic activity the ability to increase the rate of a chemical reaction

means that they increase the rates of chemical reactions compared with uncatalyzed reactions. In the context of the origin of life, catalytic molecules can facilitate the production of large numbers of complex molecules, allowing for the accumulation of such molecules. When a large group of related molecules accumulates, a complex system arises with some of the characteristics of living organisms. Such a system has a nonrandom organization, tends to reproduce itself, and competes with other systems for the simple organic molecules present in the environment.

One of the most important functions of proteins is **catalysis**, and the catalytic effectiveness of a given enzyme depends on its amino acid sequence. The specific sequence of the amino acids present ultimately determines the properties of all types of proteins, including enzymes. If not for protein catalysis, the chemical reactions that take place in our bodies would be so slow as to be useless for life processes. We discuss this point extensively in Chapters 6 and 7.

In present-day cells, the sequence of amino acids in proteins is determined by the sequence of nucleotides in nucleic acids. The process by which genetic information is translated into the amino acid sequence is very complex. DNA, one of the nucleic acids, serves as the coding material. The **genetic code** is the relationship between the nucleotide sequence in nucleic acids and the amino acid sequence in proteins. As a result of this relationship, the information for the structure and function of all living things is passed from one generation to the next. The workings of the genetic code are no longer completely mysterious, but they are far from completely understood. Theories on the origins of life consider how a coding system might have developed, and new insights in this area could shine some light on the present-day genetic code.

catalysis the process of increasing the rate of chemical reactions

genetic code the information for the structure and function of all living organisms

1-3C Molecules to Cells

▶ Which came first—the catalysts or the hereditary molecules?

Until recently, our understanding of biochemistry led to a "chicken versus the egg" conundrum when we tried to figure out how life originated. If the nucleic acids RNA and DNA are genetic materials that convey the information of heredity and proteins are the molecules that act as catalysts for biochemical reactions, then how did life ever start? Which molecule came first, and how did the other develop?

A discovery with profound implications for discussions of the origin of life is that RNA is capable of catalyzing its own processing. Until this discovery, catalytic activity was associated exclusively with proteins. RNA, rather than DNA, is now considered by many scientists to have been the original coding material, and it still serves this function in some viruses. The idea that catalysis and coding both occur in one molecule has provided a point of departure for more research on the origins of life. The "RNA world" is the current conventional wisdom, but many unanswered questions exist regarding this point of view.

According to the RNA-world theory, the appearance of a form of RNA capable of coding for its own replication was the pivotal point in the origin of life. Polynucleotides can direct the formation of molecules whose sequence is an exact copy of the original. This process depends on a template mechanism (Figure 1.11), which is highly effective in producing exact copies but is a relatively slow process. A catalyst is required, which can be a polynucleotide, even the original molecule itself. Polypeptides, however, are more efficient catalysts than polynucleotides, but there is still the question of whether they can direct the formation of exact copies of themselves. Recall that in present-day cells, the genetic code is based on nucleic acids, and catalysis relies primarily on proteins. How did nucleic acid synthesis (which requires many protein enzymes) and protein synthesis (which requires the genetic code to specify the order of amino acids) come to be?

According to this hypothesis, RNA (or a system of related kinds of RNA) originally played both roles, catalyzing and encoding its own replication. Eventually, the system evolved to the point of being able to encode the synthesis of more effective catalysts, namely proteins (Figure 1.12). Even later, DNA took

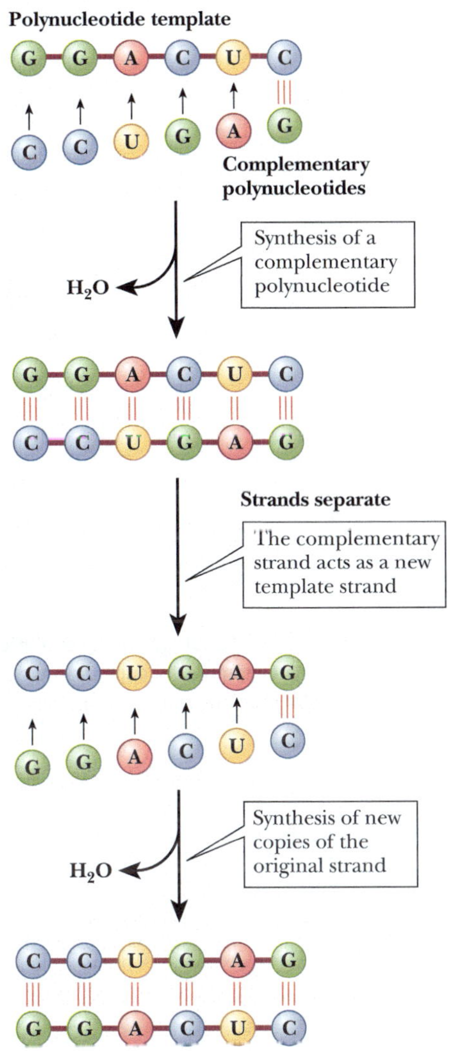

Figure 1.11 The role of templates in the synthesis of polynucleotides. Polynucleotides use a template mechanism to produce exact copies of themselves: G pairs with C and A pairs with U by a relatively weak interaction. The original strand acts as a template to direct the synthesis of a complementary strand. The complementary strand then acts as a template for the production of copies of the original strand. The original strand can be a template for a number of complementary strands, each of which can, in turn, produce a number of copies of the original strand. This process gives rise to a many-fold amplification of the original sequence. *(Copyright © 1994 from* The Molecular Biology of the Cell, *3rd edition, by A. Alberts, D. Bray, J. Lewis, M. Raff, K. Roberts, and J. D. Watson.)*

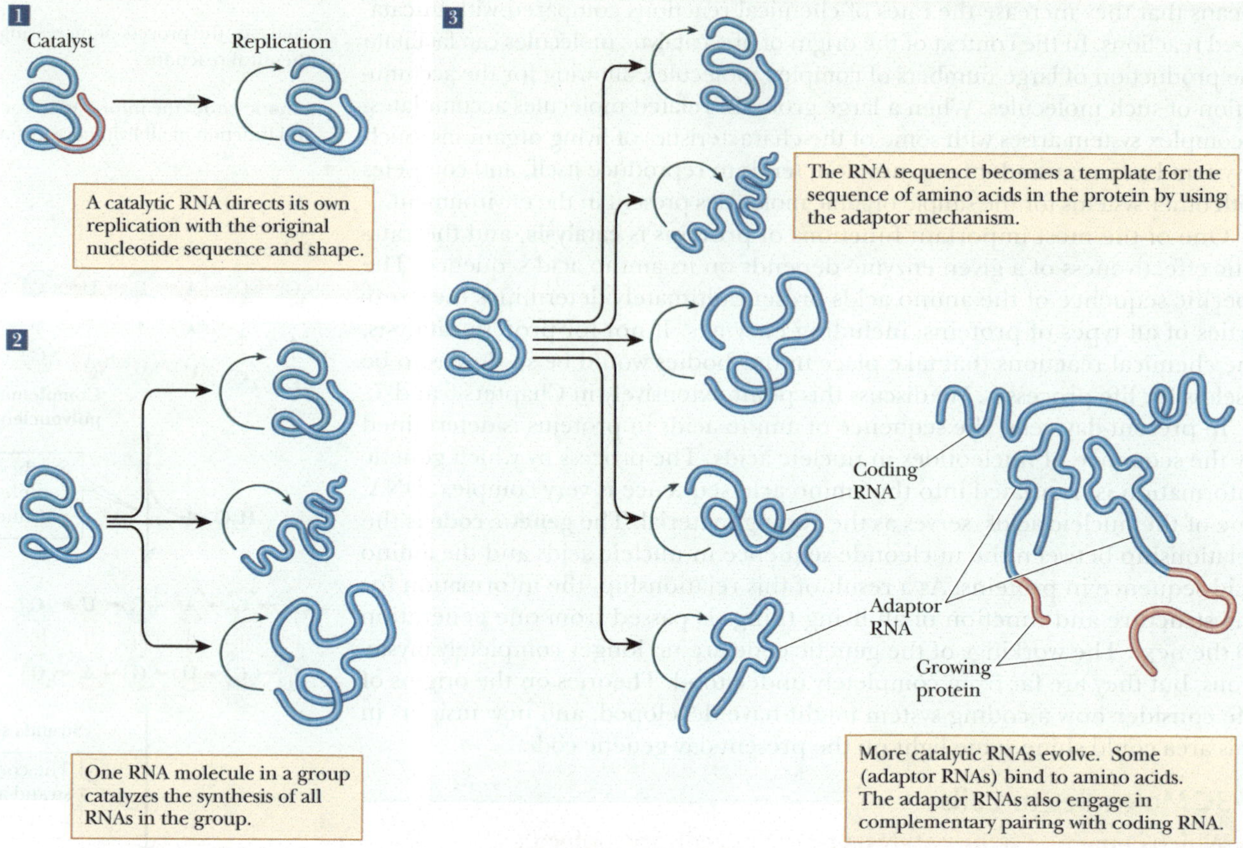

Figure 1.12 **Stages in the evolution of a system of self-replicating RNA molecules according to the RNA-world theory.** At each stage, greater complexity appears in the group of RNAs, leading eventually to the synthesis of proteins as more effective catalysts. *(Copyright © 1994 from* The Molecular Biology of the Cell, *3rd Edition, by A. Alberts, D. Bray, J. Lewis, M. Raff, K. Roberts, and J. D. Watson.)*

over as the primary genetic material, relegating the more versatile RNA to an intermediary role in directing the synthesis of proteins under the direction of the genetic code residing in DNA. A certain amount of controversy surrounds this theory, but it has attracted considerable attention.

Another key point in the development of living cells is the formation of membranes that separate cells from their environment. The clustering of coding and catalytic molecules in a separate compartment brings molecules into closer contact with each other and excludes extraneous material. For reasons we explore in detail in Chapters 2 and 8, lipids are perfectly suited to form cell membranes (Figure 1.13).

Recently, attempts have been made to combine several lines of reasoning about the origin of life into a *double-origin theory*. According to this line of thought, the development of catalysis and the development of a coding system came about separately, and the combination of the two produced life as we know it. The rise of aggregates of molecules capable of catalyzing reactions was one origin of life, and the rise of a nucleic acid–based coding system was another origin.

A theory that life began on clay particles is a form of the double-origin theory. According to this point of view, coding arose first, but the coding material was the surface of naturally occurring clay. The pattern of ions on the clay surface is thought to have served as the code, and the process of crystal growth is thought to have been responsible for replication. Nucleotides, then RNA molecules, formed on the clay surface. The RNA molecules thus formed were released from the clay surface and enclosed in lipid sacs, forming protocells. In this scenario, protocells exist in a pond with a warm side and a cold side. Double-stranded polynucleotides are formed on the cold side of the pond on a single-stranded template (Figure 1.14). The protocell moves to the warm side of the pond, where the strands separate.

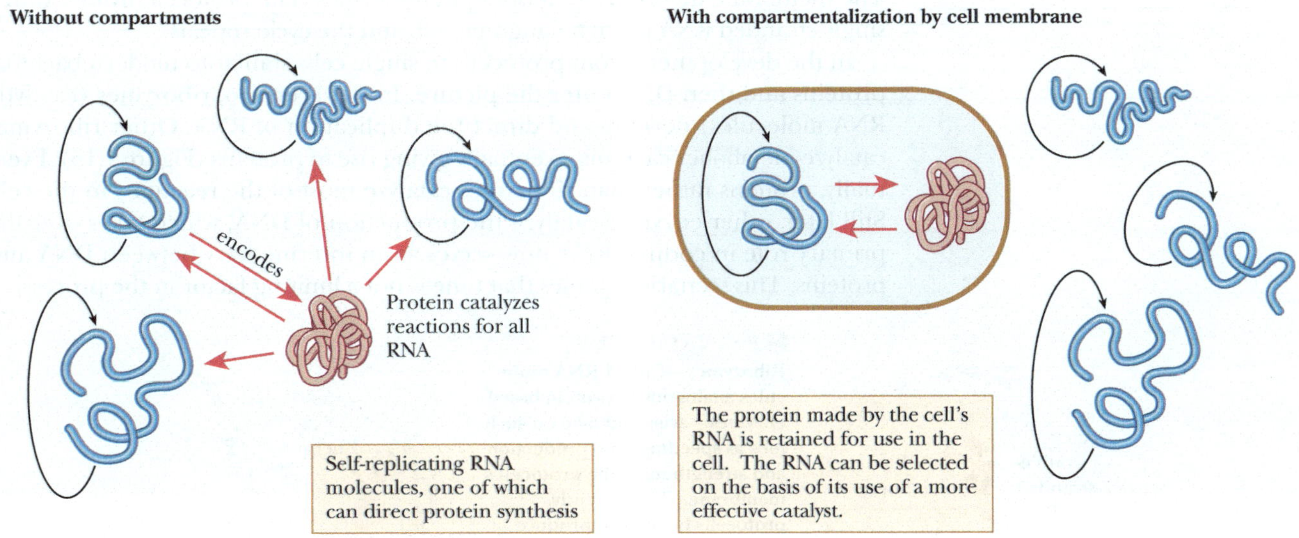

Figure 1.13 The vital importance of a cell membrane in the origin of life. Without compartments, groups of RNA molecules must compete with others in their environment for the proteins they synthesize. With compartments, the RNAs have exclusive access to the more effective catalysts and are closer to each other, making it easier for reactions to take place. *(Copyright © 1994 from* The Molecular Biology of the Cell, *3rd Edition, by A. Alberts, D. Bray, J. Lewis, M. Raff, K. Roberts, and J. D. Watson.)*

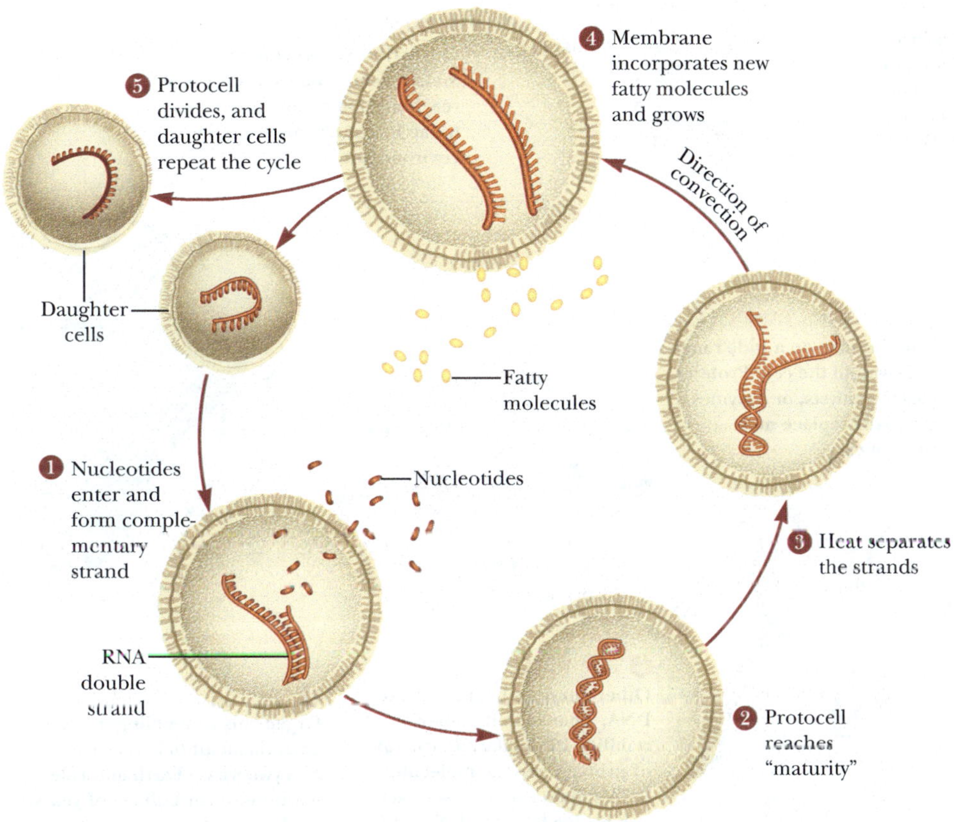

Figure 1.14 Hypothetical beginning of replication. Simple strings of nucleotides could have formed, perhaps initially lined up on the surface of clay particles. (1) On the cold side of the pond, RNA strands become surrounded by simple membranes. Nucleotides enter and form complementary strands by base pairing. (2) Over time, these protocells gain more molecules and complexity. (3) Cells find their way to the warm side of the pond, where the heat allows the RNA molecules to separate. (4) The protocell grows and gains more components. (5) Finally, the cell divides, produces daughter cells, and the process repeats. *(Based on Scientific American, a division of Nature America, Inc.)*

The membrane incorporates more lipid molecules. The protocell divides, with a single-stranded RNA in each daughter cell, and the cycle repeats.

In the development from protocells to single cells similar to modern bacteria, proteins and then DNA enter the picture. In this scenario, ribozymes (catalytic RNA molecules) develop and direct the duplication of RNA. Other ribozymes catalyze metabolic reactions, eventually giving rise to proteins (Figure 1.15). Eventually, proteins rather than ribozymes catalyze most of the reactions in the cell. Still later, other enzymes catalyze the production of DNA, which takes over the primary role in coding. RNA now serves as an intermediary between DNA and proteins. This scenario assumes that time is not a limiting factor in the process.

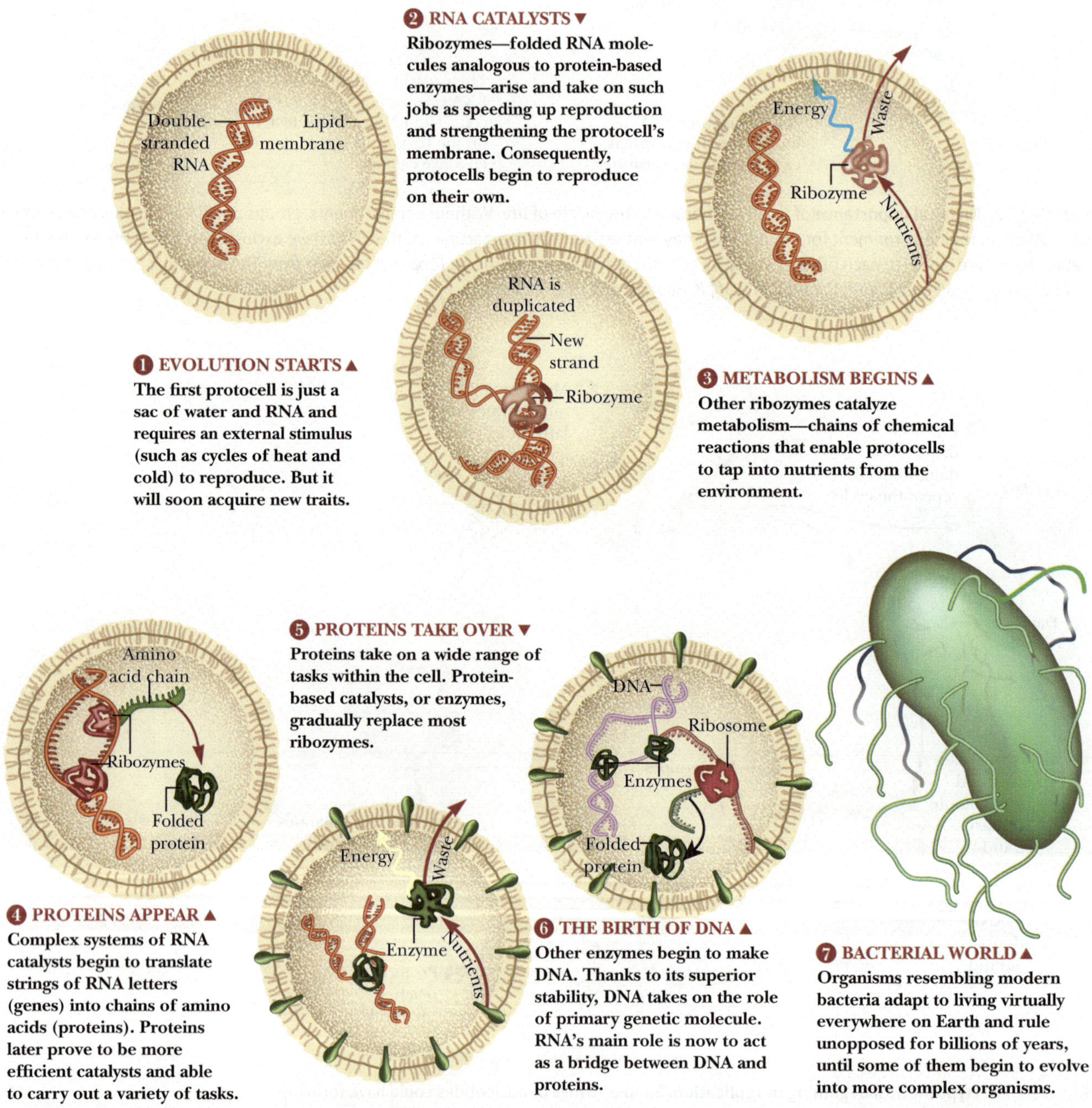

Figure 1.15 Hypothetical evolution of simple protocells to more complex cells. (1) Protocells were just a simple sac containing simple RNA molecules. They could not even replicate without cycling between cold and warm temperatures. (2) In time, the cells started replicating on their own using ribozymes. (3) With increasing complexity of RNA molecules, ribozymes begin to catalyze metabolic pathways. (4) As metabolism grows in complexity, RNA begins to be translated into proteins. Proteins prove to be more efficient catalysts. (5) Proteins gradually take over metabolism, replacing most of the functions of ribozymes. (6) New enzymes start producing DNA, which due to its superior stability, replaces RNA as the primary heredity material. (7) Organisms resembling bacteria evolve all over Earth and rule for a billion years before evolution works to create more complex organisms. *(Based on Scientific American, a division of Nature America, Inc.)*

In an attempt to study the origins of life, scientists have also attempted to combine the best properties of proteins and nucleic acids and have created peptide nucleic acids (PNA). Evidence shows that the building blocks of these hybrids could also have formed in the primordial world, and some theorize that PNA may have been the original molecule that allowed life to form. Currently, scientists are attempting to create artificial living cells based on PNA. The goal is to demonstrate that under the conditions of the "primordial soup," simple molecules could form complex molecules possessing the critical functions of catalysis and replication and that these could then form cells capable of dividing.

As of this writing, none of the theories of the origin of life is definitely established, and none is definitely disproved. The topic is still under active investigation. It seems highly unlikely that we will ever know with certainty how life originated on this planet, but these conjectures allow us to ask some of the important questions, such as those about catalysis and coding, which we are going to see many times in this text.

1-4 The Biggest Biological Distinction: Prokaryotes and Eukaryotes

Learning Objectives

After reading this section, students will be able to:

LO 1.4.1 Compare the parts of a typical animal cell, a typical plant cell, and a typical prokaryotic cell.

LO 1.4.2 Describe the functions of the following organelles: nucleus, mitochondrion, chloroplast, endoplasmic reticulum, Golgi apparatus, lysosomes, peroxisomes, cell membrane, cell wall, central vacuole.

All cells (except mammalian red blood cells) contain DNA. The complete set of genetic material in a cell is called the **genome**. **Genes** are individual units of heredity within a genome. They control individual traits by coding for a functional protein or RNA.

genome the complete set of genetic material in a cell

genes individual DNA units of inheritance

The earliest cells must have been very simple, having the minimum apparatus necessary for life processes. The types of organisms living today that probably most resemble the earliest cells are the **prokaryotes**. This word, of Greek derivation (*karyon*, "kernel, nut"), literally means "before the nucleus." Prokaryotes include *bacteria* and *cyanobacteria.* (Cyanobacteria were formerly called blue-green algae; as the newer name indicates, they are more closely related to bacteria.) Prokaryotes are single-celled organisms, but groups of them can exist in association, forming colonies with some differentiation of cellular functions.

prokaryotes microorganisms that lack a distinct nucleus and membrane-enclosed organelles

▶ What is the difference between a prokaryote and a eukaryote?

The word *eukaryote* means "true nucleus." **Eukaryotes** are more complex organisms and can be multicellular or single celled. A well-defined nucleus, set off from the rest of the cell by a membrane, is one of the chief features distinguishing a eukaryote from a prokaryote. A growing body of fossil evidence indicates that eukaryotes appeared about 1.5 billion (1.5×10^9) years ago, about 2 billion years after life first appeared on Earth. Examples of single-celled eukaryotes include yeasts and *Paramecium* (an organism frequently discussed in beginning biology courses); all multicellular organisms (e.g., animals and plants) are eukaryotes.

eukaryotes organisms whose cells have a well-defined nucleus and membrane-enclosed organelles

Eukaryotic cells are more complex and usually much larger than prokaryotic cells. The diameter of a typical prokaryotic cell is on the order of 1 to 3 μm (1×10^{-6} to 3×10^{-6} m), whereas that of a typical eukaryotic cell is about 10 to 100 μm. The distinction between prokaryotes and eukaryotes is so basic that it is now a key point in the classification of living organisms; it is far more important than the distinction between plants and animals.

The main difference between prokaryotic and eukaryotic cells is the existence of organelles, especially the nucleus, in eukaryotes. An **organelle** is a part of the

organelle a membrane-enclosed portion within a eukaryotic cell with a specific function

cell that has a distinct function; it is surrounded by its own membrane within the cell. In contrast, the structure of a prokaryotic cell is relatively simple, lacking membrane-enclosed organelles. Like a eukaryotic cell, however, a prokaryotic cell has a cell membrane, or plasma membrane, separating it from the outside world. The plasma membrane is the only membrane found in the prokaryotic cell. In both prokaryotes and eukaryotes, the cell membrane consists of a double layer (bilayer) of lipid molecules with a variety of proteins embedded in it.

Organelles have specific functions. A typical eukaryotic cell has a *nucleus* with a nuclear membrane. *Mitochondria* (respiratory organelles) and an internal membrane system known as the *endoplasmic reticulum* are also common to all eukaryotic cells. Energy-yielding oxidation reactions take place in eukaryotic mitochondria. In prokaryotes, similar reactions occur on the plasma membrane. *Ribosomes* (particles consisting of RNA and protein), which are the sites of protein synthesis in all living organisms, are frequently bound to the endoplasmic reticulum in eukaryotes. In prokaryotes, ribosomes are found free in the cytosol. The cytosol is distinct from the cytoplasm: *Cytoplasm* refers to the portion of the cell outside the *nucleus*, whereas the *cytosol* is the aqueous portion of the cell that lies outside the *membrane-bounded organelles*. *Chloroplasts*, organelles in which photosynthesis takes place, are found in plant cells and green algae. In prokaryotes that are capable of photosynthesis, the reactions take place in layers called *chromatophores*, which are extensions of the plasma membrane, rather than in chloroplasts.

Table 1.3 summarizes the basic differences between prokaryotic and eukaryotic cells.

Table 1.3 A Comparison of Prokaryotes and Eukaryotes

Organelle	Prokaryotes	Eukaryotes
Nucleus	No definite nucleus; DNA present but not separate from rest of cell	Present
Cell membrane (plasma membrane)	Present	Present
Mitochondria	None; enzymes for oxidation reactions located on plasma membrane	Present
Endoplasmic reticulum	None	Present
Ribosomes	Present	Present
Chloroplasts	None; photosynthesis (if present) is localized in chromatophores	Present in green plants

1-4A Prokaryotic Cells

Although no well-defined nucleus is present in prokaryotes, the DNA of the cell is concentrated in the **nuclear region**. This part of the cell directs the workings of the cell, much as the eukaryotic nucleus does.

nuclear region the portion of a prokaryotic cell that contains the DNA

▶ How is prokaryotic DNA organized without a nucleus?

The DNA of prokaryotes is not complexed with proteins in extensive arrays with specified architecture, as is the DNA of eukaryotes. In general, there is only a single, closed, circular molecule of DNA in prokaryotes. This circle of DNA, which is the genome, is attached to the cell membrane. Before a prokaryotic cell divides, the DNA replicates itself, and both DNA circles are bound to the plasma membrane. The cell then divides, and each of the two daughter cells receives one copy of the DNA (Figure 1.16).

In a prokaryotic cell, the cytosol (the fluid portion of the cell outside the nuclear region) frequently has a slightly granular appearance because of the presence of **ribosomes**. Because ribosomes consist of RNA and protein, they are

ribosomes the sites of protein synthesis in all organisms, consisting of RNA and protein

also called *ribonucleoprotein particles*; they are the sites of protein synthesis in all organisms. The presence of ribosomes is the main visible feature of prokaryotic cytosol. (Membrane-bound organelles, characteristic of eukaryotes, are not found in prokaryotes.)

Every cell is separated from the outside world by a **cell membrane**, or plasma membrane, an assemblage of lipid molecules and proteins. In addition to the cell membrane and external to it, a prokaryotic bacterial cell has a **cell wall**, which is made up mostly of polysaccharide material, a feature it shares with eukaryotic plant cells. The chemical natures of prokaryotic and eukaryotic cell walls differ somewhat, but a common feature is that the polymerization of sugars produces the polysaccharides found in both. Because the cell wall is made up of rigid material, it presumably serves as protection for the cell.

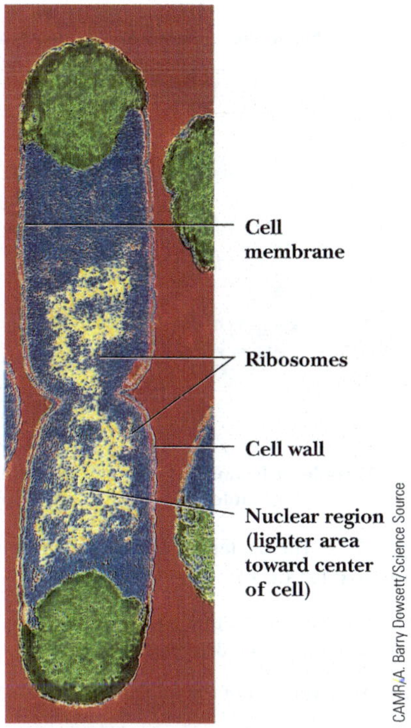

Figure 1.16 Electron micrograph of a bacterium (magnified 16,500×).
A colored electron microscope image of the bacterium *Escherichia coli*, a typical prokaryote. The pair in the center shows that division into two cells is nearly complete.

cell membrane the outer membrane of the cell that separates it from the outside world

cell wall the outer coating of bacterial and plant cells

chloroplasts organelles that are the sites of photosynthesis in green plants

1-4B Eukaryotic Cells

Multicellular plants and animals are eukaryotes, as are protista and fungi, but differences exist among them. These differences are reflected on the cellular level. One of the biggest differences between eukaryotes and prokaryotes is the presence of subcellular organelles.

Three of the most important organelles in eukaryotic cells are the nucleus, the mitochondrion, and the chloroplast. Each is separated from the rest of the cell by a double membrane. The nucleus contains most of the DNA of the cell and is the site of RNA synthesis. The mitochondria contain enzymes that catalyze important energy-yielding reactions.

Chloroplasts, which are found in green plants and green algae, are the sites of photosynthesis. Both mitochondria and chloroplasts contain DNA that differs from that found in the nucleus, and both carry out transcription and protein synthesis distinct from that directed by the nucleus.

Plant cells, like bacteria, have cell walls. A plant cell wall is mostly made up of the polysaccharide cellulose, giving the cell its shape and mechanical stability. **Chloroplasts**, the photosynthetic organelles, are found in green plants and algae. Animal cells have neither cell walls nor chloroplasts; the same is true of some protists. Figure 1.17 shows some of the important differences between typical plant cells, typical animal cells, and prokaryotes.

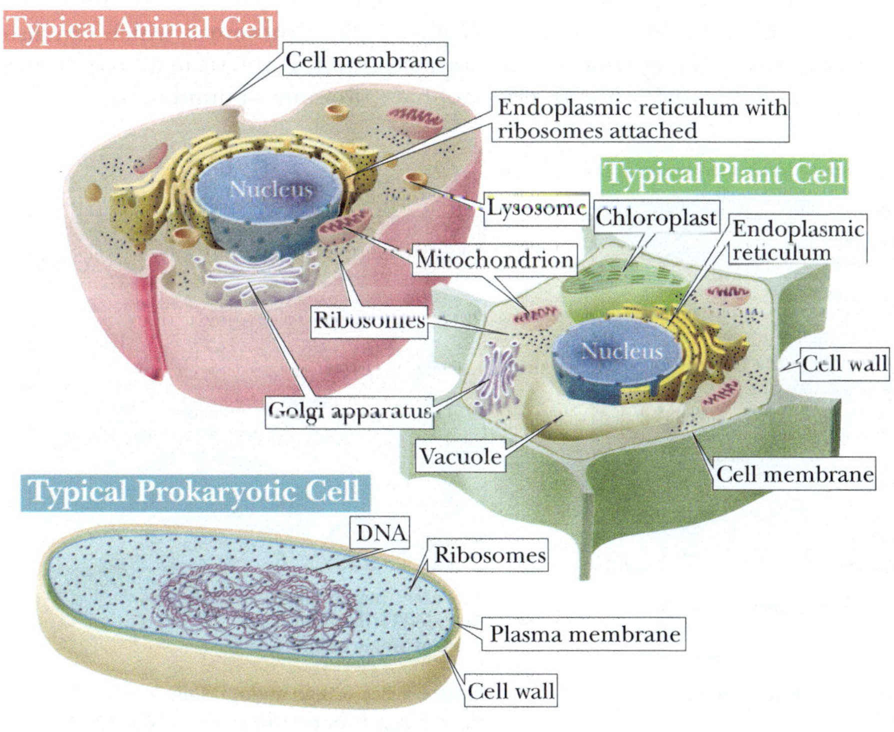

Figure 1.17 A comparison of a typical animal cell, a typical plant cell, and a typical prokaryotic cell.

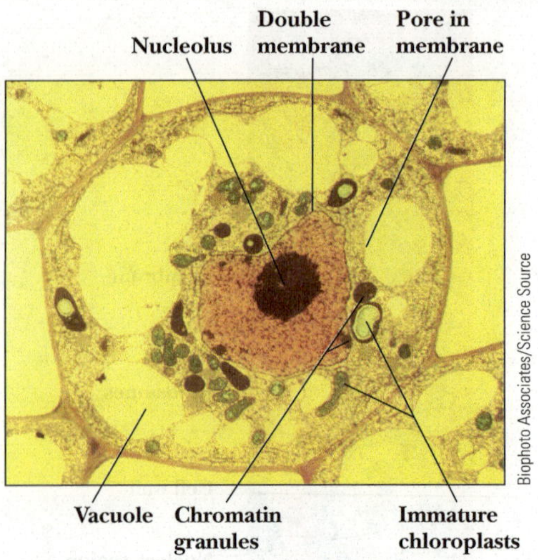

Figure 1.18 The nucleus of a tobacco leaf cell (magnified 15,000×).

nucleus the organelle that contains the main genetic apparatus in eukaryotes

nucleolus a portion of the nucleus rich in RNA

chromatin a complex of DNA and protein found in eukaryotic nuclei

chromosomes linear structures that contain the genetic material and associated proteins

mitochondrion an organelle that contains the apparatus responsible for aerobic oxidation of nutrients

cristae folds in the inner mitochondrial membrane

matrix the part of a mitochondrion enclosed within the inner mitochondrial membrane

endoplasmic reticulum a continuous single-membrane system throughout the cell

▶ What are the most important organelles?

The **nucleus** is perhaps the most important eukaryotic organelle. A typical nucleus exhibits several important structural features (Figure 1.18). It is surrounded by a *nuclear double membrane* (usually called the nuclear envelope). One of its prominent features is the **nucleolus**, which is rich in RNA. The RNA of a cell (with the exception of the small amount produced in such organelles as mitochondria and chloroplasts) is synthesized on a DNA template in the nucleolus for export to the cytoplasm through pores in the nuclear membrane. This RNA is ultimately destined for the ribosomes.

Also visible in the nucleus, frequently near the nuclear membrane, is **chromatin**, an aggregate of DNA and protein. The main eukaryotic genome (its nuclear DNA) is duplicated before cell division takes place, as in prokaryotes. In eukaryotes, both copies of DNA, which are to be equally distributed between the daughter cells, are associated with protein. When a cell is about to divide, the loosely organized strands of chromatin become tightly coiled, and the resulting **chromosomes** can be seen under a microscope. The genes, responsible for the transmission of inherited traits, are part of the DNA found in each chromosome.

A second very important eukaryotic organelle is the **mitochondrion**, which, like the nucleus, has a double membrane (Figure 1.19). The outer membrane has a fairly smooth surface, but the inner membrane exhibits many folds called **cristae**. The space within the inner membrane is called the **matrix**. Oxidation processes that occur in mitochondria yield energy for the cell. Most of the enzymes responsible for these important reactions are associated with the inner mitochondrial membrane. Other enzymes needed for oxidation reactions, as well as DNA that differs from that found in the nucleus, are found in the internal mitochondrial matrix. Mitochondria also contain ribosomes similar to those found in bacteria. Mitochondria are approximately the size of many bacteria, typically about 1 μm in diameter and 2 to 8 μm in length. They may have arisen from the absorption of aerobic bacteria by larger host cells.

The **endoplasmic reticulum** is part of a continuous single-membrane system throughout the cell; the membrane doubles back on itself to give the appearance of a double membrane in electron micrographs. The endoplasmic reticulum is attached to the cell membrane and to the nuclear membrane. It occurs in two forms: rough and smooth. The *rough endoplasmic reticulum* is studded with ribosomes bound to the membrane (Figure 1.20). Ribosomes, which can also be found free in the cytosol, are the sites of protein synthesis in all organisms. The *smooth endoplasmic reticulum* does not have ribosomes bound to it.

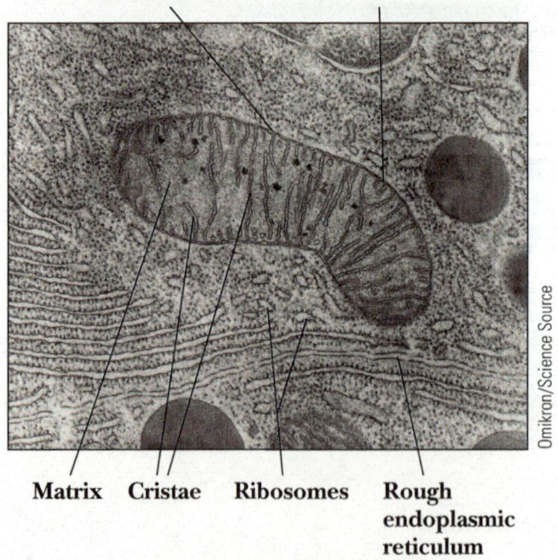

Figure 1.19 Mouse liver mitochondria (magnified 50,000×).

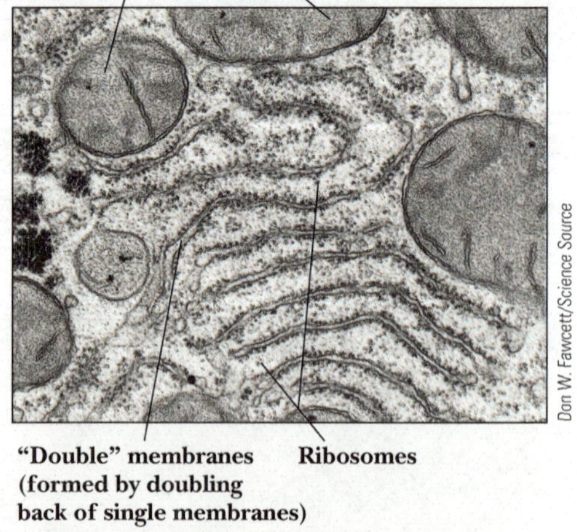

Figure 1.20 Rough endoplasmic reticulum from mouse liver cells (magnified 50,000×).

Chloroplasts are important organelles found only in green plants and green algae. Their structure includes membranes, and they are relatively large, typically up to 2 μm in diameter and 5 to 10 μm in length. The photosynthetic apparatus is found in specialized structures called *grana* (singular *granum*), membranous bodies stacked within the chloroplast. Grana are visible in the electron microscope (see Figure 1.21). Chloroplasts, like mitochondria, contain a characteristic DNA that is different from that found in the nucleus. Chloroplasts and mitochondria also contain ribosomes similar to those found in bacteria.

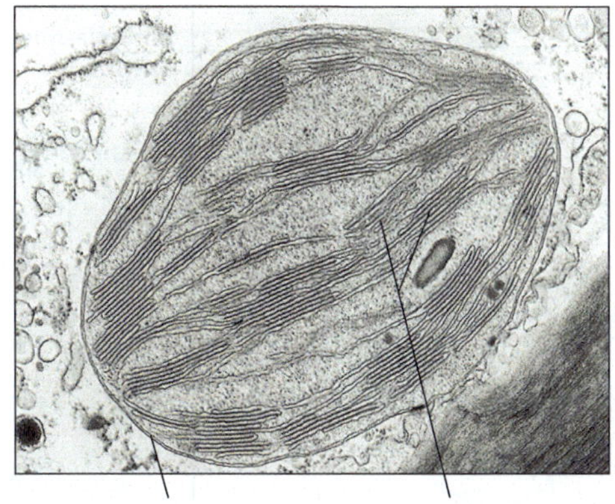

Biophoto Associates/Science Source

Figure 1.21 An electron microscope image of a chloroplast from the alga *Nitella* (magnified 60,000×).

▶ What are some other important components of cells?

Membranes are important to the structures of some less well-understood organelles. One, the **Golgi apparatus**, is separate from the endoplasmic reticulum but is frequently found close to the smooth endoplasmic reticulum. It is a series of membranous sacs (Figure 1.22). The Golgi apparatus secretes proteins from the cell, but it also appears in cells in which its primary function is not protein secretion. In particular, it is the site in the cell in which sugars are linked to other cellular components, such as proteins. The function of this organelle is still a subject of research.

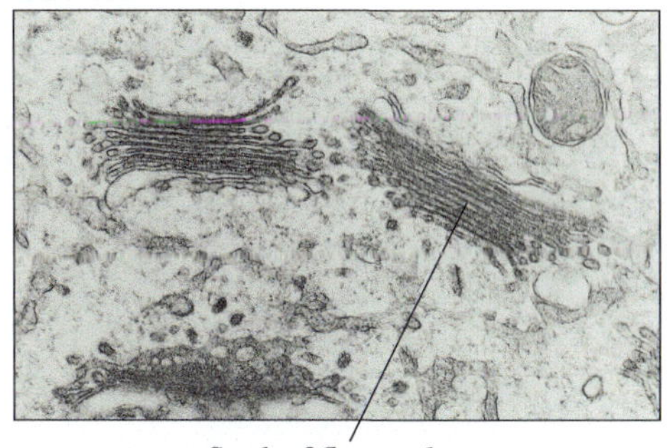

Don W. Fawcett/Science Source

Figure 1.22 Golgi apparatus from a mammalian cell (magnified 25,000×).

Other organelles in eukaryotes are similar to the Golgi apparatus in that they involve single, smooth membranes and have specialized functions. **Lysosomes**, for example, are membrane-enclosed sacs containing hydrolytic enzymes that could cause considerable damage to the cell were they not physically separated from the lipids, proteins, or nucleic acids that they are able to attack. Inside the lysosome, these enzymes break down target molecules, usually from outside sources, as a first step in processing nutrients for the cell. **Peroxisomes** are similar to lysosomes; their principal characteristic is that they contain enzymes involved in the metabolism of hydrogen peroxide (H_2O_2), which is toxic to the cell. The enzyme *catalase*, which occurs in peroxisomes, catalyzes the conversion of H_2O_2 to H_2O and O_2.

Glyoxysomes are found in plant cells only. They contain the enzymes that catalyze the *glyoxylate cycle*, a pathway that converts some lipids to carbohydrate with glyoxylic acid as an intermediate (Section 19-6).

The **cytosol** was long considered nothing more than a viscous liquid, but recent studies with electron microscopy have revealed that this part of the cell has some internal organization. The organelles are held in place by a lattice of fine strands that seem to consist mostly of protein. This **cytoskeleton (microtrabecular lattice)** is connected to all organelles (Figure 1.23). Many questions remain about its function in cellular organization, but its importance in maintaining the infrastructure of the cell is certain.

The cell membrane of eukaryotes separates the cell from the outside world. It consists of a double layer of lipids, with several types of proteins embedded in the lipid matrix. Some of the proteins transport specific substances across the membrane barrier. Transport can take place in both directions, with substances useful to the cell being taken in and others being exported.

Plant cells (and algae)—but not animal cells—have cell walls external to the plasma membrane. The cellulose that makes up plant cell walls is a major component of plant material; wood, cotton, linen, and most types of paper are mainly cellulose. Also present in plant cells are large central **vacuoles**, sacs in the cytoplasm surrounded by a single membrane. Although vacuoles sometimes appear in animal cells, those in plants are more prominent. They tend to increase in number and size as the plant cell ages. An important function of vacuoles is to isolate and dispose waste substances that are toxic to the plant.

These waste products may be unpalatable or even poisonous enough to discourage herbivores (plant-eating organisms) from ingesting them and may thus provide some protection for the plant.

Golgi apparatus a cytoplasmic organelle that consists of flattened membranous sacs, usually involved in protein secretion

lysosomes membrane-enclosed organelles that contain hydrolytic enzymes

peroxisomes membrane-bounded sacs that contain enzymes involved in the metabolism of hydrogen peroxide (H_2O_2)

glyoxysomes membrane-enclosed organelles that contain the enzymes of the glyoxylate cycle

cytosol the portion of the cell that lies outside the nucleus and the other membrane-enclosed organelles

cytoskeleton (microtrabecular lattice) a lattice of fine strands, consisting mostly of protein, that pervades the cytosol

vacuoles cavities within the cytoplasm of a cell, typically enclosed by a single membrane, that may serve secretory, excretory, or storage functions

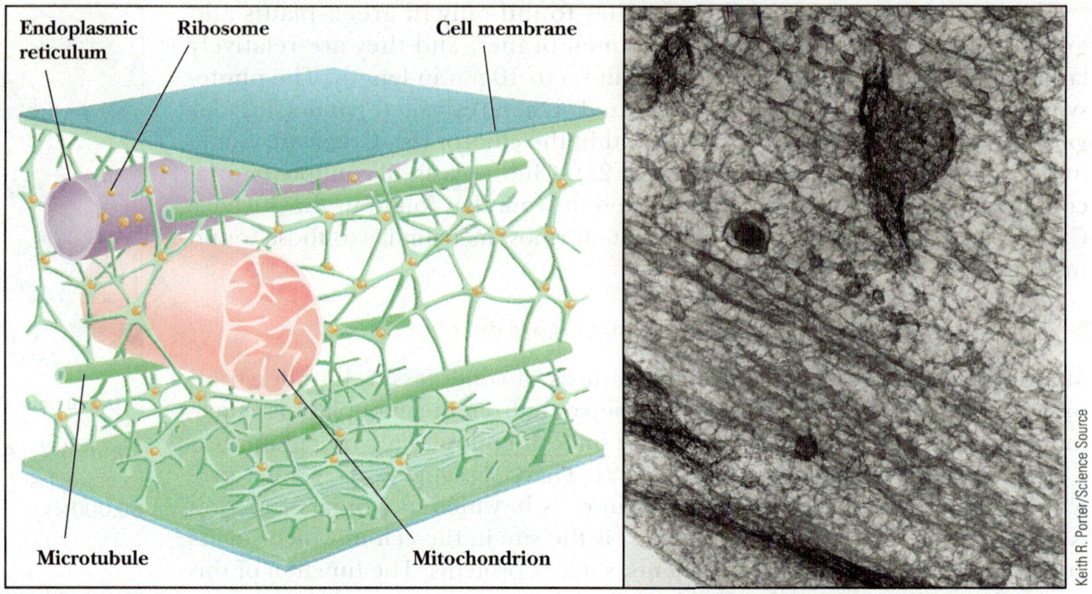

A This network of filaments, also called the cytoskeleton, pervades the cytosol. Some filaments, called microtubules, are known to consist of the protein tubulin. Organelles such as mitochondria are attached to the filaments.

B An electron micrograph of the microtrabecular lattice.

Figure 1.23 The microtrabecular lattice.

Table 1.4 summarizes organelles and their functions.

Table 1.4 A Summary of Organelles and Their Functions

Organelle	Function
Nucleus	Location of main genome; site of most DNA and RNA synthesis
Mitochondrion	Site of energy-yielding oxidation reactions; has its own DNA
Chloroplast	Site of photosynthesis in green plants and algae; has its own DNA
Endoplasmic reticulum	Continuous membrane throughout the cell; rough part studded with *ribosomes* (*the site of protein synthesis*)*
Golgi apparatus	Series of flattened membranes; involved in secretion of proteins from cells and in reactions that link sugars to other cellular components
Lysosomes	Membrane-enclosed sacs containing hydrolytic enzymes
Peroxisomes	Sacs that contain enzymes involved in the metabolism of hydrogen peroxide
Cell membrane	Separates the cell contents from the outside world; contents include organelles (held in place by the *cytoskeleton**) and the *cytosol*
Cell wall	Rigid exterior layer of plant cells
Central vacuole	Membrane-enclosed sac (plant cells) that may serve secretory, excretory, or storage functions

*Because an organelle is defined as a portion of a cell enclosed by a membrane, ribosomes are not, strictly speaking, organelles. Smooth endoplasmic reticulum does not have ribosomes attached, and ribosomes also occur free in the cytosol. The definition of *organelle* also affects the discussion of the cell membrane, cytosol, and cytoskeleton.

1-5 How We Classify Eukaryotes and Prokaryotes

Learning Objectives

After reading this section, students will be able to:

LO 1.5.1 Describe the five-kingdom classification scheme of living organisms based on the names of the five kingdoms, the relative complexities of the five kingdoms, and the relative evolutionary ages of the five kingdoms.

LO 1.5.2 Explain theories related to the development of eukaryotic cells.

Living organisms can be classified in many possible ways. The original biological classification scheme, established in the eighteenth century, divided all organisms into two kingdoms: the plants and the animals. In this scheme, plants are organisms that obtain food directly from the Sun, whereas animals are organisms that move about to search for food. Other ways to organize life include whether the organism had a cell wall or whether the organism was single celled.

With the original classification, researchers discovered that some organisms, bacteria in particular, do not have an obvious relationship to either kingdom. It has also become clear that a more fundamental division of living organisms is actually not between plants and animals but between prokaryotes and eukaryotes.

▶ How do scientists classify living organisms today?

In the twentieth century, classification schemes that divided living organisms into more than the two traditional kingdoms were introduced.

1-5A Five-Kingdom Classification System

The five-kingdom classification system (Figure 1.24) takes into account the differences between prokaryotes and eukaryotes, and it also provides classifications for eukaryotes that appear to be neither plants nor animals.

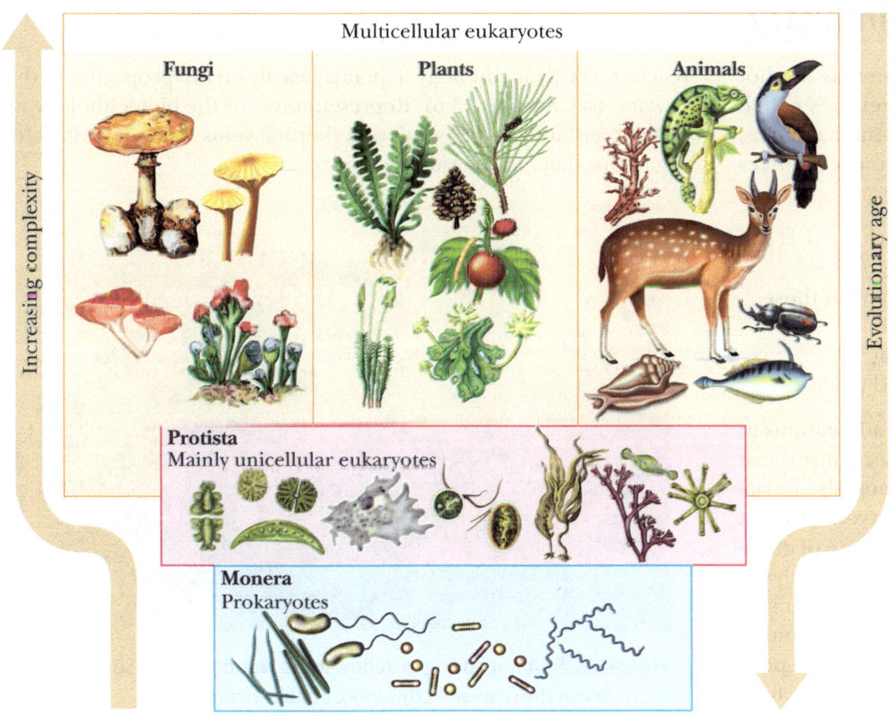

Figure 1.24 The five-kingdom classification scheme of living organisms.

Monera one of the five kingdoms used to classify living organisms; it includes prokaryotes

Protista one of the five kingdoms used to classify living organisms; it includes single-celled eukaryotes

archaebacteria prokaryotes having unique biochemical features; they live in extreme environments

eubacteria the typical bacteria of the present-day environment

The kingdom **Monera** consists only of prokaryotic organisms. Bacteria and cyanobacteria are members of this kingdom. The other four kingdoms are made up of eukaryotic organisms. The kingdom **Protista** includes unicellular organisms such as *Euglena, Volvox, Amoeba,* and *Paramecium.* Some protists, including algae, are multicellular. The three kingdoms that consist mainly of multicellular eukaryotes (with a few unicellular eukaryotes) are Fungi, Plantae, and Animalia. The kingdom Fungi includes yeasts, molds, and mushrooms. In theory, Fungi, plants, and animals evolved from simpler eukaryotic ancestors, but the major evolutionary change was the development of eukaryotes from prokaryotes.

One group of organisms can be classified as prokaryotes in the sense that the organisms lack a well-defined nucleus. These organisms are called **archaebacteria** (early bacteria) to distinguish them from **eubacteria** (true bacteria) because there are marked differences between the two kinds of organisms. Archaebacteria are found in extreme environments (see Biochemical Connections 1A) and, for this reason, are also called extremophiles. Most of the differences between archaebacteria and other organisms are biochemical features, such as the molecular structure of the cell walls, membranes, and some types of RNA.

1-5B Three-Domain Classification System

The three-domain system for classifying organisms is just an alternative way of classifying all life based on looking at the evolutionary relationships and when different life forms diverged. Specifically, this system uses the sequence of ribosomal RNA of each species. In this system, there are two divisions of prokaryotes—bacteria and archaea—and then all eukaryotes. Bacteria are the most common and are found in normal habitats, like soils, surface waters, and tissues of other organisms. Archaea are prokaryotes that inhabit extreme environments, like

1A Biochemical Connections • Biotechnology

Extremophiles: The Toast of the Industry

Archaebacteria live in extreme environments, such as the hot spring shown in Figure 1.25, so they are sometimes called extremophiles. The three groups of archaebacteria—methanogens, halophiles, and thermoacidophiles—have specific preferences about the precise nature of their environment:

- *Methanogens* are strict anaerobes that produce methane (CH_4) from carbon dioxide (CO_2) and hydrogen (H_2).
- *Halophiles* require very high salt concentrations, such as those found in the Dead Sea, for growth.
- *Thermoacidophiles* require high temperatures and acidic conditions for growth—typically, 80°C to 90°C and pH 2.

These growth requirements may have resulted from adaptations to harsh conditions on the early Earth. Because these organisms can tolerate these conditions, the enzymes they produce must also be stable. Most enzymes isolated from eubacteria and eukaryotes are not stable under such conditions. Some of the reactions that are of greatest importance to the biotechnology industry are enzyme catalyzed and carried out under conditions that cause most enzymes to lose their catalytic ability in a short time. This difficulty can be avoided by using enzymes from extremophiles. An example is the DNA polymerase from *Thermus aquaticus* (Taq polymerase). Polymerase chain reaction (PCR) technology depends heavily on the properties of this enzyme (see Section 13-6). Representatives of the biotechnology industry constantly search undersea thermal vents and hot springs for organisms that can provide such enzymes. ◗

iStockphoto.com/Stephan Hoerold

Figure 1.25 **A hot spring in Yellowstone National Park.** Some bacteria can thrive even in this inhospitable environment.

deep sea vents and salt lakes. In the three-domain classification system, eukaryotes are everything else that has a true nucleus and all the other subcellular organelles.

Nigel Cattlin/Alamy Stock Photo

Figure 1.26 The root system of a leguminous plant. Leguminous plants live symbiotically with nitrogen-fixing bacteria in their root systems.

1-5C Eukaryotic Origins

Given the complexity of eukaryotes, how did such cells arise from simpler progenitors? Symbiosis plays a large role in current theories of the rise of eukaryotes: The symbiotic association between two organisms is seen as giving rise to a new organism that combines characteristics of both the original ones. The type of symbiosis called *mutualism* is a relationship that benefits both species involved, as opposed to *parasitic symbiosis,* in which one species gains at the other's expense. A classic example of mutualism (although it has been questioned from time to time) is the lichen, which consists of a fungus and an alga. The fungus provides water and protection for the alga, whereas the alga is photosynthetic and provides food for both partners. Another example is the root-nodule system formed by a leguminous plant, such as alfalfa or beans, and anaerobic nitrogen-fixing bacteria (Figure 1.26). The plant gains useful compounds of nitrogen, and the bacteria are protected from oxygen, which is harmful to them.

Yet another example of mutualistic symbiosis, of great practical interest, is that between humans and bacteria, such as *Escherichia coli,* that live in the intestinal tract. The bacteria receive nutrients and protection from their immediate environment. In return, they aid our digestive process. Without beneficial intestinal bacteria, we would soon develop dysentery and other intestinal disorders. These bacteria are also a source of certain vitamins for us, because they can synthesize these vitamins and we cannot. The disease-causing strains of *E. coli* that appear in the news from time to time differ markedly from the ones that naturally inhabit the intestinal tract.

▶ Did symbiosis play a role in the development of eukaryotes?

endosymbiosis a symbiotic relationship in which a smaller organism is completely contained within a larger organism

In hereditary symbiosis, a larger host cell contains a genetically determined number of smaller organisms. An example is the protist *Cyanophora paradoxa,* a eukaryotic host that contains a genetically determined number of cyanobacteria (blue-green algae). This relationship is an example of **endosymbiosis,** because the cyanobacteria are contained within the host organism. The cyanobacteria are aerobic prokaryotes and are capable of photosynthesis (Figure 1.27). The host cell gains the products of photosynthesis; in return, the cyanobacteria are protected from the environment and still have access to oxygen and sunlight because of the host's small size. In this model, over the passage of many generations, the cyanobacteria would have gradually lost the ability to exist independently and would have become organelles within a new and more complex type of cell. Such a situation in the past may well have given rise to chloroplasts, which are not capable of independent existence. Their autonomous DNA and their apparatus for synthesizing ribosomal proteins can no longer meet all their needs, but the fact that these organelles have their own DNA and are capable of protein synthesis suggests that they may have existed as independent organisms in the distant past.

A similar model can be proposed for the origin of mitochondria. Imagine a large anaerobic host cell assimilating a number of smaller aerobic bacteria. The larger cell protects the smaller ones and provides them with nutrients. As in the example we used for the development of chloroplasts, the smaller cells still have access to oxygen. The larger cell is not itself capable of aerobic oxidation of nutrients, but some of the end products of its anaerobic oxidation can be further oxidized by the more efficient aerobic metabolism of the smaller cells. As a result, the larger cell can get more energy out of a given amount of

iStockphoto.com/Jouke van der Meer

Figure 1.27 Stromatolite fossils. Stromatolites are large, stony, cushion like masses composed of numerous layers of cyanobacteria (blue-green algae) that have been preserved because of their ability to secrete calcium carbonate. They are among the oldest organic remains to have been found. This specimen dates from around 2400 million years ago. Stromatolite formation reached a peak during the late Precambrian period (4600 million to 570 million years ago) but is still occurring today. This specimen was found in Argentina.

food than it could without the bacteria. In time, the two associated organisms could evolve to form a new aerobic organism, which contains mitochondria derived from the original aerobic bacteria.

The fact that both mitochondria and chloroplasts have their own DNA is an important piece of biochemical evidence in favor of this model. Additionally, both mitochondria and chloroplasts have their own apparatus for the synthesis of RNA and proteins. The genetic code in mitochondria differs slightly from that found in the nucleus, which supports the idea of an independent origin. Thus, the remains of these systems for the synthesis of RNA and protein could reflect the organelles' former existence as free-living cells. It is possible to conclude that large unicellular organisms that assimilated aerobic bacteria went on to evolve mitochondria from the bacteria and eventually gave rise to animal cells. Other types of unicellular organisms might have assimilated both aerobic bacteria and cyanobacteria and evolved both mitochondria and chloroplasts; such organisms could have given rise to green plants.

The proposed connections between prokaryotes and eukaryotes are not established with complete certainty, and they leave a number of questions unanswered. Still, they provide an interesting frame of reference from which to consider evolution and the origins of the reactions that take place in cells.

1-6 Biochemical Energetics

Learning Objectives

After reading this section, students will be able to:

LO 1.6.1 Distinguish between *spontaneous* and *nonspontaneous thermodynamic processes*.

LO 1.6.2 Relate the terms *exergonic*, *endergonic*, and *equilibrium* to changes in Gibbs energy (ΔG) values.

LO 1.6.3 Distinguish between *change in enthalpy* (ΔH) and *change in entropy* (ΔS).

▶ What is the source of energy in life processes?

reduction the gain of electrons

oxidation the loss of electrons

All cells require energy to survive. Many reactions that take place in the cell, particularly those involving the synthesis of large molecules, cannot take place unless energy is supplied. The Sun is the ultimate source of energy for all life on Earth. Photosynthetic organisms trap light energy and use it to drive the energy-requiring reactions that convert carbon dioxide and water to carbohydrates and oxygen. These reactions involve the chemical process of **reduction**. Nonphotosynthetic organisms, such as animals that consume these carbohydrates, use them as energy sources. The reactions that release energy involve the chemical process of **oxidation**. We discuss the roles that oxidation and reduction reactions play in cellular processes in Chapter 15, and we introduce many examples of such reactions in subsequent chapters. For the moment, it is useful and sufficient to recall from general chemistry that *oxidation is the loss of electrons and reduction is the gain of electrons.*

1-6A Thermodynamic Principles

thermodynamics the study of chemical transformations and transfer of energy

One of the most important questions about any process is whether the principle governing energy changes favors the process. **Thermodynamics** is the branch of science that deals with this question. The key point is that *processes that release energy are favored.* Conversely, processes that require energy are disfavored. The change in energy depends only on the state of the molecules present at the start of the process and the state of those present at the end of the process. This is true whether the process in question is the formation or breaking of a bond, the formation or disruption of an intermolecular interaction, or any possible process that requires or can release energy. We discuss these points in some detail when we look at protein folding in Chapter 4 and at energy considerations in metabolism in Chapter 15. This material is of central importance, and what we say about thermodynamics now should help you apply it in later chapters.

The hydrolysis of ATP, shown in Figure 1.28, accompanies many biochemical processes. This reaction releases energy (30.5 kJ/mol ATP = 7.3 kcal/mol ATP). More to the point, the energy released by this reaction allows energy-requiring reactions to proceed. Many ways are available to express energy transfer. One of the most common is the Gibbs energy, *G*, which is discussed in general chemistry. Also recall from general chemistry that a lowering (release) of energy leads to a more stable state of the system under consideration. The lowering of energy is frequently shown in pictorial form as being analogous to an object rolling down a hill (Figure 1.29) or over a waterfall. This representation calls on common experience and aids understanding.

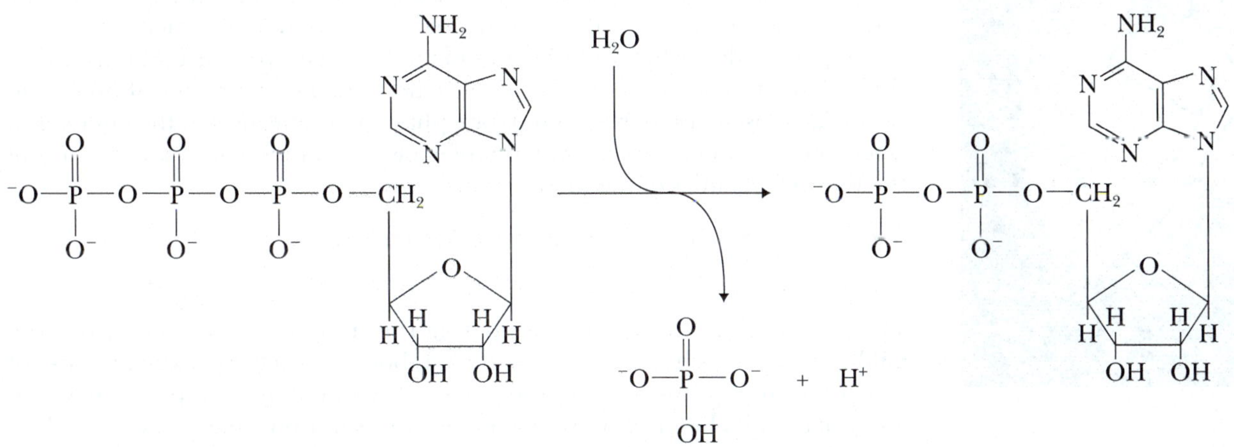

ATP (adenosine triphosphate) **Inorganic phosphate (P_i)** **ADP (adenosine diphosphate)**

Figure 1.28 Hydrolysis of adenosine triphosphate.

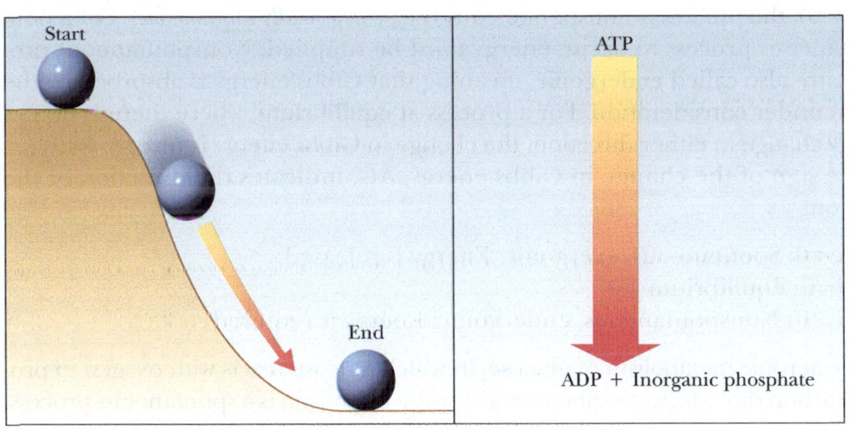

A A ball rolls down a hill, releasing potential energy.

B Adenosine triphosphate (ATP) is hydrolyzed to produce adenosine diphosphate (ADP) and inorganic phosphate, releasing energy. The release of energy when a ball rolls down a hill is analogous to the release

Figure 1.29 A schematic representation of the lowering of energy.

stephan kerkhofs/Shutterstock.com

A

Visuals Unlimited/Corbis

B

Figure 1.30 Two examples of transformations of energy in biological systems. (A) This electric ray (a marine fish in the family Torpedinidae) converts chemical energy to electrical energy, and (B) the bioluminescent dinoflagellate Pyrodinium converts chemical energy into light energy.

1-6B Energy Changes

▶ What kinds of energy changes take place in living cells?

Energy can take several forms, and it can be converted from one form to another, but it cannot be destroyed. All living organisms require and use energy in varied forms; for example, motion involves mechanical energy, and the maintenance of body temperature uses thermal energy. Photosynthesis requires light energy from the Sun. Some organisms, such as several species of fish, are striking examples of the use of chemical energy to produce electrical energy (Figure 1.30). The formation and breakdown of biomolecules involve changes in chemical energy.

Any process that takes place with no outside intervention is **spontaneous** in the specialized sense of the term used in thermodynamics. *Spontaneous does not mean "fast"; some spontaneous processes can take a long time to occur.* We can also use the term *energetically favorable* to indicate spontaneous processes. The laws of thermodynamics can be used to predict whether any change involving transformations of energy will take place. An example of such a change is a chemical reaction in which covalent bonds are broken and new ones are formed. Another example is the formation of noncovalent interactions, such as hydrogen bonds, or hydrophobic interactions, when proteins fold to produce their characteristic three-dimensional structures. The tendency of polar and nonpolar substances to exist in separate phases is a reflection of the energies of interaction between the individual molecules—in other words, a reflection of the thermodynamics of the interaction.

1-6C Spontaneity in Biochemical Reactions

▶ How can we predict what reactions will happen in living cells?

The most useful criterion for predicting the spontaneity of a process is the **Gibbs energy**, which is indicated by the symbol G. (Strictly speaking, the use of this criterion requires conditions of constant temperature and pressure, which are usual in biochemical thermodynamics.) It is not possible to measure absolute values of energy; only the *changes* in energy that occur during a process can be measured. The value of the change in Gibbs energy, ΔG (where the symbol Δ indicates change), gives the needed information about the spontaneity of the process under consideration.

The Gibbs energy of a system decreases in a spontaneous (energy-releasing) process, so ΔG is negative ($\Delta G < 0$). Such a process is called **exergonic**, meaning that Gibbs energy is released. When the change in Gibbs energy is positive ($\Delta G > 0$), the process is nonspontaneous (i.e., *energetically unfavorable*). For a nonspontaneous process to occur, energy must be supplied. Nonspontaneous processes are also called **endergonic**, meaning that Gibbs energy is absorbed by the system under consideration. For a process at **equilibrium**, where there is no net energy change in either direction, the change in Gibbs energy is zero ($\Delta G = 0$).

The sign of the change in Gibbs energy, ΔG, indicates the direction of the reaction:

- $\Delta G < 0$: Spontaneous, exergonic: Energy is released
- $\Delta G = 0$: Equilibrium
- $\Delta G > 0$: Nonspontaneous, endergonic: Energy is required

The aerobic metabolism of glucose, in which glucose reacts with oxygen to produce carbon dioxide, water, and energy for the organism, is a spontaneous process.

$$\text{Glucose} + 6\,O_2 \rightarrow 6\,CO_2 + 6\,H_2O \qquad \Delta G < 0$$

The hydrolysis of adenosine triphosphate (ATP) to adenosine diphosphate (ADP), phosphate ion, and H^+ (Figure 1.28) is spontaneous, too. As a result, the reverse reaction, in which ADP combines with phosphate ion and H^+ to give ATP, is nonspontaneous (Figure 1.31). This reaction can take place in living organisms because metabolic processes supply energy.

spontaneous in thermodynamics, characteristic of a reaction or process that takes place without outside intervention

Gibbs energy a thermodynamic quantity; diagnostic for the spontaneity of a reaction at constant temperature

exergonic a process that releases Gibbs energy

endergonic a process that absorbs Gibbs energy

equilibrium the state in which a forward process and a reverse process occur at the same rate

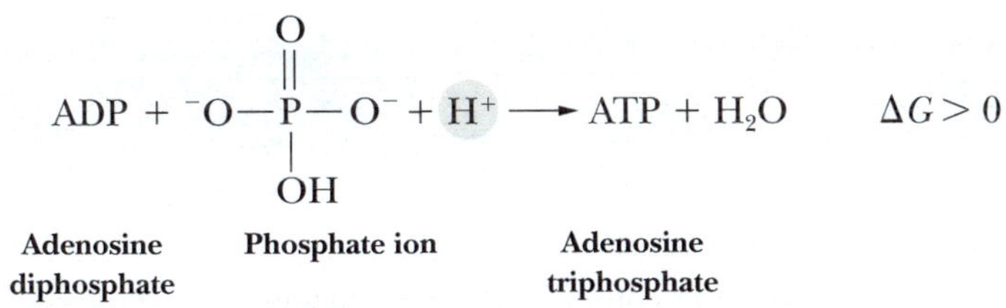

Figure 1.31 The phosphorylation of adenosine diphosphate to give adenosine triphosphate is an endergonic reaction.

1-6D Life and Thermodynamics

Does the existence of living things violate the laws of thermodynamics, specifically the second law? We now take a look at the laws of thermodynamics to clarify whether life is thermodynamically possible and to increase our understanding of this important topic.

▶ Is life thermodynamically possible?

The laws of thermodynamics can be stated in several ways. According to one formulation, the first law is "You can't get something for nothing" and the second is "You can't break even." Put less flippantly, the first law states that it is impossible to convert energy from one form to another at greater than 100% efficiency. In other words, the first law of thermodynamics is the law of conservation of energy. The second law states that even 100% efficiency in energy transfer is impossible.

The two laws of thermodynamics can be related to the Gibbs energy by means of the following equation:

$$\Delta G = \Delta H - T\Delta S$$

In this equation, G is the Gibbs energy, as before; H is the **enthalpy**; and S is the **entropy**. Discussions of the first law focus on the change in enthalpy, ΔH, which is the *heat of a reaction at constant pressure*. This quantity is relatively easy to measure. Enthalpy changes for many important reactions have been determined and are available online and in tables in general chemistry textbooks. Discussions of the second law focus on changes in entropy, ΔS, a concept that is less easily described and measured than changes in enthalpy. Entropy changes are particularly important in biochemistry.

enthalpy a thermodynamic quantity measured as the heat of a reaction at constant pressure

entropy a thermodynamic quantity measured as the number of ways a system can be arranged; it is often taken to be a measure of disorder

One of the most useful definitions of entropy arises from statistical considerations. From a statistical point of view, an increase in the entropy of a system (the substance or substances under consideration) represents an increase in the number of possible arrangements of objects, such as individual molecules. The entropy of a library's reading room is higher when books are scattered around the room than when they are in their proper places on the shelves. Scattered books are in a more dispersed state than books on shelves. The natural tendency of the Universe is in the direction of increasing dispersion of energy, and living organisms put a lot of energy into maintaining order against this tendency. As all parents know, they can spend hours cleaning up a 2-year-old's room, but the child can undo it all in seconds. Similarly, cells use a lot of energy to fight the natural tendency toward dispersion into many different arrangements and to keep the cell structure intact.

Another statement of the second law is this: *In any spontaneous process, the entropy of the Universe increases* ($\Delta S_{univ} > 0$). This statement is general, and it applies to any set of conditions. It is not confined—similar to the statement that the Gibbs energy decreases in a spontaneous process—to the special case of constant temperature and pressure. *Entropy changes are particularly important in determining the energetics of protein folding.* See **Biochemical Connections 1B**.

1B Biochemical Connections • Thermodynamics

Predicting Reactions

To illustrate the concept of entropy, imagine four molecules in a container (Figure 1.32). There is an equal chance that each molecule will be on the left or on the right side of the container. Mathematically stated, the *probability* of finding a given molecule on one side is 1/2. We can express any probability as a fraction ranging from 0 (impossible) to 1 (completely certain). There are 16 possible ways to arrange the four molecules in the container. In only one of these will all four molecules lie on the left side, but six possible arrangements exist with the four molecules evenly distributed between the two sides. *A less ordered (more dispersed) arrangement is more probable than a highly ordered arrangement.* Entropy is defined in terms of the number of possible arrangements of molecules.

Boltzmann's equation for entropy S is

$$S = k \ln W$$

In this equation, the W represents the number of possible arrangements of molecules (i.e., 16 for the system in Figure 1.32), ln is the logarithm to the base e, and k is Boltzmann's constant, which is equal to R/N, where R is the universal gas constant and N is Avogadro's number (6.02×10^{23}), the number of molecules in a mole. ◗

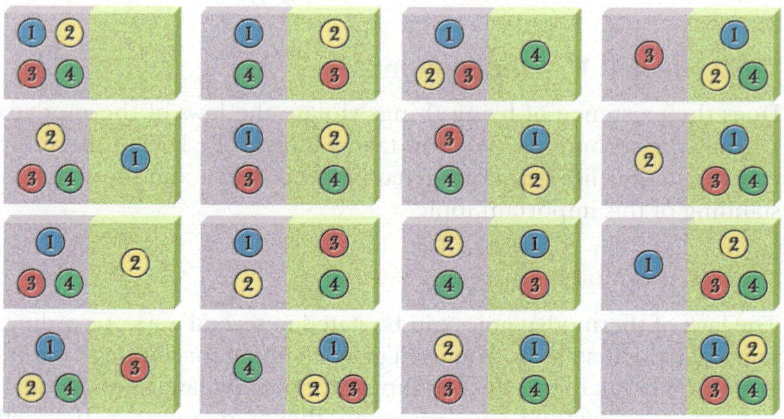

Figure 1.32 Sixteen ways to order four molecules on two halves of a box.

Summary

Biochemistry and Life

- Biochemistry is a multidisciplinary field that asks and tries to answer questions about the molecular nature of life processes.
- Many chemical reactions take place simultaneously in living cells.

Origin of Life on Earth

- The fundamental biochemical similarities observed in all living organisms have engendered speculation about the origins of life.

Chemical Foundations of Biochemistry

- The four basic molecules discussed in this text are amino acids, carbohydrates, nucleic acids, and lipids.

Functional Groups Important in Biochemistry

- Both organic chemistry and biochemistry base their approaches on the behavior of functional groups, but their emphases differ because some functional groups important to organic chemistry do not play a role in biochemistry, and vice versa.
- Important functional groups in biochemistry include carbonyl groups, hydroxyl groups, carboxyl groups, amines, amides, and esters; derivatives of phosphoric acid such as esters and anhydrides are also important.

Earth and Its Age

- Earth, along with the rest of the solar system, was formed about 4 to 5 billion years ago from elements produced by first-generation stars.

Biomolecules

- Important biomolecules can be produced under abiotic (nonliving) conditions from simple compounds postulated to have been present in the atmosphere of the early Earth.
- These simple biomolecules can polymerize, also under abiotic conditions, to give rise to compounds resembling proteins and others having a less marked resemblance to nucleic acids.

Molecules to Cells

- All cellular activity depends on the presence of catalysts, which increase the rates of chemical reactions, and on the genetic code, which directs the synthesis of the catalysts.

- In present-day cells, catalytic activity is associated with proteins, and transmission of the genetic code is associated with nucleic acids, particularly with DNA.
- Both of these functions may once have been carried out by a single biomolecule, RNA.
- It has been postulated that RNA was the original coding material, and RNA has been shown to have catalytic activity as well.
- The formation of bonds in protein biosynthesis is catalyzed by the RNA portions of the ribosome.

Prokaryotic Cells

- In prokaryotes, the cell lacks a well-defined nucleus and internal membrane; it has only a nuclear region, the portion of the cell that contains DNA, and a cell membrane that separates it from the outside world.
- The other principal feature of a prokaryotic cell's interior is the presence of ribosomes, the sites of protein synthesis.

Eukaryotic Cells

- A eukaryotic cell has a well-defined nucleus, both internal membranes and a cell membrane, and a considerably more complex internal structure than a prokaryote.
- In eukaryotes, the nucleus is separated from the rest of the cell by a double membrane.
- Eukaryotic DNA in the nucleus is associated with proteins, particularly a class of proteins called histones.
- There is a continuous membrane system, called the endoplasmic reticulum, throughout the cell. Eukaryotic ribosomes are frequently bound to the endoplasmic reticulum, but some are also free in the cytosol.
- Membrane-enclosed organelles are characteristic of eukaryotic cells.
- Two of the most important organelles are mitochondria, the sites of energy-yielding reactions, and chloroplasts, the sites of photosynthesis.
- Other components of eukaryotic cells include such features as the Golgi apparatus (involved in secretion of proteins from the cell), lysosomes (containers for hydrolytic enzymes), and the cytoskeleton (a framework for organization of various organelles).

Five-Kingdom Classification System

- One important way of classifying organisms depends on the distinction between prokaryotes and eukaryotes. In the five-kingdom scheme, prokaryotes occupy the kingdom Monera.
- The other four kingdoms consist of eukaryotes: Protista, Fungi, Plantae, and Animalia.

Eukaryotic Origins

- A good deal of research has gone into the question of how eukaryotes may have arisen from prokaryotes.
- Much of the thinking depends on the idea of endosymbiosis, in which larger cells may have absorbed aerobic bacteria, eventually giving rise to mitochondria, or photosynthetic bacteria, eventually giving rise to chloroplasts.

Thermodynamic Principles

- All cells require energy to carry out life processes. The Sun is the ultimate source of energy on Earth.
- Photosynthetic organisms trap light energy from the Sun as the chemical energy of the carbohydrates they produce. These carbohydrates then serve as energy sources for other organisms.

Energy Changes

- It is possible to measure the amounts of energy released or absorbed in a process to see whether it is likely to occur.
- Reactions that release energy are energetically favored, whereas those that require energy are disfavored.
- Energy can take several forms, which can be converted from one to another in cells but cannot be destroyed.
- Thermodynamics deals with the changes in energy that determine whether a process will take place.
- A process that will take place without outside intervention is called spontaneous.

Spontaneity in Biochemical Reactions

- In a spontaneous process, the Gibbs energy decreases ($\Delta G < 0$). In a nonspontaneous process, the Gibbs energy increases ($\Delta G > 0$).

Life and Thermodynamics

- In addition to the Gibbs energy, entropy is an important quantity in thermodynamics.
- The entropy of the Universe increases in any spontaneous process.
- Local decreases in entropy can take place within an overall increase in entropy.
- Living organisms represent local decreases in entropy.

Review Exercises

The exercises at the end of each chapter are divided into two or more categories to provide the benefit of more than one approach to reviewing the material.

Recall questions will allow you to test yourself about having important facts readily available to you. In some chapters, the material lends itself to quantitative calculations, and in those chapters you will see a *Mathematical* category of questions. *Reflect and Apply* questions ask you to put those facts to use via questions that require using the concepts in the chapter in moderately creative ways. Lastly, questions that relate specifically to Biochemical Connections boxes are labeled *Biochemical Connections.*

1-1 Basic Themes

1. **Recall** List the six simple molecules that may have given rise to early molecules and cells.
2. **Recall** List and briefly describe the eight levels of structural organization in the human body.

1-2 Chemical Foundations of Biochemistry

3. **Recall** Organic chemistry focuses on the study of compounds that
   (a) contain oxygen and nitrogen.
   (b) contain carbon and hydrogen.
   (c) are derived from petroleum.
   (d) are insoluble in water.
   (e) contain ionic bonds.
   (f) contain metals.
   (g) are derived from living organisms.
4. **Recall** Describe the characteristics of the four basic classes of small molecules found in cells.
5. **Recall** Which of the following statements defines a functional group as it relates to organic chemistry?
   (a) A type of chemical bond between carbon and hydrogen atoms
   (b) A method for naming organic compounds
   (c) A specific arrangement of atoms within a molecule that determines its chemical properties
   (d) A classification of compounds based on their molecular weight
6. **Recall** Match each entry in Column A with one in Column B; Column A lists some important functional groups, whereas Column B shows their structures.

Column A	Column B
Amino group	CH_3SH
Carbonyl group (ketone)	$CH_3CH{=}CHCH_3$
Hydroxyl group	$CH_3CH_2\overset{\overset{O}{\|}}{C}H$
Carboxyl group	$CH_3CH_2NH_2$
Carbonyl group (aldehyde)	$CH_3\overset{\overset{O}{\|}}{C}OCH_2CH_3$
Thiol group	$CH_3CH_2OCH_2CH_3$
Ester linkage	$CH_3\overset{\overset{O}{\|}}{C}CH_3$
Double bond	$CH_3\overset{\overset{O}{\|}}{C}OH$
Amide linkage	CH_3OH
Ether	$CH_3\overset{\overset{O}{\|}}{C}N(CH_3)_2$

7. **Recall** Identify the functional groups in the following compounds.

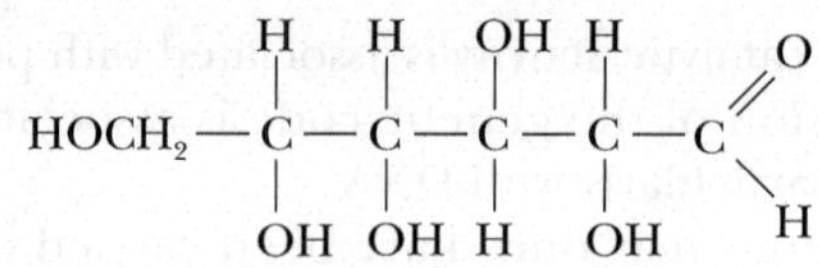

Glucose

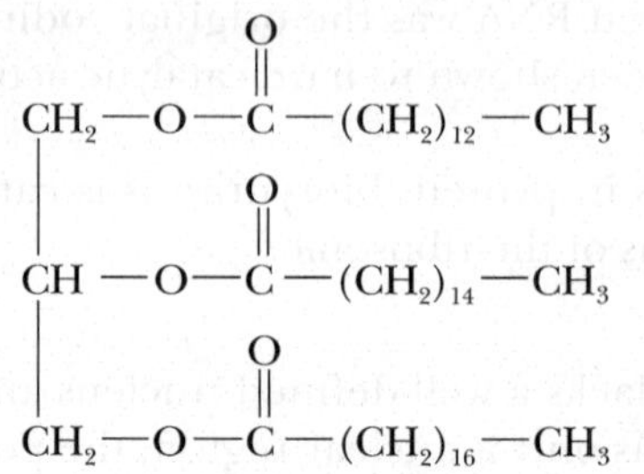

A triglyceride

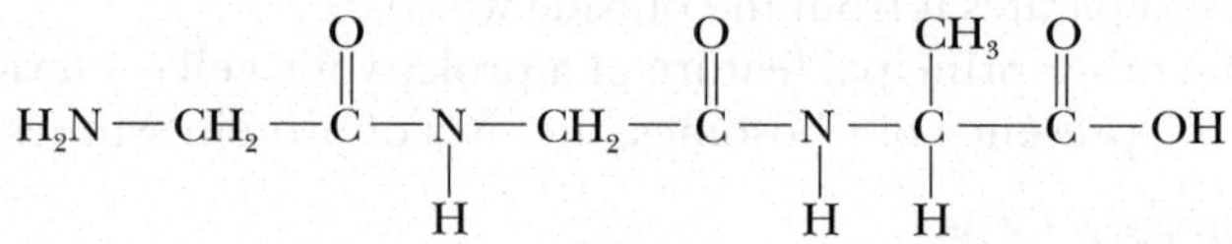

A peptide

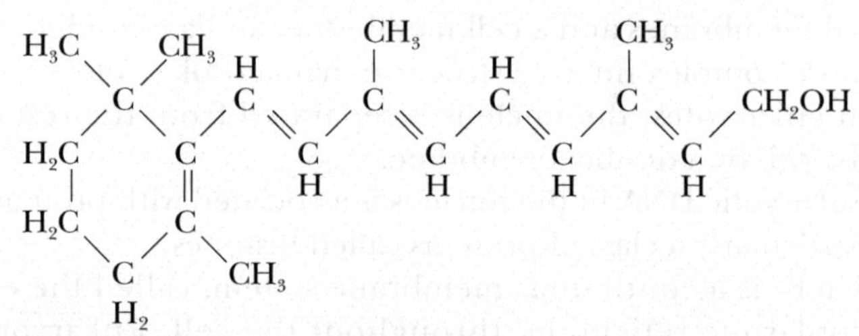

Vitamin A

8. **Reflect and Apply** Which of the following compounds would be expected to undergo chemical reactions like those of ethanol (CH_3CH_2OH)? Select all that apply.

Compound	Structure
Acetic acid	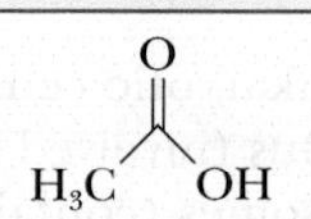
Serine	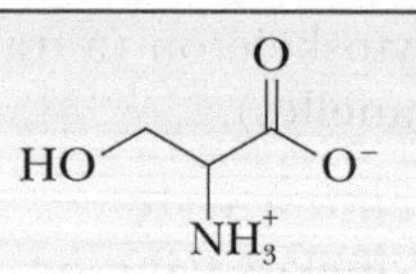
Acetone	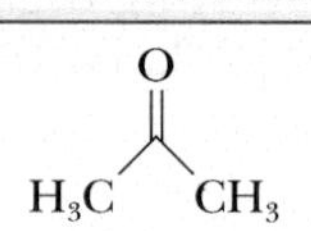

Cysteine

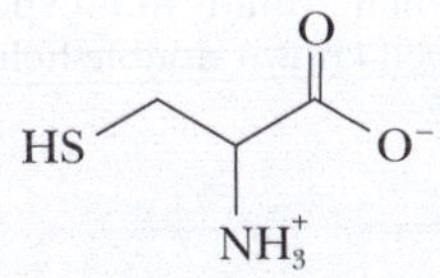

Ribose

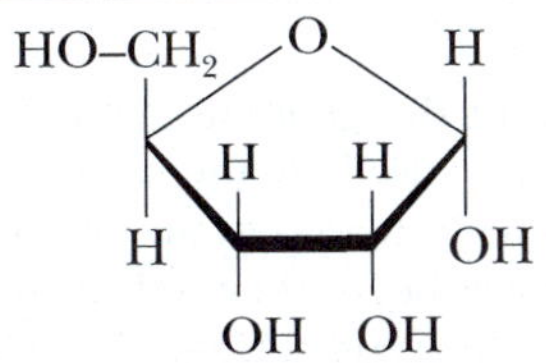

Cholesterol

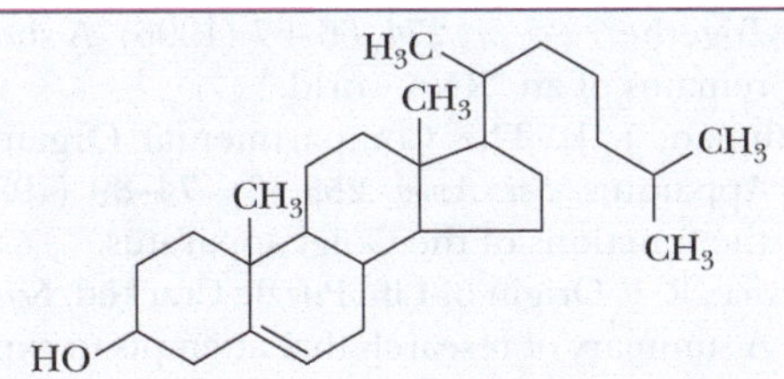

Tyrosine

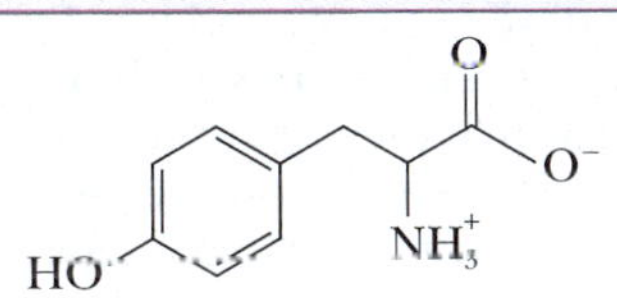

1-3 The Beginnings of Biology

9. **Recall** Define the following important biochemical terms: polymer, protein, nucleic acid, catalysis, and genetic code.
10. **Recall** Explain the directionality that is found in proteins, polysaccharides, and nucleic acids.
11. **Reflect and Apply** Nucleic acids are polymers of just four different monomers in a linear arrangement. How many different sequences are available if you make a nucleic acid polymer with only 40 monomers? How does this number compare with Avogadro's number?
12. **Reflect and Apply** RNA is often characterized as being the first "biologically active" molecule. What two properties or activities does RNA display that might be important to the evolution of life? *Hint:* Neither proteins nor DNA has *both* of these properties.
13. **Reflect and Apply** Do you consider it a reasonable conjecture that cells could have arisen as bare cytoplasm without a cell membrane?

1-4 The Biggest Biological Distinction: Prokaryotes and Eukaryotes

14. **Recall** List five differences between prokaryotes and eukaryotes.
15. **Recall** Which of the following is a constituent of the cytoplasm but not part of the cytosol?
    (a) Ribosomes
    (b) Water
    (c) Dissolved salts
    (d) Water-soluble organic compounds
16. **Recall** Do the sites of protein synthesis differ in prokaryotes versus eukaryotes?
17. **Reflect and Apply** Suppose that a scientist claims to have discovered mitochondria in bacteria. Is such a claim likely to prove valid?
18. **Recall** Draw an idealized animal cell, and identify the parts by name and function.
19. **Recall** Draw an idealized plant cell, and identify the parts by name and function.
20. **Recall** What are the differences between the photosynthetic apparatus of green plants and photosynthetic bacteria?
21. **Recall** Which organelles are surrounded by a double membrane?
22. **Recall** Which organelles contain DNA?
23. **Recall** Which organelles are the sites of energy-yielding reactions?
24. **Recall** State how the structure and function of the following organelles differ from each other: Golgi apparatus, lysosomes, peroxisomes, glyoxysomes. How do they resemble each other?
25. **Recall** How does the size of a gene compare to the size of a genome?
    (a) Genes are larger than genomes.
    (b) Genes are smaller than genomes.
    (c) Genes and genomes are approximately the same size.
    (d) Genes and genomes have no specific size comparison.

1-5 How We Classify Eukaryotes and Prokaryotes

26. **Recall** List the five kingdoms into which living organisms are divided, and give at least one example of an organism belonging to each kingdom.
27. **Recall** Which of the five kingdoms consist of prokaryotes? Which consist of eukaryotes?
28. **Reflect and Apply** How are eukaryotic cells more versatile than prokaryotic cells?
29. **Reflect and Apply** Mitochondria and chloroplasts contain some DNA, which more closely resembles prokaryotic DNA than (eukaryotic) nuclear DNA. Use this information to suggest how eukaryotes may have originated.
30. **Reflect and Apply** Fossil evidence indicates that prokaryotes have been around for about 3.5 billion years, whereas the origin of eukaryotes has been dated at only about 1.5 billion years ago. Suggest why, in spite of the lesser time for evolution, eukaryotes are much more diverse (have a much larger number of species) than prokaryotes.

1-6 Biochemical Energetics

31. **Recall** Define the following related thermodynamic terms: exergonic, endergonic, and equilibrium.
32. **Recall** Which processes are favored: those that require energy or those that release energy?
33. **Recall** Does the thermodynamic term *spontaneous* refer to a process that takes place quickly?
34. **Reflect and Apply** In which of the following processes does entropy increase? In each case, explain why it does or does not increase.
    (a) A bottle of ammonia is opened. The odor of ammonia is soon apparent throughout the room.
    (b) Sodium chloride dissolves in water.
    (c) A protein is completely hydrolyzed to the component amino acids.
35. **Reflect and Apply** Why is it necessary to specify the temperature when making a table listing ΔG values?
36. **Reflect and Apply** Why does the entropy of a system depend on temperature?
37. **Reflect and Apply** A reaction at 23°C has $\Delta G = 1$ kJ/mol. Why might this reaction become spontaneous at 37°C?

Hint: For Questions 38 through 40, consider the equation $\Delta G = \Delta H - T(\Delta S)$.

38. **Reflect and Apply** Urea dissolves very readily in water, but the solution becomes very cold as the urea dissolves. How is this possible?
39. **Reflect and Apply** Would you expect the reaction ATP $\rightarrow$ ADP + P_i to be accompanied by a decrease in entropy or an increase? Why?

40. **Reflect and Apply** The existence of organelles in eukaryotic cells represents a higher degree of organization than that found in prokaryotes. How does this affect the entropy of the Universe?

41. **Reflect and Apply** Which would you expect to have a higher entropy: DNA in its well-known double-helical form or DNA with the strands separated?

Further Reading

Chen, I. The Emergence of Cells During the Origin of Life. *Science* **314**, 1558–1559 (2006). A particularly clear summary of current theories.

de Duve, C. The Birth of Complex Cells. *Sci. Amer.* **274** (4), 50–57 (1996). A Nobel laureate summarizes endosymbiosis and other aspects of cellular structure and function.

Duke, R., D. Ojcius, and J. Young. Cell Suicide in Health and Disease. *Sci. Amer.* **275** (6), 80–87 (1996). An article on cell death as a normal process in healthy organisms and the lack of it in cancer cells.

Madigan, M., and B. Marrs. Extremophiles. *Sci. Amer.* **276** (4), 82–87 (1997). An account of various kinds of archaebacteria that live under extreme conditions and some of the useful enzymes that can be extracted from these organisms.

Morell, V. Life's Last Domain. *Science* **273**, 1043–1045 (1996). A Research News article about the genome of the archaebacterium *Methanococcus jannaschii*. This is the first genome sequence to be obtained for archaebacteria. Read in conjunction with the research article on pp. 1058–1073 of the same issue.

Pennisi, E. Laboratory Workhorse Decoded: Microbial Genomes Come Tumbling In. *Science* **277**, 1432–1434 (1997). A Research News article about the genome of the bacterium *E. coli*. This organism is widely used in the research laboratory, making its genome particularly important among the dozen bacterial genomes that have been obtained. Read in conjunction with the research article on pp. 1453–1474 of the same issue.

Robertson, H. How Did Replicating and Coding RNAs First Get Together? *Science* **274**, 66–67 (1996). A short review on possible remains of an "RNA world."

Rothman, J. E. The Compartmental Organization of the Golgi Apparatus. *Sci. Amer.* **253** (3), 74–89 (1985). A description of the functions of the Golgi apparatus.

Service, R. F. Origin-of-Life Puzzle Cracked. *Science* **347**, 1298 (2015). A summary of research that attempts to explain how the primary types of biomolecules could have evolved at the same time.

Waldrop, M. Goodbye to the Warm Little Pond? *Science* **250**, 1078–1079 (1990). Facts and theories on the role of meteorite impacts on the early Earth in the origin and development of life.

Weber, K., and M. Osborn. The Molecules of the Cell Matrix. *Sci. Amer.* **253** (4), 100–120 (1985). An extensive description of the cytoskeleton.

Life processes depend on the properties of water.

Miljan Zivkovic/Shutterstock.com

Water: The Solvent for Biochemical Reactions 2

Chapter Outline

2-1 Water and Polarity

Learning Objectives

After reading this section, students will be able to:

LO 2.1.1 Distinguish between *covalent bonds*, *polar covalent bonds*, and *nonpolar covalent bonds*.

LO 2.1.2 Distinguish between *ionic bonds*, *salt bridges*, and *ion–dipole interactions*.

LO 2.1.3 Distinguish between three types of van der Waals forces.

LO 2.1.4 Sketch an example of a micelle formed by amphipathic molecules in aqueous solution.

Water is the principal component of most cells. The geometry of the water molecule and its properties as a solvent play major roles in determining the structure and function of living systems.

A **covalent bond** involves the sharing of a pair of electrons between atoms. The tendency of an atom to attract electrons to itself in a covalent bond (i.e., to become partially negative) is called **electronegativity**. Atoms of the same element share electrons equally in a bond—that is, they have equal electronegativity—but different elements do not necessarily have the same electronegativity. Oxygen and nitrogen are both highly electronegative, much more so than carbon and hydrogen (Table 2.1).

Table 2.1 Electronegativities of Selected Elements

Element	Electronegativity*
Oxygen	3.5
Nitrogen	3.0
Sulfur	2.6
Carbon	2.5
Phosphorus	2.2
Hydrogen	2.1

*Electronegativity values are relative and are chosen to be positive numbers ranging from less than 1 for some metals to 4 for fluorine.

2-1A Polar Bonds

▶ What is polarity?

When two atoms with the same electronegativity form a bond, the electrons are shared equally between the two atoms. However, if atoms with differing electronegativity form a bond, the electrons are not shared equally and more of the negative charge is found closer to one of the atoms. In the O—H bonds in water, oxygen is more electronegative than hydrogen, so there is a higher probability that the bonding electrons are closer to the oxygen. The difference in electronegativity between oxygen and hydrogen gives rise to a *partial* positive and *partial* negative charge, usually pictured as δ^+ and δ^-, respectively (Figure 2.1). Bonds such as this are called **polar covalent bonds.** In situations in which the electronegativity difference is quite small, such as in the C—H bond in methane (CH_4), the sharing of electrons in the bond is very nearly equal, and the bond is called a **nonpolar covalent bond.**

covalent bond a chemical bond between two atoms that involves the sharing of electrons

electronegativity measure of the tendency of an atom to attract electrons to it in a chemical bond

polar covalent bonds bonds in which two atoms have an unequal share in the bonding electrons

nonpolar covalent bond a bond in which two atoms share the bonding electrons evenly

The bonds in a molecule may be polar, but the molecule itself can still be nonpolar because of its geometry. For example, the two C—O bonds in carbon dioxide are polar, but because the CO_2 molecule is linear, the attraction of the oxygen for the electrons in one bond is cancelled out by the equal and opposite attraction for the electrons by the oxygen on the other side of the molecule.

$$O{=}C{=}O$$

Water has a bent molecular geometry with a bond angle of 104.5° (Figure 2.1). As a result, the uneven sharing of electrons in the two bonds is *not* cancelled out as it is in CO_2. The bonding electrons are more likely to be found at the oxygen end of the molecule than at the hydrogen end. Molecules with positive and negative ends are called **dipoles.**

dipoles molecules with positive and negative ends due to an uneven distribution of electrons in bonds

van der Waals radius the distance between an atom's nucleus and its effective electronic surface

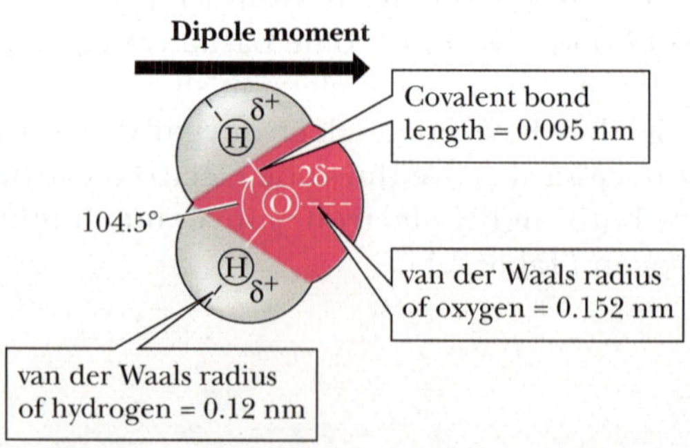

Figure 2.1 The structure of water. Oxygen has a partial negative charge, and the hydrogens have a partial positive charge. The uneven distribution of charge gives rise to the large dipole moment of water. The **van der Waals radius** is the effective distance between an atom's nucleus and its electronic surface, whether the atom is alone or part of a molecule.

2-1B Solvent Properties of Water

▶ Why do some chemical compounds dissolve in water while others do not?

The polar nature of water largely determines its solvent properties. *Ionic* compounds with full charges, such as potassium chloride (KCl; K^+ and Cl^- in solution), and *polar* compounds with partial charges (i.e., dipoles), such as ethyl alcohol (C_2H_5OH) or acetone [$(CH_3)_2C{=}O$], tend to dissolve in water, whereas less polar molecules do not dissolve as readily in water, if at all. The underlying physical principle is electrostatic attraction between unlike charges, but there are different types of bonds with different strengths depending on these electrostatic attractions. To understand why biochemicals behave the way they do in solution, we must first understand the nature of some important noncovalent interactions.

2-1C Ionic Bonds

In a crystal of salt (Figure 2.2), the positive and negative ions are held together by **ionic bonds**. Ionic bonds and covalent bonds are the strongest bonds. Most of the behavior of biomolecules, however, is due to weaker noncovalent, intermolecular forces, which we discuss next.

ionic bonds chemical bonds that involve attractions between positive and negative ions

2-1D Salt Bridges

Similar to the ionic bonds found in salt crystals, biomolecules often have ionizable groups on them. The side chain of the amino acid aspartic acid (Section 3-2) forms a carboxylate anion, $—COO^-$. The side chain of the amino acid lysine forms an amino group cation $—NH_3^+$. The attraction of these two side chains for each other in solution often controls how protein molecules fold in solution. When these oppositely charged molecules are close enough to form a hydrogen bond (Section 2-2), we call the resulting interaction a **salt bridge**.

salt bridge an attraction between oppositely charged groups where the two groups are close enough to form a hydrogen bond

2-1E Ion–Dipole Interactions

Ions in solution can also interact with molecules that have dipoles to form **ion–dipole interactions**, as shown in Figures 2.2 and 2.3A. In both these cases, water

ion–dipole interactions attractions between an ion and an opposite partial charge on a dipole

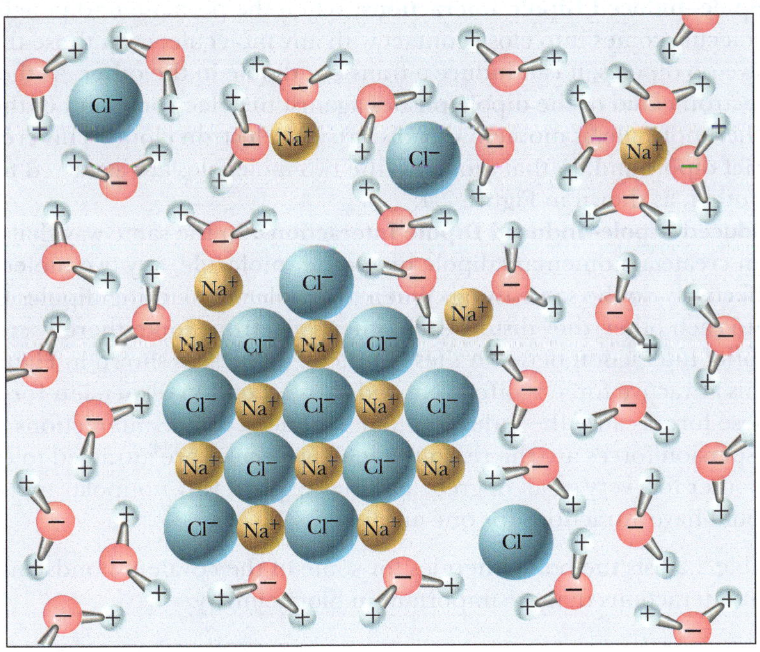

Figure 2.2 Ionic bonds become replaced by ion–dipole interactions. In ionic solids, ionic bonds hold the cations and anions together. In aqueous solution, these ionic bonds are replaced by ion–dipole interactions. The negatively charged chloride ions are attracted to the partial positive charges (the + signs on the white spheres) on water. The positively charged sodium ions are attracted to the partial negative charges (the – signs on the pink spheres) on the water. Water-surrounding ions of this type are called hydration shells.

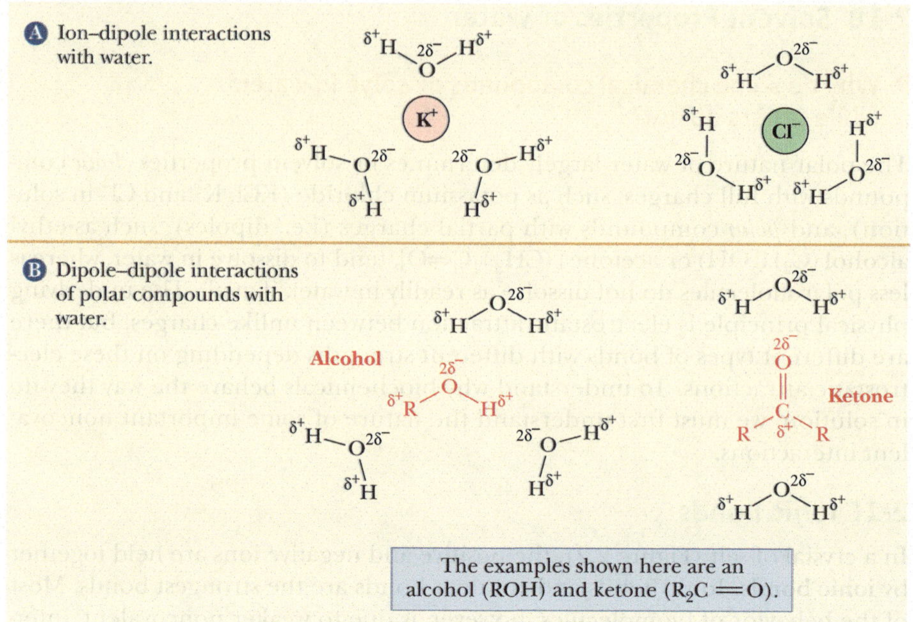

Figure 2.3 Ion–dipole and dipole–dipole interactions. Ion–dipole and dipole–dipole interactions help ionic and polar compounds dissolve in water.

is shown as the molecule with the dipole. A charged ion will interact with the corresponding opposite partial charge on the water.

2-1F van der Waals Forces

van der Waals forces noncovalent associations based on the weak attraction of transient dipoles for one another; also called a *van der Waals interaction* or *van der Waals bonds*

dipole–dipole interactions interactions where the partial positive side of one dipole attracts the partial negative side of another dipole

dipole–induced dipole interactions interactions where a dipole induces a transient dipole in a neighboring nonpolar group; one side of the dipole then attracts the partial charge of opposite sign on the induced dipole

induced dipole–induced dipole interactions interactions where two nonpolar groups bump into each other and distort each other's electron clouds, forming induced dipoles; the opposite charges on the induced dipoles then attract to form the interaction

London dispersion force an attraction between transient induced dipoles

There are several types of weak noncovalent associations that are collectively called **van der Waals forces,** after the Dutch scientist Johannes Diderik van der Waals. These forces include dipole–dipole interactions, dipole–induced dipole interactions, and induced dipole–induced dipole interactions, none of which involve an electrostatic interaction of a fully charged ion.

- **Dipole–Dipole Interactions: These forces occur between molecules that are dipoles, with the partial positive side of one molecule attracting the partial negative side of another molecule, as shown in Figure 2.3B.** There is a special type of dipole–dipole interaction called a hydrogen bond (Section 2-2).
- **Dipole–Induced Dipole Interactions:** When the permanent dipole in a molecule comes into close contact with any molecule, even those that have no dipoles, it can induce a transient dipole in the other. As the electron cloud of the dipole pushes against the electron cloud of the other molecule, it momentarily distorts the electron cloud. This creates a brief dipole and, in that moment, the two molecules are attracted to one another, as shown in Figure 2.4.
- **Induced Dipole–Induced Dipole Interactions:** In the same way that a dipole can create a momentary dipole in another molecule, any two molecules can effectively do the same thing. When two molecules lacking dipoles bump into each other, they distort each other's electron cloud, thereby creating a brief interaction between these induced dipoles, as shown in Figure 2.5. This attractive force is often referred to as a **London dispersion force**, and these forces cause the induced dipole–induced dipole interactions. London dispersion forces are the reason that all molecules are attracted to one another to a very small degree, and they explain why nonpolar molecules would have attractions for one another.

Table 2.2 lists the bond energies for some of the covalent bonds and noncovalent interactions that are important in biochemistry.

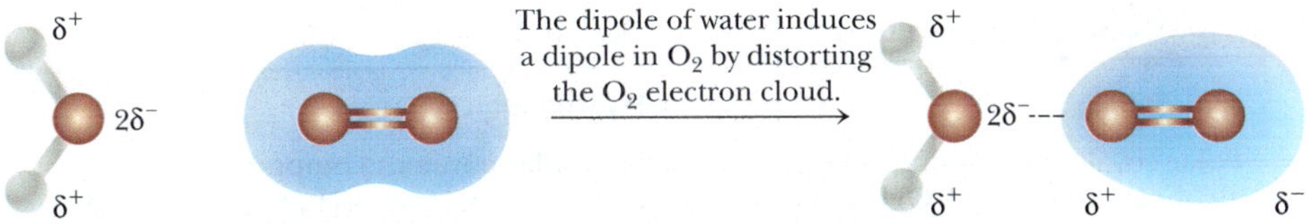

Figure 2.4 Dipole–induced dipole bonds. Polar water is able to distort the electron cloud of a nonpolar molecule, like oxygen, creating a momentary dipole. Once the dipole is created, water is attracted to it. (*J. A. Dean:* Lange's Handbook of Chemistry, *14th ed., Figure 13.2, page 584. New York, McGraw-Hill, 1992.*)

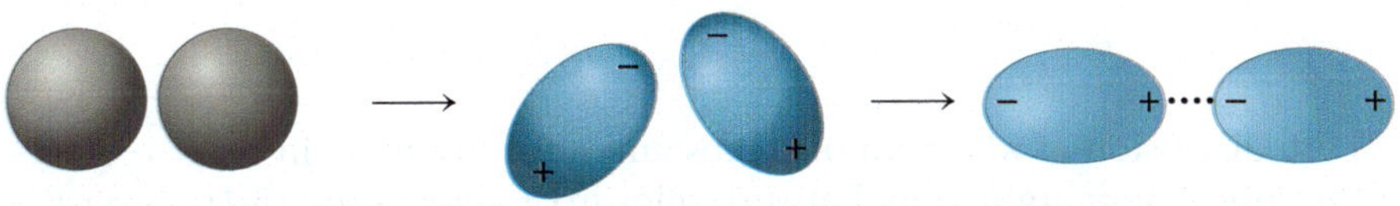

Figure 2.5 Induced dipole–induced dipole interactions. Also called London dispersion forces, these attractions arise when otherwise nonpolar molecules bump into each other and distort each other's electron shells, thereby inducing dipoles. Once the dipoles are formed, the molecules are momentarily attracted to one another. (*J. A. Dean:* Lange's Handbook of Chemistry, *14th ed., Figure 13.13. New York, McGraw-Hill, 1992.*)

Table 2.2 Strengths of Covalent Bonds and Noncovalent Interactions Found in Biochemistry

Bond Type	Strength (kcal/mol)	Strength (kJ/mol)
Covalent C—H	105	413
Covalent (O—H)	110	460
Ionic interactions	1–20	4–80
Ion–dipole	5	20
Hydrogen bonds	5	20
Van der Waals interactions	1	4

2-1G Interactions of Various Substances with Water

Why do some substances dissolve readily in water while others do not? Water is a very polar molecule, and it easily makes ion–dipole bonds with ionic compounds, or dipole–dipole bonds with other polar compounds. The situation depicted in Figure 2.2 is a prime example. Ionic bonds are very strong. Why, then, would a crystal of NaCl dissolve in water? We might think that the ionic bonds between the Na^+ and the Cl^- would make the salt crystal just stay as it was, essentially "ignoring" the water. However, when the salt dissolves and the ions become hydrated, they are then surrounded by water molecules, each one of which can make ion–dipole bonds with it. We can think of solids dissolving in water as an energetic tug-of-war. If the bonds formed between the dissolving ions and water are more favorable than the bonds holding the solid together, it will dissolve. NaCl dissolves readily. Other solids do not. Ionic and polar substances tend to dissolve in water, so they are referred to as **hydrophilic** ("water-loving," from the Greek) for this tendency (see Table 2.3).

hydrophilic tending to dissolve in water

Hydrocarbons (compounds that contain only carbon and hydrogen) are nonpolar. The favorable ion–dipole and dipole–dipole interactions responsible for the solubility of ionic and polar compounds do not occur for nonpolar compounds, so these compounds tend not to dissolve in water.

Table 2.3 Examples of Hydrophobic and Hydrophilic Substances

Hydrophilic	Hydrophobic
Polar covalent compounds (e.g., alcohols such as C_2H_5OH [ethanol] and ketones such as $(CH_3)_2C{=}O$ [acetone])	Nonpolar covalent compounds (e.g., hydrocarbons such as C_6H_{14} [hexane])
Sugars	Fatty acids, cholesterol
Ionic compounds (e.g., KCl)	
Amino acids, phosphate esters	

A full discussion of why nonpolar substances are insoluble in water requires the thermodynamic arguments that we develop in Chapters 4 and 15. However, the points made here about intermolecular interactions provide useful background information for that discussion. For the moment, it is enough to know that it is less favorable thermodynamically for water molecules to be associated with nonpolar molecules than with other water molecules. As a result, nonpolar molecules do not dissolve in water and are referred to as **hydrophobic** ("water-fearing," from the Greek). Hydrocarbons in particular tend to sequester themselves from an aqueous environment. A nonpolar solid leaves undissolved material in water, whereas a nonpolar liquid forms a two-layer system with water (an everyday example is an oil slick).

hydrophobic tending not to dissolve in water

Amphipathic molecules (Figure 2.6) have both polar (hydrophilic) and nonpolar (hydrophobic) portions. Long-chain fatty acids are amphipathic because they have a polar carboxylic acid group and a long nonpolar hydrocarbon portion. The carboxylic acid group, the "head" group, contains two oxygen atoms in addition to carbon and hydrogen; it is very polar and can form a carboxylate anion at neutral pH. The rest of the molecule, the "tail," contains only carbon and hydrogen and is thus nonpolar. A compound such as this in the presence of water tends to form structures called **micelles**, in which the polar head groups are in contact with the aqueous environment and the nonpolar tails are sequestered from the water (Figure 2.7).

amphipathic molecule a molecule that has one end with a polar, water-soluble group and another end with a nonpolar hydrocarbon group that is insoluble in water

micelles an aggregate of amphipathic molecules dispersed in a liquid

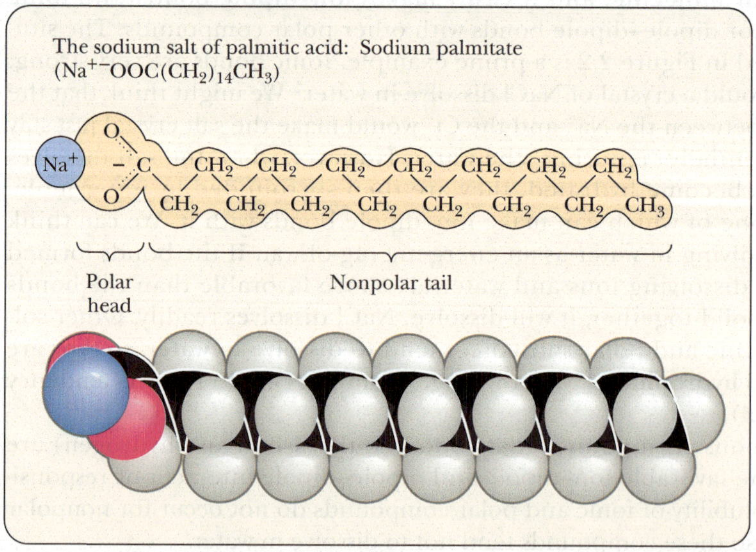

Figure 2.6 Sodium palmitate, an amphipathic molecule. Amphiphilic molecules are frequently symbolized by a ball and zigzag line structure (●〰〰), where the ball represents the hydrophilic polar head and the zigzag line represents the nonpolar hydrophobic hydrocarbon tail.

Careers in Chemistry • Justin Scott

Laboratory Director & Principal Toxicologist at Cove Environmental

Ellie Piper/Oklahoma State University

Education:
Oklahoma State University, PhD, Integrative Biology

Justin Scott, PhD, is an ecotoxicologist who measures environmental toxins and studies their impact on organisms and human health. He serves as the Laboratory Director and Principal Toxicologist at Cove Environmental, where his team analyzes wastewater treatment across various industries. Currently, they are researching how micro- and nanoplastics accumulate and interact in tissues over time, including the potential harmful effects of both the plastics and their additives. What Scott enjoys most is using scientific investigation to narrow down answers to complex questions.

Scott applies biochemistry daily to measure chemical concentrations and assess their toxicological effects. He explains that understanding how toxicants work requires examining molecular and cellular processes. His study of biochemistry deepened his knowledge of specific cell receptors and transporters and how they interact with toxicants, influencing outcomes from individual organisms to entire ecosystems.

Scott considers himself a nontraditional student, having taken time away from college to support his family. During that period, he developed valuable technical skills and research proficiency while working in a private lab. His advice to students is to not be afraid to make mistakes. "There are quite a few discoveries that were made by accident," he says "don't worry if the results aren't making sense at first. The important thing is to be honest and let the data speak for itself."

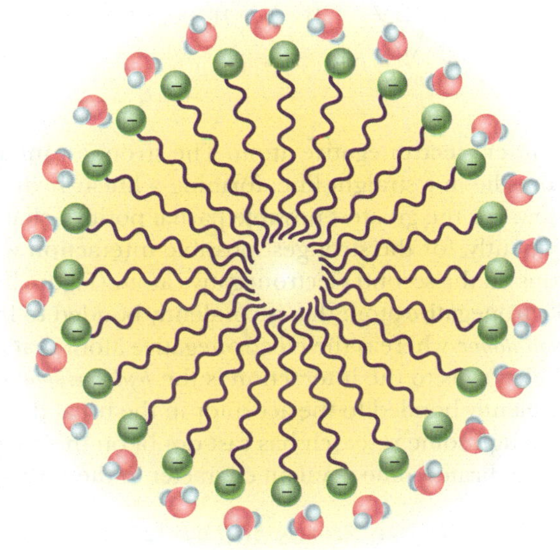

Figure 2.7 Micelle formation by amphipathic molecules in aqueous solution. When micelles form, the ionized polar head groups are in contact with water molecules, whereas the nonpolar tails of the amphipathic molecules are protected from contact with water.

2-2 Hydrogen Bonds

Learning Objectives

After reading this section, students will be able to:

LO 2.1.1 Draw examples of linear and nonlinear hydrogen bonds.

LO 2.1.2 Draw examples of hydrogen bonding in HF, H_2O, and NH_3, including acceptor and donor labels.

Recall from Section 2-1 that **hydrogen bonds** are a special type of dipole–dipole interaction. When hydrogen is covalently bonded to a very electronegative atom, such as fluorine, oxygen, or nitrogen, it develops a partial positive charge due to the polar bond. This situation does *not* occur when hydrogen is covalently bonded to carbon. The partial positive charge on hydrogen can interact with an unshared (nonbonding) pair of electrons (a source of negative charge) on

hydrogen bond a noncovalent interaction formed between a hydrogen atom covalently bonded to one electronegative atom and a lone pair of electrons on another electronegative atom

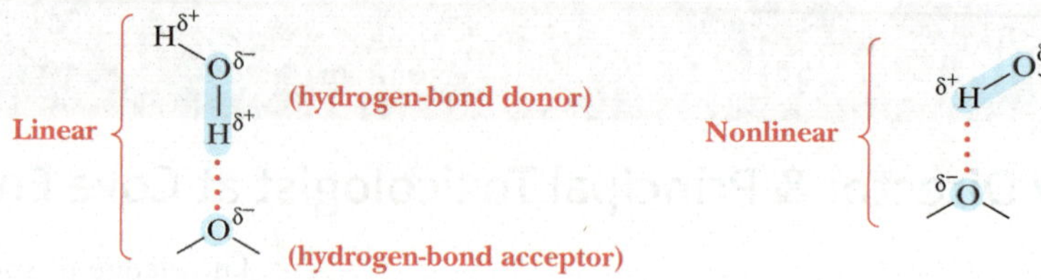

Figure 2.8 A comparison of linear and nonlinear hydrogen bonds. Nonlinear bonds are weaker than bonds in which all three atoms lie in a straight line.

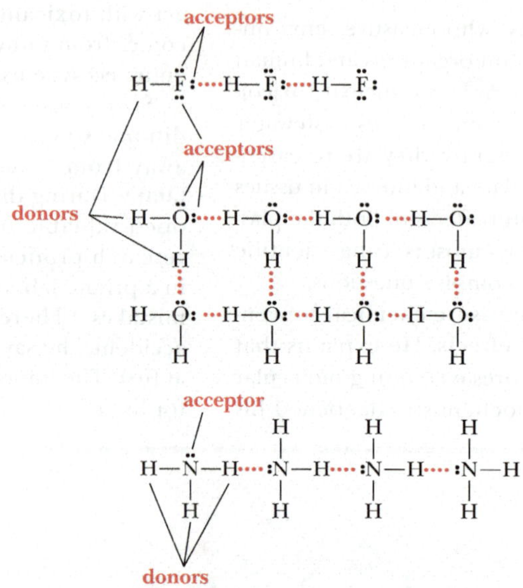

Figure 2.9 Hydrogen-bonding sites. A comparison of the numbers of hydrogen-bonding sites in HF, H_2O, and NH_3. (Actual geometries are not shown.) Each HF molecule has one hydrogen-bond donor and three hydrogen-bond acceptors. Each H_2O molecule has two donors and two acceptors. Each NH_3 molecule has three donors and one acceptor.

another electronegative atom. The strongest interactions occur when all three atoms lie in a straight line, forming a linear hydrogen bond. This arrangement allows for the greatest possible partial positive charge on the hydrogen and, consequently, for the strongest possible interaction with the unshared pair of electrons on the second electronegative atom (Figure 2.8). The group comprising the electronegative atom that is covalently bonded to hydrogen is called the *hydrogen-bond donor*, whereas the electronegative atom that contributes the unshared pair of electrons to the interaction is the *hydrogen-bond acceptor*. The hydrogen is not covalently bonded to the acceptor in the usual description of hydrogen bonding, although some research has cast doubt on this view, with experimental evidence that indicates some covalent character in the hydrogen bond.

2-2A Hydrogen Bonds and Water

▶ **Why does water have such interesting and unique properties?**

Figure 2.9 compares the number of hydrogen-bonding sites in HF, H_2O, and NH_3 Water has two hydrogens to enter into hydrogen bonds and two unshared pairs of electrons on the oxygen to which other water molecules can be hydrogen bonded. Thus, each water molecule can be involved in four hydrogen bonds—two as a donor and two as an acceptor. Compared to HF and NH_3, this is the optimal situation. Hydrogen fluoride has only one hydrogen to enter into a hydrogen bond as a donor, but it has three unshared pairs of electrons on the fluorine that could bond to other hydrogens. Ammonia has three hydrogens to donate to a hydrogen bond but only one unshared pair of electrons on the nitrogen.

The geometric arrangement of hydrogen-bonded water molecules has important implications for the properties of water as a solvent. The bond angle in water is 104.3°, as shown in Figure 2.1, and the angle between the unshared pairs of electrons is similar. The result is a tetrahedral arrangement of water molecules. Liquid water consists of hydrogen-bonded arrays that resemble ice crystals; each of these arrays can contain up to 100 water molecules. The hydrogen bonding between water molecules can be seen

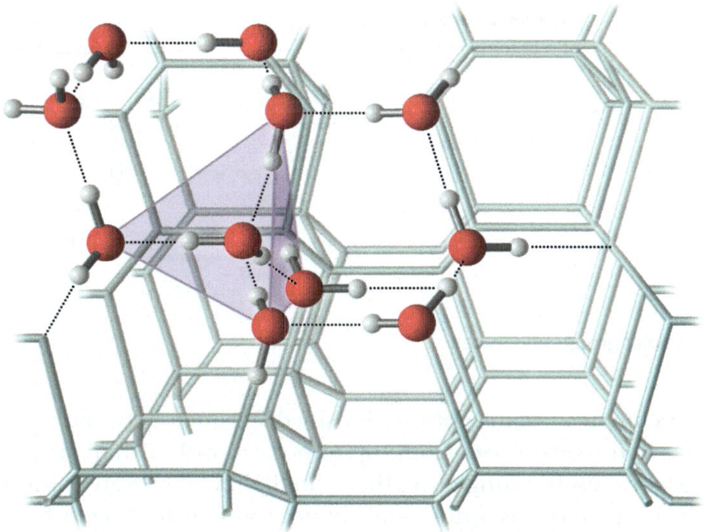

Figure 2.10 Tetrahedral hydrogen bonding in H_2O. In an array of H_2O molecules in an ice crystal, each H_2O molecule is hydrogen bonded to four others.

Table 2.4 Comparison of Properties of Water, Ammonia, and Methane

Substance	Molecular Weight	Melting Point (°C)	Boiling Point (°C)
Water (H_2O)	18.02	0.0	100.0
Ammonia (NH_3)	17.03	−77.7	−33.4
Methane (CH_4)	16.04	−182.5	−161.5

more clearly in the regular lattice structure of the ice crystal (Figure 2.10). There are several differences, however, between hydrogen-bonded arrays of this type in liquid water and the structure of ice crystals. In liquid water, hydrogen bonds are constantly breaking and new ones are constantly forming, with some molecules breaking off and others joining the cluster. A cluster can break up and re-form in 10^{-10} to 10^{-11} seconds in water at 25°C. An ice crystal, in contrast, has a more or less stable arrangement of hydrogen bonds and its number of molecules is many orders of magnitude greater than 100.

Hydrogen bonds are much weaker than normal covalent bonds. Whereas the energy required to break the O—H covalent bond is 460 kJ/mol (110 kcal/mol), the energy of hydrogen bonds in water is about 20 kJ/mol (5 kcal/mol) (Table 2.2). Even this comparatively small amount of energy is enough to affect the properties of water drastically, especially its melting point, its boiling point, and its density relative to the density of ice. Both the melting point and the boiling point of water are significantly higher than would be predicted for a molecule of its molecular weight (Table 2.4). Ammonia and methane, which have about the same molecular weight, have much lower melting points and boiling points. The forces of attraction between the molecules of these substances are weaker than the attraction between water molecules because of the number and strength of water's hydrogen bonds. The energy of this attraction must be overcome to melt ice or boil water.

Ice has a lower density than liquid water because the fully hydrogen-bonded array in an ice crystal is less densely packed than that in liquid water. Liquid water is less extensively hydrogen bonded and thus is denser than ice. This is why ice cubes and icebergs float. Most substances contract when they freeze, but the opposite is true of water. In cold weather, the cooling systems of cars require antifreeze to prevent freezing and expansion of the water, which could crack the engine block. In laboratory procedures for cell fractionation, the same principle is used in a method of disrupting cells with several cycles of freezing and thawing. Finally, aquatic organisms can survive in

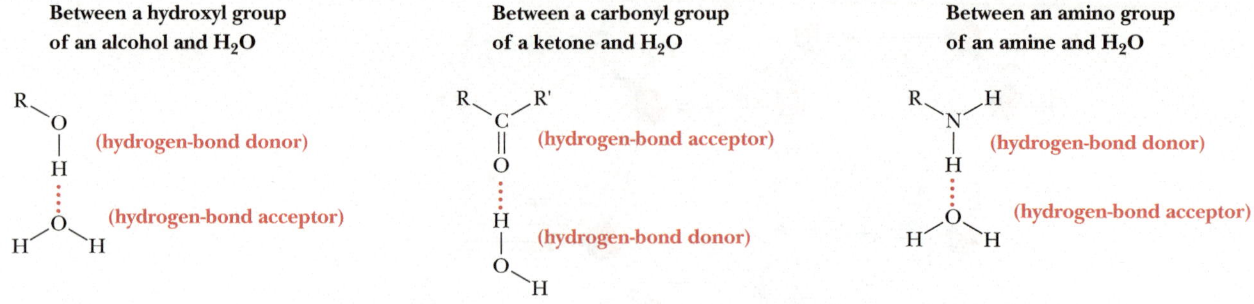

Figure 2.11 Hydrogen bonding between polar groups and water.

cold climates because of the density difference between ice and liquid water; lakes and rivers freeze from top to bottom rather than vice versa.

Hydrogen bonding also affects the behavior of water as a solvent. If a polar solute can serve as a donor or an acceptor of hydrogen bonds, not only can it form hydrogen bonds with water, but it can also be involved in nonspecific dipole–dipole interactions. Figure 2.11 shows some examples. Alcohols, amines, carboxylic acids, and esters, as well as aldehydes and ketones, can all form hydrogen bonds with water, so they are soluble in water. It is difficult to overstate the importance of water to the existence of life on Earth, and it is difficult to imagine life based on another solvent. **Biochemical Connections 2A** explores some of the implications of this statement.

2A Biochemical Connections • Chemistry

How Basic Chemistry Affects Life: The Importance of the Hydrogen Bond

Many noted biochemists have speculated that the hydrogen bond is essential to the evolution of life. Just like carbon, polymers, and stereochemistry, it is one of the criteria that can be used to search for extraterrestrial life. Even though each individual hydrogen bond (H bond) is weak, the fact that so many H bonds can form means that collectively they can exert a *very* strong force. Virtually all the unique properties of water, from its high melting point and boiling point, to the greater density of liquid water versus ice, to its solvent potency, are a result of its ability to form up to four hydrogen bonds per molecule.

Water is attracted to simple ions like Na^+ or Cl^- by polarity. Other water molecules—typically 20 or more per dissolved ion—then form H bonds with those surrounding water molecules. When we consider a simple biomolecule such as glyceraldehyde, the H bonds start at the molecule itself. At least 8 water molecules bind directly to the glyceraldehyde molecule, and then more water molecules bind to those 8 molecules.

The orderly and repetitive arrangement of hydrogen bonds in polymers determines their shape, as seen in Figure 2.12. The extended structures of cellulose and of peptides in a β-sheet allow for the formation of strong fibers through intrachain H bonding. Single helices (as in starch) and the α-helices of proteins are stabilized by intrachain H bonds. Double and triple helices, as in DNA and collagen, involve H bonds between the two or three respective strands. Collagen contains several special amino acids that have an extra hydroxyl group, which allow for additional hydrogen bonds, thereby providing stability.

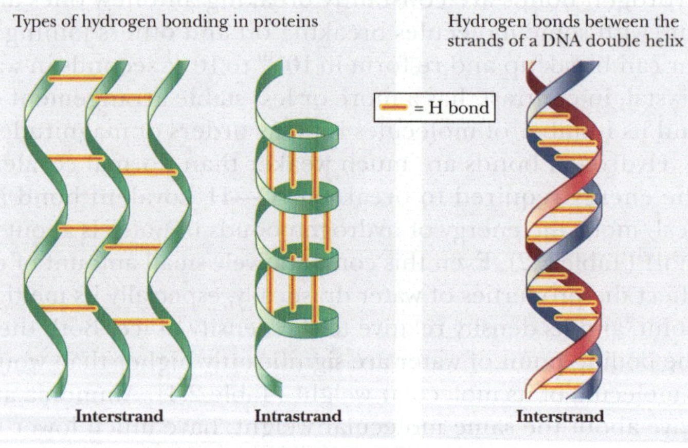

Figure 2.12 Hydrogen bonding in proteins and DNA.

Hydrogen bonding is also fundamental to the specificity of the transfer of genetic information. The complementary nature of the DNA double helix is assured by hydrogen bonds. The genetic code, both its specificity and its allowable variation, is a result of H bonds. Indeed, many compounds that cause genetic mutations work by altering the patterns of H bonding. For example, dentists often prescribe fluorouracil for cold sores (viral sores of the lip and mouth) because it causes mutations in the herpes simplex virus that causes the sores. ◗

2-2B Other Biologically Important Hydrogen Bonds

Hydrogen bonds have a vital involvement in stabilizing the three-dimensional structures of biologically important molecules, including DNA, RNA, and proteins. The hydrogen bonds between complementary bases are one of the most striking characteristics of the double-helical structure of DNA (Section 9-3). Transfer RNA also has a complex three-dimensional structure characterized by hydrogen-bonded regions (Section 9-5). Hydrogen bonding in proteins gives rise to two important structures: the α-helix and β-pleated sheet conformations. Both types of conformation are widely encountered in proteins (Section 4-3). Table 2.5 summarizes some of the most important kinds of hydrogen bonds in biomolecules.

Table 2.5 Examples of Hydrogen Bonding between Groups in Protein and Nucleic Acid Structures

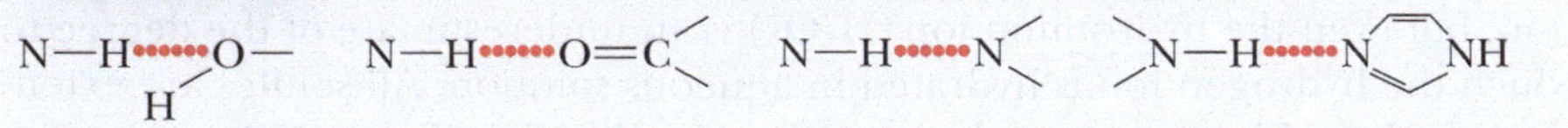

2-3 Acids, Bases, and pH

The biochemical behavior of many important compounds depends on their acid–base properties.

Learning Objectives

After reading this section, students will be able to:

LO 2.3.1 Write acid dissociation equations and acid dissociation constant expressions for a generic acid HA.

LO 2.3.2 Use the pH equation to calculate pH from $[H^+]$ and $[H^+]$ from pH.

LO 2.3.3 Describe the relationships between pH, pK_a, and the relative concentrations of an acid and its conjugate base found in the Henderson–Hasselbalch equation.

2-3A Acids and Bases

An **acid** is a molecule that acts as a proton (hydrogen ion) donor. Because there is more than one definition of an acid, this is also known as a Brønsted acid. A **base** (more specifically, a Brønsted base) is a proton acceptor. How readily acids or bases lose or gain protons depends on the chemical nature of the compounds under consideration. The degree of dissociation of acids in water, for example, ranges from essentially complete dissociation for a strong acid to practically no dissociation for a very weak acid. Any intermediate value is possible, too.

acid a molecule or group that acts as a proton (hydrogen ion) donor

base a molecule or group that acts as a proton (hydrogen ion) acceptor

It is possible to derive a numerical measure of **acid strength**, which is the amount of hydrogen ion released when a given amount of acid is dissolved in water. Such an expression, called the **acid dissociation constant**, or K_a, can be written for any acid, HA, that reacts according to the equation

acid strength the tendency of an acid to dissociate to a hydrogen ion and its conjugate base

acid dissociation constant a number that characterizes the strength of an acid

$$HA \rightleftharpoons H^+ + A^-$$

$$K_a = \frac{[H^+][A^-]}{[HA]}$$

In the K_a expression, the square brackets refer to molar concentration—that is, the concentration in moles per liter; HA refers to the acid; and A^- refers to the conjugate base of HA. For each acid, the quantity K_a has a fixed numerical value at a given temperature. This value is larger for more completely dissociated acids; thus, the greater the K_a, the stronger the acid.

Strictly speaking, the preceding acid–base reaction is a proton-transfer reaction in which water acts as a base as well as the solvent. A more correct way to write the equation is as follows:

$$HA(aq) + H_2O(\ell) \rightleftharpoons H_3O^+(aq) + A^-(aq)$$

Acid	Base	Conjugate acid of H_2O	Conjugate base of HA

The notation (aq) refers to solutes in aqueous solution, whereas (ℓ) refers to the liquid state. There are no "naked protons" (free hydrogen ions) in solution, but even the hydronium ion (H_3O^+) is an underestimate of the degree to which the hydrogen ion is hydrated in aqueous solution. All solutes are extensively hydrated in aqueous solution. We write the short form of equations for acid dissociation in the interest of simplicity, but the role of water should be kept in mind throughout our discussion.

2-3B pH

The acid–base properties of water play an important part in biological processes because of the central role of water as a solvent. The extent to which water self-dissociates to hydrogen ion and hydroxide ion,

$$H_2O \rightleftharpoons H^+ + OH^-$$

is small, but the fact that it takes place at all determines important properties of many solutes (Figure 2.13). Both the hydrogen ion (H^+) and the hydroxide ion (OH^-) are associated with several water molecules, as are all ions in aqueous solution, and the water molecule in the equation is itself part of a cluster of such molecules (Figure 2.14). It is especially important to have a quantitative estimate of the degree of dissociation of water. We can start with the expression

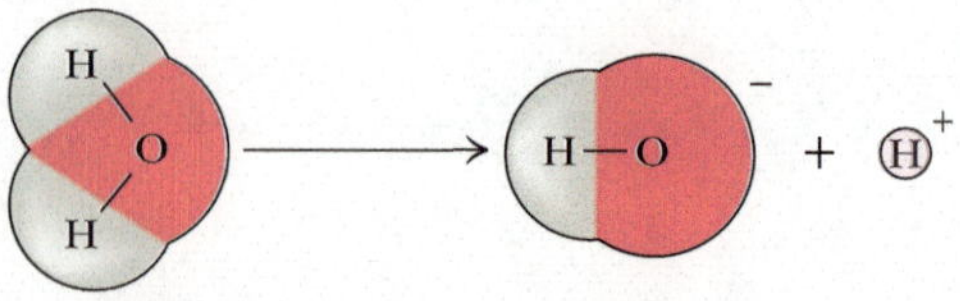

Figure 2.13 **The ionization of water.**

$$K_a = \frac{[H^+][OH^-]}{[H_2O]}$$

The molar concentration of pure water, $[H_2O]$, is quite large compared to any possible concentrations of solutes, so it can be considered a constant. (The numerical value is 55.5 *M*, which can be obtained by dividing the number of grams of water in 1 L, 1000 g, by the molecular weight of water, 18 g/mol; 1000/18 = 55.5 *M*.) Thus,

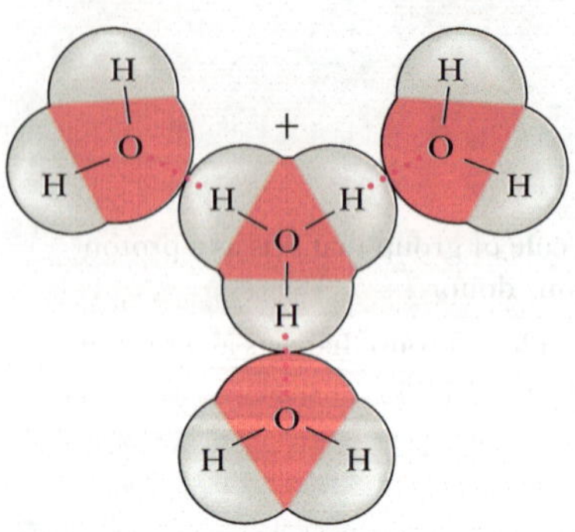

Figure 2.14 **The hydration of hydrogen ion in water.** A new constant, K_w, called the **ion product constant for water**, has just been defined, where the concentration of water has been included in its value.

ion product constant for water a measure of the tendency of water to dissociate to give hydrogen ion and hydroxide ion

$$K_a = \frac{[H^+][OH^-]}{[55.5]}$$

$$K_a \times 55.5 = [H^+][OH^-] = K_w$$

The numerical value of K_w can be determined experimentally by measuring the hydrogen ion concentration of pure water. The hydrogen ion concentration is also equal, by definition, to the hydroxide ion concentration because water is a monoprotic acid (one that releases a single proton per molecule). At 25°C in *pure* water,

$$[H^+] = 10^{-7}\ M = [OH^-]$$

Thus, at 25°C, the numerical value of K_w is given by the expression

$$K_w = [H^+][OH^-] = (10^{-7})(10^{-7}) = 10^{-14}$$

This relationship, which we have derived for pure water, is valid for *any* aqueous solution, whether neutral, acidic, or basic.

The wide range of possible hydrogen ion and hydroxide ion concentrations in aqueous solution makes it desirable to define a quantity for expressing these concentrations more conveniently than by exponential notation. This quantity is called *pH* and is defined as

$$\text{pH} = -\log_{10} [H^+]$$

with the logarithm taken to the base 10. Because of the logarithms involved, a difference of 1 pH unit implies a 10-fold difference in hydrogen ion concentration, $[H^+]$. The pH values of some typical aqueous samples can be determined by a simple calculation, which you can practice in **Apply Your Knowledge 2.1**. These practice sessions appear from time to time in the chapters. They give you the opportunity to immediately practice what was just learned.

2.1 Apply Your Knowledge

pH Calculations

Calculate the pH of the following aqueous solutions:

a. 1×10^{-3} *M* HCl
b. 1×10^{-4} *M* NaOH

Assume that the self-ionization of water makes a negligible contribution to the concentrations of hydronium ions and of hydroxide ions, which will typically be true unless the solutions are extremely dilute.

Solution

The key points in the approach to this problem are the definition of pH, which needs to be used in both parts, and the self-dissociation of water, which is needed in the second part.

a. For 1×10^{-3} *M* HCl, $[H_3O^+] = 1 \times 10^{-3}$ *M*; therefore, pH = 3.

b. For 1×10^{-4} *M* NaOH, $[OH^-] = 1 \times 10^{-4}$ *M*. Because $[OH^-][H_3O^+] = 1 \times 10^{-14}$, $[H_3O^+] = 1 \times 10^{-10}$ *M*; therefore, pH = 10.0.

When a solution has a pH of 7, it is said to be *neutral*, like pure water. Acidic solutions have pH values lower than 7, whereas basic solutions have pH values higher than 7.

Most of the acids encountered in biochemistry are weak acids. These have a K_a well below 1. To avoid having to use numbers with large, negative exponents, a similar quantity, pK_a, has been defined by analogy with the definition of pH:

$$pK_a = -\log_{10} K_a$$

The pK_a is a more convenient numerical measure of acid strength. The smaller its value, the stronger the acid. This is the reverse of the situation with K_a, where larger values imply stronger acids (Table 2.6).

Table 2.6 Dissociation Constants of Some Acids

Acid	HA	A^-	K_a	pK_a
Pyruvic acid	$CH_3COCOOH$	$CH_3C—COO^-$	3.16×10^{-3}	2.50
Formic acid	HCOOH	$HCOO^-$	1.78×10^{-4}	3.75
Lactic acid	$CH_3CHOHCOOH$	$CH_3CHOHCOO^-$	1.38×10^{-4}	3.86
Benzoic acid	C_6H_5COOH	$C_6H_5COO^-$	6.46×10^{-5}	4.19
Acetic **acid**	CH_3COOH	CH_3COO^-	1.76×10^{-5}	4.76
Ammonium ion	NH_4^+	NH_3	5.6×10^{-10}	9.25
Oxalic acid (1)	HOOC—COOH	$HOOC—COO^-$	5.9×10^{-2}	1.23
Oxalic acid (2)	$HOOC—COO^-$	$^-OOC—COO^-$	6.4×10^{-5}	4.19
Malonic acid (1)	$HOOC—CH_2—COOH$	$HOOC—CH_2—COO^-$	1.49×10^{-3}	2.83
Malonic acid (2)	$HOOC—CH_2—COO^-$	$^-OOC—CH_2—COO^-$	2.03×10^{-6}	5.69
Malic acid (1)	$HOOC—CH_2—CHOH—COOH$	$HOOC—CH_2—CHOH—COO^-$	3.98×10^{-4}	3.40
Malic acid (2)	$HOOC—CH_2—CHOH—COO^-$	$^-OOC—CH_2—CHOH—COO^-$	5.5×10^{-6}	5.26
Succinic acid (1)	$HOOC—CH_2—CH_2—COOH$	$HOOC—CH_2—CH_2—COO^-$	6.17×10^{-5}	4.21
Succinic acid (2)	$HOOC—CH_2—CH_2—COO^-$	$^-OOC—CH_2—CH_2—COO^-$	2.3×10^{-6}	5.63
Carbonic acid (1)	H_2CO_3	HCO_3^-	4.3×10^{-7}	6.37
Carbonic acid (2)	HCO_3^-	CO_3^{2-}	5.6×10^{-11}	10.20
Citric acid (1)	$HOOC—CH_2—C(OH)(COOH)$ $\vert$ $CH_2—COOH$	$HOOC—CH_2—C(OH)(COOH)$ $\vert$ $CH_2—COO^-$	8.14×10^{-4}	3.09
Citric acid (2)	$HOOC—CH_2—C(OH)(COOH)$ $\vert$ $CH_2—COO^-$	$^-OOC—CH_2—C(OH)(COOH)$ $\vert$ $CH_2—COO^-$	1.78×10^{-5}	4.75
Citric acid (3)	$^-OOC—CH_2—C(OH)(COOH)$ $\vert$ $CH_2—COO^-$	$^-OOC—CH_2—C(OH)(COO^-)$ $\vert$ $CH_2—COO^-$	3.9×10^{-6}	5.41
Phosphoric acid (1)	H_3PO_4	$H_2PO_4^-$	7.25×10^{-3}	2.14
Phosphoric acid (2)	$H_2PO_4^-$	HPO_4^{2-}	6.31×10^{-8}	7.20
Phosphoric acid (3)	HPO_4^{2-}	PO_4^{3-}	3.98×10^{-13}	12.40

▶ Why do we want to know the pH?

An equation connects the K_a of any weak acid with the pH of a solution containing both that acid and its conjugate base. This relationship has wide use in biochemistry, especially where it is necessary to control pH for optimum reaction conditions. Some reactions cannot take place if the pH varies from the optimum value. Important biological macromolecules lose activity at extremes of pH. Figure 2.15 shows how the activities of three enzymes are affected by pH. Each enzyme has a peak activity that falls off rapidly as the pH is changed from the optimum. Also, some drastic physiological consequences can result from pH fluctuations in the body. Section 2-5 has more information about how

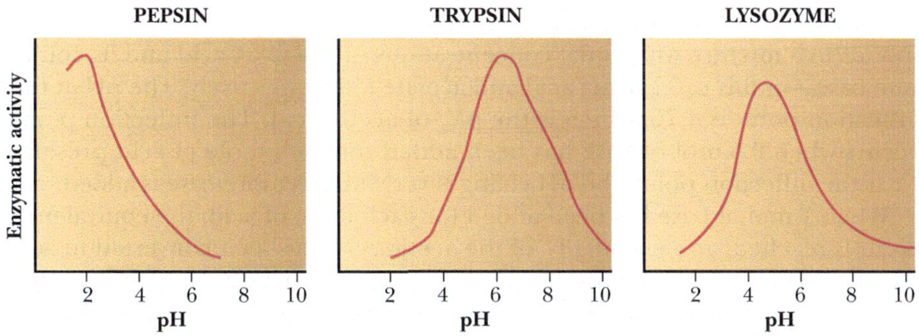

Figure 2.15 **pH versus enzymatic activity.** Pepsin, trypsin, and lysozyme all have steep pH optimum curves. Pepsin has maximum activity under very acidic conditions, as might be expected for a digestive enzyme that is found in the stomach. Lysozyme has its maximum activity near pH 5, while trypsin is most active near pH 6.

pH can be controlled. To derive the equation, we must first take the logarithm of both sides of the K_a equation:

$$K_a = \frac{[H^+][A^-]}{[HA]}$$

$$\log K_a = \log[H^+] + \log\frac{[A^-]}{[HA]}$$

$$-\log[H^+] = -\log K_a + \log\frac{[A^-]}{[HA]}$$

We then use the definitions of pH and pK_a:

$$pH = pK_a + \log\frac{[A^-]}{[HA]}$$

This relationship, known as the **Henderson–Hasselbalch equation**, is useful in predicting the properties of buffer solutions used to control the pH of reaction mixtures. When buffers are discussed in Section 2-5, we will be interested in the situation in which the concentration of acid, [HA], and the concentration of the conjugate base, $[A^-]$, are equal ($[HA] = [A^-]$). The ratio $[A^-]/[HA]$ is then equal to 1, and the logarithm of 1 is equal to zero. Therefore, when a solution contains equal concentrations of a weak acid and its conjugate base, the pH of that solution equals the pK_a value of the weak acid.

Henderson–Hasselbalch equation a mathematical relationship between the pK_a of an acid and the pH of a solution containing the acid and its conjugate base

2-4 Titration Curves

Learning Objectives

After reading this section, students will be able to:

LO 2.4.1 Draw a labeled pH titration curve for acetic acid.

LO 2.4.2 Distinguish between *monoprotic*, *diprotic*, *triprotic*, and *polyprotic acids*.

LO 2.4.3 Describe general relationships between pH, pK_a, and protonation states.

When base is added to a sample of acid, the pH of the solution changes. A **titration** is an experiment in which measured amounts of base are added to a measured amount of acid. The course of the reaction can be monitored with a pH meter. The point in the titration at which the acid is exactly neutralized is called the **equivalence point**.

titration an experiment in which a measured amount of base is added to an acid

equivalence point the point in a titration where an acid is exactly neutralized

If the pH is monitored as base is added to a sample of acetic acid in the course of a titration, an inflection point in the titration curve is reached when the pH equals the pK_a of acetic acid (Figure 2.16). As we saw in our discussion of the Henderson–Hasselbalch equation, a pH value equal to the pK_a corresponds to a mixture with equal concentrations of the weak acid and its conjugate base—in this case, acetic acid and acetate ion, respectively. The pH at the inflection point is 4.76, which is the pK_a of acetic acid. The inflection point occurs when 0.5 mol of base has been added for each mole of acid present. Near the inflection point, the pH changes very little as more base is added.

When 1 mol of base has been added for each mole of acid, the equivalence point is reached, and essentially all the acetic acid has been converted to acetate ion. Figure 2.16 also plots the relative abundance of acetic acid and acetate ion with increasing additions of NaOH. The percentage of acetic acid plus the percentage of acetate ion adds up to 100%. The acid (acetic acid) is progressively converted to its conjugate base (acetate ion) as more NaOH is added and the titration proceeds. Keeping track of the percentages of a conjugate acid and base in this way can help reveal the full significance of the reaction taking place in a titration. The form of the curves in Figure 2.16 represents the behavior of any monoprotic weak acid, but the value of the pK_a for each individual acid determines the pH values at the inflection point and at the equivalence point. See **Apply Your Knowledge 2.2.**

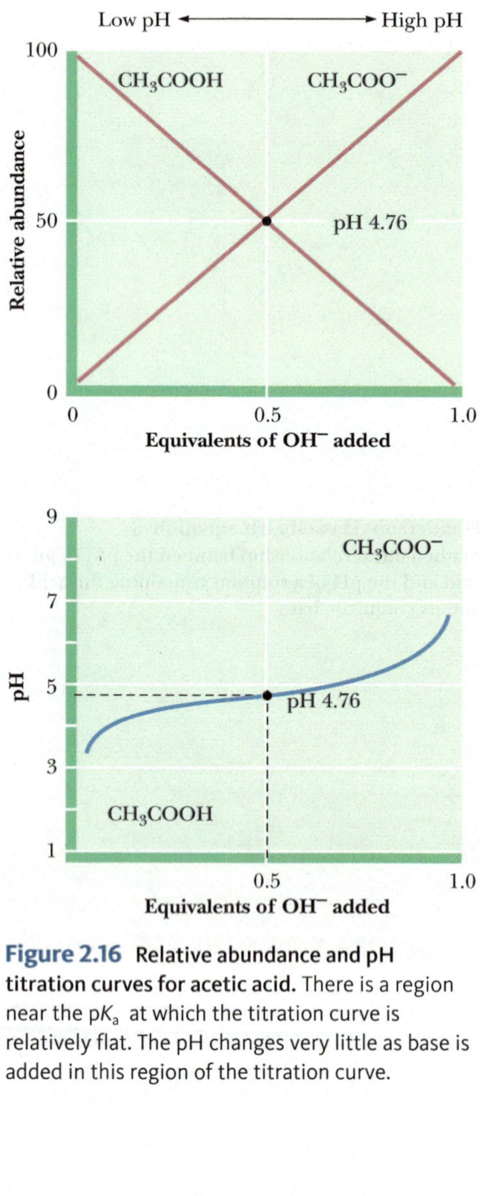

Figure 2.16 Relative abundance and pH titration curves for acetic acid. There is a region near the pK_a at which the titration curve is relatively flat. The pH changes very little as base is added in this region of the titration curve.

2.2 Apply Your Knowledge

Calculating pH Values for Weak Acid Titrations with Strong Base

Calculate the relative amounts of acetic acid and acetate ion present at the following points when 1 mol of acetic acid is titrated with sodium hydroxide. Also use the Henderson–Hasselbalch equation to calculate the values of the pH at these points.

a. 0.1 mol of NaOH is added
b. 0.3 mol of NaOH is added
c. 0.5 mol of NaOH is added
d. 0.7 mol of NaOH is added
e. 0.9 mol of NaOH is added

Solution

We approach this problem as an exercise in stoichiometry. There is a 1:1 ratio of moles of acid reacted to moles of base added. The difference between the original number of moles of acid and the number reacted is the number of moles of acid remaining. These are the values to be used in the numerator and denominator, respectively, of the Henderson–Hasselbalch equation.

a. When 0.1 mol of NaOH is added, 0.1 mol of acetic acid reacts with it to form 0.1 mol of acetate ion, leaving 0.9 mol of acetic acid. The composition is 90% acetic acid and 10% acetate ion.

$$\text{pH} = \text{p}K_a + \log\frac{0.1}{0.9}$$

$$\text{pH} = 4.76 + \log\frac{0.1}{0.9}$$

$$\text{pH} = 4.76 - 0.95$$

$$\text{pH} = 3.81$$

b. When 0.3 mol of NaOH is added, 0.3 mol of acetic acid reacts with it to form 0.3 mol of acetate ion, leaving 0.7 mol of acetic acid. The composition is 70% acetic acid and 30% acetate ion.

$$\text{pH} = \text{p}K_a + \log\frac{0.3}{0.7}$$

$$\text{pH} = 4.39$$

c. When 0.5 mol of NaOH is added, 0.5 mol of acetic acid reacts with it to form 0.5 mol of acetate ion, leaving 0.5 mol of acetic acid. The composition is 50% acetic acid and 50% acetate ion.

$$\text{pH} = \text{p}K_a + \log\frac{0.5}{0.5}$$

$$\text{pH} = 4.76$$

This one is possible to solve without doing much math, because the pH = pK_a when [HA] = [A^-]. Therefore, adding 0.5 mol of NaOH to 1 mol of acetic acid was enough NaOH to convert half of the acid to the conjugate base form. Therefore, the pH has to be equal to the pK_a.

d. When 0.7 mol of NaOH is added, 0.7 mol of acetic acid reacts with it to form 0.7 mol of acetate ion, leaving 0.3 mol of acetic acid. The composition is 30% acetic acid and 70% acetate ion.

$$\text{pH} = \text{p}K_a + \log\frac{0.7}{0.3}$$

$$\text{pH} = 5.13$$

e. When 0.9 mol of NaOH is added, 0.9 mol of acetic acid reacts with it to form 0.9 mol of acetate ion, leaving 0.1 mol of acetic acid. The composition is 10% acetic acid and 90% acetate ion.

$$\text{pH} = \text{p}K_a + \log\frac{0.9}{0.1}$$

$$\text{pH} = 5.71$$

Table 2.6 lists values for the acid dissociation constant, K_a, and for the pK_a for a number of acids. These acids are categorized into three groups: monoprotic, diprotic, and triprotic. Monoprotic acids release 1 hydrogen ion and have a single K_a and pK_a. Diprotic acids can release 2 hydrogen ions and have 2 K_a values and 2 pK_a values. Finally, triprotic acids can release 3 hydrogen ions and have 3 K_a values and 3 pK_a values. Acids that can release more than 1 hydrogen ion are also called polyprotic acids.

Is there a way to predict when the protonated versus the deprotonated forms of acids and their conjugate bases predominate? When the pH of a solution is less than the pK_a of an acid, the protonated (acid) form predominates. (Remember that the definition of pH includes a negative logarithm.) When the pH of a solution is greater than the pK_a of an acid, the deprotonated (conjugate base) form predominates (see Figure 2.16).

$$\text{pH} < \text{p}K_a$$

H^+ on, substance predominantly protonated

$$\text{pH} > \text{p}K_a$$

H^+ off, substance predominantly deprotonated

This concept can be particularly useful when we study polyprotic amino acids and peptides in Chapter 3.

2-5 Buffers

Learning Objectives

After reading this section, students will be able to:

LO 2.5.1 Explain how a buffer works upon addition of a strong acid or a strong base.

LO 2.5.2 Describe how to make a buffer solution.

buffer solution a solution that resists a change in pH upon addition of moderate amounts of strong acid or strong base

A **buffer solution** tends to resist change in pH when small to moderate amounts of a strong acid or strong base are added. A buffer solution consists of a mixture of a weak acid and its conjugate base.

▶ How do buffers work?

Let us compare the changes in pH that occur when equal amounts of strong acid or strong base are added to pure water at pH 7 and to a buffer solution at pH 7. If 1.0 mL of 0.1 *M* HCl is added to 99.0 mL of pure water, the pH drops drastically. If the same experiment is conducted with 0.1 *M* NaOH instead of 0.1 *M* HCl, the pH rises drastically (Figure 2.17).

Let's start by calculating what happens when we add 1 mL of 0.1 *M* HCl to 99.0 mL of pure water. Because HCl is a strong acid, we will assume that 0.1 *M* HCl dissociates completely to give 0.1 *M* H_3O^+. If we have 1 mL of the acid, we calculate the amount of H_3O^+ as follows:

$$1 \text{ mL} = 1 \times 10^{-3} \text{ L}$$

$$1 \times 10^{-3} \text{ L} \times 0.1 \text{ mol/L} = 1 \times 10^{-4} \text{ mol } H_3O^+$$

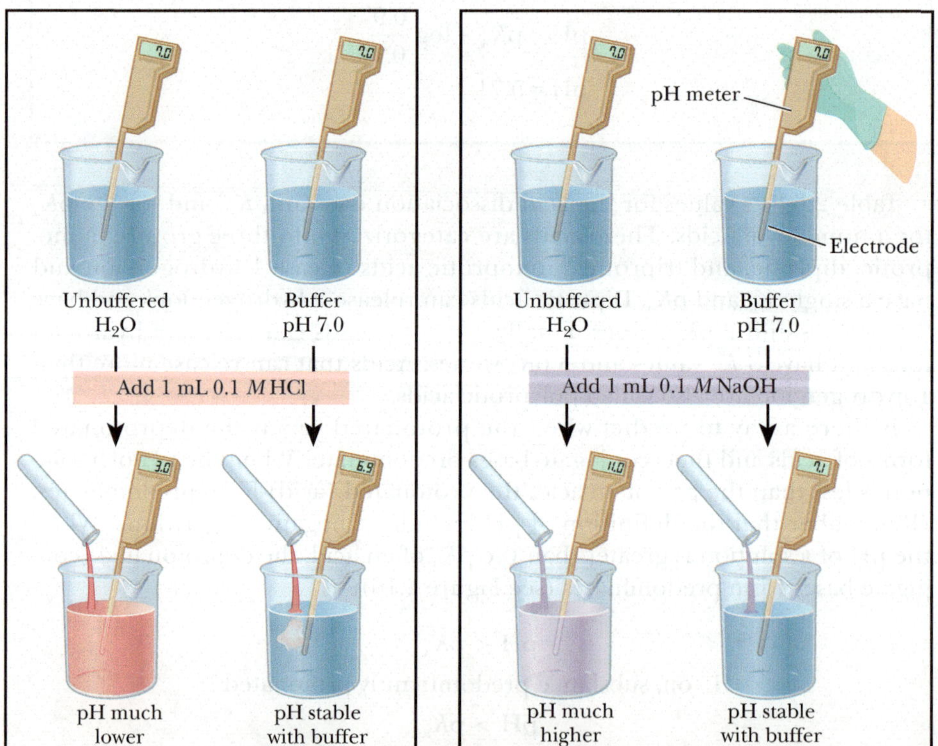

Figure 2.17 Buffering. Acid is added to the two beakers on the left. The pH of unbuffered H_2O drops dramatically while that of the buffer remains stable. Base is added to the two beakers on the right. The pH of the unbuffered water rises drastically while that of the buffer remains stable.

Therefore, 1×10^{-4} mol of H_3O^+ is diluted to a final volume of 100 mL or 0.1 L, because 1 mL was added to 99 mL. The final concentration of H_3O^+ is calculated as follows:

$$1 \times 10^{-4} \text{ mol } H_3O^+/0.1 \text{ L} = 1 \times 10^{-3}\ M$$

The pH is then calculated based on the definition:

$$\text{pH} = -\log [H_3O^+] = -\log (1 \times 10^{-3}) = 3$$

If we add 1 mL of 0.1 *M* NaOH, the calculations are done similarly to generate the concentration of $[OH^-]$, which also equals $1 \times 10^{-3}\ M$ because we used the same concentration and same volume of base.

The $[H_3O^+]$ is then calculated using the concentration of OH^- and the water equation:

$$[OH^-][H_3O^+] = 1 \times 10^{-14}$$

$$[H_3O^+] = 1 \times 10^{-14}/[OH^-] = 1 \times 10^{-14}/1 \times 10^{-3} = 1 \times 10^{-11}$$

Finally, the pH is calculated:

$$\text{pH} = -\log (1 \times 10^{-11}) = 11$$

The results are different when 99.0 mL of buffer solution is used instead of pure water. A solution that contains the hydrogen phosphate and dihydrogen phosphate ions, HPO_4^{2-} and $H_2PO_4^-$, in suitable proportions can serve as such a buffer. The Henderson–Hasselbalch equation can be used to calculate the $HPO_4^{2-}/H_2PO_4^-$ ratio that corresponds to pH 7.0, as shown in **Apply Your Knowledge 2.3.**

2.3 Apply Your Knowledge

Using the Henderson–Hasselbalch Equation

Convince yourself that the proper ratio of A^-/HA for pH 7.00 is 0.63 parts HPO_4^{2-} to 1 part $H_2PO_4^-$ by doing the calculation.

Solution

Use the Henderson–Hasselbalch equation with pH = 7.00 and pK_a = 7.20.

$$\text{pH} = pK_a + \log \frac{[A^-]}{[HA]}$$

$$7.00 = 7.20 + \log \frac{[HPO_4^{2-}]}{[H_2PO_4^-]}$$

$$-0.20 = \log \frac{[HPO_4^{2-}]}{[H_2PO_4]}$$

$$\frac{[HPO_4^{2-}]}{[H_2PO_4^-]} = 10^{(-0.20)} = 0.63$$

For purposes of illustration, let us consider a solution in which the concentrations are $[HPO_4^{2-}] = 0.063\ M$ and $[H_2PO_4^-] = 0.10\ M$; this gives the conjugate base–weak acid ratio of 0.63 seen previously. If 1.0 mL of 0.10 M HCl is added to 99.0 mL of the buffer, the reaction

$$HPO_4^{2-} + H^+ \rightleftharpoons H_2PO_4^-$$

takes place, and almost all the added H^+ will be used up. The concentrations of $[HPO_4^{2-}]$ and $[H_2PO_4^-]$ will change, and the new concentrations can be calculated.

Concentrations (mol/L)			
	$[HPO_4^{2-}]$	$[H^+]$	$[H_2PO_4^-]$
Before addition of HCl	0.063	1×10^{-7}	0.10
HCl added—no reaction yet	0.063	1×10^{-3}	0.10
After HCl reacts with HPO_4^{2-}	0.062	To be found	0.101

The new pH can then be calculated using the Henderson–Hasselbalch equation and the phosphate ion concentrations. The appropriate pK_a is 7.20 (Table 2.6).

$$pH = pK_a + \log\frac{[HPO_4^{2-}]}{[H_2PO_4^-]}$$

$$pH = 7.20 + \log\frac{0.062}{0.101}$$

$$pH = 6.99$$

The new pH is 6.99, a much smaller change than in the unbuffered pure water (Figure 2.17). Similarly, if 1.0 mL of 0.1 M NaOH is used, the same reaction takes place as in a titration:

$$H_2PO_4^- + OH^- \rightleftharpoons HPO_4^{2-}$$

Almost all the added OH^- is used up, but a small amount remains. Because this buffer is an aqueous solution, it is still true that $K_w = [H^+][OH^-]$. The increase in hydroxide ion concentration implies that the hydrogen ion concentration decreases and that the pH increases. Use the Henderson–Hasselbalch equation to calculate the new pH and to convince yourself that the result is pH = 7.01, again a much smaller change in pH than the one that took place in pure water (Figure 2.17).

Buffers work because they obey Le Châtelier's principle. This principle states that if stress is applied to a system in equilibrium, the equilibrium will shift in the direction that relieves the stress. Therefore, if we add hydrogen ion to a buffer system, we have added stress to the reaction:

$$HA \rightleftharpoons H^+ + A^-$$

To relieve the stress, the H^+ reacts with the A^- to maintain the equilibrium. We will see this effect many times in our studies. Many biological reactions will not take place unless the pH remains within fairly narrow limits, and, as a result, buffers have great practical importance in the biochemistry laboratory. See **Apply Your Knowledge 2.4**.

2.4 Apply Your Knowledge

How Do Buffers Work?

Calculate the pH if you added 3 mL of 0.1 *M* HCl to (a) 97 mL of pure water at pH 7.0 or (b) 100 mL of the same phosphate buffer at pH 7.0 from above.

Solution

a. 3 mL = 0.003 L of 0.1 *M* HCl.

$0.003\ \text{L} \times 0.1\ \text{mol/L} = 0.0003\ \text{moles}\ H^+$

The 0.0003 moles H^+ are in a final volume of 100 mL or 0.1 L:

$0.0003\ \text{moles}/0.1\ \text{L} = 0.003\ M\ H^+$

$\text{pH} = -\log 0.003 = 2.52$

b. From the description previously, we know that the phosphate buffer at pH 7.0 is composed of $HPO_4^{2-} = 0.063\ M$ and $H_2PO_4^- = 0.1\ M$. If there are 100 mL of the buffer, then the moles HA = 0.01 and the moles of $A^- - 0.0063$.

When we add 0.0003 moles of H^+ then the following reaction occurs: $HPO_4^{2-} + H^+ \rightarrow H_2PO_4^-$, until one of the reagents runs out. In this case, it will be the H^+.

The new amount of $H_2PO_4^-$ will be $0.01 + 0.0003 = 0.0103$.

The new amount of HPO_4^{2-} will be $0.0063 - 0.0003 = 0.0060$.

The pH is calculated using the Henderson–Hasselbalch equation:

$\text{pH} = 7.2 + \log(0.0060/0.0103) = 6.97$

2-5A Buffer Selection

Looking at a titration curve, such as the one in Figure 2.18A, can provide insight into how buffers work. The pH of a sample being titrated changes very little in the vicinity of the inflection point of a titration curve. Also, at the inflection point, half the amount of acid originally present has been converted to the conjugate base. The second stage of ionization of phosphoric acid,

$$H_2PO_4^- \rightleftharpoons H^+ + HPO_4^{2-}$$

was the basis of the buffer just used as an example. The pH at the inflection point of the titration is 7.20, a value numerically equal to the pK_a of the dihydrogen phosphate ion. At this pH, the solution contains equal concentrations of the dihydrogen phosphate ions and monohydrogen phosphate ions, the acid and base forms. Using the Henderson–Hasselbalch equation, we can calculate the ratio of the conjugate base form to the conjugate acid form for any

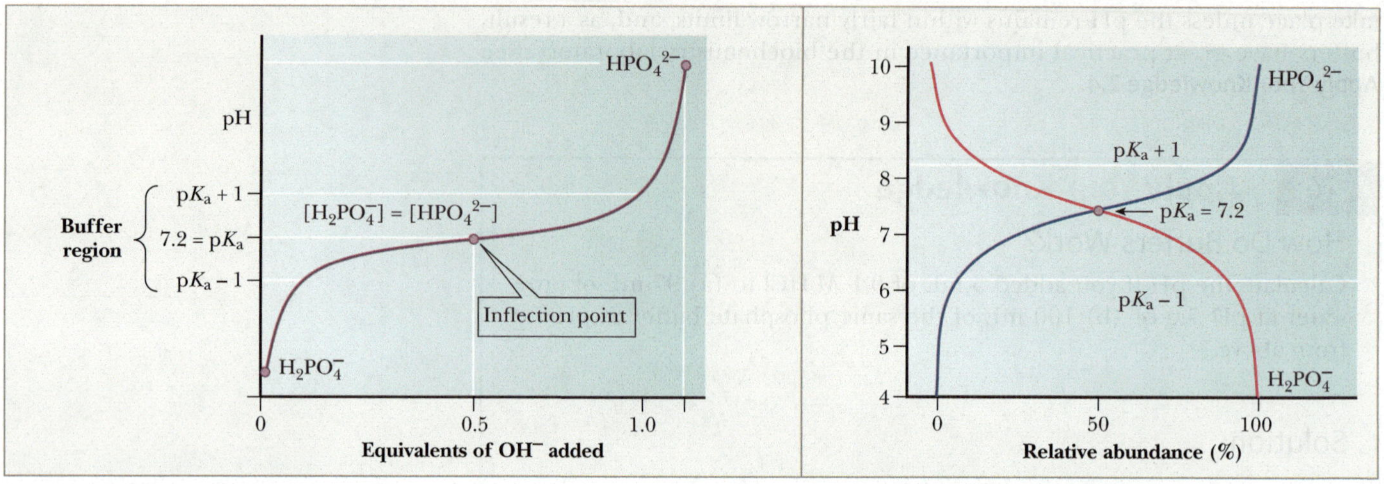

A The titration curve of $H_2PO_4^-$, showing the buffer region for the $H_2PO_4^-/HPO_4^{2-}$ pair.

B Relative abundance of $H_2PO_4^-$ and HPO_4^{2-}.

Figure 2.18 The relationship between the titration curve and buffering action in $H_2PO_4^-$.

Table 2.7 pH Values and Base–Acid Ratios for Buffers

If the pH equals	The ratio of base form to acid form equals
$pK_a - 3$	1/1000
$pK_a - 2$	1/100
$pK_a - 1$	1/10
pK_a	1/1
$pK_a + 1$	10/1
$pK_a + 2$	100/1
$pK_a + 3$	1000/1

pH when we know the pK_a. For example, if we choose a pH of 8.2 for a buffer composed of $H_2PO_4^-$ and HPO_4^{2-}, we can solve for the ratio:

$$\text{pH} = pK_a + \log\frac{[HPO_4^{2-}]}{[H_2PO_4^-]}$$

$$8.2 = 7.2 + \log\frac{[HPO_4^{2-}]}{[H_2PO_4^-]}$$

$$1.0 = \log\frac{[HPO_4^{2-}]}{[H_2PO_4^-]}$$

$$\frac{[HPO_4^{2-}]}{[H_2PO_4^-]} = 10$$

Thus, when the pH is 1 unit higher than the pK_a, the ratio of the conjugate base form to the conjugate acid form is 10. When the pH is 2 units higher than the pK_a, the ratio is 100, and so on. Table 2.7 shows this relationship for several increments of pH value.

A buffer solution can maintain the pH at a relatively constant value because of the presence of appreciable amounts of both the acid and its conjugate base.

This condition is met at pH values at or near the pK_a of the acid. If OH^- is added, an appreciable amount of the acid form of the buffer is present in solution to react with the added base. If H^+ is added, an appreciable amount of the basic form of the buffer also is present to react with the added acid.

The $H_2PO_4^-/HPO_4^{2-}$ pair is suitable as a buffer near pH 7.2, and the CH_3COOH/CH_3COO^- pair is suitable as a buffer near pH 4.76. At pH values below the pK_a, the acid form predominates, and at pH values above the pK_a, the basic form predominates. The plateau region in a titration curve, where the pH does not change rapidly, covers a pH range extending approximately 1 pH unit on each side of the pK_a. Thus, the buffer is effective within a range of about 2 pH units (Figure 2.18B).

In many biochemical studies, a strict pH range must be maintained for the experiment to be successful. Using our knowledge of the range of an effective buffer compared to its pK_a, we can select an appropriate buffer. If we were doing an experiment and needed the pH to be 7.2, we might select the $H_2PO_4^-/HPO_4^{2-}$ pair to be our buffer. If we wanted a pH near 9.0, we would look at tables of buffers to find one with a pK_a close to 9. **Biochemical Connections 2B** goes into greater detail on buffer selection.

2B Biochemical Connections • Buffer Chemistry

Buffer Selection

Much of biochemistry is studied by carrying out enzymatic reactions in a test tube or in vitro (literally, in glass). Such reactions are usually buffered to maintain a constant pH. Similarly, virtually all methods for enzyme isolation, and even for the growth of cells in tissue culture, use buffered solutions. The following criteria are typical for selecting a buffer for a biochemical reaction:

1. Suitable pK_a for the buffer
2. No interference with the reaction or with the assay
3. Suitable ionic strength of the buffer
4. No precipitation of reactants or products due to presence of the buffer
5. Nonbiological nature of the buffer

The rule of thumb is that the pK_a should be ±1 pH unit from the pH of the reaction; ±1/2 pH unit is even better. Although the perfect generic buffer would have a pH equal to its pK_a, if the reaction is known to produce an acidic product, it is advantageous if the pK_a is below the reaction pH, because then the buffer capacity increases as the reaction proceeds.

Sometimes a buffer can interfere with a reaction or with the assay method. For example, a reaction that requires or produces phosphate or CO_2 may be inhibited if too much phosphate or carbonate is present in the reaction mixture. Even the counterion may be important. Typically a phosphate or carbonate buffer is prepared from the Na^+ or K^+ salt. Because many enzymes that react with nucleic acids are activated by one of these two ions and inhibited by the other, the choice of Na^+ or K^+ for a counterion could be critical. A buffer can also affect the spectrophotometric determination of a colored assay product.

If a buffer has a poor buffering capacity at the desired pH, its efficiency can often be increased by increasing the concentration; however, many enzymes are sensitive to high salt concentration. Beginning students in biochemistry often have difficulty with enzyme isolations and assays because they fail to appreciate the sensitivity of many enzymes. Fortunately, to minimize this problem, most beginning biochemistry laboratory manuals call for the use of enzymes that are very stable.

A buffer may cause precipitation of an enzyme or even of a metallic ion that may be a cofactor for the reaction. For example, many phosphate salts of divalent cations are only marginally soluble.

Finally, it is often desirable to use a buffer that has no biological activity at all, so it can never interfere with the system being studied. TRIS [tris(hydroxymethyl)aminomethane] is a very desirable buffer, because it rarely interferes with a reaction. Special buffers, such as HEPES [4-(2-hydroxyethyl)-1-piperazineethanesulfonic acid] and PIPES [piperazine-*N*,*N'*-bis(2-ethanesulfonate)] (Table 2.8), have been developed for growing cells in tissue culture. ◗

The condition that a buffer contains appreciable amounts of both a weak acid and its conjugate base applies both to the ratio of the two forms and to the absolute amount of each present in a given solution. If a buffer solution contained a suitable ratio of acid to base, but very low concentrations of both, it would take very little added acid to use up all of the base form, and vice versa. A buffer solution with low concentrations of both the acid and the base forms is said to have a low **buffering capacity**. A buffer that contains greater amounts of both acid and base has a higher buffering capacity.

buffering capacity a measure of the amount of acid or base that can be absorbed by a given buffer solution

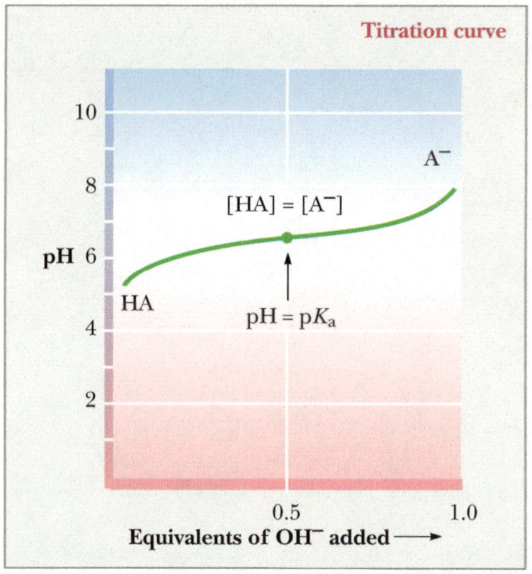

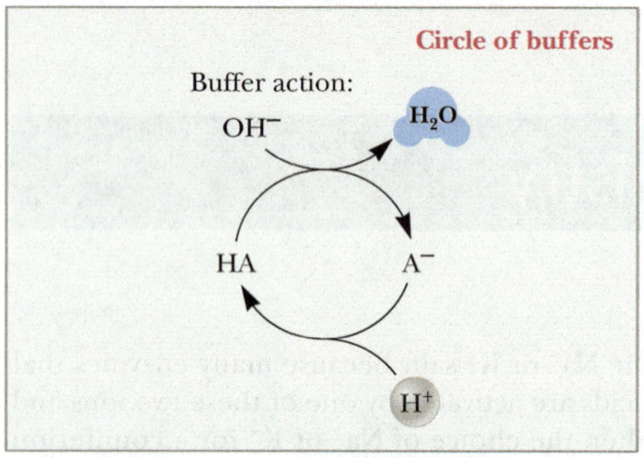

Figure 2.19 Two ways of looking at buffers. In the titration curve, the pH varies only slightly near the region in which [HA] = [A$^-$]. In the circle of buffers, adding OH$^-$ to the buffer converts HA to A$^-$, whereas adding H$^+$ converts A$^-$ to HA.

▶ How do we make buffers in the laboratory?

When we study buffers in theory, we often use the Henderson–Hasselbalch equation and do many calculations concerning ratios of conjugate base form to weak acid form. In practice, however, making a buffer is much easier. To have a buffer, all that is necessary are the two forms of the buffer present in the solution at reasonable quantities. This situation can be obtained by adding predetermined amounts of the conjugate base form (A^-) to the acid form (HA), or we could start with one and create the other, which is how it is usually done in practice. Remember that HA and A^- are interconverted by adding strong acid or strong base (Figure 2.19). To make a buffer, we could start with the HA form and add NaOH until the pH is correct, as determined by a pH meter. We could also start with A^- and add HCl until the pH is correct. Depending on the relationship of the pH we desire to the pK_a of the buffer, it may be more convenient to start with one than the other. For example, if we are making an acetic acid/acetate buffer at pH 5.7, it would make more sense to start with the A^- form and to add a small amount of HCl to bring the pH down to 5.7, rather than to start with HA and to add much more NaOH to bring the pH up past the pK_a.

2-5B Biological Buffers

Up until now, we have been considering buffers from the perspective of a chemist trying to control an experiment. However, the real importance of buffers is that they are critical to life. Buffer systems in living organisms and in the laboratory are based on many types of compounds. Because physiological pH in most organisms stays around 7, it might be expected that the phosphate buffer system would be widely used in living organisms. This is the case where phosphate ion concentrations are high enough for the buffer to be effective, as in most intracellular fluids.

▶ What naturally occurring pH buffers are present in living organisms?

The $H_2PO_4^-/HPO_4^{2-}$ pair is the principal buffer in cells. In blood, phosphate ion levels are inadequate for buffering, and a different system operates.

The buffering system in blood is based on the dissociation of carbonic acid (H_2CO_3):

$$H_2CO_3 \rightleftharpoons H^+ + HCO_3^-$$

where the pK_a of H_2CO_3 is 6.37. The pH of human blood, 7.4, is near the end of the buffering range of this system, but another factor enters into the situation.

Carbon dioxide can dissolve in water and in water-based fluids, such as blood. The dissolved carbon dioxide forms carbonic acid, which reacts, in turn, to produce bicarbonate ion. The conversion of aqueous CO_2 and water to carbonic acid is catalyzed by the enzyme carbonic anhydrase. This is one of the most efficient enzymes in biochemistry, capable of converting a million molecules of CO_2 to carbonic acid per second. Enzyme efficiency is discussed in more detail in Chapter 6.

$$CO_2(g) \rightleftharpoons CO_2(aq)$$
$$CO_2(aq) + H_2O(\ell) \rightleftharpoons H_2CO_3(aq)$$
$$H_2CO_3(aq) \rightleftharpoons H^+(aq) + HCO_3^-(aq)$$
$$\text{Net equation: } CO_2(g) + H_2O(\ell) \rightleftharpoons H^+(aq) + HCO_3^-(aq)$$

At the pH of blood, which is about 1 unit higher than the pK_a of carbonic acid, most of the dissolved CO_2 is present as HCO_3^-. The CO_2 being transported to the lungs to be expired takes the form of bicarbonate ion. A direct relationship exists between the pH of the blood and the pressure of carbon dioxide gas in the lungs. The properties of hemoglobin, the oxygen-carrying protein in the blood, also enter into the situation.

The phosphate buffer system is common in the laboratory (in vitro) as well as in living organisms (in vivo). The buffer system based on TRIS is also widely used in vitro. Other buffers that have come into wide use more recently are **zwitterions**, which are compounds that have both a positive charge and a negative charge. Zwitterions are usually considered less likely to interfere with biochemical reactions than some of the earlier buffers (Table 2.8).

Most living systems operate at pH levels close to 7. The pK_a values of many functional groups, such as the carboxyl and amino groups, are well above or well below this value. As a result, under physiological conditions, many important biomolecules exist as charged species to one extent or another. The practical consequences of this fact are explored in **Biochemical Connections 2C** and **Biochemical Connections 2D**.

zwitterions molecules that have both a positive and a negative charge

acidosis a condition in which blood pH drops below 7.35

alkalosis a condition in which blood pH rises above 7.45

Table 2.8 Acid and Base Forms of Some Useful Biochemical Buffers

Name	Acid Form		Base Form	pK_a
N-*tris*(hydroxymethyl)aminomethane (TRIS)	TRIS-H⁺ (protonated form) $(HOCH_2)_3CNH_3^+$	⇌	TRIS (free amine) $(HOCH_2)_3CNH_2$	8.3
N-*tris*(hydroxymethyl)methyl-2-aminoethane sulfonate (TES)	$^-$TES-H⁺ (zwitterionic form) $(HOCH_2)_3C\overset{+}{N}H_2CH_2CH_2SO_3^-$	⇌	$^-$TES (anionic form) $(HOCH_2)_3CNHCH_2CH_2SO_3^-$	7.55
N-2-hydroxyethylpiperazine-*N*′-2-ethane sulfonate (HEPES)	HEPES-H⁺ (zwitterionic form) 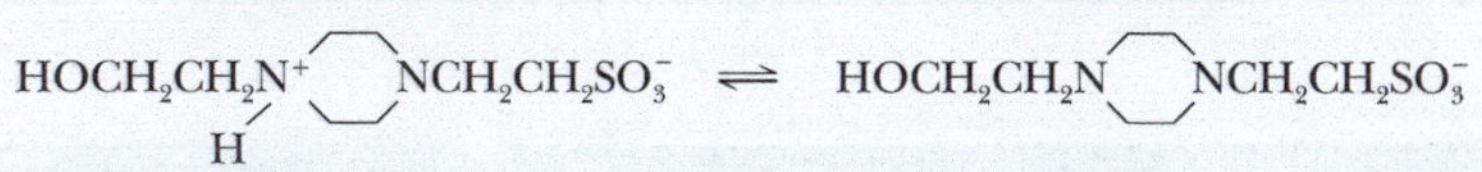		$^-$HEPES (anionic form)	7.55
3-(*N*-morpholino)propane-sulfonic acid (MOPS)	$^-$MOPS-H⁺ (zwitterionic form) 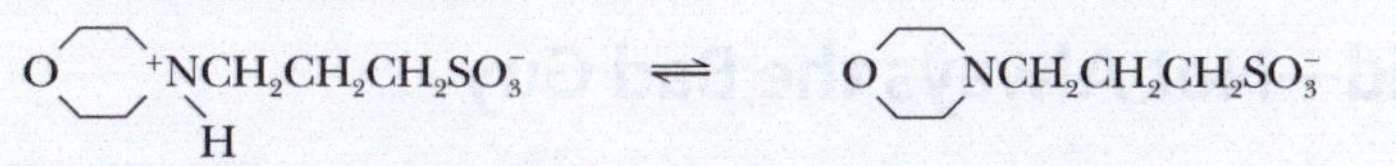		$^-$MOPS (anionic form)	7.2
Piperazine-*N*,*N*′-bis(2-ethanesulfonic acid) (PIPES)	$^{2-}$PIPES-H⁺ (protonated dianion) 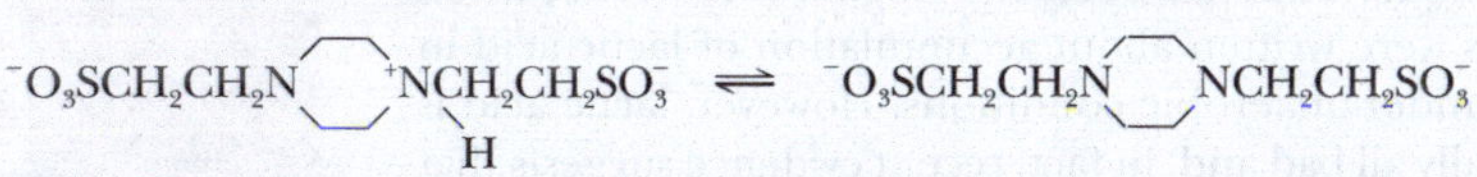		$^{2-}$PIPES (dianion)	6.8

2C Biochemical Connections • Chemistry of Blood

Some Physiological Consequences of Blood Buffering

The maintenance of proper pH is an essential function in all living organisms. The hydrogen ion concentration must be kept in a very narrow range. Normal human blood, for example, has a pH of 7.4. It may go up or down by about 0.05 depending on the concentration of metabolic waste products. Some disease states can cause pH changes that can be deadly. When human blood falls below pH 7.35, the condition is called **acidosis**. This can come from excessive metabolic production of acid or a failure of the kidneys to remove metabolic acid byproducts. This may occur during uncontrolled diabetes or during starvation. If the pH of the blood drops below 7, coma and death can occur. If the pH rises above 7.45, it is called **alkalosis**, which could be caused by excessive vomiting or taking in large amounts of alkaline drugs. If left unchecked, respiratory arrest may occur.

The process of respiration plays an important role in the buffering of blood. In particular, an increase in H^+ concentration can be dealt with by raising the rate of respiration. Initially, the added hydrogen ion binds to bicarbonate ion, forming carbonic acid.

$$H^+(aq) + HCO_3^-(aq) \rightleftharpoons H_2CO_3(aq)$$

An increased level of carbonic acid raises the levels of dissolved carbon dioxide and, ultimately, gaseous carbon dioxide in the lungs.

$$H_2CO_3(aq) \rightleftharpoons CO_2(aq) + H_2O(\ell)$$
$$CO_2(aq) \rightleftharpoons CO_2(g)$$

A high respiration rate removes this excess carbon dioxide from the lungs, starting a shift in the equilibrium positions of all the foregoing reactions. The removal of gaseous CO_2 decreases the amount of dissolved CO_2. Hydrogen ion reacts with HCO_3, and, in the process, lowers the H^+ concentration of blood back to its original level. In this way, the blood pH is kept constant.

In contrast, *hyperventilation* (excessively deep and rapid breathing) removes such large amounts of carbon dioxide from the lungs that it raises the pH of blood, sometimes to dangerously high levels that bring on weakness and fainting. Athletes, however, have learned how to use the increase in blood pH caused by hyperventilation. Short bursts of strenuous exercise produce high levels of lactic acid in the blood as a result of the anaerobic breakdown of glycogen. The presence of so much lactic acid tends to lower the pH of the blood, but a brief (30-second) period of hyperventilation before a short-distance event (say, a 400-m dash, 100-m swim, 1-km bicycle race, or any event that lasts between 30 seconds and about a minute) counteracts the effects of the added lactic acid and maintains the pH balance.

An increase in H^+ in blood can be caused by large amounts of any acid entering the bloodstream. Aspirin, like lactic acid, is an acid, and extreme acidity resulting from the ingestion of large doses of aspirin can cause *aspirin poisoning*. Exposure to *high altitudes* has an effect similar to hyperventilation at sea level. In response to the tenuous atmosphere, the rate of respiration increases. As with hyperventilation, more carbon dioxide is expired from the lungs, ultimately lowering the H^+ level in blood and raising the pH. When people who normally live at sea level are suddenly placed at a high elevation, their blood pH rises temporarily, until they become acclimated. ▶

2D Biochemical Connections • Acids and Sports

Lactic Acid—Not Always the Bad Guy

If you ask anybody who knows something of sports about lactic acid, you will likely hear that it is the acid that causes muscle pain and muscle fatigue. This has been the dogma since 1929, when the first papers were written about accumulation of lactic acid in muscle tissue under anaerobic conditions. However, lactic acid is not biochemically all bad and, in fact, recent evidence suggests that there are some benefits we did not previously know about.

In the first place, we should distinguish between the hydrogen ion dissociated from the lactic acid and the conjugate base, lactate. The H^+ ion released is the ion that is the reactive species and most likely accounts for the pain we associate with lactic acid buildup in the muscles. The lowering of the cellular pH would have effects on a variety of enzyme and muscular systems. However, the conjugate base, lactate, is removed by the blood and makes its way to the liver. Once in the liver, it is converted to glucose via a process called gluconeogenesis, which we discuss in Chapter 18. Hospital patients are often given intravenous lactate solutions to indirectly help keep their blood glucose levels up.

SHAUN BEST/Reuters

Figure 2.20 Athletes often collapse after building up too much lactic acid.

Until recently, any athlete would have proclaimed the evils of lactic acid on muscle performance. However, recent evidence suggests that lactic acid actually has a positive effect on fatiguing muscles. A recent study suggested that in a situation where muscles were becoming fatigued, the lactic acid actually maintained the muscle membrane's ability to depolarize and repolarize longer, allowing the muscles to continue to contract even though they were fatigued. When isolated muscle cells were stripped of lactic acid, they fatigued even faster. The same evidence was seen with cases of humans that suffered from a disease where they lacked the enzyme that breaks down muscle glycogen. Without the muscle glycogen breakdown, anaerobic metabolism is impossible and no lactic acid accumulates. As a result, these patients' muscle fibers fatigued even faster.

Some of the "knowledge" about the undesirable effect of lactic acid on muscle fatigue may be explained by the fact that muscle pain seems linked to decreased muscle performance. Lactic acid, as the putative cause of the muscle pain, was therefore assumed to also be the cause of the fatigue. This is an active area of research, and we still have a lot to learn about it. Despite decades of study and popular myth, we still do not really know exactly what causes muscle fatigue, although the athlete shown in Figure 2.20 would most likely tell you it was the lactic acid.

Summary

Water and Polarity

- When two atoms with the same electronegativity form a bond, the electrons are shared equally between the two atoms, leading to a nonpolar covalent bond.
- When atoms with significantly different electronegativities form a bond, the electrons are not shared equally. The more electronegative atom pulls the shared electrons closer, resulting in a partial negative charge and creating a polar covalent bond.
- The polar nature of water largely determines its solvent properties.
- Ionic compounds with full charges and polar compounds with partial charges tend to dissolve in water.
- The underlying physical principle is electrostatic attraction between unlike charges. The negative end of a water dipole attracts either a positive ion or the positive end of another dipole, whereas the positive end of a water molecule attracts either a negative ion or the negative end of another dipole.
- Some molecules are amphipathic—having both polar (hydrophilic) heads and nonpolar (hydrophobic) tail portions.
- When amphipathic molecules and water separate into layers, the polar head groups of the amphipathic molecules are in contact with the aqueous environment, whereas the nonpolar tails are sequestered from the water. London dispersion forces between the nonpolar tails provide the energetic basis for this spontaneous molecular arrangement.

Hydrogen Bonds

- Water has unique properties for a molecule with its molecular weight, such as a very high boiling point and melting point.
- This is due to the extensive hydrogen bonding possible between water molecules.
- Each water molecule has two sources of partial positive charge and two of partial negative charge. This allows water to form an array in its solid form and to bond with many other water molecules in its liquid form.
- The extensive hydrogen bonding requires large amounts of energy to disrupt, so water melts and boils at higher temperatures than other molecules of comparable molecular weight.

Acids, Bases, and pH

- Acids are compounds that release hydrogen ions (protons) when dissolved in aqueous solution. Acids are proton donors.
- Bases are compounds that accept protons.
- The mathematical definition of pH is the negative of the logarithm of the hydrogen ion concentration: $pH = -\log [H^+]$.
- pH is a measure of the acidity of the solution. The lower the pH, the more acidic the solution.
- Because of the log term, a pH change of 1 unit means a 10-fold change in hydrogen ion concentration.
- Knowing the pH and being able to control it are both important because many biological reactions require a very tight range of pH values. For example, an enzyme that is active at pH 7.0 may be completely inactive at pH 8.0.
- Solutions used in science often must have their pH controlled in order to have an experiment function correctly.
- Although local variations in pH may occur in certain subcellular organelles, a cell must maintain a pH near neutrality in order to stay alive.

Buffers

- Buffers consist of weak acids and their conjugate bases.
- If a source of extra hydrogen ion is added to a buffer solution, it reacts with the conjugate base to form the weak acid.
- If a source of hydroxide ion is added to the buffer, it reacts with the weak acid to form water and the conjugate base.
- In this way, either added H^+ or OH^- is "used up" by the buffer. This keeps the pH much more stable than if the same acid or base were added to an unbuffered system.
- We choose a buffer primarily by knowing the pH that we wish to maintain. For example, if we are performing an experiment and we want the solution to stay at pH 7.5, we look for a buffer that has a pK_a of 7.5 because buffers are most effective when the pH is close to the buffer pK_a.
- The most efficient way to make a buffer in the laboratory is to place either the weak acid form or the weak base form of the buffer compound in a container, add water, and then measure the pH with a pH meter. We then add strong acid or strong base until the pH is the desired buffer pH. Finally, we bring the solution up to the final volume so that concentration is correct.
- Buffers are not just an artificial system used in the laboratory. Living systems are buffered by naturally occurring compounds.
- Naturally occurring phosphate and carbonate buffers help maintain physiological pH near 7.0.

Review Exercises

2-1 Water and Polarity

1. **Recall** What is a van der Waals force?
2. **Recall** State whether each of the following is an example of a van der Waals force: ionic bonds, salt bridges, ion–dipole interactions, dipole–dipole interactions, hydrogen bonds, dipole–induced dipole interactions, induced dipole–induced dipole interactions (London dispersion forces).
3. **Recall** What is an induced dipole?
4. **Recall** What is a salt bridge?
5. **Recall** Under what circumstance is a molecule that has a dipole not a polar molecule?
6. **Reflect and Apply** Which would you think would be a stronger interaction and why: an interaction between a sodium ion and the partial negative charge on the oxygen in ethanol (CH_3CH_2OH) or the interaction between two ethanol molecules?
7. **Recall** List the three types of van der Waals forces from weakest to strongest.
8. **Recall** List the seven types of intermolecular attractive forces from weakest to strongest.
9. **Recall** State whether each of the following molecules is capable of forming a micelle: cholesterol, sodium lauryl sulfate, lecithin (a phosphatidyl choline), glucose, sodium palmitate, urea.

2-2 Hydrogen Bonds

10. **Recall** A hydrogen bond is a special case of what type of intermolecular force?
11. **Recall** What are two macromolecules that have hydrogen bonds as a part of their structures?
12. **Reflect and Apply** Explain why hydrogen bonding is not observed between CH_4 molecules.
13. **Reflect and Apply** Give three types of small molecules that can form hydrogen bonds; at least one of the molecules must contain nitrogen.
14. **Recall** What are the requirements for molecules to form hydrogen bonds? (What atoms must be present and involved in such bonds?)
15. **Reflect and Apply** Many properties of acetic acid can be rationalized in terms of a hydrogen-bonded dimer. Propose a structure for such a dimer.

2-3 Acids, Bases, and pH

16. **Recall** Identify the conjugate acids and bases in the following pairs of substances:

    (a) $(CH_3)_3NH^+/(CH_3)_3N$
    (b) $^+H_3N{-}CH_2COOH/^+H_3N{-}CH_2{-}COO^-$
    (c) $H_2N{-}CH_2{-}COO^-/^+H_3N{-}CH_2{-}COO^-$
    (d) $^-OOC{-}CH_2{-}COO^-/^-OOC{-}CH_2{-}COOH$
    (e) $HOOC{-}CH_2{-}COOH/^-OOC{-}CH_2{-}COOH$

17. **Recall** Identify the conjugate acids and bases in the following pairs of substances:

    (a) $(HOCH_2)_3CNH_2$ $(HOCH_2)_3CNH_3^+$

    (b) 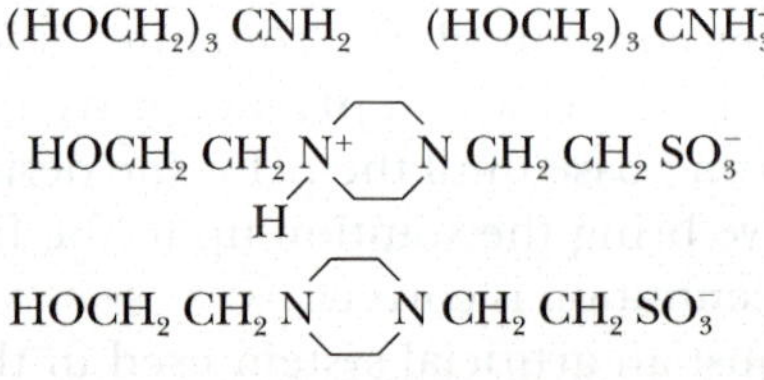

    (c) 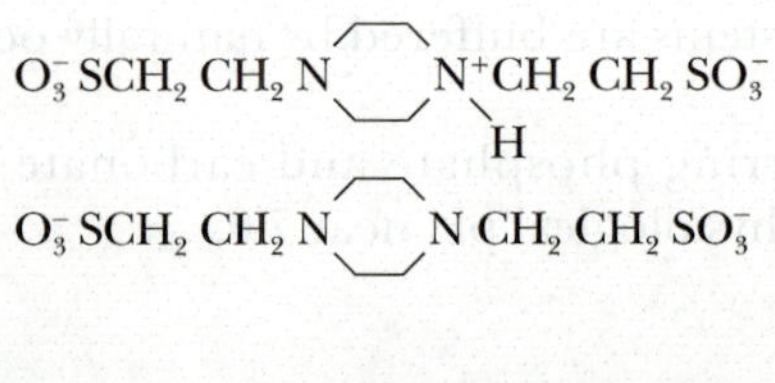

18. **Recall** Why does the pH change by 1 unit if the hydrogen ion concentration changes by a factor of 10?
19. **Mathematical** Calculate the hydrogen ion concentration, $[H^+]$, for each of the following substances:
    (a) Blood plasma, pH 7.4
    (b) Orange juice, pH 3.5
    (c) Human urine, pH 6.2
    (d) Household ammonia, pH 11.5
    (e) Gastric juice, pH 1.8
20. **Mathematical** Calculate the hydrogen ion concentration, $[H^+]$, for each of the following substances:
    (a) Saliva, pH 6.5
    (b) Intracellular fluid of liver, pH 6.9
    (c) Tomato juice, pH 4.3
    (d) Grapefruit juice, pH 3.2
21. **Mathematical** Calculate the hydroxide ion concentration, $[OH^-]$, for each of the substances listed in Question 20.
22. **Recall** What is the pH of a 0.200 *M* solution of lactic acid?

2-4 Titration Curves

23. **Recall** Define the following:
    (a) Buffering capacity
    (b) Equivalence point
    (c) Titration
24. **Recall** State whether each of the following molecules will be ionizable in aqueous solution between pH 2 and pH 12: benzoic acid, indole, imidazole, benzyl alcohol, phenol, benzamide.
25. **Reflect and Apply** Aspirin is an acid with a pK_a of 3.5; its structure includes a carboxyl group like that seen in the structure of acetic acid. To be absorbed into the bloodstream, it must pass through the membrane lining the stomach and the small intestine. Electrically neutral molecules can pass through a membrane more easily than can charged molecules. Would you expect more aspirin to be absorbed in the stomach, where the pH of gastric juice is about 1, or in the small intestine, where the pH is about 6? Explain your answer.
26. **Recall** Draw the structures of the major conjugate acid and conjugate base pair that would be present in a malate buffer at pH 5.0.
27. **Recall** You dissolve 7.684 grams of citric acid (MW 192.12 g/mol) in water to make a 200.0 mL solution. You then add 30 mL of 1.00 *M* NaOH to the citric acid solution.
    (a) Is the resulting solution a buffer?
    (b) After the addition of sodium hydroxide, at approximately what location on the following titration curve (A–G) would you be?

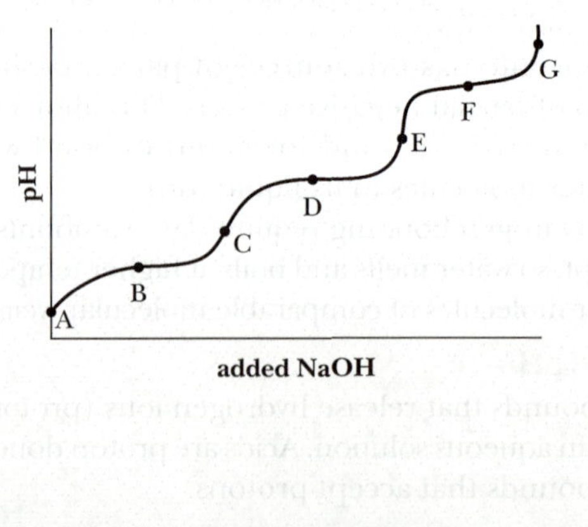

2-5 Buffers

28. **Biochemical Connections** List two criteria used to select a buffer for a biochemical reaction.
29. **Biochemical Connections** What is the relationship between pK_a and the useful range of a buffer?
30. **Mathematical** What is the $[CH_3COO^-]/[CH_3COOH]$ ratio in an acetate buffer at pH 5.00?
31. **Mathematical** What is the $[CH_3COO^-]/[CH_3COOH]$ ratio in an acetate buffer at pH 4.00?
32. **Mathematical** What is the ratio of $[TRIS]/[TRIS\text{-}H^+]$ in a TRIS buffer at pH 8.7?
33. **Mathematical** What is the ratio of $[HEPES]/[HEPES\text{-}H^+]$ in a HEPES buffer at pH 7.9?
34. **Mathematical** How would you prepare 1 L of a 0.050 *M* phosphate buffer at pH 7.5 using crystalline K_2HPO_4 and a solution of 1.0 *M* HCl?
35. **Mathematical** The buffer needed for Question 34 can also be prepared using crystalline NaH_2PO_4 and a solution of 1.0 *M* NaOH. How would you do this?
36. **Mathematical** Calculate the pH of a buffer solution prepared by mixing 75 mL of 1.0 *M* lactic acid (see Table 2.6) and 25 mL of 1.0 *M* sodium lactate.
37. **Mathematical** Calculate the pH of a buffer solution prepared by mixing 25 mL of 1.0 *M* lactic acid and 75 mL of 1.0 *M* sodium lactate.
38. **Mathematical** Calculate the pH of a buffer solution that contains 0.10 *M* acetic acid (Table 2.6) and 0.25 *M* sodium acetate.
39. **Mathematical** If you mix equal volumes of 0.1 *M* HCl and 0.20 *M* TRIS (free amine form; see Table 2.8), is the resulting solution a buffer? Why or why not?
40. **Mathematical** What would be the pH of the solution described in Question 39?
41. **Mathematical** If you have 100 mL of a 0.10 *M* TRIS buffer at pH 8.3 (Table 2.8) and you add 3.0 mL of 1 *M* HCl, what will be the new pH?
42. **Mathematical** What would be the pH of the solution in Question 41 if you were to add 3.0 mL more of 1 *M* HCl?
43. **Mathematical** What is the ratio of concentrations of acetate ion and undissociated acetic acid in a solution that has a pH of 5.12?
44. **Biochemical Connections** You need to carry out an enzymatic reaction at pH 7.5. A friend suggests a weak acid with a pK_a of 3.9 as the basis of a buffer. Will this substance and its conjugate base make a suitable buffer? Why or why not?
45. **Mathematical** If the buffer suggested in Question 44 were made, what would be the ratio of the [conjugate base]/[conjugate acid]?
46. **Biochemical Connections** Suggest a suitable buffer range for each of the following substances:
    (a) Lactic acid (pK_a = 3.86) and its sodium salt
    (b) Acetic acid (pK_a = 4.76) and its sodium salt
    (c) TRIS (pK_a = 8.3; see Table 2.8) in its protonated form and its free amine form
    (d) HEPES (pK_a = 7.55; see Table 2.8) in its zwitterionic form and its anionic form
47. **Biochemical Connections** Which of the buffers shown in Table 2.8 would you choose to make a buffer with a pH of 7.3? Explain why.
48. **Mathematical** Define buffering capacity. How do the following buffers differ in buffering capacity? How do they differ in pH?
    Buffer A: 0.01 *M* Na_2HPO_4 and 0.01 *M* NaH_2PO_4
    Buffer B: 0.10 *M* Na_2HPO_4 and 0.10 *M* NaH_2PO_4
    Buffer C: 1.0 *M* Na_2HPO_4 and 1.0 *M* NaH_2PO_4
49. **Biochemical Connections** If you wanted to make a HEPES buffer at pH 8.3 and you had both HEPES acid and HEPES base available, which would you start with, and why?
50. **Biochemical Connections** A frequently recommended treatment for hiccups is to hold your breath. The resulting condition, hypoventilation, causes a buildup of carbon dioxide in the lungs. What effect will this have on the pH of your blood?

Further Reading

Allen, D., and H. Westerblad. Lactic Acid—The Latest Performance-Enhancing Drug. *Science* **305**, 1112–1113 (2004). An article refuting certain assumptions about lactic acid buildup and muscle fatigue.

Barrow, G. M. *Physical Chemistry for the Life Sciences*, 2nd ed. New York: McGraw-Hill, 1981. Acid–base reactions are discussed in Chapter 4, with titration curves treated in great detail.

Fasman G. D., ed. *Handbook of Biochemistry and Molecular Biology: Physical and Chemical Data Section*, 2 vols., 3rd ed. Cleveland, OH: Chemical Rubber Company, 1976. Includes a section on buffers and directions for preparation of buffer solutions (vol. 1, pp. 353–378). Other sections cover all important types of biomolecules.

Ferguson, W. J., and N. E. Good. Hydrogen Ion Buffers. *Anal. Biochem.* **104**, 300–310 (1980). A description of useful zwitterionic buffers.

Gerstein, M., and M. Levitt. Simulating Water and the Molecules of Life. *Sci. Amer.* **279** (5), 101–105 (1998). A description of computer modeling as a tool to investigate the interaction of water molecules with proteins and DNA.

Hellmans, A. Getting to the Bottom of Water. *Science* **283**, 614–615 (1999). Recent research indicates that the hydrogen bond may have some covalent character, affecting the properties of water.

Jeffrey, G. A. *An Introduction to Hydrogen Bonding.* New York: Oxford University Press, 1997. An advanced, book-length treatment of hydrogen bonding. Chapter 10 is devoted to hydrogen bonding in biological molecules.

Olson, A., and D. Goodsell. Visualizing Biological Molecules. *Sci. Amer.* **268** (6), 62–68 (1993). An account of how computer graphics can be used to represent molecular structure and properties.

Pauling, L. *The Nature of the Chemical* Bond, 3rd ed. Ithaca, NY: Cornell University Press, 1960. A classic. Chapter 12 is devoted to hydrogen bonding.

Pedersen, T. H., O. B. Nielsen, G. D. Lamb, and D. G. Stephenson. Intracellular Acidosis Enhances the Excitability of Working Muscle. *Science* 305, 1144–1147 (2004). Primary article on lactic acid and its effects on muscle contraction.

Rand, R. Raising Water to New Heights. *Science* **256**, 618 (1992). A brief perspective on the contribution of hydration to molecular assembly and protein catalysis.

Westhof, E., ed. *Water and Biological Macromolecules.* Boca Raton, FL: CRC Press, 1993. A series of articles about the role of water in hydration of biological macromolecules and the forces involved in macromolecular complexation and cell–cell interactions.

Amino acids are the building blocks from which all proteins are made.

3 Amino Acids and Peptides

Chapter Outline

3-1 Amino Acids are Three Dimensional

Learning Objectives

After reading this section, students will be able to:

LO 3.1.1 Describe the general structure of an amino acid as it would appear at pH 7.

LO 3.1.2 Distinguish between *chiral* and *achiral*.

Among all the possible amino acids, only 20 are usually found in proteins. The general structure of amino acids includes an **amino group** and a **carboxyl group**, both of which are bonded to the alpha-carbon (**α-carbon**), which is the carbon next to the carboxyl group. The α-carbon is also bonded to a hydrogen and to the **side chain group**, which is represented by the letter R (Figure 3.1). The side chain group determines the identity of the particular amino acid, and there are 20 different side chain groups found among the commonly occurring amino acids.

One of the most important properties of amino acids is their three-dimensional shape, or **stereochemistry**. Every object has a mirror image. Many pairs of objects that are mirror images can be superimposed on each other; two identical coffee mugs are an example. In other cases, the mirror-image objects cannot be superimposed on one another but are related to each other as the right hand is to the left. Chemical structures that are nonsuperimposable on their mirror image are said to be **chiral** (from the Greek *cheir*, "hand"). A molecule is chiral if it contains a **chiral center**, which is an atom

Pioneers in Biochemistry • Marguerite Davis

Marguerite Davis (1887–1967) was the co-discoverer of vitamins A and B and an early contributor to the field of nutrition science. The daughter of a physician turned botanist and the granddaughter of a women's rights activist, Davis had the natural gifts and the family support she needed to make a major contribution to science.

Davis worked in the lab of Elmer McCollum at the University of Wisconsin, where they were trying to identify which compounds in food were essential to maintaining health. Davis was responsible for feeding and collecting data on the colony of rats in the lab. She worked unpaid in McCollum's lab for five years before he was finally able to obtain a salary for her.

The researchers first observed that rats fed on milk were healthier than those fed olive oil or lard. Then in 1912, they found that fat-soluble compounds could be extracted from milk and fed to the oil-fed rats to improve their health. These compounds were named vitamins A and B. This initial discovery introduced a new period of interest in research on vitamins and nutrition. Understanding the importance of vitamins to health gave doctors new tools to combat illness, and it helped policymakers confront malnutrition around the world.

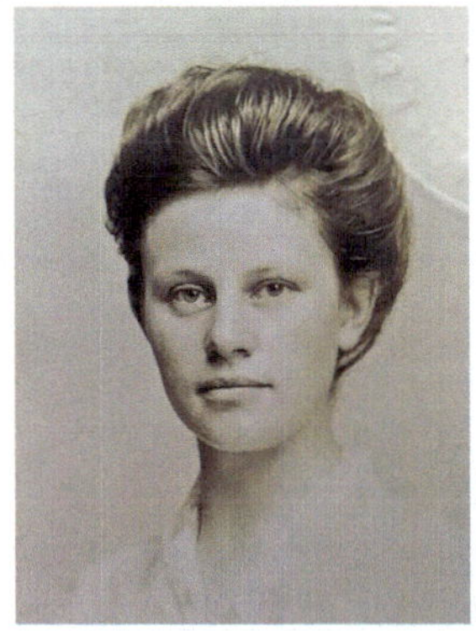
Marguerite Davis

that has four different groups bonded to it. As seen in Section 3-2, the α-carbon of glycine is not bonded to four different groups, making it **achiral**. In all the other commonly occurring amino acids, the α-carbon has four different groups bonded to it, giving rise to two nonsuperimposable mirror-image forms. Figure 3.2 shows perspective drawings of these two possibilities, or **stereoisomers**, for alanine, where the R group is —CH_3. The dashed wedges represent bonds directed away from the observer, whereas the solid wedges represent bonds directed out of the plane of the paper in the direction of the observer.

The two possible stereoisomers of another chiral compound, L- and D-glyceraldehyde, are shown for comparison with the corresponding forms of alanine. These two forms of glyceraldehyde are the basis of the classification of amino acids into L and D forms. The terminology comes from the Latin words *laevus* and *dexter*, meaning "left" and "right," respectively, which comes from the ability of optically active compounds to rotate polarized light to the left (the counterclockwise direction) or to the right (the clockwise direction). The two stereoisomers of each amino acid are designated as **L- and D-amino acids** on the basis of their similarity to the glyceraldehyde standard. When drawn in a certain orientation, the L form of glyceraldehyde has the hydroxyl group on the left side of the molecule, and the D form has it on the right side, as shown in perspective in Figure 3.2 (a Fischer projection). To determine the L or D designation for an amino acid, it is drawn as shown. The position of the amino group on the left or right side of the α-carbon determines the L or D designation. The amino acids that occur in proteins are all of the L form. Although D-amino acids occur in nature, most often in bacterial cell walls and in some antibiotics, they are not found in proteins.

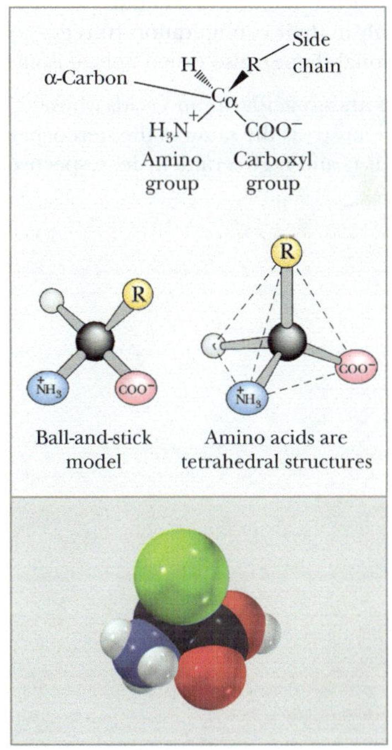

Figure 3.1 The three-dimensional nature of amino acids, showing the ionic forms that predominate at pH 7.

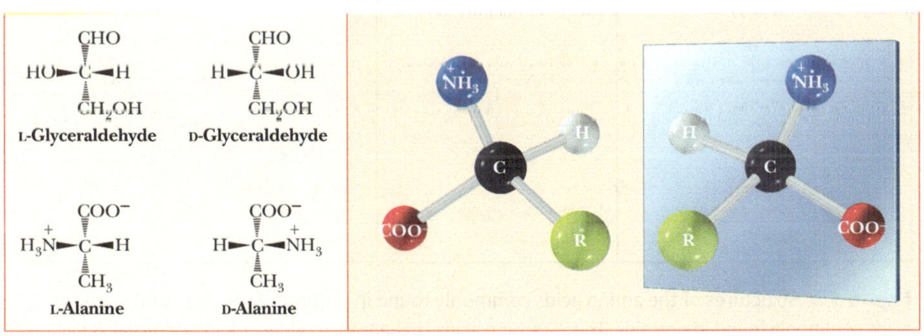

Figure 3.2 Stereochemistry of alanine and glyceraldehyde. The amino acids found in proteins have the same chirality as L-glyceraldehyde, which is opposite to that of D-glyceraldehyde.

amino group the —NH_2 functional group that can protonate to give —NH_3^+

carboxyl group the —COOH functional group that can deprotonate to give the carboxylate anion, —COO^-, and a hydrogen ion

α-carbon the carbon in an amino acid that is bonded to both the amino group and the carboxyl group

side chain group the portion of an amino acid attached to the α-carbon that determines its identity

stereochemistry the branch of chemistry that deals with the three-dimensional shape of molecules

chiral an object that is not superimposable on its mirror image

3-2 Structures and Properties of Amino Acids

Learning Objectives

After reading this section, students will be able to:

LO 3.2.1 Write the three-letter and one-letter codes for each amino acid given the full names of each amino acid.

LO 3.2.2 Explain, using structural details, why each amino acid is polar or nonpolar.

LO 3.2.3 Calculate the ratio of uncharged to charged amino acid side chains at certain pH values using the Henderson–Hasselbalch equation and given side chain pK_a values.

chiral center an atom that has four different groups bonded to it

achiral an object that is superimposable on its mirror image

stereoisomers molecules that differ from each other only in their configuration (three-dimensional shape); also called optical isomers

L- and D-amino acids amino acids whose stereochemistry is the same as the stereochemical standards L- and D-glyceraldehyde, respectively

The individual amino acids are classified according to several criteria, two of which are particularly important. The first of these is the polar or nonpolar nature of the side chain. The second depends on the presence of an acidic or basic group in the side chain. Other useful criteria include the presence of functional groups other than acidic or basic ones in the side chains and the nature of those groups.

▶ Why are amino acid side chains so important?

The side chain of the simplest amino acid, glycine, is a hydrogen atom, so in this case alone two hydrogen atoms are bonded to the α-carbon. In all other amino acids, the side chain is larger and more complex. Figure 3.3 shows the

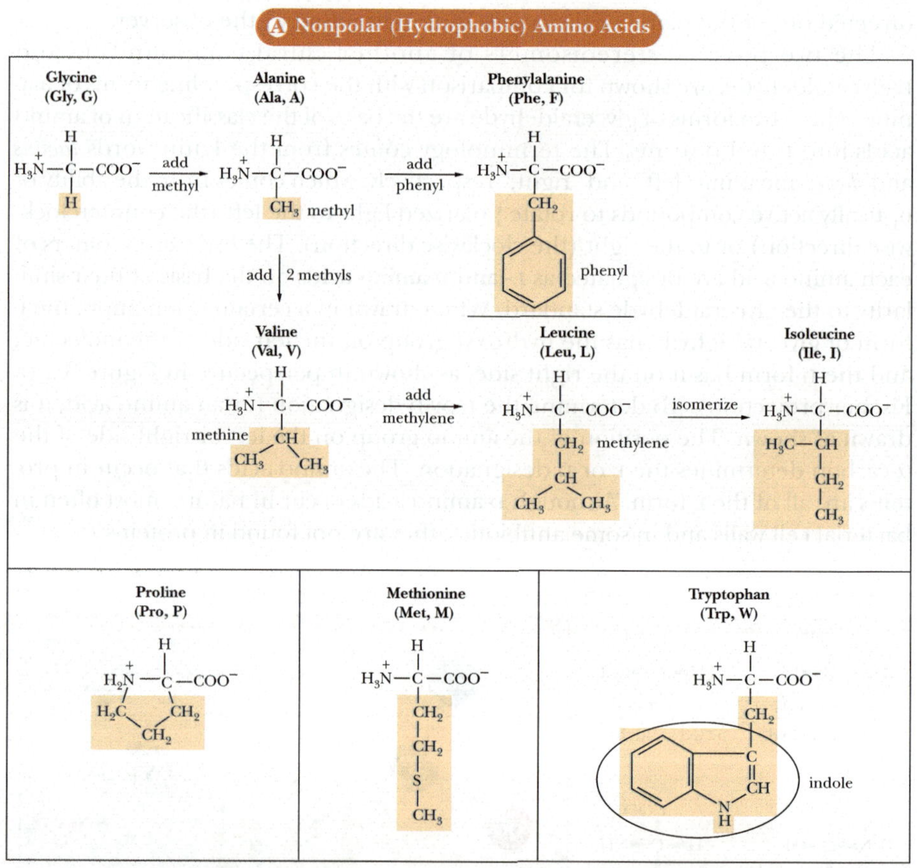

Figure 3.3 **Structures of the amino acids commonly found in proteins.** Structures of the amino acids commonly found in proteins. The 20 amino acids that are the building blocks of proteins can be classified as (A) nonpolar (hydrophobic) or (B) polar. The polar amino acids can be further classified as polar uncharged (Asn, Gln, Ser, Cys, Thr, Tyr), acidic (Asp, Glu), or basic (Lys, Arg, His). Also shown are the three-letter and one-letter abbreviations used to denote the amino acids.

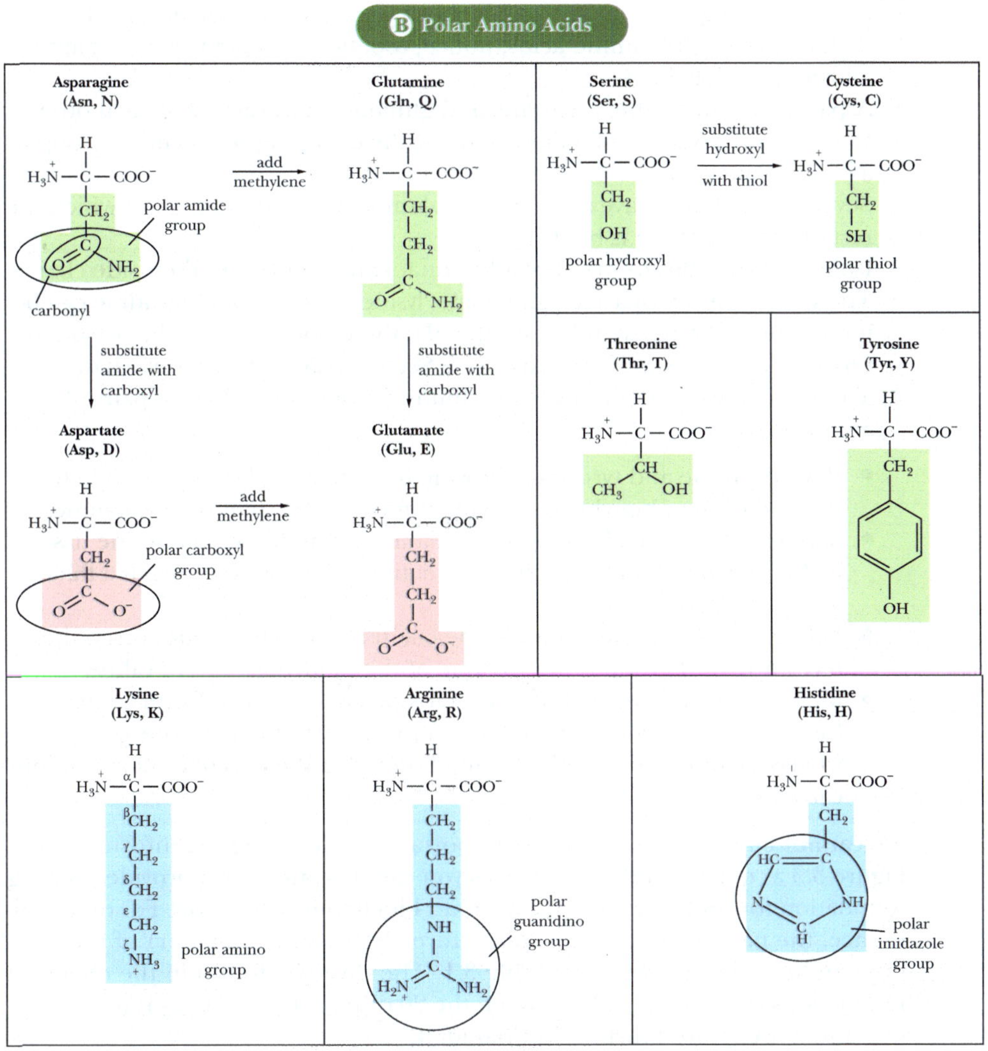

Figure 3.3—cont'd

names, three-letter codes, one-letter codes, and structures (at pH 7) of the 20 commonly occurring amino acids in a way that highlights organic chemistry nomenclature as well as relationships that connect certain amino acids to each other. (The arrows seen in Figure 3.3 are not meant to imply reactions that generate the amino acids. They are meant to show structural connections between them.) In addition to the helpful notes in Figure 3.3, here are a few extra tips to consider as you learn the structures of the amino acids:

- Carbon usually forms four covalent bonds and nitrogen usually forms three. Oxygen and sulfur usually form two covalent bonds. Once you learn the **skeletal structures** of the side chains, then you can use this information to fill in hydrogens and complete the structures. Test this with the APPLY YOUR KNOWLEDGE 3.1 exercise and continue to apply the knowledge to reduce what you have to memorize regarding the amino acid structures as well as many other biochemical structures.
- When nitrogen forms four covalent bonds, it has a positive charge. Notice how this affects the number of hydrogens on the amino group of proline at pH 7.

skeletal structure a series of atoms bonded together that form the essential structure of the compound

- Starting from glycine, a path can be built all the way to isoleucine, which includes four other amino acids; notice, too, how isoleucine's side chain is an **isomer** of leucine's.

isomer each of two or more compounds with the same chemical formula but a different arrangement of atoms in the molecule

- Asparagine, glutamine, aspartate, and glutamate are related to each other by either the addition of a methylene or the changing of an amide group to a carboxylate group.
- Several side chains have a —CH_2— ("alanine") foundation, off of which the rest of the side chain is built.
- Isoleucine and threonine have additional chiral centers in their side chains.
- The side chains of aspartate, glutamate, lysine, arginine, and histidine can be ionized or neutral, depending on the pH; these polar amino acids are often further classified as acidic or basic (see Section 3-2C and Section 3-2D).
- There are helpful similarities to be found among the following pairs of amino acids:
    - Phenylalanine and tyrosine: Tyrosine has a hydroxyl group bonded to the end of the phenyl ring, making it more polar than phenylalanine.
    - Serine and cysteine: Both have an "alanine" foundation, but serine has a **hydroxyl** group bonded to that foundation, whereas cysteine has a **thiol** group.
    - Valine and threonine: Valine's side chain has a V shape, and threonine has a hydroxyl group replacing one of the methyl groups in valine.
    - Methionine and lysine: Both of their side chains start off with chains of four groups, but methionine has a sulfur in that chain of four groups, whereas lysine has an amino group bonded at the end of its chain of four carbons.

hydroxyl an —OH group

thiol group an —SH group

methyl group a —CH_3 group

methylene group a —CH_2— group

methine group a —CH— group

phenyl group a —C_6H_5 group derived from benzene by the removal of a hydrogen atom

carbonyl group a C=O group

amide group a carbonyl group bonded to an amino group, C(=O)NH_2

We cannot emphasize enough the importance of learning the information in Figure 3.3 as quickly and accurately as you can, because it will provide a strong foundation for understanding many other biochemical concepts. For quick reference, the names and abbreviations (three-letter and one-letter) of the common amino acids are listed in Table 3.1. The diversity found in the chemical interactions that are possible between the 20 different side chains is truly amazing, and it is explored further in Chapter 4.

3.1 Apply Your Knowledge

Amino Acid Side Chain Structure Trends

1. Complete the following skeletal structures by adding any necessary hydrogen atoms.

A Ile

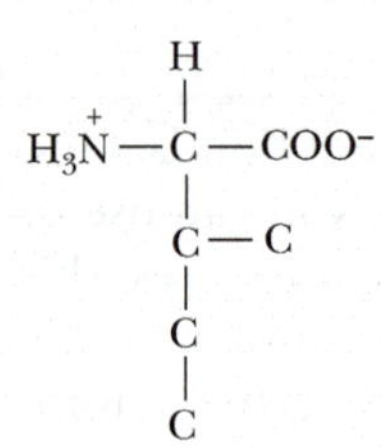

B Arg

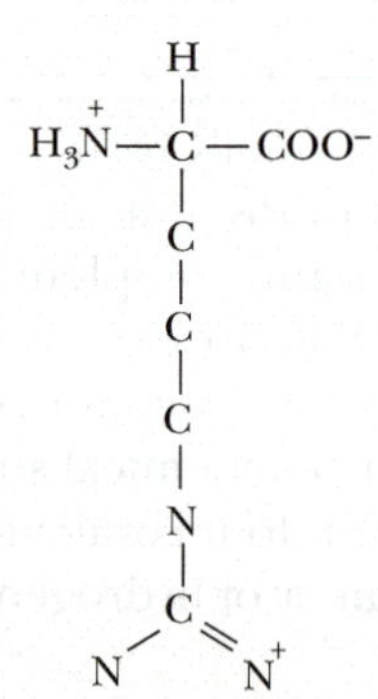

Use your knowledge of the number of covalent bonds that carbon and nitrogen usually form.

Solution

Carbon usually forms four covalent bonds, whereas nitrogen usually forms three. Thus, you should arrive at the complete structures of the side chain groups for isoleucine and arginine shown in Figure 3.3. One of the nitrogen atoms in the side chain of Arg is positively charged, however, so this is a hint that it needs *four* covalent bonds.

Notice in Figure 3.3 that the side-chain atoms are designated with letters of the Greek alphabet, counting from the α-carbon. These atoms are, in turn, the beta (β-), gamma (γ-), delta (δ-), epsilon (ε-), and zeta (ζ-) atoms (see lysine in Figure 3.3). This Greek-letter notation is based on the distance from the carbonyl group of the α-carboxyl group. Analogous notation is used in other biochemical concepts, so don't be surprised if you see it again. Finally, we frequently refer to amino acids by the three-letter or one-letter abbreviations of their names, so begin learning them now.

3-2A Nonpolar Amino Acids

The side chains of glycine, alanine, valine, leucine, isoleucine, proline, phenylalanine, tryptophan, and methionine are mainly hydrocarbon groups. As a result, these nine amino acids are nonpolar. The side chain of glycine is a hydrogen atom, but each side chain of alanine, valine, leucine, and isoleucine is an aliphatic hydrocarbon group. (In organic chemistry, the term *aliphatic* refers to the absence of a benzene ring or related structure.)

Proline has an aliphatic cyclic structure, and the nitrogen is bonded to two carbon atoms. In the terminology of organic chemistry, the amino group of proline is a secondary amine, and proline is often called an *imino acid.* In contrast, the amino groups of all the other common amino acids are primary amines. In phenylalanine, the hydrocarbon group is aromatic (it contains a cyclic group similar to a benzene ring) rather than aliphatic. In tryptophan, the side chain contains an indole ring, which is also aromatic. In methionine, the side chain contains a sulfur atom in addition to aliphatic hydrocarbon groupings (see Figure 3.3).

3-2B Polar-Neutral Amino Acids

Another group of amino acids has polar side chains that are electrically neutral (uncharged) at neutral pH. This group includes serine, threonine, tyrosine, cysteine, glutamine, and asparagine. Glycine is sometimes included here for convenience because it lacks a nonpolar side chain.

In serine and threonine, the polar group is a hydroxyl (—OH) bonded to aliphatic hydrocarbon groups. The hydroxyl group in tyrosine is bonded to an aromatic hydrocarbon group, which eventually loses a proton at higher pH. (The hydroxyl group in tyrosine is a phenol, which is a stronger acid than an aliphatic alcohol. As a result, the side chain of tyrosine can lose a proton in a titration, whereas those of serine and threonine would require such a high pH that pK_a values are not normally listed for these side chains.)

In cysteine, the polar side chain consists of a thiol group (—SH), which can react with other cysteine thiol groups to form disulfide (—S—S—) bridges in proteins in an oxidation reaction (Section 4-4A). The thiol group can also lose a proton. The amino acids glutamine and asparagine have amide groups, which are derived from carboxyl groups, in their side chains. Amide bonds do

Table 3.1 Names and Abbreviations of the Common Amino Acids

Amino Acid	Three-Letter Abbreviation	One-Letter Abbreviation
Alanine	Ala	A
Arginine	Arg	R
Asparagine	Asn	N
Aspartic acid	Asp	D
Cysteine	Cys	C
Glutamic acid	Glu	E
Glutamine	Gln	Q
Glycine	Gly	G
Histidine	His	H
Isoleucine	Ile	I
Leucine	Leu	L
Lysine	Lys	K
Methionine	Met	M
Phenylalanine	Phe	F
Proline	Pro	P
Serine	Ser	S
Threonine	Thr	T
Tryptophan	Trp	W
Tyrosine	Tyr	Y
Valine	Val	V

Note: One-letter abbreviations start with the same letter as the name of the amino acid where this is possible. When the names of several amino acids start with the same letter, phonetic names (occasionally facetious ones) are used, such as Rginine, asparDic, Fenylalanine, tWyptophan. Where two or more amino acids start with the same letter, it is the smallest one whose one-letter abbreviation matches its first letter.

not ionize in the pH range usually encountered in biochemistry. Glutamine and asparagine can be considered derivatives of glutamic acid and aspartic acid, respectively.

3-2C Acidic Amino Acids

Two amino acids, glutamic acid and aspartic acid, have ionizable carboxyl groups in their side chains in addition to the one present in all amino acids. A carboxyl group can lose a proton, forming the corresponding carboxylate anion (Section 2-5)—glutamate and aspartate, respectively, in the case of these two amino acids. Because of the presence of the carboxylate, the side chain of each of these two amino acids is negatively charged at neutral pH, but they can be neutral at low pH (Figure 3.4A). Also, the knowledge that molecules with an *-ate* ending in their name contain a carboxylate group will come in handy for understanding metabolite structures discussed in later chapters.

3-2D Basic Amino Acids

Three amino acids—histidine, lysine, and arginine—have basic side chains, and the side chain in all three is positively charged at or near neutral pH. Similar to the acidic amino acids, these basic amino acids have side chain groups that are also referred to as ionizable. In lysine, the side-chain amino group is attached to an aliphatic hydrocarbon tail. In arginine, the side-chain basic group, the guanidino group, is more complex in structure than the amino group, but it is also bonded to an aliphatic hydrocarbon tail. In free histidine, the pK_a of the side-chain imidazole group is 6.0, which is the only pK_a value close to physiological pH.

The pK_a values for amino acids depend on the environment and can change significantly within the confines of a protein. Histidine can be found in the protonated or unprotonated forms in proteins, and the properties of many proteins depend on whether individual histidine residues are or are not charged. Figure 3.4B shows the ionization of arginine. Notice how the protonated forms of basic groups are positively charged (Figure 3.4B), while the protonated forms of acidic groups are neutral (Figure 3.4A). Test yourself with the APPLY YOUR KNOWLEDGE 3.2 exercises.

3.2 Apply Your Knowledge

Structures and Properties of Amino Acids

1. In the following group, identify the amino acids with nonpolar side chains and those with basic side chains: alanine, serine, arginine, lysine, leucine, and phenylalanine.
2. The pK_a of the side-chain imidazole group of histidine is 6.0. What is the ratio of uncharged to charged side chains at pH 7.0?

Solution

In the first part of this exercise, you are asked to do a fact check on material from this chapter, and in the second part, you are asked to recall and apply concepts from an earlier chapter.

1. See Figure 3.3. Nonpolar: alanine, leucine, and phenylalanine; basic: arginine and lysine. Serine is not in either category because it has a polar side chain.
2. The ratio is 10:1 because the pH is 1 unit higher than the pK_a.

Step 1: Write the side chain group ionization chemistry for histidine.

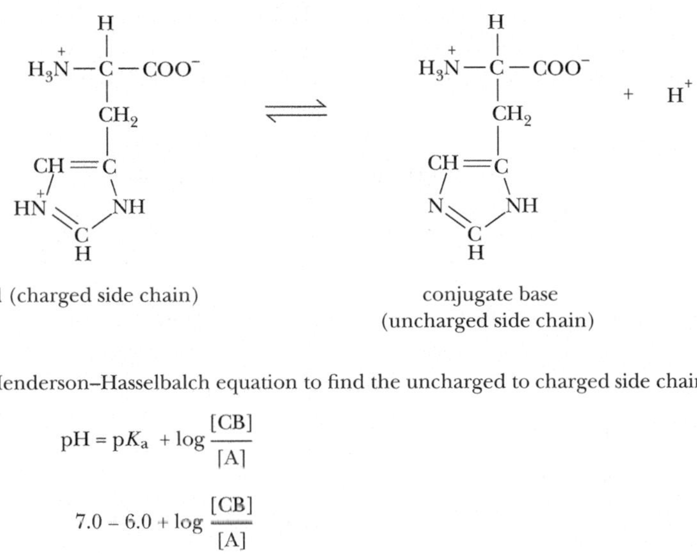

acid (charged side chain)

conjugate base (uncharged side chain)

Step 2: Use the Henderson–Hasselbalch equation to find the uncharged to charged side chain ratio.

$$pH = pK_a + \log \frac{[CB]}{[A]}$$

$$7.0 = 6.0 + \log \frac{[CB]}{[A]}$$

$$\log \frac{[CB]}{[A]} = 1.0$$

uncharged ([CB]), charged ([A]):

$$\frac{[CB]}{[A]} = 10^{1.0} = 10$$

⟶ Therefore, the uncharged to charged side chain ratio is 10:1.

Be careful when dealing with acidic group ionization chemistry because the acid form of acidic groups is uncharged while the conjugate base form of acidic groups is charged. See Figure 3.5A for an example.

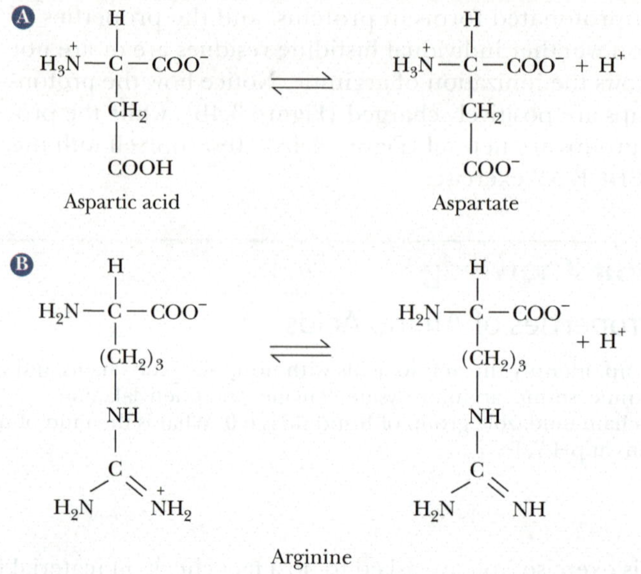

Figure 3.4 Comparison of ionization reactions for acidic and basic side-chain groups. (A) Ionization reaction for the acidic side chain of aspartic acid. (B) Ionization reaction for the basic side chain of arginine.

3-2E Uncommon Amino Acids

Many other amino acids, in addition to the ones listed here, are known to exist. They occur in some, but by no means all, proteins. Figure 3.5 shows some examples of the many possibilities. They are derived from the common amino acids and are produced by modification of the parent amino acid after the protein is synthesized by the organism in a process called posttranslational modification.

Hydroxyproline and hydroxylysine differ from the parent amino acids in that they have hydroxyl groups on their side chains; they are found only in a few connective-tissue proteins, such as collagen. Thyroxine differs from tyrosine in

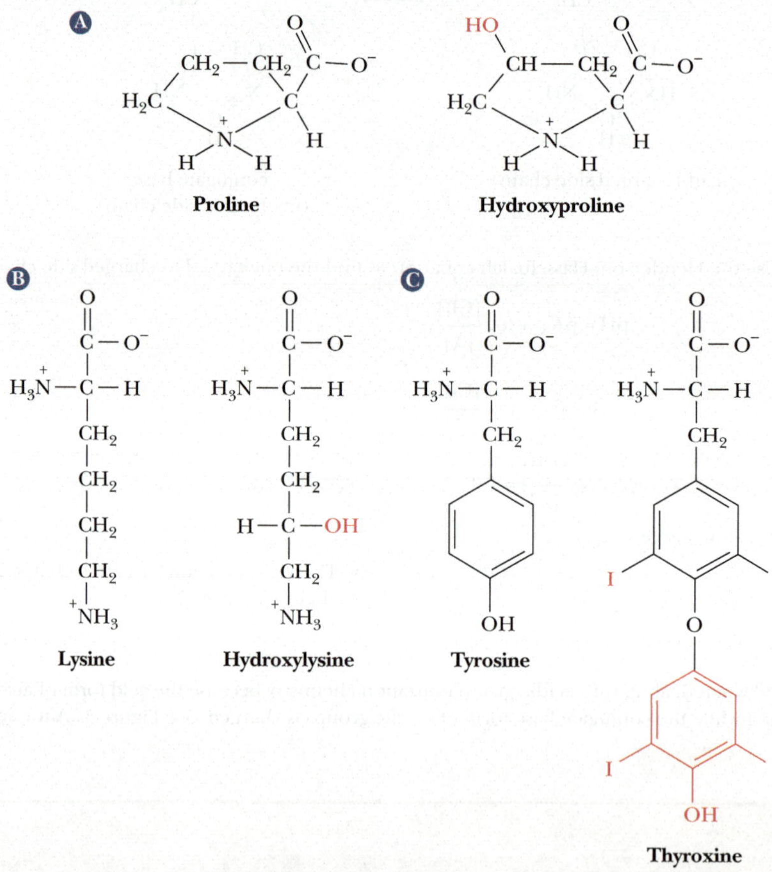

Figure 3.5 Structures of hydroxyproline, hydroxylysine, and thyroxine. The structures of the parent amino acids—(A) proline for hydroxyproline, (B) lysine for hydroxylysine, and (C) tyrosine for thyroxine—are shown for comparison. All amino acids are shown in their predominant ionic forms at pH 7.

that it has an extra iodine-containing aromatic group on the side chain; it is produced only in the thyroid gland, formed by posttranslational modification of tyrosine residues in the protein thyroglobulin. Thyroxine is then released as a hormone by proteolysis of thyroglobulin.

3-3 Amino Acids Can Act as Both Acids and Bases

Learning Objectives

After reading this section, students will be able to:

LO 3.3.1 Write deprotonation equilibria reactions for each amino acid, including all relevant details similar to those seen in Figures 3.7 and 3.9.

LO 3.3.2 Draw titration curves for each amino acid, including all relevant details similar to those seen in Figures 3.8 and 3.10.

LO 3.3.3 Calculate the isoelectric point for an amino acid given its pK_a values.

In a free amino acid, the carboxyl group and amino group of the general structure are charged at neutral pH—the carboxylate portion negatively and the amino group positively. Amino acids without charged groups on their side chains exist in neutral solution as *zwitterions* with no net charge. A **zwitterion** has equal positive and negative charges; in solution, it is electrically neutral.

zwitterion an ion that has equal positive and negative charges, making it electrically neutral (i.e., it has no net charge)

3-3A Titration of Amino Acids

Recall from Section 2-4 that acids and bases can be titrated to determine their acid or base strength.

▶ What happens when we titrate an amino acid?

When an amino acid is titrated, its titration curve indicates the reaction of each functional group with a hydroxide ion. In alanine, the carboxyl and amino groups are the two titratable groups. At very low pH, alanine has a protonated (and thus uncharged) carboxyl group and a positively charged amino group that is also protonated. Under these conditions, the alanine has a net charge of +1. As base is added, the carboxyl group loses its proton to become a negatively charged carboxylate group (Figure 3.6), and the pH of the solution increases.

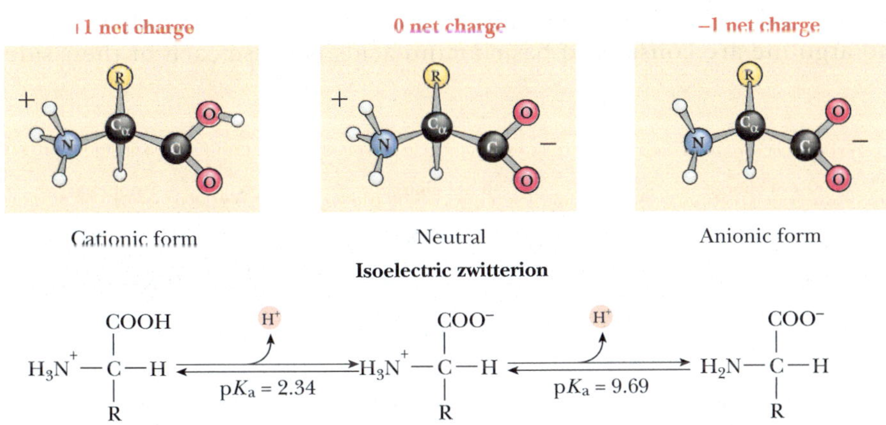

The ionic forms of the amino acids, shown without consideration of any ionizations on the side chain. The cationic form is the low-pH form, and the titration of the cationic species with base yields the zwitterions and finally the anionic form.

Figure 3.6 The ionization of amino acids lacking ionizable side chains. The pK_a values indicated are those for alanine, and Table 3.2 contains values for other amino acids.

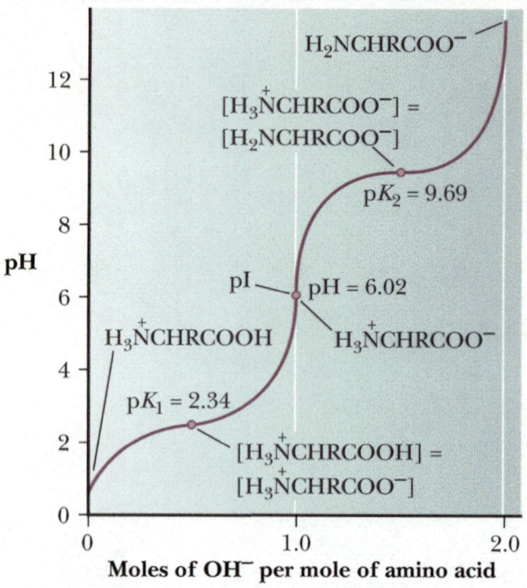

Figure 3.7 The titration curve of alanine.

The pK_a (i.e., the pH at the midpoint) for the deprotonation of the carboxyl group is 2.34, which is consistent with carboxyl groups being acidic. Alanine now has no net charge. As the pH increases still further with the addition of more base, the protonated amino group loses its proton, and the alanine molecule now has a negative charge of 1. The pK_a (i.e., the pH at the midpoint) for the deprotonation of the amino group is 9.69, which is consistent with amino groups being basic. Neutral amino acids do not exist in the form NH_2—CHR—COOH, because the amino group does not deprotonate before the carboxyl group.

While Figure 3.6 shows the structural details of the deprotonation equilibria for amino acids lacking ionizable side chains, let's now take a look at the corresponding titration curve. The titration curve of alanine (Figure 3.7) is that of a diprotic acid. The different species that are present at the various points on the curve, starting at low pH and ending at high pH, correspond to the deprotonation equilibria seen in Figure 3.6. The significance of the pI value that is indicated on the titration curve is explained at the end of this section.

Let's now take a look at titration details for histidine, which is an amino acid that has an ionizable side chain. In histidine, the imidazole side chain also contributes a titratable group. At very low pH values, the histidine molecule has a net charge of +2, because both the imidazole and the amino groups have positive charges (Figure 3.8). As base is added and the pH increases, the carboxyl group loses a proton to become a carboxylate as before, and the histidine now has a net charge of +1 (Figure 3.8). As still more base is added, the charged imidazole group loses its proton, and this is the point at which the histidine has no net charge (Figure 3.8). At still higher values of pH, the amino group loses its proton, and the histidine molecule now has a net charge of –1 (Figure 3.8).

The titratable groups in any molecule deprotonate in numerical order of their pK_a values. This is why the imidazole side chain group of histidine deprotonates before its amino group. The titration curve of histidine is that of a triprotic acid (Figure 3.9). After learning the details in Figures 3.8 and 3.9, you are now prepared to draw deprotonation equilibria structures and titration curves for any amino acid with or without an ionizable side chain group.

Like the acids we discussed in Chapter 2, the titratable groups of each of the amino acids have characteristic pK_a values. The pK_a values of α-carboxyl groups are fairly low, around 2. The pK_a values of amino groups are much higher, with values ranging from 9 to 10.5. The pK_a values of side-chain groups, including side-chain carboxyl and amino groups, depend on the groups' chemical nature. Table 3.2 lists the pK_a values of the titratable groups of the amino acids.

The classification of an amino acid as acidic or basic depends on the pK_a of the side chain, as well as the chemical nature of the group. Histidine, lysine, and arginine are considered basic amino acids because each of their side

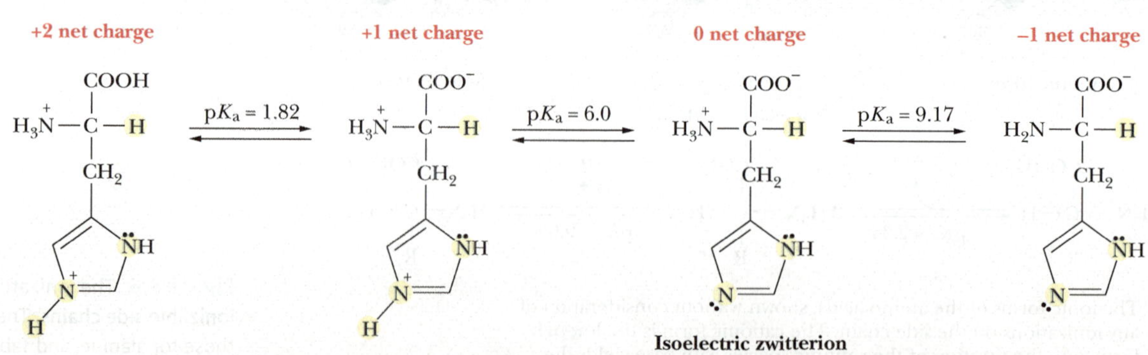

Figure 3.8 The ionization of amino acids having ionizable side chains, using histidine as the example.

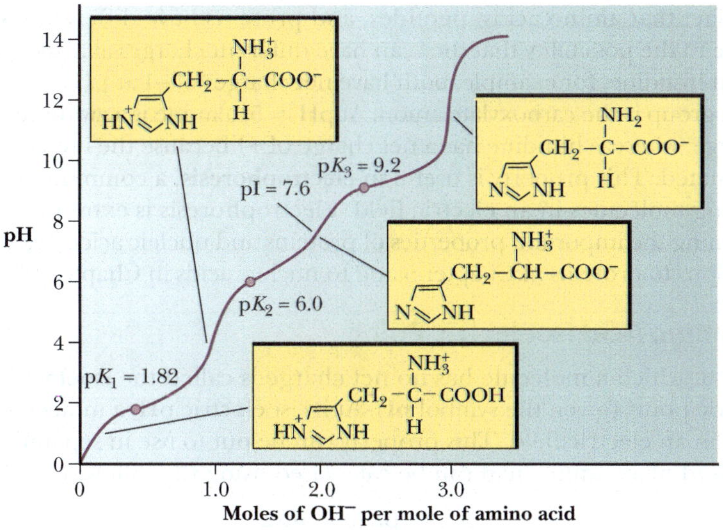

Figure 3.9 The titration curve of histidine. The isoelectric pH (pI) is the value at which positive and negative charges are the same. The molecule has no net charge.

chains has a nitrogen-containing group that can exist in either a protonated or a deprotonated form. However, histidine has a pK_a in the acidic range.

Aspartic acid and glutamic acid are considered acidic because each has a carboxylic acid side chain with a low pK_a value. These groups can still be titrated after the amino acid is incorporated into a peptide or protein, but the pK_a of the titratable group on the side chain is not necessarily the same in a protein as it is in a free amino acid. In fact, it can be very different. For example, a pK_a of 9 has been reported for an aspartate side chain in the protein thioredoxin.

Table 3.2 **pK_a Values of Common Amino Acids**

Acid	α-COOH	α-NH_3^+	RH or RH^+
Gly	2.34	9.60	
Ala	2.34	9.69	
Val	2.32	9.62	
Leu	2.36	9.68	
Ile	2.36	9.68	
Ser	2.21	9.15	
Thr	2.63	10.43	
Met	2.28	9.21	
Phe	1.83	9.13	
Trp	2.38	9.39	
Asn	2.02	8.80	
Gln	2.17	9.13	
Pro	1.99	10.6	
Asp	2.09	9.82	3.86*
Glu	2.19	9.67	4.25*
His	1.82	9.17	6.0*
Cys	1.71	10.78	8.33*
Tyr	2.20	9.11	10.07
Lys	2.18	8.95	10.53
Arg	2.17	9.04	12.48

*For these amino acids, the R group ionization occurs before the α-NH_3^+ ionization.

The fact that amino acids, peptides, and proteins have different pK_a values gives rise to the possibility that they can have different charges at a given pH. Alanine and histidine, for example, both have net charges of −1 at pH > 10; the only charged group is the carboxylate anion. At pH ≈ 5, alanine is a zwitterion with no net charge, whereas histidine has a net charge of +1 because the imidazole group is protonated. This property is useful in **electrophoresis**, a common method for separating molecules in an electric field. Electrophoresis is extremely useful in determining the important properties of proteins and nucleic acids. We discuss its applications to proteins in Chapter 5 and to nucleic acids in Chapter 13.

electrophoresis a method for separating molecules on the basis of the ratio of charge to size

3-3B Amino Acid Isoelectric Point

isoelectric pH (pI) the pH at which a molecule has no net charge; also known as the *isoelectric point*

The pH at which a molecule has no net charge is called the **isoelectric pH**, or isoelectric point (given the symbol **pI**). At its isoelectric pH, a molecule will not migrate in an electric field. This property can be put to use in separation methods. The pI of an amino acid can be calculated using the following equation:

$$pI = \frac{pK_{a1} + pK_{a2}}{2}$$

Most of the amino acids have only two pK_a values, so the pK_a values in Table 3.2 can be used in this equation to calculate the pI. For amino acids with ionizable side chains, however, it is helpful to write out the deprotonation equilibria and then choose the pK_a *before* the zwitterion as pK_a and the pK_a *after* the zwitterion as pK_{a2}. Revisit Figure 3.9 to see how this method leads to a pI value of 7.6 for histidine, because 6.0 and 9.17 are its pK_a values before and after the zwitterion, respectively. Test yourself with the exercises in APPLY YOUR KNOWLEDGE 3.3 to see how well you can use these concepts.

3.3 Apply Your Knowledge

Amino Acid Titrations

1. Which of the following amino acids has a net charge of +2 at low pH? Which has a net charge of −2 at high pH? Aspartic acid, alanine, arginine, glutamic acid, leucine, lysine.
2. What is the pI for lysine?

Solution

The first part of this exercise deals only with the qualitative description of the successive loss of protons by the titratable groups on the individual amino acids. In the second part, you need to refer to the titration curve as well to do a numerical calculation of pH values.

1. Arginine and lysine have net charges of +2 at low pH because of their basic side chains; aspartic acid and glutamic acid have net charges of −2 at high pH because of their carboxylic acid side chains. Alanine and leucine do not fall into either category because they do not have titratable side chains.
2. Step 1: Write the deprotonation equilibria using pK_a values for Lys from Table 3.2.

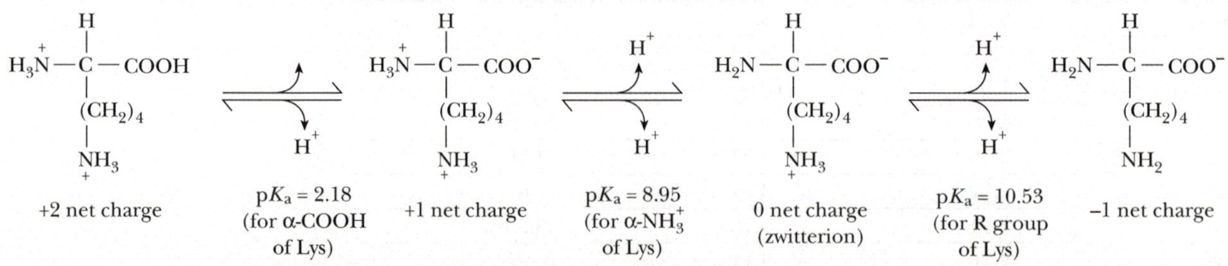

Step 2: Use the pK_a values before and after the zwitterion in the pI equation.

$$pI = \frac{pK_{a_1} + pK_{a_2}}{2} = \frac{8.95 + 10.53}{2} = \boxed{9.74}$$

3-4 The Peptide Bond

Learning Objectives

After reading this section, students will be able to:

LO 3.4.1 Label the N-terminus, C-terminus, N-to-C directionality, peptide bonds, backbone atoms, and the polarity of the backbone atoms when given a peptide structure.

LO 3.4.2 Identify positively and negatively charged residues in peptides at different pH values.

Individual amino acids can be linked by forming covalent bonds. The bond is formed between the α-carboxyl group of one amino acid and the α-amino group of the next one. Water is eliminated in the process, and the linked amino acid **residues** remain after water is eliminated (Figure 3.10). A bond formed in this way is called a **peptide bond.**

Peptides are compounds formed by linking small numbers of amino acids, ranging from two to several dozen. In a protein, many amino acids (usually more than a hundred) are linked by peptide bonds to form a **polypeptide chain** (Figure 3.11). Another name for a compound formed by the reaction between an amino group and a carboxyl group is an *amide.* Compare the amide groups shown in the peptide bonds of Figure 3.11 to the side chain amide groups of asparagine and glutamine (Figure 3.3B).

The generic peptide chain in Figure 3.11 illustrates some important general features of all peptide and polypeptide chains:

- Peptide chains have an N-terminal residue (also called the N-terminus) and a C-terminal residue (also called the C-terminus), which gives the chain a directionality (N to C).
- The repeating —C(=O)—NH— groups of atoms represent the peptide bond.

residues portions of monomer units included in polymers after splitting out water between the linked monomers

peptide bond an amide bond between amino acids in a protein

peptides molecules formed by linking two to several dozen amino acids by amide bonds

polypeptide chain the backbone of a protein; it is formed by linking amino acids by peptide (amide) bonds

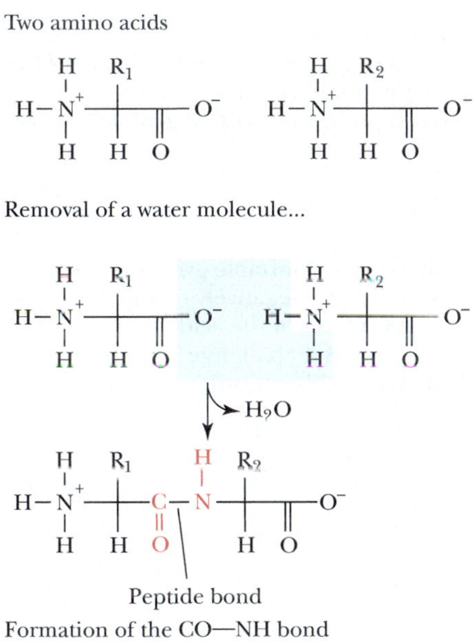

Figure 3.10 **Formation of the peptide bond.**
Garrett/Grisham, Biochemistry, 4th Edition. 2009 Cengage Learning

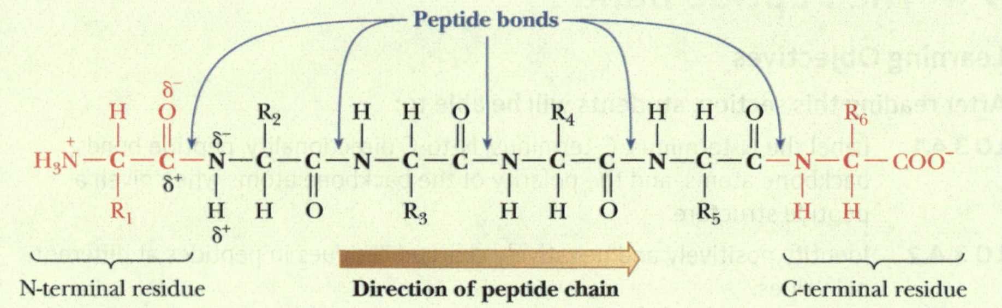

Figure 3.11 **A small peptide showing the direction of the peptide chain (N-terminal to C-terminal).** Peptide backbone atoms are in bold.

peptide backbone atoms the repeating —C(=O)—NH— group of atoms in a polypeptide chain

protein a molecule formed by linking more than several dozen amino acids by amide bonds

- The polarity of the atoms involved in the peptide bond is shown for the first two amino acids, but it applies to each of these groups in the peptide.
- These repeating atoms are frequently referred to as **peptide backbone atoms**. As we explain in Chapter 4 and beyond, they play an important role in peptide/**protein** structure.
- The repeating —C(=O)—NH— groups arise from the removal of water molecules when the carboxyl (—COO$^-$) group of one amino acid reacts with the amino (—NH_3^+) group of the next amino acid. The water molecules consist of O from the —COO$^-$ group and two H's from the —NH_3^+ group (Figure 3.10). Keeping this in mind will help you learn to correctly draw peptide and protein structures. Moreover, this process of forming a bond with the removal of water is a common way that biochemical macromolecules (e.g., carbohydrates and DNA) are linked together.

3.4 Apply Your Knowledge

Identifying Charged Groups on Peptides at Different pH Values

1. Consider the following peptide: Asp-Ala-Met
   Which residue(s) have a negatively charged group at pH 7?
2. Consider the following peptide: Gly-His-Lys-Val
   Which residue(s) have a positively charged group at pH 8?

Solution

1. Compare the pK_a values of the ionizable groups of Asp, Ala, and Met relative to pH 7. Asp and Met would have negatively charged groups at pH 7 because the pH is higher than the pK_a values of the acidic groups. Identifying Met as having a negatively charged group is a challenge here because the N- and C-termini often get overlooked as ionizable groups.

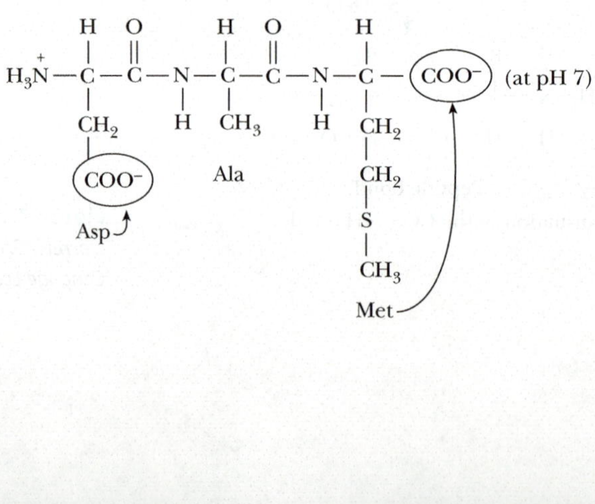

2. Compare the pK_a values of the ionizable groups of Gly, His, Lys, and Val relative to pH 8. Gly and Lys have positively charged groups at pH 8. Histidine is deprotonated and neutral because a pH of 8 is much higher than the pK_a of the imidazole group.

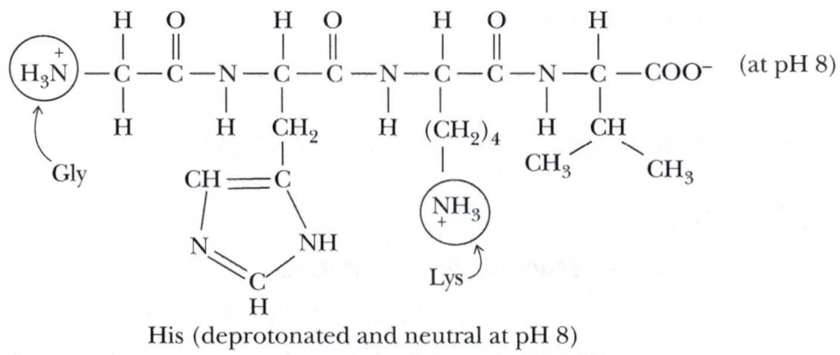

3-4A The Planar Nature of the Peptide Bond

The carbon–nitrogen bond formed when two amino acids are linked in a peptide bond is usually written as a single bond, with one pair of electrons shared between the two atoms. With a simple shift in the position of a pair of electrons, this bond can be written as a double bond. This shifting of electrons is well known in organic chemistry and results in **resonance structures**—structures that differ from one another only in the positioning of electrons. The positions of double and single bonds in one resonance structure are different from their positions in another resonance structure of the same compound. No single resonance structure actually represents the bonding in the compound; instead, all resonance structures contribute to the bonding situation.

resonance structures structural formulas that differ from each other only in the position of electrons

The peptide bond can be written as a resonance hybrid of two structures (Figure 3.12A), one with a single bond between the carbon and nitrogen and the other with a double bond between the carbon and nitrogen. As a result, the peptide bond has partial double bond character. The resonance among the peptide backbone atoms leads to three important implications for peptide and protein structure:

1. The peptide backbone atoms form a plane (see the "Amide plane" in Figure 3.12B).
2. The peptide bond is stronger than an ordinary single bond because of this resonance stabilization.

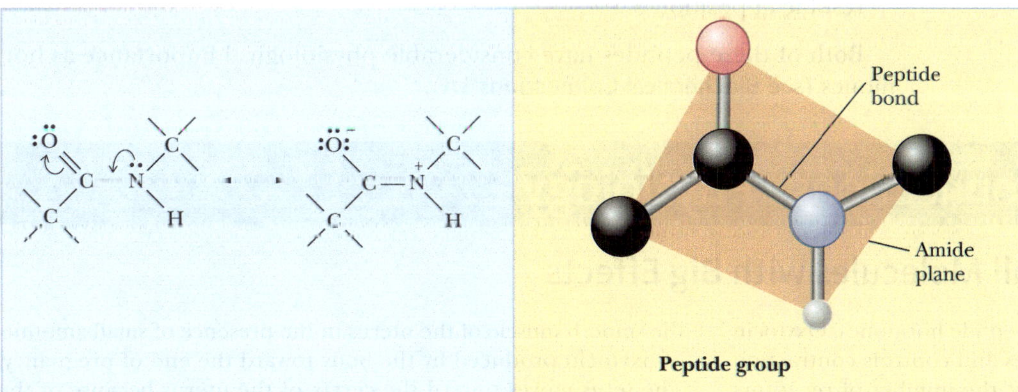

A Resonance structures of the peptide group.

B The planar peptide group.

Figure 3.12 The resonance structures of the peptide bond lead to a planar group. (*Illustration, Irving Geis. Rights owned by Howard Hughes Medical Institute.*)

3. There is free rotation around the bonds between the α-carbon of a given amino acid residue and the amino nitrogen and carbonyl carbon of that residue, but there is no significant rotation around the peptide bond. This stereochemical constraint plays an important role in determining how the protein backbone can fold.

3-5 Small Peptides with Physiological Activity

Learning Objectives

After reading this section, students will be able to:

LO 3.5.1 Compare the structures of oxytocin and vasopressin.

LO 3.5.2 Describe the physiological importance of the oxytocin and vasopressin hormones.

Chapter 4 describes the structure of proteins, and Chapters 6 and 7 discuss their action as enzymes. However, peptides also play a number of biochemical roles.

▶ What are some biological functions of small peptides?

One of the most important of their roles is that of hormones, and we present many examples of hormonal control as we discuss biochemical reactions in later chapters. Other classes of compounds, particularly steroids, can be hormones. The notoriety of steroid hormones in doping scandals in sports can obscure the fact that peptides also can be hormones. We take a quick look at peptide hormones now to prepare the way for later discussions.

Some important peptide hormones have cyclic structures. Two well-known examples with many structural features in common are oxytocin and vasopressin (Figure 3.13). We can make three comparisons between these two peptide hormones:

1. In each, there is a disulfide (—S—S—) bond responsible for the cyclic structure.
2. Each of these peptides contains nine amino acid residues, each has an amide group (rather than a free carboxyl group) at the C-terminal end, and each has a disulfide link between cysteine residues at positions 1 and 6.
3. The difference between these two peptides is that oxytocin has an isoleucine residue at position 3 and a leucine residue at position 8, whereas vasopressin has a phenylalanine residue at position 3 and an arginine residue at position 8.

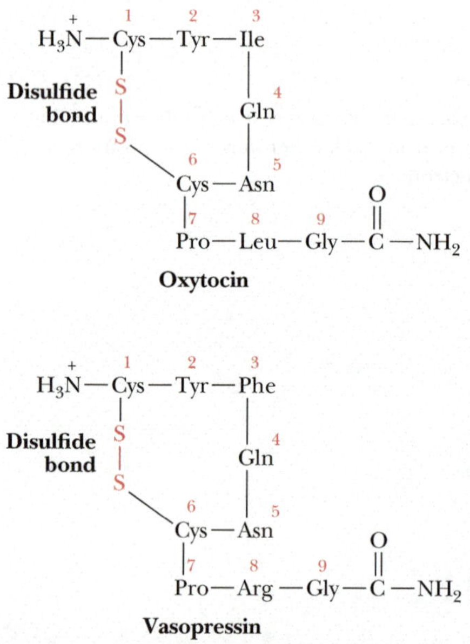

Figure 3.13 Structures of oxytocin and vasopressin.

Both of these peptides have considerable physiological importance as hormones (see **Biochemical Connections 3A**).

3A Biochemical Connections • Physiology

Peptide Hormones: Small Molecules with Big Effects

Both oxytocin and vasopressin are peptide hormones. Oxytocin induces labor in pregnant females and controls contraction of uterine muscle. During pregnancy, the number of receptors for oxytocin in the uterine wall increases. At term, the number of receptors for oxytocin is great enough to cause contraction of the smooth muscle of the uterus in the presence of small amounts of oxytocin produced by the body toward the end of pregnancy. The fetus moves toward the cervix of the uterus because of the strength and frequency of the uterine contractions. The cervix stretches, sending nerve impulses to the hypothalamus. When the impulses reach this part of the brain, positive feedback leads to the release of still more oxytocin by the posterior pituitary gland. The presence of more oxytocin leads to stronger contractions of the uterus so that the fetus is forced through the cervix and the baby is born.

Oxytocin also plays a role in stimulating the flow of milk in a nursing mother (Figure 3.14). The process of suckling sends nerve signals to the hypothalamus of the mother's brain. Oxytocin is released and carried by the blood to the mammary glands. The presence of oxytocin causes the smooth muscle in the mammary glands to contract, forcing out the milk that is in them. As suckling continues, more hormone is released, producing still more milk.

Vasopressin plays a role in the control of blood pressure by regulating contraction of smooth muscle. Like oxytocin, vasopressin is released by the action of the hypothalamus on the posterior pituitary and is transported by the blood to specific receptors. Vasopressin stimulates reabsorption of water by the kidneys, thus having an antidiuretic effect. More water is retained, and the blood pressure increases.

Although the relationship between oxytocin, childbirth, and lactation and that between vasopressin and blood pressure have been known for decades, recently even more interesting effects of these simple peptides have been discovered. In mammals, including humans, both are neuropeptides that affect behavior, especially social interactions. Abnormal social behaviors, such as pathological trusting found in Williams–Beuren syndrome, social isolation in depression, and diminished social cognition in autism negatively affect the lives of those who experience these diseases. Until recently, sorting out the myriad effects of neurochemistry on actual behavior was too daunting. Now, with techniques we discuss in later chapters, it is becoming possible to track behavior to specific genes.

Neuropeptides, including oxytocin and vasopressin, have moved to the front of the hormones thought to be involved in social behaviors. These peptides affect personality, trust, altruism, social bonding, and our ability to recognize and understand the facial expressions and feelings of others. In many species, including humans, oxytocin influences social and sexual behaviors, including intercourse, parental attachment (Figure 3.14), and pair bonding. Vasopressin has an impact on sexual functions, aggression, territoriality, and pair bonding.

Figure 3.14 Nursing stimulates the release of oxytocin. Oxytocin stimulates milk production and is also involved in emotional bonding.

Studies can involve many techniques. In nonhuman mammals, manipulation of hormone levels or hormone receptor levels can allow the study of behavioral effects. In rats, for example, infusion of oxytocin into the brain stimulates maternal behavior in virgin rats that would normally ignore or attack pups. Manipulations that blocked oxytocin receptors in the rat brain reduced maternal behaviors, such as mother–infant bonding.

Only 3% to 5% of mammals are socially monogamous or even have a distinct preference for a mate. Administration of oxytocin has been shown to introduce pair bonding and mate preference in mammals not traditionally monogamous, whereas blockage of oxytocin receptors has the opposite effect in those monogamous species studied. In male prairie voles, administration of vasopressin increases monogamy-related behaviors, such as paternal care, mate guarding, and selective mate preference. The study has gone to the genetic level, where scientists have isolated specific genes for vasopressin receptors and subsequently have been able to correlate behavior to variations in these genes.

With humans we cannot just inject peptides into brains and block receptors so easily. Other techniques have to be employed. One of those involved nasal inhalation of these neuropeptides. But how can we check actual behavior in a controlled setting? One way used a game involving economic gains and losses by individuals and teams. Participants were playing a game and making decisions that would better their results and those of their team, but what was really being tested was how their behaviors changed when they inhaled the peptides. Several conclusions were drawn—oxytocin tends to create trust between humans. Those who inhaled oxytocin were more likely to make decisions during the game that benefited others, especially in their team.

In people assigned male at birth, the hormone was a double-edged sword, however. It did tend to stimulate trust and altruistic behaviors, but only toward the males' specific group that they identified with. It tended to make the same males less trusting of outsiders. Even when there was a betrayal by a partner during the game, those who inhaled oxytocin were more likely to be forgiving and not punish the offender in the next round. On the other hand, those who did *not* inhale the oxytocin were more likely to seek revenge in the next round.

As of the spring of 2024, 1 in 36 children in the United States can be characterized as having Autism Spectrum Disorder (ASD), yet, at present, no drugs exist to treat the social impacts found. Oxytocin has been the focus of much research in the last few years and is seen as one of the more hopeful prospects for a treatment. In individuals with ASD, nasal inhalation of oxytocin has, in some studies, shown temporary improvements in social cognition, empathy, and reciprocity. Unfortunately, results are still inconclusive because other studies show no effects. Despite the lack of clarity, oxytocin can be found sold on the Internet as "liquid trust," where it is marketed as a romance enhancer—perhaps the first "love potion" actually based on some science.

Recent studies show that humans and dogs release oxytocin when they gaze into each other's eyes, perhaps accounting for why

Continued on next page

tanatat/Shutterstock.com

Arto Hakola/Shutterstock.com

Figure 3.15 Oxytocin release in humans, dogs, and wolves. (A) Both humans and dogs release oxytocin when they gaze into each other's eyes. (B) The same effect is not seen in wolves.

many humans form bonds with their animals that seem as strong as those with other humans (Figure 3.15). The same studies noted that this may have been the departure between wolves and domestic dogs, because wolves do not display this behavior and have much less interest in staring lovingly into a human's eyes.

Although much research is still needed, the results to date also bring up some interesting ethical questions. Would it be fair, for example, for a salesperson to wear a cologne or fragrance that gave off oxytocin right before they tried to sell you a car? Would people in the dating and mating game have an unfair advantage if they also wore an oxytocin-laced fragrance? One thing is clear, however—oxytocin and vasopressin are much more than nine simple amino acids. ◗

Summary

Amino Acids Are Three Dimensional

- The amino acids that are the monomer units of proteins have a general structure in common, with an amino group and a carboxyl group bonded to the same carbon atom.
- The nature of the side chains, which are referred to as R groups, is the basis of the differences among amino acids.
- Except for glycine, amino acids can exist in two forms, designated L and D. These two stereoisomers are nonsuperimposable mirror images of each other.
- The amino acids found in proteins are of the L form, but some D-amino acids occur in nature.

Structures and Properties of Amino Acids

- A classification scheme for amino acids can be based on the properties of their side chains.
- Two particularly important criteria are the polar or nonpolar nature of the side chain and the presence of an acidic or basic group in the side chain.

Nonpolar Amino Acids

- One group of amino acids has nonpolar side chains.
- The side chains are mostly aliphatic or aromatic hydrocarbons or their derivatives.

Polar-Neutral Amino Acids

- A second group of amino acids has side chains that contain electronegative elements such as oxygen, nitrogen, and sulfur.

Acidic Amino Acids

- Two amino acids, glutamic acid and aspartic acid, have ionizable carboxyl groups in their side chains.

Basic Amino Acids

- Three amino acids—histidine, lysine, and arginine—have ionizable basic side chains.

Uncommon Amino Acids

- Some amino acids are found only in a few proteins.
- They are formed from the common ones after the protein has been synthesized in the cell.

Titration of Amino Acids

- In free amino acids at neutral pH, the carboxylate group is negatively charged (acidic) and the amino group is positively charged (basic).

- Amino acids without ionizable groups on their side chains exist in neutral solution as zwitterions—that is, with no net charge.
- Titration curves of amino acids indicate the pH ranges in which titratable groups gain or lose a proton.
- Side chains of amino acids can also contribute titratable groups; the charge (if any) on the side chain must be taken into consideration in determining the net charge on the amino acid.

The Peptide Bond

- Peptides are formed by reacting the carboxyl group of one amino acid with the amino group of another amino acid to form a covalent peptide (amide) bond. The other product in this reaction is a water molecule.
- Proteins consist of polypeptide chains; the number of amino acids in a protein is usually 100 or more.
- The peptide group is planar; this stereochemical constraint plays an important role in determining the three-dimensional structures of peptides and proteins.

Small Peptides with Physiological Activity

- Small peptides, containing two to several dozen amino acid residues, can have marked physiological effects in organisms.
- Oxytocin and vasopressin each have nine amino acids and only differ in two positions.
- They both have marked biological effects in vertebrates.
- Oxytocin stimulates uterine contractions, lactation, mother–child bonding, pair bonding, feelings of trust, and altruistic behavior.
- Vasopressin is involved in the control of blood pressure, as well as being implicated in male behavior of territoriality, aggression, pair bonding, and social group defense.

Review Exercises

3-1 Amino Acids Are Three Dimensional

1. **Recall** How do D-amino acids differ from L-amino acids? What biological roles are played by peptides that contain D-amino acids?
2. **Recall** Draw the general structure of an L-amino acid as it would exist in solution at pH 7.0. Show stereochemistry as appropriate. Also consider the protonation state of the α-amino and α-carboxyl groups at pH 7.0.
3. **Reflect and Apply** Which of the following structures correctly represents the structure of an L-amino acid at pH 7.0?

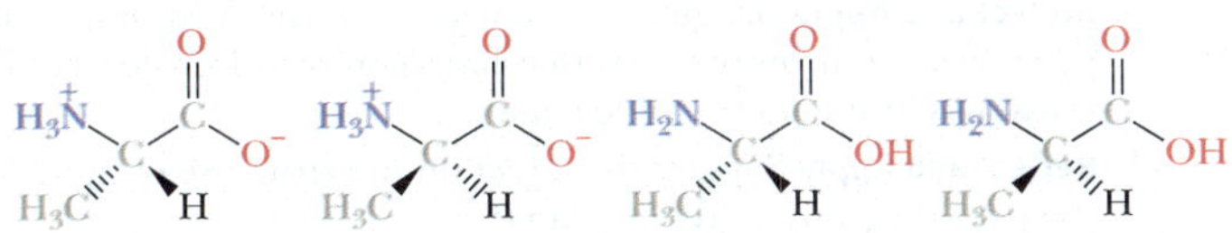

4. **Reflect and Apply** Which of the following structures represents the structure of a D-amino acid at pH 7.0?

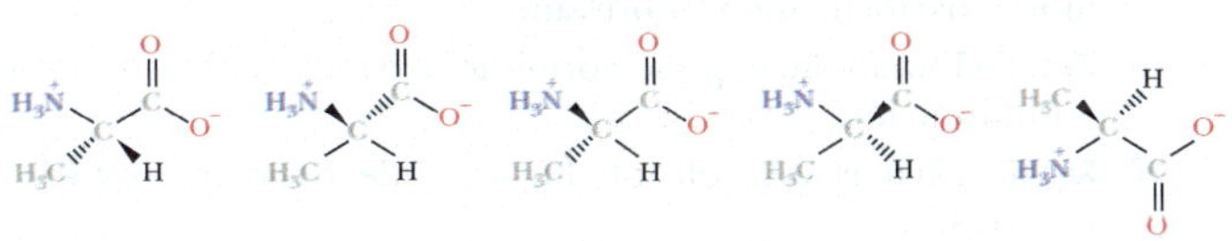

5. **Reflect and Apply** Which of the following structures represents the structure of an amino acid in solution at pH 7.0? (You do not need to consider stereochemistry.)

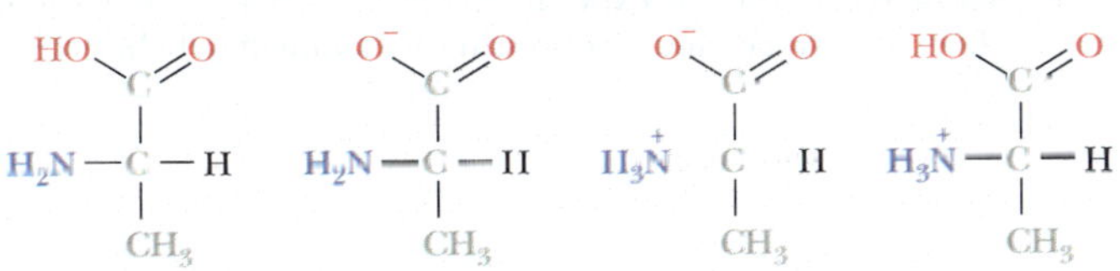

6. **Reflect and Apply** Determine whether each of the following amino acids is chiral or achiral.

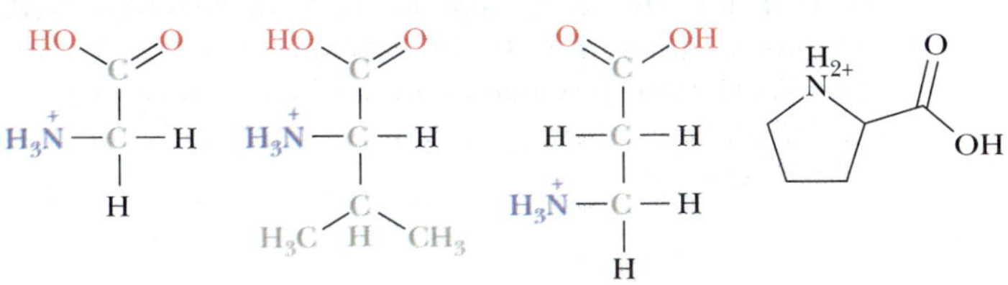

7. **Reflect and Apply** Draw the structure of a chiral α-amino acid with the formula $C_4H_9NO_2$. Draw it in the form that it would exist in solution at pH 7.0. You do not need to show the L- or D-configuration.
8. **Reflect and Apply** Draw the structure of an achiral α-amino acid with the formula $C_4H_9NO_2$. Draw it in the form that it would exist in solution at pH 7.0.

3-2 Structures and Properties of Amino Acids

9. **Recall** For each of the following, name an amino acid in which the R group contains it: a hydroxyl group, a sulfur atom, a second chiral carbon atom, an amino group, an amide group, an acid group, an aromatic ring, and a branched side chain.
10. **Recall** Provide the three-letter and one-letter codes for each of the following amino acids.
    Glycine
    Glutamic acid
    Glutamine
    Tyrosine
    Tryptophan
    Lysine
    Leucine
    Aspartic acid
11. **Recall** Identify the polar amino acids, the aromatic amino acids, and the sulfur-containing amino acids, given a peptide with the following amino acid sequence:

    Val—Met—Ser—Ile—Phe—Arg—Cys—Tyr—Leu

12. **Recall** Identify the nonpolar amino acids and the acidic amino acids in the following peptide:

    Glu—Thr—Val—Asp—Ile—Ser—Ala

13. **Recall** Which of the following structural features determines whether an α-amino acid is classified as polar or nonpolar?
    (a) The ionization state of the α-carboxyl group of the amino acid.
    (b) The ionization state of the α-amino group of the amino acid.
    (c) The stereochemical configuration of the α-carbon of the amino acid.
    (d) The nature of the side chain of the amino acid.

3-3 Amino Acids Can Act as Both Acids and Bases

14. **Mathematical** Predict the predominant ionized forms of the following amino acids at pH 7: glutamic acid, leucine, threonine, histidine, and arginine.
15. **Mathematical** Draw structures of the following amino acids, indicating the charged form that exists at pH 4: histidine, asparagine, tryptophan, proline, and tyrosine.
16. **Mathematical** Predict the predominant forms of the amino acids from Question 14 at pH 10.
17. **Mathematical** Calculate the isoelectric point of each of the following amino acids: glutamic acid, serine, tyrosine, and arginine.
18. **Mathematical** Sketch a titration curve for the amino acid cysteine, and indicate the pK_a values for all titratable groups. Also indicate the pH at which this amino acid has no net charge.
19. **Mathematical** Indicate the pH range in which the conjugate acid–base pair consisting of Asp with a +1 net charge and Asp with a 0 net charge will act as a buffer.
20. **Reflect and Apply** Suggest a reason why amino acids are usually more soluble at pH extremes than they are at neutral pH. (This does not mean that they are insoluble at neutral pH.)
21. **Reflect and Apply** Write equations to show the ionic dissociation reactions of the following amino acids: aspartic acid, valine, histidine, serine, and lysine.
22. **Reflect and Apply** Based on the information in Table 3.2, is there any amino acid that could serve as a buffer at pH 8? If so, which one?
23. **Reflect and Apply** Identify the charged groups in the peptide shown in Question 11 at pH 1 and at pH 7. What is the net charge of this peptide at these two pH values?
24. **Reflect and Apply** Consider the following peptides: Phe—Glu—Ser—Met and Val—Trp—Cys—Leu. Do these peptides have different net charges at pH 1? At pH 7? Indicate the charges at both pH values.
25. **Reflect and Apply** In each of the following two groups of amino acids, which amino acid would be the easiest to distinguish from the other two amino acids in the group, based on a titration?
    (a) Gly, Leu, Lys
    (b) Glu, Asp, Ser
26. **Reflect and Apply** Could the amino acid glycine serve as the basis of a buffer system? If so, in what two pH ranges would it be useful?
27. **Reflect and Apply** Draw the structure(s) of glutamic acid as it would appear after titrating the fully protonated (low pH) form of glutamic acid with 1.5 equivalents of NaOH. What would be the approximate charge of glutamic acid at this point?
28. **Reflect and Apply** What would be the approximate pH after titrating the fully protonated (low pH) form of glutamic acid with 1.5 equivalents of NaOH? At approximately which position on the titration curve would you be at this point?

3-4 The Peptide Bond

29. **Recall** Sketch resonance structures for the peptide group.
30. **Recall** How do the resonance structures of the peptide group contribute to the planar arrangement of this group of atoms?
31. **Reflect and Apply** Would the peptide group be planar if the amino group of amino acids were bonded to the β carbon of the amino acid, rather than to the α carbon?
32. **Reflect and Apply** Can a peptide act as a buffer? If so, why?
33. **Reflect and Apply** How do the peptides Ser—Glu—Gly—His—Ala and Gly—His—Ala—Glu—Ser differ?
34. **Reflect and Apply** What are the sequences of all the possible tripeptides that contain the amino acids aspartic acid, leucine, and phenylalanine? Use the three-letter abbreviations to express your answer.
35. **Reflect and Apply** Answer Question 34 using one-letter designations for the amino acids.
36. **Reflect and Apply** If the amino acids alanine and glycine react to form a peptide bond, is there more than one reaction product? If yes, what products?
37. **Reflect and Apply** Why do we not indicate the side chains of the amino acids that form the peptide bond (see Figure 3.12)?
38. **Reflect and Apply** Would the presence of a chiral center in an amino acid side chain affect the formation of a peptide bond?
39. **Reflect and Apply** What might you infer (or know) about the stability of amino acids when compared with that of other building-block units of biopolymers (sugars, nucleotides, fatty acids, etc.)?
40. **Reflect and Apply** Consider the peptides Gly—Pro—Ser—Glu—Thr Gly—Pro—Ser—Glu—Thr and Gly—Pro—Ser—Glu—Thr. Would a cyclic peptide bond linking the threonines and the glycines be the same chemically as the individual peptides?
41. **Reflect and Apply** Suggest a reason why the biosynthesis of amino acids and of proteins would eventually cease in an organism with carbohydrates (molecules containing only carbon, hydrogen, and oxygen) as its only food source.
42. **Reflect and Apply** You are studying with a friend who draws the structure of alanine at pH 7. It has a carboxyl group (—COOH) and an amino group (—NH_2). What suggestions would you make?
43. **Reflect and Apply** Suggest a reason (or reasons) why amino acids polymerize to form proteins that have comparatively few covalent crosslinks in the polypeptide chain.
44. **Reflect and Apply** Suggest the effect on the structure of peptides if the peptide group were not planar.

3-5 Small Peptides with Physiological Activity

45. **Recall** What are the structural differences between the peptide hormones oxytocin and vasopressin?
46. **Recall** How do the peptide hormones oxytocin and vasopressin differ in function?
47. **Recall** What is the role of the disulfide bond in oxytocin and vasopressin?
48. **Recall** Is it possible to form cyclic peptides without bonds between side chains of the component amino acids?
49. **Recall** How has oxytocin been implicated in human behavior?
50. **Recall** What types of experiments led to evidence regarding the effect of oxytocin and vasopressin on mammalian behavior?

Further Reading

Barrett, G. C., ed. *Chemistry and Biochemistry of the Amino Acids.* New York: Chapman & Hall, 1985. Wide coverage of many aspects of the reactions of amino acids.

De Dreu, C. K. W., L. L. Greer, M. J. J. Handgraaf, S. Shalvi, G. A. Van Kleef, M. Baas, F. S. Ten Velden, E. Van Dijk, and S. W. W. Feith. The Neuropeptide Oxytocin Regulates Parochial Altruism in Intergroup Conflict Among Humans. *Science,* **328,** 1408–1411 (2010). A study showing that oxytocin plays a role in human beings' nature to care for family and friends as well as defend them.

Donaldson, Z. R., and L. J. Young. Oxytocin, Vasopressin, and the Neurogenetics of Sociality. *Science,* **322**, 900–904 (2008). A review of two major small peptides and their effects on mammalian behavior.

Javitt, D. C., and J. T. Coyle. Decoding Schizophrenia. *Sci. Amer.* **290** (1), 48–55 (2004).

Larsson, A., ed. *Functions of Glutathione: Biochemical, Physiological, Toxicological, and Clinical Aspects.* New York: Raven Press, 1983. A collection of articles on the many roles of a ubiquitous peptide.

McKenna, K. W., and V. Pantic, eds. *Hormonally Active Brain Peptides: Structure and Function.* New York: Plenum Press, 1986. A discussion of the chemistry of enkephalins and related peptides.

Miller, G. The Prickly Side of Oxytocin. *Science.* **328**, 1343 (2010). A summary of research regarding effects of oxytocin on mammals.

Siddle, K., and J. C. Hutton. *Peptide Hormone Action—A Practical Approach.* Oxford, England: Oxford University Press, 1990. A book that concentrates on experimental methods for studying the actions of peptide hormones.

Stegink, L. D., and L. J. Filer, Jr. *Aspartame—Physiology and Biochemistry.* New York: Marcel Dekker, 1984. A comprehensive treatment of metabolism, sensory and dietary aspects, preclinical studies, and issues relating to human consumption (including ingestion by people with phenylketonuria and consumption during pregnancy).

Wilson, N., E. Barbar, J. Fuchs, and C. Woodward. Aspartic Acid in Reduced *Escherichia coli* Thioredoxin Has a pK_a 9. *Biochem.* **34**, 8931–8939 (1995). A research report on a remarkably high pK_a value for a specific amino acid in a protein.

Wold, F. In Vivo Chemical Modification of Proteins (Post-Translational Modification). *Ann. Rev. Biochem.* **50**, 788–814 (1981). A review article on the modified amino acids found in proteins.

Young, L. J., and C. E. Barrett. Can Oxytocin Treat Autism? *Science,* **347**, 825–826 (2015). An article about a possible treatment for autism involving oxytocin.

Zak, P. J. The Neurobiology of Trust. *Sci. Amer.* **306** (6), 88–95 (2008). A review of how oxytocin affects trust in mammals.

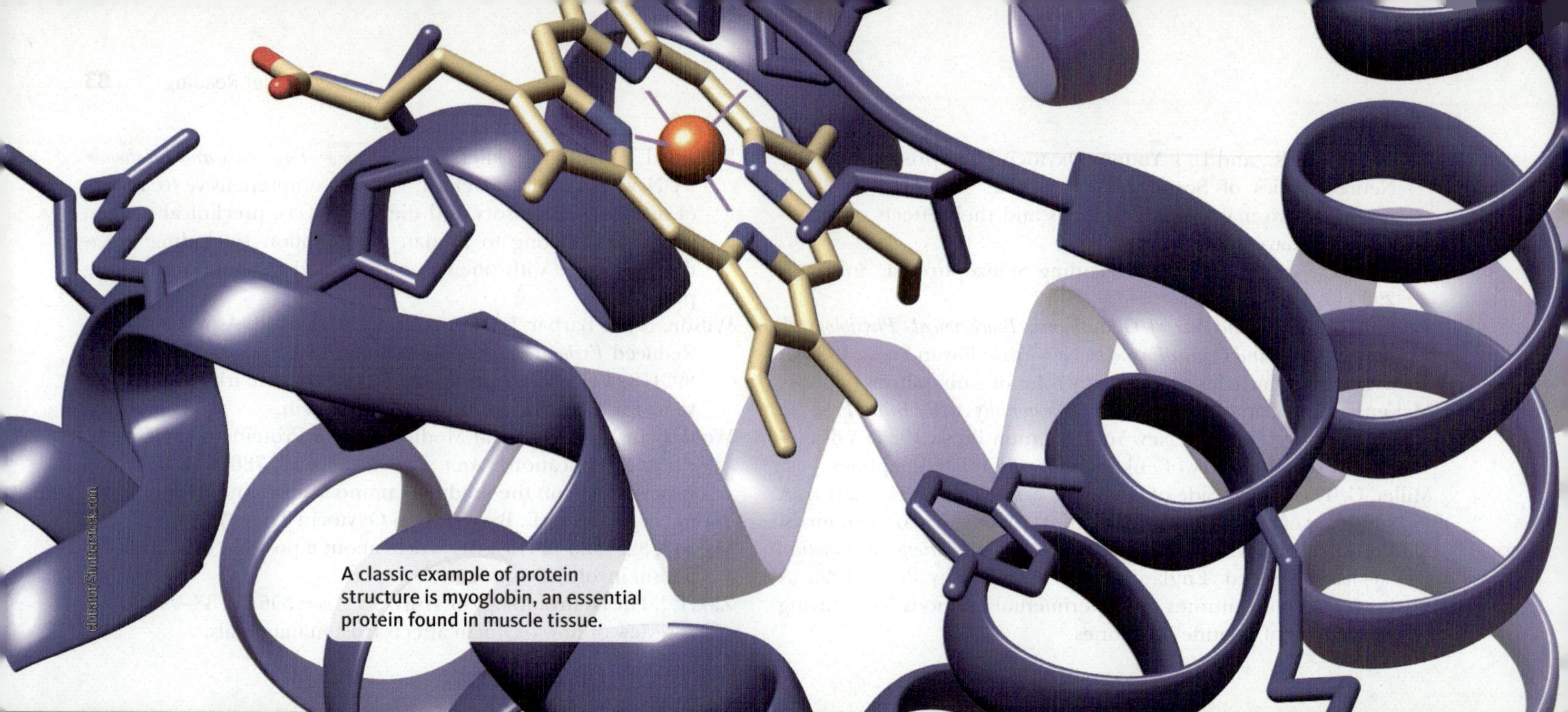
A classic example of protein structure is myoglobin, an essential protein found in muscle tissue.

4 The Three-Dimensional Structure of Proteins

Chapter Outline

4-1 Protein Structure and Function

Learning Objectives

After reading this section, students will be able to:

LO 4.1.1 Describe the importance of native conformations of proteins.

LO 4.1.2 Distinguish between the four levels of protein structure.

Biologically active proteins are polymers consisting of amino acids linked by covalent peptide bonds (Section 3-4). Many different conformations (three-dimensional structures) are possible for a molecule as large as a protein. Of these many structures, one or (at most) a few have biological activity; these are called the **native conformations**. These native conformations are needed for the proper function of each protein. Protein structures are complex, and they are defined in terms of four levels of structure, called primary, secondary, tertiary, and quaternary.

▶ What are the four levels of protein structure?

Primary structure is the N-terminal to C-terminal order in which the amino acids are covalently linked together (to review the N-terminus and C-terminus of a polypeptide, see Figure 3.11). Therefore, the polypeptide Leu—Gly—Thr—Val—Arg—Asp—His has a different primary structure than the polypeptide Val—His—Asp—Leu—Gly—Arg—Thr, even though both have the same number and kinds of amino acids. The primary structure is the one-dimensional first step in specifying the three-dimensional structure of a protein. Even though

Careers in Chemistry • Valerie George

CEO at Simply Formulas

Education:
Kent State University, BS, Biochemistry

Valerie George is a cosmetic chemist and CEO of Simply Formulas, a leading research and development lab. In an industry constantly shaped by evolving consumer trends and innovations, George formulates hair care products, researches cutting-edge ingredients, and oversees manufacturing processes. Her work spans from lab bench experiments to attending industry conferences and collaborating with marketing and education teams.

George's interest in science began at a young age, nurtured by her parents' encouragement of her curiosity. Her journey started as a high school science teacher, and it deepened her scientific foundation and love for education. While exploring research opportunities, she discovered cosmetic chemistry and was immediately drawn to the idea of turning science into everyday products like lipstick.

Valerie George

With a degree in biochemistry, George's unique blend of chemistry and biology expertise gives her a distinct advantage in understanding clinical data and formulating effective, science-backed products. She encourages students interested in biochemistry to stay curious, explore broadly, and actively build connections within the professional community. Organizations like the Society of Cosmetic Chemists, she notes, offer valuable resources, mentorship, and career development. She shares, "There are so many exciting careers with a degree in biochemistry. The opportunities are endless."

"structure" is in its name, the primary structure of a protein refers only to its amino acid sequence, not its three-dimensional conformation. However, the sequence of a protein does determine its three-dimensional structure.

native conformations the three-dimensional shapes of proteins

primary structure the N-terminal to C-terminal order in which the amino acids in a protein are linked by peptide bonds

Two three-dimensional aspects of a single polypeptide chain, called the secondary and tertiary structure, can be considered separately. **Secondary structure** is the hydrogen-bonded arrangement in space of the *peptide backbone atoms*. Recall from Section 3-4 that these are the repeating —C(=O)—NH— groups of atoms that form the peptide bonds (see Figure 3.11). The α-helix and β-pleated sheet arrangements are two different types of secondary structure (Sections 4-3A and 4-3B). Secondary structures have repetitive interactions resulting from hydrogen bonding between the amide N—H and the carbonyl groups of the peptide backbone.

secondary structure the hydrogen-bonded arrangement in space of the backbone atoms in a polypeptide chain

The conformations of the side chains of the amino acids are not part of the secondary structure, but side chain atoms are important in determining the **tertiary structure**, which is the overall structure of a protein chain. Tertiary structure includes the three-dimensional arrangement of all the atoms in the protein.

tertiary structure the arrangement in space of all the atoms in a protein

A single protein can also consist of multiple polypeptide chains called **subunits**. Each subunit has its own N-terminus and C-terminus with many amino acids linked together in between these termini. The arrangement of subunits with respect to one another is the **quaternary structure**. Interaction between subunits is mediated by noncovalent interactions such as hydrogen bonds, electrostatic attractions, and hydrophobic interactions.

subunits the individual parts of a larger molecule (e.g., the individual polypeptide chains that make up a complete protein)

quaternary structure the interaction of several polypeptide chains in a multisubunit protein

4-2 Primary Structure of Proteins

Learning Objectives

After reading this section, students will be able to:

LO 4.2.1 Describe the primary structure of proteins in terms of connectivity or spatial orientation.

LO 4.2.2 Describe the importance of protein primary structure.

The amino acid sequence (the primary structure) of a protein determines its three-dimensional structure, which, in turn, determines its function and other properties. In every protein, the correct three-dimensional structure is needed for correct functioning.

▶ Why is the primary structure of a protein important?

One of the most striking demonstrations of the importance of primary structure is found in the hemoglobin associated with *sickle-cell disease.* In this genetic disease, red blood cells cannot bind oxygen efficiently. The red blood cells also assume a characteristic sickle shape, giving the disease its name. The sickled cells tend to become trapped in small blood vessels, cutting off circulation and thereby causing organ damage. These drastic consequences stem from a change in one amino acid **residue** in the sequence of the primary structure (see Biochemical Connections 4A).

residue a specific monomer within the polymeric chain of a protein, nucleic acid, or polysaccharide

Considerable research is being done to determine the effects of changes in primary structure on the functions of proteins. Using molecular-biology techniques, such as site-directed mutagenesis, it is possible to replace any chosen amino acid residue in a protein with another specific amino acid residue. The new characteristics of the altered protein, including effects on its biological activity, can then be determined. The effects of such amino acid substitutions range from negligible to significant, depending on the protein and the nature of the residue alteration.

Determining the sequence of amino acids in a protein is a routine, but not trivial, operation in classical biochemistry. It consists of several steps, which must be carried out carefully to obtain accurate results (Section 5-4).

4-3 Secondary Structure of Proteins

Learning Objectives

After reading this section, students will be able to:

LO 4.3.1 Identify secondary and supersecondary structures in proteins.
LO 4.3.2 Distinguish between different secondary structures that can exist in proteins.
LO 4.3.3 Describe the factors that affect secondary structure formation.
LO 4.3.4 Distinguish between fibrous and globular proteins.

The secondary structure of proteins is the hydrogen-bonded arrangement of the backbone atoms of the protein chain. The nature of the bonds in the peptide backbone plays an important role here. Within each amino acid residue are two bonds with reasonably free rotation: (1) the bond between the α-carbon and the amino nitrogen of that residue and (2) the bond between the α-carbon and the carboxyl carbon of that residue. The combination of the planar peptide group and the two freely rotating bonds has important implications for the three-dimensional conformations of peptides and proteins.

A protein backbone can be visualized as a series of playing cards, each card representing a planar peptide group. The cards are linked at opposite corners by swivels, representing the bonds about which there is considerable freedom of rotation (Figure 4.1). The side-chain atoms play a vital role in determining the three-dimensional shape of a protein, but the backbone atoms play the main role in determining protein secondary structure. The angles ϕ (phi) and ψ (psi), frequently called Ramachandran angles (after the researcher who named them, G. N. Ramachandran), are used to designate rotations around the C—N and C—C bonds, respectively. The conformation of a protein backbone can be described by specifying the values of ϕ and ψ for each residue (−180° to 180°).

α-helix a folding pattern in the protein backbone that is stabilized by hydrogen bonds parallel to the helix axis

β-sheet a folding pattern in which the protein backbone is almost fully extended with hydrogen bonding between adjacent strands

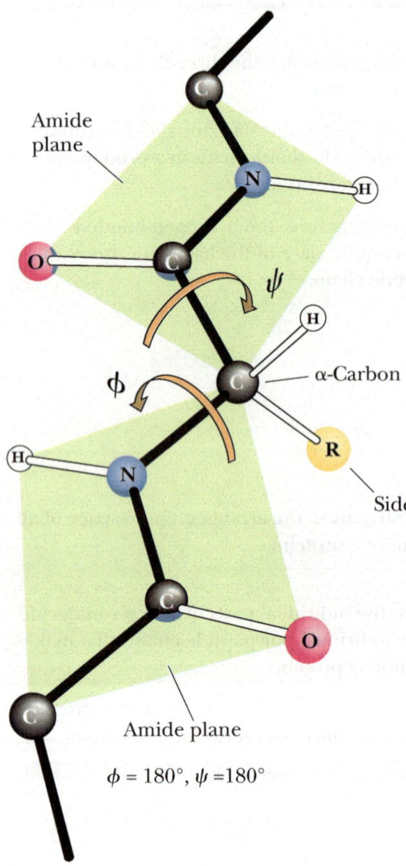

Figure 4.1 Protein conformation is described by phi and psi angles. The peptide bonds exist in a plane, and the bond between the carbonyl group and the amide group does not rotate. Two bonds coming from the alpha carbon can rotate, and secondary structure can be described by the two angles, phi and psi, shown. By definition, when the two planes are parallel, both angles are assigned the value of 180 degrees. *(Illustration: Irving Geis. Rights owned by Howard Hughes Medical Institute.)*

Two kinds of secondary structures that occur frequently in proteins are the repeating **α-helix** and **β-sheet** (or β-pleated sheet) hydrogen-bonded structures. The ϕ and ψ angles repeat themselves in contiguous amino acids in regular secondary structures. The α-helix and β-pleated sheet are not the only possible secondary structures, but they are by far the most important and deserve a closer look.

4-3A The α-Helix

The α-helix is stabilized by hydrogen bonds parallel to the helix axis within the backbone of a single polypeptide chain. Counting from the N-terminal end of the helix, the C—O group of each amino acid residue is hydrogen-bonded to the N—H group of the amino acid four residues away from it in the covalently bonded sequence (Figure 4.2). The helical conformation allows a linear arrangement of the atoms involved in the hydrogen bonds (Section 2-2), which gives the bonds maximum strength and thus makes the helical conformation very stable. There are 3.6 residues for each turn of the helix, and the *pitch* of the helix (the linear distance between corresponding points on successive turns) is 5.4 Å (Figure 4.2).

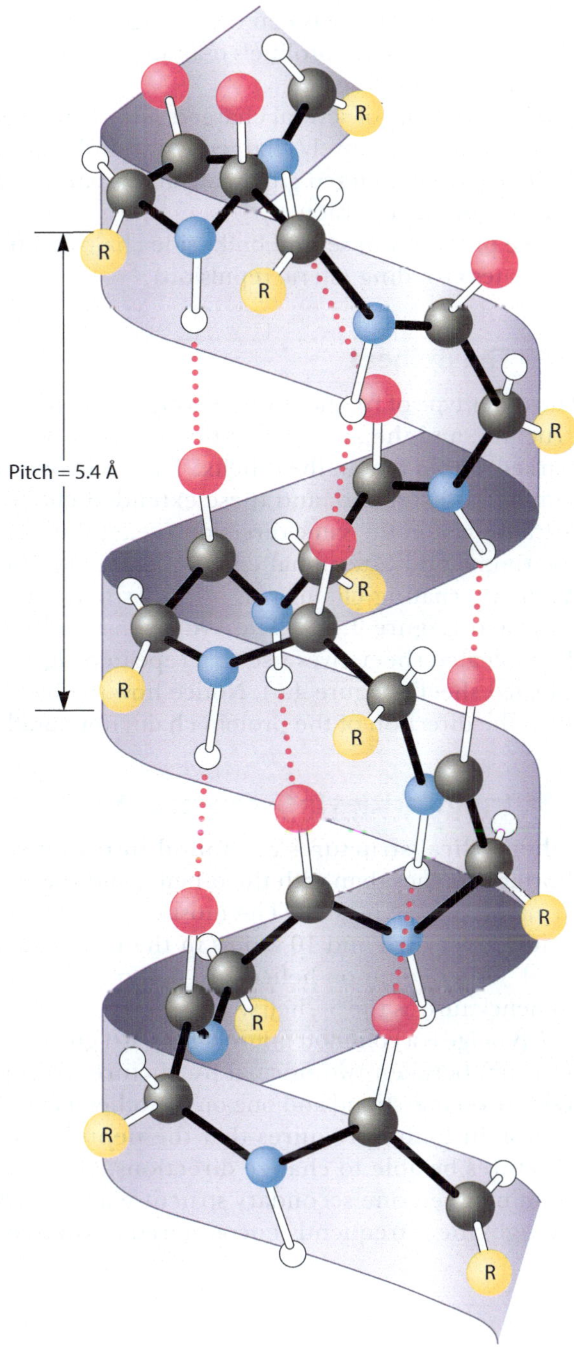

Figure 4.2 Graphic representation of the α-helix. A hybrid of a ribbon diagram and a ball-and-stick figure that shows the path of the peptide backbone, with hydrogen bonds shown as red dotted lines. R represents side chain groups of each amino acid residue.

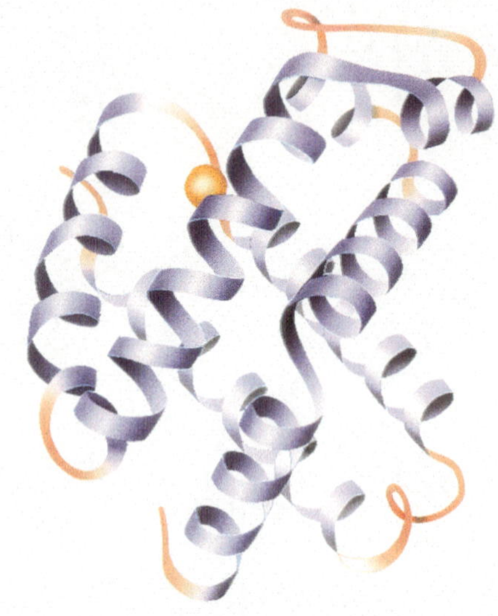

β-Hemoglobin subunit

Myohemerythrin

Figure 4.3 Three-dimensional structure of two proteins with substantial amounts of α-helix in their structures. The helices are represented by the regularly coiled sections of the ribbon diagram. β-Hemoglobin and myohemerythrin are oxygen-carrying proteins.

pitch of an α-helix the vertical distance between consecutive turns of the helix

β-strands the extended backbone segments of a protein chain that hydrogen-bond to each other to form a β-sheet

β-bulge a common nonrepetitive irregularity in β-sheets

reverse turns (or β-turns) parts of proteins where the polypeptide chain folds back on itself

The angstrom unit, $1\ \text{Å} = 10^{-10}\ \text{m} = 0.1\ \text{nm}$, is convenient for interatomic distances in molecules, but it is not a Système International (SI) unit. Nanometers ($1\ \text{nm} = 10^{-9}\ \text{m}$) and picometers ($1\ \text{pm} = 10^{-12}\ \text{m}$) are the SI units used for interatomic distances. In SI units, the **pitch of an α-helix** is 0.54 nm or 540 pm.

Notice in Figure 4.2 how the side chain groups of the α-helix (represented generically by green spheres labeled "R") point outward from the helix. They do so because there is insufficient room in the interior of the α-helix for the side chain groups. This is markedly different than in DNA (Section 9-3A), where the nitrogenous bases line the interior of the double helix.

Figure 4.3 shows the structures of two proteins with a high degree of α-helical content. Proteins have varying amounts of α-helical structures, varying from a few percent to nearly 100%. The following factors can disrupt the α-helix:

- The amino acid proline creates a bend in the backbone because of its *cyclic* structure (Figure 3.3A). It cannot fit into the α-helix because (1) rotation around the bond between the nitrogen and the α-carbon is severely restricted, and (2) proline's α-amino group cannot participate in intrachain hydrogen bonding.
- The proximity of several charged side-chain groups of the same sign, such as groups of positively charged lysine and arginine residues or groups of negatively charged glutamate and aspartate residues, creates strong electrostatic repulsion.
- The proximity of several bulky side chains on the outer surface of the helix creates crowding (steric repulsion).

4-3B The β-Sheet

The other type of frequently occurring secondary structure in proteins is β-sheet structure, and the arrangement of atoms in the β-sheet conformation differs markedly from that in the α-helix. The backbone chains in a β-sheet are almost completely extended, and these extended chains are called **β-strands**. If the β-strands run in the same direction (i.e., if they are all aligned in terms of their N-terminal and C-terminal ends), a parallel β-sheet is formed (Figure 4.4A). When alternating chains run in opposite directions, an antiparallel β-sheet is formed (Figure 4.4B). The hydrogen bonding between peptide chains in the β-pleated sheet gives rise to a repeated zigzag structure; hence, the name "pleated sheet" (Figure 4.5). Notice how the hydrogen bonds are perpendicular to the direction of the protein chain, not parallel to it as in the α-helix.

4-3C Irregularities in Secondary Structures

Other helical structures are found in proteins. These are often found in shorter stretches than with the α-helix, and they sometimes break up the regular nature of the α-helix. The most common is the 3_{10} helix, which has three residues per turn and 10 atoms in the ring formed by making the hydrogen bond. Other common helices are designated 2_7 and 4.4_{16}, following the same nomenclature as the 3_{10} helix.

A **β-bulge** is a common nonrepetitive irregularity found in antiparallel β-sheets. It occurs between two normal β-structure hydrogen bonds and involves two residues on one strand and one on the other. Figure 4.6 shows typical β-bulges.

Protein folding requires that the peptide backbones and the secondary structures be able to change directions. Often, a reverse turn marks a transition between one secondary structure and another. For steric (spatial) reasons, glycine is frequently encountered in **reverse turns (or β-turns)**, at which

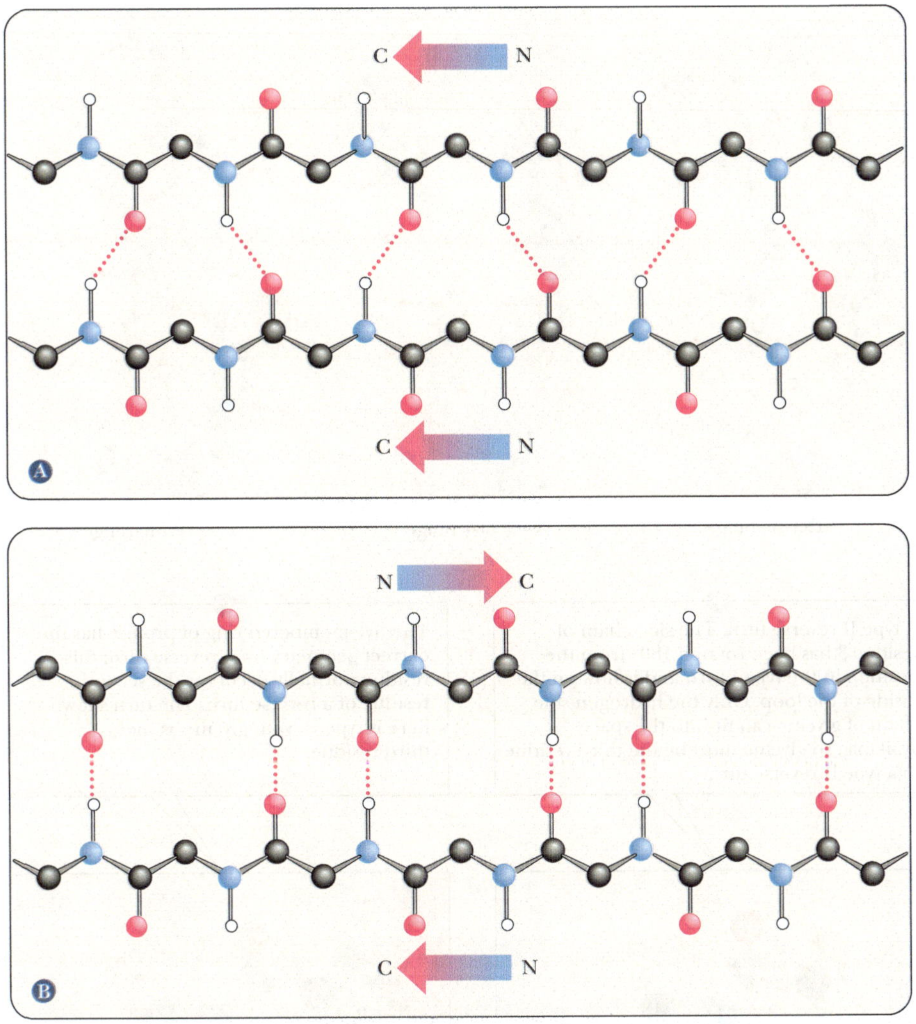

Figure 4.4 Hydrogen bonding in β-sheets. Ball-and-stick diagram showing the arrangement of hydrogen bonds in (A) parallel and (B) antiparallel β-sheets.

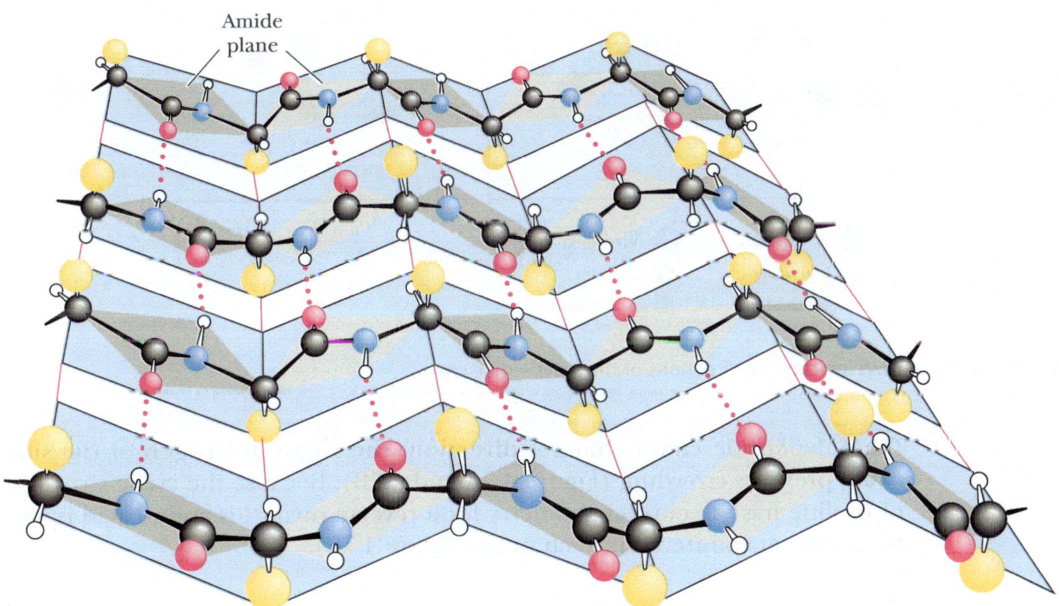

Figure 4.5 The three-dimensional form of the antiparallel β-pleated sheet arrangement. The antiparallel orientation allows for maximum hydrogen bonding because the oxygen, nitrogen, and hydrogen are in a straight line. Yellow spheres represent side chain groups. Amide planes (Figure 4.1) are also indicated. *(Illustration: Irving Geis. Rights owned by Howard Hughes Medical Institute.)*

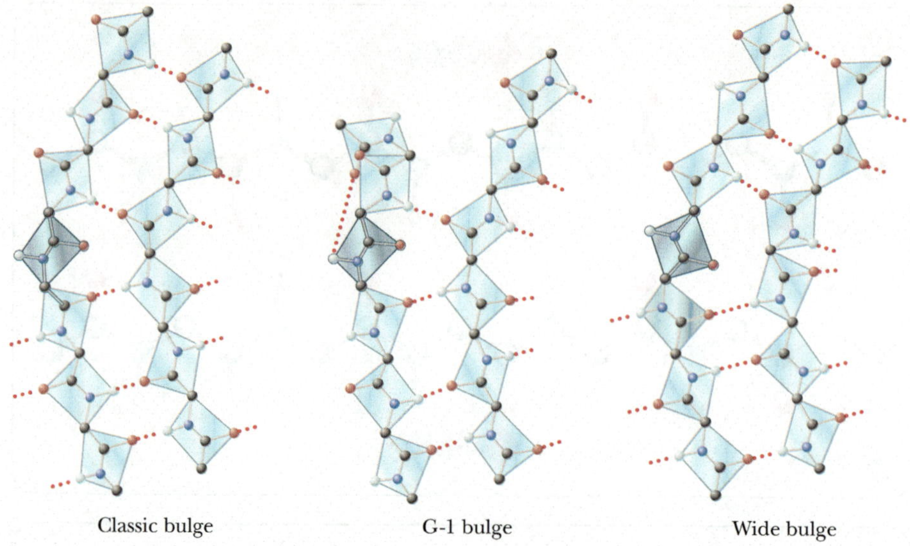

Figure 4.6 **β-Bulges.** Ball-and-stick models of three different β-bulge structures. Hydrogen bonds are shown as red dots.

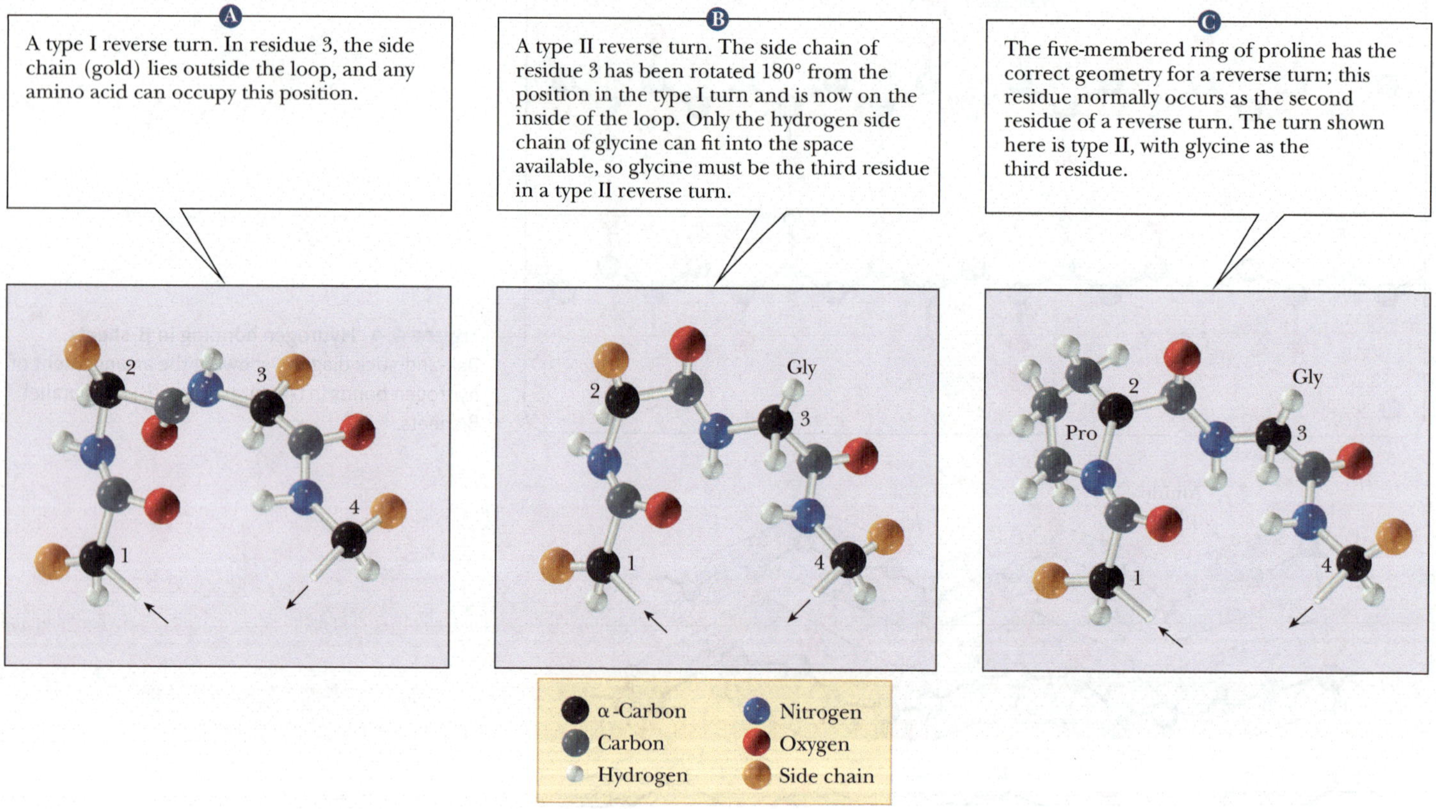

Figure 4.7 Structures of reverse turns (or β-turns). Arrows indicate the directions of the polypeptide chains.

the polypeptide chain changes direction; the single hydrogen of the side chain prevents crowding (Figure 4.7A and 4.7B). Because the cyclic structure of proline has the correct geometry for a reverse turn, this amino acid is also frequently encountered in such turns (Figure 4.7C).

4-3D Supersecondary Structures

supersecondary structures (or structural motifs) parts of proteins that consist of several adjacent elements of secondary structure

The α-helix, β-pleated sheet, and other secondary structures are combined in many ways as the polypeptide chain folds back on itself in a protein. Combinations of secondary structures produce various kinds of **supersecondary structures (or structural motifs)** in proteins. The most common feature of

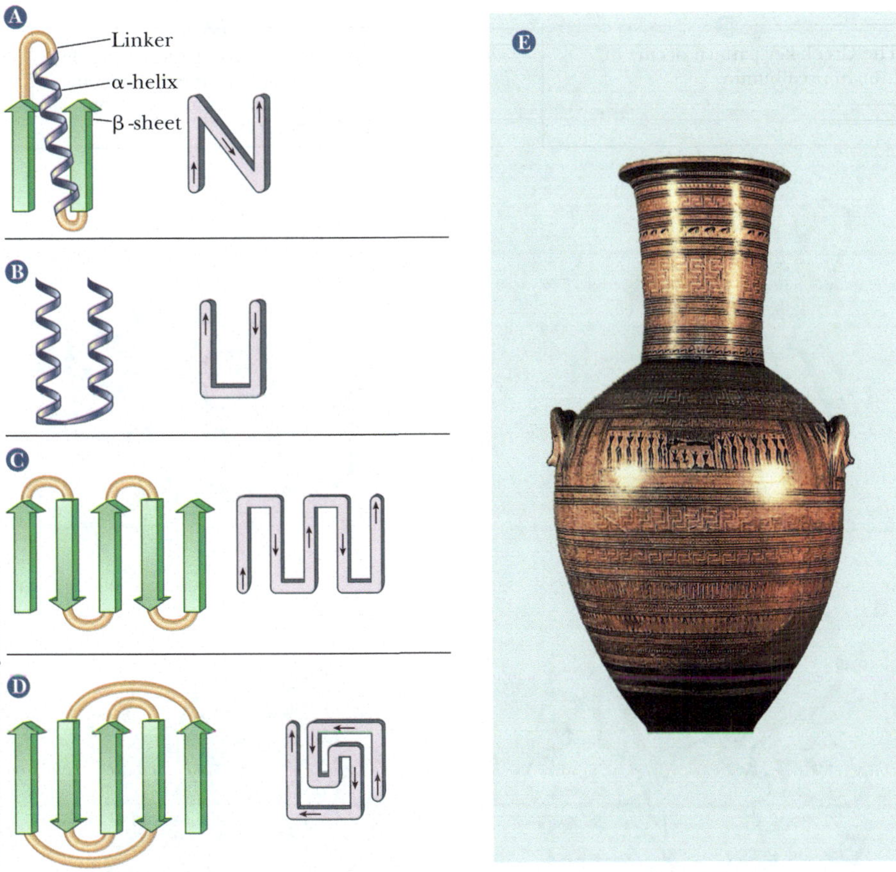

The Dipylon Amphora in the geometric style, c. 750 BC/National Archaeological Museum, Athens, Greece/The Bridgeman Art Library.

Leemage/Corbis Historical/Getty Images

Figure 4.8 Schematic diagrams of supersecondary structures. Arrows indicate the directions of the polypeptide chains. (A) A βαβ unit, (B) an αα unit, (C) a β-meander, and (D) the Greek key. (E) The Greek key motif in protein structure resembles the geometric patterns on this ancient Greek vase, giving rise to the name.

this sort is the *βαβ unit,* in which two parallel strands of β-sheet are connected by a stretch of α-helix (Figure 4.8A). An *αα unit* (helix-turn-helix) consists of two antiparallel α-helices (Figure 4.8B). In such an arrangement, energetically favorable contacts exist between the side chains in the two stretches of helix. In a *β-meander,* an antiparallel sheet is formed by a series of tight reverse turns connecting stretches of the polypeptide chain (Figure 4.8C).

Another kind of antiparallel sheet is formed when the polypeptide chain doubles back on itself in a pattern known as the *Greek key,* named for a decorative design found on pottery from the classical period (Figure 4.8E). Protein sequences that allow for a β-meander or Greek key can often be found arranged into a β-barrel in the tertiary structure of the protein (Figure 4.9). Structural motifs are important and tell us much about the folding of proteins. However, these motifs do not allow us to predict anything about the biological function of the protein because they are found in proteins and enzymes with very dissimilar functions.

4-3E The Collagen Triple Helix

Collagen, a major structural protein in connective tissues such as bone, cartilage, and tendons, is the most abundant protein in vertebrates. It is organized in water insoluble fibers of great strength. A collagen fiber consists of three polypeptide chains wrapped around each other in a ropelike twist, or triple helix, and the polypeptide chains of collagen also contain two **non-proteinogenic amino acids**—hydroxylysine and hydroxyproline (Figure 4.10). Each of the three chains has, within limits, a repeating sequence of three amino acid residues: X—Pro—Gly or X—Hyp—Gly, where Hyp stands for hydroxyproline, and any amino acid can occupy the first position, designated by X.

non-proteinogenic amino acids amino acids that are not incorporated into proteins biosynthetically

Proline and hydroxyproline can constitute up to 30% of the residues in collagen. Hydroxyproline (Figure 4.10) is formed from proline by a specific hydroxylating enzyme after the amino acids are linked together. In the amino acid sequence of collagen, every third position must be occupied by glycine. The triple

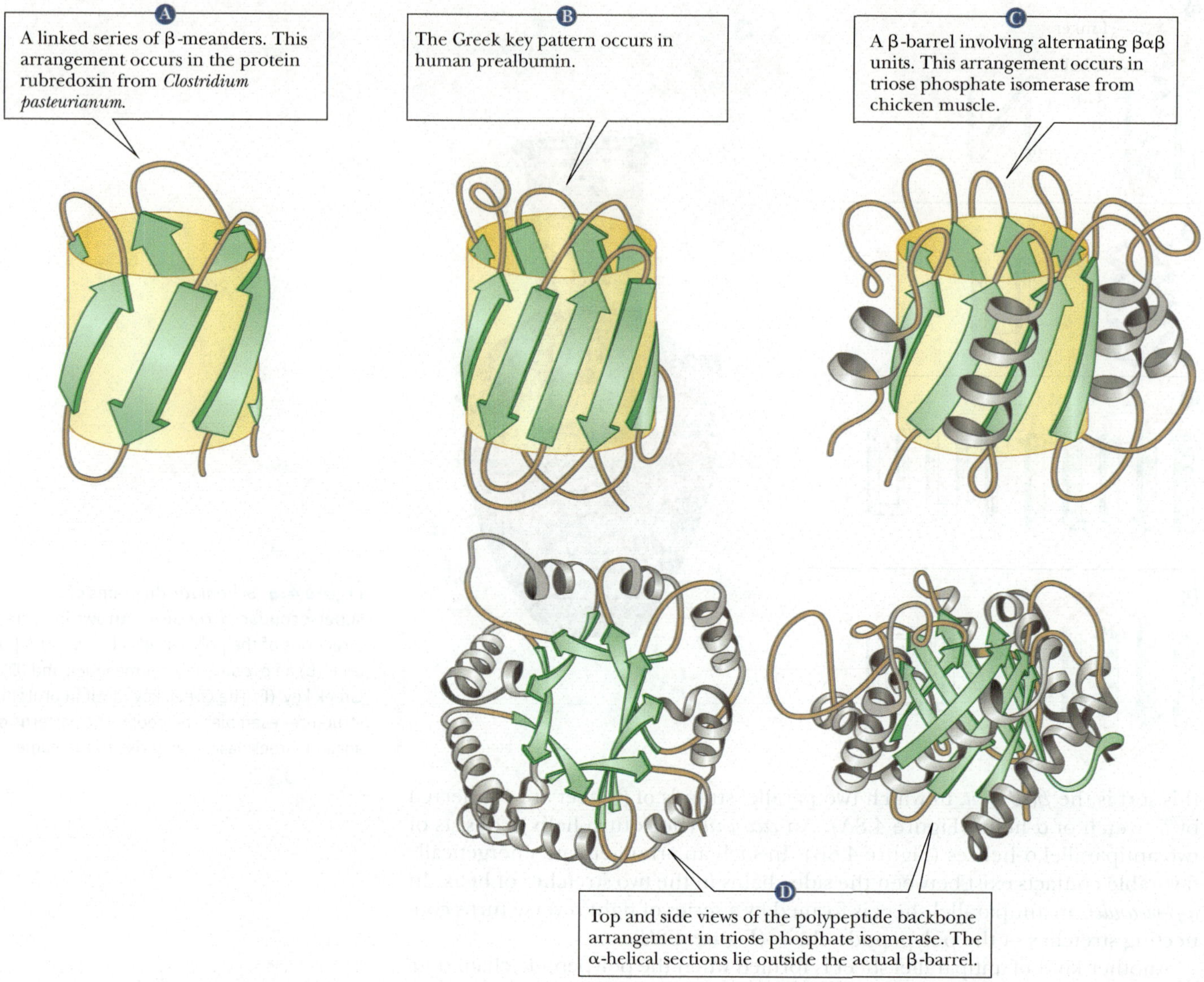

Figure 4.9 Some β-barrel arrangements.

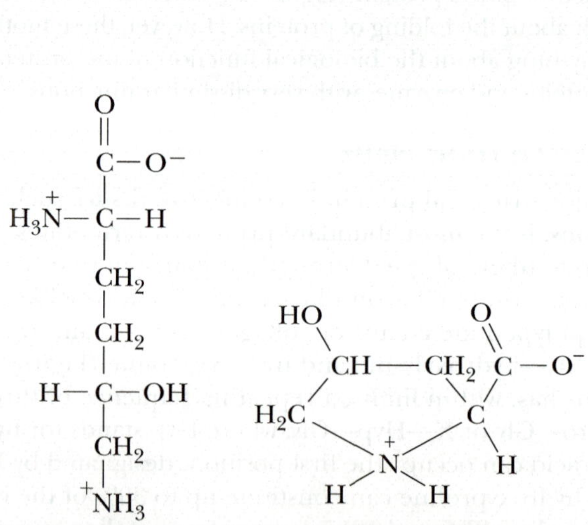

Hydroxylysine **Hydroxyproline**

Figure 4.10 Hydroxylysine and hydroxyproline.

helix is arranged so that every third residue on each chain is inside the helix. Only glycine is small enough to fit into the space available (Figure 4.11).

The three individual collagen chains are themselves helices that differ from the α-helix. They are twisted around each other in a superhelical arrangement to form a stiff rod. This triple-helical molecule is called *tropocollagen*; it is 300 nm (3000 Å) long and 1.5 nm (15 Å) in diameter. The three strands are held together by hydrogen bonds involving the hydroxyproline and hydroxylysine residues. The molecular weight of the triple-stranded array is about 300,000; each strand contains about 800 amino acid residues. Collagen is both intramolecularly and intermolecularly linked by covalent bonds formed by reactions between lysine and hydroxylysine residues. The modification that can occur to the lysine and hydroxylysine residues is an oxidative deamination that changes their side chain amino groups to an aldehyde group that can form covalent bonds to the amino groups of unmodified lysine or hydroxylysine residues or to the aldehyde groups of other modified lysine or hydroxylysine residues. The amount of cross-linking in a tissue increases with age. That is why meat from older animals is tougher than meat from younger animals.

Collagen in which the proline is not hydroxylated to hydroxyproline to the usual extent is less stable than normal collagen. Symptoms of scurvy, such as bleeding gums and skin discoloration, are the results of fragile collagen. The enzyme that hydroxylates proline and thus maintains the normal state of collagen requires ascorbic acid (vitamin C) to remain active. Scurvy is ultimately caused by a dietary deficiency of vitamin C. See **BIOCHEMICAL CONNECTIONS 16B.**

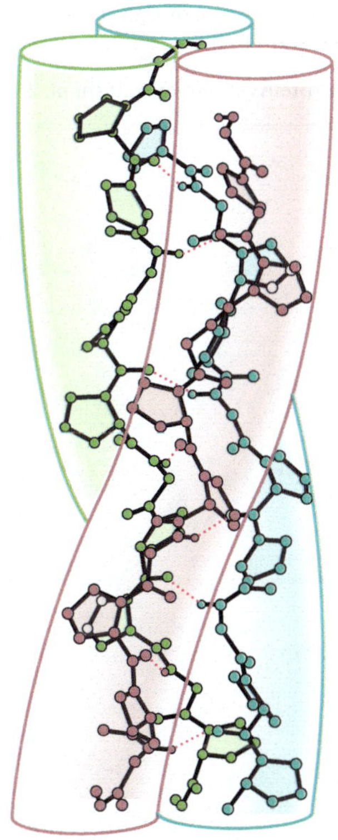

Figure 4.11 A triple helix. This figure depicts the repeating structure Poly (Gly-Pro-Pro), which is a collagen-like structure showing how three left-handed helices wrap around each other as a right-handed triple helix. *(Adapted from M. H. Miller and H. A. Scheraga, 1976, Calculation of the Structures of Collagen Models. Role of Interchain Interactions in Determining the Triple-Helical Coiled-Coil Conformations. I. Poly(glycyl-prolyl-prolyl).* Journal of Polymer Science Symposium *54:171–200. © 1976 John Wiley & Sons, Inc.)*

4-3F Two Types of Protein Conformations: Fibrous and Globular

It is difficult to draw a clear separation between secondary and tertiary structures. The nature of the side chains in a protein (part of the tertiary structure) can influence the folding of the backbone (the secondary structure). Comparing collagen with silk and wool fibers can be illuminating. Silk fibers consist largely of the protein fibroin, which, like collagen, has a fibrous structure but which, unlike collagen, consists largely of β-sheets. Fibers of wool consist largely of the protein keratin, which consists largely of α-helices. The amino acids of which collagen, fibroin, and keratin are composed determine which conformation they will adopt, but all are **fibrous proteins** (Figure 4.12A).

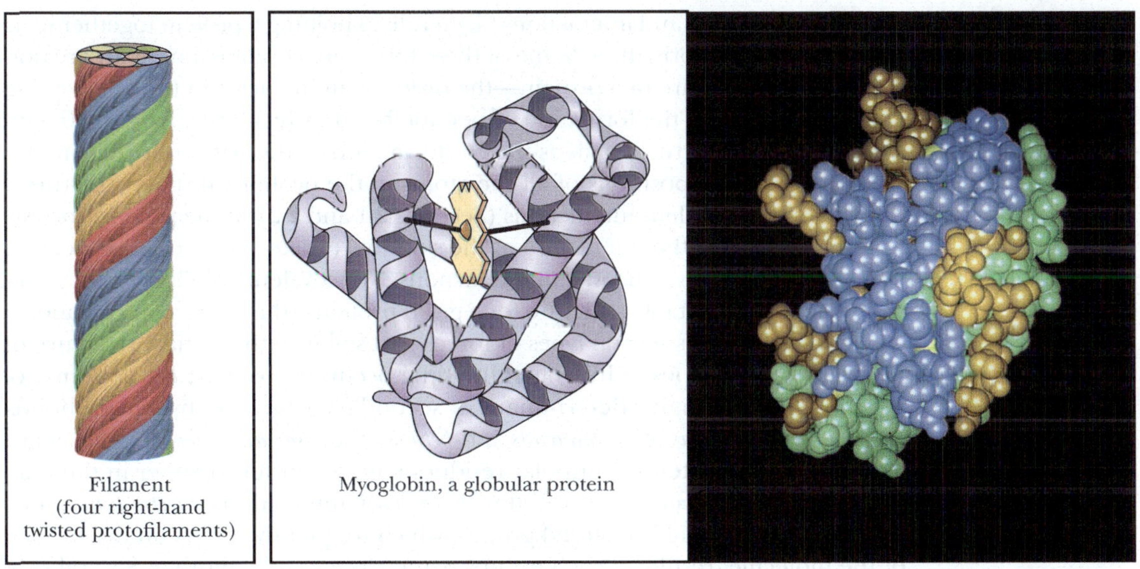

A Schematic diagrams of a portion of a fibrous protein

B Two models of globular proteins. In the model on the right, α-helices are shown in blue, β-sheets are green, and random coil (Section 4-4A) is gold.

Figure 4.12 A comparison of the shapes of fibrous and globular proteins.

fibrous proteins proteins whose overall shape is that of a long, narrow rod

globular proteins proteins whose overall shape is more or less spherical

In other proteins, the backbone folds back on itself to produce a more or less spherical shape. These are called **globular proteins** (Figure 4.12B), and we discuss many examples of them throughout the text. Globular proteins, unlike fibrous proteins, are water soluble and have compact structures; their tertiary and quaternary structures can be quite complex.

4-4 Tertiary Structure of Proteins

Learning Objectives

After reading this section, students will be able to:

LO 4.4.1 Identify tertiary structures in proteins.

LO 4.4.2 Describe the forces involved in stabilizing the tertiary structures of proteins.

LO 4.4.3 Describe methods for deducing the tertiary structure of a protein.

LO 4.4.4 Describe structural characteristics of myoglobin, including properties of its heme prosthetic group.

The tertiary structure of a protein is the three-dimensional arrangement of all the atoms in the molecule, including the conformations of the side chains and the arrangement of α-helix and β-sheet sections with respect to one another. In a fibrous protein, the overall shape of which is a long rod, the secondary structure also provides much of the information about the tertiary structure. The helical backbone of the protein does not fold back on itself, and the only important aspect of the tertiary structure that is not specified by the secondary structure is the arrangement of the atoms of the side chains.

For a globular protein, considerably more information is needed. It is necessary to determine the way in which the helical and pleated-sheet sections fold back on each other, in addition to the positions of the side-chain atoms. The interactions between the side chains play an important role in the folding of proteins. The folding pattern frequently brings residues that are separated in the amino acid sequence into proximity in the tertiary structure of the native protein.

4-4A Forces Involved in Tertiary Structures

Many types of forces and interactions play a role in holding a protein together in its correct, native conformation. Some of these forces are covalent, but many are not. The primary structure of a protein—the order of amino acids in the polypeptide chain—depends on the formation of peptide bonds, which are covalent. Higher-order levels of structure, such as the conformation of the backbone (secondary structure) and the positions of all the atoms in the protein (tertiary structure), depend on noncovalent interactions (Sections 2-1 and 2-2). If the protein consists of several subunits, the interaction of the subunits (quaternary structure, Section 4-5) also depends on noncovalent interactions. Noncovalent stabilizing forces contribute to the most stable structure for a given protein—the native conformation.

Figure 4.13 shows several types of forces that stabilize the tertiary structure of proteins. Several types of hydrogen bonding occur in proteins. *Backbone* hydrogen bonding is a major determinant of secondary structure; hydrogen bonds *between the side chains of amino acids* and *between backbone and side-chain atoms* are also possible in proteins. Nonpolar residues tend to cluster together in the interior of protein molecules as a result of *hydrophobic* interactions. *Electrostatic* attraction between oppositely charged groups, which frequently occurs on the surface of the molecule, results in such groups being close to one another. Several side chains can be *complexed* to a single metal ion.

In addition to these noncovalent interactions, *disulfide bonds* form covalent links between the side chains of cysteines. When such bonds form, they restrict

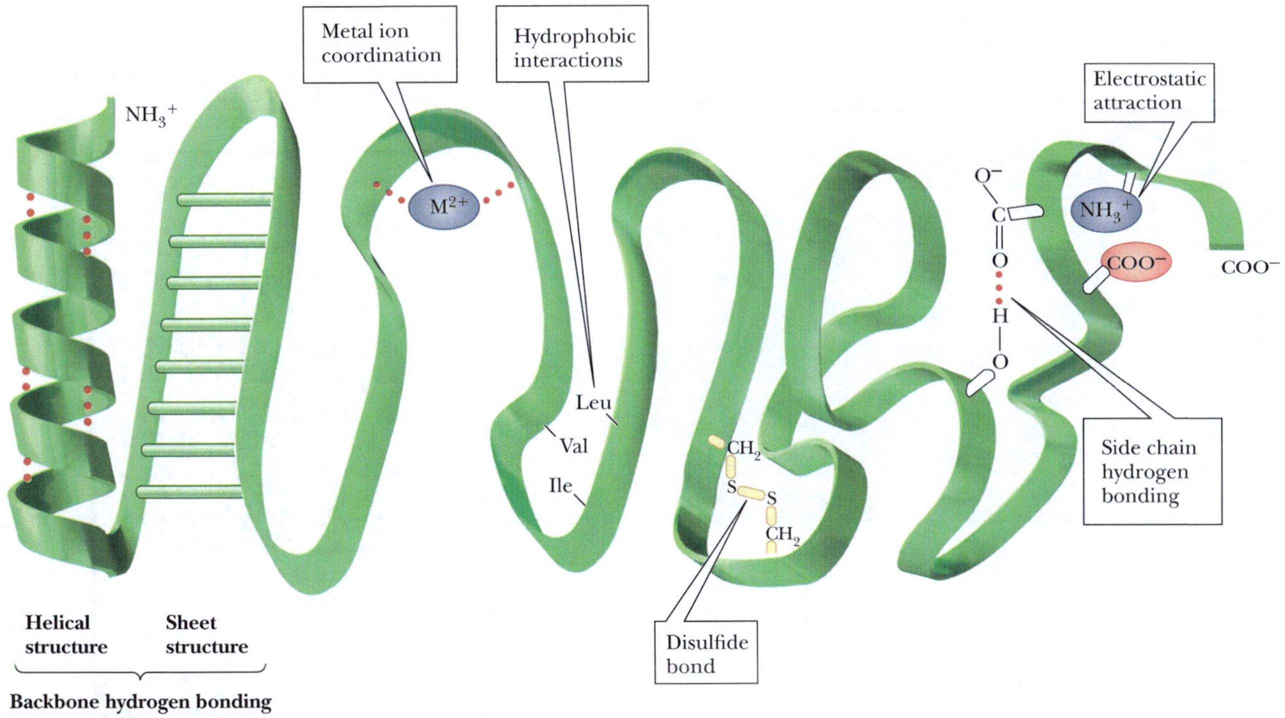

Figure 4.13 Forces that stabilize the tertiary structure of proteins. The helical structure and sheet structure are two kinds of backbone hydrogen bonding. Although backbone hydrogen bonding is part of secondary structure, the conformation of the backbone constrains the possible arrangement of the side chains.

the folding patterns available to polypeptide chains. Not every protein necessarily exhibits all possible structural features of the kinds just described. Hydrogen bonds, electrostatic interactions, and hydrophobic interactions occur in most proteins.

The three-dimensional conformation of a protein is the result of the interplay of all the stabilizing forces. It is known, for example, that proline does not fit into an α-helix and that its presence can cause a polypeptide chain to turn a corner, ending an α-helical segment. The presence of proline is not, however, a *requirement* for a turn in a polypeptide chain. Glycine is also routinely encountered at bends in polypeptide chains because its small side chain group reduces steric repulsion in a turn. The segments of proteins at bends in the polypeptide chain and in other portions of the protein that are not involved in helical or pleated-sheet structures are frequently referred to as **random coil**. In reality, the forces that stabilize each protein are responsible for its conformation.

random coil a segment of the tertiary structure of a protein that is *not* involved in secondary structure elements

▶ How can the three-dimensional structure of a protein be determined?

One type of experimental technique used to determine the tertiary structure of a protein is **X-ray crystallography**. Perfect crystals of some proteins can be grown under carefully controlled conditions. In such a crystal, all the individual protein molecules have the same three-dimensional conformation and the same orientation. Crystals of this quality can be formed only from proteins of very high purity.

X-ray crystallography an experimental method for determining the three dimensional structure of molecules, such as the tertiary or quaternary structure of proteins

When a suitably pure crystal is exposed to a beam of X-rays, a *diffraction pattern* is produced on a photographic plate (Figure 4.14A) or a radiation counter. The pattern is produced when the electrons in each atom in the molecule scatter the X-rays. The number of electrons in the atom determines the intensity of its scattering of X-rays; heavier atoms scatter more effectively than lighter atoms. The scattered X-rays from the individual atoms can reinforce each other or cancel each other (i.e., set up constructive or destructive interference),

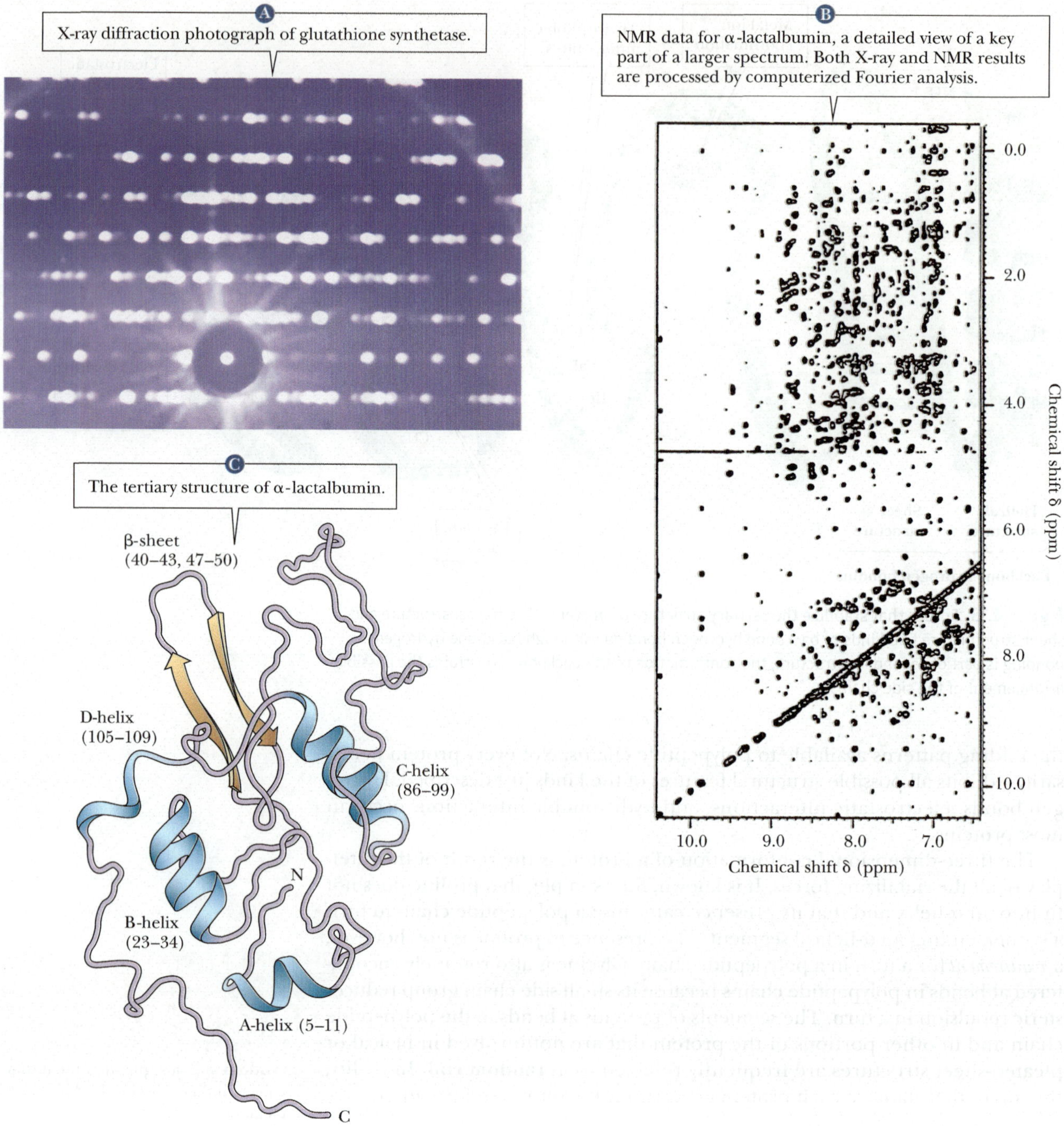

Figure 4.14 Large numbers of data points are needed to determine the tertiary structure of a protein. (See Figure 4.15 for the structure of myoglobin as determined by X-ray crystallography.) *(A, Alfred Pasieka/Science Source; B, courtesy of Professor C. M. Dobson, University of Oxford.)*

giving rise to the characteristic pattern for each type of molecule. A series of diffraction patterns taken from several angles contains the information needed to determine the tertiary structure. The information is extracted from the diffraction patterns through a mathematical analysis known as a *Fourier series.* Many thousands of such calculations are required to determine the structure of a protein. Improving the calculation procedure is a subject of active research.

nuclear magnetic resonance (NMR) spectroscopy a method for determining the three-dimensional shape of proteins in solution

Other techniques that supplement the results of X-ray diffraction are forms of **nuclear magnetic resonance (NMR) spectroscopy.** In these particular applications of NMR, called *multi-dimensional NMR,* large collections of data points are subjected to computer analysis (Figure 4.14B). Like X-ray diffraction, this method uses a Fourier series to analyze results. It is similar to X-ray diffraction in other ways—namely, it requires considerable amounts of computing power and milligram quantities of protein. Figure 4.14C shows the protein structure determined from the NMR data in Figure 4.14B.

An important way in which 2-D NMR differs from X-ray diffraction is that it uses protein samples in aqueous solution rather than crystals. This environment is closer to that of proteins in cells, and thus it is one of the main advantages of the method. Even though most proteins have a single native conformation, native conformations can still fluctuate a bit, and NMR is able to detect these fluctuations because the protein sample is not fixed in a crystal. The NMR method is undergoing constant improvement and is being applied to larger proteins as these improvements progress.

4-4B Myoglobin: An Example of Protein Structure

In many ways, myoglobin is the classic example of a globular protein. We use it here as a case study in tertiary structure. Myoglobin was the first protein for which the complete tertiary structure was determined by X-ray crystallography in 1958. The complete myoglobin molecule consists of a single polypeptide chain of 153 amino acid residues and includes a **prosthetic group**—a **heme** group in myoglobin's case—which also occurs in hemoglobin (Section 4-5A). The myoglobin molecule (including the heme group) has a compact structure, with the interior atoms very close to each other. This structure provides examples of many of the forces responsible for the three-dimensional shapes of proteins.

prosthetic group a portion of a protein that does not consist of amino acids

heme an iron-containing cyclic compound found in cytochromes, hemoglobin, and myoglobin

Figure 4.15 shows the structure of myoglobin. Myoglobin has eight α-helical regions and no β-sheet regions. Approximately 75% of the residues in myoglobin are found in these helical regions, which are designated by the letters A through H. Hydrogen bonding in the polypeptide backbone stabilizes the

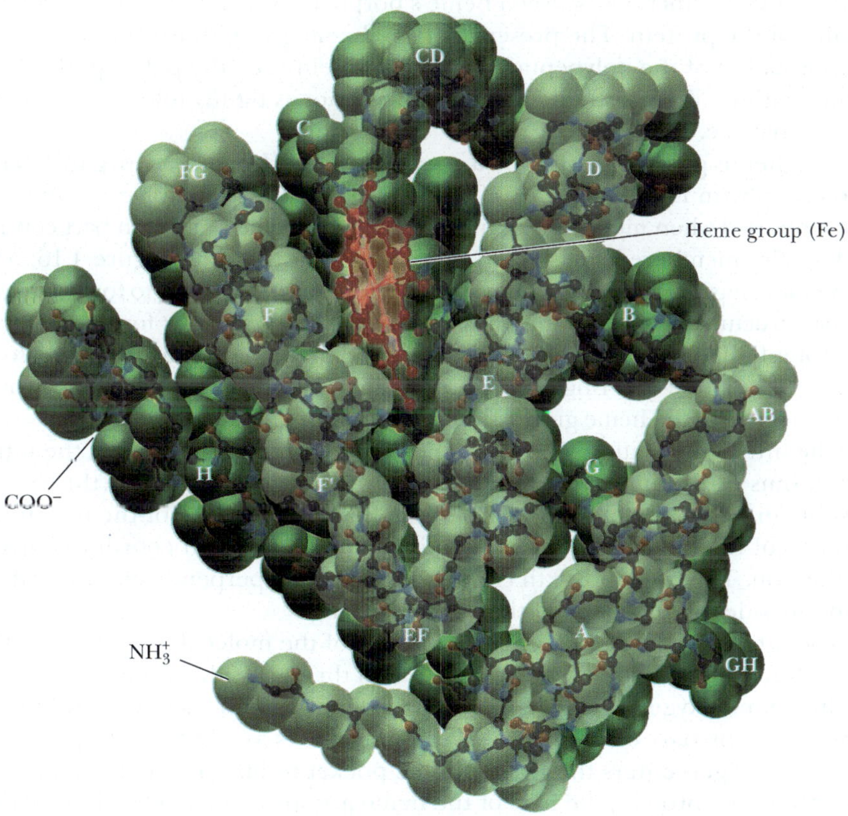

Figure 4.15 The structure of myoglobin. The peptide backbone and the heme group (the cluster of red atoms) are shown overlaid on the space-filling model. The eight helical segments are designated by the letters A through H. The AB, CD, etc. segments designate random coil sections that occur between the helical segments. "NH_3^+" and "COO^-" indicate the N-terminal and C-terminal ends, respectively.

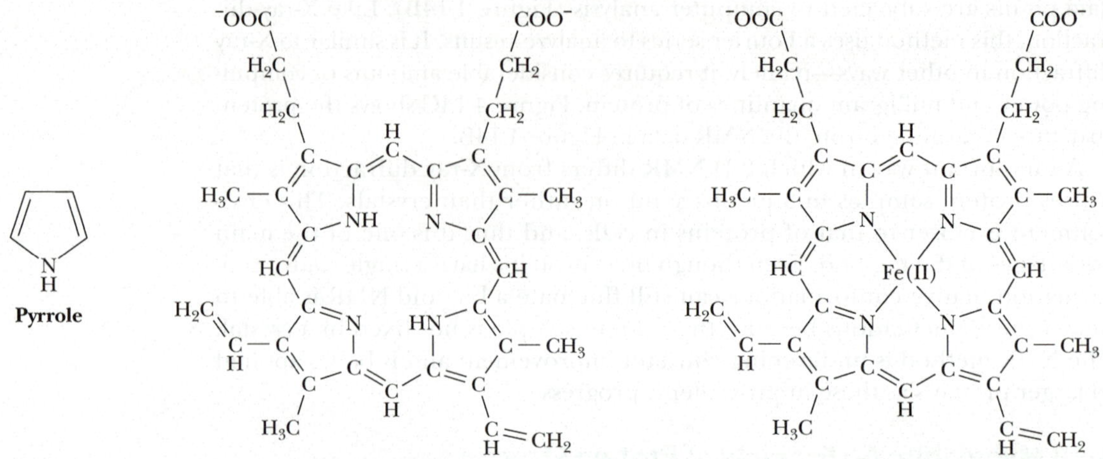

Pyrrole

Protoporphyrin IX

Heme
(Fe-protoporphyrin IX)

Figure 4.16 The structure of the heme group. Four pyrrole rings are linked by bridging groups to form a planar porphyrin ring. Several isomeric porphyrin rings are possible, depending on the nature and arrangement of the side chains. The porphyrin isomer found in heme is protoporphyrin IX. Addition of iron to protoporphyrin IX produces the heme group.

α-helical regions; amino acid side chains are also involved in hydrogen bonds. The polar residues are on the exterior of the molecule, whereas the interior of the protein contains almost exclusively nonpolar amino acid residues. Two polar histidine residues are found in the interior; they are involved in interactions with the heme group and bound oxygen and thus play an important role in the function of the molecule. The planar heme group fits into a hydrophobic pocket in the protein portion of the molecule and is held in position by hydrophobic attractions between heme's porphyrin ring and the nonpolar side chains of the protein. The presence of the heme group drastically affects the conformation of the polypeptide: The apoprotein (i.e., the polypeptide chain alone, without the prosthetic heme group) is not as tightly folded as the complete molecule.

The heme group consists of a metal ion, Fe(II), and an organic part, protoporphyrin IX (Figure 4.16). [The Fe(II) notation is equivalent to the Fe^{2+} notation used when metal ions occur in complexes.] The porphyrin part consists of four five-membered rings based on the pyrrole structure (Figure 4.16, *left*); these four rings are linked by bridging methine (—CH═) groups to form a square planar structure. The Fe(II) ion has six coordination sites, and it forms six metal–ion complexation bonds. Four of the six sites are occupied by the nitrogen atoms of the four pyrrole-type rings of the porphyrin to give the complete heme group. The presence of the heme group is required for myoglobin to bind oxygen.

The fifth coordination site of the Fe(II) ion is occupied by one of the nitrogen atoms of the imidazole side chain of histidine residue F8 (the eighth residue in helical segment F). This histidine residue is one of the two in the interior of the molecule. The oxygen is bound at the sixth coordination site of the iron. The fifth and sixth coordination sites lie perpendicular to, and on opposite sides of, the plane of the porphyrin ring.

The other histidine residue in the interior of the molecule, residue E7 (the seventh residue in helical segment E), lies on the same side of the heme group as the bound oxygen (Figure 4.17). This second histidine is not bound to the iron or to any part of the heme group, but it acts as a gate that opens and closes as oxygen enters the hydrophobic pocket to bind to the heme. In the absence of the protein, the iron of the heme group can be oxidized to Fe(III);

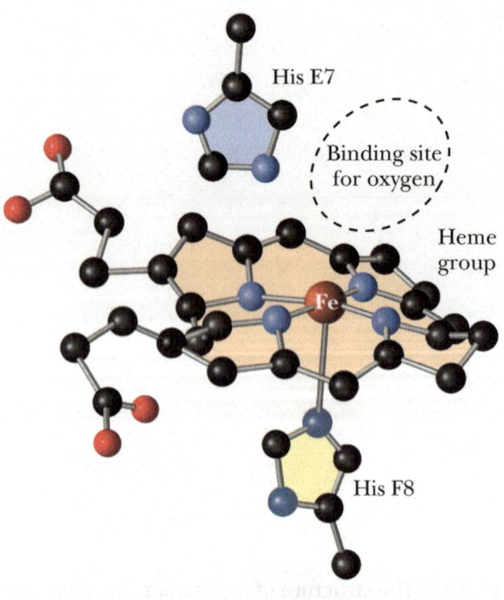

Figure 4.17 The oxygen-binding site of myoglobin. The porphyrin ring occupies four of the six coordination sites of Fe(II). Histidine F8 (His F8) occupies the fifth coordination site of the iron (see text). Oxygen is bound at the sixth coordination site of the iron, and histidine E7 lies close to the oxygen. *(Leonard Lessin/Waldo Feng/Mt. Sinai CORE.)*

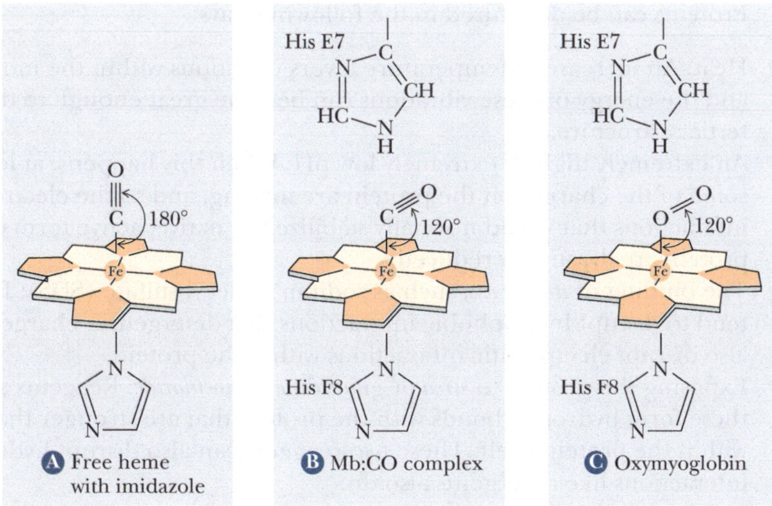

Figure 4.18 Oxygen and carbon monoxide binding to the heme group of myoglobin. The presence of the E7 histidine forces a 120° angle to the oxygen or CO.

the oxidized heme will not bind oxygen. Thus, the combination of both heme and protein is needed to oxygen storage.

▶ Why does oxygen have imperfect binding to the heme group?

The E7 histidine in myoglobin sterically inhibits oxygen from binding perpendicularly to the heme plane, with biologically important ramifications. At first, it would seem counterintuitive that oxygen would bind imperfectly to the heme group. After all, the job of both myoglobin and hemoglobin is to bind to oxygen. Wouldn't it make sense that oxygen should bind strongly? The answer lies in the fact that more than one molecule can bind to heme. Besides oxygen, carbon monoxide also binds to heme. The affinity of free heme for carbon monoxide (CO) is 25,000 times greater than its affinity for oxygen.

When carbon monoxide is forced to bind at an angle in myoglobin because of the steric block by His E7, its advantage over oxygen drops by two orders of magnitude (Figure 4.18). This guards against the possibility that traces of CO produced during metabolism would occupy all the oxygen-binding sites on the hemes. Nevertheless, CO is a potent poison in larger quantities because of its effect both on oxygen binding to hemoglobin and on the final step of the electron transport chain (Section 20-3). It is also important to remember that although our metabolism requires that hemoglobin and myoglobin bind oxygen, it would be equally disastrous if the heme never let the oxygen go. Thus, binding that is too perfect would defeat the purpose of having the oxygen-carrying proteins.

denaturation the unraveling of the three-dimensional structure of a macromolecule caused by the breakdown of noncovalent interactions

4-4C Protein Denaturation and Refolding

The noncovalent interactions that maintain the three-dimensional structure of a protein are weak, so they can be disrupted easily. The unfolding of a protein (i.e., the disruption of the tertiary structure) is called **denaturation** (Figure 4.19). Reduction of disulfide bonds (Section 4-4A) leads to even more extensive unraveling of the tertiary structure. Denaturation and reduction of disulfide bonds are frequently combined when complete disruption of the tertiary structure of proteins is desired. Under proper experimental conditions, the disrupted structure can then be completely recovered. This process of denaturation and refolding is a dramatic demonstration of the relationship between the primary structure of the protein and the forces that determine the tertiary structure. For many proteins, various other factors are needed for complete refolding, but the important point is that the primary structure determines the tertiary structure.

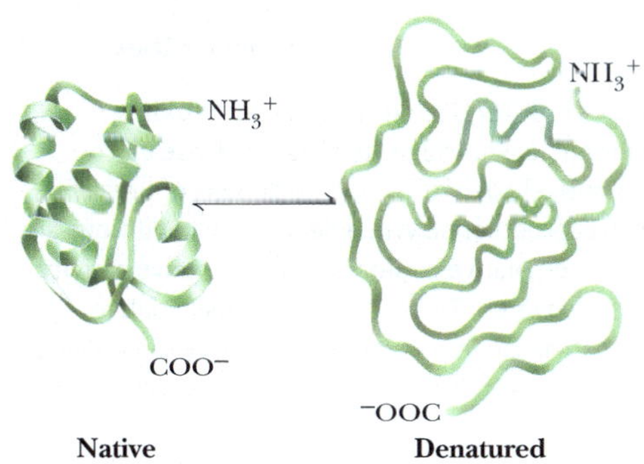

Figure 4.19 Denaturation of a protein. The native conformation can be recovered when denaturing conditions are removed.

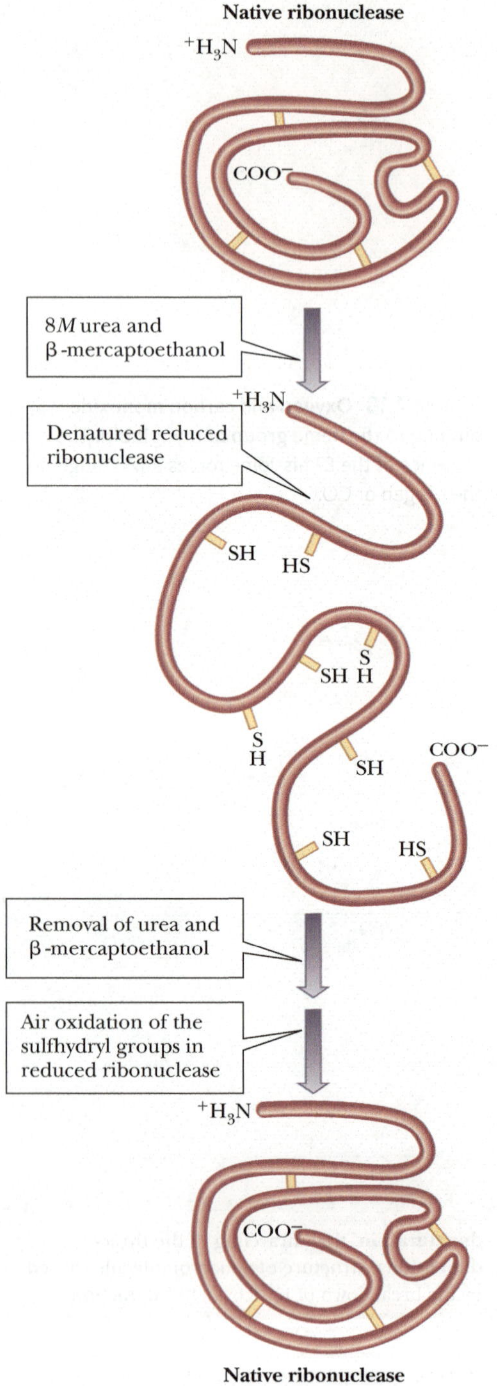

Figure 4.20 Denaturation and refolding in ribonuclease. The protein ribonuclease can be completely denatured by the actions of urea and mercaptoethanol. When denaturing conditions are removed, native conformation and native activity are recovered. The yellow segments in native ribonuclease represent disulfide (—S—S—) bridges between cysteine residues.

Proteins can be denatured in the following ways:

- Heat: An increase in temperature favors vibrations within the molecule, and the energy of these vibrations can become great enough to disrupt the tertiary structure.
- An extremely high or extremely low pH: When this happens, at least some of the charges on the protein are missing, and so the electrostatic interactions that would normally stabilize the native, active form of the protein are drastically reduced.
- The binding of *detergents,* such as sodium dodecyl sulfate (SDS): Detergents tend to disrupt hydrophobic interactions. If a detergent is charged, it can also disrupt electrostatic interactions within the protein.
- Exposing the protein to *urea* or *guanidine hydrochloride*: Reagents such as these form hydrogen bonds with the protein that are stronger than those within the protein itself. These two reagents can also disrupt hydrophobic interactions like detergents also do.

β-Mercaptoethanol ($HS—CH_2—CH_2—OH$) is frequently used to reduce disulfide (—S—S—) bridges to two sulfhydryl (—SH) groups. Urea is usually added to the reaction mixture to facilitate unfolding of the protein and to increase the accessibility of the disulfides to the reducing agent. If experimental conditions are properly chosen, the native conformation of the protein can be recovered when both β-mercaptoethanol and urea are removed (Figure 4.20). Experiments of this type provide some of the strongest evidence that the amino acid sequence of the protein contains all the information required to produce the complete three-dimensional structure. Protein researchers are pursuing with some interest the conditions under which a protein can be denatured—including reduction of disulfides—and its native conformation later recovered.

4-5 Quaternary Structure of Proteins

Learning Objectives

After reading this section, students will be able to:

LO 4.5.1 Identify quaternary structures in proteins.

LO 4.5.1 Describe the structural and functional characteristics of hemoglobin.

Quaternary structure is the fourth and final level of protein structure. It pertains only to proteins that consist of more than one polypeptide chain. Each chain is called a *subunit.* The number of chains can range from two to more than a dozen, and the chains may be identical or different. Commonly occurring examples are **dimers**, **trimers**, and **tetramers**, consisting of two, three, and four polypeptide chains, respectively. (The generic term for such a molecule, made up of a small number of subunits, is a **multimeric protein**.) Similar to tertiary structure, in quaternary structure, the subunits interact with one another noncovalently via electrostatic attractions, hydrogen bonds, and hydrophobic interactions. Although less common, covalent bonds provide strong stabilization in quaternary structure.

As a result of these noncovalent interactions, subtle changes in structure at one site on a protein molecule may cause drastic changes in properties at a distant site. Proteins that exhibit this property are called **allosteric**. Not all multisubunit proteins exhibit allosteric effects, but many do.

A classic illustration of the quaternary structure and its effect on protein properties comes from a comparison of hemoglobin, an allosteric protein, with myoglobin, which consists of a single polypeptide chain.

4-5A Hemoglobin

Hemoglobin is a tetramer consisting of four polypeptide chains: two alpha chains and two beta chains (Figure 4.21). In multimeric proteins, the types of polypeptide chains are designated with Greek letters. In this case, the terms "alpha" and "beta" have nothing to do with the α-helix or β-sheet conformations; rather, they just refer to two different polypeptide chain subunits.

The two alpha chains of hemoglobin are identical, as are the two beta chains. The overall structure of hemoglobin is denoted as $\alpha_2\beta_2$ in Greek-letter notation. Both the alpha and beta chains of hemoglobin are very similar to the myoglobin chain. The alpha chain is 141 residues long, whereas the beta chain is 146 residues long; for comparison, the myoglobin chain is 153 residues long. Many of the amino acids of the alpha chain, the beta chain, and myoglobin are *homologous*; that is, the same amino acid residues are in the same positions. The heme group is the same in myoglobin and hemoglobin.

We have previously seen that one molecule of myoglobin binds *one* oxygen molecule. In an analogous fashion, one hemoglobin molecule can bind *four* molecules of oxygen. Both hemoglobin and myoglobin bind oxygen reversibly, but the binding of oxygen to hemoglobin exhibits **positive cooperativity**, whereas oxygen binding to myoglobin does not. Positive cooperativity means that when one oxygen molecule is bound, it becomes easier for the next to bind. A graph of the oxygen-binding properties of hemoglobin and myoglobin is one of the best ways to illustrate this point (Figure 4.22).

When the degree of saturation of myoglobin with oxygen is plotted against oxygen pressure, a steady rise is observed until complete saturation is approached and the curve levels off. The oxygen-binding curve of myoglobin is thus said to be **hyperbolic**. In contrast, the shape of the oxygen-binding curve for hemoglobin is **sigmoidal**. This shape indicates the positive cooperativity seen in hemoglobin, where the binding of the first oxygen molecule facilitates the binding of the second oxygen, which facilitates the binding of the third,

dimers molecules consisting of two subunits

trimers molecules consisting of three subunits

tetramers molecules consisting of four subunits

multimeric protein an aggregate of several smaller units (monomers); bonding may be covalent or noncovalent

allosteric the property of multisubunit proteins such that a conformational change in one subunit induces a change in another subunit

positive cooperativity a cooperative effect whereby binding of the first ligand to an enzyme or protein makes it easier for the next ligand to bind

hyperbolic a characteristic of a curve on a graph such that it rises quickly and then levels off

sigmoidal referring to an S-shaped curve on a graph, characteristic of cooperative interactions

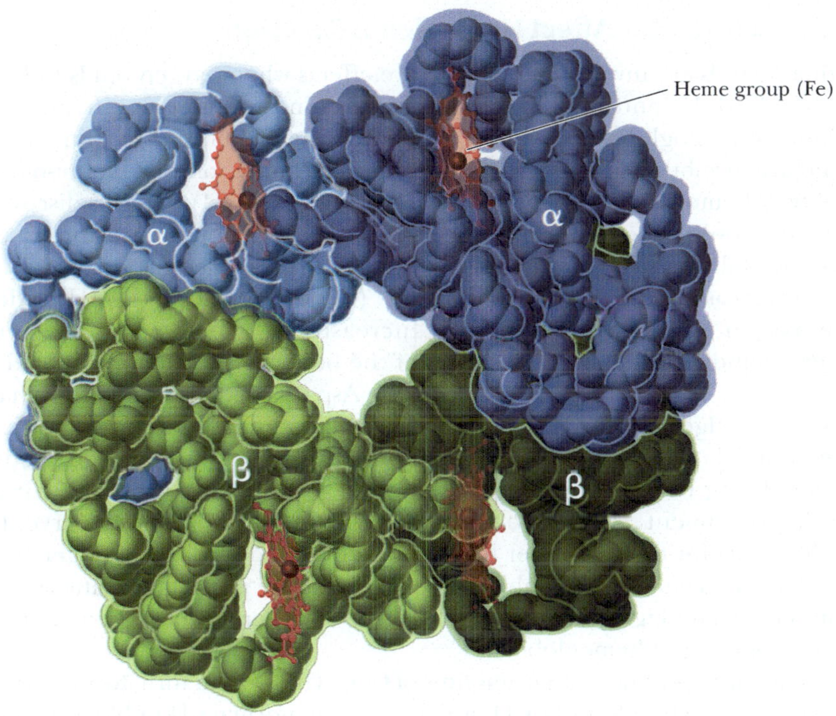

Figure 4.21 The structure of hemoglobin. Hemoglobin ($\alpha_2\beta_2$) is a tetramer consisting of four polypeptide chains (two alpha chains and two beta chains).

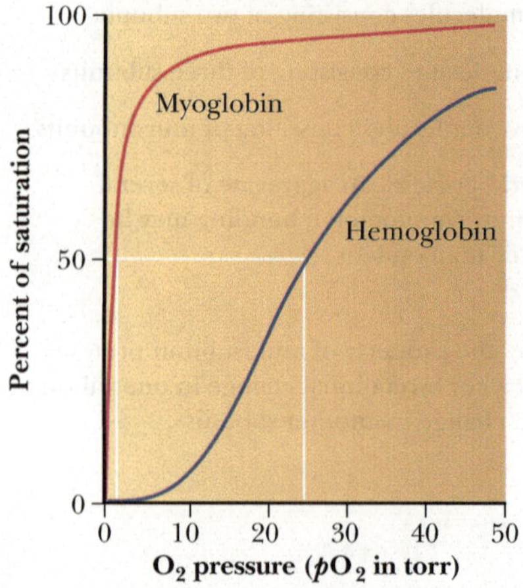

Figure 4.22 A comparison of the oxygen-binding behavior of myoglobin and hemoglobin at pH 7.4. The oxygen-binding curve of myoglobin is hyperbolic, whereas that of hemoglobin is sigmoidal. Myoglobin is 50% saturated with oxygen at 1 torr partial pressure, whereas hemoglobin does not reach 50% saturation until the partial pressure of oxygen reaches 26 torr.

torr a unit of pressure equal to that exerted by a column of mercury 1 mm high at 0°C

which in turn facilitates the binding of the fourth. However, even though cooperative binding means that binding of each subsequent oxygen is easier than the previous one, the oxygen binding of hemoglobin is still lower than that of myoglobin at any oxygen pressure. In other words, at any oxygen pressure, myoglobin will have a higher percentage of saturation than hemoglobin. This becomes particularly important when oxygen is passed from hemoglobin in the blood to myoglobin in muscle tissue.

▶ How does hemoglobin work?

The two different types of behavior exhibited by myoglobin and hemoglobin are related to the different functions of these proteins. Myoglobin *stores* oxygen in muscle, so it must bind strongly to oxygen at very low pressures. For example, it is 50% saturated at 1 torr partial pressure of oxygen. (The **torr** is a widely used unit of pressure, but it is not an SI unit. One torr is the pressure exerted by a column of mercury 1 mm high at 0°C. One atmosphere is equal to 760 torr.)

Hemoglobin, on the other hand, *transports* oxygen, so it must be able to both bind strongly to oxygen and release it easily, depending on conditions. In the alveoli of lungs (where hemoglobin must bind oxygen for transport to the tissues), the oxygen pressure is 100 torr. At this pressure, hemoglobin is 100% saturated with oxygen. In the capillaries of active muscles, the pressure of oxygen is 20 torr, corresponding to less than 50% saturation of hemoglobin, which occurs at 26 torr. In other words, hemoglobin gives up oxygen easily in capillaries, where the need for oxygen is great.

Structural changes during the binding of small molecules are characteristic of allosteric proteins such as hemoglobin. Hemoglobin has different quaternary structures in the bound (oxygenated) and unbound (deoxygenated) forms (Figure 4.23). The two beta chains are much closer to each other in oxygenated hemoglobin than in deoxygenated hemoglobin. The change is so marked that the two forms of hemoglobin have different X-ray crystal structures.

4-5B Factors That Affect Hemoglobin Function

Other ligands are involved in cooperative effects when oxygen binds to hemoglobin. Both H^+ and CO_2, which themselves bind to hemoglobin, affect the affinity of hemoglobin for oxygen by altering the protein's three-dimensional structure in subtle but important ways. The effect of H^+ on the oxygen-binding ability of hemoglobin is called the *Bohr effect* (Figure 4.24), after its discoverer, Christian Bohr (the father of physicist Niels Bohr). In contrast, the oxygen-binding ability of myoglobin is unaffected by the presence of H^+ or of CO_2.

An increase in the concentration of H^+ (i.e., a lowering of the pH) reduces the oxygen affinity of hemoglobin. Increasing $[H^+]$ protonates key amino acids, including the His 146 residues of the beta chains. The protonated histidines then become more attracted to the Asp 94 residues of the beta chains, and salt bridges are formed between these histidine and aspartate residues (Section 2-1). This favors the deoxygenated form of hemoglobin. Actively metabolizing tissue, which requires oxygen, releases H^+, thus acidifying its local environment. Hemoglobin has a lower affinity (the binding curve shifts to the right) for oxygen under these conditions, and it releases oxygen where it is needed (Figure 4.24). Table 4.1 summarizes the important features of the Bohr effect, and Figure 4.25 shows the effects of different pH environments on oxygen binding in hemoglobin.

Metabolism produces large amounts of CO_2. The CO_2, in turn, forms carbonic acid (H_2CO_3). The pK_a of H_2CO_3 is 6.35, and the normal pH of blood is 7.4. As a result, about 90% of dissolved CO_2 is present as the bicarbonate ion (HCO_3^-), releasing H^+. (Use the Henderson–Hasselbalch equation to confirm this point.)

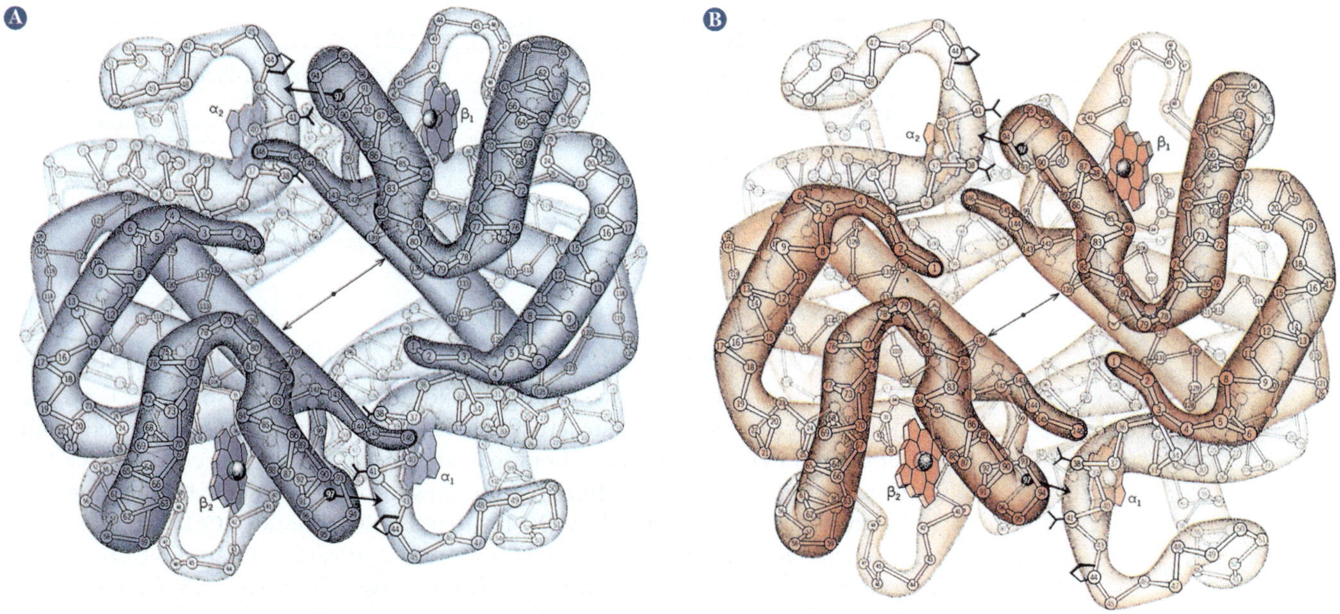

Figure 4.23 The structures of (A) deoxyhemoglobin and (B) oxyhemoglobin. The arrows in the center show that there is much less room between the subunits in oxyhemoglobin compared to deoxyhemoglobin. *(Illustration: Irving Geis. Rights owned by Howard Hughes Medical Institute. Not to be reproduced without permission.)*

$$HbO_2 + H^+ + CO_2 \underset{\text{Alveoli of lungs}}{\overset{\text{Actively metabolizing tissue (such as muscle)}}{\rightleftharpoons}} O_2 + Hb\begin{cases} CO_2 \\ H^+ \end{cases}$$

Figure 4.24 The general features of the Bohr effect. In actively metabolizing tissue, hemoglobin releases oxygen and binds both CO_2 and H^+. In the lungs, hemoglobin releases both CO_2 and H^+ and binds oxygen.

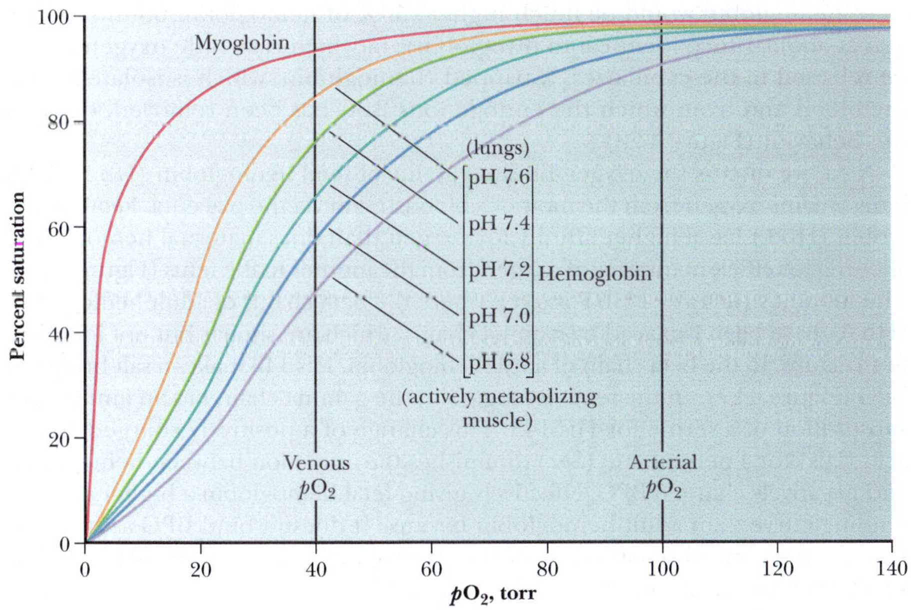

Figure 4.25 The oxygen saturation curves for myoglobin and hemoglobin at five different pH values. Hemoglobin oxygen binding is lower at the low pH values found in actively metabolizing muscle, whereas hemoglobin oxygen binding is higher at the high pH values found in the lungs.

Table 4.1 A Summary of the Bohr Effect

Lungs	Actively Metabolizing Muscle
Higher pH than actively metabolizing tissue	Lower pH due to production of H^+
Hemoglobin binds O_2	Hemoglobin releases O_2
Hemoglobin releases H^+	Hemoglobin binds H^+

The in vivo buffer system involving H_2CO_3 and HCO_3^- in blood was discussed in Section 2-5. The presence of larger amounts of H^+ as a result of CO_2 production favors the quaternary structure that is characteristic of deoxygenated hemoglobin. Hence, the affinity of hemoglobin for oxygen is lowered. The HCO_3^- is transported to the lungs, where it combines with H^+ released when hemoglobin is oxygenated, producing H_2CO_3. In turn, H_2CO_3 liberates CO_2, which is then exhaled. Hemoglobin also transports some CO_2 directly. When the CO_2 concentration is high, it combines with the α-amino groups to form carbamate groups:

$$R—NH_2 + CO_2 \rightleftharpoons R—NH—COO^- + H^+$$

This reaction turns the α-amino termini into anions, which can then interact with alpha chain Arg 141 residues, also stabilizing the deoxygenated form.

Thus, in the presence of large amounts of H^+ and CO_2, as in respiring tissue, hemoglobin releases oxygen. The presence of large amounts of oxygen and a higher pH environment in the lungs reverse the process, causing hemoglobin to bind O_2. The oxygenated hemoglobin can then transport oxygen to the tissues. The process is complex, and it allows for fine-tuning of oxygen binding according to pH and levels of oxygen and CO_2.

Another factor that affects hemoglobin function is when it binds a ligand called *2,3-bisphosphoglycerate (BPG)* (Figure 4.26). The binding of BPG to hemoglobin is electrostatic; specific interactions take place between the negative charges on BPG and positive charges on the protein (Figure 4.27). In the presence of BPG, the oxygen-binding ability of hemoglobin is lower. Recall from Figure 4.22 that the partial pressure at which 50% of hemoglobin is bound to oxygen is 26 torr. If BPG were *not* present in blood, the oxygen-binding capacity of hemoglobin would be much higher (50% of hemoglobin bound to oxygen at about 1 torr, identical to myoglobin), in which case little oxygen would be released in the capillaries. "Stripped" hemoglobin, which is isolated from the blood and from which the endogenous BPG has been removed, displays this behavior (Figure 4.28).

Next, we discuss the oxygen-binding ability of fetal hemoglobin (Hb F). The fetus obtains oxygen from the mother's bloodstream via the placenta. Fetal hemoglobin (Hb F) has a higher affinity for oxygen than does maternal hemoglobin, allowing for efficient transfer of oxygen from the mother to the fetus (Figure 4.29). The subunit structure of Hb F is $\alpha_2\gamma_2$, where the beta chains of adult hemoglobin (Hb A) have been replaced by gamma chains, which are similar but not identical in structure. In the beta chain of adult hemoglobin, His 143 makes a salt bridge to BPG (Figure 4.27). In the fetal hemoglobin, the gamma chain has an amino acid substitution of a serine for His 143. This change of a positively charged amino acid (His) for a neutral one (Ser) diminishes the attraction between hemoglobin and negatively charged BPG, effectively giving fetal hemoglobin a higher oxygen-binding curve than adult hemoglobin because it doesn't bind BPG as strongly. You can test your understanding of how hemoglobin responds to pH by trying **Apply Your Knowledge 4.1**. **Biochemical Connections 4A** discusses another type of hemoglobin chain, Hb S, which causes sickle-cell disease.

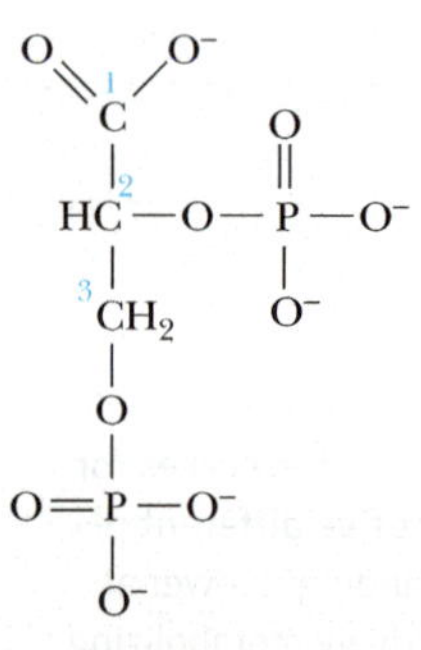

Figure 4.26 The structure of 2,3-bisphosphoglycerate (BPG). BPG is an important allosteric effector of hemoglobin.

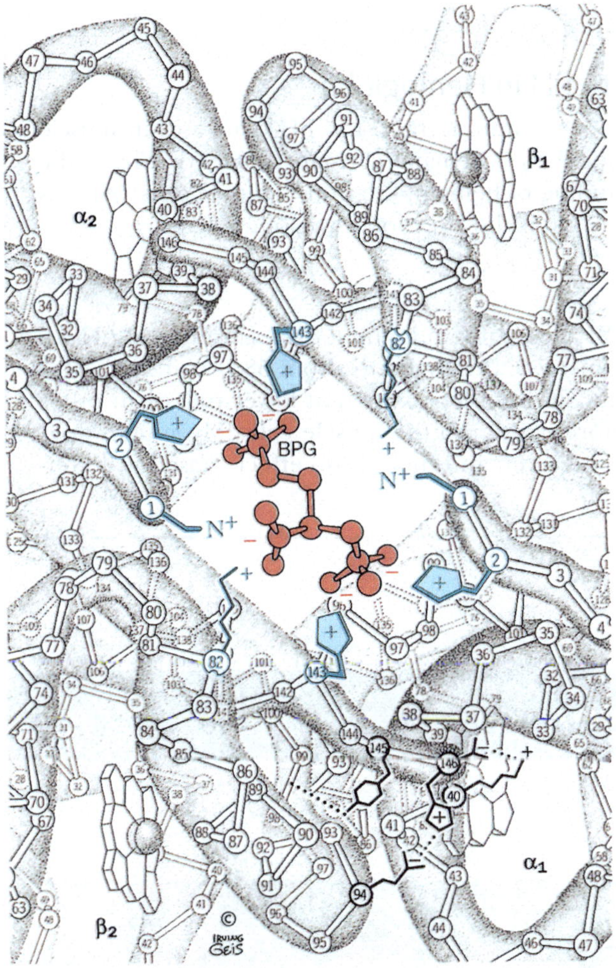

Figure 4.27 The binding of 2,3-bisphosphoglycerate (BPG) to deoxyhemoglobin. Electrostatic attractions occur between BPG and positive groups on the protein. *(Illustration: Irving Geis. Rights owned by Howard Hughes Medical Institute. Not to be reproduced without permission.)*

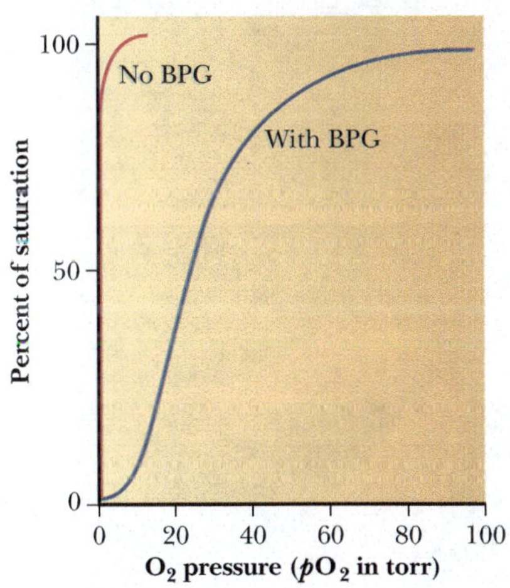

Figure 4.28 A comparison of the oxygen-binding properties of hemoglobin in the presence and absence of 2,3-bisphosphoglycerate (BPG). The presence of the BPG markedly *decreases* the affinity of hemoglobin for oxygen.

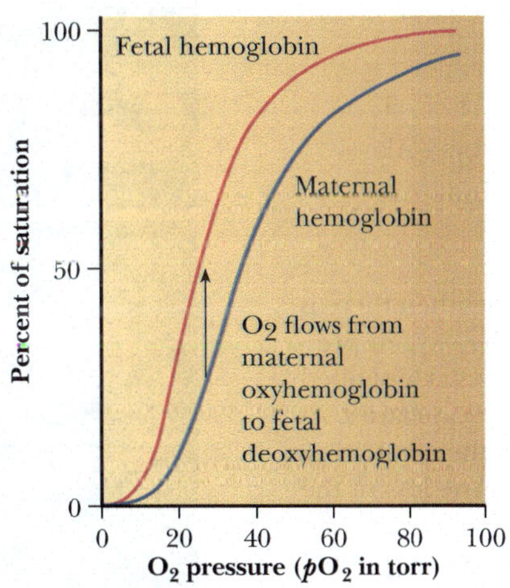

Figure 4.29 A comparison of the oxygen-binding capacity of fetal and maternal hemoglobin. Fetal hemoglobin binds less strongly to 2,3-bisphosphoglycerate and consequently has a greater affinity for oxygen than does maternal hemoglobin.

4.1 Apply Your Knowledge

Oxygen Response to pH in Hemoglobin

Assume that during a 400-m running race, the pH decreases in muscle cells from 7.6 to 7.0, while the pO_2 remains constant at 40 torr. What is the effect on the oxygen binding of hemoglobin in the muscle cells? What is the implication of this? What is the effect on myoglobin?

Solution

According to Figure 4.25, hemoglobin is about 82% saturated at pH 7.6 and a pO_2 of 40 torr. At pH 7.0, this drops to about 58% saturated. This means that hemoglobin will bind less oxygen at lower pH, or in other words, it will release more oxygen to the muscle cells. Myoglobin does not have a Bohr effect, however, and there is no effect caused by lowering the pH.

4A Biochemical Connections • Medicine

Sickle-Cell Disease

Normal adult human hemoglobin (Hb A) has two alpha chains and two beta chains. Some people, however, have a slightly different kind of hemoglobin in their blood. This hemoglobin (called Hb S) differs from the normal type only in the beta chains and only in one position on these two chains: The polar glutamic acid in the sixth position of the normal Hb beta chain is replaced by a hydrophobic valine residue in the beta chains of Hb S.

This change affects only two positions in a molecule containing 574 amino acid residues, yet it is enough to produce sickle-cell disease, a very serious medical condition.

	4	5	6	7	8	9
Normal Hb (Hb A)	—Thr—	Pro—	Glu—	Glu—	Lys—	Ala—
Sickle-cell Hb (Hb S)	—Thr—	Pro—	Val—	Glu—	Lys—	Ala—

Red blood cells carrying Hb S have a normal shape when there is an ample oxygen supply (Figure 4.30, *top*). When the oxygen pressure decreases, however, the red blood cells become sickle-shaped (Figure 4.30, *bottom*). This malformation occurs in the capillaries. As a result of this change in shape, the cells may clog the capillaries. The body's defenses destroy the clogging cells, and the loss of the blood cells causes anemia.

This change at only a single position of a chain consisting of 146 amino acids is severe enough to cause a high death rate. A child who inherits two genes programmed to produce sickle-cell hemoglobin (a homozygote) has an 80% lower chance of surviving to adulthood than a child with only one such gene (a heterozygote) or a child with two normal genes. Despite the high mortality of homozygotes, the genetic trait survives. In central Africa, 40% of the population in malaria-ridden areas carry the sickle-cell gene, and 4% are

Omikron/Science Source

Figure 4.30 Normal-shaped red blood cells (top) and sickle cells (bottom).

homozygotes. It seems that the sickle-cell genes help to acquire immunity against malaria in early childhood so that in malaria-ridden areas, the transmission of these genes is advantageous.

There is no absolute cure for sickle-cell disease, but several approaches are being looked at that will improve chances of a normal life. The U.S. Food and Drug Administration approved hydroxyurea (Figure 4.31) (sold under the name Droxia) to treat and control the symptoms of the disease.

Hydroxyurea prompts the bone marrow to manufacture fetal hemoglobin (Hb F), which does not have beta chains, which is where the mutation occurs. Instead, it has gamma chains. Thus, red blood cells containing Hb F do not sickle and do not clog the capillaries. With hydroxyurea therapy, the bone marrow still manufactures mutated Hb S, but the presence of cells with fetal hemoglobin dilutes the concentration of the sickle cells, thereby relieving the symptoms of the disease. Scientists are also using CRISPR technology (see Section 13-4C) to treat sickle-cell disease by precisely editing genes in blood stem cells to either correct the sickle mutation or boost production of fetal hemoglobin.

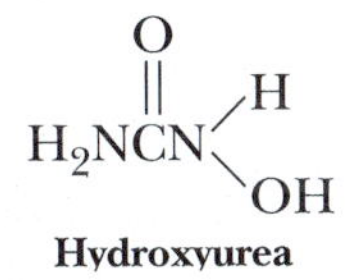

Figure 4.31 Structure of hydroxyurea.

4-6 Protein-Folding Dynamics

Learning Objectives

After reading this section, students will be able to:

LO 4.6.1 Describe how we can relate a protein's primary structure to its ability to adopt more complex structures by folding.

LO 4.6.2 Describe the thermodynamics of hydrophobic interactions in protein structure.

LO 4.6.3 Describe causes and consequences of incomplete or improper folding.

We know that the sequence of amino acids ultimately determines the three-dimensional structure of a protein. We also know that proteins can spontaneously adopt their native conformations, be denatured, and be renatured back into their native conformations, as was shown in Figure 4.20.

4-6A Primary Structure Leads to Tertiary Structure and Other Protein Properties

Modern computing techniques are highly effective at predicting protein structure from amino acid sequences alone due to the increasing power of computers and artificial intelligence (AI). The encounter of biochemistry and computing has also led to the expanding field of **bioinformatics**. The Expasy (Expert Protein Analysis System) server (**http://www.expasy.org**) has many excellent bioinformatics resources.

bioinformatics the application of computer methods to processing large amounts of information in biochemistry

Predicting protein structure is one of the principal applications of bioinformatics. Another important application is the comparison of base sequences in nucleic acids, a topic we discuss in Chapter 13, along with other methods for working with nucleic acids. Using AI methods, scientists are also making progress in engineering new functional proteins in the field of protein design.

A considerable amount of information about protein sequences and architecture is available via credible internet sources. One of the most important resources is the Protein Data Bank operated under the auspices of the Research Collaboratory for Structural Bioinformatics (RCSB). Its URL is http://www.rcsb.org. This site is the single repository of structural information about large molecules. It includes material about nucleic acids as well as proteins. Its homepage has links specifically geared toward educational applications.

4-6B Hydrophobic Interactions: A Case Study in Thermodynamics

We briefly introduced the notion of hydrophobic interactions in Section 4-4. Hydrophobic interactions have important consequences in biochemistry and play a major role in protein folding. They are thermodynamically favorable because they are spontaneous processes. It is known experimentally that proteins tend to be folded so that the nonpolar hydrophobic side chains are sequestered from water in the interior of the protein, whereas the polar hydrophilic side chains lie on the exterior of the molecule and are accessible to the aqueous environment (Figure 4.32).

The entropy of the universe increases when hydrophobic interactions occur.

$$\Delta S_{\text{univ}} > 0$$

As an example, let us assume that we have tried to mix the liquid hydrocarbon hexane (C_6H_{14}) with water and have obtained not a solution but a two-layer system, where the top layer is hexane and the bottom layer is water. Formation of a mixed solution is nonspontaneous, whereas the formation of two layers is spontaneous. Unfavorable entropy terms enter into the picture if solution formation requires the creation of ordered arrays of solvent—in this case, water (Figure 4.33).

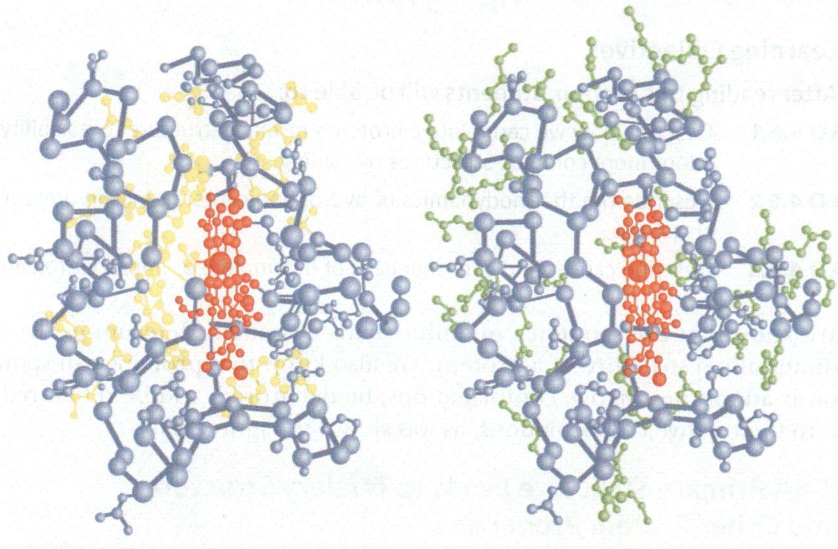

Figure 4.32 **Three-dimensional structure of the protein cytochrome c.** The interior of the protein has a porphyrin ring (red) similar to myoglobin and hemoglobin. The hydrophobic amino acids (yellow) are clustered in the interior away from water, whereas the hydrophilic ones predominate on the exterior (green). *(Illustration: Irving Geis. Rights owned by Howard Hughes Medical Institute.)*

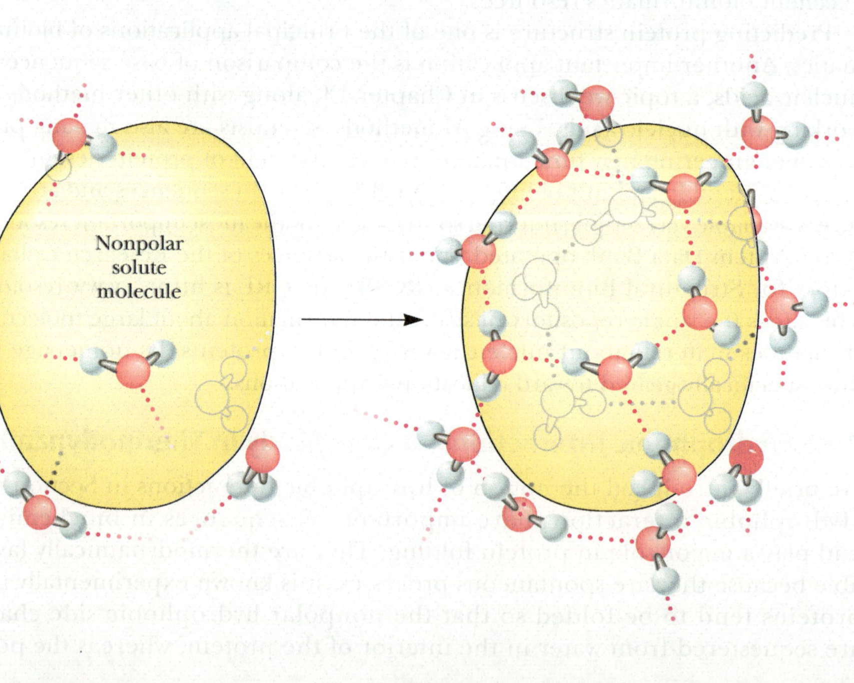

Figure 4.33 When water molecules surround a hydrophobic molecule, they are less able to interact with many more water molecules, thereby reducing entropy.

The water molecules surrounding the nonpolar molecules can hydrogen-bond with each other, but they have fewer possible orientations than if they were surrounded by other water molecules on all sides. This introduces a higher degree of order, preventing the dispersion of energy—more like the lattice of ice than liquid water—and thus a lower entropy. The required decrease in entropy is too large for the process to take place. Therefore, nonpolar substances do not dissolve in water; rather, nonpolar molecules associate with one another by hydrophobic interactions and are excluded from water. Another way to think of it is that each droplet of nonpolar material disrupts water's ability to hydrogen bond to more water molecules. The hydrogen bonding of fewer water molecules is disrupted by having just one big droplet instead of lots of little droplets.

Let us now carefully consider the thermodynamics of hydrophobic interactions between amino acids in the formation of protein tertiary structure. For example, Figure 4.13 shows hydrophobic interactions between leucine, valine, and isoleucine. We might conclude, then, that hydrophobic interactions refer only to dispersion forces between these three amino acids. However, we now know that it is not so much the attraction of the nonpolar amino acids for each other, but rather it is more that they are forced together so that water can avoid having to interact with them.

4-6C In Vivo Protein Folding Considerations

The primary structure conveys all the information necessary to produce the correct tertiary structure, but the folding process in vivo can be a bit trickier. In the protein-dense environment of the cell, proteins may begin to fold incorrectly as they are produced, or they may begin to associate with other proteins before completing their folding process. In eukaryotes, proteins may need to remain unfolded long enough to be transported across the membrane of a subcellular organelle.

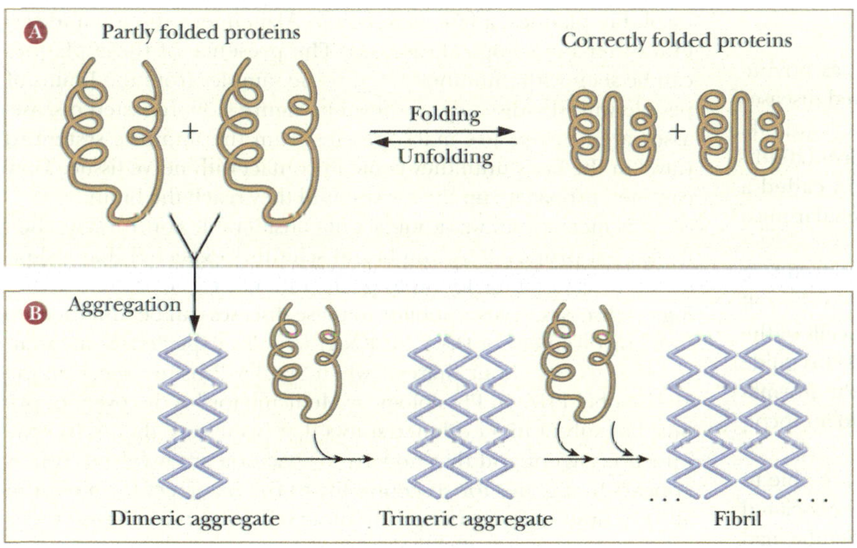

Figure 4.34 The problem of protein aggregation. (A) Partly folded polypeptide chains, released from ribosomes (the protein-synthesizing machines), normally form correctly folded, functional proteins. (B) However, partly folded proteins may sometimes associate with similar chains to form aggregates. Both soluble and insoluble aggregates can be toxic to cells. *(Reprinted by permission from Macmillan Publishers Ltd: Ellis, R. J., and Pinheiro, I. J. T. (2002). Danger—Misfolding Proteins.* Nature *416, 483–484.)*

Correctly folded proteins are usually soluble in the aqueous cell environment, or they are correctly attached to membranes (Figure 4.34A). However, when proteins do not fold correctly, they may interact with other proteins and form aggregates, as shown in Figure 4.34B. This occurs because hydrophobic regions that should be buried inside the protein remain exposed and interact with other hydrophobic regions on other molecules. Several neurodegenerative disorders, such as Alzheimer's, Parkinson's, and Huntington's diseases, and prion diseases such as Creutzfeldt-Jakob disease, are caused by the accumulation of protein deposits from such aggregates.

molecular chaperones proteins that assist in the regulated folding or unfolding of other proteins

To help prevent protein misfolding, special proteins called **molecular chaperones** aid in the correct and timely folding of many proteins. They either prevent a protein from associating with another protein with which it should not associate or keep it from associating with itself in inappropriate ways. The first such proteins discovered were a family of molecular chaperones called hsp70 (for *70,000 MW heat-shock protein*), which are proteins produced in *E. coli* cells when they are grown above optimal temperatures. Molecular chaperones exist in organisms from prokaryotes through humans, and their mechanisms of action are currently being studied. It is becoming more and more evident that protein-folding dynamics are crucial to protein function in vivo.

Protein folding is a very hot topic in biochemistry today. **Biochemical Connections 4B** describes particularly striking examples of the importance of protein folding.

4B Biochemical Connections • Medicine

Protein-Folding Diseases

Several well-known diseases are caused by misfolded proteins, including Creutzfeldt-Jakob disease, Alzheimer's disease, Parkinson's disease, and Huntington's disease. We discuss prion diseases here.

Prion Diseases

The causative agent of mad-cow disease (also known as bovine spongiform encephalopathy or BSE), as well as the related diseases scrapie in sheep, chronic wasting disease (CWD) in deer and elk, and human spongiform encephalopathy (kuru and Creutzfeldt-Jakob disease) in humans, is a small (28-kDa) protein called a prion. (Biochemists tend to call the unit of protein molecular mass the *Dalton,* abbreviated Da.)

Prions are natural glycoproteins found in the cell membranes of nerve tissue. Recently, the prion protein has been found in the cell membrane of hematopoietic stem cells, precursors to the cells of the bloodstream, and there is some evidence that the prion helps guide cell maturation. The disease state comes about when the normal form of the prion protein, PrP^{C} (Figure 4.35A), folds into an incorrect form called PrP^{SC} (Figure 4.35B).

These disease-associated forms of the prion protein are able to convert the commonly occurring forms into the disease-associated forms. This change can be propagated in nervous tissue. Scrapie had been known for years, but it had not been known to cross species barriers. Then an outbreak of mad-cow disease was shown to have followed the inclusion of sheep remains in cattle feed. It is now known that eating tainted beef from animals with mad-cow disease can cause spongiform encephalopathy, now known as new variant Creutzfeldt-Jakob disease (vCJD) in humans.

The commonly occurring prions have a large percentage of α-helix, but the disease-associated forms have more β-pleated sheets. In this case, the same protein (a single, well-defined sequence) can exist in alternative forms. These β-pleated sheets in the disease-associated proteins interact between protein molecules and form insoluble plaques, a fate also seen in Alzheimer's disease and several other neurological diseases. The presence of these plaques can be seen with immunostained tissue samples from the brains of people with the diseases, as shown in Figure 4.36. Ingested disease-associated prions use macrophages from the immune system to travel in the body until they come in contact with nerve tissue. They can then propagate up the nerves until they reach the brain.

This mechanism was a subject of considerable controversy when it was first proposed. A number of scientists expected that a slow-acting virus would be found to be the ultimate cause of these neurological diseases. A susceptibility to these diseases can be inherited, so some involvement of DNA (or RNA) was also expected. Some went so far as to talk about "heresy" when Stanley Prusiner received the 1997 Nobel Prize in Physiology or Medicine for his discovery of prions, but substantial evidence shows that prions are themselves the infectious agent and that no virus or bacteria are involved. It now appears that genes for susceptibility to the incorrect form exist in all vertebrates, giving rise to the observed pattern of disease transmission, but many individuals with the genetic susceptibility never develop the disease if they do not come into contact with disease-associated prions from another source. This combination of genetic predisposition combined with transmission by an infectious agent makes prion diseases unique. ◗

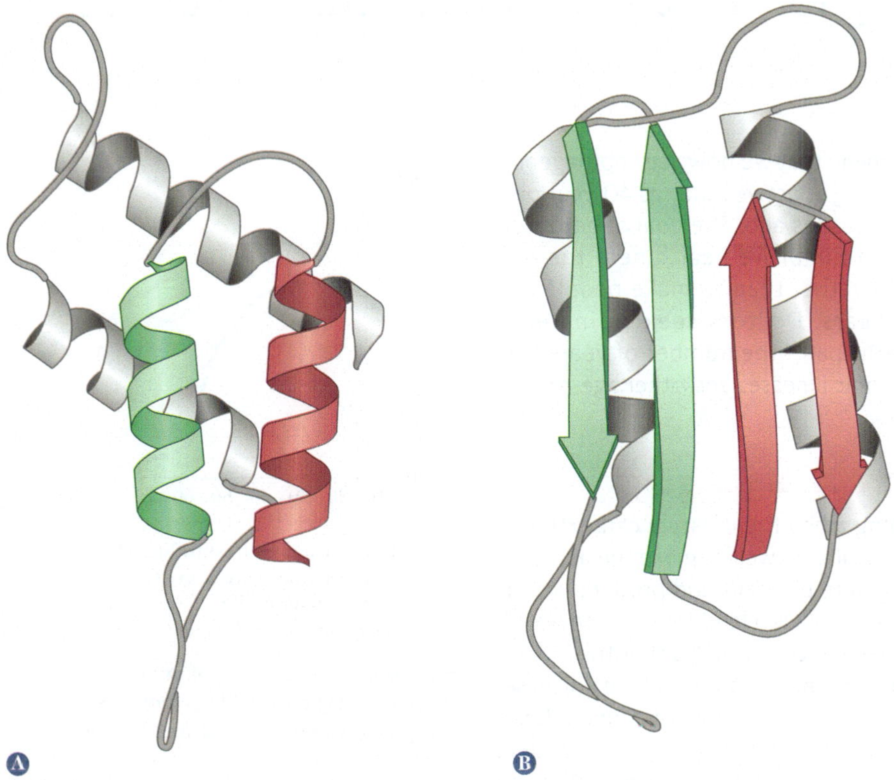

Figure 4.35 Structural differences between (A) a commonly occurring prion (PrP^{c}) and (B) a disease-associated one (PrP^{sc}).

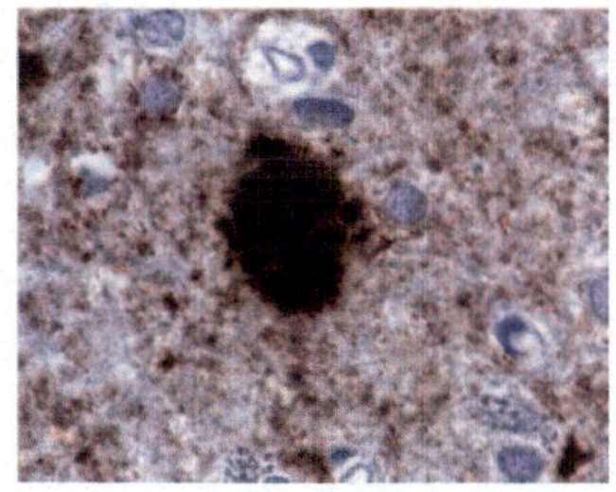

Figure 4.36 **Image of brain plaques from a patient with Creutzfeld-Jakob disease.** *(From Games Played by Rogue Proteins in Prion Disorders and Alzheimer's Disease by Adrian Aguzzi and Christian Haass (31 October 2003),* Science *302 (5646), 814. Reprinted with permission from AAAS.)*

Hot Topic

Aging—Looking for the Biochemical Fountain of Youth

Since the beginning of humankind, people have been obsessed with finding ways to turn back the clock, or at least to stop it from advancing. In this book, we provide many practical applications of biochemistry that can lead to a higher quality of life and perhaps even a longer one. Most of these are intuitive, such as maintaining a healthy lifestyle through diet, exercise, and the avoidance of negative factors like smoking, but others can be revealed through a deeper understanding of biochemical processes. Although the exact causes of aging are still not clear, it is believed to be caused by the gradual wearing out over time of the body's natural ability to maintain itself and repair damage.

When We Think and Study Aging, What Are We Really Concerned About?

There are three major issues—namely, maximum lifespan, average life expectancy, and quality of life. Maximum lifespan is the maximum time that any member of a species has lived. For humans, this is about 120 years. Average life expectancy has been increasing dramatically for humans for the last hundred years, although it is slowing considerably now. Average life expectancy has gone up over 30 years since the early 1900s, mainly due to modern medicine and public health initiatives. Diseases that killed people early in life and complications in childbirth led to many deaths at a young age, which brought down the average life expectancy. So the first is how long you would live if nothing killed you. The second is how long you really are likely to live given all environmental factors. The third consideration is the quality of life. Whether maximum life expectancy goes up or not, people today are experiencing a much higher quality of life as they age. When people say, "70 is the new 40," they mean that we now see septuagenarians being as active today as 40-year-olds were decades ago. The goal of gerontology is to improve health near the end of life, rather than to make people live 300 years. Doctors seek to increase "health span," the number of years free of chronic illnesses and other age-related issues.

Exercise and Aging

Leading a healthy lifestyle can extend a person's life, as well as make the available years more pleasant and productive. Fitness and diet can lead to a person avoiding many of the diseases that the elderly often succumb to, such as heart disease, stroke, and some forms of cancer. Being physically fit can slow the general decline we experience as we age. The earlier the fitness begins, the better. Figure 4.37 shows how different types of fitness levels are related to the deterioration with aging.

Figure 4.37 shows that the earlier physically active lifestyles begin, the greater the person's ability to do physical work and the slower the decline is. The people who were active from age 10 years had a much higher overall work capacity. Equally important, at 85, they had the same work capacity as a 25-year-old sedentary person, and they made it past 90 before the most serious decline into death began. For many people, that is the most important statistic. When people wonder whether they would really want to live for a hundred years, their answer would undoubtedly depend on what those years looked like. As a comedian once said, "Do I really want another 20 years of wearing adult diapers?" Most people, if given the choice, would like to live a long time and would probably choose the green (top) curve in the figure, where they are relatively healthy well into their old age, and then a quick decline rather than a slow and lingering one.

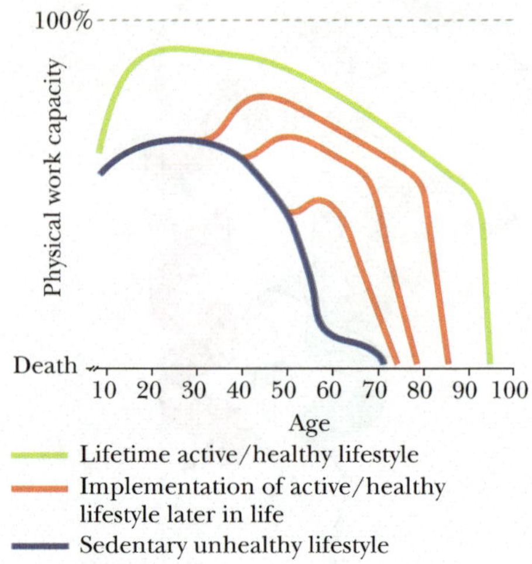

Figure 4.37 The effects of exercise on lifespan and quality of life. *(From Hoeger/Hoeger, Fitness and Wellness, 10E. © 2009 Cengage Learning.)*

Can Longevity Be Increased with Chemistry?

What if we could have increased longevity *and* quality of life? We discovered more than 70 years ago that calorie restriction (CR) is associated with increased longevity in life forms as varied as yeast and rodents. Recently, primates were added to that list. In some species, restricting caloric intake by 30% compared to normal levels was shown to increase life span by 30% or more. In addition to the life span extension, CR leads to a higher quality of life and forestalls many diseases, such as cancer, diabetes, inflammation, and even neurodegenerative diseases.

Many mechanisms for this longevity increase have been suggested, including general health benefits of weight reduction and specific improvements in DNA stability due to lower levels of oxidative compounds that are created as by-products of metabolism. However, about 15 years ago, researchers began to pinpoint a family of genes in the yeast *Saccharomyces cerevisiae* that seemed to be at the center of these increases in longevity due to CR. The best characterized of these genes is *SIR2* in yeast. *SIR2* is a member of a family of genes called the sirtuin genes, and evidence indicates that they are key regulators of the longevity mechanism. Their mode of action is based on fundamental changes in the organism's metabolism, especially the insulin-signaling pathways. In yeast and in roundworms, genetic manipulations that doubled the number of *SIR2* genes increased life span by 50%!

Humans have seven sirtuin genes and corresponding proteins, with the SIRT1 human protein being the one most closely related to the SIR2 protein in yeast. It is a stress-induced protein deacetylase that depends on NAD^+. It regulates cell survival, replicative senescence, inflammation, and metabolism via the deacetylation of histones (Section 11-5C). CR is a biological stressor like natural food scarcity. SIRT1 seems to be at the center of a generalized response to stress that primes mammals for survival. As Figure 4.38 shows, SIRT1 in mammals occupies a pivotal role in longevity through improved DNA stability, increased repair and defense, prolonged cell survival, enhanced energy production and use, and other coordinated stress responses. It has been implicated in age-related diseases, including cancer, type 2 diabetes, and Alzheimer's disease.

Mice that have been engineered to lack SIRT1 do not show the longevity increase associated with CR. Furthermore, doubling the number of *SIRT1* genes in an organism renders it unresponsive to CR. Thus, it is now generally accepted that CR promotes longevity by activating sirtuins in general and SIRT1 in particular.

Humans generally choose not to live a life of deprivation in order to reap the benefits of life-span extension, so the search for a stimulator of SIRT1 began. One of the first compounds found that is a natural activator of sirtuins is a small molecule called resveratrol (Figure 4.39), a polyphenol that is present in red wine and made by many plants when stressed. Resveratrol is a member of a class of compounds called STACs (for sirtuin-activating compounds), which are now known to be allosteric regulators of SIRT1.

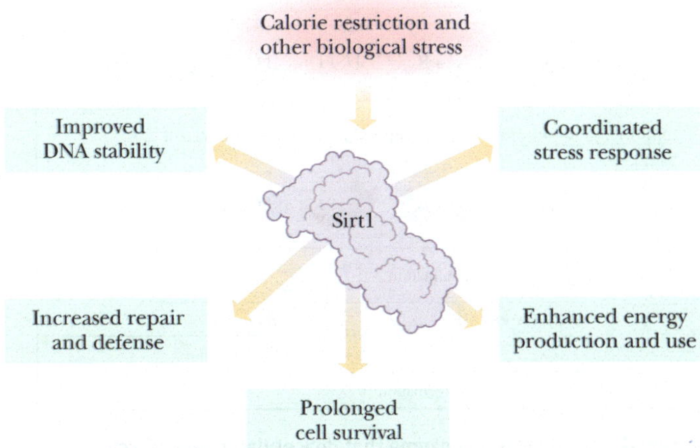

Figure 4.38 SIRT1 and its putative relationship to health and longevity. The SIRT1 enzyme appears to be responsible for the health and longevity-enhancing effects of CR in mammals. Food scarcity and other biological stressors trigger increased activity by SIRT1, which in turn alters activities within cells. By boosting manufacture of certain signaling molecules, such as insulin, SIRT1 may also coordinate the stress response throughout the body. *("Potential Alzheimer's Drug Spurs Protein Recycling" by Ken Garber)*

Feeding resveratrol to yeast, worms, or flies, or placing them on a CR diet, extends their life spans by about 30%, but only if they possess the *SIR2* gene. Resveratrol was found to increase SIRT1 activity 13-fold. As far as a small molecule that might increase longevity, nothing could be more attractive than resveratrol, especially to wine drinkers. Increased levels of SIRT1 in mice and rats allow some of the animals' cells to survive in the face of stress that would normally trigger their programmed cell death.

It does this by regulating several other key cellular proteins, such as p53 (Section 14-C), NF-κB, HSF-1, FOXO1, FOXO3, FOXO4, and PGC-1α (Figure 4.40). In addition, SIRT1 is stimulated by increased ratios

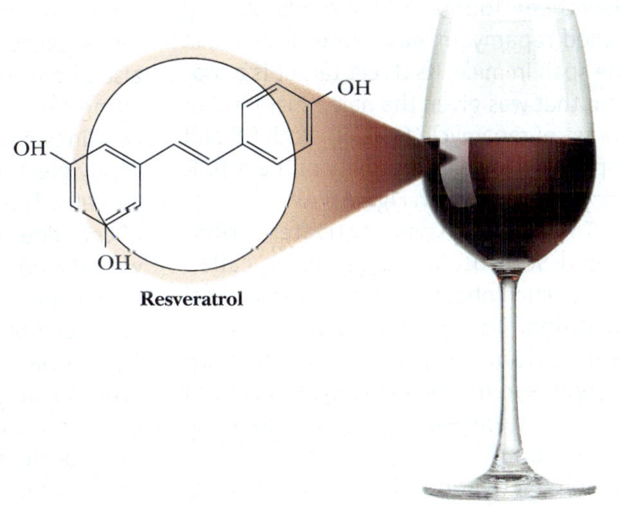

Figure 4.39 Resveratrol is an organic molecule found in red wine grapes and is believed to slow the effects of aging by activating sirtuin proteins. *(Based on Scientific American, "Unlocking the Secrets of Longevity Genes" by David A. Sinclair and Lenny Guarente, March 2006.)*

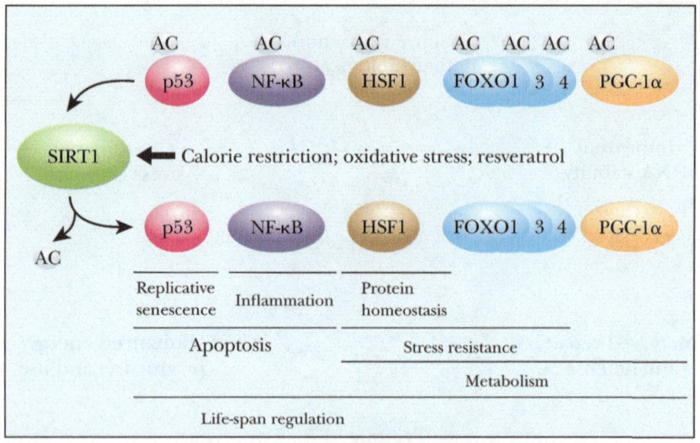

Figure 4.40 SIRT1 is an enzyme that deacetylates several key transcription factors that affect metabolism and aging. *(Based on Saunders and Verdin (2009), Stress Response to Aging.* Science *323, 1021.)*

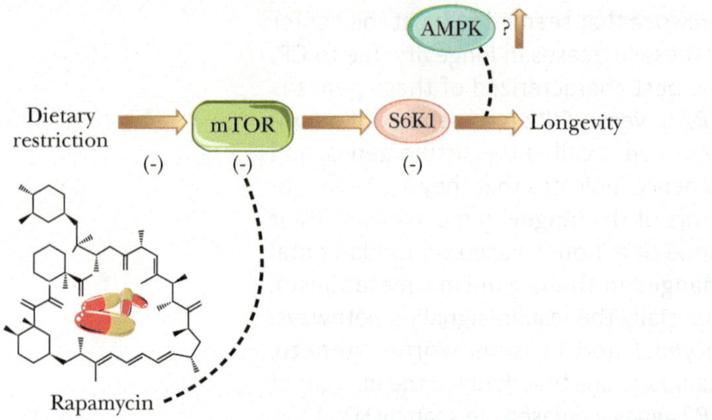

Figure 4.41 Chemical basis for longevity. Both dietary restriction and the drug rapamycin inhibit the protein mTOR. When mTOR is inhibited, activation of S6K1 is also inhibited, which leads to increased longevity. In a poorly understood mechanism, the protein AMPK is stimulated by the same process. *(Based on Kaeberlein and Kapahi (2009), Aging Is a RSKy Business.* Science *326, p. 55.)*

of $NAD^+/NADH$, a situation that arises when respiration is increased, as happens with fasting. Thus, SIRT1 is believed to act as both a sensor of nutrient availability and a regulator of the liver's response. SIRT1 has also been linked to the regulation of insulin and insulin-like growth factor. Insulin is known to play an important role in the general metabolic state of the organism.

The discovery of the sirtuins and of the effect of CR and resveratrol led to further research into aging and longevity.

Several important signaling pathways have been found to play a role. A drug called rapamycin was found to increase life span in mice. Its direct target is a protein that was given the name mammalian target of rapamycin (mTOR). Both CR and rapamycin lower the activity of the mTOR enzyme, as shown in Figure 4.41.

The mTOR enzyme activates a ribosomal S6 protein kinase (RSK), called S6K1, which phosphorylates S6 ribosomal proteins. The RSKs modulate mRNA translation and protein synthesis in response to mTOR signaling. It has been shown that longevity is increased by inhibiting the mTOR enzyme, which in turn inhibits activation of the S6K1 enzyme. Another protein kinase, AMPK, appears to be stimulated by the process.

Although we may never see a true longevity pill, the studies referenced here indicate promise that such a compound can be found. As is often the case, it should be much easier to find the treasure when we are sure the treasure exists. Both mTOR and S6K1 can be modified by small molecules, as we have seen in the case of rapamycin. Rapamycin has been shown to reduce adiposity in mice, at least in the short term. Why, then, have we not seen rapamycin on the shelves at our local pharmacy?

The reason is that we have a long way to go before we truly understand this process. For one thing, scientists are concerned about side effects. A known side effect of rapamycin when used long term is immune suppression. Furthermore, evidence exists that attempts to prolong life also often have the consequence of stimulating cancers. Several studies of cancer cell lines have shown they have significantly greater levels of sirtuins than regular cells. Thus, stimulating the longevity of cells does not work if we stimulate the wrong kind of cells. Rapamycin has also been implicated in mice and humans with glucose intolerance and insulin resistance. These side effects, if let go long enough, could outweigh the benefits to longevity.

Scientists are currently looking for ways to uncouple the positive effects on longevity with the negative effects on glucose homeostasis that rapamycin produces. The answer may lie in the details of what the mTOR protein does. It has been found that mTOR is involved in two different protein complexes, called mTORC1 and mTORC2 (Figure 4.42).

mTORC1 regulates pathways involved in autophagy, mRNA translation, and other cellular pathways, whereas mTORC2 regulates insulin signaling. Inhibiting mTOR via rapamycin inhibits both pathways, although with different effects—increasing longevity when mTORC1 is inhibited and impairing glucose metabolism when mTORC2 is inhibited. Thus, scientists continue to look for a small molecule that can have the same effect on mTORC1 as rapamycin but will not inhibit mTORC2. A human trial study took place in 2013 where males in their late 80s and 90s took rapamycin. Early results were encouraging because some members of the test group did show measurable signs of improvement in mobility studies.

Even modest successes spur increases in research in this field, human motives being what they are. One approach has been to focus on stem cells and their ability to regenerate adult cells as a potential difference between the old and young. One study from 2015 focused on SIRT7, another of the sirtuins, which is a nutrient-sensing protein in the same family as SIRT1. Under conditions of low nutrients, SIRT7 alters transcription and reduces oxidative cell metabolism and cell growth, thereby favoring cell survival under low metabolic conditions. SIRT7 was reduced in aging hematopoietic stem cells, and restoring normal expression levels of this protein caused them to revert to their younger state.

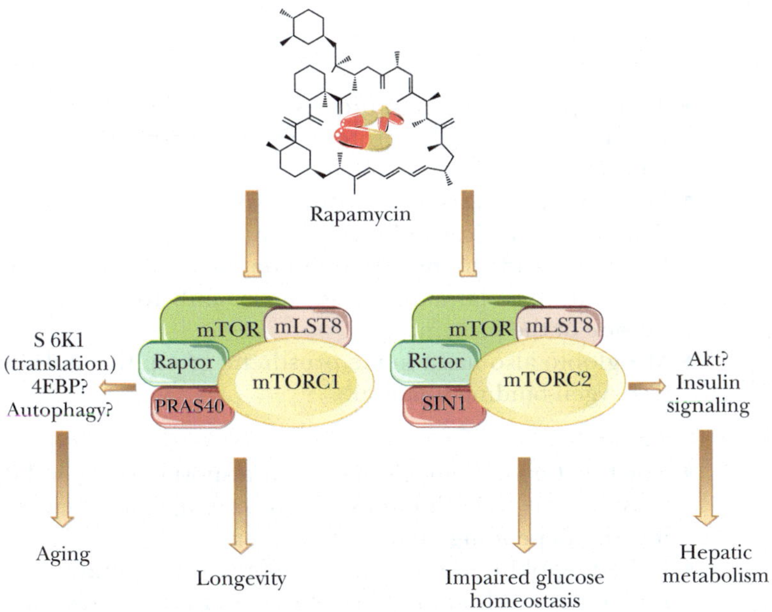

Figure 4.42 Rapamycin affects two different pathways. *(Adapted from Hughes and Kennedy (2012). Rapamycin Paradox Resolved. Science 335, 1578.)*

Some of the most exciting, yet confusing, research has come from a very old technique called parabiosis, first reported in 1864 by a French zoologist named Paul Bert. In this technique, the circulatory systems of two animals are allowed to fuse together at the site of an injury to both animals. After the fusion, the two animals share their circulatory systems and pass all blood-borne chemicals from one to the other. In the 1950s, a Cornell University researcher named Clive McKay noticed that fusing the circulatory systems of an old mouse to a young mouse had rejuvenating effects on the old mouse. However, due to a lack of analytical techniques to really figure out what was going on, it would be decades before this technique could really be studied. Since 2005, several laboratories have studied this phenomenon, and about the only conclusion that is universally shared is that there is something in young blood that stimulates a reversal of age-related deterioration in older animals, as shown in Figure 4.43.

The effect could also be seen just by injecting plasma (blood with the cells removed) from young mice into old mice, so the full technique of fusing the circulatory systems was unnecessary. Studies from the lab of Dr. Amy Wagers isolated a protein called growth differentiation factor 11 (GDF11), a 25-kDa protein that is a member of a family of well-known signaling molecules. Her research showed that injecting GDF11 into the hearts of old mice had the same age-reversal effect as the "young blood" effect. She also found that it improved mental function and the formation of new neurons in the brain. So why is not everyone getting injections of GDF11? Because the story is not over yet, and it is certainly not clear.

As the young blood phenomenon and GDF11 were being heralded as breakthroughs in gerontology, others were having trouble reconciling some questions and duplicating the results. In May 2015, a research group at the Novartis Institutes for BioMedical Research in Massachusetts challenged the GDF11 results. They noted that GDF11 is structurally similar to another type of protein called a myostatin, a protein that controls muscle growth. They wondered how two proteins that were so similar could have opposite effects. In their labs, they showed that levels of GDF11 did not go down with age. In fact, they found the opposite. They also found that increasing levels of GDF11 actually hampered repair of muscle injuries. Therefore, the debate goes on as to whether GDF11 is the key in a process that could reverse aging, is just an accidental player, or is actually unrelated.

In summary, life expectancy can be increased by maintaining a healthy lifestyle and avoiding dangerous activities. However, the knowledge gained from the studies on sirtuins, mTOR, and GDF11 has been promising.

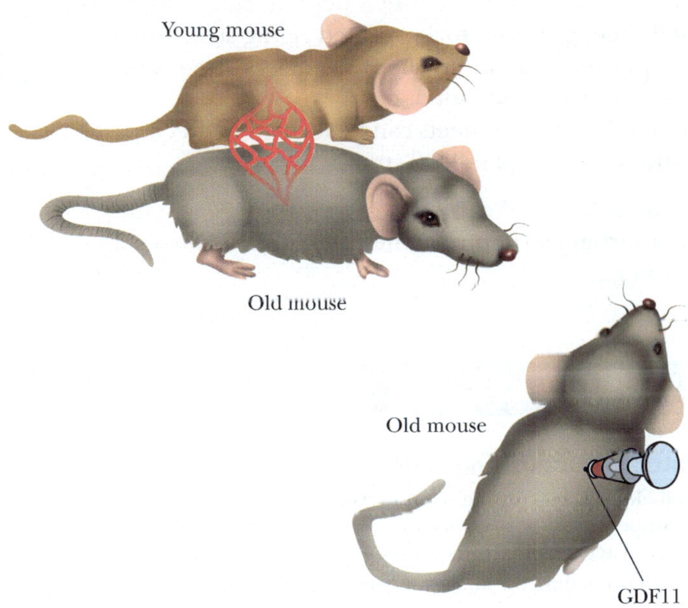

Figure 4.43 Old mice were found to be rejuvenated by two different processes. One is linking the circulatory system of the old mouse to a young mouse. The other is injecting the old mouse with a protein called growth differentiation factor 11 (GF11), which is found in larger quantities in young mice. Both processes reversed signs of aging in both muscle and brain. *(Based on "Rejuvenation Factor" in Blood Turns Back the Clock in Old Mice, Science, 344, 470–571 (2014).)*

Summary

Protein Structure and Function

- There are four levels of protein structure: primary, secondary, tertiary, and quaternary. Only proteins with multiple polypeptide chains have quaternary structure.

Primary Structure of Proteins

- Primary structure is the N-terminal to C-terminal order in which the amino acids are covalently linked.
- The amino acid sequence (the primary structure) of a protein determines its three-dimensional structure, which in turn determines its properties.
- A striking example of the importance of primary structure is sickle-cell disease, a condition caused by a change in one amino acid in each of two of the four chains of hemoglobin.

Secondary Structure of Proteins

- Protein secondary structure consists of patterns in hydrogen-bonded polypeptide backbone structure.
- The α-helix is stabilized by hydrogen bonds parallel to the helix axis. The hydrogen bonds occur between the backbone atoms of a single polypeptide chain.
- The helical conformation allows a linear arrangement of the atoms involved in the hydrogen bonds, which gives the bonds maximum strength and thus makes the helical conformation very stable.
- The β-sheet (or β-pleated sheet) is stabilized by backbone hydrogen bonds perpendicular to the direction of the protein chain, not parallel to it as in the α-helix. The hydrogen bonding between peptide chains in the β-pleated sheet gives rise to a repeated zigzag structure.
- The peptide backbone in the β-strands of the β-sheet is almost completely extended.
- Hydrogen bonds in β-sheets can be formed between parallel β-strands or antiparallel β-strands.

Tertiary Structure of Proteins

- Protein tertiary structure is the three-dimensional shape of a protein.
- Two experimental techniques used to determine the tertiary structure of a protein are X-ray crystallography and NMR.
- Myoglobin is an oxygen-storing protein. Its structure is like that of a typical globular protein, where its hydrophobic amino acids are mainly found on the inside of the protein, and its polar amino acids are mainly found on the outside of the protein.
- Myoglobin also has a heme prosthetic group that aids in its oxygen-binding function.

Hemoglobin

- The function of hemoglobin is to transport oxygen, and it must be able to both bind strongly to oxygen and release it easily, depending on conditions.
- In hemoglobin, the binding of oxygen is cooperative (as each oxygen is bound, it becomes easier for the next one to bind) and is modulated by ligands such as H^+, CO_2, and BPG.

Primary Structure Leads to Tertiary Structure and Other Protein Properties

- Using artificial intelligence methods, it is becoming increasingly more possible to predict the three-dimensional structure of a protein from its amino acid sequence. The field of protein design is rapidly developing as well.

Hydrophobic Interactions: A Case Study in Thermodynamics

- Hydrophobic interactions are spontaneous processes.
- The entropy of the universe increases when hydrophobic interactions occur.
- Hydrophobic interactions, which depend on the unfavorable entropy of the water of hydration surrounding nonpolar solutes, are particularly important determinants of protein folding.

Review Exercises

4-1 Protein Structure and Function

1. **Recall** Match the following statements about protein structure with the proper levels of organization. (i) Primary structure. (ii) Secondary structure. (iii) Tertiary structure. (iv) Quaternary structure.
   (a) The three-dimensional arrangement of all atoms
   (b) The order of amino acid residues in the polypeptide chain
   (c) The interaction between subunits in proteins that consist of more than one polypeptide chain
   (d) The hydrogen-bonded arrangement of the polypeptide backbone.
2. **Reflect and Apply** The structure of a recently studied angiotensin-converting enzyme 2 (ACE2) is described as trimeric. To which level of protein structure does the term "trimeric" refer: primary, secondary, tertiary, or quaternary structure?
3. **Recall** If a mutation occurs in a protein that alters the amino acid sequence, which level of protein structure is directly affected by this change: primary, secondary, tertiary, or quaternary structure?
4. **Recall** Define denaturation in terms of the effects on primary, secondary, tertiary, and quaternary structure.
5. **Recall** What is the nature of random-coil structure in proteins?

4-2 Primary Structure of Proteins

6. **Reflect and Apply** Rationalize the following observation: Replacements such as Lys to Arg or Leu to Ile usually have very little effect on protein structure and function.

7. **Reflect and Apply** A mutation that changes an alanine residue in a protein to an isoleucine can lead to a loss of activity. Activity could be regained when a further mutation at the same site changes the isoleucine to a glycine. Why?

4-3 Secondary Structure of Proteins

8. **Recall** List three major differences between fibrous and globular proteins.
9. **Recall** What type of noncovalent interaction is most important in stabilizing the structure of an alpha helix or beta sheet in proteins: disulfide bonds, hydrogen bonding interactions between side chains, hydrogen bonding interactions between atoms of the peptide backbone, electrostatic interactions, or van der Waals interactions?
10. **Recall** What are Ramachandran angles?
11. **Recall** What is a β-bulge?
12. **Recall** What is a reverse turn? Draw two types of reverse turns.
13. **Recall** List some of the differences between the α-helix and β-sheet forms of secondary structure.
14. **Recall** List some of the possible combinations of α-helices and β-sheets in supersecondary structures.
15. **Recall** Why is proline frequently encountered at the places in the myoglobin and hemoglobin molecules where the polypeptide chain turns a corner?
16. **Recall** Why must glycine be found at regular intervals in the collagen triple helix?
17. **Reflect and Apply** Woolen clothing shrinks when washed in hot water, but items made of silk do not. Suggest a reason based on information from this chapter.

4-4 Tertiary Structure of Proteins

18. **Recall** Draw two hydrogen bonds, one that is part of secondary structure and another that is part of tertiary structure.
19. **Recall** Draw a possible electrostatic interaction between two amino acids in a polypeptide chain.
20. **Recall** Draw a disulfide bridge between two cysteines in a polypeptide chain.
21. **Recall** Draw a region of a polypeptide chain showing a hydrophobic pocket containing nonpolar side chains.
22. **Recall** Which type of protein, globular or fibrous, is most likely to be used in structural components like hair and skin?
23. **Recall** Which of the following is a common example of a fibrous protein: hemoglobin, collagen, enzymes, or antibodies?
24. **Reflect and Apply** Which of the following is NOT a primary characteristic of globular proteins?
    (a) They are typically soluble in water
    (b) They have a compact structure.
    (c) They are usually involved in structural functions.
    (d) They often have complex tertiary and quaternary structures.
25. **Recall** Which level(s) of protein structure does myoglobin have?
26. **Recall** Which component of the oxygen binding site in myoglobin is most directly involved in decreasing the affinity of myoglobin for carbon monoxide (CO) by two orders of magnitude?
    (a) the heme group
    (b) the Fe(II) atom
    (c) the E7 histidine residue
    (d) the carboxylate groups of protoporphyrin IX
27. **Reflect and Apply** What is the highest level of protein structure found in collagen?

4-5 Quaternary Structure of Proteins

28. **Recall** List two similarities and two differences between hemoglobin and myoglobin.
29. **Recall** Which level of protein structure best explains why hemoglobin displays positive cooperativity with respect to oxygen binding curve?
30. **Recall** What are the two critical amino acids near the heme group in both myoglobin and hemoglobin?
31. **Recall** What is the highest level of organization in myoglobin? In hemoglobin?
32. **Recall** Suggest a way in which the difference between the functions of hemoglobin and myoglobin is reflected in the shapes of their respective oxygen-binding curves.
33. **Recall** Describe the Bohr effect.
34. **Recall** Describe the effect of 2,3-bisphosphoglycerate on the binding of oxygen by hemoglobin.
35. **Recall** How does the oxygen-binding curve of fetal hemoglobin (Hb F) differ from that of adult hemoglobin (Hb A)?
36. **Recall** What is the critical amino acid difference between the beta chain and the gamma chain of hemoglobin?
37. **Reflect and Apply** In oxygenated hemoglobin, $pK_a = 6.6$ for the histidines at position 146 on the beta chains. In deoxygenated hemoglobin, the pK_a of these residues is 8.2. How can this piece of information be correlated with the Bohr effect?
38. **Reflect and Apply** How does the difference between the beta chain and the gamma chain of hemoglobin explain the differences in oxygen binding between adult hemoglobin (Hb A) and fetal hemoglobin (Hb F)?
39. **Reflect and Apply** Does a fetus homozygous for sickle-cell hemoglobin (Hb S) have normal fetal hemoglobin (Hb F)?
40. **Biochemical Connections** Why might you expect to find some Hb F in adults who have sickle-cell disease?
41. **Biochemical Connections** What is the direct cause of sickle-cell disease?
42. **Biochemical Connections** What is the effect of the altered amino acid sequence in Hb S that causes the cells to form sickle shapes?
43. **Biochemical Connections** Why do scientists believe that the sickle-cell trait has not evolved out of the human population despite how deadly it is to be homozygous for the sickle-cell gene?
44. **Biochemical Connections** What is the purpose of treating a sickle-cell patient with hydroxyurea?

4-6 Protein-Folding Dynamics

45. **Reflect and Apply** Analyze the energetics of protein folding in light of the information in this chapter.
46. **Recall** According to the explanation for the hydrophobic effect, what is the primary reason why nonpolar amino acid residues associate together?
47. **Recall** What is a molecular chaperone?
48. **Biochemical Connections** What is a prion?
49. **Biochemical Connections** What are the protein secondary structures that differ between a normal prion and an infectious one?
50. **Biochemical Connections** What are some diseases caused by misfolded proteins?
51. **Biochemical Connections** What causes protein aggregates to form?
52. **Biochemical Connections** What aspects of the transmission of scrapie or other spongiform encephalopathies act like genetic diseases? What aspects act like transmissible diseases?
53. **Biochemical Connections** An isolated strain of sheep was found in New Zealand. Most of these sheep carried the gene for predisposition to scrapie, yet none of them ever came down with the disease. How do these facts relate to the transmission of prion diseases?

Further Reading

Aguzzi, A., and C. Haass. Games Played by Rogue Proteins in Prion Disorders and Alzheimer's Disease. *Science* **302**, 814–818 (2003).

Couzin, J. The Prion Protein Has a Good Side? You Bet. *Science* **311**, 1091 (2006).

Ellis, R. J., and T. J. T. Pinheiro. Danger—Misfolding Proteins. *Nature* **416**, 483–484 (2002).

Ensrink, M. After the Crisis: More Questions About Prions. *Science* **310**, 1756–1758 (2005).

Ferguson, N. M., A. C. Ghan, C. A. Donnelly, T. J. Hagenaars, and R. M. Anderson. Estimating the Human Health Risk from Possible BSE Infection of the British Sheep Flock. *Nature* **415**, 420–424 (2002).

Garber, K. Potential Alzheimer's Drug Spurs Protein Recycling, *Science*, **344**, 351 (2014).

Gibbons, A., and M. Hoffman. New 3-D Protein Structures Revealed. *Science* **253**, 382–383 (1991). Examples of the use of X-ray crystallography to determine protein structure.

Gierasch, L. M., and J. King, eds. *Protein Folding: Deciphering the Second Half of the Genetic Code.* Waldorf, MD: AAAS Books, 1990. A collection of articles on recent discoveries about the processes involved in protein folding. Experimental methods for studying protein folding are emphasized.

Glabe, C. Avoiding Collateral Damage in Alzheimer's Disease Treatment. *Science* **314**, 602–603 (2006).

Hall, S. Protein Images Update Natural History. *Science* **267**, 620–624 (1995). Combining X-ray crystallography and computer software to produce images of protein structure.

Hauptmann, H. The Direct Methods of X-Ray Crystallography. *Science* **233**, 178–183 (1986). A discussion of improvements in methods of doing the calculations involved in determining protein structure; based on a Nobel Prize address. This article should be read in connection with the one by Karle, and it provides an interesting contrast with the articles by Perutz, both of which describe early milestones in protein crystallography.

Helfand, S. L. Chaperones Take Flight. *Science* **295**, 809–810 (2002). An article about using chaperones to combat Parkinson's disease.

Holm, L., and C. Sander. Mapping the Protein Universe. *Science* **273**, 595–602 (1996). An article on searching databases on protein structure to predict the three-dimensional structure of proteins. Part of a series of articles on computers in biology.

Karle, J. Phase Information from Intensity Data. *Science* **232**, 837–843 (1986). A Nobel Prize address on the subject of X-ray crystallography. See remarks on the article by Hauptmann.

Kasha, K. J. Biotechnology and the World Food Supply. *Genome* **42** (4), 642–645 (1999). Proteins are frequently in short supply in the diet of many people in the world, but biotechnology can help improve the situation.

Legname, G., I. V. Baskakov, H. B. Nguyen, D. Riesner, F. E. Cohen, S. J. DeArmond, and S. B. Prusiner. Synthetic Mammalian Prions. *Science* **305**, 673–676 (2004).

Luzzatto, L., and R. Notaro. Haemoglobin's Chaperone. *Nature* **417**, 703–705 (2002).

Miller, G. Could They All Be Prion Diseases? *Science* **326**, 1337–1339 (2009).

Mitten, D. D., R. MacDonald, and D. Klonus. Regulation of Foods Derived from Genetically Engineered Crops. *Curr. Opin. Biotechnol.* **10**, 298–302 (1999). How genetic engineering can affect the food supply, especially that of proteins.

O'Quinn, P. R., J. L. Nelssen, R. D. Goodband, D. A. Knabe, J. C. Woodworth, M. D. Tokach, and T. T. Lohrmann. Nutritional Value of a Genetically Improved High-Lysine, High-Oil Corn for Young Pigs. *J. Anim. Sci.* **78** (8), 2144–2149 (2000). The availability of amino acids affects the proteins formed.

Peretz, D., R. A. Williamson, K. Kaneko, J. Vergara, E. Leclerc, G. Schmitt-Ulms, I. R. Mehlhorn, G. Legname, M. R. Wormald, P. M. Rudd, R. A. Dwek, D. R. Burton, and S. B. Prusiner. Antibodies Inhibit Prion Propagation and Clear Cell Cultures of Prion Infectivity. *Nature* **412**, 739–742 (2001). Description of a possible treatment for prion diseases.

Perutz, M. The Hemoglobin Molecule. *Sci. Am.* **211** (5), 64–76 (1964). A description of work that led to a Nobel Prize.

Perutz, M. The Hemoglobin Molecule and Respiratory Transport. *Sci. Am.* **239** (6), 92–125 (1978). The relationship between molecular structure and cooperative binding of oxygen.

Ruibal-Mendieta, N. L., and F. A. Lints. Novel and Transgenic Food Crops: Overview of Scientific Versus Public Perception. *Transgenic Res.* **7** (5), 379–386 (1998). A practical application of protein structure research.

Styx, G. Alzheimer's: Forestalling the Darkness. *Sci. Am.*, 51–57 (June 2010). Review of methods used to treat Alzheimer's before symptoms appear.

Willem, M., A. N. Garratt, B. Novak, M. Citron, S. Kaufmann, A. Rittger, B. DeStrooper, P. Saftig, C. Birchmeier, and C. Haass. Control of Peripheral Nerve Myelination by the ß-Secretase BACE1. *Science* **314**, 664–666 (2006).

Wolfe, M. S. Shutting Down Alzheimer's. *Sci. Am.* **294** (5), 73–74 (2006). A summary of treatment options for Alzheimer's disease.

Yam, P. Mad Cow Disease's Human Toll. *Sci. Am.* **284** (5), 12–13 (2001). An overview of mad-cow disease and how it has crossed over to infect people.

Hot Topic

Couzin-Franken, J. Aging Genes: The Sirtuin Story Unravels. *Science* **334**, 1194–1198 (2011).

Hall, S. S. In Vino Vitalis? Compounds Activate Life-Extending Genes. *Science* **301**, 1165 (2003).

Hall, S. S. Young Blood: Young Animals' Blood Holds Rejuvenating Powers. Amy Wagers Wants to Know Why. *Science,* **345**, 1234–1237 (2014).

Hughes, K. J., and B. K. Kennedy. Rapamycin Paradox Resolved. *Science* **335**, 1578–1579 (2012).

Kaeberlein, M., and Kapahi, P. Aging Is a RSKy Business. *Science* **326**, 55–56 (2009).

Kaiser, J. "Rejuvenation Factor" in Blood Turns Back the Clock in Old Mice. *Science,* **344**, 570–571.

Kaiser, J. Rejuvenating Protein Doubted. *Science* **348**, 849 (2015).

Kirkwood, T. Why Can't We Live Forever? *Sci. Am.* 42–50 (September 2010).

Leslie, M. A. Putative Antiaging Drug Takes a Step from Mice to Men. *Science,* **342**, 789 (2013).

Mohrin, M., J. Shin, Y. Liu, K. Brown, H. Luo, Y. Xi, C. M. Haynes, and D. A. Chen. Mitochondrial UPR-Mediated Metabolic Checkpoint Regulates Hematopoietic Stem Cell Aging. *Science,* **347**, 1374–1377 (2015).

Ocampo, A. and J. C. Izpisua Belmonte. Holding Your Breath for Longevity. *Science,* **347**, 1319–1320.

Saunders, L. R., and E. Verdin. Stress Response and Aging. *Science* **323**, 1021–1022 (2009).

Sinclair, D. A., and L. Guarente. Unlocking the Secrets of Longevity Genes. *Sci. Am.* 48–50 (March 2006).

Stipp, D. A New Path to Longevity. *Sci. Am.* 33–39 (January 2012).

Topisirovic, I., and N. Sonenberg. Burn Out or Fade Away? *Science* **327**, 1210–1211 (2010).

Yuan, H., and R. Marmorstein, Red Wine, Toast of the Town (Again). *Science,* **339**, 1156–1157 (2013).

Column chromatography is widely used to purify proteins.

Jochen Tack/imageBROKER.com/Alamy Stock Photo

5 Protein Purification and Characterization Techniques

Chapter Outline

5-1 Extracting Pure Proteins from Cells

Learning Objectives

After reading this section, students will be able to:

LO 5.1.1 Distinguish between different techniques to extract proteins from a cell.

LO 5.1.2 Describe the steps involved in differential centrifugation.

LO 5.1.3 Describe how to salt out proteins.

Many different proteins exist in a single cell, but a detailed study of the properties of any one protein requires a homogeneous sample consisting of only that protein. The separation and isolation, or purification, of proteins constitutes an essential first step to further experimentation. In general, separation techniques focus on size, charge, and polarity—the sources of differences between molecules. Many techniques are performed to eliminate contaminants and to arrive at a pure sample of the protein of interest. As the purification steps are followed, we make a table of the recovery and purity of the protein to gauge our success.

Table 5.1 presents typical purification steps for an enzyme. The fraction column lists purification processes that are discussed in more detail in Chapter 5. The **specific activity** column is a measure of the activity of the enzyme of interest in the fraction, divided by the total weight of all proteins in the sample, and its value should go up with each purification step. The **percent recovery** column tracks how much of the protein of interest has been retained at each step.

Table 5.1 Example of a Protein Purification Scheme: Purification of the Enzyme Xanthine Dehydrogenase from a Fungus

Fraction	Volume (mL)	Total Protein (mg)	Total Enzyme Activity (μmol product/min)	Specific Activity (μmol product/min/mg protein)	Percent Recovery
1. Crude extract	3,800	22,800	2,460	0.108	100
2. Salt precipitation	165	2,800	1,190	0.425	48
3. Ion-exchange chromatography	65	100	720	7.20	29
4. Size exclusion chromatography	40	14.5	555	38.3	23
5. Immunoaffinity chromatography	6	1.8	275	152	11

This number usually drops steadily after each purification step, and we hope that by the time the protein is pure, a sufficient sample remains for characterization.

specific activity a measurement of the amount of activity of the enzyme of interest at a certain purification step, divided by the total weight of all proteins in the sample

percent recovery a measurement of the amount of an enzyme recovered at each step of a purification experiment

5-1A Homogenization

► How do we get the proteins out of the cells?

Before the real purification steps can begin, the protein must be released from the cells and subcellular organelles. The first step, called **homogenization**, involves breaking open the cells. This can be done with a wide variety of techniques. The simplest approach is grinding the tissue in a blender with a suitable buffer. The cells are broken open, releasing soluble proteins. This process also breaks many of the subcellular organelles such as mitochondria, peroxisomes, and endoplasmic reticulum. A gentler technique is to use a Potter–Elvehjem homogenizer, a thick-walled test tube through which a tight-fitting plunger is passed. The squeezing of the homogenate around the plunger breaks open cells, but it leaves many of the organelles intact. Another technique, called sonication, involves using sound waves to break open the cells. Cells can also be ruptured by cycles of freezing and thawing. If the protein of interest is solidly attached to a membrane, detergents may have to be added to detach the proteins.

homogenization the process of breaking open cells to release the organelles

5-1B Differential Centrifugation

After the cells are homogenized, they are subjected to **differential centrifugation** (Figure 5.1). Spinning the sample at 600 times the force of gravity ($600 \times g$) results in a pellet of unbroken cells and nuclei. If the protein of interest is not found in the nuclei, this precipitate is discarded. The supernatant can then be centrifuged at higher speed, such as $15{,}000 \times g$, to bring down the mitochondria. Further centrifugation at $100{,}000 \times g$ brings down the microsomal fraction, consisting of ribosomes and membrane fragments. If the protein of interest is soluble, the supernatant from this spin will be collected and will already be partially purified because the nuclei, mitochondria, and other cellular components will have been removed.

differential centrifugation a process in which ruptured cells are centrifuged several times, increasing the force of gravity each time, to separate organelles and other subcellular particles

5-1C Salting Out

After the soluble proteins are separated from the organelles and other subcellular particles, the resulting crude extract is often subjected to a purification step called **salting out**, which is based on solubility. Ammonium sulfate is the most common reagent used for this step. Proteins have varying solubilities in polar and ionic compounds. Proteins remain soluble because of their interactions with water. When ammonium sulfate is added to a protein solution, some

salting out a purification technique for proteins based on differential solubility in salt solutions

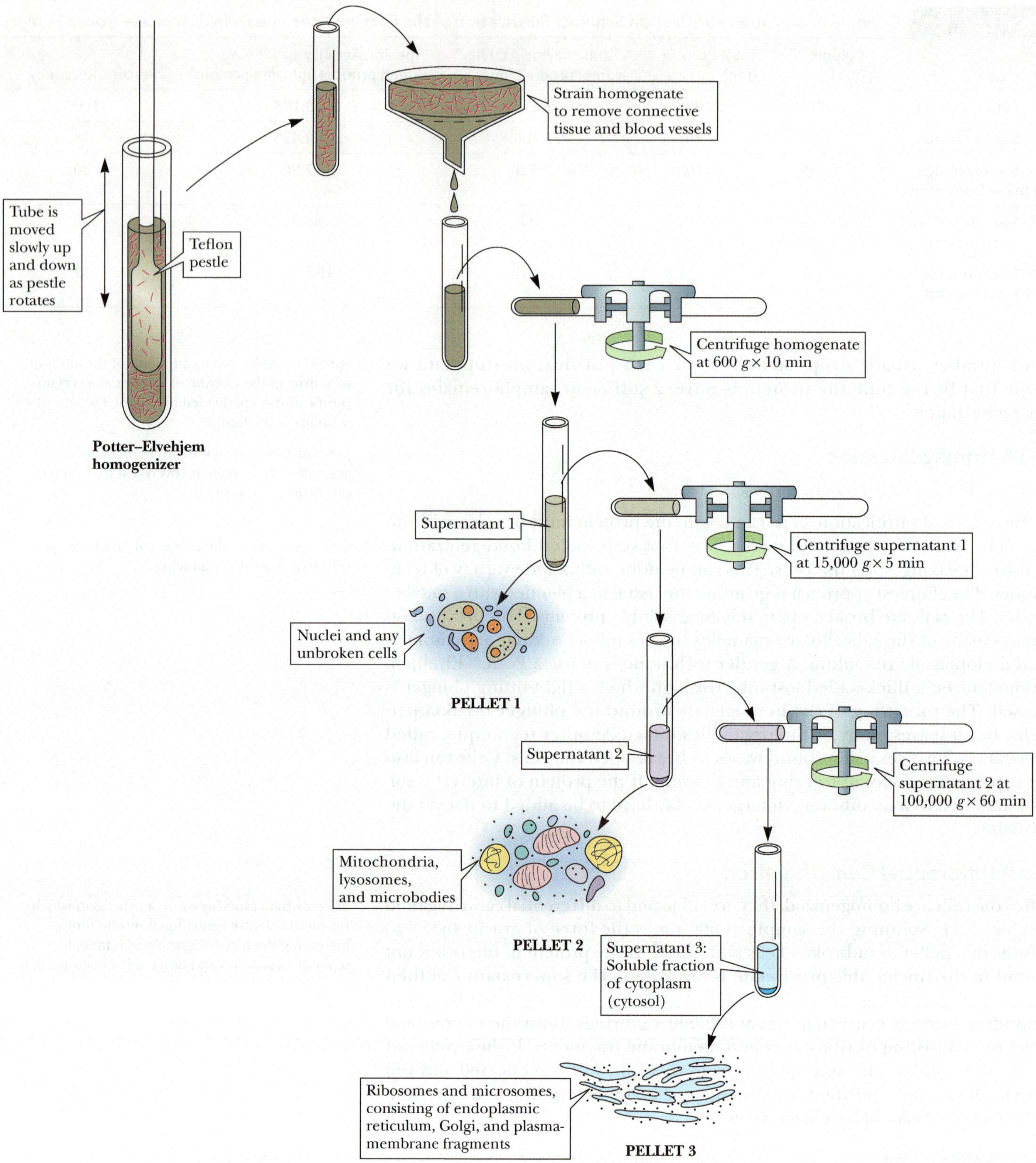

Figure 5.1 Differential centrifugation. Differential centrifugation is used to separate cell components. As a cell homogenate is subjected to increasing *g*-forces, smaller and smaller cell components end up in the pellet.

Careers in Chemistry • Liuba Pauline Williams

Education:
University of Nevada, Las Vegas, PhD (in progress), Radiochemistry
University of Nevada, Las Vegas, BS, Biochemistry

When Liuba Pauline Williams began college, she believed the only path for a biochemistry major was medical school. It wasn't until her final year that she discovered the broad range of careers in which she could use her degree to make a meaningful impact.

Williams credits her major in biochemistry for igniting her love for forensic chemistry, strengthening her critical thinking skills, and refining her laboratory techniques. Today, Williams is pursuing a doctorate in radiochemistry, where her focus is nuclear forensics. As part of her research, she simulates nuclear fireball plasmas using an arc welder and creates surrogate bomb debris. As an intern at the Nevada National Security Site, part of her job is to train the U.S. Army to identify and collect radioactive debris.

Renée Olney

As a biracial scientist—African American and Russian—Williams brings a deeply inclusive perspective to her work. Growing up in two cultures taught her the value of multiple viewpoints and the importance of open-mindedness in both science and life. "Science is for everyone," she says. "We must respect all who choose to study it with us."

Though she works in a field still where women and people of color are largely underrepresented, Williams uses her identity as a source of strength rather than limitation. "Don't let the lack of representation stop you from living your dreams," she says. "If you've never seen someone like yourself doing what you love, become that person for someone else."

of the water is taken away from the protein to make ion–dipole bonds with the salt. With less water available to hydrate the proteins, they begin to interact with each other through hydrophobic bonds.

How much ammonium sulfate should be added? The quantity of ammonium sulfate is usually measured in comparison with a 100% saturated solution. A common procedure involves bringing the solution to around 40% saturation, where a precipitate that contains contaminating proteins usually forms. These proteins are centrifuged down and discarded. Then more ammonium sulfate is added to the supernatant, often to a level of 60% to 70% saturation, and a different set of proteins, which usually contains the protein of interest, precipitates. This precipitate is collected by centrifugation, resuspended in a suitable buffer, and saved for further purification steps. These preliminary techniques do not generally give a sample that is very pure, but they do eliminate many of the contaminating proteins, thereby refining the crude homogenate extract for the more effective chromatography procedures that follow.

5-2 Column Chromatography

Learning Objectives

After reading this section, students will be able to:

LO 5.2.1 Describe the general features of column chromatography.

LO 5.2.2 Distinguish between size-exclusion chromatography, affinity chromatography, ion-exchange chromatography, and high-performance liquid chromatography.

The word *chromatography* comes from the Greek *chroma*, "color," and *graphein*, "to write." The technique was first used around the beginning of the twentieth century to separate plant pigments with easily visible colors, and it has long since been possible to separate colorless compounds as well. In chromatography, different compounds distribute themselves to varying extents between two phases—the **stationary phase** and the **mobile phase**. The mobile phase flows over the stationary material and carries the sample to be separated along with it. The components of the sample interact with the stationary phase to different extents. Some components interact relatively strongly with the stationary phase and are therefore carried along more slowly by the mobile phase than those that interact less strongly. The differing mobilities of the components are the basis of the separation.

stationary phase in chromatography, the substance that affects the separation of compounds in a homogeneous mixture

mobile phase in chromatography, the portion of the system in which the homogeneous mixture to be separated moves

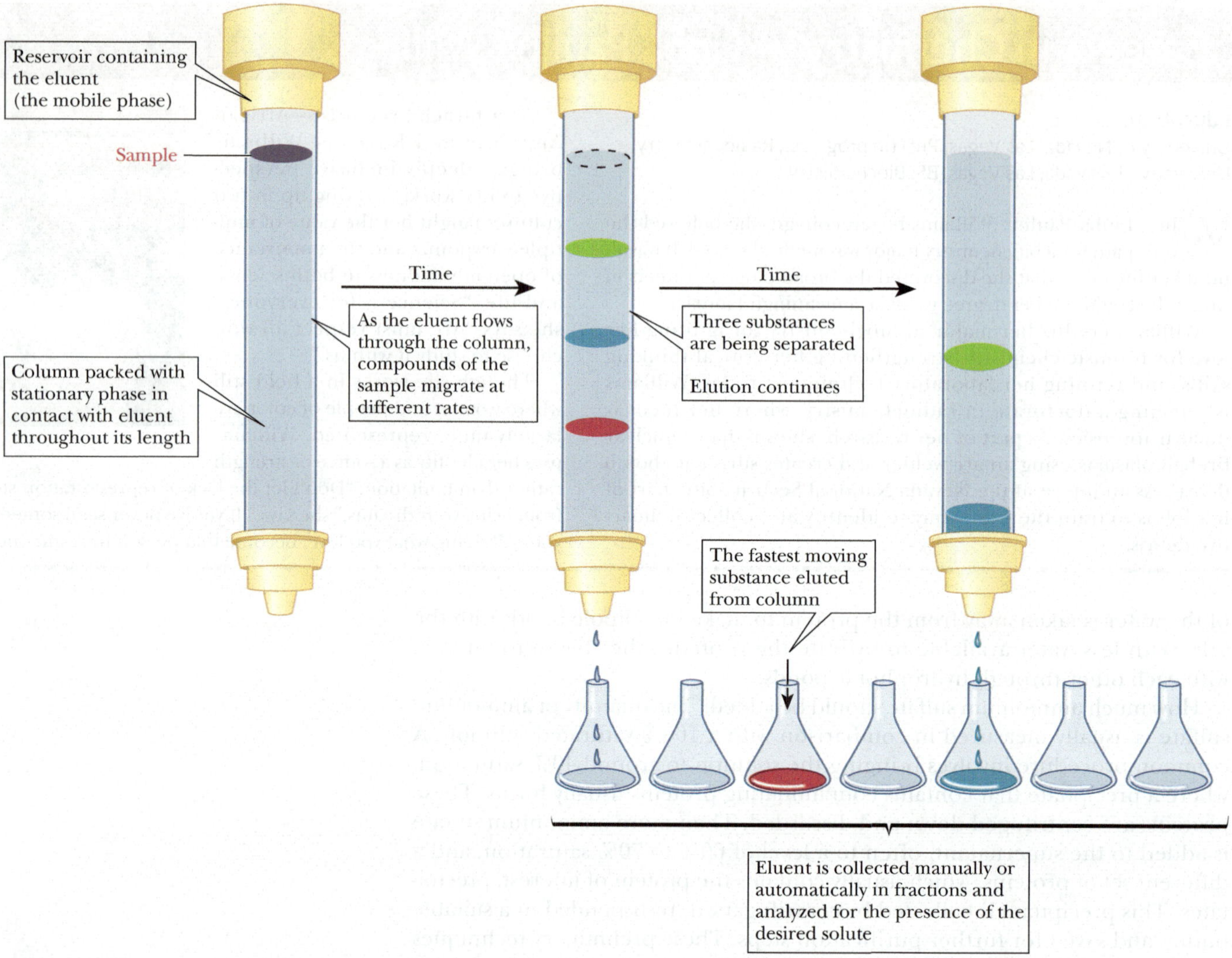

Figure 5.2 **Column chromatography.** A sample containing several components is applied to the column. The various components travel at different rates and can be collected individually.

column chromatography a form of chromatography in which the stationary phase is packed in a column

fractions in chromatography, these are the vessels that gradually capture fluid as it elutes from the column

gel-filtration chromatography a type of column chromatography in which the molecules are separated according to size as they pass through the column

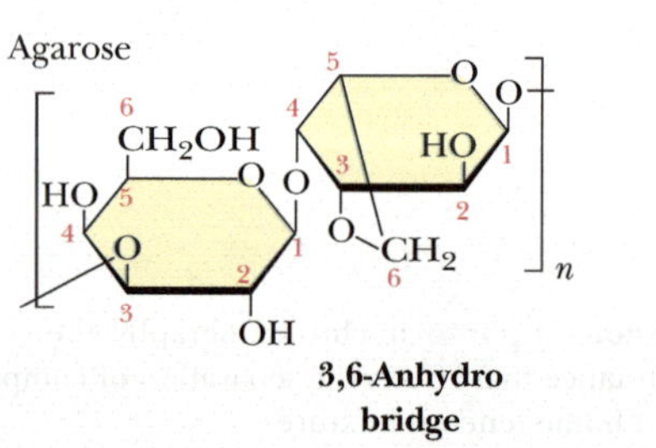

Figure 5.3 The repeating disaccharide unit of agarose that is used for column chromatography.

Many chromatographic techniques used for research on proteins are forms of **column chromatography**, in which the material that makes up the stationary phase is packed in a column (Figure 5.2). The sample is a small volume of concentrated solution that is applied to the top of the column. Then the mobile phase, called the *eluent,* is passed through the column. The entire sample eventually comes off the column, but the different compounds in the sample end up in different **fractions** depending on their differing mobilities through the stationary phase.

▶ **What are the different types of chromatography?**

5-2A Size-Exclusion Chromatography

Gel-filtration chromatography, also called size-exclusion chromatography, separates molecules on the basis of size, making it a useful way to sort proteins of varied molecular weights. It is a form of column chromatography in which the stationary phase consists of matrices consisting of **dextrans, agarose** (Figure 5.3), **polyacrylamide** (Figure 5.4), or combinations of these. The structures of these polymers produce pores in the material, and the extent of cross-linking can be controlled to select a desired pore size. When a sample is applied to the column, smaller molecules, which are able to enter the pores, tend to be delayed in their progress down the column, unlike the larger molecules. As a result, the larger molecules are eluted first, followed later by the smaller ones after they escape from the pores.

Gel-filtration chromatography is represented schematically in Figure 5.5A. The advantages of this type of chromatography are (1) its convenience as a way to separate molecules on the basis of size and (2) the fact that it can be used to estimate molecular weight by comparing the sample with a set of standards. Each type of gel used has a specific range of sizes that separate linearly with the log of the molecular weight. A typical gel-filtration elution profile is shown in Figure 5.5B. Each gel has an **exclusion limit**, which indicates a size of protein that is too large to enter the pores. All proteins of that size or larger elute first and simultaneously in the **void volume**.

5-2B Affinity Chromatography

Affinity chromatography uses the specific binding properties of proteins (Figure 5.6). It is another form of column chromatography with a polymeric material used as the stationary phase, but in this case, the polymer is not cross-linked and does not separate based on size. The distinguishing feature of affinity

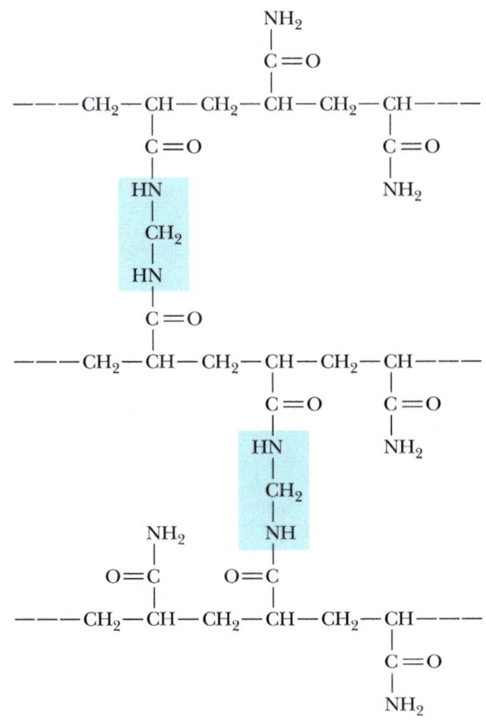

Figure 5.4 The structure of cross-linked polyacrylamide.

dextran a complex polysaccharide that is often used in column chromatography resins

agarose a complex polysaccharide used to make up resins for use in electrophoresis and column chromatography

polyacrylamide a synthetic polymer containing amide substituents on alternating carbons that can serve as a sieve or a supporting medium

exclusion limit a value for a size-exclusion resin that indicates a size of protein that is too large to enter the pores of the resin

void volume in gel-filtration chromatography, the volume of mobile phase required to elute molecules that do not enter the pores of the resin

affinity chromatography a powerful column separation procedure based on specific binding of molecules to a ligand

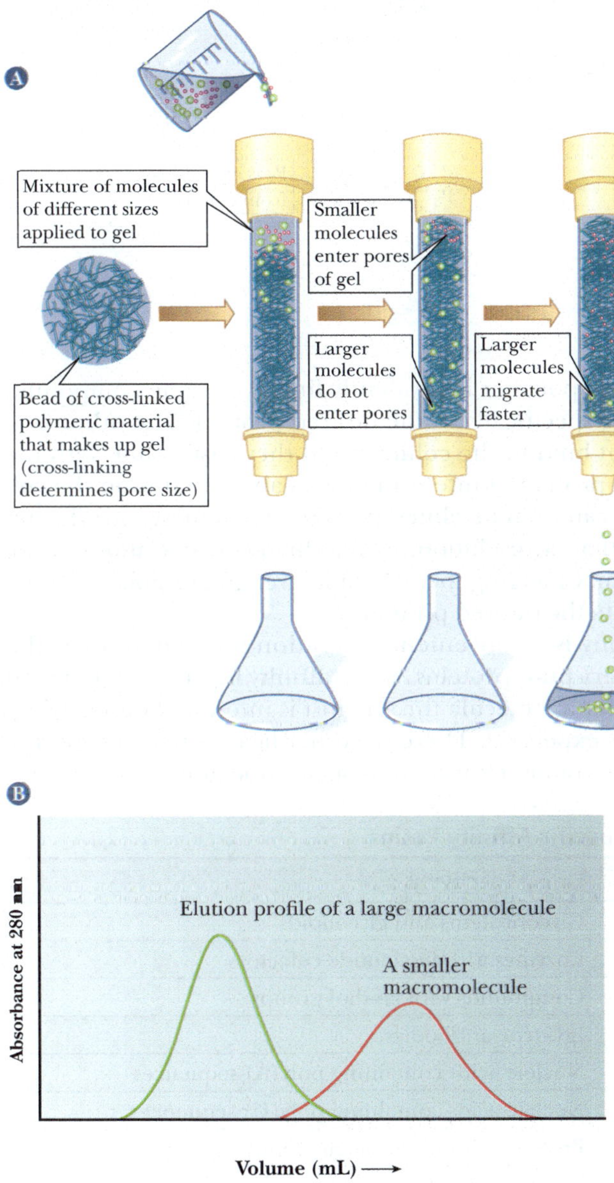

Figure 5.5 **Gel-filtration chromatography.** (A) Larger molecules are excluded from the gel and move more quickly through the column. Small molecules have access to the interior of the gel beads, so they take a longer time to elute. (B) In column chromatography used to separate proteins, the protein concentration is usually measured by UV absorption at 280 nm as the samples elute from the column.

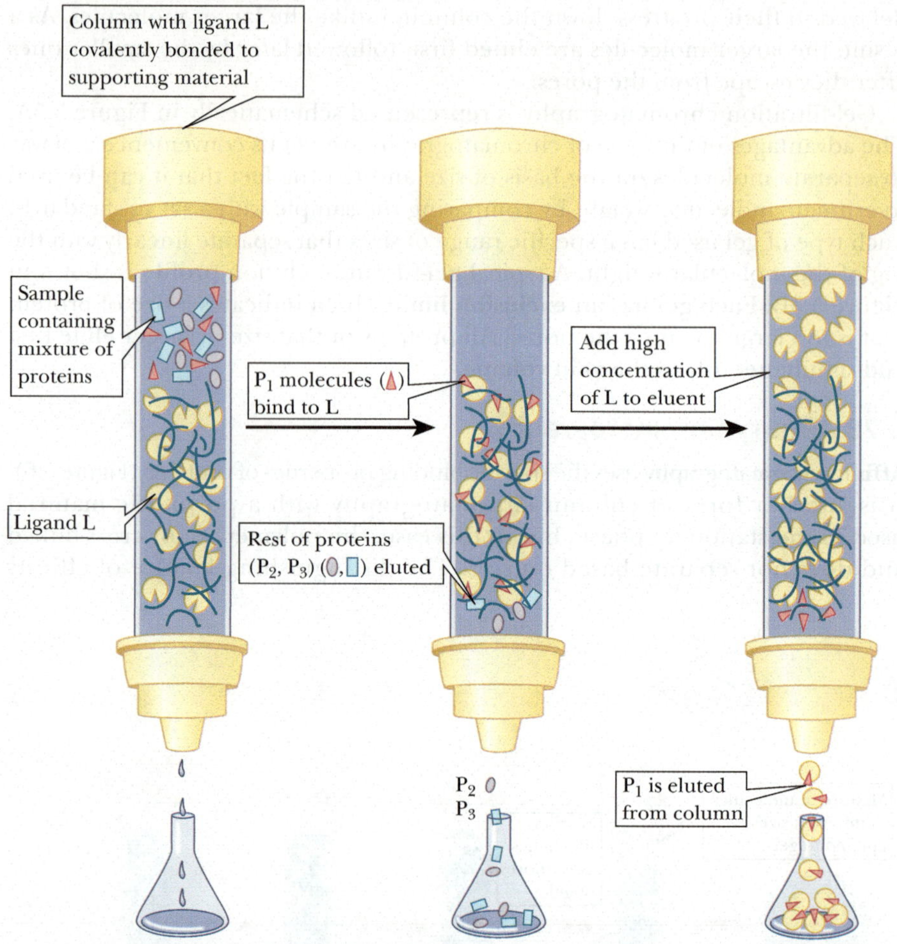

Figure 5.6 The principle of affinity chromatography. In a mixture of proteins, only one (designated P_1) binds to a ligand (L). The ligand is covalently attached to the column matrix. Once the other proteins (P_2 and P_3) have eluted, P_1 can be eluted by adding free L.

ligand a substance that binds to a biomolecule to serve a biological purpose

chromatography is that the polymer is covalently linked to some compound, called a **ligand**, that binds specifically to the desired protein. The other proteins in the sample do not bind to the column, and they easily elute with the mobile phase buffer. The protein of interest binds to and remains on the column. The bound protein can then be eluted from the column by introducing mobile-phase conditions that favor elution, such as high concentrations of the ligand in free, soluble form. Changing the pH, ionic strength, or polarity of the mobile phase can also elute the desired protein.

Affinity chromatography is a convenient separation method and has the advantage of producing very pure proteins. Some affinity ligands are designed to be completely specific for a molecule the scientist is interested in purifying. However, this is often very expensive. There are other ligands that are specific for groups of compounds. Table 5.2 lists some group-specific affinity resins.

Table 5.2 Group-Specific Affinity Resins

Group-Specific Adsorbent	Group Specificity
Concanavalin A–agarose	Glycoproteins and glycolipids
Cibacron Blue–agarose	Enzymes with nucleotide cofactors
Boronic acid–agarose	Compounds with cis-diol groups
Protein A–agarose	IgG-type antibodies
Poly(U)–agarose	Nucleic acids containing poly(A) sequences
Poly(A)–agarose	Nucleic acids containing poly(U) sequences
Iminodiacetate–agarose	Proteins with heavy metal affinity
AMP–agarose	Enzymes with NAD^+ cofactors, ATP-dependent kinases

BIOCHEMICAL CONNECTIONS 13C describes an interesting way in which affinity chromatography can be combined with molecular biological techniques to offer a one-step purification of a protein.

5-2C Ion-Exchange Chromatography

Ion-exchange chromatography is logistically similar to affinity chromatography. Both use a column resin that binds the protein of interest. With ion-exchange chromatography, however, the interaction is less specific, and separation is based on net charge. Ion-exchange resins have groups with a positive charge or a negative charge (Figure 5.7). **Cation-exchange resins** are *negatively* charged so they can exchange *positively* charged species, whereas **anion-exchange resins** are *positively* charged so they can exchange *negatively* charged species.

Figure 5.8 shows an example of cation-exchange chromatography at pH 7. The resin initially contains negatively charged—SO_3^- groups and Na^+ counterions (Figure 5.8A). Figure 5.8B shows a mixture of aspartate (Asp), serine (Ser), and lysine (Lys) amino acids added to the column (to review the structures of these amino acids, see Figure 3.3). Each amino acid has a positively charged amino group, so each can bind to the negatively charged cation-exchange resin and displace the Na^+ counterions. To elute and separate the amino acids from each other, the concentration of Na^+ can gradually be increased with an NaCl **gradient**. Because Asp is the least positively charged amino acid in the mixture, it will elute at low concentrations of Na^+, followed by Ser at higher concentrations of Na^+ (Figures 5.8C and 5.8D).

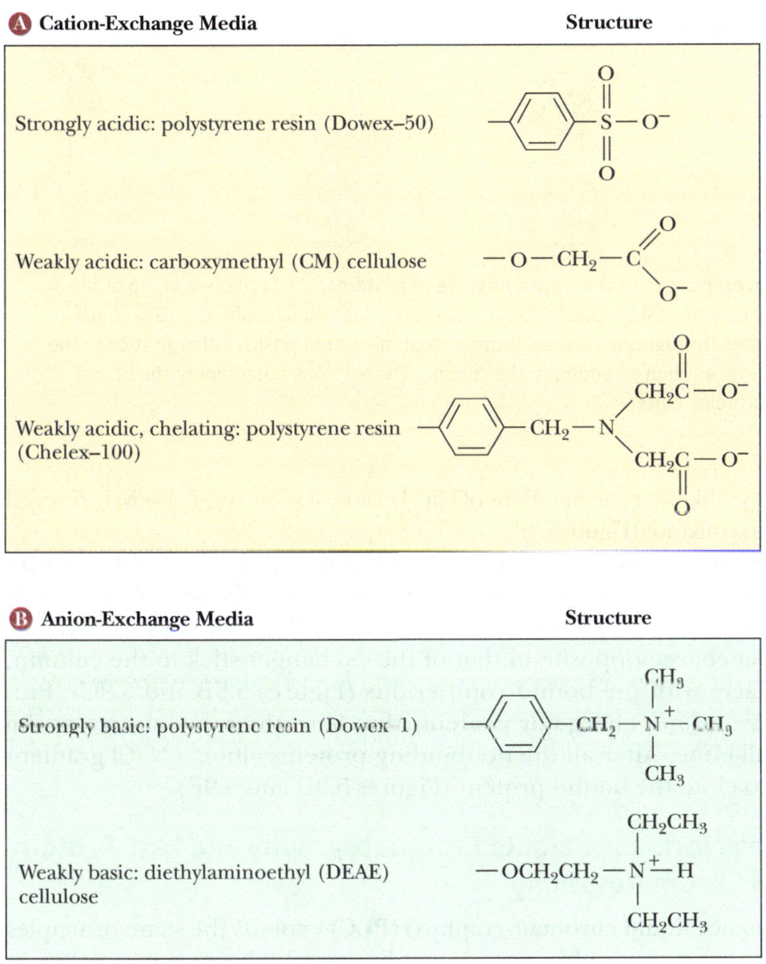

Figure 5.7 **Resins used in ion-exchange chromatography.** (A) Cation-exchange resins and (B) anion-exchange resins commonly used for biochemical separations.

ion-exchange chromatography a method for separating substances on the basis of charge

cation-exchange resin a type of ion-exchange resin that has a negative charge and binds to positively charged molecules flowing through the column

anion-exchange resin a type of ion-exchange resin that has a positive charge and binds to negatively charged molecules flowing through it

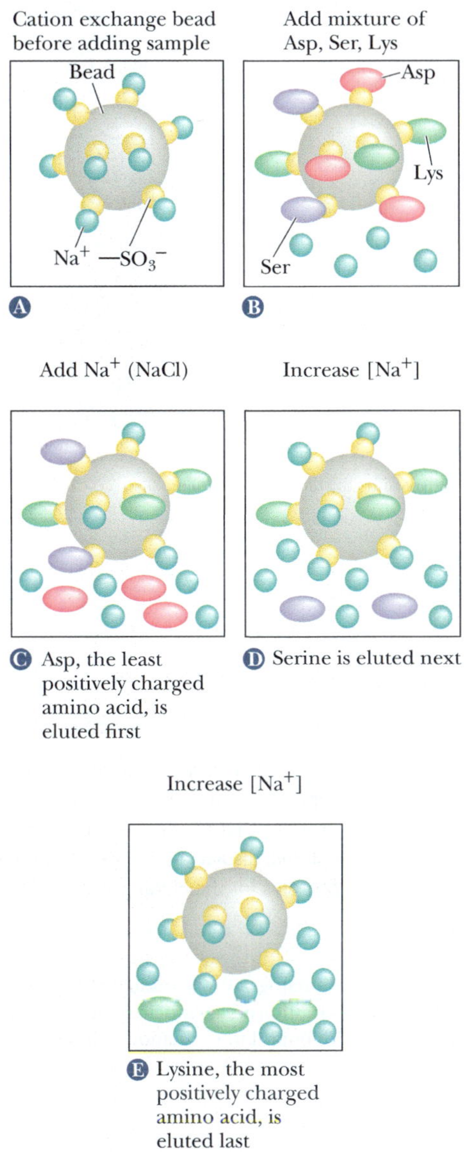

Figure 5.8 **Operation of a cation-exchange column at pH 7 to separate a mixture of aspartate, serine, and lysine amino acids.** (A) The cation-exchange resin before the sample is added. (B) A mixture of aspartate, serine, and lysine is added to the column, and the amino acids displace the Na^+ counterions. (C) A gradient of the eluting salt (e.g., NaCl) is added to the column. Aspartate, the least positively charged amino acid, is eluted first. (D) As the salt concentration increases, serine is eluted. (E) As the salt concentration is increased further, lysine, the most positively charged of the three amino acids, is eluted last.

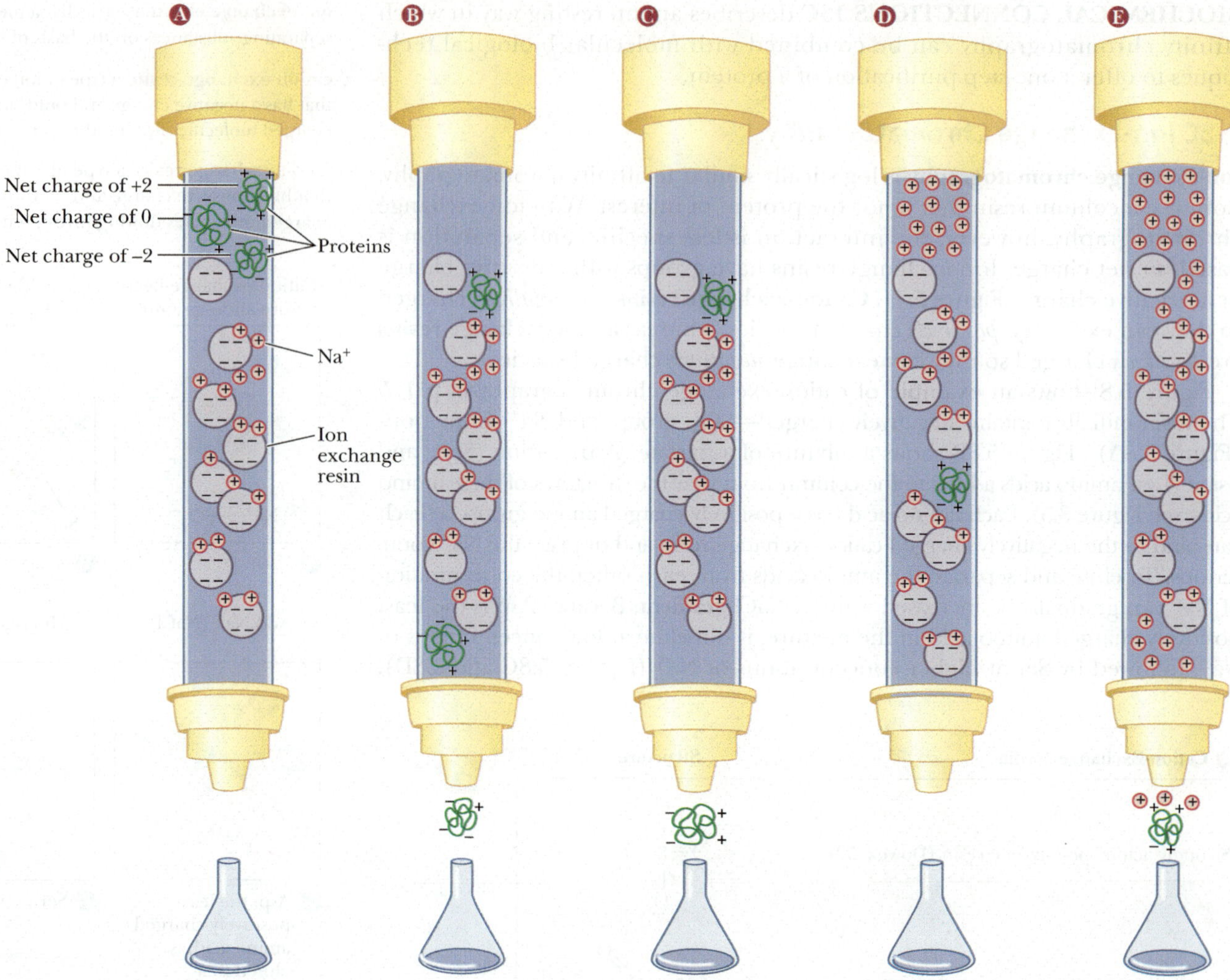

Figure 5.9 Ion-exchange chromatography using a cation exchanger to separate a mixture of proteins. (A) At the beginning of the separation, various proteins are applied to the column. The column resin is bound to Na^+ counterions (small red spheres). (B) and (C) Proteins that have no net charge or a net negative charge pass through the column. Proteins that have a net positive charge stick to the column, displacing the Na^+. (D) and (E) A gradient of Na^+ ions can then be added to the column. The Na^+ ions outcompete the bound proteins for the binding sites on the resin, and the bound proteins elute.

gradient in chromatography, a gradual change in mobile-phase conditions that causes the stepwise release of species bound to the stationary phase

Lys will elute at yet higher concentrations of Na^+ because it is the most positively charged amino acid in the mixture (Figure 5.8E).

Figure 5.9 shows how cation-exchange chromatography would separate a mixture of proteins with different net charges. The mixture of proteins is loaded on the column and allowed to flow through it (Figures 5.9A–C). Proteins that have a net charge opposite to that of the exchanger stick to the column, *exchanging* places with the bound counterions (Figures 5.9B and 5.9C). Proteins that have no net charge or proteins that have the same charge as the exchanger will elute. After all the nonbinding proteins elute, a NaCl gradient can be used to elute the bound protein (Figures 5.9D and 5.9E).

5-2D High-Performance Liquid Chromatography and Fast-Protein Liquid Chromatography

high-performance liquid chromatography (HPLC) a sophisticated chromatography technique that uses high pressure and high-resolution columns to achieve fast and clean purifications

High-performance liquid chromatography (HPLC) exploits the same principles seen in other chromatographic techniques, but very high-resolution columns that can be run under high pressures are used. High-resolution separations of small biochemical compounds can be effected very quickly using automated instrumentation. A separation that might take hours on a standard column can be done in minutes with HPLC. The separation achieved with **reverse-phase HPLC**, the most widely used HPLC technique, is based on the different hydrophobicities of the compounds in the sample. In reverse-phase HPLC, a nonpolar stationary phase and a more polar mobile phase are used to effect the separation. The less hydrophobic molecules elute first, while molecules that are more hydrophobic interact with the hydrophobic resin and elute more slowly.

reverse-phase HPLC a form of high-performance liquid chromatography in which the stationary phase is nonpolar and the mobile phase is a more polar liquid

Fast-protein liquid chromatography (FPLC) can also be used for fast separations of larger biochemical molecules such as proteins. Size-exclusion, ion-exchange, and affinity columns are commonly used in FPLC. With FPLC, all of the components are polar and, therefore, safer to use with proteins. APPLY YOUR KNOWLEDGE 5.1 shows a purification process involving several steps.

fast-protein liquid chromatography (FPLC) a specialized chromatography technique that uses lower pressures than HPLC but uses broader ranges of columns for protein purifications

5.1 Apply Your Knowledge

Protein Purification

The following table shows some typical results for a protein purification. The protein being purified is the enzyme lactate dehydrogenase, which catalyzes a reaction between lactic acid and NAD^+ to give NADH and pyruvate. The purification of an enzyme is monitored by comparing the specific activity of the enzyme at various points in the purification. The specific activity is a measure of the activity of the enzyme divided by the weight of protein in the sample. The higher the number, the purer the sample.

Purification Step	Total Activity of Enzyme (μmol product/min)	Specific Activity of Enzyme (μmol product/min/mg protein)
Crude homogenate	100,000	0.15
20,000 × *g* supernatant	75,000	0.24
Salt precipitation	36,000	0.75
Ion-exchange chromatography	12,000	3.4
Cibacron blue agarose affinity chromatography	6,000	42
Sephadex gel-filtration chromatography	500	90

Which step was the most effective in purifying the enzyme? Which step was the most costly in terms of enzyme recovery?

Solution

If we divide the specific activity of any purification step (fraction) by the one before it, we get what is called the "fold purification" for that step.

Continued on next page

The larger the number, the more effective that step was. Adding that calculation to the table gives the following results:

Purification Step	Total Activity of Enzyme (μmol product/min)	Specific Activity of Enzyme(μmol product/min/mg protein)	Fold Purification
Crude homogenate	100,000	0.15	n/a
20,000 × *g* supernatant	75,000	0.24	1.6
Salt precipitation	36,000	0.75	3.1
Ion-exchange chromatography	12,000	3.4	4.5
Cibacron blue agarose affinity chromatography	6,000	42	12.4
Sephadex gel-filtration chromatography	500	90	2.1

The results show that the affinity chromatography step gave the highest single purification of 12.4 for a single step. This is often the case due to the power of this technique.

Similar calculations can show which step was the most costly by comparing the total activity of each fraction. The last step of the purification, the gel-filtration chromatography, caused the loss of over 90% of the activity that was applied to it. This was the most costly single step. Scientists doing a purification must weigh the benefits and costs of each step, comparing the extent of purification achieved versus the amount of product lost.

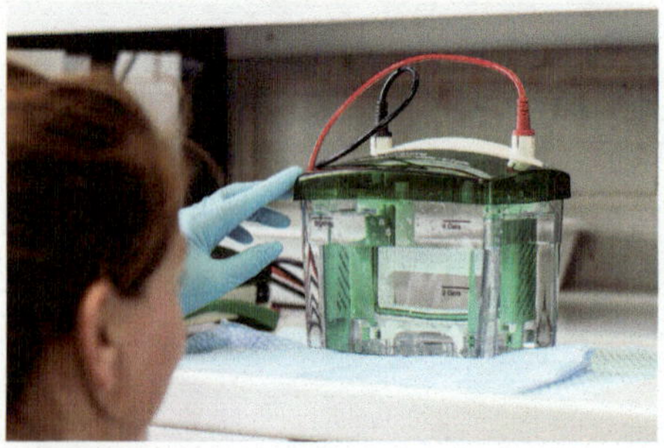

Figure 5.10 An experimental setup for protein electrophoresis. The samples are placed on the top of the gel. When the current is applied, the negatively charged proteins migrate toward the positive electrode at the bottom.

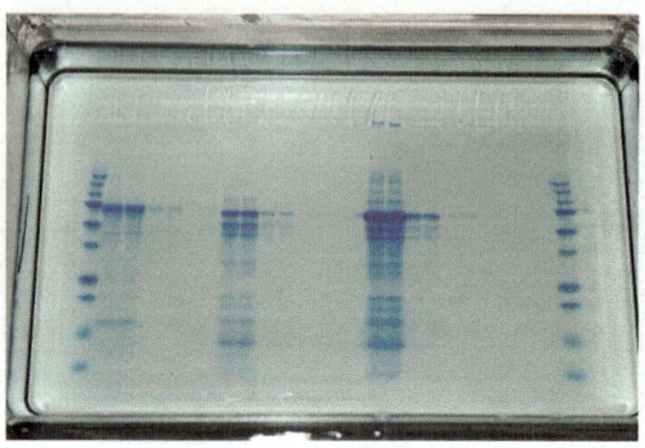

Figure 5.11 Separation of proteins by gel electrophoresis. Each band seen in the gel represents a different protein. In the SDS–PAGE technique, the sample is treated with detergent before being applied to the gel. After running the gel, the gel and the proteins in it are stained with Coomassie Blue.

Once a protein is purified, a variety of techniques are used to analyze it. Several of these techniques are discussed in Sections 5-3 through 5-5.

5-3 Polyacrylamide Gel Electrophoresis

Learning Objectives

After reading this section, students will be able to:

LO 5.3.1 Demonstrate when to use the appropriate electrophoresis support media based on knowledge of structural differences in the support media.

LO 5.3.2 Distinguish between SDS–PAGE, native gel, isoelectric focusing, and 2-D gel electrophoresis.

electrophoresis a method for separating molecules on the basis of the ratio of charge to size

Electrophoresis is a separation method based on the motion of charged particles in an electric field toward an electrode of opposite charge (Figure 5.10). Macromolecules have differing mobilities in electrophoresis based on their charge, shape, and size. Although many supporting media have been used for electrophoresis, including paper and liquids, the most common support is a polymer of agarose or acrylamide that is similar to those used for column chromatography. A sample is applied to wells that are formed in the supporting medium. An electric current is passed through the medium at a controlled voltage to achieve the desired separation. After the proteins are separated on the gel, the gel is stained to reveal the protein locations, as shown in Figure 5.11.

Agarose-based gels are most often used to separate nucleic acids and are discussed in Chapter 13. For proteins, the most common electrophoretic support is polyacrylamide (Figure 5.4). A polyacrylamide gel is prepared and cast as a continuous cross-linked matrix rather than being produced in the bead form employed in column chromatography.

5-3A SDS–PAGE Versus Native Gel Electrophoresis

In one variation of polyacrylamide gel electrophoresis, called **SDS–polyacrylamide gel electrophoresis (SDS–PAGE)**, the protein sample is treated with the detergent sodium dodecyl sulfate (SDS) before it is applied to the gel. The formula of SDS is $CH_3(CH_2)_{10}CH_2OSO_3Na$, so it contains a dodecyl sulfate anion and an Na^+ cation. The dodecyl sulfate anion binds strongly to proteins via nonspecific adsorption. The larger the protein, the more of the anion it adsorbs. SDS completely denatures proteins, breaking all the noncovalent interactions that determine tertiary and quaternary structure. This means that multisubunit proteins can be analyzed as the component polypeptide chains. All the proteins in a sample have a negative charge as a result of adsorption of the anionic $—OSO_3^-$ group. The proteins have roughly the same shape, too, which is a random coil.

SDS–polyacrylamide gel electrophoresis (SDS–PAGE) an electrophoretic technique that separates proteins on the basis of size

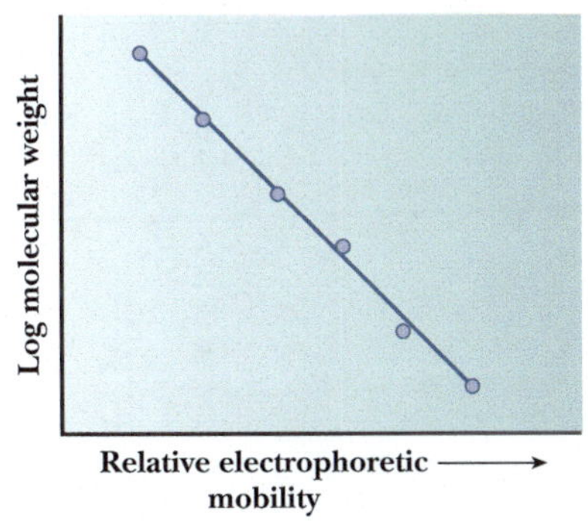

Figure 5.12 Relationship between molecular weight and mobility. A plot of the relative electrophoretic mobility of proteins in SDS–PAGE versus the log of the molecular weights of the individual polypeptides approximates a straight line.

In SDS–polyacrylamide-gel electrophoresis (SDS–PAGE), the acrylamide offers more resistance to large molecules than to small molecules. Because the shape and charge-to-size ratio are approximately the same for all the proteins in the sample, the size of the protein becomes the determining factor in the separation: Small proteins move faster than large ones. Like molecular-sieve chromatography, SDS–PAGE can be used to estimate the molecular weights of proteins by comparing the sample with standard samples. For most proteins, the log of the molecular weight is linearly related to their mobility on SDS–PAGE, as shown in Figure 5.12.

Proteins can also be separated on acrylamide without the SDS, in which case the gel is called a **native gel**. This is useful for times when the study calls for a protein in its native conformation instead of a random coil. In this case, however, the mobility is not correlated with size specifically because three variables control the movement down the gel—size, shape, and charge.

native gel a polyacrylamide electrophoresis gel without SDS or any other compound that would denature the proteins being separated

5-3B Isoelectric Focusing

Isoelectric focusing is a variation of native gel electrophoresis. Because different proteins have different numbers and types of titratable residues, they also have different isoelectric points. Recall from Section 3-3B that the isoelectric point (pI) is the pH at which a protein (or amino acid or peptide) has no net charge. At the pI, the number of positive charges exactly balances the number of negative charges. In an isoelectric focusing experiment, a native gel is prepared with a pH gradient that parallels the electric field gradient. As proteins migrate through the gel under the influence of the electric field, they encounter regions of different pH, so the charge on the protein changes. Eventually each protein reaches the point at which it has no net charge—its isoelectric point—and it no longer migrates. Each protein remains at the position on the gel corresponding to its pI, allowing for an effective method of separation.

isoelectric focusing a method for separating substances on the basis of their isoelectric points

A combination known as two-dimensional gel electrophoresis (2-D gels) allows for enhanced separation by first using isoelectric focusing in one dimension and then using SDS–PAGE run at 90° to the first gel (Figure 5.13).

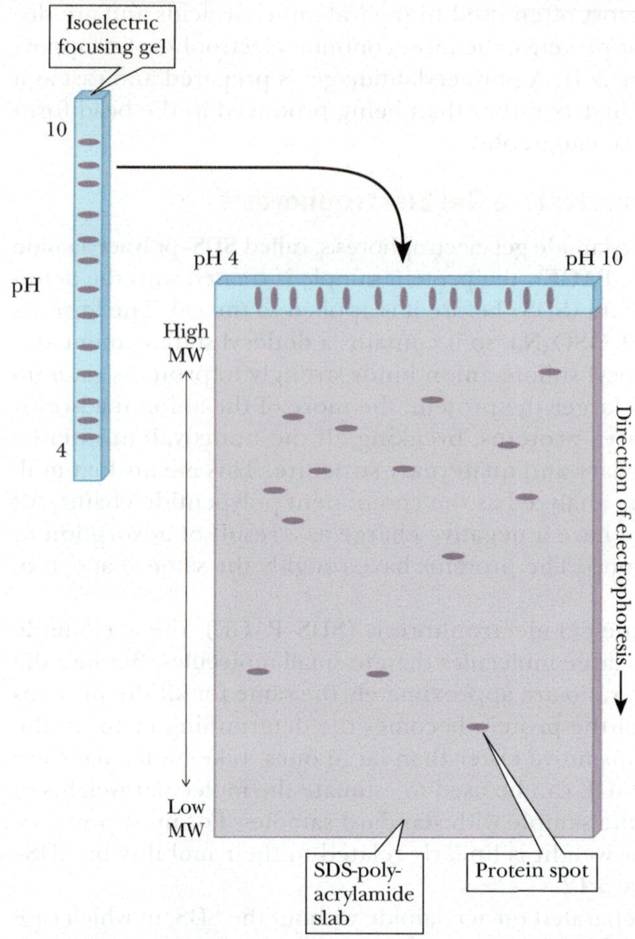

Figure 5.13 Two-dimensional electrophoresis. A mixture of proteins is separated by isoelectric focusing in one direction. The focused proteins are then run using SDS–PAGE perpendicular to the direction of the isoelectric focusing. Thus, the bands that appear on the gel have been separated first by their isoelectric points and then by size.

5-4 Determining the Primary Structure of a Protein

Learning Objectives

After reading this section, students will be able to:

LO 5.4.1 Describe the steps involved in determining the primary structure of a protein.

LO 5.4.2 Distinguish between the different proteolytic enzymes and reagents that can be used to cleave a protein into fragments.

LO 5.4.3 Describe the steps in the Edman degradation method.

Determining the primary structure, or sequence, of amino acids in a protein is a routine, but not trivial, operation in classical biochemistry. Its several parts must be carried out carefully to obtain accurate results (Figure 5.14).

5-4A Overview of the Steps Involved in Determining the Primary Structure of a Protein

Step 1 in determining the primary structure of a protein is to establish which amino acids are present and in what proportions. Breaking a protein down into its component amino acids is relatively easy: Heat a solution of the protein in acid, usually 6 *M* HCl, at 100°C to 110°C for 12 to 36 hours to hydrolyze the peptide bonds. Separation and identification of the products are somewhat

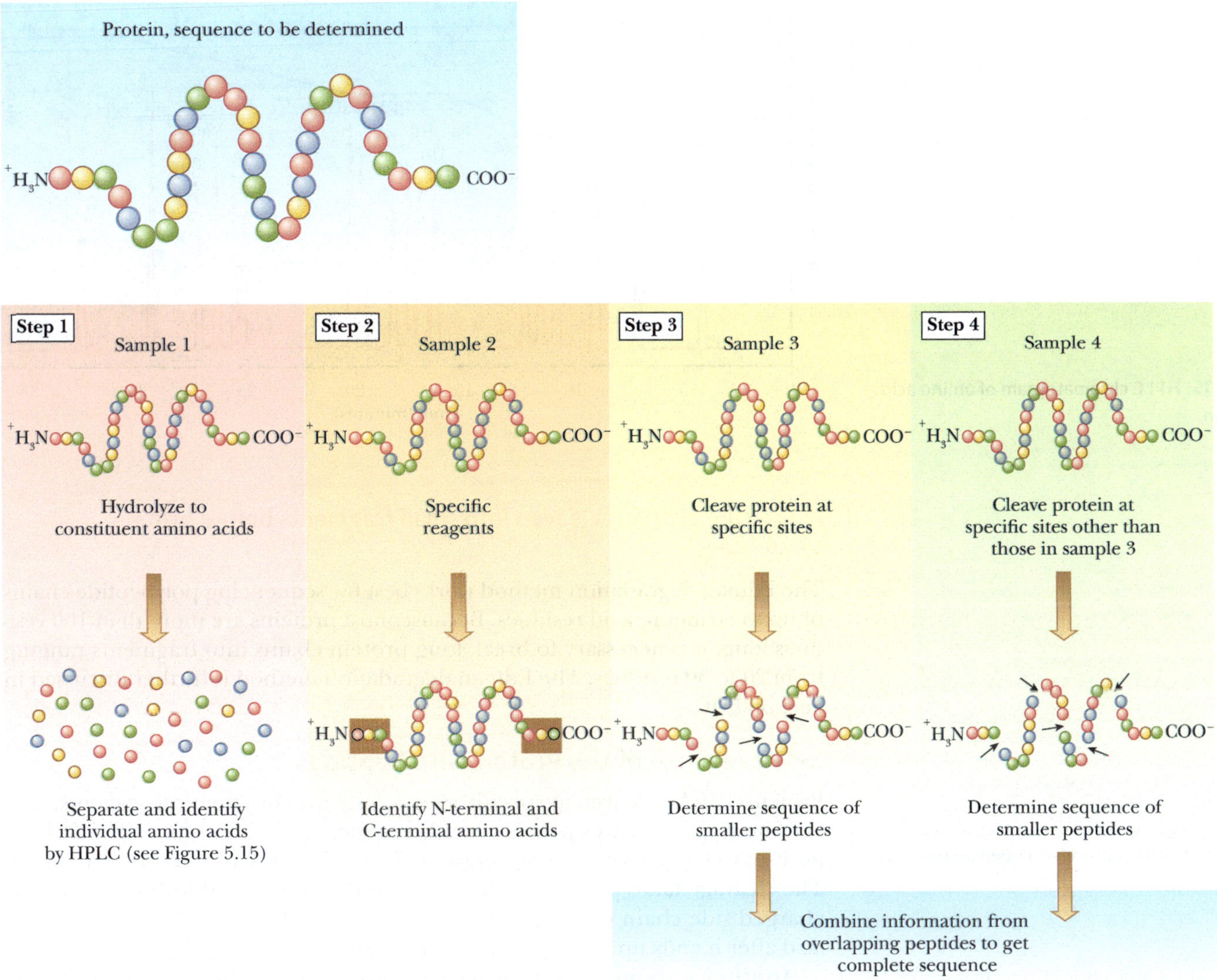

Figure 5.14 The strategy for determining the primary structure of a given protein. The amino acid can be determined by four different analyses performed on four separate samples of the same protein.

more difficult and are best done by an amino acid analyzer. This automated instrument gives both qualitative information about the identities of the amino acids present and quantitative information about the relative amounts of those amino acids. Not only does it analyze amino acids, but it also allows informed decisions to be made about which procedures to choose later in the sequencing (see Steps 3 and 4 in Figure 5.14). An amino acid analyzer separates the mixture of amino acids by either ion-exchange chromatography or HPLC. Figure 5.15 shows a typical result of amino acid separation with this technique.

In Step 2, the identities of the N-terminal and C-terminal amino acids in a protein sequence are determined. This procedure is becoming less and less necessary as the sequencing of individual peptides improves, but it can be used to check whether a protein consists of one or two polypeptide chains.

In Steps 3 and 4, the protein is cleaved into smaller fragments (further discussed in Section 5-4B), and the amino acid sequence is determined. Automated instruments can perform a stepwise modification starting from the N-terminal end, followed by cleavage of each amino acid in the sequence and the subsequent identification of each modified amino acid as it is removed. This process is called the **Edman degradation**.

Edman degradation a method for determining the amino acid sequence of peptides and proteins

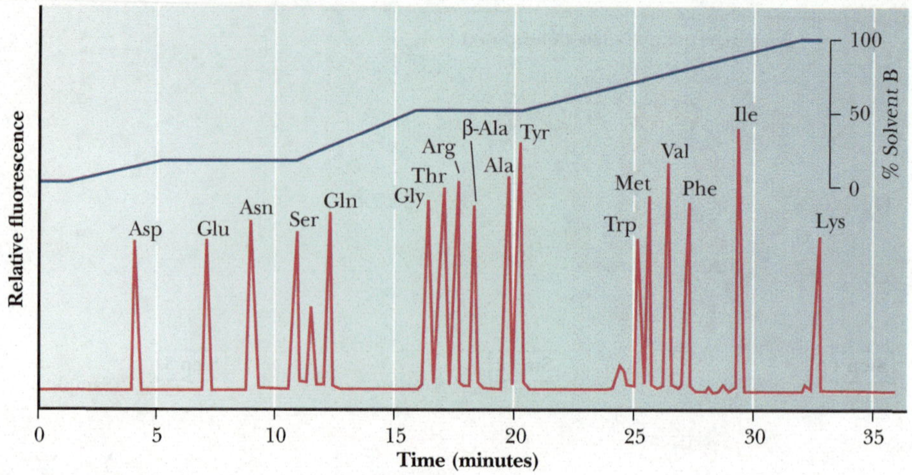

Figure 5.15 HPLC chromatogram of amino acid separation.

▶ Why are proteins cleaved into small fragments before protein sequencing?

The Edman degradation method works best for sequencing polypeptide chains of up to 50 amino acid residues. Because most proteins are more than 100 residues long, it is necessary to break long protein chains into fragments ranging from 20 to 50 residues. The Edman degradation method is further discussed in Section 5-4C.

5-4B Cleavage of the Protein into Peptides

trypsin a proteolytic enzyme that hydrolyzes peptide bonds after positively charged amino acid residues

chymotrypsin a proteolytic enzyme that hydrolyzes peptide bonds after aromatic amino acid residues

cyanogen bromide a reagent that cleaves peptide bonds after methionine residues

Proteins can be cleaved at specific sites by enzymes or chemical reagents. The enzyme **trypsin** cleaves peptide bonds that occur after amino acids that have positively charged side chain groups, such as lysine and arginine (Figure 5.16). The cleavage takes place in such a way that the amino acid with the positively charged side chain ends up with a new C-terminal end, and the next amino acid after it ends up with a new N-terminal end.

Another enzyme, **chymotrypsin**, cleaves peptide bonds preferentially after aromatic amino acids such as tyrosine, tryptophan, and phenylalanine (Figure 5.17). New termini are produced as a result of the cleavage, as seen in the trypsin reaction.

In the case of the chemical reagent **cyanogen bromide** (CNBr), the sites of cleavage are peptide bonds after methionine residues (Figure 5.18). The sulfur of the methionine reacts with the carbon of the cyanogen bromide to produce a homoserine lactone at the C-terminal end of the fragment.

The cleavage of a protein by any of these reagents (trypsin, chymotrypsin, or cyanogen bromide) produces a mixture of peptides, which are then separated by HPLC. The use of several such reagents on different samples of a protein to be sequenced produces different peptide mixtures. The sequences of a set of peptides produced by one reagent overlap with the sequences produced by another reagent (Figure 5.19). As a result, the peptides can be arranged in the proper order after their own sequences have been determined.

5-4C Edman Degradation

The actual sequencing of each peptide produced by specific cleavage of a protein is accomplished by repeated application of the Edman degradation method (Figure 5.20). The Edman reagent, *phenyl isothiocyanate,* reacts with the peptide's N-terminal residue. The modified amino acid can be cleaved off, *leaving the rest of the peptide intact,* and can be detected as the phenylthiohydantoin (PTH) derivative of the amino acid. The second amino acid of the original peptide

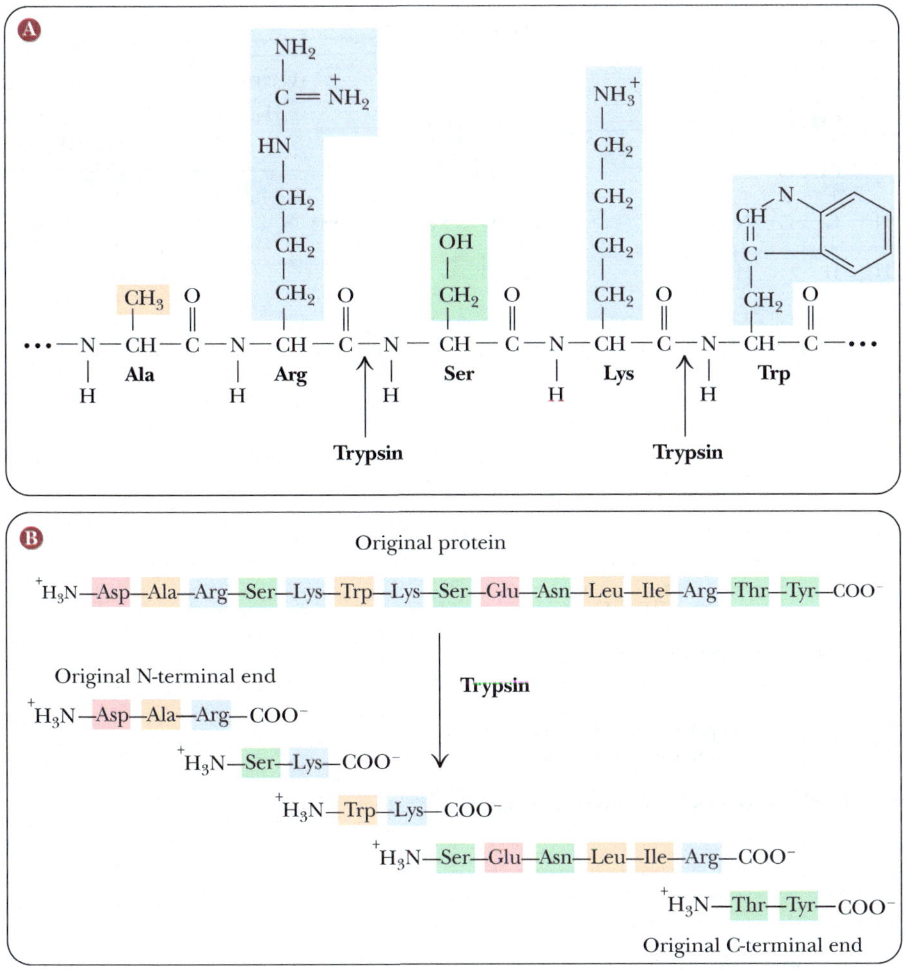

Figure 5.16 Peptide digestion with trypsin. (A) Trypsin is a proteolytic enzyme, or protease, that specifically cleaves peptide bonds after positively charged arginine or lysine residues. (B) The products of the reaction are a mixture of peptide fragments with C-terminal Arg or Lys residues and a single peptide derived from the original polypeptide's C-terminal end.

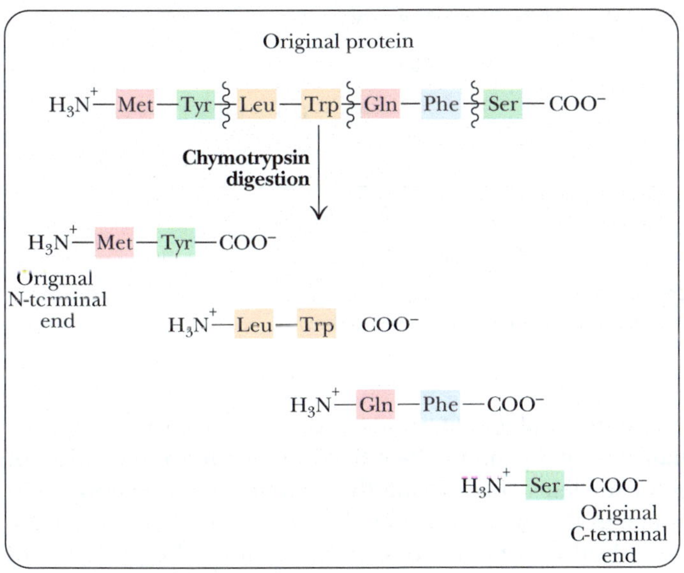

Figure 5.17 Cleavage of proteins by chymotrypsin. Chymotrypsin hydrolyzes peptide bonds after aromatic amino acids.

can then be treated in the same way, as can the third, and so on. Using an automated instrument, the process is repeated until the whole peptide is sequenced.

Another sequencing method uses the fact that the amino acid sequence of a protein reflects the base sequence of the DNA in the gene that coded for

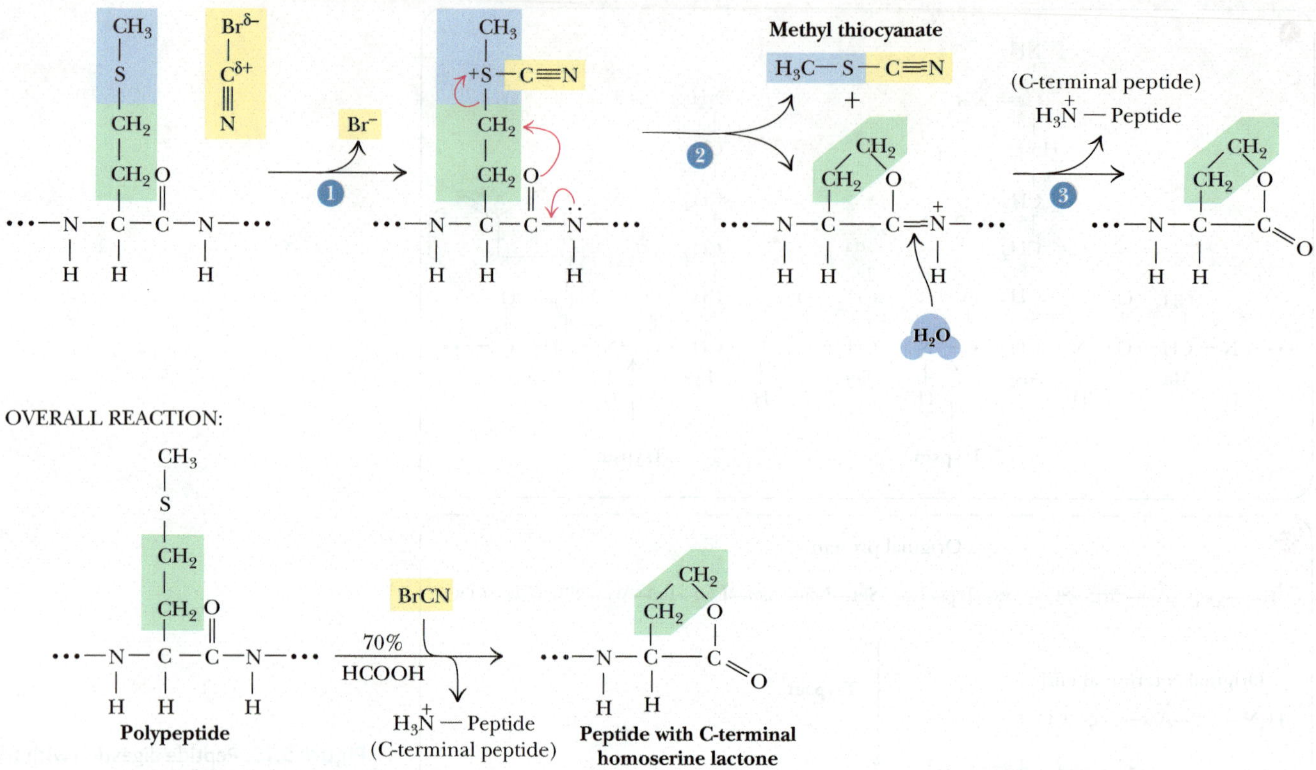

Figure 5.18 **Cleavage of proteins at internal methionine residues by cyanogen bromide.**

Chymotrypsin	$H_3\overset{+}{N}$—Leu—Asn—Asp—Phe
Cyanogen bromide	$H_3\overset{+}{N}$—Leu—Asn—Asp—Phe—His—Met
Chymotrypsin	His—Met—Thr—Met—Ala—Trp
Cyanogen bromide	Thr—Met
Cyanogen bromide	Ala—Trp—Val—Lys—COO^-
Chymotrypsin	Val—Lys—COO^-
Overall sequence	$H_3\overset{+}{N}$—Leu—Asn—Asp—Phe—His—Met—Thr—Met—Ala—Trp—Val—Lys—COO^-

Figure 5.19 **Use of overlapping sequences to determine protein sequence.** Partial digestion was effected using chymotrypsin and cyanogen bromide. For clarity, only the original N-terminus and C-terminus of the complete peptide are shown.

that protein. Using currently available methods, it is much easier to obtain the sequence of DNA than that of a protein. (See Section 13-8 for a discussion of sequencing methods for nucleic acids.) Using the genetic code (Section 12-2), the amino acid sequence of the protein can be determined immediately. Convenient though this method may be, it does not determine the positions of disulfide bonds or detect amino acids such as hydroxyproline that are modified after translation, nor does it take into account the extensive processing that occurs with eukaryotic genomes before the final protein is synthesized (Chapters 11 and 12). APPLY YOUR KNOWLEDGE 5.2 gives you an opportunity to practice analyzing peptide fragments and reconstructing the original peptide.

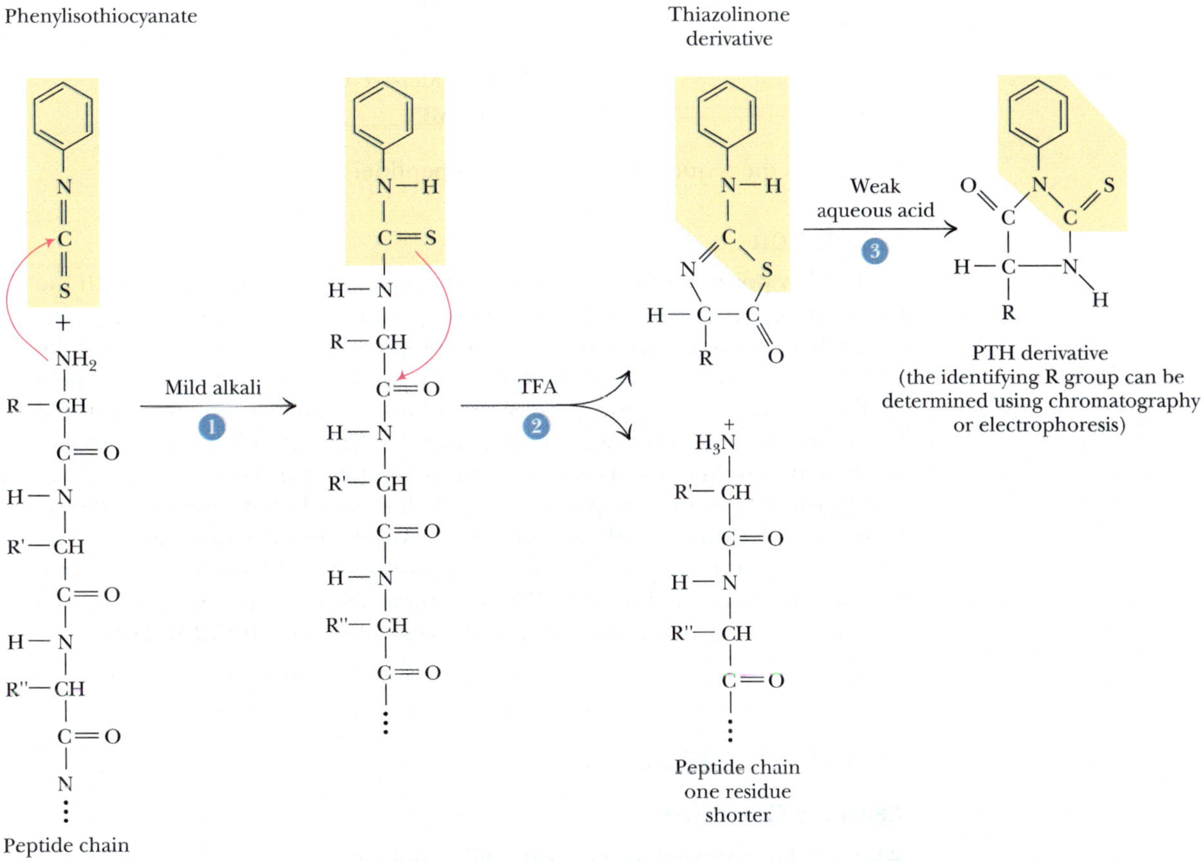

Figure 5.20 Sequencing of peptides by the Edman degradation method. (1) Phenylisothiocyanate combines with the N-terminus of a peptide under mildly alkaline conditions to form a phenylthiocarbamoyl substitution. (2) Upon treatment with trifluoroacetic acid (TFA), this cyclizes to release the N-terminal amino acid residue as a thiazolinone derivative, but the other peptide bonds are not hydrolyzed. (3) Organic extraction and treatment with aqueous acid yield the N-terminal amino acid as a phenylthiohydantoin (PTH) derivative, which can be detected in order to identify the amino acid. The process is repeated with the remainder of the peptide chain to determine the N-terminus exposed at each stage until the entire peptide is sequenced.

5.2 Apply Your Knowledge

Peptide Sequencing

A solution of a peptide of unknown sequence was divided into two samples. One sample was treated with trypsin, and the other was treated with chymotrypsin. The smaller peptides obtained by trypsin treatment had the following sequences:

Leu—Ser—Tyr—Ala—Ile—Arg
LSYAIR

and

Asp—Gly—Met—Phe—Val—Lys
DGMFVK

The smaller peptides obtained by chymotrypsin treatment had the following sequences:

Val—Lys—Leu—Ser—Tyr
VKLSY

Ala—Ile—arg
AIR

Continued on next page

and

Asp—Gly—Met—Phe
DGMF

What is the sequence of the original peptide?

Solution

The key point here is that the fragments produced by treatment with the two different enzymes have overlapping sequences. These overlapping sequences can be compared to give the complete sequence. The results of the trypsin treatment indicate that there are two basic amino acids in the peptide: arginine and lysine. One of them must be the C-terminal amino acid because no fragment was generated with a C-terminal amino acid other than these two. If there had been an amino acid other than a basic residue at the C-terminal position, trypsin treatment alone would have provided the sequence. Treatment with chymotrypsin gives the information needed. The sequence of the peptide Val—Lys—Leu—Ser—Tyr (VKLSY) indicates that lysine is an internal residue. The complete sequence is Asp—Gly—Met—Phe—Val—Lys—Leu—Ser—Tyr—Ala—Ile—Arg (DGMFVKLSYAIR).

5-5 Proteomics

Learning Objectives

After reading this section, students will be able to:

LO 5.5.1 Distinguish between different proteomics techniques such as mass spectrometry, ELISAs, western blots, and microarrays.

LO 5.5.2 Distinguish between different mass spectrometry techniques.

proteomics the field of study that involves the separation and analysis of proteins in an organism or specific parts of an organism

mass spectrometry (MS) a technique that separates molecular fragments according to their mass-to-charge ratio

electrospray ionization MS (ESI-MS) a form of mass spectrometry in which the sample is converted into fine droplets by spraying

Proteomics is the field of study that involves the separation and analysis of proteins in an organism or in specific parts of an organism being studied. **BIOCHEMICAL CONNECTIONS 5A** describes **mass spectrometry (MS)**, one of the most powerful, yet expensive, techniques for analyzing protein structure and dynamics. Mass spectrometry determines protein structure atom by atom.

5A Biochemical Connections • Instrumentation

The Power of Mass Spectrometry

In mass spectrometry (MS), the difference in the mass-to-charge ratio (*m*/*z*) of ionized atoms or molecules is used to separate them from each other. The *m*/*z* ratio is such a characteristic property that it can be used to obtain structural and chemical information about the molecules and identify them.

When the charged particles are separated on the basis of their *m*/*z* ratio, they arrive at the detector at different times. The original application of MS in the early twentieth century led to the discovery of isotopes. The isotopes of the noble gas argon were detected using apparatus that would seem quite simple now. For many years, the detection methods were based on having the substance to be analyzed in gaseous form or one that was easy to volatilize. As time went on, MS methods were developed that allow molecules as large as proteins, which are usually thought of as nonvolatile, to be analyzed.

One common type of MS is **electrospray ionization MS (ESI-MS)**. A solution of macromolecules is sprayed in the form of droplets from a capillary under a strong electric field. The droplets pick up positive charges as they leave the capillary. Evaporation of the solvent leaves multiple charged molecules. A typical 20 kDa protein will pick up 10 to 30 positive charges. The MS spectrum of this protein

(Figure 5.21) reveals all of the differently charged species as a series of sharp peaks whose consecutive m/z values differ by the charge and mass of a single proton: Decreasing m/z values indicate an increasing number of charges per molecule. **Tandem mass spectrometry** uses another spectrometer downstream from the ESI source that can analyze complex protein mixtures, such as tryptic digests or proteins emerging from an HPLC column.

In another type of MS, called **matrix-assisted laser desorption ionization-time of flight MS (MALDI-TOF MS)**, a protein sample is mixed with a chemical matrix that includes a light-absorbing substance. A laser pulse is used to excite the chemical matrix, creating a microplasma that transfers the energy to protein molecules in the sample, ionizing them and ejecting them into the gas phase. Among the products are protein molecules that have picked up a single proton. These positively charged species can be selected by the MS for mass analysis. MALDI-TOF MS is very sensitive and very accurate. Attomole (10^{-18}) quantities of a molecule can be detected.

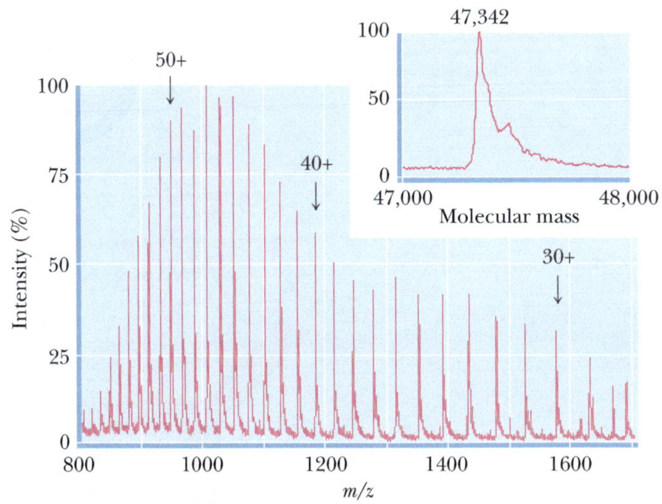

Figure 5.21 Mass spectrometry results.

5-5A Enzyme-Linked Immunosorbent Assay (ELISA)

One of the first protein identification techniques designed was the enzyme-linked immunosorbent assay or **ELISA**. This technique is based on reactions between proteins and antibodies. Antibodies are discussed further in Chapter 14, but the basic idea is that vertebrates produce antibody proteins when foreign molecules, including foreign proteins, are encountered. The antibodies bind specifically to the proteins that elicited their creation. Researchers quickly learned to harness this ability to create antibodies to recognize a wide variety of proteins. Once created, these antibodies make excellent "tags" that will seek out and bind to their target protein. By finding where these tags end up, we can know where the protein is.

Figure 5.22 shows a diagram of a basic design for this process. Step 1 shows a series of proteins that coats a well in a microtiter plate (Figure 5.23). One of the proteins, the triangle, is a protein for which an antibody has been created. This antibody, called the **primary antibody**, is shown as a Y-shaped molecule binding to its target protein in Step 2 (Section 14-3F discusses the reason we depict the antibody as a Y-shape). Unfortunately, the binding of this antibody does not make it visible, so we still don't know where the target protein is. To remedy that, a **secondary antibody** is used to locate the primary antibody (Step 3). This secondary antibody has some type of marker (Step 4) on it that catalyzes a reaction that can be observed. One such reaction involves the oxidation of 4-chloro-1-naphthol as shown here:

$$\text{4-chloro-1-naphthol}_{red} + H_2O_2 \rightarrow \text{4-chloro-1-naphthol}_{ox} + H_2O + \frac{1}{2}O_2$$

The oxidized form of this compound is purple, so wherever a purple color appears, we know we have the target protein (Figure 5.23). There are two other common ways to spot the secondary antibody. One is to have its tag be a fluorescent molecule. A second is to have the tag be radioactive. A fluorometer or a radioisotope counter can then be used, respectively, to find the target proteins.

An ELISA is usually performed in a plastic tray full of tiny wells called a microtiter plate, as shown in Figure 5.23. The protein samples being studied

tandem mass spectrometry a technique in which the output from one spectrometer is subjected to an additional reaction and separation before being analyzed in a second spectrometer

matrix-assisted laser desorption ionization-time of flight MS (MALDI-TOF MS) a technique that uses a laser to ionize the protein sample for MS

ELISA enzyme-linked immunosorbent assay

primary antibody an antibody that will react with the protein being studied or sought during an experiment

secondary antibody an antibody that will react with the primary antibody used in an ELISA or western blot experiment

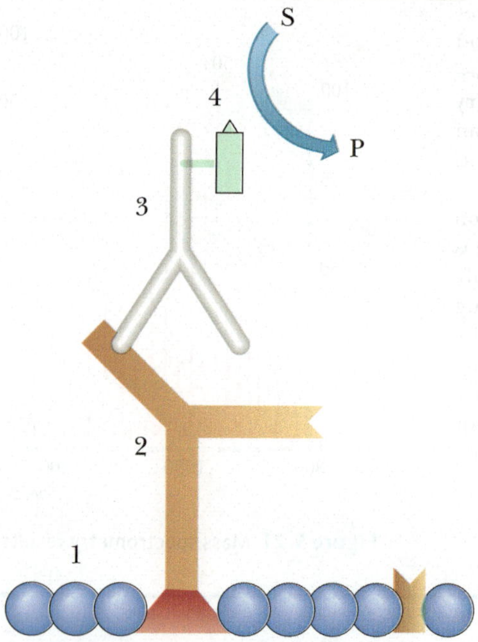

Figure 5.22 Use of primary and secondary antibodies in the ELISA method to visualize a protein. *(Farrell/Taylor, Experiments in Biochemistry, 2E. © 2006 Cengage Learning.)*

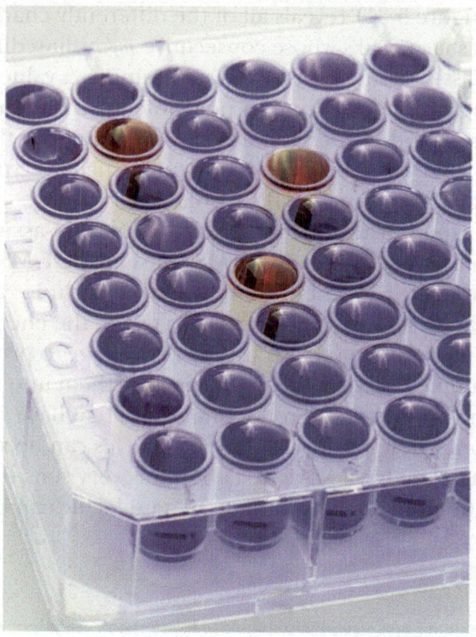

Jens Goepfert/Shutterstock.com

Figure 5.23 A microtiter plate showing the result of an ELISA.

are immobilized in the wells of the plate. Many different protein samples can be studied at one time in this way. The presence of the color in the plate tells us that the target protein is present. In quantitative analysis, the intensity of the color tells us the relative amount of the target protein.

5-5B Western Blot

western blot a technique where proteins are first separated using gel electrophoresis and then transferred to a nitrocellulose membrane for analysis and identification

A **western blot** refers to the transfer of proteins from an electrophoresis gel, usually SDS–PAGE, onto a thin membrane of nitrocellulose or some other absorbent material. Once transferred, the proteins are identified in the same manner as with an ELISA. The name, western blot, comes from a "tongue in cheek" derivation of all blotting techniques. The original blotting technique was developed for the transfer of DNA from a gel onto nitrocellulose. It was developed by a researcher named Southern, so his technique was called a Southern blot (Section 13-7). The next blot to be designed transferred RNA, and, almost as a joke, it was called a northern blot. The third blotting technique was for proteins, so scientists, many of whom like a good pun, called it a western blot.

Figure 5.24 shows the basic design of a western blot experiment. In Step 1, proteins are separated on a gel via electrophoresis, as described in Section 5-3. The gel is then sandwiched between filter paper and nitrocellulose (Step 2) and put into a second electrophoresis chamber that will drive the bands off the gel and onto the nitrocellulose membrane (Step 3). The nitrocellulose is then treated with primary and secondary antibodies, as outlined in Figure 5.22, and the relevant bands are visualized. In many cases, a comparison is made between the original gel, stained with a general protein stain like Coomassie Blue, and the antibody-labeled protein bands on the blotted membrane, as shown in Figure 5.25.

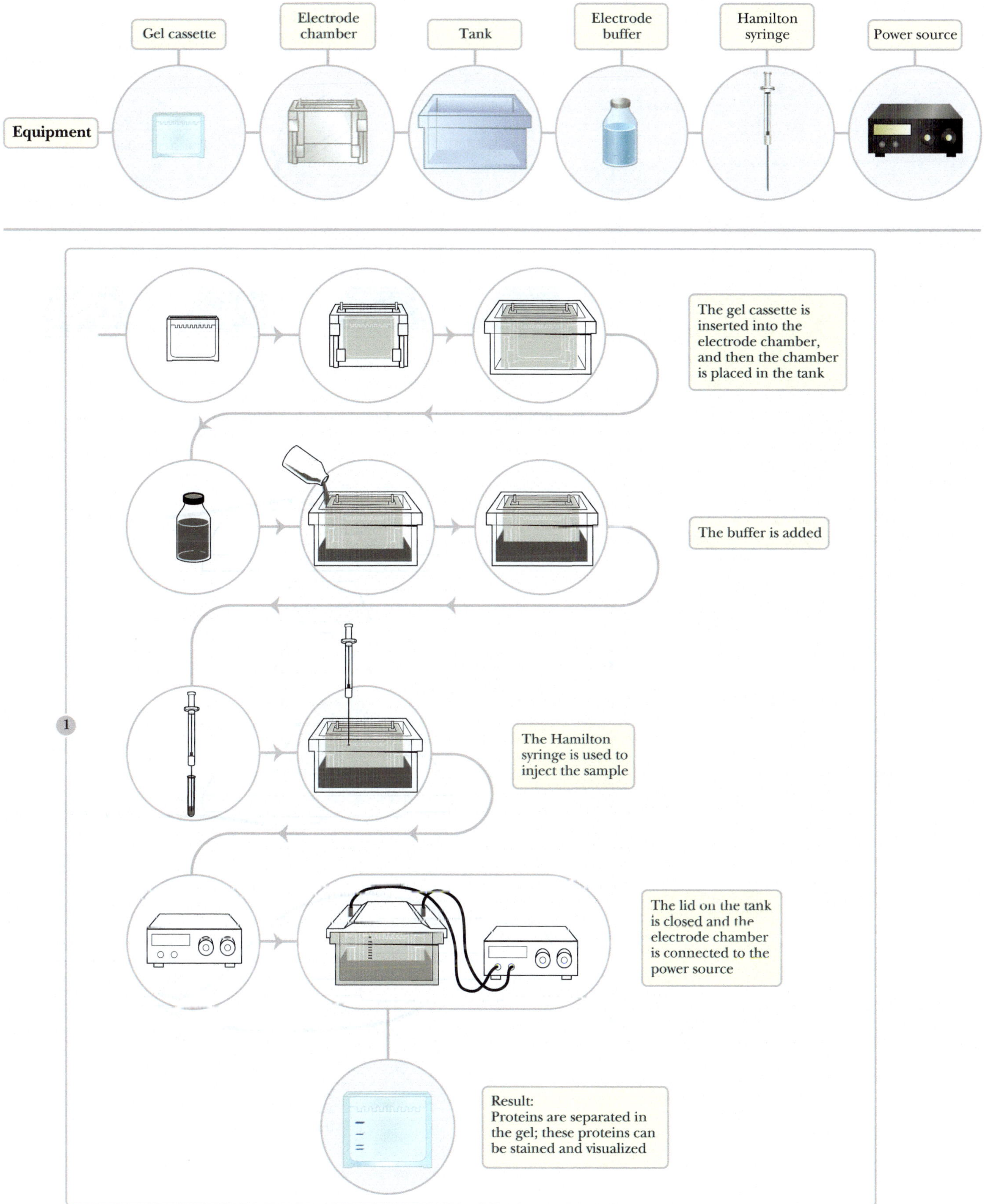

Figure 5.24 The basic design of a western blot experiment.

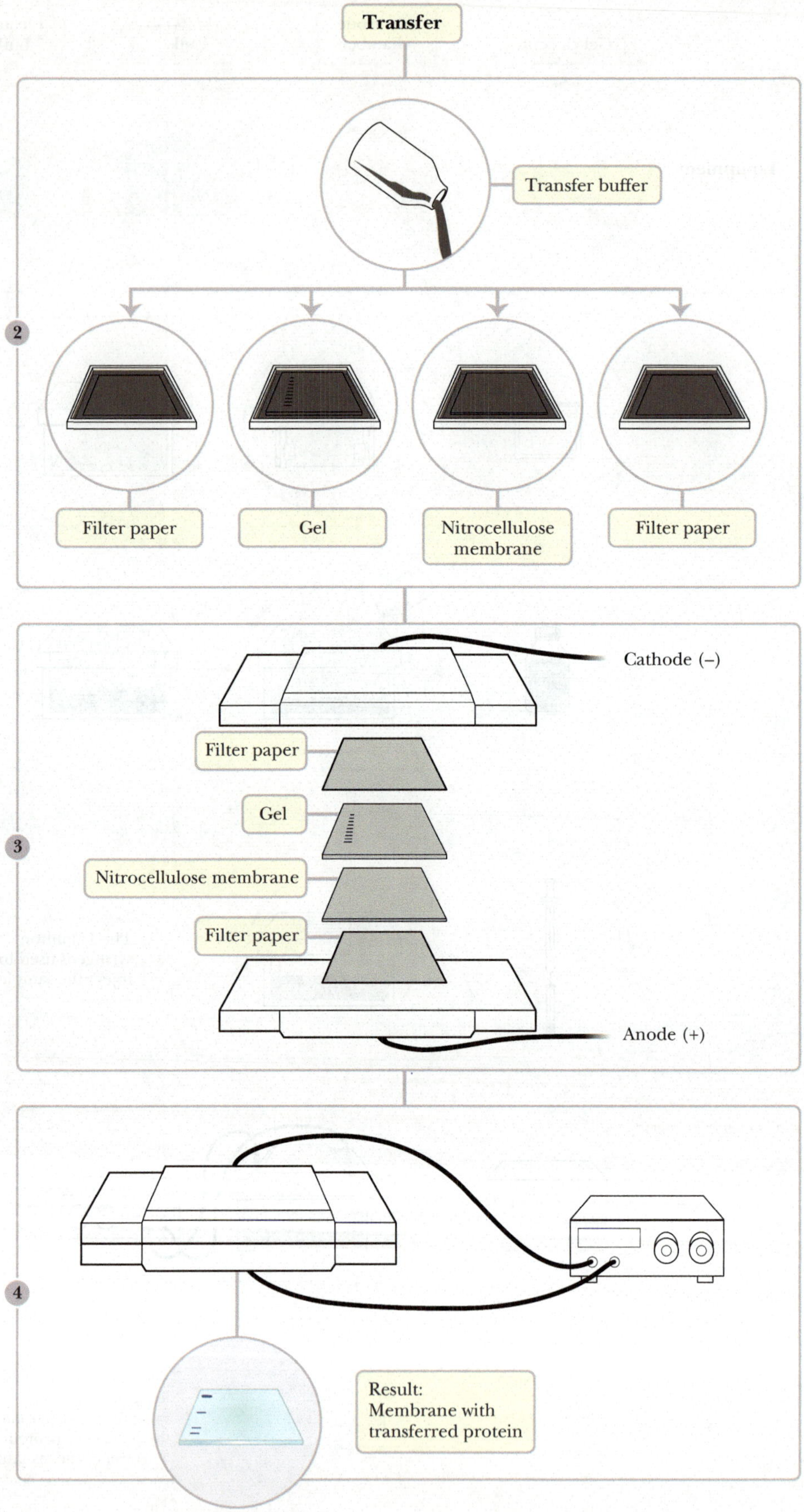

Figure 5.24—cont'd

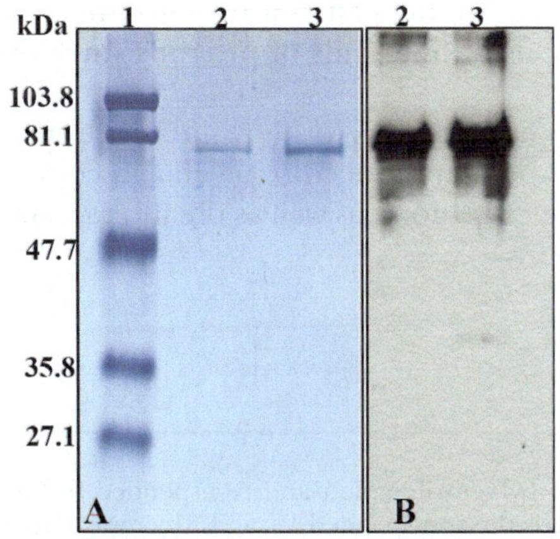

Figure 5.25 **Comparison of a Coomassie Blue-stained gel (A) and a western blot made from the same gel (B).** By using specific antibodies and markers that allow them to be seen, the locations of specific proteins can be determined on the western blot.
(A, Gustoimages/Science Source/Photo Researchers; B, 2009 Azza et al., licensee BioMed Central Ltd.)

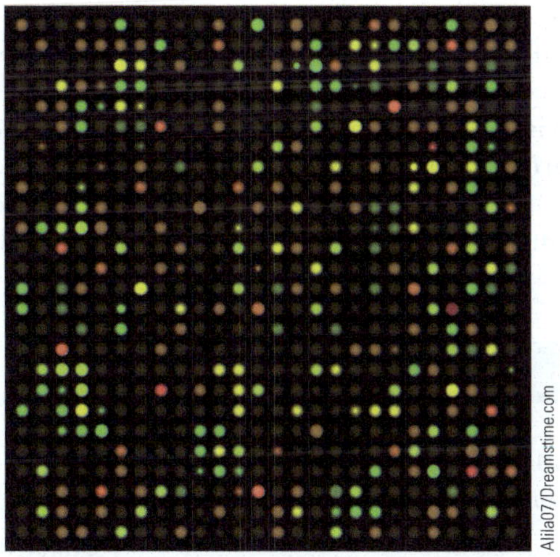

Figure 5.26 **A protein chip.** The relative colors and intensities tell the researcher which proteins are found in the various spots and how much of each is present.

5-5C Protein Microarrays

Protein microarrays, or protein chips, take the same concepts we have seen with the ELISA and western blot techniques, and they add the power of tremendous throughput. For example, a typical ELISA experiment uses a 96-well plate, so each experiment can look at 100 samples. A protein chip, on the other hand, may have 30,000 separate samples stuck on a chip a few centimeters on a side. Fluorescence is the most common way to see the results, as shown in Figure 5.26.

protein microarrays also called protein chips; small plates of a few centimeters on a side that can have tens of thousands of proteins implanted

Summary

Extracting Pure Proteins from Cells

- Disruption of cells is the first step in protein purification. The various parts of the cells can then be separated by centrifugation. This is a useful step because proteins tend to occur in given organelles.
- High salt concentrations precipitate groups of proteins, which are then further separated by chromatography and analyzed by electrophoresis.

Column Chromatography

- Gel-filtration chromatography separates proteins based on size.
- Ion-exchange chromatography separates proteins based on net charge.
- Affinity chromatography separates proteins based on their affinity for specific ligands.
- To purify a protein, many techniques are used, and often several different chromatography steps are used.

Polyacrylamide Gel Electrophoresis

- Agarose gel electrophoresis is mainly used for separating nucleic acids, and acrylamide is the usual medium for protein separation.
- When acrylamide gels are run with the chemical SDS, the proteins separate based on size alone. In the absence of SDS, proteins separate based on charge, shape, and size.

Determining the Primary Structure of a Protein

- The Edman degradation method can be used to determine the sequence of proteins that are first cut into small fragments using enzymes and chemicals.

Proteomics

- There are several ways to identify and analyze proteins.
- Mass spectrometry measures the charge-to-mass ratios of atoms in molecules and can identify a protein down to its atomic level.

- Enzyme-linked immunosorbent assay (ELISA) and western blots both exploit the specific binding of antibodies to a protein with subsequent visualization of the antibody–protein complex. In ELISA, this is carried out in a microtiter plate. With western blots, proteins are first separated by gel electrophoresis and then the bands are transferred to a membrane, such as nitrocellulose.
- A highly powerful technique, called a protein microarray or protein chip, uses thousands of proteins stuck on a slide.
- Proteomics refers to our attempts to study the entire complement of proteins being produced by a cell at a certain time under specified conditions, as well as the interactions between these proteins.

Review Exercises

5-1 Extracting Pure Proteins from Cells

1. **Recall** What types of homogenization techniques are available for solubilizing a protein?
2. **Recall** When would you choose to use a Potter–Elvehjem homogenizer instead of a blender?
3. **Recall** What is meant by "salting out" as it relates to protein purification? How does it work?
4. **Recall** What differences between proteins are responsible for their differential solubility in ammonium sulfate?
5. **Recall** How could you isolate mitochondria from liver cells using differential centrifugation?
6. **Recall** Can you separate mitochondria from peroxisomes using only differential centrifugation?
7. **Recall** Give an example of a scenario in which you could partially isolate a protein with differential centrifugation using only one spin.
8. **Recall** Describe a procedure for isolating a protein that is strongly embedded in the mitochondrial membrane.
9. **Reflect and Apply**
   (a) A yeast alcohol dehydrogenase was purified using the steps indicated in the table provided. After each step, 1.00 mL of the protein sample was used in the enzyme activity assay. Use the observed experimental data presented in the table to complete columns 4 and 5 by calculating the specific activity after each step and the fold purification at each step.

1. Purification Step	2. Total Activity of Enzyme (micromoles product/min)	3. Total Protein Concentration of Sample (mg/mL)	4. Specific Activity (micromole product/ (min mg protein))	5. Fold Purification
Crude homogenate	150,000	56		
Salt precipitation	76,000	25		
Ion exchange chromatography	58,000	8.4		
Gel filtration chromatography	37,000	3.2		
Affinity chromatography	29,000	0.4		

   (b) Which purification step yielded the highest fold of purification?
   (c) Which single purification step showed the greatest loss of activity?
10. **MATHEMATICAL** An assay was developed to detect the activity of a protein of interest. To begin purifying the protein, 20 mL of 100% saturated ammonium sulfate was added to 30 mL of a solution containing the protein of interest along with many other proteins. The precipitate that resulted (P1) was separated from the supernatant (S1) by centrifugation and resuspended in buffer. After assaying both the resuspended P1 and S1, it was found that activity remained in the supernatant. Another 20 mL of 100% saturated ammonium sulfate was added to approximately 50 mL of S1. This time, the specific activity was found to be highest in the tube containing the second resuspended precipitate (P2). What concentration of ammonium sulfate (expressed as % saturated) was used to precipitate the protein of interest?

5-2 Column Chromatography

11. **Recall** What is the basis for the separation of proteins by the following techniques?
    (a) gel-filtration chromatography
    (b) affinity chromatography
    (c) ion-exchange chromatography
    (d) reverse-phase HPLC
12. **Recall** What is the order of elution of proteins in a gel-filtration column? Why is this so?
13. **Recall** What are two ways in which a compound can be eluted from an affinity column? What could be the advantages or disadvantages of each?
14. **Recall** What are two ways that a compound can be eluted from an ion-exchange column? What could be the advantages or disadvantages of each?
15. **Recall** What are four types of compounds that make up the resin for column chromatography?
16. **Recall** Draw an example of a compound that would serve as a cation exchanger. Draw one for an anion exchanger.
17. **Recall** How can gel-filtration chromatography be used to arrive at an estimate of the molecular weight (MW) of a protein?
18. **Reflect and Apply** Sephadex G-75 has an exclusion limit of 80,000 molecular weight for globular proteins. If you tried to use this column material to separate alcohol dehydrogenase (MW 150,000) from β-amylase (MW 200,000), what would happen?
19. **Reflect and Apply** Referring to Question 18, could you separate β-amylase from bovine serum albumin (MW 66,000) using this column?
20. **Recall** What is the main difference between reverse-phase HPLC and standard ion-exchange or gel-filtration chromatography?
21. **Recall** How does HPLC differ from ion-exchange chromatography?
22. **Reflect and Apply** Design an experiment to purify protein X on an anion-exchange column. Protein X has an isoelectric point of 7.0.
23. **Reflect and Apply** Referring to Question 22, how would you purify protein X using ion-exchange chromatography if it turns out that the protein is only stable at a pH between 6 and 6.5?
24. **Reflect and Apply** An amino acid mixture consisting of lysine, leucine, and glutamic acid is to be separated by ion-exchange chromatography, using a cation-exchange resin at pH 3.5, with the eluting buffer at the same pH. Which of these amino acids will be eluted from the column first? Will any other treatment be needed to elute one of these amino acids from the column?

25. **Reflect and Apply** An amino acid mixture consisting of phenylalanine, glycine, and glutamic acid is to be separated by reverse-phase HPLC. Which of these amino acids will move fastest? Which one will move slowest?

26. **Reflect and Apply** Gel-filtration chromatography is a useful method for removing salts, such as ammonium sulfate, from protein solutions. Describe how such a separation is accomplished.

27. **Reflect and Apply** Suggest a suitable type(s) of stationary phase to separate A and B, A and C, or B and D proteins by column chromatography. (More than one method may be suitable.) Based on your selected method, indicate which of the two proteins would be retained on the column the longest.

Protein	pI	MW (kDa)	Other feature
A	7.4	105	glycoprotein
B	7.4	100	
C	7.4	50	
D	9.6	45	

28. **Reflect and Apply** To facilitate purification, recombinant proteins are often engineered with a polyhistidine tag, comprised of several adjacent histidine residues, usually at the N-terminal or C-terminal end of the protein. The nitrogen-containing imidazole side chains of the histidine residues act as a very good ligand to coordinate with nickel (Ni^{2+})
   (a) Considering this structural feature, what type of chromatographic stationary phase would be most useful to specifically purify these His-tagged proteins?
   (b) How would you change the mobile phase to release the His-tagged proteins from the column?

5-3 Polyacrylamide Gel Electrophoresis

29. **Recall** What physical parameters of a protein control its migration on electrophoresis?

30. **Recall** What types of compounds make up the gels used in electrophoresis?

31. **Recall** What types of macromolecules are usually separated on agarose electrophoresis gels?

32. **Recall** If you had a mixture of proteins with different sizes, shapes, and charges, and you separated them with electrophoresis, which proteins would move fastest toward the anode (positive electrode)?

33. **Recall** What does SDS–PAGE stand for? What is the benefit of doing SDS–PAGE?

34. **Recall** How does the addition of sodium dodecyl sulfate to proteins affect the basis of separation on electrophoresis?

35. **Recall** What electrophoresis technique involving isoelectric focusing would work best to separate proteins with
   (a) different pI values?
   (b) different molecular weights?
   (c) different quaternary structures?
   (d) different polarity?

36. **Recall** Why is the order of separation based on size opposite for gel filtration and gel electrophoresis, even though they often use the same compound to form the matrix?

37. **Recall** The accompanying figure is from an electrophoresis experiment using SDS–PAGE. The left lane has the following standards: bovine serum albumin (MW 66,000), ovalbumin (MW 45,000), glyceraldehyde 3-phosphate dehydrogenase (MW 36,000), carbonic anhydrase (MW 24,000), and trypsinogen (MW 20,000). The right lane is an unknown. Calculate the MW of the unknown.

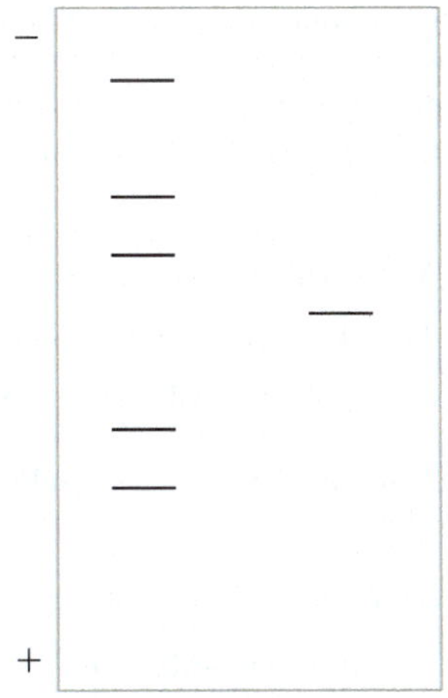

38. **Recall** Which of the following biomolecules extracted from a plant source would be best separated and analyzed by electrophoresis using an agarose gel rather than a polyacrylamide gel?
   (a) A sample of a protein
   (b) A sample of DNA
   (c) A short segment of DNA from the PCR analysis of a DNA extract
   (d) A peptide such as insulin

39. **Reflect and Apply** Gel-filtration chromatography of a protein indicates that it has a molecular weight of 84 kDa. Native gel electrophoresis of the column chromatography fraction containing this protein shows a single band. However, after SDS–PAGE analysis, two bands are observed; one band corresponds to a MW of 56 kDa, whereas the other corresponds to a MW of 14 kDa. Which of the following statements do you think is most accurate?
   (a) The 84 kDa protein has no quaternary structure.
   (b) The 84 kDa protein is a homodimer.
   (c) The 84 kDa protein is a heterodimer.
   (d) The 84 kDa protein is a homotrimer.
   (e) The 84 kDa protein is a heterotrimer.
   (f) The 84 kDa protein is a homotetramer.
   (g) The 84 kDa protein is a heterotetramer.
   (h) The 84 kDa protein is not pure.

5-4 Determining the Primary Structure of a Protein

40. **Reflect and Apply** Show, by a series of equations (with structures), the first stage of the Edman method applied to a peptide that has leucine as its N terminal residue.

41. **Reflect and Apply** A sample of an unknown peptide was divided into two aliquots. One aliquot was treated with trypsin; the other was treated with cyanogen bromide. Given the following sequences (N-terminal to C-terminal) of the resulting fragments, deduce the sequence of the original peptide.

Trypsin treatment

Asn—Thr—Trp—Met—Ile—Lys

Gly—Tyr—Met—Gln—Phe

Val—Leu—Gly—Met—Ser— Arg

Cyanogen bromide treatment

Gln—Phe

Val—Leu—Gly—Met

Ile—Lys—Gly—Tyr—Met

Ser—Arg—Asn—Thr—Trp—Met

42. **Reflect and Apply** A sample of a peptide of unknown sequence was treated with trypsin; another sample of the same peptide was treated with chymotrypsin. The sequences (N-terminal to C-terminal) of the smaller peptides produced by trypsin digestion were as follows:

Met—Val—Ser—Thr—Lys

Val—Ile—Trp—Thr—Leu—Met—Ile

Leu—Phe—Asn—Glu—Ser—Arg

The sequences of the smaller peptides produced by chymotrypsin digestion were as follows:

Asn—Glu—Ser—Arg—Val—Ile—Trp

Thr—Leu—Met—Ile

Met—Val—Ser—Thr—Lys—Leu—Phe

Deduce the sequence of the original peptide.

43. **Reflect and Apply** You are in the process of determining the amino acid sequence of a protein and must reconcile contradictory results. In one trial, you determine a sequence with glycine as the N-terminal amino acid and asparagine as the C-terminal amino acid. In another trial, your results indicate phenylalanine as the N-terminal amino acid and alanine as the C-terminal amino acid. How do you reconcile this apparent contradiction?

44. **Reflect and Apply** You are in the process of determining the amino acid sequence of a peptide. After trypsin digestion followed by the Edman degradation, you obtain the following peptide fragments:

Leu—Gly—Arg

Gly—Ser—Phe—Tyr—Asn—His

Ser—Glu—Asp—Met—Cys—Lys

Thr—Tyr—Glu—Val—Cys—Met—His

What is abnormal about these results? What might have been the problem that caused them?

45. **Reflect and Apply** Amino acid compositions can be determined by heating a protein in 6 *M* HCl and running the hydrolysate through an ion-exchange column. If you were going to do an amino acid sequencing experiment, why would you want to get an amino acid composition first?

46. **Reflect and Apply** Assume that you are getting ready to do an amino acid sequencing experiment on a protein containing 100 amino acids, and an amino acid analysis shows the following data:

Amino Acid	Number of Residues
Ala	7
Arg	23.7
Asn	5.6
Asp	4.1
Cys	4.7
Gln	4.5
Glu	2.2
Gly	3.7
His	3.7
Ile	1.1
Leu	1.7
Lys	11.4
Met	0
Phe	2.4
Pro	4.5
Ser	8.2
Thr	4.7
Trp	0
Tyr	2.0
Val	5.1

Which of the chemicals or enzymes normally used for cutting proteins into fragments would be the least useful to you?

47. **Reflect and Apply** Which enzymes or chemicals would you choose to use to cut the protein in Question 46? Why?

5-5 Proteomics

48. **Biochemical Connections** What are the two principal types of mass spectrometry?
49. **Biochemical Connections** What is the advantage of MALDI-TOF MS?
50. **Recall** What is the basis for the technique called ELISA?
51. **Recall** What is the difference between a primary antibody and a secondary antibody?
52. **Recall** What are some ways that the antibody–protein complex can be visualized in ELISA or western blots?
53. **Recall** What are the main procedures involved in a western blot?
54. **Recall** Where did western blot get its name?
55. **Reflect and Apply** What is the advantage of transferring protein bands from a gel to nitrocellulose during the western blot?
56. **Reflect and Apply** What is the point of labeling a secondary antibody with a marker that can be visualized instead of just labeling the primary antibody?
57. **Recall** What is proteomics?

Further Reading

Aebersold, R., and M. Mann. Mass Spectrometry-Based Proteomics. *Nature* **422**, 198–207 (2003).

Ahern, H. Chromatography, Rooted in Chemistry, Is a Boon for Life Scientists. *The Scientist* **10** (5), 17–19 (1996). A general treatise on chromatography.

Boyer, R. F. *Modern Experimental Biochemistry.* Boston: Addison-Wesley, 1993. A textbook specializing in biochemical techniques.

Dayhoff, M. O., ed. *Atlas of Protein Sequence and Structure.* Washington, DC: National Biomedical Research Foundation, 1978. A listing of all known amino acid sequences, updated periodically.

Deutscher, M. P., ed. *Guide to Protein Purification.* Vol. **182**, *Methods in Enzymology.* San Diego: Academic Press, 1990. The standard reference for all aspects of research on proteins.

Dickerson, R. E., and I. Geis. *The Structure and Action of Proteins,* 2nd ed. Menlo Park, CA: Benjamin Cummings, 1981. A well-written and particularly well-illustrated general introduction to protein chemistry.

Farrell, S. O., and L. Taylor. *Experiments in Biochemistry: A Hands-On Approach.* Menlo Park, CA: Thomson Learning, 2005.

A laboratory manual for undergraduates that focuses on protein purification techniques.

Kumar, A. and M. Snyder. Protein Complexes Take the Bait. *Nature* **415**, 123–124 (2002). An article that pulls together several protein purification techniques and shows how they can be used to answer real questions in protein biochemistry.

Robyt, J. F., and B. J. White. *Biochemical Techniques Theory and Practice.* Monterey, CA: Brooks/Cole, 1987. An all-purpose review of techniques.

Service, R. F. Proteomics Ponders Prime Time. *Science* **321**, 1758–1761 (2008).

Tyers, M., and M. Mann. From Genomics to Proteomics. *Nature* **422**, 193–197 (2003).

Whitaker, J. R. Determination of Molecular Weights of Proteins by Gel Filtration on Sephadex®. *Analytical Chemistry* **35** (12), 1950–1953 (1963). A classic paper describing gel filtration as an analytical tool.

Traveling over a mountain pass is an analogy frequently used to describe the progress of a chemical reaction. Enzyme catalysts speed up biochemical reactions.

6 The Behavior of Proteins: Enzyme Kinetics and Inhibition

Chapter Outline

6-1 Enzyme Kinetics versus Thermodynamics

Learning Objectives

After reading this section, students will be able to:

LO 6.1.1 Draw energy profiles for uncatalyzed and catalyzed reactions.

LO 6.1.2 Explain how energy profiles illustrate the connections and differences between enzyme thermodynamic and kinetics principles.

LO 6.1.3 Explain how temperature affects enzymatic reaction kinetics.

Of all the functions of proteins, **catalysis** is probably the most important. In the absence of catalysis, most reactions in biological systems would take place far too slowly to provide products at an adequate pace for a metabolizing organism. The catalysts that serve this function in organisms are called **enzymes**. With the exception of some RNAs (ribozymes) that have catalytic activity, all other enzymes are globular proteins (Section 4-3). Enzymes are the most efficient catalysts known; they can increase the rate of a reaction by a factor of up to 10^{20} over uncatalyzed reactions. Nonenzymatic catalysts, in contrast, typically enhance the rate of reaction by factors of 10^2 to 10^4.

Enzymes are highly specific, even to the point of being able to distinguish between stereoisomers of a given compound, and they greatly increase the speed of a reaction. In many cases, the actions of enzymes are fine-tuned by regulatory processes.

Pioneers in Biochemistry • Ruby Hirose

Ruby Hirose (1904–1960) was a Japanese American biochemist and bacteriologist. Having earned undergraduate and master's degrees in pharmacy and pharmacology, respectively, she went on to pursue a doctorate in biochemistry from the University of Cincinnati. Hirose is believed to be among the first Japanese Americans to receive a PhD. During her career as a teacher and researcher, she made contributions to the development of the polio vaccine, as well as to understanding the mechanisms involved in blood clotting, pollen allergies, and cancer.

In 1940, Hirose was one of only 10 women members of the American Chemical Society. At a time when it was difficult for women to find acceptance as scientists, Hirose faced the additional challenge of prejudice against Japanese Americans. Born outside the United States, her parents were legally prevented from purchasing property in Washington state. Later, several family members were sent to internment camps during World War II. Living in Ohio at the time, Hirose herself avoided internment.

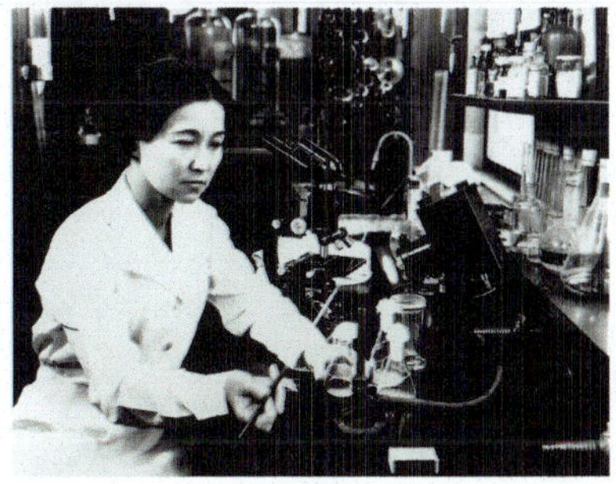

Science History Images / Alamy Stock Photo

Hirose died at age 56 from acute myeloid leukemia.

6-1A Connecting Thermodynamic and Kinetic Principles

The rate of a reaction and its thermodynamic favorability are two different topics, although they are closely related as seen in **energy profiles** (Figure 6.1). This is true of all reactions, whether or not a catalyst is involved. The difference between the energies of the reactants (the initial energy state) and the energies of the products (the final energy state) of a reaction gives the energy change for that reaction, expressed as the **standard Gibbs energy change** or ΔG° (Figure 6-1A). In Figure 6-1A, the *x*-coordinate shows the extent to which the reaction has taken place, whereas the *y*-coordinate indicates energy at various stages of the reaction. Energy changes can be described by several related thermodynamic quantities. We use standard Gibbs energy changes for our discussion; recall that the question of whether a reaction is favored depends on ΔG° (see Sections 1-6 and 15-1). Enzymes, like all catalysts, speed up reactions, but they do not alter the equilibrium constant or the Gibbs energy change. The reaction rate depends on the Gibbs energy of activation or **activation energy** ($\Delta G^{\circ\ddagger}$), the energy input required to initiate the reaction by bringing the energy of the reactants to the **transition state** energy. The activation energy for an uncatalyzed reaction is higher than that for a catalyzed reaction; in other words, an uncatalyzed reaction requires more energy to get started (Figure 6-1B). For this reason, its rate is slower than that of a catalyzed reaction.

Considerable effort has gone into elucidating the intermediate stages in reactions of interest to chemists and biochemists and determining the pathway or reaction mechanism that lies between the initial and final states. Reaction dynamics, the study of the intermediate stages of reaction mechanisms, is currently a very active field of research.

The biochemical reaction in which hydrogen peroxide (H_2O_2) is converted to water and oxygen provides an example of the effect of catalysts on activation energy.

$$2\,H_2O_2 \rightarrow 2\,H_2O + O_2$$

The activation energy of this reaction is lowered if the reaction is allowed to proceed on platinum surfaces, but it is lowered even more by the enzyme catalase. Table 6.1 summarizes the energies involved.

Chapter Outline (continued)

catalysis the process of increasing the rate of chemical reactions

enzymes biological catalysts, usually globular proteins, with some RNA molecules serving as exceptions

energy profiles a theoretical representation of a chemical reaction as an energetic pathway as the reactants become products

standard Gibbs energy change the difference between the energies of reactants and products under standard conditions

activation energy the energy input required to start a reaction by bringing the energy of the reactants to the transition state energy

transition state the intermediate state of molecules in a reaction in which old bonds are broken and new bonds are formed

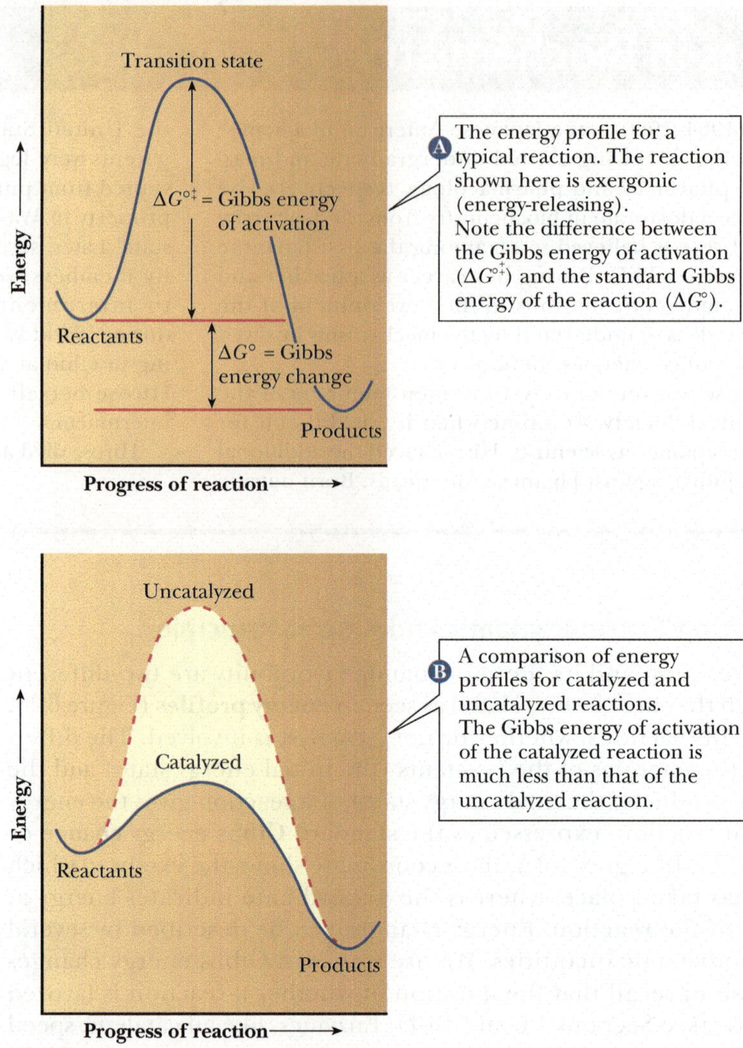

Figure 6.1 **Energy profiles.**

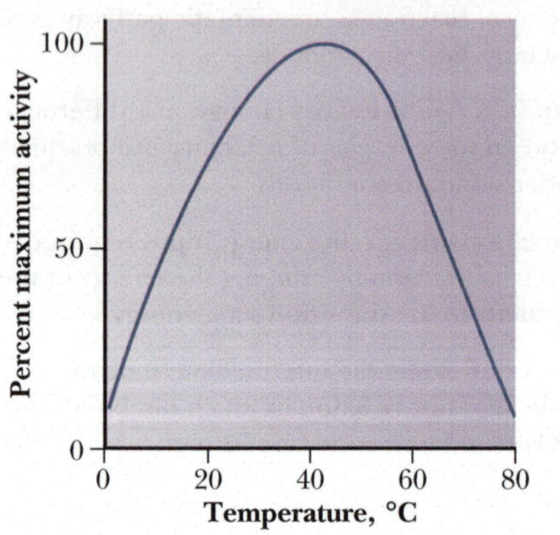

Figure 6.2 **The effect of temperature on enzyme activity.** The decrease in enzymatic activity above 50°C is due to thermal denaturation.

▶ Will a reaction go faster if you raise the temperature?

Raising the temperature of a reaction mixture increases the energy available to the reactants to reach the transition state. Consequently, the rate of a *chemical* reaction increases with temperature, but is this universally true for *biochemical* reactions? In fact, the rate of biochemical reactions increases with temperature only to a limited extent. It is helpful to raise the temperature at first, but eventually there comes a point at which heat denatures the enzyme (Section 4-4C). Above this temperature, adding more heat denatures more enzyme and slows down the reaction. Figure 6.2 shows a typical curve of the temperature effect on an enzyme-catalyzed reaction. **Biochemical Connections 6A** describes another way in which the specificity of enzymes is of great use.

Table 6.1 **Lowering of the Activation Energy of Hydrogen Peroxide Decomposition by Catalysts**

	Activation Energy		
Reaction Conditions	kJ mol^{-1}	kcal mol^{-1}	Relative Rate
No catalyst	75.2	18.0	1
Platinum surface	48.9	11.7	2.77×10^4
Catalase	23.0	5.5	6.51×10^8

Rates are given in arbitrary units relative to a value of 1 for the uncatalyzed reaction at 37°C.

6A Biochemical Connections • Health Sciences

Enzymes as Markers for Disease

Some enzymes are found only in specific tissues or in a limited number of such tissues. Figure 6.3 shows that the enzyme lactate dehydrogenase (LDH) is a tetramer that can be composed of two types of subunits (M and H). The M LDH subunit is primarily found in skeletal muscle, whereas the H LDH subunit is primarily found in heart muscle. The two different subunits differ slightly in amino acid composition. Because LDH is a tetramer of four subunits, and because the H and M subunits can combine in all possible combinations, LDH can exist in five different forms, called **protein isoforms (or isozymes)**. An increase in any form of LDH in the blood indicates some kind of tissue damage.

Similarly, there are different forms of another enzyme called creatine kinase (CK) that occur in brain, heart, and skeletal muscle tissue. Appearance of the brain type can indicate a stroke or a brain tumor, whereas the heart type indicates a heart attack. After a heart attack, CK shows up more rapidly in the blood than LDH. Monitoring the presence of both enzymes improves the diagnosis of heart attacks because elevated levels of the CK isozyme from heart muscle in the blood are a definite indication of heart tissue damage.

Additionally, researchers studying epilepsy found a link between LDH activity in the brain and epileptic seizures. Inhibiting LDH reduced the number of seizures, leading scientists to be hopeful that they have found a new and effective drug to combat this illness. Enzyme inhibition is discussed in Section 6-6.

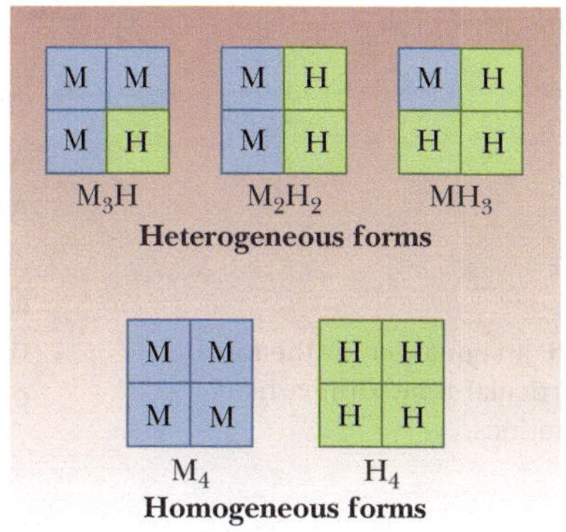

Figure 6.3 The five isozymes of lactate dehydrogenase (LDH). The symbol M refers to the form of LDH subunit that predominates in skeletal muscle, whereas the symbol H refers to the form of LDH subunit that predominates in heart muscle.

In conclusion, various enzymes can be detected using enzyme markers that rely on enzyme specificity, enhancing the diagnosis and treatment of disease.

6-2 Rates of Enzyme-Catalyzed Reactions

protein isoforms (or isozymes) multiple forms of an enzyme that catalyze the same overall reaction but have subtle physical and kinetic differences

Learning Objectives

After reading this section, students will be able to:

LO 6.2.1 Use the average rate equation to describe how reaction rates can be determined experimentally.

LO 6.2.2 Describe the principles related to zero-order enzyme kinetics.

LO 6.2.3 Determine the order of a reaction with respect to a given reactant and the overall order of the reaction using the rate law equation.

Let us now discuss how to calculate rates of enzyme-catalyzed reactions, which are additional pieces of information that biochemists can use to better understand enzymes. The rate of a chemical reaction is usually expressed in terms of a change in the concentration of a reactant or of a product in a given time interval. Any convenient experimental method can be used to monitor changes in concentration. In a general reaction of the form $A + B \rightarrow P$, where A and B are reactants and P is the product, an average rate of the reaction can be expressed either in terms of the rate of disappearance of one of the reactants or in terms of the rate of appearance of the product. The rate of disappearance of A is $-\Delta[A]/\Delta t$, where Δ symbolizes change, [A] is the molar concentration of A, and t is time. Likewise, the rate of disappearance of B is $-\Delta[B]/\Delta t$, and the rate of appearance of P is $\Delta[P]/\Delta t$. The rate of the reaction can be expressed in terms of any of these changes because the rates of appearance of the product and disappearance of the reactant are related by the stoichiometric equation for the reaction.

$$\text{Rate} = \frac{-\Delta[A]}{\Delta t} = \frac{-\Delta[B]}{\Delta t} = \frac{\Delta[P]}{\Delta t}$$

The negative signs for the changes in concentration of A and B indicate that A and B are being used up in the reaction, while the implied positive sign for $\Delta[P]$ means P is being produced.

In addition to the average rate equations just given, the rate of a reaction at a particular time under particular reactant concentration conditions can be determined using an equation of the following form, generally called a **rate law equation**:

rate law equation an equation for the rate of a reaction at a particular time with particular reactant concentrations

$$\text{Rate} = k[A]^f[B]^g$$

where k is a proportionality constant called the **rate constant**, and the other terms in the equation are the molar concentrations of the reactants raised to appropriate powers. The exponents f and g *must be determined experimentally*. They are *not necessarily* equal to the coefficients of the balanced equation, but frequently they are. When the exponents in the rate equation have been determined experimentally, a mechanism for the reaction—a description of the detailed steps along the path between reactants and products—can be proposed.

rate constant a proportionality constant that describes the rate of a reaction

The exponents in the rate equation are usually small whole numbers such as 1 or 2. (There are also some cases in which the exponent 0 occurs.) The values of the exponents are related to the number of molecules involved in the detailed steps that constitute the mechanism. The *overall order* of a reaction is the sum of all the exponents. If, for example, the rate of a reaction A → P is given by the rate equation

$$\text{Rate} = k[A]^1 = k[A] \tag{6.1}$$

where k is the rate constant and the exponent for the concentration of A is 1, then the reaction is **first order** with respect to A and first order overall.

first order describes a reaction whose rate depends on the first power of the concentration of a single reactant

If the rate law equation of a general reaction A + B → C + D is given by

$$\text{Rate} = k[A]^1[B]^1 = k[A][B] \tag{6.2}$$

where k is the rate constant, the exponent for the concentration of A is 1, and the exponent for the concentration of B is 1, then the reaction is said to be first order with respect to A, first order with respect to B, and **second order** overall. Many common reactions are first or second order. After the order of the reaction is determined experimentally, a mechanism for the reaction can be proposed.

second order describes a reaction whose rate depends on either the product of the concentrations of two reactants or the second power of the concentration of one reactant

▶ Is the rate of a reaction always based on the concentration of reactants?

Exponents in a rate law equation may be equal to zero. For example, consider a rate equation for a general reaction A → B given by

$$\text{Rate} = k[A]^0 = k \tag{6.3}$$

The order for such a reaction is called **zero order**, and the reaction rate, which is constant, does not depend on the concentrations of the reactants. Enzyme-catalyzed reactions can exhibit zero-order kinetics when the concentrations of reactants are so high that the enzyme is completely saturated with reactant molecules. This point is discussed in more detail in Section 6-4.

zero order refers to a reaction that proceeds at a constant rate independent of the concentrations of reactants

6-3 Enzyme–Substrate Binding

Learning Objectives

After reading this section, students will be able to:

LO 6.3.1 Distinguish between different models of enzyme–substrate binding.

LO 6.3.2 Draw an energy profile for an enzymatic reaction.

In an enzyme-catalyzed reaction, the enzyme binds to the **substrate** (one or more of the reactants) to form an enzyme–substrate (ES) complex. The formation of the complex leads to the formation of the transition-state species, which then forms the product. The nature of transition states in enzymatic reactions is a large field of research in itself, but some general statements can be made on the subject. A substrate binds, usually by noncovalent interactions, to a small portion of the enzyme called the **active site**, frequently situated in a cleft or crevice in the protein and consisting of certain amino acids that are essential for enzymatic activity (Figure 6.4). The catalyzed reaction takes place at the active site, usually in several steps.

substrate a reactant in an enzyme-catalyzed reaction

active site the part of an enzyme to which the substrate binds and at which certain amino acids are present to catalyze the reaction

The first step is the binding of substrate to the enzyme, which occurs because of highly specific interactions between the substrate and the side chains and backbone groups of the amino acids making up the active site. Two important models—the *lock-and-key model* and the *induced-fit model*—have been developed to describe the binding process (Figure 6.4).

In the **lock-and-key model**, there is a high degree of similarity between the shape of the substrate and the geometry of the binding site on the enzyme (Figure 6.4A). The substrate binds to a site whose shape complements its own, like a key in a lock. This model has intuitive appeal but is now largely of

lock-and-key model a description of the binding of a substrate to an enzyme such that the active site and the substrate exactly match each other in shape

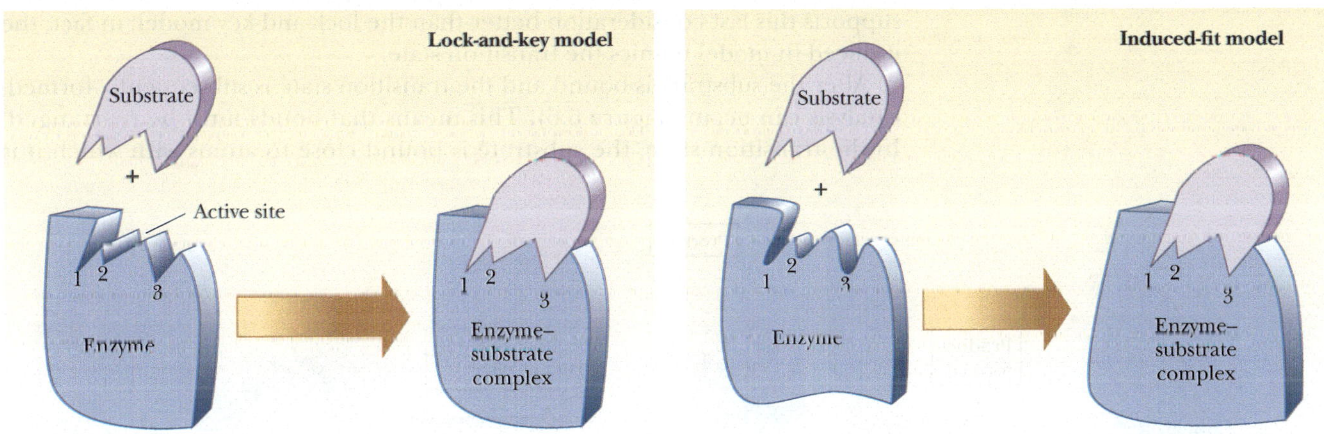

A In the lock-and-key model, the shape of the substrate and the conformation of the active site are complementary to one another.

B In the induced-fit model, the enzyme undergoes a conformational change upon binding to substrate. The shape of the active site becomes complementary to the shape of the substrate only after the substrate binds to the enzyme.

Figure 6.4 Two models for the binding of a substrate to an enzyme. The numbers in the active site represent different amino acids that are present to help catalyze the reaction.

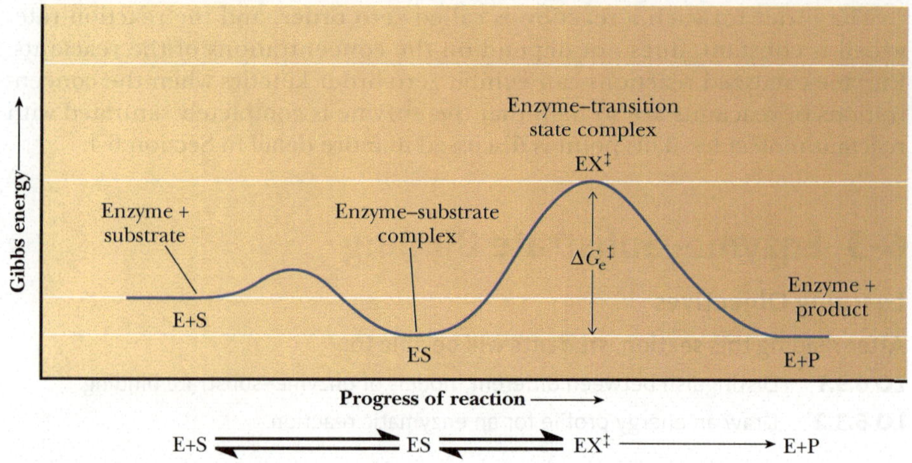

Figure 6.5 The energy profile of a reaction showing binding of the substrate to the enzyme to form an enzyme–substrate complex, formation of the enzyme–transition state complex, and formation of product.

historical interest because it does not take into account an important property of proteins—namely, their conformational flexibility.

induced-fit model a description of substrate binding to an enzyme such that the conformations of the enzyme and substrate change slightly upon their binding

In the **induced-fit model**, the binding of the substrate induces slight conformational changes in both the enzyme and substrate that result in a complementary fit after the substrate is bound (Figure 6.4B). Thus, this model takes into account the fact that proteins and their substrates have some three-dimensional flexibility.

The induced-fit model is also more attractive when we consider the nature of the transition state and the lowered activation energy that occurs with an enzyme-catalyzed reaction. The enzyme and substrate must bind to form the ES complex before anything else can happen. What would happen if this binding were too perfect? Figure 6.5 shows what happens when E and S bind. An attraction must exist between E and S for them to bind. This attraction causes the ES complex to be lower on an energy diagram than the E + S at the start. Then the bound ES must attain the conformation of the transition state $EX^{\ddagger}$. If the binding of E and S to form ES were a perfect fit, the ES would be at such a low energy that the difference between ES and $EX^{\ddagger}$ would be very large. This would slow down the rate of reaction. Many studies have shown that enzymes increase the rate of reaction by lowering the energy of the transition state, $EX^{\ddagger}$, while raising the energy of the ES complex. The induced-fit model certainly supports this last consideration better than the lock-and-key model; in fact, the induced-fit model mimics the transition state.

After the substrate is bound and the transition state is subsequently formed, catalysis can occur (Figure 6.6). This means that bonds must be rearranged. In the transition state, the substrate is bound close to atoms with which it is

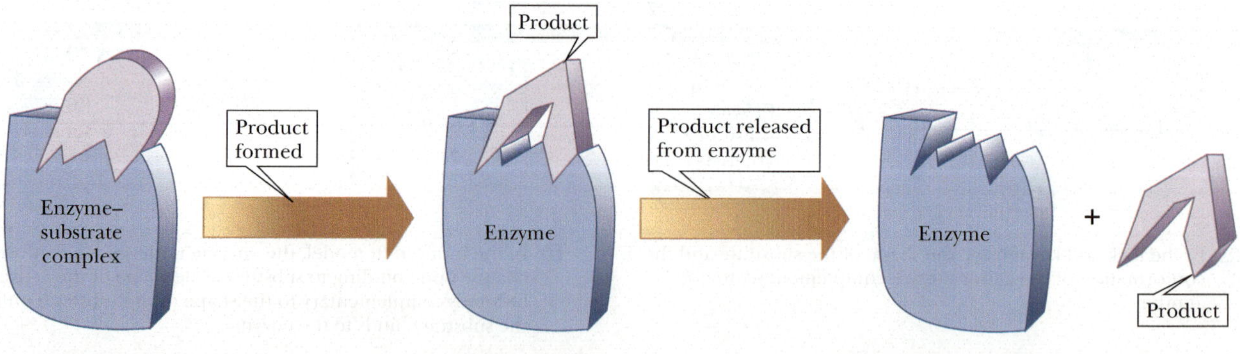

Figure 6.6 Formation of product from substrate (bound to the enzyme), followed by release of the product. *(Garrett/Grisham, Biochemistry, 4th Edition. Cengage Learning.)*

to react. Furthermore, the substrate is placed in the correct orientation with respect to those atoms. Both effects, proximity and orientation, speed up the reaction. As bonds are broken and new bonds are formed, the substrate is transformed into the product. The product is released from the enzyme, which can then catalyze the reaction of more substrate to form more product (Figure 6.6). Each enzyme has its own unique mode of catalysis, which is not surprising in view of enzymes' great specificity. Next, we discuss some kinetic parameters for describing and comparing enzymatic reactions.

6-4 The Michaelis–Menten Approach to Enzyme Kinetics

Learning Objectives

After reading this section, students will be able to:

LO 6.4.1 Calculate the enzyme turnover number from Michaelis–Menten kinetic parameters.

LO 6.4.2 Distinguish between the different possible mechanisms for enzymes with multiple substrates.

LO 6.4.3 Generate a Lineweaver–Burk double-reciprocal plot to determine V_{max} and K_M.

LO 6.4.4 Determine Michaelis–Menten model kinetic parameters for enzymes with multiple substrates.

LO 6.4.5 Use Michaelis–Menten model parameters to compare the kinetics between reactions.

A particularly useful model for the kinetics of enzyme-catalyzed reactions was devised in 1913 by Leonor Michaelis and Maud Menten. It is still the basic model for many enzymes and is widely used, even though it has undergone many modifications.

A simple enzymatic reaction is the conversion of a single substrate, S, to a product, P. The stoichiometric equation for the reaction is

$$S \rightarrow P$$

The mechanism for such a reaction can be summarized as shown in Equation 6.4:

$$E + S \underset{k_{-1}}{\overset{k_1}{\rightleftharpoons}} ES \overset{k_2}{\rightarrow} E + P \tag{6.4}$$

In this equation, k_1 is the rate constant for the formation of the ES complex, from the enzyme, E, and the substrate, S; k_{-1} is the rate constant for the reverse reaction, the dissociation of the ES complex into free enzyme and substrate; and k_2 is the rate constant for the conversion of the ES complex to product P and the subsequent release of the product from the enzyme. An assumption here is that the product is not converted to substrate to any appreciable extent. The enzyme appears explicitly in the mechanism, and the concentrations of both free enzyme, E, and ES complex, therefore, appear in the rate equations. Catalysts characteristically are regenerated at the end of the reaction, and this is true of enzymes.

When we measure the rate (also called the velocity) of an enzymatic reaction at varying substrate concentrations, we see that the rate depends on the substrate concentration, [S]. We measure the initial rate of the reaction (the rate measured immediately after the enzyme and substrate are mixed) so that we can be certain that the product is not converted to substrate to any appreciable extent. This initial velocity is sometimes written V_{init} or V_0, and it is important to remember that all the calculations involved in enzyme kinetics assume that the velocity measured is the initial velocity.

Figure 6.7 shows a graph of a theoretical enzyme kinetics experiment. In the lower region of the curve (at low levels of substrate), the reaction is first order (Section 6-3), implying that the velocity, *V*, depends on substrate concentration [S]. In the upper portion of the curve (at higher levels of substrate), the reaction is zero order—that is, the rate is independent of substrate concentration. In zero-order conditions, the active sites of all the enzyme molecules are saturated. At infinite substrate concentration, the reaction would proceed at its maximum velocity, written V_{max}.

V_{max} the theoretical rate of an enzymatic reaction at which the enzyme shows the fastest activity, also called the highest turnover

The substrate concentration at which the reaction proceeds at one-half its maximum velocity is called the **Michaelis constant (K_M)** (Figure 6.7), which can be considered an inverse measure of the affinity of the enzyme for the substrate. The lower the K_M, the higher the affinity.

Michaelis constant (K_M) an inverse measure of the strength of binding of a substrate to an enzyme

Figure 6.7 shows the effect of increasing substrate concentration on the observed rate. In such an experiment, the reaction is run at several substrate concentrations, and the rate is determined by following the disappearance of reactant, or the appearance of product, by way of any convenient method. At low-substrate concentrations, first-order kinetics are observed. At higher substrate concentrations (well beyond $10 \times K_M$), when the enzyme is saturated, the constant reaction rate characteristic of zero-order kinetics is observed. This constant rate, when the enzyme is saturated with substrate, is the V_{max} for the enzyme, a value that can be roughly estimated from the graph. The value of K_M can also be estimated from the graph. When the rate of the reaction is half its maximum value, the substrate concentration is equal to the Michaelis constant (Figure 6.7). This fact is the basis of the graphical determination of K_M.

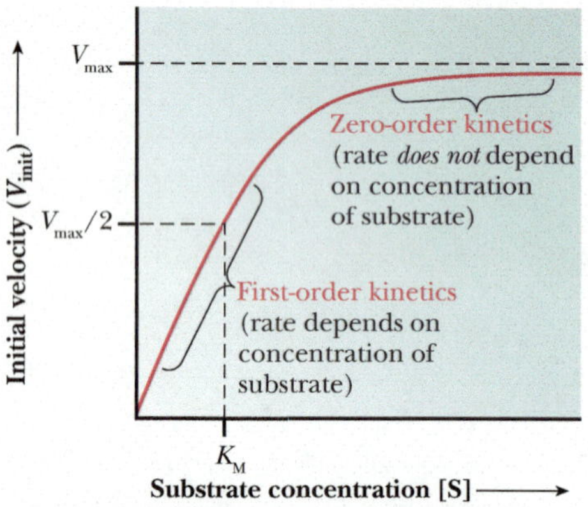

Figure 6.7 The rate and observed kinetic parameters of an enzymatic reaction depend on substrate concentration. The concentration of enzyme, [E], is constant.

6-4A Enzymes with Multiple Substrates

What happens if there is more than one substrate? The reaction used to generate the Michaelis–Menten equation was the simplest enzyme equation possible—that with a single substrate going to a single product. Most enzymes catalyze reactions containing two or more substrates. For such enzymes, one substrate is studied at a time. Consider, for example, the enzyme-catalyzed reaction

$$A + B \rightarrow P + Q$$

If we hold A at saturating levels and then vary the amount of B over a broad range, the curve of velocity versus [B] will be a hyperbola, and we can calculate the K_M for B. Conversely, we could hold the level of B at saturating levels and vary the amount of A to determine the K_M for A.

The existence of multiple substrates involved in an enzyme-catalyzed reaction leads to a variety in the order of events that can occur. With two substrates, the most common mechanisms are called the *ordered mechanism*, the *random mechanism*, and the *ping-pong mechanism*.

With an **ordered mechanism**, there is a specified order that the substrates must bind and the products must be released, as shown in Equation 6.5:

ordered mechanism an enzyme mechanism where the substrates have to bind to the enzyme in a specific order

$$E + A \rightleftharpoons EA + B \rightleftharpoons EAB \rightarrow EPQ \rightleftharpoons P + EQ \rightleftharpoons E + Q \qquad (6.5)$$

With a **random mechanism**, either one of the substrates can bind first and either of the products can be released first, as shown in Equation 6.6:

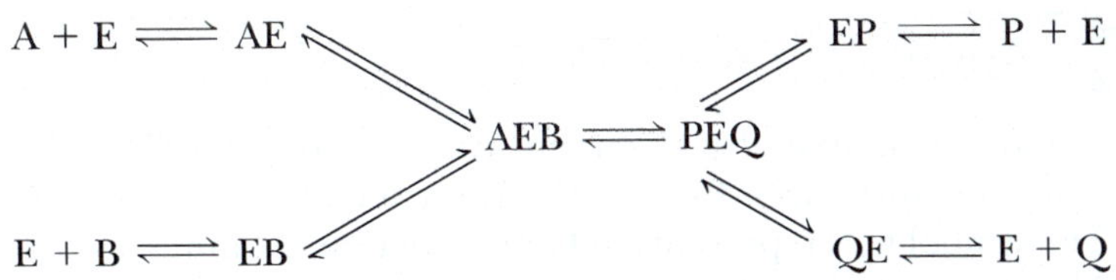

(6.6)

(From Garrett/Grisham, Biochemistry, 4th Edition. © 2009 Cengage Learning.)

random mechanism an enzyme mechanism where the substrates can bind to the enzyme in any order

In a **ping-pong mechanism**, one substrate binds to the enzyme and reacts, creating a modified version of the enzyme (E′) plus the first product, P. The product is released, and then the second substrate binds, as shown in Equation 6.7:

$$E + A \longrightarrow EA \longrightarrow E'P \xrightarrow{\downarrow P} E' \xrightarrow{\uparrow B} E'B \longrightarrow EQ \longrightarrow E + Q \qquad (6.7)$$

ping-pong mechanism an enzyme mechanism where a substrate binds to the enzyme and releases a product before the second substrate binds to the enzyme

6-4B Determining K_M and V_{max}

The curve that describes the rate of a nonallosteric enzymatic reaction is hyperbolic (Figure 6.7). From such a curve, it is quite difficult to estimate V_{max} because it is an asymptote, and the value is never reached with any finite substrate concentration that we could use in the lab. This, in turn, makes it difficult to accurately determine the K_M of the enzyme. However, the hyperbolic data can be transformed into linear data by plotting the reciprocal velocity values versus the reciprocal substrate concentration values (Figure 6.8). The equation seen at the top of Figure 6.8 can be fitted to the data, and it has the form of a straight line, $y = mx + b$, where $1/V$ takes the place of the y coordinate and $1/[S]$ takes the place of the x coordinate. The slope of the line, m, is K_M/V_{max}, and the y intercept, b, is $1/V_{max}$. When enzyme kinetics data are represented this way, it is known as a **Lineweaver–Burk double-reciprocal plot**.

Lineweaver–Burk double-reciprocal plot a graphical method for analyzing the kinetics of enzyme-catalyzed reactions where reciprocal velocity values are plotted versus reciprocal substrate concentration values

$$\frac{1}{V} = \frac{K_M}{V_{max}}\left(\frac{1}{[S]}\right) + \frac{1}{V_{max}}$$

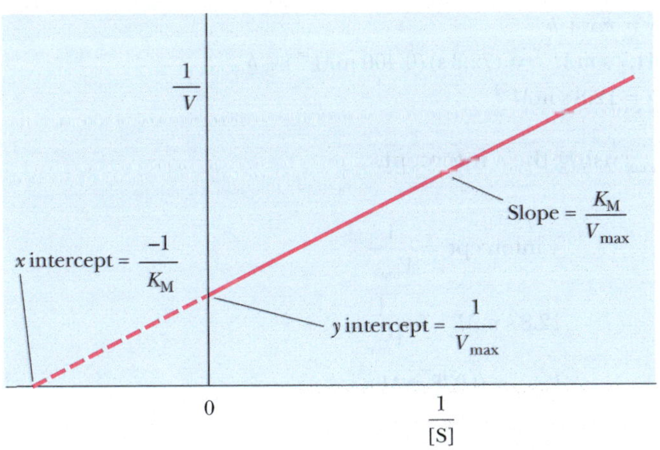

Figure 6.8 A Lineweaver–Burk double-reciprocal plot of enzyme kinetics. The reciprocal of reaction velocity, $1/V$, is plotted against the reciprocal of the substrate concentration, $1/[S]$. The slope of the line is K_M/V_{max}, and the y intercept is $1/V_{max}$. The x intercept is $-1/K_M$.

Apply Your Knowledge 6.1 gives you the opportunity to practice analyzing enzyme kinetic data to determine K_M and V_{max}.

6.1 Apply Your Knowledge

The following data describe an enzyme-catalyzed reaction. Plot these results using the Lineweaver–Burk method and determine values for K_M and V_{max}. The symbol m*M* represents millimoles per liter; 1 m*M* = 1×10^{-3} mol L^{-1}. (The concentration of the enzyme is the same in all experiments.)

Substrate Concentration (m*M*)	Velocity (m*M* s^{-1})
2.5	0.024
5.0	0.036
10.0	0.053
15.0	0.060
20.0	0.061

Solution

Step 1: Taking the reciprocal of substrate concentration and velocity gives the following results:

1/[S] (mM^{-1})	1/*V* (s mM^{-1})
0.400	41.7
0.200	27.8
0.100	18.9
0.067	16.7
0.050	16.4

Step 2: The linear regression method mentioned at the end of Step 5 would give the most accurate results, but for a rough calculation that can be done by hand, you can start by calculating the slope using the first and last data points.

$$\text{Slope} = \frac{y_2 - y_1}{x_2 - x_1} = \frac{41.7 \text{ s m}M^{-1} - 16.4 \text{ s m}M^{-1}}{0.400 \text{ m}M^{-1} - 0.05 \text{ m}M^{-1}} = 72.3 \text{ s}$$

Step 3: Calculate the *y* intercept by plugging in the slope and one data point.

$$y = mx + b$$
$$41.7 \text{ s m}M^{-1} = (72.3 \text{ s})(0.400 \text{ m}M^{-1}) + b$$
$$b = 12.8 \text{ s m}M^{-1}$$

Step 4: Calculate V_{max} using the *y* intercept.

$$y \text{ intercept} = \frac{1}{V_{max}}$$
$$12.8 \text{ s m}M^{-1} = \frac{1}{V_{max}}$$
$$V_{max} = 0.078 \text{ m}M \text{ s}^{-1}$$

Step 5: Calculate K_M using the slope and V_{max} values.

$$\text{Slope} = \frac{K_M}{V_{max}}$$

$$72.3 \text{ s} = \frac{K_M}{0.078 \text{ m}M \text{ s}^{-1}}$$

$$K_M = 5.6 \text{ m}M$$

Alternatively, you could use linear regression in Microsoft Excel, a graphing calculator, or a website to plug in the reciprocal data and determine the equation of the line as being $y = 74.4x + 12.1$. Calculating K_M and V_{max} from this linear regression equation using the techniques shown here gives the following values: $K_M = 6.1$ mM and $V_{max} = 0.083$ mM s^{-1}.

6-4C Significance of K_M and V_{max}

Recall from the beginning of this section that the Michaelis constant, K_M, can be interpreted as an inverse measure of the affinity of an enzyme for its substrate. V_{max} is related to how fast an enzyme works, but by itself V_{max} cannot be used to compare how fast one enzyme works compared to another. To get an estimate of how fast an enzyme works, the **turnover number** of an enzyme, or a quantity equal to the catalytic constant, k_2, can be used. This catalytic constant is also referred to as k_{cat} or k_p:

turnover number the number of moles of substrate that react per second per mole of enzyme

$$\frac{V}{[E_T]} = \text{turnover number} = k_{cat}$$

In this equation, $[E_T]$ is the total enzyme concentration in the experiment, and the turnover number is the number of moles of substrate that react to form product per mole of enzyme per second. This statement assumes that the enzyme is fully saturated with substrate so that the reaction is proceeding at the maximum rate. Table 6.2 lists turnover numbers for typical enzymes, where the units are *per second.*

Table 6.2 Turnover Numbers and K_M for Some Typical Enzymes

Enzyme	Function	k_{cat} = Turnover Number*	K_M**
Catalase	Conversion of H_2O_2 to H_2O and O_2	4×10^7	25
Carbonic anhydrase	Hydration of CO_2	1×10^6	12
Acetylcholinesterase	Breaks down the neurotransmitter acetylcholine into acetate and choline	1.4×10^4	9.5×10^{-2}
Chymotrypsin	Proteolytic enzyme	1.9×10^2	6.6×10^{-1}
Lysozyme	Degrades bacterial cell wall polysaccharides	0.5	6×10^{-3}

*The definition of turnover number is the moles of substrate converted to product per mole of enzyme per second. The units are s^{-1}.

**The units of K_M are millimolar.

Turnover numbers are a particularly dramatic illustration of the efficiency of enzymatic catalysis. Catalase is a particularly efficient enzyme. In Section 6-1, we encountered catalase in its role in converting hydrogen peroxide to water and oxygen. As Table 6.2 indicates, it can transform 40 million moles of substrate to product every second. **Biochemical Connections 6B** gives an interesting example of the importance of carbonic anhydrase. **Biochemical Connections 6C** gives an idea of the importance of the kinetic constants that we looked at in this section.

6B Biochemical Connections • Neuroscience

The Enzyme That Lets You Enjoy Champagne

As shown in Table 6.2, the enzyme carbonic anhydrase acts very fast, turning over a million product molecules per second per enzyme molecule. This enzyme is very important physiologically because it is responsible for how we transport CO_2 to and from the lungs as part of our metabolism. It catalyzes the reaction:

$$CO_2(g) + H_2O(l) \rightarrow H_2CO_3(aq) \rightarrow HCO_3^-(aq) + H^+(aq)$$

Carbon dioxide is common to many metabolic processes, but it travels in the blood in the form of the much more soluble carbonic acid. Blood is unable to carry enough dissolved CO_2 directly to support our metabolism, making carbonic anhydrase critical. Carbonic acid is also in equilibrium with bicarbonate (HCO_3^-) and H^+, which helps maintain blood pH.

Carbon dioxide is also what gives carbonated beverages their fizz (see Figure 6.9), which greatly affects our experience with the drink. Nobody likes flat beer or soda. However, for a long time, nobody really understood why. A clue to the puzzle came when two physicians climbed a high mountain while taking the drug acetazolamide, commonly used to prevent altitude sickness. They brought along a six-pack of beer in anticipation of celebrating reaching the top. Unfortunately, the beer was flat and tasted terrible. Further investigation showed that using the drug ruined the taste of soda and champagne, but not whisky or other noncarbonated beverages.

In 2009, a team of neuroscientists led by Dr. Charles Zuker did studies to explain this phenomenon. They identified taste receptor cells on the tongue that respond to CO_2. These cells also respond to sour taste. He determined that the molecular sensor of these cells was, in fact, a type of carbonic anhydrase called carbonic anhydrase 4. Carbonic anhydrase is inhibited by acetazolamide. This work showed that carbonic anhydrase is responsible for how we perceive carbonated beverages. In the past, people had thought that our taste perception of carbonation was due to the popping of the bubbles, which triggered mechanoreceptors in the mouth. ◗

Henrik Weis/DigitalVision/Getty Images

Figure 6.9 Champagne bubbles because of dissolved CO_2.

6C Biochemical Connections • Physical Organic Chemistry

Practical Information from Kinetic Data

The mathematics of enzyme kinetics can certainly look challenging. In fact, an understanding of kinetic parameters can often provide key information about the role of an enzyme within a living organism. Many of the ways of doing kinetic plots of this sort were developed by physical organic chemists, who then went on to propose mechanisms for reactions of all sorts based on kinetic data (see Section 7-6). Two aspects are useful: comparison of K_M and comparison of k_{cat}/K_M ratios.

Comparison of K_M

Let us start by comparing the K_M values for hexokinase and glucokinase, two enzymes that catalyze an early step in the breakdown of sugars. Both enzymes catalyze the formation of a phosphate ester linkage to a hydroxyl group of a sugar. Hexokinase can use any one of several six-carbon sugars, including glucose and fructose, the two components of sucrose (common table sugar), as substrates.

The K_M for hexokinase is 0.15 mM for glucose and 1.5 mM for fructose, which suggests that glucose is preferred over fructose as a nutrient.

Furthermore, the K_M for glucokinase, a liver-specific version of hexokinase, is 20 mM for glucose. We shall use the expression K_M here, even though some hexokinases studied do not follow Michaelis–Menten kinetics, and the term $[S]_{0.5}$ might be more appropriate. Not all enzymes have a K_M, but they do all have a substrate concentration that gives rise to $V_{max}/2$. Because the resting level for blood glucose is about 5 mM, hexokinase would be expected to be fully active for all body cells. The liver would not be competing with the other cells for glucose. However, after a carbohydrate-rich meal, blood glucose levels often exceed 10 mM, and, at that concentration, the liver glucokinase would have reasonable activity. Furthermore, because the enzyme is found only in the liver, the excess glucose will be preferentially taken into the liver, where it can be stored as glycogen until it is needed.

Comparison of k_{cat}/K_M

Even though k_{cat} alone is indicative of the maximal catalytic rate under saturating substrate conditions, [S] is rarely saturating under physiological conditions for many enzymes. The in vivo ratio of $[S]/K_M$ is often in the range of 0.01 to 1, meaning that active sites are not filled with substrate. Under these conditions, the level of substrate is small, and the amount of free enzyme approximates the level of total enzyme because most of it is not bound to substrate. The Michaelis–Menten equation can be rewritten in the following form:

$$V = \frac{V_{max}[S]}{K_M + [S]} = \frac{k_{cat}[E_T][S]}{K_M + [S]}$$

If we then replace E_T with E and assume that [S] is negligible compared with K_M, we can rewrite the equation as follows:

$$V = (k_{cat}/K_M)[E][S]$$

Thus, under these conditions, the ratio of k_{cat} to K_M is a second-order rate constant and provides a measure of the catalytic efficiency of the enzyme under nonsaturating conditions. The ratio of k_{cat} to K_M is much more constant between different enzymes than either K_M or k_{cat} alone. Looking at the first three enzymes in Table 6.2, we can see that the k_{cat} values vary over a range of about 7.3 orders of magnitude. The K_M values vary over a range of about 3.6 orders of magnitude. When the ratio of k_{cat} to K_M is compared, however, the range varies much less. The upper limit of a second-order rate constant depends on the diffusion-controlled limit of how fast the E and S can come together. The diffusion limit in an aqueous environment is in the range of 10^8 to 10^9. Many enzymes have evolved to have k_{cat} to K_M ratios that allow reactions to proceed at these limiting rates. This is referred to as being catalytically perfect.

6-5 Examples of Enzyme-Catalyzed Reactions

Learning Objectives

After reading this section, students will be able to:

LO 6.5.1 Distinguish between allosteric and nonallosteric enzymes using observed kinetics data.

LO 6.5.2 Describe the characteristics of allosteric enzymes.

Chymotrypsin is an enzyme that cleaves peptide bonds preferentially after aromatic amino acids such as tyrosine, tryptophan, and phenylalanine (Section 5-4C). Chymotrypsin also cleaves peptide bonds at other sites, such as leucine, histidine, and glutamine, but with a lower frequency than at aromatic amino acid residues. It also catalyzes the hydrolysis of ester bonds, as shown in Figure 6.10.

chymotrypsin a proteolytic enzyme that hydrolyzes peptide bonds after aromatic amino acid residues

Although ester hydrolysis is not important to the physiological role of chymotrypsin in the digestion of proteins, it is a convenient model system for investigating the enzyme's catalysis of hydrolysis reactions. The usual laboratory procedure is to use *p*-nitrophenyl esters as the substrate and to monitor the progress of the reaction by the appearance of a yellow color in the reaction mixture caused by the production of *p*-nitrophenolate ion.

In a typical reaction in which a *p*-nitrophenyl ester is hydrolyzed by chymotrypsin, the experimental rate of the reaction depends on the concentration of the substrate—in this case, the *p*-nitrophenyl ester—as shown in Figure 6.11.

The kinetics data for the reaction catalyzed by the enzyme **aspartate transcarbamoylase (ATCase)** are shown in Figure 6.12. This reaction is the first step in a pathway leading to the formation of cytidine triphosphate (CTP) and uridine triphosphate (UTP), which are ultimately needed for the biosynthesis of DNA and RNA, respectively. In this reaction, carbamoyl phosphate reacts with aspartate to produce carbamoyl aspartate and phosphate ion.

aspartate transcarbamoylase (ATCase) a classic example of an allosteric enzyme that catalyzes an early reaction in pyrimidine biosynthesis

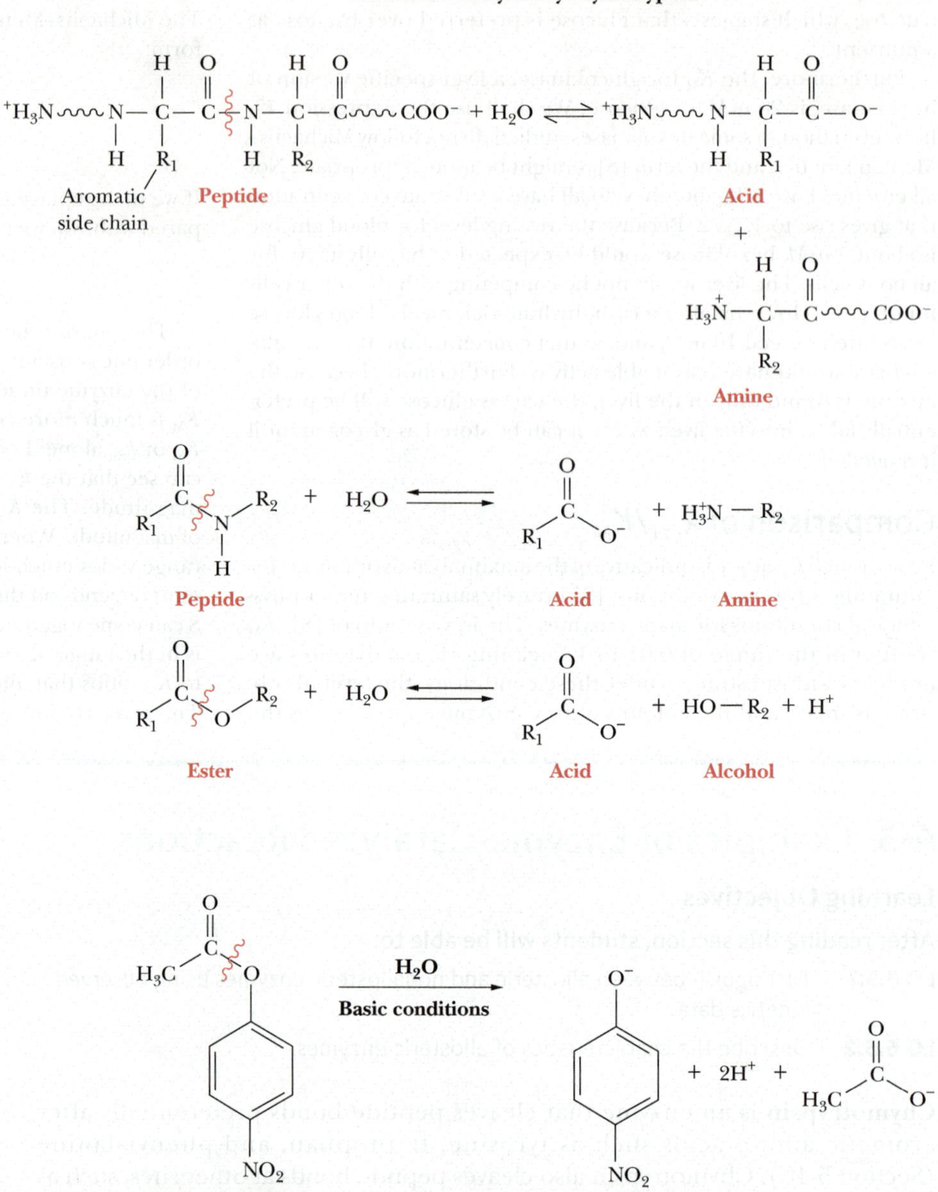

Figure 6.10 Reactions catalyzed by chymotrypsin.

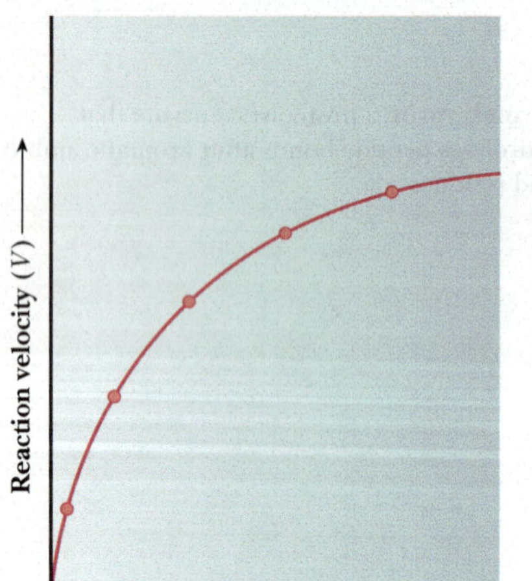

Figure 6.11 Dependence of reaction velocity, *V*, on *p*-nitrophenylacetate concentration, [S], in a reaction catalyzed by chymotrypsin. The shape of the curve for chymotrypsin is hyperbolic.

$$\text{Carbamoyl phosphate} + \text{Aspartate} \rightarrow \text{Carbamoyl aspartate} + HPO_4^{2-}$$

Reaction catalyzed by aspartate transcarbamoylase

Experimental results show that, once again, the rate of the reaction depends on substrate concentration at low and moderate concentrations, and, once again, a maximum rate is reached at high substrate concentrations. There is, however, one very important difference. For this reaction, the graph showing the dependence of reaction rate on substrate concentration has a sigmoidal shape (Figure 6.12) rather than the hyperbolic shape seen for chymotrypsin (Figure 6.11).

▶ Why do chymotrypsin and ATCase have different velocity curves?

The results of experiments on the reaction kinetics of chymotrypsin and ATCase are representative of experimental results obtained with many enzymes. The overall kinetic behavior of many enzymes resembles that of chymotrypsin, while other enzymes behave similarly to ATCase. We can use this information to draw some general conclusions about the behavior of enzymes. The comparison between the kinetic behaviors of chymotrypsin and ATCase is reminiscent of the relationship between the oxygen-binding behaviors of myoglobin and hemoglobin discussed in Chapter 4. ATCase and hemoglobin are allosteric proteins with multiple subunits, while chymotrypsin and myoglobin are proteins with one subunit and are not allosteric proteins. (Recall from Section 4-5 that allosteric proteins are the ones in which subtle changes at one site affect structure and function at another site. Cooperative effects, such as the fact that the binding of the first oxygen molecule to hemoglobin makes it easier for other oxygen molecules to bind, are a hallmark of allosteric proteins.) The differences in behavior between allosteric and nonallosteric proteins can be understood in terms of models based on structural differences between the two kinds of proteins.

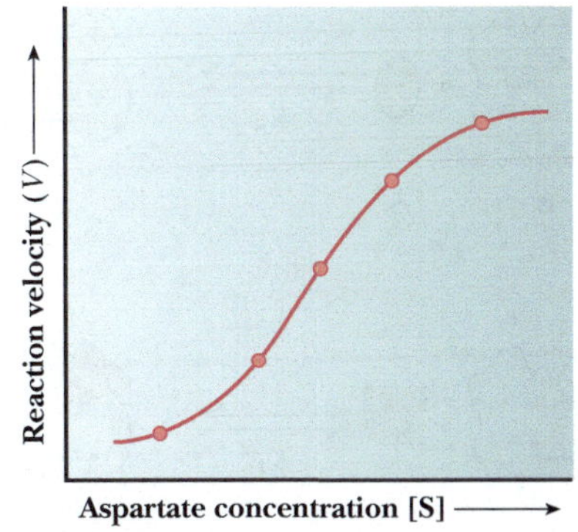

Figure 6.12 Dependence of reaction velocity, *V*, on aspartate concentration, [S], in a reaction catalyzed by aspartate transcarbamoylase (ATCase). The concentration of the other substrate for ATCase (carbamoyl phosphate) is kept constant and at saturating conditions. The shape of the curve for ATCase is sigmoidal.

6-6 Enzyme Inhibition

Learning Objectives

After reading this section, students will be able to:

LO 6.6.1 Distinguish between reversible and irreversible enzyme inhibition.

LO 6.6.2 Distinguish between the four classes of reversible enzyme inhibition.

An **inhibitor** is a substance that interferes with the action of an enzyme and slows the rate of a reaction. A good deal of information about enzymatic reactions can be obtained by observing the changes in the reaction caused by the presence of inhibitors. Inhibitors can affect an enzymatic reaction in two ways. A reversible inhibitor can bind to the enzyme and subsequently be released, leaving the enzyme in its original condition. An irreversible inhibitor reacts with the enzyme to produce a protein that is not enzymatically active and from which the original enzyme cannot be regenerated. Four major classes of reversible inhibitors can be distinguished on the basis of the sites on the enzyme to which they bind and how the binding of the inhibitor affects the binding of the substrate. These four are competitive, noncompetitive, mixed noncompetitive, and uncompetitive inhibitors.

inhibitor a substance that decreases the rate of an enzyme-catalyzed reaction

6-6A Competitive Inhibition

One class of reversible inhibitors consists of compounds very similar in structure to the substrate. In this case, the inhibitor can bind to the active site and block the substrate's access to it. This mode of action is called **competitive inhibition** because the inhibitor competes with the substrate for the active site on the enzyme. With a competitive inhibitor, we can write two different equations, both of which are mutually exclusive:

competitive inhibition a decrease in enzymatic activity caused by active-site binding of a compound that is similar in structure to the substrate

$$E + S \rightleftharpoons ES \quad \text{or} \quad E + I \rightleftharpoons EI \tag{6.8}$$

An equilibrium expression can also be determined for the breakdown of the enzyme–inhibitor complex:

$$K_I = \frac{[E][I]}{[EI]} \tag{6.9}$$

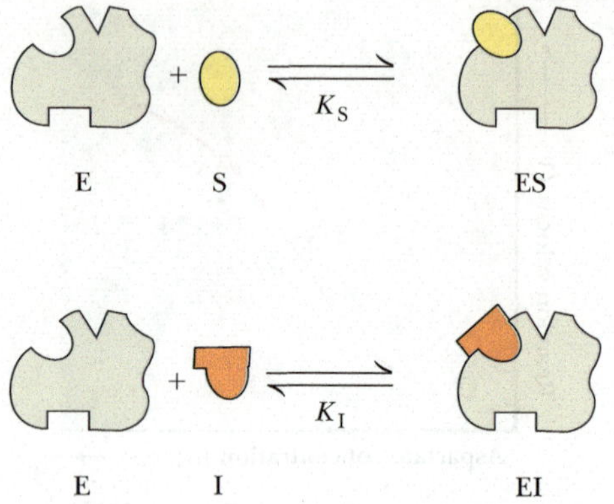

Figure 6.13 **Two mutually exclusive possibilities for substrate or inhibitor binding to an enzyme in the case of competitive inhibition.** *From Campbell/Farrell, Biochemistry, 7E. 2012 Cengage Learning.*

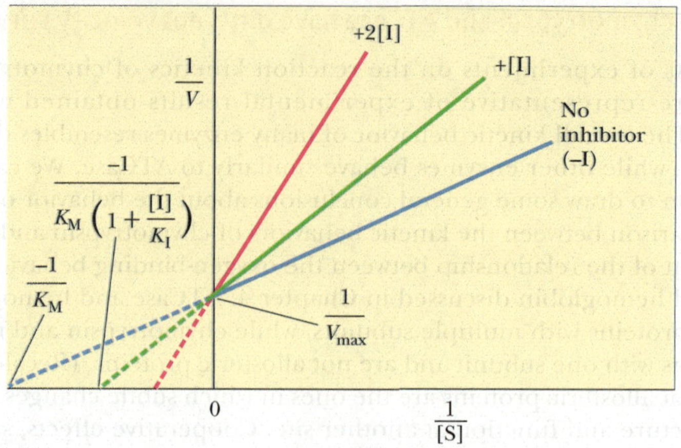

Figure 6.14 **A Lineweaver–Burk double-reciprocal plot of enzyme kinetics for competitive inhibition.** *(From Campbell/Farrell, Biochemistry, 7th Edition. 2012. Cengage Learning.)*

The competition for the active site is shown pictorially in Figure 6.13. Starting with free enzyme, the substrate (yellow) can bind to the active site or the inhibitor (red) can, but they both occupy the same space, so only one can bind at a time.

▶ How can we identify a competitive inhibitor?

The effect on the Lineweaver–Burk plot for a competitive inhibitor is shown in Figure 6.14. As seen in the plot, the *y* intercept remains the same and the *x* intercept increases in the presence of a competitive inhibitor. Because the *y* intercept remains unchanged, this means that the V_{max} stays the same (recall that the *y* intercept of these plots equals $1/V_{max}$). V_{max} stays the same in the presence of a competitive inhibitor because theoretically the inhibitor could be "outcompeted" by infinite substrate concentration. On the other hand, the increase in the *x* intercept means that the K_M value increases in the presence of a competitive inhibitor (recall that the *x* intercept of these plots equals $-1/K_M$). An increase in the K_M value reflects an apparent decrease in the affinity of the enzyme for the substrate because the substrate has to compete with the inhibitor for binding to the active site of the enzyme.

noncompetitive inhibition a form of enzyme inactivation in which a substance binds to a place other than the active site

6-6B Noncompetitive Inhibition

Another class of reversible inhibitors binds to the enzyme somewhere other than the active site and inhibits the activity of the enzyme. This mode of inhibition is called **noncompetitive inhibition**. The enzyme can bind the substrate, the inhibitor, or a combination of the two as shown in Equation 6.10.

$$\begin{array}{ccccc} & \overset{+S}{} & & & \\ E & \rightleftharpoons & ES & \rightarrow & E + P \\ +I \upharpoonleft\!\downharpoonright & & \upharpoonleft\!\downharpoonright + I & & \\ EI & \rightleftharpoons & ESI & & \\ & \underset{+S}{} & & & \end{array} \qquad (6.10)$$

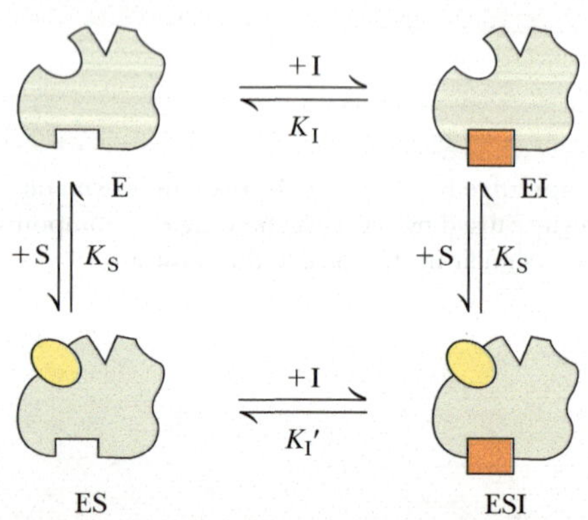

Figure 6.15 **Nature of substrate and inhibitor binding in noncompetitive inhibition.**

Noncompetitive inhibition is shown pictorially in Figure 6.15. There are two distinct binding sites, one for substrate and one for inhibitor. However, the inhibitor affects the rate of catalysis when it is bound. A limiting case of noncompetitive inhibition is called *pure noncompetitive inhibition*. In this type of

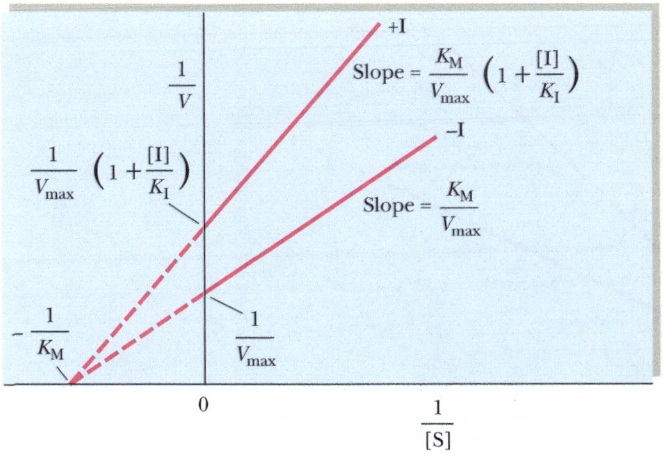

Figure 6.16 A Lineweaver–Burk plot of enzyme kinetics for noncompetitive inhibition.

inhibition, the binding of the inhibitor has no effect on the binding of the substrate. In other words, the substrate can bind just as well to E as it can to EI. The same is true of the inhibitor. It can bind just as well to E as it can to ES. The only difference is that when the inhibitor is bound, catalysis cannot occur, as shown in Equation 6.10.

▶ How can we identify a noncompetitive inhibitor?

The effect on the Lineweaver–Burk plot for a noncompetitive inhibitor is shown in Figure 6.16. Notice that the *y* intercept increases but the *x* intercept stays the same in the presence of a noncompetitive inhibitor. Because the *y* intercept increases, V_{max} decreases. V_{max} decreases in the presence of a noncompetitive inhibitor because the inhibitor could not be "outcompeted" by infinite substrate concentration, given that the inhibitor and substrate are binding to different sites on the enzyme. However, because the *x* intercept stays the same in noncompetitive inhibition, the K_M value stays the same. An unchanged K_M value means there is no change in the affinity of the enzyme for the substrate. There is no change because the substrate does not have to compete with the inhibitor for binding to the enzyme's active site.

6-6C Mixed Noncompetitive Inhibition

Pure noncompetitive inhibition is relatively rare due to its extreme parameters of two molecules being able to bind an enzyme without affecting each other's binding. A more common type of noncompetitive inhibition is called **mixed noncompetitive inhibition**. With this type of inhibition, Equation 6.10 and Figures 6.15 and 6.16 still apply, but the binding of inhibitor (I) does affect the binding of substrate (S). The dissociation constants K_I and K_I' seen in Figure 6.15 are not identical.

mixed noncompetitive inhibition a type of inhibition where the inhibitor binds to a place other than the active site and the binding of inhibitor also distorts the active site, thereby affecting the binding of substrate

▶ How can we distinguish between pure and mixed noncompetitive inhibition?

Because any amount of inhibitor will knock down the activity, the V_{max} is still reduced with a mixed noncompetitive inhibitor, just as it was with a pure noncompetitive inhibitor. However, because the binding of the inhibitor does affect the binding of the substrate, the K_M is also increased. If we look at the inhibition pattern with a Lineweaver–Burk plot, we will see that the lines do not intersect on either of the axes. Instead, they intersect somewhere in the left quadrant, as shown in Figure 6.17.

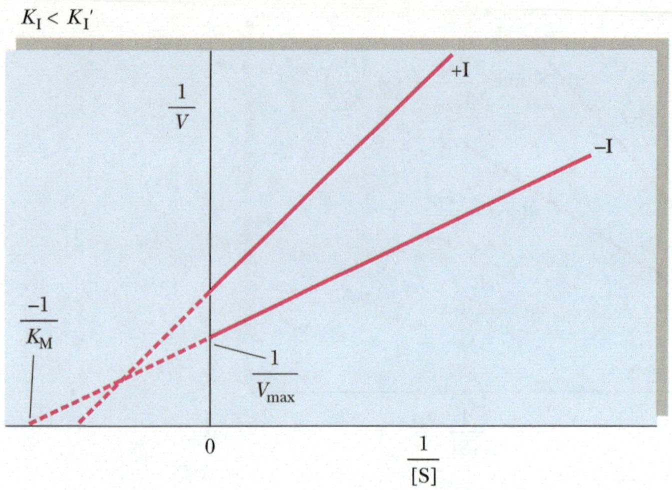

Figure 6.17 A Lineweaver–Burk double-reciprocal plot of enzyme kinetics for mixed noncompetitive inhibition. *(From Campbell/Farrell,* Biochemistry, *7E. 2012 Cengage Learning.)*

6-6D Uncompetitive Inhibition

uncompetitive inhibition a type of inhibition where the inhibitor can only bind to the enzyme–substrate (ES) complex but not to the free enzyme (E)

To complete our look at the basic types of reversible inhibition, we now turn to **uncompetitive inhibition**. An uncompetitive inhibitor only binds to the ES complex, as shown in Equation 6.11:

$$\begin{array}{c} E + S \rightleftharpoons ES \longrightarrow E + P \\ + \\ I \\ \Updownarrow \\ EIS \end{array} \quad (6.11)$$

There is a big difference between uncompetitive inhibition and noncompetitive inhibition, even though the words sound like they mean the same thing. To help in remembering the difference, consider the following mnemonic: an **un**competitive inhibitor is "**unly**" able to bind the ES complex.

▶ How can we identify an uncompetitive inhibitor?

In the Lineweaver–Burk plot for an uncompetitive inhibitor, the lines are parallel, as shown in Figure 6.18. The V_{max} is decreased as we might expect because

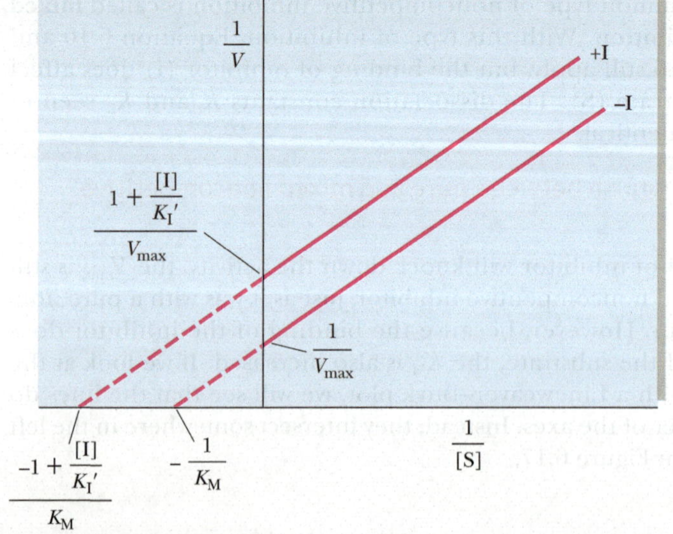

Figure 6.18 A Lineweaver–Burk double-reciprocal plot of enzyme kinetics for uncompetitive inhibition. *(Garrett/Grisham,* Biochemistry, *4th Edition. Cengage Learning.)*

any amount of inhibitor will tie up some of the ES complex into the inactive EIS form. However, the apparent K_M actually decreases, making it appear that the substrate binds to the enzyme with greater apparent affinity. This is because the presence of the inhibitor removes some of the ES complex. By Le Châtelier's principle, this forces the equation to readjust to the right, stimulating the formation of the ES complex.

Apply Your Knowledge 6.2 gives you the opportunity to practice graphing and analyzing enzyme kinetic data for an uninhibited and inhibited system.

6.2 Apply Your Knowledge

Sucrose (common table sugar) is hydrolyzed to glucose and fructose in a classic experiment in kinetics. The reaction is catalyzed by the enzyme invertase. Using the following data and the Lineweaver–Burk method to determine K_M and V_{max} values, determine the type of inhibitor that is present.

Sucrose Concentration (M)	V, No Inhibitor ($M\,s^{-1}$)	V, Inhibitor Present ($M\,s^{-1}$)
0.0292	0.182	0.083
0.0584	0.265	0.119
0.0876	0.311	0.154
0.117	0.330	0.167
0.175	0.372	0.192

Solution

Step 1: Taking the reciprocal of the substrate concentrations and of the velocity values (first and last data points) gives the following results:

1/Sucrose Concentration (M^{-1})	1/V, No Inhibitor ($s\,M^{-1}$)	1/V, Inhibitor Present ($s\,M^{-1}$)
34.2	5.49	12.0
5.71	2.69	5.21

Step 2: Calculate K_M and V_{max} when no inhibitor is present.

$$\text{Slope} = \frac{5.49\text{ s }M^{-1} - 2.69\text{ s }M^{-1}}{34.2\ M^{-1} - 5.71\ M^{-1}} = 0.0983\text{ s}$$

Calculating the y intercept and V_{max} :

$$5.49\text{ s }M^{-1} = (0.0983\text{ s})(34.2\ M^{-1}) + b$$

$$b = 2.13\text{ s }M^{-1} = \frac{1}{V_{max}}$$

$$V_{max} = 0.469\ M\text{ s}^{-1}$$

Calculating K_M :

$$\text{Slope} = \frac{K_M}{V_{max}}$$

$$0.0983\text{ s} = \frac{K_M}{0.469\ M\text{ s}^{-1}}$$

$$K_M = 0.0461\ M$$

Continued on next page

Step 3: Calculate K_M and V_{max} when inhibitor is present.

$$\text{Slope} = \frac{12.0\ \text{s}\ M^{-1} - 5.21\ \text{s}\ M^{-1}}{34.2\ M^{-1} - 5.71\ M^{-1}} = 0.238\ \text{s}$$

Calculating the y intercept and V_{max} :

$$12.0\ \text{s}\ M^{-1} = (0.238\ \text{s})(34.2\ M^{-1}) + b$$

$$b = 3.86\ \text{s}\ M^{-1} = \frac{1}{V_{max}}$$

$$V_{max} = 0.259\ M\text{s}^{-1}$$

Calculating K_M :

$$\text{Slope} = \frac{K_M}{V_{max}}$$

$$0.238\ \text{s} = \frac{K_M}{0.259\ M\ \text{s}^{-1}}$$

$$K_M = 0.0616\ \text{M}$$

Alternatively, you could use linear regression in Microsoft Excel, a graphing calculator, or a website to plug in all of the reciprocal data (not just the first and last data points) and determine the equation of the line when no inhibitor is present as $y = 0.0976x + 2.14$. Calculating K_M and V_{max} from this linear regression equation using the techniques shown here gives the following values when no inhibitor is present: $K_M = 0.0456\ M$ and $V_{max} = 0.467\ M\ \text{s}^{-1}$. When the inhibitor is present, linear regression determines the equation of the line as being $y = 0.241x + 3.92$. Calculating K_M and V_{max} from this linear regression equation using the techniques shown here gives the following values when the inhibitor is present: $K_M = 0.0615\ M$ and $V_{max} = 0.255\ M\ \text{s}^{-1}$. Because the inhibitor causes the V_{max} to decrease and the K_M to increase, it is a mixed noncompetitive inhibitor.

6-6E Irreversible Inhibition

irreversible inhibition covalent binding of an inhibitor to an enzyme, thereby permanently inactivating the enzyme

Finally, the other basic type of inhibition is **irreversible inhibition**. In this type of inhibition, the inhibitor binds to the enzyme permanently and inactivates it. Graphically, this looks most like noncompetitive inhibition, because any amount of the inhibitor will reduce the activity of the enzyme.

To conclude our exploration of enzyme inhibition, **Biochemical Connections 6D** looks at how inhibitors are used in the treatment of AIDS.

6D Biochemical Connections • Medicine

Enzyme Inhibition in the Treatment of AIDS

A key strategy in the treatment of acquired immunodeficiency syndrome (AIDS) has been to develop specific inhibitors that selectively block the actions of enzymes unique to the human immunodeficiency virus (HIV), which causes AIDS. Many laboratories are working on this approach to the development of therapeutic agents. Figure 6.19 shows the process of HIV infection. Three key enzymes are current targets for AIDS therapy—reverse transcriptase, integrase, and protease.

One of the most important target enzymes is HIV protease, an enzyme essential to the production of new virus particles in infected

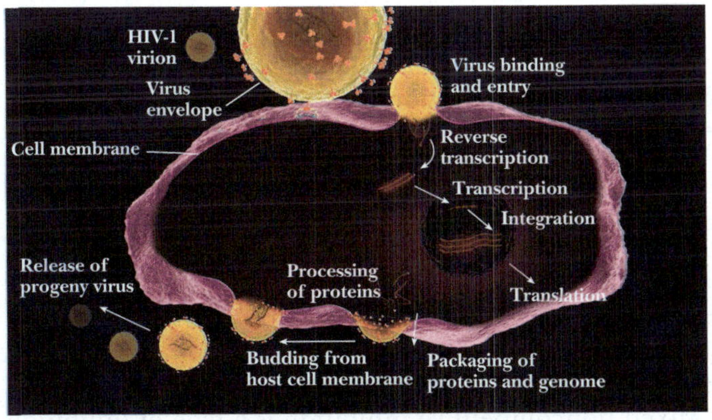

Louise Martin-Carpenter/GlaxoSmithKline

Figure 6.19 From the time a cell is infected with HIV, three key enzymes are involved in the replication of the virus: reverse transcriptase, integrase, and protease. *(Reprinted by permission from* Science *311, 943 (2006)).*

cells. HIV protease is unique to this virus. It catalyzes the processing of viral proteins in an infected cell. Without these proteins, viable virus particles cannot be released to cause further infection.

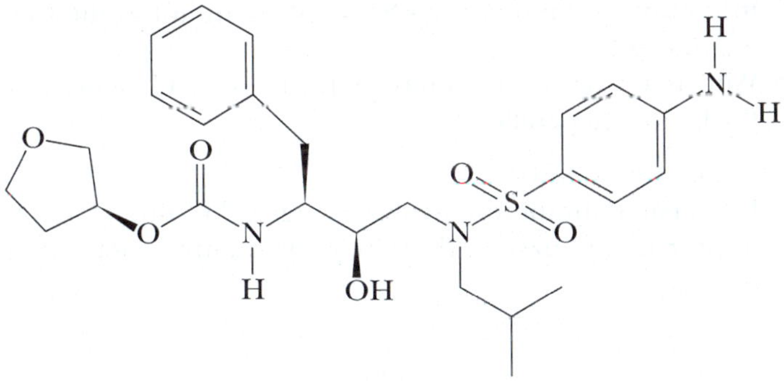

Figure 6.20A The structure of amprenavir (VX-478), an HIV protease inhibitor developed by Vertex Pharmaceuticals. *(Vertex Pharmaceuticals, Inc.)*

The structure of HIV protease, including its active site, was determined using X-ray crystallography (Section 4-4A). Using this crystal structure, scientists designed and synthesized compounds to bind to the active site. Improvements were made in the drug design by obtaining structures of a series of inhibitors bound to the active site of HIV protease. These structures were also elucidated by X-ray crystallography. This process eventually led to several compounds marketed by different pharmaceutical companies. These HIV protease inhibitors include saquinavir from Hoffman-LaRoche, ritonavir from Abbott Laboratories, indinavir from Merck, Viracept from Pfizer, and amprenavir (shown in Figure 6.20) from Vertex Pharmaceuticals.

Another target is the viral enzyme called integrase, which is needed for the virus to copy itself in the host cell. A recent drug made by Merck, called MK-0518, inhibits the integrase enzyme. Treatment of AIDS is most effective when a combination of drug therapies is used, and HIV protease, integrase, and reverse transcriptase inhibitors play an important role. The use of multiple inhibitors for the key viral enzymes allows levels of each to remain below levels that would be toxic to the cell.

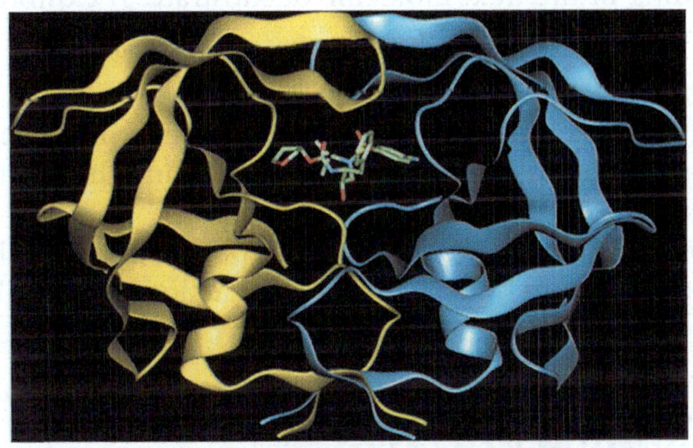

Figure 6.20B Active site of HIV-1 protease complexed with VX-478. *(Vertex Pharmaceuticals, Inc.)*

Summary

Enzyme Kinetics versus Thermodynamics

- Thermodynamic spontaneity cannot tell us whether a reaction will be fast.
- The speed of an enzyme-catalyzed reaction is a kinetic property controlled by the nature of the energy state of the ES complex and the transition state.

The Transition State

- Enzymes speed up the reaction rate by creating a situation where the Gibbs energy of activation is reduced.
- A chemical reaction may go faster at higher temperatures. However, when the reaction is catalyzed by an enzyme, this is true only for a specific range of temperatures. If the temperature is raised too much, it denatures the enzyme and the rate of reaction is reduced significantly, perhaps to zero.

Rates of Enzyme-Catalyzed Reactions

- In many situations, the concentration of the substrate influences the rate of an enzyme-catalyzed reaction. However, if there is a saturating amount of substrate, then all of the enzyme molecules are bound to substrate. Adding more substrate under these conditions will not increase the rate of reaction. When this happens, the enzyme is working at its V_{max} and is exhibiting zero-order kinetics.

Enzyme–Substrate Binding

- Enzymes and substrates are attracted to each other via noncovalent interactions, and the substrate binds to the active site of the enzyme. The enzyme's active site has amino acids in a specific orientation where they can bind to the substrate and catalyze the reaction.

Determining K_M and V_{max}

- K_M and V_{max} can be estimated by plotting the velocity versus [S]. However, a more accurate way is to make a Lineweaver–Burk plot of $1/V$ versus $1/[S]$. With such a graph, the *y* intercept yields $1/V_{max}$, which can then be converted to V_{max}. The *x* intercept is $-1/K_M$, which can be converted to K_M.

Significance of K_M and V_{max}

- Mathematically, K_M is equal to the substrate concentration that yields a velocity of $V_{max}/2$. It is an inverse measure of the affinity between the enzyme and substrate, where a low K_M indicates a high affinity. V_{max} tells us how fast the enzyme can generate product under saturating substrate conditions.

Examples of Enzyme-Catalyzed Reactions

- Chymotrypsin and ATCase exhibit different types of kinetics.
- Chymotrypsin is a nonallosteric enzyme and exhibits hyperbolic kinetics.
- ATCase is an allosteric enzyme with multiple subunits, where the binding of one molecule of substrate affects the binding of the next molecule of substrate. It exhibits sigmoidal kinetics.

Competitive Inhibition

- Competitive inhibition is a decrease in enzymatic activity caused by active-site binding of a compound that is similar in structure to the substrate.
- By comparing a Lineweaver–Burk plot of an uninhibited reaction to one for an inhibited reaction, the inhibitor can be identified as competitive if the curves intersect on the *y*-axis.

Noncompetitive Inhibition

- Noncompetitive inhibition is a form of enzyme inactivation in which a substance binds to a place other than the active site.
- With a pure noncompetitive inhibitor, where the binding of inhibitor does *not* affect the binding of substrate, a Lineweaver–Burk plot shows lines that intersect on the *x*-axis.

Mixed Noncompetitive Inhibition

- Mixed noncompetitive inhibition is a type of inhibition where the inhibitor binds to a place other than the active site and where the binding of inhibitor also distorts the active site and affects the binding of substrate.
- With a mixed noncompetitive inhibitor, the lines on a Lineweaver–Burk plot intersect somewhere in the left quadrant and not on either axis.

Uncompetitive Inhibition

- Uncompetitive inhibition is a type of inhibition where the inhibitor can bind to the ES complex but not to the free enzyme (E).
- With an uncompetitive inhibitor, the lines on a Lineweaver–Burk plot are parallel.

Irreversible Inhibition

- Irreversible inhibition involves covalent binding of an inhibitor to an enzyme, thereby permanently inactivating the enzyme.

Review Exercises

6-1 Enzyme Kinetics versus Thermodynamics

1. **Recall** How does the catalytic effectiveness of enzymes compare with that of nonenzymatic catalysts?
2. **Reflect and Apply** For the reaction of glucose with oxygen to produce carbon dioxide and water,

$$\text{Glucose} + 6\ O_2 \rightarrow 6\ CO_2 + 6\ H_2O$$

$\Delta G^\circ = -2880$ kJ mol^{-1}, making it a strongly exergonic reaction. Explain how it is possible for a sample of glucose to be maintained indefinitely in an oxygen-containing atmosphere.

3. **Reflect and Apply** Given what you know about enzyme structure, why does heating a solution containing an enzyme to above 50°C decrease enzyme activity?
4. **Reflect and Apply** Draw energy profiles for an exergonic reaction in the presence and absence of a catalyst. Use the energy profile to explain why the standard Gibbs energy change of a chemical reaction is not affected by the presence of a catalyst.
5. **Reflect and Apply** Consider a reverse reaction where the products are being converted back to substrate. What effect does a catalyst have on the activation energy of the reverse reaction?
6. **Reflect and Apply** Can the presence of a catalyst increase the amount of product obtained in a reaction?

6-2 Rates of Enzyme-Catalyzed Reactions

7. **Recall** For the hypothetical reaction

$$3A + 2B \rightarrow 2C + 3D$$

the rate was experimentally determined to be

$$\text{Rate} = k[A][B]$$

What is the order of the reaction with respect to A? With respect to B? What is the overall order of the reaction?

8. **Recall** While assaying an enzyme-catalyzed reaction, you observe that the rate of the reaction is the same even when you double the substrate concentration. What order is this reaction with respect to the substrate you are varying?
9. **Reflect and Apply** The enzyme lactate dehydrogenase catalyzes the reaction

$$\text{Pyruvate} + \text{NADH} + H^+ \rightarrow \text{lactate} + \text{NAD}^+$$

NADH absorbs light at 340 nm in the near-ultraviolet region of the electromagnetic spectrum, but NAD^+ does not. Suggest an experimental method for following the rate of this reaction, assuming that you have a spectrophotometer capable of measuring light at this wavelength.

10. **Reflect and Apply** Why is it critical to monitor the rate of the lactate dehydrogenase reaction in Question 9 in a buffered solution?

6-3 Enzyme–Substrate Binding

11. **Recall** Compare and contrast the lock-and-key and induced-fit models for binding a substrate to an enzyme.
12. **Recall** Describe the effect that a perfect fit of the enzyme–substrate complex would have on the activation energy of the catalyzed reaction. Reference the energy of the ES complex and the energy diagram in your answer.
13. **Recall** Use the energy profile for an enzymatic reaction to explain why enzymes catalyze both the forward and reverse reactions by filling in the blanks:

    The enzyme forms favorable bonds with the ______________ to decrease the activation energy, and the ______________ is the same whether the reaction proceeds from left to right or right to left.
14. **Reflect and Apply** Amino acids that are far apart in the amino acid sequence (primary structure) of an enzyme can be essential for its catalytic activity. Use your understanding of tertiary structure and the lock-and-key model to suggest a role for these amino acids. How might your answer vary if you consider the induced fit model instead?

6-4 The Michaelis–Menten Approach to Enzyme Kinetics

15. **Recall** Why is a Lineweaver–Burk plot useful in analyzing kinetic data from enzymatic reactions?
16. **Reflect and Apply** Compare the Michaelis–Menten kinetic parameters (K_M and V_{max}) of an enzyme-catalyzed reaction measured at two different enzyme concentrations.
17. **Recall** How is the turnover number of an enzyme related to V_{max}?
18. **Mathematical** Determine the values of K_M and V_{max} for the decarboxylation of a β-keto acid given the following data.

Substrate Concentration (mol L⁻¹)	Velocity (mM min⁻¹)
2.500	0.588
1.000	0.500
0.714	0.417
0.526	0.370
0.250	0.256

19. **Mathematical** The kinetic data in the following table were obtained for the reaction of carbon dioxide and water to produce bicarbonate and hydrogen ion catalyzed by carbonic anhydrase:

$$CO_2 + H_2O \rightarrow HCO_3^- + H^+$$

[H. De Voe and G. B. Kistiakowsky, *J. Am. Chem. Soc.* **83,** 274 (1961)]. From these data, determine K_M and V_{max} for the reaction.

Carbon Dioxide Concentration (mmol L⁻¹)	1/Velocity (M^{-1} s)
1.25	36×10^3
2.5	20×10^3
5.0	12×10^3
20.0	6×10^3

20. **Mathematical** For the V_{max} obtained in Question 19, calculate the turnover number (catalytic rate constant) assuming that 1×10^{-4} M of enzyme were used.
21. **Mathematical** You do an enzyme kinetic experiment and calculate a V_{max} of 100 μM of product produced per minute. If each assay contained an enzyme concentration of 0.2 mg/mL, what would be the turnover number if the enzyme had a molecular weight of 128,000 g/mol?
22. **Reflect and Apply** The enzyme D-amino acid oxidase has a very high turnover number because the D-amino acids are potentially toxic. The K_M for the enzyme is in the range of 1 to 2 mM for the aromatic amino acids and in the range of 15 to 20 mM for such amino acids as serine, alanine, and the acidic amino acids. Which of these amino acids are the preferred substrates for the enzyme?
23. **Reflect and Apply** You have the choice between two enzymes (Y and Z) that catalyze the same reaction. Enzyme Y has a lower K_M than enzyme Z, but enzyme Z has a higher V_{max} than enzyme Y. Which enzyme would produce product the fastest if you could maintain high substrate concentrations in the reaction? Assume substrate concentrations are 10 times higher than the K_M of Z.
24. **BIOCHEMICAL CONNECTIONS** When does the k_{cat}/K_M value approximate the catalytic efficiency of an enzyme?
25. **Recall** What are the three most common mechanisms for enzyme-catalyzed reactions that have two substrates?
26. **Recall** In which mechanism of multi-substrate binding will a product be released before the second substrate is bound?
27. **Recall** How do scientists determine the K_M of a substrate that is part of an ordered reaction with two substrates?

6-5 Examples of Enzyme-Catalyzed Reactions

28. **Recall** Show graphically the dependence of initial velocity on substrate concentration for an enzyme that follows Michaelis–Menten kinetics and for an allosteric enzyme.
29. **Recall** Do all enzymes display kinetics that obey the Michaelis–Menten equation? If not, which ones do not?
30. **Recall** How can you recognize an enzyme that does not display Michaelis–Menten kinetics?
31. **Recall** What does it mean to say an enzyme like aspartate transcarbamoylase (ATCase) exhibits cooperativity? Your answer should include a description of subunit behavior.
32. **Reflect and Apply** You performed an enzyme kinetics assay and generated a Lineweaver–Burk double-reciprocal plot of the data you collected only to find that that data don't form a straight line. Is your enzyme behaving allosterically? What does this mean about the number of subunits in your enzyme?

6-6 Enzyme Inhibition

33. **Recall** How can competitive and pure noncompetitive inhibition be distinguished in terms of changes to K_M?
34. **Recall** Why does a competitive inhibitor not change V_{max}?
35. **Recall** Why does a pure noncompetitive inhibitor not change the observed K_M?
36. **Recall** Distinguish between the molecular mechanisms of competitive and noncompetitive inhibition.
37. **Recall** Explain one mechanism of enzyme inhibition that cannot be reversed and a different situation that would lead to reversal of an inhibited enzyme.
38. **Recall** Where do lines intersect on a Lineweaver–Burk plot showing competitive inhibition? On a Lineweaver–Burk plot showing noncompetitive inhibition?
39. **Recall** What is the difference between pure and mixed noncompetitive inhibition?
40. **Reflect and Apply** Why do the enzyme kinetics in the presence of a pure noncompetitive inhibitor resemble the same reaction but with less enzyme present?
41. **Reflect and Apply** When we compared the binding of a mixed noncompetitive inhibitor, I, and of S to the enzyme, we assumed that the binding of I decreased the affinity of the enzyme for S. What would the Lineweaver–Burk plot look like if instead binding of I increased the affinity for S? Would this still be classified as "mixed noncompetitive inhibition"?

42. **Recall** Why does the apparent K_M decrease in the presence of an uncompetitive inhibitor?

43. **Mathematical** Draw Lineweaver–Burk plots for the behavior of an enzyme for which the following experimental data are available.

[S] (m*M*)	*V*, No Inhibitor (mmol min^{-1})	*V*, Inhibitor Present (mmol min^{-1})
3.0	4.58	3.66
5.0	6.40	5.12
7.0	7.72	6.18
9.0	8.72	6.98
11.0	9.50	7.60

Is the inhibitor competitive or noncompetitive? How can you tell?

44. **Mathematical** Describe the mechanism of inhibition for hydroxymethylaspartate in the following aspartase reaction.

[S] (molarity)	*V*, No Inhibitor (arbitrary units)	*V*, Inhibitor Present (same arbitrary units)
1×10^{-4}	0.026	0.010
5×10^{-4}	0.092	0.040
1.5×10^{-3}	0.136	0.086
2.5×10^{-3}	0.150	0.120
5×10^{-3}	0.165	0.142

45. **Reflect and Apply** Noncompetitive inhibition is a limiting case in which the effect of binding inhibitor has no effect on the affinity for the substrate and vice versa. Suggest what a Lineweaver–Burk plot would look like for an inhibitor that had a reaction scheme similar to that for a noncompetitive inhibition reaction (Section 6-6B) but where inhibitor binding lowered the affinity of EI for the substrate. What type of inhibitor would this be considered?

46. **Reflect and Apply** Would you expect an irreversible inhibitor of an enzyme to be bound by covalent or noncovalent interactions? Why?

47. **Reflect and Apply** Would you expect the structure of a noncompetitive inhibitor of a given enzyme to be similar to that of its substrate? Why or why not?

48. **Reflect and Apply** Two inhibitors (A and B) have a similar affinity for the enzyme in the absence of substrate. Inhibitor A is a noncompetitive inhibitor, while inhibitor B is a competitive inhibitor. Which inhibitor would produce the lowest initial velocity (V_o) at very high concentrations of substrate?

49. **Recall** Amprenavir (VX-478) is an HIV drug that binds to the active site of the HIV protease. What type of inhibitor is amprenavir?

Further Reading

Althaus, I., J. Chou, A. Gonzales, M. Deibel, K. Chou, F. Kezdy, D. Romero, J. Palmer, R. Thomas, P. Aristoff, W. Tarpley, and F. Reusser. Kinetic Studies with the Non-nucleoside HIV-1 Reverse Transcriptase Inhibitor U-88204E. *Biochemistry* **32**, 6548–6554 (1993). How enzyme kinetics can play a role in AIDS research.

Bachmair, A., D. Finley, and A. Varshavsky. *In Vivo* Half-Life of a Protein Is a Function of Its Amino Terminal Residue. *Science* **234**, 179–186(1986). A particularly striking example of the relationship between structure and stability in proteins.

Bender, M. L., R. L. Bergeron, and M. Komiyama. *The Bioorganic Chemistry of Enzymatic Catalysis.* New York: Wiley, 1984. A discussion of mechanisms in enzymatic reactions.

Binkley, A., Zimmerman, M., and Maguire, C. Expanding Treatment Opportunities: Reviewing the Current State of Injectable Antiretrovirals for Treatment of HIV. *Inf. Dis. and Ther.* **13**, 2475–2488 (2024). A summary of progress in the fight against AIDS.

Danishefsky, S. Catalytic Antibodies and Disfavored Reactions. *Science* **259**, 469–470 (1993). A short review of chemists' use of antibodies as the basis of "tailor-made" catalysts for specific reactions.

Dressler, D., and H. Potter. *Discovering Enzymes.* New York: Scientific American Library, 1991. A well-illustrated book that introduces important concepts of enzyme structure and function.

Dugas, H., and C. Penney. *Bioorganic Chemistry: A Chemical Approach to Enzyme Action.* New York: Springer-Verlag, 1981. Discusses model systems as well as enzymes.

Fersht, A. *Enzyme Structure and Mechanism,* 2nd ed. New York: Freeman, 1985. A thorough coverage of enzyme action.

Hammes, G. *Enzyme Catalysis and Regulation.* New York: Academic Press, 1982. A good basic text on enzyme mechanisms.

Kraut, J. How Do Enzymes Work? *Science* **242**, 533–540 (1988). A discussion of the role of transition states in enzymatic catalysis.

Lerner, R., S. Benkovic, and P. Schultz. At the Crossroads of Chemistry and Immunology: Catalytic Antibodies. *Science* **252**, 659–667 (1991). A review of how antibodies can bind to almost any molecule of interest and then catalyze some reaction of that molecule.

Marcus, R. Skiing the Reaction Rate Slopes. *Science* **256**, 1523–1524 (1992). A brief look at reaction transition states.

Miller, G. Enzyme Keeps Old Memories Alive. *Science* **317**, 887 (2007). An article that discusses how the enzyme called protein kinase M zeta plays a crucial role in maintaining long-term memories by sustaining synaptic strength.

Miller, G. Enzyme Lets You Enjoy the Bubbly. *Science* **326**, 349 (2009). This article highlights research on the enzyme carbonic anhydrase IV, which helps detect carbonation in beverages by converting CO_2 into carbonic acid.

Moore, J. W., and R. G. Pearson. *Kinetics and Mechanism,* 3rd ed. New York: Wiley Interscience, 1980. A classic treatment of the use of kinetic data to determine mechanisms.

Rini, J., U. Schulze-Gahmen, and I. Wilson. Structural Evidence for Induced Fit as a Mechanism for Antibody–Antigen Recognition. *Science* **255**, 959–965 (1992). The results of structure determination by X-ray crystallography.

Sigman, D., ed. *The Enzymes,* Vol. 20, *Mechanisms of Catalysis.* San Diego: Academic Press, 1992. Part of a definitive series on enzymes and their structures and functions.

Sigman, D., and P. Boyer, eds. *The Enzymes,* Vol. 19, *Mechanisms of Catalysis.* San Diego: Academic Press, 1990. Part of a definitive series on enzymes and their structures and functions.

Signals regulate the flow of traffic in much the same fashion as control mechanisms in chemical reactions.

The Behavior of Enzymes: Allostery, Mechanisms, and Control 7

Chapter Outline

7-1 The Behavior of Allosteric Enzymes

Learning Objectives

After reading this section, students will be able to:

LO 7.1.1 Describe the structure, reaction, regulation, and kinetics of the allosteric enzyme called aspartate transcarbamoylase (ATCase).

LO 7.1.2 Distinguish between different classes of allosteric enzymes and allosteric effectors.

The behavior of many well-known enzymes can be described quite adequately by the Michaelis–Menten model, but allosteric enzymes behave very differently. In Chapter 6, we saw similarities between the reaction kinetics of an enzyme such as chymotrypsin, which does not display allosteric behavior (Section 6-5), and the binding of oxygen by myoglobin, which is also an example of nonallosteric behavior (Section 4-5A). The analogy extends to show the similarity in the kinetic behavior of an allosteric enzyme such as aspartate transcarbamoylase (ATCase) and the binding of oxygen by hemoglobin (Section 6-5 and Section 4-5A, respectively). Both ATCase and hemoglobin are allosteric proteins; the behaviors of both exhibit cooperative effects caused by subtle changes in quaternary structure. (Recall that *quaternary structure* is the arrangement in space that results from the interaction of subunits (most frequently through noncovalent forces), and that *positive cooperativity* occurs when the binding of low levels of substrate facilitates the action of the protein at higher levels of substrate, whether the action is catalytic or some other kind of binding.)

Pioneers in Biochemistry • Edwin Krebs

Edwin Krebs (1918–2009) chose to pursue a scientific education only because he hoped it would lead to a financially secure career during the uncertainty of the Great Depression. Then he attended medical school hoping to contribute to the war effort. Serving as a medical officer in the U.S. Navy near the end of World War II turned out to be the only period of his life when he practiced clinical medicine. Unable to find a hospital position right away after being discharged from the Navy, Krebs took up the study of biochemistry, imagining it would complement his career in internal medicine. As it turned out, biochemistry research and teaching were the focus of the rest of his career.

As a researcher at the University of Washington, Krebs worked with Edmond Fischer to explain the molecular mechanism by which interconversion of the two forms of phosphorylase takes place. It was not understood until much later how fundamental this process is in human physiology and disease, controlling many metabolic processes and contributing to cell division, shape, and motility. Fischer and Krebs were jointly awarded Nobel Prize in Physiology or Medicine in 1992 for describing how reversible phosphorylation works as a switch to activate proteins and regulate various cellular processes. Krebs, who was hearing-impaired, tells the story of being surprised to find out that he had won the Nobel Prize. Having removed his hearing aids to sleep, Krebs awoke in the morning to find 75 voicemails and reporters knocking on his door.

Journal of Biological Chemistry

Chapter Outline (continued)

feedback inhibition the process by which the final product of a series of reactions inhibits the first reaction in the series

In addition to displaying cooperative kinetics, allosteric enzymes have a different response to the presence of inhibitors from that of nonallosteric enzymes.

7-1A Feedback Inhibition

ATCase catalyzes the first step in a series of reactions in which the end product is cytidine triphosphate (CTP) (Figure 7.1), a nucleoside triphosphate needed to make RNA and DNA (Chapter 9). The pathways that produce nucleotides are energetically costly and involve many steps. The reaction catalyzed by ATCase is a good example of how such a pathway is controlled to avoid overproduction of these compounds. For DNA and RNA synthesis, the levels of several nucleoside triphosphates are controlled. CTP is an inhibitor of ATCase, the enzyme that catalyzes the first reaction in the pathway. This behavior is an example of **feedback inhibition** (also called end-product inhibition), in which the end product of the sequence of reactions inhibits the first reaction in the series (Figure 7.2). Feedback inhibition is an efficient control mechanism because the entire series of reactions can be shut down when an excess of the final product exists, thus preventing the accumulation of intermediates in the pathway. Feedback inhibition is a general feature of metabolism and is not confined to allosteric enzymes. However, the observed kinetics of the ATCase reaction, including the mode of inhibition, are typical of allosteric enzymes.

7-1B Cooperative Behavior of Allosteric Enzymes

When ATCase catalyzes the condensation of aspartate and carbamoyl phosphate to form carbamoyl aspartate, the graphical representation of the rate as a function of increasing substrate concentration (aspartate) is a sigmoidal curve, rather than the hyperbola obtained with nonallosteric enzymes (Figure 7.3A). The sigmoidal curve indicates the cooperative behavior of allosteric enzymes. In this two-substrate reaction, aspartate is the substrate for which the concentration is varied, while the concentration of carbamoyl phosphate is kept constant at high levels.

Figure 7.3B compares the rate of the uninhibited reaction of ATCase with the reaction rate in the presence of CTP. In the latter case, a sigmoidal curve

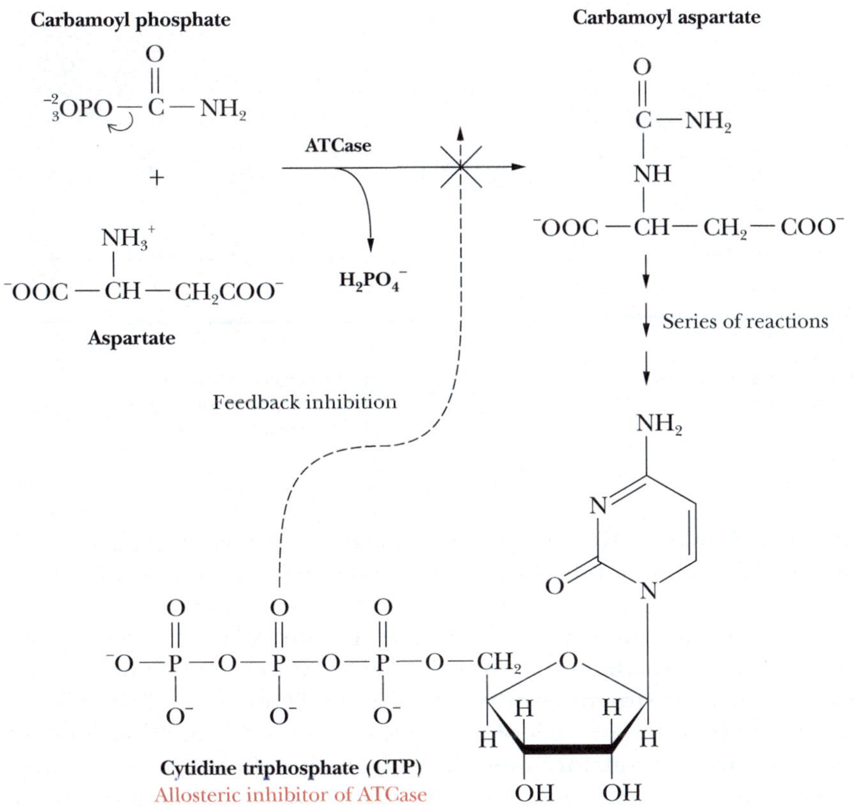

Figure 7.1 The series of reactions that produces cytidine triphosphate (CTP). Aspartate transcarbamoylase (ATCase) catalyzes the first reaction in the pathway, and high levels of CTP can feedback and inhibit ATCase.

still describes the rate behavior of the enzyme, but the curve is shifted to higher substrate levels; a higher concentration of aspartate is needed for the enzyme to achieve the same rate of reaction. At high substrate concentrations, the same maximal rate, V_{max}, is observed in the presence and absence of inhibitor. (Recall this from Section 6-6.) In the Michaelis–Menten scheme, V_{max} changes when a reaction takes place in the presence of a noncompetitive inhibitor, so noncompetitive inhibition cannot be the case here.

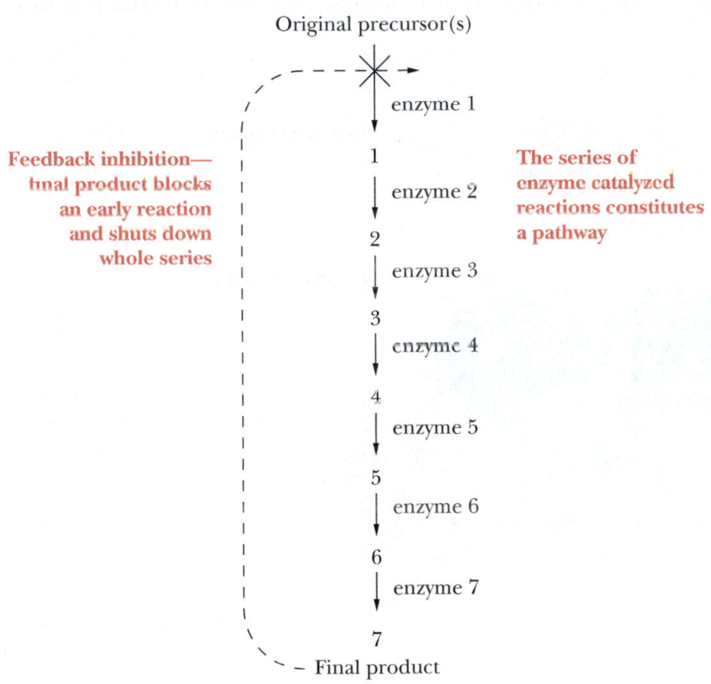

Figure 7.2 Schematic representation of a pathway, showing feedback inhibition.

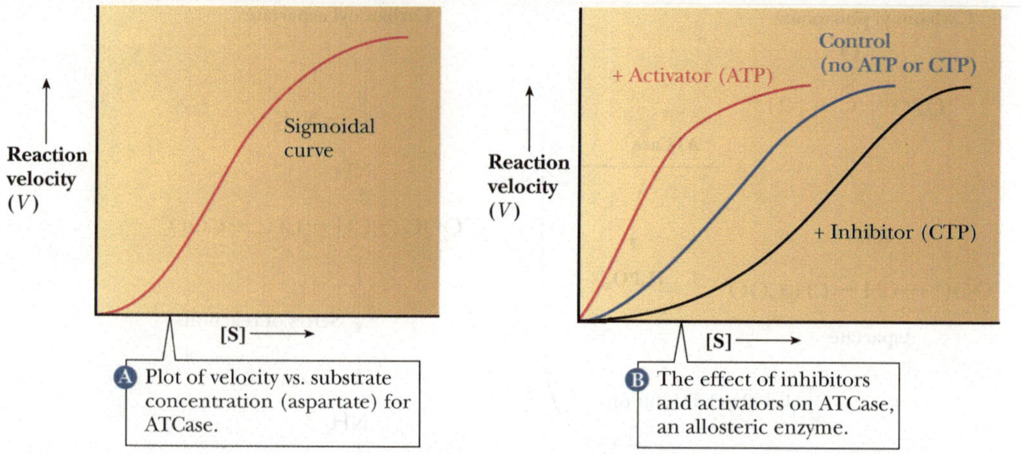

Figure 7.3 **Enzyme kinetics of aspartate transcarbamoylase (ATCase), an allosteric enzyme.**

The same Michaelis–Menten model associates this sort of behavior with competitive inhibition, but that part of the model still does not provide a reasonable picture. Competitive inhibitors bind to the same site as the substrate because they have very similar structures. The structure of CTP is very *different* from the substrate, aspartate, and it is bound to a different site on the ATCase molecule (review the structures of aspartate and CTP in Figure 7.1). ATCase is made up of two different types of subunits: a catalytic subunit that contains the active site and a regulatory subunit that contains effector (activator and inhibitor) binding sites. Six of the catalytic subunits organize into two trimers, while six of the regulatory subunits organize into three dimers (Figure 7.4). Thus, 12 protein subunits in ATCase come together in such a way as to allow for enzymatic catalysis and enzymatic regulation (activation or inhibition). Several more examples of complex enzyme structure and regulation are featured in this and other chapters of this book.

ATCase regulation becomes even more interesting when the ATCase reaction takes place not in the presence of CTP, a pyrimidine nucleoside triphosphate, but in the presence of adenosine triphosphate (ATP), a purine nucleoside triphosphate (Figure 7.5). CTP (Figure 7.1) and ATP are similar structurally, but ATP is *not* a product of the pathway that includes the reaction

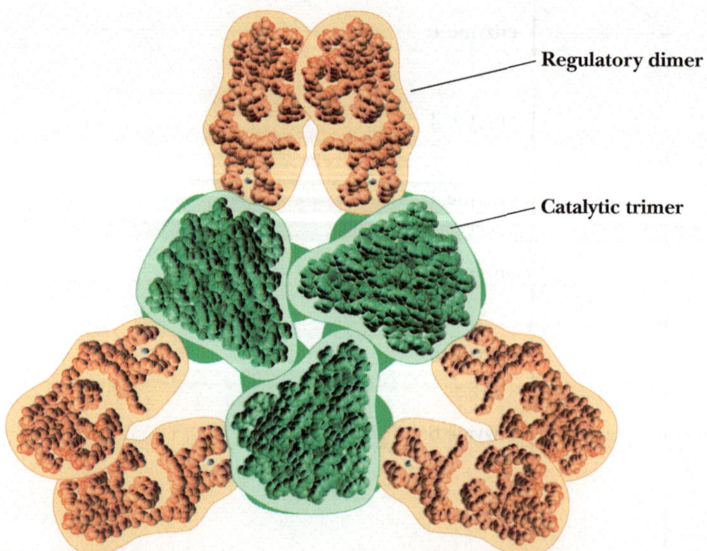

Figure 7.4 **Organization of aspartate transcarbamoylase.** There are two catalytic trimers and the three regulatory dimers.

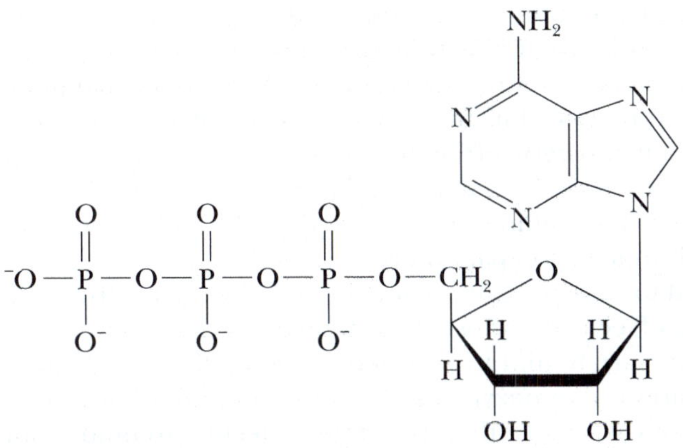

Figure 7.5 Adenosine triphosphate (ATP). ATP is a purine nucleotide and an activator of ATCase.

of ATCase and produces CTP. Both ATP and CTP are needed for the synthesis of RNA and DNA. The relative proportions of ATP and CTP are specified by the needs of the organism. If there is not enough CTP relative to the amount of ATP, the enzyme requires a signal to produce more. In the presence of ATP, the rate of the enzymatic reaction is increased at lower levels of aspartate, and the shape of the rate curve becomes less sigmoidal and more hyperbolic (Figure 7.3B). In other words, there is less cooperativity in the reaction. The binding site for ATP on the enzyme molecule is the same as that for CTP (which is not surprising in view of their structural similarity), but ATP is an activator rather than an inhibitor like CTP.

In summary, when CTP is in short supply in an organism, the ATCase reaction is not inhibited, and the binding of ATP increases the activity of the enzyme still more. Thus, slight differences in the structure of a ligand (CTP versus ATP, in this case) cause drastic changes in the activity of an enzyme. Other examples of this are discussed in other chapters, and once again, we see an opportunity to marvel at the exquisite design and control that is present in biochemical systems. The website Proteopedia offers interactive views of many enzymes. We encourage you to visit the ATCase page (and the pages of other enzymes discussed in this book) on the Proteopedia site, and click on the various links on each page to see these biochemical systems come to life in three dimensions.

Even though it is tempting to consider inhibition of allosteric enzymes in the same fashion as nonallosteric enzymes, much of the terminology is not appropriate. *Competitive inhibition* and *noncompetitive inhibition* are terms reserved for the enzymes that behave in line with Michaelis–Menten kinetics. With allosteric enzymes, the situation is more complex.

7-1c Types of Allosteric Enzymes and Allosteric Effectors

In general, two types of allosteric enzyme systems exist, called **K systems** and **V systems**. A K system is an allosteric enzyme for which the substrate concentration that yields one-half V_{max} is altered by the presence of inhibitors or activators. ATCase is a K system. Because we are not dealing with a Michaelis–Menten type of enzyme, the term K_M is not applicable. For an allosteric enzyme, the substrate level at one-half V_{max} is called the $\boldsymbol{K_{0.5}}$. In a V system, the effect of inhibitors and activators changes the V_{max} but not the $K_{0.5}$.

The key to allosteric behavior, including cooperativity and modifications of cooperativity, is the existence of multiple forms for the quaternary structures of allosteric proteins. The word *allosteric* is derived from *allo,* "other," and *steric,* "shape," referring to the fact that the possible conformations affect the

K systems combinations of allosteric enzymes and inhibitors or activators, in which the presence of the inhibitor/activator changes the substrate concentration that yields one-half V_{max}

V systems combinations of allosteric enzymes and inhibitors or activators in which the presence of the inhibitor/activator changes the maximal velocity of the enzyme but not the substrate level that yields one-half V_{max}

$\boldsymbol{K_{0.5}}$ the substrate level at one-half V_{max} for an allosteric enzyme

behavior of the protein. The binding of substrates, inhibitors, and activators changes the quaternary structure of allosteric proteins, and the changes in structure are reflected in the behavior of those proteins. A substance that modifies the quaternary structure, and thus the behavior, of an allosteric protein by binding to it is called an **allosteric effector**. The term *effector* can apply to substrates, inhibitors, or activators. Several models for the behavior of allosteric enzymes have been proposed, including the concerted model and the sequential model, and it is worthwhile to compare them.

allosteric effector a substance—substrate, inhibitor, or activator—that binds to an allosteric enzyme and affects its activity

homotropic effects allosteric effects that occur when several identical molecules are bound to a protein

heterotropic effects allosteric effects that occur when substances other than the substrate are bound to a protein

First, however, we need to explain what homotropic effects and heterotropic effects are. **Homotropic effects** are allosteric interactions that occur when several identical molecules are bound to a protein. The binding of substrate molecules to different sites on an enzyme, such as the binding of aspartate to ATCase, which makes it easier for other aspartate molecules to bind, is an example of a homotropic effect. **Heterotropic effects** are allosteric interactions that occur when different substances (such as an inhibitor and substrate) are bound to the protein. In the ATCase reaction, inhibition by CTP and activation by ATP are both heterotropic effects.

7-2 The Concerted and Sequential Models for Allosteric Enzymes

Learning Objectives

After reading this section, students will be able to:

LO 7.2.1 Distinguish between the concerted and sequential models of allosteric enzyme behavior.

LO 7.2.2 Describe negative cooperativity in the sequential model of allosteric enzyme behavior.

The two principal models for the behavior of allosteric enzymes are the concerted model (Section 7-2A) and the sequential model (Section 7-2B). They were proposed in 1965 and 1966, respectively, and both are currently used as a basis for interpreting experimental results. The concerted model has the advantage of comparative simplicity, and it describes the behavior of some enzyme systems very well.

The sequential model sacrifices a certain amount of simplicity for a more realistic picture of the structure and behavior of proteins; it also deals very well with the behavior of some enzyme systems.

7-2A The Concerted Model for Allosteric Enzymes

In 1965, Jacques Monod, Jeffries Wyman, and Jean-Pierre Changeux proposed the **concerted model** for the behavior of allosteric proteins in a paper that has become a classic in the biochemical literature. In this model, the protein has two conformations: the active R (relaxed) conformation, which binds substrate tightly, and the inactive T (tight, also called taut) conformation, which binds substrate less tightly. The distinguishing feature of this model is that the conformations of *all* subunits change simultaneously.

concerted model a description of allosteric activity in which the conformations of all subunits change simultaneously

Figure 7.6A shows a hypothetical protein with two subunits. Both subunits change conformation from the inactive T conformation to the active R conformation at the same time; that is, a concerted change of conformation occurs. The equilibrium ratio of the T/R forms is called L and is assumed to be high—that is, more enzyme is present in the unbound T form than in the unbound R form. The binding of substrate to either form can be described by the dissociation constant of the enzyme and substrate, *K*, with the affinity for substrate higher in the R form than in the T form. Thus, $K_R << K_T$. The ratio of K_R/K_T is

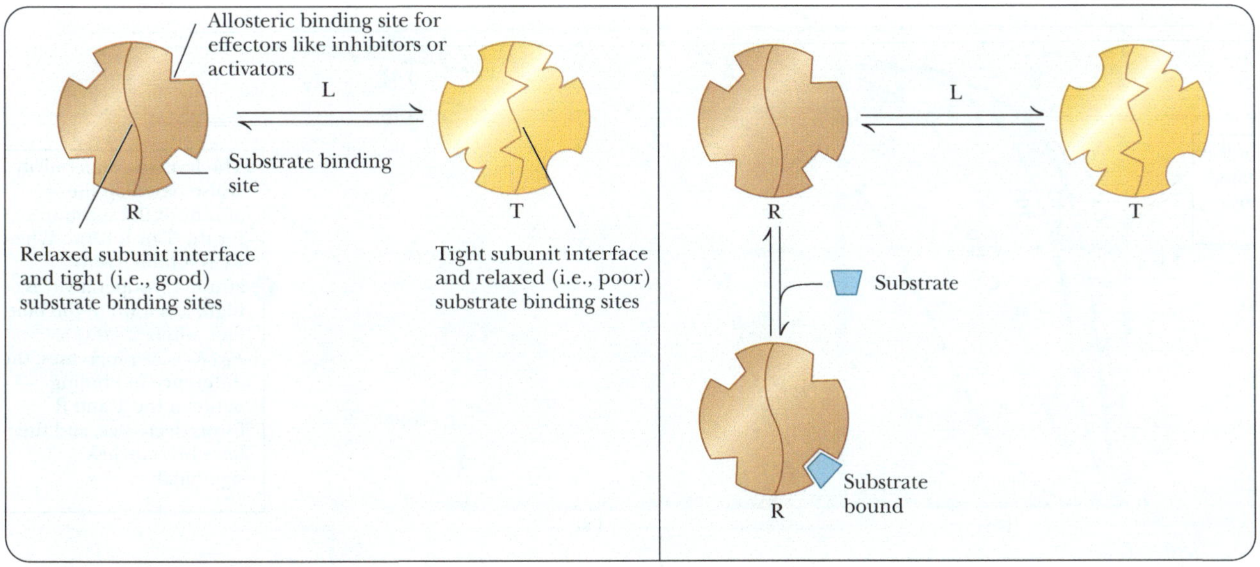

A A dimeric protein can exist in either of two conformational states at equilibrium, the T (taut) form or the R (relaxed) form. L is the ratio of the T form to the R form. With most allosteric systems, L is large, so there is more enzyme present in the T form than in the R form.

B By Le Châtelier's principle, substrate binding shifts the equilibrium in favor of the relaxed state (R) by removing unbound R. The dissociation constant for the enzyme–substrate complex is K_R for the relaxed form and K_T for the taut form. $K_R < K_T$, so the substrate binds better to the relaxed form. The ratio of K_R/K_T is called c. This figure shows a limiting case in which the taut form does not bind substrate at all, in which case K_T is infinite and $c = 0$.

Figure 7.6 Monod–Wyman–Changeux (MWC) model for allosteric transitions. It is also called the concerted model.

called c. Figure 7.6B shows a limiting case in which K_T is infinitely greater than K_R ($c = 0$). In other words, substrate will not bind to the T form at all.

The allosteric effect is explained by this model based on perturbing the equilibrium between the T and R forms. Although initially the amount of enzyme in the R form is small, when substrate binds to the R form, it removes free R form. This causes the production of more R form to reestablish the equilibrium, which makes binding more substrate possible. This shifting of the equilibrium is responsible for the observed allosteric effects. The Monod–Wyman–Changeux model has been shown mathematically to explain the sigmoidal effects seen with allosteric enzymes. The shape of the curve will be based on the L and c values. As L increases (that is, as the free T form becomes more highly favored), the shape becomes more sigmoidal (Figure 7.7). As the value for c decreases (that is, as the affinity between substrate and R form increases), the shape also becomes more sigmoidal.

In the concerted model, the effects of inhibitors and activators can also be considered in terms of shifting the equilibrium between the T and R forms of the enzyme. The binding of inhibitors to allosteric enzymes is cooperative; allosteric inhibitors bind to and stabilize the T form of the enzyme. The binding of activators to allosteric enzymes is also cooperative; allosteric activators bind to and stabilize the R form of the enzyme. When an activator, A, is present, the cooperative binding of A shifts the equilibrium between the T and R forms, with the R form favored (Figure 7.8). As a result, there is less need for substrate, S, to shift the equilibrium in favor of the R form, and less cooperativity in the binding of S is seen.

When an inhibitor, I, is present, the cooperative binding of I also shifts the equilibrium between the T and R forms, but this time the T form is favored. More substrate is needed to shift the T-to-R equilibrium in favor of the R form. A greater degree of cooperativity is seen in the binding of S.

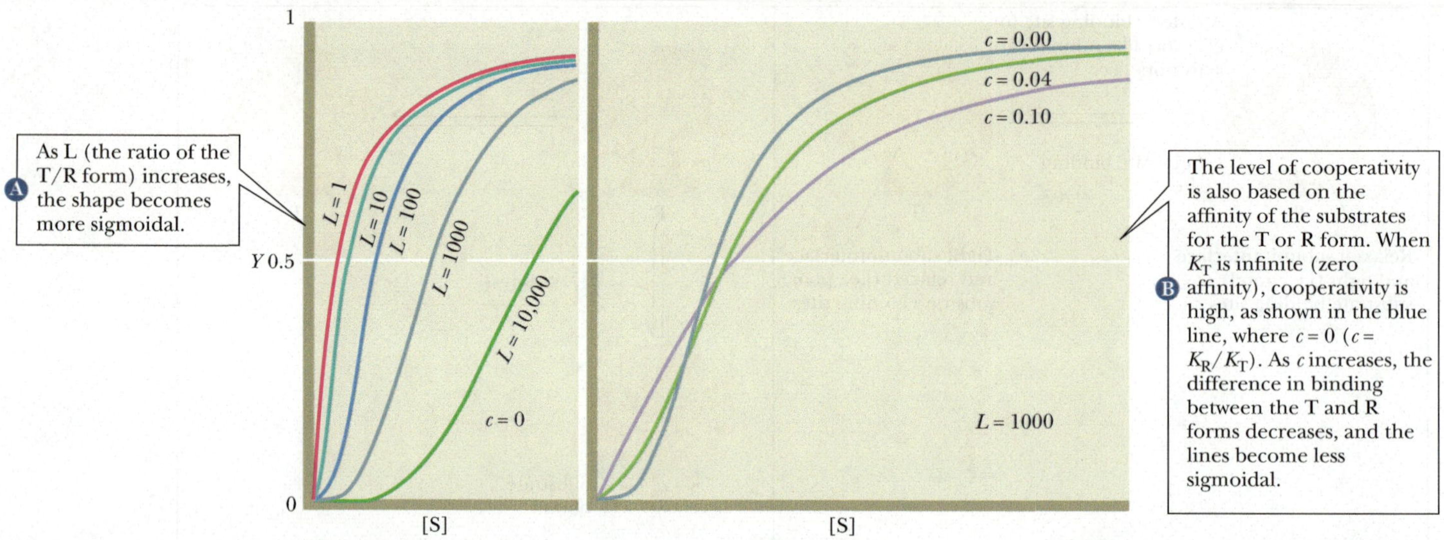

Figure 7.7 **The Monod–Wyman–Changeux (or concerted) model.** *(Adapted from Monod, J., Wyman, J., and Changeux, J. P. (1965). On the nature of allosteric transitions: A plausible model. Journal of Molecular Biology 12(92).)*

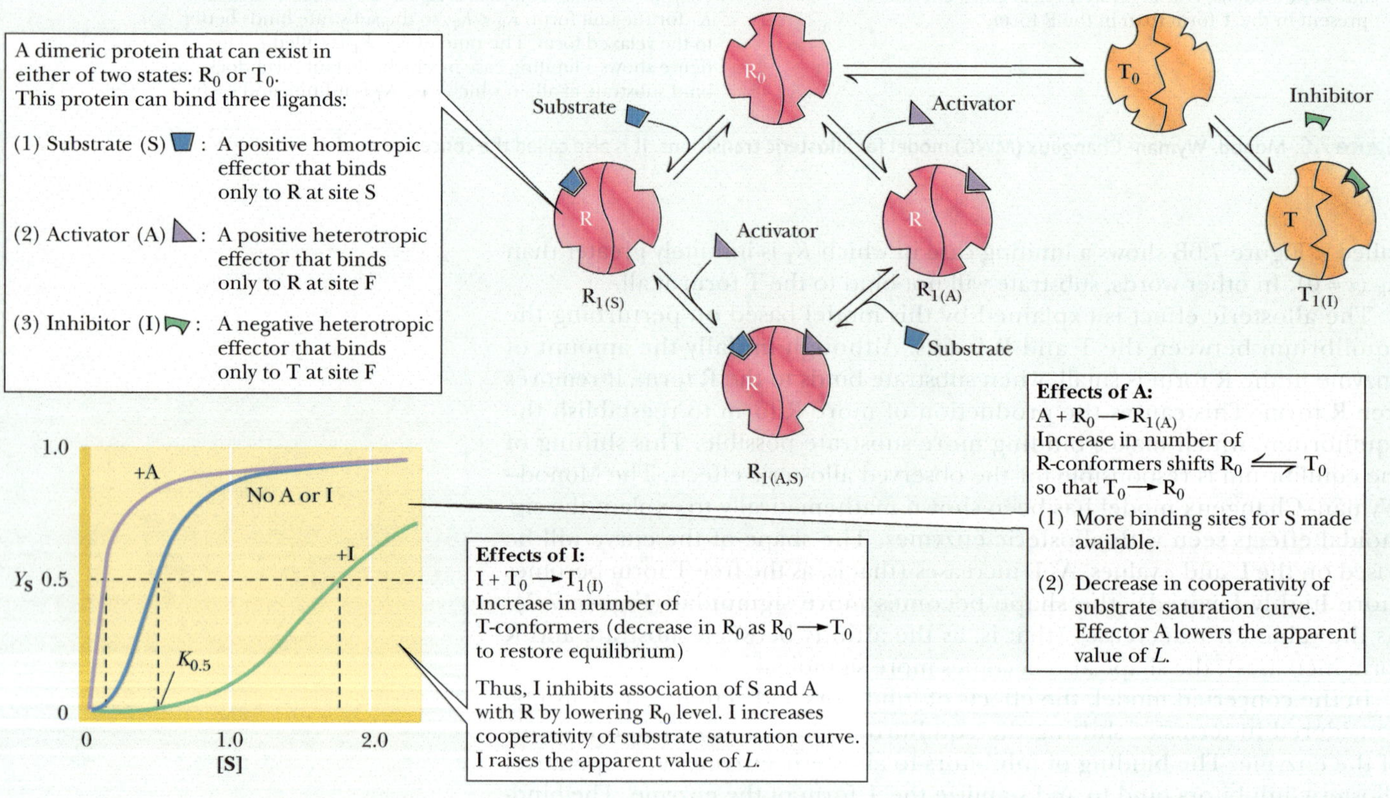

Figure 7.8 **Effects of binding activators and inhibitors with the concerted model.** An activator is a molecule that stabilizes the R form. An inhibitor stabilizes the T form.

7-2B The Sequential Model for Allosteric Enzymes

sequential model a description of the action of allosteric proteins in which a conformational change in one subunit is passed along within that subunit and then passed along to the other subunits

Daniel Koshland proposed the direct **sequential model** of allosteric behavior. As seen in Figure 7.9A, the distinguishing feature of this model is that the binding of substrate induces the conformational change from the T form to the R form—the type of behavior postulated by the induced-fit theory of substrate binding. Substrate binding to one subunit causes a conformational change at

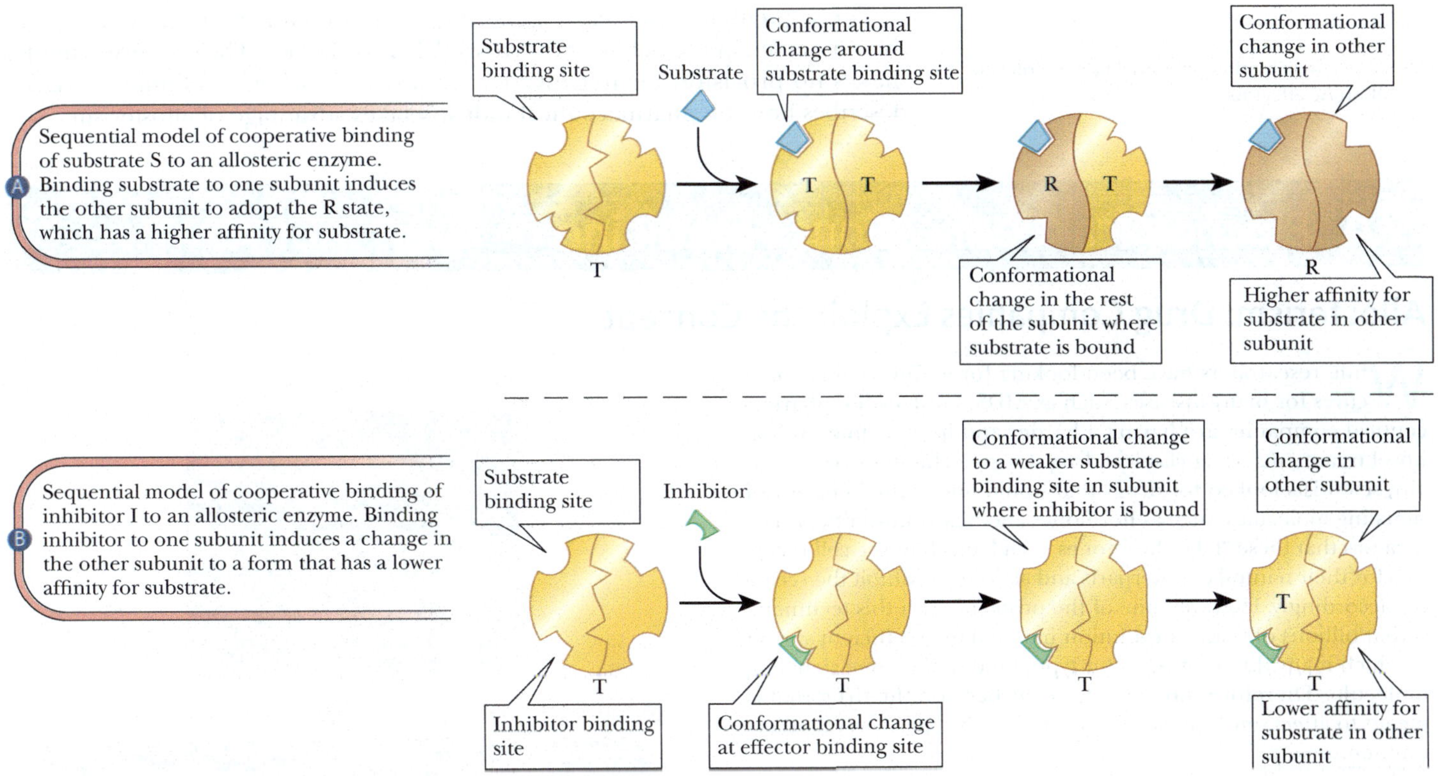

Figure 7.9 Sequential model of cooperative binding.

the active site, which causes a conformational change in the rest of that subunit, including any allosteric effector sites. After this subunit has changed to the R form, a conformational change from T to R in the other subunit becomes easier, and this is the form in which cooperative binding is expressed in this model.

In the sequential model, the binding of activators and inhibitors also takes place by the induced-fit mechanism (Figure 7.9B). The conformational change that begins with binding of inhibitor or activator to one subunit affects the conformation of the rest of that subunit and then that of other subunits in a similar manner as described in Figure 7.9A. The net result is to favor the R state when activator is present and to favor the T form when inhibitor, I, is present (Figure 7.9B). Binding I to one subunit causes a conformational change such that the T form is even less likely to bind substrate than before. This conformational change is passed along to other subunits, making them also more likely to bind inhibitor and less likely to bind substrate. This is an example of cooperative behavior that leads to more inhibition of the enzyme. Likewise, binding an activator in one subunit causes a conformational change that favors substrate binding in that subunit, and this effect is passed from one subunit to another.

7-2C Negative Cooperativity

The sequential model for binding effectors of all types, including substrates, to allosteric enzymes has a unique feature not seen in the concerted model. The conformational changes thus induced can make the enzyme less likely to bind more molecules of the same type. This phenomenon, called **negative cooperativity**, has been observed in a few enzymes. One is tyrosyl tRNA synthetase, which plays a role in protein synthesis. In the reaction catalyzed by this enzyme, the amino acid tyrosine forms a covalent bond to a molecule of transfer RNA (tRNA). In subsequent steps, the tyrosine is passed along to its place in the sequence of the growing protein. The tyrosyl tRNA synthetase consists of two subunits. Binding of the first molecule of substrate to one of the subunits inhibits binding of a second molecule to the other subunit.

negative cooperativity a cooperative effect whereby binding of the first ligand to an enzyme or protein causes the affinity for the next ligand to be lower

allosteric sites binding sites on a target molecule for allosteric effectors

The sequential model has successfully accounted for the negative cooperativity observed in the behavior of tyrosyl tRNA synthetase. The concerted model makes no provision for negative cooperativity. **Biochemical Connections 7A** describes how the pharmaceutical industry takes advantage of allosterism.

7A Biochemical Connections • Medicine

Allosterism: Drug Companies Exploit the Concept

While researchers have been looking for definitive medicinal cures for many diseases, such as AIDS, for decades, pharmaceutical companies are having a harder and harder time finding novel treatments using classical drug design techniques. Traditionally, scientists looked for drugs that would mimic the behavior of signaling molecules, such as hormones and neurotransmitters. The idea was that these "fake" hormones would attach to the cell receptors for their natural counterparts and activate or inhibit the receptor accordingly. However, one of the problems with this technique is that billions of years of evolution have led to an efficient system in which many different receptor types bind to the same signaling molecule. Therefore, side effects occur because the drug that is meant to affect one type of receptor will likely affect several others unintentionally.

More recently, pharmaceutical companies have moved beyond attempting to mimic the exact signaling molecule and are even moving away from making drugs that bind to the receptor's active site at all. Instead, they are designing drugs that bind to other sites on the receptor, called **allosteric sites** (Figure 7.10).

Allosteric drugs have several advantages over their more traditional "orthosteric (same site)" counterparts. First, binding a molecule to the actual receptor active site will give rise to an on/off or yes/no response. As we have seen in this chapter, allosteric effectors modulate the response in a more subtle way, like a dimmer switch opposed to a strict on/off switch.

Second, using allosteric drugs allows the drug to be more specific for one or a few receptor types. If there are 50 different receptors that bind the neurotransmitter glutamate, for example, evolution will have made the binding site very similar for all of the receptors; otherwise, they could not all bind glutamate. However, there would not be evolutionary pressure to make the rest of the receptors so similar. This allows scientists to find allosteric binding sites that are much more specific.

Third, allosteric drugs can be safer because they have no effect at all unless the natural ligand is present. As an example, we can compare the effects of two central nervous system depressants—phenobarbital and Valium. Phenobarbital binds to the active site of a neurotransmitter receptor and is deadly if taken in large quantities. Valium is now known to be an allosteric drug that binds to a different site on the receptors for γ-aminobutyric acid (GABA), the body's principal inhibitory neurotransmitter. Valium turns up the response of the receptor for the GABA. When Valium is bound, the response of the receptor to GABA goes up severalfold, but taking too much Valium is not as deadly as taking too much phenobarbital, because the Valium is not causing a direct effect.

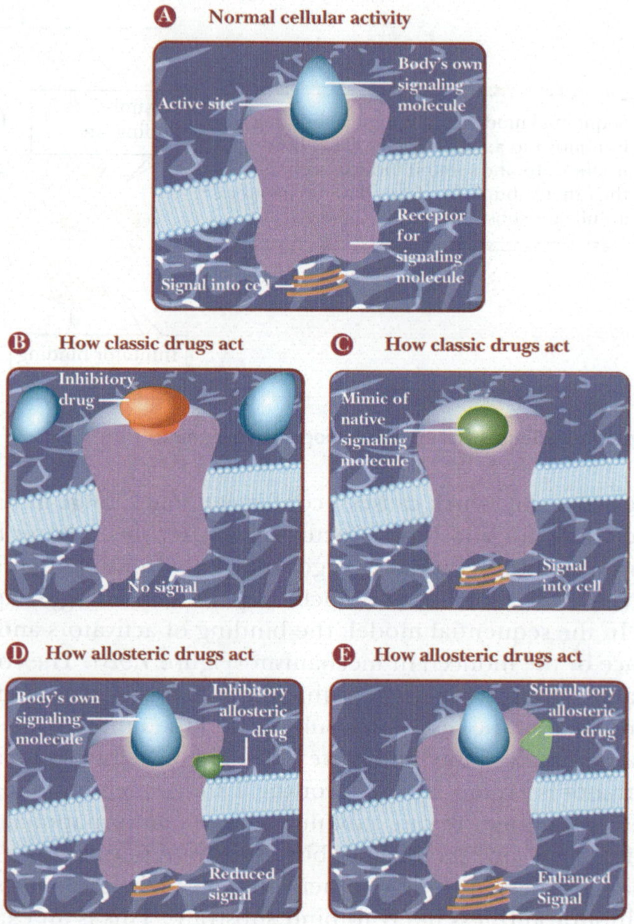

Figure 7.10 **Mechanism of drug action.** Many drugs work by altering signal cascades. Part A shows a typical signaling complex, with a membrane-bound receptor, a natural signaling molecule, and a biological effect transmitted into the cell. Classical drugs work by either blocking the access of the signaling molecule (B) or by binding and mimicking its action (C). New drug design is based on having a drug that does not bind to the active site of the receptor. Rather, it binds to an allosteric site and either (D) reduces the effect of the natural signal or (E) increases it. *(Figure on page 72 from A New Kind of Drug Target, by Melinda Wenner, Scientific American, August 2009.)*

In addition, several drugs are in or will be beginning clinical trials, including drugs that will combat Alzheimer's disease, gastroesophageal reflux disease, schizophrenia, and Parkinson's disease.

7-3 Control of Enzyme Activity by Phosphorylation

Learning Objectives

After reading this section, students will be able to:

LO 7.3.1 Describe how protein phosphorylation occurs.

LO 7.3.2 Explain an example of how protein phosphorylation controls enzyme activity.

One of the most common control mechanisms for enzymes is a form of covalent modification called phosphorylation, which involves the addition of a phosphate group to an enzyme. The side-chain hydroxyl groups of serine, threonine, and tyrosine can all form phosphate esters, as shown in Figure 7.11. The sodium–potassium ion pump, which moves potassium into the cell and sodium out, provides an important example of enzyme control by phosphorylation (Figure 7.12). The source of the phosphate group for the protein component of the sodium–potassium ion pump and for many enzyme phosphorylations is the ubiquitous ATP molecule. When ATP is hydrolyzed to adenosine diphosphate (ADP), enough energy is released to allow a number of otherwise energetically unfavorable reactions to take place. In the case of the Na^+/K^+ pump, a **protein kinase** catalyzes the transfer of a phosphate group from ATP to aspartate 369, causing a conformational change in the enzyme. *Kinase* refers to an enzyme that catalyzes the transfer of a phosphate group, almost always from ATP, to some substrate.

protein kinase a class of enzymes that covalently modify a protein by transferring a phosphate group from ATP to the protein

Many examples appear in processes involved in generating energy, as is the case in carbohydrate metabolism. Glycogen phosphorylase, which catalyzes the

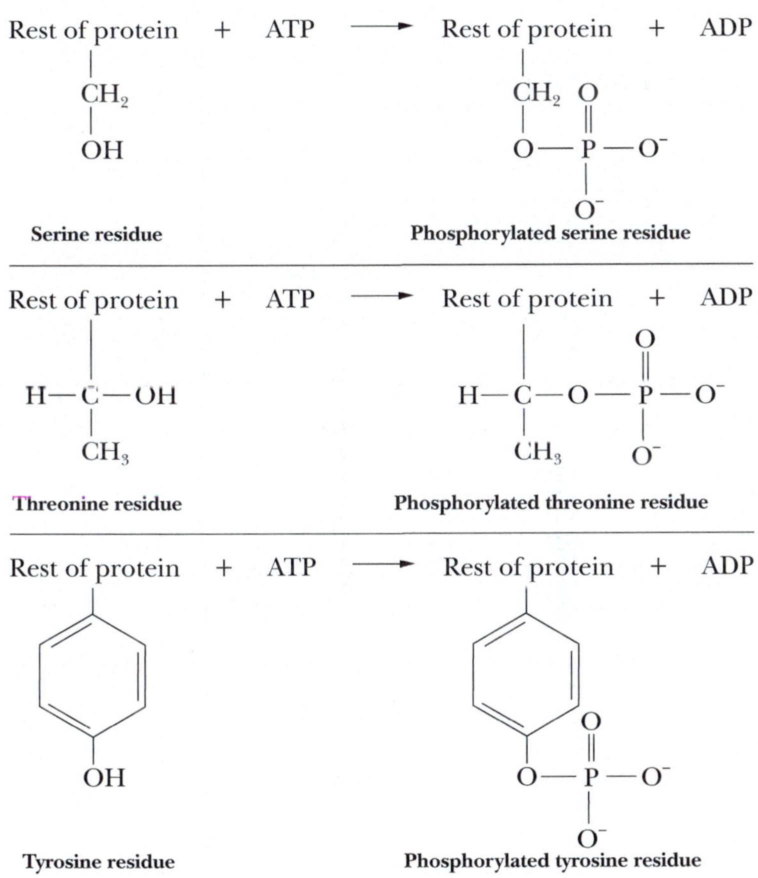

Figure 7.11 Phosphorylation of serine, threonine, and tyrosine protein residues.

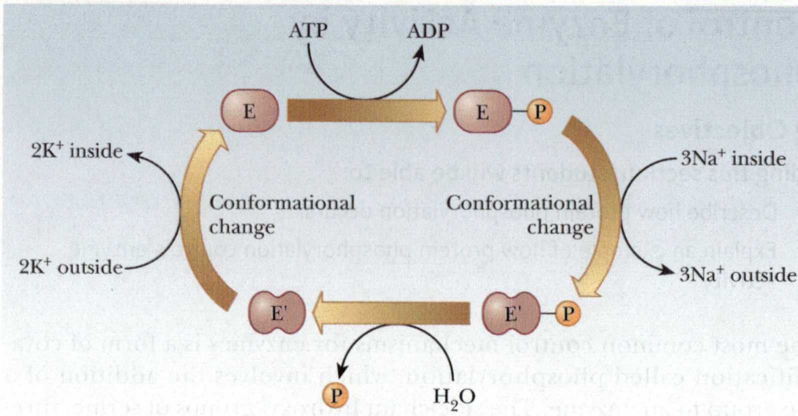

Figure 7.12 Phosphorylation of the sodium–potassium pump is involved in cycling the membrane protein between the form that binds to sodium and the form that binds to potassium. Here, E refers to the sodium–potassium pump enzyme.

initial step in the breakdown of stored glycogen (Section 18-1), exists in two forms—the phosphorylated glycogen phosphorylase *a* and the dephosphorylated glycogen phosphorylase *b* (Figure 7.13). The *a* form is more active than the *b* form, and the two forms of the enzyme respond to different allosteric effectors, depending on tissue type. Glycogen phosphorylase is thus subject to two kinds of control—allosteric regulation and covalent modification. The net result is that the *a* form is more abundant and active when phosphorylase is needed to break down glycogen to provide energy.

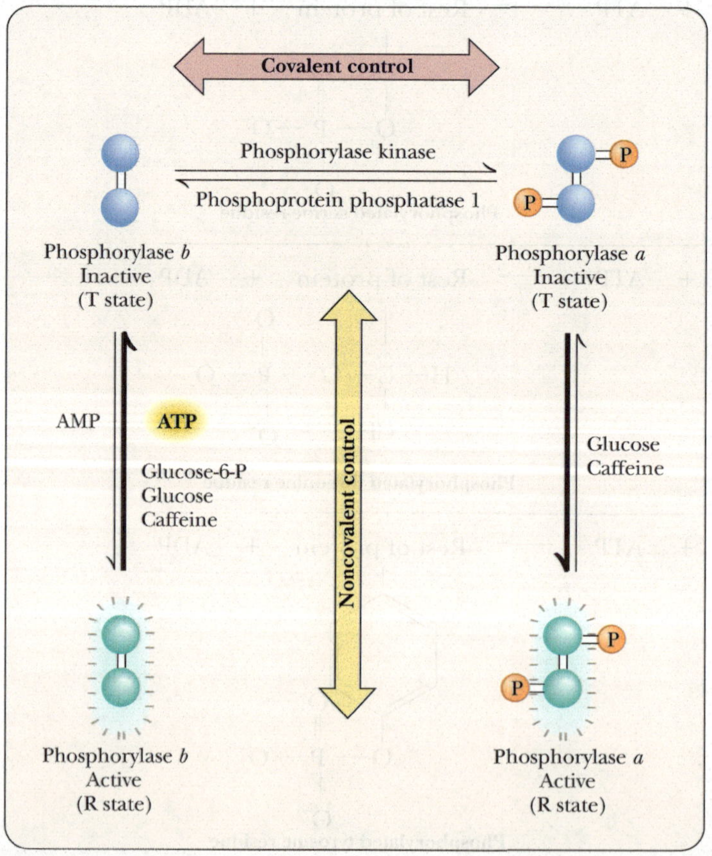

Figure 7.13 Glycogen phosphorylase activity is subject to allosteric control and covalent modification via phosphorylation. The phosphorylated form is more active. The enzyme that puts a phosphate group on phosphorylase is called phosphorylase kinase.

▶ Does phosphorylation always increase enzyme activity?

Although it would be convenient to have a model in which phosphorylation always increases the activity of an enzyme, biochemistry is not so kind to us. In reality, we cannot predict whether phosphorylation will increase or decrease the activity of an enzyme. In some systems, the effects on two opposing enzymes are coordinated. For example, a key enzyme in a catabolic pathway may be activated by phosphorylation, whereas its counterpart in an anabolic opposing pathway is inhibited by phosphorylation.

Biochemical Connections 7B describes how the medicinal effects of aspirin are related to phosphorylation and protein kinases.

7B Biochemical Connections • Medicine

An Ancient Drug Works by Stimulating a Protein Kinase

Many different enzymes are protein kinases, and researchers continue to learn more about how they control metabolism. Adenosine monophosphate-activated protein kinase (AMPK) demonstrates control of a protein kinase that is seen in many metabolic pathways. It was found that the enzyme is stimulated by salicylate, the dissociated form of salicylic acid, which is a natural compound from the bark of the willow tree that has been known to have therapeutic properties since ancient times (Figure 7.14). It is also the active ingredient in aspirin, acetylsalicylate.

Besides its properties as a painkiller, aspirin has long been used to treat a variety of health issues. For example, patients at risk of heart attacks are often advised to take small daily doses of aspirin. Researchers have shown that aspirin irreversibly acetylates and inhibits cyclooxygenase-1 (COX-1) in platelets, blocking thromboxane A_2 synthesis and thereby reducing clot formation by reducing platelet activation and aggregation. Because platelets cannot resynthesize COX-1, this sustained antiplatelet effect lowers the likelihood of clot formation in coronary arteries, reducing heart attack risk. Salicylate also shows promise in treating type 2 diabetes. ◗

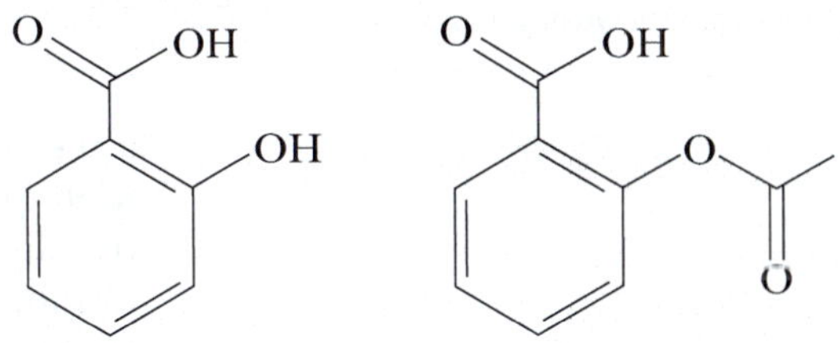

Salicylic acid   Acetylsalicyclic acid (aspirin)

Figure 7.14 Structure of aspirin and its precursor.

7-4 Zymogens

Learning Objectives

After reading this section, students will be able to:

LO 7.4.1 Describe zymogens using chymotrypsinogen as an example of zymogen activation.

Allosteric interactions control the behavior of proteins through reversible changes in quaternary structure, but this mechanism, effective though it may be, is not the only one available. A **zymogen**, an inactive precursor of an enzyme, can be irreversibly transformed into an active enzyme by cleavage of covalent bonds.

zymogen an inactive protein that can be activated by hydrolysis of specific peptide bonds

The proteolytic enzymes trypsin (Section 5-4C) and chymotrypsin provide a classic example of zymogens and their activation. Their inactive precursor molecules, trypsinogen and chymotrypsinogen, respectively, are formed in the pancreas, where they would do damage if they were in an active form. In the small intestine, where their digestive properties are needed, they are activated by cleavage of specific peptide bonds. The conversion of chymotrypsinogen to chymotrypsin is catalyzed by trypsin, which, in turn, arises from trypsinogen as

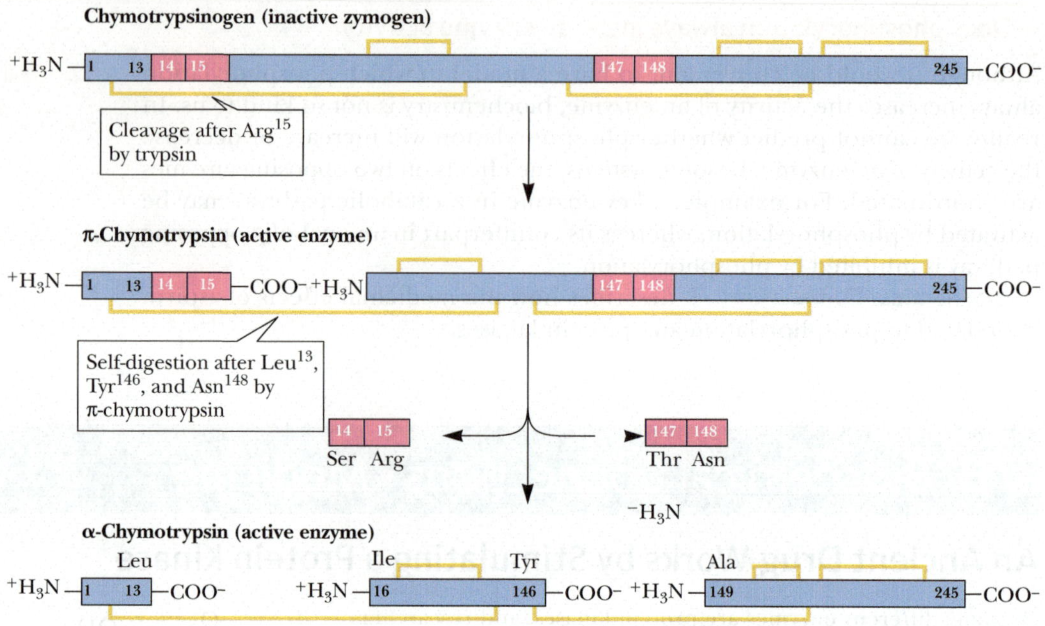

Figure 7.15 **The proteolytic activation of chymotrypsinogen.** Disulfide bonds are shown in yellow.

a result of a cleavage reaction catalyzed by the enzyme enteropeptidase. Chymotrypsinogen consists of a single polypeptide chain 245 residues long, with five disulfide (—S—S—) bonds (Figure 7.15). Recall from Section 4-4A that disulfide bonds are covalent bonds in proteins that can form between cysteine residue side-chain groups.

When chymotrypsinogen is secreted into the small intestine, trypsin, present in the digestive system, cleaves the peptide bond between arginine 15 and isoleucine 16, counting from the N-terminal end of the chymotrypsinogen sequence. The cleavage produces active π-chymotrypsin. The 15-residue fragment remains bound to the rest of the protein by a disulfide bond. Although π-chymotrypsin is fully active, it is not the end product of this series of reactions. It acts on itself to remove two dipeptide fragments, producing α-chymotrypsin, which is also fully active. The two dipeptide fragments cleaved off are Ser 14—Arg 15 and Thr 147—Asn 148; the final form of the enzyme, α-chymotrypsin, has three polypeptide chains held together by two of the five original, and still intact, disulfide bonds. (The other three disulfide bonds remain intact as well; they link portions of single polypeptide chains.) When the term *chymotrypsin* is used without specifying the α or the π form, the final α form is meant.

The changes in primary structure that accompany the conversion of chymotrypsinogen to α-chymotrypsin bring about changes in the tertiary structure, and the changes in tertiary structure allow for the necessary spatial arrangement of catalytic residues in the active site to foster catalysis. The zymogen is inactive because its tertiary structure involves a spatial arrangement of the catalytic residues that does *not* allow them to catalyze the reaction.

7-5 The Nature of the Active Site

Learning Objectives

After reading this section, students will be able to:

LO 7.5.1 Describe the techniques used to determine the identity and arrangement of critical active-site residues.

LO 7.5.2 Explain the general principles of catalytic mechanisms, using chymotrypsin as an example of an enzyme mechanism.

We now look at the specific mechanism by which an enzyme is able to increase the rate of a chemical reaction. This mechanism is based on the exact three-dimensional arrangement of the amino acids in the active site. We can ask several questions about the mode of action of an enzyme, such as the following:

1. Which amino acid residues on the enzyme are in the active site (recall this term from Section 6-3) and catalyze the reaction? In other words, which are the critical amino acid residues for the enzyme's activity?
2. What is the spatial relationship of the critical amino acid residues in the active site?
3. What is the mechanism by which the critical amino acid residues catalyze the reaction?

Answers to these questions are available for chymotrypsin, so we use its mechanism as an example of enzyme action. Information on well-known systems, such as chymotrypsin, can lead to general principles that are applicable to all enzymes.

Enzymes catalyze chemical reactions in many ways, but all reactions have in common the requirement that some reactive groups on the enzyme interact with the substrate. The side-chain reactive groups are the ones mostly involved in the action of enzymes. Hydrocarbon side chains (Figure 3.3A) do not contain reactive groups and are not involved in the process. Functional groups that can play a catalytic role include the imidazole group of histidine, the hydroxyl group of serine, the carboxyl side chains of aspartate and glutamate, the sulfhydryl group of cysteine, the amino side chain of lysine, and the phenol group of tyrosine (Figure 3.3B). If the C-terminal α-carboxyl or the N-terminal α-amino group of the peptide chain is positioned in the active site, then they, too, can play a role.

Chymotrypsin is an enzyme that catalyzes the hydrolysis of peptide bonds after aromatic amino acid residues in a protein substrate. It can cause hydrolysis after other residues as well, but this occurs at a lower frequency. In addition, chymotrypsin catalyzes the hydrolysis of esters in model studies in the laboratory (Figure 7.16). The use of model systems is common in biochemistry because a model provides the essential features of a reaction in a simple form that is easier to work with than the one found in nature. The amide (peptide) bond and the ester bond are similar enough that the enzyme can accept both types of compounds as substrates. Model systems based on the hydrolysis of esters are frequently used to study the peptide hydrolysis reaction.

A typical model compound is *p*-nitrophenyl acetate, which is hydrolyzed in two stages (Figure 7.16). The acetyl group is covalently attached to the enzyme

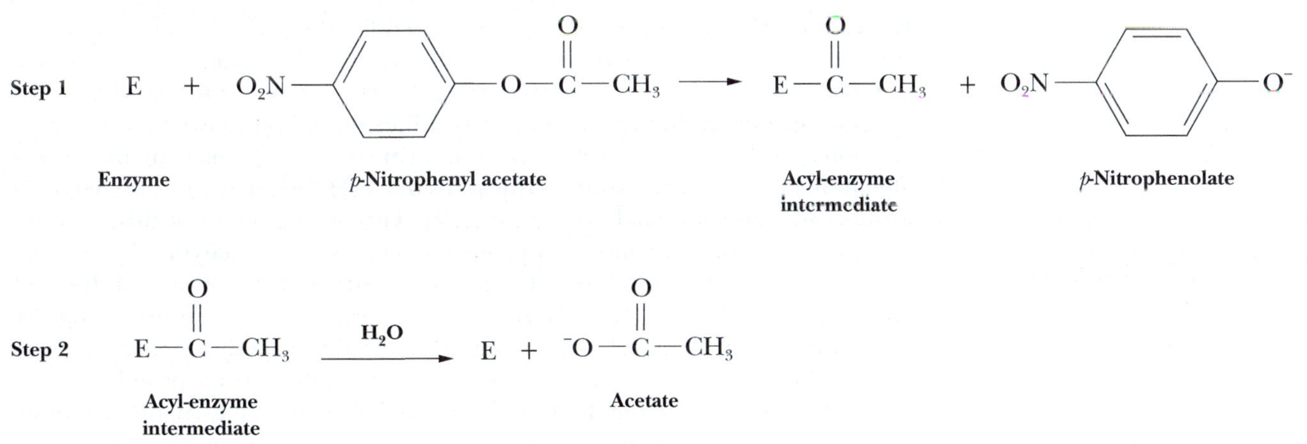

Figure 7.16 Chymotrypsin-catalyzed reaction of *p*-nitrophenyl acetate.

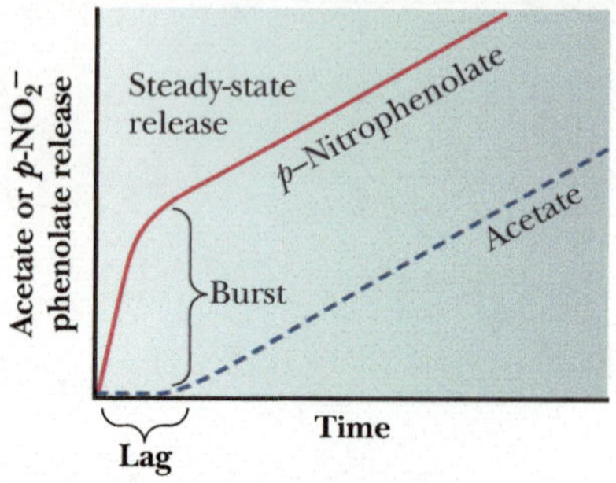

Figure 7.17 The kinetics observed in the chymotrypsin reaction. An initial burst of *p*-nitrophenolate is seen, followed by a slower, steady-state release that matches the appearance of the other product, acetate.

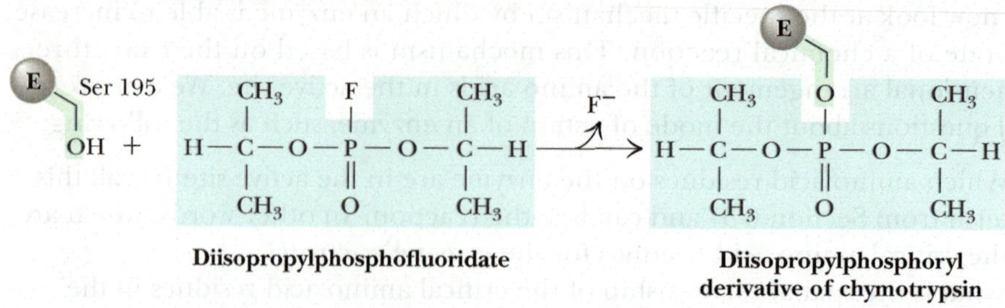

Figure 7.18 Diisopropylphosphofluoridate chemical modification of the active-site serine of chymotrypsin. Here, E refers to the chymotrypsin enzyme. The modification of this enzymatically critical serine residue inactivates the enzyme.

at the end of the first stage (step 1) of the reaction, but the *p*-nitrophenolate ion is released. In the second stage (step 2), the acyl-enzyme intermediate is hydrolyzed, releasing acetate and regenerating the free enzyme. The kinetics observed when *p*-nitrophenyl acetate is first mixed with chymotrypsin shows an initial burst and then a slower phase (Figure 7.17). This reaction is consistent with an enzyme that has two phases, one often forming an acylated enzyme intermediate.

7-5A Determining the Critical Amino Acids in the Active Site

The serine residue at position 195 (Ser 195) of chymotrypsin is required for the enzymatic activity of chymotrypsin; in this respect, chymotrypsin is typical of a class of enzymes known as **serine proteases**. Specific details of Ser 195's involvement in chymotrypsin catalysis are discussed in Section 7.5B.

serine proteases a class of proteolytic enzymes in which a serine hydroxyl plays a critical role in catalysis

▶ How can the amino acid residues that are critical for enzymatic activity be determined?

Chymotrypsin is completely inactivated when Ser 195 reacts with diisopropylphosphofluoridate (DIPF), forming a covalent bond that links the serine side chain with DIPF (Figure 7.18). This is but one example of how chemical modification of enzymes can be used in laboratory studies to probe enzymatic activity; there are several other types of chemical modification that are commonly used. The other serine residues of chymotrypsin are far less reactive and are not chemically modified by DIPF. Histidine 57 (His 57) is another critical amino acid residue in chymotrypsin. Chemical modification by a different reagent provides evidence that this residue is involved in the activity of chymotrypsin.

▶ How does the architecture of the active site affect catalysis?

Both serine 195 and histidine 57 are required for the activity of chymotrypsin; therefore, they must be close to each other in the active site. The determination of the three-dimensional structure of the enzyme by X-ray crystallography provides evidence that the active-site residues do indeed have a close spatial relationship. The folding of the chymotrypsin backbone, mostly in an antiparallel β-sheet array, positions the critical residues (His 57, Asp 102, and Ser 195) around an active-site pocket (Figure 7.19). This arrangement of three critical residues found in many different proteases is known as a **catalytic triad**. Only a few residues are directly involved in the active site, but the other hundreds of amino acid residues of chymotrypsin play an important role in providing the correct three-dimensional arrangement for the three catalytic triad residues. To play this role, these other residues participate in the various protein folding forces discussed in Section 4-4A, and the α-helix, β-sheet, β-turn, and random-coil arrangements they make are seen in Figure 7.19A. Enzyme activity is an exquisitely coordinated team effort.

catalytic triad an arrangement of three enzymatically critical amino acid residues found in the active sites of many enzymes

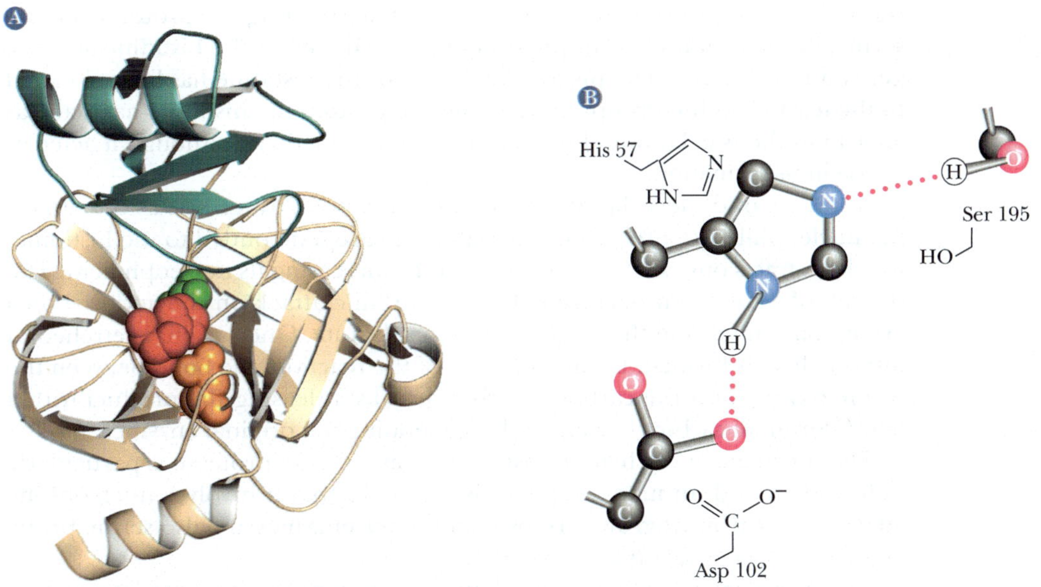

Figure 7.19 **(A) Structure of chymotrypsin (pink) in a complex with eglin C (blue ribbon structure), a target protein.** The residues of the catalytic triad (His 57, Asp 102, and Ser 195) are highlighted. His 57 (red) is flanked by Asp 102 (orange) and by Ser 195 (green). The catalytic site is filled by a peptide segment of eglin. Ser 195 is close to the peptide bond that is cleaved in the chymotrypsin reaction. His 57 and Asp 102 play roles as well. (B) The catalytic triad of the chymotrypsin active site. *(Garrett/Grisham, Biochemistry, 4th Edition. Cengage Learning.)*

Other important pieces of information about the three-dimensional structure of the active site emerge when a complex is formed between chymotrypsin and a substrate analog. When one such substrate analog, formyl-L-tryptophan (Figure 7.20), is bound to the enzyme, the tryptophan side chain fits into a hydrophobic pocket near serine 195. This type of binding is not surprising in view of the specificity of the enzyme for aromatic amino acid residues at the cleavage site.

The results of X-ray crystallography show, in addition to the binding site for aromatic amino acid side chains of substrate molecules, a definite arrangement of the amino acid side chains that are responsible for the catalytic activity of the enzyme.

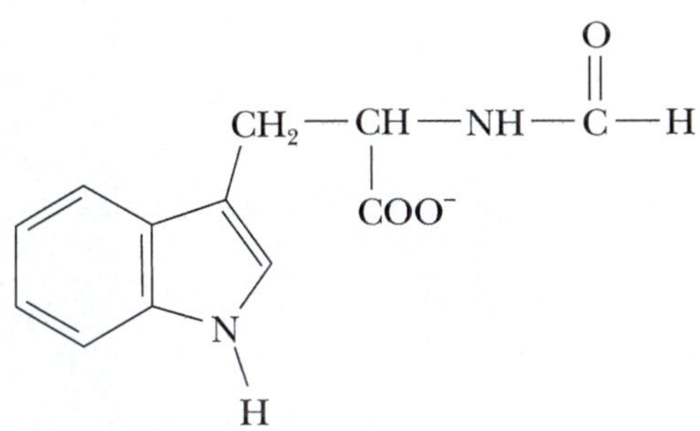

Figure 7.20 Formyl-L-tryptophan.

7-5B Mechanism of Chymotrypsin Catalysis

Any postulated reaction mechanism must be modified or discarded if it is inconsistent with experimental results. There is consensus, but not total agreement, on the main features of the mechanism discussed in this section.

The critical amino acid residues, Ser 195 and His 57, are involved in the mechanism of catalytic action. In the terminology of organic chemistry, the oxygen of the serine side chain is a **nucleophile** (that is, a nucleus-seeking substance). A nucleophile tends to bond to sites of positive charge or polarization (electron-poor sites). The nucleophilic oxygen of the serine has its hydrogen abstracted by His 57, increasing its nucleophilicity, and attacks the carbonyl carbon of the aromatic residue whose side chain is held by a hydrophobic binding pocket (Figure 7.21). The carbon now has four single bonds, so a tetrahedral intermediate has formed in which the original—C=O bond has become a single bond, and the carbonyl oxygen has become an oxyanion. The acyl-enzyme intermediate is then formed from the tetrahedral species, the peptide bond is cleaved, and the protein fragment with a new N-terminus leaves. In this part of the reaction, the amino portion of the original peptide group hydrogen bonds to the imidazole portion of the histidine. Notice that the imidazole is already protonated and that the proton came from the hydroxyl group of

nucleophile an electron-rich substance that tends to react with sites of positive charge or polarization the serine. The histidine behaves as a base in abstracting the proton from the serine; in the terminology of physical organic chemistry, the histidine acts as a general base catalyst. The proton abstracted by the histidine has been donated to the leaving amino group. In donating the proton, the histidine has acted as an acid in the breakdown of the tetrahedral intermediate, although it acted as a base in its formation.

In the second, deacylation phase of the reaction, water acts as the attacking nucleophile (Figure 7.21). The water is hydrogen-bonded to the histidine, which abstracts one of its hydrogens, thereby increasing its nucleophilicity. The oxygen of water then performs the nucleophilic attack on the acyl carbonyl group that came from the original protein substrate. Once again, a tetrahedral intermediate is formed. In the final step of the reaction, the bond between the serine oxygen and the carbonyl carbon breaks, releasing the product with a new C-terminal carboxyl group and regenerating the original enzyme.

The mechanism of chymotrypsin action as a serine protease is particularly well studied and, in many respects, is typical in that it involves nucleophilic attacks on a carbonyl group. Trypsin and other enzymes are also serine proteases (see **Biochemical Connections 7C**).

Numerous types of reaction mechanisms for enzyme action are known, and we discuss them in the contexts of the reactions catalyzed by the enzymes in question. To lay the groundwork, it is useful to discuss some general types of catalytic mechanisms and how they affect the specificity of enzymatic reactions.

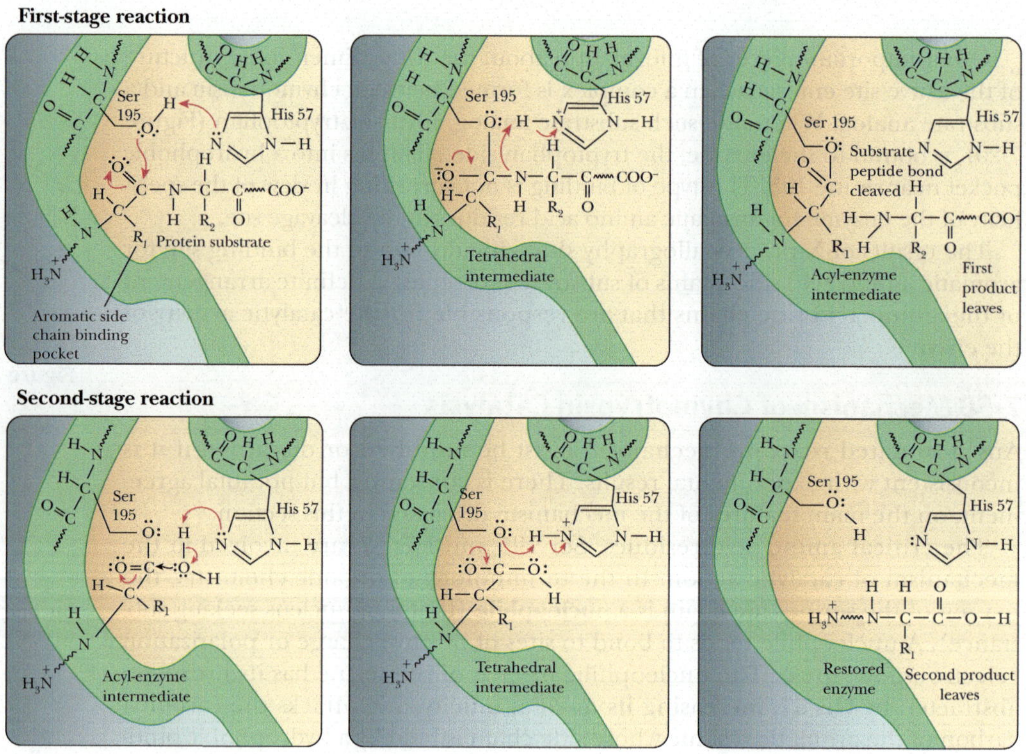

Figure 7.21 The mechanism of chymotrypsin action. In the first stage of the reaction, the nucleophile serine 195 attacks the carbonyl carbon of the substrate. In the second stage, water is the nucleophile that attacks the acyl-enzyme intermediate. Histidine 57 is involved in both stages of the reaction. R_1 refers to an aromatic side-chain group, R_2 refers to any side chain group, and the * in the first panel indicates the peptide bond that will be cleaved by chymotrypsin.

7C Biochemical Connections • Allied Health

Families of Enzymes: Proteases

Large numbers of enzymes catalyze similar functions. Many oxidation–reduction reactions take place, catalyzed by dehydrogenases. We have already seen that kinases transfer phosphate groups. Still other enzymes catalyze hydrolytic reactions. Enzymes that have similar functions may have widely varying structures. The important feature that they have in common is an active site that can catalyze the reaction in question.

A number of different enzymes catalyze the hydrolysis of proteins. Chymotrypsin is one example of the class of serine proteases, but many others are known, including elastase, which catalyzes the degradation of the connective tissue protein elastin, and the digestive enzyme trypsin (Figure 7.22). (Recall from Section 5-4C that trypsin plays a role in protein sequencing.) All these serine protease enzymes are similar in structure.

Other proteases employ reactive amino acid residues other than serine as the nucleophile in the active site. Papain, the basis of commercial meat tenderizers, is a proteolytic enzyme derived from papayas. However, it has a cysteine, rather than a serine, as the nucleophile in its active site (Figure 7.23).

Aspartic proteases differ still more widely in structure from the common serine proteases. A pair of aspartate side chains, sometimes on different subunits, participates in the reaction mechanism. A number of aspartic proteases, such as the digestive enzyme pepsin, are known. However, the most notorious aspartic protease is HIV-1 protease (Figure 7.24), the one necessary for the maturation of the human immunodeficiency virus. ◗

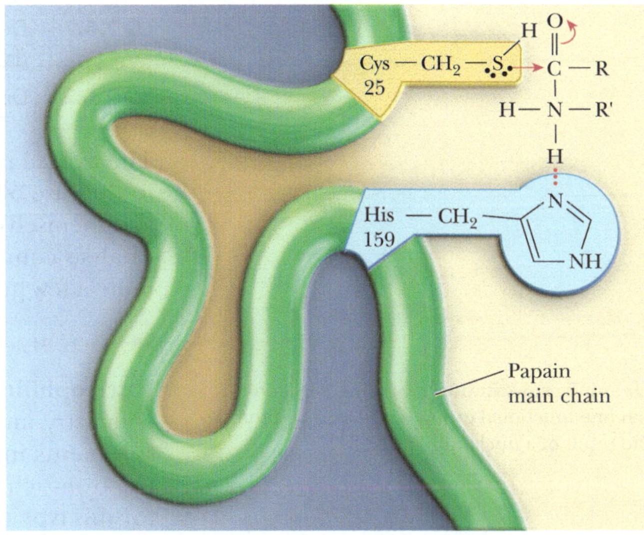

Figure 7.23 Papain is a cysteine protease. A critical cysteine residue is involved in the nucleophilic attack on the peptide bonds it hydrolyzes.

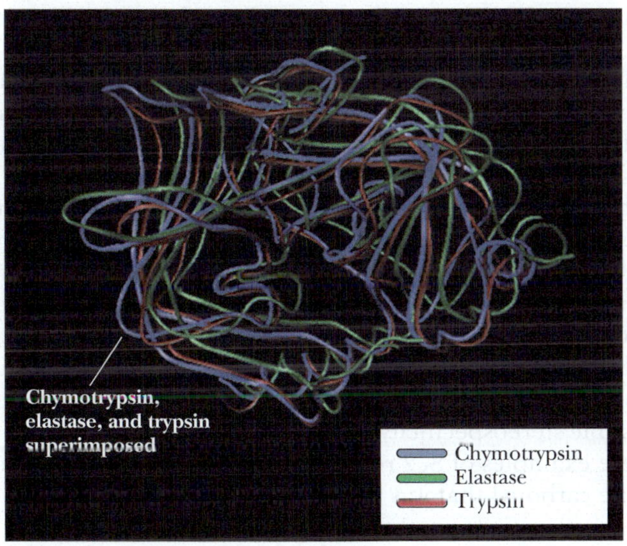

Figure 7.22 Chymotrypsin, elastin, and trypsin are serine proteases and have similar structures.

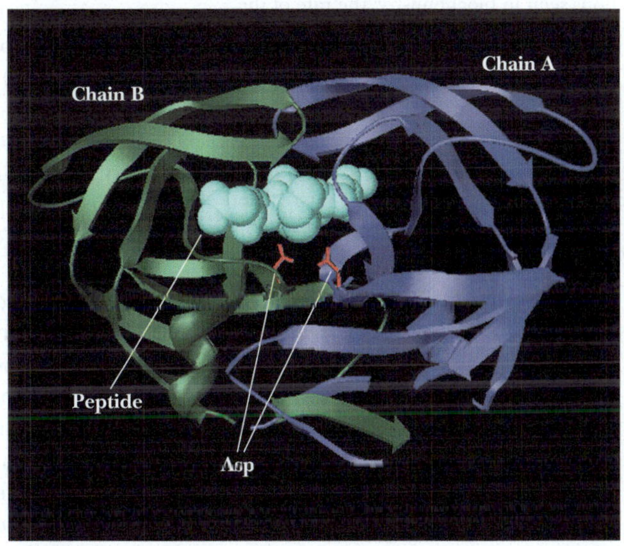

Figure 7.24 HIV-1 protease is a member of the class of enzymes called the aspartic proteases. Two aspartates are involved in the reaction.

7-6 Chemical Reactions Involved in Enzyme Mechanisms

Learning Objectives

After reading this section, students will be able to:

LO 7.6.1 Describe the general principles of metal-ion catalysis using carboxypeptidase A as an example of metal-ion catalysis.

LO 7.6.2 Distinguish between the types of nucleophilic substitution reactions in catalysis.

The overall mechanism for a reaction may be fairly complex, as we have seen in the case of chymotrypsin, but the individual parts of a complex mechanism can themselves be fairly simple. Concepts such as nucleophilic attack and acid catalysis commonly enter into discussions of enzymatic reactions. We can draw quite a few general conclusions from these two general descriptions.

▶ What are the most common types of reactions?

nucleophilic substitution reactions reactions in which one functional group is replaced by another as the result of a nucleophilic attack

Nucleophilic substitution reactions play a large role in the study of organic chemistry, and they are excellent illustrations of the importance of kinetic measurements in determining the mechanism of a reaction. A nucleophile is an electron-rich atom that attacks an electron-deficient atom. A general equation for this type of reaction is

$$R{:}X + {:}Z \rightarrow R{:}Z + X$$

where :Z is the nucleophile, and X is called a *leaving group*. In biochemistry, the carbon of a carbonyl group (C=O) is often the atom attacked by the nucleophile. Common nucleophiles are the oxygens of serine, threonine, and tyrosine.

S_N1 a unimolecular nucleophilic substitution reaction; one of the most common types of organic reactions seen in biochemistry; the rate of the reaction follows first-order kinetics

If the rate of the reaction shown here is found to depend solely on the concentration of the R:X, then the nucleophilic reaction is called an **S_N1** (substitution nucleophilic unimolecular). In an S_N1 mechanism, the slow part of the reaction is the breaking of the bond between R and X. Once that happens, the nucleophile Z adds very quickly. An S_N1 reaction follows first-order kinetics (Chapter 6).

S_N2 a bimolecular nucleophilic substitution reaction; an important type of organic reaction seen in biochemistry; the rate of the reaction follows second-order kinetics

If the nucleophile attacks the R:X while the X is still attached, then both the concentration of R:X and the concentration of :Z will be important. This reaction will follow second-order kinetics and is called an **S_N2** reaction (substitution nucleophilic, bimolecular).

The difference between S_N1 and S_N2 is very important to biochemists because it explains much about the stereospecificity of the products formed. An S_N1 reaction often leads to loss of stereospecificity. Because the leaving group is gone before the attacking group enters, the attacking group can often end up in one of two orientations, although the specificity of the active site can also limit this. With an S_N2 reaction, the fact that the leaving group is still attached forces the nucleophile to attack from a particular side of the bond, leading to only one possible stereospecificity in the product. The chymotrypsin nucleophilic attacks were examples of S_N2 reactions, although no stereochemistry is noted because the carbonyl that was attacked became a carbonyl group again at the end of the reaction and was, therefore, not chiral.

general acid–base catalysis a form of catalysis that depends on the transfer of protons

Before we discuss acid–base catalysis, recall that a Brønsted–Lowry acid is a proton donor, whereas a Brønsted–Lowry base is a proton acceptor. The concept of **general acid–base catalysis** depends on the donation and acceptance of protons by groups such as the imidazole, hydroxyl, carboxyl, sulfhydryl, amino, and phenolic side chains of amino acids; all these functional groups can act as acids or bases. The donation and acceptance of protons give rise to the bond breaking and bond re-formation that constitute the enzymatic reaction.

If the enzyme mechanism involves an amino acid donating a hydrogen ion, as in the reaction

$$R—H^+ + R—O^- \rightarrow R + R—OH$$

then that part of the mechanism would be called general acid catalysis. If an amino acid takes a hydrogen ion from one of the substrates, such as in the reaction

$$R + R—OH \rightarrow R—H^+ + R—O^-$$

then that part is called general base catalysis. The amino acid histidine often takes part in both reactions because it has a reactive hydrogen on the imidazole side chain that dissociates near physiological pH. In the chymotrypsin mechanism (Figure 7.21), histidine was involved in both acid and base catalysis.

A second form of acid–base catalysis reflects another, more general definition of acids and bases. In the Lewis formulation, an acid is an electron-pair acceptor, whereas a base is an electron-pair donor. Metal ions, including such biologically important ones as Mn^{2+}, Mg^{2+}, and Zn^{2+}, are Lewis acids. Thus, they can participate in **metal–ion catalysis** (also called Lewis acid–base catalysis). The involvement of Zn^{2+} in the enzymatic activity of carboxypeptidase A is an example of this type of behavior (Figure 7.25). This enzyme catalyzes the hydrolysis of C-terminal peptide bonds of proteins. The Zn^{2+}, which is required for the activity of the enzyme, forms a **coordination complex** to the imidazole side chains of histidines 69 and 196 and to the carboxylate side chain of glutamate 72. The zinc ion is also complexed to the substrate.

metal–ion catalysis (Lewis acid–base catalysis) a form of catalysis that depends on the Lewis definition of an acid as an electron-pair acceptor and a base as an electron-pair donor

coordination complex a chemical group that usually consists, in biochemistry, of a central metal ion and surrounding groups that donate electrons to the metal ion

The type of binding involved in the complex is similar to the binding that links iron to the large ring involved in the heme group. Binding the substrate to the zinc ion increases the polarity of the substrate's penultimate carbonyl group when the carbonyl oxygen donates electrons to the zinc ion, making the carbonyl carbon susceptible to attack by an oxygen atom of water and allowing the hydrolysis to proceed more rapidly than it does in the uncatalyzed reaction, as shown in Figure 7.25.

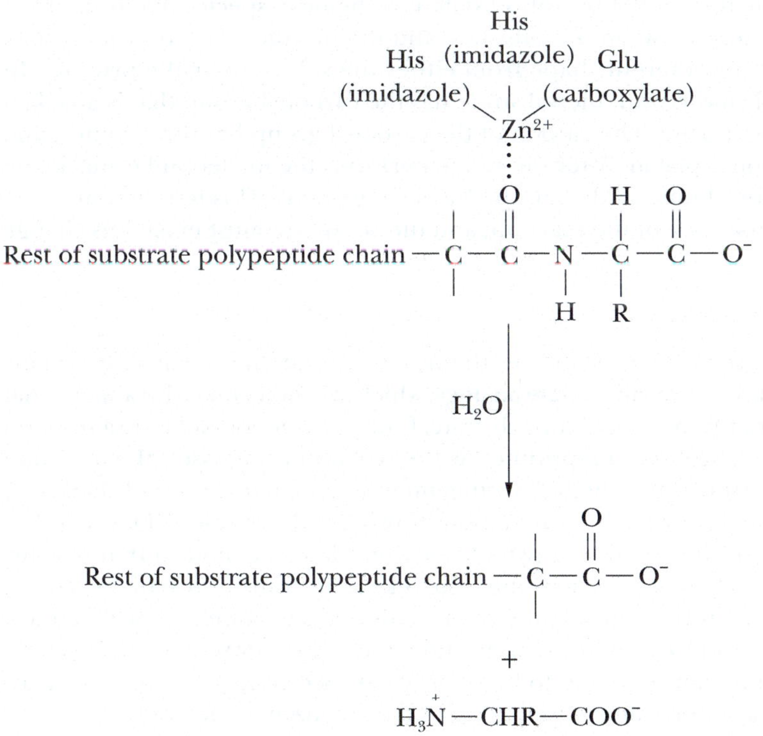

Figure 7.25 The role of zinc in metal-ion catalysis by carboxypeptidase A. A zinc ion is complexed to three side chain groups of carboxypeptidase A and to the penultimate carbonyl group of the substrate. This promotes hydrolysis and removal of the C-terminal amino acid residue from the protein substrate.

A definite connection exists between the concepts of acids and bases and the idea of nucleophiles and their complementary substances, electrophiles. A Lewis acid is an electrophile, whereas a Lewis base is a nucleophile. Catalysis by enzymes, including their remarkable specificity, is based on these well-known chemical principles operating in a complex environment.

The nature of the active site plays a particularly important role in the specificity of enzymes. An enzyme that displays *absolute specificity,* catalyzing the reaction of one, and only one, substrate to a particular product, is likely to have a fairly rigid active site that is best described by the lock-and-key model of substrate binding. The many enzymes that display *relative specificity,* catalyzing the reactions of structurally related substrates to related products, apparently have more flexibility in their active sites and are better characterized by the induced-fit model of enzyme–substrate binding; chymotrypsin is a good example.

Finally, there are *stereospecific* enzymes with specificity in which optical activity plays a role. The binding site itself must be asymmetric in this situation (Figure 7.26). If the enzyme is to bind specifically to an optically active substrate, the binding site must have the shape of the substrate and not its mirror image. There are even enzymes that introduce a center of optical activity into the product. The substrate itself is not optically active in this case. There is only one product, which is one of two possible isomers, not a mixture of optical isomers.

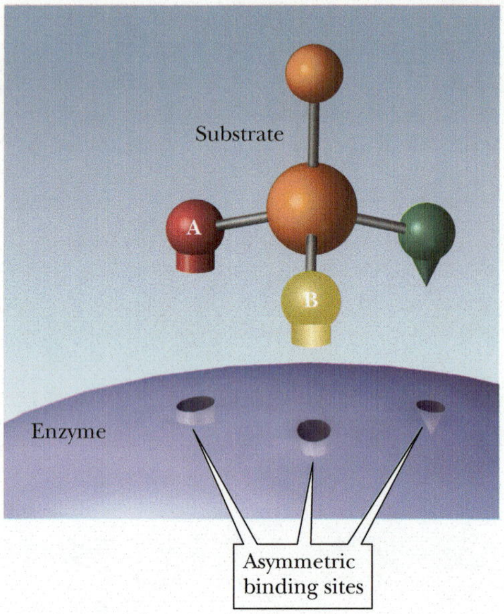

Figure 7.26 An asymmetric binding site on an enzyme can distinguish between identical groups such as A and B. The binding site consists of three parts, giving rise to asymmetric binding because one part is different from the other two.

7-7 The Active Site and Transition States

Learning Objectives

After reading this section, students will be able to:

LO 7.7.1 Analyze a transition state analog that can be used to test a proposed catalytic mechanism or act as a competitive inhibitor.

Now that we better understand mechanisms and the active site, it is worth revisiting the nature of enzyme catalysis. Recall that an enzyme lowers the activation energy by lowering the energy necessary to reach the transition state (Figure 6.1).

The true nature of the transition state is a chemical species that is intermediate in structure between the substrate and the product. This transition state often has a very different shape from either the substrate or the product. In the case of chymotrypsin, the substrate has the carbonyl group that is attacked by the reactive serine. The carbon of the carbonyl group has three bonds, and the orientation is planar. After the serine performs the nucleophilic attack, the carbon has four bonds and a tetrahedral arrangement. This tetrahedral shape is the transition state of the reaction, and the active site must make this change more likely.

▶ How do we determine the nature of the transition state?

The fact that the enzyme stabilizes the transition state has been shown many times by the use of **transition-state analogs**, which are molecules with a shape that mimics the transition state of the substrate. Proline racemase catalyzes a reaction that converts L-proline to D-proline. As the reaction progresses, the α-carbon must change from a tetrahedral arrangement to a planar form and then back to tetrahedral, but with the orientation of two bonds reversed (Figure 7.27). An inhibitor of the reaction is pyrrole-2-carboxylate, a chemical that is structurally similar to what proline would look like at its transition state because it is always planar at the equivalent carbon. This inhibitor binds to proline racemase 160 times more strongly than proline does. Transition-state analogs have been used with many enzymes to help verify a suspected mechanism and structure of the transition state, as well as to inhibit an enzyme selectively.

transition-state analogs synthesized compounds that mimic the form of the transition state of an enzyme reaction

Proline racemase reaction

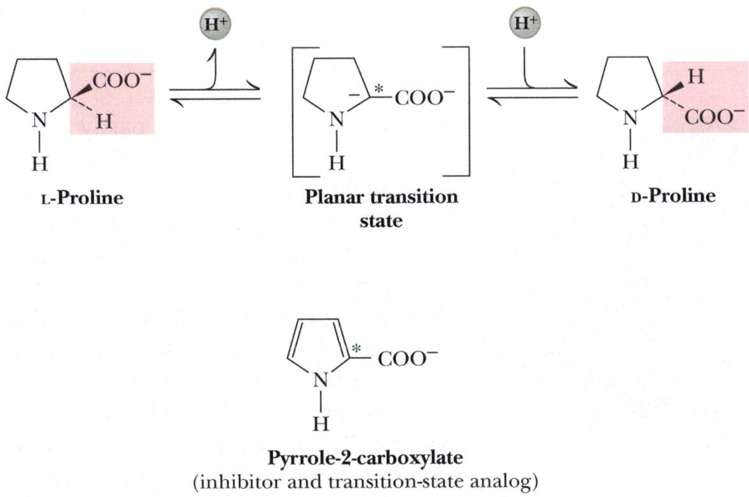

Figure 7.27 **The proline racemase reaction.** Pyrrole-2-carboxylate and Δ-1-pyrroline-2-carboxylate mimic the planar transition state of the reaction. The asterisks here denote carbons with trigonal planar geometry.

7-8 Coenzymes

Learning Objectives

After reading this section, students will be able to:

LO 7.8.1 Distinguish between cofactors and coenzymes.

Cofactors are nonprotein substances that take part in enzymatic reactions and are regenerated for further reaction. Metal ions frequently play such a role (recall metal-ion catalysis discussed in Section 7.6), and they make up one of two important classes of cofactors. The other important class (**coenzymes**) is a mixed bag of organic compounds; many of them are vitamins or are metabolically related to vitamins.

cofactors nonprotein substances that take part in enzymatic reactions and are regenerated at the end of the reaction

coenzymes nonprotein, nonmetallic substances that take part in enzymatic reactions and are regenerated at the end of the reaction

Because metal ions are Lewis acids (electron-pair acceptors), they can act as Lewis acid–base catalysts. They can also form coordination compounds by behaving as Lewis acids, whereas the groups to which they bind act as Lewis bases. Coordination compounds are an important part of the chemistry of metal ions in biological systems, as shown by Zn^{2+} in carboxypeptidase and by Fe^{2+} in hemoglobin. The coordination compounds formed by metal ions tend to have quite specific geometries, which aid in positioning the groups involved in a reaction for optimum catalysis.

Some of the most important organic coenzymes are vitamins and their derivatives, especially B vitamins. Many of these coenzymes are involved in oxidation–reduction reactions, which provide energy for the organism. Others serve as group-transfer agents in metabolic processes (Table 7.1). These coenzymes are discussed in more detail in other chapters, where the specific reactions in which they are involved are presented.

Table 7.1 **Coenzymes, Their Reactions, and Their Vitamin Precursors**

Coenzyme	Reaction Type	Vitamin Precursors	See Section
Biotin	Carboxylation	Biotin	18-2, 21-6
Coenzyme A	Acyl transfer	Pantothenic acid	15-5, 19-3, 21-6
Flavin coenzymes	Oxidation–reduction	Riboflavin (B_2)	15-5, 19-3
Lipoic acid	Acyl transfer	—	19-3
Nicotinamide adenine coenzymes	Oxidation–reduction	Niacin	15-5, 17-3, 19-3
Pyridoxal phosphate	Transamination	Pyridoxine (B_6)	23-4
Tetrahydrofolic acid	Transfer of one-carbon units	Folic acid	23-4
Thiamine pyrophosphate	Aldehyde transfer	Thiamin (B_1)	17-4, 18-4

Hot Topic

Alzheimer's Disease

Direct prion diseases like bovine spongiform encephalopathy (BSE) or Creutzfeldt–Jakob disease (Chapter 4) are not the only diseases to involve prion-like pathologies. Although BSE is transmissible, diseases such as Alzheimer's, Parkinson's, and Huntington's are not. They do, however, have pathologies involving plaque-forming proteins, and they do affect brain functioning, often with similar effects.

Basics of Alzheimer's Disease

In 1906, at the 37th annual meeting of the Society of German Psychiatrists in Tubingen, Germany, Alois Alzheimer presented a paper on the clinical and pathological characteristics of a disease that later would be named after him. In 1901, he had a 51-year-old patient with a rare condition of progressive memory loss, delusions, and hallucinations. When she died five years later, her brain was sent to a laboratory, where Alzheimer used a silver stain to localize neuritic plaques and neurofiber tangles.

Alzheimer's disease (AD) is a neurodegenerative disease generally associated with older adults. Before the term was popular, many cases of the disease would have been called dementia or just being senile, although we now know that several diseases can lead to the symptoms formerly lumped into those monikers. The chances of getting AD double every five years after the age of 65, until about a third of people in their mid-80s will show signs of the disease. The disease's observable symptoms include loss of memory, usually starting with more recent ones and proceeding in later stages to remove long-held memories, as well as personality changes and delusions. Eventually, patients don't recognize their closest family members. It is a disease that is devastating not only to the individual, but also to their family,

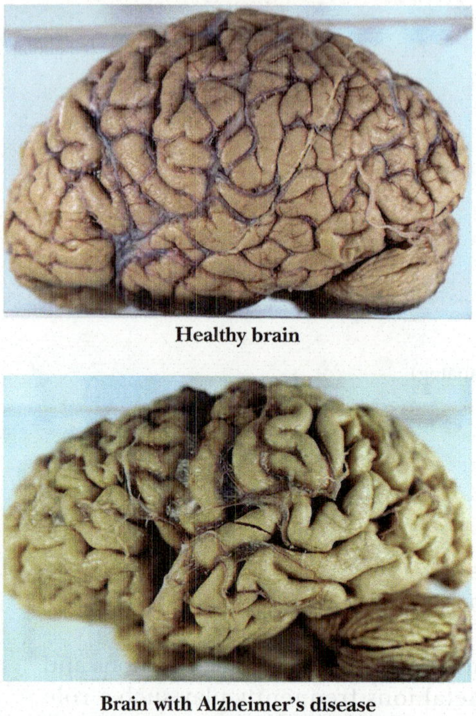

SCIENCE SOURCE/Science Source Stock Photos & Video

Figure 7.28 The brains of people with Alzheimer's disease shrink and lose some of the folding of the various lobes.

because the loss of the memories essentially erases who the person was.

One of the earliest reported pathologies of the disease was the fact that people with AD had brains that had shrunk in size and lost some of their folds, as shown in Figure 7.28. It is also now known that the brain can shrink years before there are any outward signs of the clinical symptoms.

B-Amyloid Plaques

At the heart of the disease is the destruction of some of the 100 billion neurons in the brain. The destruction is believed to be caused by plaques formed by two different proteins, one called amyloid β, or Aβ, and the other one called tau. Aβ is a short peptide of 40 or 42 amino acids, first isolated in 1984. It is derived from a larger protein called amyloid beta precursor protein (APP). APP is a membrane-bound protein, partly found inside a neuron and partly outside it. Two protein-cutting enzymes called β-secretase and γ-secretase cut out Aβ from APP, a normal process that occurs in most cells of the body. Scientists believe that the metabolism of the Aβ is what causes the disease. At high enough concentrations, the Aβ proteins form insoluble fibers and form plaques on the nerve cells. However, studies have also shown that, in tissue samples purified from AD brains, there are smaller forms of Aβ that are soluble and have not yet formed insoluble fibers. In animal models, these have been shown to be able to transmit the disease to healthy brain tissue, similarly to how prion proteins can propagate. It is now believed that cutting the Aβ into the 42–amino acid version is likely a key factor in the development of Aβ plaques. Both the soluble Aβ and insoluble fibers are toxic to neurons grown in petri dishes, and they have been shown to interfere with processes critical to learning and memory in mice.

Continuing the comparison with prion diseases, the brains of AD patients or AD test animals look remarkably similar to those with prion diseases. Figure 7.29 shows a stained section of such a brain. The brown regions are the Aβ plaques. The blue spots are formed by the other culprit, tau tangles.

Mutations in the genes that produce γ-secretase have been shown to cause

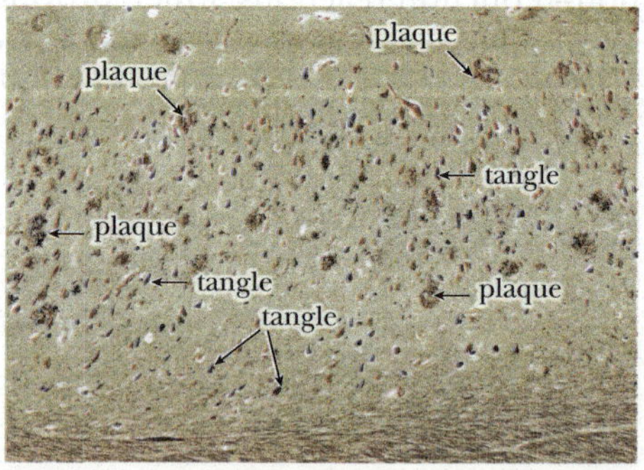

THOMAS DEERINCK, NCMIR/Science Source

Figure 7.29 Plaques and tau tangles in an AD brain.

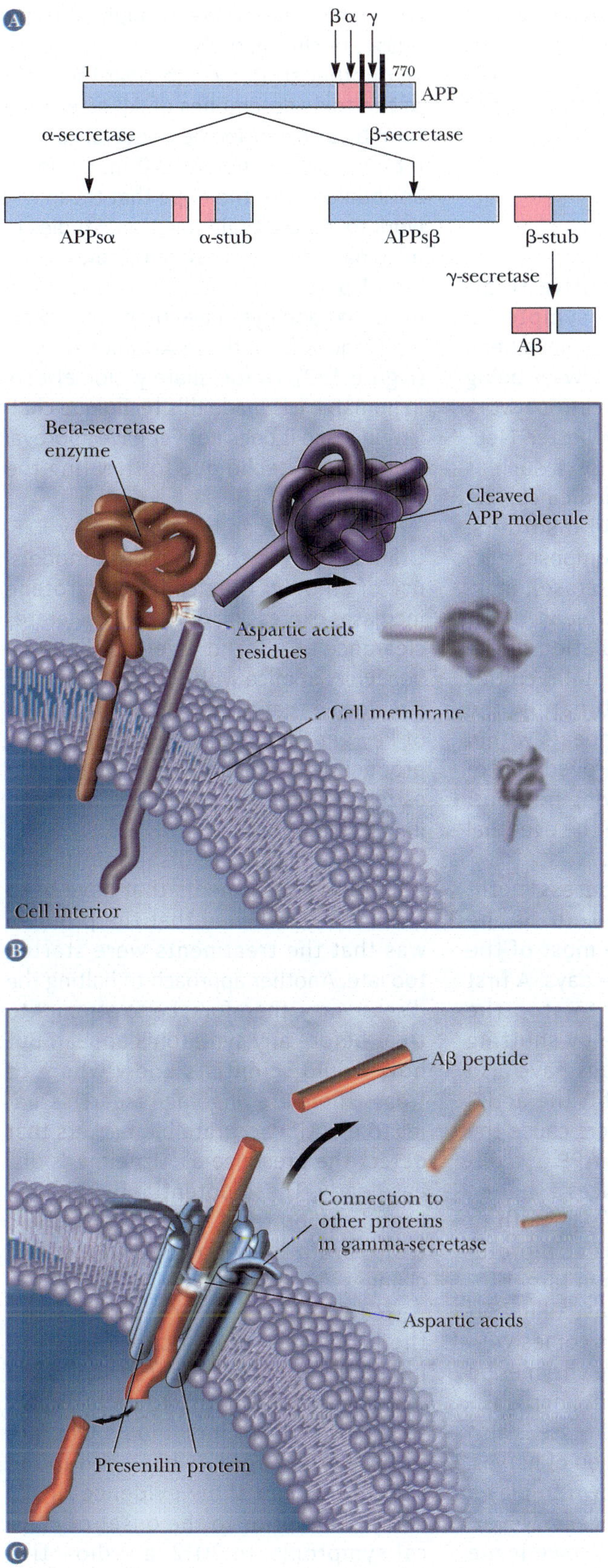

Figure 7.30 **Production of amyloid-β. In the first step, the enzyme β-secretase cuts APP.** Then the γ-secretase enzyme cuts the remaining portion of the APP inside the membrane, releasing Aβ. *(Biochem Connection,* Biochemistry, *8e, p103.)*

an early and aggressive form of AD. Other mutations of APP itself are also implicated. Combining the evidence of the importance of the Aβ peptide and the enzymes that create it has led to the "amyloid-cascade hypothesis." According to this hypothesis, AD begins with the buildup of Aβ, which is cut from the APP (Figure 7.30). In the first step, the enzyme β-secretase uses a pair of Asp residues to cut APP, and this occurs outside the cell membrane. Then the γ-secretase enzyme, itself located in the membrane, cuts the remaining portion of the APP inside the membrane, releasing Aβ. There is a third secretase, α-secretase. It cuts inside the Aβ fragment, so when it cuts, there is no possibility of getting the Aβ fragment.

Work in 2014 led to another theory on what controls the level of Aβ in the cell. The researchers found evidence that much of the Aβ is produced in a subcellular organelle called the endosome, as shown in Figure 7.31.

The research involved the discovery of a molecular complex called a retromer, which they considered to be a kind of recycling bin for the cell. Their idea is that the retromers take molecules from the endosome and carry them to the Golgi apparatus. From the Golgi, the APP proteins go to the cell membrane. They considered that if the retromers were not working fast enough, too many APP proteins would be cleaved to Aβ before they had a chance to make it to the Golgi. They had found that patients with AD had a much lower level of the retromers in their hippocampus, a key brain region for memory. Their work is now focusing on finding a drug to stabilize the retromers so that they continue to do their recycling.

Tau

Tau is a protein involved in the formation of microtubules. In the human brain, there are six isoforms of tau formed from a single gene through alternative splicing of mRNA (Section 11-8D). The role of tau in Alzheimer's disease is less clear. All six isoforms are present in patients with AD in the same proportions as in healthy individuals. Tau deposits inside the nerve cells are prevalent not just in AD, but in a wide range of neurodegenerative diseases. Because the protein tau is

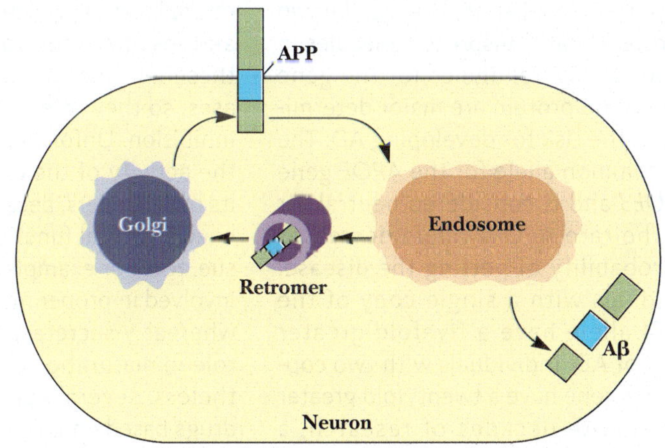

Figure 7.31 **Possible role of the retromer and the endosome in the processing of amyloid-β.** *(Science, April 25, 2014, "Potential Alzheimer's Drug Spurs Protein Recycling" by Ken Garber.)*

implicated, AD is also one of the diseases referred to as a tauopathy.

Another tauopathy that has been extensively studied is another common form of dementia called frontotemporal dementia (FTD). The tau found in tau tangles is hyperphosphorylated, which at least gives a potential target for a treatment if someone were able to find a way to inhibit the kinase that is putting the phosphates on the protein or to activate a phosphatase to remove them. Many researchers now believe that the phosphorylation of the tau proteins is somehow triggered by the Aβ plaques, and this then disrupts the microtubule assemblies in the neurons and kills them.

Tauopathies may take years to develop, and some patients have been under the care of a psychiatrist for years for behavioral issues before anyone discovers the true nature of their disease. Aggregates of tau proteins in the brain can cause inappropriate social behavior, depression, and insomnia, and eventually can lead to suicidal tendencies. Recent studies have also linked head traumas in contact athletes and posttraumatic stress disorders in soldiers with FTD and tau proteins. Head trauma in professional and amateur sports is a huge concern, and the symptoms of repeated head trauma look very similar to those of AD and other neurodegenerative diseases. Brain scans show shrinkage and loss of folds as well in such cases.

ApoE

Apolipoprotein E (ApoE) is a well-known lipoprotein that transports molecules in the blood. Several alleles for the gene for the ApoE protein are major determinants in the risk for developing AD. The most common allele for the *APOE* gene is *APOE3* and is considered neutral for risk. The rare *APOE2* allele decreases the probability of getting the disease. Individuals with a single copy of the *APOE4* allele have a fivefold greater chance of AD. Individuals with two copies of the gene have a twentyfold greater risk. Despite decades of research, a definitive answer as to why these different alleles lead to the different risks is still unknown. It has always been assumed that ApoE must play a role in transport of the Aβ peptide.

Treating Alzheimer's Disease

There have been no definitive treatments that eliminate the symptoms, but many different strategies have been tried, with new techniques always being developed. Some AD medications work by helping neurons to work better. Neurons release neurotransmitters, such as acetylcholine. Some AD medicines, like medicines for other brain disorders, block the enzyme that decomposes acetylcholine (acetylcholinesterase), making the neurotransmitter released last longer. This leads to stimulation of the neurons and improved mental faculties in those patients. Unfortunately, the improvement usually disappears within six months. Even with improved neurotransmission of the surviving neurons, this treatment cannot stop the eventual destruction of the neurons.

A true therapy must address methods to stop the destruction of the neurons, so that is where most of the effort is being placed these days. A first strategy was to attempt to stop the production of the peptide by shutting down the β-secretase. Drugs have also been developed that inhibit the activity of the γ-secretase. Others cause the β-secretase enzyme to cut APP at a different location, which produces a shorter, and less harmful, form of Aβ. Both of these enzymes are aspartic acid proteases, so they offer a common target for inhibition. Unfortunately, just killing off the activity of the enzymes comes with its own dangers, because the enzymes do have a natural function within nerve tissue, too. For example, β-secretase is also involved in proper myelination of nerves, whereas γ-secretase is known to play a role in maturation of stem cells. Nevertheless, several companies have made drugs based on inhibiting the secretases, all of which have gone through different phases of clinical trials.

Another strategy is to clear the brain of the toxic assemblies of Aβ after they form by using antibodies. Several companies are producing anti-Aβ monoclonal antibodies, with the hope that these will work to restore brain function. Studies in mice have shown these antibodies capable of preventing plaque formation in mice that don't yet have them, and clearing plaques from those AD mice that do (Figure 7.32). Unfortunately, side effects in human trials made this technique challenging, and it does not yet have approval as a general technique for treating the disease.

In 2012, research published by P. E. Cramer showed that in mouse models, use of an older drug called bexarotene increases expression of ApoE, accelerates clearance of Aβ, and reverses abnormal behaviors in mice with AD. Bexarotene is a molecule that affects the transcription of key genes in the nucleus, ultimately producing more ApoE, which is hypothesized to clear the Aβ peptides, as shown in Figure 7.33.

With many of the disappointments in potential treatments that have been seen, many believed that the problem was that the treatments were started too late. Another approach to halting the disease may, therefore, be to attempt to treat before any symptoms appear. But how would scientists know whom to treat? There are ongoing research studies to identify accurate biomarkers that detect the presence of Alzheimer's disease early. These potential biomarkers include blood tests, cerebrospinal fluid tests, brain imaging techniques, and changes in the eye or vision.

Whether current research is on the brink of finding a cure for AD or not, some of the research has been extremely useful, if only to show us how to identify those individuals who are in the early stages. It has been found that brains can show evidence of AD many years prior to the onset of clinical symptoms. In 2012, a radioactive drug was approved that binds to Aβ

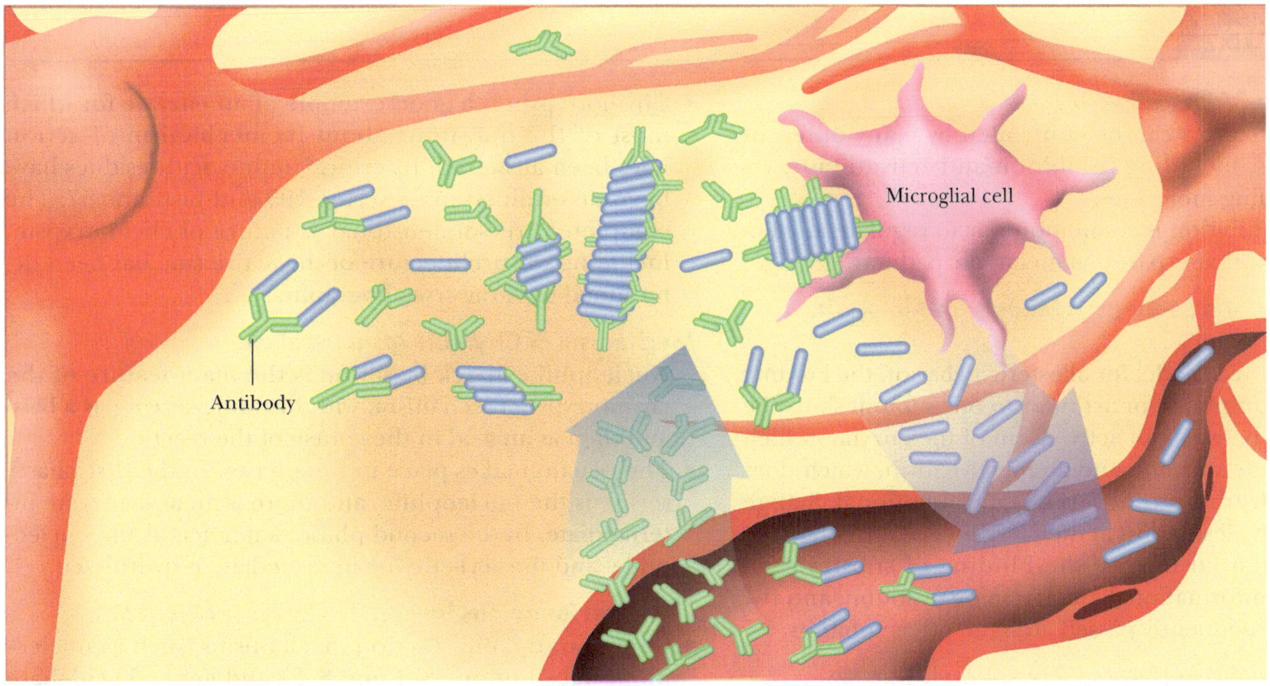

Figure 7.32 **Anti-Aβ antibodies can bind to and remove Aβ and eliminate plaques.** *(Scientific American May 2006.)*

in the brain and makes it possible to visualize plaques via positron emission tomography (PET) scans. This will make it much easier to test a patient to see if they are already developing the disease prior to the first bout of memory loss, which is usually the first warning. Accumulation of these plaques can take place between 5 and 20 years before the presence of symptoms. Tau buildup takes place from 1 to 5 years before symptoms, and brain shrinkage starts 1 to 3 years before symptoms, the latter of which can be seen with magnetic resonance imaging (MRI).

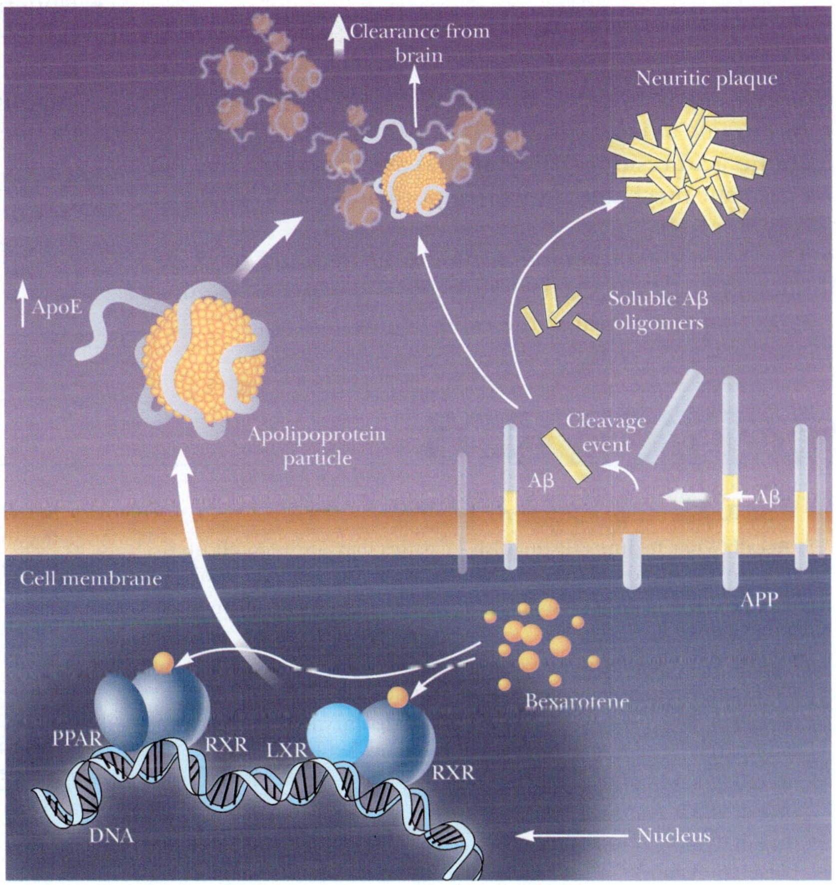

Figure 7.33 **Clearing Aβ.** Stimulating nuclear receptors increases ApoE expression and the clearance of Aβ from the brain by a mechanism requiring ApoE. *(1447, Science, 335, Old Drug, New Hope for Alzheimer's Disease by W.J. Strittmatter.)*

Summary

The Behavior of Allosteric Enzymes

- Allosteric enzymes can be controlled by many different mechanisms, including inhibition and activation by reversibly binding molecules.
- Feedback inhibition is a common way to regulate an allosteric enzyme that is part of a metabolic pathway.

The Concerted and Sequential Models for Allosteric Enzymes

- In the concerted model for allosteric behavior, the binding of substrate, inhibitor, or activator to one subunit shifts the equilibrium between an active form of the enzyme, which binds substrate strongly, and an inactive form, which does not bind substrate strongly. The conformational change takes place in all subunits at the same time.
- In the sequential model, the binding of substrate induces the conformational change in one subunit, and the change is subsequently passed along to other subunits.

Control of Enzyme Activity by Phosphorylation

- Some enzymes are activated or inactivated depending on the presence or absence of phosphate groups. This kind of covalent modification can be combined with allosteric interactions to allow for a high degree of control over enzymatic pathways.

Determining the Critical Amino Acids in the Active Site

- In enzyme studies, the nature of the critical active-site amino acid residues, their spatial arrangement, and the enzymatic mechanism are all important areas of biochemistry research.
- The use of chemical modification reagents and X-ray crystallography allows the determination of active-site amino acids that are critical to the catalytic mechanism.
- Chymotrypsin is a good example of an enzyme for which most of the questions about its mechanism of action have been answered. Its critical amino acid residues have been determined to be serine 195 and histidine 57. The complete three-dimensional structure of chymotrypsin, including the architecture of the active site, has been determined by X-ray crystallography.

Mechanism of Chymotrypsin Catalysis

- Nucleophilic attack by serine is the main feature of the chymotrypsin mechanism, with histidine serving as a base and then as an acid in the course of the reaction.
- The reaction takes place in two phases. In the first phase, serine is the nucleophile, and there is an acyl-enzyme intermediate. In the second phase, water acts as the nucleophile, and the acyl-enzyme intermediate is hydrolyzed.

Chemical Reactions Involved in Enzyme Mechanisms

- Common organic reaction mechanisms, such as nucleophilic substitution (S_N1 and S_N2) and general acid–base catalysis, are known to play roles in enzymatic catalysis.

Nature of the Transition State

- Understanding of the nature of catalysis has been aided by the use of transition-state analogs—molecules that mimic the transition state. These compounds usually bind to the enzyme better than the natural substrate and help verify the mechanism.

Coenzymes

- Nonprotein substances called cofactors are present in many enzymes to assist with catalysis. There are two classes of cofactors: metal ions and organic compounds called coenzymes.

Review Exercises

7-1 The Behavior of Allosteric Enzymes

1. **Recall** What features distinguish enzymes that undergo allosteric control from those that obey the Michaelis–Menten equation?
2. **Recall** What is the metabolic role of aspartate transcarbamoylase (ATCase)?
3. **Recall** What molecule acts as a positive effector (activator) of ATCase? What molecule acts as an inhibitor?
4. **Recall** Is the term K_M used with allosteric enzymes? What about competitive and noncompetitive inhibition? Explain.
5. **Recall** What is a K system?
6. **Recall** What is a V system?
7. **Recall** What is a homotropic effect? What is a heterotropic effect? Give an example of molecules that would show each effect when binding to ATCase.
8. **Recall** Describe the protein subunit composition of ATCase.
9. **Recall** How is the cooperative behavior of allosteric enzymes reflected in a plot of reaction rate against substrate concentration?
10. **Recall** Does the behavior of ATCase become more or less cooperative in the presence of CTP?
11. **Recall** Does the behavior of ATCase become more or less cooperative in the presence of ATP?
12. **Recall** Explain what is meant by $K_{0.5}$.
13. **Recall** Match each nonallosteric enzyme kinetic term with the equivalent term that describes allosteric enzymes.

Nonallosteric enzyme kinetic term	Similar term used for allosteric enzymes
K_M	V_{max}
V_{max}	K system
Competitive inhibition	V system
Noncompetitive inhibition	$K_{0.5}$

14. **Reflect and Apply** Phosphofructokinase is a key regulatory step of glucose metabolism. It catalyzes the conversion of ATP and fructose 6-phosphate to ADP and fructose 1,6-bisphosphate. ATP also binds to an allosteric site and leads to a decrease in V_{max}. Use three different kinetic terms that all accurately describe the interaction of ATP with phosphofructokinase.

7-2 The Concerted and Sequential Models for Allosteric Enzymes

15. **Recall** Describe how the subunit conformations vary in the concerted and sequential models of allosteric enzymes.
16. **Recall** Which allosteric model can explain negative cooperativity? Explain why.
17. **Recall** With the concerted model, what conditions favor greater cooperativity?
18. **Recall** With respect to the concerted model, what is the L value? What is the *c* value?
19. **Reflect and Apply** Thymidylate synthase is a homodimer that catalyzes conversion of the DNA damaging molecule, dUMP, into dTMP. Researchers found that in the absence of dUMP only one active site in the dimer is available. However, binding of substrate to the one available site changes the conformation of the second subunit to expose the second active site. Does thymidylate follow the concerted or sequential model for allostery? Explain your choice.

7-3 Control of Enzyme Activity by Phosphorylation

20. **Recall** Describe the general reaction catalyzed by a protein kinase.
21. **Recall** What amino acids are often phosphorylated by kinases?
22. **Recall** Explain how phosphorylation is involved in the function of the sodium–potassium pump.
23. **Reflect and Apply** Nonhydrolyzable analogs of ATP have a stable bond to the terminal phosphate so that it cannot be easily removed. This class of molecules is a notorious inhibitor of protein kinases. If you observed the kinetics of a cooperative multi-subunit protein kinase in the presence of an ATP analog, which type of inhibition would you most likely see? Why?

7-4 Zymogens

24. **Reflect and Apply** Why is it necessary for the body to produce a zymogen for chymotrypsin instead of making it in the active form directly?
25. **Recall** Protein cleavage is frequently associated with the loss of protein function, but what do you call the category of enzymes that are actually activated by proteolytic cleavage? Describe an example of one such enzyme.

7-5 The Nature of the Active Site

26. **Recall** What are the two critical amino acids in the active site of chymotrypsin?
27. **Recall** Explain the techniques used to determine the identity and arrangement of critical active-site residues.
28. **Recall** Draw and describe the two stages of the chymotrypsin reaction mechanism.
29. **Reflect and Apply** Briefly describe the role of nucleophilic catalysis in the mechanism of the chymotrypsin reaction.
30. **Reflect and Apply** Explain the function of histidine 57 in the mechanism of chymotrypsin.
31. **Reflect and Apply** Histidine, with a pK_a near physiological pH, is frequently found in active sites. Use chymotrypsin to explain why this pK_a is useful in catalysis.

7-6 Chemical Reactions Involved in Enzyme Mechanisms

32. **Reflect and Apply** Describe the property of metal ions that makes them useful cofactors, and describe the role of the zinc ion in the enzymatic activity of carboxypeptidase A.
33. **Recall** In biological macromolecules, what functional group is often attacked by a nucleophile?
34. **Reflect and Apply** Categorize the following examples from biochemical reactions as functional groups acting as either Lewis acids, Lewis bases, Brønsted–Lowry acids, or Brønsted–Lowry bases: 1) the sulfhydryl of cysteine donates a hydrogen; 2) a zinc ion accepts electrons; 3) the oxygen of a protein backbone donates a lone pair of electrons; 4) the imidazole of histidine is protonated.
35. **Reflect and Apply** What is meant by general acid catalysis with respect to enzyme mechanisms?
36. **Reflect and Apply** Explain the difference between an S_N1 reaction mechanism and an S_N2 reaction mechanism.
37. **Reflect and Apply** You are monitoring a reaction and find that it is not stereospecific. Do you predict the reaction takes place using an S_N1 or S_N2 reaction mechanism? Why?

7-7 The Active Site and Transition States

38. **Reflect and Apply** What would be the characteristics of a transition-state analog for the chymotrypsin reaction?
39. **Reflect and Apply** Describe the molecule you would use as a high-affinity K system inhibitor of carboxypeptidase A.
40. **Reflect and Apply** Why does a high affinity to a transition-state analog help verify a reaction mechanism? Reference the induced-fit model in your explanation.

7-8 Coenzymes

41. **Recall** Provide an example of a coenzyme and an inorganic cofactor.
42. **Recall** Fill in the blanks with either "coenzymes" or "cofactors" to make the following statement true: __________ are all types of __________, but not all __________ are __________.
43. **Reflect and Apply** Would a disease that affects vitamin absorption have more effect on enzymes that use coenzymes or inorganic cofactors? Why?

Further Reading

Collins, T. J., and C. Walter. Little Green Molecules. *Sci. Amer.* **294** (3): 82–90 (2006). An article describing constructed molecules with enzyme-like properties that clean up pollution.

Danishefsky, S. Catalytic Antibodies and Disfavored Reactions. *Science* **259**, 469–470 (1993). A short review of chemists' use of antibodies as the basis of "tailor-made" catalysts for specific reactions.

Dressler, D., and H. Potter. *Discovering Enzymes.* New York: Sci. Amer. Library, 1991. A well-illustrated book that introduces important concepts of enzyme structure and function.

Koshland, D., G. Nemethy, and D. Filmer. Comparison of Experimental Binding Data and Theoretical Models in Proteins Containing Subunits. *Biochem.* **5**, 365–385 (1966).

Kraut, J. How Do Enzymes Work? *Science* **242**, 533–540 (1988). An advanced discussion of the role of transition states in enzymatic catalysis.

Lerner, R., S. Benkovic, and P. Schultz. At the Crossroads of Chemistry and Immunology: Catalytic Antibodies. *Science* **252**, 659–667 (1991). A review of how antibodies can bind to almost any molecule of interest and then catalyze some reaction of that molecule.

Marcus, R. Skiing the Reaction Rate Slopes. *Science* **256**, 1523–1524 (1992). A brief, advanced-level look at reaction transition states.

Monod, J., J. Wyman, and J.-P. Changeux. On the Nature of Allosteric Transitions: A Plausible Model. *J. Mol. Biol.* **12**, 88–118 (1965).

Sigman, D., ed. *The Enzymes,* Vol. 20, *Mechanisms of Catalysis.* San Diego: Academic Press, 1992. Part of a definitive series on enzymes and their structures and functions.

Sigman, D., and P. Boyer, eds. *The Enzymes,* Vol. 19, *Mechanisms of Catalysis.* San Diego: Academic Press, 1990. Part of a definitive series on enzymes and their structures and functions.

Wenner, M. A New Kind of Drug Target. *Sci. Amer.* **301** (2), 70–76 (2009).

Wolan, D. W., J. A. Zorn, D. C. Gray, and J. A. Wells. Small-Molecule Activators of a Proenzyme. *Science* **326**, 853–858 (2009).

Hot Topic

Garber, K. Potential Alzheimer's Drug Spurs Protein Recycling. *Science,* **344**, 351 (2014)

Glabe, C. Avoiding Collateral Damage in Alzheimer's Disease Treatment. *Science,* **314**, 602–603 (2006).

Goedert, M. and M. G. Spillantini. A Century of Alzheimer's Disease. *Science,* **314**, 777–781 (2006).

Miller, G. Could They All Be Prion Diseases? *Science,* **326**, 1337–1339 (2009).

Miller, G. Stopping Alzheimer's Before It Starts. *Science,* **337**, 790–792 (2012).

Stix, G. Alzheimer's: Forestalling the Darkness. *Scientific American,* 51–57 (June 2010).

Strittmatter, W. J. Old Drug, New Hope for Alzheimer's Disease. *Science,* **335**, 1447–1448 (2012).

Walker, L. C. and M. Jucker. Seeds of Dementia. *Scientific American,* 53–57 (May 2013).

Wolfe, M. S. Shutting Down Alzheimer's. *Scientific American,* 73–79 (May 2006).

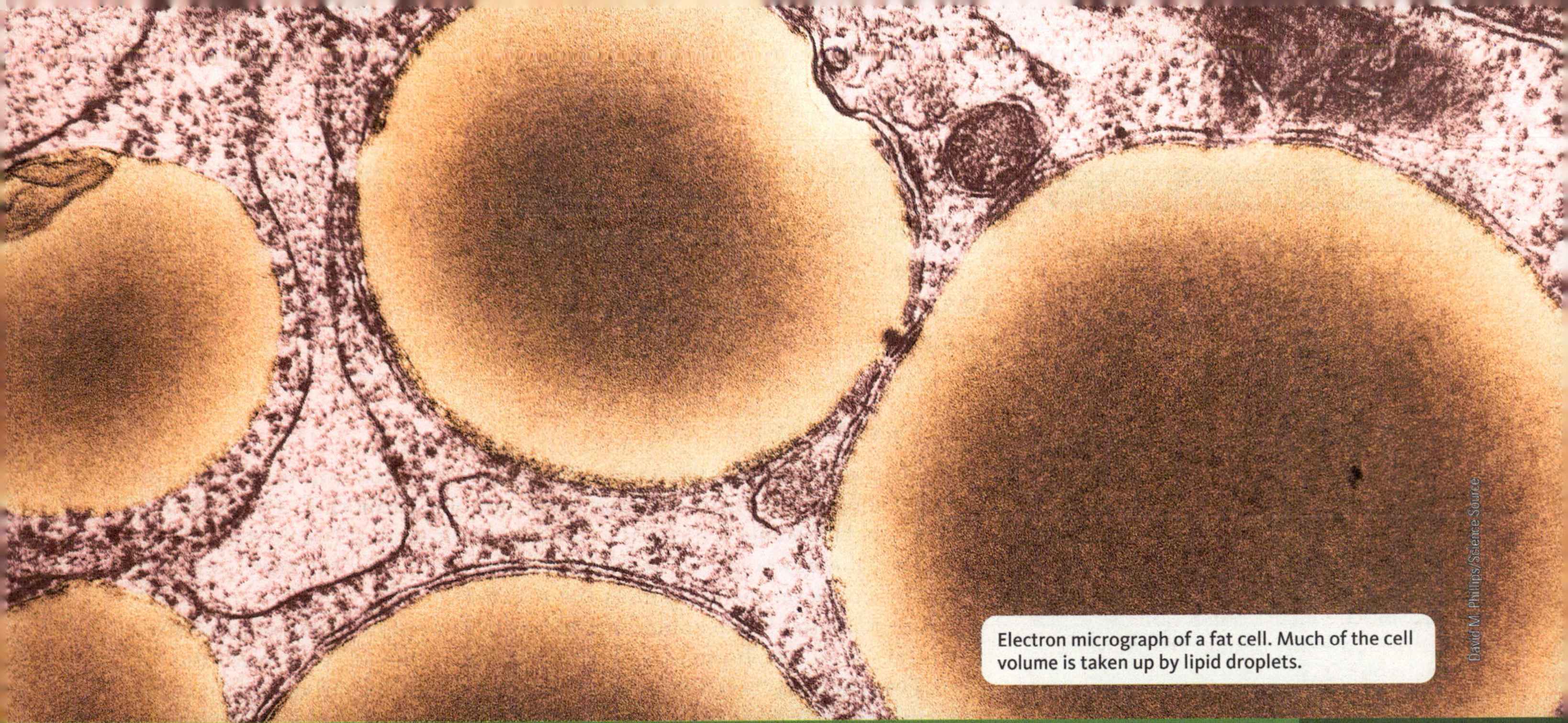

Electron micrograph of a fat cell. Much of the cell volume is taken up by lipid droplets.

David M. Phillips/Science Source

Lipids and Biological Membranes 8

Chapter Outline

8-1 The Definition of a Lipid

Learning Objectives

After reading this section, students will be able to:

LO 8.1.1 Distinguish between the characteristics of lipids and other macromolecules.

LO 8.1.2 Distinguish between the two main groups of lipids.

So far, we have discussed many biochemical aspects of proteins, and in this chapter, we will discuss a different class of biochemical compounds called **lipids**. They are found in many biological areas and are an important component of biological membranes. In terms of chemistry, lipids are a mixed bag of organic compounds that share some properties based on structural similarities—mainly a preponderance of nonpolar groups. Because of this, lipids are marginally soluble (at best) in water but readily soluble in organic solvents such as chloroform or acetone.

Classified according to their chemical nature, lipids fall into two main groups. One group, which consists of open-chain compounds with polar head groups and long nonpolar tails, includes *fatty acids, triacylglycerols, sphingolipids, phosphoacylglycerols,* and *glycolipids*. The second major group consists of fused-ring compounds, the *steroids*; an important representative of this group is cholesterol.

Chapter Outline (continued)

8-2 The Chemical Natures of the Lipid Types

Learning Objectives

After reading this section, students will be able to:

LO 8.2.1 Distinguish between six different types of open-chain lipids.

LO 8.2.2 Distinguish between different types of steroids.

8-2A Fatty Acids

A fatty acid is an **amphipathic** molecule with a polar head group and a nonpolar tail, giving it both hydrophilic and hydrophobic properties. The polar end of a fatty acid molecule is the acidic **carbonyl group** derivative called a **carboxyl group**, and the nonpolar end is a hydrocarbon chain. The carboxyl group can ionize under the proper conditions (Section 2-4).

Figure 8.1 shows some common fatty acids. For palmitic acid, a condensed structural formula is shown as well as a line-angle formula. As we study the different kinds of lipids here in Chapter 8, keep in mind that carbon usually forms four covalent bonds and oxygen usually forms two covalent bonds, as seen explicitly in the condensed structural formula for palmitic acid. Fatty acids normally contain an even number of carbon atoms, and the hydrocarbon chain is usually unbranched. If there are carbon–carbon double bonds in the chain, the fatty acid is *unsaturated*; if there are only single bonds, the fatty acid is *saturated*. To distinguish between saturated and unsaturated fatty acids, remember that the hydrocarbon tails of saturated fatty acids are completely filled (i.e., saturated) with hydrogens while unsaturated fatty acids are not. Tables 8.1 and 8.2 list a few examples of these two classes of fatty acids.

In unsaturated fatty acids, the stereochemistry at the double bond is usually a **cis double bond** rather than a **trans double bond**. The difference between cis and trans fatty acids is very important to their overall shape, as seen by comparing the structures of oleic acid and elaidic acid in Figure 8.1. A cis double bond puts a kink in the long-chain hydrocarbon tail, whereas the shape of a trans fatty acid is like that of a saturated fatty acid in its fully extended conformation. The notation used in Table 8.2 to name fatty acids indicates the number of carbon atoms and the number of double bonds. In this system, 18:0 denotes an

lipids a diverse group of organic compounds whose structures contain mostly nonpolar groups and a few polar groups

amphipathic a molecule that has one end with a polar, water-soluble group and another end with a nonpolar group that is insoluble in water

carbonyl group a functional group composed of a carbon atom double-bonded to an oxygen atom

carboxyl group the —COOH functional group containing a carbonyl group that is bonded to a hydroxyl (—OH) group

cis double bond a double bond where the heavier groups attached to the double bond are on the same side of the double bond

Careers in Chemistry • Tonya Lane

CEO and Cosmetic Chemist at Curly Chemistry

Education:
North Carolina Central University, MS, Chemistry
North Carolina Central University, BS, Chemistry

Tonya Lane is the CEO and founder of Curly Chemistry, a platform that empowers the natural hair community with the science behind natural hair care. As a cosmetic chemist, Lane uses her expertise in biochemistry to create innovative solutions for textured hair. By understanding how ingredients interact with skin and hair at the molecular level, she develops beauty products that are both safe and effective."

Lane's passion for science was first ignited in high school by an inspiring chemistry teacher who helped her see the science behind everyday life. "When I chose to major in chemistry in college, I realized that this same principle applies to beauty products, which I've always loved." Her early career began immediately after earning her master's degree, stepping directly into the beauty industry. Her understanding of biochemical principles like protein structure and lipid behavior gives her a competitive edge in creating formulas that truly perform.

Tonya Lane

Lane's journey wasn't always easy. "To be transparent, I wasn't a straight-A student," she shares. "I often struggled to fully understand the material." What helped her overcome academic challenges was persistence—committing to learning, asking for help, and never giving up. These experiences taught her the value of resilience, a lesson she now brings into her professional life as well.

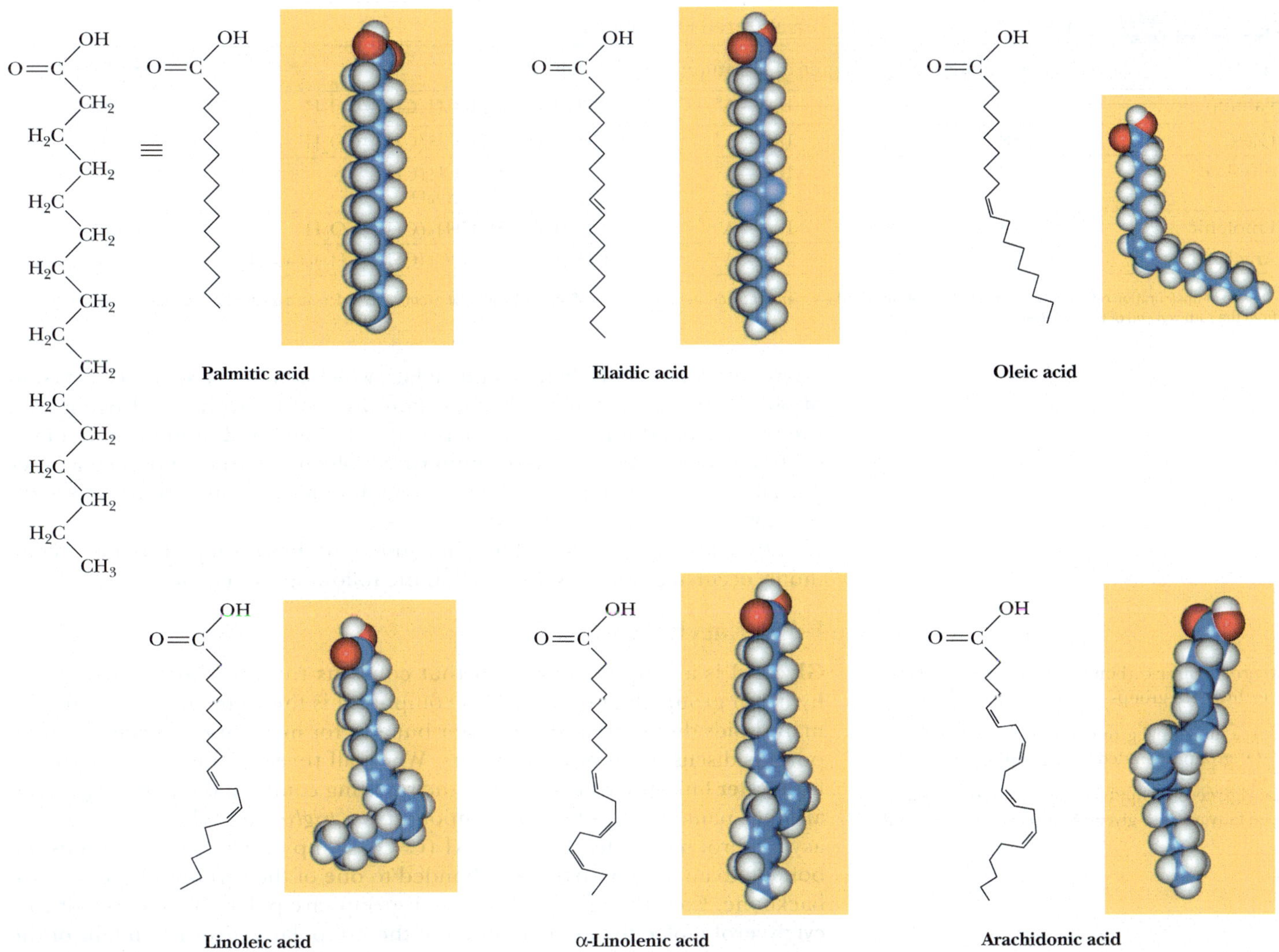

Figure 8.1 **The structures of some typical fatty acids.** Most naturally occurring fatty acids contain even numbers of carbon atoms. The double bonds are nearly always cis and rarely conjugated.

Table 8.1 **Typical Naturally Occurring Saturated Fatty Acids**

Acid	Number of Carbon Atoms	Formula	Melting Point (°C)
Lauric	12	$CH_3(CH_2)_{10}CO_2H$	44
Myristic	14	$CH_3(CH_2)_{12}CO_2H$	58
Palmitic	16	$CH_3(CH_2)_{14}CO_2H$	63
Stearic	18	$CH_3(CH_2)_{16}CO_2H$	71
Arachidic	20	$CH_3(CH_2)_{18}CO_2H$	77

trans double bond a double bond where the heavier groups attached to the double bond are on opposite sides of the double bond

18 carbon saturated fatty acid with no double bonds, whereas 18:1 denotes an 18-carbon fatty acid with one double bond.

The double bonds in unsaturated fatty acids usually start at carbon 9 (except for arachidonic acid), and they also get a Δ notation, where the superscript numbers after the Δ indicate where the double bonds are. When there are multiple double bonds in a fatty acid, the double bonds are spaced three carbons apart. Section 8-7 discusses the importance of arachidonic acid and its derivatives.

Unsaturated fatty acids have lower melting points than saturated ones. Plant oils are liquid at room temperature because they have higher proportions of

Table 8.2 Typical Naturally Occurring Unsaturated Fatty Acids

Acid	Number of Carbon Atoms	Degree of Unsaturation*	Formula	Melting Point (°C)
Palmitoleic	16	16:1—Δ^9	$CH_3(CH_2)_5CH{=}CH(CH_2)_7CO_2H$	−0.5
Oleic	18	18:1—Δ^9	$CH_3(CH_2)_7CH{=}CH(CH_2)_7CO_2H$	16
Linoleic	18	18:2—$\Delta^{9,12}$	$CH_3(CH_2)_4CH{=}CH(CH_2)$ $CH{=}CH(CH_2)_7CO_2H$	−5
Linolenic	18	18:3—$\Delta^{9,12,15}$	$CH_3(CH_2CH{=}CH)_3(CH_2)_7CO_2H$	−11
Arachidonic	20	20:4—$\Delta^{5,8,11,14}$	$CH_3(CH_2)_4(CH{=}CHCH_2)_4CH_2)_2CO_2H$	−50

*Degree of unsaturation refers to the number of double bonds. The superscript indicates the position of double bonds. For example, Δ^9 refers to a double bond at the ninth carbon atom from the carboxyl end of the molecule.

unsaturated fatty acids than do animal fats, which tend to be solids. Conversion of oils to fats is a commercially important process. It involves hydrogenation, the process of adding hydrogen across the double bond of unsaturated fatty acids to produce the saturated counterpart. Oleomargarine, in particular, uses partially hydrogenated vegetable oils, which tend to include trans fatty acids (see Biochemical Connections 8A).

Fatty acids are rarely found free in nature, but they form parts of many commonly occurring lipids, as discussed in the following subsections.

8-2B Triacylglycerols

glycerol a three-carbon compound that contains three hydroxyl groups, one bound to each carbon

ester a functional group in which a carbonyl (C=O) group is bonded to an —OR group

triacylglycerol a lipid formed by esterification of three fatty acids to glycerol; also called a *triglyceride*

Glycerol is a simple compound that contains three carbons, each with a hydroxyl group (Figure 8.2). This compound is the foundation not only for many lipids discussed in this chapter but also for many of the metabolic compounds discussed in other chapters. When all three of the hydroxyl groups form **ester** linkages with fatty acids, the resulting compound is a **triacylglycerol**; another name for this type of compound is a *triglyceride*. When drawing a triacylglycerol molecule, the carbonyl (C=O) group of each fatty acid must be bonded to an oxygen that is also bonded to one of the carbons of the glycerol backbone. Even though the three ester groups are polar, the majority of triacylglycerol molecules are composed of the nonpolar hydrocarbon tails of the esterified fatty acids. It is common for three different fatty acids to be esterified to the hydroxyl groups of the same glycerol molecule. Triacylglycerols accumulate in adipose tissue (primarily fat cells) and provide a means of storing fatty acids, particularly for metabolic energy production (Section 21-3).

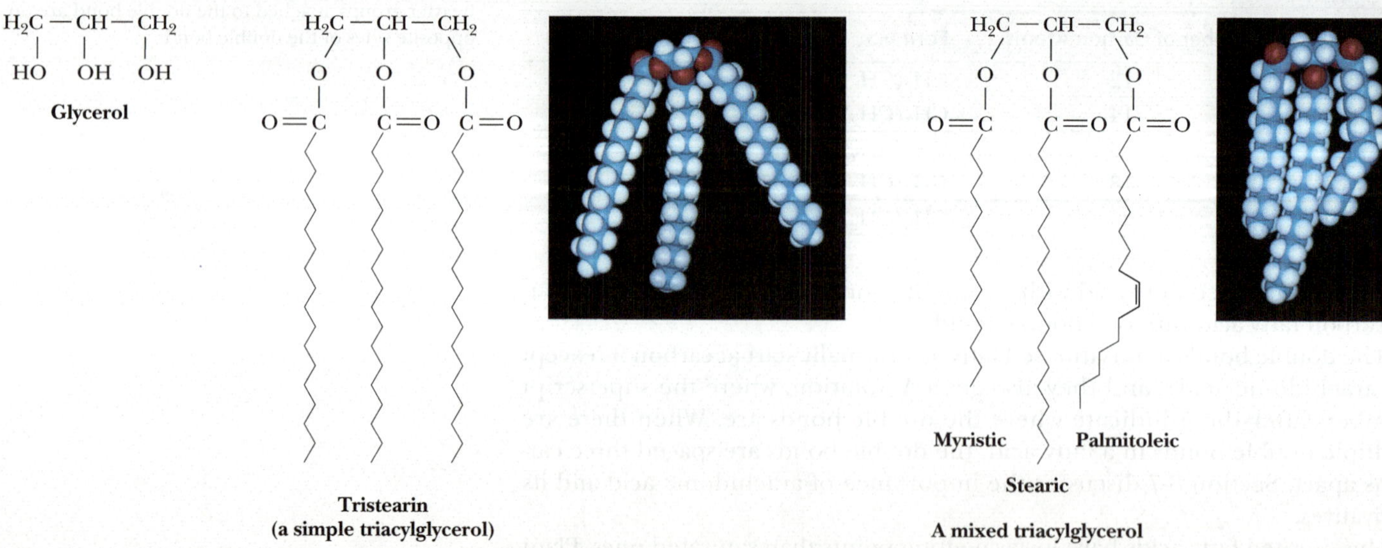

Figure 8.2 Triacylglycerols are formed from esterification reactions between glycerol and fatty acids.

When an organism needs free fatty acids, the ester linkages of triacylglycerols can be hydrolyzed by enzymes called **lipases** (Figure 8.3). A similar reaction called saponification can be performed in the lab using aqueous sodium hydroxide (NaOH). In both reactions, the products are glycerol and the fatty acids. In saponification, soaps (sodium salts of the fatty acids) are formed. When soaps are used with hard water, the calcium and magnesium ions in the water react with the fatty acids to form a precipitate—the characteristic scum left on the insides of sinks and bathtubs. The other commercial product of saponification, glycerol, is used in creams and lotions as well as in the manufacture of the explosive nitroglycerine.

lipases enzymes that hydrolyze lipids

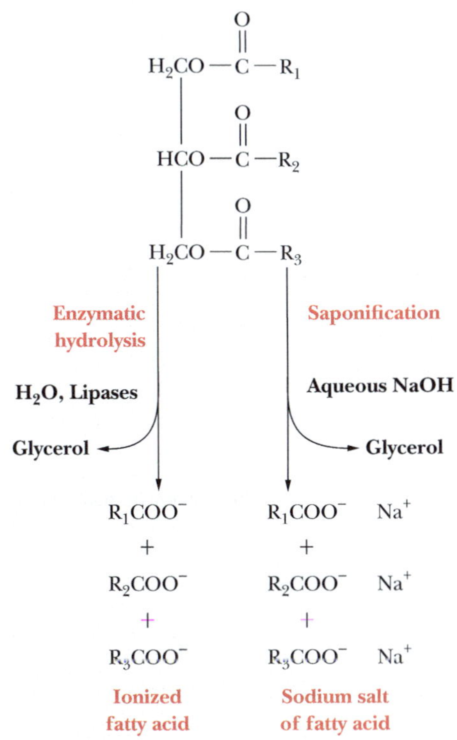

Figure 8.3 Release of fatty acids from triacylglycerols. Fatty acids can be released from a triacylglycerol either by enzymatic hydrolysis or by saponification. Glycerol is also a product of these reactions. R_1, R_2, and R_3 here represent long hydrocarbon chains.

8-2C Phosphoacylglycerols

In a class of compounds called **phosphatidic acids**, the primary (first) hydroxyl group of glycerol is esterified to a phosphoric acid molecule, and the other two carbons of the glycerol are esterified to fatty acids (Figure 8.4A). When the phosphate group of a phosphatidic acid molecule forms another ester group to another **alcohol**, a **phosphoacylglycerol** is formed (Figure 8.4B). The natures of the fatty acids attached to phosphoacylglycerols vary widely, as they do in triacylglycerols. As a result, the names of these types of lipids must be considered generic names.

Some of the most important phosphoacylglycerols are phosphatidylethanolamine (PE), phosphatidylserine (PS), phosphatidylcholine (PC), phosphatidylinositol (PI), and phosphatidylglycerol (PG) (Figure 8.5). In each of these types of compounds, the nature of the fatty acids in the molecule can vary widely. All these compounds have long, nonpolar, hydrophobic tails and polar, highly hydrophilic head groups and thus are markedly amphipathic. (We have already seen this characteristic in fatty acids.) In a phosphoacylglycerol, the polar head group is charged because the phosphate group is ionized at neutral pH. A positively charged amino group is also frequently contributed by an amino alcohol esterified to the phosphoric acid. Phosphoacylglycerols are important components of lipid bilayer biological membranes and are discussed further in Section 8-3. These lipid bilayers are frequently drawn in cartoon form in biology courses, and here, we get to form higher resolution images in our minds of the biochemical groups that are involved in forming these lipid bilayers.

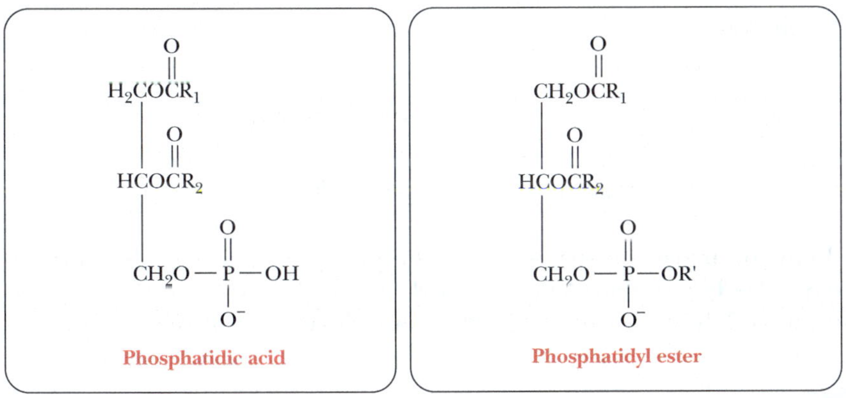

A A **phosphatidic acid**, in which glycerol is esterified to phosphoric acid and to two different carboxylic acids. R_1 and R_2 represent the hydrocarbon chains of the esterified fatty acids.

B A phosphatidyl ester (**phosphoacylglycerol**). Glycerol is esterfied to two carboxylic acids, stearic acid and linoleic acid, as well as to phosphoric acid. Phosphoric acid, in turn, is esterified to a second alcohol, R'OH.

Figure 8.4 The molecular architecture of phosphoacylglycerols.

phosphatidic acids a compound in which two fatty acids and phosphoric acid are esterified to the three hydroxyl groups of glycerol

alcohol an organic compound with one or more hydroxyl (—OH) groups attached

phosphoacylglycerol a phosphatidic acid molecule with another alcohol esterified to the phosphoric acid group

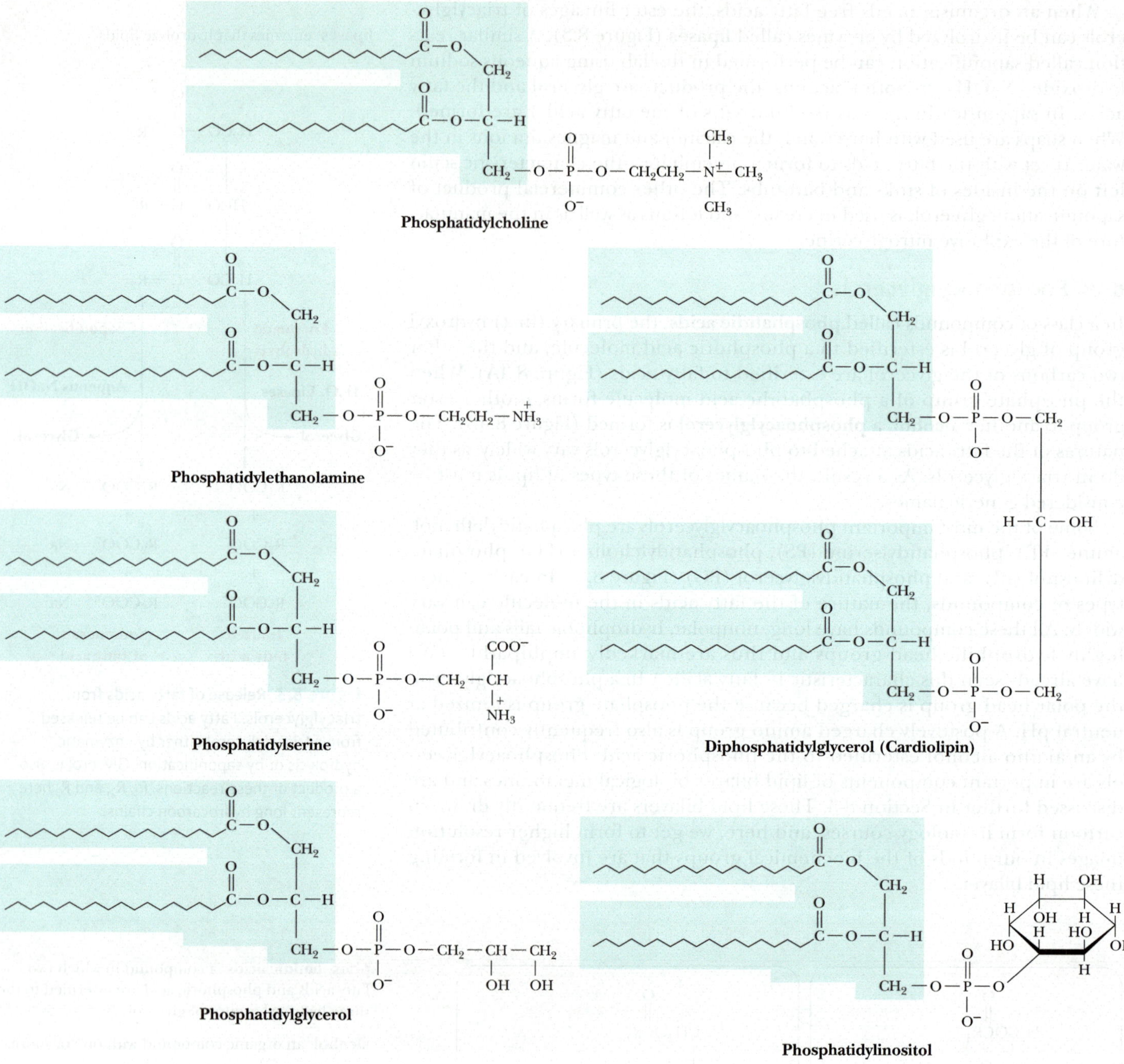

Figure 8.5 Structures of some important phosphoacylglycerols. The esterified fatty acids in these phosphoacylglycerol structures represent simple examples, but in reality, the hydrocarbon tails of the esterified fatty acids can vary in length and in the extent to which they are unsaturated.

To help you learn the structures of the phosphoacylglycerols shown in Figure 8.5, **Apply Your Knowledge 8.1** works through how the different parts of phosphoacylglycerols can come together to form the entire molecule.

8.1 Apply Your Knowledge

Phosphoacylglycerol Structures

Given the following phosphoglycerol molecule, use condensation reactions (recall Figure 3.10) to build a PE molecule. Show how two 16:0 fatty acids could get esterified to the glycerol hydroxyls and how one ethanolamine molecule could get esterified to the phosphoric acid group.

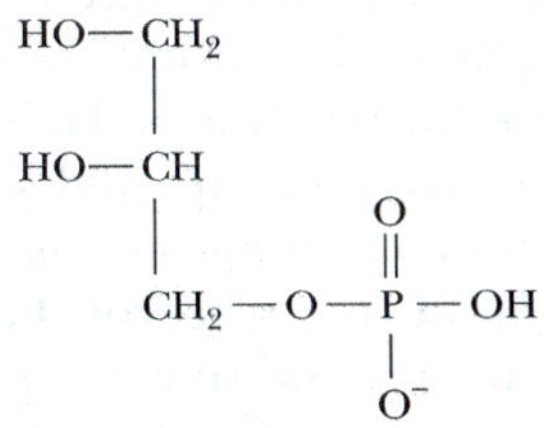

Solution

Building a phosphoacylglycerol molecule from its parts can help you understand how to draw them. We encourage you to try similar exercises using choline, serine (recall from Figure 3.3B), and glycerol as the alcohols that get esterified to the phosphoric acid group to form PC, PS, and PG, respectively.

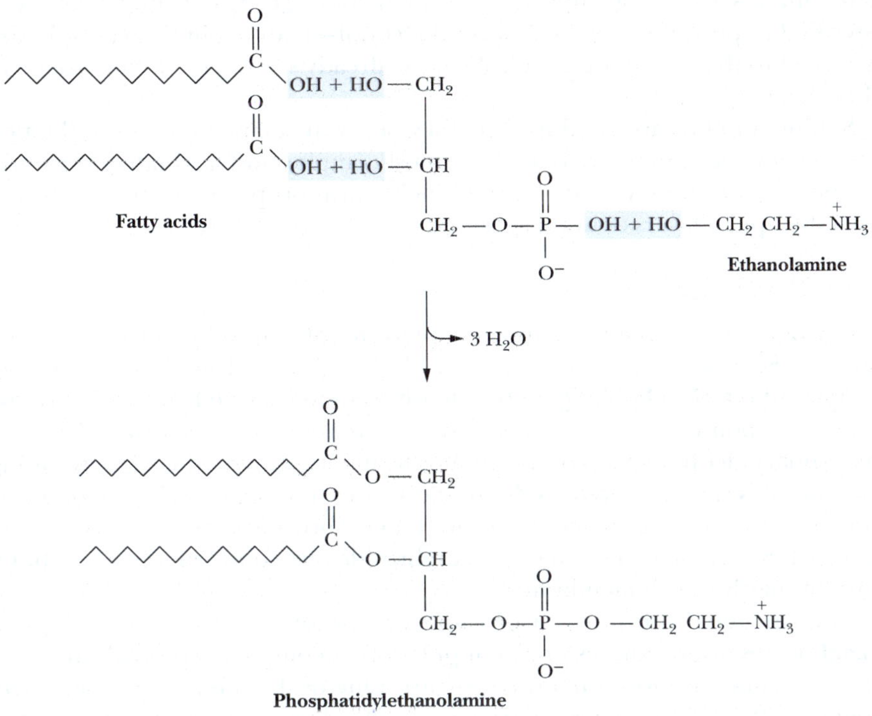

Waxes mixtures of esters of long-chain carboxylic acids and long-chain alcohols

8-2D Waxes

Waxes are complex mixtures of esters of long-chain carboxylic acids and long-chain alcohols. They frequently serve as protective coatings for both plants and animals. In plants, they coat stems, leaves, and fruit; in animals, they are found on fur, feathers, and skin. Myricyl cerotate (Figure 8.6), the principal component of carnauba wax, is produced by the Brazilian wax palm. Carnauba wax is extensively used in floor wax and automobile wax. The principal component of spermaceti, a wax produced by whales, is cetyl palmitate (Figure 8.6). The use of spermaceti as a component of cosmetics made it one of the most highly prized products of nineteenth-century whaling efforts.

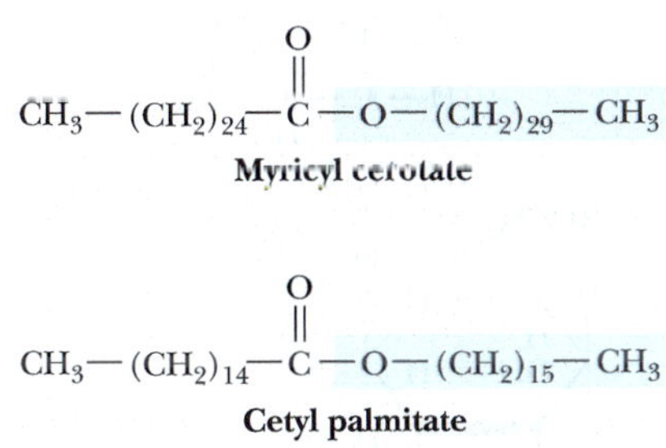

Figure 8.6 Structures of some waxes. The shaded portions represent the long-chain alcohols that are esterified to the unshaded fatty acid portions.

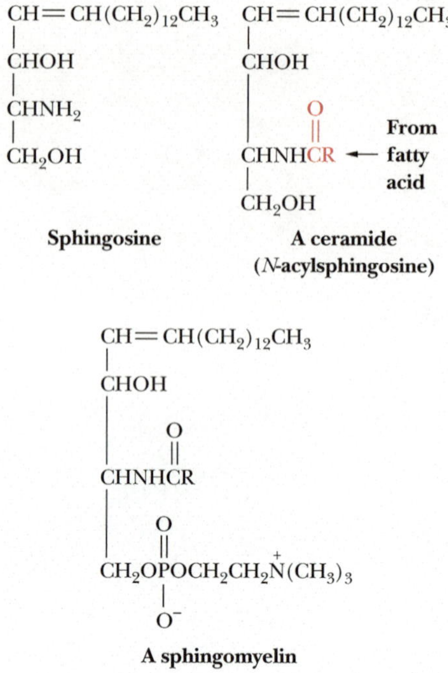

Figure 8.7 Structures of sphingosine and related sphingolipids.

8-2E Sphingolipids

Sphingolipids contain the long-chain amino alcohol sphingosine, from which this class of compounds takes its name (Figure 8.7). To help you understand and draw the structure of sphingosine, notice how the bottom three carbons are similar to glycerol except for two things. First, the middle carbon has an amino group instead of a hydroxyl group, and second, a long hydrocarbon chain (hence the categorization of sphingosine as a long-chain amino alcohol) is attached to the top carbon of the "glycerol" backbone. Sphingolipids are found in both plants and animals, and they are particularly abundant in the nervous system. The simplest compounds of this class are the **ceramides**, which consist of one fatty acid linked to the amino group of sphingosine by an amide bond (see Figure 8.7, but also review Figure 3.10, which shows the formation of an amide bond).

In **sphingomyelins**, the primary alcohol group of sphingosine is esterified to phosphoric acid, which, in turn, is esterified to another amino alcohol, choline (Figure 8.7). Notice the structural similarities between sphingomyelin and other phospholipids. Two long hydrocarbon chains are attached to a backbone that contains alcohol groups. One of the alcohol groups of the backbone is esterified to phosphoric acid. A second alcohol—choline, in this case—is also esterified to the phosphoric acid. We have already seen that choline occurs in phosphoacylglycerols.

Sphingomyelins are amphipathic; they occur in cell membranes in the nervous system. Our brains are one of the most complex and fascinating organs in our bodies, and here we can see the biochemical properties of the lipids that form the lipid bilayers of brain cells.

sphingolipids lipids whose structure is based on sphingosine, a long-chain amino alcohol

ceramides lipids that contain one fatty acid linked to sphingosine by an amide bond

sphingomyelins compounds in which the primary alcohol of sphingosine is esterified to phosphoric acid, which is also esterified to another amino alcohol

glycolipid a lipid that has a sugar group covalently attached

cerebroside a glycolipid that contains a ceramide that has formed a covalent bond with a sugar group

gangliosides glycolipids with a complex carbohydrate structure that contains more than three sugars; one of the sugars is always a sialic acid

8-2F Glycolipids

If a carbohydrate is bound to an alcohol group of a lipid by a glycosidic linkage (see Section 16-2 for a discussion of glycosidic linkages), the resulting compound is a **glycolipid**. Quite frequently, ceramides (see Figure 8.7) are the parent compounds for glycolipids, and the glycosidic bond is formed between the primary alcohol group of the ceramide and a sugar residue. The resulting compound is called a **cerebroside**. In most cases, the sugar is glucose or galactose; for example, a glucocerebroside is a cerebroside that contains glucose (Figure 8.8). As the name indicates, cerebrosides are found in nerve and brain cells, primarily in cell membranes.

The carbohydrate portion of these compounds can be very complex. **Gangliosides** are examples of glycolipids with a complex carbohydrate structure that contains more than three sugars. One of them is always a sialic acid (Figure 8.9). These compounds are also referred to as acidic glycosphingolipids because of their net negative charge at neutral pH. Glycolipids are often found as markers on cell membranes and play a large role in tissue and organ specificity. Gangliosides are also present in large quantities in nerve tissues. Their biosynthesis and breakdown are discussed in Section 21-7.

Before we move on to the steroids, the other major class of lipids, we encourage you to build understanding of the lipids seen in the preceding subsections (i.e., waxes, sphingolipids, and glycolipids) by naming and drawing their individual parts. Then, use what you learned in the **Apply Your Knowledge 8.1** exercise to propose the condensation reactions that could build the entire structures of these molecules.

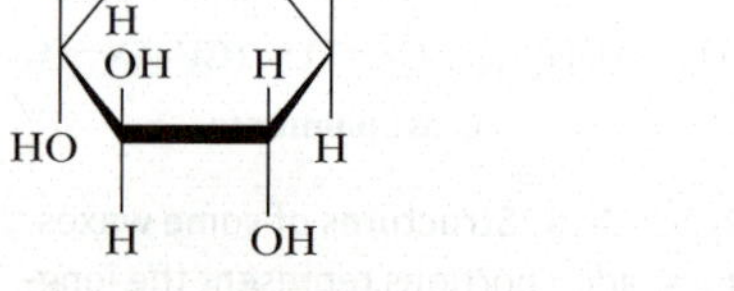

Figure 8.8 Structure of a glucocerebroside.

8-2G Steroids

Many compounds of widely differing functions are classified as **steroids** because they have the same general structure: a fused-ring system consisting of three six-membered rings (the A, B, and C rings) and one five-membered

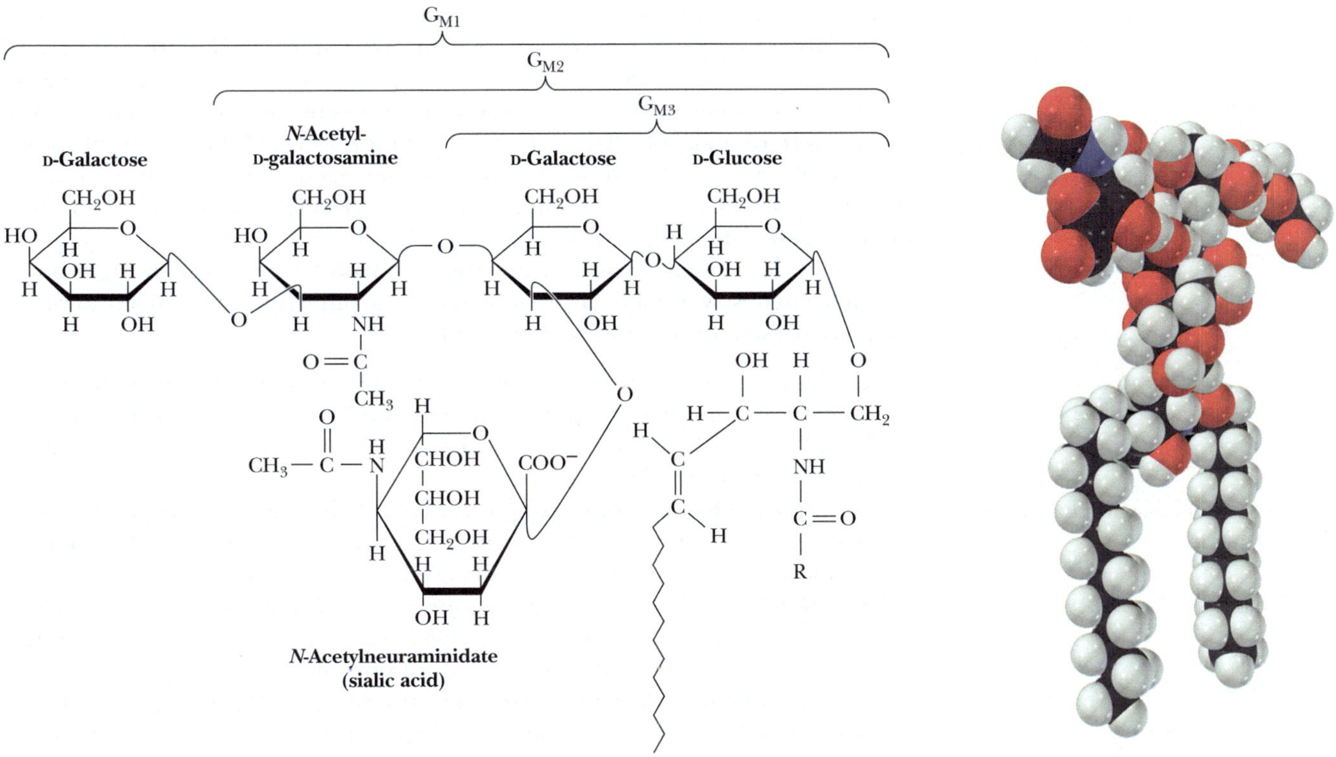

Gangliosides G_{M1}, G_{M2}, and G_{M3}

Gangliosides

Figure 8.9 The structures of several important gangliosides. Also shown is a space-filling model of ganglioside G_{M1}.

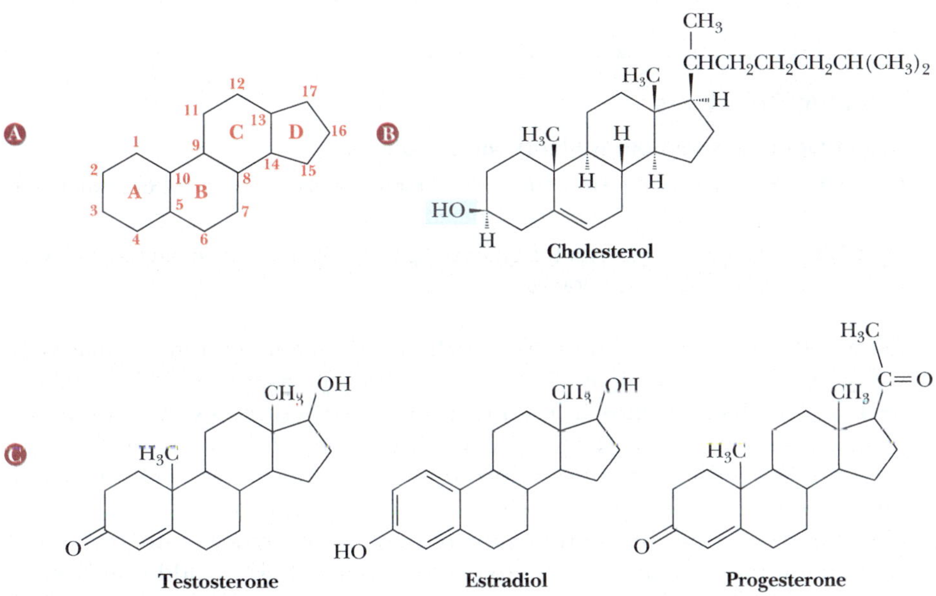

Figure 8.10 Structures of some steroids. (A) The fused-ring structure of steroids. (B) Cholesterol. (C) Three steroid sex hormones.

ring (the D ring) (Figure 8.10A). The steroid that is of most interest in our discussion of membranes is **cholesterol** (Figure 8.10B). The only hydrophilic group in the cholesterol structure is the single hydroxyl group. As a result, the molecule is highly hydrophobic. Cholesterol is widespread in biological membranes, especially in animals, but it does not occur in prokaryotic cell membranes. The presence of cholesterol in membranes affects membrane fluidity

steroids lipids with a characteristic fused-ring structure that have a wide range of biological functions

cholesterol a steroid that occurs in cell membranes; also serves as the precursor of other steroids and vitamins

(Section 8-3B). Cholesterol has a number of other important biological functions, including its role as a precursor of other steroids (Figure 8.10C) and of vitamin D_3 (Section 8-6B). However, cholesterol is best known for its harmful effects on health when it is present in excess in the blood. It plays a role in the development of *atherosclerosis,* a condition in which lipid deposits block the blood vessels and lead to heart disease (see Section 21-8). There are many other important steroids, including the sex hormones testosterone, estradiol, and progesterone (Figure 8.10C).

Finally, let us learn how to draw these steroid structures by using a bit of organic chemistry nomenclature to name the groups (mostly from left to right) that are present in these steroid molecules:

- Cholesterol has a hydroxyl group (hence the "-ol" in the name cholesterol) on carbon 3, methyl groups at carbons 10 and 13, a double bond at carbon 5, and a "2-methylheptane" group connected between carbon 17 and the sixth carbon of the heptane chain.
- Testosterone has a **ketone** group on carbon 3 (hence the "-one" in the name testosterone), a double bond at carbon 4, methyl groups at carbons 10 and 13, and a hydroxyl group at carbon 17.
- Estradiol has two hydroxyl groups (one at carbon 1 and one at carbon 17; hence the "-diol" in its name), a methyl at carbon 13, and a phenyl ring A, which means a methyl at carbon 10 is not possible.
- Progesterone has a ketone at carbon 3 (just like testosterone), a double bond at carbon 4, methyl groups at carbons 10 and 13, and an **acetyl** group at carbon 17. An acetyl group is part of acetic acid (Section 2-4), and it is part of many metabolic compounds seen in other chapters.

ketone a carbonyl (C=O) group that is bonded to two other carbon atoms

acetyl a two-carbon group where a carbonyl (C=O) group is bonded to a methyl (—CH_3) group

8-3 Biological Membranes

Learning Objectives

After reading this section, students will be able to:

LO 8.3.1 Explain how the structures of membrane lipids and proteins determine the characteristics of biological membranes.

LO 8.3.2 Distinguish between the types of lipids found in the inner and outer layers of biological membranes.

Every cell has a cell membrane (also called a plasma membrane); eukaryotic cells also have membrane-enclosed organelles such as nuclei and mitochondria. The molecular basis of the membrane's structure lies in its lipid and protein components. In Section 8-3, we discuss how the interaction between the lipid bilayer and membrane proteins determines membrane function. Membranes not only separate cells from the external environment but also play important roles in the transport of specific substances into and out of cells and in the cellular response to extracellular signals. In addition, a number of important enzymes are found in membranes and depend on this environment for their function. First, we take a look at the lipid component of membranes.

8-3A Lipid Bilayers

Phosphoacylglycerols, such as those seen in Figure 8.4, are the principal lipid components of membranes, and they form a **lipid bilayer** (Figure 8.11). Biological membranes contain glycolipids in addition to phosphoacylglycerols as part of the lipid component. Steroids are also present in the membranes of eukaryotes—cholesterol in animal membranes and similar compounds called

lipid bilayer an aggregate of a lipid molecule in which the polar head groups are in contact with water and the hydrophobic tails are not

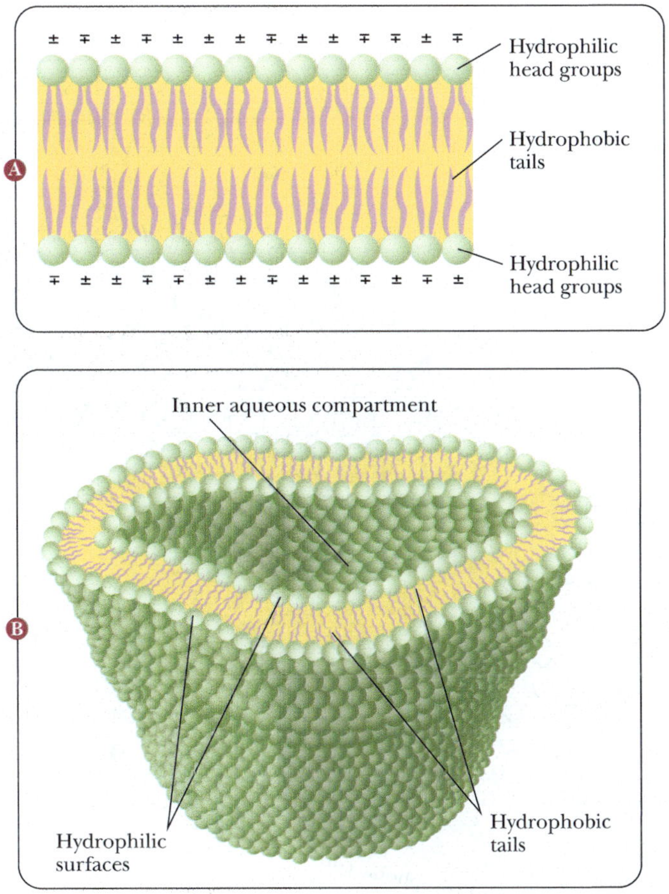

Figure 8.11 Lipid bilayers. (A) Schematic drawing of a portion of a bilayer consisting of phospholipids. The polar surfaces of the bilayer contain charged groups. The hydrocarbon "tails" lie in the interior of the bilayer. (B) Cutaway view of a lipid bilayer vesicle. Notice the aqueous inner compartment and the fact that the inner layer is more tightly packed than the outer layer. *(From Bretscher, M. S. The Molecules of the Cell Membrane.* Scientific American, *October 1985, p.* 103. *Art by Dana Burns-Pizer.)*

phytosterols in plants. In the lipid bilayer part of the membrane (Figure 8.11), the polar head groups are in contact with water, whereas the nonpolar tails lie in the interior of the membrane. The whole bilayer arrangement is held together by noncovalent interactions such as van der Waals and hydrophobic interactions (Section 2-1). The surfaces of the bilayer are polar and contain charged groups. The nonpolar hydrocarbon interior of the bilayer consists of the saturated and unsaturated hydrocarbon chains of fatty acids and the nonpolar portions of cholesterol (Figure 8.12).

Both the inner and outer layers of the bilayer contain mixtures of lipids, but their compositions differ and can be used to distinguish the inner and outer layers from each other (Figure 8.12). Bulkier molecules tend to occur in the outer layer, whereas smaller molecules tend to occur in the inner layer. For simplicity, the polar head groups in Figures 8.11 and 8.12 are shown as spheres, but their full structural details are shown in Figure 8.5 for phosphoacylglycerols, Figure 8.7 for sphingomyelins, Figure 8.8 for cerebrosides, and Figure 8.9 for gangliosides.

Table 8.3 lists the lipid composition of various types of membranes in rat liver cells. The distribution of major lipid types such as PC, PE, and cholesterol varies widely.

8-3B Membrane Fluidity

The arrangement of the hydrocarbon interior of the bilayer can be ordered and rigid or disordered and fluid. The bilayer's fluidity depends on its composition. In saturated fatty acids, a linear arrangement of the hydrocarbon chains leads to close packing of the molecules in the bilayer and thus to rigidity. Unsaturated fatty acids have a kink in the hydrocarbon chain that does not exist in

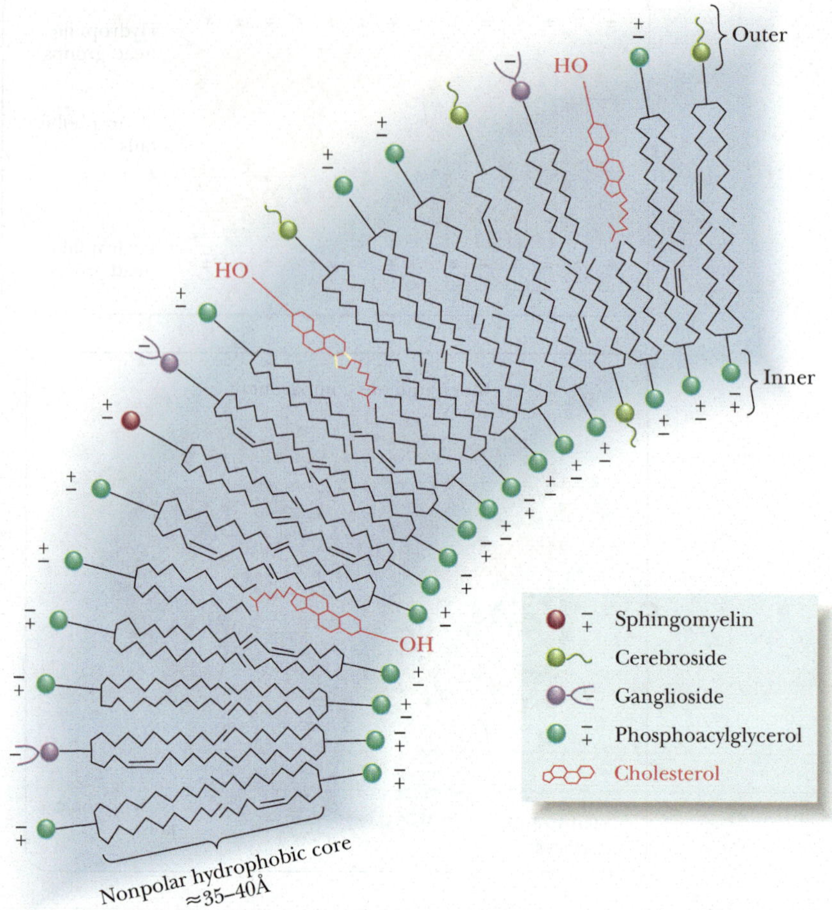

Figure 8.12 Lipid bilayer asymmetry. The compositions of the outer and inner layers differ; the concentration of bulky molecules is higher in the outer layer, which has more room. Complete structural examples of the polar head groups are shown in Section 8-2E for sphingomyelins, Section 8-2F for cerebrosides and gangliosides, and Section 8-2C for phosphoacylglycerols.

Table 8.3 The Lipid Composition of Membranes in Rat Liver Cells, in Weight Percent

	Type of Membrane				
Type of Lipid	Nuclear Membrane	Golgi Apparatus	Mitochondria	Lysosomes	Plasma Membrane
Phosphatidylcholine	49	42	38	27	28
Phosphatidylethanolamine	13	17	34	9	16
Sphingolipids	3	7	0	13	12
Phosphatidylinositol	10	10	5	3	6
Phosphatidylserine	3	5	0	0	6
Cardiolipin	3	0	17	0	0
Minor lipids	4	3	0	0	0
Cholesterol	15	17	4	33	28

saturated fatty acids (Figure 8.13). The kinks cause disorder in the packing of the chains, which makes for a more open structure than would be possible for straight saturated chains (Figure 8.14). In turn, the disordered structure caused by the presence of unsaturated fatty acids with cis double bonds (and therefore kinks) in their hydrocarbon chains causes greater fluidity in the bilayer. The lipid components of a bilayer are always in motion, to a greater extent in more fluid bilayers and to a lesser extent in more rigid ones.

The presence of cholesterol may also enhance order and rigidity. The fused-ring structure of cholesterol is itself quite rigid, and the presence of cholesterol stabilizes the extended straight-chain arrangement of saturated fatty acids by van der Waals interactions (Figure 8.15). The lipid portion of a plant membrane has a higher percentage of unsaturated fatty acids, especially polyunsaturated (containing two or more double bonds) fatty acids, than does the lipid portion of an animal membrane. In contrast, cholesterol is a characteristic of animal membranes that is largely absent from plant membranes. As a result, animal membranes tend to be less fluid (more rigid) than plant membranes, while prokaryotic membranes which lack steroids, are the most fluid of all. With heat, ordered bilayers become less ordered; bilayers that are comparatively disordered become even more disordered. **Biochemical Connections 8A** looks at some connections between the fatty acid composition of bilayers and membranes and how they behave at different temperatures.

Recall that the distribution of lipids is not the same in the inner and outer portions of the bilayer. Because the bilayer is curved, the molecules of the inner layer are more tightly packed (Figure 8.12). Bulkier molecules, such as cerebrosides (see Section 8-2F), tend to be located in the outer layer. There is very little tendency for "flip-flop" migration of lipid molecules from one layer of the bilayer to another, but it does occur occasionally. Lateral motion of lipid molecules within one of the two layers frequently takes place, however, especially in more fluid bilayers.

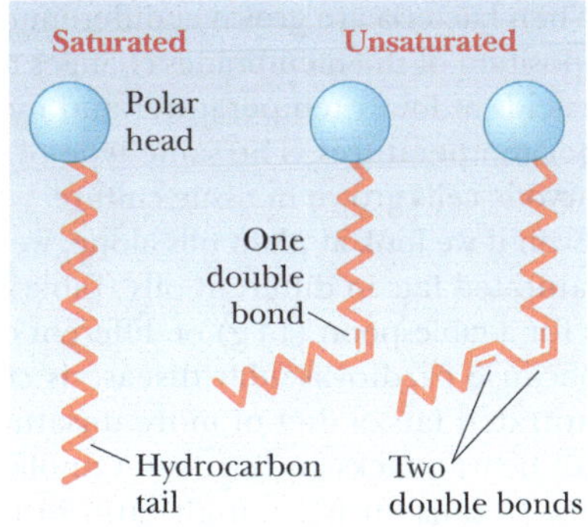

Figure 8.13 The effect of double bonds on the conformations of the hydrocarbon tails of fatty acids. Unsaturated fatty acids have kinks in their tails.

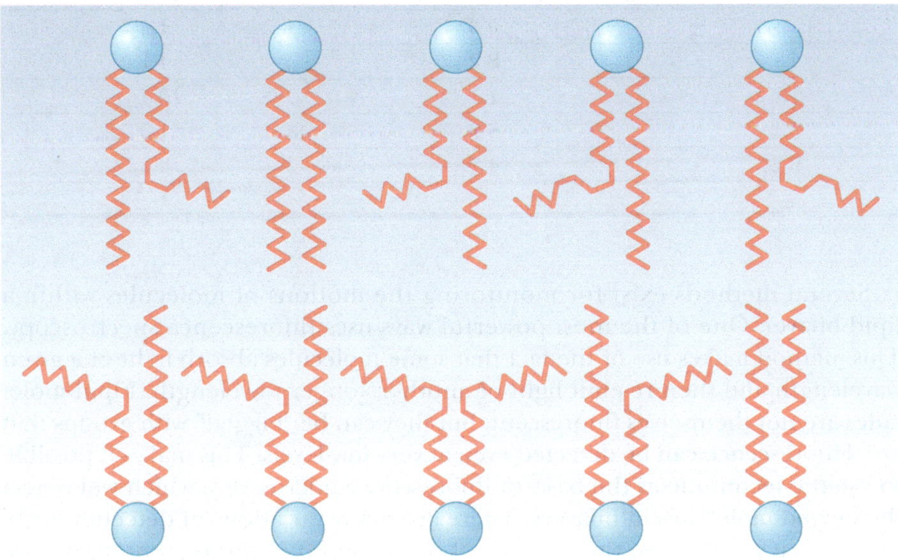

Figure 8.14 Schematic drawing of a portion of a highly fluid phospholipid bilayer. The kinks in the unsaturated side chains prevent close packing of the hydrocarbon portions of the phospholipids.

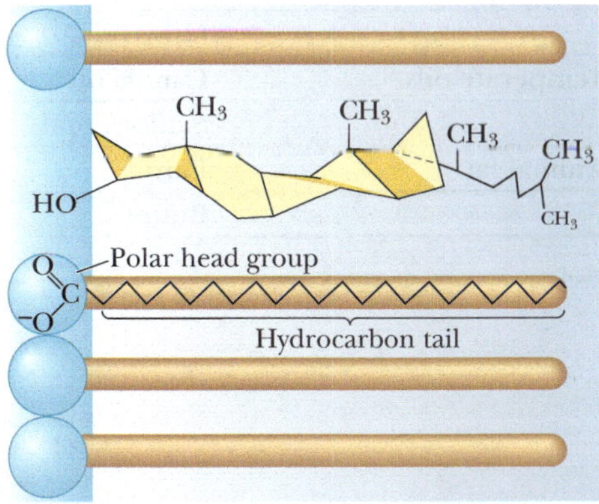

Figure 8.15 Stiffening of the lipid bilayer by cholesterol. The presence of cholesterol in a membrane reduces fluidity by stabilizing extended chain conformations of the hydrocarbon tails of fatty acids as a result of van der Waals interactions.

8A Biochemical Connections • Nutrition

Butter Versus Margarine—Which Is Healthier?

We use the terms animal "fats" and plant "oils" because of the solid and fluid nature of these two groups of lipids, respectively. The major difference between fats and oils is the percentage of unsaturated fatty acids in the triglycerides and the phosphoglycerides of membranes. This difference is far more important than the fact that the length of the fatty acid chain can affect the melting points. Butter is an exception; it has a high proportion of short-chain fatty acids and thus can "melt in your mouth." Membranes must maintain a certain degree of fluidity to be functional. Consequently, unsaturated fats are distributed in varying proportions in different parts of the body. The membranes of internal organs of warm-blooded mammals have a higher percentage of saturated fats than do the membranes of skin tissues, which helps keep the membrane more solid at the higher temperature of the internal organ. An extreme example of this is found in the legs and the body of reindeer, where marked differences exist in the percentages of saturated fatty acids.

Continued on next page

When bacteria are grown at different temperatures, the fatty acid composition of the membranes changes to reflect more unsaturated fatty acids at lower temperatures and more saturated fatty acids at higher temperatures. The same type of difference can be seen in eukaryotic cells grown in tissue culture.

Even if we look at plant oils alone, we find different proportions of saturated fats in different oils. Table 8.4 compares the distribution for a tablespoon (14 g) of different oils.

Because cardiovascular disease is correlated with diets high in saturated fats, a diet of more unsaturated fats may reduce the risk of heart attacks and strokes. Canola oil is an attractive dietary choice because it has a high ratio of unsaturated fatty acids to saturated fatty acids. Since the 1960s, we have known that foods higher in polyunsaturated fats are healthier. Thus, companies began to market butter substitutes that were based on unsaturated fatty acids but that would also have the physical characteristics of butter such as being solid at room temperature. They accomplished this task by partially hydrogenating the double bonds in the unsaturated fatty acids making up the oils. The irony here is that, to avoid eating the saturated fatty acids in butter, butter substitutes were created from polyunsaturated oils by removing some of the double bonds, thus making them more saturated. In addition, many of the soft spreads that are marketed as being healthy (safflower oil spread and canola oil spread) may indeed pose new health risks. In the hydrogenation process, some double bonds are converted to the trans form. Studies now show that trans fatty acids raise the ratio of low-density lipoprotein (LDL) cholesterol compared with high-density lipoprotein (HDL) cholesterol, a positive correlator of heart disease. Thus, the effects of trans fatty acids are similar to those of saturated fatty acids. However, there are butter substitutes that have been developed and marketed as having "no trans fatty acids."

Table 8.4 Fatty Acid Composition of Some Common Fats and Oils

Types of Oil or Fat	Example	Saturated (g)	Monounsaturated (g)	Polyunsaturated (g)
Tropical oils	Coconut oil	13	0.7	0.3
Semitropical oils	Peanut oil	2.4	6.5	4.5
	Olive oil		10.3	1.3
Temperate oils	Canola oil	1	8.2	4.1
	Safflower oil	1.3	1.7	10.4
Animal fat	Lard	5.1	5.9	1.5
	Butter	9.2	4.2	0.6

Several methods exist for monitoring the motions of molecules within a lipid bilayer. One of the most powerful ways uses fluorescence spectroscopy. This method makes use of the fact that some molecules absorb light of a given wavelength and then re-emit light of another, longer wavelength. Lipid molecules are not themselves fluorescent, but they can be "tagged" with groups that are. Fluorescence can be detected even at very low levels. This makes it possible to use the technique as the basis of fluorescence microscopy, which will detect the tagged molecules in bilayers. There are many variations in detection techniques, but in all cases, they are based on the emitted fluorescent light. The use of fluorescence can be expanded in studies on actual membranes rather than bilayers. Membranes contain proteins in addition to the lipid bilayer. The side chains of tryptophan and tyrosine have intrinsic fluorescence, and this property can be used to obtain information about the protein portion of the membrane. **Biochemical Connections 8B** discusses another application of the physical properties of membranes.

8B Biochemical Connections • Biotechnology

Membranes in Drug Delivery

Getting a pharmaceutical drug to a place where it can be most effective is of great importance, with enormous commercial ramifications. A number of technologies have been applied to the task. They include injections, topical application to the skin, and implants. Most importantly for our purposes, they take advantage of the properties of membranes and model systems for membranes. We have seen how lipid molecules form layered structures, and these structures are the key point in developing a delivery system. In Chapter 2, we saw how amphipathic molecules such as fatty acids can form micelles, which are assemblages with a single layer of fatty acids, with the nonpolar tails in the middle sequestered from water and the polar heads in contact with water (Figure 2.7). Nonpolar substances can be packaged in micelles for delivery where desired. Recall that lipid bilayers (Figure 8.11) have polar surfaces (exterior and interior) and nonpolar portions of the component molecules sandwiched in between.

Because the driving force behind the formation of lipid bilayers is the exclusion of water from the hydrophobic region of lipids and not some enzymatic process, artificial membranes can be created in the lab. **Liposomes** are stable structures based on a lipid bilayer that form a spherical vesicle. These vesicles can be prepared with polar therapeutic agents packaged inside the aqueous interior of the vesicle. The vesicle can then be used to deliver the agent to a target tissue.

8-4 Membrane Proteins

liposomes spherical aggregates of lipids arranged so that the polar head groups are in contact with water and the nonpolar tails are sequestered from water

Learning Objectives

After reading this section, students will be able to:

LO 8.4.1 Describe characteristics of two categories of membrane proteins.

LO 8.4.2 Use the fluid mosaic model to explain the arrangement of lipids and proteins in the membrane.

8-4A Membrane Protein Types

Proteins in a biological membrane can be associated with the lipid bilayer in either of two ways—as **peripheral proteins** on the surface of the membrane or as **integral proteins** embedded in and traversing the lipid bilayer (Figure 8.16). The integral protein rhodopsin (shown in blue) consists mostly of helical portions that span the membrane. The peripheral G protein is a trimer. The three different subunits are shown in green (α), purple (β), and red (γ). Peripheral proteins can also associate with the charged head groups of the lipid bilayer. Peripheral proteins interact with integral membrane proteins or with membrane lipids by polar interactions, electrostatic interactions, or both. They can be removed by such mild treatment as raising the ionic strength of the medium. The relatively numerous charged particles present in a medium of higher

peripheral proteins proteins loosely bound to the outside of a membrane

integral proteins proteins embedded in a membrane

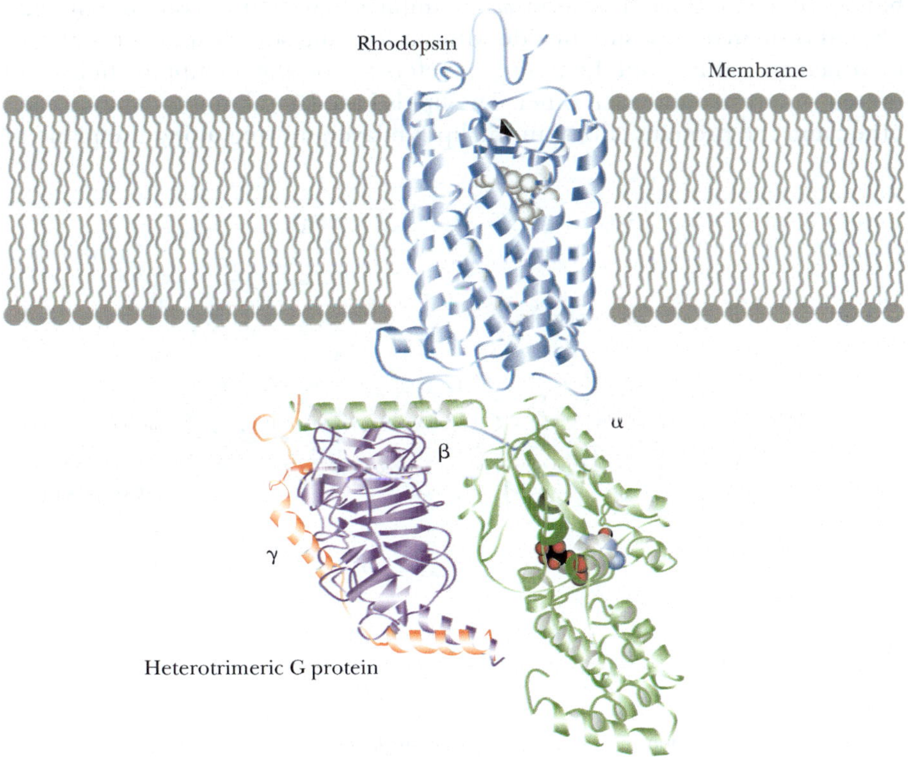

Figure 8.16 Integral and peripheral proteins. The integral protein rhodopsin spans the membrane. The heterotrimeric G protein is a peripheral protein. The three different subunits are marked alpha, beta, and gamma. *(Adapted from figure in Hamm, H. E. & Gilchrist, A.: Current Opinion in Cell Biology 1996 8: 189–196, with permission from Elsevier.)*

ionic strength undergo more electrostatic interactions with the lipid and with the protein, "swamping out" the comparatively fewer electrostatic interactions between the protein and the lipid.

Removing integral proteins from membranes is much more difficult. Harsh conditions, such as treatment with detergents or extensive sonication (exposure to ultrasonic vibrations), are usually required. Such measures frequently denature the protein, which often remains bound to lipids in spite of all efforts to obtain it in pure form. The denatured protein is inactive, whether or not it remains bound to lipids. Fortunately, nuclear magnetic resonance techniques enable researchers to study proteins of this sort in living tissue or in reconstituted membranes. The structural integrity of the whole membrane system appears to be necessary for the activities of most membrane proteins.

Integral membrane proteins traverse the membrane in a variety of ways. When a protein completely spans the membrane, it is often in the form of an α-helix or β-sheet. These structures minimize contact of the polar parts of the peptide backbone with the nonpolar lipids in the interior of the bilayer. Proteins can also be anchored to the lipids via covalent bonds from cysteines or free amino groups on the protein to one of several lipid anchors. Myristoyl and palmitoyl groups are common anchors (Figure 8.17).

transport proteins protein components of a membrane that mediate the entry of specific substances into a cell

receptor proteins proteins on a cell membrane with specific binding sites for extracellular signaling molecules

Membrane proteins have a variety of functions. Most, but not all, of the important functions of the membrane are carried out by protein components of the membrane. **Transport proteins** help move substances in and out of the cell, whereas **receptor proteins** are important in the transfer of extracellular signals, such as those carried by hormones or neurotransmitters, into the cell. In addition, some enzymes are tightly bound to membranes; examples include many of the enzymes responsible for aerobic oxidation reactions, which are found in specific parts of mitochondrial membranes. Some of these enzymes are on the inner surface of the membrane, and some are on the outer surface.

8-4B The Fluid-Mosaic Model of Membrane Structure

fluid-mosaic model the model for membrane structure in which proteins and a lipid bilayer exist side by side without covalent bonds between the proteins and lipids

How do the lipid and protein components interact in a biological membrane? Currently, the **fluid-mosaic model** is the most widely accepted description of biological membranes. The term *mosaic* implies that the two components (lipids and proteins) exist side by side without forming some other substance of intermediate nature, and the term *fluid* refers to how the membrane lipids and proteins move around each other. The lipids have lateral motion among each other, and the proteins "float" in the lipid bilayer, moving along the plane of

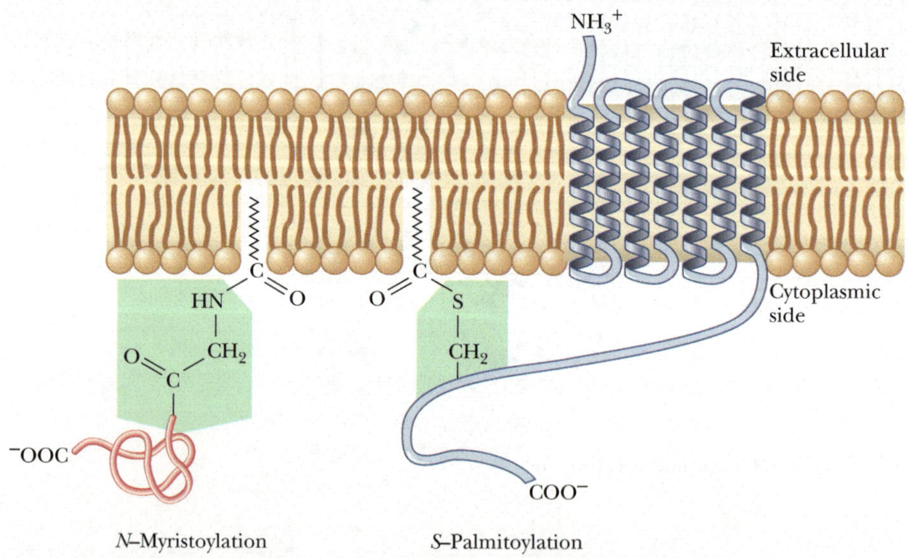

Figure 8.17 Certain proteins are anchored to biological membranes by lipid anchors. Particularly common are the *N*-myristoyl and *S*-palmitoyl anchoring motifs shown here. *N*-Myristoylation always occurs at an *N*-terminal glycine residue, whereas thioester linkages occur at cysteine residues within the polypeptide chain. *G*-Protein–coupled receptors, with seven transmembrane segments, may contain one (and sometimes two) palmitoyl anchors in thioester linkage to cysteine residues in the C-terminal segment of the protein.

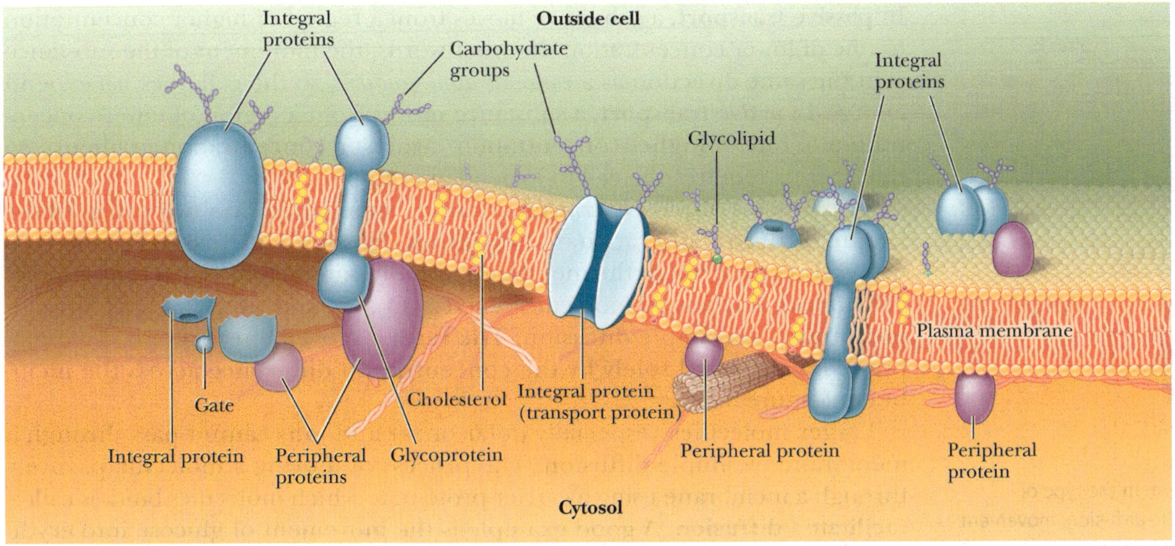

Figure 8.18 Fluid-mosaic model of membrane structure. Integral membrane proteins span the membrane, but peripheral proteins are associated with one side or the other. *(Russell/Hertz/Mcmillan, Biology, 3rd Edition, Cengage Learning.)*

the membrane. The basic structure of biological membranes is that of the various molecules of the lipid bilayer, with integral membrane proteins embedded in the bilayer structure and peripheral membrane proteins interacting with the polar surfaces of the membrane lipids or integral proteins (Figure 8.18).

8-5 The Functions of Membranes

Learning Objectives

After reading this section, students will be able to:

LO 8.5.1 Describe mechanisms for the transport of molecules across membranes.

LO 8.5.2 Use human growth hormone signaling to explain how messages cross membranes.

Three important functions—*transport, catalysis,* and the *receptor property*—take place in or on membranes (in addition to the structural role of membranes as the boundaries and containers of all cells and of the organelles within eukaryotic cells).

- Transport: Membranes are semipermeable barriers to the flow of substances into and out of cells and organelles. Transport through the membrane can involve the lipid bilayer as well as membrane proteins.
- Catalysis: Enzymes bound to the membrane can catalyze reactions on the membrane.
- Receptor property: In this property, proteins bind specific extracellular substances that trigger biochemical responses in the cell.

These three functions primarily involve membrane proteins. We discuss membrane catalysis in subsequent chapters (especially in our treatment of aerobic oxidation reactions in Chapters 19 and 20). Membrane transport and membrane receptors, on the other hand, we now consider in turn.

8-5A Membrane Transport

The most important question about the transport of substances across biological membranes is whether the process requires the cell to expend energy.

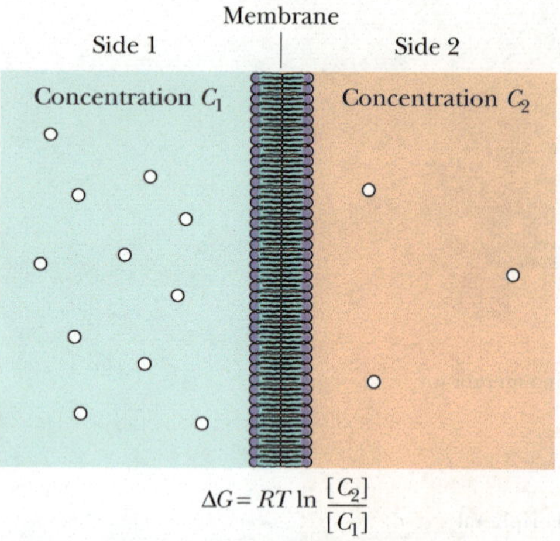

Figure 8.19 Simple diffusion. In the type of passive transport called simple diffusion, movement of an uncharged species across a membrane depends only on the concentrations (C_1 and C_2) on the two sides of the membrane. Here, $C_2 < C_1$, so the uncharged species diffuses from side 1 to side 2.

passive transport the process by which a substance moves across a membrane without an expenditure of energy by the cell

active transport the energy-requiring process of moving substances across a membrane against a concentration gradient

simple diffusion the process of passing through a membrane without requiring a carrier or the expenditure of energy

facilitated diffusion a process by which substances enter a cell by binding to a carrier protein; this process does not require energy

primary active transport a process that uses ATP hydrolysis to move molecules or ions across a cell membrane

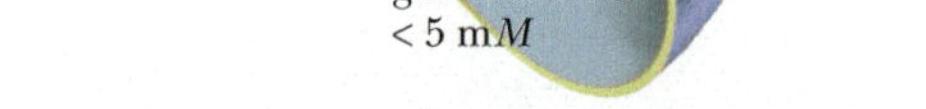

Figure 8.20 Facilitated diffusion. Glucose passes into an erythrocyte via glucose permease by facilitated diffusion. Glucose flows using its concentration gradient via passive transport. *(From Principles of Biochemistry, by Albert L. Lehninger, David L. Nelson and Michael M. Cox. Copyright © 1993 by W. H. Freeman & Co. Used with permission.)*

In **passive transport**, a substance moves from a region of higher concentration to one of lower concentration. In other words, the movement of the substance is in the same direction as a *concentration gradient,* so the cell does *not* expend energy. In **active transport**, a substance moves from a region of lower concentration to one of higher concentration (against a concentration gradient), so this process requires the cell to expend energy.

The process of passive transport can be subdivided into two categories—simple diffusion and facilitated diffusion. In **simple diffusion**, a molecule moves directly through the membrane without interacting with another molecule. Small, uncharged molecules, such as O_2, N_2, and CO_2, can pass through membranes via simple diffusion. The rate of movement through the membrane is controlled solely by the concentration difference across the membrane (Figure 8.19).

Larger molecules (especially polar ones) and ions cannot pass through a membrane by simple diffusion. The process of moving a molecule passively through a membrane using a carrier protein, to which molecules bind, is called **facilitated diffusion**. A good example is the movement of glucose into erythrocytes. The concentration of glucose in the blood is about 5 mM, whereas the glucose concentration in the erythrocyte is less than 5 mM. Glucose passes through a carrier protein called glucose permease (Figure 8.20). This process is categorized as facilitated diffusion because no energy is expended and a carrier protein is used. In addition, facilitated diffusion is identified by the fact that the rate of transport, when plotted against the concentration of the molecule being transported, gives a hyperbolic curve similar to that seen in Michaelis–Menten enzyme kinetics (Figure 8.21).

Carrier proteins create a pore in the membrane to allow transport. To create such pores, these proteins often have several α-helical portions that span the membrane; in other carrier proteins, a β-barrel (Figure 4.9) forms the pore. The exterior portions of these proteins are in contact with the lipid bilayer, so they contain hydrophobic amino acid residues (Figure 3.3A). On the other hand, the interior portions, through which ions or polar molecules pass, contain hydrophilic amino acid residues (Figure 3.3B). This orientation is the inverse of that observed in water-soluble globular proteins.

Active transport requires moving substances against a concentration gradient. It is identified by the presence of a carrier protein and the need for an energy source to move solutes against a gradient. In **primary active transport**, the movement of molecules against a gradient is directly linked to the hydrolysis of a high-energy molecule such as ATP. The situation is so markedly similar to pumping water uphill that one of the most extensively studied examples of active transport, moving potassium ions into a cell and simultaneously moving sodium ions out of the cell, is referred to as the sodium–potassium pump (or Na^+/K^+ pump).

Figure 8.22 illustrates the action of the Na^+/K^+ pump. Under normal circumstances, the concentration of K^+ is higher inside a cell than in extracellular fluids ($[K^+]_{inside} > [K^+]_{outside}$), but the concentration of Na^+ is lower inside the cell than out ($[Na^+]_{inside} < [Na^+]_{outside}$). The energy required to move these ions against their gradients comes from an exergonic (energy-releasing) reaction, the hydrolysis of ATP to ADP and P_i (phosphate ion). There can be no transport of these ions without the energy provided by the hydrolysis of ATP. The same protein appears to serve both as the enzyme that hydrolyzes the ATP (the ATPase) and as the transport protein; it consists of several subunits. The reactants and products of this hydrolysis reaction—ATP, ADP, and P_i—remain within the cell, and the phosphate becomes covalently bonded to the transport protein for part of the process (review Figure 7.12, which shows how phosphorylation controls the Na^+/K^+ pump).

The Na^+/K^+ pump operates in several steps (Figure 8.22). One subunit of the protein hydrolyzes the ATP and transfers the phosphate group to an aspartate side chain on another subunit (Step 1). (The bond formed here is a mixed anhydride; see Section 1-2.) Simultaneously, binding of three Na^+ ions from the interior of the cell takes place. The phosphorylation of one subunit causes a conformational change in the protein, which opens a channel or pore through which the three Na^+ ions can be released to the extracellular fluid (Step 2). Outside the cell, two K^+ ions bind to the pump enzyme, which is still phosphorylated (Step 3). Another conformational change occurs when the bond between the enzyme and the phosphate group is hydrolyzed. This second conformational change regenerates the original form of the enzyme and allows the two K^+ ions to enter the cell (Step 4). The pumping process transports three Na^+ ions out of the cell for every two K^+ ions transported into the cell. A mechanism for the sodium–potassium pump is shown in Figure 8.23.

The operation of the pump can be reversed when there is no K^+ and a high concentration of Na^+ in the extracellular medium; in this case, ATP is produced by the phosphorylation of ADP. The actual operation of the Na^+/K^+ pump is not completely understood and probably is even more complicated than we now know. There is also a calcium ion (Ca^{2+}) pump, which is a subject of equally active investigation. Unanswered questions about the detailed mechanism of active transport provide opportunities for future research.

Another type of transport is called **secondary active transport**. An example is the galactoside permease, also called lactose permease, in bacteria (Figure 8.24).

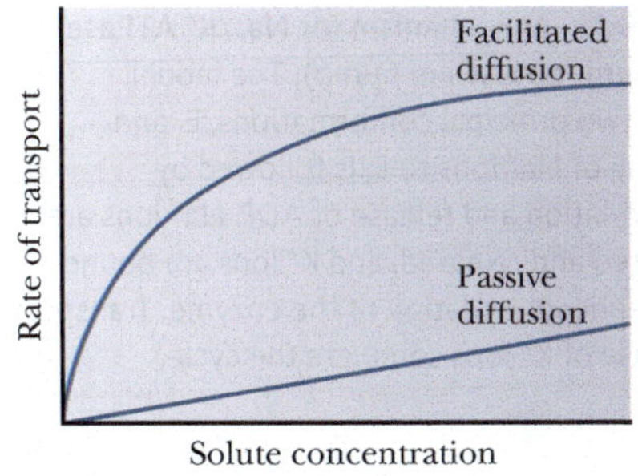

Figure 8.21 Passive diffusion and facilitated diffusion may be distinguished graphically. The plots for facilitated diffusion are similar to kinetics plots of enzyme-catalyzed reactions (Figure 6.11), and they display enzyme saturation behavior.

secondary active transport a type of active transport where the transporter uses an electrochemical gradient created by a proton pump to move a substance against its concentration gradient

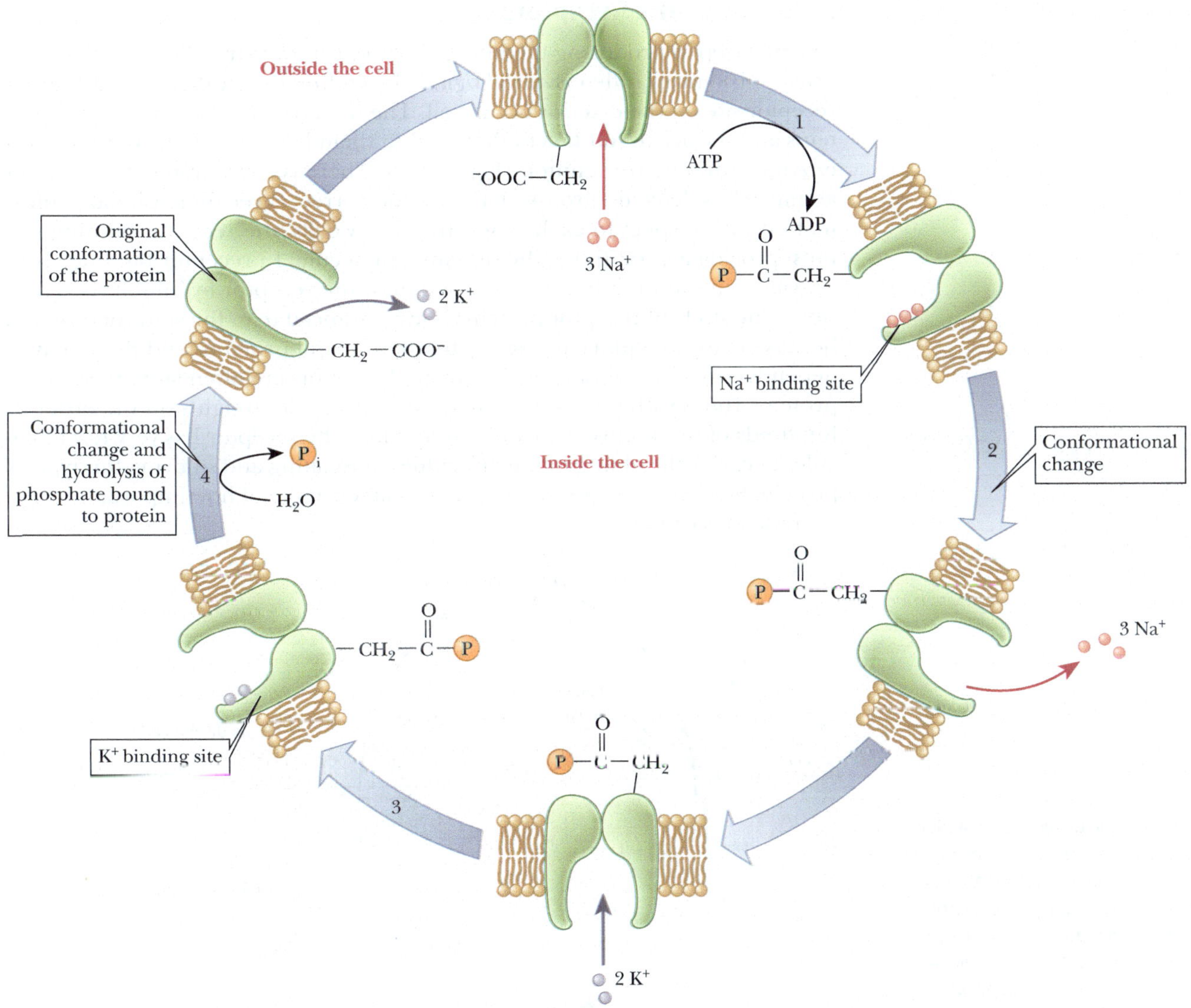

Figure 8.22 The sodium–potassium ion pump.

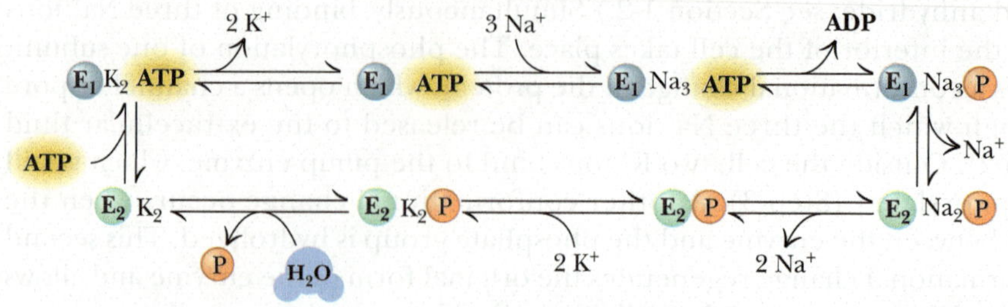

Figure 8.23 A mechanism for Na^+/K^+ ATPase (the sodium–potassium pump). The model assumes two principal conformations, E_1 and E_2. Binding of Na^+ ions to E_1 is followed by phosphorylation and release of ADP. Na^+ ions are transported and released, and K^+ ions are bound before dephosphorylation of the enzyme. Transport and release of K^+ ions complete the cycle.

The lactose concentration inside the bacterial cell is higher than the concentration outside, so moving lactose into the cell requires energy. The galactoside permease does not directly hydrolyze ATP, however. Instead, it harnesses the energy by letting hydrogen ions flow through the permease into the cell with their concentration gradient. As long as more energy is available allowing the hydrogen ions to flow ($-\Delta G$) than is required to concentrate the lactose ($+\Delta G$), the process is possible. However, to arrive at a situation in which there is a higher concentration of hydrogen ions on the outside than on the inside, some other primary active transporter must establish the hydrogen ion gradient. Active transporters that create hydrogen ion gradients are called **proton pumps**.

proton pumps integral membrane proteins that create a hydrogen ion gradient across the membrane

8-5B Membrane Receptors

The first step in producing the intracellular effects of extracellular, biologically active substances, called *chemical signals*, is binding the substance to a protein receptor site on the exterior of the cell. The interaction between receptor proteins and the active substances that bind to them has features in common with enzyme–substrate recognition (Figure 6.4). There is a requirement for certain amino acid side chain groups that have the correct three-dimensional conformation with respect to each other. In other words, the receptor binding site must provide a good fit for the signaling molecule. In receptor binding, as in enzyme behavior, inhibition of the action of the receptor by inhibitors is possible. The study of receptor proteins is less advanced than the study of enzymes because many receptors are tightly bound integral proteins, and their activity depends on the membrane environment. Receptors are often large multimeric proteins (ones with several subunits), with molecular weights on the order of hundreds of thousands. Also, quite frequently, the receptor has very few molecules in each cell, adding to the difficulties of isolating and studying this type of protein. Biochemical Connections 8C gives an example of how important receptor functions can be.

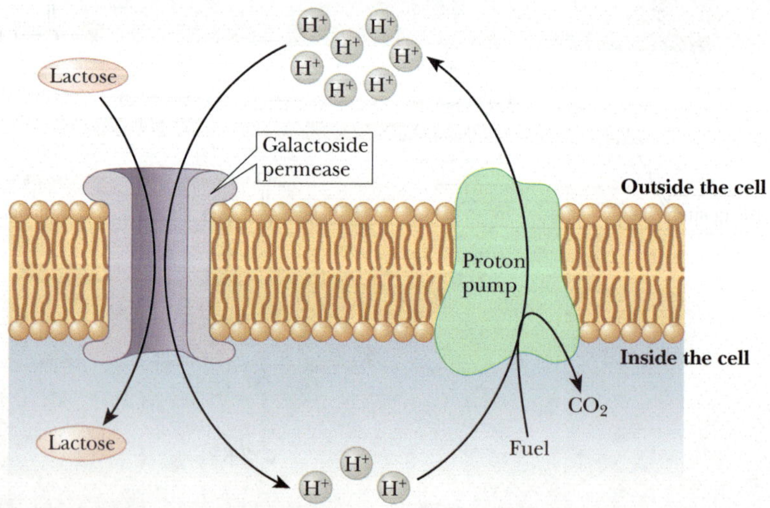

Figure 8.24 An example of secondary active transport. Galactoside permease uses the higher concentration of H^+ outside the cell created by a proton pump to drive the transport of lactose against its concentration gradient into the cell. *(Principles of Biochemistry, by Albert L. Lehninger, David L. Nelson and Michael M. Cox. Copyright © 1993 by W. H. Freeman & Co.)*

8C Biochemical Connections • Physiology

Lipid Droplets Are Not Just Great Balls of Fat

In electron micrographs of fat cells, large **lipid droplets**, also called **adiposomes**, are easily visible. Very visible are the large lipid droplets. For decades, these structures have been thought of as great balls of fat, a convenient way to store triacylglycerols for consumption. However, these droplets are surrounded by a phospholipid membrane that contains many membrane proteins with widely varying activities. On the negative side, they may also be involved in several lipid diseases, cardiovascular diseases, and diabetes. These lipid droplets are now being thought of as a subcellular organelle in their own right.

One of the first clues that lipid droplets were more than a simple pool of fat came in the early 1990s from research by Constantine Londos. He and his colleagues identified a protein called *perilipin* on the membrane of the lipid droplets in fat cells. They discovered that when cells are stimulated to metabolize the fatty acids in the lipid droplets, this protein is phosphorylated. This suggests a more complicated mechanism for controlling lipid digestion in fat cells than previously imagined. Many more proteins have since been identified on the membrane of lipid droplets.

It is now believed that perilipin guards the fat resources of the lipid droplet. When not phosphorylated, the protein does not allow fat-digesting enzymes access to the triacylglycerols. When phosphorylated, the protein shifts conformation and allows access. Studies with mutant mice lacking perilipin have shown that such mice eat much more than their wild-type counterparts yet burn off two-thirds of the extra calories consumed. Figure 8.25 compares three different strains of mice. The mouse on the left is normal. The mouse on the right is an obese mouse from a strain that lacks the ability to respond to an appetite-suppressing hormone called *leptin*. The mouse in the middle is a double mutant that cannot respond to leptin but also lacks perilipin. In this case, the extra fat burning that takes place almost makes up for the increased appetite. Further research on such proteins in the membranes of lipid droplets could lead to useful antiobesity therapies. Semaglutides, moreover, which are peptides that mimic hormones like leptin, are seeing success in the treatment of type 2 diabetes and as weight-loss medications.

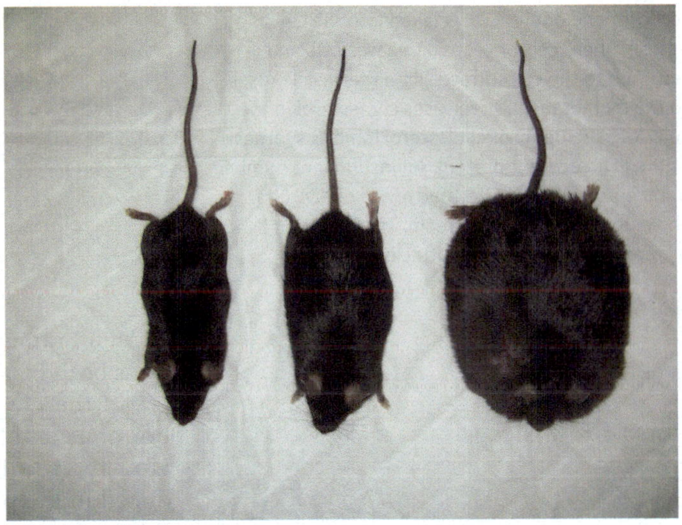

Figure 8.25 Three different strains of mice. Mouse on the right: an obese mouse that does not produce leptin because of a genetic mutation in the leptin gene. Mouse in the middle: a mouse that does not produce leptin but maintains relatively normal body weight because of its enhanced fat metabolism caused by the lack of perilipin. Mouse on the left: a wild-type, normal mouse that produces leptin as well as perilipin.

Receptor proteins operate in various ways, giving rise to different receptor actions. Many examples are discussed in context in later chapters on metabolism and its control. Receptors can be very specific in their activity, and we can use one example now as a case study for receptor activity.

LDL receptors are important types of receptors that bind LDL, the principal carrier of cholesterol in the bloodstream. LDL is a particle that consists of various lipids—in particular cholesterol and phosphoglycerides—as well as a protein. The protein portion of the LDL particle binds to the LDL receptor of a cell (Figure 8.26). The complex formed between the LDL and the receptor is pinched off into the cell in a process called *endocytosis*. The receptor protein is then recycled back to the surface of the cell.

The cholesterol portion of the LDL is used in the cell, but an oversupply of cholesterol causes problems. Excess of cholesterol inhibits the synthesis of LDL receptor. If there are too few receptors for LDL, the level of cholesterol in the bloodstream increases. Eventually, the excess cholesterol is deposited in the arteries, blocking them severely. This blocking of arteries, called atherosclerosis, can eventually lead to heart attacks and strokes. In many industrialized countries, typical blood cholesterol levels are high, and the incidence of heart attacks and strokes is correspondingly high.

lipid droplets (adiposomes) lipid-rich organelles that regulate the storage and release of fatty acids

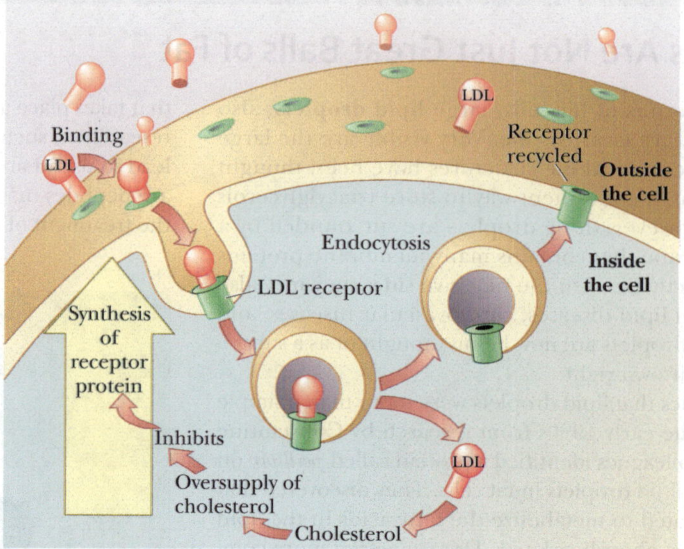

Figure 8.26 The mode of action of the LDL receptor. A portion of the membrane with LDL receptor and bound LDL is taken into the cell as a vesicle. The receptor protein releases LDL and is returned to the cell surface when the vesicle fuses to the membrane. LDL releases cholesterol in the cell. An oversupply of cholesterol inhibits synthesis of the LDL receptor protein. An insufficient number of receptors leads to elevated levels of LDL and cholesterol in the bloodstream. This situation increases the risk of heart attack.

Membrane receptors combine protein chemistry, lipid chemistry, interactions both inside and outside the cell, and metabolic pathways. A good example of the elegance of such interactions is the receptor for human growth hormone and how information from outside the cell is transmitted to the cell. Human growth hormone (hGH) is a 191-amino-acid peptide hormone produced in the pituitary gland, which stimulates growth, cell reproduction, and cell regeneration.

The hGH receptor has two identical protein subunits that extend from the cell membrane outward, as shown in Figure 8.27. This is a common design for cell receptors, where signal binding to a dimeric receptor changes the conformation of the dimer and causes either activation or inactivation of processes inside the cell. For example, a single molecule of hGH binds sequentially to the outer domain of one subunit and then the other. Binding of hGH causes movement of the dimer, and this movement is transmitted through

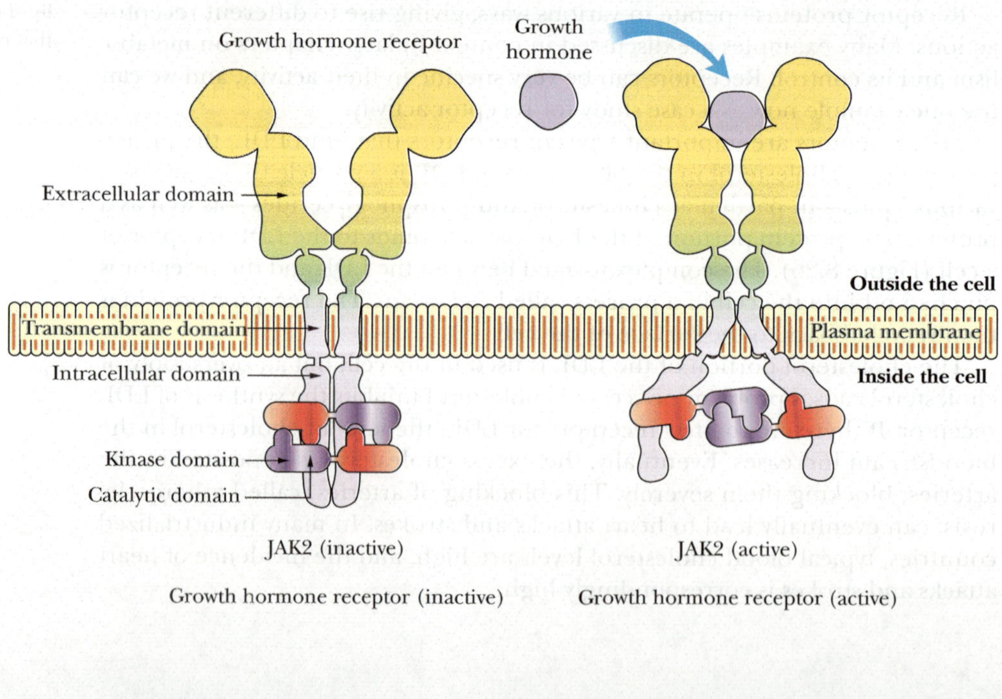

Figure 8.27 Mechanism of action of human growth hormone and its receptor. Binding of the human growth hormone to the receptor outside the cell is transmitted into action inside the cell by movement of the transmembrane protein activating Janus kinase 2 inside the cell. *(Science, 344, p. 703.)*

the transmembrane domain to the intracellular domain of the receptor. The receptor is associated with a kinase enzyme called Janus kinase-2 (JAK2), which is inactive initially. The movement of the receptor caused by hormone binding changes the orientation of the catalytic site of JAK2, thereby activating the enzyme. As we explain in other chapters, activation of protein kinases is a common and critical part of cell signaling. Frequently, activating a kinase starts a cascade of phosphorylation events that ultimately lead to the observed effect of the hormone.

8-6 Lipid-Soluble Vitamins and Their Functions

Learning Objectives

After reading this section, students will be able to:

LO 8.6.1 Describe the characteristics of four lipid-soluble vitamins.

LO 8.6.2 Distinguish between the functions of four lipid-soluble vitamins.

Some vitamins, having a variety of functions, are of interest in this chapter because they are soluble in lipids. These lipid-soluble vitamins are hydrophobic, which accounts for their solubility characteristics (Table 8.5).

8-6A Vitamin A

The extensively unsaturated hydrocarbon **β-carotene** is the precursor of **vitamin A**, which is also known as **retinol**. As the name suggests, β-carotene is abundant in carrots, but it also occurs in other yellow and orange vegetables. When an organism requires vitamin A, β-carotene is cleaved into two molecules of vitamin A (Figure 8.28A). To draw the structure of vitamin A, keep in mind the following features:

- It has a 15-carbon chain that starts with a 6-membered ring.
- It has two methyl groups on carbon 1.
- It has methyl groups on carbons 5, 9, and 13.
- It has double bonds on every odd-numbered carbon starting at carbon 5.
- Finally, it has a hydroxyl group (hence the "-ol" in the name retinol) on carbon 15.

β-carotene an unsaturated hydrocarbon; the precursor of vitamin A

vitamin A (retinol) the lipid-soluble compound responsible for the primary photochemical event in vision

A derivative of vitamin A plays a crucial role in vision when it is bound to a protein called *opsin*. The cone cells in the retina of the eye contain several types of opsin and are responsible for vision in bright light and for color vision. The rod cells in the retina contain only one type of opsin; they are responsible for vision in dim light. The chemistry of vision has been more extensively studied in rod cells than in cone cells, so we discuss events that take place in rod cells.

Vitamin A gets oxidized to 11-trans-retinal, where the "-al" in the molecule name indicates the presence of an **aldehyde** group (Figure 8.28B). This oxidation of vitamin A is catalyzed by the enzyme retinol dehydrogenase. **Dehydrogenase** enzymes oxidize molecules by reducing an electron acceptor like NAD^+.

aldehyde a functional group where a carbonyl (C=O) is bonded to at least one hydrogen atom

dehydrogenase an enzyme that oxidizes a substrate by reducing an electron acceptor, usually NAD^+

Table 8.5 Lipid-Soluble Vitamins and Their Functions

Vitamin	Function
Vitamin A	Serves as the site of the primary photochemical reaction in vision
Vitamin D	Regulates calcium (and phosphorus) metabolism
Vitamin E	Serves as an antioxidant; necessary for reproduction in rats and may be necessary for reproduction in humans
Vitamin K	Has a regulatory function in blood clotting

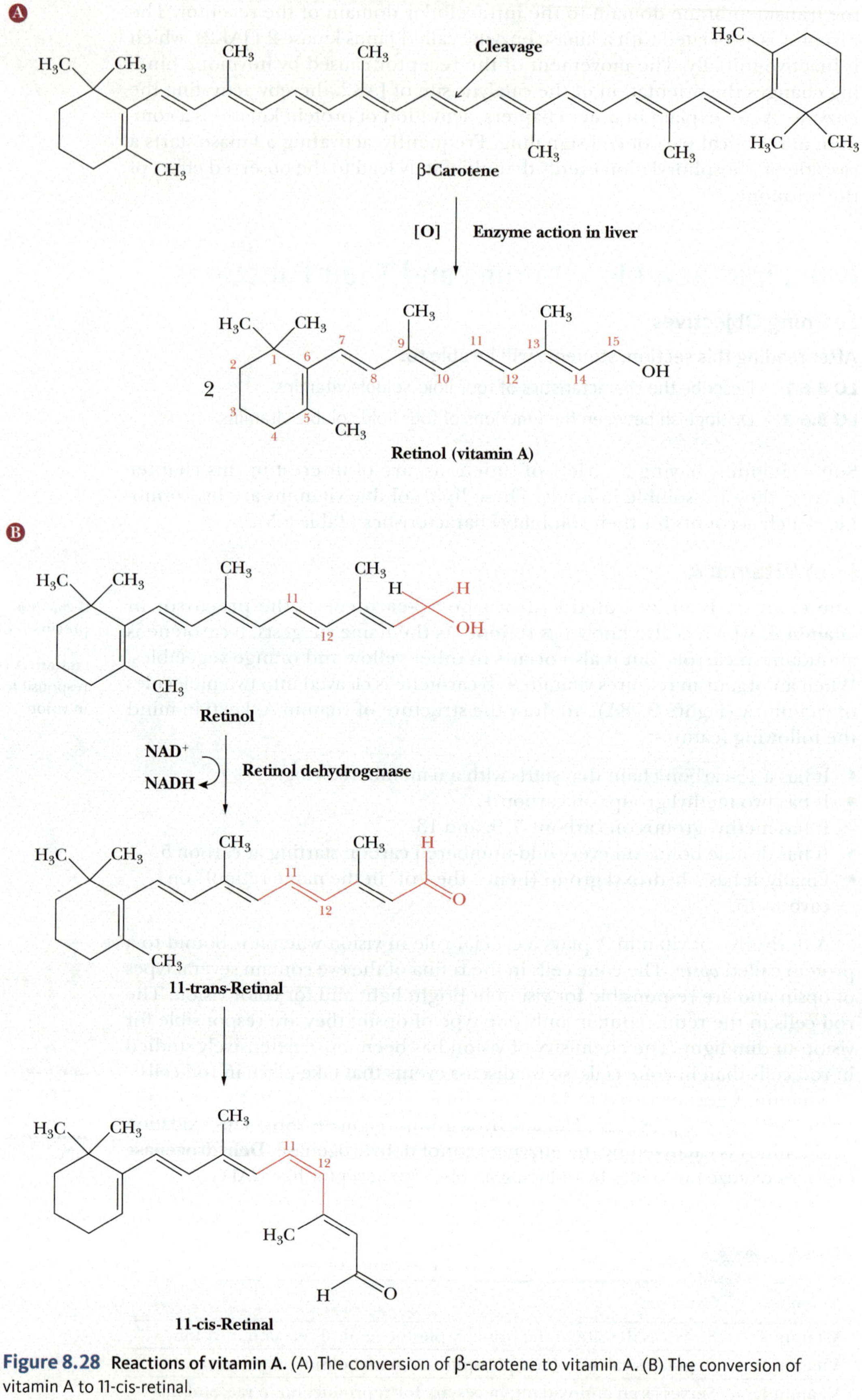

Figure 8.28 **Reactions of vitamin A.** (A) The conversion of β-carotene to vitamin A. (B) The conversion of vitamin A to 11-cis-retinal.

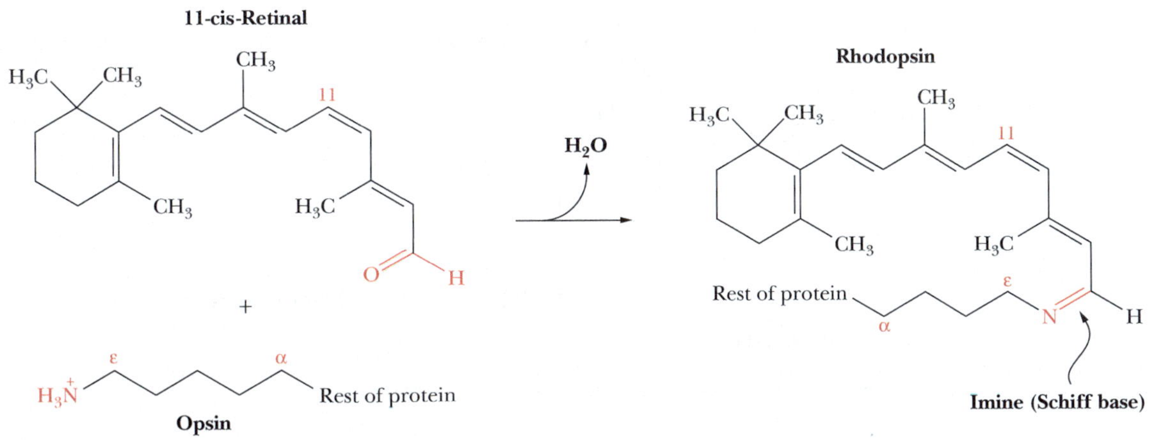

Figure 8.29 The formation of rhodopsin from 11-cis-retinal and opsin.

The next reaction, catalyzed by retinal isomerase, involves a cis–trans isomerization around the double bond at carbon 11. The aldehyde group of 11-cis-retinal forms an imine (also called a Schiff base) with the side-chain amino group of a lysine residue in rod-cell opsin to form **rhodopsin** (Figure 8.29). For more details about how rhodopsin plays a role in vision, see **Biochemical Connections 8D**.

rhodopsin a protein crucial to vision; it is formed by the reaction of 11-cis-retinal and opsin

8-6B Vitamin D

The several forms of **vitamin D** play a major role in the regulation of calcium and phosphorus metabolism. One of the most important of these compounds, vitamin D_3 (cholecalciferol), is formed from cholesterol by the action of ultraviolet radiation from the Sun (Figure 8.31). Vitamin D_3 is further processed in the body to form hydroxylated derivatives, which are the metabolically active form of this vitamin. The presence of vitamin D_3 leads to increased synthesis of a Ca^{2+}-binding protein, which increases the absorption of dietary calcium in the intestines. This process results in calcium uptake by the bones.

vitamin D a lipid-soluble compound that regulates calcium and phosphorus metabolism

A deficiency of vitamin D can lead to *rickets,* a condition in which the bones of growing children become soft, resulting in skeletal deformities. Children, especially infants, have higher requirements for vitamin D than do adults.

8D Biochemical Connections • Neuroscience

Vision Has Great Chemistry

The primary chemical reaction in vision, the one responsible for generating an impulse in the optic nerve, involves a cis–trans isomerization around one of the double bonds in the retinal portion of rhodopsin. When rhodopsin is active (that is, when it can respond to visible light), the double bond between carbon atoms 11 and 12 of the retinal (11-cis-retinal) has the cis orientation (Figure 8.30). Under the influence of light, an isomerization reaction occurs at this double bond, producing all-trans-retinal. Because the all-trans form of retinal cannot bind to opsin, all-trans-retinal and free opsin are released. As a result of this reaction, an electrical impulse is generated in the optic nerve and is transmitted to the brain to be processed as a visual event. The active form of rhodopsin is regenerated by enzymatic isomerization of the all-trans-retinal back to the 11 cis form and subsequent re-formation of the rhodopsin.

Vitamin A deficiency can have drastic consequences, as would be predicted from its importance in vision. Night blindness—and even total blindness—can result, especially in children. On the other hand, an excess of vitamin A can have harmful effects such as bone fragility. Lipid-soluble compounds are not excreted as readily as water-soluble substances, and excessive amounts of lipid-soluble vitamins can accumulate in adipose tissue.

Continued on next page

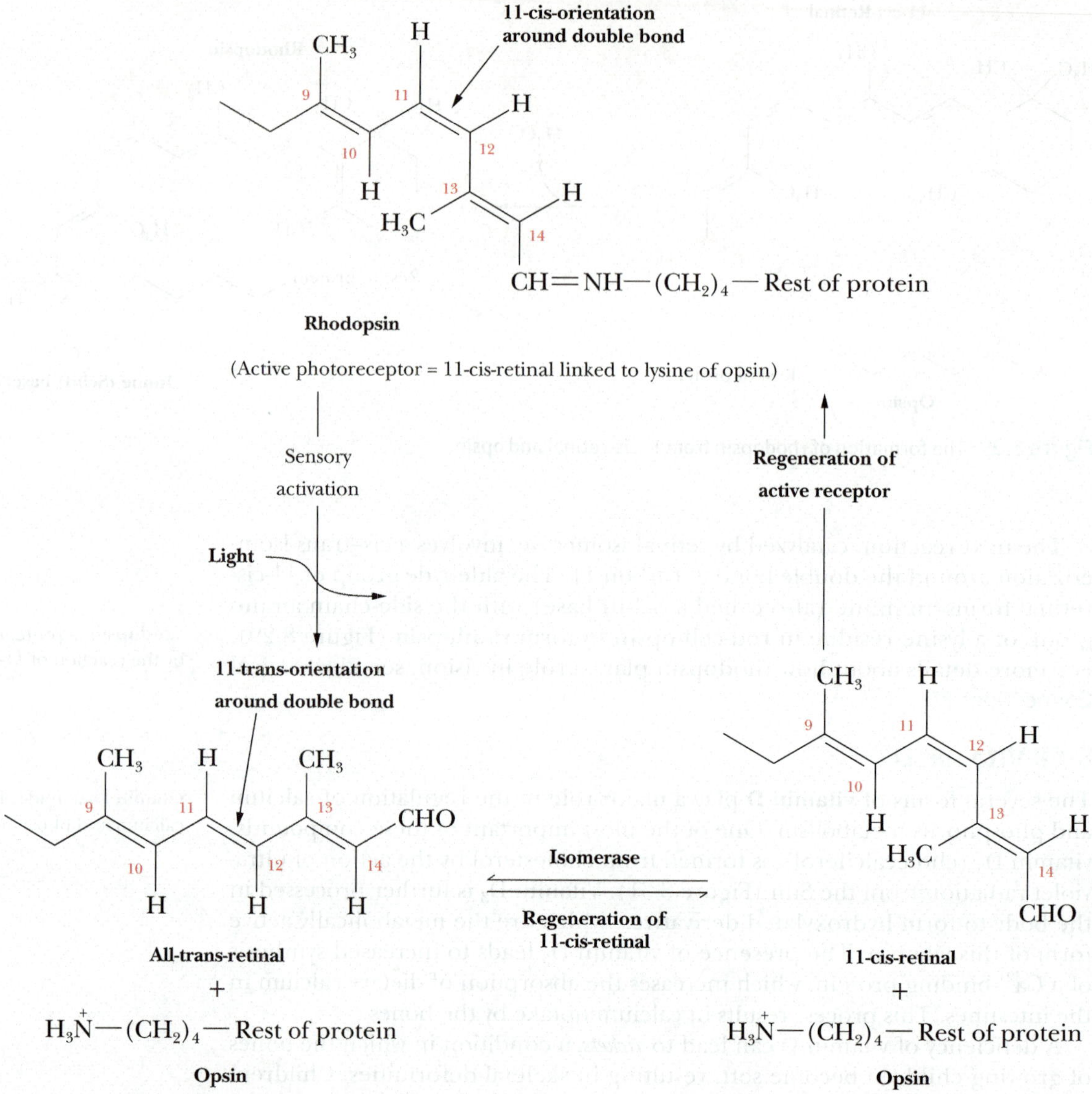

Figure 8.30 The primary chemical reaction of vision.

Milk with vitamin D supplements is available to most children. Adults who are exposed to normal amounts of sunlight do not usually require vitamin D supplements.

vitamin E a lipid-soluble antioxidant

antioxidant a reducing agent that removes potentially damaging oxidizing agents

8-6C Vitamin E

Vitamin E is an **antioxidant**, and its most active form is α-tocopherol (Figure 8.32). Antioxidants are reducing agents that react with oxidizing agents before they can potentially damage other biomolecules. In rats, vitamin E is required for reproduction and for prevention of the disease *muscular dystrophy*. It is not known whether this requirement exists in humans. Vitamin E has been shown to protect important compounds, including vitamin A, from degradation in the laboratory; it probably also serves this function in organisms.

Recent research has shown that the interaction of vitamin E with membranes enhances its effectiveness as an antioxidant. Another function of antioxidants such as vitamin E is to react with, and thus to remove, the very reactive and

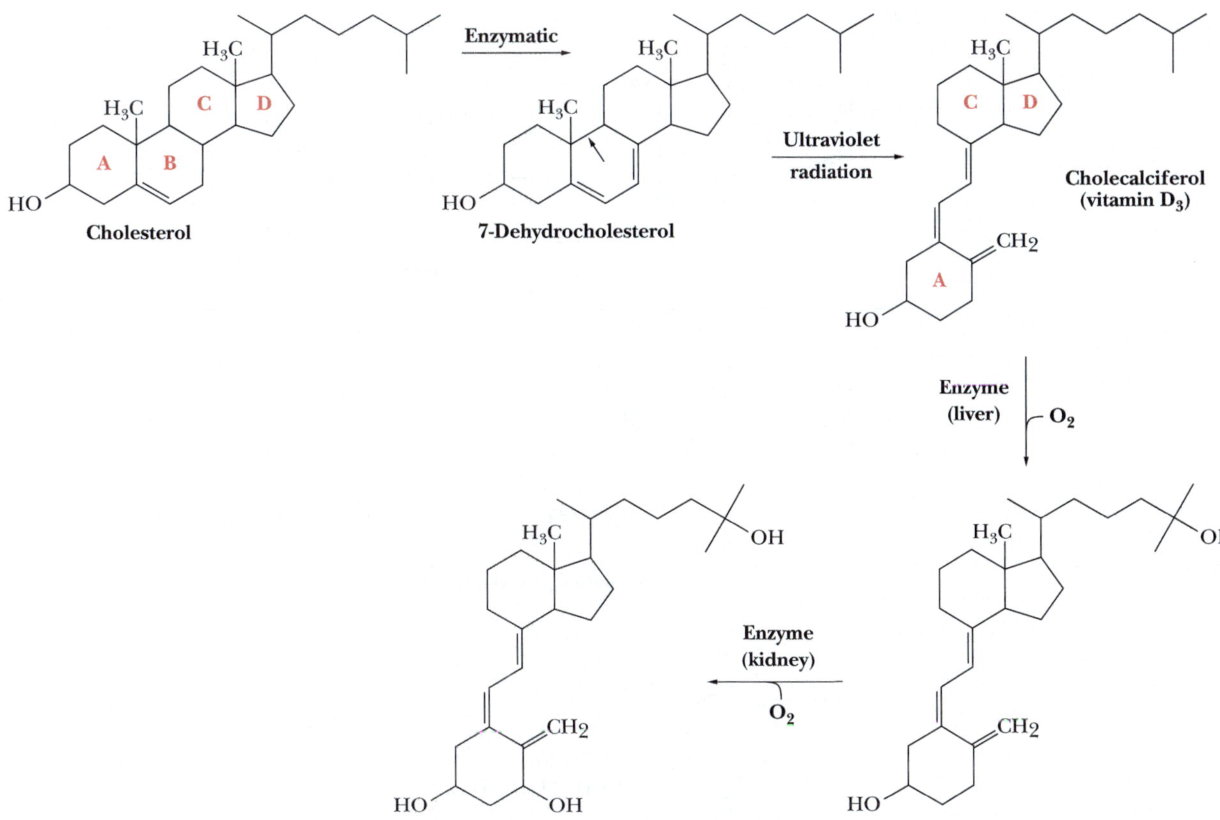

Figure 8.31 **Reactions of vitamin D.** The photochemical cleavage occurs at the bond shown by the arrow; electron rearrangements after the cleavage produce vitamin D_3. The final product, 1,25-dihydrocholecalciferol, is the form of the vitamin that is most active in stimulating the intestinal absorption of calcium and phosphate and in mobilizing calcium for bone development.

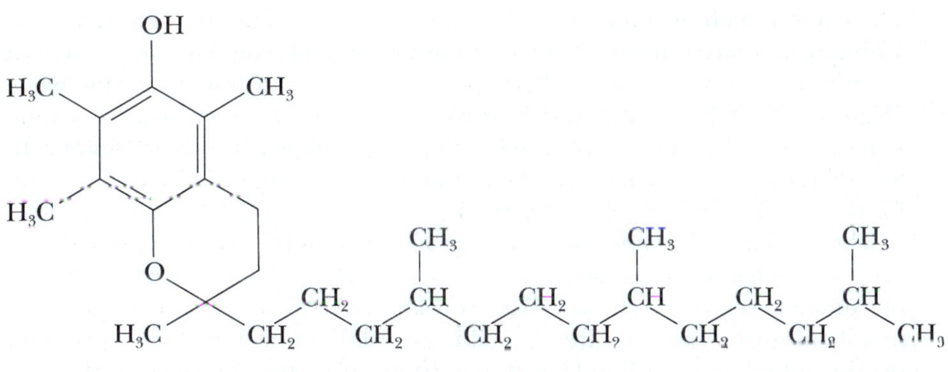

Figure 8.32 **The most active form of vitamin E is α-tocopherol.**

highly dangerous substances known as **free radicals**. A free radical has at least one unpaired electron, which accounts for its high degree of reactivity. Free radicals may play a part in the development of cancer and in the aging process.

free radicals highly reactive molecules that have at least one unpaired electron

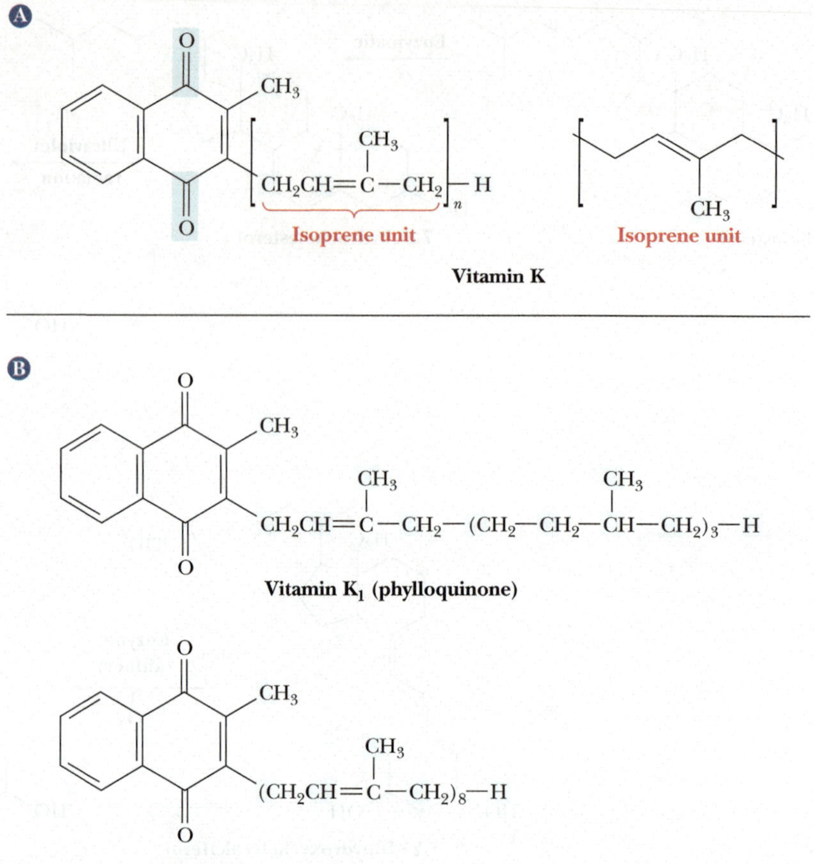

Figure 8.33 Vitamin K. (A) The general structure of vitamin K, which is required for blood clotting. The value of *n* is variable, but it is usually <10. (B) Vitamin K_1 has one unsaturated isoprene unit; the rest are saturated. Vitamin K_2 has eight unsaturated isoprene units.

8-6D Vitamin K

vitamin K lipid-soluble compound that plays an important role in blood clotting

The name **vitamin K** comes from the Danish *koagulation* because this vitamin is an important factor in the blood-clotting process. The bicyclic ring system of vitamin K contains two carbonyl groups, the only polar groups on the molecule (Figure 8.33). A long unsaturated hydrocarbon side chain consists of repeating *isoprene* units, the number of which determines the exact form of vitamin K. Several forms of this vitamin can be found in a single organism, but the reason for this variation is not well understood.

The presence of vitamin K is required in the complex process of blood clotting, which involves many steps and many proteins and has stimulated numerous unanswered research questions. It is known that vitamin K is required to modify prothrombin and other proteins involved in the clotting process. With prothrombin, vitamin K aids in the addition of carboxyl groups to the side chains of several glutamate residues (Figure 8.34). This modification of glutamate produces γ-carboxyglutamate residues (recall the Greek-letter notation for amino acids discussed in Section 3-2 and seen in Figure 3.3). The two carboxyl groups form a *bidentate* ("two teeth") *ligand,* which can bind calcium ion (Ca^{2+}). If prothrombin is not modified in this way, it does not bind Ca^{2+}. Even though there is a lot more to be learned about blood clotting and the role of vitamin K in the process, this point, at least, is well established because Ca^{2+} is required for blood clotting.

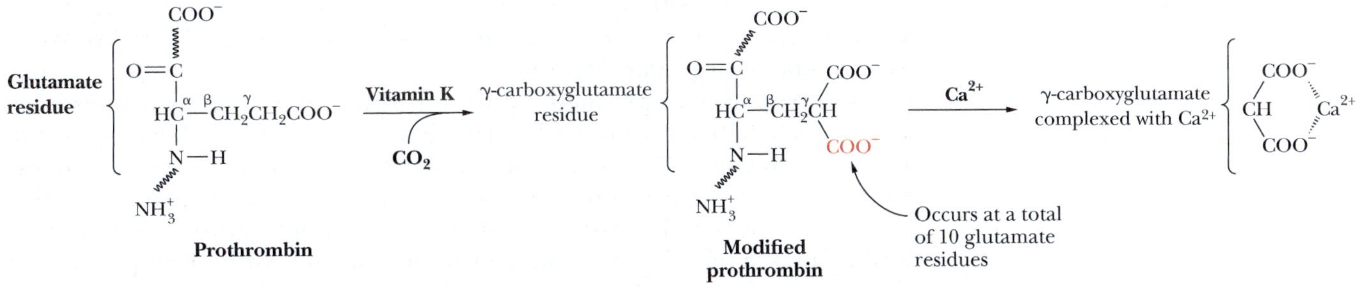

Figure 8.34 The role of vitamin K in the modification of glutamate residues in prothrombin.

8-7 Arachidonic Acid Derivatives

Learning Objectives

After reading this section, students will be able to:

LO 8.7.1 Describe the characteristics of three types of arachidonic acid derivatives.

LO 8.7.2 Distinguish between the functions of three types of arachidonic acid derivatives.

8-7A Prostaglandins

A group of compounds derived from fatty acids has a wide range of physiological activities; they are called **prostaglandins** because they were first detected in seminal fluid, which is produced by the prostate gland. It has since been shown that they are widely distributed in a variety of tissues. The metabolic precursor of all prostaglandins is arachidonic acid, a fatty acid that contains 20 carbon atoms and four double bonds (Table 8.2). The production of the prostaglandins from arachidonic acid takes place in several steps, which are catalyzed by enzymes.

prostaglandins derivatives of arachidonic acid that contain a five-membered ring and are involved in a wide range of biological functions

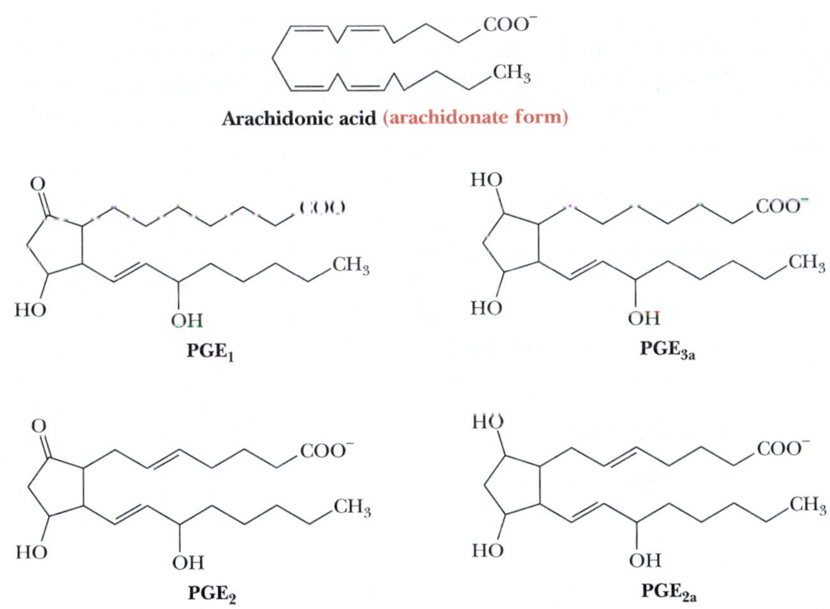

Figure 8.35 Arachidonic acid and some prostaglandins.

The prostaglandins themselves each have a five-membered ring; they differ from one another in the numbers and positions of double bonds and hydroxyl/ketone functional groups (Figure 8.35).

The structures of prostaglandins and their laboratory syntheses have been topics of great interest to organic chemists, largely because of the many physiological effects of these compounds and their possible usefulness in the pharmaceutical industry. Some of the functions of prostaglandins are control of blood pressure, stimulation of smooth muscle contraction, and induction of inflammation. Aspirin inhibits the synthesis of prostaglandins, particularly in blood platelets, a property that accounts for its anti-inflammatory and fever-reducing properties. Cortisone and other steroids also have anti-inflammatory effects because of their inhibition of prostaglandin synthesis.

Prostaglandins are known to inhibit the aggregation of platelets. They may thus be of therapeutic value by preventing the formation of blood clots, which can cut off the blood supply to the brain or the heart and cause certain types of strokes and heart attacks. Even if this behavior were the only useful property of prostaglandins, it would justify considerable research effort. Heart attacks and strokes are two of the leading causes of death in industrialized countries. The study of prostaglandins has also been of great interest because of their possible antitumor and antiviral activity.

8-7B Leukotrienes

leukotrienes substances derived from leukocytes (white blood cells) that have three conjugated double bonds and are of pharmaceutical importance

Leukotrienes are compounds that, like prostaglandins, are derived from arachidonic acid. They are found in leukocytes (white blood cells) and have three conjugated double bonds; these two facts account for the "-tri-" and the "-ene" parts of the name *leukotriene.* Leukotriene C (Figure 8.36) is a typical member of this group. The 20 carbon atoms in the carboxylic acid backbone are what relate this compound structurally to arachidonic acid. (The 20-carbon prostaglandins and leukotrienes are also called eicosanoids.) An important property of leukotrienes is their constriction of smooth muscle, especially in the lungs. Asthma attacks may result from this constricting action because the synthesis of leukotriene C appears to be facilitated by allergic reactions such as a reaction to pollen. Drugs that inhibit the synthesis of leukotriene C are now being used in the treatment of asthma, as are other drugs designed to block leukotriene receptors. In the United States, the incidence of asthma has increased drastically since 1980, providing considerable incentive to find new treatments. Leukotrienes may also have inflammatory properties and may be involved in rheumatoid arthritis.

8-7C Thromboxanes

ethers functional groups where an oxygen atom is bonded to two alkyl carbon atoms

Thromboxanes are a third class of arachidonic acid derivatives. They contain cyclic **ethers** as part of their structures. The most widely studied member of the group, thromboxane A_2 (TxA_2) (Figure 8.37), is known to induce platelet aggregation and smooth muscle contraction.

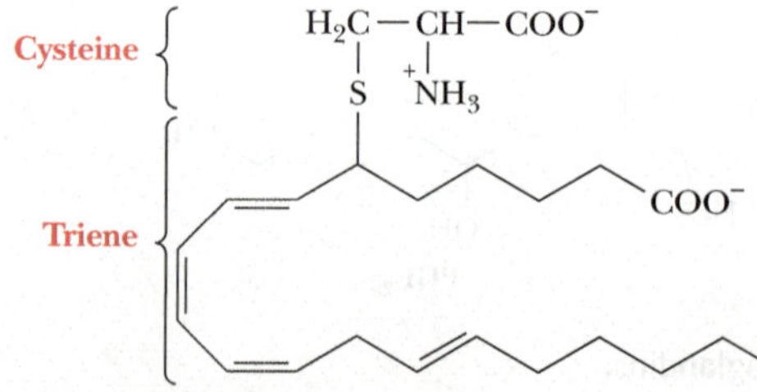

Figure 8.36 Leukotriene C.

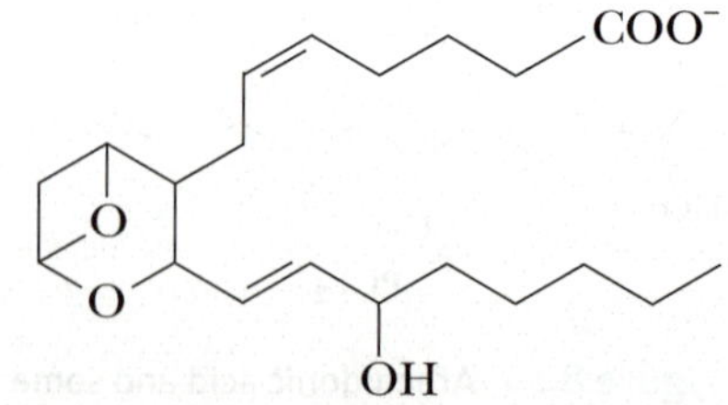

Figure 8.37 Thromboxane A_2 (TxA_2).

Biochemical Connections 8E discusses some nutrition-related connections among the topics in Chapter 8.

8E Biochemical Connections • Nutrition

Why Should We Eat More Salmon?

Platelets are elements in the blood that initiate blood clotting and tissue repair by releasing clotting factors and platelet-derived growth factor (PDGF). Turbulence in the bloodstream may cause platelets to rupture. Fat deposits and bifurcations of arteries lead to such turbulence, so platelets and PDGF are implicated in blood clotting and growth of atherosclerotic plaque. Furthermore, the anaerobic conditions that exist under a large plaque deposit may lead to weakness and dead cells in the arterial wall, aggravating the problem.

Among peoples who depend on fish as a major food source, including some Inuit tribes, very little heart disease is diagnosed, even though people in these groups eat high-fat diets and have high levels of blood cholesterol. Analysis of their diet led to the discovery that certain highly unsaturated fatty acids are found in the oils of fish and diving mammals (Figure 8.38). One class of these fatty acids is called omega-3 (ω_3), an example of which is eicosapentenoic acid (EPA):

$$CH_3CH_2(CH{=}CHCH_2)_5(CH_2)_2COOH$$

Eicosapentenoic acid (EPA)

There is a double bond at the third carbon atom from the end of the hydrocarbon tail. The omega system of nomenclature is based on numbering the double bonds from the last carbon in the fatty acid instead of from the carbonyl group [the delta (Δ) system]. Omega is the last letter in the Greek alphabet.

The omega-3 fatty acids inhibit the formation of certain prostaglandins and thromboxane A, which is similar in structure to prostaglandins. Thromboxane released by ruptured arteries causes other platelets to clump in the immediate area and to increase the size of the blood clot. Any disruption in thromboxane synthesis results in a lower tendency to form blood clots and, thus, in a lower potential for artery damage.

Nuts such as almonds, pistachios, and walnuts are also healthy sources of omega-3 fatty acids.

Figure 8.38 Salmon is a tasty source of important lipids.

Hot Topic The Science of Happiness and Depression

Happiness is an important issue in our lives, and many books and research studies have been published on factors that affect happiness. In the United States, we have always thought about the fundamental importance of happiness, which is why our nation's founders included "life, liberty and the pursuit of happiness" in the Declaration of Independence.

The Brain and Pleasure

To begin to understand the biochemical nature of happiness, we must first be able to understand how the brain works with respect to the emotions that we associate with happiness. For decades scientists have been trying to determine where in the brain the "happiness centers" are and how they work. In the 1950s, a psychiatrist named Robert Heath did some of the pioneering work in mapping the brain by implanting electrodes into the brains of patients who suffered from a wide variety of conditions, including epilepsy, schizophrenia, and depression. He was trying to find the biological seat of the disorders by stimulating regions in the brain, and he made some remarkable discoveries. Patients who were catatonic with despair would smile, talk, and laugh while areas of their brains were stimulated, but once the stimulation ended, they returned to their catatonic states.

To continue the study, Heath fitted a handful of patients with electrodes so they could control their own stimulation. One patient self-stimulated 1,500 times in a three-hour session. These experiments helped to define structures in the brain that would be called the "pleasure center," and the theories from these experiments lasted for decades. In the years that followed, many researchers used animal models to demonstrate similar results. Although it might not be intuitive that it's possible to tell if a rat is depressed or not, certain behaviors are common to all mammals, and researchers use these cues to determine whether animals other than humans are depressed.

An underlying assumption in the early work, however, was that repetition implied that the individual was deriving pleasure from the response. What if this were not always the case? In the early 2000s, Morten Kringelbach and Kent Berridge began to question this underlying assumption. Many known addictive behaviors can be found where an individual repeats an action obsessively but does not seem to truly enjoy the object of their obsession. Using animal models, they began to study the "like versus want" conundrum. The regions of the brain identified in the 1950s are positioned at the front of the brain and are activated by the neurotransmitter dopamine (Figure 8.39), released by neurons that originate near the brainstem.

Kringelbach and Berridge hypothesized that if these areas of the brain were truly the pleasure centers, then flooding them with dopamine or removing dopamine receptors would alter an animal's response to a pleasurable stimulus, but that is not what they found. They used knockout mice that lacked a protein that would remove dopamine once it had been released. Thus, these animals would have more dopamine in their systems. If the dopamine was leading to pleasure, they should be "happier" mice when faced with pleasurable stimuli. However, what they found was that the mice would seek pleasurable stimuli, such as sweet foods, more actively than their normal counterparts, but once they attained the reward, they did not seem to enjoy it more. Rodents that are depleted of dopamine show no interest in sugary treats at all. They will starve to death unless actively nursed. Dopamine, therefore, does not seem to be the single neurotransmitter of pleasure but is highly linked to motivation to attain pleasure—a subtle distinction.

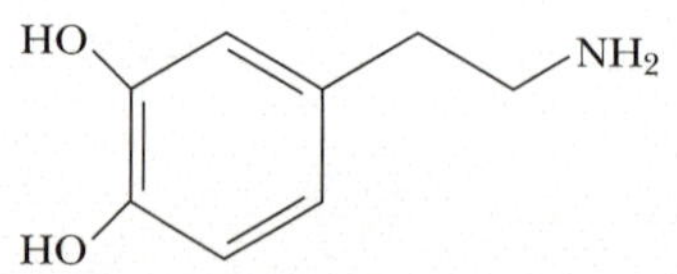

Figure 8.39 Dopamine.

In human studies, dopamine also seems more related to a person's perception of how much they want something rather than how much they actually enjoy it. This was also confirmed by studies of people addicted to recreational drugs, many of which act by flooding the brain with dopamine. The addiction persists long after the pleasurable sensation associated with the drug subsides. Persons with alcohol-use and drug-use disorders often "have to" ingest their particular poison even if they don't want to. Similar addiction effects have been seen with social media use and dopamine release in adults and teens, and warnings and advice for curbing social media use continue to grow.

The original work from the 1950s focused on an area of the brain near the brainstem, now called the ventral tegmental area, as shown in Figure 8.40. This area communicates with several other areas of the brain as shown, but it is now believed to be related to the wanting part of pleasure rather than the enjoying part. Wanting is a part of pleasure as well. For example, receiving a gift you really wanted is often more rewarding than receiving one you did not.

Without discounting the importance of wanting, scientists looked to find other areas of the brain that were related to pleasure, and these have become known as hedonic hotspots. One of these is found in a subregion of the nucleus accumbens (shown in red in Figure 8.40). Another is found near the ventral pallidum (also shown in red). Kringelbach and Berridge identified these by finding areas in the brains of rodents that, when stimulated, would increase the outward signs of enjoyment. The pathways that are related to feelings of pleasure in these structures are not based on dopamine. Rather, they are based on the following pentapeptide hormones:

Tyr—Gly—Gly—Phe—Met

or

Tyr—Gly—Gly—Phe—Leu

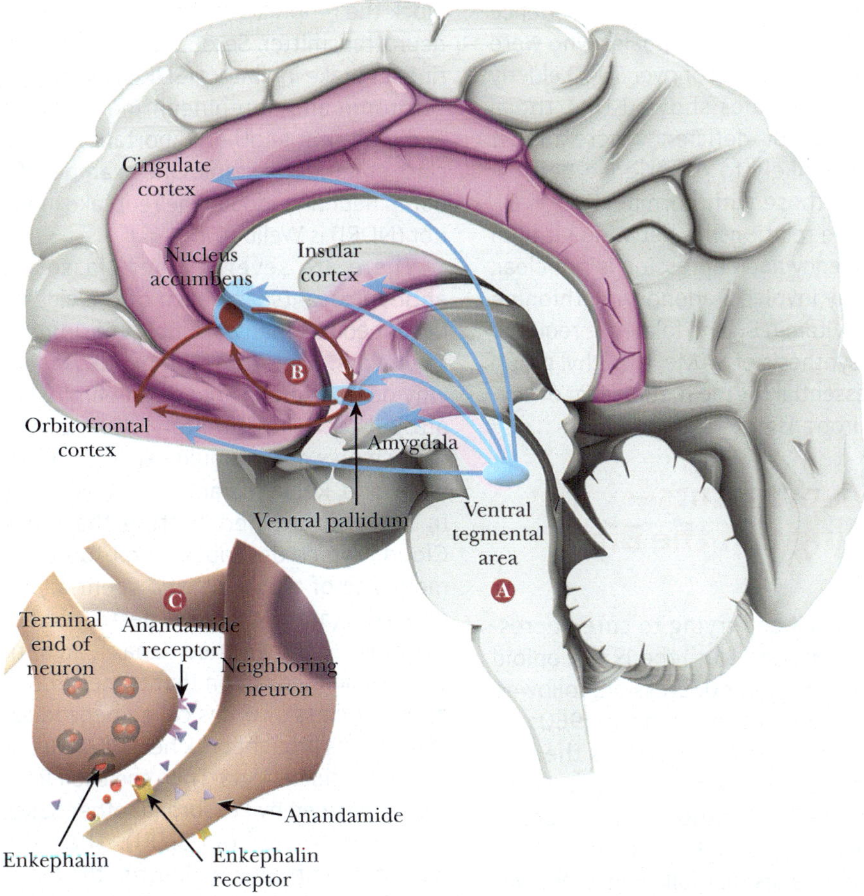

Figure 8.40 Wanting and liking. Evidence shows that there is a difference between wanting and liking, although they are related. In (A), several neural pathways associated with wanting begin in the ventral tegmental area and are propagated to other parts of the brain (blue arrows). These pathways of wanting are often associated with release of the neurotransmitter dopamine. The pathways were once thought to be the only pathways related to pleasure. More recently, researchers focused on other areas that are related to liking (B). These involve hedonic hotspots (shown in red), such as the nucleus accumbens and the ventral pallidum. The chemistry involved in these hedonic hotspots is shown in (C). A pleasurable stimulus, like the taste of chocolate, prompts the terminal end of a neuron to release enkephalins, which migrate to a neighboring neuron where they bind to enkephalin receptors. This releases another neurotransmitter, anandamide, which is the body's natural version of THC, the active ingredient of marijuana. The anandamide moves back to the anandamide receptors on the first neuron, setting up a feedback loop that extends the feeling of pleasure. *(AXS Biomedical Animation Studio Inc.)*

Enkephalins bind to opioid receptors in nearby neurons. These neurons then release another hormone called anandamide (Figure 8.41). Anandamide is the body's natural version of the THC molecule found in marijuana, and it is thought to regulate and propagate the feeling of pleasure initiated by the enkephalins.

Although each of the hedonic hot spots is small, as shown in Figure 8.40, they make an interrelated complex with many parts of the brain. It is quite complicated and undoubtedly will be studied for years to come, but the separation of wanting and liking itself was a milestone for psychology.

Don't Worry—Be Happy

The causes of depression vary from person to person, but unfortunately, millions of Americans experience major depression. It is important to recognize that major depression is not the same as being sad. Over the course of a lifetime, everyone has moments of sadness. These moments of sadness, even intense ones, are natural and healthy. Major depression can be

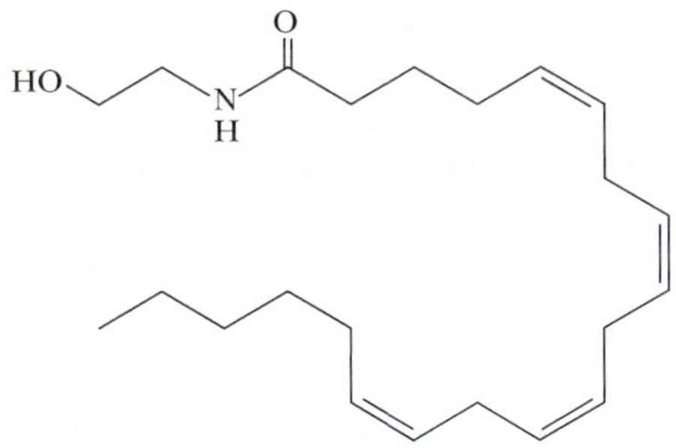

Figure 8.41 **Anandamine.**

a chronic or acute condition characterized by sustained sorrowful moods, guilty feelings, feelings of worthlessness, loss of motivation, and loss of interest in the activities of life. It impairs thinking, sleep, and appetite. It can also be fatal. An estimated 15% of patients with major depression die by suicide. There are currently many different means to treat depression, so let's look at some of the research.

Fatty Acids

In early 2011, researchers reported online in the journal *Nature Neuroscience* that deficiencies in omega-3 polyunsaturated fatty acids alter functioning of the endocannabinoid system, a group of lipids and their receptors that are involved in mood, pain sensations, and other processes. They noted that mice subjected to a diet low in omega-3 polyunsaturated fatty acids have lower omega-3 levels in the brain, and this is associated with an alteration in the functioning of the endocannabinoid system, specifically a deficit in the signaling of the CB1 cannabinoid receptor in the prefrontal cortex of the brain. The cannabinoid receptors are a class of cell membrane receptors under the G-protein–coupled receptor superfamily. Cannabinoid receptors are activated by three major groups of ligands: endocannabinoids (produced by the mammalian body), plant cannabinoids (such as THC, produced by the cannabis plant), and synthetic cannabinoids. All of the endocannabinoids and plant cannabinoids are lipophilic—that is, all are fat-soluble compounds.

Studies followed over 600 cases of people with depression and showed that patients who had diets higher in trans fatty acids were almost 50% more likely to experience depression. While much research remains to be done, this could indicate that eating salmon and other sources of omega-3 fatty acids, as well as diets low in trans fatty acids, can help your head as well as your heart.

Vitamin D

In research studies, a clear association has been found between low vitamin D levels and depression. However, in studies where vitamin D was given as a supplement to treat depression, results have been mixed. Therefore, vitamin D supplements may help as one part of an overall health plan for those who are vitamin D deficient. Furthermore, type 2 diabetes is on the rise in America, and some evidence exists that several factors, including diet and vitamin deficiencies, can lead to this disease, as well as the associated depression.

Vitamin B_{12}

Vitamin B_{12} has the largest and most complex chemical structure of all the vitamins. It is unique among vitamins in that it contains a metal ion, cobalt. *Cobalamin* is the term used to refer to compounds having vitamin B_{12} activity. Methylcobalamin and 5-deoxyadenosylcobalamin are the forms of vitamin B_{12} used in the human body. Most supplements contain cyanocobalamin, which is readily converted to 5-deoxyadenosyl and methylcobalamin. In mammals, cobalamin is a cofactor for only two enzymes, methionine synthase and L-methylmalonyl-CoA mutase.

Methylcobalamin is required for the function of the folate-dependent enzyme, methionine synthase. This enzyme is required for the synthesis of the amino acid, methionine, from homocysteine. Methionine in turn is required for the synthesis of *S*-adenosylmethionine, a methyl group donor used in many biological methylation reactions, including the methylation of a number of sites within DNA and RNA. Vitamin B_{12} deficiency is also believed to contribute to depression. Studies have found that up to 30% of patients hospitalized for depression are deficient in vitamin B_{12}. A study of several hundred females over the age of 65 with physical disabilities found that females deficient in vitamin B_{12} were more likely to experience severe depression than females who were not deficient. A study of over 3000 elderly males and females showed that those with vitamin B_{12} deficiency were almost 70% more likely to experience depression than those with normal vitamin B_{12} levels. The relationship between vitamin B_{12} deficiency and depression is not clear, but it may involve *S*-adenosylmethionine (SAMe). Vitamin B_{12} and folate are required for the synthesis of SAMe, a methyl group donor essential for the metabolism of several neurotransmitters.

Antidepressants—Bringing Out the Big Guns?

People have been trying to cure depression for centuries. Until the 1950s, opioid drugs were used for depression, followed by amphetamines through the 1960s. In 1952, psychiatrist Max Luri was the first to use a dedicated antidepressant, Isoniazid (Figure 8.42), although it had previously been used to fight tuberculosis.

In the decades that followed, there was an explosion in the number and types of drugs marketed to fight this poorly understood disease. A few of the major classes of antidepressant and how they function are shown in the following. Most of them inhibit the reuptake of the neurotransmitters that are associated with happiness and depression.

Reuptake inhibitors are antidepressants that block the reuptake of certain neurotransmitters. Selective serotonin reuptake inhibitors (SSRIs) are currently one of the most popular antidepressants. They block the reabsorption of the neurotransmitter, serotonin. Common examples of these are Prozac™, Zoloft™, and Lexapro™. Norepinephrine reuptake inhibitors (NRIs) block the reuptake of the norepinephrine neurotransmitter. Serotonin–norepinephrine reuptake inhibitors (SNRIs) block both serotonin and norepinephrine reuptake and represent another important modern class of antidepressant. One example of a norepinephrine–dopamine reuptake inhibitor (NDRI) is Wellbutrin™.

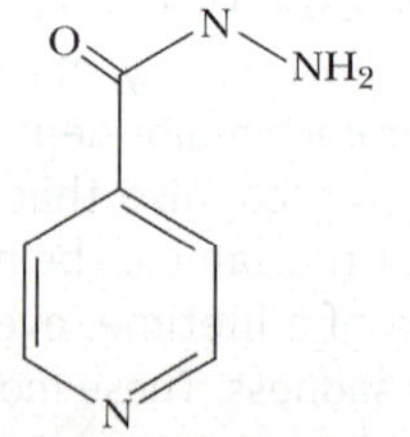

Figure 8.42 Isoniazid.

There are several other classes of antidepressants, and each of the classes described has many examples of drugs in their class. Therefore, you might wonder why depression has not been eliminated altogether. Unfortunately, none of the drugs works as we would expect in a perfect world. Robin Marantz Henig describes in her article titled "Lifting the Black Cloud," which describes an all-too-common case of a patient who spent a year on Paxil™ (an SSRI), but it destroyed her sex drive. She then tried Xanax™, an anti-depression medication that comes from a different class of drugs called benzodiazepines. This brought back her sex drive but had other side effects. Then she tried Paxil again, followed by Lexapro (another SSRI), and then Pristiq™ (an SNRI). Then she went on Zoloft and Wellbutrin, the latter of which was supposed to offset the side effects of the Zoloft. Unfortunately, this trial-and-error approach is common with the prescription of antidepressants. They do not work for everyone and they don't work the same in everyone they affect. The most common and popular since the 1980s and 1990s are the SSRIs and SNRIs, but they fail to work in 30% of cases. They also take on the order of weeks to start working, which is a very long time if you're experiencing depression. During those weeks, the chances of suicide are up to five times higher. Given these issues and others that are seen with current medications, pharmaceutical companies continue to look for better, faster-acting medicines. Such medicines would eliminate the dangerous waiting period as well as just make it more efficient to determine which one is right for a given individual. One route that pharmaceutical companies have tried involved animal models using drugs that are known to be very fast. One such drug is ketamine (Figure 8.43).

Ketamine is an analgesic and a street drug called special K. In large doses it causes hallucinations, and in rodents it

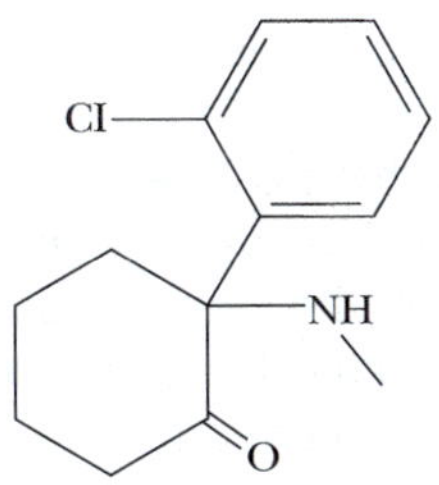

Figure 8.43 Ketamine.

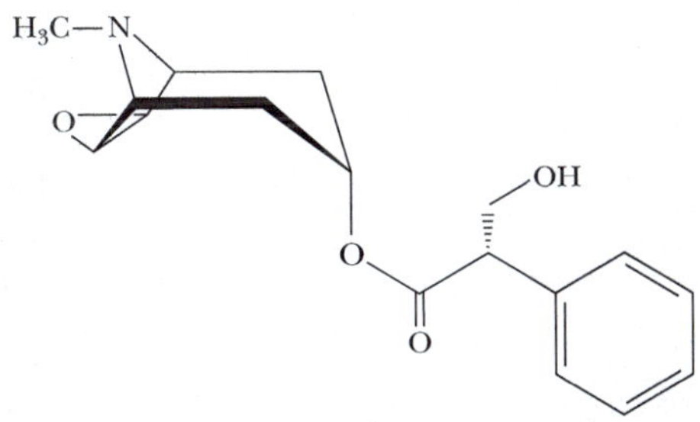

Figure 8.45 Scopolamine.

can be toxic to nerve cells, making it less than an ideal antidepressant for humans. Ketamine has been shown to work very quickly and does almost immediately alleviate symptoms, but it is a serious drug and extreme care must be taken if used. Studies in rats have shown that the drug causes rats to start forming proteins needed to build new synapses between neurons in the prefrontal cortex, an area of the brain that behaves abnormally in depressed animals. Within 24 hours, the rats began to show new synaptic spines along the dendrites, which are projections that receive impulses from other neurons. In several studies, the more spines, the quicker the nerve impulses are transmitted and the less the animal shows signs of depression. In depression, there is atrophy in the prefrontal cortex and the hippocampus. Creation of the synaptic spines is one way that ketamine or other similar drugs are thought to help combat depression. Figure 8.44 shows how dendrites sprout more spines when ketamine is administered.

Further research on ketamine has demonstrated that the neuronal growth is mediated through activation of an enzyme called mTOR, the same enzyme seen in the Chapter 4 Hot Topic on aging. Because many consider ketamine to be too risky

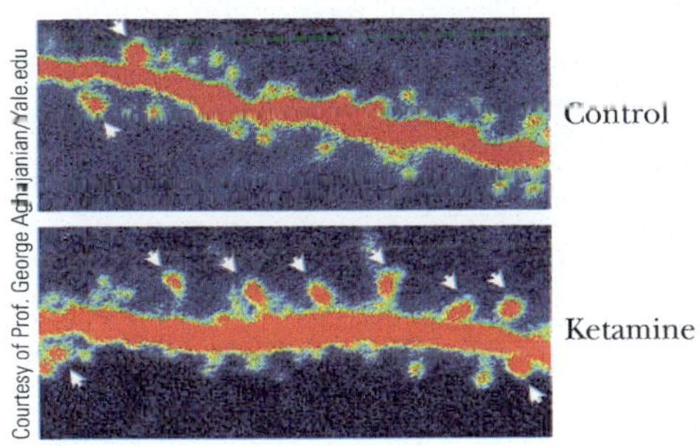

Figure 8.44 In the presence of ketamine, neuronal dendrites are shown to sprout more spines than controls do.

for an antidepressant, scientists are looking for other drugs that activate mTOR. Ketamine stimulates mTOR by preventing glutamate, a principal brain neurotransmitter, from docking to its receptor, called an NMDA (*N*-methyl-D-aspartate) receptor, so finding drugs that block that site is an active area of research.

Scopolamine (Figure 8.45) is another drug that acts quickly and is already in use for another purpose—namely motion sickness.

Scopolamine influences a different system. It blocks the binding of the neurotransmitter acetylcholine to a type of receptor called a muscarinic receptor. In studies with over 40 subjects, intravenous injections of scopolamine were shown to relieve symptoms of depression within three days, with some patients reporting feeling better the next day. However, one challenge is the need for intravenous injections. This makes it not as practical as a pill or a skin patch, but these two techniques have not proved nearly as useful.

Besides the problem of how slowly many antidepressants work, researchers are also trying to tackle the issue that the known drugs do not work for all people. Some companies are targeting other receptor systems, such as the nicotinic receptors, so named because they respond to nicotine as well as acetylcholine, their natural substrate. However, some of the most exciting work involves different approaches instead of just coming up with a drug that will bind a receptor. One is methods to increase neurogenesis in the hippocampus, because it was shown that one of the issues seen with patients experiencing depression was atrophy of the neurons in the brain. It had also been known that the SSRIs and SNRIs already in use act not only by affecting serotonin levels but also by increasing neuronal growth. Additionally, chronic stress has been shown to reduce neurogenesis, while aerobic exercise has been shown to stimulate neurogenesis.

Coming full circle, some unique studies are focusing on serotonin but with a completely different approach. They aim to increase serotonin receptors using gene therapy. The gene being looked at is *p11*, which codes for a protein instrumental in the movement of serotonin receptors to the cell surface. Paul Greengard and colleagues at the Rockefeller Institute showed that in both rodents and humans, individuals with depression-like behaviors also showed low levels of *p11*. Using an adenovirus carrier, Greengard put *p11* directly into the nucleus accumbens of deficient mice and their depressed behavior diminished. Currently studies are underway to see if a similar approach will work in other animal models and eventually in human clinical trials.

Unfortunately, in an estimated 20% of cases of people with depression, none of the methods, ranging from psychotherapy, to medication, to electroconvulsive therapy, are effective. Such cases are referred to as treatment-resistant depression (TRD). Patients with TRD may be candidates for another type of treatment, called deep brain stimulation.

Deep brain stimulation will never be a patient's first choice of treatment because it requires doctors to drill holes in the skull and implant electrodes deep in the brain, in an area called the subcallosal cingulate area, or Brodmann area after a German neuroanatomist who mapped the human brain in 1909. The Brodmann area is connected to many other parts of the brain that have been implicated in depression. As shown in Figure 8.46, the treatment involves sending electrical pulses via electrodes to this area of the brain. Although still in its infancy as a treatment for depression, similar techniques are more widely established for treatment of disorders like Parkinson's disease, where it has been used for more than 100,000 patients.

In many studies, patients immediately noted a lifting of their mental burden once the electrodes were turned on. Furthermore, long-term stimulation seemed to lead to longer-lasting benefits. The unit is

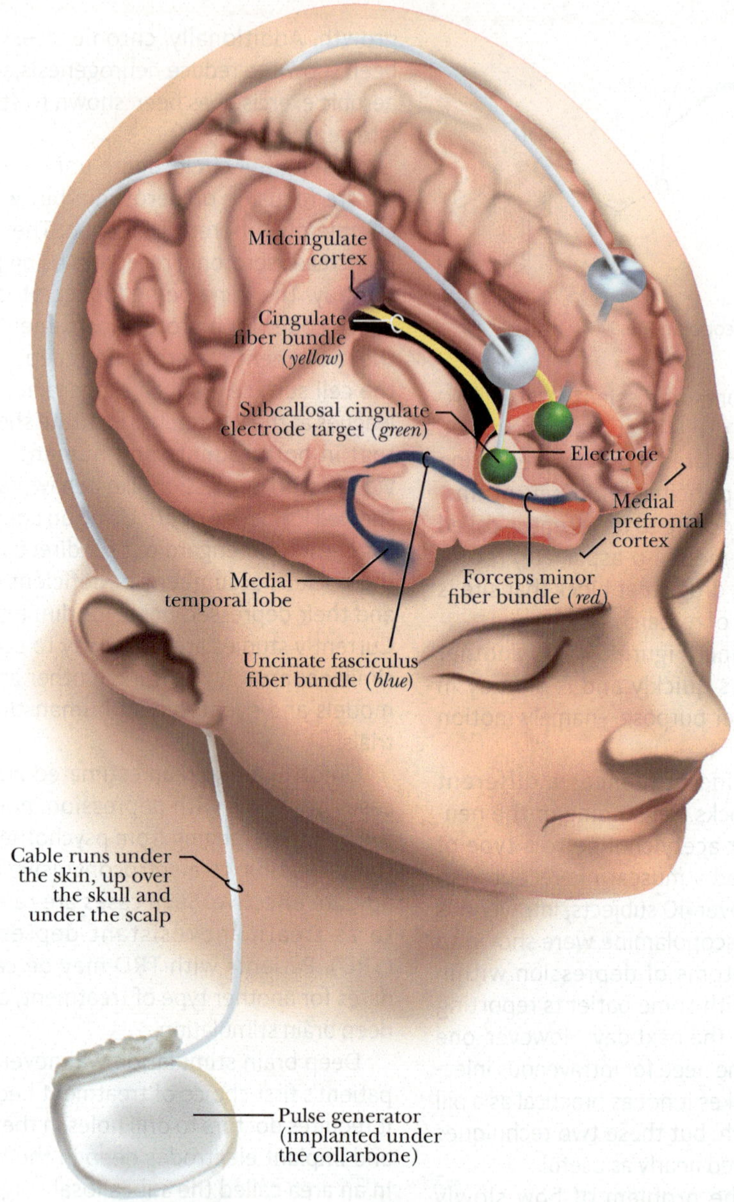

Figure 8.46 Deep brain stimulation. Deep brain stimulation of the subcallosal cingulate area is being used to treat certain cases of depression. A battery is implanted beneath the collarbone, and wires under the skin connect the battery to electrodes in the brain. *(Science, page 71, February 2015.)*

eventually powered by the brain equivalent of a pacemaker.

As seen with other treatments, there is a tremendous range in success with this technique. In different studies, the range of success, which is measured by a clinical response of at least a 50% improvement, varied between 40% and 70%. Researchers are currently trying to establish why there is such a range, but the technique still offers hope to many patients who experience depression, as well as other mental health conditions such as bipolar disorder, schizophrenia, Alzheimer's disease, and even anorexia.

The use of psychedelic drugs such as psilocybin and MDMA is also being extensively researched for use in the treatment of TRD. One way these drugs are thought to work in treating depression is by enhancing neuroplasticity in the brain. They are believed to cause sustained brain reorganization and formation of new neural connections. However, significant work remains to be done in the area psychedelics regarding their safe use, regulation, and cultural acceptance.

This article just scratches the surface of the body of science around happiness and depression, but it should be clear that the brain is a very complicated organ, as is the chemistry associated with it. It is almost ironic that a branch of science that is looking for ways to extend lifespan stumbled on an enzyme, mTOR, the inhibition of which correlates with longer lives, while another branch looking to cure depression found that activating the same enzyme correlates with reducing depression. We can only hope that we are not really left with a choice to make of either long life or happiness.

Summary

The Definition of a Lipid

- Lipids are compounds that are insoluble in water but soluble in nonpolar organic solvents.
- One group of lipids consists of open-chain compounds, each with a polar head group and a long nonpolar tail; this group includes fatty acids, triacylglycerols, phosphoacylglycerols, sphingolipids, and glycolipids.
- Glycerol, fatty acids, and phosphoric acid are frequently obtained as degradation products of lipids.
- A second major group of lipids consists of fused-ring compounds, the steroids.

Fatty Acids

- Fatty acids are long-chain carboxylic acids that may or may not have double bonds in their hydrocarbon portion.

Triacylglycerols

- Triacylglycerols are the storage forms of fatty acids in which fatty acids are esterified to glycerol.

Phosphoacylglycerols

- Phosphoacylglycerols differ from triacylglycerols in having a phosphorus-containing portion esterified to glycerol. These compounds are important components of biological membranes.

Waxes and Sphingolipids

- Waxes are esters of fatty acids and long-chain alcohols.
- Sphingolipids do not contain glycerol but have a long-chain alcohol called sphingosine as part of their structure instead.

Glycolipids

- In glycolipids, a carbohydrate portion is covalently bonded to the lipid.

Steroids

- Steroids are lipids that have a characteristic fused-ring structure. Different steroids vary from each other in the types of functional groups that are bonded to the fused-ring structure.

Lipid Bilayers

- A biological membrane consists of lipids and membrane proteins.
- The lipid part is a bilayer, with the polar head groups in contact with the aqueous interior and exterior of the cell, and the nonpolar portions of the lipid in the interior of the membrane.

Membrane Fluidity

- The presence of unsaturated fatty acids in membrane components leads to greater fluidity than with a preponderance of saturated fatty acids.
- Conversely, the presence of saturated fatty acids and cholesterol tends to stiffen the bilayer.
- Lateral motion of lipid molecules within one layer of a membrane occurs frequently.

Membrane Proteins

- The proteins that occur in membranes can be peripheral proteins, which are found on the surface of the membrane, or integral proteins, which span the lipid bilayer.
- Various structural motifs, such as bundles of α-helices or β-barrels, occur in proteins that span membranes.
- Peripheral proteins are loosely attached to one surface of the membrane by hydrogen bonds or electrostatic attractions, whereas integral proteins are embedded solidly in the membrane.

The Fluid-Mosaic Model of Membrane Structure

- The fluid-mosaic model describes the interaction of lipids and proteins in biological membranes. The membrane proteins and lipids "float" around each other in the lipid bilayer.

The Functions of Membranes

- Three important functions take place in or on membranes—namely transport, catalysis, and receptor properties.
- Transport across the membrane can involve the lipid bilayer as well as the membrane proteins.
- Catalysis is carried out by enzymes bound to the membrane.
- Receptor proteins bind extracellular signals that trigger responses inside the cell.

Membrane Transport

- The most important question about transport of substances across biological membranes is whether the process requires expenditure of energy by the cell.
- In passive transport, a substance moves from a region of higher concentration to one of lower concentration, requiring no expenditure of energy by the cell.
- Active transport requires the energy of ATP hydrolysis to move substances against a concentration gradient. Carrier proteins are also required for active transport.
- The sodium–potassium ion pump is an example of active transport.

Membrane Receptors

- Receptor proteins in the membrane bind biologically important, extracellular substances that trigger a biochemical response in the cell. The interaction between receptor proteins and the active substances to which they bind is very similar to enzyme–substrate recognition.
- The action of a receptor frequently depends on a conformational change in the receptor protein, which causes activation or inactivation of a process inside the cell.

Lipid-Soluble Vitamins and Their Functions

- Lipid-soluble vitamins are hydrophobic, accounting for their solubility properties.
- A derivative of vitamin A plays a crucial role in vision.
- Vitamin D controls calcium and phosphorus metabolism, affecting the structural integrity of bones.
- Vitamin E is an antioxidant.
- The presence of vitamin K is required in the blood-clotting process.

Arachidonic Acid Derivatives

- The unsaturated fatty acid arachidonic acid is the precursor of prostaglandins, leukotrienes, and thromboxanes.
- These compounds have a wide range of physiological activities, such as stimulation of smooth-muscle contraction, induction of inflammation, control of blood pressure, and inhibition of blood-platelet aggregation.

Review Exercises

8-1 The Definition of a Lipid

1. **Recall** Proteins, nucleic acids, and carbohydrates are grouped by common structural features found within their group. What is the basis for grouping substances as lipids?
2. **Reflect and Apply** Name two atomic elements that are found in all lipids and three additional elements that are found in some, but not all, lipids.

8-2 The Chemical Natures of the Lipid Types

3. **Recall** What structural features do a triacylglycerol and phosphatidylethanolamine (PE) have in common? How do the structures of these two types of lipids differ?
4. **Recall** Draw the structure of a phosphoacylglycerol that contains glycerol, oleic acid, stearic acid, and choline.
5. **Recall** What structural features do a sphingomyelin and phosphatidylcholine (PC) have in common? How do the structures of these two types of lipids differ?
6. **Reflect and Apply** What do the structures of phosphatidylinositol and a ganglioside have in common? How do the structures of these two types of lipids differ?
7. **Recall** You have just isolated a pure lipid that contains only sphingosine and a fatty acid. To what class of lipids does it belong?
8. **Recall** What structural features does a sphingolipid have in common with proteins?
9. **Recall** Write the structural formula for a triacylglycerol and name the component parts.
10. **Recall** How does the structure of steroids differ from that of the other lipids discussed here in Chapter 8?
11. **Recall** What are the structural features of waxes? What are some common uses of compounds of this type?
12. **Reflect and Apply** Compare and contrast the structure of *glycoproteins* with *glycolipids*.
13. **Reflect and Apply** Which is more hydrophilic, cholesterol or phospholipids? Defend your answer.
14. **Reflect and Apply** Write a reaction, with structural formulas, for the saponification of a triacylglycerol.
15. **Reflect and Apply** Succulent plants from arid regions generally have waxy surface coatings. Suggest why such a coating is valuable for the survival of the plant.

8-3 Biological Membranes

16. **Recall** Which of the following lipids is *not* found in animal membranes?
    (a) Phosphoacylglycerols
    (b) Cholesterol
    (c) Triacylglycerols
    (d) Glycolipids
    (e) Sphingolipids
17. **Recall** Which of the following statements is (are) consistent with what is known about membranes?
    (a) A membrane consists of a layer of proteins sandwiched between two layers of lipids.
    (b) The compositions of the inner and outer lipid layers are the same in any individual membrane.
    (c) Membranes contain glycolipids and glycoproteins.
    (d) Lipid bilayers are an important component of membranes.
    (e) Most membrane proteins are covalently linked to a lipid.
18. **Recall** Explain two ways that cholesterol influences membrane fluidity in animal cell membranes.
19. **Reflect and Apply** In lipid bilayers, there is an order–disorder transition similar to the melting of a crystal. In a lipid bilayer in which most of the fatty acids are unsaturated, would you expect this transition to occur at a higher temperature, a lower temperature, or the same temperature as it would in a lipid bilayer in which most of the fatty acids are saturated? Why?
20. **Reflect and Apply** Suggest a reason why the cell membranes of bacteria grown at 20°C tend to have a higher proportion of unsaturated fatty acids than the membranes of bacteria of the same species grown at 37°C. In other words, the bacteria grown at 37°C have a higher proportion of saturated fatty acids in their cell membranes.
21. **Reflect and Apply** What is the energetic driving force for the formation of phospholipid bilayers?

8-4 Membrane Proteins

22. **Recall** How can fluorescence techniques be used to monitor the motion of lipids and proteins in membranes?
23. **Recall** Describe two categories of membrane proteins that do *not* span across both lipid bilayers of the membrane.
24. **Recall** Transmembrane proteins and lipid-anchored proteins both interact with lipids in the membrane. What is the difference in the type of bonds formed with lipids for the two categories of membrane proteins?
25. **Biochemical Connections** Perilipin is a protein that binds to the phospholipid monolayer on the surface of lipid droplets. A conserved patch of charged and polar residues is critical for this interaction. Do you predict that perilipin is an integral or peripheral membrane protein? Explain your prediction.
26. **Reflect and Apply** A membrane consists of 50% protein by weight and 50% phosphoglycerides by weight. The average molecular weight of the lipids is 800 Da, and the average molecular weight of the proteins is 50,000 Da. Calculate the molar ratio of lipid to protein.
27. **Reflect and Apply** Suppose that you are studying an integral membrane protein involved in transporting ions in and out of cells.

Would you expect to find the nonpolar residues of the protein in the interior surface where the ions travel or the exterior surface of the protein? Why? Would you expect to find the polar residues of the protein in the interior or the exterior surface of the protein? Why?

28. **Reflect and Apply** Researchers attach a green fluorescent molecule to a sphingolipid and a red fluorescent molecule to phosphatidylinositol and incorporate the molecules within mammalian cells. They report that if they use a laser to destroy the fluorescence of a spot of red signal that the bleached spot rapidly turns red again, but when the laser is applied to an area with green signal, it takes longer for fresh molecules carrying the green signal to return to the area. Explain what you can conclude regarding the fluidity of the cell membranes that contain phosphatidylinositol and sphingolipid.
29. **Reflect and Apply** Which statements are consistent with the fluid-mosaic model of membranes?
    (a) All membrane proteins are bound to the interior of the membrane.
    (b) Both proteins and lipids undergo transverse (flip-flop) diffusion from the inside to the outside of the membrane.
    (c) Some proteins and lipids undergo lateral diffusion along the inner or outer surface of the membrane.
    (d) Carbohydrates are covalently bonded to the outside of the membrane.
    (e) The term *mosaic* refers to the arrangement of just the lipids in the membrane.

8-5 The Functions of Membranes

30. **Reflect and Apply** Suggest a reason why inorganic ions, such as K^+, Na^+, Ca^{2+}, and Mg^{2+}, do not cross biological membranes by simple diffusion.
31. **Recall** Which statements accurately describe how membrane transport works?
    (a) Active transport moves a substance from a region in which its concentration is lower to one in which its concentration is higher.
    (b) All transport requires an input of energy.
    (c) Transport proteins may be involved in bringing substances into cells.
32. **Recall** What happens to the human growth hormone receptor when it binds hGH?
33. **Recall** What is JAK2 and why is it important to the effect of hGH?

8-6 Lipid-Soluble Vitamins and Their Functions

34. **Recall** What is the structural relationship between vitamin D_3 and cholesterol?
35. **Recall** Describe an important chemical property of vitamin E.
36. **Recall** What are isoprene units? What do they have to do with lipids?
37. **Recall** List the fat-soluble vitamins and give a physiological role for each.
38. **Biochemical Connections** What is the role in vision of the cis–trans isomerization of retinal?
39. **Reflect and Apply** List four vitamins that can enter the cell through the plasma membrane using simple diffusion. Explain your reasoning.
40. **Reflect and Apply** Why can some vitamin-K antagonists act as anticoagulants?

8-7 Arachidonic Acid Derivatives

41. **Biochemical Connections** Describe the structure of an omega-3 fatty acid.
42. **Recall** What are the main structural features of leukotrienes?
43. **Recall** What are the main structural features of prostaglandins?
44. **Reflect and Apply** List three classes of compounds derived from arachidonic acid. Suggest some reasons for the amount of biomedical research devoted to these compounds.
45. **Biochemical Connections** Given what you know about the function of prostaglandins, explain why drugs have been developed to control their synthesis.

Further Reading

Barinaga, M. Forging a Path to Cell Death. *Science* **273**, 735–737 (1996). A Research News article describing a process apparently missing in cancer cells, and that depends on interactions among receptor proteins on cell surfaces.

Bayley, H. Building Doors into Cells. *Sci. Amer.* **277** (3), 62–67 (1997). Protein engineering can create artificial pores in membranes for drug delivery.

Beckman, M. Great Balls of Fat. *Science* **311**, 1232–1234 (2006). An article about lipid droplets and their organelle nature.

Bretscher, M. S. The Molecules of the Cell Membrane. *Sci. Amer.* **253** (4), 100–108 (1985). A particularly well-illustrated description of the roles of lipids and proteins in cell membranes.

Brown, M. S., and J. L. Goldstein. A Receptor-Mediated Pathway for Cholesterol Homeostasis. *Science* **232**, 34–47 (1986). A description of the role of cholesterol in heart disease.

Dautry-Varsat, A., and H. F. Lodish. How Receptors Bring Proteins and Particles into Cells. *Sci. Amer.* **250** (5), 52–58 (1984). A detailed description of endocytosis.

Engelman, D. Crossing the Hydrophobic Barrier: Insertion of Membrane Proteins. *Science* **274**, 1850–1851 (1996). A short review of the processes by which transmembrane proteins become associated with lipid bilayers.

Hajjar, D., and A. Nicholson. Atherosclerosis. *Amer. Scientist* **83**, 460–467 (1995). The cellular and molecular basis of lipid deposition in arteries.

Karow, J. Skin So Fixed. *Sci. Amer.* **284** (3), 21 (2001). A discussion of liposomes used to deliver DNA-repair enzymes to skin cells.

Keuhl, F. A., and R. W. Egan. Prostaglandins, Arachidonic Acid, and Inflammation. *Science* **210**, 978–984 (1980). A discussion of the chemistry of these compounds and their physiological effects.

Leonard, T.A., M. Loose, and S. Martens. The Membrane Surface as a Platform that Organizes Cellular and Biochemical Processes. *Dev. Cell* **58**, 1315–1332 (2023). This review summarizes current understanding of the biochemistry of membrane-localized reactions, with a focus on how membranes organize and regulate various cellular and viral processes.

Wells, J.A. and A. A. Kossiakoff. New Tricks for an Old Dimer. *Science*, **344**, 703–704 (2014).

Wood, R. D., M. Mitchell, J. Sgouros, and T. Lindahl. Human DNA Repair Genes. *Science* **291** (5507), 1284–1289 (2001).

Youness, R.A., A. Dawoud, O. ElTahtawy, and M. A. Farag. Fat-Soluble Vitamins: Updated Review of their Role and Orchestration in Human Nutrition Throughout Life Cycle with Sex Differences. *Nut. & Met.* **19**, Article Number: 60 (2022). This review highlights the roles of fat-soluble vitamins in physiological processes such as immune regulation, vision, bone, and mental health, and discusses their potential for preventing various diseases.

Hot Topic

Lozano, A. M., and H. S. Maybert. Treating Depression at the Source: Electrical Stimulation Deep Within the Brain May Alleviate Devastating Mood Disorders. *Sci. Amer.*, 68–73 (February, 2015).

Marantz Henig, R. Lifting the Black Cloud. *Sci. Amer.* 66–71 (March 2012).

Kringelbach, M. L., and K. C. Berridge. The Joyful Mind. *Sci. Amer.* 40–45 (August 2012).

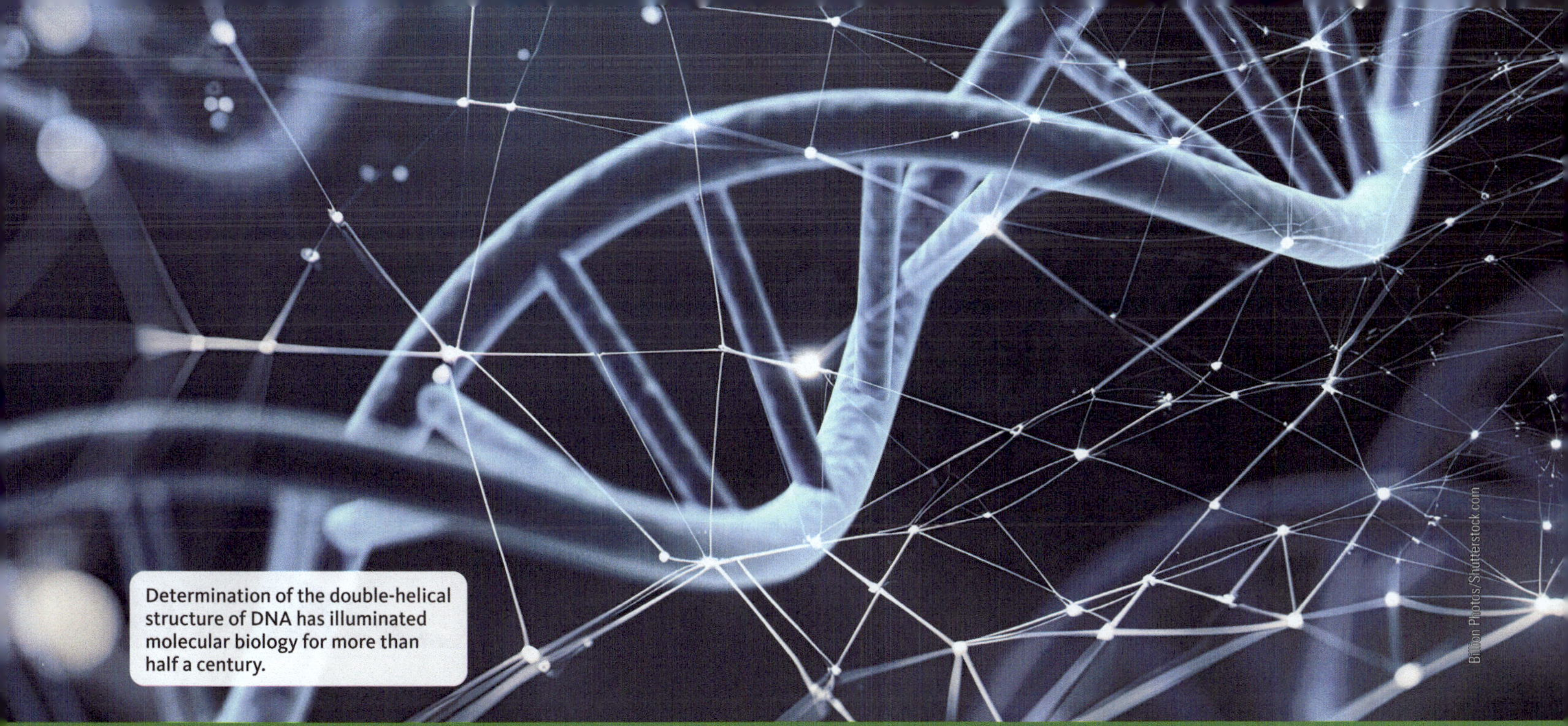

Determination of the double-helical structure of DNA has illuminated molecular biology for more than half a century.

Billion Photos/Shutterstock.com

Nucleic Acids: How Structure Conveys Information 9

9-1 Levels of Structure in Nucleic Acids

Learning Objectives

After reading this section, students will be able to:

LO 9.1.1 Describe the four levels of structure in nucleic acids.

LO 9.1.2 Distinguish between DNA and RNA.

In Chapter 4, we identified four levels of structure—primary, secondary, tertiary, and quaternary—in proteins. **Nucleic acids** can be viewed as having the same levels of structure, although the specific details are different. The *primary structure* of nucleic acids is the order of bases in the polynucleotide sequence, and the *secondary structure* is its base-pairing features. As seen with proteins, the *tertiary structure* of nucleic acids is their three-dimensional shape. The *quaternary structure* is the interactions between nucleic acids or between nucleic acids and proteins.

There are two principal types of nucleic acids: deoxyribonucleic acid (DNA) and ribonucleic acid (RNA). The main difference between the two types is one fewer hydroxyl group on the sugar moieties in DNA (hence the deoxy- prefix), and this is seen in Section 9.2. Other important differences between DNA and RNA appear in their secondary and tertiary structures, so we describe these structural features separately for DNA and for RNA.

Chapter Outline

Pioneers in Biochemistry • Libin T. Cheng

Libin T. Cheng (1900–2010) was a Chinese scientist who pioneered the study of biochemistry and nutrition in China. He did his graduate studies in American universities, where he met many of the early luminaries of biochemistry. Later he was instrumental in establishing biochemistry as a field of study in China, where he founded a biochemistry research institute in the Central Medical School and made biochemistry a major at Nanjing University.

Cheng focused his postgraduate research on the biochemical properties of soybeans, one of the key sources of protein in the Chinese diet. He went on to apply this research to the question of how to improve the nutrition of the nation. He also used an experimental approach to counter long-held folk beliefs about food, notably disproving the idea that certain foods could be poisonous in combination.

In his 70s, Cheng turned his attention to the biochemistry of aging. He investigated how aging brought changes to enzymes, nucleic acids, blood cells, and organs. The result of his cellular and molecular research was a theory of a metabolic basis for the aging mechanism. Cheng must have learned something significant from his research into aging, given that he lived to the age of 110.

National Library of Medicine

9-2 The Covalent Structure of Polynucleotides

Learning Objectives

After reading this section, students will be able to:

LO 9.2.1 Distinguish between the three components of nucleotides.

LO 9.2.2 Draw segments of RNA and DNA.

Polymers are molecular chains of small units called monomers. The monomers of nucleic acids are **nucleotides** (recall from Section 3-4 that the monomers of peptides and proteins are amino acids). An individual nucleotide consists of three parts—a nitrogenous base, a sugar, and a phosphate—all of which are covalently bonded together.

The order of bases in the nucleic acids of DNA contains the information necessary to produce the correct sequence of bases in RNA and the correct amino acid sequence in the cell's proteins.

Nucleic acids organic substances in living cells, especially DNA or RNA, whose molecules consist of nucleotides linked together

Polymers molecular chains of small units called monomers

nucleotides purine or pyrimidine bases bonded to sugars (ribose or deoxyribose), which are bonded, in turn, to phosphate groups

9-2A Components of Nucleotides

The first component of a nucleotide is the **nucleobase**, and there are two types—*pyrimidines* and *purines* (Figure 9.1). In this case, the word *base* does not refer to an alkaline compound, such as NaOH; rather, it refers to a one- or two-ring nitrogenous aromatic compound. Three **pyrimidine bases** (single-ring aromatic compounds)—*cytosine, thymine,* and *uracil*—commonly occur. Cytosine is found both in RNA and in DNA. Uracil occurs only in RNA. In DNA, thymine is substituted for uracil; thymine is also found to a small extent in some forms of RNA. The common **purine bases** (double-ring aromatic compounds) are *adenine* and *guanine,* both of which are found in RNA and in DNA (Figure 9.1). To help you distinguish pyrimidines from purines, keep in mind that *pyrimidine* is the longer name but the smaller base (one ring), whereas *purine* is the shorter name but the larger base (two rings). To learn the structures of each of the five bases in Figure 9.1, recall that carbon usually forms four covalent bonds, nitrogen three, and oxygen two. Finally, the chemical names of the bases are as follows:

nucleobase nitrogen-containing aromatic compounds that make up the coding portion of nucleic acids

pyrimidine bases nitrogen-containing aromatic compounds that contain a six-membered ring

purine bases nitrogen-containing aromatic compounds that contain a six-membered ring fused to a five-membered ring

- Cytosine: 2-oxy-4-aminopyrimidine
- Thymine: 2,4-dioxy-5-methylpyrimidine

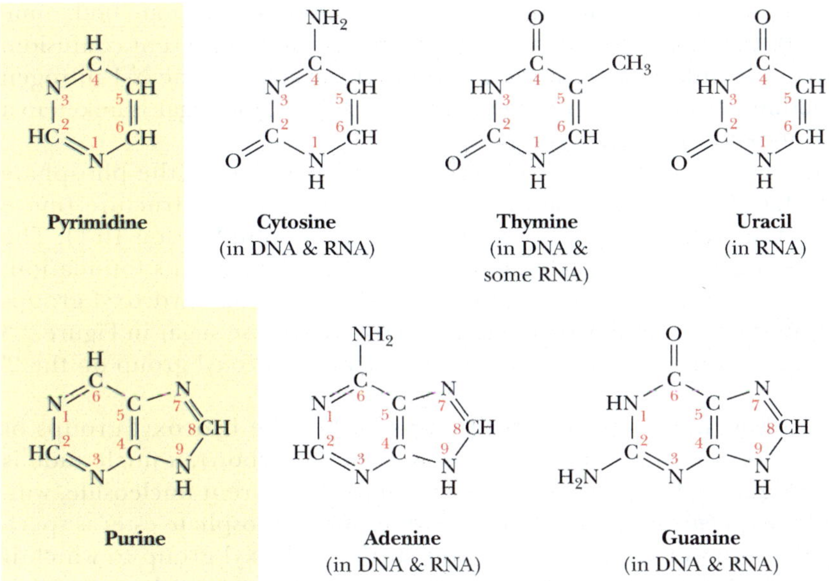

Figure 9.1 Structures of the common nucleobases. The structures of pyrimidine and purine are shown for comparison.

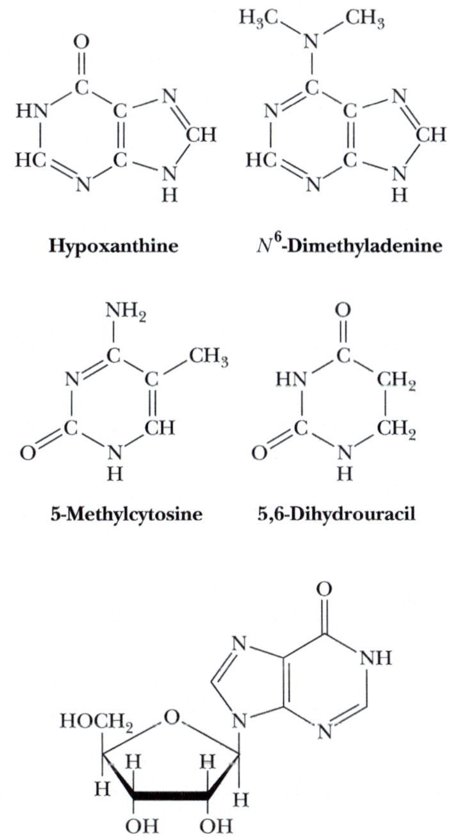

Inosine, an uncommon nucleoside

Figure 9.2 Structures of some of the less common nucleic acid bases. When hypoxanthine is bonded to a sugar, the corresponding compound is called inosine.

- Uracil: 2,4-dioxypyrimidine
- Adenine: 6-aminopurine
- Guanine: 2-amino-6-oxypurine

In addition to these five commonly occurring bases, there are "unusual" bases, with slightly different structures, that are found principally—but not exclusively—in transfer RNA (Figure 9.2). In many cases, the base is modified by methylation.

The next component of nucleotides is sugar moieties (ribose in RNA nucleotides and deoxyribose in DNA nucleotides). When one of these sugars forms a glycosidic linkage to a nucleobase, a **nucleoside** is formed. Glycosidic linkages and the stereochemistry of sugars are discussed in detail in Section 16-2. For now, it is sufficient to say that a *glycosidic bond* links a sugar and some other moiety. When the sugar is β-D-ribose, the resulting compound is a **ribonucleoside**; when the sugar is β-D-deoxyribose, the resulting compound is a **deoxyribonucleoside** (Figure 9.3).

nucleoside a structure in which a purine or pyrimidine base has formed a glycosidic bond to a sugar (ribose or deoxyribose)

ribonucleoside a compound formed when a nucleobase forms a glycosidic bond with ribose

deoxyribonucleoside a compound formed when a nucleobase forms a glycosidic bond with deoxyribose

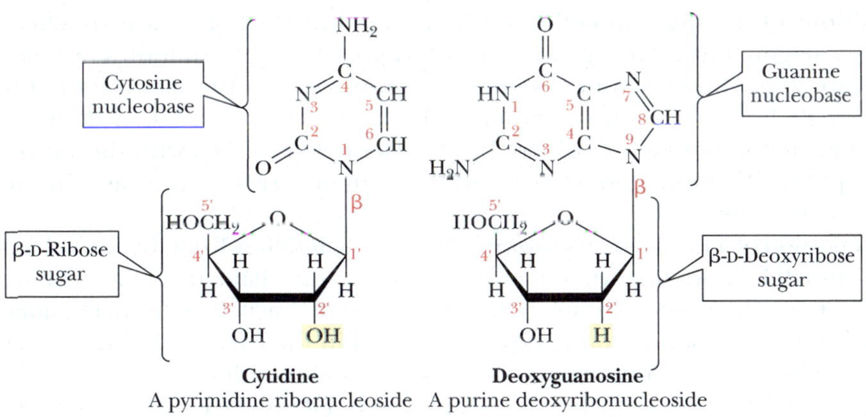

Figure 9.3 A comparison of the structures of a ribonucleoside and a deoxyribonucleoside. A nucleoside consists of a sugar and a base; it does not have a phosphate group in its structure.

The ring atoms of the base and the carbon atoms of the sugar are both numbered, with the numbering of the sugar atoms primed to prevent confusion. The glycosidic linkage is from the 1' carbon of the sugar to the N-1 nitrogen of pyrimidines or to the N-9 nitrogen of purines. Thus, the sugar is linked to a nitrogen in both cases (an *N*-glycosidic bond).

Before moving to the third component of nucleotides (the phosphate groups), let's look more closely at the ribose sugar moiety structure (more details of the cyclic structure of sugars are presented in Section 16-1). The β-D-ribose sugar in Figure 9.3 has a five-membered ring as its foundation, where an oxygen atom is part of the ring. There are also hydroxyl groups present on the 2', 3', and 5' carbons. The β-D-deoxyribose sugar in Figure 9.3 is very similar, the difference being that it lacks a hydroxyl group on the 2' carbon.

When phosphoric acid is esterified to one of the hydroxyl groups of the sugar portion of a nucleoside (usually the 5' carbon), a nucleotide is formed (Figure 9.4). A nucleotide is named for the parent nucleoside, with the suffix *-monophosphate* added; the position of the phosphate ester is specified by the number of the carbon atom at the hydroxyl group to which it is esterified—for instance, adenosine 3'-monophosphate or deoxycytidine 5'-monophosphate.

The 5' nucleotides are most commonly encountered in nature. If additional phosphate groups form anhydride linkages to the first phosphate, the corresponding nucleoside diphosphates and triphosphates are formed. Recall this point from Section 1-2. These compounds are also nucleotides.

9-2B Polymerization of Nucleotides

3',5'-phosphodiester bond a covalent linkage in which phosphoric acid is esterified to the 3' hydroxyl of one nucleoside and the 5' hydroxyl of another nucleoside; this type of bond forms the backbone of nucleic acids

The polymerization of nucleotides gives rise to nucleic acids. The linkage between monomers in nucleic acids involves the formation of two ester bonds by phosphoric acid. The hydroxyl groups to which the phosphoric acid is esterified are those bonded to the 3' and 5' carbons on adjacent residues. The resulting repeated linkage is a **3',5'-phosphodiester bond**. The nucleotide residues of nucleic acids are numbered from the 5' end, which normally carries a phosphate group, to the 3' end, which normally has a free hydroxyl group.

Figure 9.5 shows the structure of a fragment of an RNA chain. The *sugar–phosphate backbone* repeats itself down the length of the chain. Therefore, the most important feature of the structure of nucleic acids is the sequence of the bases. Abbreviated forms of the structure can be written to convey this essential information. In one system of notation, single letters, such as A, G, C, U, and T, represent the individual bases. Vertical lines show the positions of the sugar moieties to which the individual bases are attached, and a diagonal line through the letter P represents a phosphodiester bond (Figure 9.5). However, an even more common system of notation uses only the single letters to show the order of the bases. The sequence of the RNA oligonucleotide in Figure 9.5 can be represented as ACGU, with the sugar–phosphate backbone understood and the sequence read left to right in the 5' to 3' direction.

A portion of a DNA chain is shown in Figure 9.6. It differs from the RNA chain in Figure 9.5 in two important ways: the sugars in the chain are 2'-deoxyribose rather than ribose, and thymine is present instead of uracil. In abbreviated notation, the deoxyribonucleotide can be specified in the usual manner. A letter d could be added to indicate a deoxyribonucleotide residue; for example, dT could be substituted for T, and the DNA chain in Figure 9.6 could be specified as d(TGCA). However, given that the sequence must refer to DNA because of the presence of thymine, the sequence TGCA is unambiguous and works well to represent this DNA chain.

A Ribonucleotides

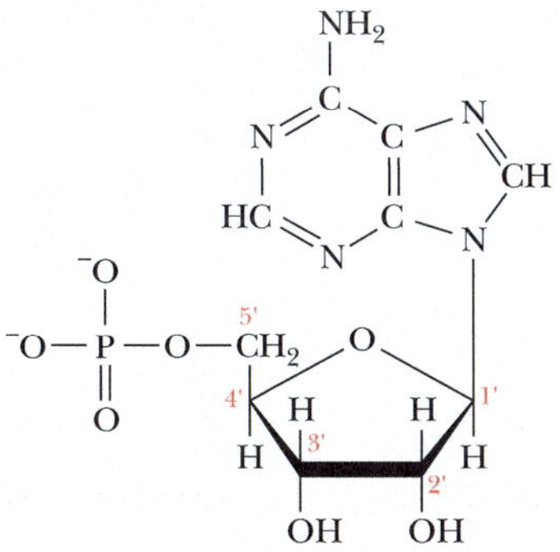

Adenosine 5'-monophosphate

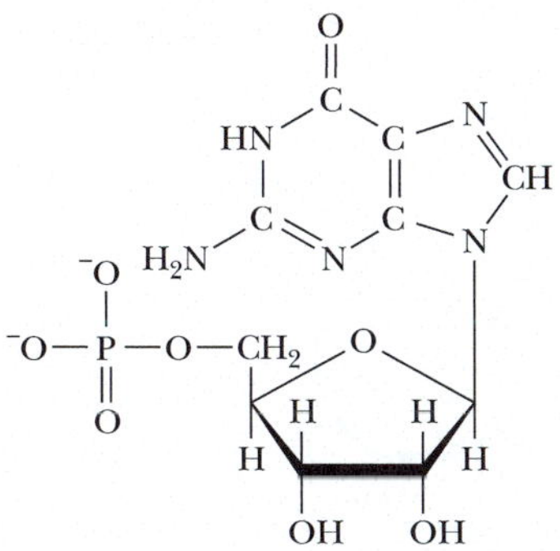

Guanosine 5'-monophosphate

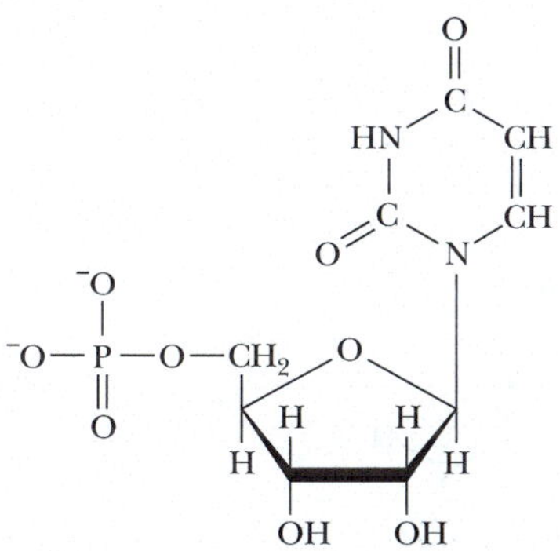

Uridine 5'-monophosphate

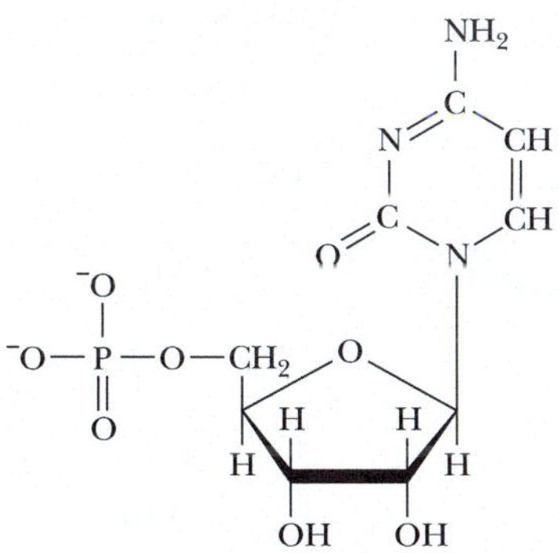

Cytidine 5'-monophosphate

B Deoxyribonucleotides

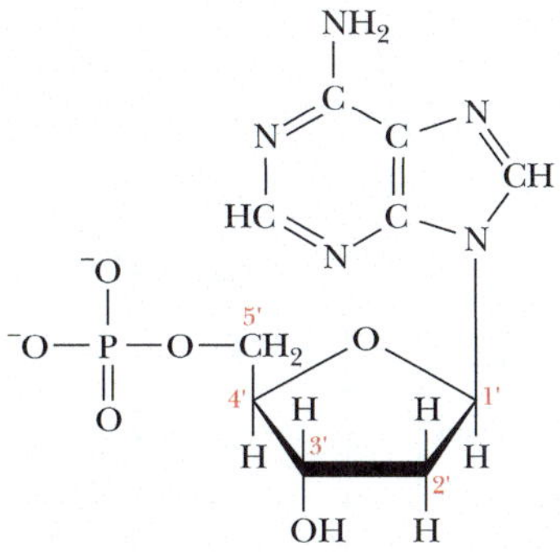

Deoxyadenosine 5'-monophosphate

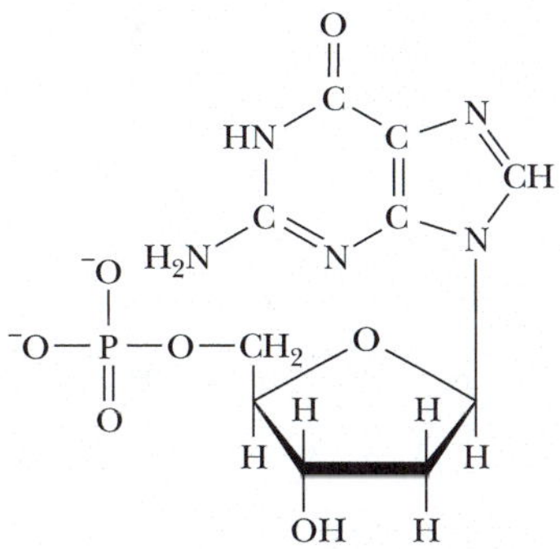

Deoxyguanosine 5'-monophosphate

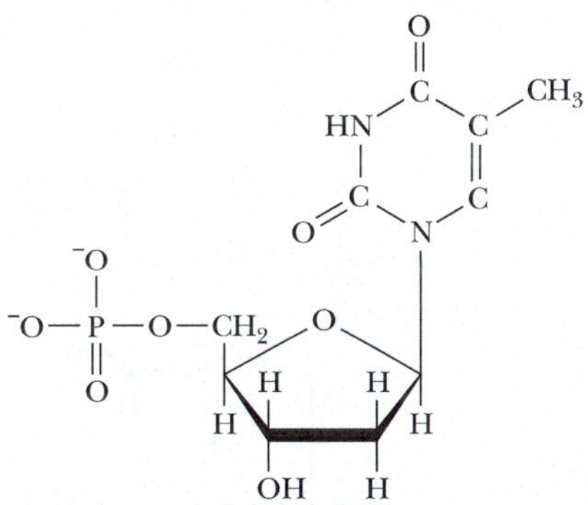

Deoxythymidine 5'-monophosphate

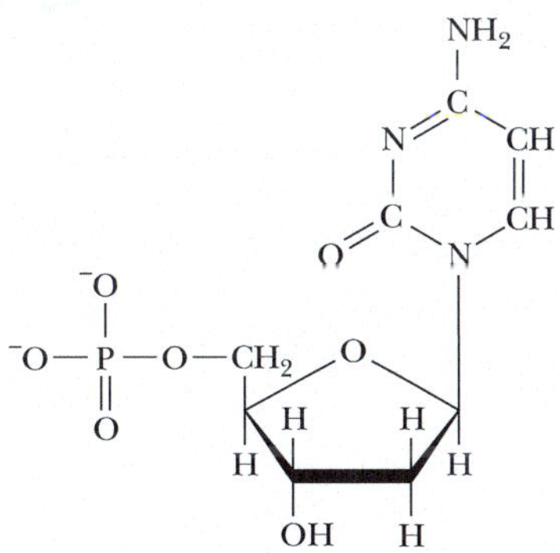

Deoxycytidine 5'-monophosphate

Figure 9.4 The structures and names of the commonly occurring nucleotides. Each nucleotide has a phosphate group in its structure. All structures are shown in the forms that exist at pH 7.

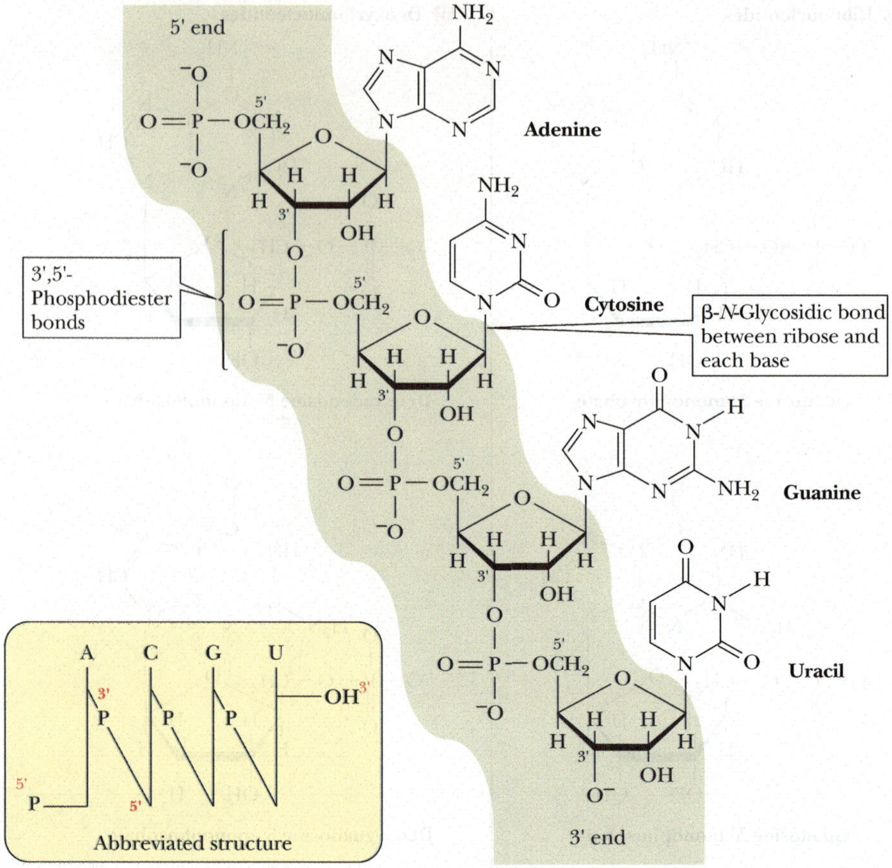

Figure 9.5 A fragment of an RNA chain. The sequence of nucleotides, read from the 5' end (top) to the 3' end (bottom), is ACGU.

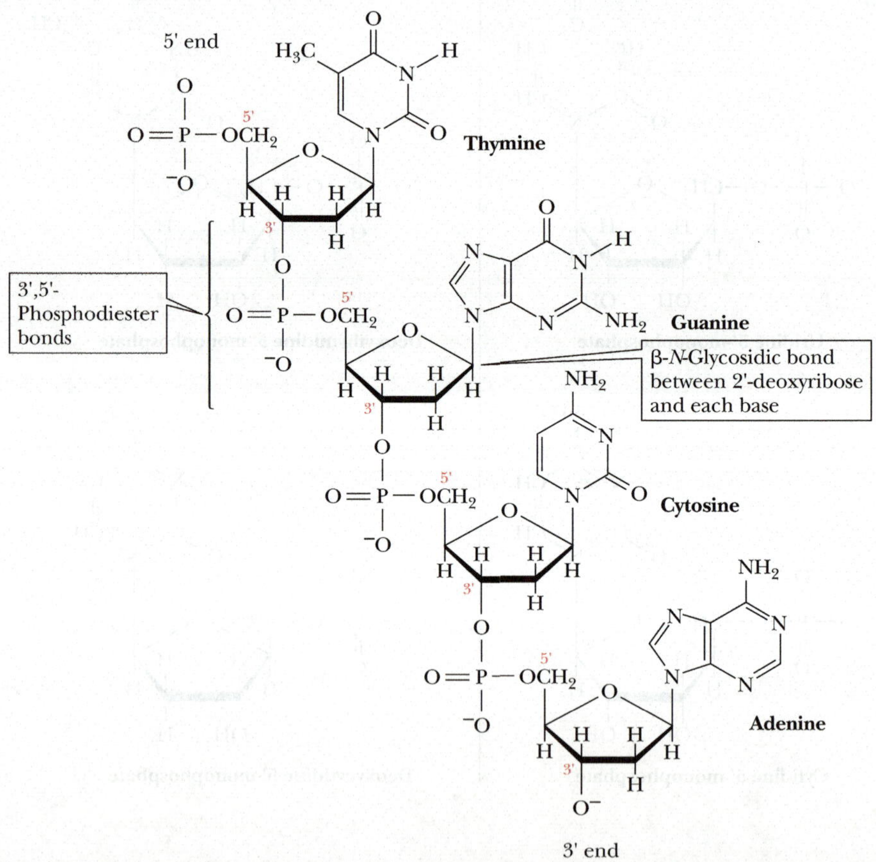

Figure 9.6 A portion of a DNA chain. The sequence of nucleotides, read from the 5' end (top) to the 3' end (bottom), is TGCA.

Biochemical Connections 9A describes an interesting problem between science and law when it comes to the "ownership" of genes.

9A Biochemical Connections • Law

Who Owns Genes?

Following the explosion in information that came from the Human Genome Project (see Biochemical Connections 9B), commercial firms, universities, and even government agencies began to look for patents on genes, which began a long philosophical and legal battle. Human cells have about 24,000 genes, which are the blueprint for the 100 trillion cells in our body. At one point, about 20% of the human genome had been patented.

How could a company patent a biological entity? Well, they were not able to actually patent you or your genes—at least not the ones you carry around. What was patentable was purified DNA containing the sequence of the gene and techniques that allow the study of the genes. The idea of patenting information began with a landmark case in 1972, when Ananda M. Chakrabarty, a General Electric engineer, filed for a patent on a strain of *Pseudomonas* bacteria that could break down oil slicks more efficiently. He experimented with the bacteria, getting them to take up DNA from plasmids that conferred the cleanup ability. The patent office rejected the patent on the grounds that products of nature and live organisms cannot be patented. However, the battle was not over, and in 1980, the U.S. Supreme Court heard the appeal in the same year that the techniques of molecular biology and recombinant DNA technology really began to take off. Chief Justice Warren Burger declared arguments against patenting life irrelevant by stating "anything under the sun that is made by man" could be patented. The ruling was close—only 5–4 in favor of Chakrabarty. A patent on a cloned gene or the protein it produces gave the owner exclusivity in marketing the protein, such as insulin or erythropoietin. Figure 9.7 shows how a chromosome map of patented genes looked at one point.

Many issues stir the controversy. Proponents for the patent system point out that it takes money to drive research. Companies will not want to invest hundreds of thousands to millions of dollars in research if they cannot get a tangible gain. Allowing them to patent a product means they can eventually recover their investment. Opponents believe a patent on what amounts to information stifles more research and even prevents the advancement of medicine. If a company holds the patent to a gene known to be involved in a disease, then others cannot study it effectively and perhaps come up with better or cheaper treatments. The latter point of view came under intense scrutiny because patents on diagnostic genes inhibit both research and clinical medicine. At the heart of the conflict were patents for two genes related to breast cancer, *BRCA1* and *BRCA2*, both owned by Myriad Genetics, Inc., of Salt Lake City at one point. In 2009, a group of patients, doctors, and research professionals brought a suit to invalidate those patents. They argued that the two genes are "products of nature" and should never have been patented in the first place.

Opponents of gene patents claimed a big victory in June 2013, when the U.S. Supreme Court ruled against Myriad Genetics in the suit of *BRCA1* and *BRCA2*, stating that human genes cannot be patented. So, who owns our genes? At the moment, we do.

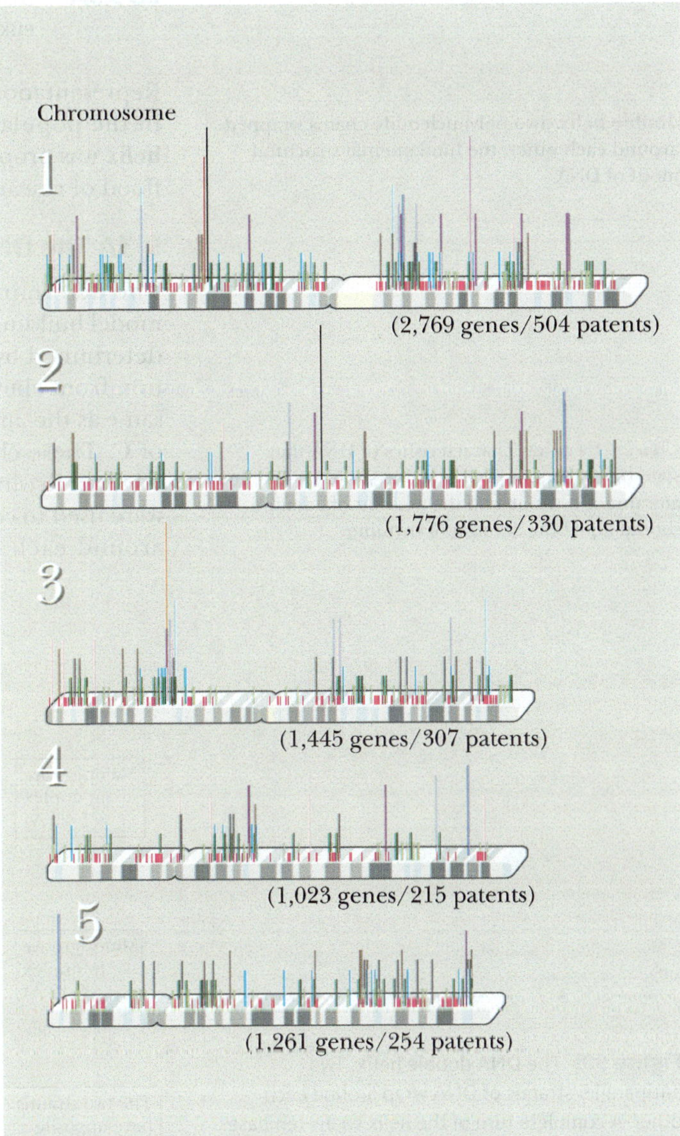

Figure 9.7 This chromosome map offers an indication of how often genes had been patented in the United States. Each colored bar represents the number of patents in a given segment of a chromosome, which can contain several genes. Previous patents could claim multiple genes, and one gene used to be able to receive multiple patents. *(Laurie Grace)*

9-3 The Structure of DNA

Learning Objectives

After reading this section, students will be able to:

LO 9.3.1 Describe characteristics of the DNA double helix.

LO 9.3.2 Describe hydrogen bonding in standard Watson–Crick base pairs.

LO 9.3.3 Distinguish between three different conformations of DNA.

LO 9.3.4 Distinguish between DNA supercoiling characteristics in prokaryotes and eukaryotes.

double helix two polynucleotide chains wrapped around each other; the fundamental structural motif of DNA

Representations of the double-helical structure of DNA have become common in the popular press as well as in the scientific literature. When the **double helix** was proposed by James Watson and Francis Crick in 1953, it touched off a flood of research activity, leading to great advances in molecular biology.

9-3A The DNA Double Helix

Chargaff's rules characteristics of DNA that state that the amount of guanine should equal the amount of cytosine and the amount of adenine should equal the amount of thymine

The determination of the double-helical structure was based primarily on model building and X-ray diffraction patterns. Information from X-ray patterns determined by Rosalind Franklin and Maurice Wilkins was added to information from chemical analyses that showed that the amount of A was always the same as the amount of T, and that the amount of G always equaled the amount of C. These characteristics are known as **Chargaff's rules**, after the Austrian scientist Erwin Chargaff, who proposed them. Both of these lines of evidence were used to conclude that DNA consists of two polynucleotide chains wrapped around each other to form a double helix (Figure 9.8). Hydrogen bonds

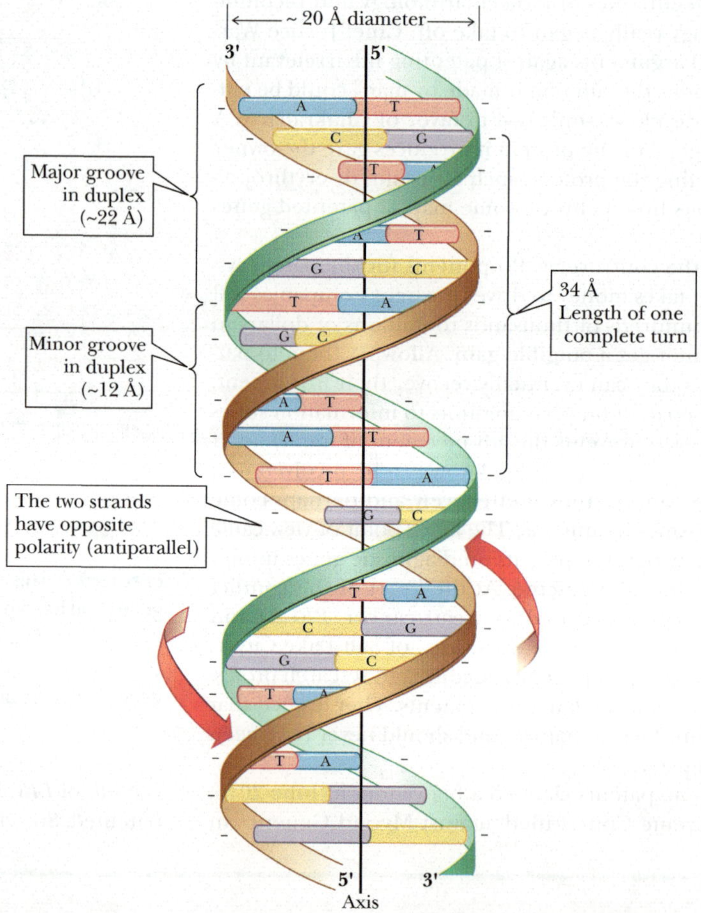

Figure 9.8 The DNA double helix. Two antiparallel strands of DNA wrap around each other. A complete turn of the helix spans ten base pairs, covering a distance of 34 Å (3.4 nm). The individual base pairs are spaced 3.4 Å (0.34 nm) apart. The places where the strands cross hide base pairs that extend perpendicular to the viewer. The inside diameter is 11 Å (1.1 nm), and the outside diameter is 20 Å (2.0 nm). Within the cylindrical outline of the double helix are two grooves: a small one (called the minor groove) and a large one (called the major groove). Both are large enough to accommodate polypeptide chains. The minus signs alongside the strands represent the many negatively charged phosphate groups along the entire length of each strand.

between bases on opposite chains determine the alignment of the helix, with the paired bases lying in planes perpendicular to the helix axis. The sugar–phosphate backbone is the outer part of the helix. The chains run in antiparallel directions, one 3' to 5' and the other 5' to 3'.

The X-ray diffraction pattern of DNA demonstrated the helical structure and the diameter. The combination of evidence from X-ray diffraction and chemical analysis led to the conclusion that the base pairing is *complementary,* meaning that adenine pairs with thymine and that guanine pairs with cytosine. Because complementary base pairing occurs along the entire double helix, the two chains are also referred to as *complementary strands.* By 1953, studies of the base composition of DNA from many species had already shown that, to within experimental error, the mole percentages of adenine and thymine (moles of these substances as percentages of the total) were equal; the same was found to be the case with guanine and cytosine. An adenine–thymine (A–T) base pair forms two hydrogen bonds between the bases, whereas a guanine–cytosine (G–C) base pair forms three (Figure 9.9).

The inside diameter of the sugar–phosphate backbone of the double helix is about 11 Å (1.1 nm). The distance between the points of attachment of the bases to the two strands of the sugar–phosphate backbone is the same for the two base pairs (A–T and G–C), about 11 Å (1.1 nm), which allows for a double helix with a smooth backbone and no overt bulges. Base pairs other than A–T and G–C are possible, such as A–C or G–T, but they do not have the correct hydrogen-bonding pattern or the right dimensions (purine–purine or pyrimidine–pyrimidine pairs) to allow for a smooth double helix.

The outside diameter of the helix is 20 Å (2 nm). The length of one complete turn of the helix along its axis is 34 Å (3.4 nm) and contains 10 base pairs. The atoms that make up the two polynucleotide chains of the double helix do not completely fill an imaginary cylinder around the double helix;

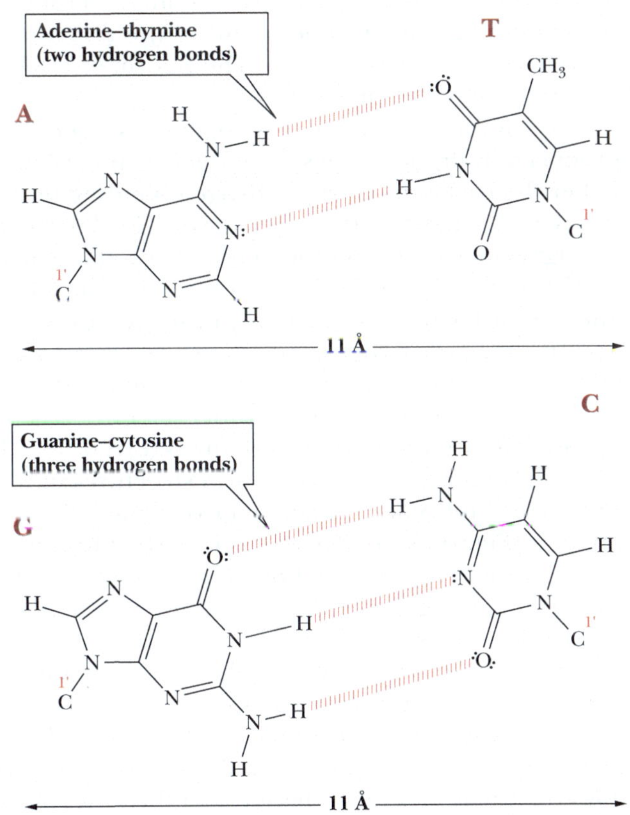

Figure 9.9 Base pairing. The adenine–thymine (A–T) base pair has two hydrogen bonds, whereas the guanine–cytosine (G–C) base pair has three hydrogen bonds. Hydrogen bonds are shown as red vertical lines.

major groove the larger of two empty spaces in an imaginary cylinder that encloses the DNA double helix

minor groove the smaller of two empty spaces in an imaginary cylinder that encloses the DNA double helix

they leave empty spaces known as grooves. There is a large **major groove** and a smaller **minor groove** in the double helix; both can be sites at which drugs or polypeptides bind to DNA (see Figure 9.8). At neutral, physiological pH, each phosphate group of the backbone carries a negative charge. Positively charged ions, such as Na^+ or Mg^{2+}, and polypeptides with positively charged side chains must be associated with DNA in order to neutralize the negative charges. Eukaryotic DNA, for example, is complexed with histones, which are positively charged proteins that have an abundance of lysine and arginine residues (Figure 3.3B).

9-3B DNA Conformations

B-DNA the most common form of the DNA double helix

A-DNA a form of a DNA double helix characterized by having 11 base pairs per turn and major and minor grooves with dimensions that are more similar to each other than those of B-DNA

Z-DNA a form of DNA that is a left-handed helix; it occurs naturally under certain circumstances

The form of DNA that we have been discussing so far is called **B-DNA**. It is thought to be the principal form that occurs in nature. However, other conformations can occur, depending on conditions such as the nature of the positive ion associated with the DNA and the specific sequence of bases. One of those other forms is **A-DNA**, which has 11 base pairs for each turn of the helix. Its base pairs are not perpendicular to the helix axis, but lie at an angle of about 20° to the perpendicular (Figure 9.10). An important shared feature of A-DNA and B-DNA is that both are right-handed helices; that is, the helix winds upward in the direction in which the fingers of the right hand curl when the thumb is pointing upward (Figure 9.11). The A form of DNA was originally found in dehydrated DNA samples, and many researchers believed that the A form was an artifact of DNA preparation. DNA:RNA hybrids, where one strand is DNA and the other is RNA, can adopt an A formation because the 2'-hydroxyl on the ribose prevents an RNA helix from adopting the B form. RNA:RNA hybrids may also be found in the A form.

Another variant form of the double helix, **Z-DNA**, is left-handed; it winds in the direction of the fingers of the left hand (Figure 9.11). Z-DNA is known to occur in nature, most often when there is a sequence of alternating purine–pyrimidine pairs, such as CGCGCG. Sequences with 5-methylcytosine (cytosine methylated at the number 5 position of the pyrimidine ring; see Figure 9.2) can also be found in the Z form. It may play a role in the regulation of gene expression. The Z form of DNA is also a subject of active research among biochemists.

The Z form of DNA can be thought of as a derivative of the B form of DNA, produced by flipping one side of the backbone 180° without having to break either the backbone or the hydrogen bonding of the complementary bases. Figure 9.12 shows how this might occur. The Z form of DNA gets its name from the zigzag look of the phosphodiester backbone when viewed from the side.

base stacking intrastrand interactions between bases that are next to each other in a DNA chain

Although it is easy to focus completely on the interstrand base pairing and the order of bases in DNA, other features of DNA structure are just as important. The ring portions of the DNA bases are very hydrophobic, and they can form intrastrand interactions with each other via hydrophobic bonding (van der Waals interactions) of their pi-cloud electrons. These intrastrand interactions are usually referred to as **base stacking**, and even single-stranded DNA tends to form structures in which the bases can stack. In standard B-DNA, each base pair is rotated 32° with respect to the preceding one (Figure 9.13). This form is perfect for maximal base pairing, but it is not optimal for maximal overlap of the bases. In addition, the edges of the bases that are exposed to the minor groove must come into contact with water in this form.

Many of the bases twist in a characteristic way, called *propeller twist* (Figure 9.14). In this form, the base-pairing distances are less optimal, but the base stacking is more optimal, and water is eliminated from the minor-groove contacts with the bases. Besides twisting, bases also slide sideways, allowing them to interact better with the bases above and below them. The

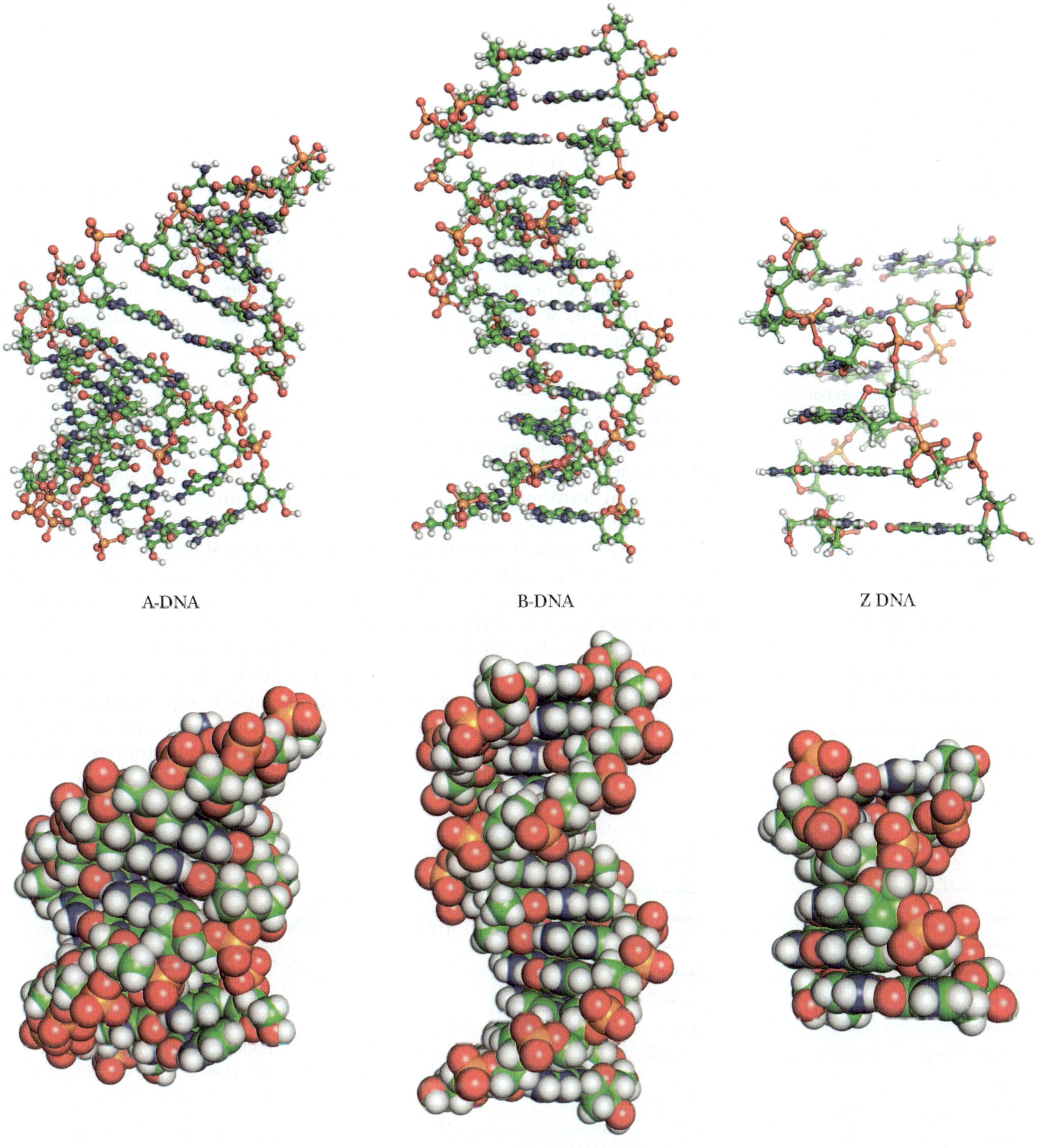

Figure 9.10 Comparison of three forms of DNA. Ball-and-stick models (top) and space-filling models (bottom) for A-, B-, and Z-DNA, respectively.

twist and slide depend on which bases are present, and researchers have identified that a basic unit for studying DNA structure is actually a dinucleotide with its complementary pairs. This is called a *step* in the nomenclature of DNA structure. For example, Figure 9.14 depicts an AG/CT step, which tends to adopt a different structure from a GC/GC step.

As more and more is learned about DNA structure, it is evident that the standard B-DNA structure, while a good model, does not really describe local

Figure 9.11 **Left- and right-handed helices.** In each case, the helix turns the same direction as the fingers of a left or right hand would curl toward the thumb.

regions of DNA very well. Many DNA-binding proteins recognize the overall structure of a sequence of DNA, which depends upon the sequence but is not the DNA sequence itself.

9-3C DNA Supercoils

The DNA molecule has a length considerably greater than its diameter; it is not completely stiff and can fold back on itself in a manner similar to that of proteins as they fold into their tertiary structures. The double helix we have discussed so far is relaxed, which means that it has no twists in it other than the helical twists themselves. Pictures of relaxed DNA make it easier to understand the basics of DNA structure, but in reality, most DNA double helices have more or less twisting than the relaxed form. These deviations from normal twisting, which are called **DNA supercoils**, determine the amount of strain in the DNA.

▶ How does prokaryotic DNA supercoil into its tertiary structure?

Prokaryotic DNA is circular, and it forms supercoils. If the strands are underwound, they form **negative supercoils**. If they are overwound, they form **positive supercoils** (Figure 9.15). Underwound duplex DNA has fewer than the normal number of turns, whereas overwound DNA has more. DNA supercoiling is analogous to twisting or untwisting a rope so that it is torsionally stressed. Negative supercoiling introduces a torsional stress that favors unwinding of the right-handed B-DNA double helix, whereas positive supercoiling overwinds such a helix. Both forms of supercoiling compact the DNA, like a rubber band wound up into a ball.

Enzymes that affect the supercoiling of DNA have been isolated from a variety of organisms. Naturally occurring circular DNA is negatively supercoiled except during replication, when it becomes positively supercoiled. Cellular regulation of this process is critical. Enzymes that are involved in changing the supercoiled state of DNA are called **topoisomerases**, and they fall into two

DNA supercoils deviations from relaxed double helix twisting in DNA

negative supercoils circular DNA with *fewer* than the normal number of turns of the helix

positive supercoils circular DNA with *more* than the normal number of turns of the helix

topoisomerases enzymes that relax supercoiling in closed circular DNA

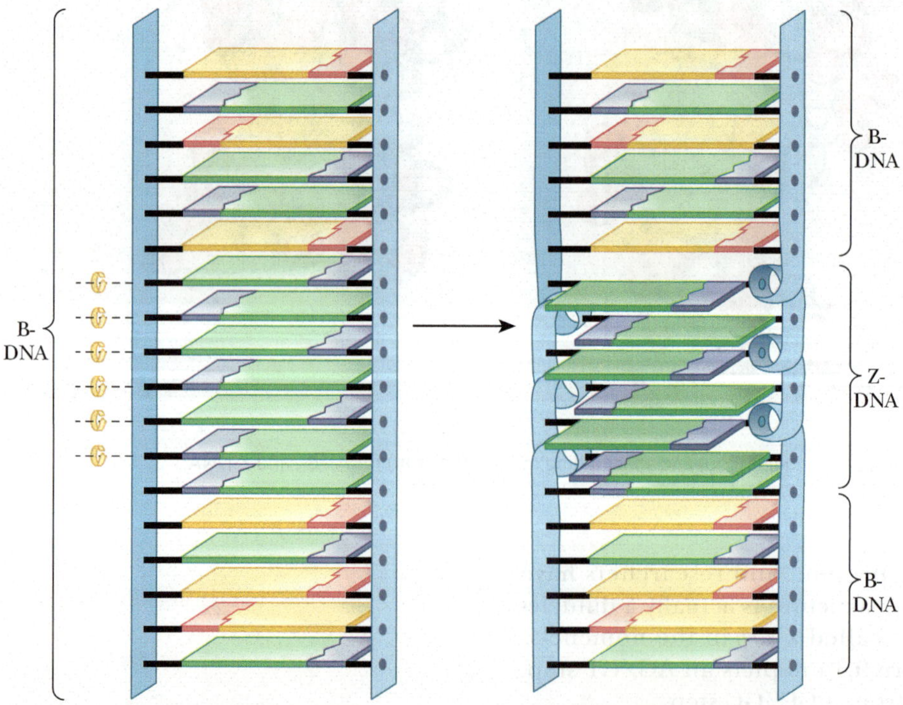

Figure 9.12 **Formation of Z-DNA.** A Z-DNA section can form in the middle of a section of B-DNA by rotation of the base pairs, as indicated by the curved arrows.

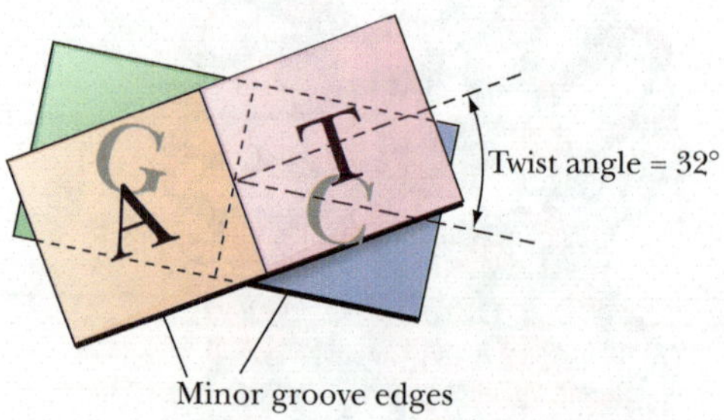

Figure 9.13 **Helical twists.** Two base pairs with 32° of right-handed helical twist; the minor-groove edges are drawn with heavy shading.

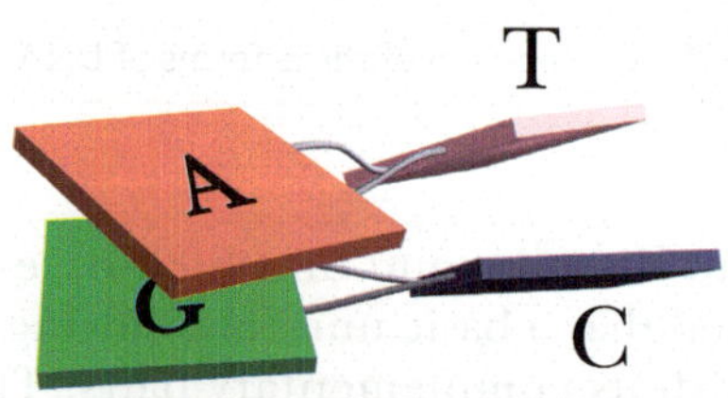

Figure 9.14 **Propeller-twisted base pairs.** The hydrogen bonds between bases are distorted by this motion yet remain intact. The minor-groove edges of the bases are shaded.

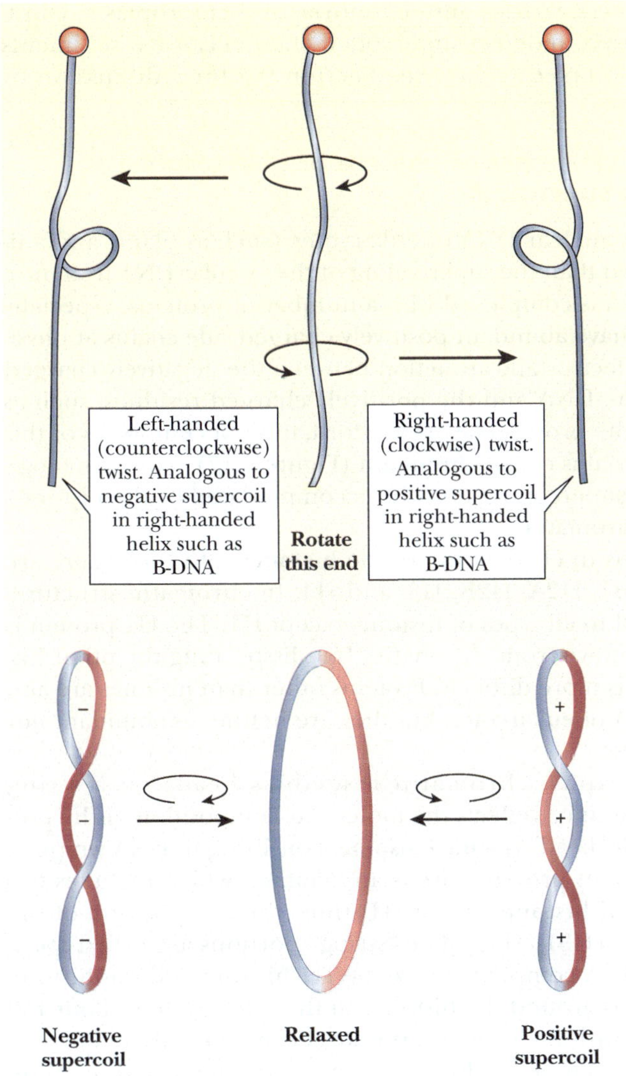

Figure 9.15 Supercoiled DNA topology. The DNA double helix can be approximated as a two-stranded, right-handed coiled rope. If one end of the rope is rotated counterclockwise, the strands begin to separate (negative supercoiling). If the rope is twisted clockwise (in a right-handed fashion), the rope becomes overwound (positive supercoiling). Get a piece of right-handed multistrand rope and carry out these operations to convince yourself. *(Based on Pennisi, E. (2006).* Science *312(5779), 1467–68.)*

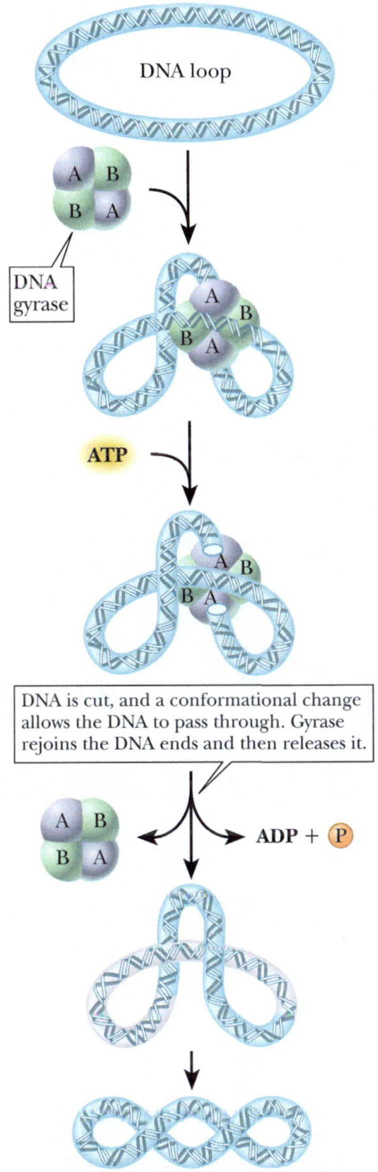

Figure 9.16 A model for the action of bacterial DNA gyrase (topoisomerase II).

classes. Class I topoisomerases cut the phosphodiester backbone of one strand of DNA, pass the other end through, and then reseal the backbone. Class II topoisomerases cut both strands of DNA, pass some of the remaining DNA helix between the cut ends, and then reseal. In either case, supercoils can be added or removed. As we explain in upcoming chapters, these enzymes play an important role in replication and transcription, where separation of the helix strands causes supercoiling. **DNA gyrase** is a bacterial topoisomerase that introduces negative supercoils into DNA. The mechanism is shown in Figure 9.16. The enzyme is a tetramer. It cuts both strands of DNA, so it is a class II topoisomerase.

Supercoiling has been observed experimentally in naturally occurring DNA. Particularly strong evidence has come from electron micrographs that clearly show coiled structures in circular DNA from a number of different

DNA gyrase an enzyme that introduces supercoiling into closed circular DNA

sources, including bacteria, viruses, mitochondria, and chloroplasts. Ultracentrifugation can be used to detect supercoiled DNA because it sediments more rapidly than the relaxed form. (See Section 9-5 for a discussion of ultracentrifugation.)

▶ How does supercoiling take place in eukaryotic DNA to form its tertiary and quaternary structures?

chromatin a complex of DNA and protein found in eukaryotic nuclei

histones proteins that have an abundance of positively charged residues and form complexes with eukaryotic DNA

nucleosome a globular structure in chromatin in which DNA is wrapped around an aggregate of histone molecules

The supercoiling of the nuclear DNA of eukaryotes (such as plants and animals) is more complicated than the supercoiling of the circular DNA from prokaryotes. Eukaryotic DNA is complexed with a number of proteins, especially with basic proteins that have abundant positively charged side chains at physiological (neutral) pH. Electrostatic attraction between the negatively charged phosphate groups on the DNA and the positively charged residues, such as lysine and arginine, on the proteins favors the formation of complexes of this sort. The resulting material is called **chromatin** (Figure 9.17). Thus, topological changes induced by supercoiling must be accommodated by the histone–protein component of chromatin.

The principal proteins in chromatin are the **histones**, of which there are five main types, called H1, H2A, H2B, H3, and H4. In chromatin structure, the DNA is tightly bound to all types of histone except H1. The H1 protein is comparatively easy to remove from chromatin, but dissociating the other histones from the complex is more difficult. Proteins other than histones are also complexed with the DNA of eukaryotes, but they are neither as abundant nor as well studied as histones.

In electron micrographs, chromatin resembles beads on a string (Figure 9.17). This appearance reflects the molecular composition of the protein–DNA complex. Each "bead" is a **nucleosome**, consisting of DNA wrapped around a histone core. This protein core is an octamer, which includes two molecules of each type of histone except H1; thus, the composition of the octamer is $(H2A)_2(H2B)_2(H3)_2(H4)_2$. The "string" portions are called *spacer regions*; they consist of DNA complexed to some H1 histone and nonhistone proteins. As the DNA coils around the histones in the nucleosome, about 150 base pairs are in contact with the proteins; the spacer region is about 30 to 50 base pairs long. Histones can be modified by acetylation, methylation, phosphorylation, and ubiquitination. Ubiquitin is a protein involved in the degradation of other proteins. It is discussed further in Chapter 12. Modifying histones changes their DNA-binding characteristics, and how these changes affect replication and transcription is a subject of active research (Chapters 10 and 11, respectively).

The actual structure and spacing of the nucleosomes appear to be important in their own right. While most scientists focus on the actual sequence of DNA, there is evidence that the chromatin structure is important for other reasons, too. When scientists studied the rate of genetic variation in the DNA of the Japanese killifish, *Oryzias latipes*, they found that the variation was cyclical and corresponded to a chromatin structure. However, what surprised the researchers was that the cyclic nature was different depending on what type of mutation rate they looked at (Figure 9.18). In the linker region, the rate of insertion or deletion mutations peaked. These are mutations where an extra nucleotide is added or deleted, respectively. In this same region, the rate of mutation by substitution, where one nucleotide is substituted for another, was at its low point. How these facts relate to mutation rates and natural selection is a mystery, but clearly the structure of chromatin and the spacing of the nucleosomes must be very important to both because the rate of mutation is not random and is affected by those structures. **Biochemical Connections 9B** describes the Human Genome Project, the attempt to sequence the entire human genome.

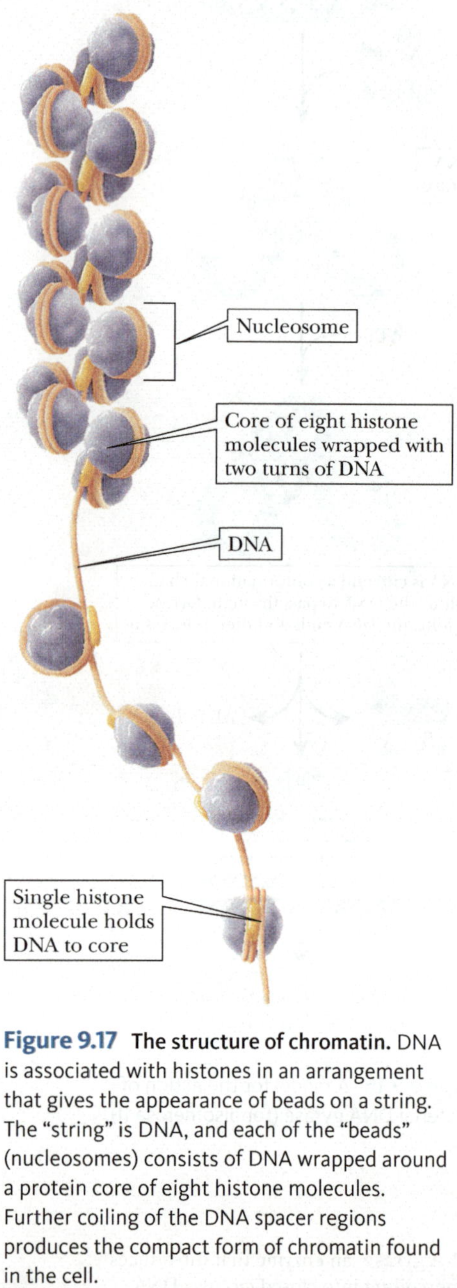

Figure 9.17 The structure of chromatin. DNA is associated with histones in an arrangement that gives the appearance of beads on a string. The "string" is DNA, and each of the "beads" (nucleosomes) consists of DNA wrapped around a protein core of eight histone molecules. Further coiling of the DNA spacer regions produces the compact form of chromatin found in the cell.

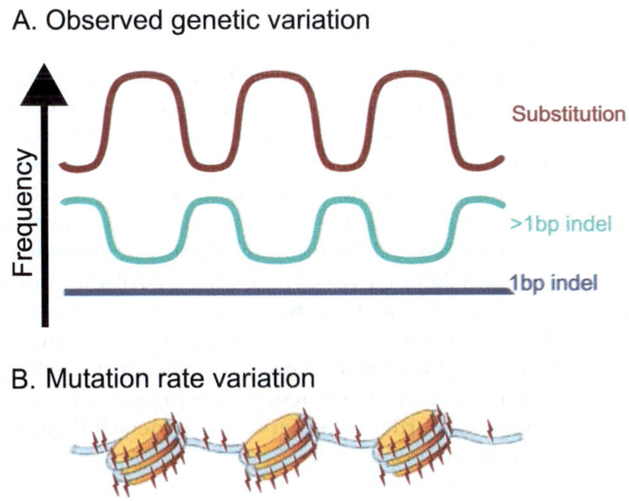

Figure 9.18 **The rate of mutation surrounding nucleosomes is not constant.** Substitution mutations peak in the nucleosomes themselves and are at a low point in the linker regions. Conversely, insertion and deletion mutations (indel in the figure) are the opposite. Researchers are attempting to establish why this is. *(From Semple, C. A., and Taylor, M. S. (2009). Molecular Biology: The Structure of Change.* Science *323, 347–348. Reprinted with permission from AAAS.)*

9B Biochemical Connections • Genetics

The Human Genome Project: Treasure or Pandora's Box?

The Human Genome Project was a massive attempt to sequence the entire human genome, some 3.3 billion base pairs spread over 23 pairs of chromosomes. This project, started formally in 1990, was a worldwide effort driven forward by two groups. One was a private company called Celera Genomics, and its preliminary results were published in *Science* in February 2001. The other was a publicly funded group of researchers called the International Human Genome Sequencing Consortium. Their preliminary results were published in *Nature* in February 2001. Researchers were surprised to find only about 30,000 genes in the human genome. This figure has since dwindled to fewer than 25,000. This is similar to the number of genes found in other eukaryotes, including some organisms as simple as the roundworm *Caenorhabditis elegans.*

What do we do with the information? From this information, we will eventually be able to identify all human genes and to determine which sets of genes are likely to be involved in all human genetic traits, including diseases that have a genetic basis. There is an elaborate interplay of genes, so it may never be possible to say that a defect in a given gene will ensure that the individual will develop a particular disease. Nevertheless, some forms of genetic screening will certainly become a routine part of medical testing in the future. It would be beneficial, for example, if someone more susceptible to heart disease than the average person were to have this information at an early age. This person could then decide on some minor adjustments in lifestyle and diet that might make heart disease much less likely to develop.

Many people are concerned that the availability of genetic information could lead to genetic discrimination. For that reason, the Human Genome Project is a rare example of a scientific project in which definite percentages of financial support and research effort have been devoted to the ethical, legal, and social implications of the research. The question is often posed in this form: Who has a right to know your genetic information? You? Your doctor? Your potential spouse or employer? An insurance company? These questions are not trivial, but they have not yet been answered definitively. Many citizens have expressed concern that genetic screening would lead to a new type of prejudice and bigotry aimed against "genetically challenged" people. Many people have suggested that there is no point in screening for potentially disastrous genes if there is no meaningful therapy for the disease they may "cause." However, couples often want to know in advance if they are likely to pass on a potentially lethal disease to their children.

Two specific examples are pertinent here:

1. There is no advantage in testing for the breast cancer gene if a woman is *not* in a family at high risk for the disease. The presence of a "normal" gene in such a low-risk individual tells nothing about whether a mutation might occur in the future. The risk of breast cancer is not changed if a low-risk person has the normal gene, so mammograms and monthly self-examination are in order.
2. The presence of a gene has not always predicted the development of the disease. Some individuals who have been shown to be carriers of the gene for Huntington's disease have lived

Continued on next page

to old age without developing the disease. Some males who are functionally sterile have been found to have cystic fibrosis, which carries a side effect of sterility due to the improper chloride-channel function that is a feature of that disease (see Section 13-7). They learn this when they go to a clinic to assess the nature of their fertility problem, even though they may never have shown true symptoms of the disease as a child—other than perhaps a high occurrence of respiratory ailments.

The development of genetic tests has become very controversial in recent years. Many scientists balk at the high costs, and many question whether the tests are worth it. For example, so much is still unknown about cancer that it is unclear whether knowing that a person has one or more genetic traits that might predispose them to cancer would be useful information, because some traits might indicate only a small percentage change in the chance of getting the disease.

Since the Human Genome Project was finished, scientists have been eager to sequence other genomes to compare with human DNA. Such studies could show conserved sequences and shed light on those genes that are key to the survival of similar biological species, as well as those that indicate changes that have occurred during the evolutionary process. Therefore, the National Human Genome Research Institute (NHGRI) assembled a list of 32 mammals and 24 other vertebrates whose DNA they would like to analyze. At an international symposium of geneticists, they determined that freezers around the world already contained DNA samples from over 16,000 different species. Shortly thereafter, the Genome 10K Project was launched, which proposes to sequence 10,000 genomes in order to preserve and understand genetic diversity in animals on our planet. ◗

9-4 Denaturation of DNA

Learning Objectives

After reading this section, students will be able to:

LO 9.4.1 Describe a DNA denaturation experiment.

LO 9.4.2 Describe the factors that lead to different DNA denaturation results.

The hydrogen bonds between base pairs are an important factor in holding the double helix together. The amount of stabilizing energy associated with the hydrogen bonds is not great, but the hydrogen bonds hold the two polynucleotide chains in the proper alignment. However, the stacking of the bases in the native conformation of DNA contributes the largest part of the stabilization energy. Energy must be added to a sample of DNA to break the hydrogen bonds and to disrupt the stacking interactions. This is usually carried out by heating the DNA in solution.

▶ How can we monitor DNA denaturation?

The heat denaturation of DNA, also called *melting*, can be monitored experimentally by observing the absorption of ultraviolet light. The bases absorb light in the 260-nm-wavelength region. As the DNA is heated and the strands separate, the wavelength of absorption does not change, but the amount of light absorbed increases (Figure 9.19). This effect is called *hyperchromicity*. It is based on the fact that the bases, which are stacked on top of one another in native DNA, become unstacked as the DNA is denatured.

Under a given set of conditions, there is a characteristic midpoint of the melting curve (the transition temperature, or melting temperature, written T_m) for DNA from each distinct source. The underlying reason for this property is that each type of DNA has a given, well-defined base composition. A G–C base pair forms three hydrogen bonds, so it is held together more strongly than an A–T base pair, which forms only two hydrogen bonds (Figure 9.9). Therefore, the melting temperature of a DNA molecule increases as the percentage of G–C base pairs increases.

Renaturation of denatured DNA is possible on slow cooling (Figure 9.20). The separated strands can recombine and form the same base pairs responsible for maintaining the double helix. The renaturation is also called annealing and is a very important step in many techniques, such as polymerase chain reaction (PCR) (Section 13-6), which are employed to study molecular biology.

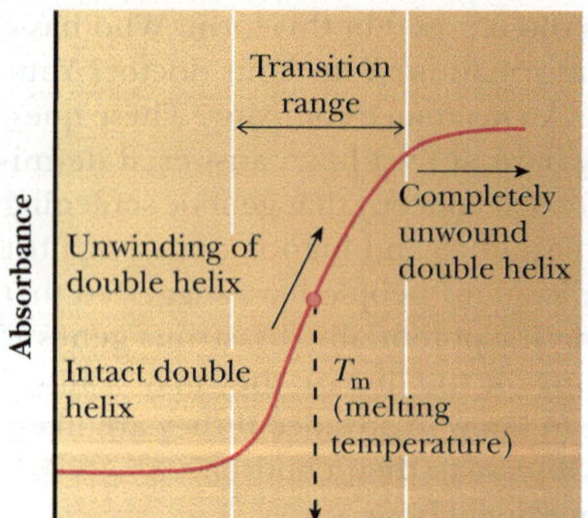

Figure 9.19 The experimental determination of DNA denaturation. This is a typical melting-curve profile of DNA, depicting the hyperchromic effect observed on heating. The transition (melting) temperature, T_m, increases as the guanine and cytosine (the G–C content) increase. Therefore, the entire curve would be shifted to the right for a DNA with higher G–C content and to the left for a DNA with lower G–C content.

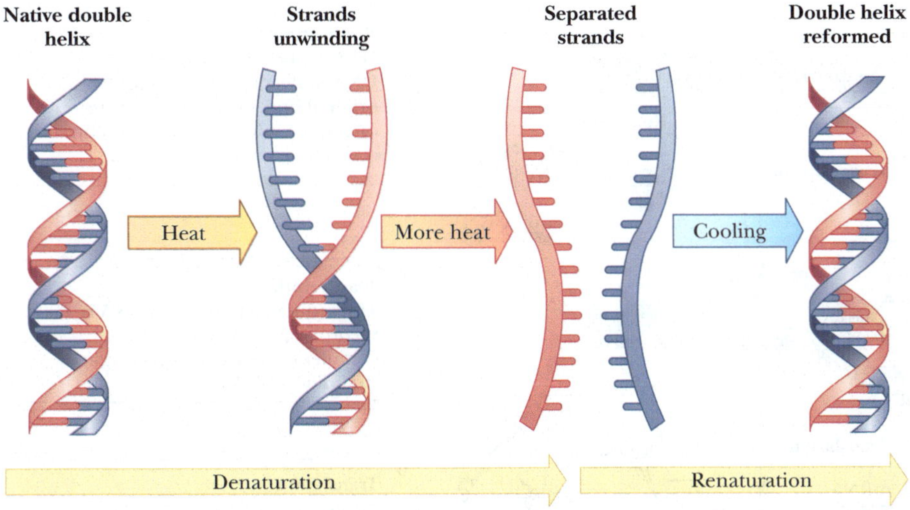

Figure 9.20 Helix unwinding in DNA denaturation. The double helix unwinds when DNA is denatured, then the strands eventually separate. The double helix re-forms on renaturation with slow cooling and annealing.

9-5 The Types of RNA

Learning Objectives

After reading this section, students will be able to:

LO 9.5.1 Distinguish between *DNA replication*, *transcription*, and *translation.*

LO 9.5.2 Distinguish between seven types of RNA.

There are many types of RNA molecules, with more being discovered every day. We discuss a few of them here in Chapter 9 and then explore them in more depth in Chapter 11.

9-5A The Three Primary RNA Types

Figure 9.21 shows the process of information transfer in cells. Other than the process of **DNA replication**, where the DNA is copied, various kinds of RNA participate in the synthesis of proteins in a series of reactions ultimately directed by the base sequence of the cell's DNA (its primary structure). *The base sequences of all types of RNA are determined by that of DNA.* The process by which the order of bases is passed from DNA to RNA is called **transcription** (Chapter 11).

Ribosomes, in which ribosomal RNA (rRNA) is associated with ribosomal proteins, bind messenger RNA (mRNA) and transfer RNA (tRNA) for the assembly of a polypeptide chain during **translation**, the process of protein synthesis (Figure 9.21). Amino acids are brought to the assembly site covalently bonded to tRNA, as aminoacyl-tRNAs. The order of bases in mRNA specifies the order of amino acids in the growing protein. A sequence of three bases in mRNA directs the incorporation of a particular amino acid into the growing protein chain. More details of protein synthesis are discussed in Chapter 12.

To make sense of the terms *transcription* and *translation,* think of the everyday meanings of the words. For example, a transcriptionist makes a written or digital copy of something that is said, using the same language but in a different format—just as DNA and RNA use the same language of nucleotides but in a different format. On the other hand, a translator converts one language to another. In the cellular process of translation, the language of nucleotides in the mRNA is converted to the language of amino acids in the protein that gets synthesized.

The details of transcription and translation differ in prokaryotes and in eukaryotes (Figure 9.22). In prokaryotes, there is no nuclear membrane, so mRNA can direct the synthesis of proteins while it is still being transcribed. Eukaryotic mRNA, on the other hand, undergoes considerable processing. One of the most important

DNA replication the process where an exact copy of DNA is made before cell division

transcription the process where a sequence of DNA bases is recorded as a sequence of RNA bases

ribosomes particles consisting of ribosomal RNA and ribosomal proteins that bind messenger RNA and transfer RNA in order to synthesize proteins

translation the process of protein synthesis in which the amino acid sequence of the protein reflects the sequence of bases in the gene that codes for that protein

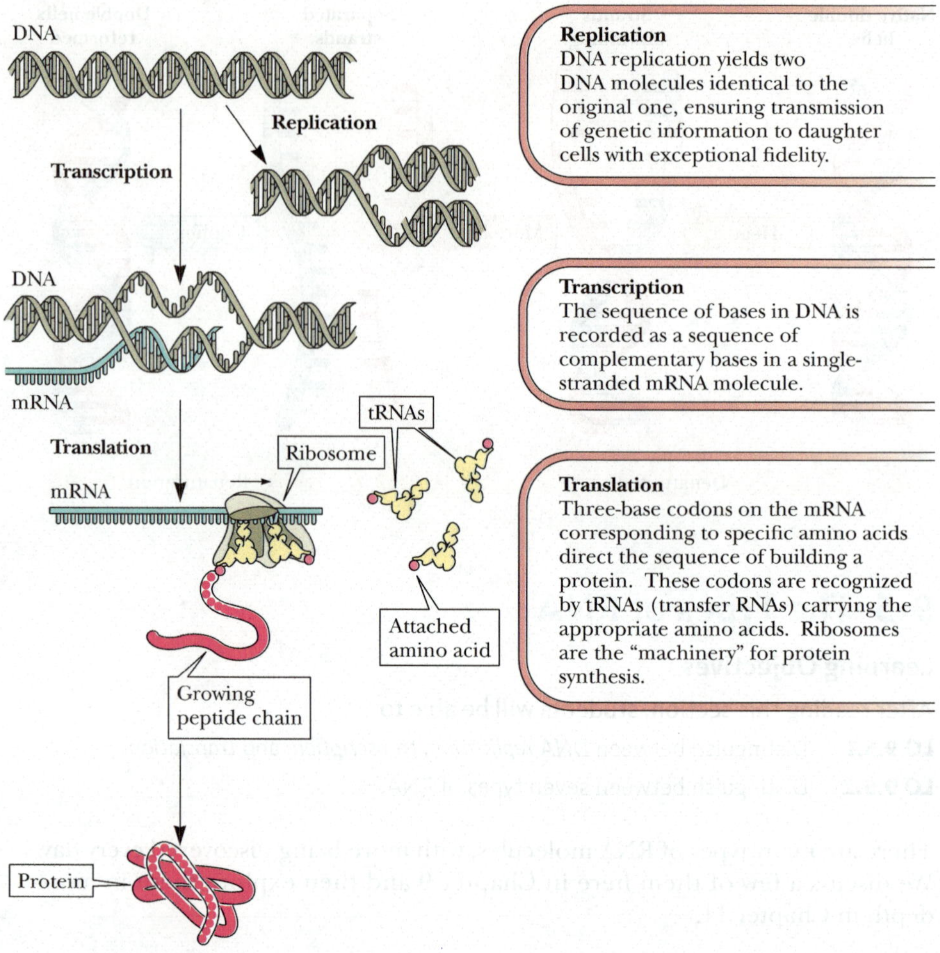

Figure 9.21 The fundamental process of information transfer in cells. This information transfer flows in one direction and is encapsulated in the central dogma of molecular biology: DNA → RNA → protein.

parts of the process is splicing out intervening sequences (introns) so that the parts of the mRNA that will be expressed (exons) are contiguous to each other.

9-5B Other RNA Types

It seems that every year a new type of RNA is discovered. We discuss a few of them here, but this is not a comprehensive list.

Small nuclear RNA (snRNA) is found only in the nucleus of eukaryotic cells, and it is distinct from the other RNA types. snRNA molecules are involved in the processing of initial mRNA transcription products to a mature form suitable for export from the nucleus to the cytoplasm for translation. Micro RNA (miRNA) and small interfering RNA (siRNA) are relatively recent discoveries. They are the main players in **RNA interference (RNAi)**, a process that was first discovered in plants and later in mammals, including humans. In RNAi, miRNAs or siRNAs bind to other RNA or DNA segments and regulate **gene expression.** RNAi is also being used extensively by scientists who wish to eliminate the effect of a gene to help discover its function (see Chapter 13).

RNA interference (RNAi) a process where short pieces of RNA, such as small interfering RNA and micro RNA, affect gene expression

gene expression the process by which information encoded in DNA is used to direct the assembly of a protein

Long noncoding RNA (lncRNA) is around 200 bases or longer and previously may have been mistaken for mRNA. These RNA molecules do not code for protein, and they lack the traditional sequences indicating that they are translated into protein. Estimates vary between 10,000 and 200,000 of these in a typical mammalian genome. Although only about 1% of the mammalian genome is transcribed into protein-coding RNA, upward of 70% may be transcribed into lncRNA, and many of these are thought to be overlapping sequences. They have been linked to development, such as silencing of the second X chromosome in females, as well as certain disease states. Table 9.1 summarizes the types of RNA.

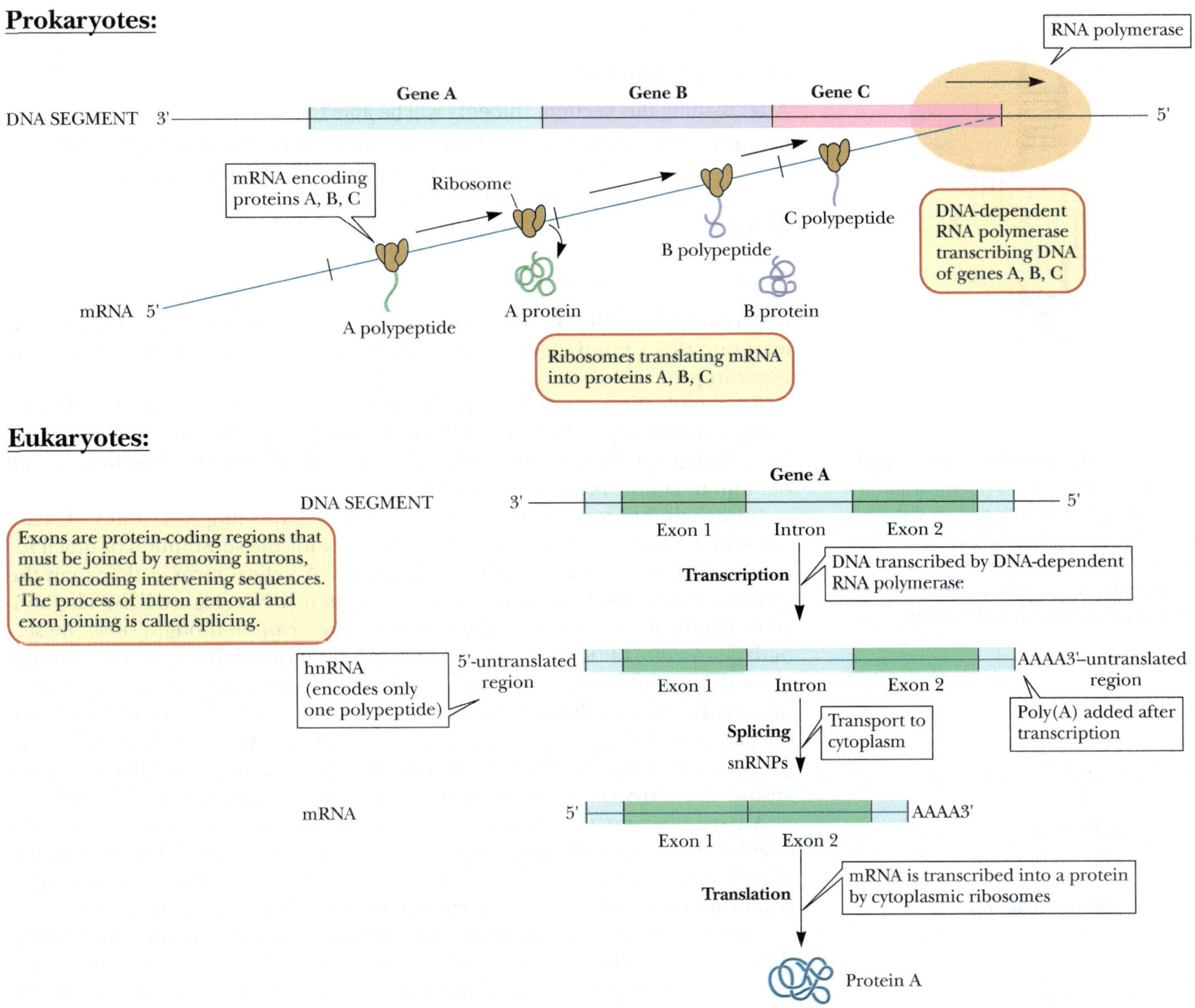

Figure 9.22 **The role of mRNA in transcription.** The properties of mRNA molecules in prokaryotic versus eukaryotic cells during transcription and translation.

Table 9.1 The Roles of Different Kinds of RNA

RNA Type	Size	Function
Transfer RNA (tRNA)	Small	Transports amino acids to the site of protein synthesis
Ribosomal RNA (rRNA)	Several kinds—variable in size	Combines with proteins to form ribosomes, the sites of protein synthesis
Messenger RNA (mRNA)	Variable (long)	Directs amino acid sequence of proteins
Small nuclear RNA (snRNA)	Small	Processes initial mRNA to its mature form in eukaryotes
Small interfering RNA (siRNA)	Small	Affects gene expression; used by scientists to knock out a gene being studied
Micro RNA (miRNA)	Small	Affects gene expression; important in growth and development
Long noncoding RNA (lncRNA)	Variable (long)	Does not code for protein but appears to affect development and be related to certain disease states

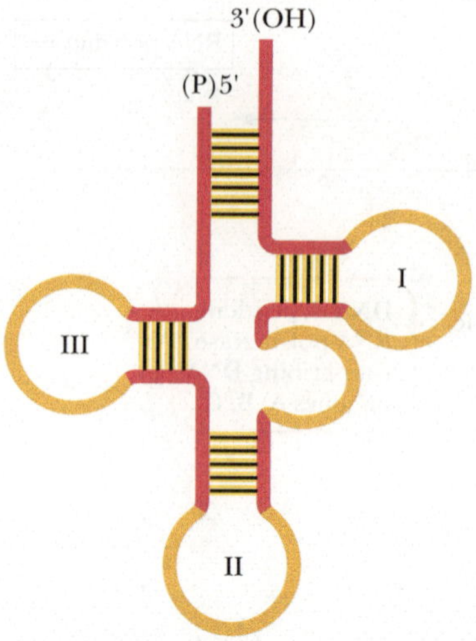

Figure 9.23 The cloverleaf depiction of transfer RNA. Double-stranded regions (shown in red) are formed by folding the molecule and are stabilized by hydrogen bonds between complementary base pairs. Peripheral loops are shown in yellow. There are three major loops (numbered I–III) and one minor loop of variable size (not numbered).

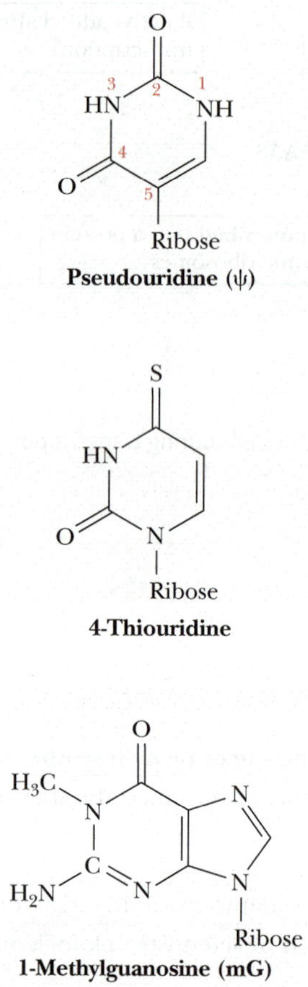

Figure 9.24 Structures of some modified bases found in transfer RNA. The pyrimidine in pseudouridine is linked to ribose at C-5 rather than at the usual N-1.

9-6 Roles for RNA

Learning Objectives

After reading this section, students will be able to:

LO 9.6.1 Describe the roles and structural details of the following types of RNA: transfer RNA, ribosomal RNA, messenger RNA, small nuclear RNA, small interfering RNA, and micro RNA

LO 9.6.2 Explain ribosome structure.

9-6A Transfer RNA

Different types of tRNA molecules can be found in every living cell because at least one tRNA bonds specifically to each of the amino acids that commonly occur in proteins.

Frequently there are several tRNA molecules for each amino acid. A tRNA is a single-stranded polynucleotide chain, between 73 and 94 nucleotide residues long, that generally has a molecular mass of about 25,000 Da. (Biochemists call the unit of atomic mass the *Dalton,* abbreviated Da.)

Intrachain hydrogen bonding occurs in tRNA, forming A–U and G–C base pairs similar to those that occur in DNA (except for the substitution of uracil for thymine). The duplexes thus formed have the A-helical form, rather than the B-helical form, which is the predominant form in DNA (Figure 9.8). The molecule can be drawn as a *cloverleaf structure,* which can be thought of as the secondary structure of tRNA because it shows the hydrogen bonding between certain bases (Figure 9.23). The hydrogen-bonded portions of the molecule are called *stems,* and the non-hydrogen-bonded portions are called *loops.* Some of these loops contain modified bases (Figure 9.24). During protein synthesis, both tRNA and mRNA are bound to the ribosome in a definite spatial arrangement that ultimately ensures the correct order of the amino acids in the growing polypeptide chain.

A particular tertiary structure is necessary for tRNA to interact with the enzyme that covalently attaches the amino acid to the 2' or 3' hydroxyl group of its 3' end. To produce this tertiary structure, the tRNA folds into an L-shaped conformation that has been determined by X-ray diffraction (Figure 9.25).

After the tRNA is transcribed from DNA, a specific enzyme, ATP(CTP): tRNA nucleotidyltransferase adds the sequence CCA to the 3' end. This is a necessary step prior to the linking of the correct amino acid to the tRNA. Recently, this enzyme was found to play a role in quality control of tRNAs. If there is a structural defect in the tRNA or if the molecule is a small tRNA-like molecule but not true tRNA, then the enzyme puts a CCACCA at the 3' end instead. This activates a rapid decay pathway of the tRNA molecule.

9-6B Ribosomal RNA

In contrast with tRNA, rRNA molecules tend to be quite large, and only a few types of rRNA exist in a cell. Because of the intimate association between rRNA and proteins, a useful approach to understanding the structure of rRNA is to investigate ribosomes themselves.

The RNA portion of a ribosome accounts for 60%–65% of the total weight, and the protein portion constitutes the remaining 40%–35% of the weight. Dissociation of ribosomes into their components has proved to be a useful way of studying their structure and properties. A particularly important endeavor has been to determine both the number and the kind of RNA and protein molecules that make up ribosomes. This approach has helped elucidate the role of ribosomes in protein synthesis.

In both prokaryotes and eukaryotes, a ribosome consists of two subunits: a small subunit and a large subunit. The small subunit consists of one large rRNA molecule and about 20 different ribosomal proteins; the large subunit consists of two rRNA molecules in prokaryotes (three in eukaryotes) and about 35 different ribosomal proteins in prokaryotes (about 50 in eukaryotes). The subunits are easily dissociated from one another in the laboratory by lowering the Mg^{2+} concentration of the medium. Raising the Mg^{2+} concentration to its original level reverses the process, and active ribosomes can be reconstituted by this method.

A technique called *analytical ultracentrifugation* has proved very useful for studying ribosomes. Figure 9.26 shows a schematic of an analytical ultracentrifuge. We need not consider all the details of this technique, as long as it is clear that its basic aim is the observation of the motion of ribosomes, RNA, or protein in a centrifuge. The motion of the particle is characterized by a *sedimentation coefficient,* expressed in *Svedberg units* (S), which are named after Theodor Svedberg, the Swedish scientist who invented the ultracentrifuge. The S value increases with the molecular weight of the sedimenting particle, but it is not directly proportional to it because the particle's shape also affects its sedimentation rate.

Ribosomes and rRNA have been studied extensively via sedimentation coefficients. Most research on prokaryotic systems has been done with the bacterium *Escherichia coli,* which we use as an example here. An *E. coli* ribosome typically has a sedimentation coefficient of 70S. When an intact 70S bacterial ribosome dissociates, it produces a light 30S subunit and a heavy 50S subunit. Notice that the values of sedimentation coefficients are not additive, thereby demonstrating the dependence of the S value on the shape of the particle. The 30S subunit contains a 16S rRNA and 21 different proteins. The 50S subunit contains a 5S rRNA, a 23S rRNA, and 34 different proteins (Figure 9.27). For comparison, eukaryotic ribosomes have a sedimentation coefficient of 80S, and the small and large subunits are 40S and 60S, respectively. The small subunit of eukaryotes contains an 18S rRNA, and the large subunit contains three types of rRNA molecules: 5S, 5.8S, and 28S.

The 5S rRNA has been isolated from many different types of bacteria, and the nucleotide sequences have been determined. A typical 5S rRNA is about 120 nucleotide residues long and has a molecular mass of about 40,000 Da. Some sequences have also been determined for the 16S and 23S rRNA molecules. These larger molecules are about 1500 and 2500 nucleotide residues long, respectively. The molecular mass of 16S rRNA is about 500,000 Da, and that of 23S rRNA is about one million Da. The degrees of secondary and tertiary structure in the larger RNA molecules appear to be substantial.

A secondary structure has been proposed for 16S rRNA (Figure 9.28), and suggestions have been made about the way in which the proteins associate with the RNA to form the 30S subunit.

The *self-assembly of ribosomes* takes place in the living cell, but the process can be duplicated in the laboratory. The elucidation of ribosomal structure was truly a formidable scientific challenge, and the 2009 Nobel Prize in Chemistry was awarded to Ada Yonath, Venkatraman Ramakrishnan, and Thomas Steitz for their innovations in ribosomal X-ray crystallography. The binding of antibiotics to bacterial ribosomal subunits so as to prevent self-assembly of the ribosome is one focus of the investigation. The structure of ribosomes is also one of the points used to compare and contrast eukaryotes, eubacteria, and archaebacteria (Chapter 1). The study of RNA became much more exciting in 1986, when Thomas Cech showed that certain RNA molecules exhibited catalytic activity (Section 11-9). Equally exciting was the discovery that the rRNA—and not ribosomal protein—is the part of a ribosome that catalyzes the formation of peptide bonds in bacteria (Section 12-4C).

9-6C Messenger RNA

The least abundant of the three main types of RNA is mRNA. In most cells, it constitutes no more than 5%–10% of the total cellular RNA. The sequences

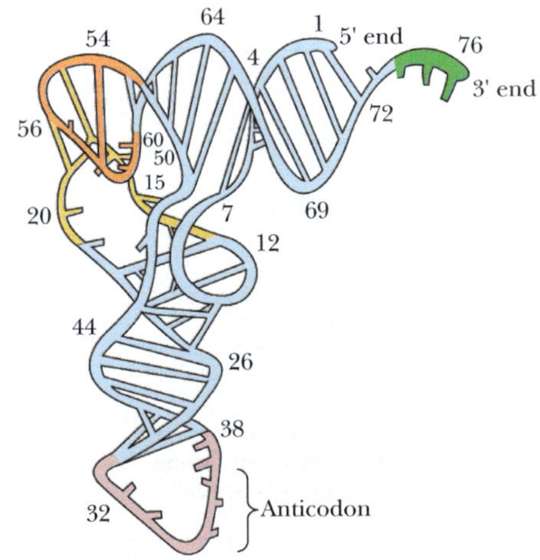

Figure 9.25 The three-dimensional structure of yeast phenylalanine tRNA as deduced from X-ray diffraction studies of its crystals. The tertiary folding is illustrated, and the ribose–phosphate backbone is presented as a continuous ribbon; hydrogen bonds are indicated by crossbars. Unpaired bases are shown as short, unconnected rods. The anticodon loop that binds the mRNA codon is at the bottom, and the —CCA 3'—OH acceptor end is at the top right.

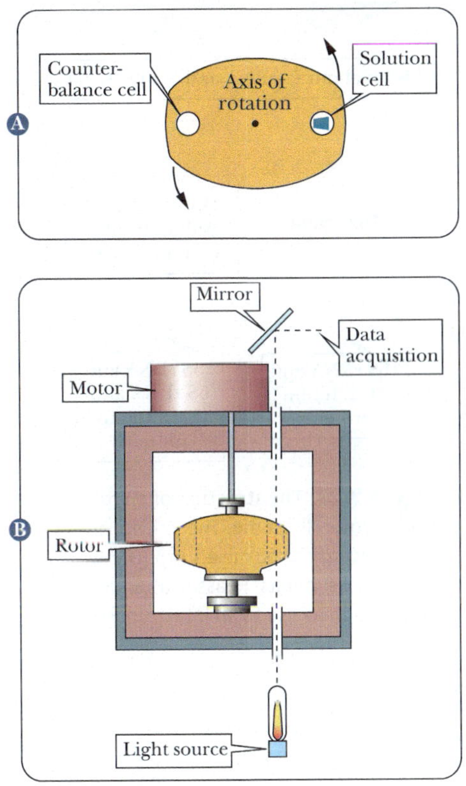

Figure 9.26 The analytical ultracentrifuge. (A) Top view of an ultracentrifuge rotor. The solution cell has optical windows; the cell passes through a light path once each revolution. (B) Side view of an ultracentrifuge rotor. The optical measurement taken as the solution cell passes through the light path makes it possible to monitor the motion of sedimenting particles.

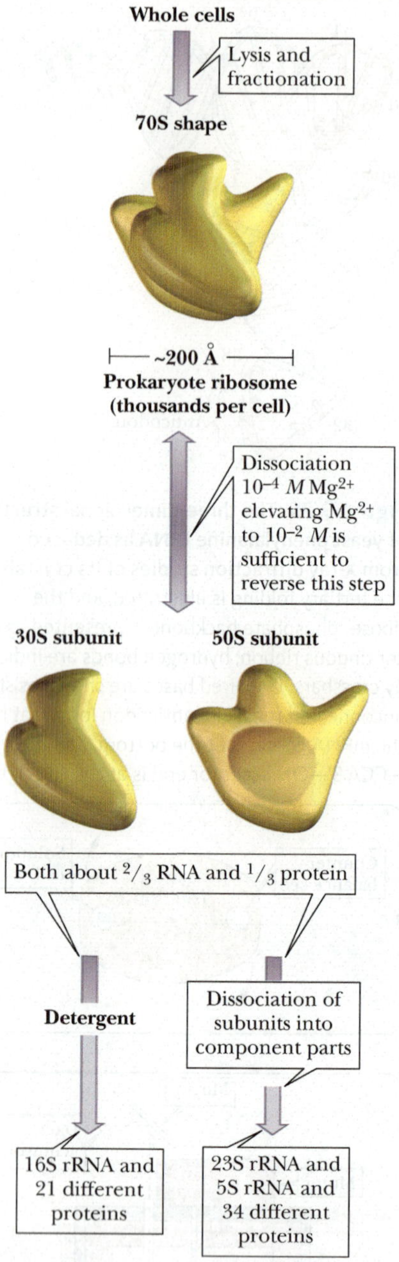

Figure 9.27 The structure of a typical prokaryotic ribosome. The individual components can be mixed, producing functional subunits. Reassociation of subunits gives rise to an intact ribosome.

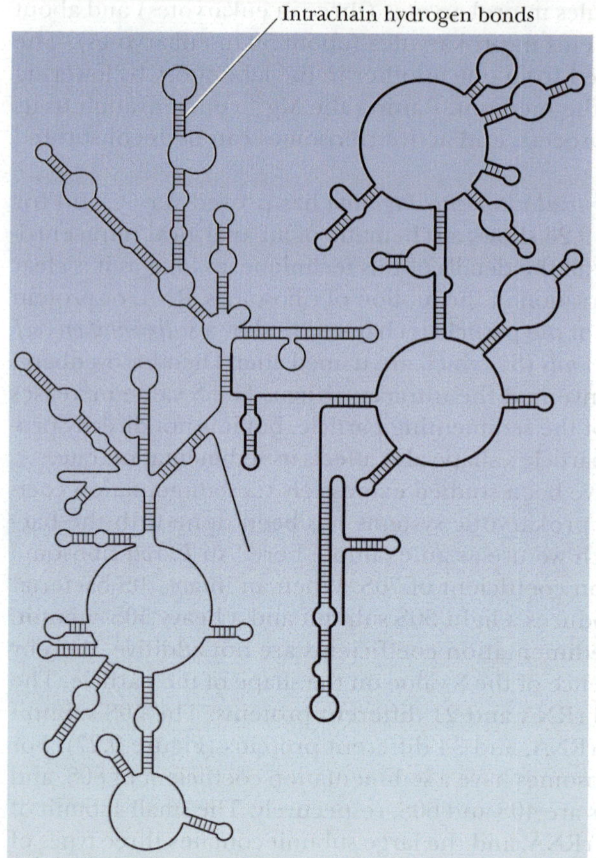

Figure 9.28 A schematic drawing of a proposed secondary structure for 16S rRNA. The intrachain folding pattern includes loops and double-stranded regions. Notice the extensive intrachain hydrogen bonding.

of bases in mRNA specify the order of the amino acids in proteins. In rapidly growing cells, many different proteins are needed within a short time interval. Fast turnover in protein synthesis becomes essential. Consequently, mRNA is formed when it is needed, it directs the synthesis of proteins, and then it is degraded so that the nucleotides can be recycled. Of the main types of RNA, mRNA is the one that usually turns over most rapidly in the cell. Both tRNA and rRNA (as well as ribosomes themselves) can be recycled intact for many rounds of protein synthesis.

The sequence of mRNA bases that directs the synthesis of a protein reflects the sequence of DNA bases in the gene that codes for that protein, although this mRNA sequence is often altered after it is produced from the DNA. Messenger RNA molecules are heterogeneous in size, as are the proteins whose sequences they specify. Less is known about possible intrachain folding in mRNA, with the exception of folding that occurs during the termination of transcription (Chapter 11). It is also likely that several ribosomes are associated with a single mRNA molecule at some time during the course of protein synthesis. In eukaryotes, mRNA is initially formed as a larger precursor molecule called **heterogeneous nuclear RNA (hnRNA)**. These contain lengthy portions of intervening sequences called **introns** that do not encode a protein. These introns are removed by posttranscriptional splicing. In addition, protective units called *5′-caps* and *3′ poly(A) tails* are added before the mRNA is complete (Section 11-7).

heterogeneous nuclear RNA (hnRNA) eukaryotic RNA that is initially produced by transcribing DNA; it contains intervening sequences that do not code for any proteins

introns intervening sequences in DNA that do not appear in the final sequence of mRNA

9-6D Small Nuclear RNA

Small nuclear RNA (snRNA) is found in the nucleus of eukaryotic cells, and it is about 100–200 nucleotides long. These RNA molecules form complexes with proteins to generate **small nuclear ribonucleoproteins** (abbreviated as **snRNPs** but pronounced as "snurps"). Their function is to help with the processing of the initial mRNA transcribed from DNA into a mature form that is ready for export out of the nucleus. The processing events involve splicing the RNA, and many researchers are working on elucidating RNA splicing mechanisms.

small nuclear ribonucleoproteins protein–RNA complexes found in the nucleus that aid in processing RNA molecules for export to the cytosol

9-6E RNA Interference

The process called RNA interference (RNAi) is a gene regulation and protection mechanism in many species. It uses siRNAs and miRNAs to eliminate or suppress the expression of a gene such as one that is causing uncontrolled cell growth or a gene that came from a virus. These small RNAs (20–30 nucleotides long) have been found to have an enormous control over gene expression. As a result, scientists use them to study gene expression. In what has become an explosion of new biotechnology, many companies have been created to produce and market designer siRNA to knock out hundreds of known genes. The biotech applications of RNA interference are discussed more in Chapter 13.

9-7 RNA and Medical Applications

Learning Objectives

After reading this section, students will be able to:

LO 9.7.1 Describe CRISPR technology.

LO 9.7.2 Explain mRNA vaccines.

9-7A CRISPR

CRISPR, which stands for *c*lustered *r*egularly *i*nterspaced *s*hort *p*alindromic *r*epeats, are repetitive stretches of DNA found in bacteria and archaea. These sequences interact with proteins known as CRISPR-associated proteins (Cas proteins). The combination is the basis of a protection mechanism that the bacteria use against foreign DNA. The proteins are guided to specific DNA sequences and then cut the DNA in two. They are guided by a short strand of RNA that they take from the invading organism and then use it to target the invader's DNA. Researchers quickly realized that if they could provide their own designer RNA, then they could control where the Cas proteins attack. They took the natural bacterial system and used it to produce a specific gene-cutting tool. Additional steps are also able to repair and replace the DNA in specific ways. Several biotech firms are currently working to use CRISPR/Cas gene editing to develop treatments for cystic fibrosis and sickle-cell disease. This technology is discussed in greater detail in Chapter 13.

9-7B mRNA Vaccines

When the world shut down in March 2020 due to the COVID-19 pandemic, a race began to develop a vaccine for the virus. This included Operation Warp Speed in the United States. In a relatively short amount of time, several vaccines were developed for COVID-19, including **mRNA vaccines**. This vaccine technology is also being used in the treatment of cancer. These vaccines use a copy of an mRNA molecule that produces a viral protein or a cancer cell protein. When the vaccine is delivered, the patient's immune system can use the viral or cancer cell protein produced from the vaccine to learn how to identify and destroy virus particles or cancer cells.

mRNA vaccines vaccines that use mRNA to produce an immune response against a virus or cancer cell

epigenetics the study of how DNA is changed and modified in a way that affects gene expression but without an actual change in the base sequence

Biochemical Connections 9C describes a few more interesting details regarding DNA and information transfer in cells.

9C Biochemical Connections • Genetics

Why Identical Twins Are Not Identical

Much has been learned about the differences between nature and nurture by studying twins. Often twins separated at birth are later studied to see how different they have become. Their differences and similarities give us insight into how much of our physiology and our behavior is governed by genetics. Sometimes, however, twins raised together under seemingly identical circumstances can turn out very different. While the actual DNA sequence of identical twins is the same, two twins can be very different in other ways. The study of **epigenetics** sheds light on the causes of the differences in identical twins. Epigenetics refers to changes in DNA that are not reflected in the actual base sequence. Epigenetic modifications of DNA act like switches that turn on or off certain genes. If these modifications are not the same in each of two twins, then the twins will no longer be identical.

The best-known example of an epigenetic mechanism is DNA methylation, where a cytosine is tagged with a methyl group, as shown in Figure 9.29. This is usually associated with shutting off the expression of the gene. Another epigenetic mechanism is chromatin remodeling. The histone proteins discussed in Section 9.3 can be modified by the addition of methyl, acetyl, or phosphate groups. This influences, in turn, the activity of adjacent genes. Acetylation generally switches on the expression of the genes, while methylation usually silences the expression, as shown in Figure 9.30.

Because certain disease states can be linked to epigenetic states, it is possible for one individual to develop a disease while their identical twin does not. Susceptibility to diseases is often a family trait, but the actual mechanism of getting the disease may require epigenetic changes in the DNA of a cell. Epigenetic changes are known to be important in the field of cancer research, but only recently have scientists begun to study the relationship between epigenetic state and other diseases such as schizophrenia, immune deficiencies, obesity, diabetes, and heart disease. ◗

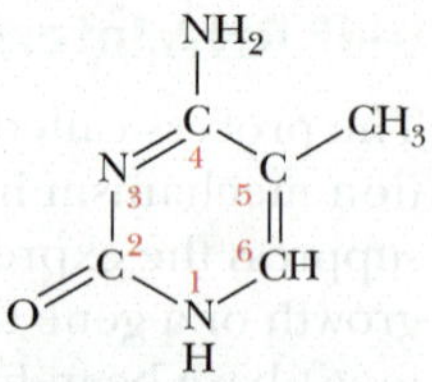

Figure 9.29 Structure of 5-methylcytosine.

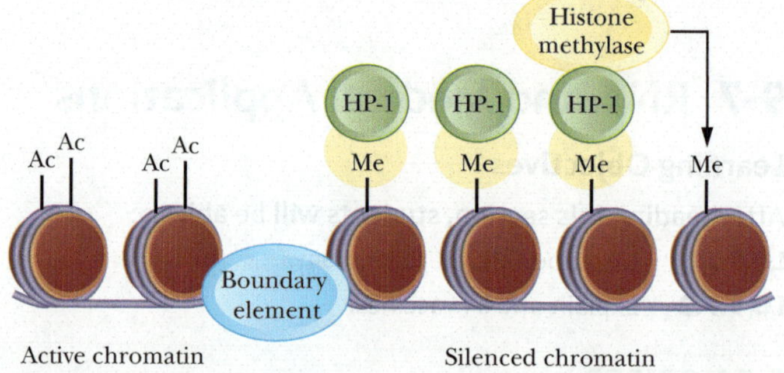

Figure 9.30 Methylation and acetylation of histones regulate expression of genes. *(Reprinted by permission from Macmillan Publishers Ltd: Bannister, A., et al. (2001). Selective Recognition of Methylated Lysine 9 on Histone H3 by the HP1 Chromo Domain.* Nature **410,** *120–124.)*

Hot Topic

The Genetics of Breast Cancer

An enormous amount of research has been devoted to breast cancer. In October 2012, the Cancer Genome Atlas Network published results that represented the culmination of decades of work. They called their results "comprehensive molecular portraits of human breast tumors." Let us look at how this story developed.

One of the first major studies focused on the genetic aspects of breast cancer. In 1990, a group of researchers in California studied 23 extended families with 146 cases of breast cancer. They focused on the number of affected individuals in a given family, especially those with early onset of the disease. The research method combined classic genetic analysis with DNA studies. Statistical analysis of cancer incidence in the family trees indicated that these cancers were not sporadic cases that arise from somatic mutations in individuals but were genetically linked (Figure 9.31). In two of these families, males developed breast cancer. It is very unusual for males to do so, but it does occur. DNA was obtained from blood samples taken from 329 members of the families being studied. Southern blots (Section 13-7) revealed information about the homology of DNA samples among various family members. The change in DNA linked to the occurrence of breast cancer was localized to chromosome 17, characterized by a variable number of tandem repeats of specific DNA sequences. This result was the first milestone in determining the genetics of breast cancer, and the gene was designated *BRCA1*.

At about the same time, researchers in Utah were carrying out similar investigations on the genetics of breast cancer. They used genetic and cytologic analysis to study the relationship between proliferative breast disease (PBD) and breast cancer. PBD is a term used to describe several kinds of nonmalignant breast lesions. They found that a high incidence of PBD, which was detected cytologically, strongly correlated with a high risk for breast cancer due to family history. PBD may represent a precancerous state. Their results led them to consider the possibility that the development of breast cancer involves several genes.

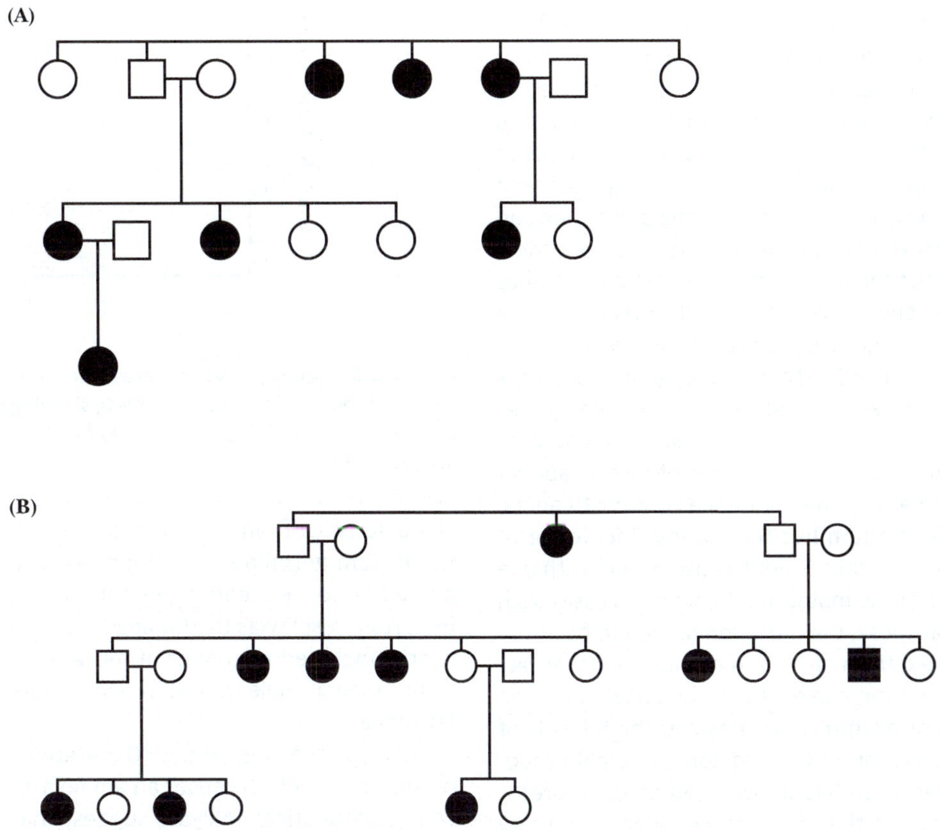

Figure 9.31 Family trees showing the genetic distribution of breast cancer in two families. Each horizontal row represents a generation. Males are shown as squares and females as circles. A solid symbol represents an individual who developed breast cancer, and an open symbol represents nonoccurrence. As seen here, males can also develop breast cancer, although it is uncommon.

In 1994, an International group of researchers announced the discovery of a second gene for breast cancer susceptibility—one that was not linked to *BRCA1*. This second gene, designated *BRCA2*, is located on chromosome 13. Some significant differences between the two genes were readily apparent. Both confer a high degree of susceptibility to breast cancer, but they differ in the degree of risk of ovarian cancer. The presence of *BRCA1* substantially increases the risk of ovarian cancer, but *BRCA2* does not do so. In addition, male breast cancer does not appear to be linked to *BRCA1*, but there is a small risk for men who carry the *BRCA2* gene. For females who carry these genes, the probability of breast cancer at some point in their life is over 80%. Some females who have this genetic predisposition have opted for double mastectomies as a preventive measure.

In May 2013, the use of double mastectomies as a preventive measure received wide publicity when the actress Angelina Jolie announced that she had undergone this surgery. DNA testing had shown that she carried the mutation in the *BRCA1* gene that drastically increases the probability of developing breast cancer. In her contribution to the Opinion Pages of the *New York Times*, entitled "My Medical Choice," Jolie indicated that she was strongly motivated to see that her children did not lose their mother at an early age. She made her choice public in the hope that other women could benefit from her experience. This news did indeed lead to worldwide news coverage and discussion of all aspects of breast cancer. A number of females chose to share their experience of mastectomies, both for the prevention and treatment of cancer. These stories show that the management of breast cancer has changed markedly over the years, in large part because of advances in research.

The *BRCA1* and *BRCA2* genes are cancer suppressors, and mutations in these genes in affected families increase the probability of cancer. A number of factors appear to affect the incidence of breast cancer. Some families have a low incidence of breast cancer until a case occurs with one of these mutations. In nearly all cases such as these, the mutation proved to be inherited from fathers. A number of other factors can and do affect the onset of cancer. For example, females with high levels of physical activity and normal weight in adolescence had a lower tendency to breast cancer than females who were sedentary or experienced obesity in adolescence. It also appears that risk is increasing with time. Females born before 1940 in affected families have high risk, but it is higher still for females born after 1940. Figure 9.32 compares the incidence of breast cancer, particularly at early ages, for females born before and after 1940.

The success of the Human Genome Project provided access to the sequence of portions of the genome affected by these cancer-predisposing mutations. In 2006, extensively studied protein-encoding genes were sequenced using tissue samples from patients with cancer. Breast tumors and colorectal tumors were used in the study because they are among the most common cancers. As Figure 9.33 shows, several rounds of polymerase chain reaction (PCR) amplification and subsequent sequencing were needed to accomplish this task. Sequences of the same genes in normal tissue were used as controls. Another part of the task was to select the somatic mutations that are characteristic of cancer and to distinguish them from germline mutations that are passed on to succeeding generations. An important result was the finding that each cancer included a number of mutations, rather than a single mutation responsible for cancer.

In a typical cancer, about 80 mutations occur, each of which causes an amino acid change. Statistical analyses suggest that not all of these mutations lead to cancer. The number of mutations that trigger or maintain cancer is probably fewer than 15. However, identifying the mutations is only a beginning. It does not provide information about *how* the mutation leads to the development of a tumor. Identifying the proteins encoded by the mutant genes is a step forward because it indicates the pathways that may be affected. A few small changes in the efficiency of several reactions may well combine to be the triggering

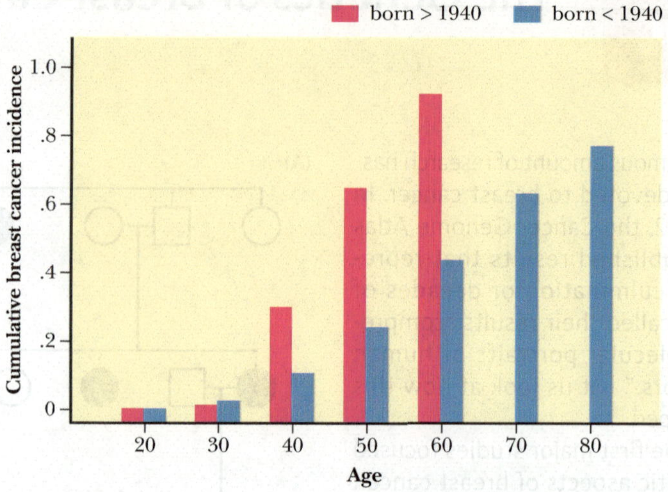

Figure 9.32 A comparison of breast cancer incidence for females with *BRCA1* and *BRCA2* mutations born before and after 1940, showing increasing risk with time. *(Based on King, M., et al. (2003). Breast and Ovarian Cancer Risks Due to Inherited Mutations in BRCA1 and BRCA2.* Science *302(5645), 645.)*

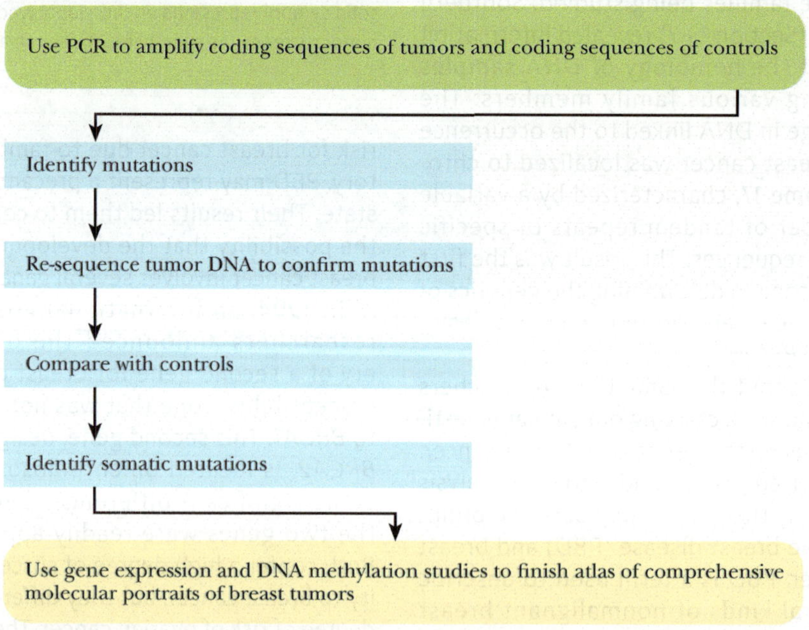

Figure 9.33 Schematic diagram of the screening procedure used to identify cancer-associated mutations. *(Adapted from Sjöblom, T., et al. (2006). The Consensus Coding Sequences of Human Breast and Colorectal Cancers.* Science ***314, 269.)***

event in developing cancer. Insight into structural changes in proteins that catalyze reactions in these pathways can suggest therapeutic approaches. In this study, mutations were found in a number of enzyme genes, including the gene *GALNT5* that codes for a glycosylation enzyme. Figure 9.34 shows the known crystal structure of this enzyme with bound substrate. The purple spheres are the mutated amino acids, which lie close to the active site, superimposed on the structure. This information does not suggest any therapeutic advance, but it does suggest approaches to develop new therapies.

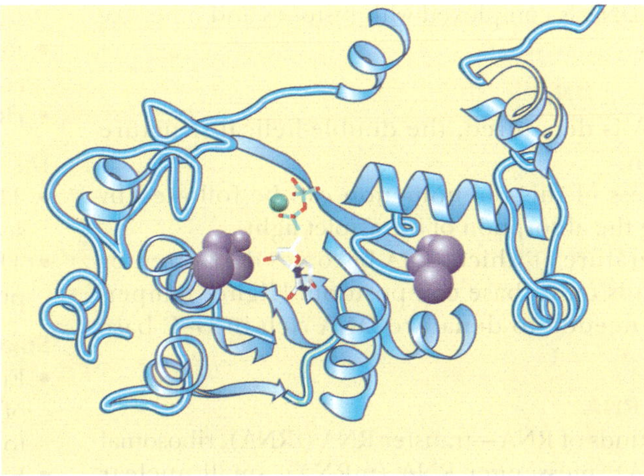

Figure 9.34 Structure of the enzyme coded by the GALNT5 gene. The substrate is shown as a stick model, and the purple spheres represent amino acid mutations near the active site that are found in a tumor. *(Adapted from Macmillan Publishers Ltd: The Cancer Genome Atlas Network (2012). Comprehensive Molecular Portraits of Human Breast Tumours.* Nature, *490, 62.)*

The work of the Cancer Genome Atlas Network built upon all this previous research, culminating in publication of a wide-ranging picture of the genetic changes in breast cancer. Results from a number of methods went into the analysis. An analysis of anomalies in DNA methylation was one technique, and sequencing of exons was another. Gene expression analyses with both mRNA and miRNA also played a role. Gene expression by DNA was yet another method, as was analysis of single nucleotide polymorphisms (single-base changes in DNA). Even with automated techniques, this work was an enormous undertaking. Refer back to Figure 9.33 for an outline of the method.

These results have some correlation with clinical features, but the correspondence is incomplete. We know, for example, that the chemotherapeutic agent Herceptin is effective only in the kinds of cancer where the gene for human epidermal growth factor (HER2) is overexpressed. It is possible to determine whether this is the case in individual patients and to plan treatment accordingly. Ideally, the time will come when we can correlate all mutations with specific clinical features and plan treatment accordingly. That time is not here yet, but this kind of work brings it closer.

Summary

Levels of Structure in Nucleic Acids

- DNA (deoxyribonucleic acid) and RNA (ribonucleic acid) are the two kinds of nucleic acids. DNA contains the sugar deoxyribose, whereas RNA contains the sugar ribose.
- The primary structure of nucleic acids is the order of bases in the polynucleotide sequence, and the secondary structure is its base pairing features. As seen with proteins, the tertiary structure of nucleic acids is their three-dimensional shape. The quaternary structure is the interactions between nucleic acids or between nucleic acids and proteins.

The Covalent Structure of Polynucleotides

- The monomers of nucleic acids are nucleotides. An individual nucleotide consists of three parts—a nitrogenous base, a sugar, and a phosphoric acid residue—all of which are covalently bonded together. Nucleotides are linked by phosphodiester bonds to form the phosphate–sugar backbone of nucleic acids.

The Structure of DNA

- The double helix originally proposed by Watson and Crick is the most striking feature of DNA structure.
- The two coiled strands run in antiparallel directions with hydrogen bonds between complementary bases.
- Adenine pairs with thymine, whereas guanine pairs with cytosine.
- Some variations on the usual representation of the double helix (B-DNA) are known to exist. In A-DNA, the base pairs lie at an angle to the helix axis, and in Z-DNA, the helix is left-handed, rather than the more usual right-handed form of B-DNA.

DNA Supercoils

- Supercoiling is a feature of DNA structure in both prokaryotes and eukaryotes.
- Prokaryotic DNA is normally circular and is twisted into a supercoiled form before the circle is closed.
- The form of supercoiling plays a role in DNA replication.

- Eukaryotic DNA is complexed with histones and other basic proteins.

Denaturation of DNA

- When DNA is denatured, the double-helical structure breaks down.
- The progress of this phenomenon can be followed by monitoring the absorption of ultraviolet light.
- The temperature at which DNA becomes denatured by heat depends on its base composition. Higher temperatures are needed to denature DNA rich in G–C base pairs.

The Types of RNA

- The many kinds of RNA—transfer RNA (tRNA), ribosomal RNA (rRNA), messenger RNA (mRNA), small nuclear RNA (snRNA), micro RNA (miRNA), small interfering RNA (siRNA), and long noncoding RNA (lncRNA)—differ in structure and function.

Transfer RNA

- Transfer RNA is relatively small, about 80 nucleotides long.
- tRNA exhibits extensive intrachain hydrogen bonding, represented in two dimensions by a cloverleaf structure.
- Amino acids are brought to the site of protein synthesis bonded to tRNAs.

Ribosomal RNA

- Ribosomal RNA molecules tend to be quite large and are complexed with proteins to form ribosomal subunits.
- rRNA also exhibits extensive internal hydrogen bonding.

Messenger RNA

- The sequence of bases in a given mRNA determines the sequence of amino acids in a specified protein.
- The size of mRNA molecules varies with the size of the protein.

Small Nuclear RNA

- Eukaryotic mRNA is processed in the nucleus by a fourth type of RNA, small nuclear RNA, which is complexed with proteins to give small nuclear ribonuclear protein particles (snRNPs).
- Eukaryotic mRNA is initially produced in an immature form that must be processed by snRNPs, which remove introns and add protective units at the 5' and 3' ends.

RNA interference

- Micro RNA and small interfering RNA are both very small, about 20 to 30 bases long. They function in the control of gene expression.

RNA and Medical Applications

- Several technologies, including CRISPR/Cas and mRNA vaccines, are being studied for their potential to treat and cure diseases.

Review Exercises

9-1 Levels of Structure in Nucleic Acids

1. **Reflect and Apply** Answer the following questions using the concept of levels of structure (primary, secondary, tertiary, quaternary) as defined for proteins.
   (a) What level is shown by double-stranded DNA?
   (b) What level is shown by tRNA?
   (c) What level is shown by mRNA?
2. **Reflect and Apply** Use examples to describe why you think the following statement is true or false: RNA does not adopt quaternary structures.

9-2 The Covalent Structure of Polynucleotides

3. **Recall** What is the structural difference between thymine and uracil?
4. **Recall** What is the structural difference between adenine and hypoxanthine?
5. **Recall** Give the name of the base, the ribonucleoside and/or deoxyribonucleoside, and the ribonucleoside triphosphate for A, G, C, T, and U.
6. **Recall** How are the structures of ATP and dATP different?
7. **Recall** Draw an RNA strand with the following sequence: 5' ACG 3'.
8. **Reflect and Apply** (a) Is it biologically advantageous that DNA is stable? Why or why not? (b) Is it biologically advantageous that mRNA is unstable? Why or why not?

9-3 The Structure of DNA

9. **Recall** In what naturally occurring nucleic acids would you expect to find A-form helices, B-form helices, Z-form helices, nucleosomes, and circular DNA?
10. **Recall** Draw a G–C base pair. Draw an A–T base pair.
11. **Reflect and Apply** Give the sequence on the opposite strand for (a) ACGTAT, (b) AGATCT, and (c) CTGGTA. All read 5' → 3' and your answer should as well.
12. **Recall** Are the sequences shown in Question 11 those of RNA or DNA? How can you tell?
13. **Recall** Which of the following statements is (are) true?
    (a) Bacterial ribosomes consist of 40S and 60S subunits.
    (b) Prokaryotic DNA is normally complexed with histones.
    (c) Prokaryotic DNA normally exists as a closed circle.
    (d) Circular DNA is supercoiled.
14. **Recall** How do the major and minor grooves in B-DNA compare to those in A-DNA?
15. **Recall** Which of the following statements is (are) true?
    (a) The two strands of DNA run parallel from their 5' end to their 3' end.
    (b) An adenine–thymine base pair contains three hydrogen bonds.
    (c) Positively charged counterions are associated with DNA.
    (d) DNA base pairs are always perpendicular to the helix axis.
16. **Recall** Define *supercoiling, positive supercoil, topoisomerase,* and *negative supercoil.*
17. **Recall** What is propeller twist?
18. **Recall** What is an AG/CT step?
19. **Recall** Why does propeller-twist occur?
20. **Recall** What is the difference between B-DNA and Z-DNA?
21. **Recall** If circular B-DNA is positively supercoiled, will these supercoils be left-handed or right-handed?

22. **Recall** Which biomolecules make up chromatin?
23. **Reflect and Apply** Would you expect to find adenine–cytosine or guanine–thymine base pairs in DNA? Use your understanding of hydrogen bonding to explain why or why not.
24. **Reflect and Apply** List three mechanisms that relax the twisting stress in helical DNA molecules.
25. **Reflect and Apply** Explain how DNA gyrase works.
26. **Reflect and Apply** Explain, and draw a diagram to show, how acetylation or phosphorylation of histones could change the binding affinity between DNA and histones.
27. **Reflect and Apply** Would you expect to find adenine–guanine or cytosine–thymine base pairs in DNA? Why or why not?
28. **Reflect and Apply** One of the original structures proposed for DNA had all the phosphate groups positioned in close proximity at the center of a long fiber. Give a reason why this proposal was rejected.
29. **Reflect and Apply** What is the complete base composition of a double-stranded eukaryotic DNA that contains 22% guanine?
30. **Reflect and Apply** Why was it necessary to specify that the DNA in Question 29 is double-stranded?

9-4 Denaturation of DNA

31. **Recall** Fill in the blanks to accurately describe a standard DNA denaturation experiment. Increasing ______________ is applied to the DNA sample while ______________ at 260 nm wavelength is measured to quantify DNA base unstacking. As DNA denatures, the measured intensity will ______________ (increase/decrease/shift).
32. **Recall** Why does DNA with a high A–T content have a lower transition temperature, T_m, than DNA with a high G–C content?
33. **Reflect and Apply** Polyethylene glycol is hydrophobic, so its presence in solution decreases the hydrophobic forces that promote base stacking. What effect would you expect this to have on the melting temperature (T_m) of double-stranded DNA? Explain.

9-5 The Types of RNA

34. **Recall** What are the three types of RNA needed for protein synthesis?
35. **Recall** What determines the base sequence of all types of RNA?
36. **Reflect and Apply** You discovered a gene that you hypothesize affects cell growth and division. Which type of RNA will you add to the cell to decrease the expression of your gene of interest and test your hypothesis?
37. **Recall** What is the name of the process that produces RNA from a DNA template?
38. **Recall** What is the basic purpose of tRNA?
39. **Recall** The base sequence of which type of RNA is responsible for determining the order of amino acids in a protein?
40. **Recall** What is the name of the process by which amino acids are linked together in the correct order based on an RNA sequence?
41. **Recall** What are some key differences in how mRNA is processed in prokaryotes versus eukaryotes?
42. **Recall** What are small nuclear RNAs?
43. **Recall** What are micro RNAs?

9-6 Roles for RNA

44. **Recall** Sketch a typical cloverleaf structure for transfer RNA. Point out any similarities between the cloverleaf pattern and the proposed structures of ribosomal RNA.
45. **Reflect and Apply** A friend tells you that only four different kinds of bases are found in RNA. What would you say in reply?
46. **Recall** What is the purpose of small nuclear RNA? What is an snRNP?
47. **Recall** Which type of RNA is the biggest? Which is the smallest?
48. **Recall** Which type of RNA has the least amount of secondary structure?
49. **Recall** What is the role of ATP(CTP): tRNA nucleotidyletransferase?
50. **Recall** What is RNA interference?
51. **Reflect and Apply** Would you expect tRNA or mRNA to be more extensively hydrogen bonded? Why?
52. **Reflect and Apply** The structures of tRNAs contain several unusual bases in addition to the typical four. Suggest a function for the unusual bases.
53. **Reflect and Apply** Would you expect mRNA or rRNA to be degraded more quickly in the cell? Why?
54. **Reflect and Apply** Which would be more harmful to a cell, an error in DNA replication or an error in transcription? Why?
55. **Reflect and Apply** Explain briefly what happens to eukaryotic mRNA before mRNA is translated to protein. Where do these events take place in the cell?
56. **Reflect and Apply** Explain why a 50S ribosomal subunit and a 30S ribosomal subunit combine to form a 70S subunit, instead of an 80S subunit.

9-7 RNA and Medical Applications

57. **Recall** Why do researchers believe they can use CRISPR/Cas to treat disease?
58. **Recall** Briefly describe how mRNA vaccines work.
59. **Reflect and Apply** Most mRNA vaccines need to be injected the day they are defrosted, while medications made from proteins like insulin are stable at room temperature for 28 days. Use your understanding of mRNA and protein structure to explain these differences in storage requirements.

Further Reading

Baltimore, D. Our Genome Unveiled. *Nature* **409**, 814–816 (2001). A Nobel Prize winner's guide to the special issue describing human genome sequencing.

Berg, P., and M. Singer. *Dealing with Genes: The Language of Heredity.* Mill Valley, CA: University Science Books, 1992. Two leading biochemists have produced an eminently readable book on molecular genetics; highly recommended.

Cech, T. R. The Ribosome Is a Ribozyme. *Science* **289**, 878–879 (2000). The title says it all.

Church, G. M. Genomes for All. *Sci. Amer.* **295** (1), 47–54 (2006). An excellent summary of how human genomes are determined.

Claverie, J. M. Fewer Genes, More Noncoding RNA. *Science* **309**, 1529–1530 (2005). As more human genome information is discovered, the number of genes coding for proteins is decreasing and the amount of noncoding RNA products is increasing.

Claverie, J. M. What If There Are Only 30,000 Human Genes? *Science* **252**, 1255–1257 (2001). Implications of the low gene number for human molecular biology.

Collins, F., et al. (International Human Genome Sequencing Consortium). Initial Sequencing and Analysis of the Human Genome. *Nature* **409**, 860–921 (2001). One of two simultaneous publications of the sequence of the human genome.

Couzin, J. DNA Test for Breast Cancer Draws Criticism. *Science* **322**, 357 (2008).

Couzin, J. Mini RNA Molecules Shield Mouse Liver from Hepatitis. *Science* **299**, 995 (2003). An example of RNA interference.

Couzin, J. Small RNAs Make Big Splash. *Science* **298**, 2296–2297 (2002). A description of the small, recently discovered forms of RNA.

Couzin, J. The Twists and Turns in BRCA's Path. *Science* **302**, 591–593 (2003). Genes involved in breast cancer have given researchers some big surprises and continue to do so.

Dennis, C. Altered States. *Nature* **421**, 686–688 (2003). Epigenetic states can control disease states and explain why identical twins are not so identical.

Gitlin, L., S. Karelsky, and R. Andino. Short Interfering RNA Confers Intracellular Antiviral Immunity in Human Cells. *Nature* **418**, 430–434 (2002). An example of RNA interference.

Gorman, C. and D. F. Maron. The RNA Revolution. *Sci. Amer.*, 53–59 (April 2014) A summary of RNA-based therapeutics.

Haussecker, D. and M. A. Kay. Drugging RNAi. *Science*, **346**, 1069–1070 (2015) An article about advances in the creation and delivery of RNA interference drugs.

Jeffords, J. M., and T. Daschle. Political Issues in the Genome Era. *Science* **252**, 1249–1251 (2001). Comments on the Human Genome Project by two members of the U.S. Senate.

Jenuwein, T., and C. D. Allis. Translating the Histone Code. *Science* **293**, 1074–1079 (2001). An in-depth article about chromatin, histones, and methylation.

Lake, J. A. Evolving Ribosome Structure: Domains in Archaebacteria, Eubacteria, Eocytes and Eukaryotes. *Ann. Rev. Biochem.* **54**, 507–530 (1985). A review of the evolutionary implications of ribosome structure.

Lake, J. A. The Ribosome. *Sci. Amer.* **245** (2), 84–97 (1981). A look at some of the complexities of ribosome structure.

Lau, N. C., and D. P. Bartel. Censors of the Genome. *Sci. Amer.* **289** (2), 34–41 (2003). An article primarily about RNA interference.

Lee, J. T. Epigenetic Regulation by Long Noncoding RNAs. *Science*, **338**, 1435–1439 (2012) A thorough description of the roles of long noncoding RNA in the epigenetic mechanism.

Levy-Lahad, E., and S. E. Plon. A Risky Business—Assessing Breast Cancer Risk. *Science* **302**, 574–575 (2003). A discussion of risk factors and probabilities for *BRCA* gene carriers.

Marshall, E. Lawsuit Challenges Legal Basis for Patenting Human Genes. *Science* **324**, 1000–1001 (2009).

Moffat, A. Triplex DNA Finally Comes of Age. *Science* **252**, 1374–1375 (1991). Triple helices as "molecular scissors."

Paabo, S. The Human Genome and Our View of Ourselves. *Science* **252**, 1219–1220 (2001). A look at human DNA and its comparison with the DNA of other species.

Peltonen, L., and V. A. McKusick. Dissecting Human Disease in the Postgenomic Era. *Science* **252**, 1224–1229 (2001). How diseases may be studied in the genomic era.

Pennisi, E. DNA's Molecular Gymnastics. *Science* **312**, 1467–1468 (2006). A review of the alternative forms of DNA.

Pennisi, E. The Evolution of Epigenetics. *Science* **293**, 1063–1105 (2001). A mini-symposium on epigenetics.

Pennisi, E. Lengthy RNAs Earn Respect as Cellular Players. *Science*, **344**, 1072 (2014). An article about long noncoding RNAs.

Pennisi, E. No Genome Left Behind. *Science* **326**, 794-795 (2009).

Saw, P.E. and E. Song. Advancements in Clinical RNA Therapeutics: Present Developments and Prospective Outlooks. *Cell Rep. Med.* **5**, Article number: 101555 (2024). This review discusses RNA therapeutics, highlighting the progress from the approval of siRNA drugs to the development of mRNA-based therapeutics for various diseases.

Scovell, W. M. Supercoiled DNA. *J. Chem. Ed.* **63**, 562–565 (1986). A discussion focused mainly on the topology of circular DNA.

Semple, C. A. M., and M. S. Taylor. The Structure of Change. *Science* **323**, 347 (2009).

Stix, G. Owning the Stuff of Life. *Sci. Amer.* **294** (2), 76–83 (2009).

Venter, J. C., et al. The Sequence of the Human Genome. *Science* **291**, 1304–1351 (2001). One of two simultaneous publications of the sequence of the human genome.

Watson, J. D., and F. H. C. Crick. Molecular Structure of Nucleic Acid. A Structure for Deoxyribose Nucleic Acid. *Nature* **171**, 737–738 (1953). The original article describing the double helix.

Wolfsberg, T., J. McEntyre, and G. Schuler. Guide to the Draft Human Genome. *Nature* **409**, 824–826 (2001). How to analyze the results of the Human Genome Project.

Hot Topic

The Cancer Genome Atlas Network. Comprehensive Molecular Portraits of Human Breast Tumours. *Nature* **490**, 61–70 (2012). Hundreds of authors from all over the world compile genetic information about breast tumors.

Hall, J., Lee, M., Newman, B., Morrow, J., Anderson, L., Huey, B., and M. King. Linkage of Early-Onset Familial Breast Cancer to Chromosome 17q21. *Science* **250**, 1684–1689 (1990). Discovery of a breast cancer gene (*BRCA1*).

King, M., J. Marks, and J. Mardell. for the New York Breast Cancer Study Group. Breast and Ovarian Cancer Risks Due to Inherited Mutations in *BRCA1* and *BRCA2*. *Science* **302**, 643–646 (2003). Factors that affect the development of cancer in females with these mutations.

Sjoblom, T., et al. The Consensus Coding Sequences of Human Breast and Colorectal Cancers. *Science* **314**, 268–274 (2006). The use of molecular biology techniques to obtain sequence information on frequently occurring tumor mutations.

Skolnick, M., et al. Inheritance of Proliferative Breast Disease in Breast Cancer Kindreds. *Science* **250**, 1715–1720 (1990). A report about high incidence of proliferative breast disease in families with a high incidence of breast cancer.

Wood, L., et al. The Genomic Landscapes of Human Breast and Colorectal Cancers. *Science* **318**, 1008–1013 (2007). An extensive discussion of cancer genetics.

Wooster, R., et al. Localization of Breast Cancer Susceptibility Gene, *BRCA2*, to Chromosome 13q 12–13. *Science* **265**, 2088–2090 (1994). A report on the second major breast cancer susceptibility gene.

The biochemical process of DNA replication comes very close to producing an exact copy of a cell's genetic material before the cell divides.

AYO Production/Shutterstock.com

Biosynthesis of Nucleic Acids: Replication 10

Chapter Outline

10-1 The Flow of Genetic Information in the Cell

Learning Objectives

After reading this section, students will be able to:

LO 10.1.1 Describe the central dogma of molecular biology.

LO 10.1.2 Explain exceptions to the central dogma of molecular biology.

The sequence of bases in DNA encodes genetic information. The duplication of DNA— giving rise to a new DNA molecule with the same base sequence as the original—is necessary whenever a cell divides to produce daughter cells. This duplication process is called **replication**. The actual formation of gene products requires RNA; the production of RNA using a DNA template is called **transcription**, which is the topic of Chapter 11. The base sequence of DNA is reflected in the base sequence of RNA.

Three kinds of RNA are involved in the biosynthesis of proteins. Of the three, messenger RNA (mRNA) is particularly important. A sequence of three bases in mRNA specifies the identity of one amino acid in a manner directed by the genetic code. The process by which the base sequence directs the amino acid sequence is called **translation**, which is the topic of Chapter 12.

In nearly all organisms, the flow of genetic information is DNA $\rightarrow$ RNA $\rightarrow$ protein. The only major exceptions are some viruses (called retroviruses) in which RNA, rather than DNA, is the genetic material.

In those viruses, RNA can direct its own synthesis as well as that of DNA. The enzyme **reverse transcriptase** catalyzes this process. Not all viruses in which RNA is the genetic material are retroviruses, but all retroviruses have a reverse transcriptase. In fact, that is the origin of the term *retrovirus,* referring to the reverse of the usual situation with transcription. In cases of infection by retroviruses, such as HIV, reverse transcriptase is a target for drug design. Figure 10.1 shows ways in which information is transferred in the cell. This scheme has been called the **central dogma of molecular biology**.

10-2 Replication of DNA

Learning Objectives

After reading this section, students will be able to:

LO 10.2.1 Summarize three important challenges in DNA replication.

LO 10.2.2 Analyze the Meselson–Stahl experiment to determine how it supports the semiconservative model of DNA replication.

replication the process that duplicates DNA

transcription the process that uses a DNA template to form RNA

translation the process of protein synthesis in which the amino acid sequence of the protein reflects the sequence of bases in the gene that codes for that protein

reverse transcriptase the enzyme that directs the synthesis of DNA on an RNA template

central dogma of molecular biology a theory about information flow in cells from DNA to RNA to protein

Naturally occurring DNA exists in many forms. Single- and double-stranded DNAs are known, and both can exist in linear and circular forms. As a result, it is difficult to generalize about all possible cases of DNA replication. Because many DNAs are double-stranded, we can present some general features of the replication of double-stranded DNA, features that apply both to linear and to circular DNA. Most of the details of the process that we discuss here were first investigated in prokaryotes—particularly in the bacterium *Escherichia coli* (*E. coli*). We use information obtained by experiments on this organism for most of our discussion of the topic. Section 10-7 discusses differences between prokaryotic and eukaryotic replication.

The process by which one double-helical DNA molecule is duplicated to produce two such double-stranded molecules is complex. The very complexity

Pioneers in Biochemistry • Dorothy Crowfoot Hodgkin

Dorothy Crowfoot Hodgkin (1910–1994) discovered a fascination with chemistry and crystals at age 10 with the encouragement of her mother, a botanist. There were few women chemists at the time, but Hodgkin pursued her study of chemistry with the support of financial sponsors and academic mentors, many of them women. At her secondary school, she was one of only two girls allowed to study chemistry along with the boys. After graduating college with a chemistry degree, Hodgkin was awarded a PhD from Cambridge University in 1937 for research on X-ray crystallography and the chemistry of sterols.

Hodgkin devoted her career at Oxford University to investigating biochemical structures through X-ray crystallography. One of Hodgkin's early achievements was confirming the structure of penicillin in 1946. She went on to determine the structure of vitamin B_{12}, which has one of the most complex structures of all vitamins. In 1964, Hodgkin was just the third woman to be awarded the Nobel Prize in Chemistry for these and other achievements.

Despite suffering from crippling rheumatoid arthritis, Hodgkin continued to refine crystallography techniques to enable her to analyze complex molecules like insulin, a biologically significant molecule that had fascinated her for decades. She and her team finally uncovered the structure of insulin in 1969, which was instrumental in making possible the mass production of insulin to treat diabetes.

Hulton Deutsch/Corbis Historical/Getty Images

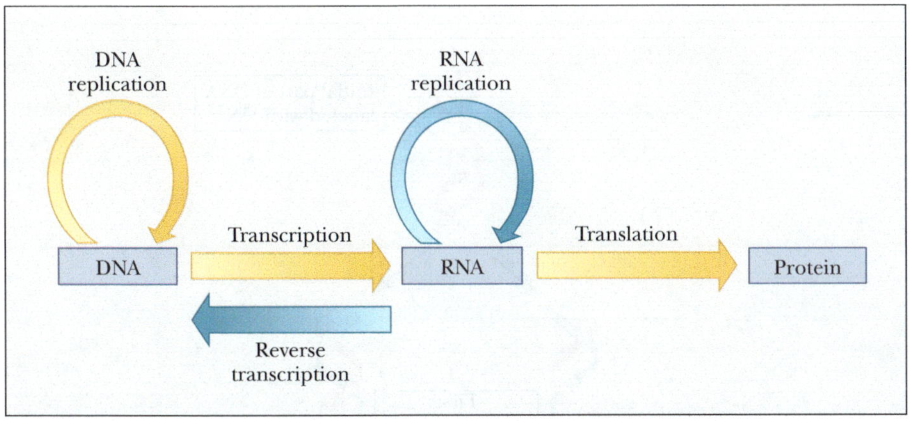

Figure 10.1 Mechanisms for the transfer of information in the cell. The yellow arrows represent typical processes, whereas the blue arrows represent special cases (processes that occur mostly in RNA viruses).

allows for a high degree of fine-tuning, which ensures, in turn, considerable fidelity in replication. The cell faces three important challenges in carrying out the necessary steps:

- *Separating the two DNA strands:* The two strands of DNA are wound around each other in such a way that they must be unwound if they are to be separated. In addition to achieving continuous unwinding of the double helix, the cell also must protect the unwound portions of DNA from the action of **nucleases,** which preferentially attack single-stranded DNA.
- *Synthesizing of DNA from the 5′ to the 3′ end:* The DNA synthesis is catalyzed by **DNA polymerases**, and two antiparallel strands must be synthesized in the same direction on antiparallel templates (see Section 10-3A).
- *Guarding against errors in replication:* The correct base must be added to the growing polynucleotide chain.

nucleases enzymes that hydrolyze a nucleic acid; specific for DNA or RNA

DNA polymerases enzymes that catalyze the synthesis of new DNA strands

More details about these challenges are discussed here in Section 10-2 and in Sections 10-3, 10-4, and 10-5.

10-2A Semiconservative Replication

DNA replication involves the separation of the two original strands and the production of two new strands with the original strands used as templates. Each new DNA molecule contains one strand from the original DNA and one newly synthesized strand. This situation is called **semiconservative replication** (Figure 10.2). The details of the process differ in prokaryotes and eukaryotes, but semiconservative replication is observed in all organisms.

semiconservative replication the mode in which DNA is reproduced such that one strand comes from parent DNA and the other strand is newly formed

▶ How did scientists figure out that replication is semiconservative?

Semiconservative replication of DNA was established unequivocally in the late 1950s by experiments performed by Matthew Meselson and Franklin Stahl. *E. coli* bacteria were grown with $^{15}NH_4Cl$ as the sole nitrogen source. The usual isotope of nitrogen is ^{14}N, so ^{15}N is a heavy isotope of nitrogen. In such a medium, all newly formed nitrogen compounds, including purine and pyrimidine nucleobases, become labeled with ^{15}N. The ^{15}N-labeled DNA has a higher density than unlabeled DNA, which contains ^{14}N. In this experiment, the ^{15}N-labeled cells were then transferred to a medium that contained only ^{14}N. The cells continued to grow in the new medium. With every new generation of growth, a sample of DNA was extracted and analyzed by the technique of **density-gradient centrifugation** (Figure 10.3). This technique depends on the fact that heavy ^{15}N DNA (DNA that contains ^{15}N alone) forms a band at the

density-gradient centrifugation the technique of separating substances in an ultracentrifuge by applying the sample to the top of a tube that contains a solution of varying densities

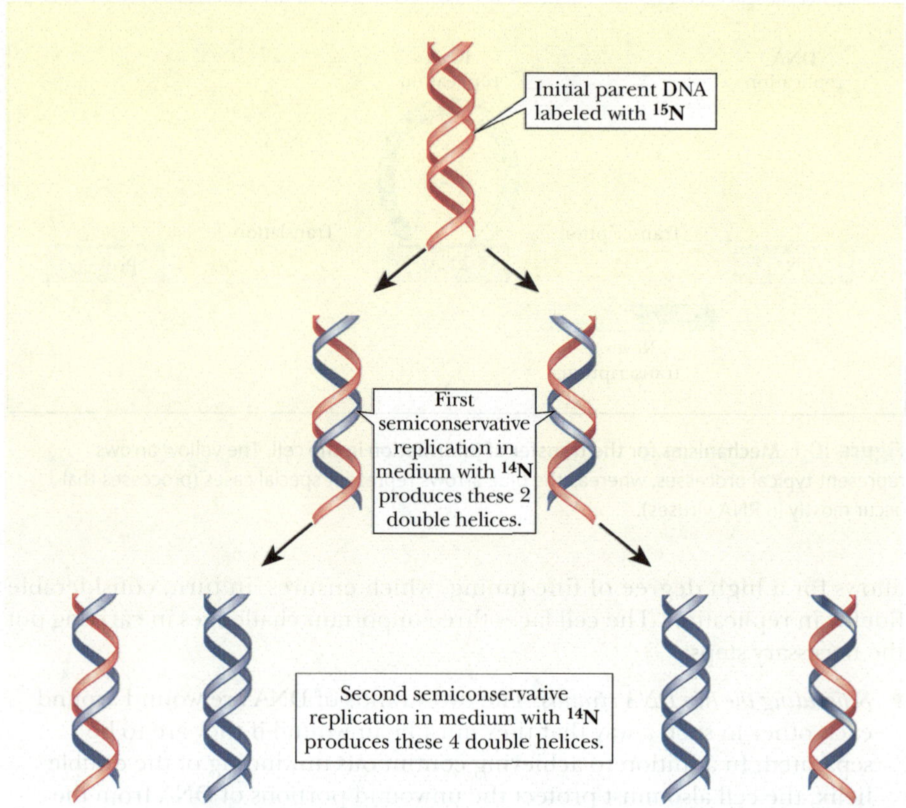

Figure 10.2 The production of new DNA strands in semiconservative replication. Red indicates the original ^{15}N strands, whereas blue indicates the new ^{14}N strands.

bottom of the tube, whereas light ^{14}N DNA (containing ^{14}N alone) appears at the top of the tube. DNA containing a 50–50 mixture of ^{14}N and ^{15}N appears at a position halfway between the two bands. In the actual experiment, this 50–50 hybrid DNA was observed after one generation, a result to be expected with semiconservative replication. After two generations in the lighter medium, half of the DNA in the cells should be the 50–50 hybrid and half should be the lighter ^{14}N DNA, a result also expected in semiconservative replication.

10-2B Direction of Replication

origin of replication the point at which the DNA double helix begins to unwind at the start of replication

replication forks in DNA replication, the points at which new DNA strands are formed

As seen in Figure 10.4, the DNA double helix unwinds during replication at a specific point called the **origin of replication** (OriC in *E. coli*). New polynucleotide chains are synthesized using each of the exposed strands as a template. Two possibilities exist for the growth of the new strands: synthesis can take place in both directions from the origin of replication or in one direction only. It has been established that DNA synthesis is bidirectional in most organisms, with the exception of a few viruses and plasmids. (Plasmids are rings of DNA that are found in bacteria and that replicate independently of the regular bacterial genome. They are discussed in Section 13-3.) For each origin of replication, there are two points **(replication forks)** at which new polynucleotide chains are formed. A "bubble" (also called an "eye") of newly synthesized DNA between regions of the original DNA is a manifestation of the advance of the two replication forks in opposite directions. This feature is also called a θ structure because of its resemblance to the lowercase Greek letter theta.

One such bubble (and one origin of replication) exists in the circular DNA of prokaryotes (Figure 10.4A). In eukaryotes, several origins of replication—and thus several bubbles—exist (Figure 10.4B). The bubbles grow larger and eventually merge, giving rise to two complete daughter DNAs. This bidirectional growth with respect to the origin of both new polynucleotide chains represents *net chain growth.* However, both new polynucleotide chains are always

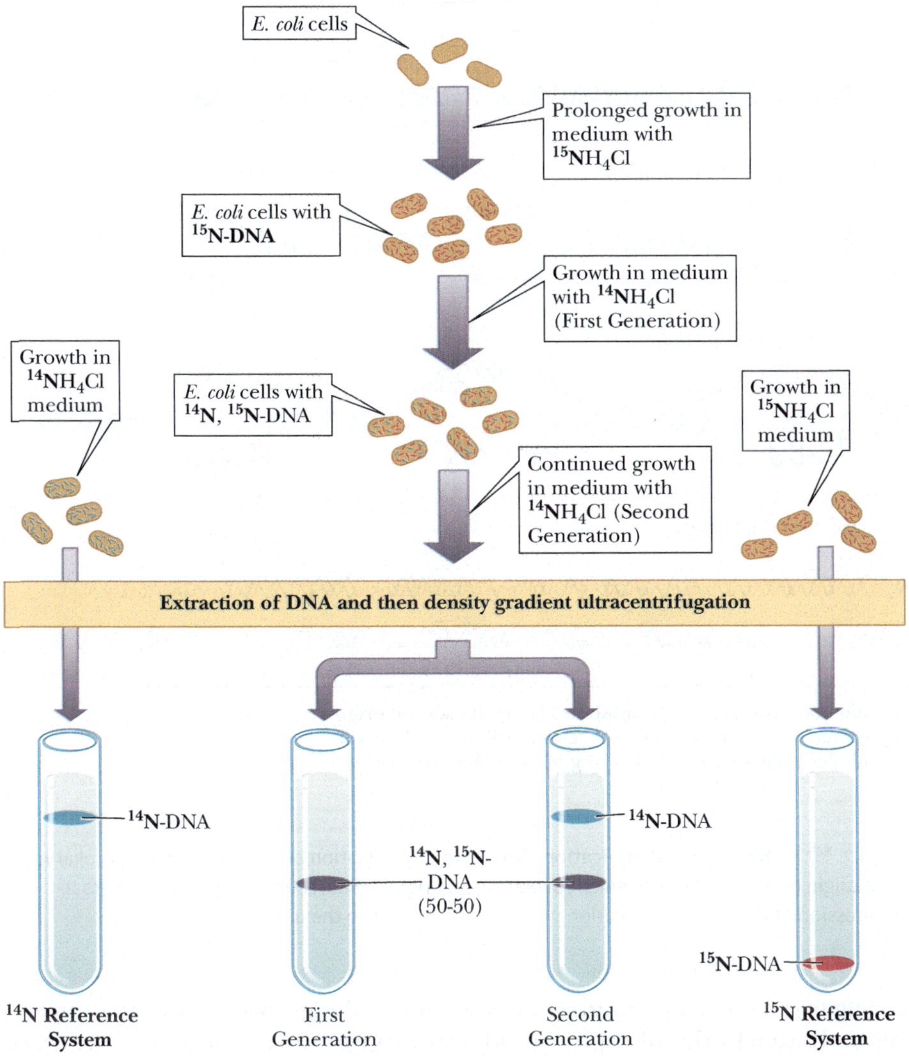

Figure 10.3 The experimental evidence for semiconservative replication. Heavy DNA labeled with ^{15}N forms a band at the bottom of the tube, whereas light DNA with ^{14}N forms a band at the top (see the reference systems on the left and right). DNA that forms a band at an intermediate position has one heavy strand and one light strand.

synthesized in the 5' → 3' direction, and the ways in which this is accomplished are discussed in Section 10-3.

10-3 DNA Polymerase

Learning Objectives

After reading this section, students will be able to:

LO 10.3.1 Describe the reaction mechanism by which nucleotides are added to a growing DNA chain.

LO 10.3.2 Justify the biological necessity of discontinuous replication using structural features of DNA.

LO 10.3.3 Distinguish between three DNA polymerases of *E. coli*.

10-3A Discontinuous DNA Replication

All synthesis of nucleotide chains occurs in the 5' → 3' direction from the perspective of the chain being synthesized. This is due to the nature of the reaction of DNA synthesis. The last nucleotide added to a growing chain has a 3'-hydroxyl on the sugar. The incoming nucleotide has a 5'-triphosphate on

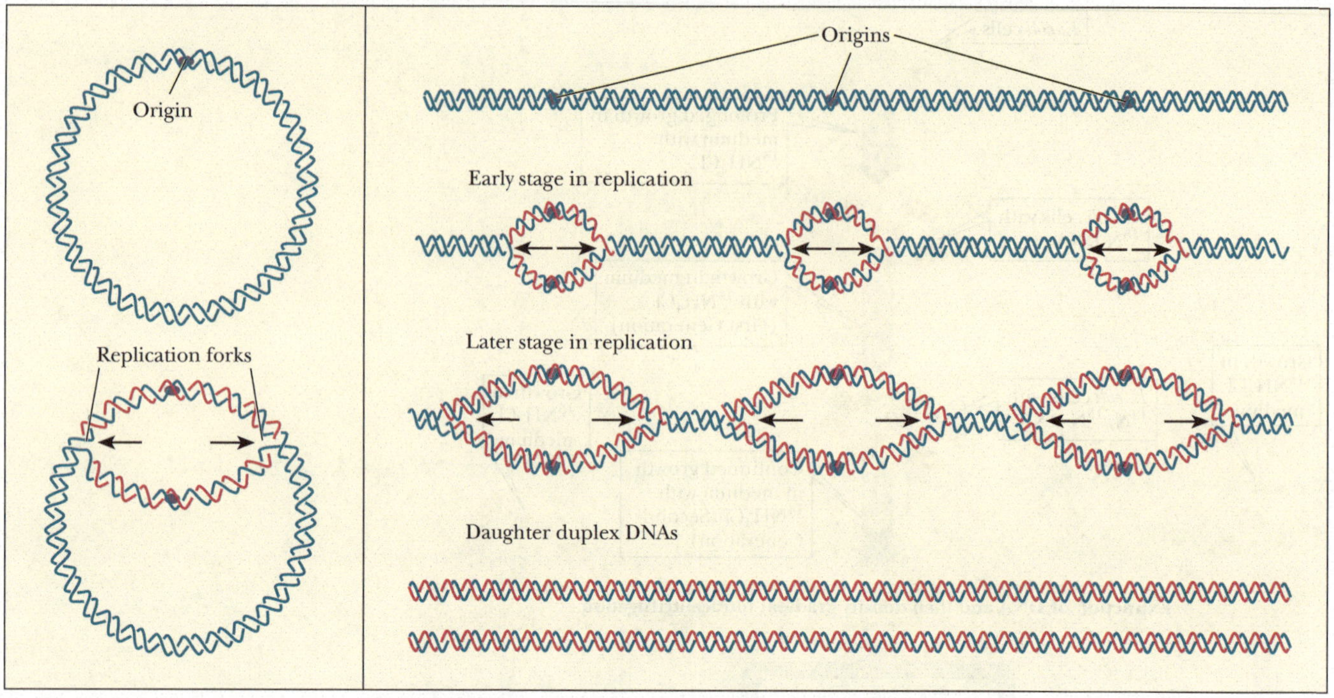

A Replication of the chromosome of *E. coli*, a typical prokaryote. There is one origin of replication, and there are two replication forks.

B Replication of a eukaryotic chromosome. There are several origins of replication, and there are two replication forks for each origin. The "bubbles" that arise from each origin eventually coalesce.

Figure 10.4 Bidirectional replication. Bidirectional replication of DNA is shown for prokaryotes (one origin of replication) and eukaryotes (several origins). Bidirectional replication refers to progression of the overall replication process with respect to the origin(s).

pyrophosphate a phosphorous oxyanion consisting of two phosphates bonded together

its sugar. The 3'-hydroxyl group at the end of the growing chain is a nucleophile. It attacks the phosphorus adjacent to the sugar in the nucleotide to be added to the growing chain, leading to the elimination of **pyrophosphate** and the formation of a new phosphodiester bond (Figure 10.5). We discussed nucleophilic attack by a hydroxyl group at length in the case of chymotrypsin (Section 7-5B); here we see another instance of this kind of mechanism, and several more instances of nucleophilic attacks by hydroxyl groups are discussed in other chapters. Therefore, it is helpful to keep this mechanism in mind. The further in depth we study DNA, the more the directionality of 5' → 3' can lead to confusion over which strand of DNA we are discussing. If you always remember that all synthesis of nucleotides occurs in the 5' → 3' direction from the perspective of the growing chain, it will be much easier to understand the processes to come.

This universal nature of synthesis presents a problem for the cell because as the DNA synthesis proceeds along a replication fork, the two strands are going in opposite directions.

▶ How can replication machinery proceed along the DNA in one direction if the two parent strands are antiparallel and if the new strands must be synthesized in the 5' to 3' direction?

In the discontinuous model of DNA replication, two different modes of polymerization occur for the two growing strands (Figure 10.6). One newly formed strand (the leading strand) is formed continuously from its 5' end to its 3' end at the replication fork on the exposed 3' to 5' template strand. The other strand (the lagging strand) is formed discontinuously in small fragments (typically

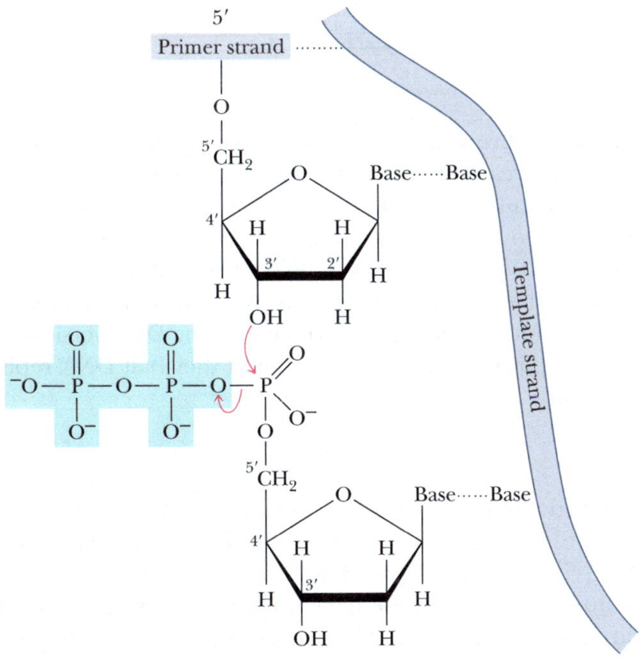

Figure 10.5 The addition of a nucleotide to a growing DNA chain. The 3'-hydroxyl group at the end of the growing DNA chain is a nucleophile. It attacks at the phosphorus adjacent to the sugar in the next nucleotide to be added to the growing chain. Pyrophosphate (shaded) is eliminated, and a new phosphodiester bond is formed.

1000–2000 nucleotides long), sometimes called **Okazaki fragments** after the scientists who first studied them. The fragments of the lagging strand are then linked together by an enzyme called **DNA ligase**.

Apply Your Knowledge 10.1 looks at how the chemistry of DNA synthesis can explain why two drugs work to help fight acquired immune deficiency syndrome (AIDS).

Okazaki fragments short sequences of DNA that are synthesized discontinuously in the lagging strand during DNA replication

DNA ligase the enzyme that links separate stretches of DNA

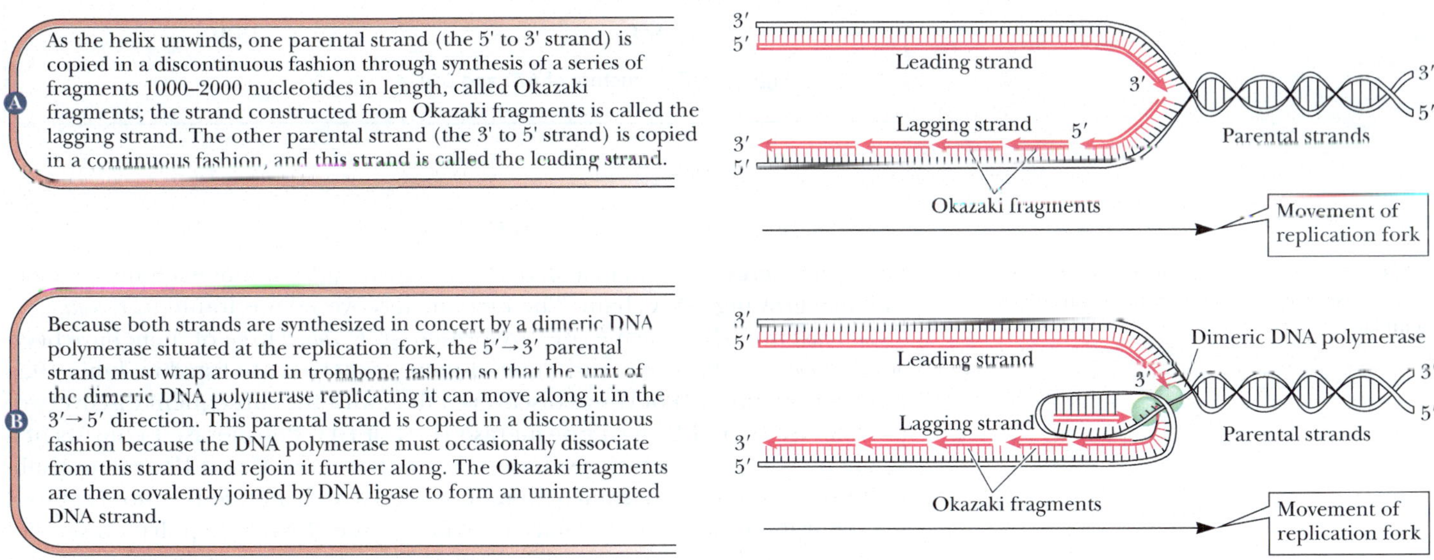

Figure 10.6 The discontinuous model for DNA replication. Newly synthesized DNA is shown in red. Because DNA polymerases only polymerize nucleotides 5' → 3', both strands must be synthesized in the 5' → 3' direction.

10.1 Apply Your Knowledge

DNA Structure

A nucleoside derivative called 3'-azido-3'-deoxythymidine (AZT) has been widely used in the treatment of AIDS, caused by human immunodeficiency virus (HIV), as has another nucleoside derivative called 2'-3'-dideoxyinosine (DDI). These compounds, shown in Figure 10.7, inhibit the reverse transcriptase enzyme that HIV uses to create a DNA version of its RNA genome. Propose a reason for the effectiveness of these two compounds. *Hint:* Assuming that these two compounds get added to the growing DNA version of the HIV genome in the replication process, how might they halt DNA replication of the viral genome?

Solution

Both compounds lack a hydroxyl group at the 3'-position of the sugar moiety that would be used in a nucleophilic attack on the next nucleotide to be added (recall Figure 10.5). Therefore, they cannot form the phosphodiester linkages found in the backbone of DNA, so they interfere with the replication of the AIDS virus by preventing nucleic acid synthesis.

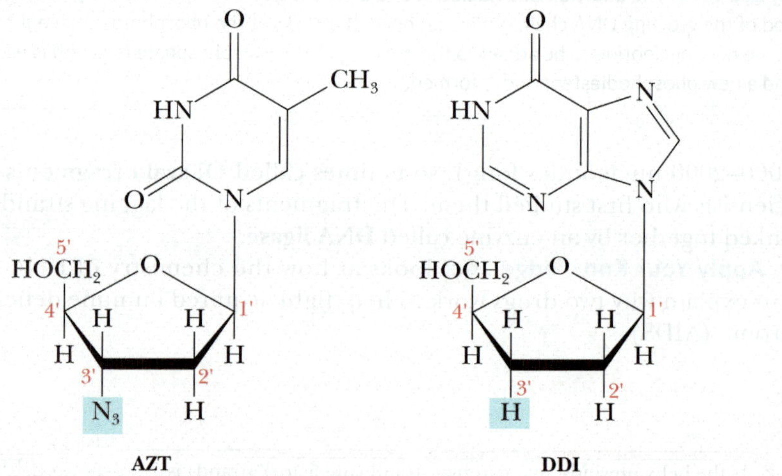

Figure 10.7 Structure of AZT and DDI.

10-3B DNA Polymerase from *E. coli*

DNA polymerase the enzyme that forms new DNA strands using deoxyribonucleotides and a DNA template

DNA polymerase enzymes catalyze the successive addition of each new nucleotide to a growing DNA chain. The first one discovered was found in *E. coli.*

At least five DNA polymerases are present in *E. coli.* Three of them have been studied more extensively, and some of their properties are listed in Table 10.1. DNA polymerase I (Pol I) was discovered first, with the subsequent discovery of polymerase II (Pol II) and polymerase III (Pol III). Polymerase I consists of a single polypeptide chain (or subunit), but polymerases II and III are multisubunit proteins that share some common subunits. Polymerase II is not required for replication; rather, it is strictly a repair enzyme. Two more polymerases, Pol IV and Pol V, have also been discovered. They, too, are repair enzymes, and both are involved in a unique repair mechanism called the SOS response (see **Biochemical Connections 10B**). Two important considerations regarding the polymerases are the speed of the synthetic reaction (turnover number) and the

Table 10.1 **Properties of Some DNA Polymerases of *E. coli***

Property	Pol I	Pol II	Pol III
Mass (kDa)	103	90	830
Turnover number (min^{-1})	600	30	1200
Processivity	200	1500	≥500,000
Number of subunits	1	≥4	≥10
Structural gene	*polA*	*polB**	*polC**
Polymerization 5' → 3'	Yes	Yes	Yes
Exonuclease 5' → 3'	Yes	No	No
Exonuclease 3' → 5'	Yes	Yes	Yes

*Polymerization subunit only. These enzymes have multiple subunits, and some of them are shared between both enzymes.

processivity, which is the number of nucleotides joined before the enzyme dissociates from the template (Table 10.1).

Polymerase III consists of a core enzyme responsible for the polymerization and 3' **exonuclease activity**—consisting of α-, ε-, and θ-subunits—and a number of other subunits, including a dimer of α-subunits responsible for DNA binding, and the γ-complex—consisting of γ-, δ-, δ'-, χ-, and ψ-subunits (also known as the clamp loader)—which allows the β-subunits to form a clamp that surrounds the DNA and slides along it as polymerization proceeds (Figure 10.8). Table 10.2 lists the subunit composition of the DNA Pol III complex. All of these polymerases add nucleotides to a growing polynucleotide chain but have different roles in the overall replication process. As can be seen in Table 10.1, DNA Pol III has the highest turnover number and a huge processivity compared to polymerases I and II.

If DNA polymerases are added to a single-stranded DNA template with all the deoxyribonucleoside triphosphates (dNTPs) necessary to make a strand of DNA, no reaction occurs. It was discovered that DNA polymerases cannot catalyze de novo synthesis. All three enzymes require the presence of a **primer**, a short oligonucleotide strand to which the growing polynucleotide chain is covalently attached in the early stages of replication. In essence, DNA polymerases must have

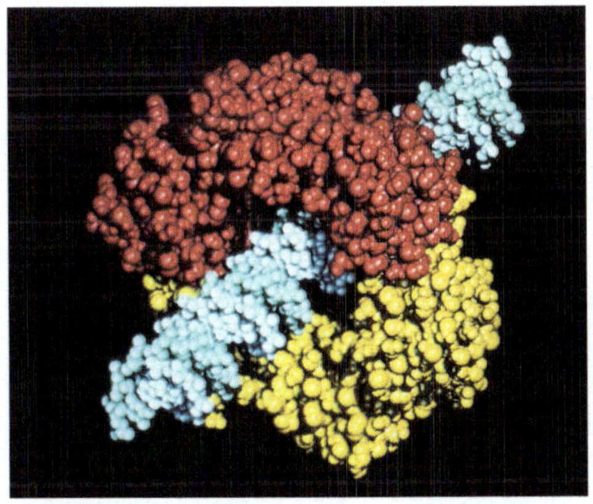

Courtesy of John Kuriyan, University of California, Berkeley

Figure 10.8 **The dimer of β-subunits of DNA polymerase III bound to DNA.** One monomer is shown in yellow, the other in red. The dimer forms a closed loop around the DNA (shown in blue). The rest of the polymerase III holoenzyme is not shown. The remainder of the holoenzyme consists of the core enzyme responsible for the polymerization and the 3' exonuclease activity (α-, ε-, and θ-subunits) and the γ-complex (consisting of γ-, δ-, δ', χ-, and ψ-subunits, also known as the clamp loader), which allows the β-subunits to form a clamp that surrounds the DNA and slides along it as polymerization proceeds. *(Adapted from Kong, X. P., et al. Three-Dimensional Structure of the Subunit of E. Coli DNA Polymerase Holoenzyme: A Sliding DNA Clamp. Cell* 69, *425–437 (1992).)*

processivity the number of nucleotides incorporated in a growing DNA chain before the DNA polymerase dissociates from the template DNA

exonuclease activity in DNA replication, removal of nucleotides from the end of a nucleic acid molecule

primer in DNA replication, a short stretch of RNA hydrogen bonded to the template DNA to which the growing DNA strand is bonded at the start of synthesis

Table 10.2 **The Subunits of *E. coli* DNA Polymerase III Holoenzyme**

Subunit	Mass (kDa)	Structural Gene	Function
α	130.5	*polC* (*dnaE*)	Polymerase
ε	27.5	*dnaQ*	3'-exonuclease
θ	8.6	*holE*	α, ε assembly?
τ	71	*dnaX*	Assembly of holoenzyme on DNA (clamp loader)
β	41	*dnaN*	Sliding clamp, processivity
γ	47.5	*dnaX* (*Z*)	Part of the γ-complex* (clamp loader)
δ	39	*holA*	Part of the γ-complex*
δ'	37	*holB*	Part of the γ-complex*
χ	17	*holC*	Part of the γ-complex*
ψ	15	*holD*	Part of the γ-complex*

*Subunits γ, δ, δ', χ, and ψ form the so-called γ-complex, which is responsible for the placement of the β-subunits (the sliding clamp) on the DNA. The γ-complex is referred to as the clamp loader. The δ- and τ-subunits are encoded by the same gene.

a nucleotide with a free 3'-hydroxyl already in place so that they can add the first nucleotide as part of the growing chain. In natural replication, this primer is RNA.

The DNA polymerase reaction requires all four dNTPs—dTTP, dATP, dGTP, and dCTP. The dNTPs serve two roles: they are the building blocks of the newly synthesized strands, and they provide the energy to drive the reaction when the high-energy phosphate bonds are hydrolyzed:

$$\text{dNTP} \rightarrow \text{inserted dNMP} + \text{PP}_i$$

Mg^{2+} and a DNA template are also necessary. Because of the requirement for an RNA primer, all four ribonucleoside triphosphates—ATP, UTP, GTP, and CTP—are needed as well; they are incorporated into the primer. The primer (RNA) is hydrogen bonded to the template (DNA); the primer provides a stable framework on which the nascent chain can start to grow. The newly synthesized DNA strand begins to grow by forming a covalent linkage to the free 3'-hydroxyl group of the primer.

It is now known that DNA Pol I has a specialized function in replication—repairing and "patching" DNA—and that DNA Pol III is the enzyme primarily responsible for the polymerization of the newly formed DNA strand. The major function of DNA polymerases II, IV, and V is as repair enzymes. The exonuclease activities listed in Table 10.1 are part of the proofreading and repair functions of DNA polymerases, a process by which incorrect nucleotides are removed from the polynucleotide so that the correct nucleotides can be incorporated. The 3' → 5' exonuclease activity, which all three polymerases possess, is part of the **proofreading** function in which incorrect nucleotides are removed in the course of replication and are replaced by the correct ones. Proofreading is done one nucleotide at a time. The 5' → 3' exonuclease activity clears away short stretches of nucleotides during **repair**, usually involving several nucleotides at a time. This is also how the RNA primers are removed.

proofreading the process of removing incorrect nucleotides when DNA replication is in progress

repair the enzymatic removal of incorrect nucleotides from DNA and their replacement by correct ones

10-4 Proteins Required for DNA Replication

Learning Objectives

After reading this section, students will be able to:

LO 10.4.1 Explain the functions of key proteins involved in the DNA replication process.

LO 10.4.2 Describe the common structural features and functions of DNA polymerases.

DNA replication is carried out in all organisms by a multiprotein complex called the **replisome**. In bacteria, the replisome consists of 13 different proteins that work together to synthesize DNA on the leading and lagging strands.

replisome a complex of DNA polymerases, primase, gyrase, helicase, ligase, and other proteins at the replication fork

Two issues arise in separating the two strands of the original DNA so that it can be replicated. The first is how to achieve continuous unwinding of the double helix. To complicate matters, prokaryotic DNA exists in a supercoiled, closed-circular form (see Section 9-3C). The second problem is how to protect single-stranded stretches of DNA that are exposed to intracellular nucleases as a result of the unwinding.

10-4A Supercoiling and Replication

An enzyme called **DNA gyrase** (a class II topoisomerase) catalyzes the conversion of relaxed, circular DNA with a nick in one strand to the supercoiled form—with the nick sealed—that is found in normal prokaryotic DNA (Figure 10.9). A slight unwinding of the helix before the nick is sealed introduces the supercoiling. The energy required for the process is supplied by the hydrolysis of ATP. Some

DNA gyrase an enzyme that introduces supercoiling into closed circular DNA

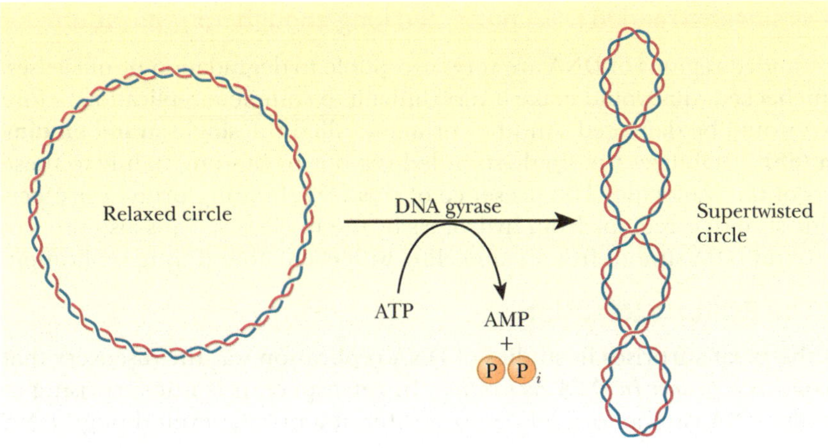

Figure 10.9 **DNA gyrase introduces supertwisting in circular DNA.**

evidence exists that DNA gyrase causes a double-strand break (DSB) in DNA in the process of converting the relaxed, circular form to the supercoiled form.

▶ How does replication work with supercoiled DNA?

In replication, the role of the gyrase is somewhat different. The prokaryotic DNA is negatively supercoiled in its natural state; however, opening the helix during replication would introduce positive supercoils ahead of the replication fork. If the replication fork continued to move, the torsional strain of the positive supercoils would eventually make further replication impossible. DNA gyrase fights these positive supercoils by putting negative supercoils ahead of the replication fork (Figure 10.10). A helix-destabilizing protein, called a **helicase**, promotes unwinding by binding at the replication fork. A number of helicases are known, including the *dnaB protein* and the *rep protein*.

helicase a protein that unwinds the double helix of DNA in the process of replication

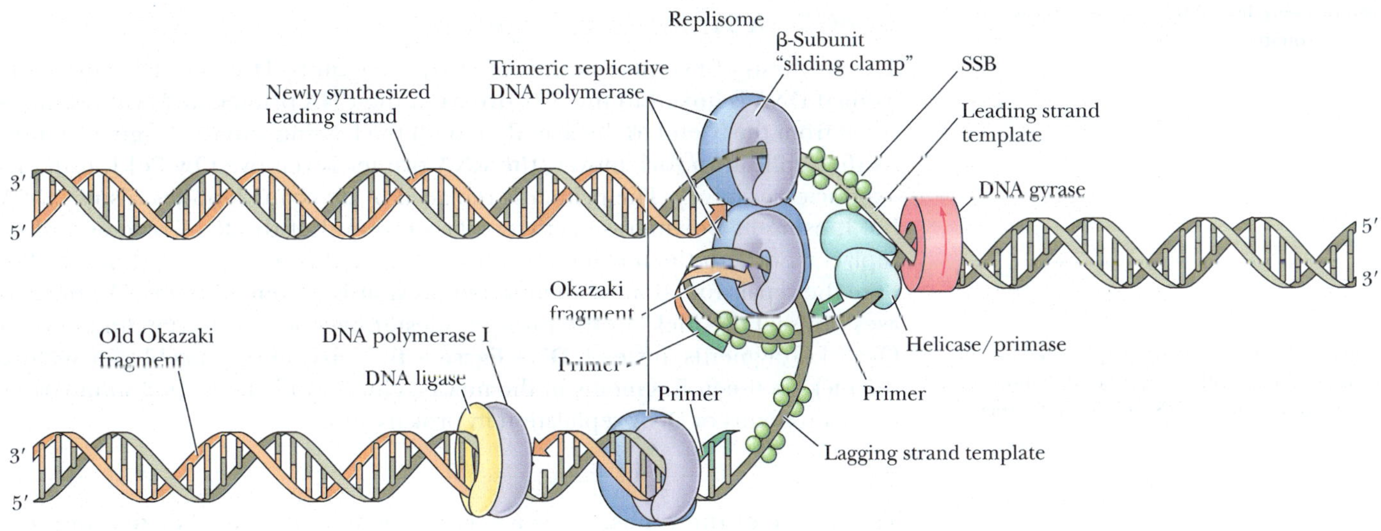

Figure 10.10 **General features of a replication fork.** The DNA duplex is unwound by the action of DNA gyrase and helicase, and the single strands are coated with single-strand binding proteins (SSBs). Primase periodically generates primers on the lagging strand. Each half of the dimeric replicative polymerase is a holoenzyme bound to its template strand by a β-subunit sliding clamp. DNA Pol I and DNA ligase act downstream on the lagging strand to remove RNA primers, replace them with DNA, and ligate the Okazaki fragments.

▶ How is single-stranded DNA protected long enough for replication?

single-strand binding protein (SSB) in DNA replication, a protein that protects exposed single-strand sections of DNA from nucleases; it also prevents the unwound DNA strands from reannealing with each other during replication

Single-stranded regions of DNA are very susceptible to degradation by nucleases. If left unchecked, this would make it very difficult to complete replication before the DNA would be damaged. Another protein, called the **single-strand binding protein (SSB)**, stabilizes the single-stranded regions by binding tightly to these portions of the molecule. The presence of this DNA-binding protein protects the single-stranded regions from hydrolysis by the nucleases. SSBs also prevent the unwound DNA strands from reannealing with each other during replication.

10-4B The Primase Reaction

One of the great surprises in studies of DNA replication was the discovery that *RNA serves as a primer in DNA replication.* In retrospect, it is not surprising at all, because RNA can be formed de novo without a primer, even though DNA synthesis requires a primer. This finding lends support to theories of the origin of life in which RNA, rather than DNA, was the original genetic material. The fact that RNA has been shown to have catalytic ability in several cases has added support to that theory (Chapter 11). A primer is needed in replication because the main replicative DNA polymerases can only add nucleotides to an existing strand of nucleotides. A primer in DNA replication must have a free 3'-hydroxyl to which the growing chain can attach, and both RNA and DNA can provide this group. The primer activity of RNA was first observed in vivo. In some of the original in vitro experiments, DNA was used as a primer because a primer consisting of DNA was expected. Living organisms are far more complex than isolated molecular systems and, as a result, can be full of surprises for researchers.

▶ Where does the primer come from?

primase the enzyme that makes a short section of RNA to act as a primer for DNA synthesis

primosomes the complexes at replication forks that are responsible for creating the RNA primers that are used to start DNA synthesis; they consist of seven proteins, including primase, helicase, and five other proteins

It has subsequently been found that a separate enzyme, called **primase**, is responsible for copying a short stretch of the DNA template strand to produce the RNA primer sequence. The first primase was discovered in *E. coli.* The enzyme consists of a single polypeptide chain, with a molecular weight of about 60,000 Da (i.e., 60,000 g/mol). There are 50–100 molecules of primase in a typical *E. coli* cell, and they are complexed with other proteins to form **primosomes**. The general features of DNA replication, including the use of an RNA primer, appear to be common to all prokaryotes.

10-4C DNA Synthesis and Ligation

DNA ligase the enzyme that catalyzes the formation of the phosphodiester bonds between Okazaki fragments in the DNA replication process

The synthesis of two new strands of DNA is begun by DNA Pol III. The newly formed DNA is linked to the 3'-hydroxyl of the RNA primer, and synthesis proceeds from the 5' end to the 3' end on both the leading and the lagging strands. As the replication fork moves, the RNA primer is removed by Pol I, using its exonuclease activity. The primer is replaced by deoxynucleotides, also by DNA Pol I, using its polymerase activity. (The removal of the RNA primer and its replacement with the missing portions of the newly formed DNA strand by Pol I are the repair function we mentioned previously.) None of the DNA polymerases can seal the nicks in the phosphodiester backbone that remain between Okazaki fragments; instead, **DNA ligase** is the enzyme responsible for sealing the nicks between fragments in the new lagging strand. Table 10.3 summarizes the main points of DNA replication in prokaryotes.

▶ What is the nature of the replisome?

The nature of the replisome is a very active area of study. For decades, the dogma has been that the replisome contains two molecules of Pol III, one for the leading strand and one for the lagging strand. While this was intuitive, studies from 2010 to 2012 have shown, instead, that three Pol III enzymes are involved, as shown in Figure 10.10. By labeling specific molecules at the replication fork with fluorescent markers and watching the location of bound

Table 10.3 A Summary of DNA Replication in Prokaryotes

1. DNA synthesis is bidirectional. Two replication forks advance in opposite directions from an origin of replication.
2. The direction of DNA synthesis is from the 5' end to the 3' end of the newly formed strand. One strand (the leading strand) is formed continuously, while the other strand (the lagging strand) is formed discontinuously as Okazaki fragments.
3. Five DNA polymerases have been found in *E. coli.* Polymerase III is primarily responsible for the synthesis of new strands. The first polymerase enzyme discovered, Pol I, is involved in synthesis, proofreading, and repair. Polymerases II, IV, and V function as repair enzymes under unique conditions.
4. DNA gyrase introduces a swivel point in advance of the movement of the replication fork. A helix-destabilizing protein, a helicase, binds at the replication fork and promotes unwinding. The exposed single-stranded regions of the template DNA are stabilized by single-strand binding proteins (SSBs).
5. Primase catalyzes the synthesis of RNA primers that help start DNA replication.
6. The primers are removed by Pol I, which also replaces the primers with deoxyribonucleotides. DNA ligase seals the remaining nicks between the Okazaki fragments.

proteins, researchers were able to show that three molecules of Pol III are bound at or near the replication fork. By comparing the locations of Pol III and single-stranded binding proteins, they concluded that a new Pol III is used for each Okazaki fragment, and that the replisome has one Pol III dedicated to leading strand synthesis and two dedicated to lagging strand synthesis.

▶ How do clamp loaders work?

As shown in Figure 10.8, *E. coli* DNA Pol III has a dimer of β-subunits that forms a closed ring around the DNA, called a sliding clamp. This is a complication for the entire process of replication, because somehow this ring has to get around the DNA. The clamp cannot spontaneously surround the DNA chain because the dimer is a closed circle. Instead, the part of the Pol III enzyme that is called the clamp loader opens the sliding clamp and inserts the DNA chain. In all species, these clamp loaders evolved independently to be very similar. In *E. coli*, the clamp loader is the γ- and τ-complexes, with the γ-complex being composed of several subunits, as described in Table 10.2. All clamp loaders are pentameric enzymes that are members of a family of ATPases called the AAA+ superfamily.

10-5 Proofreading and Repair in Replication

Learning Objectives

After reading this section, students will be able to:

LO 10.5.1 Explain proofreading processes in DNA replication.

LO 10.5.2 Distinguish between the mismatch repair, base excision repair, nucleotide excision repair, and double-strand break repair mechanisms of DNA repair.

DNA replication takes place only once each generation in each cell, unlike other processes, such as RNA and protein synthesis, which occur many times. It is essential that the fidelity of the replication process be as high as possible to prevent **mutations**, which are errors in replication. Mutations are frequently harmful, even lethal, to organisms. Nature has devised several ways to ensure that the nucleobase sequence of DNA is copied faithfully.

mutations changes in DNA sequence causing subsequent changes in an organism that can be transmitted genetically

10-5A Proofreading in Replication

Proofreading refers to the identification and removal of incorrect nucleotides immediately after they are added to the growing DNA during the replication process. The proofreading process in cells is so accurate that it prevents errors in replication from happening more than once every 10^9 to 10^{10} base pairs. DNA Pol I has three active sites, as demonstrated by Hans Klenow in 1970. Pol I can be cleaved into two major fragments. One of them (the Klenow fragment) contains the polymerase activity and the proofreading activity. The other fragment contains the 5' → 3' exonuclease activity.

Figure 10.11 shows the proofreading activity of Pol I. Errors in hydrogen bonding lead to the incorporation of an incorrect nucleotide into a growing DNA chain once in every 10^4 to 10^5 base pairs. However, Pol I uses its 3' exonuclease activity to remove the incorrect nucleotide, and replication resumes when the correct nucleotide is added, also by Pol I. Although the specificity of hydrogen-bonded base pairing accounts for one error in every 10^4 to 10^5 base pairs, the proofreading function of DNA polymerase improves the fidelity of replication to one error in every 10^9 to 10^{10} base pairs.

10-5B Mutagens

mutagens agents that bring about a mutation; examples include radiation and chemical substances that alter DNA

In addition to experiencing those spontaneous mutations caused by misreading the genetic code, organisms are frequently exposed to **mutagens**, agents that produce mutations. Common mutagens include ultraviolet light, ionizing radiation (radioactivity), and various chemical agents, all of which lead to changes in DNA over and above those produced by spontaneous mutation. The most common effect of ultraviolet light is the creation of pyrimidine dimers (Figure 10.12). The π electrons from two carbons on each of two pyrimidines form a cyclobutyl ring, which distorts the normal shape of the DNA and interferes with replication and transcription. Alternatively, chemical damage, which is often caused by free radical oxidation, can lead to a break in the phosphodiester backbone of the DNA strand (Figure 10.13). This is one of the primary reasons that antioxidants are so popular as dietary supplements.

10-5C Mismatch Repair

mismatch repair (MMR) a type of DNA repair that begins when repair enzymes find two bases that are incorrectly paired

When damage has managed to escape the normal exonuclease activities of DNA polymerases I and III, prokaryotes have a variety of other repair mechanisms at their disposal. In **mismatch repair (MMR)**, a class of enzymes called "Mut" proteins recognize that two bases are incorrectly paired (Figure 10.14). The area with the mismatch is removed, and DNA polymerases replicate the area again. For example, Figure 10.14A shows a DNA replication mistake involving a G–T mismatch. The repair enzymes take advantage of the lack of DNA methylation

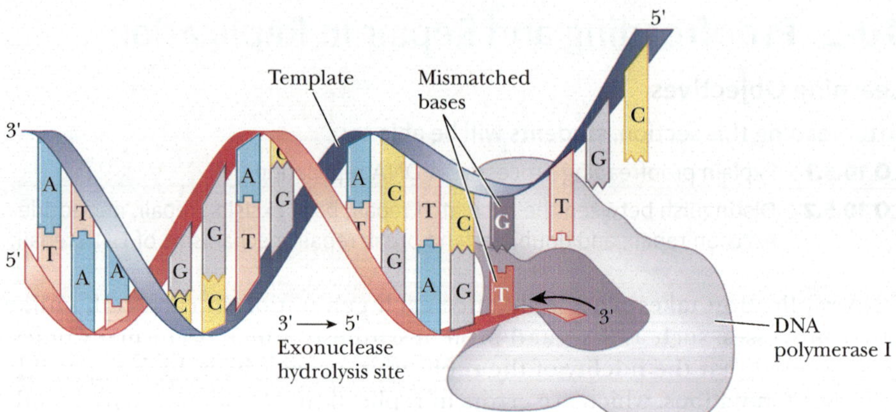

Figure 10.11 DNA polymerase proofreading. The 3' → 5' exonuclease activity of DNA polymerase I removes nucleotides from the 3' end of the growing DNA chain.

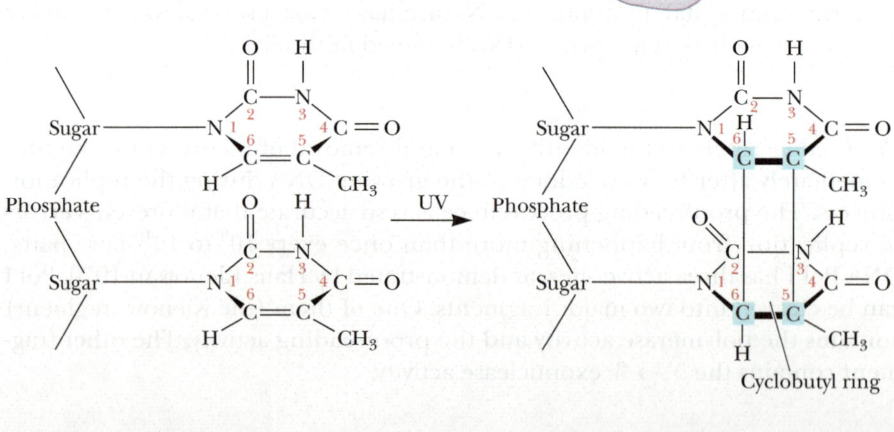

Figure 10.12 UV irradiation causes dimerization of adjacent thymine bases. A cyclobutyl ring is formed when two covalent bonds are formed between adjacent pyrimidine rings. One covalent bond forms between the carbons numbered 5, and the other forms between the carbons numbered 6. Normal base pairing is disrupted by the presence of such dimers.

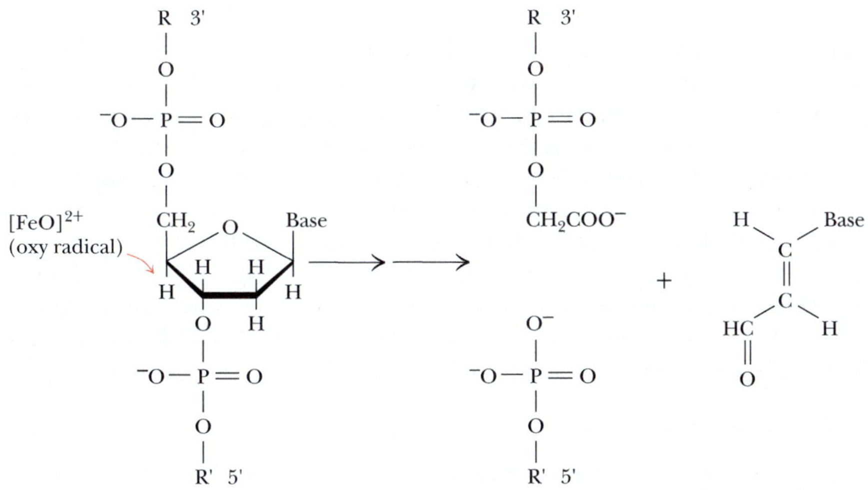

Figure 10.13 Oxidation damage. Oxygen radicals, in the presence of metal ions such as Fe^{2+}, can destroy sugar rings in DNA, breaking the strand.

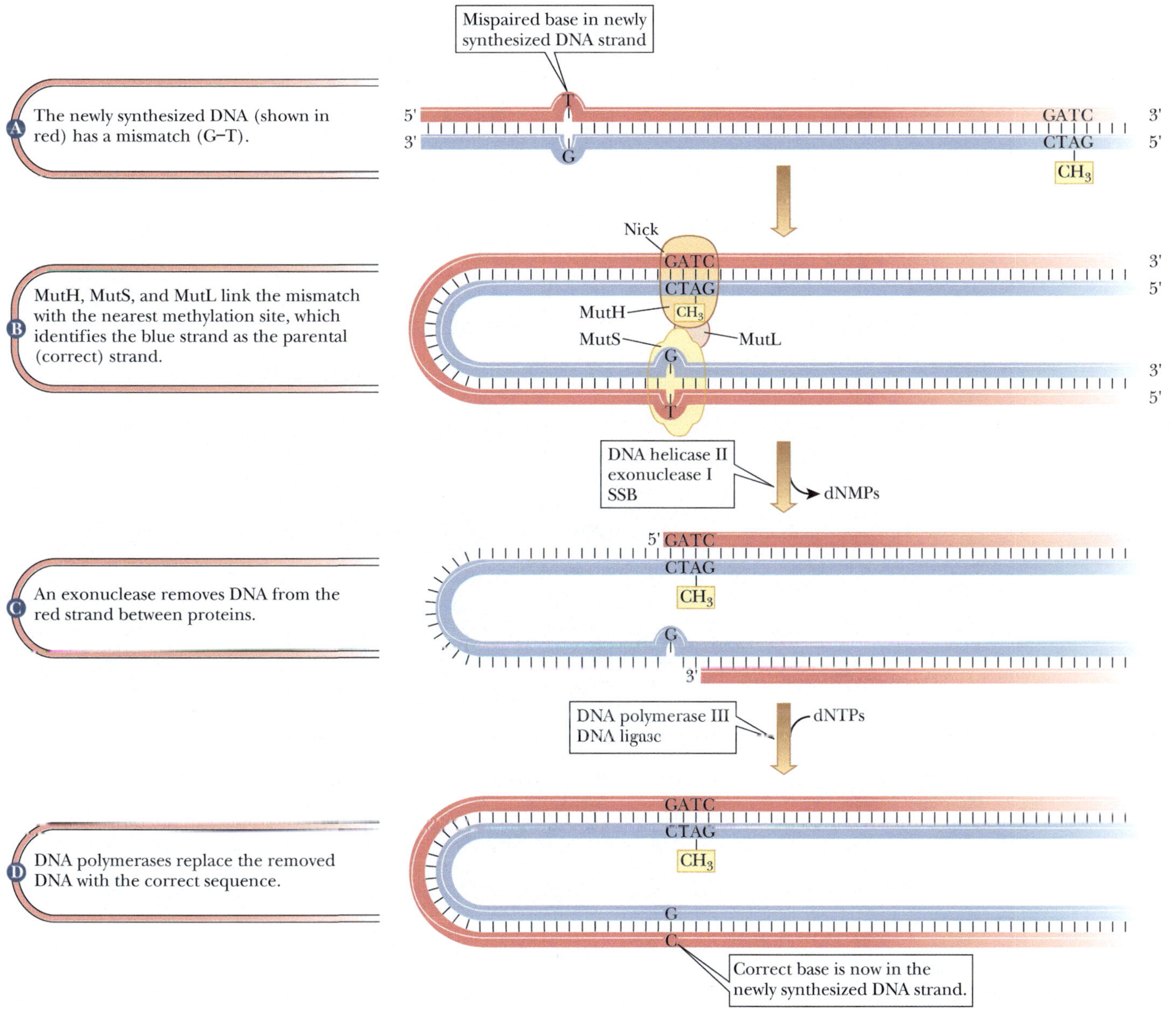

Figure 10.14 Mismatch repair in *E. coli. (From* Principles of Biochemistry, *by Albert L. Lehninger, David L. Nelson, and Michael M. Cox. Copyright © 1993 by W. H. Freeman & Co.)*

on new DNA strands to identify which nucleobase is the incorrect one. Because the parental strands contain methylated adenines, the enzymes can distinguish the parental strands from newly synthesized daughter strands without the methylated bases. Thus, in Figure 10.14A, the T is the mistake and not the G because the strand with the G has a methylated adenine base seen on the right-hand side of the figure. Several proteins are then involved in the repair process. In Figure 10.14B, *MutH, MutS,* and *MutL* proteins form a loop between the mistake and a methylation site. DNA helicase II helps unwind the DNA. *Exonuclease I* removes the section of DNA containing the mistake (Figure 10.14C). Single-stranded binding proteins protect the template (blue) strand from degradation. DNA Pol III then fills in the missing nucleotides, and DNA ligase seals the nick in the repaired daughter strand (Figure 10.14D).

base excision repair (BER) a type of DNA repair for small, non-helix distorting damage that begins with an enzyme removing a damaged base, followed by removal and replacement of the rest of the nucleotide

10-5D Base Excision Repair

Another repair system is called **base excision repair (BER)** (Figure 10.15). This repair system is used by the cell when the damage is small and non-helix

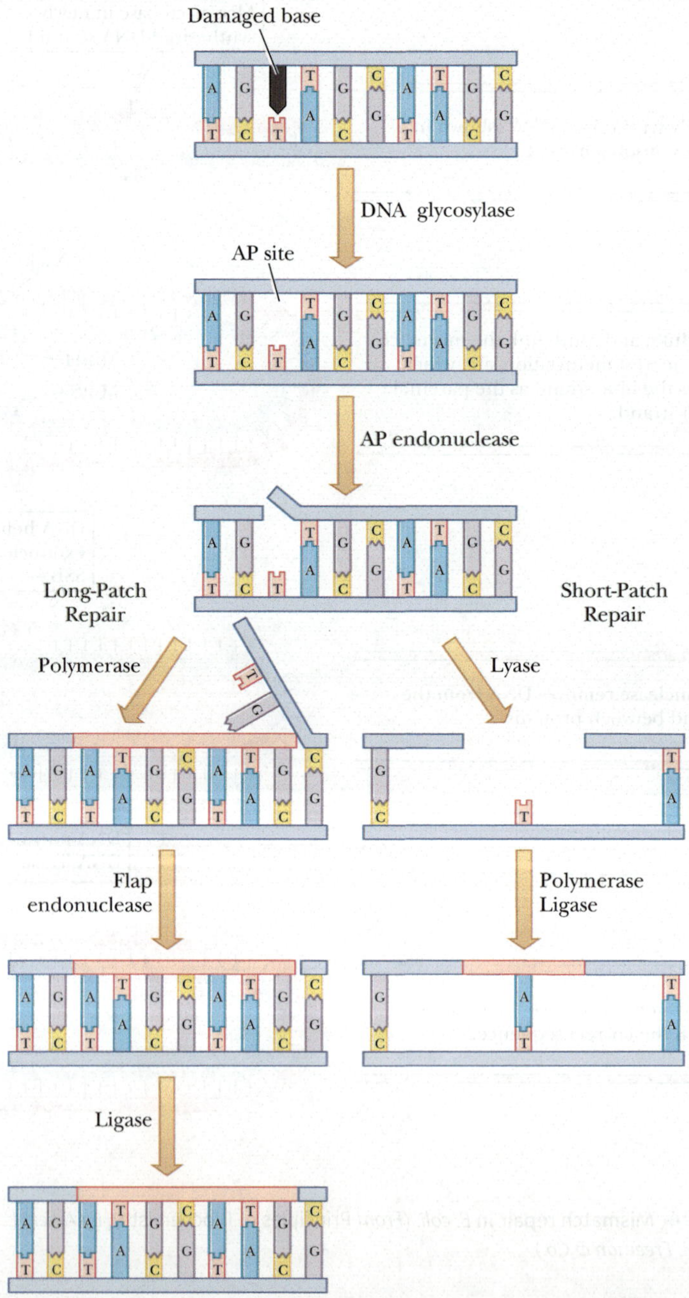

Figure 10.15 **Base excision repair (BER).** A damaged base (▮) is excised from the sugar–phosphate backbone by DNA glycosylase, and an AP endonuclease creates a break at the AP site. The repair process then follows either long-patch BER or short-patch BER.

distorting. A nucleobase that has been damaged by oxidation or other chemical modification is removed by *DNA glycosylase,* leaving an *AP site,* so called because it is apurinic or apyrimidinic (without a purine or pyrimidine nucleobase), but the site still retains the sugar-phosphate backbone. An *apurinic/apyrimidinic (AP) endonuclease* then creates a break in the sugar–phosphate backbone. At this point, two possible routes can be taken: long-patch BER or short-patch BER. The mechanisms behind which route the cell chooses are still being investigated. In long-patch BER, a DNA polymerase not only replaces the damaged AP site but also adds a few more nucleotides, displacing the AP site as well as a few other nucleotides and creating a flap. A flap endonuclease then digests the flap, and DNA ligase seals the nick. In short-patch BER, after the break is created by the AP endonuclease, a lyase removes only the phosphate and sugar backbone moieties of the AP site. Then, DNA polymerase adds the one correct nucleotide, and DNA ligase seals the nick.

10-5E Nucleotide Excision Repair

Nucleotide excision repair (NER) is used by cells for DNA damage caused by ultraviolet radiation or chemical means that lead to deformed DNA structures. Figure 10.16 shows how a large section of DNA containing the lesion is removed by *ABC excinuclease* (excinucleases, or excision nucleases, remove segments of nucleotides during DNA repair). DNA polymerase I and DNA ligase then work to fill in the gap. This type of repair is also the most common repair for ultraviolet radiation damage in mammals. Defects in NER mechanisms can have drastic consequences. One of the most remarkable examples is the disease *xeroderma pigmentosum.* Individuals with this condition develop numerous skin cancers at an early age because they do not have the repair system to correct damage caused by ultraviolet light. The endonuclease that nicks the damaged portion of the DNA is probably the missing enzyme. The repair enzyme that recognizes the lesion has been named XPA protein after the disease. The cancerous lesions eventually spread throughout the body, causing death. **Biochemical Connections 10A** describes another aspect of DNA replication fidelity.

nucleotide excision repair (NER) a type of DNA repair in which damage that leads to deformed DNA structure is repaired by removal of a section of DNA containing the damage

10-5F Double-Strand Break Repair

When both strands of DNA are broken, the result is called a *double-strand break (DSB),* and these are a big threat to the stability of the genome. One repair mechanism that exists to handle a DSB is called **nonhomologous end-joining (NHEJ)**, as shown

nonhomologous end-joining (NHEJ) a type of DNA repair used for double-strand breaks in DNA

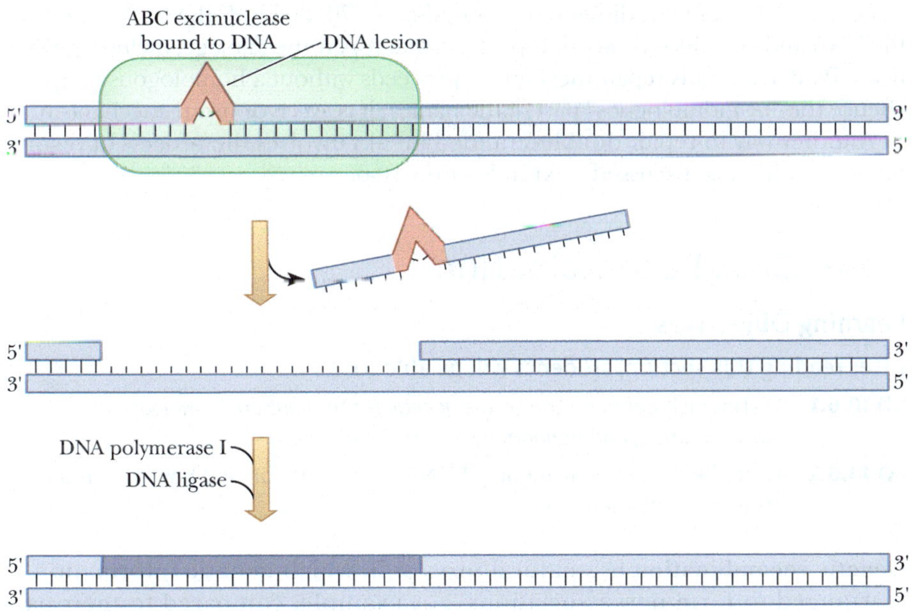

Figure 10.16 Nucleotide excision repair. When a serious lesion, such as a pyrimidine dimer, is detected, ABC excinuclease binds to the region and cuts out a large piece of DNA, including the lesion. DNA polymerase I and DNA ligase then resynthesize and seal the DNA.
From Principles of Biochemistry, *by Albert L. Lehninger, David L. Nelson, and Michael M. Cox. Copyright © 1993 by W. H. Freeman & Co.)*

10A Biochemical Connections • Genetics

Why Does DNA Contain Thymine and Not Uracil?

Given that both uracil and thymine base-pair with adenine, why does RNA contain uracil and DNA contain thymine? Scientists now believe that RNA was the original hereditary molecule and that DNA developed later. If we compare the structure of uracil and thymine (Figure 10.17), the only difference is the presence of a methyl group at C-5 of thymine. This group is not on the side of the molecule involved in base pairing. Because carbon sources and energy are required to methylate a molecule, there must be a reason for DNA developing with a base that does the same thing as uracil but that requires more energy to produce. The answer is that thymine helps guarantee replication fidelity. One of the most common spontaneous mutations of bases is the natural deamination of cytosine, as shown in Figure 10.18.

At any moment, a small but finite number of cytosines lose their amino groups to become uracil. Imagine that during replication, a C–G base pair separates. If at that moment the C deaminates to U, it would tend to base-pair to A instead of to G. If U were a natural base in DNA, the DNA polymerases would just line up an adenine across from the uracil, and there would be no way to know that the uracil was a mistake. This would lead to a much higher level of mutation during replication. Because uracil is an unnatural base in DNA, DNA polymerases can recognize it as a mistake and can replace it. Thus, the incorporation of thymine into DNA, though energetically more costly, helps ensure that the DNA is replicated faithfully. ◗

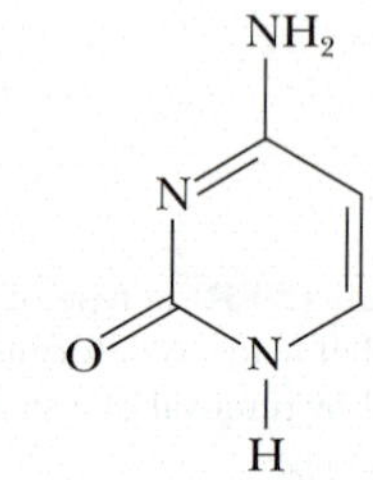

Cytosine
(2-oxy-4-amino pyrimidine)

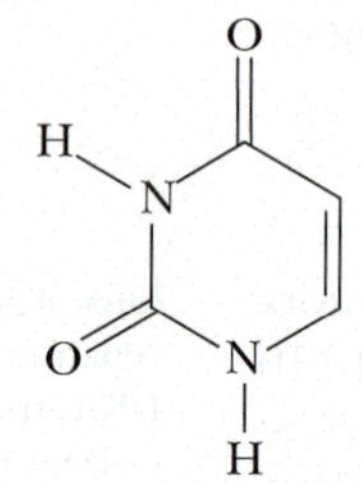

Uracil
(2-oxy-4-oxy pyrimidine)

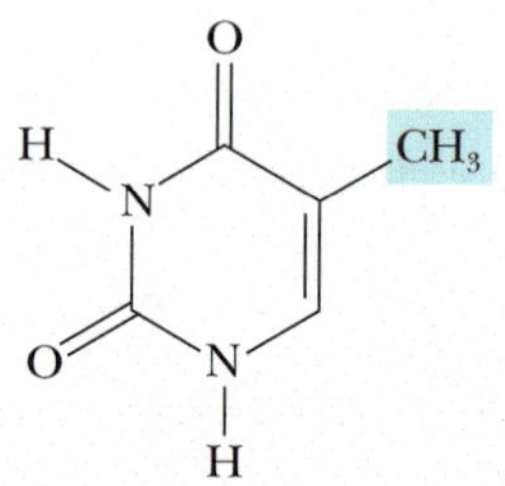

Thymine
(2-oxy-4-oxy 5-methyl pyrimidine)

Figure 10.17 Structures of cytosine, uracil, and thymine.

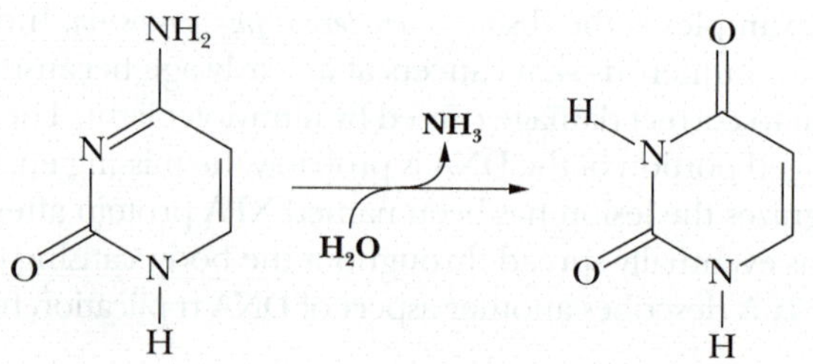

Cytosine **Uracil**

Figure 10.18 Deamination of cytosine.

in Figure 10.19. A heterodimeric protein called Ku70/80 binds the broken ends of the DNA and recruits several other proteins that repair the damage, including DNA ligase IV. Because this repair mechanism proceeds without a homologous template (hence the "*non*homologous" part of the name), it is an error-prone mechanism.

Another way to repair double-stranded breaks involves the process of recombination, which is discussed next in Section 10-6.

10-6 DNA Recombination

Learning Objectives

After reading this section, students will be able to:

LO 10.6.1 Distinguish between the terms *genetic recombination*, *homologous recombination*, and *nonhomologous recombination*.

LO 10.6.2 Describe the Holliday model of DNA recombination and how to demonstrate it experimentally.

genetic recombination general term for several processes whereby genetic information is rearranged

Genetic recombination is a natural process in which genetic information is rearranged to form new associations. For example, compared to their parents, progeny may have new combinations of traits because of recombination.

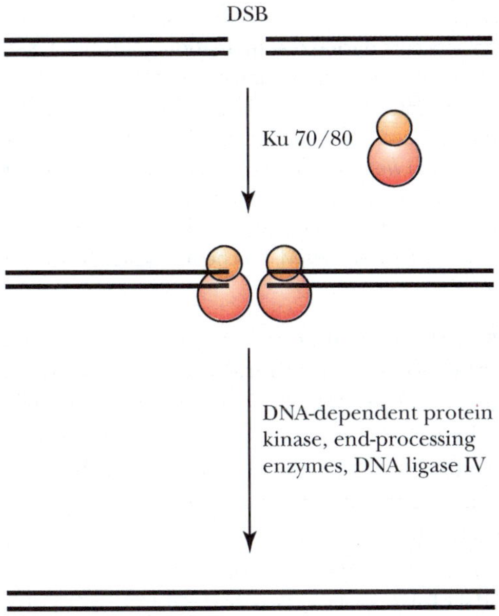

Figure 10.19 Nonhomologous end-joining (NHEJ): one of the mechanisms for repairing a double-strand break (DSB) in DNA. *(From GARRETT/GRISHAM,* **Biochemistry,** *4E. Cengage Learning, 2009.)*

At the molecular level, genetic recombination is the exchange of one DNA sequence with another or the incorporation of a DNA sequence into another. If the recombination involves a reaction between homologous sequences, then the process is called **homologous recombination**. When very different nucleotide sequences recombine, it is **nonhomologous recombination**.

homologous recombination genetic recombination between homologous DNA sequences

nonhomologous recombination genetic recombination between very different nucleotide sequences

The process involved in homologous recombination is also termed *general recombination* because the enzymes that mediate the exchange can use essentially any pair of homologous DNA sequences. It occurs in all organisms and is prevalent during the production of gametes in diploid organisms during meiosis. In higher animals, it also occurs in somatic cells and is responsible for the rearrangements within immune cells that lead to the tremendous diversity of immunoglobulins that vertebrates possess (see Chapter 14). Recombination does not occur randomly around a genome. Some areas of a genome are much more likely to show recombination. These zones are called **recombination hot spots**.

recombination hot spots areas in a genome that are likely to have recombination events

Recombination was first shown in 1961 by Meselson and Weigle, who used two different phages to infect bacteria (Figure 10.20). One of the phages had light DNA and one had heavy DNA. Without recombination, the light DNA would always package into light virus particles, and the heavy would package into heavy particles. This would lead to only two populations of phages after infection. Meselson and Weigle's results showed, however, that there were intermediate combinations that had DNA of different weights. This demonstrated that the phage DNA was recombining.

Recombination occurs by the breakage and reunion of DNA strands so that physical exchange of DNA parts takes place. The mechanism was deduced in 1964 by Robin Holliday and is referred to as the **Holliday model** (Figure 10.21). First, the two homologous DNA segments align (Figure 10.21A). In eukaryotes, this is called *chromosome pairing*. A nick occurs at the same place on two homologous strands, shown as the (−) strands in Figure 10.21B. The DNAs on the two strands then swap places, or cross over, at the nick by the process of *strand invasion*, shown in Figure 10.21C. The crossing over can then proceed down each strand of DNA, like opening a zipper and reclosing on a different one. The branch migration and two possible cleavage events lead to strand exchange between the two homologous DNA pieces (Figure 10.21E–H). These recombination events lead to the exchange of genes and traits caused by them.

Holliday model the model for how recombination occurs between homologous chromosomes

In *E. coli*, the principal molecules involved are the *RecBCD* enzyme complex, which initiates recombination; the *RecA* protein, which binds single-stranded DNA; and the *RuvA, RuvB*, and *RuvC* proteins, which drive branch migration.

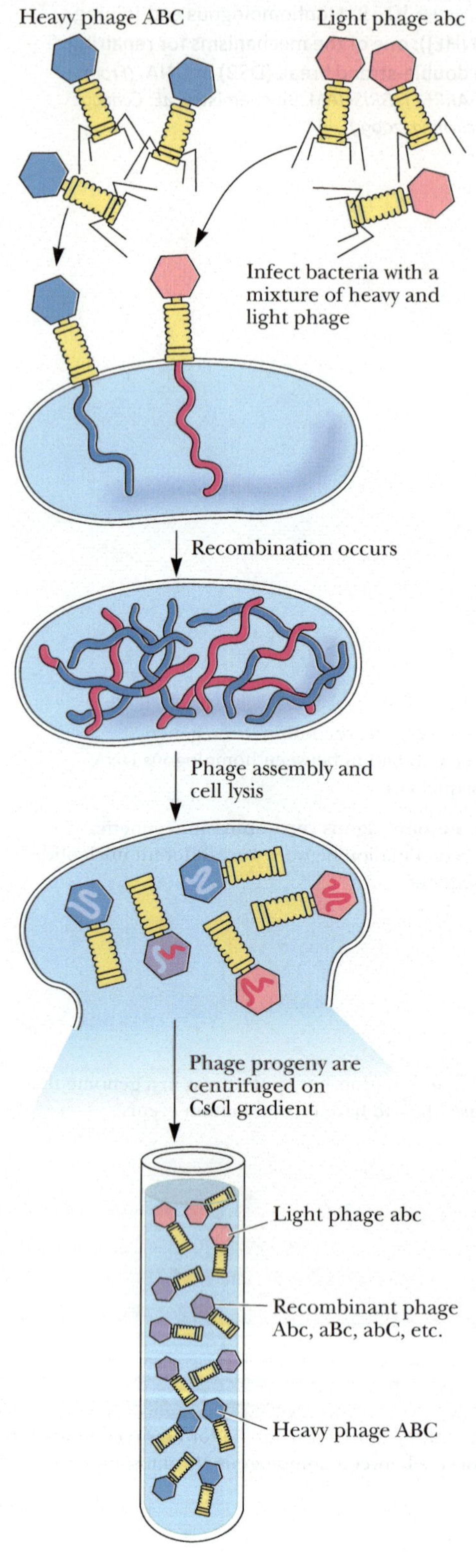

Figure 10.20 Meselson and Weigle's experiment demonstrated that a physical exchange of chromosome parts actually occurs during recombination. Density-labeled "heavy" phage, symbolized as ABC in this diagram, was used to coinfect bacteria along with "light" phage, the abc phage. The progeny from the infection were collected and subjected to CsCl density gradient centrifugation. Parental-type ABC and abc phage were well separated in the gradient, but recombinant phage (ABc, Abc, etc.) were distributed diffusely between the two parental strands because they contained viral genomes constituted from fragments of both "heavy" and "light" DNA. These recombinant genomes were formed by breakage and reunion of parental "heavy" and "light" chromosomes. *(From GARRETT/GRISHAM,* Biochemistry, *4E. © 2009 Cengage Learning.)*

Eukaryotic homologs of these prokaryotic proteins have been found, indicating that the fundamental process of general recombination is conserved.

Recombination is a critical process during meiosis. Surprisingly, the segregation of chromosomes during the formation of gametes is quite inaccurate, with estimates indicating that abnormal chromosome numbers in gametes, called **aneuploidy**, occur in 10%–25% of all conceptions. This is the leading cause of pregnancy loss and birth defects. The process of recombination is critical to the correct segregation of the chromosomes and is an active area of research for that reason. Scientists have discovered several genes involved in controlling recombination. One gene discovered, *PRDM9*, was found to be very important

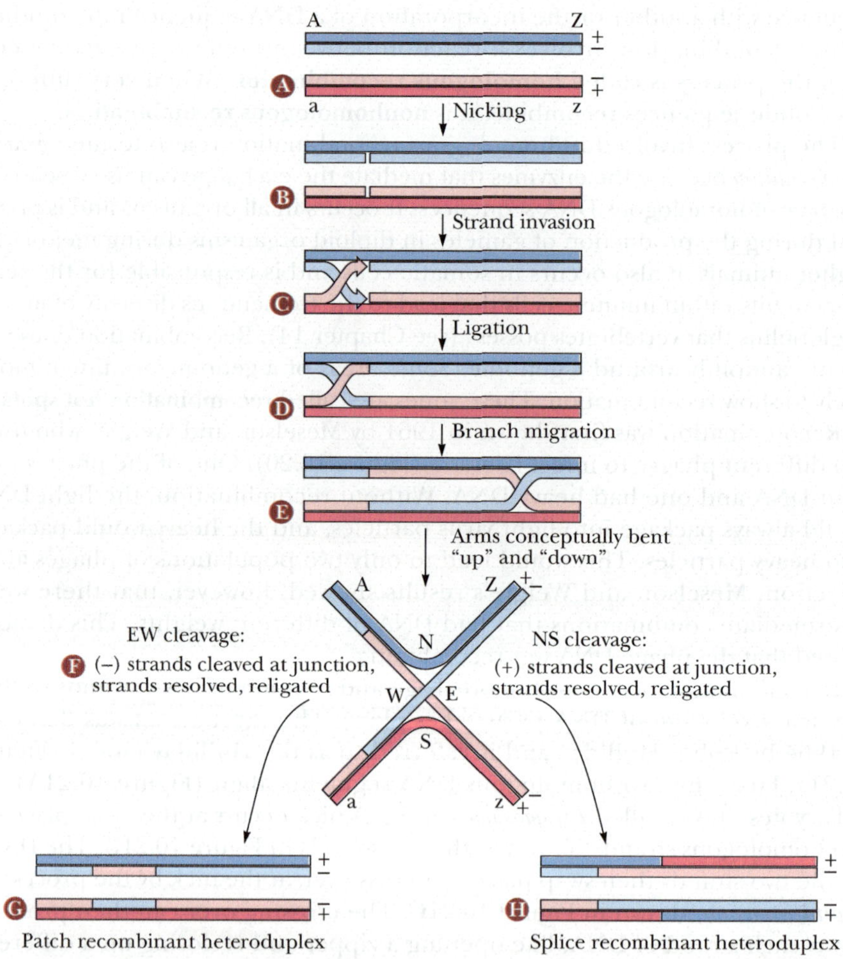

Figure 10.21 The Holliday model for homologous recombination. The + signs and − signs label strands of like polarity. For example, assume that the two strands running 5' → 3' as read left to right are labeled +, and the two strands running 3' → 5' as read left to right are labeled −. Only strands of like polarity exchange DNA during recombination. *(From GARRETT/GRISHAM,* **Biochemistry,** *4E. © 2009 Cengage Learning.)*

aneuploidy a situation where a cell has an abnormal number of chromosomes

in maintaining the recombination in known hot spots. The protein product of PRDM9 is a zinc-finger protein (Section 11-7) that acts as a histone methyltransferase. Several variations of the sequence for PRDM9 have been discovered, and researchers are studying the effects of changes in this gene on recombination and correct meiosis. **Biochemical Connections 10B** describes how repair and recombination work to protect bacteria from severe DNA damage.

10B Biochemical Connections • Microbiology

The SOS Response in *E. coli*

When bacteria are subjected to extreme conditions and a great deal of DNA damage occurs, the normal repair mechanisms are not up to the task of repairing it. However, bacteria have one last card to play, which is called, appropriately, the SOS response. At least 15 proteins are activated as part of this response, including DNA polymerases II, IV, and V. Another important protein is called *RecA*. It gets its name from the fact that it is involved in a recombination event. Homologous DNA can recombine by a variety of mechanisms (see Section 10-6).

Figure 10.22 demonstrates the mechanics of prokaryotic DNA repair under two different circumstances. The starting situation

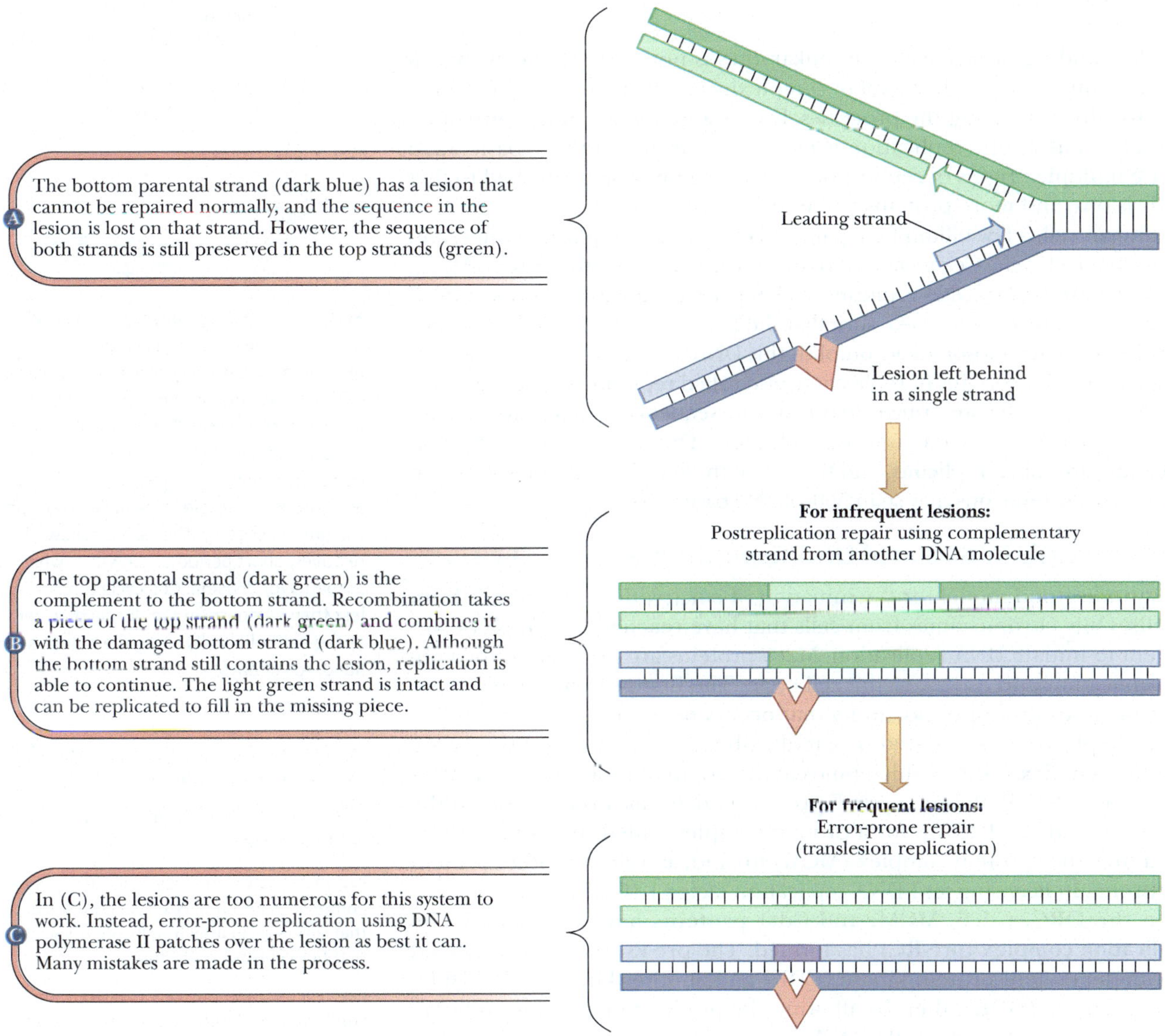

Figure 10.22 **Recombination can be used to repair infrequent lesions.** (From Principles of Biochemistry, *by Albert L. Lehninger, David L. Nelson, and Michael M. Cox. Copyright © 1993 by W. H. Freeman & Co.)*

is shown in Figure 10.22A. The parental strands are shown in dark green and dark blue. The bottom parental strand has a serious lesion (pink wedge) that cannot be fixed by the other repair mechanisms we have discussed in this section. In the bottom strand, the sequence at the lesion could be lost. If the lesions are infrequent, recombination can be used to repair the bottom strand. In Figure 10.22B, a piece of the top parental strand (dark green) is used to recombine with the bottom parental strand (dark blue). In this way, replication is able to continue. The light green strand can be replicated normally to fill in the missing piece. This type of recombination is seen when the lesions are infrequent. If the lesions are too frequent, then recombination will not work. In that case, error-prone replication occurs. The DNA polymerase just "has a good guess" over the lesion and continues. Many errors occur, and the mutation rate is high, but some of the bacteria will survive.

10-7 Eukaryotic DNA Replication

Learning Objectives

After reading this section, students will be able to:

LO 10.7.1 Distinguish between prokaryotic and eukaryotic DNA polymerases.

LO 10.7.2 Distinguish between the processes of prokaryotic and eukaryotic DNA replication.

Our understanding of replication in eukaryotes is not as extensive as that in prokaryotes, due to the higher level of complexity in eukaryotes and the consequent difficulty in studying the processes. Even though many of the principles are the same, eukaryotic replication is more complicated in three basic ways: There are multiple origins of replication, the timing must be controlled to that of cell divisions, and more proteins and enzymes are involved.

In a human cell, a few billion base pairs of DNA must be replicated once, and only once, per cell cycle. Cell growth and division are divided into phases—M, G_1, S, and G_2 (Figure 10.23). DNA replication takes place during a few hours in the S phase, and pathways exist to make sure that the DNA is replicated only once per cycle. Eukaryotic chromosomes accomplish this DNA synthesis by having replication begin at multiple origins of replication, also called **replicators**. These are specific DNA sequences that are usually between gene sequences. An average human chromosome may have several hundred replicators. The zones where replication is proceeding are called **replicons**, and the size of these varies with the species. In higher mammals, replicons may span 500–50,000 base pairs.

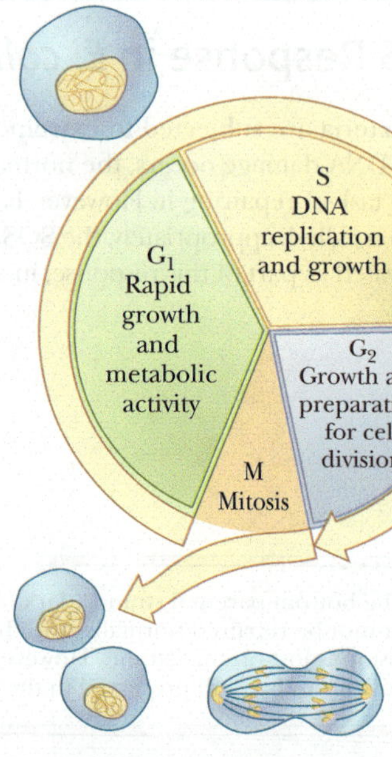

Figure 10.23 The eukaryotic cell cycle. The processes of mitosis and cytoplasm/cell membrane division define the M phase ("M" for *mitosis*). G_1 ("G" for *gap*, not *growth*) is typically the longest part of the cell cycle; G_1 is characterized by rapid growth and metabolic activity. Cells that are quiescent—that is, not growing and dividing (such as neurons)—are said to be in G_0. The S phase is the time of DNA synthesis. S is followed by G_2, a relatively short period of growth in which the cell prepares for division. Cell cycle times vary from less than 24 hours (in rapidly dividing cells, such as the epithelial cells lining the mouth and gut) to hundreds of days.

10-7A Eukaryotic Initiation of DNA Replication and Cell Division

The best understood model for the initiation of eukaryotic replication is from yeast cells. Only chromosomes from cells that have reached the G_1 phase are competent to initiate DNA replication. Many proteins are involved in the control of replication and its link to the cell cycle. The first proteins involved are seen during a window of opportunity that occurs between the early and late G_1 phase. Replication is initiated by a multisubunit protein called the **origin recognition complex (ORC)**, which binds to the origin of replication. The next protein to bind is cell division cycle 6 protein (Cdc6), and together with the ORC, they provide a platform for a protein complex called the minichromosome maintenance protein complex (MCM) to bind, as well as another protein called chromatin licensing and DNA replication factor 1 (Cdt1).

When the ORC, Cdc6, MCM, and Cdt1 proteins have assembled, the **pre-replication complex (pre-RC)** has formed. The pre-RC then gets activated by a cyclin-dependent protein kinase (CDK) and another kinase called Dbf4-dependent kinase (DDK), thereby allowing the pre-RC to transition to being an initiation complex in which the MCM proteins serve as a scaffold for other proteins (Cdc45 and GINS) to bind and unwind the DNA with helicase activity.

replicators the multiple origins of replication in eukaryotic DNA synthesis

replicons parts of chromosomes in which DNA synthesis is taking place

origin recognition complex (ORC) a protein complex bound to DNA throughout the cell cycle that serves as an attachment site for several proteins that help control replication

pre-replication complex (pre-RC) the combination of ORC, Cdc6, MCM, and Cdt1 proteins that binds DNA and makes it competent for replication in eukaryotes

Once the DNA gets unwound, DNA polymerases α and ε can then bind, as well as other proteins such as RPA and PCNA (see Section 10-7C).

One of the keys to linking replication to cell division is that some of these proteins have been found to be cytosolic. Thus, they have access to the chromosomes and their origins of replication only *when the nuclear membrane dissolves during mitosis.* Recall the typical animal cell and its nucleus (Section 1-4B), as well as the discussion of membrane lipids and proteins in Sections 8-3 and 8-4. Then, let us take a moment to marvel at the coordinated events and chemical processes that must occur for the nuclear membrane lipids and their associated membrane proteins to dissolve during mitosis and then come back together when the cell has divided. The mechanisms involved in these processes are an active area of research.

10-7B Eukaryotic DNA Polymerases

At least 19 different polymerases are present in eukaryotes, of which five have been studied more extensively (Table 10.4). The use of animals rather than plants for study avoids the complication of any DNA synthesis in chloroplasts. The five best studied polymerases are called α, β, γ, δ, and ε. The α, β, δ, and ε enzymes are found in the nucleus, whereas the γ form occurs in mitochondria.

▶ How are eukaryotic polymerases related to prokaryotic ones?

Polymerase α was the first discovered, and it has the most subunits. It also has the ability to make primers, but it lacks a 3' → 5' proofreading activity and has low processivity. After making the RNA primer, Pol α adds about 20 nucleotides and is then replaced by Pol δ and ε. Polymerase δ is the principal DNA polymerase in eukaryotes. It interacts with a special protein called *PCNA* (for *proliferating cell nuclear antigen*). PCNA is the eukaryotic equivalent of the part of Pol III that functions as the dimeric β-subunit sliding clamp shown in Figure 10.8. It is a trimer of three identical proteins that surround the DNA (Figure 10.24). The role of DNA polymerase ε is less clear, but recent evidence suggests it is involved in leading strand replication as well as nucleotide and BER. It may replace polymerase δ in lagging strand synthesis. DNA polymerase β appears to be a repair enzyme. DNA polymerase γ carries out DNA replication in mitochondria (recall from Section 1-4B that mitochondria are energy-producing organelles). Several of the DNA polymerases isolated from animals lack exonuclease activity (the

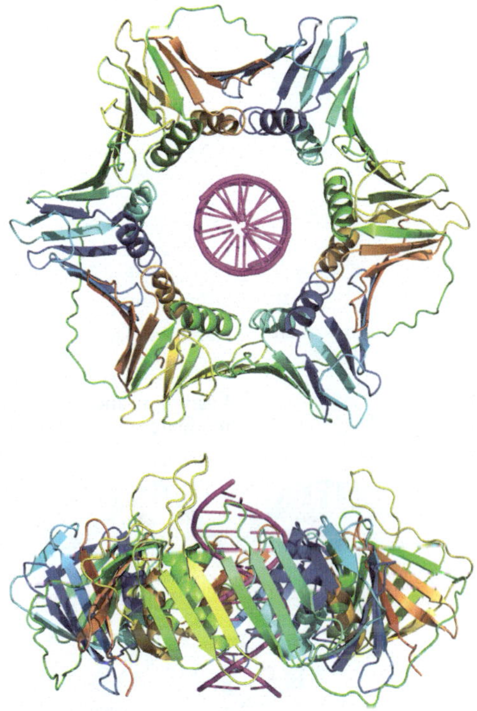

Own work/public domain

Figure 10.24 Structure of the proliferating cell nuclear antigen (PCNA) homotrimer complexed with DNA. (A) Top view of the PCNA ring (multicolored) surrounding DNA (magenta). (B) A side view. The trimeric PCNA ring of eukaryotes is remarkably similar to its prokaryotic counterpart, the dimeric β-subunit sliding clamp (Figure 10.8). (From https://commons.wikimedia.org/wiki/File:Sliding_clamp_dna_complex.png?uselang=en#Licensing (A) and https://commons.wikimedia.org/wiki/File:Sliding_clamp_dna_complex_side.png (B).)

Table 10.4 The Biochemical Properties of Eukaryotic DNA Polymerases

	α	δ	ε	β	γ
Mass (kDa)					
Native	>250	170	256	36–38	160–300
Catalytic core	165–180	125	215	36–38	125
Other subunits	70, 50, 60	48	55	None	35, 47
Location	Nucleus	Nucleus	Nucleus	Nucleus	Mitochondria
Associated functions					
3' → 5' exonuclease	No	Yes	Yes	No	Yes
Primase	Yes	No	No	No	No
Properties					
Processivity	Low	High	High	Low	High
Fidelity	High	High	High	Low	High
Replication	Yes	Yes	Yes	No	Yes
Repair	No	?	Yes	Yes	No

Source: Adapted from Kornberg, A., and Baker, T. A., 1992. *DNA Replication,* 2nd ed. New York: W. H. Freeman and Co.

Table 10.5 Differences in DNA Replication in Prokaryotes and Eukaryotes

Prokaryotes	Eukaryotes
Five polymerases (I, II, III, IV, V)	Five polymerases (α, β, γ, δ, ε)
Functions of polymerase:	Functions of polymerases:
I is involved in synthesis, proofreading, repair, and removal of RNA primers	α is a polymerizing enzyme
II is also a repair enzyme	β is a repair enzyme
III is the main polymerizing enzyme	γ is involved in mitochondrial DNA synthesis
IV, V are repair enzymes under unusual conditions	δ is the main polymerizing enzyme; ε is the leading strand replication enzyme
Polymerases are also exonucleases	Not all polymerases are exonucleases
One origin of replication	Several origins of replication
Okazaki fragments 1000–2000 residues long	Okazaki fragments 100–300 residues long
No proteins complexed to DNA	Histones complexed to DNA

α and β enzymes). In this regard, the animal enzymes differ from prokaryotic DNA polymerases. Separate exonucleolytic enzymes exist in animal cells.

10-7C The Eukaryotic Replication Fork

The general features of DNA replication in eukaryotes are similar to those in prokaryotes. Table 10.5 summarizes the differences. As with prokaryotes, DNA replication in eukaryotes is semiconservative. There is a leading strand with continuous synthesis in the 5' → 3' direction and a lagging strand with discontinuous synthesis in the 5' → 3' direction. An RNA primer is formed by a specific enzyme in eukaryotic DNA replication, as is the case with prokaryotes, but in this case the primase activity is associated with Pol α. The structures involved at the eukaryotic replication fork are shown in Figure 10.25. Compare these

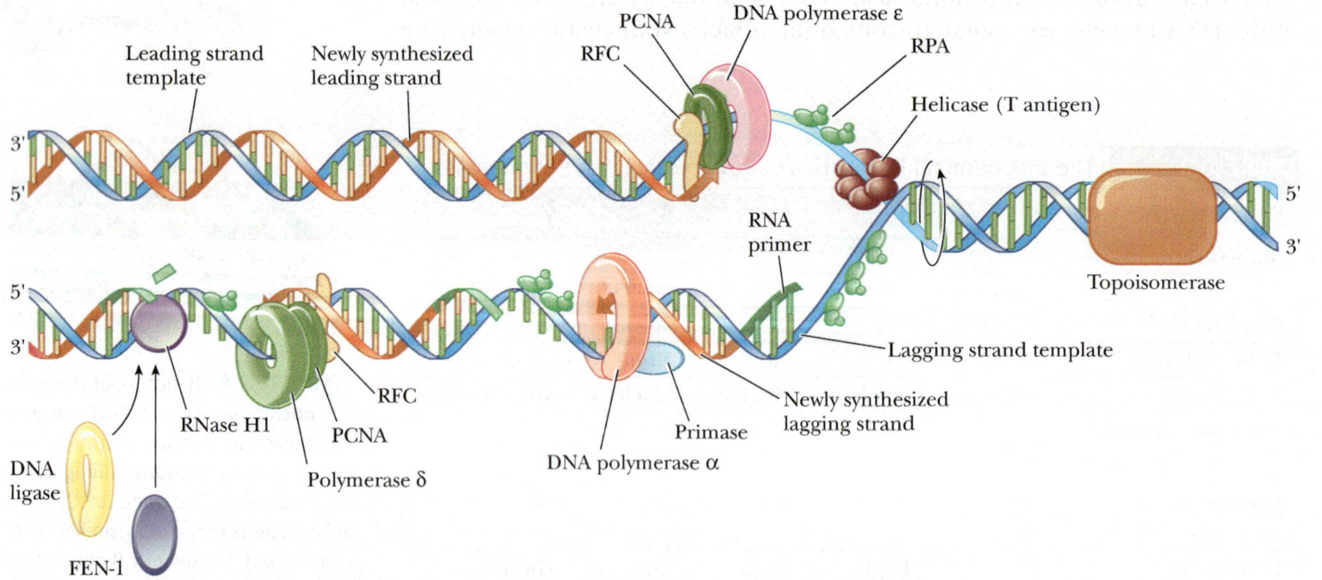

Figure 10.25 The basics of the eukaryotic replication fork. The primase activity is associated with DNA polymerase α. After a few nucleotides are incorporated, DNA polymerase δ, with its associated proteins called PCNA and RFC, binds and does the majority of the synthesis. The enzymes FEN-1 and RNase H1 degrade the RNA primers in eukaryotic replication. *(Karp, Gerald. Figure 13–22 from* **Cellular and Molecular Biology***. John Wiley & Sons, Inc., 2013. Copyright © John Wiley & Sons, Inc., 2013. Reproduced with permission of John Wiley & Sons, Inc.)*

to the structures involved at the prokaryotic replication fork shown in Figure 10.10.

The formation of Okazaki fragments (typically 150–200 nucleotides long in eukaryotes) is initiated by DNA polymerase α (Pol α). After the RNA primer is made and a few nucleotides are added by Pol α, the polymerase dissociates and is replaced by DNA polymerase δ (Pol δ) and its attached proliferating cell nuclear antigen (PCNA) clamp protein. Another protein, called RFC (replication factor C), acts as a clamp loader in attaching PCNA to the DNA and to Pol δ. The RNA primer is eventually degraded, but, in the case of eukaryotes, the polymerases do not have the 5' → 3' exonuclease activity to do it. Instead, separate enzymes, FEN-1 and RNase H1, degrade the RNA. Continued movement of Pol δ fills in the gaps made by primer removal, and DNA ligase seals the nicks between the Okazaki fragments. As mentioned in Section 10-7A, the MCM complex, Cdc45, and GINS complex proteins perform the helicase activity, unwinding the parental DNA strands. As with prokaryotic replication, topoisomerases relieve the torsional strain from unwinding the helix, and a single-strand binding protein, called RPA (replication protein A), protects the DNA from degradation. DNA polymerase ε (Pol ε) catalyzes leading strand DNA synthesis.

Another important difference between DNA replication in prokaryotes and in eukaryotes is that prokaryotic DNA is not complexed to histones, as is eukaryotic DNA. Histone biosynthesis occurs at the same time and at the same rate as DNA biosynthesis. In eukaryotic replication, histones associate with DNA as it is formed. During replication, the histones can be modified in such a way that they release the portions of DNA that are being replicated. For example, several lysine residues on the histones can be acetylated, removing their positive charge and weakening the interaction with the DNA so that the replication machinery can come in and associate with it. Although much of the early research into replication focused on the enzymes used for polymerizing DNA, as well as the proteins required to initiate replication, one thing that is seen with all of the processes involved in the flow of genetic information is that whatever starts must also stop. At the end of replication, replisomes from adjacent replication origins meet and DNA ligase seals the nicks. Quite a tangle of DNA and proteins could occur at that point, and researchers are "untangling" our understanding of this process. One thing that is known is that the protein complexes in the replisome must disassemble. This disassembly is based on modifications to the proteins in question, often by tagging the replisome with a specific protein called ubiquitin. Ubiquitin comes up again in Chapter 12 when we look at the termination of translation. An important aspect of DNA replication in eukaryotes specifically affecting human aging is described in **Biochemical Connections 10C**. To end this chapter, **Biochemical Connections 10D** explores some fascinating research involving RNA replication.

10C Biochemical Connections • Allied Health

Telomerase and Cancer

Replication of linear DNA molecules, such as human chromosomes, poses particular problems at the ends of the molecules. Remember that at the 5' end of a strand of DNA being synthesized, there is initially a short RNA primer, which must later be removed and replaced by DNA. This is never a problem with a circular template because the DNA polymerase I that is coming from the 5' side of the primer (the previous Okazaki fragment) can then patch over the RNA with DNA. With a linear chromosome, however, this is not possible. At each end is a 3' and a 5' DNA chain. The 5'-end template strand is not a problem because a DNA polymerase copying it is moving from 5' to 3' and can proceed to the end of the chromosome from the last RNA primer. The 3'-end template strand poses a problem, however (see Figure 10.26A). The RNA primer at the 5' end of the new strand does not have any way of being replaced. All DNA polymerases require a primer and, because nothing is upstream (to the 5' side), the RNA primer cannot be replaced with

Continued on next page

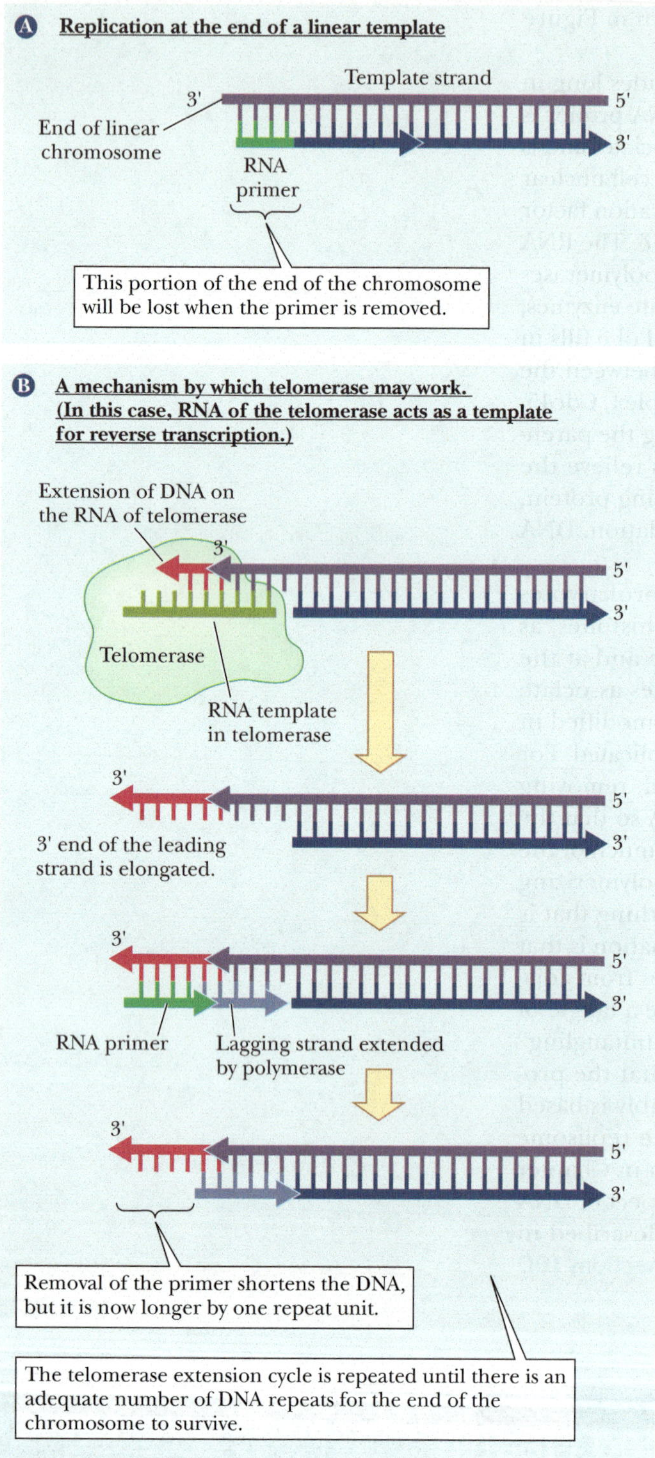

Figure 10.26 Action of telomerase.

telomeres regions of repetitive nucleotide sequences at the ends of linear chromosomes

DNA. RNA is unstable, and the RNA primer is degraded in time. In effect, unless some special mechanism is created, the linear molecule gets shorter each time it is replicated.

To help with this problem in 3'-end replication, the ends of eukaryotic chromosomes have special structures called **telomeres**, which are a series of repeated DNA sequences. In human sperm-cell and egg-cell DNA, the sequence is 5'TTAGGG3', and this sequence is repeated more than 1000 times at the end of the chromosomes. This repetitive DNA is noncoding and acts as a buffer against degradation of the DNA sequence at the ends, which would occur with each replication as the RNA primers are degraded. Some evidence shows a relationship between longevity and telomere length, and some researchers have suggested that the loss of the telomere DNA with age is part of the natural aging process. Eventually, the DNA becomes nonviable and the cell dies.

However, even with long telomeres, cells eventually die when their DNA gets shorter with each replication unless there is some compensatory mechanism. In cells, this compensatory mechanism involves an enzyme called telomerase, as shown in Figure 10.26B, which provides a mechanism for the synthesis of the telomeres. The enzyme telomerase is a ribonuclear protein, containing a section of RNA that is the complement of the telomere. In humans, this sequence is 5'CCCUAA3'. Telomerase binds to the 5' strand at the chromosome end and uses a reverse transcriptase activity to synthesize DNA (shown in red) on the 3' strand, using its own RNA as the template. This allows the template strand (shown in purple) to be elongated, effectively lengthening the telomere.

When the nature of telomerase was discovered, researchers originally believed that it was a "fountain of youth" and that, if we could figure out how to keep it going, cells (and perhaps individuals) would never die. Subsequent work has shown that even though the enzyme telomerase *does* remain active in rapidly growing tissues, such as blood cells, the intestinal lumen, skin, and others, it is *not* active in most adult tissues. Therefore, when the cells of most adult tissues divide, they cannot fully preserve their chromosome ends, or telomeres. With each round of cell division, telomeres become progressively shorter. When they reach a critically short length, the cell can no longer divide and enters senescence or dies. This gradual telomere shortening contributes to normal cellular aging and tissue decline over time.

A big surprise was the discovery that telomerase is reactivated in cancer cells, explaining in part their immortality and their ability to keep dividing rapidly. This observation has opened a new possibility for cancer therapy: If we can prevent the reactivation of telomerase in cancerous tissues, the cancer might die of natural causes. The study of telomerase is just the tip of the iceberg. Other mechanisms must exist to protect the integrity of chromosomes besides telomerase. Using techniques described in Chapter 13, researchers have genetically engineered mice to lack telomerase. These mice showed continued shortening of their telomeres with successive replication and generations, but eventually the chromosome shortening stopped, indicating that some other process also conserved the length of the chromosomes. Many research questions remain regarding the relationship between telomeres, DNA replication, and aging. ◗

10D Biochemical Connections • Evolutionary Biology

Self-Replicating RNAs

While Chapter 10 has dealt with DNA replication, RNA replication is also a hot topic in biochemistry. Proponents of the "RNA world" hypothesis believe that RNA was the original molecule of heredity—the first that took simple compounds and turned them into larger molecules having a function. To demonstrate how evolution could have started, scientists have been looking for a way to bridge the gap between the simple molecules that were available before life on Earth began and those molecules that could reproduce, metabolize, form cells, and so on. In essence, scientists are looking for ways to create life in a test tube. If scientists can create artificial life, then it becomes much easier to imagine how it could have happened naturally with billions of years of trial and error.

In 2008, Gerald Joyce and Tracey Lincoln published work showing that they had created a series of small RNA molecules that could reproduce, change, and compete for limited resources, exactly the traits that Charles Darwin envisioned at the species level when he proposed the *survival of the fittest*. In their study they started with 24 RNA variants, all of which reproduced, but at different rates. Sometimes the reproduction was not faithful, and mutations were made. Some of the mutations survived better than the originals and replaced them. They let the system run for 100 hours, during which time they saw an amplification of the replicating molecules by a factor of 10^{23}. The original molecules died out and were replaced by other molecules, essentially demonstrating chemical evolution in vitro (see Figure 10.27).

The studies are very exciting to evolutionary biologists but do not, by themselves, prove evolution. The definition of life includes many aspects, and reproductive success is only one of them. It is a big step from finding that RNA molecules can reproduce and change to finding that they can develop metabolism and form cells. The next step will be to demonstrate that as the RNA molecules evolve, they can develop new abilities and functions. This is an area of active research, with many scientists trying to find the missing pieces.

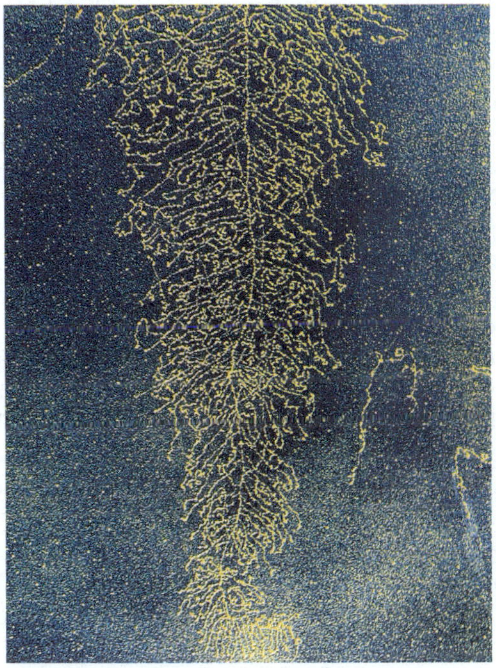

Professor Oscar L. Miller/Science Source

Figure 10.27 Reproducing molecules of RNA spread out from a single DNA spine.

Summary

Semiconservative Replication

- The semiconservative nature of replication was determined with an elegant experiment. *E. coli* bacteria were grown with $^{15}NH_4Cl$ as the sole nitrogen source. In such a medium, all newly formed nitrogen compounds, including purine and pyrimidine nucleobases, become labeled with ^{15}N. The ^{15}N-labeled cells were then transferred to a medium that contained only ^{14}N.
- With every new generation of growth, a sample of DNA was extracted and analyzed by density-gradient centrifugation. DNA containing a 50–50 mixture of ^{14}N and ^{15}N appeared at a position halfway between the two bands after one generation, a result to be expected with semiconservative replication.

Direction of Replication

- Replication goes in both directions from the origin of replication. That is to say, each strand of DNA is replicated, and from each origin, two replication forks move in opposite directions.

Discontinuous Replication

- Two different modes of polymerization exist for the two growing strands. One newly formed strand (the leading strand) is formed continuously from its 5' end to its 3' end at the replication fork on the exposed 3' to 5' template strand. The other strand (the lagging strand) is formed discontinuously in small fragments, called Okazaki fragments. The fragments of the lagging strand are linked together by DNA ligase.

Proteins Required for Replication

- DNA gyrase is used to induce negative supercoils in the DNA to compensate for the positive supercoils that form in front of the replication fork where the strands are separated.
- Specific proteins called single-strand binding proteins bind to the single-stranded regions, protecting them from nucleases and preventing the parental strands from reannealing.
- The enzyme primase uses a DNA strand as a template and creates a complementary RNA strand that is the primer for DNA synthesis.
- DNA polymerases have a very common structure that is often compared to a right hand, with domains referred to as the fingers, palm, and thumb. The active site where the polymerase reaction is catalyzed lies in the crevice within the palm domain that lies between the fingers and thumb domains. The fingers domain acts in deoxynucleotide recognition and binding, whereas the thumb is responsible for DNA binding.
- Three DNA polymerase III (Pol III) enzymes are involved in the prokaryotic replisome: One Pol III enzyme is dedicated to leading strand synthesis and two Pol III enzymes are dedicated to lagging strand synthesis.
- The polymerizing part of Pol III is a dimer of the β-subunit, and it forms a sliding clamp around the DNA chain. The part of the Pol III enzyme that is called the clamp loader opens the sliding clamp and inserts the DNA chain.
- All clamp loaders are pentameric enzymes that are members of a family of ATPases called the AAA+ superfamily.

Proofreading and Repair in Replication

- Spontaneous mutation of bases and insertion of the wrong nucleotide would normally lead to an error every 10^4 to 10^5 bases. However, the proofreading capability of the DNA polymerases allows mismatched nucleotides to be removed, reducing the errors to one in 10^9 to 10^{10}.
- Mismatch repair (MMR), base excision repair (BER), and nucleotide excision repair (NER) are different mechanisms that cells use to fix mistakes and DNA damage.

Eukaryotic Replication and Cell Division

- Eukaryotic replication is tied to cell division by several proteins, including the origin recognition complex, replication activator protein, and replication licensing factors.
- The process is controlled by cyclins, proteins produced during the G_1 and S phases that bind to cyclin-dependent kinases and activate replication.

Eukaryotic DNA Polymerases

- Eukaryotic DNA polymerases are labeled α through ε.
- The various polymerases vary in size, complexity, processivity, and their levels of exonuclease and repair activities. The γ version is found only in mitochondria, but the others are found in the nucleus.

Review Exercises

10-1 The Flow of Genetic Information in the Cell

1. **Recall** Define *replication, transcription,* and *translation.*
2. **Recall** What is the name of the dominant theory that describes the flow of genetic information in cells?
3. **Reflect and Apply** Is the following statement true or false? Why? "The flow of genetic information is always DNA → RNA → protein."
4. **Reflect and Apply** You want to carry out an experiment to show that DNA replication is a semiconservative process, but you are out of ^{15}N-enriched growth medium. However, you do have two growth media that are enriched with either ^{31}P or ^{32}P. Will this media substitution work? Why or why not?
5. **Reflect and Apply** Why is it more important for DNA to be replicated accurately than transcribed accurately?

10-2 Replication of DNA

6. **Recall** Why is the replication of DNA referred to as a semiconservative process? What is the experimental evidence for the semiconservative nature of the process? What experimental results would you expect if replication of DNA were a conservative process?
7. **Recall** What is a replication fork? Why is it important in replication?
8. **Recall** Describe the structural features of an origin of replication.

10-3 DNA Polymerase

9. **Recall** Do DNA polymerase enzymes also function as exonucleases?
10. **Recall** Compare and contrast the properties of the enzymes DNA polymerase I and polymerase III from *E. coli.*
11. **Reflect and Apply** Define *processivity* and indicate the importance of this concept in DNA replication.
12. **Reflect and Apply** Comment on the dual role of the monomeric reactants in replication.
13. **Reflect and Apply** DNA synthesis always takes place from the 5' to the 3' end. The template strands have opposite directions. How does nature deal with this situation?
14. **Reflect and Apply** What would happen to the replication process if the growing DNA chain did not have a hydroxyl group at the 3' end?
15. **Reflect and Apply** Explain how the discontinuous model of DNA replication leads to differences in the distance between the primers and the DNA polymerases replicating the two different strands. Which polymerase is in closest proximity to the replication primers, the polymerase(s) acting on the leading strand or lagging strand? Why?

10-4 Proteins Required for DNA Replication

16. **Recall** List the substances required for the replication of DNA catalyzed by DNA polymerase.
17. **Recall** Describe the discontinuous synthesis of the lagging strand in DNA replication.
18. **Recall** What are the functions of the gyrase, primase, and ligase enzymes in DNA replication?
19. **Recall** Single-stranded regions of DNA are attacked by nucleases in the cell, yet portions of DNA are in a single-stranded form during the replication process. Explain.
20. **Recall** What is the primer in DNA replication?
21. **Reflect and Apply** Why is a short RNA primer needed for replication?
22. **Recall** What are the common features of all DNA polymerases?
23. **Reflect and Apply** You discover a bacterial DNA polymerase that has decreased rates of DNA synthesis compared to previously characterized enzymes but still binds the DNA template and deoxynucleotides effectively. Which domain of the polymerase would most likely contain the mutations affecting catalytic activity?
24. **Recall** Why is a clamp loader necessary in replication?

10-5 Proofreading and Repair in Replication

25. **Recall** How does proofreading take place in the process of DNA replication?
26. **Recall** Describe the excision repair process in DNA, using the excision of thymine dimers as an example.
27. **Biochemical Connections** Of what benefit is it for DNA to have thymine rather than uracil?
28. **Reflect and Apply** Can methylation of nucleotides play a role in DNA replication? If so, what sort of role?
29. **Biochemical Connections** Can prokaryotes deal with drastic DNA damage in ways that are not available to eukaryotes?
30. **Recall** What is a direct way of repairing double-stranded DNA breaks?
31. **Recall** What proteins are used in NHEJ?
32. **Recall** Why is NHEJ an error-prone mechanism?
33. **Reflect and Apply** Cisplatin is a chemotherapy drug that creates crosslinks between nitrogenous bases on opposite strands. Enzymatic removal of this lesion generates a double-stranded break that must be repaired by recombination or NHEJ. Why can't these crosslinked nucleotides be repaired using base excision repair or nucleotide excision repair?
34. **Recall** What is the role of Ku70/80 in direct repair of DSBs?

10-6 DNA Recombination

35. **Recall** How did Meselson and Weigle demonstrate recombination?
36. **Reflect and Apply** How has the use of DNA labeled with heavy isotopes been instrumental in our understanding of replication?
37. **Reflect and Apply** Predict the relationship between the frequency of chromosomal hot spots and the genetic diversity observed within a species. Use your understanding of the Holliday model to explain your prediction.
38. **Recall** What is the Holliday model?

10-7 Eukaryotic DNA Replication

39. **Recall** Compared to prokaryotes, do eukaryotes have fewer, more, or the same number of origins of replication?
40. **Recall** How does DNA replication in eukaryotes differ from the process in prokaryotes?
41. **Recall** What role do histones play in eukaryotic DNA replication?
42. **Reflect and Apply** How do the DNA polymerases of eukaryotes differ from those of prokaryotes?
43. **Reflect and Apply** What is the relationship between the control of DNA synthesis in eukaryotes and the stages of the cell cycle?
44. **Biochemical Connections** What would be the effect on DNA synthesis if the telomerase enzyme were inactivated?
45. **Reflect and Apply** Would it be advantageous to a eukaryotic cell to have histone synthesis take place at a faster rate than DNA synthesis?
46. **Biochemical Connections** Name an important difference in the replication of circular DNA versus linear double-stranded DNA.
47. **Reflect and Apply** Eukaryotic cells have mechanisms to avoid signaling for homologous recombination when DNA is damaged during the first growth phase of the cell, G_1, but use homologous recombination preferentially during the second growth phase, G_2, after DNA has been replicated in S phase. Explain why this is.
48. **Reflect and Apply** Why is it reasonable that eukaryotes have a DNA polymerase (Pol γ) that operates only in mitochondria?
49. **Biochemical Connections** What is meant by the "RNA world"?

Further Reading

Bell, S. P. Terminating the Replisome. *Science*, **346**, 418–419 (2014). An article discussing the way the replisome is disassembled when replication is done.

Botchan, M. Coordinating DNA Replication with Cell Division: Current Status of the Licensing Concept. *Proc. Nat. Acad. Sci.* **93**, 9997–10,000 (1996). An article about control of replication in eukaryotes.

Buratowski, S. DNA Repair and Transcription: The Helicase Connection. *Science* **260**, 37–38 (1993). How repair and transcription are coupled.

Cheung, V. G., S. L. Sherman, and E. Feingold Genetic Control of Hotspots. *Science* **327**, 791–792 (2010).

Gibbs, W. W. Evolution in a Bottle. *Sci. Amer.* **300** (4), 18–21 (2009).

Gilbert, D. M. Making Sense of Eukaryotic DNA Replication Origins. *Science* **294**, 96–100 (2001). Information on replication origins in eukaryotes.

Kornberg, A., and T. Baker. *DNA Replication,* 2nd ed. New York: Freeman, 1991. Most aspects of DNA biosynthesis are covered. The first author received a Nobel Prize for his work in this field.

Kucherlapati, R., and R. A. DePinho. Telomerase Meets Its Mismatch. *Nature* **411**, 647–648 (2001). An article about a possible relationship between telomerase and mismatch repair.

Radman, M., and R. Wagner. The High Fidelity of DNA Duplication. *Sci. Amer.* **259** (1), 40–46 (1988). A description of replication, concentrating on the mechanisms for minimizing errors.

Stillman, B. Cell Cycle Control of DNA Replication. *Science* **274**, 1659–1663 (1996). A description of how eukaryotic replication is controlled and linked to cell division.

Varmus, H. Reverse Transcription. *Sci. Amer.* **257** (3), 56–64 (1987). A description of RNA-directed DNA synthesis. The author was one of the recipients of the 1989 Nobel Prize in Physiology or Medicine for his work on the role of reverse transcription in cancer.

Wu, L., and D. Hickson. DNA Ends RecQ-uire Attention. *Science* **292**, 229–230 (2001). An article describing various ways that the ends of chromosomes are protected.

A medical transcriptionist converts language spoken by doctors to a text form of the same language. In the biochemical process of transcription, the nucleic acid language of DNA is converted to the RNA form of nucleic acid language.

Keith Brofsky/Photodisc/Getty Images

Transcription of the Genetic Code: The Biosynthesis of RNA 11

Chapter Outline

11-1 Overview of Transcription

Learning Objectives

After reading this section, students should be able to:

LO 11.1.1 Distinguish between *transcription* and *gene expression*.

LO 11.1.2 Compare features of RNA synthesis in prokaryotes and eukaryotes.

As we explained in Chapter 10, the central dogma of molecular biology is that DNA is transcribed into RNA, and RNA is translated into protein. The process of making RNA from DNA is called **transcription**, and it is the major control point in **gene expression** and the production of proteins.

In transcription, one of the strands of the double-stranded DNA molecule is used as a template to synthesize a complementary sequence of RNA. The RNA sequence differs from DNA in one respect: The DNA base thymine (T) is replaced by the RNA base uracil (U). The backbone structure of RNA also differs from the backbone structure of DNA in one respect: The backbone sugars in RNA strands are ribose instead of deoxyribose (see Figures 9.5 and 9.6 for sample RNA strand and DNA strand structures, respectively). Of all the DNA in a cell, only some is transcribed. Transcription produces all the types of RNA—mRNA, tRNA, rRNA, snRNA, miRNA, and siRNA. New types of RNA and new functions for it are found every year.

transcription the process of formation of RNA using a DNA template

gene expression the process by which the information in a gene is used to make a protein or noncoding RNA

The details of RNA transcription differ somewhat in prokaryotes and eukaryotes. For example, the process is much more complicated in eukaryotes, involving a large number of proteins called transcription factors (Section 11-4C). Most of the research on the subject has been done in prokaryotes, especially *Escherichia coli* (*E. coli*), but some general features are found in all organisms, except in the case of cells infected by RNA viruses.

▶ **What are the basics common to all transcription?**

Although many differences exist between transcription in prokaryotes and eukaryotes, and even differences between the transcription of different types of RNA in eukaryotes, some aspects are constant. Table 11.1 summarizes the main features of the process.

11-2 Transcription in Prokaryotes

Learning Objectives

After reading this section, students should be able to:

LO 11.2.1 Describe the features of prokaryotic RNA polymerase.

LO 11.2.2 Explain the components of prokaryotic transcription promoters.

LO 11.2.3 Distinguish between events in the initiation, elongation, and termination phases of prokaryotic transcription.

11-2A RNA Polymerase in *Escherichia coli*

The enzyme that synthesizes RNA is called **DNA-dependent RNA polymerase (RNAP)**, and the most extensively studied RNAP is from *E. coli*. The molecular weight of this enzyme is about 470,000 Da, and it has a multisubunit structure.

Careers in Chemistry • Stephen Liu

Medical Director/Chairman, Department of Emergency Medicine, Adventist Health White Memorial, Emergency Physician at US Acute Care Solutions (USACS)

Education:
The University of North Carolina–Chapel Hill, MD, Medicine
The University of North Carolina–Chapel Hill, BA, Economics with Honors and Chemistry

Dr. Stephen Liu is an emergency physician at Adventist Health White Memorial medical center in Los Angeles and his daily work centers around treating patients who may be experiencing one of the worst days of their lives. In this fast-paced environment, no two days are alike. Dr. Liu manages everything from stabilizing patients to diagnosing common conditions, to reassuring patients when an emergent condition is discovered. Each shift brings a wide range of cases, making adaptability and clinical reasoning essential.

Dr. Liu's journey to this career of "controlled chaos" started in college. His love for science and desire to understand why and how things work led him to pursue a degree in chemistry; however, he soon discovered that simply memorizing content was not enough to make the language of chemistry truly click. After college, he spent 4 years in medical school at the University of North Carolina–Chapel Hill and then completed his residency training at the Stanford University Medical Center.

Stephen Liu

To this day, Dr. Liu still considers himself a scientist, even if his work is a few steps removed from traditional research and basic science. He sees medicine as a direct application of scientific principles—understanding how the body functions, how it responds to illness, and how treatments and medications create change. His advice to biochemistry students who are considering a career in medicine: "You don't need to be a star biochemist to pursue medicine—trust me on this. It certainly helps, but it's not a requirement. I've come to really appreciate the foundation biochemistry gave me, especially when it comes to understanding pharmacology and how medications affect the body. But it's important to know that you can absolutely build a successful career in medicine without being an expert in biochemistry."

Table 11.1 General Features of RNA Synthesis

1. RNA is initially synthesized using a DNA template in the process called transcription; the enzyme that catalyzes the process is DNA-dependent RNA polymerase (RNAP).
2. All four ribonucleoside triphosphates (ATP, GTP, CTP, and UTP) are required, as is Mg^{2+}.
3. A primer is not needed in RNA synthesis.
4. As is the case with DNA biosynthesis, the RNA chain grows from the 5′ to the 3′ end. The nucleotide at the 5′ end of the chain retains its triphosphate group (abbreviated ppp).
5. The enzyme uses one strand of the DNA as the template for RNA synthesis. The enzyme binds to the template strand and moves along it in the 3′ to 5′ direction.
6. The DNA template is unchanged in transcription, and the base sequence of the DNA contains signals for initiation and termination of RNA synthesis.

Five different types of subunits, designated α, ω, β, β', and σ, have been identified. The actual composition of the enzyme is $\alpha_2\omega\beta\beta'\sigma$. The σ subunit (or σ factor) is rather loosely bound to the rest of the RNAP enzyme (the $\alpha_2\omega\beta\beta'$ portion), which is called the **core enzyme**. The **holoenzyme** consists of all the RNAP subunits, including the σ subunit.

DNA-dependent RNA polymerase (RNAP) the enzyme that catalyzes the production of RNA using a DNA template

core enzyme (of RNA polymerase) the RNAP enzyme lacking the σ subunit

holoenzyme in general, a complex enzyme that has all component parts, including coenzymes and all subunits

▶ What do the subunits of RNA polymerase do?

The σ subunit of RNA polymerase (RNAP) is involved in the recognition of specific promoters (Section 11-2B), whereas the β, β', α, and ω subunits combine to make the active site for RNA synthesis. The mechanism of the polymerization reaction of RNAP is an active area of research.

▶ Which of the DNA strands is used in transcription?

Figure 11.1 shows the basics of information transfer from DNA to protein. Of the two strands of DNA, one of them is used as the template for the synthesis of a particular RNA product. RNA polymerase reads this **template strand** from 3′ to 5′, so the new RNA strand is synthesized in the 5′ to 3′ direction. The template strand has several other names, such as the antisense strand, because its code is the complement of the RNA that is produced. It is sometimes called the (–) strand by convention.

template strand (antisense, or [–] strand) the DNA strand that is used as a template for RNA synthesis (strand)

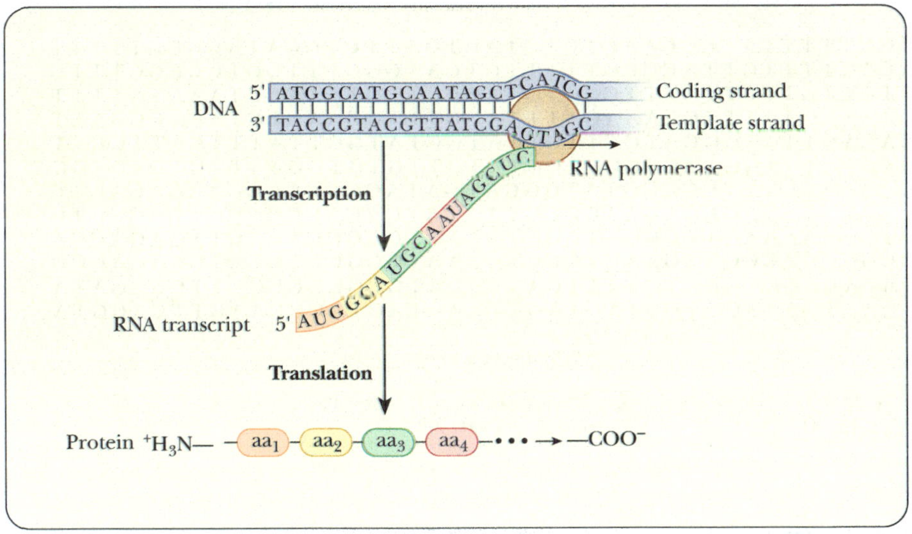

Figure 11.1 The basics of transcription. RNA polymerase uses the template strand of DNA to make an RNA transcript that has the same sequence as the nontemplate DNA strand (the coding strand), with the exception that T is replaced by U. If this RNA is mRNA, it can later be translated to protein.

coding strand (sense, [+], or nontemplate strand) the DNA strand that has the same sequence as the RNA that is synthesized from the template (strand)

The other strand is called the **coding strand** (or the sense strand) because its sequence of DNA will be the same as the RNA sequence that is produced (with the exception of U replacing T). Although the coding strand is not used directly in transcription, its sequence contains the *code* for the protein amino acid sequence that is eventually produced. It is also called the (+) strand by convention, or even the nontemplate strand. For our purposes, we use the terms *template strand* and *coding strand* throughout. Because the DNA in the coding strand has the same sequence as the RNA that is produced, it is used when discussing the sequence of genes for proteins or for promoters and controlling elements on the DNA.

The core enzyme of RNA polymerase, which *lacks* the σ subunit, is catalytically active but lacks specificity in terms of which strand of the DNA to use as the template strand. The holoenzyme of RNA polymerase, which *has* the σ subunit, binds to specific DNA sequences, and it binds specifically to the template strand and produces the correct RNA strand. The essential role of the σ subunit is recognition of the **promoter** (a DNA sequence that signals the start of RNA transcription; see Section 11-3). The loosely bound σ subunit is released after transcription begins and about 10 nucleotides have been added to the new RNA strand. Prokaryotes have more than one type of σ subunit, and the different types of σ subunits direct RNA polymerases to different promoters and cause the transcription of various genes depending on the metabolic needs of the bacterial cell.

promoter the portion of DNA to which RNA polymerase binds at the start of transcription

11-2B Promoter Structure

Even the simplest organisms contain a great deal of DNA that is not transcribed. RNA polymerase must have a way of knowing which of the two DNA strands is the template strand and where the first nucleotide of the gene to be transcribed is located.

▶ How does RNA polymerase know where to begin transcription?

Promoters are DNA sequences that provide direction for RNA polymerase. The promoter region to which RNA polymerase binds is located *upstream* (farther to the 5′ side of the coding strand) from the gene for the RNA to be synthesized (Figure 11.2, *top*). The RNA is formed in the 5′ to 3′ direction, so the

← Upstream

Gene		−35 region		Pribnow box (−10 region)		Transcription start site (TSS) (+1)	
araBAD	GGATCCT	ACCTGACGCTTT	TTATCGCAACTCTC	TACTGT	TTCTCCAT	A	CCCGTTTTT
araC	GCCGTGAT	TATAGACACTTT	TGTTACGCGTTTT	TGTCAT	GGCTTTG	G	TCCCGCTTTG
bioA	TTCCAAAACG	TGTTTTTTGTTG	TTAATTCGGTG	TAGACT	TGTAAA	A	CCTAAATCTTTT
bioB	CATAATCG	ACTTGTAAACCA	AATTGAAAAGATT	TAGGTT	TACAAGTC	T	ACACCGAAT
galP2	ATTTATTCC	ATGTCACACTTT	TCGCATCTTTGT	TATGCT	ATGGTT	A	TTTCATACCAT
lac	ACCCCAG	GCTTTACACTTT	ATGCTTCCGGCTCG	TATGTT	GTGTG	GA	ATTGTGAGCGG
lacI	CCATCGAA	TGGCGCAAAACC	TTTCGCGGTATGG	CATGAT	AGCGCCC	G	GAAGAGAGTC
rrnA1	AAAATAAAT	GCTTGACTCTGT	AGCGGGAAGGCG	TATTAT	CACACC	CCC	GCGCCGCTG
rrnD1	CAAAAAAAT	ACTTGTGCAAAA	AATTGGGATCCC	TATAAT	GCGCCTCC	G	TTGAGACGA
rrnE1	CAATTTTTC	TATTGCGGCCTG	CGGAGAACTCCC	TATAAT	GCGCCTCC	A	TCGACACGG
*t*RNA^{Tyr}	CAACGTAAC	ACTTTACAGCGG	CGCGTCATTTGA	TATGAT	GCGCCCC	G	CTTCCCGATA
trp	AAATGAGC	TGTTGACAATTA	ATCATCGAACTAG	TTAACT	AGTACGCA	A	GTTCACGTA

	−35 region										Pribnow box						
Consensus sequence:	T	C	T	T	G	A	C	A	T	···[11–15 bp]···	T	A	T	A	A	T	···[5–8 bp]···
% occurrence of indicated base	42	38	82	84	79	64	53	45	41		79	95	44	59	51	96	

Figure 11.2 Sequences of 12 representative promoters from *E. coli*. By convention, these promoter sequences are given as the sequence that is found on the coding strand, going from left to right in the 5′ to 3′ direction. Promoter consensus sequence bases for *E. coli* are shown in the bottom row. The numbers below the consensus sequence bases indicate the percentage of the time that a certain position is occupied by the indicated nucleotide.

polymerase moves along the template strand in its 3′ to 5′ direction. By convention, all *control sequences* (DNA sequences that either activate or repress transcription) are given in the coding strand, which runs 5′ to 3′. It is important to remember the correct orientations of the template strand and the coding strand. The promoter sequence is given based on the coding strand, even though the RNA polymerase is actually binding to the template strand. Promoters are upstream, which means to the 5′ side of the coding strand and to the 3′ side of the template strand.

Most bacterial promoters have at least three components. Figure 11.2 shows some typical promoter sequences for 12 well-studied *E. coli* genes. For example, the *araBAD* row in the figure shows the promoter sequence for the *araBAD* operon, which contains genes that code for proteins that metabolize the sugar L-arabinose in *E. coli* (operons are discussed in Section 11-3C). In general, the promoter component closest to the first nucleotide to be incorporated is about 10 bases upstream, so it is called the −10 region or the **Pribnow box** after its discoverer. Also by convention, the first base to be incorporated into the RNA chain is said to be at position +1 and is called the **transcription start site (TSS)**, which is another of the three components of bacterial promoters. All the nucleotides upstream from this start site are given negative numbers.

Upstream from the Pribnow box, there are 16–18 bases that are completely variable. Moving further upstream, the next promoter element is about 35 bases upstream of the TSS and is simply called the **−35 region (−35 element)**. An element is a general term for a DNA sequence that is important in controlling transcription. The area from the −35 element to the Pribnow box to the TSS is called the **core promoter**. Upstream of the core promoter, **UP elements** can be present, which enhance the binding of RNA polymerase. UP elements usually extend from the −40 to −60 region. The region from the 5′ end of the UP element to the TSS is known as the **extended promoter**.

The base sequence of promoter regions has been determined for a number of prokaryotic genes, and a striking feature is that they contain many bases in common. **Consensus sequences**, in transcription, are nucleotide sequences that are most frequently found at each position in a sequence alignment (Figure 11.2, *bottom*). Figure 11.2 shows sequence alignments for 12 *E. coli* genes. Notice how the Pribnow boxes and −35 regions are A–T rich, with two hydrogen bonds per base pair. Consequently, they are more easily unwound than G–C-rich regions, which have three hydrogen bonds per base pair. This is important because the DNA strands must be unwound for RNA polymerase to bind to the template strand and produce an RNA strand (Figure 11.1). Figure 11.2 shows the consensus sequences for the Pribnow box and −35 regions. The consensus sequence for the Pribnow box contains only A and T bases, which means that A and T bases are most frequently found at each position in Pribnow box sequences.

Even though the −10 and −35 regions of many genes are similar, there are also some significant variations that are important to the metabolism of the organism. Besides directing the RNA polymerase to the correct gene, the promoter base sequence controls the frequency with which the gene is transcribed. Some promoters are strong, and others are weak. A strong promoter binds RNA polymerase tightly, so the gene is transcribed more often. In general, as a promoter sequence varies from the consensus sequence, the binding of RNA polymerase becomes weaker. Therefore, genes with weaker promoters get transcribed less often.

Pribnow box an A–T-rich DNA base sequence that is part of a prokaryotic promoter; it is located 10 bases before the TSS

transcription start site (TSS) the location on the template DNA strand where the first ribonucleotide is used to initiate RNA synthesis

−35 region (−35 element) a portion of DNA that is 35 base pairs upstream from the start of RNA transcription; it usually has a TTGACA sequence and is important in the control of RNA synthesis in bacteria

core promoter in prokaryotic transcription, the portion of the DNA from the transcription start site to the Pribnow box to the −35 region

UP elements prokaryotic promoter elements that are 40–60 bases upstream of the transcription start site and enhance the binding of RNA polymerase to DNA

extended promoter in prokaryotic transcription, the DNA from the transcription start site to the UP element

consensus sequences (in transcription) DNA sequences to which RNA polymerase binds; they contain nucleotides that are most frequently found at each position in a sequence alignment

11-2C Initiation of Transcription in Prokaryotes

The initiation of transcription in prokaryotes is the part of their transcription stages that has been studied the most. It is also the part that is the most controlled. Prokaryotic transcription initiation begins when the σ subunit of RNA

closed complex the complex that initially forms between RNA polymerase and DNA before transcription begins, and the DNA strands are not separated

open complex the form of the complex of RNA polymerase and DNA that occurs during transcription when the DNA strands get separated

polymerase binds to the promoter and forms what is called the **closed complex** (Figure 11.3). The σ subunit directs the polymerase to the promoter. It bridges the −10 and −35 regions of the promoter to the RNA polymerase core via a flexible "flap" in the σ subunit. Core enzymes lacking the σ subunit bind to areas of DNA that lack promoters. The holoenzyme may bind to "promoter-less" DNA, but it dissociates without transcribing.

Next, the **open complex** forms (Figure 11.3, *Step 3*). Studies show that a portion of the β′ and the σ subunits initiates strand separation, melting about 14 base pairs surrounding the TSS. A purine ribonucleoside triphosphate is the first base in RNA, and it binds to its complementary DNA base at position +1. Of the purines, A tends to occur more often than G. This first residue retains its 5′-triphosphate group (indicated by ppp in Figure 11.3).

After the strands have separated, a transcription bubble of about 17 base pairs moves down the DNA sequence to be transcribed (Figure 11.3, *Steps 4 and 5*), and RNA polymerase catalyzes the formation of the phosphodiester bonds between the incorporated ribonucleotides. When about 10 nucleotides have been incorporated, the σ subunit dissociates and is later recycled to bind to another RNA polymerase core enzyme (Figure 11.3, *Step 6*).

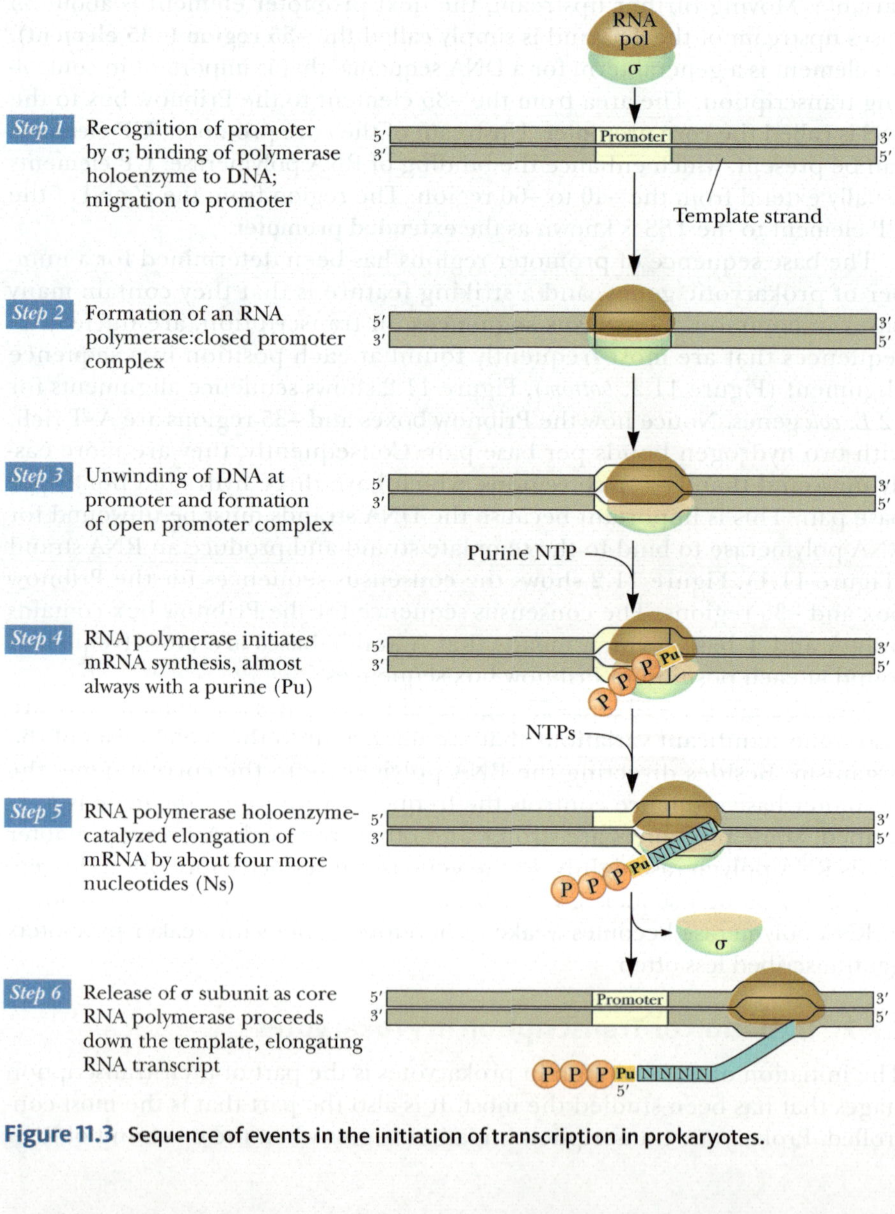

Figure 11.3 Sequence of events in the initiation of transcription in prokaryotes.

11-2D Chain Elongation

The transcription process supercoils DNA, with negative supercoiling upstream of the transcription bubble and positive supercoiling downstream, as shown in Figure 11.4. Topoisomerases relax the supercoils in front of and behind the advancing transcription bubble. The rate of chain elongation is not constant. The RNA polymerase moves quickly through some DNA regions and slowly through others. It may pause for as long as one minute before continuing.

Instead of finishing every RNA chain, RNA polymerase actually releases most chains near the beginning of the process, after about 5–10 nucleotides have been assembled, in a process called *abortive transcription*. The cause of abortive transcription is the failure of RNA polymerase to break its own bonds to the promoter via the σ subunit. Figure 11.5 shows a model of the mechanism of the transition between the open complex and transcription elongation. In order for chain elongation to occur, the RNA polymerase must be able to launch itself off the promoter. Given the tight binding between the σ subunit and the promoter, this requires substantial energy. The RNA polymerase is bound tightly to the DNA promoter. It "scrunches" the DNA into itself, causing torsional strain in the separated DNA strands. Like a bow loading up with potential energy as the bowstring is pulled, this provides the energy to allow the polymerase to break free.

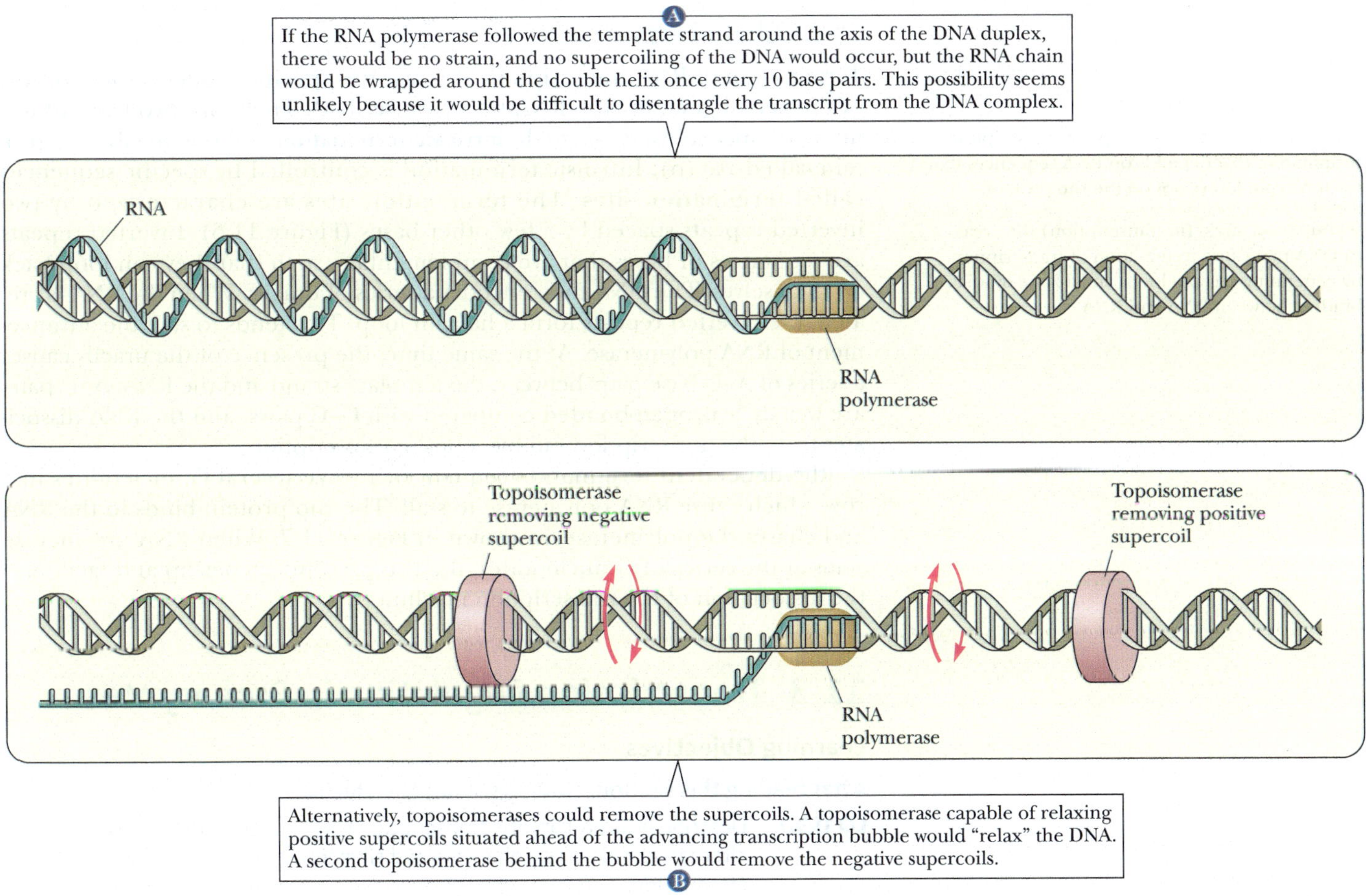

Figure 11.4 Two models for transcription elongation. *(Adapted from Futcher, B. (1988). Supercoiling and transcription, or vice versa? Trends in Genetics 4, 271–272, with permission from Elsevier.)*

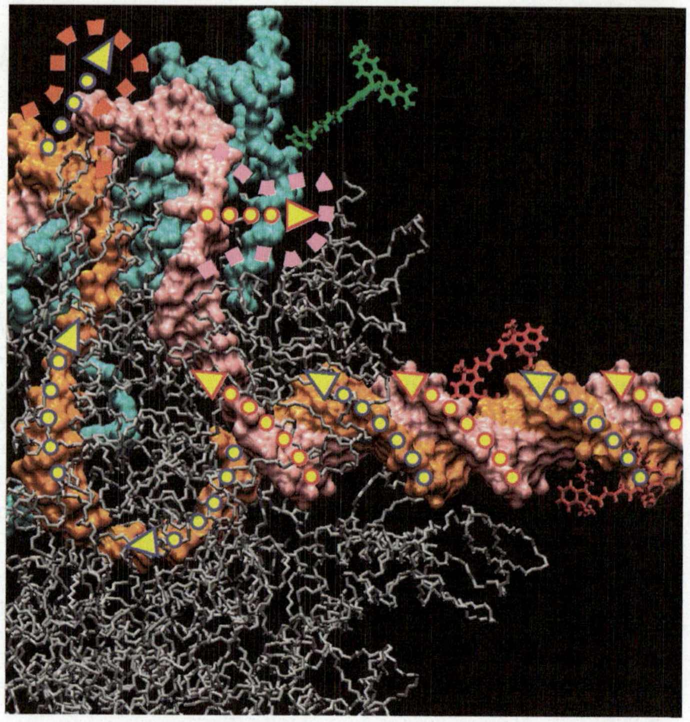

Figure 11.5 Monitoring events in prokaryotic transcription. The model shows an open promoter complex of bacterial RNA polymerase poised to begin RNA synthesis, with arrows designating motions of DNA segments that occur as the first nucleotides are combined to form an RNA chain. Downstream DNA (on the right) rotates inward and separates in the active-site channel of the polymerase. Template (orange) and coding (pink) DNA strands follow the indicated paths, moving as the RNA chain (not shown) is polymerized on the template DNA strand. Sites of extrusion of single-stranded DNA are shown by outlined arrows at the top. σ^{70} (blue); RNA polymerase core chains (gray). Fluorescent tags, shown in red and green, were used to monitor the movements that demonstrated the "scrunching" phenomenon. *(Courtesy of Achillefs Kapandis [University of Oxford], Shimon Weiss [University of California, Los Angeles], and Richard H. Ebright [Rutgers University].)*

11-2E Chain Termination

intrinsic termination the type of transcription termination that depends on RNA sequences that form hairpin loops, not on the rho protein

termination sites (in transcription) the areas in DNA that cause termination of transcription by generating hairpin loops and a zone of weak binding between DNA and RNA

Termination of RNA transcription also involves specific sequences *downstream* of the actual gene for the RNA to be transcribed. There are two types of termination mechanisms—namely, **intrinsic termination** and one involving a protein called *rho* (ρ). Intrinsic termination is controlled by specific sequences called **termination sites**. The termination sites are characterized by two inverted repeats spaced by a few other bases (Figure 11.6). Inverted repeats are sequences of bases that are complementary, such that they can loop back on themselves. The DNA then encodes a series of uracils. When the RNA is created, the inverted repeats form a hairpin loop. This tends to stall the advancement of RNA polymerase. At the same time, the presence of the uracils causes a series of A–U base pairs between the template strand and the RNA. A–U pairs are weakly hydrogen-bonded compared with G–C pairs, and the RNA dissociates from the transcription bubble, ending transcription.

Rho-dependent termination sequences involve several G nucleotides in a row, which cause RNA polymerase to stall. The rho protein binds to the RNA and chases the polymerase, as shown in Figure 11.7. When RNA polymerase stalls at the series of G nucleotides, the rho protein catches up and facilitates the dissociation of the transcription machinery.

11-3 Transcription Regulation in Prokaryotes

Learning Objectives

After reading this section, students should be able to:

LO 11.3.1 Describe how prokaryotic transcription is regulated in the following ways: alternative σ factors, enhancers and silencers, operons, and transcription attenuation.

LO 11.3.2 Distinguish between the following gene expression control mechanisms: induction, repression, positive control, and negative control.

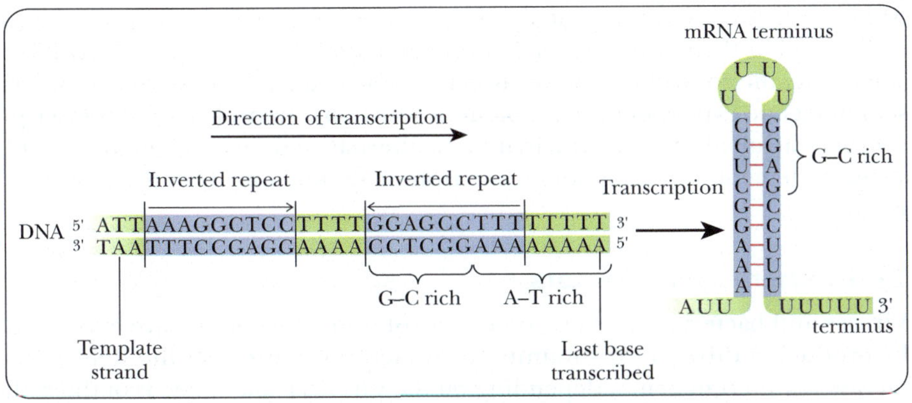

Figure 11.6 Inverted repeats terminate transcription. Inverted repeats in the DNA sequence being transcribed can lead to an mRNA molecule that forms a hairpin loop. This is often used to terminate transcription.

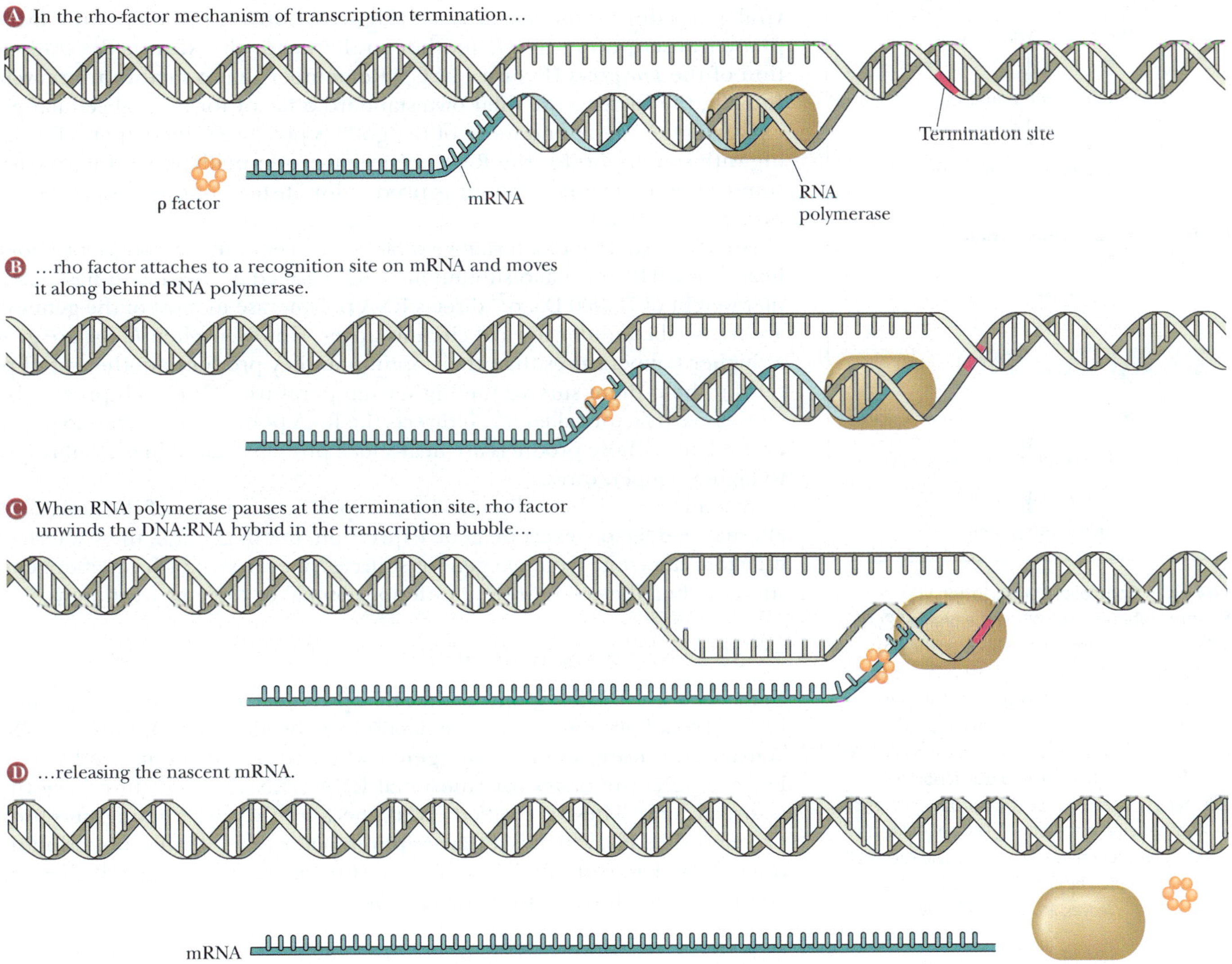

Figure 11.7 The rho-dependent mechanism of transcription termination. Rho (ρ) factor (A) attaches to a recognition site on mRNA and (B) moves along it behind RNA polymerase. (C) When RNA polymerase pauses at the termination site (a series of G nucleotides), rho factor unwinds the DNA:RNA hybrid in the transcription bubble, releasing the nascent mRNA (D).

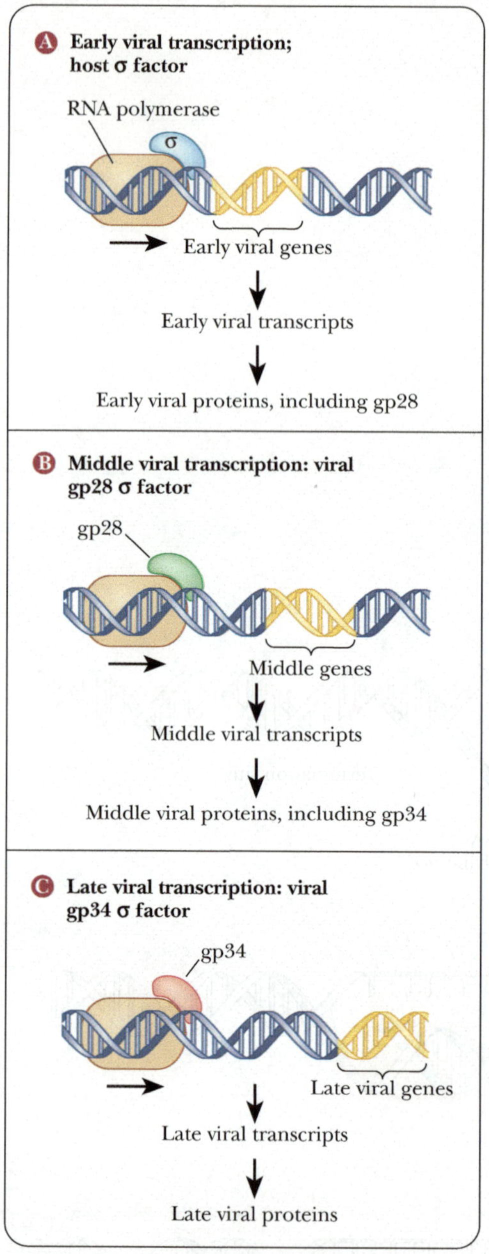

Figure 11.8 Control of transcription via different σ subunits. (A) When the phage SPO1 (or Okubovirus) infects bacterial cells, the host RNA polymerase (tan) and the host σ factor (blue) transcribe the early genes of the infecting viral DNA. One of the early gene products is gp28 (green), an alternative viral σ factor. (B) The gp28 directs the RNA polymerase to transcribe the middle genes, where one of their products is gp34 (red), another viral σ factor. (C) The gp34 directs the host's RNA polymerase to transcribe the viral late genes. *(Adapted by permission from* Molecular Biology, *by R. F. Weaver, McGraw-Hill, 1999.)*

enhancers DNA sequences that lie upstream of extended promoters, where transcription factors can bind to help increase the rate of transcription of downstream genes

Although many RNAs and proteins are produced in even a simple prokaryotic cell, not all of them are produced at the same time or in the same quantities. Therefore, the control of transcription is an important area of study for understanding gene expression (i.e., protein production). In prokaryotes, transcription is controlled in four principal ways: alternative σ factors, enhancer and silencer elements, operons, and transcription attenuation. They are discussed in turn.

11-3A Alternative σ Factors

Viruses and bacteria can exert some control over which genes are expressed by producing different σ subunits (or σ factors) that direct the RNA polymerase to different genes, depending on the various protein needs of the cell. A classic example of how this works is the action of phage SPO1, a virus that infects various bacterial cells. This virus has been renamed several times, but it has gone by the name Okubovirus since 2018. The virus has a set of genes called the *early genes,* which are transcribed by the host's RNA polymerase using the regular host σ subunit (Figure 11.8A). One of the viral early genes codes for a protein called *gp28.* This protein is a viral σ factor, which directs the bacterial host RNA polymerase to preferentially transcribe more of the viral genes during the *middle phase* (Figure 11.8B). A product of the middle phase transcription is *gp34,* another viral σ factor that directs the transcription of the *late genes* (Figure 11.8C). Remember that σ factors are recycled. Initially, the bacteria use their own standard σ factor for normal cellular gene expression. As more and more of the gp28 is produced, it competes for binding with standard σ for the RNA polymerase, eventually subverting the cell's transcription machinery for virus production instead of the normal bacterial protein production.

Another example of alternative σ factors is seen in the response of *E. coli* to heat shock. The normal σ subunit in *E. coli* is called σ^{70} because it has a molecular weight of 70,000 Da. σ^{70} directs RNA polymerase to most of the genes that are transcribed under normal circumstances. However, when *E. coli* are grown at higher temperatures than their optimum, they produce another set of proteins to help them survive the higher temperatures. One such protein is an alternative σ factor called σ^{32}. It directs the RNA polymerase to bind to promoters for genes whose products are heat-shock proteins that help the cell respond to higher temperatures.

In summary, we have discussed a few specific examples of the control that alternative σ factors exert on gene expression, but in general, these alternative σ factors can come from exogenous sources, like viruses, or from endogenous sources when the cell is dealing with nonoptimal or extreme conditions.

11-3B Enhancers and Silencers

Enhancers are DNA sequences that lie upstream of the extended promoter region (recall promoter structure details from Section 11-2B). Enhancers help activate transcription of certain genes when certain proteins bind to them. For example, the genes for ribosomal RNA (rRNA) production have three upstream sites. These are called *Fis sites* because they are binding sites for the protein called Fis, which is a **transcription factor** that elevates transcription of the rRNAs (Figure 11.9). These sites start from the end of the UP element at –60 and extend further upstream to –150.

▶ What is the difference between an enhancer and a promoter?

When a DNA sequence is labeled a promoter, it implies that the RNA polymerase binds to that region of DNA. An enhancer, on the other hand, is a DNA sequence that is usually upstream of the promoter. The polymerase does

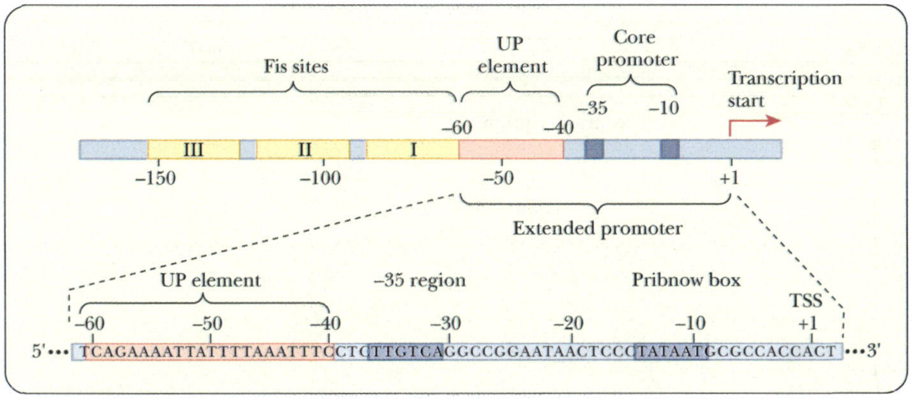

Figure 11.9 Elements of a bacterial promoter. The core promoter includes the -10 and -35 regions. The extended promoter includes the UP element. Upstream of the UP element, there may be enhancers, such as the Fis sites seen in the promoters for genes that code for ribosomal RNA in *E. coli*. The protein Fis, which binds to Fis sites, is a transcription factor that elevates rRNA transcription levels. *(Adapted by permission from* Molecular Biology, *by R. F. Weaver, McGraw-Hill, 1999.)*

transcription factors proteins or other complexes that bind to DNA sequences and alter basal levels of transcription

not bind to enhancers, but transcription factors do. When enhancers allow a response to changing metabolic conditions, such as temperature shock, they are usually referred to as **response elements**. When binding the transcription factor increases the level of transcription, the element is said to be an enhancer. When binding the transcription factor decreases transcription, the element is said to be a **silencer**. The position and orientation of enhancers are less important than for sequences that are part of the promoter. Molecular biologists can study the nature of control elements such as promoters, enhancers, and silencers by making changes to them. For example, when enhancer sequences are moved from one place on the DNA to another or have their sequences reversed, they still function as enhancers.

response elements DNA sequences that lie upstream of extended promoters, where transcription factors can bind to help *increase* the rate of transcription of genes whose protein products respond to specific stress conditions

silencer a DNA sequence that lies upstream of extended promoters, where transcription factors can bind to help *decrease* the rate of transcription of downstream genes

11-3C Operons

In prokaryotes, genes that encode enzymes of certain metabolic pathways are often controlled as a group, with the genes encoding the proteins of the pathway being close together and under the control of a common promoter. Such a group of genes is called an **operon**. Usually, these genes are not transcribed all the time. Rather, the production of these proteins can be triggered by the presence of a suitable substance called an **inducer**. This phenomenon is called **induction**. Most inducers are small molecules or proteins. A particularly well-studied example of an operon is the lactose operon (the *lac* operon) in *E. coli*.

operon a group of DNA sequences including an operator, promoter, regulatory genes, and structural genes that work together in specific metabolic pathways

inducer a molecule that turns on the transcription of a gene

induction (of enzyme synthesis) the triggering of the production of an enzyme by the presence of a specific inducer

The *lac* operon contains genes that regulate the transport and metabolism of the disaccharide lactose (Section 16-3). One of the genes in the operon is *lacZ*, which codes for the β-galactosidase enzyme (Figure 11.10, *bottom*). β-Galactosidase hydrolyzes the glycosidic linkage between galactose and glucose, the two monosaccharides that combine to make lactose. *E. coli* can survive with lactose as its sole carbon source. To do so, the bacterium needs β-galactosidase to catalyze the first step in lactose degradation.

The production of β-galactosidase takes place only in the presence of lactose, not in the presence of other carbon sources such as glucose. A metabolite of lactose, allolactose, is the actual inducer, and β-galactosidase is an *inducible enzyme*.

Operons contain **structural genes**, and they also often contain **regulatory genes**. Structural genes encode the protein products that are involved in the biochemical pathway of the operon, whereas regulatory genes encode protein products that can activate or repress transcription of the structural genes in the operon. Besides *lacZ* (Figure 11.10), two other structural genes, *lacY* and

structural gene an operon gene that directs the synthesis of a metabolic pathway protein

regulatory gene an operon gene that directs the synthesis of a protein that can activate or repress expression of structural genes in the operon

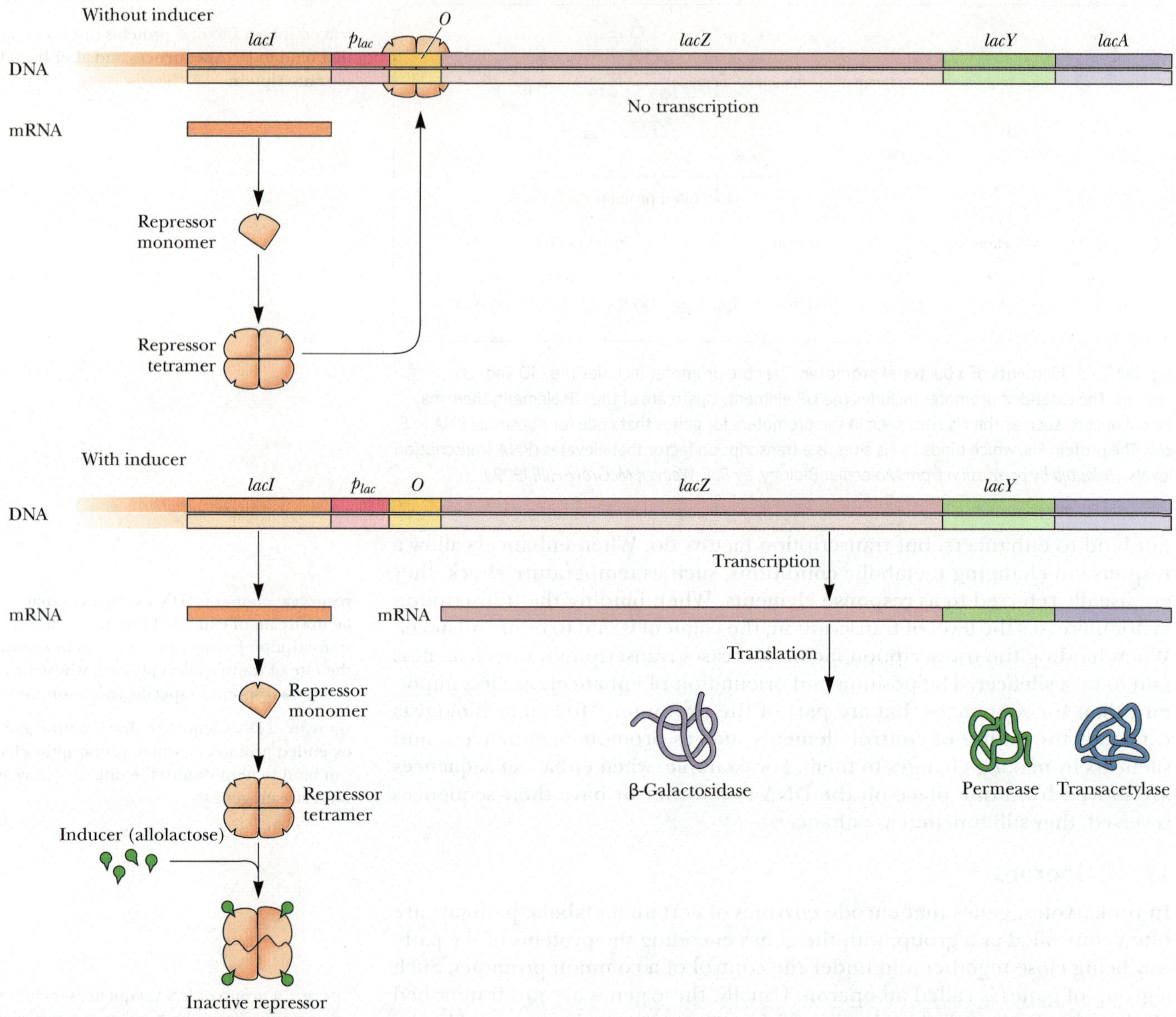

Figure 11.10 The lactose operon (*lac* operon). The *lacI* gene produces a repressor protein that forms a tetramer and represses the *lac* operon by binding to the operator (*O*). When the repressor is bound to the operator, RNA polymerase cannot bind the promoter (p_{lac}) and transcribe the structural genes (*lacZ*, *lacY*, and *lacA*). In the presence of an inducer (allolactose), the repressor cannot bind to the operator, so the operon genes can be transcribed when RNA polymerase binds to the promoter. The products of these genes (β-galactosidase, permease, and transacetylase) help transport and metabolize lactose in order to produce energy for the cell.

lacA, are part of the *lac* operon. The *lacY* gene encodes the enzyme lactose permease, which allows lactose to enter the cell. The *lacA* gene encodes an enzyme called β-galactoside transacetylase. The function of this last enzyme is not exactly known, but some hypothesize that its role is to acetylate certain sugars that the cell cannot metabolize in order to prevent their entry into the cell.

The expression of these structural genes is under the control of a regulatory gene called *lacI*. The *lacI* regulatory gene is responsible for the production of a repressor protein. As seen in the top part of Figure 11.10, the repressor protein forms a tetramer that binds to the **operator**, preventing the binding of RNA polymerase and thereby preventing the expression of the structural genes *lacZ*, *lacY*, and *lacA*. In the presence of the inducer, allolactose, the repressor cannot bind the operator, so the structural gene is no longer prevented from being expressed. Consequently, in the presence of the inducer, the β-galactosidase, permease, and transacetylase proteins are produced, which means that lactose

operator the DNA element to which a repressor of protein synthesis binds

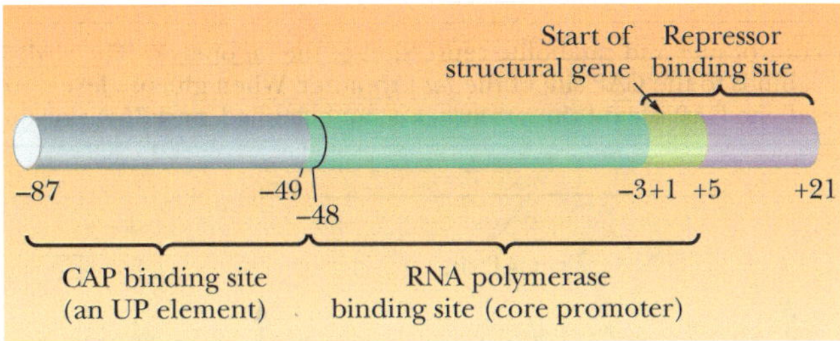

Figure 11.11 Extended promoter details in the *lac* operon. The numbers refer to base pairs. Negative numbers are assigned to base pairs in the regulatory sites. Positive numbers indicate the structural gene, starting with base pair +1. The catabolite activator protein (CAP) binding site is seen next to the RNA polymerase binding site.

can enter the cell, lactose can be broken down into glucose and galactose, and then the glucose can enter the glycolysis pathway to produce energy for the cell (Chapter 17).

The lactose operon is an example of *negative regulation* because the *lac* operon is turned on unless something is present to turn it off, which is the repressor protein in this case. The *lacI* repressor gene is adjacent to the structural genes in the *lac* operon, but this need not be the case. Many operons are known in which the regulatory gene is far removed from the structural genes.

▶ What are the different types of regulation in the *lac* operon?

We just saw how the repressor tetramer protein can negatively regulate expression of the structural genes in the *lac* operon, but the *lac* operon can also be positively regulated in a complex process called **catabolite repression**. It may be a bit confusing that a process called catabolite repression is a way in which the *lac* operon gets positively regulated, but let's carefully consider the details. In general, a catabolite is a molecule that is found in a metabolic pathway. In biochemistry, therefore, there are many, many catabolites. In this case, however, the catabolite involved in the repression of the *lac* operon is glucose.

catabolite repression (in the *lac* operon) repression of the synthesis of *lac* proteins by the presence of the glucose catabolite

catabolite activator protein (CAP) a protein that can bind to a promoter when complexed with cAMP, allowing RNA polymerase to bind to its entry site on the same promoter

Let's begin with what happens when glucose levels are low before exploring the repression that happens when glucose levels are high. The mechanism by which *E. coli* recognizes low levels of glucose involves the promoter and a regulatory protein that binds the promoter to help recruit RNA polymerase. The promoter has two regions. One is the binding site for RNA polymerase, and the other is the binding site for a positive regulatory protein (in contrast to the negative regulatory repressor protein). The positive regulatory protein for the *lac* operon is called the **catabolite activator protein (CAP)**. Figure 11.11 shows the extended promoter of the *lac* operon, including the CAP binding site, which is an UP element (Section 11-2B). Figure 11.12 shows a structural determination of the catabolite activator protein bound to the CAP DNA site. When CAP binds to the CAP site, it helps recruit RNA polymerase to the *lac* operon promoter so that it can express the *lac* genes.

Figure 11.13 shows how the binding of CAP to the promoter depends on the presence or absence of 3′,5′-cyclic AMP (cAMP). The structure of cAMP is shown in Figure 11.14. Notice how the phosphate forms a bond to the 3′ carbon of the ribose sugar as well as to the 5′ carbon, creating the cyclic structure (compare this to the structure of AMP in Figure 9.4). When glucose is not present, cAMP is formed, serving as a "hunger signal" for the cell. CAP forms a complex with cAMP. The complex binds to the CAP site in the promoter region. When the complex is bound to the CAP site on the promoter, the RNA polymerase can bind at the binding site available to it and proceed with transcription (Figure 11.13). The *lac* promoter (p_{lac}, Figure 11.10) is a weak promoter, meaning that RNA polymerase binding is minimal in the absence of the CAP–cAMP complex bound to the CAP site (recall the promoter strength discussion at the end of Section 11-2B). The CAP site is an example of an enhancer element, and the CAP–cAMP complex is a transcription factor. The modulation of transcription by CAP is a type of *positive regulation*. The *lac* operon is induced

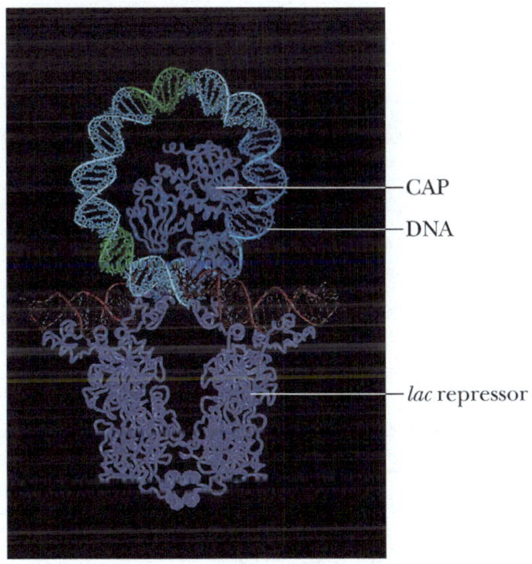

Figure 11.12 Structures of the *lac* repressor protein and catabolite activator protein (CAP) bound to DNA. *(Republished with permission of American Association for the Advancement of Science. Used with permission from Mitchell Lewis et al. (1 March 1996) Science 271 (5253), 1247. Permission conveyed through Copyright Clearance Center, Inc.)*

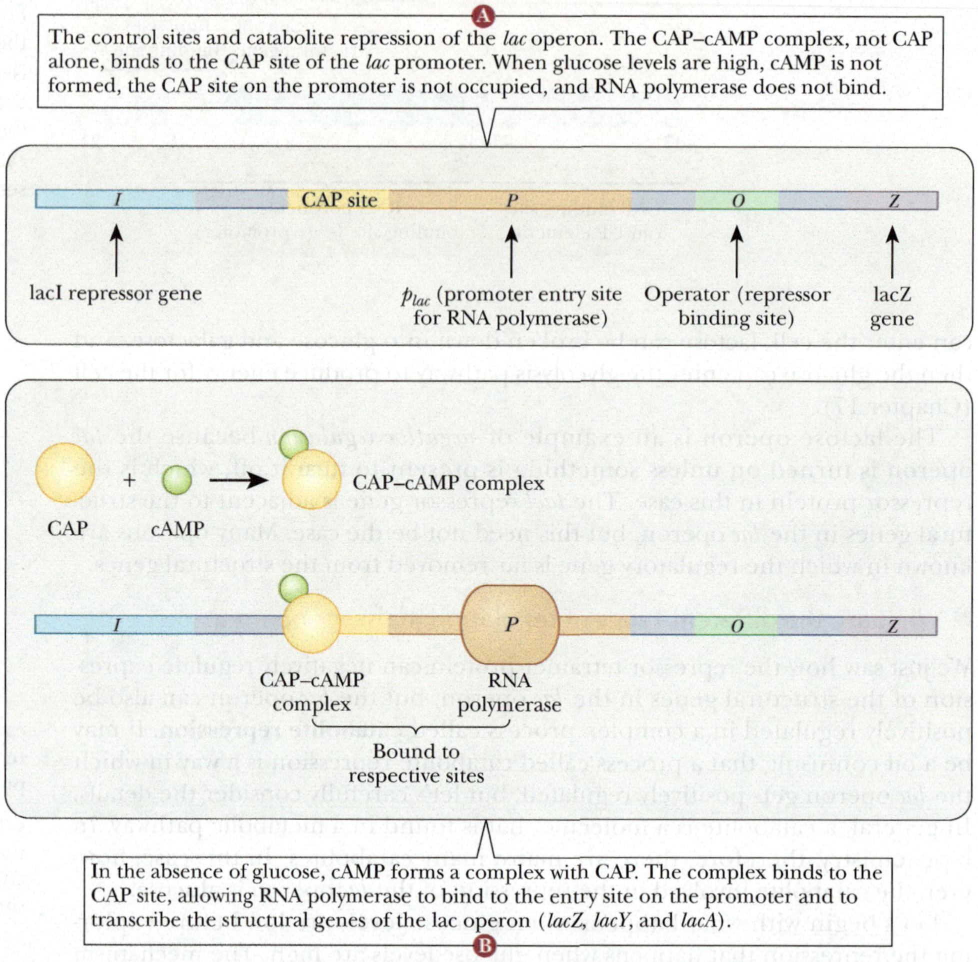

Figure 11.13 Catabolite repression and positive regulation of the *lac* operon.

when *E. coli* has lactose and no glucose available to it as a carbon source. When both glucose and lactose are present, the cell does not make the *lac* operon proteins. The repression of the synthesis of the *lac* operon proteins by the presence of glucose is called catabolite repression.

The preceding details involved the positive regulation processes that happen when there are low levels of glucose in bacteria. When bacterial cells have an adequate supply of glucose, catabolite repression processes happen. High levels of the glucose catabolite cause levels of cAMP to be low. When cAMP levels are low, the CAP–cAMP complex is not formed. Therefore, the complex does not bind the CAP site and recruit RNA polymerase, so expression of the *lac* genes is turned off. Thus, if the cell has enough glucose, then it doesn't need the *lac* proteins to produce more glucose. The combination of positive and negative regulation with the *lac* operon means that the presence of lactose is necessary, but not sufficient, for transcription of the operon structural genes. It takes the presence of lactose (to remove the repressor) *and* the absence of glucose (to produce cAMP and the CAP–cAMP complex) for the operon to be active. As we explain in other chapters, many transcription factors and response elements involve the use of cAMP, a common messenger in the cell.

In summary, operons can be controlled by positive or negative regulation mechanisms. They can also be classified as *inducible* or *repressible.* Therefore, there are four general possibilities for the control of gene expression, as shown in Figure 11.15. The top left figure shows a negative control system with induction. It is negative control because a repressor protein stops transcription when

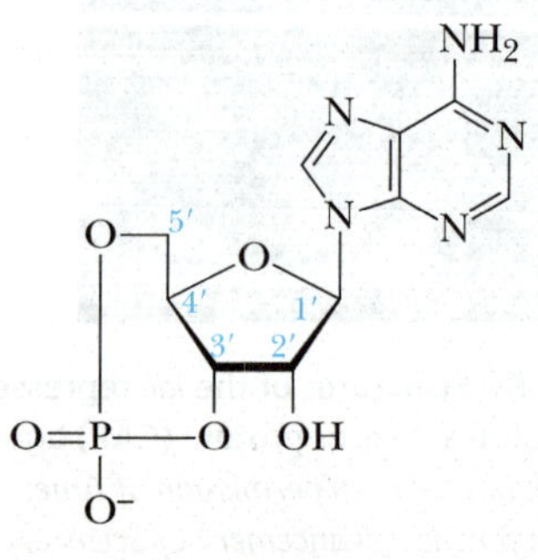

Figure 11.14 The structure of 3′,5′-cyclic adenosine monophosphate.

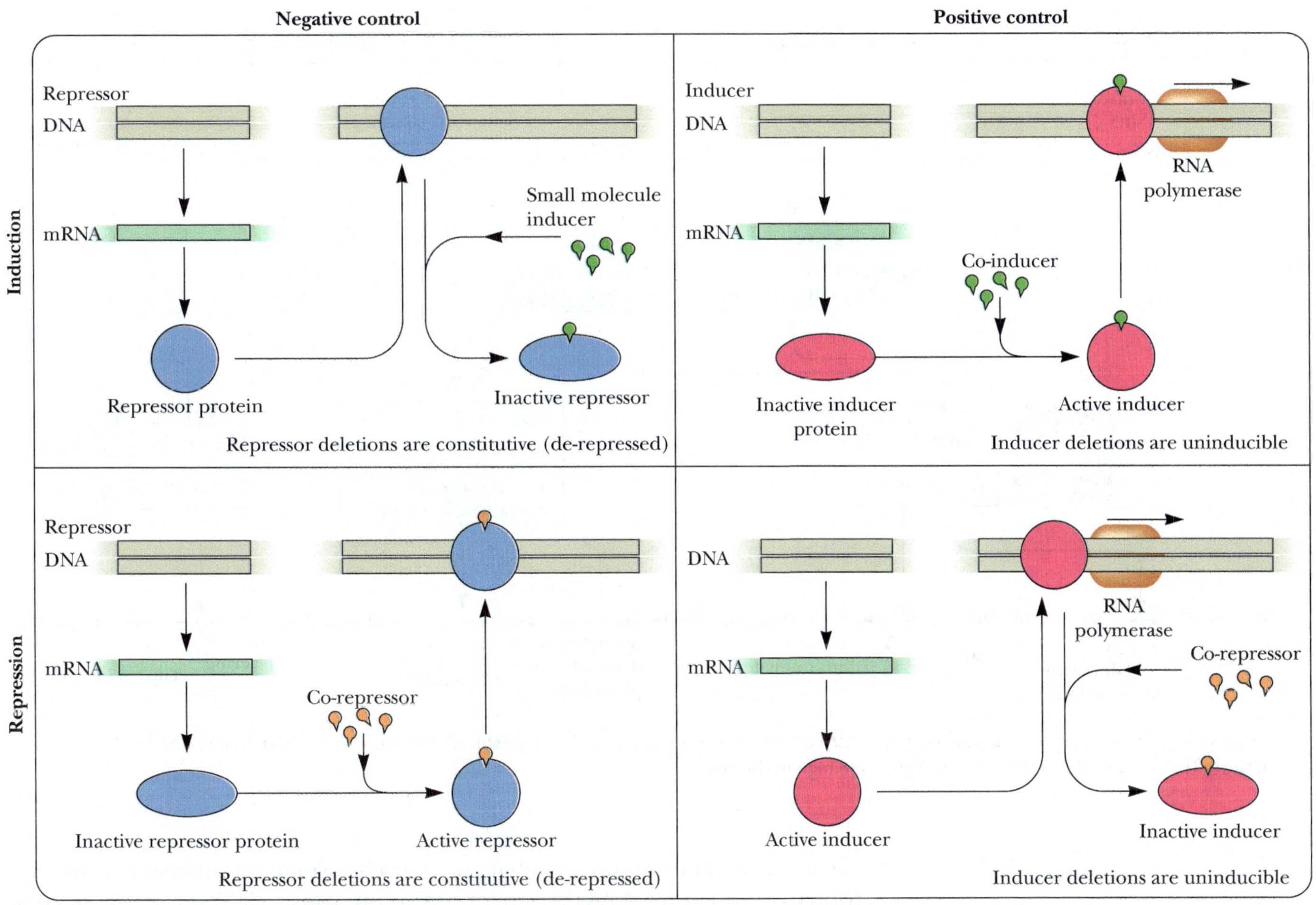

Figure 11.15 Basic control mechanisms seen in the control of genes. They may be inducible or repressible, and they may be positively or negatively controlled. Inducers and co-inducers can either stimulate inducers or inhibit repressors, whereas co-repressors can either activate repressors or inactivate inducers.

it binds to the promoter. It is an inducible system because the presence of the inducer releases the repression, as we saw with the *lac* operon. Negative control systems can be identified by the fact that if the gene for the repressor is mutated in some way that stops the expression of the repressor, the operon is always expressed. Genes that are always expressed are called **constitutive**.

constitutive refers to the transcription and expression of genes that are not controlled by anything other than the inherent binding of the RNA polymerase to the promoter

The top right of Figure 11.15 shows a positively controlled inducible system. The controlling protein is an inducer that binds to the promoter, stimulating transcription, but it works only when bound to its *co-inducer*. This is what is seen with the CAP–cAMP complex in the positive regulation of the *lac* operon. Such positively controlled systems can be identified by the fact that, if the gene for the inducer is mutated, it cannot be expressed—that is, it is *uninducible*.

The bottom left of Figure 11.15 shows a negatively controlled repressible system. A repressor stops transcription, but this repressor functions only in the presence of a *co-repressor*. Like an inducer, a co-repressor could be a protein or a small molecule. Finally, the bottom right of Figure 11.15 shows a positively controlled repressible system. An inducer protein binds to the promoter, stimulating transcription; but in the presence of the co-repressor, the inducer is inactivated.

The *trp* operon of *E. coli* (Figure 11.16) is an example of a repressible operon. The five structural genes (*trpE* through *trpA*) code for polypeptide chains that make four different enzymes (shown in the three boxes near the bottom of the figure). The bottom of Figure 11.16 shows how these enzymes

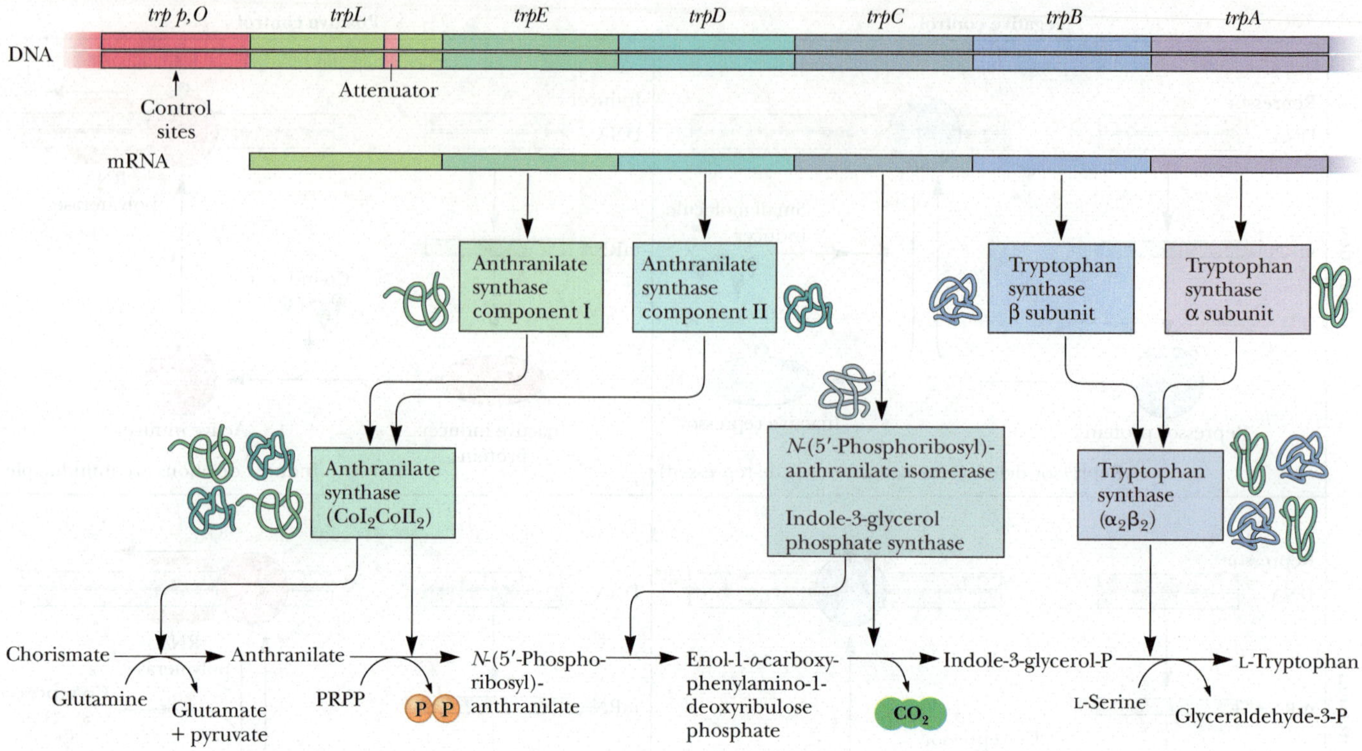

Figure 11.16 **The *trp* operon of *E. coli*.** *trp p* is the *trp* operon promoter and *O* is the operator, where a repressor–Trp complex can bind to inactivate the operon when levels of tryptophan are high in the cell.

catalyze the multistep process that converts chorismate to the amino acid tryptophan (see Figure 3.3A). Control of the operon is via a repressor protein that binds to two molecules of tryptophan. When tryptophan is plentiful, this repressor–tryptophan complex binds to the *trp* operator that is next to the *trp* promoter. This binding prevents the binding of RNA polymerase, so the operon is not transcribed. When tryptophan levels are reduced, the repression is lifted because the repressor will not bind to the operator in the absence of the co-repressor, tryptophan. This is an example of a system that is repressible and under negative regulation, as shown in Figure 11.15. The *trp* repressor protein is itself produced by the *trpR* operon and also represses that operon. It is an example of **autoregulation** because the product of the *trpR* operon regulates its own production.

autoregulation (of an operon) a system in which the product of an operon regulates its own production

11-3D Transcription Attenuation

Similar to how the *lac* operon is regulated in multiple ways, the *trp* operon is regulated by **transcription attenuation** processes in addition to repression processes. This control mechanism works by altering transcription *after* it has begun via transcription termination or pausing. Prokaryotes have no separation of transcription and translation as eukaryotes do, so the ribosomes are attached to the mRNA while it is being transcribed. The *trp* operon's first gene is the *trpL* sequence that codes for a leader transcript. The transcript of the leader sequence has four regions that can bind to each other in different ways to form different hairpin structures (Figure 11.17).

transcription attenuation a type of transcription control in which the transcription is controlled after it has begun via pausing and early release of incomplete RNA sequences

One possible secondary structure that the leader sequence transcript can form has binding between regions 1 and 2 and between regions 3 and 4 (Figure 11.17, *left*), but let's focus on conditions that affect the binding between regions 3 and 4. We first consider conditions where levels of tryptophan (Trp)

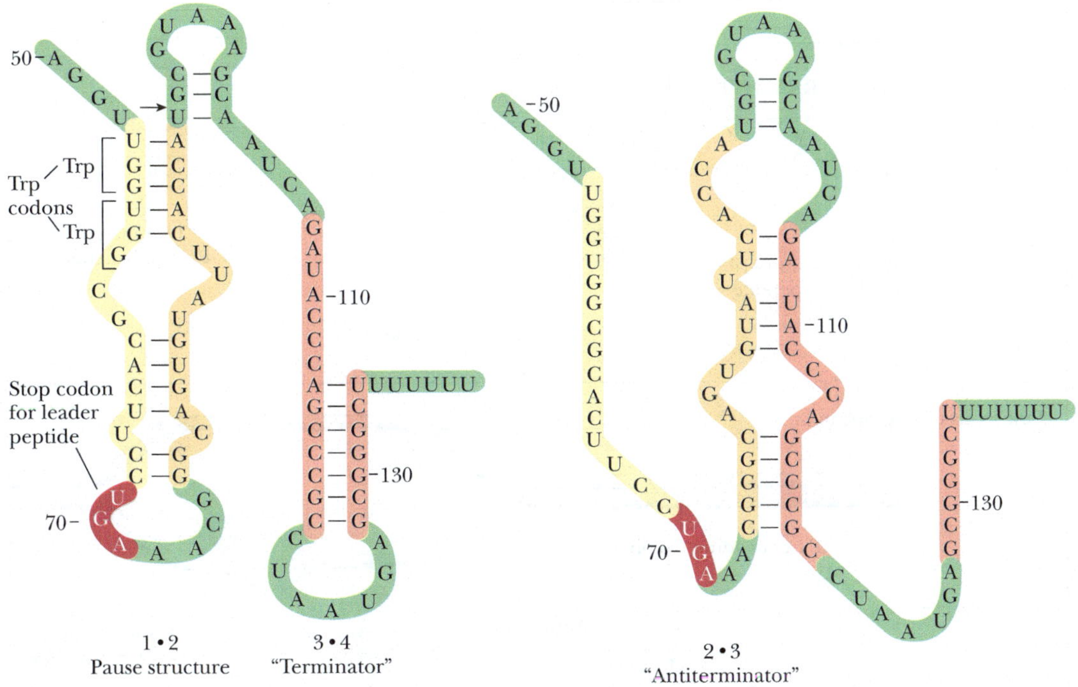

Figure 11.17 Alternative secondary structures can form in the mRNA of the *trpL* leader sequence of the *trp* operon. Binding between regions 1 and 2 (yellow and tan) is called a pause structure. Regions 3 and 4 (purple) then form a terminator hairpin loop. Alternative binding between regions 2 and 3 forms an antiterminator structure.

are low in the cell. We assume that the leader sequence is being transcribed by RNA polymerase and that a ribosome is following the polymerase and translating the transcript as well. The leader sequence transcript has two Trp codons in region 1, so if Trp levels are low, then the ribosome will have to wait longer for tRNA molecules carrying Trp to arrive (Section 9-6A). This pausing of translation will cause region 1 to be occupied by the ribosome, meaning that regions 2 and 3 of the leader transcript can bind to each other (Figure 11.17, *right*). In this case (low levels of Trp), the "terminator" hairpin between regions 3 and 4 did not form, so RNA polymerase remains bound and proceeds to transcribe the rest of the *trp* operon genes, which leads to the eventual production of Trp (Figure 11.16, *bottom*).

On the other hand, when Trp is plentiful, the ribosome does *not* stall at the Trp codons. Instead, it stalls when it reaches the UGA codon (a stop codon). When the ribosome stalls at this stop codon, it occupies parts of regions 1 and 2 of the transcript, which allows the "terminator" hairpin between regions 3 and 4 to form (Figure 11.17, *left*). This hairpin causes RNA polymerase to dissociate, preventing further expression of the *trp* operon genes. Figure 11.18 shows an overview of the attenuation processes involving the ribosome, the hairpin, and RNA polymerase when there are high levels of tryptophan in *E. coli* compared to when there are low levels of tryptophan.

In summary, whereas the repression processes control transcription of the *trp* operon based on levels of tryptophan in the cell, the attenuation processes control transcription of the operon based indirectly on levels of tryptophan in the form of tryptophan aminoacyl-tRNA molecules. The attenuation mechanism seen in the biosynthesis of tryptophan is also seen in similar ways in the biosynthesis of histidine, phenylalanine, and threonine.

Biochemical Connections 11A discusses the related topic of riboswitches.

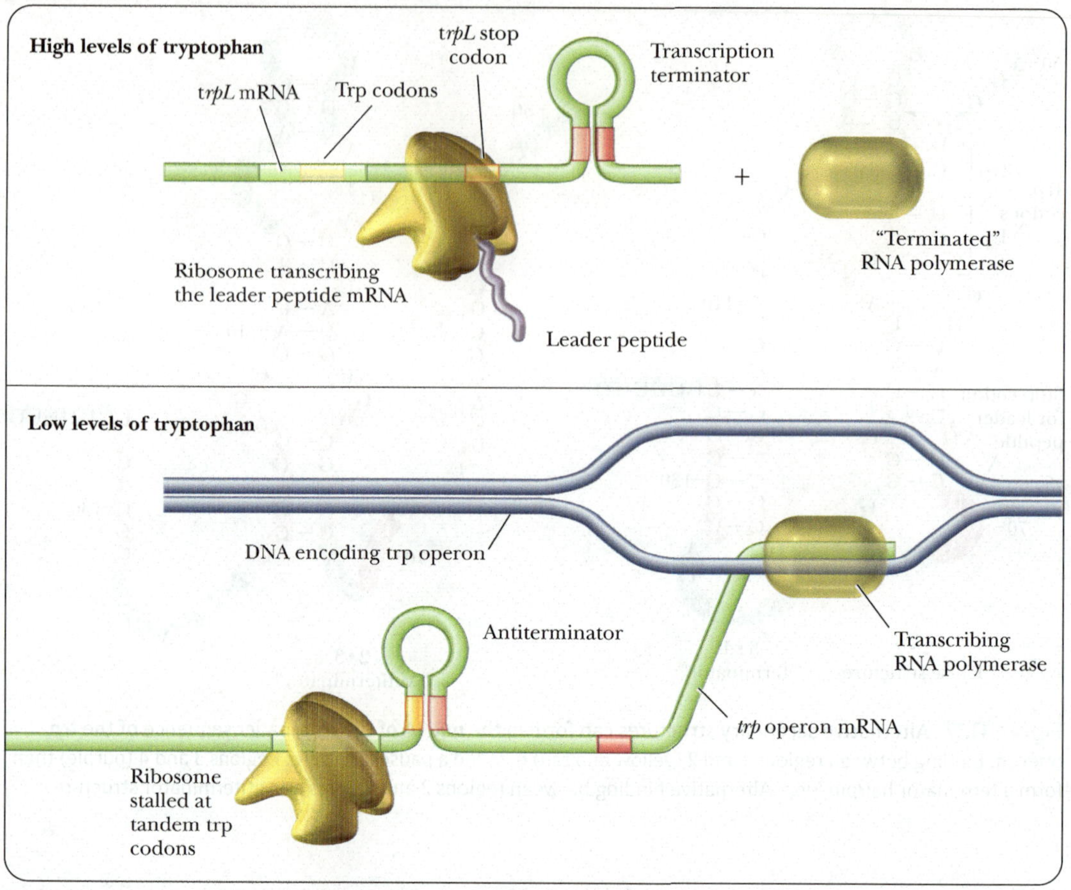

Figure 11.18 The attenuation mechanism in the trp operon. Top: When tryptophan levels are high, the ribosome quickly translates the leader peptide because Trp-carrying tRNAs are readily available when the ribosome translates the Trp codons. This causes premature abortion of the transcript as the terminator loop is allowed to form. Bottom: When tryptophan levels are low, the ribosome stalls at the trp codons, allowing the antiterminator loop to form, and transcription continues.

11A Biochemical Connections • Bacteriology

Riboswitches Provide Another Weapon Against Pathogens

Besides the transcription controls discussed in the previous section, prokaryotes have another control mechanism based on the mRNA transcript itself. While studying the metabolism of certain vitamins and cofactors, scientists discovered that the mRNA of pathways involved in making these vitamins could bind the vitamins themselves. These mRNAs have two functions—sensing and decision-making. Together, these functions are called *riboswitches.* A riboswitch is made up of two parts. At the 5′ end is a sensing domain called an *aptamer,* which can bind signal molecules, and the other part is the *expression platform,* whose structure can directly influence gene expression.

Figure 11.19A shows how one of the known riboswitches binds to thiamine pyrophosphate (TPP). The aptamer portion of the mRNA is able to bind to TPP and therefore acts as a sensor for this vitamin. If the riboswitch senses that the vitamin is present, then it can respond in a way that prevents the translation of the mRNA, thereby avoiding a pathway that would produce more of a molecule that is not needed. In the example of the TPP sensor, translation is prevented when a hairpin loop forms in the expression platform of the riboswitch that blocks the translation initiation site (Figure 11.19B).

Other translation-halting processes have also been discovered. One riboswitch recognizes another vitamin metabolite, flavin mononucleotide (FMN). When FMN is bound to the aptamer, a terminator hairpin is created similar to the one that forms during transcription attenuation (Figure 11.18C). In another example, a riboswitch that recognizes the sugar glucosamine-6-phosphate (GlcN6P) becomes susceptible to faster degradation in the presence of that sugar (Figure 11.18D).

More than a dozen human pathogens have riboswitches as part of their metabolism. Researchers are hoping to find molecules that can act like a competitive inhibitor and fool the riboswitch into acting as though the natural substrate were present. If the riboswitch controls a vital process, then shutting off the riboswitch would kill the pathogen. Riboswitch targets have been identified in many different bacteria that can be harmful to humans, including *Bacillus anthracis, Hemophilus influenza, Helicobacter pylori, Salmonella enterica,* and *Streptococcus pneumoniae.*

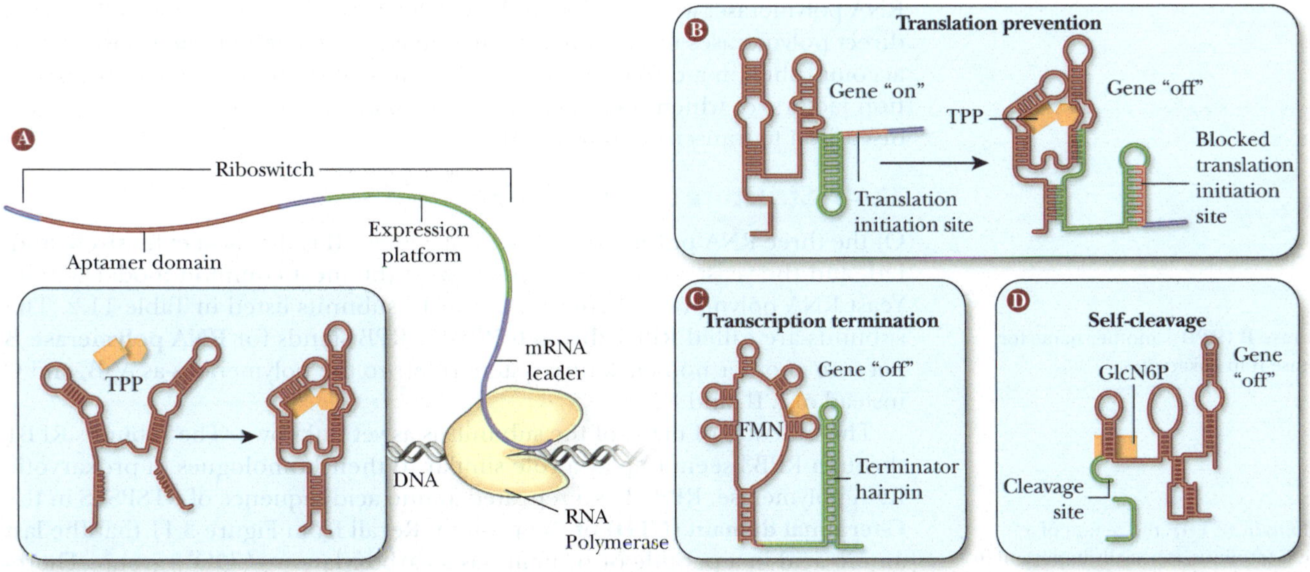

Figure 11.19 Metabolite sensing. (A) An aptamer for the coenzyme thiamine pyrophosphate (TPP) assumes a defined shape (left) as it exits the polymerase. When TPP is present, the aptamer binds to it, grasping the molecule tightly (right). Riboswitch responses: Riboswitches employ a variety of strategies to control protein manufacture. When TPP is absent, for example, the expression platform can leave a translation initiation site open to ribosomes, allowing expression of the gene's instructions to remain "on" (B, left). When TPP is bound by the aptamer, the expression platform can form a hairpin that blocks translation, turning the gene "off" (B, right). A riboswitch sensing the coenzyme flavin mononucleotide (FMN) forms a terminator hairpin that halts transcription of its message by polymerase (C). Another ribozyme triggered by glucosamine-6-phosphate (GlcN6P) self-destructs by cleaving itself (D). *(©Tolpa Studios, Inc.)*

11-4 Transcription in Eukaryotes

Learning Objectives

After reading this section, students should be able to:

LO 11.4.1 Describe the features of eukaryotic RNA polymerase.

LO 11.4.2 Explain the components of eukaryotic transcription promoters.

LO 11.4.3 Distinguish between events in the initiation, elongation, and termination phases of eukaryotic transcription.

We have seen that prokaryotes have a single RNA polymerase that is responsible for the synthesis of all three kinds of prokaryotic RNA–mRNA, tRNA, and rRNA. The polymerase can switch σ factors to interact with different promoters, but the core polymerase stays the same. The transcription process is predictably more complex in eukaryotes than in prokaryotes. Three primary RNA polymerases with different activities are known to exist. Each one transcribes a different set of genes and recognizes a different set of promoters:

1. RNA polymerase I is found in the nucleolus (Section 1-4B) and synthesizes precursors of most, but not all, ribosomal RNAs.
2. RNA polymerase II is found in the nucleus and synthesizes mRNA, miRNA, and snRNA.
3. RNA polymerase III is found in the nucleus and synthesizes the tRNAs and precursors of 5S ribosomal RNA.

All three of these eukaryotic RNA polymerases are large (500–700 kDa), complex proteins consisting of 10 or more subunits. Their overall structures differ, but they all have a few subunits in common. They all have two larger subunits that share sequence homology with the β and β′ subunits of prokaryotic RNA polymerase, which make up the catalytic unit. There are no σ subunits to direct polymerases to promoters. The detection of a gene to be transcribed is accomplished in a different way in eukaryotes, and the presence of transcription factors, of which there are hundreds, plays a larger role. We restrict our discussion to transcription by Pol II.

11-4A Structure of RNA Polymerase II

Of the three RNA polymerases, RNA polymerase II is the most extensively studied, and the yeast *Saccharomyces cerevisiae* is the most common model system. Yeast RNA polymerase II consists of the 12 subunits listed in Table 11.2. The subunits are called RPB1 through RPB12. RPB stands for **RNA polymerase B** because another nomenclature system refers to the polymerases as A, B, and C instead of I, II, and III.

RNA polymerase B (RPB) another name for RNA polymerase II in eukaryotes

The function of many of the subunits is as yet unknown. The subunits RPB1 through RPB3 seem to play a role similar to their homologues in prokaryotic RNA polymerase. RPB1 has a repeated amino acid sequence of PTSPSYS in the **C-terminal domain (CTD)** of the protein. Recall from Figure 3.11 that the last amino acid in a peptide or protein has a carboxylate (—COO^-) group. Therefore, the CTD of a protein includes the very last amino acid in the polypeptide chain and several other amino acids that precede the last one. Threonine, serine, and tyrosine are all substrates for phosphorylation (Section 7-3), which is important in the control of transcription initiation.

C-terminal domain (CTD) the region of a protein at the C-terminus, especially important in eukaryotic RNA polymerase B

X-ray crystallography has been used to determine the structure of RNA polymerase II (Figure 11.20). Notable features include a pair of jaws formed by subunits RPB1, RPB5, and RPB9, which appear to grip the DNA downstream of the active site. A clamp near the active site is formed by RPB1, RPB2, and RPB6, which may be involved in locking the DNA:RNA hybrid to the polymerase, increasing the stability of the transcription unit.

Table 11.2 Yeast RNA Polymerase II Subunits

Subunit	Size (kDa)	Features	*E. coli* Homologue
RPB1	191.6	DNA binding-domain and polymerase activity	β′
RPB2	138.8	Contacts the DNA template and the new RNA strand	β
RPB3	35.3	Core subunit that interacts with several other RPB subunits	α
RPB4	25.4	Binds NTPs, ssDNA, and new RNA strand	σ
RPB5	25.1	Core subunit that interacts with several other RPB subunits	
RPB6	17.9	Core subunit that interacts with several other RPB subunits	
RPB7	19.1	May regulate polymerase activity	
RPB8	16.5	Core subunit that interacts with several other RPB subunits	
RPB9	14.3	Binds DNA template along with RPB1	
RPB10	8.3	Core subunit that interacts with several other RPB subunits	
RPB11	13.6	Interacts with RPB3	
RPB12	7.7	Interacts with RPB1 and RPB2	

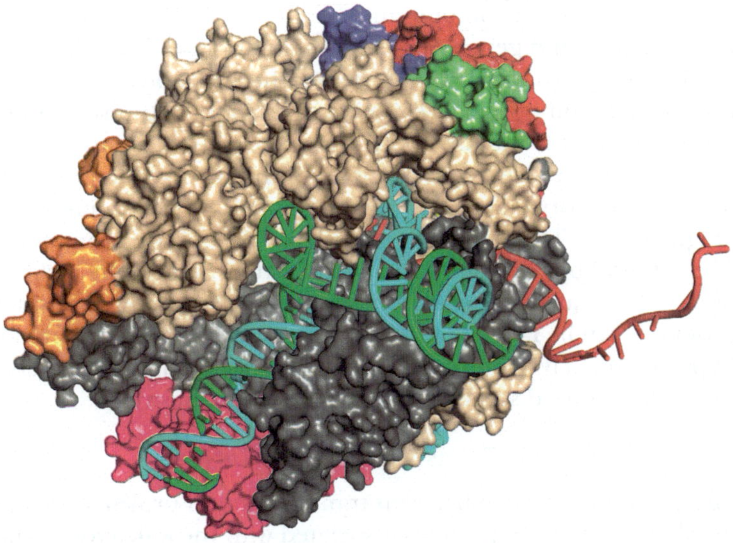

Figure 11.20 **Architecture of yeast RNA polymerase II.** Transcription of DNA (helical structure) into RNA (red) is shown. The template strand of DNA is shown in blue, and the coding strand in green. Transcription takes place in the clamp region of the active site, shown at center right. The jaws that hold DNA downstream of the active site are shown at lower left. *(Courtesy of Roger Kornberg/Stanford University.)*

Structural work on RNA polymerases from prokaryotes and eukaryotes has led some to conclude that because extensive homology exists between the core regions of RNA polymerases from bacteria, yeast, and humans, it is possible that RNA polymerase evolved eons ago, at a time when only prokaryotes existed. The theories speculate that as more complex organisms developed, layers of other subunits were added to the core polymerase to reflect the more complicated metabolism and compartmentalization of eukaryotes.

11-4B Pol II Promoters

▶ How does Pol II recognize the correct DNA to transcribe?

Pol II promoters have four elements (Figure 11.21). The first includes a variety of *upstream elements*, which act as enhancers or silencers (Section 11-3B). Specific binding proteins either activate transcription above basal levels, in the case of enhancers, or suppress it, in the case of silencers. Two other upstream elements that are close to the core promoter are the GC box (–40), which has a consensus sequence of GGGCGG, and the CAAT box (extending to –110), which has a consensus sequence of GGCCAATCT.

The second eukaryotic promoter element, found at position –25, is the **TATA box**, which has a consensus sequence of TATAAA. This is very similar to the Pribnow box element in prokaryotic core promoters (Figure 11.2).

TATA box a promoter element found in eukaryotic transcription that is located 25 bases upstream of the transcription start site

The third eukaryotic promoter element includes the TSS at position +1, but in the case of eukaryotes, it is surrounded by a sequence called the **initiator element** (*Inr*). This sequence is not well conserved. For instance, the sequence for a particular gene type may be $_{-3}$YYCAYYYYY$_{+6}$, in which Y indicates either pyrimidine (i.e., C or T), and A is the purine at the TSS.

initiator element a loosely conserved DNA sequence surrounding the transcription start site in eukaryotic promoters

The fourth eukaryotic promoter element is a possible downstream regulator, although these are rarer than upstream regulators. Many natural promoters lack at least one of the four elements. The initiator plus the TATA box

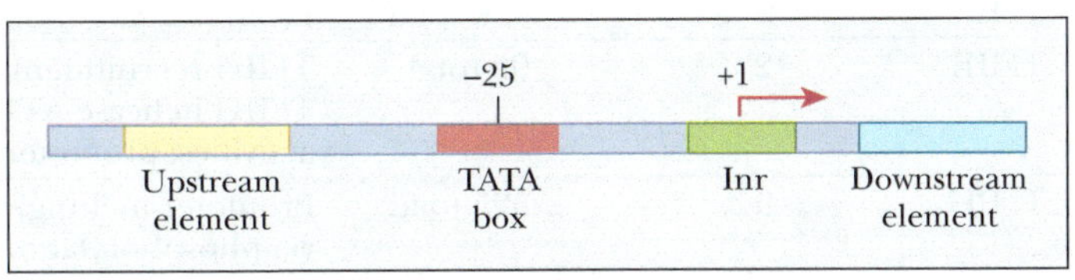

Figure 11.21 **Four elements of Pol II promoters.**

make up the core promoter and are the two most consistent parts across different species and genes. Some genes do not have TATA boxes; they are called "TATA-less" promoters. In some genes, the TATA box is necessary for transcription, so deletion of the TATA box causes a loss of transcription. In others, the TATA box orients the RNA polymerase correctly. Elimination of the TATA box in these genes causes transcription at random starting points. Whether a particular regulatory element is considered part of the promoter or not is often a judgment call. Those that are considered part of the promoter are close to the TSS (50–200 bp) and show specificity with regard to the distance and orientation of the sequence. Regulatory sequences that are not considered part of the promoter can be far removed from the TSS, and their orientation is irrelevant. Experiments have shown that such sequences still work, even when they are reversed or moved several thousand base pairs upstream.

11-4C Eukaryotic Transcription: Initiation

The biggest difference between transcription in prokaryotes and eukaryotes is the sheer number of proteins associated with the eukaryotic version of the process. Any protein that regulates transcription but is not itself a subunit of RNA polymerase is a **transcription factor**. There are many transcription factors for eukaryotic transcription. The molecular mass of the entire complex of Pol II and all of the associated factors exceeds 2.5 million Da.

transcription factor any protein that regulates transcription but is not itself a subunit of RNA polymerase

preinitiation complex (PIC) in eukaryotic transcription, the phase where RNA polymerase and about 100 other proteins, including the general transcription factors, bind to the DNA

Eukaryotic transcription initiation begins by the formation of a **preinitiation complex (PIC)**, and most of the control of transcription occurs at this step. This complex normally contains RNA polymerase II and six *general transcription factors (GTFs)*: TFIIA, TFIIB, TFIID, TFIIE, TFIIF, and TFIIH.

▶ What do eukaryotic transcription factors do?

The general transcription factors are required for all eukaryotic promoters. Much work is still going on to determine the structure and function of each of the parts of the preinitiation complex. Each GTF has a specific function, and each is added to the complex in a defined order. Table 11.3 summarizes the components of the preinitiation complex.

Figure 11.22 shows the sequence of events in eukaryotic Pol II transcription. The first step in the formation of the preinitiation complex is the recognition of the TATA box by TFIID. This transcription factor is actually a combination of

Table 11.3 General Transcription Initiation Factors

Factor	Number of Subunits	Size (kDa)	Function
TFIID-TBP	1	27	TATA box recognition, positioning of TATA box DNA around TFIIB and Pol II
TFIID-TAF$_{II}$s	14	15–250	Core promoter recognition (non-TATA elements), positive and negative regulation
TFIIA	3	12, 19, 35	Stabilization of TBP binding; stabilization of TAF–DNA binding
TFIIB	1	38	Recruitment of Pol II and TFIIF; start-site recognition for Pol II
TFIIF	3	156 total	Promoter targeting of Pol II
TFIIE	2	92 total	TFIIH recruitment; modulation of TFIIH helicase ATPase and kinase activities; promoter melting
TFIIH	9	525 total	Promoter melting; promoter clearance via phosphorylation of CTD

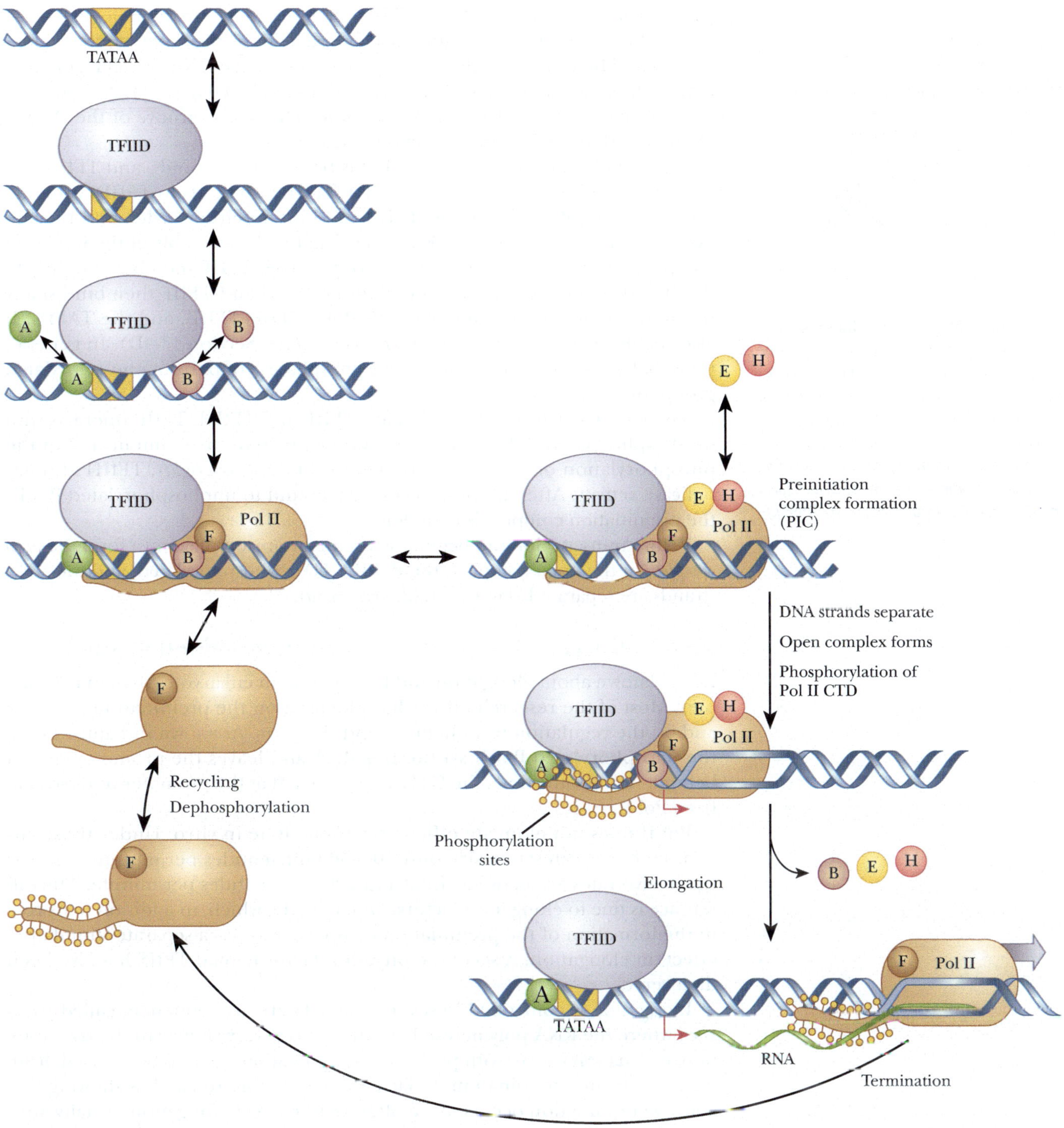

Figure 11.22 A schematic representation of the order of events of eukaryotic transcription. TFIID (which contains the TATA-box binding protein, TBP) binds to the TATA box. TFIIA and TFIIB then bind, followed by recruitment of RNA polymerase II and TFIIF. TFIIH and TFIIF then bind to form the preinitiation complex (PIC). Kinases phosphorylate the CTD of Pol II, leading to the open complex, in which the DNA strands are separated. RNA is produced during elongation as Pol II and TFIIF leave the promoter and the other GTFs behind. Pol II dissociates during the termination phase, and the CTD is dephosphorylated. Pol II/TFIIF is then recycled to bind to another promoter.

several proteins. The primary protein is called *TATA-binding protein (TBP)*. Associated with TBP are many other proteins called *TBP-associated factors (TAFIIs)*. Because TBP is also present and required for Pol I and Pol III, it is a universal transcription factor. TBP is highly conserved. From species as different as yeast,

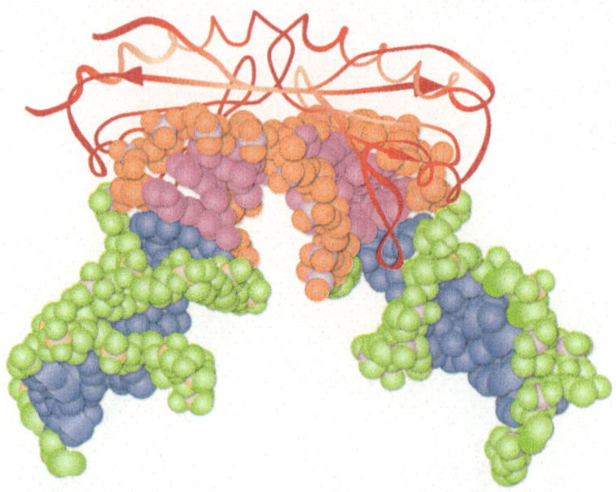

Figure 11.23 Model of yeast TATA-binding protein (TBP) binding to DNA. The DNA backbone of the TATA box is shown in orange, and the TATA bases are shown in pink. Adjacent DNA sequences are shown in green. The TBP, which is shown in red, sits on the minor groove of the DNA like a saddle. *(Reprinted by permission from Crystal Structure of a Yeast TBP/TATA-Box Complex by Kim, Y., Geiger, J. H., Hahn, S., and Sigler, P. B.,* Nature *365,* p. 512 *[1993].)*

plants, fruit flies, and humans, the TBPs are more than 80% homologous—that is, they have more than 80% identical amino acids in the same sequence positions. The TBP protein binds to the minor groove of the DNA (recall Figure 9.8) at the TATA box via the last 180 amino acids of its CTD. Figure 11.23 shows how the TBP sits on the TATA box like a saddle. The minor groove of the DNA is opened, and the DNA is bent to an 80° angle.

Figure 11.22 shows that once TFIID is bound, TFIIA binds, and TFIIA also interacts with both the DNA and TFIID. TFIIB also binds to TFIID, bridging the TBP and Pol II. TFIIA and TFIIB can bind in either order, and they do not interact with each other. TFIIB is critical for the assembly of the initiation complex and for the location of the correct TSS. TFIIF then binds tightly to Pol II and suppresses nonspecific binding. Pol II and TFIIF then bind stably to the promoter. TFIIF interacts with Pol II, TBP, TFIIB, and the TAFIIs. It also regulates the activity of the *CTD phosphatase* (Section 11-4D). In general, phosphatases are proteins that remove phosphate groups from phosphorylated groups or molecules.

The last two factors to be added are TFIIE and TFIIH. TFIIE interacts with unphosphorylated Pol II, and these two factors have been implicated in the phosphorylation of polymerase II (Figure 11.22, *lower right*). TFIIH also has helicase activity. After all these GTFs have bound to unphosphorylated Pol II, the preinitiation complex is complete.

Before transcription can begin, the preinitiation complex must form the *open complex.* In the open complex, the Pol II CTD is phosphorylated, and the DNA strands are separated (Figure 11.22, *lower right*).

11-4D Eukaryotic Transcription: Elongation and Termination

Less is known about elongation and termination in eukaryotes than in prokaryotes. Most of the research efforts have focused on the preinitiation complex and on the regulation by enhancers and silencers. As shown in Figure 11.22, the phosphorylated Pol II synthesizes RNA and leaves the promoter region behind. At the same time, the GTFs either are left at the promoter or dissociate from Pol II.

Pol II does not elongate efficiently when alone in vitro. Under those circumstances, it can synthesize only 100–300 nucleotides per minute, whereas the in vivo rates are between 1500 and 2000 nucleotides per minute. The difference is due to elongation factors. One is TFIIF, which, in addition to its role in the formation of the preinitiation complex, also has a separate stimulatory effect on elongation. A second elongation factor named TFIIS has also been discovered.

Elongation is controlled in several ways. There are sequences called *pause sites,* where the RNA polymerase hesitates. This is very similar to the transcription attenuation we saw with prokaryotes. Elongation can also be aborted, leading to premature termination. In fact, more and more research is showing that abortive termination occurs more often than correct elongation, usually after just a few nucleotides have been linked. Finally, elongation can proceed past the normal termination point. This is called *antitermination.* The TFIIF class of elongation factors promotes a rapid read-through of pause sites, perhaps locking the Pol II into an elongation-competent form that does not pause and dissociate.

The TFIIS class of elongation factors is called *arrest release factors.* They help the RNA polymerase move again after it has paused. A third class of elongation factors consists of the *P-TEF* and *NELF* proteins (*positive-transcription elongation factor* and *negative elongation factor*). They increase the productive form of transcription and decrease the abortive form or vice versa.

Termination begins by stopping the RNA polymerase. There is a eukaryotic consensus sequence for termination, which is AAUAAA. This sequence may be 100–1000 bases away from the actual end of the mRNA. After termination occurs, the transcript is released, and the Pol II open form (phosphorylated) is released from the DNA. The phosphates are removed by phosphatases, and the Pol II/TFIIF complex is recycled for another round of transcription (Figure 11.22, *lower left*).

11-5 Transcription Regulation in Eukaryotes

Learning Objectives

After reading this section, students should be able to:

LO 11.5.1 Describe how eukaryotic transcription is regulated by the mediator complex.

LO 11.5.2 Explain how eukaryotic transcription is regulated by chromatin remodeling complexes, covalent modifications of histones, and response elements.

In Section 11-4, we saw how the general transcription machinery, consisting of the RNA polymerase and GTFs, functions to initiate transcription. This is the general case that is consistent for all transcription of mRNA. However, this machinery alone produces only a low level of transcription called the **basal level**. The actual transcription level of some genes may be many times above or below the basal level. The differences in transcription levels are caused by gene-specific transcription factors called *activators* or *repressors*. Recall that eukaryotic DNA is complexed to histone proteins in chromatin. The DNA is wound tightly around the histone proteins, and many of the promoters and other regulatory DNA sequences may be inaccessible much of the time.

basal level (of transcription) the level of transcription that occurs due solely to RNA polymerase and the GTFs

11-5A The Role of Mediator in Eukaryotic Transcription Activation and Repression

General transcription activation requires a protein called **mediator**, which serves to bridge the promoter, RNA polymerase, and general transcription machinery with specific remote enhancers and silencers. Mediator is a giant complex with a mass of over 1 million daltons, comprising over 20 distinct subunits in yeast and more than 30 subunits in humans, referred to as the MED proteins. Mediator is a crescent-shaped protein with a head, middle, and tail. Eight MED proteins are found in the head and middle regions, and four make up the tail. The tail section binds to co-activators or co-repressors that are themselves bound to enhancer and silencer elements, as shown in Figure 11.24. The head and middle portions bind to the CTD of RNA polymerase II. The middle region also binds to the GTF TFIIF. Mediator bridges the promoter region with the enhancer region to activate transcription, as shown in Figure 11.24A. Or, in the opposite case, it binds to the silencer element but does not recruit RNA Pol II to the promoter, as shown in Figure 11.24B. Mediator is the ultimate regulator of eukaryotic transcription, essential to the transcription of almost every RNA polymerase II–dependent gene. Mediator is also required for basal transcription of these genes.

mediator a giant protein complex that bridges the promoter, GTFs, and remote silencers and enhancers

▶ How does transcription work around the nucleosomes?

One of the more complicating differences between prokaryotic and eukaryotic transcription is the state the DNA is found in with the latter. As seen in Section 9-3C and Figure 9.17, the DNA in eukaryotes is wound tightly around the nucleosomes, a core of basic histone proteins. In its tightly packed state, RNA polymerase II has no access to the promoter regions, and transcription cannot occur. Essentially, the presence of nucleosomes represses transcription.

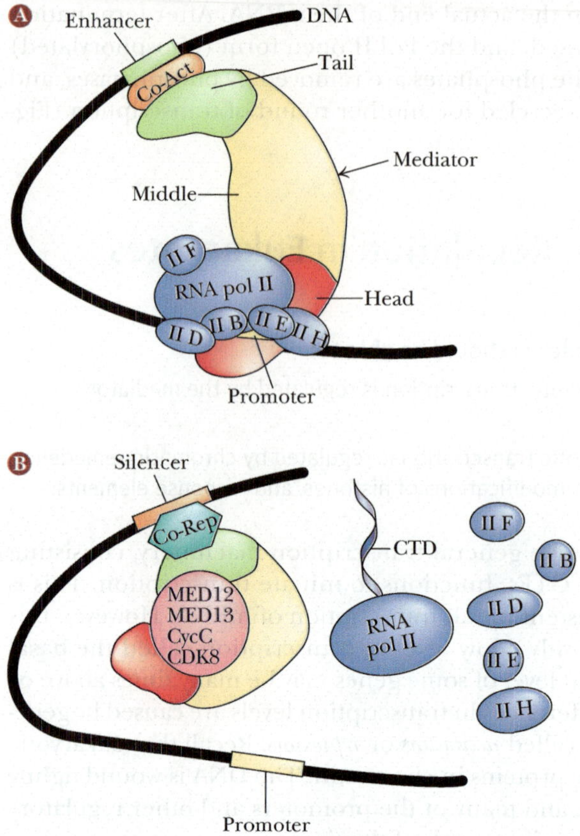

Figure 11.24 **The role of mediator in the regulation of eukaryotic gene expression.** (A) The multiprotein complex mediator is a transcription activator. It bridges the promoter region to an enhancer, increasing the transcription of a gene. (B) Mediator can also help repress transcription. When it binds to a repressor region, including the repressive complex MED12/MED13/CycC/CDK8, it then fails to recruit RNA polymerase to the promoter region.

chromatin remodeling complexes enzyme complexes that mediate ATP-dependent conformational changes in nucleosome structure that lead to transcription

histone-modifying enzymes enzymes that make covalent modifications to the histone core octamer in order to regulate histone–DNA binding

Activation of eukaryotic transcription depends on the following circumstances: the interaction of RNA polymerase with the promoter and transcription machinery, and the relief of repression caused by the chromatin structure. This relief depends on factors that can reorganize the chromatin and alter the nucleosomes so that the promoters in the DNA become accessible to the transcription machinery. Therefore, two sets of protein factors are important in the activation of eukaryotic transcription: **chromatin remodeling complexes** that mediate ATP-dependent conformational changes in nucleosome structure, and **histone-modifying enzymes** that introduce covalent modifications into the N-terminal tails of the histone core octamer (Section 11-5C). These two processes are closely linked.

11-5B Chromatin Remodeling Complexes

Chromatin remodeling complexes are huge (1 megadalton = 1,000,000 Da) assemblies containing ATP-dependent enzymes that loosen the DNA:protein interactions in nucleosomes by a variety of mechanisms involving sliding, ejecting, inserting, and otherwise restructuring the core octamers (Figure 11.25). There are several families of remodeling complexes. The most well studied are the SNF/SWI and RSC families of remodelers. The ultimate goal of the remodelers is to create space between the nucleosomes and to expose the DNA so that RNA polymerase can be recruited to the promoter.

11-5C Covalent Modification of Histones

Another important part of transcription activation via modification of chromatin involves the covalent modification of histone proteins. In the transcriptionally inactive state, the negative charges on the phosphates of the DNA backbone are tightly bound to the positive charges on the basic proteins of the

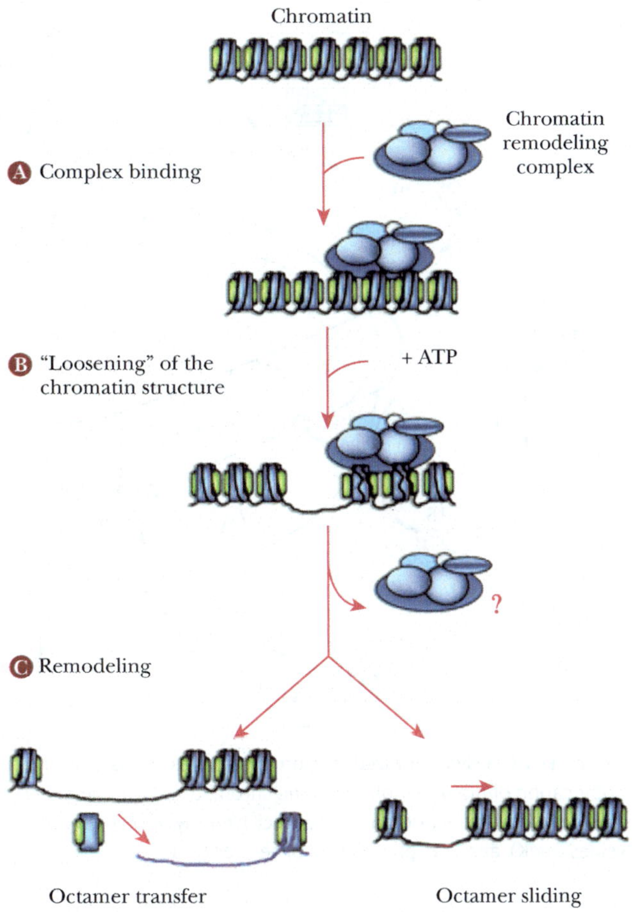

Figure 11.25 **Chromatin remodeling.**
(A) The chromatin remodeling complex binds to the chromatin in an ATP-independent manner. (B) ATP is hydrolyzed, and the chromatin structure is loosened. The remodeling complex is released in a manner that is not well understood as indicated by the question mark. (C) Remodeling can take multiple forms, such as sliding of the octamers or their removal and transfer. The end result is that the loosening of the chromatin allows the transcription machinery to access the DNA.

histone. To activate transcription, this tight binding must be relaxed. The most important modification of the histones is the acetylation of the ε-amino groups of lysine on the histone tails by **histone acetyltransferases (HATs)**. Acetylating the lysine removes the positive charge and loosens the binding of the DNA. The weakened interaction between histones and DNA opens chromatin structure and allows transcription factors and RNA polymerase access to the DNA in order to transcribe genes. This process can be reversed by **histone deacetylases (HDACs)**. Other modifications also play a role in transcription regulation through histones, including phosphorylation of serine residues and methylation of lysine and arginine residues. As shown in Figure 11.26, there are many sites for the covalent modification of histones.

In an attempt to understand the complex mechanisms of how modification of histones and remodeling of chromatin affect transcription, scientists have formed a **histone code** hypothesis to describe all of the variables that control eukaryotic transcription activation by these processes. The main concept of the hypothesis is that the covalent modifications of the histones not only serve to affect the binding between DNA and the histones but also recruit proteins that recognize the histone modifications. These proteins then hypothetically catalyze chromatin structure alterations that promote transcription.

Unfortunately for those who study transcription, the situation is too complex to draw simple conclusions about the effects that each type of histone modification causes. For example, there are many sites where histone methylation occurs (Figure 11.26), and methylating some of the histone residues leads to transcription activation, while methylating others leads to transcription repression. In histone H3, methylation of lysine 4 is associated with active transcription, but methylation of lysine 9 is associated with repressed transcription.

histone acetyltransferases (HATs) enzymes that acetylate lysine residues on histone proteins

histone deacetylases (HDACs) enzymes that remove acetyl groups from acetylated lysine residues on histone proteins

histone code a hypothesis regarding the combination of chromatin remodeling processes that control eukaryotic transcription

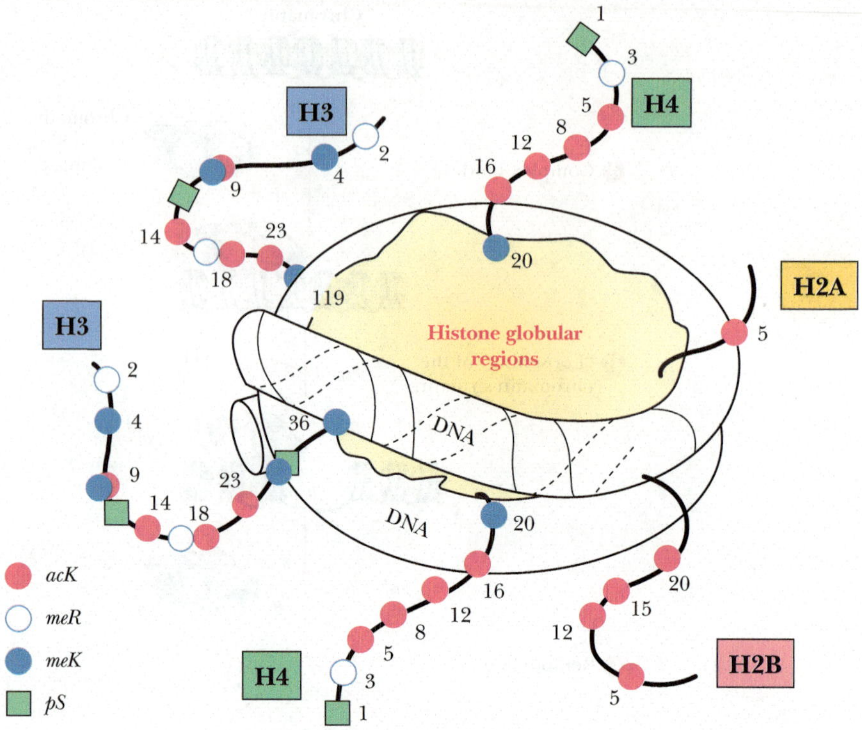

Figure 11.26 Covalent modification of histones. Transcription is regulated by modification of histone proteins in the chromatin. There are many sites for modification, as shown, such as acetylation of lysines (acK), methylation of arginines (meR), methylation of lysines (meK), and phosphorylation of serines (pS).

Acetylation and phosphorylation of histone residues also have similar differences in modification state versus their effect on transcription.

11-5D Eukaryotic Response Elements

Some eukaryotic transcription control mechanisms can be categorized based on a common response to certain metabolic factors. Similar to prokaryotic transcription control mechanisms seen in Section 11-3B, eukaryotes also have response elements, which are short DNA sequences where transcription factors can bind and regulate transcription. Examples include the heat-shock element (HSE), the glucocorticoid-response element (GRE), the metal-response element (MRE), and the cyclic-AMP-response element (CRE).

▶ How do response elements work?

These response elements all bind transcription factor proteins that are produced under certain cellular conditions, causing the expression of genes whose protein products help the cell respond to nonoptimal conditions. Eukaryotic transcription control is not the same as in prokaryotic operons because eukaryotic genes are not linked in sequence and are not controlled by a single promoter. Several different genes, all with unique promoters, may all be affected by the same transcription factor binding the response element.

In the case of HSE, elevated temperatures lead to the production of specific heat-shock transcription factors that activate the associated genes. Glucocorticoid hormones bind to a steroid receptor. Once bound, this becomes the transcription factor that binds to the GRE.

We now look more closely at the CRE as an example of eukaryotic control of transcription. Hundreds of research papers deal with this topic, as more and more genes are found to have this response element as part of their control.

Remember that 3′,5′-cyclic adenosine monophosphate (cAMP; see Figure 11.14) was also involved in the control of prokaryotic operons via its complexing with the CAP protein and the binding of the CAP–cAMP complex to the CAP enhancer element (Figure 11.13).

Cyclic AMP is produced as a second messenger from several hormones such as epinephrine and glucagon (Section 24-3B). When the levels of cAMP rise, the activity of **cAMP-dependent protein kinase** (protein kinase A) is stimulated. This enzyme phosphorylates many other proteins and enzymes inside the cell and is usually associated with switching the cell to a catabolic mode, in which macromolecules are broken down for energy. Protein kinase A phosphorylates a protein called **cyclic-AMP-response-element-binding protein (CREB)**, which binds to the cyclic-AMP-response element and activates the associated genes. The CREB protein does not directly contact the basal transcription machinery (RNA polymerase and GTFs), however, and the activation requires another protein. **CREB-binding protein (CBP)** binds to CREB after it has been phosphorylated and bridges the response element and the promoter region, as shown in Figure 11.27. After this bridge is made, transcription is activated above basal levels. Many abbreviations are used in the language of transcription, and Table 11.4 summarizes the more important ones.

cAMP-dependent protein kinase an important enzyme in the control of metabolism; it is stimulated by cAMP and when stimulated phosphorylates key proteins

cyclic-AMP-response-element-binding protein (CREB) an important transcription factor in eukaryotes that binds to the CRE and activates transcription

CREB-binding protein (CBP) an important mediator of transcription; it links the basal transcription machinery to CREB

The CBP protein and a similar one called p300 are transcription factors that play central roles in the response to several different hormone signals, as can be seen in Figure 11.28. Several hormones that act through cAMP cause the phosphorylation and binding of CREB to CBP. Similarly, steroid and thyroid hormones, and some others, act on receptors in the nucleus to bind to CBP/p300. Finally, growth factors and stress signals cause *mitogen-activated protein kinase (MAPK)* to phosphorylate transcription factors *AP-1 (activating protein 1)* and *Sap-1a,* both of which bind to CBP. Binding of these response proteins to the CBP or p300 proteins leads to activation of genes whose protein products respond to the hormone signals.

CBP/p300 are known to have histone acetyltransferase activity, so they also stimulate transcription via loosening of the nucleosomes. They are also thought to recruit mediator to the promoter region (Figure 11.27, *lower right*).

Researchers have linked several known human diseases to the reduction of active levels of CBP, including Huntington's disease, spinocerebellar ataxia, and fragile X syndrome, all of which are characterized by mutations that increase levels of a trinucleotide repeat of CAG.

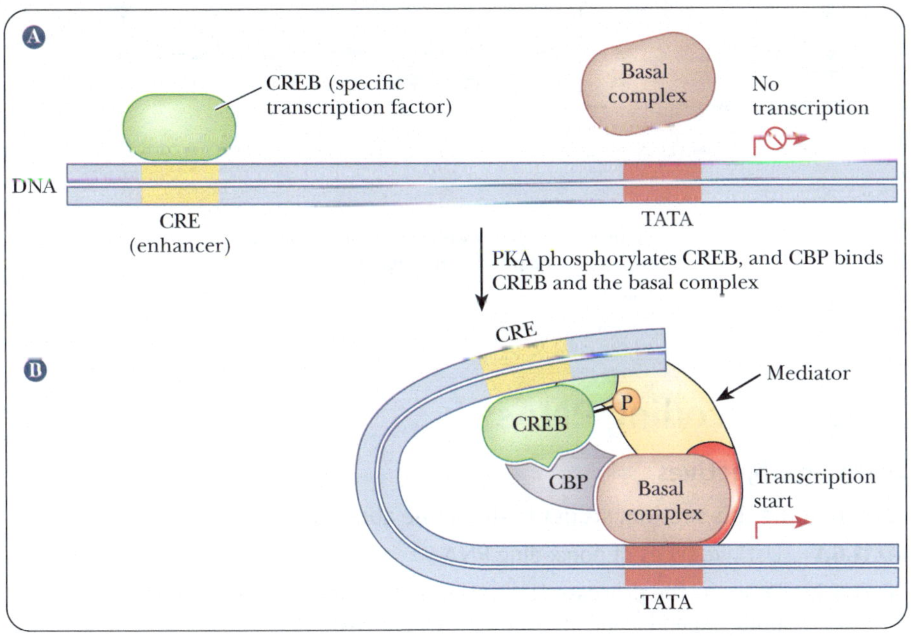

Figure 11.27 Activation of transcription via cyclic-AMP-response-element-binding protein (CREB) and CREB-binding protein (CBP). (A) Unphosphorylated CREB does not bind to CBP, and no transcription occurs. (B) Phosphorylation of CREB causes binding of CREB to CBP, which forms a complex with the basal transcription complex (RNA polymerase and general transcription factors), thereby activating transcription. The mediator protein complex (Section 11-5A) is also involved in transcription activation. *(Adapted by permission from* Molecular Biology, *by R. F. Weaver, McGraw-Hill, 1999.)*

Table 11.4 Abbreviations Used in Transcription

bZIP	Basic-region leucine zipper	NTD	N-terminal domain
CAP	Catabolite activator protein	N-TEF	Negative transcription elongation factor
CBP	CREB-binding protein	Pol II	RNA polymerase II
CRE	Cyclic-AMP-response element	P-TEF	Positive-transcription elongation factor
CREB	Cyclic-AMP-response-element-binding protein	RPB	RNA polymerase B (Pol II)
CREM	Cyclic-AMP-response-element modulating protein	RNP	Ribonucleoprotein particle
CTD	C-terminal domain	snRNP	Small nuclear ribonucleoprotein particle ("snurp")
GRE	Glucocorticoid-response element	TAF	TBP-associated factor
GTF	General transcription factor	TATA	Consensus promoter element in eukaryotes
HSE	Heat-shock-response element	TBP	TATA-box binding protein
HTH	Helix–turn–helix	TCR	Transcription-coupled repair
Inr	Initiator element	TF	Transcription factor
MRE	Metal-response element	TSS	Transcription start site
MAPK	Mitogen-activated protein kinase	XP	Xeroderma pigmentosum

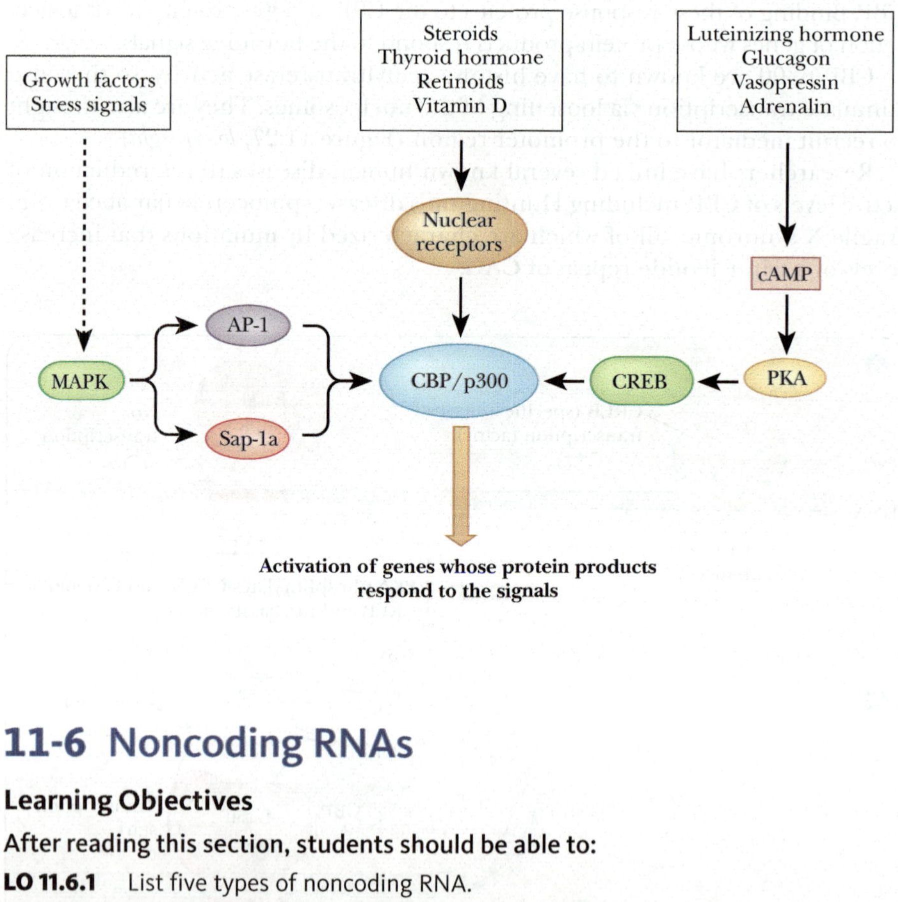

Figure 11.28 Multiple ways in which CREB-binding protein (CBP) and p300 are involved in activating gene expression. MAPK is mitogen-activated protein kinase. It acts on two other transcription factors, AP-1 and Sap-1a, which bind to CBP. Steroid hormones affect nuclear receptors, which then bind to CBP. Other hormones activate a cAMP cascade, leading to phosphorylation of CREB, which then binds to CBP. *(Adapted by permission from* Molecular Biology, *by R. F. Weaver, McGraw-Hill, 1999.)*

11-6 Noncoding RNAs

Learning Objectives

After reading this section, students should be able to:

LO 11.6.1 List five types of noncoding RNA.

LO 11.6.2 Describe the processes that occur in RNA interference by micro RNA (miRNA) and small interfering RNA (siRNA).

While so far we have focused on the traditional transcription of DNA that leads to protein-producing mRNA, the reality is that most of the RNA transcribed in our genome does not make protein (e.g., tRNAs, rRNAs, siRNAs, miRNAs, and snRNAs). Scientists believe that most of the transcription of RNA leads to gene regulation and explains much of the complexity of higher organisms, as well as the many differences between higher organisms that have very similar DNA. Some estimates suggest that as much as 98% of the transcriptional output of our genomes is comprised of noncoding RNAs (ncRNAs). These ncRNAs have been linked to many processes, including regulating transcription, gene silencing, replication, RNA processing, RNA modification, translation, protein stabilization, and protein translocation. While the number and type of RNAs seem to grow daily, the main focus of research is on two types—namely, **micro RNA (miRNA)** and **small interfering RNA (siRNA)**, both of which were introduced in Section 9-6E and are discussed in more detail in Section 11-6A.

micro RNAs (miRNA) short stretches of RNA that bind nonspecifically to mRNAs (imperfect match) and prevent gene expression

small interfering RNA (siRNA) short stretches of RNA that bind specifically to mRNAs (perfect match) and prevent gene expression

11-6A Micro RNAs and Small Interfering RNAs

Micro RNA (miRNA) and small interfering RNA (siRNA) are small double-stranded RNAs (dsRNA) that are involved in the control of gene expression via several related mechanisms. At first, siRNAs were thought to come from exogenous sources (i.e., originating *outside* the cell), whereas miRNAs were thought to be endogenous (i.e., originating *inside* the cell). However, endogenous siRNAs have been discovered, and scientists have started using exogenous miRNAs in their gene expression research. The first miRNA genes, called lin-4 and let-7, were found in the roundworm *C. elegans* in 1993 and 2000, respectively. Although they were initially thought to be peculiar to the roundworm, it was soon discovered that let-7 was present in many species, including humans. Exogenous siRNA sources include those from a viral infection, a transposable element, or a molecule supplied by a scientist attempting to knock out the function of a gene.

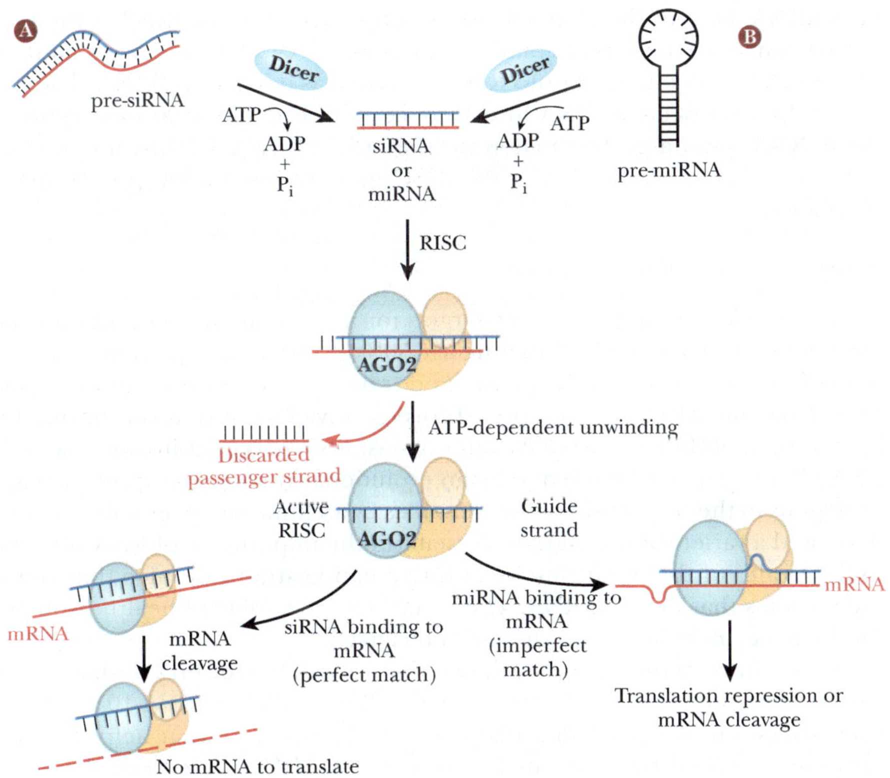

Figure 11.29 RNA interference. (A) Double-stranded RNA from exogenous sources, such as a virus or a synthetic RNA created in the lab, is cleaved by the enzyme Dicer to yield siRNA. (B) Pre-miRNA from endogenous sources, like a gene or transposon, is cleaved by Dicer to yield miRNA. miRNA and siRNA are bound by the RISC protein complex, including the Argonaute2 (AGO2) subunit. The passenger strand of the dsRNA is discarded. Active RISC is then recruited to the mRNA that matches the sequence of the antisense strand. When the binding is perfect, AGO2 cleaves the mRNA, so there is nothing to translate (*left side bottom*). When the binding is imperfect, hairpin loops form and the RISC/RNA complex represses translation or causes mRNA cleavage (*right side bottom*).

Figure 11.29 shows mechanisms of RNA interference that start with the Dicer protein. Dicer is in the family of endonucleases known as RNase III. It takes larger dsRNAs and cuts them to their characteristic small size, leaving a two-nucleotide-long 3′ overhang. Once the Dicer has cleaved the RNA, it passes the RNA to a protein complex called the RNA-induced silencing complex (RISC). In an ATP-dependent process, RISC unwinds the double-stranded RNA and selects the antisense strand, which is referred to as the guide strand. The other strand, referred to as the passenger strand, is discarded. One of the components of RISC is always a nuclease protein in the family of proteins known as the Argonautes. Argonaute proteins, like the AGO2 protein seen in Figure 11.29, guide the RISC complex to the targeted mRNA and facilitate binding and mRNA cleavage.

The effects of miRNA and siRNA are similar because both diminish the expression of specific genes, but in different ways, depending on the complementary nature of the miRNA or the siRNA. The extent of the matching of the sequences largely controls the ultimate effects, which collectively are referred to as **RNA interference (RNAi)**.

RNA interference (RNAi) the most understood form of RNA silencing, whereby small RNA molecules bind to mRNA, either causing its destruction so it cannot be translated or blocking its translation directly

Figure 11.29 shows the basics of these processes. In part A, the cell detects an exogenous dsRNA. This might be from a scientist-supplied molecule or from a virus. The enzyme Dicer binds to the dsRNA, ATP is hydrolyzed, and the dsRNA is cut into 21- to 27-nucleotide fragments of siRNA. This is then loaded onto the RNA-induced silencing complex (RISC), including a member of the Argonaute family (AGO2). In another ATP-dependent reaction, the siRNA is unwound, and the passenger strand is discarded, leaving active RISC bound to the antisense strand. RISC is then recruited to the mRNA with the target sequence. In the case of siRNA shown here, the matching of the antisense strand to the mRNA is perfect. When that happens, the Argonaute protein, also known as Slicer, cleaves the mRNA. In this way, the gene that made the mRNA is silenced because its mRNA is destroyed before it can be translated to protein.

MiRNA action follows a process similar process to siRNA action and uses most of the same enzymes. In part B of Figure 11.29, the starting material is pre-miRNA. One of the characteristics of this type of pre-miRNA is the presence of a hairpin loop. Dicer binds to the pre-miRNA and cleaves it to miRNA. The miRNA is then loaded onto RISC, the same as it was with siRNA. The process is then the same as RNA interference, with one important exception—the miRNA is not a perfect match for the mRNA target. In this case, there is either cleavage of the mRNA or the RISC sticks to the mRNA, preventing its translation.

► Why did RNA silencing develop?

Genomes of higher organisms are targets for invasion by viruses and transposable elements. Estimates are that nearly 50% of the human genome is a remnant of previous invasions by viruses and transposons over evolutionary time. In essence, our DNA has to fight off foreign invaders, just as our antibodies have to fight off foreign proteins and organisms such as parasites and bacteria. RNA silencing is now believed to be an evolutionarily conserved process that is analogous to the protection of our genomes by an immune system. Researchers have used a variety of techniques to establish the importance of RNA silencing to the health of the organism, including creating strains of mice that lacked the proteins that make miRNA. The results were a variety of health problems for the mice, including heart disease and cancer.

Studies in humans have also shown the importance of miRNA-based processes. For example, loss of a particular miRNA, called miRNA-101, leads to overexpression of a particular histone methyltransferase that helps the progression of several types of cancer. Studies with HIV-infected cells have also shown that miRNA is important to controlling the spread of the disease and that HIV actively suppresses the production of the cell's miRNAs that would otherwise be able to fight it. Micro RNAs have been implicated in various heart conditions through their influence on the production of the α and β myosin heavy chain, the predominant contractile proteins in the heart.

In late 2006, a California-based research team claimed to have found that small RNA molecules can also be found to work as activators. Researchers were attempting to use RNAi to block transcription of the human tumor suppressor gene, E-cadherin. When they added synthetic RNAs designed to target the gene's DNA, they found instead that production of the tumor suppressor went up instead of down. This process is referred to as RNA activation (RNAa). This process has been another powerful tool in the scientist's arsenal because it has expanded ways to manipulate genes and approach the fight against genetic diseases. **Biochemical Connections 11B** describes how a micro RNA helps repair nerves after an injury.

11B Biochemical Connections • Medicine

A Micro RNA Helps Regenerate Nerve Synapses After Injury

Another role for miRNA has been discovered. An miRNA called miRNA-206 (miR-206) was found to be involved in reinnervating the neuromuscular junction after injury. It was also found to improve survival in a mouse model of the disease amyotrophic lateral sclerosis (ALS), otherwise known as Lou Gehrig's disease. It has been known for some time that miRNAs play a role in muscle response to stress, and researchers used this as a basis for their hypothesis that diseases of muscle tissue might be due to loss of particular miRNAs. In normal mice, sciatic nerve injury results in loss of nerve function in the muscle and leads to an increase in miR-206. Using a line of mice that had ALS and inactivated miR-206, the time from onset of the disease was shortened, indicating this miRNA has a protective effect on nerves in the muscle. Further research indicated that the direct effect of the micro RNA is to repress the expression of histone deacetylase 4 (HDAC4) in the muscle cells. HDAC4 normally inhibits reinnervation by blocking expression of fibroblast growth factor binding protein 1 (FGFBP1), which acts on other fibroblast growth factors (FGF).

As can be seen in Figure 11.30, miR-206 binds to the mRNA that expresses the histone deacetylase, blocking its translation. This allows the production of the FGFBP1 in the muscle fiber, which then passes into the neuromuscular junction and stimulates FGF. This growth factor moves into the motor neuron and promotes reinnervation. This pathway was shown to work both with damaged nerves and with the diseased nerves from the ALS mice. This finding supports a growing body of evidence for the importance of miRNA to neurological function and susceptibility to disease. Micro RNA networks have been implicated in Parkinson's disease, Huntington's disease, and Alzheimer's disease. ▶

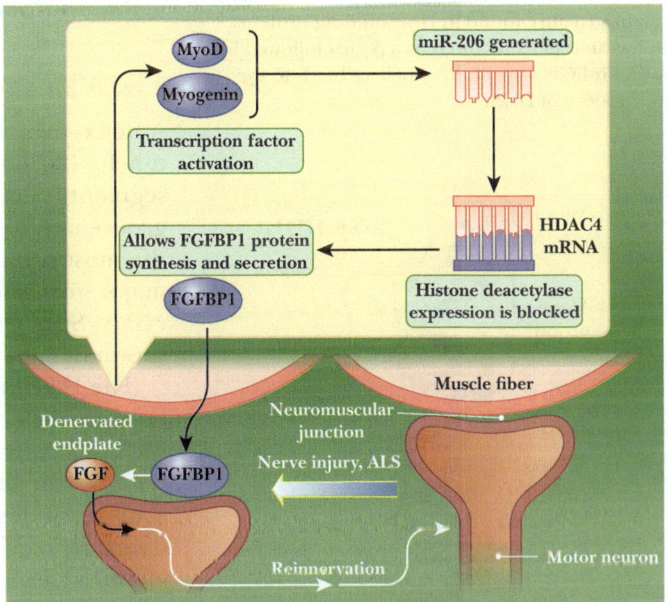

Neuromuscular junction	miR-206	HDAC4	FGFBP1	FGF	Reinnervation
Denervate	Enhance	Reduce	Enhance	Enhance	Enhance
	Inactivate	Enhance	Reduce	Reduce	Reduce
		Inactivate	Enhance	Enhance	Enhance

Figure 11.30 From injury to repair. A damaged adult neuromuscular junction triggers a signaling cascade in the muscle fiber that involves the miRNA called miR-206. This increases the effect of the growth factor FGF on the neuron and promotes reinnervation of the muscle. Reinnervation can be enhanced or diminished by altering upstream events. *(From Brown, R. H. (2009). A reinnervating microRNA.* Science *326, 1494–1495. Used with permission from AAAS)*

11-7 Structural Motifs in DNA-Binding Proteins

Learning Objectives

After reading this section, students should be able to:

LO 11.7.1 Describe the features of the following DNA-binding motifs of DNA-binding proteins, as well as features of the interactions between these motifs and DNA: helix–turn–helix, zinc finger, and basic leucine zipper motifs.

LO 11.7.2 Distinguish between the four categories of transactivation domains found in eukaryotic transcription factors.

So far, we have discussed many examples of proteins that bind to DNA, such as repressor proteins, transcription factors, and polymerases. The DNA-binding domains of DNA-binding proteins are the parts of the proteins that form direct contacts with the DNA. These domains that form contacts with the DNA do so by the same types of interactions that bind protein structures together—namely hydrogen bonding, electrostatic attractions, and hydrophobic interactions (Section 4-4A).

structural motifs short, three-dimensional parts of a protein's structure that appear in a variety of different proteins

The DNA-binding domains of DNA-binding proteins contain **structural motifs**, which are common structures that appear in a variety of different proteins. Many different DNA-binding motifs have been discovered, and the following subsections discuss three of them: the helix–turn–helix (HTH), zinc finger, and basic-region leucine zipper (bZIP) motifs. These motifs involve interactions with DNA in either the major or minor groove, with the major groove being more common.

helix–turn–helix (HTH) a common DNA-binding domain structure found in DNA-binding proteins that has an α-helix followed by a β-turn followed by another α-helix; one of the α-helices binds to the major groove of DNA

11-7A Helix–Turn–Helix Motifs

A **helix–turn–helix (HTH)** motif is defined by the presence of a segment of α-helix (Section 4-3A) that fits into the DNA major groove followed by a β-turn segment followed by another α-helix (Figure 11.31). The width of the major groove and the α-helix are similar, so the protein helix can fit snugly. This is the most common motif because the standard form of DNA, B-DNA, has the major groove of the correct size, and no alterations in its topology are necessary. Such binding proteins are often dimers with two regions of HTH, as shown in Figure 11.31.

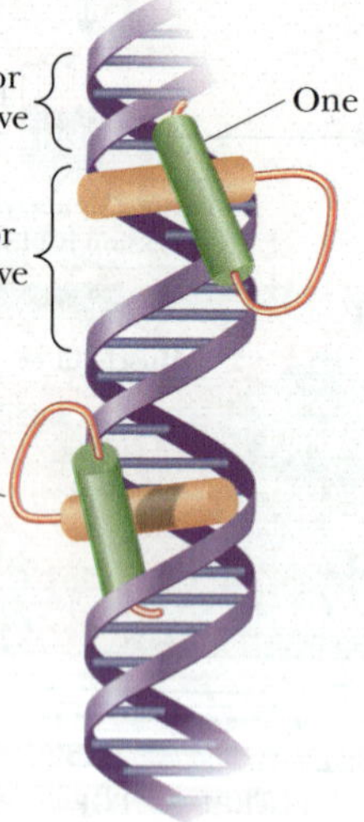

Figure 11.31 The helix–turn–helix (HTH) motif. The cylinders represent α-helices, whereas the regions connecting the α-helices represent the β-turns of the HTH motif. Proteins containing the HTH motif bind to DNA via the major groove, and they often occur as dimers that provide two regions of HTH. (*Annual Review of Biochemistry. Vol. 58 by Richardson, Charles C. Reproduced with permission of ANNUAL REVIEWS INC in the format Republish in other published product via Copyright Clearance Center. © 1989 by Annual Reviews.)*

The HTH motif is a sequence of 20 amino acids that is relatively conserved in many different DNA-binding proteins. Table 11.5 shows the sequence for the HTH region of several transcription factors. The first helical region is composed of the first eight residues of the region. A β-turn sequence of four amino acids separates it from the second helical region (residues 13–20 of the HTH domain). Position 9 of HTH domains, located inside the turn, is almost always a glycine residue because glycine's tiny hydrogen side chain makes it easier for the polypeptide chain to turn back on itself and change direction (Section 4-3C).

Proteins that recognize DNA with specific base sequences are more likely to bind to the major groove. The orientation of the bases in the standard base pairings puts more of the unique structure into the major groove. For example, Figure 11.32 shows how glutamine and arginine can interact favorably with adenine and guanine, respectively. Interactions like these are just a few examples of the many possible types of interactions between amino acid residues of DNA-binding proteins and the major groove bases of DNA. Some interactions, however, including many in the minor groove, only read the DNA indirectly. As discussed in Section 9-3B, the B form of DNA is not as constant as was once thought. Local variations in helix structure occur based on the actual sequence, especially when there are A–T-rich areas. The bases undergo extensive propeller twist. Many proteins bind to the edges of the bases that protrude into the minor groove. Studies have shown that artificial molecules that mimic

Table 11.5 Amino Acid Sequences in the HTH Regions of Selected Transcription Regulatory Proteins

434 *Rep* and *Cro* are bacteriophage 434 proteins; *Lam Rep* and *Cro* are bacteriophage λ proteins; CAP, *trp Rep,* and *Lac Rep* are catabolite activator protein, *Trp* repressor, and *lac* repressor of *E. coli,* respectively. *Antp* is the homeodomain protein of the *Antennapedia* gene of the fruit fly *Drosophila melanogaster.* The numbers in each sequence indicate the location of the HTH within the amino acid sequences of the various polypeptides.

				Helix							Turn							Helix		
	1	2	3	4	5	6	7	8	9	10	11	12	13	14	15	16	17	18	19	20
434 Rep	17-Gln	Ala	Glu	**Leu**	**Ala**	Gln	Lys	**Val**	**Gly**	Thr	Thr	Gln	Gln	Ser	**Ile**	Glu	Gln	**Leu**	Glu	Asn-36
434 Cro	17-Gln	Thr	Glu	**Leu**	**Ala**	Thr	Lys	**Ala**	**Gly**	Val	Lys	Gln	Gln	Ser	**Ile**	Gln	Leu	**Ile**	Glu	Ala-36
Lam Rep	33-Gln	Glu	Ser	**Val**	**Ala**	Asp	Lys	**Met**	**Gly**	Met	Gly	Gln	Ser	Gly	**Val**	Gly	Ala	**Leu**	Phe	Asn-52
Lam Cro	16-Gln	Thr	Lys	**Thr**	**Ala**	Lys	Asp	**Leu**	**Gly**	Val	Tyr	Gln	Ser	Ala	**Ile**	Asn	Lys	**Ala**	Ile	His-35
CAP	169-Arg	Gln	Glu	**Ile**	**Gly**	Glu	Ile	**Val**	**Gly**	Cys	Ser	Arg	Glu	Thr	**Val**	Gly	Arg	**Ile**	Leu	Lys-18
Trp Rep	68-Gln	Arg	Glu	**Leu**	**Lys**	Asn	Glu	**Leu**	**Gly**	Ala	Gly	Ile	Ala	Thr	**Ile**	Thr	Arg	**Gly**	Ser	Asn-87
Lac Rep	6-Leu	Tyr	Asp	**Val**	**Ala**	Arg	Leu	**Ala**	**Gly**	Val	Ser	Tyr	Gln	Thr	**Val**	Ser	Arg	**Val**	Val	Asn-25
Antp	31-Arg	Ile	Glu	**Ile**	**Ala**	His	Ala	**Leu**	**Cys**	Leu	Thr	Glu	Arg	Gln	**Ile**	Lys	Ile	**Trp**	Phe	Gln-50

Source: Adapted from Harrison, S. C., and Aggarwal, A. K., 1990, DNA recognition by proteins with the helix–turn–helix motif. *Annual Review of Biochemistry* 59, 933–969.

the base protrusions into the minor groove are equally capable of binding to many transcription factors.

Thus, although the base pairing is clearly important to DNA, sometimes the overall shape of the bases is important for other reasons. A particular binding protein might be recognizing not the part of the base that is involved in hydrogen bonds, but rather a part that is protruding into the grooves.

11-7B Zinc-Finger Domains

In 1985, it was discovered that a transcription factor of RNA polymerase III, TFIIIA, had 9 repeating structures of 30 amino acids each. Each repeat contained 3 closely spaced cysteines and 2 closely spaced histidines 12 amino acids later. It was also found that this factor had enough associated zinc ions to bind to each of the repeats. This led to the discovery of the **zinc-finger** motif

zinc finger a common DNA-binding domain structure found in DNA-binding proteins that has an antiparallel β-sheet followed by an α-helix; the fold is stabilized by the coordination of a zinc ion by two cysteine and two histidine residues

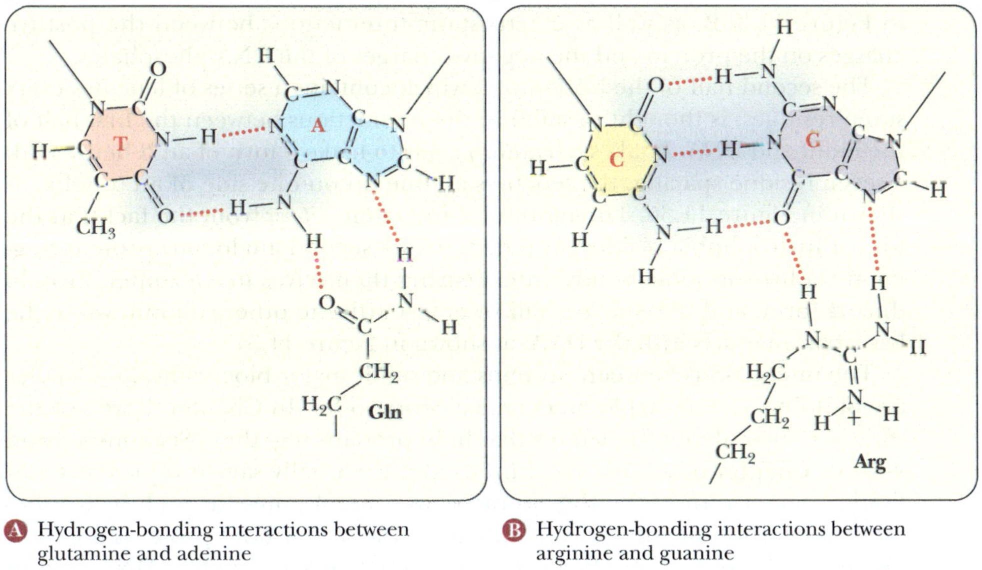

Ⓐ Hydrogen-bonding interactions between glutamine and adenine

Ⓑ Hydrogen-bonding interactions between arginine and guanine

Figure 11.32 A few examples of DNA–amino acid interactions. Amino acid residues in DNA-binding proteins can make hydrogen-bonding interactions like these with DNA bases.

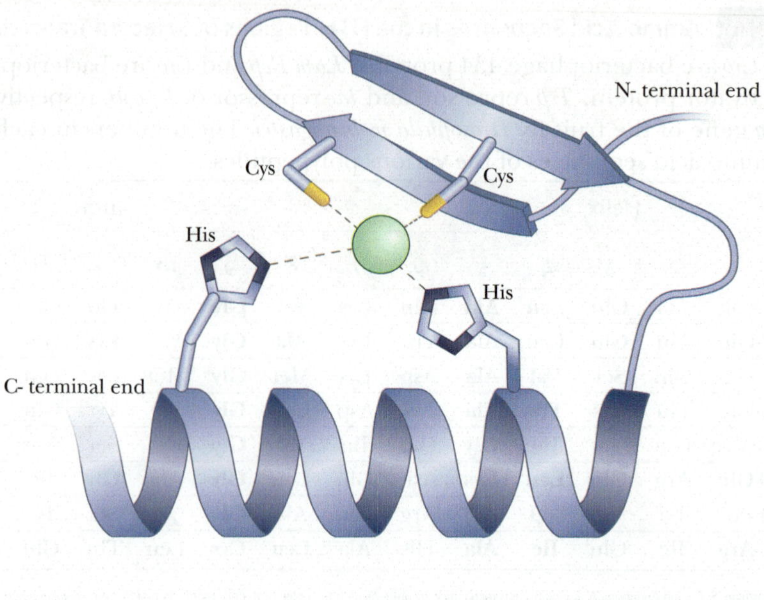

Figure 11.33 **The Cys_2His_2 zinc-finger motif.** The zinc-finger motif consists of an antiparallel β-sheet followed by an α-helix. The zinc ion (green) is coordinated by nitrogen atoms (dark blue) from two histidine residues and sulfur atoms (yellow) from two cysteine residues. *(Wikipedia Contributors, Zinc Finger, Wikipedia, The Free Encyclopedia. https://en.wikipedia.org/wiki/Zinc_finger (accessed [July 29, 2025].)*

in DNA-binding proteins, which is represented in Figure 11.33. The coordination of the cysteines and histidines to the zinc ion stabilizes the motif structure, which includes an α-helix that can bind to the major groove of DNA using hydrogen-bonding interactions like those seen in Figure 11.32. The zinc ion coordination also stabilizes an antiparallel β-sheet (Section 4-3B) that precedes the α-helix (Figure 11.33). When TFIIIA (or any other zinc-finger-containing protein) binds to DNA, the repeated zinc fingers follow the major groove around the DNA, as shown in Figure 11.34.

basic-region leucine zipper (bZIP) a common DNA-binding domain structure found in DNA-binding proteins that has leucines on two α-helices that interdigitate with each other; each α-helix has positively charged residues that bind to the major groove of DNA

11-7C Basic-Region Leucine Zipper Motifs

A third major class of DNA-binding domain motifs found in DNA-binding proteins is called the **basic-region leucine zipper (bZIP)** motif. Many transcription factors are known to contain this motif, including CREB. Figure 11.35 shows the sequence homology of several bZIP-containing transcription factors. Half of the bZIP motif is composed of basic regions A and B, which contain many conserved residues of lysine, arginine, and histidine. These residues bind the major groove of DNA using hydrogen-bonding interactions such as those seen in Figure 11.32B, as well as electrostatic interactions between the positive charges on the protein and the negative charges of the DNA phosphates.

The second half of the bZIP motif, which contains a series of leucines every seven residues, is thought to stabilize the interactions between the first half of the motif and DNA. It takes 3.6 amino acids to make a turn of an α-helix. With a seven-residue spacing, the leucines all line up on one side of an α-helix, as shown in Figure 11.36. The motif gets the name *zipper* from the fact that the line of hydrophobic residues interacts with a second analogous protein fragment via hydrophobic bonds, interweaving themselves like a zipper. Protein dimers form, and the leucine half interacts with the other subunit, while the basic part interacts with the DNA, as shown in Figure 11.37.

The interactions between proteins and other major biochemical molecules are building a remarkable biochemical story so far. In Chapter 4, we saw the various noncovalent interactions that hold proteins together (Sections 4-3 and 4-4). In Chapter 8, we discussed lipids and eventually saw how proteins and lipids interact with each other in biological membranes through hydrophobic and hydrophilic interactions similar to those that hold proteins together (Section 8-4). In Chapter 9, we discussed nucleic acids such as DNA, and in Section 11-7, we are seeing specific hydrophilic interactions that bring proteins and DNA together. There are so many possible combinations of these

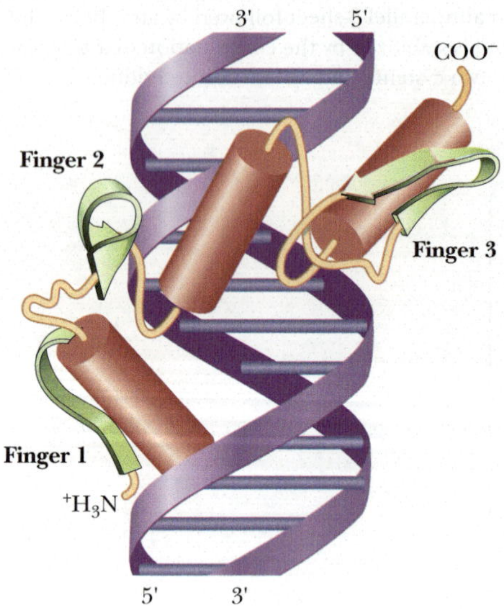

Figure 11.34 **Zinc-finger proteins follow the major groove of DNA.** The cylinders represent α-helices that are stabilized by the coordination of cysteines and histidines to zinc ions. These α-helices can then form hydrogen-bonding interactions with major groove bases. *(Adapted from Pavletich, N., and Pabo, C. O. (1991).* Science *252, 809, Figure 2. Used with permission from AAAS.)*

Protein	Sequence (Basic region A; Basic region B; Leucine zipper)
C/EBP	278–DKNSNEYRVRRERNNIAVRKSHDKAKQRNVETQQKVLELTSDNDRLRKRVEQLSRELDTLRG–341
Jun	257–SQERIKAERKRMRNRIAASKCHKRKLERIARLEEKVKTLKAQNSELASTANMLTEQVAQLKO–320
Fos	233–EERRRIRRIRRERNKMAAAKCRNRRRELTDTLQAETDQLEDKKSALQTEIANLLKEKEKLEF–296
GCN4	221–PESSDPAALKRARNTEAARRSRARKLQRMKOLEDKVEELLSKNYHLENEVARLKKLVGER–COOH
YAP1	60–DLDPETKQKRTAQNRAAQRAFHERKERKMKELEKKVQSLESIQQQNEVEATFLRDQLITLVN–123
CREB	279–EEAARKREVRLMKNREAARECRRKKKEYVKCLENRVAVLENQNKTLIEELKALKDLYCHKSD–342
Cys-3	95–ASRLAAEEDKRKRNTAASARFRIKKKQREQALEKSAKEMSEKVTQLEGRIQALETENKYLKG–148
CPC1	211–EDPSDVVAMKRARNTLAARKSBERKAQRLEELEAKIEELIAERDRYKNLALAHGASTE–COOH
HBP1	176–WDERELKKQKRLSNRESARRSRLRKQAECEELGQRAEALKSENSSLRIELDRIKKEYEELLS–239
TGA1	68–SKPVEKVLRRLAQRNEAARKSRLRKKAYVQQLENSKLKLIQLEQELERARKQGMCVGGGVDA–131
Opaque2	223–MPTEERVRKRKESNRESARRSRYRKAAHLKELEDQVAQLKAENSCLLRRIAALNQKYNDANV–286

Figure 11.35 **Amino acid sequences of DNA-binding proteins with basic-region leucine zippers.** *(Adapted from Vinson, C. R., Sigler, P. B., and McKnight, S. L. (1989).* Science *246, 912, Figure 1.)*

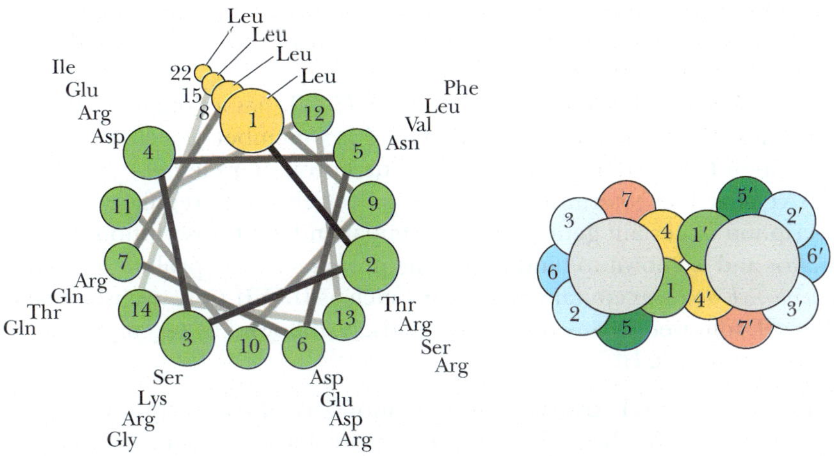

(a) (b)

Figure 11.36 **Schematic representations of basic-region leucine zipper motifs.** (A) A helical wheel representation of one α-helix of a basic-region leucine zipper. The amino acids listed show the progression down the helix. The leucines line up along one side, forming a hydrophobic spine. (B) A top-view schematic of two α-helices in a leucine zipper. If 1 and 1′ are leucines that are analogous to the residue 1 Leu in (A), then the 8th residues from these first leucines (not shown) would also be interdigitating leucines between the two helices. The same would apply to the 15th, 22nd, and 29th residues from the first leucines in each helix (as seen in Figure 11.35 sequences), forming the leucine zipper. *(Part A is based on Landschulz, W. H., Johnson, P. F., and McKnight, S. L. (1988).* Science *240, 1759–1764, Figure 1. Wikipedia Contributors, Leucine Zipper, Wikipedia, The Free Encyclopedia. https://en.wikipedia.org/wiki/Leucine_zipper [accessed July 29, 2025].)*

fundamental interactions (both within biochemical molecules and between them) that are essential to life, and there are many more of them to learn about in subsequent chapters.

11-7D Transactivation Domains

These three motifs—HTH, zinc finger, and basic-region leucine zipper—are commonly involved when transcription factors bind to DNA. However, transcription factors often also bind to other proteins called *co-activators* or *co-repressors*, and they do so through parts of their structures called **transactivation domains**. The motifs whereby transcription factors recognize other proteins can be broken down into four categories:

transactivation domains the parts of transcription factor structures that contain binding sites for other proteins that partner with the transcription factors to coregulate transcription

1. *Acidic domains* are regions rich in Asp and Glu acidic amino acid residues. *Gal4* is a transcription factor in yeast that activates the genes for metabolizing galactose. It has a domain of 49 amino acids, 11 of which are acidic.

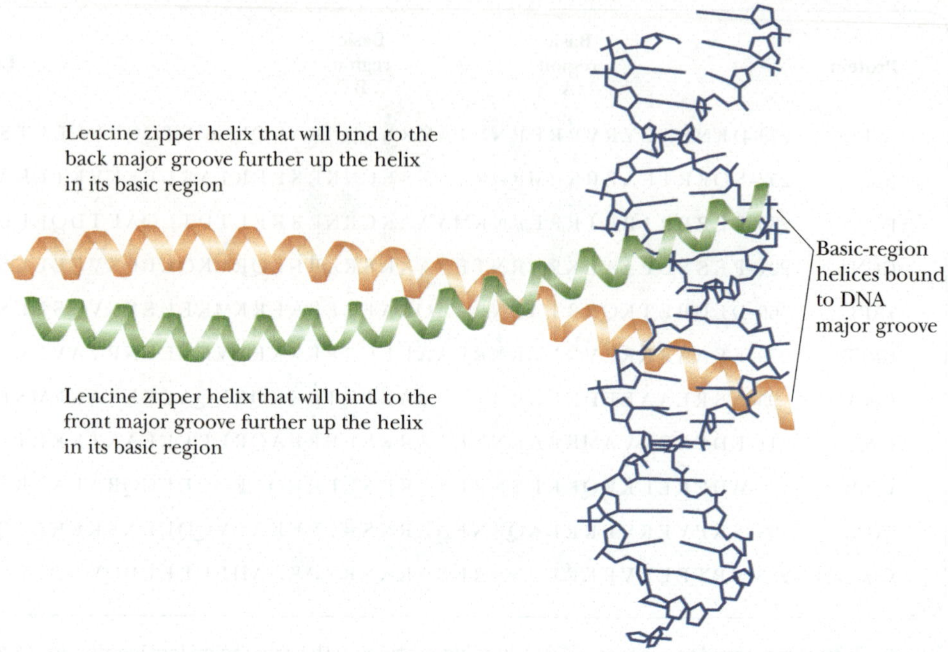

Figure 11.37 Structure of the bZIP transcription factor. Crystal structure of the bZIP transcription factor c-Fos:c-Jun bound to a DNA oligomer containing the AP-1 consensus target sequence TGACTCA. The basic-region binds to the DNA (the green helix is bound in the *front* major groove, whereas the orange helix is bound in the *back* major groove), while the leucine regions of the two helices bind to form a leucine zipper via hydrophobic interactions. *(Reprinted by permission from Crystal structure of the heterodimeric bZIP transcription factor c-Fos:c-Jun bound to DNA by Glover, J. N. M., and Harrison, S. C.,* Nature *373, 257–261 [1995].)*

2. *Glutamine-rich domains* are seen in several transcription factors. *Sp1* is an upstream transcription factor that activates transcription in the presence of an additional promoter element called a *GC box*. It has 2 glutamine-rich domains, one of which contains 39 glutamines in 143 amino acids. CREB and CREM also have this domain.
3. A *proline-rich domain* is seen in the activator *CTF-1*. It has a domain of 84 amino acids, 19 of which are prolines. CTF-1 is a member of a class of transcription factors that bind to an extended promoter element called a CCAAT box. The N-terminal domain has been shown to regulate transcription of certain genes. The C-terminal end is a transcription regulator and is known to bind to histone proteins via the proline repeats.
4. *Isoleucine-rich domains* contain repetitions such as IIXXII (i.e., two isoleucine residues, I, followed by any two amino acids, X, followed by two more isoleucine residues, I).

Despite the seemingly overwhelming complexity of transcription factors, understanding them has been made more manageable by the similarities in the motifs described here in Section 11-7. For example, if a new protein is discovered, evidence of its role as a transcription factor can be determined by locating one or more of the DNA-binding protein motifs discussed in this section.

11-8 Posttranscriptional RNA Modifications

Learning Objectives

After reading this section, students should be able to:

LO 11.8.1 Explain the posttranscriptional modifications that are made to transfer RNA (tRNA), ribosomal RNA (rRNA), and messenger RNA (mRNA) molecules.

LO 11.8.2 Describe the events that occur in the splicing of mRNA precursor molecules.

The three principal kinds of RNA—tRNA, rRNA, and mRNA—are all modified enzymatically after transcription to give rise to the functional form of the RNA in question. The type of processing in prokaryotes can differ greatly from that in eukaryotes, especially in the case of mRNA. The initial size of the RNA transcripts is greater than the final size because of leader sequences at the 5′ end

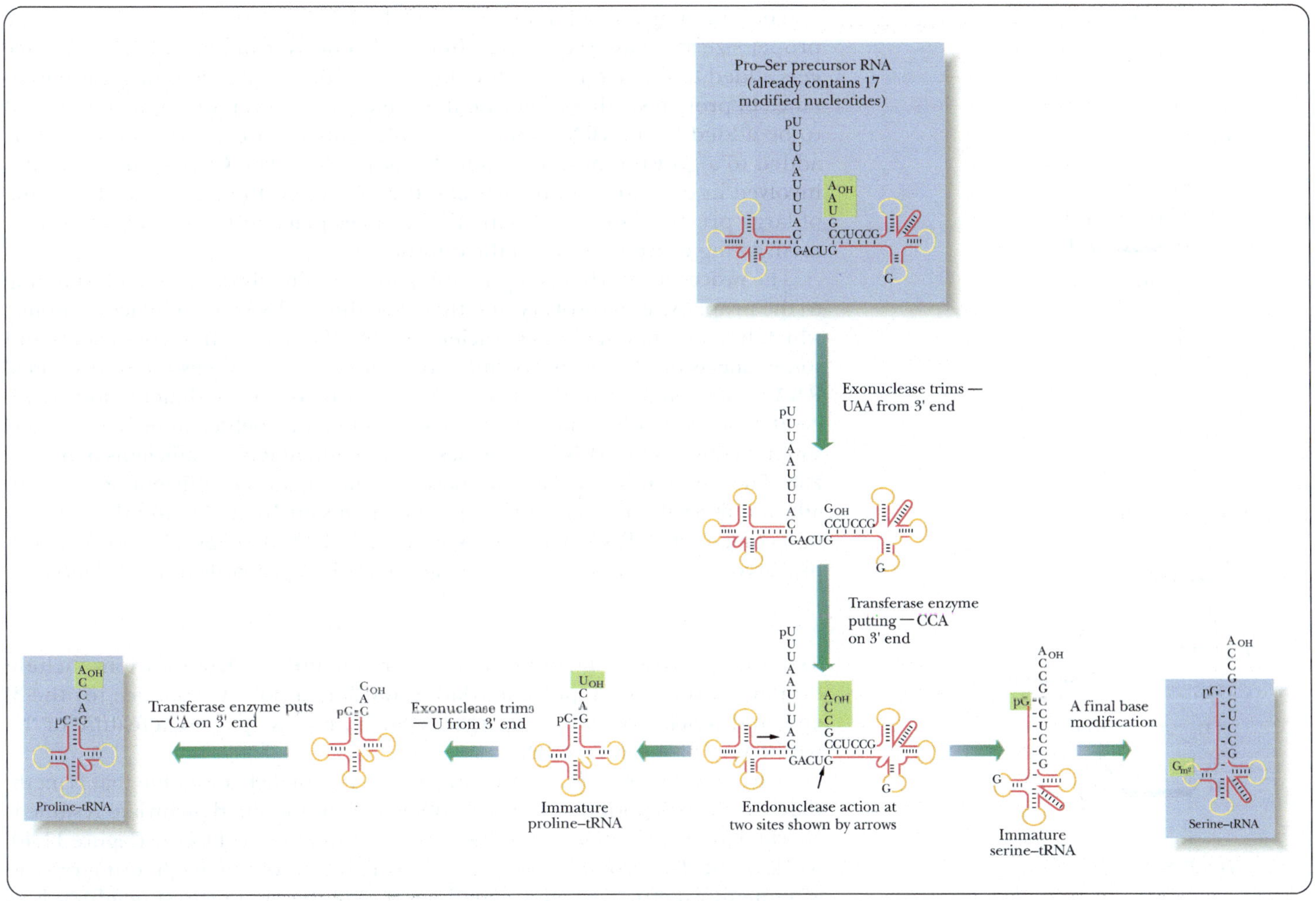

Figure 11.38 **Posttranscriptional modifications of a tRNA precursor.** Dashes represent hydrogen-bonded base pairs. The symbols G_{OH}, C_{OH}, A_{OH}, and U_{OH} refer to 3′ ends, which have free hydroxyl groups; $G^2{}_m$ is a methylated guanine; pU, pG, and pC refer to 5′ ends that have free phosphate groups.

and trailer sequences at the 3′ end. The leader and trailer sequences must be removed, and other forms of *trimming* by nucleases are also possible. Additionally, *terminal sequences* can be added after transcription, *intron removal* can occur, and *base modification* is frequently observed, especially in tRNA.

11-8A Transfer RNA and Ribosomal RNA Posttranscriptional Modifications

In eukaryotes, the precursor of several tRNA molecules (pre-tRNA) is frequently transcribed in one long polynucleotide sequence. Three types of modification—trimming, addition of terminal sequences, and base modification—always take place in the transformation of the initial tRNA transcripts into mature tRNAs (Figure 11.38). Introns are not always present in pre-tRNA molecules, so intron removal is not always necessary. The enzyme responsible for trimming the 5′ ends of all *E. coli* tRNAs, *RNase P*, consists of both RNA and protein. The RNA moiety of RNase P is responsible for its catalytic activity. This was one of the first examples of catalytic RNA (Section 11-9). Some tRNA base modifications take place before trimming, and some occur after. Methylation and substitution of sulfur for oxygen are two of the more usual types of base modification. (Refer to Section 9-2 and the discussion of transfer RNA in Section 9-5 for the structures of some of the modified bases.) One type of methylated nucleotide found only in eukaryotes contains a 2′-*O*-methylribosyl group (Figure 11.39).

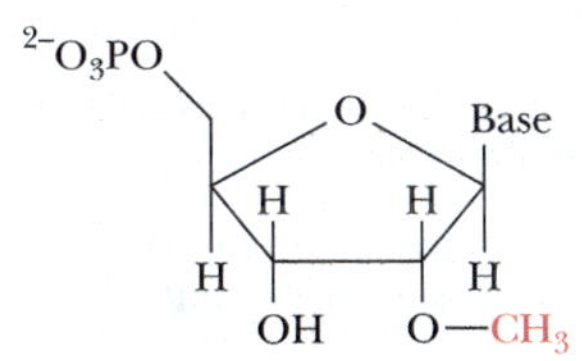

2′-*O*-Methylribosyl moiety

Figure 11.39 **The structure of a nucleotide containing a 2′-*O*-methylribosyl group.**

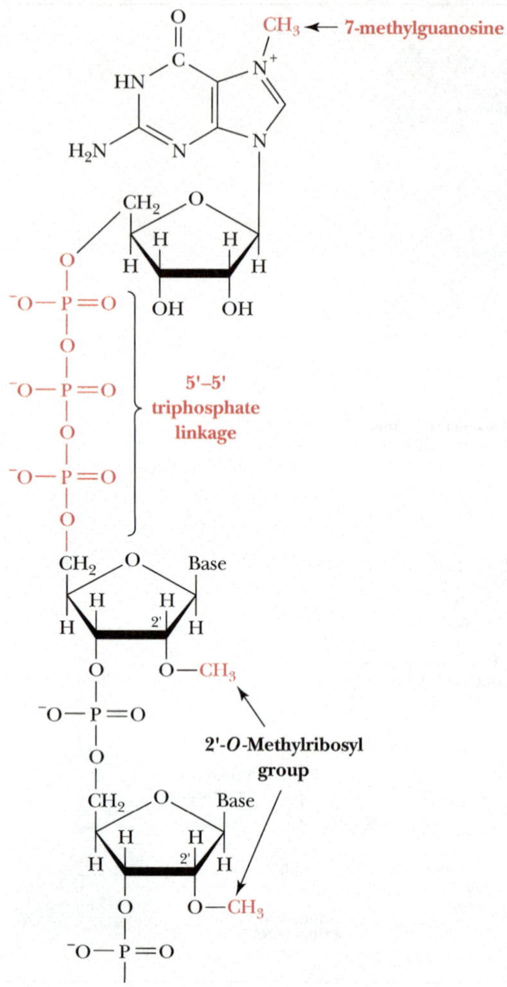

Figure 11.40 The structure of an mRNA cap.

The trimming and addition of terminal nucleotides produce tRNAs with the proper size and base sequence. After the 3′ end is trimmed, a CCA sequence gets added to the 3′ end, and the presence of this sequence is of great importance in protein synthesis because the 3′ end is the acceptor for an amino acid to be added to the tRNA (Section 12-3). Only then can the amino acid be added to a growing protein chain (Section 12-4). The CCA sequence is also involved in the interactions between the tRNA and the ribosome. Trimming of large precursors of eukaryotic tRNAs takes place in the nucleus, but most methylating enzymes occur in the cytosol.

The processing of rRNAs is primarily a matter of methylation and of trimming to the proper size. In prokaryotes, there are three rRNAs in an intact ribosome, which has a sedimentation coefficient of 70S. (Sedimentation coefficients and some aspects of ribosomal structure are reviewed in the discussion of ribosomal RNA in Section 9-5.) In the smaller subunit, which has a sedimentation coefficient of 30S, one RNA molecule has a sedimentation coefficient of 16S. The 50S subunit contains two rRNA molecules, with sedimentation coefficients of 5S and 23S. The ribosomes of eukaryotes have a sedimentation coefficient of 80S, with 40S and 60S subunits. The 40S subunit contains an 18S RNA, and the 60S subunit contains a 5S RNA, a 5.8S RNA, and a 28S RNA. Base modifications in both prokaryotic and eukaryotic rRNA are accomplished primarily by methylation.

11-8B Messenger RNA Posttranscriptional Modifications

Extensive processing takes place in eukaryotic mRNA. Modifications include capping of the 5′ end, polyadenylating (adding a poly-A sequence to) the 3′ end, and **splicing** of coding sequences. Such processing is not a feature of the synthesis of prokaryotic mRNA.

splicing the process whereby noncoding sequences of mRNA are removed and the coding sequences are linked together

The cap at the 5′ end of eukaryotic mRNA is a guanylate residue that is methylated at the N-7 position (Figure 11.40, *top*). This modified guanylate residue is attached to the neighboring residue by a 5′–5′ triphosphate linkage (Figure 11.40, *middle*). The 2′-hydroxyl group of the ribosyl portion of the neighboring residue is frequently methylated, and sometimes that of the next nearest neighbor is as well (Figure 11.40, *lower right*). The **polyadenylate (poly-A) tail** (abbreviated *poly-A* or *poly[r(A)$_n$]*) at the 3′ end of a message (typically 100–200 nucleotides long) is added before the mRNA leaves the nucleus. It is thought that the presence of the tail protects the mRNA from nucleases and phosphatases, which would degrade it. According to this point of view, the adenylate residues would be cleaved off before the portion of the molecule that contains the actual message is attacked. The presence of the 5′ cap also protects the mRNA from exonuclease degradation.

polyadenylate (poly-A) tail a long sequence of adenosine residues at the 3′ end of eukaryotic mRNA

The presence of the poly-A tail has been very fortuitous for researchers. By designing an affinity chromatography column (Chapters 5 and 13) with a poly-T tail (or poly[d(T)] tail), the isolation of mRNA from a cell lysate can be quickly accomplished. This enables the study of transcription by looking at which genes are being transcribed at a particular time under various cellular conditions.

The genes of prokaryotes are continuous; every base pair in a continuous prokaryotic gene is reflected in the base sequence of the mRNA. The genes of eukaryotes are not necessarily continuous; eukaryotic genes frequently contain intervening sequences that do not appear in the final base sequence of the mRNA for that gene product. The DNA sequences that are in the final mRNA product are called **exons.** The intervening sequences, which are spliced out and not expressed, are called **introns**. One way to distinguish exons from introns is to consider that ***ex***ons are the sequences that are retained and ***ex***ported from the nucleus for translation into protein. Such genes with introns are often referred to as **split genes.** The expression of a eukaryotic gene involves not just its transcription but also the processing of the primary transcript into its final form. Figure 11.41 shows how a split gene might be processed. When the gene is transcribed, the mRNA transcript contains regions at the 5′ and 3′ ends that are not translated, and several introns shown in green. Splicing events then remove the introns and link the

exons DNA sequences that are expressed in the final sequence of mRNA

introns intervening sequences in DNA that do not appear in the final sequence of mRNA

split genes genes that contain intervening sequences not present in the mature RNA

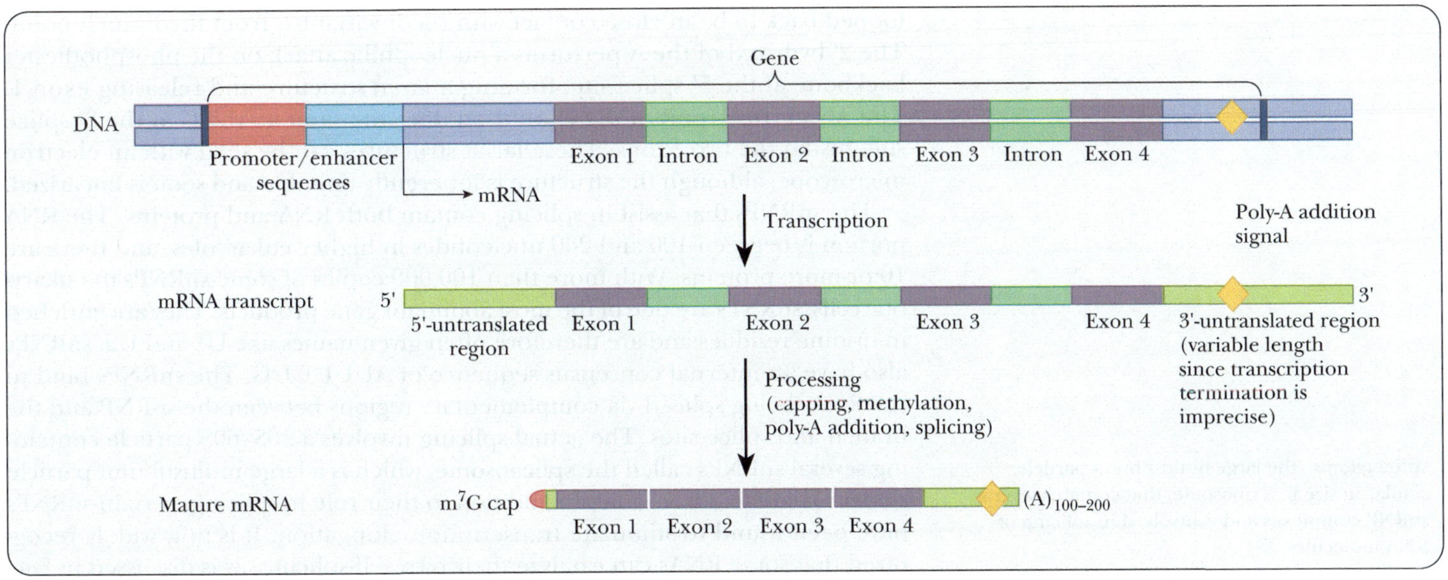

Figure 11.41 The transcription and processing of a eukaryotic split gene in eukaryotes. The mature mRNA molecule has a 7-methylguanosine (m^7G) cap and a poly-A ($A_{100-200}$) tail.

exons together. The 3′ end is modified by adding a poly-A tail, and a 7-methylguanosine cap is added to the 5′ end to yield the mature mRNA.

Some genes have very few introns, while others have many. There is one intron in the gene for the muscle protein actin; there are 2 in both the α- and β-chains of hemoglobin, 3 in lysozyme, and so on, up to as many as 50 introns in a single gene. The pro α-2 collagen gene in chickens is about 40,000 base pairs long, but the actual coding regions amount to only 5000 base pairs spread out over 51 exons. With so much splicing needed, the splicing mechanisms must be very accurate. Splicing is a little easier because the genes have the exons in the correct order, even if they are separated by introns. Also, the primary transcript is usually spliced in the same positions in all tissues of the organism.

A major exception to this is the splicing that occurs with immunoglobulins, in which antibody diversity is maintained by having multiple ways of splicing mRNA. In the last few years, more eukaryotic proteins that are the products of alternative splicing have been discovered. The need for this was also demonstrated by the data from the Human Genome Project. Differential splicing was necessary to explain the fact that the known number of proteins exceeds the number of human genes found.

▶ **How are introns spliced out to make mature RNA?**

11-8C The Splicing Reaction: Lariats and Snurps

The removal of intervening intron sequences from pre-mRNA takes place in the nucleus, where it is performed by snRNA and proteins that form small nuclear ribonucleoproteins (snRNPs, Section 9-6D). These snRNPs interact with RNA as it is formed, keeping it in a form that can be accessed by other proteins and enzymes. The substrate for splicing is the capped, polyadenylated pre-mRNA. Splicing requires cleavage at the 5′ and 3′ ends of introns and the joining of the two ends. This process must be done with great precision to avoid shifting the sequence of the mRNA product.

Figure 11.42 shows how splicing occurs. Specific sequences make up the *splice sites* for the process, with GU at the 5′ end and AG at the 3′ end of the introns in higher eukaryotes. A *branch site* within the intron also has a conserved sequence. This site is found 18–40 bases upstream from the 3′ splice site. The branch site sequence in higher eukaryotes is rich in pyrimidines (Figure 11.42, *top*). With the help of snRNPs, the G that is always present on the 5′ end of the intron gets looped back to be in close contact with the invariant A from the branch point. The 2′ hydroxyl of the A performs a nucleophilic attack on the phosphodiester backbone at the 5′ splice site, forming a *lariat* structure and releasing exon 1. The AG at the 3′ end of the exon then does the same to the G at the 3′ splice site, fusing the two exons. These lariat structures can be seen with an electron microscope, although the structure is inherently unstable and soon is linearized.

The snRNPs that assist in splicing contain both RNA and proteins. The RNA portion is between 100 and 200 nucleotides in higher eukaryotes, and there are 10 or more proteins. With more than 100,000 copies of some snRNPs in eukaryotic cells, snRNPs are one of the most abundant gene products. They are enriched in uridine residues and are therefore often given names like U1 and U2. snRNPs also have an internal consensus sequence of AUUUUUG. The snRNPs bind to the RNAs being spliced via complementary regions between the snRNP and the branch and splice sites. The actual splicing involves a 50S–60S particle containing several snRNPs called the **spliceosome**, which is a large multisubunit particle similar in size to a ribosome. In addition to their role in splicing, certain snRNPs have been found to stimulate transcription elongation. It is now widely recognized that some RNAs can catalyze their own self-splicing, as is discussed in Section 11-9. An important similarity between the two processes is that both proceed via a lariat mechanism by which the splice sites are brought together.

spliceosome the large multisubunit particle, similar in size to a ribosome, that contains several snRNP complexes and is involved in splicing of RNA molecules

11-8D Alternative RNA Splicing

Gene expression can also be controlled at the level of mRNA splicing. The mRNA transcripts of many proteins are always spliced in the same way, but some mRNA transcripts can be spliced in different ways to produce **isoforms** of a protein. In humans, 5% of the proteins produced have isoforms based on alternative splicing. These differences may occur by having two forms of the mRNA in the same cell, or there might be only one form in one tissue and a different form in another tissue. Regulatory proteins can affect the recognition of splice sites and direct the alternative splicing.

isoforms (in RNA splicing) different forms of a protein produced by alternative mRNA splicing reactions

It has been found that a protein called Tau accumulates in the brain of people with Alzheimer's disease. This protein has six isoforms generated by differential splicing, with the forms appearing during specific developmental stages.

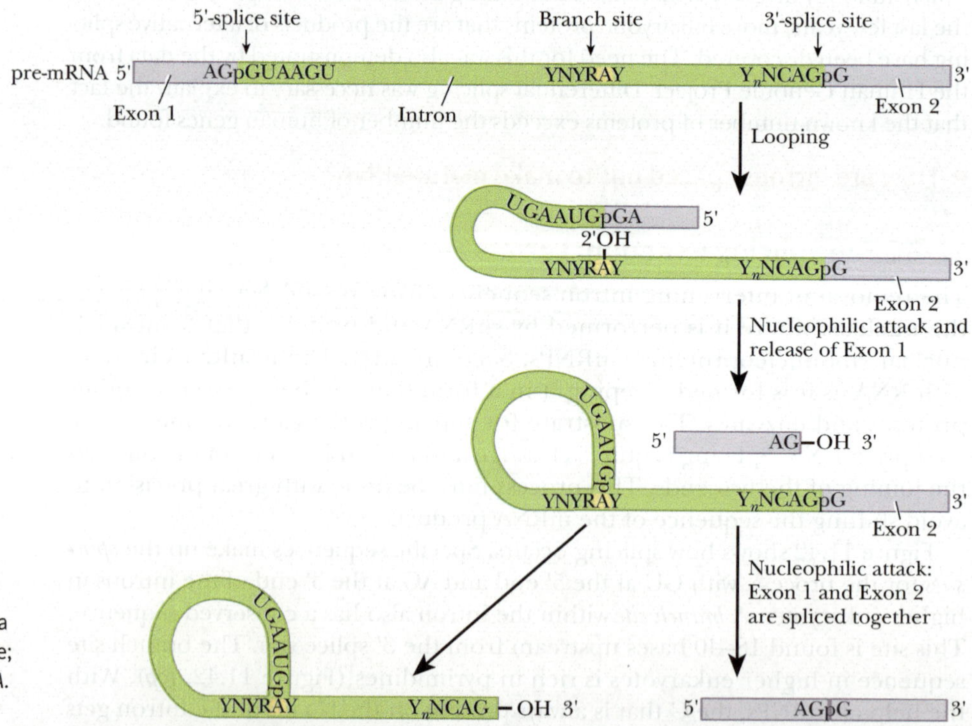

Figure 11.42 Splicing of mRNA precursors. Exons 1 and 2 are separated by the intervening sequence (intron) shown in green. In the splicing together of the two exons, a lariat forms in the intron. The branch site is rich in pyrimidines (Y is a pyrimidine ribonucleotide; N is any ribonucleotide; R is a purine nucleotide). *(Adapted from Sharp, P. A. (1987).* Science *235, 766, Figure 1.)*

ribozymes RNA molecules that have catalytic ability

11-9 Ribozymes

Learning Objectives

After reading this section, students should be able to:

LO 11.9.1 Define *ribozyme.*

LO 11.9.2 Distinguish between Group I intron and Group II intron ribozymes.

There was a time when proteins were considered the only biological macromolecules capable of catalysis. The discovery of the catalytic activity of RNA has thus had a profound impact on the way biochemists think. A few enzymes with RNA components had been discovered, such as telomerase (Biochemical Connections 10C). It was later discovered that RNase P (Section 11-8A) is an enzyme that is composed only of RNA. Enzymes in which RNA performs the catalytic activity are called **ribozymes**. One exciting discovery in ribozyme research was RNA that can catalyze its own self-splicing. It has also been shown that RNAs can catalyze reactions involved in protein synthesis, as is explained further in Chapter 12. The catalytic efficiency of catalytic RNAs is less than that of protein enzymes, and the catalytic efficiency of currently existing RNA systems is greatly enhanced by the presence of protein subunits in addition to the RNA.

▶ What are the characteristics of a ribozyme?

Several classes of ribozymes are known to exist. *Group I intron* ribozymes catalyze their own excision from mRNA, tRNA, and rRNA precursors. An example is the self splicing that takes place in pre-rRNA of the ciliate protist *Tetrahymena* (Figure 11.43). The transesterification (of phosphoric acid esters) that takes place here releases one end of the intron. The free 3′-OH end of the exon attacks the 5′ end of the other exon, splicing the two exons and releasing the intron. The free 3′-OH end of the intron then attacks a nucleotide 15 residues from the 5′ end, cyclizing the intron and releasing a 5′ terminal sequence. The precision of this sequence of reactions depends on the folded conformation of the RNA, which remains internally hydrogen-bonded throughout the process. In vitro, this catalytic RNA can act many times, being regenerated in the usual way for a true catalyst. In vivo, however, it appears to act only once by splicing itself out.

Group II intron ribozymes display a lariat mechanism of operation similar to the mechanism seen in Section 11-8 that was facilitated by snRNPs. There is no requirement for an external GTP nucleotide; the 2′-OH of an internal adenosine attacks the phosphate at the 5′ splice site. However, Group II intron ribozymes require assistance from proteins for splicing in vivo. DNA cannot self-splice in this fashion because it does not have a 2′-OH.

The folding of the RNA is crucial to its catalytic activity, as is the case with protein catalysts. A divalent cation (Mg^{2+} or Mn^{2+}) is required; it is quite likely that metal ions stabilize the folded structure by neutralizing some of the negative charges on the phosphate groups of the RNA. A divalent cation is essential for the functioning of the hammerhead ribozymes, which can be catalytically active with as few as 43 nucleotides. The folding of RNA is such that large scale conformational changes can take place with great precision. Similar large-scale changes take place in the ribosome in protein synthesis and in the spliceosome in the processing of mRNA.

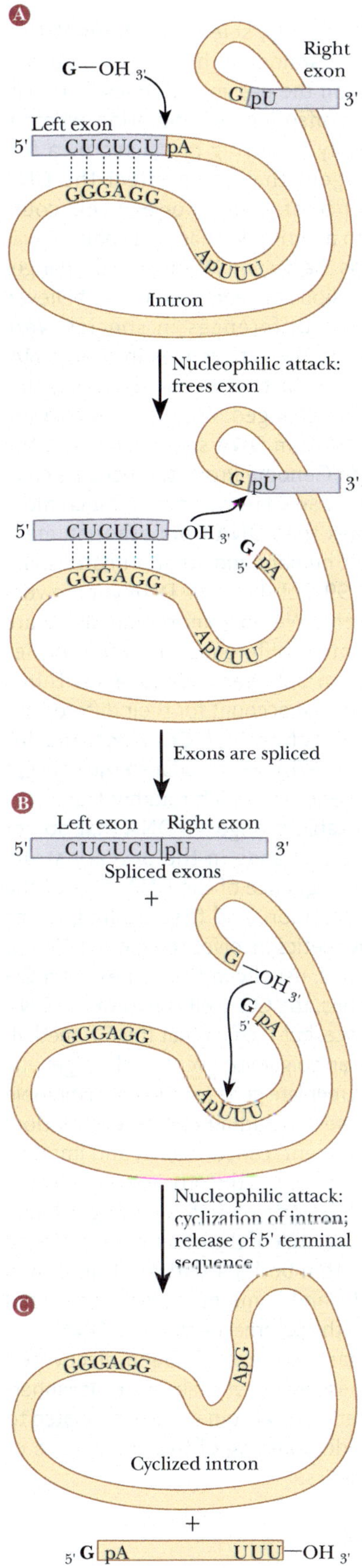

Figure 11.43 The self-splicing of pre-rRNA of the ciliate protist *Tetrahymena*, a Group I intron ribozyme. (A) A guanine nucleotide attacks at the splice site of the exon on the left, giving a free 3′-OH end. (B) The free 3′-OH end of the exon attacks the 5′ end of the exon on the right, splicing the two exons and releasing the intron. (C) The free 3′-OH end of the intron then attacks a nucleoside 15 residues from the 5′ end, cyclizing the intron and releasing a 5′ terminal sequence.

Hot Topic Epigenetics

Initially, scientists thought that mammalian heredity was based solely on the passing of genes from parents to offspring. Any mutation, whether naturally occurring or caused by a foreign agent that could affect the DNA sequence of a sperm or egg cell, could lead to a heritable change. Evolution was seen as the accumulation of such changes over billions of years, and it was believed that the differences in species were based on the differences in their DNA. However, in the recent decades, this view has changed. With the information gleaned from DNA sequencing and the Human Genome Project, it became clear that human DNA was not so much more complex than DNA from other animals. Indeed, humans and other primates share about 99% of the same DNA yet are very different. We now know that there are many other differences between species and even between individuals within a species that account for their differences.

In **Biochemical Connections 9C** (Section 9-7B), we introduced the concept of epigenetics, which roughly translates to heritable changes in DNA that do not involve a change in the primary structure or sequence of the DNA. One of the most common, and first studied, forms of epigenetics involves the methylation of cytosines, usually in the context of a CG pair. Thus, the CG itself is part of the DNA sequence, but the fact that the C is methylated is an epigenetic event. Such epigenetic events regulate gene expression in multiple processes, including cell differentiation, development, cancer, aging, and imprinting. Imprinting is the process whereby the DNA from one parent is distinguishable from the DNA of the other parent. Not all genes from both parents are transcribed equally, and some processes are biased toward the paternal or maternal DNA.

It has also recently been discovered that these epigenetic labels are also inherited. Let's look at some concepts related to our understanding of the importance of epigenetics.

Windows into the Past

Although we are a long way from being able to create a dinosaur out of blood taken from a mosquito trapped in amber, as was described in the *Jurassic Park* movies, we are learning a lot from ancient DNA. Until recently, studies were only able to document DNA *sequences* from old specimens, but nowadays, we can also look at the *epigenetic state* of DNA collected. A study by Pedersen et al. looked at the sequence of a 4000-year-old pre-Inuit that revealed methylation patterns that were similar to those found in modern humans of today. One known methylation site is methylated so predictably with age that it is called a clock site. Using this methylation site, Pedersen was able to determine that this ancient human died in their 50.

Other studies have been able to analyze samples taken from ancient hominids that lived between 50,000 and 130,000 years ago. Regions of DNA have been identified that have differential methylation of genes involved in limb development compared to modern humans. This is very exciting news for those who want to be able to study evolution and track the movement from pre-*Homo sapiens* to modern humans.

Another startling discovery is that heritable epigenetic changes accompany environmental, climatic, and social changes. An example is noted epigenetic markers that can be tracked to a famine in Holland in 1944 and 1945. Subsequent studies of data from people born in 1890, 1905, and 1920 in Sweden, along with their descendants, have shown a correlation between feast/famine cycles and the tendency for female descendants of women who suffered one of the cycles to die from cardiovascular disease. Such studies indicate that biological stressors can lead to heritable changes that are passed on to the next generation. They also indicate a relationship to imprinting because some of these inheritance patterns are biased for one gender or another. If such biological stressors can create a heritable epigenetic change, then the ramifications for human development are significant. Societal changes, like the change from a hunter/gatherer society to industrialization, could have influenced human epigenetics in a tangible way. Modern changes in diet and lifestyle could be doing the same thing right now.

Non-CG Methylation

Although still in its infancy, recent work has shown that methylation of DNA is not confined to CG pairs. In the mammalian brain and in embryonic stem cells, there is also a high level of methylation of cytosines that is not in the traditional CG context, and this form of methylation is highly correlated with a reduction in expression of the gene. Evidence that this type of epigenetic marking of the DNA is related to brain development has been shown by studies that demonstrate that newborn mice, and humans for that matter, have very little non-CG methylation in the neurons of their brains, but as they develop, more and more of the DNA is so methylated. This change in methylation state corresponds strongly to the period of rapid brain development and formation of neural synapses. Proteins have been isolated that react with and bind to non-CG methylation sites. One such protein is called methyl-CpG binding protein 2 (MECP2), and disruption of MECP2 binding is known to cause the neurological disorder called Rett's syndrome.

Epigenetics and Environmental Toxicology

Work by Skinner et al. has shown that pollutants—namely agricultural chemicals, some common plastics, and even jet fuel—can induce epigenetic modifications. What was more striking was that these modifications, along with their concomitant health risks, were able to be passed on to future generations. In the past, toxicologists would study the potential side effects of chemicals to ascertain if they were safe. One of the biggest threats was that a drug or other chemical would act as a mutagen, something that would change the sequence of DNA, otherwise known as a mutation. Such changes can lead to cancer. However, it appears that just checking the DNA sequence of individuals exposed to the chemical is not enough.

Ironically, much of the information that Skinner determined was the result of an accident. His lab was studying the effects of certain toxins on rats. They had determined that some of them caused reproductive problems for the animals exposed. A graduate student later accidentally crossed descendants of exposed animals together and found that their pups suffered from the same problems in many cases. The crossed parents had not been exposed as adults or even in the womb, so there was no chance that the negative effects were caused by the toxin directly. Subsequent studies showed that the descendants of the exposed animals showed the same anomalous epigenetic markers as the exposed animals, showing that altered epigenetic states could be inherited and that such inheritance could lead to health issues. Thus, when considering whether an environmental influence is safe or not, its ability to mutate DNA is important, but equally important is its ability to cause **epimutations**—that is, heritable changes to DNA structure but not to the DNA sequence.

To look further into such possibilities, researchers are studying whether similar inheritance patterns have been seen in humans. In 1976, an explosion at a dioxin plant in Italy exposed nearby residents to the highest level of dioxin ever recorded. They checked the blood levels of 1000 residents and followed their health. They discovered a correlation between dioxin exposure and relative infertility. They also found that females younger than 13 at the time of the exposure had double the normal risk of developing metabolic syndrome, a collection of problems that make it more likely a person will get diabetes or suffer from heart disease later on. They have since found that granddaughters of the exposed women have increased levels of thyroid problems.

The Epigenome and Histone Code

One of the main focuses of epigenetics is the study of the histone code, which describes all of the modifications to histones that take place and attempts to map them against their effects on cellular processes such as replication, transcription, DNA repair, and control of the cell cycle. In the last few years, science has uncovered some striking effects of changes in epigenetic states on our health.

The **epigenome** is the totality of chemical changes to DNA and chromatin, and more is learned about it every day. Studies have shown that many of the chemical changes can be acquired by interactions with oncogenic compounds or DNA damage. Many compounds have been discovered to act as **chromatin remodelers (CRs)**. Mutations in histone modifiers, histone proteins, nucleosome remodelers, DNA methylators, and other regulators have been found to cause epimutations in several cancers.

A major focus in medicine will be identifying how these mutations in CRs lead to cancer, as well as how pharmacology can stop the process. Epigenetics has also been implicated in cognitive aging, as shown in Figure 11.44. Age-related cognitive decline begins in the late 40 and is most pronounced in declarative memory, the ability to recall facts and experiences. Research has led to a hypothesis that much of the decline can be tracked to the epigenetic state of the cells in the prefrontal cortex and the hippocampus of the brain. It is known that chromatin remodeling in the hippocampus is necessary to stabilize long-term memories. A 2010 study by Peleg et al. in *Science* found that aged mice showed a disruption of experience-dependent epigenetic modification of the acetylation site of lysine 12 on histone 4 (H4K12). This was associated with a concomitant loss of normal memory-associated transcription in the hippocampus. They also discovered a memory-associated gene, *Formin 2*, which is required for normal memory. The transcription of this gene was found to be disrupted in aging mice. To tie the theory together, they showed that when they infused the mice's hippocampus with an inhibitor of histone deacetylase (HDAC), it increased the acetylation of H4K12, restored memory-associated transcription, and restored behavioral memory function.

Their work ties together three related processes: chromatin regulation, memory-associated transcription regulation, and a molecular basis for the decline in cognitive memory with aging.

epimutations heritable changes to DNA where the structure of DNA is changed (base methylation state, for example), but the sequence of DNA is not changed

epigenome the total chemical changes to DNA and chromatin that are known in an organism that are not associated with actual base sequence changes

chromatin remodelers (CRs) compounds that alter gene expression by modifying the form of chromatin

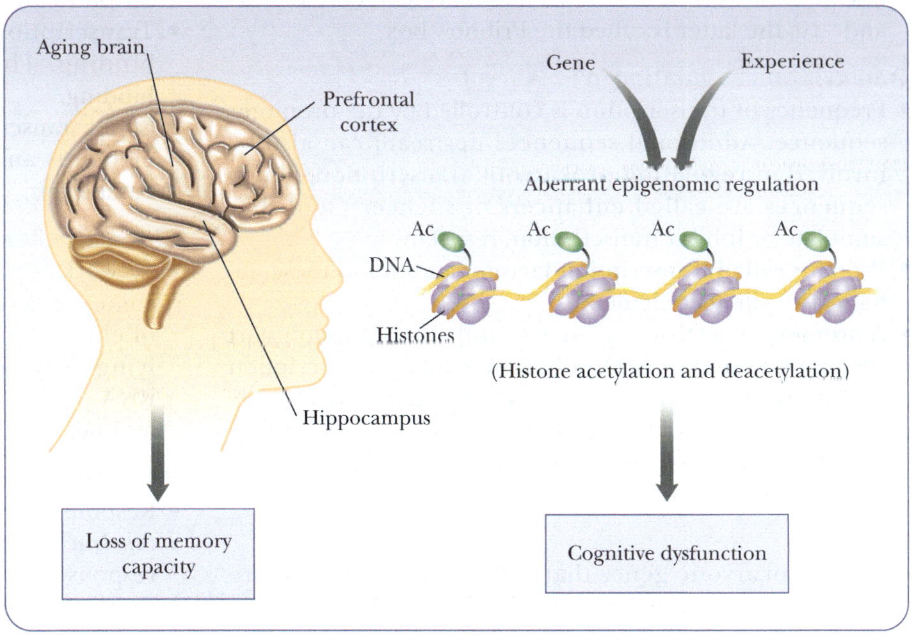

Figure 11.44 **Age-related epigenetic changes have been shown to play a role in cognitive function.** (*From Sweatt, J. D. [2010].* Science *328, 701. Reprinted with permission from AAAS.*)

Summary

Overview of Transcription

- RNA synthesis is the transcription of the base sequence of DNA to that of RNA.
- All RNAs are synthesized using a DNA template.

Transcription in Prokaryotes

- The enzyme that catalyzes transcription is DNA-dependent RNA polymerase. All four ribonucleoside triphosphates—ATP, GTP, CTP, and UTP—are required, as is Mg^{2+}. There is no need for a primer in RNA synthesis.
- As is the case with DNA biosynthesis, the RNA chain grows from the 5′ to the 3′ end.
- The enzyme uses one strand of the DNA (the antisense strand or template strand) as the template for RNA synthesis. The RNA product has a sequence that matches the other strand of DNA, the coding strand.

RNA polymerase in *E. coli*

- In prokaryotes, the subunits of RNA polymerase—α, β, β′, and ω—make up the core of the enzyme and are responsible for the enzymatic activity that catalyzes nucleotide incorporation.
- The σ subunit of RNA polymerase is used for promoter recognition.
- RNA polymerase moves along the template strand from 3′ to 5′ and produces the RNA from 5′ to 3′.

Promoter Structure

- In prokaryotic transcription, RNA polymerase is directed to the gene to be transcribed by the interactions between the polymerase's σ subunit and sequences of DNA near the start site called promoters.
- Consensus sequences have been established for prokaryotic promoters, and the key elements are sequences at −35 and −10; the latter is called the Pribnow box.

Transcription Regulation in Prokaryotes

- Frequency of transcription is controlled by the promoter sequence. Additional sequences upstream can also be involved in regulating prokaryotic transcription. These sequences are called enhancers or silencers, and they stimulate or inhibit transcription, respectively.
- Proteins called transcription factors can bind to these enhancer or silencer elements.
- A promoter is a DNA sequence near the transcription start site that is bound by RNA polymerase during transcription initiation. Its position and orientation are critical to its function. Enhancers are DNA sequences that are farther away from the start site. Their position and orientation are not as important.

Operons

- Many prokaryotic genes that produce proteins that are part of a pathway are controlled in groups called operons.
- The regulatory gene of the *lac* operon, *lacI*, produces a protein called the repressor. This protein monomer, once transcribed and translated, combines to form a tetramer.
- The tetramer is the active repressor, and it binds to the operator portion of the *lac* promoter.
- In the presence of the inducer (lactose), the repressor no longer binds, and the repression is released.

Transcription Attenuation

- In prokaryotes, transcription and translation are linked.
- With attenuation, the RNA being transcribed is also being translated.
- Depending on the speed of simultaneous translation, the RNA produced can form different hairpin loop structures.
- In one orientation, the hairpin loop acts as a terminator and aborts the transcription before the actual proteins can be translated, thereby attenuating, or reducing, the amount of expression of certain genes.
- In another orientation, the transcription is allowed to proceed.

Transcription in Eukaryotes

- The most well-studied eukaryotic promoter region is called the TATA box.
- Although some genes lack TATA boxes, they are the most consistent part of eukaryotic promoters that RNA polymerase recognizes.

Eukaryotic Transcription: Initiation

- There are a large number of eukaryotic transcription factors. Some are called GTFs, and they are involved in transcription initiation.
- Transcription factors aid in promoter recognition and binding. They have a specific order and location of binding.
- Other transcription factors bind to enhancers or response elements and increase the rate of transcription above basal levels.

Chromatin Remodeling Complexes

- To work around the tight winding of DNA in the nucleosomes, eukaryotic transcription is activated by the use of chromatin remodeling complexes and histone modifying enzymes. The combination of the two allows the RNA polymerase to have access to the promoter region and begin transcription.

Response Elements

- Response elements are DNA sequences similar to enhancers, but they are involved in a bigger picture of metabolic responses.
- Common response elements are the HSE and the CRE.

Micro RNAs and Small Interfering RNAs

- Micro RNA (miRNA) and small interfering RNA (siRNA) are small double-stranded RNAs that are involved in the control of gene expression.

RNA Interference

- RNA interference is a process involving miRNA and siRNA, whereby they diminish the expression of specific genes. The RNA silencing processes involving miRNA are almost the same as those for RNA silencing with siRNA. The biggest difference is that the match of the sequences in miRNA silencing is not perfect, whereas the matching of sequences in siRNA silencing is.
- RNA silencing protects our DNA from invading viruses or transposons. It is also used by researchers to study the effects of silencing certain genes in cells.

DNA-Binding Domains

- There are some common and easily recognizable motifs that occur in the DNA-binding domains of transcription factors and other DNA-binding proteins. These motifs include the helix–turn–helix, zinc finger, and basic-region leucine zipper motifs. These sections of the protein allow easy binding to the DNA.
- In addition to binding DNA, transcription factors often bind other proteins such as co-activators or co-repressors. The sites of protein–protein interactions can also be identified by common motifs such as acidic domains, glutamine-rich domains, proline-rich domains, and isoleucine-rich domains.

Posttranscriptional Modification of RNA

- Messenger RNA is modified in several ways in eukaryotes.
- Two modifications believed to be protection mechanisms are 5′ capping and 3′ polyadenylation.
- The 5′ cap uses a unique 5′–5′ bond that standard nucleases would not be able to degrade.
- The other major modification is the removal of introns by splicing to yield the correct final mRNA.

The Splicing Reaction: Lariats and Snurps

- The splicing reaction often involves ribonucleoprotein molecules called snRNPs ("snurps") that contain small nuclear RNA (snRNA) and proteins. Other times, the RNA being spliced catalyzes the reaction itself.
- The general splicing mechanism involves a 3′ splice point, a 5′ splice point, and a branch point. An intermediate in the process has a lariat shape.

Ribozymes

- Ribozymes are RNA molecules with catalytic ability.
- Some ribozymes are a combination of RNA and protein, and others are just RNA.
- To be a ribozyme, the RNA portion must be involved in the catalysis of the reaction.
- Scientists believe that RNA could have been the first molecule that combined the ability to carry genetic information for replication and the ability to catalyze reactions.

Review Exercises

11-2 Transcription in Prokaryotes

1. **Recall** What is the difference in the requirement for a primer in RNA transcription compared to DNA replication?
2. **Recall** List three important properties of RNA polymerase from *E. coli*.
3. **Recall** What is the subunit composition of *E. coli* RNA polymerase?
4. **Recall** What is the difference between the core enzyme and the holoenzyme?
5. **Recall** During transcription in prokaryotes, what specific event occurs after the RNA polymerase holoenzyme binds to the promoter but before elongation begins?
6. **Recall** What are the different terms used to describe the two strands of DNA involved in transcription?
7. **Recall** Define the *promoter region* and list three of its properties.
8. **Recall** Put the following in 5′ to 3′ order on the coding strand: UP element, Pribnow box, TSS, –35 region, Fis site.
9. **Recall** Distinguish between rho-dependent termination and intrinsic termination.
10. **Reflect and Apply** Diagram a section of DNA being transcribed. Give the various names for the two strands of DNA.

11-3 Transcription Regulation in Prokaryotes

11. **Recall** Define *inducer* and *repressor*.
12. **Reflect and Apply** Explain how the presence or absence of allolactose and glucose affects transcription of the *lac* operon in *E. coli*.
13. **Recall** What is a σ factor? Why is it important in transcription?
14. **Recall** What is the difference between σ^{70} and σ^{32}?
15. **Recall** What is the function of the catabolite activator protein?
16. **Recall** What is transcription attenuation?
17. **Reflect and Apply** What role does an operon play in the synthesis of enzymes in prokaryotes?
18. **Reflect and Apply** Diagram a termination of transcription, showing how inverted repeats can be involved in releasing the RNA transcript.
19. **Reflect and Apply** Give an example of a viral system in which alternative σ factors can control which genes are transcribed. Explain how this works.
20. **Reflect and Apply** Explain, with mRNA diagrams, how transcription attenuation works in the *trp* operon.

11-4 Transcription in Eukaryotes

21. **Biochemical Connections** What is an aptamer?
22. **Biochemical Connections** What is a riboswitch?
23. **Biochemical Connections** What are various ways that a riboswitch shuts off translation when it binds to its target molecule?
24. **Biochemical Connections** How is the discovery of riboswitches relevant to bacterial pathology?
25. **Recall** Define *exon* and *intron.*
26. **Recall** What are some of the main differences between transcription in prokaryotes and in eukaryotes?
27. **Recall** What are the products of the reactions of the three eukaryotic RNA polymerases?
28. **Recall** List the components of eukaryotic Pol II promoters.
29. **Recall** List the Pol II general transcription factors.
30. **Reflect and Apply** What are the functions of TFIIH?

11-5 Transcription Regulation in Eukaryotes

31. **Recall** What specific histone modification is commonly associated with increased transcriptional activity in eukaryotic cells, and how does it promote gene expression?
32. **Recall** Describe the function of three eukaryotic response elements.
33. **Recall** What is the purpose of CREB?
34. **Reflect and Apply** How does regulation of transcription in eukaryotes differ from regulation of transcription in prokaryotes?
35. **Reflect and Apply** How do the roles of enhancers and silencers differ from each other?
36. **Reflect and Apply** How do response elements modulate RNA transcription?
37. **Reflect and Apply** Diagram a gene that is affected by CRE and CREB, showing which proteins and nucleic acids contact each other.
38. **Reflect and Apply** Explain the relationship between TFIID, TBP, and TAFs.
39. **Reflect and Apply** Explain the different ways in which eukaryotic transcription elongation is controlled.
40. **Reflect and Apply** Explain the importance of CREB, giving examples of genes activated by it.
41. **Reflect and Apply** Give examples of structural motifs found in transcription factors that interact with other proteins instead of DNA.
42. **Recall** What is a mediator and how does it work?
43. **Recall** What is the difference in how a mediator works with activation versus repression of transcription?
44. **Recall** How do nucleosomes add to the complexity of eukaryotic transcription?
45. **Recall** What are the two main circumstances necessary for active transcription in eukaryotes?
46. **Recall** What two sets of protein factors are important in the activation of eukaryotic transcription?
47. **Recall** What are chromatin remodeling complexes?
48. **Recall** Why are histone-modifying enzymes important?
49. **Recall** What are the major types of covalent modification of histones?
50. **Recall** What enzymes are involved in histone acetylation and deacetylation?

11-6 Noncoding RNAs

51. **Recall** What are micro RNAs?
52. **Recall** What are small interfering RNAs?
53. **Recall** In what processes are noncoding RNAs important?
54. **Recall** What is RNA interference?
55. **Reflect and Apply** What is a potential link between several types of cancer and miRNA-101?
56. **Biochemical Connections** How is miRNA-206 beneficial to an organism?
57. **Biochemical Connections** Micro RNAs play a role in physical nerve damage. What other neurological diseases are they associated with?
58. **Recall** What enzymes are involved in the production of both miRNA and siRNA?
59. **Recall** What are the guide and passenger strands in the production of small RNAs?
60. **Recall** What is the principal difference between the binding of siRNA and the binding of miRNA to mRNA?
61. **Recall** What is the difference between how miRNA stops gene expression and how siRNA does?

11-7 Structural Motifs in DNA-Binding Proteins

62. **Recall** List three important structural motifs in DNA-binding proteins.
63. **Reflect and Apply** Draw examples of the major structural motifs in DNA-binding proteins and explain how they bind.

11-8 Posttranscriptional RNA Modifications

64. **Recall** List several ways in which mRNA is processed after transcription.
65. **Reflect and Apply** Which posttranscriptional modification is unique to tRNA compared to rRNA and mRNA, and why is the modification important?
66. **Reflect and Apply** Why is a trimming process important in converting precursors of tRNA and rRNA to the active forms?
67. **Reflect and Apply** What are snRNPs? What is their role in the processing of eukaryotic mRNAs?
68. **Reflect and Apply** What roles can RNA play other than the transmission of the genetic message?
69. **Reflect and Apply** Diagram the formation of a lariat in RNA processing.
70. **Reflect and Apply** Explain how differential splicing of RNA is thought to be relevant to the information gathered from the Human Genome Project.

11-9 Ribozymes

71. **Recall** What is a ribozyme? List some examples of ribozymes.
72. **Reflect and Apply** Outline a mechanism by which RNA can catalyze its own self-splicing.
73. **Reflect and Apply** Why are proteins more effective catalysts than RNA molecules?
74. **Biochemical Connections** What does "epigenetics" mean?
75. **Biochemical Connections** What is an "epimutation"?
76. **Biochemical Connections** What is the link between epigenetics and cancer?

Further Reading

Barinaga, M. Ribozymes: Killing the Messenger. *Science* **262**, 1512–1514 (1993). A report on research designed to use ribozymes to attack the RNA genome of HIV.

Barrick, J. E., and R. R. Breaker The Power of Riboswitches. *Sci. Amer.* 296 (1), 50–57 (2007).

Bentley, D. RNA Processing: A Tale of Two Tails. *Nature* **395**, 21–22 (1998). The relationship between RNA processing and the structure of the RNA polymerase that made it.

Brivanlou, A. H., and J. E. Darnell. Signal Transduction and the Control of Gene Expression. *Science* **295**, 813–818 (2002). A comprehensive review of eukaryotic transcription factors.

Brown, R. H. A Reinnervating MicroRNA. *Science* **326**, 1494–1495 (2009).

Bushnell, D. A., K. D. Westover, R. E. Davis, and R. D. Kornberg. Structural Basis of Transcription: An RNA Polymerase II-TFIIB Cocrystal at 4.5 Angstroms. *Science* **303**, 983–988 (2004). A study of enzyme structure using X-ray crystallography.

Cammarota, M., et al. Cyclic AMP-Responsive Element Binding Protein in Brain Mitochondria. *J. Neurochem.* **72** (6),2272–2277 (1999). An article about the suspected relationship of CREB to memory.

Cech, T. R. RNA as an Enzyme. *Sci. Amer.* **255** (5), 64–75 (1986). A description of the discovery that some RNAs can catalyze their own self-splicing. The author was a recipient of the 1989 Nobel Prize in Chemistry for this work.

Cramer, P. Self-Correcting Messages. *Science* **313**, 447–448 (2006). This paper identifies proofreading capability in transcription.

Cramer, P., et al. Architecture of RNA Polymerase II and Implications for the Transcription Mechanism. *Science* **288**, 640–649 (2000). Good overview of RNA polymerase structure and function.

De Cesare, D., and P. Sassone-Corsi. Transcriptional Regulation by Cyclic AMP-Responsive Factors. *Prog. Nucleic Acid Res. Mol. Biol.* **64**, 343–369 (2000). A review of cAMP response elements.

deHaseth, P. L., and T. W. Nilsen. When a Part Is as Good as the Whole. *Science* **303**, 1307–1308 (2004). An article describing the structure and function of RNA polymerase.

Doudna, J. A. RNA Structure: A Molecular Contortionist. *Nature* **388**, 830–831 (1997). An article about RNA structure and its relationship to transcription.

Erhard, K. F., J. L, Stonaker., S. E, Parkinson., J. P, Lim., C. J, Hale., and J. B. Hollick, RNA Polymerase IV Functions in Paramutation in Zea Mays. *Science* **323**, 1201–1204 (2009).

Fong, Y. W., and Q. Zhou. Stimulatory Effect of Splicing Factors on Transcriptional Elongation. *Nature* 414, 929–933 (2001). An article about the link between transcription and splicing.

Garber, K. Small RNAs Reveal an Activating Side. *Science* **314**, 741–742 (2006). Preliminary information about how small RNAs might also regulate transcription by activation as well as interference.

Grant, P. A., and J. L. Workman. Transcription: A Lesson in Sharing? *Nature* **396**, 410–411 (1998). A review of TBP and TAFs that discusses the requirements for transcription.

Hanawalt, P. C. DNA Repair: The Bases for Cockayne Syndrome. *Nature* **405**, 415 (2000). The relationship between DNA repair and transcription factors.

Kapanidis, A. N., Margaret., E. Ho., S. O. Kortkhonjia, E., Weiss, S., Ebright, R. H. Initial Transcription by RNA Polymerase Proceeds Through a DNA-Scrunching Mechanism. *Science* **314**, 1144–1147 (2006). Primary research article about using fluorescence to determine the mechanism of polymerase action.

Kuras, L., and K. Struhl. Binding of TBP to Promoters *In Vivo* Is Stimulated by Activators and Requires Pol II Holoenzyme. *Nature* **399**, 609–613 (1999). The nature of the TATA box and binding proteins.

Kuznedelov, K., L. Minakhin, A. Niedzuda-Majka, S. L. Dove, D. Rogulja, B. E. Nickels, A. Hochschild, T. Heyduk, and K. Severinov. A Role for Interaction of the RNA Polymerase Flap Domain with the σ Subunit in Promoter Recognition. *Science* **295**, 855–857 (2002). The most recent structural information on bacterial RNA polymerase binding.

Mandelkow, E. Alzheimer's Disease: The Tangled Tale of Tau. *Nature* **402**, 588–589 (1999). Phosphorylation of a neuron protein leads to neurofibrillary tangles that are associated with Alzheimer's dementia.

Montminy, M. Transcriptional Activation: Something New to Hang Your HAT On. *Nature* **387**, 654–655 (1997). Transcription in eukaryotes requires opening the DNA/histone complex, which can be controlled by acetylation.

Plasterk, R. H. A. RNA Silencing: The Genome's Immune System. *Science* **296**, 1263–1265 (2002).

Revyakin, A., C. Liu, R. H. Ebright, and T. R. Strick. Abortive Initiation and Productive Initiation by RNA Polymerase Involve DNA Scrunching. *Science* **314**, 1139–1143 (2006). Primary research article about using fluorescence to determine the mechanism of polymerase action.

Rhodes, D., and A. Klug. Zinc Fingers. *Sci. Amer.* **268** (2), 56–65 (1993). How the structure of these zinc-containing proteins enables them to play a role in regulating the activity of genes.

Riccio, A., et al. Mediation by a CREB Family Transcription Factor of NGF-Dependent Survival of Sympathetic Neurons. *Science* **286**, 2358–2361 (1999). How CREB may help protect neurons in times of stress.

Roberts, J. W. RNA Polymerase, a Scrunching Machine. *Science* **314**, 1097–1098 (2006). Overview of two *Science* articles regarding the mechanism of RNA polymerase movement in chain elongation.

Steitz, J. A. Snurps. *Sci. Amer.* **258** (6), 56–63 (1988). A discussion of the role of small nuclear ribonucleoproteins, or snRNPs, in the removal of introns from mRNA.

Storz, G. An Expanding Universe of Noncoding RNAs. *Science* **296**, 1260–1262 (2002).

Struhl, K. A. Paradigm for Precision. *Science* **293**, 1054–1055 (2001). This article discusses how transcription factors, coactivators, and histone acetyltransferases work together to enhance transcription.

Tupler, R., G. Perini, and M. R. Green. Expressing the Human Genome. *Nature* **409**, 832–833 (2001). An excellent review of transcription and the use of data from the Human Genome Project to search for transcription factors.

Westover, K. D., D. A. Bushnell, and R. D. Kornberg. Structural Basis of Transcription: Separation of RNA from DNA by RNA

Polymerase II. *Science* **303**, 1014–1016 (2004). An article describing the mechanism of separation of the DNA and RNA during transcription.

Young, B. A., T. M. Gruber, and C. A. Gross. Minimal Machinery of RNA Polymerase Holoenzyme Sufficient for Promoter Melting. *Science* **303**, 1382–1384 (2004). Research article about RNA polymerase structure and function.

Zenkin, N., Y. Yuzenkova, and K. Severinov. Transcript-Assisted Transcriptional Proofreading. *Science* **313**, 518–520 (2006). Research on possible proofreading in transcription based on secondary structures formed by the RNA transcript.

Hot Topic

Roberts, J. W. Molecular Basis of Transcription Pausing. *Science,* 344, 1226–1227 (2014). Description of molecular interactions between RNA polymerase and transcription pausing sequences.

Luo, C. and J. R. Ecker Exceptional Epigenetics in the Brain. *Science,* 348, 1094–1095 (2015). An article about new evidence for epigenetic control of brain development in mammals.

Orlando, L. and E. Willerslev An Epigenetic Window into the Past? *Science,* 345, 511–512 (2014). A description of new technologies for creating epigenetic maps that can help us understand human evolution.

Skinner, M. A New Kind of Inheritance. *Sci. Amer.,* 45–51 (August, 2014). Comprehensive article about epigenetic changes and their inheritance.

Pedersen, J. S. Genome-Wide Nucleosome Map and Cytosine Methylation Levels of an Ancient Human Genome. Genome Res. (2014). An article describing the methylation patterns and techniques for determining them for a 4000-year-old Inuit.

Sweatt, J. D. Epigenetics and Cognitive Aging. *Science,* 328, 701–702 (2010). An article describing how epigenetic changes in the brain accompany aging.

A medical translator translates medical information from one language to another language. In the biochemical process of translation, the nucleic acid language of RNA is converted to the amino acid language of proteins.

Brian Eichhorn/Shutterstock.com

Protein Synthesis: Translation of the Genetic Message 12

Chapter Outline

12-1 Translating the Genetic Message

Learning Objectives

After reading this section, students will be able to:

LO 12.1.1 Distinguish between the different types of RNA that are involved in protein biosynthesis.

LO 12.1.2 List the steps in protein biosynthesis.

Translation (i.e., protein biosynthesis) is a complex process requiring ribosomes, messenger RNA (mRNA), transfer RNA (tRNA), and a number of protein factors. The ribosome is the site of protein synthesis. The mRNA and tRNA, which are bound to the ribosome in the course of protein synthesis, provide the code and bring the right building blocks, respectively, for synthesizing proteins with their amino acids in the correct order.

Figure 12.1 shows an overview of the steps involved in translation. Before an amino acid can be incorporated into a growing protein chain, it must first be *activated* and covalently attached to a tRNA molecule to form an aminoacyl-tRNA (Section 12-3). The actual formation of the polypeptide chain occurs in three steps:

1. *Chain initiation*: The first aminoacyl-tRNA is bound to the mRNA at the site that encodes the start of polypeptide synthesis. In this complex, the mRNA and the ribosome are bound to each other. The next aminoacyl-tRNA forms a complex with the ribosome

Chapter Outline (continued)

and with mRNA. The binding site for the second aminoacyl-tRNA is next to that for the first aminoacyl-tRNA.

2. *Chain elongation*: A peptide bond is formed between the amino acids, and the ribosome moves one codon down the mRNA. The chain elongation process repeats itself until the polypeptide chain is complete.
3. *Chain termination*: The completed polypeptide chain is released from the ribosome.

Each of these phases has many distinguishing features, but before we look at each of them in detail, we must discuss the genetic code.

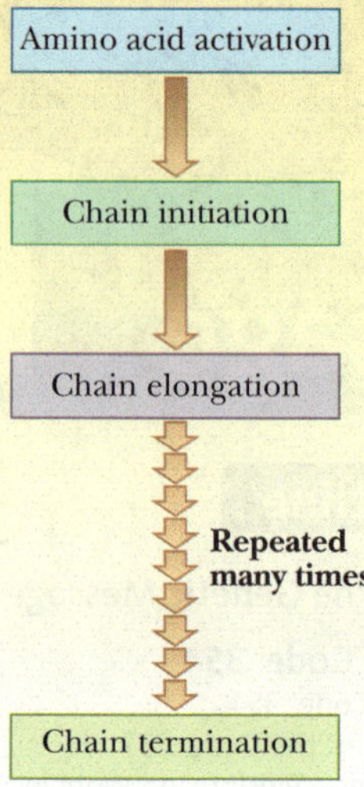

Figure 12.1 The steps in protein biosynthesis.

12-2 The Genetic Code

Learning Objectives

After reading this section, students will be able to:

LO 12.2.1 Distinguish between the terms *triplet, nonoverlapping, commaless, degenerate,* and *universal* as they apply to the genetic code.

LO 12.2.2 Describe how the filter-binding assay was used to determine which trinucleotide codons code for which amino acids.

Some of the most important features of the genetic code can be specified by saying that the genetic message is contained in a *triplet, nonoverlapping, nonpunctuated, commaless, degenerate, nearly universal* code. Each of these terms has a definite meaning that describes the way in which the code is translated.

12-2A Nature of the Code

A **triplet code** means that a sequence of three bases (called a **codon**) is needed to specify one amino acid. Cells must translate the genetic code of DNA, which contains four bases (A, G, C, and T), into the language of the 20 common amino acids that are found in proteins. With a sequence of three bases and with four possible bases at each position, there are $4 \times 4 \times 4 = 64$ possibilities, or 64 possible codons, which is more than enough to encode the 20 amino acids. The term *nonoverlapping* indicates that no bases are shared between consecutive codons; the ribosome moves along the mRNA three bases at a time, rather than one or two at a time (Figure 12.2A). This

triplet code (codon) a sequence of three bases (a triplet) in mRNA that specifies one amino acid in a protein

Pioneers in Biochemistry • Dorothy Andersen

Dorothy Andersen (1901–1963) was the pioneering physician who first identified cystic fibrosis (CF). The daughter of a Danish immigrant, Andersen was born in North Carolina and later moved to Vermont after her father's death.

As an assistant pathologist, Andersen performed an autopsy on a child and noticed a specific group of abnormalities that she later identified in the case histories of 49 other patients. Just a few years later, in 1938, she published the article in which she first described and named cystic fibrosis.

Although Andersen pursued several lines of research over her career, she continued to develop her understanding of CF. In the 1940s she was involved in demonstrating the genetic basis for the disease, and she later contributed to several iterations of diagnostic testing. Later, her research contributed to the development of the noninvasive "sweat test," which is still used to diagnose CF today. Andersen also developed the first treatments for CF, including a special diet, vitamin supplements, and replacement therapy for pancreatic enzymes.

As has often been the case for female scientists, Andersen's contributions to CF research were undervalued during her lifetime. Today, science historians look back and recognize how Andersen's perseverance and meticulous research methods shaped our understanding of the disease.

Science History Images/Alamy Stock Photo

is important because it ensures that the ribosome translates mRNA in a consistent reading frame, producing the correct sequence of amino acids. If the code were overlapping, a single mutation could affect multiple amino acids, increasing the chance of harmful errors. If the ribosome moved along the mRNA more than three bases at a time, this situation would be referred to as "a punctuated code." Because no intervening bases exist between codons, the code is *commaless* (Figure 12.2B). In a *degenerate* code, more than one triplet can encode the same amino acid. Sixty-four (4 × 4 × 4) possible triplets of the four bases occur in RNA, and all are used to encode the 20 amino acids or one of the three stop signals. There is a significant difference between a degenerate code and an ambiguous one. Each amino acid may have more than one codon, so the genetic code is a little redundant, but no codon can encode more than one amino acid. If it did, the code would be ambiguous, and the protein-synthesizing machinery would not know which amino acid should be inserted in the sequence. All 64 codons have been assigned meanings, with 61 of them coding for amino acids and the remaining 3 serving as the termination signals (Table 12.1).

Two amino acids, tryptophan and methionine, have only one codon each, but the rest have more than one. Multiple codons for a single amino acid are not randomly distributed in Table 12.1 but have one or two bases in common. The bases that are common in codons that code for the same amino acid are usually the first and second bases, with more room for variation in the third base, which is called the "wobble" base. This degeneracy (or redundancy) of the code acts as a buffer against deleterious mutations. In the codons for Val, Ser, Pro, Thr, Ala, and Gly, for example, the third base is completely irrelevant. Thus, any mutation in the third base of these codons would not change the amino acid at that location. A mutation in the DNA that does not lead to a change in the amino acid translated is called a *silent mutation.*

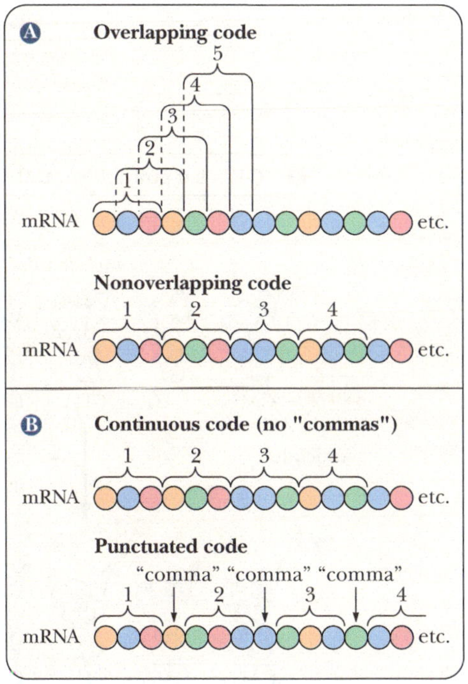

Figure 12.2 Theoretically possible genetic codes. (A) An overlapping versus a nonoverlapping code. (B) A continuous versus a punctuated code.

Table 12.1 The Genetic Code

First Position (5′-end)	Second Position				Third Position (3′-end)
	U	C	A	G	
U	UUU Phe	UCU Ser	UAU Tyr	UGU Cys	U
	UUC Phe	UCC Ser	UAC Tyr	UGC Cys	C
	UUA Leu	UCA Ser	UAA Stop	UGA Stop	A
	UUG Leu	UCG Ser	UAG Stop	UGG Trp	G
C	CUU Leu	CCU Pro	CAU His	CGU Arg	U
	CUC Leu	CCC Pro	CAC His	CGC Arg	C
	CUA Leu	CCA Pro	CAA Gln	CGA Arg	A
	CUG Leu	CCG Pro	CAG Gln	CGG Arg	G
A	AUU Ile	ACU Thr	AAU Asn	AGU Ser	U
	AUC Ile	ACC Thr	AAC Asn	AGC Ser	C
	AUA Ile	ACA Thr	AAA Lys	AGA Arg	A
	AUG Met*	ACG Thr	AAG Lys	AGG Arg	G
G	GUU Val	GCU Ala	GAU Asp	GGU Gly	U
	GUC Val	GCC Ala	GAC Asp	GGC Gly	C
	GUA Val	GCA Ala	GAA Glu	GGA Gly	A
	GUG Val	GCG Ala	GAG Glu	GGG Gly	G

*AUG signals translation initiation as well as coding for Met residues.

Third-Base Degeneracy Is Color-Coded		
Third-Base Relationship	Third Bases with Same Meaning	Number of Codons
Third base irrelevant	U, C, A, G	32 (8 families)
Purines	A or G	12 (6 pairs)
Pyrimidines	U or C	14 (7 pairs)
Three out of four	U, C, A	3 (AUX = Ile)
Unique definitions	G only	2 (AUG = Met) (UGG = Trp)
Unique definition	A only	1 (UGA = Stop)

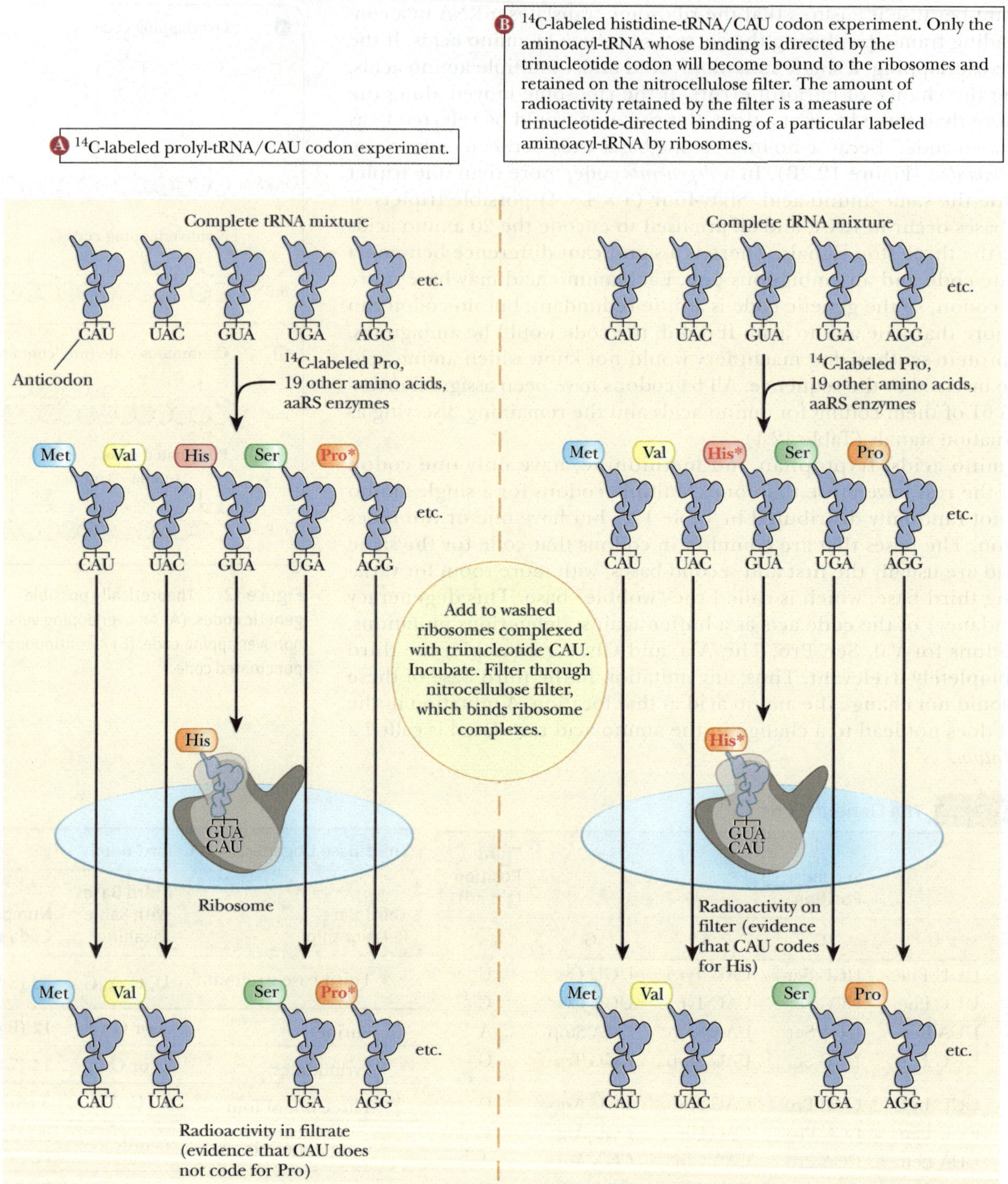

Figure 12.3 The filter-binding assay for elucidation of the genetic code. A reaction mixture combines washed ribosomes, Mg^{2+}, a particular trinucleotide, and all possible aminoacyl-tRNAs, one of which carries a radioactively (^{14}C) labeled amino acid (indicated by an asterisk). Aminoacyl-tRNA synthetase (aaRS) enzymes are used to generate the aminoacyl-tRNAs. Use of this binding assay to test the 64 possible codon trinucleotides against the 20 different amino acids quickly enabled researchers to assign most of the triplet code sequences to individual amino acids. *(Based on Nirenberg, M. W., and Leder, P. (1964). RNA codewords and protein synthesis.* Science *Vol. 145, pp.* 1399-1407.*)*

In addition, the second base of the codon appears to be very important for determining the type of amino acid (i.e., polar or nonpolar—see Figure 3.3). For example, when the second base is U, all the amino acids generated from the codon possibilities are nonpolar. Thus, if the first or third base were mutated, the mutation would not be silent, but the damage would not be as great because one hydrophobic amino acid would be replaced with another. Codons sharing the same first letter often code for amino acids that are products of one another or precursors of one another. These features of the genetic code provide a certain amount of protection for organisms against DNA mutations.

12-2B Determination of the Code

The assignment of triplets in the genetic code was based on several types of experiments. One of the most significant experiments involved the use of synthetic polyribonucleotides as messengers. When homopolynucleotides (i.e., polyribonucleotides that contain only one type of base) are used as a *synthetic mRNA* for polypeptide synthesis in laboratory systems, homopolypeptides (i.e., polypeptides that contain only one kind of amino acid) are produced. When poly U is the messenger, the product is polyphenylalanine because UUU codes for phenylalanine. With poly A as the messenger, polylysine is formed. The product for poly C is polyproline, and the product for poly G is polyglycine. This procedure was used to establish the code for the four possible homopolymers quickly.

When an alternating copolymer (i.e., a polymer with an alternating sequence of two bases) is the messenger, the product is an alternating polypeptide (a polypeptide with an alternating sequence of two amino acids). For example, when the sequence of the polynucleotide is –ACACACACACACACACACACAC–, the polypeptide produced has alternating threonines and histidines. There are two types of coding triplets in this polynucleotide, ACA and CAC, but this experiment cannot establish which one codes for threonine and which one codes for histidine.

Although more information is needed for an unambiguous assignment, this result does prove that the code is a triplet code. If it were a doublet code, the product would be a mixture of two homopolymers, one specified by the codon AC and the other by the codon CA. The terminology for the different ways of reading this message as a doublet is to say that they have different **reading frames**, /AC/AC/ and /CA/CA/. In a triplet code, only one reading frame is possible—namely /ACA/CAC/ACA/CAC/—which gives rise to an alternating polypeptide. Use of other synthetic polynucleotides can yield other coding assignments, but, as in our example here, many questions remain.

reading frames the starting points for reading genetic messages

Other methods are needed to answer the remaining questions about the codon assignments. One of the most useful methods is the **filter-binding assay** (Figure 12.3). In this technique, a complete set of aminoacyl-tRNA molecules, one of which has an amino acid that is radioactively labeled with carbon-14 (^{14}C), is mixed with ribosomes and synthetic trinucleotides that are bound to a filter (each synthetic trinucleotide mimics an mRNA codon). The mixture of tRNAs is passed through the filter, and some bind to the trinucleotide while others pass through (recall from Figure 9.25 that tRNAs have an **anticodon** sequence that binds to mRNA codon sequences). If the radioactive label is detected on the filter, then it is known that the particular tRNA did bind to the trinucleotide, and this binding is evidence that the trinucleotide codes for the amino acid that is radioactively labeled in the experiment. If the radioactive label is found in the solution that flowed through the filter, then the tRNA did *not* bind to the trinucleotide, in which case the trinucleotide does *not* code for the amino acid that is radioactively labeled in the experiment.

filter-binding assay a method used to determine the base sequence of many mRNA codons

anticodon the sequence of three bases (triplet) in tRNA that hydrogen bonds with the mRNA triplet that specifies a given amino acid

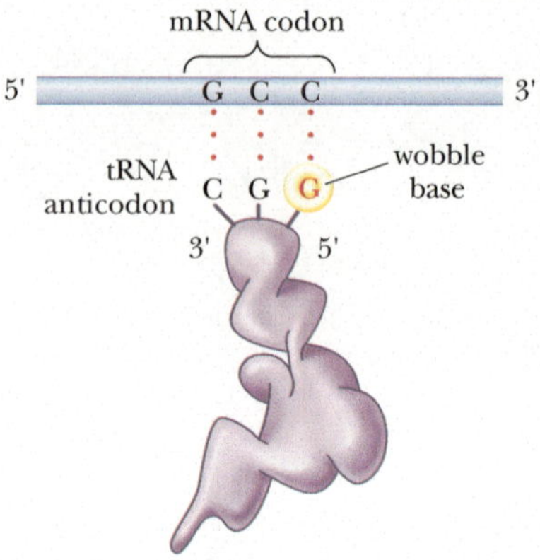

Figure 12.4 Wobble base pairing. The wobble base of the anticodon is the one at the 5′ end of the anticodon; it forms hydrogen bonds with the last base of the mRNA codon, the one at the 3′ end of the codon. *(Adapted from Crick, F. H. C. (1966). Codon–anticodon pairing: The wobble hypothesis.* Journal of Molecular Biology *19, 548–555, with permission from Elsevier.)*

wobble base pairing the possible variations in the binding of the third base of a codon to the first base of an anticodon that are allowed by several acceptable forms of base pairing between mRNA and tRNA

This technique depends on the fact that aminoacyl-tRNAs bind strongly to ribosomes in the presence of the correct trinucleotide. In this situation, the trinucleotide plays the role of an mRNA codon, and binding assays are repeated with each of the 64 possible trinucleotide sequences. For example, if the aminoacyl-tRNA for ^{14}C-labeled histidine binds to the ribosome in the presence of the trinucleotide CAU, the sequence CAU is established as a codon for histidine (Figure 12.2B). About 50 of the 64 codons were identified by this method.

12-2C Codon–Anticodon Pairing and Wobble

Figure 12.3 shows that an mRNA codon base pairs with a complementary anticodon of a tRNA when an amino acid is incorporated during protein synthesis. Because there are 64 possible codons, it is reasonable to assume that there are 64 types of tRNA. In fact, the number is less than 64 in all cells.

▶ If there are 64 codons, how can there be fewer than 64 tRNA molecules?

Some tRNAs bond to one codon exclusively, but many of them can recognize more than one codon because of variations in the allowed pattern of hydrogen bonding. This variation, called **wobble base pairing** (Figure 12.4), applies to the first base of an anticodon—the one at the 5′ end—but not to the second or the third base. Because mRNA is read from the 5′ to the 3′ end, the first base of the anticodon can display wobble base pairing to the third base of the codon—the one at the 3′ end of the codon.

As seen in Table 12.2, the base in the wobble position of the anticodon can base-pair with several different bases in the codon, not just the base specified by Watson–Crick base pairing. For example, when the wobble base of the anticodon is uracil, it can base-pair not only with adenine, as expected, but also with guanine, the other purine base. When the wobble base is guanine, it can base-pair with cytosine, as expected, and also with uracil, the other pyrimidine base. The purine base hypoxanthine frequently occurs in the wobble position in many tRNAs, and it can base-pair with adenine, cytosine, and uracil in the codon (Figure 12.5). Adenine and cytosine do not form any base pairs other than the expected ones with uracil and guanine, respectively (Table 12.2). To summarize, when the wobble position is occupied by I (from inosine, the nucleoside made up of ribose and hypoxanthine), G, or U, variations in hydrogen bonding are allowed; when the wobble position is occupied by A or C, these variations do not occur.

The wobble model provides insight into some aspects of the degeneracy of the code. In many cases, the degenerate codons for a given amino acid differ in the third base, the one that pairs with the wobble base of the anticodon. Fewer different tRNAs are needed because a given tRNA can base-pair with

Table 12.2 Base-Pairing Combinations in the Wobble Scheme

Base at 5′ End of Anticodon	Base at 3′ End of Codon
I*	A, C, or U
G	C or U
U	A or G
A	U
C	G

*I = hypoxanthine.

There are no variations in base pairing when the wobble position is occupied by A or C.

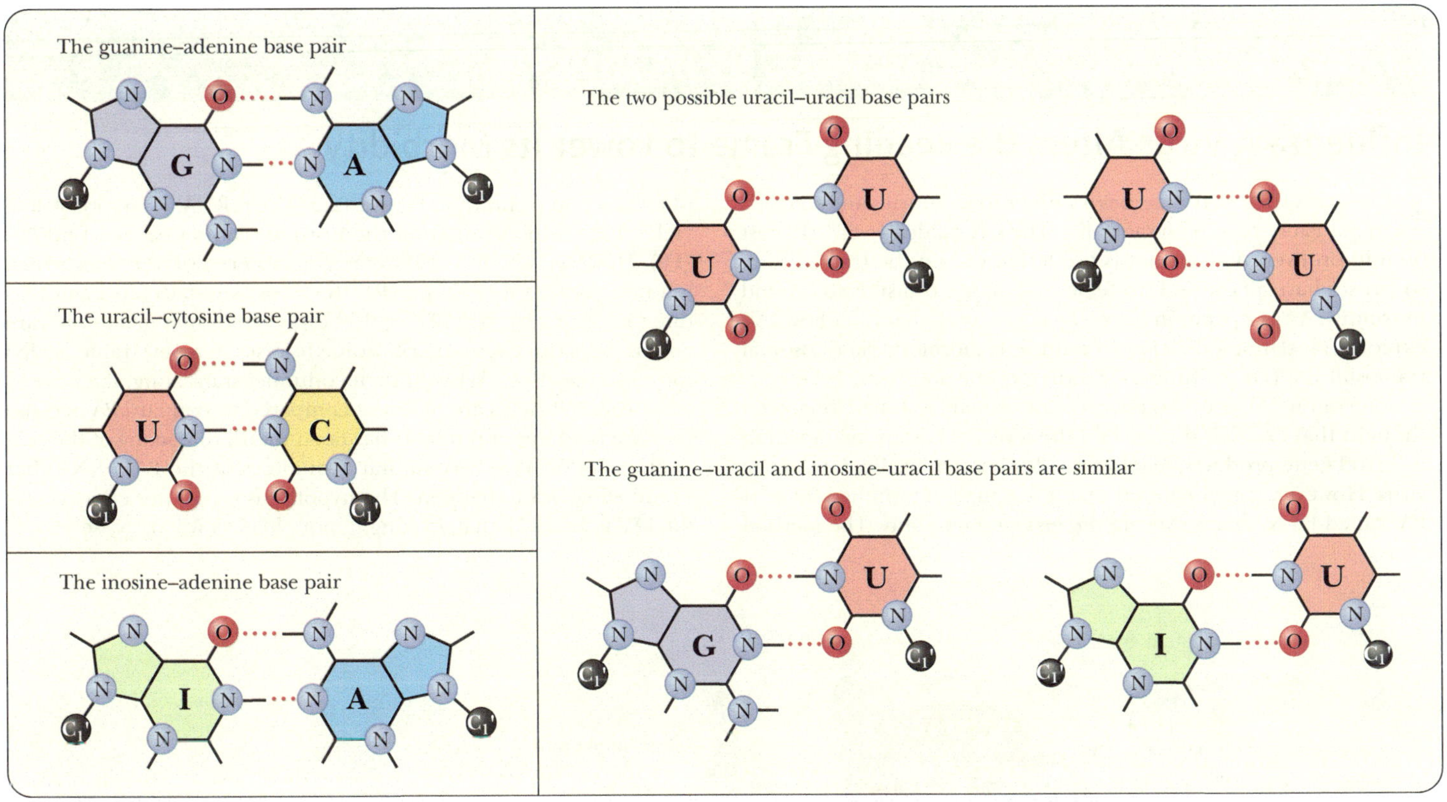

Figure 12.5 Various base-pairing alternatives. G:A is unlikely because the 2-NH_2 of G cannot form one of its H bonds; even water is sterically excluded. U:C may be possible, even though the two C=O are juxtaposed. Two U:U arrangements are feasible. G:U and I:U are both possible and somewhat similar. The purine pair I:A is also possible. *(Adapted from Crick, F. H. C. (1966). Codon–anticodon pairing: The wobble hypothesis.* Journal of Molecular Biology *19, 548–555, with permission from Elsevier.)*

several codons. As a result, a cell would have to invest less energy in the synthesis of needed tRNAs. The existence of wobble also minimizes the damage that can be caused by a misreading of the code. If, for example, a leucine codon, CUU, were to be misread as CUC, CUA, or CUG during transcription of mRNA, this codon would still be translated as leucine during protein synthesis; no damage to the organism would occur.

A *universal* code is one that is the same in all organisms. The universality of the code has been observed in viruses, prokaryotes, and eukaryotes. However, there are some exceptions. Some codons seen in mitochondria are different from those seen in the nucleus. There are also at least 16 organisms that have code variations. For example, the marine alga *Acetabularia* translates the standard stop codons, UAG and UAA, as a glycine rather than as a stop. Fungi of the genus *Candida* translate the codon CUG as a serine, where that codon would specify leucine in most organisms. The evolutionary origin of these differences is not known, but many researchers believe that understanding these code variations is important to understanding evolution. Additionally, even though the code is not universal across all organisms, it is universal in all tissues within a multicellular organism.

Although there are different codons that specify some amino acids, not all codons are represented equally. Some are common, and some are rare. This has implications for the speed of translation because rare codons often have low quantities of the matching tRNAs. Biochemical Connections 12A describes an interesting effect based on codon frequency that is exploited by the influenza A virus (IAV).

12A Biochemical Connections • Virology

Influenza A Virus Alters the Reading Frame to Lower Its Morbidity

Influenza A virus (IAV) has been with us for centuries and remains a major cause of human mortality. The IAV epidemic of 1918 cost 50 million lives. A virus that is too lethal will not be transmitted, so viruses have all evolved strategies to balance transmissibility and morbidity. A 2012 paper in *Science* by Jagger et al. describes how IAV exploits translation efficiency to reduce its morbidity via ribosomal frameshifting (i.e., shifting the reading frame on the mRNA).

One of the IAV proteins is a protease, called PA. It had long been thought that PA itself diminished the levels of certain mammalian cell viral gene products, leading to a lessened mortality rate of the virus. However, Jagger showed that it is actually a modified form of PA, called PA-X, that is causing the lessened mortality. The method of production is shown in Figure 12.6. The mRNA, when translated, yields 191 amino acids from the normal reading frame of mRNA (F0). However, when the ribosome encounters a CGU codon, which is a rare codon for arginine (R), the ribosome stalls on the mRNA due to the relative lack of Arg-tRNA molecules that match the rare codon. This can cause the ribosome to reset to a new frame, called the +1 frame (F+1). When it then continues translating, the next 61 amino acids it yields are different compared to standard PA because they are read out of frame. It has been found that many of the sub-strains of IAV have very similar sequences of the PA mRNA that would allow this to happen. The hypothesis is that this effect allows the IAV to be less lethal, making it more likely it will be passed on. ◗

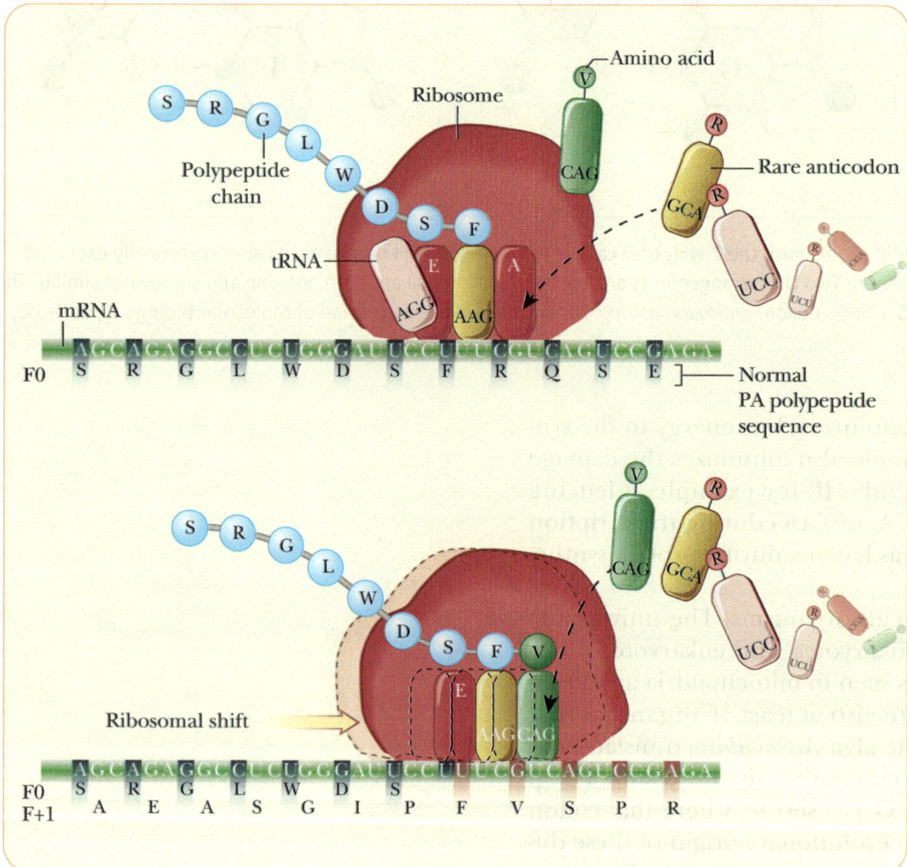

Figure 12.6 A rare codon stalls the ribosome while it translates the PA protein, producing the PA-X variant of the protein. Top: The normal reading frame (F0) produces the normal PA polypeptide sequence. Bottom: A stall due to a rare Arg codon can cause the ribosome to shift to the +1 frame (F+1) and begin translating a new sequence, producing the PA-X variant (the alternate residues shaded in red are synthesized after the shift). *(Based on Yewdell, J. W., and Ince, W. L. (2012). Frameshifting to PA-X influenza.* Science *337, 164.)*

12-3 Aminoacyl-tRNA Formation

Learning Objectives

After reading this section, students will be able to:

LO 12.3.1 Describe how aminoacyl-tRNAs are formed.

LO 12.3.2 Explain what is meant by the term *second genetic code*.

Before an amino acid can be added to a tRNA molecule to form an aminoacyl-tRNA molecule that binds to mRNA and the ribosome during translation, the amino acid must first be *activated.* Figure 12.7 shows how the activation of the

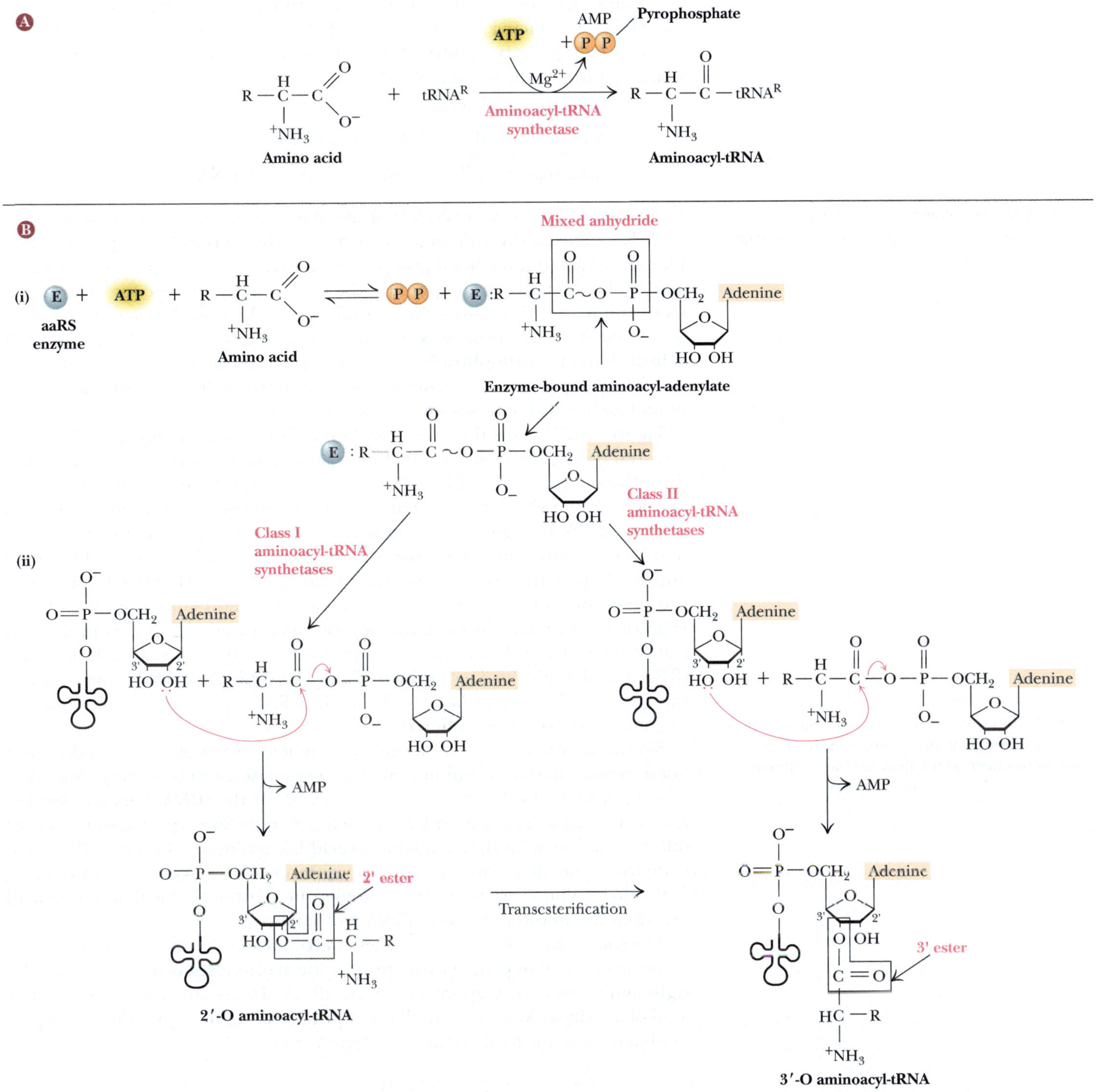

Figure 12.7 The aminoacyl-tRNA synthetase (aaRS) reaction. (A) The overall aaRS reaction. $tRNA^R$ represents a generic tRNA molecule that is designed to carry an amino acid with side-chain "R." Ever-present pyrophosphatases in cells quickly hydrolyze the PP_i produced in the aaRS reaction, rendering aminoacyl-tRNA synthesis thermodynamically favorable and essentially irreversible. (B) The two steps in the aaRS reaction: (i) formation of an aminoacyl-adenylate (the "~" character, a so-called squiggle bond, is used to indicate a high-energy bond—in this case, the high-energy mixed anhydride bond) and (ii) nucleophilic attack on the activated amino acid moiety of the mixed anhydride by either the 2′-OH (class I aaRSs) or 3′-OH (class II aaRSs) of the ribose on the terminal adenylic acid at the 3′ end common to all tRNAs. Those aminoacyl-tRNAs formed as 2′ esters undergo a transesterification that moves the aminoacyl group to the 3′ carbon of the tRNA 3′ end. Only the 3′ ester form of aminoacyl-tRNAs is used in protein synthesis at the ribosome.

synthetase an enzyme that uses energy (in the form of ATP hydrolysis) to catalyze the formation of a new covalent bond that links two molecules together

amino acid and the formation of the aminoacyl-tRNA take place in two separate steps, both of which are catalyzed by an aaRS enzyme. **Synthetase** enzymes are discussed in several other chapters. In general, they use energy (in the form of ATP hydrolysis) to catalyze the formation of a new covalent bond that links two molecules together. First, the amino acid forms a covalent bond to an adenine nucleotide, producing an aminoacyl-AMP. The free energy of hydrolysis of ATP provides energy for bond formation. The aminoacyl moiety is then transferred to tRNA, forming an aminoacyl-tRNA.

$$\text{Amino acid} + \text{ATP} \rightarrow \text{aminoacyl-AMP} + \text{PP}_i$$
$$\text{Aminoacyl-AMP} + \text{tRNA} \rightarrow \text{aminoacyl-tRNA} + \text{AMP}$$
$$\overline{\text{Amino acid} + \text{ATP} + \text{tRNA} \rightarrow \text{aminoacyl-tRNA} + \text{AMP} + \text{PP}_i}$$

anhydride a functional group that has two carbonyl (C=O) groups bonded to the same oxygen atom

Aminoacyl-AMP is a mixed **anhydride** of a carboxylic acid and a phosphoric acid. The mixed anhydride in this case has a carbonyl (C=O) group and a phosphorus–oxygen double bond group bonded to the same oxygen atom. Because anhydrides are reactive compounds, the Gibbs energy change for the hydrolysis of aminoacyl-AMP favors the second step of the overall reaction. Another point that favors the process is the energy released when pyrophosphate (PP_i) is hydrolyzed to orthophosphate (P_i) to replenish the phosphate pool in the cell. The *pyr-* prefix comes from Greek, and it means "fire." It is used in a few other biochemical names seen in other chapters.

Figure 12.7A shows the overall aaRS reaction, whereas Figure 12.7B shows the step-by-step information in the reaction. In the first part of the reaction, the enzyme uses ATP and the amino acid to form an enzyme-bound aminoacyl-adenylate species that has a high-energy mixed anhydride bond [Figure 12.7B, (*i*)]. In the second part of the reaction, the aaRS enzyme fosters a nucleophilic attack on the activated amino acid moiety of the mixed anhydride by either the 2′-OH (class I aaRSs) or 3′-OH (class II aaRSs) of the ribose on the terminal adenylic acid at the 3′ end common to all tRNAs [Figure 12.7B, (*ii*)]. Those aminoacyl-tRNAs formed as 2′ esters undergo a transesterification that moves the aminoacyl group to the 3′ carbon of the tRNA 3′ end. Only the 3′ ester form of aminoacyl-tRNAs is used in protein synthesis at the ribosome. Class I and class II aaRSs appear to be unrelated and suggest a **convergent evolution** event.

convergent evolution (of enzymes) when enzymes appear to have evolved to catalyze the same or similar biochemical reactions but have different structures and amino acid sequences

Several tRNAs can exist for each amino acid, but a given tRNA does not bond to more than one amino acid. The synthetase enzyme requires Mg^{2+} and is highly specific both for the amino acid and for the tRNA. Separate synthetases exist for each amino acid, so there are 20 different synthetases for the 20 different amino acids. When an amino acid has multiple different codons that code for it and therefore has multiple different tRNAs (with corresponding anticodons) that carry it to the ribosome, the synthetase for that amino acid functions for all those different tRNA molecules.

The specificity of the aaRS enzymes contributes to the accuracy of the translation process. The primary function of the synthetases is to assure that the right amino acid pairs up with the right tRNA. The synthetases have another level of activity as well. An extra level of proofreading by the synthetases is part of what is sometimes called the *second genetic code.*

▶ What is the second genetic code?

The two-stage aaRS reaction (Figure 12.7B) allows for selectivity to operate at two levels: that of the amino acid and that of the tRNA. These two levels of selectivity allow the synthetase enzymes to accurately pair tRNA molecules with the correct amino acids, and these are the aspects that play a role in the concept of the second genetic code. The specificity of the first stage uses the fact

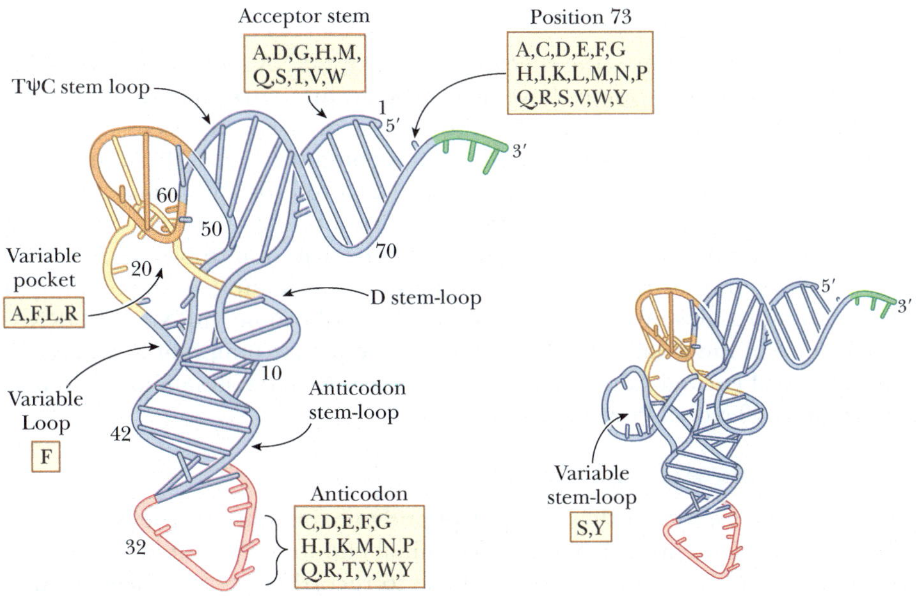

Figure 12.8 Ribbon diagram of the tRNA tertiary structure and tRNA recognition sites for various aminoacyl-tRNA synthetases (aaRSs). Numbers represent nucleotide positions. The locations of nucleotides recognized by the various aaRSs are indicated; shown within the boxes are one-letter codes of the amino acids whose respective aaRSs interact at the discriminator base (position 73), the acceptor stem, the variable pocket, the variable loop, and/or the anticodon. The inset shows additional recognition sites in those tRNAs having a variable stem-loop structure. *(Adapted from Saks, M. E., Sampson, J. R., and Abelson, J. N. (1994). The transfer RNA problem: A search for rules.* Science *263, 191–197, Figure 2. Used with permission from AAAS.)*

that the aminoacyl-AMP remains bound to the enzyme. For example, isoleucyl-tRNA synthetase can form an aminoacyl-AMP of isoleucine or the structurally similar valine. If the valyl moiety is then transferred to the tRNA for isoleucine, it is detected by an editing site in the tRNA synthetase, which then hydrolyzes the incorrectly acylated aminoacyl-tRNA. Thus, the selectivity resides in the tRNA, not in the amino acid.

The second aspect of selectivity depends on the selective recognition of tRNAs by aaRSs. Specific binding sites on tRNA are recognized by aaRSs. The exact position of the recognition site varies with different synthetases, and this feature, in and of itself, is a source of greater specificity. Contrary to what you might expect, the anticodon is not always the part of the tRNA that is recognized by the aaRS, although it is frequently involved. Figure 12.8 shows the locations of the recognition sites on the tRNAs for various amino acids.

The recognition of the correct tRNA by each synthetase is vital to the fidelity of translation because most of the final proofreading occurs at this step.

12-4 Prokaryotic Translation

Learning Objectives

After reading this section, students will be able to:

LO 12.4.1 Describe the architecture of the prokaryotic ribosome.

LO 12.4.2 Explain the events that happen during the initiation, elongation, and termination phases of prokaryotic translation.

LO 12.4.3 Describe how the rare amino acids selenocysteine and pyrrolysine get incorporated into proteins.

As seen in the processes of replication and transcription, the details of the events in translation differ in prokaryotes and eukaryotes. In our discussion of prokaryotic translation, we use *Escherichia coli* (*E. coli*) as our principal example because all aspects of protein synthesis have been most extensively studied in this bacterium. As was the case with replication and transcription, translation can be divided into stages—namely, initiation, elongation, and termination.

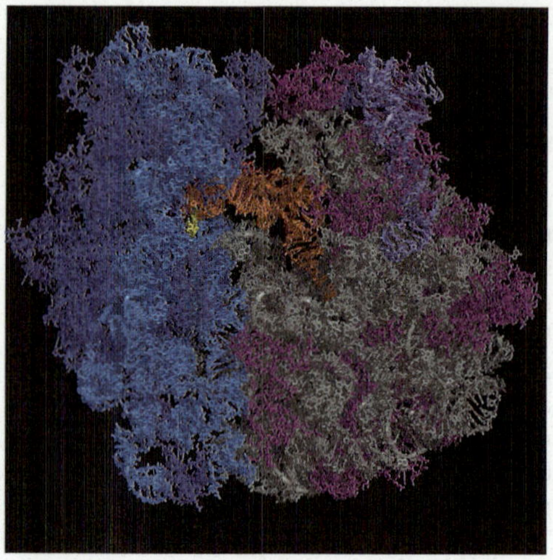

Figure 12.9 Ribosome structure as determined by X-ray crystallography. A tRNA is shown in orange. The portion of the mRNA hydrogen bonded to the tRNA can be seen on the left (gold). The peptidyl transferase is shown in gray on the right.

12-4A Prokaryotic Ribosomal Architecture

Protein synthesis requires the specific binding of mRNA and aminoacyl-tRNAs to the ribosome. Ribosomes have a specific architecture that facilitates the binding. The details of prokaryotic ribosome structure were discussed in Section 9-6B. Recall that prokaryotic ribosomes contain a small 30S subunit that contains a 16S ribosomal RNA (rRNA) and 21 proteins and a large 50S subunit that contains a 5S rRNA, a 23S rRNA, and up to 34 proteins. The small and large subunits combine to form the 70S prokaryotic ribosome complex. Figure 12.9 shows a structural representation of an example of the binding of the main molecules involved in translation: the ribosome, mRNA, and tRNA. Elucidation of the details of ribosomal structure was a triumph of X-ray crystallography and earned Ada Yonath, Venkatraman Ramakrishnan, and Thomas Steitz the 2009 Nobel Prize in Chemistry.

12-4B Prokaryotic Initiation of Translation

In all organisms, the synthesis of polypeptide chains starts at the N-terminal end, and the chain grows from the N-terminal end to the C-terminal end. This is one of the reasons that scientists chose to record DNA sequences from 5′ to 3′ and to focus on the coding strand of DNA and the mRNA. The coding strand DNA sequences are read from 5′ to 3′, and the mRNA is read from 5′ to 3′, corresponding to protein synthesis from the N-terminus to the C-terminus.

In prokaryotes, the initial N-terminal amino acid of all proteins is *N*-formylmethionine (fMet) (Figure 12.10). However, this residue is often removed by posttranslational processing after the polypeptide chain is synthesized (Section 12-6). Therefore, there are two different tRNAs for methionine in *E. coli*—one for unmodified methionine (Figure 3.3B) and one for *N*-formylmethionine. These two tRNAs are called $tRNA^{Met}$ and $tRNA^{fMet}$, respectively (recall from Section 12-3 that each tRNA is specific for a certain amino acid, and the superscript in tRNA nomenclature identifies the tRNA). The aminoacyl-tRNAs that they form with methionine are called $Met\text{-}tRNA^{Met}$ and $Met\text{-}tRNA^{fMet}$, respectively (the prefixes in tRNA nomenclature indicate a bound amino acid). In the case of the *N*-formyl-Met-$tRNA^{fMet}$ (fMet-$tRNA^{fMet}$) formation seen in Figure 12.10, a formylation reaction takes place after methionine is bonded to the tRNA. The source of the formyl (–CH=O) group is N^{10}-formyltetrahydrofolate (Section 23-4C). The enzyme that catalyzes the reaction is the aptly named Met-$tRNA^{fMet}$ formyltransferase, and it uses a cofactor called tetrahydrofolate (THF) to transfer a formyl group to the methionine. THF, which is used in a variety of biochemical one-carbon group transfer reactions, is discussed in more detail in Section 23-4C.

$tRNA^{Met}$ is a different tRNA molecule than $tRNA^{fMet}$, and the methionine bound to $tRNA^{Met}$ does not get formylated. Both tRNAs ($tRNA^{Met}$ and $tRNA^{fMet}$) contain the anticodon sequence 3′-UAC-5′, which base-pairs with the codon mRNA sequence 5′-AUG-3′. The $tRNA^{fMet}$ anticodon recognizes the AUG codon, which is the **start codon** when it occurs at the beginning of the mRNA sequence

start codon the mRNA codon that begins the sequence that provides the code for polypeptide synthesis

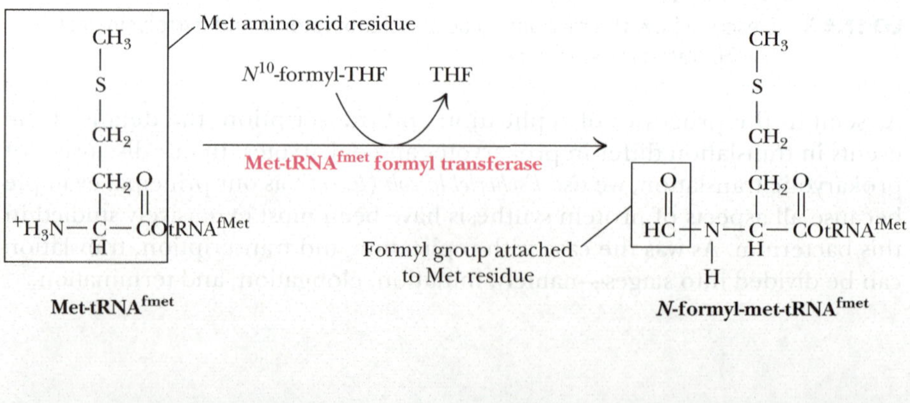

Figure 12.10 Formation of the *N*-formylmethionine-$tRNA^{fMet}$ (fMet-$tRNA^{fMet}$) molecule. Methionine must be bound to $tRNA^{fMet}$ to be formylated. The methionine and modified methionine groups are connected to the tRNA by a 3′ ester connection (Figure 12.7B, *lower right*). THF is a cofactor that helps transfer one-carbon groups; a formyl group is transferred in this reaction.

that directs the synthesis of the polypeptide ("school *starts* in *AUG*ust" may be a helpful mnemonic here). The same 3′-UAC-5′ anticodon in the normal methionine tRNA ($tRNA^{Met}$) recognizes the AUG codon when it is found in an internal position of the mRNA sequence, as opposed to when the AUG codon is at the start position.

The start of polypeptide synthesis requires the formation of an **initiation complex** (Figure 12.11). At least eight components enter into the formation of the initiation complex, including mRNA, the 30S ribosomal subunit, fMet-$tRNA^{fMet}$, guanosine triphosphate (GTP), and three protein initiation factors, called IF-1, IF-2, and IF-3. The IF-3 protein facilitates the binding of

initiation complex (in translation) the aggregate of mRNA, the first aminoacyl-tRNA, ribosomal subunits, and initiation factors needed at the start of protein synthesis

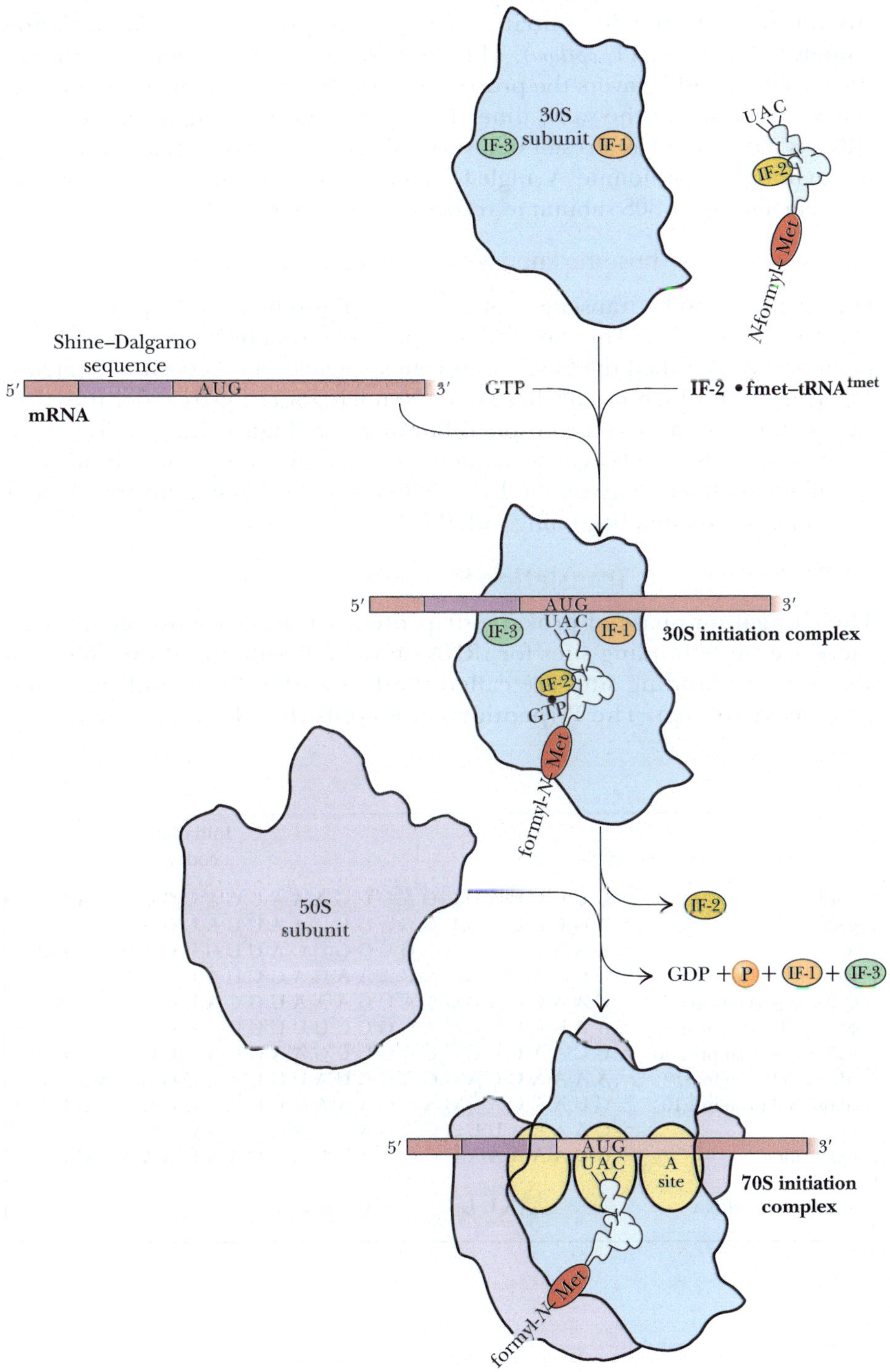

Figure 12.11 The formation of an initiation complex. The 30S ribosomal subunit binds to mRNA and fMet-$tRNA^{fMet}$ in the presence of GTP and the three initiation factors (IF-1, IF-2, and IF-3), forming the 30S initiation complex. The 50S ribosomal subunit is added, forming the 70S initiation complex. The oval in the ribosome that the tRNA is occupying is the P site.

mRNA to the 30S ribosomal subunit. It also appears to prevent the premature binding of the 50S subunit, which takes place in a subsequent step of the initiation process. IF-2 binds GTP and aids in the selection of the initiator tRNA (fMet-$tRNA^{fMet}$) from all the other aminoacylated tRNAs available.

IF-1 binds in the area that eventually becomes the A of the ribosome, preventing another aminoacyl-tRNA from binding to the second mRNA codon until the whole ribosome is assembled. IF-1 also appears to bind to IF-3 and to IF-2, and it facilitates the action of both. It also catalyzes the separation of the 30S and the 50S ribosomal subunits being recycled for another round of translation.

30S initiation complex the combination of mRNA, aminoacyl-tRNA, and 30S ribosomal subunit

70S initiation complex a 30S initiation complex plus a 50S ribosomal subunit

The resulting combination of mRNA, the 30S ribosomal subunit, and fMet-$tRNA^{fMet}$ is the **30S initiation complex** (Figure 12.11, *middle*). A 50S ribosomal subunit binds to the 30S initiation complex to produce the **70S initiation complex** (Figure 12.11, *bottom*). The hydrolysis of GTP to guanosine diphosphate (GDP) and P_i favors the process by providing energy; the initiation factors are released at the same time. The correct positioning of the initiator tRNA is maintained as a result of a small difference between it and the tRNA for an internal methionine. A single C–A mismatched base pair near the acceptor stem allows the 30S subunit to recognize the initiator tRNA.

▶ How does the ribosome know where to start translating?

Shine–Dalgarno sequence a prokaryotic mRNA ribosomal binding site that is located upstream of the start codon; it assists in the process of translation initiation

For the mRNA to be translated correctly, the ribosome must be placed at the correct start location. The start codon is preceded by a purine-rich leader segment of mRNA, called the **Shine–Dalgarno sequence** (5′-GGAGGU-3′) (Figure 12.11, *top left*), which usually lies about 10 nucleotides upstream of the AUG start codon and acts as a ribosomal binding site. Figure 12.12 shows some characteristic Shine–Dalgarno sequences. This purine-rich area binds to a pyrimidine-rich sequence on the 16S rRNA part of the 30S subunit and aligns it for proper translation beginning with the AUG start codon.

12-4C Prokaryotic Translation Elongation

P (peptidyl) site the ribosome binding site on a ribosome for the tRNA that carries the growing peptide chain

The elongation phase of prokaryotic protein synthesis is possible because there are three binding sites for tRNA on the 50S subunit of the 70S ribosome. These binding sites are called the E site, the P site, and the A site (Figure 12.13, *top*). The **P (peptidyl) site** binds the tRNA that carries the

	Sequence (Initiation codon highlighted: AUG/GUG)
araB	– U U U G G A U G G A G U G A A A C G A U G G C G A U U –
galE	– A G C C U A A U G G A G C G A A U U A U G A G A G U U –
lacI	– C A A U U C A G G G U G G U G A U U G U G A A A C C A –
lacZ	– U U C A C A C A G G A A A C A G C U A U G A C C A U G –
Q β phage replicase	– U A A C U A A G G A U G A A A U G C A U G U C U A A G –
ϕX174 phage A protein	– A A U C U U G G A G G C U U U U U U A U G G U U C G U –
R17 phage coat protein	– U C A A C C G G G G U U U G A A G C A U G G C U U C U –
ribosomal protein S12	– A A A A C C A G G A G C U A U U U A A U G G C A A C A –
ribosomal protein L10	– C U A C C A G G A G C A A A G C U A A U G G C U U U A –
trpE	– C A A A A U U A G A G A A U A A C A A U G C A A A C A –
trpL leader	– G U A A A A A G G G U A U C G A C A A U G A A A G C A –
3′-end of 16S rRNA	3′ HO A U U C C U C C A C U A G – 5′

Figure 12.12 Various Shine–Dalgarno mRNA sequences recognized by *E. coli* ribosomes. These sequences lie about 10 nucleotides upstream from their respective AUG initiation codon and are complementary to the UCCU core sequence element of *E. coli* 16S rRNA (bottom). G:U as well as canonical G:C and A:U base pairs are involved here.

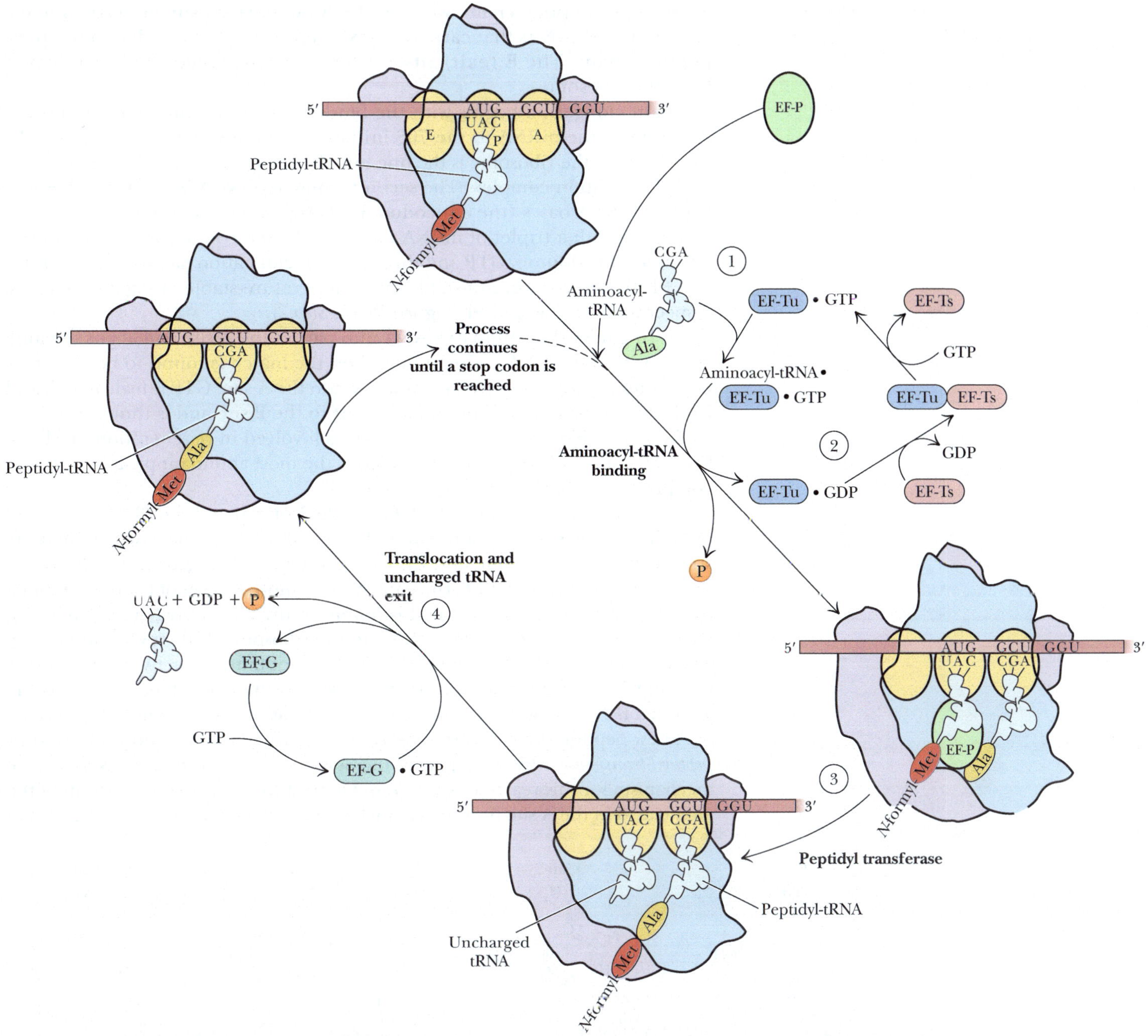

Figure 12.13 A summary of the steps in the elongation phase of prokaryotic translation. Step 1: An aminoacyl-tRNA binds to the A site on the ribosome. Elongation factor EF-Tu (Tu) and GTP are required. The P site on the ribosome is already occupied. Step 2: Elongation factor EF-Tu is released from the ribosome and regenerated in a process requiring elongation factor EF-Ts (Ts) and GTP. Step 3: The polypeptide chain is transferred to the A site amino acid residue, forming a new covalent bond and extending the length of the polypeptide chain by one amino acid, this also leaves an uncharged tRNA at the P site. Step 4: In the translocation step, the ribosome moves one codon down the mRNA, which moves the peptidyl-tRNA in the A site to the P site, leaving an empty A site for the next aminoacyl-tRNA to bind. The uncharged tRNA is moved to the E site and subsequently released. Elongation factor EF-G and GTP are required.

A (aminoacyl) site the ribosome binding site for the aminoacyl-tRNA that is carrying the next amino acid to be added to the growing peptide chain

E (exit) site the ribosome site where an uncharged tRNA is released from the ribosome

uncharged tRNA a tRNA molecule that is not covalently bonded to an amino acid

growing polypeptide chain, whereas the **A (aminoacyl) site** binds the incoming aminoacyl-tRNAs that carry the next amino acid to be added to the polypeptide chain. The **E (exit) site** is where an **uncharged tRNA** is released from the ribosome.

Chain elongation begins with the binding of the second aminoacyl-tRNA specified by the mRNA to the 70S initiation complex (Figure 12.13, *Step 1*). The P site on the ribosome is the one initially occupied by the fMet-$tRNA^{fMet}$ in the 70S initiation complex. The second aminoacyl-tRNA binds at the A site. A triplet of tRNA bases (the anticodon 3′-CGA-5′ in our example) forms hydrogen bonds with a triplet of mRNA bases (GCU; the codon for alanine, in this example). In addition, GTP and three protein elongation factors—EF-P, EF-Tu, and EF-Ts (temperature-unstable and temperature-stable elongation factors, respectively)—are required (Figure 12.13, *Step 2*).

EF-Tu guides the aminoacyl-tRNA into part of the A site and aligns the anticodon with the mRNA codon. Only when the match is found to be correct is the aminoacyl-tRNA inserted completely into the A site. GTP is hydrolyzed, and EF-Tu dissociates. EF-P is bound adjacent to the P site and is thought to help facilitate peptide bond formation. EF-Ts is involved in regeneration of EF-Tu-GTP. This small EF-Tu protein (43 kDa) is the most abundant protein in *E. coli,* comprising 5% of the dry weight of the cell.

The exact mechanism by which GTP is hydrolyzed and the tRNA is released by EF-Tu is an active area of research. In late 2009, researchers determined the structure of the ribosome and EF-Tu complex to 3.6-angstrom resolution. It has also been shown that EF-Tu is involved in another level of translation fidelity. When the correct amino acid is bound to the correct tRNA, EF-Tu is efficient at delivering the activated tRNA to the ribosome. If the tRNA and amino acid are mismatched, then either EF-Tu does not bind the activated tRNA very well, in which case it does not deliver it well to the ribosome, or it binds the activated tRNA too well, in which case it does not release it from the ribosome.

A new peptide bond (Section 3-4) is then formed in a reaction catalyzed by *peptidyl transferase,* which is a part of the 50S subunit (Figure 12.13, *Step 3*). The mechanism for this reaction is shown in Figure 12.14. The α-amino group of the amino acid in the A site performs a nucleophilic attack on the carbonyl group of

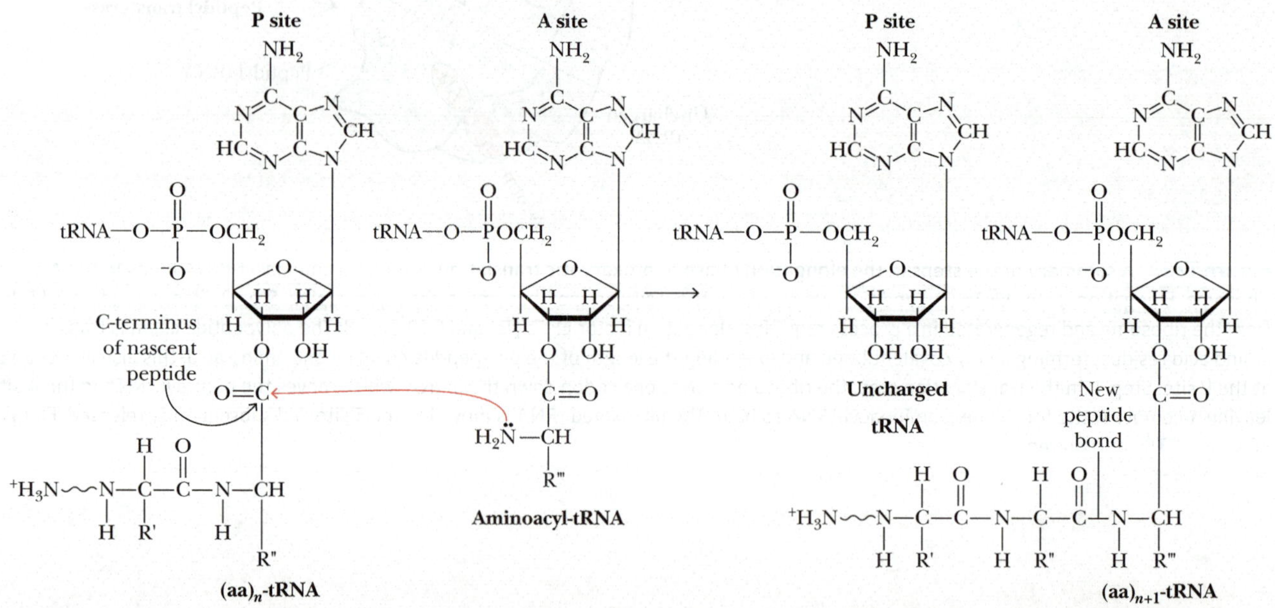

Figure 12.14 The peptidyl transferase mechanism and peptide bond formation in protein synthesis. A nucleophilic attack by the α-amino group of the amino acid at the A site on the carbonyl carbon of the last amino acid of the growing polypeptide chain at the P site produces a new peptide bond and extends the growing polypeptide chain by one amino acid. $(aa)_n$ refers to a polypeptide that is *n* amino acids long.

the last amino acid of the growing polypeptide chain that is linked to the tRNA in the P site. The growing polypeptide is now at the A site, and it is one amino acid longer. An uncharged tRNA with no amino acid attached is now at the P site.

A **translocation** step then takes place before another amino acid can be added to the growing chain (Step 4 in Figure 12.13). In the process, the uncharged tRNA moves from the P site to the E site, from which it is subsequently released; the peptidyl-tRNA moves from the A site to the vacated P site. In addition, the mRNA moves with respect to the ribosome. Another elongation factor, EF-G—also a protein—is required at this point, and, once again, GTP is hydrolyzed to GDP and P_i.

translocation in protein synthesis, the motion of the ribosome along the mRNA as the genetic message is being read

The three steps of the chain elongation process are aminoacyl-tRNA binding, peptide bond formation, and translocation (Steps 1, 3, and 4 in Figure 12.13). They are repeated for each amino acid specified by the genetic message of the mRNA until the stop signal is reached. Step 2 in Figure 12.13 shows the regeneration of EF-Tu-GTP.

12-4D Prokaryotic Termination of Translation

A stop signal is required for the termination of protein synthesis. The codons UAA, UAG, and UGA are the stop signals (Table 12.1). These codons are not recognized by any tRNAs, but they are recognized by proteins called release factors (RFs) (Figure 12.15). One of two protein RFs (RF-1 or RF-2) is required, as is GTP, which is bound to a third RF, RF-3. RF-1 binds to UAA and UAG, and RF-2 binds to UAA and UGA. RF-3 does not bind to any codon, but it does facilitate the activity of the other two RFs. Either RF-1 or RF-2 is bound near the A site of the ribosome when one of the termination codons is reached. Studies have shown that a conserved sequence of Pro-X-Thr in RF1 and Ser-Pro-Phe in RF2 controls their codon specificities. RF binding not only blocks the binding of a new aminoacyl-tRNA but also affects the activity of the peptidyl transferase so that the bond between the carboxyl end of the peptide and the tRNA is hydrolyzed. A conserved sequence of Gly–Gly–Gln is essential for the hydrolysis reaction of the RF. Then GTP is hydrolyzed, and the whole complex dissociates, setting free the RFs, tRNA, mRNA, and the 30S and 50S ribosomal subunits. All these components can be reused in further protein synthesis. Table 12.3 summarizes the steps in protein synthesis and the components required for each step.

12-4E The Rare 21st and 22nd Amino Acids in the Genetic Code

There are many amino acids, such as citrulline and ornithine (Section 23-6C), that are *not* building blocks of proteins. Other nonstandard amino acids, such as hydroxyproline (Section 3-2E), are formed after translation by posttranslational modification. When discussing amino acids and translation, the magic number used to be 20; that is, the 20 standard amino acids and only the standard 20 were put onto tRNA molecules for protein synthesis. In the late 1980s, however, another amino acid was found in proteins from eukaryotes and prokaryotes alike, including humans, and it was given the name selenocysteine (Sec). It was designated as the 21st **proteinogenic amino acid**. Sec is a cysteine residue (Figure 3.3B) in which the sulfur atom has been replaced by a selenium atom (Se), as shown in Figure 12.16A. Its importance in life was demonstrated when strains of mice lacking the ability to make Sec were not viable. It has been found in the active sites of enzymes involved in clearing reactive oxidative molecules during thyroid hormone activation.

proteinogenic amino acid an amino acid that can be used by cellular translation machinery in protein synthesis

The codon for Sec is UGA, which normally is a translation stop codon. During translation of an mRNA that produces a Sec-bearing protein, the UGA is interpreted differently due to interactions with the specialized elongation factors called SelB in bacteria and EFsec in humans. The first step in Sec synthesis

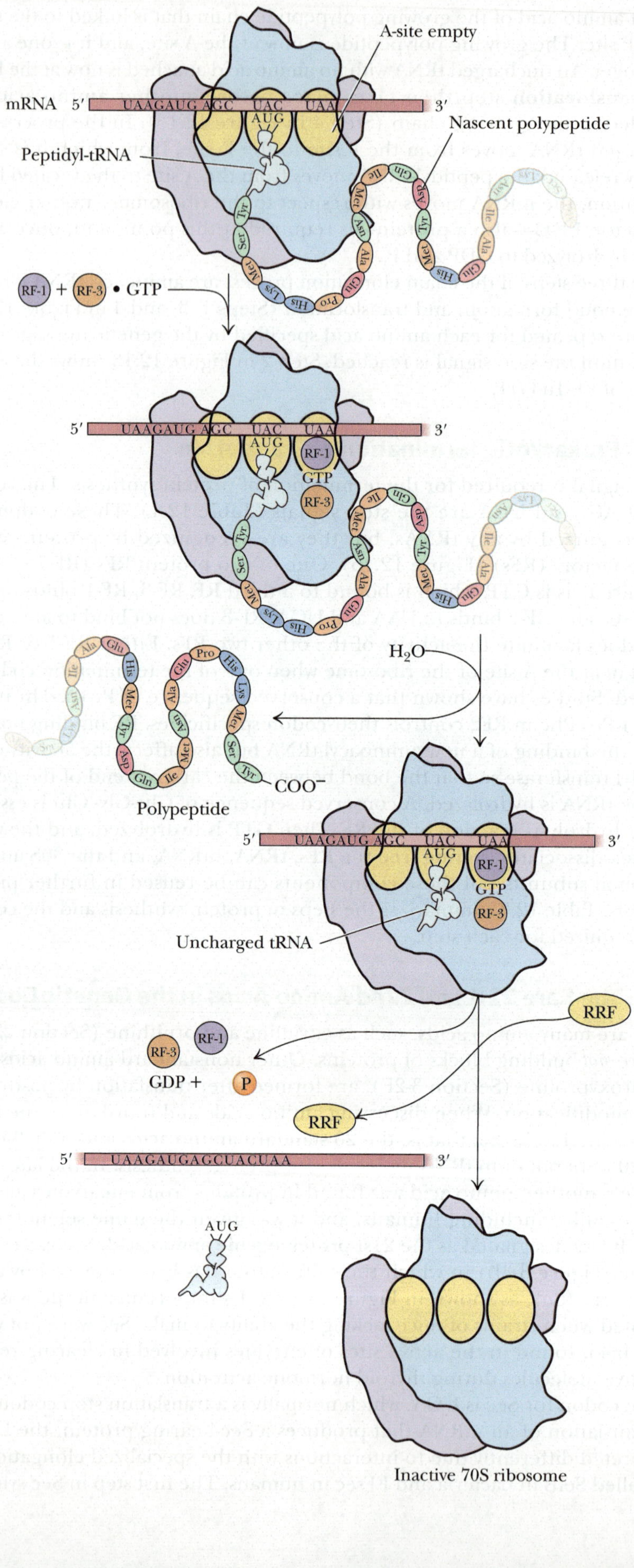

Figure 12.15 The events in peptide chain termination. When a stop codon is encountered on the mRNA, release factor (RF) proteins bind and catalyze the termination of translation. A ribosome release factor (RRF) also plays a role in the final step of translation termination.

Table 12.3 Components Required for Each Step of Protein Synthesis in *Escherichia coli*

Step	Components
Amino acid activation	Amino acids
	tRNAs
	Aminoacyl-tRNA synthetases
	ATP, Mg^{2+}
Chain initiation	fMet-$tRNA^{fMet}$
	Shine–Dalgarno sequence and AUG start codon on mRNA
	30S ribosomal subunit
	50S ribosomal subunit
	Initiation factors (IF-1, IF-2, and IF-3)
	GTP, Mg^{2+}
Chain elongation	70S ribosome
	mRNA codons
	Aminoacyl-tRNAs
	Elongation factors (EF-Tu, EF-Ts, EF-P, and EF-G)
	GTP, Mg^{2+}
Chain termination	70S ribosome
	Termination codon (UAA, UAG, or UGA) on mRNA
	Release factors (RF-1, RF-2, and RF-3)
	GTP, Mg^{2+}

is the acylation of $tRNA^{Sec}$ by seryl-tRNA synthetase, the synthetase that normally uses serine, to yield Ser-$tRNA^{Sec}$. (Recall from Section 12-3 that the prefix indicates the attached amino acid and the superscript identifies the type of tRNA molecule.) The serine residue bound to the $tRNA^{Sec}$ molecule is then converted to Sec by a series of reactions, producing Sec-$tRNA^{Sec}$, which is recognized by special translation elongation factors. Then the β-hydroxyl group of Ser-$tRNA^{Sec}$ is phosphorylated by *O*-phosphoseryl-tRNA kinase to give *O*-phosphoseryl-$tRNA^{Sec}$, which, in turn, gives the substrate used for the final reaction that catalyzes the conversion of the phosphoseryl group into the selenocysteinyl group.

Pyrrolysine (Pyl) is another rare amino acid (the 22nd proteinogenic amino acid) that is used in the biosynthesis of some bacterial proteins (it is not found in humans). Figure 12.16B shows the structure of Pyl. It contains a pyrrole ring (Figure 4.16, *left*) that is linked to the side-chain amino group of lysine via a carbonyl group. The pyrrole ring of Pyl also has a methyl group substituent. The codon for Pyl is UAG, which is also normally a stop codon. However, this stop codon can be translated as Pyl in certain bacterial genes.

12-4F The Ribosome Is a Ribozyme

In 2000, the complete structure of the large ribosomal subunit was determined by X-ray crystallography to 2.4-Å (0.24-nm) resolution (Figure 12.17). Ribosomes had been studied for 40 years, but the complete structure had been elusive. When the active sites for peptidyl transferase were looked at, it turned out that there is no protein in the vicinity of the new peptide bond, suggesting this was another example of an RNA molecule having catalytic ability. For example, the inset of Figure 12.17 shows another view of the peptidyl transferase mechanism shown first in Figure 12.14. One additional feature of the mechanism seen in Figure 12.17 is the role of an rRNA adenine

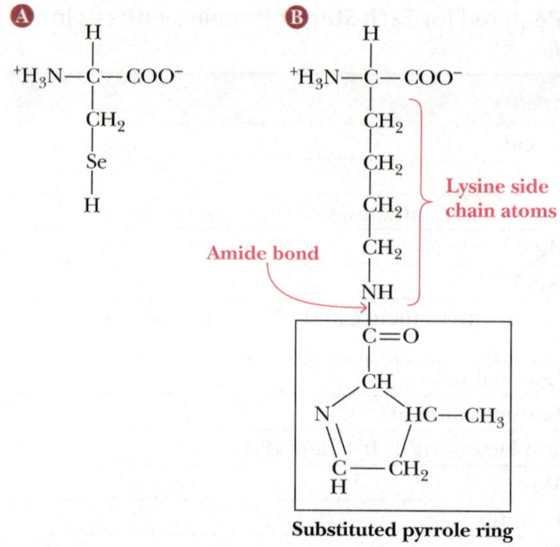

Figure 12.16 The structures of the rare proteinogenic amino acids. (A) Selenocysteine and (B) pyrrolysine (Pyl). In Pyl, the carboxyl substituent (–COO⁻) of the substituted pyrrole ring forms an amide bond with the side-chain amino group ($-NH_3^+$) of lysine.

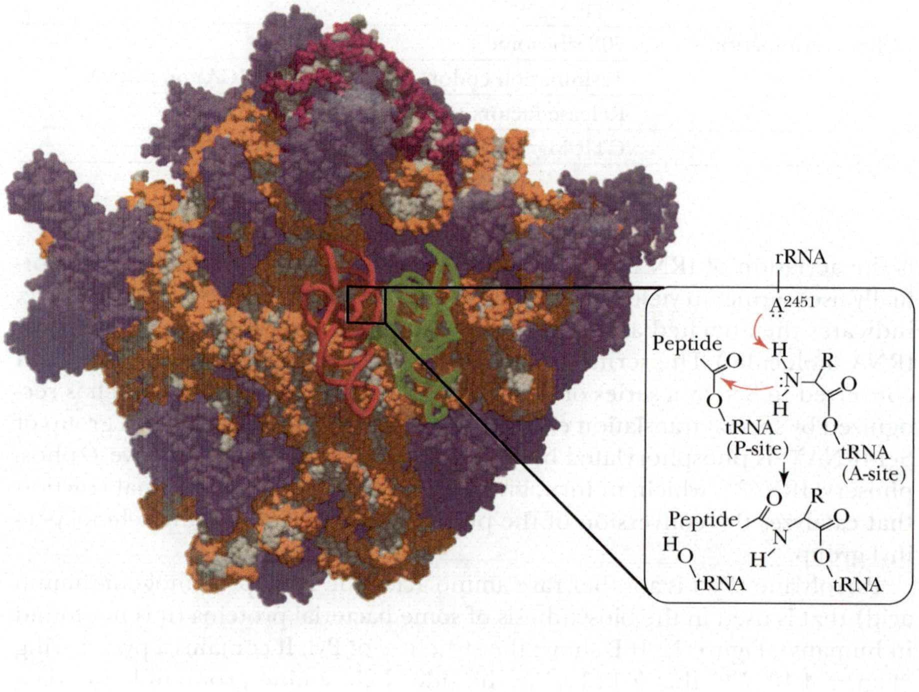

Figure 12.17 Ribosome structure. The large subunit of the ribosome seen from the viewpoint of the small subunit, with proteins in purple, 23S rRNA in orange and gray, 5S rRNA (at the top) in burgundy and gray, and A-site tRNA (green) and P-site tRNA (red) docked. In the box, the peptidyl transfer mechanism is catalyzed by RNA. The general base (adenine 2451 in *E. coli* 23S rRNA) is rendered unusually basic by its environment within the folded structure; it could abstract a proton at any of several steps, one of which is shown here. *(Republished with permission of American Association for the Advancement of Science. Used with permission from Cech, T. (2000). The ribosome is a ribozyme.* Science *289, 878. Permission conveyed through Copyright Clearance Center, Inc.)*

moiety (A^{2451}) in abstracting (removing) a proton from the α-amino group of the amino acid in the A site, which helps the amino group perform the nucleophilic attack.

12-4G Polysomes

In our description of protein synthesis, we have considered, up to now, the reactions that take place at one ribosome. It is, however, not only possible, but quite usual, for several ribosomes to be attached to the same mRNA. Each of these ribosomes bears a polypeptide in one of various stages of completion, depending on the position of the ribosome as it moves along the mRNA (Figure 12.18). This complex of an mRNA molecule with several ribosomes bound to it is called a **polysome**; an alternative name is *polyribosome.* In prokaryotes, translation

polysome the assemblage of several ribosomes bound to one mRNA

begins very soon after mRNA transcription. It is possible for a molecule of mRNA that is still being transcribed to have a number of ribosomes attached to it that are in various stages of translating that mRNA. It is also possible for DNA to be in various stages of being transcribed. In this situation, several molecules of RNA polymerase are attached to a single gene, giving rise to several mRNA molecules, each of which has a number of ribosomes attached to it. Therefore, prokaryotic genes can be simultaneously transcribed and translated. This process, which is called *coupled translation* (Figure 12.19), is possible in prokaryotes

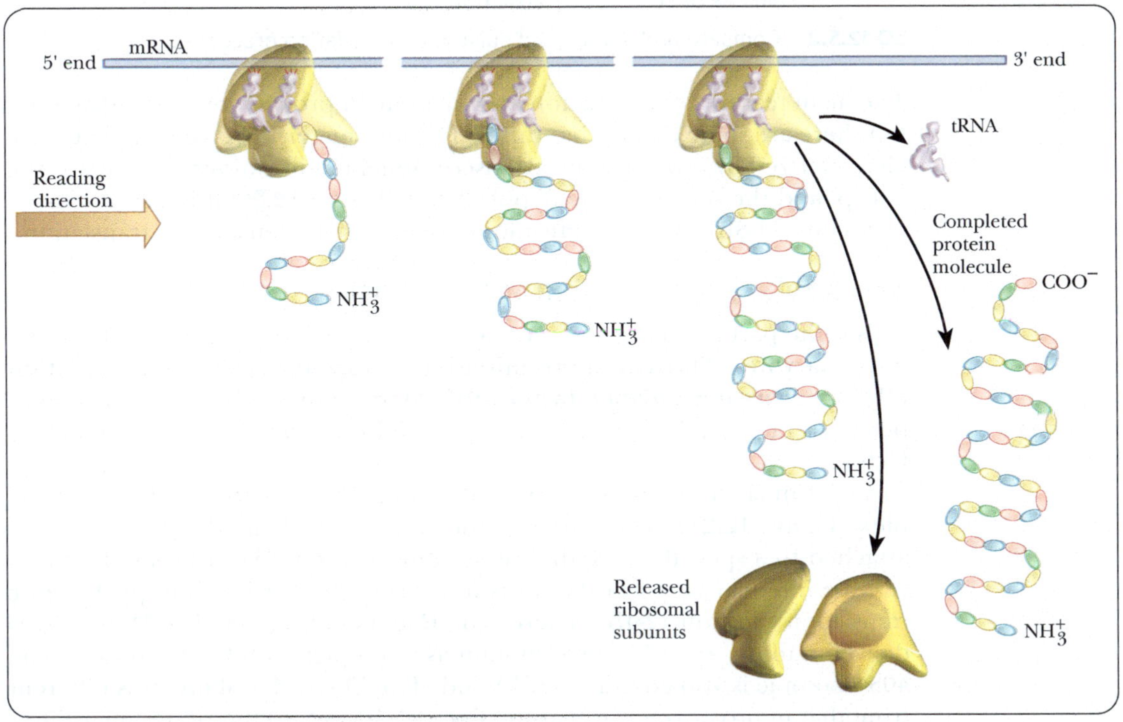

Figure 12.18 Simultaneous protein synthesis on polysomes. A single mRNA molecule is translated by several ribosomes simultaneously. Each ribosome produces one copy of the polypeptide chain specified by the mRNA. When the protein has been completed, the ribosome dissociates into subunits that are used in further rounds of protein synthesis.

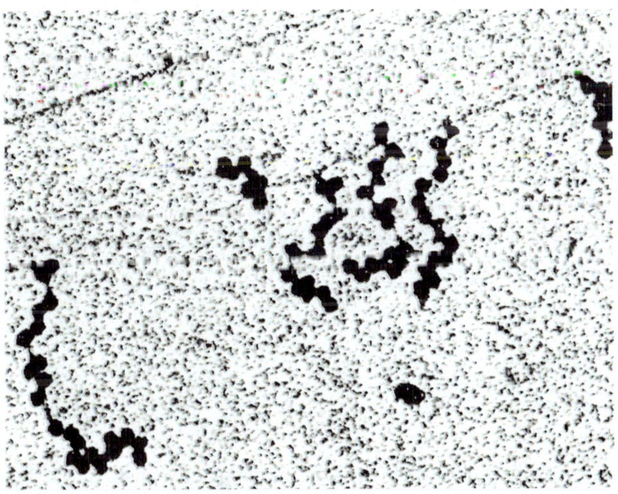

Figure 12.19 Electron micrograph showing coupled translation. The dark spots are ribosomes, arranged in clusters on a strand of mRNA. Several mRNAs have been transcribed from one strand of DNA (diagonal line from center left to upper right). *From Visualization of Bacterial Genes in Action by O. L. Miller, Jr., B.A. Hamkalo, and C. A. Thomas, Jr. (24 July 1970) Science 169 (3943), 392. Used with permission of AAAS.*

because of the lack of cell compartmentalization. In eukaryotes, mRNA is produced in the nucleus, whereas most protein synthesis takes place in the cytosol, as discussed next in Section 12-5.

12-5 Eukaryotic Translation

Learning Objectives

After reading this section, students will be able to:

LO 12.5.1 Describe events that happen in the initiation, elongation, and termination phases of eukaryotic translation.

LO 12.5.2 Compare prokaryotic and eukaryotic translation processes.

The main features of translation are the same in prokaryotes and eukaryotes, but there are some details that differ. The messenger RNAs of eukaryotes are characterized by two major posttranscriptional modifications. The first is the 5′ cap, and the second is the 3′ poly A tail (Figure 12.20, it is also discussed in Section 11-8B). Both modifications are essential to eukaryotic translation.

12-5A Eukaryotic Initiation of Translation

eukaryotic initiation factor (eIF) a protein involved in the initiation of translation in eukaryotes

This is the part of eukaryotic translation that is the most different from that in prokaryotes. Thirteen more initiation factors are given the designation eIF, for **eukaryotic initiation factor (eIF)**. Many of these eIFs are multisubunit proteins. Table 12.4 summarizes pertinent information about these initiation factors.

Step 1 in chain initiation involves the assembly of a 43S preinitiation complex (Figure 12.21). The initial amino acid is usually methionine, which is attached to a special $tRNA_i$ that serves only as the initiator tRNA. This first residue in eukaryotic translation is not formylated, as it was in prokaryotic translation (Section 12-4B). There is no fMet in eukaryotes. The Met-$tRNA_i$ is delivered to the 40S ribosomal subunit as a complex with GTP and eIF2. The 40S ribosome is also bound to eIF1A and eIF3. This order of events is different from that in prokaryotes in that the first tRNA binds to the ribosome without the presence of the mRNA.

In Step 2, the mRNA is recruited. There is no Shine–Dalgarno sequence to locate the start codon. The 5′ cap orients the ribosome to the correct AUG via what is called a *scanning mechanism,* which is driven by ATP hydrolysis.

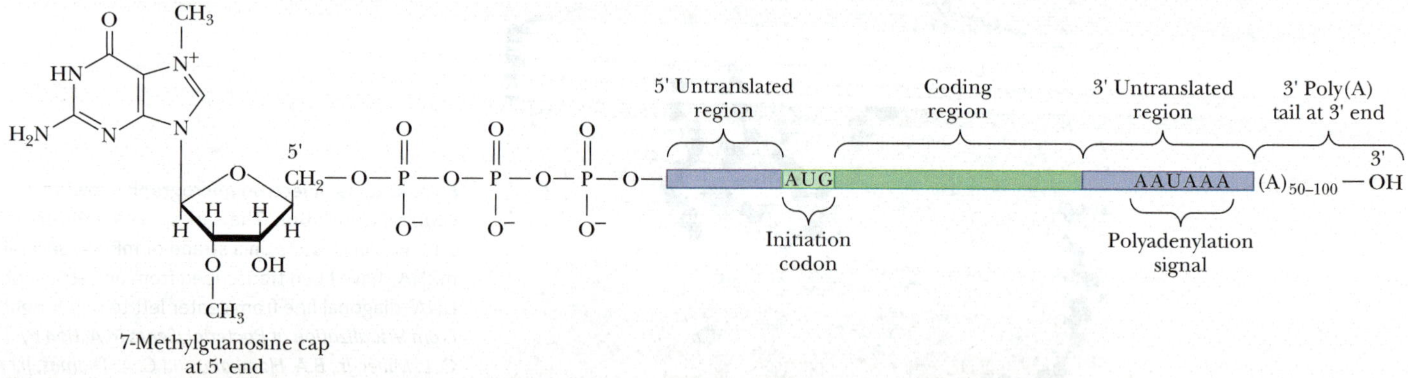

Figure 12.20 The characteristic structure of eukaryotic mRNAs. Untranslated regions ranging between 40 and 150 bases in length occur at both the 5′ and 3′ ends of the mature mRNA. An initiation codon at the 5′ end, invariably AUG, signals the translation start site.

Table 12.4 Properties of Eukaryotic Translation Initiation Factors

Factor	Subunit	Size (kDa)	Function
eIF1		15	Enhances initiation complex formation
eIF1A		17	Stabilizes Met-tRNA_i binding of 40S ribosomes
eIF2		125	GTP-dependent Met-tRNA_i binding to 40S ribosomes
	α	36	Regulated by phosphorylation
	β	50	Binds mRNA and Met-tRNA_i
	γ	55	Binds GTP, Met-tRNA_i
eIF2B		270	Promotes guanine nucleotide exchange on eIF2
	α	26	Binds GTP
	β	39	Binds ATP
	γ	58	Binds ATP
	δ	67	Regulated by phosphorylation
	ε	82	
eIF2C		94	Stabilizes ternary complex in the presence of RNA
eIF3		800	13-subunit complex that promotes Met-tRNA_i and mRNA binding and nearly all other steps of translation initiation
eIF4A		46	Binds RNA; ATPase; RNA helicase; promotes mRNA binding to 40S ribosomes
eIF4B		80	Binds mRNA; promotes RNA helicase activity and mRNA binding to 40S ribosomes
eIF4E		25	Promotes ribosome binding to mRNA caps
eIF4F			Complex binds to mRNA caps; RNA helicase activity; promotes mRNA binding to 40S
eIF4G		153.4	Component of the eIF4F cap-binding complex
eIF5		48.9	Promotes hydrolysis of GTP and recruitment of 60S ribosome
eIF6			Promotes association of 40S and 60S ribosomes

Adapted from Clark, B. F. C., et al., eds. (1996). Prokaryotic and eukaryotic translation factors. *Biochimie 78*, 1119–1122.

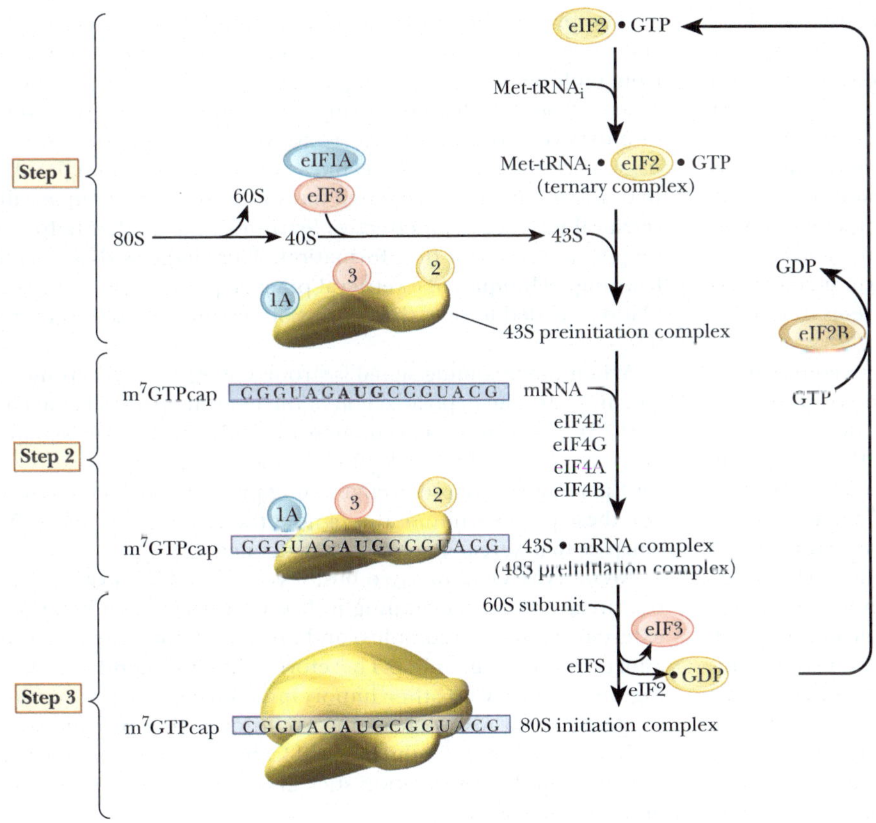

Figure 12.21 The three stages in the initiation of translation in eukaryotic cells. See Table 12.4 for a description of the functions of the eukaryotic initiation factors (eIFs).

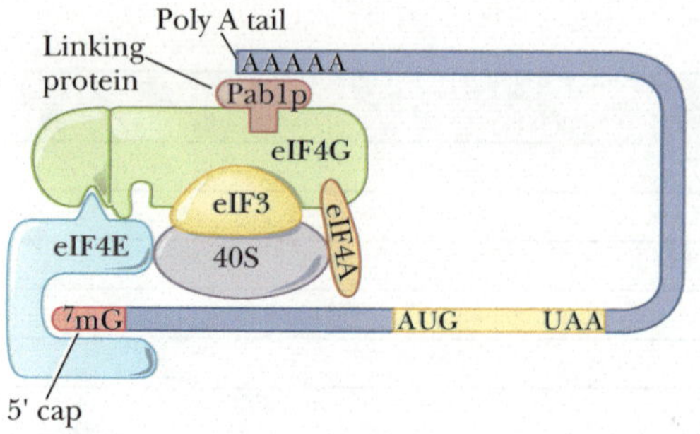

Figure 12.22 mRNA looping in eukaryotic translation initiation. Initiation factor eIF4G serves as a multipurpose adapter to engage the ^{7}methyl-G cap:eIF4E complex, the Pab1p:poly(A) tract, and the 40S ribosomal subunit in eukaryotic translation initiation. *(Adapted from Heutze, H. W. (1997). eIF4G: A multipurpose ribosome adapter?* Science *275, 500–501. Reprinted with permission from AAAS.)*

Kozak sequence the base sequence surrounding a eukaryotic start codon that helps the ribosome recognize where to start translation

eIF4E is also a cap-binding protein, which forms a complex with several other eIFs. A *poly A binding protein (Pab1p)* links the poly A tail to eIF4G. The eIF-40S complex is initially positioned upstream of the start codon (Figure 12.22). It moves downstream until it encounters the first AUG in the correct context. The context is determined by a few bases surrounding the start codon, called the **Kozak sequence.** It is characterized by the consensus sequence $_{-3}$ACC**AUG**G$_{+4}$. The ribosome may skip the first AUG it finds if the next one has the Kozak sequence, although the AUG closest to the 5′ end of the mRNA is usually the start codon. Another factor is the presence of mRNA secondary structure. If hairpin loops form downstream of an AUG, an earlier AUG may be chosen. The mRNA and the seven eIFs constitute the 48S preinitiation complex.

In Step 3, the 60S ribosome is recruited, forming the 80S initiation complex. GTP is hydrolyzed, and the initiation factors are released. **Biochemical Connections 12B** describes important links between translation initiation factors, protein synthesis, and memory.

12B Biochemical Connections • Neurology

Protein Synthesis Makes Memories

Memories are either short term or long term. Short-term memories last from seconds to minutes, while long-term memories last days, months, or even a lifetime. Neuroscientists have long been fascinated by what makes one memory stick while another does not. When you meet someone at a party and they tell you their name, you might forget it in seconds. However, your best friend's name is converted to long-term memory and might last your whole life. While there are tremendous individual variations in memory capability, one thing that is known for sure is that making long-term memories relies on new protein synthesis. Animals given drugs that block protein synthesis cannot form new long-term memories, yet their ability to make short-term memories is preserved. Therefore, making information stick seems to require making new brain proteins.

Both long- and short-term memories come from connections between neurons, called synapses, where one neuron emits a signal from its axon that is received by the next neuron's dendrite. Longer-term memories are made when the synapse is made stronger, or "sensitized" to further signals. In order for this strengthening to occur, genes in the nucleus of the neuron must be activated and proteins produced. The central paradox for neuroscientists studying memory has always been this: How does a gene activated in a neuron's nucleus *know* when to strengthen the synapse permanently? In searching for that answer, they set off to find out what proteins were involved in the process. By the mid-1990s, researchers had determined that a critical role is played by the cyclic AMP response element binding protein (CREB) transcription factor (Section 11-5D) in turning short-term memories into long-term ones. Researchers are searching for other proteins, especially ones that specifically strengthen the synapse. The entire process can be envisioned as shown in Figure 12.23.

The process begins when strong stimulation depolarizes the cell membrane of the nerve (1). This stimulation could come from multiple firings involving the action potentials of a single synapse or from simultaneous firings from multiple synapses. Not all impulses received cause a nerve to fire its own impulse, thereby passing on the impulse. Only ones that are strong enough will do that. Learning science research has shown that a technique called retrieval practice is an effective learning strategy that helps with long-term retention of information. This suggests that effective learning techniques like retrieval practice provide the strong stimulation needed to start the process of making synapse-strengthening proteins.

When the incoming signal is strong enough, the receiving neuron fires (2). The depolarization of the neuron opens calcium channels (3). Calcium enters the neuron and activates key signaling enzymes, which activate CREB (4). CREB activates gene expression for the putative synapse-strengthening proteins (5). In the last step (6), these proteins diffuse throughout the cell but only affect those synapses that were temporarily strengthened.

Memory is a complicated business, and we have just scratched the surface of understanding it. It is clear that a combination of external factors (the stimulus) and internal factors (the chemistry) affects how we remember. Therefore, when we spend time learning biochemistry or other important subjects, research strongly suggests that if we use learning techniques that provide strong brain stimulation, then the internal factors that happen biochemically to help the information stick are several processes involving protein synthesis.

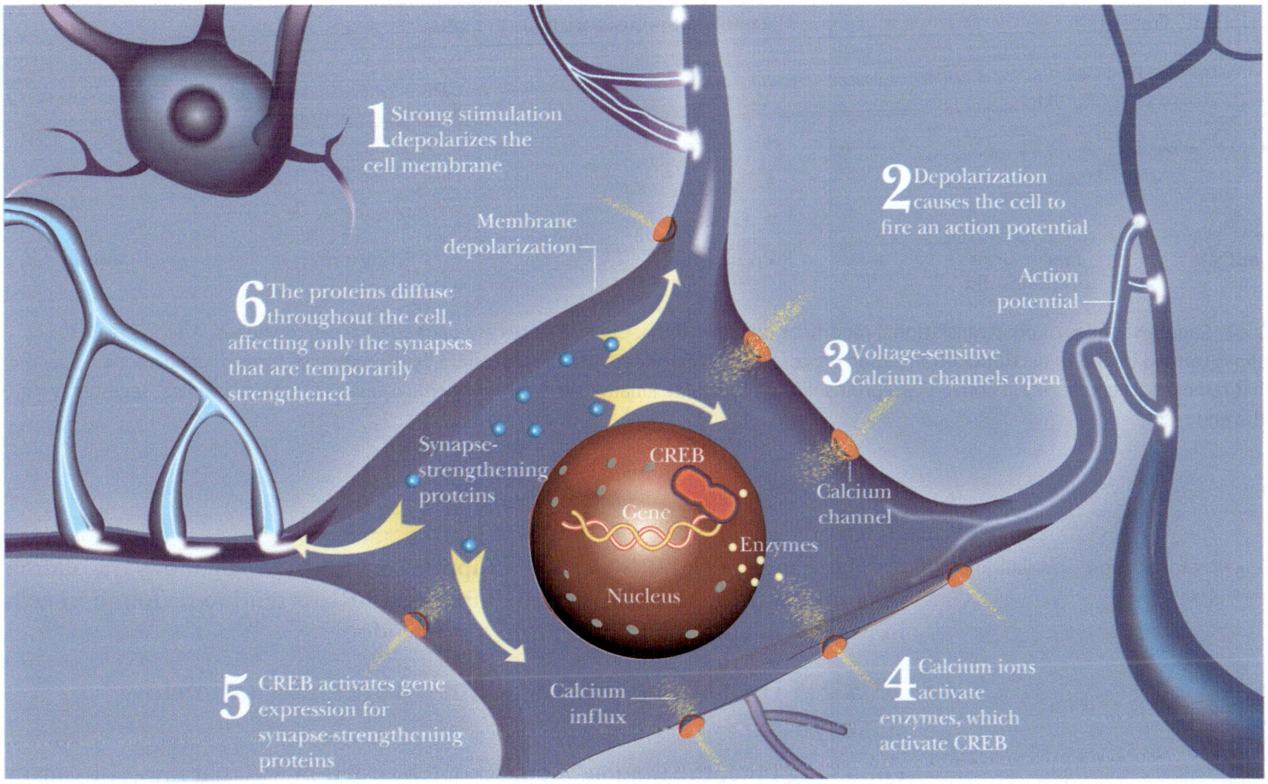

Figure 12.23 **Protein synthesis makes memories.** *(From Making Memories Stick by R. Douglas Fields,* Scientific American, *February 2005, page* 80.*)*

12-5B Eukaryotic Translation Elongation

Polypeptide chain elongation in eukaryotes is very similar to that of prokaryotes, as seen in Figure 12.13. The same mechanism of peptidyl transferase and ribosome translocation is observed. The structure of the eukaryotic ribosome is different because there is no E site—only the A and P sites. There are two eukaryotic elongation factors: eEF1 and eEF2. The eEF1 consists of two subunits, eEF1A and eEF1B. The 1A subunit is the counterpart of EF-Tu in prokaryotes. The 1B subunit is the equivalent of the EF-Ts in prokaryotes. The eEF2 protein is the counterpart of the prokaryotic EF-G, which causes translocation.

Many of the differences between translation in prokaryotes and eukaryotes can be seen in response to inhibitors of protein synthesis and to toxins. The antibiotic chloramphenicol (a trade name is Chloromycetin) binds to the A site and inhibits peptidyl transferase activity in prokaryotes but not in eukaryotes. This property has made chloramphenicol useful in treating bacterial infections. In eukaryotes, diphtheria toxin is a protein that interferes with protein synthesis by decreasing the activity of the eukaryotic elongation factor eEF2.

12-5C Eukaryotic Termination of Translation

As in prokaryotic termination, the ribosome encounters a stop codon—UAG, UAA, or UGA, which are not recognized by a tRNA molecule. In prokaryotes, three different RFs—RF1, RF2, and RF3—were used, with two of them alternating, depending on which stop codon was found (Section 12-4D). In eukaryotes, only one RF binds to all three stop codons and catalyzes the hydrolysis of the bond between the C-terminal amino acid and the tRNA.

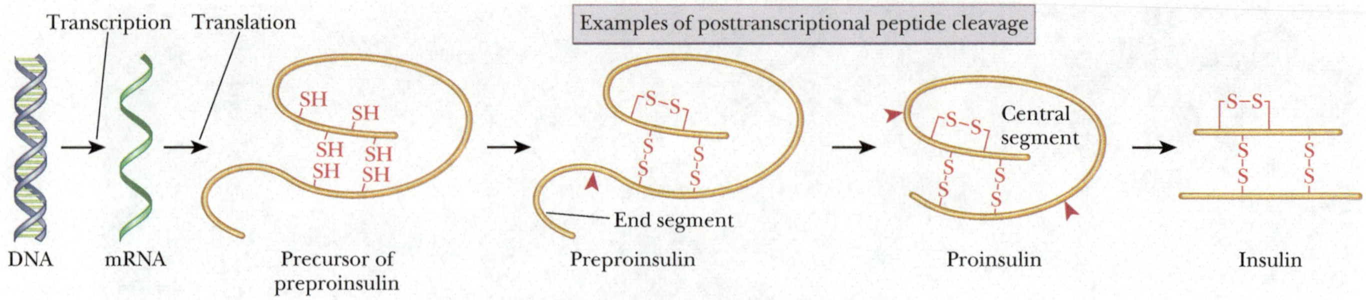

Figure 12.24 Some examples of posttranslational modification of proteins. Cleavage sites are marked by red arrowheads. After a precursor of preproinsulin is formed by the transcription–translation process, it is transformed into preproinsulin by formation of three disulfide bonds. Specific cleavage that removes an end segment converts preproinsulin to proinsulin. Finally, two further specific cleavages remove a central segment, with insulin as the end result.

suppressor tRNA a tRNA that allows translation to continue through a stop codon

There is a special tRNA called a **suppressor tRNA**, which allows translation to continue through a stop codon. Suppressor tRNAs tend to be found in cells in which a mutation has introduced a stop codon.

12-5D Is There Coupled Transcription and Translation in Eukaryotes?

For a long time, the dogma of eukaryotic translation was that it was physically separated from transcription. Transcription occurred in the nucleus, and mRNA was then exported to the cytosol for translation. Although this system is accepted as the normal process, evidence has been found that the nucleus has all of the components (mRNA, ribosomes, and protein factors) necessary for translation. In addition, evidence shows that, in isolated test systems, proteins are translated in the nucleus. Some researchers have suggested that 10% to 15% of the cell's protein synthesis occurs in the nucleus.

12-6 Posttranslational Modification of Proteins

Learning Objectives

After reading this section, students will be able to:

LO 12.6.1 Describe several ways that proteins can be modified after they are synthesized.

LO 12.6.2 Explain the role that the ribosome can play in protein folding.

Newly synthesized polypeptides are frequently processed before they reach the form in which they have biological activity. Recall the discussion of zymogens (Section 7-4) and how, in prokaryotes, *N*-formylmethionine is cleaved off after translation (Section 12-4B). Specific bonds in precursors can be hydrolyzed, as in the cleavage of preproinsulin to proinsulin and of proinsulin to insulin (Figure 12.24). Proteins destined for export to specific parts of the cell or from the cell have leader sequences at their N-terminal ends. These leader sequences, which direct the proteins to their proper destination, are recognized and removed by specific PAs associated with the *endoplasmic reticulum.* The finished protein then enters the *Golgi apparatus,* which directs it to its final destination.

In addition to the processing of proteins by breaking bonds, other substances can be linked to the newly formed polypeptide. Various cofactors, such as heme groups, can be added, and some amino acid residues can be covalently modified, as in the conversion of proline to hydroxyproline. Other covalent modifications can take place, such as the addition of carbohydrates or lipids to yield an active final form of the protein in question. Proteins can also be methylated, phosphorylated, arginylated, and ubiquitinylated (Section 12-7).

▶ Once translated, do proteins always have the correct three-dimensional structure?

A highly important question concerns the proper folding of the newly synthesized protein. In principle, the primary structure of the protein conveys enough information to specify its three-dimensional structure. In the cell, the complexity of the process and the number of possible conformations make it less likely that a protein would spontaneously fold into the correct conformation.

Besides decoding the genetic information and synthesizing the nascent peptides, research also indicates that the ribosomes themselves are involved in protein folding. For example, T4 lysozyme synthesis was studied on a single ribosome, and the folding of the peptide was compared to how it would fold in a translation-free system. While translation efficiency from a ribosome has many variables, such as tRNA abundance, codon order, and mRNA secondary structure, it now appears that the presence of the ribosome itself is able to confer the ability for a protein to fold correctly, perhaps in concert with the other variables. **Biochemical Connections 12C** gives an example of how silent mutations can affect translation and protein folding, and **Biochemical Connections 12D** describes other processes involved in protein folding in vivo.

12C Biochemical Connections • Genetics

Silent Mutations Are Not Always Silent

A silent mutation is a mutation that changes DNA but not the amino acid incorporated. For example, if the DNA coding strand has a UUC, it codes for phenylalanine. If a mutation in the DNA changes the sequence to UUU, then the DNA has undergone a silent mutation because both UUU and UUC code for Phe. At least, that is what scientists believed for decades. However, studies have shown that this is not always true. Researchers at the National Cancer Institute were studying a gene called *MDR1,* which is named for its association with multiple drug resistance in tumor cells. They had sequences of this gene and knew that there were some common silent mutations. Interestingly, they discovered that there was a phenotypic response to silent mutations of this gene that influenced patients' response to certain drugs. This was striking because a silent mutation is supposed to have no effect on the final product.

Apparently, not all codons are translated equally. Different codons may require alternate versions of the tRNA for a particular amino acid. Even though the amino acid incorporated is the same, the pace at which the ribosome is able to incorporate the amino acid differs depending on which codon it is. This is a situation reminiscent of the transcription attenuation that we discussed in Section 11-3D. As shown in Figure 12.25, translation kinetics can affect the form of the final protein. If the wild-type codon is used, then translation proceeds normally and produces the normal conformation of the protein. However, if a silent mutation changes the pace of the movement of the ribosome, then an abnormal protein conformation is created due to folding differences.

Other examples of phenotypic effects due to silent mutations have also been discovered. Besides the protein-folding differences shown in Figure 12.25, there are other processes based on synonymous codon differences. One of the most frequent is based on differences in the processing of the mRNA after transcription (Section 11-8). Research has shown that the mRNA exons not only contain the code that should lead to production of the correct amino acids, but they also contain the information necessary for the correct removal of introns. There are specific sequences, called *exonic splicing enhancers (ESEs),* that tell the splicing machinery where to splice out introns. The codons GGA and GGG both encode glycine and both can occur in ESEs. However, GGA is a much more potent splicing enhancer. As shown in Figure 12.26, a silent mutation change in an ESE could lead to the incorrect removal of introns and an entire exon being left out of the final mRNA.

Another process involving silent mutations is caused by secondary structures of mRNA. While textbooks often depict mRNA as a linear molecule for simplicity, mRNA can also fold up into many different structures. Changing any nucleotide can potentially change the

Continued on next page

way the mRNA folds, and such changes can affect the speed with which the mRNA is translated—if it can be translated at all. Researchers studying a gene involved in pain tolerance discovered that silent mutations in an intron of the gene *COMT* (catechol-*O*-methyltransferase) are responsible for different levels of an important enzyme that affects how we perceive pain (Figure 12.27). They showed that the different mutations led to different folding patterns of the mRNA, which translated to different levels of the enzyme in the cell. While Figure 12.27 does not show mRNA secondary structures, it does show the silent mutations that are believed to be responsible for the different effects on pain sensitivity.

Researchers continue to study this phenomenon because it is now known that more than 50 human diseases are caused by silent mutations, including Marfan syndrome, androgen-insensitivity syndrome, cholesteryl ester storage disease, McArdle disease, and phenylketonuria. ◗

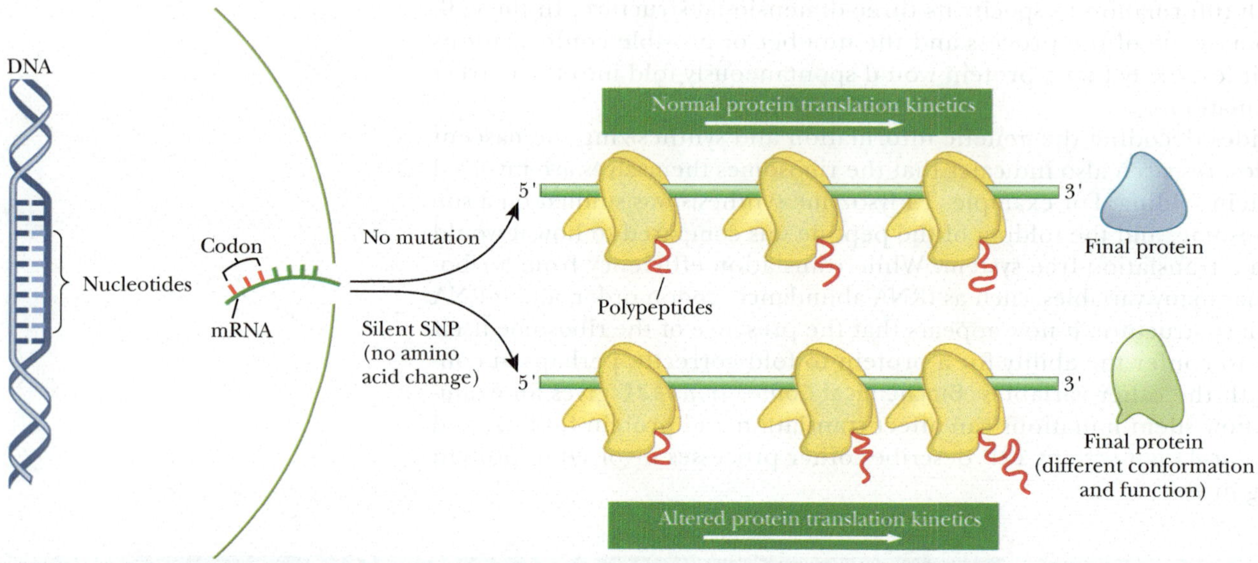

Figure 12.25 Translation kinetics and protein folding. Unaffected translation kinetics result in a correctly folded protein. Abnormal kinetics, caused by the ribosome moving faster or slower through certain mRNA regions, can produce a different final protein conformation. Abnormal kinetics may arise from a silent single nucleotide polymorphism (SNP) in a gene that creates a codon synonymous with the wild-type codon. However, this synonymous codon substitution may lead to different kinetics of mRNA translation, thus yielding a protein with a different final structure and function. *(Based on Komar, A. A. (2007). SNPs, silent but not invisible.* Science *Vol. 315, pp. 466-467.)*

NORMAL SPLICING
When eukaryotic DNA is transcribed, the nascent mRNA has sequences that code for amino acids, which are called exons, and other sequences that are noncoding, called introns. The exons contain short sequences called exonic splicing enhancers (ESEs). Ser- and Arg-rich proteins (SR) bind to the ESE regions and create the spliceosome. The introns are removed, leaving just the exons joined together to form the final mRNA.

EXON SKIPPING
When a silent mutation occurs in the ESE region, it can cause the splice site to be missed by the SR machinery. In that case, one of the exons may be missed as it gets spliced out with the two adjoining introns, leading to an incorrect mRNA product.

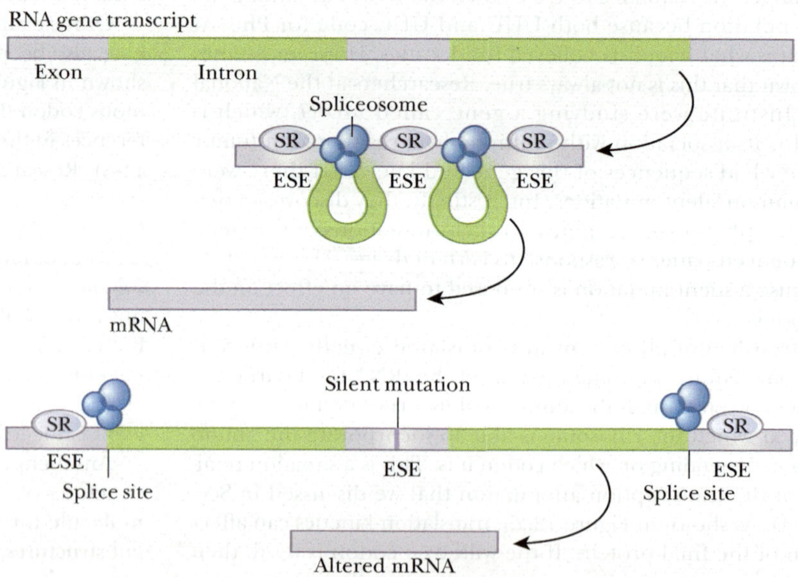

Figure 12.26 Silent mutations in an exonic splicing enhancer region of an exon can lead to altered splicing and an altered mRNA product.

Changes in the sequence inside an intron can lead to different phenotypic variations. This can be seen with a gene for pain tolerance called COMT, for catechol-*O*-methyltransferase. There are three common variations, including one for low pain tolerance, one for average pain tolerance, and one for high pain tolerance. There are three different sites involved in the mutations that lead to altered pain sensitivity, involving both synonymous and nonsynonymous mutations.

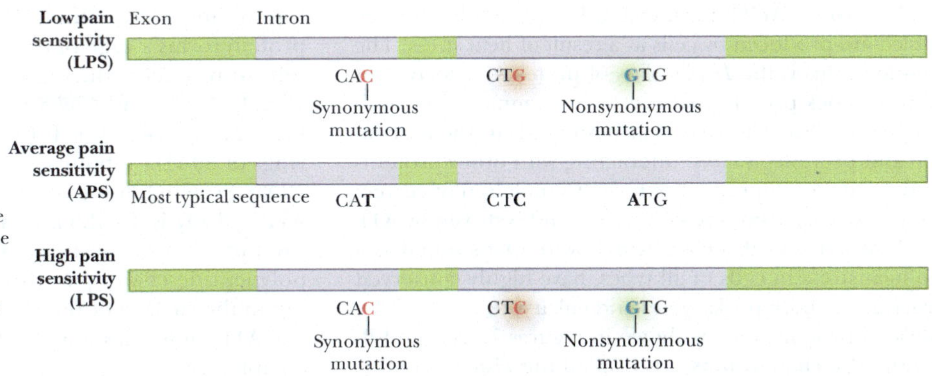

Figure 12.27 **Different sequences in the introns in a gene can lead to different phenotypes, as shown for a gene for pain tolerance.**

12D Biochemical Connections • Biophysical Chemistry

Chaperones: Preventing Unsuitable Associations

It is sometimes said that a chaperone's task is to prevent unsuitable associations. The class of proteins known as **molecular chaperones** operates in this way by preventing aggregation of newly formed proteins until they fold into their active forms. The information necessary for protein folding is present in the amino acid sequence, and many proteins fold correctly without any outside help, as shown in Figure 12.28A. However, some proteins may form aggregates with other proteins, or may fold with

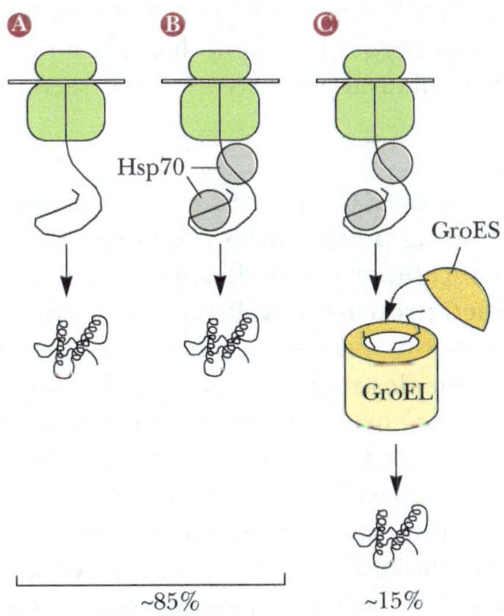

Figure 12.28 **Protein-folding pathways.** (A) Chaperone-independent folding. (B) Folding facilitated by a chaperone (gray)—in this case, the Hsp70 protein. About 85% of proteins fold by one of the two mechanisms shown in (A) and (B). (C) Folding facilitated by chaperonins—in this case, GroEL and GroES. *(Adapted from Netzer, W. J., and Hartl, F. U. (1998). Protein folding in the cytosol: Chaperonin-dependent and -independent mechanisms.* Trends in Biochemical Sciences *23, 68–73, Figure 2, with permission from Elsevier.)*

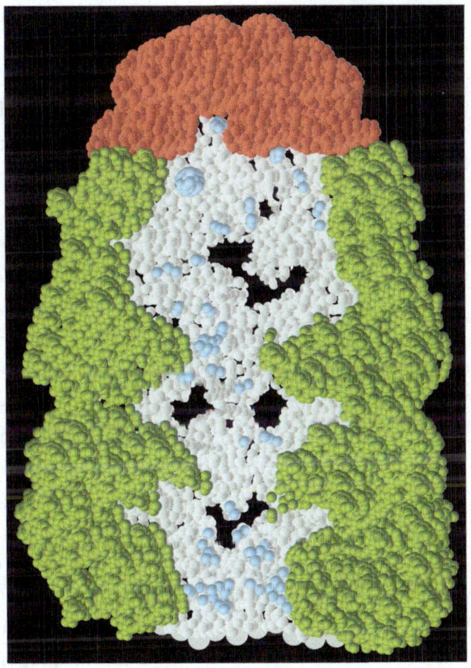

Figure 12.29 **Structure of the GroEL–GroES complex, emphasizing the central cavity.** GroEL is pictured in green and GroES in orange. *(Adapted by permission from Macmillan Publishers Ltd: Xu, Z., et al. (1997). The crystal structure of the asymmetric GroEL–GroES-(ADP)7 chaperonin complex.* Nature *388, 741–750, Fig. 1. Molecular graphics courtesy of Paul B. Sigler, Yale University.)*

Continued on next page

incorrect secondary and tertiary structures, unless they interact first with a chaperone. Well-known examples include heat-shock proteins, which are produced by cells as a result of heat stress. The prime example of this is the *Hsp70* class of proteins, named after the 70-kDa heat-shock protein that occurs in mammalian cytosol, shown in Figure 12.28B. The Hsp70 protein binds to the nascent polypeptide and prevents it from interacting with other proteins or from folding into an unproductive form. Completion of correct folding requires release from the chaperone and is driven by ATP hydrolysis. All proteins in this class, which were first studied as a response to heat stress in cells of all types, have highly conserved primary structures in both prokaryotes and eukaryotes.

About 85% of proteins fold as shown in Figures 12.28A and B. Another group, the **chaperonins** (also called the *Hsp60* proteins after their 60-kDa molecular weight) is known to be involved in the folding of the other 15% of proteins. A large multisubunit protein forms a cage of 60-kDa subunits around the nascent protein to protect it during the folding process, shown in Figure 12.28C. *GroEL* and *GroES* are the best-characterized chaperonins from *E. coli.* GroEL is formed by two stacked seven-membered rings of 60-kDa subunits with a central cavity, as shown in Figure 12.29. Protein folding occurs in the central cavity and depends on ATP hydrolysis. GroES is a single seven-membered ring of 10-kDa subunits that sits on top of GroEL. During protein folding, the polypeptide chain goes through cycles of binding and unbinding to the surface of the central cavity. In some cases, more than 100 ATP molecules must be hydrolyzed before protein folding is complete. ◗

molecular chaperones proteins that help other proteins fold correctly and arrive at their correct location in the cell

chaperonins proteins that form a barrel structure with a lid that mediate the folding of newly synthesized proteins

12-7 Protein Degradation

Learning Objectives

After reading this section, students will be able to:

LO 12.7.1 Explain how proteins are marked for cellular degradation.

LO 12.7.2 Describe events that happen in the cellular degradation of proteins.

One of the most often overlooked controls of gene expression occurs at the level of the degradation of proteins. Proteins are in a dynamic state in which they are turned over often. Athletes are painfully aware of this because it means that they must work very hard to get in shape, but then they get out of shape very quickly. Some classes of proteins experience a 50% turnover every three days. In addition, abnormal proteins that were formed from errors in either transcription or translation are degraded quickly. It is believed that a single break in the peptide backbone of a protein is enough to trigger the rapid degradation of the pieces, because breakdown products from natural proteins are rarely seen in vivo.

▶ How does the cell know which proteins to degrade?

proteasomes multisubunit complexes of proteins that mediate the degradation of other, suitably tagged proteins

If proteins are degraded so quickly, then the process must be heavily controlled to avoid the destruction of the wrong polypeptides. The degradation pathways are restricted to degradative subcellular organelles, such as lysosomes, or to macromolecular structures called **proteasomes.** Proteins are directed to lysosomes by specific signal sequences, often added in a posttranslational modification step. Once in the lysosome, the destruction is nonspecific, and it is carried out by acidic hydrolases in the acidic environment inside the organelle. Proteasomes are found in both prokaryotes and eukaryotes, and specific pathways exist to target a protein so that it complexes with a proteasome and is degraded.

ubiquitinylation the process of binding the polypeptide ubiquitin to a protein to mark it for degradation

In eukaryotes, the most common mechanism for targeting proteins for destruction in a proteasome is by **ubiquitinylation.** Ubiquitin is a small polypeptide (76 amino acids) that is highly conserved in eukaryotes. It is aptly named because it is found ubiquitously in most tissues of eukaryotic organisms. There is a high degree of homology between the ubiquitin amino acid sequences in species as widespread as yeast and humans. When ubiquitin is linked to another protein, it condemns that protein to destruction in a proteasome.

Figure 12.30 shows the mechanism of ubiquitinylation. Three enzymes are involved—*ubiquitin-activating enzyme (E1), ubiquitin-conjugating enzyme (E2),* and *ubiquitin ligase (E3).* E1 activates ubiquitin using ATP, similar to how an amino acid gets activated (see part *i* of Figure 12.7B). Next, E1

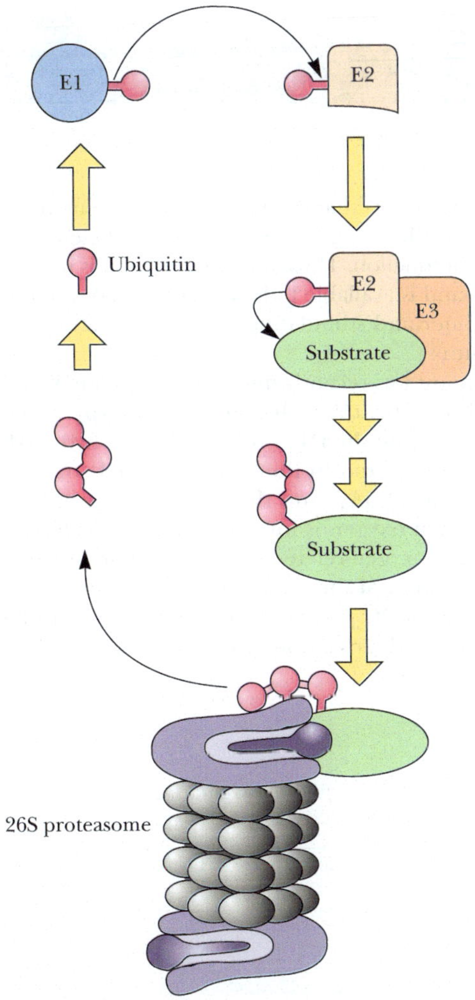

Figure 12.30 **Diagram of the ubiquitin–proteasome degradation pathway.** Pink "lollipop" structures symbolize ubiquitin molecules. *(Adapted from Hilt, W., and Wolf, D. H. (1996). Proteasomes: Destruction as a program.* Trends in Biochemical Science *21, 96–102, Figure 1, with permission from Elsevier.)*

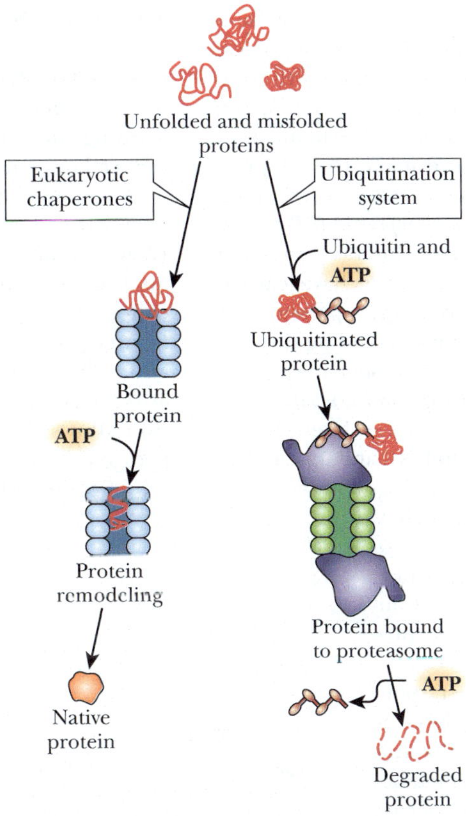

Figure 12.31 **Misfolded proteins can be repaired and folded correctly by eukaryotic chaperones, or they can be targeted for destruction by ubiquitination.** *(From GARRETT/GRISHAM,* Biochemistry, *4E.)*

catalyzes the transfer of the activated ubiquitin to an E1 active site cysteine, followed by another transfer of the ubiquitin, this time to an E2 active site cysteine, which is catalyzed by the E2 enzyme (Figure 12.30, *top*). E3 then transfers the ubiquitin to a free amino group on the targeted protein, either the N-terminus or a lysine side chain, using the C-terminal glycine carboxyl group of ubiquitin to make the peptide bond connection.

Proteins can also be polyubiquitinylated, where the extra ubiquitin moieties get attached through their C-terminal glycine carboxyl group and a lysine side chain on a ubiquitin that is already attached to the protein that is set to be degraded (Figure 12.30, *lower right*).

If we combine what we know about the chaperone (**Biochemical Connections 12D**) and protein degradation processes, we obtain a good picture of how the cell monitors quality control of its proteins. Cells start with proteins that may be unfolded or misfolded. If they are unfolded, they go through the chaperone process and are folded correctly (see Figure 12.31). If they are misfolded, then they go through the ubiquitination system and are degraded. **Biochemical Connections 12E** discusses an interesting example of how transcription regulation and protein degradation work together to control the process of acclimation to high altitude.

12E Biochemical Connections • Physiology

How Do We Adapt to High Altitude?

Those of us who live at low altitudes are quite aware of the sensations of oxygen deprivation and the associated physiological changes that occur with prolonged exposure to high altitudes. Physiologists have studied this phenomenon for many years, and biochemists have tried to find the mechanism that explains how cells sense the oxygen partial pressure and make the adaptive changes. Two of the major changes are the increase in red blood cells, which is stimulated by the hormone *erythropoietin (EPO)*, and *angiogenesis*, the formation of new capillaries, which is stimulated by *vascular endothelial growth factor (VEGF)*. Researchers have learned a great deal about how cells respond to low oxygen pressure, or *hypoxia*, and the results have many applications, including the production of drugs to treat inflammation, heart disease, and cancer.

A family of transcription factors called *hypoxia-inducible factors (HIFs)* is the key to these processes (Figure 12.32). Heterodimers composed of HIFα and HIFβ subunits bind to DNA and up-regulate a variety of genes when the partial pressure of oxygen in the blood is low. Oxygen can be low for many reasons, such as a person being at high altitude or a tissue having to work unusually hard. During a heart attack or stroke, the oxygen partial pressure can drop, and these transcription factors can help express genes whose protein products reduce the damage.

The genes that are controlled by HIFs are responsible for angiogenesis and the production of EPO, as well as the production of glycolytic enzymes that can provide energy for the cells when aerobic metabolism is compromised. In addition, many types of cancers are found associated with elevated levels of HIFs. HIFs may be involved in the tumor's ability to grow, which requires a large oxygen supply. As it turns out, if cell growth is uncontrolled, the dimerization of the two HIF subunits is unchecked, which would lead to constant expression of the adaptive genes. When this happens, overgrowth of endothelial cells occurs, leading to tumors. Any given cell maintains relatively constant levels of the HIFβ subunit, but the level of the HIFα subunit is regulated.

The system is mainly controlled by the degradation of the HIFα subunit (Figure 12.32, *bottom right*). Proline 564 in the HIFα subunit can be hydroxylated by an enzyme called *proline hydroxylase (PH)*. After HIFα is hydroxylated, it binds to a protein called von Hippel–Lindau protein (pVHL), which was first discovered to be a tumor suppressor (see Section 14-8 for more on tumor suppressors). After the pVHL is bound, it stimulates the organization of a complex with *ubiquitin ligase (UL)*, which ubiquitinylates the HIFα subunit. Recall from Section 12-7 that when ubiquitin protein is attached to another protein, it targets the protein for transport to a proteasome, where the protein is degraded. Searching for how this pathway is related to the body's ability to sense the oxygen partial pressure is currently an area of active research. Evidence exists that the PH has a requirement for iron and oxygen. It is possible that, in the absence of sufficient oxygen, the HIF PH cannot function. Thus, the HIFα subunit is not targeted for destruction and is available to bind to the HIFβ subunit, forming the active dimer and stimulating the adaptive effects of reduced oxygen pressure.

A second control point was recently discovered to be mediated by another hydroxylase. This time, the target is an asparagine (Asn) residue on HIFα. When the HIFα subunit has the hydroxylated Asn residue, it cannot bind to the transcriptional cofactor called mediator (Section 11-5A), and therefore cannot induce transcription. Thus, when oxygen is low, two different hydroxylations that would reduce oxygen-related gene expression proceed at reduced levels. Reduction of the asparagine hydroxylase reaction allows the HIFα to do its job, and reduction of the prolyl hydroxylase reaction stops degradation of the HIFα by the ubiquitin-linked pathway.

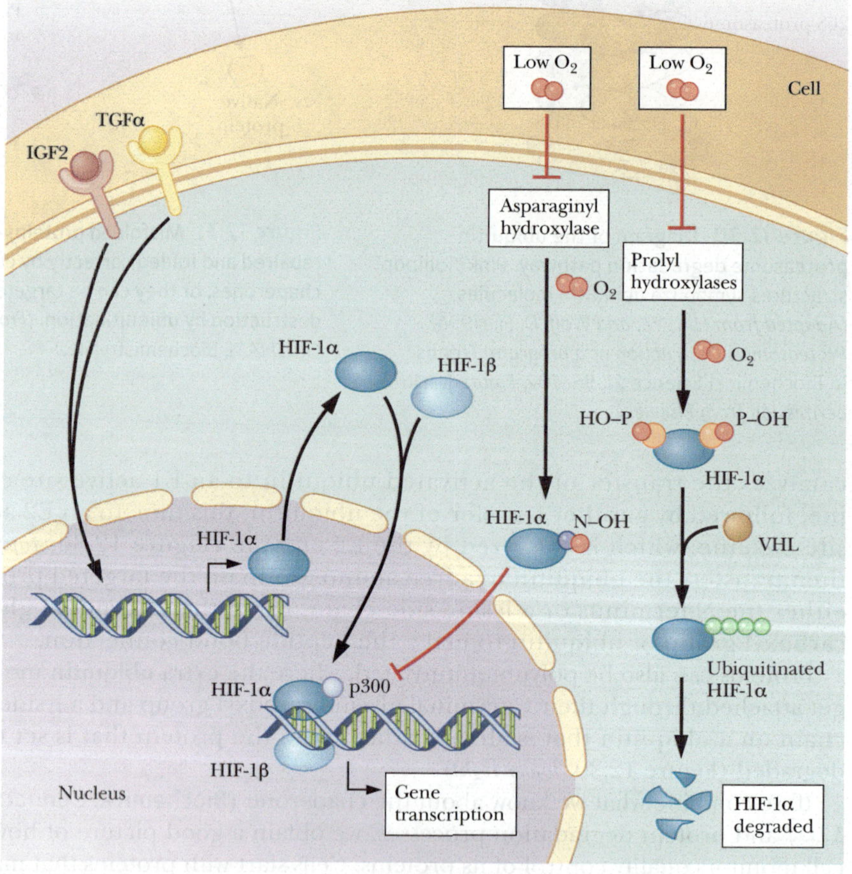

Figure 12.32 **The control of the level and function of hypoxia-inducible factor α by the level of oxygen.** *(Adapted from Marx, J. (2004). How cells endure low oxygen. Science 303, 1454–1456. Illustration by Carin Cain. Reprinted with permission from AAAS.)*

HIF is also controlled in a positive way by *insulin growth factor 2 (IGF2)* and by *transforming growth factor α (TGFα)* (Figure 12.32, *top left*). These are common growth factors that are active in several growth and differentiation pathways, and they are not related to oxygen availability.

Researchers are looking at control of HIF as a potential therapy against cancer because so many cancers are found to be associated with high HIF levels. Because tumors need oxygen to grow, stopping the tumor's activation of HIF-mediated transcription could effectively choke out the tumor before it progresses.

Summary

The Genetic Code

- The genetic code for living organisms is contained in a triplet, nonoverlapping, commaless, degenerate, and universal code.
- A variety of techniques were used to determine the genetic code. The most comprehensive experiment was the filter-binding experiment of Nirenberg, which used specific trinucleotides bound to a filter to see which aminoacyl-tRNA molecules would bind.

Codon–Anticodon Pairing and Wobble

- There are only about half as many tRNA molecules as there are codons for amino acids. This occurs because of wobble in the third base (5′ end) of the anticodon on the tRNA.
- The wobble base can break Watson–Crick base-pairing rules under certain circumstances. For example, if the wobble base is a U, then it can bind to an A or a G of the mRNA codon. As a result, fewer tRNA molecules are needed in translation.

Aminoacyl-tRNA Formation

- Before an amino acid is attached to a tRNA molecule to form an aminoacyl-tRNA, the amino acid is first activated in an ATP-dependent manner. The amino acid can then be transferred to the tRNA and brought to the ribosome as a building block for protein synthesis.
- The second genetic code refers to the specificity with which the aaRSs put the correct amino acid onto the correct tRNA. There is only minimal proofreading past that point, so the correct loading of amino acids is critical. If the wrong amino acid is loaded on the tRNA, it is quickly hydrolyzed by the synthetase to avoid such errors.

Prokaryotic Initiation of Translation

- In prokaryotic translation, the correct mRNA AUG start codon is identified by its proximity to a consensus sequence called the Shine–Dalgarno sequence. The Shine–Dalgarno sequence is complementary to a sequence on the small subunit of the prokaryotic ribosome. The ribosome is initially positioned on the Shine–Dalgarno sequence, which aligns it for correct translation initiation at the start codon. Initiation factor proteins and the fMet-$tRNA^{fMet}$ molecule are also involved.

Prokaryotic Translation Elongation

- The E site, P site, and A site on the ribosome bind exiting uncharged tRNAs, the peptidyl-tRNAs, and incoming aminoacyl-tRNAs, respectively, during translation elongation steps that build the new protein chain. Elongation factor proteins also play roles in the elongation phase of translation.

Prokaryotic Termination of Translation

- When a stop codon (UAA, UAG, or UGA) is encountered on the mRNA, RF proteins bind and terminate translation.

Eukaryotic Translation

- In eukaryotes, the main features of translation are the same as in prokaryotes, but the process differs in certain details.
- Eukaryotic mRNA has a 5′ cap and a 3′ poly A tail, both of which are involved in the formation of the initiation complex.
- There is no Shine–Dalgarno sequence, but there is a Kozak sequence surrounding the correct AUG start codon.
- The number of initiation factors and elongation factors is much higher in eukaryotes than in prokaryotes.
- Stop codons in eukaryotes are sometimes recognized by suppressor tRNAs, which allow the insertion of rare amino acids, such as Sec and Pyl.

Posttranslational Modification of Proteins

- Proteins are often processed after they are synthesized. In prokaryotes, the N-terminal methionine is often removed. Disulfide bridges may form. Cofactors and prosthetic groups, like heme, may be added. Enzymes may also cut off pieces of the nascent peptide.
- In theory, the primary structure of the protein determines its three-dimensional structure. In reality, however, proteins often need the help of a chaperone to arrive at the correct structure.

Protein Degradation

- Proteins are degraded in subcellular organelles, such as lysosomes, or in macromolecular structures called proteasomes.
- Many proteins are targeted for destruction by being bound to a protein called ubiquitin.
- Damaged proteins are degraded very quickly.

Review Exercises

12-1 Translating the Genetic Message

1. **Recall** Prepare a flow chart showing the stages of protein synthesis.

12-2 The Genetic Code

2. **Reflect and Apply** A genetic code in which two bases encode a single amino acid is not adequate for protein synthesis. Give a reason why.
3. **Recall** Define *degenerate code.*
4. **Reflect and Apply** The genetic code is described as *nonoverlapping.* What does this mean, and why is it important for protein synthesis?
5. **Recall** How can the binding assay technique be used to assign coding triplets to the corresponding amino acids?
6. **Recall** Which nucleotides break the rules of Watson–Crick base pairing when they are found at the wobble position of the anticodon? Which ones do not?
7. **Recall** Describe the role of stop codons in the termination of protein synthesis.
8. **Reflect and Apply** Consider a three-base sequence in the template strand (not the coding strand) of DNA: 5′ … 123 … 3′, in which 1, 2, and 3 refer to the relative positions of deoxyribonucleotides. Comment on the probable effect on the resulting protein if the following point mutations (one-base substitutions) occurred:
   (a) changing one purine for another in position 1
   (b) changing one pyrimidine for another in position 2
   (c) changing a purine to a pyrimidine in position 2
   (d) changing one purine for another in position 3
9. **Reflect and Apply** Hypoxanthine, as part of the inosine monophosphate nucleotide, frequently occurs as the first base in tRNA anticodons that pair with the third base of mRNA codons. Why is this?
10. **Reflect and Apply** Is it reasonable that codons for the same amino acid have one or two nucleotides in common? Why or why not?
11. **Reflect and Apply** How would protein synthesis be affected if a single codon could specify the incorporation of more than one amino acid (an ambiguous code)?
12. **Reflect and Apply** Comment on the evolutionary implications of the differences in the genetic code observed in mitochondria.
13. **Biochemical Connections** Why would influenza A virus (IAV) evolve to be less lethal?
14. **Biochemical Connections** Explain how a frameshift in reading mRNA affects the translation of the mRNA for the IAV protease.

12-3 Aminoacyl-tRNA Formation

15. **Recall** Describe the two-step process by which an aminoacyl-tRNA synthetase charges a tRNA with its corresponding amino acid.
16. **Recall** What is the role of ATP in amino acid activation?
17. **Recall** Outline the proofreading processes in aminoacyl-tRNA formation.
18. **Recall** What ensures fidelity in protein synthesis? How does this compare with the fidelity of replication and transcription?
19. **Recall** Can the same enzyme esterify more than one amino acid to its corresponding tRNA?
20. **Reflect and Apply** A friend tells you that they are starting a research project on aminoacyl esters. Your friend asks you to describe the biological role of this class of compounds. What do you tell them?
21. **Reflect and Apply** Is aminoacyl-tRNA formation energetically favored? Why or why not?

12-4 Prokaryotic Translation

22. **Recall** Describe the subunit composition of the prokaryotic ribosome, including the sizes of each subunit, the molecules that are present in each subunit, and the total size of the ribosome.
23. **Recall** Identify the following by describing their functions: EF-G, EF-Tu, EF-Ts, EF-P, and peptidyl transferase.
24. **Recall** What are the components of the prokaryotic translation initiation complex? How do they interact with one another?
25. **Recall** What is the role of the 50S ribosomal subunit in prokaryotic protein synthesis?
26. **Recall** What are the A site and the P site? How are their roles in protein synthesis similar? How do they differ? What is the E site?
27. **Recall** Describe the role of the stop signals in protein synthesis.
28. **Recall** What is the Shine–Dalgarno sequence? What role does it play in protein synthesis?
29. **Reflect and Apply** You are studying with a friend who says that the hydrogen-bonded portions of tRNA play no important role in its function. What is your reply?
30. **Reflect and Apply** *E. coli* has two tRNAs for methionine. What is the basis for the distinction between the two?
31. **Reflect and Apply** In prokaryotic protein synthesis, *N*-formylmethionine (fMet) is the first amino acid incorporated, whereas (normal) methionine is incorporated in eukaryotes. The same codon (AUG) serves both. What prevents methionine from being inserted at the beginning and *N*-formylmethionine in the interior?
32. **Reflect and Apply** Describe the recognition process by which the tRNA for *N*-formylmethionine interacts with the portion of mRNA that specifies the start of transcription.
33. **Reflect and Apply**
    (a) How many amino acid activation cycles are needed for a protein with 150 amino acids?
    (b) How many initiation cycles are needed for a protein with 150 amino acids?
    (c) How many elongation cycles are needed for a protein with 150 amino acids?
    (d) How many termination cycles are needed for a protein with 150 amino acids?
34. **Reflect and Apply** What is the energy cost per amino acid in prokaryotic protein synthesis? Relate this to entropy and fidelity.
35. **Reflect and Apply** Would it be possible to calculate the energy cost of protein synthesis, including the cost of making mRNA and DNA?
36. **Reflect and Apply** Suggest a possible conclusion from the fact that peptidyl transferase is one of the most conserved sequences in all of biology.
37. **Reflect and Apply** Would you expect electron microscopy to give detailed information about ribosomal structure? *Hint:* Compare Figure 12.17 to Figure 12.19.
38. **Reflect and Apply** How does it improve the efficiency of protein synthesis to have several binding sites for tRNA close to each other on the ribosome?
39. **Reflect and Apply** A virus does not contain ribosomes. How does it manage to ensure the synthesis of its proteins?

12-5 Eukaryotic Translation

40. **Recall** Why is selenocysteine called the 21st amino acid when there are many more amino acids found than the 20 basic ones coded for by the genetic code?
41. **Recall** What is unique about selenocysteine?
42. **Recall** What conserved sequences are important to codon recognition in prokaryotic release factors?
43. **Recall** What sequence in the prokaryotic release factors is important for the hydrolysis of the peptide bond?
44. **Recall** What are two major similarities between protein synthesis in bacteria and in eukaryotes? What are two major differences?
45. **Reflect and Apply** Protein synthesis takes place much more slowly in eukaryotes than in prokaryotes. Suggest a reason why this is so.
46. **Reflect and Apply** Why is it advantageous to have a mechanism to override the effect of stop codons in protein synthesis?
47. **Biochemical Connections** What process distinguishes a short-term memory from a long-term one?
48. **Biochemical Connections** What transcription factor in the nucleus of a neuron is known to play a role in creating long-term memories?
49. **Biochemical Connections** What is one experiment done to show that protein synthesis is required to make long-term memories?
50. **Biochemical Connections** How does the strength of an incoming nerve impulse affect memory?

12-6 Posttranslational Modification of Proteins

51. **Recall** Describe the key posttranslational modifications required to produce mature, active insulin.
52. **Reflect and Apply** The amino acid hydroxyproline is found in collagen. There is no codon for hydroxyproline. Explain the occurrence of this amino acid in a common protein.
53. **Biochemical Connections** What are chaperones?
54. **Biochemical Connections** How would protein synthesis be different if there were no chaperones?
55. **Biochemical Connections** How were chaperones discovered?
56. **Reflect and Apply** Are chaperones the only proteins that aid in correct protein folding?

12-7 Protein Degradation

57. **Recall** What role does ubiquitin play in the degradation of proteins?
58. **Reflect and Apply** Consider protein degradation in the absence of ubiquitinylation. Is the process likely to be more or less efficient?
59. **Reflect and Apply** Compare the roles of lysosomes and proteasomes in protein degradation. How do they differ in terms of the types of proteins they degrade and the processes involved?
60. **Reflect and Apply** Is it reasonable to expect that protein degradation can take place at any location in a cell?
61. **Biochemical Connections** What are exonic splicing enhancers, and how are they important?
62. **Biochemical Connection** What is a silent mutation? Why is the name "silent mutation" a bit of a misnomer?
63. **Biochemical Connections** How can a silent mutation lead to "exon skipping"?
64. **Biochemical Connections** What process in mRNA folding leads to sensitivity to pain due to an effect on the COMT gene?
65. **Biochemical Connections** What diseases are known to be caused by silent mutations?
66. **Biochemical Connections** What transcription factors are important to how we adapt to high altitude?
67. **Biochemical Connections** How does protein degradation play a role in how we adapt to high altitude?

Further Reading

Ban, N. The Complete Atomic Structure of the Large Ribosomal Subunit at 2.4 Ångstrom Resolution. *Science* **289**, 905–920. (2000). Current information on the structure of the ribosome.

Blaha, G., R. E. Stanley, and T. A. Steitz, Formation of the First Peptide Bond: The Structure of EF-P Bound to the 70S Ribosome. *Science* **325**, 966–969 (2009).

Cech, T. R. The Ribosome Is a Ribozyme. *Science* **289**, 878–879 (2000). A classic paper describing the RNA-based catalysis of the ribosome.

Chamary, J. V. and L. D. Hurst, The Price of Silent Mutations. *Sci. Amer.* **300** (6), 46–53 (2009).

Fabrega, C., M. A. Farrow, B. Mukhopadhyay, V. de Crecy-Lagard, A. R. Ortiz, and P. Schimmel. An Aminoacyl tRNA Synthetase Whose Sequence Fits into Neither of the Two Known Classes. *Nature* **411**, 110–114 (2001). An article describing a tRNA synthetase discovered to not belong to one of the two known classes.

Freeland, S. J. and L. D. Hurst. Evolution Encoded. *Sci. Amer.* **290** (4), 84–91 (2004). An article on the evolutionary fitness of the genetic code.

Goldberg, A. Functions of the Proteasome: The Lysis at the End of the Tunnel. *Science* **268**, 522–523 (1995). A perspective view of the multisubunit proteins involved in protein degradation.

Hartl, F. Molecular Chaperones in Cellular Protein Folding. *Nature* **381**, 571–579 (1996). A review article on the various classes of molecular chaperones.

Hentze, M. eIF4G: A Multipurpose Ribosome Adapter? *Science* **275**, 500–501 (1997). A short review of translation initiation in eukaryotes.

Hentze, M. W. Believe It or Not—Translation in the Nucleus. *Science* **293**, 1058–1059 (2001). An article introducing the research that showed nuclear translation.

Ibba, M. Discriminating Right from Wrong. *Science* **294**, 70–71 (2001). A summary of the research showing a type of proofreading ability by EF-Tu.

Iborra, F. J., D. A. Jackson, and P. R. Cook. Coupled Transcription and Translation within Nuclei of Mammalian Cells. *Science* **293**, 1139–1142 (2001). The primary research showing translation in the nucleus.

Komar, A. A. SNPs, Silent But Not Invisible. *Science* **315**, 466–467 (2007). An article describing the nature of translation kinetics changes based on silent mutations.

LaRiviere, F. J., A. D. Wolfson, and O. C. Uhlenbeck. Uniform Binding of Aminoacyl-tRNAs to Elongation Factor Tu by Thermodynamic Compensation. *Science* **294**, 154–168 (2001). An in-depth article about the evidence that EF-Tu provides another level of fidelity in translation.

Liljas, A. Getting Close to Termination. *Science* **322**, 863–865 (2008).

Palioura, S., R. L. Sherrer, T. A. Steitz, D. Soll, and M. Simonovic. The Human SepSepS-tRNA(sec) Complex Reveals the Mechanism of Selenocysteine Formation. *Science* **325**, 321–325 (2009).

Polacek, N., M. Gaynor, A. Yassin, and A. S. Mankin. Ribosomal Peptidyl Transferase Can Withstand Mutations at the Putative Catalytic Nucleotide. *Nature* **411**, 498–501 (2001). An article questioning the ribozyme theory of peptidyl transferase.

Schmeing, T. M., R. M. Voorhees, A. C. Kelley, Y. G. Gao, F. V. Murphy, J. R. Weir, and V. Ramakrishnan. The Crystal Structure of the Ribosome Bound to EF-Tu and Aminoacyle-tRNA. *Science* **326**, 677–678 (2009).

Soares, C. Codon Spell Check. *Sci. Amer.* **296** (5), 23–24 (2007). An introduction to how silent mutations are not always silent.

Zhu, H. and H. F. Bunn. How Do Cells Sense Oxygen? *Science* **292** (5516), 449–451 (2001). An article about protein degradation and gene expression in the system that allows adaptation to high altitude.

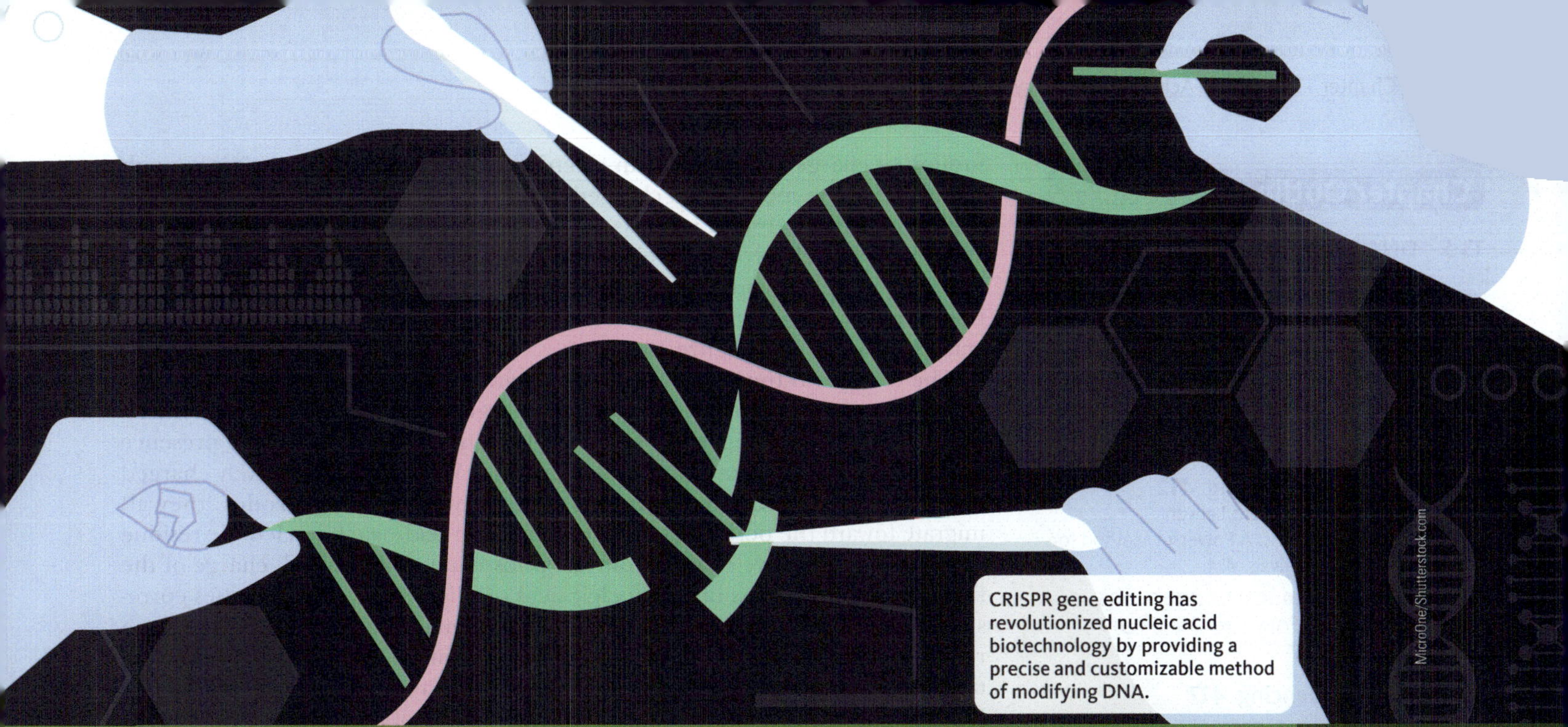

MicroOne/Shutterstock.com

CRISPR gene editing has revolutionized nucleic acid biotechnology by providing a precise and customizable method of modifying DNA.

Nucleic Acid Biotechnology Techniques 13

Chapter Outline

13-1 Purification and Detection of Nucleic Acids

Learning Objectives

After reading this section, students will be able to:

LO 13.1.1 Describe methods for separating nucleic acids.

LO 13.1.2 Explain methods for detecting nucleic acids.

Striking examples of the power of biotechnology techniques for manipulating DNA and RNA have sparked enormous amounts of discussion over the years. In Chapter 13, we focus our attention on some of the most important methods used in nucleic acid biotechnology.

Experiments on nucleic acids frequently involve extremely small quantities of materials of widely varying molecular sizes. Two of the primary necessities in these experiments are to separate the components of a nucleic acid mixture and to detect the components after separation. Fortunately, powerful methods exist for accomplishing both goals.

13-1A Nucleic Acid Separation Techniques

Any separation method depends on the differences between the items to be separated. Charge and size are two properties of molecules that are frequently used for separation. One of the most widely used techniques in molecular biology, **gel electrophoresis**, uses both of these properties (Section 5-3). Electrophoresis is based on the

Chapter Outline (continued)

gel electrophoresis a technique for separating molecules on the basis of size, charge, and shape using a gel as a support and sieving material

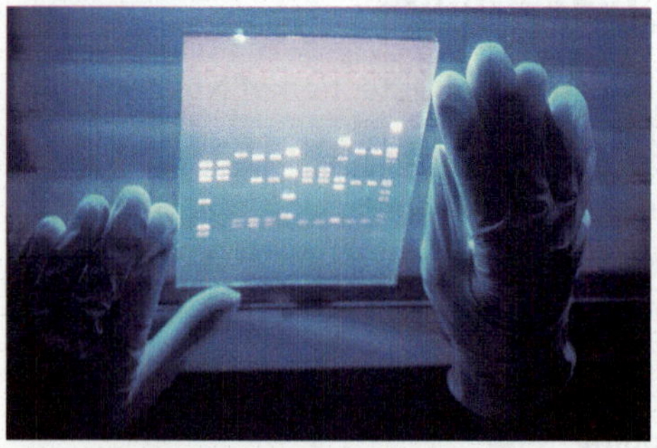

Figure 13.1 Separation of nucleic acids by gel electrophoresis. Each band seen in the gel represents a different nucleic acid molecule.

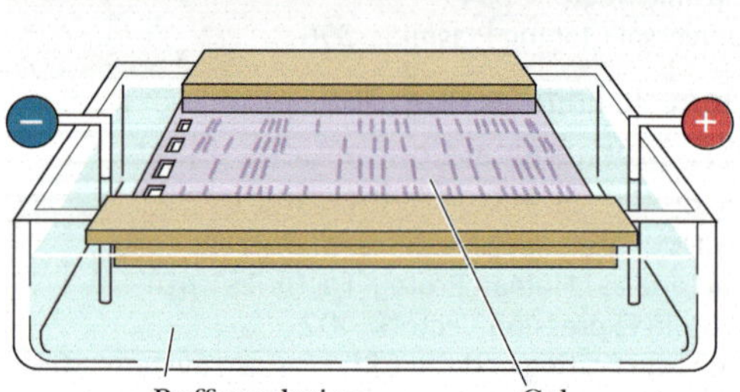

Figure 13.2 An experimental setup for nucleic acid gel electrophoresis. The samples are placed in wells (see the left side of the gel). When the current is applied, the negatively charged oligonucleotides migrate toward the positive electrode.

nucleases enzymes that hydrolyze nucleic acids (DNA or RNA)

motion of charged particles in an electric field. For our purposes, it is enough to know that the motion of a charged molecule in an electric field depends on the ratio of its charge to its mass. A sample is applied to a support medium. With the use of electrodes, an electric current is passed through the medium to move the sample through the support medium and achieve the desired separation. A polymeric gel called agarose is used as the support medium for nucleic acid electrophoresis (Figure 13.1).

The charge on the molecules to be separated leads them to move through the gel toward an electrode of opposite charge. Nucleic acids and oligonucleotide fragments are negatively charged at neutral pH because of the presence of the phosphate groups (Figures 9.5 and 9.6). When these negatively charged molecules are placed in an electric field between two electrodes, they all migrate toward the positive electrode. In nucleic acids, each nucleotide residue contributes a negative charge from the phosphate to the overall charge of the fragment, but the mass of the nucleic acid or oligonucleotide increases correspondingly. Thus, the ratio of charge to mass remains approximately the same regardless of the size of the molecule in question. As a result, the separation takes place simply on the basis of size and is due to the sieving action of the gel. In a given amount of time, with a sample consisting of a mixture of oligonucleotides, a smaller oligonucleotide moves farther than a larger one in an electrophoretic separation. The oligonucleotides move in the electric field because of their charges; the distances they move in a given time depend on their sizes.

Most separations are done with an agarose gel in a horizontal position, called a *submarine gel* because it is underneath the buffer in the chamber. Many different samples can be separated on a single gel. Each sample is loaded at a given place (a distinct well) at the negative-electrode end of the gel, and an electric current flows until the separation is complete (Figure 13.2).

13-1B Nucleic Acid Detection Methods

After the DNA pieces have been separated, they must be treated in some way that allows them to be seen. One detection method involves treating the sample with the compound *ethidium bromide* (Figure 13.1). Its molecular structure includes a planar portion that can slip between the bases of DNA, giving ethidium bromide fluorescent properties when it binds to DNA that differ from those observed when it is free in solution. An ethidium bromide solution is added when the agarose gel is prepared, and ethidium bromide binds to the DNA fragments when they move through the gel. After running the gel, the DNA fragments in the sample can be seen as bands by shining ultraviolet light on the gel (Figure 13.1).

Ethidium bromide is a strong carcinogen, and great care must be taken in its use and disposal. Because of this, newer fluorescent dyes have been developed, such as SyBr Green and SyBr Gold, which are not as harmful and can be disposed of more easily.

13-2 Restriction Endonucleases

Learning Objectives

After reading this section, students will be able to:

LO 13.2.1 Distinguish between *nucleases* and *restriction endonucleases*.

LO 13.2.2 Identify DNA or RNA sequences that can be cleaved by a restriction endonuclease.

A class of enzymes, called **nucleases**, catalyzes the hydrolysis of the phosphodiester backbones of nucleic acids. Some nucleases are specific for DNA; others

Careers in Chemistry • Emmanuelle Charpentier

Scientific and Managing Director at Max Planck Unit for the Science of Pathogens

Education

University Pierre & Marie Curie (now Sorbonne University), PhD, Microbiology

University Pierre & Marie Curie, Master Studies, Biochemistry, Microbiology, and Genetics

University Pierre & Marie Curie, Bachelor Studies, Life Sciences, Biochemistry, Microbiology, and Genetics

Emmanuelle Charpentier's foundation in biochemistry has been central to her scientific career. Her research shaped one of the most transformative discoveries in modern biology: the CRISPR-Cas9 genome editing system. Charpentier's work revealed how bacteria defend themselves from viruses and how this natural defense mechanism can be harnessed to edit the genomes of other organisms, including human cells. Today, she serves as the Scientific and Managing Director of the Max Planck Unit for the Science of Pathogens in Berlin, where she works closely with the next generation of researchers, sharing with them her deep passion for science.

Charpentier credits the discipline and perseverance she developed as a young student of classical piano and ballet with preparing her for the demands of scientific research. From her parents and mentors, she learned to think independently and challenge conventional wisdom. Her experience navigating scientific environments as a woman in STEM also highlighted the importance of diversity and inclusion within research teams. Having worked and studied alongside scientists from a range of cultural and national backgrounds, she deeply values how such diversity can enrich scientific perspectives.

Hallbauer & Fioretti

To students, Charpentier offers this advice: "Science is often about exploring the unknown, and sometimes the greatest advances come from unexpected observations." She emphasizes the importance of interdisciplinary thinking and reminds aspiring scientists that "science is a global endeavor. Be open to collaborating with people from different cultures and backgrounds. Your path may not always be linear, but that's part of the adventure."

are specific for RNA. Cleavage from the ends of the molecule (by *exo*nucleases) is possible, as is cleavage in the middle of the chain (by *endo*nucleases). Some enzymes are specific for single-stranded nucleic acids, whereas others cleave double-stranded ones. One group of nucleases, called **restriction endonucleases**, has played a crucial role in the development of recombinant DNA technology.

restriction endonucleases enzymes that hydrolyze double-stranded DNA at specific sequences of bases

Restriction enzymes were discovered in the course of genetic investigations of bacteria and *bacteriophages* (*phages* for short; from the Greek *phagein*, "to eat"), the viruses that infect bacteria. The researchers found that bacteria use the restriction endonucleases as a protection mechanism that produces double-chain breaks at the unmethylated specific sequences in phage DNA, thus *restricting* the replication of the viral DNA (Figure 13.3). The cells' own corresponding DNA sequences, in which methylated bases occur, are not attacked. These cleavage enzymes consequently degrade DNA from any source *but* the host cell. DNA from any source can be cleaved by a restriction enzyme if it contains the enzyme's target sequence. Thousands of restriction endonucleases have been discovered in a variety of bacterial species. Table 13.1 shows target DNA sequences for several restriction enzymes.

13-2A Many Restriction Endonucleases Produce Sticky Ends

Each restriction endonuclease hydrolyzes only a specific bond of a specific sequence in DNA. The sequences recognized by restriction endonucleases—their sites of action—read the same from left to right as they do from right

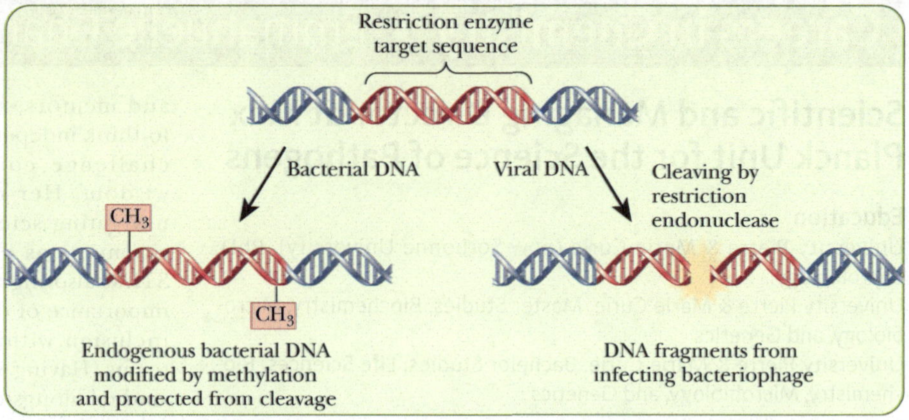

Figure 13.3 Restriction enzymes protect bacteria from bacteriophage viruses.

Table 13.1 Restriction Endonucleases and Their Cleavage Sites

Enzyme*	Recognition and Cleavage Site	Enzyme*	Recognition and Cleavage Site
*Bam*HI	↓ 5′-GGATCC-3′ 3′-CCTAGG-5′ ↑	*Hpa*II	↓ 5′-CCGG-3′ 3′-GGCC-5′ ↑
*Eco*RI	↓ 5′-GAATTC-3′ 3′-CTTAAG-5′ ↑	*Not*I	↓ 5′-GCGGCCGC-3′ 3′-CGCCGGCG-5′ ↑
*Hae*III	↓ 5′-GGCC-3′ 3′-CCGG-5′ ↑	Pst	↓ 5′-CTGCAG-3′ 3′-GACGTC-5′ ↑
*Hind*III	↓ 5′-AAGCTT-3′ 3′-TTCGAA-5′ ↑		

Arrows indicate the positions where phosphodiester bonds are cleaved by the restriction endonucleases.
*The name of the restriction endonuclease consists of a three-letter abbreviation of the bacterial species from which it is derived—for example, Eco for *Escherichia coli.*

palindrome a message that reads the same backward or forward

to left (on the complementary strand). The term for such a sequence is a **palindrome**. ("Taco cat" and "Race car" are examples of linguistic palindromes.)

A restriction endonuclease called *Eco*RI, which is commonly used in biochemical studies, was discovered in *E. coli.* Each restriction endonuclease is designated by an abbreviation of the name of the organism in which it was found, and the abbreviation usually consists of the first letter of the name of the genus and the first two letters of the name of the species. The *Eco*RI site in DNA is 5′-GAATTC-3′, where the base sequence on the other strand is 3′-CTTAAG-5′. The sequence from left to right on one strand is the same as the sequence from right to left on the other strand. The phosphodiester bond between G and A is the one hydrolyzed (Figure 13.4). This same break is made on both strands of the DNA.

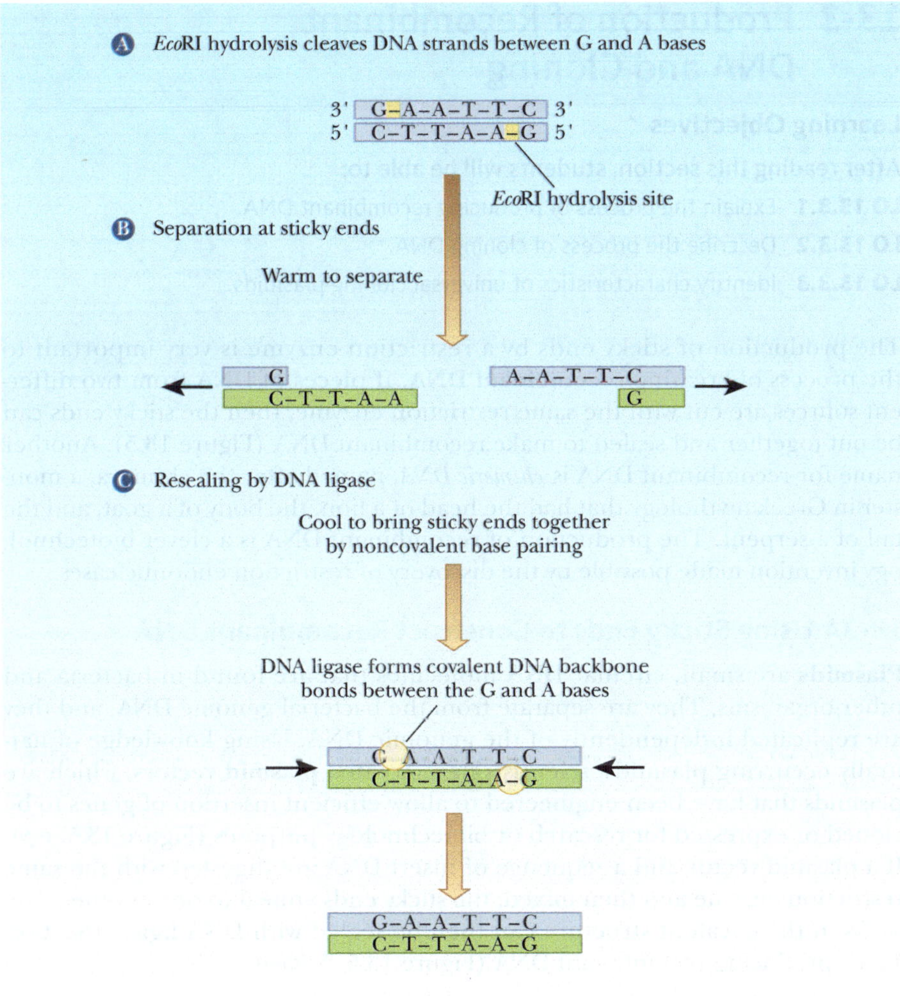

Figure 13.4 Hydrolysis of DNA by restriction endonucleases and the resealing of the DNA backbone by DNA ligase. Sometimes, restriction enzymes can also cut nonspecifically or at sites with sequences other than their usual recognition sequences. This is called *star (*) activity* and can be seen under such conditions as when the enzyme concentration was too high or the enzyme was incubated with the DNA too long. Also, different enzymes from different species may recognize the same DNA sequence and have the same specificity for cut sites. Such enzymes are called *isoschizomers* (the iso part meaning "equal" and the schiz part, also seen in schism, meaning "split").

There are four nucleotide residues—two adenines and two thymines in each strand—between the two breaks on opposite strands, leaving **sticky ends**, which can still be joined by hydrogen bonding interactions between the complementary bases, even though the DNA backbone has been hydrolyzed at the cleavage sites. With the ends held in place by the hydrogen bonds, the two breaks can then be resealed covalently by the action of DNA ligases (Section 10-4). If no ligase is present, the ends can remain separated, and the hydrogen bonding at the sticky ends holds the molecule together until gentle warming or vigorous stirring effects a separation. Some restriction enzymes, however, cut in a way that leaves a blunt end. *Hae*III is one such restriction enzyme:

sticky ends short, single-stranded stretches at the ends of cleaved double-stranded DNA; they can provide sites to which other DNA molecules with the same sticky-end sequence can be linked

$$\begin{array}{ccc} 5' - GGCC - 3' & \xrightarrow{\textbf{\textit{Hae}III}} & 5' - GG \quad\quad CC - 3' \\ 3' - CCGG - 5' & & 3' - CC \quad\quad GG - 5' \end{array}$$

Blunt-end cut

13-3 Production of Recombinant DNA and Cloning

Learning Objectives

After reading this section, students will be able to:

LO 13.3.1 Explain the process of producing recombinant DNA.

LO 13.3.2 Describe the process of cloning DNA.

LO 13.3.3 Identify characteristics of universal cloning plasmids.

The production of sticky ends by a restriction enzyme is very important to the process of creating **recombinant DNA**. If pieces of DNA from two different sources are cut with the same restriction enzyme, then the sticky ends can be put together and sealed to make recombinant DNA (Figure 13.5). Another name for recombinant DNA is *chimeric DNA*, named after the chimera, a monster in Greek mythology that had the head of a lion, the body of a goat, and the tail of a serpent. The production of recombinant DNA is a clever biotechnology invention made possible by the discovery of restriction endonucleases.

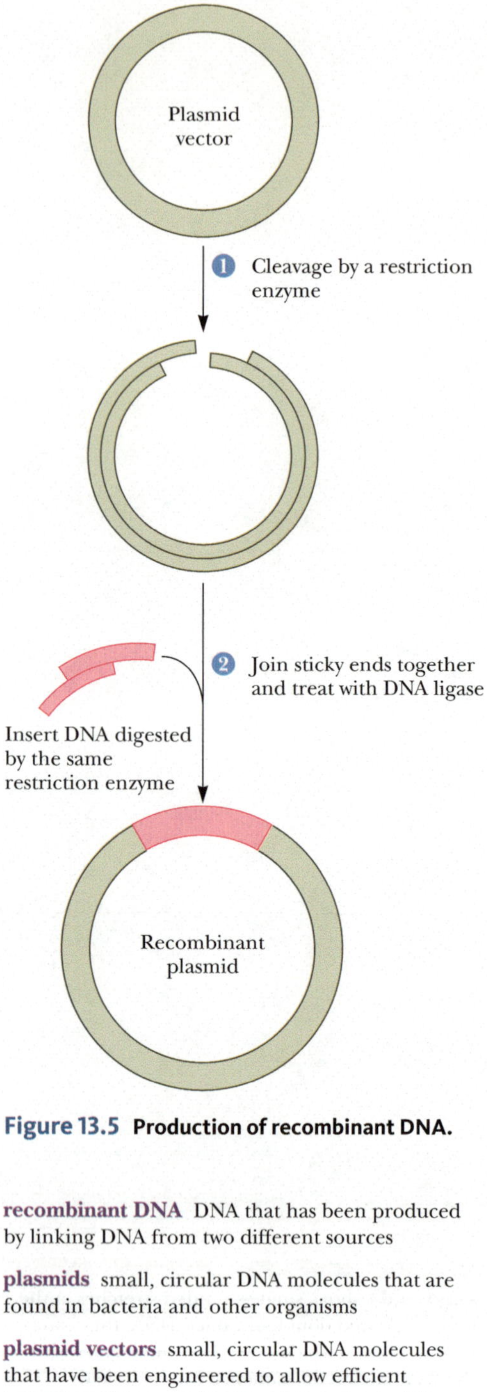

Figure 13.5 Production of recombinant DNA.

recombinant DNA DNA that has been produced by linking DNA from two different sources

plasmids small, circular DNA molecules that are found in bacteria and other organisms

plasmid vectors small, circular DNA molecules that have been engineered to allow efficient insertion of genes to be cloned or expressed for research or biotechnology purposes

DNA cloning a molecular biology technique that involves making many copies of a sequence of DNA

transformation a process by which external DNA is introduced into a cell

13-3A Using Sticky Ends to Construct Recombinant DNA

Plasmids are small, circular DNA molecules that are found in bacteria and other organisms. They are separate from the bacterial genome DNA, and they are replicated independently of the genomic DNA. Using knowledge of naturally occurring plasmids, scientists have created **plasmid vectors**, which are plasmids that have been engineered to allow efficient insertion of genes to be cloned or expressed for research or biotechnology purposes (Figure 13.5, *top*). If a plasmid vector and a sequence of insert DNA are digested with the same restriction enzyme and then mixed, the sticky ends anneal to one another. The nicks in the covalent structure can then be sealed with DNA ligase (Section 10-4), producing recombinant DNA (Figure 13.5, *bottom*).

13-3B DNA Cloning

DNA cloning is an important molecular biology technique in which a sequence of DNA is first inserted into a plasmid vector. Then that recombinant plasmid is introduced into bacterial cells using a process called **transformation**, where it is replicated, producing many copies of that DNA sequence. The term *clone* refers to a genetically identical population, whether of organisms, cells, viruses, or DNA molecules. Every member of the population is derived from a single cell, virus, or DNA molecule. Bacteriophage viruses and bacterial cells can produce large numbers of progeny relatively easily in a research laboratory.

To clone individual cells, whether from a bacterial or a eukaryotic source, a small number of cells is spread thinly over a suitable growth medium in a dish. Spreading the cells thinly ensures that each cell will multiply in isolation from the others. Each colony of cells that appears on the dish will then be a clone derived from a single cell (Figure 13.6). Because large quantities of bacteria can be grown in short time intervals under laboratory conditions, it is useful to introduce DNA from a larger, slower-growing organism into bacteria and to produce more of the desired DNA by inserting it into a plasmid vector and then cloning the recombinant DNA in bacterial cells.

13-3C Plasmid Vectors

When a recombinant plasmid (Figure 13.5, *bottom*) is taken up by bacteria using transformation techniques, the bacteria that contain the DNA insert can then be grown in fermentation tanks under conditions that allow them to divide rapidly, amplifying the inserted DNA many thousandfold (Figure 13.7).

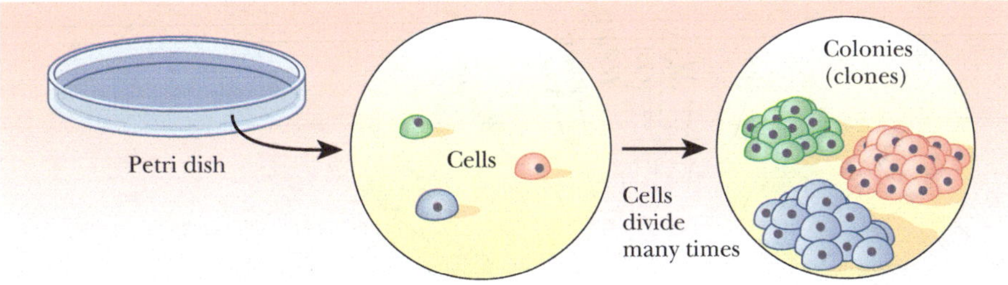

Figure 13.6 The cloning of cells. Each individual cell divides many times, producing a colony of progeny. Each colony is a clone containing many identical cells. *(Adapted with permission from* Dealing with Genes: The Language of Heredity, *by Paul Berg and Maxine Singer, ©1992 by University Science Books.)*

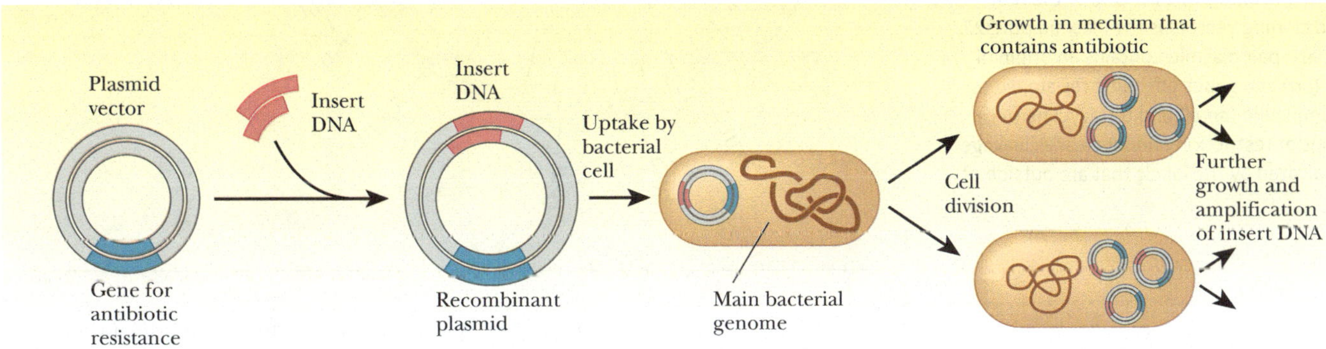

Figure 13.7 Selecting for and cloning recombinant DNA. The plasmid vector contains a gene for antibiotic resistance. When bacteria are grown in a medium that contains the antibiotic, only those cells that have the recombinant plasmid can grow. Bacteria without a plasmid cannot grow in this medium. *(Adapted with permission from* Dealing with Genes: The Language of Heredity, *by Paul Berg and Maxine Singer, ©1992 by University Science Books.)*

How are we to know that the bacteria being grown are ones that have the recombinant DNA containing the insert DNA? Plasmid vectors have some type of *selectable marker* that lets us know that the growing bacterial colonies contain the plasmid. These markers are usually genes that confer resistance to antibiotics. After transformation, the bacteria are plated on a medium containing the antibiotic to which the plasmid carries resistance. In that way, only the bacteria that took up the plasmids will grow (Figure 13.7). One of the first plasmids that was engineered for cloning was pBR322 (Figure 13.8). This simple plasmid was created from a naturally occurring one found in *Escherichia coli (E. coli)*. Like all plasmids, it has an origin of replication, so it can replicate independently of the rest of the genome. It has genes that confer resistance to two antibiotics, tetracycline and ampicillin. The genes are labeled as *tet*^{r} and *amp*^{r} in Figure 13.8. The pBR322 plasmid also has several restriction enzyme sites. The selection of a restriction site is very important to a cloning experiment. For example, if the restriction site chosen is inside one of the selection markers, the resistance to the antibiotic is lost upon insertion of the insert DNA.

One of the early stumbling blocks to cloning was finding the right plasmid—namely, one that had restriction sites that matched the enzymes needed to cut the insert DNA from its original source. As the technology to design plasmids improved, they were created with regions that had many different restriction sites in a small space. This region is called a **multiple cloning site (MCS)** or a *polylinker* (Figure 13.9).

multiple cloning site (MCS) a region of a bacterial plasmid with many unique restriction sites; an example of a polylinker.

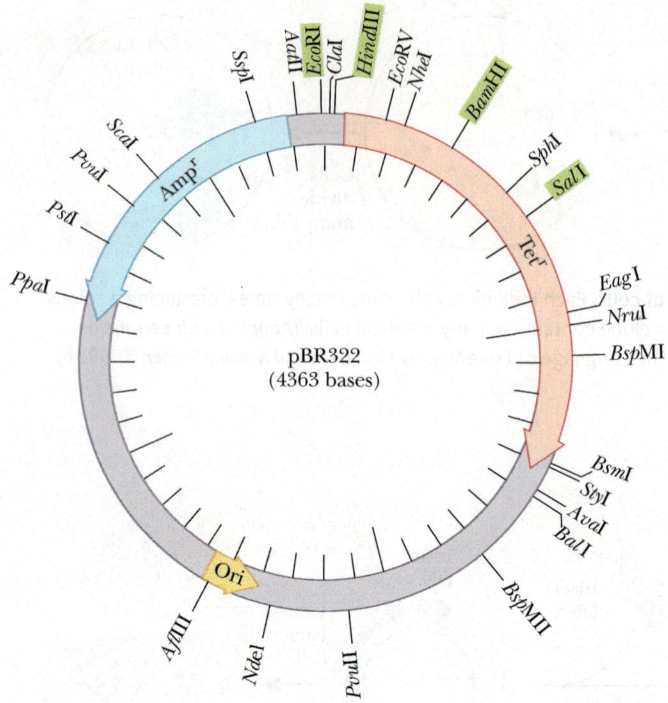

Figure 13.8 **Plasmid pBR322.** One of the first widely used cloning vectors is the plasmid pBR322. This 4363-base-pair plasmid contains an origin of replication (*ori*) and genes encoding resistance to the drugs ampicillin (*amp*ʳ) and tetracycline (*tet*ʳ). The locations of restriction endonuclease cleavage sites are indicated by the labels that are outside of the circle.

1	2	3	4	5	1	2	3	4	5	6	7	8	9	10	11	12	13	14	6
Met	Thr	Met	Ile	Thr	Asn	Ser	Pro	Asp	Pro	Ser	Thr	Cys	Arg	Ser	Thr	Asp	Pro	Gly	Asn	Ser
ATG	ACC	ATG	ATT	ACG	AAT	TCC	CCG	GAT	CCG	TCG	ACC	TGC	AGG	TCG	ACG	GAT	CCG	GGG	AAT	TCA

*Eco*RI *Bam*HI *Sal*I *Acc*I *Hinc*II *Pst*I *Sal*I *Acc*I *Hinc*II *Bam*HI *Eco*RI

Figure 13.9 **A vector cloning site containing multiple restriction sites.** This is called a polylinker or multiple cloning site (MCS). The colored amino acids are from the *lacZ* gene that is part of the plasmid. The MCS does not disrupt the normal reading frame of this sequence, so this plasmid can be used for blue/white screening (Section 13-3D) *(Adapted from Figure 1.14.2 in* Current Protocols in Molecular Biology. *Edited by Frederick M. Ausubel. John Wiley & Sons, Inc., 1995.)*

13-3D Universal Cloning Plasmids

A popular cloning vector series is based on the pUC plasmids (Figure 13.10). The acronym *pUC* stands for *u*niversal *c*loning *p*lasmid. Each of these cloning vectors has an extensive multiple cloning site (MCS), which helps solve another problem with cloning—the directionality of the inserted DNA. Depending on what is to be done with the cloned DNA, controlling its orientation in the vector may be important. If only one restriction enzyme is used, such as *Bam*H1, then the insert DNA can enter the plasmid in either of two directions. However, if the insert DNA is cut out of its source at one end with *Bam*H1 and at the other end with *Hind*III, for example, and if these same two restriction enzymes are used to open up the plasmid, then the ends match in only one direction (Figure 13.10).

The use of the pUC plasmids also aids in the selection procedure. The pUC plasmids have a *lacZ* gene that allows for a selection technique called **blue/white screening.** Recall that the *lacZ* gene codes for the α-subunit of the enzyme β-galactosidase, which is used to cleave disaccharides such as lactose (Section 11-3C). The MCS in pUC plasmids is located inside the *lacZ*

blue/white screening a method for determining whether bacterial cells have incorporated a plasmid that includes insert DNA

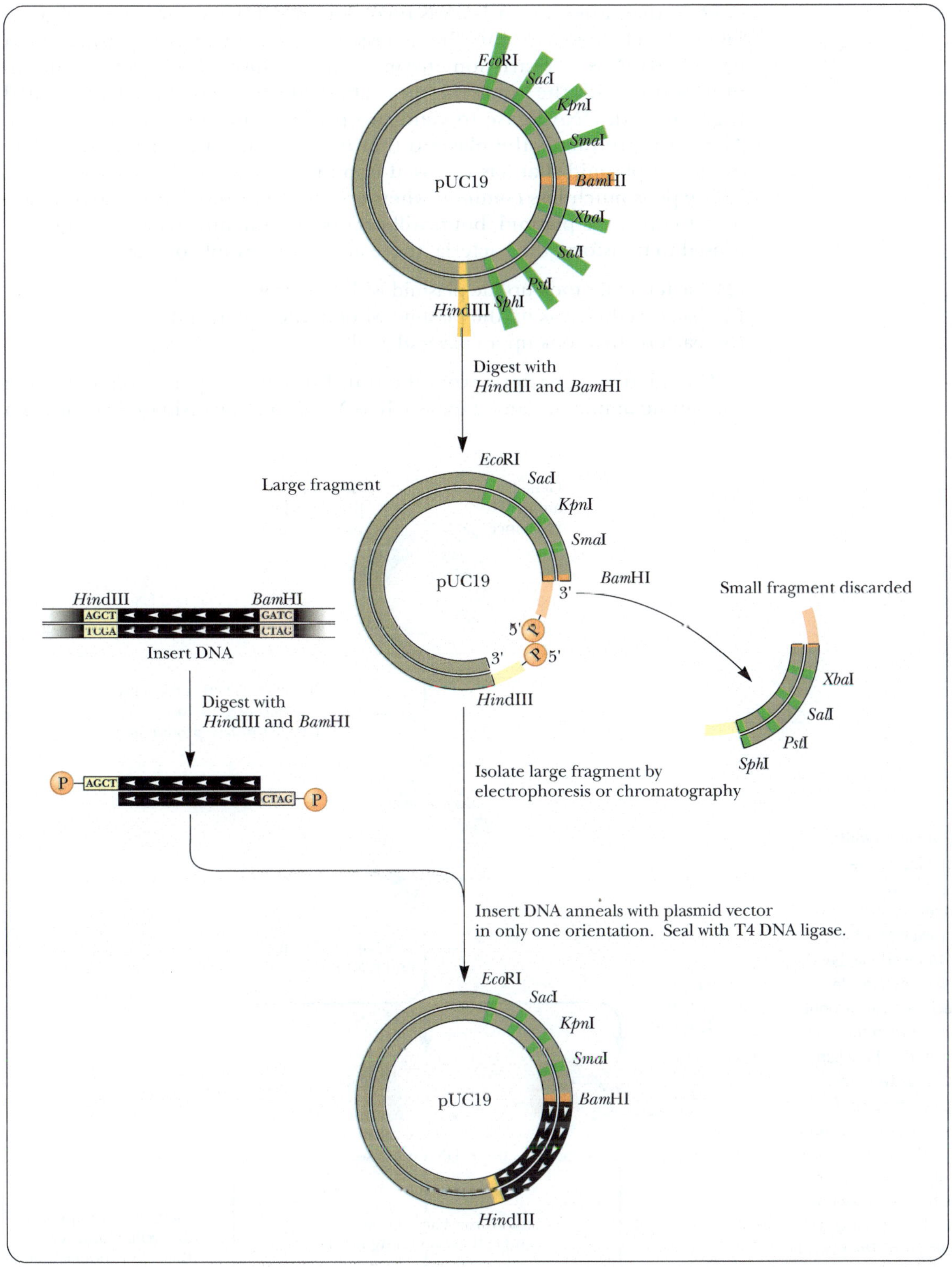

Figure 13.10 Cloning with pUC plasmids. The pUC series of plasmids is very popular. They have extensive multiple cloning sites. Here we see an example of directional cloning. Two different restriction enzymes are used to cut open the MCS and to cut out a piece of DNA to be cloned. As a result, the DNA that is to be inserted can be incorporated in only one orientation.

gene so that, when insert DNA is incorporated, it inactivates the *lacZ* gene. Figure 13.11 shows how this characteristic is useful. The pUC plasmid is cut open in its MCS by restriction enzymes, and the insert DNA (red) is cut out of its source with the same enzymes. These are then combined and joined together with DNA ligase to yield two products in the ligation reaction. The main product is the plasmid that now contains the insert DNA. The other is a plasmid that has reclosed upon itself without the inserted DNA. This type is much less common when two different restriction enzymes are used to open the plasmid, but it still occurs infrequently. When this mixture is used to transform the bacteria, there are three possible outcomes:

(1) bacteria that took up the plasmid with the insert,
(2) bacteria that took up the plasmid without the insert, and
(3) bacteria that took up no plasmid at all.

The mixture of bacteria from the transformation is plated on a medium containing ampicillin and a dye such as X-gal. β-Galactosidase hydrolyzes a

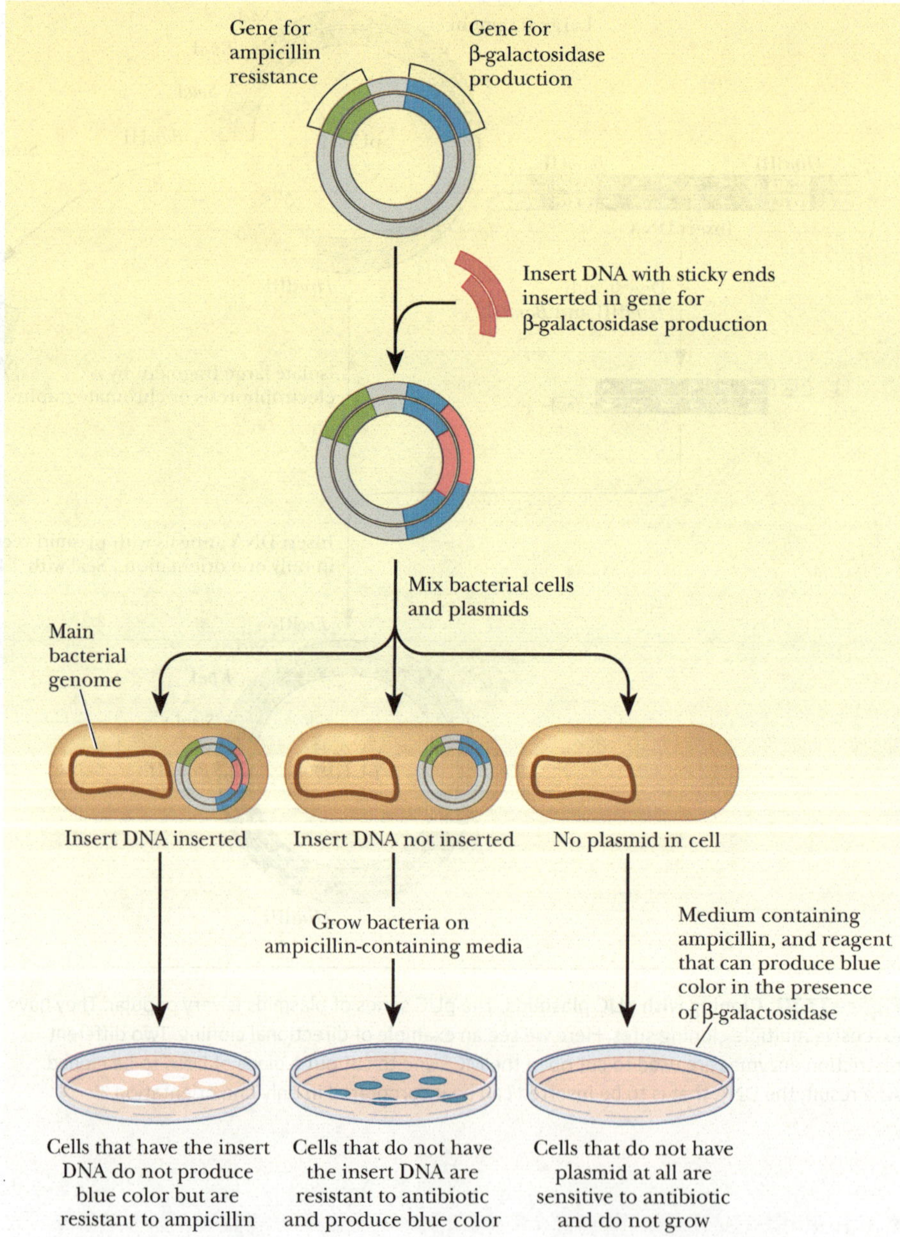

Figure 13.11 Clone selection via blue/white screening. A *lacZ* gene is located inside the multiple cloning site of a pUC plasmid. When the plasmid and the DNA to be cloned are cut with a restriction enzyme and then mixed together, two possibilities result. The DNA insert can be incorporated as shown (red insert seen inside plasmid), or the plasmid can recircularize without the insert (blue gene seen intact). When this mixture is used to transform bacteria, there can be three products. (*left*) The bacteria take up a plasmid that has the insert. This plasmid confers ampicillin resistance to the cells, but the *lacZ* gene is inactivated by the presence of the insert. These cells grow and are the normal off-white color of bacterial colonies. (*middle*) The bacteria take up the recircularized plasmid. This plasmid confers ampicillin resistance, so the cells grow. The plasmid makes the α-subunit of β-galactosidase. The β-galactosidase cleaves the X-gal, causing the dye to turn blue, so these cells grow with a blue color. (*right*) Bacteria take up no plasmid at all. These cells do not grow because of their sensitivity to ampicillin. *(Adapted with permission from* Dealing with Genes: The Language of Heredity, *by Paul Berg and Maxine Singer, ©1992 by University Science Books.)*

bond in the X-gal molecule, turning it blue. If the bacteria take up no plasmid at all, they lack the ampicillin-resistance gene and do not grow (Figure 13.11, *bottom right*). If the bacteria take up a plasmid lacking the insert, they have a functional *lacZ* gene and produce the α-subunit of β-galactosidase. These colonies produce active β-galactosidase, which cleaves the dye, X-gal. These colonies grow with a blue color (Figure 13.11, *bottom middle*). If the bacteria take up a plasmid that contains the insert, the *lacZ* gene is inactivated. These colonies are the off-white color of normal bacterial colonies on agar (Figure 13.11, *bottom left*).

13-4 Genetic Engineering

Learning Objectives

After reading this section, students will be able to:

LO 13.4.1 Explain examples of genetic engineering.

LO 13.4.2 Describe how genetic engineering can be used for large-scale production of human proteins.

LO 13.4.3 Explain how the genome editing technique called CRISPR-Cas9 works.

The previous sections dealt with how DNA of interest could be inserted into a vector and amplified by cloning. One of the most important purposes for doing this is to be able to produce the gene product in larger quantities than could be acquired by other means. When an organism is intentionally changed at the molecular level so that it exhibits different traits, we say it has been *genetically engineered.*

In a sense, **genetic engineering** on an organismal level has been around since humans first started to use selective breeding on plants and animals. This procedure did not deal directly with the molecular nature of genetic material, nor was the appearance of traits under human control. Breeders had to cope with changes that arose spontaneously, and the only choice was whether to breed for a trait or to let it die out. An understanding of the molecular nature of heredity and the ability to manipulate those molecules in the laboratory have added to our ability to control the appearance of these traits.

genetic engineering the process of manipulating the genome of an organism to achieve a desired end

The practice of selective alteration of organisms for both agricultural and medical purposes has profited greatly from recombinant DNA methods. Genetic engineering of crop plants is an active field of research. Genes for increased yields, frost resistance, and resistance to pests are introduced into commercially important plants such as strawberries, tomatoes, and corn. Similarly, animals of commercial importance—mostly mammals, but also including fish—are also genetically altered. Genetic variations introduced in animals have medical implications. Mice with altered genetic makeup are used in the research laboratory to enhance our understanding of human diseases such as cancer, Parkinson's, and Alzheimer's. In all cases, the focus of the research is to introduce *traits that can be inherited* by the descendants of the treated organisms. In the treatment of human genetic disease, however, the aim is not to produce heritable changes. Serious ethical questions arise with the manipulation of human genetics; consequently, the focus of research has been on forms of **gene therapy** in which cells of specific tissues in a living person are altered in a way that alleviates the effects of the disease. Examples of diseases that may someday be treated in this way include cystic fibrosis (CF), hemophilia, Duchenne muscular dystrophy, and severe combined immune deficiency (SCID).

gene therapy a method for treating a genetic disease by introducing a good copy of a defective gene

Since the advent of recombinant DNA technology, it is possible (within limits) to change specific genes, and even to change specific DNA sequences within those genes, to alter the inherited characteristics of organisms. Bacteria can be altered to produce large amounts of medically and economically important proteins. Animals can be manipulated to cure, or to alleviate the symptoms of, their genetic diseases, and agriculturally important plants can be made to produce greater crop yields or be given increased resistance to pests. **Biochemical Connections 13A** gives some examples of agricultural applications of genetic engineering.

13A Biochemical Connections • Plant Science

Genetic Engineering in Agriculture

The idea of genetic engineering in humans often upsets people because they are afraid of "playing God." Genetic manipulations in plants are less controversial, although many people are still concerned about the practice. Nevertheless, many types of modifications have been made, and some have been introduced with little fanfare and have shown some signs of success. Several examples are listed here. Many of the modifications that have been made using genetic engineering are just controlled versions of the selective breeding used for centuries to improve crop and animal production.

1. *Disease resistance.* Because most high-production crops involve specialized strains, many are more susceptible to fungal disease and insect damage. As a consequence, many herbicides and insecticides are applied liberally during the growing season. In many cases, other plants have natural resistance to these pests. When the gene that produces the resistance can be isolated, it can be transferred into other plants. There has been limited success in transferring such resistance to crop species. In 2000, the public became more aware of genetically modified (GM) food crops because of news about Bt corn, a crop carrying a bacterial gene that produces a toxin poisonous to certain caterpillars. The Bt gene, from *Bacillus thuringiensis,* has been put into corn and cotton to increase crop yields by killing the caterpillars that would otherwise eat those crops. It also reduces the amount of pesticides needed to grow the crops. On the negative side is the potential harm to other species. For example, environmental groups opposed the planting of Bt corn because of its effect on the monarch butterfly in initial lab tests. However, in 2022, the United States Department of Agriculture (USDA) published a study that suggested that Bt corn has little impact on nontarget insects and other organisms.
2. *Nitrogen fixation.* Nitrogen fixation is most easily accomplished by bacteria that grow in nodules on the roots of certain legumes such as beans, peas, and alfalfa. We now realize that the genes of the nitrogen-fixing bacteria actually become shared with the genes of the host plant. Much research is going on to determine whether these genes could be incorporated into other plants, which would reduce the amount of nitrogen fertilizer needed for maximal plant growth and crop production.
3. *Frost-free plants.* Many marine organisms, such as fish found in the Antarctic, produce a so-called antifreeze protein—a protein characterized by its hydrophobic surface—which prevents the formation of ice crystals at low temperatures. In plants exposed to freezing temperatures, ice crystals in their tissues actually cause the frost damage. Insertion of the gene for the antifreeze protein into strawberries and potatoes, for example, has resulted in crops that are stable during late spring frosts or in areas with very short growing seasons. Much controversy has ensued over these crops because, in each case, a foreign gene from a nonplant species was introduced into a plant. This type of cross-species gene modification causes widespread concern, even though the taste and texture of the modified product are indistinguishable from those of foods produced by unmodified plants.
4. *Tomatoes with a long shelf life.* The first genetically modified food granted a license for human consumption was the Flavr Savr tomato, first created by a California biotech firm called Calgene. Polygalacturonase (PG), which degrades pectin in the cell walls, is an enzyme that hastens the softening and eventual rotting of fruit. Calgene worked to reduce the effects of this enzyme so that tomatoes would not soften as quickly, theoretically allowing them to ripen longer and then be picked while they were still firm. To attack PG, they inserted a second PG gene but in the antisense direction. When the antisense gene is expressed, it interferes with the expression of the normal gene. The tomatoes were first produced in 1994. Unfortunately for Calgene, their tomatoes were never a financial success and were withdrawn from the market in 1997. Calgene was later purchased by Monsanto.

   Three companies, including Monsanto, then took a different approach, which was to reduce the amount of naturally produced ethylene, a ripening agent, in the fruit. Their technique was to inhibit the production of 1-aminocyclopropane-1-carboxylic acid (ACC), a precursor of ethylene. The Monsanto version included a bacterial version of ACC deaminase, which degrades ACC. While the techniques for modifying tomatoes were successful in the sense that the fruit did ripen slower, most were not commercially successful. There is only one genetically modified tomato that is approved for consumption. It is called the Sicilian Rouge tomato, and it was approved in 2021 in Japan. The Sicilian Rouge tomato has the improved nutritional characteristic of higher levels of gamma-aminobutyric acid (GABA), a neurotransmitter that has a variety of health benefits. It is also the first CRISPR-edited (Section 13-4C) food to enter the market.

5. *Bruise resistant and healthier potato.* In 2014, the J. R. Simplot Company gained USDA approval for a genetically modified potato, called the Innate potato. The desired attributes of this potato include resistance to blemishes caused by bruising, reduction in sugar content, and lower levels of the suspected carcinogen acrylamide, which can be generated by making French fries. Unlike many controversial modifications introduced by genetic engineering, the innate potato was created by splicing genes with desirable attributes from wild potatoes into potato DNA. Thus, much of the criticism associated with nonnatural manipulation of the genome with foreign genes was averted.
6. *Good predator attraction.* Researchers in the Netherlands were studying the mustard plant, *Arabidopsis thaliana*, which is very susceptible to attack by herbivorous spider mite predators. They introduced a strawberry gene into the mustard plants that produces a chemical attractant for predatory mites that eat the herbivorous spider mites (Figure 13.12).

Sigrid Gombert/Science Source
iStock.com/Tomasz Klejdysz

Figure 13.12 The mustard plant *Arabidopsis thaliana* has undergone transgenic alteration to adopt favorable characteristics that enhance productivity.

13-4A Bacteria as "Human Protein Factories"

We can use the reproductive power of bacteria to express large quantities of a mammalian protein of interest; however, the process is often more complicated than it might seem because most mammalian proteins are heavily processed after their initial transcription and translation (Section 12-7). Because bacteria have little posttranslational modification of their proteins, they lack the enzymes necessary for this processing.

▶ How can human proteins be made by bacteria?

An application of genetic engineering that is of considerable practical importance is the production of human insulin by *E. coli*. This was one of the first human proteins produced through genetic engineering, and its production eliminated the problems related to harvesting insulin from large numbers of laboratory animals and giving humans a peptide from another species. The process is far from straightforward, however. A significant problem is that the insulin gene is split (Figure 13.13). It contains an *intron*—a DNA sequence that codes for RNA that is eventually deleted in the processing of the mRNA that directs the synthesis of the protein (see Section 11-8B). Only the RNA transcribed from DNA sequences called *exons* appears in mature mRNA that is exported from the nucleus for expression. (Recall that one way to remember the difference between introns and exons is that ***ex***ons are the sequences that are ***ex***ported.) Bacteria do not have the cellular apparatus for splicing introns out of RNA transcripts to give functional mRNA. Another problem is that the polypeptide encoded by this mRNA contains an end peptide and a central peptide, which need to be removed from it by further processing in insulin-producing cells to yield two polypeptide chains, designated A and B (Figure 13.13, *bottom*). Bacteria lack the enzymes needed to do these peptide processing events.

The approach to this problem is to use two synthetic DNAs, one encoding the A chain of insulin and the other encoding the B chain (Figure 13.14). These synthetic DNAs are produced in the laboratory using methods that were developed by synthetic organic chemists. Each DNA is inserted into a separate plasmid vector. Two different batches of *E. coli* are then transformed and cloned, each with one of the two recombinant plasmids (Figure 13.14, *middle*). Each group of bacteria produces one of the two polypeptide chains of insulin. The A and B polypeptide chains are extracted, purified, and mixed, finally

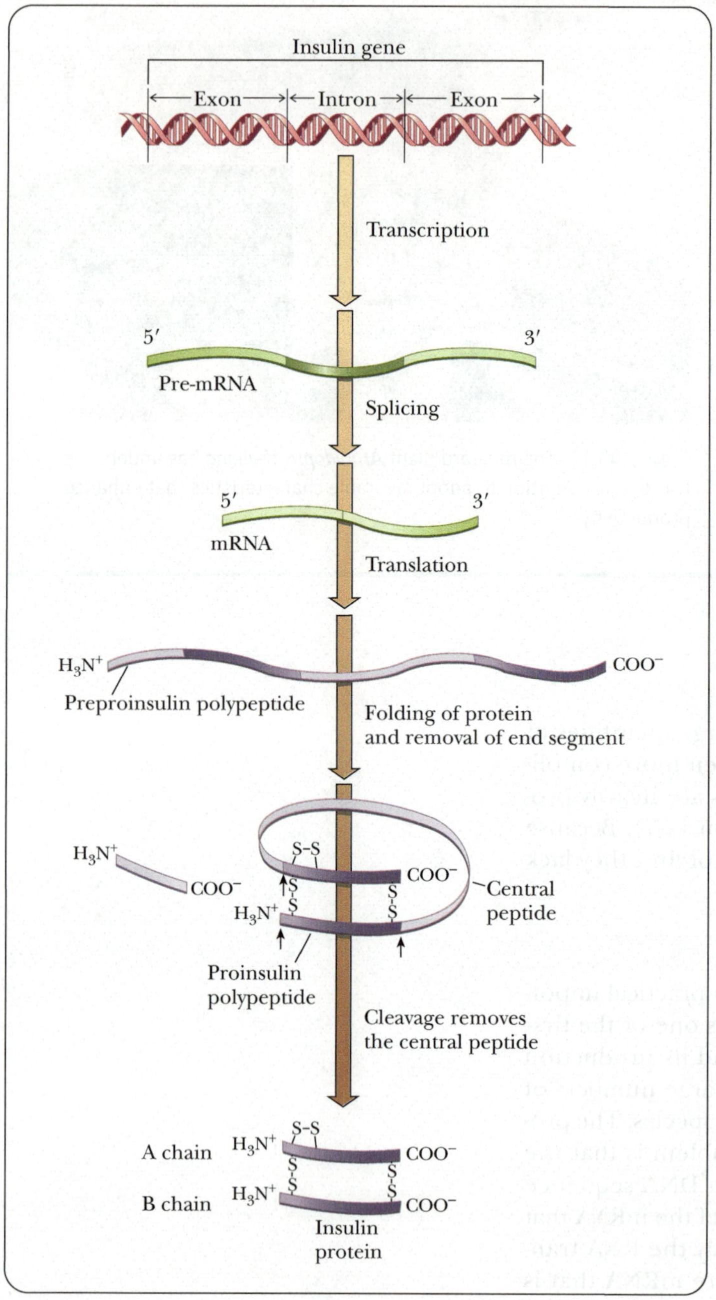

Figure 13.13 Synthesis of insulin in humans. The insulin gene is a split gene. The intervening sequence (intron) is spliced out of the mRNA. Only the portions of the gene called exons are reflected in the base sequence of mRNA. Once protein synthesis takes place, the polypeptide is folded, cut, and spliced. The end product, active insulin, has two polypeptide chains as a result. *(Adapted with permission from* Dealing with Genes: The Language of Heredity, *by Paul Berg and Maxine Singer, ©1992 by University Science Books.)*

producing functional human insulin. To review protein extraction and purification methods, see Sections 5-1 and 5-2, respectively.

13-4B Protein Expression Vectors

The plasmid vectors pBR322 and pUC (Section 13-3C) are referred to as *cloning vectors*. They are used to incorporate insert DNA and amplify it. However, if the goal is to produce recombinant protein product from insert DNA, they are unsuitable—an **expression vector** is needed instead.

▶ What is an expression vector?

An expression vector has many of the same attributes as a cloning vector such as the origin of replication, a multiple cloning site (MCS), and at least one selectable marker (Figure 13.15). In addition, it must be able to be transcribed by the genetic machinery of the bacteria into which it is transformed. This means that it must have a promoter for RNA polymerase (Section 11-2B), and the RNA transcribed must have a ribosomal binding site so that it can be translated (Section 12-4B). It must also have a transcription termination sequence (Section 11-2E); otherwise, the entire plasmid is transcribed instead of just the inserted gene. Figure 13.15 shows a schematic of an expression vector. Upstream of the insert DNA sequence is the transcription promoter. Often, this is engineered to be the promoter sequence for a viral RNA polymerase called *T7 RNA polymerase*. A key feature of the T7 RNA polymerase promoter is that it allows a high level of transcription of the gene of interest, and it can be regulated to control the timing and amount of protein expression. There is also a T7 terminator at the other end of the MCS.

After the insert is successfully ligated, the plasmid is transformed into an expression strain of bacteria, such as *E. coli* JM109 DE3. What makes this strain unique is that it has a gene that produces T7 RNA polymerase, but the gene is under the control of the *lac* operon (Section 11-3C). Once the bacteria are growing well with the plasmid, the cells are given a lactose analog called isopropylthiogalactoside (IPTG). This removes the lac repressor protein, thereby stimulating the *lac* operon and expression of T7 RNA polymerase, which then binds to the T7 promoter on the protein expression vector and transcribes the gene for the protein of interest. The bacterial cells then translate the mRNA into protein. This selective control of the expression is important because many foreign proteins are toxic to the cells; expression must be timed carefully by waiting for the bacteria to grow sufficiently before adding IPTG. **Biochemical Connections 13B** gives an example of how protein expression can be linked to a novel purification scheme, and **Biochemical Connections 13C** gives a practical application for genetic engineering to create a protein that can be easily purified.

expression vector a plasmid containing DNA elements that can direct the synthesis of a desired protein

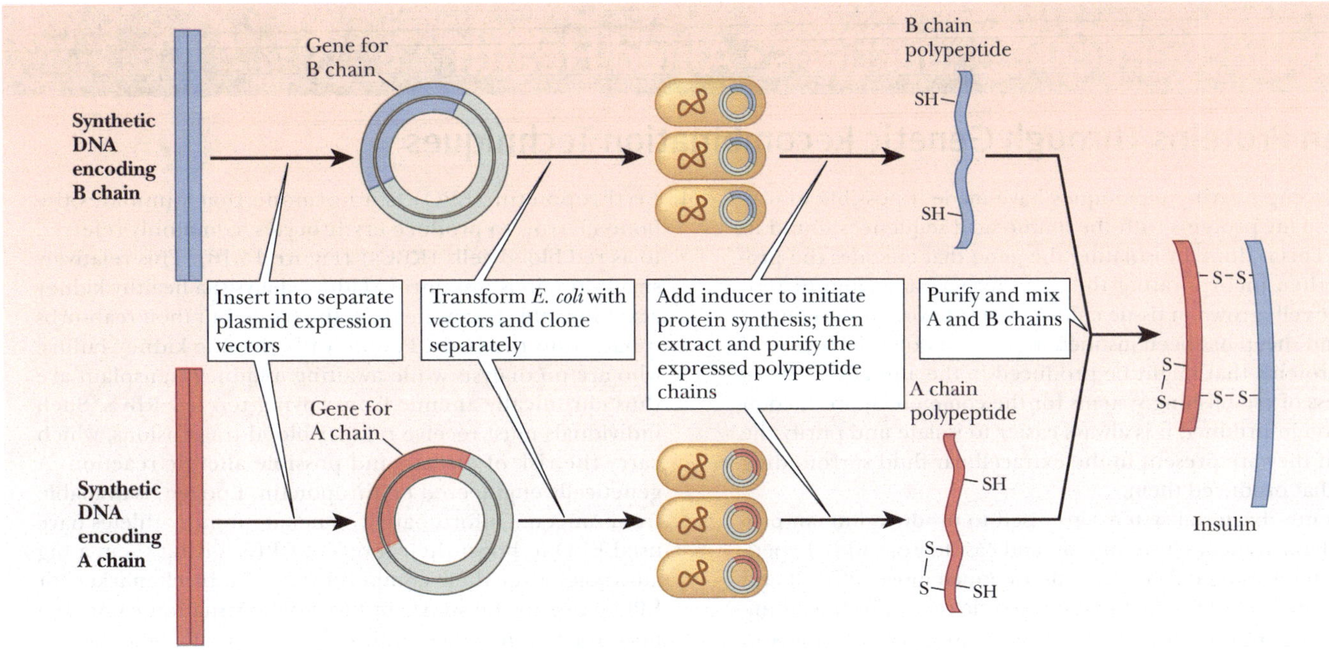

Figure 13.14 Production of recombinant human insulin. Active human insulin can be produced in bacteria by the use of two separate batches of *E. coli*. Each batch produces one of the two chains, the A chain or the B chain. The two chains are mixed to produce active insulin. *(Adapted with permission from* Dealing with Genes: The Language of Heredity, *by Paul Berg and Maxine Singer, ©1992 by University Science Books.)*

13-4C Genome Editing and the CRISPR-Cas9 System

genome editing a type of genetic engineering in which the genome of an organism is altered by insertions, deletions, or other modifications

Genome editing is a type of genetic engineering in which the genome of an organism is altered by insertions, deletions, or other modifications. It is a very promising set of tools that must be used carefully because it can make precise edits at site-specific locations. These changes get passed on every time a cell divides, so they carry the risk of unforeseen harmful effects.

The most widely used genome editing technique is CRISPR-Cas9, which is often referred to simply as CRISPR. CRISPR stands for clustered, regularly interspaced, short palindromic repeat, and the technology uses the Cas9 RNA-guided endonuclease for genetic engineering. The Cas9 nuclease is used in combination with a user-designed guide RNA to selectively cut target sequences of DNA. The Cas9 guide RNA can be read by the repair machinery of the cell, resulting in deletion, addition, or modification of the host organism's DNA. Gene editing performed with CRISPR-Cas9 has been shown to work in a wide range of organisms, including plants, animals, and humans, and some of its promises include disease treatment and prevention, personalized medicine, and agricultural improvements.

CRISPR was first described in 1987, and fundamental research was performed on the genome editing approach up until a major breakthrough in 2012, when it was demonstrated that the CRISPR/Cas9 RNA-guided DNA endonuclease could be applied to mediate site-specific genome engineering in eukaryotic cells, including human cells. Since 2012, thousands of studies have been published, and fundamental research on in vitro model systems has shifted to in vivo mammalian testing.

To understand how genome editing with the CRISPR-Cas9 system works, let's discuss the four key components of the system: CRISPR RNA

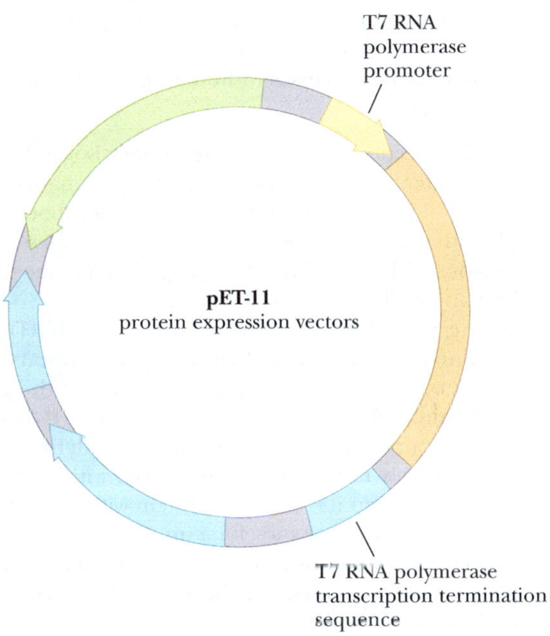

Figure 13.15 pET-11, a pET expression vector. These plasmids have the usual DNA components such as an origin of replication, a multiple cloning site (MCS), and an antibiotic resistance gene that provides resistance to ampicillin. In addition, their MCS is between the promoter for binding T7 RNA polymerase and a termination site for T7 RNA polymerase. When these vectors carry inserts, the insert DNA can be transcribed in the cell by T7 RNA polymerase. The cell then translates the mRNA into protein. *(Courtesy of Stratagene.)*

13B Biochemical Connections • Allied Health

Human Proteins Through Genetic Recombination Techniques

Genetic engineering techniques have made it possible to prepare many proteins with the amino acid sequences found in humans. This is done by isolating the gene that encodes the protein and then incorporating the gene into a bacterium or into eukaryotic cells grown in tissue culture. Recent work in the cloning of cows and sheep has been justified in part as a potential source of human proteins that might be produced in the animals' milk. In the process of establishing systems for the commercial production of any protein product, it is always easier to isolate and purify the proteins if they are present in the extracellular fluid surrounding the cells that produced them.

Frequently, bacterial systems are used to produce human proteins. Bacteria are very cheap to grow and easy to work with. Experimental systems using eukaryotic cells are much more difficult and expensive, in spite of the fact that eukaryotic cells offer the advantage of being able to add sugar residues to glycoproteins and perhaps help the proteins fold into their biologically active conformations. Still, using bacteria and recombinant DNA for protein production has its complications. Growing proteins in bacteria requires the addition of the Shine–Dalgarno sequence (Section 12-4) to the mRNA to ensure its binding to the ribosome. The genetic manipulations necessary to make the product in a foreign host often result in a protein that is slightly modified, usually by having a few extra amino acids on the N-terminus. The presence of the extra amino acids complicates the process of approval by the FDA because there is more concern about side effects.

Some notable successes in human protein production include the following:

1. Insulin (Section 13-4A) was traditionally isolated from horses and swine, and it was used to treat type I diabetes mellitus. However, it was discovered that over time, about 5% of people with diabetes develop a severe allergy to these nonhuman versions of the protein. These people can now be treated with human insulin produced in bacterial cells.
2. In the past, human growth hormone (HGH) has been obtained only by extraction from the pituitary glands of cadavers, a practice that carries the risk of the cells being contaminated with HIV or other diseases. HGH is used for therapy in people with growth hormone deficiencies. HGH is a relatively large protein hormone, with more than 300 amino acids, but it is a simple protein with no sugar residues, so it was relatively easy to clone into bacteria. The availability of increased amounts of safe HGH has resulted in the emergence of a black market in the sale of the hormone for muscle building in athletes.
3. Two proteins, tissue plasminogen activator (TPA) and enterokinase (EK), are known to dissolve blood clots. If they are injected into the body within a critical period after a heart attack or stroke, either of these two proteins can prevent or minimize the disastrous effects of blood clots in the heart or brain. Without genetic recombination biotechnology techniques, enough of these proteins would never be available for this treatment to be of practical use.
4. Erythropoietin (EPO) is a hormone that stimulates the bone marrow to produce erythrocytes, commonly referred to as red blood cells (RBCs) (Figure 13.16). This relatively small protein is lost during kidney dialysis; a healthy kidney also filters this hormone from the blood but then reabsorbs it back into the body. People with chronic kidney failure who are on dialysis while awaiting a kidney transplant are thus chronically anemic from having too few RBCs. Such individuals must receive regular blood transfusions, which carry the risk of disease and possible allergic reaction. A genetically engineered erythropoietin, Epogen, is available from Amgen. Unfortunately, some endurance athletes have used EPO to boost their levels of RBCs, giving them a big advantage over their competitors, and a black market for EPO exists in the sports industry. The situation came to a head during the 1998 Tour de France when a team doctor was arrested crossing the border into France while carrying countless vials of EPO. The EPO was eventually traced to a theft from a hospital. The team was ejected from the race, and the international governing body of cycling has invested hundreds of thousands of dollars to combat EPO usage, including developing assays for the detection of recombinant EPO. ◗

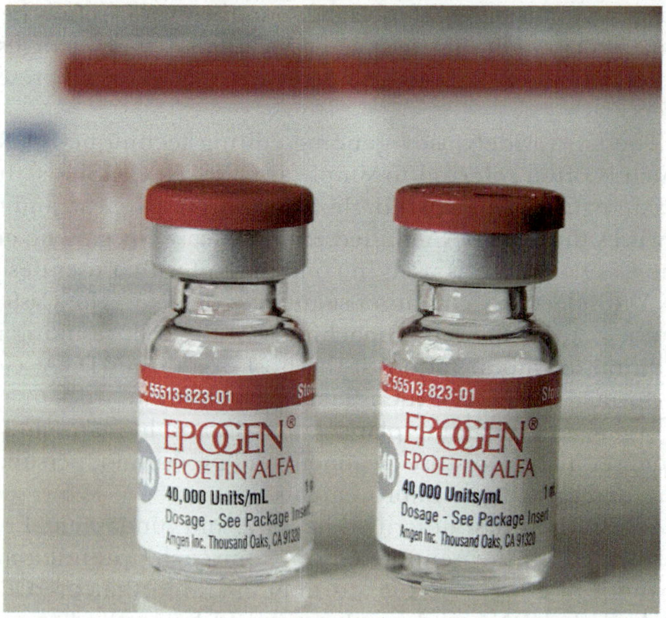

Figure 13.16 **Epogen, a recombinant erythropoietin human peptide hormone produced in bacteria through cloning and expression.** This drug is used to boost red blood cells in patients who have lost blood through surgery or kidney dialysis. It has, unfortunately, also become one of the drugs of choice for illegally enhancing athletic performance.

13C Biochemical Connections • Analytical Chemistry (Chromatography)

Fusion Proteins and Fast Purifications

Affinity chromatography (Section 5-2B) is a powerful technique for protein purification. Molecular biologists have taken the idea one step further and have incorporated affinity chromatography ligand-binding sites directly into a protein to be expressed. A protein is created that contains not only the amino acid sequence of the desired polypeptide but also some extra amino acids at the N-terminus or C-terminus. These new proteins are called **fusion proteins**. Figure 13.17 shows how this might work.

An expression vector that has a promoter for T7 polymerase, followed by an **ATG start codon**, is used. A *His-tag* sequence that follows the ATG codes for six histidine residues. Following the His-tag is a sequence that is specific for a proteolytic enzyme called *enterokinase.* Finally comes the MCS, where the gene of interest can be cloned. After the desired gene is cloned into this vector, it is transformed into bacteria and expressed. The fusion proteins that are translated have the initial methionine, six histidines, the enterokinase-specific amino acid sequence, and then the desired protein.

Recall from Section 5-2B that an affinity chromatography resin uses a ligand that binds specifically to a protein of interest. A nickel affinity column is set up, which is very specific for histidine residues. The cells are lysed and passed over the column. All the proteins pass through except the fusion protein, which binds tightly to the nickel column. The fusion protein can be eluted with imidazole, a histidine analog. Enterokinase is then added to cleave off the His-tag, leaving the desired protein. Under ideal circumstances, this can be an almost perfect, one-step purification of the protein.

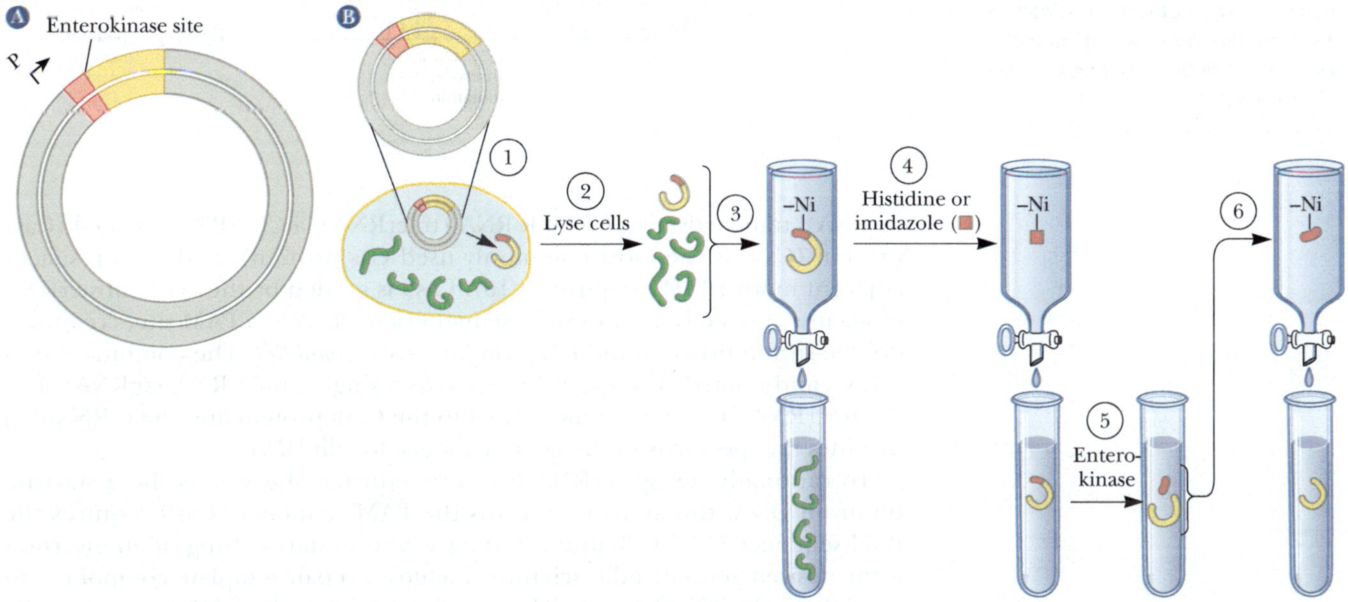

Figure 13.17 Affinity chromatography and fusion proteins combine to purify a protein efficiently. (A) An expression vector is used that inserts six histidines (the His-tag) at the N-terminus just after the ATG start site. Following the His-tag (red) is a sequence that codes for an enterokinase site, after which is the protein you are trying to purify (yellow). (B) In Step 1, bacterial cells are transformed with plasmid, and the fusion protein (red and yellow) is expressed along with other cell proteins (green). In Step 2, the cells are lysed, and in Step 3, the lysate is run over a nickel affinity column. Nickel attracts the His-tag (red) on the fusion protein, and the other cell components are washed out. In Step 4, high concentrations of histidine or imidazole (red rectangles) are used to elute the fusion protein from the nickel affinity column. In Step 5, enterokinase is added to the purified fusion protein in order to make a cut that separates the His-tag from the desired protein. In Step 6, the digested fusion protein components are run over the nickel affinity column, leaving only the desired protein. *(Adapted from* Molecular Biology *by R. F. Weaver, McGraw-Hill, 1999.)*

fusion proteins proteins made via recombinant DNA techniques that include the desired protein sequence as well as a leader sequence that will aid in quick separation and purification of the protein

ATG start codon the DNA sequence (or its RNA equivalent AUG) that initiates protein synthesis

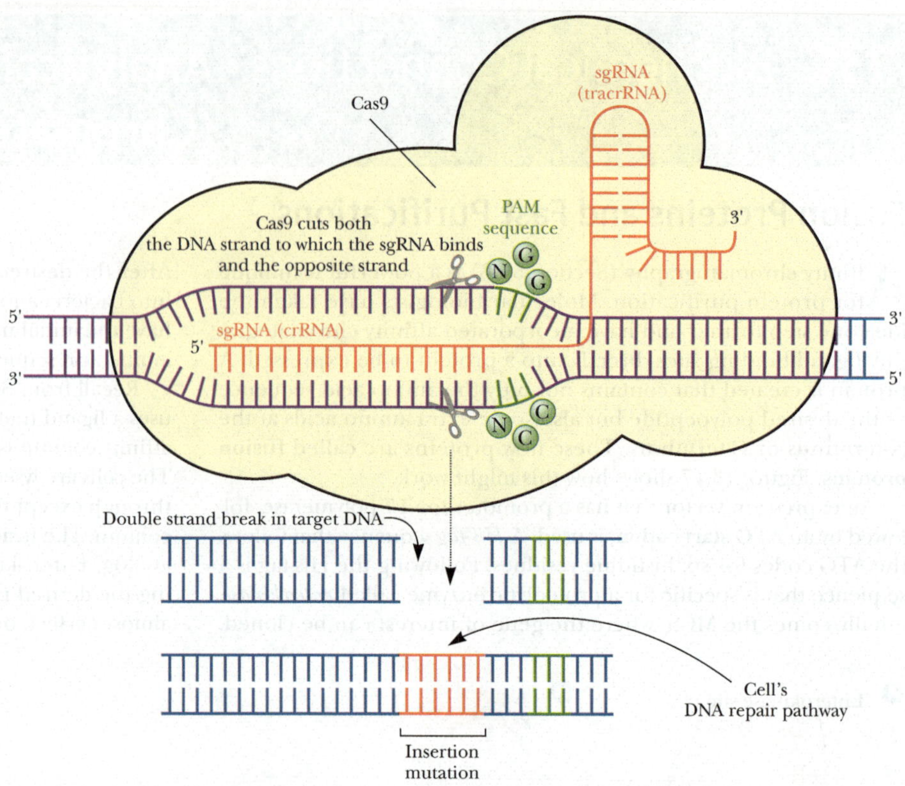

Figure 13.18 A CRISPR cut resulting in the introduction of an insertion mutation into the gene sequence. *(Storrs; A CRISPR Fore-Cas-t; (2014)* The Scientist Magazine; *http://www.thescientist.com/?articles.view/articleNo/39239title/A-CRISPR-Fore-Cas-t/ (accessed 27 November 2015).)*

(crRNA), trans-activating CRISPR RNA (tracrRNA), a CRISPR-associated (Cas) protein (Cas9 is the most commonly used Cas protein), and a protospacer adjacent motif (PAM) (Figure 13.18). Cas9 is guided by the crRNA:tracrRNA molecule to selectively cut target sequences of DNA at a PAM site, creating a double-strand break in the DNA (Figure 13.18, *middle*). The combination of crRNA and tracrRNA is also referred to as a single guide RNA (sgRNA). It is the tracrRNA that anchors the sgRNA to the Cas9 protein and the crRNA that provides the specificity of the gene sequence to edit DNA.

To rationally design crRNA for gene editing, there must be a starting point on DNA; this starting point is the PAM sequence. Cas9 requires the PAM sequence 5′-NGG-3′ in the coding strand to start cutting. If an insertion is the desired genome edit, scientists include a repair template component to the delivered CRISPR system. The template contains the DNA sequence to be inserted and sequences on both ends that match either side of the double-stranded break. The cell's DNA repair mechanisms then use the repair template to repair the cut, resulting in an insertion of the desired DNA sequence (Figure 3.18, *bottom*). For example, the inserted DNA sequence could be a functional version of a gene that is defective or missing in a patient. There are several ways to deliver CRISPR components into a cell, such as viral vectors, electroporation, microinjection, lipid nanoparticles, plasmid transfection, and ribonucleoprotein complexes.

An advantage of the CRISPR/Cas9 system over other gene-editing techniques is that it is very specific because it requires an exact identity match between complementary strands of crRNA and target DNA that is adjacent to a PAM site. PAM sites are abundant on DNA, with genes containing many PAM sites. Thus, selectivity and control over genome editing are easily attainable, given the accuracy of modern sequencing and molecular biology methods.

CRISPR/Cas9 has revolutionized human gene therapy, plant and animal agriculture, synthetic biology, viral gene disruption, and myriad other yet-to-be-contrived applications. There remain many challenges to the implementation of gene therapy in humans, but there is significant excitement in the scientific and medical fields at the emergence and development of CRISPR/Cas9 genome editing strategies.

13-5 DNA Libraries

Learning Objectives

After reading this section, students will be able to:

LO 13.5.1 Describe how DNA libraries and cDNA libraries are created.

LO 13.5.2 Explain how an individual clone in a DNA library can be found.

LO 13.5.3 Distinguish between the purposes of a DNA library compared to a cDNA library.

Because methods exist for selecting regions of DNA from the genome of an organism of interest and cloning those regions in suitable vectors, a question that immediately comes to mind is whether we can take *all* the DNA of an organism (the total genome) and clone it in chunks of reasonable size. The answer is that we can do this, and the result is a **DNA library (genomic library)**.

DNA library (genomic library) a collection of cloned DNA fragments that includes the total genome of an organism

Figure 13.19 shows the steps involved in the construction of a DNA library. Using methods discussed in Section 13-3, DNA fragments extracted from an

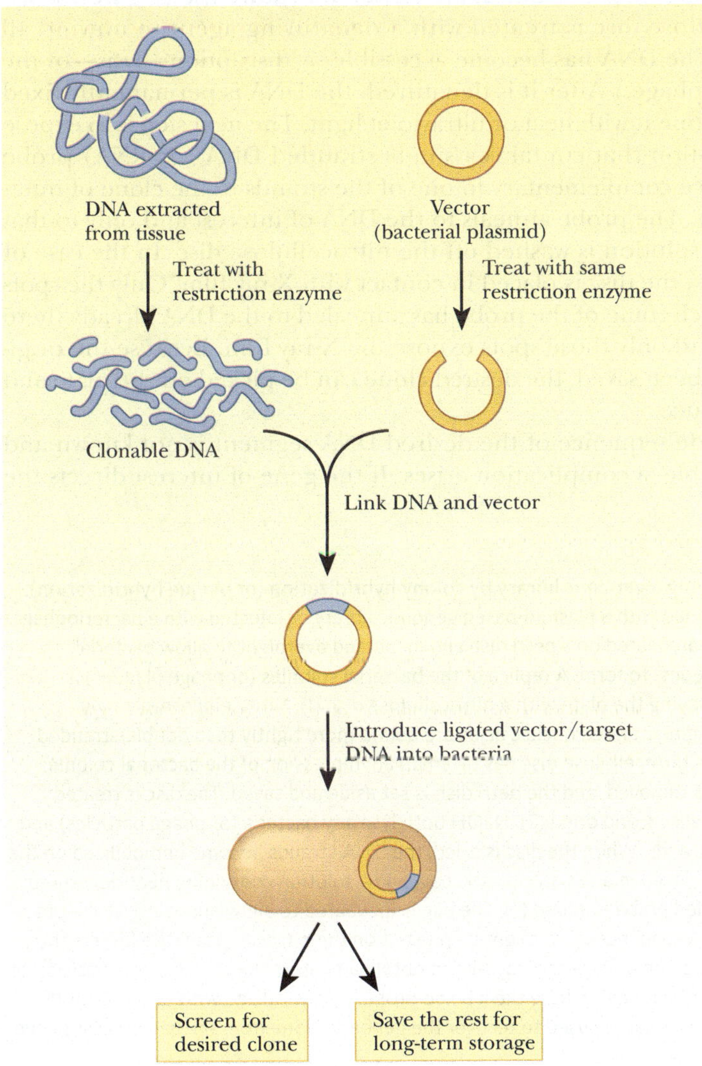

Figure 13.19 Steps involved in the construction of a DNA library. *(Adapted with permission from* Dealing with Genes: The Language of Heredity, *by Paul Berg and Maxine Singer, ©1992 by University Science Books.)*

organism are recombined with a plasmid vector. Bacterial cells are then transformed under conditions where each cell is likely to only take up one recombined plasmid, each one containing a different fragment of the organism's genome. The bacterial cells can then be stored and screened later to select a cell that has a fragment of interest (Figure 13.19, *bottom*). The screening process is discussed further in Section 13-5A.

The process of constructing a DNA library can be quite laborious, leading many researchers to obtain previously constructed libraries from other laboratories or from commercial sources. Some journals require that libraries and individual clones that have been discussed in articles they publish be freely available to other laboratories.

13-5A Finding an Individual Clone in a DNA Library

Imagine that, after a DNA library has been constructed, someone wants to find a single clone (e.g., one that contains a gene responsible for an inherited disease) out of the hundreds of thousands, or possibly millions, in the library. This degree of selectivity requires specialized techniques. One of the most useful of these techniques depends on separating and annealing complementary strands. An imprint is taken of the petri dish on which the bacterial colonies (or phage plaques) have grown. A nitrocellulose disc is placed on the dish and then removed. Some of each colony or plaque is transferred to the disc, and *the position of each is the same as it was on the dish.* The rest of the original colonies or plaques remain on the dish and can be stored for future use (Figure 13.20).

The nitrocellulose disc is treated with a denaturing agent to unwind all the DNA on it. (The DNA has become accessible by disruption—lysis—of the bacterial cells or phage.) After it is denatured, the DNA is permanently fixed to the disc by treatment with heat or ultraviolet light. The next step is to expose the disc to a solution that contains a single-stranded DNA (or RNA) probe that has a sequence complementary to one of the strands in the clone of interest (Figure 13.20). The probe anneals to the DNA of interest and only to that DNA. Any excess solution is washed off the nitrocellulose disc. In the case of radioactive probes, the disc is placed in contact with X-ray film. Only the spots on the disc in which some of the probe has annealed to the DNA already there are radioactive, and only those spots expose the X-ray film. Because the original Petri dish has been saved, the desired clone can be picked off the plate and allowed to reproduce.

If the nucleotide sequence of the desired DNA segment is not known and no probe is available, a complication arises. If the gene of interest directs the

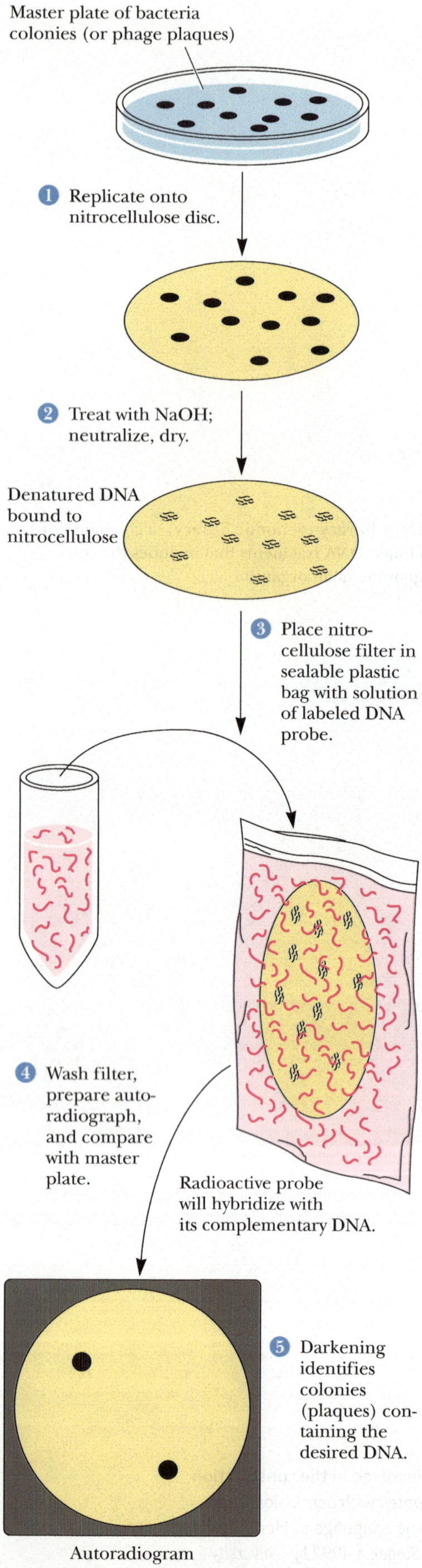

Figure 13.20 Screening a genomic library by colony hybridization (or plaque hybridization). Host bacteria transformed with a plasmid-based genomic library or infected with a bacteriophage-based genomic library are plated on a petri dish and incubated overnight to allow bacterial colonies (or phage plaques) to form. A replica of the bacterial colonies (or phage plaques) is then obtained by overlaying the plate with a nitrocellulose disc (1). Nitrocellulose strongly binds nucleic acids; single-stranded nucleic acids are bound more tightly than double-stranded nucleic acids. Once the nitrocellulose disc has taken up an impression of the bacterial colonies (or phage plaques), it is removed, and the petri dish is set aside and saved. The disc is treated with 2 M NaOH, neutralized, and dried (2). NaOH both lyse any bacteria (or phage particles) and dissociates the DNA strands. When the disc is dried, the DNA strands become immobilized on the filter. The dried disc is placed in a sealable plastic bag, and a solution containing heat-denatured (single-stranded), labeled probe is added (3). The bag is incubated to allow annealing of the probe DNA to any target DNA sequences that might be present on the nitrocellulose. The filter is then washed, dried, and placed on a piece of X-ray film to obtain an autoradiograph (4). The position of any spots on the X-ray film reveals where the labeled probe has hybridized with target DNA (5). The location of these spots can be used to recover the genomic clone from the bacteria (or phage plaques) on the original petri dish.

synthesis of a given protein, a vector is chosen that allows cloned genes to be transcribed and translated. If the presence of the desired protein can be detected by its function, that serves as the basis for detecting it. Alternatively, labeled antibodies can be used as a basis for protein detection.

13-5B cDNA Libraries

In some cases, scientists are interested in studying genes that are being actively transcribed in an organism or in a specific tissue, so instead of a genomic library, a **cDNA library** can be constructed from the existing RNA molecules in the organism or specific tissue (Figure 13.21). The RNA of interest (usually mRNA) is used as the template for the synthesis of complementary DNA (cDNA) in a reaction catalyzed by reverse transcriptase. The cDNA is then incorporated into a vector (Figure 13.21, *bottom*). From this point on, the process of producing a cDNA library is virtually identical to that for constructing a genomic DNA library. For a given organism, a genomic DNA library is the same no matter what the tissue source, and the DNA represents both expressed and unexpressed DNA. In contrast, a cDNA library is different depending on the tissue conditions used and the gene expression profile of the cells.

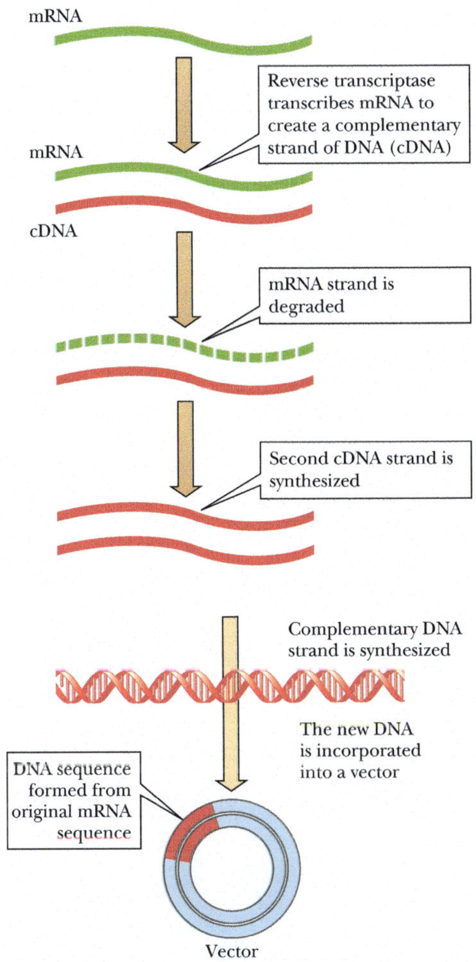

Figure 13.21 **Formation of a cDNA library.** *(Adapted with permission from* Dealing with Genes: The Language of Heredity, *by Paul Berg and Maxine Singer, ©1992 by University Science Books.)*

13-6 The Polymerase Chain Reaction

Learning Objectives

After reading this section, students will be able to:

LO 13.6.1 List the steps involved in amplifying a sample of DNA using the polymerase chain reaction (PCR).

LO 13.6.2 List the steps involved in quantifying the amount of DNA in a sample using quantitative PCR (qPCR).

LO 13.6.3 Distinguish between *cloning*, *PCR*, and *qPCR*.

In many cases, researchers want to increase the amount of a given DNA sequence many times over without cloning that DNA, and the method that makes this amplification possible is the **polymerase chain reaction (PCR)**. Any chosen DNA, ranging from a few hundred to a few thousand base pairs, can be amplified, and it need not be separated from the rest of the DNA in a sample before the procedure is applied. PCR is an automated process that copies both complementary strands of the desired DNA sequence.

Scientists had long wished for a cell-free, automated way of synthesizing DNA, but any system that could be automated would need to function at high temperatures so that the DNA strands could be separated physically without the need for the many enzymes found in DNA replication, such as topoisomerases and helicases. Unfortunately, such temperatures (around 90°C) would denature and inactivate the DNA polymerases that would be making the DNA strands. What made the process possible was the discovery of bacteria that live around deep sea hydrothermal vents, under extreme pressures, and at temperatures higher than 100°C. If bacteria can live under those conditions, then their enzymes must be able to function at those temperatures. The bacterium, *Thermus aquaticus,* from which a heat-resistant polymerase is extracted, is one of those that live in hot springs, and its DNA polymerase is called *Taq* polymerase.

Figure 13.22 shows three cycles of a PCR reaction. Before starting the reaction, several ingredients are combined in a PCR tube—namely, source DNA, designed primers, dNTPs, and a thermostable DNA polymerase such

cDNA library a library of actively transcribed genes that is constructed from complementary DNA obtained from the mRNA of an organism or a specific tissue

polymerase chain reaction (PCR) a rapid in vitro method for making millions of copies of a sequence of DNA

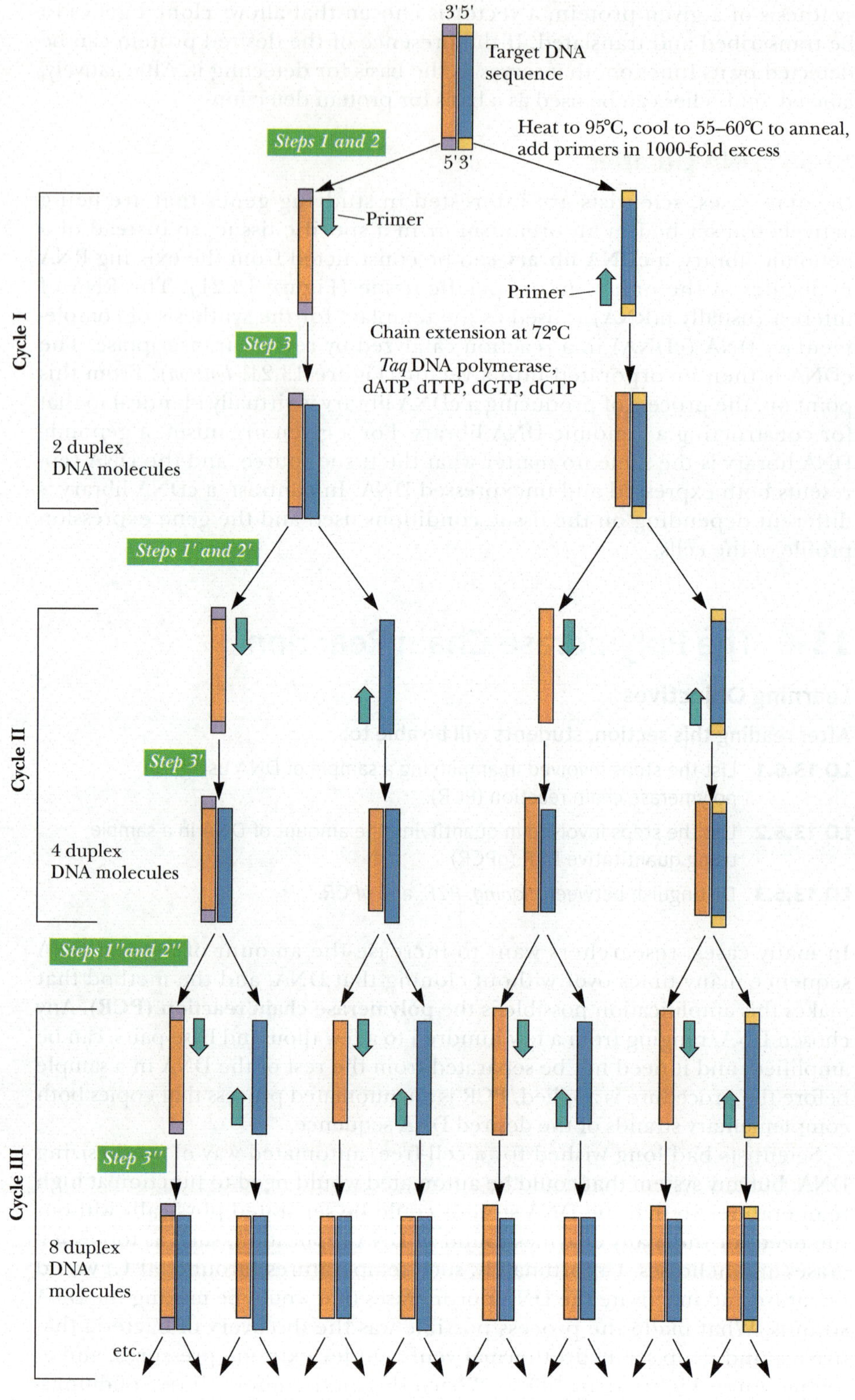

Figure 13.22 Polymerase chain reaction (PCR). Oligonucleotides complementary to a given DNA sequence prime the synthesis of only that sequence. Heat-stable *Taq* DNA polymerase survives many cycles of heating. Theoretically, the amount of the specific primed sequence is doubled in each cycle.

as *Taq* polymerase. The tube is then placed in a *thermocycler* (or *thermal cycler*) machine that automates 20–35 cycles of temperature increases and decreases described next. At the start of the process, the two DNA strands are separated by heating to 95°C. The thermocycler then cools to an annealing temperature typically between 55 and 60°C so that the short oligonucleotide primers can anneal to the DNA strands. These primers are designed by the researchers to be complementary to the ends of the DNA chosen for amplification and serve the same purpose as the RNA primers in normal replication (Section 10-3B).

Once the primers have annealed to the DNA, the temperature is raised to the optimal temperature for the activity of *Taq* polymerase, which is 72°C. The polymerase then begins synthesizing the new DNA from the 3′ end of the primer. The two complementary strands grow in the 5′ to 3′ direction, and the *Taq* polymerase is allowed to work until the desired length of DNA has been synthesized (Figure 13.22, *end of Cycle I*). This first cycle doubles the amount of the DNA target sequence. These processes of unwinding the DNA strands, annealing the primers, and extending from the primers are repeated, bringing about a second doubling of the DNA target sequence. It is not necessary to add more primer because it is present in large excess. The amount of DNA continues to double after each PCR cycle, and after 20–35 cycles, millions to hundreds of millions of copies of the DNA target sequence are obtained.

With good primer design, other DNA sequences are not amplified and do not interfere with the reaction or subsequent use of the amplified DNA. PCR primers have to be sufficiently long to be specific for the target sequence but not too long because nonspecific annealing and other problems may occur. Usually, the primers are 18–30 bases long. They must also have optimal binding properties, such as an amount of G and C that is sufficient to allow them to anneal before the entire DNA renatures. In addition, the two primers should contain similar amounts of G and C so that they have similar melting temperatures (Section 9-4). The sequence of the primer must not lead to secondary structures within a primer or between the two different primers; otherwise, the primers bind to themselves instead of the DNA being amplified. For example, if a primer had the sequence AAAAATTTTT, it would form a hairpin loop with itself and would not be available to bind the DNA.

▶ What are the advantages of PCR?

Amplification of the amounts of DNA in extremely small samples has made it possible to obtain accurate analyses that were not possible earlier. Forensic applications of the technique have resulted in positive identifications of crime victims and suspects. Even minuscule amounts of ancient DNA, such as those available from Egyptian mummies, can now be researched after amplification.

13-6A Quantitative PCR Allows Sensitive Measurement of DNA Samples

One innovation in biotechnology expands on the PCR technique to give quantitative results. Standard PCR is used for producing large quantities of a DNA target sequence, but a newer technique called **quantitative PCR (qPCR)** allows the PCR reaction to generate time-point data that can be used to determine how much of the DNA target sequence was in the cell or sample originally. Figure 13.23 shows how this works. PCR is run on multiple samples containing the DNA sequence of interest. As the number of cycles increases, a threshold value is reached where the amount of DNA can be measured. For this purpose, the DNA is labeled with fluorescent markers. The more DNA that was in the original sample, the earlier in the process the results can be seen (shown by the different colored lines crossing the orange horizontal threshold line). Scientists also use qPCR to quantify levels of gene expression by first using reverse transcriptase to convert RNA transcripts into DNA molecules that then get amplified in the qPCR technique.

quantitative PCR (qPCR) a polymerase chain reaction technique that is used to determine the amount of a target DNA or RNA sequence initially present

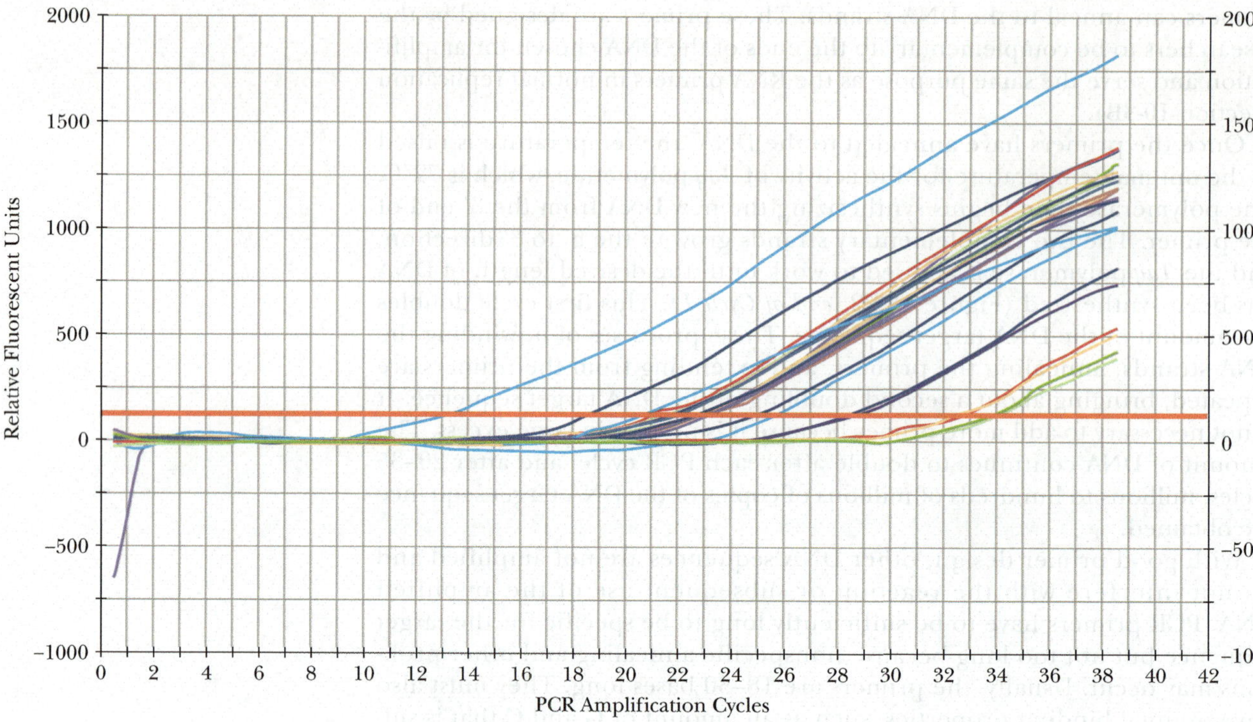

Figure 13.23 **Quantitative PCR (qPCR).** Samples with different amounts of initial target DNA are shown in different colors. The horizontal orange line is a threshold value used to compare the unknown samples to known standard samples.

13-7 DNA Fingerprinting

Learning Objectives

After reading this section, students will be able to:

LO 13.7.1 Describe how restriction fragment length polymorphism (RFLP) analysis is performed.

LO 13.7.2 Explain how RFLP analysis can be used in paternity testing.

Southern blot a technique, invented by E. M. Southern, used for transferring DNA from an agarose gel after electrophoresis onto a membrane, such as one made of nitrocellulose

The unique DNA characteristics of individuals can be studied and compared in forensic analyses using a technique called *DNA fingerprinting.* One version of this technique is called *restriction fragment length polymorphism (RFLP) analysis.* In RFLP analysis, DNA is extracted from a sample of an individual's cells, and then the DNA is digested with restriction enzymes (Section 13-2A). After that, a **Southern blot** is performed as shown in Figure 13.24. The digested DNA is run on an agarose gel (Figure 13.24, *Step 2*). The DNA fragments can be seen directly on the gel if it is soaked in ethidium bromide and viewed under ultraviolet light (Section 13-1). As shown in Figure 13.24, Step 3, electrophoresis yields bands of varying sizes, depending on the nature of the DNA and the restriction enzymes used.

If greater sensitivity is needed, or if the number of fragments would be too great to distinguish the bands, this technique can be modified to visualize only selected DNA sequences. The first step in visualizing only selected DNA sequences would be to transfer the DNA to a nitrocellulose membrane using the Southern blot technique (Figure 13.24, *Step 4*). The agarose gel is soaked in NaOH to denature the DNA because only single-stranded DNA binds to the nitrocellulose. The membrane is placed on the agarose gel, which is on top of a filter-paper wick placed in buffer. Dry absorbent paper is placed on top of the nitrocellulose filter. Wicking action carries the buffer from the buffer chamber up through the gel and nitrocellulose into the dry paper. The DNA bands move out of the gel and stick to the nitrocellulose (Figure 13.24, *Step 5*). The next

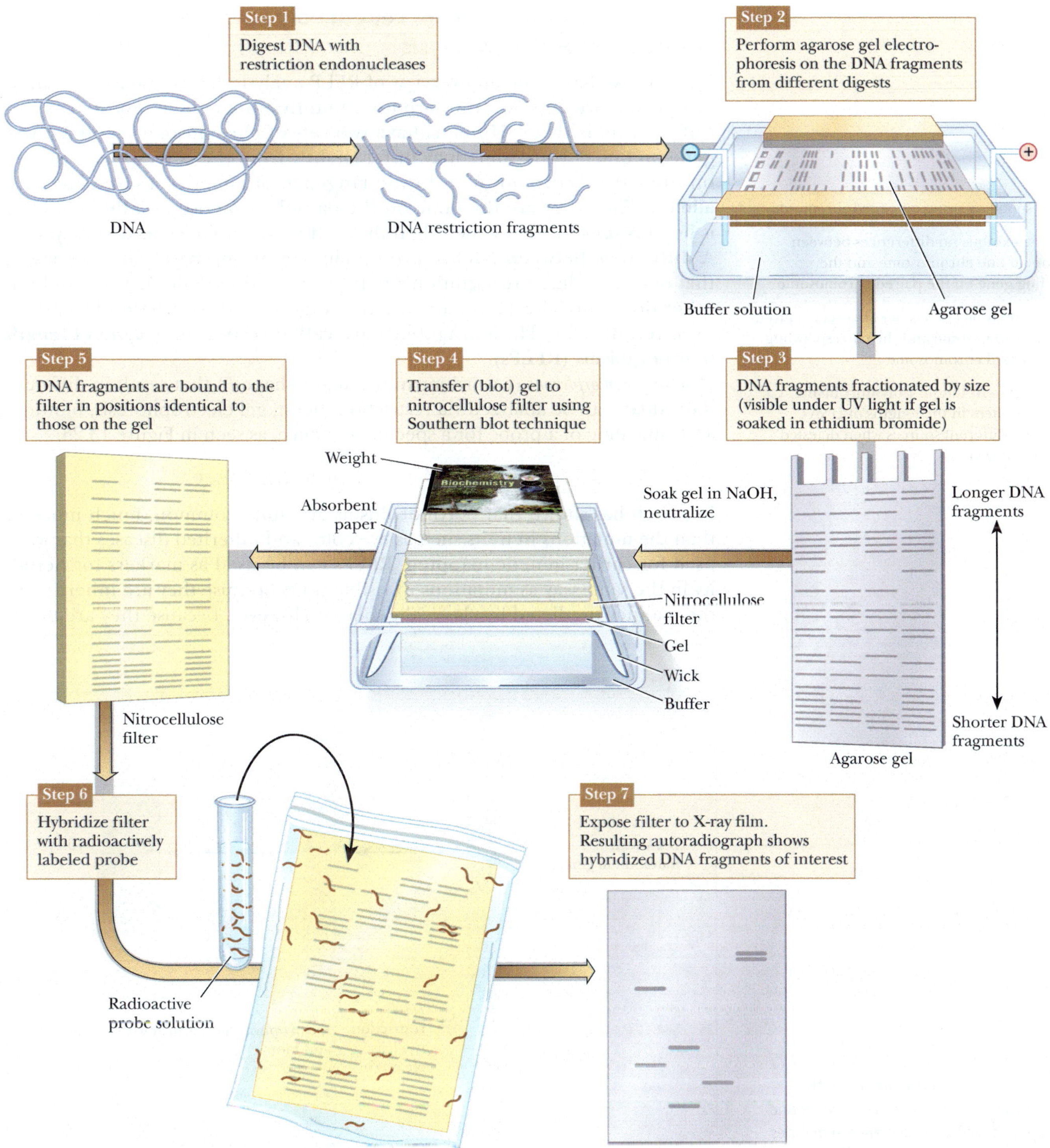

Figure 13.24 Restriction fragment length polymorphism (RFLP) analysis and the Southern blot technique. Electrophoretically separated DNA restriction fragments are transferred to a nitrocellulose sheet. A radioactively labeled probe for a DNA sequence of interest is bound to the nitrocellulose, and bands representing the presence of DNA sequences of interest are visualized with an autoradiograph.

step is to visualize the bands on the nitrocellulose. A specific DNA probe that is labeled with ^{32}P is incubated with the nitrocellulose membrane (Figure 13.24, *Step 6*). The DNA probe binds to DNA fragments that are complementary. The membrane is then placed on photographic paper to produce an autoradiograph that shows the presence of any DNA fragments that are of interest. The use of the specific probe greatly reduces the number of bands seen and isolates desired DNA sequences.

13-7A Restriction-Fragment Length Polymorphisms: A Powerful Method for Forensic Analysis

Now that we have seen an overview of RFLP analysis, let us discuss what polymorphisms are and how they are used to differentiate between the DNA of different individuals. Humans have two sets of chromosomes, and a given gene on one chromosome may differ slightly from the corresponding gene on the paired chromosome. In the language of genetics, these genes are **alleles**. When they are the same on the paired chromosomes, the organism is **homozygous** for that gene; when they differ, the organism is **heterozygous**. A difference between alleles, even a change in one base pair, can mean that one allele has a recognition site for a restriction endonuclease and the other does not. Therefore, a restriction digest yields restriction fragments of different sizes. These fragments are called **restriction-fragment length polymorphisms (RFLPs)**.

alleles corresponding genes on paired chromosomes

homozygous exhibits no differences between a given gene on one chromosome and the corresponding gene on the paired chromosome

heterozygous exhibits differences between a given gene on one chromosome and the corresponding gene on the paired chromosome

restriction-fragment length polymorphisms (RFLPs) differences in the lengths of DNA fragments from different sources when digested with restriction enzymes

The *polymorphisms* (a word meaning "many shapes") that result in different individuals can be analyzed via restriction digest, gel electrophoresis, blotting, and annealing of a probe for a specific sequence, as seen in Figure 13.25.

▶ How can differences in DNA from individuals be seen?

Research has shown that polymorphisms are quite common, much more so than the mutations in traits such as eye color and inherited diseases that were used for earlier genetic mapping. RFLPs can be used as markers for heredity in the same way as mutations in visible traits because they are inherited in the manner predicted by classical genetics. However, because they are much

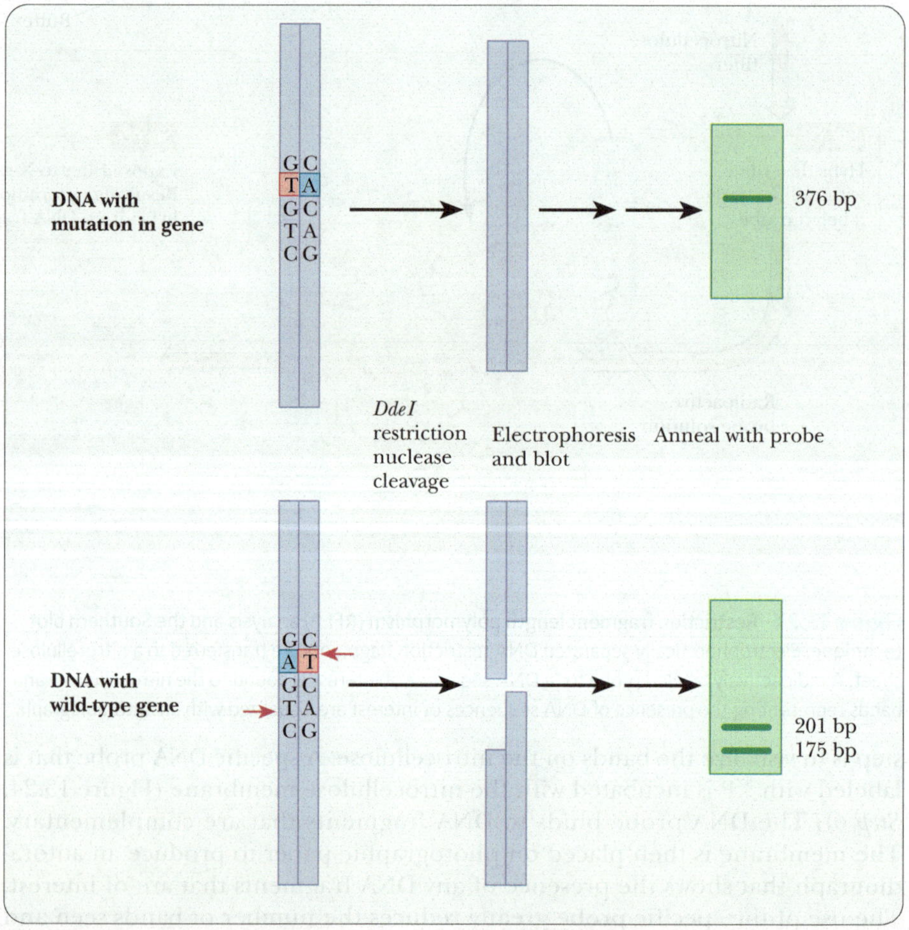

Figure 13.25 Restriction fragment length polymorphisms. In this example, a change of one base pair seen in a wild-type sequence eliminates a restriction nuclease cleavage site in the mutated sequence. (The red arrows in the wild-type sequence indicate the cleavage sites for the *DdeI* restriction enzyme.) The corresponding DNA with a mutation does not have this cleavage site. The difference can be detected by electrophoresis, followed by blotting and the annealing of a probe specific for this fragment. (The abbreviation *bp* stands for *base pairs*.) *(Adapted with permission from* Dealing with Genes: The Language of Heredity, *by Paul Berg and Maxine Singer, ©1992 by University Science Books.)*

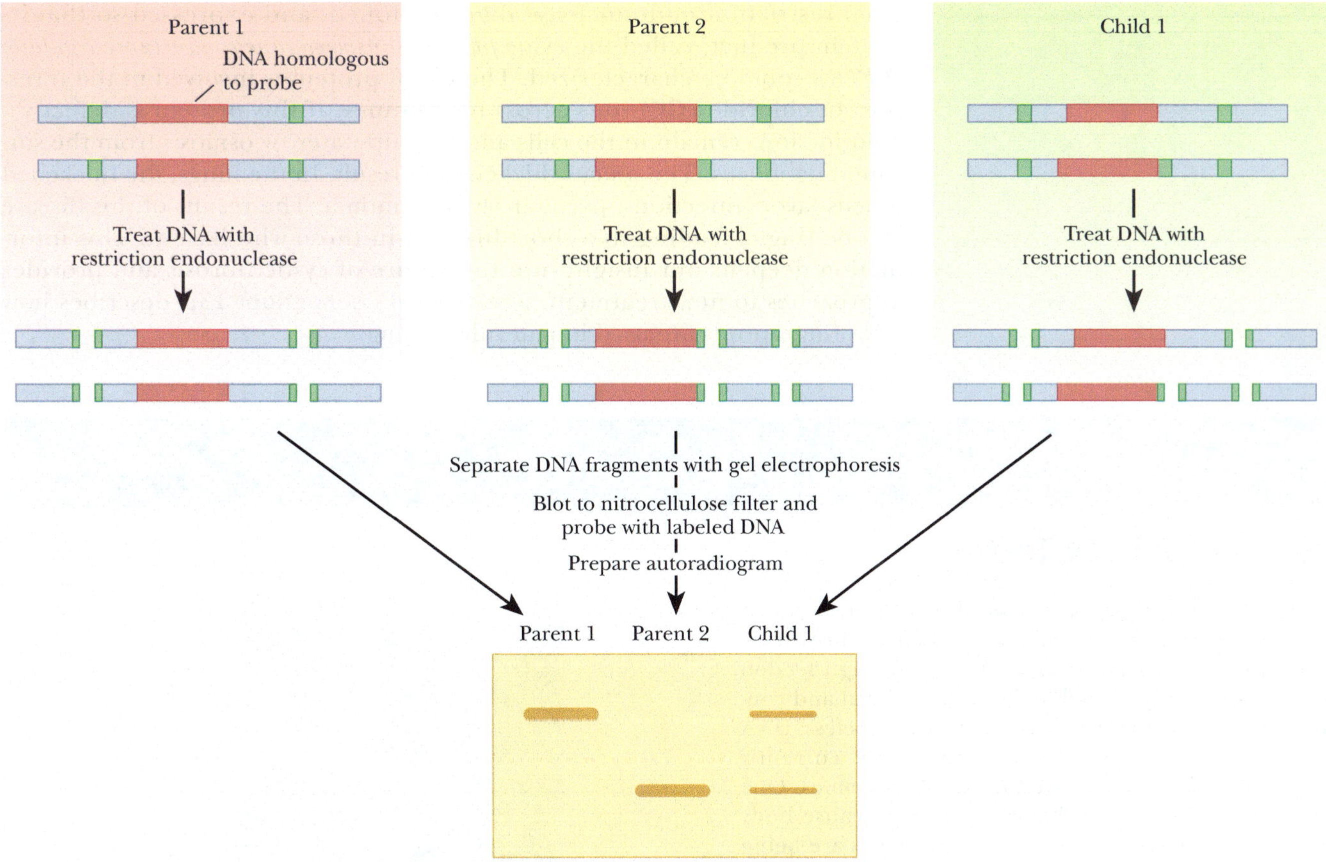

Figure 13.26 Paternity testing. Restriction fragment length polymorphisms (RFLPs) can be detected by probes for homologous DNA in paternity testing. The green blocks indicate restriction sites. Parent 1 has DNA with two restriction sites near the DNA that is complementary to the probe. Parent 2 has a different sequence of DNA where there are three restriction sites in the same chromosome area, which results in a smaller restriction fragment that the probe will bind to. A single smaller restriction fragment is detected by the homologous probe for Parent 2. Their child has inherited one copy of each DNA strand from each parent. Therefore, the child's RFLP analysis produces one fragment of each size, indicating a positive paternity test in this case. *(Adapted with permission from* Dealing with Genes: The Language of Heredity, *by Paul Berg and Maxine Singer, ©1992 by University Science Books.)*

more abundant than mutations that lead to phenotypic variations, they have provided many more markers for detailed genetic mapping.

Figure 13.26 shows how RFLP analysis can be used in paternity testing. A child gets one of each allele from each parent, so every fragment that the child has must also be present in one of the parents. Thus, it would be easy to eliminate a prospective father undergoing a test for paternity. Figure 13.26 shows an example of results that suggest a match in a paternity test, but if the child had an RFLP band that the prospective father did not have, then the prospective father would be excluded as a match. The same kinds of analyses are done with evidence found at crime scenes. Suspects in criminal cases may be exonerated if their DNA samples do not match those found at the scene.

RFLP analysis was also used extensively in the process of locating the altered gene that causes cystic fibrosis, a prevalent genetic disease. Once the gene was located on chromosome 7, a series of RFLP markers was used to help map its exact position (Figure 13.27). The CF gene was then isolated

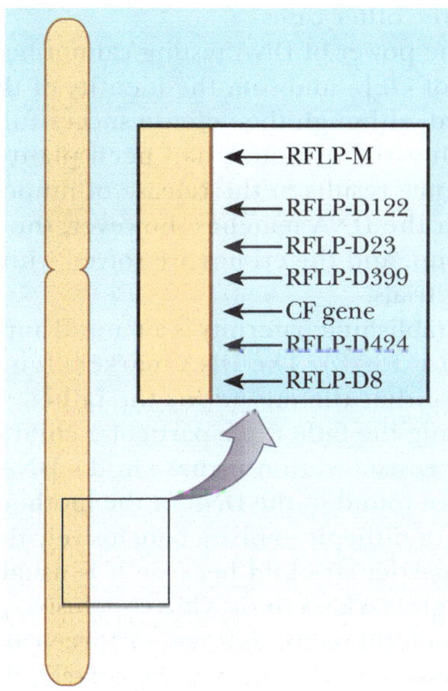

Figure 13.27 Localization of the gene associated with cystic fibrosis (CF) on human chromosome 7. The RFLP markers are given arbitrary names. The location of the CF gene is given relative to the RFLP markers. *(Adapted with permission from* Dealing with Genes: The Language of Heredity, *by Paul Berg and Maxine Singer, ©1992 by University Science Books.)*

from restriction endonuclease digests, cloned, and expressed so that its protein product, called the *cystic fibrosis transmembrane conductance regulator (CFTR)*, could be characterized. The CFTR protein is involved in the transport of chloride (Cl^-) ions across membranes. If this protein is defective, chloride ions remain in the cells and take up water by osmosis from the surrounding mucus. The mucus thickens as a result. In the lungs, the thickened mucus favors infections, particularly pneumonia. The results of this disease can be tragic, leading to a short lifespan in those who have it. This information deepens our insight into the nature of cystic fibrosis and provides approaches to new treatment. **Biochemical Connections 13D** describes how DNA fingerprints are used in paternity testing.

13D Biochemical Connections • Forensics

Forensic Uses of DNA Testing

It has been suggested that tissue samples should be taken from all convicted felons to make it possible to identify their presence at future crime sites by DNA fingerprinting (Section 13-7). This may seem very desirable, but some ethical and constitutional questions must first be answered. Nevertheless, DNA testing has been used to determine whether people currently in custody might have been involved in unsolved crimes. One crime frequently examined in this way is rape because body fluids are usually left behind. Many unsolved crimes are being solved, and some convicted prisoners are being found innocent. In at least one bizarre case, a felon was found innocent of the crime for which he was in jail, only to be rearrested in a few weeks because his DNA matched samples found on rape victims in three other cases.

The power of DNA testing cannot be overemphasized. In the trial of O. J. Simpson, the identity of the DNA was never questioned, although the defense successfully raised questions as to whether the evidence had been planted. In most cases, DNA evidence results in the release of innocent suspects in a crime. When the DNA matches, however, the rate of plea bargaining goes up, and the crimes are solved without long and expensive court trials.

Establishing paternity is a natural application for DNA testing (Figure 13.28). The DNA markers found in a child must arise from either the mother or the father. A man who is suspected of being the father of a particular child is immediately excluded from consideration if that child's DNA contains markers that are not found in the DNA of the mother or the suspect father. It is more difficult to prove conclusively that a person is the father of a particular child because it is usually too expensive to test enough markers to provide conclusive proof. However, in cases in which there are only two or three candidate fathers, it is usually possible to determine the correct one. ◗

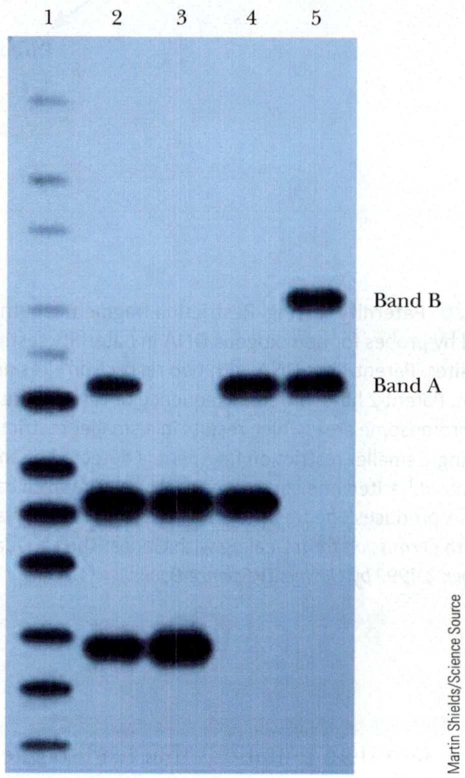

Figure 13.28 Paternity test DNA fingerprint. This gel shows possible results of a DNA fingerprint from a paternity test. Lane 1 shows a DNA size ladder used as a control. Lane 2 is a DNA pattern from a mother. Lanes 3 and 4 are DNA patterns from possible fathers. Lane 5 is the DNA pattern from the child. Every band in the child's DNA must have a corresponding band in either the mother's or the father's DNA. Band A in the child's DNA matches a band in the mother's DNA and in the prospective father's DNA in lane 4. However, band B in the child's DNA is not seen in the mother's DNA, nor is it seen in either of the prospective father's DNA. This excludes both possible fathers from being the father of the child.

13-8 DNA Sequencing

Learning Objectives

After reading this section, students will be able to:

LO 13.8.1 Describe how DNA sequencing can be performed.

The primary structure of a protein determines its secondary and tertiary structures. The same is true of nucleic acids; the nature and order of monomer units determine the properties of the whole molecule. Base pairing in both RNA and DNA depends on a series of complementary bases, whether these bases are on different polynucleotide strands, as in DNA, or on the same strand, as is frequently the case in RNA. Sequencing of nucleic acids is now fairly routine, and this relative ease would have amazed the scientists of the 1950s and 1960s.

The method devised by Sanger and Coulson for determining the base sequences of nucleic acids depends on selective interruption of oligonucleotide synthesis. A single-stranded DNA fragment whose sequence is to be determined is used as a template for the synthesis of a complementary strand. The new strand grows from the 5′ end to the 3′ end. This unique direction of growth is true for all nucleic acid synthesis. The synthesis is interrupted at every possible site in a population of molecules. The interruption of synthesis depends on the presence of 2′,3′-dideoxyribonucleoside triphosphates (ddNTPs) (Figure 13.29). These ddNTPs can be incorporated into a growing DNA chain, but they lack a 3′ hydroxyl group to form a bond with another nucleoside triphosphate (recall the mechanism of nucleotide addition to a growing DNA chain seen in Figure 10.5). The incorporation of a ddNTP into the growing chain causes termination at that point. The presence of small amounts of ddNTPs combined with normal dNTP building blocks in a replicating mixture causes random terminations of chain growth.

Figure 13.30 shows the Sanger–Coulson method for sequencing DNA. The DNA to be sequenced is mixed with a short oligonucleotide that serves as a primer for synthesis of the complementary strand. The primer is hydrogen-bonded toward the 3′ end of the DNA to be sequenced. The DNA with primer is divided into four separate reaction mixtures. Each reaction mixture contains all four deoxyribonucleoside triphosphates (dNTPs), one of which is labeled to allow the newly synthesized fragments to be visualized by fluorescence. In addition, each of the reaction mixtures contains one of the four ddNTPs. Synthesis of the chain is allowed to proceed in each of the four reaction mixtures. In each mixture, chain termination occurs at all possible sites for that nucleotide. When gel electrophoresis is performed on each reaction mixture, a band corresponding to each position of chain termination appears. The sequence of the newly formed strand, which is complementary to that of the template DNA, can be "read" directly from the sequencing gel (Figure 13.30, *bottom*).

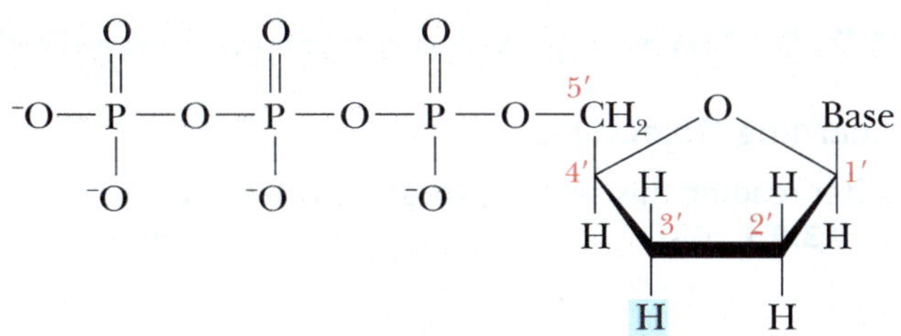

Figure 13.29 Structure of a 2′,3′-dideoxyribonucleoside triphosphate (ddNTP). The lack of a hydroxyl group on the 3′ carbon of ddNTPs prevents another nucleotide from being added to the end of a nucleic acid chain that has just had a ddNTP added to it.

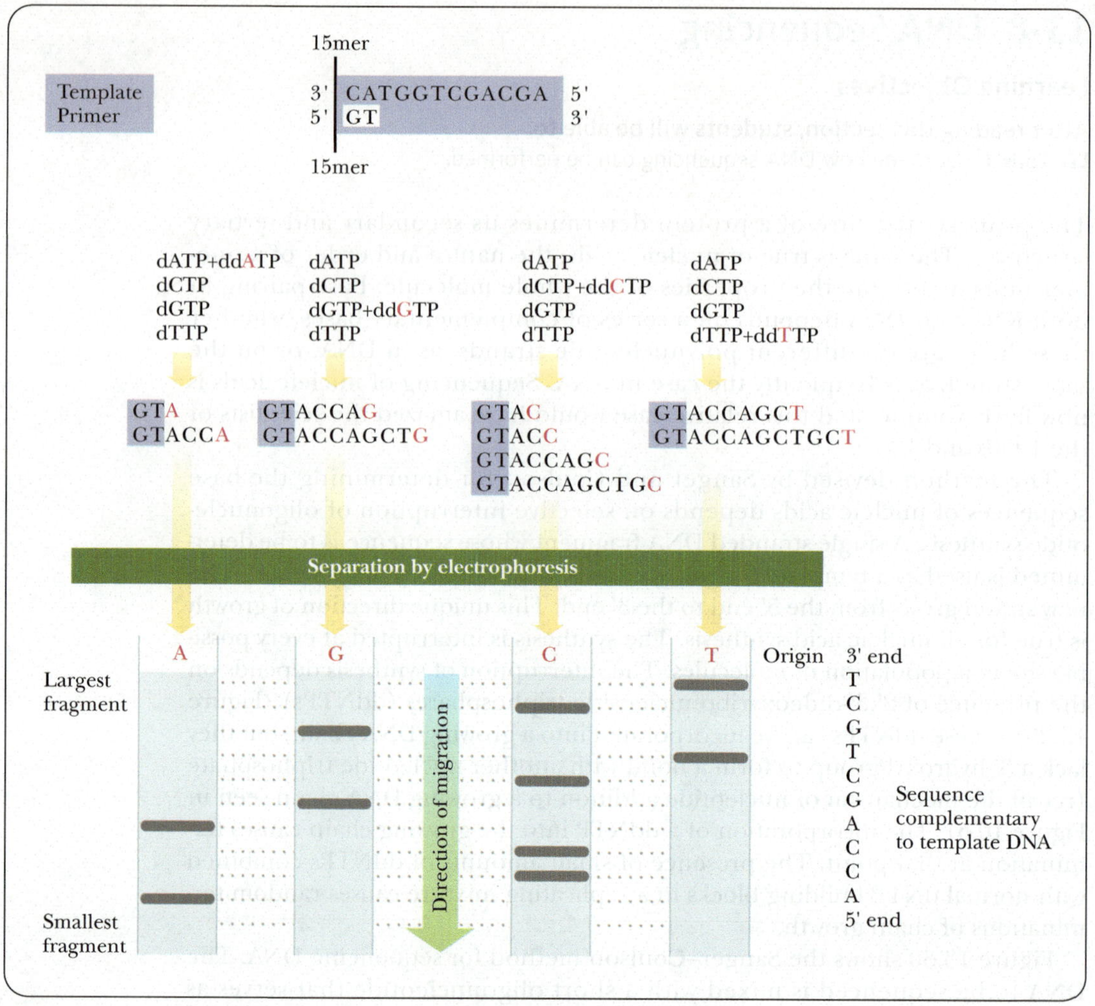

Figure 13.30 **The Sanger–Coulson method for sequencing DNA.** A primer at least 15 residues long is hydrogen-bonded to the 3′ end of the DNA to be sequenced. Four reaction mixtures are prepared; each contains the four dNTPs and one of the four possible ddNTPs. In each reaction mixture, synthesis takes place, but in a given population of molecules, synthesis is interrupted at every possible site. A mixture of oligonucleotides of varying lengths is produced. The components of the mixture are separated by gel electrophoresis.

A variation on this method, called dye-terminator sequencing, uses a single reaction mixture with a different fluorescent label on each of the four ddNTPs. Each fluorescent label can be detected by its characteristic spectrum, requiring only a single gel electrophoresis experiment. The use of fluorescent labels makes it possible to automate DNA sequencing, with the whole process under computer control. Commercial kits are available for these sequencing methods (Figure 13.31).

When RNA is to be sequenced, the method of choice is not to analyze the RNA itself but to use the methods of DNA sequencing on a DNA complementary (cDNA) to the RNA in question. The cDNA, in turn, is generated by using the enzyme reverse transcriptase, which catalyzes the synthesis of DNA from an RNA template.

13-9 Genomics and Other "-omics"

Learning Objectives

After reading this section, students will be able to:

LO 13.9.1 Distinguish between *genomics*, *proteomics*, *transcriptomics*, *metabolomics*, and *kinomics*.

With more and more full-genome DNA sequences becoming available, it is tempting to compare genomes to see whether patterns emerge from genes

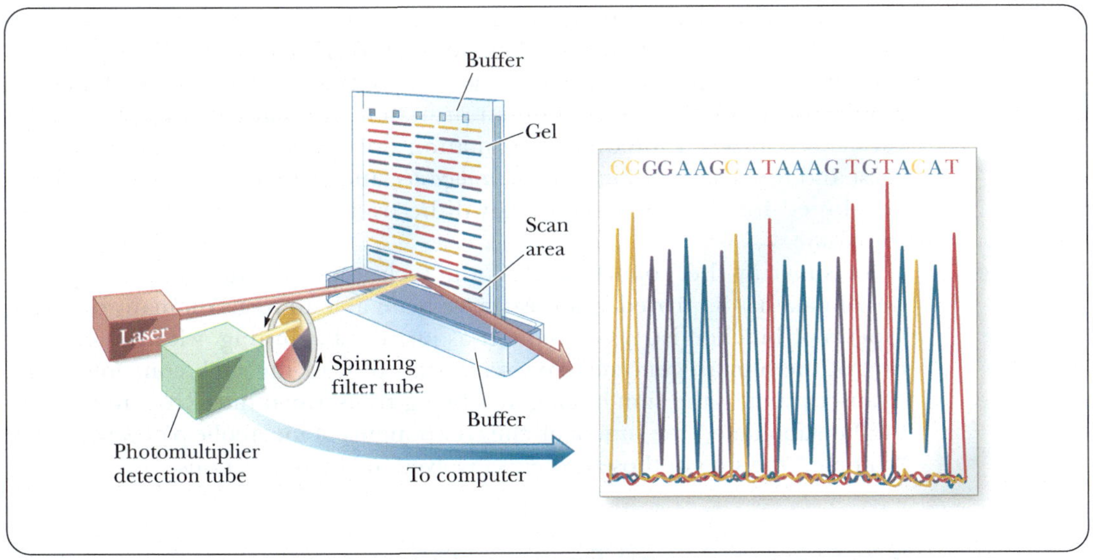

Figure 13.31 Dye-terminator sequencing of DNA. Four reactions are set up, one for each base, and the primer in each is end-labeled with one of four different fluorescent dyes; the dyes color-code the base-specific sequencing protocol (a unique dye is used in each dideoxynucleotide reaction). The four reaction mixtures are then combined and run in one lane. Thus, each lane in the gel represents a different sequencing experiment. As the differently sized fragments pass down the gel, a laser beam excites the dye in the scan area. The emitted energy passes through a rotating color filter and is detected by a fluorometer. The color of the emitted light identifies the final base in the fragment. *(Applied Biosystems, Inc., Foster City, CA.)*

that encode proteins with similar functions. The enormous amount of data makes the use of a computer essential for the process. Databases on genome and protein sequences are so extensive that they require information technology at its best to solve problems. Knowing the full DNA sequence of the human genome, for example, allows us to better address the causes of disease. That prospect was one of the main incentives for undertaking the Human Genome Project.

A number of genomes are available online, along with software for sequence comparisons. An example is the material available from the Sanger Institute. In November 2003, the researchers at this institute announced that they had sequenced 2 billion bases from the DNA of several organisms (human, mouse, zebrafish, yeast, and the roundworm *Caenorhabditis elegans,* among others). If this amount of DNA were the size of a spiral staircase, it would reach from Earth to the Moon.

A question implicit in the determination of the genome of any organism is that of the assignment of sequences to the chromosome to which they belong. This is a challenging task, but computer algorithms make it possible. Using these algorithms, scientists can compare genomes to see what changes have occurred in the DNA of complex organisms compared with those of simpler ones.

Beyond this application lie numerous medical applications, which are leading to a number of surprises. Two closely related genes (*BRCA1* and *BRCA2*) involved in the development of breast cancer interact with other genes and proteins, and this is a topic of feverish research. The connection between these genes and a number of seemingly unrelated cancers is only starting to be unraveled. There is a need to determine not only the genetic blueprint but the manner in which an organism puts it into action.

The **proteome** is the protein version of the genome. In all organisms for which sequence information is available, **proteomics** (the study of interactions among all the proteins in a cell) is assuming an important place in the life sciences. If the genome is the script, the proteome puts the play on

proteome the total protein content of the cell

proteomics study of interactions among all the proteins of the cell

stage. The genetically determined amino acid sequence of proteins determines their structure and how they interact with each other. Those interactions determine how they behave in a living organism. The potential medical applications of knowing the human proteome are many, but most have not yet been realized. Proteomic information does exist for eukaryotes such as yeast and the fruit fly *Drosophila melanogaster,* and the methods that have been developed for those experiments will be useful in unraveling the human proteome.

This is truly the "-omics" era! As the technology to study massive quantities of data has improved, scientists have been able to focus on many different aspects of biochemistry. Genomics was just the beginning, with the elucidation of the human genome. Now scientists are studying such concepts as the "transcriptome," the sum of all genes being transcribed into RNAs in the cell, the "metabolome," the sum of all the components of metabolic pathways, or the "kinome," the sum of those proteins involved in kinase reactions.

13-10 DNA and Protein Microarrays

Learning Objectives

After reading this section, students will be able to:

LO 13.10.1 Describe how DNA microarrays are performed.

LO 13.10.2 Describe how protein microarrays are performed.

LO 13.10.3 Distinguish between the purposes of a DNA microarray compared to a protein microarray.

DNA microarray (DNA chip or gene chip) an arrangement of DNA samples on a single tiny computer chip, on which many samples can be examined simultaneously

transcriptome the group of genes that are being transcribed at a given time

Thousands of genes and their products (i.e., RNA and proteins) in a given living organism function in complicated and harmonious ways, and microarrays are used to study large numbers of these functions and interactions at once. A **DNA microarray (DNA chip or gene chip)** allows for the analysis of an entire genome in one experiment and is used to study gene expression—the transcription rates of the genome in vivo. The genes that are being transcribed at any particular time are known as the **transcriptome.**

The principle behind the microarray is the placement of specific nucleotide sequences in an ordered array, which then base-pair with complementary sequences of DNA or RNA that have been labeled with fluorescent markers of different colors. The locations where binding occurred and the colors observed are then used to quantify the amount of DNA or RNA bound. Microarray chips are manufactured by high-speed robotics, which can put thousands of samples on a glass slide with an area of about 1 cm^2. The diameter of an individual sample might be 200 μm or less. Several different methods are used for implanting the DNA to be studied on the chip, and many companies make microarray chips.

▶ How do DNA microarrays work?

Figure 13.32 shows an example of how microarrays could be used to determine whether a potential new drug would be harmful to liver cells. In Step 1, a microarray is purchased or constructed that has single-stranded DNA representing thousands of different genes, each applied at a specific spot on the microarray chip. In Step 2, different samples of liver cells are collected—one sample is treated with the potential drug and the other is untreated. The mRNAs being transcribed in these cells are then collected, and the gene expression in the treated cells is theoretically different from that of the untreated cells.

In Step 3, the mRNAs are converted to cDNAs using reverse transcriptase and fluorescently labeled dNTPs. For example, green fluorescent nucleotides can be added to the cDNA from the untreated cells, and red fluorescent

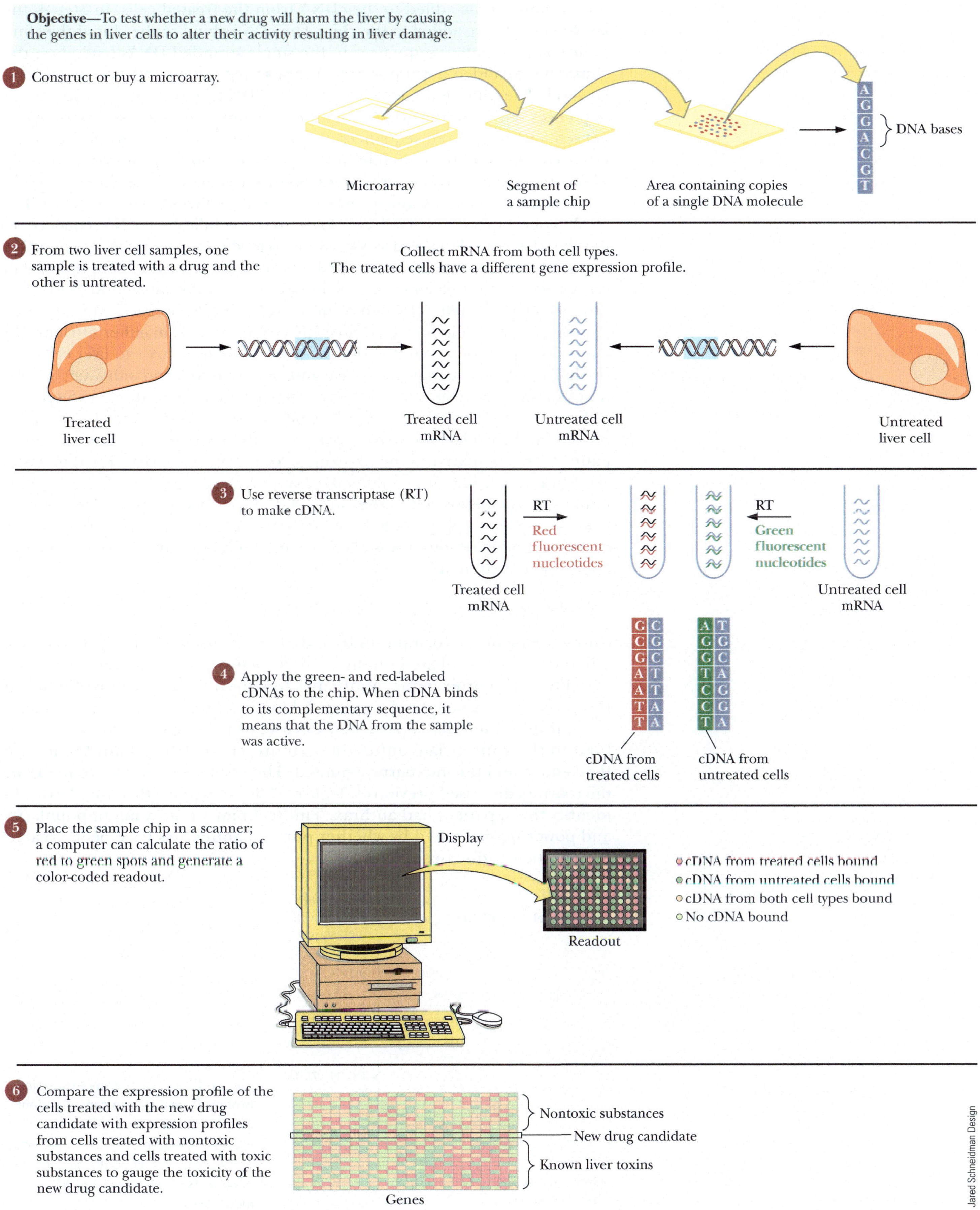

Figure 13.32 **How microarrays work.** In Step 2, the blue screens over the DNA molecules indicate different areas of active gene expression.

nucleotides can be added to the cDNA from the treated cells. In Step 4, the labeled cDNAs are added to the chip. The cDNAs bind to the chip if they find their complementary sequences in the single-stranded DNAs loaded on the chip. The expanded portion of Step 4 shows what is happening at the molecular level. The blue sequences represent the DNA bound to the chip. Green sequences represent the cDNA from the untreated cells that bind to their target sequences. The red sequences are cDNA from treated cells. Some of the DNA sequences on the chip bind nothing. Some bind only the red sequences, while others bind only the green ones. Some sequences on the chip bind both.

In Step 5, the chip is scanned, and a computer analyzes the fluorescence. The results appear as a series of colored dots. A red dot indicates a DNA sequence on the chip that bound to the cDNA from the treated cells. This indicates an mRNA that was being expressed in treated cells. A green dot indicates RNA produced in untreated but not treated cells. A yellow dot would indicate an mRNA that was produced equally well in treated or untreated cells. Black spaces indicate DNA sequences on the chip for which no mRNA was produced in either situation. To answer the question about whether the potential drug is toxic to liver cells, the results from the microarray would be compared to controls run with liver cells and drugs known to be toxic versus those known to be nontoxic, as shown in Step 6.

Figure 13.33 shows the results of a study designed to scan cells from cancer patients and correlate microarray patterns with prognoses. The four different patterns are compared to the percentage of patients who developed metastases. Information like this could be critical to cancer treatment. Doctors often have to choose between different strategies. If they had access to data like this from their patients, they would be able to predict the likelihood of the patient's developing more serious forms of cancer and thereby be able to choose a more appropriate treatment.

13-10A Protein Microarrays

Another type of microchip uses bound proteins instead of DNA. These protein microarrays are based on interactions between proteins and antibodies (Chapter 14). For example, antibodies to known diseases can be bound to the microarray. A sample of a patient's blood can then be put on the microarray. If the patient has a particular disease, proteins specific to that disease bind to the appropriate antibodies. Fluorescently labeled antibodies are then added and the microarray scanned. The results look similar to the DNA microarrays discussed previously. Figure 13.34 shows how this would work to identify that a patient had anthrax. This technique is growing in popularity and power but is limited by whether purified antibodies have been created for a particular disease.

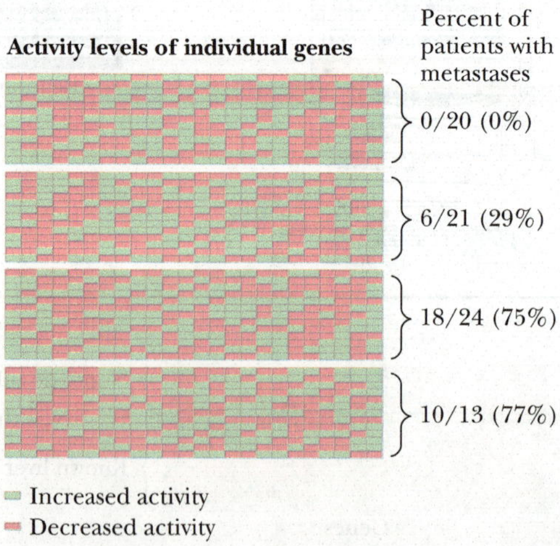

Figure 13.33 Microarray expression profiles from different cancer patients. Four microarrays show different expression patterns for cancer patients. These are compared to the percentage of patients who later developed metastases.

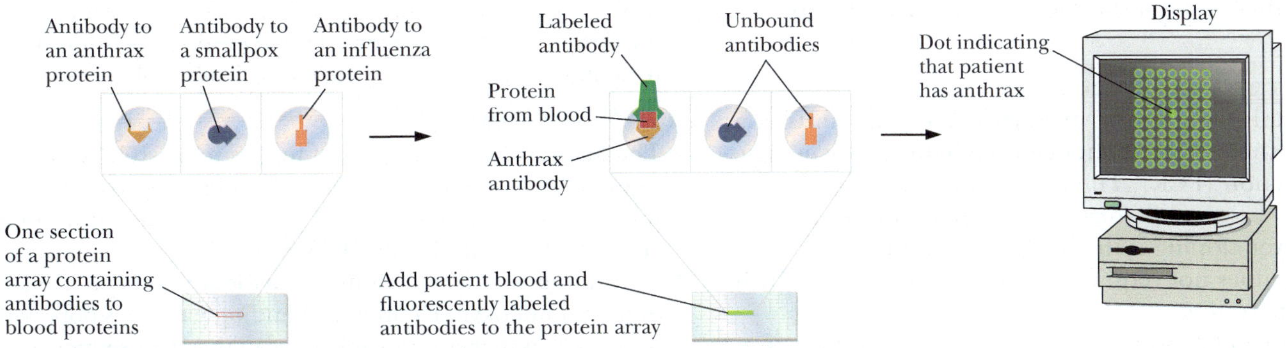

Figure 13.34 **Protein microarray analysis.**

Summary

Nucleic Acid Separation Techniques

- Two of the primary necessities for successful experiments with nucleic acids are to separate the components of a mixture and to detect the presence of nucleic acids. Agarose gels prepared with ethidium bromide are commonly used for these purposes.

Nucleic Acid Detection Methods

- DNA can be visualized with a variety of dyes. Ethidium bromide is most commonly used, but it is a carcinogen. Safer dyes, such as SyBr Green and SyBr Gold, can also be used.
- Bands of dyed DNA fluoresce under UV light and can be seen easily.

Restriction Endonucleases

- Restriction endonucleases play a large role in the manipulation of DNA.
- These enzymes produce short, single-stranded stretches, called sticky ends, at the ends of cleaved DNA. The sticky ends provide a way to link DNA strands from different sources.

Production of Recombinant DNA and Cloning

- DNA sequences of interest can be selectively cut by using restriction endonucleases and then spliced into a bacterial plasmid using DNA ligase to produce recombinant DNA plasmids.
- The recombinant plasmids can be inserted into bacterial cells. Growing the bacteria then produces large amounts of the recombinant DNA by the process of cloning.

Plasmid Vectors

- Plasmids that are used in cloning are small circular pieces of bacterial DNA that are engineered to carry several useful genes, such as an antibiotic resistance gene. Once inside a bacterium, they replicate when the bacterial DNA replicates.

Universal Cloning Plasmids

- The universal cloning plasmid (pUC) contains the ampicillin-resistance gene and a *lacZ* gene, which produces β-galactosidase. When cells are grown on an agar plate containing ampicillin and a dye that will turn blue in the presence of β-galactosidase, only those cells that have the pUC plasmid containing both genes will grow, and they will appear blue in color.
- The pUC plasmids are widely used for producing recombinant DNA because they have a multiple cloning site (MCS) region in the *lacZ* gene. The MCS region has many different restriction enzyme sites, and these sites can be used to cut and ligate pUC plasmids and DNA insert sequences. The recombinant pUC plasmids with insert DNA have inactive *lacZ* genes, so cells with these plasmids grown on an agar plate containing ampicillin and a dye that will turn blue in the presence of β-galactosidase will appear as the normal white color of bacterial colonies.

Genetic Engineering

- Scientists use genetic engineering to modify the characteristics of an organism by manipulating its DNA.
- If a human gene is inserted into a plasmid and the plasmid is transformed into a bacterial cell, then the plasmid is replicated when the cell divides and replicates. If the plasmid is an expression vector, then it can also transcribe and translate the protein from the human gene being carried by the plasmid. In this way, bacteria can produce human proteins and other proteins relatively quickly and cheaply for science and medicine.
- CRISPR-Cas9 is a widely used genome editing technique. It uses a designed RNA guide molecule to selectively cut and modify target sequences of genomic DNA.

DNA Libraries

- DNA libraries are created by digesting genomic DNA into segments that are suitable for insertion into a plasmid, followed by cloning.
- To find individual clones in a series of DNA library plates, colonies are transferred to filter discs. These discs can be reacted with specific DNA probes for a sequence of interest. The probe has ^{32}P in it, so when the probe binds, its location can be determined with autoradiography. The

colony at that position can then be isolated and amplified for study.

The Polymerase Chain Reaction

- An alternative method of producing large amounts of a relatively short DNA sequence is called the polymerase chain reaction (PCR). The procedure is automated and relies on a heat-stable form of DNA polymerase. The advantage over cloning is that PCR is faster, produces more DNA, and does not require plasmids or bacterial growth to generate results.
- Quantitative PCR (qPCR) allows for sensitive measurements of the original amount of target DNA that was in a sample.

DNA Fingerprinting

- The ability to analyze DNA fragments is important for basic research and for forensic science.
- DNA fingerprinting makes it possible to identify individuals from their DNA samples. In this technique, DNA is digested with restriction enzymes, and a banding pattern is seen upon electrophoresis of the digest. No two individuals have the same pattern, just as no two people have the same fingerprints. This technique is often used for paternity tests and for the identification of criminals.

DNA Sequencing

- The Sanger–Coulson method makes it possible to determine a DNA sequence by using labeled dideoxynucleotides. These nucleotides lack 3′ hydroxyl groups and cause chain termination during DNA synthesis.
- By running parallel reactions with a dideoxy version of each of the four deoxynucleotides, a banding pattern is obtained on a gel that makes it possible to read the DNA sequence.

DNA and Protein Microarrays

- Robotic technology is used to load thousands of samples of DNA or protein onto a microchip. Biological samples are then overlaid on the chip, and the binding of molecules in the sample to molecules on the chip is evaluated through the use of fluorescence.
- Researchers use microarrays to analyze gene expression or the presence of certain proteins under different conditions.

Review Exercises

13-1 Purification and Detection of Nucleic Acids

1. **Reflect and Apply** When proteins are separated using native gel electrophoresis, size, shape, and charge control their rate of migration on the gel. Why does DNA separate based on size, and why do we not worry much about shape or charge?

13-2 Restriction Endonucleases

2. **Reflect and Apply** What type of DNA sequences do restriction endonucleases typically recognize, and how are restriction endonucleases used in biochemistry?
3. **Recall** How does the use of restriction endonucleases of different specificities aid in the sequencing of DNA?
4. **Recall** What is the importance of methylation in the activity of restriction endonucleases?
5. **Recall** Why don't restriction endonucleases hydrolyze DNA from the organism that produces them?
6. **Recall** Where did restriction endonucleases get their name?
7. **Recall** Give three examples of DNA palindromes.
8. **Reflect and Apply** What are three differences between the sites recognized by *Hae*III and those recognized by *Bam*HI?
9. **Recall** What are sticky ends? What is their importance in recombinant DNA technology?
10. **Recall** What would be an advantage of using *Hae*III for a cloning experiment? What would be a disadvantage?

13-3 Production of Recombinant DNA and Cloning

11. **Recall** Describe the key steps involved in producing recombinant DNA, including the enzymes used and how the recombinant DNA is introduced into bacterial cells.
12. **Recall** Describe the cloning of DNA.
13. **Recall** Describe the methods you would use to test for the uptake of a plasmid with a DNA insert.
14. **Recall** What is blue/white screening? What is the key feature of a plasmid that is used for it?
15. **Recall and Apply** What are some general "requirements" for recombinant DNA technology?

13-4 Genetic Engineering

16. **Recall** What are the purposes of genetic engineering in agriculture?
17. **Biochemical Connections** You go for a drive in the country with some friends and pass a cornfield with a sign. Your friends do not understand the cryptic message on the sign, with the letters "Bt" followed by some numbers. How would you be able to enlighten them based on information from this chapter?
18. **Reflect and Apply** Using information we have seen about lactate dehydrogenase, how could you clone and express human lactate dehydrogenase 3 (LDH 3) in bacteria? Recall that LDH 3 is a tetramer that has the subunit composition H_2M_2.
19. **Recall** What are the requirements for a recombinant protein expression vector?
20. **Reflect and Apply** What is a distinguishing feature of a recombinant protein expression vector compared to other vectors, such as a cloning vector, that enables high-level production of a recombinant protein, and why is it important?
21. **Reflect and Apply** The genes for both the α- and β-globin chains of hemoglobin contain introns (i.e., they are split genes). How would this fact affect your plans if you wanted to introduce the gene for α-globin into a bacterial plasmid and have the bacteria produce α-globin?
22. **Recall** Explain how CRISPR-Cas9 technology targets and edits specific DNA sequences, and describe why this technology is important.

13-5 DNA Libraries

23. **Reflect and Apply** What are the differences between a DNA library and a cDNA library?
24. **Recall** Why do some journals require the authors of articles describing DNA libraries to make those libraries available to other researchers?

25. **Biochemical Connections** What is a fusion protein? How are fusion proteins involved in cloning and expression?

13-6 The Polymerase Chain Reaction

26. **Recall** Why is temperature control so important in the polymerase chain reaction?
27. **Recall** Why is the use of temperature-stable DNA polymerase an important factor in the polymerase chain reaction?
28. **Recall** What are the criteria for "good" primers in a PCR reaction?
29. **Reflect and Apply** What difficulties arise in the polymerase chain reaction if there is contamination of the DNA that is to be copied?
30. **Reflect and Apply** Each of the following pairs of primers has a problem with it. Tell why the primers would not work well.
    (a) Forward primer 5′ GCCTCCGGAGACCCATTGG 3′
    Reverse primer 5′ TTCTAAGAAACTGTTAAGG 3′
    (b) Forward primer 5′ GGGGCCCCTCACTCGGGGCCCC 3′
    Reverse primer 5′ TCGGCGGCCGTGGCCGAGGCAG 3′
    (c) Forward primer 5′ TCGAATTGCCAATGAAGGTCCG 3′
    Reverse primer 5′ CGGACCTTCATTGGCAATTCGA 3′
31. **Recall** What is qPCR?
32. **Recall** What is the functional difference between regular PCR and qPCCR?

13-7 DNA Fingerprinting

33. **Reflect and Apply** Suppose that you are a prosecuting attorney. How has the introduction of the polymerase chain reaction changed your job?
34. **Reflect and Apply** Why is DNA evidence more useful as exclusionary evidence than for the positive identification of a suspect?

13-8 DNA Sequencing

35. **Reflect and Apply** Write the DNA sequence for the template strand that gives rise to the following sequencing gel, prepared using the Sanger method with a radioactive label at the 5′ end of the primer.

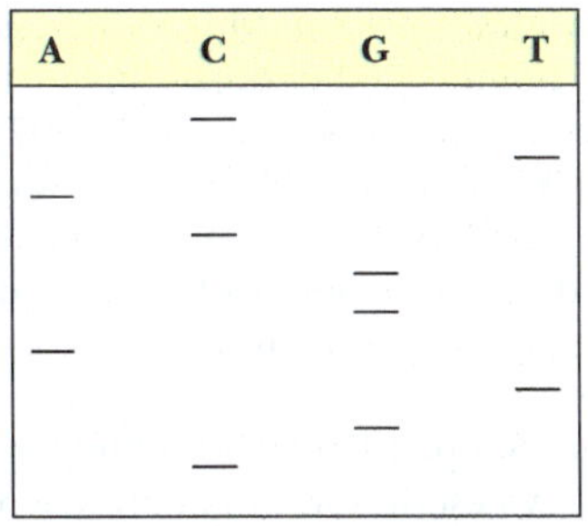

36. **Reflect and Apply** Although techniques are available for determining the sequences of amino acids in proteins, it is becoming more and more common to sequence proteins indirectly by determining the base sequence of the gene for the protein and then inferring the amino acid sequence from the genetic-code relationships. Suggest why the latter technique is being used for proteins.
37. **Reflect and Apply** Sometimes knowing the DNA sequence of a gene that codes for a protein does not tell you the amino acid sequence. Suggest several reasons why this is so.

13-9 Genomics and Other "-omics"

38. **Recall** What is the difference between the genome and the proteome?
39. **Recall** Has proteomic analysis been done on multicellular eukaryotes?

13-10 DNA and Protein Microarrays

40. **Recall** Explain how a DNA microarray works.
41. **Reflect and Apply** If you wanted to study the nature of transcription in yeast under aerobic versus anaerobic conditions, how could you use DNA microarrays to accomplish this?
42. **Recall** How are DNA microarrays used to screen a patient's cells for a cancer prognosis?
43. **Recall** What are the key differences between DNA microarrays and protein microarrays, and how are they used in research?

Further Reading

Anderson, N. L., and G. Valkers. Protein Arrays—A New Option. *Sci. Amer.* **286** (2), 51 (2002). A good overview of protein arrays and how they work.

Berg, P., and M. Singer. *Dealing with Genes: The Language of Heredity.* Mill Valley, CA: University Science Books, 1992. Two leading biochemists have produced an eminently readable book on molecular genetics. Highly recommended.

Brown, K. Seeds of Concern. *Sci. Amer.* **284** (4), 52–57 (2000). Discussion of potential environmental problems associated with genetically modified foods.

DeRisi, J. L. Exploring the Metabolic and Genetic Control of Gene Expression on a Genomic Scale. *Science* **278**, 680–686 (1997). Comprehensive discussion of microarray use for mapping transcription changes in yeast under anaerobic or aerobic conditions.

Editors of *Science* et al. Genome Issue. *Science* **274**, 533–567 (1996). A series of articles on genomes of organisms from yeast to humans, including a map of the human genome. The articles include an editorial about policy issues. A web feature is also associated with this issue.

Editors of *Science* et al. Genome Issue. *Science* **302**, 587–608 (2003). A series of articles on progress in applying genomic information to medicine. The article by Couzin on the *BRCA* genes appears on pp. 591–593.

Friend, S. H., and R. B. Stoughton. The Magic of Microarrays. *Sci. Amer.* **286** (2), 44–49 (2002). An in-depth article about microarray technology.

Hannon, G. J. RNA Interference. *Nature* **418**, 244–251 (2002). A description of one of the most important topics in nucleic acid manipulation.

Li, T., C. Y. Chang, D. Y. Jin, P. J. Lin, A. Khvorova, and D. W. Stafford. Identification of the Gene for Vitamin K Epoxide Reductase. *Nature* **427**, 541–544 (2004). A report on an important use of RNA interference.

MacBeath, G. Printing Proteins as Microarrays for High-Throughput Function Determination. *Science* **289**, 1760–1763 (2000). A research article on protein microchips.

Marx, J. DNA Arrays Reveal Cancer in Its Many Forms. *Science* **289**, 1670–1672 (2000). The use of DNA chips in the study of common cancers.

O'Brien, S., and M. Dean. In Search of AIDS-Resistance Genes. *Sci. Amer.* **277** (3), 44–51 (1997). Genetic resistance to HIV infection may provide the basis of new approaches to prevention and therapy of AIDS.

Pennisi, E. Laboratory Workhorse Decoded: Microbial Genomes Come Tumbling In. *Science* **277**, 1432–1434 (1997). A Research News article on the determination of the complete genome of the bacterium *Escherichia coli,* with a discussion of information available on genomes of other organisms. To be read in conjunction with the research article on pp. 1453–1474 of the same issue.

Pollack, A. U.S.D.A. Approves Modified Potato. Next Up: French Fry Fans. *The New York Times.* N.p., 7 Nov. 2014, Web. 20 Sept. 2015. Article describes benefits of genetic modification to create the innate potato leading to increased value and decreased toxicity.

Reichhardt, T. Will Souped Up Salmon Sink or Swim. *Nature* **406**, 10–12 (2000). A description of genetically modified oversized salmon.

Rodriguez, A. V., C. Rodriguez-Oramas, E. S. Velazquez, A. Hardisson de la Torre, C. R. Armendariz, and C. C. Iruzubieta. Myths and Realities about Genetically Modified Food: A Risk-Benefit Analysis. *Appl. Sci.* **12** (6), 2861–2877 (2022). Reflections on possible problems with GM foods.

Ronald, P. Making Rice Disease-Resistant. *Sci. Amer.* **277** (5), 100–105 (1997). Genetic engineering to improve yields of one of the world's most important food crops.

Sander, J. D., and J. K. Joung. CRISPR-Cas Systems for Editing, Regulating and Targeting Genomes. *Nat. Biotechnol.* **32** (4), 347–355 (2014). Gene editing can be used to insert new genes or suppress harmful genes using RNA-based technology.

Service, R. F. Protein Arrays Step Out of DNA's Shadow. *Science* **289**, 1673 (2000). A single-page review of protein microchips.

Tabashnik, B. E., Brevault, T., and Carriere, Y. Insect Resistance to Bt Crops: Lessons from the First Billion Acres. *Nat. Biotechnol.* **31**, 510–521 (2013). An article about the global adoption of crops engineered to kill specific pests while being safe for humans and non-target organisms.

Windbichler, N., M. Menichelli, P. A. Papathanos, S. B. Thyme, H. Li, U. Y. Ulge, B. T. Hovde, D. Baker, R. J. Monnat Jr., A. Burt, A. Crisanti. A Synthetic Homing Endonuclease-Based Gene Drive System in the Human Malaria Mosquito. *Nature* **473**, 212–215 (2011). An application of gene editing technology may be to reduce mosquito populations, leading to fewer incidents of malaria.

Hot Topic

Barrangou, R. "Cas9 Targeting and the CRISPR Revolution." *Science* **344**, 707–708 (2014).

Doudna, J. A. and E. Charpentier. "The New Frontier of Genome Engineering with CRISPR-Cas9." *Science* **346**, issue 6213 (2014).

Gantz, V. M., N. Jasinskiene, O. Tatarenkova, A. Fazekas, V. M. Macias, E. Bier, and A. A. James. "Highly Efficient Cas9-Mediated Gene Drive for Population Modification of the Malaria Vector Mosquito *Anopheles stephensi.*" *PNAS,* DOI 10.1073/pnas.1521077112 (2015).

Long, C., J. R. McAnally, J. M. Shelton, A. A. Mireault, R. Bassel-Duby, and E. N. Olson. "Prevention of Muscular Dystrophy in Mice by CRISPR/CAS9-Mediated Editing of Germline DNA." *Science* **345**, 1184–1188 (2014).

Sander, J. D. and J. K. Joung. "CRISPR-Cas Systems for Editing, Regulating and Targeting Genomes." *Nat. Biotechnol.* 32(4), 347–355 (2014).

Windbichler, N., M. Menichelli, P. A. Papathanos, S. B. Thyme, H. Li, U. Y. Ulge, B. T. Hovde, D. Baker, R. J. Jr. Monnat, A. Burt, and A. Crisanti. "A Synthetic Homing Endonuclease-Based Gene Drive System in the Human Malaria Mosquito." *Nature* **473**, 212–215 (2011).

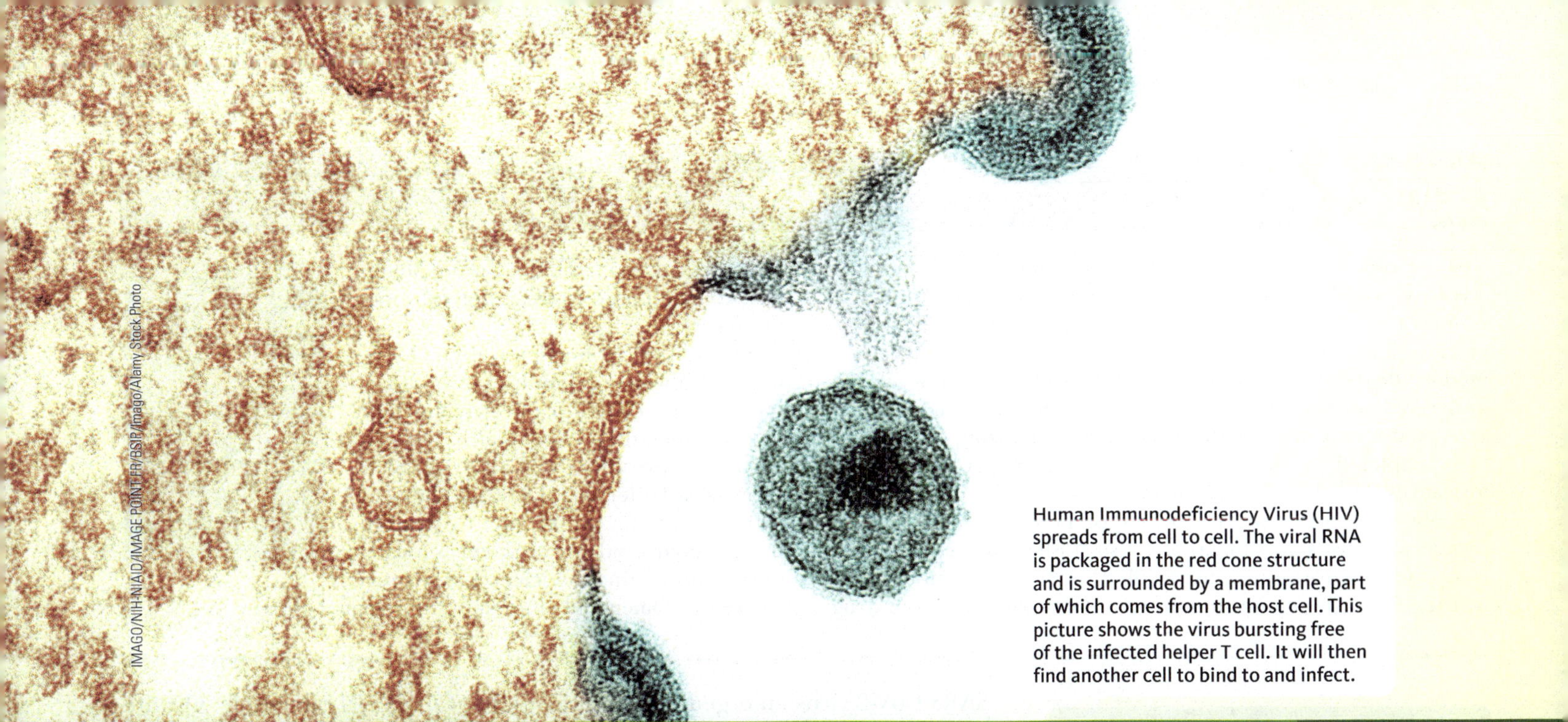

Human Immunodeficiency Virus (HIV) spreads from cell to cell. The viral RNA is packaged in the red cone structure and is surrounded by a membrane, part of which comes from the host cell. This picture shows the virus bursting free of the infected helper T cell. It will then find another cell to bind to and infect.

Viruses, Cancer, and Immunology 14

Chapter Outline

14-1 Viruses

Learning Objectives

After reading this section, students should be able to:

LO 14.1.1 Describe the structure of a typical virus.

LO 14.1.2 Explain events in the life cycle of a virus.

Viruses have always been difficult to classify according to normal taxonomy. Many have argued over whether they should be considered living things. They cannot reproduce independently, and they cannot make proteins or generate energy independently, so they do not meet all of the requirements for life as we have traditionally defined it. But if they are not life forms, then what are they? A simple definition of a **virus** is that it is an entity consisting of a small amount of genetic material, packaged in and assisted by viral proteins, that interacts with host cells. Most viruses have only one type of nucleic acid, either DNA or RNA. Depending on the virus, this nucleic acid could be either single-stranded or double-stranded.

14-1A Common Viruses

Viruses are pathogens of bacteria, plants, and animals, and they are known for the diseases they cause. Some viruses are deadly, such as the fast-acting Ebola virus, which can have a mortality rate of more than 85%. Others can cause serious, lifelong conditions, such as the slow-acting human immunodeficiency virus (HIV), which causes acquired immunodeficiency syndrome (AIDS). In late 2019, the

Pioneers in Biochemistry • Jewel Plummer Cobb

Jewel Plummer Cobb (1924–2017) was a cancer researcher, university president, and advocate for increasing the number of women and people of color in STEM fields. Cobb's father, whose grandfather was enslaved, was the first Black man to receive a medical degree from Cornell University. Cobb herself began her undergraduate studies at the University of Michigan, but left in her sophomore year because of the segregated housing on campus. She completed her degree in biology at Talladega College, a historically Black college in Alabama.

Cobb's scientific research focused on how melanin in the skin protects against skin cancer. She also studied how cell division is affected by hormones, ultraviolet light, and chemotherapy. In the 1960s, she made the important discovery that the drug methotrexate can treat skin cancer. Methotrexate is still used to treat a broad range of cancers.

Cal State Fullerton

In addition to her scientific achievements, Cobb was committed to broadening access to scientific careers as a mentor, teacher, and administrator, all while serving as the president of California State University, Fullerton. She was recognized with a Lifetime Achievement Award by the National Academy of Sciences for her work in promoting the sciences as a career choice for women and other underrepresented groups.

Chapter Outline (continued)

SARS-CoV-2 virus emerged, causing the global COVID-19 pandemic. On the other hand, some viruses are simply annoying, such as rhinovirus, which causes common colds.

▶ What is the structure of a virion?

The infectious virus particles that exist outside of host cells are called **virions**, and they can be composed of nucleic acid, **coat proteins (CPs)**, spike proteins, and lipids (Figure 14.1). At the center of the virion is the nucleic acid. Surrounding this is the **capsid**, which is a protein coat. The combination of the nucleic acid and the capsid is called the *nucleocapsid*. For some viruses, such as the rhinovirus, that is the extent of the particle. Many other viruses, including HIV, have a *membrane envelope* surrounding the nucleocapsid. Many viruses also have *protein spikes* that help them attach to their host cell. Figure 14.1 shows the main features of a virus.

The overall shape of a virus varies. The classic viral shape most often seen in the literature has an icosahedral shape, which appears hexagonal in profile, with a rod sticking out of it that attaches to the host cell and acts like a syringe to inject the nucleic acid (Figure 14.2).

virus an entity consisting of a small amount of genetic material, packaged in and assisted by viral proteins, that interacts with host cells

virions virus particles that exist outside of host cells that can consist of nucleic acid, coat proteins, spike proteins, and lipids

coat proteins (CPs) the proteins that make up the capsid or protein coat that surrounds the genetic material of a virus

capsid the protein coat of a virus

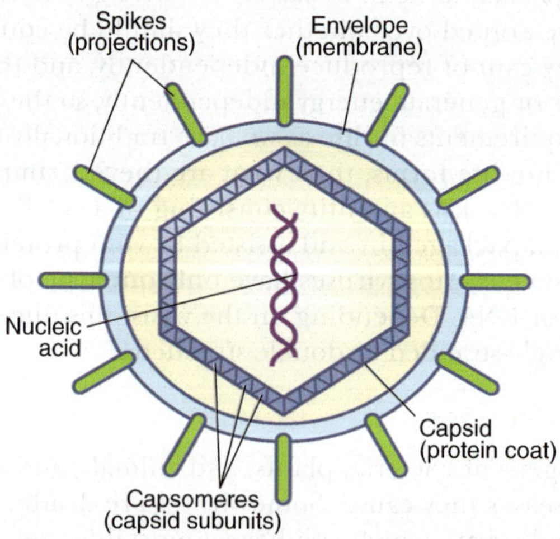

Figure 14.1 A schematic diagram of a typical virion. The nucleic acid is in the middle, surrounded by a protein coat called the capsid. Many viruses also have an envelope membrane that is usually covered with protein spikes.

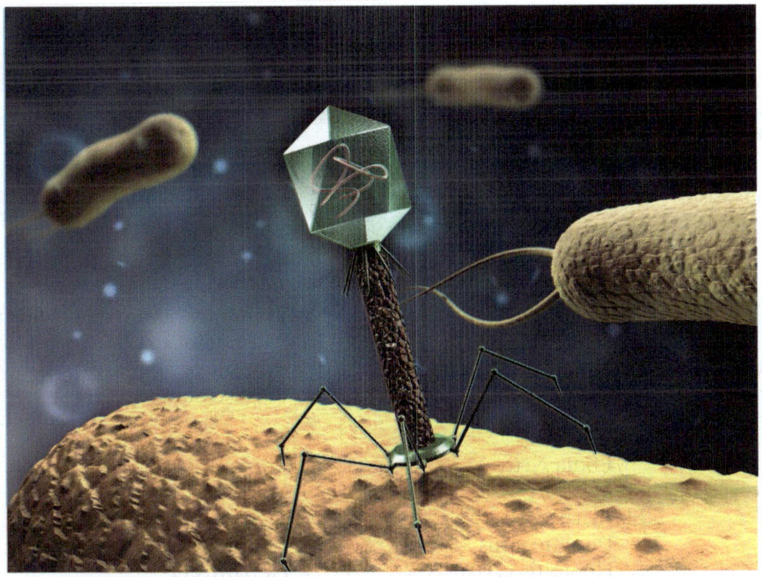

Figure 14.2 A digital illustration of a bacteriophage T4 virion infecting a bacterial cell.

14-1B Families of Viruses

Although many characteristics distinguish viruses, most are organized by whether they have a genome of DNA or of RNA and whether they have an envelope. In addition, the nature of the nucleic acid (linear versus circular, small versus large, single-stranded versus double-stranded) and the mode of incorporation (nucleic acid remains separate versus nucleic acid joins with host chromosome) distinguish the different virus types. Table 14.1 lists some of the known viral diseases and the families of viruses that cause them.

14-1C Virus Life Cycles

Most viruses cannot survive for long periods outside cells, so they must quickly gain access to a cell. There are several mechanisms for gaining access, and preventing access to the cell has been a major focus of pharmaceutical companies trying to develop antiviral drugs. Figure 14.3 shows a generic example of a virus infecting a cell. The virus binds to the cell membrane and releases its DNA into the cell. The viral DNA is then replicated by host DNA polymerases and transcribed by host RNA polymerases, producing viral mRNA. Translation of the viral mRNA leads to production of spike proteins and the coat proteins of the capsid. New virions are produced and then released from the cell. This cycle of viral reproduction is called the **lytic cycle** because the host cells are lysed by this process. Viruses do not always lyse their host cells, however. A separate process, called the **lysogenic cycle**, involves the incorporation of the viral DNA into the host chromosome, but the cell is not destroyed. Viruses can switch from one pathway to another based on the condition of the host cells. If the host is healthy, there is sufficient material to allow the virus to replicate and produce new virions. If the host cell is starved or unhealthy, there may be insufficient energy and material to do so. In this case, lysogeny allows the DNA to incorporate into the host cell, where it can wait until the cell's health improves.

lytic cycle in virus reproduction, the cycle that involves the virus infecting a cell, using cellular machinery to produce many virions, and then lysing the cell to release the new virions

lysogenic cycle a viral life cycle pathway in which the viral DNA integrates into the host DNA but does not lyse the host cell

Simian virus 40 (SV40) is a DNA virus (Figure 14.4A). It appears to be spherical, but it is actually an icosahedron, a geometric shape with 20 faces that are equilateral triangles (Figure 14.4B). The genome of this virus is a closed circle of

Table 14.1 Vertebrate Viruses and Diseases They Cause

DNA-Containing Viruses, Nonenveloped	
Adenoviruses	Respiratory and gastrointestinal diseases
Circoviruses	Anemia in chickens
Iridoviruses	Various diseases of insects, fish, and frogs
Papovaviruses	
Papillomaviruses	Warts
Human papillomavirus	Cervical cancer
Parvoviruses	
Human parvovirus B19	Fifth disease (a childhood rash)
Canine parvovirus	Viral gastroenteritis in dogs

DNA-Containing Viruses, Enveloped	
African swine fever virus	African swine fever (rarely in humans)
Hepadnaviruses	Hepatitis B
Herpesviruses	
Cytomegalovirus	Birth defects
Epstein–Barr virus (EBV)	Infectious mononucleosis and Burkitt's lymphoma
Herpes simplex virus (HSV) type 1	Cold sores
Herpes simplex virus (HSV) type 2	Genital herpes
Varicella-zoster virus (herpes zoster virus)	Chickenpox and shingles
Poxviruses	
Monkeypox virus	Monkeypox
Variola major virus	Smallpox

RNA-Containing Viruses, Nonenveloped	
Astroviruses	Gastroenteritis
Birnaviruses	Various diseases of birds, fish, and insects
Calciviruses	Norwalk gastroenteritis
Picornaviruses	
Hepatitis A virus	Hepatitis A
Polioviruses	Poliomyelitis
Rhinoviruses	Common cold
Reoviruses	
Rotaviruses	Infantile gastroenteritis

RNA-Containing, Enveloped	
Arenaviruses	
Lassa virus	Lassa fever
Arteriviruses	
Equine arteritis virus	Equine viral arteritis
Bunyaviruses	
California encephalitis virus	California encephalitis
Hantaviruses	Epidemic hemorrhagic fever or pneumonia
Coronaviruses	Respiratory disease, possibly gastroenteritis
Filoviruses	
Ebola virus	Ebola disease
Marburg virus	Marburg disease
Flaviviruses	
Dengue virus	Dengue fever
Hepatitis C virus	Hepatitis C
St. Louis encephalitis virus	St. Louis encephalitis
Yellow fever virus	Yellow fever
Orthomyxoviruses	Influenza
Paramyxoviruses	
Measles virus	Measles
Mumps virus	Mumps
Respiratory syncytial virus (RSV)	Pneumonia, bronchitis
Retroviruses	
Human T-cell lymphotropic viruses (HTLVs)	Leukemia, lymphoma
Human immunodeficiency virus (HIV)	Acquired immunodeficiency syndrome (AIDS)
Rhabdoviruses	
Rabies virus	Rabies
Togaviruses	
Rubella virus	Rubella (German measles)
Eastern equine encephalomyelitis (EEE) virus	Encephalomyelitis

double-stranded DNA, with genes that encode the amino acid sequences of five proteins. Three of the five proteins are coat proteins. Of the remaining two proteins, one, the large-T protein, is involved in the development of the virus when it infects a cell. The function of the fifth protein, the small-T protein, is to aid in cell transformation by inhibiting the host's protein phosphatase 2 activity.

The outcome of infection by SV40 depends on the organism infected (Figure 14.5). When simian cells are infected, the virus enters the cell and loses its protein coat. The viral DNA is expressed first as mRNAs and then as proteins. The large-T protein is the first one made (Figure 14.5, *middle*), triggering the replication of viral DNA, followed by viral coat proteins. The virus takes over the cellular machinery for both replication of DNA and protein synthesis. New virus particles are assembled, and eventually the infected cell bursts, releasing the new virus particles to infect other cells.

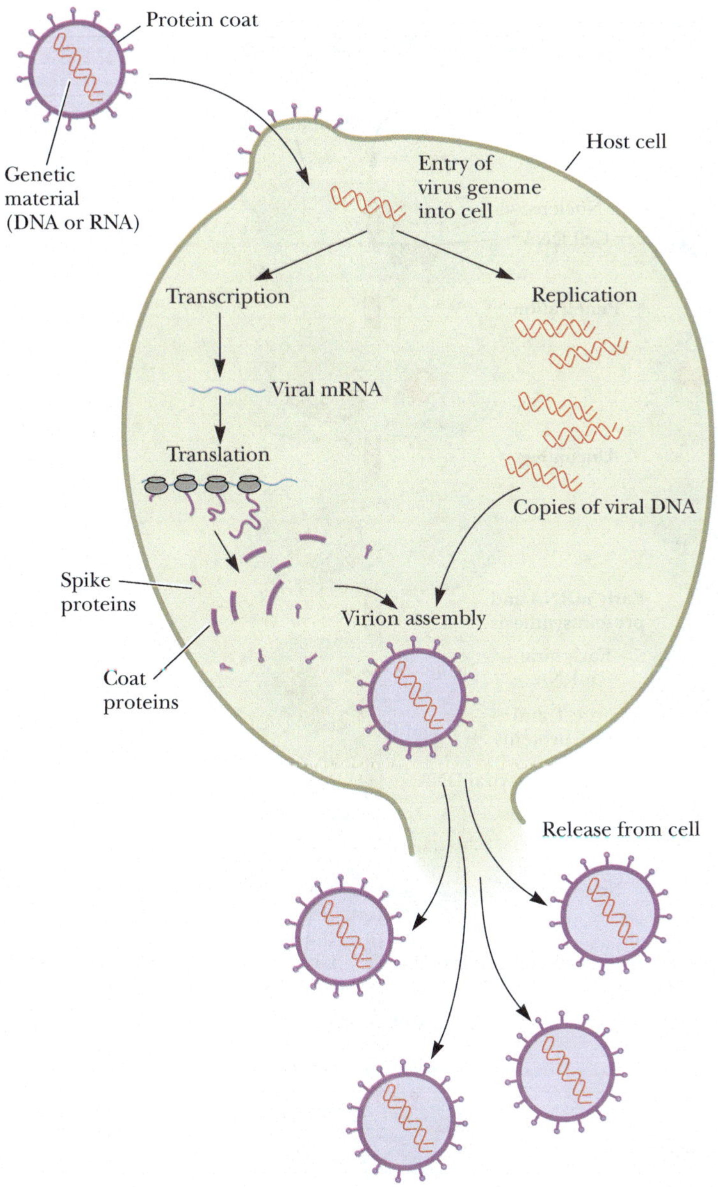

Figure 14.3 The lytic virus life cycle. Viruses are mobile bits of genetic information encapsulated in a protein coat. The genetic material may be either DNA or RNA. Once this genetic material gains entry into the host cell, it takes over the host machinery for macromolecular synthesis and subverts it to the synthesis of viral-specific nucleic acids and proteins. These virus components are assembled into mature virus particles, which are then released from the cell. Often, this parasitic cycle of virus infection leads to cell death and disease.

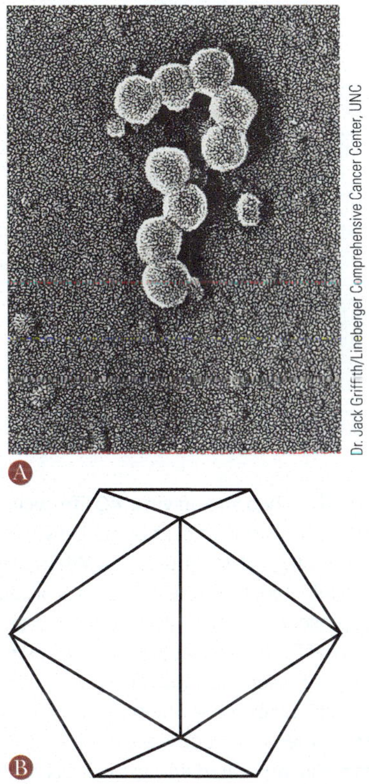

Figure 14.4 The architecture of simian virus 40 (SV40). (A) Virus particles appear almost spherical in electron micrographs, but, on closer examination, they can be seen to have an icosahedral shape. (B) The geometry of an icosahedron. This regular polyhedron has 20 faces, all of which are equilateral triangles of identical size.

The results are different when SV40 infects rodent cells. The process is the same as far as the production of the large-T protein, but replication of the viral DNA does not take place. The SV40 DNA already present in the cell can be lost or can be integrated into the DNA of the host cell. If the SV40 DNA is lost, the infection has no apparent result. If it becomes integrated into the DNA of the host cell, the infected cell loses control of its own growth. As a result of the accumulation of large-T protein, the infected cell behaves like a cancer cell. The large-T gene is an **oncogene**—one that causes cancer. Its mechanism is a subject of active research. The relationship between viruses and cancer is discussed in more detail in Section 14-4.

oncogene a gene that causes cancer when a triggering event takes place

▶ How does a virus infect a cell?

A virus must attach to a host cell before it can penetrate, which is why so much research is involved in studying the exact mechanisms of viral attachment. A common method of attachment involves the binding of one of the spike proteins on the envelope of the virus to a specific receptor on the host cell. Figure 14.6 shows an example of HIV attachment. A specific spike protein called *gp120* binds

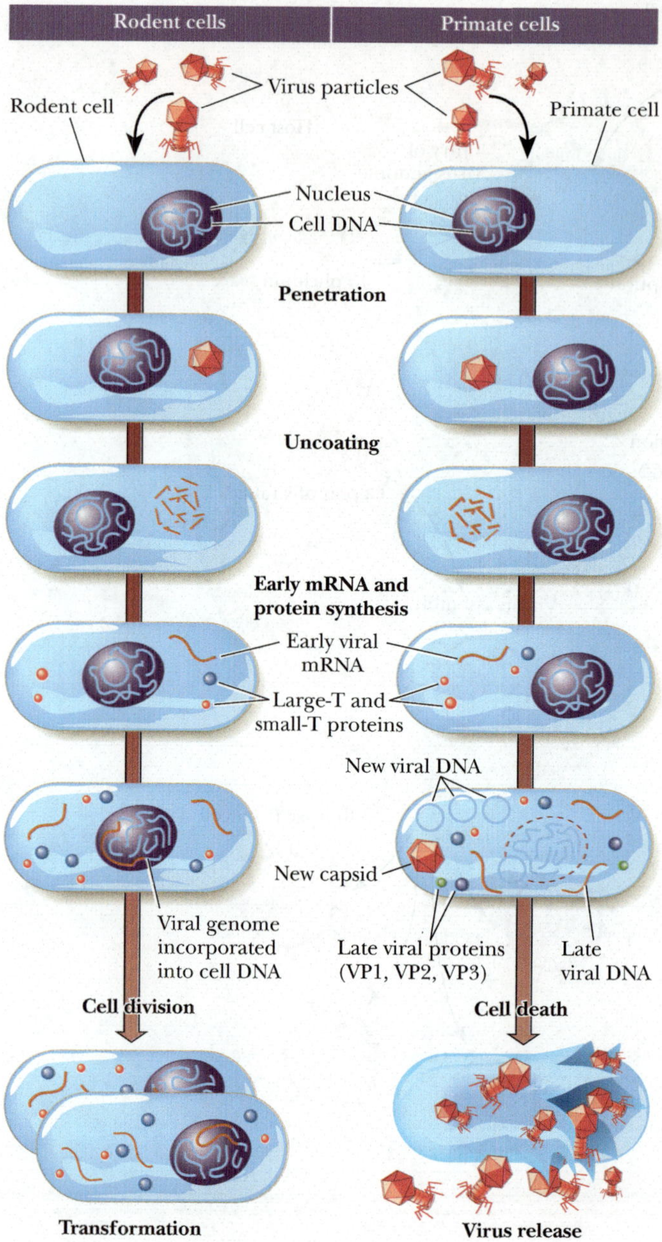

Figure 14.5 Two simian virus 40 life cycles. The outcome of infection of cells by simian virus 40 depends on the nature of the cells. When primate cells are infected, the large-T protein is produced, and its presence ultimately leads to the production of new viral DNA and coat proteins. New virus particles are assembled and released; the death of the host cell takes place when the new virions are released. When rodent cells are infected, the viral genome is incorporated into the cell DNA. *(Based on* Dealing with Genes: The Language of Heredity *by Paul Berg and Maxine Singer, ©1992 by University Science Books.)*

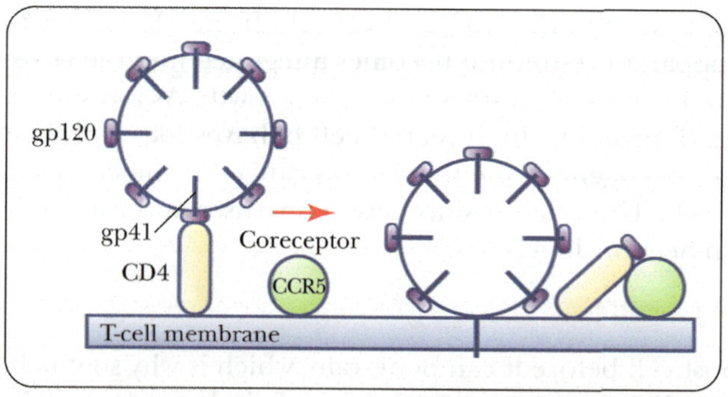

Figure 14.6 HIV attachment to a helper T cell. A specific spike protein called gp120 binds to a CD4 receptor on helper T cells. After this happens, a coreceptor, such as CCR5, complexes with CD4 and gp120. Another spike protein, gp41, then punctures the cell so that the capsid can enter.

to a CD4 receptor on helper T cells. After this happens, a coreceptor complexes with CD4 and gp120. Another spike protein, *gp41,* then punctures the cell so that the capsid can enter.

14-2 Retroviruses

Learning Objectives

After reading this section, students will be able to:

LO 14.2.1 Distinguish the life cycle of retroviruses from the life cycle of other viruses.

LO 14.2.2 Explain areas of retrovirus research.

14-2A Replication of Retroviruses

A retrovirus gets its name from the fact that its replication is backward compared to the central dogma of molecular biology: It makes DNA from RNA. The genome of a retrovirus is single-stranded RNA. Once it infects the cell, this RNA is used as a template to make a double-stranded DNA. The enzyme that does this is a virally encoded **reverse transcriptase (RT)**.

reverse transcriptase (RT) the enzyme that directs the synthesis of DNA on an RNA template

One of the unique features of the retrovirus life cycle is that the DNA produced by reverse transcription must be incorporated into the host DNA. This occurs because the ends of the DNA produced contain long terminal repeats (LTRs). LTRs are well known in DNA recombination events, and they allow the viral DNA to combine with the host's DNA. Figure 14.7 shows some aspects of the life cycle of a retrovirus.

14-2B Retroviruses: They Are Not All Bad

▶ Why are retroviruses important?

Retroviruses are the subject of extensive research in virology for several reasons:

1. Retroviruses have been linked to cancer, and more aspects of the relationship between viruses and cancer are discovered every day.
2. HIV is a retrovirus. HIV is the causative agent of the disease AIDS. AIDS treatments and a definitive cure are among the primary goals of retroviral research (see Section 14-5).
3. Retroviruses can be used in gene therapy, as described in **Biochemical Connections 14A**.

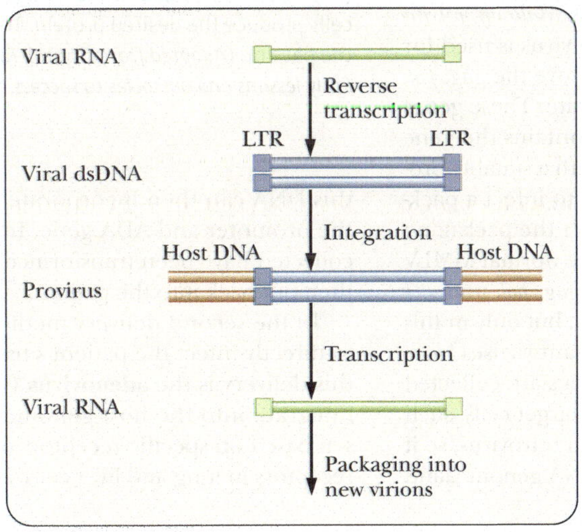

Figure 14.7 Some aspects of the life cycle of a retrovirus. Viral RNA is released into the host cell, where the viral reverse transcriptase (RT) synthesizes double-stranded DNA from it. The DNA then incorporates into the host's DNA via recombination using the long terminal repeat (LTR) sequences. Eventually, the DNA is transcribed to RNA, which is then packaged into new virions.

14A Biochemical Connections • Medicine

Viruses Are Used for Gene Therapy

Although viruses have usually been seen as problems for humans, there are areas where they are beneficial. For example, viruses can be used to make alterations in somatic cells, whereby a genetic disease is treated by the introduction, removal, or modification of genetic material in a patient's cells. This is called **gene therapy**. A successful form of gene therapy involves the gene for *adenosine deaminase (ADA)*, an enzyme involved in purine catabolism (Section 23-8). If this enzyme is missing, dATP builds up in tissues, inhibiting the action of the enzyme ribonucleotide reductase. This results in a deficiency of the other three deoxyribonucleoside triphosphates (dNTPs). The dATP (in excess) and the other three dNTPs (deficient) are precursors for DNA synthesis. This imbalance particularly affects DNA synthesis in lymphocytes, on which much of the immune response depends.

Individuals who are homozygous for ADA deficiency develop *severe combined immune deficiency (SCID)*, the "bubble-boy" syndrome. These individuals are highly prone to infection because of their highly compromised immune systems. The ultimate goal of the gene therapy in these situations is to take bone marrow cells from affected individuals, introduce the gene for ADA into the cells using a virus as a vector, and then reintroduce the bone marrow cells into the body, where they will produce the desired enzyme. The first clinical trials for ADA-SCID were simple enzyme-replacement therapies begun in 1982. The patients were given injections of ADA. Later clinical trials focused on correction of the gene in mature T cells. In 1990, transformed T cells were given to recipients via transfusions.

In trials at the National Institutes of Health (NIH), two girls, ages four and nine at the start of treatment, showed improvement to the extent that they could attend regular public schools and have no more than the average number of infections. Administration of bone marrow stem cells in addition to T cells was the next step; clinical trials of this procedure were undertaken with two infants, ages four months and eight months, in 2000. After 10 months, the children were healthy and had restored immune systems.

There are two types of delivery methods in human gene therapy. The first, called *ex vivo,* is the type used to combat SCID. Ex vivo delivery means that somatic cells are removed from the patient, altered with the gene therapy, and then given back to the patient. The most common vector for this is the retrovirus *Maloney murine leukemia virus (MMLV)*. Figure 14.8 shows how the virus is used for gene therapy. Some of the MMLV is altered to remove the *gag, pol,* and *env* genes, rendering the virus unable to replicate. These genes are replaced with an **expression cassette**, which contains the gene being administered, such as the ADA gene, along with a suitable promoter (Section 11-2B). This mutated virus is used to infect a packaging cell line. Normal MMLV is also used to infect the packaging cell line, which is not susceptible to the MMLV. The normal MMLV does not replicate in the packaging cell line. Its *gag, pol,* and *env* genes restore the mutated virus's ability to replicate, but only in this cell line. These controls are necessary to keep mutant viruses from escaping to other tissues. The mutated virus particles are collected from the packaging cell line and used to infect the target cells, such as bone marrow cells in the case of SCID. MMLV is a retrovirus, so it infects the target cell and produces DNA from its RNA genome, and

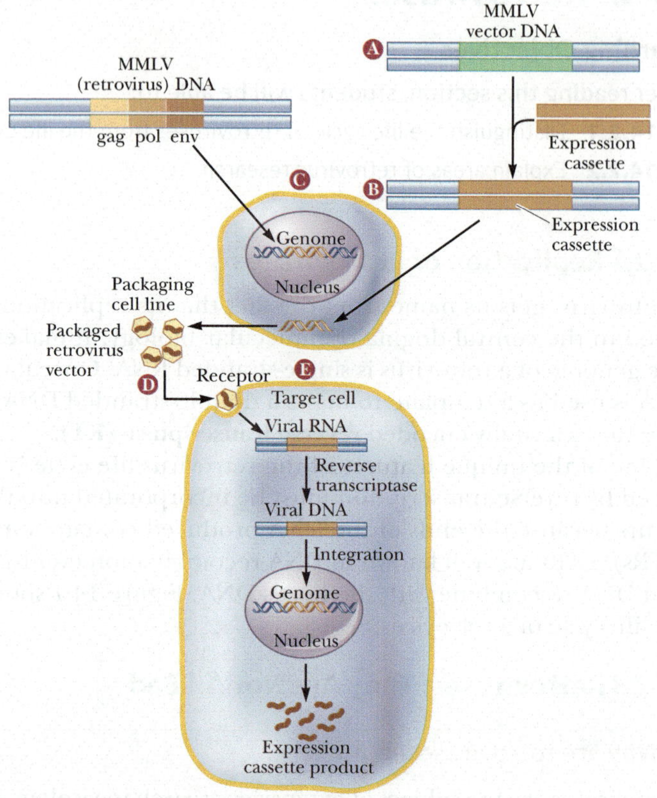

Figure 14.8 Gene therapy via retroviruses. The Maloney murine leukemia virus (MMLV) is used for ex vivo gene therapy. (A) Essential genes (*gag, pol, env*) are removed from the virus and (B) replaced with an expression cassette containing the gene being replaced with gene therapy. Removal of the essential viral genes renders the viruses unable to replicate. (C) The altered virus is then grown in a packaging cell line that allows replication. (D) Viruses are collected and used to infect cultured target cells from the patient needing the gene therapy. (E) The altered virus produces RNA, which then produces DNA via reverse transcriptase (RT). The DNA integrates into the patient's cells' genome, and the patient's cells produce the desired protein. The cultured cells are then given back to the patient. *(Adapted from Crystal, R. G. (1995). Transfer of genes to humans: Early lessons and obstacles to success.* Science *270, 404, Figure 1.)*

this DNA can then incorporate into the host genome, along with the promoter and ADA gene. In this way, the target cells that were collected have been transformed and produce ADA. These cells are then put back into the patient.

In the second delivery method, called in vivo, the virus is used to directly infect the patient's tissues. The most common vector for this delivery is the **adenovirus** (which is a DNA virus that does not integrate into the host genome). A particular vector can be chosen based on specific receptors on the target tissue. Adenovirus has receptors in lung and liver cells, and it has been used in clinical trials for gene therapy of cystic fibrosis (CF) and ornithine transcarbamoylase deficiency.

Clinical trials using gene therapy to combat CF and certain tumors in humans began in the 1990s, and the list of diseases that are being treated by this technique continues to grow. In 2009, a fatal demyelinating brain disease called X-linked adrenoleukodystrophy (ALD) was shown to respond well to gene therapy. In mice, gene therapy has been successful in fighting diabetes. The field of gene therapy is exciting and full of promise, but there are many obstacles to success in humans. There are also many risks, such as the risk of a dangerous immunological response to the vector carrying the gene or the danger of a gene becoming incorporated into the host chromosome at a location that activates a cancer-causing gene. This possibility is discussed further in Section 14-4.

Returning to the case of CF, the disease has been a great frustration to researchers. After much hope (and hype) surrounding the potential for a gene therapy cure, the reality has not matched the promise. Gene therapy simply has not worked for a variety of reasons. Although a cure has remained elusive, the prognosis for CF patients is much better today than it was when gene therapy clinical trials started in the 1990s. New drugs and treatments have improved the quality of life for CF patients. ◗

All retroviruses have certain genes in common. There is a gene for proteins of the nucleocapsid, often called CPs. They all have a gene for RT, and they all have genes for **envelope proteins (EPs)**. Figure 14.9 shows a schematic of the RNA genomes of common retroviruses. In the case of the Rous sarcoma virus, the genome also contains an oncogene that causes tumors (see Section 14-4).

gene therapy a method for treating a genetic disease by introducing, removing, or modifying genetic material in a patient's cells

expression cassette in gene therapy, the assemblage that contains the gene being transferred

adenovirus a simple DNA virus that is often used in the in vivo gene therapy delivery method; it does not integrate into the host genome

envelope proteins (EPs) proteins found in the outer membrane of a virus

14-3 The Immune System

Learning Objectives

After reading this section, students will be able to:

LO 14.3.1 Compare the innate immune system to the adaptive immune system.

LO 14.3.2 Describe the structure and function of antibodies.

LO 14.3.3 Explain how monoclonal antibodies are produced.

The immune system is a general term for many cellular and enzymatic processes that allow organisms to defend themselves from viruses, bacteria, and other parasites. All vertebrates have an immune system, and we discuss several aspects of this critical system here in Section 14-3.

14-3A Immunity and Vaccines

One distinctive characteristic of the immune system is its ability to *distinguish self from nonself.* This ability enables the cells and molecules responsible for immunity to recognize and destroy pathogens (disease-causing agents, such as viruses and bacteria) when they invade the body—or even one's own cells when they become cancerous. Because infectious diseases can be fatal, the operation of the immune system can be a matter of life and death. Striking confirmation

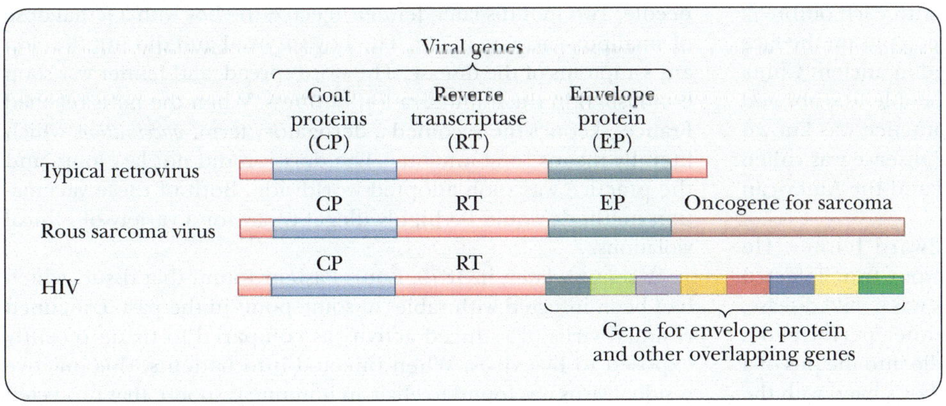

Figure 14.9 Typical retrovirus genes. The RNA genomes of all retroviruses have genes for coat proteins (CPs), for reverse transcriptase (RT), and for envelope protein (EP). In addition to these essential genes, the Rous sarcoma virus carries the sarcoma oncogene. The HIV genome is more complex, with a number of overlapping genes for EPs and other proteins. *(Adapted from Dealing with Genes: The Language of Heredity by Paul Berg and Maxine Singer, ©1992 by University Science Books.)*

Figure 14.10 Allergic reactions arise when the immune system attacks innocuous substances. Allergies to plant pollens are common, producing well-known symptoms such as sneezing.

autoimmune disease a disease in which the immune system attacks the body's own tissues

vaccine a weakened or killed form of an infectious agent that is injected into an organism so that it will make antibodies against the true infectious agent

of this last point is apparent in the lives of those who have AIDS. This disease so weakens the immune system that those who have it become prey to infections that proceed unchecked, with ultimately fatal consequences.

Conversely, suppression of the immune system can save lives as well as take them. For example, the development of drugs that suppress the immune system has made *organ transplants* possible. Recipients of hearts, lungs, kidneys, or livers tolerate the transplanted organs without rejecting them because these drugs prevent the immune system from attacking the grafts. However, the immune suppression also makes transplant recipients more susceptible to infections.

▶ What happens when the immune system doesn't work?

The immune system can also go awry in distinguishing self from nonself. The result is **autoimmune disease**, in which the immune system attacks the body's own tissues. Examples include rheumatoid arthritis, insulin-dependent diabetes, and multiple sclerosis. A significant portion of research on the immune system is directed toward developing approaches for treating these diseases. Allergies are another example of improper functioning of the immune system (Figure 14.10). Millions have asthma as a result of allergies to plant pollens and other allergens (substances that trigger allergic attacks). Food allergies can evoke violent reactions that may be life-threatening.

Over the years, researchers have unraveled some of the mysteries of the immune system and have used its properties as a therapeutic aid. Edward Jenner developed the first **vaccine**—against smallpox—in 1796. Since that time, it has been used so effectively as a preventive measure that smallpox has been eradicated. The action of vaccines of this sort depends on exposure to the infectious agent in a weakened form. The immune system mounts an attack, and *the immune system retains "memory" of the exposure.* In subsequent encounters with the same pathogen, the immune system can mount a quick and effective defense. This ability to retain "memory" is another major characteristic of the immune system. It is hoped that current research can be carried to the point of developing vaccines that can treat AIDS in people who are already infected. Other strategies are directed at finding treatments for autoimmune diseases. Still others are attempting to use the immune system to attack and destroy cancer cells. **Biochemical Connections 14B** describes the early work that led to the discovery of vaccines.

14B Biochemical Connections • Medicine

The First Vaccine: Bad Science Gone Good

Smallpox was a scourge over many centuries, with each outbreak leaving many people dead and others with deep pits on the face and body. A form of immunization was practiced in ancient China and the Middle East by intentionally exposing people to scabs and fluids of lesions of victims of smallpox. This practice was known as variolation in the Western world, where the disease was called variola. Variolation was introduced to England and the American colonies in 1721.

Figure 14.11 shows English physician Edward Jenner. He observed that milkmaids who had contracted cowpox from infected cows seemed to be immune to smallpox. Cowpox was a mild disease, whereas smallpox could be lethal. In 1796, Jenner performed a potentially deadly experiment: He dipped a needle into the pus of a cowpox-infected milkmaid and then scratched a boy's hand with the needle. Two months later, Jenner injected the boy with a lethal dose of smallpox-carrying agent. The boy survived and did not develop any symptoms of the disease. The word spread, and Jenner was soon established in the immunization business. When the news reached France, skeptics there coined a derogatory term, *vaccination*, which literally means "encowment." The derision did not last long, and the practice was soon adopted worldwide. Both of these vaccination methods would be highly illegal today for a variety of ethical violations.

A century later, in 1879, Louis Pasteur found that tissue, which had been infected with rabies at some point in the past, contained residual virus of reduced activity as compared to tissue recently exposed to live virus. When injected into patients, this inactive residual virus was found to elicit an immune response that protected patients against rabies. Pasteur named these attenuated, protective antigens *vaccines* in honor of Jenner's work. Today, immunization and vaccination are synonymous.

Vaccines are available for several diseases, including polio, measles, and smallpox, to name a few. A vaccine may be made of either dead viruses or bacteria or of weakened ones. For example, the Salk polio vaccine is a polio virus that has been made harmless by treatment with formaldehyde; it is given by intramuscular injection. In contrast, the Sabin polio vaccine is a mutated form of the wild virus; the mutation makes the virus sensitive to temperature. The mutated live virus is taken orally. The body temperature and the gastric juices render it harmless before it penetrates the bloodstream.

In an unprecedented response to the Ebola outbreak of 2014, researchers expedited vaccine development, which traditionally takes ten or more years, to less than one year. By July 2015, more than 27,000 people in Guinea, Liberia, and Sierra Leone had contracted Ebola, with death rates ranging from 25%–90% of infected patients. Phase III trials of an experimental vaccine administered to patients in Guinea yielded favorable responses in 75% to 100% of participants. The vaccine was administered by a "ring vaccination" method that offered treatment to individuals who had come in contact with infected family members. The global response and efficacy of the vaccine development effort have set a new standard for dealing with the world's viral challenges. The efficacy of the vaccine and the time required to move from animal studies to patients benefited from significant fundamental research on the topic and the slow rate of mutation of the Ebola virus.

Many cancers have specific carbohydrate, cell-surface tumor markers. If such antigens could be introduced by injection without endangering the individual, they could provide an ideal vaccine. Although you cannot use live or even attenuated cancer cells for vaccination, a synthetic analog of a natural tumor marker would elicit the same immune reaction that the original cancer-surface markers do. Thus, injecting such innocuous synthetic compounds may prompt the body to produce immunoglobulins that can provide a cure for the cancer—or at least prevent the occurrence of it. A compound called 12:13 dEpoB, a derivative of the macrolide epothilon B, is currently being explored for its potential to become an anticancer vaccine.

Figure 14.11 Edward Jenner developed the world's first vaccine in 1796. It was a safe and effective way to prevent smallpox and has led to eradication of this disease. His methodology, however, was anything but safe.

Vaccines change lymphocytes into plasma cells that produce large quantities of antibodies to fight any invading antigens. This is only the immediate, short-term response, however. Some lymphocytes become memory cells rather than plasma cells. These memory cells do not secrete antibodies but instead store them to serve as a detecting device for future invasion of the same foreign cells.

Smallpox, which was once one of humanity's worst scourges, has been totally wiped out, and smallpox vaccination is no longer required. Because smallpox is a potential weapon of bioterrorism, the U.S. government continues to enhance its smallpox vaccine production capabilities.

▶ How does the immune system work?

There are two important aspects to the immunity process: those that operate on the cellular level and those that operate on the molecular level. In addition, we compare the **innate immune system** to the **adaptive immune system**.

A major component of the immune system is the class of cells called **leukocytes**, otherwise known as white blood cells. Like all blood cells, they arise from common precursor cells (stem cells) in the bone marrow. Unlike other blood cells, however, they can leave the blood vessels and circulate in the lymphatic system. Lymphoid tissues (such as lymph nodes, the spleen, and, above all, the thymus gland) play important roles in the workings of the immune system.

innate immune system the first lines of defense in the immune system that act against pathogens in nonspecific ways

adaptive immune system the parts of the immune system that act against pathogens in specialized ways that can also be stored in the immune system's "memory" the next time the pathogen is encountered

leukocytes white blood cells that play an important role in the functioning of the immune system

14-3B The Innate Immune System

Considering the tremendous numbers of bacteria, viruses, parasites, and toxins that our bodies have to deal with, why aren't we continually sick? There are many weapons in the human immune defense system, and in this section, we discuss the innate immune system, which provides the first lines of defense.

dendritic cells innate immunity cells that are the first to encounter viruses or bacteria

macrophages cells of the innate immune system that digest cellular debris and pathogens

natural killer (NK) cells cells of the innate immune system that kill cells that have been infected or are cancerous

major histocompatibility complex (MHC) a protein complex that displays an antigen on the surface of cells of the immune system

T cells cells of the adaptive immune system that are involved in recognizing and responding to antigens presented by MHCs

helper T cells (T_H cells) cells of the immune system that activate other immune cells

antigen-presenting cells (APCs) cells of the immune system that display antigens on their surface through their MHC complexes and activate the adaptive immunity system

cytokines soluble proteins that serve as signaling molecules for the immune system

killer T cells cells of the adaptive immune system that kill infected or cancerous cells

B cells white blood cells that produce immune system antibodies

The innate immune system has several parts. One part includes *physical barriers* such as skin, mucus, and tears. All of these hinder penetration by pathogens and do not require specialized cells to fight the pathogen. However, if a pathogen (i.e., a bacterium, virus, or other parasite) breaches this outer layer of defense, the *cellular warriors* of the innate system come into play. The cells of the innate immune system that we discuss are **dendritic cells, macrophages,** and **natural killer (NK) cells.**

One of the first and most important cells to join the fight are the dendritic cells, so called because of their dendrites, which are long, tentacle-like projections (Figure 14.12). Dendritic cells are found in the skin, the mucous membranes, the lungs, and the spleen, and are the first cells of the innate system to have a crack at any virus or bacterium that wanders across their path. Using suction-cup-like receptors, they grab onto invaders and then engulf them by endocytosis. These cells then chop up the devoured pathogens and bring parts of their proteins to the surface. Here the protein fragments are displayed on a protein called a **major histocompatibility complex (MHC)**. The dendritic cells travel through the lymph to the spleen, where they present these antigens to a class of **T cells** of the adaptive immune system, called **helper T cells (TH cells)**.

Dendritic cells are members of a class of cells referred to as **antigen-presenting cells (APCs)**, and they are the starting point in most of the responses that are traditionally associated with the immune system. Once the dendritic cells present their antigens to the helper T cells, the latter release chemicals called **cytokines** that stimulate other members of the immune system, such as **killer T cells** (also called cytotoxic T cells or T_C cells) and **B cells**. Figure 14.13 shows an interaction between a dendritic cell of the innate immune system and a T cell of the adaptive immune system.

There are two classes of MHCs (I and II), based on their structure and on what they bind to. MHC I complexes bind to killer T cells, while MHC II complexes bind to helper T cells. The term "MHC complex" may seem a bit redundant because the "C" of "MHC" stands for "complex," but MHCs present protein fragments to other immune system cells. Therefore, the term *MHC complex* refers to the entity consisting of the MHC proteins complexed to the pathogen peptides. Besides the link between the two cells based on the identity of the MHC complex, another link (or perhaps two) is always necessary before cell proliferation occurs. The double signal is a trademark of most

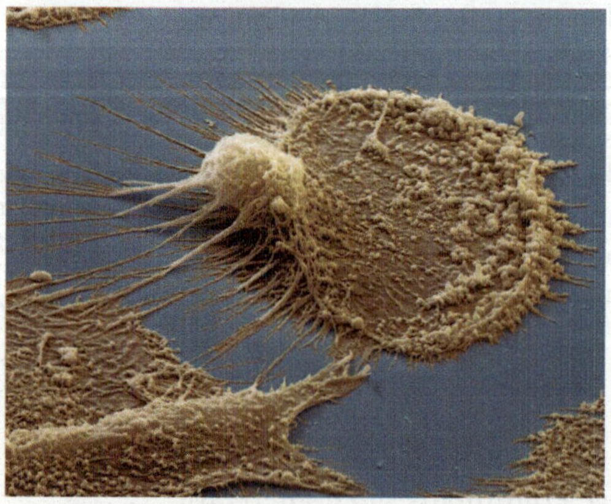

Eye of Science/Science Source

Figure 14.12 A human dendritic cell. Dendritic cells get their name from their tentacle-like arms.

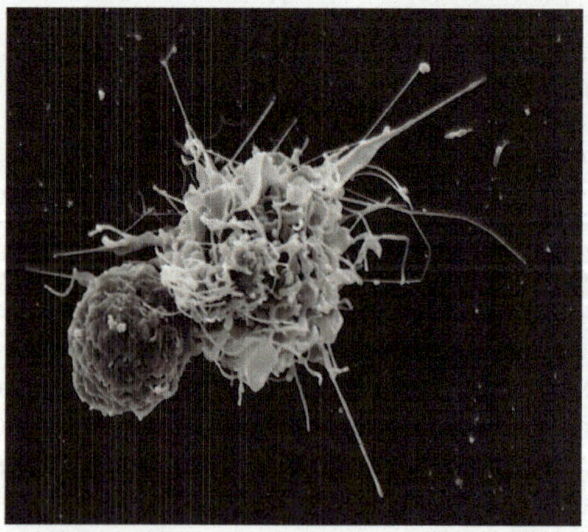

Courtesy of Steve Haley, William Bowers, and Richard Hunt, South Carolina Medical School

Figure 14.13 A dendritic cell interacting with a T cell. This figure shows a rat dendritic cell (*right*) interacting with a T cell (*left*). Through these interactions, dendritic cells of the innate immune system teach T cells of the adaptive immune system what to attack.

immune-cell responses, and it is thought to be a mechanism for making sure that the immune system is not activated in error.

Another important cell type in the innate immunity system is the NK cell (Figure 14.14), which is a member of a class of leukocytes called **lymphocytes** because it is derived from a type of stem cell called a *lymphoid stem cell.* NK cells kill off cells that have been infected by viruses or that are cancerous, and they secrete cytokines that call up other cells, such as macrophages, another innate immunity cell type that destroys microbes. They also work with dendritic cells, in a sense. If an infection is small, the NK cells may end up killing off the infected dendritic cells before the rest of the immune system is activated. Thus, NK cells help decide whether the adaptive immune system needs to be activated.

lymphocytes a major class of white blood cells of the immune system to which T cells, B cells, and NK cells belong

Interferons are a family of signaling proteins produced primarily by innate immune cells in response to viral infection. Once dendritic cells release interferons, the interferons trigger antiviral defenses in neighboring cells, enhance antigen presentation, and play a crucial role in activating NK cells. Because interferons are normally produced in small quantities and are rapidly degraded, obtaining enough of them for research or clinical use was historically difficult. However, recombinant DNA technology made it possible to clone interferon genes and express them, allowing large-scale production of these important immune system regulators.

14-3C The Adaptive Immune System

Adaptive immunity depends on two other types of lymphocytes: *T cells* and *B cells.* T cells develop primarily in the thymus gland, and B cells develop primarily in the bone marrow, accounting for their names (Figure 14.15). Much of the *cellular aspect* of adaptive immunity is the province of the T cells, whereas much of the *molecular aspect* depends on the activities of the B cells.

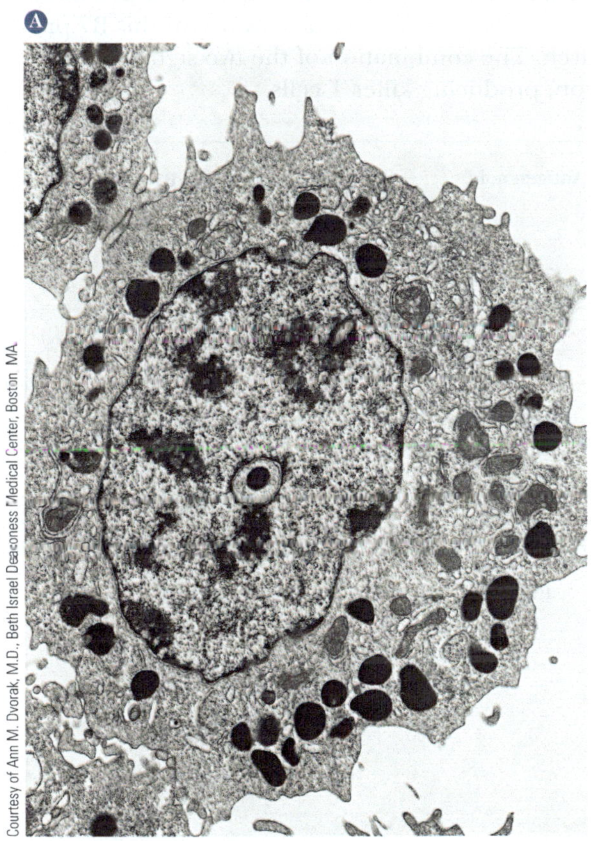

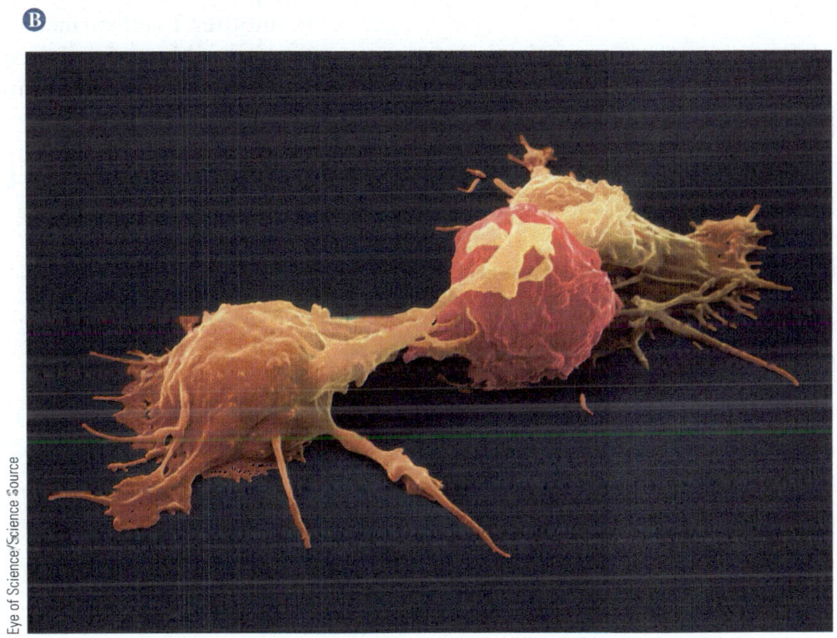

Figure 14.14 Natural killer (NK) cells are among the first cells to be involved in an immune response. They are nonphagocytic cells that can interact with and destroy other cells, such as those infected with a virus or cancer cells. (A) An electron micrograph. (B) A high-resolution photomicrograph.

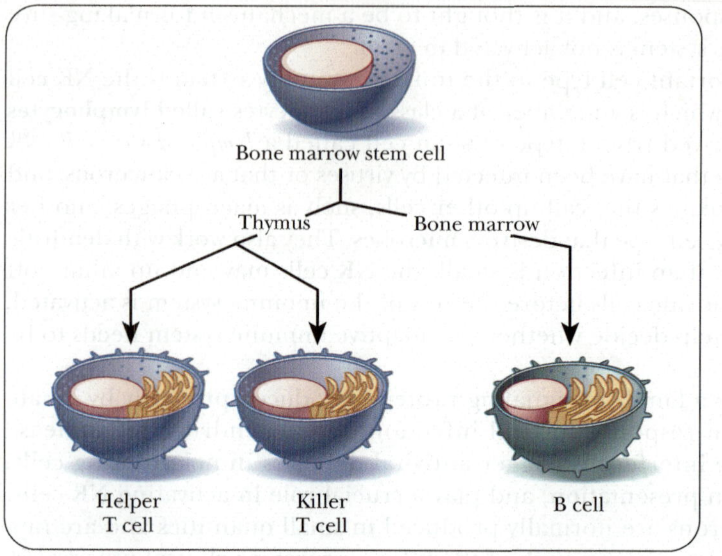

Figure 14.15 The development of lymphocytes. All lymphocytes are ultimately derived from the stem cells of the bone marrow. Some of these cells travel to the thymus, where two kinds of T cells develop: helper T cells and killer T cells. B cells develop in the bone marrow.

▶ What do T cells do?

T cells can have a number of functions. As T cells differentiate, each becomes specialized for one of the possible functions. The first of these possibilities, that of the killer T cells, involves *T-cell receptors (TCRs)* on their surfaces that recognize and bind to **antigens**, the foreign substances that trigger the immune response. The antigens are presented to the T cells by antigen-presenting cells (APCs), such as macrophages and dendritic cells. The APCs ingest and process antigens and then present them to the T cells. The processed antigen takes the form of a short peptide bound to an MHC I protein on the surface of the APC. Figure 14.16 shows how this works for macrophages. The macrophage also presents another molecule, a protein of a family known as B7, which binds to another T-cell-surface protein called CD28; the exact nature of this B7 protein is a subject of active research. The combination of the two signals leads to T-cell growth and differentiation, producing killer T cells.

antigens substances that trigger an immune response

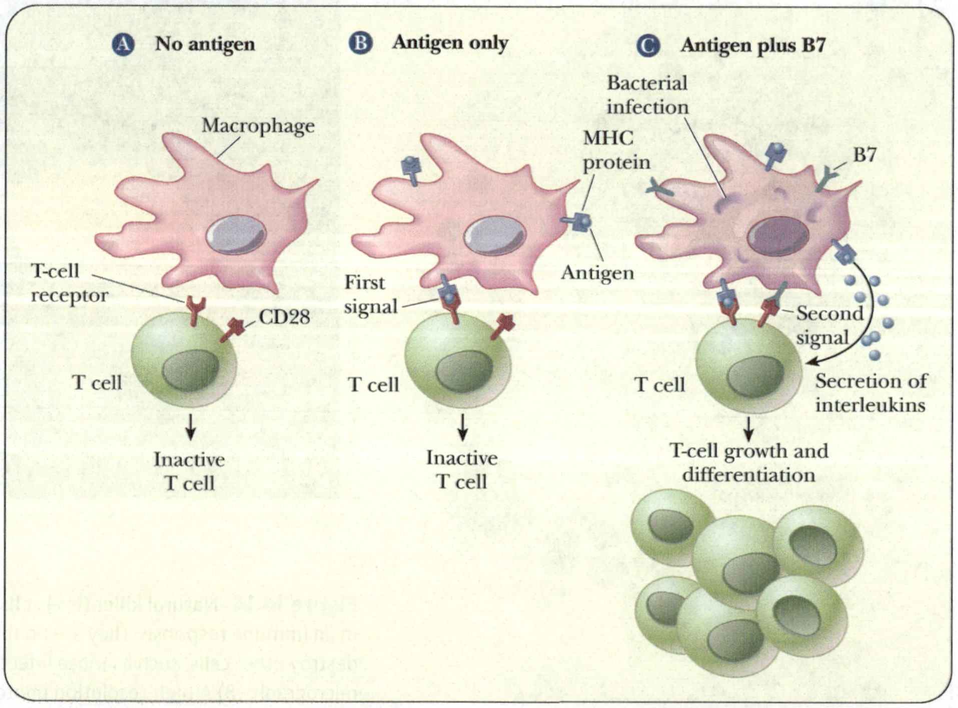

Figure 14.16 A two-stage process leads to the growth and differentiation of T cells. (A) In the absence of antigen, T cells do not proliferate. (B) In the presence of antigen alone, the T-cell receptor (TCR) binds to antigen presented on the surface of a macrophage cell by the MHC protein. There is still no proliferation of T cells because the second signal is missing. In this way, the body can avoid an inappropriate response to its own antigens. (C) When an infection takes place, a B7 protein is produced in response to the infection. The B7 protein on the surface of the infected cell binds to a CD28 protein on the surface of the immature T cell, giving the second signal that allows it to grow and proliferate. *(Adapted from "How the Immune System Recognizes Invaders" by Charles A. Janeway, Jr.; illustration by Ian Warpole.* Sci. Amer. *September 1993.)*

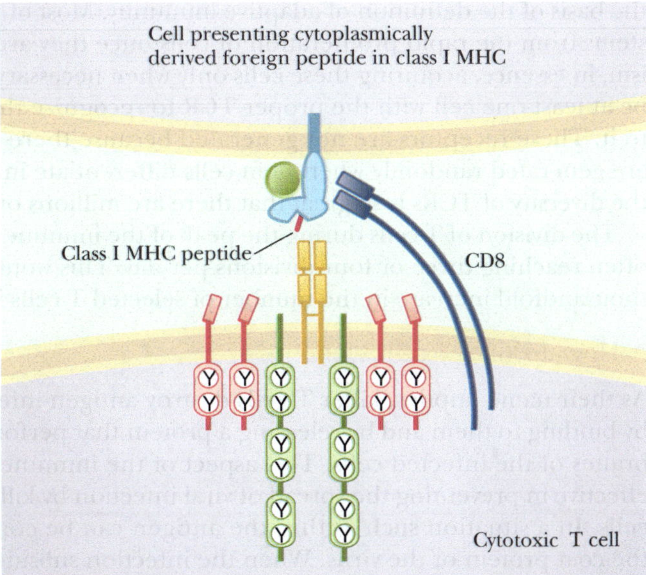

Figure 14.17 Interaction between cytotoxic T cells (killer T cells) and antigen-presenting cells. Foreign peptides that are derived from the cytoplasm of infected cells are displayed on the surface by major histocompatibility complex class I (MHC I) proteins. These bind to the T-cell receptor (TCR) of a killer T cell. A docking protein called CD8 helps link the two cells.

Proliferation of killer T cells is also triggered when macrophages bound to T cells produce small cytokine proteins called **interleukins** (recall from Section 14-3B that, in general, cytokines are signaling molecules of the immune system). The T cells make an interleukin-receptor protein as long as they are bound to the macrophage but do not do so when they are no longer bound. In this way, T cells do not proliferate in an uncontrolled fashion. A killer T cell also has another membrane protein called CD8, which helps it dock to the MHC of the APC, as shown in Figure 14.17. In fact, the CD8 protein is such a distinguishing characteristic that many researchers use the term *CD8 cells* instead of *killer T cells.*

interleukins cytokine proteins that regulate the development, activation, and behavior of immune cells

T cells that bind to a given antigen, and *only to that antigen,* grow when these conditions are fulfilled, highlighting the specificity of which the immune system is capable. Many substances, including ones that do not exist in nature, can be antigens. The remarkable adaptability of the immune system in dealing with so many possible challenges is another of its main features. The process by which only the cells that respond to a given antigen grow in preference to other T cells is called **clonal selection** (Figure 14.18). The immune system can thus be versatile in its responses to the challenges it meets. This clonal selection is

clonal selection the process by which the immune system responds selectively to antibodies actually present in an organism

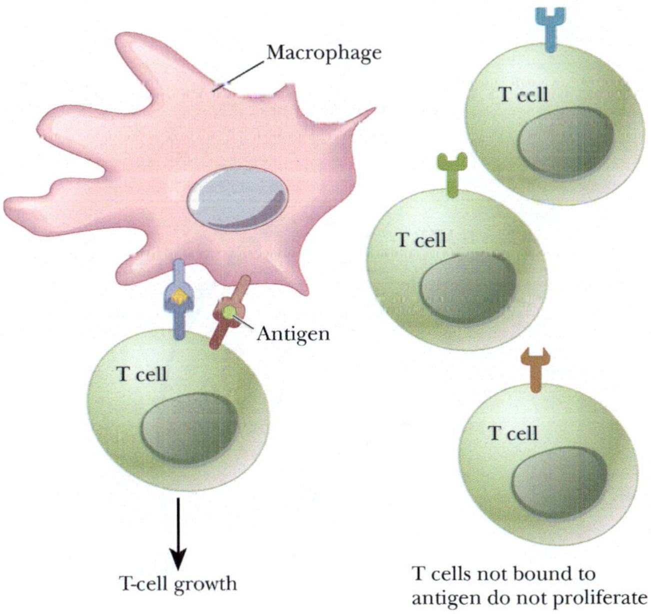

Figure 14.18 Clonal selection. Clonal selection allows the immune system to be both versatile and efficient in responding to a wide range of possible antigens. Many different types of cells can be produced by the immune system, allowing it to deal with almost any possible challenge. Only cells that respond to an antigen that is actually present are produced in quantity; this is an efficient use of resources.

the basis of the definition of adaptive immunity. Most of the response of T cells stems from the rapid proliferation of cells once they are selected—the organism, in essence, acquiring these cells only when necessary. However, there must be at least one cell with the proper TCR to recognize the antigen and to bind to it. These receptors are not generated because there is a need; rather, they are generated randomly when stem cells differentiate into T cells. Fortunately, the diversity of TCRs is so great that there are millions of TCR specificities.

The division of T cells during the peak of the immune response is very rapid, often reaching three or four divisions per day. This would lead to more than a thousandfold increase in the number of selected T cells in a few days.

▶ How do killer T cells work?

As their name implies, killer T cells destroy antigen-infected cells. They do so by binding to them and by releasing a protein that perforates the plasma membranes of the infected cells. This aspect of the immune system is particularly effective in preventing the spread of viral infection by killing virus-infected host cells. In a situation such as this, the antigen can be considered all or part of the coat protein of the virus. When the infection subsides, some memory cells remain, conferring immunity against later attacks from the same virus.

▶ What is the role of helper T cells?

T cells play another role in the immune system when they develop receptors for a different group of antigen-presenting MHC proteins—in this case, MHC II. These become *helper T cells,* which develop in much the same way as killer T cells. Helper T cells are also referred to as *CD4+ cells* because of the presence of that unique membrane protein. CD4 helps the cell dock to the MHC of the antigen-presenting cell (APC), as shown in Figure 14.19. In the following paragraphs, we use "CD4+" sometimes and just "CD4" other times. "CD4" refers to the CD4 membrane protein itself, whereas "CD4+ cell" specifies that the T cell being referred to expresses CD4 on its surface.

The function of helper T cells is primarily to aid in the stimulation of B cells. Maturing B cells display the MHC II protein, with processed antigen, on their surfaces. Here, again, the MHC proteins play a key role in the immune system. This property has led to a considerable amount of research to determine their structure, including determination by X-ray crystallography. The MHC II of the B cells is the binding site for helper T cells. The binding of helper T cells to immature B cells releases interleukins IL-2 and IL-4 (Figure 14.20A). These

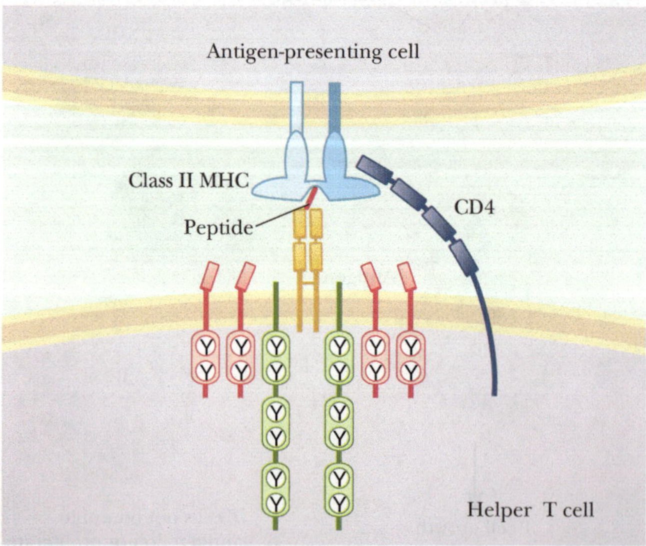

Figure 14.19 Interaction between helper T cells and antigen-presenting cells. Foreign peptides are displayed on the surface by MHC II proteins. These bind to the T-cell receptor of a helper T cell. A docking protein called CD4 helps link the two cells.

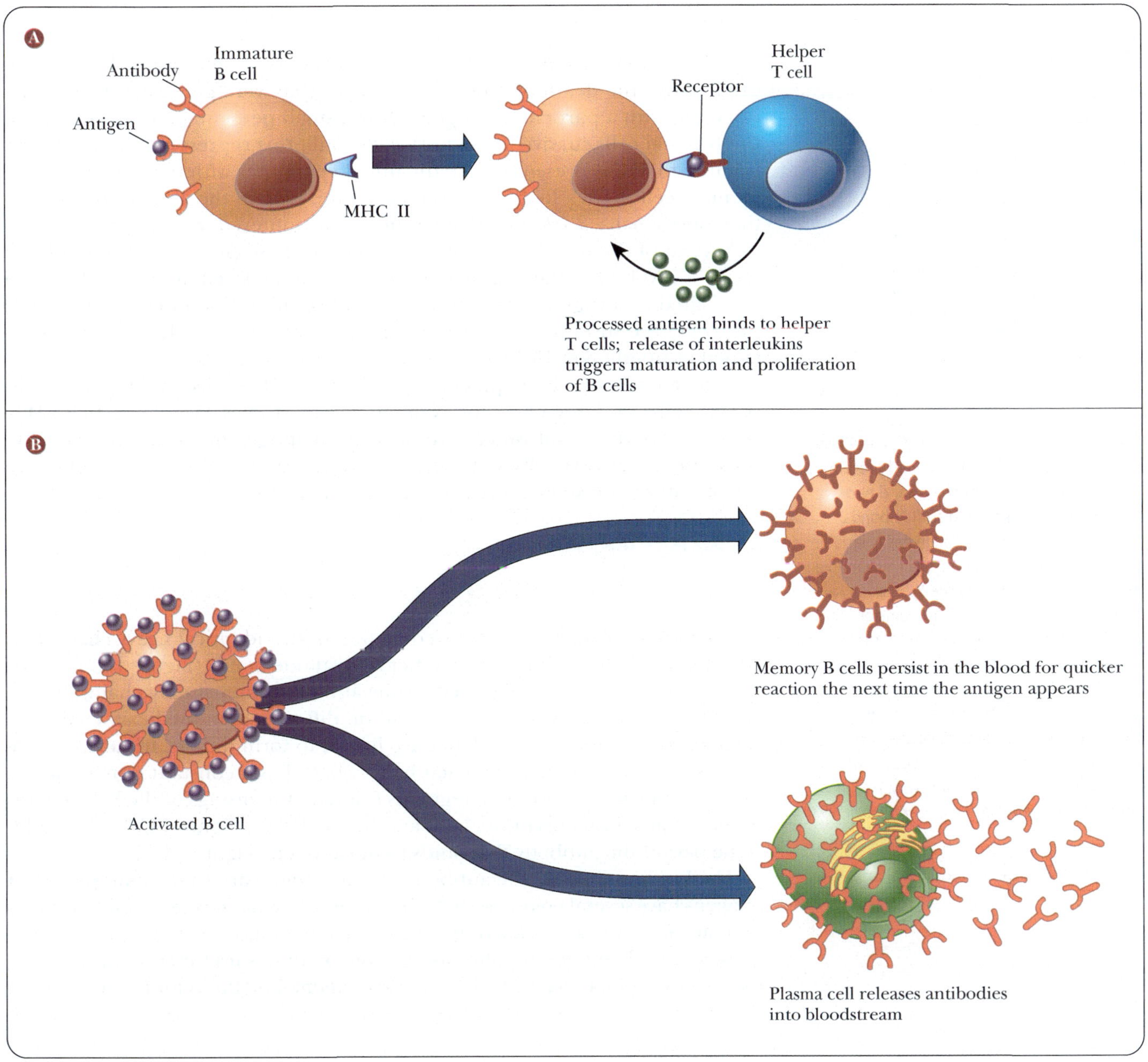

Figure 14.20 Helper T cells aid in the development of B cells. (A) A helper T cell has a receptor for the MHC II protein on the surface of immature B cells. When the helper T cells bind to the processed antigen presented by the MHC II protein, they release interleukins and trigger the maturation and proliferation of B cells. (B) The activated B cells can then differentiate into memory B cells or plasma cells. *(Based on Jared Schneidman Design.)*

interleukins trigger the maturation and proliferation of B cells. Activated B cells can then differentiate into memory B cells (Figure 14.20B, *top right*) or into plasma cells (Figure 14.20B, *lower right*).

Both memory B cells and plasma cells produce **antibodies** (also known as immunoglobulins), the proteins that we devote most of our time to as we discuss the molecular aspects of the immune response. Memory B cells display antibodies on their surfaces in addition to the MHC II proteins, and they persist after an infection in order to respond faster the next time an antigen appears. Plasma cells release circulating antibodies into the bloodstream, where they bind to antigens, marking them for destruction by the immune system. Helper T cells also help stimulate killer T cells and APCs via the release of interleukins.

antibodies (immunoglobulins) glycoproteins that bind to and immobilize a substance that the cell recognizes as foreign

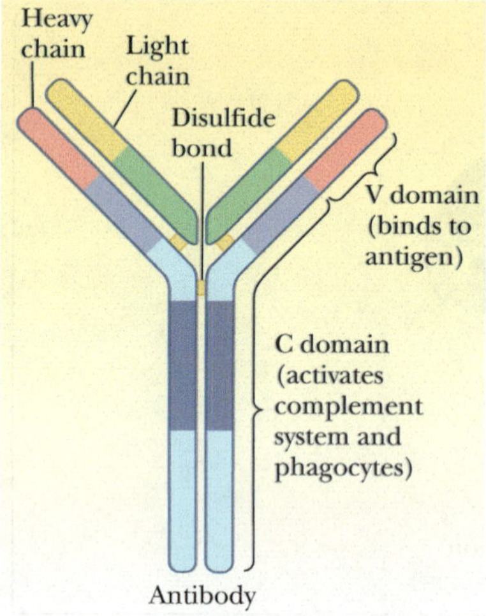

Figure 14.21 A schematic diagram of an antibody. A typical antibody molecule is a Y-shaped protein consisting of two identical light chains and two identical heavy chains linked by disulfide bonds. Each light chain and each heavy chain has a variable (V) domain and a constant (C) domain. The variable region, which is at the prongs of the Y, binds to antigen. It consists of the variable domains of the light and heavy chains. The constant region, toward the stem of the Y, activates the complement system (the parts of the immune system that destroy antibody-bound antigens) and phagocytes. It consists of the constant domains of the heavy chains. *(Adapted from "How the Immune System Recognizes Invaders" by Charles A. Janeway, Jr.; illustration by Ian Warpole.* Sci. Amer. *September 1993.)*

▶ How do T cells develop memory?

T cells can develop memory as well. The process of generating T-cell memory involves the death of most of the T cells that were generated by the first infection with a particular antigen. Only a small percentage (5% to 10%) of the original cells survive as memory cells. Still, this represents a much larger number than was present before the initial encounter with the antigen. These memory cells have a higher reproductive rate, even in the absence of antigen, than does a naive *T cell* (one that has never encountered the antigen).

Several of the interleukins play key roles in these processes. Interleukin 7 (IL-7) is involved in the maintenance of naive killer T cells at low levels. When stimulated by antigens, the maturation and proliferation of cytotoxic T cells (T_C cells) are stimulated by interleukin 2 (IL-2). Memory T_C cells are maintained by interleukin 15 (IL-15).

In T-cell memory development, killer T cells (CD8+ cells) and helper T cells (CD4+ cells) work together. For example, several studies have shown that CD8+ cells expand when confronted with the correct antigen in the absence of CD4+ cells. However, CD8+ cells that were clonally expanded without CD4+ cells were unable later to make memory cells that were as active.

14-3D Antibodies

▶ What are some characteristics of antibody structure?

Antibodies are Y-shaped proteins consisting of two identical heavy chains and two identical light chains held together by disulfide bonds (Figure 14.21). They are glycoproteins, with oligosaccharides linked to their heavy chains. There are different classes of antibodies based on differences in the heavy chains. In some of these classes, heavy chains are linked to form dimers, trimers, or pentamers. Each light chain and each heavy chain has a constant domain and a variable domain. The variable region is found at the prongs of the Y. It consists of the variable domains of the light and heavy chains (Figure 14.21, *top*), and it is the part of the antibody that binds to the antigen (Figure 14.22).

epitopes binding sites for antibodies on an antigen

The binding sites for the antibody on the antigen are called **epitopes**. Most antigens have several epitopes, so the immune system has several possible avenues of attack for naturally occurring antigens. Each antibody can bind to two antigens, and each antigen usually has multiple binding sites for antibodies, giving rise to a precipitate that is the basis of experimental methods for immunological

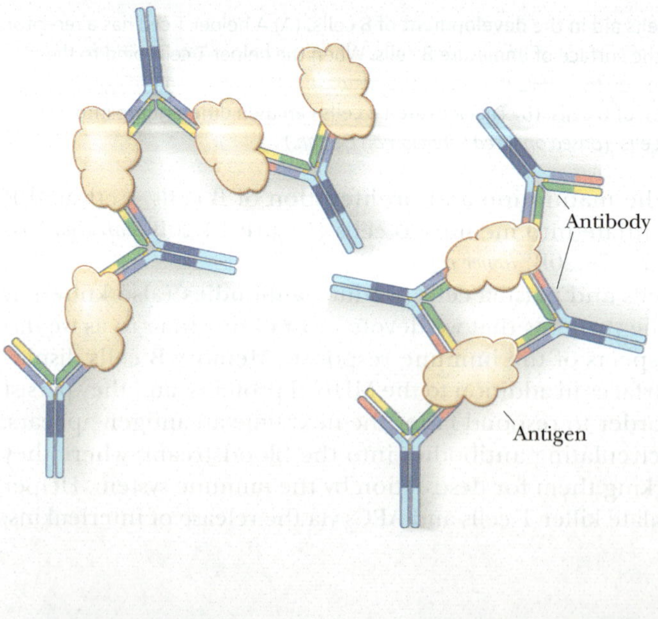

Figure 14.22 An antigen–antibody reaction forms a precipitate. An antigen, such as a bacterium or virus, typically has several binding sites for antibodies. Each variable region of an antibody (each prong of the Y) can bind to a different antigen, so aggregates form. The aggregates then form a precipitate that is attacked by phagocytes and the complement system.

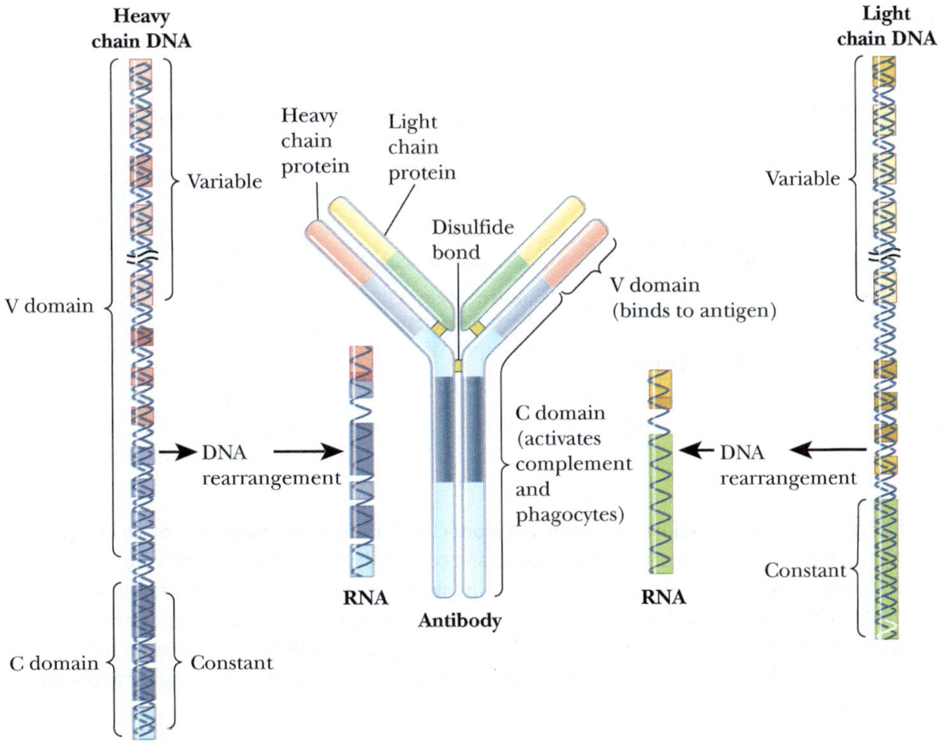

Figure 14.23 The formation of antibody heavy and light chains. The heavy and light chains of antibodies are encoded by genes that consist of a number of DNA segments. These segments rearrange and, in the process, give rise to different chains in each B cell. Because the joining is highly variable, comparatively few gene segments can give rise to millions of distinct antibodies. *(Adapted from "How the Immune System Recognizes Invaders" by Charles A. Janeway, Jr.; illustration by Ian Warpole.* Sci. Amer. *September 1993.)*

research. The constant region consists of the constant domains of the heavy chains, and it is located at the hinge and the stem of the Y (Figure 14.21, *bottom*); this part of the antibody is recognized by the *complement system* (the portion of the immune system that destroys antibody-bound antigen) and by *phagocytes* (the immune cells that engulf and digest foreign pathogens).

How does the body produce so many highly diverse antibodies to respond to essentially any possible antigen? Similar to how, in the English language, the letters of the alphabet can be arranged in countless ways to give a variety of words, the same possibility for enormous numbers of rearrangements exists with the gene segments that code for the variable portions of antibody chains. Antibody genes are inherited as small fragments that join to form a complete gene in individual B cells as they develop (Figure 14.23). When gene segments are joined, the enzymes that catalyze the process add random DNA bases to the ends of segments being spliced, allowing for the wide variety of antibodies that can be produced. This rearrangement process takes place in the genes for both the light and the heavy chains. (The exon splicing and mRNA processing that we discussed in Section 11-8 still take place as well.) In addition to these factors, B lymphocytes have a particularly high rate of somatic mutation, in which changes in the base sequence of DNA occur as the cell develops. Changes outside the germ cells apply only to the organism in which they take place and are not passed on to succeeding generations.

▶ How are monoclonal antibodies produced?

Each B cell (and each progeny plasma cell) produces only one kind of antibody. In principle, each such cell should be a source of homogeneous antibodies by cloning. This is not possible in practice because lymphocytes do not grow continuously in culture. In the late 1970s, Georges Köhler and César Milstein developed a method to circumvent this problem, a feat for which they received the Nobel Prize in Physiology or Medicine in 1984. The technique requires fusing lymphocytes that make the desired antibody with mouse myeloma cells. The resulting **hybridoma** (hybrid myeloma), like all cancer cells, can be cloned

hybridoma a kind of cell produced by fusing lymphocytes and mouse myeloma cells

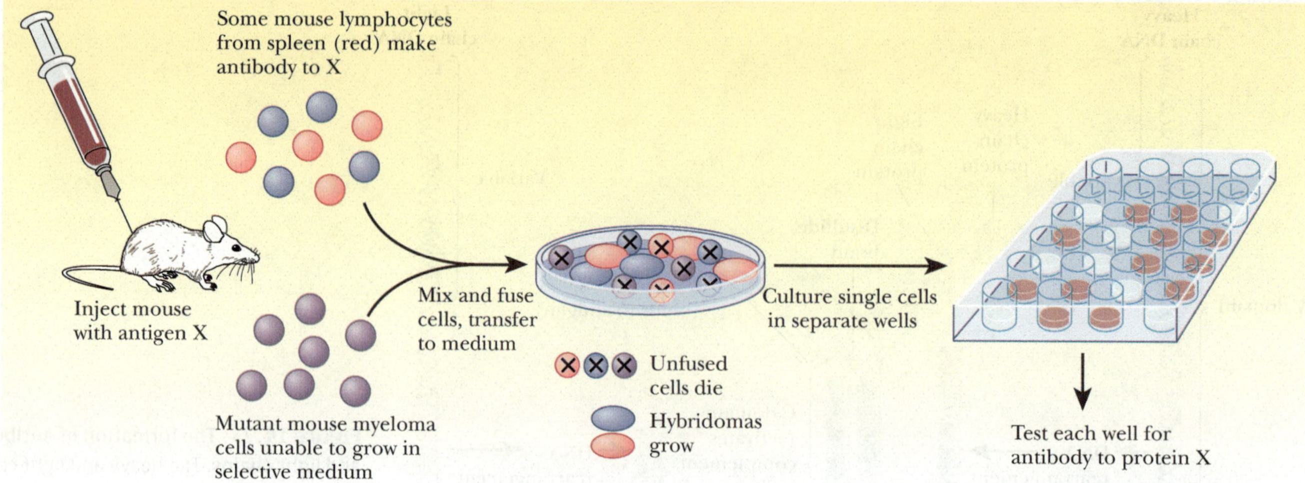

Figure 14.24 A procedure for producing monoclonal antibodies against a protein antigen X. A mouse is immunized against the antigen X, and some of its spleen lymphocytes produce antibodies. The lymphocytes are fused with mutant myeloma cells that cannot grow in a given medium because they lack an enzyme found in the lymphocytes. Unfused cells die because lymphocytes cannot grow in culture, and the mutant myeloma cells cannot survive in this medium. The individual cells are grown in culture in separate wells and are tested for antibodies to antigen X.

monoclonal antibodies antibodies produced from the progeny of a single cell and specific for a single antigen

in culture (Figure 14.24) and produces the desired antibody. Because the clones are the progeny of a single cell, they produce homogeneous **monoclonal antibodies**. Using this method, it is possible to produce antibodies to almost any antigen in quantity. Monoclonal antibodies can be used to assay for biological substances that can act as antigens. A striking example of their usefulness is in testing blood for the presence of HIV; this procedure has become routine to protect the public blood supply.

Scientists have also expanded the principle of "one antigen–one antibody" and have created designer antibodies that recognize two different proteins. These "two-in-one" antibodies can be useful tools against diseases that do not respond well to single treatments. For example, cancer and AIDS are often attacked with multiple treatments anyway, and creating such designer antibodies adds another weapon to the arsenal.

14-3E Distinguishing Self from Nonself

With all the power the immune system has to attack foreign invaders, it must also do so with discretion because we have our own cells that display proteins and other macromolecules on their surfaces. How the immune system knows not to attack these cells is a complicated and fascinating topic. When the body makes a mistake and attacks one of its own cells, the result is an autoimmune disease, examples of which are rheumatoid arthritis, lupus, multiple sclerosis, scleroderma, Crohn's disease, and some forms of diabetes.

T cells and B cells have a wide variety of receptors on their surfaces. The affinities for a given antigen vary greatly. Below a certain threshold, an encounter between a lymphocyte receptor and an antigen is insufficient to trigger that cell to become active and begin to multiply. These same cells also have stages of development. They mature in the bone marrow or the thymus and go through an early stage in which receptors first begin to appear on their surfaces.

In the case of T cells, a precursor form called a *double-positive (DP) cell* has both the CD4 and the CD8 proteins (Figure 14.25). This cell is the turning point for the fate of its progeny. If the receptors of the DP cell do not recognize anything, including self-antigens or self-MHC proteins, then it dies by neglect

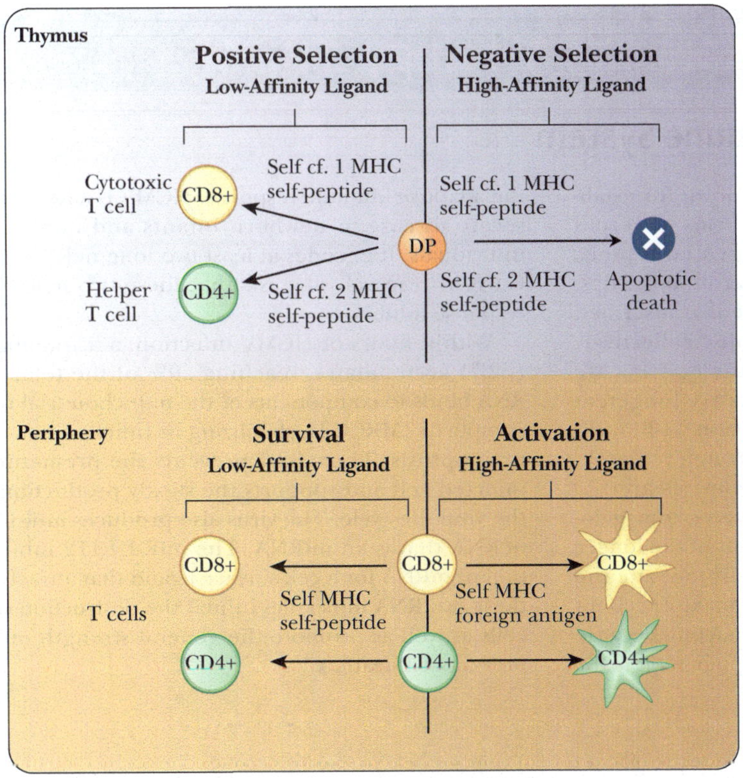

Figure 14.25 T cell development and selection. Top: In the thymus, a precursor to T cells called a double-positive (DP) cell is the turning point in the fate of T-cell progeny. If the DP cell reacts to nothing, including self-peptides or MHCs, then it dies by neglect (not shown). If it recognizes self-peptides or MHCs with high affinity, then it is programmed for apoptosis to avoid an autoimmune response. If it recognizes self-peptides or MHCs with low affinity, it differentiates into killer T cells and helper T cells. Bottom: After leaving the thymus for peripheral tissues, the CD8+ and CD4+ cells survive using low-affinity interactions with self-peptides. Alternatively, when they interact with high affinity to a foreign antigen, they become activated. *(Based on Werlen, G., et al. (2003). Signaling life and death in the thymus; Timing is everything.* Science *299, 1859–1863.)*

(not shown in the figure). If the receptors recognize self-antigens or MHC but with low affinity, then the cell undergoes *positive selection* and differentiates into a killer T cell (CD8+ cell) or a helper T cell (CD4+ cell) (Figure 14.25, *top left*). On the other hand, if a DP cell's receptors encounter self-antigens that are recognized with high affinity, it undergoes a process called *negative selection* and is programmed for apoptosis, or cell death (Figure 14.25, *top right*). After leaving the thymus for peripheral tissues, the CD8+ and CD4+ cells survive using low-affinity interactions with self-antigens (Figure 14.25, *bottom left*). Alternatively, when they interact with high affinity to a foreign antigen, they become activated (Figure 14.25, *bottom right*).

By the time the lymphocytes leave their tissue of origin, they have therefore already been stripped of the most dangerous individual cells that would tend to react to self-antigens. Some individual cells will still have a receptor with very low affinity for a self-antigen. If these slip out of the bone marrow or thymus, they do not initiate an immune response because their affinity is below the minimum threshold, and there is always the requirement for a secondary signal. They would need to have another cell, such as a macrophage, also present them with an antigen. Besides binding an antigen to its receptor, a B cell would need to receive an interleukin 2 from a helper T cell that had also been stimulated by the same antigen.

All of these safeguards lead to the delicate balance that must be maintained by the immune system, a system that simultaneously has the diversity to bind to almost any molecule but does not react to the myriad proteins that are recognized as self. Autoimmune diseases have a wide range of effects on the lives of those who experience them, but treatments for these diseases continue to improve.

Our immune systems have coevolved with many invading organisms and viruses. It is a constant battle, with each side constantly adapting. **Biochemical Connections 14C** gives one example of how viral RNAs try to outwit the immune system.

14C Biochemical Connections • Virology

Viral RNAs Can Outwit the Immune System

Herpesviruses are pathogenic DNA viruses known for establishing long-term latent infections. Some forms of herpes are sexually transmitted diseases and lead to genital lesions. Others are known for leading to "fever blisters" around the lips. Once the virus infects the cell, its viral genome is transcribed into viral mRNA and other smaller RNA molecules, collectively known as noncoding RNA (ncRNA). Some of the ncRNAs are very small, micro RNAs (miRNAs), while others are longer, at about 100 nucleotides. As we discussed in Section 11-6, miRNAs can selectively inhibit gene expression. Through research, the function of the longer viral ncRNAs has become clearer.

Researchers studying a herpesvirus called human cytomegalovirus (HCMV) have shown that one of the functions of the longer ncRNA from the virus is to allow the infected cells to evade the innate immunity response (Figure 14.26). RNAs make excellent weapons against the immune system. They are fast-acting because they do not need to be translated, and they are poor targets for the adaptive immune response. HCMV is a herpesvirus that causes severe disease in newborn infants and immunocompromised individuals. It encodes at least two long ncRNAs and 11 miRNAs. Current research suggests that these two ncRNAs are immune-response inhibitors.

Within hours of HCMV infection, a 2.7-kilobase viral ncRNA (β2.7) accumulates, reaching 20% of the total viral RNA. This RNA binds to components of the mitochondrial respiratory chain complex I (MRCC-I), stabilizing its function as a countermeasure to apoptosis. Thus, β2.7 prevents the premature death of the infected cell and supports the steady production of ATP during the viral life cycle. The virus also produces miR-UL112, a shorter ncRNA that is an miRNA. The miR-UL112 inhibits the production of mRNA for a cell-surface ligand that attracts NK cells. Thus, these two RNA molecules inhibit the destruction of the cell by NK cells as well as promote the general strength of the host cell to prevent cell death. ◗

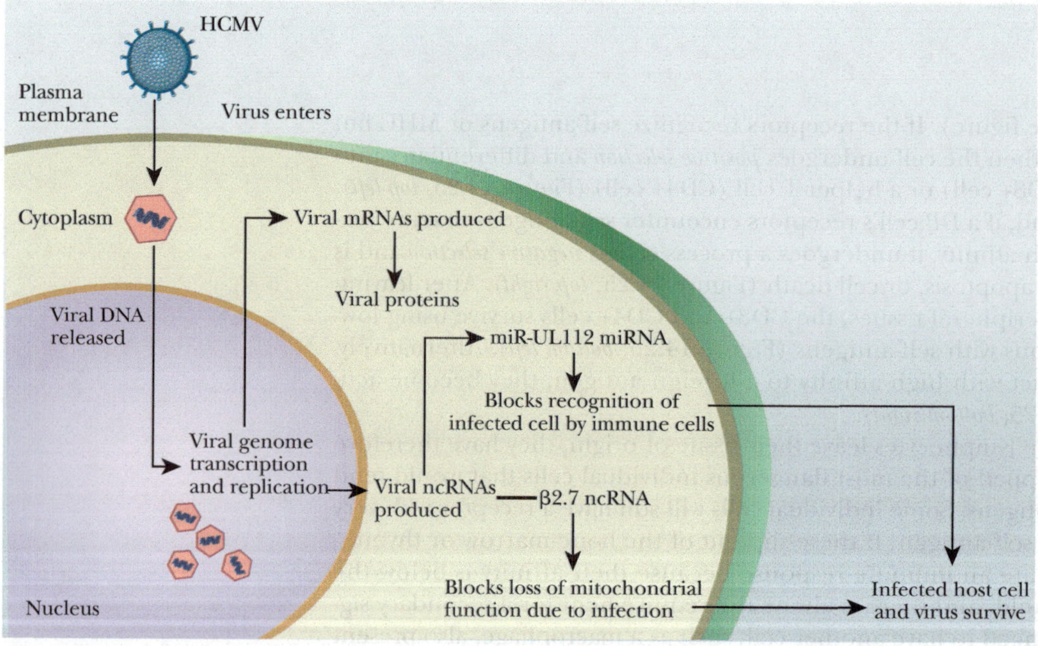

Figure 14.26 Viral RNA shuts down the immune response. Infection by HCMV induces a state of metabolic stress that would normally trigger host immune responses, including cell lysis or apoptosis (cell death). HCMV prevents these immune responses using two viral ncRNAs, a micro RNA called miR-UL112 and a noncoding RNA called β2.7 ncRNA. The β2.7 ncRNA binds to components of the mitochondrial respiratory chain complex I (MRCC-I) and stabilizes mitochondrial energy production. The miR-UL112 miRNA blocks the expression of cell-surface ligand, MICB, which attracts natural killer (NK) cells and hence prevents lysis by these immune cells that would recognize MICB. *(Based on Cullen, B. R. (2007). Outwitted by viral RNAs.* Science *317, p. 329.)*

14-4 Cancer

Learning Objectives

After reading this section, students will be able to:

LO 14.4.1 Describe the characteristics of cancer cells.

LO 14.4.2 Distinguish between oncogenes and tumor-suppressor genes.

LO 14.4.3 Contrast different ways of fighting cancer.

Cancer is one of the leading causes of death in humans. It is characterized by cells that grow and divide out of control, often spreading to other tissues and causing them to become cancerous. Some estimates suggest that a third of all humans will get cancer during their lifetimes, so it is a disease that is important for everyone to understand. The older a person gets, the more likely they are to get cancer. A 70-year-old is about 100 times more likely to get cancer than a 20-year-old.

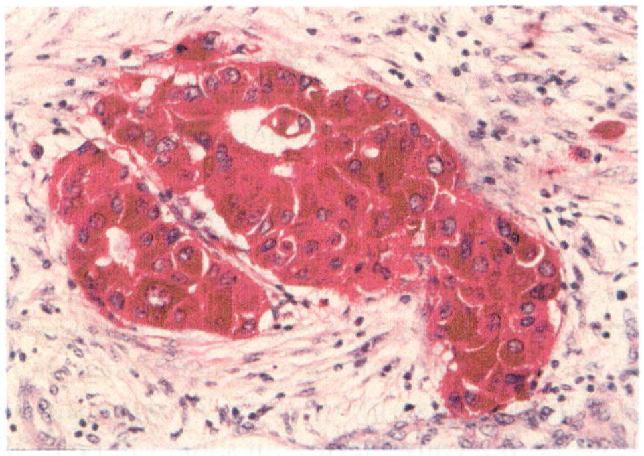

Michael Abbey/Science Source

Figure 14.27 Tumor cells. Normal cells stop growing when they are squeezed by other cells. This tumor cell continues dividing and growing even though it is being squeezed by the adjacent tissue.

14-4A Cancer Cell Characteristics

Multiple problems must occur in a cell before it becomes cancerous. That may be why, even though cancer is common, most people grow to old age and do not get cancer. All life-threatening cancers have at least six characteristics in common:

1. Cancer cells continue to grow and divide in situations in which normal cells do not. Most cells grow only when they receive a growth-chemical signal, but cancer cells manage to keep growing without such signals.
2. Cancer cells continue to grow even when the neighboring cells send out "stop-growth" signals. For example, normal cells stop growing when compressed by other cells. Somehow, tumors manage to avoid this. (Figure 14.27 shows a tumor cell expanding and squeezing against neighboring tissue.)
3. Cancer cells manage to keep going and avoid a "self-destruct" signal that usually occurs when DNA damage has occurred.
4. Cancer cells can co-opt the body's vascular system, causing the growth of new blood vessels to supply the cancerous cells with nutrients.
5. Cancer cells are essentially immortal. Normal cells can divide only a finite number of times, usually in the 50–70 range. However, cancer cells and tumors can divide far more than that.
6. Cells that exhibit the first five characteristics can be a problem, but cancer cells have the ability to break loose, travel to other parts of the body, and create new tumors. This process, called **metastasis**, makes cancer cells particularly lethal. Stationary tumors can often be removed by surgery. However, once a cancer starts spreading, it is almost impossible to stop. Of every ten deaths due to cancer—including a high percentage of lung, colon, and breast cancers—approximately nine of them are due to cancers that metastasized. **Biochemical Connections 14D** looks at some pivotal research in the search for cancer's origins.

metastasis the spread of cancer throughout the body

14D Biochemical Connections • Genetics

Cancer: The Dark Side of the Human Genome

In a very real sense, cancer is the most common genetic disease. Somatic mutations (ones that do not affect the germ cells passed on to the following generations) constantly take place in the human body, and they accumulate as a person ages. When these mutations reach a critical number in essential locations on the genome, cancer develops. Figure 14.28 shows an electron micrograph of a cancer cell. The number of such mutations can easily reach well into the thousands.

By December 2009, scientists at the Wellcome Trust Sanger Institute in the United Kingdom had sequenced the genome of melanoma (skin cancer) and lung cancer. They found more than 30,000 mutations in the melanoma genome and more than 23,000 in

Continued on next page

lung cancer. The genomes of many other types of cancer cells have been determined, thanks to initiatives like The Cancer Genome Atlas (TCGA) and the International Cancer Genome Consortium (ICGC). These projects help researchers better understand the molecular basis of cancer.

With the sequencing of cancer cell genomes, it is possible to determine the factors that led to the development of the cancer. As an example, most of the mutations associated with melanoma arise from excessive exposure to the sun. Likewise, smoking causes most of the DNA errors in lung cancer. Researchers estimate that a new mutation occurs with every 15 cigarettes smoked. Most of the mutations do not cause harm, but some will be in DNA sequences where alterations lead to cancer.

This knowledge is helping researchers understand cancer development and improve early detection methods. Even partial knowledge of the breast cancer genome led to the drug Herceptin, a chemotherapeutic agent that deals with the results of a gene that is overexpressed. The gene in question, HER2, encodes a receptor protein on the cell surface. In normal cells, the receptor binds to a signaling compound to activate processes within a cell. In cancer, activation takes place without a signal, leading to the uncontrolled cell proliferation characteristic of cancer. Herceptin has no effect on cancers in which the gene in question is not overexpressed. It can even be harmful because of its toxicity, especially to the heart. This drug is but one example of the benefits that will come from knowing the cancer genome. Many more breakthroughs can be expected as this research continues.

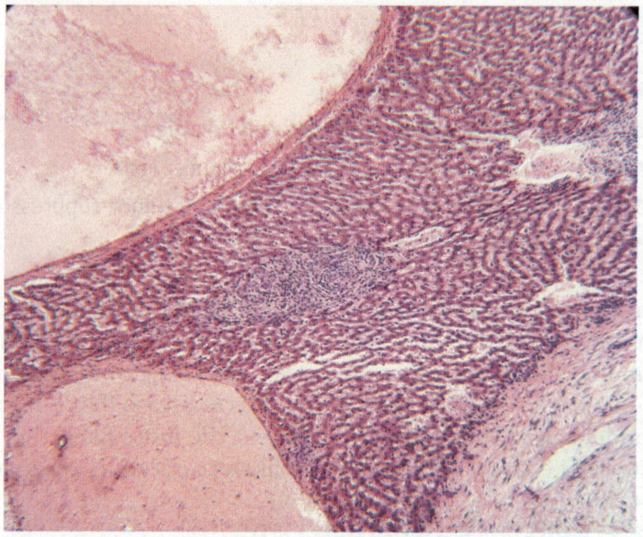

Jubal Harshaw/Shutterstock.com

Figure 14.28 Electron micrograph of a cancer cell.

14-4B Oncogenes and Tumor-Suppressor Genes

▶ What causes cancer?

You may have heard that many things, such as smoking or eating grilled meat, cause cancer. Or exposure to radiation or asbestos causes cancer. These things may play a role, but they cannot truly be the ultimate cause. The real cause may be a combination of insults to the cell that turns it malignant. Cancer is ultimately a DNA disease. It has its roots in changes in the DNA inside a cell. Somehow these changes cause the loss of control of cell division and the other characteristics described so far.

The changes in the DNA cause changes to specific proteins that are responsible for controlling the cell cycle. Most cancer-causing mutations of DNA affect two types of genes, called proto-oncogenes and tumor suppressors. A **proto-oncogene** is a gene whose protein product stimulates growth and cell division. Mutations in proto-oncogenes can cause them to become **oncogenes**, which are permanently active. A **tumor suppressor** is a gene that makes a protein that restricts the cell's ability to divide. If a mutation damages the gene for a tumor suppressor, then the cell has lost its brakes and divides out of control. A third type of gene, called a *DNA repair gene*, is sometimes involved in cancer. Mutations to DNA repair genes can prevent the repair of mutations in proto-oncogenes and tumor-suppressor genes. Most cancers result from multiple mutations across these gene categories.

proto-oncogene normal eukaryotic genes that stimulate cell growth or cell division

oncogenes a mutated proto-oncogene that is permanently active in stimulating cell growth or cell division

tumor suppressor a cancer-related gene that encodes a protein that inhibits cell division

▶ What are some examples of oncogenes?

Historically, an oncogene was a gene that was found to be implicated in cancer. The root word, *onco,* means "cancer." In 1911, a scientist named Peyton Rous demonstrated that solutions taken from chicken carcinomas could infect other cells. This was the first discovery of tumor viruses, and Rous was given the Nobel Prize in Physiology or Medicine in 1966 for his discovery. The virus was called the **Rous sarcoma virus**, and it was the first retrovirus shown to cause

Rous sarcoma virus the first virus to be shown to cause cancer

cancer. The gene that was specific to the cancer is called *v-src*, for *viral sarcoma*. This gene encodes a protein that causes transformation of the host cell into a cancer cell. Thus, the gene was given the name *oncogene*. The protein was called pp60^{src}, which stands for a phosphoprotein of 60,000-Da molecular weight from the sarcoma virus (*src*).

However, it was later found that the sequence of the *v-src* gene was very similar to that of a normal gene in eukaryotes called cellular *src* (*c-src*). *c-src* and other normal cell-growth genes and cell-division genes were later called *proto-oncogenes*. Many proto-oncogenes are normal and necessary for proper growth and development in eukaryotic cells. However, some transforming event causes the proto-oncogene to lose control. Sometimes this is due to a viral infection. In other cases, the event that causes a proto-oncogene to become an oncogene is not known. Table 14.2 lists some proto-oncogenes implicated in human tumors. Many of these genes are involved in signal-transduction pathways that affect the transcription of genes that speed up cell division. Figure 14.29 shows an example of one of these pathways. At the bottom of Figure 14.29 is the CREB-binding protein (CBP). Recall from Section 11-5D that we discussed the control of transcription in eukaryotes and noted that many signaling pathways were routed through the CBP/p300 coactivator. One of these pathways involves **mitogen-activated protein kinase (MAPK)** and a transcription factor called AP-1. To understand the roles of many of the proto-oncogenes listed in Table 14.2, let us explore more deeply the pathway in Figure 14.29.

mitogen-activated protein kinase (MAPK) an enzyme that responds to cell growth and stress signals and phosphorylates key proteins that act as transcription factors

The process starts when an extracellular signal, such as a growth factor, binds to a receptor on the cell membrane (Figure 14.29, *top left*). This receptor is a *tyrosine kinase* that dimerizes, and then each tyrosine kinase monomer phosphorylates several of its own tyrosine residues in a process called *autophosphorylation*. Once phosphorylated, the receptors are bound by an adaptor molecule, a protein called *GRB2* (pronounced "*grab two*"), which has a phosphotyrosine-binding domain that is very similar to a domain found in the pp60^{src} protein. The other end of GRB2 binds to a protein called *Sos*.

At this point, there is an interaction with a very important 21-kDa protein called p21^{ras} or just *Ras*. (kDa refers to the kilodalton unit that is a measure of protein size, where 1 kDa = 1000 daltons, and 1 dalton = 1 g/mol.) Ras is involved in about 30% of human tumors. The designation *Ras* comes from *Ra*t *s*arcoma, the original tissue in which it was discovered. The Ras family of proteins is GTP-binding proteins. In their resting state, they are bound to GDP. After the cell signal, the Sos replaces the GDP with GTP. Intrinsic hydrolysis of the GTP returns the protein to its inactive state, but this process is slow. Proteins known as GTPase-activating proteins (GAPs) speed up this hydrolysis and are involved in the control of the Ras proteins. GAPs inactivate Ras

Table 14.2 A Representative List of Proto-oncogenes Implicated in Human Tumors

Proto-oncogene	Neoplasm(s)
abl	Chronic myelogenous leukemia
*erb*B-1	Squamous cell carcinoma; astrocytoma
*erb*B-2 (*neu*)	Adenocarcinoma of breast, ovary, and stomach
myc	Burkitt's lymphoma; carcinoma of lung, breast, and cervix
H-*ras*	Carcinoma of colon, lung, and pancreas; melanoma
N-*ras*	Carcinoma of genitourinary tract and thyroid; melanoma
ros	Astrocytoma
src	Carcinoma of colon
jun, *fos*	Several

Adapted from Bishop, J. M. (1991). Molecular themes in oncogenesis, *Cell* 64: 235–248.

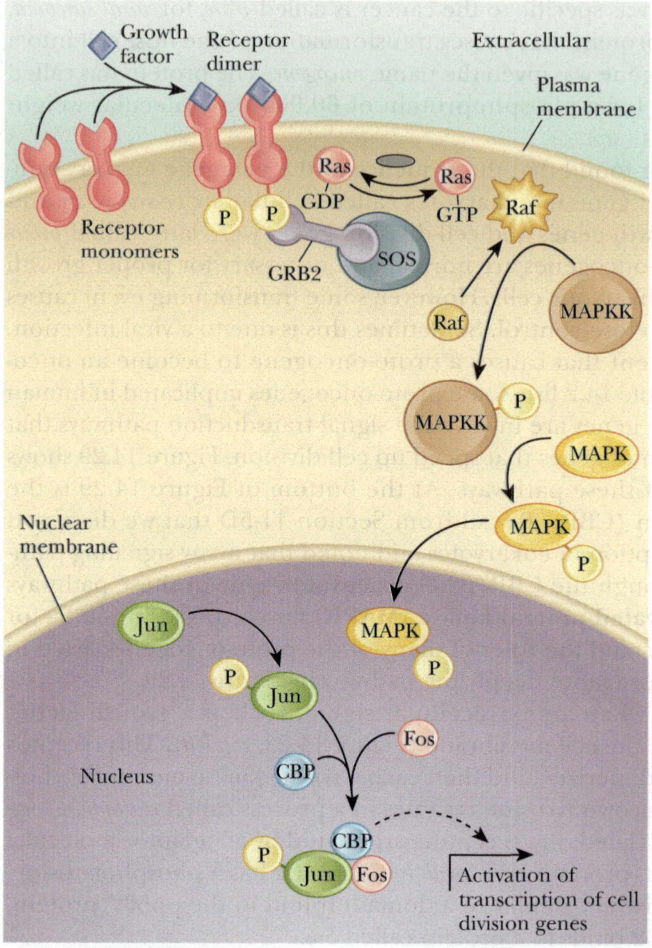

Figure 14.29 A mitogen-activated protein kinase (MAPK) signal-transduction pathway. Signal transduction starts when a growth factor binds to receptor monomers on the cell membrane. The receptor monomers are tyrosine kinases, which then dimerize, and each monomer phosphorylates several of its own tyrosine residues. The phosphorylated receptor is then recognized by GRB2, which binds to the Ras exchanger SOS. SOS is activated to exchange GDP for GTP on Ras, activating it. Ras moves Raf to the cell membrane, where it becomes active. Raf phosphorylates mitogen-activated protein kinase kinase (MAPKK), which then phosphorylates mitogen-activated protein kinase (MAPK). MAPK enters the nucleus and phosphorylates Jun. Phosphorylated Jun binds to Fos and CBP, and transcription of cell-division genes is activated. *(Adapted by permission from* Molecular Biology *by R. F. Weaver, McGraw-Hill, 1999, p. 375.)*

by accelerating the hydrolysis of GTP. Oncogenic forms of Ras have impaired GTPase activity and are insensitive to GAPs, thus leaving them bound to GTP, which causes them to stimulate cell division continually.

Although Ras mutations have been some of the most-studied mutations leading to cancer, Ras is found rather early in the process that ultimately leads to cell division. Activated Ras attracts another protein called Raf, which then phosphorylates serines and threonines on the enzyme *mitogen-activated protein kinase kinase (MAPKK).* MAPKK then phosphorylates the enzyme mitogen-activated protein kinase (MAPK). MAPK then enters the nucleus and phosphorylates a transcription factor called *Jun.* Jun binds to another transcription factor called *Fos.* Together, Jun and Fos make up the transcription factor called *AP-1* in Section 11-5D. AP-1 binds to CREB-binding protein (CBP) and stimulates the transcription of genes that lead to rapid cell division. The *jun* and *fos* proto-oncogenes code for these proteins (see Table 14.2). Also, in 2002, researchers screening 20 different genes in 378 cancer samples found that the Raf gene (*c-Raf*) was mutated in 70% of the malignant melanoma samples.

▶ What are some examples of tumor-suppressor genes?

In addition to the cell-growth and cell-division proteins produced by proto-oncogenes, many human genes produce proteins called *tumor suppressors.* These human genes are called *tumor-suppressor genes.* Tumor-suppressor proteins inhibit the transcription of genes that would cause increased replication. When a mutation occurs in any of these suppressors, replication and division become uncontrolled, and tumors result. Table 14.3 lists some human tumor-suppressor genes.

Table 14.3 A Representative List of Tumor-Suppressor Genes Implicated in Human Tumors

Tumor-Suppressor Gene	Neoplasm(s)
RBI	Retinoblastoma; osteosarcoma; carcinoma of breast, bladder, and lung
p53	Astrocytoma; carcinoma of breast, colon, and lung; osteosarcoma
WT1	Wilms' tumor
DCC	Carcinoma of colon
NF1	Neurofibromatosis type 1
FAP	Carcinoma of colon
MEN-1	Tumors of parathyroid, pancreas, pituitary, and adrenal cortex

Adapted from Bishop, J. M. (1991). Molecular themes in oncogenesis. *Cell* 64, 235–248.

Mutations in the gene that codes for a 53-kDa protein designated *p53* are found in more than half of all human cancers. When the gene is operating normally, it acts as a tumor suppressor; when it is mutated, it is involved in a wide variety of cancers. By the end of 1993, mutations in the *p53* gene had been found in 51 types of human tumors. The role of p53 is to slow down cell division and to promote cell death (apoptosis) under certain circumstances, including when DNA is damaged or when cells are infected by viruses.

p53 is a transcription factor that regulates the expression of genes involved in key cellular processes. When cancer-causing mutations occur in p53, it can no longer bind to DNA in a normal fashion. The mode of action of p53 as a tumor suppressor is twofold. As shown in Figure 14.30, it is an activator of RNA transcription; it "turns on" the transcription and translation of several genes. One of them, *Pic1,* encodes a 21-kDa protein, *p21,* that is a key regulator of DNA synthesis and thus of cell division. The p21 protein, which is present in normal cells but is missing from (or mutated in) cancer cells, binds to the enzymes known as *cyclin-dependent kinases (CDKs).* CDKs, as their name implies, become active only when they associate with proteins called cyclins. Recall from Section 10-7A that cell division depends on the activity of CDKs. Some of the oncogenes discussed so far work in such a way that the result is an overproduction of the CDK proteins, which keeps the cells dividing continuously. Normal levels of p53 protein cannot turn these genes off in cancer cells, but they could do this in normal cells. In normal cells, the result is that the cell cycle remains in the state between mitosis (in which cells divide) and the replication of DNA for the next cell division. DNA repair can take place at this stage. If the attempts at DNA repair fail, the p53 protein may trigger apoptosis, the programmed cell death characteristic of normal cells, but not of cancer cells.

The important point is that two different mechanisms are operating here. One is analogous to the brakes failing in your car (inadequate or defective p53 protein), and the other (overproduction of CDKs) is equivalent to the accelerator sticking in the open position—two opposite mechanisms with the same result: the car crashes.

A number of factors come together in explaining the variety of diseases we call cancer. Mutations of DNA lead to changes in the proteins that control cell growth by either directly causing cell division or allowing it to occur by default. Still other mutations interfere with DNA repair. The possibility of finding new cancer therapies—and perhaps even cancer cures—is enhanced by understanding these contributing factors and how they affect each other.

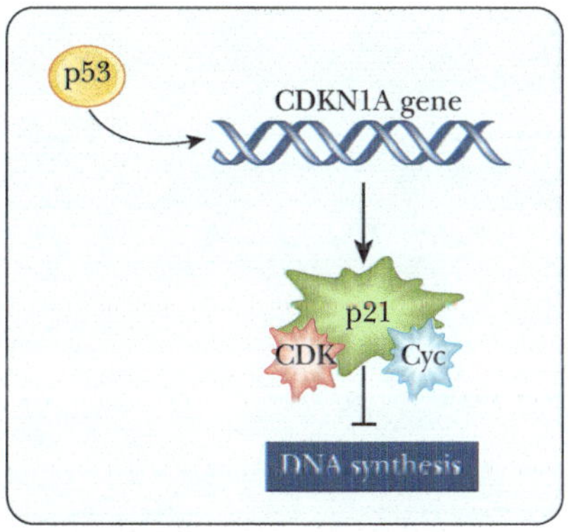

Figure 14.30 An important role of p53. The 53-kDa protein called p53 is a transcription factor. One of its roles is to bind to the CDKN1A gene and turn on the expression of its gene product, a 21-kDa protein called p21. p21 then binds to complexes of cyclin-dependent kinases (CDKs) and cyclins. The direct result of this binding is inhibition of DNA synthesis with the indirect result being the inhibition of cell division. *(Based on Sutliff, K. (1993).* Science *262, 1644, Figure 1.)*

14-4C Viruses and Cancer

The work by Rous mentioned in Section 14-4B showed how viruses could cause cancer in certain situations. The close homology between the oncogene sequence found in some viruses and the proto-oncogene sequences in the mammalian genome has led many researchers to theorize that the oncogenes may have been of mammalian origin. In the course of repeated infections and travels, the virus may pick up pieces of DNA from a host and deliver another piece of DNA to a host. In the course of the rapid mutation that occurs in retroviruses, these proto-oncogenes could be mutated to a form that is oncogenic.

Retroviruses that cause cancer in humans are known; some forms of leukemia (caused by HTLV-I and HTLV-II, which infect T cells of the immune system) are well-known examples, as well as cervical cancer caused by cervical papillomavirus. Theoretically, any retrovirus that inserts its DNA into the host chromosome could accidentally disable a tumor-suppressor gene or enable an oncogene by insertion of a strong promoter sequence near a proto-oncogene. One of the biggest fears of using in vivo delivery techniques for human gene therapy is that the viral DNA inserted into a human chromosome might become incorporated into an otherwise healthy tumor-suppressor gene. This would potentially cause an even greater problem while trying to solve one of the patient's problems by providing a functional gene they were missing. This happened, unfortunately, in 2003, when researchers in France were using viral gene therapy to treat patients with X-linked SCID (see **Biochemical Connections 14A**). In 9 of 11 cases, the viral gene therapy restored the immune systems of the patients. However, two patients developed leukemia. It was later found that the virus had inserted itself, in each case, near a gene that has been found to be a leukemia oncogene. This was a tragic setback in viral gene therapy, but it remains a powerful tool that continues to be refined.

14-4D Fighting Cancer

Cancer can be treated in a variety of ways:

- Traditional approaches such as surgeries to remove tumors or radiation and chemotherapy to kill cancerous cells.
- Targeted therapies such as treatment with monoclonal antibodies or hormone therapy.
- Immunotherapy approaches such as T cell modifications to recognize and kill cancer cells.
- Other treatments such as CRISPR-Cas9 gene editing (Section 13-4C) or oncolytic virus therapy techniques (see below).

Despite this wide range of treatments, scientists and doctors must be very careful when tinkering with processes involved in cell growth.

▶ How can viruses help fight cancer?

As we have discovered here in Chapter 14, viruses come in many types and cause many diseases. Viruses can be very specific to a single cell type because they rely on a protein receptor on the cell to gain entry. Liver cells display receptors that nerve cells do not, and vice versa. Oncologists (doctors who treat cancer) have treated cancer for years with techniques such as radiation therapy and chemotherapy. These techniques attempt to target cancer cells but, in the end, destroy other cells as well. In some sense, the goal of chemotherapy is to kill the cancer before the treatment kills the patient. If doctors could come up with a treatment that would be completely specific to cancer cells, it would go a long way both toward stopping the cancer and toward making the patient's life more comfortable during the treatment. This was another chance for researchers to find something helpful about viruses.

In the 1990s, a new type of treatment for cancer, called **oncolytic virus therapy**, was begun. This technique was designed to target human tumor cells grafted onto mice, and the treatment eliminated the human tumors using an adenovirus (Table 14.1). We now discuss two of the strategies used in oncolytic virus therapy. One is to use the virus to attack and kill the cancer cell directly. The second is to have the virus ferry in a gene to the cancer cell that makes the cell more susceptible to a chemotherapy agent.

oncolytic virus therapy a form of cancer treatment that uses viruses

One of the biggest challenges in virotherapy is to make sure the virus specifically targets the cancer cell. The common adenovirus is not specific for cancer cells, so in order to use it for virotherapy, other techniques must also be employed. One of these is called **transductional targeting**. In this technique, antibodies are attached to the virus. These antibodies are created so that they target the cancer cell (Figure 14.31). In this way, the normally indiscriminant adenovirus attacks only cancer cells. Once inside, the virus reproduces and eventually lyses the cell.

transductional targeting a cancer-fighting technique in which antibodies are attached to a virus so that it will infect a cancer cell

Another approach is called **transcriptional targeting**, shown in Figure 14.34. With this technique, the replication genes for adenovirus are placed after a promoter that is specific to a cancer cell. For example, skin cells make much more of the pigment melanin than other cells. Therefore, the genes for enzymes that make melanin are turned on more often in skin cells than in other cells (Figure 14.34). Adenovirus can be engineered to have the promoter for the melanin-producing enzyme near the genes for virus replication. In cancerous skin cells, these promoters are triggered more often, so adenovirus replicates

transcriptional targeting a cancer-fighting technique in which a virus is engineered to respond to specific transcription factors or promoters that are active only in cancer cells

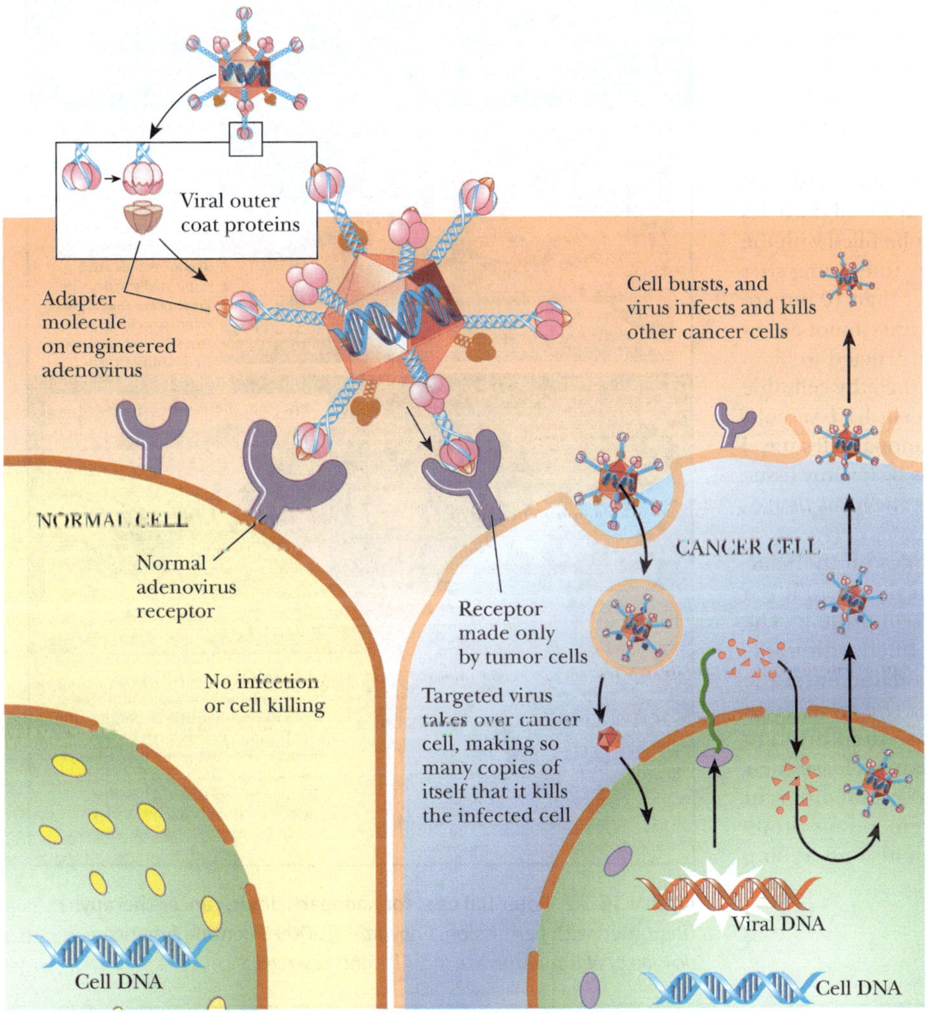

Figure 14.31 Transductional targeting in virotherapy. Viruses, such as adenovirus, are used to infect and destroy cancer cells selectively. Spikes on the adenovirus are mutated so that they recognize unique receptors on cancer cells. The virus selectively infects and lyses the cancer cells. *(Based on Terese Winslow.)*

much quicker in skin-cancer cells, killing them specifically. Similar techniques have been used to target liver-cancer cells and prostate-cancer cells.

The other basic strategy is to have the virus ferry in a gene that makes the cancer cell more susceptible to chemotherapy. One such system uses a virus that targets rapidly dividing cells. Inside these cells, and only in these cells, the gene carried by the virus converts an innocuous prodrug into an anticancer drug. These viruses are sometimes called "smart viruses" for their ability to select only the cancer cells. They then allow drugs to be used that are not harmful to normal cells.

nanoparticles artificial particles 10–100 nm in diameter that can be used in medicine to carry specific compounds into cells

Biochemical Connections 14E describes the use of nanotechnology to fight cancer. To end this section on cancer, **Biochemical Connections 14F** looks at cancer and inflammation.

14E Biochemical Connections • Biotechnology

Nanotechnology Tackles Cancer

The explosion of technology in the last 10 years has made it possible to envision medical procedures that are faster, cheaper, and more effective than more traditional treatments. One of the goals of cancer treatments is to minimize the damage done to healthy tissues and cells while being able to target the cancer cells. With traditional radiation or chemotherapy, this has been difficult. As described at the end of this section, using viruses to deliver treatment to specific targets is one approach that is very promising.

Another approach that is being studied and is in clinical trials for several cancers is the use of **nanoparticles**. These artificial particles can be made in various sizes and materials and can be filled with the specific molecule to be delivered to the target. The interesting sizes are those between 10 and 100 nm. Particles smaller than 10 nm are rapidly removed from the body. Particles over 100 nm cannot access the cells we wish to target. However, particles constructed to be in between in size can access cancer cells better than healthy cells due to the nature of cancer. Most tumors and cancer cells have vasculature that is damaged and porous. Therefore, as nanoparticles travel through the bloodstream, they will continually pass by healthy tissue but then can leak out through the pores into the cancerous tissues, as shown in Figure 14.32.

The nanoparticles are created to have molecules on their surface that will be attracted to the cancer cells. In this case, it is a polyethylene glycol molecule linked to a transferrin protein. The particle is made up of a synthetic polymer and filled with a specific siRNA. The particles move through the blood and enter the cancer cells. The cancer cells have receptors for the transferrin protein and bind to them. The nanoparticles are then brought inside by endocytosis, whereupon they degrade and release their siRNA contents. Clinical trials are underway with several different nanoparticle types, and this technology shows tremendous promise for improving the quality of life of cancer patients undergoing treatment. ◗

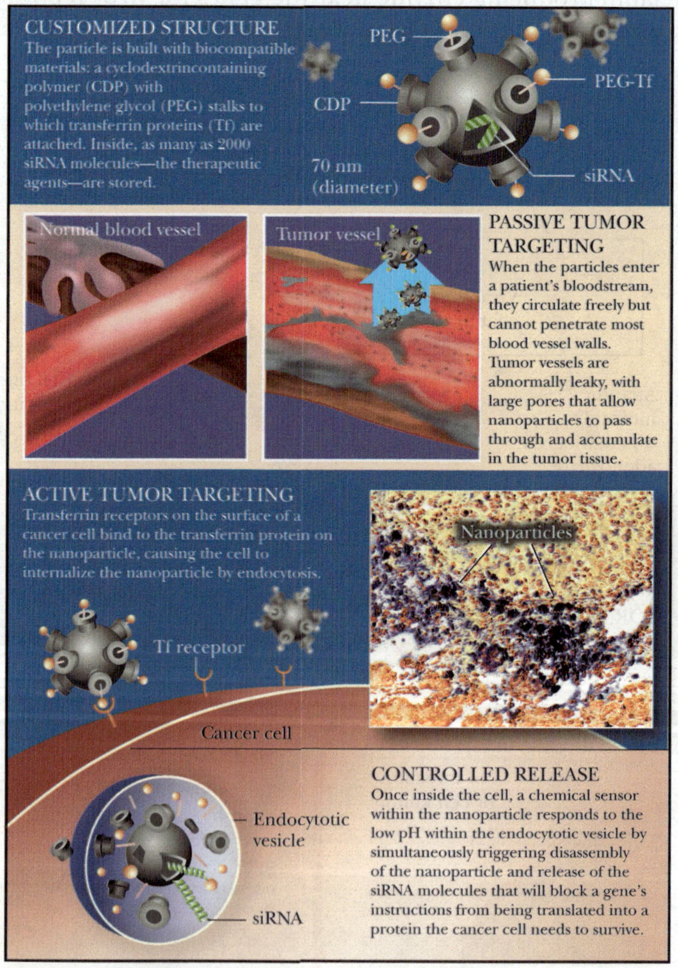

NCI/Science Source

Figure 14.32 Potential uses for nanoparticles in cancer therapy. (Reprinted with permission. *Copyright* ©2009 Scientific American, *a division of Nature America, Inc. All rights reserved.)*

14F Biochemical Connections • Immunology and Oncology

The Relationship Between the Immune System and Cancer

In Section 14-3B, we looked at innate immunity. Many of the processes involving macrophages and other innate immunity cell types lead to inflammation. Inflammation is known to be the underlying contributor to almost every chronic human disease known, including arthritis, Crohn's disease, diabetes, heart disease, Alzheimer's disease, and stroke. It has recently been shown that inflammation is also linked to cancer.

Cancer begins with a series of genetic changes that lead to cells overreplicating. It can then progress to the metastasis stage, where cancer cells break free from their original location and set up new cancerous colonies at other sites. However, the stages in development are controlled by many factors, and researchers are now realizing that invasive cancers use cells of the innate immune system. These cells normally come to our rescue to protect against damage and disease, but in cancer they can be subverted for the good of the tumor and the ill of the patient. As shown in Figure 14.33, cells of the innate system can be both beneficial and detrimental to the fight against cancer. As antigen-presenting cells (APCs), dendritic cells, or macrophages stimulate T cells and B cells, which lead to an attack on a tumor cell. However, they also lead to an inflammatory response that leads to tumor growth through the production of cytokines and growth factors. The evidence indicates that the inflammation surrounding macrophages encourages the conversion of precancerous tissue to full malignancy.

This realization has led researchers to broaden the search for cancer prevention and management strategies. If doctors can treat the inflammation, then they might be able to stop the progression of the disease, essentially turning cancer into a manageable long-term disease. Patients rarely die from the primary cancer; rather, they die from metastasis. Current research is looking at drug therapies, including the simple use of nonsteroidal anti-inflammatory drugs (NSAIDs) such as aspirin, more selective inhibition of prostaglandin E_2, and specific inhibition of cytokines such as IL-6 and IL-8. ◗

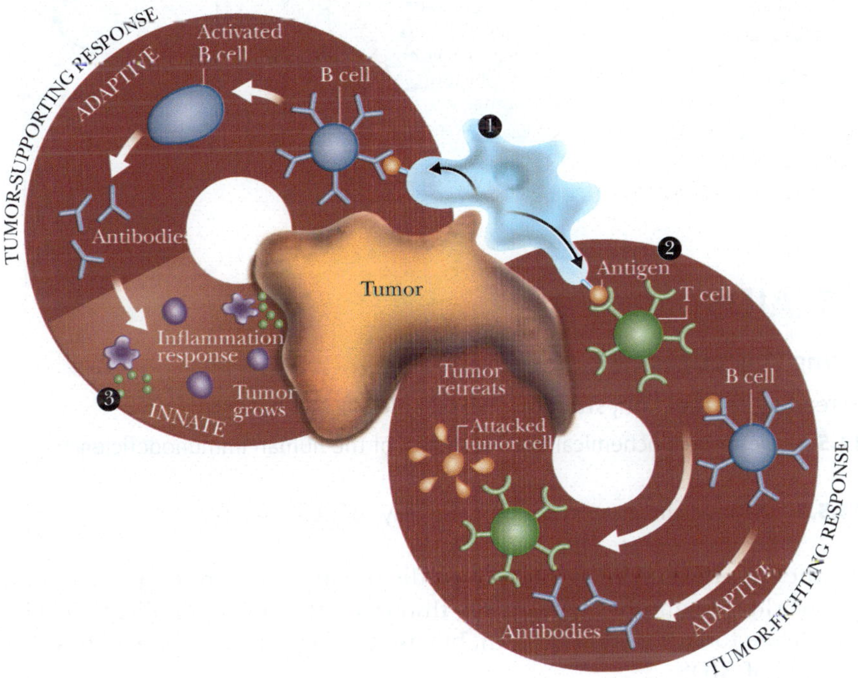

❶ A dendritic or other cell (the exact one has not been identified experimentally) "displays" a piece of the tumor to adaptive T cells and B cells.

❷ In some instances, T cells and B cell-produced antibodies identify and destroy tumor cells, as they would a pathogen.

❸ But when B cells receive certain yet to be delineated signals, they produce antibodies that activate the innate immune system to help tumor cells to survive and flourish.

Figure 14.33 The immune paradox. Two arms of the immune system—the innate and the adaptive—are exquisitely well adapted for fighting pathogens, but their role in combating cancer is more paradoxical. The innate system furnishes an initial inflammatory response to a microbial insult by attacking any invading pathogen indiscriminately, whereas adaptive immunity furnishes a delayed response that hones in on a particular pathogen. In cancer, both systems may sometimes attack tumor cells. But a tumor protects itself by recruiting the innate system to enhance its development. *(Based on Scientific American, "A Malignant Flame" by Gary Stix,* Scientific American, *297[1], 60–67 [2007].)*

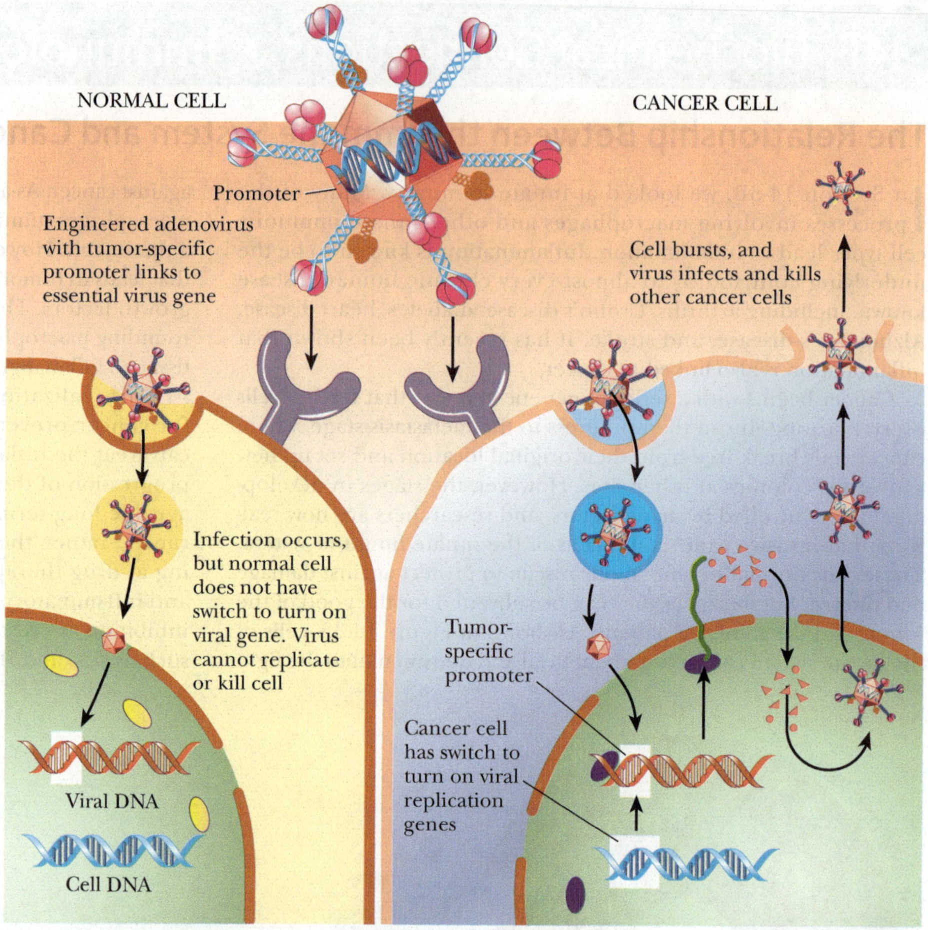

Figure 14.34 Transcriptional targeting in virotherapy. A tumor-specific promoter is engineered into the adenovirus. The adenovirus infects many cells, but it is activated to replicate only in the cancer cell. *(Based on Terese Winslow.)*

14-5 AIDS

Learning Objectives

After reading this section, students will be able to:

LO 14.5.1 Describe biochemical characteristics of the human immunodeficiency virus (HIV).

LO 14.5.2 Distinguish between the different ways of fighting HIV.

HIV is the causative agent of AIDS. This disease affects more than 40 million people worldwide and has continually thwarted attempts to eradicate it. The best medicine today can slow it down, but as of today, only one person has ever been cured of AIDS.

14-5A HIV: A Model Retrovirus

The HIV genome is a single-stranded RNA that has a number of proteins packed around it, including the virus-specific reverse transcriptase (RT), integrase, and protease. A protein coat surrounds the RNA–protein assemblage, giving the overall shape of a truncated cone. Finally, a membrane envelope encloses the protein coat. The envelope consists of a phospholipid bilayer formed from the plasma membrane of cells infected earlier in the life cycle of the virus, as well as some specific glycoproteins, such as gp41 and gp120, as shown in Figure 14.35.

HIV offers a classic example of the mode of operation of retroviruses. The HIV infection begins when the virus particle binds to receptors on the surface

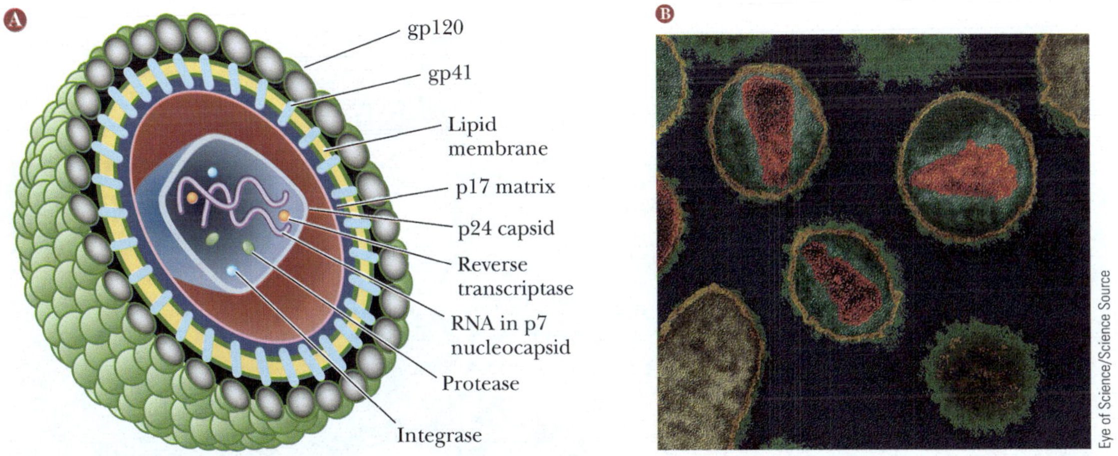

Figure 14.35 The architecture of HIV. (A) The RNA genome is surrounded by p7 nucleocapsid proteins and by several viral enzymes—namely reverse transcriptase (RT), integrase, and protease. The truncated cone consists of p24 capsid protein subunits. The p17 matrix (another layer of protein) lies inside the envelope, which consists of a lipid bilayer and glycoproteins such as gp41 and gp120. (B) An electron micrograph shows both mature virus particles, in which the core (the truncated cone) is visible, and immature virus particles, in which it is not. *(Based on Bettelheim/Brown/Campbell/Farrell/Torres,* Introduction to General, Organic, and Biochemistry, *10E. ©2013 Cengage Learning.)*

of a cell (Figure 14.36, *Step 1*). The viral core is inserted into the cell and partially disintegrates. The RT catalyzes the production of DNA from the viral RNA. The viral DNA is integrated into the DNA of the host cell. The DNA, including the integrated viral DNA, is transcribed to RNA. Smaller RNAs are produced first, specifying the amino acid sequences of viral regulatory proteins. Larger RNAs, which specify the amino acid sequences of viral enzymes and coat proteins (CPs), are made next. The viral protease assumes particular importance in the budding of new virus particles. Both the viral RNA and viral proteins are included in the budding virus, as is some of the membrane of the infected cell.

▶ How does HIV confound the immune system?

Why was HIV so deadly and so hard to stop back in the 1980s and 1990s? Many viruses, such as adenovirus, cause nothing more than the common cold; others, such as the virus that causes severe acute respiratory syndrome (SARS), are deadly. HIV has several characteristics that prevented effective treatment for many years. Ultimately, it is tricky because of its targets, the helper T cells. The immune system is under constant attack by the virus, and millions of helper T cells and killer T cells are called up to fight billions of virus particles. Through degradation of the T-cell membrane via budding and the activation of enzymes that lead to cell death, the T-cell count diminishes to the point that the infected person is no longer able to mount a suitable immune response. As a result, the individual eventually succumbs to pneumonia or another opportunistic disease.

There are many reasons that the disease is so persistent. For example, it is slow-acting. Other viruses have been eradicated quickly because they were quick to act, making it easy to find infected people before they had a chance to spread the disease. In contrast, HIV-infected individuals can go years before they become aware that they have the disease. There is a small percentage of people, called controllers, who manage to live to old age with HIV but never develop AIDS, although they are carriers and can pass it to someone else. However, this is only a small part of what makes HIV so difficult to kill.

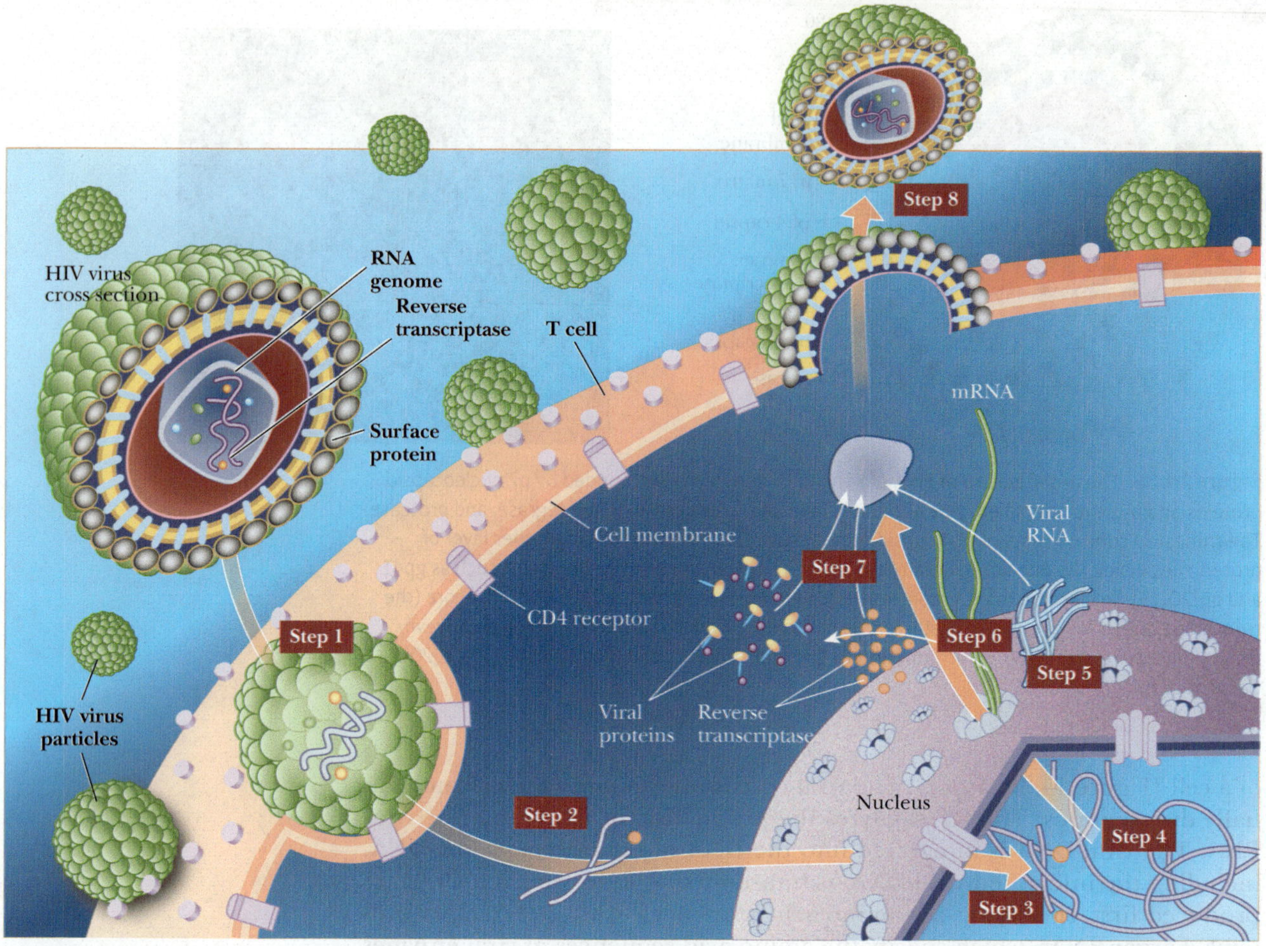

Figure 14.36 **The HIV life cycle.** HIV infection begins when the virus particle binds to CD4 receptors on the surface of a cell (Step 1). The viral core is inserted into the cell and partially disintegrates (Step 2). The reverse transcriptase catalyzes the production of DNA from the viral RNA. The viral DNA is integrated into the DNA of the host cell (Step 3). The DNA, including the integrated viral DNA, is transcribed to RNA (Step 4). Smaller RNAs are produced first, specifying the amino acid sequence of viral regulatory proteins (Step 5). Larger RNAs, which specify the amino acid sequences of viral enzymes and coat proteins, are made next (Step 6). The viral protease assumes particular importance in the budding of new virus particles (Step 7). Both the viral RNA and viral proteins are included in the budding virus, as is some of the membrane of the infected cell (Step 8). *(Based on Bettelheim/Brown/Campbell/Farrell/Torres,* Introduction to General, Organic, and Biochemistry, *10E. ©2013 Cengage Learning.)*

HIV was difficult to treat because of the following reasons:

- HIV was difficult to find. For an immune system to fight a virus, it needs to locate specific macromolecules that can be bound to antibodies or T-cell receptors (TCRs). The reverse transcriptase (RT) of HIV is very inaccurate in replication. As a result, HIV mutates rapidly, a situation that presents a considerable challenge to those who want to devise treatments for AIDS. The virus mutates so rapidly that multiple strains of HIV may be present in a single individual.
- Another trick the virus plays is a conformational change of the gp120 protein when it binds to the CD4 receptor on a T cell. The normal shape of the gp120 monomer may elicit an antibody response, but these antibodies are largely ineffective. The gp120 forms a complex

with gp41 and changes shape when it binds to CD4. It also binds to a secondary site on the T cell that normally binds to a cytokine. This change exposes a part of gp120 that was previously hidden and, therefore, cannot elicit antibodies.

- HIV is also adept at evading the innate immunity system. Natural killer cells attempt to attack the virus, but HIV binds a particular cell protein, called cyclophilin, to its capsid, which blocks the antiviral agent restriction factor-1. Another of HIV's proteins blocks the viral inhibitor called CEM-15, which normally disrupts the viral life cycle.
- Lastly, HIV hides from the immune system by cloaking its outer membrane in sugars that are very similar to the natural sugars found on most of its host's cells, making it harder for the immune system to recognize it.

14-5B Fighting HIV and AIDS

Science has been fighting a frustrating battle with HIV and AIDS since it was discovered in the early 1980s. So far, the fight has been carried on two fronts: the search for a vaccine and antiviral drugs. A vaccine has yet to be found for several of the reasons listed at the end of Section 14-5A. However, one strategy for using a vaccine to stimulate the body's immunity to HIV is shown in Figure 14.37. DNA for a unique HIV gene, such as the *gag* gene, is injected into muscle. The *gag* gene leads to the Gag protein, which is taken up by antigen-presenting cells and displayed on their cell surfaces. This then elicits the cellular immune response, stimulating killer and helper T cells. It also stimulates the humoral immune response, spurring production of antibodies. Figure 14.37 also shows a second phase of the treatment, a booster shot of an altered adenovirus that carries the *gag* gene.

Unfortunately, most attempts at making antibodies have proved unsuccessful to date, although there is still much active research in this area, especially when combined with other techniques.

▶ What is antiretroviral therapy (ART)?

While the search for an effective vaccine has continued with little to minor success, pharmaceutical companies have succeeded by designing drugs that inhibit retroviral proteins and capsid formation. Collectively, these drugs are used in *antiretroviral therapy (ART)*, and they can reduce the virus to undetectable levels, preventing disease progression and transmission. There are several classes of antiviral HIV drugs:

- Early entry inhibitors that prevent HIV from entering cells.
- Reverse transcriptase inhibitors that prevent the creation of DNA from HIV RNA.
- Integrase inhibitors that prevent HIV DNA from integrating into host DNA.
- Protease inhibitors that prevent the processing of viral proteins needed for new HIV particles.
- Capsid inhibitors that prevent capsid formation and virion assembly.

▶ Is there hope for a cure for HIV?

Several approaches are currently being explored in the elusive search for an AIDS cure, and most experts agree that a widely available HIV cure will be developed by the year 2035. CRISPR-Cas9 gene editing (Section 13-4C) and immune-based therapies such as broadly neutralizing antibodies (bNAbs) that block multiple HIV strains appear to be the most promising approaches to finding a cure.

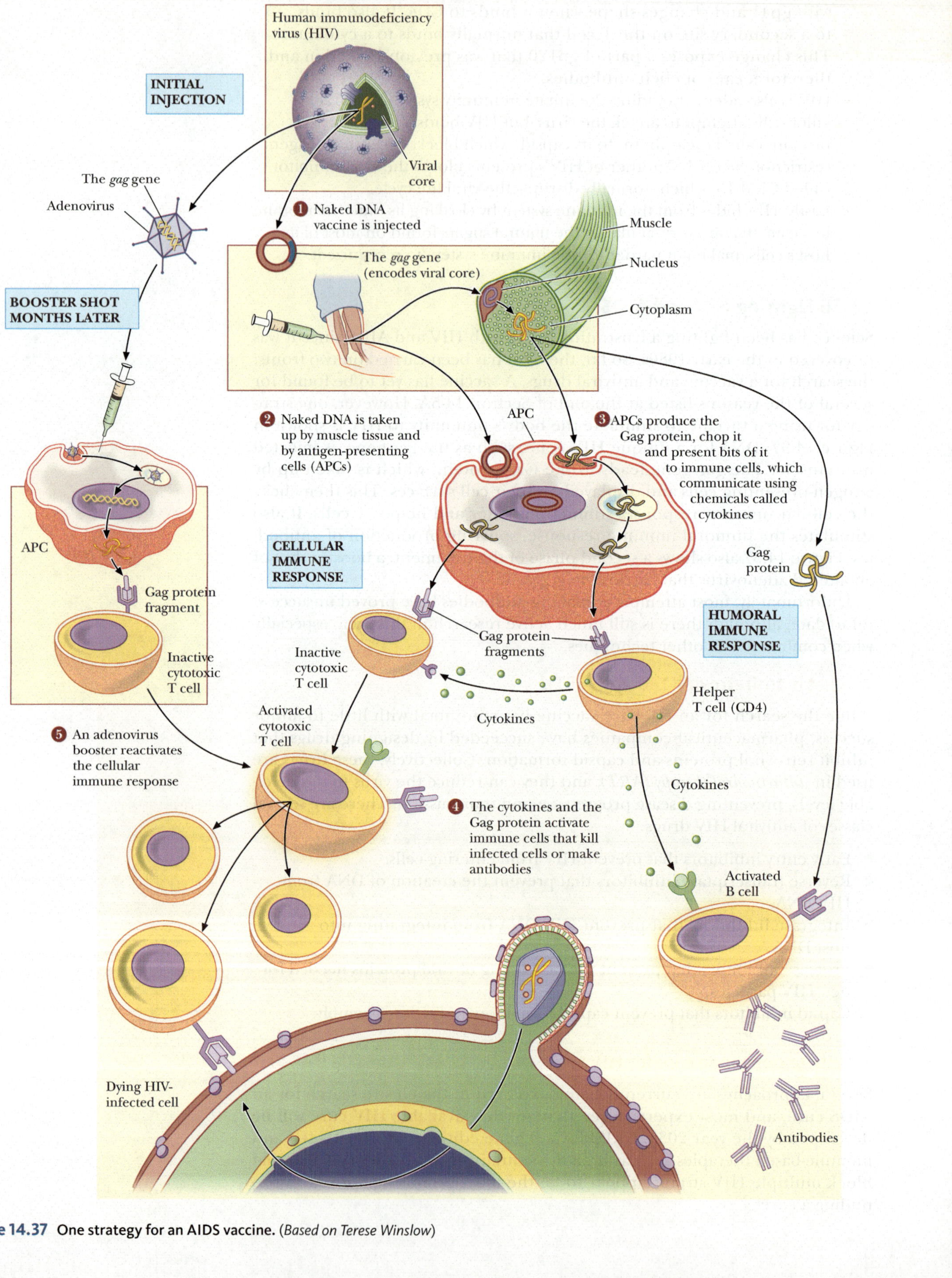

Figure 14.37 One strategy for an AIDS vaccine. (*Based on Terese Winslow*)

Hot Topic

Stem Cells: Science and Politics

Stem cells are the precursors of all other cell types. They are undifferentiated cells that have the ability to form any cell type as well as to replicate into more stem cells. Stem cells are often called progenitor cells because of their ability to differentiate into many cell types. A pluripotent stem cell is one that can give rise to all cell types in an embryo or in an adult. Some cells are called multipotent because they can differentiate into more than one cell type, but not into all cell types. The further from a zygote a cell is in the course of development, the less the potency of the cell type. The use of stem cells is an exciting field of research that really took off in the late 1990s.

History of Stem Cell Research

Stem cell research began in the 1970s with studies on teratocarcinoma cells, which are found in testicular cancers. These cells are bizarre blends of differentiated and undifferentiated cells. They were referred to as embryonal carcinoma (EC) cells. They were found to be pluripotent, which led to the idea of using them for therapy. However, such research was suspended because the cells had come from tumors, which made their use dangerous, and because they were aneuploid, which means they had the wrong number of chromosomes. The possibility of cells becoming cancerous is one of the major hurdles to overcome when we consider using stem cells for tissue therapy.

Early work with embryonic stem cells (ESCs) came from cells that were grown in culture after being taken from embryos. Researchers found that these stem cells could be grown in culture and maintained for long periods. Most differentiated cells, on the other hand, will not grow for extended periods in culture. Stem cells are maintained in culture by the addition of certain factors, such as leukemia inhibitory factor or feeder cells (nonmitotic cells such as fibroblasts). Once released from these controls, ESCs can differentiate into all kinds of cells, as shown in Figure 14.38. However, research with ESCs raised ethical concerns because it required the destruction of embryos. Under the administration of George W. Bush, funding restrictions on ESC research were put in place that were then reversed under the administration of Barack Obama. A major breakthrough in 2006 was the discovery of induced pluripotent stem cells (iPSCs), but it took a while before work with iPSCs reached the level of work that could be achieved with ESCs. Today, iPSCs are widely used in stem cell research, and they are discussed in more detail in this Hot Topic feature.

Stem Cells Offer Hope

Stem cells placed into a particular tissue, such as blood, differentiate and grow into blood cells. Others placed into brain tissue grow into brain cells. This was a very exciting discovery. Previously, it had been believed that there was little hope for patients with spinal cord and other nerve damage because nerve cells do not normally regenerate. In theory, neurons could be produced to treat neurodegenerative diseases, such as Alzheimer's disease or Parkinson's disease. Muscle cells could be produced to treat muscular dystrophies and heart disease. In one study, mouse stem cells were injected into a mouse heart that had undergone a myocardial infarction. The cells spread from an unaffected region into the infarcted zone and began to grow new heart tissue. Human pluripotent stem cells have been used to regenerate nerve tissue in rats with nerve injuries and have been shown to improve motor and cognitive ability in rats that suffered strokes. (See the articles by Sussman, by Aldhous, and by Donovan and Gearhart cited in the bibliography.) Results such as these led some scientists to claim that stem cell technology will be the most important advancement since cloning.

Truly pluripotent stem cells have been harvested primarily from embryonic tissue, and these cells show the greatest ability to differentiate into various tissues and to

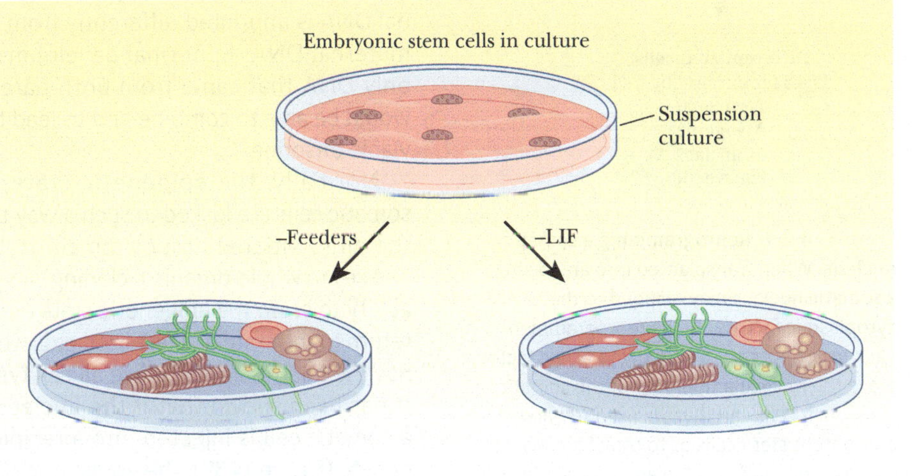

Figure 14.38 Pluripotent embryonic stem cells can be grown in cell culture. They can be maintained in an undifferentiated state by growing them on certain feeder cells, such as fibroblasts, or by using leukemia inhibitory factor (LIF). When removed from the feeder cells or when the LIF is removed, they begin to differentiate into a wide variety of tissue types, which could then be harvested and grown for tissue therapy. *(Taken from Donovan, P. J., and Gearhart, J. (2001). The end of the beginning for pluripotent stem cells.* Nature *414, 92–97.)*

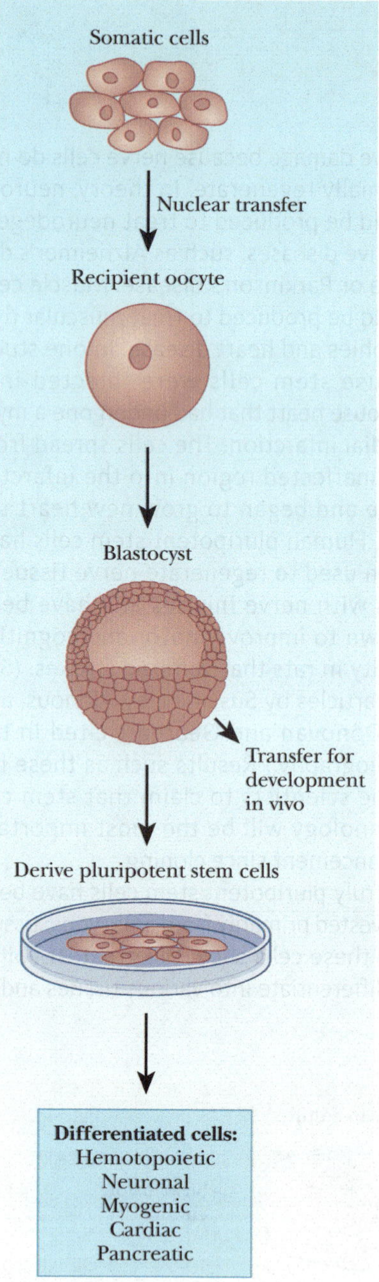

Figure 14.39 Reprogramming a somatic nucleus. When transplanted into an oocyte, a somatic nucleus may respond to the cytoplasmic factors and be reprogrammed back to totipotency. These cytoplasmic factors erase the molecular memory of the somatic cells. Such cells can then be used to harvest pluripotent stem cells or to transfer a blastocyst into a carrier and develop an organism in vivo. *(Taken from Surani, M. A. (2001). Reprogramming of genome function through epigenetic inheritance.* Nature *414, 122–128.)*

reproduce in cell culture. Stem cells have also been taken from adult tissues because organisms always contain some stem cells even at the adult stage. These cells are usually multipotent—that is, they can form several different cell types, but they are not as versatile as ESCs. For this reason, ESCs were the gold standard for tissue therapy for several years. The acquisition and use of stem cells can also be related to a technique called cell reprogramming, which is a necessary component of whole-mammal cloning, such as the cloning that produced the sheep named Dolly.

Most somatic cells in an organism contain the same genes, but the cells develop as different tissues with extremely different patterns of gene expression. A mechanism that alters the expression of genes without changing the actual DNA sequence is called an epigenetic mechanism. An epigenetic state of the DNA in a cell is a heritable trait that allows a "molecular memory" to exist in the cells. In essence, a liver cell remembers where it came from and continues to divide and remain a liver cell. These epigenetic states involve methylation of cytosine–guanine dinucleotides and interactions with proteins of chromatin. Mammalian genes have an additional level of epigenetic information called imprinting, which allows the DNA to retain a molecular memory of its germline origin. The paternal DNA is imprinted differently from the maternal DNA. In normal development, only DNA that came from both parents would be able to combine and to lead to a viable offspring.

Normally, the epigenetic states of somatic cells are locked in such a way that the differentiated tissues remain stable. The key to whole-organism cloning was the ability to erase the epigenetic state and to return to the state of a fertilized egg, which has the potential to produce all cell types. It has been shown that if the nucleus of a somatic cell is injected into a recipient oocyte (Figure 14.39), the epigenetic state of the DNA can be reprogrammed or at least partially reprogrammed. The molecular memory is erased, and the cell begins to behave like a true zygote. This can be used to derive pluripotent stem cells or to transfer a blastocyst into a mother carrier for growth and development. In November 2001, the first cloned human blastocyst was created in this way, with the aim of growing enough cells to harvest pluripotent stem cells for research.

The Search for Less Controversial Stem Cells

In 2006, Japanese researcher Shinya Yamanaka made a major breakthrough in the search for less controversial stem cells. While many other researchers were taking embryonic stem cells and working on controlling how to correctly differentiate them into various target tissues, he took an opposite approach and began looking for ways of creating stem cells from regular somatic cells. He hypothesized that specific proteins would be found in embryonic cells but not in differentiated cells. He thought that if he could introduce the genes for these proteins into differentiated cells, he might be able to convert them back to a pluripotent state.

After four years of research, he had uncovered 24 factors that would transform skin fibroblast cells into pluripotent cells in mice and found that these cells were almost identical to stem cells. He then found that by introducing four specific genes into the fibroblast cells, he could accomplish the same thing. In 2006, he published his landmark article in which he identified these four critical genes as *Oct3/4, Sox2, c-Myc*, and *Klf4*. The cells derived from this process are called induced pluripotent stem cells (iPSCs). Yamanaka and other researchers have derived iPSCs from many tissues, including liver, stomach, and brain. These cells show some of the same abilities as true stem cells and have been turned into skin, muscle, cartilage, and nerve cells. In 2007, researchers in the United States extended the technique to create human iPSCs in work recognized as the first runner-up in *Science* magazine's *Breakthrough of the Year.*

Cutting Out the Middleman

Whether using embryonic stem cells or induced pluripotent stem cells, the original research that headed toward tissue therapy involved several changes to cell type. First, a cell had to be turned into a pluripotent cell. Then the pluripotent cell had to be converted to the desired tissue type. Finally, a new process that involves avoiding this transition and changing one cell type directly into another one was discovered. In late 2008, researchers at Harvard University used cell reprogramming techniques to change one type of pancreatic cell in mice into the beta cells that produce insulin. This technique allows the direct programming of one cell type into another. After sifting through 1000 transcription factors, the researchers found that by turning on only three genes in the exocrine cells of the mouse pancreas, they could change exocrine cells to insulin-producing beta cells. A similar technique, called lineage switching, involves another way to change cell types, in this case by moving the cell back toward a less differentiated cell until a branch point and then re-differentiating it into something else.

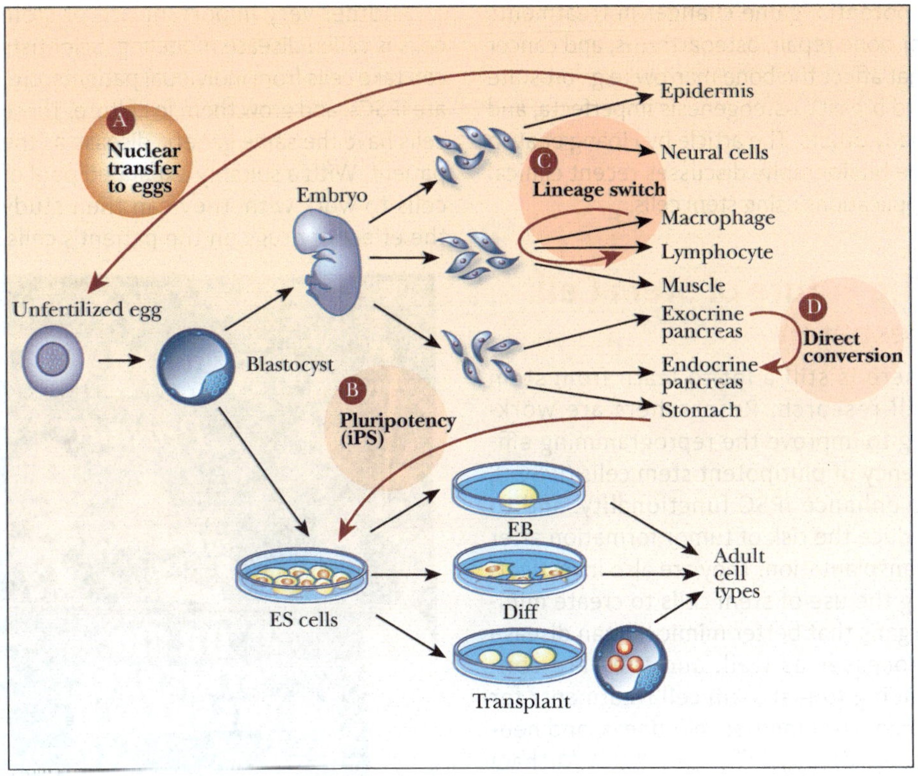

Figure 14.40 **Many paths to reprogramming.** See the text for details. *(From Gurdon, J. B., and Melton, D. A. [2008]. Nuclear reprogramming in cells.* Science *322 (5909), 1811. Reprinted with permission from AAAS.)*

Many Paths to Reprogramming

Researchers interested in studying paths to cell reprogramming have a variety of paths to choose from these days, as shown in Figure 14.40. Path A shows nuclear transfer to an unfertilized egg, such as the process that led to the cloned sheep, Dolly. Path B shows how a differentiated cell from the stomach can be used to regress to an undifferentiated state in the form of an induced pluripotent cell. Path C shows a lineage switch where a macrophage becomes a lymphocyte. Path D shows the direct conversion of an exocrine cell to an endocrine beta cell.

In 2015, Chan et al. mapped skeletal stem cell differentiation in mice that led to the production of bone, cartilage, and stromal tissue. They discovered what they referred to as niche factors, which can be induced in specified combinations to activate skeletal or even nonskeletal tissues to form cartilage or bone and bone marrow stroma. Figure 14.41 shows a schematic of the steps and niche factors involved in the differentiation of mouse skeletal stem cells into cartilage, bone, and stroma tissue. This research represents

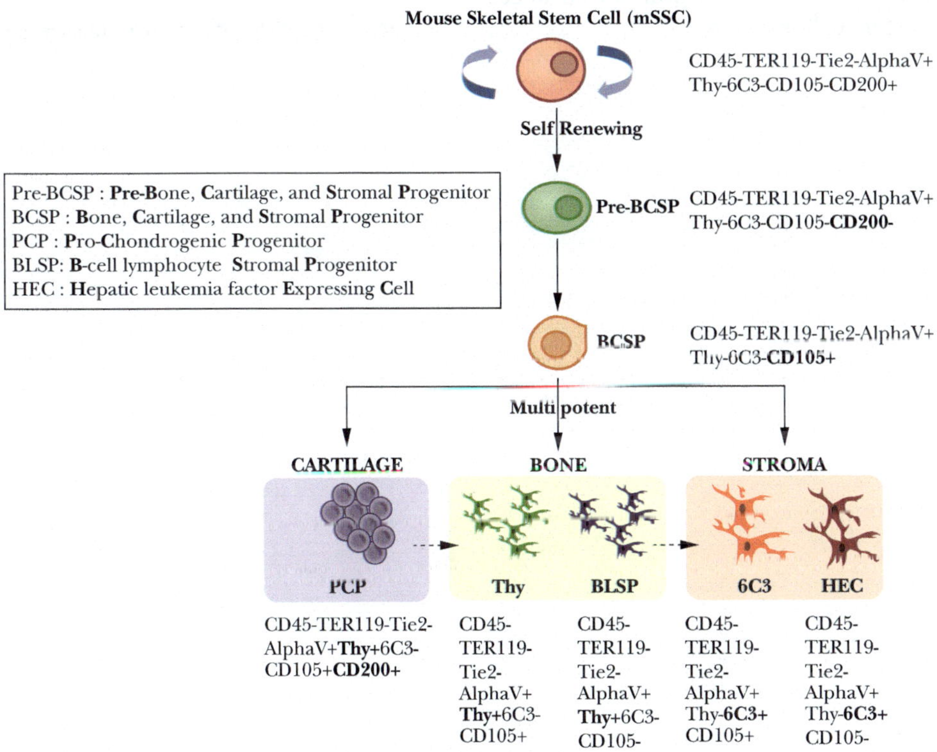

Figure 14.41 **Mouse skeletal stem cell differentiation pathway and niche factors involved in the formation of cartilage, bone, and stroma tissue.** *(Chan, C.K.F., et al. Identification and Specification of the Mouse Skeletal Stem Cell.* Cell, *160, 285–298 [2015].)*

a potential game changer in treatments for bone repair, osteoarthritis, and cancer that affect the bone marrow (e.g., prostate and breast), osteogenesis imperfecta, and many others. The article by Hoang et al. in the bibliography discusses recent clinical applications using stem cells.

The Future of Stem Cell Research

There is still a lot to learn from stem cell research. Researchers are working to improve the reprogramming efficiency of pluripotent stem cells (iPSCs), to enhance iPSC functionality, and to reduce the risk of tumor formation after transplantation. They are also investigating the use of stem cells to create mini-organs that better mimic human disease processes as well, and they are continuing to test stem cell treatments for spinal cord injuries, blindness, and neurodegenerative diseases. Actor Michael J. Fox (Figure 14.42), who has Parkinson's disease, is the most vocal proponent of stem cell research, and his Michael J. Fox Foundation for Parkinson's Research continues to fund and promote advances in stem cell research.

Another very important use of stem cells is called disease modeling. Scientists can take cells from individual patients, create iPSCs, and grow them in culture. These cells have the same genetic disease as the patient. With a suitably expanded pool of cells to work with, they can then study the effect of drugs on the patient's cells, essentially allowing direct experimentation on a virtual "patient in a petri dish." While not as flashy as the potential cure of a severed spinal cord via tissue therapy, in the short term, disease modeling will allow uncontroversial use of a patient's own cells to help treat a disease (see the article by Vogel in the bibliography).

Scott J. Ferrell/Contributor/CQ-Roll Call, Inc./Getty Images

Figure 14.42 **Celebrity activists.** Michael J. Fox has been a vocal proponent for stem cell research.

Summary

Viruses

- Viruses are known to cause many diseases, and they may be specific to a particular species and cell type.
- Viruses enter the cell by binding to specific receptors on the cell.
- Once inside the cell, the virus may replicate, form new viruses, and burst the cell.
- The virus may also hide its DNA by incorporating it into the host's DNA.

Common Viruses

- Viruses have various shapes. Some are rod-shaped, like the tobacco mosaic virus (TMV). Others have a hexagonal shape, like the bacteriophage T2 virus.
- At the center of a virus is its nucleic acid genetic material, which is surrounded by a protein coat called a capsid. The combination of the nucleic acid and the capsid is called the nucleocapsid.
- Many viruses also have a membrane envelope surrounding the nucleocapsid. Some also have protein spikes that help the virus attach to a host cell.

Families of Viruses

- Vertebrate viruses can be split into four families based on whether their genome contains DNA or RNA and whether or not they contain a protein coat.

Virus Life Cycles

- Simian virus 40 (SV40) can infect either human or rodent cells, but its life cycle differs depending on which of those two kinds of cells it is infecting.
- Human cells infected by the virus produce large-T protein, which leads to the production of new viral DNA and coat proteins. The new virions aggregate until they are excreted from the cell, often accompanied by the rupture and destruction of the host cell.
- Rodent cells infected by simian virus 40 will incorporate the viral genome into their own cellular DNA and propagate the virus by cellular division.
- In order for the virus to infect either a human or a rodent cell, a common method of attachment involves the binding of one of the spike proteins on the envelope of the virus to a specific receptor on the host cell.

Retroviruses

- Retroviruses have a genome based on RNA.
- When they infect a cell, their RNA is turned into DNA.
- The DNA is then incorporated into the host's DNA genome as part of the replication cycle for the virus.
- HIV, the causative agent of the disease AIDS, is a retrovirus.
- Retroviruses have also been linked to certain cancers, but they are also used in several forms of gene therapy.
- A retrovirus contains single-stranded RNA that gets incorporated into the DNA of a host cell by the viral reverse transcriptase and viral integrase proteins, resulting in host cell DNA that contains the viral RNA code.

Immunity and Vaccines

- The immune system consists of the cells and molecules that recognize and destroy pathogens that have invaded the human body.
- Vaccines consist of dead or weakened virus or bacteria that, once introduced into the human body, allow the immune system to produce its own biodefense against infection.

The Innate Immune System

- One type of immunity, called innate immunity, consists of physical barriers, such as skin, and cellular warriors, such as dendritic cells.
- This system is always present and waiting to attack invading organisms or even cancerous cells.
- Another type of immunity, called adaptive immunity, is based on two types of T cells (killer T cells and helper T cells) and on B cells. These cells are generated randomly with receptors that can be specific for an unimaginable number of antigens.
- When these cells encounter their specific antigens, they are stimulated to multiply, exponentially increasing the number of cells that can fight the invading organism.
- Adaptive immune cells also leave behind memory cells so that if the same pathogen is seen again, the body is faster to eliminate it.
- Immune cells must also be able to recognize self from nonself. T cells and B cells are conditioned in their early stages of development not to recognize proteins from that individual. In some cases, this system breaks down, and a person may be attacked by their own immune system, which may lead to an autoimmune disease.

The Adaptive Immune System

- Adaptive immunity depends on T cells and B cells.
- T cells produced in the thymus gland are responsible for cellular aspects of adaptive immunity. As they differentiate, they form killer T cells or helper T cells. Killer T cells bind to antigen-presenting cells (APCs), such as macrophages. This causes the T cell to proliferate. The T cell also secretes chemicals that destroy the APC. Helper T cells also bind to APCs, but instead of destroying these cells, they stimulate B cells.
- B cells produced in the bone marrow are responsible for producing antibodies that bind foreign antigens and mark them for destruction.
- Another characteristic of the adaptive immune system is memory. After infection, a small percentage of memory cells survive, and they are able to reproduce rapidly if the antigen appears again.

Antibodies

- Antibodies are Y-shaped proteins composed of two light chains and two heavy chains held together by disulfide bonds.

- Each light chain and heavy chain has a constant domain and a variable domain. When the chains come together to form the antibody, there is a constant region and a variable region. The variable region is the part that binds specifically to an antigen.
- Antibodies bind to antigens, creating a reaction that forms a precipitate of the antibody–antigen complexes. This precipitate is then attacked by phagocytes and the complement system.

Distinguishing Self from Nonself

- The body maintains a delicate balance between maintaining an arsenal of antibodies and controlling cellular destruction to avoid autoimmune diseases.

Cancer Cell Characteristics

- In cancer cells, mutations have occurred that cause the cells to grow and divide uncontrollably.
- Cancer cells, which amplify external growth signals or produce their own, are insensitive to antigrowth signals emitted by other cells.
- Cancer cells can avoid apoptosis when other cells would be triggered to self-destruct, and they can replicate indefinitely; normal cells can replicate 50 to 70 times before dying.
- Cancer cells also emit chemical signals that promote blood vessel growth in their area to deliver oxygen to the cancerous tissues. Moreover, they can defy signals that normally hold cells in place, thus traveling and thriving in other locations in the body.

Oncogenes and Tumor-Suppressor Genes

- Cancer is a genetic disease, and the ultimate cause is mutations in DNA, whether this is caused by an inherited condition or tissue damage from chemicals or radiation.
- Most cancers have been linked to specific genes called oncogenes or tumor-suppressor genes. When these genes are mutated, the cell loses the ability to control its replication.
- Although DNA damage starts the cancer, evidence suggests that inflammation associated with the innate immune system triggers many cancers to progress.

Viruses and Cancer

- There are examples of retroviruses that cause cancer in humans, such as some forms of leukemia and cervical cancer. However, scientists are also using viruses to treat cancer.
- Oncolytic virus therapy is a method used to treat certain types of cancer. It either uses a virus to attack and kill a cancer cell directly or attacks cancer indirectly by activating the immune system.

Fighting Cancer

- There are many classical ways to fight cancer, such as radiation therapy and chemotherapy. Both of these are very hard on healthy cells and, therefore, on the patient.
- Other techniques using viruses are now being tried to target cancer cells more directly, and some of these are showing tremendous promise.
- Because more than half of the known cancers involve a mutated and nonfunctional p53 gene, restoring the activity of p53 is a focus of cancer research.

HIV: A Model Retrovirus

- The HIV is a virus that significantly affects millions of people worldwide.
- It attacks the host's helper T cells, compromising the person's ability to mount a suitable immune response to disease.
- HIV has many tricks in its bag that confounded our ability to fight it for many years. It mutates so rapidly that a given person may have several versions, making it difficult to attack. It also hides in tissues, lies dormant for years, and finds ways of thwarting the best antibodies the body can muster.

Fighting HIV and AIDS

- Antiretroviral drugs are the most effective method to treat HIV. These drugs target specific HIV enzymes, like the HIV reverse transcriptase. Although not a cure, antiretroviral drugs can effectively delay viral progression.
- Scientists continue to search for a vaccine for HIV.

Review Exercises

14-1 Viruses

1. **Recall** What is the genetic material of a virus?
2. **Recall** Define the following:
   (a) virion
   (b) capsid
   (c) nucleocapsid
   (d) protein spike
3. **Recall** What are three factors that determine how a virus is classified into a family?
4. **Recall** How does a virus infect a cell?
5. **Recall** What is the difference between the lytic pathway and the lysogenic pathway?
6. **Reflect and Apply** If you were going to design a drug to fight a virus, what would be likely targets for the drug design?
7. **Reflect and Apply** Some viruses can undergo lysis or lysogeny even in the same host. What might be a reason for this? Under what conditions might the virus favor one strategy over the other?
8. **Reflect and Apply** What might be the characteristics of cells of a human who is immune to HIV infection?

14-2 Retroviruses

9. **Recall** What is unique about the life cycle of a retrovirus?
10. **Recall** What enzyme is responsible for the production of viral DNA from a retrovirus?

11. **Recall** What are three reasons that retroviruses are studied heavily?
12. **Recall** What is meant by gene therapy?
13. **Recall** What are two types of viruses used for gene therapy, and how are they manipulated to make them useful?
14. **Recall** What are two potential hazards of gene therapy?

14-3 The Immune System

15. **Recall** What is innate immunity? What is adaptive immunity?
16. **Recall** What are the components of innate immunity?
17. **Recall** What are the components of adaptive immunity?
18. **Recall** Where do T cells mature, where do B cells mature, and from where and from what type of cell do they both originate?
19. **Recall** What is the purpose of major histocompatibility complexes (MHCs)?
20. **Recall** What is clonal selection?
21. **Reflect and Apply** Describe the relationship between the innate immune system and the adaptive immune system.
22. **Reflect and Apply** One of the first human proteins cloned was interferon. Why is it important to be able to produce interferon in a lab?
23. **Reflect and Apply** Describe how the cells of the adaptive immune system develop so that they do not recognize self-antigens but do recognize foreign antigens.
24. **Recall** What part of the immune system has been linked to the progression of cancer?
25. **Recall** Describe the polypeptide chains that make up an antibody and describe how they are arranged.
26. **Biochemical Connections** What type of viral RNAs from herpesviruses confound the immune system?
27. **Biochemical Connections** Explain the mode of action of the herpes viral RNAs and how they confound the immune system.

14-4 Cancer

28. **Recall** What are the characteristics of cancer cells?
29. **Recall** What is a tumor suppressor? What is an oncogene?
30. **Recall** Why are the proteins called p53 and Ras studied heavily?
31. **Recall** How are viruses related to cancer?
32. **Reflect and Apply** What is virotherapy?
33. **Reflect and Apply** Describe the difference between a tumor suppressor and an oncogene with respect to the actual causes of cancer.
34. **Recall** Briefly explain how activation of a receptor tyrosine kinase can lead to activation of the MAP kinase pathway.
35. **Reflect and Apply** Describe the relationships between *Ras*, *jun*, and *fos*.
36. **Reflect and Apply** Describe the nature of p53 reactivation as a cancer-fighting strategy.
37. **Reflect and Apply** Describe different techniques that might restore p53 to a cell lacking it.
38. **Biochemical Connections** Describe the positive and negative effects of the innate immune system on cancer cells.
39. **Biochemical Connections** What were the results of sequencing the human melanoma and lung cancer genomes?
40. **Biochemical Connections** What causes the majority of the mutations found in human melanomas and lung cancer?

14-5 AIDS

41. **Recall** Which cells does HIV attack?
42. **Recall** How does HIV gain entry into the cells it attacks?
43. **Recall** How does HIV confound the human immune system?
44. **Recall** What two types of therapy are used to fight AIDS?
45. **Reflect and Apply** Does the HIV virus kill a patient directly?
46. **Reflect and Apply** How is the gp120 protein important to HIV and its ability to confound the immune system?
47. **Recall** What is ART?

Further Reading

Bakker, T. C. M., and M. Zbinden. Counting on Immunity. *Nature* **414**, 262–263 (2001). An article about how, in some species, individuals select mates that have major histocompatibility proteins as dissimilar to their own as possible.

Banchereau, J. The Long Arm of the Immune System. *Sci. Amer.* **187** (5), 52–59 (2002). An in-depth article about dendritic cells.

Batzing, B. *Microbiology: An Introduction.* Pacific Grove, CA: Brooks/Cole, 2002. Basic textbook in microbiology, including chapters on viruses.

Cartier, N., et al. Hematopoietic Stem Cell Gene Therapy with a Lentiviral Vector in X-Linked Adrenoleukodystrophy. *Science* **326**, 818–822 (2009). An article that demonstrated that stem cells genetically modified with a lentiviral vector expressing ABCD1 could safely halt cerebral demyelination in patients with X linked adrenoleukodystrophy who lacked matched donors.

Check, E. Back to Plan A. *Nature* **423**, 912–914 (2003). A discussion of the different strategies for using antibodies in the fight against AIDS.

Check, E. Trial Suggests Vaccines Could Aid HIV Therapy. *Nature* **422**, 650 (2003). An article about the efficacy of using antibodies to fight HIV.

Cohen, J. A Race Against Time to Vaccinate Against Novel H1N1 Virus. *Science* **325**, 1328–1329 (2009). An article that reports on global efforts to develop, test, and distribute a vaccine for the novel H1N1 influenza virus.

Cohen, J. Confronting the Limits of Success. *Science* **296**, 2320–2324 (2002). An article about the problems associated with finding vaccines and other treatments for AIDS.

Cohen, J. Escape Artist par Excellence. *Science* **299**, 1505–1508 (2003). An article about how HIV confounds the immune system.

Cohen, J. Straight from the Pig's Mouth: Swine Research with Swine Influenzas. *Science* **325**, 140–141 (2009). An article that discusses the critical role of swine in the emergence and transmission of novel influenza strains, emphasizing the importance of swine based studies in understanding influenza dynamics and informing public health responses.

Couzin-Frankel, J. The Promise of a Cure: 20 Years and Counting. *Science* **324**, 1504–1507 (2009). An article that reflects on the progress and challenges in cystic fibrosis research over two decades, emphasizing the significance of identifying the CFTR gene and the ongoing efforts to develop effective therapies.

Couzin-Frankel, J. Replacing an Immune System Gone Haywire. *Science* **327**, 772–774 (2010). An article that explores the potential of hematopoietic stem cell transplantation (HSCT) as a treatment for autoimmune diseases by resetting the immune system.

Cullen, B. R. Outwitted by Viral RNAs. *Science* **317**, 329–330 (2007). An article about how small RNA molecules produced by herpesviruses can confound the immune system of the host.

Enserink, M. High Hopes for Guinean Vaccine Trial. *Science* **347**, 219–220 (2015). Description of the development and delivery of an Ebola vaccine in Guinea.

Ferbeyere, G., and S. W. Lowe. The Price of Tumour Suppression. *Nature* **415**, 26–27 (2002). An article about the trade-offs between aging and tumor suppression.

Gibbs, W. W., and C. Soares. Preparing for a Pandemic. *Sci. Amer.* **293** (5), 45–52 (2005). An article that discusses the global efforts and strategies being implemented to prepare for a potential influenza pandemic.

Greene, W. AIDS and the Immune System. *Sci. Amer.* **269** (3), 98–105 (1993). A description of the HIV virus and its life cycle in T cells.

Heath, J. R., M. E. Davis, and L. Hood. Nanomedicine Targets Cancer. *Sci. Amer.* **300** (2), 44–51 (2009). An article that discusses how nanomedicine can revolutionize cancer treatment by enabling molecular-level diagnosis and therapy.

Janssen, E. M., E. E. Lemmens, T. Wolfe, U. Christen, M. G. von Herrath, and S. P. Schoenberger. CD4+ T Cells Are Required for Secondary Expansion and Memory in CD8+ T Lymphocytes. *Nature* **421**, 852–855 (2003). An in-depth article about the research that led to the understanding of the CD4/CD8 cell relationship in memory.

Jardetzky, T. Conformational Camouflage. *Nature* **420**, 623–624 (2002). An article about how HIV can hide from antibodies.

Kaech, S. M., and R. Ahmed. CD8 T Cells Remember with a Little Help. *Science* **300**, 263–265 (2003). An article about how memory develops in immune cells.

Kaiser, J. Seeking the Cause of Induced Leukemias in X-SCID Trial. *Science* **299**, 495 (2003). An article about the search for information about why two patients receiving gene therapy developed leukemia.

Landovitz, R. J., H. Scott, and S. G. Deeks. Prevention, Treatment, and Cure, of HIV Infection *Nat. Rev. Microbiol.* **21**, 657–670 (2023). This review discusses progress in HIV prevention, vaccine development, treatment strategies, and efforts toward a cure.

Leslie, M. Flu Antibodies Stir New Hope for Treatment, Vaccine. *Science* **323**, 1160 (2009). An article that discusses the discovery of antibodies that can neutralize multiple strains of the influenza virus.

Marx, J. Recruiting the Cell's Own Guardian for Cancer Therapy. *Science* **315**, 1211–1213 (2007). An article about promising research into reactivation of p53 in cancer patients.

McCune, J. M. The Dynamics of CD4+ T-Cell Depletion in HIV Disease. *Nature* **410**, 974–979 (2001). An in-depth article about T cells and how they are affected by HIV.

McMichael, A. J., and S. L. Rowland-Jones. Cellular Immune Responses to HIV. *Nature* **410**, 980–987 (2001). An in-depth review of the immune response to HIV infection.

Nettelbeck, D. M., and D. T. Curiel. Tumor-Busting Viruses. *Sci. Amer.* **289** (4), 68–75 (2003). An article describing a use for viruses as specific weapons against cancer.

Nossal, G. J. V. A Purgative Mastery. *Nature* **412**, 685–686 (2001). An article about the diversity of the immune system.

Oliveira, G., and C. J. Wu. Dynamics and Specificities of T cells in Cancer Immunotherapy. *Nat. Rev. Cancer* **23**, 295–316 (2023). This review focuses on the role of T cells in cancer immunotherapy.

Parham, P. The Unsung Heroes. *Nature* **423**, 20 (2003). An article about the innate immune system.

Parren, P. W. H. I., and D. R. Burton. Two-in-One Designer Antibodies. *Science* **323**, 1567–1568 (2009). An article that discusses the development of bispecific antibodies that can simultaneously bind two different antigens, enhancing therapeutic efficacy by targeting multiple disease pathways with a single molecule.

Reusch, T. B. H., M. A. Häberli, P. B. Aeschlimann, and M. Milinski. Female Sticklebacks Count Alleles in a Strategy of Sexual Selection Explaining MHC Polymorphism. *Nature* **414**, 300–302 (2001). An article about how female sticklebacks select mates based on diversity of MHC proteins.

Serbina, N. V., and E. G. Pamer. Giving Credit Where Credit Is Due. *Science* **301**, 1856–1857 (2003). An article about the importance of dendritic cells.

Sprent, J., and D. F. Tough. T Cell Death and Memory. *Science* **293**, 245–247 (2001). An article about how T cells are selected.

Stix, G. A Malignant Flame. *Sci. Amer.* **297** (1), 60–67 (2007). A comprehensive article about inflammation and cancer.

Straus, E. Cancer-Stalling System Accelerates Aging. *Science* **295**, 28–29 (2002). An article about the apparent link between aging and cancer protection.

Werlen, G., B. Hausmann, D. Naeher, and E. Palmer. Signaling Life and Death in the Thymus: Timing Is Everything. *Science* **299**, 1859–1863 (2003). An article about how immune cells must be selected that recognize foreign molecules but not self.

Hot Topic

Aldhous, P. Can They Rebuild Us? *Nature* **410**, 622–625 (2001). An article about how stem cells might lead to therapies to rebuild tissues.

Chan, C. K. F., et al. Identification and Specification of the Mouse Skeletal Stem Cell. *Cell*, **160**, 285–298 (2015). An article about how skeletal stem cells have been mapped to identify triggers that determine whether they will form bone, cartilage, and stromal tissue.

Gurdon, J. B., and D. A. Melton. Nuclear Reprogramming in Cells. *Science* **322**, 1811–1815 (2008). An article that discusses how transferring a differentiated cell nucleus into an enucleated egg can reprogram the cell to an embryonic state, demonstrating the potential for cellular identity reversal.

Hoang, D. M., P. T. Pham, T. Q. Bach, A. T. L. Ngo, Q. T. Nguyen, T. T. K. Phan, G. H. Nguyen, P. T. T. Le, V. T. Hoang, N. R. Forsyth, M. Heke, and L. T. Nguyen, Stem Cell-Based Therapy for Human Diseases. *Signal Transduct. Target. Ther.* **7**, Article Number: 272 (2022). This review discusses clinical applications using stem cells derived from various sources, highlighting their therapeutic potential.

Hornyak, T. Turning Back the Cellular Clock. *Sci. Amer.*, 112–115 (December 2008). A review of the initial research that led to iPS cells.

Normile, D. Recipe for Induced Pluripotent Stem Cells Just Got Clearer. *Science* **325**, 803 (2009). An article that discusses how recent research has clarified the role of the p53 gene in the reprogramming process, revealing that inhibiting p53 activity at specific stages can enhance the efficiency of generating induced pluripotent stem (iPS) cells.

Stadtfeld, M., M. Nagaya, J. Utikal, G. Weir, and K. Hochedlinger, Induced Pluripotent Stem Cells Generated without Viral Integration. *Science* **322**, 945–949 (2008). The original work showing that iPS cells can be produced using adenovirus.

Sussman, M. Cardiovascular Biology: Hearts and Bones. *Nature* **410**, 640–641 (2001). An article about how stem cells can be used to regenerate tissue.

Vogel, G. Diseases in a Dish Take Off. *Science* **330**, 1172–1173 (2010). An article that discusses how reprogramming patient cells into induced pluripotent stem cells (iPSCs) has enabled researchers to create disease-specific cell models, offering new insights into disease mechanisms and accelerating drug discovery.

Similar to how the controlled flow of water through a hydroelectric dam generates energy in the form of electricity, the controlled transfer of electrons in mitochondria generates energy in the form of ATP.

The Importance of Energy Changes and Electron Transfer in Metabolism 15

Chapter Outline

15-1 Standard States for Gibbs Energy Changes

Learning Objectives

After reading this section, students will be able to:

LO 15.1.1 Describe standard states for Gibbs energy changes.

LO 15.1.2 Distinguish between *endergonic reactions* and *exergonic reactions*.

In Section 1-6, we saw how the lowering of energy, which really means energy dispersal on the molecular level, is spontaneous in the thermodynamic sense. We have looked at some examples of energy changes, but we have not gone into detail about them. An example is the transport of substances into and out of cells across the cell membrane, such as the glucose transport system and the sodium–potassium pump described in Section 8-5. Here in Chapter 15, we examine how energy considerations apply to metabolism, and in several of the chapters that follow, we consider the energetics of reactions in various metabolic pathways.

Careers in Chemistry • Lily Rosewater

Founder of Pitch Science

Education:
Curtin University: PhD, Biomedical Science
University of Western Australia: BSc (Hons), Neuroscience

Lily Rosewater turns science into stories through her business, Pitch Science. She works with a range of science-based brands, including nonprofits, research institutes, and science education organizations, to create online content. She says, "Science communication is such a unique career, where you can be both analytical and creative. I get paid to read, learn, and talk about exciting, new research from a range of life science fields. No two days are the same, and it's challenging in the best way."

Before starting Pitch Science, Rosewater completed her PhD in Biomedical Science, assessing the vulnerability of critical brain cells to DNA damage in concussion and multiple sclerosis. She also tested the effectiveness of a new, injectable treatment for concussion, comprised of a combination of three different ion channel inhibitors. Her PhD taught her how to think critically about research and extract key findings without compromising accuracy and integrity.

Rosewater, a late-diagnosed autistic and ADHD woman, embraces her neurodivergence. When her husband was diagnosed with multiple sclerosis, she turned to her science background to clearly communicate the condition to him and his family. These experiences deepened her appreciation for her education and communication skills. When it comes to the importance of learning biochemistry, she shares, "I often find myself drawing upon biochemistry when doing science communication. Biochemistry is the building blocks of life."

15-1A Standard Gibbs Energy Change

standard states the set of substance conditions used for comparisons of chemical reactions

Standard conditions for any process can be defined and then used as the basis for comparing reactions. The choice of standard conditions is arbitrary. For a process under standard conditions, all substances involved in the reaction are in their **standard states** at 25°C (298 K), in which case they are also said to be at *unit activity*. For pure solids and pure liquids, the standard state is the pure substance itself. For gases, the standard state is usually taken as a pressure of 1.00 atmosphere of that gas. For solutes, the standard state is usually taken as 1.00 molar concentration. Strictly speaking, these definitions for gases and for solutes are approximations, but they are valid for all but the most exacting work.

▶ What do standard states have to do with Gibbs energy changes?

For any general reaction

$$a\,\mathrm{A} + b\,\mathrm{B} \rightarrow c\,\mathrm{C} + d\,\mathrm{D}$$

we can write an equation that relates the Gibbs energy change (ΔG) for the reaction under *any* conditions to the Gibbs energy change under *standard* conditions (ΔG°); the superscript ° refers to standard conditions. This equation is

$$\Delta G = \Delta G^\circ + RT \ln \frac{[\mathrm{C}]^c[\mathrm{D}]^d}{[\mathrm{A}]^a[\mathrm{B}]^b}$$

In this equation, the square brackets indicate molar concentrations, R is the gas constant (8.314 J/mol/K), and T is the absolute temperature. The notation ln refers to natural logarithms (to the base e) rather than logarithms to the base 10, for which the notation is log. This equation holds under all circumstances; the reaction does not have to be at equilibrium. Therefore, the value of ΔG under a given set of conditions depends on the value of ΔG° and on the concentration of reactants and products (given by the second term in the equation). Most biochemical reactions are described in terms of $\Delta G°$, which is the ΔG under standard conditions (1.00 M concentration for solutes). There is only one $\Delta G°$ for a reaction at a given temperature.

$\Delta G = 0$ when the reaction is at equilibrium, so

$$0 = \Delta G° + RT \ln \frac{[C]^c[D]^d}{[A]^a[B]^b}$$

$$\Delta G° = -RT \ln \frac{[C]^c[D]^d}{[A]^a[B]^b}$$

The concentrations are now equilibrium concentrations, and this equation can be rewritten

$$\Delta G° = -RT \ln K_{eq}$$

where K_{eq} is the equilibrium constant for the reaction. We now have a relationship between the equilibrium concentrations of reactants and products and the standard Gibbs energy change. After we have determined the equilibrium concentrations of reactants by any convenient method, we can calculate the equilibrium constant, K_{eq}. We can then calculate the standard Gibbs energy change, $\Delta G°$, from the equilibrium constant.

▶ How do living things use energy?

The Gibbs energy change, ΔG, is perhaps the most suitable way to measure energy changes in living systems because it measures *the energy available to do work at constant temperature and pressure,* which describes the living state. Even cold-blooded organisms are at constant temperature and pressure at any given point in time; any temperature and pressure changes are so slow that they do not affect measurements of ΔG.

15-1B Spontaneity and Reversibility

The concept of *spontaneity* can be confusing, but it merely means that a reaction can occur without added energy. This is similar to water held behind a dam at the top of a hill, which has the potential energy to flow downhill but does not do so unless someone opens the dam. Because water flows only downhill, that is the direction with a negative value of the Gibbs energy change ($-\Delta G$); pumping water uphill is nonspontaneous (requires energy) and has a positive value of the Gibbs energy change ($+\Delta G$). If the Gibbs energy change is only 1 kcal/mol (about 4 kJ/mol) in either direction, then the reaction is considered freely reversible. The reaction can readily go in either direction. If we add reactants or remove products, the reaction shifts to the right; if we remove reactants or add products, the reaction shifts to the left.

This effect is an example of Le Châtelier's principle: When a stress is applied to a system at equilibrium, the system reacts so as to minimize the effect of that stress. Here, the stress is a change in the concentration of one of the components of the reaction. Even though a reaction may appear to be energetically unfavorable to a slight extent, a change in the concentration of reactant or product can increase the amount of a reaction product compared to the amount present before the change in concentration. This is a key aspect of a number of metabolic pathways; many reactions in the middle of the pathway are likely to be freely reversible. This means that the same enzymes can be used whether the pathway is in the process of breaking down a substance or forming the substance.

In reversible metabolic pathways, often just the reactions at the ends are irreversible, and these reactions can be turned on or off to turn the whole pathway on or off, or even to reverse it.

▶ How do many of the endergonic reactions in biochemistry occur?

Several biochemical reactions are *endergonic* ($\Delta G > 0$), meaning that they require an input of energy. Many of these reactions are coupled to *exergonic* reactions ($\Delta G < 0$) to allow them to proceed efficiently. For example, this occurs when the energy-requiring phosphorylation of glucose is coupled to the energy-releasing hydrolysis of one phosphate group of adenosine triphosphate (ATP). There are not really two reactions going on; the enzyme merely transfers the phosphate from the ATP directly to the glucose. We can think of the phosphorylation of glucose and the hydrolysis of ATP as two parts of the same reaction that, when added together, result in an overall exergonic reaction.

15-1C Energy Balance

The law of conservation of energy must always be obeyed. Thus, the energy taken up by an organism is used in some way and never disappears. This is *energy balance,* which can be represented by the following equation:

$$\text{Energy intake} = \text{internal heat produced} + \text{work} + \text{storage}$$

Heat and *work* include not only the mechanical work of motion but also anabolic processes and all the activities of an organism. In organisms, *storage* refers to chemical energy, which can be converted to other forms of energy as needed. Stored chemical energy can also be accessed as needed to drive endergonic reactions, as we have just seen.

We explore the implications of these statements in subsequent parts of this chapter.

15-2 A Modified Standard State for Biochemical Applications

Learning Objectives

After reading this section, students will be able to:

LO 15.2.1 Explain the modified standard state for biochemical applications.

▶ Why do we need a modified standard state for biochemical applications?

We have just seen that the calculation of standard Gibbs energy changes includes the stipulation that all solutes be in standard states at 25°C (298 K) and a concentration of 1 *M.* However, if the hydrogen-ion concentration of a solution is 1 *M,* then the pH is zero. (Recall that the logarithm of 1 to any base is zero.) The interior of a living cell is, in many respects, an aqueous solution of the cellular components, and the pH of such a system is normally in the neutral range. Biochemical reactions in the laboratory are usually carried out in buffers that are also at or near neutral pH. For this reason, it is convenient to define, for biochemical practice, a modified standard state—one that differs from the original standard state only by the change in hydrogen-ion concentration from 1 *M* to 1×10^{-7} *M,* implying a pH of 7.

When Gibbs energy changes are calculated on the basis of this modified standard state, they are designated by the symbol $\Delta G^{\circ\prime}$ (spoken "delta G naught prime"). **APPLY YOUR KNOWLEDGE 15.1** gives you practice in doing this calculation.

15.1 Apply Your Knowledge

Using Equilibrium Constants to Determine $\Delta G^{\circ\prime}$

Let us assume that the relative concentrations of reactants have been determined for a reaction carried out at pH 7 and 25°C (298 K). Such concentrations can be used to calculate an equilibrium constant, K'_{eq}, which, in turn, can be used to determine the standard Gibbs energy change, $\Delta G^{\circ\prime}$, for the reaction. A typical reaction to which this kind of calculation can be applied is the hydrolysis of adenosine triphosphate (ATP) at pH 7, yielding adenosine diphosphate (ADP), monohydrogen phosphate ion (written as P_i), and H^+:

$$\text{ATP} + \text{H}_2\text{O} \rightleftharpoons \text{ADP} + \text{P}_i + \text{H}^+$$

$$K'_{eq} = \frac{[\text{ADP}][\text{P}_i][\text{H}^+]}{[\text{ADP}]} \quad \text{pH 7, 25°C}$$

The concentrations of the solutes are used to approximate their activities, and the activity of the water is one. The value for K'_{eq} for this reaction, determined in the laboratory, is 2.23×10^5. Once we have this information, we can determine the standard Gibbs energy change by substituting into the equation $\Delta G^{\circ\prime} = -RT \ln K'_{eq}$. The key point is to choose the correct quantities to substitute and to keep track of units. Substituting $R = 8.31$ J/mol/K, $T = 298$ K, and $\ln K'_{eq} = \ln 2.23 \times 10^5 = 12.3$,

$$\Delta G^{\circ\prime} = -RT \ln K'_{eq}$$

$$\Delta G^{\circ\prime} = -(8.31 \text{ J/mol/K})(298 \text{ K})(12.3)$$

$$\Delta G^{\circ\prime} = -3.05 \times 10^4 \text{ J/mol} = -30.5 \text{ kJ/mol} = -7.29 \text{ kcal/mol}$$

where 1 kJ = 0.239 kcal.

In addition to illustrating the usefulness of a modified standard state for biochemical work, the negative value of $\Delta G^{\circ\prime}$ indicates that the reaction of hydrolysis of ATP to ADP is a spontaneous process in which energy is released.

15-3 The Nature of Metabolism

Learning Objectives

After reading this section, students will be able to:

LO 15.3.1 Distinguish between *metabolism*, *catabolism*, and *anabolism*.

▶ What is metabolism?

Until now, we have discussed some basic chemical principles and investigated the nature of the molecules of which living cells are composed. We have yet to discuss the bulk of chemical reactions of biomolecules themselves, which constitute **metabolism**, the biochemical basis of all life processes. The molecules of carbohydrates, fats, and proteins taken into an organism are processed in a variety of ways (Figure 15.1). The breakdown of larger molecules to smaller ones is called **catabolism**. Small molecules are used as the starting points of a variety of reactions to produce larger and more complex molecules, including

metabolism the sum total of all biochemical reactions that take place in an organism

catabolism the breakdown of nutrients to provide energy

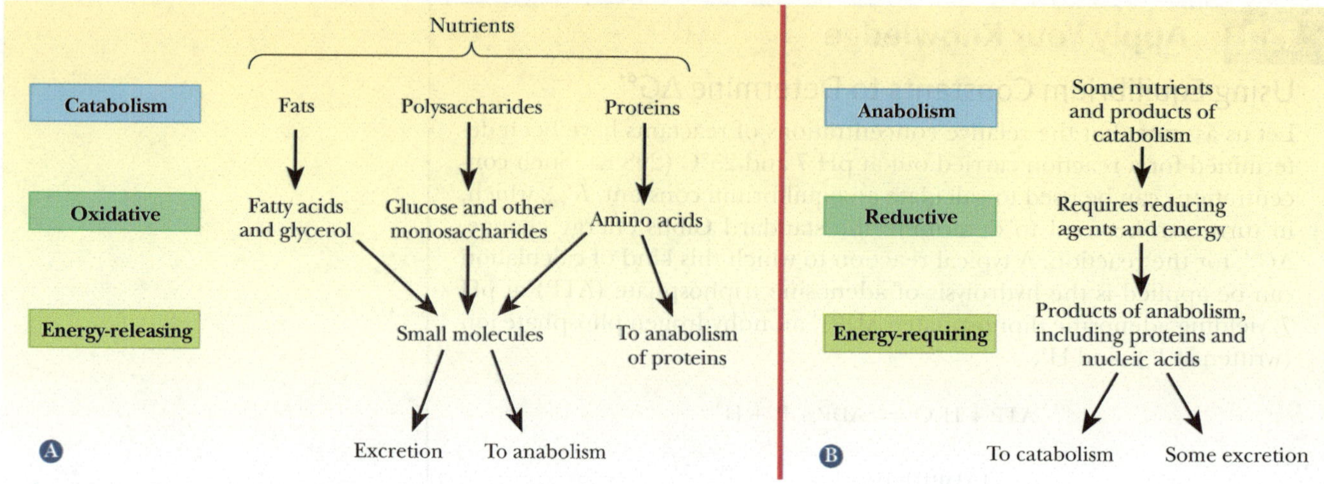

Figure 15.1 A comparison of (A) catabolism and (B) anabolism.

anabolism the synthesis of biomolecules from simpler compounds

proteins and nucleic acids; these processes are called **anabolism**. Catabolism and anabolism involve separate pathways that are not simply the reverse of each other.

Catabolism is an oxidative process that releases energy; anabolism is a reductive process that requires energy. We explore several of the implications of this statement in Chapters 17–23. Here in Chapter 15, we discuss oxidation and reduction (electron-transfer reactions) and their relation to the use of energy by living cells. **BIOCHEMICAL CONNECTIONS 15A** deals with another aspect of the unique energetics of living things.

The outline of metabolism in Figure 15.1 is very general. In the specific case of photosynthetic organisms, the energy source is radiant energy from the Sun, the ultimate energy source for Earth. Some bacteria can oxidize inorganic materials, including nitrates, sulfites, and hydrogen sulfide, producing adenosine triphosphate (ATP). In this text, we concentrate on common cases, but never forget that life has many variations and special cases.

15A Biochemical Connections • Thermodynamics

Living Things Are Unique Thermodynamic Systems

Do living organisms obey the laws of thermodynamics? They most definitely do. Most classical treatments of thermodynamics deal with closed systems at equilibrium. A closed system can exchange energy, but not matter, with its surroundings.

A living organism is *not* a closed system. Instead, it is an open system because it can exchange both matter and energy with its surroundings. Because living organisms are open systems, they cannot be at equilibrium as long as they are alive, as shown in Figure 15.2. They can, however, achieve a *steady state,* which is a stable condition. It is the state in which living things can operate at maximum thermodynamic efficiency. This point was established by Ilya Prigogine, winner of the 1977 Nobel Prize in Chemistry for his work on nonequilibrium thermodynamics (Figure 15.3). He showed that for systems not at equilibrium, ordered structures can arise from disordered ones. This treatment of thermodynamics is quite advanced and highly mathematical, but the results are more directly applicable to biological systems than those of classical thermodynamics. This approach applies not only to living organisms but also to the growth of cities and to predictions of auto traffic.

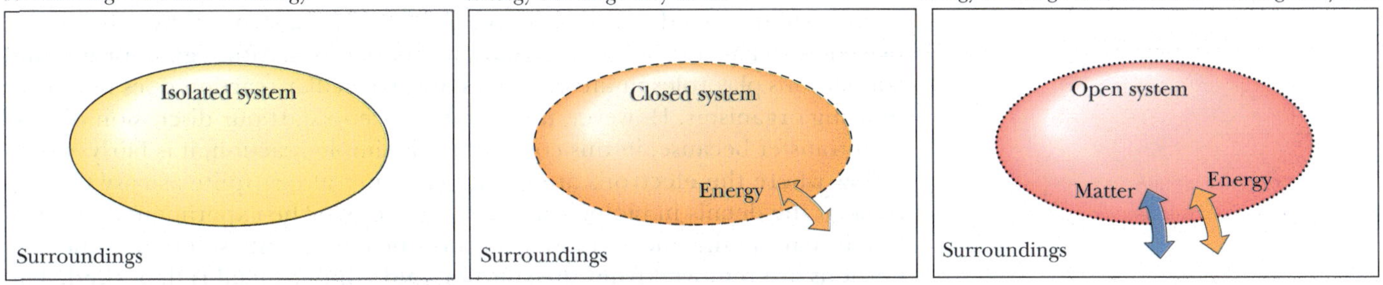

Figure 15.2 The characteristics of isolated, closed, and open systems.

AFP/AFP/Getty Images

Figure 15.3 **Ilya Prigogine (1917–2003).** Ilya Prigogine was born in Moscow in 1917. His family moved to Germany to escape the Russian Revolution and subsequently moved to Belgium. He studied at the Université Libre in Brussels and remained there as a faculty member to conduct research on nonequilibrium thermodynamics. He was also associated with the University of Texas, which marked his Nobel Prize in a unique way: a tower on the Texas campus is illuminated when one of the university's sports teams wins a championship, so the university illuminated the tower when his Nobel Prize was announced.

15-4 The Role of Oxidation and Reduction in Metabolism

Learning Objectives

After reading this section, students will be able to:

LO 15.4.1 Distinguish between *reduction*, *oxidation*, *reducing agent*, and *oxidizing agent*.

LO 15.4.2 Describe examples of oxidation and reduction in metabolism.

Oxidation–reduction reactions, also referred to as *redox* reactions, are those in which electrons are transferred from a donor to an acceptor. **Oxidation** is the loss of electrons, whereas **reduction** is the gain of electrons. The substance that loses electrons (the electron donor)—that is, the one that is oxidized—is called the **reducing agent** or reductant. The substance that gains electrons (the electron acceptor)—the one that is reduced—is called the **oxidizing agent** or oxidant. Both an oxidizing agent and a reducing agent are necessary for the transfer of electrons (an oxidation–reduction reaction) to take place. In other

oxidation–reduction reactions reactions that involve the transfer of electrons from one reactant to another

oxidation the loss of electrons

reduction the gain of electrons

reducing agent a substance that gives up electrons to other substances, thereby causing them to be reduced

oxidizing agent a substance that accepts electrons from other substances, thereby causing them to be oxidized

words, an oxidation cannot occur without a reduction, and a reduction cannot occur without an oxidation.

An oxidation–reduction reaction occurs when a strip of metallic zinc is placed in an aqueous solution containing copper ions. Although both zinc and copper ions play roles in life processes, this particular reaction does not occur in living organisms. However, it is a good place to start our discussion of electron transfer because, in this comparatively simple reaction, it is fairly easy to follow where the electrons are going. (It is not always quite as easy to keep track of the details in biological redox reactions.) The experimental observation is that the zinc metal disappears and zinc ions go into solution, while copper ions are removed from the solution and copper metal is deposited. The equation for this reaction is

$$Zn(s) + Cu^{2+}(aq) \rightarrow Zn^{2+}(aq) + Cu(s)$$

The notation (s) signifies a solid, and (aq) signifies a solute in aqueous solution.

In the reaction between zinc metal and copper ion, the Zn lost two electrons to become the Zn^{2+} ion and was oxidized. A separate equation can be written for this part of the overall reaction, and it is called the **half reaction** of oxidation:

half reaction an equation that explicitly shows the loss or gain of electrons that occur in an oxidation–reduction reaction

$$Zn \rightarrow Zn^{2+} + 2e^-$$

Zn is the reducing agent (it loses electrons; it is an electron donor; it is oxidized).

Likewise, the Cu^{2+} ion gained two electrons to form Cu and was reduced. An equation can also be written for this part of the overall reaction and is called the half reaction of reduction:

$$Cu^{2+} + 2e^- \rightarrow Cu$$

Cu^{2+} is the oxidizing agent (it gains electrons; it is an electron acceptor; it is reduced).

If the two equations for the half reactions are combined, the result is an equation for the overall reaction:

$Zn \rightarrow Zn^{2+} + 2e^-$	Oxidation
$Cu^{2+} + 2e^- \rightarrow Cu$	Reduction
$Zn + Cu^{2+} \rightarrow Zn^{2+} + Cu$	Overall reaction

This reaction is a particularly clear example of electron transfer, so keep these basic principles in mind when we examine the flow of electrons in the more complex redox reactions of aerobic metabolism. In many of the biological redox reactions we discuss, the oxidation state of a carbon atom changes. Figure 15.4 shows the changes that occur as carbon in its most reduced form (an alkane) becomes oxidized to an alkene (a carbon–carbon double bond; not shown in Figure 15.4), then hydrated to an alcohol (hydroxyl) carbon, which can be oxidized to a carbonyl carbon, which can be oxidized to a carboxyl carbon, which can ultimately be oxidized to carbon dioxide. Each of

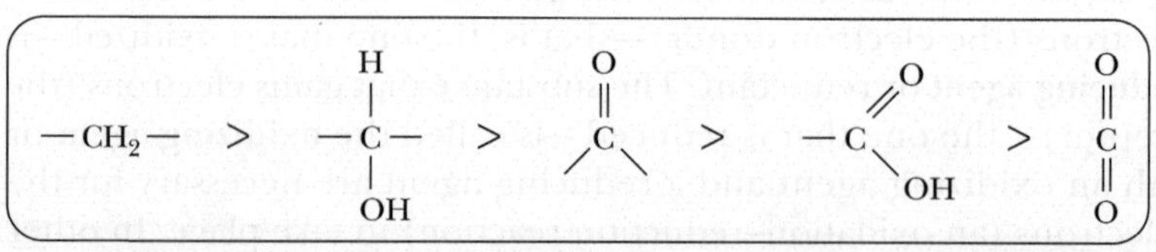

Figure 15.4 Changes to the oxidation state of carbon atoms in biomolecules. Carbon is oxidized as it changes from $—CH_2—$ (single-bond carbon) to —CHOH— (hydroxyl carbon) to C=O (carbonyl carbon) to —COOH (carboxyl carbon) to CO_2 (carbon dioxide).

these oxidations requires the loss of two electrons. Oxidations like these (or their reverse reactions) appear in several metabolic pathway reactions discussed in subsequent chapters, so you should become familiar with these redox patterns.

Oxidation–reduction reactions are discussed at length in textbooks of general and inorganic chemistry, but the oxidation of nutrients by living organisms to provide energy requires its own special treatment. The description of redox reactions in terms of oxidation numbers, which is widely used with inorganic compounds, can also be used to deal with the oxidation of carbon-containing molecules. However, our discussion is more pictorial and easier to follow if we write equations for the half reactions and then concentrate on the functional groups of the reactants and products and on the number of electrons transferred. An example is the oxidation half reaction for the conversion of ethanol to acetaldehyde, as shown in Figure 15.5.

Drawing the Lewis electron-dot structures for the functional groups involved in the reaction helps us keep track of the electrons being transferred. In the oxidation of ethanol, there are 12 electrons in the part of the ethanol molecule involved in the reaction and 10 electrons in the corresponding part of the acetaldehyde molecule; two electrons are transferred to an electron acceptor (an oxidizing agent). This type of "bookkeeping" is useful for dealing with biochemical reactions. Many biological oxidation reactions, like this example, are accompanied by the transfer of a proton (H^+). The oxidation half reaction has been written as a reversible reaction because the occurrence of oxidation or reduction depends on the other reagents present.

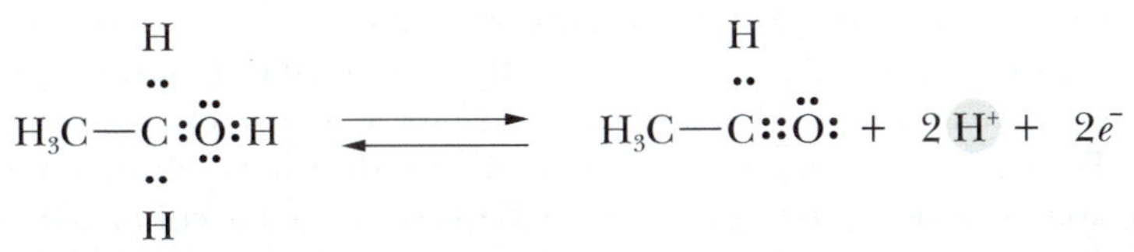

Ethanol (12 electrons in groups involved in reaction)

Acetaldehyde (10 electrons in groups involved in reaction)

Figure 15.5 The half reaction of the oxidation of ethanol to acetaldehyde.

15-5 Coenzymes in Biologically Important Oxidation–Reduction Reactions

Learning Objectives

After reading this section, students will be able to:

LO 15.5.1 Distinguish between the oxidized and reduced forms of nicotinamide adenine dinucleotide.

LO 15.5.2 Distinguish between the oxidized and reduced forms of flavin adenine dinucleotide.

▶ What are the reactions of NAD and FAD, two key oxidation–reduction coenzymes?

In many redox reactions in biochemistry, nicotinamide adenine dinucleotide (NAD) is a **coenzyme** that can serve as either an oxidizing agent in the form of NAD^+ or as a reducing agent in the form of NADH (Figure 15.6). To help you learn the structure, notice the two individual nucleotides of this coenzyme. Notice, too, that the two nucleotides are not connected with one phosphate

coenzyme an organic nonprotein substance that takes part in an enzymatic reaction

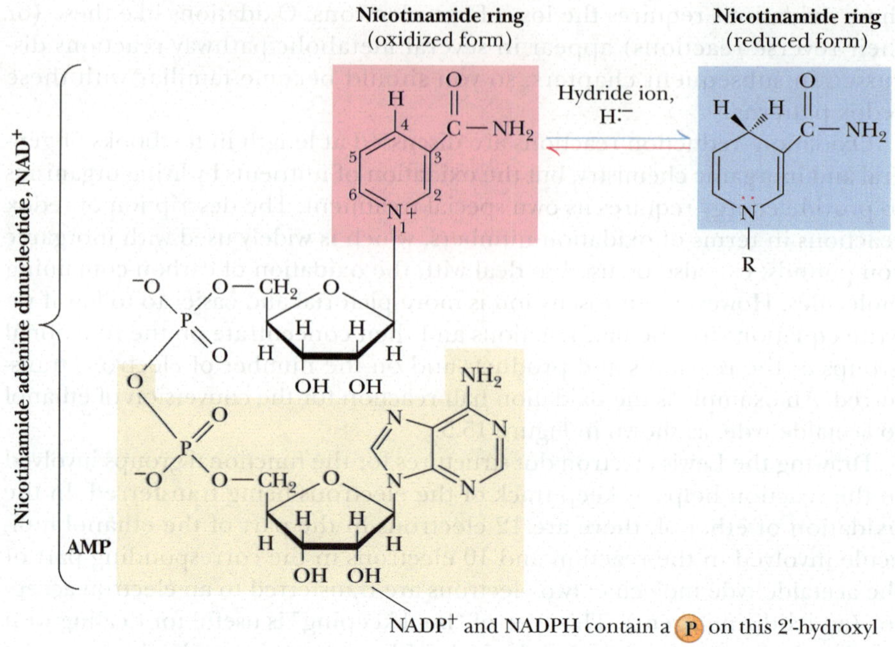

Figure 15.6 The structures and redox states of the nicotinamide adenine dinucleotide coenzyme. The adenine nucleotide (AMP) is shaded in tan. Hydride ion (a proton with two electrons) transfers to NAD^+ to produce NADH. The nitrogen of the nicotinamide ring goes from having four bonds and a positive charge to three bonds and no charge.

between the 3′ and 5′ carbons of the ribose sugars, as seen in Section 9-2B. Instead, they are connected by two phosphates between the 5′ carbons. The nicotinamide portion of the NAD coenzyme is the functional group that is involved in the redox reactions, and you can see how its structure changes as it gains electrons (Figure 15.6, *top right*). Nicotinamide is a derivative of nicotinic acid (also called niacin), one of the B-complex vitamins. A similar coenzyme is called NADPH (for which the oxidized form is $NADP^+$). It differs from NADH by having an additional phosphate group attached to the adenine ribose (Figure 15.6, *bottom*).

Because the redox chemistry occurs at the nicotinamide portion of NAD, the structure of the coenzyme is sometimes simplified by designating the rest of the compound as R (Figure 15.7, *top right*). The two electrons that are lost when NADH is converted to NAD^+ can be considered to come from the bond between carbon and the lost hydrogen, with the nitrogen lone-pair electrons becoming involved in a bond. The loss of a hydrogen and two electrons can

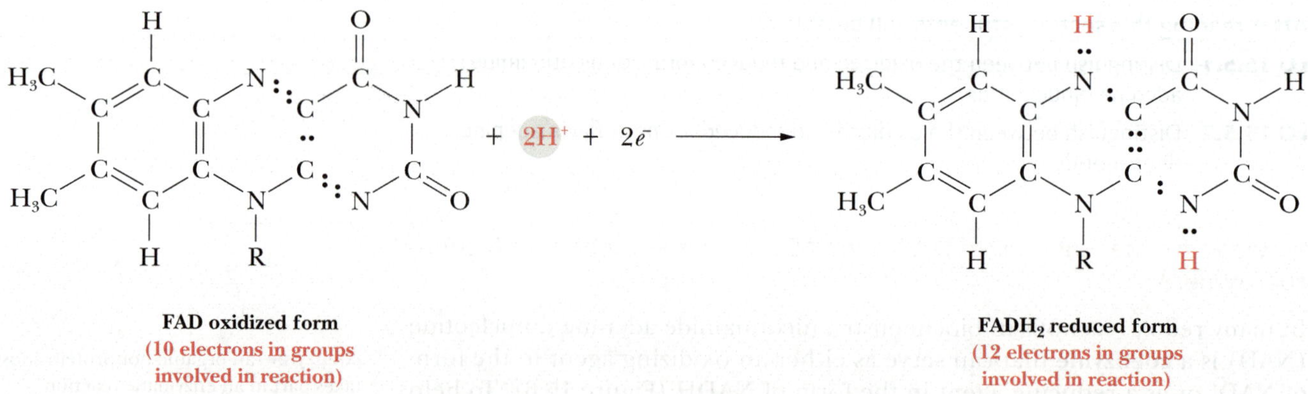

Figure 15.7 The half reaction of reduction of flavin adenine dinucleotide (FAD) to $FADH_2$. R represents the rest of the FAD coenzyme structure shown in Figure 15.8.

be considered to be the loss of a hydride ion ($H:^-$) by NADH and is sometimes written that way.

If ethanol and NADH were mixed in a test tube, no reaction could take place because there would be no electron acceptor. If, however, NADH were mixed with acetaldehyde, which is an oxidized species, a transfer of electrons could take place, producing ethanol and NAD^+. (This reaction would take place very slowly in the absence of an enzyme to catalyze it. Here we have an excellent example of the difference between the thermodynamic and kinetic aspects of reactions. The reaction is spontaneous in the thermodynamic sense but very slow in the kinetic sense.)

$$NADH \rightarrow NAD^+ + H^+ + 2e^- \qquad \text{Half reaction of oxidation}$$

$$CH_3CHO + 2H^+ + 2e^- \rightarrow CH_3CH_2OH \qquad \text{Half reaction of reduction}$$

$$\underset{\text{Acetaldehyde}}{NADH + H^+ + CH_3CHO} \rightarrow \underset{\text{Ethanol}}{NAD^+ + CH_3CH_2OH} \qquad \text{Overall reaction}$$

Such a reaction does take place in some organisms as the last step of alcoholic fermentation. The NADH is oxidized while the acetaldehyde is reduced.

Another coenzyme that can serve as an oxidizing agent is flavin adenine dinucleotide (FAD), which can be reduced to $FADH_2$ (Figure 15.7). The symbol $FADH_2$ explicitly recognizes that protons (hydrogen ions) as well as electrons are accepted by FAD. The full structure of FAD is shown in Figure 15.8. The structures shown in Figure 15.7 point out the electrons that are transferred in the reaction. Several other coenzymes contain the flavin group; they are derived from the vitamin riboflavin (vitamin B_2).

Oxidation of nutrients to provide energy for an organism cannot take place without reduction of some electron acceptor, such as NAD^+ or FAD. The ultimate electron acceptor in aerobic oxidation is oxygen. Alternatively, the

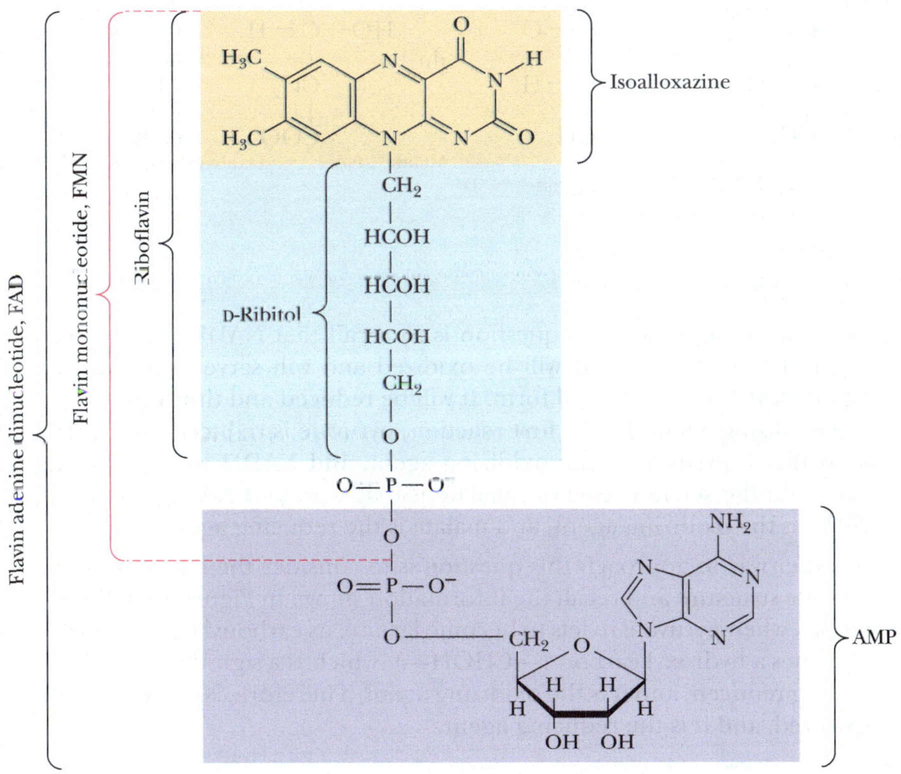

Figure 15.8 The structure of flavin adenine dinucleotide (FAD). The adenine nucleotide (AMP) is shaded in purple, and the flavin mononucleotide (FMN) is shaded in blue and tan. Even in organisms that rely on the nicotinamide coenzymes (NADH and NADPH) for many of their oxidation–reduction cycles, the flavin coenzymes fill essential roles. Flavins are stronger oxidizing agents than NAD^+ and $NADP^+$. They can be reduced by both one-electron and two-electron pathways and can be reoxidized easily by molecular oxygen. Enzymes that use flavins to carry out their reactions—flavoenzymes—are involved in many kinds of oxidation–reduction reactions.

reduction of metabolites plays a significant role in living organisms in *anabolic* processes. Important biomolecules are synthesized in organisms by many reactions in which a metabolite is reduced while the reduced form of a coenzyme, such as NADPH, is oxidized.

APPLY YOUR KNOWLEDGE 15.2 gives you practice in identifying biological oxidizing and reducing agents.

In most of the remaining chapters of this book, we present many reactions in which NAD^+ or FAD is the oxidizing agent. Biomolecules, particularly carbohydrates and lipids, are oxidized to yield energy. Many intermediate reactions occur as a sugar, such as glucose, or a lipid, such as a fatty acid, undergoes complete oxidation to carbon dioxide and water. These pathways are long and complex, with a number of reactions in addition to the actual redox reactions. The additional reactions transform the original reactant, such as glucose, into a molecule that can be oxidized easily. When an oxidation reaction takes place, the oxidizing agent, NAD^+ or FAD, is reduced, thereby gaining electrons and producing NADH and $FADH_2$.

The energy released in these reactions is used to phosphorylate adenosine diphosphate (ADP) to adenosine triphosphate (ATP) in a process in which the phosphorylation reactions are coupled to the reoxidation of NADH and $FADH_2$. Hydrolysis of ATP, in turn, releases energy that can be used to make possible reactions that require energy. We explore the reactions involved in these processes in Chapters 17, 19, and 20, and parts of Chapters 21 and 23.

15.2 Apply Your Knowledge

Oxidation and Reduction

In the following reactions, identify the substance oxidized, the substance reduced, the oxidizing agent, and the reducing agent.

(a)

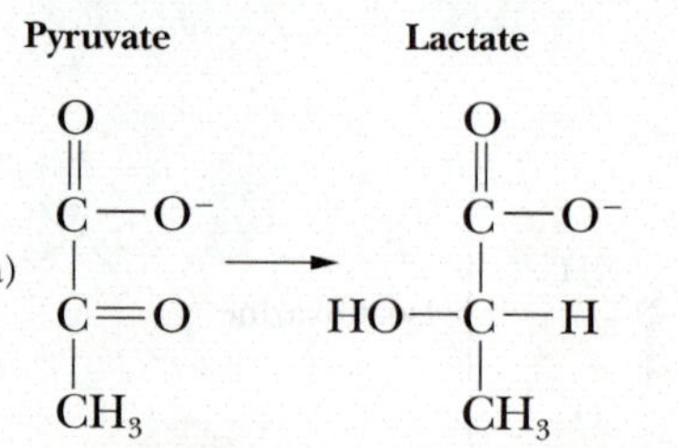

Pyruvate + NADH + H^+ ⟶ Lactate + NAD^+

(b)

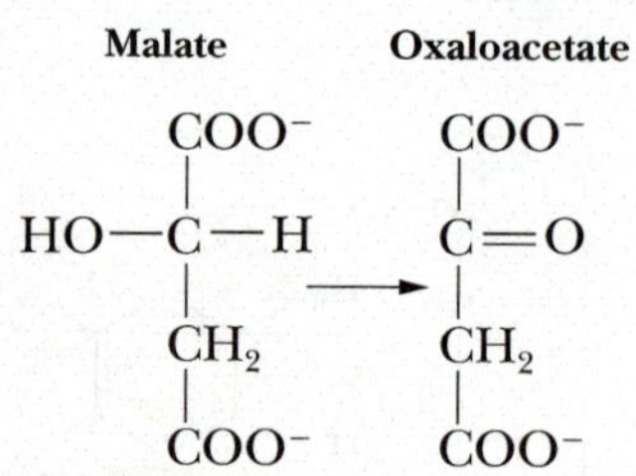

Malate + NAD^+ ⟶ Oxaloacetate + NADH + H^+

Solution

One way to approach this question is to recall that NADH is a reduced form of the coenzyme. It will be oxidized and will serve as a reducing agent. NAD^+ is the oxidized form. It will be reduced and thus will serve as the oxidizing agent. In the first reaction, pyruvate is reduced and NADH is oxidized; pyruvate is the oxidizing agent, and NADH is the reducing agent. In the second reaction, malate is oxidized, and NAD^+ is reduced; NAD^+ is the oxidizing agent, and malate is the reducing agent.

Another way to approach this question is to consider the changes in metabolite structure and recall the information shown in Figure 15.4. For example, when pyruvate reacts to become lactate, its carbonyl carbon (C=O) becomes a hydroxyl carbon (—CHOH—), which is a sign that pyruvate is getting reduced, and it is the oxidizing agent. Therefore, NADH is getting oxidized, and it is the reducing agent.

In anabolic pathways, reducing agents such as NADH and NADPH, which were produced in catabolic pathways, are oxidized, producing NAD^+ and $NADP^+$. The "reducing power" of NADH and NADPH, along with ATP produced in the oxidation reactions of catabolism, is needed for the endergonic reduction reactions of anabolism. We look at a number of these reactions in parts of Chapters 18, 21, and 22.

15-6 Coupling of Energy Production and Energy Use

Learning Objectives

After reading this section, students will be able to:

LO 15.6.1 Distinguish between energy production, energy use, and energy coupling in metabolic reactions.

LO 15.6.2 Describe ATP structure.

How is the energy released by the oxidation of nutrients trapped and used? This energy cannot be used directly; it must be shunted into an easily accessible form of chemical energy.

In Section 1-6, we saw that several phosphorus-containing compounds, such as adenosine triphosphate (ATP), can be hydrolyzed easily, and that the reaction releases energy. The formation of ATP is intimately linked with the release of energy from the oxidation of nutrients. The coupling of energy-producing reactions and energy-requiring reactions is a central feature in the metabolism of all organisms.

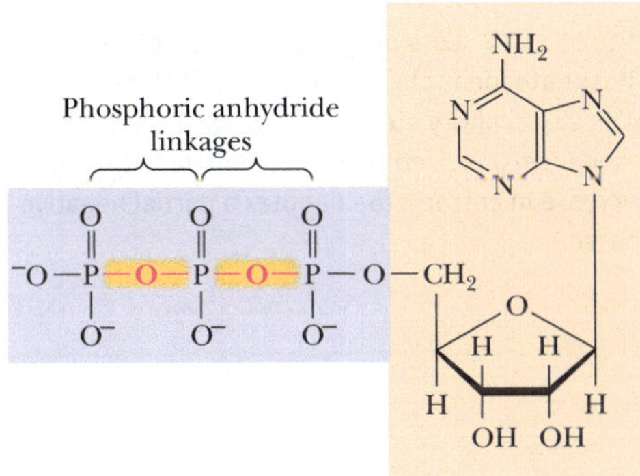

ATP

Figure 15.9 The structure of adenosine triphosphate (ATP). The phosphoric anhydride bonds in ATP are referred to as "high-energy" bonds because they require or release convenient amounts of energy, depending on the direction of the reaction.

▶ How do energy-producing reactions allow energy-requiring reactions to take place?

The phosphorylation of adenosine diphosphate (ADP) to produce ATP requires energy, which can be supplied by the oxidation of nutrients. Conversely, the hydrolysis of ATP to ADP releases energy. Figure 15.9 shows the structure of ATP and its "high-energy" bonds.

The forms of ADP and ATP shown here in Section 15-6 are in their ionization states for the physiological pH of 7.4. The symbol P_i for phosphate ion comes from its name in biochemical jargon, "inorganic phosphate." There are four negative charges on ATP and three on ADP; electrostatic repulsion makes ATP less stable than ADP. Energy must be expended to put an additional negatively charged phosphate group on ADP by forming a covalent bond to the phosphate group being added.

In addition, there is an entropy loss when ADP is phosphorylated to ATP. Inorganic phosphate can adopt multiple resonance structures, and the loss of these potential structures results in a *decrease* in entropy when the phosphate is attached to ADP. There is an *increase* in entropy when the phosphate is released from ATP (Figure 15.10). The $\Delta G°'$ for the reaction refers to the usual biochemical convention of pH 7 as the standard state for the hydrogen ion (Section 15-2). However, there is a marked decrease in electrostatic repulsion on hydrolysis of ATP to ADP (Figure 15.11).

The thermodynamics of the phosphorylation of ADP to ATP help us to see how the reverse reaction, the hydrolysis of ATP to ADP and phosphate ion, releases so much energy when it is needed (30.5 kJ/mol or 7.3 kcal/mol):

$$ATP + H_2O \rightarrow ADP + P_i + H^+$$
$$\Delta G°' = -30.5 \text{ kJ/mol} = -7.3 \text{ kcal/mol}$$

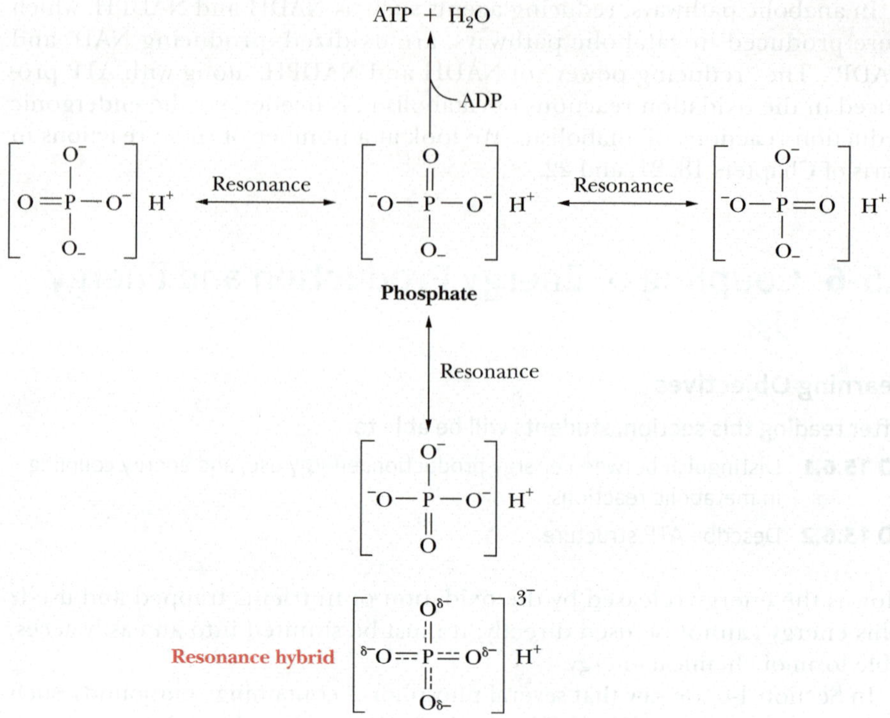

Figure 15.10 Loss of a resonance-stabilized phosphate ion in the production of ATP. When ADP is phosphorylated to ATP, there is a loss of the resonance-stabilized phosphate ion, resulting in a decrease in entropy. (δ– denotes a partial negative charge.)

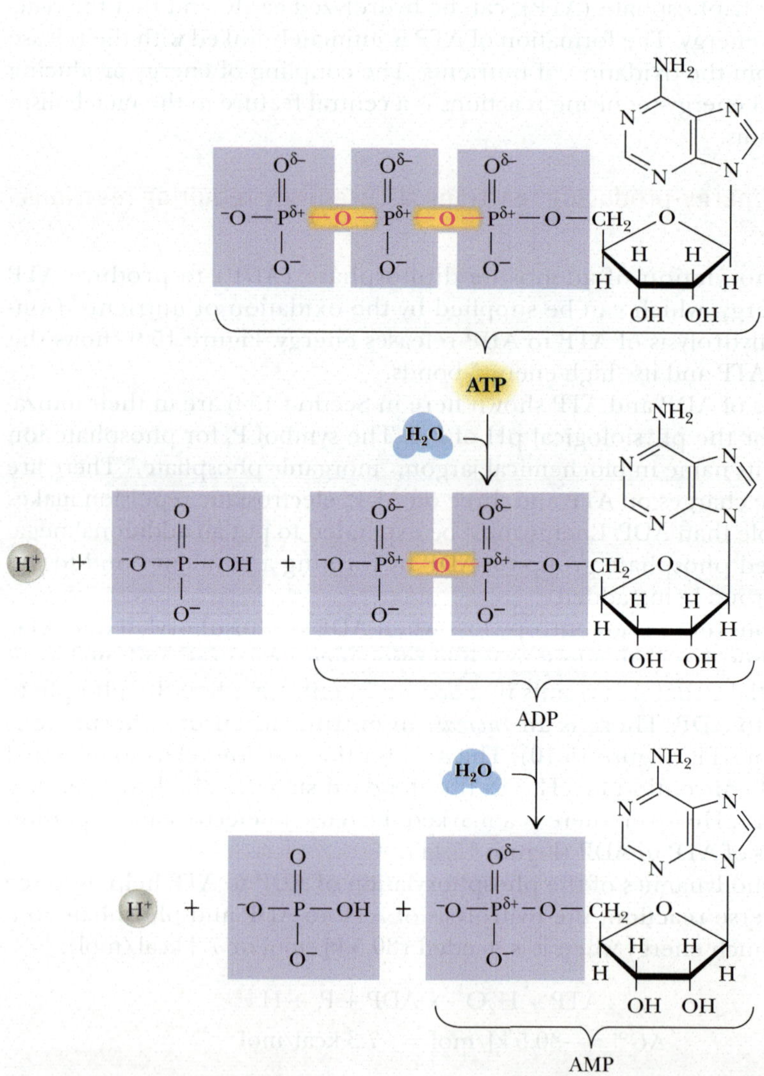

Figure 15.11 Decrease in electrostatic repulsion on the hydrolysis of ATP. Hydrolysis of ATP to ADP (and/or hydrolysis of ADP to AMP) reduces electrostatic repulsion between the negatively charged phosphate groups.

Table 15.1 Gibbs Energy Changes of the Hydrolysis of Selected Organophosphates

	$\Delta G^{\circ\prime}$	
Compound	**kJ/mol**	**kcal/mol**
Phosphoenolpyruvate	−61.9	−14.8
Carbamoyl phosphate	−51.4	−12.3
Creatine phosphate	−43.1	−10.3
Acetyl phosphate	−42.2	−10.1
ATP (to ADP)	−30.5	−7.3
Glucose-1-phosphate	−20.9	−5.0
Glucose-6-phosphate	−12.5	−3.0
Glycerol-3-phosphate	−9.7	−2.3

The bond that is hydrolyzed when this reaction takes place is sometimes called a "high-energy bond" because hydrolysis of that bond releases a useful amount of energy. Another way of indicating such a bond is ~P. Numerous organophosphate compounds with high-energy bonds play roles in metabolism, but ATP is by far the most important (Table 15.1). In some cases, the Gibbs energy of hydrolysis of organophosphates is higher than that of ATP and is thus able to drive the phosphorylation of ADP to ATP.

Phosphoenolpyruvate (PEP), a molecule involved in glycolysis (Chapter 17), tops the list. It is a very high-energy compound because of the resonance stabilization of the liberated phosphate when it is hydrolyzed (the same effect as that seen with ATP) and because the tautomerization of the enol form of pyruvate to the more stable keto form of pyruvate is possible (Figure 15.12). Both effects increase the entropy upon hydrolysis.

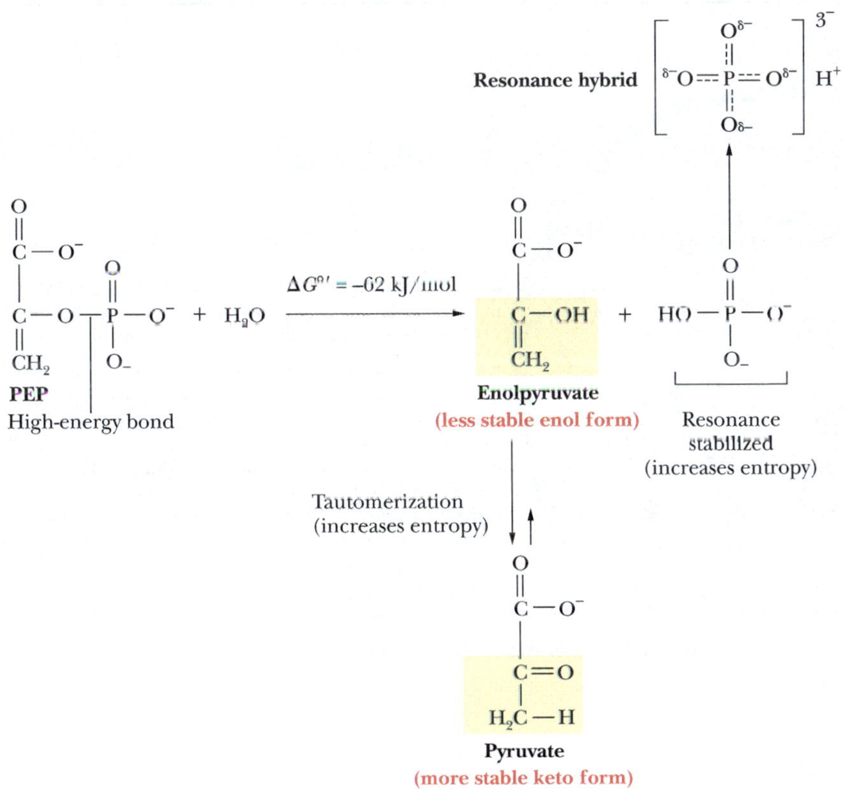

Figure 15.12 Increase in entropy on the hydrolysis of phosphoenolpyruvate (PEP). Entropy increases when phosphoenolpyruvate is hydrolyzed to pyruvate and phosphate. Both the formation of the keto form of pyruvate and the resonance structures of phosphate lead to the increase in entropy.

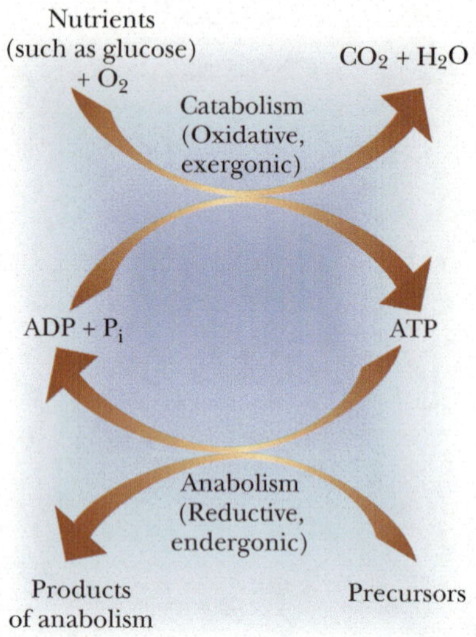

Figure 15.13 **The role of ATP as energy currency in processes that release energy and processes that use energy.**

The high energy of ATP is not stored energy, just as an electric current does not represent stored energy. Both ATP and electric current must be produced when they are needed—by organisms or by a power plant, as the case may be. The cycling of ATP and ADP in metabolic processes is a way of shunting energy from its production (by the oxidation of nutrients) to its uses (in processes such as the biosynthesis of essential compounds or muscle contraction) when it is needed. The oxidation processes take place when the organism needs the energy that can be generated by the hydrolysis of ATP.

Energy storage in an organism is usually in the form of fats and carbohydrates, which are metabolized for ATP (energy) production as needed. Certain small biomolecules, such as creatine phosphate, can also serve as vehicles for storing chemical energy. Creatine phosphate can donate its phosphate to ADP to quickly produce ATP when it is needed. The energy that must be supplied for the many endergonic reactions in life processes comes directly from the hydrolysis of ATP and indirectly from the oxidation of nutrients. The nutrients produce the energy needed to phosphorylate ADP to ATP (Figure 15.13). In addition to its key role in bioenergetics, ATP can do many other things, especially in the realm of intercellular communication. BIOCHEMICAL CONNECTIONS 15B gives some details.

Next, we examine some biological reactions that release energy and see how some of that energy is used to phosphorylate ADP to ATP. The multistep conversion of glucose to lactate ions is an exergonic and anaerobic process. Two molecules of ADP are phosphorylated to ATP for each molecule of

15B Biochemical Connections • Physiology

ATP in Cell Signaling

The role of ATP in generating and using energy within cells is summarized in Figure 15.14. Cells produce ATP in their mitochondria and use it to provide energy for their activities, such as protein synthesis. ATP is widely involved in chemical signaling between cells. For example, during the transmission of nerve signals, ATP is released along with neurotransmitters.

Enzymes hydrolyze ATP to ADP, AMP, and adenosine. Specific receptors exist on the surface of receiving cells: P2X receptors for ATP itself, P2Y receptors for ATP and ADP, and P1 receptors for AMP and adenosine.

The binding of ATP and its degradation products to their individual receptors sets off a response within the receiving cell. The effects can vary widely. For example, ATP release from neurons regulates pain perception and inflammatory response. In many cases, a short-term effect, such as hydrolysis of a chemical bond, triggers a long-term effect. Release of calcium ions from an intracellular reservoir, for example, frequently leads to the activation of key enzymes. In Chapter 24, we discuss many examples of the responses of cells to extracellular triggers as a result of receptor binding. ◗

A

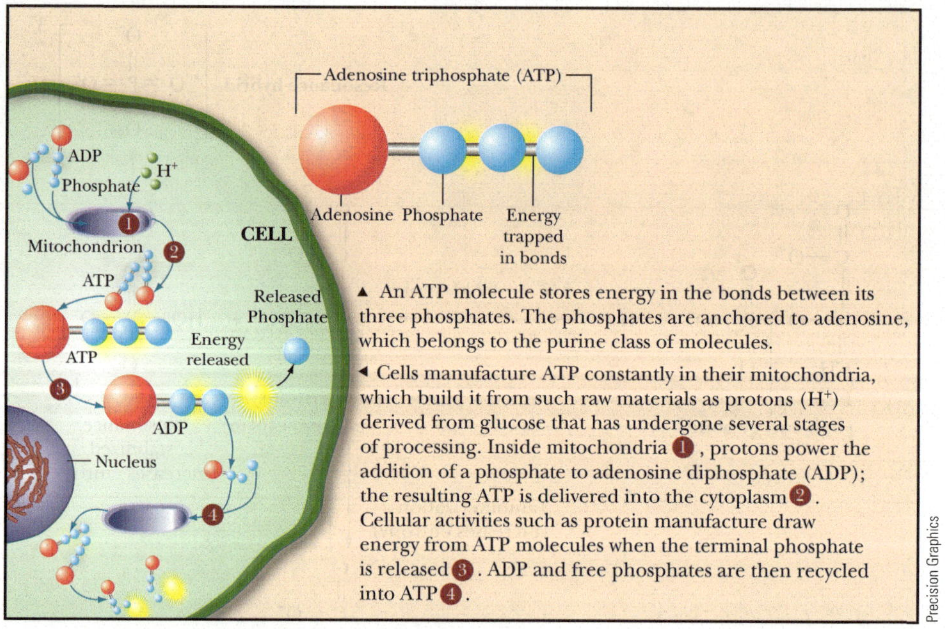

Figure 15.14A **ATP in the generation and use of energy.**

B

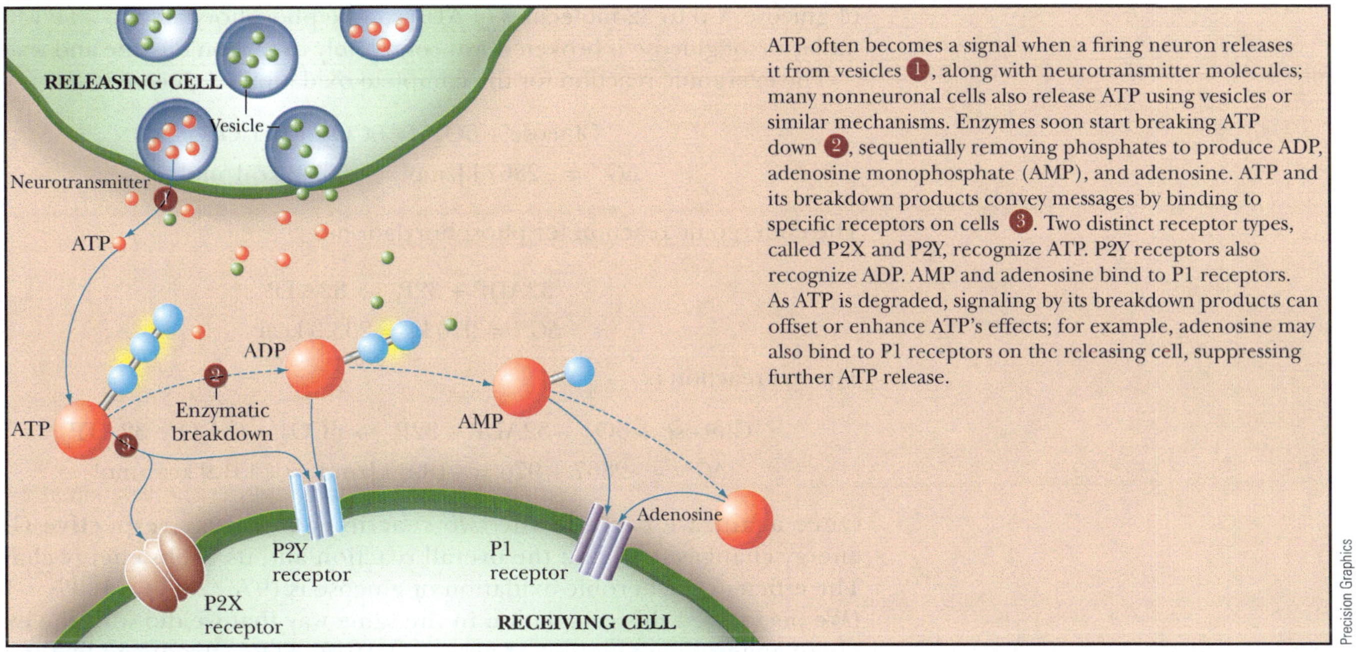

Figure 15.14B ATP in cell signaling.

glucose metabolized. The basic reactions are the production of lactate, which is exergonic,

$$\text{Glucose} \rightarrow 2\ \text{Lactate ions}\ \Delta G^{\circ\prime} = -184.5\ \text{kJ/mol} = -44.1\ \text{kcal/mol}$$

and the phosphorylation of two moles of ADP for each mole of glucose, which is endergonic:

$$2\text{ADP} + 2\text{P}_i \rightarrow 2\text{ATP}$$
$$\Delta G^{\circ\prime} = 61.0\ \text{kJ/mol} = 14.6\ \text{kcal/mol}$$

(To keep things as simple as possible, we write the equation for phosphorylation of ADP in terms of ADP, P_i, and ATP only.) The overall reaction is

$$\text{Glucose} + 2\text{ADP} + 2\text{P}_i \rightarrow 2\ \text{Lactate ions} + 2\text{ATP}$$
$$\Delta G^{\circ\prime}\ \text{overall} = -184.5 + 61.0 = -123.5\ \text{kJ/mol} = -29.5\ \text{kcal/mol}$$

Not only can we add the two chemical reactions to obtain an equation for the overall reaction, but we can also add the Gibbs energy changes for the two reactions to find the overall Gibbs energy change. We can do this because the Gibbs energy change is a **state function**; it depends only on the initial and final states of the system under consideration, not on the path between those states.

state function a function that depends only on the initial and final states of a system, not on the path between the states

The exergonic reaction provides energy, which drives the endergonic reaction. This phenomenon is called **coupling**. The percentage of the released energy that is used to phosphorylate ADP is the efficiency of energy use in anaerobic metabolism; it is $(61.0/184.5) \times 100\% \approx 33\%$. The "61.0" comes from the number of kilojoules required to phosphorylate 2 moles of ADP to ATP, whereas "184.5" is the number of kilojoules released when 1 mole of glucose is converted to 2 moles of lactate.

coupling the process by which an exergonic reaction provides energy for an endergonic one

The breakdown of glucose goes further under aerobic conditions than under anaerobic conditions. The end products of aerobic oxidation are six molecules of carbon dioxide and six molecules of water for each molecule of glucose. Up to 32 molecules of ADP can be phosphorylated to ATP when 1 molecule of glucose is broken down completely to carbon dioxide and water.

The exergonic reaction for the complete oxidation of glucose is

$$\text{Glucose} + 6O_2 \rightarrow 6CO_2 + 6H_2O$$
$$\Delta G^{\circ\prime} = -2867 \text{ kJ/mol} = -685.2 \text{ kcal/mol}$$

The endergonic reaction for phosphorylation is

$$32\text{ADP} + 32\text{P}_i \rightarrow 32\text{ATP}$$
$$\Delta G^{\circ\prime} = 976 \text{ kJ} = 233.3 \text{ kcal}$$

The net reaction is

$$\text{Glucose} + 6O_2 + 32\text{ADP} + 32\text{P}_i \rightarrow 6CO_2 + 6H_2O + 32\text{ATP}$$
$$\Delta G^{\circ\prime} = -2867 + 976 = -1891 \text{ kJ/mol} = -451.9 \text{ kcal/mol}$$

Once again, we can add the two reactions and their respective Gibbs energy changes to obtain the overall reaction and its Gibbs energy change. The efficiency of aerobic oxidation of glucose is $(976/2867) \times 100\% \approx 34\%$. (We performed this calculation in the same way that we did with the example of anaerobic oxidation of glucose.) More ATP is produced by the coupling process in aerobic oxidation of glucose than by the coupling process in anaerobic oxidation. **APPLY YOUR KNOWLEDGE 15.3** gives you a chance to practice this kind of calculation.

15.3 Apply Your Knowledge

Predicting Reactions: Calculations of Gibbs Energy Changes

Let's use the values from Table 15.1 to calculate $\Delta G^{\circ\prime}$ for the following reaction and determine whether it is spontaneous. The most important point here is that we add algebraically. In particular, we have to remember to change the sign of the $\Delta G^{\circ\prime}$ when we reverse the direction of the reaction.

$$\text{ADP} + \text{Phosphoenolpyruvate} \rightarrow \text{ATP} + \text{Pyruvate}$$

From Table 15.1,

$$\text{Phosphoenolpyruvate} + H_2O \rightarrow \text{Pyruvate} + \text{P}_i$$
$$\Delta G^{\circ\prime} = -61.9 \text{ kJ/mol} = -14.8 \text{ kcal/mol}$$

Also,

$$\text{ATP} + H_2O \rightarrow \text{ADP} + \text{P}_i \qquad \Delta G^{\circ\prime} = -30.5 \text{ kJ/mol} = -7.3 \text{ kcal/mol}$$

We want the reverse of the second reaction:

$$\text{ADP} + \text{P}_i \rightarrow \text{ATP} + H_2O \qquad \Delta G^{\circ\prime} = +30.5 \text{ kJ/mol} = +7.3 \text{ kcal/mol}$$

We now add the two reactions and their Gibbs energy changes:

$$\text{Phosphoenolpyruvate} + H_2O \rightarrow \text{Pyruvate} + \text{P}_i$$
$$\text{ADP} + \text{P}_i \rightarrow \text{ATP} + H_2O$$
$$\text{Phosphoenolpyruvate} + \text{ADP} \rightarrow \text{Pyruvate} + \text{ATP} \qquad \textbf{Net reaction}$$
$$\Delta G^{\circ\prime} = -61.9 \text{ kJ/mol} + 30.5 \text{ kJ/mol} = -31.4 \text{ kJ/mol}$$
$$\Delta G^{\circ\prime} = -14.8 \text{ kcal/mol} + 7.3 \text{ kcal/mol} = -7.5 \text{ kcal/mol}$$

The reaction is spontaneous, as indicated by the negative value of $\Delta G^{\circ\prime}$. However, even though this reaction is spontaneous, nothing would happen if you simply put these chemicals in a test tube. This is because biochemical reactions require enzymes to catalyze them. We discuss many of these metabolic enzymes in subsequent chapters.

The exergonic hydrolysis of ATP produced by the breakdown (aerobic or anaerobic) of glucose can be coupled to endergonic processes, such as muscle contraction in exercise. As any jogger or long-distance swimmer knows, aerobic metabolism involves large quantities of energy, processed in a highly efficient fashion. We have now seen two examples of the coupling of exergonic and endergonic processes—aerobic oxidation of glucose and anaerobic fermentation of glucose—involving different amounts of energy. Both examples of coupling, aerobic and anaerobic, are simplified versions of multistep pathways. Some of the intermediate steps in these pathways are energetically unfavorable, but the whole series of reactions can be considered to be coupled.

In organisms, energy is constantly transferred from the processes in which it is produced to those in which it is used. The chemical energy in the molecules being oxidized is released as needed by the reactions of metabolism. We discuss the details of the *anaerobic breakdown* of glucose in Chapter 17 and those of the *aerobic breakdown* of glucose, fatty acids, and amino acids in Chapters 19 and 20. It is of paramount importance to remember that *all the reactions of metabolism proceed simultaneously*. Metabolism in a living organism cannot take place if it is not energetically favorable. Thus, we must consider all of metabolism to be a large-scale coupled reaction.

15-7 Metabolic Activation

Learning Objectives

After reading this section, students will be able to:

LO 15.7.1 Explain coenzyme A structure.

LO 15.7.2 Describe metabolite activation using coenzyme A.

The metabolic oxidation of glucose that we saw in Section 15-6 does not take place in one step. The anaerobic breakdown of glucose requires many steps, and the complete aerobic oxidation of glucose to carbon dioxide and water has still more steps. One of the most important points about the multistep nature of all metabolic processes, including the oxidation of glucose, is that the many stages allow for the efficient production and use of energy. The electrons produced by the oxidation of glucose are passed along to oxygen, the ultimate electron acceptor, by intermediate electron acceptors. Many of the intermediate stages of the oxidation of glucose are coupled to adenosine triphosphate (ATP) production by the phosphorylation of adenosine diphosphate (ADP).

▶ How is coenzyme A involved in metabolic activation?

A step frequently encountered in metabolic pathways is the activation of a **metabolite. Metabolite activation** involves a metabolite bonding to some other molecule, such as a coenzyme or AMP, where the Gibbs energy change for breaking this new bond is negative. In other words, metabolite activation helps make the next reaction in the metabolic pathway exergonic. For example, if substance A is the metabolite that needs to react with substance B to give AB, then the following series of reactions might take place:

metabolite a compound found in a metabolic pathway

metabolite activation the process of bonding a metabolite to another molecule so that the next reaction of the metabolite is energetically favorable

A + Coenzyme → A—Coenzyme		**Activation step**
A—Coenzyme + B → AB + Coenzyme	$\Delta G^{\circ\prime} < 0$	**Exergonic reaction**

There are many examples of metabolite activation in metabolic processes. One of the most useful examples involves forming a covalent bond with a compound known as coenzyme A (CoA-SH).

The structure of coenzyme A is complex, and it consists of several smaller components linked together covalently (Figure 15.15).

- One part is 3′-P-5′-ADP, a derivative of adenosine with phosphate groups esterified to the sugar, as shown in the structure.
- Another part is derived from the vitamin pantothenic acid.
- The part of the molecule involved in activation reactions contains a thiol group (Figure 15.15, *left*).

Coenzyme A is frequently written as CoA-SH to emphasize that the thiol group is the reactive portion of the molecule. For example, carboxylic acids form thioester linkages with CoA-SH. The metabolically active form of a carboxylic acid is the corresponding acyl-CoA thioester, in which the thioester linkage is a high-energy bond (Figure 15.16). Thioesters are high-energy compounds because of the possible dissociation of the products after hydrolysis and the resonance structures of the products. For example, when acetyl-CoA is hydrolyzed, the —SH at the end of the molecule can dissociate slightly to form H^+ and CoA-S^-. The acetate released by the hydrolysis is stabilized by resonance. Acetyl-CoA is a particularly important metabolic intermediate; other acyl-CoA species figure prominently in lipid metabolism.

The important coenzymes we have discussed in this chapter—NAD^+, $NADP^+$, FAD, and coenzyme A—share an important structural feature: all contain an adenine nucleotide (AMP). In $NADP^+$, there is an additional phosphate group at the 2′ position of the ribose group of the AMP, and in CoA-SH, there is an additional phosphate group at the 3′ position.

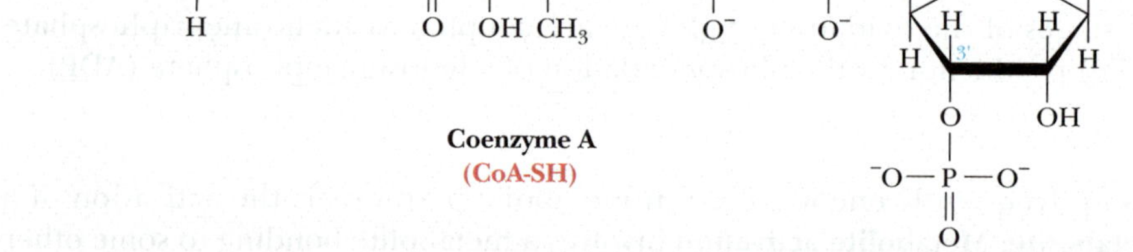

Figure 15.15 The structure of coenzyme A (CoA-SH). Several substructures in CoA-SH are also indicated, including its functional thiol group (*left*), which contributes to the activation of some metabolites in metabolic pathways.

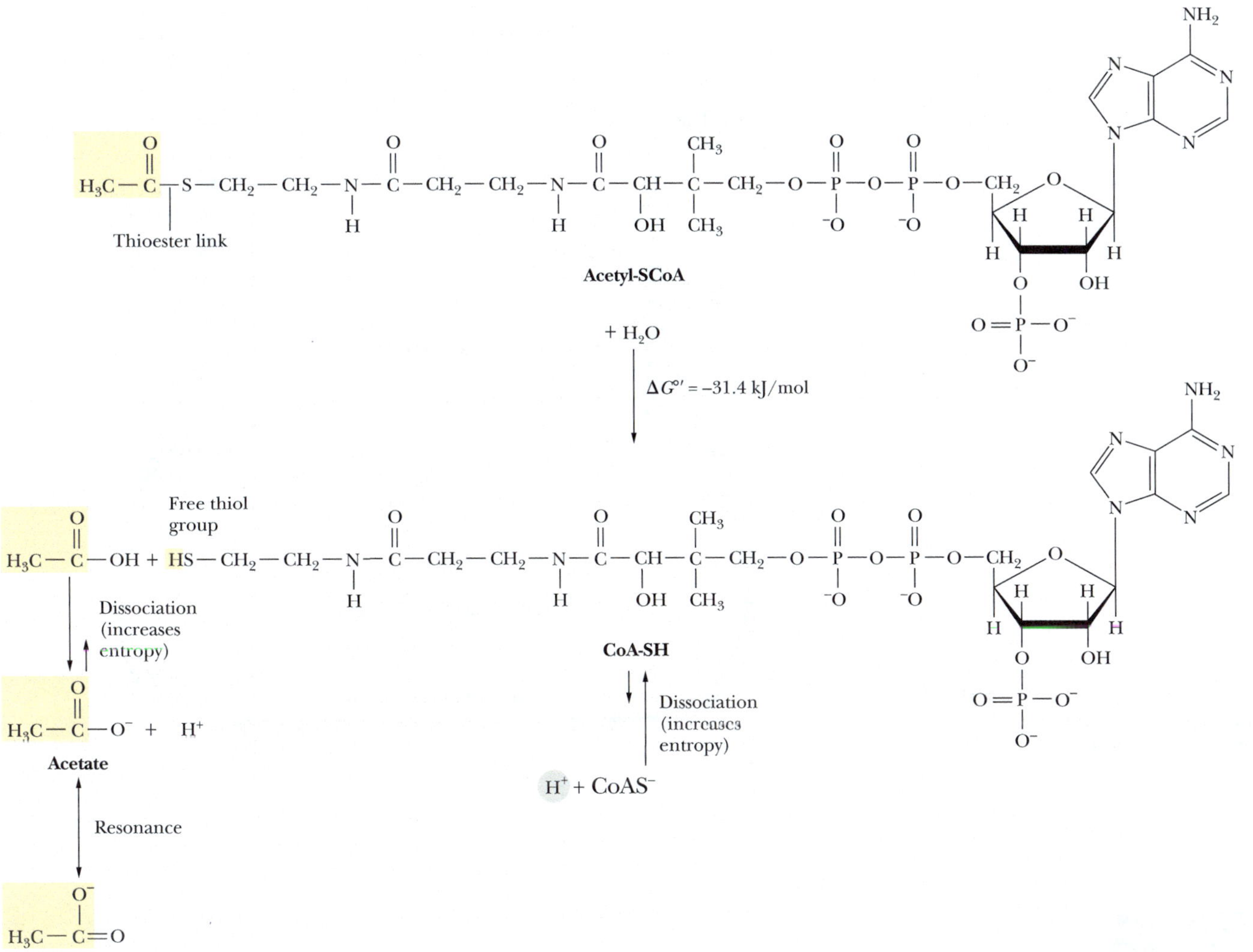

Figure 15.16 The hydrolysis of acetyl-CoA. The products are stabilized by resonance and by the increase in entropy that results from dissociation.

Like catabolism, anabolism proceeds in stages. Unlike catabolism, which releases energy, anabolism requires energy, and the ATP produced by catabolism is hydrolyzed to release the needed energy. Reactions in which metabolites are reduced are part of anabolism; they require reducing agents, such as NADH and NADPH, both of which are the reduced forms of coenzymes mentioned in this chapter. In their oxidized forms, these coenzymes serve as the intermediate oxidizing agents needed in catabolism. In their reduced form, the same coenzymes provide the "reducing power" needed for the anabolic processes of biosynthesis; in this case, the coenzymes act as reducing agents.

We are now in a position to expand on our earlier statements about the natures of anabolism and catabolism. Figure 15.17 is an outline of metabolic pathways that explicitly points out two important features of metabolism: the role of electron transfer and the role of ATP in the release and utilization of energy. Even though this outline is more extended than the one in Figure 15.1, it is still very general. The more important specific pathways have been studied in detail, and some are still the subjects of active research. We discuss some of the most important metabolic pathways in the remainder of this textbook.

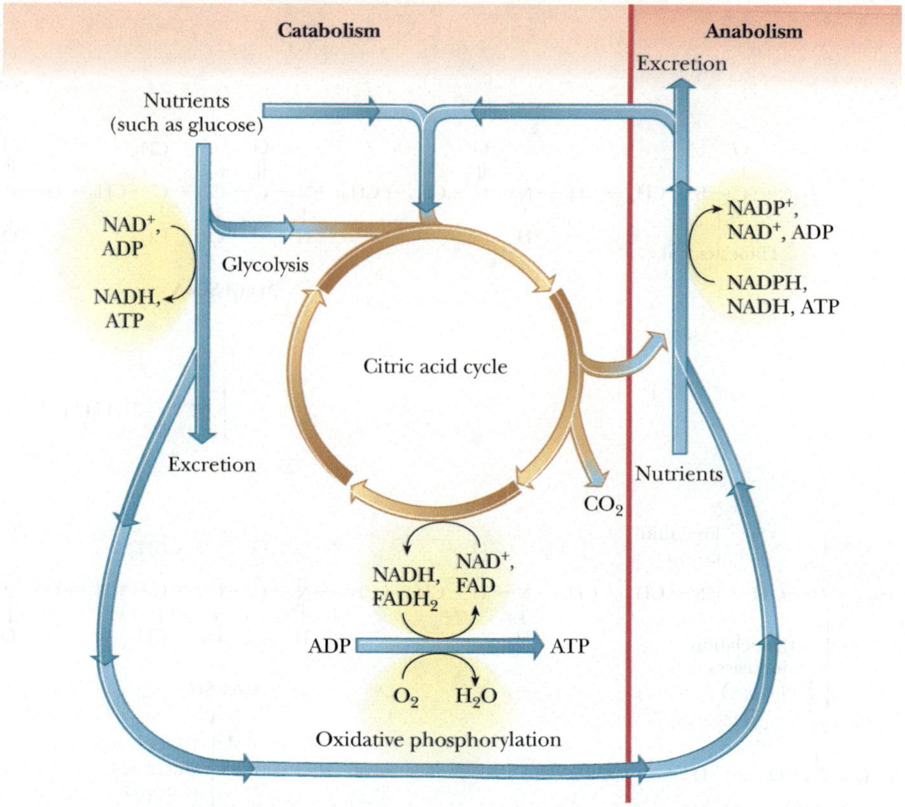

Figure 15.17 The role of electron transfer and ATP production in metabolism. NAD^+, FAD, and ATP are continuously recycled.

Summary

Standard Gibbs Energy Change

- Standard states are used to compare different reactions.
- A change in Gibbs energy under any set of conditions can be compared to the Gibbs energy change under standard conditions ($\Delta G°$), in which the concentration of all reactants in solution is set at 1 *M*.
- Gibbs energy changes under standard conditions can be related to the equilibrium constant of a reaction by the equation $\Delta G° = -RT \ln K_{eq}$.

Spontaneity and Reversibility

- A spontaneous reaction is one that will occur without added energy.
- Reactions that release energy are the ones that are most likely to occur.
- Reactions that require energy will occur if they are coupled to energy-releasing reactions such that the overall process releases energy.
- Le Châtelier's principle applies to reactions in equilibrium and states that when a stress is applied to a system at equilibrium, the system reacts so as to minimize the effect of that stress.
- Driving endergonic reactions can be achieved by reaction coupling. For example, the endergonic phosphorylation of glucose can be coupled to the relatively larger exergonic hydrolysis of a phosphate group from ATP, resulting in a net exergonic process.

Energy Balance

- The amount of energy consumed by a reaction is equivalent to the heat produced by the reaction in combination with the amount of work and chemical energy storage resulting from the reaction.

A Modified Standard State for Biochemical Applications

- $\Delta G°'$ ("delta G naught prime") represents the standard state for biochemical reactions where the pH of the system is 7 rather than 0.

The Nature of Metabolism

- The reactions of biomolecules in a cell constitute metabolism.
- Catabolism is the breakdown of larger molecules into smaller ones.
- Anabolism is the reaction of small molecules to produce larger and more complex ones.
- Catabolism and anabolism are separate pathways, not the reverse of each other.

The Role of Oxidation and Reduction in Metabolism

- Catabolism is an oxidative process that releases energy; anabolism is a reductive process that requires energy.
- Oxidation–reduction (redox) reactions are those in which electrons are transferred from a donor to an acceptor.
- Oxidation is the loss of electrons, whereas reduction is the gain of electrons.

Coenzymes in Biologically Important Oxidation–Reduction Reactions

- Many biologically important redox reactions involve coenzymes, such as nicotinamide adenine dinucleotide (NAD) and flavin adenine dinucleotide (FAD).
- These coenzymes appear in many redox reactions as electron donors or electron acceptors.

Coupling of Energy Production and Energy Use

- The coupling of energy-producing reactions and energy-requiring reactions is a central feature in the metabolism of all organisms.
- In catabolism, oxidative reactions are coupled to the endergonic production of ATP by the phosphorylation of ADP.
- Aerobic metabolism is a more efficient means of using the chemical energy of nutrients than is anaerobic metabolism.
- In anabolism, the exergonic hydrolysis of the high-energy bond of ATP releases the energy needed to drive endergonic reductive reactions.
- Metabolism proceeds in steps, and the many steps allow for the efficient production and use of energy.

Metabolic Activation

- The process of metabolite activation, which produces high-energy intermediates, occurs in many metabolic pathways.
- One example is the formation of thioester linkages by the reaction of carboxylic acids with coenzyme A.

Review Exercises

15-1 Standard States for Gibbs Energy Changes

1. **Recall** Is there a connection between the Gibbs energy change for a reaction and its equilibrium constant? If there is a connection, what is it?
2. **Reflect and Apply** What do the following indicators tell you about whether a reaction can proceed as written or how it could proceed as written?
   (a) The Gibbs energy change is positive.
   (b) The reaction is exergonic.
3. **Reflect and Apply** Consider the reaction

$$\text{Glucose-6-phosphate} + H_2O \rightarrow \text{Glucose} + P_i$$

$$K_{eq} = \frac{[\text{glucose}]\,[P_i]}{[\text{glucose-6-P}]}$$

   $K_{eq} = 122$ at pH 8.5 and 38°C. Can you determine the rate of the reaction from this information?
4. **Recall** What conditions are necessary for the Gibbs energy change to be used to predict the spontaneity of a reaction?
5. **Reflect and Apply** Why is it important that energy released by exergonic reactions can be used to provide energy for endergonic reactions?

15-2 A Modified Standard State for Biochemical Applications

6. **Recall** In biochemistry, what are the standard conditions used to define the standard Gibbs energy change ($\Delta G^{\circ\prime}$)?
7. **Recall** Why is it necessary to define a modified standard state for biochemical applications of thermodynamics?
8. **Recall** Which of the following statements is (are) true about the modified standard state for biochemistry? For each, explain why or why not.
   (a) $[H^+] = 1 \times 10^{-7}$ *M*, not 1 *M*.
   (b) The concentration of any solute is 1×10^{-7} *M*.
9. **Recall** How can you tell if the standard Gibbs energy change given for a reaction is for chemical standard states or biological standard states?
10. **Recall** Can the thermodynamic property $\Delta G^{\circ\prime}$ be used to predict the speed of a reaction in a living organism? Why or why not?
11. **Mathematical** Calculate $\Delta G^{\circ\prime}$ at 25°C for the following values of K'_{eq}: 1×10^4, 1, 1×10^{-6}.
12. **Mathematical** The $\Delta G^{\circ\prime}$ for the reaction Citrate → Isocitrate is 6.64 kJ/mol (1.59 kcal/mol). The $\Delta G^{\circ\prime}$ for the reaction Isocitrate → α-Ketoglutarate is −267 kJ/mol (−63.8 kcal/mol). What is the $\Delta G^{\circ\prime}$ for the conversion of citrate to α-ketoglutarate? Is that reaction exergonic or endergonic, and why?
13. **Mathematical** All the organophosphate compounds listed in Table 15.1 undergo hydrolysis reactions in the same way as ATP. The following equation illustrates the situation for glucose-1-phosphate.

$$\text{Glucose-1-phosphate} + H_2O \rightarrow \text{Glucose} + P_i$$
$$\Delta G^{\circ\prime} = -20.9 \text{ kJ/mol}$$

   Using the Gibbs energy values in Table 15.1, predict whether the following reactions will proceed in the direction written, and calculate the $\Delta G^{\circ\prime}$ for the reaction, assuming that the reactants are initially present in a 1:1 molar ratio.
   (a) ATP + Creatine → Creatine phosphate + ADP
   (b) ATP + Glycerol → Glycerol-3-phosphate + ADP
   (c) ATP + Pyruvate → Phosphoenolpyruvate + ADP
   (d) ATP + Glucose → Glucose-6-phosphate + ADP
14. **Reflect and Apply** Why are $\Delta G^{\circ\prime}$ values not rigorously applicable to biochemical systems?

15-3 The Nature of Metabolism

15. **Recall** Briefly explain the relationship between the terms *metabolism, catabolism,* and *anabolism.*
16. **Recall** Organize the following words into two related groups: catabolism, energy-requiring, reductive, anabolism, oxidative, energy-yielding.

17. **Biochemical Connections** Describe the difference between a closed system and an open system in terms of what they can exchange with their surroundings.

18. **Biochemical Connections** Comment on the statement that the existence of life is a violation of the second law of thermodynamics, adding concepts from this chapter to those discussed in Chapter 1.

19. **Reflect and Apply** Would you expect the production of sugars by plants in photosynthesis to be an exergonic or endergonic process? Give the reason for your answer.

20. **Reflect and Apply** Would you expect the biosynthesis of a protein from the constituent amino acids in an organism to be an exergonic or endergonic process? Give the reason for your answer.

21. **Reflect and Apply** Adult humans synthesize large amounts of ATP in the course of a day, but their body weights do not change significantly. In the same time period, the structures and compositions of their bodies also do not change appreciably. Explain this apparent contradiction.

15-4 The Role of Oxidation and Reduction in Metabolism

22. **Recall** List the key functional groups formed at each step of two progressive oxidations of an alkane carbon.

23. **Recall** Identify the molecules oxidized and reduced in the following reactions and write the half reactions.
(a) $CH_3CH_2CHO + NADH \rightarrow CH_3CH_2CH_2OH + NAD^+$
(b) $Cu^{2+}(aq) + Fe^{2+}(aq) \rightarrow Cu^+(aq) + Fe^{3+}(aq)$

24. **Recall** For each of the reactions in Exercise 23, identify the oxidizing agent and the reducing agent.

15-5 Coenzymes in Biologically Important Oxidation–Reduction Reactions

25. **Recall** Briefly describe the role of NAD^+ in biochemical oxidation–reduction reactions and how its structure enables this function.

26. **Recall** What structural feature do NAD^+, $NADP^+$, and FAD have in common?

27. **Recall** What is the structural difference between NADH and NADPH?

28. **Recall** How does the difference between NADH and NADPH affect the reactions in which they are involved?

29. **Recall** Which coenzyme, NAD^+ or NADH, is a reactant in the oxidation of a nutrient? What is the reason for your answer?

30. **Reflect and Apply** Which of the following statements are true? For each, explain why or why not.
(a) All coenzymes are electron-transfer agents.
(b) Coenzymes do not contain phosphorus or sulfur.
(c) Generating ATP is a way of storing energy.

31. **Reflect and Apply** The following half reactions play important roles in metabolism.

$$\tfrac{1}{2}\,O_2 + 2H^+ + 2e^- \rightarrow H_2O$$
$$NADH + H^+ \rightarrow NAD^+ + 2H^+ + 2e^-$$

Which of these two is a half reaction of oxidation? Which one is a half reaction of reduction? Write the equation for the overall reaction. Which reagent is the oxidizing agent (electron acceptor)? Which reagent is the reducing agent (electron donor)?

32. **Reflect and Apply** Draw the structures of NAD^+ and FAD, showing where the electrons and hydrogens go when the molecules are reduced.

33. **Reflect and Apply** There is a reaction in carbohydrate metabolism in which glucose-6-phosphate reacts with $NADP^+$ to give 6-phosphoglucono-δ-lactone and NADPH.

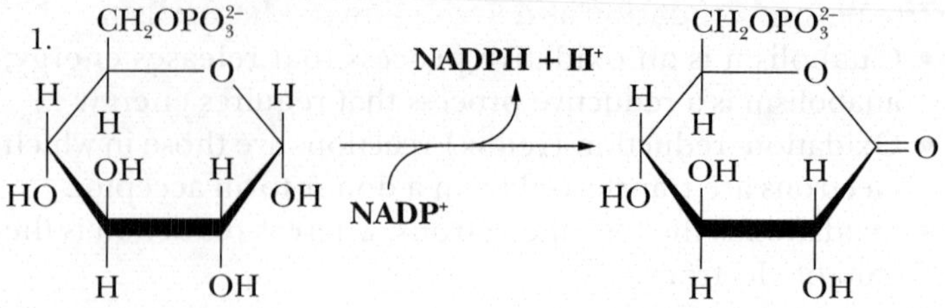

Glucose-6-phosphate 6-Phosphoglucono-δ-lactone

In this reaction, which substance is oxidized, and which is reduced? Which substance is the oxidizing agent, and which is the reducing agent?

34. **Reflect and Apply** There is a reaction in which succinate reacts with FAD to give fumarate and $FADH_2$.

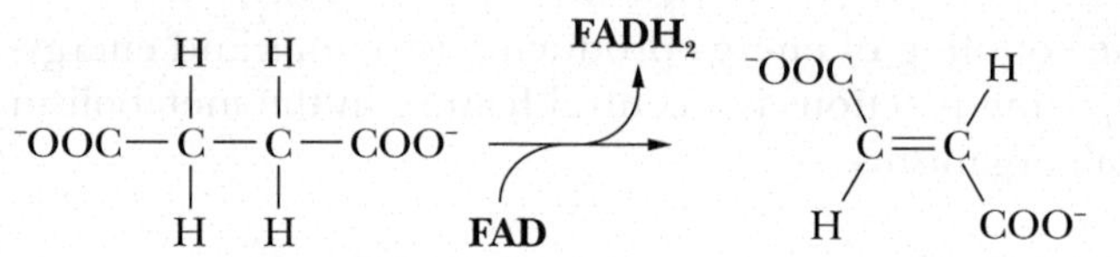

Succinate Fumarate

In this reaction, which substance is oxidized, and which is reduced? Which substance is the oxidizing agent, and which is the reducing agent?

35. **Reflect and Apply** Suggest a reason why catabolic pathways generally use NAD^+, whereas anabolic pathways generally use NADPH.

15-6 Coupling of Energy Production and Energy Use

36. **Recall** Does all the energy released by an exergonic reaction go into driving a coupled endergonic reaction? Give an example to support your answer.

37. **Reflect and Apply** How does the release of chemical energy make metabolism possible?

38. **Mathematical** Using the data in Table 15.1, calculate the value of $\Delta G°'$ for the following reaction.

Creatine phosphate + Glycerol → Creatine + Glycerol-3-phosphate

Hint: This reaction proceeds in stages. ATP is formed in the first step, and the phosphate group is transferred from ATP to glycerol in the second step.

39. **Mathematical** Using information from Table 15.1, calculate the value of $\Delta G°'$ for the following reaction.

Glucose-1-phosphate → Glucose-6-phosphate

40. **Mathematical** Show that the hydrolysis of ATP to AMP and $2P_i$ releases the same amount of energy by either of the two following pathways.

Pathway 1

$$ATP + H_2O \rightarrow ADP + P_i$$
$$ADP + H_2O \rightarrow AMP + P_i$$

Pathway 2

$$ATP + H_2O \rightarrow AMP + PP_i \text{ (Pyrophosphate)}$$
$$PP_i + H_2O \rightarrow 2P_i$$

41. **Mathematical** The standard Gibbs energy change for the reaction

$$\text{Arginine} + \text{ATP} \rightarrow \text{Phosphoarginine} + \text{ADP}$$

is +1.7 kJ/mol. From this information and the data in Table 15.1, calculate the $\Delta G^{\circ\prime}$ for the reaction

$$\text{Phosphoarginine} + H_2O \rightarrow \text{Arginine} + P_i$$

42. **Reflect and Apply** What are the usual ionic forms of ATP and ADP in typical cells? Does this information have any bearing on the Gibbs energy change for the conversion of ATP to ADP?
43. **Reflect and Apply** How does the Gibbs energy change for the hydrolysis of the phosphate bond of ATP (−30.5 kJ/mol; −7.3 kcal/mol) compare to those of other organophosphates listed in Table 15.1 (e.g., sugar phosphates, creatine phosphate).
44. **Reflect and Apply** Would you expect an increase or a decrease in entropy to accompany the hydrolysis of phosphatidylcholine to its constituent parts (glycerol, two fatty acids, phosphoric acid, and choline)? Why?
45. **Reflect and Apply** Explain why, and show how, phosphoenolpyruvate is a high-energy compound.

15-7 Metabolic Activation

46. **Reflect and Apply** Why is the process of activation a useful strategy in metabolism?
47. **Reflect and Apply** What is the molecular logic that makes a pathway with a number of comparatively small energy changes more likely than a single reaction with a large energy change?
48. **Reflect and Apply** Why are thioesters considered to be high-energy compounds?
49. **Reflect and Apply** Explain why several biochemical pathways start by putting a coenzyme A onto the molecule that initiates the pathway.

Further Reading

Two standard multivolume references cover specific aspects of metabolism in detail. One of these, the third edition of *The Enzymes* (P. D. Boyer, ed.; New York: Academic Press) is a series that has been in production since 1970. The other, *Comprehensive Biochemistry* (M. Florkin and E. H. Stotz, eds.; New York: Elsevier), has been in production since 1962.

Atkins, P. W. *The Second Law.* San Francisco: Freeman, 1984. A highly readable nonmathematical discussion of thermodynamics.

Chang, R. *Physical Chemistry with Applications to Biological Systems,* 2nd ed. New York: Macmillan, 1981. Chapter 12 contains a detailed treatment of thermodynamics.

Fasman, G. D., ed. *Handbook of Biochemistry and Molecular Biology,* 3rd ed., Sec. D, *Physical and Chemical Data.* Cleveland, OH: CRC Press, 1976. Volume 1 contains data on the Gibbs energies of hydrolysis of many important compounds, especially organophosphates.

Harold, F. M. *The Vital Force: A Study of Bioenergetics.* New York: Freeman, 1986. Energetic aspects of many important life processes.

Hinkle, P. C., and R. E. McCarty. How Cells Make ATP. *Sci. Amer.* **238** (3), 104–125 (1978). Getting old, but a particularly good treatment of energy coupling.

Kang, D., S. Kim, J. Sohn, and B. H. Sung. Insights into Enzyme Reactions with Redox Cofactors in Biological Conversion of CO_2. *J. Microbiol. Biotechnol.* **33** (11), 1403–1411 (2023). This review summarizes seven known natural CO_2 fixation pathways and describes recent studies related to enzyme reactions with redox cofactors.

Khakh, B., and G. Burnstock. The Double Life of ATP. *Sci. Amer.* **301** (6), 84–92 (2009). A description of the role of ATP in cell signaling.

Prigogine, I., and I. Stengers. *Order Out of Chaos.* Toronto: Bantam Press, 1984. A comparatively accessible treatment of the thermodynamics of biological systems. Prigogine won the 1977 Nobel Prize in Chemistry for his pioneering work on the thermodynamics of complex systems.

Lightspring/Shutterstock.com

Like bread and pasta, fruits and vegetables are sources of carbohydrates, which provide energy.

16 Carbohydrates

Chapter Outline

16-1 Simple Sugars: Their Structures and Stereochemistry

Learning Objectives

After reading this section, students will be able to:

LO 16.1.1 Distinguish between monosaccharide, oligosaccharide, and polysaccharide structures.

LO 16.1.2 Describe three functions of carbohydrates in living organisms.

LO 16.1.3 Explain stereochemistry concepts related to simple sugar structures.

When the word *carbohydrate* was coined, it originally referred to compounds of the general formula $C_n(H_2O)_n$. However, only the simple sugars, or **monosaccharides**, fit this formula exactly. The other types of carbohydrates—namely oligosaccharides and polysaccharides—contain multiple monosaccharide units and have slightly different general formulas. **Oligosaccharides** are formed when a few (Greek *oligos*) monosaccharides are linked; **polysaccharides** are formed when many (Greek *polys*) monosaccharides are bonded together. Similar to the reaction that adds amino acids together to form polypeptide chains (Section 3-4), the reaction that adds monosaccharide units to a growing carbohydrate molecule involves the loss of one H_2O for each new link formed. This loss of H_2O accounts for the deviation of oligosaccharides and polysaccharides from the general formula of the monosaccharides $[C_n(H_2O)_n]$. However, monosaccharides, oligosaccharides, and polysaccharides are all referred to as carbohydrates, while the term *sugar* is commonly used for only the sweet-tasting monosaccharide and disaccharide carbohydrates.

Careers in Chemistry • Mona Minkara

Assistant Professor at Northeastern University

Education:
University of Florida, PhD, Chemistry

Mona Minkara

Mona Minkara is an Assistant Professor and computational chemist who leads the Minkara COMBINE Lab (Computational Modeling for Biointerface Engineering) at Northeastern University. Her research investigates the molecular foundations of innate immunity and respiration, with a particular focus on how collectins interact with glycans on the surfaces of viruses and bacteria.

From an early age, Professor Minkara was curious about the natural world. At the age of seven, she was diagnosed with macular degeneration and cone-rod dystrophy and lost her eyesight. Despite skepticism from doctors and teachers about her ability to succeed in science, she refused to be discouraged. She remained committed to her dream of becoming a scientist.

Throughout her education, Professor Minkara often found herself in classrooms where teachers had little or no experience working with blind students. Undaunted by these obstacles, she developed her own learning strategies. These approaches eventually evolved into the Blind Scientist Toolkit: a growing set of tactile, auditory, and digital resources designed to make science more inclusive and accessible (https://monaminkara.com/blind-scientist-tools).

Professor Minkara encourages students to follow their own paths despite challenges. "Careers in science—like science itself—are nonlinear, shaped by curiosity, persistence, and the courage to ask for help," she says. Her message to students is both practical and empowering: "Every scientist brings a unique lens to their work. Yours matters.'"

16-1A Carbohydrate Functions

Many commonly encountered carbohydrates are polysaccharides, including glycogen, which is found in animals, and starch and cellulose, which occur in plants. Carbohydrates play a number of important roles in biochemistry. First, they are major energy sources (Chapters 17–20 are devoted to carbohydrate metabolism). Second, oligosaccharides play a key role in processes that take place on the surfaces of cells, particularly in cell–cell interactions and immune recognition. In addition, polysaccharides are essential structural components of several classes of organisms. Cellulose is a major structural component of grasses and trees, and other polysaccharides are major structural components of bacterial cell walls.

monosaccharides compounds that contain a single carbonyl group and two or more hydroxyl groups

oligosaccharides compounds that contain a few sugars linked by glycosidic bonds

polysaccharides compounds that contain many sugars linked by glycosidic bonds

16-1B Monosaccharide Structure and Stereochemistry

The building blocks of all carbohydrates are the simple sugars called monosaccharides, which contain a single carbonyl (C=O) group and two or more hydroxyl (–OH) groups. Described using other organic chemistry terminology, a monosaccharide can be a polyhydroxy aldehyde (**aldose**) or a polyhydroxy ketone (**ketose**) (Figure 16.1A). Recall from Section 1-2E that an aldehyde functional group is a carbonyl (C=O) bonded to at least one hydrogen, whereas a ketone functional group is a carbonyl (C=O) bonded to two carbons. The simplest monosaccharides contain three carbon atoms and are called trioses (*tri* meaning "three"). *Glyceraldehyde* is the aldose with three carbons

aldose a sugar that contains an aldehyde group as part of its structure

ketose a sugar that contains a ketone group as part of its structure

(an aldotriose), and *dihydroxyacetone* is the ketose with three carbon atoms (a ketotriose). In this and other chapters, it is helpful to know that if a biochemical compound name ends in "-ose," then the compound is a sugar and that structurally, the compound will contain a carbonyl group and two or more hydroxyl groups.

stereoisomers (optical isomers) molecules that have the same molecular formula and sequence of bonded atoms but differ only in their configuration (the three-dimensional orientation of their atoms)

We saw in Section 3-1 that some molecules are not superimposable on their mirror images and that these mirror images are **stereoisomers (optical isomers)** of each other. In some monosaccharides, nonsuperimposable, *non-mirror-image* stereoisomers are also possible, as shown for glyceraldehyde in Figures 16.1B and 16.1C. A chiral (asymmetric) carbon atom is the usual source of optical isomerism, as was the case with amino acids. Glyceraldehyde is the simplest carbohydrate that contains a chiral carbon. Its two isomeric forms, which are mirror images of each other, differ in the position of the hydroxyl group bonded to the central carbon. (Dihydroxyacetone [Figure 16.1A] does not contain a chiral carbon atom and does not exist in nonsuperimposable mirror-image forms.) The two forms of glyceraldehyde are designated D-glyceraldehyde and L-glyceraldehyde (more details regarding the D and L designations are discussed in Figure 16.2). Mirror-image stereoisomers are also called **enantiomers**, so D-glyceraldehyde and L-glyceraldehyde are enantiomers of each other.

enantiomers mirror-image, nonsuperimposable stereoisomers

A A comparison of glyceraldehyde (an aldotriose) and dihydroxyacetone (a ketotriose).

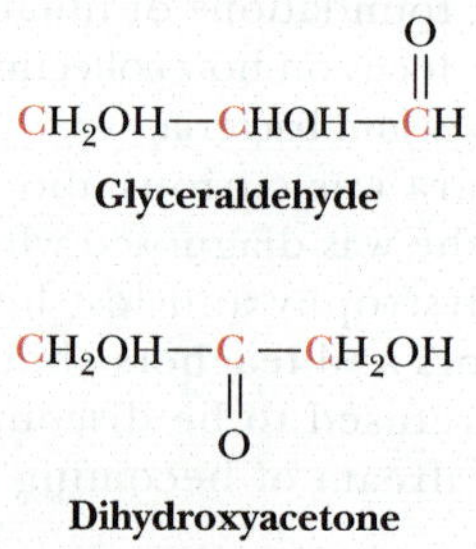

B The structure of D-glyceraldehyde and a space-filling model of D-glyceraldehyde.

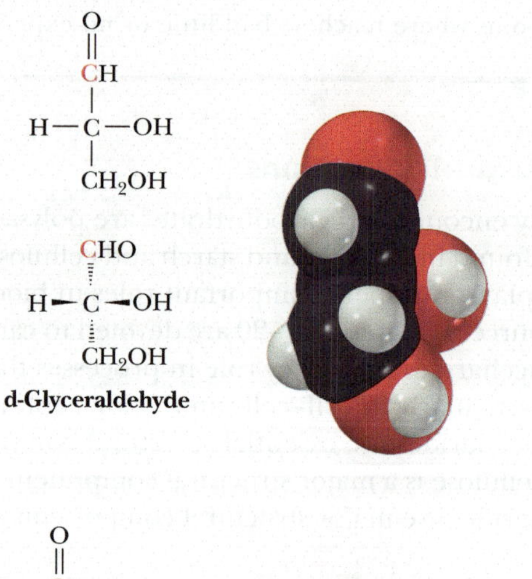

C The structure of L-glyceraldehyde and a space-filling model of L-glyceraldehyde.

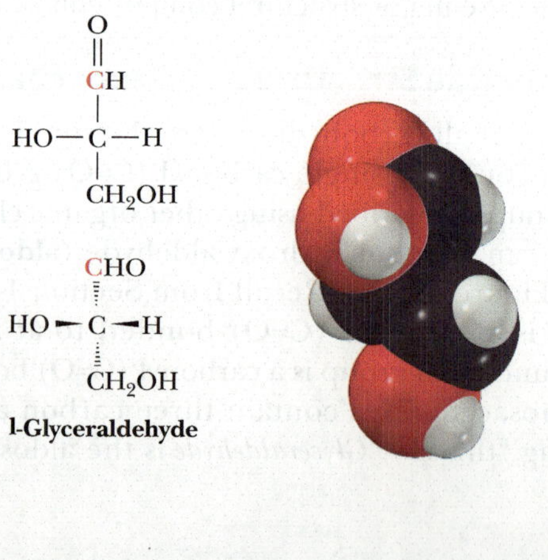

(Leonard Lessin/Waldo Feng/Mt. Sinai CORE.)

Figure 16.1 The structures of the simplest carbohydrates, the trioses.

Certain conventions are used to represent the three-dimensional structures of stereoisomers with two-dimensional drawings. The dashed wedges represent bonds directed away from the viewer, below the plane of the paper, whereas the solid wedges represent bonds directed toward the viewer, above the plane of the paper. The **configuration** is the three-dimensional arrangement of groups around a chiral carbon atom, and stereoisomers differ from each other in configuration. The D,L system to denote stereochemistry is widely used by biochemists. Organic chemists tend to use a more recent one, designated the *R,S* system, which more reliably describes the wide variety of configurational differences seen in organic compounds. There is no one-to-one correspondence between the two systems. For example, some D-isomers are *R*, and some are *S*.

configuration the three-dimensional arrangement of groups around a chiral carbon atom

The two enantiomers of glyceraldehyde are the only possible stereoisomers of three-carbon sugars, but the possibilities for stereoisomerism increase as the number of carbon atoms increases. Aldoses with four, five, six, and seven carbon atoms are called aldotetroses, aldopentoses, aldohexoses, and aldoheptoses, respectively. The corresponding ketoses are ketotetroses, ketopentoses, ketohexoses, and ketoheptoses. To show the possible stereoisomer structures of these molecules, we need to say more about the convention for a two-dimensional perspective of the molecular structure, which is called the **Fischer projection** method, after the German chemist Emil Fischer who established the structures of many sugars.

Fischer projection a two-dimensional representation of the stereochemistry of three-dimensional molecules

Figure 16.2 shows the Fischer projections of some common six-carbon sugars. In a Fischer projection, bonds written vertically on the two dimensional paper represent bonds directed *behind* the paper in three dimensions, whereas bonds written horizontally represent bonds directed *in front of* the paper in three dimensions. Figure 16.2A shows how the most highly oxidized carbon (the carbonyl carbon of monosaccharides) is written at the "top" (or closest to the "top"), and how the numbering of the carbon atoms starts there at the "top." Thus, the aldehyde carbon of an aldose is designated carbon 1 or C-1. In the ketose shown, the ketone group is at C-2, the carbon atom next to the

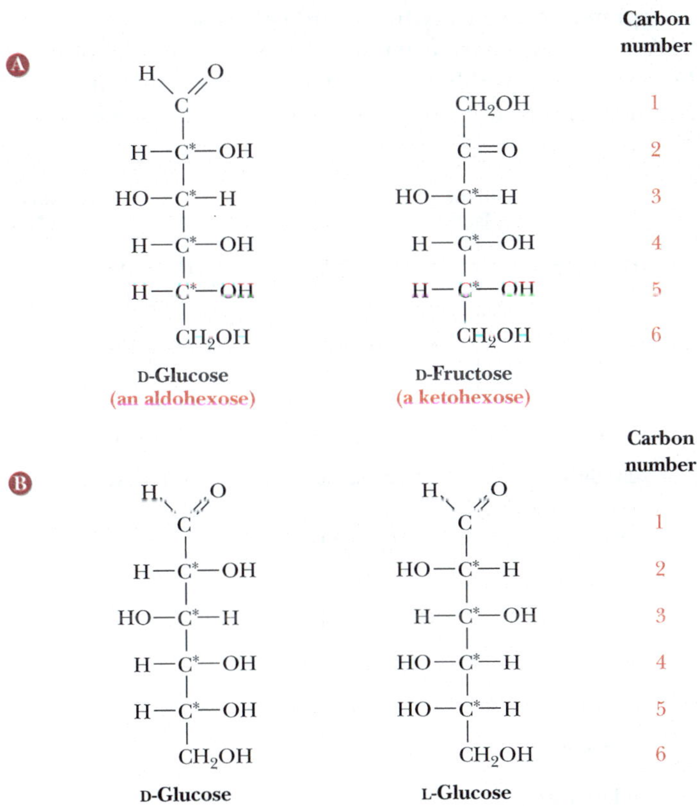

Figure 16.2 Fischer projections of monosaccharides. Chiral carbons are indicated by asterisks. (A) Examples of an aldohexose (D-glucose) and a ketohexose (D-fructose), showing the numbering of the carbon atoms, which always start at the end of the monosaccharide that is closest to the carbonyl group. (B) A comparison of the structures of the D-glucose and L-glucose enantiomers.

"top." The asterisks in Figure 16.2 indicate the *chiral carbons.* Confirm for yourself that each chiral carbon has four different groups bonded to it. Most common sugars are aldoses rather than ketoses, so our discussion focuses mainly on aldoses.

Figure 16.2B compares the structures of the D-glucose and L-glucose enantiomers. Notice how the hydroxyl groups at each chiral carbon (C-2 through C-5) point in opposite directions. At C-2, for example, the OH in D-glucose points to the right, whereas it points to the left in L-glucose. The designation of sugar configurations as L or D depends on the arrangement at the chiral carbon with the highest number, which is C-5 for glucose. In Fischer projections of the D configuration, the hydroxyl group of the highest-numbered chiral carbon points to the right, whereas it points to the left in the L configuration. Confirm for yourself that this is true for the structures in Figure 16.2.

To further explore the stereochemistry of monosaccharides, let us consider the possible aldotetrose stereoisomers (Figure 16.3). In general, the number of possible stereoisomers for a monosaccharide equals 2^n, where n is the number of chiral carbons. Because an aldotetrose has two chiral carbons (C-2 and C-3), there are 2^2, or four, possible stereoisomers (Figure 16.3A).

Figure 16.3B shows the Fischer projections of the two D aldotetrose stereoisomers, D-erythrose and D-threose. These two D-isomers have the same configuration at C-3, but they differ in configuration (arrangement of the —OH group) at the other chiral carbon, C-2. These stereoisomers are *not* superimposable on each other, but neither are they mirror images of each other. Stereoisomers that are nonsuperimposable, non-mirror-images of each other are called **diastereomers**. The two L aldotetrose isomers, L-erythrose and L-threose, are diastereomers, too.

diastereomers nonsuperimposable, non-mirror-image stereoisomers

L-Erythrose is the enantiomer (nonsuperimposable mirror image) of D-erythrose, and L-threose is the enantiomer of D-threose. L-Threose is a diastereomer of both D- and L-erythrose, and L-erythrose is a diastereomer of both D- and L-threose. **Epimers** are diastereomers that differ from each other in the configuration at only one chiral carbon. Therefore, D-erythrose and D-threose are epimers, as are L-erythrose and L-threose.

epimers stereoisomers that differ only in configuration around one chiral carbon atom

As a monosaccharide gets longer and the number of stereoisomers increases, however, it is possible to have diastereomers that are not epimers, as seen in Figure 16.4 with the aldopentose and aldohexose stereoisomers. Aldopentoses

A Enantiomers D- and L-erythrose and D- and L-threose. The carbons are numbered. The designation of D or L depends on the configuration at the highest-numbered chiral carbon atom.

B Diastereomers D-erythrose and D-threose.

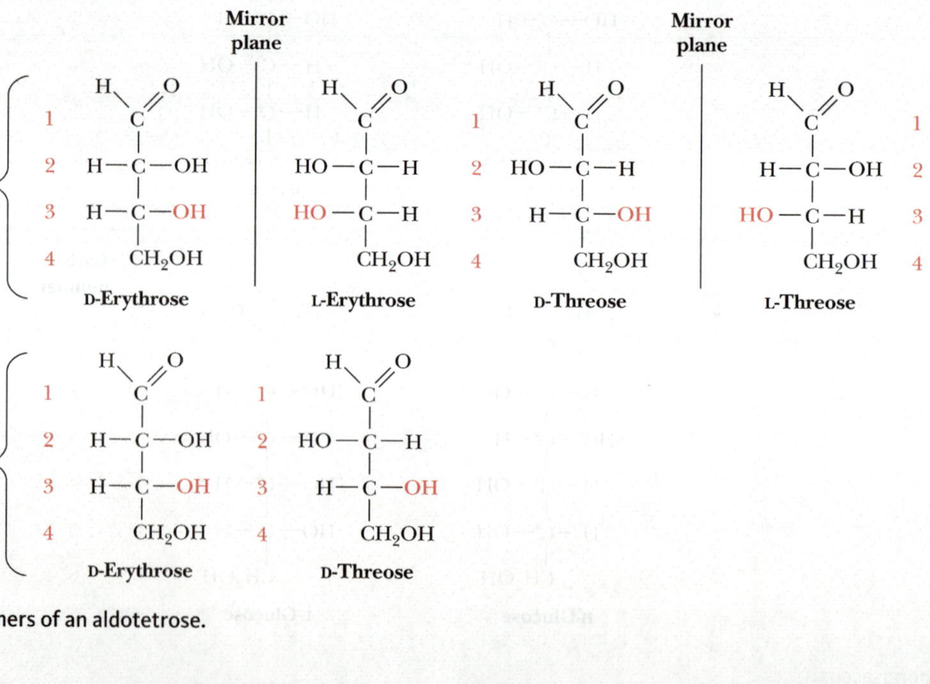

Figure 16.3 Stereoisomers of an aldotetrose.

A Aldoses containing from three to six carbon atoms, with the numbering of the carbon atom shown. Note that the figure shows only half the possible isomers. For each isomer shown, there is an enantiomer that is not shown, the L series.

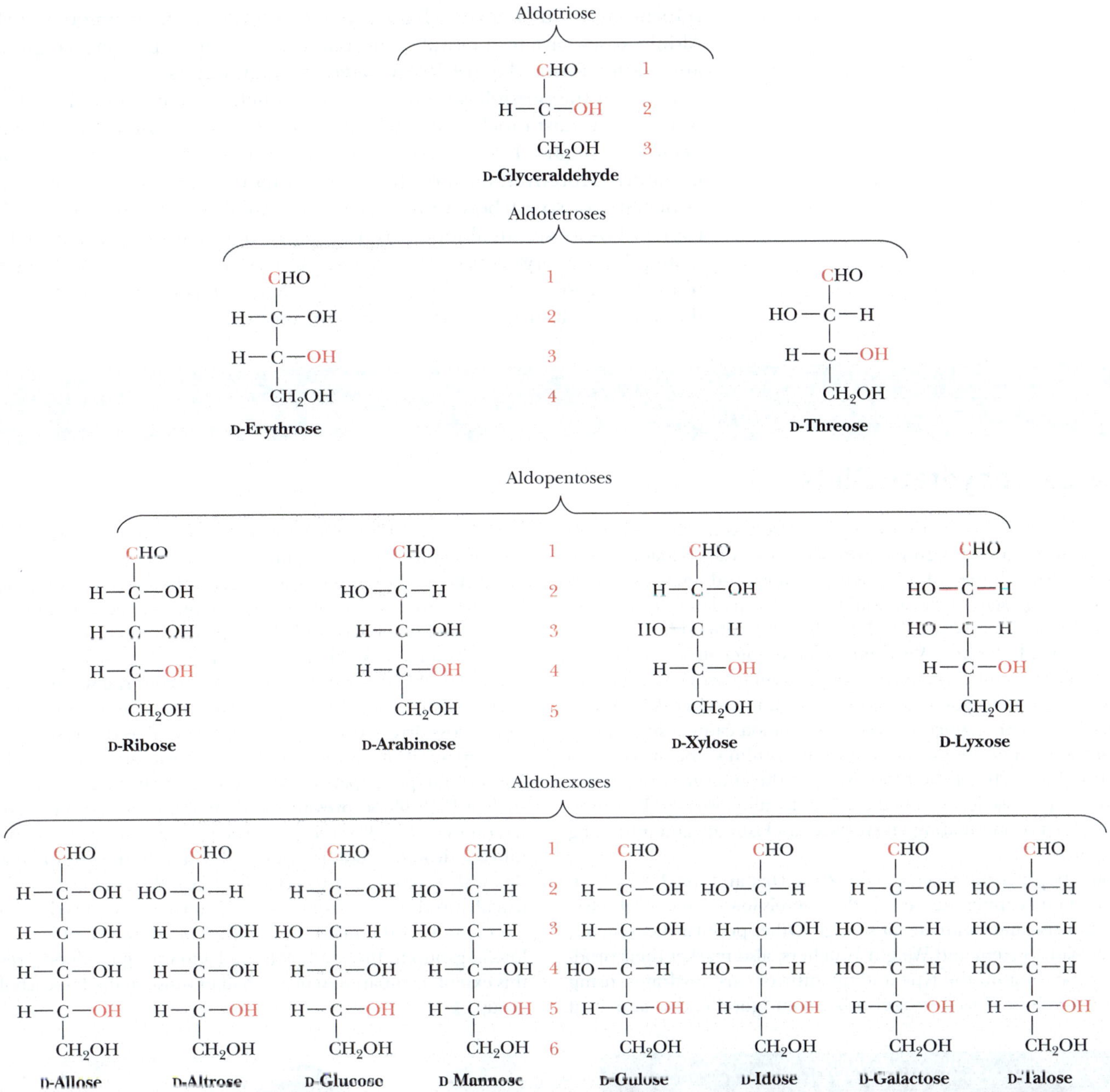

B The relationship between mirror images is of interest to mathematicians as well as to chemists. Lewis Carroll (C. L. Dodgson), the author of *Alice's Adventures in Wonderland*, was a contemporary of Emil Fischer.

Courtesy of Beverlee Tito Simboli

Figure 16.4 Stereochemical relationships among monosaccharides.

have three chiral carbons, so there are $2^3 = 8$ possible stereoisomers—four D forms and four L forms (Figure 16.4A, middle; L-enantiomers not shown). Aldohexoses have four chiral carbons and $2^4 = 16$ stereoisomers—eight D forms and eight L forms (Figure 16.4A, *bottom;* L-enantiomers not shown).

Some of the possible stereoisomers are much more common in nature than others, and most biochemical discussions center on those common, naturally occurring sugars. For example, D sugars, rather than L sugars, predominate in nature. Moreover, most of the sugars in nature, especially in foods, contain either five- or six-carbon atoms. As a result, we discuss D-glucose (an aldohexose) and D-ribose (an aldopentose) far more than any other sugars. Glucose is a ubiquitous energy source, and ribose plays an important role in the structure of nucleic acids. **Biochemical Connections 16A** discusses an important result of the wide distribution of carbohydrates in foods.

16A Biochemical Connections • Nutrition and Health

Low-Carbohydrate Diets

In the 1970s, the diets that were supposed to be the healthiest were low in fat and high in carbohydrates. "Carbo-loading" was the craze for athletes of all types, genders, and ages, as well as for the average sedentary person. Over 30 years later, things have changed so much that you can buy a burger wrapped in a piece of lettuce instead of a bun. Why did a macromolecule once thought to be healthy become something people want to avoid? The answer has to do with how glucose, the primary monosaccharide of life, is metabolized. Rising glucose levels in the blood cause a subsequent rise in the level of the hormone insulin. Insulin stimulates cells to take up glucose from the blood so that the cells get the energy, and blood-glucose levels remain stable. Insulin also has the unfortunate effect of stimulating fat synthesis and storage and inhibiting fat burning.

Some popular diets, such as the Zone Diet and the Atkins Diet, are based on keeping the carbohydrate levels low so that insulin levels do not rise and stimulate fat storage. Other popular diet systems, such as NutriSystem and Weight Watchers, also market their products by focusing on the type and quantity of carbohydrates, using a "glycemic index" to distinguish between "good carbs" and "bad carbs" (Figure 16.5). Carbohydrate sources that are high in fiber are also recognized as promoting overall health.

In the case of athletes, however, little evidence suggests that a low-carbohydrate diet is effective for athletic performance because of the extended time needed to replenish muscle and liver glycogen when the athlete is not on a high-carbohydrate diet.

Early in 2010, a study found that, for overall weight loss, a low-carbohydrate diet was as effective as a regimen that combined weight-loss drugs with a low-fat diet. Useful as this result was, it was accompanied by another finding with wide-ranging implications. Many of the participants in this study had chronic health problems, such as high blood pressure or diabetes, in addition to being overweight or obese. The group on the low-carbohydrate diet had a significant drop in blood pressure compared to the group on a low-fat diet with weight-loss drugs (Figure 16.6). Weight loss in and of itself usually produces a drop in blood pressure, with a reduction or an elimination of the need for medication. However, only 21% of the low-fat group in this study achieved a lowering of blood pressure to this extent, compared to nearly half of those in the low-carbohydrate group. ▶

Charles D. Winters

Figure 16.5 Some food products make a point of their low glycemic index.

Monkey Business Images/Shutterstock.com

Figure 16.6 Weight loss on low-carbohydrate diets can lead to a marked drop in blood pressure.

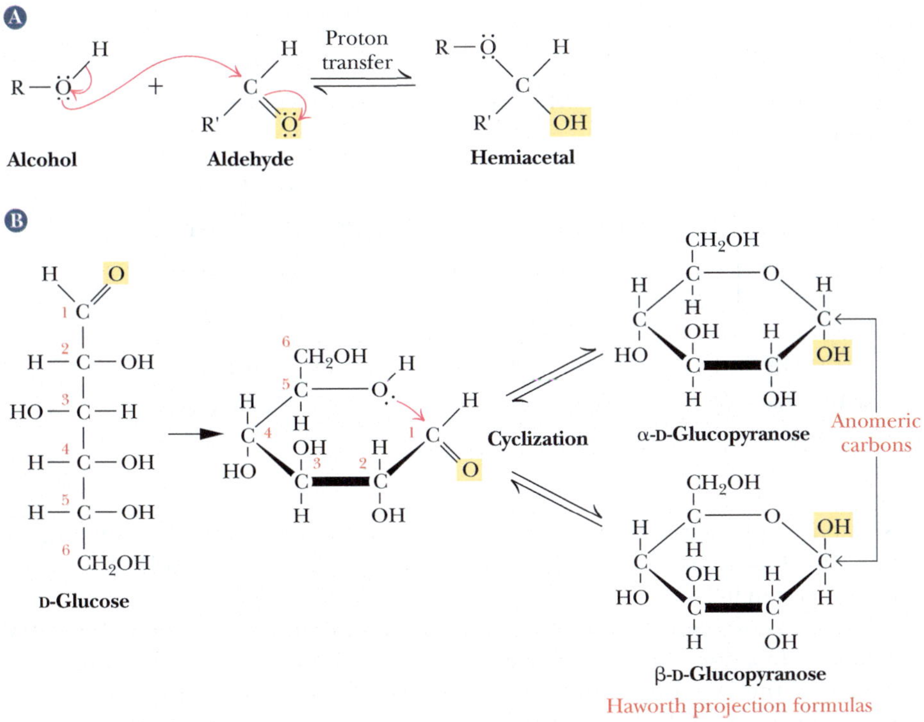

Figure 16.7 The cyclization of a monosaccharide. (A) A general mechanism of hemiacetal formation. (B) The open-chain form of D-glucose undergoes an intramolecular reaction to form a cyclic hemiacetal. Some of the mechanistic details from panel A are missing in panel B due to space constraints, but the same mechanistic details seen in panel A apply in panel B.

16-1C Cyclic Sugars

Sugars, especially those with five- or six-carbon atoms, normally exist as cyclic molecules rather than as the open-chain forms we have shown so far. When a sugar cyclizes, a **hemiacetal** functional group forms, and the hemiacetal group concept is revisited and developed further in Section 16-2B. Figure 16.7A shows a general mechanism for the formation of a hemiacetal. Compare it with the cyclization mechanism for D-glucose shown in Figure 16.7B. The cyclization takes place as a result of interaction between the functional groups on distant carbons, such as C-1 and C-5, to form a cyclic hemiacetal (in aldohexoses).

hemiacetal a compound that is formed by the reaction of an aldehyde with an alcohol and is found in the cyclic structure of sugars; its defining feature is a carbon atom bonded to an –OH group and an –OR group

In this case, the carbonyl carbon becomes a new chiral center called the **anomeric carbon**. The cyclic sugars can take either of two different forms, depending on the orientation of atoms at the anomeric carbon. These different forms are designated α and β and are called **anomers** of each other. In α-anomers of D-glucose, the hydroxyl group on the anomeric carbon points down, below the plane of the sugar ring, whereas in β-anomers of D-glucose, the hydroxyl group on the anomeric carbon points up, above the sugar ring (Figure 16.7B, *right*). A mnemonic to help you remember this involves the following baseball analogy: beta (β) sounds somewhat similar to "batter," and in baseball, a common phrase is "batter up!" Therefore, in the beta anomer, the hydroxyl group points up. The situation is the opposite for anomers of L-glucose, but in all cases, the α-anomer has its anomeric hydroxyl group on the opposite side of the ring as the hydroxymethyl group, while the β-anomer has its anomeric hydroxyl group on the same side of the ring as the hydroxymethyl group.

anomeric carbon the chiral center created when a sugar cyclizes

anomers one of the possible stereoisomers formed when a sugar assumes the cyclic form

The Fischer projection of the α-anomer of a D sugar has the anomeric hydroxyl group to the right of the anomeric carbon (C—OH), whereas the β-anomer of a D sugar has the anomeric hydroxyl group to the left of the anomeric carbon (Figure 16.8). The open-chain carbonyl species can readily form either the α- or β-anomer, and the anomers can be converted from one form to another by opening to the free carbonyl species and re-cyclizing to the other form. In some biochemical molecules, any anomer of a given sugar can be

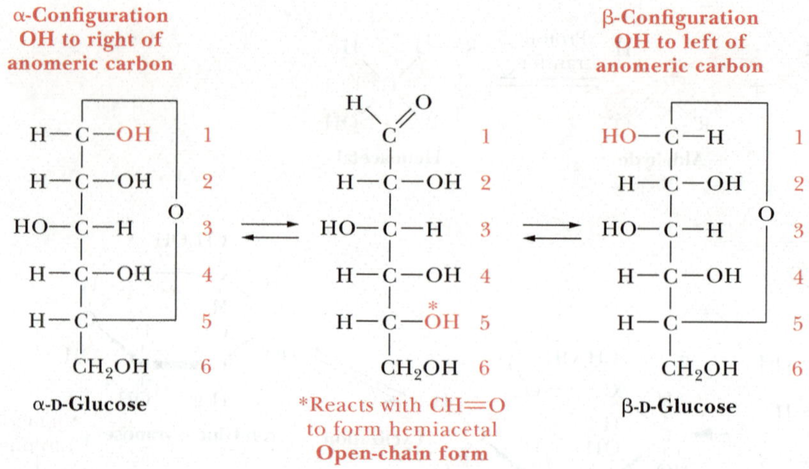

Figure 16.8 Fischer projection formulas of three forms of D-glucose. The α and β forms can be converted into each other through the open-chain form. The configuration at carbon 5 determines the D designation.

used, but in other cases, only one anomer occurs. For example, in living organisms, only β-D-ribose and β-D-deoxyribose are found in RNA and DNA, respectively (Section 9-2A).

Fischer projection formulas are useful for describing the stereochemistry of sugars, but their long bonds and right-angle bends do not give a realistic picture of the bonding situation in the cyclic forms, nor do they accurately represent the overall shape of the molecules. **Haworth projection formulas** (Figure 16.9) are more useful for those purposes. In Haworth projections, the cyclic structures of sugars are shown in perspective drawings as planar five- or six-membered rings viewed nearly edge on. A five-membered ring is called a **furanose** because of its resemblance to furan (Figure 16.9A); a six-membered ring is called a **pyranose** because of its resemblance to pyran (Figure 16.9B). Therefore, glucose is a pyranose, while ribose (Section 9-2A) is a furanose. Recall from Section 16-1B that the "-ose" suffix denotes a sugar, and sugar structures contain several hydroxyl groups and a carbonyl group.

Haworth projection formulas a perspective representation of the cyclic forms of sugars

furanose a cyclic sugar with a five-membered ring, named for its resemblance to the ring system in furan

pyranose a cyclic sugar with a six-membered ring, named for its resemblance to the ring system in pyran

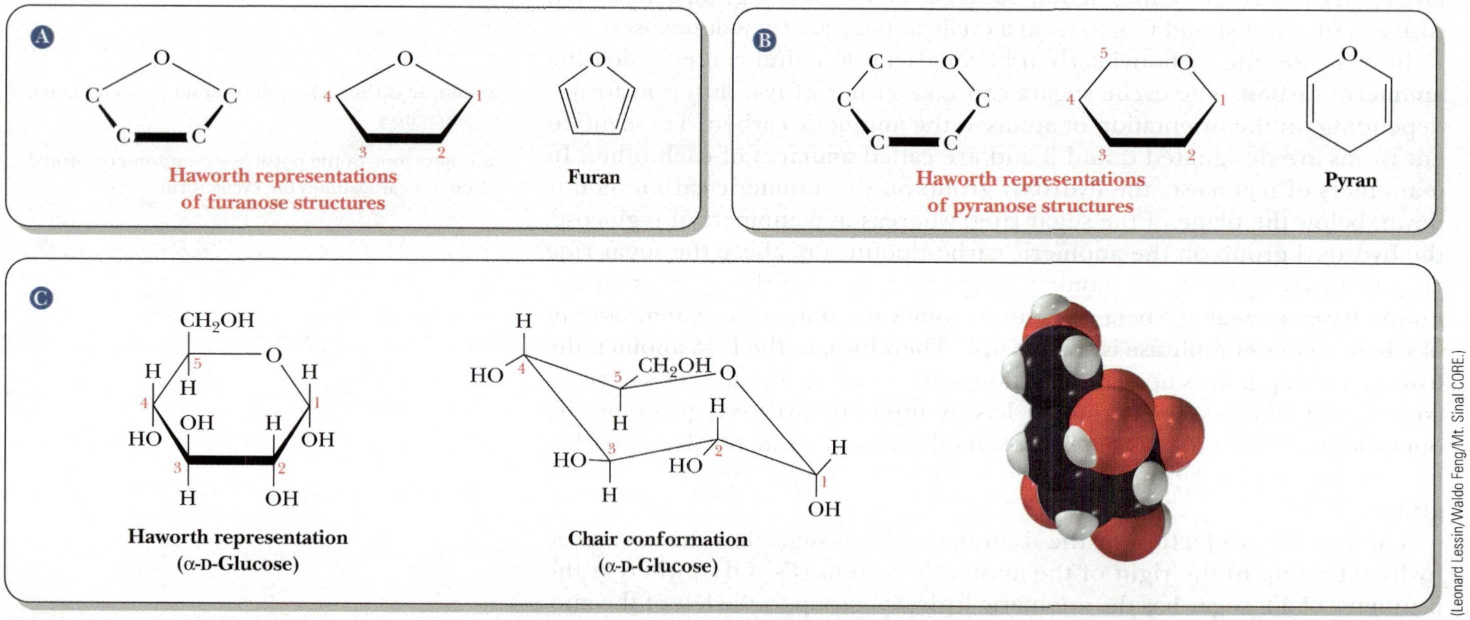

Figure 16.9 **Haworth representations of sugar structures.** (A) A comparison of the structure of furan with Haworth representations of furanoses. (B) A comparison of the structure of pyran with Haworth representations of pyranoses. (C) α-D-Glucopyranose in the Haworth representation (*left*), the chair conformation (*middle*), and as a space-filling model (*right*).

These cyclic formulas approximate the shapes of the actual molecules better for furanoses than for pyranoses. The five-membered rings of furanoses are in reality very nearly planar, whereas the six-membered rings of pyranoses actually exist in solution in the chair conformation (Figure 16.9C). The chair conformation and the Haworth projections are alternative ways of expressing the same information. Even though the Haworth formulas are approximations, they are useful shorthand for the structures of reactants and products in many reactions that we discuss. The Haworth projections represent the stereochemistry of sugars more realistically than do the Fischer projections, and the Haworth scheme is adequate for our purposes. That is why biochemists use it, even though organic chemists prefer the chair form. As biochemists, we use Haworth projections in our discussion of sugars.

Figure 16.10 compares the Fischer and Haworth methods of representing sugar structures. For a D sugar, any group that is written to the *right* of the carbon in a Fischer projection points *downward* in a Haworth projection; any group that is written to the *left* in a Fischer projection points *upward* in a Haworth projection. The terminal —CH_2OH group, which contains the carbon atom with the highest number in the numbering scheme, is shown pointing upward. The structures of α- and β-D-glucose, which are both pyranoses, and β-D-ribose, which is a furanose, illustrate this point. In the α-anomer, the hydroxyl group on the anomeric carbon points down, as mentioned previously in this section. In the β-anomer, the hydroxyl group on the anomeric carbon points up. The same convention holds for α- and β-anomers of furanoses.

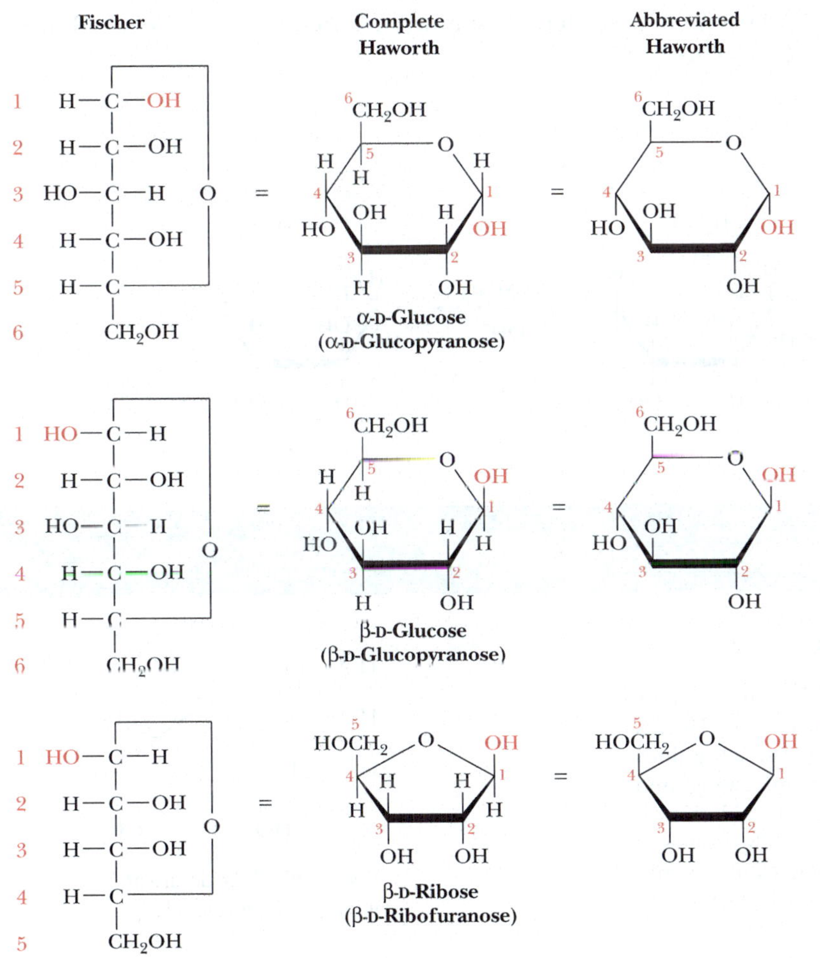

Figure 16.10 A comparison of the Fischer, complete Haworth, and abbreviated Haworth representations of α- and β-D-glucose (glucopyranose) and β-D-ribose (ribofuranose). In the Haworth representation, the α-anomer is represented with the OH group (red) downward, and the β-anomer is represented with the OH group (red) upward.

16-2 Reactions of Monosaccharides

Learning Objectives

After reading this section, students will be able to:

LO 16.2.1 Describe redox chemistry concepts related to simple sugars.

LO 16.2.2 Draw the formation of various types of glycosidic bonds between monosaccharides.

LO 16.2.3 Draw structures of amino sugars.

16-2A Redox Reactions of Simple Sugars

Oxidation and reduction reactions of sugars play key roles in biochemistry. For example, the complete oxidation of sugars in aerobic processes is coupled with a chain of electron transfer reactions that ultimately produce adenosine triphosphate (ATP), which is a nucleotide molecule that provides energy for organisms to carry out their life processes. The reverse of the complete oxidation of sugars is the reduction of CO_2 and H_2O to form sugars, a process that takes place in photosynthesis.

Several oxidation reactions of sugars can be used to identify sugars. Aldehyde groups can be oxidized to give the carboxyl group that is characteristic of acids, and this reaction is the basis of a test for the presence of aldoses. When the aldehyde is oxidized, some oxidizing agent must be reduced. Aldoses are called **reducing sugars** because of this type of reaction; ketoses can also be reducing sugars because they isomerize to aldoses. In the cyclic form, the compound produced by oxidation of an aldose is a *lactone* (a cyclic ester linking the carboxyl group and one of the sugar alcohols, as shown in Figure 16.11). A lactone of considerable importance to humans is discussed in **Biochemical Connections 16B**.

reducing sugars sugars that have a free hemiacetal group (or a free hemiketal group) that can react with an oxidizing agent

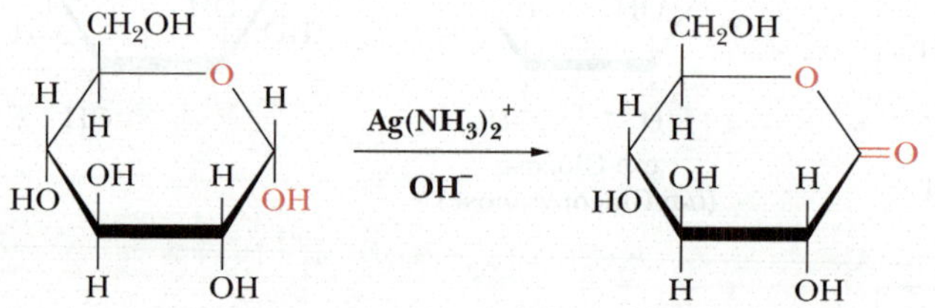

Figure 16.11 Oxidation of a sugar to a lactone. An example of an oxidation reaction of sugars: oxidation of α-D-glucose hemiacetal to give a lactone. Deposition of free silver as a silver mirror indicates that the reaction has taken place.

16B Biochemical Connections • Nutrition

Vitamin C Is Related to Sugars

Vitamin C (ascorbic acid) is an unsaturated lactone (a cyclic ester) with a five-membered ring structure (Figure 16.12). Most animals can synthesize vitamin C; the exceptions are guinea pigs and primates, including humans. As a result, guinea pigs and primates must acquire vitamin C in their diet. Air oxidation of ascorbic acid when certain foods go stale, followed by hydrolysis of the ester bond, leads to a loss of activity as a vitamin. Consequently, a lack of fresh food can cause vitamin C deficiencies, which, in

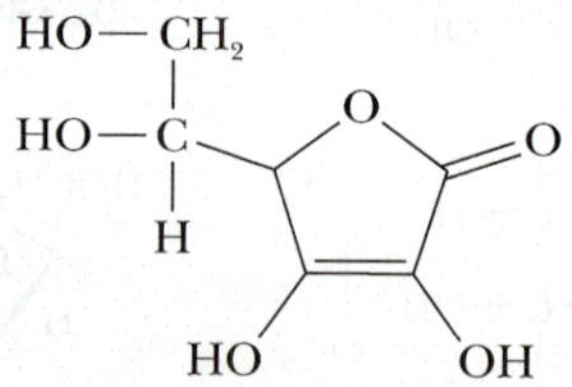

Figure 16.12 Structure of vitamin C (ascorbic acid).

turn, can lead to the disease scurvy. In this disease, defects in collagen structure cause skin lesions and fragile blood vessels. The presence of hydroxyproline is necessary for collagen stability because of hydrogen-bonded cross-links between collagen strands (Section 4-3E). Ascorbic acid is essential for the activity of prolyl hydroxylase, which converts proline residues in collagen to hydroxyproline. Therefore, a lack of ascorbic acid eventually leads to the fragile collagen responsible for the symptoms of scurvy.

The British navy introduced citrus fruit into the diet of sailors in the eighteenth century to prevent scurvy during long sea voyages, and many people still consume citrus fruit for its vitamin C.

Potatoes are another important source of vitamin C, not because potatoes contain a *high* concentration of ascorbic acid but because we eat so many potatoes. ◗

Two types of reagent are used in the laboratory to detect the presence of reducing sugars. The first of these is Tollens reagent, which uses the silver ammonia complex ion, $Ag(NH_3)_2^+$, as the oxidizing agent. A silver mirror is deposited on the wall of the reaction vessel if a reducing sugar is present, as a result of the Ag^+ in the complex ion being reduced to free silver metal (Figure 16.13). A more modern method, specific for the detection of glucose but not other reducing sugars, is based on the use of the enzyme glucose oxidase.

In addition to oxidized sugars, there are some important reduced sugars. In *deoxy sugars,* a hydrogen atom is substituted for one of the hydroxyl groups of the sugar. One of these deoxy sugars is L-fucose (L-6-deoxygalactose), which is found in the carbohydrate portions of some glycoproteins (Figure 16.14), including the ABO blood-group antigens. The name *glycoprotein* (Section 16-5) indicates that these substances are proteins that are attached to some carbohydrate group (*glykos* is Greek for "sweet"). An even more important example of a deoxy sugar is D-2-deoxyribose, the sugar found in DNA (Figure 16.14).

The carbonyl group of the sugar can be reduced to a hydroxyl group. The resulting compound is one of the polyhydroxy alcohols known as *alditols.* Two compounds of this kind, xylitol and sorbitol, are derivatives of the sugars xylulose and sorbose, respectively. They have commercial importance as sweeteners in sugarless chewing gum and candy (Figure 16.15). Even though alditols have several hydroxyl groups, they are not classified as sugars because they lack a carbonyl group. Alditols are sometimes referred to as *sugar alcohols.*

Charles D. Winters

Figure 16.13 A silver mirror produced by the oxidation of an aldehyde group. After the addition of Tollens reagent to an aldehyde, a silver mirror has been deposited on the inside of this flask.

16-2B Sugar Esters and Glycosidic Bonds

▶ What are some important esterification reactions of sugars?

The hydroxyl groups of sugars behave exactly like all other alcohols in the sense that they can react with acids and derivatives of acids to form esters. The phosphate esters are particularly important because they are the usual intermediates in the breakdown of carbohydrates to provide energy. Phosphate esters are frequently formed by the transfer of a phosphate group from ATP to

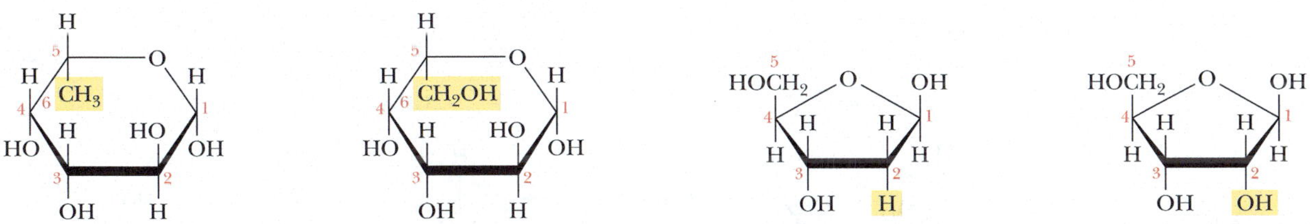

β-L-Fucose (6-Deoxy-β-L-galactose) β-L-Galactose β-D-Deoxyribose (2-Deoxy-β-D-ribose) β-D-Ribose

Figure 16.14 Structures of two deoxy sugars, β-L-fucose and β-D-deoxyribose. The structures of their parent sugars, β-L-galactose and β-D-ribose, respectively, are shown for comparison.

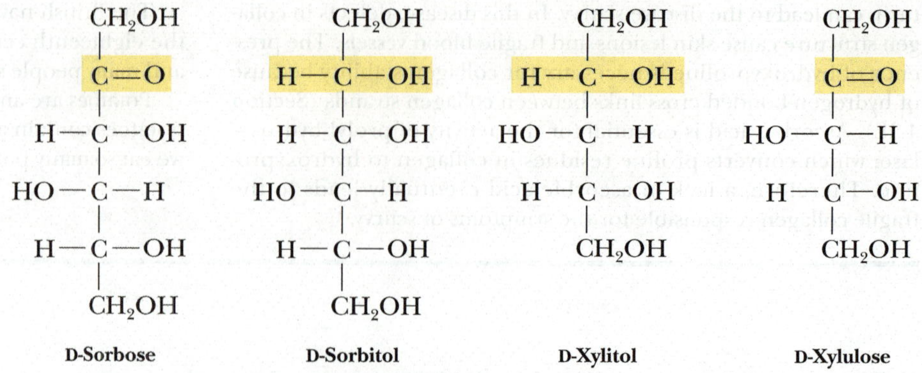

Figure 16.15 The structures of two common alditols, D-sorbitol and D-xylitol. The structures of their related sugars, D-sorbose and D-xylulose, respectively, are shown for comparison.

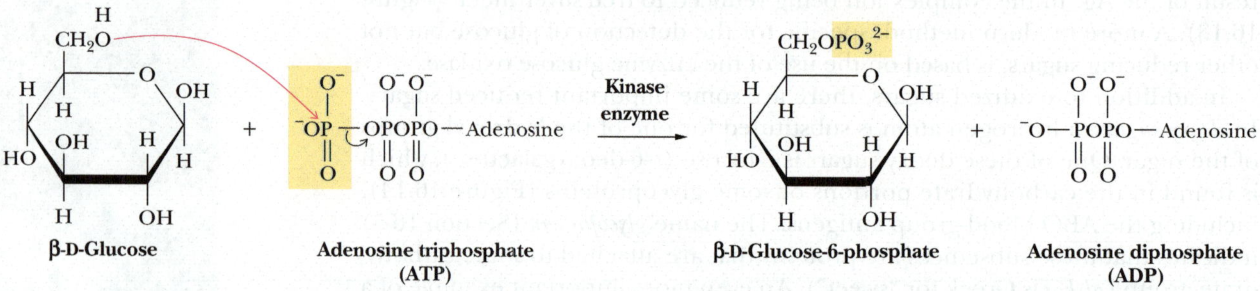

Figure 16.16 The formation of a phosphate ester of glucose. Adenosine triphosphate (ATP) is the phosphate group donor, and the kinase enzyme interacts with both ATP and the hydroxyl group on carbon 6 to specify the phosphorylation of carbon 6.

give the phosphorylated sugar and adenosine diphosphate (ADP), as shown in Figure 16.16. Such reactions play an important role in the metabolism of sugars (Section 17-2).

▶ What are glycosides, and how do they form?

acetal group a functional group consisting of a carbon bonded to two —OR groups

glycoside a compound in which a sugar is bonded to a non-sugar molecule

glycosidic bond a bond between a sugar and another molecule, which may or may not be another sugar molecule

It is also possible for an anomeric carbon hydroxyl group to react with a hydroxyl group on another molecule to form a glycosidic bond (Figure 16.17). Recall from Figure 16.7A how a hemiacetal forms when a sugar cyclizes. Here, the formation of a glycosidic linkage forms an **acetal group**, which consists of a carbon bonded to two —OR groups. This type of reaction involves the anomeric carbon of the sugar in its cyclic form. (Recall that the anomeric carbon is the carbonyl carbon of the open-chain form of the sugar and is the one that becomes a chiral center in the cyclic form.) In general, a sugar hemiacetal carbon can react with an alcohol to give a *full acetal* or **glycoside**. The newly formed bond is called a **glycosidic bond**. The glycosidic bonds discussed in this

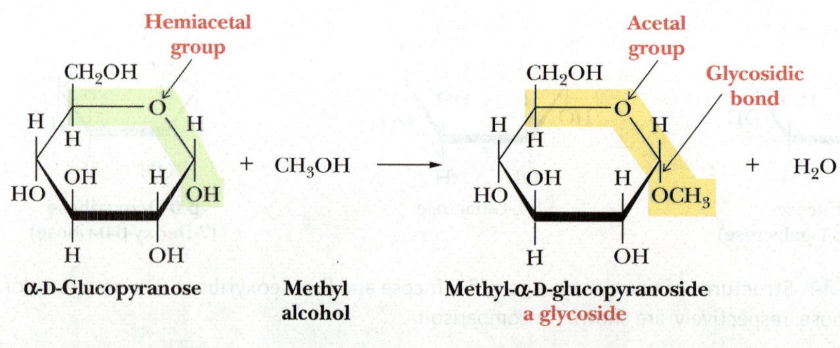

Figure 16.17 An example of the formation of a glycoside. Methyl alcohol (CH_3OH) and an α-D-glucopyranose react to form the corresponding glycoside. In these types of reactions, the hemiacetal group becomes an acetal group.

chapter are *O*-glycosidic bonds, with each sugar bonded to an oxygen atom of another molecule. (We encountered *N*-glycosidic bonds in Section 9-2A when we discussed nucleosides and nucleotides, in which the sugar is bonded to a nitrogen atom of a nucleobase.) Glycosides derived from furanoses are called **furanosides**, and those derived from pyranoses are called **pyranosides**.

furanosides glycosides involving a sugar with a five-membered ring

pyranosides glycosides involving a sugar with a six-membered ring

Now that we have introduced sugars and discussed the formation of glycosidic bonds between a sugar and a non-sugar molecule, let us next discuss glycosidic bonds between sugars that are involved in oligosaccharides and polysaccharides. Glycosidic linkages in these molecules can take various forms; the anomeric carbon of one sugar can be bonded to any one of the —OH groups on a second sugar to form an α- or β-glycosidic linkage. Many different combinations are found in nature. The —OH groups are numbered so that they can be distinguished, and the numbering scheme follows that of the carbon atoms. The notation for the glycosidic linkage between the two sugars specifies which anomeric form of the sugar is involved in the bond and also specifies which carbon atoms of the two sugars are linked together.

For example, Figure 16.18 shows two ways in which two α-D-glucose molecules can be linked together to form two different *disaccharides*. In an α(1 → 4) linkage, the α-anomeric carbon (C-1) of the first glucose molecule is joined in a glycosidic bond to the fourth carbon atom (C-4) of the second glucose molecule. In an α(1 → 6) linkage, the C-1 of the first glucose molecule is linked to the C-6 of the second glucose molecule. Figure 16.19 shows another possibility of a glycosidic bond, this time between two β-D-glucose molecules joined in a β,β(1 → 1) linkage. The anomeric forms at both C-1 carbons must be specified because the linkage is between the two anomeric carbons, each of which is C-1.

In all of the linkages shown in Figures 16.18 and 16.19, the α or β notation for each linkage only specifies the direction that the *first* bond points when going from the anomeric carbon. It does *not* specify the direction that the second bond then points when going to the other carbon. The directionality of the second bond is specified by the stereochemistry of the hydroxyl group at the other carbon. For example, many of the linkages seen in the rest of this

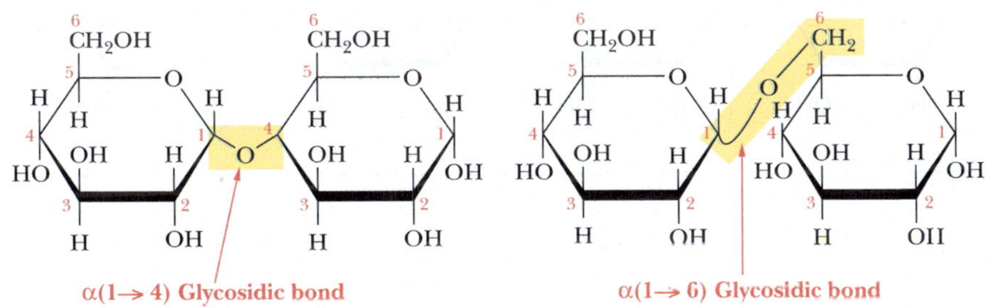

Figure 16.18 Two different disaccharides of α-D-glucose. These two chemical compounds have different properties because one has an α(1 → 4) linkage and the other has an α(1 → 6) linkage.

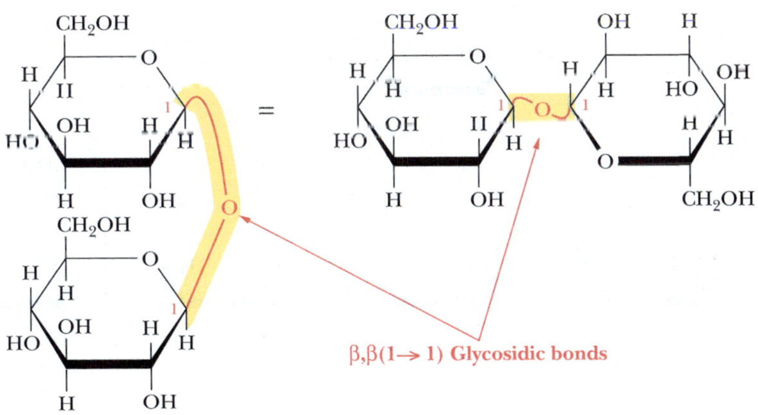

Figure 16.19 A disaccharide of β-D-glucose. Both anomeric carbons (C-1) are involved in the glycosidic linkage.

chapter go from an anomeric carbon of one glucose residue to the fourth carbon (C-4) of another glucose residue, and the hydroxyl group of C-4 in the cyclic form of glucose (as it is typically drawn) points down. Therefore, all the glycosidic bonds that we show from the fourth carbon of glucose point downward, no matter whether they are part of an α- or β-linkage. Also, some of the linkages in Figures 16.18 and 16.19 are curved instead of straight, and this is done for a variety of reasons, including artistic flexibility when drawing disaccharides and polysaccharides.

When oligosaccharides and polysaccharides form as a result of glycosidic bonding, their chemical natures depend on which monosaccharides are linked together and also on the particular glycosidic bond formed (i.e., which anomers and which carbon atoms are linked together). For example, the structural difference between cellulose and starch depends on the glycosidic bond formed between glucose monomers in each polysaccharide. Cellulose and starch differ in their glucose anomer residue composition (β and α anomer residues, respectively). This subtle difference in orientation leads to drastically different properties for cellulose compared to starch (Section 16-4). Also, both linear and branched-chain polymers can be formed in the starch polysaccharide (Figure 16.20). Amylose is a linear form of starch, and it is composed only of α(1 → 4) linkages (Figure 16.20A). Amylopectin is a branched form of starch, and it has α(1 → 6) linkages that occur periodically along the α(1 → 4) chains (Figure 16.20B).

We have seen that the anomeric carbon is involved in the glycosidic linkage and that the test for the presence of sugars—specifically for reducing sugars—requires a reaction of the group at the anomeric carbon. The internal anomeric carbons in oligosaccharides are not free to give the test for reducing sugars. Only if the end residue is a free hemiacetal rather than a glycoside

A The linear polyglucose chain occurs in amylose. All glycosidic bonds are α(1→ 4).

B The branched-chain polymer occurs in amylopectin and glycogen. Branched-polyglucose-chain glycosidic bonds are α(1→ 6) at branch points, but all glycosidic bonds along the chain are α(1→ 4).

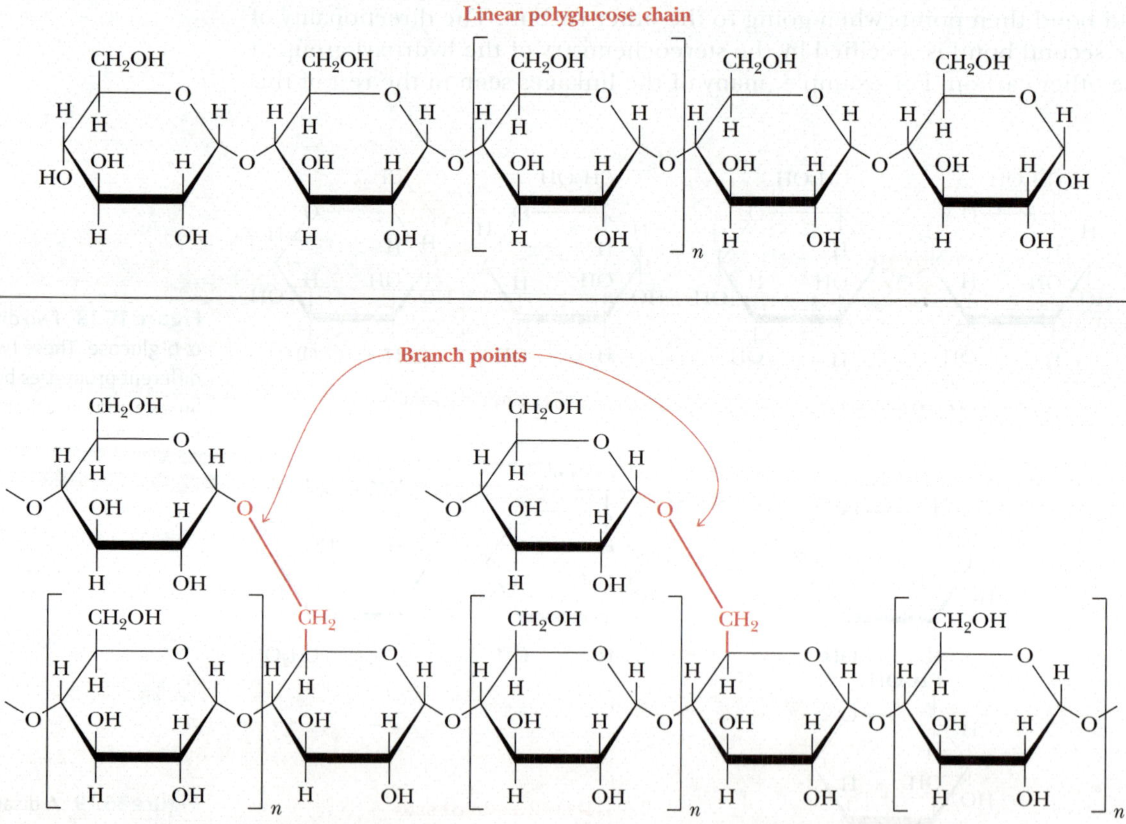

Figure 16.20 Linear and branched-chain polymers of α-D-glucose.

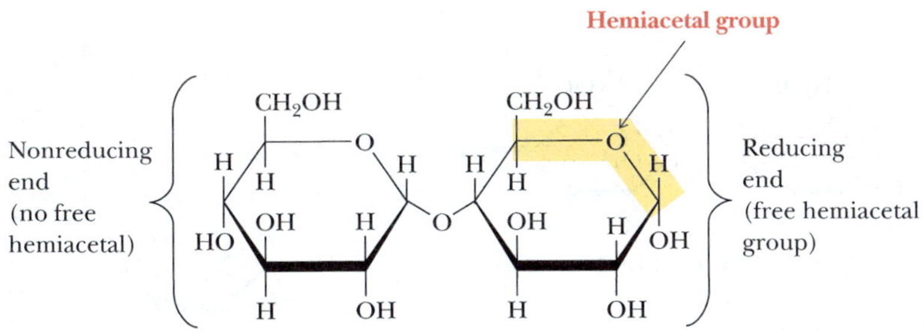

Figure 16.21 **A reducing sugar.** Any carbohydrate with a free hemiacetal end is a reducing sugar. Reducing sugars do contain a reducing end and a nonreducing end, however.

will there be a positive test for a reducing sugar (Figure 16.21). The level of detection can be important for such a test. A sample that contains molecules of a large polysaccharide, where each large molecule has only one reducing end, might well produce a negative test because there are not enough reducing ends to detect.

16-2C Amino Sugars

Amino sugars are a class of compounds related to the monosaccharides. We do not go into the chemistry of their formation, but it is useful to have some acquaintance with them when we discuss polysaccharides. In sugars of this type, an amino group (—NH_2) or one of its derivatives is substituted for a hydroxyl group of the parent sugar.

amino sugars sugars with a substituted amino group as part of their structure

In *N*-acetyl amino sugars, the amino group itself carries an *acetyl group* (CH_3—CO—) as a substituent (Figure 16.22). Two particularly important examples are *N*-acetylglucosamine (NAG) and its derivative, *N*-acetylmuramic acid (NAM), which has a carboxylic acid reacted with the hydroxyl group on the third carbon. These two compounds are components of bacterial cell walls. We did not specify whether NAM belongs to the L or D series of configurations, and we did not specify the α- or β-anomer. This type of shorthand is the usual practice with β-D-glucose and its derivatives; the D configuration and the β-anomeric form are so common that we need not specify them all the time unless we want to make a specific point. The position of the amino group is also left unspecified because the discussion of amino sugars usually centers on a few compounds whose structures are well known.

16-3 Some Important Oligosaccharides

Learning Objective

After reading this section, students will be able to:

LO 16.3.1 Distinguish between the structures of various disaccharide molecules.

Oligomers of sugars frequently occur as **disaccharides**, formed by linking two monosaccharide units by glycosidic bonds. Three of the most important examples of oligosaccharides are disaccharides. They are sucrose, lactose, and

disaccharides two monosaccharides (monomeric sugars) linked by a glycosidic bond

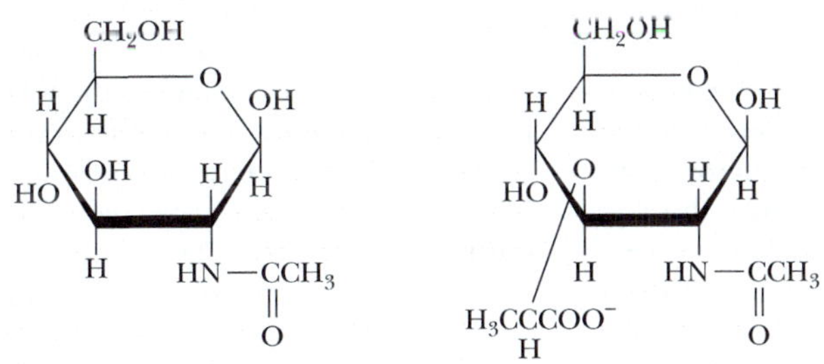

Figure 16.22 The structures of *N*-acetylglucosamine (NAG) and *N*-acetylmuramic acid (NAM).

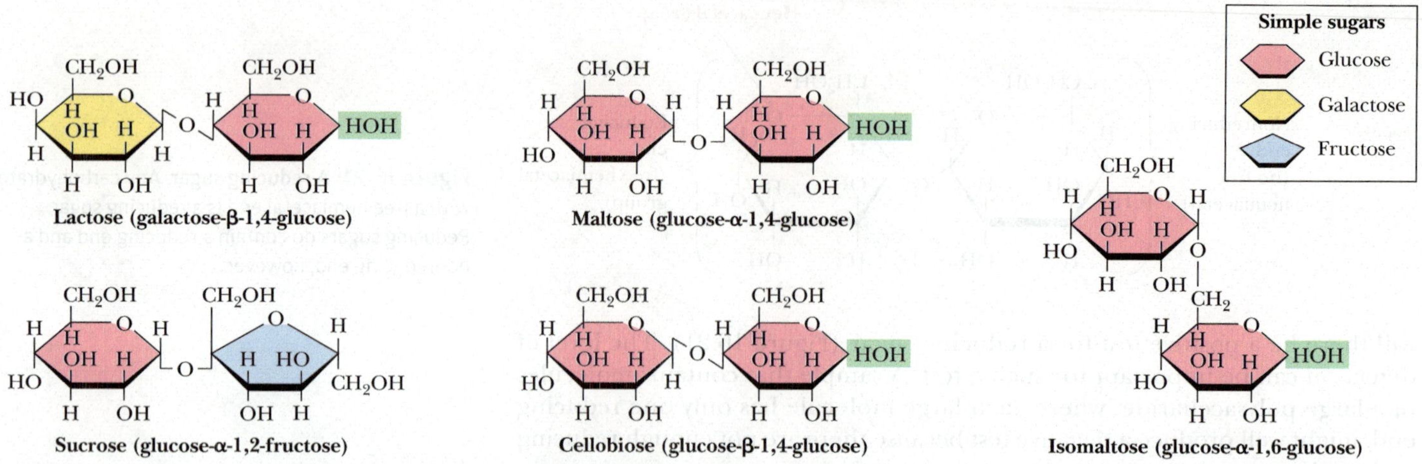

Figure 16.23 The structures of several important disaccharides. The notation —HOH at the anomeric carbons means that the configuration can be either α or β. Each of these disaccharides (except sucrose) is a reducing sugar because they have free hemiacetal groups (see the right side of each disaccharide as they are drawn here). Sucrose has no free hemiacetal, so it is a nonreducing sugar. Nevertheless, every one of the other reducing sugars in this figure does have a *nonreducing end* (see the left side of each disaccharide as they are drawn here).

maltose (Figure 16.23). Two other disaccharides, isomaltose and cellobiose, are shown for comparison.

▶ What makes sucrose an important compound?

Sucrose, which is common table sugar, is extracted from sugarcane and sugar beets. The monosaccharide units that make up sucrose are α-D-glucose and β-D-fructose. Glucose (an aldohexose) is a pyranose, whereas fructose (a ketohexose) is a furanose. The α-C-1 carbon of the glucose is linked to the β-C-2 carbon of the fructose (Figure 16.23) in a glycosidic linkage that has the notation $\alpha,\beta(1 \rightarrow 2)$. Sucrose is not a reducing sugar because both anomeric groups are involved in the glycosidic linkage, so the resulting molecule has no free hemiacetal group. Free glucose is a reducing sugar, and free fructose can also give a positive test, even though it is a ketone rather than an aldehyde in the open-chain form. Fructose and ketoses in general can act as reducing sugars because they can isomerize to aldoses in a rather complex rearrangement reaction. (We need not concern ourselves with the details of this isomerization.) The sucrose in table sugar improves the taste of many dietary items, but too much sucrose can cause health problems. The search for healthier sweeteners is described in **Biochemical Connections 16C.**

16C Biochemical Connections • Nutrition

Artificial Sweeteners

When animals consume sucrose, it is hydrolyzed to glucose and fructose, which are then degraded by metabolic processes to provide energy. Humans consume large quantities of sucrose, and excess consumption can contribute to health problems; this fact has led to a search for other sweetening agents. One that has been proposed is fructose itself. It is sweeter than sucrose; therefore, a smaller amount (by weight) of fructose than sucrose can produce the same sweetening effect with fewer calories. Consequently, high-fructose corn syrup is frequently used in food processing. The presence of fructose changes the texture of food, and the reaction to the change tends to depend on the preference of the consumer.

Artificial sweeteners have been produced in the laboratory and have frequently been suspected of having harmful side effects; the ensuing controversies bear eloquent testimony to the human craving for sweets. Saccharin, for example, has been found to cause cancer in laboratory animals, as have cyclamates, but the applicability of these results to human carcinogenesis has been questioned by some. Aspartame (NutraSweet) has been suspected of causing neurological problems, especially in individuals whose metabolism cannot tolerate phenylalanine.

Sucralose (Figure 16.24), a derivative of sucrose, is another artificial sweetener. Marketed under the trade name Splenda, it differs from sucrose in two ways:

1. Three of the hydroxyl groups in sucrose have been replaced with chlorine atoms.
2. The configuration at carbon atom 4 of the six-membered pyranose ring of glucose has been inverted, producing a galactose derivative.

The three hydroxyl groups that have been replaced by chlorine atoms are those bonded to carbon atoms 1 and 6 of the fructose moiety and to carbon atom 4 of the galactose moiety. Sucralose is not metabolized by the body, so it does not provide calories. Tests conducted so far, as well as anecdotal evidence, indicate that it is a safe sugar substitute. It is likely to find wide use in the near future. We can safely predict that the search for nonfattening sweeteners will continue and that it will be accompanied by controversy.

General acceptance of the artificial sweeteners aspartame, sucralose, and saccharin has led many dieters to question the health benefits when their use does not appear to provide calorie-saving and weight-loss benefits. Researchers may have identified the reason for this apparent disconnect. A compelling correlation has been made between the use of artificial sweeteners and an increase in the population of bacteria in the gut microbiome that is responsible for enhanced calorie extraction from food, increased fat production, and hormone production changes that positively influence appetite. The net effect is that the change to the gut microbiome appears to negatively affect glucose metabolism and may lead to glucose intolerance. These findings are consistent with the observed prevalence of obesity and type 2 diabetes mellitus among hard-core artificial sweetener consumers.

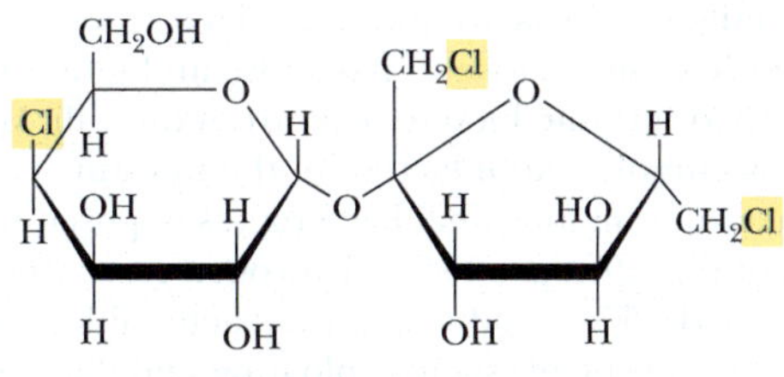

Figure 16.24 **The structure of sucralose.** Marketed under the trade name Splenda, it differs from sucrose in the substitution of chlorine for three hydroxyls.

▶ Are any other disaccharides important to us?

Lactose is a disaccharide made up of β-D-galactose and D-glucose (Figure 16.23). Galactose is the C-4 epimer of glucose. In other words, the only difference between glucose and galactose is the inversion of configuration at C-4. The glycosidic linkage is β(1 → 4), between the anomeric carbon C-1 of the β form of galactose and the C-4 carbon of glucose. Because the anomeric carbon of glucose is *not* involved in the glycosidic linkage, it can be in either the α or the β form. The two anomeric forms of lactose can be specified, and the designation refers to the glucose residue; galactose must be present as the β-anomer because the β form of galactose is required by the structure of lactose. Lactose is a reducing sugar because the group at the anomeric carbon of the glucose portion is not involved in a glycosidic linkage, so it is free to react with oxidizing agents. The manner in which lactose is metabolized by the body can have important health implications, which are described in **Biochemical Connections 16D**.

16D Biochemical Connections • Nutrition

Lactose Intolerance

Humans can be intolerant of cow's milk and cow's milk products for several reasons. Sugar intolerance results from the inability either to digest or to metabolize certain sugars. This problem differs from a food allergy, which involves an immune response (Section 14-3). A negative reaction to sugars in the diet usually involves intolerance, whereas proteins, including those found in milk, tend to cause allergies. Most sugar intolerance is due to missing or defective enzymes, so this is another example of an inborn error of metabolism.

Lactose is sometimes referred to as milk sugar because it occurs in cow's milk. In some adults, a deficiency of the enzyme lactase in the intestinal villi causes a buildup of the disaccharide

Continued on next page

when cow's milk products are ingested. This is because lactase is necessary to degrade lactose to galactose and glucose so that it can be absorbed into the bloodstream from the villi. Without the enzyme, an accumulation of lactose in the intestine can be acted on by the lactase from intestinal bacteria (as opposed to the desirable lactase of the villi), producing hydrogen gas, carbon dioxide, and organic acids. The products of the bacterial lactase reaction lead to digestive problems such as bloating and diarrhea, as does the presence of undegraded lactose. In addition, the by-products of the extra bacterial growth draw water into the intestine, thus aggravating the diarrhea. This disorder affects only about one-tenth of the white population of the United States, but it is more common among Black Americans, Asians, Native Americans, and Hispanics.

Even if the enzyme lactase is present so that lactose can be broken down by the body, a different but related problem can occur in the further metabolism of galactose. If the enzyme that catalyzes a subsequent reaction in the pathway is missing and galactose builds up, a condition known as galactosemia can result. This is a severe problem in infants because the nonmetabolized galactose accumulates within cells and is converted to the hydroxy sugar galactitol, which cannot escape. Water is drawn into these cells, and the swelling and edema cause damage. The critical tissue is the brain, which is not fully developed at birth. The swelling cells can crush the brain tissue, resulting in severe and irreversible intellectual disability. The clinical test for this disorder is inexpensive and is required by law in all states.

The dietary therapy for these two problems is quite different. Individuals with lactose intolerance must avoid lactose throughout their lives. Fortunately, tablets like Lactaid are available to add to cow's milk, as are lactose- and galactose-free formulas for feeding infants. True fermented food products such as yogurt and many cheeses (especially aged ones) have had their lactose degraded during fermentation. However, many foods are not processed in this way, so individuals with lactose intolerance need to exercise caution in their food choices.

There is no way to treat cow's milk to make it safe for people who have galactosemia, so affected individuals must avoid cow's milk during childhood. Fortunately, a galactose-free diet is easy to achieve simply by avoiding cow's milk or by selecting from a variety of lactose-free products (Figure 16.25). After puberty, the development of other metabolic pathways for galactose alleviates the problem in most afflicted individuals. For people who want to avoid cow's milk, there are plenty of milk substitutes, such as soy milk, almond milk, or rice milk. You can even get your latte or mocha made with dairy substitutes at coffee shops. ◗

Lena Bukovsky/Shutterstock.com

Figure 16.25 Dairy substitutes for lactose intolerance. Nondairy products with nut, oat, rice, or soy ingredients help those with lactose intolerance meet their calcium needs.

Maltose is a disaccharide obtained from the hydrolysis of starch. It consists of two residues of D-glucose in an $\alpha(1 \rightarrow 4)$ linkage (Figure 16.23). Maltose differs from *cellobiose,* a disaccharide that is obtained from the hydrolysis of cellulose, only in the glycosidic linkage. In cellobiose, the two residues of D-glucose are bonded together in a $\beta(1 \rightarrow 4)$ linkage (Figure 16.23). Mammals can digest maltose but not cellobiose. Yeast, specifically brewer's yeast, contains enzymes that hydrolyze the starch in sprouted barley (barley malt), first to maltose and then to glucose, which is fermented in the brewing of beer. Maltose is also used in other beverages such as malted milk.

16-4 Structures and Functions of Polysaccharides

Learning Objectives

After reading this section, students will be able to:

LO 16.4.1 Distinguish between the structures of various forms of linear and branched homopolysaccharides of glucose.

LO 16.4.2 Describe the structural and functional aspects of chitin.

LO 16.4.3 Explain the structural and functional aspects of cell wall and connective tissue molecules.

When many monosaccharides are linked together, the result is a **polysaccharide**. Polysaccharides that occur in organisms are usually composed of very few types of monosaccharide components. A polymer that consists of only one type of monosaccharide is a *homopolysaccharide;* a polymer that consists of more than one type of monosaccharide is a *heteropolysaccharide.* Glucose is the most common monomer in polysaccharides. When there is more than one type of monomer, frequently only two types of molecules occur in a repeating sequence.

polysaccharide a carbohydrate polymer consisting of many sugar residues bonded together

A complete characterization of a polysaccharide includes specifying which monomers are present and, if necessary, the sequence of monomers. The type of glycosidic linkage must also be specified because it determines the function of the polysaccharide. Cellulose and chitin are polysaccharides with β-glycosidic linkages, and both are structural materials that provide support and protection. Starch and glycogen, also polysaccharides, have α-glycosidic linkages, and they serve as carbohydrate-storage polymers in plants and animals, respectively.

16-4A Linear Homopolysaccharides of Glucose

▶ How do cellulose and amylose (in starch) differ from one another?

Cellulose is the major structural component of plants, especially of wood and plant fibers. It is a linear homopolysaccharide of β-D-glucose in which all residues are linked in β(1 → 4) glycosidic bonds (Figure 16.26A). Cellulose is made of repeating disaccharide units of β-cellobiose, which contains two β-glucose monomers that are in a flipped orientation relative to each other (Figure 16.26B). Individual cellulose polysaccharide chains are hydrogen-bonded together (Figure 16.26C), giving plant fibers their mechanical strength.

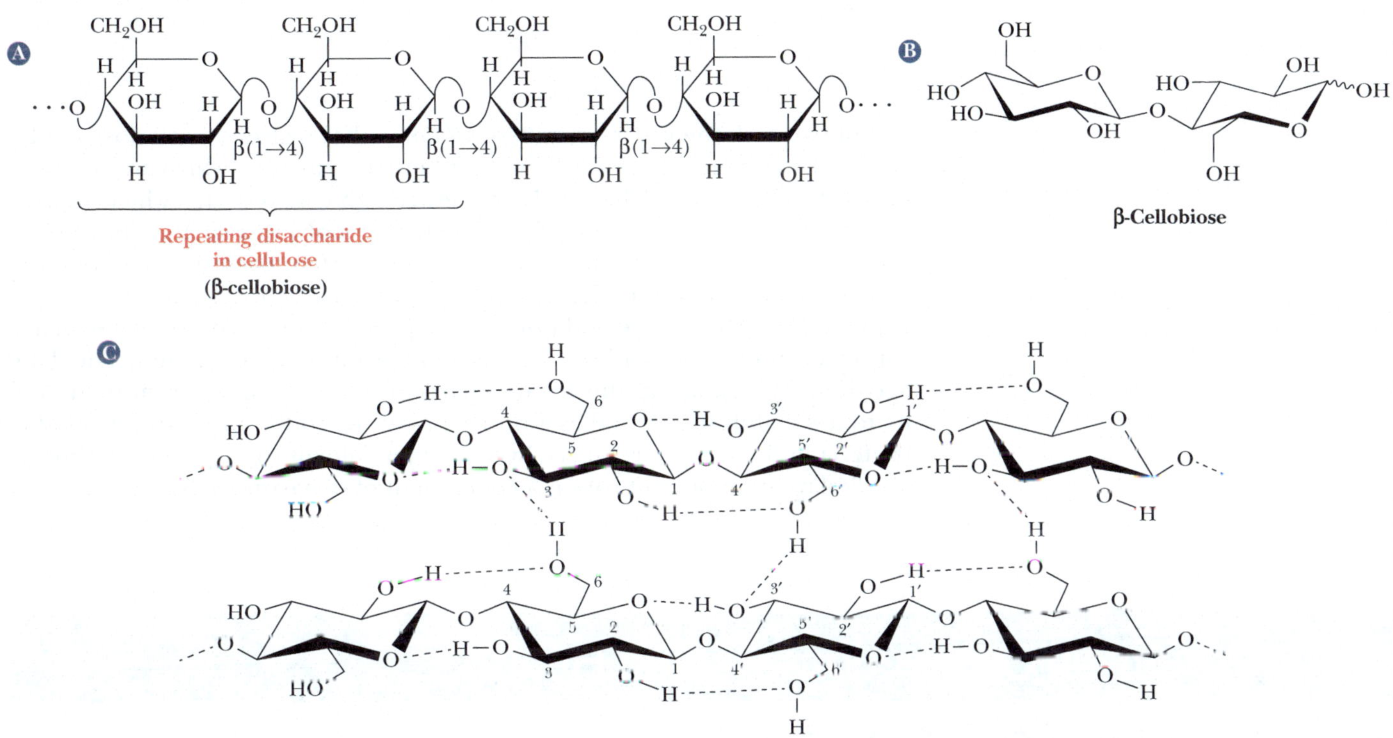

Figure 16.26 The polymeric, hydrogen-bonded structure of cellulose. (A) The polymeric structure of cellulose. The dots at the ends of the structure indicate that the linear polymer continues in the same fashion in both directions. (B) The structure of β-cellobiose, which is the repeating disaccharide unit of cellulose. The second residue in the β-cellobiose structure is "flipped," with carbon 6 pointing down. The hydroxyl on the anomeric carbon appears to point down as well, but this is just a consequence of the "flipped" orientation of the residue. (C) The polymeric structure of cellulose gives rise to long chains that can hydrogen-bond to one another.

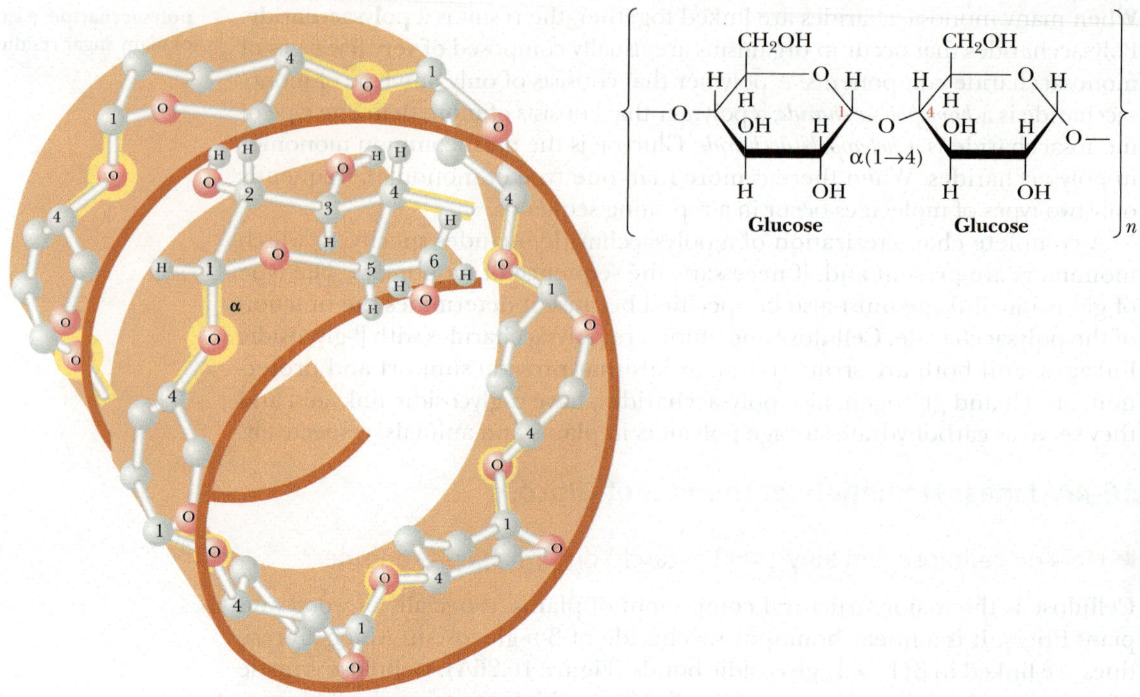

Figure 16.27 The structure of amylose in starch is based on the α-anomer of glucose. The repeating monomer of starch is the α-anomer of glucose, which gives rise to a chain that folds into a helical form. The repeating dimer has α(1 → 4) linkages throughout. The first glucose residue is shown with all its atoms labeled, but only the ring atoms are shown for the rest of the residues.

Animals lack the enzymes, called *cellulases,* that hydrolyze cellulose to glucose. Such enzymes attack the β-linkages between glucoses, which are common to structural polymers. The α-linkages between glucose residues, which animals *can* digest, are characteristic of energy-storage polymers such as amylose in starch (Figure 16.27). Cellulases are found in certain bacteria, including the bacteria that inhabit the digestive tracts of insects such as termites and grazing animals such as cattle and horses. The presence of these bacteria explains why cows and horses can live on grass and hay but humans cannot. The damage done by termites to the wooden parts of buildings arises from their ability to use cellulose in wood as a nutrient—owing to the presence of suitable bacteria in their digestive tracts. Even though animals cannot digest cellulose, it still plays an important role as a component of dietary fiber (see **Biochemical Connections 16E**).

16E Biochemical Connections • Allied Health

Why Is Dietary Fiber So Good for You?

Fiber in the diet is colloquially called roughage. It is principally made of complex carbohydrates, it may have some protein components, and it is moderately to fully insoluble. The health benefits of fiber have been widely studied. Roughage stimulates peristaltic action and thus helps move the digested food through the intestines, decreasing the transit time through the gut.

Potentially toxic substances in food and in bile fluid bind to fiber and are exported from the body, thus preventing them from damaging the lower intestine or being reabsorbed there. Statistical evidence indicates that a diet high in fiber also reduces colon and other cancers precisely because fiber binds suspected carcinogens. It is also plausible that the benefit is due to a lack of other items in the

Figure 16.28 Many breakfast cereals advertise their high-fiber content. Fruit adds to the fiber content of such a breakfast.

high-fiber diet. People on high-fiber diets also tend to take in less fat and fewer calories (Figure 16.28). Any difference in heart disease or cancer occurrence may be due to the health benefits of dietary fiber.

How does dietary fiber help reduce cholesterol? Fiber binds cholesterol-containing bile acids in the intestine, preventing their reabsorption. This causes the liver to draw cholesterol from the blood to create more bile acids. The resulting lower LDL cholesterol levels in the blood reduce the risk of plaque formation in arteries.

Fiber comes in two forms: soluble and insoluble. The most common insoluble fiber is cellulose, which is found in lettuce, carrots, bean sprouts, celery, brown rice, most other vegetables, many fruit skins, tree nuts, chia seeds, and pumpernickel bread. Insoluble fiber binds various molecules but otherwise merely forms bulk in the lower intestine. Soluble fibers include amylopectin and other pectins, as well as complex starches. Uncooked and mildly processed foods contain a higher proportion of this type of fiber.

The differences between soluble and insoluble fiber are apparent from their structure. Both kinds of fiber are polymers of the simple carbohydrates, which are monosaccharides or disaccharides. The difference between the two kinds of fiber depends on the bonding between the monomer units. In cellulose, the bonds between units are based on the β-anomeric form. These bonds cannot be hydrolyzed by enzymes in the body (Figure 16.29).

In starch, on the other hand, the bonding is based on the α-anomeric form, which *can* be hydrolyzed (Figure 16.30). As discussed in Section 16-4B, starch can exist in two forms: amylose and amylopectin. Amylose is the unbranched form of the polymer, whereas amylopectin is the branched-chain form.

Because of increased surface area, these fibers seem to be more beneficial. Good sources include bran (especially oat bran), barley, and fresh fruits (with skin), Brussels sprouts, potatoes with skin, beans, and zucchini. Soluble fiber binds water very well, increasing satiety by helping to fill the stomach.

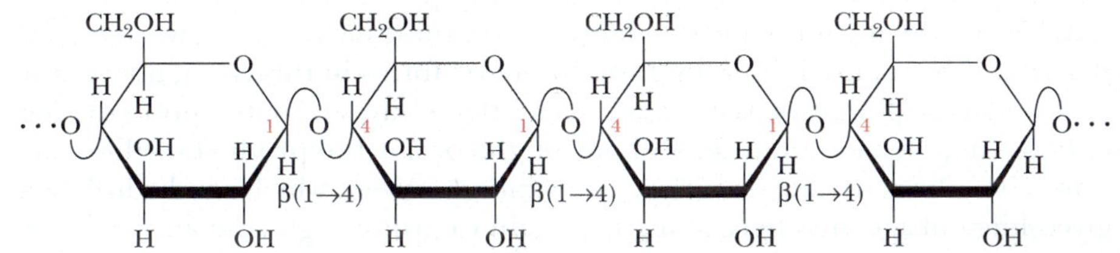

Figure 16.29 Cellulose is a good source of fiber in our diets because we cannot break down the beta glycosidic bonds.

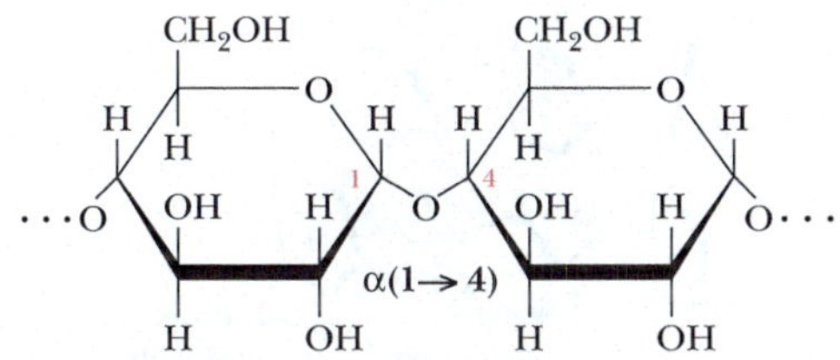

Figure 16.30 Starch contains alpha glycosidic bonds that can be hydrolyzed in humans.

16-4B Branched Homopolysaccharides of Glucose

▶ Is there more than one form of starch?

The importance of carbohydrates as energy sources suggests that there is some use for polysaccharides in metabolism. We now discuss in more detail some polysaccharides, such as starch, that serve as vehicles for storage of glucose.

Starches are polymers of α-D-glucose that occur in plant cells, usually as starch granules in the cytosol. Recall that there are α-linkages in starch, in contrast with the β-linkages in cellulose. The types of starches can be distinguished from one another by their degree of chain branching. Amylose is a linear polymer of glucose, with all the residues linked together by α(1 → 4) bonds (Figure 16.31, *top*). Amylopectin, on the other hand, is a branched-chain polymer, with the branches starting at α(1 → 6) linkages along the chain of α(1 → 4) linkages (Figure 16.31, *bottom*).

starches polymers of glucose that play an energy-storage role in plants

The most common conformation of amylose is a helix with six residues per turn. Iodine molecules can fit inside the helix to form a starch–iodine complex, which has a characteristic dark-blue color (Figure 16.32). The formation of this complex is a well-known test for the presence of the amylose form of

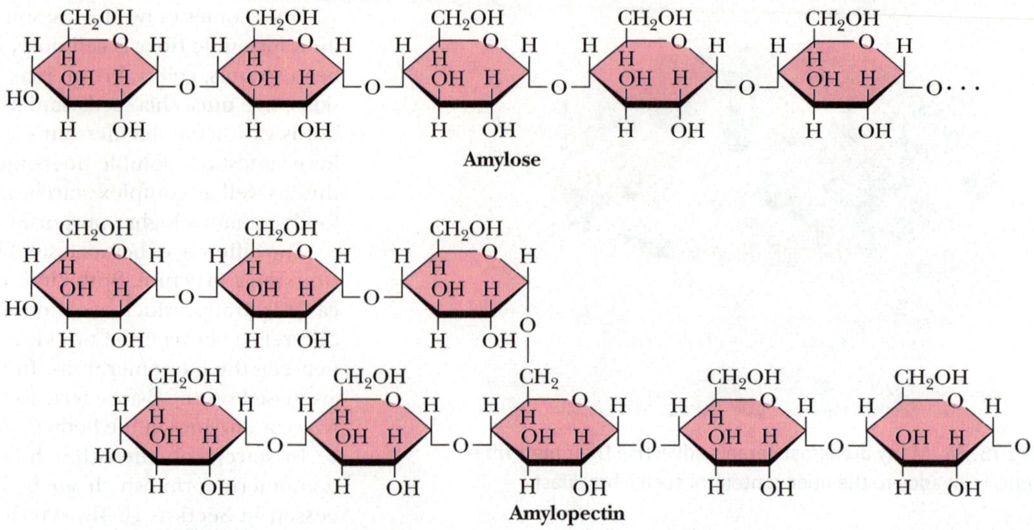

Figure 16.31 **Amylose and amylopectin are the two forms of starch.** The linear linkages are α(1 → 4) in each form of starch, but the branches in amylopectin are α(1 → 6). Branches in amylopectin occur every 24–30 residues.

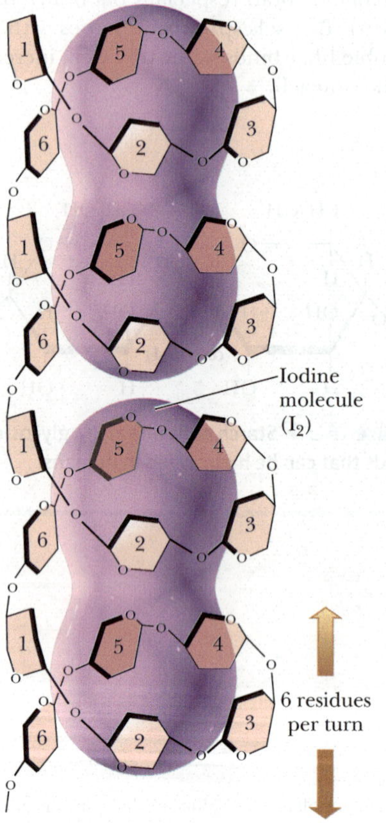

Figure 16.32 **The starch–iodine complex.** Amylose occurs as a helix with six residues per turn. In the starch–iodine complex, the iodine molecules are parallel to the long axis of the helix. Four turns of the helix are shown here. Six turns of the helix, containing 36 glycosyl residues, are required to produce the characteristic blue color of the complex.

glycogen a polymer of glucose that is an important energy-storage molecule in animals

starch. Amylopectin forms a tree-like structure instead of the helical structure of amylose, and the color of the product obtained when amylopectin and glycogen react with iodine is red-brown, not blue.

Because starches are storage molecules, both plants and animals contain enzymes that hydrolyze starches. As a result, glucose can be released from starch when the organism needs energy. Two of these enzymes, known as *α*- and *β-amylase* (the α and β do not signify anomeric forms in this case), attack α(1 → 4) linkages. β-Amylase is an *exoglycosidase* that cleaves from the nonreducing end of the polymer. Maltose, a dimer of glucose, is the product of the reaction. The other enzyme, α-amylase, is an *endoglycosidase,* which can hydrolyze a glycosidic linkage anywhere along the chain to produce glucose and maltose. Amylose is a linear chain of glucose residues, so it can be completely degraded to glucose and maltose by the two amylases. Amylopectin, on the other hand, is *not* completely degraded because the branching linkages are untouched. Nevertheless, *debranching enzymes* occur in both plants and animals that degrade the α(1 → 6) linkages at the branch points. When these enzymes are combined with the amylases, they can completely degrade both forms of starch.

▶ How is glycogen related to starch?

Although starches occur only in plants, *glycogen* is a similar carbohydrate-storage polymer in animals. **Glycogen** is a branched-chain polymer of α-D-glucose, and in this respect, it is similar to the amylopectin form of starch. Like amylopectin, glycogen consists of a chain of α(1 → 4) linkages with α(1 → 6) linkages at the branch points. The main difference between glycogen and amylopectin is that glycogen is more highly branched (Figure 16.33). Branch points occur about every 10 residues in glycogen versus about every 25 residues in amylopectin. In glycogen, therefore, the average chain length is 10 glucose residues. At the heart of every glycogen molecule is a protein called glycogenin, which is discussed in Section 18-1.

Glycogen is found in animal cells in granules similar to the starch granules in plant cells. Glycogen granules are observed in well-fed liver and muscle cells, but they are not seen in some other cell types, such as brain and heart cells under normal conditions. Some athletes, particularly endurance athletes like long-distance runners, try to build up their glycogen reserves before a race by eating large amounts of carbohydrates.

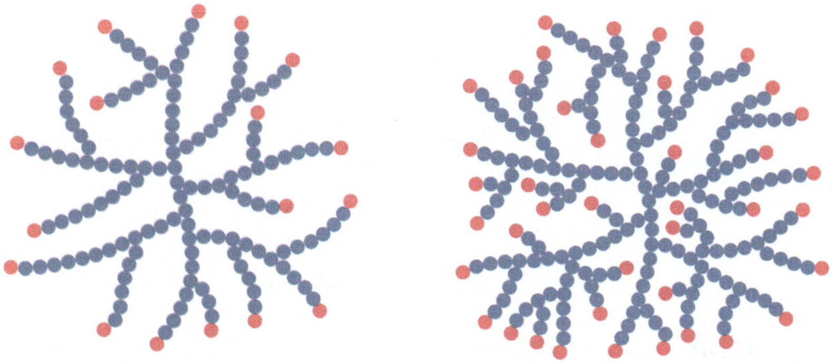

Figure 16.33 A comparison of the degrees of branching in amylopectin and glycogen. Each circle represents a glucose residue, and red circles indicate nonreducing ends of branches where glucose monomers can be removed by enzymes for energy production.

When the organism needs energy, various degradative enzymes remove glucose units (Section 18-1). Glycogen phosphorylase is one such enzyme; it cleaves one glucose at a time from the nonreducing end of a branch to produce glucose-1-phosphate, which then enters the metabolic pathways of carbohydrate breakdown. Debranching enzymes also play a role in the complete breakdown of glycogen. The number of branch points is significant for two reasons. First, a more branched polysaccharide is more water-soluble. This may not be as important for a plant, but for a mammal, the amount of glycogen in solution is. There are glycogen-storage diseases caused by lower-than-normal levels of branching enzymes. The glycogen products resemble starch and can fall out of solution, forming glycogen crystals in the muscles and liver. Second, when an organism needs energy quickly, the glycogen phosphorylase has more potential targets if there are more branches, allowing quicker mobilization of glucose. This is not as important to a plant, so amylopectin is not as highly branched as glycogen.

16-4C Linear Homopolysaccharides of Glucose Derivatives

▶ What is chitin?

Like cellulose, chitin is a linear homopolysaccharide with all the residues linked in β(1 → 4) glycosidic bonds. Whereas the monomer in cellulose is β-D-glucose, the monomer in chitin is a derivative of β-D-glucose called *N*-acetyl-β-D-glucosamine (Figure 16.34). This compound differs from glucose only in the substitution of the *N*-acetylamino group (—NH—CO—CH_3) for the hydroxyl group (—OH) on carbon C-2. Like cellulose, chitin plays a structural role and has a fair amount of mechanical strength because the individual strands are held together by hydrogen bonds. It is a major structural component of the exoskeletons of invertebrates such as insects and crustaceans (a group that includes lobsters and shrimp), and it also occurs in the cell walls of algae, fungi, and yeasts.

16-4D Heteropolysaccharides

▶ What role do polysaccharides play in the structure of cell walls?

In organisms that have cell walls, such as bacteria and plants, the walls consist largely of polysaccharides. The cell walls of bacteria and plants have biochemical differences, however.

Heteropolysaccharides, as opposed to the homopolysaccharides discussed so far, are the major components of *bacterial* cell walls. A distinguishing feature of prokaryotic cell walls is that the polysaccharides are cross-linked by peptides. The repeating unit of the polysaccharide consists of two residues held together by β(1 → 4) glycosidic links, as was the case in cellulose and chitin. One of the

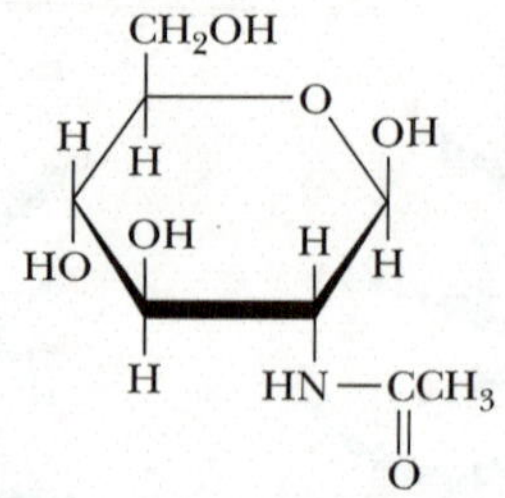

N-Acetyl-β-D-glucosamine

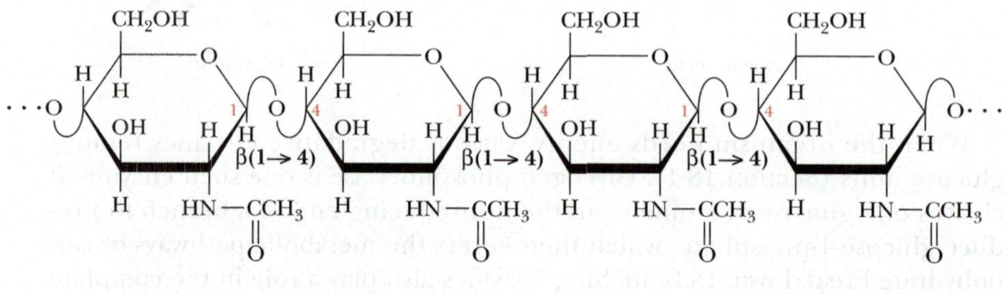

Figure 16.34 The polymeric structure of chitin. Chitin is a polysaccharide that contains *N*-acetylglucosamine (NAG) monomers linked by β(1→4) bonds.

two monomers is *N*-acetyl-D-glucosamine, which occurs in chitin, and the other monomer is *N*-acetylmuramic acid (NAM) (Figure 16.35A). The structure of NAM differs from that of *N*-acetylglucosamine (NAG) by the substitution of a lactic acid side chain [—O—CH(CH_3)—COOH] for the hydroxyl group (—OH) on carbon 3. NAM is found only in prokaryotic cell walls; it does not occur in eukaryotic cell walls.

The cross-links in bacterial cell walls consist of small peptides. One of the best-known examples is the cell wall of the bacterium *Staphylococcus aureus.* It is an oligomer of four amino acids (a tetrapeptide) bonded to *N*-acetylmuramic acid (NAM) (Figure 16.35B). The carboxyl group of the lactic acid side chain of NAM forms an amide bond with the N-terminal end of a tetrapeptide that has the sequence L-Ala–D-Gln–L-Lys–D-Ala. Recall that bacterial cell walls are one of the few places where D-amino acids occur in nature. The occurrence of D-amino acids and NAM in bacterial cell walls but not in plant cell walls represents a biochemical as well as a structural difference between prokaryotes and eukaryotes. The tetrapeptides containing the D-amino acids are themselves cross-linked to two other pentapeptides by another small peptide, in this case consisting of five glycine amino acids (Figure 16.35C).

The glycine pentamers form peptide bonds to the C-terminal end and to the side-chain ε-amino group of the lysine in the tetrapeptide, as well as the peptide bonds to other NAM-linked tetrapeptides (Figure 16.35C). This extensive cross-linking produces a three-dimensional network of considerable mechanical strength, which is why bacterial cell walls are extremely difficult to disrupt. The material that results from the cross-linking of polysaccharides by peptides is a **peptidoglycan** (Figure 16.35D), so named because it has both peptide and carbohydrate components.

peptidoglycan a polysaccharide that contains peptide cross-links; it is found in bacterial cell walls

Plant cell walls consist largely of cellulose (Section 16-4A). The other important polysaccharide component found in plant cell walls is **pectin**, a polymer made up mostly of D-galacturonic acid, a derivative of galactose in which the hydroxyl group on carbon C-6 has been oxidized to a carboxyl group (Figure 16.36).

pectin a polymer of galacturonic acid; it occurs in the cell walls of plants

Pectin is extracted from plants because it has commercial importance in the food-processing industry as a gelling agent in yogurt, fruit preserves, jams, and jellies. The major nonpolysaccharide component in plant cell walls, especially in woody plants, is *lignin* (Latin *lignum,* "wood"). Lignin is a polymer of coniferyl alcohol, and it is a very tough and durable material (Figure 16.37). Unlike bacterial cell walls, plant cell walls contain comparatively little peptide or protein.

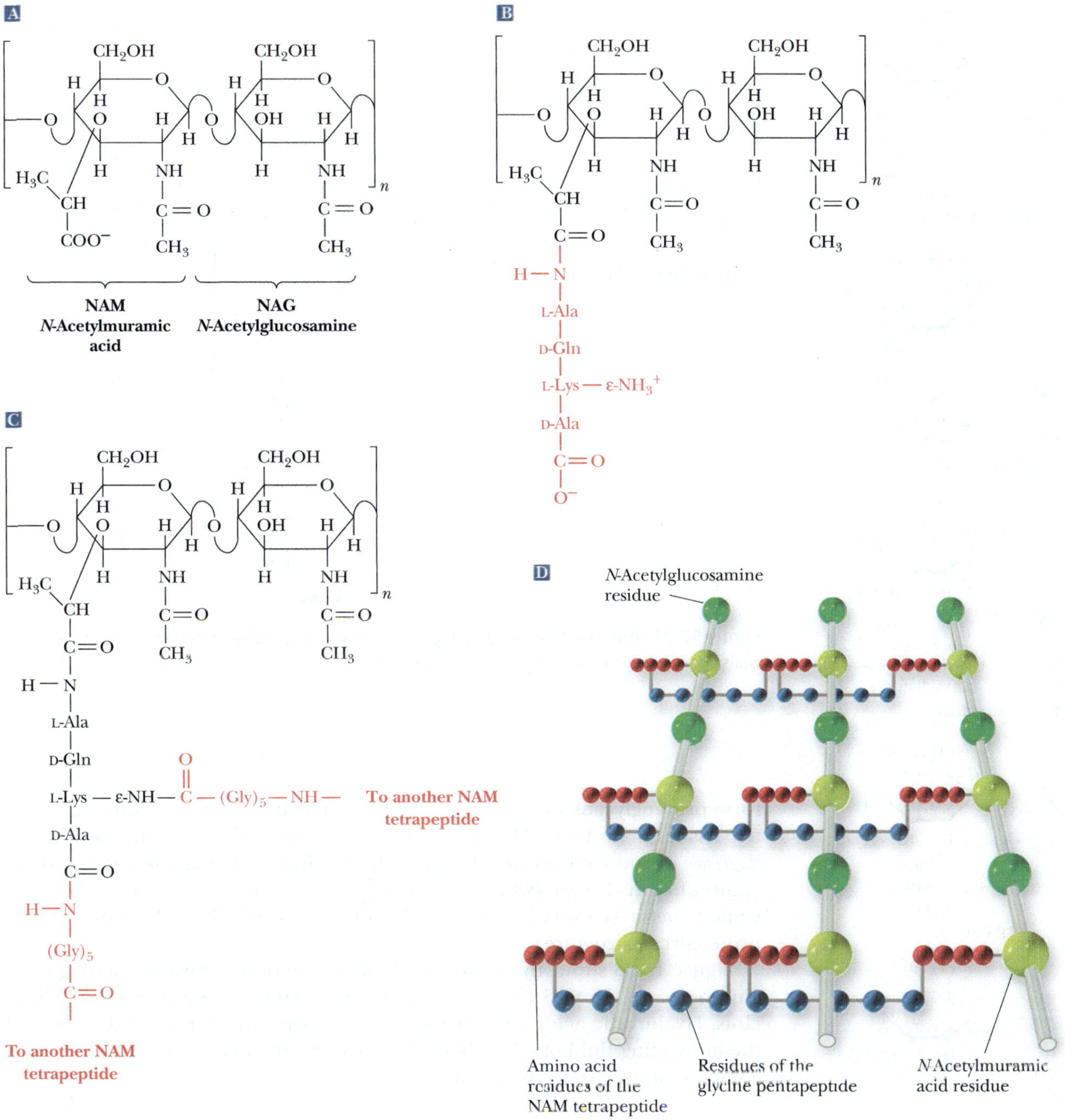

Figure 16.35 **The structure of the peptidoglycan of the bacterial cell wall of *Staphylococcus aureus*.** (A) The repeating peptidoglycan NAM-NAG disaccharide. (B) The repeating peptidoglycan disaccharide with the NAM-linked tetrapeptide (shown in red). (C) Addition of the pentaglycine cross-links (shown in red). (D) Schematic diagram of the peptidoglycan. The sugars are the larger spheres. The red spheres are the amino acid residues of the NAM-linked tetrapeptides, and the blue spheres are the glycine residues of the pentapeptide cross-links.

▶ Do polysaccharides play any specific roles in connective tissue?

Glycosaminoglycans are a type of polysaccharide based on a repeating disaccharide in which one of the sugars is an amino sugar and at least one of them has a negative charge owing to the presence of a sulfate group or a carboxyl group (Figure 16.38). These polysaccharides combine with core proteins to

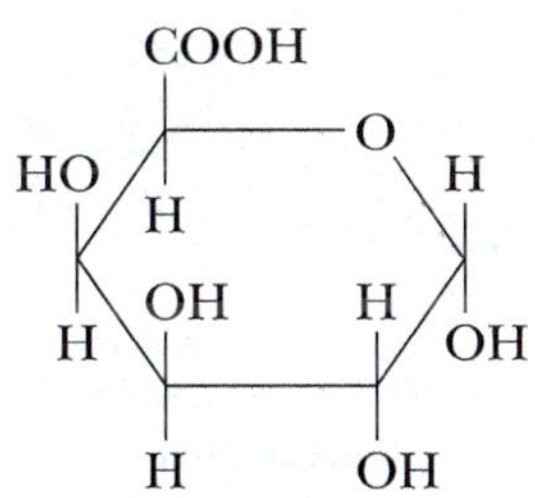

D-Galacturonic acid

Figure 16.36 **Structure of D-galacturonic acid.**

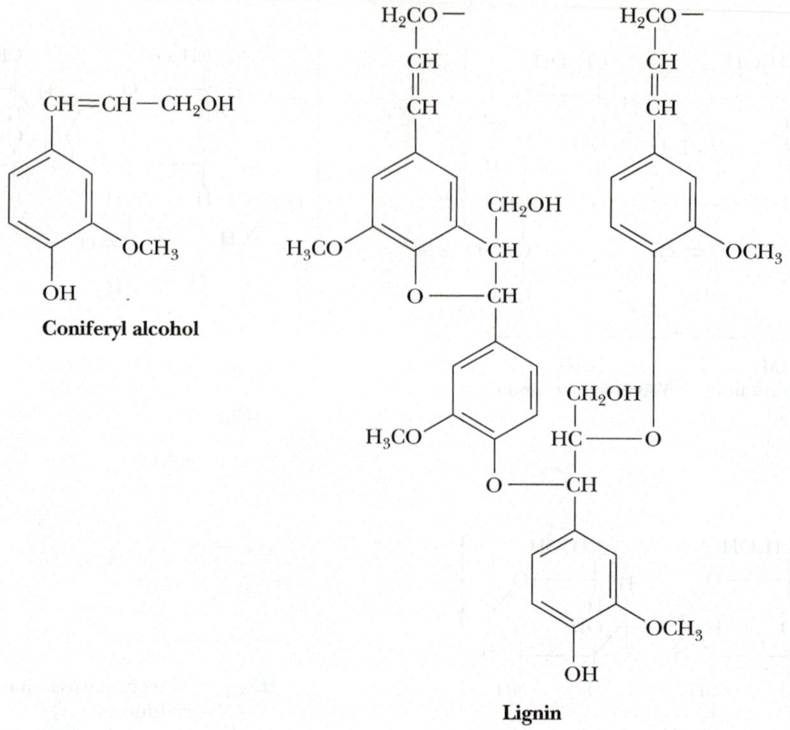

Figure 16.37 **The structure of lignin.** Lignin is a tough and durable component of plant cell walls and is a polymer of phenolic coniferyl alcohol.

form **proteoglycans** that help maintain the structure and function of connective tissues in animals. The term *proteoglycan* is very similar to the term *peptidoglycan* introduced previously. Recall from Chapter 4 that a *peptide* is a short chain of amino acids (which is what appears in a peptidoglycan structure), while a *protein* is a very long chain of amino acids (which is what appears in a proteoglycan structure).

Figure 16.38 shows the disaccharide structure of the most common glycosaminoglycans. Heparin is a natural anticoagulant that helps prevent blood clots. Hyaluronic acid is a component of the vitreous humor of the eye and the lubricating fluid of joints. The chondroitin sulfates and keratan sulfate are components of connective tissue.

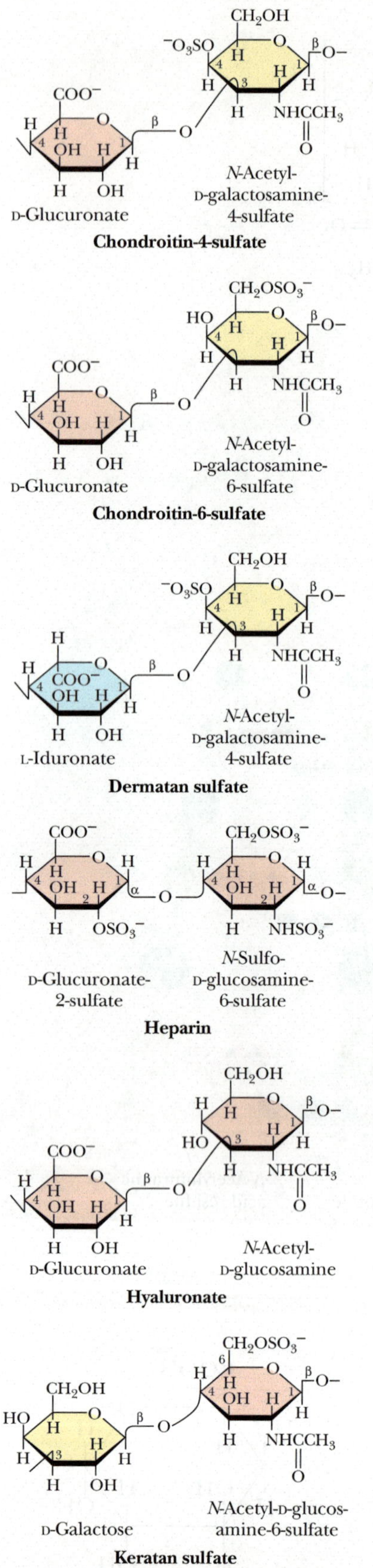

Figure 16.38 **Glycosaminoglycans are formed from repeating disaccharide units.** They often occur as components of the proteoglycans.

16-5 Glycoproteins

Learning Objectives

After reading this section, students will be able to:

LO 16.5.1 Describe the importance of carbohydrates in the immune response.

LO 16.5.2 Explain the functional aspects of glycoproteins.

Glycoproteins contain carbohydrate residues attached to the polypeptide chain. Some of the most important examples of glycoproteins are involved in the immune response; for example, *antibodies*, which bind to and immobilize *antigens* (the substances attacking the organism), are glycoproteins. Carbohydrates also play an important role as **antigenic determinants**, the portions of an antigenic molecule that antibodies recognize and to which they bind.

▶ How are carbohydrates important in the immune response?

The oligosaccharide portion of glycoproteins serves as an antigenic determinant in human blood groups. There are four human blood groups: A, B, AB, and O. The distinctions between the groups depend on the oligosaccharide portions of the glycoproteins on the surfaces of the blood cells called erythrocytes. In all blood types, the oligosaccharide portion of the molecule contains the sugar L-fucose, shown in Figure 16.14 as an example of a deoxy sugar. *N*-Acetylgalactosamine is found at the nonreducing end of the oligosaccharide in the type-A blood-group antigen (Figure 16.39). In type-B blood, α-D-galactose takes the place of *N*-acetylgalactosamine. In type-O blood, neither of these terminal residues is present, and in type-AB blood, both kinds of oligosaccharide are present. **Biochemical Connections 16F** goes into detail about blood-group antigens.

glycosaminoglycans connective tissue polysaccharides consisting of repeating disaccharide units that have amino-group derivatives bonded to one of the sugar residues of the disaccharide

proteoglycans a compound found in connective tissue that contains core proteins bonded to glycosaminoglycan groups

antigenic determinants the portions of molecules that antibodies can bind to and recognize as foreign

β-*N*-Acetylgalactosamine (1→3) β-Galactose (1→3) β-*N*-Acetylgalactosamine
↑ 2
1
α-l-Fucose
Nonreducing end
Type-A blood-group antigen

α-Galactose (1→3) β-Galactose (1→3) β-*N*-Acetylgalactosamine
↑ 2
1
α-l-Fucose
Nonreducing end
Type-B blood-group antigen

Figure 16.39 The structures of the blood-group antigenic determinants.

16F Biochemical Connections • Allied Health

Glycoproteins and Blood Transfusions

If a blood transfusion is attempted with incompatible blood types, such as when blood from a type-A donor is given to a type-B recipient, then an antigen–antibody reaction takes place because the type-B recipient has antibodies to the type-A blood. The characteristic oligosaccharide residues of type-A blood cells serve as the antigen. A cross-linking reaction occurs between antigens and antibodies, and the blood cells clump together. In the case of a transfusion of type-B blood to a type-A recipient, antibodies to type-B blood produce the same result. Type-O blood has neither antigenic determinant, so people with type-O blood are considered universal donors. However, these people have antibodies to both type-A and type-B blood, so they are not universal acceptors. Type-AB people have both antigenic determinants. As a result, they do not produce either type of antibody, so they are universal acceptors. Table 16.1 correlates blood transfusion relationships that are and are not compatible. Figure 16.40 shows schematically how the blood-group antigens are arranged on the surface of a red blood cell. ◗

Table 16.1 Transfusion Relationships

Blood Type	Makes Antibodies Against	Can Receive From	Can Donate To
O	A, B	O	O, A, B, AB
A	B	O, A	A, AB
B	A	O, B	B, AB
AB	None	O, A, B, AB	AB

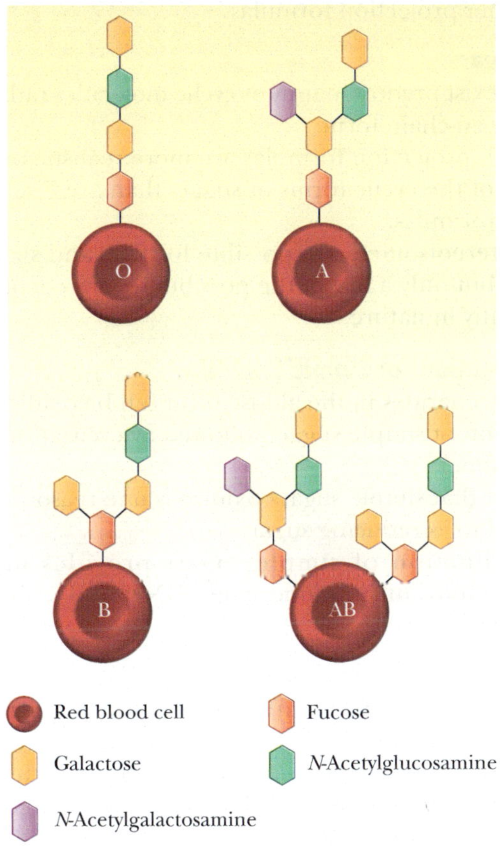

Figure 16.40 Arrangement of blood-group antigens on red blood cells.

Glycoproteins also play an important role in eukaryotic cell membranes. The sugar portions are added to the protein as it passes through the Golgi on its way to the cell surface. Those glycoproteins with an extremely high-carbohydrate content (85–95% by weight) are involved in forming connective tissue support as well as in cell signaling processes. Proteoglycans are constantly being synthesized and broken down. If there is a lack of the lysosomal enzymes that degrade them, proteoglycans accumulate with tragic consequences. One of the most striking consequences is the genetic disease known as Hurler's syndrome, in which the material that accumulates includes large amounts of amino sugars (Section 16-2). This disease leads to skeletal deformities, severe intellectual disability, and death in early childhood.

Summary

Carbohydrate Functions

- Carbohydrates serve as sources of energy, cell–cell interaction, immune recognition, and structural components.

Monosaccharide Structure and Stereochemistry

- The simplest examples of carbohydrates are monosaccharides, compounds that each contain a single carbonyl group and two or more hydroxyl groups.
- Monosaccharides frequently encountered in biochemistry are sugars that contain three to seven carbon atoms.
- Sugars contain one or more chiral centers, and the configurations of the possible stereoisomers can be represented by Fischer projection formulas.

Cyclic Sugars

- Sugars exist predominantly as cyclic molecules rather than in an open-chain form.
- Haworth projection formulas are more realistic representations of the cyclic forms of sugars than are Fischer projection formulas.
- Many stereoisomers are possible for five- and six-carbon sugars, but only a few of the possibilities are encountered frequently in nature.

Redox Reactions of Simple Sugars

- Monosaccharides in the aldose form can be oxidized.
- Oxidation of simple sugars followed by cyclization results in a lactone.
- The fact that simple sugars oxidize is one reason they are referred to as reducing sugars.
- The reduction of simple sugars provides building blocks important in blood types, DNA, and commercial sweeteners.

Sugar Esters and Glycosidic Bonds

- Esterification of sugars with phosphoric acid plays an important role in metabolism.
- The most important reaction of sugars by far is the formation of glycosidic linkages, which give rise to oligosaccharides and polysaccharides.

Amino Sugars

- Amino sugars are the basis of cell wall structures.

Some Important Oligosaccharides

- The disaccharides sucrose, lactose, and maltose are three important oligosaccharides.
- Sucrose is a common table sugar. It is a disaccharide formed when a glycosidic bond forms between glucose and fructose.
- Lactose occurs in milk, and maltose is obtained via the hydrolysis of starch.

Linear Homopolysaccharides of Glucose

- In polysaccharides, the repeating unit of the polymer is frequently limited to one or two kinds of monomer.
- Cellulose and starch differ in the anomeric form of their glycosidic bonds: the α form occurs in starch and the β form occurs in cellulose.

Branched Homopolysaccharides of Glucose

- Starch exists in two polymeric forms: the linear amylose and the branched amylopectin. Glycogen is also a branched homopolysaccharide of glucose.
- Starch, found in plants, and glycogen, which occurs in animals, differ from each other in the degree of branching in the polymer structure.

Linear Homopolysaccharides of Glucose Derivatives

- Cellulose and chitin are polymers based on single kinds of monomer units—glucose and *N*-acetyl-β-D-glucosamine, respectively. Both polymers play structural roles in organisms.

Heteropolysaccharides

- In bacterial cell walls, polysaccharides are cross-linked to peptides.
- Plant cell walls consist primarily of glucose.
- Glycosaminoglycans are a type of polysaccharide based on a repeating disaccharide in which one of the sugars is an amino sugar, and at least one of them has a negative charge due to the presence of a sulfate group or a carboxyl group.

- They play a role in joint lubrication and in the blood-clotting process.

Glycoproteins

- In glycoproteins, carbohydrate residues are covalently linked to the polypeptide chain.
- Such glycoproteins can play a role in the recognition of antigen sites.
- A common example is the ABO blood group, in which the three major blood types are distinguished by sugar molecules attached to the protein.

Review Exercises

16-1 Simple Sugars: Their Structures and Stereochemistry

1. **Recall** Define the following terms: polysaccharide, furanose, pyranose, aldose, ketose, glycosidic bond, oligosaccharide, glycoprotein.
2. Describe distinguishing features of monosaccharide structure.
3. **Recall** Name which, if any, of the following are epimers of D-glucose: D-mannose, D-galactose, and D-ribose.
4. **Reflect and Apply** What is the difference between *enantiomers*, *diastereomers*, and *epimers*?
5. **Recall** How many possible epimers of D-glucose exist?
6. **Recall** Why are furanoses and pyranoses the most common cyclic forms of sugars?
7. **Recall** How many chiral centers are there in the open-chain form of glucose? In the cyclic form?
8. **Reflect and Apply** The following are Fischer projections for a group of five-carbon sugars, all of which are aldopentoses. Identify the pairs that are enantiomers and the pairs that are epimers. (The sugars shown here are not all of the possible five-carbon sugars.)

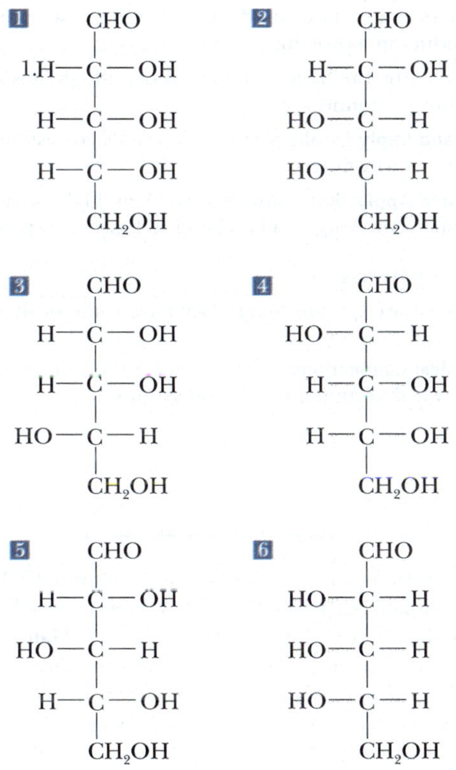

9. **Reflect and Apply** Consider the structures of arabinose and ribose. Explain why nucleotide derivatives of arabinose, such as ara-C and ara-A, are effective metabolic poisons.

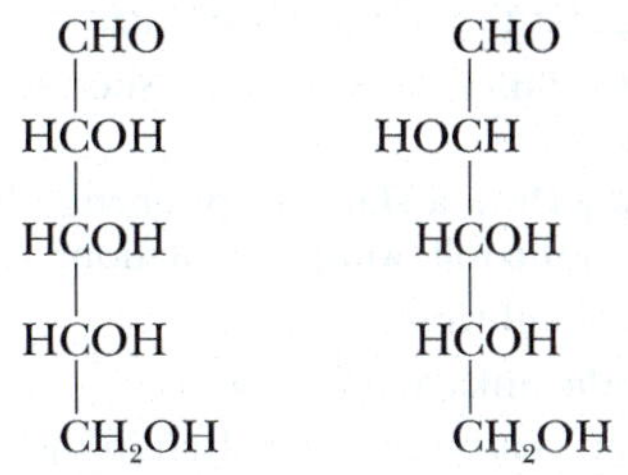

D-Ribose D-Arabinose

10. **Reflect and Apply** Two sugars are epimers of each other. Is it possible to convert one to the other without breaking covalent bonds?
11. **Reflect and Apply** How does the cyclization of sugars introduce a new chiral center?
12. **Reflect and Apply** Convert the following Haworth projections to a Fischer projection. Name the monosaccharides you have drawn.

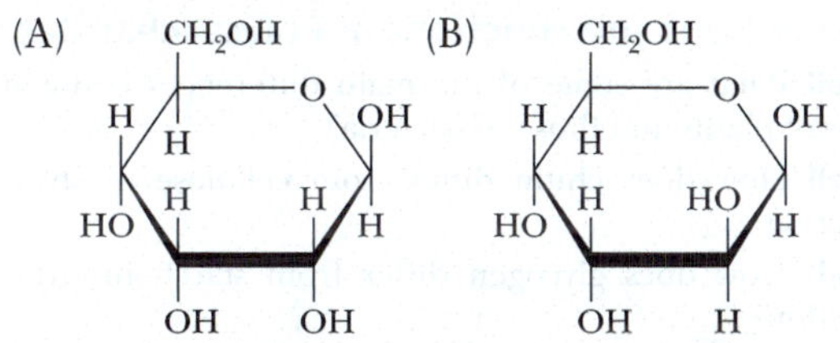

13. **Reflect and Apply** Convert each of the following chair conformations to an open-chain form and to a Fischer projection. Name the monosaccharides you have drawn.

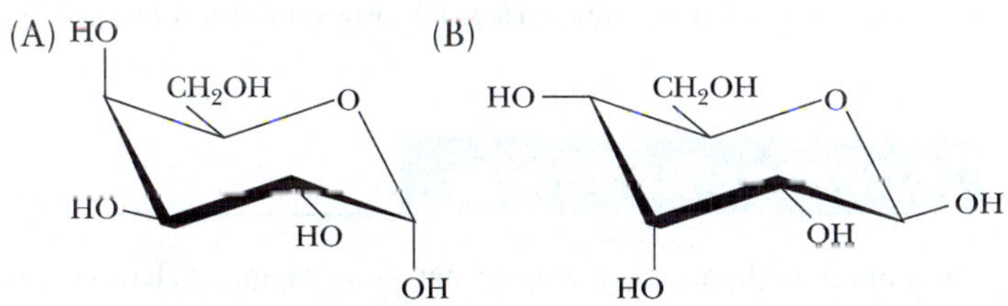

14. **Reflect and Apply** Starting with a Fischer projection of D-fructose, write equations showing the formation of α-D-fructopyranose, α-D-fructofuranose, β-D-fructopyranose, and β-D-fructofuranose.

15. **Reflect and Apply** Starting with the open-chain form of D-ribose, write equations for the cyclization reactions that form the pyranose and the furanose forms.

16-2 Reactions of Monosaccharides

16. **Recall** Describe the type of reaction involved in forming a bond between two monosaccharides, and give the name of the bond.
17. **Recall** Considering Figure 16.22, describe how the structure of *N*-acetylglucosamine (NAG) differs from that of glucose, and then explain how the structure of *N*-acetylmuramic acid (NAM) differs from that of NAG.
18. **Recall** Define the term *reducing sugar.*
19. **Biochemical Connections** What are the structural differences between vitamin C and sugars? Do these structural differences play a role in the susceptibility of this vitamin to air oxidation?

16-3 Some Important Oligosaccharides

20. **Recall** Name two differences between sucrose and lactose. Name two similarities.
21. **Reflect and Apply** Draw a Haworth projection for the disaccharide gentibiose, given the following information:
    (a) It is a dimer of glucose.
    (b) The glycosidic linkage is $\beta(1 \rightarrow 6)$.
    (c) The anomeric carbon not involved in the glycosidic linkage is in the α configuration.
22. **Biochemical Connections** What is the metabolic basis for the observation that many adults cannot ingest large quantities of cow's milk without developing gastric difficulties?
23. **Reflect and Apply** Draw Haworth projection formulas for dimers of glucose with the following types of glycosidic linkages:
    (a) A $\beta(1 \rightarrow 4)$ linkage (both molecules of glucose in the β form)
    (b) An $\alpha,\alpha(1 \rightarrow 1)$ linkage
    (c) A $\beta(1 \rightarrow 6)$ linkage (both molecules of glucose in the β form)

16-4 Structures and Functions of Polysaccharides

24. **Recall** What are some of the main differences between the cell walls of plants and those of bacteria?
25. **Recall** How does chitin differ from cellulose in structure and function?
26. **Recall** How does glycogen differ from starch in structure and function?
27. **Recall** What is the main structural difference between cellulose and starch, and how does the structural difference affect the shape of each polysaccharide?
28. **Reflect and Apply** Pectin, which occurs in plant cell walls, exists in nature as a polymer of D-galacturonic acid methylated at carbon 6 of the monomer. Draw a Haworth projection for a two-residue segment of pectin with one methylated and one unmethylated monomer unit in an $\alpha(1 \rightarrow 4)$ linkage.
29. **Reflect and Apply** All naturally occurring polysaccharides have one terminal residue, which contains a free anomeric carbon. Why do these polysaccharides *not* give a positive chemical test for a reducing sugar?
30. **Reflect and Apply** An amylose chain is 5000 glucose units long. At how many places must it be cleaved to reduce the average length to 2500 units? To 1000 units? To 200 units? What percentage of the glycosidic links is hydrolyzed in each case? (Even partial hydrolysis can drastically alter the physical properties of polysaccharides and thus affect their structural role in organisms.)
31. **Reflect and Apply** Suppose that a polymer of glucose with alternating $\alpha(1 \rightarrow 4)$ and $\beta(1 \rightarrow 4)$ glycosidic linkages has just been discovered. Draw a Haworth projection for a four-residue segment of such a polysaccharide.
32. **Reflect and Apply** Glycogen is highly branched. What advantage, if any, does this provide to an animal?
33. **Reflect and Apply** No animal can digest cellulose. Reconcile this statement with the fact that many animals are herbivores that depend heavily on cellulose as a food source.
34. **Reflect and Apply** How does the presence of α-bonds versus β-bonds influence the digestibility of glucose polymers by humans? *Hint:* There are *two* effects.
35. **Reflect and Apply** How do the sites of cleavage of starch differ from one another when the cleavage reaction is catalyzed by α-amylase and β-amylase?
36. **Biochemical Connections** What is the benefit of fiber in the diet?
37. **Reflect and Apply** How would you expect the active site of a cellulase to differ from the active site of an enzyme that degrades starch?
38. **Reflect and Apply** Compare the information in the sequence of monomers in a polysaccharide with that in the sequence of amino acid residues in a protein.
39. **Recall** Describe the basic structure of peptidoglycans and indicate where they are found.
40. **Reflect and Apply** Could bacterial cell walls consist largely of protein? Why or why not?
41. **Reflect and Apply** Some athletes eat diets high in carbohydrates before an event. Suggest a biochemical basis for this practice.

16-5 Glycoproteins

42. **Recall** What are glycoproteins? What are some of their biochemical roles?
43. **Biochemical Connections** Briefly indicate the role of glycoproteins as antigenic determinants in blood groups.

Further Reading

Most organic chemistry textbooks have one or more chapters on the structures and reactions of carbohydrates.

Kritchevsky, K., C. Bonfield, and J. Anderson, eds. *Dietary Fiber: Chemistry, Physiology, and Health Effects.* New York: Plenum Press, 1990. A topic of considerable current interest, with explicit connections to the biochemistry of plant cell walls.

Sharon, N. Carbohydrates. *Sci. Amer.* **243** (5), 90–102 (1980). A good overview of structures.

Sharon, N., and H. Lis. Carbohydrates in Cell Recognition. *Sci. Amer.* **268** (1), 82–89 (1993). The development of drugs to stop infection and inflammation by targeting cell-surface carbohydrates.

Suez, J., T. Korem, D. Zeevi, G. Zilberman-Schapira, C. A. Thaiss, O. Maza, D. Israeli, N. Zmora, S. Gilad, A. Weinberger, Y. Kuperman, A. Harmelin, I. Kolodkin-Gal, H. Shapiro, Z. Halpern, E. Segal, and E. Elinav. Artificial Sweeteners Induce Glucose Intolerance by Altering the Gut Microbiota. *Nature* **514**, 181–186 (2014).

Takahashi, N., and T. Muramatsu. *Handbook of Endoglycosidases and Glycoamidases.* Boca Raton, FL: CRC Press, 1992. A source of practical information on how to manipulate biologically important carbohydrates.

jeep2499/Shutterstock.com

For athletes, efficient use of carbohydrates can provide the margin of victory.

Glycolysis 17

Chapter Outline

17-1 The Overall Pathway of Glycolysis

Learning Objectives

After reading this section, students will be able to:

LO 17.1.1 Describe an overview of glycolysis, including the starting material, key intermediates, and an accounting of ATP consumption and production in the pathway.

LO 17.1.2 Explain three possible fates of pyruvate in glucose metabolism.

Chapters 17–20 cover carbohydrate metabolism and its central role in life processes. Chapters 17 and 18 focus on reactions of glucose, which is one of the most widely distributed sugars in nature. The subject here in Chapter 17 is glycolysis, the conversion of glucose to pyruvate, and the control of this vital pathway.

17-1A Overview of Glycolysis

The first stage of glucose metabolism in organisms from bacteria to humans is called **glycolysis**, and it was the first biochemical pathway elucidated. In glycolysis, one molecule of glucose (a six-carbon compound) is converted to fructose-1,6-bisphosphate (also a six-carbon compound), which eventually gives rise to two molecules of pyruvate (a three-carbon compound) (Figure 17.1). The glycolytic pathway is also called the Embden–Meyerhof pathway in honor of Gustav Embden and Otto Meyerhof, who were largely responsible for elucidating the pathway in the 1940s.

Glycolysis involves many steps, including the reactions in which metabolites of glucose are oxidized. Each reaction in the pathway is

Chapter Outline (continued)

catalyzed by an enzyme specific for that reaction. In each of two reactions in the pathway, one molecule of adenosine triphosphate (ATP) is hydrolyzed for each molecule of glucose metabolized; the energy released in the hydrolysis of these two ATP molecules makes coupled endergonic reactions possible. In each of two other reactions, two molecules of ATP are produced by phosphorylation of adenosine diphosphate (ADP) for each molecule of glucose, giving a total of four ATP molecules produced. Thus, there is a net gain of two ATP molecules for each molecule of glucose processed in glycolysis (Figure 17.1, *top*). Glycolysis plays a key role in the way organisms extract energy from nutrients.

Pioneers in Biochemistry • Severo Ochoa

Severo Ochoa (1905–1993) was a pioneering biochemist and Nobel Prize recipient whose discoveries were fundamental to understanding genetics. Born in Spain, Ochoa studied and worked around the world, eventually becoming a U.S. citizen in midlife.

Trained as a medical doctor, Ochoa launched his career in biochemical research in the 1930s, just as the field of biochemistry began to truly flourish. His early research focused on glycolysis in muscle tissue, which evolved into a lifelong interest in enzymatic processes. His work contributed to the understanding of the metabolism of carbohydrates and fatty acids as well as the biosynthesis of nucleic acids.

In 1959, Ochoa shared the Nobel Prize in Physiology or Medicine with Arthur Kornberg for their discoveries of the mechanisms in the biological synthesis of ribonucleic and deoxyribonucleic acids. In his acceptance speech, Ochoa described why he found biochemical research so compelling: "The study of the way in which the basic substances of life (the nucleic acids and the proteins) are perpetuated from generation to generation… and the spectacular advances of the science of genetics bring us closer to an understanding of life's most characteristic features."

Bettmann/Bettmann/Getty Images

glycolysis the anaerobic breakdown of glucose to three-carbon compounds

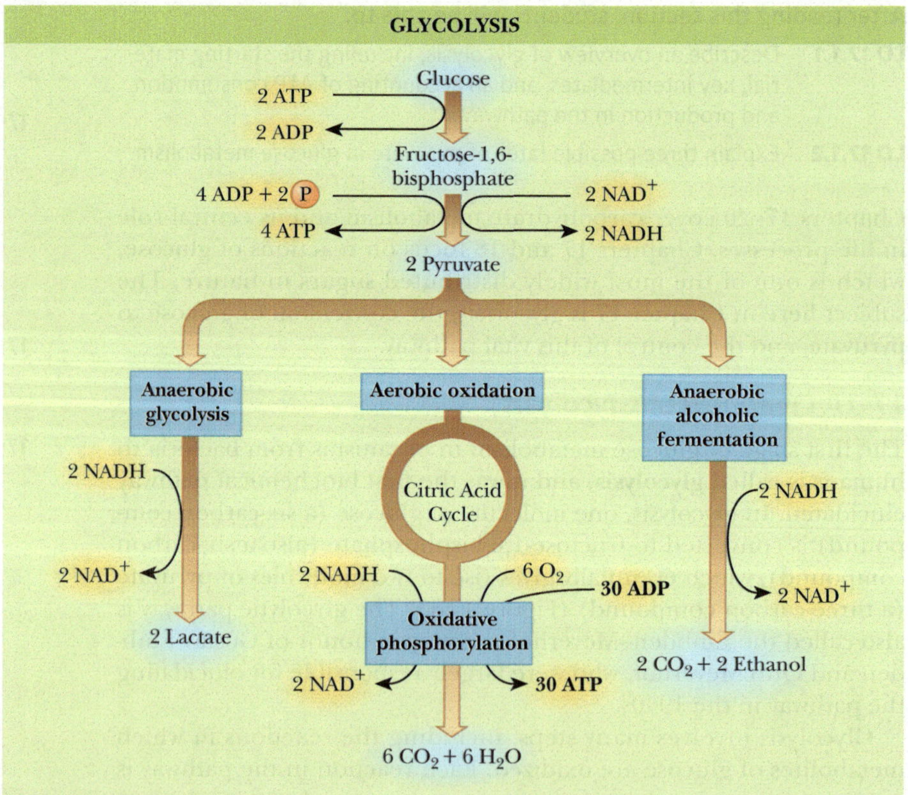

Figure 17.1 An overview of glycolysis and pyruvate metabolism. In glycolysis, one molecule of glucose is converted into two molecules of pyruvate. Under aerobic conditions, pyruvate is then oxidized to CO_2 and H_2O by the citric acid cycle (Chapter 19) and oxidative phosphorylation (Chapter 20). Under anaerobic conditions, lactate is produced, especially in muscles. Alcoholic fermentation occurs in yeast. In all of these pyruvate metabolism processes, the reduced form of nicotinamide adenine dinucleotide (NADH) produced in the conversion of glucose to pyruvate is reoxidized to NAD^+ in the subsequent reactions of pyruvate.

17-1B Pyruvate: A Key Intermediate in Glycolysis

▶ What are the possible fates of pyruvate?

When pyruvate is formed, it can have one of several fates (Figure 17.1). In aerobic metabolism (in the presence of oxygen), pyruvate loses carbon dioxide. The remaining two carbon atoms become linked to coenzyme A (Section 15-7) as an acetyl group to form acetyl-CoA, which then enters the citric acid cycle (Chapter 19). There are two fates for pyruvate in anaerobic metabolism (in the absence of oxygen). The more common fate of pyruvate in anaerobic metabolism is its reduction to lactate (Section 17-4A). This process is called **anaerobic glycolysis** to distinguish it from the conversion of glucose to pyruvate, which is simply called glycolysis. Anaerobic metabolism is the only energy source in mammalian red blood cells, as well as in several species of bacteria, such as *Lactobacillus* in sour milk and *Clostridium botulinum* in tainted canned foods. In organisms capable of alcoholic fermentation, pyruvate loses carbon dioxide, this time producing acetaldehyde, which is reduced to produce ethanol (Section 17-4B). **Biochemical Connections 17A** discusses an important practical application of fermentation.

anaerobic glycolysis the pathway for the conversion of glucose to lactate; distinguished from glycolysis, which is the conversion of glucose to pyruvate

17A Biochemical Connections • Environmental Science

Biofuels from Fermentation

The grave concern about the depletion of fossil fuels, especially petroleum-based ones, has led to interest in developing renewable energy sources. Most kinds of organic matter can be used as fuel. Wood and animal dung have been used for centuries and continue to be used in many parts of the world. Still, many kinds of engines are designed to operate with liquid fuels, such as gasoline in cars. Ethanol is a common liquid product of the fermentation of carbohydrates, and its use as a fuel has been explored widely (Figure 17.2).

Ethanol is not normally used as the sole fuel in an internal combustion engine but is usually mixed with gasoline. The specific formulation is designated with an E number, such as E10 or E25, where the number refers to the percentage of ethanol in the mixture. E10 is used in nearly all gas sold in the United States, and the use of E25 is widely implemented as part of Brazil's ethanol program.

Recent developments in biofuels produced from fermentation focus on improving feedstock diversity, enhancing fermentation processes, and scaling up the production of advanced biofuels like cellulosic ethanol, butanol, and other higher-energy biofuels. Innovations in microbial engineering, waste-to-energy solutions, and fermentation integration with carbon capture are positioning fermentation as a key technology for sustainable biofuels. As these technologies advance and become more cost-effective, fermentation-based biofuels are expected to play an increasingly important role in the global shift toward cleaner, renewable energy sources.

It is possible to use any number of carbohydrate sources to obtain ethanol. Almost any kind of plant material can be fermented to produce ethanol, which is then recovered by distillation. In 2007, the Energy Independence and Security Act was signed into law by President George W. Bush. By 2013, 16.55 billion gallons of renewable fuels were required to be blended into the U.S. fuel supply, with the goal of achieving 36 billion gallons of renewable fuel usage by 2022. Widespread implementation of E10 has been achieved, but the *E10 blend wall* has hampered adoption of E15 and higher ethanol-fuel blends. The *E10 blend wall* is the term used to describe the limit to biofuel that can be mixed with gas (10 parts biofuel to 90 parts gasoline) before harmful effects are observed in motor vehicles. The majority of vehicles in the United States are unable to accommodate ethanol at levels higher than E10, with motorcycles and small engines being particularly susceptible to damage and the violation of warranties.

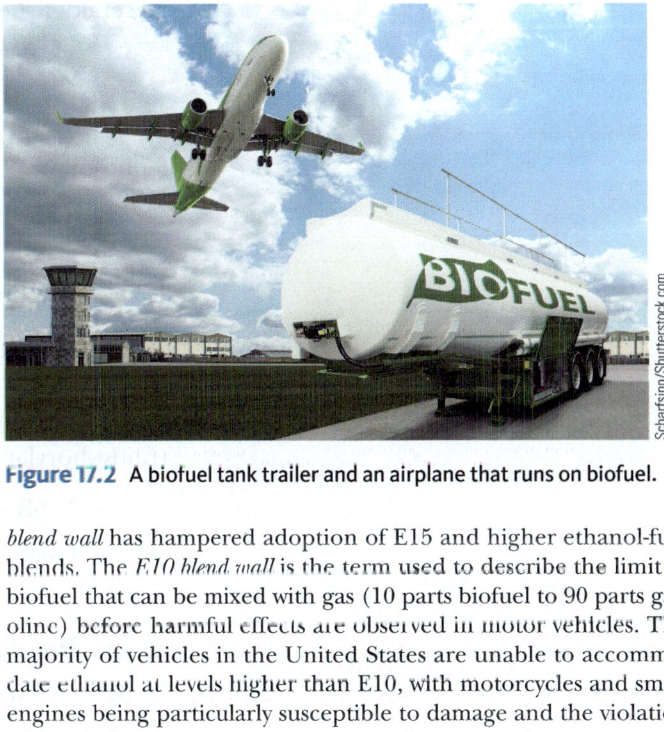

Figure 17.2 A biofuel tank trailer and an airplane that runs on biofuel.

Methyl esters of free fatty acids obtained from oil of plant origin have been widely utilized for biodiesel. Examples of plant sources that contain high oil content are algae, soy, rapeseed (canola oil), and coffee beans. Solvent extraction of oil from the plant biomass, followed by saponification of the oil to free fatty acids and subsequent esterification with methanol, provides an efficient means to produce biodiesel. ◗

metabolites substances that are produced in, or that are necessary for, metabolic pathways

Figure 17.3 shows the names of the **metabolites** in the glycolytic pathway, and Sections 17-2, 17-3, and 17-4 discuss structures of the metabolites as well as the enzymes that catalyze each reaction. The breakdown of glucose to pyruvate can be summarized as follows:

Glucose **(six carbon atoms)** → 2 Pyruvate **(three carbon atoms)**

2ATP + 4ADP + $2P_i$ → 2ADP + 4ATP **(Phosphorylation)**

Glucose + 2ADP + $2P_i$ → 2 Pyruvate + 2ATP **(Net reaction)**

All sugars in the pathway have the D configuration, a fact we assume throughout this chapter.

17-1C The 10 Reactions of Glycolysis

Step 1. ***Phosphorylation*** of glucose to give glucose-6-phosphate (ATP is the source of the phosphate group).

Glucose + ATP → Glucose-6-phosphate + ADP

Step 2. ***Isomerization*** of glucose-6-phosphate to give fructose-6-phosphate.

Glucose-6-phosphate → Fructose-6-phosphate

Step 3. ***Phosphorylation*** of fructose-6-phosphate to give fructose-1,6-bisphosphate (ATP is the source of the phosphate group).

Fructose-6-phosphate + ATP → Fructose-1,6-bisphosphate + ADP

Step 4. ***Cleavage*** of fructose-1,6-bisphosphate to give two 3-carbon fragments, glyceraldehyde-3-phosphate and dihydroxyacetone phosphate.

Fructose 1,6-bisphosphate → Glyceraldehyde-3-phosphate + Dihydroxyacetone phosphate

Step 5. ***Isomerization*** of dihydroxyacetone phosphate to give glyceraldehyde-3-phosphate.

Dihydroxyacetone phosphate → Glyceraldehyde-3-phosphate

Step 6. ***Oxidation*** (and phosphorylation) of glyceraldehyde-3-phosphate to give 1,3-bisphosphoglycerate.

Glyceraldehyde-3-phosphate + NAD^+ + P_i → NADH + 1,3-Bisphosphoglycerate + H^+

Step 7. ***Transfer of a phosphate group*** from 1,3-bisphosphoglycerate to ADP (phosphorylation of ADP to ATP) to give 3-phosphoglycerate.

1, 3-Bisphosphoglycerate + ADP → 3-Phosphoglycerate + ATP

Step 8. ***Isomerization*** of 3-phosphoglycerate to give 2-phosphoglycerate.

3-Phosphoglycerate → 2-Phosphoglycerate

Step 9. ***Dehydration*** of 2-phosphoglycerate to give phosphoenolpyruvate.

2-Phosphoglycerate → Phosphoenolpyruvate + H_2O

Step 10. ***Transfer of a phosphate group*** from phosphoenolpyruvate to ADP (phosphorylation of ADP to ATP) to give pyruvate.

Phosphoenolpyruvate + ADP → Pyruvate + ATP

Only one of the 10 steps in this pathway (Step 6) involves an electron-transfer reaction. We now look at each of these reactions in detail.

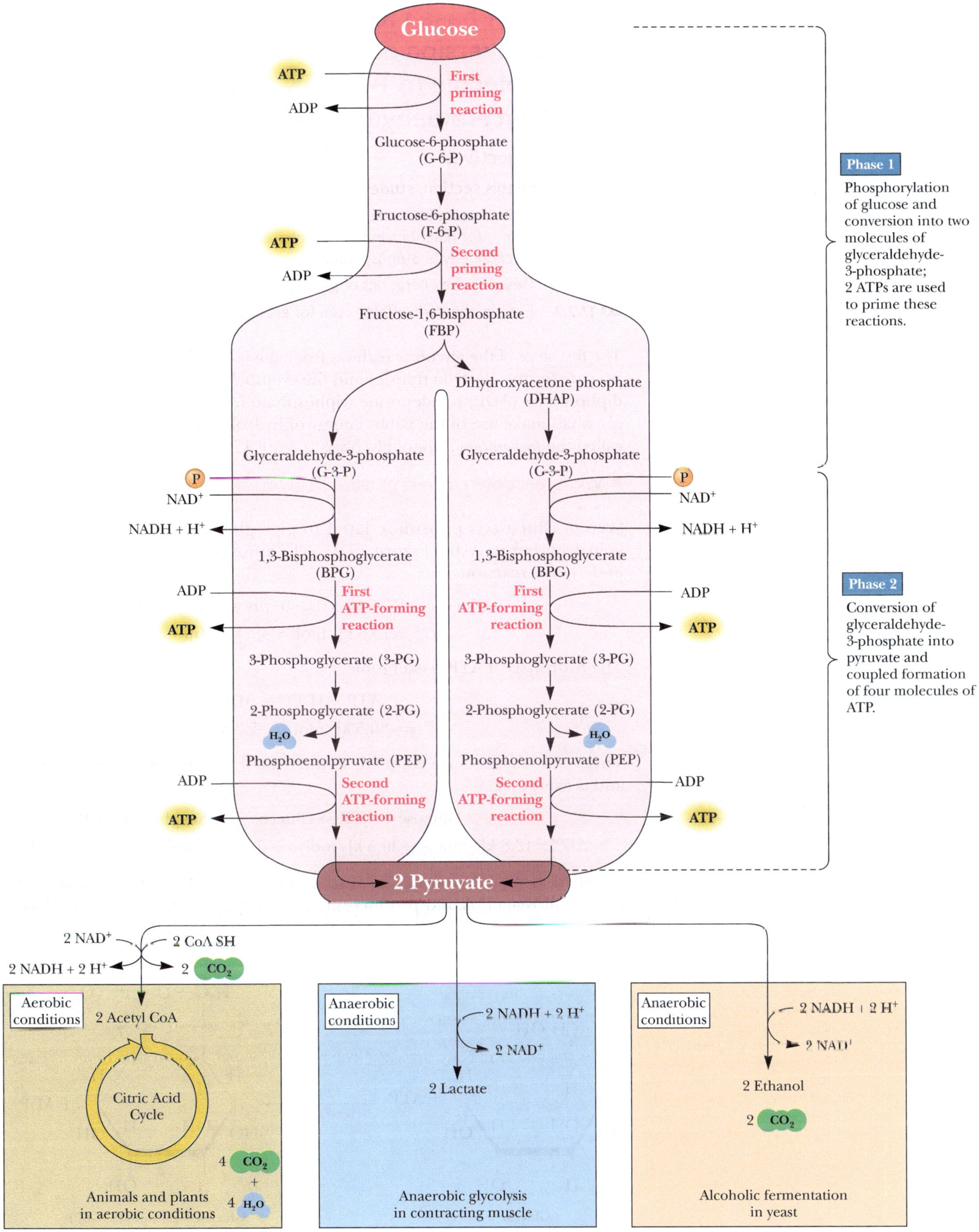

Figure 17.3 The glycolytic pathway.

17-2 The Phase 1 Reactions of Glycolysis: Conversion of a Six-Carbon Glucose Molecule to Two Three-Carbon Glyceraldehyde-3-Phosphate Molecules

Learning Objectives

After reading this section, students will be able to:

LO 17.2.1 Write metabolite names, metabolite structures, and enzyme names for reactions in the first phase of glycolysis from glucose to glyceraldehyde-3-phosphate.

LO 17.2.2 Describe the energetics of glycolysis reactions.

LO 17.2.3 Explain allosteric effects seen for enzymes in the first phase of glycolysis.

The first steps of the glycolytic pathway from glucose to glyceraldehyde-3-phosphate prepare for the electron transfer and the eventual phosphorylation of adenosine diphosphate (ADP) to adenosine triphosphate (ATP). These initial reactions of glycolysis make use of the Gibbs energy of hydrolysis of ATP, and they are often called the *preparation phase* of glycolysis (Figure 17.3, *Phase 1*).

▶ **What reactions convert glucose to glyceraldehyde-3-phosphate?**

Step 1. Glucose is phosphorylated to give glucose-6-phosphate, and ATP is hydrolyzed to ADP (Figure 17.4). The phosphorylation of glucose is an endergonic reaction.

$$\text{Glucose} + P_i \rightarrow \text{Glucose-6-phosphate} + H_2O$$
$$\Delta G^{\circ\prime} = 13.8 \text{ kJ/mol} = 3.3 \text{ kcal/mol}$$

The hydrolysis of ATP is exergonic.

$$\text{ATP} + H_2O \rightarrow \text{ADP} + P_i$$
$$\Delta G^{\circ\prime} = -30.5 \text{ kJ/mol} = -7.3 \text{ kcal/mol}$$

These two reactions are coupled, so the overall reaction is the sum of the two and is exergonic.

$$\text{Glucose} + \text{ATP} \rightarrow \text{Glucose-6-phosphate} + \text{ADP}$$
$$\Delta G^{\circ\prime} = 13.8 \text{ kJ/mol} + (-30.5 \text{ kJ/mol}) = -16.7 \text{ kJ/mol} = -4.0 \text{ kcal/mol}$$

Recall that $\Delta G^{\circ\prime}$ is calculated under standard states with the concentration of all reactants and products at 1 *M* except hydrogen ion. If we look at

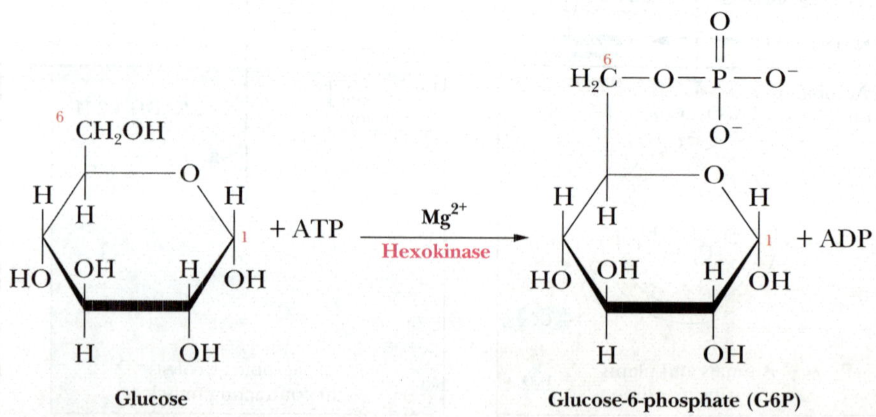

Figure 17.4 Step 1 of glycolysis: the phosphorylation of glucose.

the actual Gibbs energy change in the cell, the number varies depending on cell type and metabolic state, but a typical value for this reaction is −33.9 kJ/mol = −8.12 kcal/mol. Thus, the reaction is typically even more favorable under cellular conditions. Table 17.1 gives the $\Delta G°'$ and ΔG values for all the reactions of anaerobic glycolysis in erythrocytes.

The phosphorylation of glucose illustrates the use of chemical energy originally produced by the oxidation of nutrients and ultimately trapped by the phosphorylation of ADP to ATP. Recall from Section 15-6 that ATP does not represent stored energy, just as an electric current does not represent stored energy. The chemical energy of nutrients is released by oxidation and is made available for immediate use on demand by being trapped as ATP.

The enzyme that catalyzes this reaction is **hexokinase**. The term *kinase* is applied to the class of ATP-dependent enzymes that transfer a phosphate group from ATP to a substrate. The substrate of hexokinase is not necessarily glucose; rather, it can be any one of a number of hexoses such as glucose, fructose, and mannose. Glucose-6-phosphate is a **feedback inhibitor** of hexokinase; this is a control point in the pathway. ATP is an inhibitor of other kinase reactions in glycolysis because the cell does not need to have active glycolysis enzymatic activity occurring if there is a sufficient level of ATP available. ATP does not inhibit hexokinase, however, because the product of the hexokinase reaction (glucose-6-phosphate) is needed for other metabolic pathways discussed in Chapter 18.

hexokinase the first enzyme of glycolysis; it catalyzes the transfer of phosphate from ATP to six-carbon sugars

feedback inhibitor a molecule produced in a metabolic pathway that can reduce the activity of an enzyme found in an earlier stage of the pathway

Recall from **Biochemical Connections 6A** in Section 6-1A that isozymes are multiple forms of an enzyme that catalyze the same overall reaction but have subtle physical and kinetic differences. Some organisms or tissues contain multiple isozymes of hexokinase. One isoform of hexokinase found in the human liver, called glucokinase, lowers blood glucose levels after you've eaten a meal. Liver glucokinase requires a much higher substrate level to achieve saturation than hexokinase does. Because of this, when glucose levels are high, the liver can metabolize glucose via glycolysis preferentially over other tissues. When glucose levels are low, hexokinase is still active in all tissues. **Biochemical Connections 17B** discusses the importance of maintaining blood glucose levels in a normal range.

17B Biochemical Connections • Allied Health

Dolphins as a Model for Humans with Diabetes

Did you know that dolphins can share with humans symptoms typical of type 2 diabetes (Figure 17.5)? For a number of years, veterinarians monitored the health of 52 dolphins owned by the U.S. Navy. Blood samples revealed that fasting dolphins have high levels of blood glucose, similar to humans with diabetes. Blood glucose levels return to normal after a meal. The dolphins have been observed to maintain glucose levels on a high-protein diet. When the animals have high levels of glucose, they also have high levels of iron and triglycerides in the blood, typical of diabetes. Dolphins and humans respond to low-carbohydrate, high-protein diets with similar metabolic changes. In addition, the central nervous system of both species has a high demand for glucose, in view of the high brain-to-body-mass ratio common to both. These similarities, combined with the way the symptoms can be turned on and off with diet in dolphins, make dolphins a model for humans with diabetes.

Figure 17.5 Dolphins can share symptoms of type 2 diabetes with humans.

Consistent with the induced-fit model of enzyme–substrate binding (Section 6-3), a large conformational change takes place in hexokinase when substrate is bound (Figure 17.6). It has been shown by X-ray crystallography that, in the absence of substrate, two lobes of the enzyme that surround the binding site are quite far apart. When glucose is bound, the two lobes move closer together, and the glucose becomes almost completely surrounded by protein.

glucose-6-phosphate isomerase the enzyme that isomerizes glucose-6-phosphate to fructose-6-phosphate

Figure 17.6 A comparison of the conformations of free hexokinase and hexokinase complexed with glucose, its substrate.

Step 2. Glucose-6-phosphate isomerizes to give fructose-6-phosphate (Figure 17.7). Recall that isomers are compounds that have the same formula but a different arrangement of atoms and different properties. **Glucose-6-phosphate isomerase** is the enzyme that catalyzes Step 2 of glycolysis. The C-1 aldehyde group of glucose-6-phosphate is reduced to a hydroxyl group, and the C-2 hydroxyl group is oxidized to give the ketone group of fructose-6-phosphate, with no net oxidation or reduction. (Recall from Section 16-1 that glucose is an aldose, a sugar whose open-chain, noncyclic structure contains an aldehyde group, while fructose is a ketose, a sugar whose corresponding structure contains a ketone group.) The phosphorylated forms, glucose-6-phosphate and fructose-6-phosphate, are an aldose and a ketose, respectively.

Step 3. Fructose-6-phosphate is phosphorylated again, producing fructose-1,6-bisphosphate (Figure 17.8).

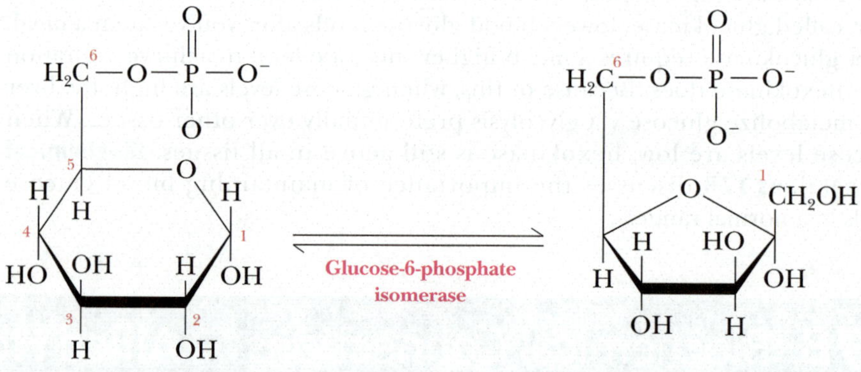

Glucose-6-phosphate (G6P) Fructose-6-phosphate (F6P)

Figure 17.7 Isomerization of glucose-6-phosphate.

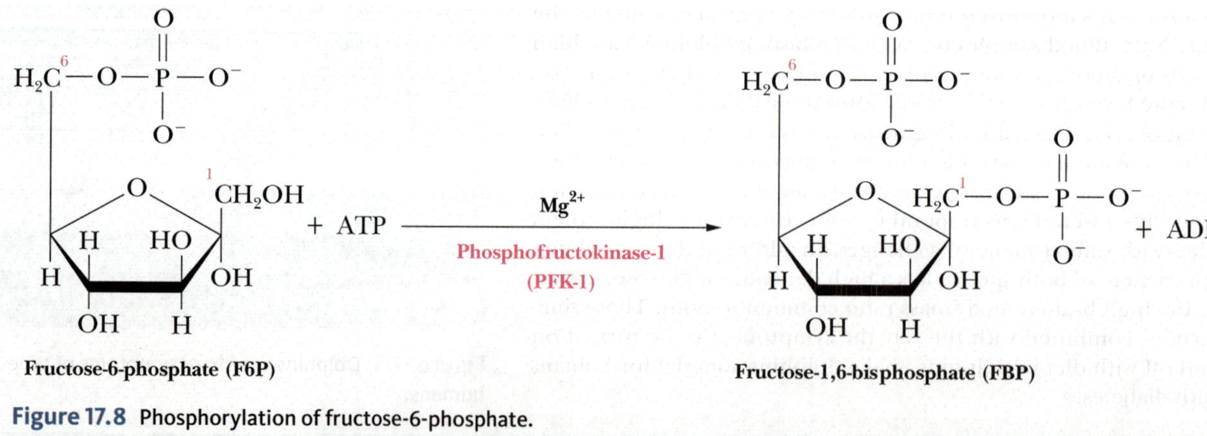

Fructose-6-phosphate (F6P) Fructose-1,6-bisphosphate (FBP)

Figure 17.8 Phosphorylation of fructose-6-phosphate.

Table 17.1 The Reactions of Glycolysis and Their Standard Gibbs Energy Changes

			$\Delta G^{\circ\prime}$*		ΔG**
Step	Reaction	Enzyme	kJ/mol	kcal/mol	kJ/mol
1	Glucose + ATP → Glucose-6-phosphate + ADP	Hexokinase/Glucokinase	−16.7	−4.0	−33.9
2	Glucose-6-phosphate → Fructose-6-phosphate	Glucose-6-phosphate isomerase	+1.67	+0.4	−2.92
3	Fructose-6-phosphate + ATP → Fructose-1,6-bisphosphate + ADP	Phosphofructokinase-1	−14.2	−3.4	−18.8
4	Fructose-1,6-bisphosphate → Dihydroxyacetone phosphate + Glyceraldehyde-3-phosphate	Aldolase	+23.9	+5.7	−0.23
5	Dihydroxyacetone phosphate → Glyceraldehyde-3-phosphate	Triose phosphate isomerase	+7.56	+1.8	+2.41
6	2(Glyceraldehyde-3-phosphate + NAD^+ + P_i → 1,3-Bisphosphoglycerate + NADH + H^+)	Glyceraldehyde-3-P dehydrogenase	2(+6.20)	2(+1.5)	2(−1.29)
7	2(1,3-Bisphosphoglycerate + ADP → 3-Phosphoglycerate + ATP)	Phosphoglycerate kinase	2(−18.8)	2(−4.5)	2(+0.1)
8	2(3-Phosphoglycerate → 2-Phosphoglycerate)	Phosphoglyceromutase	2(+4.4)	2(+1.1)	2(+0.83)
9	2(2-Phosphoglycerate → Phosphoenolpyruvate + H_2O)	Enolase	2(+1.8)	2(+0.4)	2(+1.1)
10	2(Phosphoenolpyruvate + ADP → Pyruvate + ATP)	Pyruvate kinase	2(−31.4)	2(−7.5)	2(−23.0)
Overall	Glucose + 2ADP + $2P_i$ + NAD^+ → 2 Pyruvate → 2ATP + NADH + H^+		−73.3	−17.5	−98.0
	2(Pyruvate + NADH + H^+ → Lactate + NAD^+)	Lactate dehydrogenase	2(−25.1)	2(−6.0)	2(−14.8)
	Glucose + 2ADP + $2P_i$ → 2 Lactate + 2ATP		−123.5	−29.5	−127.6

*$\Delta G^{\circ\prime}$ values are assumed to be the same at 25°C and 37°C and are calculated for standard-state conditions (1 *M* concentration of reactants and products, pH 7.0).
**ΔG values are calculated at 310 K (37°C) using steady-state concentrations of these metabolites found in erythrocytes.

As in Step 1, the endergonic reaction of the phosphorylation of fructose-6-phosphate is coupled to the exergonic reaction of the hydrolysis of ATP, and the overall reaction is exergonic (see Table 17.1).

The reaction in which fructose-6-phosphate is phosphorylated to give fructose-1,6-bisphosphate is the one in which the sugar is committed to glycolysis. Glucose-6-phosphate and fructose-6-phosphate can play roles in other pathways, but fructose-1,6-bisphosphate does not. After fructose-1,6-bisphosphate is formed from the original sugar, no other pathways are available, and the molecule must undergo the rest of the reactions of glycolysis. The phosphorylation of fructose-6-phosphate is highly exergonic and irreversible, and **phosphofructokinase-1 (PFK-1)**, the enzyme that catalyzes it, is the key regulatory enzyme in glycolysis.

phosphofructokinase-1 (PFK-1) the key allosteric control enzyme in glycolysis; it catalyzes the transfer of a phosphate from ATP to fructose-6-phosphate

PFK-1 is a tetramer that is subject to allosteric feedback regulation of the type we discussed in Section 7-1. There are two types of subunits, designated M and L. Figure 17.9 shows that they can combine into tetramers to give different isozyme permutations—namely, M_4, M_3L, M_2L_2, ML_3, and L_4. These isozymes have subtle physical and kinetic differences. The subunits differ slightly in amino acid composition, so the two isozymes can be separated from each other by electrophoresis (Section 5-3). The tetrameric form that occurs in muscle is designated M_4, while that in liver is designated L_4. In red blood cells, several of the combinations can be found. Individuals who lack the gene that directs the synthesis of the M form of the enzyme can carry on glycolysis in their livers but experience muscle weakness because they lack the enzyme in muscle.

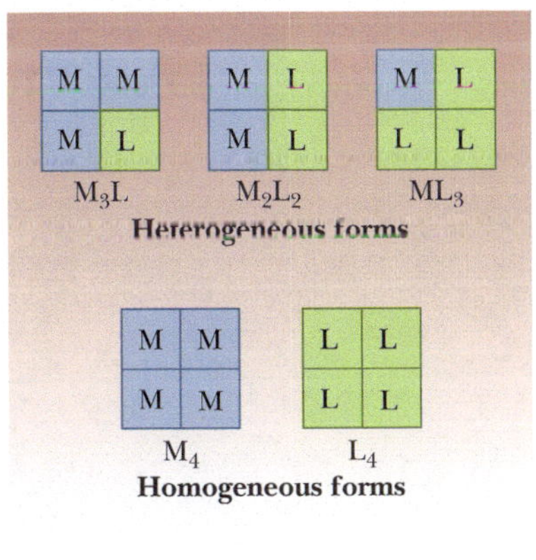

Figure 17.9 The possible isozymes of phosphofructokinase-1.

When the rate of the PFK-1 reaction is observed at varying concentrations of substrate (fructose-6-phosphate), the sigmoidal curve typical of allosteric

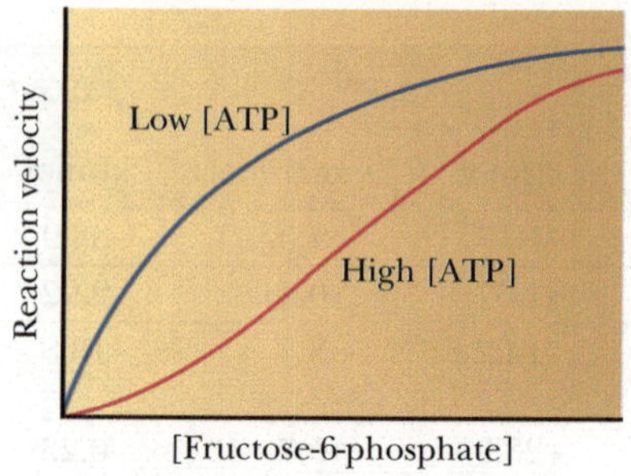

Figure 17.10 Allosteric effects on phosphofructokinase-1 (PFK-1) kinetics. At high [ATP], PFK-1 behaves cooperatively, and the plot of enzyme activity versus [fructose-6-phosphate] is sigmoidal. High [ATP] thus inhibits PFK-1, decreasing the enzyme's affinity for its substrate, fructose-6-phosphate. At low [ATP], the plot is more hyperbolic, reflecting the higher affinity of PFK-1 for its substrate.

aldolase in glycolysis, the enzyme that catalyzes the reverse of an aldol condensation of fructose-1,6-bisphosphate

triosephosphate isomerase the enzyme that isomerizes dihydroxyacetone phosphate to glyceraldehyde-3-phosphate

enzymes is obtained (Figure 17.10). ATP is an allosteric effector in the reaction. Recall that allosteric control (Section 4-5) is when binding of a ligand at one site causes changes at another binding site in a protein. In this case, binding of ATP to one site of PFK-1 causes changes at its active site that depress the rate of the reaction. Conversely, low levels of ATP stimulate the PFK-1 reaction. When there is a high level of ATP in the cell, a good deal of chemical energy is immediately available from the hydrolysis of ATP. The cell does not need to metabolize glucose for energy, so the presence of ATP inhibits the glycolytic pathway at this point. There is also another, more potent allosteric effector of PFK-1 called fructose-2,6-bisphosphate; we discuss its mode of action in Section 18-3 when we consider general control mechanisms in carbohydrate metabolism.

Step 4. Fructose-1,6-bisphosphate, a six-carbon molecule, is split into two three-carbon fragments, dihydroxyacetone phosphate and glyceraldehyde-3-phosphate (Figure 17.11). The cleavage reaction here is the reverse of an aldol condensation, and the enzyme that catalyzes it is called **aldolase**. In the enzyme isolated from most animal sources (the one from muscle is the most extensively studied), the basic side chain of an essential lysine residue plays the key role in catalyzing this reaction. The thiol group of a cysteine also acts as a base here.

Step 5. The dihydroxyacetone phosphate is isomerized to glyceraldehyde-3-phosphate (Figure 17.12).

The enzyme that catalyzes Step 5 is **triosephosphate isomerase**. (Both dihydroxyacetone phosphate and glyceraldehyde-3-phosphate are trioses that are phosphorylated.)

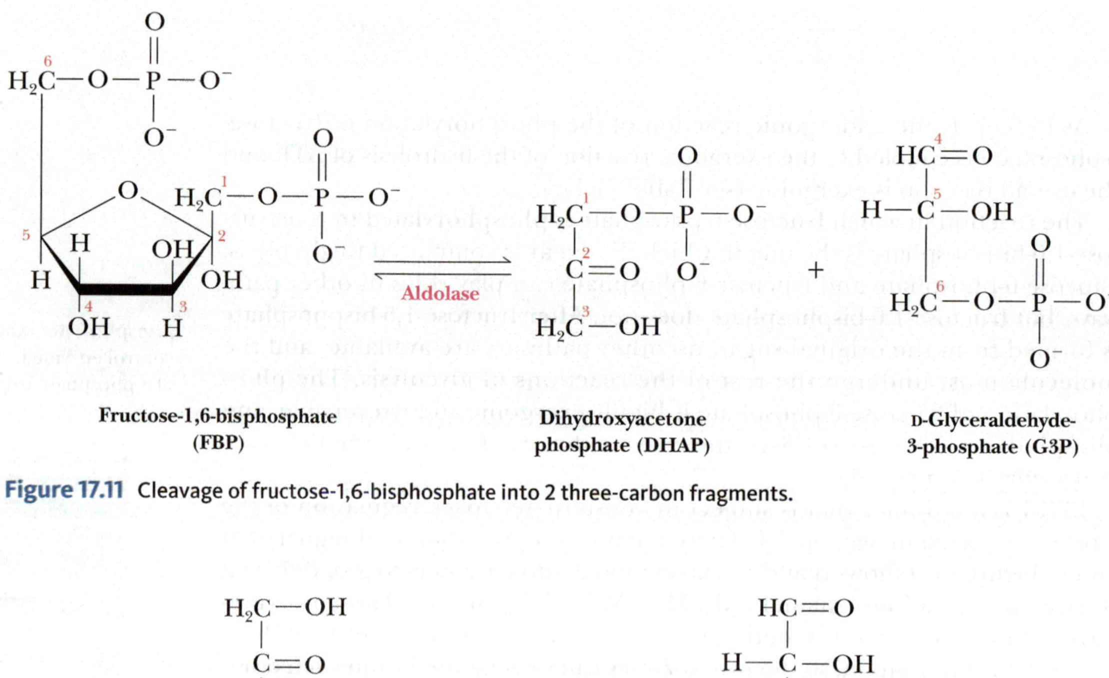

Figure 17.11 Cleavage of fructose-1,6-bisphosphate into 2 three-carbon fragments.

Figure 17.12 Isomerization of dihydroxyacetone phosphate to glyceraldehyde-3-phosphate.

One molecule of glyceraldehyde-3-phosphate has already been produced by the aldolase reaction in Step 4; Step 5 produces a second molecule of glyceraldehyde-3-phosphate. Thus, at the halfway point of the 10-step glycolytic pathway, *one* molecule of glucose has been converted into *two* molecules of glyceraldehyde-3-phosphate. The flowchart in Figure 17.13 shows these structural changes. Each molecule of glyceraldehyde-3-phosphate continues through the remaining glycolysis reactions to produce pyruvate (Figure 17.14, *Phase 2*).

The ΔG value for the triosephosphate isomerase reaction under physiological conditions is slightly positive (+2.41 kJ/mol = +0.58 kcal/mol). Does this mean that the reaction does not occur and that glycolysis is halted at this step? Just as coupled reactions involving ATP hydrolysis add their ΔG values together for the overall reaction, glycolysis is composed of many reactions that have very negative ΔG values that can drive the reaction to completion. A few reactions in glycolysis have small, positive ΔG values (see Table 17.1), but four reactions have very large, negative values, so ΔG for the whole process is negative.

17-3 The Phase 2 Reactions of Glycolysis: Glyceraldehyde-3-Phosphate Is Converted into Pyruvate

Learning Objectives

After reading this section, students will be able to:

LO 17.3.1 Write metabolite names, metabolite structures, and enzyme names for reactions in the second phase of glycolysis from glyceraldehyde-3-phosphate to pyruvate.

LO 17.3.2 Write the mechanism of a nicotinamide adenine dinucleotide–linked dehydrogenase.

LO 17.3.3 Describe allosteric effects seen for enzymes in the second phase of glycolysis.

In Steps 6–10 of glycolysis, *two* molecules of each of the three-carbon compounds take part in every reaction for each original glucose molecule. Figure 17.14 summarizes the second part of the pathway, which is often referred to as the *payoff phase* of glycolysis because adenosine triphosphate (ATP) is produced instead of used in this phase. Additionally, this phase produces two molecules of NADH, which carry electrons to the electron transport chain, thereby producing even more ATP.

▶ What reactions convert glyceraldehyde-3-phosphate to pyruvate?

Step 6. Glyceraldehyde-3-phosphate is oxidized to 1,3-bisphosphoglycerate (Figure 17.15).

Step 6 is *the* characteristic reaction of glycolysis. It involves the addition of a phosphate group to glyceraldehyde-3-phosphate as well as an electron-transfer reaction from glyceraldehyde-3-phosphate to NAD^+. We can simplify the discussion by considering the two parts separately.

In an electron-transfer reaction, one species is oxidized while another is reduced. The oxidation half reaction converts the aldehyde group (—CHO) of glyceraldehyde-3-phosphate to a carboxylic acid group (—COOH).

$$RCHO + H_2O \rightarrow RCOOH + 2H^+ + 2e^-$$

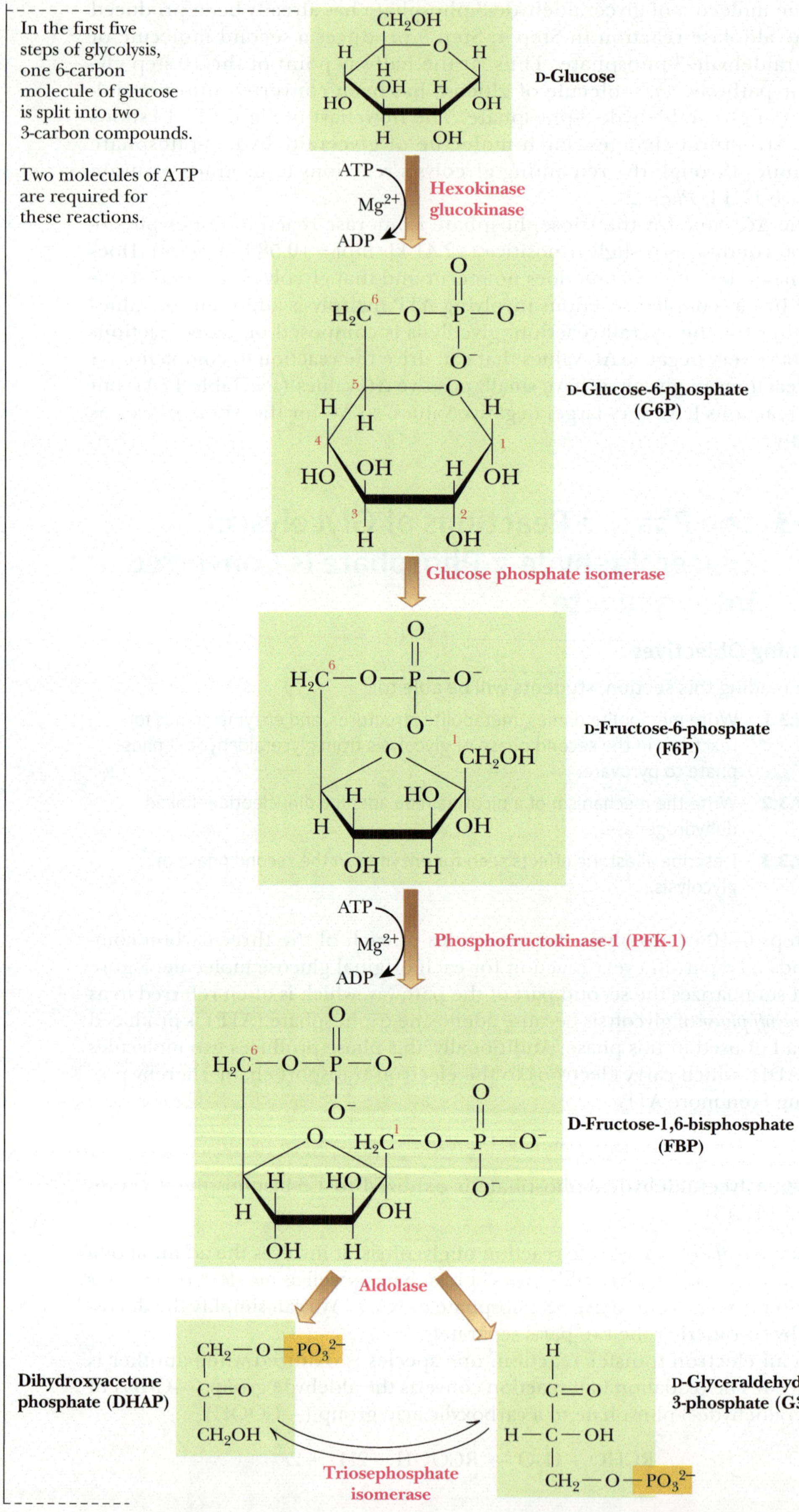

Figure 17.13 Chemical structure changes associated with the conversion of glucose to glyceraldehyde-3-phosphate.

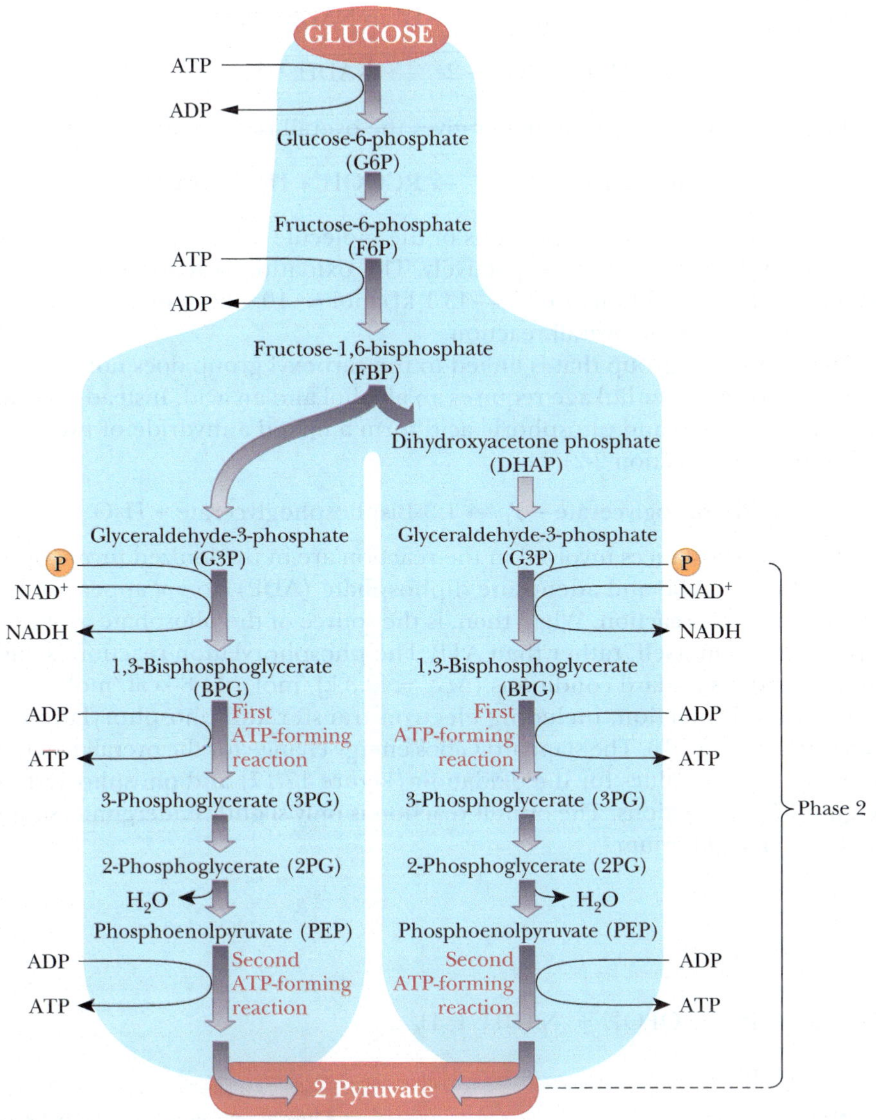

Figure 17.14 The second phase of glycolysis.

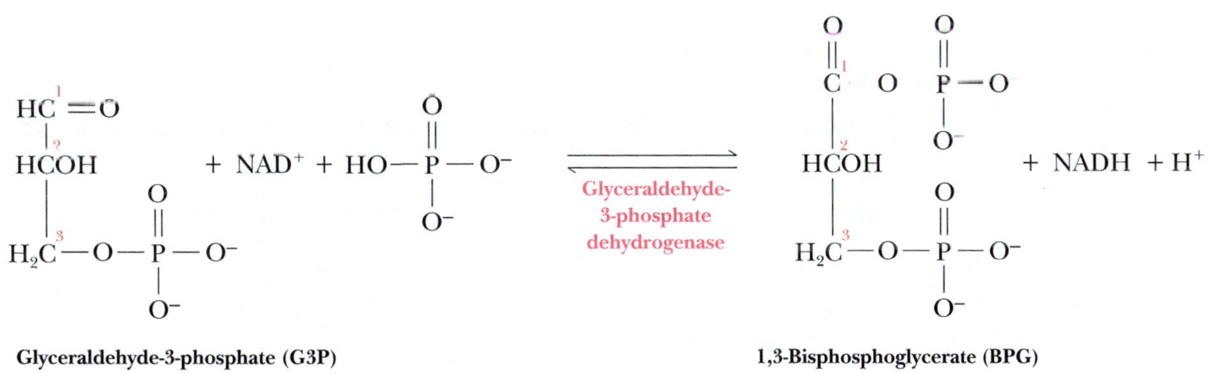

Figure 17.15 Glyceraldehyde-3-phosphate is oxidized to 1,3-bisphosphoglycerate.

The reduction half reaction converts NAD^+ to NADH (Section 15-9).

$$NAD^+ + 2H^+ + 2e^- \rightarrow NADH + H^+$$

Adding these two reactions together gives the overall redox reaction:

$$RCHO + H_2O + NAD^+ \rightarrow RCOOH + H^+ + NADH$$

The R group represents the portions of the molecule other than the aldehyde and carboxylic acid groups, respectively. The oxidation reaction is exergonic under standard conditions ($\Delta G^{\circ\prime} = -43.1$ kJ/mol = -10.3 kcal/mol), but oxidation is only part of the overall reaction.

The phosphate group that is linked to the carboxyl group does not form an ester because an ester linkage requires an alcohol and an acid. Instead, the carboxylic acid group and phosphoric acid form a mixed anhydride of two acids by loss of water (Section 2-2),

$$\text{3-Phosphoglycerate} + P_i \rightarrow \text{1,3-Bisphosphoglycerate} + H_2O$$

in which the substances involved in the reaction are in the ionized form appropriate at pH 7. ATP and adenosine diphosphate (ADP) do *not* appear in the equation for this reaction. What, then, is the source of the phosphate group? It is phosphate ion itself, rather than ATP. The phosphorylation reaction is endergonic under standard conditions ($\Delta G^{\circ\prime} = 49.3$ kJ/mol = 11.8 kcal/mol).

The overall reaction, including electron transfer and phosphorylation, is shown in Figure 17.16. The standard Gibbs energy change for the overall reaction is the sum of the values for the oxidation (Figure 17.17) and phosphorylation (Figure 17.18) reactions. The overall reaction is only slightly endergonic, so it is not far from equilibrium.

$$RCHO + HOPO_3^{2-} + NAD^+ \rightleftharpoons RC(=O){-}OPO_3^{2-} + NADH + H^+$$

or

$$\text{Glyceraldehyde-3-phosphate} + P_i + NAD^+ \xrightarrow{\text{Glyceraldehyde-3-phosphate dehydrogenase}} \text{1,3-Bisphosphoglycerate} + NADH + H^+$$

Figure 17.16 General redox reaction from an aldehyde to a phosphoester (*top*) followed by the specific redox reaction of glycolysis (*bottom*).

1. Oxidation of glyceraldehyde-3-phosphate ($\Delta G^{\circ\prime} = -43.1$ kJ mol^{-1} = -10.3 kcal mol^{-1})

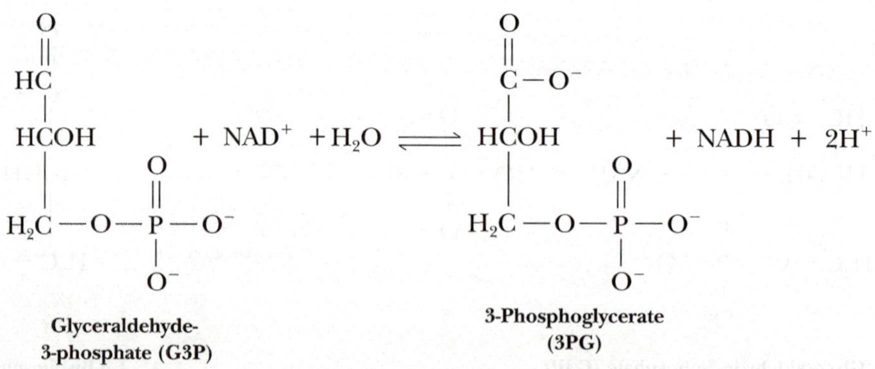

Figure 17.17 The exergonic conversion of glyceraldehyde-3-phosphate to 3-phosphoglycerate.

2. Phosphorylation of 3-phosphoglycerate ($\Delta G^{\circ\prime}$ = 49.3 kJ mol^{-1} = 11.8 kcal mol^{-1})

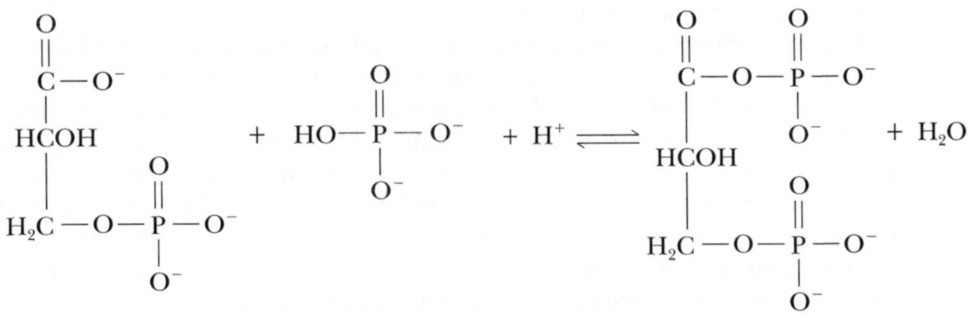

3-Phosphoglycerate (3PG)

1,3-Bisphosphoglycerate (BPG)

Figure 17.18 The endergonic conversion of 3-phosphoglycerate to 1,3-bisphosphoglycerate.

$$\begin{aligned}\Delta G^{\circ\prime}\text{ overall} &= \Delta G^{\circ\prime}\text{ oxidation} + \Delta G^{\circ\prime}\text{ phosphorylation}\\ &= (-43.1\text{ kJ/mol}) + (49.3\text{ kJ/mol})\\ &= 6.2\text{ kJ/mol} = 1.5\text{ kcal/mol}\end{aligned}$$

This value of the standard Gibbs energy change is for the reaction of *one* mole of glyceraldehyde-3-phosphate; the value must be multiplied by 2 to get the value for each mole of glucose ($\Delta G^{\circ\prime}$ = 12.4 kJ/mol = 3.0 kcal/mol). Table 17.1 shows that ΔG under cellular conditions is slightly negative (−1.29 kJ/mol or −0.31 kcal/mol).

The enzyme that catalyzes the conversion of glyceraldehyde-3-phosphate to 1,3-bisphosphoglycerate is **glyceraldehyde-3-phosphate dehydrogenase (G3PDH)**. This enzyme is one of a class of similar enzymes, the NAD-linked dehydrogenases. In these enzymes, nicotinamide adenine dinucleotide (NAD) is a coenzyme (Section 15-5) that is the oxidizing agent for the G3PDH enzyme. Each dehydrogenase enzyme we discuss here in Chapter 17 or in the other metabolism chapters catalyzes a redox reaction involving a metabolite paired with a coenzyme. Moreover, the name of the metabolite being oxidized appears before the word "dehydrogenase" in the name of each dehydrogenase enzyme. For reversible dehydrogenase reactions, this naming convention is based on the direction in which the substrate is being oxidized.

glyceraldehyde-3-phosphate dehydrogenase (G3PDH) an enzyme in glycolysis that catalyzes the oxidation of glyceraldehyde-3-phosphate and the production of NADH

The structures of a number of NAD-linked dehydrogenases have been studied via X-ray crystallography. The overall structures are not strikingly similar, but the structure of the binding site for the NAD coenzyme is quite similar in all these enzymes (Figure 17.19). The oxidizing agent is NAD^{+}; both the

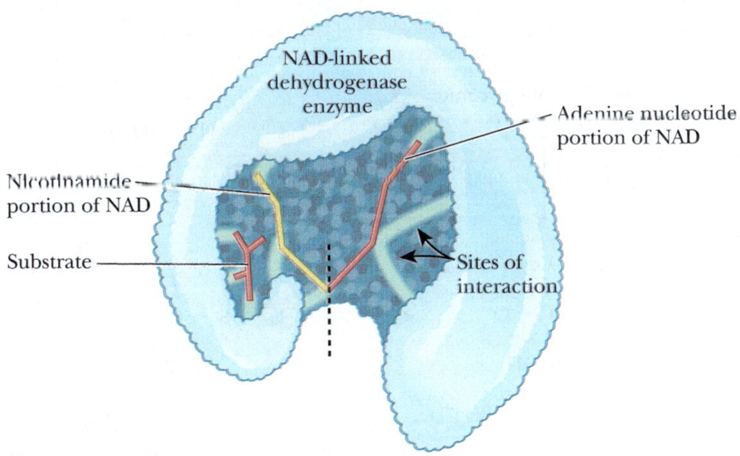

Figure 17.19 Schematic view of the binding site of an NAD-linked dehydrogenase. There are specific binding sites for the adenine nucleotide portion of the NAD coenzyme (shown in red to the right of the dashed line) and for the nicotinamide portion of the coenzyme (shown in yellow to the left of the dashed line), in addition to the binding site for the substrate. Specific interactions with amino acid residues of the enzyme hold the substrate and coenzyme in the proper positions. Sites of interaction are shown as a series of pale-green lines.

oxidized (NAD^+) and reduced (NADH) forms of the coenzyme bind to the enzyme. One portion of the binding site is specific for the nicotinamide ring, and one portion is specific for the adenine ring.

Figure 17.20 shows the mechanism of the G3PDH reaction. The G3PDH enzyme is a tetramer, consisting of four identical subunits. Each subunit binds one molecule of NAD^+, and each subunit contains an essential cysteine residue. To start the reaction, the active-site cysteine residue performs a nucleophilic attack on the carbonyl carbon of the glyceraldehyde-3-phosphate substrate (Figure 17.20, *top left*). Electron rearrangement then results in a transfer of electrons to the nicotinamide ring of NAD^+, producing NADH and a thioester intermediate (Figure 17.20, *top right* and *bottom right*). This thioester involving the cysteine residue is the key intermediate in this reaction.

In the phosphorylation step, the thioester acts as a high-energy intermediate (see Section 15-7 for a discussion of thioesters). Phosphate ion attacks the thioester, forming a mixed anhydride of the carboxylic and phosphoric acids, which is also a high-energy compound (Figure 17.20, *bottom right* and *bottom left*). This compound is 1,3-bisphosphoglycerate, the product of the reaction. Production of ATP requires a high-energy compound as the starting material. 1,3-Bisphosphoglycerate fulfills this requirement and has its phosphate group transferred to ADP in the next step in a highly exergonic reaction.

Step 7. A phosphate from 1,3-bisphosphoglycerate is transferred to ADP, producing 3-phosphoglycerate and ATP (Figure 17.21). Step 7 is one of the two reactions in which ATP is produced in glycolysis by the phosphorylation of ADP.

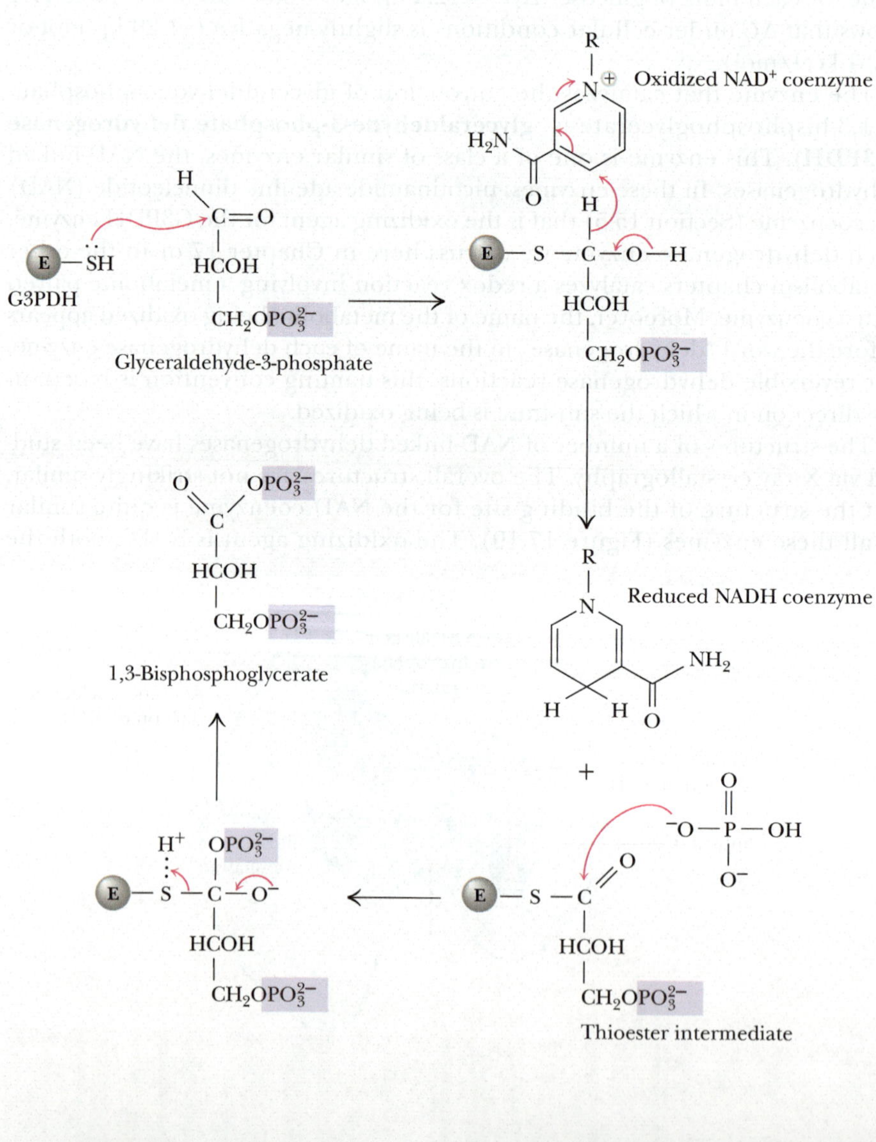

Figure 17.20 The mechanism of the glyceraldehyde-3-phosphate dehydrogenase (G3PDH) reaction in glycolysis whereby glyceraldehyde-3-phosphate is oxidized to 1,3-bisphosphoglycerate and NAD^+ is reduced to NADH. "E" represents the G3PDH enzyme, which is a tetramer, with each subunit containing many amino acid residues. For simplicity, however, only the thiol group of one active-site cysteine residue is shown. Also, only the nicotinamide portion of the NAD coenzyme is shown (see Section 15-5 for the entire structure of NAD).

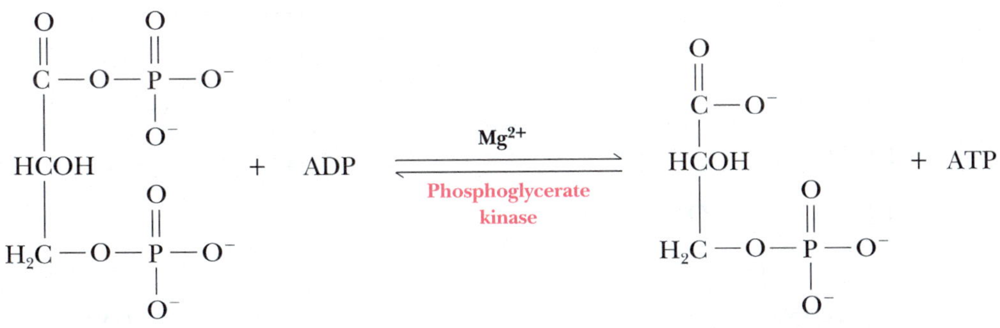

Figure 17.21 Dephosphorylation of 1,3-bisphosphoglycerate to 3-phosphoglycerate coupled with phosphorylation of ADP to ATP.

The enzyme that catalyzes Step 7 is **phosphoglycerate kinase**. Recall that *kinase* is the generic name for a class of ATP-dependent phosphate-group transfer enzymes. The most striking feature of the reaction has to do with the energetics of the phosphate-group transfer. As mentioned previously, 1,3-bisphosphoglycerate can easily transfer its phosphate group to other substances, and in this reaction, 1,3-bisphosphoglycerate transfers its phosphate group to ADP.

phosphoglycerate kinase the enzyme that catalyzes the transfer of a phosphate group from 1,3-bisphosphoglycerate to ADP

This transfer is typical of a type of phosphorylation called **substrate-level phosphorylation**. It is different from oxidative phosphorylation (Sections 20-1 through 20-5). In oxidative phosphorylation, the transfer of phosphate groups is linked to electron-transfer reactions in which oxygen is the ultimate electron acceptor. The only requirement for substrate-level phosphorylation is that the standard Gibbs energy of the hydrolysis reaction is more negative than that for the hydrolysis of the new phosphate compound being formed. The standard Gibbs energy of hydrolysis of 1,3-bisphosphoglycerate is –49.3 kJ/mol. Also, the standard Gibbs energy of hydrolysis of ATP is –30.5 kJ/mol, and we must change the sign of the Gibbs energy change when the reverse reaction occurs:

substrate-level phosphorylation a reaction in which the source of phosphorus is inorganic phosphate ion, not ATP

$$\text{ADP} + \text{P}_i + \text{H}^+ \rightarrow \text{ATP} + \text{H}_2\text{O}$$
$$\Delta G^{\circ\prime} = 30.5 \text{ kJ/mol} = 7.3 \text{ kcal/mol}$$

The net reaction is

$$\text{1,3-bisphosphoglycerate} + \text{ADP} \rightarrow \text{3-Phosphoglycerate} + \text{ATP}$$
$$\Delta G^{\circ\prime} = -49.3 \text{ kJ/mol} + 30.5 \text{ kJ/mol} = -18.8 \text{ kJ/mol} = -4.5 \text{ kcal/mol}$$

Two molecules of ATP are produced by this reaction for each molecule of glucose that enters the glycolytic pathway. In the earlier stages of the pathway, two molecules of ATP were invested to produce fructose-1,6-bisphosphate, and now they have been recovered. At this point, the balance of ATP use and production is exactly even. The next few reactions bring about the production of two more molecules of ATP for each original molecule of glucose, leading to the net gain of two ATP molecules in glycolysis.

Step 8. The phosphate group is transferred from carbon 3 to carbon 2 of the glycerate backbone (Figure 17.22), setting the stage for the reaction that follows. The enzyme that catalyzes Step 8 is **phosphoglyceromutase**. Mutases are specialized types of isomerases that catalyze the movement of functional groups within a molecule.

phosphoglyceromutase the enzyme that catalyzes the isomerization of 3-phosphoglycerate to 2-phosphoglycerate, in which the phosphate group is moved from carbon 3 to carbon 2

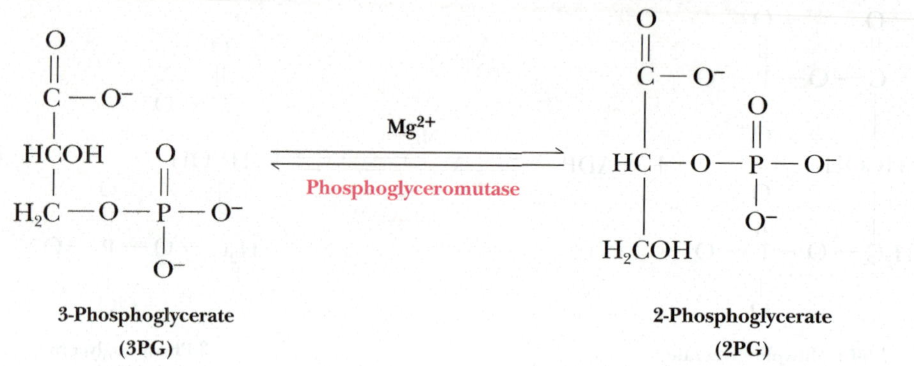

Figure 17.22 Phosphate transfer converts 3-phosphoglycerate to 2-phosphoglycerate.

enolase the enzyme that catalyzes the conversion of 2-phosphoglycerate to phosphoenolpyruvate

Step 9. 2-Phosphoglycerate loses a molecule of water, producing phosphoenolpyruvate (Figure 17.23). **Enolase**, the enzyme that catalyzes this reaction, requires Mg^{2+} as a cofactor. The water molecule that is eliminated binds to Mg^{2+} in the course of the reaction.

Step 10. The phosphate group of phosphoenolpyruvate is transferred to ADP in another example of *substrate-level phosphorylation*, producing ATP and pyruvate (Figure 17.24). The double bond shifts to the oxygen on carbon 2, and a hydrogen shifts to carbon 3. Phosphoenolpyruvate is a high-energy compound with a high phosphate-group transfer potential. The Gibbs energy of hydrolysis of this compound is more negative than that of ATP (–61.9 kJ/mol versus –30.5 kJ/mol, or –14.8 kcal/mol versus –7.3 kcal/mol). The reaction that occurs here in step 10 can be considered the sum of the hydrolysis of phosphoenolpyruvate and the phosphorylation of ADP.

$$\text{Phosphoenolpyruvate} \rightarrow \text{Pyruvate} + P_i$$
$$\Delta G^{\circ\prime} = -61.9 \text{ kJ/mol} = -14.8 \text{ kcal/mol}$$
$$\text{ADP} + P_i \rightarrow \text{ATP}$$
$$\Delta G^{\circ\prime} = 30.5 \text{ kJ/mol} = 7.3 \text{ kcal/mol}$$

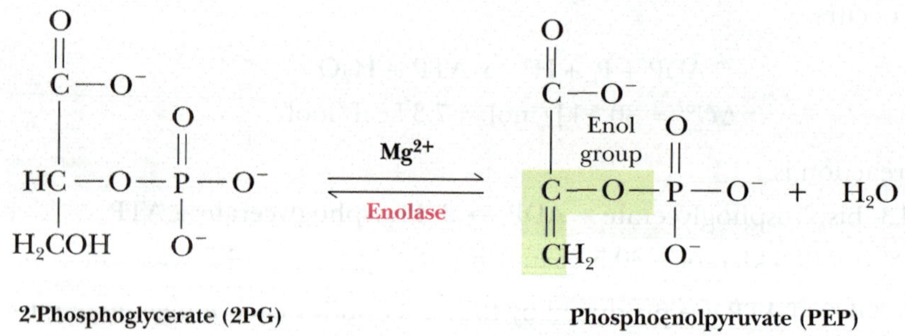

Figure 17.23 Phosphoenolpyruvate forms from the dehydration of 2-phosphoglycerate.

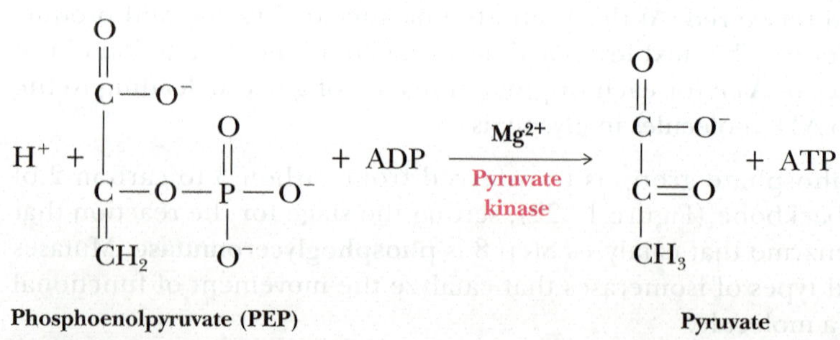

Figure 17.24 Dephosphorylation of phosphoenolpyruvate forms pyruvate and ATP.

The net reaction is

$$\text{Phosphoenolpyruvate} + \text{ADP} \rightarrow \text{Pyruvate} + \text{ATP}$$
$$\Delta G^{\circ\prime} = -31.4 \text{ kJ/mol} = -7.5 \text{ kcal/mol}$$

Because two moles of pyruvate are produced for each mole of glucose, twice as much energy is released for each mole of starting material.

Pyruvate kinase (PK) is the enzyme that catalyzes Step 10. It is an allosteric enzyme consisting of four subunits of two different types (M and L), as we saw with phosphofructokinase-1 (PFK-1) in Step 3 (see Figures 17.8 and 17.9). PK is inhibited by ATP. The conversion of phosphoenolpyruvate to pyruvate slows down when the cell has a high concentration of ATP—that is, when the cell does not have a great need for energy in the form of ATP. Because of the different isozymes of PK found in liver versus muscle, the control of glycolysis is handled differently in these two tissues, which we discuss in detail in Chapter 18. Figure 17.25 shows the structural changes in the second phase of glycolysis, from glyceraldehyde-3-phosphate to pyruvate.

pyruvate kinase (PK) the enzyme that catalyzes the final step common to all forms of glycolysis, the conversion of phosphoenolpyruvate to pyruvate and the phosphorylation of ADP to ATP

17-4 Anaerobic Metabolism of Pyruvate

Learning Objectives

After reading this section, students will be able to:

LO 17.4.1 Write metabolite names, metabolite structures, and enzyme names for reactions seen in the two possible fates for pyruvate in anaerobic metabolism: lactic acid fermentation and alcoholic fermentation.

LO 17.4.2 Explain the purpose of alcoholic fermentation and lactic acid fermentation.

LO 17.4.3 Write the mechanism of a thiamine pyrophosphate-linked decarboxylase.

17-4A Reduction of Pyruvate to Lactate

▶ How does the conversion of pyruvate to lactate take place in muscle?

The final reaction of anaerobic glycolysis is the reduction of pyruvate to lactate (Figure 17.26). This reaction is also exergonic ($\Delta G^{\circ\prime} = -25.1$ kJ/mol = -6.0 kcal/mol); as before, we need to multiply this value by 2 to find the energy yield for each molecule of glucose that enters the pathway. Lactate is a dead end in muscle metabolism, but it can be recycled in the liver to form pyruvate and even glucose by a pathway called gluconeogenesis ("creation of new glucose," read from right to left), which we discuss in Section 18-2.

Lactate dehydrogenase (LDH) is the enzyme that catalyzes this reaction. Recall from the glyceraldehyde-3-phosphate dehydrogenase (G3PDH) reaction (Step 6) of glycolysis that dehydrogenases oxidize a metabolite with a concomitant reduction of a coenzyme. These trends still hold here when you consider the reverse reaction (i.e., the conversion of lactate to pyruvate). Furthermore, remember that dehydrogenases are named by placing the metabolite being oxidized before the word "dehydrogenase."

lactate dehydrogenase an NAD-linked dehydrogenase that catalyzes the conversion of pyruvate to lactate

Like G3PDH, LDH is an NAD-linked dehydrogenase and consists of four subunits. There are two kinds of subunits, designated M and H, which vary slightly in amino acid composition. The quaternary structure of the tetramer can vary according to the relative amounts of the two kinds of subunits, yielding five possible isozymes: M_4, M_3H, M_2H_2, MH_3, and H_4. In human skeletal muscle, the homogeneous tetramer of the M_4-type predominates, and in the heart, the other homogeneous possibility, the H_4 tetramer, is the predominant form. The heterogeneous forms—M_3H, M_2H_2, and MH_3—occur in blood serum.

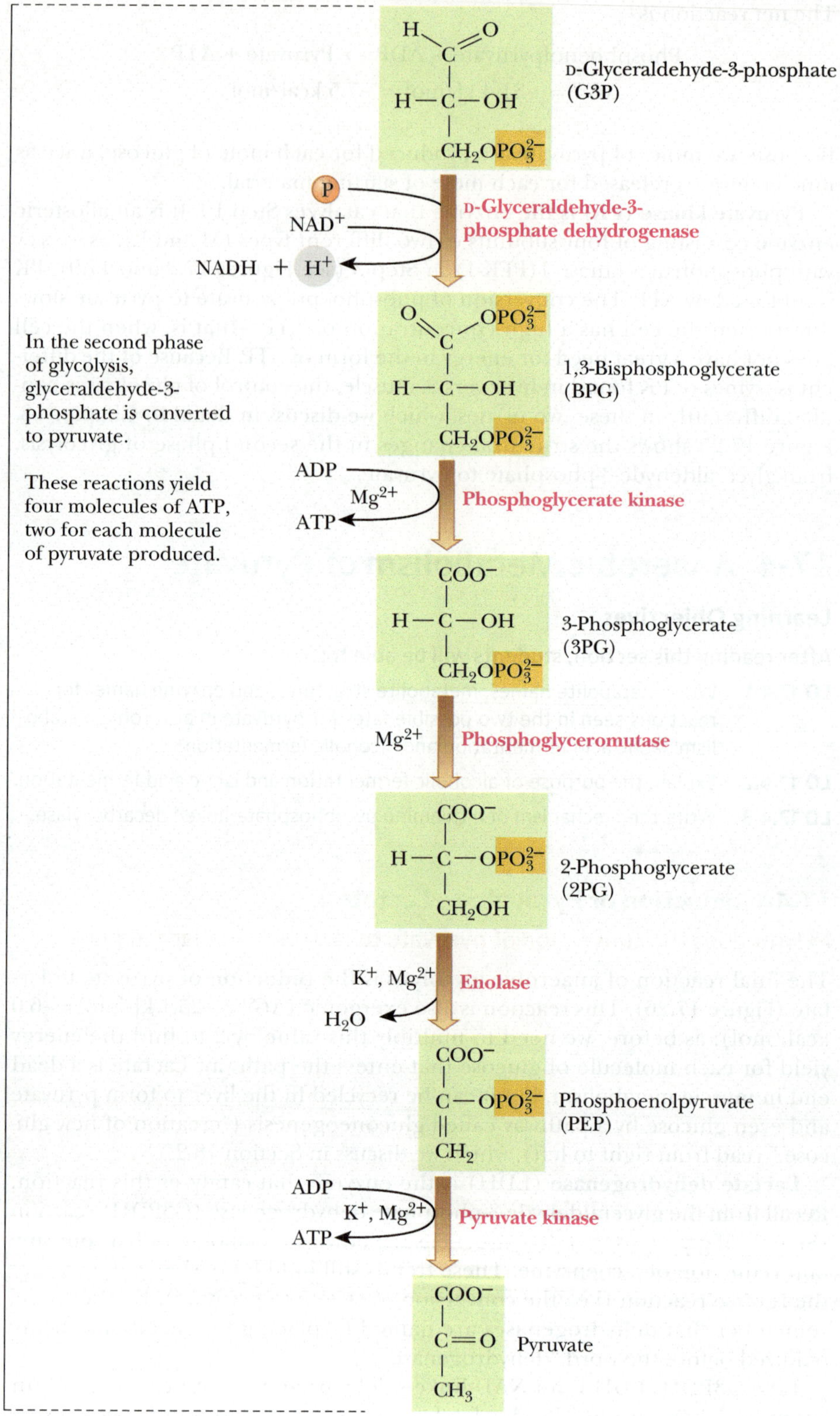

Figure 17.25 Chemical structure changes associated with the conversion of glyceraldehyde-3-phosphate to pyruvate in the second phase of glycolysis.

A very sensitive clinical test for heart disease is based on the existence of the various isozymic forms of this enzyme. The relative amounts of the H_4 and MH_3 isozymes in blood serum increase drastically after myocardial infarction (heart attack) compared with normal serum.

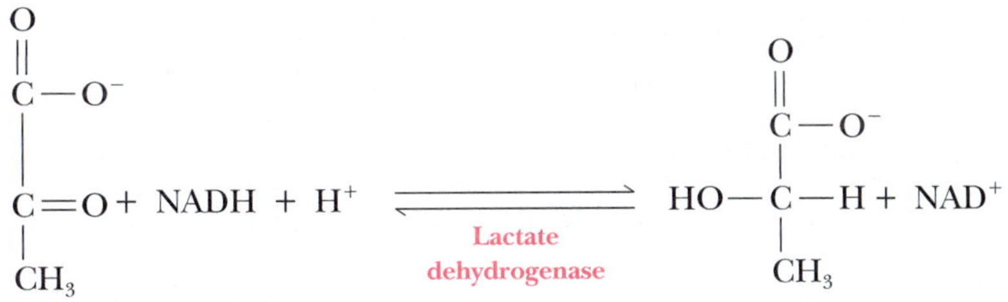

Pyruvate

Lactate

Figure 17.26 Reduction of pyruvate to lactate.

The different isozymes have slightly different kinetic properties due to their subunit compositions. The H_4 isozyme (also called LDH 1) has a higher affinity for lactate as a substrate. The M_4 isozyme (LDH 5) is allosterically inhibited by pyruvate. These differences reflect the isozymes' general roles in the different aerobic and anaerobic metabolic conditions found in various tissues. **Biochemical Connections 17C** describes another aspect of anaerobic metabolism.

Why is the reduction of pyruvate to lactate (a waste product in aerobic organisms) the last step in anaerobic glycolysis, a pathway that provides energy for the organism by the oxidation of nutrients? To understand why, we need to consider the relative amounts of NAD^+ and NADH in a cell. The reduction half reaction can be written as follows:

$$\text{Pyruvate} + 2\text{H}^+ + 2e^- \rightarrow \text{Lactate}$$

17C Biochemical Connections • Allied Health (Dentistry)

What Is the Connection Between Anaerobic Metabolism and Dental Plaque?

Dental caries, or tooth decay, is one of the most prevalent diseases in the United States and possibly in the world, although modern treatments such as fluoride and flossing have greatly reduced its incidence in young people. Contributing factors in tooth decay are a combination of a diet high in refined sugars, the development of dental plaque, and anaerobic metabolism.

The high-sugar diet allows for rapid growth of bacteria in the mouth, and sucrose is perhaps the most efficiently used sugar because the bacteria can make their polysaccharide "glue" more efficiently from this nonreducing sugar. The bacteria grow in expanding, sticky colonies, forming plaque on the tooth surface (Figure 17.27). The bacteria growing under the surface of the plaque must utilize anaerobic metabolism because oxygen does not diffuse readily through the waxy surface of dental plaque. The two predominant by-products, lactate and pyruvate, are relatively strong organic acids, and these acid products actually destroy the enamel surface. The bacteria grow rapidly in the pock holes. If the enamel is eaten all the way through, the bacteria grow even more readily in the softer dentin layer beneath the enamel.

Fluoridation results in a much harder enamel surface, and the fluoride may inhibit the metabolism of the bacteria. Daily flossing with the correct technique below the gumline disrupts the plaque, and the anaerobic conditions never get started. ▶

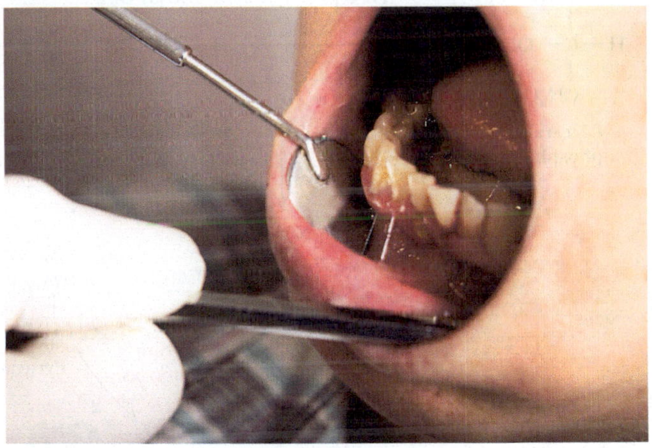

iStock.com/101dalmatians

Figure 17.27 Yellow discoloration of teeth is due to the buildup of dental plaque.

The oxidation half reaction is

$$NADH + H^+ \rightarrow NAD^+ + 2e^- + 2H^+$$

Thus, the overall reaction is

$$\text{Pyruvate} + NADH + H^+ \rightarrow \text{Lactate} + NAD^+$$

The NADH produced from NAD^+ by the earlier oxidation of glyceraldehyde-3-phosphate is used up in anaerobic glycolysis with no net change in the relative amounts of NADH and NAD^+ in the cell (Figure 17.28). The regeneration of NAD^+ from NADH is needed under anaerobic conditions in the cell so that NAD^+ will be present for further glycolysis to take place. Without this regeneration, the oxidation reactions in anaerobic organisms would soon come to a halt because of the lack of NAD^+ to serve as an oxidizing agent in fermentative processes. The production of lactate buys time for the organism experiencing anaerobic metabolism and shifts some of the load away from the muscles and onto the liver, in which gluconeogenesis can reconvert lactate to pyruvate and glucose (Chapter 18). The same considerations apply in alcoholic fermentation (which is discussed next).

On the other hand, NADH is a frequently encountered reducing agent in many reactions, and it is lost to the organism in lactate production. Aerobic metabolism makes more efficient use of reducing agents ("reducing power") such as NADH because the conversion of pyruvate to lactate does not occur in aerobic metabolism. The NADH produced in the stages of glycolysis leading to the production of pyruvate is available for use in other reactions in which a reducing agent is needed.

17-4B Pyruvate Conversion to Ethanol by Fermentation

▶ How does alcoholic fermentation take place?

Two other reactions related to the glycolytic pathway lead to the production of ethanol by *alcoholic fermentation*. This process is one of the alternative fates of pyruvate mentioned at the beginning of Section 17-1. In the first of the two

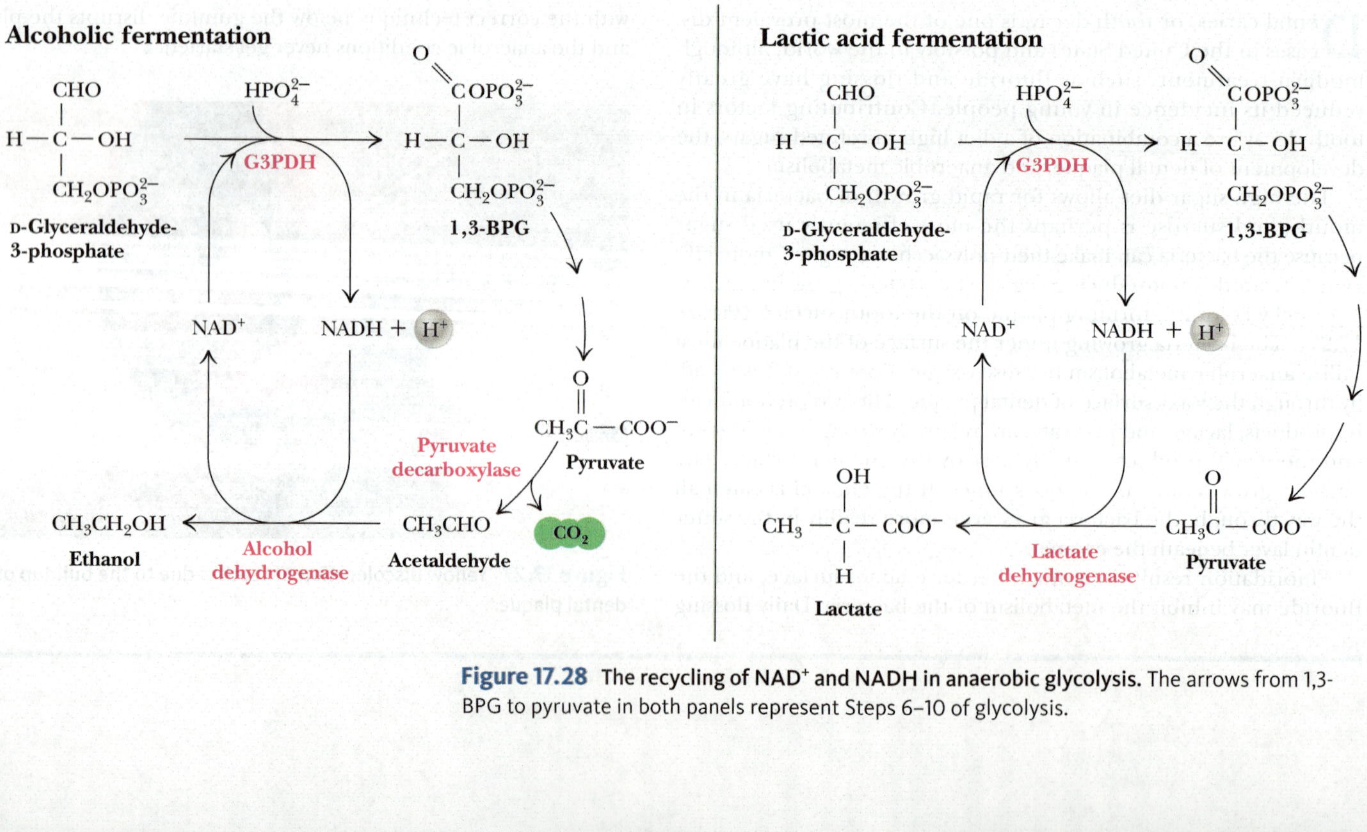

Figure 17.28 **The recycling of NAD^+ and NADH in anaerobic glycolysis.** The arrows from 1,3-BPG to pyruvate in both panels represent Steps 6–10 of glycolysis.

reactions that lead to the production of ethanol, pyruvate is decarboxylated (loses carbon dioxide) to produce acetaldehyde (Figure 17.29).

The enzyme that catalyzes this reaction is **pyruvate decarboxylase**. This enzyme requires Mg^{2+} and **thiamine pyrophosphate (TPP)**, a cofactor we have not encountered before (Figure 17.30). (Thiamine itself is vitamin B_1.) In TPP, the carbon atom between the nitrogen and the sulfur in the thiazole ring is highly reactive. It forms a carbanion (an ion with a negative charge on a carbon atom) quite easily, and the carbanion, in turn, attacks the carbonyl group of pyruvate to form an adduct (Figure 17.31).

Carbon dioxide splits off, leaving a two-carbon fragment covalently bonded to TPP. There is a shift of electrons, and the two-carbon fragment splits off, producing acetaldehyde. The two-carbon fragment bonded to TPP is sometimes called activated acetaldehyde, and TPP can be found in several reactions that are decarboxylations. The carbon dioxide produced is responsible for the bubbles in beer and sparkling wines.

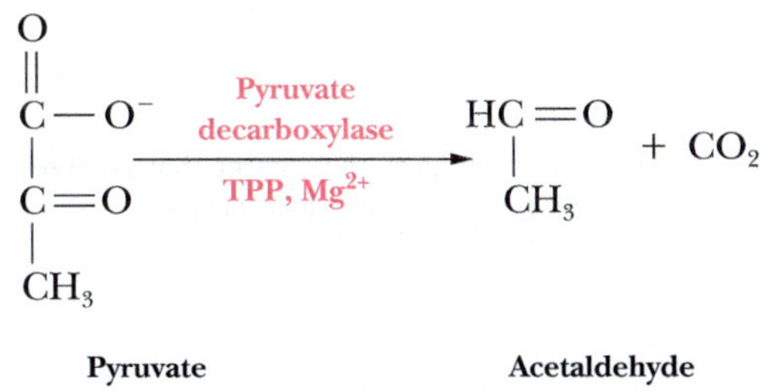

Figure 17.29 Decarboxylation of pyruvate forms acetaldehyde. The thiamine pyrophosphate (TPP) coenzyme and magnesium ion are required by the pyruvate decarboxylase enzyme.

pyruvate decarboxylase the enzyme that catalyzes the removal of the carboxyl group of pyruvate, producing acetaldehyde and CO_2

thiamine pyrophosphate (TPP) a coenzyme that stabilizes carbanion intermediates during the transfer or removal of carbon fragments

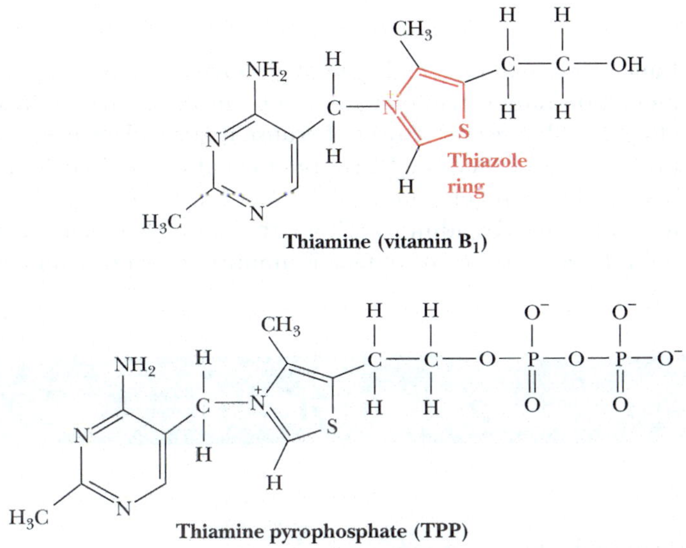

Figure 17.30 The structures of thiamine (vitamin B_1) and thiamine pyrophosphate (TPP), the active form of the coenzyme.

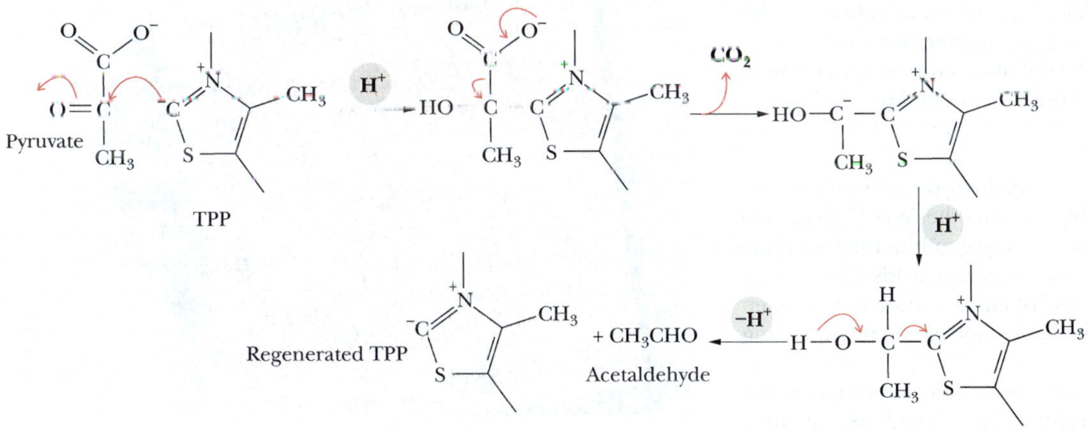

Figure 17.31 The mechanism of the pyruvate decarboxylase reaction. The carbanion form of the thiazole ring of thiamine pyrophosphate (TPP) is strongly nucleophilic. The carbanion attacks the carbonyl carbon of pyruvate to form an adduct. Carbon dioxide splits out, leaving a two-carbon fragment (activated acetaldehyde) covalently bonded to the coenzyme. A shift of electrons releases acetaldehyde, regenerating the TPP carbanion.

alcohol dehydrogenase the enzyme that catalyzes the conversion of acetaldehyde to ethanol

Acetaldehyde is then reduced to produce ethanol, and at the same time, one molecule of NADH is oxidized to NAD^+ for each molecule of ethanol produced (Figure 17.28; see the bottom of the alcoholic fermentation panel). **Alcohol dehydrogenase**, the enzyme that catalyzes the conversion of acetaldehyde to ethanol, is similar to LDH in many ways. The most striking similarity is that both are NAD-linked dehydrogenases, and both are tetramers.

The reduction reaction of alcoholic fermentation is similar to the reduction of pyruvate to lactate because it recycles NAD^+ and thus allows further anaerobic oxidation (fermentation) reactions. The net reaction for alcoholic fermentation is

$$\text{Glucose} + 2\text{ADP} + 2\text{P}_i + 2\text{H}^+ \rightarrow 2\ \text{Ethanol} + 2\text{ATP} + 2\text{CO}_2 + 2\text{H}_2\text{O}$$

NAD^+ and NADH do not appear explicitly in the net equation. It is essential that NADH is recycled to NAD^+ here, just as it is when lactate is produced, so that there can be further anaerobic oxidation. Biochemical Connections 17D describes an important physiological consequence of ethanol consumption.

17-4C Cancer Detection by Lactate

▶ **What is the connection between lactate production and cancer?**

Cancer cells exhibit a high rate of glycolysis, which is followed by lactic acid fermentation rather than oxidation of pyruvate to carbon dioxide and water. This effect is called *aerobic glycolysis* because it takes place even when oxygen is plentiful. It is also called the Warburg effect after its discoverer and has sometimes been called "cancer's molecular sweet tooth." It is used routinely in the diagnosis of cancer by administering a radioactively labeled analog of glucose that binds to hexokinase. Its uptake is monitored by positron emission tomography.

17D Biochemical Connections • Allied Health

Fetal Alcohol Syndrome

The variety of injuries to a fetus caused by maternal consumption of ethanol is called fetal alcohol syndrome. In the catabolism of ethanol by the body, the first step is conversion to acetaldehyde—the reverse of the last reaction of alcoholic fermentation (see Figure 17.28). The level of acetaldehyde in the blood of a pregnant female is the key to detecting fetal alcohol syndrome. Acetaldehyde is transferred across the placenta, where it then accumulates in the liver of the fetus. Acetaldehyde is toxic, and this is one of the most important factors in fetal alcohol syndrome.

In addition to the toxic effects of acetaldehyde, consumption of ethanol during pregnancy harms the fetus in other ways. It depresses the transfer of nutrients to the fetus, resulting in lower levels of sugars (hypoglycemia), vitamins, and essential amino acids. Lower levels of oxygen (hypoxia) also occur. This last effect is more drastic when the mother smokes during pregnancy, as well as when consuming alcohol.

The labels of alcoholic beverages now include a warning against consumption during pregnancy (Figure 17.32). The American Medical Association has issued the unequivocal warning that "there is no known safe level of alcohol during pregnancy." ◗

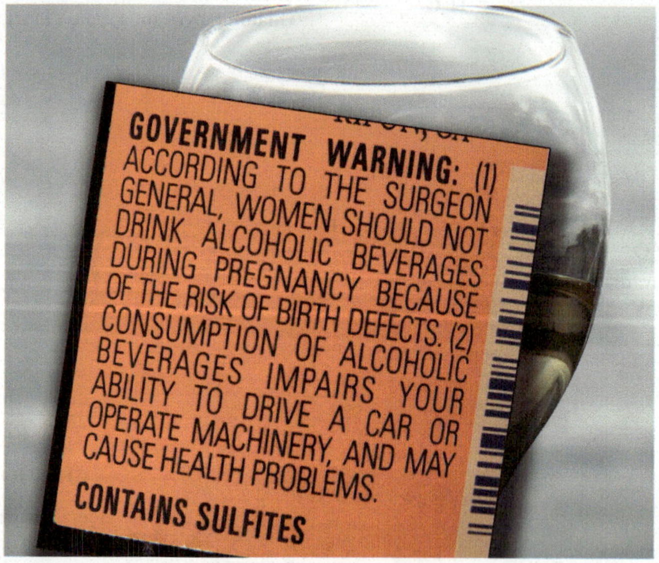

Charles D. Winters

Figure 17.32 Surgeon General warning that pregnant females should not drink alcohol.

17E Biochemical Connections • Cancer Research

Using Pyruvate Kinase Isozymes to Treat Cancer

The metabolism of cancer cells differs from that of normal cells. In carbohydrate metabolism, cancer cells rely on anaerobic metabolism, whereas normal cells tend mainly to utilize the aerobic degradation of glucose to obtain energy. Two isozymes of pyruvate kinase (PK) play a key role in this difference. The PKM1 isozyme is found in normal cells, whereas the PKM2 form predominates in cancer cells. Although PKM2 catalyzes the conversion of phosphoenolpyruvate to pyruvate less efficiently than PKM1, cancer cells compensate through increased glucose uptake and glycolytic flux. As a result, steady-state pyruvate levels are paradoxically higher –not because PKM2 is more active, but because glycolysis overall is upregulated and pyruvate consumption by mitochondria is reduced. The accumulation of most of the glycolytic intermediates allows carbon to be diverted into anabolic pathways that fuel biosynthesis and tumor growth.

Researchers at the National Institutes of Health (NIH) set out to activate the PKM2 isozyme in hopes of switching the metabolism of cancer cells to that of normal cells. They were able to achieve their goal by combining the natural allosteric activator of PK, fructose-1,6-bisphosphate, with a second synthetic activator, a member of the class of molecules known as diarylsulfonamides (Figure 17.33).

PKM2 forms a tetramer with increased activity, similar to that of PKM1, in the presence of fructose-1,6-bisphosphate and the diarylsulfonamide. The structure of this tetramer has been determined by X-ray crystallography. In Figure 17.34, two molecules of the diarylsulfonamide (red) are bound to the tetramer, in addition to one molecule of fructose-1,6-bisphosphate (black) for each subunit of the tetramer. This result is only one approach to metabolic treatment for cancer, but it may well lead the way to others.

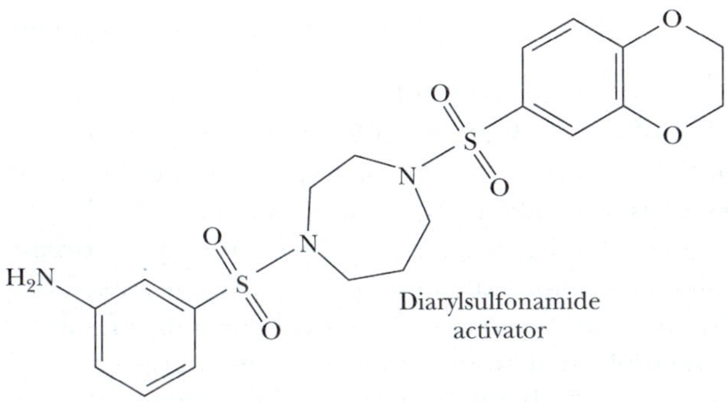

Figure 17.33 **The structure of the synthetic diarylsulfonamide activator.** (*From* Chemical and Engineering News, *February 2010, p. 46.*)

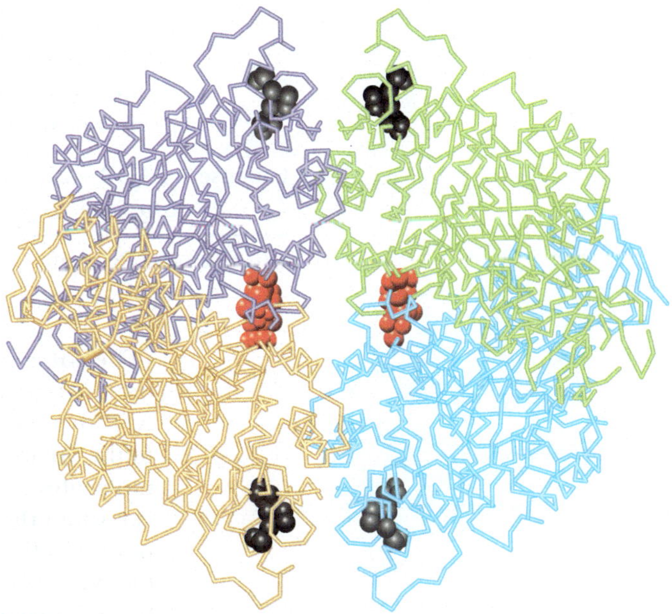

Courtesy of Bum-Soo Hong

Figure 17.34 **Tetramer of pyruvate kinase (PK) formed in the presence of the natural activator (shown in black) and the synthetic activator (shown in red).**

Research has focused on ways to modulate this effect to treat cancer. A number of oncogenes (Section 14-4B) are activated under these conditions. Genes for glycolytic enzymes are activated in turn, especially ones that play a role in reactions of pyruvate. One possible avenue for treatment is to find suitable activators or inhibitors of such enzymes to switch metabolism back to the normal aerobic pathway from glycolysis. **Biochemical Connections 17E** has a report on such research.

17-5 Energy Production in Glycolysis

Learning Objectives

After reading this section, students will be able to:

LO 17.5.1 Describe the overall energy production in glycolysis.

LO 17.5.2 Explain the differences between in vitro and in vivo Gibbs energy change values.

▶ What is the energy yield from glycolysis?

Now that we have seen the reactions of the glycolytic pathway, we can do some bookkeeping and determine the standard Gibbs energy change for the entire pathway by using the data from Table 17.1.

The overall process of glycolysis is exergonic. We can calculate $\Delta G°'$ for the entire reaction by adding up the $\Delta G°'$ values from each of the steps. Remember that all of the reactions from triose phosphate isomerase to pyruvate kinase (PK) are doubled. This gives a final figure from glucose to two pyruvates of –74.0 kJ/mol = –17.5 kcal/mol. The energy released in the exergonic phases of the process drives the endergonic reactions. The net reaction of glycolysis explicitly includes an important endergonic process, that of phosphorylation of two molecules of adenosine diphosphate (ADP), producing two molecules of adenosine triphosphate (ATP).

$$2\text{ADP} + 2\text{P}_i \rightarrow 2\text{ATP}$$

$$\Delta G°' \text{ reaction} = 61.0 \text{ kJ/mol} = 14.6 \text{ kcal/mol glucose consumed}$$

Without the production of ATP, the reaction of one molecule of glucose to produce two molecules of pyruvate would be even more exergonic. Thus, *subtracting* out the synthesis of ATP:

$$\text{Glucose} + 2\text{ADP} + 2\text{P}_i \rightarrow 2\text{ Pyruvate} + 2\text{ATP} \quad \Delta G°' = -73.4 \text{ kJ/mol} - 17.5 \text{ kcal/mol}$$

$$-(2\text{ ADP} + 2\text{P}_i \rightarrow 2\text{ ATP}) \quad \Delta G°' = -61.0 \text{ kJ/mol} - 14.6 \text{ kcal/mol}$$

$$\text{Glucose} \rightarrow 2\text{ Pyruvate} \quad \Delta G°' = -134.4 \text{ kJ/mol}$$

$$= -32.1 \text{ kcal/mol glucose consumed}$$

(The corresponding figure for the conversion of one mole of glucose to two moles of lactate is –184.6 kJ/mol = –44.1 kcal/mol.) Without the production of ATP, the energy released by the conversion of glucose to pyruvate would be lost to the organism and dissipated as heat. The energy required to produce the two molecules of ATP for each molecule of glucose can be recovered by the organism when the ATP is hydrolyzed in some metabolic process. We discussed this point briefly in Section 15-6, when we compared the thermodynamic efficiency of anaerobic and aerobic metabolism. The percentage of the energy released by the breakdown of glucose to lactate that is "captured" by the organism when ADP is phosphorylated to ATP is the efficiency of energy use in glycolysis; it is $(61.0/184.6) \times 100\% \approx 33\%$. This percentage comes from calculating the energy used to phosphorylate two moles of ATP as a percentage of the energy released by the conversion of one mole of glucose to two moles of lactate. The net release of energy in glycolysis, 123.6 kJ (29.5 kcal) for each mole of glucose converted to lactate, is dissipated as heat by the organism. Without the production of ATP to serve as a source of energy for other metabolic processes, the energy released by glycolysis would serve no purpose for the organism, except to help maintain body temperature in warm-blooded animals. A soft drink with ice can help keep you warm even on the coldest day of winter (if it is not a diet drink) because of its high-sugar content.

The Gibbs energy changes we have listed in this section are the standard values, assuming the standard conditions, such as 1 *M* concentrations of all solutes except hydrogen ion. Concentrations under physiological conditions can differ markedly from standard values. Fortunately, there are well-known methods (Section 15-2) for calculating the difference in the Gibbs energy change. Also, large changes in concentrations frequently lead to relatively small differences in the Gibbs energy change—about a few kilojoules per mole. Some of the Gibbs energy changes may be different under physiological conditions from the values listed here for standard conditions, but the underlying principles and the conclusions drawn from them remain the same.

17-6 Control of Glycolysis

Learning Objectives

After reading this section, students will be able to:

LO 17.6.1 Compare the three glycolysis control points, including enzyme names and inhibitor names.

LO 17.6.2 Describe additional allosteric and covalent modification controls for the pyruvate kinase reaction of glycolysis.

▶ Where are the control points in the glycolytic pathway?

One of the most important questions to ask about any metabolic pathway is, at which points is control exercised? Pathways can be "shut down" if an organism has no immediate need for their products, which saves energy for the organism. In glycolysis, three reactions are control points (Figure 17.35):

- Step 1: the reaction of glucose to glucose-6-phosphate, catalyzed by hexokinase;

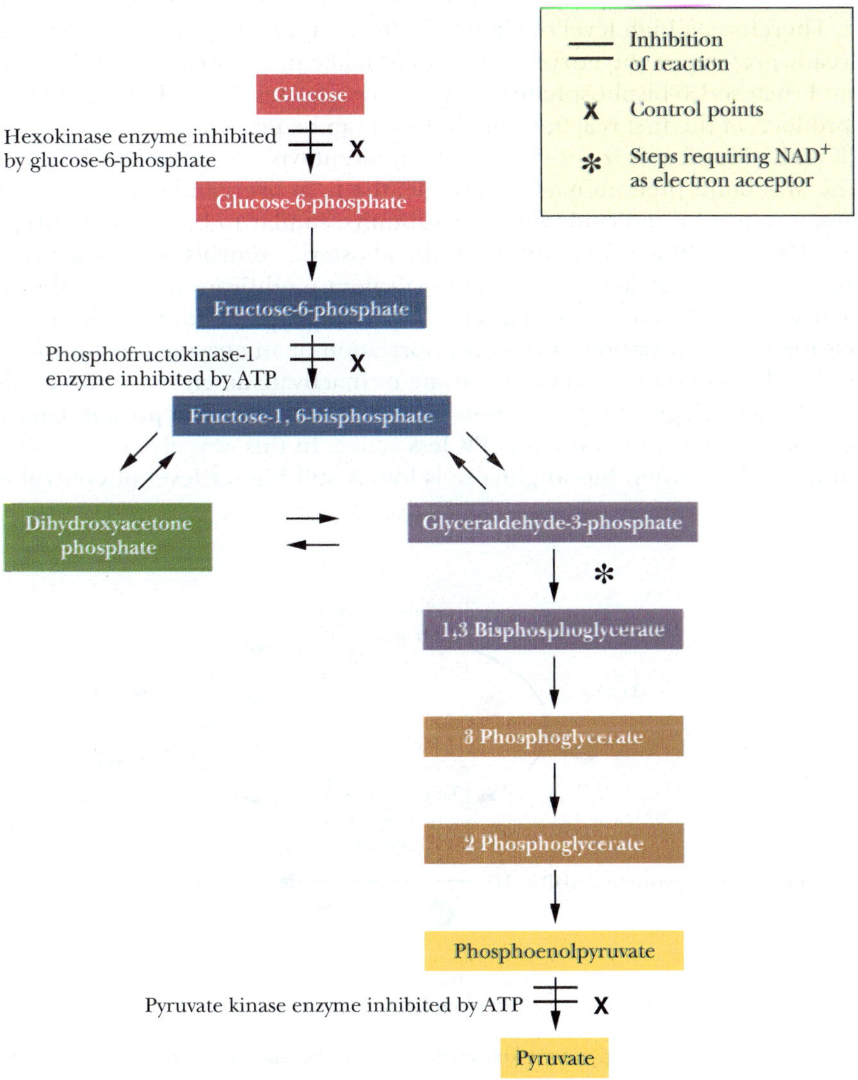

Figure 17.35 Control points in glycolysis.

- Step 3: the production of fructose-1,6-bisphosphate, catalyzed by phosphofructokinase-1 (PFK-1); and
- Step 10: the reaction of PEP to pyruvate, catalyzed by pyruvate kinase (PK).

It is frequently observed that control is exercised near the start and end of a pathway, as well as at points involving key intermediates such as fructose-1,6-bisphosphate. When we have learned more about the metabolism of carbohydrates, we can return to the role of PFK-1 and fructose-1,6-bisphosphate in the regulation of several pathways of carbohydrate metabolism (Section 18-3).

Adenosine triphosphate (ATP) inhibits enzymes at two of the three control points of glycolysis (Steps 3 and 10) because it is prudent to shut down a pathway that leads ultimately to the production of ATP if the cell has sufficient levels of ATP already. However, ATP does not inhibit the hexokinase enzyme at the first step of glycolysis, because the glucose-6-phosphate product of the first step is needed for other metabolic pathways such as glycogen synthesis (Section 18-1) and the pentose phosphate pathway (Section 18-4).

Let us next discuss the last glycolysis control point a bit further. The PK enzyme at Step 10 is allosterically affected by several compounds. ATP and alanine both inhibit it. The ATP makes sense because there would be no reason to sacrifice glucose to make more energy if there is ample ATP. The effect of alanine may be less intuitive. Alanine is the amino version of pyruvate. In other words, it is one reaction away from pyruvate via an enzyme called a transaminase. Therefore, a high level of alanine indicates that a high level of pyruvate is already present, so the enzyme that would make more pyruvate can be shut down. Fructose-1,6-bisphosphate allosterically activates PK so that the incoming products of the first reactions of glycolysis can be processed.

PK is also found as isozymes with three different types of subunits: M, L, and A. The M subunit predominates in muscle; the L, in liver; and the A, in other tissues. A native PK molecule has four subunits, similar to lactate dehydrogenase (LDH) and PFK-1. In addition to the allosteric controls mentioned previously, the liver isozymes are subject to covalent modification, too, as shown in Figure 17.36. Recall that a frequently occurring type of covalent modification is the phosphorylation or dephosphorylation of an enzyme (Section 7-3). Phosphorylation events can either activate or inactivate an enzyme. Low levels of blood sugar trigger the production of a protein kinase. The protein kinase phosphorylates PK, which renders PK less active. In this way, glycolysis is shut down in the liver when blood glucose is low. A still higher level of control is

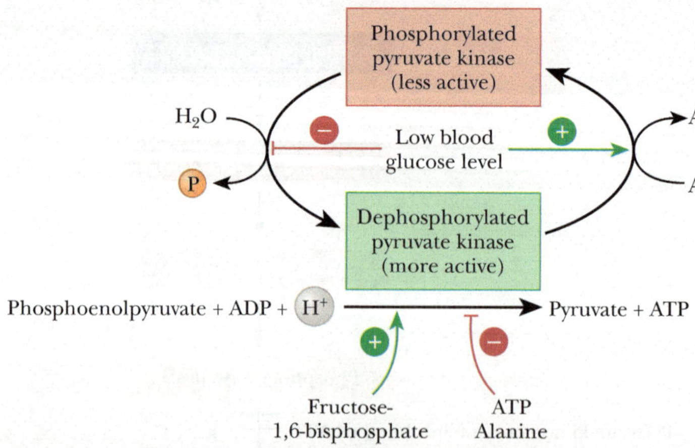

Figure 17.36 Control of liver pyruvate kinase (PK) by phosphorylation. When blood glucose is low, the phosphorylation of PK is favored. The phosphorylated form is less active, thereby slowing glycolysis and allowing pyruvate to produce glucose by gluconeogenesis.

exercised by the action of hormones. Insulin and other hormones control the level of glucose in the bloodstream, activating and deactivating metabolic pathways as needed.

Hexokinase is inhibited by high levels of its product, glucose-6-phosphate. When glycolysis is inhibited through PFK-1, glucose-6-phosphate builds up, shutting down hexokinase. This keeps glucose from being metabolized in the liver when it is needed in the blood and other tissues. However, the liver contains a second enzyme, glucokinase, that phosphorylates glucose. (See **Biochemical Connections 6C**.) Glucokinase has a higher K_M for glucose than hexokinase, so it functions only when glucose is abundant. If there is an excess of glucose in the liver, glucokinase phosphorylates it to glucose-6-phosphate. By doing so, it can eventually be polymerized into glycogen (Section 18-1) or used to make five-carbon sugars (Section 18-4).

Summary

Overview of Glycolysis

- In glycolysis, one molecule of glucose gives rise, after a series of reactions, to two molecules of pyruvate.
- Along the way, two net molecules of adenosine triphosphate (ATP) and NADH are produced.

Pyruvate: A Key Intermediate in Glycolysis

- In aerobic metabolism, pyruvate is further oxidized to carbon dioxide and water.
- In anaerobic metabolism, the product is lactate, or in organisms capable of alcoholic fermentation, it is ethanol.

The 10 Reactions of Glycolysis

- A series of 10 reactions converts one molecule of glucose into two molecules of pyruvate.
- Four of these reactions transfer a phosphate group, three are isomerizations, one is a cleavage, one is a dehydration, and only one is an oxidation.

First Phase of Glycolysis: Glucose to Glyceraldehyde-3-Phosphate

- In the first reaction of glycolysis, glucose is phosphorylated to glucose-6-phosphate, using an ATP in the process.
- Glucose-6-phosphate is isomerized to fructose-6-phosphate, which is then phosphorylated again to fructose-1,6-bisphosphate, utilizing another ATP.
- Fructose-1,6-bisphosphate is a key intermediate, and the enzyme that catalyzes its formation, phosphofructokinase-1 (PFK-1), is an important control point in the pathway.
- Fructose 1,6-bisphosphate is then split into two three-carbon compounds, glyceraldehyde-3-phosphate and dihydroxyacetone phosphate, the latter of which is also converted to glyceraldehyde-3-phosphate.
- The overall reaction in the first half of the pathway is the conversion of one molecule of glucose into two molecules of glyceraldehyde-3-phosphate at the expense of two molecules of ATP.

Second Phase of Glycolysis: Glyceraldehyde-3-Phosphate to Pyruvate

- Glyceraldehyde-3-phosphate is oxidized to 1,3-bisphosphoglycerate, and NAD^+ is reduced to NADH.
- The 1,3-bisphosphoglycerate is then converted to 3-phosphoglycerate, and ATP is produced.
- 3-Phosphoglycerate is converted in two steps to phosphoenolpyruvate, an important high-energy compound.
- Phosphoenolpyruvate is then converted to pyruvate, and ATP is produced.
- The overall reaction of the second half of the pathway is that two molecules of glyceraldehyde-3-phosphate are converted into two molecules of pyruvate, and four molecules of ATP are produced.

Fates for Pyruvate in Anaerobic Metabolism

- Pyruvate has one of two fates in anaerobic metabolism: either lactic acid fermentation or alcoholic fermentation.
- In lactic acid fermentation, pyruvate is reduced to lactate in muscle, catalyzed by the enzyme lactate dehydrogenase. NAD^+ is recycled in this process.
- In organisms capable of alcoholic fermentation, pyruvate loses carbon dioxide to produce acetaldehyde, which is reduced to produce ethanol. Thiamine pyrophosphate (TPP) is a coenzyme needed for the decarboxylation reaction, and NAD^+ is also recycled in this process.

Cancer Detection by Lactate

- Cancer cells exhibit a high rate of glycolysis, which is followed by lactic acid fermentation, even at high oxygen levels.
- This characteristic can be used to diagnose cancer.
- Research is in progress to find methods of cancer treatment by restoring the more usual patterns of metabolism.

Glycolysis Energy Output

- In each of two reactions in the pathway, one molecule of ATP is hydrolyzed for each molecule of glucose metabolized.

- In each of two other reactions, two molecules of ATP are produced by phosphorylation of adenosine diphosphate (ADP) for each molecule of glucose, giving a total of four ATP molecules produced.
- There is a net gain of two ATP molecules for each molecule of glucose processed in glycolysis.
- The anaerobic breakdown of glucose to lactate can be summarized as follows: Glucose + 2ADP + $2P_i \rightarrow$ 2 Lactate + 2ATP

Reaction	$\Delta G^{\circ\prime}$	
	kJ/mol	kcal/mol
Glucose + 2ADP + $2P_i \rightarrow$ 2 Pyruvate + 2ATP	−73.3	−17.5
2(Pyruvate + NADH + $H^+ \rightarrow$ Lactate + NAD^+)	−50.2	−12.0
Glucose + 2ADP + $2P_i \rightarrow$ 2 Lactate + 2ATP	−123.5	−29.5

- The overall process of glycolysis is exergonic.
- Without the production of ATP, glycolysis would be even more exergonic, but the energy released would be lost to the organism and dissipated as heat.

Regulation of Carbohydrate Metabolism

- There are three control points in the glycolytic pathway.
- The first is the hexokinase reaction at the beginning, where glucose is converted to glucose-6-phosphate.
- The second is the phosphofructokinase-1 reaction, which produces fructose-1,6-bisphosphate. This is the commitment step in glycolysis because fructose-1,6-bisphosphate cannot be used for anything else in the organism.
- The third is the pyruvate kinase reaction at the end of glycolysis, where phosphoenolpyruvate is converted to pyruvate.
- The second and third control points have the effect of slowing down the glycolysis pathway when energy is not needed and speeding it up when there is a need.

Review Exercises

17-1 The Overall Pathway of Glycolysis

1. **Recall** Which reaction or reactions that we have discussed in this chapter require ATP? Which reaction or reactions produce ATP? List the enzymes that catalyze the reactions that require and produce ATP.
2. **Recall** Which reaction or reactions that we have discussed in this chapter require NADH? Which reaction or reactions require NAD^+? List the enzymes that catalyze the reactions that require NADH and those that require NAD^+.
3. **Recall** What are the possible metabolic fates of pyruvate?

17-2 The Phase 1 Reactions of Glycolysis: Conversion of a Six-Carbon Glucose to Two Three-Carbon Glyceraldehyde-3-Phosphate Molecules

4. **Recall** Describe the reactions of glycolysis that convert glucose to fructose-1,6-bisphosphate.
5. **Recall** Explain the origin of the name of the enzyme aldolase.
6. **Recall** Define *isozymes* and give an example from the material discussed in this chapter.
7. **Recall** Why would enzymes be found as isozymes?
8. **Recall** Why is the formation of fructose-1,6-bisphosphate the committed step in glycolysis?
9. **Reflect and Apply** Show that the reaction Glucose $\rightarrow$ 2 Glyceraldehyde-3-phosphate is slightly endergonic; that is, it is not too far from equilibrium. Use the data in Table 17.1.
10. **Reflect and Apply** What is the metabolic advantage of having both hexokinase and glucokinase to phosphorylate glucose?
11. **Reflect and Apply** What are the metabolic effects of not being able to produce the M subunit of phosphofructokinase?
12. **Reflect and Apply** In what way is the observed mode of action of hexokinase consistent with the induced-fit theory of enzyme action?
13. **Reflect and Apply** How does ATP act as an allosteric effector in the mode of action of phosphofructokinase?

17-3 The Phase 2 Reactions of Glycolysis: Glyceraldehyde-3-Phosphate Is Converted to Pyruvate

14. **Recall** At what point in glycolysis are all the reactions considered doubled?
15. **Recall** Describe the first ATP-forming reaction of glycolysis, including how many ATP molecules are produced for each molecule of glucose.
16. **Recall** Which of the enzymes discussed in this chapter are NAD-linked dehydrogenases?
17. **Recall** Define *substrate-level phosphorylation* and give an example from the reactions discussed in this chapter.
18. **Recall** Which reactions are the control points in glycolysis?
19. **Recall** Which molecules act as inhibitors of glycolysis? Which molecules act as activators?
20. **Recall** Many NAD-linked dehydrogenases have similar active sites. Which part of glyceraldehyde-3-phosphate dehydrogenase would be the most conserved between other enzymes?
21. **Recall** Several of the enzymes of glycolysis fall into classes that occur often in metabolism. What reaction types are catalyzed by each of the following:
    (a) Kinases
    (b) Isomerases
    (c) Dehydrogenases
22. **Recall** What is the difference between an isomerase and a mutase?
23. **Reflect and Apply** Using general knowledge of dehydrogenase enzyme activity, predict the type of reaction that pyruvate dehydrogenase catalyzes.
24. **Reflect and Apply** Is the reaction of 2-phosphoglycerate to phosphoenolpyruvate a redox reaction? Give the reason for your answer.
25. **Reflect and Apply** Identify the carbon atom that changes oxidation state during the reaction catalyzed by glyceraldehyde-3-phosphate dehydrogenase. What is the functional group that changes during the reaction?

26. **Reflect and Apply** Discuss the logic of the nature of the allosteric inhibitors and activators of glycolysis. Why would these molecules be used?
27. **Reflect and Apply** Many species have a third type of LDH subunit that is found predominantly in the testes. If this subunit, called C, were expressed in other tissues and could combine with the M and H subunits, how many LDH isozymes would be possible? What would their compositions be?
28. **Reflect and Apply** The M and H subunits of lactate dehydrogenase have very similar sizes and shapes but differ in amino acid composition. If the only difference between the two were that the H subunit had a glutamic acid in a position where the M subunit had a serine, how would the five isozymes of LDH separate on native gel electrophoresis at pH 8.6? (To review the details of protein electrophoresis, see Section 5-3A.)
29. **Reflect and Apply** Most metabolic pathways are relatively long and appear to be very complex. For example, there are 10 individual chemical reactions in glycolysis, converting glucose to pyruvate. Suggest a reason for the complexity.

17-4 Anaerobic Metabolism of Pyruvate

30. **Recall** What is the metabolic purpose of lactate production?
31. **Reflect and Apply** Using the Lewis electron-dot notation, show explicitly the transfer of electrons in the following redox reactions.
    (a) Pyruvate + NADH + H^+ → Lactate + NAD^+
    (b) Acetaldehyde + NADH + H^+ → Ethanol + NAD^+
    (c) Glyceraldehyde-3-phosphate + NAD^+ → 3-Phosphoglycerate + NADH + H^+ (redox reaction only)
32. **Reflect and Apply** Briefly discuss the role of thiamine pyrophosphate (TPP) in enzymatic reactions, using material from this chapter to illustrate your points.
33. **Reflect and Apply** What is unique about TPP that makes it useful in decarboxylation reactions?
34. **Reflect and Apply** Most hunters know that meat from animals that have been run to death tastes sour. Suggest a reason for this observation.
35. **Biochemical Connections** Cancer cells grow so rapidly that they have a higher rate of anaerobic metabolism than most body tissues, especially at the center of a tumor. Can you use drugs that inhibit the enzymes of anaerobic metabolism in the treatment of cancer? Why or why not?
36. **Recall** What is the Warburg effect, and what does it have to do with the topics of this chapter?

17-5 Energy Production in Glycolysis

37. **Recall** What is the net gain of ATP molecules derived from the reactions of glycolysis?
38. **Reflect and Apply** Why do cells produce much less ATP per glucose during anaerobic glycolysis compared to aerobic respiration? What are the biological advantages and disadvantages of this lower efficiency?
39. **Reflect and Apply** What are coupled reactions in biochemical pathways? Identify which reactions in glycolysis are coupled reactions and explain why.
40. **Reflect and Apply** Identify which steps in glycolysis are physiologically irreversible and explain why.
41. **Reflect and Apply** Show, by a series of equations, the energetics of the phosphorylation of ADP by phosphoenolpyruvate.
42. **Reflect and Apply** What should be the net ATP yield for glycolysis when fructose, mannose, and galactose are used as the starting compounds? Justify your answer.
43. **Reflect and Apply** In the muscles, glycogen is broken down via the following reaction:

    $$(\text{Glucose})_n + P_i \rightarrow \text{Glucose-1-phosphate} + (\text{Glucose})_{n-1}$$

    What would be the ATP yield per molecule of glucose in the muscles if glycogen were the source of the glucose?
44. **Reflect and Apply** Using Table 17.1, predict whether the following reaction is thermodynamically possible:

    $$\text{Phosphoenolpyruvate} + P_i + 2\text{ADP} \rightarrow \text{Pyruvate} + 2\text{ATP}$$

45. **Reflect and Apply** According to Table 17.1, several reactions have very positive $\Delta G^{\circ\prime}$ values. How can this be explained, given that these reactions do occur in the cell?
46. **Reflect and Apply** According to Table 17.1, four reactions have positive ΔG values. How can this be explained?

17-6 Control of Glycolysis

47. **Reflect and Apply** Consider a cell with high levels of glucose-6-phosphate (G6P). Which enzyme would be inhibited? What would happen if glycolysis reactions began to increase? Why?
48. **Reflect and Apply** List the names of the three glycolysis enzymes that are regulated, indicate which control point is the most important for glycolysis, and explain why.
49. **Recall** Do hormones play a role in the control of metabolism? Is glycolysis a possible pathway for such control?
50. **Reflect and Apply** Why does it make sense for cells to regulate metabolic pathways at or near the beginning of a pathway?

Further Reading

Bodner, G. M. Metabolism: Part I, Glycolysis, or the Embden–Meyerhoff Pathway. *J. Chem. Ed.* **63**, 566–570 (1986). A clear, concise summary of the pathway. Part of a series on metabolism of carbohydrates and lipids.

Boyer, P. D., ed. *The Enzymes,* Vols. 5–9. New York: Academic Press, 1972. A standard reference with review articles on the glycolytic enzymes; lactate dehydrogenase and alcohol dehydrogenase appear in Volume 10.

Florkin, M., and E. H. Stotz, eds. *Comprehensive Biochemistry.* New York: Elsevier, 1967. Another standard reference. Volume 17, *Carbohydrate Metabolism,* deals with glycolysis.

Cao, J. and Y. Chen. Natural Compounds Regulate Glycolysis in Hypoxic Tumor Environment. *Biomed. Res. Int.* **2015**, Article ID 354143 (2015). This review discusses the role of natural compounds in modulating glycolysis under hypoxic conditions in tumors, offering insights into potential therapeutic approaches targeting cancer metabolism.

Karl, P. I., B. H. J. Gordon, C. S. Lieber, and S. E. Fisher. Acetaldehyde Production and Transfer by the Perfused Human Placental Cotyledon. *Science* **242**, 273–275 (1988). A report describing some of the processes involved in fetal alcohol syndrome.

Kierans, S. J. and T. C. Taylor. Glycolysis: A Multifaceted Metabolic Pathway and Signaling Hub. *J. Biol. Chem.* **300**, Article ID 107906 (2024). This review discusses the role of the glycolytic pathway as a source of signaling molecules in eukaryotic cells, emphasizing the newfound recognition of glycolysis beyond energy production.

Light, W. J. *Alcoholism and Women, Genetics, and Fetal Development.* Springfield, IL: Thomas, 1988. A book that devotes a large amount of space to fetal alcohol syndrome.

Lipmann, F. A Long Life in Times of Great Upheaval. *Ann. Rev. Biochem.* **53**, 1–33 (1984). The reminiscences of a Nobel laureate whose research contributed greatly to the understanding of carbohydrate metabolism. Very interesting reading from the standpoints of autobiography and the author's contributions to biochemistry.

Ika Rahma H/Shutterstock.com

Similar to the careful attention to detail and balance in the arrangement of a Japanese bento box, carbohydrate metabolism pathways are also compartmentalized and balanced with exquisite detail.

Release Mechanisms, Storage Mechanisms, and Control in Carbohydrate Metabolism 18

Chapter Outline

18-1 How Glycogen Is Degraded and Produced

Learning Objectives

After reading this section, students will be able to:

LO 18.1.1 Describe how glycogen is degraded and produced.

LO 18.1.2 Explain how glycogen breakdown and glycogen synthesis are regulated.

When we digest a meal high in carbohydrates, we have a supply of glucose that exceeds our immediate needs. We store glucose as glycogen (Figure 18.1). Glycogen is a polymer that is similar to the starches found in plants; it differs from starch only in the degree of chain branching. In fact, glycogen is sometimes called "animal starch" because of this similarity. A look at the metabolism of glycogen provides some insights into how glucose can be stored in this form and made available on demand. In the degradation of glycogen, several glucose residues can be released simultaneously, one from each end of a branch, rather than one at a time as would be the case in a linear polymer.

Chapter Outline (continued)

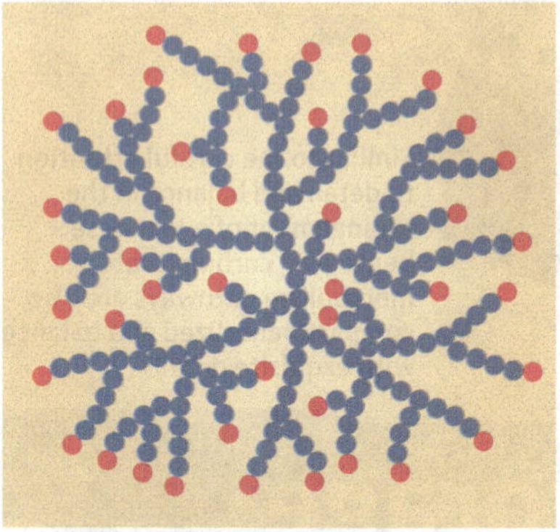

Glycogen

Figure 18.1 The branched-chain structure of glycogen. The highly branched structure of glycogen makes it possible for several glucose residues to be released at once to meet energy needs. This would not be possible with a linear polymer. The red dots indicate the terminal glucose residues that are released from glycogen. The more branch points there are, the more of these terminal residues are available at one time.

glycogenolysis the process of breaking down glycogen into free glucose

18-1A Glycogen: Optimized for Efficiency

The branched structure of glycogen makes it possible for an organism to meet its short-term demands for energy by increasing the glucose supply as quickly as possible. Mathematical modeling has shown that the structure of glycogen is *optimized* for its ability to store and deliver energy quickly and for the longest amount of time possible. The key to this optimization is the average chain length of the branches (13 residues). If the average chain length were much greater or much shorter, glycogen would not be as efficient a vehicle for energy storage and release on demand. Experimental results support the conclusions reached from the mathematical modeling.

18-1B Glycogen Breakdown

Glycogen is found primarily in the liver and muscles. When blood glucose is low, the liver responds by breaking down its glycogen stores to release glucose in a process called **glycogenolysis**. Four reactions play roles in the conversion of glycogen to glucose:

1. A glucosyl residue is removed from liver glycogen by **glycogen phosphorylase** to produce glucose-1-phosphate (Figure 18.2). This cleavage is a **phosphorolysis** reaction rather than a hydrolysis reaction. Glycogen phosphorylase cleaves the $\alpha(1 \rightarrow 4)$ linkages in glycogen.
2. Glucose-1-phosphate is isomerized by **phosphoglucomutase** to give glucose-6-phosphate (Figure 18.3).
3. Complete breakdown of glycogen also requires a reaction in which a debranching enzyme hydrolyzes the $\alpha(1 \rightarrow 6)$ glycosidic bonds at branch points in the glycogen structure.
4. Glucose-6-phosphate is dephosphorylated by **glucose-6-phosphatase** to produce glucose, which enters the blood from the liver to replenish the supply.

In general, *phosphatases* remove phosphate groups from molecules. In muscle, the glycogen breakdown process is similar, except that glucose-6-phosphate obtained from the first few steps of glycogen breakdown enters the glycolytic pathway directly for energy production.

Pioneer in Chemistry • Chi Che Wang (1894–1979)

Chi Che Wang (1894–1979) was a Chinese biochemist and educator whose work in nutrition and food science helped advance medicine, agriculture, and public health. Born in Suzhou, China, Wang was among the earliest Chinese girls sponsored by the government to study abroad. She earned her bachelor's degree from Wellesley College in 1914, followed by a master's in chemistry (1916) and a Ph.D. in chemistry and nutrition (1918) from the University of Chicago. Her dissertation examined the chemical composition of traditional Chinese foods, bridging biochemical science and cultural practices to improve local diets.

Wang worked as an assistant and later an instructor in food chemistry, becoming the first Chinese woman to join an American faculty. After postdoctoral research at Johns Hopkins University, she oversaw the Department of Blood Chemistry at the Morris Memorial Institute for Medical Research at Michael Reese Hospital from 1920 to 1930. In later decades, Wang held research and teaching roles at several medical institutions, including the Mayo Clinic, offering significant insights into metabolism, nutrition, and pediatric kidney disease.

Wang was elected a Fellow of the American Association for the Advancement of Science in 1922. She also co-founded the Chicago Chinese Women's Club, promoting education and civic engagement among Chinese women. Today, her legacy is commemorated in Chicago with Chi Che Wang Park, honoring her enduring impact as a scientist, educator, and leader.

Wellesley College, The Wellesley Legenda (1914 yearbook)

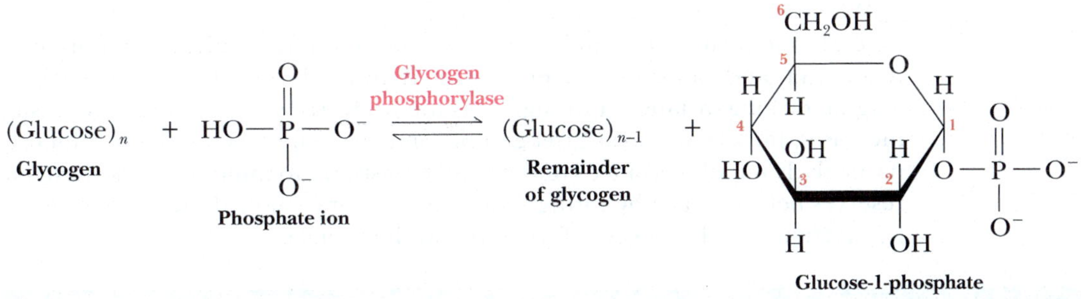

Figure 18.2 Phosphorolysis of glucose from glycogen to produce glucose-1-phosphate.

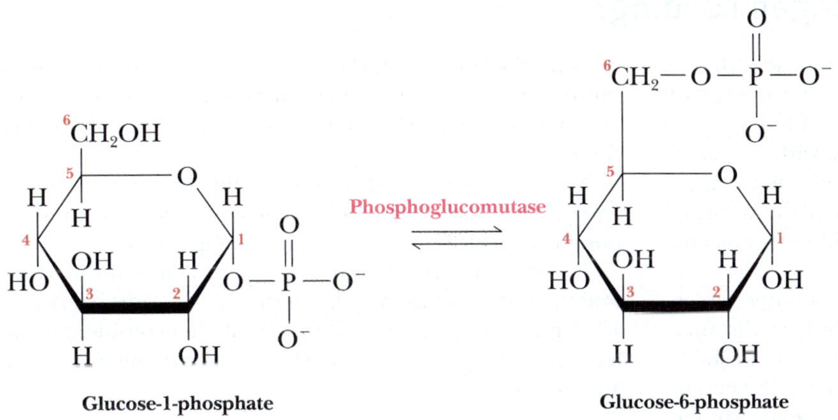

Figure 18.3 Isomerization of glucose-1-phosphate to glucose-6-phosphate.

The debranching of glycogen involves the transfer of a "limit branch" of three glucose residues to the end of another branch, where they are subsequently removed by glycogen phosphorylase. The same glycogen debranching enzyme then hydrolyzes the α(1 → 6) glycosidic bond of the last glucose residue remaining at the branch point (Figure 18.4).

glycogen phosphorylase the enzyme in the glycogen breakdown pathway that catalyzes the phosphorolysis of glycogen to give glucose-1-phosphate

phosphorolysis the addition of phosphoric acid across a bond, such as the glycosidic bond in glycogen, giving glucose phosphate and a glycogen remainder one residue shorter; it is analogous to hydrolysis (addition of water across a bond)

phosphoglucomutase the enzyme that catalyzes the isomerization of glucose-1-phosphate to glucose-6-phosphate

glucose-6-phosphatase the enzyme that removes phosphate from glucose-6-phosphate to produce glucose

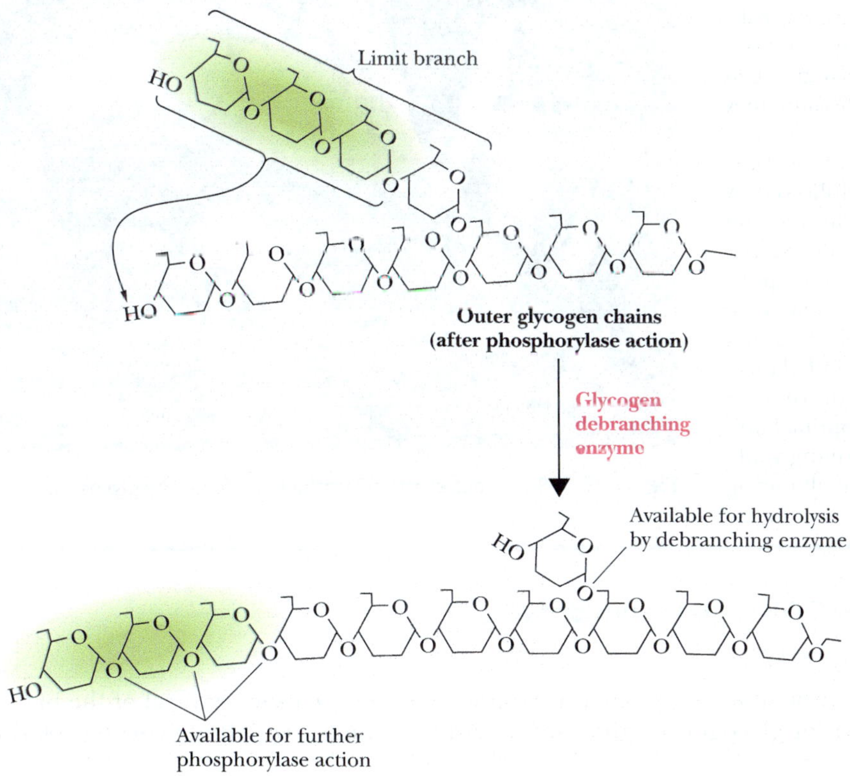

Figure 18.4 The mode of action of the debranching enzyme in glycogen breakdown. The enzyme transfers three α(1 → 4)-linked glucose residues from a limit branch to the end of another branch. The same enzyme also catalyzes the hydrolysis of the α(1 → 6)-linked residue at the branch point.

When an organism needs energy quickly, glycogen breakdown is important. Muscle tissue can mobilize energy production from glycogen more easily than fat and can do so anaerobically. With low-intensity exercise, such as jogging or long-distance running, fat is the preferred fuel, but as the intensity increases, muscle and liver glycogen become more important. Some athletes, particularly middle-distance runners and cyclists, try to build up their glycogen reserves before a race by eating large amounts of carbohydrates. **Biochemical Connections 18A** has more information on the subject.

18A Biochemical Connections • Exercise Physiology

Why Do Athletes Go in for Glycogen Loading?

Glycogen is the primary energy source for a muscle that is at rest and then starts working vigorously. The energy of adenosine triphosphate (ATP) hydrolysis derived from glycogen breakdown is initially produced *anaerobically,* with the lactic acid product being processed back to glucose in the liver. As an athlete becomes well-conditioned, the muscle cells have more mitochondria, allowing for more *aerobic* metabolism of fats and carbohydrates for energy.

The switch to aerobic metabolism takes a few minutes, which is why athletes must warm up before an event. In long-distance events, athletes rely more on fat metabolism than they do in short-distance events, but in any race, there is a final surge at the end in which the level of muscle glycogen may well determine the winner.

The idea behind glycogen loading is that if there is more available glycogen, then a person can carry out anaerobic metabolism for a longer period of time, either at the end of a distance event or for the entire event if the effort level is high enough. This is probably true, but several questions come to mind: How long does the glycogen last? What is the best way to "load" glycogen? Is it safe? Theoretical calculations estimate that it takes 8–12 minutes to use all the glycogen in the skeletal muscle, although this range varies greatly depending on the intensity level. Allowing for the loading of extra glycogen, it might last half an hour. Evidence shows that glycogen may be used more slowly in well-conditioned athletes because they exhibit higher fat utilization.

Early loading methods involved glycogen depletion for three days via a high-protein diet and extreme exercise, followed by loading from a high-carbohydrate diet and resting. This method yields a marked increase in glycogen, but some of it is then stored in the heart (which usually has little or no glycogen). The practice actually stresses the heart muscle. There is clearly some danger here. Dangers are also associated with the high-protein diet because too much protein often leads to a mineral imbalance, which also stresses the heart and the kidneys. Again, there is some danger. In addition, the training was often nonoptimal during the week because the athlete had trouble performing while on the low-carbohydrate diet and didn't train much at all during the loading phase. Simple carbohydrate loading without previous extreme glycogen depletion does increase glycogen but not as much; however, this increase does not risk potential stress to the heart.

Simple loading merely involves eating diets rich in pasta, starch, and complex carbohydrates for a few days before the strenuous exertion. It is not clear whether simple loading works.

It is certainly possible to increase the amount of glycogen in muscle, but how long it will last during vigorous exercise (Figure 18.5) remains uncertain. Ultimately, all diet considerations for athletes are very individual, and what works for one may not work for another. ◗

Maxisport/Shutterstock.com

Figure 18.5 Strenuous exertion by athletes is fueled by glycogen.

18-1C Glycogen Production

▶ **How is glycogen formed from glucose?**

The formation of glycogen from glucose is not the exact reversal of the breakdown of glycogen to glucose. It requires energy, which is provided by the

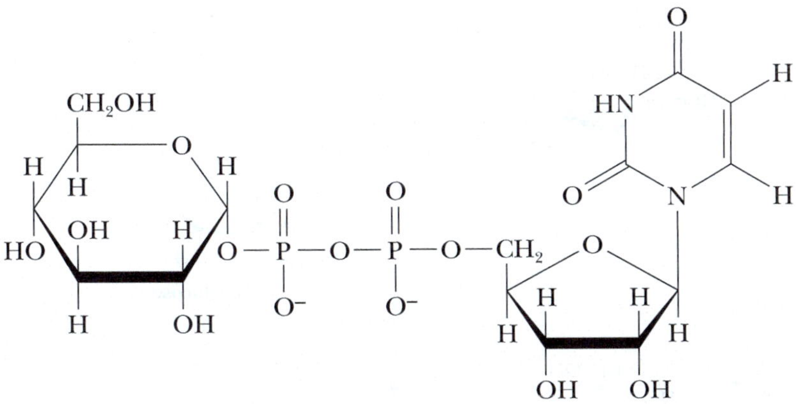

Figure 18.6 Structure of uridine diphosphate glucose (UDPG).

hydrolysis of a nucleoside triphosphate, uridine triphosphate (UTP), and it occurs in three stages. In the first stage of glycogen synthesis, glucose-1-phosphate (obtained from glucose-6-phosphate by an isomerization reaction) reacts with UTP to produce uridine diphosphate glucose (also called uridine diphosphate glucose [UDP-glucose or UDPG]; see Figure 18.6) and pyrophosphate (PP_i):

	$\Delta G^{\circ\prime}$	
	kJ mol^{-1}	kcal mol^{-1}
Glucose-1-phosphate + UTP $\rightleftarrows$ UDPG 1 PPi	0	0
$H_2O + PP_i \rightarrow 2P_i$	−30.5	−7.3
Overall: Glucose-1-phosphate + UTP $\rightarrow$ UDPG + $2P_i$	−30.5	−7.3

The enzyme that catalyzes the overall reaction in stage 1 is **UDP-glucose pyrophosphorylase**. The exchange of one phosphoric anhydride bond for another has a Gibbs energy change close to zero. The release of energy comes about when the enzyme inorganic pyrophosphatase catalyzes the hydrolysis of pyrophosphate to two phosphates, a strongly exergonic reaction.

UDP-glucose pyrophosphorylase the enzyme in the glycogen synthesis pathway that combines glucose-1-phosphate and UTP to produce uridine diphosphate glucose (also called UDP-glucose or UDPG)

It is common in biochemistry to see the energy released by the hydrolysis of pyrophosphate combined with the Gibbs energy of hydrolysis of a nucleoside triphosphate. The coupling of these two exergonic reactions to a reaction that is not energetically favorable allows an otherwise endergonic reaction to take place. The supply of UTP is replenished by an exchange reaction with ATP, which is catalyzed by **nucleoside diphosphate kinase**:

nucleoside diphosphate kinase the enzyme that catalyzes the reversible exchange of phosphate groups between nucleoside diphosphates and nucleoside triphosphates

$$\text{UDP + ATP} \rightleftarrows \text{UTP + ADP}$$

This exchange reaction makes the hydrolysis of any nucleoside triphosphate energetically equivalent to the hydrolysis of ATP.

The addition of UDPG to a growing chain of glycogen is the second stage in glycogen synthesis. It involves the formation of a new $\alpha(1 \rightarrow 4)$ glycosidic bond in a reaction catalyzed by the enzyme **glycogen synthase** (Figure 18.7). This enzyme cannot simply form a bond between two isolated glucose molecules; it must add to an existing chain with $\alpha(1 \rightarrow 4)$ glycosidic linkages. The initiation of glycogen synthesis requires a primer for this reason. The hydroxyl group of a specific tyrosine of the protein *glycogenin* serves this purpose. In the very first event of glycogen synthesis, a glucose residue is linked to this tyrosine hydroxyl, and glucose residues are successively added to this first one. The glycogenin molecule itself acts as the catalyst for the addition of glucoses until

glycogen synthase the enzyme that catalyzes the addition of glucosyl residues from uridine diphosphate glucose (UDPG) to the ends of glycogen chains

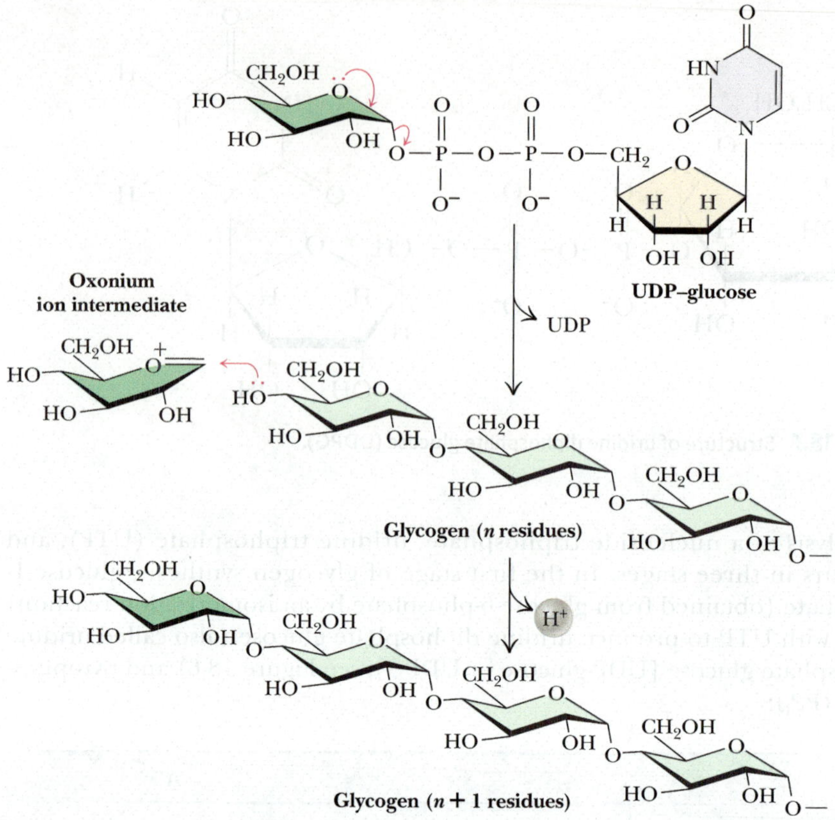

Figure 18.7 The reaction catalyzed by glycogen synthase. A glucose residue is transferred from UDPG to the growing end of a glycogen chain in an $\alpha(1\rightarrow4)$ linkage.

there are about eight of them linked together. At that point, glycogen synthase takes over.

The third stage in the synthesis of glycogen is the formation of $\alpha(1 \rightarrow 6)$ branch points along the $\alpha(1 \rightarrow 4)$ glycosidic linkages of the main chain. A *branching enzyme* accomplishes this task. It does so by transferring a segment about seven residues long from the end of a growing chain to a branch point, where it catalyzes the formation of the required $\alpha(1 \rightarrow 6)$ glycosidic linkage (Figure 18.8). This enzyme has already catalyzed the breaking of an $\alpha(1 \rightarrow 4)$ glycosidic linkage in the process of transferring the oligosaccharide segment. Each transferred segment must come from a chain at least 11 residues long, and each new branch point must be at least four residues away from the nearest existing branch point.

18-1D Balancing Glycogen Breakdown and Production

How does an organism ensure that glycogen synthesis and glycogen breakdown do not operate simultaneously? If this were to occur, the main result would be the hydrolysis of UTP, which would waste chemical energy stored in the phosphoric anhydride bonds. A major controlling factor lies in the behavior of glycogen phosphorylase. This enzyme is subject not only to *allosteric control* but also to another control feature: *covalent modification.* In allosteric control (Section 4-5), recall that the binding of a ligand at one site causes changes at another binding site in a protein. Recall, too, that a frequently occurring type of covalent modification is phosphorylation or dephosphorylation of an enzyme (Section 7-3). Phosphorylation events can either activate or inactivate an enzyme.

Figure 18.9 summarizes some of the salient allosteric and covalent modification control features that affect glycogen phosphorylase activity. The enzyme is a dimer that exists in two forms, the inactive T (taut) form and the active R

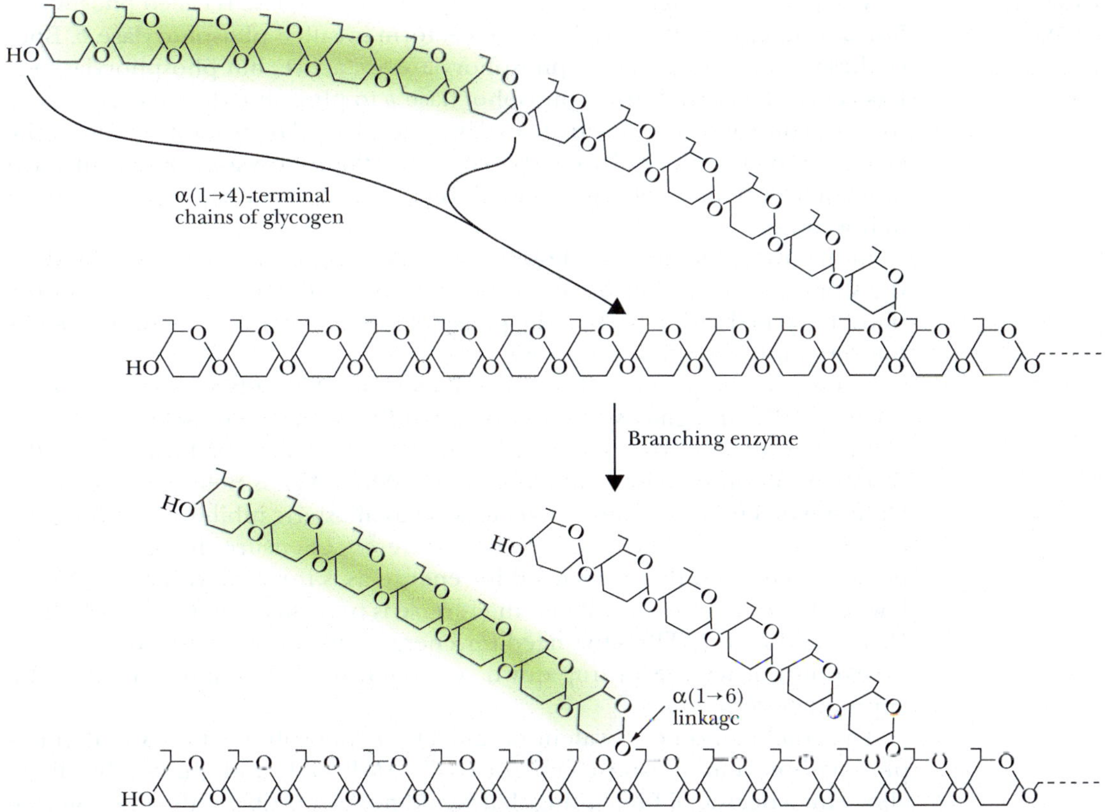

Figure 18.8 The mode of action of the branching enzyme in glycogen synthesis. A segment seven residues long is transferred from a growing branch that is at least 11 residues long to a new branch point that is at least four residues away from the nearest existing branch point.

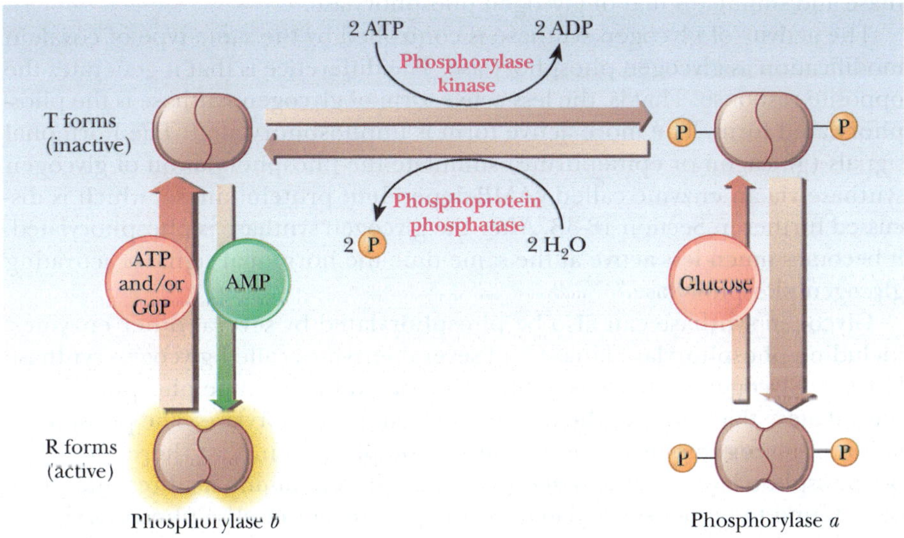

Figure 18.9 Glycogen phosphorylase activity is subject to two types of control: allosteric control and covalent modification. Glycogen phosphorylase *a* and glycogen phosphorylase *b* respond to different allosteric effectors (see the text). ATP, glucose-6-phosphate (G6P), and glucose are allosteric inhibitors that convert glycogen phosphorylase to its inactive T forms. AMP is an allosteric activator that converts the dephosphorylated T form to the dephosphorylated R form, glycogen phosphorylase *b*, which is active but less so than the phosphorylated glycogen phosphorylase *a*. Dephosphorylation of the *a* form of the enzyme converts it to the less active *b* form. Only the inactive T form is subject to phosphorylation and dephosphorylation.

(relaxed) form. (Recall the discussion of T forms and R forms of allosteric enzymes in Section 7-2A.) In the T form (and *only* in the T form), glycogen phosphorylase can be modified by phosphorylation of a specific serine residue on each of the two subunits. The esterification of the serines to phosphoric acid is catalyzed by the enzyme *phosphorylase kinase*; the dephosphorylation is catalyzed by *phosphoprotein phosphatase* (Figure 18.9, *top*).

phosphorylase *a* the phosphorylated form of glycogen phosphorylase that is more active

phosphorylase *b* the dephosphorylated form of glycogen phosphorylase that is less active.

The phosphorylated R form of glycogen phosphorylase is called **phosphorylase *a***, whereas the dephosphorylated R form is called **phosphorylase *b***. Both of these R forms of glycogen phosphorylase are active, but phosphorylase *b* is less active. The switch from phosphorylase *b* to phosphorylase *a* is the major form of control over the activity of phosphorylase. The response time of the changes is on the order of seconds to minutes. Phosphorylase is also controlled more quickly in times of urgency by allosteric effectors, with a response time of milliseconds.

In the liver, glucose is an allosteric inhibitor of phosphorylase *a*. It binds to the substrate site and favors the transition to the T state. It also exposes the phosphorylated serines so that the phosphatase can hydrolyze them. This shifts the equilibrium to phosphorylase *b*.

In muscle, the primary allosteric effectors are ATP, adenosine monophosphate (AMP), and glucose-6-phosphate (G6P). When the muscles use ATP to contract, AMP levels rise. This increase in AMP stimulates the formation of the R state of phosphorylase *b*, which is active. When ATP is plentiful or glucose-6-phosphate builds up, these molecules act as allosteric inhibitors, shifting the equilibrium back to the T form. These differences ensure that glycogen will be degraded when there is a need for energy, as is the case with high [AMP], low [G6P], and low [ATP]. When the reverse is true—namely, low [AMP], high [G6P], and high [ATP]—the need for energy, and consequently for glycogen breakdown, is less. "Shutting down" glycogen phosphorylase activity is the appropriate response.

The combination of covalent modification (phosphorylation and dephosphorylation) and allosteric control (ATP, AMP, G6P, and glucose binding) allows for a degree of fine-tuning that would not be possible with either mechanism alone. Hormonal control also plays a role. When blood glucose levels are low, glucagon is released, activating a series of events that lead to phosphorylation and activation of glycogen phosphorylase. When epinephrine is released from the adrenal gland in response to stress, it triggers a series of events, discussed more fully in Section 24-4, that suppresses the activity of glycogen synthase and stimulates that of glycogen phosphorylase.

The activity of glycogen synthase is controlled by the same type of covalent modification as glycogen phosphorylase. The difference is that it generates the opposite response. That is, the less active form of glycogen synthase is the phosphorylated form. The more active form is unphosphorylated. The hormonal signals (glucagon or epinephrine) stimulate the phosphorylation of glycogen synthase via an enzyme called cAMP-dependent protein kinase, which is discussed further in Section 18-3B. After the glycogen synthase is phosphorylated, it becomes much less active at the same time the hormonal signal is activating glycogen phosphorylase.

Glycogen synthase can also be phosphorylated by several other enzymes, including phosphorylase kinase and several enzymes called glycogen synthase kinases. Glycogen synthase is dephosphorylated by the same phosphoprotein phosphatase that removes the phosphate from phosphorylase. The phosphorylation of glycogen synthase is also more complicated in that there are multiple phosphorylation sites. As many as nine different amino acid residues have been found to be phosphorylated. As the progressive level of phosphorylation increases, the activity of the enzyme decreases.

Glycogen synthase is also under allosteric control. Glucose-6-phosphate is an allosteric activator. However, the two forms of glycogen synthase respond very differently to glucose-6-phosphate. The phosphorylated (less active) form is sometimes called *glycogen synthase D* (for "glucose-6-phosphate dependent") because it is active only under very high concentrations of glucose-6-phosphate. In fact, the level necessary to give significant activity would be beyond the physiological range. The dephosphorylated form is sometimes called *glycogen synthase I* (for "glucose-6-phosphate independent") because it is active even with low concentrations of glucose-6-phosphate. Thus, even though purified enzymes can be shown to respond to allosteric effectors, the true control over the activity of glycogen synthase is its phosphorylation state, which, in turn, is controlled by hormonal states.

The fact that two target enzymes, glycogen phosphorylase and glycogen synthase, are modified in the same way by the same enzymes links the opposing processes of synthesis and breakdown of glycogen even more intimately.

Finally, the modifying enzymes are themselves subject to covalent modification and allosteric control. This feature complicates the process considerably but adds the possibility of an amplified response to small changes in conditions. A small change in the concentration of an allosteric effector of a modifying enzyme can cause a large change in the concentration of an active, modified target enzyme; this amplification response is due to the fact that the substrate for the modifying enzyme is itself an enzyme. At this point, the situation has become very complex indeed, but it is a good example of how opposing processes of breakdown and synthesis can be controlled to the advantage of an organism. The synthesis of glucose from lactate, discussed in Section 18-2, provides another example of opposing carbohydrate metabolism pathways and how they are controlled.

18-2 Gluconeogenesis Produces Glucose from Pyruvate

Learning Objectives

After reading this section, students will be able to:

LO 18.2.1 Write metabolite names, metabolite structures, and enzyme names for the reactions in the gluconeogenesis pathway.

LO 18.2.2 Write the mechanism of a biotin carboxylase enzyme.

LO 18.2.3 Describe the direct and indirect phosphoenolpyruvate gluconeogenesis pathways as well as the cellular conditions that favor each pathway.

The conversion of pyruvate to glucose occurs by a process called **gluconeogenesis**. Broken down into its three parts (gluco-neo-genesis) and read right to left, *gluconeogenesis* means "creation of new glucose." Most of the glucose from gluconeogenesis is used to support the brain and red blood cells and to maintain proper blood glucose levels. The brain and red blood cells are the most "glucose-hungry" tissues in your body. Gluconeogenesis is not the exact reversal of glycolysis. We first met pyruvate as a product of glycolysis, but it can arise from other sources to be the starting point for the synthesis of glucose. Three of the reactions of glycolysis are irreversible, and these reactions are bypassed in gluconeogenesis. It is common that the biosynthesis and the degradation of many important biomolecules follow different pathways.

gluconeogenesis the metabolic pathway by which glucose is synthesized from lactate

What are the differences between glycolysis and gluconeogenesis at the sites of the three irreversible reactions of glycolysis (Figure 18.10, *yellow boxes*)? The first of the gluconeogenesis reactions that are different from glycolysis occurs in the conversion of pyruvate to phosphoenolpyruvate (PEP). The second is the conversion of fructose-1,6-bisphosphate to fructose-6-phosphate, and the third is the conversion of glucose-6-phosphate to glucose. The net result of gluconeogenesis includes the reversal of three irreversible glycolytic reactions, but gluconeogenesis is different at these steps, with different reactions and different enzymes.

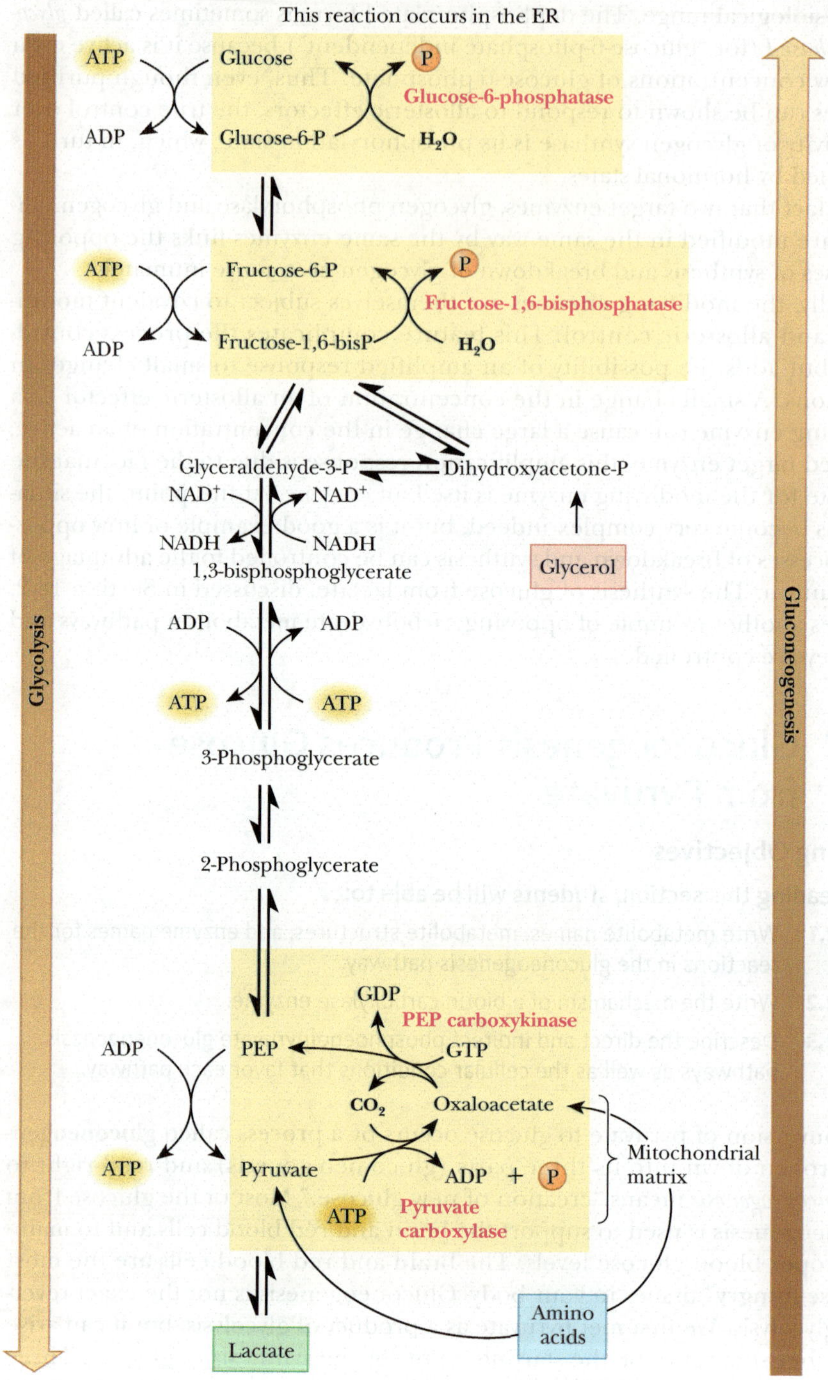

Figure 18.10 The opposing pathways of gluconeogenesis and glycolysis. The yellow-shaded boxes indicate the reactions and enzymes of gluconeogenesis that differ from glycolysis. All other reactions involve the same enzymes as those used in glycolysis. Species in blue, green, and pink-shaded boxes indicate other entry points for gluconeogenesis (in addition to pyruvate).

18-2A The Initial Reactions of Gluconeogenesis: Pyruvate to Phosphoenolpyruvate in Two Steps

▶ **Why is oxaloacetate an intermediate in gluconeogenesis?**

The conversion of pyruvate to phosphoenolpyruvate in gluconeogenesis takes place in two steps. The first step is the conversion of pyruvate to oxaloacetate (Figure 18.11). This step requires energy, which is available from the hydrolysis of ATP.

pyruvate carboxylase the first gluconeogenesis-specific enzyme that catalyzes the conversion of pyruvate to oxaloacetate

The enzyme that catalyzes this reaction is **pyruvate carboxylase**, an allosteric enzyme found in the mitochondria. Acetyl-CoA is an allosteric effector that

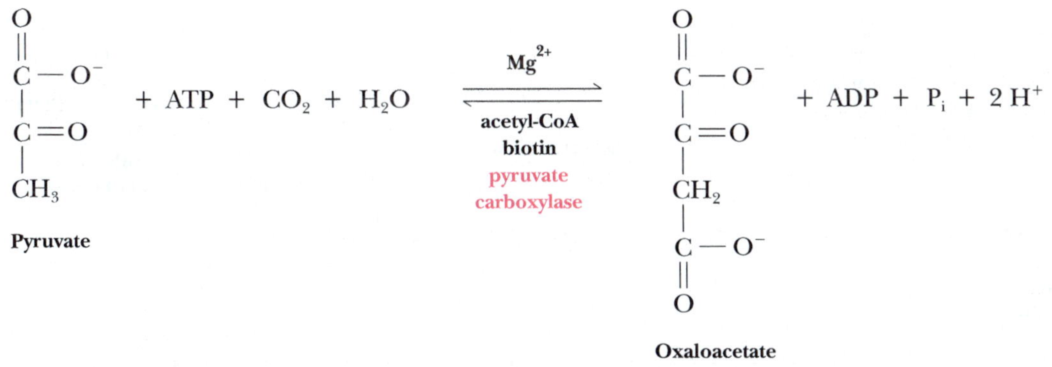

Figure 18.11 The first reaction of gluconeogenesis that differs from glycolysis: the conversion of pyruvate to oxaloacetate by pyruvate carboxylase.

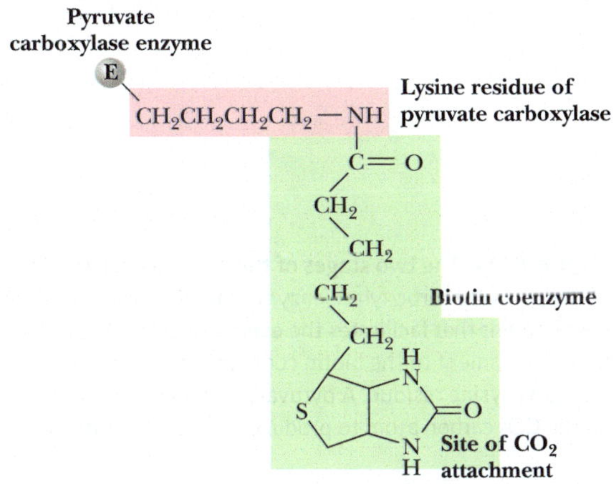

Figure 18.12 The structure of the biotin coenzyme, its site of CO_2 attachment for carboxylation reactions, and its mode of attachment to a lysine residue of the pyruvate carboxylase enzyme.

activates pyruvate carboxylase. If high levels of acetyl-CoA are present (in other words, if there is more acetyl-CoA than is needed to supply the citric acid cycle), then pyruvate (a precursor of acetyl-CoA) can be diverted to gluconeogenesis to produce glucose. Oxaloacetate from the citric acid cycle can frequently be a starting point for gluconeogenesis as well. Magnesium ion (Mg^{2+}) and biotin are also required for effective catalysis. We have seen Mg^{2+} as a cofactor before, but we have not seen biotin, which requires some discussion.

Biotin is a *coenzyme* that is a carrier of carbon dioxide for carboxylation reactions. It has a specific site for the covalent attachment of CO_2, and biotin itself forms a covalent attachment to the enzymes that it assists (Figure 18.12). The carboxyl group of the biotin forms an amide bond with the ε-amino group of a specific lysine side chain of pyruvate carboxylase. Figure 18.13 shows details of the pyruvate carboxylase mechanism and the roles of CO_2, ATP, and biotin in this enzymatic reaction. Biotin and another carboxylase enzyme are discussed in the biosynthesis of fatty acids (Section 21-6A).

biotin a coenzyme that carries CO_2 for carboxylation reactions

The second step in the conversion of pyruvate to phosphoenolpyruvate in gluconeogenesis is the conversion of oxaloacetate to phosphoenolpyruvate. This reaction is catalyzed by the enzyme **phosphoenolpyruvate carboxykinase (PEPCK)** (Figure 18.14). The PEPCK enzyme is found in the mitochondria and the cytosol. This reaction also involves the hydrolysis of a nucleoside triphosphate—guanosine triphosphate (GTP) in this case, rather than ATP.

phosphoenolpyruvate carboxykinase (PEPCK) the second gluconeogenesis-specific enzyme that catalyzes the conversion of oxaloacetate to phosphoenolpyruvate

The successive carboxylation and decarboxylation reactions are both close to equilibrium (their standard Gibbs energies have low values); as a result, the conversion of pyruvate to phosphoenolpyruvate is also close to equilibrium ($\Delta G°' = 2.1 \text{ kJ mol}^{-1} = 0.5 \text{ kcal mol}^{-1}$). A small increase in the level of

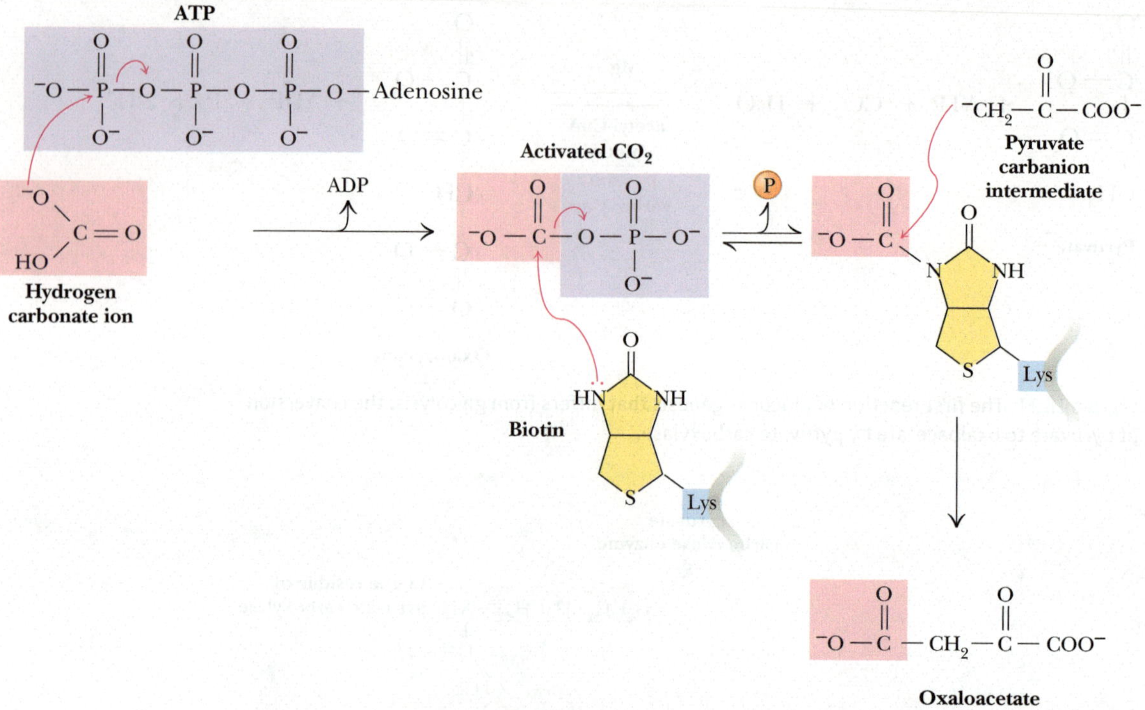

Figure 18.13 The two stages of the pyruvate carboxylase reaction mechanism. The active site of the pyruvate carboxylase enzyme contains amino acid residues that provide a biochemical environment that facilitates the events seen in this mechanism. ATP is used to activate CO_2 prior to its attachment to the biotin coenzyme that is covalently attached to pyruvate carboxylase through a lysine residue. A pyruvate carbanion intermediate then performs a nucleophilic attack on the CO_2 carbon atom to produce the oxaloacetate product.

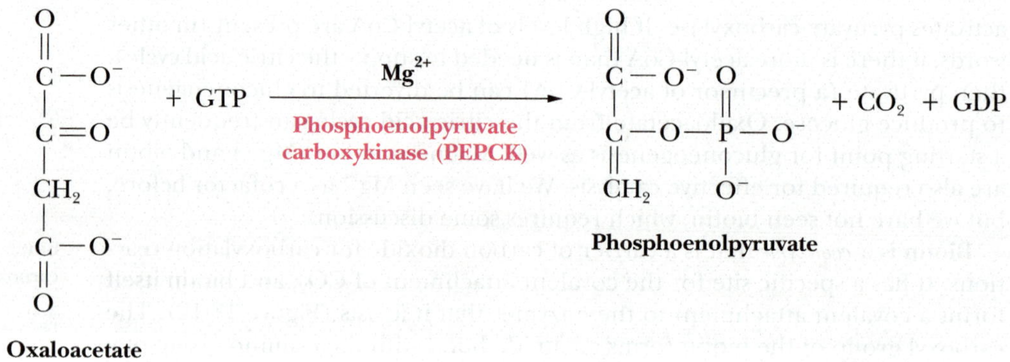

Figure 18.14 The second reaction of gluconeogenesis that differs from glycolysis: the conversion of oxaloacetate to phosphoenolpyruvate by phosphoenolpyruvate carboxykinase.

law of mass action in chemistry, the relationship between the concentrations of products and reactants in a system at equilibrium

oxaloacetate can drive the equilibrium to the right, whereas a small increase in the level of phosphoenolpyruvate can drive it to the left. The **law of mass action** relates the concentrations of reactants and products in a system at equilibrium, and *Le Châtelier's principle* explains how a system at equilibrium responds when the equilibrium is disturbed. Changing the concentration of reactants or products causes a shift to reestablish equilibrium. A reaction proceeds to the right upon the addition of reactants and to the left upon the addition of products.

$$\text{Pyruvate} + \text{ATP} + \text{GTP} \rightarrow \text{Phosphoenolpyruvate} + \text{ADP} + \text{GDP} + P_i$$

The oxaloacetate formed in the mitochondria can have two fates when it comes to continuing gluconeogenesis in the cytosol and converting the oxaloacetate to phosphoenolpyruvate (PEP) in the next step:

- Direct PEP pathway: Oxaloacetate can get converted to PEP in the mitochondria, which can then leave the mitochondria via a specific transporter to continue gluconeogenesis in the cytosol. This path is favored when lactate is the source of pyruvate.
- Indirect PEP pathway: Oxaloacetate can be turned into malate via mitochondrial malate dehydrogenase, a reaction that uses nicotinamide adenine dinucleotide (NADH), as shown near the top of Figure 18.15. Malate can then leave the mitochondria and have the reaction reversed by cytosolic malate dehydrogenase to produce oxaloacetate, a reaction that produces NADH in the cytosol (Figure 18.15, *bottom*). The oxaloacetate then gets converted to PEP to continue gluconeogenesis in the cytosol. This path is favored when NADH is needed in the cytosol.

18-2B The Final Reactions of Gluconeogenesis: Dephosphorylations of Sugar Phosphates

The previous subsection discussed the initial reactions of gluconeogenesis and ended with the production of phosphoenolpyruvate. The final gluconeogenesis reactions discussed in this subsection are *not* the *very next* reactions of gluconeogenesis after the production of phosphoenolpyruvate; those would be the conversions of phosphoenolpyruvate to fructose-1,6-bisphosphate (Figure 18.10, *middle*). The other two reactions in which gluconeogenesis *differs* from glycolysis are ones in which a phosphate-ester bond to a sugar-hydroxyl group

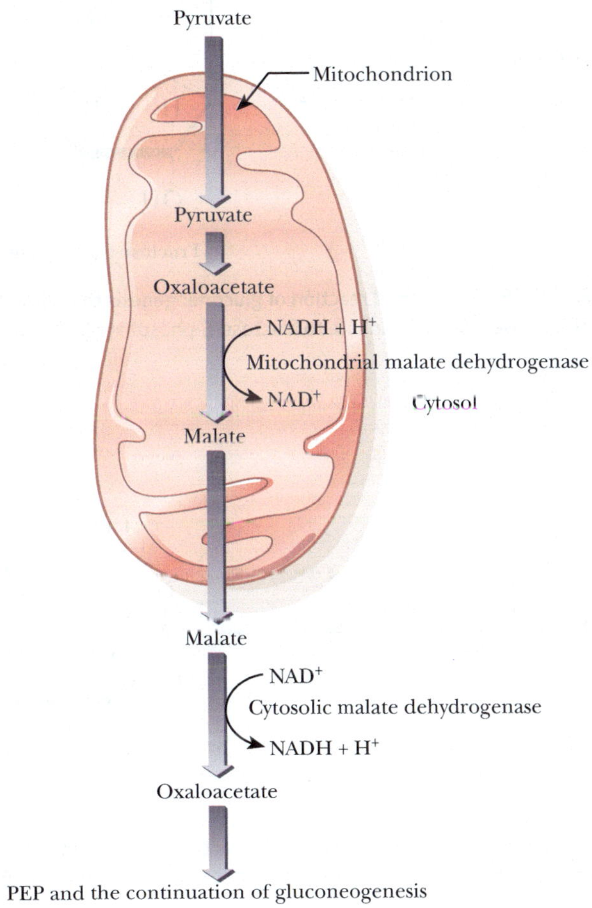

Figure 18.15 The indirect phosphoenolpyruvate (PEP) pathway is used when NADH is needed in the cytosol. Pyruvate from the cytosol is transported from the cytosol to a mitochondrion. Pyruvate is then converted to oxaloacetate and malate in the mitochondria. Malate is transported to the cytosol and then oxidized back to oxaloacetate (with the concomitant production of cytosolic NADH), which is converted to PEP in the continuation of gluconeogenesis.

is hydrolyzed. Both reactions are catalyzed by *phosphatases*, and both reactions are exergonic. The first reaction is the hydrolysis of fructose-1,6-bisphosphate to produce fructose-6-phosphate and phosphate ion ($\Delta G^{\circ\prime} = -16.7\ \text{kJ mol}^{-1} = -4.0\ \text{kcal mol}^{-1}$) (Figure 18.16).

fructose-1,6-bisphosphatase (FBPase-1) the third gluconeogenesis-specific enzyme that catalyzes the conversion of fructose-1,6-bisphosphate to fructose-6-phosphate

This reaction is catalyzed by the enzyme **fructose-1,6-bisphosphatase (FBPase-1)**, an allosteric enzyme strongly inhibited by adenosine monophosphate (AMP). High AMP levels signal that the cell needs ATP, so AMP inhibits FBPase-1 to slow down gluconeogenesis, allowing glycolysis (the breakdown of glucose for energy) to proceed.

glucose-6-phosphatase the fourth (and final) gluconeogenesis-specific enzyme that catalyzes the conversion of glucose-6-phosphate to glucose

The second reaction is the hydrolysis of glucose-6-phosphate to glucose and phosphate ion ($\Delta G^{\circ\prime} = -13.8\ \text{kJ mol}^{-1} = -3.3\ \text{kcal mol}^{-1}$) (Figure 18.17). The enzyme that catalyzes this reaction is **glucose-6-phosphatase**.

When we discussed glycolysis, we saw that both of the phosphorylation reactions, which are the reverse of these two phosphatase-catalyzed reactions, are endergonic. In glycolysis, the phosphorylation reactions must be coupled to the hydrolysis of ATP to make them exergonic and thus energetically allowed. In gluconeogenesis, the organism can make direct use of the fact that the hydrolysis reactions of the sugar phosphates are exergonic. The corresponding reactions are not the reverse of each other in the two pathways. They differ from each other in whether they require ATP and in the enzymes involved. Hydrolysis of glucose-6-phosphate to glucose occurs in the endoplasmic reticulum. This is an example of an interesting pathway that requires three cellular locations (mitochondria, cytosol, and endoplasmic reticulum).

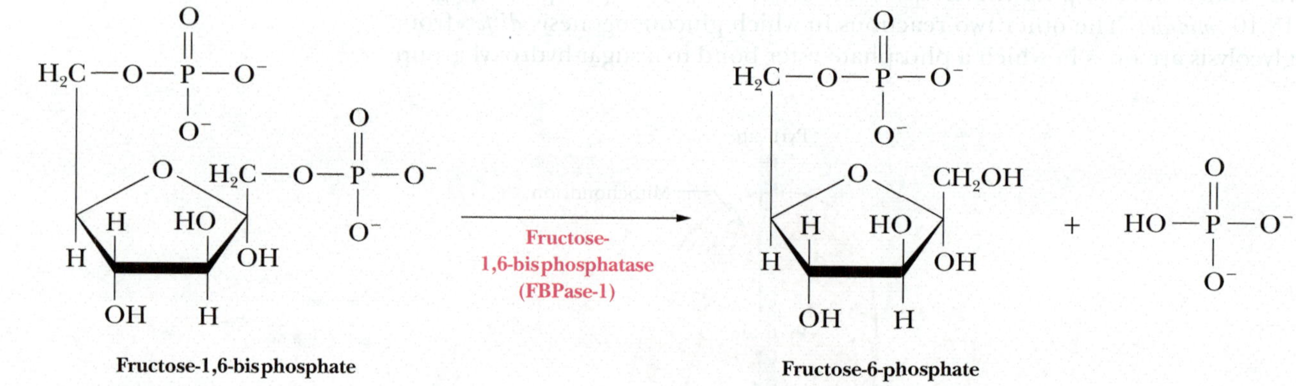

Figure 18.16 The third reaction of gluconeogenesis that differs from glycolysis: the conversion of fructose-1,6-bisphosphate to fructose-6-phosphate by fructose-1,6-bisphosphatase.

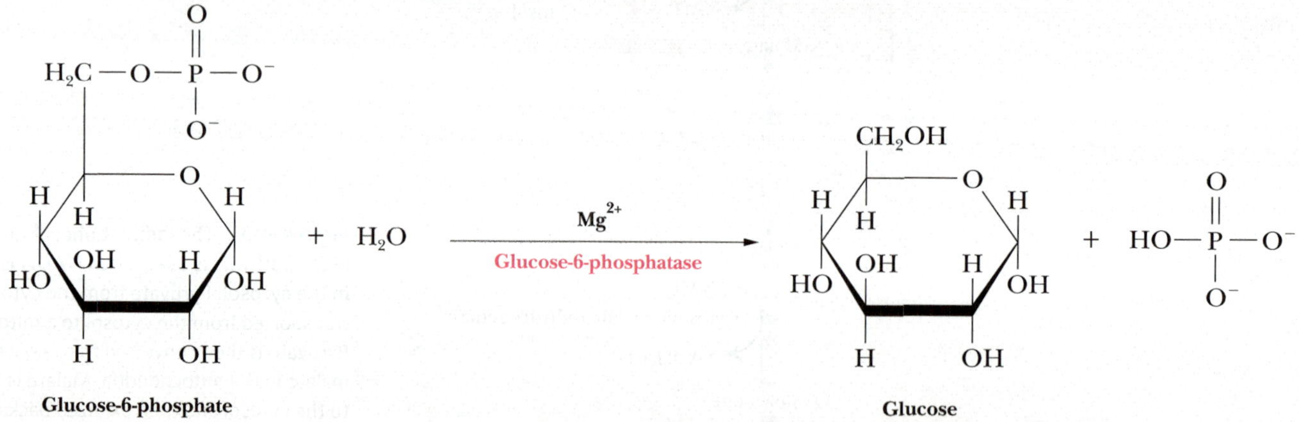

Figure 18.17 The fourth (and final) reaction of gluconeogenesis that differs from glycolysis: the conversion of glucose-6-phosphate to glucose by glucose-6-phosphatase.

18-3 Control of Carbohydrate Metabolism

Learning Objectives

After reading this section, students will be able to:

LO 18.3.1 Explain how glycolysis and gluconeogenesis are regulated.

LO 18.3.2 Describe the events in the hormonal regulation cascade pathways of glycogen breakdown and glycogen synthesis.

LO 18.3.3 Summarize and give examples of four metabolic regulation mechanisms.

LO 18.3.4 Explain details of the Cori cycle, including key intermediates, redox agent details, and nucleoside triphosphate consumption/production.

We have now seen several aspects of carbohydrate metabolism: glycolysis, the breakdown and synthesis of glycogen, and gluconeogenesis. Glucose has a central role in all these processes. It is the starting point for glycolysis, in which it is broken down to pyruvate, and for the synthesis of glycogen, in which many glucose residues combine to give the glycogen polymer. Glucose is also the product of gluconeogenesis, which has the net effect of reversing glycolysis. Glucose is also obtained from the breakdown of glycogen. Each of the opposing pathways—glycolysis and gluconeogenesis, on the one hand, and the breakdown and synthesis of glycogen, on the other hand—is not the *exact* reversal of the other, even though the net results are. In other words, a different path is used to arrive at the same place. It is time to see how all these related pathways are controlled.

18-3A Reciprocal Regulation in Glucose Metabolism

▶ Why is reciprocal regulation a key feature of glucose metabolism?

It is important to consider the **reciprocal regulation** that occurs in sets of opposing pathways. Reciprocal regulation is when a ligand has opposite control effects in opposing pathways. This can occur when the same allosteric effector molecule binds to enzymes of opposing pathways, activating an enzyme in one pathway while inhibiting an enzyme in the other pathway. It is a waste of resources for an organism to have opposing pathways operating simultaneously at a high level. Let us look at reciprocal regulation in each of the two sets of opposing pathways discussed so far in this chapter.

reciprocal regulation when a ligand has opposite control effects on enzymes in opposing pathways

The first step of glycolysis, the conversion of glucose to glucose-6-phosphate, differs from the reverse reaction in gluconeogenesis, with different enzymes in the two pathways. Glycolysis and gluconeogenesis also differ enzymatically at two other points: the conversion of fructose-6-phosphate and fructose-1,6-bisphosphate, and the interconversion of pyruvate and phosphoenolpyruvate (Figure 18.18, *middle* and *bottom*). The control mechanisms for these enzymes are ones that appear many times in metabolism: allosteric control and covalent modification. These forms of control are not a factor with the opposing hexokinase and glucose-6-phosphatase reactions, where product and substrate levels, respectively, provide control (Figure 18.18, *top*). We first compared glycolysis and gluconeogenesis in Figure 18.10, where we simply pointed out the enzymes that are unique to gluconeogenesis. In Figure 18.18, we extend the biochemical understanding by comparing the corresponding enzymes of the two pathways side by side, along with the substances that activate and inhibit them.

In gluconeogenesis, the hydrolysis of fructose-1,6-bisphosphate to fructose-6-phosphate is catalyzed by fructose-1,6-bisphosphatase. In the middle of Figure 18.18, the allosteric control effects of a molecule called fructose-2,6-bisphosphate (F2,6BP) are shown. F2,6BP is produced from fructose-6-phosphate specifically to regulate the flow of metabolites through glycolysis

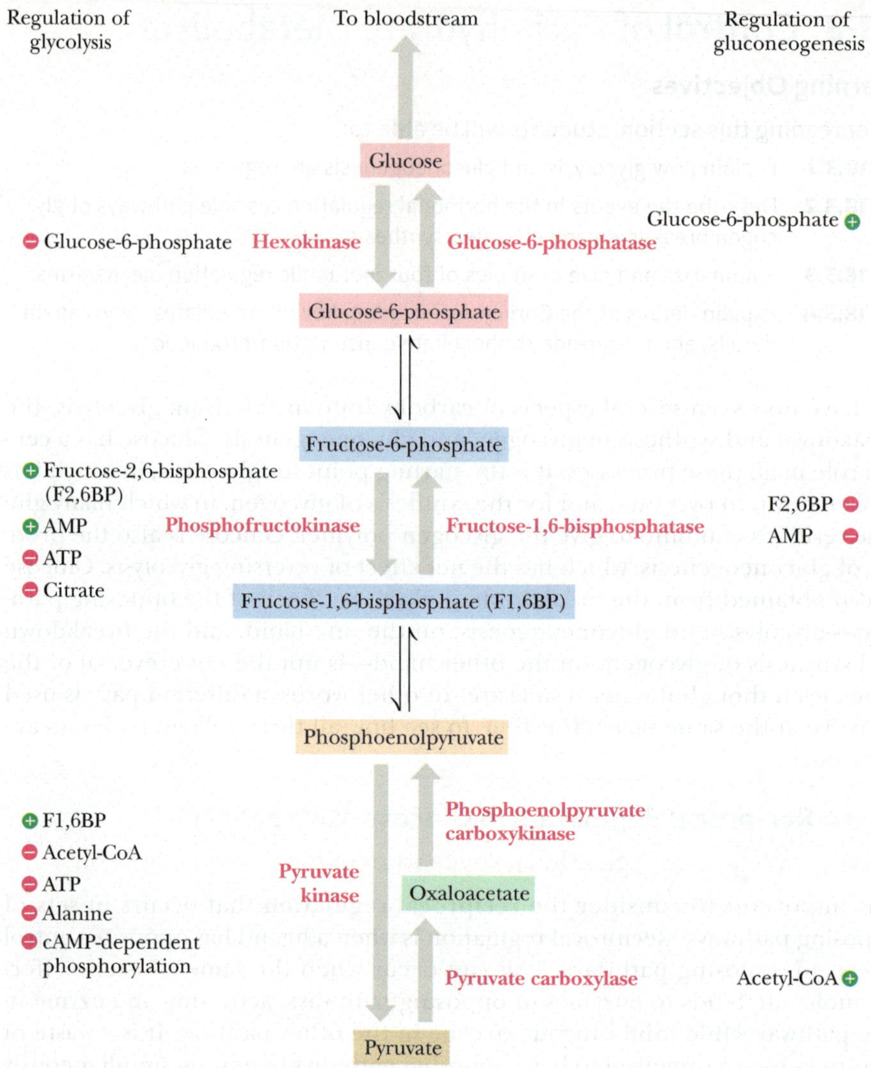

Figure 18.18 The principal regulatory mechanisms in glycolysis and gluconeogenesis. Activators are indicated by green plus signs and inhibitors by red minus signs. (*From GARRETT/GRISHAM, Biochemistry,4E. ©2009 Cengage Learning.*)

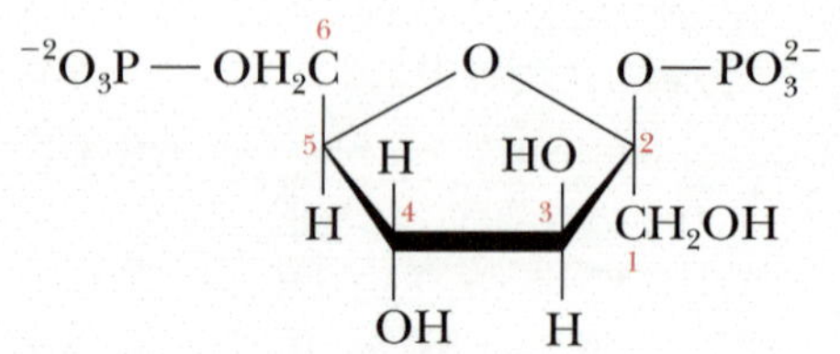

Figure 18.19 Structure of fructose-2,6-bisphosphate.

substrate cycling the control process in which opposing reactions are catalyzed by different enzymes

phosphofructokinase-2 (PFK-2) the enzyme that catalyzes the conversion of fructose-6-phosphate (F6P) to fructose-2,6-bisphosphate (F2,6BP), which is a key regulator of glycolysis and gluconeogenesis

fructose-2,6-bisphosphatase-2 (FBPase-2) the enzyme that catalyzes the conversion of fructose-2,6-bisphosphate (F2,6BP) to fructose-6-phosphate (F6P)

and gluconeogenesis based on the cell's energy needs. F2,6BP amplifies and fine-tunes the effects of ATP and AMP that are also seen in the figure. This enzyme is subject to allosteric inhibition by F2,6BP and by AMP. We saw in Section 17-2 that, in glycolysis, phosphofructokinase (PFK), the enzyme that catalyzes the phosphorylation of fructose-6-phosphate to fructose-1,6-bisphosphate, exhibits the opposite allosteric behavior. Phosphofructokinase is activated by fructose-2,6-bisphosphate and by AMP, whereas it is inhibited by ATP and citrate. The term **substrate cycling** refers to the fact that opposing reactions can be catalyzed by different enzymes. Consequently, the opposing reactions can be independently regulated and have different rates. It would not be possible to have different rates with the same enzyme because a catalyst speeds up a reaction and the reverse of the reaction to the same extent (Section 6-2).

The intracellular level of fructose-2,6-bisphosphate (F2,6BP; Figure 18.19) is a key point here. F2,6BP is an important allosteric activator of phosphofructokinase (PFK), the key enzyme of glycolysis; it is also an inhibitor of fructose bisphosphate phosphatase (FBPase), which plays a role in gluconeogenesis. A high concentration of F2,6BP stimulates glycolysis, whereas a low concentration stimulates gluconeogenesis. The concentration of F2,6BP in a cell depends on the balance between its synthesis, catalyzed by **phosphofructokinase-2 (PFK-2)**, and its breakdown, catalyzed by **fructose-2,6-bisphosphatase-2 (FBPase-2)**.

The enzymes that control the formation and breakdown of F2,6BP are themselves controlled by a phosphorylation/dephosphorylation mechanism similar to what we have already seen in the case of glycogen phosphorylase and glycogen synthase (Section 18-1D). Both enzyme activities are located on the same protein (a dimer of about 100 kDa molecular mass). Phosphorylation of the dimeric protein leads to an increase in the activity of FBPase-2 and a decrease in the concentration of F2,6BP, ultimately stimulating gluconeogenesis (Figure 18.20). Dephosphorylation of the dimeric protein leads to an increase in PFK-2 activity and an increase in the concentration of F2,6BP, ultimately stimulating glycolysis. The net result is similar to the control of glycogen synthesis and breakdown that we discussed in Section 18-1D.

Figure 18.21 shows the co-effects of fructose-2,6-bisphosphate (F2,6BP) and AMP on the activity of fructose-1,6-bisphosphatase (FBPase). The data suggest that the F2,6BP inhibitor works by itself, but its effect is greatly increased by the presence of FBPase's allosteric inhibitor, AMP.

18-3B Hormone Control of Glycogen Metabolism

▶ What is the role of hormones in the regulation of glycogen synthesis and breakdown?

The breakdown and synthesis of glycogen exhibit similar reciprocal regulation. Recall from Section 18-1D that both glycogen synthase and glycogen phosphorylase are controlled by covalent modification (phosphorylation and dephosphorylation) as well as by allosteric control. Recall also that the same enzyme system is responsible for the modification of both the synthase and phosphorylase. In addition, hormonal control is of paramount importance in glycogen synthesis and breakdown, and the PFK-2/FBPase-2 bifunctional enzyme seen in Figure 18.20 is also regulated by hormones (insulin and glucagon).

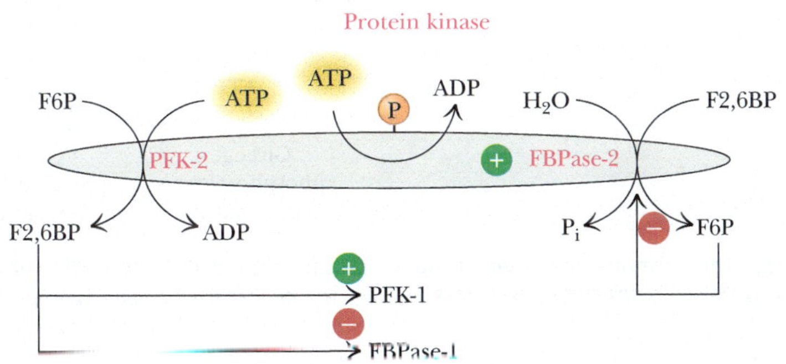

Figure 18.20 The formation and breakdown of fructose-2,6-bisphosphate (F2,6BP). These two processes are catalyzed by two enzyme activities (phosphofructokinase-2 [PFK-2] and fructose-2,6-bisphosphatase-2 [FBPase-2]) on the same bifunctional protein (represented by the gray oval). These two enzyme activities are controlled by a phosphorylation/dephosphorylation mechanism. Phosphorylation of the bifunctional enzyme by a protein kinase activates the FBPase-2 activity of the enzyme that degrades F2,6BP (green plus sign in the gray oval), whereas dephosphorylation (not shown) activates the PFK-2 activity of the enzyme that produces F2,6BP. Fructose-6-phosphate (F6P) inhibits FBPase-2 activity. Under metabolic conditions where fructose-2,6-bisphosphate (F2,6BP) is produced (*lower left*), the F2,6BP molecule goes on to activate the phosphofructokinase-1 (PFK-1) enzyme of glycolysis, and it also inhibits the fructose-1,6-bisphosphatase (FBPase-1) enzyme of gluconeogenesis.

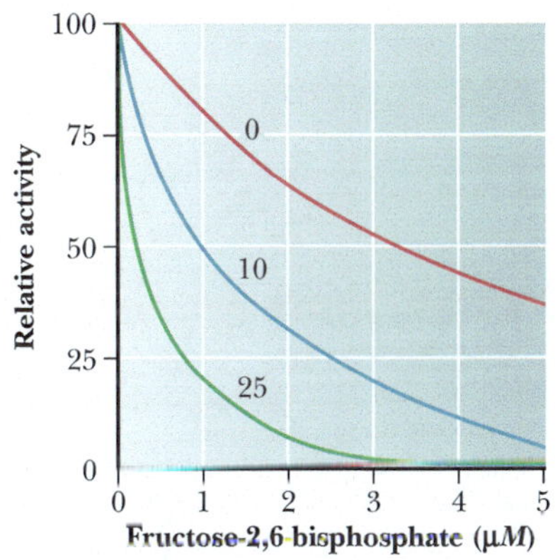

Figure 18.21 Allosteric effects in the control of fructose-1,6-bisphosphatase: the effect of AMP (0, 10, and 25 μ*M* [micromolar]) on the inhibition of fructose-1,6-bisphosphatase activity by fructose-2,6-bisphosphate. Activity was measured in the presence of 10 μ*M* fructose-1,6-bisphosphate. (*Adapted from Van Schaftingen, E., and Hers, H. G. (1981). Inhibition of fructose-1,6-bisphosphatase by fructose-2,6-bisphosphate.* Proc. Nat. Acad. Sci. *78, 2861–2863.*)

adenylyl cyclase the enzyme that catalyzes the conversion of ATP to 3′,5′-cyclic AMP (cAMP), a critical second messenger molecule

An enzymatic cascade triggers the activation of glycogen phosphorylase. Figure 18.22 shows the chain of events that ultimately leads to the activation of glycogen phosphorylase, which leads to the release of glucosyl residues from glycogen (Section 18-1B). Figure 18.23 shows the first reaction in this cascade, the **adenylyl cyclase** reaction. Adenylyl cyclase also goes by the less commonly used name "adenylate cyclase." The products of this reaction are 3′,5′-cyclic AMP (cAMP), which is an activator of the next kinase in the series. cAMP is a critical *second messenger* in many cascade pathways, with hormones being the first messengers. Pyrophosphate is the other product of this reaction. The hydrolysis of pyrophosphate releases energy to drive the reaction. The end product of the cascade seen in Figure 18.22 is the active phosphorylated form of glycogen phosphorylase *a*. We can see, by comparing Figure 18.22 with Figure 18.9, that the covalent modification—the addition of a phosphate group that increases the activity of glycogen phosphorylase—is a downstream event in the cascade that is triggered by a hormone. Recall that Figure 18.9 also shows allosteric control details in the regulation of glycogen breakdown.

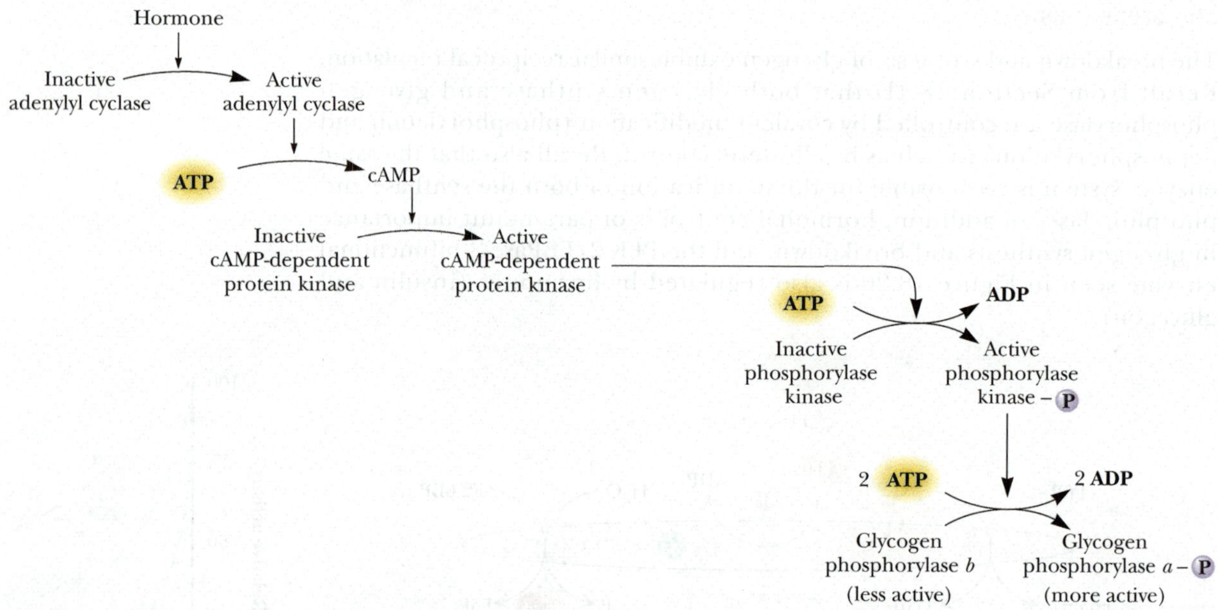

Figure 18.22 The hormone-activated enzymatic cascade that leads to activation of glycogen phosphorylase. (*From GARRETT/GRISHAM,* Biochemistry, *4E. ©2009 Cengage Learning.*)

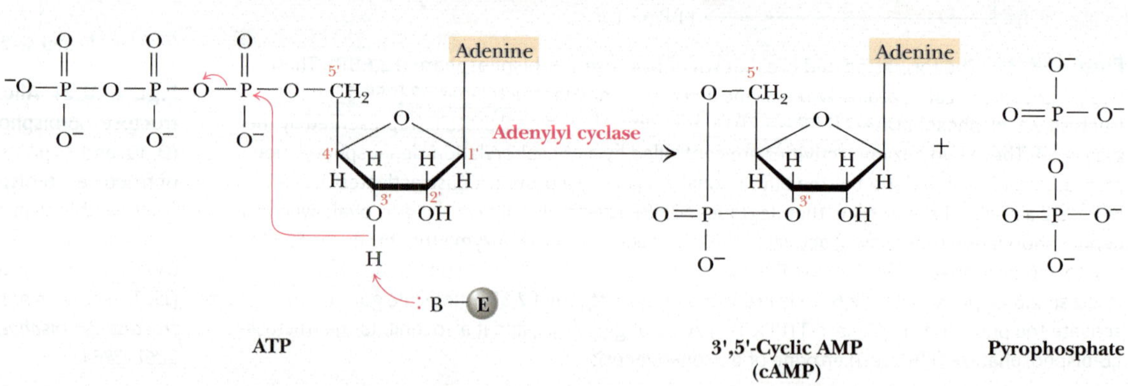

Figure 18.23 The adenylyl cyclase reaction. The reaction is driven forward by the subsequent hydrolysis of pyrophosphate by the enzyme inorganic pyrophosphatase. Unprimed numbers are used to label the carbon atoms of adenine (not shown), so primed numbers are used for the ribose carbons. (*From GARRETT/GRISHAM,* Biochemistry, *4E. ©2009 Cengage Learning.*)

In the case of glycogen synthesis, insulin (a hormone) binds to receptors on the cell surface, which triggers a series of covalent modifications, eventually activating glycogen synthase kinase. This kinase phosphorylates glycogen synthase, producing glycogen synthase D, which can be activated by the allosteric effector glucose-6-phosphate (recall this point from Section 18-1D). Phosphoprotein phosphatase, which plays a role in glycogen breakdown, dephosphorylates glycogen synthase I, which is not subject to allosteric activation by glucose-6-phosphate and which is more active than glycogen synthase D. The various aspects of glucose metabolism are connected by the many levels of reciprocal regulation.

Insulin is secreted by the β cells of the islets of Langerhans in the pancreas in response to increased blood glucose levels, such as those after a meal. Uptake of insulin by cells, as a result of insulin binding to receptors on the surface, triggers the protein kinase cascade that eventually leads to glycogen synthesis (Figure 18.24). Another protein kinase cascade triggered by insulin binding to receptors stimulates the action of GLUT4, a glucose transport protein. GLUT4 is released from vesicles within the cell to the cell surface, where it brings glucose into the cell. Inside the cell, glucose is converted to glucose-6-phosphate, which can have two fates. It can be incorporated into glycogen (after conversion to glucose-1-phosphate), or it can serve as an allosteric activator of the phosphorylated glycogen synthase.

When blood glucose levels decrease, the hormones glucagon and epinephrine play important roles. Glucagon is a 29-amino acid peptide hormone secreted by the α cells of the islets of Langerhans, whereas epinephrine (also called adrenaline) is an amino acid derivative. Binding of either of these hormones initiates a cascade that activates glycogen phosphorylase and inhibits glycogen synthase (Figure 18.25). In both cases, the phosphorylase cascade amplifies the original signal of hormone binding. The two hormones differ in the time scale of their actions. Epinephrine is important on the very short time scale of the "fight-or-flight" response. Glucagon operates with insulin over a longer time scale to stabilize blood glucose levels. The response time of a given control mechanism can be one of its most important features.

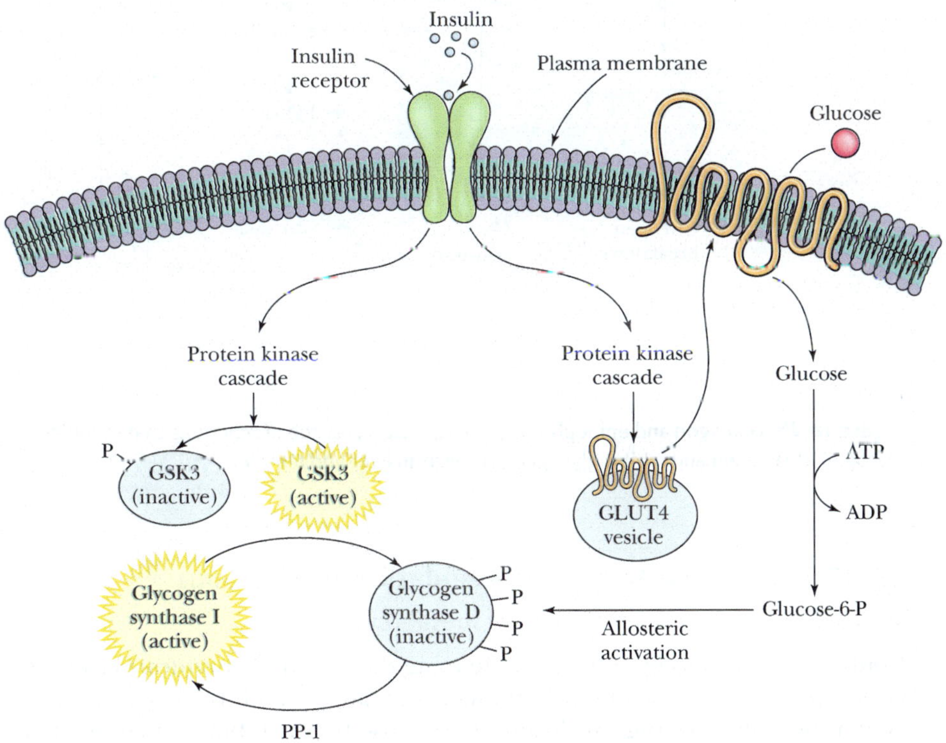

Figure 18.24 Binding of insulin to plasma membrane receptors triggers protein kinase cascades that stimulate glycogen synthesis. Glucose uptake leads to the syntheses of glycogen and glucose-6-phosphate, which allosterically activates the otherwise inactive form of glycogen synthase. *(From GARRETT/GRISHAM,* Biochemistry, *4E. ©2009 Cengage Learning.)*

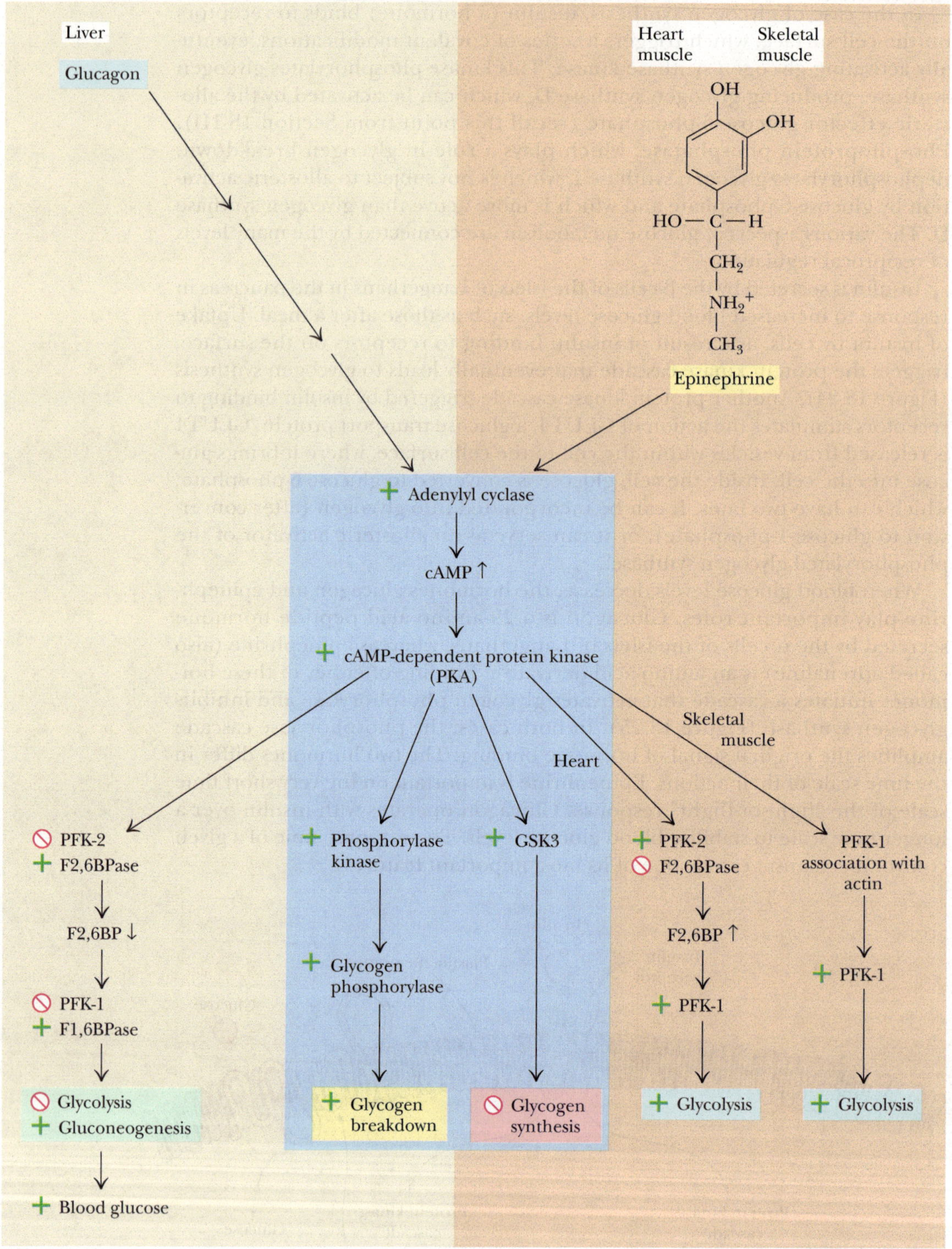

Figure 18.25 Glucagon and epinephrine each activate a cascade of reactions that stimulate glycogen breakdown and inhibit glycogen synthesis in liver and muscle.

18-3C Summary of Metabolic Regulation Mechanisms

▶ **What are the main categories of metabolic control?**

Control mechanisms operate by modulating the action of enzymes. Having several different ways of doing this allows for greater flexibility in achieving this goal, especially allowing for different response times. Table 18.1 summarizes

Table 18.1 Mechanisms of Metabolic Control

Type of Control	Mode of Operation	Examples
Allosteric	Effectors (substrates, products, or coenzymes) of a pathway inhibit or activate an enzyme. (Responds rapidly to external stimuli.)	ATCase (Section 7-1A); phosphofructokinase (Section 17-2)
Covalent modification	Inhibition or activation of an enzyme depends on the formation or breaking of a bond, frequently by phosphorylation or dephosphorylation. (Responds rapidly to external stimuli.)	Sodium–potassium pump (Section 8-5A); glycogen phosphorylase, glycogen synthase (Section 18-1D)
Substrate cycles	Two opposing reactions, such as the formation and breakdown of a given substance, are catalyzed by different enzymes, which can be activated or inhibited separately. (Responds rapidly to external stimuli.)	Glycolysis (Chapter 17) and gluconeogenesis (Section 18-2)
Genetic control	The amount of enzyme present depends on the regulation of transcription. (Longer-term control than the other mechanisms listed here.)	*Lac* operon (Section 11-3C)

important mechanisms of metabolic control. Even though we discuss them here in the context of carbohydrate metabolism, they apply to all aspects of metabolism.

18-3D Glycogen Metabolism in Different Body Organs

▶ How do different organs share carbohydrate metabolism?

Using combinations of the control mechanisms listed in Table 18.1, an organism can set up a division of labor among tissues and organs to maintain control of glucose metabolism. A particularly clear example is found in the Cori cycle (Figure 18.26). The Cori cycle is named for Gerty and Carl Cori, who first described it in 1929. In the Cori cycle, a loop of lactate production and recycling occurs between muscles and the liver, enabling efficient use of energy during intense activity.

Glycolysis in fast-twitch skeletal muscle produces lactate under conditions of oxygen debt, such as a sprint. Fast-twitch muscle has comparatively few mitochondria, so metabolism is largely anaerobic in this tissue. The buildup of lactate contributes to the muscular aches that follow strenuous exercise. Lactate then travels through the blood to the liver, where gluconeogenesis recycles the lactate into glucose. Glucose produced in the liver is transported back to skeletal muscle by the blood, where it becomes an energy source for the next burst of exercise. This is the main reason that athletes receive post-event massages and that they always cool down after the event. Cooling down keeps the blood flowing through the muscles and allows the lactate and other acids to leave the cells and enter the blood. Massages increase this movement from the cells to the blood.

Notice that we have a division of labor between two different types of organs—muscle and liver. In the same cell (of whatever type), these two metabolic pathways—glycolysis and gluconeogenesis—are not highly active simultaneously. When the cell needs ATP, glycolysis is more active; when there is little need for ATP, gluconeogenesis is more active. Because of the hydrolysis of ATP and GTP in the reactions of gluconeogenesis that differ from those

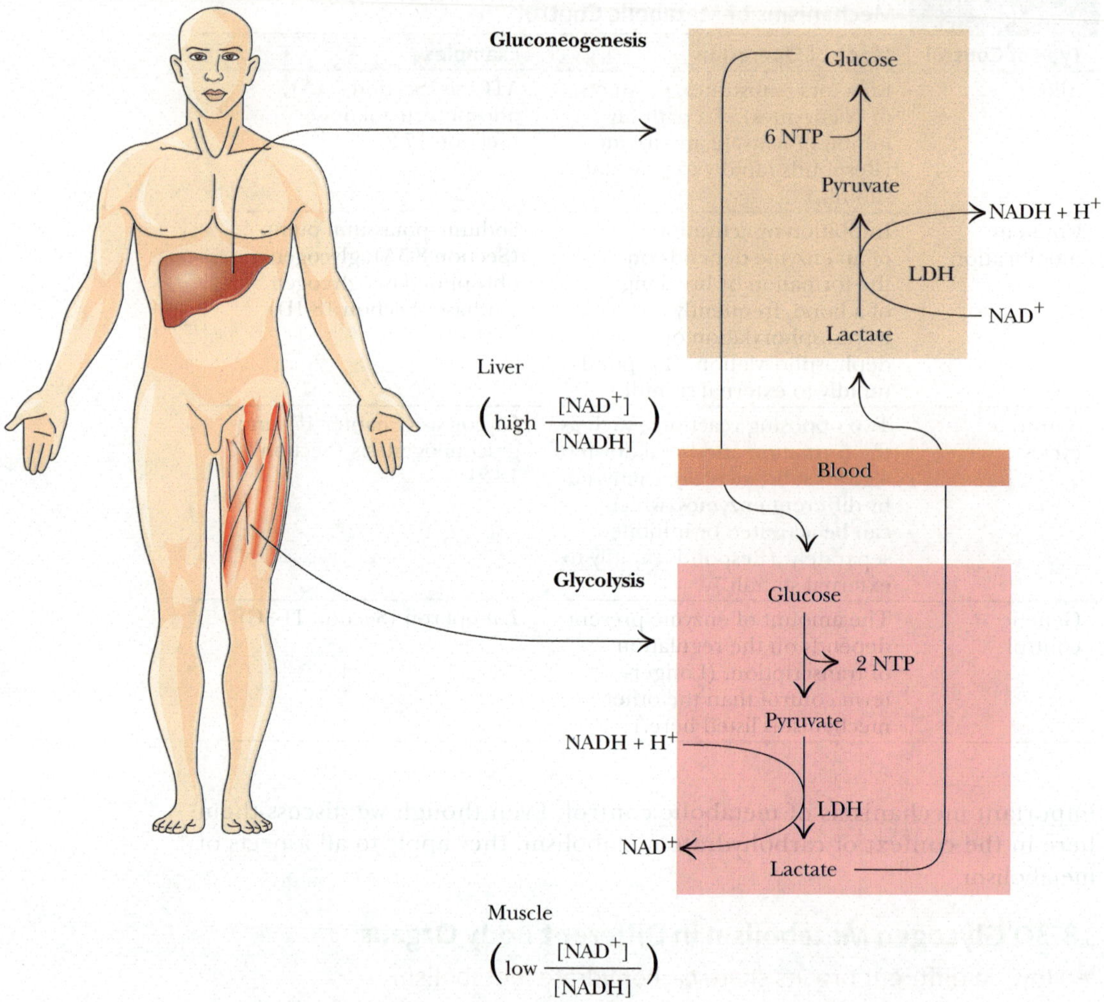

Figure 18.26 **The Cori cycle.** Lactate produced in muscles by glycolysis is transported by the blood to the liver. Gluconeogenesis in the liver converts the lactate back to glucose, which can be carried back to the muscles by the blood. Glucose can be stored as glycogen until it is degraded by glycogenolysis. (NTP stands for nucleoside triphosphate, and it represents ATP or GTP in these processes.)

of glycolysis, the overall pathway from two molecules of pyruvate back to one molecule of glucose is exergonic ($\Delta G°' = -37.6 \text{ kJ mol}^{-1} = -9.0 \text{ kcal mol}^{-1}$, for one mole of glucose). The conversion of pyruvate to lactate is exergonic, which means that the reverse reaction is endergonic. The energy released by the exergonic conversion of pyruvate to glucose by gluconeogenesis facilitates the endergonic conversion of lactate to pyruvate.

The Cori cycle requires the net hydrolysis of two ATP and two GTP. ATP is produced by the glycolytic part of the cycle, but the portion involving gluconeogenesis requires yet more ATP in addition to GTP.

Glycolysis:

$$\text{Glucose} + 2\text{NAD}^+ + 2\text{ADP} + 2\text{P}_i \rightarrow$$
$$2\text{ Pyruvate} + 2\text{NADH} + 4\text{H}^+ + \mathbf{2ATP} + 2\text{H}_2\text{O}$$

Gluconeogenesis:

$$2\text{ Pyruvate} + 2\text{NADH} + 4\text{H}^+ + \mathbf{4ATP} + \mathbf{2GTP} + 6\text{H}_2\text{O} \rightarrow$$
$$\text{Glucose} + 2\text{NAD}^+ + 4\text{ADP} + 2\text{GDP} + 6\text{P}_i$$

Overall:

$$2ATP + 2GTP + 4H_2O \rightarrow 2ADP + 2GDP + 4P_i$$

The hydrolysis of both ATP and GTP is the price of increased simultaneous control of the two opposing pathways.

18-4 Glucose Is Sometimes Diverted Through the Pentose Phosphate Pathway

Learning Objectives

After reading this section, students will be able to:

LO 18.4.1 Write metabolite names, metabolite structures, and enzyme names for the reactions in the pentose phosphate pathway.

LO 18.4.2 Describe how the pentose phosphate pathway is regulated.

The **pentose phosphate pathway** is an alternative to glycolysis and differs from it in several important ways. In glycolysis, one of our most important concerns was the production of ATP. In the pentose phosphate pathway, the production of ATP is not the crux of the matter. As the name of the pathway indicates, five-carbon sugars, including ribose, are produced from glucose. Ribose and its derivative, deoxyribose, play an important role in the synthesis of nucleotides and nucleic acids (to review the structures of these molecules, see Figures 9.4–9.6 in Section 9-2). Another important facet of the pentose phosphate pathway is the production of a coenzyme called *nicotinamide adenine dinucleotide phosphate (NADPH),* a compound that differs from nicotinamide adenine dinucleotide (NADH) by having one extra phosphate group esterified to the ribose ring of the adenine nucleotide portion of the molecule (Figure 18.27). NAD is a coenzyme involved in oxidative catabolic reactions, like the glyceraldehyde-3-phosphate dehydrogenase reaction of glycolysis (Section 17-3). In contrast, NADP is a coenzyme involved in reductive anabolic reactions. For example, NADPH plays an important role in the biosynthesis of lipids (see Chapter 21). Thus, the simple presence or absence of a phosphate group on this dinucleotide determines its role in the breakdown or synthesis of biochemical molecules.

pentose phosphate pathway a pathway in sugar metabolism that gives rise to five-carbon sugars and NADPH

The pentose phosphate pathway begins with a series of oxidation reactions that produce NADPH and a five-carbon sugar called ribulose-5-phosphate (Figure 18.28, *top*). The remainder of the pathway involves the nonoxidative reshuffling of the carbon skeletons of the sugars involved (Figure 18.28, *bottom*). The products of these nonoxidative reactions include substances such as ribose-5-phosphate, which is used in nucleic acid biosynthesis. Other products include fructose-6-phosphate and glyceraldehyde-3-phosphate, which play a role in glycolysis. Some of these reshuffling reactions also occur when sugars are produced in photosynthesis.

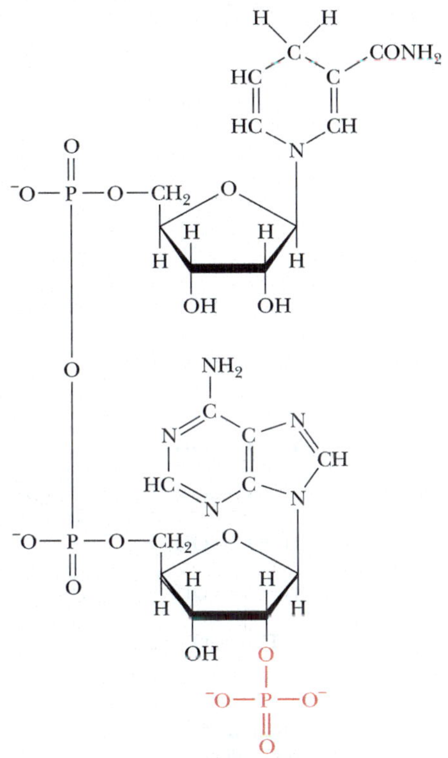

Figure 18.27 The structure of reduced nicotinamide adenine dinucleotide phosphate (NADPH).

18-4A Pentose Phosphate Pathway: Oxidative Reactions

▶ **What are the oxidative reactions of the pentose phosphate pathway?**

To keep track of the carbon backbone of the sugars and their aldehyde and ketone functional groups, we write the formulas in the open-chain form. In the first reaction of the pathway, the aldehyde group of glucose-6-phosphate is oxidized to a carboxylate group to form 6-phosphogluconate (Figure 18.28, *top left*). The enzyme that catalyzes this reaction is **glucose-6-phosphate dehydrogenase**. The NADPH produced by the reaction can be used in other metabolic pathways that are anabolic. In general, dehydrogenases catalyze the oxidation of a metabolite with a concomitant coenzyme reduction. Moreover, the name

glucose-6-phosphate dehydrogenase the first enzyme of the pentose phosphate pathway that catalyzes the oxidation of glucose-6-phosphate to 6-phosphogluconate and the reduction of $NADP^+$ to NADPH

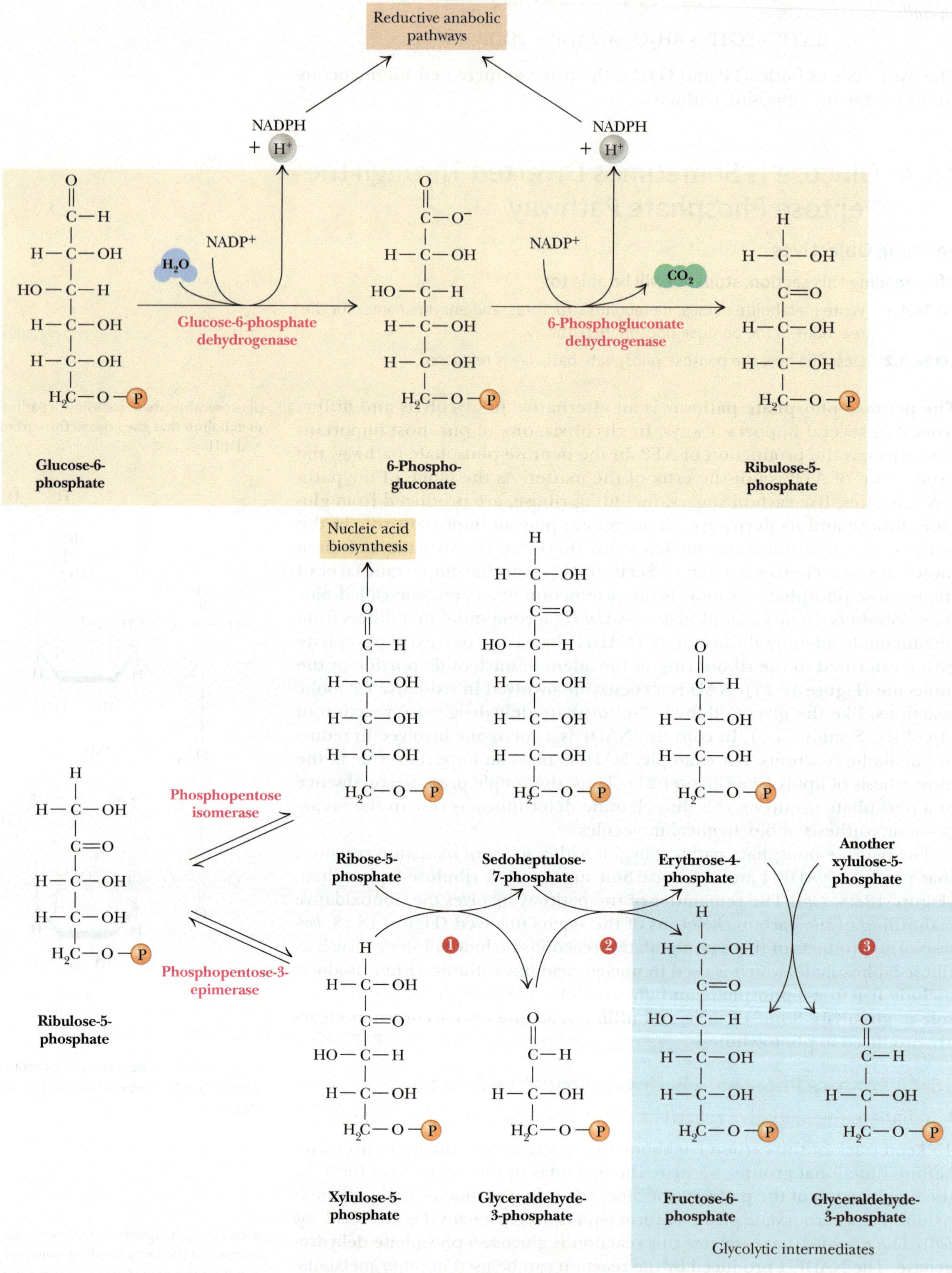

Figure 18.28 The pentose phosphate pathway. The numerals in the red circles indicate the steps discussed in the text.

of the metabolite being oxidized appears before the word "dehydrogenase" in the name of each dehydrogenase enzyme.

The next reaction is an oxidative decarboxylation, and NADPH is produced once again (Figure 18.28, *top right*). The 6-phosphogluconate molecule loses its carboxyl group, which is released as carbon dioxide, and a hydroxyl group is oxidized to a carbonyl, producing the five-carbon keto-sugar (ketose) ribulose-5-phosphate as the other product. The enzyme that catalyzes this reaction is **6-phosphogluconate dehydrogenase**. Several other oxidative decarboxylation reactions are discussed in the tricarboxylic acid (TCA) cycle in Chapter 19.

6-phosphogluconate dehydrogenase the second enzyme of the pentose phosphate pathway that catalyzes the oxidative decarboxylation of 6-phosphogluconate to ribulose-5-phosphate and the reduction of $NADP^+$ to NADPH

18-4B Pentose Phosphate Pathway: Nonoxidative Reactions

▶ What are the nonoxidative reactions of the pentose phosphate pathway, and why are they important?

In the remaining steps of the pentose phosphate pathway, several reactions transfer two- and three-carbon units. There are two different reactions in which ribulose-5-phosphate isomerizes (Figure 18.28, *bottom left*). In one of these reactions, catalyzed by phosphopentose-3-epimerase, there is an inversion of configuration around carbon atom 3. The result is an epimer of ribulose-5-phosphate called xylulose-5-phosphate, which is also a ketose. (To review epimers, see Section 16-1B.) The other isomerization reaction, catalyzed by phosphopentose isomerase, produces a sugar with an aldehyde group (an aldose) rather than a ketone. In this second reaction, ribulose-5-phosphate isomerizes to ribose-5-phosphate. Ribose-5-phosphate is a necessary building block for the synthesis of nucleotides, nucleic acids, and coenzymes such as NAD and NADP.

The group-transfer reactions that link the pentose phosphate pathway with glycolysis require the two five-carbon sugars produced by the isomerization of ribulose-5-phosphate. Two molecules of xylulose-5-phosphate and one molecule of ribose-5-phosphate rearrange to give two molecules of fructose-6-phosphate and one molecule of glyceraldehyde-3-phosphate. In other words, three molecules of pentose (with five carbon atoms each) give two molecules of hexose (with six carbon atoms each) and one molecule of a triose (with three carbon atoms). The total number of carbon atoms (15) does not change, but there is considerable structural rearrangement as a result of the group transfer.

Two enzymes, *transketolase* and *transaldolase,* are responsible for the reshuffling of the carbon atoms of sugars such as ribose-5-phosphate and xylulose-5-phosphate in the remainder of the pathway, which consists of three reactions (Figure 18.28, red numerals 1, 2, and 3). Transketolase transfers a two-carbon unit (a ketol group), and transaldolase transfers a three-carbon unit (an aldol group). Transketolase catalyzes the first and third reactions in the rearrangement process, and transaldolase catalyzes the second reaction. The results of these transfers are summarized in Table 18.2:

Table 18.2 Group-Transfer Reactions in the Pentose Phosphate Pathway

	Reactants	Enzyme	Products
Two-carbon shift	$C_5 + C_5$	$\underset{}{\overset{\text{Transketolase}}{\rightleftarrows}}$	$C_7 + C_3$
Three carbon shift	$C_7 + C_3$	$\overset{\text{Transaldolase}}{\rightleftarrows}$	$C_4 + C_6$
Two-carbon shift	$C_5 + C_4$	$\overset{\text{Transketolase}}{\rightleftarrows}$	$C_6 + C_3$
Net reaction	$3C_5$	$\rightleftarrows$	$2C_6 + C_3$

- In the first of these reactions, a two-carbon unit from xylulose-5-phosphate (five carbons) is transferred to ribose-5-phosphate (five carbons) to give sedoheptulose-7-phosphate (seven carbons) and glyceraldehyde-3-phosphate (three carbons), as also shown in Figure 18.28 (*bottom*, red numeral 1).

- In the reaction catalyzed by transaldolase, a three-carbon unit is transferred from the seven-carbon sedoheptulose-7-phosphate to the three-carbon glyceraldehyde-3-phosphate (Figure 18.28, *bottom*, red numeral 2). The products of the reaction are fructose-6-phosphate (six carbons) and erythrose-4-phosphate (four carbons).
- In the final reaction of this type in the pathway, xylulose-5-phosphate reacts with erythrose-4-phosphate. This reaction is catalyzed by transketolase. The products of the reaction are fructose-6-phosphate and glyceraldehyde-3-phosphate (Figure 18.28, *bottom*, red numeral 3).

In the pentose phosphate pathway, glucose-6-phosphate can be converted to fructose-6-phosphate and glyceraldehyde-3-phosphate by a means other than the glycolytic pathway. For this reason, the pentose phosphate pathway is also called the *hexose monophosphate shunt,* and this name is used in some texts. A major feature of the pentose phosphate pathway is the production of ribose-5-phosphate and NADPH. Next, we discuss how the control mechanisms of the pentose phosphate pathway can respond to the varying needs of organisms for either or both of these compounds.

18-4C Regulation of the Pentose Phosphate Pathway

▶ How is the pentose phosphate pathway controlled?

The reactions catalyzed by transketolase and transaldolase are reversible, which allows the pentose phosphate pathway to respond to the needs of an organism. Even though we "started" our discussion of the pentose phosphate pathway with glucose-6-phosphate, the pathway can begin in other ways, depending on whether there is a greater need for ribose-5-phosphate or for NADPH. The operation of the oxidative portion of the pathway depends strongly on the organism's requirement for NADPH. The need for ribose-5-phosphate can be met in other ways because ribose-5-phosphate can be obtained by starting with the glycolytic intermediates that we "ended" our initial discussion of the pathway with (Figure 18.28, *bottom*, red numerals). The reverse of these reactions can proceed without the oxidative reactions of the pentose phosphate pathway, and this use of glycolytic intermediates for the production of ribose-5-phosphate is also shown in Figure 18.29.

If the organism needs more NADPH than ribose-5-phosphate, the reaction series goes through the complete pathway just discussed. The oxidative reactions at the beginning of the pathway are needed to produce NADPH. The net reaction for the oxidative portion of the pathway is

$$\text{6 Glucose-6-phosphate} + 12\text{NADP}^+ + 6\text{H}_2\text{O} \rightarrow$$
$$\text{6 Ribose-5-phosphate} + 6\text{CO}_2 + 12\text{NADPH} + 12\text{H}^+$$

Biochemical Connections 18B discusses a clinical manifestation of an enzyme malfunction in the pentose phosphate pathway.

If the organism has a greater need for ribose-5-phosphate than for NADPH, then fructose-6-phosphate and glyceraldehyde-3-phosphate intermediates from glycolysis can be used in the transketolase and transaldolase pentose phosphate pathway reactions to produce ribose-5-phosphate, bypassing the oxidative portion of the pentose phosphate pathway (follow the red-shaded path in Figure 18.29 down to glyceraldehyde-3-phosphate and then go up to ribose-5-phosphate). The reactions catalyzed by transketolase and transaldolase are reversible, and this fact plays an important role in the organism's ability to adjust its metabolism to changes in conditions.

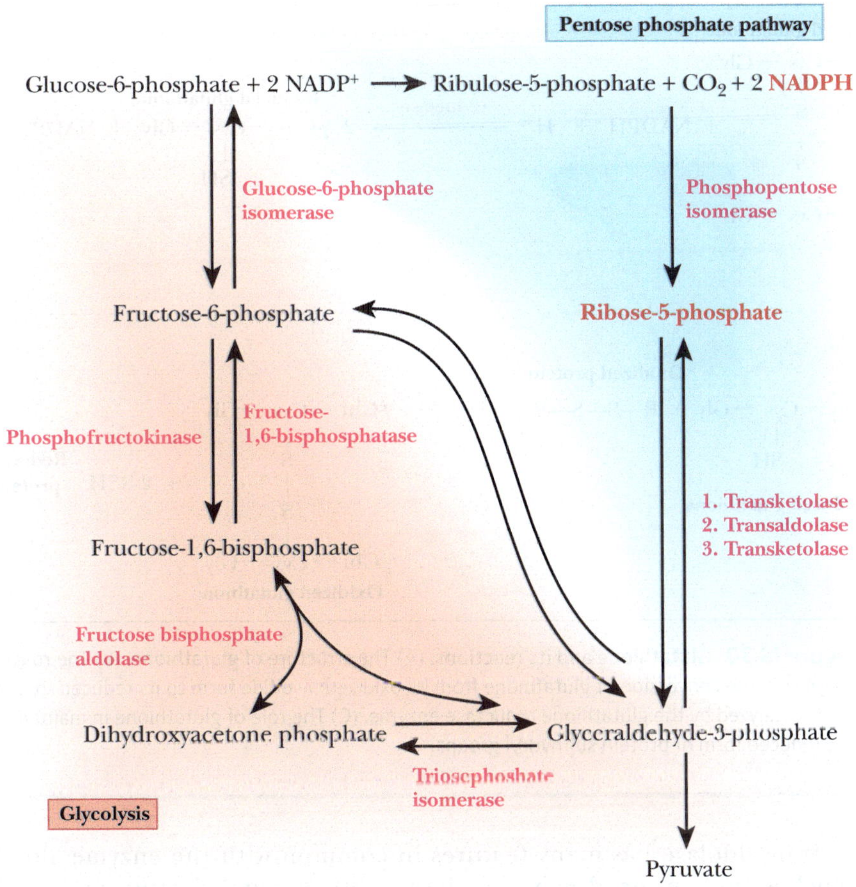

Figure 18.29 Relationships between the pentose phosphate pathway and glycolysis. If the organism needs NADPH more than ribose-5-phosphate, the entire pentose phosphate pathway is operative. If the organism needs ribose-5 phosphate more than NADPH, the nonoxidative reactions of the pentose phosphate pathway operate in reverse, using the glycolytic intermediates fructose-6-phosphate and glyceraldehyde-3-phosphate to produce ribose-5-phosphate.

18B Biochemical Connections • Allied Health

The Pentose Phosphate Pathway and Hemolytic Anemia

The pentose phosphate pathway is the only source of NADPH in red blood cells, which, as a result, are highly dependent on the proper functioning of the enzymes involved. A glucose-6-phosphate dehydrogenase deficiency leads to an NADPH deficiency, which can, in turn, lead to *hemolytic anemia* because of the wholesale destruction of red blood cells.

The relationship between NADPH deficiency and anemia is an indirect one, which involves a tripeptide called glutathione (Figure 18.30A). NADPH is required to reduce the peptide glutathione from the disulfide to the free thiol form (Figure 18.30B). Mammalian red blood cells lack mitochondria, which host many redox reactions.

Consequently, these cells are limited in the ways in which they can deal with redox balance. A substance like glutathione, which can take part in redox reactions, assumes greater importance than would be the case in cells with large numbers of mitochondria. For example, the presence of the reduced form of glutathione is necessary for the maintenance of the sulfhydryl groups of hemoglobin and other proteins in their reduced forms (Figure 18.30C), as well as for keeping the Fe(II) of hemoglobin in its reduced form.

Glutathione also maintains the integrity of red blood cells by reacting with peroxides that would otherwise degrade fatty-acid side chains in the cell membrane. About 11% of Black Americans are affected by glucose-6-phosphate dehydrogenase deficiency.

This condition, like the sickle-cell trait, leads to increased resistance to malaria, accounting for some of its persistence in the gene pool in spite of its otherwise deleterious consequences. ◗

Continued on next page

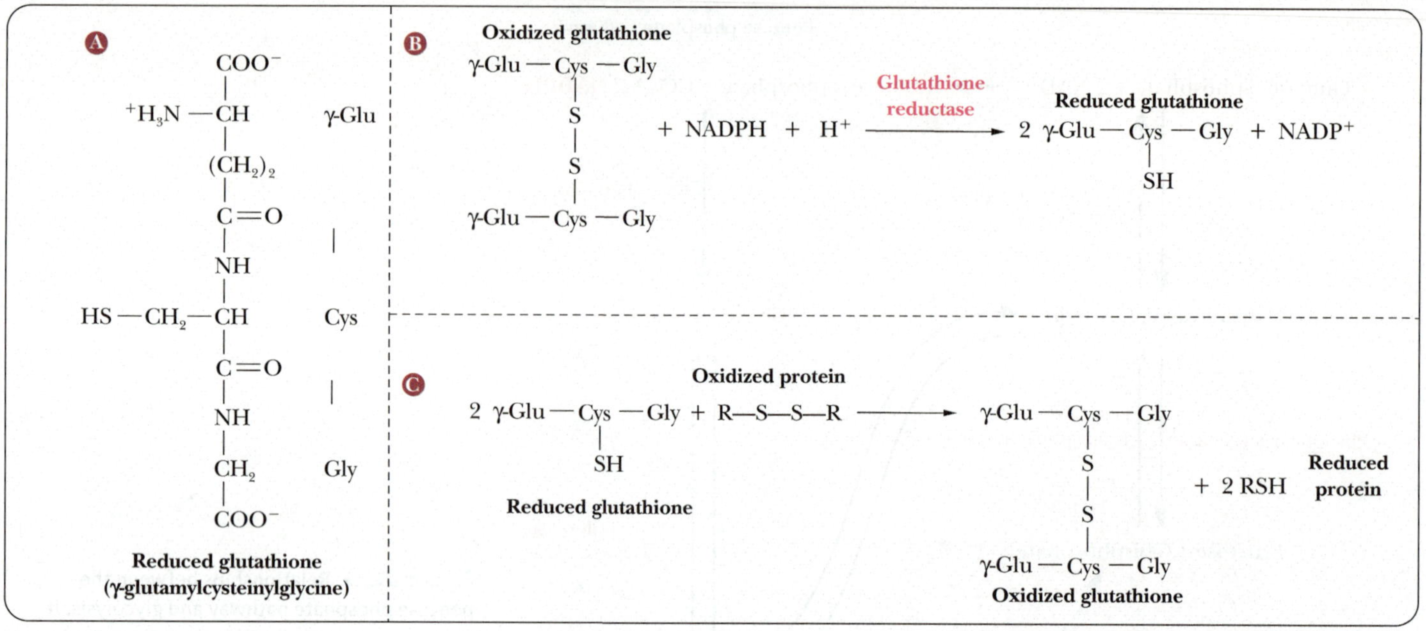

Figure 18.30 Glutathione and its reactions. (A) The structure of glutathione. (B) The role of NADPH in the conversion of glutathione from its oxidized disulfide form to its reduced free thiol form, catalyzed by the glutathione reductase enzyme. (C) The role of glutathione in maintaining the reduced form of protein sulfhydryl groups.

Transaldolase has many features in common with the enzyme aldolase, which we met in the glycolytic pathway. Both an aldol cleavage and an aldol condensation occur at different stages of the reaction. Transketolase resembles pyruvate decarboxylase, the enzyme that converts pyruvate to acetaldehyde (Section 17-4B), in that it also requires Mg^{2+} and thiamine pyrophosphate (TPP). As in the pyruvate decarboxylase reaction, a carbanion plays a crucial role in the reaction mechanism, which is similar to that of the conversion of pyruvate to acetaldehyde.

Summary

Glycogen: Optimized for Efficiency

- Glycogen can readily be broken down into glucose in response to energy needs.

Glycogen Breakdown

- Glycogen phosphorylase uses phosphate to break an $\alpha(1 \rightarrow 4)$ linkage, yielding glucose-1-phosphate and a glycogen molecule shorter by one glucose.
- Debranching enzymes aid in the degradation of glycogen around the $\alpha(1 \rightarrow 6)$ linkages.

Glycogen Production

- When an organism has an available supply of extra glucose, more than is immediately needed as a source of energy extracted in glycolysis, it forms glycogen, a polymer of glucose.
- Glycogen synthase catalyzes the reaction between a glycogen molecule and UDP-glucose to add a glucose molecule to the glycogen via an $\alpha(1 \rightarrow 4)$ linkage.
- Branching enzyme moves a section of a glycogen branch to form a new $\alpha(1 \rightarrow 6)$ branch point.

Balancing Glycogen Breakdown and Production

- Control mechanisms, such as allosteric control and covalent modification, ensure that both the formation and breakdown of glycogen are not active simultaneously, a situation that would waste energy.

The Initial Reactions of Gluconeogenesis: Pyruvate to Phosphoenolpyruvate in Two Steps

- The conversion of pyruvate (the product of glycolysis) to glucose takes place by a process called gluconeogenesis, which is not the exact reversal of glycolysis.
- Gluconeogenesis differs from glycolysis at the three irreversible steps in glycolysis. One of the irreversible steps in glycolysis is the conversion of phosphoenolpyruvate to pyruvate. The gluconeogenesis counterpart to this reaction occurs in two steps: the conversion of pyruvate to

oxaloacetate, followed by the conversion of oxaloacetate to phosphoenolpyruvate.

The Final Reactions of Gluconeogenesis: Dephosphorylations of Sugar Phosphates

- The hydrolysis of sugar phosphates by phosphatases is energetically favorable, and these steps have the effect of reversing the early energy-requiring steps of glycolysis.

Reciprocal Regulation in Glucose Metabolism

- If different enzymes catalyze opposing pathways, both pathways cannot be active at the same time, saving resources for the cell.

Hormone Control of Glycogen Metabolism

- Hormone control of glycogen metabolism can be achieved by insulin, glucagon, or epinephrine.
- The binding of insulin to receptors on the cell surface triggers the protein kinase cascade that leads to glycogen synthesis.
- The binding of glucagon and epinephrine to their receptors leads to the cascade that inhibits glycogen synthesis and activates glycogen breakdown.

Summary of Metabolic Regulation Mechanisms

- Four main categories of metabolic control are possible: allosteric control, covalent modification, substrate cycling, and genetic control.

Glycogen Metabolism in Different Body Organs

- In the same organ, glycolysis and gluconeogenesis are not highly active simultaneously. When the cell needs ATP, glycolysis is more active; when there is little need for ATP, gluconeogenesis is more active.
- The Cori cycle reveals the compartmentalization of these opposing metabolic pathways between the muscle and liver. The division of labor between liver and muscle allows glycolysis and gluconeogenesis to take place in different organs to serve the needs of an organism.

Pentose Phosphate Pathway: Oxidative Reactions

- The pentose phosphate pathway is another glucose metabolism pathway.
- In this pathway, five-carbon sugars, including ribose (a building block for nucleotides and nucleic acids), are produced from glucose.
- In the oxidative reactions of the pathway, NADPH (a reducing agent used in biosynthetic pathways) is also produced.

Pentose Phosphate Pathway: Nonoxidative Reactions

- The nonoxidative reactions of the pentose phosphate pathway produce various five-carbon sugars, particularly ribose.
- They are important when an organism has less need for NADPH but needs the sugars.

Regulation of the Pentose Phosphate Pathway

- Control of the pentose phosphate pathway allows the organism to adjust the relative levels of production of five-carbon sugars and NADPH according to its needs.
- If the organism needs NADPH more than ribose-5-phosphate, the entire pentose phosphate pathway is operative.
- If the organism needs ribose-5-phosphate more than NADPH, the nonoxidative reactions of the pentose phosphate pathway, operating in reverse, produce ribose-5-phosphate.

Review Exercises

18-1 How Glycogen Is Degraded and Produced

1. **Recall** Describe the reactions that break down glycogen into free glucose and indicate the name of the pathway.
2. **Reflect and Apply** Why is it essential that the mechanisms that activate glycogen synthesis also deactivate glycogen phosphorylase?
3. **Recall** How does phosphorolysis differ from hydrolysis?
4. **Recall** Why is it advantageous that the breakdown of glycogen gives rise to glucose-6-phosphate rather than glucose?
5. **Recall** Briefly outline the role of UDPG in glycogen biosynthesis.
6. **Recall** Name two control mechanisms that play a role in glycogen biosynthesis. Give an example of each.
7. **Recall** Describe how glycogen breakdown is regulated in the liver when blood glucose is low.
8. **Biochemical Connections** You are planning to go on a strenuous hike and are advised to eat plenty of high-carbohydrate foods, such as bread and pasta, for several days beforehand. Suggest a reason for the advice.
9. **Biochemical Connections** Would eating candy bars high in sucrose, rather than complex carbohydrates, help build up glycogen stores?
10. **Biochemical Connections** Would it be advantageous to consume a candy bar with a high refined-sugar content *immediately* before you start a strenuous hike?
11. **Reflect and Apply** What is the source of the energy needed to incorporate glucose residues into glycogen? How is it used?
12. **Reflect and Apply** Why is a primer required in glycogen synthesis?
13. **Reflect and Apply** Is the glycogen synthase reaction exergonic or endergonic? What is the reason for your answer?
14. **Reflect and Apply** How does a high cellular ATP concentration affect gluconeogenesis and glycogen synthesis in the liver? Explain your reasoning.

18-2 Gluconeogenesis Produces Glucose from Pyruvate

15. **Recall** Describe the first and last reactions of gluconeogenesis.
16. **Recall** What reactions in this chapter require acetyl-CoA or biotin?
17. **Recall** Which steps of glycolysis are irreversible? What bearing does this observation have on the reactions in which gluconeogenesis differs from glycolysis?
18. **Recall** What is the role of biotin in gluconeogenesis?
19. **Recall** How does the role of glucose-6-phosphate in gluconeogenesis differ from that in glycolysis?
20. **Reflect and Apply** Avidin, a protein found in raw egg whites, binds to biotin so strongly that biotin could become unavailable for biotin-requiring enzymes. What is the possible effect of avidin on glycogen formation? On gluconeogenesis? On the pentose

phosphate pathway? (Note: Cooking eggs denatures avidin, preventing it from binding biotin.)

21. **Recall** Which enzyme in gluconeogenesis bypasses the PFK-1 step of glycolysis, and what reaction does it catalyze?

18-3 Control of Carbohydrate Metabolism

22. **Recall** Which reaction or reactions discussed in this chapter require ATP? Which reaction or reactions produce ATP? List the enzymes that catalyze the reactions that require and produce ATP.
23. **Recall** How does fructose-2,6-bisphosphate play a role as an allosteric effector?
24. **Reflect and Apply** Describe how pyruvate carboxylase is regulated and explain why it makes sense.
25. **Recall** What is the Cori cycle?
26. **Reflect and Apply** The concentration of lactate in blood rises sharply during a sprint and declines slowly for about an hour afterward. What causes the rapid rise in lactate concentration? What causes the decline in lactate concentration after the run?
27. **Reflect and Apply** Why is it advantageous for two control mechanisms—allosteric control and covalent modification—to be involved in the metabolism of glycogen?
28. **Reflect and Apply** How do the control mechanisms in glycogen metabolism lead to amplification of response to a stimulus?
29. **Reflect and Apply** Why would you expect to see that reactions of substrate cycles involve different enzymes in different directions?
30. **Reflect and Apply** Suggest a reason or reasons why the Cori cycle takes place in the liver and in the muscle.
31. **Reflect and Apply** Explain how fructose-2,6-bisphosphate can play a role in more than one metabolic pathway.
32. **Reflect and Apply** How can the synthesis and breakdown of fructose-2,6-bisphosphate be controlled independently?
33. **Reflect and Apply** How is it advantageous for animals to convert ingested starch to glucose and then to incorporate the glucose into glycogen?
34. **Reflect and Apply** How does control of the glucose-6-phosphatase reaction differ from that of the fructose-1,6-bisphosphatase reaction?
35. **Recall** How does the action of allosteric effectors differ in the reactions catalyzed by phosphofructokinase and fructose-1,6-bisphosphatase?
36. **Recall** Name two forms of control of enzymatic action. Which of the two is more important in the control of glycogen breakdown?
37. **Recall** What role does insulin play in glycogen synthesis?
38. **Recall** What roles do glucagon and epinephrine play in glycogen breakdown?
39. **Recall** How do glucagon and epinephrine differ in chemical structure?

18-4 Glucose Is Sometimes Diverted Through the Pentose Phosphate Pathway

40. **Recall** Describe the first reaction of the pentose phosphate pathway.
41. **Recall** List three differences in structure or function between NADH and NADPH.
42. **Recall** What are four possible metabolic fates of glucose-6-phosphate?
43. **Biochemical Connections** What is the connection between the material in this chapter and hemolytic anemia?
44. **Recall** Show how the pentose phosphate pathway (PPP), which is connected to the glycolytic pathway, can do the following:
    (a) Make mostly or only NADPH
    (b) Make mostly or only pentose phosphates
45. **Recall** What is a major difference between transketolase and transaldolase?
46. **Biochemical Connections** List two ways in which glutathione functions in red blood cells.
47. **Recall** Does thiamine pyrophosphate play a role in the reactions of the pentose phosphate pathway? If so, what is that role?
48. **Reflect and Apply** Using the Lewis electron-dot notation, show explicitly the transfer of electrons in the following redox reaction.

$$\text{Glucose-6-phosphate} + \text{NADP}^+ \rightarrow$$
$$\text{6-Phosphoglucono-}\delta\text{-lactone} + \text{NADPH} + \text{H}^+$$

The lactone is a cyclic ester that is an intermediate in the production of 6-phosphogluconate.

49. **Reflect and Apply** Suggest a reason why a different reducing agent (NADPH) is used in anabolic reactions rather than NADH, which plays a role in catabolic ones.
50. **Reflect and Apply** Explain how the pentose phosphate pathway can respond to a cell's need for ATP, NADPH, and ribose-5-phosphate.
51. **Reflect and Apply** How would it affect the reactions of the pentose phosphate pathway to have only an epimerase and not both an epimerase and an isomerase to catalyze the reshuffling reactions?

Further Reading

Florkin, M., and E. H. Stotz, eds. *Comprehensive Biochemistry.* New York: Elsevier, 1969. A standard reference. Volume 17, *Carbohydrate Metabolism,* deals with glycolysis and related topics.

Horecker, B. L., in Florkin, M., and E. H. Stotz, eds. *Comprehensive Biochemistry.* New York: Elsevier, 1964. Transaldolase and Transketolase. Volume 15, *Group Transfer Reactions,* includes a review of these two enzymes and their mechanism of action.

Katz, A. A Century of Exercise Physiology: Key Concepts in Regulation of Glycogen Metabolism in Skeletal Muscle. *Eur. J. Appl. Physiol.* **122**, 1751–1772 (2022). This review discusses the mechanisms controlling glycogen metabolism in skeletal muscle, emphasizing the role of glycogen levels in regulating glycogenolysis during high-energy demand situations.

Lipmann, F. A Long Life in Times of Great Upheaval. *Ann. Rev. Biochem.* **53**, 1–33 (1984). The reminiscences of a Nobel laureate whose research contributed greatly to the understanding of carbohydrate metabolism. Very interesting reading from the standpoint of autobiography and the author's contributions to biochemistry.

Shulman, R. G., and D. L. Rothman. Enzymatic Phosphorylation of Muscle Glycogen Synthase: A Mechanism for Maintenance of Metabolic Homeostasis. *Proc. Nat. Acad. Sci.* **93**, 7491–7495 (1996). An in-depth article about metabolic flux and covalent modification of enzymes.

TeSlaa, T., M. Ralser., J. Fan., and J. D. Rabinowitz. The Pentose Phosphate Pathway in Health and Disease. *Nat. Metab.* **5**, 1275–1289 (2023). This review highlights the role of the pentose phosphate pathway in redox regulation and biosynthetic pathways, discussing its importance in various contexts, including development, differentiation, and disease states.

The citric acid cycle is the central pathway of metabolism. It plays a pivotal role in the production of energy by a cell.

The Citric Acid Cycle 19

Chapter Outline

19-1 The Central Role of the Citric Acid Cycle in Metabolism

Learning Objectives

After reading this section, students will be able to:

LO 19.1.1 Distinguish between the *citric acid cycle*, *electron transport*, and *oxidative phosphorylation*.

LO 19.1.2 Distinguish between *catabolic*, *anabolic*, and *amphibolic* as they pertain to metabolic pathways.

The evolution of aerobic metabolism, by which nutrients are oxidized to carbon dioxide and water, was an important step in the history of life on Earth. Organisms can obtain far more energy from nutrients by aerobic oxidation than by anaerobic oxidation. (Even yeast—which is usually thought of in terms of the anaerobic reactions of alcoholic fermentation and is responsible for producing bread, beer, and wine—uses the citric acid cycle and aerobically degrades glucose to carbon dioxide and water.) We saw in Chapter 17 that glycolysis produces only two molecules of adenosine triphosphate (ATP) for each molecule of glucose metabolized. First in Chapter 19 and then in Chapter 20, we explain how 30–32 molecules of ATP can be produced from each molecule of glucose in complete aerobic oxidation to carbon dioxide and water. Three processes play roles in aerobic

Chapter Outline (continued)

metabolism: the **citric acid cycle**, which we discuss here in Chapter 19, and **electron transport** and **oxidative phosphorylation**, both of which we discuss in Chapter 20 (Figure 19.1). These three processes operate together in aerobic metabolism; the separate discussion is a matter of convenience only.

Metabolism consists of catabolism, which is the oxidative breakdown of nutrients, and anabolism, which is the reductive synthesis of biomolecules. The citric acid cycle is **amphibolic**, meaning that it plays a role in both catabolism and anabolism. Although the citric acid cycle is part of the pathway of aerobic oxidation of nutrients (a catabolic pathway; see Section 19-7), some of the molecules that are included in this cycle are the starting points of biosynthetic (anabolic) pathways (see Section 19-8). Metabolic pathways operate simultaneously, even though we talk about them separately. Always keep this point in mind.

The citric acid cycle has two other common names. One is the *Krebs cycle*, after Sir Hans Krebs, who first investigated the pathway (work for which he received the Nobel Prize in Physiology or Medicine in 1953). The other name is the *tricarboxylic acid cycle* (or *TCA cycle*), from the fact that some of the molecules involved are acids with three carboxyl groups. We begin our discussion with a general overview of the pathway and then go on to discuss specific reactions.

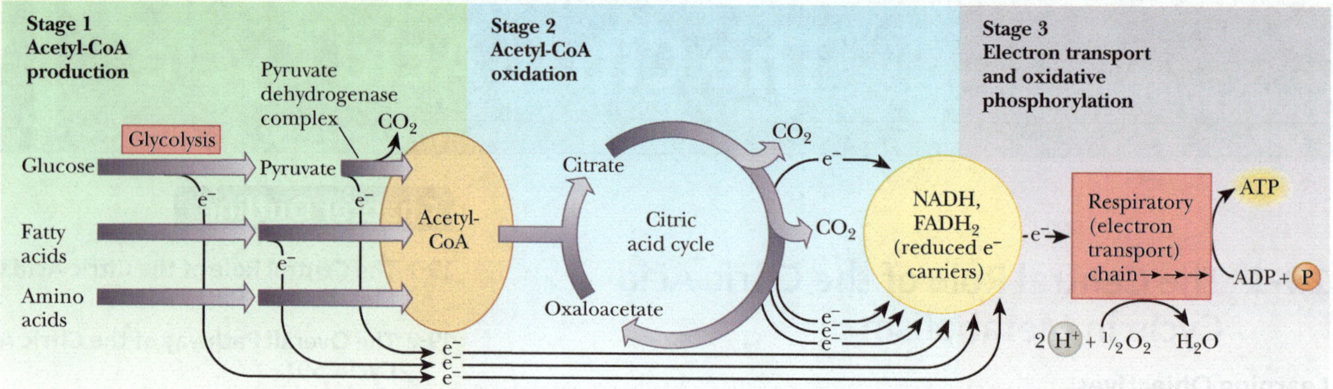

Figure 19.1 The central relationship of the citric acid cycle to catabolism. Amino acids, fatty acids, and glucose can all produce acetyl-CoA in stage 1 of catabolism. In stage 2, acetyl-CoA enters the citric acid cycle. Stages 1 and 2 produce reduced electron carriers (shown here as e^-). In stage 3, the electrons enter the electron transport chain, which then produces ATP.

Pioneers in Biochemistry • Har Gobind Khorana

Har Gobind Khorana (1922–2011) was an Indian American biochemist and Nobel laureate. Khorana was born to Hindu parents in a small, rural village in British India, where his family were among only a few literate residents. After pursuing scientific degrees in India, Khorana completed his education in England, where he began a lifelong interest in proteins and enzymes. Khorana's research took him to Switzerland, Canada, and finally the United States, where he became a naturalized citizen in the 1960s.

Khorana is best known for his research on the genetic code and nucleosynthesis.

In 1968, he was awarded the Nobel Prize in Physiology or Medicine jointly with Robert W. Holley and Marshall W. Nirenberg for research that showed how the order of nucleotides in nucleic acids control the cell's synthesis of proteins. He also demonstrated that nucleotides are always transmitted in groups of three, including specific groups that signal a start or stop to protein production.

A pioneer in his field, Khorana was among the first to pursue gene manipulation. He developed techniques to chemically synthesize oligonucleotides in the 1970s and was the first to completely synthesize a functional gene outside of a living organism. His gene synthesis techniques were foundational to the field of genetic engineering.

19-2 The Overall Pathway of the Citric Acid Cycle

Learning Objectives

After reading this section, students will be able to:

LO 19.2.1 Explain the compartmentalization of the citric acid cycle within mitochondria.

LO 19.2.2 Identify the starting material of the citric acid cycle and its intermediates, and account for the reducing agent and the CO_2 production in the pathway.

An important difference between glycolysis and the citric acid cycle is the part of the cell in which these pathways occur. In eukaryotes, glycolysis occurs in the cytosol, while the citric acid cycle takes place in mitochondria. Most of the enzymes of the citric acid cycle are present in the mitochondrial matrix.

19-2A Mitochondrial Structure

A quick review of some aspects of mitochondrial structure is in order here because we want to describe the exact location of each of the components of the citric acid cycle and the electron transport chain. Recall from Section 1-4B that a mitochondrion has an inner and an outer membrane (Figure 19.2). The region enclosed by the inner membrane is called the **mitochondrial matrix**, and an **intermembrane space** exists between the inner and outer membranes. The inner membrane is a tight barrier between the matrix and the cytosol, and very few compounds can cross this barrier without a specific transport protein (Section 8-5A). The reactions of the citric acid cycle take

citric acid cycle a central metabolic pathway; part of aerobic metabolism

electron transport (to oxygen) a series of oxidation–reduction reactions by which the electrons derived from the oxidation of nutrients are passed to oxygen

oxidative phosphorylation a process for generating ATP; it depends on the creation of a pH gradient within the mitochondrion as a result of electron transport

amphibolic able to be a part of both anabolic and catabolic pathways

mitochondrial matrix the part of the mitochondrion enclosed by the inner mitochondrial membrane

intermembrane space the region between the inner and outer mitochondrial membranes

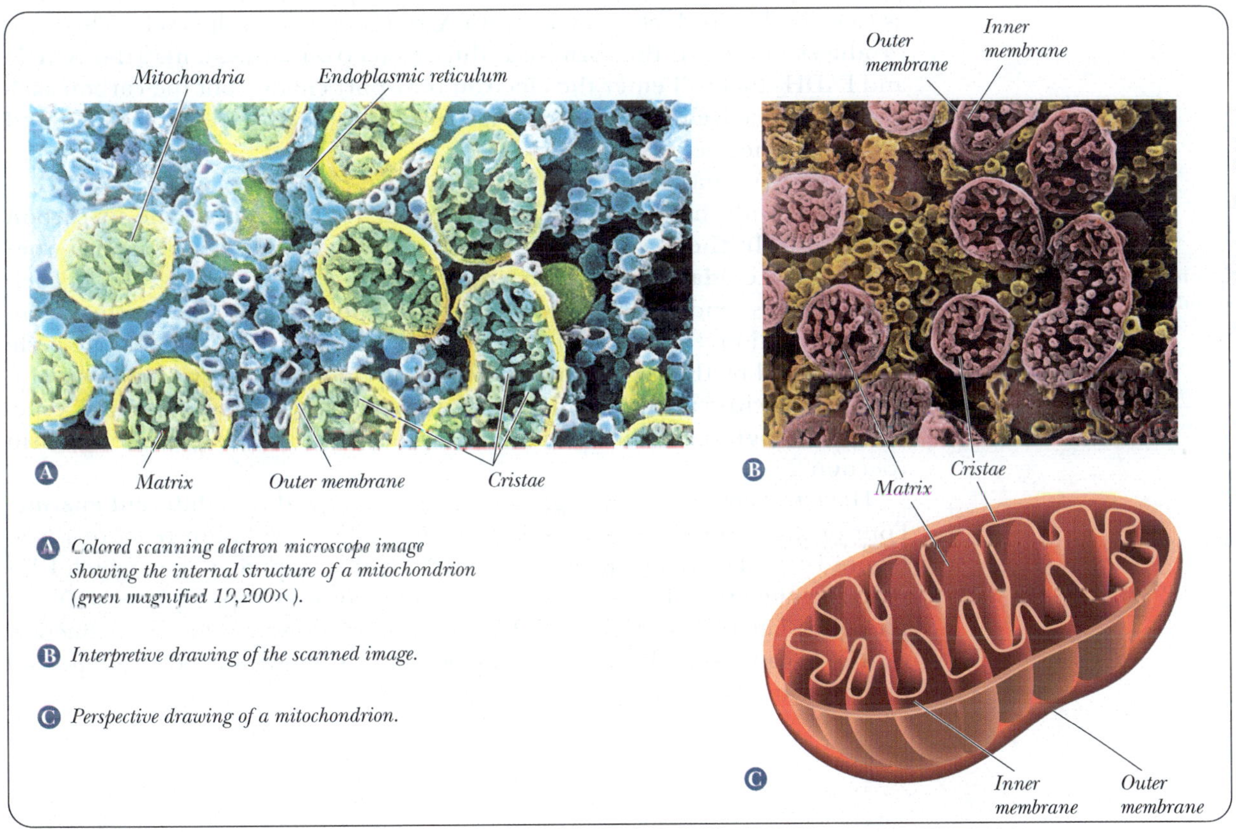

(A) *Colored scanning electron microscope image showing the internal structure of a mitochondrion (green magnified 19,200×).*

(B) *Interpretive drawing of the scanned image.*

(C) *Perspective drawing of a mitochondrion.*

(A) Professors P. M. Motta, S. Makabe and T. Naguro/Science Photo Library/Science Source; (B) Professor Pietro M. Motta/Science Source; (C) jack0m/iStock Vectors/Getty Images

Figure 19.2 The structure of a mitochondrion. (For an electron micrograph of mitochondrial structure, see Figure 1.19.)

place in the matrix, except for Step 6 of the cycle, in which the intermediate electron acceptor is flavin adenine dinucleotide (FAD). The enzyme that catalyzes the FAD-linked reaction is an integral part of the inner mitochondrial membrane and is linked directly to the electron transport chain (Section 20-3).

19-2B Overview of the Reactions in the Citric Acid Cycle

The citric acid cycle is shown in schematic form in Figure 19.3. Under aerobic conditions, pyruvate produced by glycolysis is oxidized further. Carbon dioxide and water are the final products after pyruvate and other metabolic intermediates go through the citric acid cycle, electron transport, and oxidative phosphorylation.

- First, the pyruvate is oxidized to one carbon dioxide molecule and to one acetyl group, which becomes linked to an intermediate, coenzyme A (CoA) (Section 15-7).
- Next, the acetyl-CoA enters the citric acid cycle. In the citric acid cycle, two more molecules of carbon dioxide are produced for each molecule of acetyl-CoA that enters the cycle, and electrons are transferred in the process. The immediate electron acceptor in all cases but one is NAD^+, which is reduced to NADH. In that one case, FAD is the intermediate electron acceptor. It is derived from riboflavin (vitamin B_2), and it takes up two electrons and two hydrogen ions to produce $FADH_2$.
- Finally, the electrons are passed from NADH and $FADH_2$ through several stages of an electron transport chain, with a different redox reaction at each step. The final electron acceptor is oxygen, with water as the product.

Starting from pyruvate, a three-carbon compound, three carbons are lost as CO_2 via the production of acetyl-CoA and one turn of the cycle. The cycle produces energy in the form of reduced electron equivalents (the NADH and $FADH_2$ that will enter the electron transport chain), but the carbon skeletons are effectively lost. The cycle also produces one high-energy compound directly, guanosine triphosphate (GTP).

In the first reaction of the cycle, the two-carbon group of acetyl-CoA condenses with the four-carbon oxaloacetate ion to produce the six-carbon citrate ion. In the next few steps, the citrate isomerizes, and then it goes through two **oxidative decarboxylation** reactions, where the reactant in each reaction gets oxidized and loses carbon dioxide (CO_2). Then, after a synthetase reaction, the four-carbon succinate molecule is produced. The cycle is completed by the regeneration of oxaloacetate from succinate in several steps. Many citric acid cycle intermediates take part in other pathways, especially α-ketoglutarate, which is very important in amino acid metabolism (Section 23-4).

oxidative decarboxylation a reaction in which a molecule gets oxidized and loses carbon dioxide

The citric acid cycle has eight steps, each catalyzed by a different enzyme. Four of the eight steps—Steps 3, 4, 6, and 8—are oxidation reactions (see Figure 19.3). The oxidizing agent is NAD^+ in all except Step 6, in which FAD plays the same role. In Step 5, a molecule of guanosine diphosphate (GDP) is phosphorylated to produce GTP. This reaction is equivalent to the production of ATP because the phosphate group is then easily transferred to adenosine diphosphate (ADP), producing GDP and ATP.

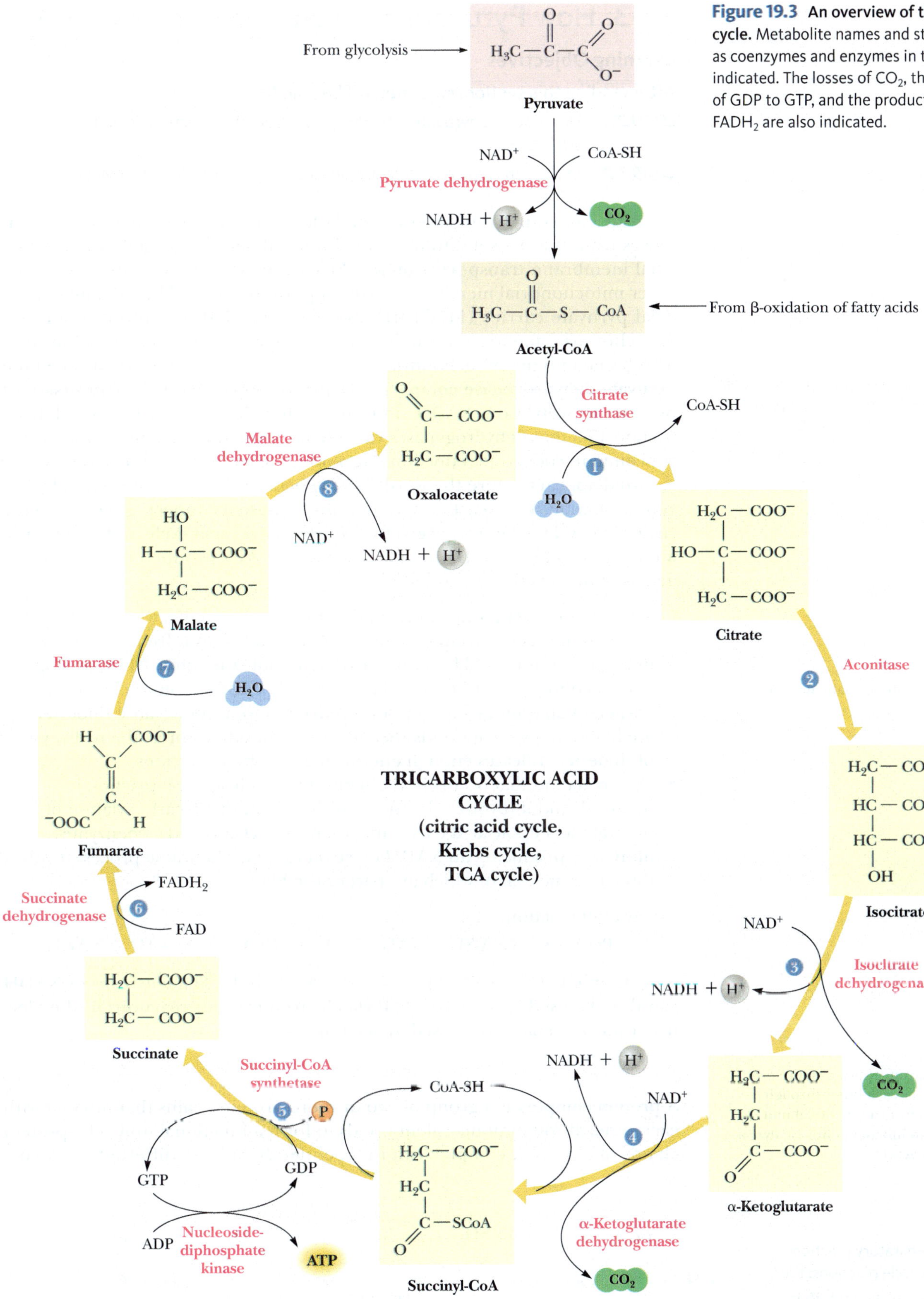

Figure 19.3 **An overview of the citric acid cycle.** Metabolite names and structures as well as coenzymes and enzymes in the pathway are indicated. The losses of CO_2, the phosphorylation of GDP to GTP, and the production of NADH and $FADH_2$ are also indicated.

19-3 How Pyruvate Is Converted to Acetyl-CoA

Learning Objectives

After reading this section, students will be able to:

LO 19.3.1 Describe the pyruvate dehydrogenase complex reaction, including structures.

LO 19.3.2 Explain the pyruvate dehydrogenase complex reaction mechanism.

pyruvate dehydrogenase complex a multienzyme complex that catalyzes the conversion of pyruvate to acetyl-CoA, NADH, and carbon dioxide

Pyruvate can come from several metabolic pathways, including glycolysis. It moves from the cytosol into the mitochondrial matrix via specific mitochondrial membrane transport proteins. From the cytosol, pyruvate crosses the outer mitochondrial membrane through porin proteins. Then the mitochondrial pyruvate carrier (MPC) transport protein that is found in the inner mitochondrial membrane carries pyruvate across the inner mitochondrial membrane into the mitochondrial matrix. There, an enzyme system called the **pyruvate dehydrogenase complex** (PDC) is responsible for the conversion of pyruvate to acetyl-CoA, NADH, and carbon dioxide (Figure 19.4). Recall from Section 17-3 that dehydrogenases catalyze the oxidation of a metabolite with a concomitant coenzyme reduction. Moreover, the name of the metabolite being oxidized appears before the word *dehydrogenase* in the name of each dehydrogenase enzyme. This reaction is important because it connects glycolysis to the citric acid cycle, produces acetyl-CoA for the citric acid cycle and other pathways, and produces NADH that contributes to ATP production. A few more details of the reaction are as follows:

thioester a sulfur-containing analogue of an ester

- There is an —SH group at one end of the CoA molecule, which is the point at which the acetyl group is attached. As a result, CoA is frequently shown in equations as CoA-SH. Because CoA is a thiol (the sulfur [thio] analog of an alcohol), acetyl-CoA is a **thioester**, with a sulfur atom replacing an oxygen of an ester group. This difference is important because thioesters are high-energy compounds (Section 15-7). In other words, the hydrolysis of thioesters releases enough energy to drive other reactions.
- The oxidation that occurs in the pyruvate dehydrogenase reaction is nuanced, and it involves electron transfers from the decarboxylation of pyruvate to a lipoic acid coenzyme, then to FAD and NAD^+ coenzymes, ultimately producing an NADH electron carrier. The whole process involves three enzymes, all of which are part of the PDC.

The overall reaction,

$$\text{Pyruvate} + \text{CoA-SH} + \text{NAD}^+ \rightarrow \text{Acetyl-CoA} + \text{CO}_2 + \text{H}^+ + \text{NADH}$$

is exergonic ($\Delta G^{\circ\prime} = -33.4$ kJ/mol $= -8.0$ kcal/mol). The electrons of NADH can then be used to generate ATP via electron transfer reactions in the electron transport chain (discussed in Section 20-3).

▶ How many enzymes are needed to convert pyruvate to acetyl-CoA?

protein complex a group of two or more protein chains that interact with each other through noncovalent bonds to form a functional unit; each protein chain can independently catalyze a different enzymatic reaction

A **protein complex** is a group of two or more protein chains that interact with each other through noncovalent bonds to form a functional unit. The protein chains in a protein complex can independently catalyze different enzymatic

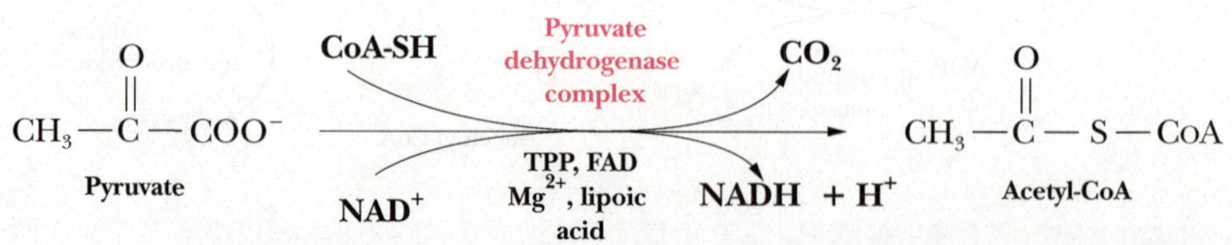

Figure 19.4 The preparatory reaction before the citric acid cycle reactions: the conversion of pyruvate to acetyl-CoA is catalyzed by the pyruvate dehydrogenase complex.

reactions. This makes a protein complex different from a multimeric protein such as ATCase (Section 7-1), which catalyzes one enzymatic reaction. Five enzymes make up the pyruvate dehydrogenase complex in mammals. They are *pyruvate dehydrogenase (PDH), dihydrolipoyl transacetylase, dihydrolipoyl dehydrogenase, pyruvate dehydrogenase kinase,* and *pyruvate dehydrogenase phosphatase.* The first three are involved in the conversion of pyruvate to acetyl-CoA. The kinase and the phosphatase are enzymes used in the control of PDH (Section 19-5) and are present on a single polypeptide. The reaction takes place in five steps. Two enzymes catalyze reactions of *lipoic acid,* a compound that has a disulfide group in its oxidized form and two sulfhydryl groups in its reduced form.

Lipoic acid can act as an oxidizing agent; the reaction involves hydrogen transfer, which frequently accompanies biological oxidation–reduction reactions (Section 15-5). Another reaction of lipoic acid is the formation of a thioester linkage with the acetyl group before it is transferred to acetyl-CoA. Lipoic acid can act simply as an oxidizing agent, or it can simultaneously take part in two reactions—a redox reaction and the shift of an acetyl group by transesterification.

The first stage in the reaction sequence that converts pyruvate to carbon dioxide and acetyl-CoA is catalyzed by *pyruvate dehydrogenase* (Figure 19.5, numeral 1). This enzyme of the complex is also referred to as *E1,* and it requires thiamine pyrophosphate (TPP; a metabolite of vitamin B_1, or thiamine) as a coenzyme. The coenzyme is not covalently bonded to the enzyme; instead, it is held together by noncovalent interactions. Mg^{2+} is also required. We saw the action of TPP as a coenzyme in the conversion of pyruvate to acetaldehyde, catalyzed by pyruvate decarboxylase (Section 17-4B). In the pyruvate

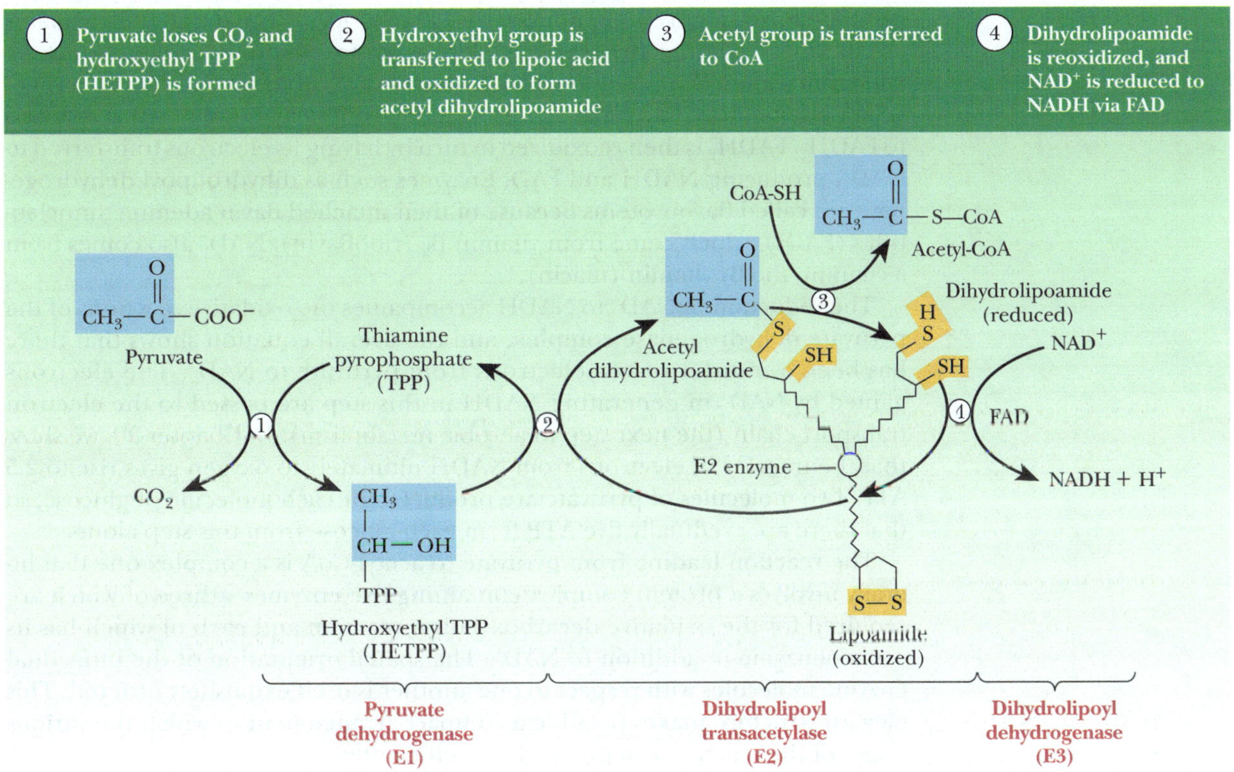

Figure 19.5 The mechanism of the pyruvate dehydrogenase complex reaction. Three of the five pyruvate dehydrogenase complex enzymes—pyruvate dehydrogenase (E1), dihydrolipoyl transacetylase (E2), and dihydrolipoyl dehydrogenase (E3)—are involved in the conversion of pyruvate to acetyl-CoA.

dehydrogenase reaction, an α-keto acid, pyruvate, loses carbon dioxide; the remaining two-carbon unit becomes covalently bonded to TPP.

The second stage of the reaction is catalyzed by *dihydrolipoyl transacetylase* (Figure 19.5, numeral 2). This enzyme is also referred to as *E2*, and it requires lipoic acid as a coenzyme. Lipoic acid is a vitamin-like cofactor consisting of an eight-carbon fatty acid (recall general fatty acid structure details from Section 8-2A) that has a five-membered disulfide ring at carbons 6 and 8 (Figure 19.5, *bottom right*). Lipoic acid is covalently bonded to the enzyme by an amide bond between the carboxyl group of lipoic acid and the ε-amino group of a lysine side chain on the dihydrolipoyl transacetylase enzyme. Once this connection occurs, the lipoic acid is then referred to as a lipoamide group. The two-carbon unit that originally came from pyruvate is transferred from TPP to one of the sulfur atoms of the lipoamide group, and in the process of breaking the disulfide ring, the hydroxyl group of the two-carbon unit is oxidized to produce an acetyl group. The disulfide group of lipoic acid is the oxidizing agent, which is itself reduced, and the product of the reaction is a thioester. In other words, the acetyl group is now covalently bonded to lipoic acid by a thioester linkage.

The third stage of the reaction is also catalyzed by dihydrolipoyl transacetylase (Figure 19.5, numeral 3). A molecule of coenzyme A (CoA-SH) attacks the thioester linkage between the acetyl group and the lipoamide group, and the acetyl group is transferred to a thioester linkage with CoA. The reaction of pyruvate and CoA-SH has now reached the stage of the products, carbon dioxide and acetyl-CoA, but the lipoamide group is in a reduced form.

The rest of the steps regenerate the oxidized form of the lipoamide group, so further reactions can be catalyzed by a regenerated dihydrolipoyl transacetylase enzyme. In the fourth stage of the overall reaction, the enzyme *dihydrolipoyl dehydrogenase* reoxidizes the reduced lipoamide from the sulfhydryl to the disulfide form (Figure 19.5, numeral 4). The lipoic acid coenzyme still remains covalently bonded to the transacetylase enzyme. The dehydrogenase uses a coenzyme called FAD (Section 15-5), which is bound to the enzyme by noncovalent interactions, to oxidize the lipoamide group. As a result, FAD is reduced to $FADH_2$. $FADH_2$ is then reoxidized in turn by having its electrons transferred to NAD^+, producing NADH and FAD. Enzymes such as dihydrolipoyl dehydrogenase are called flavoproteins because of their attached flavin adenine dinucleotides (FADs), which come from vitamin B_2 (riboflavin). NAD^+ also comes from a vitamin, the B_3 vitamin (niacin).

The reduction of NAD^+ to NADH accompanies the oxidative reactions of the pyruvate dehydrogenase complex, and the overall equation shows that there has been a transfer of two electrons from pyruvate to NAD^+. The electrons gained by NAD^+ in generating NADH in this step are passed to the electron transport chain (the next step in aerobic metabolism). In Chapter 20, we show that the transfer of electrons from NADH ultimately to oxygen gives rise to 2.5 ATP. Two molecules of pyruvate are produced for each molecule of glucose, so that there are eventually five ATP from each glucose from this step alone.

The reaction leading from pyruvate to acetyl-CoA is a complex one that literally involves a protein complex containing five enzymes—three of which are required for the oxidative decarboxylation reaction and each of which has its own coenzyme in addition to NAD^+. The spatial orientation of the individual enzyme molecules with respect to one another is itself exquisitely ordered. This elegant structure makes possible a compact arrangement in which the various stages of the reaction can proceed very efficiently.

A compact arrangement, such as the one in the pyruvate dehydrogenase multienzyme complex, has two great advantages over an arrangement in which the various components are more widely dispersed. First, the various stages of the reaction can take place more efficiently because the reactants, the enzymes, and their coenzymes are so close to each other. The role of lipoic acid is particularly important here. Recall that lipoic acid is covalently attached to the transacetylase enzyme that occupies a central position in the complex. The lipoic acid and the lysine side chain to which it is bonded are long enough to act as a "swinging arm," which can move to the site of each of the steps of the reaction (Figure 19.5, numerals 2, 3, and 4). As a result of the swinging-arm action, the lipoic acid can move to the pyruvate dehydrogenase site to accept the two-carbon unit and then transfer it to the active site of the transacetylase. The acetyl group can then be transesterified to CoA-SH from the lipoic acid. Finally, the lipoic acid can swing to the active site of the dehydrogenase so that the sulfhydryl groups can be reoxidized to a disulfide.

The second advantage of a multienzyme complex is that regulatory controls can be applied more efficiently in such a system than in a single enzyme molecule. In the case of the pyruvate dehydrogenase complex, the controlling factors are intimately associated with the multienzyme complex itself, which are the other two enzymes of the five-enzyme complex (discussed in Section 19-5B).

19-4 The Individual Reactions of the Citric Acid Cycle

Learning Objectives

After reading this section, students will be able to:

LO 19.4.1 Describe the eight individual reactions of the citric acid cycle, including structures of the reactants and products and the names of the enzymes that catalyze the reactions.

LO 19.4.2 Distinguish between *synthases* and *synthetases*.

The reactions of the citric acid cycle proper and the enzymes that catalyze them are listed in Table 19.1 (they are also shown in Figure 19.3). We now discuss each of these reactions in turn.

Step 1. ***Formation of Citrate*** The first step of the citric acid cycle is the reaction of acetyl-CoA and oxaloacetate to form citrate and coenzyme A (CoA-SH) (Figure 19.6). This reaction is called a condensation because two molecules are joined together with the release of a small byproduct—CoA-SH in

Table 19.1 The Reactions of the Citric Acid Cycle

Step	Reaction	Enzyme
1	Acetyl-CoA + Oxaloacetate + H_2O → Citrate + CoA-SH	Citrate synthase
2	Citrate → Isocitrate	Aconitase
3	Isocitrate + NAD^+ → α-Ketoglutarate + NADH + CO_2 + H^+	Isocitrate dehydrogenase
4	α-Ketoglutarate + NAD^+ + CoA-SH → Succinyl-CoA + NADH + CO_2 + H^+	α-Ketoglutarate dehydrogenase
5	Succinyl-CoA + GDP + P_i → Succinate + GTP + CoA-SH	Succinyl-CoA synthetase
6	Succinate + FAD → Fumarate + $FADH_2$	Succinate dehydrogenase
7	Fumarate + H_2O → L-Malate	Fumarate hydratase
8	L-Malate + NAD^+ → Oxaloacetate + NADH + H^+	Malate dehydrogenase

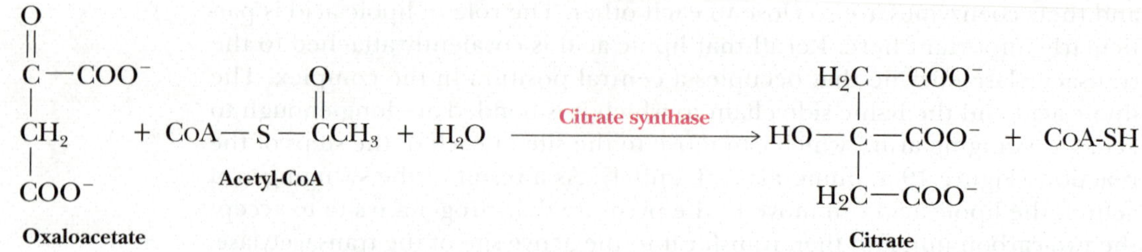

Figure 19.6 Step 1 of the citric acid cycle: the conversion of oxaloacetate and acetyl-CoA to citrate is catalyzed by the citrate synthase enzyme.

this case. The reaction starts with the joining of acetyl-CoA and oxaloacetate to form citryl-CoA, followed by the hydrolysis of citryl-CoA to give citrate and CoA-SH.

citrate synthase the enzyme that catalyzes the conversion of oxaloacetate and acetyl-CoA to citrate in Step 1 of the citric acid cycle

synthase an enzyme that makes a new covalent bond between two molecules; it does not require the energy of ATP hydrolysis

aconitase the enzyme that catalyzes the isomerization of citrate to isocitrate in Step 2 of the citric acid cycle

The reaction is catalyzed by the enzyme **citrate synthase**. In general, a **synthase** is an enzyme that makes a new covalent bond between two molecules, but it does not require the energy of ATP hydrolysis. It is an exergonic reaction ($\Delta G^{\circ\prime} = -32.8$ kJ/mol $= -7.8$ kcal/mol) because the hydrolysis of a thioester releases energy. Thioesters are considered to be high-energy compounds.

Step 2. ***Isomerization of Citrate to Isocitrate*** The second reaction of the citric acid cycle, the one catalyzed by **aconitase**, is the isomerization of citrate to isocitrate (Figure 19.7). The enzyme requires Fe^{2+}. One of the most interesting features of the reaction is that citrate, a symmetrical (achiral) compound, is converted to isocitrate, a chiral compound—a molecule that cannot be superimposed on its mirror image. **Biochemical Connections 19A** discusses some reactions of aconitase inhibitors.

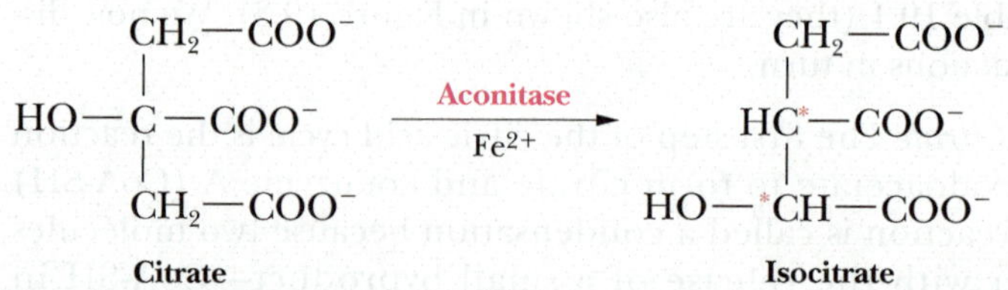

Figure 19.7 **Step 2 of the citric acid cycle: the isomerization of citrate to isocitrate is catalyzed by the aconitase enzyme.** The asterisks indicate the two chiral carbons in isocitrate.

19A Biochemical Connections • Toxicology

Fluorine Compounds and Carbohydrate Metabolism

Both ionic and covalent compounds of fluorine play roles in the metabolism of carbohydrates. In ionic form, the fluoride ion is an inhibitor of the glycolytic enzymes phosphoglyceromutase, enolase, and pyruvate kinase. Two compounds with covalent bonds, fluoroacetate and fluorocitrate, play a role in the inhibition of the citric acid cycle. Of these two, fluoroacetate is noteworthy for its role in plant development; it is a precursor of *fluoroacetyl-CoA*, which is an inhibitor of citrate synthase.

The source of the fluoroacetyl-CoA is fluoroacetate, which is found in the leaves of various types of poisonous plants, including the locoweed plant (Figure 19.8). Animals that ingest these plants form fluoroacetyl-CoA, which, in turn, is converted to fluorocitrate by their citrate synthase (Figure 19.9). Fluorocitrate, in turn, is a potent inhibitor of *aconitase*, the enzyme that catalyzes the next

spline_x/Shutterstock.com

Figure 19.8 The locoweed poison plant.

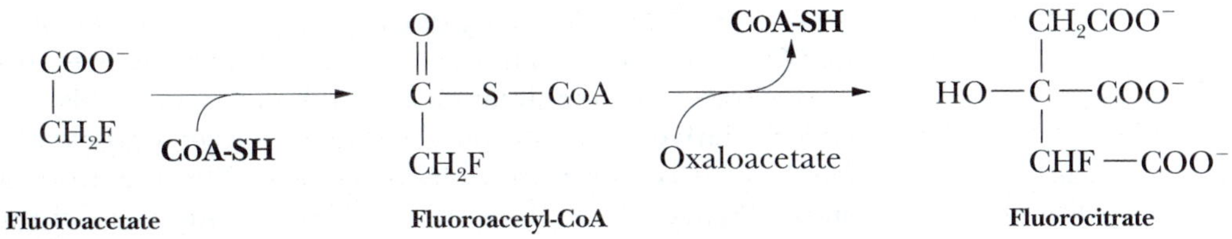

Figure 19.9 The formation of fluorocitrate from fluoroacetate. Fluoroacetate is a potent inhibitor of the aconitase enzyme in the citric acid cycle.

reaction of the citric acid cycle. These plants are poisonous because they produce a potent inhibitor of life processes.

The poison called Compound 1080 (pronounced "ten-eighty") is sodium fluoroacetate. Ranchers who want to protect their sheep from attacks by coyotes put the poison just outside the ranch fence. When the coyotes eat this poison, they die. New Zealand is a remote and isolated ecosystem that endured tens of millions of years in the absence of predatory mammals. Human explorations to the islands dating back to the tenth century have led to the current thriving predatory mammal population of bushtail possums, stoats, and rats, which have nearly decimated native wildlife, including reptiles and birds like the kokako, mohua, and kiwi. In addition, nonpredatory rabbits have raised concern due to overpopulation and their ability to carry diseases that can decimate native populations. New Zealand is the world's largest user of Compound 1080 as a controversial means of pest control. The use of Compound 1080 in New Zealand is so widespread due to the absence of native mammals; other parts of the world that allow its use are restricted due to concern for the impact on native mammal populations. The mechanism of poisoning by Compound 1080 is the same as that by plant poisons. ◗

Isocitrate has two chiral carbon atoms, so it has four possible isomers. Only one of the four, however, is produced by this reaction. (We shall not discuss the nomenclature of the isomers of isocitrate here. See the end-of-chapter question about the other isomers.) Aconitase, the enzyme that catalyzes the conversion of citrate to isocitrate, can select one end of the citrate molecule in preference to the other. This type of behavior means that the enzyme can bind a symmetrical substrate in an asymmetrical binding site. In Section 7-6, we mentioned that this possibility exists, and here we have an example of it. The enzyme forms an asymmetrical three-point attachment to the citrate molecule (Figure 19.10). The reaction proceeds by removing a water molecule from the citrate to produce *cis*-aconitate. Water is then added back to the *cis*-aconitate to give isocitrate (Figure 19.11).

The intermediate, *cis*-aconitate, remains bound to the enzyme during the course of the reaction. There is some evidence that the citrate is complexed to the Fe(II) in the active site of the enzyme in such a way that the citrate curls back on itself in a nearly circular conformation. As a result, some biochemists refer to this situation as the "ferrous wheel."

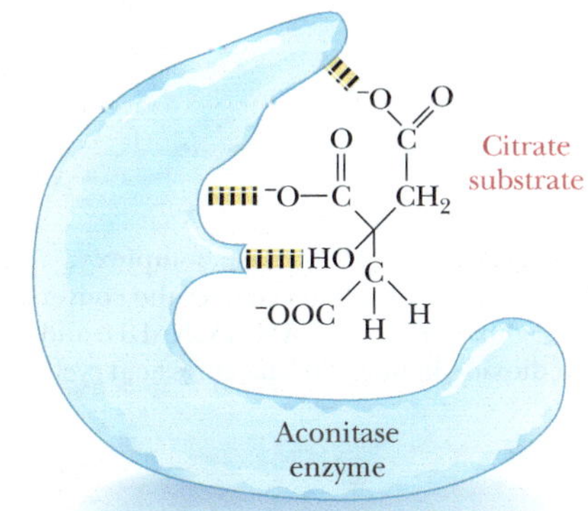

Figure 19.10 The asymmetrical three-point attachment of the citrate substrate to the aconitase enzyme. The three-point attachment makes the two $—CH_2— COO^-$ ends of citrate stereochemically nonequivalent.

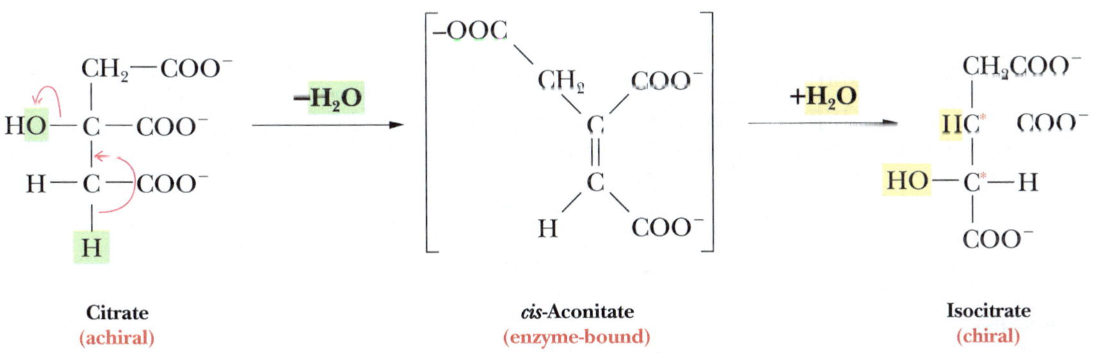

Figure 19.11 The aconitase enzyme mechanism showing steps in the conversion of citrate to isocitrate. The asterisks indicate the two chiral carbon atoms in isocitrate.

isocitrate dehydrogenase the enzyme that catalyzes the conversion of isocitrate to α-ketoglutarate in Step 3 of the citric acid cycle; this is the first of two oxidative decarboxylations in the cycle that produce NADH and CO_2

Step 3. ***Oxidative Decarboxylation of Isocitrate to Form α-Ketoglutarate, NADH, and CO_2*** The third step in the citric acid cycle is the oxidative decarboxylation of isocitrate to α-ketoglutarate and carbon dioxide. This reaction is the first of two oxidative decarboxylations in the citric acid cycle; the enzyme that catalyzes it is **isocitrate dehydrogenase**. The reaction takes place in two steps (Figure 19.12). First, the hydroxyl group of isocitrate is oxidized to a carbonyl group in an oxalosuccinate intermediate, which remains bound to the enzyme. Then, oxalosuccinate is decarboxylated, and the carbon dioxide and α-ketoglutarate are released.

This is the first reaction of the citric acid cycle proper in which NADH is produced. One molecule of NADH is produced from NAD^+ at this stage by the loss of two electrons in the oxidation. As we saw in our discussion of the pyruvate dehydrogenase complex (PDC), each NADH produced leads to the production of 2.5 ATP in later stages of aerobic metabolism. Recall also that for each original molecule of glucose, there will be two NADH produced in this step, equivalent to five ATP.

Step 4. ***Oxidative Decarboxylation of α-Ketoglutarate to Form Succinyl-CoA, NADH, and CO_2*** A second citric acid cycle oxidative decarboxylation takes place in Step 4, in which carbon dioxide and succinyl-CoA are formed from α-ketoglutarate and CoA (Figure 19.13).

This reaction is similar to the one in which acetyl-CoA is formed from pyruvate, with NADH produced from NAD^+. Once again, two molecules of NADH and eventually five ATP are produced from this step for each original molecule of glucose.

α-ketoglutarate dehydrogenase complex a multienzyme complex that catalyzes the conversion of α-ketoglutarate to succinyl-CoA, NADH, and carbon dioxide in Step 4 of the citric acid cycle

The reaction occurs in several stages and is catalyzed by an enzyme system called the **α-ketoglutarate dehydrogenase complex**, which is very similar to the pyruvate dehydrogenase complex (PDC). This multienzyme system consists of three enzymes that catalyze the overall reaction. The reaction takes place in several steps, and there is again a requirement for thiamine pyrophosphate

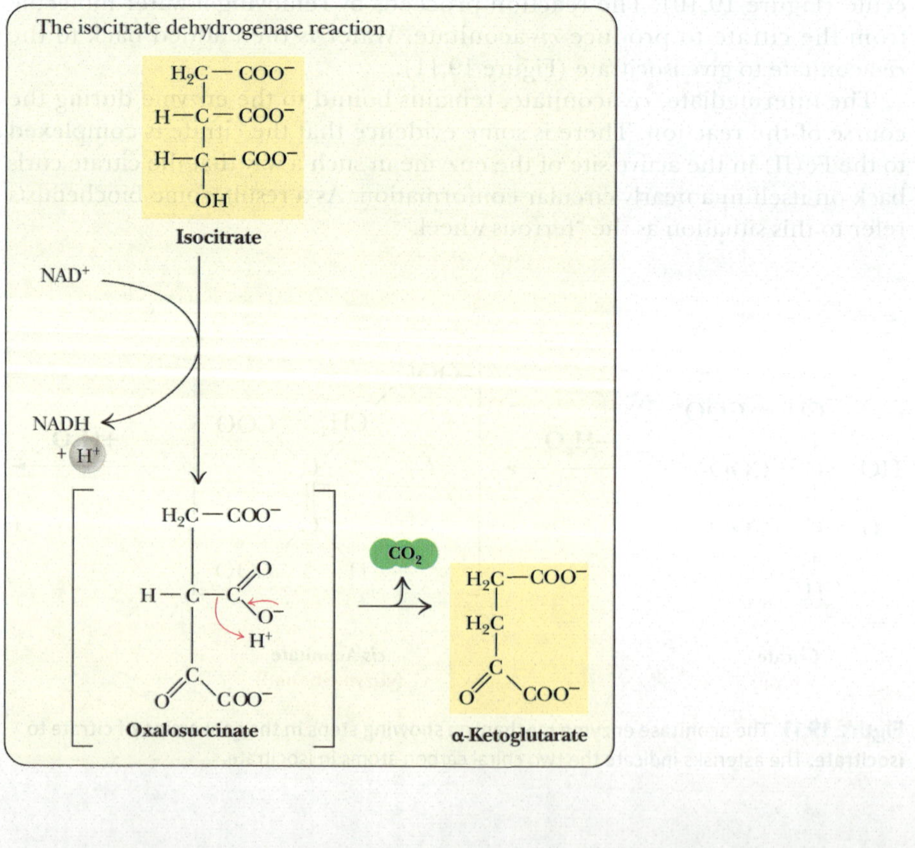

Figure 19.12 Step 3 of the citric acid cycle: the oxidative decarboxylation of isocitrate to form α-ketoglutarate, NADH, and CO_2 is catalyzed by the isocitrate dehydrogenase enzyme.

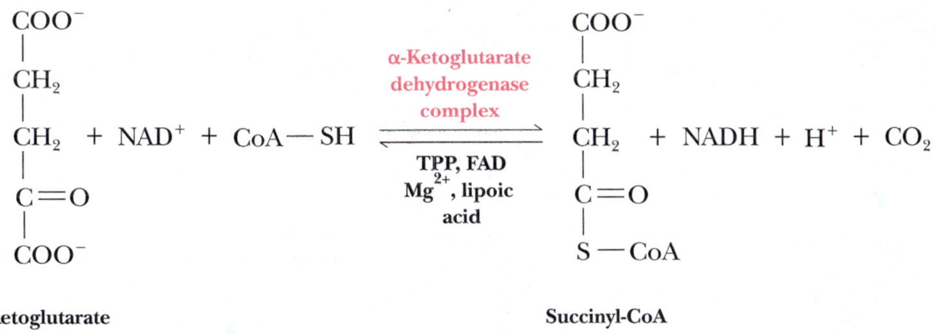

Figure 19.13 Step 4 of the citric acid cycle: the oxidative decarboxylation of α-ketoglutarate to form succinyl-CoA, NADH, and CO_2 is catalyzed by the α-ketoglutarate dehydrogenase enzyme.

(TPP), FAD, lipoic acid, and Mg^{2+} (recall the roles these coenzymes play in the PDC mechanism shown in Figure 19.5). This reaction is highly exergonic ($\Delta G°' = -33.4$ kJ/mol $= -8.0$ kcal/mol), as is the one catalyzed by pyruvate dehydrogenase.

At this point, two molecules of CO_2 have been produced by the oxidative decarboxylations of the citric acid cycle. Removal of the CO_2 makes the citric acid cycle irreversible in vivo, although in vitro each separate reaction is reversible. Do the two molecules of CO_2 come from the two carbon atoms of acetyl-CoA or from somewhere else? Labeling studies discussed in **Biochemical Connections 19B** answer this question in some detail.

The two CO_2 molecules arise from carbon atoms that were part of the oxaloacetate with which the acetyl group condensed. The carbons of this acetyl group are incorporated into the oxaloacetate that will be regenerated for the next round of the cycle. The release of the CO_2 molecules has a profound influence on mammalian physiology. The α-ketoglutarate dehydrogenase complex reaction is the third in which we have encountered an enzyme that requires TPP.

Step 5. ***Conversion of Succinyl-CoA to Succinate*** In the next step of the cycle, the thioester bond of succinyl-CoA is hydrolyzed to produce succinate and CoA-SH; an accompanying reaction is the phosphorylation of GDP to GTP

19B Biochemical Connections • Labeling Methods

What Is the Origin of the CO_2 Released by the Citric Acid Cycle?

The use of radioactive labels has shown that the two carbons that are released as CO_2 in the citric acid cycle do *not* come from the acetyl-CoA that entered the cycle. Moreover, the results of these studies have provided detailed information about how the reactions of the cycle take place.

Not only do the two CO_2 molecules released in the citric acid cycle *not* come from the carbons of acetyl-CoA that condensed with oxaloacetate at the start of the cycle, but they also take a number of turns to release both carbons. To complicate matters, the fate of the carbonyl carbon of acetyl-CoA differs from that of the methyl carbon, as shown in Figure 19.14. Figure 19.14A shows the fate of the carbonyl carbon of acetyl-CoA. That carbon is retained while two carbons of the oxaloacetate to which it condensed are lost in succession, forming succinyl-CoA. When succinyl-CoA is hydrolyzed to succinate to regenerate oxaloacetate, the label can appear in either of the carboxyl carbons. They are both lost in the second turn of the cycle.

Figure 19.14B shows the fate of the methyl carbon of the original acetyl-CoA. In the first turn of the cycle, the incoming methyl carbon becomes evenly distributed between the methylene and carbonyl carbons of oxaloacetate. These methylene and carbonyl carbons are retained in the second turn of the cycle. In the third turn, half of the carbon from the original methyl group has become one of the carboxyl carbons of oxaloacetate, and it will subsequently be lost as CO_2. In the fourth turn of the cycle, half of the remaining carbon is lost. The original methyl carbon is retained in the first two turns of the cycle, and then 50% is lost in each remaining turn (Figure 19.14, *bottom*).

Continued on next page

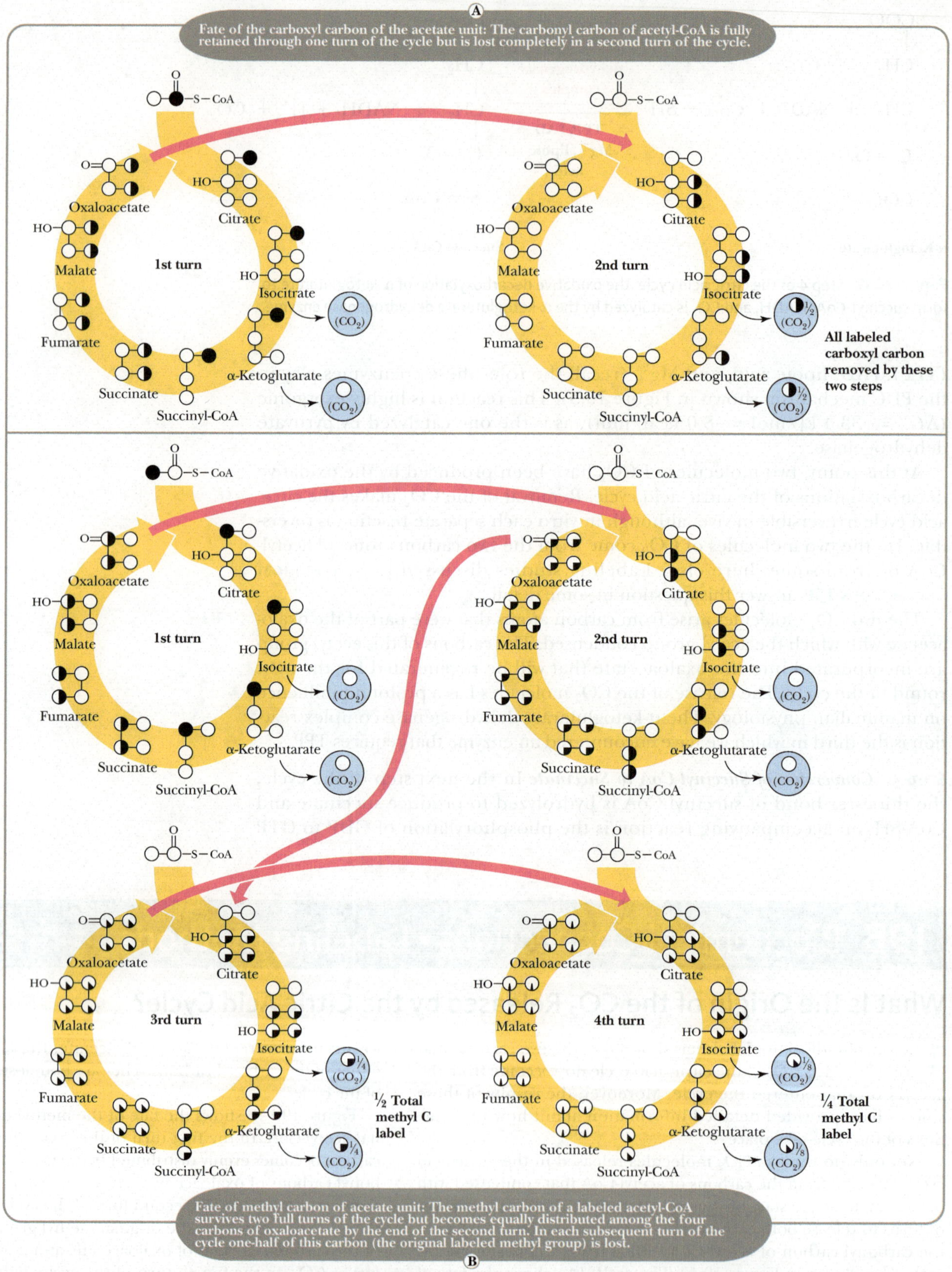

Figure 19.14 **The fate of the carbon atoms of acetate in successive TCA cycles.** Assume at the start that labeled acetate is added to cells containing unlabeled metabolites. The circles with fractional shading represent the experimental observation that the label appears at multiple carbon positions. *(From GARRETT/GRISHAM,* Biochemistry, *4E. ©2009 Cengage Learning.)*

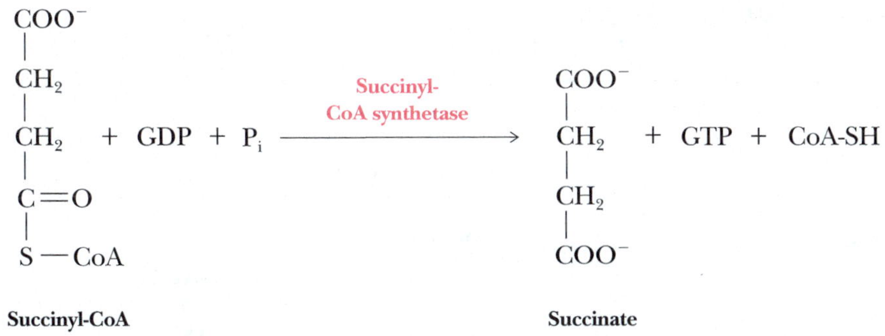

Figure 19.15 **Step 5 of the citric acid cycle: the conversion of succinyl-CoA to succinate is catalyzed by succinyl-CoA synthetase.**

(Figure 19.15). The entire reaction is catalyzed by the enzyme **succinyl-CoA synthetase**. In general, a **synthetase** is an enzyme that creates a new covalent bond between two molecules and requires the direct input of energy from a high-energy phosphate. The name of the enzyme for this reaction describes the reverse reaction. Succinyl-CoA synthetase would produce succinyl-CoA while spending an ATP or another high-energy phosphate. The actual citric acid cycle reaction is the opposite of that.

succinyl-CoA synthetase the enzyme that catalyzes the conversion of succinyl-CoA and GDP to succinate and GTP, releasing CoA-SH, in Step 5 of the citric acid cycle

synthetase an enzyme that creates a new covalent bond between two molecules and requires the direct input of energy from a high-energy phosphate

Recall that we encountered a synthase (citrate synthase) in Step 1 of the citric acid cycle (Figure 19.6). The difference between a synthase and a synthetase is that a synthase does *not* require energy from phosphate-bond hydrolysis, whereas a synthetase does. In the reaction mechanism, a phosphate group covalently bonded to the enzyme is directly transferred to GDP. The phosphorylation of GDP to GTP is endergonic, as is the corresponding ADP-to-ATP reaction ($\Delta G°' = 30.5$ kJ/mol = 7.3 kcal/mol).

The energy required for the phosphorylation of GDP to GTP is provided by the hydrolysis of succinyl-CoA to produce succinate and CoA. The Gibbs energy of hydrolysis ($\Delta G°'$) of succinyl-CoA is −33.4 kJ/mol (−8.0 kcal/mol). The overall reaction is slightly exergonic (−2.9 kJ/mol = −0.7 kcal/mol) and, as a result, does not contribute greatly to the overall production of energy by the mitochondrion.

The enzyme *nucleoside diphosphate kinase* catalyzes the transfer of a phosphate group from GTP to ADP to give GDP and ATP.

$$\text{GTP} + \text{ADP} \rightarrow \text{GDP} + \text{ATP}$$

This reaction step is called *substrate-level phosphorylation* to distinguish it from *oxidative phosphorylation*. In oxidative phosphorylation, the production of ATP is coupled to the electron transport chain (see Chapter 20). Step 7 of glycolysis, the conversion of 1,3-bisphosphoglycerate and ADP to 3-phosphoglycerate and ATP (Section 17-3), is another example of substrate-level phosphorylation. The distinction between substrate-level phosphorylation and oxidative phosphorylation is an important one. In substrate-level phosphorylation, the Gibbs energy of hydrolysis of the other compound (succinyl-CoA here) provides the energy for the phosphorylation reaction. The production of ATP in this reaction is the only place in the citric acid cycle in which chemical energy in the form of ATP is made available to the cell.

In Steps 6–8 of the citric acid cycle, the four-carbon succinate ion is converted to the oxaloacetate ion to complete the cycle (Figure 19.16):

- Step 6: Oxidation of a carbon–carbon single bond to a double bond with FAD as the oxidizing agent
- Step 7: Addition of H_2O across the double bond to introduce a hydroxyl group
- Step 8: Oxidation of the hydroxyl group to a carbonyl group with NAD^+ as the oxidizing agent

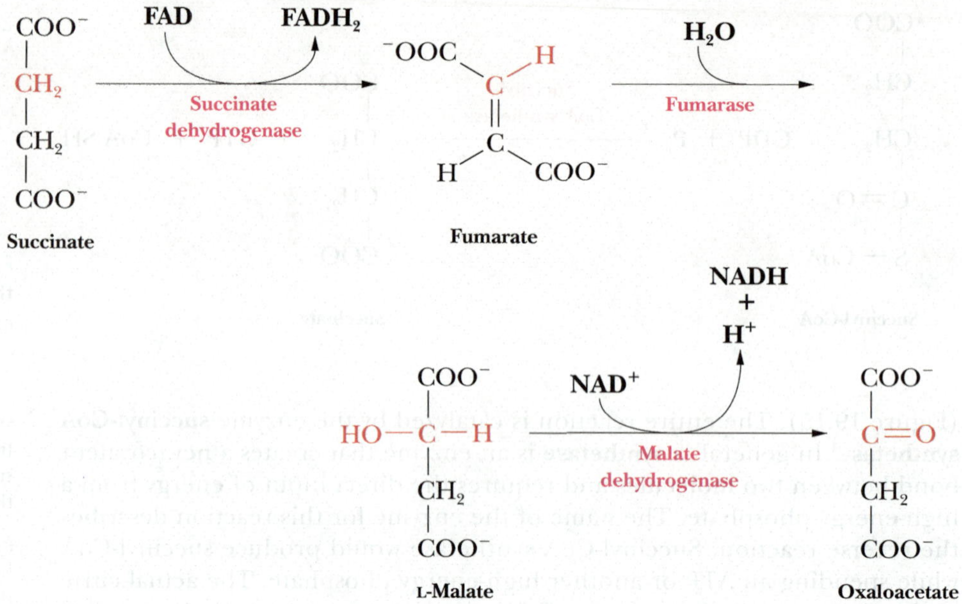

Figure 19.16 The formation of oxaloacetate from succinate.

This same three-step pattern of chemical reactions appears again in the β-oxidation pathway of fatty acid catabolism (Section 21-2B). For now, let us discuss each of these final citric acid cycle steps in turn.

Step 6. ***Oxidation of Succinate to Form Fumarate and*** $FADH_2$ The single bond between the methylene carbons of succinate is oxidized to a double bond in fumarate, and FAD is reduced to $FADH_2$. This reaction is catalyzed by the enzyme **succinate dehydrogenase** (Figure 19.17). Succinate dehydrogenase is part of a multienzyme integral membrane protein complex known as Complex II of the electron transport chain (Section 20-3A). Complex II is found in the inner mitochondrial membrane (recall that the other seven individual enzymes of the citric acid cycle are found in the mitochondrial matrix). The electron acceptor, which is FAD rather than NAD^+, is covalently bonded to the enzyme; succinate dehydrogenase is also called a flavoprotein because of the flavin moiety in FAD.

succinate dehydrogenase the enzyme that catalyzes the conversion of succinate to fumarate in Step 6 of the citric acid cycle

The overall reaction is

$$\text{Succinate} + \text{E-FAD} \rightarrow \text{Fumarate} + \text{E-FADH}_2$$

The E-FAD and E-$FADH_2$ in the equation indicate that the FAD coenzyme electron acceptor is covalently bonded to the enzyme. The $FADH_2$ group will transfer these electrons in the electron transport chain, and eventually to oxygen, and gives rise to 1.5 ATP, rather than 2.5, as is the case with NADH. Two molecules of $FADH_2$ and eventually three ATP are produced from this step for each original molecule of glucose.

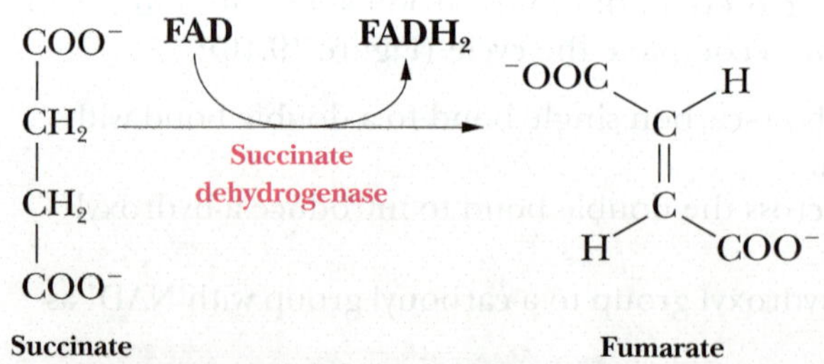

Figure 19.17 Step 6 of the citric acid cycle: the oxidation of succinate to form fumarate and $FADH_2$ is catalyzed by the succinate dehydrogenase enzyme found in Complex II of the electron transport chain.

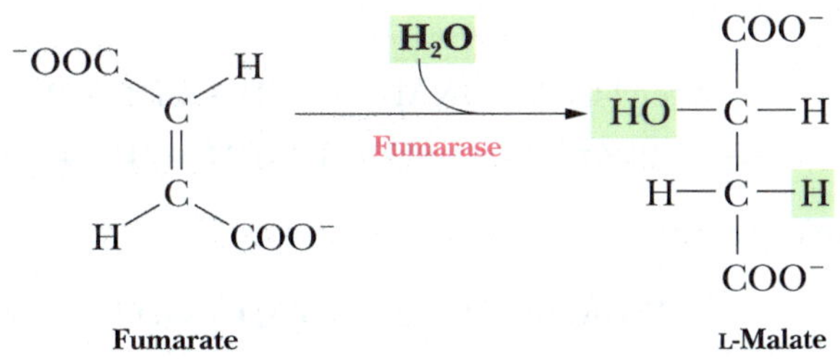

Figure 19.18 Step 7 of the citric acid cycle: the conversion of fumarate to L-malate is catalyzed by the fumarate hydratase enzyme.

Another interesting feature of succinate dehydrogenase is that it contains iron atoms but does not contain a heme group. It is referred to as a **nonheme iron protein** or an *iron–sulfur protein*. The latter name refers to the fact that the protein contains several clusters of iron and sulfur atoms. These *iron–sulfur clusters* are discussed further in Section 20-3C.

nonheme iron protein a protein that contains iron and sulfur but no heme group; also known as an iron–sulfur protein

Step 7. ***Hydration of Fumarate to Form L-Malate*** Step 7, which is catalyzed by the enzyme **fumarate hydratase**, is a hydration reaction. Water is added across the carbon–carbon double bond of fumarate to give malate (Figure 19.18). The carbon atom bonded to the hydroxyl group is chiral because it is bonded to four different groups. Thus, malate has two enantiomers, L- and D-malate. Only L-malate is produced, however, because this reaction is stereospecific.

fumarate hydratase the enzyme that catalyzes the conversion of fumarate to malate in Step 7 of the citric acid cycle

Step 8. ***Oxidation of L-Malate to Regenerate Oxaloacetate*** The hydroxyl group of malate is oxidized to a carbonyl group in oxaloacetate, and another molecule of NAD^+ is reduced to NADH (Figure 19.19). This reaction is catalyzed by the enzyme **malate dehydrogenase**. The oxaloacetate can then react with another molecule of acetyl-CoA to start another round of the cycle.

malate dehydrogenase the enzyme that catalyzes the oxidation of malate to form oxaloacetate and NADH in Step 8 of the citric acid cycle

The oxidation of pyruvate by the pyruvate dehydrogenase complex (PDC) and the citric acid cycle results in the production of three molecules of CO_2. As a result of these oxidation reactions, one molecule of GDP is phosphorylated to GTP, one molecule of FAD is reduced to $FADH_2$, and four molecules of NAD^+ are reduced to NADH. Of the four molecules of NADH produced, three come from the citric acid cycle, and one comes from the reaction of the PDC. The overall stoichiometry of the oxidation reactions is the sum of the pyruvate dehydrogenase reaction and the citric acid cycle. Only one high-energy phosphate, GTP, is produced *directly* from the citric acid cycle, but many more ATP will arise from the reoxidation of NADH and $FADH_2$ and the transfer of their electrons to the electron transport chain (Section 20-3A).

Pyruvate dehydrogenase complex:

$$\text{Pyruvate} + \text{CoA-SH} + NAD^+ \rightarrow \text{Acetyl-CoA} + \text{NADH} + CO_2 + H^+$$

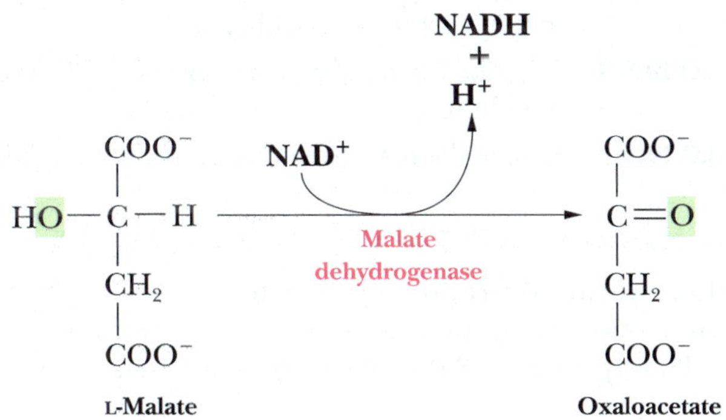

Figure 19.19 Step 8 of the citric acid cycle: the oxidation of L-malate to form oxaloacetate and NADH is catalyzed by the malate dehydrogenase enzyme.

Citric acid cycle:

$$\text{Acetyl-CoA} + 3\text{NAD}^+ + \text{FAD} + \text{GDP} + \text{P}_i + 2\text{H}_2\text{O} \rightarrow$$
$$2\text{CO}_2 + \text{CoA-SH} + 3\text{NADH} + 3\text{H}^+ + \text{FADH}_2 + \text{GTP}$$

Overall reaction:

$$\text{Pyruvate} + 4\text{NAD}^+ + \text{FAD} + \text{GDP} + \text{P}_i + 2\text{H}_2\text{O} \rightarrow$$
$$3\text{CO}_2 + 4\text{NADH} + \text{FADH}_2 + \text{GTP} + 4\text{H}^+$$

Eventual ATP production per pyruvate:

4 NADH → 10 ATP (2.5 ATP for each NADH)
1 FADH → 1.5 ATP (1.5 ATP for each $FADH_2$)
1 GTP → 1 ATP
Total 12.5 ATP per pyruvate or 25 ATP per glucose

There were also two ATP produced per glucose in glycolysis and two NADH, which will give rise to another five ATP (seven more ATP total). In Chapter 20, we have a lot more to say about the production of ATP from the complete oxidation of glucose.

At this point, we would do well to revisit Figure 19.3, which shows the entire citric acid cycle. When studying a pathway such as this, we learn many details, but it is helpful to revisit the big picture to see the metabolite names and their structures together with the enzymes and coenzymes that are involved in the entire pathway. The most important reactions can be identified by those that have important cofactors (NADH, $FADH_2$, GTP). Also important are the steps where CO_2 is released.

These important reactions of the citric acid cycle contribute significantly to our metabolism. One of its purposes is to produce energy. It does so by producing GTP directly and by producing reduced electron carriers (NADH and $FADH_2$). The three decarboxylations mean that for every three carbons entering as pyruvate, three carbons are effectively lost during the cycle, a fact that has many implications to our metabolism, including the following: waste elimination, the coupling of carbon removal to NADH and $FADH_2$ production, which leads to energy production, and the regeneration of oxaloacetate so that the citric acid cycle can continue.

19-5 Energetics and Control of the Citric Acid Cycle

Learning Objectives

After reading this section, students will be able to:

LO 19.5.1 Compare the Gibbs energy values of the reactions that convert pyruvate to CO_2 via the citric acid cycle.

LO 19.5.2 Describe the regulation that occurs in the conversion of pyruvate to acetyl-CoA.

LO 19.5.3 Explain the regulation that occurs in the citric acid cycle.

19-5A Energetics of the Citric Acid Cycle

Like a ring of fire, the citric acid cycle is strongly exergonic. As mentioned in Section 19-3, the reaction of pyruvate to acetyl-CoA is exergonic ($\Delta G°' = -33.4$ kJ/mol $= -8.0$ kcal/mol). The citric acid cycle itself is also exergonic

Table 19.2 The Energetics of Conversion of Pyruvate to CO_2 via the Citric Acid Cycle

		$\Delta G°'$	
Step	Reaction	kJ/mol	kcal/mol
	Pyruvate + CoA-SH + NAD^+ → Acetyl-CoA + NADH + CO_2	−33.4	−8.0
1	Acetyl-CoA + Oxaloacetate + H_2O → Citrate + CoA-SH + H^+	−32.2	−7.7
2	Citrate → Isocitrate	+6.3	+1.5
3	Isocitrate + NAD^+ → α-Ketoglutarate + NADH + CO_2 + H^+	−7.1	−1.7
4	α-Ketoglutarate + NAD^+ + CoA-SH → Succinyl-CoA + NADH + CO_2 + H^+	−33.4	−8.0
5	Succinyl-CoA + GDP + P_i → Succinate + GTP + CoA-SH	−3.3	−0.8
6	Succinate + FAD → Fumarate + $FADH_2$	~0	~0
7	Fumarate + H_2O → L-Malate	−3.8	−0.9
8	L-Malate + NAD^+ → Oxaloacetate + NADH + H^+	+29.2	+7.0
Overall:	Pyruvate + 4NAD^+ + FAD + GDP + P_i + 2H_2O → 3 CO_2 + 4NADH + $FADH_2$ + GTP + 4H^+	−77.7	−18.6

($\Delta G°' = -44.3$ kJ/mol = −10.6 kcal/mol), and you are asked to confirm this point in an end-of-chapter question. The standard Gibbs energy changes for the individual reactions are listed in Table 19.2. Of the individual reactions of the cycle, only Step 8, the oxidation of malate to oxaloacetate, is strongly endergonic ($\Delta G°' = +29.2$ kJ/mol = +7.0 kcal/mol). However, this endergonic reaction is coupled to Step 1, the condensation of acetyl-CoA and oxaloacetate to produce citrate and coenzyme A, which is strongly exergonic ($\Delta G°' = -32.2$ kJ/mol = −7.7 kcal/mol). (Recall that these values for the Gibbs energy changes refer to standard conditions. The effect of concentrations of metabolites in vivo can change matters drastically.) In addition to the energy released by the oxidation reactions, even more energy will be released in the electron transport chain. When the four NADH and single $FADH_2$ produced by the pyruvate dehydrogenase complex and the citric acid cycle are reoxidized by the electron transport chain, considerable quantities of ATP are produced.

Control of the citric acid cycle is exercised at three points; that is, three enzymes within the citric acid cycle play a regulatory role (Figure 19.20). In addition, Section 19-5B discusses the control of access to the cycle via pyruvate dehydrogenase.

19-5B Pre-cycle Regulation

▶ **How is access to the citric acid cycle controlled in the pyruvate dehydrogenase reaction?**

The overall citric acid cycle reaction is part of a pathway that produces energy in the form of ATP. Therefore, the pyruvate dehydrogenase complex (PDC), which produces a product that initiates the cycle, is inhibited by ATP and NADH in prokaryotic cells because both compounds are abundant when a cell has energy readily available. Conversely, the PDC is activated by ADP and AMP, which are abundant when a cell needs energy.

In mammalian cells, the PDC is regulated allosterically as mentioned above, and it is also regulated covalently by phosphorylation. PDC is inhibited when a phosphate group is covalently attached to the pyruvate dehydrogenase (E1) enzyme (recall the different enzymes of the complex shown in Figure 19.5) in a reaction catalyzed by the PDC enzyme **pyruvate dehydrogenase kinase**

pyruvate dehydrogenase kinase the enzyme of the pyruvate dehydrogenase complex that phosphorylates the E1 enzyme of the complex, thereby inhibiting its activity in the production of acetyl-CoA

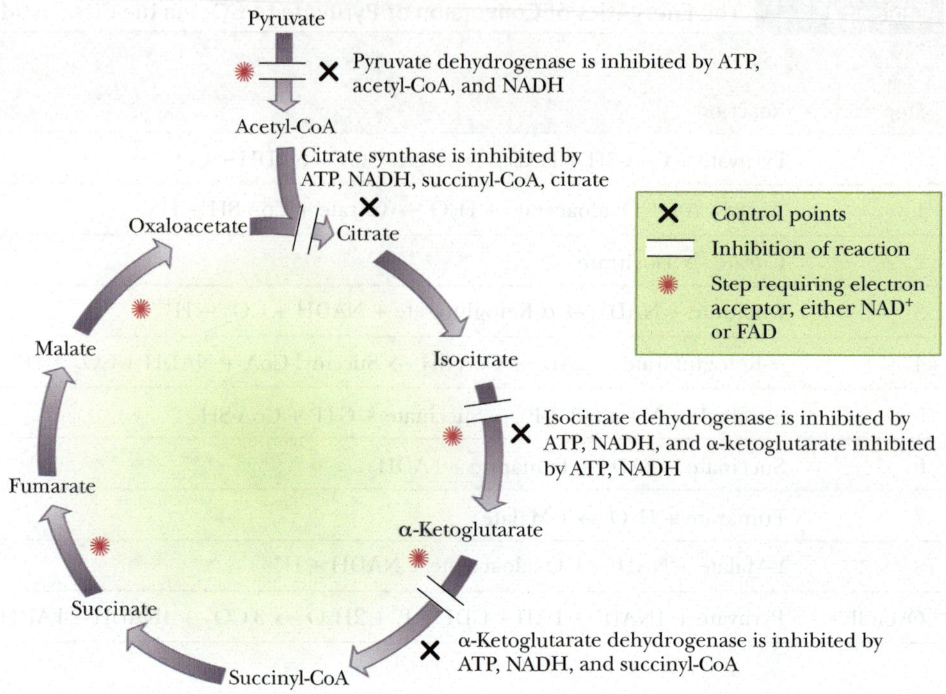

Figure 19.20 Control points in the conversion of pyruvate to acetyl-CoA and in the citric acid cycle.

(recall that the pyruvate dehydrogenase complex contains five enzymes). When the need arises for pyruvate dehydrogenase to be activated, the hydrolysis of the phosphate ester linkage (dephosphorylation) is catalyzed by another PDC enzyme, **pyruvate dehydrogenase phosphatase**. This latter enzyme is itself activated by Ca^{2+}. Both enzymes are associated with the mammalian pyruvate dehydrogenase complex, permitting effective control of the overall reaction from pyruvate to acetyl-CoA.

pyruvate dehydrogenase phosphatase the enzyme of the pyruvate dehydrogenase complex that dephosphorylates the E1 enzyme of the complex, thereby stimulating its activity in the production of acetyl-CoA

How do high levels of ATP inhibit the pyruvate dehydrogenase enzyme? High levels of ATP activate the PDH kinase, which then phosphorylates three serine residues of the E1 enzyme of the complex (recall the E1 enzyme shown in the left portion of Figure 19.5). These covalent modifications inactivate the E1 enzyme of the complex. Pyruvate dehydrogenase is also inhibited by high levels of its product, acetyl-CoA, in a feedback inhibition mechanism.

19-5C Citric Acid Cycle Regulation

▶ How is control exerted within the citric acid cycle?

Within the citric acid cycle itself, the three control points are the reactions catalyzed by citrate synthase (Step 1), isocitrate dehydrogenase (Step 3), and the α-ketoglutarate dehydrogenase complex (Step 4).

- In general, the first reaction of a metabolic pathway is one in which regulatory control appears. That is the case here with citrate synthase, which is an allosteric enzyme inhibited by ATP, NADH, succinyl-CoA, and its own product, citrate.
- The second regulatory step in the citric acid cycle is the isocitrate dehydrogenase reaction. In this case, ADP, AMP, and NAD^+ are activators of the enzyme, while ATP and two of the reaction products—NADH and α-ketoglutarate—are inhibitors.
- The α-ketoglutarate dehydrogenase complex is the third regulatory step in the citric acid cycle. As before, ATP and two of the reaction products—NADH and succinyl-CoA—are inhibitors of this reaction.

Table 19.3 Relationship Between the Metabolic State of a Cell and ATP/ADP and NADH/NAD⁺ Ratios

Cells in a resting metabolic state
Need and use comparatively little energy
High ATP, low ADP levels (high ATP/ADP ratio)
High NADH, low NAD^+ levels (high NADH/NAD^+ ratio)
Cells in a highly active metabolic state
Need and use more energy than resting cells
Low ATP, high ADP levels (low ATP/ADP ratio)
Low NADH, high NAD^+ levels (low NADH/NAD^+ ratio)

These recurring themes in the regulation of metabolism reflect the ways in which a cell can adjust to an active state that requires more energy or to a resting state that does not.

When a cell is metabolically active, it uses ATP and NADH at a great rate, producing large amounts of ADP and NAD^+ (Table 19.3). In other words, when the ATP/ADP ratio is low, the cell is using energy and needs to release more energy from stored nutrients. A low NADH/NAD^+ ratio is also characteristic of an active metabolic state. On the other hand, a resting cell has fairly high levels of ATP and NADH. The ATP/ADP ratio and the NADH/NAD^+ ratio are also high in resting cells, which do not need to maintain a high level of oxidation to produce energy.

When cells have low energy requirements (that is, when they have a high "energy charge") with high ATP/ADP and NADH/NAD^+ ratios, the presence of so much ATP and NADH serves as a signal to "shut down" the enzymes responsible for oxidative reactions. When cells have a low energy charge, characterized by low ATP/ADP and NADH/NAD^+ ratios, the need to release more energy and to generate more ATP serves as a signal to "turn on" the oxidative enzymes. This relationship of energy requirements to enzyme activity is the basis for the overall regulatory mechanism exerted at a few key control points in metabolic pathways.

19-6 The Glyoxylate Cycle: Producing Glucose from Fatty Acid Catabolism

Learning Objectives

After reading this section, students will be able to:

LO 19.6.1 Describe the purpose of the glyoxylate cycle.

LO 19.6.2 Explain the five key steps of the glyoxylate cycle.

▶ How can plants produce carbohydrates from fatty acid breakdown and acetyl-CoA, whereas animals cannot?

Animals can convert carbohydrates to fats (Section 21-6), but they cannot convert fats to carbohydrates. Conversely, in plants, fungi, and some bacteria, acetyl-CoA from fatty acid catabolism can serve as the starting material for the biosynthesis of carbohydrates in the **glyoxylate cycle**. Two enzymes (**isocitrate lyase** and **malate synthase**) are responsible for the ability of plants and bacteria to produce glucose from fatty acids. They are discussed in detail here in Section 19-6, and their reactions bypass several citric acid cycle reactions. The five key steps of the glyoxylate cycle are as follows:

- Step 1: Condensation of acetyl-CoA and oxaloacetate by citrate synthase to form citrate. The acetyl-CoA can come from fatty acid breakdown

glyoxylate cycle a pathway in plants and some bacteria whereby they can bypass several citric acid cycle reactions to produce carbohydrates from fatty acids

isocitrate lyase the glyoxylate cycle enzyme that catalyzes the cleavage of isocitrate to glyoxylate and succinate

malate synthase the glyoxylate cycle enzyme that catalyzes the reaction of glyoxylate with acetyl-CoA to produce malate

(Section 21-2B), and it can be considered to be a source of two carbons for eventual carbohydrate synthesis.

- Step 2: Isomerization of citrate to isocitrate catalyzed by aconitase.
- Step 3: Cleavage of isocitrate by isocitrate lyase to form succinate and glyoxylate (Figure 19.21A). This reaction bypasses the decarboxylation reactions of the citric acid cycle (Figure 19.22, *middle*) in order to conserve carbons for carbohydrate synthesis. The glyoxylate product is the namesake of the cycle. The succinate continues in the citric acid cycle, but the two carbons of glyoxylate combine with two more carbons from another acetyl-CoA from fatty acid breakdown in Step 4.
- Step 4: Condensation of glyoxylate and acetyl-CoA by malate synthase to form malate (Figure 19.21B). This step completes the joining of four carbons from fatty acid catabolism. The malate can continue on a gluconeogenic path toward the synthesis of glucose, as seen in the top left of Figure 19.22.
- Step 5: Oxidation of malate to oxaloacetate catalyzed by malate dehydrogenase. The oxaloacetate can then be converted to phosphoenolpyruvate (PEP) for gluconeogenesis (Section 18-2). When this oxaloacetate is diverted to gluconeogenesis, it is not removing an oxaloacetate for continuing the citric acid cycle because the succinate produced in Step 3 of the glyoxylate cycle continues in the citric acid cycle to produce oxaloacetate as well.

The net result is that carbons from two molecules of acetyl-CoA (from fatty acid breakdown) enter the glyoxylate cycle (one acetyl-CoA at Step 1 and one acetyl-CoA at Step 4), and these carbons can ultimately be used to produce glucose. The relationship between the glyoxylate cycle bypass reactions and the whole citric acid cycle is shown in Figure 19.22.

glyoxysomes membrane-enclosed organelles in plants that contain the enzymes of the glyoxylate cycle

Specialized organelles in plants, called **glyoxysomes**, are the sites of the glyoxylate cycle. This pathway is particularly important in germinating seeds. The fatty acids stored in the seeds are broken down for energy during germination. First, the fatty acids give rise to acetyl-CoA, which can enter the citric acid cycle and go on to release energy in the reactions discussed in Section 19-4. The citric acid cycle and the glyoxylate cycle can operate simultaneously. Acetyl-CoA

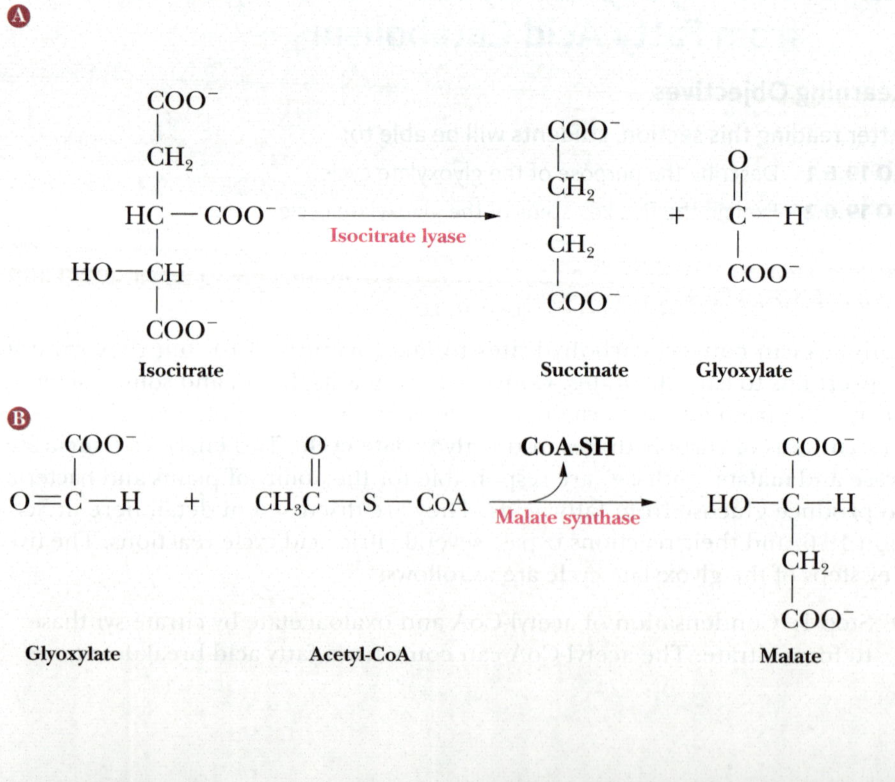

Figure 19.21 Reactions in plants and bacteria that produce carbohydrates from fats.

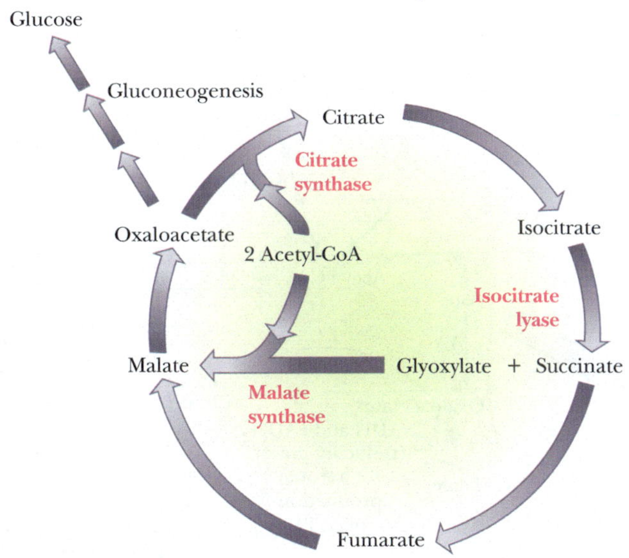

Figure 19.22 The glyoxylate cycle. This pathway results in the net conversion of two acetyl-CoA to oxaloacetate, bypassing the decarboxylation steps of the citric acid cycle and conserving the carbons for the synthesis of glucose. All the reactions are shown in purple. The unique reactions of the glyoxylate cycle are shown with a light green highlight in the center of the circle.

from fatty acid catabolism can be used for energy production in the citric acid cycle, and it can also serve as the starting point for the synthesis of glucose and other compounds needed by the growing seedling. (Recall that carbohydrates play an important structural, as well as energy-producing, role in plants. To review the structural role, see the structure of cellulose in Figure 16.26 in Section 16-4A.)

The glyoxylate cycle also occurs in bacteria, which allows many types of bacteria to live on very limited carbon sources. They have metabolic pathways that can produce all the biomolecules they need from quite simple molecules. The glyoxylate cycle is one example of how bacteria manage this feat.

19-7 The Citric Acid Cycle in Catabolism

Learning Objectives

After reading this section, students will be able to:

LO 19.7.1 Draw a labeled diagram showing the central role of the citric acid cycle in the catabolism of sugars, fatty acids, and amino acids.

▶ How is the citric acid cycle involved in the breakdown of nutrients?

Like an essential traffic circle of life, metabolic routes enter and exit the citric acid cycle, as explained in Sections 19-7 and 19-8. The nutrients taken in by an organism can include large molecules. This is especially true of animals, which ingest polysaccharides and proteins, which are polymers, as well as storage lipids (recall triacylglycerols from Section 8-2B) and membrane lipids (recall phosphoacylglycerols from Section 8-2C). Nucleic acids constitute a very small percentage of the nutrients present in foodstuffs, so we do not consider their catabolism.

The first step in the breakdown of nutrients is the degradation of large molecules to smaller ones. Polysaccharides are hydrolyzed by specific enzymes to produce sugar monomers; an example is the breakdown of starch by amylases. Lipases hydrolyze triacylglycerols to give fatty acids and glycerol (Section 8-2B). Proteins are digested by proteases, with amino acids as the end products. Sugars, fatty acids, and amino acids then enter their specific catabolic pathways.

In Sections 17-2 and 17-3, we discussed the glycolytic pathway by which sugars are converted to pyruvate, which then enters the citric acid cycle (Figure 19.3). In Section 21-2B, we explain how fatty acids are converted to acetyl-CoA

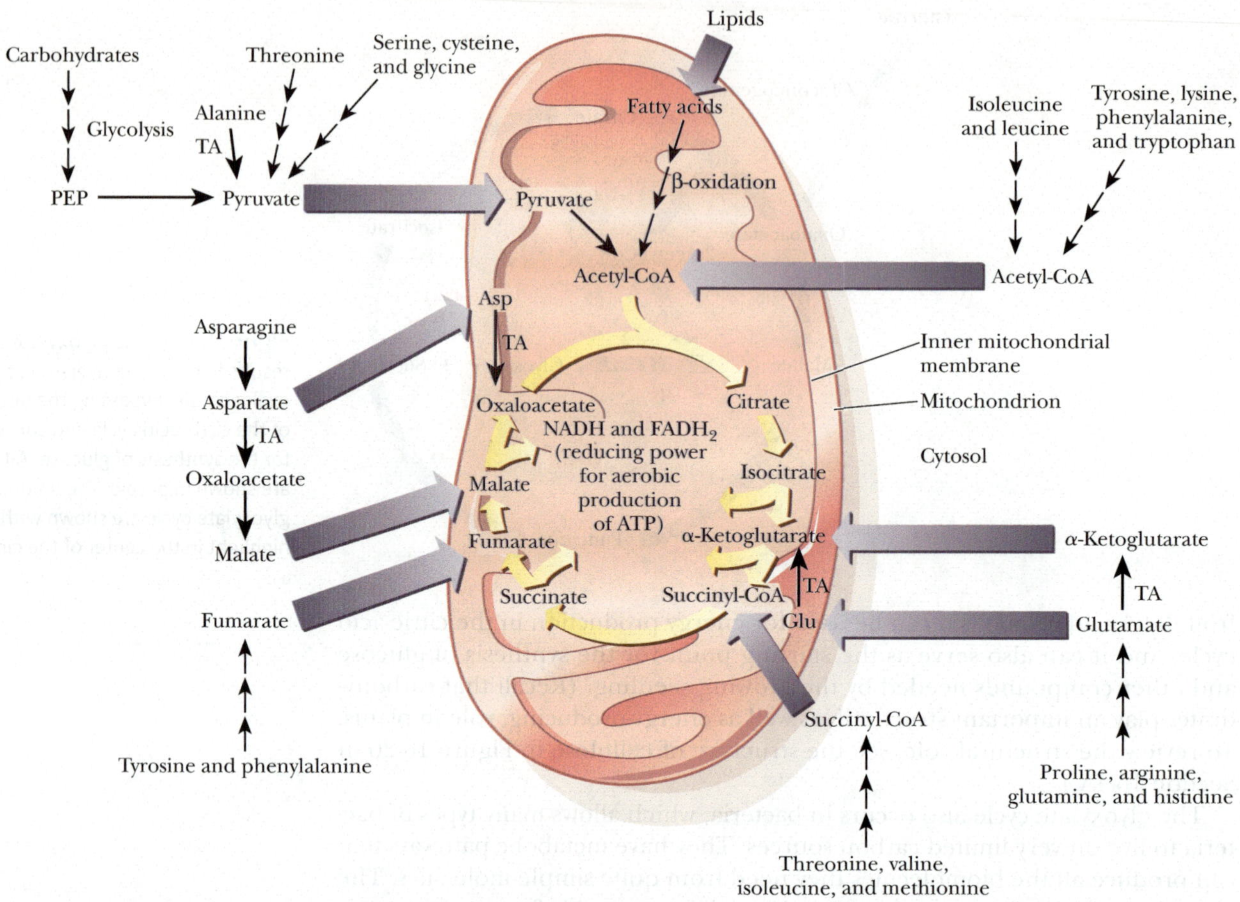

Figure 19.23 A summary showing the central role of the citric acid cycle in the catabolism of sugars, fatty acids, and amino acids. The end products of the catabolism of carbohydrates, lipids, and amino acids all appear. (PEP is the phosphoenolpyruvate; TA is the transamination; →→→ is a multistep pathway.)

in the β-oxidation pathway. Amino acids enter the cycle by various paths. We discuss the catabolic reactions of amino acids in Section 23-6.

Figure 19.23 shows schematically the various catabolic pathways by which sugars, fatty acids, and amino acids can feed into the citric acid cycle. The catabolic reactions occur in the cytosol; the citric acid cycle takes place in mitochondria. Many of the end products of catabolism cross the mitochondrial membrane and then participate in the citric acid cycle. Figure 19.23 also shows the outline of pathways by which amino acids are converted to components of the citric acid cycle. Just as "all roads lead to Rome," all of these catabolic pathways lead to the citric acid cycle.

19-8 The Citric Acid Cycle in Anabolism

Learning Objective

After reading this section, students will be able to:

LO 19.8.1 Draw a labeled diagram showing the central role of the citric acid cycle in the anabolism of sugars, fatty acids, and amino acids.

▶ How is the citric acid cycle involved in the synthesis of biomolecules?

In Section 19-7, we saw how various nutrients can be broken down into intermediates that can enter the citric acid cycle. Conversely, the citric acid cycle is also a *source* of starting materials for the biosynthesis of many important biomolecules,

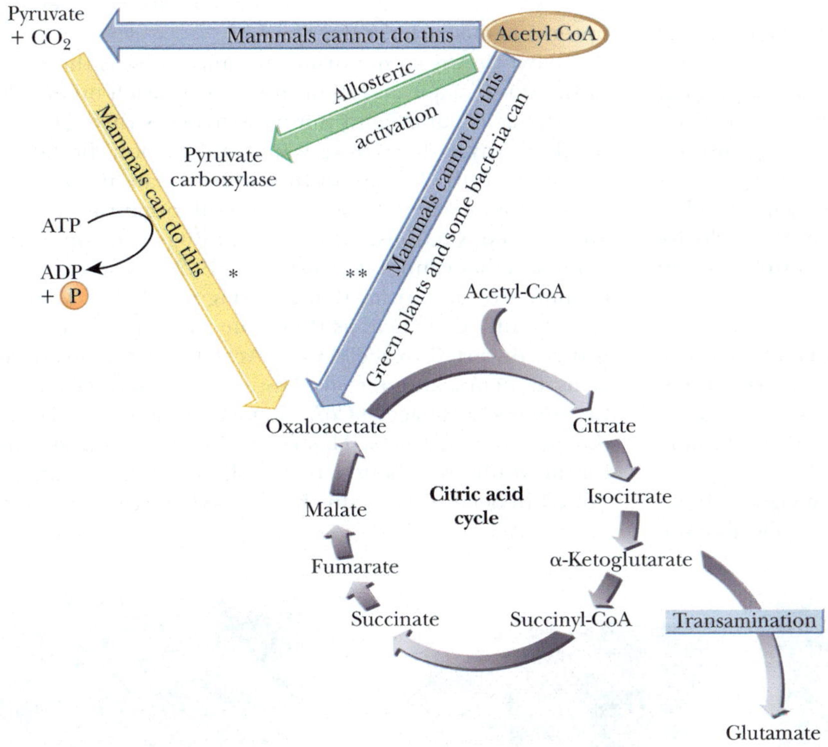

Figure 19.24 How mammals keep an adequate supply of downstream metabolic intermediates when an upstream intermediate is used for other pathways. An anabolic reaction uses a citric acid cycle intermediate (α-ketoglutarate is transaminated to glutamate in our example), competing with the rest of the cycle. The concentration of acetyl-CoA rises and signals the allosteric activation of pyruvate carboxylase to produce more oxaloacetate.

*Anaplerotic reaction.

**Part of glyoxylate pathway.

but the supply of the starting materials that are components of the cycle must be replenished if the cycle is to continue operating. See **Biochemical Connections 19C** for more details on this point.

In particular, the oxaloacetate in an organism must be maintained at a level sufficient to allow acetyl-CoA to enter the cycle, but sometimes the α-ketoglutarate intermediate that is a precursor for oxaloacetate in the citric acid cycle gets used in the anabolic pathway that synthesizes glutamate (see the bottom right of Figure 19.24; also, notice the *glutamate side chain structure* [Section 3-2] in the structure of α-ketoglutarate shown in the bottom right of Figure 19.12). In some organisms, acetyl-CoA can be converted to oxaloacetate in the glyoxylate cycle to replenish its supply (Section 19-6), but mammals cannot do this. In mammals, oxaloacetate can be supplied from pyruvate by the enzyme *pyruvate carboxylase* (Figure 19.24, *left*). We encountered this enzyme and this reaction in the context of gluconeogenesis (see Section 18-2A), and here we see another highly important role for this enzyme in the replenishing of oxaloacetate. Because the citric acid cycle supplies starting materials for anabolic pathways, the supply of oxaloacetate for the citric acid cycle would soon be depleted if there were no means of producing it from a readily available precursor like pyruvate. A reaction like the pyruvate carboxylase reaction that replenishes an important intermediate in a metabolic cycle or pathway is called an **anaplerotic reaction**.

anaplerotic reaction a reaction that ensures an adequate supply of an important intermediate in a metabolic cycle or pathway

19C Biochemical Connections • Metabolic Diversity

Animals Can't Use All the Same Energy Sources as Plants and Bacteria

The citric acid cycle is important not only as a source of energy during aerobic metabolism but also as a source of metabolites in the synthesis of important biomolecules. For example, it is a source of starting materials for the production of amino acids, carbohydrates, vitamins, nucleotides, and heme. However, if these intermediates are used for the synthesis of other molecules, then they must be

Continued on next page

replenished to maintain the catabolic nature of this cycle. The term *anaplerotic* means "filling up," and the reactions that replenish citric acid cycle intermediates are called anaplerotic reactions. One source of needed compounds, available to all organisms, is the group of amino acids that can be converted to citric acid cycle intermediates in a single reaction. A simple reaction available to all organisms is the addition of carbon dioxide to the pyruvate and phosphoenolpyruvate (PEP) generated from the metabolism of sugars. Another source, important in bacteria and plants, is the glyoxylate cycle discussed in Section 19-6. This source is vital to the ability of plants to fix carbon dioxide into carbohydrates (Figure 19.25).

Some anaerobic organisms have developed only parts of the citric acid cycle, which they use exclusively to make important precursors. These simple yet important reactions emphasize the truly connected nature of what we often artificially separate into "pathways." They also illustrate the convergence of life processes on a few key molecules and metabolic steps.

Which molecule is arguably the most important metabolic intermediate? Acetyl-CoA is perhaps *the* central molecule of metabolism. If you plot a chart of all known metabolic pathways, acetyl-CoA ends up close to the center of that chart.

Acetyl-CoA is so important because it links the metabolism of the three major classes of nutrients to each other. All sugars, all fatty acids, and many amino acids pass through acetyl-CoA on their way to becoming water and carbon dioxide. Equally important is the key use of this intermediate in the synthesis of major biomolecules. Some, but not all, organisms can carry out all these conversions. Humans and dogs provide examples of mammals that cannot do so (Figure 19.26), whereas bacteria are examples of organisms that can (Figure 19.27). Many bacteria can live off acetic acid as their sole carbon source; however, it is first converted to acetyl-CoA. Acetyl-CoA is converted to fatty acids, terpenes, and steroids. More important is the conversion of two molecules of acetyl-CoA to malate in plants and bacteria via the glyoxylate pathway. This key compound is the starting point for the synthesis of both amino acids and carbohydrates. As mentioned in Section 19-6, this key glyoxylate reaction is missing in mammals. ◗

HamsterMan/Shutterstock.com

Figure 19.25 Plants can produce carbohydrates from carbon dioxide.

georgina/Alamy Stock Photo

Figure 19.26 Dogs cannot carry out all the conversions for citric acid cycle intermediates.

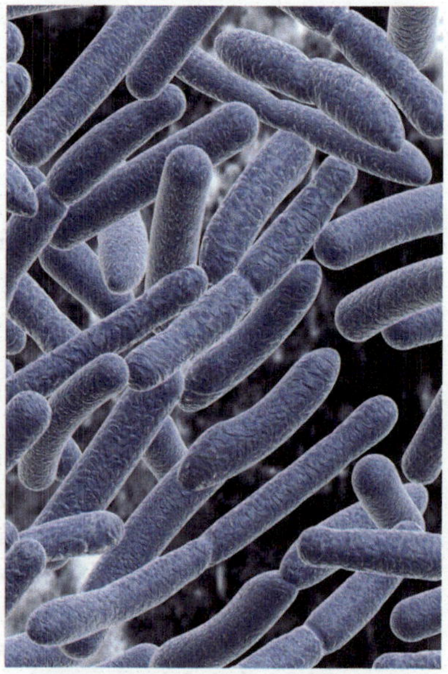

Fedorov Oleksiy/Shutterstock.com

Figure 19.27 Bacteria can carry out all the conversions for citric acid cycle intermediates.

19-8A The Roles of Pyruvate and Acetyl-CoA in Animal and Plant Metabolism

The pyruvate carboxylase reaction, which produces oxaloacetate from pyruvate (Figure 19.24, *left*), provides a connection between the amphibolic citric acid cycle and the anabolism of sugars in the gluconeogenesis pathway. On this same topic of carbohydrate anabolism, we note again that pyruvate cannot be produced from acetyl-CoA in mammals. Because acetyl-CoA is the end product of the catabolism of fatty acids, mammals cannot exist with fats or acetate as the sole carbon source. The intermediates of carbohydrate metabolism would soon be depleted. Carbohydrates are the principal energy and carbon source in animals, and glucose is especially critical in humans because it is the preferred fuel for our brain cells. Plants can carry out the conversion of acetyl-CoA to pyruvate and oxaloacetate, so they can exist without carbohydrates as a carbon source. The conversion of pyruvate to acetyl-CoA does take place in both plants and animals (see Section 19-3).

The anabolic reactions of gluconeogenesis take place in the cytosol. Oxaloacetate is not transported across the mitochondrial membrane. Two mechanisms exist for the transfer of molecules needed for gluconeogenesis from mitochondria to the cytosol (Figure 19.28):

- One mechanism takes advantage of the fact that phosphoenolpyruvate (PEP) can be formed from oxaloacetate in the mitochondrial matrix (this reaction is the next step in gluconeogenesis); PEP is then transferred to the cytosol, where the remaining gluconeogenesis reactions take place (Figure 19.28, *upper left*).
- The other mechanism relies on the fact that malate, which is another intermediate of the citric acid cycle, can be transferred to the cytosol. There is a *malate dehydrogenase* enzyme in the cytosol as well as in mitochondria, and malate can be converted to oxaloacetate in the cytosol.

$$\text{Malate} + \text{NAD}^+ \rightarrow \text{Oxaloacetate} + \text{NADH} + \text{H}^+$$

Oxaloacetate is then converted to PEP, leading to the rest of the steps of gluconeogenesis (Figure 19.28, *middle left*).

Gluconeogenesis has many steps in common with the production of glucose in photosynthesis, but photosynthesis also has many reactions in common with the pentose phosphate pathway. It is intriguing that common mechanisms that coordinate carbohydrate metabolism in all its aspects are found in different organisms.

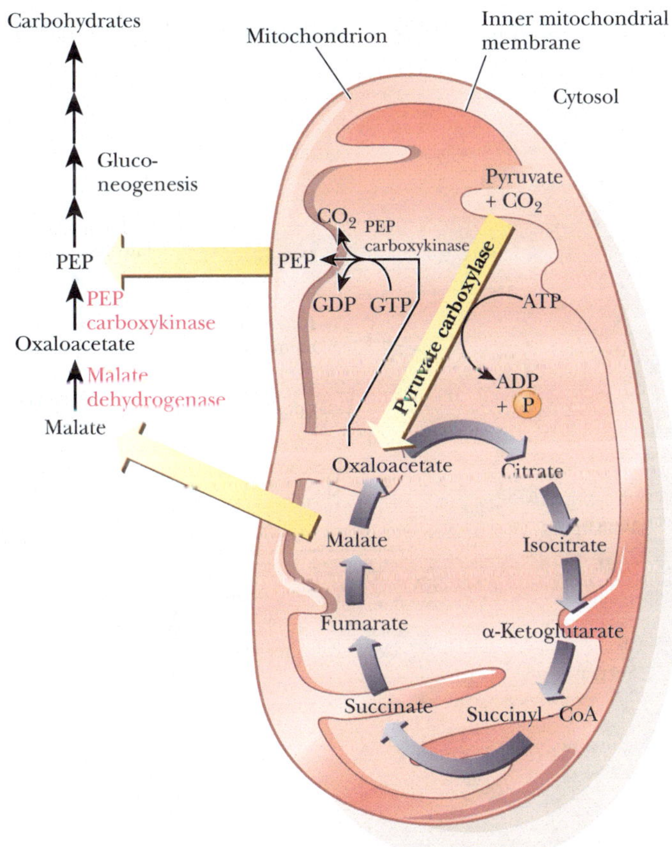

Figure 19.28 Transfer of the starting materials of gluconeogenesis from the mitochondrion to the cytosol. Phosphoenolpyruvate (PEP) can be transferred from the mitochondrion to the cytosol, as can malate. Oxaloacetate is not transported across the mitochondrial membrane.

19-8B The Citric Acid Cycle in Lipid Anabolism

► **How is lipid anabolism related to the citric acid cycle?**

The starting molecule of lipid anabolism is acetyl-CoA. The anabolic reactions of lipid metabolism, like those of carbohydrate metabolism, take place in the cytosol; these reactions are catalyzed by soluble enzymes that are not bound to membranes. Acetyl-CoA is mainly produced in mitochondria, whether from pyruvate or from the breakdown of fatty acids. An indirect transfer mechanism exists for the transfer of acetyl-CoA, in which citrate is transferred to the cytosol (Figure 19.29, *bottom right*). Citrate reacts with CoA-SH to produce citryl-CoA, which is then cleaved to yield oxaloacetate and acetyl-CoA. The enzyme that catalyzes this reaction requires ATP and is called **ATP-citrate lyase**. The overall reaction is

ATP-citrate lyase an enzyme that catalyzes the conversion of citrate to oxaloacetate in a fatty acid synthesis metabolic pathway

$$\text{Citrate} + \text{CoA-SH} + \text{ATP} \rightarrow \text{Acetyl-CoA} + \text{Oxaloacetate} + \text{ADP} + P_i$$

Acetyl-CoA is the starting point for lipid anabolism in both plants and animals. An important source of acetyl-CoA is the catabolism of carbohydrates. We have just seen that animals cannot convert lipids to carbohydrates, but they can convert carbohydrates to lipids. The efficiency of the conversion of carbohydrates to lipids in animals is discussed in **Biochemical Connections 19D**.

The right side of Figure 19.29 also shows how the oxaloacetate from the ATP-citrate lyase reaction can be used in the production of NADPH, which is also needed as a reducing agent for fatty acid synthesis. Oxaloacetate can be reduced to malate by malate dehydrogenase.

$$\text{Oxaloacetate} + \text{NADH} + \text{H}^+ \rightarrow \text{Malate} + \text{NAD}^+$$

Malate can also move out of mitochondria by active transport processes. Either way, malate can be oxidatively decarboxylated to pyruvate by **malic enzyme**, which requires $NADP^+$.

malic enzyme an enzyme that catalyzes the conversion of malate to pyruvate and produces NADPH needed for fatty acid synthesis

$$\text{Malate} + \text{NADP}^+ \rightarrow \text{Pyruvate} + CO_2 + \text{NADPH} + \text{H}^+$$

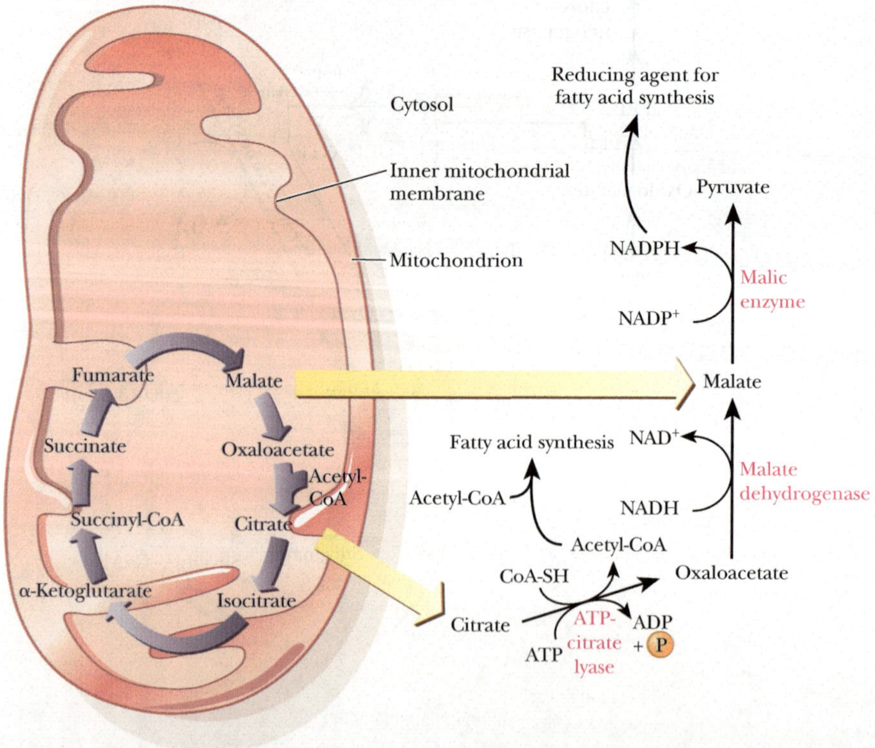

Figure 19.29 Transfer of the starting materials of lipid anabolism from the citric acid cycle in the mitochondrion to the cytosol. It is not definitely established whether acetyl-CoA is transported from the mitochondrion to the cytosol.

19D Biochemical Connections • Nutrition

Why Is It So Hard to Lose Weight?

For humans, it is far easier to gain weight than to lose it. If we had to analyze the specific chemical reactions that make this a reality, we would look very carefully at the citric acid cycle, especially the decarboxylation reactions.

All food in excess can be stored as fat. This is true for carbohydrates, proteins, and fats. In addition, these molecules can be interconverted, with the exception that fats cannot give a net yield of carbohydrates, as we saw in Section 19-6. Why is it that fats cannot yield carbohydrates? In order for a fat molecule to make glucose, it would have to enter the citric acid cycle as acetyl-CoA and then be drawn off as oxaloacetate for gluconeogenesis. Unfortunately, the two carbons that enter are effectively lost by the decarboxylations. (We have already seen that, in one round of the citric acid cycle, it isn't really these same two carbons that are lost; nevertheless, a two-carbon loss is a two-carbon loss, regardless of which two carbons they are.) This leads to an imbalance in the catabolic pathways versus the anabolic pathways.

All roads lead to fat, but fat cannot lead back to carbohydrate. Humans are very sensitive to glucose levels in the blood because so much of our metabolism is geared toward protecting our brain cells, which prefer glucose as a fuel. If we eat more carbohydrates than we need, the excess carbohydrates turn to fat. It is very easy to put on fat, especially as we age. What about the reverse? Why don't we just stop eating? Won't that reverse the process? The answer is yes and no. When we start eating less, fat stores are mobilized for energy. Fat is an excellent source of energy because it forms acetyl-CoA and gives a steady influx for the citric acid cycle. Thus, we can lose some weight by reducing our caloric intake. Unfortunately, our blood sugar also drops as soon as our glycogen stores run out. We don't have very much stored glycogen that could maintain our blood glucose levels.

After the blood glucose drops, we become depressed, sluggish, and irritable. We start having negative thoughts such as "This dieting thing is really stupid. I should eat a pint of ice cream." If we continue the diet, and given that we cannot turn fats into carbohydrates, where does the blood glucose come from? Only one source remains: proteins. Proteins are degraded to amino acids, and they are eventually converted to pyruvate for gluconeogenesis. Thus, we begin to lose muscle as well as fat.

Using our knowledge of biochemistry, we can create a combination of exercise and a diet consisting of healthy protein, carbohydrates, and fats. If you exercise correctly, you can train your body to use fats to supply acetyl-CoA for the citric acid cycle (Figure 19.30). If you maintain a healthy, complete diet, you maintain your blood glucose

Figure 19.30 Regular exercise and a proper diet keep fat from accumulating.

and do not degrade proteins for that purpose; your ingested carbohydrates are sufficient to maintain blood glucose and carbohydrate stores. With the proper balance of exercise to food intake, and the proper balance of the right types of nutrients, we can increase the breakdown of fat without sacrificing our carbohydrate stores or our proteins.

These last two reactions are also shown structurally in Figure 19.31, and they are a reduction reaction followed by an oxidation, so there is *no net oxidation.* There is, however, a *substitution of NADPH for NADH.*

This last point is important because many of the enzymes of fatty acid synthesis require NADPH. The pentose phosphate pathway (Section 18-4) is the principal source of NADPH in most organisms, but here we have another source.

The two ways of producing NADPH (through the pentose phosphate pathway or through citric acid cycle intermediates) once again illustrate how metabolic

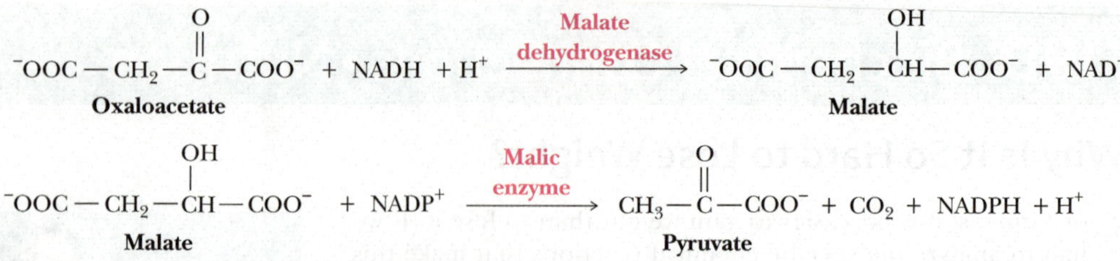

Figure 19.31 Reactions involving citric acid cycle intermediates that produce NADPH for fatty acid anabolism. These reactions take place in the cytosol.

pathways are related. The exchange reactions involving malate and citryl-CoA constitute a control mechanism in lipid anabolism, while the pentose phosphate pathway is part of carbohydrate metabolism. Both carbohydrates and lipids are important energy sources in many organisms, particularly in animals.

19-8C Amino Acids and the Citric Acid Cycle

▶ How is amino acid metabolism related to the citric acid cycle?

The anabolic reactions that produce amino acids have, as a starting point, the intermediates of the citric acid cycle that can cross the mitochondrial membrane into the cytosol. For example, malate can cross the mitochondrial membrane and give rise to oxaloacetate in the cytosol. In Chapter 23, we explain how oxaloacetate can undergo a transamination reaction to produce aspartate, and aspartate, in turn, can undergo further reactions to form not only amino acids but also other nitrogen-containing metabolites, such as pyrimidines. Similarly, isocitrate can cross the mitochondrial membrane and produce α-ketoglutarate in the cytosol. Glutamate arises from α-ketoglutarate as a result of another transamination reaction, and glutamate undergoes further reactions to form still more amino acids. Succinyl-CoA gives rise not to amino acids but to the porphyrin ring of the heme group. Another difference is that the first reaction of heme biosynthesis, the condensation of succinyl-CoA and glycine to form δ-aminolevulinic acid, takes place in the mitochondrial matrix, whereas the remainder of the pathway occurs in the cytosol.

An overall outline of anabolic reactions and their connections to the citric acid cycle is shown in Figure 19.32. We used the same type of diagram in Figure 19.23 to show an overall outline of catabolic reactions and their connections to the citric acid cycle. The similarity of the two schematic diagrams points out that catabolism and anabolism, while not exactly the same, are closely related. The operation of any metabolic pathway, anabolic or catabolic, can be "speeded up" or "slowed down" in response to the needs of an organism by control mechanisms such as feedback control. Regulation of metabolism takes place in similar ways in many different pathways.

19-9 The Link to Oxygen

Learning Objective

After reading this section, students will be able to:

LO 19.9.1 List the biochemical pathways involved in the equation for the aerobic oxidation of glucose.

The citric acid cycle is considered part of aerobic metabolism, but we have not encountered any reactions in this chapter in which oxygen takes part. However, the reactions of the citric acid cycle are intimately related to those of electron transport and oxidative phosphorylation, which do eventually lead to oxygen.

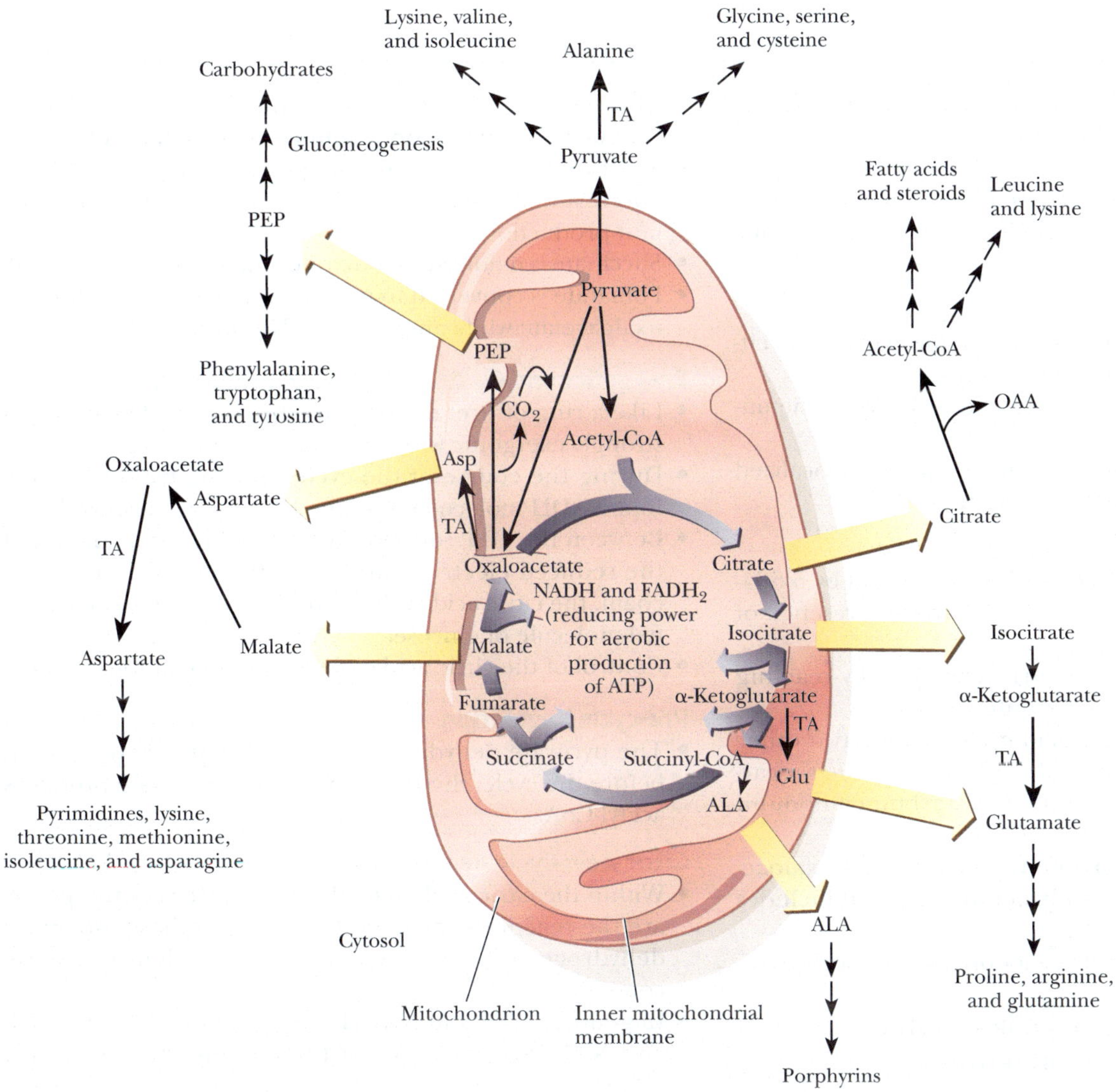

Figure 19.32 A summary of the central role of the citric acid cycle in the anabolism of sugars, fatty acids, and amino acids. Pathways for the biosynthesis of carbohydrates, lipids, and amino acids all appear. (OAA is oxaloacetate, ALA is δ-aminolevulinic acid, and PEP is phosphoenolpyruvate; TA is transamination; →→→ is a multistep pathway.)

The citric acid cycle provides a vital link between the chemical energy of nutrients and the chemical energy of ATP. We explain in Chapter 20 that many molecules of ATP can be generated as a result of electron transfer from the NADH and $FADH_2$ molecules generated in the citric acid cycle.

Recall the classic equation for the aerobic oxidation of glucose:

$$\text{Glucose} + 6O_2 \rightarrow 6H_2O + 6CO_2$$

- We saw the metabolism of glucose through glycolysis in Chapter 17.
- Here, in Chapter 19, we saw where the CO_2 comes from—namely, the three decarboxylation reactions associated with the citric acid cycle.
- In Chapter 20, we show how O_2 and H_2O get involved after events in the electron transport chain.

If insufficient oxygen is available, the electron transport chain is unable to process the reduced electron carriers from the TCA cycle, so it slows down. Continued activity under these circumstances causes the pyruvate produced by glycolysis to be processed anaerobically into lactate.

Summary

The Central Role of the Citric Acid Cycle in Metabolism

- The citric acid cycle is also known as the Krebs cycle or tricarboxylic acid (TCA) cycle.
- The citric acid cycle plays a central role in metabolism. It is the first part of aerobic metabolism, and it is amphibolic (both catabolic and anabolic).

Mitochondrial Structure

- Unlike glycolysis, which takes place in the cytosol, the citric acid cycle occurs in the mitochondria.
- Most of the enzymes of the citric acid cycle are in the mitochondrial matrix.
- Succinate dehydrogenase, the sole exception, is localized in the inner mitochondrial membrane.

Overview of Reactions in the Citric Acid Cycle

- Pyruvate produced by glycolysis is transformed by oxidative decarboxylation into acetyl-CoA in the presence of coenzyme A.
- Acetyl-CoA then enters the citric acid cycle by reacting with oxaloacetate to produce citrate.
- The reactions of the citric acid cycle include two other oxidative decarboxylations, which transform the six-carbon compound citrate into the four-carbon compound succinate.
- The cycle is completed by the regeneration of oxaloacetate from succinate in a multistep process that includes two other oxidation reactions.
- The overall citric acid cycle reaction, starting with pyruvate, is

$$\text{Pyruvate} + 4\text{NAD}^+ + \text{FAD} + \text{GDP} + \text{P}_i + 2\text{H}_2\text{O} \rightarrow$$
$$3\text{CO}_2 + 4\text{NADH} + \text{FADH}_2 + \text{GTP} + 4\text{H}^+$$

- NAD^+ and FAD are the electron acceptors in the oxidation reactions, and the cycle is strongly exergonic.

How Pyruvate Is Converted to Acetyl-CoA

- Pyruvate is produced by glycolysis in the cytosol of the cell.
- The citric acid cycle takes place in the matrix of the mitochondria, so the pyruvate must first pass through a transporter into this organelle.
- The pyruvate dehydrogenase complex (PDC) resides in the mitochondrial matrix. It is a large, multisubunit protein made up of three enzymes involved in the production of acetyl-CoA and NADH, plus two enzyme activities involved in the control of the other three enzymes of the complex.
- The PDC reaction requires several coenzymes, including FAD, lipoic acid, and TPP.

The Individual Reactions of the Citric Acid Cycle

- In the first reaction of the citric acid cycle proper, acetyl-CoA condenses with oxaloacetate to give citrate, a six-carbon compound.
- Citrate isomerizes to isocitrate, which then undergoes an oxidative decarboxylation to α-ketoglutarate, a five-carbon compound.
- This then undergoes another oxidative decarboxylation, producing succinyl-CoA, a four-carbon compound.
- The two decarboxylation steps produce CO_2 and NADH.
- Succinyl-CoA is converted to succinate with the concomitant production of GTP.
- Succinate is oxidized to fumarate, and $FADH_2$ is produced.
- Fumarate is converted to malate, which is then oxidized to oxaloacetate while another NADH is produced.

Energetics and Control of the Citric Acid Cycle

- Like a ring of fire, the overall citric acid cycle pathway is strongly exergonic with a ΔG° of −77.7 kJ/mol.
- During the course of the cycle, starting from pyruvate, four NADH molecules and one $FADH_2$ are produced.
- Between the GTP formed directly and the reoxidation of the reduced electron carriers by the electron transport chain, the citric acid cycle produces 25 ATP for each original molecule of glucose.
- Control of the citric acid cycle is exercised at four points.

Pre-cycle Regulation

- The pyruvate dehydrogenase reaction is a control point before the cycle, the reaction in which pyruvate produces acetyl-CoA.

Citric Acid Cycle Regulation

- Within the citric acid cycle, the other three control points are the reactions catalyzed by citrate synthase, isocitrate dehydrogenase, and the α-ketoglutarate dehydrogenase complex.
- In general, ATP and NADH are inhibitors, whereas ADP and NAD^+ are activators of the enzymes at the control points.

The Glyoxylate Cycle: Producing Glucose from Fatty Acid Catabolism

- In plants and bacteria, the glyoxylate cycle is a pathway related to the citric acid cycle.
- Isocitrate lyase and malate synthase are the two enzymes in plants and bacteria that allow them to convert fatty acids to glucose by bypassing the decarboxylation reactions of the citric acid cycle. These enzymes are not present in animals.
- This pathway converts fatty acids stored in seeds to carbohydrates during germination.

The Citric Acid Cycle in Catabolism

- Like an essential traffic circle of life, the citric acid cycle has many routes entering and exiting it.
- Many members of the three basic nutrient types—proteins, fats, and carbohydrates—are metabolized to smaller molecules that can cross the mitochondrial membrane and enter the citric acid cycle as one of the intermediate molecules. In this way, the cycle allows us to get energy from the food we eat.
- Carbohydrates and many amino acids can enter the cycle either as pyruvate or as acetyl-CoA.
- Lipids enter as acetyl-CoA.

- Because a transamination reaction is possible with glutamate and α-ketoglutarate, almost any amino acid can be transaminated to glutamate, thereby producing α-ketoglutarate that can enter the cycle.
- Several other pathways lead to amino acids entering the pathway as succinate, fumarate, or malate.

The Citric Acid Cycle in Anabolism

- Although the citric acid cycle takes place in the mitochondria, many anabolic reactions take place in the cytosol.
- Oxaloacetate, the starting material for gluconeogenesis, is a component of the citric acid cycle.
- Citrate from the citric acid cycle can be transported to the cytosol, where it can be converted to oxaloacetate and acetyl-CoA by the ATP-citrate lyase enzyme. The acetyl-CoA can then be used in fatty acid synthesis.
- Malate can also be transported across the mitochondrial membrane, and it plays a role in lipid anabolism. Malate is oxidatively decarboxylated to pyruvate by an enzyme that requires $NADP^+$, producing NADPH, which is a reducing agent used in fatty acid synthesis.
- Additionally, most of the citric acid cycle intermediates have anabolic pathways leading to amino acids, as well as some that lead to porphyrins or pyrimidines.

The Link to Oxygen

- Glycolysis and the citric acid cycle account for some of the overall equation for the oxidation of glucose:

$$C_6H_{12}O_6 + O_2 \rightarrow 6CO_2 + 6H_2O$$

- Glucose is the first compound in glycolysis.
- The decarboxylation steps of the citric acid cycle account for the CO_2.
- However, the oxygen and water in the equation do not appear until the last step of the electron transport chain.
- If insufficient oxygen is available, the electron transport chain is unable to process the reduced electron carriers from the TCA cycle, so it slows down.
- Continued activity under these circumstances causes the pyruvate produced by glycolysis to be processed anaerobically to lactate.

Review Exercises

19-1 The Central Role of the Citric Acid Cycle in Metabolism

1. **Recall** Which pathways are involved in the anaerobic metabolism of glucose? Which pathways are involved in the aerobic metabolism of glucose?
2. **Recall** How many ATPs can be produced from one molecule of glucose anaerobically? Aerobically?
3. **Recall** What are two other names used to describe the citric acid cycle pathway, and what is the reasoning behind these other names?
4. **Recall** The citric acid cycle is an amphibolic pathway. What does that mean?

19-2 The Overall Pathway of the Citric Acid Cycle

5. **Recall** In what part of the cell does the citric acid cycle take place, and why is this location important? Does this differ from the part of the cell where glycolysis occurs?
6. **Recall** How does pyruvate from glycolysis get to the pyruvate dehydrogenase complex (PDC)?
7. **Recall** What electron acceptors play a role in the citric acid cycle?
8. **Recall** What three molecules produced during the citric acid cycle are an indirect or direct source of high-energy compounds?

19-3 How Pyruvate Is Converted to Acetyl-CoA

9. **Recall** Describe the reaction that occurs right before the citric acid cycle and explain three reasons why it is important.
10. **Recall** How many enzymes are associated with the mammalian pyruvate dehydrogenase complex? What are their functions?
11. **Recall** Briefly describe the dual role of lipoic acid in the pyruvate dehydrogenase complex.
12. **Recall** What is the advantage to the organization of the PDH complex?
13. **Recall** In the PDH reaction, there are five vitamin or vitamin-like cofactors that play roles. What are the cofactors, and what are the vitamins?
14. **Reflect and Apply** Draw the structures of the activated carbon groups bound to thiamine pyrophosphate in the three enzymes that contain this coenzyme. *Hint:* Keto–enol tautomerism may enter into the picture.
15. **Reflect and Apply** Prepare a sketch showing how the individual reactions of the three enzymes of the pyruvate dehydrogenase complex give rise to the overall reaction.

19-4 The Individual Reactions of the Citric Acid Cycle

16. **Recall** Why is the reaction catalyzed by citrate synthase considered a condensation reaction?
17. **Biochemical Connections** What is fluoroacetate? Why is it used?
18. **Reflect and Apply** Describe the second reaction of the citric acid cycle and explain its role in the cycle.
19. **Recall** With respect to stereochemistry, what is unique about the reaction catalyzed by aconitase?
20. **Recall** In which steps of the aerobic processing of pyruvate is CO_2 produced?
21. **Recall** In which steps of the aerobic processing of pyruvate are reduced electron carriers produced?
22. **Recall** What type of reaction is catalyzed by isocitrate dehydrogenase and α-ketoglutarate dehydrogenase?
23. **Recall** What are the similarities and differences between the reactions catalyzed by the pyruvate dehydrogenase and α-ketoglutarate dehydrogenase complexes?
24. **Recall** Compare and contrast what it means when the name of an enzyme includes the word *synthase* or *synthetase.*
25. **Recall** How does substrate-level phosphorylation differ from phosphorylation linked to the electron transport chain?
26. **Recall** Give an example of substrate-level phosphorylation in a pathway other than the citric acid cycle.

27. **Recall** Why is the production of a GTP molecule equivalent to the production of an ATP molecule?
28. **Recall** What are the major differences between the oxidations in the citric acid cycle that use NAD^+ as an electron acceptor and the one that uses FAD?
29. **Reflect and Apply** ATP is a competitive inhibitor of NADH binding to malate dehydrogenase, as are ADP and AMP. Suggest a structural basis for this inhibition.
30. **Recall** Is the conversion of fumarate to malate a redox (electron transfer) reaction? Give the reason for your answer.
31. **Reflect and Apply** We have seen one of the four possible isomers of isocitrate, the one produced in the aconitase reaction. Draw the configurations of the other three.
32. **Reflect and Apply** Show, by Lewis electron-dot structures of the appropriate portions of the molecule, where electrons are lost in the following conversions:
    (a) Pyruvate to acetyl-CoA
    (b) Isocitrate to α-ketoglutarate
    (c) α-Ketoglutarate to succinyl-CoA
    (d) Succinate to fumarate
    (e) Malate to oxaloacetate

19-5 Energetics and Control of the Citric Acid Cycle

33. **Recall** Which steps of the aerobic metabolism of pyruvate through the citric acid cycle are control points?
34. **Recall** Describe how the pyruvate dehydrogenase complex (PDC) is regulated in prokaryotic cells, and explain why it makes sense.
35. **Recall** Describe the multiple ways that pyruvate dehydrogenase (PDH) is controlled.
36. **Reflect and Apply** What are the two most common inhibitors of steps of the citric acid cycle and of the reaction catalyzed by pyruvate dehydrogenase, and why does it make sense?
37. **Reflect and Apply** How does an increase in the ADP/ATP ratio affect the activity of isocitrate dehydrogenase?
38. **Reflect and Apply** How does an increase in the $NADH/NAD^+$ ratio affect the activity of pyruvate dehydrogenase?
39. **Reflect and Apply** Would you expect the citric acid cycle to be more or less active when a cell has a high ATP/ADP ratio and a high $NADH/NAD^+$ ratio? Give the reason for your answer.
40. **Reflect and Apply** Would you expect $\Delta G^{\circ\prime}$ for the hydrolysis of a thioester to be (a) large and negative, (b) large and positive, (c) small and negative, or (d) small and positive? Give the reason for your answer.
41. **Reflect and Apply** Acetyl-CoA and succinyl-CoA are both high-energy thioesters, but their chemical energy is put to different uses. Elaborate.
42. **Reflect and Apply** Some reactions of the citric acid cycle are endergonic. Show how the overall cycle is exergonic. (See Table 19.2.)
43. **Reflect and Apply** Using the information in Chapters 17–19, calculate the amount of ATP that can be produced from one molecule of lactose metabolized aerobically through glycolysis and the citric acid cycle.

19-6 The Glyoxylate Cycle: Producing Glucose from Fatty Acid Catabolism

44. **Recall** Briefly describe the purpose of the glyoxylate cycle and how it works.
45. **Recall** Indicate which enzymes of the citric acid cycle are bypassed in the glyoxylate cycle and explain why.
46. **Recall** What are the unique reactions of the glyoxylate cycle?
47. **Biochemical Connections** Why is it possible for bacteria to survive on acetic acid as a sole carbon source, but not human beings?

19-7 The Citric Acid Cycle in Catabolism

48. **Recall** Describe the various purposes of the citric acid cycle.
49. **Reflect and Apply** The intermediates of glycolysis are phosphorylated, but those of the citric acid cycle are not. Suggest a reason why.
50. **Reflect and Apply** Discuss oxidative decarboxylation, using a reaction from this chapter to illustrate your points.
51. **Reflect and Apply** Many soft drinks contain citric acid as a significant part of their flavor. Does citric acid provide nutritional value? Why or why not?

19-8 The Citric Acid Cycle in Anabolism

52. **Recall** NADH is an important coenzyme in catabolic processes, whereas NADPH appears in anabolic processes. Explain how an exchange of the two can be effected.
53. **Biochemical Connections** What are the anaplerotic reactions in mammals?
54. **Reflect and Apply** Why is acetyl-CoA considered the central molecule of metabolism?

19-9 The Link to Oxygen

55. **Reflect and Apply** Why is the citric acid cycle considered part of aerobic metabolism, even though molecular oxygen does not appear in any reaction?

Further Reading

Arnold, P. K. and L. W. S. Finley. Regulation and Function of the Mammalian Tricarboxylic Acid Cycle. *J. Biol. Chem.* **299**, Article ID 102838 (2023). This review revisits the TCA cycle, exploring its potential origins, historical discovery, and detailed regulatory mechanisms. It emphasizes the cycle's central role in energy production and biosynthesis, discussing how various factors influence its activity.

Bodner, C. M. The Tricarboxylic Acid (TCA), Citric Acid, or Krebs Cycle. *J. Chem. Ed.* **63**, 673–677 (1986). A concise and well-written summary of the citric acid cycle. Part of a series on metabolism.

Boyer, P. D., ed. *The Enzymes.* 3rd ed. New York: Academic Press, 1975. There are reviews on aconitase in Volume 5 and on dehydrogenases in Volume 11.

Kloska, S. M., K. Palczynski, T. Marciniak, T. Talaska, B. J. Wysocki, P. Davis, P., and T. A. Wysocki. Integrating Glycolysis, Citric Acid Cycle, Pentose Phosphate Pathway, and Fatty Acid Beta-Oxidation Into a Single Computational Model. *Sci. Rep.* **13**, Article ID 14484 (2023). Researchers propose a computational model that integrates multiple metabolic pathways, including the TCA cycle, using queueing theory. This model offers a quantitative framework to understand the interplay and regulation of these interconnected pathways.

Krebs, H. A. *Reminiscences and Reflections.* New York: Oxford Univ. Press, 1981. A review of the citric acid cycle, along with the autobiography.

Popjak, G. Stereospecificity of Enzyme Reactions. In Boyer, P. D., ed., *The Enzymes,* 3rd ed., Vol. 2, *Kinetics and Mechanism.* New York: Academic Press, 1970. A review of stereochemical aspects of the citric acid cycle.

See also the bibliographies for Chapters 16, 17, and 18.

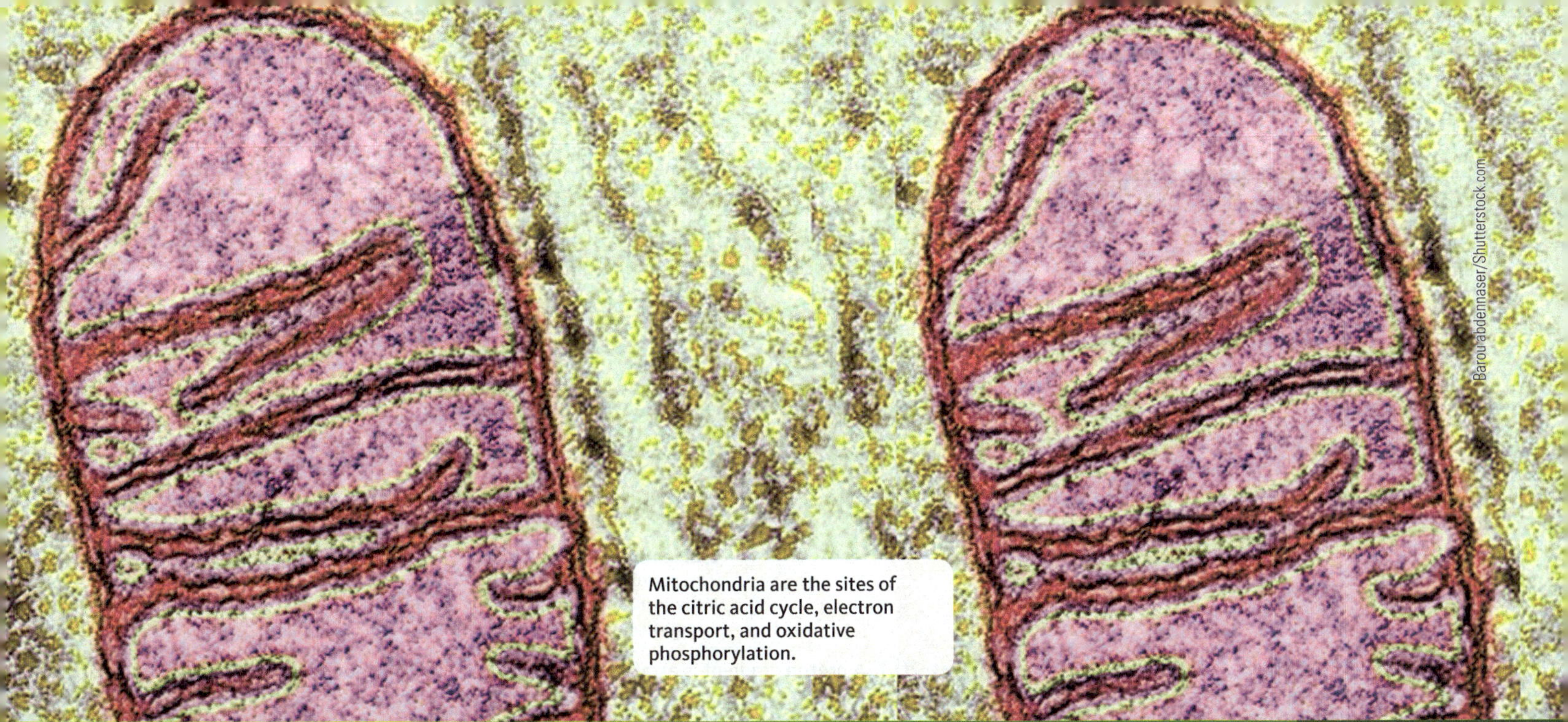

Mitochondria are the sites of the citric acid cycle, electron transport, and oxidative phosphorylation.

Electron Transport and Oxidative Phosphorylation 20

Chapter Outline

20-1 The Role of Electron Transport in Metabolism

Learning Objectives

After reading this section, students will be able to:

LO 20.1.1 Distinguish between *electron transport chain* and *oxidative phosphorylation*.

LO 20.1.2 Describe details of mitochondrial structure and proton gradients.

Aerobic metabolism is a highly efficient way for an organism to extract energy from nutrients. In eukaryotic cells, the aerobic processes (including the conversion of pyruvate to acetyl-CoA, the citric acid cycle, and electron transport) all occur in the mitochondria, while the anaerobic process, glycolysis, takes place outside the mitochondria in the cytosol. We have not yet seen any reactions in which oxygen plays a part, but here in Chapter 20, we discuss the role of oxygen in metabolism as the final acceptor of electrons in the **electron transport chain**. The oxidation–reduction (redox) reactions of the electron transport chain take place in the inner mitochondrial membrane.

Chapter Outline (continued)

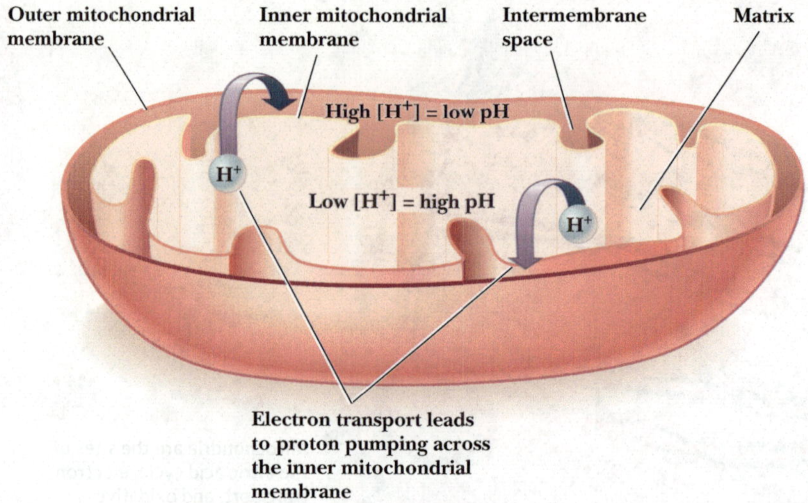

Figure 20.1 A proton gradient is established across the inner mitochondrial membrane as a result of electron transport. Transfer of electrons through the electron transport chain leads to the pumping of protons from the matrix to the intermembrane space. The proton gradient (also called the pH gradient), together with the membrane potential (a voltage across the membrane), provides the basis of the chemiosmotic coupling mechanism that drives ATP synthesis.

electron transport chain the series of intermediate carriers that transfer electrons from NADH and $FADH_2$ to oxygen

oxidative phosphorylation a process for generating ATP; it depends on the creation of a pH gradient within the mitochondrion as a result of electron transport

proton gradient the difference between the hydrogen ion concentrations in the mitochondrial matrix and in the intermembrane space, which is the basis of the coupling between oxidation and phosphorylation

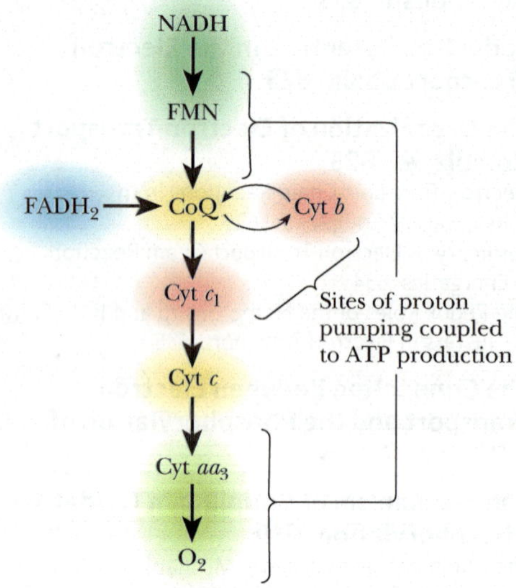

Figure 20.2 Schematic representation of the electron transport chain, showing sites of proton pumping coupled to oxidative phosphorylation. FMN is the flavin coenzyme *flavin mononucleotide*, which differs from flavin adenine dinucleotide (FAD) in not having an adenine nucleotide. CoQ is coenzyme Q. Cyt *b*, cyt c_1, cyt *c*, and cyt aa_3 are the heme-containing proteins cytochrome *b*, cytochrome c_1, cytochrome *c*, and cytochrome aa_3, respectively.

► What is the importance of mitochondrial structure in adenosine triphosphate (ATP) production?

The energy released by the oxidation of nutrients is used by organisms in the form of the chemical energy of ATP. Production of ATP in the mitochondria is the result of **oxidative phosphorylation**, in which adenosine diphosphate (ADP) is phosphorylated to give ATP. The production of ATP by oxidative phosphorylation (an endergonic process) is separate from electron transport to oxygen (an exergonic process), but the reactions of the electron transport chain are strongly linked to one another and are tightly coupled to the synthesis of ATP by phosphorylation of ADP. The operation of the electron transport chain leads to the pumping of protons (hydrogen ions) from the matrix into the intermembrane space, creating a pH gradient (also called a **proton gradient**) where the impermeable inner membrane prevents their return except through ATP synthase (Figure 20.1 and Section 20-5). This proton gradient represents stored potential energy and provides the basis of the coupling mechanism called *chemiosmotic coupling*. Oxidative phosphorylation gives rise to most of the ATP production associated with the complete oxidation of glucose.

The NADH and $FADH_2$ molecules generated in glycolysis and the citric acid cycle transfer electrons to oxygen in the series of electron transport chain reactions. The NADH and $FADH_2$ are oxidized to NAD^+ and FAD and can be used again in various metabolic pathways. Oxygen, the ultimate electron acceptor, is reduced to water, thus completing the process by which glucose is completely oxidized to carbon dioxide and water. Recall from Chapter 19 how carbon dioxide is produced from pyruvate being converted to acetyl-CoA and from other subsequent citric acid cycle reactions. In Chapter 20, we see how water is produced.

The complete series of oxidation–reduction reactions of the electron transport chain is presented in schematic form in Figure 20.2. A particularly noteworthy point about electron transport is that, on average, 2.5 moles of ATP are generated for each mole of NADH that enters the electron transport chain, whereas, on average, 1.5 moles of ATP are produced for each mole of $FADH_2$. The general outline of the process is that NADH passes electrons to coenzyme

Pioneers in Biochemistry • Mildred Cohn

Mildred Cohn (1913–2009) was a pioneering biochemist who overcame gender and religious discrimination in her distinguished career as a researcher and professor. Cohn was born in the Bronx, New York, to Jewish parents who had recently immigrated from Russia. She began her college study of chemistry at age 15, despite the advice from an instructor that it was "unladylike" for women to become chemists.

Cohn's pursuit of an advanced degree was first derailed by rules excluding women from assistantships. Later, she faced steep obstacles securing employment, as industry positions were routinely advertised exclusively for Christian men. In the end, she benefited from the mentorship of several Nobel laureates to make her way in the world of research.

Cohn is best known for her pioneering use of nuclear magnetic resonance (NMR) to investigate adenosine triphosphate (ATP) in one of her talks, she explained, "I didn't intend to be an assistant for the rest of my life, so I started a new field of research." Cohn was the first to distinguish the three phosphorus atoms of ATP spectroscopically, and her research helped elucidate the biochemistry of ATP and metabolism. Her techniques for using NMR to study metabolic processes at the molecular level were widely adopted.

By the end of her career, Cohn had been granted honorary doctorates from nine universities and had received numerous awards, including the National Medal of Science.

Bettmann/Bettmann/Getty Images

Q, as does $FADH_2$, providing an alternative mode of entry into the electron transport chain. Electrons are then passed from coenzyme Q to a series of proteins called cytochromes (which are designated by lowercase letters) and, eventually, to oxygen.

20-2 Reduction Potentials in the Electron Transport Chain

Learning Objectives

After reading this section, students will be able to:

LO 20.2.1 Describe how standard reduction potentials are measured.

LO 20.2.2 Explain how reduction potentials can be used to predict the direction of electron transport.

Up until now, most of the energy considerations we have had concerned phosphorylation potentials. In Section 15-6, we saw how the Gibbs energy change associated with the hydrolysis of adenosine triphosphate (ATP) could be used to drive otherwise endergonic reactions. The opposite is also true—when a reaction is highly exergonic, it can drive the formation of ATP. When we look closely at the energy changes in electron transport, a more useful approach is to consider the change in energy associated with the movement of electrons from one carrier to another. Each carrier in the electron transport chain can be isolated and studied, and each can exist in an oxidized or reduced form (Section 15-5).

If we had two potential electron carriers, such as NADH and coenzyme Q, for example, how would we know whether electrons would be more likely to be transferred from NADH to coenzyme Q or the other way around? This is determined by measuring a **reduction potential** for each of the carriers. A molecule with a high reduction potential tends to be reduced if it is paired with a molecule with a lower reduction potential. This is measured by making a simple

reduction potential a standard voltage that indicates the tendency of a reduction half reaction to take place

battery cell, as shown in Figure 20.3. The reference point is the half–cell on the right, where hydrogen ion is in aqueous solution in equilibrium with hydrogen gas. The reduction of hydrogen ion to hydrogen gas is the control and is considered to have a voltage (E) of zero. The sample to be tested is in the other half–cell. The electric circuit is completed by a bridge with a salt-containing agar gel.

$$2H^+ + 2e^- \rightarrow H_2$$

▶ How can reduction potentials be used to predict the direction of electron transport?

Figure 20.3A shows what happens if ethanol and acetaldehyde are put into the sample half–cell. Electrons flow away from the sample cell and toward the reference cell. This means that the hydrogen ion is being reduced to hydrogen gas and the ethanol is being oxidized to acetaldehyde. Therefore, the hydrogen/H^+ pair has a higher reduction potential than the ethanol/acetaldehyde pair. The opposite occurs in Figure 20.3B. When fumarate and succinate are put into the sample half–cell, the electrons flow in the opposite direction, meaning that fumarate is being reduced to succinate while hydrogen gas is being oxidized to H^+. The direction of electron flow and the magnitude of the observed voltage allow us to make a table like the one shown in Table 20.1. Because this is a table of standard reduction potentials, all the reactions are written as reductions. The value being measured is the standard biological voltage of each half reaction $E°'$. This value is calculated based on the compounds in the cells being at 1 M and the pH being 7 at the standard temperature of 25°C.

To interpret the data in Table 20.1 for the purpose of electron transport, we need to look at the reduction potentials of the electron carriers involved. A reaction at the top of the table tends to occur as a reduction, as written, if it is paired with a reaction that is lower down on the table. The reaction lower in the table then occurs as an oxidation—that is, it occurs in the reverse of what is written in Table 20.1. For example, we have already seen that the final step of the electron transport chain is the reduction of oxygen to water. This reaction is at the top of Table 20.1, with a reduction potential of 0.816 V, a very positive number. If this reaction were paired directly with NAD^+/NADH, what would happen? The standard reduction potential for NAD^+ forming NADH is given near the bottom of the table. Its reduction potential is –0.320 V.

$$NAD^+ + 2H^+ + 2e^- \rightarrow NADH + H^+ \qquad E°' = -0.320\ V$$

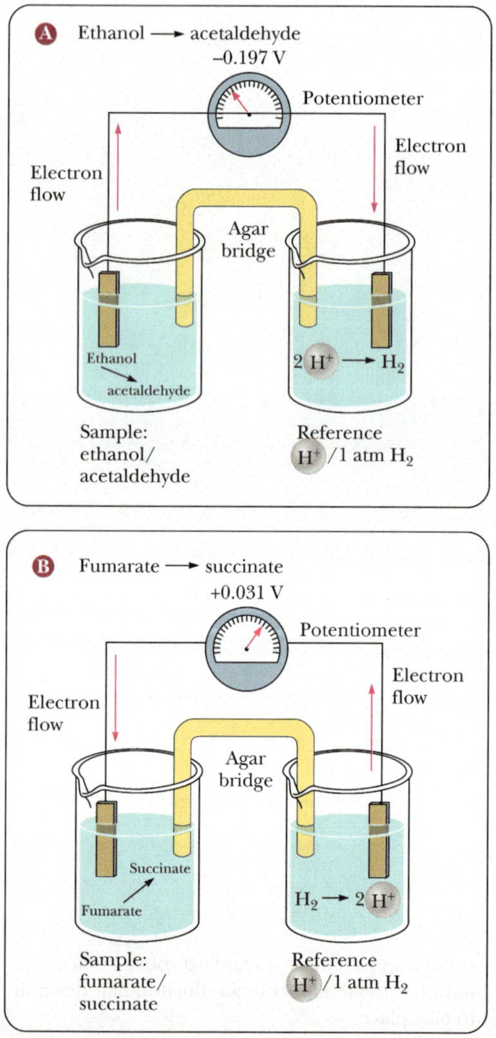

Figure 20.3 Experimental apparatus used to measure the standard reduction potential of the indicated redox couples: (A) the ethanol/acetaldehyde couple and (B) the fumarate/succinate couple. Part (A) shows a sample/reference half–cell pair for measuring the standard reduction potential of the ethanol/acetaldehyde couple. Because electrons flow toward the reference half cell and away from the sample half–cell, the standard reduction potential is negative, specifically –0.197 V. In contrast, the fumarate/succinate couple in part (B) accepts electrons from the reference half–cell; that is, reduction occurs spontaneously in the system, and the reduction potential is thus positive. For each half–cell, a *half–cell reaction* describes the reaction taking place. For the fumarate/succinate half–cell coupled to a H^+/H_2 reference half–cell (B), the reaction taking place is indeed the reduction of fumarate.

$$\text{Fumarate} + 2H^+ + 2e^- \rightarrow \text{Succinate}$$
$$E°' = +0.031\ V$$

However, the reaction occurring in the ethanol/acetaldehyde half–cell (A) is the oxidation of ethanol, which is the reverse of the reaction listed in Table 20.1.

$$\text{Ethanol} \rightarrow \text{Acetaldehyde} + 2H^+ + 2e^-$$
$$E°' = +0.197\ V$$

Table 20.1 Standard Reduction Potentials for Several Biological Reduction Half Reactions

Reduction Half Reaction	$E°'$ (V)
$\frac{1}{2}O_2 + 2H^+ + 2e^- \rightarrow H_2O$	0.816
$Fe^{3+} + e^- \rightarrow Fe^{2+}$	0.771
Cytochrome $a_3(Fe^{3+}) + e^- \rightarrow$ Cytochrome $a_3(Fe^{2+})$	0.350
Cytochrome $a(Fe^{3+}) + e^- \rightarrow$ Cytochrome $a(Fe^{2+})$	0.290
Cytochrome $c(Fe^{3+}) + e^- \rightarrow$ Cytochrome $c(Fe^{2+})$	0.254
Cytochrome $c_1(Fe^{3+}) + e^- \rightarrow$ Cytochrome $c_1(Fe^{2+})$	0.220
$CoQH^{\bullet} + H^+ + e^- \rightarrow CoQH_2$ (coenzyme Q)	0.190
$CoQ + 2H^+ + 2e^- \rightarrow CoQH_2$	0.060
Cytochrome $b_H(Fe^{3+}) + e^- \rightarrow$ Cytochrome $b_H(Fe^{2+})$	0.050
Fumarate + $2H^+ + 2e^- \rightarrow$ Succinate	0.031
$CoQ + H^+ + e^- \rightarrow CoQH^{\bullet}$	0.030
$FAD + 2H^+ + 2e^- \rightarrow FADH_2$	0.003–0.091*
Cytochrome $b_L(Fe^{3+}) + e^- \rightarrow$ Cytochrome $b_L(Fe^{2+})$	–0.100
Oxaloacetate + $2H^+ + 2e^- \rightarrow$ Malate	–0.166
Pyruvate + $2H^+ + 2e^- \rightarrow$ Lactate	–0.185
Acetaldehyde + $2H^+ + 2e^- \rightarrow$ Ethanol	–0.197
$FMN + 2H^+ + 2e^- \rightarrow FMNH_2$	–0.219
$FAD + 2H^+ + 2e^- \rightarrow FADH_2$	–0.219
1,3-Bisphosphoglycerate + $2H^+ + 2e^- \rightarrow$ Glyceraldehyde-3-phosphate + P_i	–0.290
$NAD^+ + 2H^+ + 2e^- \rightarrow NADH + H^+$	–0.320
$NADP^+ + 2H^+ + 2e^- \rightarrow NADPH + H^+$	–0.320
α-Ketoglutarate + $CO_2 + 2H^+ + 2e^- \rightarrow$ Isocitrate	–0.380
Succinate + $CO_2 + 2H^+ + 2e^- \rightarrow$ α-Ketoglutarate + H_2O	–0.670

*Typical values for reduction of bound FAD in flavoproteins such as succinate dehydrogenase. The value for reduction of free FAD is –0.219 V. We have listed a number of components of the electron transport chain individually. We are going to see them again as part of integral protein complexes. We have also included values for a number of reactions we saw in previous chapters.

This means that, if the two half reactions are paired during a redox reaction, the one for the NADH must be reversed. NADH gives up its electrons so that oxygen can be reduced to water:

$$\begin{array}{lll} & NADH + H^+ \rightarrow NAD^+ + 2H^+ + 2e^- & 0.320 \\ & \frac{1}{2}O_2 + 2H^+ + 2e^- \rightarrow H_2O & 0.816 \\ \hline \text{Sum} & NADH + \frac{1}{2}O_2 + H^+ \rightarrow NAD^+ + H_2O & 1.136 \end{array}$$

The overall voltage for this reaction is the sum of the standard reduction potentials—in this case, 0.816 V + 0.320 V = 1.136 V. We had to change the sign on the standard reduction potential for NADH because we had to reverse the direction of its reaction.

The $\Delta G^{\circ\prime}$ of a redox reaction is calculated using

$$\Delta G^{\circ\prime} = -nF\,\Delta E^{\circ\prime}$$

where n is the number moles of electrons transferred, F is the Faraday constant (96.485 kJ/V/mol), and $\Delta E^{\circ\prime}$ is the total voltage for the two half reactions. As we can see from this equation, $\Delta G^{\circ\prime}$ is negative when $\Delta E^{\circ\prime}$ is positive. Therefore, we can always calculate the direction in which a redox reaction will go under standard conditions by combining the two half reactions in the way that gives the largest positive value for $\Delta E^{\circ\prime}$. For this example, $\Delta G^{\circ\prime}$ would be calculated as follows:

$$\Delta G^{\circ\prime} = -(2)(96.485 \text{ kJ/V/mol})(1.136 \text{ V}) = -219 \text{ kJ/mol}$$

This would be a very large number if NADH reduced oxygen directly. As we explain in Section 20-3, NADH passes its electrons along a chain that eventually leads to oxygen, but it does not reduce oxygen directly.

Just as there is a difference between ΔG° and ΔG, there is a similar difference between ΔE° and ΔE. Recall from Chapter 15 that we devoted several sections to the question of standard states, including the modified standard state for biochemical reactions. The notations ΔG and ΔE refer to the Gibbs energy change and the reduction potential under any conditions, respectively. When all components of a reaction are in their standard state (1 atm pressure, 25°C, all solutes at 1 M concentration), we write ΔG° and ΔE°, respectively, for the standard Gibbs energy change and standard reduction potential. The modified standard state for biochemical reactions takes note of the fact that having all solutes at 1 M concentration includes the hydrogen ion concentration. That implies a pH equal to zero. Consequently, we define a modified standard state for biochemistry that differs from the usual one only in that pH = 7. Under these conditions, we write $\Delta G^{\circ\prime}$ and $\Delta E^{\circ\prime}$ for the standard Gibbs energy change and the standard reduction potential, respectively. The true direction of electron flow in a redox reaction is also based on the true values of the concentrations for the reactants and products because the cellular concentrations are never 1 M.

20-3 The Organization of Electron Transport Complexes

Learning Objectives

After reading this section, students will be able to:

LO 20.3.1 Identify the electron transfer reactions that occur in each of the four respiratory complexes.

LO 20.3.2 Describe the roles of heme groups and iron–sulfur clusters in electron transport.

respiratory complexes integral enzymes of the inner mitochondrial membrane that are made up of multiple subunits and are involved in the process of energy production

Intact mitochondria isolated from cells can carry out all the reactions of the electron transport chain; the electron transport apparatus can also be resolved into its component parts by a process called fractionation. Four separate **respiratory complexes** can be isolated from the inner mitochondrial membrane. These complexes are multienzyme systems. In Chapter 19, we encountered other examples of such multienzyme complexes, such as the pyruvate dehydrogenase complex (Section 19-3) and the α-ketoglutarate dehydrogenase complex (Section 19-4). Each of the respiratory complexes can carry out the reactions of a portion of the electron transport chain.

20-3A Electron Transport Chain Reactions in the Respiratory Complexes

Complex I The first complex, **NADH-CoQ oxidoreductase** (also known as **NADH dehydrogenase**), catalyzes the first steps of the electron transport chain—namely, the transfer of electrons from NADH to **coenzyme Q (CoQ).** This complex is an integral part of the inner mitochondrial membrane and includes, among other subunits, several proteins that contain an iron–sulfur cluster and the flavoprotein that oxidizes NADH. (The total number of subunits is more than 20. This complex is a subject of active research, which has proven to be a challenging task because of its complexity. It is particularly difficult to generalize about the nature of the iron–sulfur clusters because they vary from species to species.) The flavoprotein has a flavin coenzyme, called flavin mononucleotide (FMN), which differs from flavin adenine dinucleotide (FAD) in not having an adenine nucleotide (Figure 20.4).

NADH-CoQ oxidoreductase (NADH dehydrogenase) the first complex of the electron transport chain; it catalyzes the transfer of electrons from NADH to coenzyme Q

coenzyme Q (CoQ) an oxidation–reduction coenzyme in mitochondrial electron transport that is the common electron acceptor from the NADH and $FADH_2$ electron carriers; also known as ubiquinone

The Complex I reaction occurs in several steps:

- The first step is the transfer of electrons from NADH to the flavin portion of the flavoprotein:

$$NADH + H^+ + E\text{—}FMN \rightarrow NAD^+ + E\text{—}FMNH_2$$

in which the notation E—FMN indicates that the flavin is covalently bonded to the enzyme.

- In the second step, the reduced flavoprotein is reoxidized, and the oxidized form of the iron–sulfur protein is reduced. (The Fe—S notation indicates the iron–sulfur clusters.).

$$E\text{—}FMNH_2 + 2Fe\text{—}S_{oxidized} \rightarrow E\text{—}FMN + 2Fe\text{—}S_{reduced} + 2H^+$$

- Finally, the reduced iron–sulfur protein donates its electrons to coenzyme Q, which becomes reduced to $CoQH_2$.

$$2Fe\text{—}S_{reduced} + CoQ + 2H^+ \rightarrow 2Fe\text{—}S_{oxidized} + CoQH_2$$

Coenzyme Q is also called *ubiquinone,* and its oxidized and reduced forms are shown in Figure 20.5.

The overall equation for the Complex I reaction is

$$NADH + H^+ + CoQ \rightarrow NAD^+ + CoQH_2$$

This reaction is one of the three responsible for the proton pumping that creates the pH (proton) gradient (Figure 20.1). Figure 20.6 shows how the Complex I reactions fit in with the overall picture of the four respiratory complexes that are involved in electron transport to oxygen.

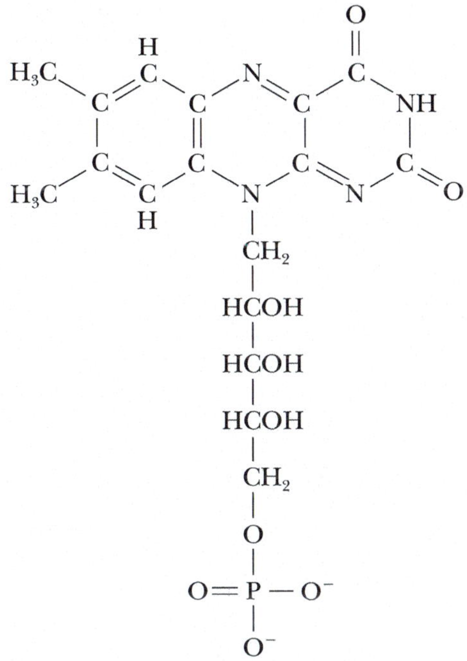

Figure 20.4 **The structure of flavin mononucleotide (FMN).**

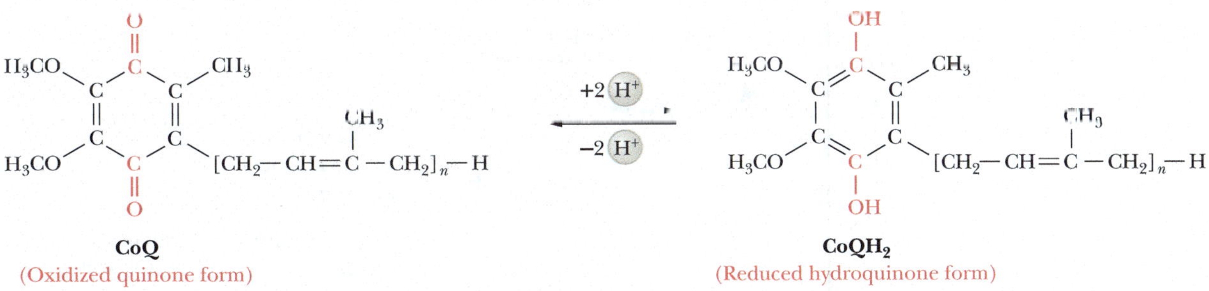

Figure 20.5 **The oxidized and reduced forms of coenzyme Q (CoQ).** Coenzyme Q is also called ubiquinone, and the value of n is usually 10 in humans and most mammals.

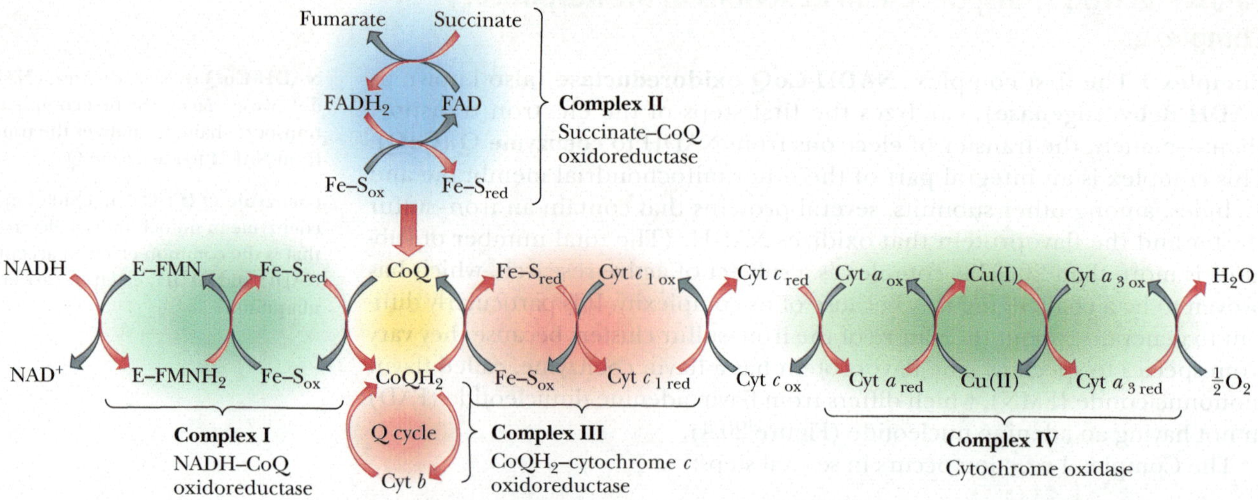

Figure 20.6 **The electron transport chain, showing the respiratory complexes.** In the reduced cytochromes, the iron is in the Fe(II) oxidation state; in the oxidized cytochromes, the iron is in the Fe(III) oxidation state.

The standard Gibbs-energy change ($\Delta G^{\circ\prime} = -81$ kJ/mol $= -19.4$ kcal/mol) for the overall Complex I reaction indicates that the reaction is strongly exergonic, releasing enough energy to drive the phosphorylation of ADP to ATP (Figure 20.7). An important consideration about proton pumping and electron transport is the subtle differences among the electron carriers. Although they can all exist in an oxidized or reduced form, they reduce each other in a certain order, as we saw in Section 20-2. In other words, reduced NADH donates its electrons to coenzyme Q, but not the other way around. Thus, there is a direction to the electron flow in the complexes discussed here.

The other important subtlety is that some carriers, such as NADH, carry electrons and hydrogens in their reduced forms; others, such as the iron–sulfur protein we just saw, can carry only electrons. This is the basis of the proton pumping that ultimately leads to ATP production. When a carrier such as NADH reduces the iron–sulfur protein, it passes along its electrons but not its hydrogens. The architecture of the inner mitochondrial membrane and the electron carriers allows the hydrogen ions to pass out on the opposite side of the membrane. We look more closely at this in Section 20-5.

The final electron receptor of Complex I, coenzyme Q, is mobile—that is to say, it is free to move in the membrane and to pass the electrons it has gained

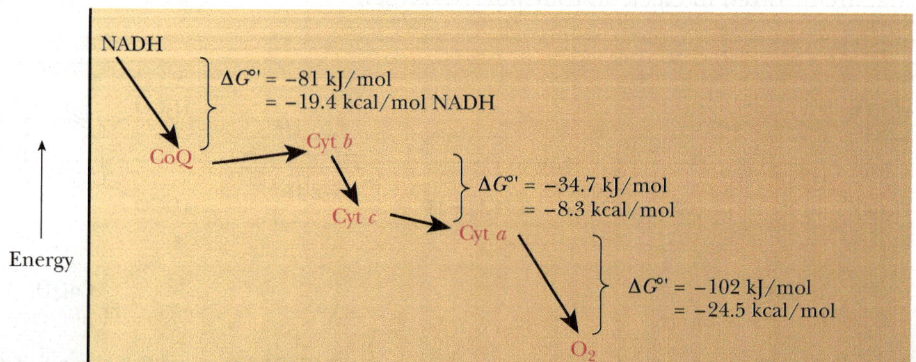

Figure 20.7 **The energetics of electron transport.**

to the third complex for further transport to oxygen. We shall now see that the second complex also transfers electrons from an oxidizable substrate to coenzyme Q.

Complex II The second of the four membrane-bound complexes, **succinate-CoQ oxidoreductase**, also catalyzes the transfer of electrons to coenzyme Q. However, its source of electrons (in other words, the substance being oxidized) differs from the oxidizable substrate (NADH) acted on by NADH-CoQ oxidoreductase. In this case, the substrate is succinate from the citric acid cycle, which is oxidized to fumarate by a flavin enzyme (Figure 20.6, *top left*).

succinate-CoQ oxidoreductase the second complex of the electron transport chain, it catalyzes the transfer of electrons from succinate to coenzyme Q

The Complex II reaction also occurs in several steps:

- The first step is the transfer of electrons from succinate to the flavin portion of the flavoprotein:

$$\text{Succinate} + \text{E—FAD} \rightarrow \text{Fumarate} + \text{E—FADH}_2$$

  The E—FAD notation indicates that the flavin portion is covalently bonded to the enzyme.

- In the second step, the flavin group is reoxidized as the iron–sulfur protein of the complex is reduced.

$$\text{E—FADH}_2 + \text{Fe—S}_{\text{oxidized}} \rightarrow \text{E—FAD} + \text{Fe—S}_{\text{reduced}}$$

- Finally, the reduced iron–sulfur protein donates its electrons to oxidized coenzyme Q, producing $CoQH_2$.

$$\text{Fe—S}_{\text{reduced}} + \text{CoQ} + 2\text{H}^+ \rightarrow \text{Fe—S}_{\text{oxidized}} + \text{CoQH}_2$$

The overall Complex II reaction is

$$\text{Succinate} + \text{CoQ} \rightarrow \text{Fumarate} + \text{CoQH}_2$$

We saw the first step of this reaction when we discussed the oxidation of succinate to fumarate as part of the citric acid cycle (Section 19-4). The enzyme traditionally called succinate dehydrogenase, which catalyzes the oxidation of succinate to fumarate, has been shown to be a part of this enzyme complex. Recall that the succinate dehydrogenase portion consists of a flavoprotein and an iron–sulfur protein. The other components of Complex II are a *b*-type cytochrome and two iron–sulfur proteins. The whole complex is also an integral enzyme complex in the inner mitochondrial membrane. The standard Gibbs energy change ($\Delta G°'$) is −13.5 kJ/mol = −3.2 kcal/mol. The overall reaction is exergonic, but there is not enough energy from this reaction to drive ATP production, and no hydrogen ions are pumped out of the matrix during this step.

In further steps of the electron transport chain, electrons are passed from coenzyme Q, which is then reoxidized, to the first of a series of very similar proteins called **cytochromes**. Each of these proteins contains a heme group, and in each heme group, the iron is successively reduced to Fe(II) and reoxidized to Fe(III). This situation differs from that of the iron in the heme group of myoglobin and hemoglobin (Sections 4-4B and 4-5A, respectively), which remains in the reduced form as Fe(II) through the entire process of oxygen binding and transport. There are also some structural differences between the heme group in hemoglobin and the heme groups in the various types of cytochromes.

cytochromes groups of heme-containing proteins in the electron transport chain

The successive oxidation–reduction reactions of the cytochromes

$$\text{Fe(III)} + e^- \rightarrow \text{Fe(II)} \quad \text{(reduction)}$$

and

$$\text{Fe(II)} \rightarrow \text{Fe(III)} + e^- \quad \text{(oxidation)}$$

differ from one another in their ΔG° values because of the influences of the various types of hemes and protein structures. Each of the proteins is slightly different in structure, and thus each protein has slightly different properties, including the tendency to participate in oxidation–reduction reactions. The different types of cytochromes are distinguished by lowercase letters (*a, b, c*); further distinctions are possible with subscripts, as in c_1.

$CoQH_2$-cytochrome *c* oxidoreductase (cytochrome bc_1 complex) the third complex of the electron transport chain, it catalyzes the transfer of electrons from reduced coenzyme Q to cytochrome *c*

Complex III The third complex, **$CoQH_2$-cytochrome *c* oxidoreductase** (also called **cytochrome bc_1 complex**), catalyzes the oxidation of reduced coenzyme Q ($CoQH_2$). The electrons produced by this oxidation reaction are passed along to cytochrome *c* in a multistep process. The overall Complex III reaction is

$$CoQH_2 + 2\text{ Cyt }c[\text{Fe(III)}] \rightarrow CoQ + 2\text{ Cyt }c[\text{Fe(II)}] + 2H^+$$

Recall that the oxidation of coenzyme Q involves two electrons, whereas the reduction of Fe(III) to Fe(II) requires only one electron. Therefore, two molecules of cytochrome *c* are required for every molecule of coenzyme Q; this is indicated by the stoichiometric coefficients seen in the overall reaction. The components of this complex include cytochrome *b* (actually two *b*-type cytochromes, cytochrome b_H and b_L), cytochrome c_1, and several iron–sulfur proteins (Figure 20.6). Cytochromes can carry electrons, but not hydrogens. This is another location where hydrogen ions leave the matrix. When reduced $CoQH_2$ is oxidized to CoQ, the hydrogen ions pass out on the other side of the membrane. Complex III is also an integral enzyme complex in the inner mitochondrial membrane.

Figure 20.8 shows how the electron transport chain reactions in Complexes I, II, and III are connected to each other, as well as the connection to Complex IV (discussed below). Complex II is not shown in Figure 20.8, but

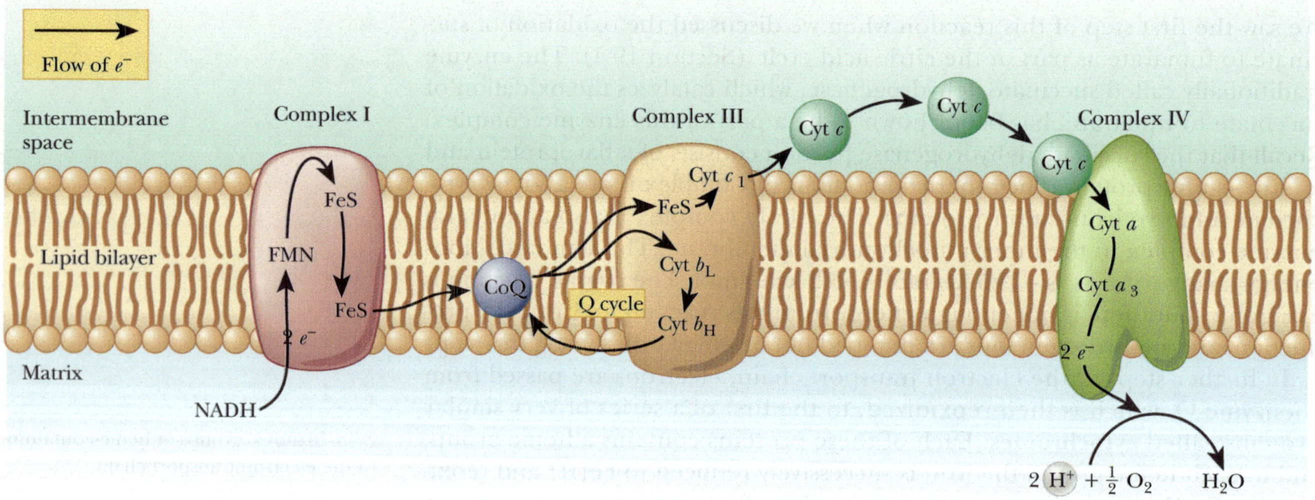

Figure 20.8 The electron transport chain reactions of the respiratory complexes in the inner mitochondrial membrane, showing the flow of electrons from NADH to O_2. NADH has accepted electrons from substrates such as pyruvate, isocitrate, α-ketoglutarate, and malate. The binding site for NADH is on the matrix side of the membrane. Coenzyme Q (CoQ) is soluble in the lipid bilayer. Complex II is not shown, but it also contributes two electrons to CoQ. Complex III contains two *b*-type cytochromes, which are involved in the Q cycle. Cytochrome *c* is loosely bound to the membrane, facing the intermembrane space. In Complex IV, the binding site for oxygen lies on the side toward the matrix.

it also contributes electrons to CoQ, which then transfers electrons to Complex III. Coenzyme Q is soluble in the lipid component of the mitochondrial membrane. It is separated from the complex in the fractionation process that resolves the electron transport apparatus into its component parts, but the coenzyme is probably close to the respiratory complexes in the intact membrane. Cytochrome *c* itself is not part of the complex but is loosely bound to the outer surface of the inner mitochondrial membrane, facing the intermembrane space.

These two important electron carriers, coenzyme Q and cytochrome *c*, are *not* part of the respiratory complexes, but they can move freely in the membrane, acting as mobile carriers of electrons between the respiratory complexes. The respiratory complexes themselves move within the membrane (recall lateral motion within membranes from Section 8-3), and it has been observed that *Complex I–III–IV supercomplexes* can form, increasing the efficiency of proton pumping and pH gradient formation.

The flow of electrons from reduced coenzyme Q to the other components of the complex does not take a simple, direct path. Instead, a cyclic flow of electrons involves coenzyme Q twice. This behavior depends on the fact that, as a quinone, coenzyme Q can exist in three forms (Figure 20.9). The semiquinone form, which is intermediate between the oxidized and reduced forms, is of crucial importance here. Because of the crucial involvement of coenzyme Q, this portion of the pathway is called the **Q cycle** (Figure 20.8, *middle*).

In part of the Q cycle, *one* electron is passed from reduced coenzyme Q to the iron–sulfur clusters and then to cytochrome c_1, leaving coenzyme Q in the semiquinone form.

$$\text{CoQH}_2 \rightarrow \text{Fe—S} \rightarrow \text{Cyt } c_1$$

The Fe—S notation indicates the iron–sulfur clusters. The overall reaction involving coenzyme Q and cytochrome c_1 can be written as

$$\text{CoQH}_2 + \text{Cyt } c_1(\text{oxidized}) \rightarrow \text{Cyt } c_1(\text{reduced}) + \text{CoQ}^- \text{ (semiquinone anion)} + 2\text{H}^+$$

The semiquinone, along with the oxidized and reduced forms of coenzyme Q, participates in a cyclic process in which the two *b* cytochromes are reduced and oxidized in turn. A second molecule of coenzyme Q is involved, transferring a second electron to cytochrome c_1, and from there to the mobile carrier cytochrome *c*. We are going to omit a number of details of the process in the interest of simplicity. Each of the two molecules of coenzyme Q involved in the Q cycle loses one electron. The net result is the same as if one molecule of CoQ had lost two electrons. One molecule of $CoQH_2$ is regenerated, and one is oxidized to CoQ, which is consistent with this picture. Most important, the Q cycle provides a mechanism for electrons to be transferred one at a time from coenzyme Q to cytochrome c_1.

Similar to the reactions of Complex I discussed previously, proton pumping, to which ATP production is coupled, occurs as a result of the reactions of Complex III. The Q cycle is implicated in the process, and the standard Gibbs-energy change ($\Delta G°'$) is –34.7 kJ = –8.3 kcal for each mole of NADH that enters the electron transport chain (Figure 20.7, *middle*). The phosphorylation of ADP requires 30.5 kJ/mol = 7.3 kcal/mol, so the reaction catalyzed by Complex III supplies enough energy to drive the production of ATP.

Complex IV The fourth complex, **cytochrome c oxidase**, catalyzes the final steps of electron transport: the transfer of electrons from cytochrome *c* to oxygen.

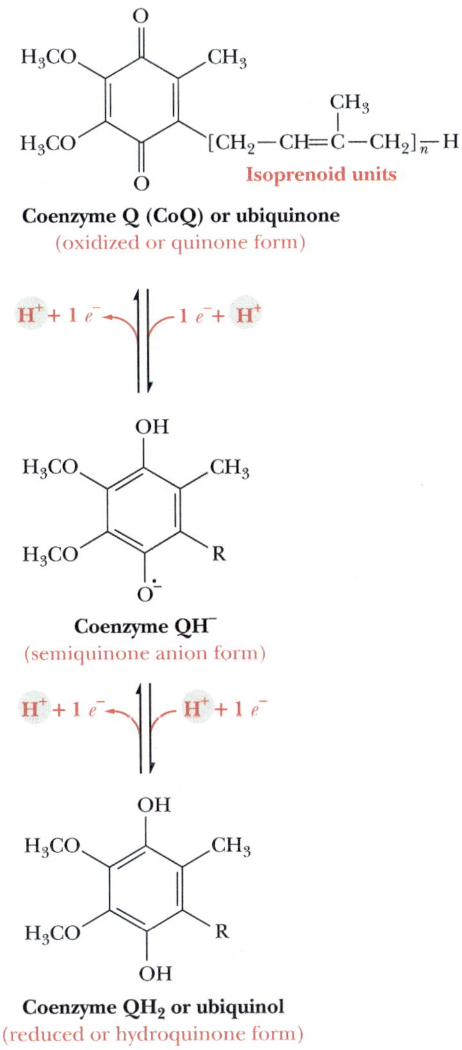

Figure 20.9 The oxidized and reduced forms of coenzyme Q, showing the intermediate semiquinone anion form involved in the Q cycle.

Q cycle a series of reactions in the electron transport chain that provides the link between two-electron transfers and one-electron transfers

cytochrome c oxidase the fourth complex of the electron transport chain, it catalyzes the transfer of electrons from cytochrome *c* to oxygen

The overall reaction is

$$2\ \text{Cyt}\ c[\text{Fe(II)}] + 2\text{H}^+ + \tfrac{1}{2}\text{O}_2 \rightarrow 2\ \text{Cyt}\ c[\text{Fe(III)}] + \text{H}_2\text{O}$$

Proton pumping also takes place as a result of this reaction. Like the other respiratory complexes, cytochrome oxidase is an integral part of the inner mitochondrial membrane and contains cytochromes *a* and a_3, as well as two Cu^{2+} ions that are involved in the electron transport process. Taken as a whole, this complex contains about 10 subunits. In the flow of electrons, the copper ions are intermediate electron acceptors that lie between the two *a*-type cytochromes in the following sequence:

$$\text{Cyt}\ c \rightarrow \text{Cu}_\text{A} \rightarrow \text{Cyt}\ a \rightarrow \text{Cyt}\ a_3\text{-Cu}_\text{B} \rightarrow \text{O}_2$$

To explain the sequence more explicitly, reduced cytochrome *c* binds to Complex IV and donates an electron to the Cu_A center of the complex. From the Cu_A center, the electron is passed to cytochrome *a*. Then the electron is passed to the cytochrome a_3-Cu_B binuclear center, which contains two metal ions: an iron in the heme group of cytochrome a_3 and a copper ion. Finally, the aforementioned electron, plus one more electron (via another round of the same process), is donated to molecular oxygen.

Note that in this final reaction we have finally seen the link to molecular oxygen in aerobic metabolism.

20-3B Summary of Electron Transport Chain Reaction Energetics

The standard Gibbs energy change ($\Delta G^{\circ\prime}$) is −110 kJ = −26.3 kcal for each mole of NADH that enters the electron transport chain (see Figure 20.7). We have now seen the three places in the respiratory chain where electron transport is coupled to ATP production by proton pumping. These three places are the NADH dehydrogenase reaction, the oxidation of cytochrome *b*, and the reaction of cytochrome oxidase with oxygen, although the mechanism for proton transfer in cytochrome oxidase remains a mystery. Table 20.2 summarizes the energetics of electron transport reactions.

The order of the respiratory complexes in the electron transport chain was established by using specialized spectroscopic techniques on intact

Table 20.2 The Energetics of Electron Transport Reactions

Reaction	$\Delta G^{\circ\prime}$ kJ/(mol NADH)	$\Delta G^{\circ\prime}$ kcal/(mol NADH)
$\text{NADH} + \text{H}^+ + \text{E—FMN} \rightarrow \text{NAD}^+ + \text{E—FMNH}_2$	−38.6	−9.2
$\text{E—FMNH}_2 + \text{CoQ} \rightarrow \text{E—FMN} + \text{CoQH}_2$	−42.5	−10.2
$\text{CoQH}_2 + 2\ \text{Cyt}\ b[\text{Fe(III)}] \rightarrow \text{CoQ} + 2\text{H}^+ + 2\ \text{Cyt}\ b[\text{Fe(II)}]$	+11.6	+2.8
$2\ \text{Cyt}\ b[\text{Fe(II)}] + 2\ \text{Cyt}\ c_1[\text{Fe(III)}] \rightarrow 2\ \text{Cyt}\ c_1[\text{Fe(II)}] + 2\ \text{Cyt}\ b[\text{Fe(III)}]$	−34.7	−8.3
$2\ \text{Cyt}\ c_1[\text{Fe(II)}] + 2\ \text{Cyt}\ c[\text{Fe(III)}] \rightarrow 2\ \text{Cyt}\ c[\text{Fe(II)}] + 2\ \text{Cyt}\ c_1[\text{Fe(III)}]$	−5.8	−1.4
$2\ \text{Cyt}\ c[\text{Fe(II)}] + 2\ \text{Cyt}\ (aa_3)[\text{Fe(III)}] \rightarrow 2\ \text{Cyt}\ (aa_3)[\text{Fe(II)}] + 2\ \text{Cyt}\ c[\text{Fe(III)}]$	−7.7	−1.8
$2\ \text{Cyt}\ (aa_3)[\text{Fe(II)}] + \tfrac{1}{2}\text{O}_2 + 2\text{H}^+ \rightarrow 2\ \text{Cyt}\ (aa_3)[\text{Fe(III)}] + \text{H}_2\text{O}$	−102.3	−24.5
Overall reaction: $\text{NADH} + \text{H}^+ + \tfrac{1}{2}\text{O}_2 \rightarrow \text{NAD}^+ + \text{H}_2\text{O}$	−220	−52.6

mitochondria capable of carrying out electron transport. When the mitochondria are exposed to compounds called respiratory inhibitors, specific electron transfer reactions are inhibited. The specific reaction inhibited depends on the nature of the inhibitor. Components of the electron transport chain before the blocked reaction accumulate in their reduced form and can be detected spectroscopically. Similarly, the oxidized forms of components after the blocked reaction in the chain accumulate, and they also can be detected spectroscopically.

20-3C The Redox Roles of the Heme Group and Iron–Sulfur Clusters in Electron Transport

▶ **What is the nature of the iron-containing proteins of electron transport?**

In contrast to the electron carriers in the early stages of electron transport, such as NADH, FMN, and CoQ, the cytochromes are macromolecules. These proteins are found in all types of organisms and are typically located in membranes. In eukaryotes, the usual site is the inner mitochondrial membrane, but cytochromes can also occur in the endoplasmic reticulum.

All cytochromes contain the heme group, which is also a part of the structure of myoglobin and hemoglobin (Sections 4-4B and 4-5A, respectively). In the cytochromes, the iron of the heme group does not bind to oxygen; instead, the iron is involved in the series of redox reactions discussed so far. There are differences in the side chains of the heme group of the cytochromes involved in the various stages of electron transport (Figure 20.10). These structural differences, combined with the variations in the polypeptide chain and in the way the polypeptide chain is attached to the heme, account for the differences in properties among the cytochromes in the electron transport chain.

Nonheme iron proteins do not contain a heme group, as their name indicates. Many of the most important proteins in this category contain sulfur, as

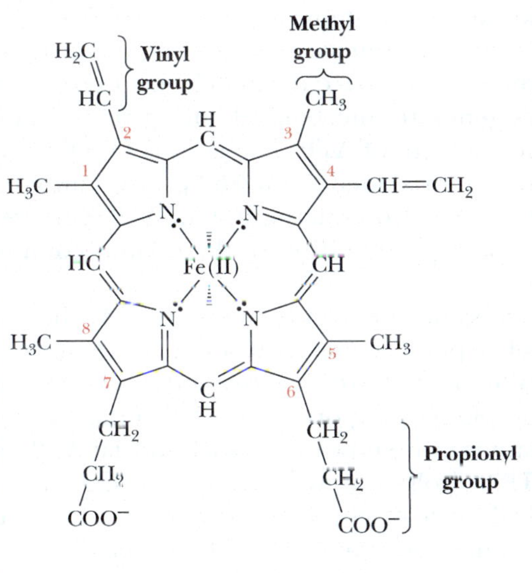

POSITION	*a* CYTOCHROMES	*c* CYTOCHROMES
1	Same	Same
2 (in *a*)	$-CH(OH)-CH_2-(CH_2-CH=C(CH_3)-CH_2)_3H$	
2 (in *c*)		$-CHCH_3$ with S—protein (Covalent attachment)
3	Same	Same
4	Same	$-CHCH_3$ with S—protein
5	Same	Same
6	Same	Same
7	Same	Same
8	$-C(H)=O$ (Formyl group)	Same

A Structures of the heme of all *b* cytochromes and of hemoglobin and myoglobin. The wedge bonds shows the fifth and sixth coordination sites of the iron atom.

B A comparison of the side chains of *a* and *c* cytochromes to those of *b* cytochromes.

Figure 20.10 The heme group of cytochromes.

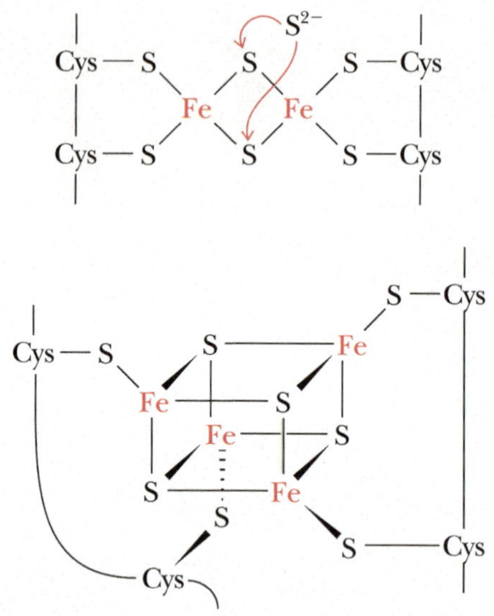

Figure 20.11 Iron–sulfur bonding in nonheme iron proteins.

is the case with the iron–sulfur proteins that are components of the respiratory complexes. The iron is usually bound to cysteine or to S^{2-} (Figure 20.11). There are still many questions about the location and mode of action of iron–sulfur proteins in mitochondria.

20-4 The Connection Between Electron Transport and the Phosphorylation of ADP to ATP

Learning Objectives

After reading this section, students will be able to:

LO 20.4.1 List the components of ATP synthase.

LO 20.4.2 Describe the concepts of uncoupling and the P/O ratio in oxidative phosphorylation.

Some of the energy released by the oxidation reactions in the electron transport chain is used to drive the phosphorylation of adenosine diphosphate (ADP). The phosphorylation of each mole of ADP requires 30.5 kJ = 7.3 kcal. As we have seen, each reaction catalyzed by three of the four respiratory complexes provides more than enough energy to drive this reaction, although it is by no means a direct usage of this energy. A common theme in metabolism is that energy to be used by cells is converted to the chemical energy of adenosine triphosphate (ATP) as needed. The energy-releasing oxidation reactions give rise to proton pumping and thus to the pH gradient across the inner mitochondrial membrane. In addition to the pH gradient, a voltage difference across the membrane is generated by the concentration differences of ions inside and outside. The energy of the electrochemical potential (voltage drop) across the membrane is converted to the chemical energy of ATP by the coupling process.

▶ What is the coupling factor in oxidative phosphorylation?

ATP synthase the enzyme responsible for the production of ATP in mitochondria

A coupling factor called **ATP synthase** is needed to link oxidation and phosphorylation. ATP synthase is a multimeric protein that is separate from the electron transport complexes. It serves this function by allowing the protons that were pumped into the intermembrane space during electron transport to flow back into the matrix. This generates mechanical energy that is then converted to chemical energy in the form of ATP. The complete ATP synthase protein spans the inner mitochondrial membrane and projects into the matrix as well. Figure 20.12 shows an electron micrograph of ATP synthase, and Figure 20.13 shows a schematic representation of the various components of ATP synthase.

The portion of ATP synthase that spans the membrane is called F_0. It consists of three different kinds of polypeptide chains (a, b, and c). The portion that projects into the matrix is called F_1; it consists of five different kinds of polypeptide chains in the ratio $\alpha_3\beta_3\gamma\delta\varepsilon$. The F_1 portion of ATP synthase is the site of ATP synthesis. ATP synthase is also known as mitochondrial ATPase because the reverse reaction of ATP hydrolysis, as well as phosphorylation, can be catalyzed by the enzyme. The hydrolytic reaction was discovered before the reaction of the synthesis of ATP, hence, the name. The 1997 Nobel Prize in Chemistry was shared by Paul Boyer of UCLA and John Walker of the Medical Research Council in Cambridge, England, for their elucidation of the structure and mechanism of this enzyme. (The other half of this prize went to Jens Skou for his work on the sodium–potassium pump [Section 8-5], which also functions as an ATPase.)

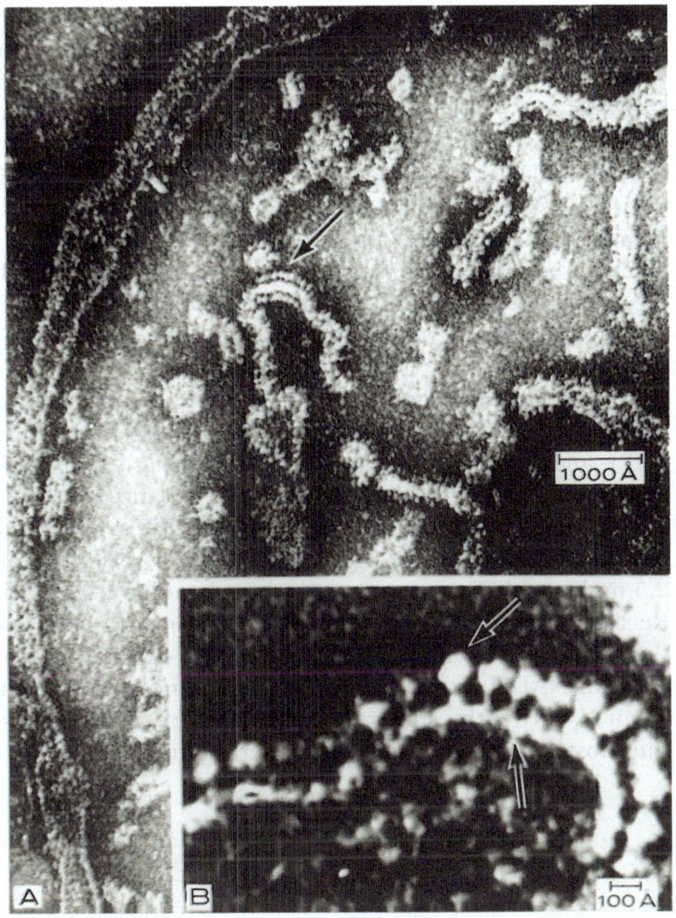

Figure 20.12 **Electron micrograph of projections of the F_1 subunit of ATP synthase into the matrix of a mitochondrion.** Notice the difference in scale between parts A and B. The top arrows indicate the matrix side and the F_1 subunit. The bottom arrow in part B indicates the intermembrane space. (*Science Source.*)

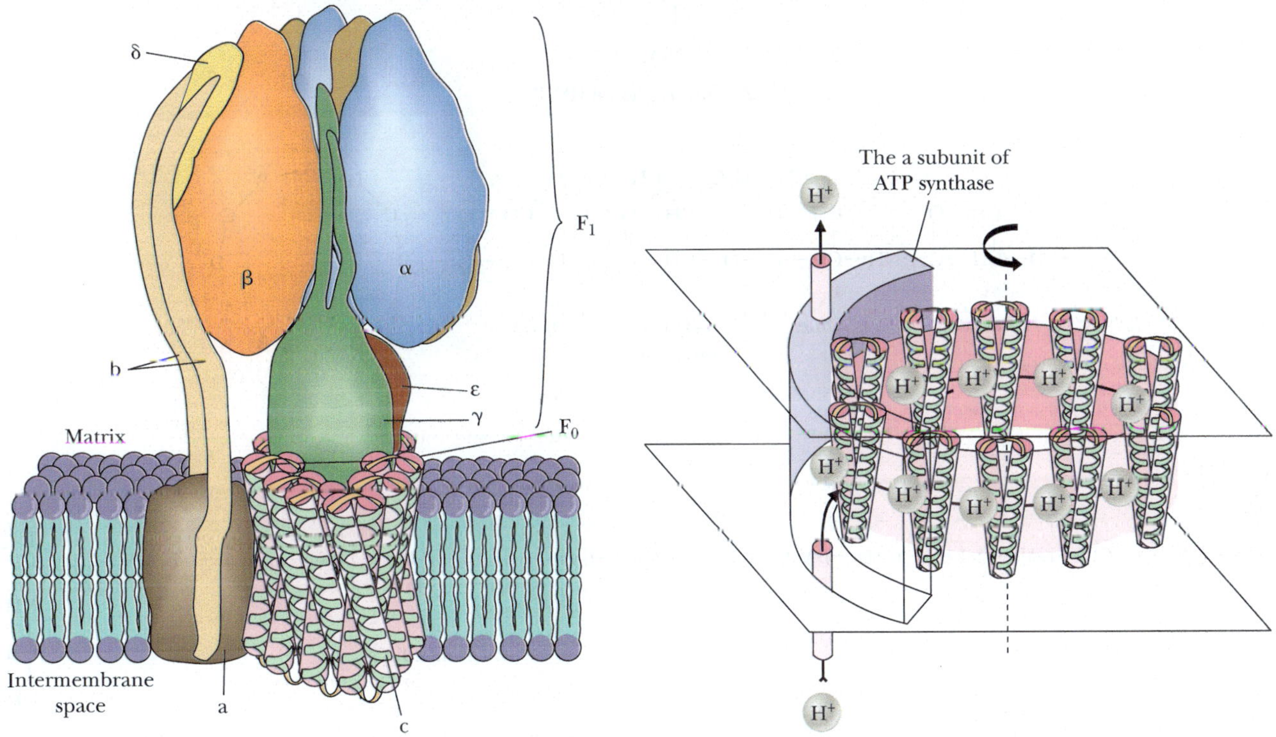

Figure 20.13 **A model of the F_1 and F_0 components of ATP synthase, a rotating molecular motor.** The a, b, α, β, and δ subunits constitute the stator (or stationary part) of the motor, whereas the c, γ, and ε subunits form the rotor (the part that rotates). Flow of protons through the structure turns the rotor and drives the cycle of conformational changes in α and β that synthesize ATP.

uncoupler a substance that reduces the proton gradient in mitochondria by allowing protons to bypass ATP synthase, allowing electron transport to proceed in the absence of phosphorylation

When mitochondrial oxidation processes are operating normally, electron transport from NADH or $FADH_2$ to oxygen results in proton pumping that establishes a proton gradient. When the protons flow back into the matrix through ATP synthase, ATP is produced. However, when a compound called an **uncoupler** is present, oxygen is still reduced to H_2O through electron transport, but ATP is not produced. Uncouplers work by reducing the proton gradient that is established by electron transport. They do this by allowing protons to bypass ATP synthase and go back into the matrix through another route. Therefore, uncouplers inhibit the phosphorylation of ADP without affecting electron transport. A well-known example of an uncoupler is *2,4-dinitrophenol.* Various antibiotics, such as *valinomycin* and *gramicidin A,* are also uncouplers (Figure 20.14).

P/O ratio the ratio of ATP produced by oxidative phosphorylation to oxygen atoms consumed in electron transport

A term called the **P/O ratio** is used to indicate the coupling of ATP production to electron transport. The "P" stands for *p*hosphorylation (of ADP to ATP), and the "O" stands for the molecular *o*xygen that is consumed by electron transport. More specifically, the P/O ratio gives the number of moles of P_i consumed in the reaction ADP + P_i → ATP for each mole of oxygen atoms consumed in the reaction $\frac{1}{2}O_2 + 2H^+ + 2e^- \rightarrow H_2O$. As mentioned previously, 2.5 moles of ATP are produced when 1 mole of NADH is oxidized to NAD^+. Recall that oxygen is the ultimate acceptor of the electrons from NADH and that $\frac{1}{2}$ mole of O_2 molecules (one mole of oxygen atoms) is reduced for each mole of NADH oxidized. The experimentally determined P/O ratio is 2.5 when NADH is the substrate oxidized. The P/O ratio is 1.5 when $FADH_2$ is the substrate oxidized (also an experimentally determined value). Originally,

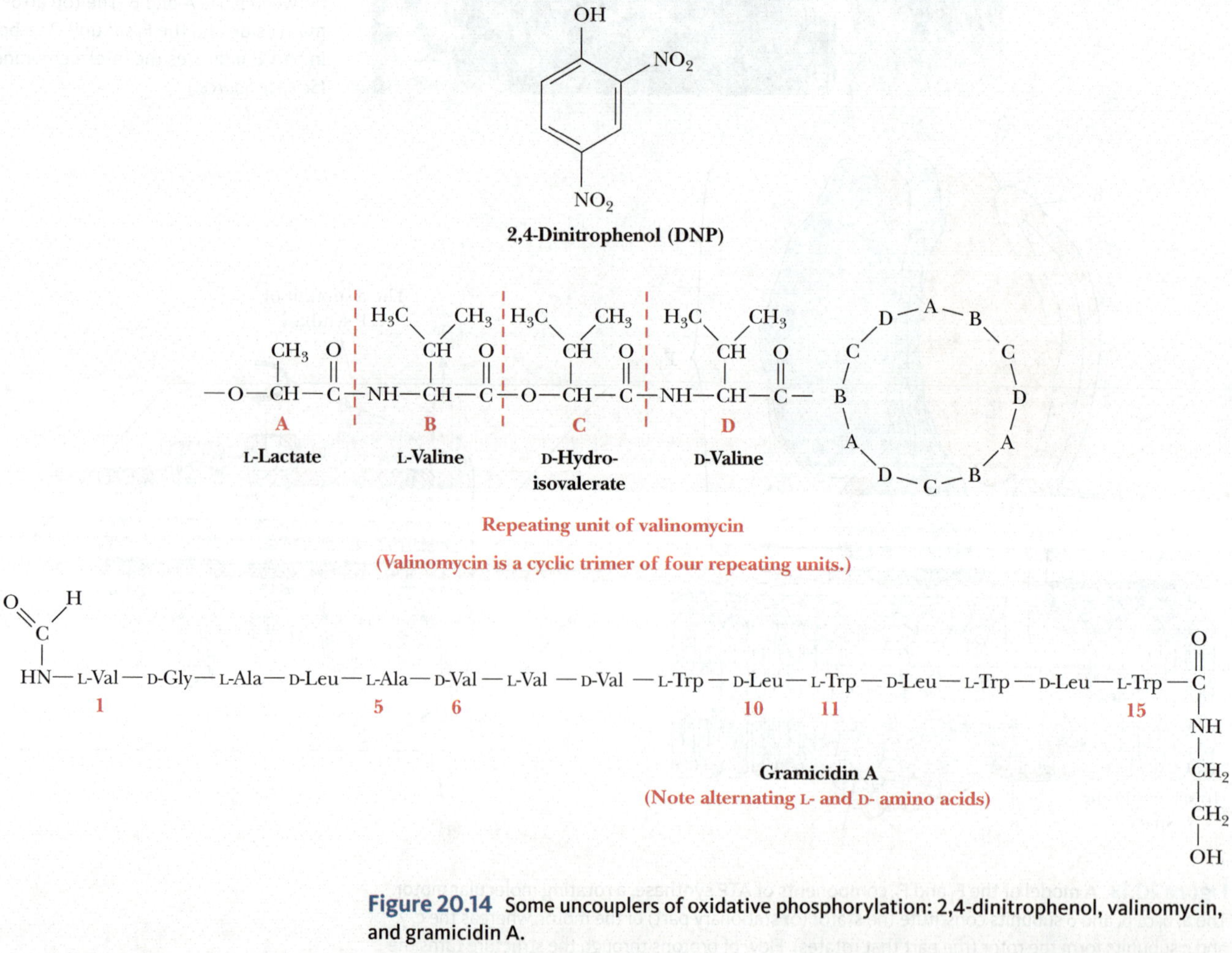

Figure 20.14 Some uncouplers of oxidative phosphorylation: 2,4-dinitrophenol, valinomycin, and gramicidin A.

biochemists had used integer values of 3 and 2 for the P/O ratios for reoxidation of NADH and $FADH_2$, respectively. The noninteger consensus values given here underscore the complexity of electron transport, oxidative phosphorylation, and the manner in which they are coupled.

20-5 The Mechanism of Coupling in Oxidative Phosphorylation

Learning Objectives

After reading this section, students will be able to:

LO 20.5.1 Cite experimental evidence for the chemiosmotic coupling mechanism in oxidative phosphorylation.

LO 20.5.2 Explain the conformational coupling mechanism in oxidative phosphorylation.

Several mechanisms have been proposed to account for the coupling of electron transport and adenosine triphosphate (ATP) production. The **chemiosmotic coupling** mechanism is widely accepted, and other mechanisms have been proposed that serve to complement or refine it. Coupling of electron transport and ATP production is still an active area of research, so our understanding of these mechanisms continues to be refined.

chemiosmotic coupling the mechanism for coupling electron transport to oxidative phosphorylation; it requires the establishment of a proton gradient across the inner mitochondrial membrane that is used to generate ATP

20-5A The Chemiosmotic Coupling Mechanism in Oxidative Phosphorylation

▶ What is chemiosmotic coupling?

As originally proposed, the chemiosmotic coupling mechanism was based entirely on the difference in proton concentration between the intermembrane space and the matrix of an actively respiring mitochondrion. In other words, the proton (hydrogen ion, H^+) gradient across the inner mitochondrial membrane is the crux of the matter. The proton gradient exists because the various proteins that serve as electron carriers in the respiratory chain are not symmetrically oriented with respect to the two sides of the inner mitochondrial membrane, nor do they react in the same way with respect to the matrix and the intermembrane space.

Figure 20.15 repeats the information found in Figure 20.8, with the addition of the flow of protons. The number of protons transported by respiratory complexes is well studied, but the exact number is not absolutely certain under all conditions (Figure 20.15 shows a consensus estimate for each complex). In the process of electron transport, the proteins of the respiratory complexes take up protons from the matrix to transfer them in redox reactions; these electron carriers subsequently release protons into the intermembrane space when they are reoxidized, creating the proton gradient. As a result, there is a higher concentration of protons in the intermembrane space than in the matrix. As a result, the intermembrane space has a lower pH than the matrix. The proton gradient, in turn, can drive the production of ATP that occurs when the protons flow back into the matrix.

Since chemiosmotic coupling was first suggested by the British scientist Peter Mitchell in 1961, a considerable body of experimental evidence has accumulated to support it, including the following:

1. A system with definite inside and outside compartments (closed vesicles) is essential for oxidative phosphorylation. The process does not occur in soluble preparations or in membrane fragments without compartmentalization.

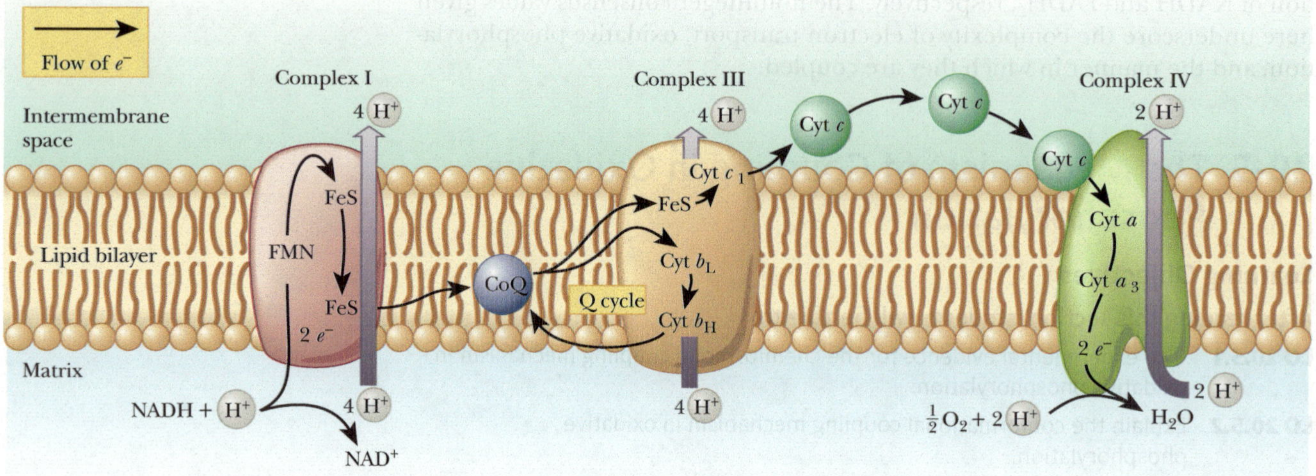

Figure 20.15 The creation of a proton gradient by the respiratory complexes. The overall effect of the electron transport reaction series is to move protons (H^+) out of the matrix into the intermembrane space, creating a difference in pH across the membrane that can be used to produce ATP. Complex II does not contribute directly to establishing the proton gradient, but it does contribute electrons to CoQ, which then contributes electrons to Complex III.

2. Submitochondrial preparations using fragments of the inner mitochondrial membrane can be produced that contain closed vesicles; such vesicles can carry out oxidative phosphorylation, and the asymmetrical orientation (i.e., proton pumping in one direction) of the respiratory complexes with respect to the membrane can be demonstrated (Figure 20.16).
3. A model system for oxidative phosphorylation can be constructed with proton pumping in the absence of electron transport. The model system consists of reconstituted membrane vesicles, mitochondrial ATP synthase, and a proton pump. The pump is bacteriorhodopsin, a protein found in the membrane of halobacteria. The proton pumping takes place when the protein is illuminated (Figure 20.17).
4. The existence of the pH gradient has been demonstrated and confirmed experimentally.

The way in which the proton gradient leads to the production of ATP depends on ion channels through the inner mitochondrial membrane; these channels are a feature of the structure of ATP synthase. Figure 20.18 shows the

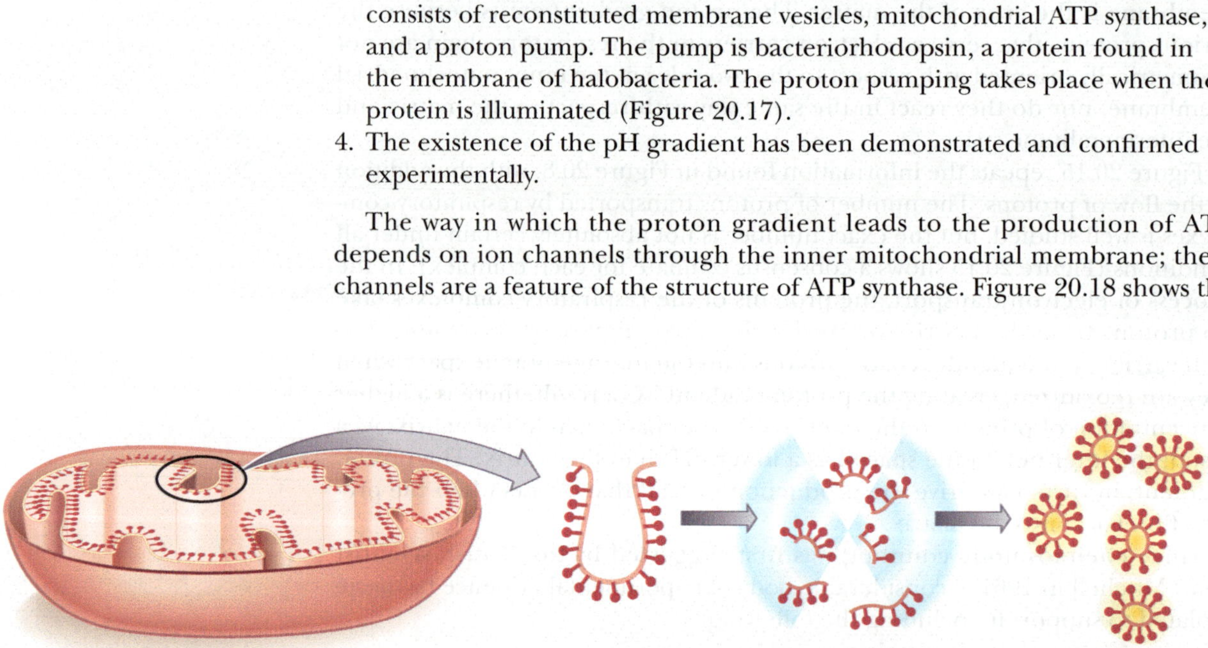

Figure 20.16 Closed vesicles prepared from fragments of the inner mitochondrial membrane can pump protons and produce ATP, supporting the chemiosmotic coupling theory.

chemiosmotic coupling mechanism. In this theory, electron transport pumps protons (H^+) into the intermembrane space, creating a higher concentration of protons there. Protons then flow back into the matrix through ion channels in the ATP synthase; the F_0 part of the protein is the proton channel (Figure 20.18, *bottom right*). The flow of protons is accompanied by the formation of ATP, which takes place in the F_1 unit (Figure 20.18, *bottom right*). The unique feature of chemiosmotic coupling is the direct linkage of the proton gradient to the phosphorylation reaction, and specific details of the process continue to be elucidated by researchers. Before moving on to the conformational coupling mechanism, which provides some explicit details, let us revisit the topic of *uncouplers.*

A reasonable mode of action for uncouplers can be proposed in light of the chemiosmotic coupling theory. For example, the dinitrophenol uncoupler is an acid; its conjugate base, the dinitrophenolate anion, acts as an uncoupler because it can react with protons in the intermembrane space, reducing the difference in proton concentration between the two sides of the inner mitochondrial membrane.

The antibiotic uncouplers, such as gramicidin A and valinomycin, are **ionophores**. Ionophores create channels through which ions such as H^+, K^+, and Na^+ can pass through the mitochondrial membrane. When they do this, the proton gradient is reduced, resulting in the uncoupling of oxidation and phosphorylation.

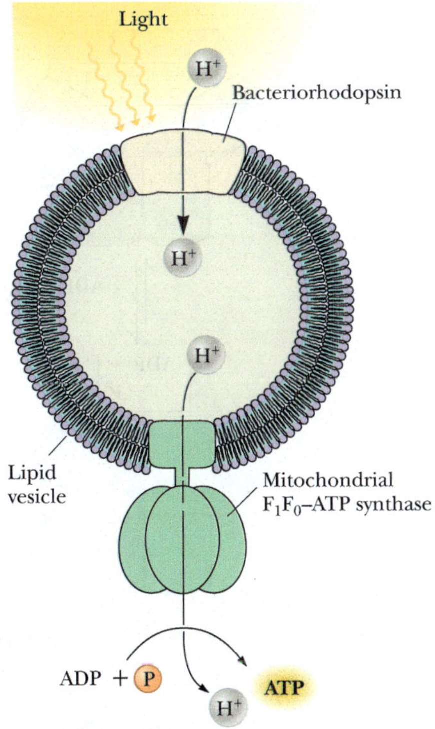

Figure 20.17 **ATP can be produced by closed vesicles with bacteriorhodopsin as a proton pump, supporting the chemiosmotic coupling theory.**

ionophores substances that create channels for ions to pass through the inner mitochondrial membrane

conformational coupling a refinement of the chemiosmotic coupling theory; it states that coupling of electron transport to oxidative phosphorylation depends on a conformational change in the ATP synthase

20-5B The Conformational Coupling Mechanism in Oxidative Phosphorylation

A **conformational coupling** mechanism has also been proposed as a refinement of the chemiosmotic coupling theory. In this theory, the proton gradient is indirectly related to ATP production. The proton gradient leads to conformational changes in a number of proteins, particularly in the ATP synthase itself.

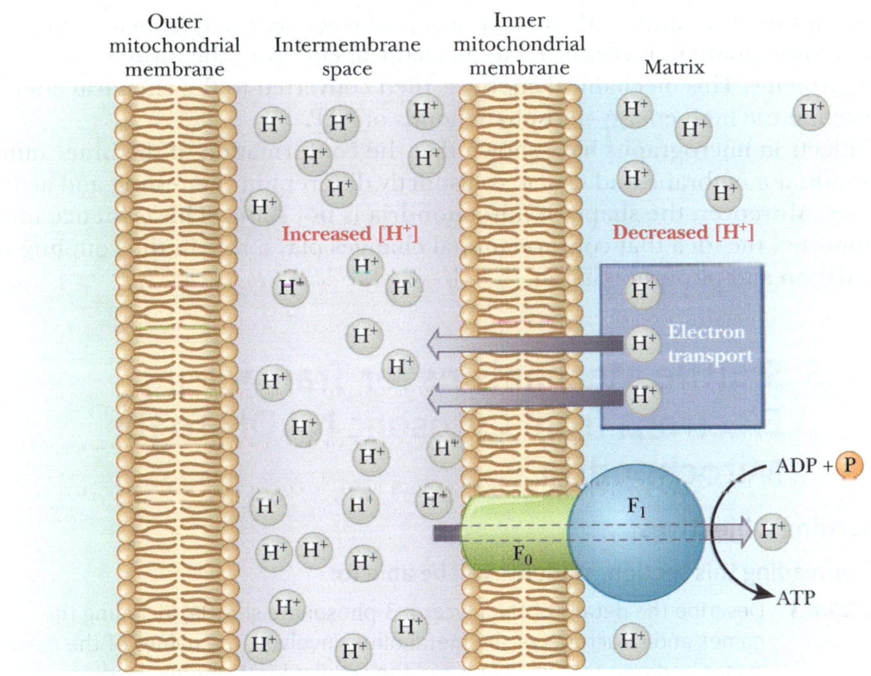

Figure 20.18 **The chemiosmotic coupling mechanism.** Electron transport pumps protons (H^+) into the intermembrane space, creating a higher concentration of protons there. Formation of ATP accompanies the flow of protons back into the mitochondrial matrix (*bottom right*).

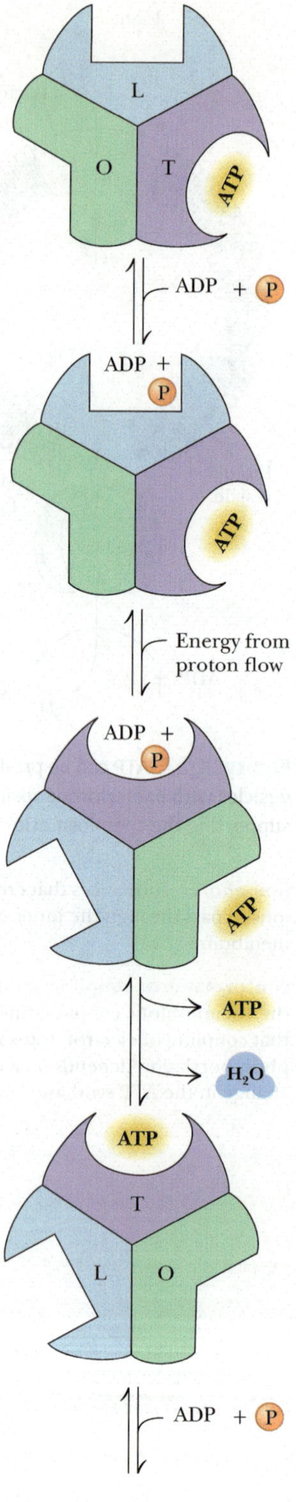

Figure 20.19 The conformational coupling mechanism in oxidative phosphorylation. According to the conformational coupling mechanism refinement of the chemiosmotic theory, the effect of proton flow is to cause a conformational change that leads to the release of already-formed ATP from ATP synthase. There are three sites for substrate on the synthase and three possible conformational states: open (O), with low affinity for substrate; loose-binding (L), which is not catalytically active; and tight-binding (T), which is catalytically active. At the top of this figure, ATP synthase has catalyzed the production of ATP at the T site. Next, more substrate (ADP and P) binds to the L site, and then proton flow provides energy for conformational changes in all three sites. The T site changes to the O conformation and releases its ATP, the L site changes to the T conformation and catalyzes the phosphorylation of ADP to ATP, and the O site changes to the L conformation and is ready to receive more substrate so that the cycle can repeat upon more proton flow aided by electron transport.

Recent evidence appears to show that the proton gradient is involved in the release of tightly bound ATP from the synthase as a result of the conformational change (Figure 20.19).

There are three sites for substrate on the synthase and three possible conformational states: open (O), with low affinity for substrate; loose-binding (L), which is not catalytically active; and tight-binding (T), which is catalytically active. At any given time, each site is in one of these three different conformational states. These states interconvert as a result of proton flow through the synthase, which provides mechanical energy for rotation of the γ subunit of ATP synthase (Figure 20.13). ATP already formed by the synthase is bound at a site in the T conformation, while adenosine diphosphate (ADP) and P_i bind at a site in the L conformation. Proton flow converts the site in the T conformation to the O conformation, releasing the ATP. The site at which ADP and P_i are bound assumes the T conformation, which can then give rise to ATP.

Next, let us discuss more details of the rotation in ATP synthase. The F_1 portion of ATP synthase acts as a rotary motor. The c, γ, and ε subunits constitute the rotor, turning within a stationary barrel of a domain consisting of the subunit in association with the $\alpha_3\beta_3$ hexamer and the a and b subunits (refer to Figure 20.13 for a detailed picture of the subunits). The γ and ε subunits constitute the rotating "shaft" that mediates the energy exchange between the proton flow at F_0 and ATP synthesis at F_1. In essence, the chemical energy of the proton gradient is converted to mechanical energy in the form of the rotating proteins. This mechanical energy is then converted to the chemical energy stored in the high-energy phosphate bonds of ATP.

Electron micrographs have shown that the conformation of the inner mitochondrial membrane and cristae is distinctly different in the resting and active states. Moreover, the shape of mitochondria is not static. This evidence long supported the idea that conformational changes play a role in the coupling of oxidation and phosphorylation.

20-6 Shuttle Mechanisms for Transferring Electrons from Cytosolic NADH to the Mitochondria

Learning Objectives

After reading this section, students will be able to:

LO 20.6.1 Describe the details of the glycerol-3-phosphate shuttle, including the names and structures of the metabolites involved, the names of the coenzymes and enzymes involved, and the cellular locations involved.

LO 20.6.2 Explain the details of the malate–aspartate shuttle, including the names and structures of the metabolites involved, the names of the coenzymes and enzymes involved, and the cellular locations involved.

The NADH that is produced in the glyceraldehyde-3-phosphate dehydrogenase step of glycolysis in the cytosol cannot cross the inner mitochondrial membrane to enter the electron transport chain. However, these electrons can be transferred to carrier molecules, such as glycerol-3-phosphate or malate, that can cross the membrane. The number of ATP molecules that is ultimately generated from each original molecule of glucose depends on the carrier molecule that is used, which varies according to the type of cell in which it occurs.

20-6A The Glycerol-3-Phosphate Shuttle Mechanism of Electron Transfer to the Matrix

One carrier system that has been extensively studied in insect flight muscle is the **glycerol-3-phosphate shuttle** (Figure 20.20). In the cytosol, dihydroxyacetone phosphate (DHAP) can be reduced to a glycerol-3-phosphate electron carrier using the NADH from glycolysis as the reducing agent. This reaction is catalyzed by cytosolic *glycerol-3-phosphate dehydrogenase* (Figure 20.20, *left*). This might appear to divert DHAP away from use in glycolysis, but the cytosolic pool of DHAP for glycolysis and this shuttle is generally sufficient for both processes to proceed in parallel. The DHAP is also recycled in this shuttle.

glycerol-3-phosphate shuttle a mechanism for transferring electrons from NADH in the cytosol to $FADH_2$ in the mitochondrion

The glycerol-3-phosphate then diffuses through one of the various **porins** that are present in the outer mitochondrial membrane. Once inside the intermembrane space, an FAD-dependent enzyme called mitochondrial glycerol phosphate dehydrogenase, which is on the outer face of the inner mitochondrial membrane, oxidizes the glycerol-3-phosphate back to DHAP, and FAD gets reduced to $FADH_2$ (Figure 20.20, *right*). The $FADH_2$ (which has now ultimately received electrons from cytosolic NADH) then passes electrons through the electron transport chain, leading to the production of 1.5 moles of ATP for each mole of cytosolic NADH, and the DHAP diffuses back to the cytosol for more shuttling of cytosolic NADH electrons. This mechanism has also been observed in mammalian muscle and brain.

porin membrane proteins that allow small, water-soluble molecules to passively diffuse across the membrane

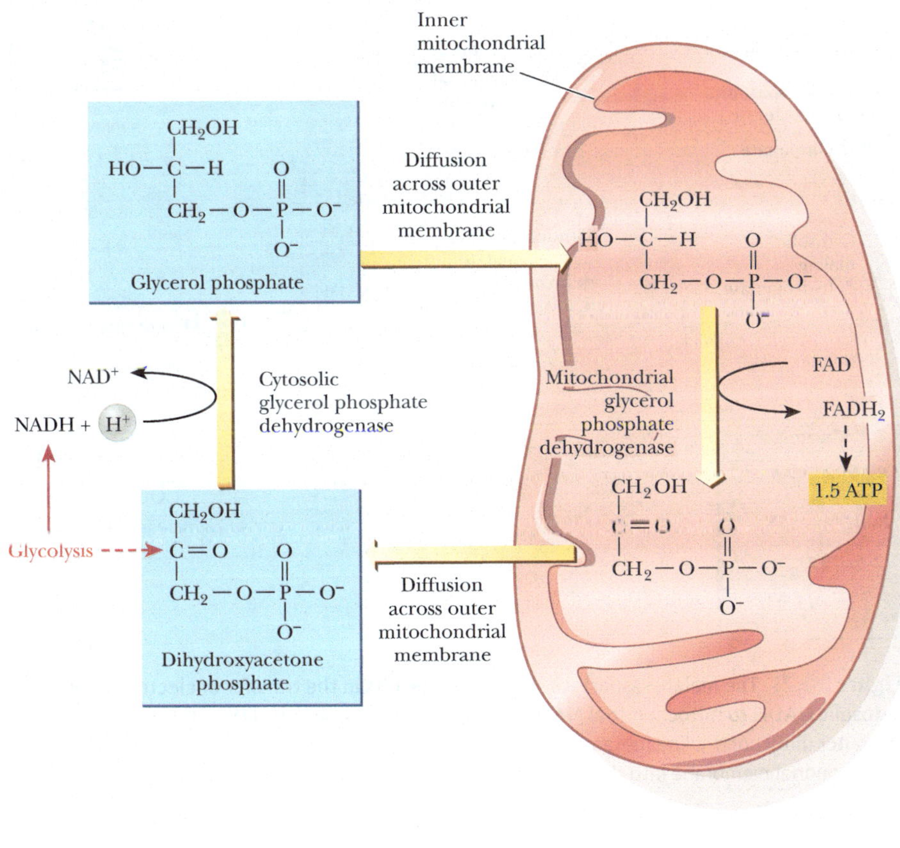

Figure 20.20 The glycerol-3-phosphate shuttle that results in the transfer of electrons from cytosolic NADH to mitochondrial $FADH_2$. The mitochondrial reaction occurs in the intermembrane space on the inner membrane surface, not in the matrix.

malate–aspartate shuttle a mechanism for transferring electrons from NADH in the cytosol to NADH in the mitochondrion

20-6B The Malate–Aspartate Shuttle Mechanism of Electron Transfer to the Matrix

A more complex and more efficient shuttle mechanism is the **malate–aspartate shuttle**, which has been found in mammalian kidney, liver, and heart tissue (Figure 20.21). This shuttle uses the fact that malate can cross the mitochondrial membrane, while oxaloacetate cannot. The noteworthy point about this shuttle mechanism is that the transfer of electrons from NADH in the cytosol produces NADH in the mitochondrion. In the cytosol, oxaloacetate is reduced to malate by cytosolic malate dehydrogenase, accompanied by the oxidation of cytosolic NADH to NAD^+. The malate then crosses the mitochondrial membrane. In the mitochondrion, the conversion of malate back to oxaloacetate is catalyzed by the mitochondrial malate dehydrogenase (one of the enzymes of the citric acid cycle). Oxaloacetate is converted to aspartate, which can also cross the mitochondrial membrane. Aspartate is converted to oxaloacetate in the cytosol, completing the cycle of reactions. Additionally, this shuttle involves reactions catalyzed by enzymes called *aminotransferases,* which are discussed in Section 23-4B.

The NADH that is produced in the mitochondrion thus passes electrons to the electron transport chain. With the malate–aspartate shuttle, 2.5 moles of ATP are produced for each mole of cytosolic NADH, rather than 1.5 moles of ATP in the glycerol-3-phosphate shuttle, which uses $FADH_2$ as a carrier. **Biochemical Connections 20A** discusses some practical applications of our understanding of catabolic pathways.

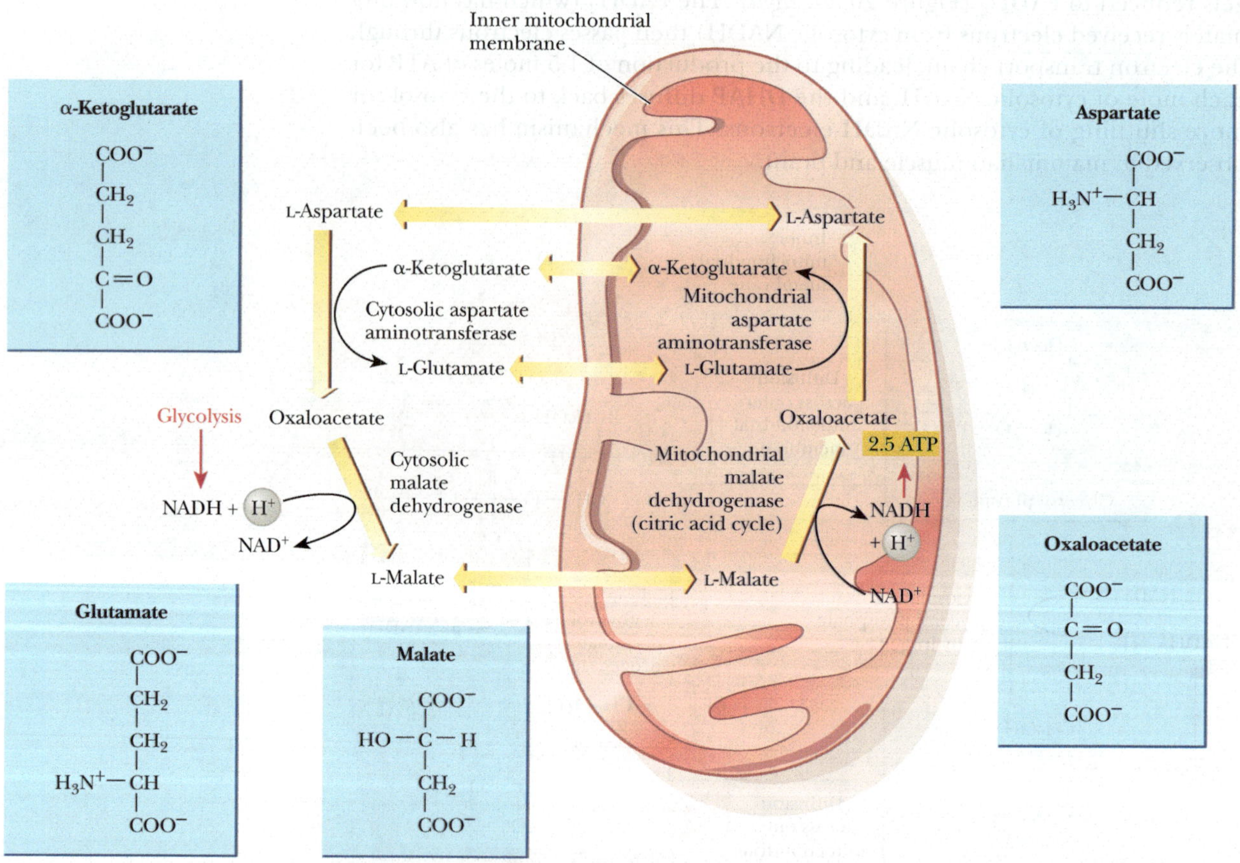

Figure 20.21 The malate–aspartate shuttle that results in the transfer of electrons from cytosolic NADH to mitochondrial NADH. Metabolites in the shuttle diffuse through porins in the outer mitochondrial membrane. Transport proteins then transport them across the inner mitochondrial membrane into the mitochondrial matrix, where they are involved in reactions.

20A Biochemical Connections • Allied Health

Sports and Metabolism

Genetic endowment and training are important to the success of an athlete, especially at the elite level, but a keen understanding of physiology and metabolism is equally important. To plan nutrition for performance, a serious athlete must understand the nature of metabolism as it relates to their chosen sport. A working muscle has four different sources of energy available after a period of rest:

1. Creatine phosphate, which reacts directly with ADP in substrate-level phosphorylation to produce ATP.
2. Glucose from the glycogen of muscle stores, initially consumed by anaerobic metabolism.
3. Glucose from the liver, from both its glycogen stores and the gluconeogenesis from lactic acid produced in the muscle (the Cori cycle), again initially consumed by anaerobic metabolism.
4. Aerobic metabolism in the muscle mitochondria.

Initially, all four energy sources are available to the muscle. When the creatine phosphate runs out, only the other sources are left. When muscle glycogen runs out, the anaerobic boost it provided slows down, and when the liver glycogen is gone, only aerobic metabolism to carbon dioxide and water is left.

It is difficult to accurately calculate how much each of these nutrients might supply to a rapidly working muscle, but simple calculations are consistent with there being less than a one-minute supply of creatine phosphate. This figure correlates with the length of time for sprint events, which is typically less than a minute. Creatine supplements for athletes are sold in health food stores, and results suggest that for powerlifting or short sprints, such as the 100-meter dash, this supplementation is effective.

There is about 10–30 minutes' worth of glycogen in the muscle cells, with this figure varying dramatically based on the intensity of the exercise. Performance in running events ranging in distance from 1500 meters up to 10 kilometers can be heavily influenced by the muscle glycogen levels at the start of the event. Glycogen loading (Biochemical Connections 18A) could affect this last figure significantly. One reason for the difficulty in making these calculations is the uncertainty of what proportion of the liver glycogen is metabolized only to lactic acid and how much is metabolized in the liver. One rate-limiting step for aerobic metabolism is the shuttling of both NADH and pyruvate from the cytoplasm into the mitochondrion.

Well-conditioned and well-trained athletes actually have a higher number of mitochondria in their muscle cells. For long-distance events, such as the marathon or road-cycling events, aerobic metabolism certainly comes into play. *Fat burning* is the term frequently used, and it reflects a metabolic fact. Fatty acids are degraded to acetyl-CoA, which then enters the citric acid cycle; marathoners and cyclists are known for their notably lean frames, with a minimal amount of body fat. Running a marathon, which takes between two and three hours for very fit runners, uses more fatty acids and is done at a lower level of oxygen uptake than riding in a professional road-cycling event that may take up to seven hours. Thus, there are differences in metabolism for sports even within the category known as endurance events.

Underscoring the importance of the electron transport chain and mitochondria to the athlete is the story of the cyclist Greg LeMond, who is the only American to win the Tour de France. He won it a total of three times (Figure 20.22). Shortly after his first Tour de France victory in 1986, he was shot in the back with buckshot while hunting. Although he recovered and won two more Tours de France, he never really felt well again. He later commented that even in his final Tour de France victory in 1990, something was definitely wrong. During the next few years, he never made it back into the top places in a race. He seemed to be putting on weight and didn't respond to training. Finally, in desperation, he underwent some painful muscle biopsies and discovered that he had a rare condition called mitochondrial myopathy. When he trained hard, his mitochondria began to disappear. He was essentially an aerobic athlete without the ability to process fuels aerobically. He retired from competition shortly thereafter. While the hunting accident itself did not cause LeMond's mitochondrial myopathy, the physical trauma and prolonged recovery may have exacerbated his symptoms. ◗

Figure 20.22 Winner of the 1986, 1989, and 1990 Tour de France, Greg LeMond is the only American to win the world's most prestigious bicycle race.

20-7 The ATP Yield from the Complete Oxidation of Glucose

Learning Objective

After reading this section, students will be able to:

LO 20.7.1 Calculate the net ATP yield from glucose oxidation, as well as the ATP yields from various steps in glycolysis, the citric acid cycle, and oxidative phosphorylation when the glycerol-3-phosphate shuttle is used compared to the malate–aspartate shuttle.

In Chapters 17–20, we have discussed many aspects of the complete oxidation of glucose to carbon dioxide and water. At this point, it is useful to do some bookkeeping to see how many molecules of adenosine triphosphate (ATP) are produced for each molecule of glucose oxidized. Recall that some ATP is produced in glycolysis, but that far more ATP is produced by aerobic metabolism. Table 20.3 summarizes ATP production and also follows the reoxidation of NADH and $FADH_2$.

Table 20.3 Yield of ATP from Glucose Oxidation

Pathway	ATP Yield per Glucose: Glycerol-3-Phosphate Shuttle	ATP Yield per Glucose: Malate–Aspartate Shuttle	NADH	$FADH_2$
Glycolysis: glucose to pyruvate (cytosol)				
Phosphorylation of glucose	−1	−1		
Phosphorylation of fructose-6-phosphate	−1	−1		
Dephosphorylation of 2 molecules of 1,3-BPG	+2	+2		
Dephosphorylation of 2 molecules of PEP	+2	+2		
Oxidation of 2 molecules of glyceraldehyde-3-phosphate yields 2 NADH			+2	
Pyruvate conversion to acetyl-CoA (mitochondria)				
2 NADH produced			+2	
Citric acid cycle (mitochondria)				
2 molecules of GTP from 2 molecules of succinyl-CoA	+2	+2		
Oxidation of 2 molecules each of isocitrate, α-ketoglutarate, and malate yields 6 NADH			+6	
Oxidation of 2 molecules of succinate yields 2 $FADH_2$				+2
Oxidative phosphorylation (mitochondria)				
2 NADH from glycolysis yield 1.5 ATP each if NADH is oxidized by glycerol–phosphate shuttle; 2.5 ATP by malate–aspartate shuttle	+3	+5	−2	
Oxidative decarboxylation of 2 pyruvate to 2 acetyl-CoA: 2 NADH produce 2.5 ATP each	+5	+5	−2	
2 $FADH_2$ from each citric acid cycle produce 1.5 ATP each	+3	+3		−2
6 NADH from citric acid cycle produce 2.5 ATP each	+15	+15	−6	
Net yield	+30	+32	0	0

Note: These P/O ratios of 2.5 and 1.5 for mitochondrial oxidation of NADH and $FADH_2$ are "consensus values." Because they may not reflect actual values, and because these ratios may change depending on metabolic conditions, these estimates of ATP yield from glucose oxidation are approximate.

Hot Topic

What Does Brown Adipose Tissue Have to Do with Obesity?

When electron transport generates a proton gradient, some of the energy produced takes the form of heat. Dissipation of energy as heat is useful to organisms in two situations: cold-induced non-shivering thermogenesis (the production of heat) and diet-induced thermogenesis. Cold-induced non-shivering thermogenesis enables animals to survive in the cold once they have become adapted to such conditions, and diet-induced thermogenesis prevents the development of obesity in spite of prolonged overeating. (Energy is dissipated as heat as food molecules are metabolized instead of being stored as fat.) These two processes may be the same biochemically. They occur principally, if not exclusively, in brown adipose tissue, which is rich in mitochondria. (Brown fat takes its color from the large number of mitochondria present in it, unlike white fat cells, which have far fewer mitochondria.)

The key to this "inefficient" use of energy in brown adipose tissue appears to be a mitochondrial protein called thermogenin, also referred to as the *uncoupling protein or UCP1* (Palou et al. 1998). When this membrane-bound protein is activated in thermogenesis, it serves as a proton channel through the inner mitochondrial membrane. Like all other uncouplers, it "punches a hole" in the mitochondrial membrane and decreases the effect of the proton gradient. Protons flow back into the matrix through thermogenin, bypassing the ATP synthase complex.

Since 2009, an increasing number of studies have been conducted on the physiology of brown adipose tissue in humans. Traditionally, this work on both obesity and adaptation to cold stress has been done on small mammals such as rats, mice, and hamsters (Figure 20.23). Medical advances in imaging technology, including positron-emission tomography and computed tomography (PET-CT), magnetic resonance imaging (MRI), and thermal imaging, have made it possible for researchers to locate and study the activity of brown fat cells in humans. The results of this research have led to a broader understanding of how brown fat in adults may contribute to triglyceride uptake from blood, improved carbohydrate regulation, and reduced white fat accumulation (Moyer 2014). The implications are tremendous because they relate to potential pathways to combat heart disease, diabetes, and obesity, respectively.

Figure 20.23 Hamsters serve as a model to study obesity and cold stress.

Brown adipose tissue and white adipose tissue are the two common types of adipose tissues in the human body (Saito et al. 2014). When we gain weight, the extra mass is due to the accumulation of white adipose tissue around our midsection and rump. White adipose tissue is our body's way of storing energy for later use. In contrast, brown adipose tissue is most abundant in infants and functions to dissipate energy as heat through non-shivering thermogenesis. The ability of mitochondria-rich brown fat to generate heat is the reason this tissue is of such great interest to scientists. The fuel for the mitochondria to create heat is in the form of carbohydrates and fatty acids that have been oxidized. Researchers are actively investigating the driving force for brown fat metabolism as it relates to carbohydrate regulation and triglyceride uptake from blood. Figure 20.24 shows a PET-CT image that contrasts two human body types and ages using ^{18}F-labeled fluorodeoxyglucose as an agent to see brown fat as dark splotches.

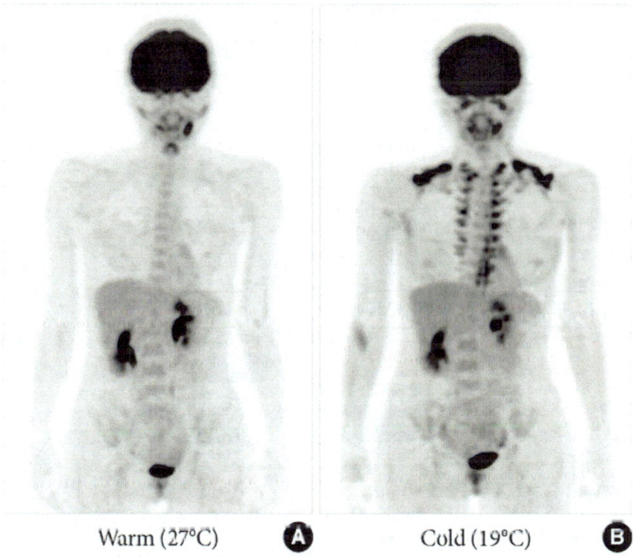

Figure 20.24 PET scan of old (*left*) versus young (*right*) humans showing brown fat deposits as dark splotches. FDG represents the radiolabeled glucose analog ^{18}F-fluorodeoxyglucose, and BAT is brown adipose tissue. Factors besides age that contribute to increased brown fat levels are cold temperature, lean body type, and diet (especially capsinoid consumption).

Figure 20.24 correlates differences in brown fat deposits due to factors including temperature, age, and percent body fat. Yoneshiro and colleagues (2013) explored the hypothesis that brown fat disappearance with aging results in increased incidents of obesity that can be reduced by cold exposure and diet. Specifically, they demonstrated that subjects who spent two hours each day in a depressed temperature environment (~19°C) were able to increase their brown fat deposits within six weeks. A similar increase in brown fat was observed when the subjects consumed capsinoids (i.e., hot peppers). This study provided evidence for "brown-type" adipose tissue called beige fat.

Beige fat was originally thought to be white adipose tissue that changed color due to an increase in the number of mitochondria initiated by either cold stress or capsinoid consumption. The beige adipose tissue is capable of *UCP1* expression, thus providing the benefits of brown fat. More detailed studies to definitively link beige fat exclusively to white adipose tissue demonstrated that an alternative pathway must exist that is not entirely reliant on white fat. The hormone irisin has been associated with the "browning of white adipose tissue" in skeletal muscle (Harms and Seale 2013). Figure 20.25 shows the two pathways by which beige adipose tissue can form, in addition to the roles of white and brown fat in response to cold and capsinoid activation.

Figure 20.25 provides an overview of the influence of cold and capsinoids on activating transient receptor potential channels (TRP). This stimulus releases noradrenalin, which in turn binds to β-adrenergic receptors (βAR). βAR binding by noradrenalin induces hydrolysis of triglycerides (TG) through secondary signaling. Thermogenesis is fueled by oxidized glucose and oxidized fatty acids that have been released through triglyceride hydrolysis, in addition to stimulus of uncoupling protein 1 (UCP1). Prolonged exposure to cold and/or capsinoid consumption results in the accumulation of brown fat and the appearance of beige fat from white fat. The net result of sympathetically activated thermogenesis is a decrease in white adipose tissue and an increase in the ability of the body to burn energy. These are two very beneficial attributes in the fight against type II diabetes and obesity.

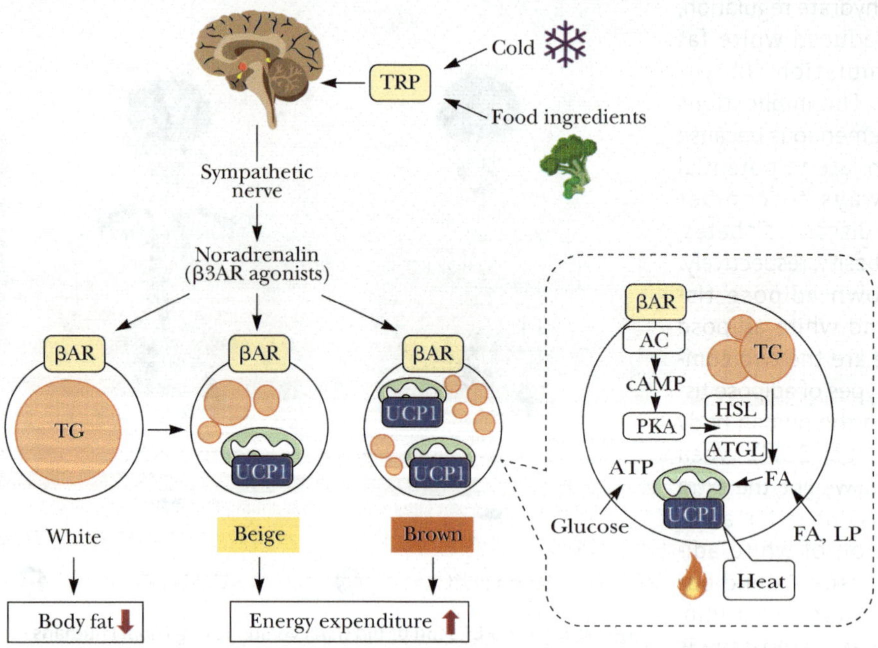

Figure 20.25 Cold and capsinoid activation of thermogenesis and its effect on fat tissue types. Acronyms used in the figure include transient receptor potential channels (TRP), β-adrenergic receptors (βAR), triglycerides (TG), uncoupling protein 1 (UCP1), adenylate cyclase (AC), adipose triglyceride lipase (ATGL), cAMP-dependent protein kinase (PKA), hormone-sensitive lipase (HSL), lipoprotein (LP), and fatty acid (FA). *Figure 1 from Saito, M. Human brown adipose tissue: Regulation and anti-obesity potential. Endocrine J. 61, 409–416 (2014).*

Additional factors linked to the management of obesity that may be associated with brown fat regulation include exercise and sleep. Researchers have evidence to suggest that exercise, especially in cold conditions, can enhance metabolic rates beyond just the benefit of calories burned. Sanchez-Delgado and colleagues (2015) report that exercise may lead to brown fat activation independent of the sympathetic nervous system. The molecules proposed to activate brown fat through exercise are cardiac natriuretic peptides, irisin, interleukin-6 (IL-6), β-aminoisobutyric acid (BAIBA), and fibroblast growth factor 21 (FGF21). Figure 20.26 provides an overview of how researchers propose these activators influence brown and white fat through exercise. Their findings suggest that exercise-induced thermogenesis by brown fat activation result in favorable body weight regulation.

The final factor associated with the health benefits of brown fat is sleep. Kapás and Szentirmai (2014) have correlated the role of brown adipose tissue with restorative sleep following periods of sleeplessness. Observational research suggests that sleep on the order of eight hours per night leads to a healthier metabolism, while humans who obtain five or fewer hours of sleep per night are at higher risk for obesity. The molecular basis for the correlation between sleep and obesity has been elusive, but Kapás and Szentirmai propose that impaired brown adipose tissue function may be the culprit. They have studied the sleep patterns of *UCP1*-deficient mice and associated the influence of sleeplessness with weight gain. Their current hypothesis is based on the monitoring of thermogenesis and its role in temperature regulation and metabolism during sleep/awake cycles. A restorative sleep, on the order of eight hours per night for humans, correlates with brown fat metabolism, thermogenesis, body temperature regulation, and resistance to obesity. While there remains much research to be conducted on brown adipose tissue, the potential health benefits as they relate to type II diabetes (carbohydrate regulation), obesity (thermogenesis), and heart disease (triglyceride uptake) are becoming increasingly clear.

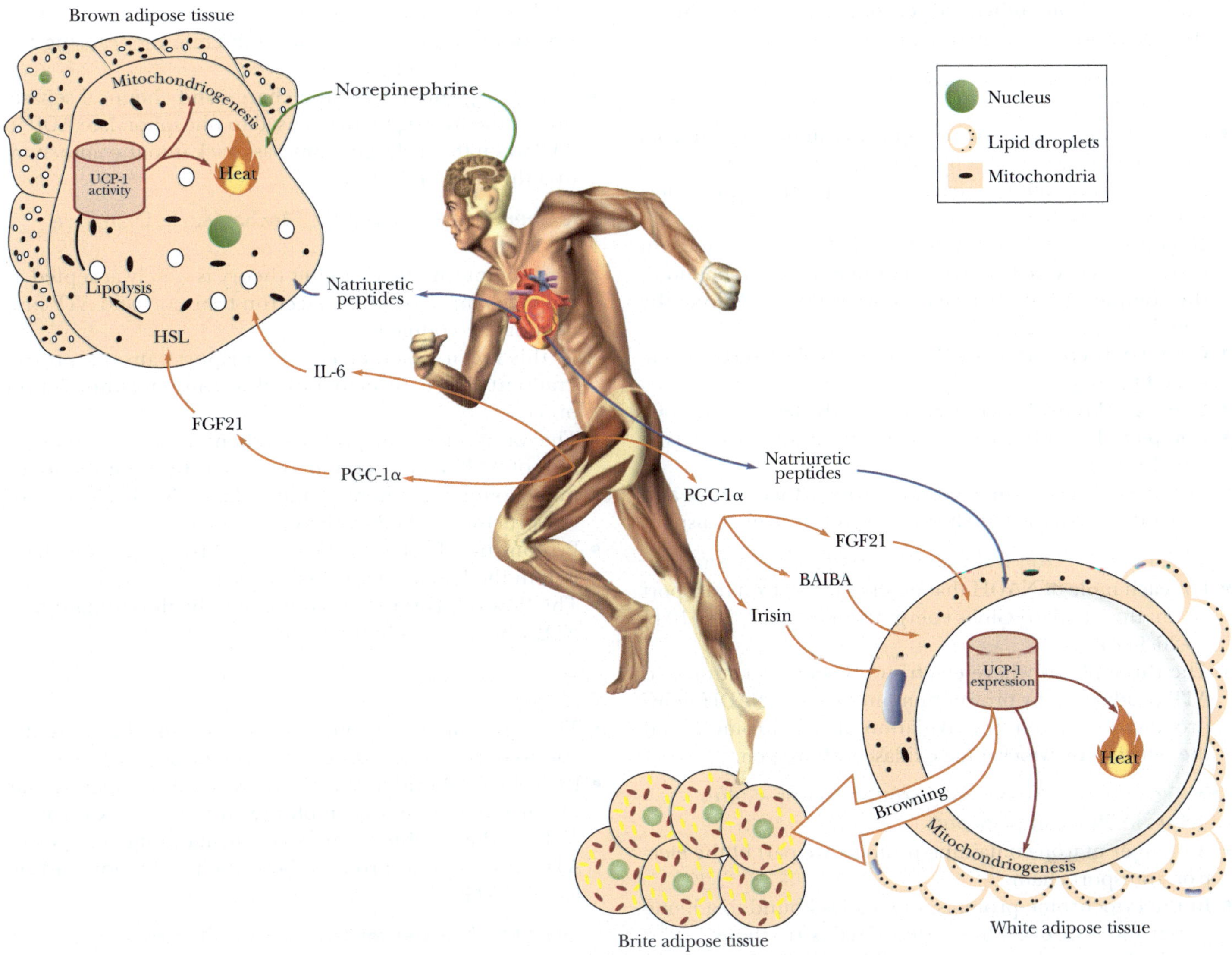

Figure 20.26 **Proposed activation of brown adipose tissue through exercise.** *Note*: BRITE and beige adipose tissue are used synonymously in the chemical literature. *Figure 1 from Sanchez-Delgado, G., B. Martinez-Tellez, J. Olza, C. M. Aguilera, Á. Gil, J. R. Ruiz. Role of exercise in the activation of brown adipose tissue. Ann. Nutr. Metab. 67, 21–32 (2015).*

Summary

The Role of Electron Transport in Metabolism

- In the final stages of aerobic metabolism, electrons are transferred from NADH and $FADH_2$ to oxygen (the ultimate electron acceptor) in a series of oxidation–reduction reactions known as the electron transport chain.
- In the process, protons are pumped across the inner mitochondrial membrane.
- This series of events depends on the presence of oxygen in the final step.
- This pathway allows for the reoxidation of the reduced electron carriers produced in glycolysis, the citric acid cycle, and several other catabolic pathways and is the true source of the ATPs produced by catabolism.
- Phosphorylation depends on the compartmentalized structure of mitochondria.

Reduction Potentials in the Electron Transport Chain

- The overall reaction of the electron transport chain shows a very large, negative $\Delta G^{\circ\prime}$ due to the large differences in reduction potentials between the reactions involving NADH and those involving oxygen.
- If NADH were to reduce oxygen directly, the $\Delta E^{\circ\prime}$ would be more than 1 V.
- In reality, many redox reactions occur between NADH and oxygen, and the correct order of events in the electron transport chain was predicted by comparing the reduction

potentials of the individual reactions long before the order was established experimentally.

Electron Transport Chain Reactions in the Respiratory Complexes

- Four separate respiratory complexes can be isolated from the inner mitochondrial membrane.
- Each can carry out the reactions of a portion of the electron transport chain.
- In addition to the respiratory complexes, two electron carriers, coenzyme Q and cytochrome *c*, are not bound to the complexes but are free to move within and along the membrane, respectively.
- Complex I reoxidizes NADH and sends electrons to coenzyme Q (CoQ).
- Complex II reoxidizes $FADH_2$ and sends electrons to CoQ.
- Complex III involves the Q cycle and shuttles electrons to cytochrome *c*.
- Complex IV takes the electrons from cytochrome *c* and passes them to oxygen in the final step of electron transport.

Summary of Electron Transport Chain Reaction Energetics

- For each mole of NADH that enters the electron transport chain, the standard Gibbs energy change ($\Delta G°'$) is –110 kJ = –26.3 kcal.
- The three places where electron transport is coupled to ATP production by proton pumping are the NADH dehydrogenase reaction, the oxidation of cytochrome *b*, and the reaction of cytochrome oxidase with oxygen.

The Redox Roles of the Heme Group and Iron–Sulfur Clusters in Electron Transport

- A number of iron-containing proteins are part of the electron transport chain.
- In the cytochrome proteins, the iron is bound to a heme group and is involved in a series of redox reactions.
- Unlike hemoglobin and myoglobin, which also contain heme groups, the iron in cytochromes does not bind oxygen.
- In other nonheme iron proteins, the iron is bound to the protein along with sulfur in iron–sulfur clusters.

The Connection Between Electron Transport and the Phosphorylation of ADP to ATP

- A multimeric protein called ATP synthase is a coupling factor that links oxidation and phosphorylation.
- The complete protein spans the inner mitochondrial membrane and projects into the matrix as well.
- The portion of the protein that spans the membrane is called F_0; it consists of three different kinds of polypeptide chains (a, b, and c).
- The portion that projects into the matrix is called F_1; it consists of five different kinds of polypeptide chains (α, β, γ, δ, and ε, in the ratio $\alpha_3\beta_3\gamma\delta\varepsilon$).
- The F_1 sphere is the site of ATP synthesis.
- ATP synthase is also known as mitochondrial ATPase.
- During the process of electron transport, several reactions occur in which reduced carriers that have both electrons and protons to donate are linked to carriers that can only accept electrons.
- At these points, hydrogen ions are released to the other side of the inner mitochondrial membrane, causing the formation of a pH gradient.
- The energy inherent in the charge and chemical separation of the hydrogen ions is used to phosphorylate ADP to ATP when the hydrogen ions pass back into the mitochondria through ATP synthase.

The Chemiosmotic Coupling Mechanism in Oxidative Phosphorylation

- The chemiosmotic coupling theory is widely accepted as the mechanism by which electron transport and ATP production are coupled.
- In this mechanism, electron transport causes a proton gradient that is directly linked to the phosphorylation process.
- The way in which the proton gradient leads to the production of ATP depends on ion channels through the inner mitochondrial membrane; these channels are a feature of the structure of ATP synthase.
- Protons flow back into the matrix through proton channels in the F_0 part of the ATP synthase.
- The flow of protons is accompanied by the formation of ATP, which occurs in the F_1 unit.

The Conformational Coupling Mechanism in Oxidative Phosphorylation

- The conformational coupling mechanism adds more details to our understanding of the chemiosmotic theory.
- Experimental evidence for this mechanism suggests that the proton gradient promotes proton flow back through ATP synthase, which causes conformational changes in ATP synthase that promote both the production and release of ATP.

Shuttle Mechanisms for Transferring Electrons from Cytosolic NADH to the Mitochondria

- Two shuttle mechanisms—the glycerol-3-phosphate shuttle and the malate–aspartate shuttle—transfer electrons from the NADH molecules produced in cytosolic reactions into the mitochondrion.
- In the glycerol-3-phosphate shuttle, which is found in muscle and brain, the electrons are transferred to FAD; in the malate–aspartate shuttle, which is found in kidney, liver, and heart, the electrons are transferred to NAD^+.
- With the malate–aspartate shuttle, 2.5 molecules of ATP are produced for each molecule of cytosolic NADH, rather than 1.5 ATP in the glycerol-3-phosphate shuttle, a point that affects the overall yield of ATP in these tissues.
- Approximately 2.5 molecules of ATP are generated for each molecule of NADH that enters the electron transport chain and approximately 1.5 molecules of ATP for each molecule of $FADH_2$.

The ATP Yield from the Complete Oxidation of Glucose

- When the pyruvate generated from glycolysis can enter the citric acid cycle, and the resulting NADH and $FADH_2$ molecules are reoxidized through the electron transport chain, a total of 30 or 32 ATPs are produced, with the difference due to the two possible shuttles.

Review Exercises

20-1 The Role of Electron Transport in Metabolism

1. **Recall** In terms of mitochondrial structure, describe the proton pumping that occurs to establish the proton gradient and explain why the gradient is important for producing energy in the form of ATP.
2. **Recall** Briefly summarize the steps in the electron transport chain from NADH to oxygen.
3. **Recall** Are electron transport and oxidative phosphorylation the same process? Why or why not?
4. **Reflect and Apply** How does mitochondrial structure contribute to aerobic metabolism, particularly in the integration of the citric acid cycle and electron transport?

20-2 Reduction Potentials in the Electron Transport Chain

5. **Reflect and Apply** Why is it reasonable to compare the electron transport process to a battery?
6. **Recall** Why are all the reactions in Table 20.1 written as reduction reactions?
7. **Recall** In terms of reduction potentials, describe the direction of electron flow between electron carriers in the mitochondrial electron transport chain.
8. **Mathematical** Calculate $E^{\circ\prime}$ for the following reaction:

$$\text{NADH} + \text{H}^+ + \tfrac{1}{2}\text{O}_2 \rightarrow \text{NAD}^+ + \text{H}_2\text{O}$$

9. **Mathematical** Calculate $E^{\circ\prime}$ for the following reaction:

$$\text{NADH} + \text{H}^+ + \text{Pyruvate} \rightarrow \text{NAD}^+ + \text{Lactate}$$

10. **Mathematical** Calculate $E^{\circ\prime}$ for the following reaction:

$$\text{Succinate} + \tfrac{1}{2}\text{O}_2 \rightarrow \text{Fumarate} + \text{H}_2\text{O}$$

11. **Mathematical** For the following reaction, identify the electron donor and the electron acceptor, and calculate $E^{\circ\prime}$.

$$\text{FAD} + 2\ \text{Cyt}\ c\ (\text{Fe}^{2+}) + 2\text{H}^+ \rightarrow \text{FADH}_2 + 2\ \text{Cyt}\ c\ (\text{Fe}^{3+})$$

12. **Mathematical** Which is more favorable energetically, the oxidation of succinate to fumarate by NAD^+ or by free FAD? Give the reason for your answer.

20-3 The Organization of Electron Transport Complexes

13. **Recall** Describe the electron transport reactions of Complex I and give the other name for this enzyme complex.
14. **Reflect and Apply** List the reactions of electron transport that liberate enough energy to drive the phosphorylation of ADP to ATP.
15. **Reflect and Apply** Show how the reactions of the electron transport chain differ from those in Question 14 when $FADH_2$ is the starting point for electron transport. Show how the reactions that liberate enough energy to drive the phosphorylation of ADP to ATP differ from the pathway when NADH is the starting point.
16. **Recall** What do cytochromes have in common with hemoglobin or myoglobin?
17. **Recall** How does the chemical activity of the cytochromes differ from that of hemoglobin and myoglobin?
18. **Recall** Which of the following is not bound in respiratory complexes: cytochromes, flavoproteins, iron–sulfur proteins, or coenzyme Q?
19. **Recall** Do any of the respiratory complexes play a role in the citric acid cycle? If so, what is that role?
20. **Recall** Do all the respiratory complexes generate enough energy to phosphorylate ADP to ATP? Explain your reasoning.
21. **Reflect and Apply** Two biochemistry students are about to use mitochondria isolated from rat liver for an experiment on oxidative phosphorylation. The directions for the experiment specify the addition of purified cytochrome *c* from any source to the reaction mixture. Why is the added cytochrome *c* needed? Why can the source be different from that of the mitochondria?
22. **Reflect and Apply** What are two advantages of the components of the electron transport chain being embedded in the inner mitochondrial membrane?
23. **Reflect and Apply** Experimental evidence strongly suggests that the protein portions of cytochromes have evolved more slowly (as judged by the number of changes in amino acids per million years) than the protein portions of hemoglobin and myoglobin and even more slowly than hydrolytic enzymes. Suggest a reason why.
24. **Reflect and Apply** What is the advantage of having mobile electron carriers in addition to large membrane-bound complexes of carriers?
25. **Reflect and Apply** What is the advantage of having a Q cycle in electron transport despite its complexity?
26. **Reflect and Apply** Why do the electron-transfer reactions of the cytochromes differ in standard reduction potential, even though all the reactions involve the same oxidation–reduction reaction of iron?
27. **Reflect and Apply** What is the underlying reason for the differences in spectroscopic properties among the cytochromes?
28. **Reflect and Apply** What would be some of the challenges involved in removing respiratory complexes from the inner mitochondrial membrane to study their properties?
29. **Mathematical** Using the information in Table 20.2, calculate $\Delta G^{\circ\prime}$ (in kJ/mol) for the following reaction:

$$2\ \text{Cyt}\ aa_3\ [\text{oxidized; Fe(III)}] + 2\ \text{Cyt}\ b\ [\text{reduced; Fe(II)}] \rightarrow$$
$$2\ \text{Cyt}\ aa_3\ [\text{reduced; Fe(II)}] + 2\ \text{Cyt}\ b\ [\text{oxidized; Fe(III)}]$$

20-4 The Connection Between Electron Transport and the Phosphorylation of ADP to ATP

30. **Recall** Describe the role of the F_1 portion of ATP synthase in oxidative phosphorylation.
31. **Recall** Is mitochondrial ATP synthase an integral membrane protein?
32. **Recall** Define *P/O ratio* and indicate why it is important.
33. **Recall** In what sense is mitochondrial ATP synthase a motor protein?
34. **Reflect and Apply** What is the approximate P/O ratio that can be expected if intact mitochondria are incubated in the presence of oxygen, along with added succinate?
35. **Reflect and Apply** Why is it difficult to determine an exact number for P/O ratios?
36. **Reflect and Apply** What are some of the difficulties in determining the exact number of protons pumped across the inner mitochondrial membrane by the respiratory complexes?

20-5 The Mechanism of Coupling in Oxidative Phosphorylation

37. **Recall** Briefly summarize the main concepts of the chemiosmotic coupling mechanism.
38. **Recall** Why does ATP production require an intact mitochondrial membrane?
39. **Recall** Briefly describe the role of uncouplers in oxidative phosphorylation.
40. **Recall** Describe the direction of proton movement during ATP synthesis in mitochondria and describe how this movement drives the production of ATP.
41. **Reflect and Apply** Can an uncoupler of oxidative phosphorylation inhibit electron transport from one component of the electron transport chain to another? Why or why not?
42. **Recall** What is the experimental evidence that mitochondria can change conformation during respiration?
43. **Recall** Is it possible to have proton pumping in the absence of electron transport?

20-6 Shuttle Mechanisms for Transferring Electrons from Cytosolic NADH to the Mitochondria

44. **Reflect and Apply** Describe the processes that occur in the glycerol-3-phosphate shuttle and explain its purpose.
45. **Recall** How does the yield of ATP from the complete oxidation of one molecule of glucose in muscle and brain differ from that in liver, heart, and kidney? What is the underlying reason for this difference?
46. **Reflect and Apply** The malate–aspartate shuttle yields about 2.5 moles of ATP for each mole of cytosolic NADH. Why does nature use the glycerol-3-phosphate shuttle, which yields only about 1.5 moles of ATP?

20-7 The ATP Yield from the Complete Oxidation of Glucose

47. **Mathematical** Using Table 20.3, what yield of ATP can be expected from the complete oxidation (using the malate–aspartate shuttle) of each of the following substrates by the reactions of glycolysis, the citric acid cycle, and oxidative phosphorylation?
(a) Fructose-1,6-bisphosphate
(b) Glucose
(c) Phosphoenolpyruvate
(d) Glyceraldehyde-3-phosphate
(e) NADH
(f) Pyruvate

Further Reading

Dickerson, R. E. Cytochrome *c* and the Evolution of Energy Metabolism. *Sci. Amer.* **242** (3), 136–152 (1980). An account of the evolutionary implications of cytochrome *c* structure.

Fillingame, R. The Proton-Translocating Pumps of Oxidative Phosphorylation. *Ann. Rev. Biochem.* **49**, 1079–1114 (1980). A review of chemiosmotic coupling.

Fillingame, R. H. Molecular Rotary Motors. *Science* **286**, 1687–1688 (1999). A review of research on ATP synthase.

Hatefi, Y. The Mitochondrial Electron Transport and Oxidative Phosphorylation System. *Ann. Rev. Biochem.* **54**, 1015–1069 (1985). A review that emphasizes the coupling between oxidation and phosphorylation.

Hinkle, P. C., and R. E. McCarty. How Cells Make ATP. *Sci. Amer.* **238** (3), 104–123 (1978). Chemiosmotic coupling and the mode of action of uncouplers.

Juhaszova, M., E. Kobrinsky, D. B. Zorov, M. A. Aon, S. Cortassa, and S. Sollott Setting the Record Straight: A New Twist on the Chemiosmotic Mechanism of Oxidative Phosphorylation. *Function (Oxf)* **3**, zqac018 (2022). This opinion article revisits the original proposal of chemiosmotic theory, discussing its foundational role in understanding mitochondrial energy metabolism and exploring recent insights into oxidative phosphorylation mechanisms.

Lane, M. D., P. L. Pedersen, and A. S. Mildvan. The Mitochondrion Updated. *Science* **234**, 526–527 (1986). A report on an international conference on bioenergetics and energy coupling.

Mitchell, P. Keilin's Respiratory Chain Concept and Its Chemiosmotic Consequences. *Science* **206**, 1148–1159 (1979). A Nobel Prize lecture by the scientist who first proposed the chemiosmotic coupling hypothesis.

Moser, C. C., et al. Nature of Biological Electron Transfer. *Nature* **355**, 796–802 (1992). An advanced treatment of electron transfer in biological systems.

Stock, D., A. G. W. Leslie, and J. F. Walker. Molecular Architecture of the Rotary Motor in ATP Synthase. *Science* **286**, 1700–1705 (1999). An article about the structure and function of ATP synthase.

Trumpower, B. The Protonmotive Q Cycle: Energy Transduction by Coupling of Proton Translocation to Electron Transfer by the Cytochrome *bc*1 Complex. *J. Biol. Chem.* **265**, 11409–11412 (1990). An advanced article that goes into detail about the Q cycle.

Vercellino, I., and L. A. Sazanov. The Assembly, Regulation, and Function of the Mitochondrial Respiratory Chain. *Nat. Rev. Mol. Cell. Biol.* **23**, 141–161 (2022). This review discusses the composition and function of the mitochondrial oxidative phosphorylation system.

Vignais, P. V., and J. Lunardi. Chemical Probes of the Mitochondrial ATP Synthesis and Translocation. *Ann. Rev. Biochem.* **54**, 977–1014 (1985). A review about the synthesis and use of ATP.

Xu, H., S. Z. DeLuca, and P. H. O'Farrell. Manipulating the Metazoan Mitochondrial Genome with Targeted Restriction Enzymes. *Science* **321**, 575–577 (2008). An article about using mutations in *Drosophila* mitochondrial DNA as models for human disease.

Hot Topic

Kapás, L. and E. Szentirmai. Brown Adipose Tissue at the Intersection of Sleep and Temperature Regulation. *Temperature* **1**, 16–17 (2014). Proposed link between brown adipose tissue and restorative sleep and body temperature regulation.

Harms, M. and P. Seale. Brown and Beige Fat: Development, Function, and Therapeutic Potential. *Nature Med.*, **19**, 1252–1263 (2013). Review of genes and pathways that regulate brown and beige adipocyte biology used to identify therapeutic targets for metabolic diseases such as type II diabetes and obesity.

Moyer, M. W. Supercharging Brown Fat to Battle Obesity. *Sci. Amer.*, **311**, 30–32 (2014). Article describing the application of state-of-the-art technology to monitor, detect, and understand hormone regulation of fat.

Palou, A., C. Picó, M. L. Bonet, and P. Oliver. The Uncoupling Protein, Thermogenin. *Int. J. Biochem. Cell Biol.* **30**, 7–11 (1998). Article describing the role of thermogenin in heat production in adipose tissue in response to cold exposure or overeating.

Saito, M. Human Brown Adipose Tissue: Regulation and Anti-obesity Potential. *Endocrine J.* **61**, 409–416 (2014). Review providing evidence that brown adipose tissue can be stimulated in adult humans by cold exposure and consumption of capsinoids.

Sanchez-Delgado, G., B. Martinez-Tellez, J. Olza, C. M. Aguilera, Á. Gil, and J. R. Ruiz. Role of Exercise in the Activation of Brown Adipose Tissue. *Ann. Nutr. Metab.* **67**, 21–32 (2015). Article that proposes the molecular link between exercise and the activation of brown adipose tissue.

Yoneshiro, T., S. Aita, M. Matsushita, T. Kayahara, T. Kameya, Y. Kawai, T. Iwanaga, and M. Saito. Recruited Brown Adipose Tissue as an Antiobesity Agent in Humans. *J. Clin. Invest.* **123**, 3404–3408 (2013). Research paper providing evidence for temperature and diet to recover brown adipose tissue deposits in adults.

Lipid metabolism allows polar bears to thrive in arctic climates and to endure months of hibernation.

Lipid Metabolism 21

Chapter Outline

21-1 Lipids Are Involved in the Generation and Storage of Energy

Learning Objective

After reading this section, students will be able to:

LO 21.1.1 Identify three molecules produced by the oxidation of lipids that go on to produce energy.

In the past few chapters, we have seen how energy can be released by the catabolic breakdown of carbohydrates in aerobic and anaerobic processes. In Chapter 16, we saw that there are carbohydrate polymers (such as starch in plants and glycogen in animals) that represent stored energy, in the sense that these carbohydrates can be hydrolyzed to monomers and then oxidized to provide energy in response to the needs of an organism. In this chapter, we describe how the metabolic oxidation of lipids releases large quantities of energy through the production of acetyl-CoA, NADH, and $FADH_2$ and how lipids represent an even more efficient way of storing chemical energy.

21-2 Catabolism of Lipids

Learning Objectives

After reading this section, students will be able to:

LO 21.2.1 Describe how fatty acids are released from triacylglycerols and phosphoacylglycerols.

Chapter Outline (continued)

LO 21.2.2 Explain how fatty acyl groups are transferred from the cytosol to the mitochondrial matrix.

LO 21.2.3 Distinguish between the reactions in the β-oxidation pathway for saturated fatty acids.

The oxidation of fatty acids is the chief source of energy in the catabolism of lipids; in fact, lipids that are sterols (steroids that have a hydroxyl group as part of their structure; Section 8-2G) are not catabolized as a source of energy but are excreted. Both triacylglycerols (Section 8-2B), which are the main storage form of the chemical energy of lipids, and phosphoacylglycerols (PAGs) (Section 8-2C), which are important components of biological membranes, have fatty acids as part of their covalently bonded structures. In both types of compounds, the bond between the fatty acid and the rest of the molecule can be hydrolyzed, with the reaction catalyzed by suitable groups of enzymes—**lipases** in the case of triacylglycerols (Figure 21.1, *left*) and **phospholipases** in the case of PAGs (Figure 21.1, *right*).

lipases enzymes that hydrolyze lipids

phospholipases enzymes that hydrolyze phospholipids

Several different phospholipases can be distinguished on the basis of the site at which they hydrolyze phospholipids (Figure 21.2). Phospholipase A_2 is widely distributed in nature; biochemists are also actively studying its structure and mode of action, which involves the hydrolysis of phospholipids at the surface of micelles (Section 2-1). Phospholipase D occurs in spider venom and is responsible for the tissue damage that accompanies spider bites. Snake venoms also contain phospholipases; the concentration of phospholipases is particularly high in venoms with comparatively low concentrations of the toxins (usually small peptides) that are characteristic of some kinds of venom. The lipid products of hydrolysis lyse red blood cells, preventing clot formation. The snakebite victim then bleeds to death.

Combining what we have just seen in Figures 21.1 and 21.2, let us consider the products of individual lipases or phospholipases that would catalyze the release of fatty acids. Lipases cleave one fatty acid at a time from an acylglycerol molecule. For example, the products of a triacylglycerol lipase reaction would produce a diacylglycerol and one fatty acid. Phospholipase A_1 or phospholipase A_2 activity would also release one fatty acid, but the products in each case (assuming a phosphoacylglycerol as substrate) would be a **lysophospholipid** and one fatty acid. A lysophospholipid is a phosphoacylglycerol that has only one fatty acyl chain instead of the usual two.

lysophospholipid a phosphoacylglycerol that has only one fatty acyl chain instead of the usual two

lipolysis the metabolic pathway by which triacylglycerols are hydrolyzed to glycerol and three free fatty acid molecules

In contrast to the release of fatty acids from PAGs seen in Figure 21.2, the release of fatty acids from triacylglycerols (a process called **lipolysis**) in

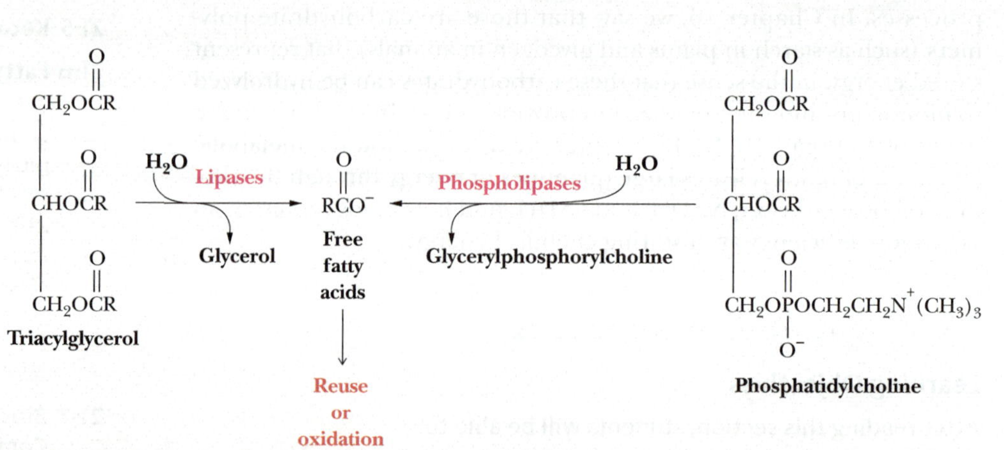

Figure 21.1 The release of fatty acids for future use. The source of fatty acids can be a triacylglycerol *(left)* or a phosphoacylglycerol such as phosphatidylcholine *(right).* Each R group represents a 16–20 carbon-long hydrocarbon chain.

Pioneers in Biochemistry • Marie Maynard Daly

Marie Maynard Daly (1921–2003) was a pioneering biochemist and the first Black woman to receive a PhD in chemistry. As many male students were diverted from higher education to serve in World War II, universities began to admit more women and other underrepresented groups. Daly took advantage of these new opportunities to attend Queens College as one of the few Black students and later NYU, where she earned a master's degree. At Columbia, where she received her PhD, Daly joined the lab of the university's only female chemistry professor, who became an important mentor for her. Although some assumed that Black and female scientists would serve only as placeholders for the white men who had gone to war, Daly proved otherwise through a lifetime of contributions to biochemistry research.

Daly began her career in genetics, working with a research team that made discoveries that were foundational to the field, including the amino acid composition of histones and the role of RNA in protein production. After this, Daly pivoted to medical research, where her discoveries were fundamental in establishing the relationship of cholesterol in the diet with hypertension and heart attack.

A pioneer in her field, Daly was committed to increasing diversity in scientific institutions. As a professor of biochemistry, she worked to recruit, train, and support Black students and later founded a scholarship specifically for minority students in STEM.

adipocytes is controlled by hormones (Figure 21.3). The top left of Figure 21.3 shows a hormone, such as epinephrine or glucagon, binding to a receptor on the plasma membrane of an adipose cell. This hormone binding activates adenylate cyclase, which produces 3′,5′-cyclic AMP (cAMP). cAMP activates protein kinase A (cAMP-dependent protein kinase). Recall that cAMP is a key second messenger molecule seen in many biochemical processes and that we first discussed it in Section 11-3C. Protein kinase A phosphorylates and activates other enzymes that are involved in lipolysis. The first enzyme directly involved in the lipolysis is *adipose triglyceride lipase (ATGL)*, which cleaves a fatty acid from the triacylglycerol, forming a diacylglycerol molecule (Figure 21.3, *middle right*). Then, *hormone-sensitive lipase (HSL)* cleaves a fatty acid from the diacylglycerol to produce a monoacylglycerol, which can then be split into glycerol and the last fatty acid by *monoacylglycerol lipase (MGL)* (Figure 21.3, *bottom right*).

21-2A Fatty Acid Transport into Mitochondria

Fatty acid oxidation and breakdown begin with activation of the fatty acid molecule (Figure 21.4). Recall from Section 15-7 that we discussed metabolite activation as a process that thermodynamically helps the next step in a metabolic pathway occur. Fatty acid activation involves the formation of a thioester bond between the carboxyl group of the fatty acid and the thiol group of coenzyme A (CoA-SH). The activated form of the fatty acid is an acyl-CoA, the exact nature of which depends on the nature of the fatty acid itself. Keep in mind throughout this discussion that all acyl-CoA molecules are thioesters because the fatty acid is esterified to the thiol group of CoA.

The enzyme that catalyzes the formation of the ester bond, *fatty acyl CoA synthetase*, requires adenosine triphosphate (ATP) for its action. In the course of the reaction, an acyl adenylate intermediate is formed (Figure 21.4, *Step 1*). This is an intermediate activation step that helps the next step of acyl group transfer to coenzyme A (CoA-SH). ATP is converted to adenosine monophosphate (AMP) and PP_i, rather than to adenosine diphosphate (ADP) and P_i. The PP_i is hydrolyzed to two P_i; the hydrolysis of two high-energy phosphate bonds provides energy for the activation of the fatty acid and is equivalent to the use of two ATP. The formation of the acyl-CoA is endergonic without the energy provided by the hydrolysis of the two high-energy bonds. The hydrolysis

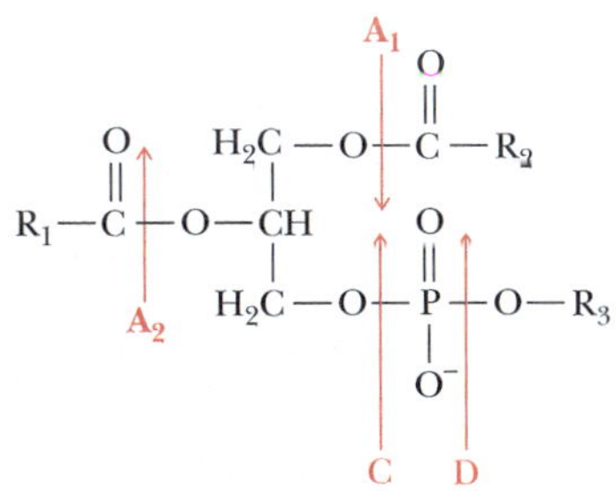

A phosphoacylglycerol

Figure 21.2 Several phospholipases hydrolyze phosphoacylglycerols. They are designated A_1, A_2, C, and D, depending on their site of action as shown.

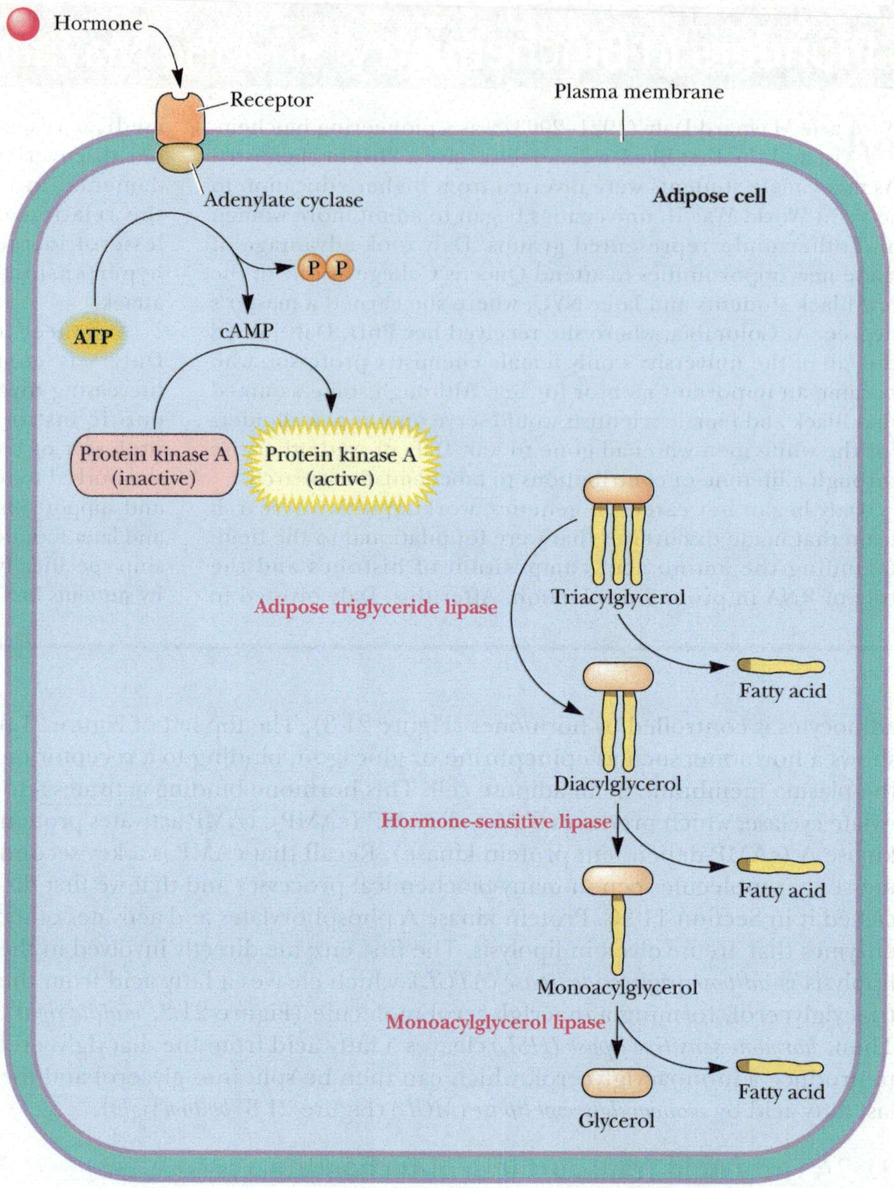

Figure 21.3 Hormone-activated liberation of fatty acids from triacylglycerols in adipose tissue. Hormone binding to a receptor on an adipose cell activates adenylate cyclase and cAMP production. cAMP activates protein kinase A, which then helps activate the other enzymes involved in releasing fatty acid molecules from the triacylglycerol.

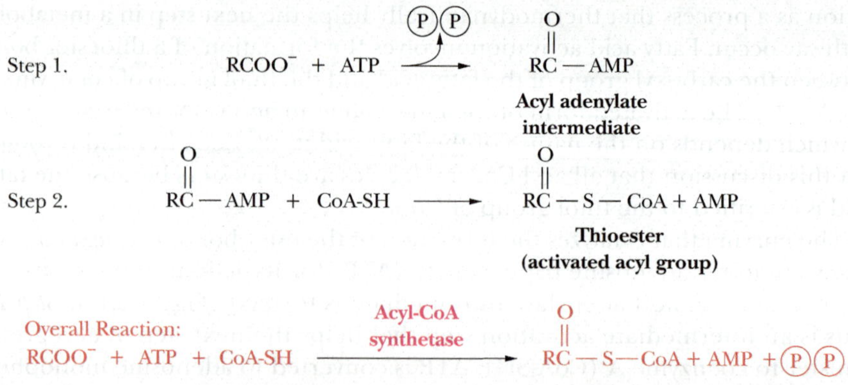

Figure 21.4 Fatty acid activation: the formation of fatty acyl-CoA by fatty acyl-CoA synthetase. In the first step, fatty acyl-CoA synthetase binds a free fatty acid and uses ATP to form an acyl adenylate intermediate. This can be considered an intermediate activation step that helps the next activation step of transferring the fatty acyl group to coenzyme A (CoA-SH). The R group represents a 16–20-carbon-long hydrocarbon chain, and $RCOO^-$ represents a fatty acid.

of ATP to AMP and two P_i represents an increase in entropy, too, which makes this process even more thermodynamically favored.

There are several synthetase enzymes of this type, some specific for longer-chain fatty acids and some for shorter-chain fatty acids. Both saturated and unsaturated fatty acids can serve as substrates for these enzymes. The esterification takes place in the cytosol, but the rest of the reactions of fatty acid oxidation occur in the mitochondrial matrix. Therefore, the activated fatty acid must be transported into a mitochondrion, where the rest of the fatty acid oxidation and breakdown process can proceed.

Figure 21.5 shows how the activated fatty acid is transported from the cytosol into the mitochondrial matrix by the **carnitine shuttle**. This process begins on the cytosolic surface of the outer mitochondrial membrane with carnitine

carnitine shuttle the process used in fatty acid metabolism to shuttle fatty acyl groups from the cytosol to the mitochondrial matrix

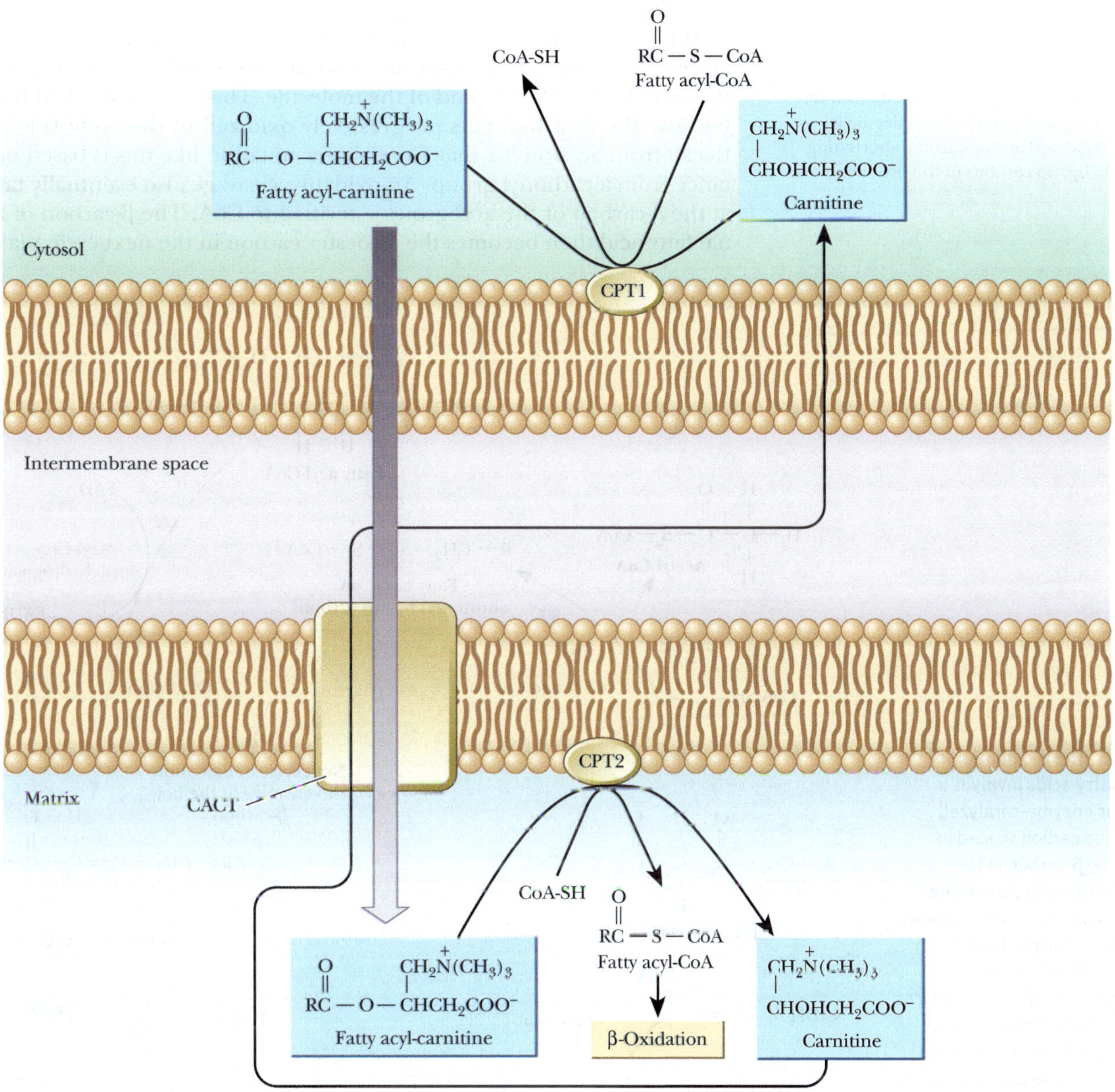

Figure 21.5 The carnitine shuttle pathway that transports fatty acyl-CoA molecules from the cytosol to the mitochondrial matrix. Carnitine palmitoyltransferase 1 (CPT1) catalyzes the transfer of the fatty acyl group from coenzyme A to carnitine, forming fatty acyl-carnitine, which travels through the outer mitochondrial membrane into the intermembrane space. Carnitine-acylcarnitine translocase (CACT) then transports fatty acyl-carnitine across the inner mitochondrial membrane into the mitochondrial matrix, where carnitine palmitoyltransferase 2 (CPT2) catalyzes transfer of the fatty acyl group from carnitine back to coenzyme A, forming a fatty acyl-CoA molecule that is ready for oxidation and breakdown. Finally, the carnitine in the matrix moves back to the cytosol so that it can shuttle the next fatty acyl group to the matrix.

palmitoyltransferase 1 (CPT1) catalyzing the transfer of the fatty acyl group from coenzyme A to carnitine, thereby forming an acyl-carnitine molecule that crosses the outer mitochondrial membrane into the intermembrane space (Figure 21.5, *top*). The fatty acyl-carnitine molecule is then transported across the inner mitochondrial membrane by an integral membrane protein called CACT (Figure 21.5, *middle*). Once inside the mitochondrial matrix, carnitine palmitoyltransferase 2 (CPT2) catalyzes transfer of the fatty acyl group from carnitine back to coenzyme A (Figure 21.5, *bottom*), and the carnitine group moves back across CACT and through the outer mitochondrial membrane to the cytosol so that it can shuttle the next fatty acyl group to the matrix. The fatty acyl-CoA molecule can now undergo its oxidation and breakdown process in the mitochondrial matrix.

21-2B Oxidation of Saturated Fatty Acids

β-oxidation the main pathway of catabolism for fatty acids, where a fatty acyl-CoA molecule is oxidized and cleaved at its β-carbon, shortening the molecule by two carbons in the production of acetyl-CoA

In the mitochondrial matrix, a repeated sequence of reactions successively cleaves two-carbon units from the fatty acyl-CoA molecule, starting from the thioester [—C(=O)—S] end of the molecule. This process is called **β-oxidation** because the β-carbon gets progressively oxidized in the cycle (Figure 21.6). Recall from Section 3-2 that Greek-letter notation like this is based on the distance from a carbonyl group. An oxidative cleavage also eventually takes place at the β-carbon of the acyl group esterified to CoA. The β-carbon of the original fatty acid then becomes the thioester carbon in the next cycle that oxidizes

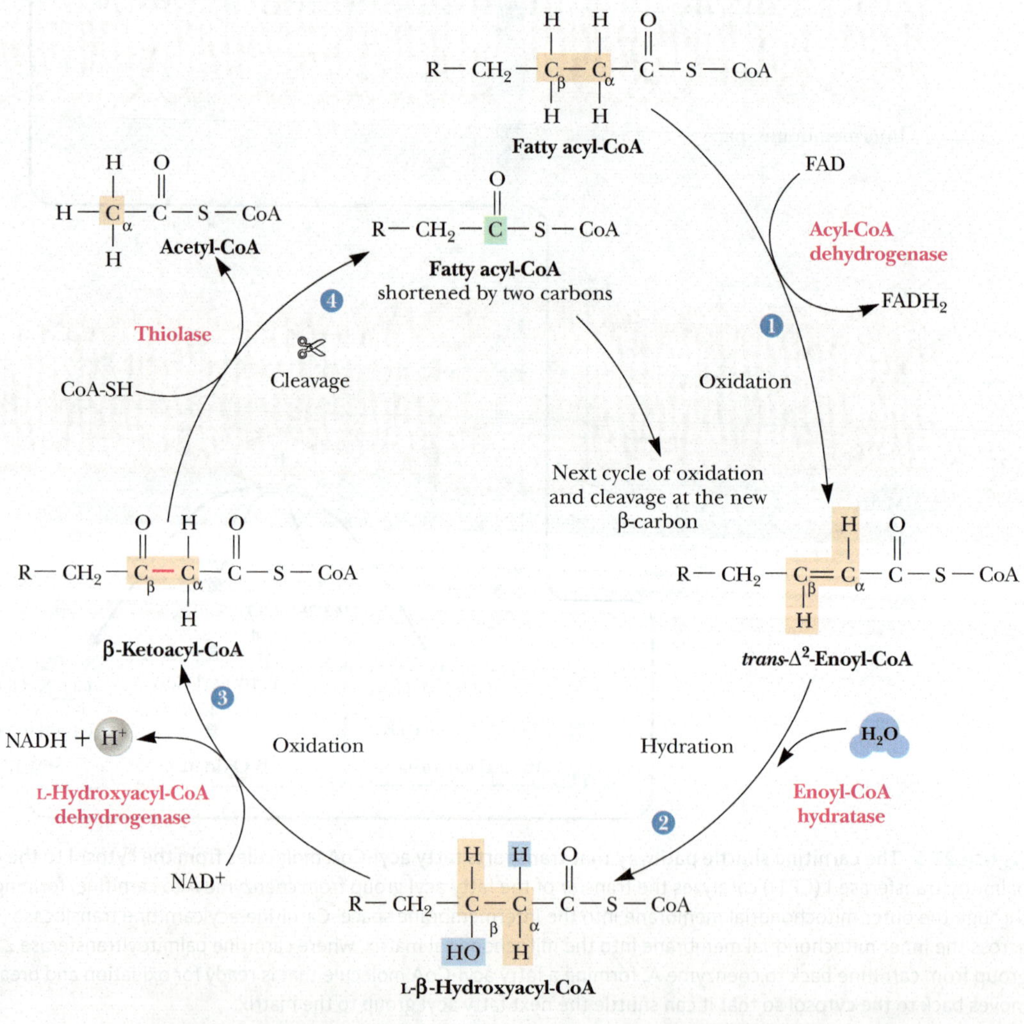

Figure 21.6 The β-oxidation of saturated fatty acids involves a cycle of four enzyme-catalyzed reactions. The carbon shaded in green was the β-carbon at the beginning of the cycle, but at the end of the cycle, there is now a new β-carbon that is two carbons down the acyl chain from the carbon shaded in green. Each cycle produces one FADH$_2$ and one NADH, and it liberates acetyl-CoA, resulting in a fatty acid that is two carbons shorter. Recall from Section 8-2A that the Δ symbol represents a double bond, and the number associated with it is the location of the double bond (based on counting the carbonyl group as carbon 1).

and cleaves at a new β-carbon. The whole cycle requires the following four reactions:

Step 1. The fatty acyl-CoA is *oxidized* to an α,β unsaturated acyl-CoA (also called a β-enoyl-CoA). The product has a *trans* arrangement at the double bond. This reaction is catalyzed by an FAD-dependent acyl-CoA dehydrogenase. Recall from Section 19-4 that an FAD-dependent dehydrogenase (succinate dehydrogenase) also catalyzed the oxidation of a carbon–carbon single bond to a carbon–carbon double bond in the citric acid cycle.

Step 2. The unsaturated acyl-CoA is *hydrated* to produce a β-hydroxyacyl-CoA. This reaction is catalyzed by the enzyme enoyl-CoA hydratase.

Step 3. A second *oxidation* reaction is catalyzed by β-hydroxyacyl-CoA dehydrogenase, an NAD^+-dependent enzyme. The product is a β-ketoacyl-CoA.

Step 4. The enzyme thiolase catalyzes the *cleavage* of the β-ketoacyl-CoA; a molecule of CoA is required for the reaction. The products are acetyl-CoA and an acyl-CoA that is two carbons shorter than the original molecule that entered the β-oxidation cycle. The CoA is needed in this reaction to form the new thioester bond in the smaller acyl-CoA molecule. This smaller molecule then undergoes another round of the β-oxidation cycle at the new β-carbon of the fatty acyl-CoA molecule that is now two carbons shorter.

Fatty acids with even numbers of carbon atoms are the ones normally found in nature. When a fatty acid with an even number of carbon atoms undergoes many successive rounds of the β-oxidation cycle, the final products are two molecules of acetyl-CoA. The total number of molecules of acetyl-CoA produced in breaking down a fatty acid is equal to half the number of carbon atoms in the original fatty acid. For example, stearic acid contains 18 carbon atoms, so it gives rise to 9 molecules of acetyl-CoA. The conversion of one 18-carbon stearic acid molecule to nine 2-carbon acetyl units requires eight, not nine, cycles of β-oxidation (Figure 21.7). Each acetyl-CoA that is produced in fatty acid breakdown enters the citric acid cycle to produce carbon dioxide and more NADH and $FADH_2$ (Section 19-4). Each NADH and $FADH_2$ produced will then carry electrons to the electron transport chain for oxidative phosphorylation and the production of ATP (Sections 20-3 and 20-4).

Recall that most of the enzymes of the citric acid cycle are located in the mitochondrial matrix, and we have just seen that the β-oxidation cycle takes place in the matrix as well. In addition to mitochondria, other sites of β-oxidation are known. Peroxisomes and glyoxysomes (Section 1-5), organelles that carry out oxidation reactions, are also sites of β-oxidation, albeit to a far lesser extent than the mitochondria. Certain hypolipidemic drugs, called fibrates, are used in an attempt to control obesity by stimulating β-oxidation.

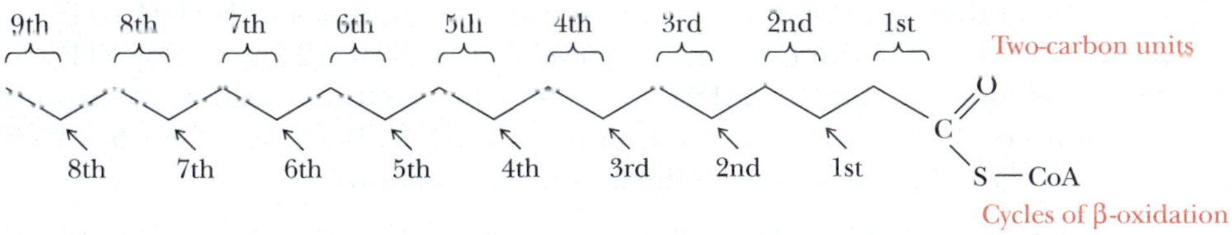

Figure 21.7 Stearic acid (18 carbons) gives rise to nine 2-carbon units after eight cycles of β-oxidation. The ninth two-carbon unit remains esterified to CoA after eight cycles of β-oxidation have removed eight successive two-carbon units, starting at the thioester end on the right. Thus, it takes only eight rounds of β-oxidation to completely process an 18-carbon fatty acid to acetyl-CoA.

21-3 The Energy Yield from the Oxidation of Fatty Acids

Learning Objective

After reading this section, students will be able to:

LO 21.3.1 Compare the energy yield from the complete oxidation of fatty acids to the energy yield from the complete oxidation of glucose.

In carbohydrate metabolism, the energy released by oxidation reactions is used to drive the production of ATP, with most of the ATP produced in aerobic processes. In the same aerobic processes—namely, the citric acid cycle and oxidative phosphorylation—the energy released by the oxidation of acetyl-CoA formed by β-oxidation of fatty acids can also be used to produce ATP. There are two sources of ATP to keep in mind when calculating the overall yield of ATP. The first source is the reoxidation of the NADH and $FADH_2$ produced by the β-oxidation of the fatty acid to acetyl-CoA. The second source is ATP production from the processing of the acetyl-CoA through the citric acid cycle and oxidative phosphorylation.

We use the oxidation of stearic acid, which contains 18 carbon atoms, to calculate the energy yield from the β-oxidation of fatty acids. Eight cycles of β-oxidation are required to convert one mole of stearic acid to nine moles of acetyl-CoA; in the process, eight moles of FAD are reduced to $FADH_2$, and eight moles of NAD^+ are reduced to NADH.

$$CH_3(CH_2)_{16}\overset{\overset{\displaystyle O}{\|}}{C}-S-CoA + 8\,FAD + 8\,NAD^+ + 8\,H_2O + 8\,CoA\text{-}SH \longrightarrow$$

$$9\,CH_3-\overset{\overset{\displaystyle O}{\|}}{C}-S-CoA + 8\,FADH_2 + 8\,NADH + 8\,H^+$$

The nine moles of acetyl-CoA produced from each mole of stearic acid enter the citric acid cycle. One mole of $FADH_2$ and three moles of NADH are produced for each mole of acetyl-CoA that enters the citric acid cycle. At the same time, one mole of GDP is phosphorylated to produce GTP for each turn of the citric acid cycle.

$$9\,CH_3\overset{\overset{\displaystyle O}{\|}}{C}-S-CoA + 9\,FAD + 27\,NAD^+ + 9\,GDP + 9\,P_i + 27\,H_2O \longrightarrow$$

$$18\,CO_2 + 9\,CoA\text{-}SH + 9\,FADH_2 + 27\,NADH + 9\,GTP + 27\,H^+$$

The $FADH_2$ and NADH produced by β-oxidation and by the citric acid cycle enter the electron transport chain, and ATP is produced by oxidative phosphorylation. In our example, there are 17 moles of $FADH_2$ (8 from β-oxidation and 9 from the citric acid cycle); there are also 35 moles of NADH (8 from β-oxidation and 27 from the citric acid cycle). Recall that 2.5 moles of ATP are produced for each mole of NADH that enters the electron transport chain, and 1.5 moles of ATP result from each mole of $FADH_2$. Because $17 \times 1.5 = 25.5$ and $35 \times 2.5 = 87.5$, we can write the following equations:

$$17FADH_2 + 8.5O_2 + 25.5ADP + 25.5P_i \rightarrow 17FAD + 25.5ATP + 17H_2O$$

$$35NADH + 35H^+ + 17.5O_2 + 87.5ADP + 87.5P_i \rightarrow 35NAD^+ + 87.5ATP + 35H_2O$$

The overall yield of ATP from the oxidation of stearic acid can be obtained by adding the equations for β-oxidation, the citric acid cycle, and oxidative phosphorylation. In this calculation, we take GDP as equivalent to ADP and GTP as equivalent to ATP, which means that the equivalent of nine ATP must be added to those produced in the reoxidation of $FADH_2$ and NADH. There are 9 ATP equivalent to the 9 GTP from the citric acid cycle, 25.5 ATP from the reoxidation of $FADH_2$, and 87.5 ATP from the reoxidation of NADH, for a grand total of 122 ATP.

$$CH_3(CH_2)_{16}\overset{\overset{\Large O}{\|}}{C}—S—CoA + 26\ O_2 + 122\ ADP + 122\ P_i \rightarrow$$

$$18\ CO_2 + 17\ H_2O + 122\ ATP + CoA\text{-}SH$$

The activation step in which stearyl-CoA was formed is not included in this calculation, and we must subtract the ATP that was required for that step. Even though only one ATP was required, two high-energy phosphate bonds are lost because of the production of AMP and PP_i. The pyrophosphate must be hydrolyzed to phosphate (P_i) before it can be recycled in metabolic intermediates. As a result, we must subtract the equivalent of two ATP for the activation step. The net yield of ATP becomes 120 moles of ATP for each mole of stearic acid that is completely oxidized. See Table 21.1 for a balance sheet. Keep in mind that these values are theoretical consensus values that not all cells attain.

As a comparison, 32 moles of ATP can be obtained from the complete oxidation of one mole of glucose, but glucose contains 6, rather than 18, carbon atoms. Three glucose molecules contain 18 carbon atoms, so a more relevant comparison is the ATP yield from the oxidation of three glucose molecules, which is 3 × 32 = 96 ATP for the same number of carbon atoms. The yield of ATP from the oxidation of the lipid is still higher than that from the carbohydrate, even for the same number of carbon atoms.

Why is there this difference of 120 – 96 = 24 ATP molecules? A fatty acid is all hydrocarbon except for the carboxyl group—that is, it exists in a highly reduced state. A sugar is already partly oxidized because of the presence of its oxygen-containing groups. Because the oxidation of a fuel leads to the reduced electron carriers used in the electron transport chain, a more reduced fuel, such as a fatty acid, can be oxidized further than a partially oxidized fuel, such as a carbohydrate.

Water is produced in the oxidation of fatty acids, just as water is produced in the complete oxidation of carbohydrates. The production of **metabolic water** is a common feature of aerobic metabolism. This process can be a source of water for organisms that live in desert environments. Camels are a well-known

metabolic water the water produced as a result of the complete oxidation of nutrients; sometimes it is the only water source of desert-dwelling organisms

Table 21.1 The Balance Sheet for Oxidation of One Molecule of Stearic Acid

Reaction	NADH Molecules	$FADH_2$ Molecules	ATP Molecules
1. Stearic acid → Stearyl-CoA (activation step)	+8	+8	–2
2. Stearyl-CoA → 9 Acetyl-CoA (8 cycles of β-oxidation)	+27	+9	
3. 9 Acetyl-CoA → 18 CO_2 (citric acid cycle); GDP → GTP (9 molecules)			+9
4. Reoxidation of NADH from β-oxidation cycle	–8		+20
5. Reoxidation of NADH from citric acid cycle	–27		+67.5
6. Reoxidation of $FADH_2$ from β-oxidation cycle		–8	+12
7. Reoxidation of $FADH_2$ from citric acid cycle		–9	+13.5
Net	0	0	+120

There is no net change in the number of molecules of NADH or $FADH_2$.

example; the stored lipids in their humps are a source of both energy and water during long trips through the desert. Kangaroo rats provide an even more striking example of adaptation to an arid environment. These animals have been observed to live indefinitely without having to drink water. They live on a diet of seeds, which are rich in lipids but contain little water. The metabolic water that kangaroo rats produce is adequate for all their water needs. This metabolic response to arid conditions is usually accompanied by a reduced output of urine.

21-4 Catabolism of Unsaturated Fatty Acids and Odd-Carbon Fatty Acids

Learning Objective

After reading this section, students will be able to:

LO 21.4.1 Distinguish between the catabolism of odd-carbon fatty acids, monounsaturated fatty acids, and polyunsaturated fatty acids.

Fatty acids with odd numbers of carbon atoms are not as frequently encountered in nature as those with even numbers of carbon atoms. When odd-numbered fatty acids undergo β-oxidation, the last cycle of β-oxidation produces a three-carbon propionyl-CoA molecule (Figure 21.8, *top left*). To utilize propionyl-CoA for energy production, an enzymatic pathway exists to convert propionyl-CoA to succinyl-CoA, which then enters the citric acid cycle (Figure 21.8, *middle*). In this pathway, propionyl-CoA is first carboxylated to methylmalonyl-CoA in a reaction catalyzed by propionyl-CoA carboxylase, which then undergoes rearrangement to form succinyl-CoA. Because propionyl-CoA is also a product of the catabolism of several amino acids, the conversion of propionyl-CoA to succinyl-CoA also plays a role in amino acid metabolism (see Section 23-6A). The conversion of methylmalonyl-CoA to succinyl-CoA requires vitamin B_{12} (cyanocobalamin), which has a cobalt(III) ion in its active state.

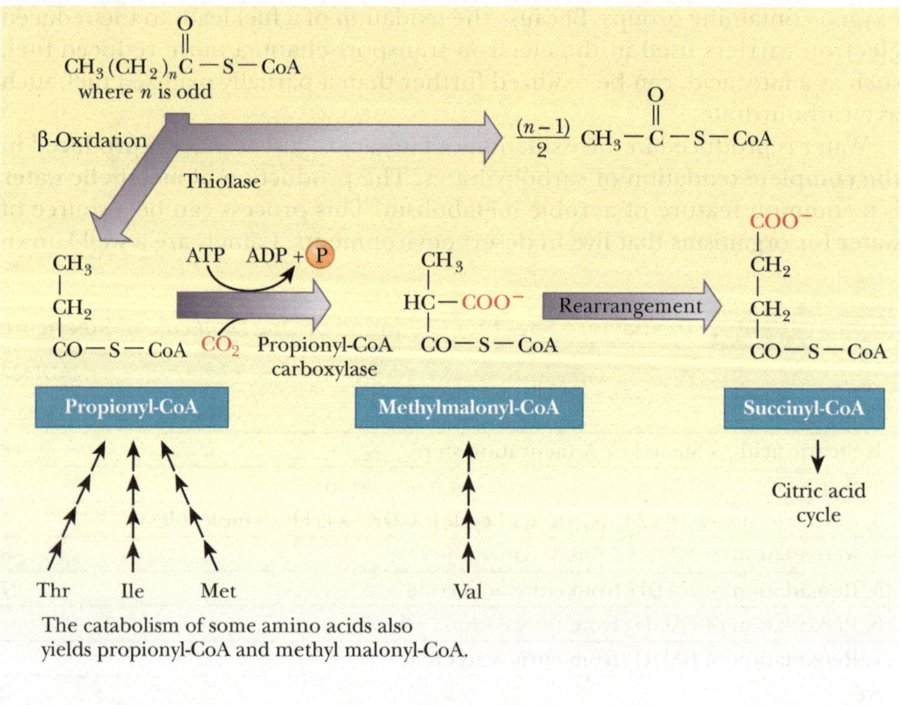

Figure 21.8 The oxidation of a fatty acid containing an odd number of carbon atoms. The last cycle of β-oxidation of a fatty acid with an odd number of carbons produces a three-carbon propionyl-CoA molecule. To utilize this molecule for energy production, propionyl-CoA undergoes a few reactions to produce succinyl-CoA, which can enter the citric acid cycle. The rearrangement of methylmalonyl-CoA to succinyl-CoA is catalyzed in two steps by a racemase enzyme, followed by a mutase enzyme.

21-4A Oxidation of Monounsaturated Fatty Acids

▶ **How does the oxidation of unsaturated fatty acids differ from that of saturated fatty acids?**

The conversion of a monounsaturated fatty acid to acetyl-CoA requires a cis–trans isomerization (Figure 21.9, *middle*)—a reaction that is not encountered in the oxidation of saturated acids. Successive rounds of β-oxidation of oleic acid (18:1) provide an example of these reactions. β-Oxidation gives rise to unsaturated fatty acids in which the double bond is in the trans arrangement, whereas the double bonds in most naturally occurring fatty acids are cis. In the case of oleic acid, there is a cis double bond between carbons 9 and 10. Three rounds of β-oxidation produce a 12-carbon unsaturated fatty acid with a cis double bond between carbons 3 and 4 (Figure 21.9, *top*). The hydratase of the β-oxidation cycle requires a trans double bond between carbon atoms 2 and 3 as a substrate. A cis–trans isomerase produces a trans double bond between carbons 2 and 3 from the cis double bond between carbons 3 and 4. From this point forward, the fatty acid is metabolized in the same way as saturated fatty acids (Figure 21.9, *bottom*). When oleic acid is β-oxidized, the first step (fatty acyl-CoA dehydrogenase) is skipped, and the isomerase deals with the cis double bond, putting it into the proper position and orientation to continue the pathway.

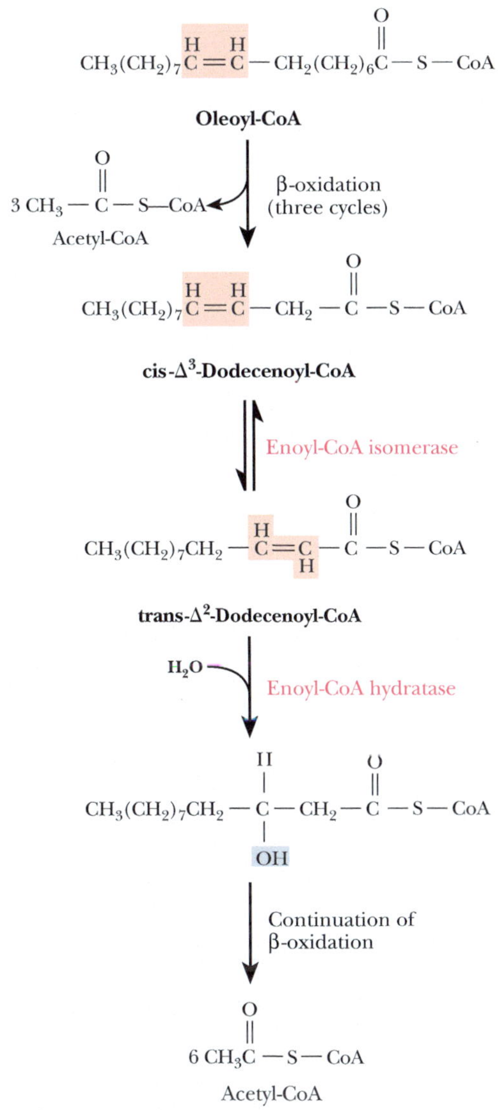

Figure 21.9 β-Oxidation of unsaturated fatty acids. In the case of oleoyl-CoA, three β-oxidation cycles produce three molecules of acetyl-CoA and leave cis-Δ^3-dodecenoyl-CoA. Rearrangement of the cis-Δ^3-enoyl-CoA species by enoyl-CoA isomerase gives the trans-Δ^2 species, which then proceeds normally through the β-oxidation pathway.

21-4B Oxidation of Polyunsaturated Fatty Acids

When polyunsaturated fatty acids are β-oxidized, another enzyme is needed to handle the second double bond. Let's consider how linoleic acid (18:2) would be metabolized (Figure 21.10). This fatty acid has cis double bonds at positions 9 and 12, which are indicated as cis-Δ^9 and cis-Δ^{12} (Figure 21.10, *top*). Three normal cycles of β-oxidation occur, as in our example with oleic acid, before the isomerase must switch the position and orientation of the double bond. The cycle of β-oxidation continues until a 10-carbon fatty acyl-CoA is attained that has one cis double bond on its carbon 4 (cis-Δ^4). Then the first step of β-oxidation occurs, putting in a trans double bond between carbons 2 and 3 (α and β). Normal β-oxidation cannot continue at this point because the fatty acid with the two double bonds so close together is a poor substrate for the hydratase. Therefore, a second new enzyme, *2,4-dienoyl-CoA reductase*, uses NADPH to reduce this intermediate. The result is a fatty acyl-CoA with a trans double bond between carbons 3 and 4. The isomerase then switches the trans double bond from carbon 3 to carbon 2, and β-oxidation continues.

A molecule with three double bonds, such as linolenic acid (18:3), would use the same two enzymes to handle the double bonds. The first double bond requires the isomerase. The second one requires the reductase and the isomerase, and the third requires the isomerase. For practice, you can diagram the β-oxidation of an 18-carbon molecule with cis double bonds at positions 9, 12, and 15 to see that this is true. Unsaturated fatty acids make up a large enough portion of the fatty acids in storage fat (40% for oleic acid alone) to make the reactions of the cis–trans isomerase and the reductase of particular importance.

The oxidation of unsaturated fatty acids does not generate as many ATPs as it would for a saturated fatty acid with the same number of carbons. This is because the presence of a double bond means that the acyl-CoA dehydrogenase step will be skipped. Thus, fewer $FADH_2$ will be produced.

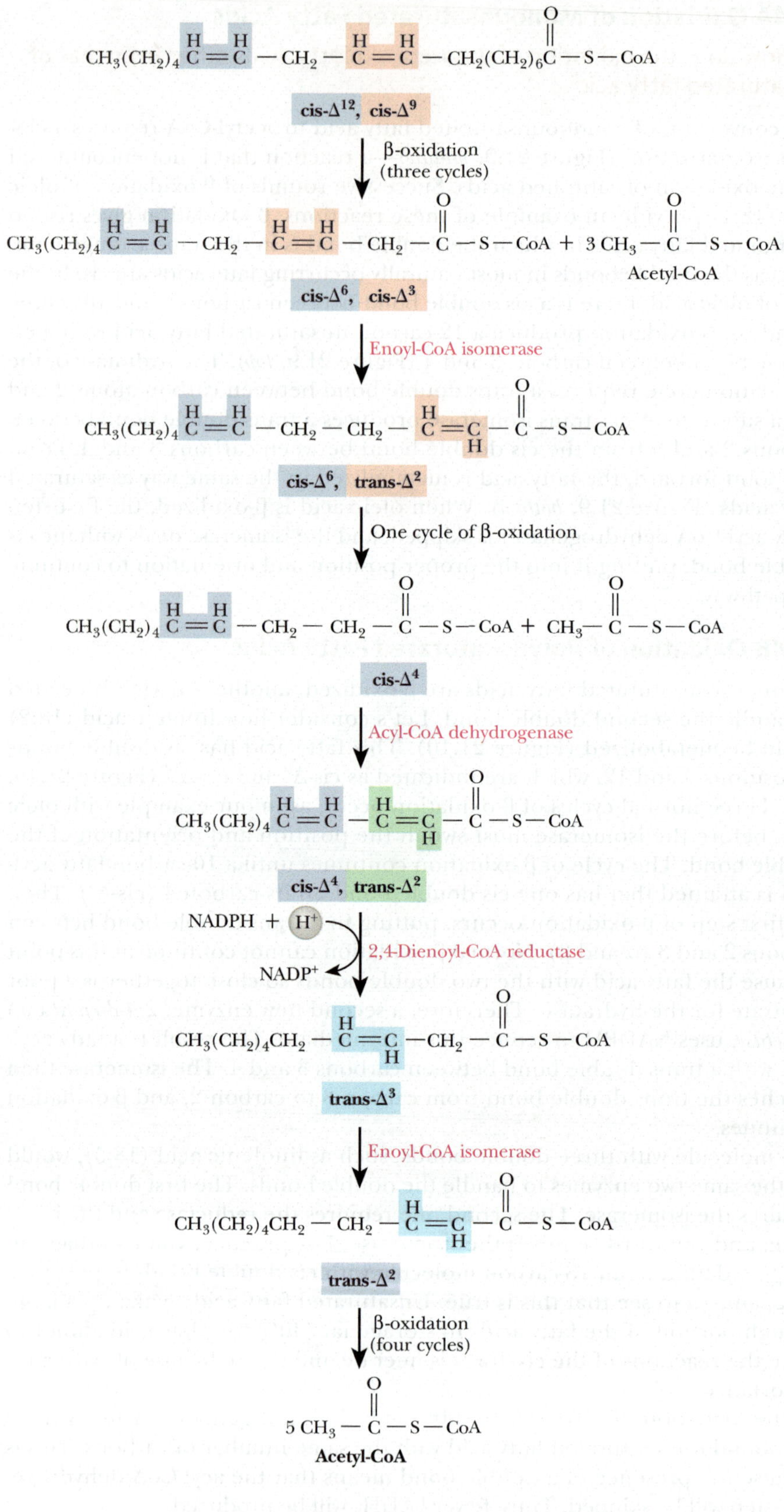

Figure 21.10 The oxidation pathway for polyunsaturated fatty acids, illustrated for linoleic acid. Three cycles of β-oxidation on linoleoyl-CoA yield three molecules of acetyl-CoA and a cis-Δ^3, cis-Δ^6 fatty acyl-CoA intermediate, which is converted to a trans-Δ^2, *cis*-Δ^6 intermediate. An additional round of β-oxidation gives cis-Δ^4 enoyl-CoA, which is oxidized to the trans-Δ^2, *cis*-Δ^4 species by acyl-CoA dehydrogenase. The subsequent action of 2,4-dienoyl-CoA reductase yields the trans-Δ^3 product, which is converted by enoyl-CoA isomerase to the trans-Δ^2 form. Normal β-oxidation then produces five additional molecules of acetyl-CoA.

21-5 Ketone Bodies

Learning Objectives

After reading this section, students will be able to:

LO 21.5.1 Describe the physiological conditions that result in the production of ketone bodies.

LO 21.5.2 Explain the reactions that produce ketone bodies.

ketone bodies ketone-based molecules produced in the liver during overutilization of fatty acids when carbohydrates are limited

Ketone bodies are substances related to acetone that are produced in the liver when an excess of acetyl-CoA arises from β-oxidation. This condition occurs when not enough oxaloacetate is available to react with the large amounts of acetyl-CoA that could enter the citric acid cycle. A situation like this can come about when an organism has a high intake of lipids and a low intake of carbohydrates, but there are also other possible causes such as starvation and diabetes. Starvation conditions cause an organism to break down fats for energy, leading to the production of large amounts of acetyl-CoA by β-oxidation. The amount of acetyl-CoA is excessive by comparison with the amount of oxaloacetate available to react with it. In the case of people with diabetes mellitus, the cause of the imbalance is not an inadequate intake of carbohydrates but rather the inability to metabolize them.

▶ Do acetone and acetyl-CoA have a connection in lipid metabolism?

The reactions that result in the three primary ketone bodies (acetoacetate, *β*-hydroxybutyrate, and acetone) start with the condensation of two molecules of acetyl-CoA to produce acetoacetyl-CoA. This reaction is catalyzed by thiolase and is followed by the formation of β-hydroxy-β-methylglutaryl-CoA (HMG-CoA), catalyzed by HMG-CoA synthase (Figure 21.11, *top*). *Acetoacetate* is produced when HMG-CoA lyase catalyzes the cleavage of an acetyl-CoA molecule from HMG-CoA, as well as the oxidation of the hydroxyl group of HMG-CoA to a ketone (Figure 21.11, *middle*).

Acetoacetate can then have two fates. A reduction reaction can produce β-*hydroxybutyrate* from acetoacetate. The other possible reaction is the spontaneous decarboxylation of acetoacetate to give *acetone*. The odor of acetone can frequently be detected on the breath of people with diabetes whose disease is not controlled by suitable treatment. The excess of acetoacetate, and consequently of acetone, is a pathological condition known as *ketosis*. Because acetoacetate and β-hydroxybutyrate are acidic, their presence at high concentration overwhelms the buffering capacity of the blood. The body deals with the consequent lowering of blood pH (ketoacidosis) by excreting H^+ into the urine, accompanied by excretion of Na^+, K^+, and water. Severe dehydration can result (excessive thirst is a classic symptom of diabetes); diabetic coma is another possible danger.

The principal site of synthesis of ketone bodies is the liver mitochondria, but they are then transported to other tissues for energy production pathways that extract acetyl-CoA back out of the ketone bodies. It is easy to transport ketone bodies in the bloodstream because, unlike fatty acids, they are water-soluble and do not need to be bound to proteins such as serum albumin. Even though glucose is the usual fuel in most tissues and organs, acetoacetate can be used as a fuel in the heart muscle and the renal cortex in order to spare glucose for the brain.

Even in organs such as the brain, in which glucose is the preferred fuel, starvation conditions can lead to the use of acetoacetate for energy. In this situation, acetoacetate is converted to two molecules of acetyl-CoA, which can then enter the citric acid cycle. The key point here is that starvation gives rise to long-term, rather than short-term, regulation over a period of hours to days,

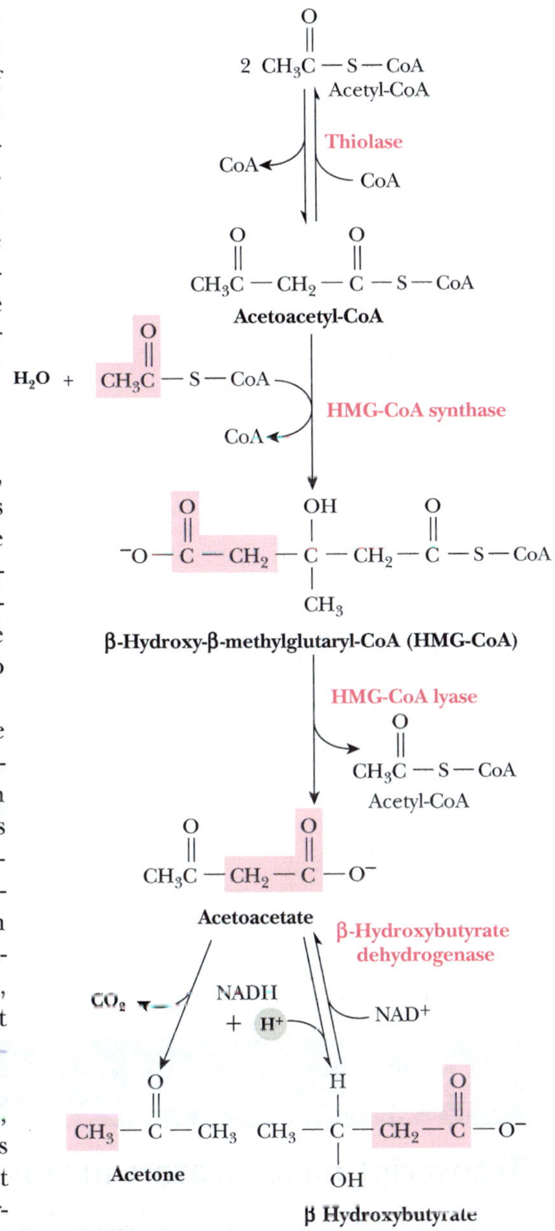

Figure 21.11 The formation of ketone bodies from excess acetyl-CoA, synthesized primarily in the liver. The pink shading helps us see how the atoms of excess acetyl-CoA end up in three primary ketone bodies: acetoacetate, β-hydroxybutyrate, and acetone. The acetoacetate to acetone reaction happens spontaneously under physiological conditions, but it can also be catalyzed by acetoacetate decarboxylase.

rather than minutes. The decreased level of glucose in the blood over a period of days changes the hormone balance in the body, particularly involving insulin and glucagon (see Section 24-5). (Short-term regulation, such as allosteric interactions or covalent modification, can occur in a matter of minutes.) The rates of protein synthesis and breakdown are subject to change under these conditions. The specific enzymes involved are those involved in fatty acid oxidation (increase in levels) and those for lipid biosynthesis (decrease in levels).

21-6 Fatty Acid Biosynthesis

Learning Objectives

After reading this section, students will be able to:

LO 21.6.1 Explain the reactions involved in fatty acid biosynthesis.

LO 21.6.2 Compare the structures of coenzyme A and the phosphopantetheine group of the acyl carrier protein.

The anabolism, or biosynthesis, of fatty acids is not simply a reversal of the reactions of β-oxidation. Recall that, in general, anabolic and catabolic pathway counterparts are not the exact reverse of each other; for instance, gluconeogenesis (Section 18-2) is not simply a reversal of the reactions of glycolysis (Sections 17-2 and 17-3). In the case of fatty acids, the anabolic reactions take place in the cytosol, whereas the catabolic reactions of β-oxidation take place in the mitochondrial matrix.

The first step in fatty acid biosynthesis is the transport of acetyl-CoA to the cytosol. Acetyl-CoA can be formed either by β-oxidation of fatty acids or by decarboxylation of pyruvate. (Degradation of certain amino acids also produces acetyl-CoA; see Section 23-6.) Most of these reactions take place in the mitochondria, requiring a transport mechanism to export acetyl-CoA to the cytosol for fatty acid biosynthesis. The transport mechanism is based on the fact that citrate can cross the mitochondrial membrane. Acetyl-CoA condenses with oxaloacetate, which cannot cross the mitochondrial membrane, to form citrate (recall that this is the first reaction of the citric acid cycle). **Biochemical Connections 21A** describes another form of control in lipid biosynthesis.

21A Biochemical Connections • Gene Expression

Transcription Activators in Lipid Biosynthesis

Transcription factors can do double duty in cells, and an example that directly deals with the material in this chapter is the action of the transcription factor XBP1 in mouse liver. XBP1 regulates genes that deal with improperly folded proteins and genes that control lipid synthesis (Figure 21.12). It was already well established that other transcription factors, called SREBPs, played a role in regulating gene expression leading to lipid synthesis. The signal for regulation by SREBPs is low cholesterol in the cytoplasm, with the signal being passed into the endoplasmic reticulum and then into the Golgi apparatus. Processing of the proteins involved takes place in the Golgi before the proteins enter the nucleus. The role of XBP1 in regulating lipid biosynthesis follows a different pathway. As shown in Figure 21.12, stress factors trigger a response by binding to IRE-1α in the endoplasmic reticulum, leading to the release of XBP1 into the Golgi. Another transcription factor, ATF6, enters the Golgi in a similar fashion. After processing, all these transcription factors play roles in regulating lipid synthesis. What is new in this picture is the hitherto unknown involvement of XBP1 in lipid metabolism. ▶

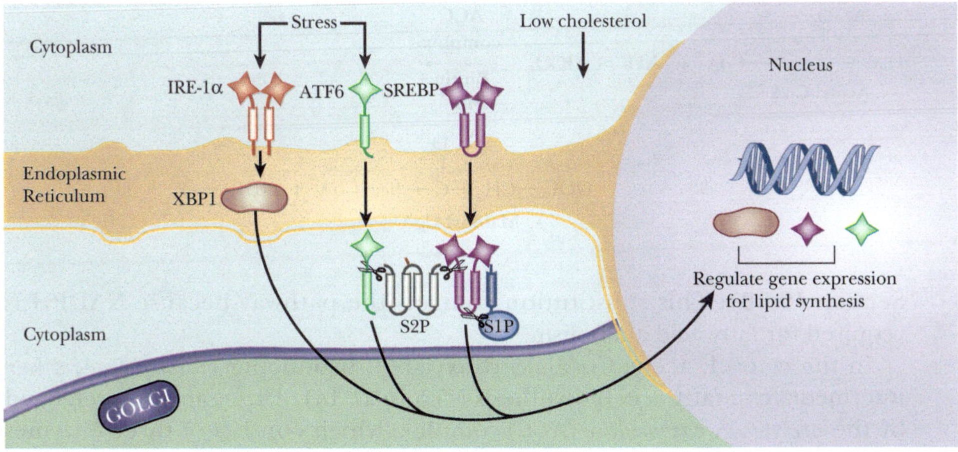

Figure 21.12 Transcription factors in lipid synthesis. XBP1, ATF6, and SREBP all undergo processing in the Golgi before they enter the nucleus. ATF6 activates XBP1. In turn, XBP1 and SREBPs regulate the expression of the genes for lipid synthesis. The regulation of SREBPs by cholesterol is a different process from the activation of ATF6 and XBP1 in response to stress.

21-6A First Steps in Fatty Acid Biosynthesis

► **How do the first steps of fatty acid synthesis take place?**

The citrate that is exported to the cytosol can undergo the reverse citrate synthase reaction (Section 19-4), producing oxaloacetate and acetyl-CoA (Figure 21.13, *middle right*). Acetyl-CoA then enters the pathway for fatty acid biosynthesis, while oxaloacetate undergoes a series of reactions in which NADPH is substituted for NADH (see the discussion of lipid anabolism in

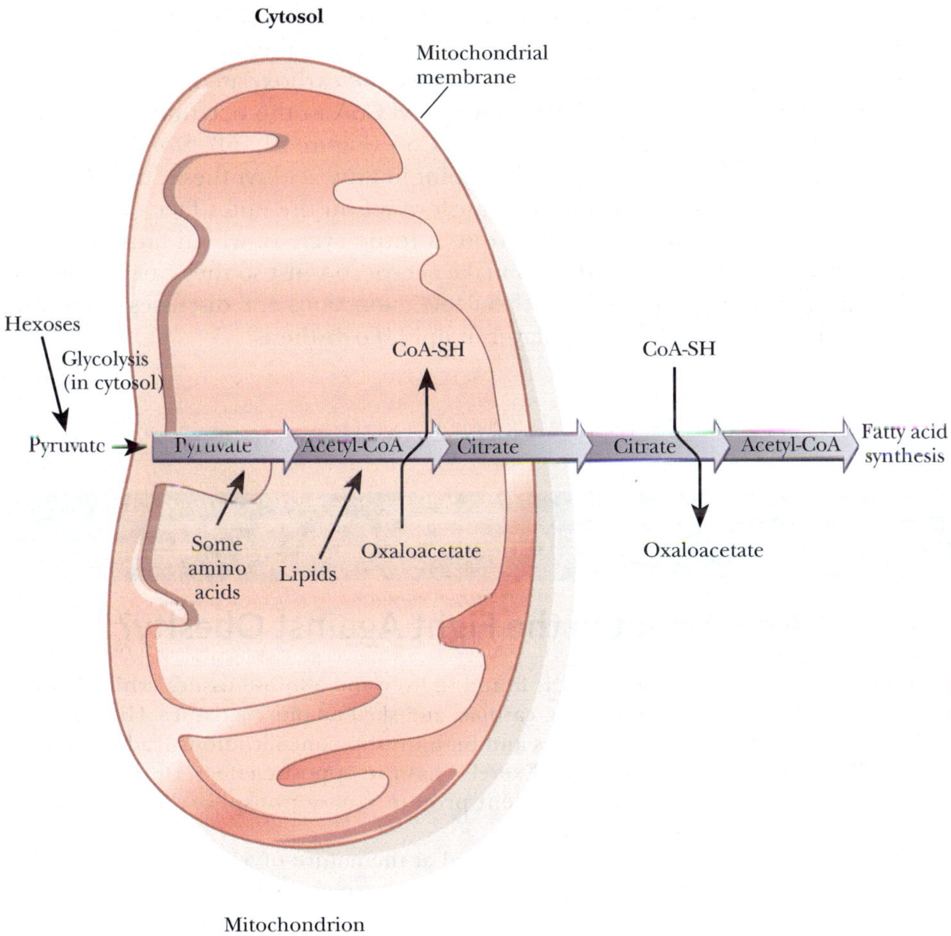

Figure 21.13 The transport of acetyl groups from the mitochondrion to the cytosol. Citrate synthase catalyzes the formation of citrate from acetyl-CoA and oxaloacetate in the mitochondrial matrix. Citrate is then transported to the cytosol by a citrate transporter. In the cytosol, the ATP-citrate lyase enzyme cleaves citrate into acetyl-CoA and oxaloacetate, and the acetyl-CoA can then be used in the fatty acid synthesis pathway.

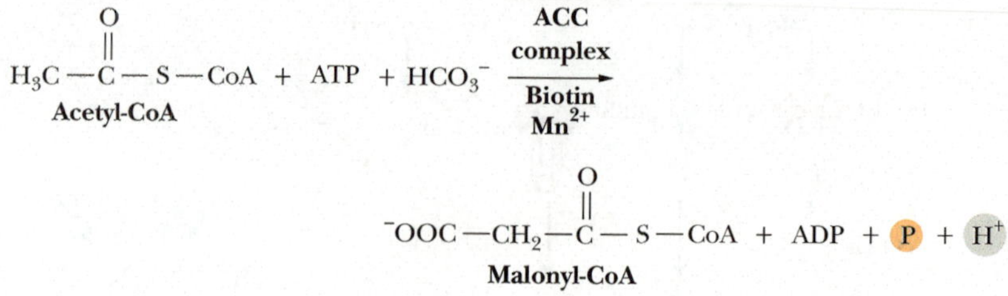

Figure 21.14 The formation of malonyl-CoA from acetyl-CoA, catalyzed by the acetyl-CoA carboxylase (ACC) complex. This is one of the first steps in the fatty acid biosynthesis pathway.

Section 19-8B). This substitution controls the pathway because NADPH is required for fatty acid anabolism.

In the cytosol, acetyl-CoA is carboxylated, producing *malonyl-CoA,* a key intermediate in fatty acid biosynthesis (Figure 21.14). This reaction is catalyzed by the *acetyl-CoA carboxylase (ACC)* complex, which consists of three enzymes and requires Mn^{2+}, biotin, and adenosine triphosphate (ATP) for activity. Biotin is a coenzyme that serves as a covalently bound cofactor that carries carboxyl groups (CO_2). Recall that we discussed more details of biotin attachment to enzymes, as well as another example of biotin in action in the pyruvate carboxylase reaction of the gluconeogenesis pathway (Section 18-2A). We have also previously seen enzyme complexes that consist of several separate protein molecules with different enzymatic activities (recall the details of the pyruvate dehydrogenase complex discussed in Section 19-3).

In this case, ACC consists of the three proteins: *biotin carboxylase,* the *biotin carrier protein,* and *carboxyl transferase.* Biotin carboxylase catalyzes the transfer of the carboxyl group to biotin (Figure 21.15B, *Step 1*). The "activated CO_2" (the carboxyl group derived from the bicarbonate ion HCO_3^-) is covalently bonded to biotin. Biotin (whether carboxylated or not) is bonded to the biotin carrier protein by an amide linkage to the ε-amino group of a lysine side chain. The amide linkage to the side chain that bonds biotin to the biotin carrier protein is long enough and flexible enough to move the carboxylated biotin into position to transfer the carboxyl group to acetyl-CoA in the reaction catalyzed by carboxyl transferase, producing malonyl-CoA (Figure 21.15B, *Step 2*).

In addition to its role as a starting point in fatty acid synthesis, malonyl-CoA strongly inhibits the carnitine acyltransferase I on the outer face of the outer mitochondrial membrane. This avoids a futile cycle in which fatty acids are β-oxidized in the mitochondria to make acetyl-CoA just so they can be remade into fatty acids in the cytosol. **Biochemical Connections 21B** discusses other ways in which malonyl-CoA is a key player in lipid biosynthesis.

21B Biochemical Connections • Nutrition

Acetyl-CoA Carboxylase Complex—A New Target in the Fight Against Obesity?

Malonyl-CoA has two very important functions in metabolism. First, it is the committed intermediate in fatty acid synthesis. Second, it strongly inhibits carnitine palmitoyltransferase I and, therefore, fatty acid oxidation. The level of malonyl-CoA in the cytosol can determine whether the cell will be oxidizing fats or storing fats. The enzyme that produces malonyl-CoA is the acetyl-CoA carboxylase (ACC) complex. There are two isoforms of this enzyme encoded by separate genes. ACC1 is found in liver and adipose tissues, while ACC2 is found in liver, cardiac, and skeletal muscle tissues. High glucose concentrations and high insulin concentrations lead to stimulation of ACC2. Exercise has the opposite effect. During exercise, an AMP-dependent protein kinase phosphorylates ACC2 and inactivates it.

Some studies looked at the nature of weight gain and weight loss with respect to ACC2 (see papers by Ruderman and Flier and by Abu-Elheiga et al. cited in Further Reading). The investigators created a strain of mice lacking the gene for ACC2. These mice ate more than their wild-type counterparts but had significantly lower stores of lipids (30% to 40% less in skeletal muscle and 10% less in cardiac muscle) (Figure 21.16). Even the adipose tissue, which still had ACC1, showed a reduction in stored triacylglycerols of up to 50%. The mice showed no other abnormalities. They grew and reproduced normally and had normal life spans. The investigators concluded that reduced pools of malonyl-CoA due to the lack of ACC2 result in increased β-oxidation via removal of the block on carnitine palmitoyltransferase I and a decrease in fatty acid synthesis. They speculated that ACC2 would be a good target for drugs used to combat obesity. While ACC2 is still a target in obesity research, challenges remain with unwanted side effects that accompany drugs that target ACC2. ◗

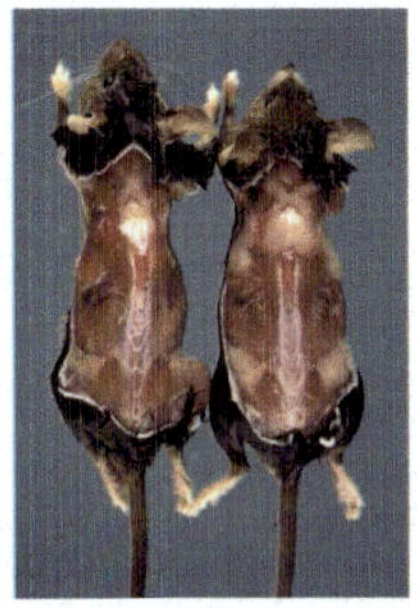

Figure 21.16 The amount of white fat under the skin of the mouse on the left, which lacks the gene for ACC2, is less than that of the mouse on the right, which has the gene. (*From Continuous Fatty Acid Oxidation and Reduced Fat Storage in Mice Lacking Acetyl-CoA Carboxylase 2 by Lufti Abu-Elheiga, et al., [March 30, 2001].* Science *291 (5513): 2613. Used with permission of AAAS.*)

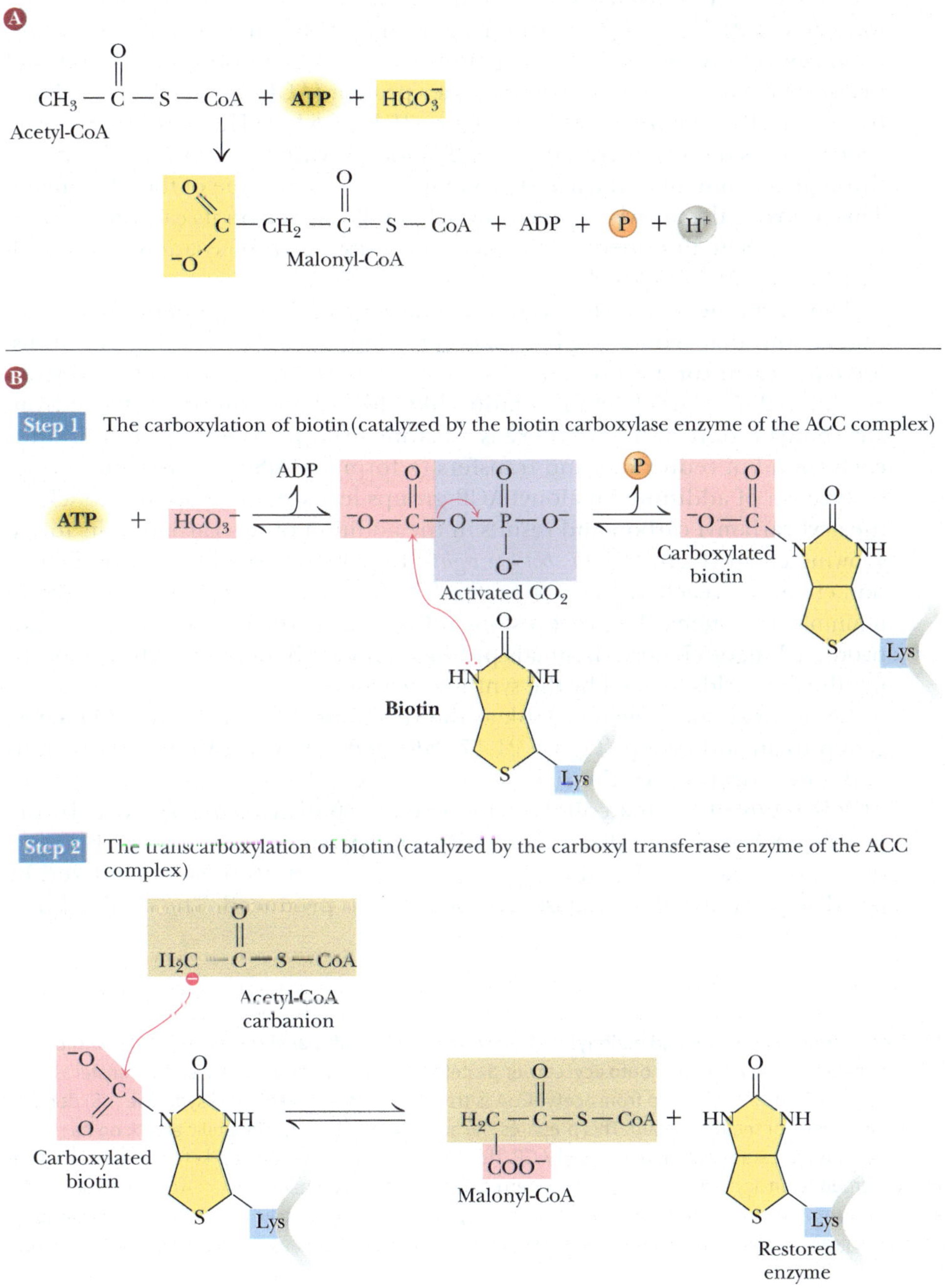

Figure 21.15 The acetyl-CoA carboxylase (ACC) complex reaction mechanism. (A) The overall ACC reaction produces malonyl-CoA for fatty acid synthesis. (B) A mechanism for the ACC reaction. In step 1, bicarbonate is activated for carboxylation reactions by the formation of a bond with a phosphate from ATP. Then, a nucleophilic attack from a nitrogen of the biotin coenzyme on the carboxyl carbon of the activated CO_2 forms *N*-carboxybiotin (carboxylated biotin). The biotin-lysine connection is abbreviated in this figure (see Figure 18.12 for the complete connection). In step 2, a typical biotin-dependent reaction, nucleophilic attack by the acetyl-CoA carbanion on the carboxyl carbon of *N*-carboxybiotin—a transcarboxylation—yields the carboxylated malonyl-CoA product.

21-6B The Fatty Acid Synthase Complex

▶ What is the mode of action of enzymes in the fatty acid synthase complex?

fatty acid synthase a giant enzyme complex that makes long-chain fatty acids from acetyl-CoA

acyl carrier protein (ACP) a protein that functions in fatty acid synthesis to carry activated carbon groups

The biosynthesis of fatty acids involves the successive addition of two-carbon units to the growing chain. Two of the three carbon atoms of the malonyl group of malonyl-CoA are added to the growing fatty acid chain with each cycle of the biosynthetic reaction. This reaction, like the formation of the malonyl-CoA itself, requires a multienzyme complex located in the cytosol and not attached to any membrane. The complex, made up of the individual enzymes, is called **fatty acid synthase** (FAS).

The usual product of fatty acid anabolism is *palmitate,* the 16-carbon saturated fatty acid (Figure 21.17). All except two of the 16 carbons come from the acetyl group of acetyl-CoA via malonyl-CoA, with the two carbon-atom exceptions coming from acetyl-CoA directly at the very beginning of the pathway (Figure 21.17, *top left*). Recall from Section 21-6A that malonyl-CoA arises from acetyl-CoA. In the initial reactions of the FAS complex, acetyl-CoA and malonyl-CoA are loaded onto **acyl carrier protein (ACP)** in reactions catalyzed by the transferase enzymes of FAS (Figure 21.17, *top left*). The acetyl group from acetyl-CoA is then transferred to the β-ketoacyl synthase (KS) domain of FAS through attachment of the acetyl group to a cysteine residue of the KS domain. This prepares the acetyl group for a nucleophilic attack on its carbonyl carbon from an enolate intermediate that is formed after decarboxylation of malonyl-ACP (Figure 21.17, *top middle*).

This forms acetoacetyl-ACP (a β-ketoacyl group) that undergoes a series of reactions that reduce the β-ketoacyl group to an acyl group that gets transferred back to the KS domain of FAS (Figure 21.17, *top right*). Recall from Section 3-2 that Greek-letter notation like this is based on the distance from a carbonyl group, and here there is a ketone group at the beta carbon. After each round of reductions and transfers to form acyl-KS intermediates, decarboxylation of additional malonyl-ACP groups leads to nucleophilic attack on the acyl carbonyl carbon and results in the addition of two-carbon units to the growing chain (Figure 21.17, *bottom right*). In palmitate synthesis, once the carbon chain has reached 16 carbons, the free fatty acid is released from ACP. In mammalian systems, the process stops at C_{16} because the FAS complex does not produce longer chains. Mammals produce longer-chain fatty acids by modifying the fatty acids formed by the synthase reaction.

Let us next take a deeper look at the reactions that reduce the β-ketoacyl group to an acyl group (Figure 21.17, *top right*). They involve two reductions and a dehydration. In the first reduction, the β-keto group of acetoacetyl-ACP is reduced to an alcohol, giving rise to D-β-hydroxybutyryl-ACP. In the process, NADPH is oxidized to $NADP^+$. The enzyme that catalyzes this reaction is β-ketoacyl-ACP reductase. Next, in a dehydration step catalyzed by β-hydroxyacyl-ACP dehydratase, crotonyl-ACP is produced. The double bond

Figure 21.17 The pathway of palmitate fatty acid synthesis from acetyl-CoA and malonyl-CoA, catalyzed by the fatty acid synthase (FAS) complex. In the initial reactions of the FAS complex, acetyl-CoA and malonyl-CoA are loaded onto acyl carrier protein (ACP) in reactions catalyzed by transferase enzymes of FAS. Before the condensation to form acetoacetyl-ACP, the acetyl group from acetyl-CoA is transferred to the β-ketoacyl synthase (KS) domain of FAS through attachment of the acetyl group to a cysteine residue of the KS domain. This prepares the acetyl group for a nucleophilic attack on its carbonyl carbon from an enolate intermediate that is formed after decarboxylation of malonyl-ACP. This forms acetoacetyl-ACP (a β-ketoacyl group), which undergoes a series of reactions that reduce the β-ketoacyl group to an acyl group that gets transferred back to the KS domain of FAS. After each round of reductions and transfers to form acyl-KS intermediates, decarboxylation of additional malonyl-ACP groups leads to nucleophilic attack on the acyl carbonyl carbon and results in the addition of two-carbon units to the growing chain. In palmitate synthesis, once the carbon chain has reached 16 carbons, the free fatty acid is released from ACP.

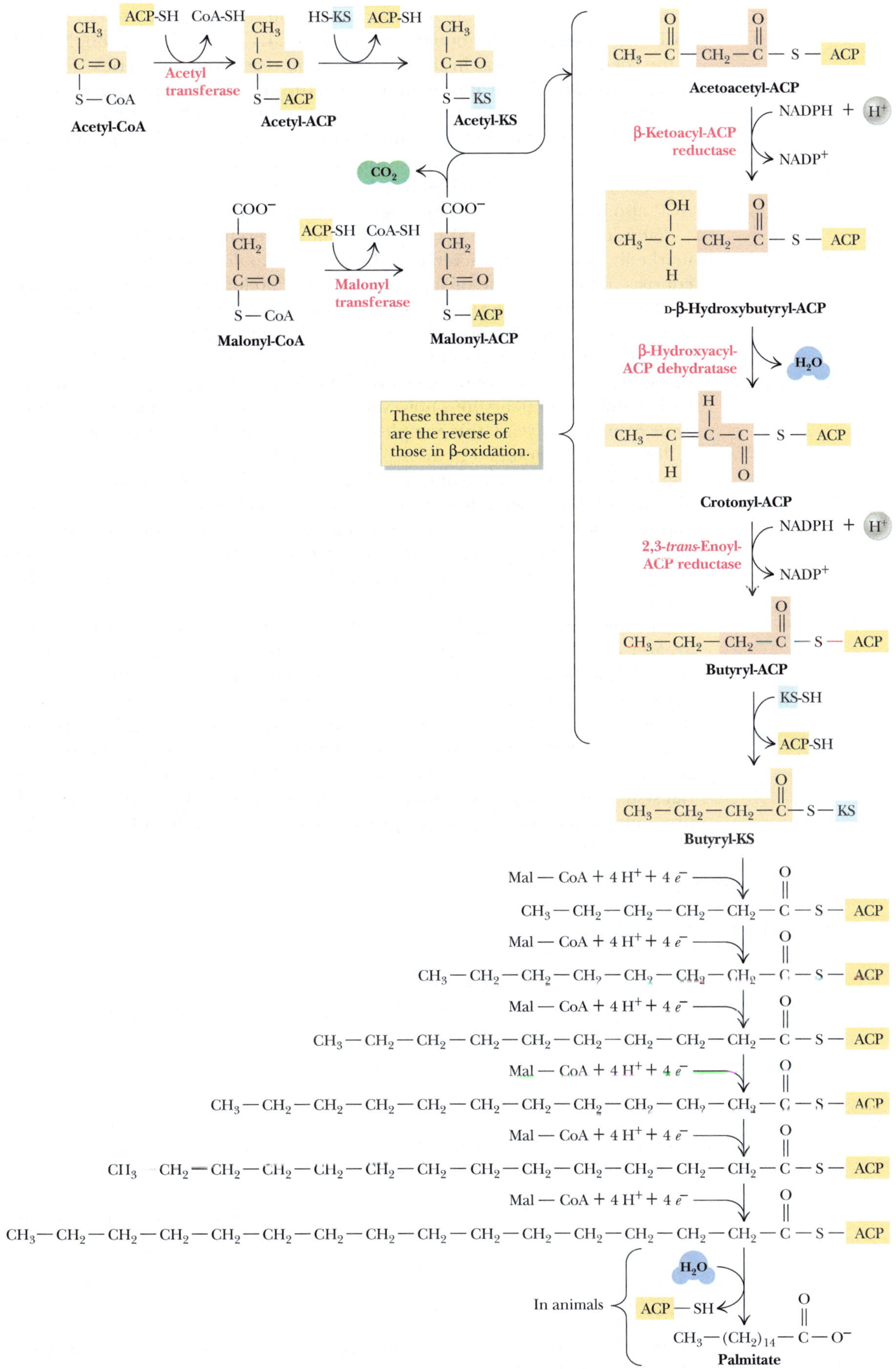

Acetyl-CoA
Acetyl transferase
Acetyl-ACP
Acetyl-KS
CO_2
Malonyl-CoA
Malonyl transferase
Malonyl-ACP
Acetoacetyl-ACP
NADPH + H^+
β-Ketoacyl-ACP reductase
$NADP^+$
D-β-Hydroxybutyryl-ACP
β-Hydroxyacyl-ACP dehydratase
H_2O
These three steps are the reverse of those in β-oxidation.
Crotonyl-ACP
2,3-*trans*-Enoyl-ACP reductase
Butyryl-ACP
KS-SH
ACP-SH
Butyryl-KS
Mal — CoA + 4 H^+ + 4 e^-
In animals
Palmitate

is in the trans configuration. A second reduction reaction, catalyzed by β-enoyl-ACP reductase, produces butyryl-ACP (Figure 21.17). In this reaction, NADPH is the coenzyme, as it was in the first reduction reaction in this series.

The group on the ACP protein to which the growing acyl group is bonded is a 4-phosphopantetheine group, which in turn is bonded to a serine side chain of ACP. As seen in Figure 21.18, this group is structurally similar to coenzyme A (CoA-SH) itself. At the very beginning of fatty acid synthesis, the acetyl group is transferred from CoA-SH, to which it is bound by a thioester linkage, to the ACP, also by a thioester linkage (Figure 21.17, *top left*).

Fatty acid synthase (FAS) complexes from different types of organisms have markedly different characteristics. In *Escherichia coli,* the multienzyme system consists of an aggregate of separate enzymes, including a separate ACP. The ACP is of primary importance to the complex and is thought to occupy a central position in it. The phosphopantetheine group plays the role of a "swing-ing arm," much like that of biotin, which was discussed in Section 21-6A. This bacterial system has been extensively studied and is a typical example of a FAS complex. In eukaryotes, however, fatty acid synthesis occurs on a multienzyme complex. In yeast, this complex consists of two different types of subunits, called α and β, arranged in an $\alpha_6\beta_6$ complex. In mammals, the FAS complex contains only one type of subunit, but the active enzyme is a dimer of this single subunit. It has been determined by X-ray crystallography that the mammalian synthase contains two reaction chambers in which the various components are held in proximity to each other as the reaction proceeds. Each of the subunits is a *multifunctional enzyme* that catalyzes reactions requiring several different proteins in the *E. coli* system.

The structure of fungal FAS complex was elucidated in the mid-2000s in greater detail by X-ray crystallography (Figure 21.19). The results demonstrated that the multiple active sites for the synthase reactions in this type of eukaryote are arranged in the reaction chamber so that a circular movement of the ACP-bound substrates can deliver the substrate to each specific active site. The growing fatty acid chain swings back and forth between enzyme activities contained on different subunits, using ACP as a "swinging arm." Like the bacterial system, this eukaryotic system keeps all the components of the reaction in proximity to one another, which provides another example of the advantages of multienzyme complexes.

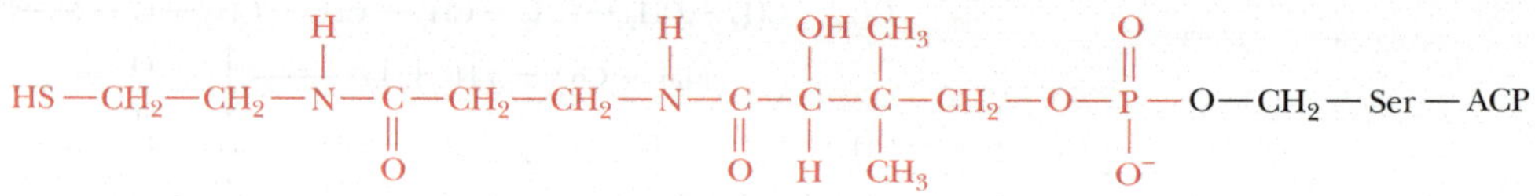

Phosphopantetheine group of ACP

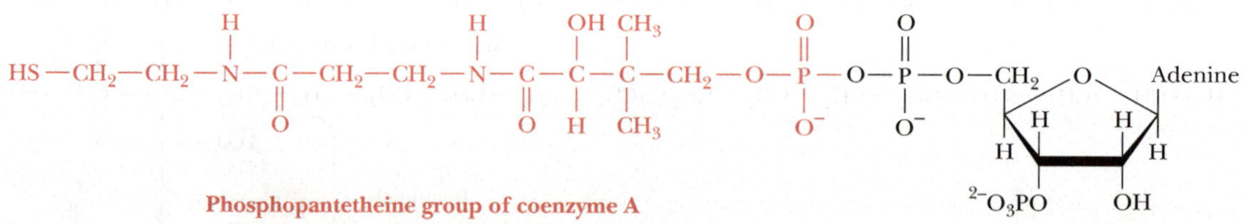

Phosphopantetheine group of coenzyme A

Figure 21.18 Structural similarities between coenzyme A and the phosphopantetheine group of acyl carrier protein (ACP).

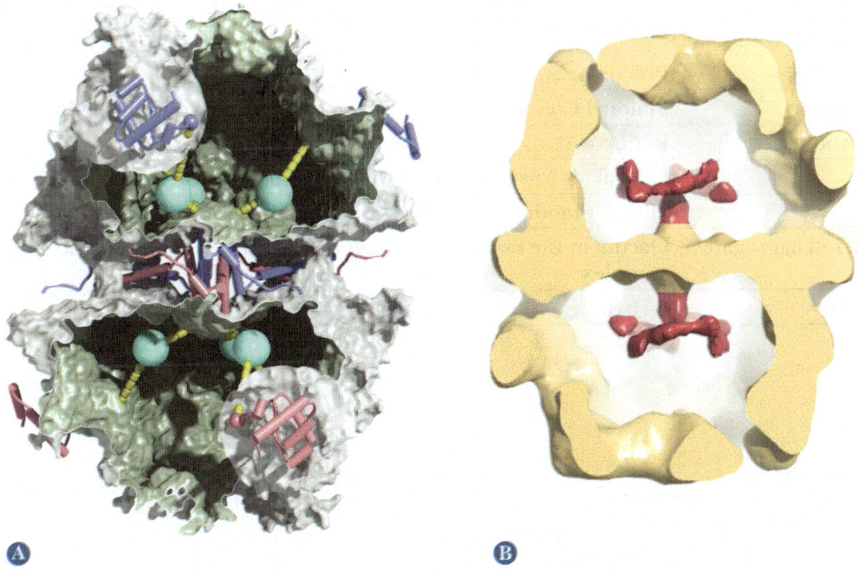

Figure 21.19 The architecture of fungal fatty acid synthase. (A) Side view of two reaction chambers, one on top of the other. The ACP portions are shown as cyan spheres and yellow lines. (B) A comparison structure based on electron microscopy. *(Adapted from Jenni, S., et al. (2007). Structure of Fungal Fatty Acid Synthase and Implications for Iterative Substrate Shuttling.* Science *316, 258.)*

Several additional reactions are required for the elongation of fatty acid chains past 16 carbons and for the introduction of double bonds. When mammals produce fatty acids with longer chains than that of palmitate, the reaction does not involve cytosolic FAS complexes. There are two sites for the chain-lengthening reactions: the endoplasmic reticulum (ER) and the mitochondrion. In the chain-lengthening reactions in the mitochondrion, the intermediates are of the acyl-CoA type rather than the acyl-ACP type. In other words, the chain-lengthening reactions in the mitochondrion are the reverse of the catabolic reactions of fatty acids, with acetyl-CoA as the source of added carbon atoms; this is a difference between the main pathway of fatty acid biosynthesis seen in Figure 21.17 and these modification reactions. In the ER, the source of additional carbon atoms is malonyl-CoA. The modification reactions in the ER also differ from the biosynthesis of palmitate in that, like the mitochondrial reactions, no intermediates are bound to ACP.

Reactions in which a double bond is introduced in fatty acids mainly take place on the ER. The insertion of the double bond is catalyzed by a mixed-function oxidase that requires molecular oxygen (O_2) and NAD(P)H. During the reaction, both NAD(P)H and the fatty acid are oxidized, while oxygen is reduced to water. Reactions linked to molecular oxygen are comparatively rare. Mammals cannot introduce a double bond beyond carbon atom 9 (counting from the carboxyl end) of the fatty acid chain. As a result, linoleate [$CH_3—(CH_2)_4—CH=CH—CH_2—CH=CH—(CH_2)_7—COO^-$], with two double bonds, and linolenate [$CH_3—(CH_2)_4—CH=CH\quad CH_2CH=CH—CH_2—CH=CH—(CH_2)_4COO^-$], with three double bonds, must be included in the diets of mammals. They are **essential fatty acids** because they cannot be synthesized in our bodies and must be obtained from our diet. Good sources of essential fatty acids are salmon, seeds, tree nuts, and eggs. They are precursors of other lipids, including prostaglandins.

essential fatty acids the polyunsaturated fatty acids (such as linoleic acid) that the body cannot synthesize; they must be obtained from dietary sources

Even though both the anabolism and the catabolism of fatty acids require successive reactions of two-carbon units, the two pathways are not the exact reversal of each other. The differences between the two pathways are summarized in Table 21.2. Also, the sites in the cell in which various anabolic and catabolic reactions take place are shown in Figure 21.20. We conclude Section 21-6 with **Biochemical Connections 21C**, which describes the results of excessive lipid biosynthesis.

Table 21.2 A Comparison of Fatty Acid Degradation and Biosynthesis

Degradation	Biosynthesis
1. Product is acetyl-CoA	Precursor is acetyl-CoA
2. Malonyl-CoA is not involved; no requirement for biotin	Malonyl-CoA is the source of two-carbon units; biotin required
3. Oxidative process; requires NAD^+ and FAD and produces ATP	Reductive process; requires NADPH and ATP
4. Fatty acids form thioesters with CoA-SH	Fatty acids form thioesters with acyl carrier proteins (ACP-SH)
5. Starts at thioester end	Starts at methyl end (CH_3CH_2—)
6. Occurs in the mitochondrial matrix, with no ordered aggregate of enzymes	Occurs in the cytosol, catalyzed by an ordered multienzyme complex
7. β-Hydroxyacyl intermediates have the L configuration	β-Hydroxyacyl intermediates have the D configuration

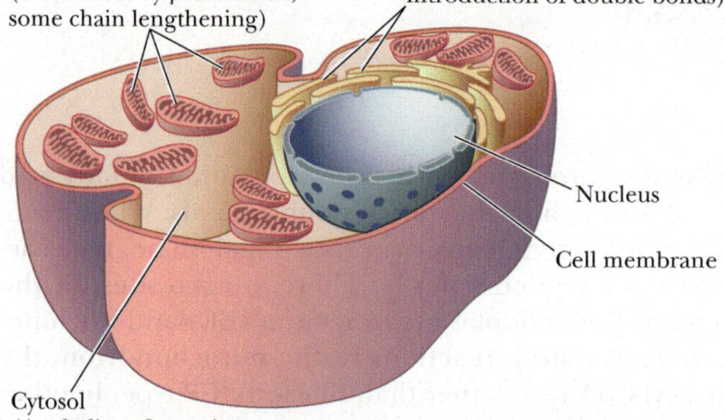

Figure 21.20 A portion of an animal cell, showing the sites of various aspects of fatty acid metabolism. The cytosol is the site of fatty acid anabolism. It is also the site of formation of acyl-CoA, which is transported to the mitochondrion for catabolism by the β-oxidation process. Some chain-lengthening reactions (beyond C_{16}) take place in the mitochondria. Other chain-lengthening reactions take place in the endoplasmic reticulum (ER), as do reactions that introduce double bonds.

21C Biochemical Connections • Genetics

A Gene for Obesity

Obesity has long been associated with several known disease states, such as diabetes and even cancer. Researchers have identified several genes that are linked to obesity (Figure 21.21). For example, *FTO*, the fat mass and obesity-associated gene, is a gene that shows a clear relationship to a tendency toward obesity. *FTO* encodes an enzyme that modifies RNA, affecting the expression of genes involved in lipid metabolism. In early research on *FTO*, a British team of scientists studied samples from more than 4,000 individuals and identified a relationship between *FTO* and body mass index (BMI). A specific variant of *FTO* with a single nucleotide change was found. Individuals who had two copies of the variant were 1.67 times more likely to be obese than individuals lacking any copies of the variant. Further research has suggested that while *FTO* variants increase obesity risk, lifestyle factors like diet, exercise, and environment still play major roles. ◗

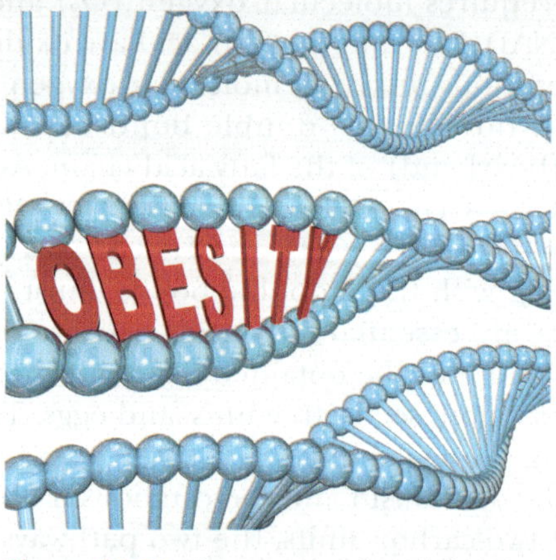

Figure 21.21 The commonality of obesity in modern society has been linked to several genes, including the *FTO* gene.

21-7 Biosynthesis of Acylglycerols and Compound Lipids

Learning Objective

After reading this section, students will be able to:

LO 21.7.1 Distinguish between the biosynthetic reactions that produce triacylglycerols, phosphoacylglycerols, and sphingolipids.

Other lipids, including triacylglycerols, phosphoacylglycerols (PAGs), and steroids, are derived from fatty acids and metabolites of fatty acids, such as acetoacetyl-CoA. Free fatty acids do not occur in the cell to any great extent; they are normally found incorporated in triacylglycerols and PAGs. The biosynthesis of these two types of compounds takes place principally on the ER of liver cells or fat cells (adipocytes).

21-7A Triacylglycerol Biosynthesis

The glycerol portion of lipids is derived from glycerol-3-phosphate, a compound available from reduction of the dihydroxyacetone phosphate, a glycolysis metabolite (Figure 21.22, *top left*). In the liver and kidney, another source is glycerol released by the degradation of acylglycerols followed by phosphorylation (Figure 21.22, *top middle*). Once glycerol-3-phosphate is made by either of these two pathways, an acyl group of a fatty acid is transferred from an acyl-CoA. The products of this reaction are CoA-SH and a *lysophosphatidate* (a monoacylglycerol phosphate), and the reaction is catalyzed by a transferase enzyme (Figure 21.22, *middle left*). The acyl group is shown as esterified at carbon atom 2 (C-2) in this series of equations, but it is equally likely that it is esterified at C-1. A second acylation reaction takes place, catalyzed by the same transferase enzyme, producing a *phosphatidate* (a diacylglycerol phosphate) (Figure 21.22, *top right*). Phosphatidates occur in membranes and are precursors of other phospholipids, but in triacylglycerol synthesis, the phosphate group of the phosphatidate is removed by hydrolysis, producing a *diacylglycerol* (Figure 21.22, *middle right*). A third acyl group is added in a reaction catalyzed by another transferase (Figure 21.22, *bottom right*).

21-7B Phosphoacylglycerol Biosynthesis

Phosphoacylglycerols (PAGs) are based on phosphatidates, with the phosphate group esterified to another alcohol, frequently a nitrogen-containing alcohol such as ethanolamine (see the discussion of phosphoacylglycerols [phosphoglycerides] and how they come together to form the lipid bilayers of cell membranes in Section 8-2C). The conversion of phosphatidates to other phospholipids frequently requires the presence of nucleoside triphosphates, particularly *cytidine triphosphate* (CTP). The role of CTP depends on the type of organism because the details of the biosynthetic pathway are not the same in mammals and bacteria. We shall use a comparison of the synthesis of phosphatidylethanolamine (PE) in mammals and in bacteria (Figure 21.23) as a case study of the kinds of reactions commonly encountered in phosphoacylglycerol biosynthesis.

In bacteria, CTP reacts with phosphatidate to produce cytidine diphosphodiacylglycerol (a CDP diglyceride). The CDP diglyceride reacts with serine to form *phosphatidylserine (PS)*, which is also a phosphoacylglycerol. Phosphatidylserine is then decarboxylated to give *PE*. In eukaryotes, the synthesis of PE requires two separate pathways whose metabolites eventually join to form the PE molecule (Figure 21.24). The first of these two pathways involves the removal of the phosphate group of the phosphatidate, producing a

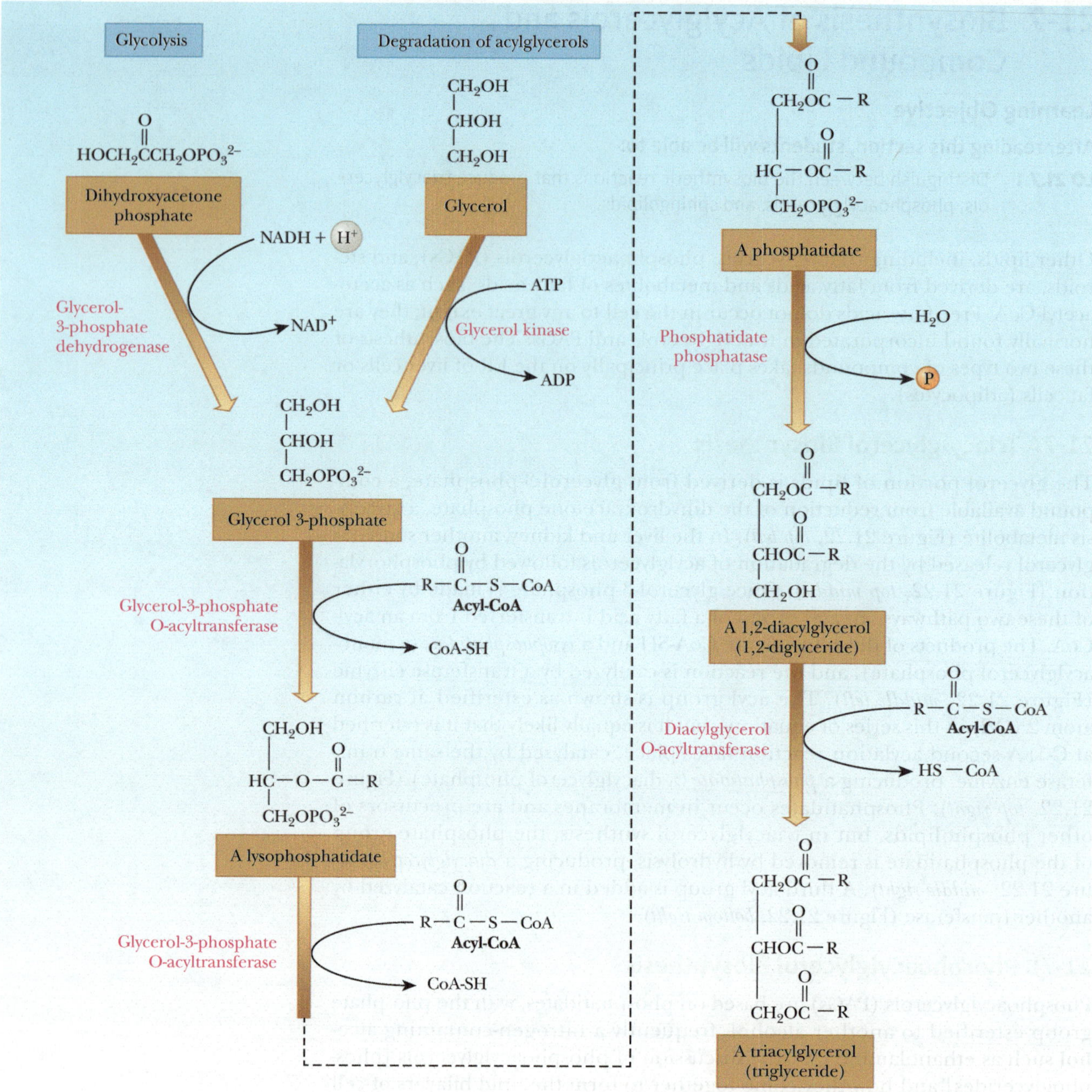

Figure 21.22 **Pathways for the biosynthesis of triacylglycerols.** The glycerol-3-phosphate metabolite (middle left) can be obtained from glycolysis or from the degradation of acylglycerols. From there, a series of acyl transfer reactions produce a triacylglycerol (bottom right).

diacylglycerol (Figure 21.24, *right*); the second pathway involves the reaction of phosphoethanolamine with CTP to produce pyrophosphate (PP_i) and cytidine diphosphate ethanolamine (CDP-ethanolamine). The CDP-ethanolamine and diacylglycerol from the two pathways then react to form PE (Figure 21.24, *bottom left*).

Figure 21.24 depicts PE biosynthesis in eukaryotes generally, but in mammals specifically, there is an important interconversion between PE and phosphatidylserine (PS) that helps maintain proper lipid balance across different organelles (Figure 21.25). Alcohol exchange from ethanolamine to serine catalyzed by phosphatidylserine synthase 2 (PSS2) in the ER allows the conversion

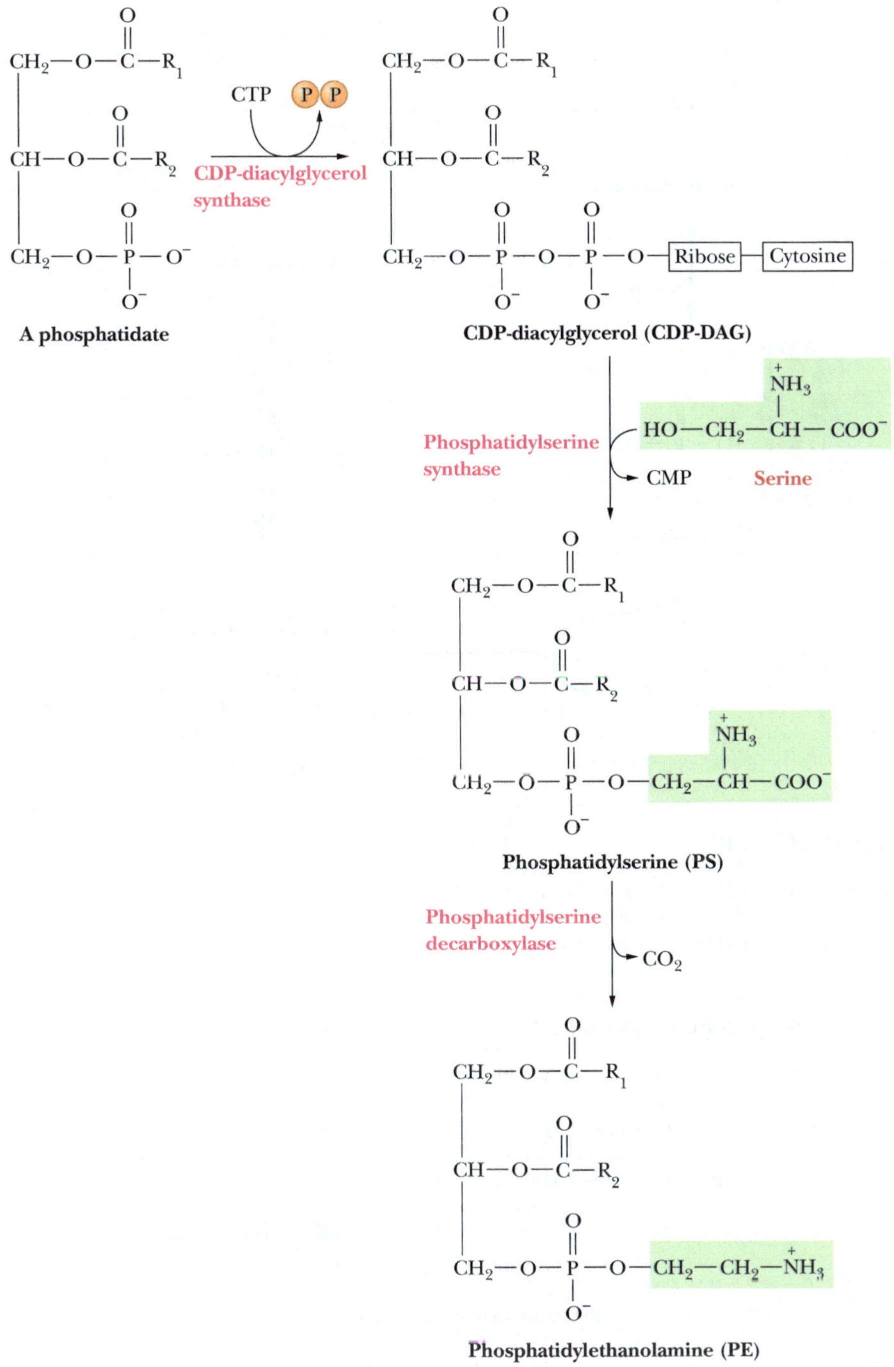

Figure 21.23 The biosynthesis of the phosphoacylglycerol called phosphatidylethanolamine (PE) in bacteria. The phosphatidate at the beginning of this pathway can come from either of the two pathways seen on the left side of Figure 21.22. Also, notice how this pathway can be used for the production of phosphatidylserine, another phosphoacylglycerol.

of PE to PS (Figure 21.25, *right*), while decarboxylation of PS to PE catalyzed by phosphatidylserine decarboxylase occurs in the mitochondria (Figure 21.25, *left*).

21-7C Sphingolipid Biosynthesis

The structural backbone of sphingolipids is not glycerol but *sphingosine*, a long-chain amine (see the discussion of sphingolipids in Section 8-2E). The precursors of the sphingosine backbone are palmitoyl-CoA and the amino acid serine, which react to produce 3-ketosphinganine in a reaction catalyzed by serine palmitoyltransferase (Figure 21.26, *top*). The carboxyl group of the serine is lost

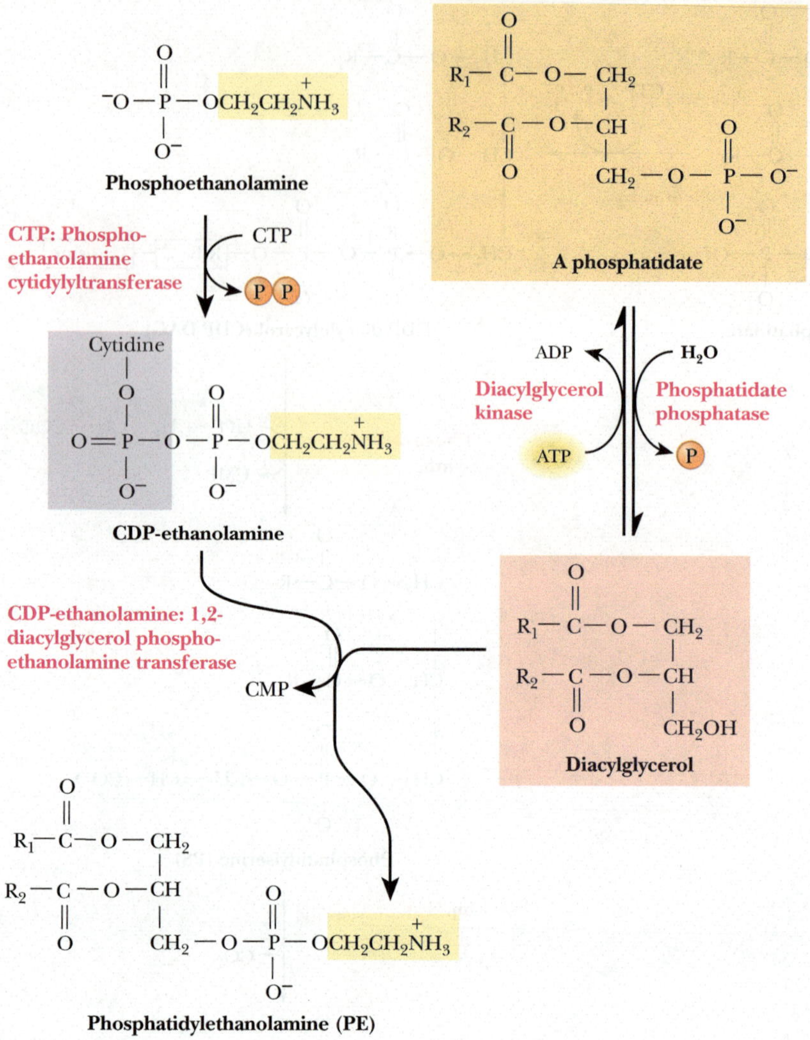

Figure 21.24 The biosynthesis of phosphatidylethanolamine (PE) in eukaryotes. Comparing the details here with the details in Figure 21.23 helps us see the differences in PE synthesis in eukaryotic cells versus bacterial cells.

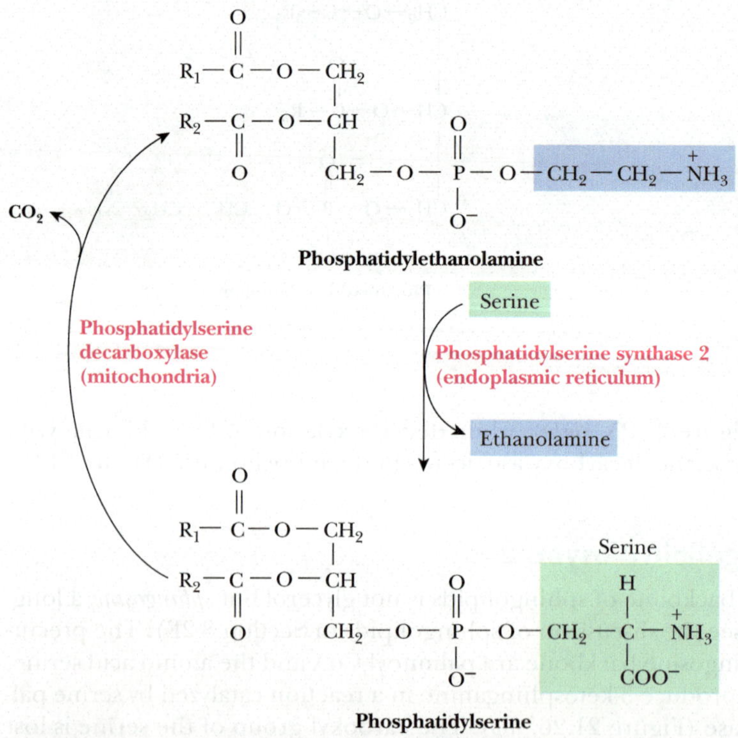

Figure 21.25 The interconversion of phosphatidylethanolamine and phosphatidylserine in mammals. These reactions help maintain proper lipid balance across different organelles in mammalian cells.

Figure 21.26 Formation of a ceramide from palmitoyl-CoA. The "dihydro-" of dihydroceramide refers to the two hydrogens on carbons 4 and 5 that have been removed to form the carbon–carbon double bond in the ceramide. R represents a hydrocarbon chain of at least 16 carbons. Therefore, the ceramide product has two fatty acyl tail groups. This type of molecule plays crucial roles in skin barrier protection and cell signaling.

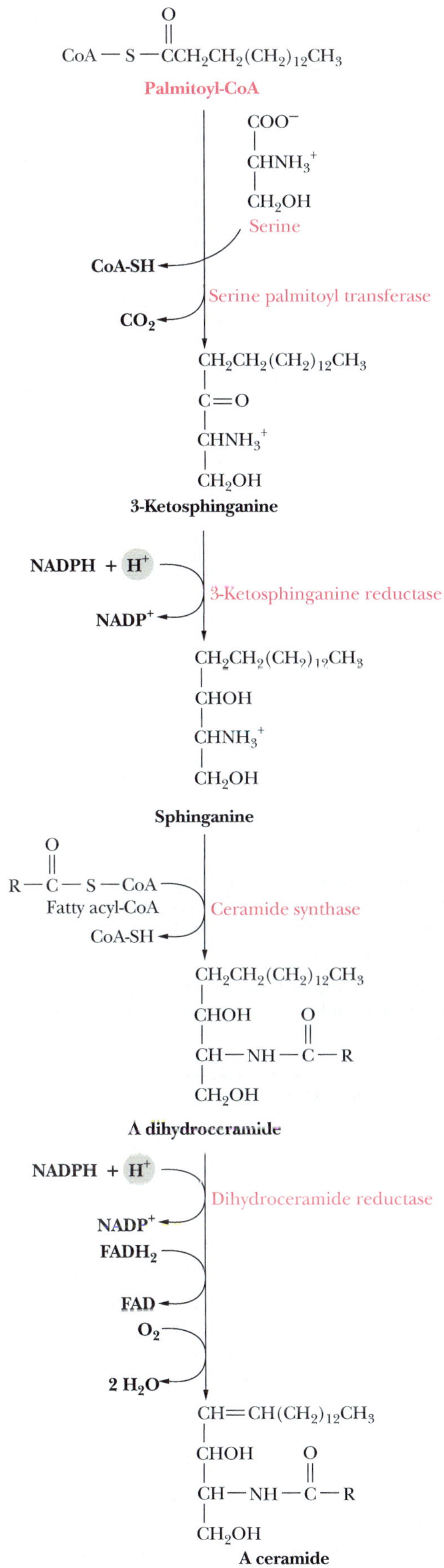

as CO_2 in the course of this reaction. The ketone group is then reduced to a hydroxyl group in a reaction catalyzed by 3-ketosphinganine reductase, forming *sphinganine.* Next, in a reaction catalyzed by ceramide synthase, the amino group of sphinganine reacts with another acyl-CoA to form an amide bond in a *dihydroceramide.* The acyl groups that can be added by ceramide synthase can vary widely in length, and they can be saturated or unsaturated, depending on the specific role needed for the ceramide being synthesized.

Finally, an oxidation reaction catalyzed by dihydroceramide reductase introduces a double bond, resulting in a **ceramide**. Ceramides play crucial roles in skin barrier protection and cell signaling, and they are the parent compounds of sphingomyelins, cerebrosides, and gangliosides (Section 8-2E). Attachment of phosphorylcholine to the primary alcohol group of a ceramide produces a *sphingomyelin,* whereas attachment of sugars such as glucose at the same site produces *cerebrosides. Gangliosides* are formed from ceramides by attachment of oligosaccharides that contain a sialic acid residue, also at the primary alcohol group. See the discussion of sphingolipids in Section 8-2E for the structures of these compounds.

21-8 Steroid Biosynthesis

Learning Objectives

After reading this section, students will be able to:

LO 21.8.1 Describe the reactions involved in the biosynthesis of cholesterol.

LO 21.8.2 Explain the biosynthetic reactions that use cholesterol as a precursor to produce bile acids and steroids other than cholesterol.

LO 21.8.3 Distinguish between the different types of lipoproteins.

21-8A Cholesterol Biosynthesis

The ultimate precursor of all the carbon atoms in **cholesterol** and in the other steroids that are derived from cholesterol is the acetyl group of acetyl-CoA. There are many steps in the biosynthesis of steroids. Figure 21.27 outlines some of the major intermediates in the process. The condensation of three acetyl groups produces mevalonate, which contains six carbons (Figure 21.27, *top left*). Decarboxylation of mevalonate produces the five-carbon isoprene unit frequently encountered in the structure of lipids (Figure 21.27, *top right*). The involvement of **isoprene units** is a key point in the biosynthesis of steroids and many other compounds that have the generic name *terpenes.* Some important details are as follows:

- Vitamins A, E, and K come from reactions involving terpenes that humans cannot carry out. That is why we must consume these vitamins in our diets; vitamin D, the remaining lipid-soluble vitamin, is derived from cholesterol (Section 8-6).
- Isoprene units are involved in the biosynthesis of ubiquinone (coenzyme Q) (Section 20-3A) and of derivatives of proteins and tRNA with specific five-carbon units attached.
- Isoprene units are also often added to proteins to act as anchors when the protein is attached to a membrane.

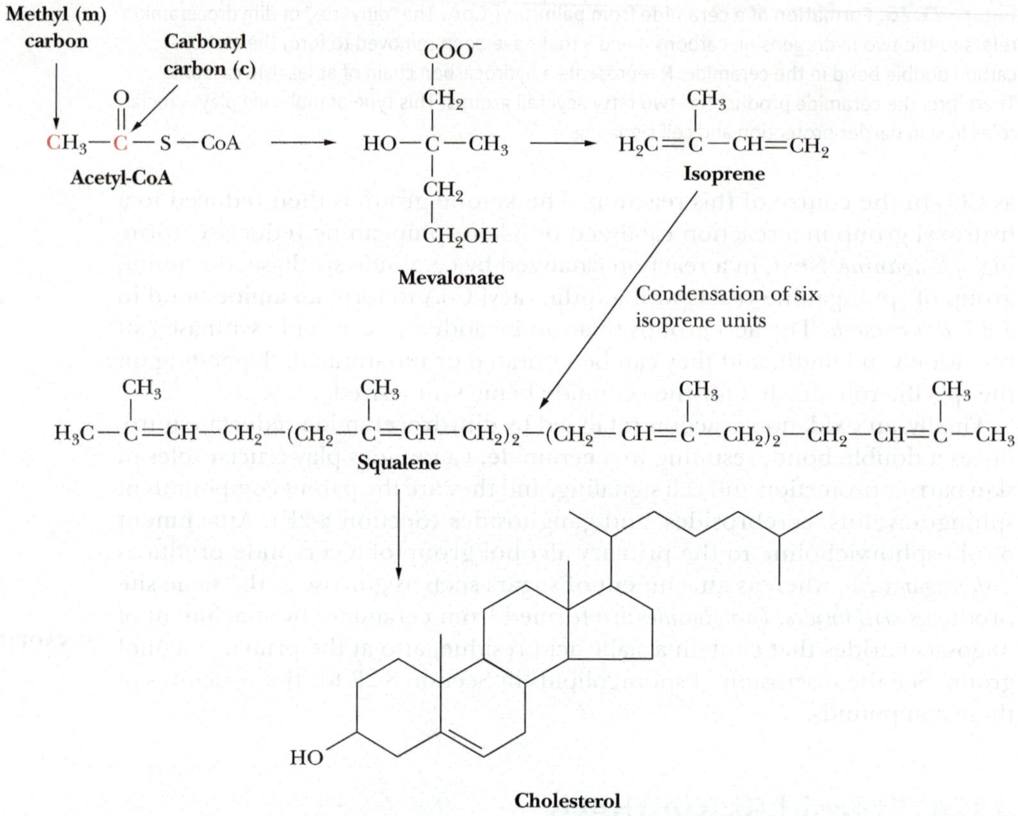

Figure 21.27 The biosynthesis of cholesterol. The enzymes that catalyze these conversions, as well as the names of other intermediates involved, are discussed in more detail in Figures 21.29, 21.31, and 21.32. Also, the final destinations of the "m" and "c" carbons of acetyl-CoA in the final cholesterol structure are shown in Figure 21.28.

ceramide a lipid that contains one fatty acid linked to a sphingosine backbone by an amide bond; it plays crucial roles in skin barrier protection and cell signaling

cholesterol a steroid that occurs in cell membranes and that also serves as the precursor of other steroids

isoprene units five-carbon groups that are used in the biosynthesis of steroids

Continuing the discussion of cholesterol biosynthesis, six isoprene units condense to form squalene, which contains 30 carbon atoms (Figure 21.27, *middle*). Finally, squalene is converted to cholesterol, which contains 27 carbon atoms (Figure 21.27, *bottom*); squalene can also be converted to other sterols.

$$\text{Acetyl - CoA} \rightarrow \text{Mevalonate} \rightarrow \text{Isoprene} \rightarrow \text{Squalene} \rightarrow \text{Cholesterol}$$
$$C_2 \qquad C_6 \qquad C_5 \qquad C_{30} \qquad C_{27}$$

Twelve of the carbon atoms of cholesterol arise from the thioester carbon of the acetyl group; these are the carbon atoms labeled "c" in Figure 21.28. The other 15 carbon atoms arise from the methyl carbon of the acetyl group; these are the carbon atoms labeled "m." Let's now look at the individual steps of the process in more detail.

The conversion of three acetyl groups of acetyl-CoA to *mevalonate* takes place in several steps (Figure 21.29). We saw the first of these steps, the production of acetoacetyl-CoA from two molecules of acetyl-CoA, when we discussed the formation of ketone bodies and the anabolism of fatty acids (Figure 21.11, *top*). A third molecule of acetyl-CoA condenses with acetoacetyl-CoA to produce *β-hydroxy-β-methylglutaryl-CoA* (also called HMG-CoA or 3-hydroxy-3-methylglutaryl-CoA) (Figure 21.29, *middle*). Recall that HMG-CoA is also an intermediate in the production of ketone bodies (Section 21-5).

This reaction is catalyzed by the enzyme hydroxymethylglutaryl-CoA synthase; one molecule of CoA-SH is released in the process. The next reaction, the production of mevalonate from hydroxymethylglutaryl-CoA, is catalyzed by the enzyme hydroxymethylglutaryl-CoA reductase (HMG-CoA reductase) (Figure 21.29, *bottom*). A carbonyl group, the one esterified to CoA-SH, is

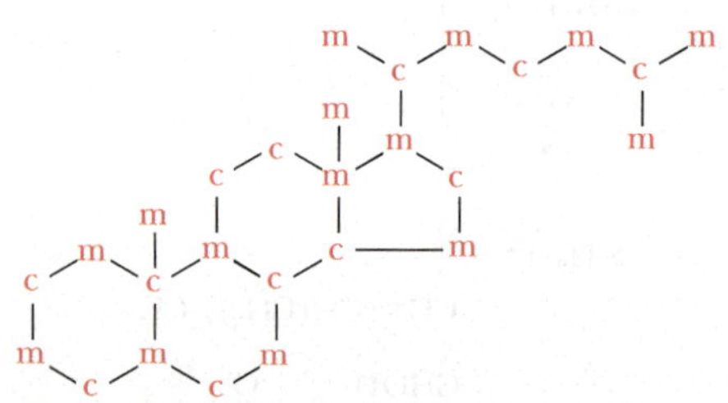

Figure 21.28 The labeling pattern of cholesterol. Each letter "m" indicates a methyl carbon, and each letter "c" indicates a carbonyl carbon, all of which come from acetyl-CoA.

reduced to a hydroxyl group, and the CoA-SH is released. This step is inhibited by high levels of cholesterol and is the major control point of cholesterol synthesis. It is also a target for drugs to lower cholesterol levels in the body. Drugs such as *lovastatin* are inhibitors of hydroxymethyl-CoA reductase and are widely prescribed to lower blood cholesterol levels (Figure 21.30A). The drug is metabolized to mevinolinic acid (Figure 21.30B), which is a transition-state analog of a tetrahedral intermediate in the reaction catalyzed by HMG-CoA reductase (Figure 21.30C).

Continuing the details of cholesterol biosynthesis, mevalonate is then converted to an isoprenoid unit by a combination of phosphorylation, decarboxylation, and dephosphorylation reactions (Figure 21.31). To start, three successive reactions, each of which is catalyzed by an enzyme that requires ATP, give rise to *isopentenyl pyrophosphate,* a five-carbon isoprenoid derivative (Figure 21.31, *top half*). Isopentenyl pyrophosphate is converted to *dimethylallyl pyrophosphate,* another isoprenoid derivative, in a rearrangement reaction catalyzed by the enzyme *isopentenyl pyrophosphate isomerase.*

To finish the bottom half of Figure 21.31, two condensation reactions involving additions of two more isopentenyl pyrophosphate groups occur, catalyzed by a synthetase in the first condensation and a synthase in the second condensation. As a result, *farnesyl pyrophosphate,* a 15-carbon compound, is produced. In the final reaction, two molecules of farnesyl pyrophosphate condense to form *squalene,* a 30-carbon compound. The reaction is catalyzed by squalene synthase, and NADPH is required for the reaction.

To complete the biosynthesis of cholesterol, Figure 21.32 shows the conversion of squalene to cholesterol. The details of this conversion are far from simple. First, squalene is converted to *squalene epoxide* in a reaction catalyzed by squalene monooxygenase that requires both NADPH and molecular oxygen (O_2) (Figure 21.32, *top*). Squalene epoxide then undergoes a remarkably complex cyclization reaction catalyzed by squalene epoxide cyclase to form *lanosterol.* The mechanism of the reaction is a concerted reaction—that is, one in which each part is essential for any other part to take place. No portion of a concerted reaction can be left out or changed because it all takes place simultaneously rather than in a sequence of steps. The conversion of lanosterol to cholesterol is also a complex process. There are many steps and two possible routes that remove three methyl groups and move double bonds (Figure 21.32, *middle*), but we do not discuss the specific details of these processes.

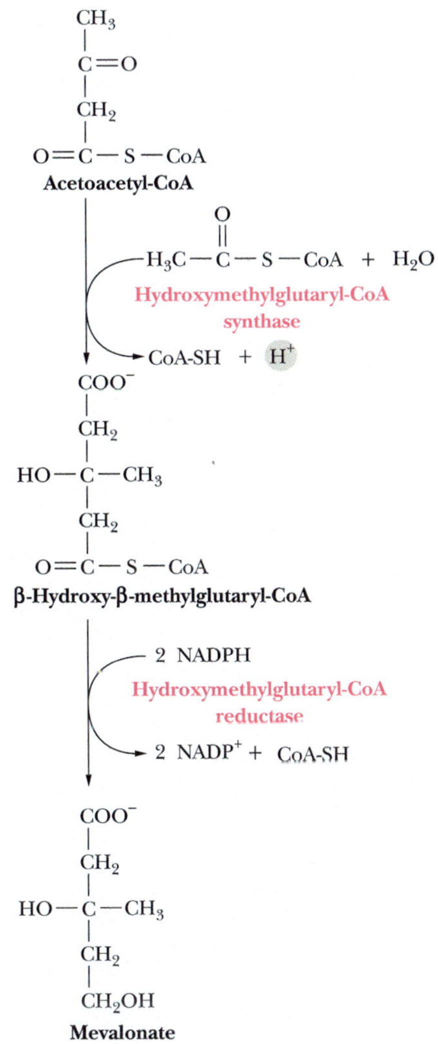

Figure 21.29 The biosynthesis of mevalonate. The acetoacetyl-CoA in this figure is produced from two molecules of acetyl-CoA; see Figure 21.11 for details.

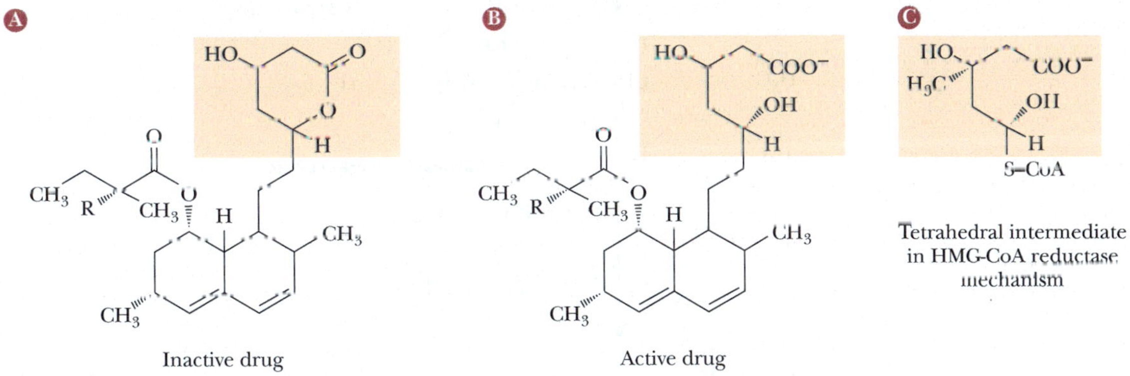

R = H: Mevinolin (Lovastatin, MEVACOR®)
R = CH_3: Synvinolin (Simnastatin, ZOCOR®)

Figure 21.30 The structures of (A) the inactive forms of lovastatin and synvinolin, (B) the active form of mevinolinic acid, and (C) the natural tetrahedral intermediate in the HMG-CoA reductase mechanism. Compare the active form of the drugs with the natural intermediate to see how the drugs are designed to be competitive inhibitors of HMG-CoA reductase.

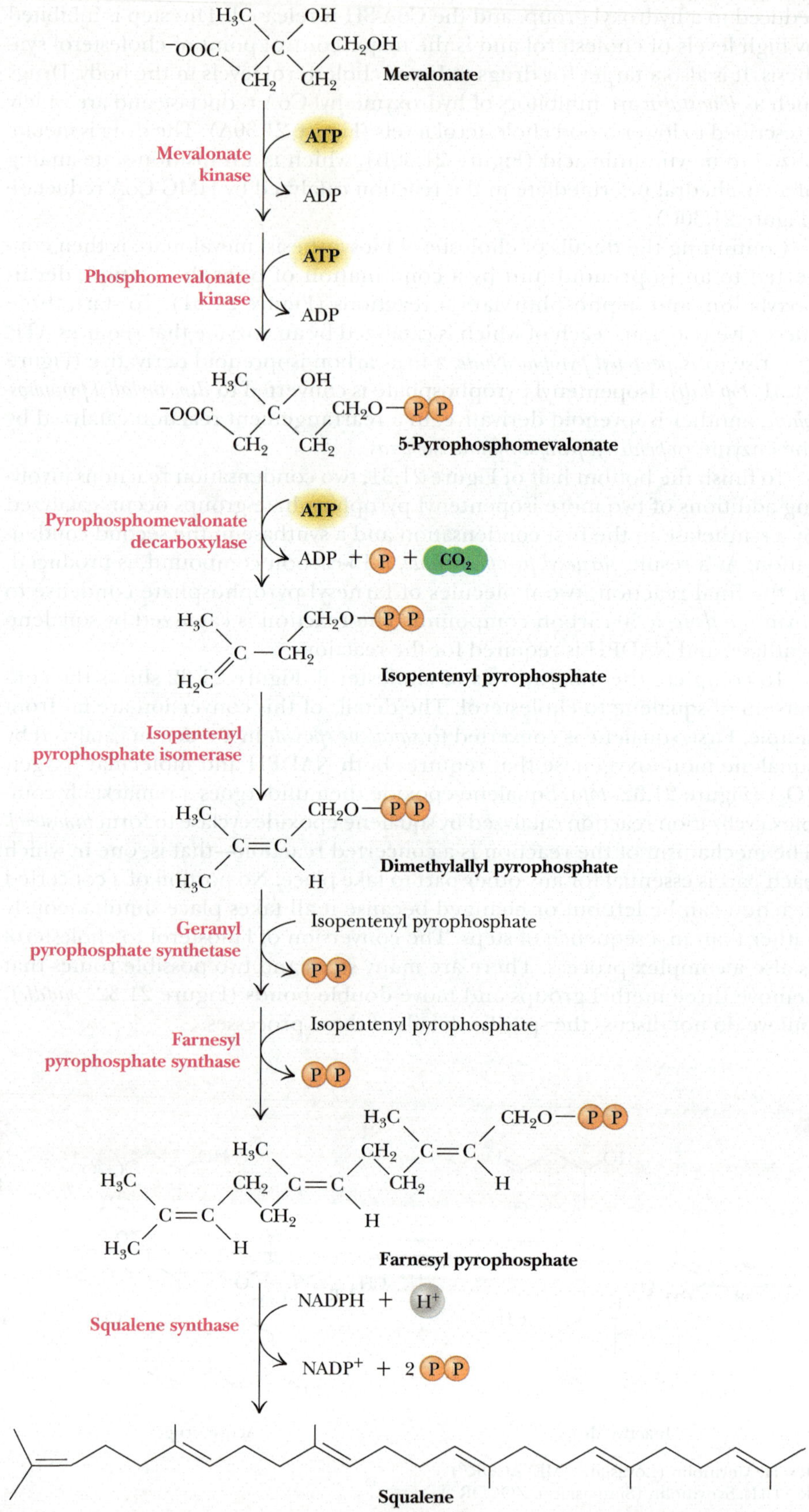

Figure 21.31 The conversion of mevalonate to squalene. In the final reaction, catalyzed by squalene synthase, two 15-carbon molecules of farnesyl pyrophosphate condense to form the 30-carbon squalene molecule.

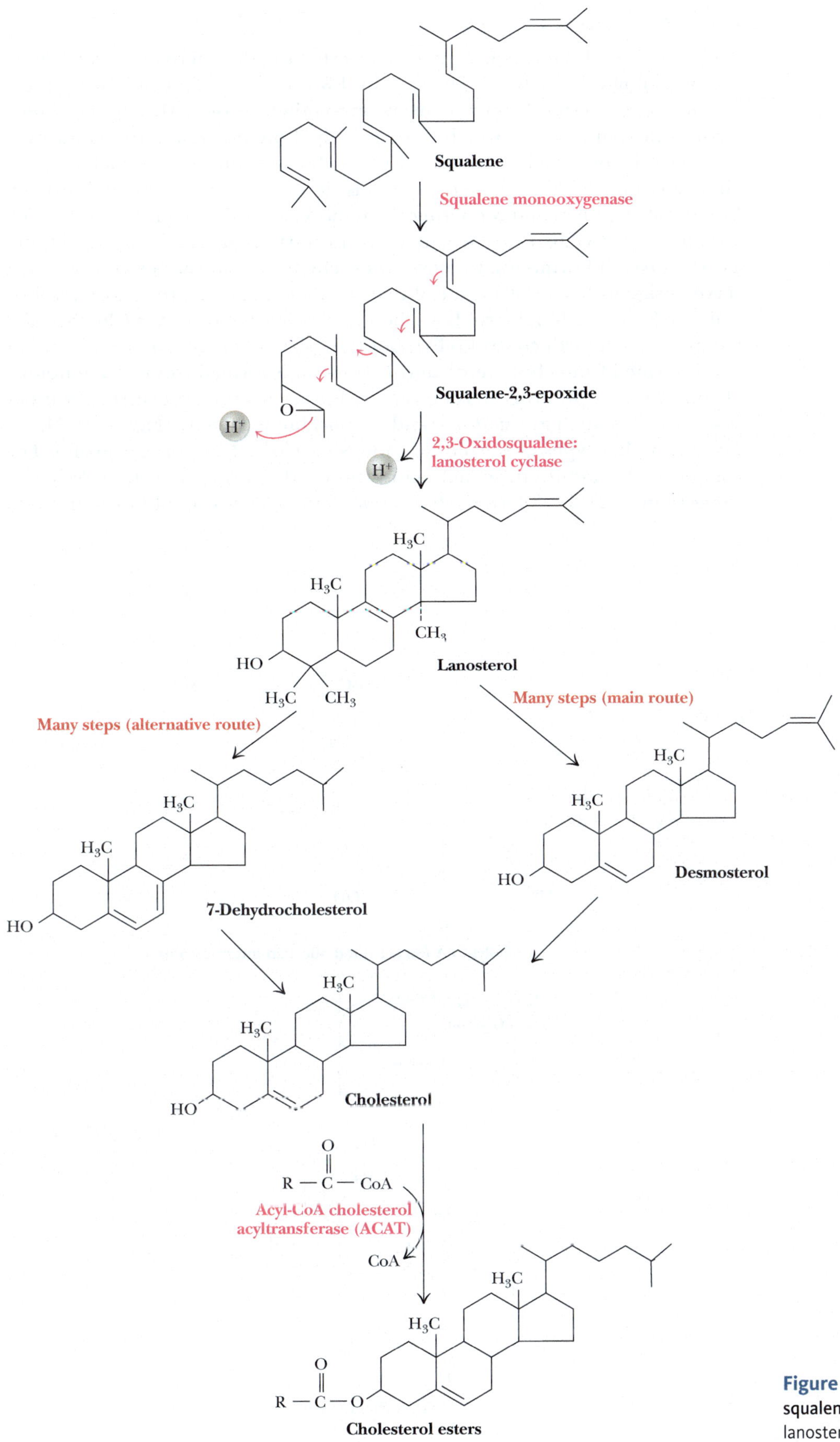

Figure 21.32 Cholesterol is synthesized from squalene via lanosterol. There are two routes from lanosterol to cholesterol.

21-8B Cholesterol as a Precursor to Other Steroids

After cholesterol is formed, it can be converted to other steroids of widely varying physiological functions. The smooth ER is an important site for both the synthesis of cholesterol and its conversion to other steroids. Most of the cholesterol formed in the liver, which is the principal site of cholesterol synthesis in mammals, is converted to *bile acids,* such as cholate, and *bile salts,* such as glycocholate (Figure 21.33). These compounds aid in the digestion of lipid droplets by emulsifying them and rendering them more accessible to enzymatic attack.

Cholesterol is also the precursor of important *steroid hormones* (Figure 21.34). Like all hormones, whatever their chemical nature, steroid hormones serve as signals from outside a cell that regulate metabolic processes within a cell (see Section 24-3). Steroids are best known as sex hormones (they are also components of birth-control pills), but they play other roles as well. *Pregnenolone* is formed from cholesterol, and *progesterone* is formed from pregnenolone (Figure 21.34, *top*). Progesterone is a sex hormone and a precursor for other sex hormones, such as *testosterone* and *estradiol* (an estrogen) (Figure 21.34, *bottom left*). Other types of steroid hormones also arise from progesterone. The role of sex hormones in sexual maturation is discussed in Section 24-3. *Cortisone* (Figure 21.34, *bottom right*) is a *glucocorticoid,* a group of hormones that

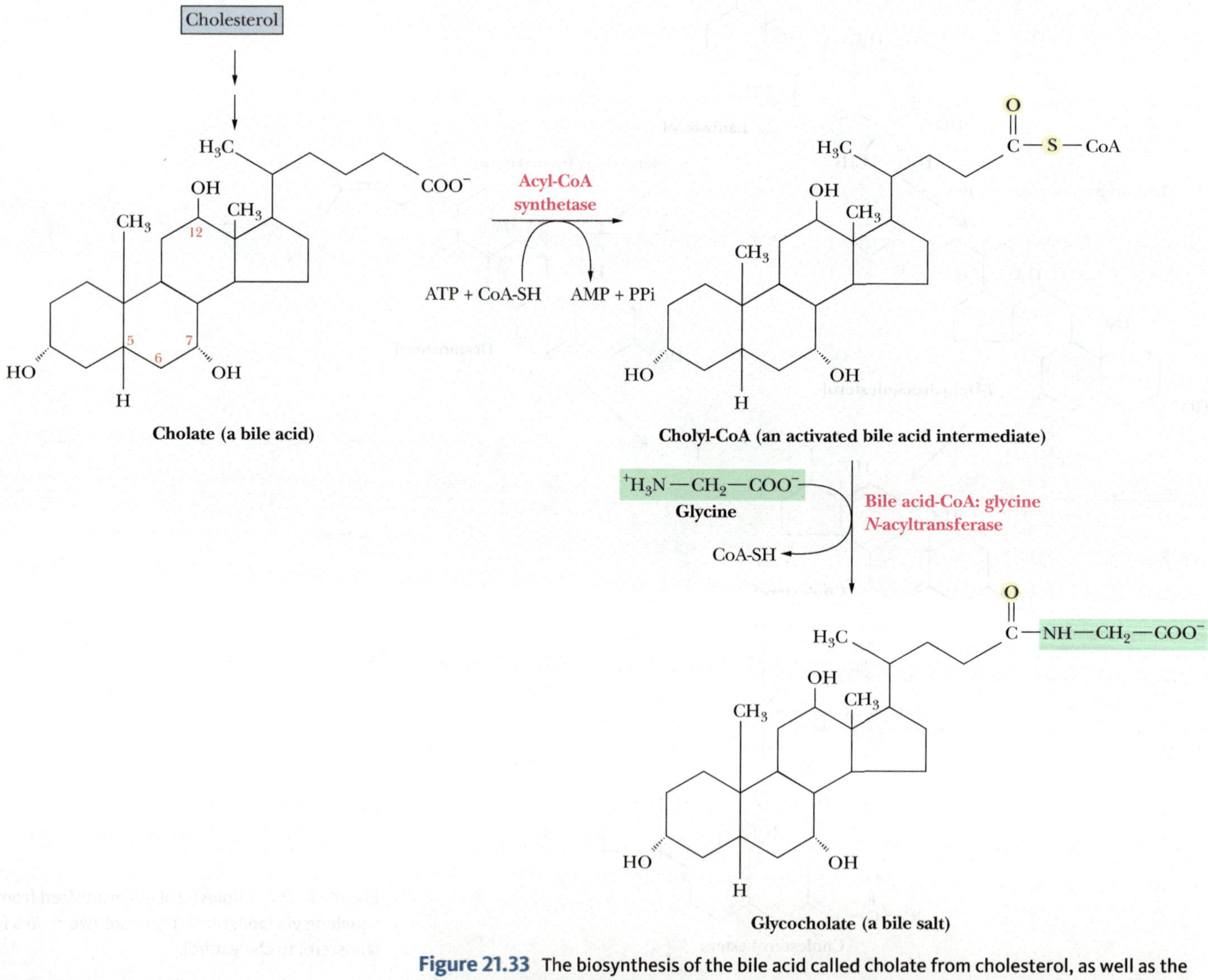

Figure 21.33 The biosynthesis of the bile acid called cholate from cholesterol, as well as the bile salt called glycocholate. The several steps from cholesterol to cholate occur in the liver. The "glyco-" prefix of glycocholate refers to the glycine group.

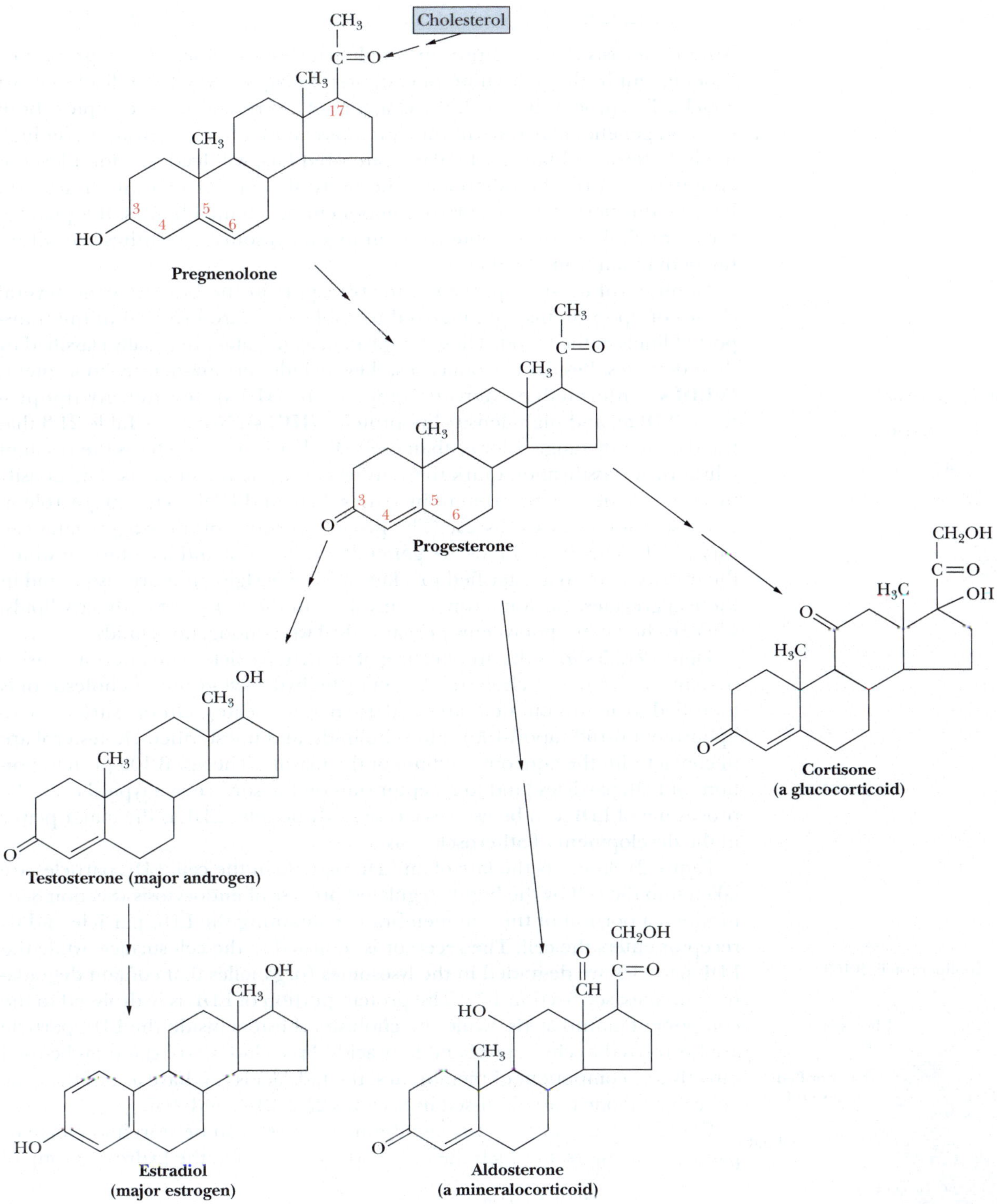

Figure 21.34 **The synthesis of steroid hormones from cholesterol.** Steroid hormones serve as extracellular signals that bind to receptors and regulate metabolic processes within a cell.

play a role in carbohydrate metabolism, as well as in the metabolism of proteins and fatty acids. *Mineralocorticoids* constitute another class of hormones that are involved in the metabolism of electrolytes, including metal ions ("minerals") and water. *Aldosterone* (Figure 21.34, *bottom middle*) is a mineralocorticoid. In cells in which cholesterol is converted to steroid hormones, an enlarged smooth ER is frequently observed, providing a site for the process to take place.

21-8C The Role of Cholesterol in Heart Disease: LDL and HDL

Atherosclerosis is a condition in which arteries are blocked to a greater or lesser extent by the deposition of cholesterol plaques, which can lead to heart attacks. The process by which the clogging of arteries occurs is complex. Both diet and genetics play roles in the development of atherosclerosis. A diet high in cholesterol and fats, particularly saturated fats, will lead to a high level of cholesterol in the bloodstream. The body also makes its own cholesterol because this steroid is a necessary component of cell membranes. It is possible for more cholesterol to come from endogenous sources (synthesized within the body) than from the diet.

Cholesterol must be packaged for transport in the bloodstream; several classes of lipoproteins (summarized in Table 21.3) are involved in the transport of lipids in the blood. These lipoprotein aggregates are usually classified by their densities. Besides chylomicrons, they include very-low-density lipoproteins **(VLDLs)**, intermediate-density lipoproteins **(IDLs)**, low-density lipoproteins **(LDLs)**, and high-density lipoproteins **(HDLs)**. Notice in Table 21.3 that the IDL density range is lower than the LDL density range. This is the result of a historical classification quirk that can be confusing for students. The density increases as the protein content increases. LDL and HDL play a major role in our discussion of heart disease. The protein portions of these aggregates can vary widely. The major lipids are generally cholesterol and its esters, in which the hydroxyl group is esterified to a fatty acid; triacylglycerols are also found in these aggregates. Chylomicrons are involved in the transport of dietary lipids, whereas the other lipoproteins primarily deal with endogenous lipids.

VLDLs very-low-density lipoproteins

IDLs intermediate-density lipoproteins

LDLs low-density lipoproteins

HDLs high-density lipoproteins

Figure 21.35 shows the architecture of an LDL particle. The interior consists of many molecules of cholesteryl esters (the hydroxyl group of cholesterol is esterified to an unsaturated fatty acid, such as linoleate). On the surface, apolipoprotein B-100 (apoB-100), phospholipids, and unesterified cholesterol are in contact with the aqueous medium of the plasma. The apoB-100 protein portions of LDL particles bind to receptor sites on the surface of a typical cell. The processing of LDL can be used as a case study because LDL is the major player in the development of atherosclerosis.

Figure 21.36 shows the fate of an LDL particle in the cell. LDL particles are taken into the cell by the highly regulated process of endocytosis (Section 8-5), in which a portion of the cell membrane containing the LDL particle and its receptor enters the cell. The receptor is returned to the cell surface, while the LDL particles are degraded in the lysosomes (organelles that contain degradative enzymes; see Section 1-5). The protein portion of LDL is hydrolyzed to the component amino acids, while the cholesterol esters inside the LDL particle are hydrolyzed to cholesterol and fatty acids. Free cholesterol can then be used directly as a component of membranes; the fatty acids can have any of the catabolic or anabolic fates discussed in Sections 21-2, 21-4, and 21-7.

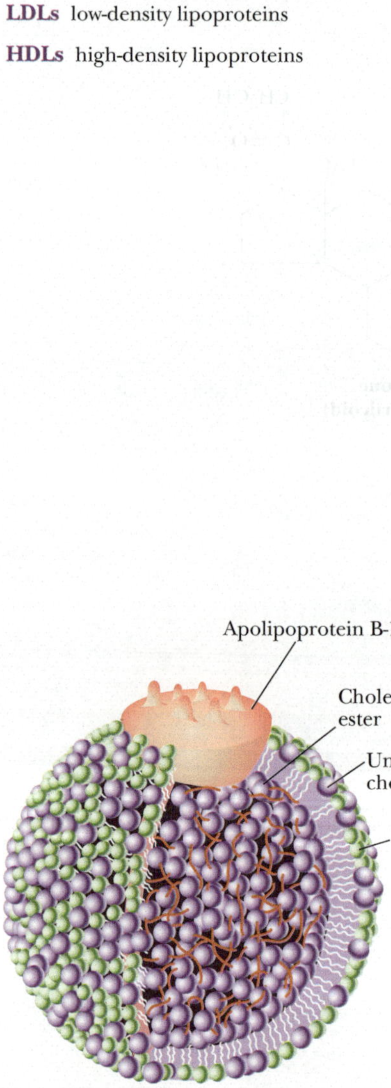

Figure 21.35 Schematic diagram of a low-density lipoprotein (LDL) particle. *(Based on M. S. Brown and J. L. Goldstein, 1984, How LDL Receptors Influence Cholesterol and Atherosclerosis,* Sci. Amer. *251 (5), 58–66.)*

Cholesterol not needed for membrane synthesis can be stored as oleate or palmitoleate esters, in which the fatty acid is esterified to the hydroxyl group of

Table 21.3 Major Classes of Lipoproteins in Human Plasma

Lipoprotein class	Density (g mL^{-1})
Chylomicrons	<0.95
VLDL	0.95–1.006
IDL	1.006–1.019
LDL	1.019–1.063
HDL	1.063–1.210

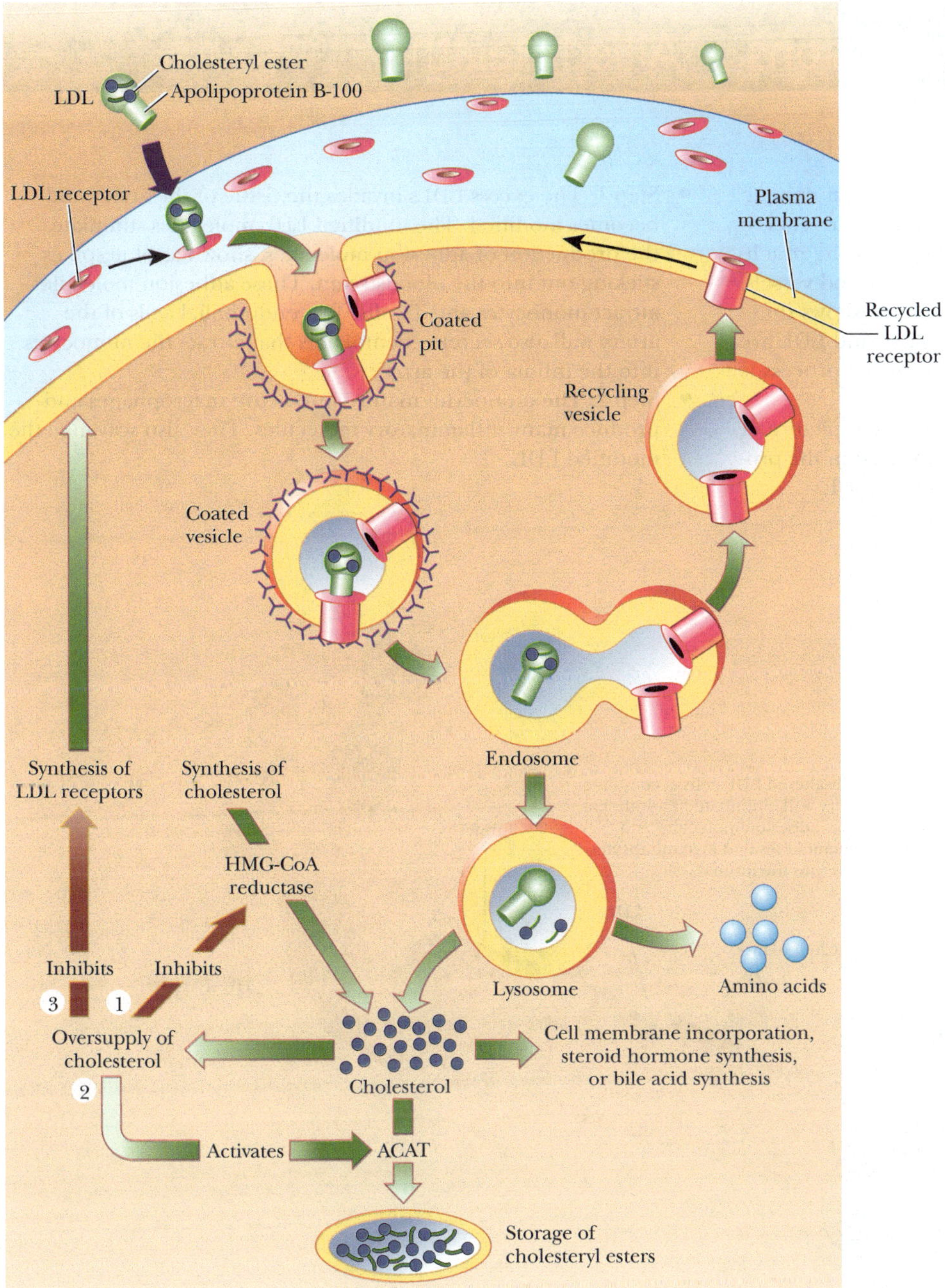

Figure 21.36 The intracellular fate of the cholesterol in a low-density lipoprotein (LDL) particle. Cholesterol esters from the LDL particle are hydrolyzed to cholesterol in lysosomes, while the apolipoprotein B-100 proteins are hydrolyzed to their constituent amino acids. An oversupply of cholesterol in cells inhibits LDL receptor synthesis, and this leads to a buildup of LDL in arteries and an increased risk of heart disease. Acyl-CoA:cholesterol acyltransferase (ACAT) is the enzyme that esterifies cholesterol to fatty acids for storage. (*Based on M. S. Brown and J. L. Goldstein,1984, How LDL Receptors Influence Cholesterol and Atherosclerosis,* Sci. Amer. *251 (5), 58–66.*)

cholesterol. The production of these esters is catalyzed by acyl-CoA:cholesterol acyltransferase (ACAT), and the presence of free cholesterol increases the enzymatic activity of ACAT. In addition, cholesterol inhibits both the synthesis and the activity of the enzyme hydroxymethylglutaryl-CoA reductase (HMG-CoA reductase). This enzyme catalyzes the production of mevalonate (Figure 21.29), the reaction that is the committed step in cholesterol biosynthesis. This point has important implications. Dietary cholesterol suppresses the synthesis of cholesterol by the body, especially in tissues other than the liver. An effect of the presence of free cholesterol in the cell is inhibition of synthesis of LDL receptors. As a result of reduction in the number of receptors, cellular uptake of cholesterol is inhibited, and the level of LDL in the blood increases, leading to the deposition of atherosclerotic plaques. **Biochemical Connections 21D** goes into more detail about the process.

21D Biochemical Connections • Allied Health

Atherosclerosis

Atherosclerosis causes more deaths every year than cancer. It causes chest pain, heart attacks, and strokes. For years, it was described with simple models that described how the much-maligned LDL particles would lay down plaques in blood vessels, leading to blockage. However, more recent evidence shows these old models to be overly simplistic. Lipid metabolism and LDL are certainly involved, but a more important villain in the process is inflammation.

Figure 21.37 shows a more realistic and complicated picture of the formation of atherosclerosis. Excess LDL is involved in the process, but it is not caused by a simple laying down of the LDL.

- Step 1: The excess LDLs invades the tissue of the artery and becomes modified. The modified LDL molecules stimulate the production of adhesion molecules, shown as blue spikes sticking out into the bloodstream. These adhesion molecules attract monocytes and T cells. The endothelial cells of the artery wall also secrete chemokines that attract the monocytes into the intima of the artery.
- Step 2: The monocytes mature into active macrophages and produce many inflammatory molecules. They also scavenge the modified LDL.

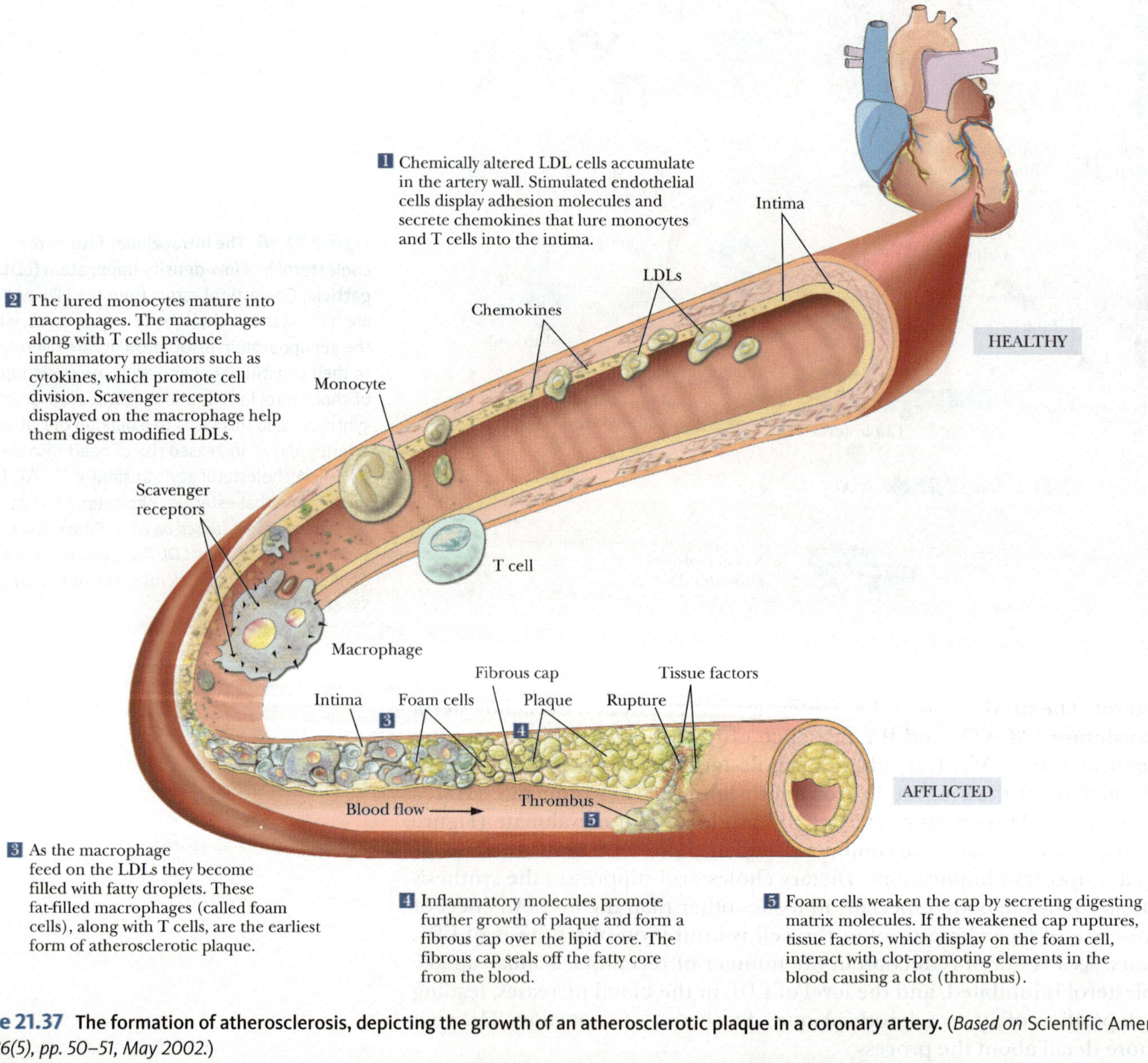

Figure 21.37 The formation of atherosclerosis, depicting the growth of an atherosclerotic plaque in a coronary artery. (*Based on* Scientific American, *vol. 286(5), pp. 50–51, May 2002.*)

- Step 3: The macrophages continue to digest the LDL and become filled with lipid droplets, becoming what are called foam cells. These cells form a fatty streak that is the first outward signal of atherosclerosis.
- Step 4: Inflammation promotes growth of the plaque and formation of a fibrous cap over the lipids. The cap protrudes into the bloodstream but also protects the blood from the deposits.
- Step 5: More inflammatory molecules can cause the cap to rupture. The foam cells also release tissue factor, a potent blood-clotting agent. The ruptured plaque leads to the formation of a thrombus, or blood clot, in the artery. If the clot is big enough, it can cause a heart attack—the death of cardiac cells.

Research has shown a direct connection between inflammation and high lipid levels in the blood. Cytokines play a key role in regulating the levels of enzymes that control lipid metabolism. In mice deficient in LDL receptors (these mice cannot control lipid levels in their blood), inhibition of enzymes produced by cytokines led to a lowering of LDL levels. This result strongly suggests that the immune system affects lipid metabolism and provides indicators as to the direction of future research on the subject.

From a practical standpoint, the take-home message may still be the same. Practices that reduce LDL are still considered necessary for a healthy heart. However, when scientists try to come up with medical countermeasures, this full understanding of all the chemical factors involved in atherosclerosis is critical to the development of cures. ◗

The crucial role of LDL receptors in maintaining healthy levels of cholesterol in the bloodstream is especially clear in the case of *familial hypercholesterolemia,* which results from a defect in the gene that codes for the active receptors. An individual who has one gene that codes for the active receptor and one defective gene is heterozygous for this trait. Heterozygotes have blood cholesterol levels that are above average; therefore, they are at a higher risk for heart disease than the general population. An individual with two defective genes, and thus with no active LDL receptor, is homozygous for the trait. Homozygotes have very high blood cholesterol levels from birth, and there are recorded cases of heart attacks in two-year-olds with this condition. Patients who are homozygous for familial hypercholesterolemia usually die before age 20. Another genetic abnormality involved in hypercholesterolemia is the one that gives rise to a faulty apolipoprotein E, a component of IDL and VLDL, which is involved in the uptake of lipids by the cell. The unfortunate result is the same.

Before we leave this discussion, we should mention HDL, the "good" cholesterol. Unlike LDL, which transports cholesterol from the liver to the rest of the body, HDL transports it back to the liver for degradation into bile acids. It is desirable to have low levels of cholesterol and LDL in the bloodstream, but it is also desirable to have as high a proportion of total cholesterol as possible in the form of HDL. High levels of LDL and low levels of HDL are correlated with the development of heart disease. Factors that are known to increase HDL levels, such as regular strenuous exercise, decrease the probability of heart disease. Smoking reduces the level of HDL and is highly correlated with heart disease.

Summary

Lipids Are Involved in the Generation and Storage of Energy

- We previously saw how carbohydrates are processed, and lipids are another class of nutrients that are processed catabolically and anabolically.
- The catabolic oxidation of lipids releases large quantities of energy, whereas the anabolic formation of lipids represents an efficient way of storing chemical energy.

Fatty Acid Transport into Mitochondria

- After an initial activation step in the cytosol, with the formation of a fatty acyl-CoA corresponding to each fatty acid, each acyl group is transesterified to carnitine for transport into the mitochondrial matrix.
- Once inside the matrix, the acyl group is again transesterified to form a fatty acyl-CoA that then undergoes β-oxidation.

Oxidation of Saturated Fatty Acids

- The oxidation of fatty acids, which takes place in the mitochondrial matrix, is the chief source of energy in the catabolism of lipids.
- In this process, two-carbon units are successively removed from the thioester end of the fatty acyl-CoA to produce

acetyl-CoA, which subsequently enters the citric acid cycle for energy production.

- The reactions that liberate the acetyl-CoA units from a fatty acyl-CoA also produce NADH and $FADH_2$, which eventually produce ATP via the electron transport chain.

The Energy Yield from the Oxidation of Fatty Acids

- There is a net yield of 120 ATP molecules for each molecule of stearic acid (an 18-carbon compound) that is completely oxidized to carbon dioxide and water.
- The source of these ATP molecules is the production of NADH and $FADH_2$ in the β-oxidation pathway, as well as the NADH, $FADH_2$, and GTP produced when the acetyl-CoA molecules are processed through the citric acid cycle and the electron transport chain.

Oxidation of Monounsaturated Fatty Acids

- The pathway of catabolism of fatty acids includes reactions in which unsaturated and saturated fatty acids can be metabolized.
- For monounsaturated fatty acids to be broken down to acetyl-CoA, enoyl-CoA isomerase catalyzes a cis–trans isomerization, which allows enoyl-CoA hydratase to hydrate the trans double bond, allowing for the continuation of the β-oxidation and breakdown of the fatty acyl-CoA molecule.

Oxidation of Polyunsaturated Fatty Acids

- For fatty acids containing multiple double bonds to be catabolized, the enoyl-CoA isomerase is required for the first double bond, and 2,4-dienoyl-CoA reductase is required for subsequent double bonds.
- The 2,4-dienoyl-CoA reductase converts two conjugated pi bonds into a single pi bond that can be isomerized by enoyl-CoA isomerase, which allows β-oxidation to continue.

Ketone Bodies

- Ketone bodies are substances related to acetone that are produced when an excess of acetyl-CoA results from β-oxidation.
- This situation can arise from a large intake of lipids and a low intake of carbohydrates or can occur in diabetes, in which the inability to metabolize carbohydrates causes an imbalance in the breakdown products of carbohydrates and lipids.

First Steps in Fatty Acid Biosynthesis

- The anabolism of fatty acids proceeds by a different pathway from the catabolism of fatty acids in β-oxidation.
- First, acetyl-CoA is transported to the cytosol, where it is converted to malonyl-CoA.

The Fatty Acid Synthase Complex

- Two carbons of malonyl-CoA then get added to the growing fatty acyl chain in a series of reactions catalyzed by an ordered multienzyme complex called fatty acid synthase.
- Some of the most important differences between the fatty acid anabolism and catabolism are the requirement for biotin in anabolism, but not in catabolism, and the requirement for NADPH in anabolism, rather than the NAD^+ required in catabolism.

Triacylglycerol Biosynthesis

- Most compound lipids, such as triacylglycerols, phosphoacylglycerols, and sphingolipids, have fatty acids as precursors.
- Triacylglycerols consist of a glycerol moiety derived from glycerol-3-phosphate and acyl groups of fatty acids that are transferred from acyl-CoA.

Phosphoacylglycerol Biosynthesis

- In the case of phosphoacylglycerols, two fatty acids and phosphoric acid are added to a glycerol backbone.
- The addition of remaining groups requires nucleoside triphosphates and differs between mammals and bacteria.

Sphingolipid Biosynthesis

- Sphingolipid biosynthesis starts when palmitoyl-CoA combines with serine to form a sphingosine backbone precursor. Then, in a series of reactions, fatty acids are linked to the sphingosine backbone, producing ceramides.
- Other moieties, including sugars, can be added to ceramides, producing gangliosides and other compounds.

Cholesterol Biosynthesis

- The starting material for steroid biosynthesis is acetyl-CoA.
- Isoprene units are formed from acetyl-CoA in the early stages of a lengthy process that ultimately leads to cholesterol.
- HMG-CoA is a key intermediate in cholesterol biosynthesis, and its formation is a target of cholesterol-lowering drugs.

Cholesterol as a Precursor to Other Steroids

- Cholesterol can be converted to other steroids, including bile acids, sex hormones, glucocorticoids, and mineralocorticoids.

The Role of Cholesterol in Heart Disease: LDL and HDL

- Cholesterol must be packaged for transport in the bloodstream, and several classes of lipoprotein particles are involved.
- One class of lipoprotein particles is LDL (low-density lipoprotein or "bad cholesterol"), whereas another class is HDL (high-density lipoprotein or "good cholesterol").
- Both dietary cholesterol and genetic factors influence the role of cholesterol in heart disease.

Review Exercises

21-1 Lipids Are Involved in the Generation and Storage of Energy

1. **Reflect and Apply** (a) The major energy storage compound of animals is fats (except in muscles). Why would this be advantageous? (b) Why don't plants use fats or oils as their *major* energy storage compound?

21-2 Catabolism of Lipids

2. **Recall** What is the difference between phospholipase A_1 and A_2?
3. **Reflect and Apply** Compare the products of phospholipase A_1 or phospholipase A_2 with those of a lipase. Assume a phosphoacylglycerol as the substrate for the phospholipases and a triacylglycerol as the substrate for the lipase.
4. **Recall** How are lipases activated hormonally?
5. **Recall** What is the metabolic purpose of linking a fatty acid to coenzyme A (CoA) to produce acyl-CoA molecules?
6. **Recall** Outline the role of carnitine in the transport of fatty acyl groups into the mitochondrion. How many enzymes are involved? What are they called?
7. **Recall** Describe the first reaction of the β-oxidation pathway.
8. **Recall** What is the difference between the type of oxidation catalyzed by acyl-CoA dehydrogenase and that catalyzed by β-hydroxy-CoA dehydrogenase?
9. **Recall** Draw a six-carbon saturated fatty-acyl CoA molecule and show where the double bond is created during the first step of β-oxidation. What is the orientation of this bond?
10. **Reflect and Apply** Why does the degradation of palmitic acid to eight molecules of acetyl-CoA require seven, rather than eight, rounds of the β-oxidation process?
11. **Reflect and Apply** Given the nature of the hormonal activation of lipases, what carbohydrate pathways would be activated or inhibited under the same conditions?

21-3 The Energy Yield from the Oxidation of Fatty Acids

12. **Recall** Compare the energy yields from the oxidative metabolism of glucose and stearic acid. To be fair, calculate it on the basis of ATP equivalents per carbon and also ATP equivalents per gram.
13. **Reflect and Apply** Which generates more ATP—the processing of the reduced electron equivalents formed during β-oxidation through the electron transport chain or the processing of the acetyl-CoA generated from β-oxidation through the citric acid cycle and the electron transport chain? Explain.
14. **Mathematical** Calculate the ATP yield for the complete oxidation of one molecule of palmitic acid (16 carbons). How does this figure differ from that obtained for stearic acid (18 carbons)? Consider the β-oxidation steps, the processing of acetyl-CoA through the citric acid cycle, and the electron transport chain.
15. **Reflect and Apply** It is frequently said that camels store water in their humps for long desert journeys. How would you modify this statement based on the information in this chapter?

21-4 Catabolism of Unsaturated Fatty Acids and Odd-Carbon Fatty Acids

16. **Recall** Describe briefly how β-oxidation of an odd-chain fatty acid is different from that of an even-chain fatty acid.
17. **Reflect and Apply** State the number and identity of molecules that enter the citric acid cycle from the complete oxidation of a fatty acid with 17 carbons.
18. **Reflect and Apply** You hear a fellow student say that the oxidation of unsaturated fatty acids requires the same group of enzymes as the oxidation of saturated fatty acids. Is the statement true or false? Why?
19. **Recall** Compared to saturated fatty acids, what are the unique enzymes needed to β-oxidize a monounsaturated fatty acid?
20. **Recall** Compared to saturated fatty acids, what are the unique enzymes needed to β-oxidize a polyunsaturated fatty acid?
21. **Mathematical** Calculate the net ATP yield from the complete processing of a saturated fatty acid containing 17 carbons. Consider the β-oxidation steps, processing of acetyl-CoA through the citric acid cycle, and electron transport.
22. **Mathematical** Calculate the net ATP yield from oleic acid (18:1 Δ^9). *Hint:* Remember the step that bypasses acyl-CoA dehydrogenase.
23. **Mathematical** Calculate the net ATP yield from linoleic acid (18:2 $\Delta^{9,12}$). For this calculation, assume that the loss of an NADPH is the same as the loss of an NADH.
24. **Reflect and Apply** How many cycles of β-oxidation are required to process a fatty acid with 17 carbons?
25. **Reflect and Apply** Fatty acids cannot yield a *net* gain in carbohydrates. Why can odd-chain fatty acids be thought to break this rule to a small extent?

21-5 Ketone Bodies

26. **Recall** Describe what ketone bodies are, where they are produced, and the conditions that promote their production.
27. **Recall** Briefly outline the reactions involved in ketone body production.
28. **Reflect and Apply** Why might a doctor smell the breath of a person known to have diabetes who has just passed out?
29. **Reflect and Apply** Why might a person who chronically consumes large amounts of alcohol have a "fatty liver"?
30. **Reflect and Apply** A person following a very low-carbohydrate diet notices a metallic taste in the morning. What biochemical process could explain this symptom?

21-6 Fatty Acid Biosynthesis

31. **Recall** Describe the first step of fatty acid biosynthesis.
32. **Reflect and Apply** Compare and contrast the pathways of fatty acid breakdown and biosynthesis. What features do these two pathways have in common? How do they differ?
33. **Recall** Outline the steps in the acetyl-CoA carboxylase mechanism.
34. **Reflect and Apply** What is the metabolic importance of malonyl-CoA?
35. **Recall** How are the two redox reactions of β-oxidation different from their counterparts in fatty acid synthesis?
36. **Recall** How is ACP similar to coenzyme A? How is it different?
37. **Recall** What is the purpose of having ACP as a distinct activating group for fatty acid synthesis?
38. **Recall** Why are linoleate and linolenate considered essential fatty acids? What step in the production of polyunsaturated fatty acids are mammals unable to perform?
39. **Reflect and Apply** Is it possible to convert fatty acids to other lipids without acyl-CoA intermediates?
40. **Reflect and Apply** What is the role of citrate in the transport of acetyl groups from the mitochondrion to the cytosol?
41. **Reflect and Apply** In the mitochondrion, a short-chain carnitine acyltransferase can take acetyl groups from acetyl-CoA and transfer them to carnitine. How might this be related to lipid biosynthesis?
42. **Reflect and Apply** In fatty acid synthesis, malonyl-CoA, rather than acetyl-CoA, is used as a "condensing group." Suggest a reason for this.

43. **Reflect and Apply** (a) Where in an earlier chapter have we encountered something comparable to the action of the acyl carrier protein (ACP) of fatty acid synthesis? (b) What is a critical feature of the action of the ACP?

21-7 Biosynthesis of Acylglycerols and Compound Lipids

44. **Recall** What is the source of the glycerol in triacylglycerol synthesis?
45. **Recall** What is the activating group used in the formation of phosphoacylglycerols?
46. **Recall** What are the differences between the synthesis of phosphatidylethanolamine in prokaryotes and eukaryotes?

21-8 Steroid Biosynthesis

47. **Recall** How are isoprene units important in cholesterol biosynthesis and other biochemical pathways?
48. **Recall** A cholesterol sample is prepared using acetyl-CoA labeled with ^{14}C at the carbonyl group as precursor. Which carbon atoms of cholesterol are labeled?
49. **Recall** Which molecules have cholesterol as a precursor?
50. **Reflect and Apply** What structural feature do all steroids have in common? What are the biosynthetic implications of this common feature?
51. **Reflect and Apply** In steroid synthesis, squalene is oxidized to squalene epoxide. This reaction is somewhat unusual, in that both a reducing agent (NADPH) and an oxidizing agent (O_2) are required. Why are both needed?
52. **Reflect and Apply** Why must cholesterol be packaged for transport rather than occurring freely in the bloodstream?
53. **Reflect and Apply** A drug that reduces blood cholesterol has the effect of stimulating the production of bile salts. How might this result in lower blood cholesterol? *Hint:* There are two ways.

Further Reading

Abu-Elheiga, L., M. M. Matzuk, K. A. H. Abo-Hashema, and S. J. Wakil. Continuous Fatty Acid Oxidation and Reduced Fat Storage in Mice Lacking ACC2. *Science* **291**, 2613–2616 (2001). An article about metabolic effects seen in mice lacking one of the isoforms of acetyl-CoA carboxylase.

Bodner, C. M. Lipids. *J. Chem. Ed.* **63**, 772–775 (1986). Part of a series of concise and clearly written articles on metabolism.

Brown, M. S., and J. L. Goldstein. How LDL Receptors Influence Cholesterol and Atherosclerosis. *Sci. Amer.* **251** (5), 58–66 (1984). A description of the role of cholesterol in heart disease by the winners of the 1985 Nobel Prize in Physiology or Medicine.

Horton, J. Unfolding Lipid Metabolism. *Science* **320**, 1433–1434 (2008). The dual role of a transcription factor that regulates genes for response to improperly folded proteins and genes for lipid synthesis.

Jenni, S., M. Leibundgut, D. Boehringer, C. Frick, B. Mikolasek, B., and N. Ban. Structure of Fungal Fatty Acid Synthase and Implications for Iterative Substrate Shuttling. *Science* **316**, 254–261 (2007). An in-depth look at the reaction sites for a multiple-subunit enzyme.

Kaiser, J. Mysterious, Widespread Obesity Gene Found Through Diabetes Study. *Science* **316**, 185 (2007). A single-page article that describes the discovery of a gene highly correlated with obesity.

Krutch, J. W. *The Voice of the Desert.* New York: Morrow, 1975. Chapter 7, "The Mouse That Never Drinks," is a description, primarily from a naturalist's point of view, of the kangaroo rat, but it does make the point that metabolic water is this animal's only source of water.

Lawn, R. Lipoprotein (a) in Heart Disease. *Sci. Amer.* **266** (6), 54–60 (1992). Relates the properties of lipids and protein structure to the blockage of arteries characteristic of heart disease.

Libby, P. Atherosclerosis: The New View. *Sci. Amer.* **286** (5), 47–55 (2002). The role of inflammation in the development of heart disease.

Lo, J., Y. Wang, A. Tumanov, M. Bamji, Z. Yao, C. Reardon, G. Getz, and Y. Fu. Lymphotoxin (beta) Receptor-Dependent Control of Lipid Homeostasis. *Science* **316**, 285–288 (2007). The role of immune cells in liver in controlling lipid levels in blood.

Maier, T., S. Jenni, and N. Ban. Architecture of Mammalian Fatty Acid Synthase at 4.5 Å Resolution. *Science* **311**, 1258–1262 (2006). A pictorial view of the active site of a large enzyme.

Ruderman, N., and J. S. Flier. Chewing the Fat—ACC and Energy Balance. *Science* **291**, 2558–2561 (2001). A summary of information about acetyl-CoA carboxylase and lipid metabolism.

Wakil, S. J., and E. M. Barnes. Fatty Acid Metabolism. *Compr. Biochem.* **18**, 57–104 (1971). Extensive coverage of the topic.

Wang, X.,M. Gan, Y. Wang, S. Wang, Y. Lei, K. Wang, X. Zhang, L. Chen, Y. Zhao, L. Niu, S. Zhang, L. Zhu, and L. Shen. Comprehensive Review on Lipid Metabolism and RNA Methylation: Biological Mechanisms, Perspectives and Challenges. *Int. J. Biol. Macromol.* **270**, 132057 (2024). This review article explores the relationship between RNA methylation and lipid metabolism, highlighting how RNA modifications influence adipogenesis and lipid utilization across various organs.

Yoon, H., J. L. Shaw, M. C. Haigis, and A. Greka. Lipid Metabolism in Sickness and in Health: Emerging Regulators of Lipotoxicity. *Mol. Cell* **81**, 3708–3730 (2021). This review article discusses advances in understanding the roles of specific lipid species in lipid metabolism and in metabolic disorders.

Lush rain forest vegetation. Photosynthesis linked to oxygen plays an essential role in all plant and animal life.

Photosynthesis 22

Chapter Outline

22-1 Chloroplasts Are the Site of Photosynthesis

Learning Objectives

After reading this section, students will be able to:

LO 22.1.1 Write the net equation of photosynthesis.

LO 22.1.2 Describe the structure of chloroplasts.

LO 22.1.3 Distinguish between the structures of different types of chlorophyll.

22-1A The Net Reaction Equation of Photosynthesis and Chloroplast Structure

Photosynthetic organisms, such as green plants, convert carbon dioxide (CO_2) and water into carbohydrates such as glucose (written here as $C_6H_{12}O_6$) and molecular oxygen (O_2).

$$6CO_2 + 6H_2O \rightarrow C_6H_{12}O_6 + 6O_2$$

The equation actually represents two pathways. One pathway, called the **light-dependent** (or **light**) **reactions of photosynthesis**, involves the oxidation of water to produce oxygen and the conversion of light energy from the Sun into chemical energy in the form of adenosine triphosphate (ATP) and nicotinamide adenine dinucleotide phosphate (NADPH). In prokaryotes and eukaryotes, solar energy is absorbed by **chlorophyll** to supply the energy needed in the light reactions. The NADPH generated by the light reactions is the reducing agent needed in the second photosynthetic pathway, called the **light-independent (Calvin cycle) reactions of photosynthesis**. In this pathway, the fixation of CO_2 into glucose and other sugars does not use solar energy directly but rather uses it indirectly in the form of the ATP and NADPH produced in the course of the light reactions.

In prokaryotes such as cyanobacteria, photosynthesis takes place in granules bound to the plasma membrane. The site of photosynthesis in eukaryotes such as green plants and green algae is the **chloroplast** (Figure 22.1), a membrane-enclosed organelle that we discussed in Section 1-5. Like the mitochondrion, the chloroplast has inner and outer membranes and an intermembrane space. In addition, within the chloroplast are bodies called **grana**, which consist of

Chapter Outline (continued)

light-dependent (light) reactions of photosynthesis the first stage of photosynthesis that involves the oxidation of water to produce oxygen and the conversion of light energy from the Sun into chemical energy in the form of ATP and NADPH

chlorophyll the principal photosynthetic pigment responsible for trapping light energy from the Sun

light-independent (Calvin cycle) reactions of photosynthesis the second stage of photosynthesis that uses ATP and NADPH as chemical energy sources to fix CO_2 into glucose and other sugars

chloroplast the organelle that is the site of photosynthesis in green plants

grana bodies within the chloroplast that contain the thylakoid disks, the site of photosynthesis

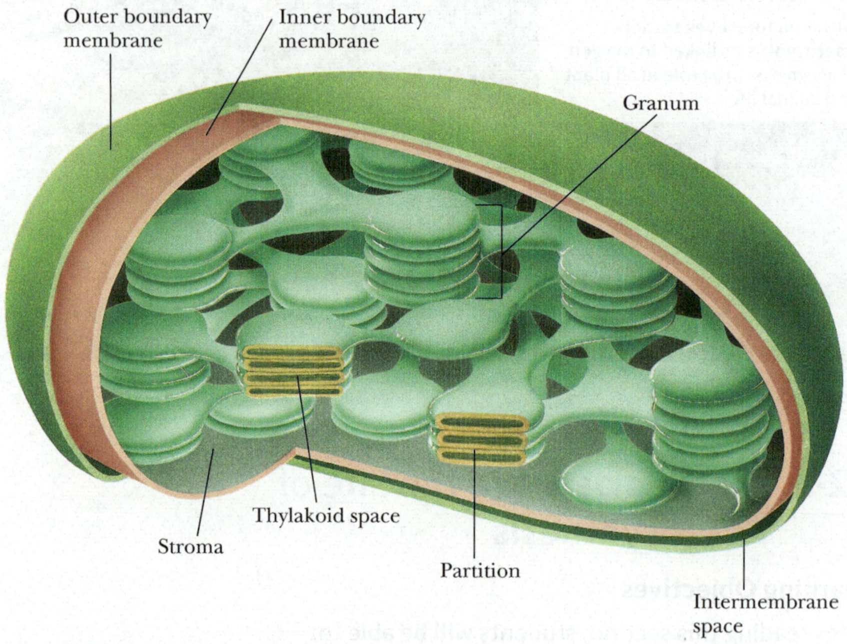

Figure 22.1 A schematic diagram of a chloroplast. Chloroplasts are organelles where photosynthesis occurs.

Pioneers in Biochemistry • Alice Ball

Alice Ball (1892–1916) was a chemist who developed a groundbreaking treatment for leprosy, which today is known as the Ball Method.

Ball began—and ended—her brief career at the University of Hawai'i (then the College of Hawai'i), where she was the first Black woman to receive a master's degree and become a professor of chemistry. At that time in Hawai'i, patients with leprosy (now known as Hansen's disease) were exiled to an island and left to await their deaths. This horrific fate was primarily suffered by Native Hawaiians, who lacked acquired immunity to the disease. Although there was a traditional, plant-based treatment in use, it was unreliable, and there was no effective method to administer it.

Using her knowledge of the chemical makeup of plants, Ball developed a method to process oil from a tree into a solution that maintained the oil's therapeutic properties but was also stable and able to be injected. The injectable form of the oil produced by the Ball Method remained the only effective treatment for leprosy until the 1940s.

Sadly, Ball died at the age of 24 without knowing the long-lasting impact of her research. Although official records cite the cause of death as tuberculosis, some conjecture that she died from chemical exposure in her teaching lab.

FLHC JN8 / Alamy Stock Photo

stacks of flattened membranes called **thylakoid disks**. The grana are connected by membranes called *intergranal lamellae*. The thylakoid disks are formed by the folding of a third membrane within the chloroplast.

thylakoid disks sites of light-trapping reactions in chloroplasts

The folding of the thylakoid membrane creates two spaces in the chloroplast in addition to the intermembrane space. The **stroma** lies within the inner membrane and outside the thylakoid membrane. In addition to the stroma, there is a **thylakoid space** within the thylakoid disks themselves. The trapping of light and the production of oxygen take place in the thylakoid disks. The Calvin cycle reactions, in which CO_2 is fixed into carbohydrates, take place in the stroma. Figure 22.2 shows an overview of the events that happen in chloroplasts during the light-dependent and light-independent reactions of photosynthesis.

stroma in a chloroplast, the portion of the organelle that is equivalent to the mitochondrial matrix; the site of sugar production in photosynthesis

thylakoid space the portion of the chloroplast between the thylakoid disks

22-1B Chlorophyll Structure and Properties

The primary event in photosynthesis is the absorption of light by chlorophyll. The high-energy states (excited states) of chlorophyll are useful in photosynthesis because the light energy can be passed along and converted into chemical energy in the light reactions. There are two principal types of chlorophyll, *chlorophyll* a and *chlorophyll* b. Eukaryotes such as green plants and green algae contain both chlorophyll *a* and chlorophyll *b*.

Prokaryotes such as cyanobacteria (formerly called blue-green algae) contain only chlorophyll *a*. Photosynthetic bacteria other than cyanobacteria have bacteriochlorophylls, with *bacteriochlorophyll* a being the most common. Organisms such as green and purple sulfur bacteria, which contain bacteriochlorophylls, do not use water as the ultimate source of electrons for the redox reactions of photosynthesis, nor do they produce oxygen. Instead, they use other electron sources such as H_2S, which produces elemental sulfur instead of oxygen. Organisms that contain bacteriochlorophyll are anaerobic and have only one photosystem, whereas green plants have two different photosystems.

The structure of chlorophyll is similar to that of the heme group (Section 4-4B) found in myoglobin, hemoglobin, and the cytochromes, in that it is based on the tetrapyrrole ring of porphyrins (Figure 22.3). In chlorophyll, the metal ion bound to the tetrapyrrole ring is Mg^{2+}, rather than the iron ion that occurs in heme. Another difference between chlorophyll and heme is the presence of a cyclopentanone ring fused to the tetrapyrrole ring. A long hydrophobic side chain, the phytol group, contains four isoprenoid units (five-carbon units that are basic building blocks in many lipids; Section 21-8A) and anchors

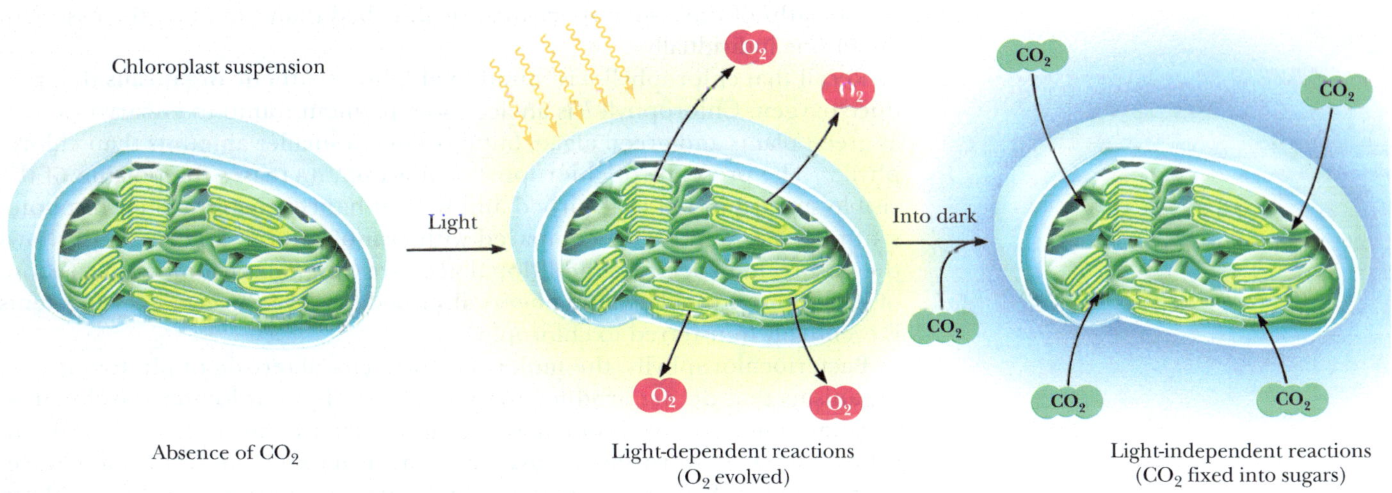

Figure 22.2 The light-dependent and light-independent reactions of photosynthesis. Light-dependent reactions are associated with the thylakoid membranes, whereas light-independent reactions are associated with the stroma.

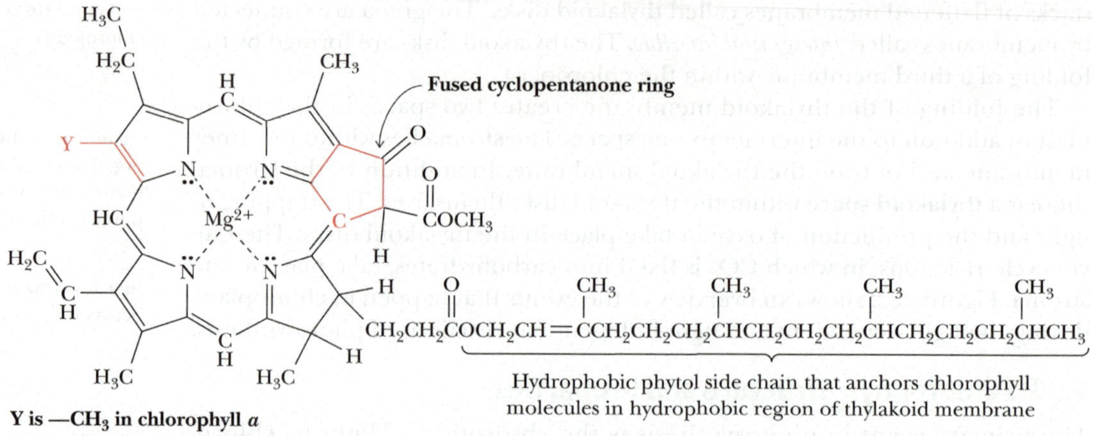

Y is —CH_3 in chlorophyll *a*
Y is —CHO in chlorophyll *b*
Y is —CH_3 in bacteriochlorophyll *a*
(and highlighted bond is saturated)

Figure 22.3 Molecular structures of chlorophyll *a*, chlorophyll *b*, and bacteriochlorophyll *a*.

the chlorophyll to the thylakoid membrane through hydrophobic interactions. The phytol group is covalently bonded to the rest of the chlorophyll molecule by an ester linkage between the alcohol group of the phytol and a propionic acid side chain on the porphyrin ring.

The difference between chlorophyll *a* and chlorophyll *b* lies in the substitution of an aldehyde group for a methyl group on the porphyrin ring. The difference between bacteriochlorophyll *a* and chlorophyll *a* is that a double bond in the porphyrin ring of chlorophyll *a* is saturated in bacteriochlorophyll *a*. The lack of a conjugated system (alternating double and single bonds) in the porphyrin ring of bacteriochlorophylls causes a significant difference in the absorption of light by bacteriochlorophyll *a* compared with chlorophylls *a* and *b*.

In organisms such as green plants and algae that utilize multiple chlorophyll types, chlorophyll *a* is the central pigment to which all absorbed light energy is transferred. Chlorophyll *b* and other pigments that help absorb light energy are called **accessory pigments**. The absorption spectra of chlorophyll *a* and chlorophyll *b* differ slightly (Figure 22.4A). Both absorb light in the red and blue portions of the visible spectrum (600 to 700 nm and 400 to 500 nm, respectively), and the presence of both types of chlorophyll guarantees that more wavelengths of the visible spectrum are absorbed than would be the case with either one individually.

accessory pigments plant pigments other than chlorophyll *a* that play a role in absorbing light energy for photosynthesis

Recall that chlorophyll *a* is found in all photosynthetic organisms that produce oxygen. Chlorophyll *b* is an accessory pigment found in eukaryotes such as green plants and green algae, but it occurs in smaller amounts than chlorophyll *a*. The presence of chlorophyll *b*, however, increases the portion of the visible spectrum that is absorbed and thus enhances the efficiency of photosynthesis in green plants compared with cyanobacteria. Figure 22.4B shows the absorption spectra of additional accessory pigments compared to those of the chlorophylls. The light energy absorbed by all of the accessory pigments is eventually transferred to chlorophyll *a*.

Bacteriochlorophylls, the molecular form characteristic of photosynthetic organisms that do *not* produce oxygen, absorb light at longer wavelengths. The wavelength of maximum absorption of bacteriochlorophyll *a* is 780 nm; other bacteriochlorophylls have absorption maxima at still longer wavelengths such as 870 nm or 1050 nm. Light of wavelength longer than 800 nm is part of the infrared rather than the visible, region of the spectrum. The wavelength of light absorbed plays a critical role in the light reactions of photosynthesis because the energy of light is inversely related to wavelength (see **Biochemical Connections 22A**).

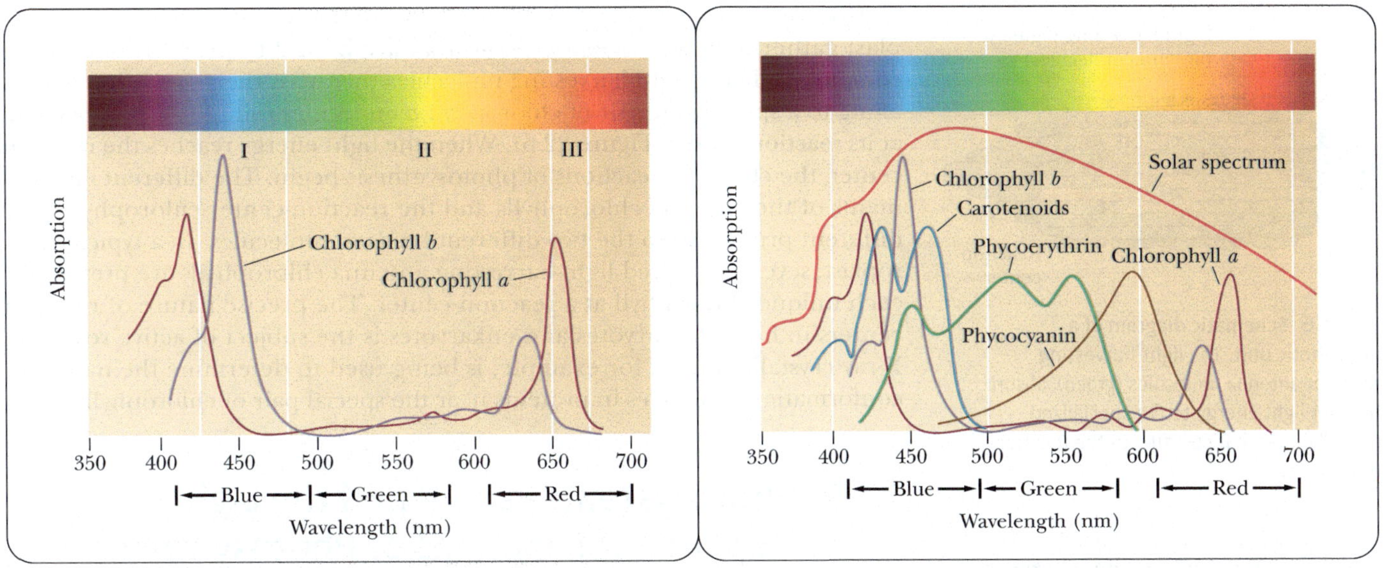

A The absorption of visible light by chlorophylls *a* and *b*. The areas marked I, II, and III are regions of the spectrum that give rise to chloroplast activity. There is greater activity in regions I and III, which are close to major absorption peaks. There are high levels of O_2 production when light from regions I and III is absorbed by chloroplasts. Lower (but measurable) activity is seen in region II, where some of the accessory pigments absorb.

B The absorption of light by accessory pigments compared to that of chlorophyll *a*. The accessory pigments absorb light of wavelengths across the solar spectrum that chlorophyll *a* does not absorb, and they transfer their energy to chlorophyll *a*.

Figure 22.4 **The visible absorption spectrum of chlorophyll *a* compared with those of various accessory pigments.**

22A Biochemical Connections • Physics

The Relationship Between Wavelength and Energy of Light

The equation that relates the wavelength and energy of light is of crucial importance for our purposes. Early in the twentieth century, Max Planck established that the energy of light is directly proportional to its frequency.

$$E = hv$$

where E is energy, h is a constant (the Planck constant), and v is the frequency of the light. Because the wavelength of light is related to the frequency by the following equation,

$$v = \frac{c}{\lambda}$$

where λ is wavelength, v is frequency, and c is the velocity of light, we can rewrite the expression for the energy of light in terms of wavelength rather than frequency.

$$E = hv = \frac{hc}{\lambda}$$

Therefore, light of shorter wavelength (higher frequency) is higher in energy than light of longer wavelength (lower frequency; Figure 22.5). ◗

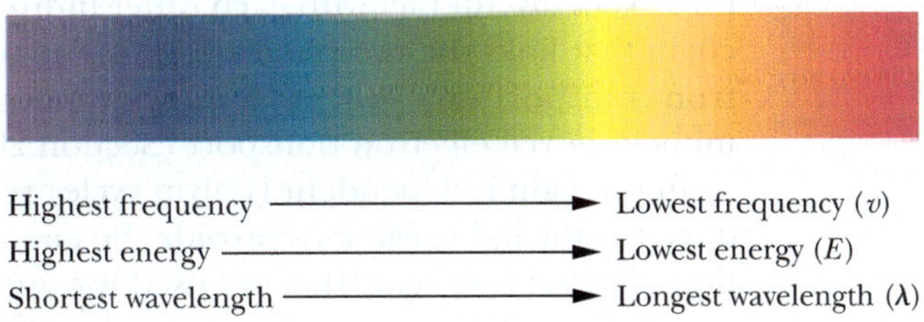

Figure 22.5 **The relationships among frequency, energy, and wavelength of light.** In the visible spectrum, violet light has a shorter wavelength (λ), higher frequency (v), and higher energy (E) than red light. Therefore, smaller amounts of energy are absorbed for colors other than violet in the visible spectrum.

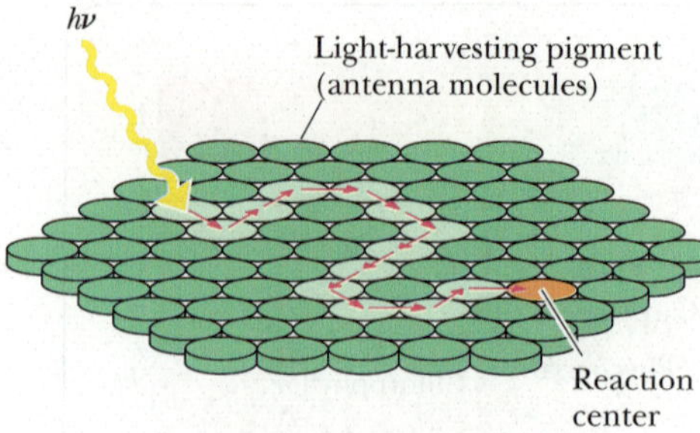

Figure 22.6 Schematic diagram of a photosynthetic unit. The light-harvesting pigments, or antenna molecules (green), absorb and transfer light energy to the specialized chlorophyll dimer that constitutes the reaction center (orange).

antenna complexes membrane-bound pigment and protein complexes that capture light energy and transfer it to special pairs of chlorophyll *a* molecules

photosystem a membrane-bound protein complex that carries out the light reactions of photosynthesis

reaction center the site in a photosystem where a special pair of chlorophyll *a* molecules converts light energy from the Sun into chemical energy

Most of the chlorophyll *a* and accessory pigment molecules in a chloroplast gather light as *antenna pigments* that are bound to proteins in **antenna complexes**. These light-harvesting molecules then pass their excitation energy along to a specialized pair of chlorophyll *a* molecules found in a **photosystem** at its **reaction center** (Figure 22.6). When the light energy reaches the reaction center, the chemical reactions of photosynthesis begin. The different environments of the antenna chlorophylls and the reaction-center chlorophylls give different properties to the two different kinds of molecules. In a typical chloroplast, several hundred light-harvesting antenna chlorophylls are present for each unique chlorophyll at a reaction center. The precise nature of reaction centers in both prokaryotes and eukaryotes is the subject of active research. X-ray crystallography, for example, is being used to determine the nature of conformational changes in proteins near the special pair of chlorophylls.

22-2 Photosystems I and II and the Light-Dependent Reactions of Photosynthesis

Learning Objectives

After reading this section, students will be able to:

LO 22.2.1 Describe the electron transfer reactions and energy-harnessing reactions of the light-dependent reactions of photosynthesis.

LO 22.2.2 Explain the S-state cycle in photosystem II.

LO 22.2.3 Discuss cyclic photophosphorylation in photosystem I.

LO 22.2.4 Distinguish between anoxygenic photosystems and oxygenic photosystems.

22-2A Overview of the Light-Dependent Reactions in Photosynthesis

photophosphorylation the process in which light energy is used to phosphorylate ADP to ATP

In the light-dependent reactions of photosynthesis, light energy is used to oxidize water to oxygen, and $NADP^+$ is reduced to NADPH. The series of redox reactions is coupled to the phosphorylation of adenosine diphosphate (ADP) to adenosine triphosphate (ATP) in a process called **photophosphorylation**.

$$H_2O + NADP^+ \rightarrow NADPH + H^+ + O_2$$
$$ADP + P_i \rightarrow ATP$$

photosystem I (PSI) the portion of the photosynthetic apparatus responsible for the production of NADPH

photosystem II (PSII) the portion of the photosynthetic apparatus responsible for the splitting of water to produce oxygen

The light reactions consist of two parts, accomplished by two distinct but related photosystems. One part of the reaction is the reduction of $NADP^+$ to NADPH, carried out by **photosystem I (PSI)**. The second part of the reaction is the oxidation of water to produce oxygen, carried out by **photosystem II (PSII)**. Both photosystems carry out redox (electron transfer) reactions, and the two photosystems interact with each other indirectly through an electron transport chain that links the two photosystems. The production of ATP is linked to electron transport in a process similar to that seen in the production of ATP by mitochondrial electron transport (Section 20-4).

In the light-independent (Calvin cycle) reactions, the ATP and NADPH produced in the light reactions provide the energy and reducing power for the fixation of CO_2 into sugars that are used for organism growth, energy storage, and metabolism (discussed further in Section 22-5). The Calvin cycle reactions also constitute a redox process because the carbon in carbohydrates is in a more reduced state than the highly oxidized carbon in CO_2. The light-dependent and light-independent reactions do not take place separately, but they are separated for purposes of discussion only.

The net electron transport reaction of the two photosystems taken together is, except for the substitution of NADPH for NADH, the reverse of mitochondrial electron transport. The half-reaction of reduction is that of $NADP^+$ to NADPH, whereas the half-reaction of oxidation is that of water to oxygen.

$$NADP^+ + 2H^+ + 2e^- \rightarrow NADPH + H^+$$
$$H_2O \rightarrow \frac{1}{2}O_2 + 2H^+ + 2e^-$$
$$\overline{NADP^+ + H_2O \rightarrow NADPH + H^+ + \frac{1}{2}O_2}$$

This is an endergonic reaction with a positive $\Delta G°' = +220$ kJ/mol = +52.6 kcal/mol. The light energy absorbed by the chlorophylls in both photosystems provides the energy that allows this endergonic reaction to take place. A series of electron carriers embedded in the thylakoid membrane links these reactions. The electron carriers have an organization very similar to that of the carriers in the electron transport chain.

Photosystem I can be excited by light of wavelengths shorter than 700 nm, but photosystem II requires light of wavelengths shorter than 680 nm for excitation. Both photosystems must operate for the chloroplast to produce NADPH, ATP, and O_2 because the two photosystems are connected by the electron transport chain. The two systems are, however, structurally distinct in the chloroplast; photosystem I can be released preferentially from the thylakoid membrane by treatment with detergents. The reaction centers of the two photosystems provide different environments for the unique chlorophylls involved.

22-2B The Z-Scheme of Photosynthesis and Photophosphorylation of ADP to ATP

The **Z-scheme of photosynthesis** describes the energy flow of electrons through photosystem II (PSII) and photosystem I (PSI) during the light-dependent reactions of photosynthesis (Figure 22.7A). P_{680} is a specialized pair of chlorophyll *a* molecules in the reaction center of PSII (Figure 22.7A, *bottom left*). The reaction center of PSII is designated P_{680} because the longest wavelength of absorbed light that initiates its reactions is 680 nm. When P_{680} absorbs light energy, an electron is excited to a higher energy state (represented as P_{680}*), and the excited electron is transferred to an electron transport chain. P_{680}* is a strong oxidizing agent that pulls electrons from water in the **oxygen-evolving complex (OEC)** of PSII, producing oxygen and reducing P680 so that an electron can be excited again by light energy. The electron donor of the OEC is labeled "D" in the lower left portion of Figure 22.7A. This oxidation of water also releases protons into the lumen, which helps create a proton gradient for ATP production.

Z-scheme of photosynthesis a diagram of the energy flow of electrons through photosystem II and photosystem I during the light-dependent reactions of photosynthesis

oxygen-evolving complex (OEC) the part of photosystem II where water is oxidized, producing oxygen

We now return to the electron transport chain reactions that occur after P_{680}* transfers its excited electron:

- **Pheophytin (pheo)** is a modified form of chlorophyll that is referred to as a *primary electron acceptor* because it captures the electron from P_{680}* and passes it on to a series of other acceptors.
- The first of these acceptors is Q_A, which is attached to PSII, and the next is Q_B, which is also attached to PSII.
- Q_A and Q_B are bound plastoquinones, and Q_B is a binding site for a mobile plastoquinone called **plastoquinone (PQ)**. PQ accepts two electrons to become PQH_2 (after two rounds of PSII excitation) and carries the electrons to the cytochrome b_6f complex (Figure 22.7A, *middle*).
- Electron transfer reactions in the cytochrome b_6f complex cause protons from PQH_2 oxidation to be pumped into the lumen to help create a proton gradient for ATP production. Then, one electron moves forward to another

pheophytin (pheo) a photosynthetic pigment that differs from chlorophyll only in having two hydrogen atoms in place of a magnesium ion

plastoquinone (PQ) a substance similar to coenzyme Q; it is part of the electron transport chain that links the two photosystems in photosynthesis

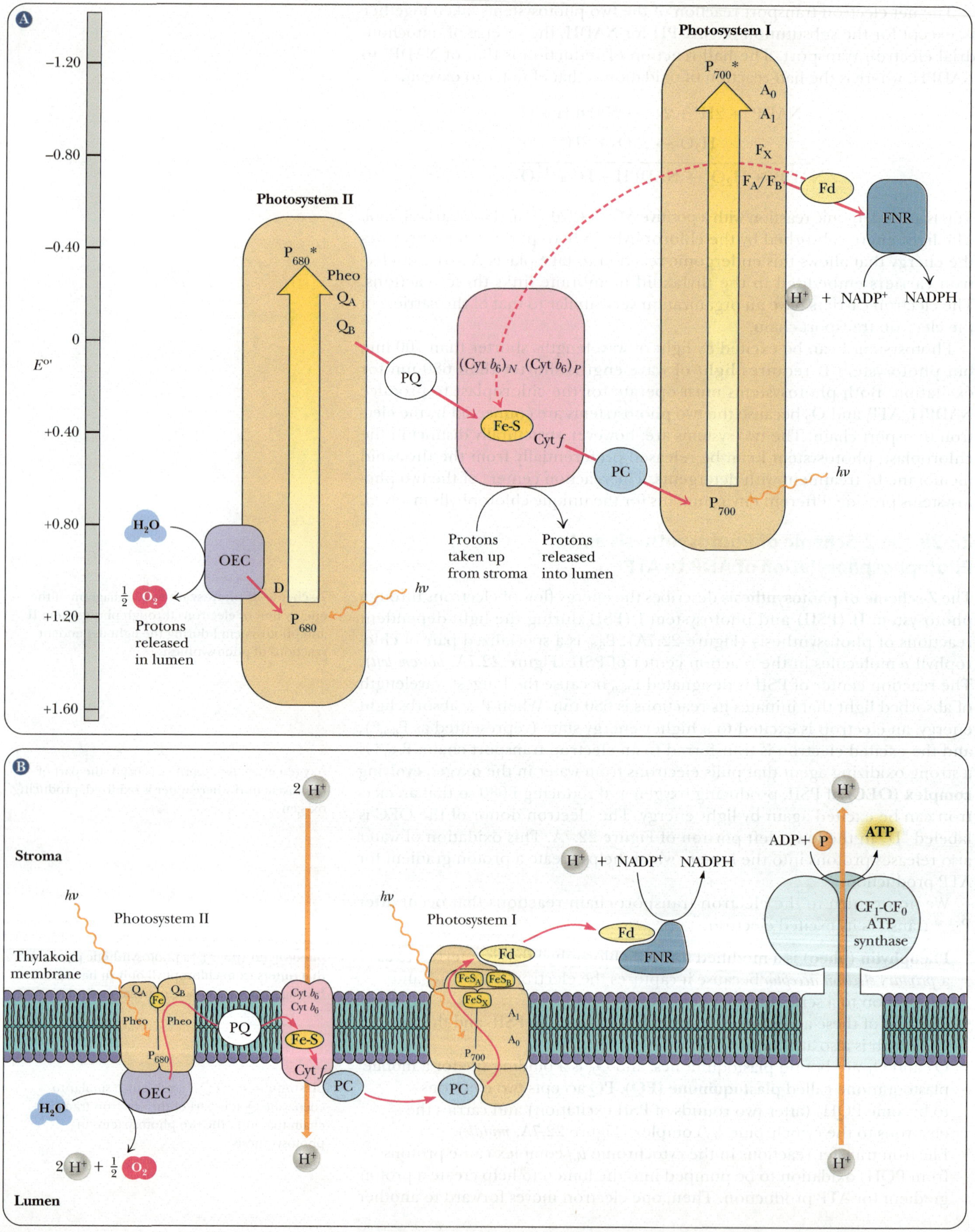
A
−1.20
−0.80
−0.40
0
+0.40
+0.80
+1.20
+1.60
E°'
Photosystem II
P680*
Pheo
QA
QB
D
P680
OEC
H2O
1/2 O2
Protons released in lumen
hν
PQ
(Cyt b6)N (Cyt b6)P
Fe-S
Cyt f
PC
Protons taken up from stroma
Protons released into lumen
Photosystem I
P700*
A0
A1
FX
FA/FB
P700
hν
Fd
FNR
H+ + NADP+
NADPH
B
Stroma
Thylakoid membrane
Lumen
hν
Photosystem II
QA
Fe
QB
Pheo
Pheo
P680
OEC
H2O
2 H+ + 1/2 O2
PQ
2 H+
Cyt b6
Cyt b6
Fe-S
Cyt f
PC
H+
hν
Photosystem I
Fd
FeSA
FeSB
FeSX
A1
A0
P700
PC
Fd
FNR
H+ + NADP+ NADPH
H+
ADP + P
ATP
CF1-CF0 ATP synthase
H+

Figure 22.7 The Z-scheme of photosynthesis. (A) The Z-scheme is a diagrammatic representation of photosynthetic electron flow from H_2O to $NADP^+$. The energy relationships can be derived from the $E^{\circ\prime}$ scale beside the Z diagram, with lower standard potentials and therefore greater energy as you go from bottom to top. Energy input as light is indicated by two broad arrows, one photon appearing in P_{680} and the other in P_{700}. P_{680}* and P_{700}* represent excited states. Electron loss from P_{680}* and P_{700}* creates P_{680} and P_{700}. The representative components of the three supramolecular complexes (photosystem II, the cytochrome b_6f complex, and photosystem I) are shown in the larger shaded boxes enclosed by solid black lines. A number of components are represented by letters of the alphabet—chlorophylls and quinones by *A* and *Q*, respectively, and iron–sulfur clusters by *F*—and are further distinguished by subscripts. OEC, PQ, PC, Fd, and FNR refer to the oxygen-evolving complex, plastoquinone, plastocyanin, ferredoxin, and the ferredoxin–$NADP^+$ reductase complex, respectively. Proton translocations that establish the proton-motive force driving ATP synthesis are illustrated as well. (B) This figure shows the functional relationships among PSII, the cytochrome b_6f complex, PSI, and the photosynthetic chloroplast (CF_1-CF_0) ATP synthase complex within the thylakoid membrane. The electron acceptors Q_A (for PSII) and A_1 (for PSI) are at the stromal side of the thylakoid membrane, whereas the electron donors to P_{680} and P_{700} are situated at the lumenal side of the membrane. The consequence is charge separation (stroma, lumen) across the membrane. Also, protons are translocated into the thylakoid lumen, giving rise to a chemiosmotic gradient that is the driving force for ATP synthesis by CF_1CF_0–ATP synthase.

mobile carrier called **plastocyanin (PC)**, which transfers the electron to the reaction center of PSI called P_{700}.

- P_{700} is a specialized pair of chlorophyll *a* molecules, and it is designated P_{700} because the longest wavelength of absorbed light that initiates its reactions is 700 nm. The electron that is now on P_{700} is ready for another energy boost from light absorption that converts P_{700} to P_{700}*.
- P_{700}* then transfers the excited electron to another *primary electron acceptor* that is a specialized chlorophyll *a* molecule called A_0.
- Another electron transport chain is then initiated and continues from A_0 through a series of electron acceptors to ferredoxin (Fd).
- Fd carries the electron to the ferredoxin–$NADP^+$ reductase complex (FNR), which catalyzes the reduction of $NADP^+$ to NADPH. NADPH is then used in the light-independent (Calvin cycle) reactions of photosynthesis to fix CO_2 into sugars (discussed in Section 22-5).

plastocyanin (PC) a copper-containing protein; it is part of the electron transport chain that links the two photosystems in photosynthesis

The Z-scheme helps us see that there are two places in the light-dependent reactions of photosynthesis (in PSII and in PSI) where the absorption of light supplies energy to make endergonic electron transfer and endergonic proton release/pumping reactions take place. The "Z" of the scheme is rather lopsided and lies on its side. The energy differences in these energy absorption and release processes are shown on the vertical axis of Figure 22.7A.

Now that we have finished an initial discussion of the Z-scheme of the light-dependent reactions of photosynthesis, let us connect these reactions to the production of ATP, which, along with the NADPH produced, will be used in the Calvin cycle reactions of photosynthesis to fix CO_2 into sugars. Panel B of Figure 22.7 shows functional relationships between PSII, the cytochrome b_6f complex, PSI, and FNR in the thylakoid membrane, as well as the connection to ATP production by the *chloroplast (CF_1-CF_0) ATP synthase complex.* In the lower left portion of Figure 22.7B, notice that the production of a proton gradient in the lumen is caused

- by the release of protons in the oxidation of water catalyzed by the OEC and
- by the protons that are pumped by the cytochrome b_6f complex.

These protons flow through the CF_1-CF_0 ATP synthase complex back into the stroma, resulting in the photophosphorylation of ADP to ATP (Figure 22.7B, *right*).

▶ What are some additional details related to photosystem II?

As seen in Figure 22.7A, the path of electron excitation and transfer starts with the reactions in PSII rather than in PSI. The reason for the nomenclature is that PSI was studied extensively at an earlier date than PSII because it is easier to extract PSI from the thylakoid membrane than it is to extract PSII.

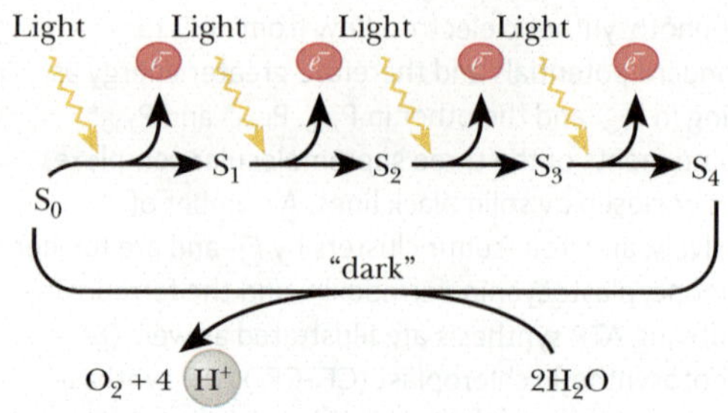

Figure 22.8 The Kok cycle—also called the S-state cycle in photosystem II (PSII). The PSII reaction center passes through five different oxidation states, S_0 through S_4, in the course of oxygen evolution.

The oxidation of water by PSII to produce oxygen is the ultimate source of electrons in photosynthesis. These electrons are subsequently passed from PSII to PSI by the electron transport chain. The electrons from water are needed to "fill the hole" that is left when the absorption of one photon of light leads to the donation of an electron from PSII to the electron transport chain.

The electrons released by the oxidation of water are first passed to P_{680}, which is reduced. There are intermediate steps in this reaction because four electrons are required for the oxidation of water, and P_{680}* can accept only one electron at a time. A manganese-containing protein complex and several other protein components are required. The OEC of PSII passes through a series of five oxidation states (designated as S_0 through S_4) in the transfer of four electrons in the process of evolving oxygen (Figure 22.8). This happens in a cycle known as the *Kok cycle* or the *S-state cycle.*

In the Kok cycle, one electron is passed from water to PSII for each quantum of light. In the process, the components of the reaction center go successively through oxidation states S_1 through S_4. The S_4 state decays spontaneously to the S_0 state and, in the process, oxidizes two water molecules to one oxygen molecule, and four protons are released simultaneously. The immediate electron donor to the P_{680} chlorophyll, shown as D in Figure 22.7, is a tyrosine residue of one of the protein components that does not contain manganese. Several quinones serve as intermediate electron transfer agents to accommodate the four electrons donated by one water molecule. Redox reactions of manganese also play a role here. Even this mechanism is an oversimplification. Attempts to observe the direct production of oxygen by the S_4 state imply that some intermediate (S_4') directly produces oxygen after deprotonation of S_3 and loss of an electron by S_4. The main point is that the oxygen-evolving complex is very complex indeed.

In PSII, the primary (or first) electron acceptor is a molecule of pheophytin, one of the accessory pigments of the photosynthetic apparatus (Figure 22.7A, *top left*). The structure of pheophytin differs from that of chlorophyll only in the substitution of two hydrogens for magnesium (to review the chlorophyll structure, see Figure 22.3). The transfer of electrons is mediated by events that take place at the reaction center. The next electron acceptor is plastoquinone (PQ). The structure of plastoquinone (Figure 22.9) is similar to that of coenzyme Q (ubiquinone), a part of the respiratory electron transport chain (Section 20-3), and plastoquinone serves a very similar purpose in the photosynthetic transfer of electrons and hydrogen ions. When PQ accepts two electrons, the carbonyl groups become hydroxyl groups to form PQH_2, the reduced form of PQ.

▶ What are some additional details related to the cytochrome b_6f complex and the electron transport chain that connects the two photosystems?

The electron transport chain that links the two photosystems consists of pheophytin, plastoquinone, a complex of plant cytochromes (the cytochrome b_6f complex), a copper-containing protein called plastocyanin (PC), and the oxidized form of P_{700} (see Figure 22.7A, *middle*). The b_6f complex of plant cytochromes consists of two *b*-type cytochromes (cytochrome b_6) and a *c*-type cytochrome (cytochrome *f*). This complex is similar in structure to the cytochrome bc_1 complex in mitochondria and occupies a similar central position in an electron transport chain. This part of the photosynthetic apparatus is the subject of active research. A Q-cycle (recall this from Section 20-3) operates here as well. In plastocyanin, the copper ion is the actual electron carrier; the copper ion exists as Cu(II) and Cu(I) in the oxidized and reduced forms, respectively. This photosynthetic electron transport chain has another similarity to the respiratory one in mitochondria—that of coupling to ATP generation.

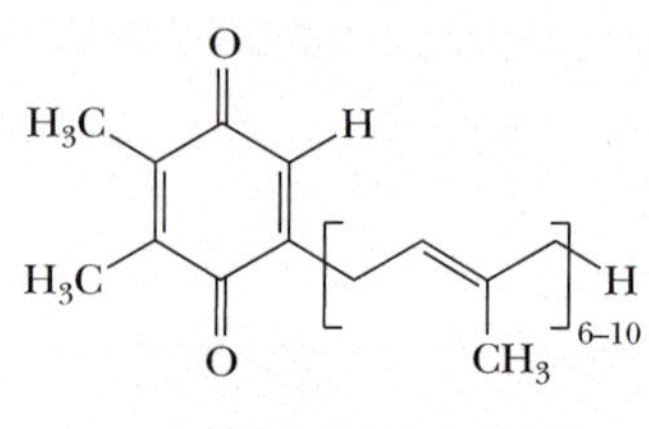

Figure 22.9 The structure of plastoquinone (PQ). When PQ accepts two electrons, the carbonyl groups become hydroxyl groups to form PQH_2, the reduced form of PQ. The length of the aliphatic side chain varies in different organisms.

▶ What are some additional details related to photosystem I?

When the oxidized special chlorophyll *a* pair of P_{700} in the reaction center of PSI accepts electrons from the electron transport chain, it is reduced and ready to absorb a photon of light. Absorption of light by PSII does not raise the electrons to a high enough energy level to reduce $NADP^+$, so the additional photon absorbed by PSI provides the remaining needed energy.

The absorption of light by P_{700} then leads to the series of electron transfer reactions of PSI (Figure 22.7A, *top right*). The F_X and F_A/F_B electron acceptors that occur after the A_1 acceptor in the series are iron–sulfur clusters that are bound to PSI. Recall that we saw iron–sulfur clusters in respiratory electron transport in Section 20-3C. The F_A/F_B notation represents two iron–sulfur clusters that are so close together on one of the PSI subunits that the electron can pass through either one of them on the way to the mobile ferredoxin (Fd) electron carrier. Soluble ferredoxin in turn reduces an FAD-containing enzyme called the ferredoxin–$NADP^+$ reductase complex (FNR). The FAD portion of the FNR enzyme reduces $NADP^+$ to NADPH (Figure 22.7A, *top right*). We can summarize the main features of the PSI processes in two equations, in which the notation ferredoxin refers to the soluble form of the protein.

$$P_{700}^* + \text{Ferredoxin}_{\text{oxidized}} \rightarrow P_{700} + \text{Ferredoxin}_{\text{reduced}}$$

$$2\text{Ferredoxin}_{\text{reduced}} + H^+ + NADP^+ \rightarrow 2\text{Ferredoxin}_{\text{oxidized}} + NADPH$$

P_{700}^* donates one electron to ferredoxin, but the electron transfer reactions of FAD and $NADP^+$ involve two electrons. Thus, an electron from each of two ferredoxins is required for the production of NADPH.

The net reaction for the two photosystems together is the flow of electrons from H_2O (Figure 22.7A, *bottom left*) to $NADP^+$ (Figure 22.7A, *top right*).

$$2H_2O + 2NADP^+ \rightarrow O_2 + 2NADPH + 2H^+$$

22-2C Cyclic Photophosphorylation in Photosystem I

In addition to the electron transfer reactions just described, it is also possible for excited electrons from PSI to cycle back to the cytochrome b_6f complex using soluble ferredoxin (Fd) as the electron ferry in a process called **cyclic photophosphorylation** (see the dashed line in Figure 22.7A that goes from Fd back to the cytochrome b_6f complex, as well as details in Figure 22.10). This process boosts ATP production due to the protons that are pumped by the cytochrome b_6f complex. This boost in ATP production happens without generating NADPH (Figure 22.7A, *top right*) or splitting water to produce oxygen (Figure 22.7A, *bottom left*) because the electrons are rerouted from going to the ferredoxin–$NADP^+$ reductase complex (FNR), and they do not originate from PSII. Cyclic photophosphorylation takes place when there is a high NADPH/$NADP^+$ ratio in the cell (i.e., when there is not enough $NADP^+$ present in the cell to accept all the electrons generated by the excitation of P_{700}). Researchers are still trying to elucidate the exact details of this process. For example, a mysterious membrane protein that is not a part of the cytochrome b_6f complex somehow transfers the electrons from the soluble ferredoxin protein to the mobile membrane carrier called plastoquinone (PQ). The mystery protein is currently referred to as *ferredoxin–plastoquinone reductase (FQR)* (not shown in Figure 22.10).

cyclic photophosphorylation a light-driven process in photosynthesis in which excited electrons from photosystem I are cycled back to the cytochrome b_6f complex, contributing to ATP production without generating NADPH or oxygen

22-2D Anoxygenic Photosystems

The molecular structure and reactions of photosystems are a subject of intense interest to biochemists. Early on in photosynthesis studies, the most extensively studied system was that from anaerobic phototrophic bacteria of the genus

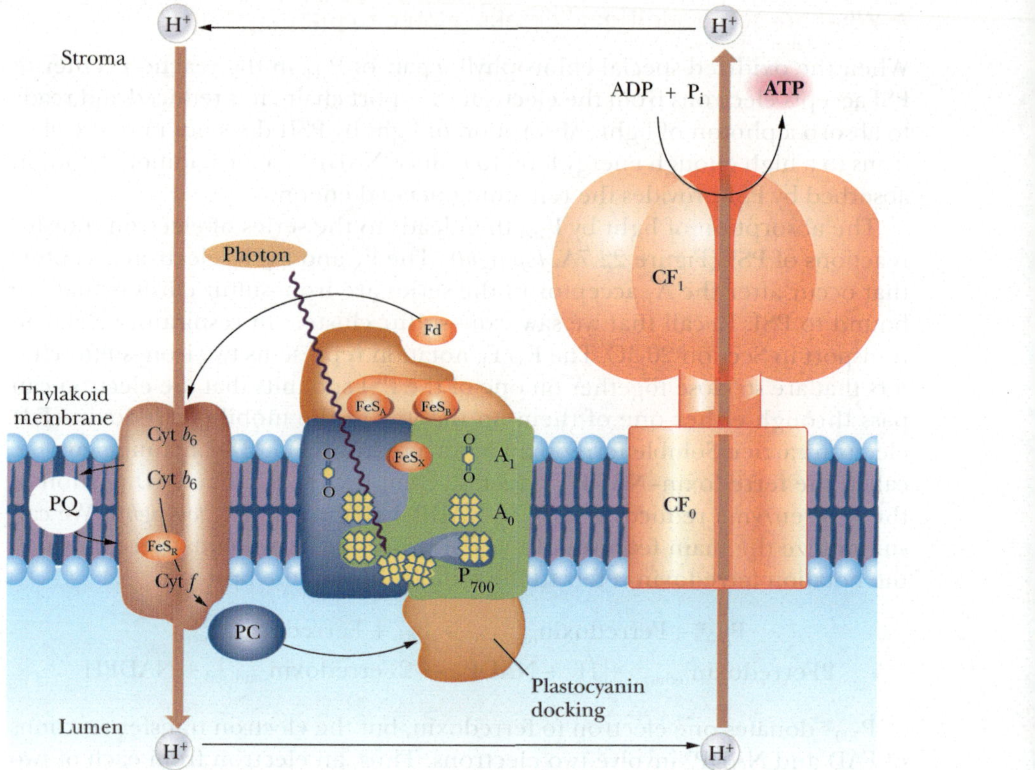

Figure 22.10 The cyclic photophosphorylation pathway that boosts ATP production. The mobile electron carrier ferredoxin (Fd) carries excited electrons from photosystem I (*center*) to the cytochrome b_6f complex (*left*), where electron transfer reactions that lead back to photosystem I pump protons into the lumen, which can then flow through the chloroplast (CF_1-CF_0) ATP synthase complex to phosphorylate ADP to ATP. Water is not split into oxygen, and no NADPH is produced when this pathway is used. *(Adapted from Arnon, D. I. (1984). The discovery of photosynthetic phosphorylation.* Trends Biochem. Science *9, 258–262, with permission from Elsevier.)*

Rhodopseudomonas. These bacteria do not produce molecular oxygen as a result of their photosynthetic activities, but enough similarities exist between the photosynthetic reactions of *Rhodopseudomonas* and photosynthesis linked to oxygen to lead scientists to draw conclusions about the nature of reaction centers in all organisms. Because the structure of **anoxygenic photosystems**—like the one from *Rhodopseudomonas viridis* (*R. viridis*) modeled in Figure 22.11A—has been elucidated by X-ray crystallography, the structures of PSI and PSII in *oxygenic photosynthesis* have also been determined and have been shown to be markedly similar. The detailed process that goes on at the reaction center of the photosystem from *R. viridis* in the absence of oxygen gives us a more well-rounded view of the diversity of photosynthetic processes found in different organisms.

anoxygenic photosystems photosystems used by certain types of photosynthetic organisms (such as some bacteria) that do not produce oxygen during photosynthesis

In the reaction center of the *R. viridis* photosystem, there is a pair of bacteriochlorophyll *a* molecules (designated P_{870} because light of 870 nm is the maximum excitation wavelength). This critical pair of chlorophylls is embedded in a protein complex that is, in turn, an integral part of the photosynthetic membrane. (We refer to the bacteriochlorophylls simply as chlorophylls in order to simplify the discussion.)

Accessory pigments, which also play a role in the light-trapping process, have specific positions close to the special pair of chlorophylls (P_{870}). The absorption of light by the special pair of chlorophylls raises one of their electrons to a higher energy level (Figure 22.11A, *middle*). This electron is passed on to a series of accessory pigments. The first of these accessory pigments is a bacteriochlorophyll *a* monomer called B_A, which is located in the L polypeptide of the reaction center (there is another bacteriochlorophyll *a* monomer called B_B that is located in the M polypeptide of the reaction center, but it is not used under normal conditions). Then the electron is passed to a bacteriopheophytin *a* (H_A) molecule, which is structurally similar to chlorophyll, differing only in having two hydrogens in place of the magnesium. The next electron acceptor is a quinone (Q_A), which is located in the M polypeptide of the reaction center.

The final electron acceptor is another quinone (Q_B), which is located near the L polypeptide of the reaction center. An iron atom helps to facilitate the electron transfer between Q_A and Q_B, converting Q_B into a quinol (QH_2). Figure 22.12 shows the structures of two quinones (menaquinone and ubiquinone); the exact identities of the quinones that are used as Q_A and Q_B in anoxygenic reaction centers vary depending on the organism. In the structure of QH_2, the ketone groups of the quinone that is used as Q_B are reduced to hydroxyl groups. QH_2 then enters a Q-cycle that eventually passes the electron to the cytochrome complex, which transfers the electron back to P870, thereby completing a cyclic electron flow that pumps protons for ATP synthesis. The cytochrome complex is bound to the membrane by a diacylglycerol moiety (Figure 22.11B).

The separation of charge is equivalent to a battery, a form of stored energy. The reaction center has acted as a transducer, converting light energy to a form usable by the cell to carry out the energy-requiring reactions of photosynthesis. These processes that take place in the reaction center of the *R. viridis* photosystem serve as a model for reaction centers in photosynthesis linked to oxygen. Research has suggested that a specific tyrosine residue in polypeptide L moves when the special pair of chlorophylls is activated by light. This conformational change is important in stabilizing the charge separation that provides energy for the reactions that require energy.

22-3 Photosynthesis and ATP Production

Learning Objectives

After reading this section, students will be able to:

LO 22.3.1 Describe how photosynthesis and ATP production are coupled.

LO 22.3.2 Explain experimental evidence suggesting the coupling of a proton gradient and ATP production in chloroplasts.

In Section 20-4, we saw that a proton gradient across the inner mitochondrial membrane drives the phosphorylation of ADP in respiration. In photosynthesis, the mechanism of photophosphorylation is essentially the same as that of the production of ATP in the respiratory electron transport chain. In fact, some of the strongest evidence for the chemiosmotic coupling of phosphorylation to electron transport has been obtained from experiments on chloroplasts rather than mitochondria.

Chloroplasts can synthesize ATP from ADP and P_i *in the dark* if they are provided with a pH (or proton) gradient (Figure 22.13). If isolated chloroplasts are allowed to equilibrate in a pH 4 buffer for several hours, their internal pH will be equal to 4. If the pH of the buffer is raised rapidly to 8 and ADP and P_i are added simultaneously, ATP will be produced. The production of ATP does not require the presence of light; the proton gradient produced by the pH difference supplies the driving force for phosphorylation. This experiment provides solid evidence for the chemiosmotic coupling mechanism.

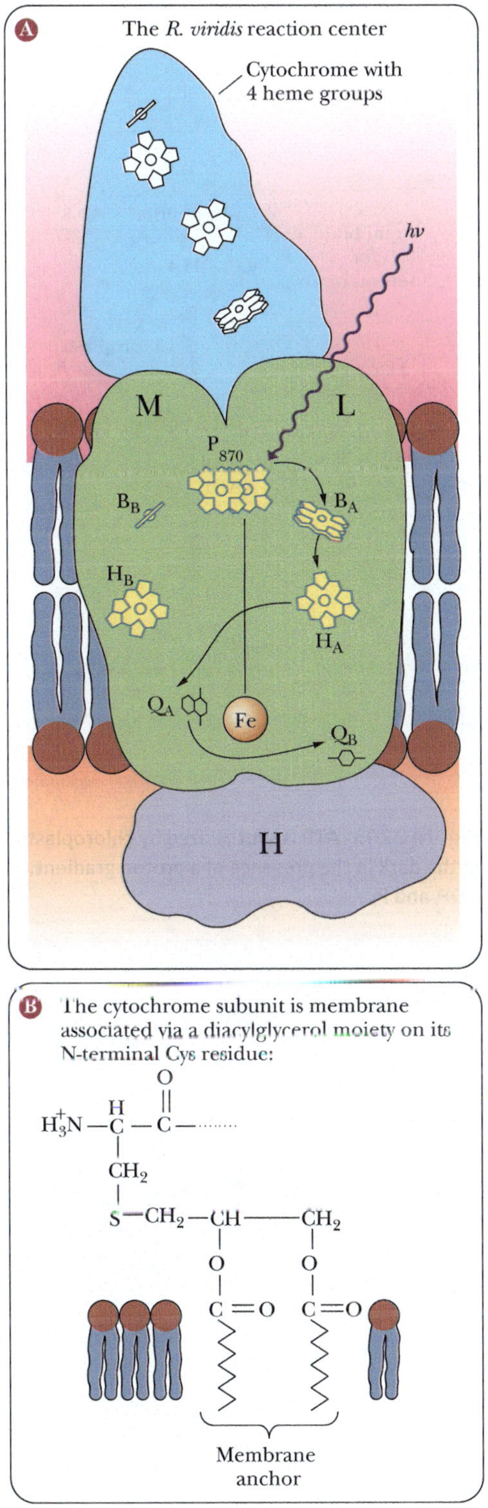

Figure 22.11 Model of the structure and activity of the *Rhodopseudomonas viridis* reaction center. Four polypeptides (designated cytochrome, M, L, and H) make up the reaction center, an integral membrane complex. The M and L polypeptides both consist of five membrane-spanning α-helices, and the H polypeptide has a single membrane-spanning α-helix. The prosthetic groups are spatially situated so that rapid electron transfer from P_{870}* to the Q_B quinone is facilitated. Photoexcitation of P_{870} leads to the reduction of the L-branch B_A only. P_{870} is re-reduced via an electron provided through the heme groups of the cytochrome.

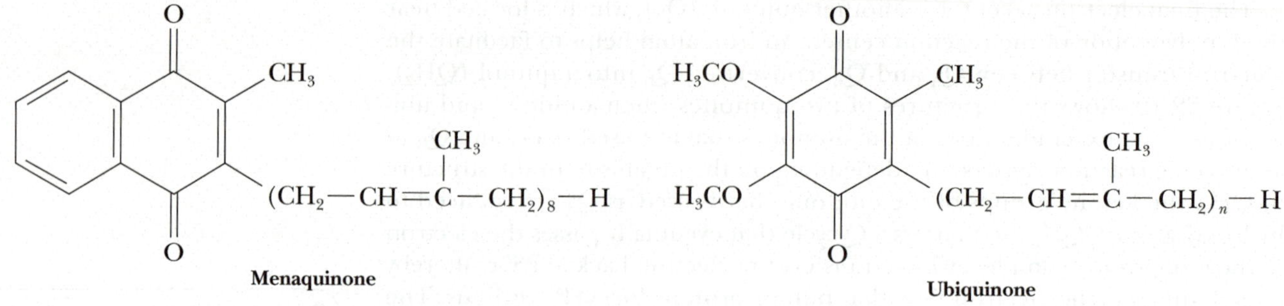

Figure 22.12 The structures of two kinds of quinones that are used as electron acceptors in anoxygenic reaction centers.

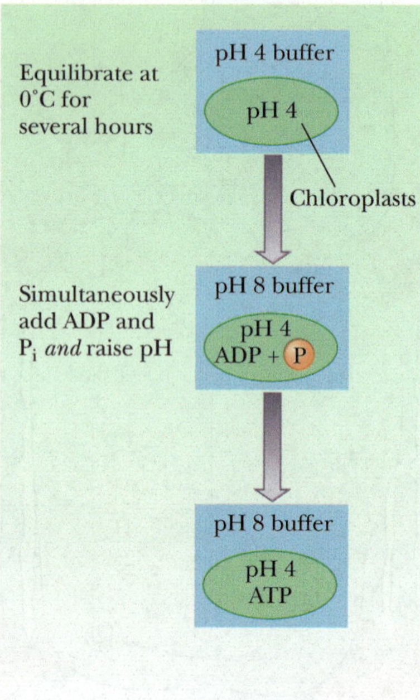

Figure 22.13 ATP is synthesized by chloroplasts in the dark in the presence of a proton gradient, ADP, and P_i.

▶ How does ATP production in chloroplasts resemble the process in mitochondria?

Several reactions contribute to the generation of a proton gradient and ATP production in chloroplasts in an actively photosynthesizing cell (Figure 22.14). The oxidation of water releases H^+ into the thylakoid space. Electron transport from photosystem II (PSII) and photosystem I (PSI) also helps create the proton gradient by involving plastoquinone and cytochromes in the process. Then PSI reduces $NADP^+$ by using H^+ in the stroma to produce NADPH. As a result, the pH of the thylakoid space is lower than that of the stroma. We saw a similar situation in Chapter 20 when we discussed the pumping of protons from the mitochondrial matrix into the intermembrane space. The ATP synthase in chloroplasts is similar to the mitochondrial enzyme; in particular, it consists of two parts, CF_1 and CF_0, where the C stands for "chloroplast" and serves to distinguish them from their mitochondrial counterparts, F_1 and F_0, respectively. Evidence exists that the components of the electron transport chain in chloroplasts are arranged asymmetrically in the thylakoid membrane, as is the case in mitochondria. An important consequence of this asymmetrical arrangement is the release of ATP and NADPH produced by the light reactions into the stroma, where they provide energy and reducing power for the Calvin cycle reactions of photosynthesis.

In mitochondrial electron transport, four respiratory complexes are connected by soluble electron carriers. Similarly, the electron transport apparatus of the thylakoid membrane consists of several large membrane-bound complexes: PSII, the cytochrome b_6f complex, and PSI. As in mitochondrial electron transport, several soluble electron carriers form the connection between the protein complexes. In the thylakoid membrane, the soluble carriers are plastoquinone and plastocyanin, which have a role similar to that of coenzyme Q and cytochrome *c* in mitochondria (Figure 22.14). The proton gradient created by electron transport drives the synthesis of ATP in chloroplasts, as in mitochondria.

22-4 Evolutionary Implications of Photosynthesis With and Without Oxygen

Learning Objectives

After reading this section, students will be able to:

LO 22.4.1 Compare and contrast photosynthesis with and without oxygen.

LO 22.4.2 Write a photosynthesis reaction with a hydrogen donor other than water.

LO 22.4.3 Write a photosynthesis reaction with a hydrogen acceptor other than CO_2.

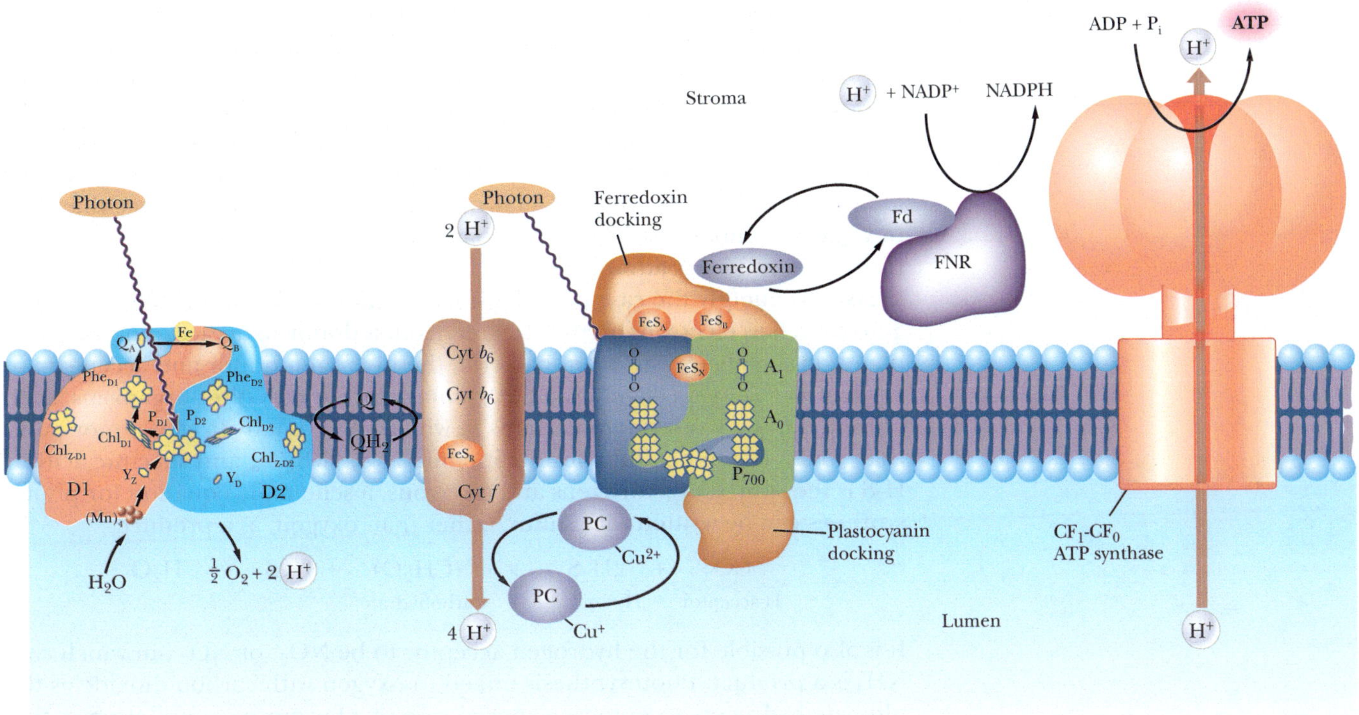

Figure 22.14 **The mechanism of photophosphorylation in chloroplasts.** Photosynthetic electron transport establishes a proton gradient that is used by the CF_1-CF_0 ATP synthase to drive ATP synthesis. Critical to this mechanism is the fact that the membrane-bound components of light-induced electron transport and ATP synthesis are asymmetric with respect to the thylakoid membrane so that directional discharge and uptake of H^+ ensue, generating the proton-motive force. The number of protons pumped through the ATP synthase varies by species and is the subject of active research. ATP and NADPH produced by these processes are used in the Calvin cycle reactions of photosynthesis (see Section 22-5).

Photosynthetic prokaryotes other than cyanobacteria have only one photosystem and do not produce oxygen (recall the discussion of the *R. viridis* reaction center in Section 22-2D and Figure 22.11). The chlorophyll in these organisms is different from that found in photosystems linked to oxygen. Anaerobic photosynthesis is not as efficient as photosynthesis linked to oxygen, but the anaerobic version of the process appears to be an evolutionary way station. Anaerobic photosynthesis is a means for organisms to use solar energy to satisfy their needs for food and energy. Although it produces ATP efficiently, it is less efficient than that of aerobic photosynthesis for carbon fixation.

▶ Is it possible to have photosynthesis without producing oxygen?

A possible scenario for the development of photosynthesis starts with heterotrophic bacteria that contain some form of chlorophyll, probably bacteriochlorophyll. (*Heterotrophs* are organisms that depend on their environment for organic nutrients and energy.) In such organisms, the light energy absorbed by chlorophyll can be trapped in the form of ATP and NADPH. The important point about such a series of reactions is that photophosphorylation takes place, ensuring an independent supply of ATP for the organism. In addition, the supply of NADPH facilitates the synthesis of biomolecules from simple sources such as CO_2. Under conditions of limited food supply, organisms that can synthesize their own nutrients have a selective advantage.

Organisms of this sort are *autotrophs* (not dependent on an external source of biomolecules) but are also anaerobes. The ultimate electron source that they use is not water but some more easily oxidized substance, such as H_2S, as is the case with present-day green sulfur bacteria (and purple sulfur bacteria), or various organic compounds, as is the case with present-day purple nonsulfur bacteria. These organisms do not possess an oxidizing agent powerful enough to split water, which is a far more abundant electron source than H_2S or organic compounds. The ability to use water as an electron source confers a further evolutionary advantage.

As is frequently the case in biological oxidation–reduction reactions, hydrogens as well as electrons are transferred from a donor to an acceptor. In green plants, green algae, and cyanobacteria, the hydrogen donor and acceptor are H_2O and CO_2, respectively, with oxygen as a product. Other organisms, such as bacteria, carry out photosynthesis in which there is a hydrogen donor other than water. Some possible donors include H_2S, $H_2S_2O_3$, and succinic acid. If H_2S is the source of hydrogens and electrons, a schematic equation for photosynthesis can be written with sulfur, rather than oxygen, as a product.

$$\underset{\text{H-acceptor}}{CO_2} + \underset{\text{H-donor}}{2H_2S} \rightarrow \underset{\text{Carbohydrate}}{(CH_2O)} + 2S + H_2O$$

It is also possible for the hydrogen acceptor to be NO_2^- or NO_3^-, in which case NH_3 is a product. Photosynthesis linked to oxygen with carbon dioxide as the ultimate hydrogen acceptor is a special case of a far more general process that is widely distributed among many different organisms.

Cyanobacteria were apparently the first organisms to have the ability to use water as the ultimate reducing agent in photosynthesis. As we have seen, this ability requires a second photosystem as well as a new variety of chlorophyll—namely, chlorophyll *a* rather than bacteriochlorophyll in this case. Chlorophyll *b* had not yet appeared on the scene because it occurs only in eukaryotes. Therefore, the basic system of aerobic photosynthesis was in place with cyanobacteria. As a result of aerobic photosynthesis by cyanobacteria, Earth acquired its present atmosphere with its high levels of oxygen. The existence of all other aerobic organisms depended ultimately on the activities of cyanobacteria.

Plants can carry out a number of reactions that animals cannot. **Biochemical Connections 22B** describes the product of another series of reactions unique to one species of plant. This product is of great practical importance.

22B Biochemical Connections • Applied Genetics

Improving the Yield of Antimalarial Plants

A plant called *Artemisia annua* has long been used in Chinese folk medicine. As is the case with many folk remedies, this plant contains a pharmacologically active agent. This compound, which has been named "artemisinin," contains a peroxide group as part of a ring system (Figure 22.15). It is highly effective in treating malaria, which affects hundreds of millions of people worldwide each year, causing hundreds of thousands of deaths. Unfortunately, artemisinin is sometimes in short supply, depending on global demand.

The genetic map of this plant has been determined, and it was used to identify genetic loci that can be used to improve the yields of artemisinin significantly. Additionally, efforts have been made to synthesize artemisinin and related compounds in the laboratory. Other attempts have been made to produce it in yeast and bacteria using recombinant DNA technology. More recently, genetically engineered yeast cells have been used to produce a precursor of artemisinin, which is then converted to the desired compound by one of several methods. In 2014, synthetic biology applied to the engineering of microorganisms allowed scientists to produce a precursor of artemisinin that can be converted to artemisinin in a semi-synthetic procedure (see Paddon & Keasling, 2014). ▶

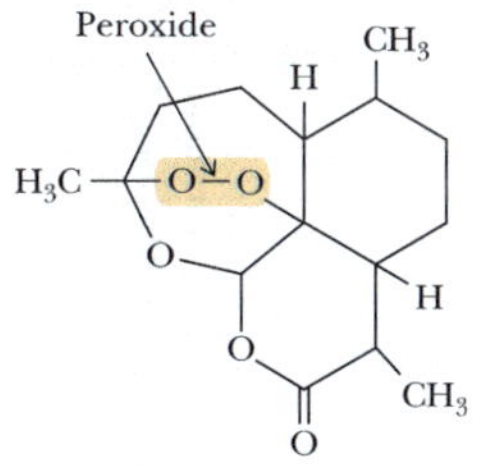

Figure 22.15 Structure of the antimalarial compound artemisinin. The genetic map of the plant from which it is extracted, *Artemisia annua*, is known, including loci that will allow for the development of high-yielding plants. *(Based on Graham, I. A., et al. (2010). The genetic map of Artemisia annua L. identifies loci affecting yield of the antimalarial drug artemisinin.* Science *327 (5963), 328. Copyright © 2015 Cengage Learning®.)*

22-5 The Light-Independent (Calvin Cycle) Reactions of Photosynthesis

Learning Objectives

After reading this section, students will be able to:

LO 22.5.1 Describe the light-independent reactions of photosynthesis.

LO 22.5.2 Explain how nuclear gene products and chloroplast gene products work together in the ribulose-1,5-bisphosphate carboxylase/oxygenase (rubisco) enzyme.

The actual storage form of the carbohydrates produced from carbon dioxide by photosynthesis is not glucose but disaccharides (e.g., sucrose in sugarcane and sugar beets) and polysaccharides (starch and cellulose). However, it is customary and convenient to write the carbohydrate product as glucose, and we follow this time-honored practice. **Biochemical Connections 22C** discusses the energetic aspects of carbohydrate production by plants in detail.

22C Biochemical Connections • Agriculture

Plants Feed Animals—Plants Need Energy—Plants Can Produce Energy

All living things on Earth depend on photosynthetic plants for food, and genetic engineering can address desirable agricultural goals. These goals include improving the nutrient content of seeds and edible plant parts as well as increasing drought resistance. Strengthening defenses against plant pests also makes use of added genes. If plants can be engineered to carry out nitrogen fixation, this modification can lead to decreased use of fertilizers.

The energy required to grow food crops, especially those used for livestock feed, is another important consideration. Figure 22.16 shows the results of a Swedish study on energy requirements for producing food. Energy amounts are given in megajoules. Note the amounts of energy that go into livestock production.

In addition to nutrients, plants produce large quantities of cellulose, which humans cannot digest but which can be a source of biofuels. Biofuels of plant origin are widely discussed as alternatives to petroleum-based energy sources. Although plants like corn are needed as food by humans and animals, they contain parts that are not. For example, the leftover leaves, stalks, and cobs from corn production can provide an important source of cellulose. Wood chips and waste from paper mills are another possible source. Plants such as sorghum, which are not an important food source, are yet another possible supply.

The next step is to turn cellulose into a usable fuel, preferably in liquid form. The process is fundamentally that of breaking up an oxygen-rich polymer to obtain smaller molecules that mostly contain carbon and hydrogen. The goal is achieved by catalytic fast pyrolysis. The method works well and shows promise. A big question in further development is cost. Biofuels need to compete with petroleum-based fuels. Fluctuations in the price of a barrel of oil will play a large role in the future of biofuels.

Continued on next page

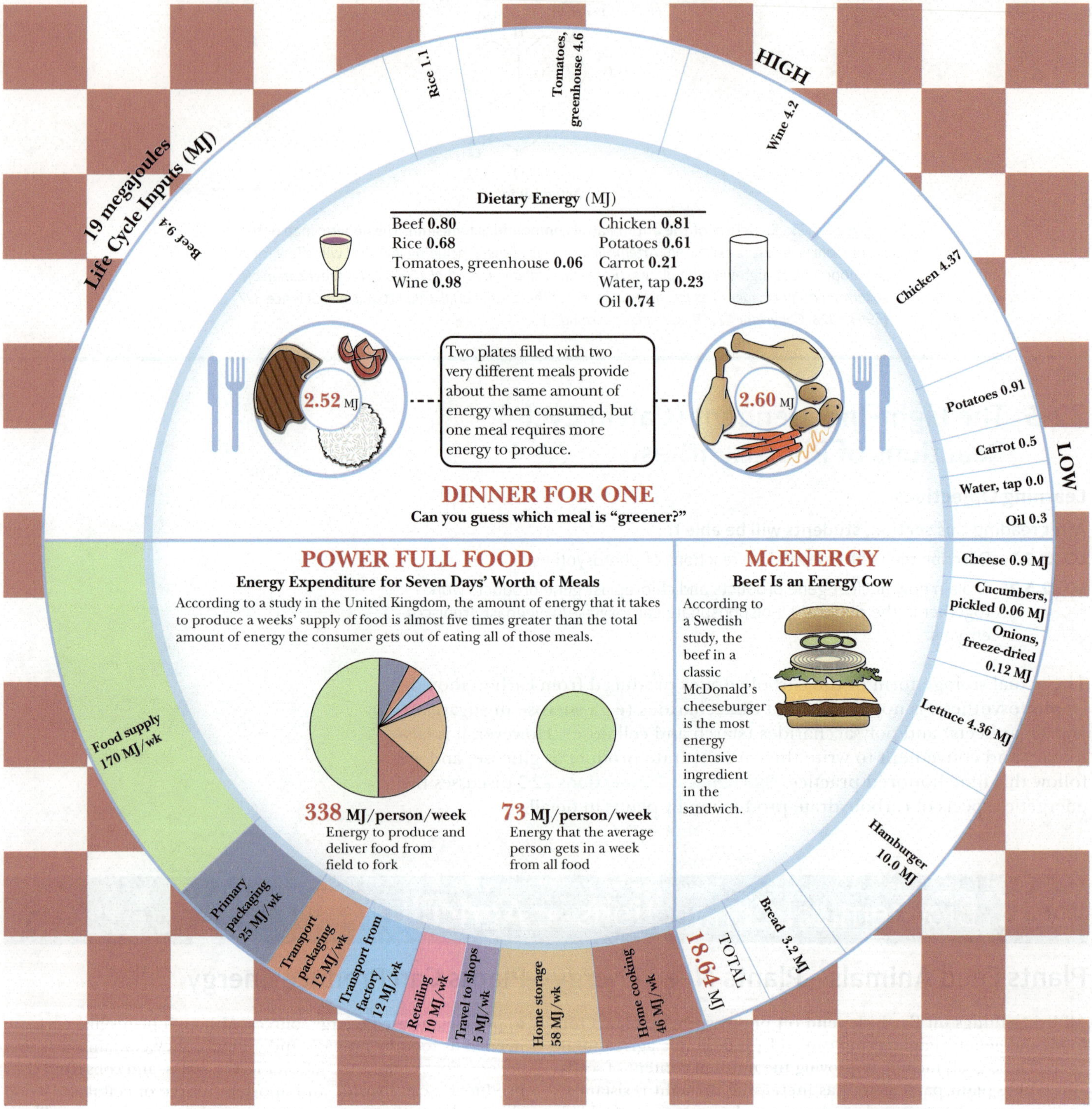

Figure 22.16 **Energy requirements for producing food.** *Tester M, Langridge P. Breeding technologies to increase crop production in a changing world. Science. 2010 Feb 12;327(5967):818-22. doi: 10.1126/science.1183700. PMID: 20150489*

In catalytic fast pyrolysis, cellulose is introduced into a chamber maintained at 500°C. The cellulose polymer breaks down within a second. The three-dimensional catalyst removes oxygen from the fragments and induces cyclization reactions. The final products are aromatic compounds similar to those found in gasoline. Figure 22.17 shows the catalytic fast pyrolysis process by which cellulose is converted into a usable fuel. ◗

Continued on next page

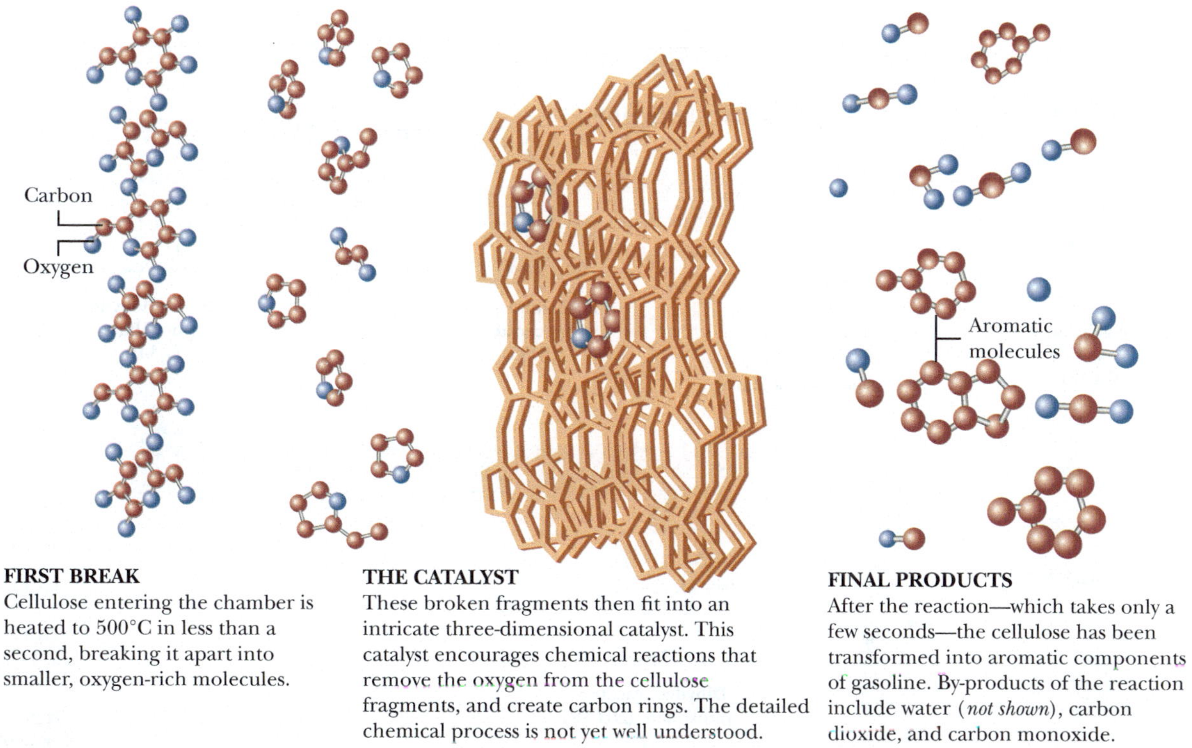

Figure 22.17 **Catalytic fast pyrolysis process by which cellulose is converted into a usable fuel.** *(From* Scientific American, *July 2009, vol. 301, issue 1, p. 55.)*

Carbon dioxide fixation takes place in the stroma. The equation for the overall reaction, like all equations for photosynthetic processes, is deceptively simple.

$$6CO_2 + 12NADPH + 18ATP \xrightarrow{\text{Enzymes}} C_6H_{12}O_6 + 12NADP^+ + 18ADP + 18P_i$$

The actual reaction pathway has some features in common with glycolysis (Sections 17-2 and 17-3) and some in common with the pentose phosphate pathway (Section 18-4).

The overall reaction pathway is cyclic and is called the *Calvin cycle* (Figure 22.18), after Melvin Calvin, the scientist who first investigated it. His research earned him a share of the 1961 Nobel Prize in Chemistry. The net reaction of 6 molecules of carbon dioxide to produce one molecule of glucose requires the carboxylation of 6 molecules of a five-carbon key intermediate, ribulose-1,5-bisphosphate, to form 6 molecules of an unstable six-carbon intermediate, which then splits to give 12 molecules of 3-phosphoglycerate. Of these, two molecules of 3-phosphoglycerate react in turn, ultimately producing glucose. The remaining 10 molecules of 3-phosphoglycerate are used to regenerate the 6 molecules of ribulose-1,5-bisphosphate.

22-5A The Calvin Cycle

The first reaction of the Calvin cycle is the condensation of ribulose-1,5-bisphosphate with carbon dioxide to form a six-carbon intermediate, 2-carboxy-3-ketoribitol-1,5-bisphosphate, which quickly hydrolyzes to give two molecules

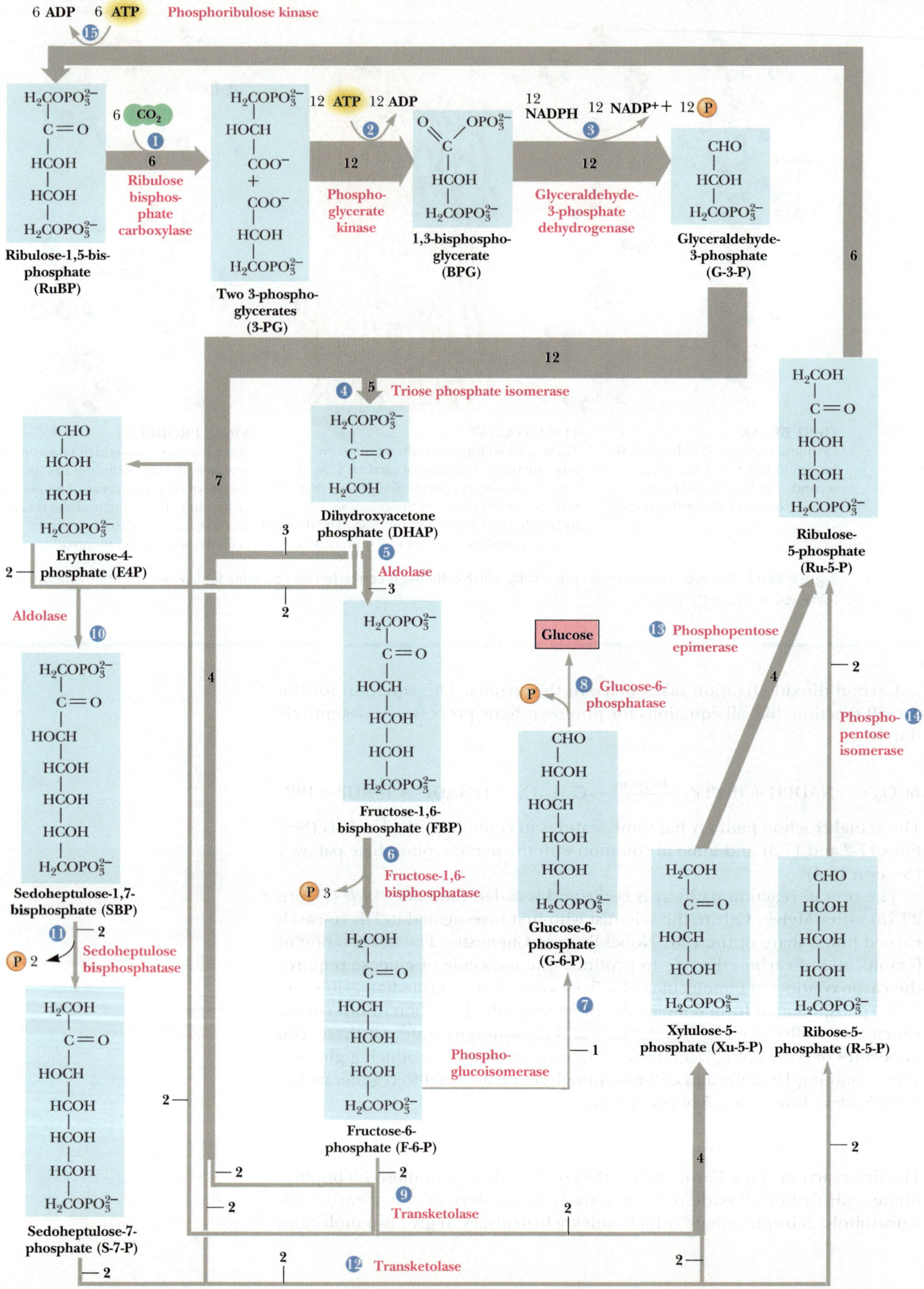

Figure 22.18 **The Calvin cycle of reactions.** The number associated with the arrow at each step indicates the number of molecules reacting in a turn of the cycle that produces one molecule of glucose. Reactions are numbered as in Table 22.1.

of 3-phosphoglycerate (Figure 22.19). The reaction is catalyzed by the enzyme **ribulose-1,5-bisphosphate carboxylase/oxygenase (rubisco)**. Rubisco is located on the stromal side of the thylakoid membrane and is one of the most abundant proteins in nature: It accounts for about 15% of the total protein in chloroplasts. The molecular weight of rubisco is about 560,000 Da, and it consists of eight large subunits (molecular weight 55,000 Da) and eight small subunits (molecular weight 15,000 Da) (Figure 22.20). Recall that chloroplasts in a plant cell have their own DNA in addition to the DNA found in the nucleus of the plant cell. The sequence of the large subunit of rubisco is encoded by a chloroplast gene, and that of the small subunit is encoded by a nuclear gene. The endosymbiotic theory for the development of eukaryotes is consistent with the idea of independent genetic material in organelles. The large rubisco subunit (expressed from a chloroplast gene) is catalytic, whereas the small rubisco subunit (expressed from a nuclear gene) plays a regulatory role, an observation that is consistent with an endosymbiotic origin for organelles such as chloroplasts.

ribulose-1,5-bisphosphate carboxylase/oxygenase (rubisco) the enzyme that catalyzes the first step in carbon dioxide fixation in photosynthesis, the condensation of ribulose-1,5-bisphosphate with carbon dioxide

The incorporation of CO_2 into 3-phosphoglycerate represents the actual carbon fixation process; the remaining reactions are those of carbon shuffling to produce various carbohydrates. The next two reactions after the rubisco reaction lead to the reduction of 3-phosphoglycerate to form glyceraldehyde-3-phosphate (Figure 22.18, *top*). The reduction takes place in the same fashion as in gluconeogenesis, except for one unique feature: The reactions in chloroplasts require NADPH rather than NADH for the reduction of 1,3-bisphosphoglycerate to glyceraldehyde-3-phosphate. When glyceraldehyde-3-phosphate is formed, it can have two alternative fates. One is the production of six- and seven-carbon sugars (see all the "-ose" metabolites in Figure 22.18), and the other is the regeneration of ribulose-1,5-bisphosphate, which completes the "cycle" of the Calvin cycle. Table 22.1 summarizes the reactions that take place and indicates their stoichiometry.

The formation of glucose from glyceraldehyde-3-phosphate takes place in the same manner as in gluconeogenesis (reactions 4 through 8 of Figure 22.18 and Table 22.1). These reactions were also discussed in more detail in Section 18-2. The conversion of glyceraldehyde-3-phosphate to dihydroxyacetone phosphate is catalyzed by triosephosphate isomerase. Dihydroxyacetone phosphate reacts, in turn, with glyceraldehyde-3-phosphate in a reaction catalyzed by aldolase, giving rise to fructose-1,6-bisphosphate. Fructose-1,6-bisphosphate is dephosphorylated by fructose-1,6-bisphosphatase to give fructose-6-phosphate, which is then isomerized by phosphoglucose isomerase to produce glucose-6-phosphate. Glucose-6-phosphate is dephosphorylated by glucose-6-phosphatase to produce glucose.

22-5B Regeneration of Starting Material in the Calvin Cycle

The Calvin cycle starts with ribulose-1,5-bisphosphate (Figure 22.18, *top left*), and the cyclic process of regenerating ribulose-1,5-bisphosphate can be divided into four phases: *preparation, reshuffling, isomerization,* and *phosphorylation*. The preparation phase involves reactions 4 through 6 of Figure 22.18 and Table 22.1, and it begins with the conversion of some of the glyceraldehyde-3-phosphate to dihydroxyacetone phosphate (catalyzed by triosephosphate isomerase). Portions of both the glyceraldehyde-3-phosphate and the dihydroxyacetone phosphate are then condensed to form fructose-1,6-bisphosphate (catalyzed by aldolase). Fructose-1,6-bisphosphate is hydrolyzed to fructose-6-phosphate (catalyzed by fructose-1,6-bisphosphatase). With a supply of glyceraldehyde-3-phosphate, dihydroxyacetone phosphate, and fructose-6-phosphate now available, the reshuffling phase can begin.

Most of the reactions of the reshuffling process are the same as the ones we have seen as part of the pentose phosphate pathway (Section 18-4). Consequently, we concentrate just on the main outline of the process summarized in

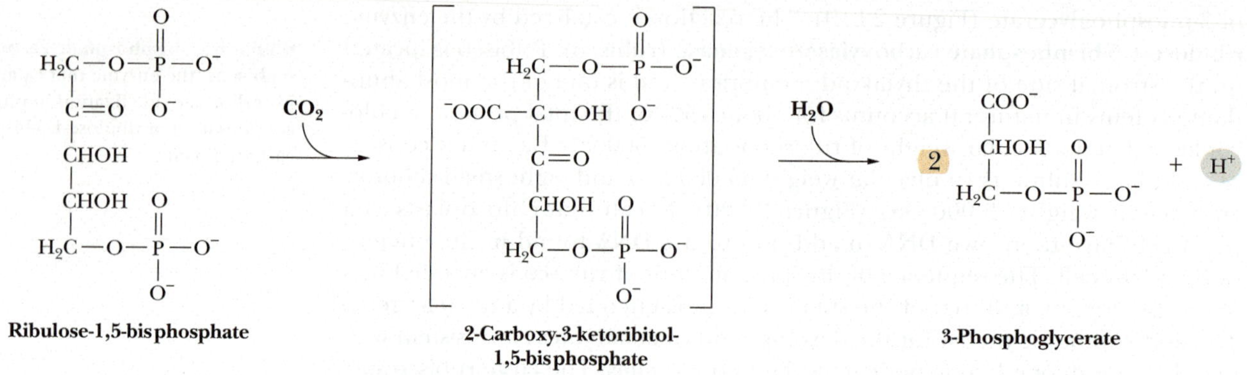

Figure 22.19 **The reaction of ribulose-1,5-bisphosphate with CO_2, catalyzed by ribulose-1,5-bisphosphate carboxylase/oxygenase (rubisco), ultimately produces two molecules of 3-phosphoglycerate.**

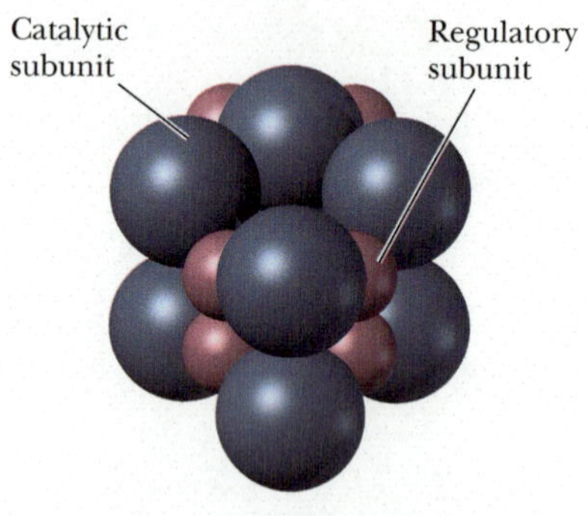

Figure 22.20 **A schematic of the overall structure and composition of ribulose-1,5-bisphosphate carboxylase (rubisco).** In rubisco, the eight catalytic subunits are 55,000 Da in size, and the eight regulatory subunits are 15,000 Da in size.

reactions 9 through 12 of Figure 22.18 and Table 22.1. These reactions, catalyzed in turn by *transketolase, aldolase,* and *sedoheptulose bisphosphatase,* rearrange the carbon skeletons in the reshuffling phase of the Calvin cycle.

The isomerization phase (reactions 13 and 14 in Figure 22.18 and Table 22.1) involves the conversion of both ribose-5-phosphate and xylulose-5-phosphate to ribulose-5-phosphate. *Ribose-5-phosphate isomerase* catalyzes the conversion of ribose-5-phosphate to ribulose-5-phosphate, and *xylulose-5-phosphate epimerase* catalyzes the conversion of xylulose-5-phosphate to ribulose-5-phosphate. The reverse of both of these reactions takes place in the pentose phosphate pathway, catalyzed by the same enzymes.

In the final phosphorylation step (reaction 15 in Figure 22.18 and Table 22.1), ribulose-1,5-bisphosphate is regenerated by the phosphorylation

Table 22.1 **The Calvin Cycle Series of Reactions**

Reactions 1 through 15 constitute the cycle that leads to the formation of one equivalent of glucose. The enzyme catalyzing each step, a concise reaction, and the overall carbon balance are given. Numbers in parentheses show the number of carbon atoms in the substrate and product molecules. Prefix numbers indicate in a stoichiometric fashion how many times each step is carried out in order to provide a balanced net reaction.

1. Ribulose-1,5-bisphosphate carboxylase/oxygenase: 6 CO_2 + 6 H_2O + 6 RuBP → 12 3-PG	6(1) + 6(5) → 12(3)
2. 3-Phosphoglycerate kinase: 12 3-PG + 12 ATP → 12 1,3-BPG + 12 ADP	12(3) → 12(3)
3. NADP-glyceraldehyde-3-phosphate dehydrogenase: 12 1,3-BPG + 12 NADPH → 12 NADP + 12 G-3-P + 12 P_i	12(3) → 12(3)
4. Triosephosphate isomerase: 5 G-3-P → 5 DHAP	5(3) → 5(3)
5. Aldolase: 3 G-3-P + 3 DHAP → 3 FBP	3(3) +3(3) → 3(6)
6. Fructose-1,6-bisphosphatase: 3 FBP + 3 H_2O → 3 F-6-P + 3 P_i	3(6) → 3(6)
7. Phosphoglucose isomerase: 1 F-6-P → 1 G-6-P	1(6) → 1(6)
8. Glucose-6-phosphatase: 1 G-6-P + 1 H_2O → 1 Glucose + 1 P_i	1(6) → 1(6)
The remainder of the pathway involves regenerating six RuBP acceptors (30 C) from the leftover two F-6-P (12 C), four G-3-P (12 C), and two DHAP (6 C).	
9. Transketolase: 2 F-6-P + 2 G-3-P → 2 Xu-5-P + 2 E4P	2(6) + 2(3) → 2(5) + 2(4)
10. Aldolase: 2 E4P + 2 DHAP → 2 SBP	2(4) + 2(3) → 2(7)
11. Sedoheptulose bisphosphatase: 2 SBP + 2 H_2O → 2 S-7-P + 2 P_i	2(7) → 2(7)
12. Transketolase: 2 S-7-P + 2 G-3-P → 2 Xu-5-P + 2 R-5-P	2(7) + 2(3) → 4(5)
13. Phosphopentose epimerase: 4 Xu-5-P → 4 Ru-5-P	4(5) → 4(5)
14. Phosphopentose isomerase: 2 R-5-P → 2 Ru-5-P	2(5) → 2(5)
15. Phosphoribulose kinase: 6 Ru-5-P + 6 ATP → 6 RuBP + 6 ADP	6(5) → 6(5)
Net: 6 CO_2 + 18 ATP + 12 NADPH + 12 H^+ + 12 H_2O → Glucose + 18 ADP + 18 P_i + 12 $NADP^+$	6(1) → 1(6)

of ribulose-5-phosphate. This reaction requires ATP and is catalyzed by the enzyme *phosphoribulose kinase*. The reactions leading to the regeneration of ribulose-1,5-bisphosphate summarized in Table 22.1 give the following *net* equation for the fixation of CO_2 into glucose in photosynthesis:

$$6CO_2 + 18ATP + 12NADPH + 12H^+ + 12H_2O \rightarrow$$
$$\text{Glucose} + 12NADP^+ + 18ADP + 18P_i$$

The efficiency of energy use in photosynthesis can be calculated as follows. The $\Delta G°'$ for the reduction of CO_2 to glucose is +478 kJ (+114 kcal) for each mole of CO_2, and the energy of light of 600-nm wavelength is 1593 kJ/mol (381 kcal/mol). We do not explain in detail here how this figure for the energy of the light is obtained, but it comes ultimately from the equation $E = h\nu$. Light of wavelength 680 nm or 700 nm has lower energy than light at 600 nm. Thus, the efficiency of photosynthesis is at least (478/1593) × 100% = 30%. **Biochemical Connections 22D** goes into another aspect of the unique features of plant biochemistry.

22D Biochemical Connections • Genetics

Chloroplast Genes

Chloroplasts, like mitochondria, have their own DNA. Scientists speculate that these organelles were originally cyanobacteria that were engulfed by a eukaryotic cell through endosymbiosis (Section 1-5). Some interesting, even elaborate, interactions have been discovered between nuclear and chloroplast gene products. There are about 3,000 chloroplast proteins, and 95% of them are encoded by nuclear genes.

One of the interesting interactions between nuclear and chloroplast gene products involves ribulose-1,5-bisphosphate carboxylase/oxygenase (rubisco), the principal enzyme of CO_2 fixation. The large subunit of this enzyme is coded by a chloroplast gene, and the small subunit by a nuclear gene. The mechanisms that coordinate the two syntheses to ensure equimolar production of both subunits are still not completely understood.

The nuclear gene for the small subunit of rubisco is translated in the cytoplasm, and the protein is then transported to the chloroplast, protected by a chaperonin, via targeting mechanisms (see Section 12-6). Special targeting sequences are used to direct various nuclear products to the appropriate chloroplast location, using reactions that require ATP hydrolysis. The chaperonin aids in the formation of the final, active complex.

The chloroplast encodes its own RNA polymerase, ribosomal and transfer RNAs, and about one-third of the ribosomal proteins. The DNA polymerase, aminoacyl synthetases, and the rest of the ribosomal proteins in chloroplasts come from nuclear genes. Different nuclear genes may be used for chloroplasts in different, specialized tissues of the plant. In different classes of plants, different genes are coded by the nucleus, even though all the same proteins constitute the chloroplast. The sequences of many of these enzymes encoded in the nucleus resemble those found in bacteria more than the sequences of proteins encoded by other nuclear genes. The chloroplast rRNAs likewise resemble bacterial rRNAs. The four subunits of the chloroplast-coded RNA polymerase are homologous to the four subunits of bacterial RNA polymerase. Furthermore, the chloroplast mRNA uses a Shine–Dalgarno sequence to bind to the ribosome, and it does not have a cap or a poly-A tail. These observations are consistent with the idea that the chloroplast (and mitochondrion) originated from bacteria-like organisms, symbiotically taken into early non-photosynthetic eukaryotic cells, with some subsequent transfer of organelle genes to the nucleus.

The separation of genetic material between the nucleus and the chloroplast requires coordination of the transcription of genes in two different locations. How plant cells are able to accomplish this is an active research area. It is now known that there are communications in both directions between the two organelles. A signal proceeding from the nucleus to the chloroplast is considered the more "normal" pathway and is referred to as *anterograde signaling*. When the signal goes from another organelle to the nucleus, it is called *retrograde signaling*. About seven different retrograde signaling processes have been discovered in algae and higher plants.

Figure 22.21 shows three examples of retrograde signaling processes. One of them involves Mg-protoporphyrin IX, a tetrapyrrole generated during chlorophyll biosynthesis. Also studied are signals arising from the expression of chloroplast genes and those from the photosynthetic electron transport (PET) chain. Stress signals and developmental impairment from the cytosol increase the levels of another set of related genomes-uncoupled (GUN) proteins, GUN4 and GUN5. These three retrograde systems have been shown to act through the GUN1 protein, which is a member of a large family of enigmatic transcription factors in higher plants (Figure 22.21, *top right*). It has been shown that all three signaling pathways affect the level of GUN1 in the chloroplast. It is currently unknown how the GUN1 signal makes its way to the nucleus. However, in the nucleus, the effect is to increase a transcription factor called ABI4 (abscisic acid-insensitive 4), which then inhibits transcription of targeted nuclear genes involved in chloroplast metabolism. This common

Continued on next page

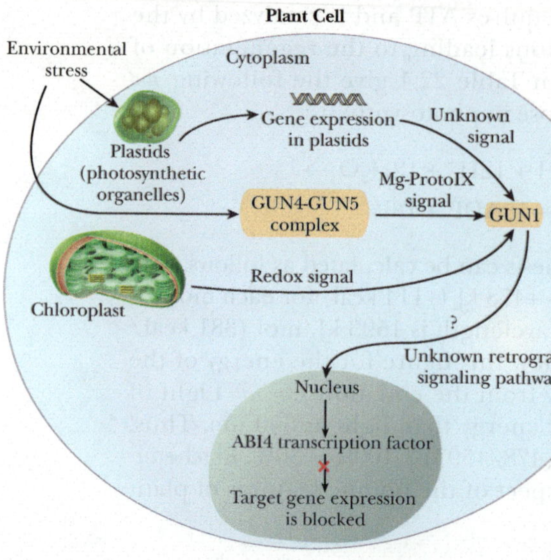

Figure 22.21 Three different chloroplast signals are focused down to one common signal molecule, the genomes-uncoupled 1 (GUN1) protein. By an unknown mechanism, the GUN1 signal protein leaves the chloroplast and affects the transcription of nuclear genes by the nuclear transcription factor ABI4. (*Based on Zhang, D. P., et al. (2007) Signaling to the nucleus with a loaded GUN.* Science *316 (5825), 700-701.*)

pathway focused on the GUN1 protein allows the smooth coordination of the nuclear part of chloroplast metabolism in response to many environmental signals. Epigenetic mechanisms, such as DNA methylation, play a highly important role in gene expression in plants. In fact, plants display significantly more epigenetic gene expression than animals.

22-6 CO_2 Fixation into Sugars in Tropical Plants

Learning Objectives

After reading this section, students will be able to:

LO 22.6.1 Describe the reactions and physiological locations of the reactions in the C_4 pathway of CO_2 fixation into sugars in tropical plants.

LO 22.6.2 Explain the activity of the ribulose-1,5-bisphosphate carboxylase/oxygenase (rubisco) enzyme when levels of O_2 are high compared to CO_2.

C_4 pathway an alternative way to fix CO_2 found in tropical plants

In tropical plants, there is a **C_4 pathway**, so named because it involves four-carbon compounds (Figure 22.22). The operation of this pathway (also called the *Hatch–Slack pathway*) ultimately leads to the C_3 (based on 3-phosphoglycerate) pathway of the Calvin cycle. (There are other C_4 pathways, but this one is the most widely studied. Corn—maize—is an important example of a C_4 plant, and it is certainly not confined to the tropics.)

▶ What is different about CO_2 fixation in tropical plants?

When CO_2 enters the leaf through pores in the outer mesophyll cells, it first reacts with phosphoenolpyruvate to produce oxaloacetate and P_i in a reaction catalyzed by phosphoenolpyruvate carboxylase. Oxaloacetate is reduced to malate in a reaction catalyzed by malate dehydrogenase, with the concomitant oxidation of NADPH. Malate is then transported to the bundle-sheath cells (the next layer) through channels that connect the two kinds of cells.

In the bundle-sheath cells, malate is decarboxylated to give pyruvate and CO_2. In the process, $NADP^+$ is reduced to NADPH. The CO_2 reacts with ribulose-1,5-bisphosphate to enter the Calvin cycle. Pyruvate is transported back to

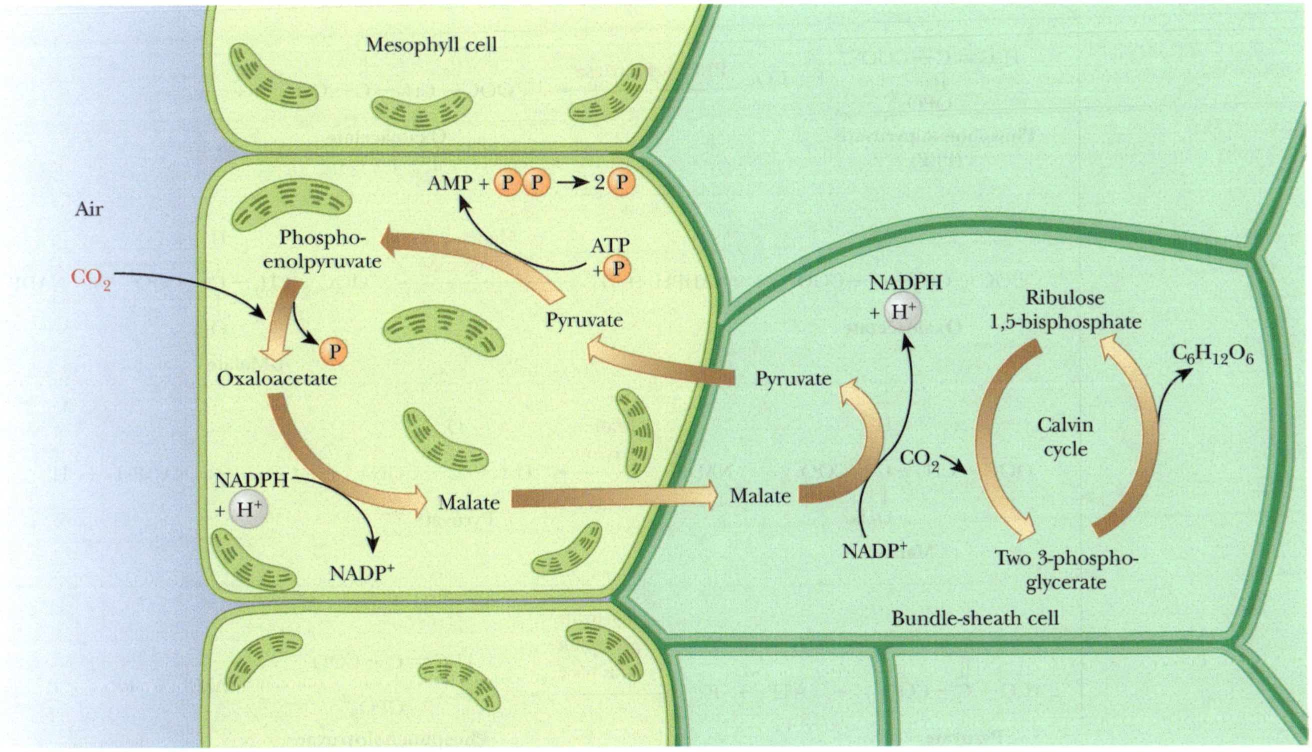

Figure 22.22 The C_4 pathway, also called the Hatch–Slack pathway, of CO_2 fixation into sugars in tropical plants. Atmospheric CO_2 enters the mesophyll cells of tropical plants and first combines with phosphoenolpyruvate to form oxaloacetate, which is then reduced to malate. The malate then donates CO_2 to the Calvin cycle in bundle-sheath cells, ultimately giving rise to the production of glucose ($C_6H_{12}O_6$). The malate becomes pyruvate, which is shuttled back to the mesophyll cells to be converted into phosphoenolpyruvate to begin the CO_2 fixation process again.

the mesophyll cells, where it is phosphorylated to phosphoenolpyruvate, which can react with CO_2 to start another round of the C_4 pathway. When pyruvate is phosphorylated, ATP is hydrolyzed to AMP and PP_i. This situation represents a loss of two high-energy phosphate bonds, equivalent to the use of two ATP. Consequently, the C_4 pathway requires two more ATP equivalents than the Calvin cycle alone for each CO_2 incorporated into glucose. Even though more ATP is required for the C_4 pathway than for the Calvin cycle, there is abundant light in tropical areas to produce the extra ATP by the light reactions of photosynthesis. Figure 22.23 shows the metabolite structures involved in these reactions of the C_4 pathway, as well as the names of the enzymes that catalyze each reaction.

The C_4 pathway fixes CO_2 in the mesophyll cells only to unfix and refix it in the bundle sheath cells, where CO_2 then enters the C_3 pathway of the Calvin cycle. This observation raises the question of the advantage to tropical plants of using the C_4 pathway. The conventional wisdom on the subject focuses on the role of CO_2, but there is more to the situation than that. According to the conventional view, the point of the C_4 pathway is that it concentrates CO_2 and, as a result, accelerates the process of photosynthesis. Leaves of tropical plants have small pores to minimize water loss, and these small pores decrease CO_2 entry into the plant. Another point to consider is that the K_M for CO_2 of phosphoenolpyruvate carboxylase is lower than that of rubisco, allowing the outer mesophyll cells to fix CO_2 at a lower concentration. This also increases the concentration gradient of CO_2 across the leaf and facilitates the movement of

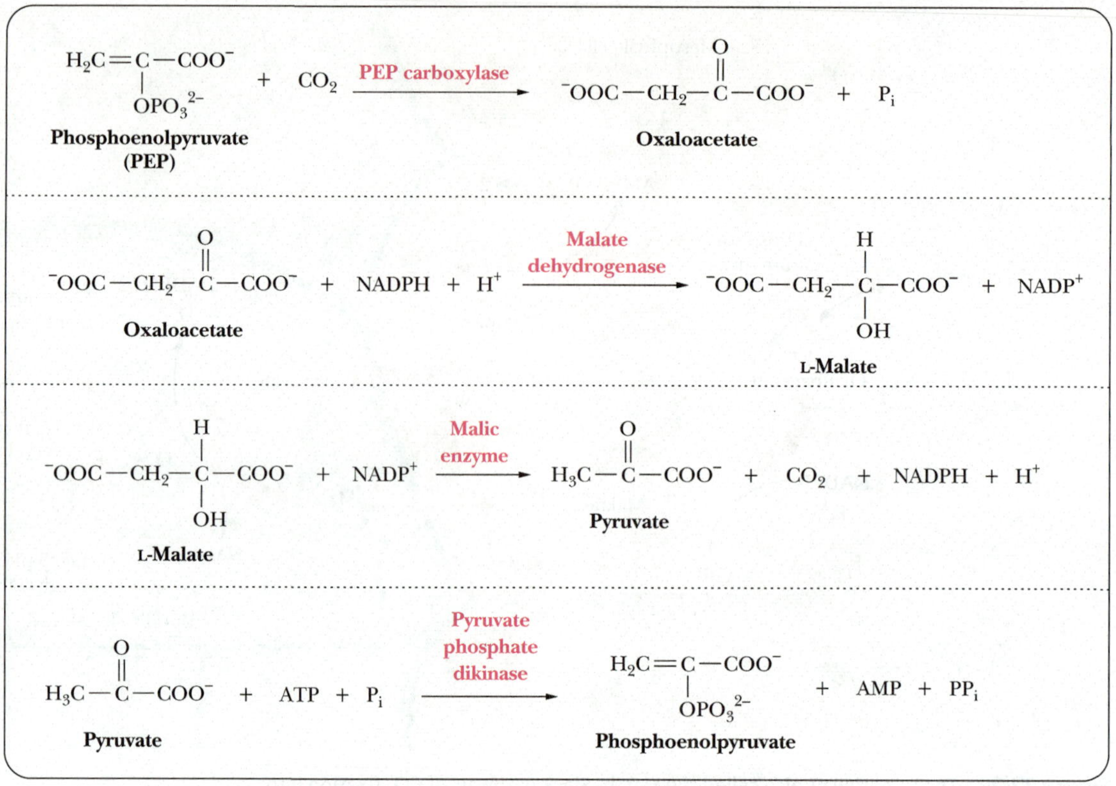

Figure 22.23 The characteristic reactions of the C_4 pathway, also called the Hatch–Slack pathway, of CO_2 fixation into sugars in tropical plants. The physiological locations of these reactions are shown in Figure 22.22.

photorespiration the process by which rubisco fixes oxygen instead of CO_2 under conditions of high O_2 or low CO_2 levels

CO_2 into the leaf through the pores. In tropical areas, where there is abundant light, the amount of CO_2 available to plants controls the rate of photosynthesis.

The C_4 pathway allows tropical plants to grow more quickly in these conditions and to produce more biomass per unit of leaf area than plants that use the C_3 pathway. A more comprehensive view of the subject includes a consideration of the high levels of oxygen compared with CO_2 and the process of **photorespiration**, in which oxygen is used instead of CO_2 during the reaction catalyzed by rubisco.

Although the actual biological role of photorespiration is not known, several points are well established. The oxygenase activity appears to be an unavoidable, wasteful activity of rubisco. Photorespiration is a salvage pathway that saves some of the carbon that would be lost because of the oxygenase activity of rubisco. In fact, photorespiration is essential to plants even though the plant pays the price in loss of ATP and reducing power; mutations that affect this pathway can be lethal. The principal substrate oxidized in photorespiration is *glycolate* (Figure 22.24). The product of the oxidation reaction, which takes place in peroxisomes of leaf cells (Section 1-4), is *glyoxylate.* (Photorespiration is localized in peroxisomes.) Glycolate arises ultimately from the oxidative breakdown of ribulose-1,5-bisphosphate. The enzyme that catalyzes this reaction is ribulose-1,5-bisphosphate carboxylase/oxygenase, acting as an oxygenase (linked to O_2) rather than as a carboxylase (linked to CO_2) that fixes CO_2 into 3-phosphoglycerate.

When levels of O_2 are high compared with those of CO_2, ribulose-1,5-bisphosphate is oxygenated to produce phosphoglycolate (which gives rise to glycolate) and 3-phosphoglycerate by photorespiration, rather than the two molecules of 3-phosphoglycerate that arise from carboxylation reaction. This

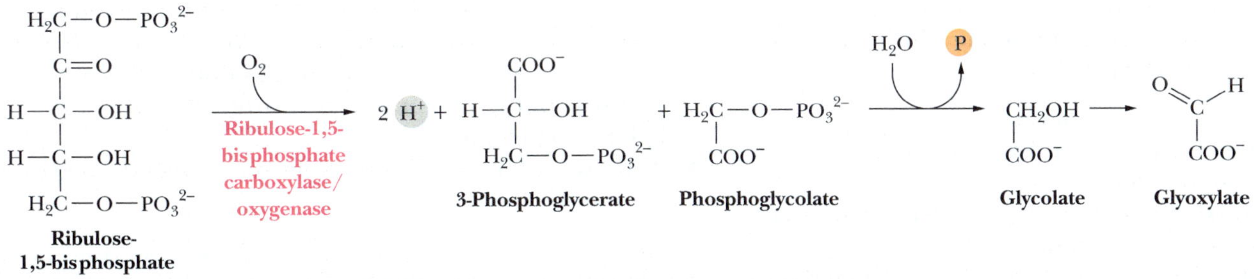

Figure 22.24 The characteristic reactions of photorespiration, which occurs in tropical conditions. When the levels of oxygen are relatively high compared with those of CO_2, ribulose-1,5-bisphosphate carboxylase/oxygenase uses oxygen to catalyze the formation of 3-phosphoglycerate and phosphoglycolate instead of using CO_2 to catalyze the formation of two molecules of 3-phosphoglycerate.

situation occurs in C_3 plants. In C_4 plants, the small pores decrease the entry not only of CO_2 but also of O_2 into the leaves. The ratio of CO_2 to O_2 in the bundle-sheath cells is relatively high as a result of the operation of the C_4 pathway, favoring the carboxylation reaction. C_4 plants have reduced oxygenase activity through compartmentation and thus have less need for photorespiration. This is an advantage in the hot climates in which C_4 plants are principally found.

Summary

The Net Reaction Equation of Photosynthesis and Chloroplast Structure

- The equation for photosynthesis actually represents two processes:

$$6CO_2 + 6H_2O \rightarrow C_6H_{12}O_6 + 6O_2$$

- One process, the oxidation of water to produce oxygen, requires light energy from the Sun, and the other process, the fixation of CO_2 to give sugars, uses solar energy indirectly.
- Chloroplasts are complex organelles where photosynthesis occurs.

Chlorophyll Structure and Properties

- In eukaryotes, the light reactions of photosynthesis take place in the thylakoid membranes of chloroplasts.
- A series of membrane-bound electron carriers and pigments is able to harness the light energy of the Sun. Chlorophylls play key roles in harnessing light energy, and they contain a porphyrin ring with a central Mg^{2+} ion.
- The trapping of light takes place at a reaction center within the chloroplast, and the process requires a special pair of chlorophyll *a* molecules in a unique environment.

Overview of the Light-Dependent Reactions in Photosynthesis

- In the light reactions, water is oxidized to produce oxygen, accompanied by the reduction of $NADP^+$ to NADPH.
- The light reactions consist of two parts, each carried out by a separate photosystem.

The Z-Scheme of Photosynthesis and Photophosphorylation of ADP to ATP

- When photosystem II transfers electrons from water to an electron transport chain, oxygen is produced.
- The electrons generated by photosystem II and carried by the electron transport chain reduce $NADP^+$ to NADPH in the reactions of photosystem I.
- Both photosystems harness light energy that is needed for the reduction of $NADP^+$ and for the phosphorylation of ADP to ATP.

Cyclic Photophosphorylation in Photosystem I

- Under conditions of a high $NADPH/NADP^+$ ratio, cyclic electron transport in photosystem I can be coupled to the production of ATP without the production of either NADPH or O_2.

Anoxygenic Photosystems

- Early research on the molecular structure of photosystems and their reaction centers was carried out on anoxygenic photosystems from anaerobic phototrophic bacteria that do not produce oxygen as a result of their photosynthetic processes.
- In anoxygenic reaction centers, specific chlorophylls, accessory pigments, and quinones are associated with membrane-spanning polypeptides, and the whole assemblage

is arranged to facilitate the transfer of electrons to carry out the reactions of photosynthesis and the creation of a proton gradient for ATP production.

Photosynthesis and ATP Production

- ATP production in chloroplasts has some commonality to the same process in mitochondria.
- The two photosystems are linked by an electron transport chain coupled to the production of ATP.
- A proton gradient drives the production of ATP in photosynthesis, as it does in mitochondrial respiration.

Evolutionary Implications of Photosynthesis With and Without Oxygen

- Some forms of bacteria also perform photosynthesis, although they often have simpler systems involving only a single photosystem.
- Early photosynthetic bacteria probably used an electron donor other than water and did not produce oxygen.
- Bacteria that appeared later, as well as eukaryotes, eventually developed the dual photosystems and the ability to produce oxygen from water, which led to Earth having an oxygen-rich atmosphere.

The Calvin Cycle

- The overall reaction pathway of sugar production is cyclic and is called the Calvin cycle.
- The Calvin cycle reactions of photosynthesis involve the net synthesis of one molecule of glucose from six molecules of CO_2.
- The net reaction of 6 molecules of CO_2 to produce 1 molecule of glucose requires the carboxylation of 6 molecules of a five-carbon key intermediate, ribulose-1,5-bisphosphate, ultimately forming 12 molecules of 3-phosphoglycerate.
- Of these, two molecules of 3-phosphoglycerate react to give rise to glucose.

Regeneration of Starting Material in the Calvin Cycle

- The remaining 10 molecules of 3-phosphoglycerate are used to regenerate the 6 molecules of ribulose-1,5-bisphosphate through a series of reactions that include the reshuffling of carbon skeletons and isomerizations.

CO_2 Fixation into Sugars in Tropical Plants

- In addition to the Calvin cycle, there is an alternative pathway for CO_2 fixation in tropical plants.
- It is called the C_4 pathway because it involves four-carbon compounds. It is also called the Hatch–Slack pathway.
- In this pathway, CO_2 reacts in the outer (mesophyll) cells with phosphoenolpyruvate to produce oxaloacetate and P_i.
- Oxaloacetate in turn is reduced to malate.
- Malate is transported from mesophyll cells, where it is produced, to inner (bundle-sheath) cells, where it passes CO_2 to the Calvin cycle for glucose production.
- Plants in which the C_4 pathway operates grow more quickly and produce more biomass per unit of leaf area than C_3 plants, in which only the Calvin cycle operates.

Review Exercises

22-1 Chloroplasts Are the Site of Photosynthesis

1. **Recall** Why does chlorophyll appear green to the human eye?
2. **Recall** What are the principal metal ions used in electron transfer in chloroplasts? Compare them with the ions found in mitochondria.
3. **Recall** How is the structure of chloroplasts like that of mitochondria? How does it differ?
4. **Recall** List three ways in which the structure of chlorophyll differs from that of heme.
5. **Recall** Why do plants contain accessory pigments in addition to chlorophylls *a* and *b*?
6. **Reflect and Apply** The first amino acid in protein synthesis in the chloroplast is *N*-formylmethionine. What is the significance of this fact?

22-2 Photosystems I and II and the Light-Dependent Reactions of Photosynthesis

7. **Recall** Is it fair to say that the synthesis of NADPH in chloroplasts is merely the reverse of NADH oxidation in mitochondria? Explain your answer.
8. **Recall** Outline the events that take place at the photosynthetic reaction center in *Rhodopseudomonas*.
9. **Recall** What are the two places where light energy is required in the light reaction of photosynthesis? Why must energy be supplied at precisely these points?
10. **Recall** Do all the chlorophyll molecules in a photosynthetic reaction center play the same roles in the light reactions of photosynthesis? Explain your answer.
11. **Recall** Describe some similarities between the electron transport chains in chloroplasts and mitochondria.
12. **Reflect and Apply** Which is likely to have evolved first, the electron transport chain in chloroplasts or mitochondria? Explain your answer.
13. **Reflect and Apply** Uncouplers of oxidative phosphorylation in mitochondria also uncouple photoelectron transport and ATP synthesis in chloroplasts. Give an explanation for this observation.
14. **Reflect and Apply** A larger proton gradient is required to form a single ATP in chloroplasts than in mitochondria. Suggest a reason why. *Hint:* Ions can move across the thylakoid membrane more easily than across the inner mitochondrial membrane.
15. **Reflect and Apply** Albert Szent-Gyorgi, a pioneer in early photosynthesis research, stated, "What drives life is a little electric current, kept up by the sunshine." What did he mean by this?
16. **Biochemical Connections** What is implied about the energy requirements of photosystems I and II by the fact that there is a difference in the minimum wavelength of light needed for them to operate (700 nm for photosystem I and 680 nm for photosystem II)?
17. **Reflect and Apply** Why is a photosynthetic reaction center comparable to a battery?

18. **Reflect and Apply** Antimycin A (an inhibitor of electron transport) is an inhibitor of photosynthesis in chloroplasts. Why is this likely to be so?
19. **Reflect and Apply** Would you expect H_2O or CO_2 to be the source of the oxygen produced in photosynthesis? Give the reason for your answer.
20. **Reflect and Apply** Why do we describe the path of electrons in photosynthesis as starting in photosystem II and ending in photosystem I? In other words, why is the nomenclature "backward"?
21. **Reflect and Apply** It has taken considerable research to establish the number of protons pumped across the mitochondrial membrane at the various stages of electron transport. Would you expect to encounter difficulties in determining the number of protons pumped in electron transport across the thylakoid membrane? Why or why not?
22. **Reflect and Apply** The oxidation of water requires four electrons, but chlorophyll molecules can transfer only one electron at a time. Describe how these two statements can be reconciled.
23. **Reflect and Apply** Why does a loosely bound cytochrome play a unique role in the reaction-center events in *Rhodopseudomonas?*

22-3 Photosynthesis and ATP Production

24. **Recall** In cyclic photophosphorylation in photosystem I, ATP is produced even though water is not split. Explain how the process takes place.
25. **Recall** What are the major similarities and differences between ATP synthesis in chloroplasts and mitochondria?
26. **Recall** How can a proton gradient be created in cyclic photophosphorylation in photosystem I?
27. **Reflect and Apply** Can ATP production take place in chloroplasts in the absence of light? Give the reason for your answer.
28. **Reflect and Apply** What is the advantage to plants of having the option of both cyclic and noncyclic pathways for photophosphorylation?

22-4 Evolutionary Implications of Photosynthesis With and Without Oxygen

29. **Recall** Is water the only possible electron donor in photosynthesis? Why or why not?
30. **Reflect and Apply** Suppose a prokaryotic organism that contains both chlorophyll *a* and chlorophyll *b* has been discovered. Comment on the evolutionary implications of such a discovery.

22-5 The Light-Independent (Calvin Cycle) Reactions of Photosynthesis

31. **Recall** Why is rubisco likely to be the most abundant protein in nature?
32. **Biochemical Connections** Is the sequence of amino acids in rubisco encoded by nuclear genes or not? Explain.
33. **Reflect and Apply** Name some other metabolic pathways that have reactions similar to those of the Calvin cycle reactions of photosynthesis.
34. **Reflect and Apply** Using information from Sections 15-2 and 15-6, show how the $\Delta G°'$ of 478 kJ (114 kcal) is obtained for each mole of CO_2 fixed in photosynthesis. The reaction in question is $6CO_2 + 6H_2O \rightarrow \text{Glucose} + 6O_2$.
35. **Reflect And Apply** If photosynthesizing plants are grown in the presence of $^{14}CO_2$ for six Calvin cycles, will every carbon atom of the glucose produced be labeled with the radioactive carbon? Why or why not?
36. **Biochemical Connections** What key aspects of chloroplasts (and mitochondria) are consistent with the theory that they may have once been bacteria? List three specific features.
37. **Reflect And Apply** Explain why the evolution of the pathway for the regeneration of ribulose-1,5-bisphosphate from glyceraldehyde-3-phosphate might have required relatively few enzymatic adaptations.
38. **Reflect And Apply** Rubisco is thought to have evolved from a CO_2-fixing enzyme that operated in a low-oxygen environment. Why would oxygenase activity in rubisco's ancestral enzyme have been a problem after photosynthetic organisms appeared?
39. **Reflect And Apply** Does the whole Calvin cycle represent carbon dioxide fixation? Why or why not?
40. **Reflect And Apply** Why do we refer to the conversion of six molecules of carbon dioxide (six carbon atoms) to one molecule of glucose (also six carbon atoms) as a *net* reaction?

22-6 CO_2 Fixation into Sugars in Tropical Plants

41. **Recall** How does photosynthesis in tropical plants differ from the process in non-tropical plants?
42. **Recall** Describe what photorespiration is and compare it with the Calvin cycle.
43. **Reflect And Apply** Why is it advantageous for tropical plants to use the C_4 rather than the C_3 fixation pathway?
44. **Reflect And Apply** What would be the effect on plants if photorespiration did not exist?

Further Reading

Bering, C. L. Energy Interconversions in Photosynthesis. *J. Chem. Ed.* **62**, 659–664 (1985). A discussion of basic concepts of photosynthesis, concentrating on the light reactions and photosystems.

Bishop, M. B., and C. B. Bishop. Photosynthesis and Carbon Dioxide Fixation. *J. Chem. Ed.* **64**, 302–305 (1987). Concentrates on the Calvin cycle.

Danks, S. M., E. H. Evans, and P. A. Whittaker. *Photosynthetic Systems: Structure, Function, and Assembly.* New York: Wiley, 1983. A short book with excellent electron micrographs of chloroplasts and related structures in Chapter 1.

Deisenhofer, J., and H. Michel. The Photosynthetic Reaction Center from the Purple Bacterium *Rhodopseudomonas viridis. Science* **245**, 1463–1473 (1989). The authors' Nobel Prize address describing their work on the structure of the reaction center.

Deisenhofer, J., H. Michel, and R. Huber. The Structural Basis of Photosynthetic Light Reactions in Bacteria. *Trends Biochem. Science* **10**, 243–248 (1985). A discussion of the photosynthetic reaction center in bacteria.

Dennis, D. T. *The Biochemistry of Energy Utilization in Plants.* New York: Chapman and Hall, 1987. A short book on plant biochemistry.

Editors of *Science* and several authors. Food Security. *Science* **327**, 797–834 (2010). A special section on a number of important topics in agriculture.

Govindjee and W. J. Coleman. How Plants Make Oxygen. *Sci. Amer.* **262** (2), 50–58 (1990). This article focuses on the water-oxidizing apparatus of photosystem II.

Halliwell, B. *Chloroplast Metabolism: The Structure and Function of Chloroplasts in Green Leaf Cells.* New York: Oxford University Press, 1981. A detailed description of chloroplast activity.

Hathway, D. *Molecular Mechanisms of Herbicide Selectivity.* New York: Oxford University Press, 1989. A short book primarily devoted to the differences in enzyme activity in weeds and desirable plants.

Haumann, M., P. Liebisch, C. Muller, M. Barra, M. Grabolle, and H. Dau. Photosynthetic O_2 Formation by Time-Resolved X-Ray Experiments. *Science* **310**, 1019–1021. (2007). Experimental observation of the oxygen-producing state and possible intermediates in oxygen evolution.

Hipkins, M. F., and N. R. Baker, eds. *Photosynthesis: Energy Transduction: A Practical Approach.* Oxford, England: IRL Press, 1986. A collection of articles about research methods used to study photosynthesis.

Huber, G., and B. Dale. Grassoline at the Pump. *Sci. Amer.* **301** (1), 52–59 (2009). An article about plant waste and weeds as possible sources of biofuels.

Jiang, F., F. H. Kleiner, and M. Aubin-Tam. Harnessing Photosynthesis for Materials, Devices, and Environmental Technologies. *Curr. Op. Biotech.* **92**, Article 103265 (2025). This review article highlights the utilization of photosynthetic organisms in developing sustainable materials and devices.

Karplus, P., M. Daniels, and J. Herriott. Atomic Structure of Ferredoxin-NADP+ Reductase: Prototype for a Structurally Novel Flavoenzyme Family. *Science* **251**, 60–66 (1991). The structure of a key enzyme involved in nitrogen and sulfur metabolism, as well as in photosynthesis.

Koussevitzky, S., A. Nott, T. C. Mockler, F. Hong, G. Sachetto-Martins, M. Surpin, J. Lim, R. Mittler, and J. Chory. Signals from Chloroplasts Converge to Regulate Nuclear Gene Expression. *Science* **316**, 715–718 (2007). A research article about how chloroplast and nuclear gene expression is coordinated.

Liu, J., D. Zhang, L. Bao, T. Mao, L. Zhao, C. Liu, G. Huang, and Y. Zhai Plant Nanobiotechnology: A New Strategy to Enhance Crop Photosynthesis. *Env. Sci: Nano* **11**, 4655–4665 (2024). This review article explores how nanobiotechnology can be leveraged to improve photosynthetic efficiency in crops.

Margulis, L. *Early Life.* Boston: Science Books International, 1982. Chapters 2 and 3 discuss the evolutionary development of photosynthesis.

Milhous, W., and P. Weina. The Botanical Solution for Malaria. *Science* **327**, 279–280 (2010). A report on a plant-derived drug for the treatment of malaria.

Paddon, C. J., and J. D. Keasling. Semi-synthetic Artemisinin: A Model for the Use of Synthetic Biology in Pharmaceutical Development. *Nature Reviews Microbiology* **12**, 355–367 (2014). A review describing advances in synthetic biology applied to providing a commercial route for the production of artemisinin.

Wohri, A., et al. Light-Induced Structural Changes in a Photosynthetic Reaction Center Caught by Laue Diffraction. *Science* **328**, 630–633 (2010). A report on light-induced conformational change in a photosynthetic reaction center.

Youvan, D. C., and B. L. Marrs. Molecular Mechanisms of Photosynthesis. *Sci. Amer.* **256** (6), 42–48 (1987). A detailed description of a bacterial photosynthetic reaction center and the molecular events that take place there.

Zhang, D. Signaling to the Nucleus with a Loaded GUN. *Science* **316**, 700–701 (2007). A prelude to the Koussevitzky article regarding coordination of nuclear and chloroplast gene expression.

Zuber, H. Structure of Light-Harvesting Antenna Complexes of Photosynthetic Bacteria, Cyanobacteria and Red Algae. *Trends Biochem. Sci.* **11**, 414–419 (1986). This article concentrates on the protein portion of the photosynthetic reaction center.

Root nodules of leguminous plants play a pivotal role in nitrogen fixation.

The Metabolism of Nitrogen 23

Chapter Outline

23-1 Nitrogen Metabolism: An Overview

Learning Objectives

After reading this section, students will be able to:

LO 23.1.1 Distinguish between major classes of biochemical compounds that have nitrogen in them and those that do not.

LO 23.1.2 Summarize the nitrogen reactions that occur between the atmosphere and the biosphere.

We have seen the structures of many types of compounds that contain nitrogen, including amino acids, porphyrins, and nucleotides, but we have not discussed their metabolism. The metabolic pathways we have dealt with up to now have mainly involved compounds of carbon, hydrogen, and oxygen, such as carbohydrates and lipids.

Several important topics can be included in our discussion of the metabolism of nitrogen (see Figure 23.1 for an overview). The first of these is **nitrogen fixation**, the process by which inorganic molecular nitrogen from the atmosphere (N_2), which is unusable by organisms, is incorporated first into the ammonium ion (NH_4^+) (Figure 23.1, *right*) and then into organic compounds that organisms can use. Nitrate ion (NO_3^-), another kind of inorganic nitrogen, is the form in which nitrogen is found in the soil, and many fertilizers contain nitrates—frequently potassium nitrate. **Nitrification** is the conversion of the ammonium ion to nitrate. It is accomplished by

Chapter Outline (continued)

nitrogen fixation the conversion of molecular nitrogen to ammonium ion

nitrification the conversion of ammonium ion to nitrate ion

assimilatory nitrate reduction the process by which organisms convert nitrate to the ammonium ion as a nitrogen source for nitrogen metabolic pathways

denitrification the process by which nitrates and nitrites are broken down into molecular nitrogen

nitrifying bacteria. **Assimilatory nitrate reduction** (Figure 23.1, *bottom* and *middle*), in which nitrate is converted back to the ammonium ion, is another way for organisms to obtain nitrogen. Nitrate ion and nitrite ion (NO_2^-) are also involved in **denitrification** reactions, in which *denitrifying bacteria* return nitrogen to the atmosphere (Figure 23.1, *middle*).

Ammonium ion formed by either nitrogen fixation or assimilatory nitrate reduction enters the biosphere, where organisms like plants, algae, and some bacteria use it as a nitrogen source for their nitrogen-containing compounds. The ammonium ion nitrogen then becomes *organic nitrogen*, and this organic nitrogen can be passed to animals through food chains (Figure 23.1, *middle*). Finally, animal waste products such as urea are excreted and degraded to the ammonium ion by microorganisms. The word *ammonium ion* comes from *sal ammoniac* (ammonium chloride), which was first prepared from the dung of camels at the temple of Jupiter Ammon in North Africa. The process of death and decay releases the ammonium ion in both plants and animals. Denitrifying bacteria reverse the conversion of the ammonium ion to nitrate and then recycle NO_3^- as free N_2 (Figure 23.1, *middle*).

The topic of nitrogen metabolism includes the biosynthesis and breakdown of *amino acids, purines,* and *pyrimidines*; the metabolism of *porphyrins* is also related to that of amino acids. Many of these pathways, particularly the anabolic ones, are long and complex. In discussing pathways in which the amount of material is large and highly detailed, we concentrate on the most important points. Specifically, we concentrate on overall patterns and on interesting reactions of wide applicability. We are also interested in health-related aspects of this material.

Nitrogen from the atmosphere is converted to the ammonium ion by nitrogen-fixing bacteria, and then that nitrogen can be passed to plants and animals. A huge chemistry breakthrough occurred when Fritz Haber and Carl Bosch learned to fix atmospheric nitrogen into fertilizer, revolutionizing agriculture (Figure 23.1, *top left*).

Careers in Chemistry • Annemarie Ross

Associate Professor at the National Technical Institute for the Deaf/Rochester Institute of Technology

Education:
University at Buffalo: PhD, Curriculum, Instruction, and the Science of Learning
Rochester Institute of Technology: MS, Interdisciplinary Degree in Biotechnology/Chemistry
Rochester Institute of Technology: BS, Biochemistry

Dr. Annemarie Ross is an Associate Professor at the National Technical Institute for the Deaf at the Rochester Institute of Technology. Her work focuses on understanding how Deaf individuals learn and developing strategies to better prepare them for scientific tasks. Her goal is to empower Deaf students to pursue careers as scientists. "I make sure to emphasize to my own students, as well as colleagues, that Deafness is not a barrier to working in the laboratory sciences," she shares.

Dr. Ross went through a nontraditional career path to get where she is today. After earning her BS in biochemistry, she worked as an assistant chemical coordinator at IBM. Finding herself eager to learn more biochemistry, she went back to school to pursue an MS degree and worked as a graduate teaching assistant to support her studies. It was there that she fell in love with teaching and decided to make that her primary focus.

Dr. Ross's passion for supporting the Deaf community is deeply rooted in her own experiences. As a Deaf student, she was fortunate to receive strong support—from her parents' advocacy to mentors who valued her unique perspective in scientific research. She acknowledges that Deaf students often face challenges but believes these can be overcome with careful preparation and practical strategies, such as comparing notes with peers and self-advocacy skills. "It's a passion of mine to help my own community," she says, "and one of the best ways I do that is by researching how Deaf students learn best. I integrate those skills into my science courses so they can become great scientists."

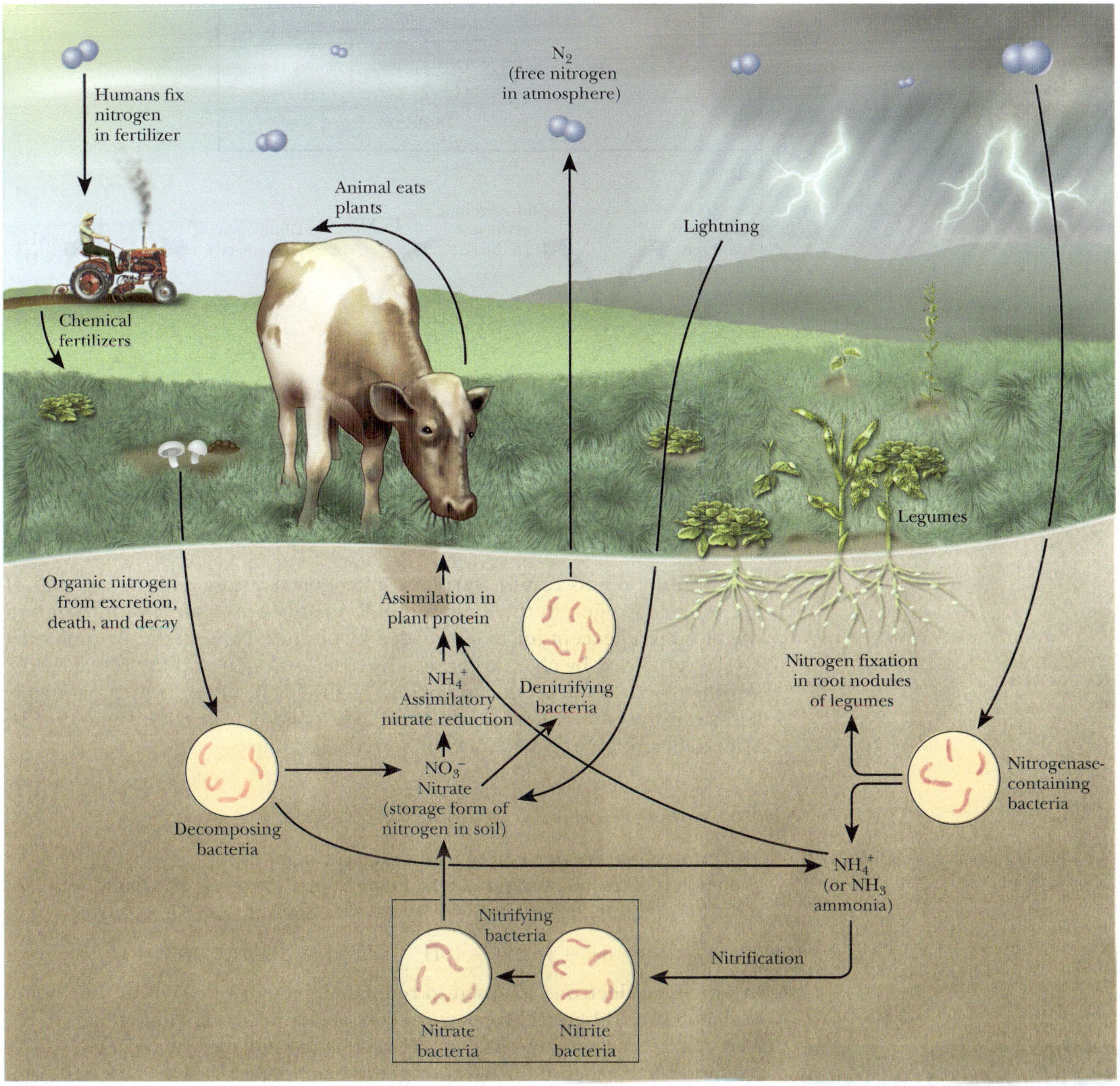

Figure 23.1 The flow of nitrogen between the atmosphere and the biosphere.

23-2 Nitrogen Fixation

Learning Objectives

After reading this section, students will be able to:

LO 23.2.1 Describe the reactions that occur in the nitrogenase complex.

LO 23.2.2 Explain the symbiotic relationship between nitrogen-fixing bacteria and legumes.

Bacteria are responsible for the reduction of N_2 to the ammonium ion (NH_4^+). Typical nitrogen-fixing bacteria are symbiotic organisms that form nodules on the roots of leguminous plants such as beans and alfalfa (Figure 23.1, *middle right*). Many free-living microbes and some cyanobacteria also fix nitrogen.

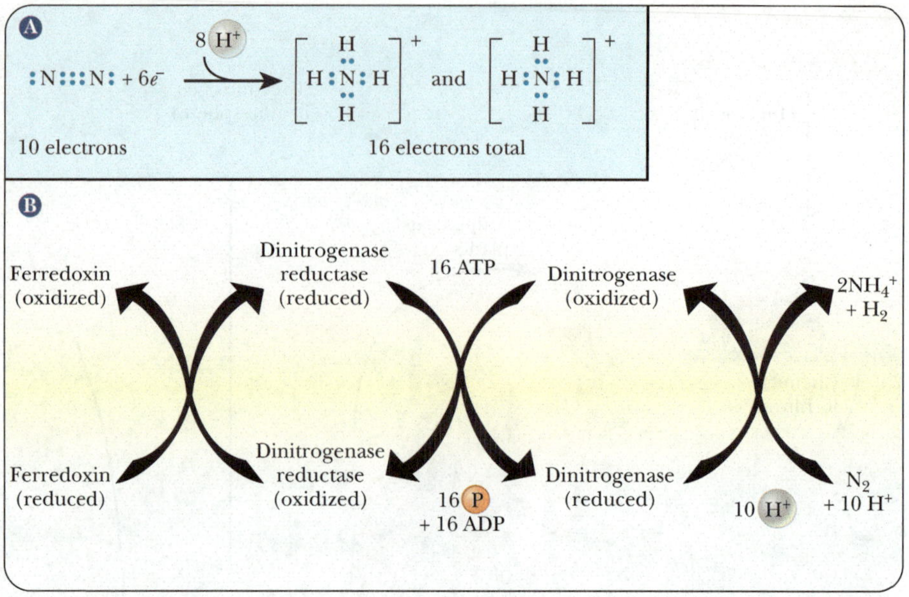

Figure 23.2 Some aspects of the nitrogenase reaction. (A) The reduction of N_2 to $2NH_4^+$. (B) The path of electrons from ferredoxin to N_2.

Plants and animals cannot carry out nitrogen fixation. This conversion of molecular nitrogen to the ammonium ion is the only source of nitrogen in the biosphere, except for that provided by nitrates. Parenthetically, ammonia (NH_3), obtained by chemical synthesis from nitrogen and hydrogen, is the starting point for the production of many synthetic fertilizers, which frequently contain nitrates. Fertilizer production, developed between 1909 and 1913 by Fritz Haber and Carl Bosch, revolutionized agriculture.

▶ How is nitrogen from the atmosphere incorporated into the ammonium ion?

nitrogenase the enzyme complex that catalyzes nitrogen fixation from atmospheric nitrogen to ammonium ion

The **nitrogenase** enzyme complex found in nitrogen-fixing bacteria catalyzes the production of the ammonium ion from molecular nitrogen. The half-reaction of reduction (Figure 23.2A) is

$$N_2 + 8e^- + 16ATP + 10H^+ \rightarrow 2NH_4^+ + 16ADP + 16P_i + H_2$$

in which six electrons are used to reduce molecular nitrogen to the ammonium ion. An additional two electrons are used to reduce hydrogen ions to H_2. The total reaction catalyzed by nitrogenase is an eight-electron reduction. The oxidation half-reaction in the nitrogenase complex varies because different organisms use different substances to supply electrons (i.e., to be oxidized).

Two protein components make up the nitrogenase complex (Figure 23.2B). Ferredoxin (Fd) is not part of the complex, but it is a soluble electron carrier that delivers electrons to the nitrogenase complex (recall that Fd also plays an important role in electron transfer in photosynthesis; Section 22-2B). The two proteins specific to the nitrogenase reaction are an iron–sulfur (Fe–S) protein called *dinitrogenase reductase*, and an iron–molybdenum (Fe–Mo) protein called *dinitrogenase*. The flow of electrons is from ferredoxin to dinitrogenase reductase to dinitrogenase (with ATP hydrolysis) to nitrogen.

The nature of the nitrogenase complex is a subject of active research. Significant progress was made in this work with the determination by X-ray crystallography of the three-dimensional structure of both the Fe protein and the Fe–Mo protein from *Azotobacter vinelandii* (Figure 23.3). The Fe protein is a dimer ("the iron butterfly"), with the iron–sulfur cluster located at the butterfly's head. The nitrogenase is even more complicated, with several types of subunits arranged into tetramers.

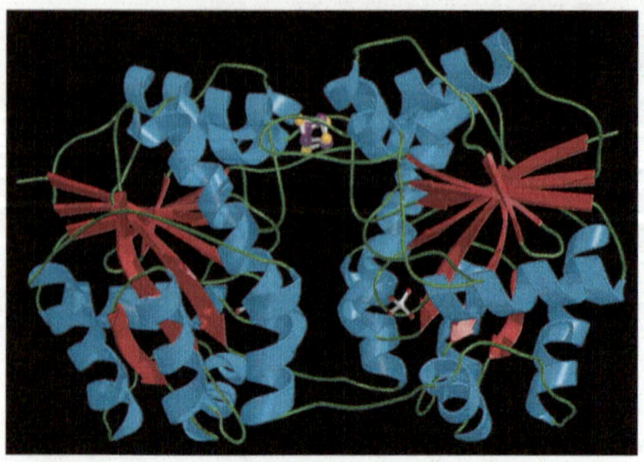

Figure 23.3 The X-ray structure of the *Azotobacter vinelandii* Fe–protein dimer. *(From Crystallographic structure of the nitrogenase iron protein from Azotobacter vinelandii. MM Georgiadis, H Komiya, P Chakrabarti, D Woo, JJ Kornuc, and DC Rees. (September 18, 1992)* Science *257 (5077), 1653. Reprinted with permission from AAAS.)*

23A Biochemical Connections • Plant Science

Why Is the Nitrogen Content of Fertilizers So Important?

Crop production increased dramatically as a result of extensive use of fertilizers, especially those that supply nitrogen in a form that plants can use readily. Both ammonium and nitrate ions are used in fertilizers; even ammonia gas can be pumped into the ground if enough water is available in the soil to dissolve it.

Ammonia is toxic to animals, so it often comes as a surprise that ammonia gas itself can be used for fertilization. Plants can assimilate the ammonium ion rapidly, but they usually never get the chance to do so because the nitrifying soil bacteria, especially *Nitrosomonas* and *Nitrobacter*, rapidly convert the ammonium ion first to nitrite and then to nitrate. The final nitrate product is easily converted back to the ammonium ion, but the process requires energy. Figure 23.4 shows nodules formed on the roots of alfalfa by nitrogen-fixing bacteria that naturally produce the ammonium ion.

Ammonium ion is especially useful as a fertilizer in the early spring and for germinating plants. In the spring, the soil is usually damp enough to dissolve the ammonium ion so that it can move to the plants. Because light is less available in the early spring, the young plants do not have enough energy to convert the nitrate back to the ammonium ion until their chloroplasts develop fully. Fortunately, because of the condition of the soil, the ammonium ion goes directly to the plants rather than to the soil bacteria.

The genes for the nitrogenase complex enzymes for nitrogen fixation have been studied extensively. Much research is under way to determine whether these genes can be incorporated into crop plants, which would reduce the amount of nitrogen fertilizer needed for maximal plant growth and crop production.

Two other sources of nitrogen fixation are often overlooked. The first is the chemical synthesis of ammonia from H_2 and N_2, called the Haber–Bosch process, after its discoverers, the German chemists Fritz Haber and Carl Bosch. This reaction is very important for the formation of chemical fertilizers, and it is responsible for a great deal of the organic nitrogen currently found in the biosphere. The second source of fixed nitrogen is that produced by lightning. The intense energy of a lightning strike breaks down nitrogen molecules, allowing them to react with oxygen to form nitrate ions. ◗

Figure 23.4 Nitrogen-fixing bacteria form nodules on alfalfa roots.

Other studies of the Fe–Mo protein by X-ray crystallography have revealed details that may shed light on the mechanism of the nitrogenase reaction. A single-carbon unit is inserted at the active site of the enzyme, and that insertion requires *S*-adenosylmethionine (SAM), a one-carbon transfer agent that we discuss in Section 23-4 in its role in amino acid biosynthesis. The removal of hydrogens from a methyl group inserted into the active site may play an important role in the mechanism of the nitrogenase reaction. Ferredoxin, dinitrogenase reductase, and dinitrogenase combine to perform a series of single-electron transfers, eventually transferring the eight electrons necessary to complete the reduction of N_2 to NH_4^+.

The reactions of nitrogen fixation consume a great deal of energy (Figure 23.2B, *middle*). It is estimated that about half of the ATP produced from photosynthesis in legumes is used to fix nitrogen. BIOCHEMICAL CONNECTIONS 23A discusses some important consequences of nitrogen fixation.

23-3 Feedback Inhibition in Nitrogen Metabolism

Learning Objectives

After reading this section, students will be able to:

LO 23.3.1 Write the reaction in which glutamate is converted to glutamine.

LO 23.3.2 Describe how glutamine-precursor products participate in feedback inhibition.

feedback inhibition the process by which the final product of a series of reactions inhibits the first reaction in the series

The biosynthetic pathways that produce amino acids and the bases of nucleotides (purines and pyrimidines) are long and complex, requiring a large investment of energy by the organism. If there is a high level of some end product, such as an amino acid or a nucleotide, the cell saves energy by not producing that compound. However, the cell needs a signal to tell it to stop producing more of that particular compound. The signal is frequently part of a **feedback inhibition** mechanism, in which the end product of a metabolic pathway inhibits an enzyme at the beginning of the pathway. We saw an example of such a control mechanism when we discussed the allosteric enzyme aspartate transcarbamoylase in Section 7-1. This enzyme plays a role in nitrogen metabolism by catalyzing one of the early stages of pyrimidine nucleotide biosynthesis. It is inhibited by the end product of that pathway—namely, the nitrogenous base-containing nucleotide called cytidine triphosphate (CTP).

Feedback inhibition is frequently encountered in the biosynthesis of other nitrogen-containing compounds. Another prime example of allosteric regulation by feedback inhibition is found in the activity of the enzyme glutamine synthetase, one of the key enzymes in amino acid biosynthesis (Figure 23.5). Nine allosteric inhibitors are involved in the regulation of

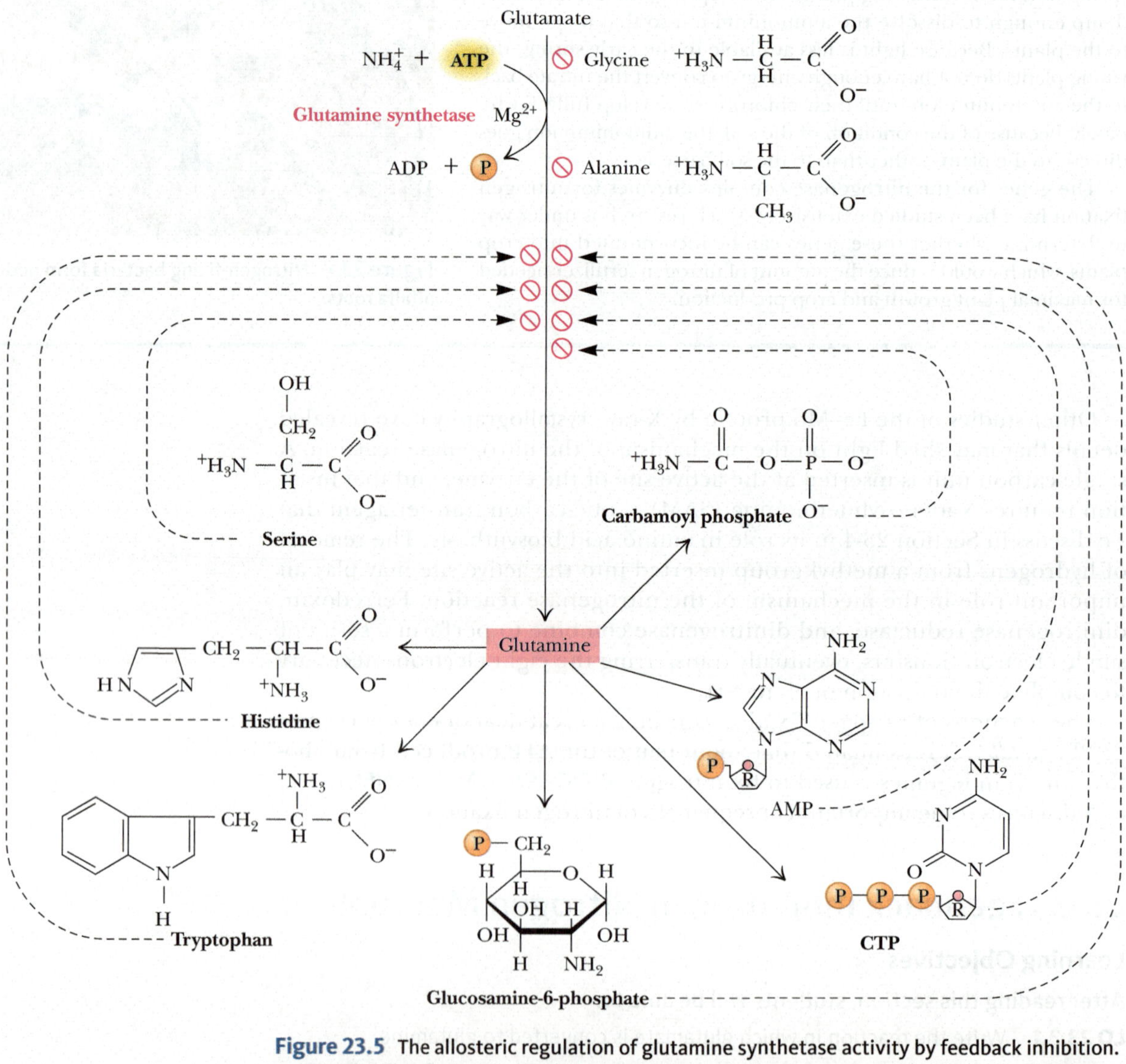

Figure 23.5 The allosteric regulation of glutamine synthetase activity by feedback inhibition. Metabolites that derive from glutamine (*middle*) feedback and inhibit the glutamine synthetase enzyme that produces their glutamine precursor.

glutamine synthetase: glycine, alanine, serine, histidine, tryptophan, CTP, AMP, carbamoyl phosphate, and glucosamine-6-phosphate. Glycine, alanine, and serine are key indicators of amino acid metabolism in the cell. Each of the other six compounds represents an end product of a biosynthetic pathway that depends on glutamine (Figure 23.5, *middle*). Feedback inhibition is very effective because a single product molecule can inhibit an enzyme that affects the synthesis of many other product molecules.

23-4 Amino Acid Biosynthesis

Learning Objectives

After reading this section, students will be able to:

LO 23.4.1 Write reactions for the biosynthesis of glutamate and glutamine.

LO 23.4.2 Explain transamination reactions and one-carbon transfer reactions in amino acid metabolism.

LO 23.4.3 Classify amino acid families based on common carbon-based precursors.

Ammonium ion is toxic in high concentrations, and so it must be incorporated into biologically useful compounds when it is formed by the reactions of nitrogen fixation discussed earlier in this chapter. The amino acids glutamate and glutamine are of central importance in the process. Glutamate is formed from NH_4^+ and α-ketoglutarate in a **reductive amination** reaction that requires NADPH (Figure 23.6A). This reaction is reversible and is catalyzed by *glutamate dehydrogenase* (GDH). Glutamate is a major donor of amino groups in reactions, and α-ketoglutarate is a major acceptor of amino groups. The reducing power in the form of NADPH is provided by the pentose phosphate pathway (Section 18-4). The conversion of glutamate to glutamine is catalyzed by *glutamine synthetase* (GS) in an **amidation** reaction that requires ATP (Figure 23.6B).

reductive amination a chemical reaction in which a carbonyl group (in an aldehyde or ketone) is converted into an amine group

amidation a chemical reaction in which an amine group is connected to a carbonyl group, forming an amide group

These reactions fix inorganic nitrogen (NH_4^+), forming organic (carbon-containing) nitrogen compounds, such as amino acids, but they frequently do not operate in this sequential fashion. In fact, the combination of GDH and GS is responsible for most of the assimilation of the ammonium ion into organic compounds, especially in organisms that are rich in nitrogen sources. However, the K_M of GS is considerably lower than that of GDH. When nitrogen is limiting, as is frequently the case in plants, the conversion of glutamate to glutamine is the preferred mode of nitrogen assimilation. This means that the supply of glutamate becomes depleted unless there is some way to replenish it. The reductive amination of α-ketoglutarate with the amide nitrogen of glutamine as the nitrogen source is how this is done.

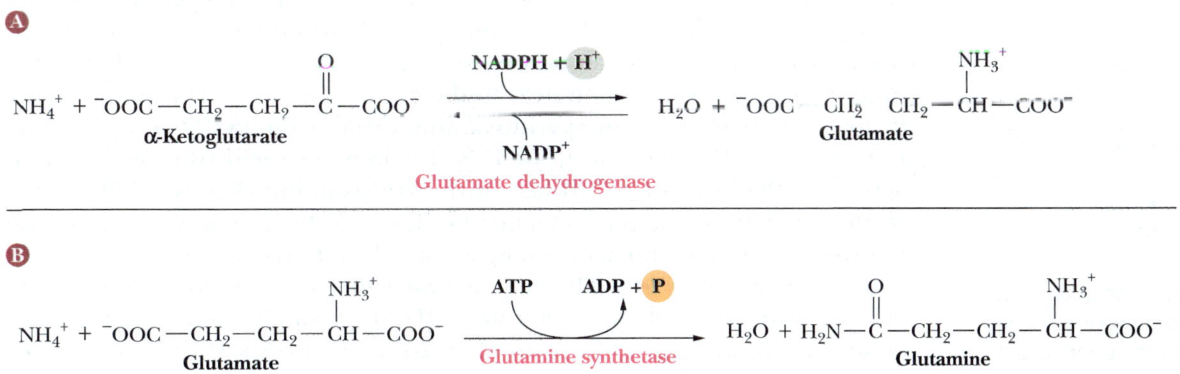

Figure 23.6 Biosynthesis of glutamate and glutamine. (A) The production of glutamate from α-ketoglutarate. (B) The production of glutamine from glutamate.

$$\text{Reductant} + \alpha\text{-Ketoglutarate} + \text{Glutamine} \rightarrow 2\ \text{Glutamate} + \text{Oxidized reductant}$$

The reductant can be NADH, NADPH (in yeast and bacteria), or reduced ferredoxin (in plants). The enzyme that catalyzes this reaction is *glutamate synthase*; it is also known as glutamate:oxoglutarate aminotransferase (GOGAT). A GS/GOGAT complex exists in plants and allows them to cope with conditions of limited nitrogen availability.

transamination the transfer of amino groups from one molecule to another; an important process in the anabolism and catabolism of amino acids

In other reactions of amino acid anabolism, the α-amino group of glutamate and the side-chain amino group of glutamine are transferred to other compounds in **transamination** reactions (discussed in Section 23-4B).

23-4A Common Features in Amino Acid Metabolism

The biosynthesis of amino acids involves a common set of reactions. In addition to transamination reactions, the transfer of one-carbon units, such as formyl or methyl groups, occurs frequently. We are not going to discuss all the details of the reactions that give rise to amino acids. We can, however, organize this material by grouping amino acids into families based on common precursors (Figure 23.7). The reactions of some of individual families of amino acids provide good examples of reactions that are of general importance, such as transamination reactions and one-carbon transfer reactions.

We can also make some generalizations about amino acid metabolism in terms of the relationship of the carbon skeleton to the citric acid cycle and the related reactions of pyruvate and acetyl-CoA (Figure 23.8). The citric acid cycle is amphibolic; it plays a part in both catabolism and anabolism. The anabolic aspect of the citric acid cycle contributes to amino acid biosynthesis, while the catabolic aspect is involved in the breakdown of amino acids.

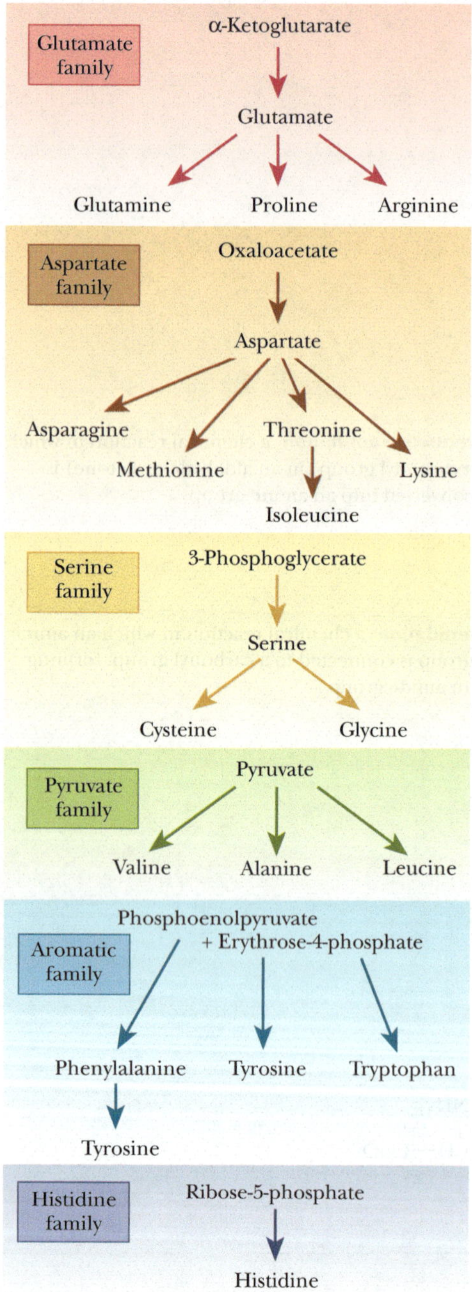

Figure 23.7 Families of amino acids based on biosynthetic pathways. The nitrogen-containing amino acids in each family have a common carbon-based precursor.

23-4B Transamination Reactions

▶ How are transamination reactions important in amino acid biosynthesis?

Enzymes that catalyze transamination reactions, in which an amino group is transferred from one molecule to another, are called *aminotransferases*, and they require pyridoxal phosphate (PyrP) as a coenzyme (Figure 23.9A). We discussed this compound in Section 7-8 as a typical example of a coenzyme, and here we can see its mode of action in context. Recall from that section and Table 7.1 that pyridoxine is also known as vitamin B_6.

In transamination reactions, PyrP forms a Schiff base with the amino group of Substrate I (the amino-group donor) (Figure 23.9B, *top left*). The next stage is a rearrangement followed by hydrolysis, which removes Product I (an α-keto acid conjugate of Substrate I) (Figure 23.9B, *top right*). Recall from Section 3-2 that the Greek-letter notation indicates the distance from a carbonyl group. Therefore, an α-keto acid has a ketone group (R—C=O—R′) on the α-carbon (the carbon next to the carbonyl group). The amino group is now being carried as the pyridoxamine form of the PyrP coenzyme (Figure 23.9B, *middle right*). Substrate II is also an α-keto acid (the amino-group acceptor) that forms a Schiff base with pyridoxamine (Figure 23.9B, *bottom right*). Again, there is a rearrangement followed by hydrolysis, which gives rise to Product II (an amino acid conjugate of Substrate II) and regenerates PyrP (Figure 23.9B, *bottom left*). The net reaction is that an α-amino acid (Substrate I) reacts with an α-keto acid (Substrate II) in a transamination reaction (an amino-group transfer) to form another α-keto acid (Product I) and another α-amino acid (Product II).

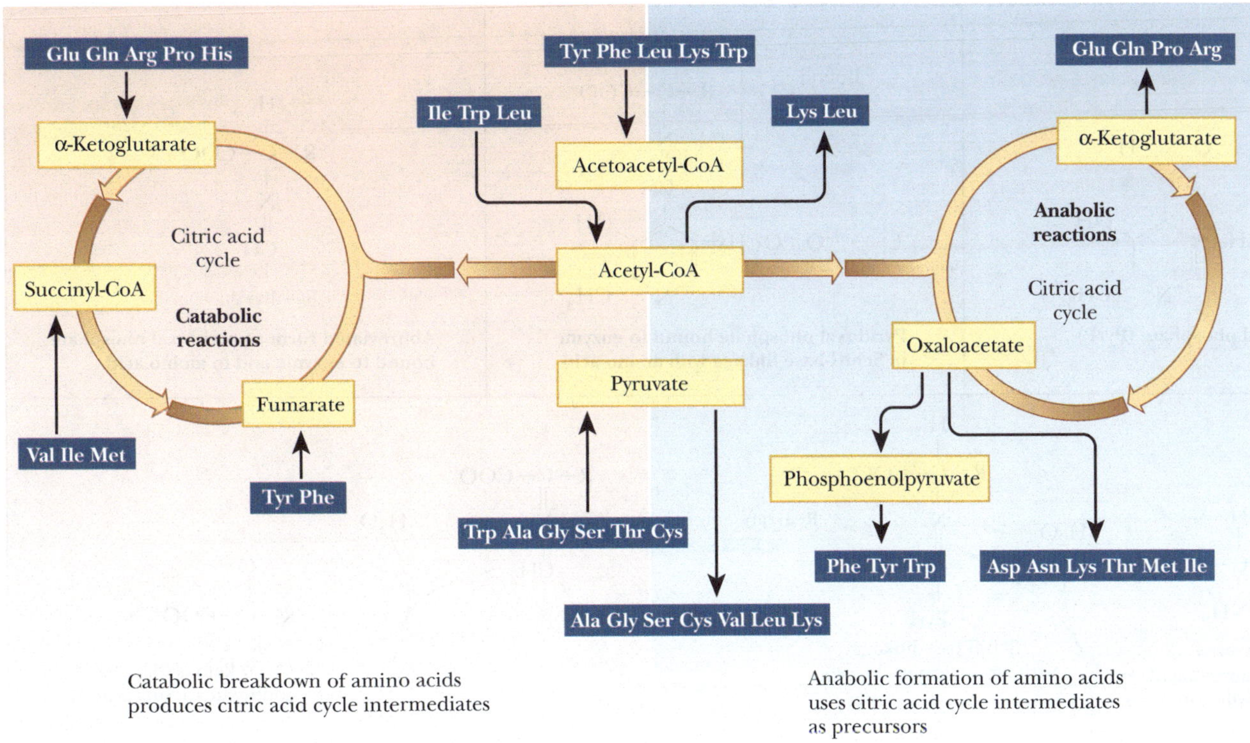

Figure 23.8 The relationship between amino acid metabolism and the citric acid cycle. Amino acid catabolism relationships with the citric acid cycle are shown on the left, and amino acid anabolism relationships with the citric acid cycle are shown on the right.

Figure 23.10A shows a general transamination reaction, and Figure 23.10B shows a specific example that leads to the biosynthesis of aspartate. In amino acid biosynthesis transamination reactions, glutamate is usually the amino-group donor. It becomes α-ketoglutarate after the reaction. When not involved with one of the substrates, the pyridoxal group is bound in a Schiff-base linkage to an active site ε-NH_2 group of lysine. Pyridoxal phosphate is a versatile coenzyme that is also involved in other reactions, including decarboxylations, racemizations, and the movement of hydroxymethyl groups, as in the conversion of serine to glycine (see Section 23-4C).

23-4C One-Carbon Transfer Reactions

▶ What is the importance of one-carbon transfer reactions in amino acid metabolism?

In addition to transamination reactions, one-carbon transfer reactions are crucial in amino acid biosynthesis. A good example of a one-carbon transfer is found in the reactions that produce the amino acids of the serine family. This family also includes glycine and cysteine (Figure 23.7, *middle*). Serine and glycine are frequently precursors in other biosynthetic pathways. A discussion of the synthesis of cysteine also provides some insight into the metabolism of sulfur, as well as that of nitrogen.

The ultimate precursor of serine is 3-phosphoglycerate (3PG), which is obtainable from Step 7 of the glycolysis pathway (Section 17-3). Other than continuing in the glycolysis pathway by having its phosphate group moved to carbon 2 (see Step 8 of glycolysis in Section 17-3), 3PG can be used in serine

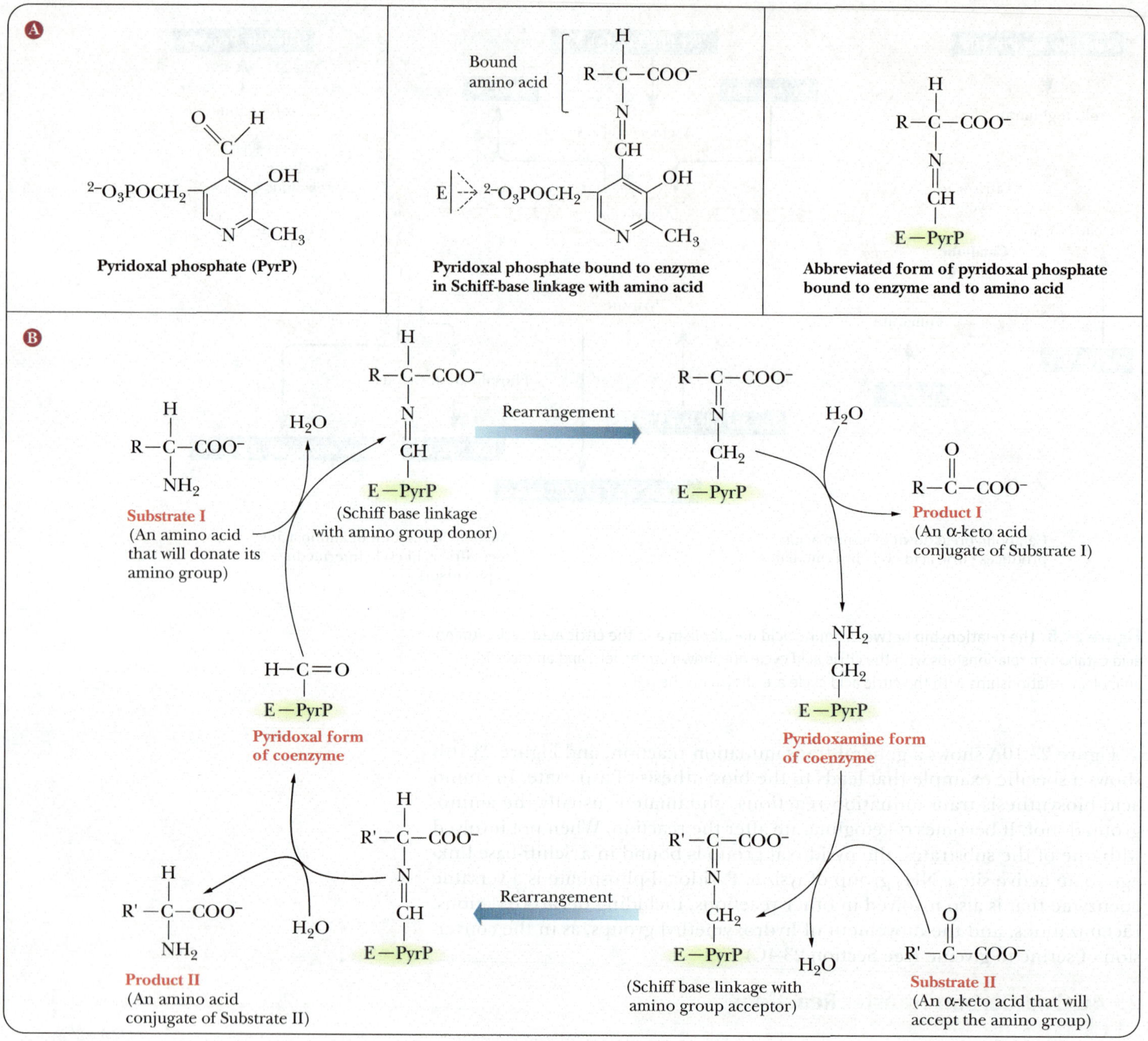

Figure 23.9 The role of pyridoxal phosphate in transamination reactions. (A) The mode of binding of pyridoxal phosphate (PyrP) to an aminotransferase enzyme (E) and to the substrate amino acid. The Schiff-base linkage is also involved when PyrP is bound to the product amino acid. (B) The transamination reaction itself. The original substrate, an amino acid, is deaminated, while an α-keto acid is aminated to form an amino acid. The net reaction is one of amino-group transfer. The PyrP coenzyme is regenerated at the end of the reaction.

biosynthesis, as seen in Figure 23.11. The hydroxyl group on carbon 2 of 3PG can be oxidized to a keto group in a reaction catalyzed by 3-phosphoglycerate dehydrogenase, giving an α-keto acid called 3-phosphohydroxypyruvate. A transamination reaction catalyzed by phosphoserine aminotransferase, in which glutamate is the nitrogen donor, produces 3-phosphoserine and α-ketoglutarate. Hydrolysis of the phosphate group of 3-phosphoserine by phosphoserine phosphatase then gives rise to serine (Figure 23.11, *bottom*).

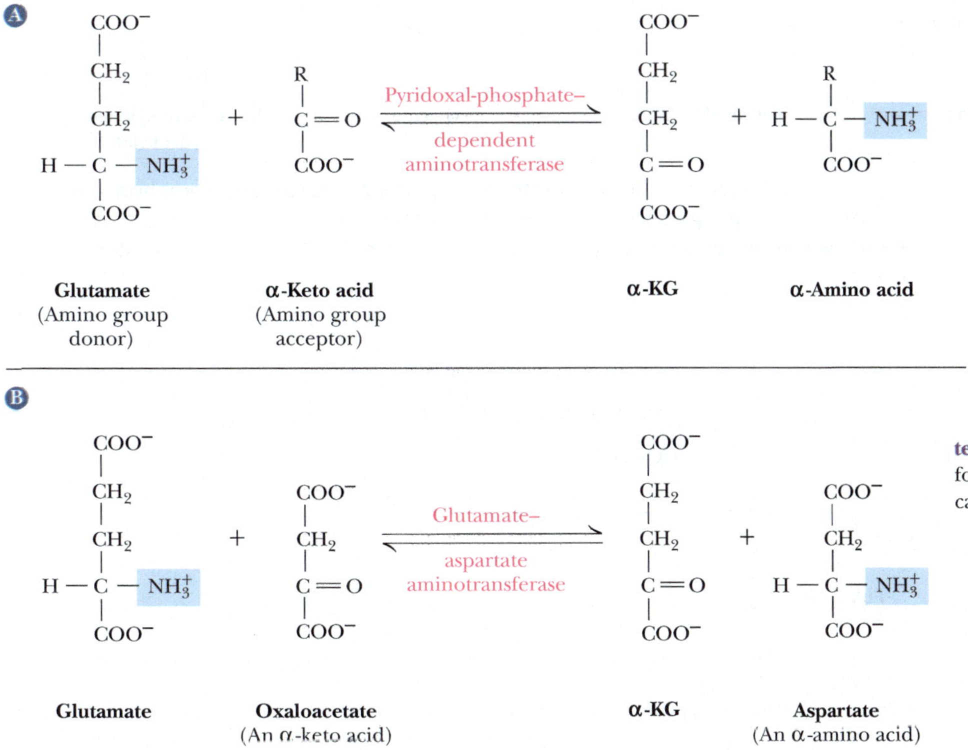

Figure 23.10 Transamination reactions switch an amino group from glutamate to an α-keto acid to produce various other amino acids. Glutamate and α-ketoglutarate (α-KG) are one donor/acceptor pair. (A) A general example of a transamination reaction. (B) A specific example of a transamination reaction, in which the other donor/acceptor pair is aspartate and oxaloacetate. See Figure 23.9B for specific details of how the pyridoxal phosphate (PyrP) coenzyme aids these reactions.

For the biosynthesis of glycine, serine is usually used in a one-carbon transfer reaction catalyzed by serine hydroxymethyltransferase (SHMT). SHMT uses **tetrahydrofolate (THF)** as a coenzyme that carries a one-carbon unit, and it also uses pyridoxal phosphate (Figure 23.12). Tetrahydrofolate is a derivative of folic acid, and it is a frequently encountered carrier of one-carbon units in metabolic pathways. Its structure has three parts: a substituted pteridine ring, *p*-aminobenzoic acid, and glutamic acid (Figure 23.13A, *top*). In addition to the N^5,N^{10}-methylenetetrahydrofolate version of THF that is involved in the conversion of serine to glycine, Figure 23.13A also shows five other ways that THF can carry a one-carbon unit for one-carbon transfer reactions. Figure 23.13B shows a schematic of the reactions that these different forms of carbon-carrying THF molecules are involved in. Folic acid is a vitamin that has been identified as essential in preventing birth defects; consequently, it is now a recommended supplement for all females of childbearing age. There is also some evidence that folic acid may prevent heart disease in both males and females over the age of 50.

Tetrahydrofolate is not the only carrier of one-carbon units. We have previously encountered biotin, a carrier of CO_2, and we have discussed the role that biotin plays in gluconeogenesis (Section 18-2) and in the anabolism of fatty acids (Section 21-6). *S*-adenosylmethionine (SAM) is another one-carbon carrier.

The conversion of serine to cysteine involves some interesting reactions. The source of the sulfur in animals differs from that in plants and bacteria. In plants and bacteria, serine is acetylated to form *O*-acetylserine. This reaction is catalyzed by *serine acetyltransferase (SAT),* with acetyl-CoA as the acetyl-group

tetrahydrofolate (THF) the metabolically active form of the vitamin folic acid; a carrier of one-carbon groups

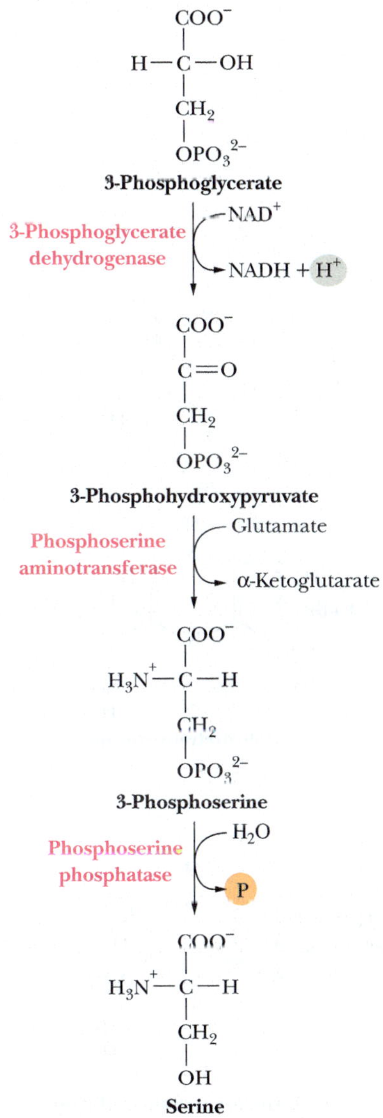

Figure 23.11 The biosynthesis of serine. 3-Phosphoglycerate, a metabolite from the glycolysis pathway, can be used in this metabolic pathway that produces the serine amino acid.

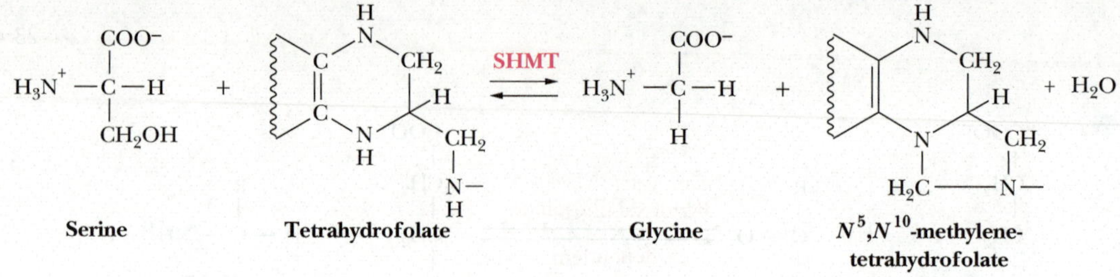

Figure 23.12 The conversion of serine to glycine, showing the role of the tetrahydrofolate coenzyme in accepting a one-carbon unit. This one-carbon transfer reaction is catalyzed by serine hydroxymethyltransferase (SHMT). It also uses pyridoxal phosphate as an additional coenzyme.

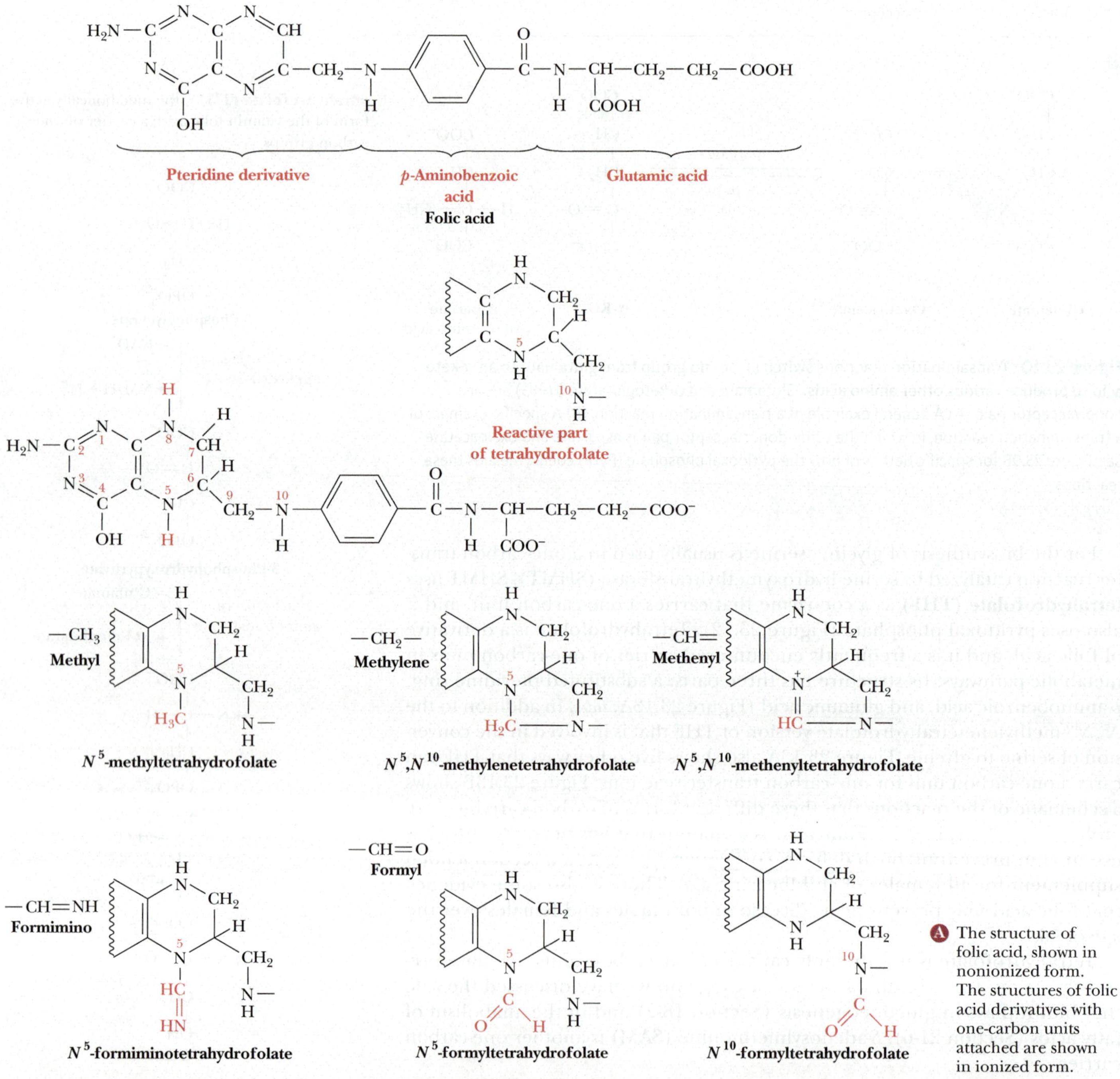

Figure 23.13 Structure and reactions of folic acid. (A) The structure of folic acid, shown in non-ionized form. (B) The reactions that introduce one-carbon units into tetrahydrofolate (THF) link seven different folate intermediates that carry one-carbon units in three different oxidation states (−2, 0, and +2). *(Adapted from T. Brody et al., in L. J. Machlin,* Handbook of Vitamins. *New York: Marcel Dekker, 1984.)*

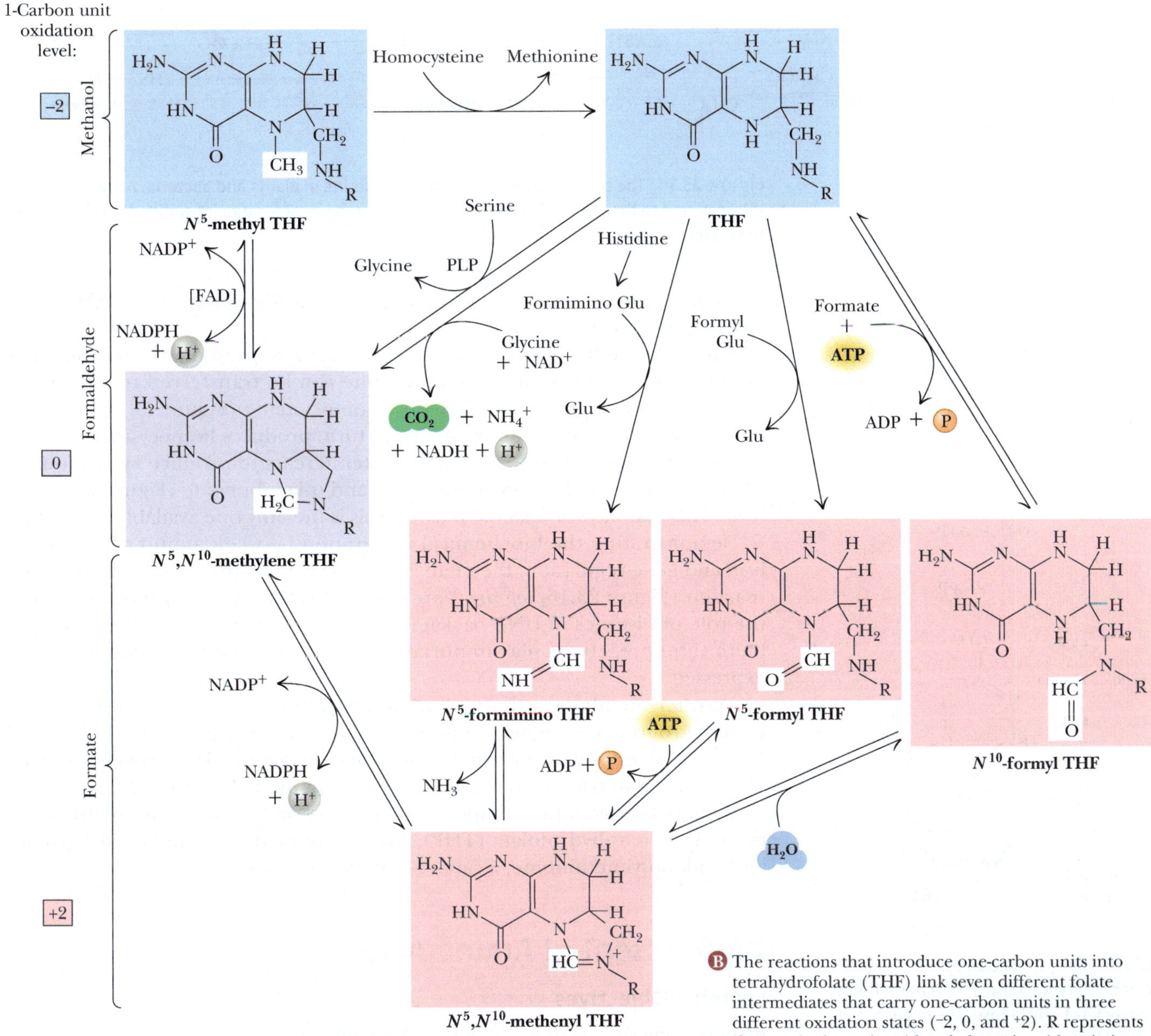

B The reactions that introduce one-carbon units into tetrahydrofolate (THF) link seven different folate intermediates that carry one-carbon units in three different oxidation states (−2, 0, and +2). R represents the *p*-aminobenzoic acid and glutamic acid moieties that form the rest of these THF-derivative molecules.

Figure 23.13 continued

donor (Figure 23.14, *left*). Conversion of *O*-acetylserine to cysteine is catalyzed by *O*-acetylserine(thiol)lyase (OASTL). It uses sulfide (S^{2-}) produced in the *assimilatory sulfate reduction* pathway, which uses sulfate (SO_4^{2-}) taken up from the environment by plants and bacteria.

Animals form cysteine from serine by a different pathway because they do not have the enzymes to carry out the sulfate-to-sulfide conversion seen in plants and bacteria. Instead, the cysteine biosynthesis reaction sequence in animals involves the amino acid methionine.

Methionine, which is produced by reactions of the aspartate family in bacteria and plants, cannot be produced by animals. It must be obtained from dietary sources. It is an **essential amino acid** because it cannot be synthesized by the body. The ingested methionine reacts with ATP to form ***S*-adenosylmethionine (SAM)**,

essential amino acid an amino acid that cannot be synthesized by the body and must therefore be obtained in the diet

***S*-adenosylmethionine (SAM)** an important carrier molecule of methyl groups in amino acid metabolism

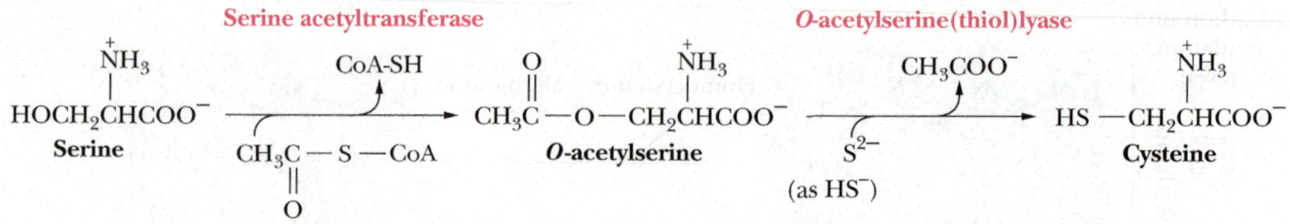

Figure 23.14 The biosynthesis of cysteine from serine in plants and bacteria. An acetyl group is transferred to the side chain of serine, followed by the incorporation of sulfide as HS^-, forming the cysteine amino acid.

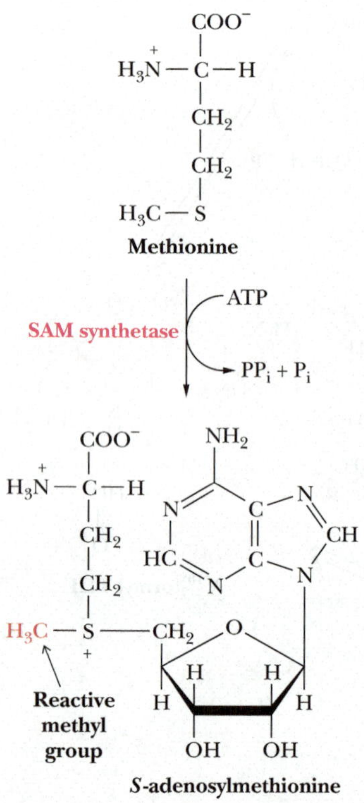

Figure 23.15 The production of S-adenosylmethionine (SAM) from methionine, catalyzed by SAM synthetase. Methionine is an essential amino acid that must come from dietary sources.

which has a highly reactive methyl group, in a reaction catalyzed by SAM synthetase (Figure 23.15).

S-adenosylmethionine is a carrier of methyl groups in many reactions. The methyl group from *S*-adenosylmethionine can be transferred to any one of several acceptors, producing *S*-adenosylhomocysteine (Figure 23.16, *top right*). Hydrolysis of *S*-adenosylhomocysteine, in turn, produces homocysteine (Figure 23.16, *bottom right*). Serine and homocysteine react to produce cystathionine, which hydrolyzes to form cysteine, NH_4^+, and α-ketobutyrate (Figure 23.16, *bottom*). This pathway for cysteine biosynthesis is the only one available to animals.

Demonstrating the biochemical integration between amino acid metabolism and gene regulation, the methyl group acceptors in the methyltransferase reaction (Figure 23.16, *top right*) are sometimes DNA bases or histones (recall the role of histones in DNA packaging from Section 9-3C). DNA and histone methylation reactions play important roles in determining which genes are expressed.

Another metabolic role for *S*-adenosylmethionine has also been discovered. A group of researchers in Boston showed that threonine metabolism and *S*-adenosylmethionine can regulate histone methylation. This process has been observed to give rise to pluripotent stem cells in mouse embryos.

We have now seen three important carriers of one-carbon units: biotin, a carrier of CO_2; tetrahydrofolate (THF), a carrier of methylene and formyl groups; and *S*-adenosylmethionine, a carrier of methyl groups.

23-5 Essential Amino Acids

Learning Objectives

After reading this section, students will be able to:

LO 23.5.1 Describe the importance of essential amino acids.

LO 23.5.2 Identify the essential amino acids in human nutrition.

The biosynthesis of proteins requires the presence of all the constituent amino acids. If one of the 20 amino acids is missing or in short supply, protein biosynthesis is inhibited. Some organisms, such as *Escherichia coli,* can synthesize all the amino acids they need. Other species, including humans, must obtain essential amino acids from dietary sources. The essential amino acids in human nutrition are listed in Table 23.1. The body can synthesize some of these amino acids, but not in sufficient quantities for its needs, especially in the case of growing children. This last point applies particularly to children's requirements for arginine and histidine. Amino acids are not stored (except in proteins), and dietary sources of essential amino acids are needed at regular intervals. Protein deficiency, especially a prolonged deficiency in sources that contain essential amino acids, leads to the disease *kwashiorkor.* The problem in this disease, particularly severe in growing children, is not only starvation but also the breakdown of the body's own proteins.

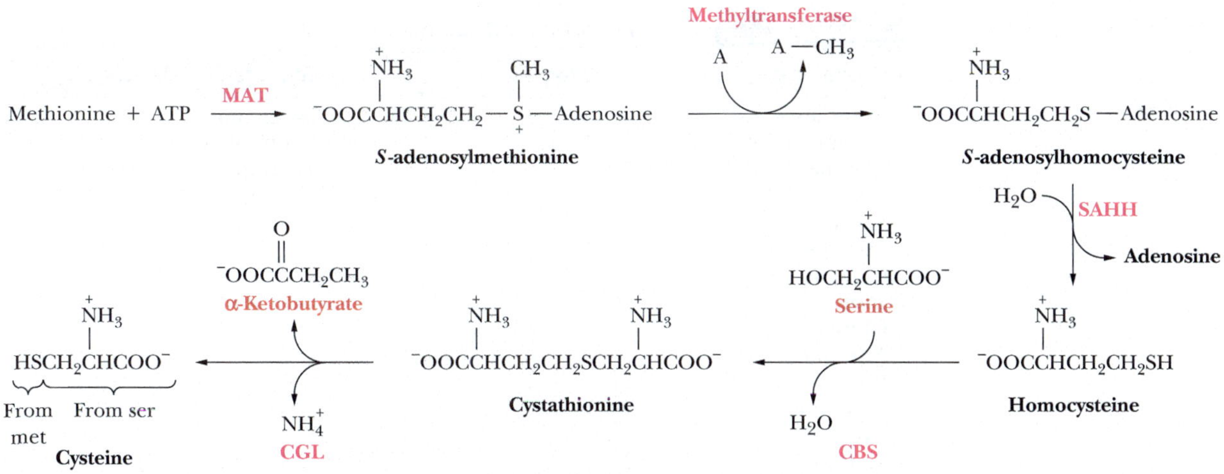

Figure 23.16 The biosynthesis of cysteine in animals. MAT stands for methionine adenosyltransferase. "A" in the methyltransferase reaction stands for *acceptor*, and acceptors of the methyl group can be molecules such as DNA, histones, RNA, and various metabolites. SAHH stands for *S*-adenosylhomocysteine hydrolase, CBS stands for cystathionine β-synthase, and CGL stands for cystathionine γ-lyase.

Table 23.1 Amino Acid Requirements in Humans

Essential	Nonessential
Arginine*	Alanine
Histidine†	Asparagine
Isoleucine	Aspartate
Leucine	Cysteine
Lysine	Glutamate
Methionine	Glutamine
Phenylalanine	Glycine
Threonine	Proline
Tryptophan	Serine
Valine	Tyrosine

*Mammals synthesize arginine but cleave most of it to urea (Section 23-6).
†Essential for children but not necessarily for adults.

23-6 Amino Acid Catabolism and Nitrogen Excretion

Learning Objectives

After reading this section, students will be able to:

LO 23.6.1 Distinguish between *glucogenic amino acids* and *ketogenic amino acids.*

LO 23.6.2 Identify the different compounds in which nitrogen is excreted.

LO 23.6.3 Describe the reactions of the urea cycle.

When we specifically focus on the catabolism of amino acids, the first step we consider is the removal of nitrogen by transamination (see the reverse reaction in Figure 23.10A). In these transamination reactions, the amino nitrogen of the amino acid that is to be degraded is transferred to α-ketoglutarate to produce glutamate, leaving behind the carbon skeleton of the amino acid in the form of an α-keto acid. The fates of the carbon skeleton and of the nitrogen that is now attached to glutamate are considered separately.

Table 23.2 Glucogenic and Ketogenic Amino Acids

Glucogenic	Ketogenic	Glucogenic and Ketogenic
Aspartate	Leucine	Isoleucine
Asparagine	Lysine	Phenylalanine
Alanine		Tryptophan
Glycine		Tyrosine
Serine		
Threonine		
Cysteine		
Glutamate		
Glutamine		
Arginine		
Proline		
Histidine		
Valine		
Methionine		

23-6A Pathways for Amino Acid Carbon Skeletons

glucogenic amino acid an amino acid that has pyruvate or oxaloacetate as a catabolic breakdown product

ketogenic amino acid an amino acid that has acetyl-CoA or acetoacetyl-CoA as a catabolic breakdown product

Metabolism of the carbon skeletons of amino acids follows two general pathways, depending on the type of amino acid (Table 23.2). A **glucogenic amino acid** yields pyruvate or oxaloacetate upon degradation. Pyruvate and oxaloacetate are starting materials for the production of glucose by gluconeogenesis (Section 18-2A). A **ketogenic amino acid** breaks down to acetyl-CoA or acetoacetyl-CoA, leading to the formation of ketone bodies (Section 21-5). The carbon skeletons of amino acids give rise to metabolic intermediates such as pyruvate, acetyl-CoA, acetoacetyl-CoA, α-ketoglutarate, succinyl-CoA, fumarate, and oxaloacetate (Figure 23.8, *left*). Oxaloacetate is a key intermediate in the breakdown of the carbon skeletons of amino acids because of its dual role in the citric acid cycle and in gluconeogenesis. The amino acids degraded to acetyl-CoA and acetoacetyl-CoA are used in the citric acid cycle, but mammals cannot synthesize glucose from acetyl-CoA. This fact is the basis of the distinction between glucogenic and ketogenic amino acids. Glucogenic amino acids can be converted to glucose, with oxaloacetate as an intermediate, whereas ketogenic amino acids cannot be converted to glucose. Some amino acids have more than one pathway for catabolism, which explains why four of the amino acids are listed as both glucogenic and ketogenic.

23-6B Excretion of Excess Nitrogen

Excess nitrogen from the breakdown of nitrogen-containing molecules is excreted in one of three forms: *ammonium ions, urea,* or *uric acid* (Figure 23.17). Animals, such as fish and amphibians, that live in an aquatic environment excrete nitrogen as ammonium ions; they are protected from the toxic effects of high concentrations of ammonium ions not only by the removal of ammonium ions from their bodies but also by rapid dilution of the excreted ammonium ions by the water in the environment. The principal waste product of nitrogen metabolism in mammals and amphibians (in their terrestrial stages) is urea (a water-soluble compound). Birds and reptiles excrete nitrogen in the form of uric acid, which is insoluble in water. This is advantageous for birds because they do not have to carry the excess weight of water, which could hamper flight, to rid themselves of waste products. This is also advantageous for reptiles, which often live in dry environments, for water conservation. **BIOCHEMICAL CONNECTIONS 23B** provides further details on this point.

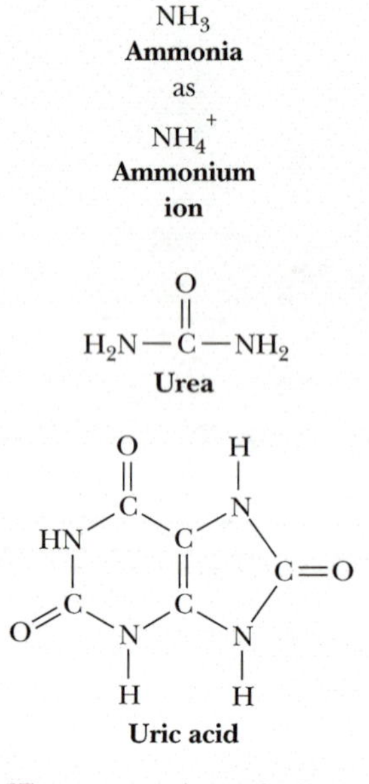

Figure 23.17 Compounds in which excess nitrogen is excreted. Fish and amphibians in water-rich environments excrete excess nitrogen as ammonium ions. Mammals and amphibians (in their terrestrial stages) excrete excess nitrogen mostly as urea. Birds and reptiles excrete nitrogen in the form of uric acid.

23B Biochemical Connections • Physiology

Water and the Disposal of Nitrogen Wastes

Ammonia gas is toxic to most organisms and must usually be disposed of rapidly. The mechanism of nitrogen-waste disposal generally depends on the amount of water available to the organism in question. For example, bacteria and fish, which live in "infinite" water supplies, usually simply release ammonium ions into the medium, where organisms that are lower on the evolutionary scale can use them. Fish sometimes produce trimethylamine, another highly water-soluble compound, which is the characteristic "fish odor." Most terrestrial animals do not have "infinite" water supplies, but mammals, which have urinary bladders, usually live in conditions where adequate water is available. Their mechanism for disposing of most toxins is to produce urea, a water-soluble compound, and then excrete it through the urine. Thus, urea becomes a major by-product of nitrogen metabolism in mammals.

Reptiles in the desert do not usually have much water available (Figure 23.18), and birds cannot afford to carry the weight of a fluid-filled bladder. These animals do not make urea; instead, they convert all their waste nitrogen to uric acid (Figure 23.17), the concentrated white solid so familiar in bird droppings.

Uric acid, the typical waste product from purines, can cause problems in primates due to its marginal water solubility. Deposits of uric acid in the joints and extremities cause gout (Section 23-8). Other mammals do not have a problem with uric acid because they convert it to allantoin, which is very water-soluble (Figure 23.19).

Figure 23.18 Desert-dwelling spiny-tailed iguanas convert most of their waste nitrogen to uric acid to save water.

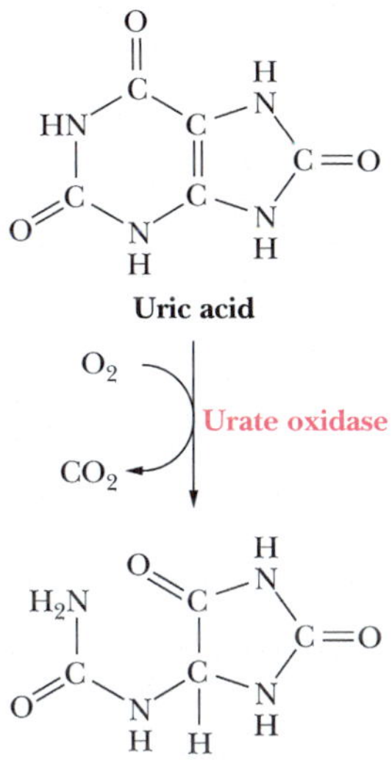

Figure 23.19 Catabolism of uric acid to allantoin. Most mammals other than primates use this reaction, catalyzed by urate oxidase, to convert water-insoluble uric acid from purine catabolism to water-soluble allantoin, which is excreted in urine.

23-6C Urea Cycle in Nitrogen Excretion

▶ **What is the role of the urea cycle in amino acid breakdown?**

A central pathway in nitrogen metabolism is the **urea cycle** (Figure 23.20). The nitrogens that enter the urea cycle come from several sources. One of the nitrogens of urea is added in the mitochondria, and its immediate precursor is glutamate, which releases ammonium ions via glutamate dehydrogenase (see the reverse reaction in Figure 23.6A). Recall that the ammonium ion nitrogens of glutamate ultimately come from the catabolism of amino acids as a result of transamination reactions in which α-ketoglutarate is the amino-group acceptor (see the reverse reaction in Figure 23.10A). Another source of free ammonium ions that can enter the urea cycle is a reaction catalyzed by *mitochondrial glutaminase* in which glutamine is broken down to glutamate and ammonium ions.

urea cycle a pathway that leads to the excretion of the waste products of nitrogen metabolism, especially those of amino acids

A preparatory condensation reaction catalyzed by *carbamoyl phosphate synthetase I* (CPS-I) between ammonium ions (from amino acid breakdown) and

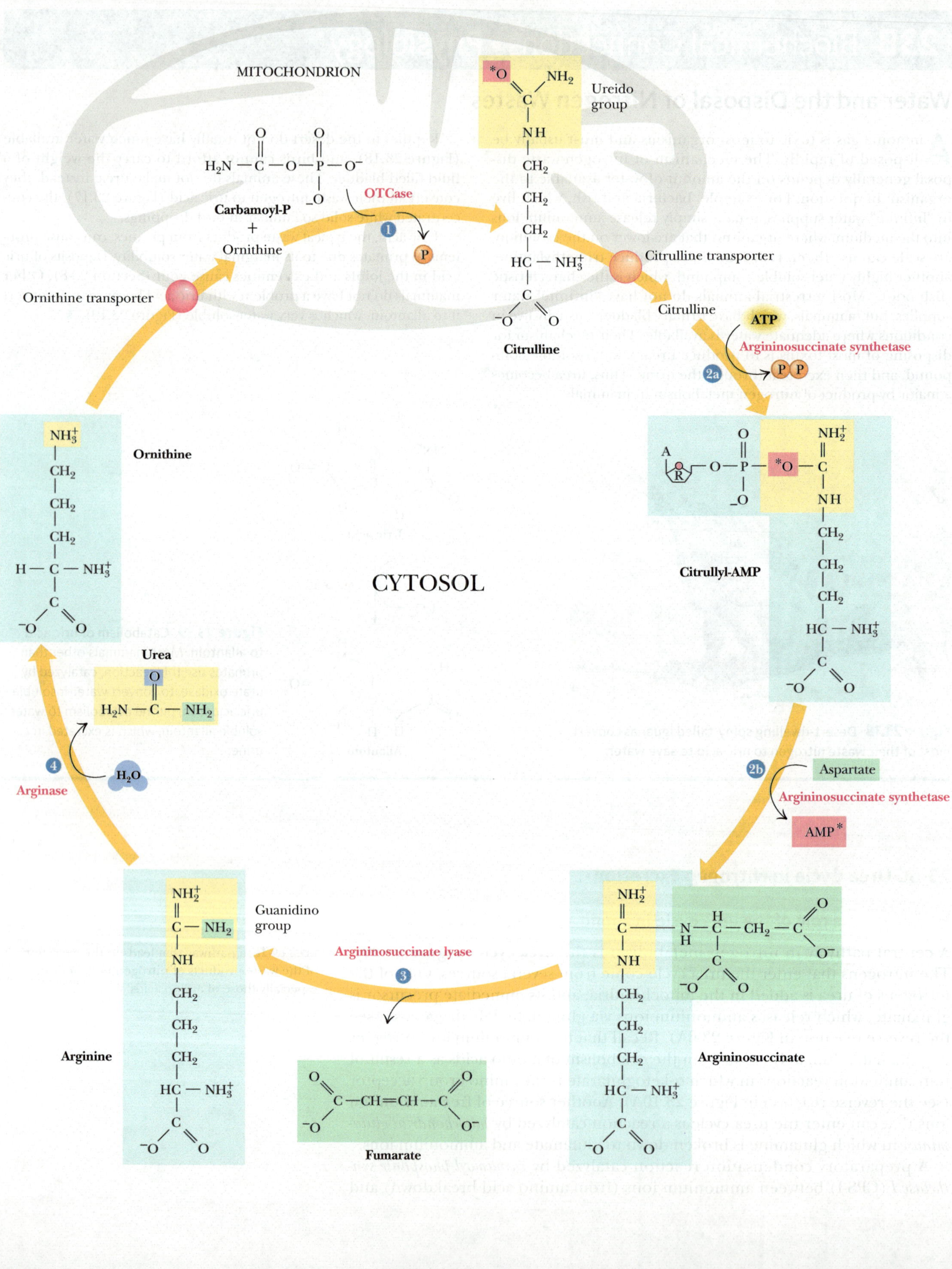
MITOCHONDRION
Carbamoyl-P
+
Ornithine
OTCase
1
Ureido group
Citrulline
Citrulline transporter
Ornithine transporter
ATP
Argininosuccinate synthetase
2a
Citrullyl-AMP
Ornithine
CYTOSOL
Urea
4
Arginase
2b
Aspartate
Argininosuccinate synthetase
AMP*
Guanidino group
Argininosuccinate lyase
3
Arginine
Fumarate
Argininosuccinate

Figure 23.20 The urea cycle series of reactions. A preparatory condensation reaction catalyzed by carbamoyl phosphate synthetase I (CPS-I) between ammonium ions (from amino acid breakdown) and carbon dioxide produces carbamoyl phosphate (carbamoyl-P) in a reaction not shown. Transfer of the carbamoyl group of carbamoyl-P to ornithine by ornithine transcarbamoylase (OTCase, reaction 1) yields citrulline. The citrulline ureido group is then activated by reaction with ATP to give a citrullyl–AMP intermediate (reaction 2a); AMP is then displaced by aspartate, which is linked to the carbon framework of citrulline via its α-amino group (reaction 2b). The course of reaction 2 was verified using ^{18}O-labeled citrulline. The ^{18}O label (indicated by the asterisk, *) was recovered in AMP. Citrulline and AMP are joined via the ureido *O atom. The product of this reaction is argininosuccinate; the enzyme catalyzing the two steps of reaction 2 is argininosuccinate synthetase. The next step (reaction 3) is carried out by argininosuccinase, which catalyzes the nonhydrolytic removal of fumarate from argininosuccinate to give arginine. Hydrolysis of arginine by arginase (reaction 4) yields urea and ornithine, completing the urea cycle.

carbon dioxide produces carbamoyl phosphate in a reaction that occurs before the cyclic steps of the urea cycle. It requires the hydrolysis of two molecules of ATP for each molecule of carbamoyl phosphate. After the CPS-I reaction, the following steps occur in the urea cycle:

Step 1. Carbamoyl phosphate reacts with *ornithine* to form *citrulline* in a reaction catalyzed by *ornithine transcarbamoylase (OTCase)*. Citrulline is then transported to the cytosol.

Step 2. A second nitrogen enters the urea cycle when aspartate reacts with citrulline to form *argininosuccinate* in a reaction catalyzed by *argininosuccinate synthetase*; this reaction requires ATP (AMP and PP_i are produced). The amino group of the aspartate is the source of the second nitrogen in the urea that will be formed in this series of reactions.

Step 3. Argininosuccinate is split by *argininosuccinate lyase* to produce *arginine* and *fumarate*. **Lyases** are enzymes that remove or add groups to molecules, often forming or breaking double bonds, without the involvement of water or redox chemistry.

lyases enzymes that remove or add groups to molecules, often forming or breaking double bonds, without the involvement of water or redox chemistry

Step 4. Arginine is hydrolyzed by arginase to give *urea* and regenerate ornithine, which is transported back to the mitochondrion to start another round of the cycle.

Ornithine is an amino acid that is structurally very similar to lysine (Section 3-2), and it gets its name from the Greek word *ornithos*, meaning "bird." It was first isolated in 1877 from the analysis of bird excrement. As previously mentioned, birds excrete nitrogen primarily as uric acid, but arginine metabolism in birds produces ornithine. Hence, it was named "ornithine" in honor of its avian origin.

Figure 23.21 shows some ways that the urea cycle and the citric acid cycle are connected. The production of fumarate in Step 3 of the urea cycle is a link between the urea cycle and the citric acid cycle. Fumarate is an intermediate of the citric acid cycle, and it can be converted to oxaloacetate (Figure 23.21, *bottom*). The oxaloacetate metabolite produced in the citric acid cycle can also contribute to the urea cycle when it is used in a transamination reaction to produce the aspartate needed for Step 2 of the urea cycle (Figure 23.21, *bottom right*). In this reaction, oxaloacetate serves as an amino-group acceptor in a direct route of amino acid catabolism into the urea cycle. However, recall that the usual amino-group acceptor in amino acid catabolism is α-ketoglutarate, as mentioned at the beginning of Section 23-6. The urea cycle and citric acid cycle pathways were both discovered by Hans Krebs.

In humans, urea production in the urea cycle is used to excrete excess nitrogen, such as would be found after consuming a high-protein meal. The pathway is confined to the liver. Arginine, the immediate precursor to urea, is the

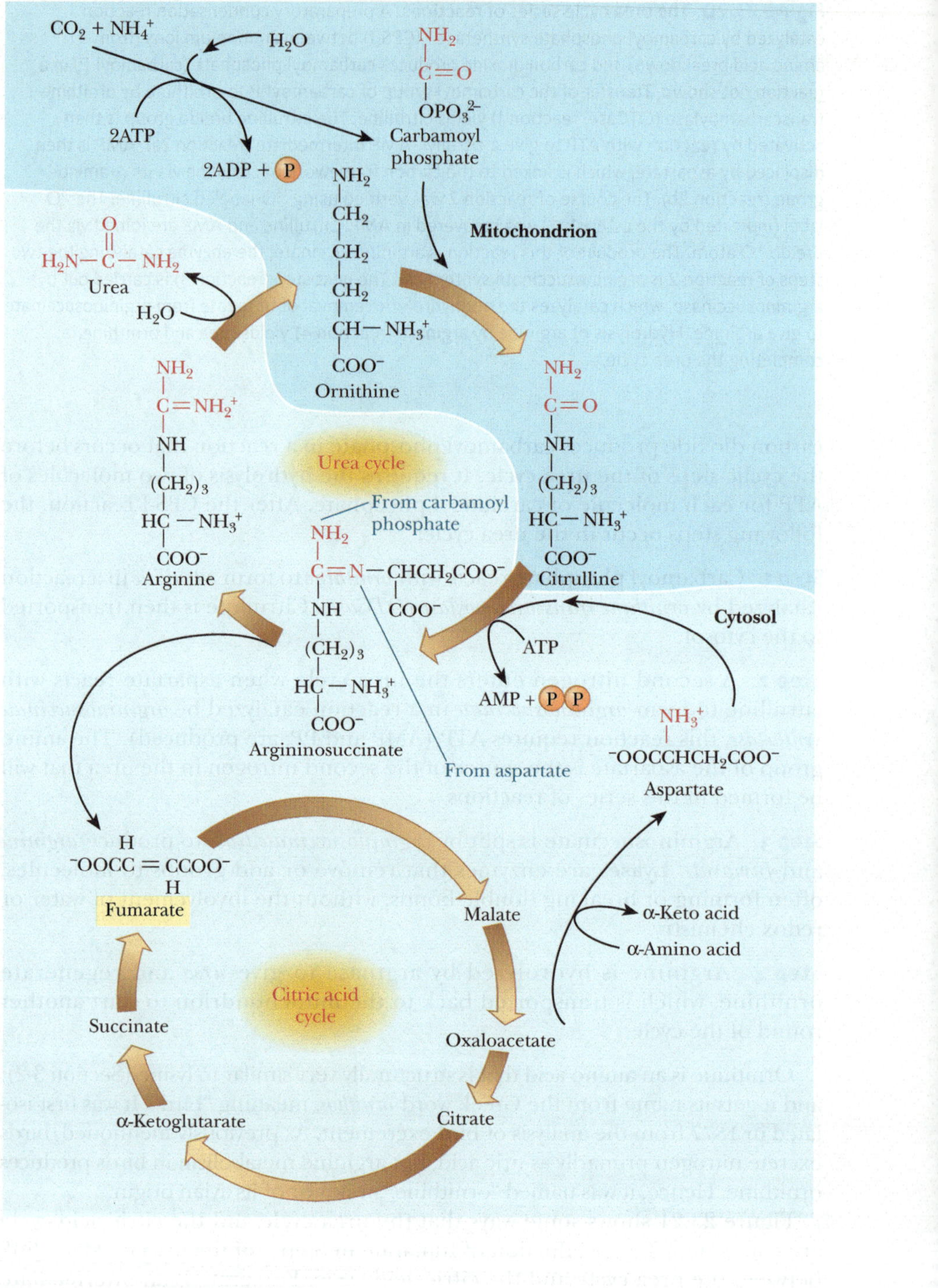

Figure 23.21 The urea cycle and some of its links to the citric acid cycle. Part of the urea cycle takes place in the mitochondrion and part in the cytosol. In the cytosol, fumarate produced by the urea cycle can be used in the citric acid cycle for energy production. In another link between the two cycles, oxaloacetate that is produced in the citric acid cycle can contribute to the urea cycle when it is used in a transamination reaction to produce the aspartate needed for converting citrulline to argininosuccinate.

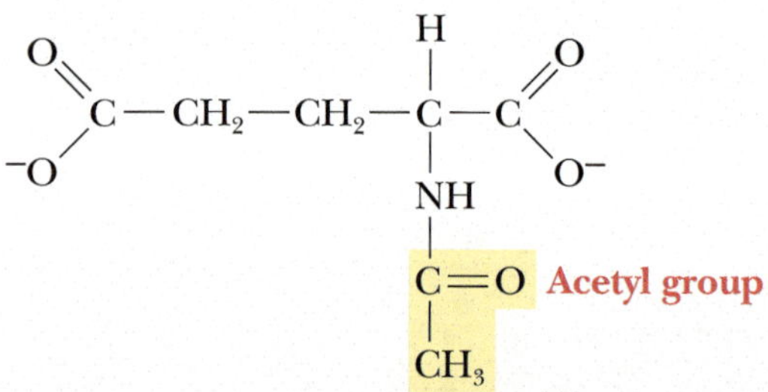

Figure 23.22 Structure of *N*-acetylglutamate. In this molecule, an acetyl group is attached to the amino group of glutamate.

most nitrogen-rich amino acid, but the source of the nitrogens in arginine can be derived from any of the other 20 amino acids.

The major control point in the urea cycle is the preparatory mitochondrial carbamoyl phosphate synthetase I (CPS-I) reaction, and the formation of carbamoyl phosphate is the committed step in the urea cycle. CPS-I is allosterically activated by *N*-acetylglutamate (Figure 23.22). *N*-acetylglutamate is formed by a reaction between glutamate and acetyl-CoA, catalyzed by *N*-acetylglutamate synthase. This enzyme is activated by increased concentrations of arginine. Thus, when amino acid catabolism is high, large amounts of glutamate are present from the degradation of glutamine, from synthesis via glutamate dehydrogenase, and from transamination reactions. Increased glutamate levels lead to increased levels of *N*-acetylglutamate, followed by increased activity of the urea cycle.

In addition, any time arginine builds up, either because of protein catabolism or because ornithine is building up due to a low level of CPS-I activity, the arginine stimulates synthesis of *N*-acetylglutamate and therefore increases CPS-I activity.

23-7 Purine Biosynthesis and Catabolism

Learning Objectives

After reading this section, students will be able to:

LO 23.7.1 Describe purine biosynthesis reactions.

LO 23.7.2 Explain purine catabolism reactions.

We discussed the formation of ribose-5-phosphate as part of the pentose phosphate pathway in Section 18-4, and the biosynthetic pathways for both purine and pyrimidine nucleotides use this preformed ribose-5-phosphate. Purines and pyrimidines are metabolized in different ways, and we consider them separately (purine metabolism in Section 23-7 and pyrimidine metabolism in Section 23-8).

23-7A Conversion of Inosine Monophosphate to Adenosine Monophosphate and Guanosine Monophosphate

In the synthesis of purine nucleotides [adenosine monophosphate (AMP) and guanosine monophosphate (GMP)], the growing double-ring system is bonded to ribose-5-phosphate while the purine skeleton is being assembled. Figure 23.23 shows how the five-membered ring is synthesized first, and then the six-membered ring. All four nitrogen atoms of the purine double-ring system are derived from amino acids: two from glutamine, one from aspartate, and one from glycine. Two of the five carbon atoms (adjacent to the glycine nitrogen) also come from glycine, two more come from tetrahydrofolate derivatives, and the fifth comes from CO_2.

Although ribose-5-phosphate is not shown in Figure 23.23, it is indeed attached while the purine double-ring system is being formed, and at the end of the purine assembly reactions, inosine-5'-monophosphate (IMP) is produced (Figure 23.24, *top*). IMP is the precursor of both AMP and GMP. The conversion of IMP to AMP takes place in two stages (Figure 23.24, *left*). The first step is the reaction of aspartate with IMP to form adenylosuccinate. This reaction is catalyzed by adenylosuccinate synthetase and requires GTP, not ATP, as an energy source. The cleavage of fumarate from adenylosuccinate to produce AMP is catalyzed by adenylosuccinase. This enzyme also functions in the synthesis of the six-membered ring of IMP. The chemistry in these two AMP synthesis reactions is similar to the chemistry seen in Steps 2 and 3 of the urea cycle—namely, a synthetase-catalyzed reaction involving aspartate, followed by a lyase catalyzed reaction producing fumarate.

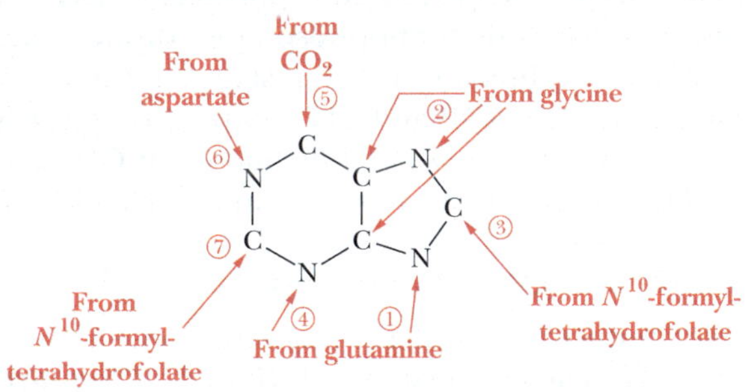

Figure 23.23 Sources of the atoms in the purine double-ring system in purine nucleotide biosynthesis. The numbering system indicates the order in which each atom or group of atoms is added. See the IMP structure in Figure 23.24 for the standard numbering convention that is used for the atoms of the purine double-ring system.

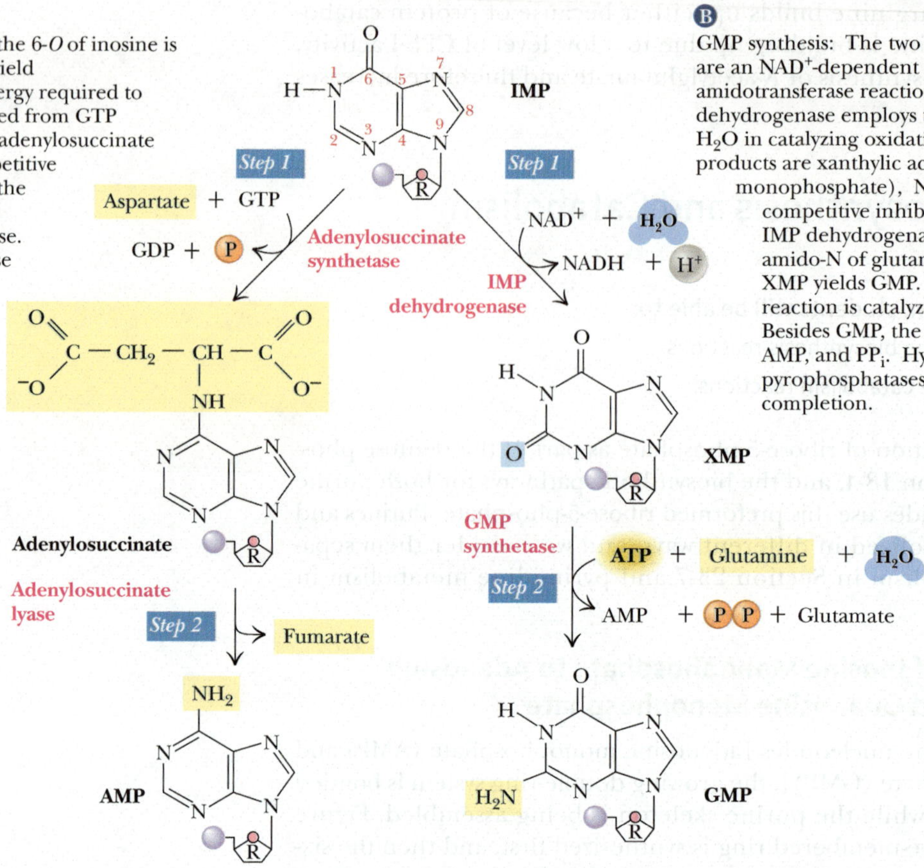

A AMP synthesis: In Step 1, the 6-*O* of inosine is displaced by aspartate to yield adenylosuccinate. The energy required to drive this reaction is derived from GTP hydrolysis. The enzyme is adenylosuccinate synthetase. AMP is a competitive inhibitor (with respect to the substrate IMP) of adenylosuccinate synthetase. In Step 2, adenylosuccinase (also known as adenylosuccinate lyase) carries out the nonhydrolytic removal of fumarate from adenylosuccinate, leaving AMP.

B GMP synthesis: The two reactions of GMP synthesis are an NAD^+-dependent oxidation followed by an amidotransferase reaction. In Step 1, IMP dehydrogenase employs the substrates NAD^+ and H_2O in catalyzing oxidation of IMP at C-2. The products are xanthylic acid (XMP or xanthosine monophosphate), NADH, and H^+. GMP is a competitive inhibitor (with respect to IMP) of IMP dehydrogenase. In Step 2, transfer of the amido-N of glutamine to the C-2 position of XMP yields GMP. This ATP-dependent reaction is catalyzed by GMP synthetase. Besides GMP, the products are glutamate, AMP, and PP_i. Hydrolysis of PP_i to 2 P_i by pyrophosphatases drives this reaction to completion.

Figure 23.24 **The synthesis of adenosine monophosphate (AMP) and guanosine monophosphate (GMP) from inosine monophosphate (IMP).**

The conversion of IMP to GMP also takes place in two stages (Figure 23.24, *right*). The first of the two steps is an oxidation in which the C—H group at the C-2 position is converted into a keto group. The oxidizing agent in the reaction is NAD^+, and the enzyme involved is IMP dehydrogenase. The nucleotide formed by the oxidation reaction is xanthosine-5′-phosphate (XMP). Next, an amino group from the side chain of glutamine replaces the C-2 keto group of XMP to produce GMP. This reaction is catalyzed by GMP synthetase; ATP is hydrolyzed to AMP and PP_i in the process. Some control is exerted, depending on the relative levels of purine nucleotides. For example, GTP is needed for the synthesis of adenine nucleotides (Figure 23.24A, Step 1), whereas ATP is required for the synthesis of guanine nucleotides (Figure 23.24B, Step 2). Each of the purine nucleotides must occur at a reasonably high level for the other to be synthesized.

Subsequent phosphorylation reactions produce purine nucleoside diphosphates (ADP and GDP) and triphosphates (ATP and GTP) (Figure 23.25). The purine nucleoside monophosphates, diphosphates, and triphosphates are all feedback inhibitors of the first stages of their own biosynthesis (Figure 23.25, *top*). Also, AMP, ADP, and ATP inhibit the conversion of IMP to adenine nucleotides (Figure 23.25, *middle left*), and GMP, GDP, and GTP inhibit the conversion of IMP to xanthylate and to guanine nucleotides (Figure 23.25, *middle right*).

23-7B Energy Requirements for the Production of AMP and GMP

The production of IMP starting with ribose-5-phosphate requires the equivalent of 7 ATP. The conversion of IMP to AMP requires hydrolysis of an additional

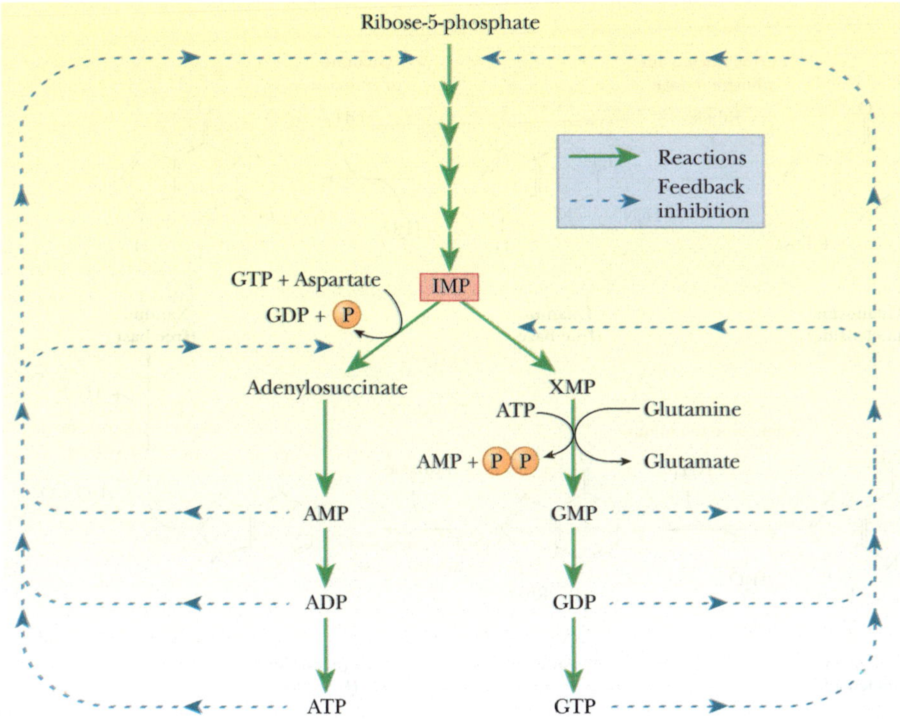

Figure 23.25 The role of feedback inhibition in the regulation of purine nucleotide biosynthesis. Each of the six purine nucleotides (AMP, ADP, ATP, GMP, GDP, and GTP) can feed back and inhibit enzymes at the first stages of their biosynthesis (*top*), or they can inhibit the enzymes where the inosine monophosphate precursor is a substrate (*middle*).

high-energy bond—in this case, that of GTP. Therefore, in the formation of AMP from ribose-5-phosphate, the equivalent of 8 ATP is needed. The conversion of IMP to GMP requires two high-energy bonds, given that a reaction occurs in which ATP is hydrolyzed to AMP and PP_i. Therefore, for the production of GMP from ribose-5-phosphate, the equivalent of 9 ATP is necessary.

The anaerobic oxidation of glucose produces only 2 ATP for each molecule of glucose (Section 17-1). Therefore, *anaerobic organisms* require four molecules of glucose (which produce 8 ATP) for each AMP they form, or five molecules of glucose (which produce 10 ATP) for each GMP. The process is more efficient for *aerobic organisms*. Because 30 or 32 ATP result from each molecule of glucose, depending on the type of tissue, aerobic organisms can optimally produce four AMP (requiring 32 ATP) or three GMP (requiring 36 ATP) for each molecule of glucose oxidized. A mechanism for the reuse of purines, rather than complete turnover and new synthesis, saves energy for organisms.

23-7C Purine Catabolism

Recall the structure of RNA discussed in Section 9-2B. RNA hydrolysis produces individual pyrimidine (CMP and UMP) and purine (AMP and GMP) nucleotides. The catabolism of GMP proceeds by removal of the phosphate by *5′-nucleotidase* to produce guanosine, followed by removal of the ribose sugar by *purine nucleoside phosphorylase (PNP)* to produce guanine. This is followed by removal of the amino group by *guanine deaminase* to produce xanthine (Figure 23.26A, *top*).

The catabolism of AMP proceeds by removal of the phosphate by 5'-nucleotidase to produce guanosine, followed by removal of the ribose sugar by PNP to produce guanine. This is followed by the removal of the amino group by guanine deaminase to produce xanthine (Figure 23.26A, *bottom*). The bottom of Figure 23.26A also shows how AMP can be converted to IMP, and IMP can be converted to inosine, but this is not the main route of AMP catabolism in every tissue.

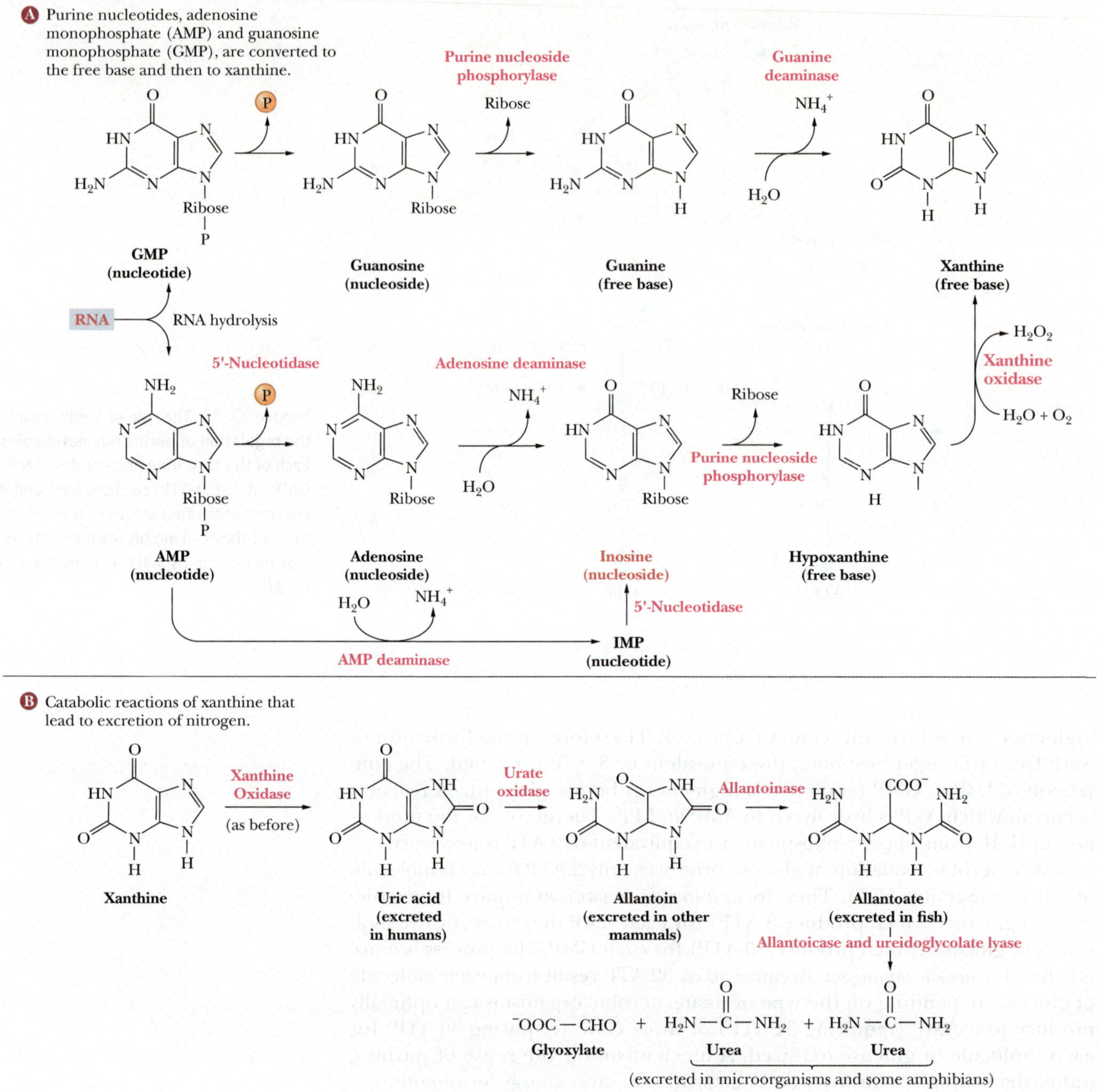

Figure 23.26 The reactions of purine catabolism.

The xanthine base is a common degradation product of both adenine and guanine. Xanthine is oxidized in turn to uric acid by xanthine oxidase (Figure 23.26B). In birds, some reptiles, insects, Dalmatian dogs, and primates (including humans), uric acid is the end product of purine metabolism and is excreted. In all other terrestrial animals, including other mammals, uric acid is converted to allantoin by urate oxidase, and allantoin is the product excreted. In fish, allantoin is converted to allantoate by allantoinase and then excreted. Allantoate is further degraded by two enzymes (allantoicase and ureidoglycolate lyase) to glyoxylate and urea by microorganisms and some amphibians (Figure 23.26B, *bottom right*).

Gout is a disease in humans that is caused by the overproduction of uric acid. Deposits of uric acid (which is barely soluble in water) accumulate in the joints of the hands and feet. Allopurinol is a compound used to treat gout; it inhibits the degradation of hypoxanthine to xanthine and of xanthine to uric acid, preventing the buildup of uric acid deposits.

Salvage reactions are important in the metabolism of purine nucleotides because of the amount of energy required for the synthesis of the purine bases. A free purine base that has been cleaved from a nucleotide can produce the corresponding nucleotide by reacting with the compound phosphoribosylpyrophosphate (PRPP) in a reaction catalyzed by adenine phosphoribosyltransferase (Figure 27.27A). The PRPP for this reaction is formed by the transfer of a pyrophosphate group from ATP to ribose-5-phosphate in a reaction catalyzed by PRPP synthetase (Figure 23.27B). The synthesis of purines requires energy, but their breakdown does not release energy. It is very useful for an organism to recycle as much of the carbon and nitrogen of purines as possible. This salvage also avoids some of the problems caused by the accumulation of breakdown products such as uric acid.

salvage reactions reactions that reuse compounds, such as purines, that require a large amount of energy to produce

Two different enzymes with different specificities with respect to the purine base catalyze salvage reactions. The reaction

$$\text{Adenine} + \text{PRPP} \rightarrow \text{AMP} + \text{PP}_i$$

is catalyzed by adenine phosphoribosyltransferase (APRT). The corresponding reactions of guanine and hypoxanthine,

$$\text{Hypoxanthine} + \text{PRPP} \xrightarrow{\text{HGPRT}} \text{IMP} + \text{PP}_i$$

$$\text{Guanine} + \text{PRPPHGPRT} \xrightarrow{\text{HGPRT}} \text{GMP} + \text{PP}_i$$

are catalyzed by hypoxanthine-guanine phosphoribosyltransferase (HGPRT) (Figure 23.28).

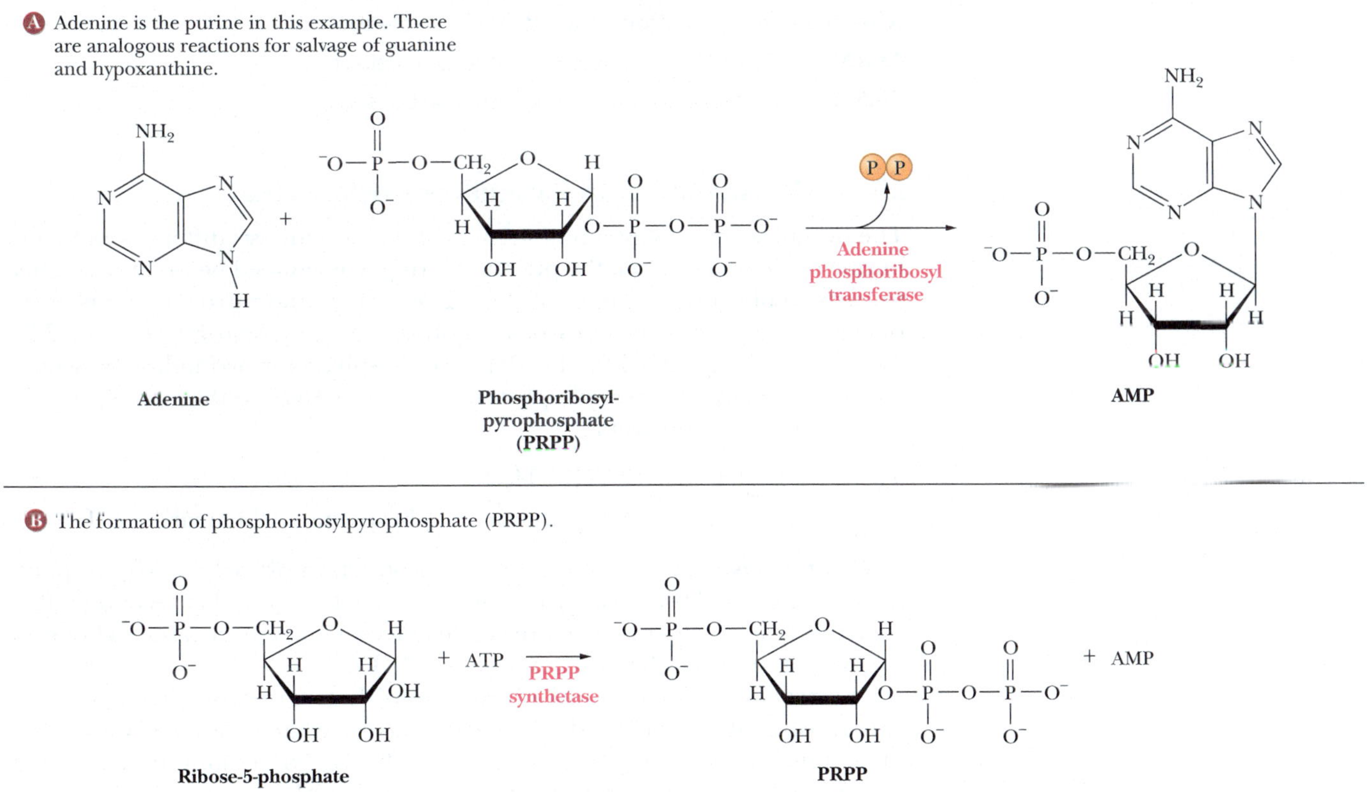

Figure 23.27 Purine salvage reactions. These reactions reuse purine rings to save energy.

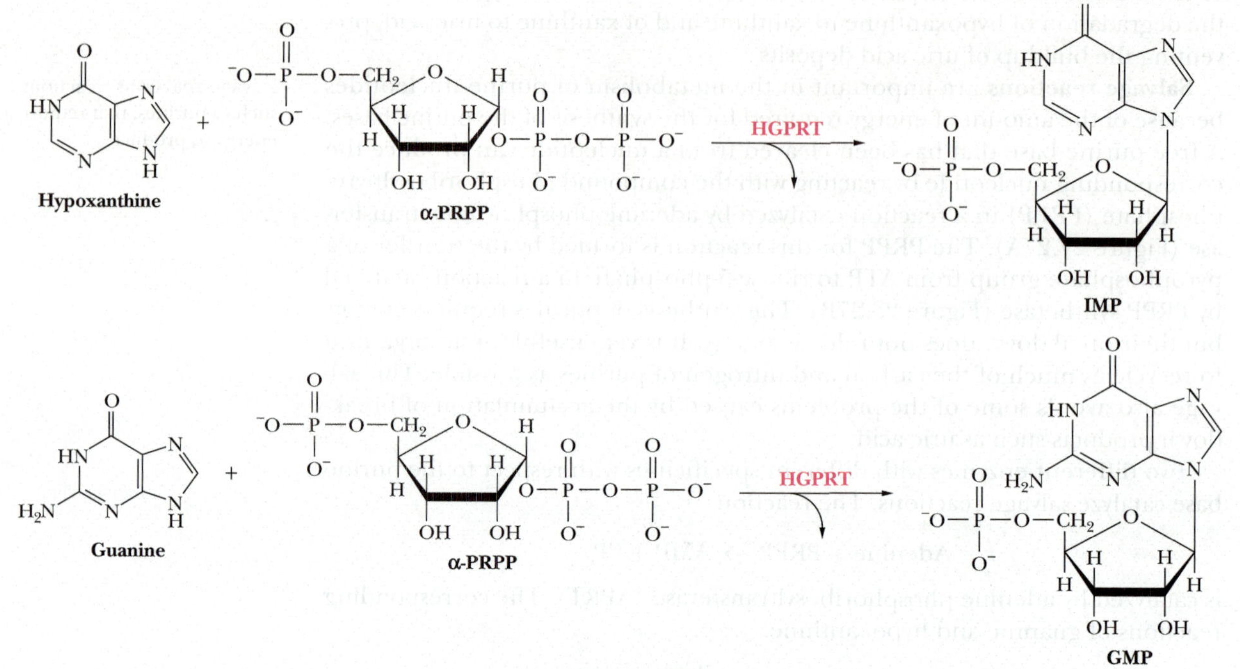

Figure 23.28 **Purine salvage by the HGPRT reaction.** These reactions reuse purine rings (*left*) to produce purine nucleotides (*right*), thereby saving energy for the cell. The compound phosphoribosylpyrophosphate (PRPP) is used in these reactions.

23-8 Pyrimidine Biosynthesis and Catabolism

Learning Objectives

After reading this section, students will be able to:

23.8.1 Describe pyrimidine biosynthesis reactions.

23.8.2 Explain pyrimidine catabolism reactions.

23-8A The Anabolism of Pyrimidine Nucleotides

The overall scheme of pyrimidine nucleotide biosynthesis differs from that of purine nucleotides in that the pyrimidine ring is assembled before it is attached to ribose-5-phosphate (Figure 23.29). The carbon and nitrogen atoms of the pyrimidine ring come from carbamoyl phosphate and aspartate (Figure 23.29, Steps 1–3). The production of carbamoyl phosphate for pyrimidine biosynthesis is catalyzed by *carbamoyl phosphate synthetase II (CPS-II)* in the cytosol, and the nitrogen donor is glutamine.

$$HCO_3^- + \text{Glutamine} + 2\text{ATP} + H_2O \rightarrow$$
$$\text{Carbamoyl phosphate} + \text{Glutamate} + 2\text{ADP} + P_i$$

We previously saw a reaction for the production of carbamoyl phosphate when we discussed the urea cycle in Section 23-6. However, that reaction differs from this one in two ways: The urea cycle reaction takes place in mitochondria, and in that reaction, the nitrogen donor is NH_4^+.

The reaction of carbamoyl phosphate with aspartate to produce *N*-carbamoylaspartate is catalyzed by aspartate transcarbamoylase (ATCase). It is the committed step in pyrimidine biosynthesis. The compounds involved in reactions up to this point in the pathway can play other roles in metabolism; after this point, *N*-carbamoylaspartate can be used only to produce

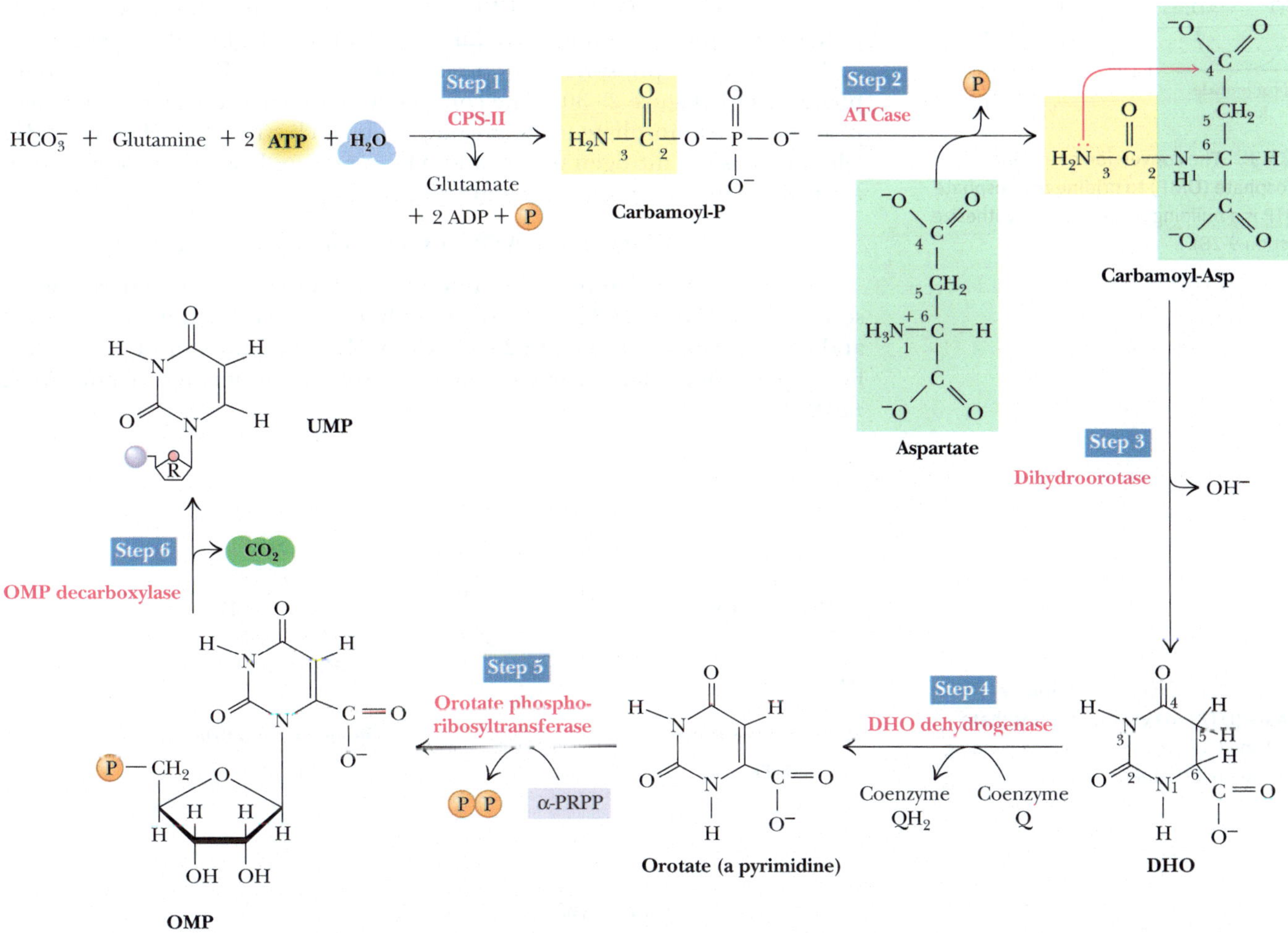

Figure 23.29 The pyrimidine nucleotide biosynthetic pathway. Step 1: Carbamoyl phosphate (carbamoyl-P) synthesis is catalyzed by carbamoyl phosphate synthetase II (CPS-II). Step 2: Condensation of carbamoyl phosphate and aspartate to yield carbamoyl aspartate (carbamoyl-Asp) is catalyzed by aspartate transcarbamoylase (ATCase). Step 3: An intramolecular condensation catalyzed by dihydroorotase gives the six-membered heterocyclic ring characteristic of pyrimidines. The product is dihydroorotate (DHO). Step 4: The oxidation of DHO by dihydroorotate dehydrogenase gives orotate. (In bacteria, NAD^+ is the electron acceptor from DHO.) Step 5: Phosphoribosylpyrophosphate (PRPP) provides the ribose-5-P moiety that transforms orotate into orotidine-5'-monophosphate (OMP), a pyrimidine nucleotide. Orotate phosphoribosyltransferase joins N-1 of the pyrimidine to the ribosyl group in the appropriate β-configuration. PP_i hydrolysis renders this reaction thermodynamically favorable. Step 6: Decarboxylation of OMP by OMP decarboxylase yields uridine monophosphate (UMP), a pyrimidine nucleotide.

pyrimidines—thus the term "committed step." The next step, the conversion of *N*-carbamoylaspartate to dihydroorotate, takes place in a reaction that involves an intramolecular dehydration (loss of water) as well as cyclization, catalyzed by dihydroorotase (Figure 23.29, Step 3). Dihydroorotate is oxidized to orotate by dihydroorotate dehydrogenase (DHO dehydrogenase), with the concomitant reduction of NAD^+ to NADH (Figure 23.29, Step 4). A pyrimidine nucleotide is now formed by the reaction of orotate with phosphoribosylpyrophosphate (PRPP) to give orotidine-5'-monophosphate (OMP) (Figure 23.29, Step 5). This reaction, catalyzed by orotate phosphoribosyltransferase, is similar to the one that takes place in purine salvage (Figure 23.27). Finally, orotidine-5'-monophosphate decarboxylase (OMP decarboxylase) catalyzes

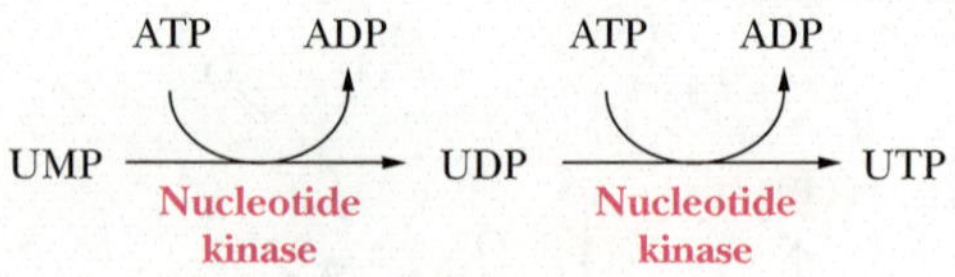

Figure 23.30 The conversion of uridine monophosphate (UMP) to uridine triphosphate (UTP). UTP is a building block used to synthesize RNA (Section 9-2B).

the conversion of OMP to UMP (uridine-5'-monophosphate), which is the precursor of the remaining pyrimidine nucleotides (Figure 23.29, Step 6).

Two successive phosphorylation reactions convert UMP to uridine triphosphate (UTP) (Figure 23.30). The conversion of uracil to cytosine, which takes place in the triphosphate form, is catalyzed by CTP synthetase (Figure 23.31). Glutamine is the nitrogen donor, and ATP is required, as we saw previously in the GMP synthetase reaction (Figure 23.24B, Step 2).

$$\text{UTP} + \text{Glutamine} + \text{ATP} \rightarrow \text{CTP} + \text{Glutamate} + \text{ADP} + \text{P}_i$$

Feedback inhibition in pyrimidine nucleotide biosynthesis takes place in several ways (Figure 23.32). CTP is an inhibitor of aspartate transcarbamoylase and of CTP synthetase (Figure 23.32, *left*). UMP is an inhibitor of an even earlier step, the one catalyzed by carbamoyl phosphate synthetase II (Figure 23.32, *right*).

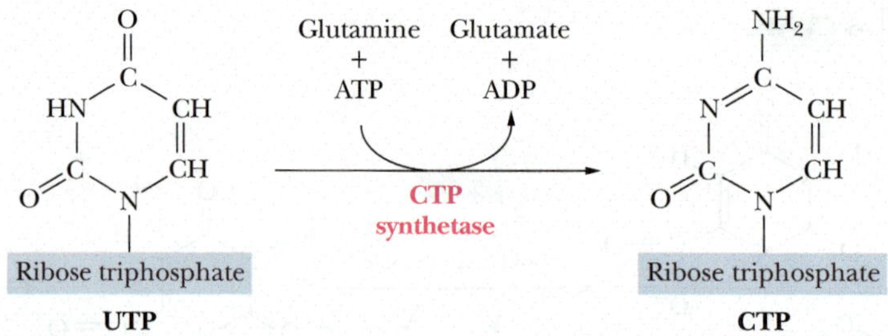

Figure 23.31 The conversion of uridine triphosphate (UTP) to cytosine triphosphate (CTP). UTP and CTP are the pyrimidine building blocks used in RNA synthesis.

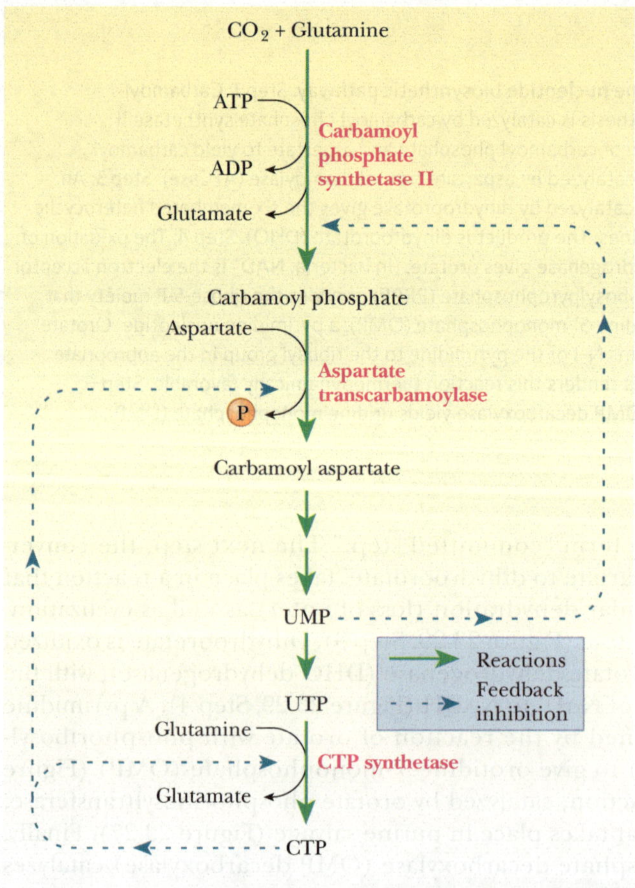

Figure 23.32 The role of feedback inhibition in the regulation of pyrimidine nucleotide biosynthesis. CTP is an inhibitor of aspartate transcarbamoylase and of CTP synthetase (*left*). UMP is an inhibitor of carbamoyl phosphate synthetase II (*right*).

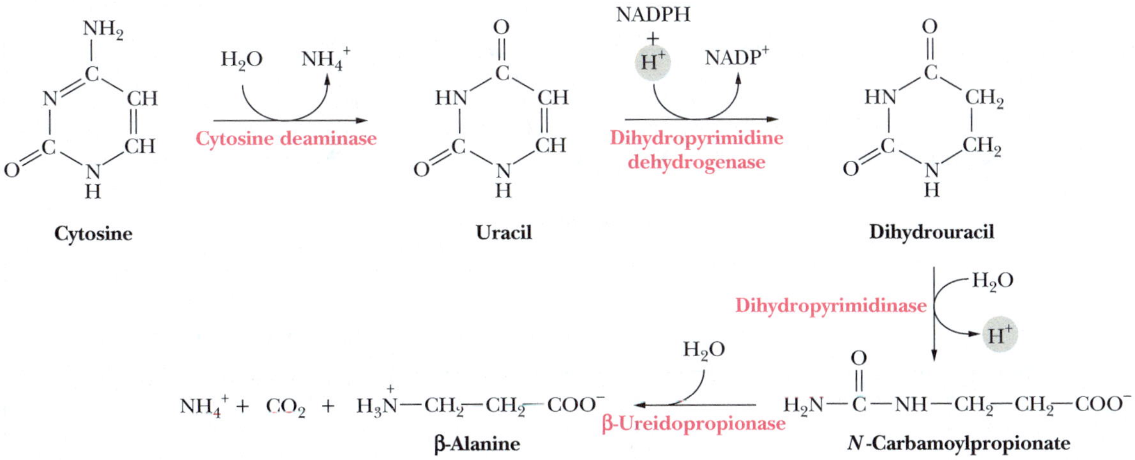

Figure 23.33 The catabolism of pyrimidine bases. The end products of these reactions are β-alanine, CO_2, and NH_4^+. The β-alanine can be incorporated into coenzyme A (Section 15-7) or other metabolites, the CO_2 is exhaled, and the NH_4^+ is converted to urea in the urea cycle (Section 23-6C).

23-8B Pyrimidine Catabolism

Figure 23.33 shows the catabolism of pyrimidines. Pyrimidine nucleotides are broken down first to the nucleoside and then to the base, as purine nucleotides are. Cytosine can be deaminated to uracil by *cytosine deaminase*, and the double bond of the uracil ring is reduced by *dihydropyrimidine dehydrogenase* to produce dihydrouracil. The dihydrouracil ring is then opened by *dihydropyrimidinase* to produce *N*-carbamoylpropionate, which in turn is broken down by *β-ureidopropionase* to NH_4^+ (converted to urea—see Section 23-6C), CO_2 (exhaled), and β-alanine (incorporated into coenzyme A—see Section 15-7, and other metabolites) (Figure 23.33, *bottom*).

23-9 Conversion of Ribonucleotides to Deoxyribonucleotides

Learning Objectives

After reading this section, students will be able to:

LO 23.9.1 Describe how ribonucleotides are converted to deoxyribonucleotides.

LO 23.9.2 Write the reactions by which deoxythymidine triphosphate is produced from deoxyuridine diphosphate.

23-9A Conversion of Ribonucleoside Diphosphates to Deoxyribonucleoside Diphosphates

Recall from Section 9-2A that the difference between the ribonucleotides used in RNA synthesis and the deoxyribonucleotides used in DNA synthesis is the absence of the 2' hydroxyl group on the ribose moiety of deoxyribonucleotides. For the biosynthesis of deoxyribonucleotides, ribonucleoside diphosphates (NDPs), whose biosynthesis reactions have been discussed in Sections 23-7 and 23-8, are reduced to 2'-deoxyribonucleoside diphosphates (dNDPs) in all organisms (Figure 23.34A); NADPH is the reducing agent.

$$\text{Ribonucleoside diphosphate} + \text{NADPH} + \text{H}^+ \rightarrow \text{Deoxyribonucleoside diphosphate} + \text{NADP}^+ + \text{H}_2\text{O}$$

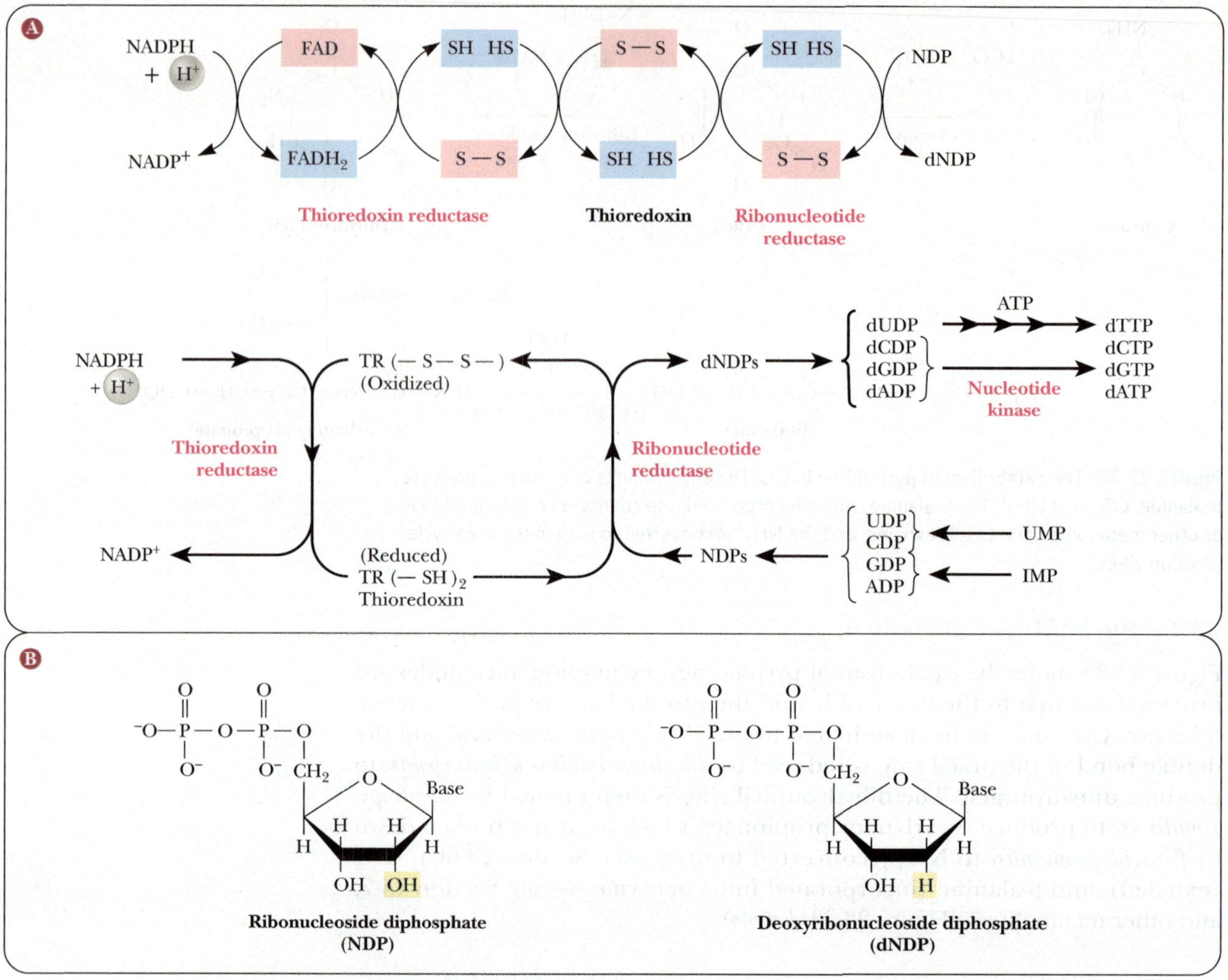

Figure 23.34 Conversion of ribonucleoside diphosphates (NDPs) to deoxyribonucleoside diphosphates (dNDPs). (A) The (—S—S—)/(—SH HS—) oxidation–reduction cycle involving ribonucleotide reductase, thioredoxin, thioredoxin reductase, and NADPH. (B) General structures of NDP and dNDP nucleotides.

The actual process, which is catalyzed by the *ribonucleotide reductase complex,* is more complicated than the preceding equation would indicate and involves some intermediate electron carriers. The ribonucleotide reductase complex from *E. coli* has been extensively studied, and its mode of action gives some clues to the nature of the process. Two other proteins, thioredoxin and thioredoxin reductase, are required. *Thioredoxin* contains a disulfide (S—S) group in its oxidized form and two sulfhydryl (—SH) groups in its reduced form. NADPH reduces thioredoxin in a reaction catalyzed by *thioredoxin reductase.* The reduced thioredoxin in turn reduces a ribonucleoside diphosphate to a deoxyribonucleoside diphosphate, and this reaction is actually catalyzed by ribonucleotide reductase. General structures of an NDP and a dNDP are shown in Figure 23.24B, where you should confirm for yourself that the 2' hydroxyl group is missing in the dNDP structure. This reaction produces dADP, dGDP, dCDP, and dUDP nucleotides. The first three are phosphorylated to give the corresponding deoxyribonucleoside triphosphates (dNTPs), which are substrates for the synthesis of DNA. The fourth required substrate for DNA synthesis is deoxythymidine triphosphate (dTTP), and we now see how dTTP is produced from dUDP.

23-9B Conversion of Deoxyuridine Diphosphate to Deoxythymidine Triphosphate

dTTP is the final DNA building block whose biosynthesis we need to discuss. For its biosynthesis, a one-carbon transfer is required to convert the uracil base of deoxyuridine diphosphate (dUDP) to a thymine base by attaching the methyl group that distinguishes thymine from uracil. The one-carbon transfer is catalyzed by *thymidylate synthase* (Figure 23.35, *middle*). The source of the one-carbon unit is N^5,N^{10}-methylenetetrahydrofolate (recall from Section 23-4C that tetrahydrofolate derivatives are commonly used in one-carbon transfer reactions), which is converted to dihydrofolate (DHF) in the process. The DHF coenzyme gets converted back to the metabolically active THF form in a reaction catalyzed by *dihydrofolate reductase* that requires NADPH (Figure 23.36, *bottom right*). Once THF is regenerated, a one-carbon group gets added back to it by *serine hydroxymethyltransferase* in a reaction requiring serine (Figure 23.36, *bottom left*).

Because a supply of dTTP is necessary for DNA synthesis, inhibition of enzymes that catalyze its production of dTTP inhibits the growth of rapidly dividing cells. Cancer cells, like all fast-growing cells, depend on continued DNA synthesis for growth. Inhibitors of thymidylate synthetase, such as fluorouracil, and inhibitors of dihydrofolate reductase, such as aminopterin and methotrexate (structural analogues of folate), have been used in cancer chemotherapy. The intent of such therapy is to inhibit the formation of dTTP and thus of DNA in cancer cells, causing the death of the cancer cells with minimal effect on normal cells, which grow more slowly. Chemotherapy has adverse side effects because of the highly toxic nature of most of the drugs involved; normal cells are affected to some extent, although less than the cancer cells. Enormous amounts of research are focused on finding safe and effective forms of cancer treatment. The specific role of folic acid and its derivatives in medicine is the subject of **BIOCHEMICAL CONNECTIONS 23C**.

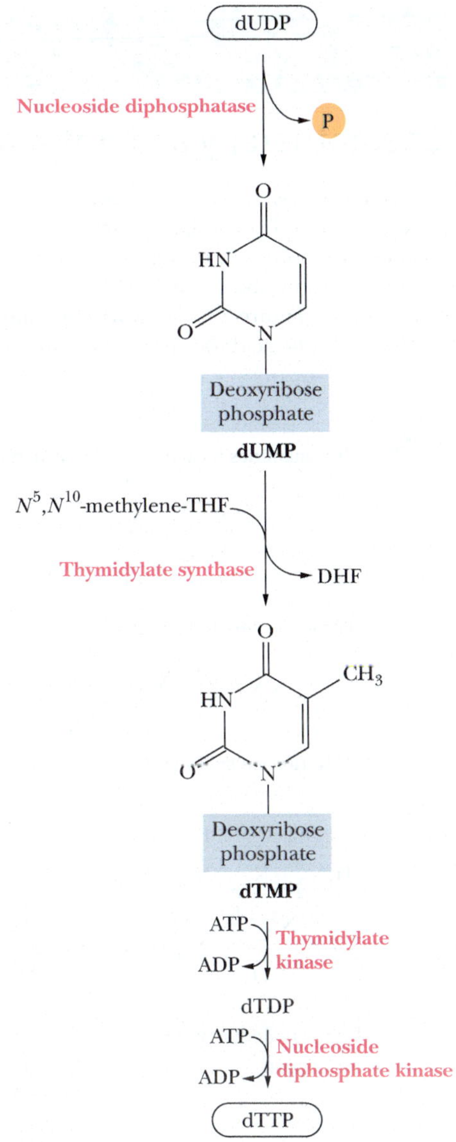

Figure 23.35 The conversion of deoxyuridine diphosphate (dUDP) to deoxythymidine triphosphate (dTTP). dTTP is a building block used in DNA synthesis. The other three DNA building blocks are produced as shown in Figure 23.34A. THF is tetrahydrofolate, and DHF is dihydrofolate.

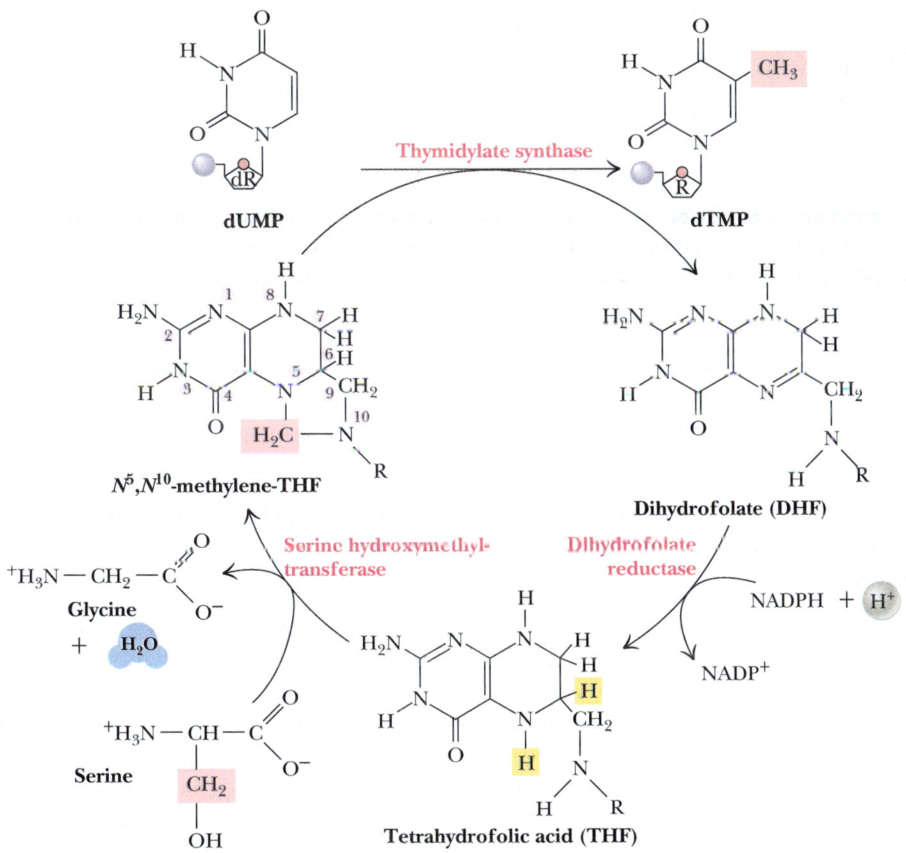

Figure 23.36 The thymidylate synthase reaction. This reaction produces deoxythymidine monophosphate (dTMP) and regenerates the N^5,N^{10}-methylenetetrahydrofolate coenzyme that is required for the thymidylate synthase reaction. The one-carbon methyl (—CH_3) group that gets added to the uracil base to produce the thymine base is ultimately derived from the β-carbon of serine.

23C Biochemical Connections • Medicine

Chemotherapy and Antibiotics—Taking Advantage of the Need for Folic Acid

We have already seen the importance of folic acid and its derivative, tetrahydrofolate, in several reactions. This importance has been exploited in human medicine. Bacteria synthesize folic acid from *p*-aminobutyric acid (PABA). A type of antibiotic called a sulfonamide (Figure 23.37A) works by competing with PABA in the synthesis of folic acid. Because folic acid is critical to the formation of purines and pyrimidines, antagonists of folic acid metabolism are used to inhibit nucleic acid synthesis and cell growth. Rapidly dividing cells, such as those found in cancers and tumors, are more susceptible to these antagonists. Several related compounds, such as methotrexate (Figure 23.37B), are used in chemotherapy to inhibit cancer cell growth. ◗

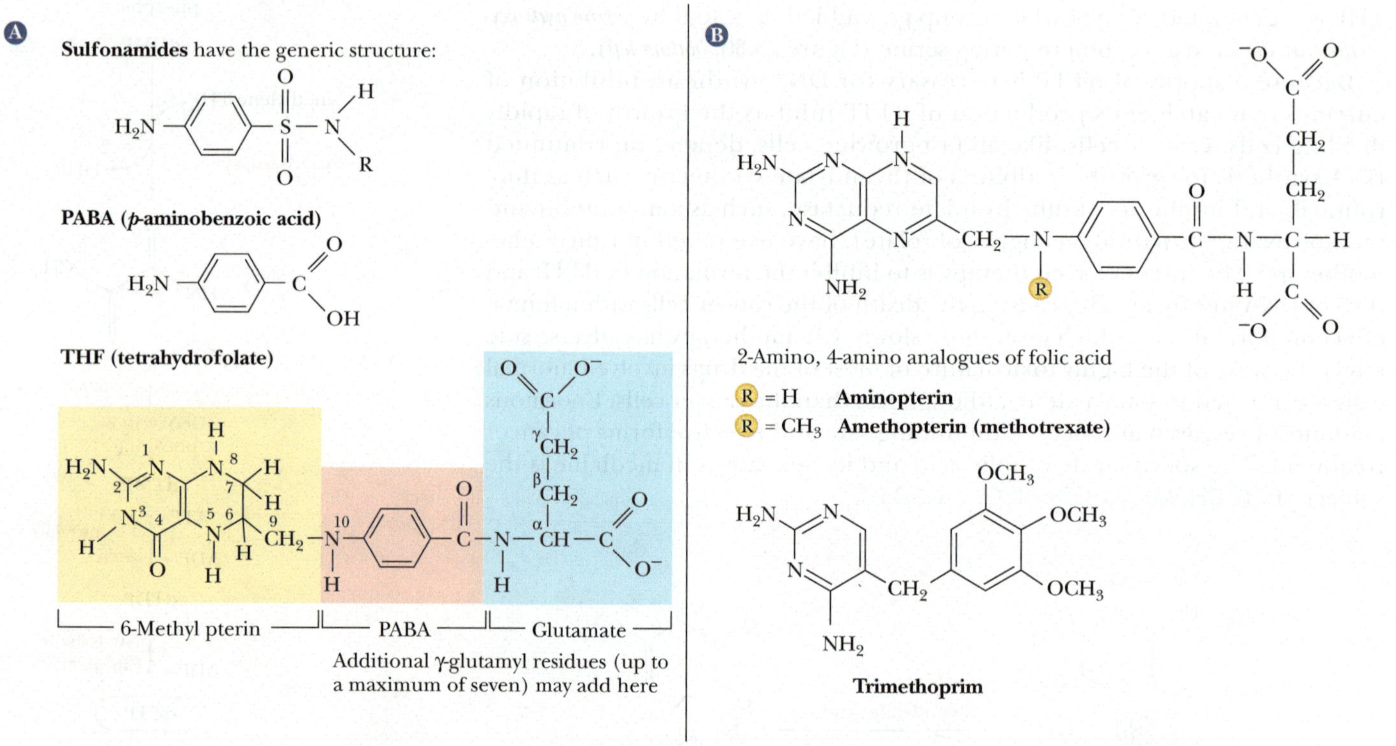

Figure 23.37 (A) Sulfa drugs (sulfonamides) act as antibiotics because of their similarity to p-aminobenzoic acid (PABA), a precursor in folic acid synthesis. Sulfa drugs compete with PABA and stop folic acid synthesis in bacteria. (B) Three compounds that are used in chemotherapy because they interfere with folic acid metabolism. They are nearly irreversible inhibitors of dihydrofolate reductase, having a 1000-fold greater affinity than dihydrofolate.

Summary

Nitrogen Metabolism: An Overview

- The metabolism of nitrogen encompasses a number of topics, including the anabolism and catabolism of amino acids, porphyrins, and nucleotides.
- Atmospheric nitrogen is the ultimate source of this element in biomolecules.

Nitrogen Fixation

- Nitrogen fixation is the process by which molecular nitrogen from the atmosphere is made available to organisms in the form of ammonium ions.
- Nitrification reactions convert NH_4^+ to NO_3^- and provide another source of nitrogen.

Feedback Inhibition in Nitrogen Metabolism

- Feedback inhibition control mechanisms are a unifying factor in biosynthetic pathways involving nitrogen compounds.
- Most of the nitrogen metabolism pathways are long and complicated and use a great deal of energy.
- Shutting off these processes when enough of the final product has built up is important to the energy flux of the cell.

Common Features in Amino Acid Metabolism

- In the anabolism of amino acids, transamination reactions play an important role.
- Glutamate and glutamine are frequently the amino-group donors.

Transamination Reactions

- The enzymes that catalyze transamination reactions are called aminotransferases, and they frequently require pyridoxal phosphate as a coenzyme.

One-Carbon Transfer Reactions

- One-carbon transfer reactions also operate in the anabolism of amino acids.
- Carriers are required for the one-carbon groups transferred.
- Tetrahydrofolate is a carrier of one-carbon groups, and *S*-adenosylmethionine is a carrier of methyl groups.

Essential Amino Acids

- Some species, including humans, cannot synthesize all the amino acids required for protein synthesis and must therefore obtain these essential amino acids from dietary sources.
- About half of the standard 20 amino acids are essential in humans, including arginine, histidine, isoleucine, leucine, lysine, methionine, phenylalanine, threonine, tryptophan, and valine.

Pathways for Amino Acid Carbon Skeletons

- The catabolism of amino acids has two parts: the fate of the nitrogen of the amino group and the fate of the carbon skeleton.
- The carbon skeleton is converted to pyruvate or oxaloacetate in the case of glucogenic amino acids, or to acetyl-CoA or acetoacetyl-CoA in the case of ketogenic amino acids.

Excretion of Excess Nitrogen

- Nitrogen from the breakdown of nitrogen-containing compounds is excreted by organisms in one of three forms: ammonium ion, urea, or uric acid.

Urea Cycle in Nitrogen Excretion

- In the urea cycle, nitrogen released by the catabolism of amino acids is converted to urea.
- The urea cycle and the citric acid cycle are linked in two ways: Fumarate produced in the urea cycle can be used in the citric acid cycle, and oxaloacetate from the citric acid cycle can accept an amino group from amino acid catabolism to become aspartate for nitrogen removal via the urea cycle.

Conversion of Inosine Monophosphate to Adenosine Monophosphate and Guanosine Monophosphate

- The anabolic pathway of nucleotide synthesis involving purines differs from that involving pyrimidines.
- Both pathways use preformed ribose-5-phosphate but differ with regard to the point in the pathway at which the sugar phosphate is attached to the base.
- In the case of purine nucleotides, the growing base is attached to the sugar phosphate during the synthesis, ultimately giving rise to inosine monophosphate (IMP).
- IMP is converted to AMP and GMP in reactions subject to a high degree of feedback control.

Energy Requirements for the Production of AMP and GMP

- The energy requirements for the production of AMP and GMP are high—more than 30 ATP equivalents in both cases.

Purine Catabolism

- In catabolism, purine bases are frequently salvaged and reattached to sugar phosphates.
- Otherwise, purines are broken down into uric acid.

The Anabolism of Pyrimidine Nucleotides

- In pyrimidine biosynthesis, the base is formed first and then attached to the sugar phosphate.

Pyrimidine Catabolism

- Pyrimidines are degraded to β-alanine, CO_2, and NH_4^+. NH_4^+ is then excreted via the urea cycle.

Conversion of Ribonucleotides to Deoxyribonucleotides

- Deoxyribonucleotides for DNA synthesis are produced by the reduction of ribonucleoside diphosphates to deoxyribonucleoside diphosphates.

Conversion of Deoxyuridine Diphosphate to Deoxythymidine Triphosphate

- Another reaction specifically needed to produce substrates for DNA synthesis is the conversion of uracil to thymine.
- This pathway, which requires a tetrahydrofolate derivative as the carrier for one-carbon transfer, is a target for cancer chemotherapy.

Review Exercises

23-1 Nitrogen Metabolism: An Overview

1. **Recall** What is nitrogen fixation?
2. **Recall** What kinds of organisms can fix nitrogen? Which ones cannot?

23-2 Nitrogen Fixation

3. **Recall** How is nitrogen fixed (i.e., how is N_2 converted to NH_4^+)? How is it subsequently assimilated into organic compounds?
4. **Biochemical Connections** What is the Haber–Bosch process?

5. **Recall** Write the overall reaction for the fixation of nitrogen via the nitrogenase complex.
6. **Recall** Describe the nitrogenase complex. How is the enzyme organized? What are its unique components?

23-3 Feedback Inhibition in Nitrogen Metabolism

7. **Reflect and Apply** Why are nitrogen-utilizing pathways controlled by feedback inhibition?
8. **Reflect and Apply** Metabolic cycles are rather common (Calvin cycle, citric acid cycle, urea cycle). Why are cycles useful to organisms?

23-4 Amino Acid Biosynthesis

9. **Recall** What is the relationship between α-ketoglutarate, glutamate, and glutamine in amino acid anabolism?
10. **Recall** Draw a transamination reaction between α-ketoglutarate and alanine.
11. **Recall** Diagram the reactions involving glutamate dehydrogenase and glutamine synthetase that produce glutamine from the ammonium ion and α-ketoglutarate.
12. **Reflect and Apply** What is the difference between glutamine synthetase and glutaminase?
13. **Recall** Draw the mechanism of transamination with pyridoxal phosphate.
14. **Recall** What cofactors are involved in one-carbon transfer reactions of amino acid anabolism?
15. **Recall** Draw the structures of (a) folic acid and (b) tetrahydrofolate (THF). (c) Show how THF serves as a carrier of one-carbon groups.
16. **Recall** Why is there no net gain of methionine if homocysteine is converted to methionine with *S*-adenosylmethionine as the methyl donor?
17. **Recall** Write a general equation that illustrates how glutamate plays a central role in the biosynthesis of amino acids.
18. **Recall** By means of a structural formula, show how *S*-adenosylmethionine is a carrier of methyl groups.
19. **Reflect and Apply** Sulfanilamide and related sulfa drugs were widely used to treat diseases of bacterial origin before penicillin and more advanced drugs were readily available. The inhibitory effect of sulfanilamide on bacterial growth can be reversed by *p*-aminobenzoate. Suggest a mode of action for sulfanilamide.

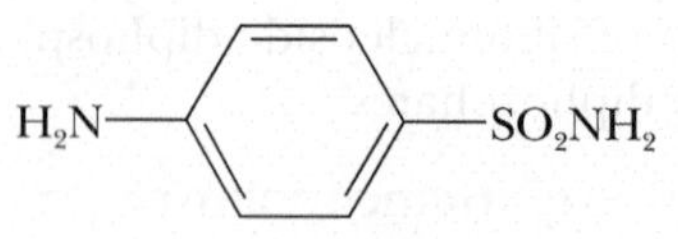

Sulfanilamide

20. **Reflect and Apply** Proteins contain methionine but not α-amino-*n*-hexanoic acid. The only structural difference is the substitution of $—CH_2—$ for —S—. Both groups are similar in size and hydrophobic character. Why is methionine more advantageous than α-amino-*n*-hexanoic acid?

23-5 Essential Amino Acids

21. **Recall** Some amino acids are classified as essential in the human diet. What does it mean for an amino acid to be essential, and why must these amino acids be obtained from food?
22. **Recall** In general, what categories of amino acids are essential in humans?
23. **Recall** What is the additional essential amino acid for individuals with phenylketonuria? Why?

23-6 Amino Acid Catabolism and Nitrogen Excretion

24. **Recall** What defines an amino acid as an α-amino acid? How many α-amino acids participate directly in the urea cycle? Of these, how many can be used for protein synthesis?
25. **Recall** Write an equation for the net reaction of the urea cycle. Show how the urea cycle is linked to the citric acid cycle.
26. **Reflect and Apply** Compare the structures of citrulline and ornithine to those of two of the 20 standard amino acids.
27. **Recall** Which amino acids in the urea cycle are the links to the citric acid cycle? Show how these links occur.
28. **Recall** How many ATPs are required for one round of the urea cycle? Where do these ATPs get used?
29. **Recall** How is carbamoyl phosphate synthetase I (CPS-I) controlled?
30. **Recall** What is the logic behind high levels of arginine positively regulating *N*-acetylglutamate synthase?
31. **Recall** How does the level of glutamate affect the urea cycle?
32. **Recall** When amino acids are catabolized, what are the end products of the carbon skeletons for glucogenic amino acids? For ketogenic amino acids?
33. **Recall** Will an amino acid be glucogenic or ketogenic if it is catabolized to the following molecules?
   (a) Phosphoenolpyruvate
   (b) α-Ketoglutarate
   (c) Succinyl-CoA
   (d) Acetyl-CoA
   (e) Oxaloacetate
   (f) Acetoacetate
34. **Biochemical Connections** What species excrete excess nitrogen as ammonium ion? Which ones excrete it as uric acid, and which ones excrete it as urea?
35. **Biochemical Connections** Would you expect an ostrich to excrete excess nitrogen as uric acid, urea, or ammonium ion? Make an argument for your answer.
36. **Reflect and Apply** Why is arginine an essential amino acid when it is made in the urea cycle?
37. **Reflect and Apply** People on high-protein diets are advised to drink lots of water. Why?
38. **Reflect and Apply** Why is it better, when running a marathon, to drink a beverage with sugar for energy rather than one with amino acids?

23-7 Purine Biosynthesis and Catabolism

39. **Biochemical Connections** How is the importance of folic acid related to chemotherapy?
40. **Recall** What are the sources of the carbons and nitrogens in the purine bases?
41. **Recall** What is the structural difference between inosine and adenosine?
42. **Recall** How is tetrahydrofolate important to purine synthesis?
43. **Recall** Does the conversion of IMP to GMP use or produce ATP, either directly or indirectly? Justify your answer.
44. **Recall** Discuss the role of feedback inhibition in the anabolism of purine-containing nucleotides.
45. **Recall** How many high-energy phosphate bonds must be hydrolyzed in the pathway that produces GMP from guanine and PRPP by the PRPP salvage reaction, compared with the number of such bonds hydrolyzed in the pathway leading to IMP and then to GMP?

46. **Reflect and Apply** Why do most mammals, other than primates, not suffer from gout?

23-8 Pyrimidine Biosynthesis and Catabolism

47. **Recall** What is an important difference between the biosynthesis of purine nucleotides and that of pyrimidine nucleotides?
48. **Recall** Compare the fates of the products of purine and pyrimidine catabolism.

23-9 Conversion of Ribonucleotides to Deoxyribonucleotides

49. **Recall** What roles do thioredoxin and thioredoxin reductase play in the metabolism of nucleotides?
50. **Recall** Suggest a mode of action for fluorouracil in cancer chemotherapy.
51. **Reflect and Apply** Chemotherapy patients receiving cytotoxic (cell-killing) agents such as FdUMP (the UMP analogue that contains fluorouracil) and methotrexate temporarily go bald. Why does this take place?

Further Reading

Anand, S. K., Governale, T. Governale, X. Zhang, B. Razani, A. Yurdagul, C. B. Pattillo, and O. Rom. Amino Acid Metabolism and Atherosclerotic Cardiovascular Disease. *Am. J. Path.* **194**, 510–524 (2024). This review article examines the role of amino acid metabolism in atherosclerotic cardiovascular disease.

Bender, D. A. *Amino Acid Metabolism,* 2nd ed. New York: John Wiley, 1985. A general treatment of the topic, with a particularly good section on tryptophan metabolism.

Benkovic, S. On the Mechanism of Action of Folate- and Biopterin-Requiring Enzymes. *Ann. Rev. Biochem.* **49**, 227–254 (1980). A review of one-carbon transfers.

Braunstein, A. E. Amino Group Transfer. In Boyer, P. D., ed. *The Enzymes,* Vol. 9, 3rd ed. New York: Academic Press, 1973. A dated, but standard, reference.

Karplus, P., M. Daniels, and J. Herriott. Atomic Structure of Ferredoxin-NADP$^+$ Reductase: Prototype for a Structurally Novel Flavoenzyme Family. *Science* **251**, 60–66 (1991). The structure of a key enzyme involved in nitrogen and sulfur metabolism, as well as in photosynthesis.

Kim, J., and D. Rees. Crystallographic Structure and Functional Implications of the Nitrogenase Molybdenum-Iron Protein from *Azotobacter vinelandii. Nature* **360**, 553–560 (1992). X-ray crystallography makes an important contribution to understanding the structure of a key protein involved in the nitrogen fixation process.

Leslie, M. Internal Affairs. *Science* **326**, 929–931 (2009). A description of cytoplasmic responses to gout and other diseases.

Meyer, E., N. Leonard, B. Bhat, J. Stubbe, and J. Smith. Purification and Characterization of the *pur* E, *pur* K, and *pur* C Gene Products: Identification of a Previously Unrecognized Energy Requirement in the Purine Biosynthetic Pathway. *Biochem.* **31**, 5022–5032 (1992). The discovery of a hitherto unsuspected requirement for additional ATP in the biosynthesis of purines.

Mullen, N. J., and P. K. Singh. Nucleotide Metabolism: A Pan-Cancer Metabolic Dependency. *Nat. Rev. Cancer* **23**, 275–294 (2023). This review article explores the mechanisms by which cancer cells rely on nucleotide biosynthesis and discusses potential combination therapies targeting these pathways.

Orme-Johnson, W. Nitrogenase Structure: Where to Now? *Science* **257**, 1639–1640 (1992). Thoughts about nitrogen fixation based on the determination of the structure of nitrogenase by X-ray crystallography.

Stadtman, E. R. Mechanisms of Enzyme Regulation in Metabolism. In Boyer, P. D., ed. *The Enzymes,* Vol. 1, 3rd ed. New York: Academic Press, 1970. A review dealing with the importance of feedback control mechanisms.

In a relay race, a baton is passed from one runner to the next, ultimately leading to the final product of race completion. In organisms, molecules like hormones, neurotransmitters, and other signaling effectors get passed between proteins, ultimately leading to a specific cellular response.

Koto Images/Shutterstock.com

24 Integration of Metabolism: Cellular Signaling

Chapter Outline

24-1 Connections Between Metabolic Pathways

Learning Objectives

After reading this section, students will be able to:

LO 24.1.1 Describe the relationships between photosynthesis and other metabolic pathways.

LO 24.1.2 Explain the relationship between the citric acid cycle and other metabolic pathways.

In the preceding chapters, we learned about a number of individual metabolic pathways. Some metabolites, such as pyruvate, oxaloacetate, and acetyl-CoA, appear in more than one pathway. Furthermore, reactions of metabolism can take place simultaneously, and it is important to consider control mechanisms by which some reactions and pathways are turned on and off.

All metabolism is ultimately linked to photosynthesis and the energy from the Sun (Figure 24.1). The light-dependent (or light) reactions of photosynthesis produce adenosine triphosphate (ATP) and nicotinamide adenine dinucleotide phosphate (NADPH), which are then used to make carbohydrates in the light-independent (or dark) reactions. These carbohydrates are the source of nutrients for other organisms. ATP and NADPH are the two consistent links between different forms of metabolism. Besides linking the light and dark reactions of photosynthesis, they are the most direct link between catabolism and anabolism (Figure 24.1). Other common molecules, such as sugars, PEP, pyruvate, and acetyl-CoA, also form a bridge between catabolic and anabolic processes. We now focus on some of the relationships among pathways by considering some of the physiological responses to biochemical events.

The *citric acid cycle* (Section 19-4) plays a central role in metabolism. Three main points can be considered in assigning a central role to the citric acid cycle. The first of these is its part in the catabolism of nutrients of the main types: carbohydrates, lipids, and proteins (Section 19-7). The second is the function of the citric acid cycle in the anabolism of carbohydrates, lipids, and amino acids (Section 19-8). The third and final point is the relationship between individual metabolic pathways and the citric acid cycle. When we discuss these broader considerations, we can and should address questions that involve more than individual cells and the reactions that go on in them, such as questions of what goes on in tissues and in whole organs. Here in Chapter 24, we look at three such topics—nutrition, hormonal control, and the wide-reaching effects of signaling pathways. **Biochemical Connections 24A** describes the way in which one compound can affect an entire organism.

Chapter Outline (continued)

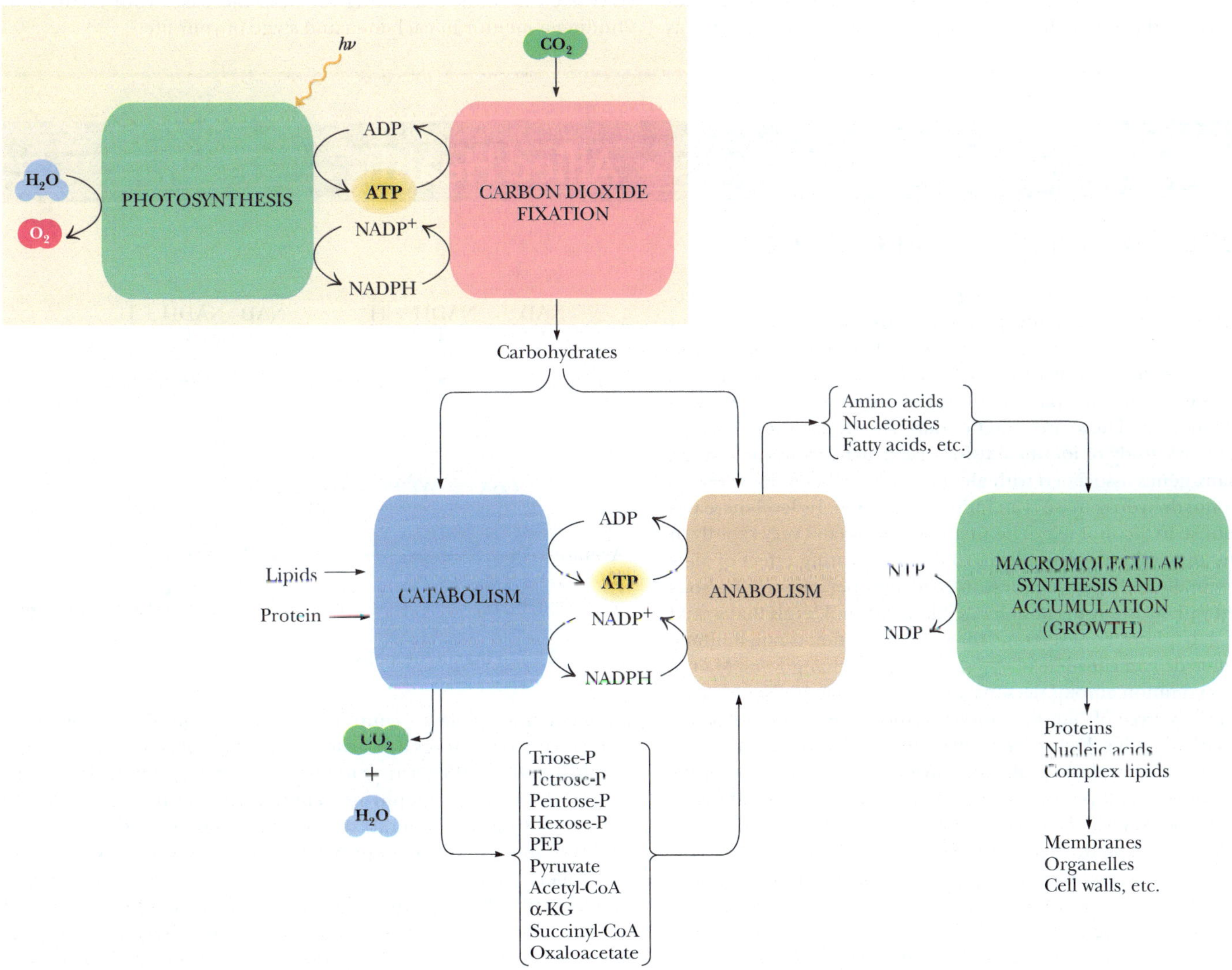

Figure 24.1 All metabolism is connected to photosynthesis and energy from the Sun. This block diagram of intermediary metabolism shows the relationship between anabolic and catabolic processes and the common metabolites seen in many pathways.

Careers in Chemistry • Saul Treviño

Professor of Chemistry at Houston Christian University
Author of *Biochemistry*, 10E (Cengage)

Education:
Texas A&M University System Health Science Center: PhD, Medical Sciences (emphasis in protein chemistry)
Texas A&M University: BS, Biochemistry

Saul Treviño is a dedicated Professor of Chemistry at Houston Christian University and the author of this very *Biochemistry* textbook. He has built a fulfilling career through teaching, advising, and mentoring young scientists in their research. He shares, "What I enjoy most about my work is not only sharing the wonders of chemistry and biochemistry with new sets of students each semester but also learning from each new set of students as they ask questions and develop as learners."

Professor Treviño originally pursued a career in medicine; however, partway through college, he discovered a deep joy in teaching. A pivotal experience was working in the biochemistry labs of Drs. Nick Pace and Marty Scholtz, where he developed expertise in molecular biology and protein chemistry. He continued in the same lab as a research assistant and later as a graduate student. These formative experiences, along with mentorship from individuals of diverse backgrounds, inspired his commitment to teaching and mentoring the next generation of learners.

Joanie Trammell/ProShots Photography

Professor Treviño's path was not without its challenges. After graduate school, he was initially turned down for a full-time teaching position. However, a part-time adjunct role gave him the chance to hone his teaching skills and develop his own course materials. He shares a valuable piece of advice from one of his mentors, "Each person that you ask for help may not end up being a trusted mentor, but asking for help increases your chances of finding a mentor in each area and stage of your life."

24A Biochemical Connections • Allied Health

Alcohol Consumption and Addiction

Alcohol is the most abused drug in the United States, and alcoholism is among the most common diseases. Statistics about deaths due to drunk driving are available, but no one knows how many other accidental deaths may be indirectly caused by alcohol. Many believe that some particular biochemistry must be associated with alcoholism. There are genetic traits, shown most forcefully in a benchmark study of identical twins raised apart from each other, and many genes associated with alcoholism have been discovered.

Alcohol dehydrogenase is an inducible enzyme. Its level increases in response to alcohol use. The first reaction occurs very rapidly in persons affected with alcoholism, so the intoxicating effect of alcohol is actually reduced (i.e., less intoxication per ounce). Persons affected with alcoholism can tolerate blood alcohol levels that would be lethal to others. For everyone, the second reaction is rate-limiting. Acetaldehyde can cause headaches, nausea, and hangovers. Malnutrition is common among persons affected with alcoholism because alcohol is a source of "empty calories" without important nutrients, particularly vitamins. Figure 24.2 shows the catabolism of ethanol.

Biochemical, psychological, and nutritional effects of alcohol are not the same for all people. The studies of twins indicate the possibility of a person who is "born" affected with alcoholism who could be totally hooked on the first drink. Fetal alcohol syndrome (see **Biochemical Connections 17D**) is of particular concern to women and their children. Ethanol is a teratogen; there is no "safe" level of alcohol during pregnancy. Fetal alcohol syndrome occurs in up to 5 of every 1,000 births. Indicators include stunted growth, dysfunction of the central nervous system, and a characteristic facial shape.

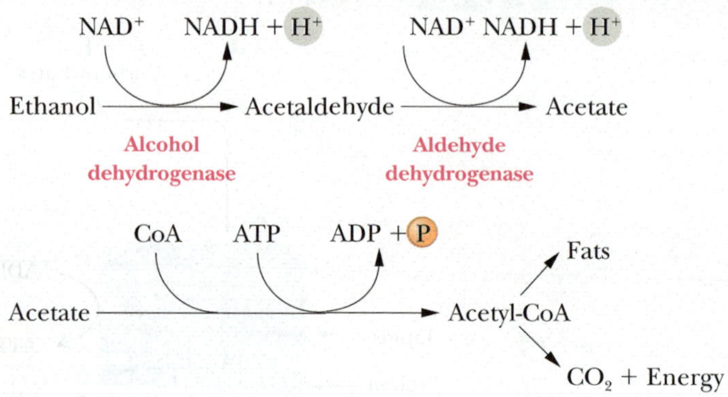

Figure 24.2 Ethanol catabolism.

Biochemistry plays a central role in addiction. Because many of the psychoactive drugs are structural analogs of serotonin and epinephrine, their effects are enhanced, or they compete with those of serotonin and epinephrine. There is increasing interest in the effects of drugs (in general) on the production of endorphins and enkephalins, the short peptides that are the brain's own opiate painkillers. In people who are not affected with alcoholism ethanol inhibits enkephalin synthesis because the pleasant effect of alcohol replaces the need for enkephalins. Part of the misery of a hangover is caused by the lack of enkephalins; the hangover usually lasts until the level of these compounds returns to normal.

24-2 Biochemistry and Nutrition

Learning Objectives

After reading this section, students will be able to:

LO 24.2.1 Distinguish between macronutrients and micronutrients.

LO 24.2.2 Describe three important points regarding dietary protein.

LO 24.2.3 Compare the 1992 Food Guide Pyramid to the MyPlate food guide.

LO 24.2.4 Explain factors that contribute to obesity.

The molecules that we process by catabolic reactions ultimately come from outside the body because we are heterotrophic organisms (dependent on external food sources). We devote this section to a brief look at how the foods we eat are sources of substrates for catabolic reactions. We should also bear in mind that nutrition is related to physiology as well as to biochemistry. This last point is certainly appropriate because many early biochemists were physiologists by training.

24-2A Required Nutrients

In humans, the catabolism of **macronutrients** (carbohydrates, fats, and proteins) to supply energy is an important aspect of nutrition. In the United States, most diets provide more than an adequate number of nutritional calories. The typical American diet is high enough in fat that essential fatty acids (Section 21-6) are seldom, if ever, deficient. The only concern is that the diet contains an adequate supply of protein. If the intake of protein is sufficient, the supply of essential amino acids (Section 23-5) is normally also sufficient. Packaging on food items frequently lists the protein content in terms of both the number of grams of protein and the percentage of the daily value (DV) suggested by the Food and Nutrition Board under the auspices of the National Research Council of the National Academy of Sciences (see Table 24.1). Daily values have replaced the recommended daily allowances (RDAs) formerly seen on food packaging.

macronutrients ones needed in large amounts, such as proteins, carbohydrates, or fats

There are some key biochemical concepts to remember when analyzing a diet for protein content. First, there is no storage form for proteins. This means that proteins eaten in excess do a person no good in terms of satisfying that person's future protein requirements. All protein consumed in excess of what is needed is turned into carbohydrate or fat, and the nitrogen from the amino group must be eliminated through the urea cycle (Section 23-6). Ingesting too much protein can therefore be stressful on the liver and kidneys because of the overproduction of ammonia that must be eliminated. This is the same risk faced by certain athletes who take creatine to build muscle because creatine is a highly nitrogenated compound.

Second, the essential amino acids must be consumed daily in order for proteins to be made. It would be difficult to find a protein that does not have at least one residue of each of the common 20 amino acids. Half of these amino acids are essential, and if the diet is lacking or low in even one of these essential amino acids, then protein synthesis is not possible. Not all proteins are created equal. The protein efficiency ratio (PER) is an indication of how complete a protein is. However, mixing proteins correctly is very important, especially for vegetarians and vegans. A protein that is very low in lysine would have a low PER value. If a second protein had a low PER because it was low in tryptophan, it could be combined with the low-lysine protein to give a combination with a high PER. However, this would work only if the two were eaten together.

Third, proteins are always being produced and degraded in the body. Because of that, even if a person does not seem to be doing any activities that

Table 24.1 Recommended Nutrient Daily Values for the Average Male and Female, Aged 19 to 22

Nutrient	Male	Female
Protein	56 g	44 g
Lipid-soluble vitamins		
Vitamin A	1 mg RE*	8 mg RE*
Vitamin D	7.5 μg†	7.5 μg†
Vitamin E	10 mg α-TE‡	8 mg α-TE‡
Water-soluble vitamins		
Vitamin C	60 mg	60 mg
Thiamine (vitamin B_1)	1.5 mg	1.1 mg
Riboflavin (vitamin B_2)	1.7 mg	1.3 mg
Vitamin B_6	3 μg	3 μg
Vitamin B_{12}	3 μg	3 μg
Niacin	3 μg	3 μg
Folic acid	19 mg	14 mg
Pantothenic acid (estimate)	10 mg	10 mg
Biotin (estimate)	0.3 mg	0.3 mg
Minerals		
Calcium	800 mg	800 mg
Phosphorus	800 mg	800 mg
Magnesium	350 mg	300 mg
Zinc	15 mg	15 mg
Iron	10 mg	18 mg
Copper (estimate)	3 mg	3 mg
Iodine	150 μg	150 μg

*RE = retinol equivalent, where 1 retinol equivalent = 1 μg retinol or 6 μg β-carotene. See Section 8-6.
†As cholecalciferol. See Section 8-6.
‡α-TE = α-tocopherol equivalent, where 1 α-TE = 1 μg D-α-tocopherol. See Section 8-6. Data from the Food and Nutrition Board, National Academy of Sciences—National Research Council, Washington, DC, 1988.

would tend to require protein replenishment, there is a constant need for quality protein in order to maintain the body's structures. Athletes are painfully aware of that fact. They must train constantly, and they get out of shape quickly when they stop, the effect being even more pronounced as the athlete reaches middle age.

24-2B Dietary Vitamins

micronutrients vitamins and minerals that are needed in small amounts

Micronutrients (vitamins and minerals) are also listed on food packaging. The vitamins we require are compounds that are necessary for metabolic processes; either our bodies cannot synthesize them, or they cannot synthesize them in amounts sufficient for our needs. As a result, we must obtain vitamins from dietary sources. In Table 24.1, recommended daily values are listed for the fat-soluble vitamins—vitamins A, D, and E (Section 8-6)—but care must be taken to avoid overdoses of these vitamins. Excesses can be toxic when large amounts of fat-soluble vitamins accumulate in adipose tissue. Excess vitamin A is especially toxic. With water-soluble vitamins, turnover is frequent enough that the danger of excess is not normally a problem.

The water-soluble vitamins with listed DVs are vitamin C, which is necessary for the prevention of scurvy, and the B vitamins—niacin, pantothenic acid, vitamin B_6, riboflavin, thiamine, folic acid, biotin, and vitamin B_{12}. The B vitamins are the precursors of the metabolically important coenzymes listed in Table 7.1, where references to the reactions in which the coenzymes play a role are given. We have seen many pathways in which NADH, NADPH, FAD, TPP, biotin, pyridoxal phosphate, and coenzyme A are found, all of which come from vitamins. A summary of vitamins and their metabolic roles is given in Table 24.2. Frequently, the actual biochemical role is played by a metabolite of the vitamin rather than by the vitamin itself, but this point does not affect the dietary requirement.

24-2C Dietary Minerals

minerals in nutrition, inorganic substances required as ions or free elements

Minerals, in the nutritional sense, are inorganic substances required in the ionic or free-element form for life processes. The macrominerals (those needed in the largest amounts) are sodium, potassium, chloride, magnesium, phosphorus, and calcium. The required amounts of all these minerals, except calcium, can easily be satisfied by a normal diet. Deficiencies of calcium can, and frequently do, occur. Calcium deficiencies can lead to bone fragility, with a concomitant risk of fracture, which is a problem especially for elderly females. Calcium supplements are indicated in such cases.

Requirements for some microminerals (trace minerals) are not always clear. It is known, for example, from biochemical evidence that chromium is necessary for glucose metabolism (a role that has recently been suggested for chromium picolinate) and that manganese is necessary for bone formation, but no deficiencies of these elements have been recorded.

Requirements have been established for iron, copper, zinc, iodide, and fluoride; there are DVs for all of these minerals except fluoride. In the case of copper and zinc, needs are easily met by dietary sources, and overdoses can be toxic. A deficiency of iodide, leading to an enlarged thyroid gland (Section 24-3), has been a problem in some parts of the United States for many years. Iodized salt is used to prevent this deficiency, and it has become unusual to find table salt without an iodine supplement. Fluoride is administered to prevent tooth decay in children and, with that end in mind, has been added to water supplies, sometimes causing considerable controversy. Iron is important because it is part of the structure of the ubiquitous heme proteins. Females

Table 24.2 Vitamins: Chemical and Biochemical Facts

Vitamin	Metabolic Function	Reference
Water-Soluble Vitamins		
B_1 (thiamine)	Aldehyde transfer, decarboxylation in alcoholic fermentation and citric acid cycle	Sections 17-4, 19-3
B_2 (riboflavin)	Oxidation–reduction reactions, especially in citric acid cycle and electron transport	Section 19-2
B_6 (pyridoxine)	Transamination reactions, especially of amino acids	Section 23-4
Niacin (nicotinic acid)	Oxidation–reduction reactions, found in many metabolic processes	Section 19-3
Biotin	Carboxylation reactions in carbohydrate and lipid metabolism	Sections 18-2, 21-6
Pantothenic acid	Acyl transfer in many metabolic processes	Section 15-7
Folic acid	One-carbon group transfer, especially in nitrogen containing compounds	Sections 23-6, 23-11
C (ascorbic acid)	Hydroxylates collagen	Biochemical Connections 16B
Lipoic acid (?)	Acyl transfer, oxidation–reduction	Section 19-3
(It has been questioned whether lipoic acid is a vitamin.)		
Lipid-Soluble Vitamins		
A	Isomerization mediates visual process	Section 8-6
D	Regulates calcium and phosphorus metabolism, especially in bone	Section 8-6
E	Antioxidant	Section 8-6
K	Mediates protein modification required for blood clotting	Section 8-6

of childbearing age are more susceptible to iron deficiencies than are other segments of the population, and in some cases, supplements are advised. These recommended levels vary with the age of the individual and are subject to adjustment for the level of activity. **Biochemical Connections 24B** explores dietary iron in more detail.

24B Biochemical Connections • Nutrition

Iron: An Example of a Mineral Requirement

Iron, whether in the form Fe(II) or Fe(III), is usually found in the body in association with proteins. Little or no iron can be found "free" in the blood. Because iron-containing proteins are ubiquitous, there is a dietary requirement for this mineral. Severe deficiencies can lead to iron-deficiency anemia.

Iron usually occurs in the Fe(III) form in food. This is also the form released from iron pots when food is cooked in them. However, iron must be in the Fe(II) state to be absorbed, and dietary iron supplements can be purchased in this state (Figure 24.3). Reduction from Fe(III) to Fe(II) can be accomplished by ascorbate (vitamin C) or by succinate. Factors that affect absorption include the solubility of a given compound of iron, the presence of antacids in the digestive tract, and the source of the iron. To give some examples, iron may form insoluble complexes with phosphate or oxalate, and the presence of antacids in the digestive tract may decrease iron absorption. Iron from meats is more easily absorbed than iron from plant sources.

Requirements for iron vary according to age and gender. Infants and adult males need 10 mg per day; infants are born with a three- to six-month supply. Children and females (ages 16 through 50) need 15 to 18 mg per day. Females lose 20 to 23 mg of iron during each menstrual period. Pregnant and lactating females need more than 18 mg per day. After a blood loss, anyone, regardless of age or gender, needs more than these amounts. Distance runners, particularly marathoners, are also at risk of becoming anemic because of the loss of blood cells in the feet caused by the pounding of thousands of footfalls that occur during a long run. People with iron deficiencies may experience a craving for nonfood items such as clay, chalk, and ice.

While iron is an essential nutrient, too much iron can also have consequences. Iron normally is bound to proteins such as hemoglobin, myoglobin, and transferrin. If a person has too much iron, it will overload the proteins it should bind to. In that case, free iron will increase. Free iron is a very reactive ion and causes oxidative damage to cells. Iron toxicity can come from genetic disorders where a person's ability to manage their iron content is defective, or it can come from taking in far too much iron. As is the case with many compounds, more is not always better, at least not once you surpass the amount you need. ◗

Figure 24.3 A ferrous iron, Fe(II), dietary supplement.

24-2D Food Guides

In 1992, one approach to publicizing healthful food selection was the development of the Food Guide Pyramid, a graphic display that focuses on a diet sufficient in nutrients but without excesses (Figure 24.4). The goal was to use a well-chosen diet to promote good health. To avoid confusion, the development of this scheme had to take into account the fact that many people were familiar with the older recommendations about food groups. The newer recommendations pay particular attention to increasing the amount of fiber and decreasing the amount of fat in the typical diet. Variety and moderation are key concepts of the graphic presentation. From the biochemical point of view, these recommendations translate into a diet based primarily on carbohydrates, with enough protein to meet the needs for essential amino acids (Section 23-5).

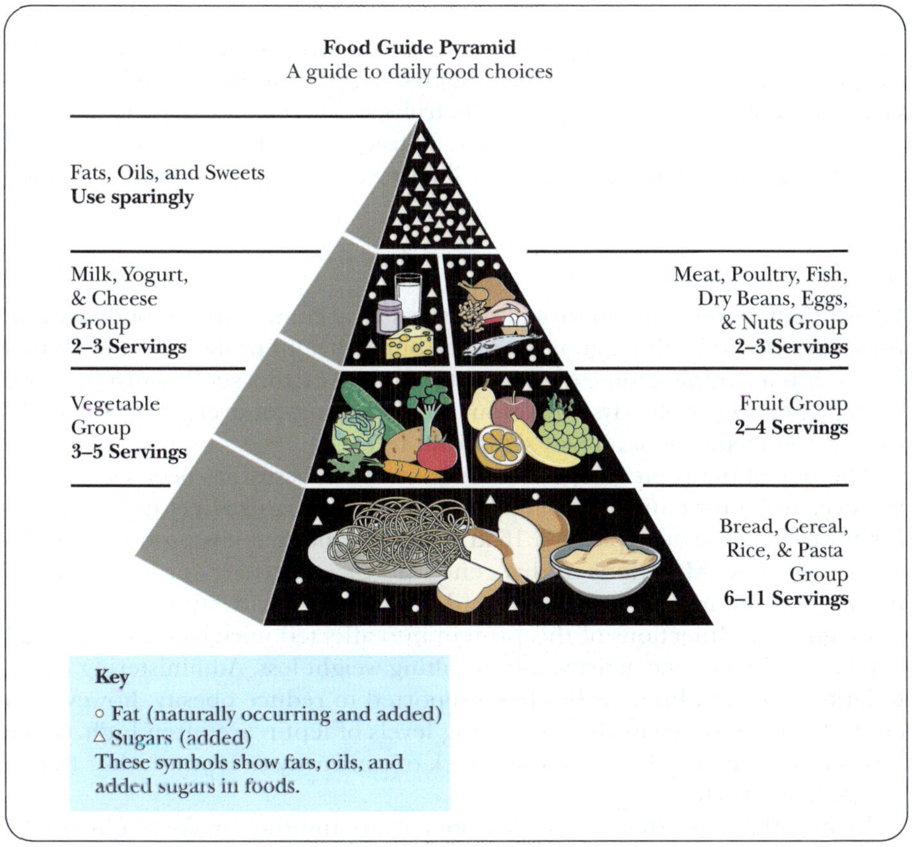

Figure 24.4 The 1992 Food Guide Pyramid (USDA). In 1992, the USDA recommended choices that reflected a diet based primarily on carbohydrates. Smaller amounts of proteins and lipids were considered sufficient to meet the body's needs.

Notice in Figure 24.4 that carbohydrates are the base, with the correct amount suggested to be 6 to 11 servings of foods rich in complex carbohydrates, such as bread, cereal, rice, or pasta. Lipids should not contribute more than 30% of daily calories, but the typical American diet currently is about 45% fat. High-fat diets have been linked to heart disease and to some kinds of cancer, so the recommendation about lipid intake is of considerable importance.

▶ How has nutritional advice shifted since the 1992 Food Guide Pyramid?

In the years since the development of the 1992 Food Guide Pyramid, nutritional advice has shifted from a one-size-fits-all, low-fat, high-carbohydrate guide to a personalized MyPlate guide, which was introduced in 2011 (Figure 24.5). Instead of a pyramid visual, MyPlate depicts a plate and glass with five food groups

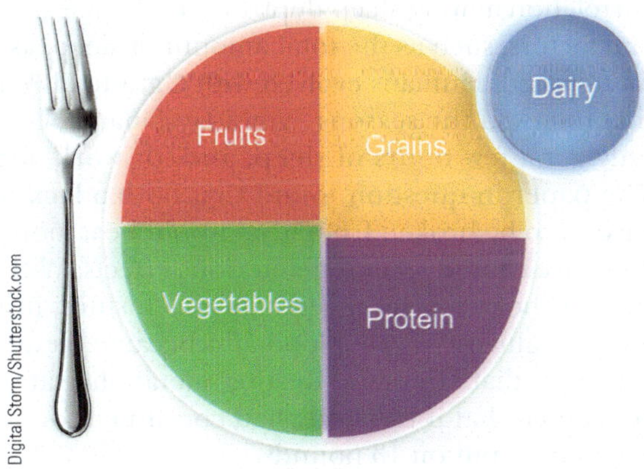

Figure 24.5 The MyPlate Food Guide. The MyPlate website (myplate.gov) and mobile app (Start Simple with MyPlate) are designed to help people build balanced, healthy meals based on proper portions of five food groups (vegetables, fruits, grains, protein, and dairy).

represented: vegetables, fruits, grains, protein, and dairy. The MyPlate website (myplate.gov) and mobile app (Start Simple with MyPlate) are designed to help people build balanced, healthy meals. Key messages of the MyPlate guide include choosing variety, making half your plate fruits and vegetables, making half your grains whole grains, moving to low-fat or fat-free dairy, varying your protein, and prioritizing nutrient-dense foods while limiting added sugars, sodium, and saturated fat.

24-2E Obesity

Obesity is defined as an abnormal or excessive accumulation of body fat that presents a risk to health, and it is a major public health problem in the United States. It is a complex topic of great interest, and factors such as insulin resistance, leptin or ghrelin dysregulation, the psychology of overeating, sleep disorders, and medications all play significant roles.

The role of the protein leptin in the control of obesity has been established in mice, and information is just appearing concerning its effect in humans. It is known that in mice, leptin is a 16-kDa protein and that it is produced by the *obesity (ob)* gene. Mutations in this gene lead to a deficiency of leptin, which in turn leads to increased appetite and decreased activity, ultimately leading to weight gain. Injections of this protein into affected mice lead to decreased appetite and increased activity, with resulting weight loss. Administering leptin to leptin-deficient humans has been reported to reduce obesity; however, in clinically obese subjects, the circulating levels of leptin are often high. Some forms of obesity may be caused by a lack of sensitivity to leptin rather than a lack of leptin itself.

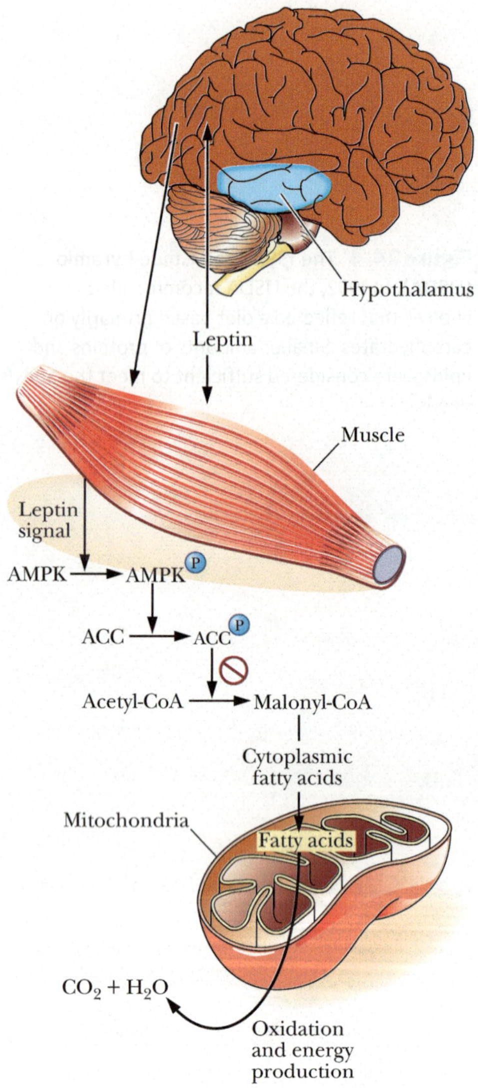

Figure 24.6 Leptin has multiple effects on metabolism. It affects the brain, lowering appetite. It also inactivates acetyl-CoA carboxylase (ACC). Reduced activity of ACC leads to a reduction in malonyl-CoA, which stimulates fatty acid oxidation and reduces fatty acid synthesis. *(Reprinted by permission from Macmillan Publishers Ltd: Friedman, J. (2002). Diabetes: Fat in all the wrong places.* Nature *415, 268, Fig. 1.)*

Leptin stimulates the oxidation of fatty acids and the uptake of glucose by muscle cells. It does so by stimulating AMP-activated protein kinase, which phosphorylates an isoform of acetyl-CoA carboxylase (ACC) in muscle cells, rendering it less active (Figure 24.6). Recall from Section 21-6 that ACC plays a pivotal role in fat metabolism. When ACC activity is decreased, malonyl-CoA levels decrease and the mitochondria can take up and oxidize fatty acids. Leptin also inhibits the production of the mRNA for hepatic stearoyl-CoA desaturase, an enzyme that adds double bonds to saturated fatty acids, leading to less lipid synthesis.

Leptin also works directly on the nervous system. Both leptin and insulin (Section 24-5) are long-term regulators of appetite. They circulate in the blood at concentrations roughly proportional to body-fat mass. They inhibit appetite by inhibiting specific neurons in the hypothalamus. Several laboratories have shown interest in using this information to develop treatments for human obesity.

While it would be very convenient if scientists could find one catch-all solution to obesity, the problem is far too complex for that, and new factors come to light every day. A paper in *Science* gave the results of a study that showed a correlation among sleep deprivation, diabetes, and obesity. They looked at several factors, such as the total amount of sleep, as well as when the sleep cycle is interrupted. Humans evolved with a gradually changing night-and-day pattern that followed the seasons. Nowadays, many people get up when it is still dark, only get a few hours of sleep, and/or work odd hours, such as night shifts. The paper in question found that both a lack of sleep and forcing the sleep rhythm to be broken from what would be a more normal 24-hour cycle of light and dark caused several deleterious effects. With just a week of reduced sleep, subjects began to show resistance to insulin, just like in type II diabetes. The levels of glucose in the blood also became elevated to a range that would indicate a prediabetic state. Resting metabolic rate also dropped 8%. The paper concluded that if the study had continued for a year at that level, the subjects would have put on 13 pounds.

This phenomenon may be counterintuitive. A person would be tempted to think that if they got less sleep, they would have a higher metabolic rate because they would have more total hours of activity. This study indicates just the opposite. A traditional day/night cycle with seven to eight hours of sleep is actually healthier in terms of maintaining a good metabolic rate and not putting on weight.

A major breakthrough in the fight against obesity occurred in 2021 with the release of clinical trial results for a medication called Wegovy. Wegovy is a GLP-1 (glucagon-like peptide-1) receptor agonist, also called *semaglutide*, that mimics the natural GLP-1 hormone. Since Wegovy's release, other formulations of semaglutide (Ozempic and Rybelsus) have been produced for treating type II diabetes. When semaglutide binds to GLP-1 receptors, it activates a stimulatory G protein inside the cell, which leads to a cascade of effects that reduce appetite (see Section 24-3B for more details on the metabolic effects of hormone binding).

In 2022 and 2023, more progress in the fight against obesity and type II diabetes occurred with the development of *tirzepatide* in the Mounjaro (2022 FDA approval) and Zepbound (2023 FDA approval) formulations. Tirzepatide is a dual GIP/GLP-1 receptor agonist (GIP stands for glucose-dependent insulinotropic polypeptide, a hormone involved in regulating insulin and metabolism), and it has shown even greater effectiveness than semaglutide in treating obesity.

24-3 Hormones and Second Messengers

Learning Objectives

After reading this section, students will be able to:

LO 24.3.1 Describe the role of hormones.

LO 24.3.2 Explain the control of the production of hormones.

LO 24.3.3 Summarize the events leading to the production of the second messengers cAMP, Ca^{2+}, IP_3, and DAG.

24-3A Hormones

The metabolic processes within a given cell are frequently regulated by signals from outside the cell. A usual means of intercellular communication takes place through the workings of the **endocrine system**, in which the ductless glands produce **hormones** as intercellular messengers.

endocrine system the system of ductless glands that release hormones into the bloodstream

hormones substances produced by endocrine glands and delivered through the bloodstream to target cells, producing regulatory effects on metabolic processes

▶ What are hormones?

Hormones are transported from the sites of their synthesis to the sites of action by the bloodstream (Figure 24.7). Typical hormones include steroids, such as estrogens, androgens, and mineralocorticoids; polypeptides, such as insulin and endorphins; and amino acid derivatives, such as epinephrine and norepinephrine (Table 24.3).

Hormones have several important functions in the body. They help maintain **homeostasis**, the balance of biological activities in the body. The effect of insulin in keeping the blood glucose level within narrow limits is an example of this function. The operation of epinephrine and norepinephrine in the "fight-or-flight" response is an example of the way in which hormones mediate responses to external stimuli. Finally, hormones play roles in growth and development, as seen in the roles of growth hormone and the sex hormones. The methods and insights of biochemistry and physiology alike have helped illuminate the workings of the endocrine system.

homeostasis the balance of biological activities in the body

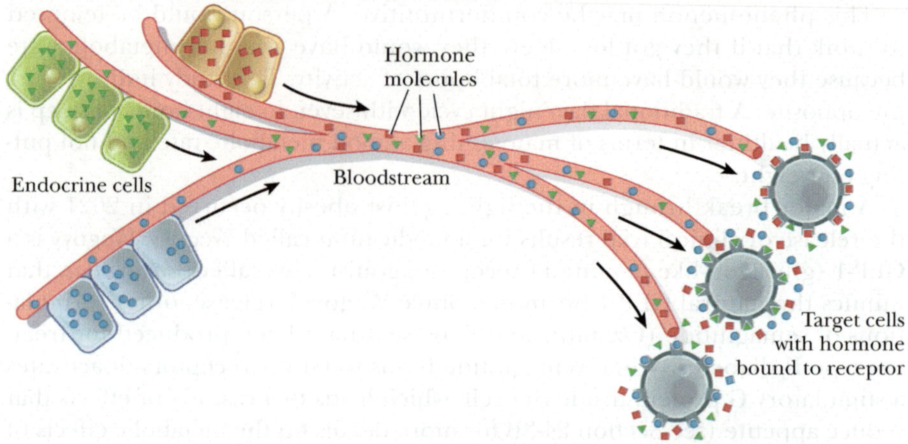

Figure 24.7 **Blood carries many hormones.** Endocrine cells secrete hormones into the bloodstream, which transports them to target cells. Target cells have specific receptors that bind to particular hormones, thereby generating their metabolic effects.

Table 24.3 Selected Human Hormones

Hormone	Source	Major Effects
Polypeptides		
Corticotropin-releasing factor (CRF)	Hypothalamus	Stimulates release of ACTH
Gonadotropin-releasing factor (GnRF)	Hypothalamus	Stimulates release of FSH and LH
Thyrotropin-releasing factor (TRF)	Hypothalamus	Stimulates release of TSH
Growth hormone-releasing factor (GRF)	Hypothalamus	Stimulates release of growth hormone
Adrenocorticotropic hormone (ACTH)	Anterior pituitary	Stimulates release of adrenocorticosteroids
Thyrotropin (TSH)	Anterior pituitary	Stimulates release of thyroxine
Follicle-stimulating hormone (FSH)	Anterior pituitary	In ovaries, stimulates ovulation and estrogen synthesis; in testes, stimulates spermatogenesis
Luteinizing hormone (LH)	Anterior pituitary	In ovaries, stimulates estrogen and progesterone synthesis; in testes, stimulates androgen synthesis
Met-enkephalin	Anterior pituitary	Has opioid effects on central nervous system
Leu-enkephalin	Anterior pituitary	Has opioid effects on central nervous system
β-Endorphin	Anterior pituitary	Has opioid effects on central nervous system
Vasopressin	Posterior pituitary	Stimulates water resorption by kidneys and raises blood pressure
Oxytocin	Posterior pituitary	Stimulates uterine contractions and flow of milk
Insulin	Pancreas (β-cells of islets of Langerhans)	Stimulates uptake of glucose from bloodstream
Glucagon	Pancreas (α-cells of islets of Langerhans)	Stimulates release of glucose to bloodstream
Steroids		
Glucocorticoids	Adrenal cortex	Decrease inflammation, increase resistance to stress
Mineralocorticoids	Adrenal cortex	Maintain salt and water balance
Estrogens	Gonads and adrenal cortex	Stimulate development of secondary sex characteristics, particularly in females
Androgens	Gonads and adrenal cortex	Stimulate development of secondary sex characteristics, particularly in males
Amino acid derivatives		
Epinephrine	Adrenal medulla	Increases heart rate and blood pressure
Norepinephrine	Adrenal medulla	Decreases peripheral circulation, stimulates lipolysis in adipose tissue
Thyroxine	Thyroid	Stimulates metabolism generally

The release of hormones exerts control on the cells of target organs; other control mechanisms, however, determine the workings of the endocrine gland that releases the hormone in question. Simple negative feedback mechanisms, such as the one illustrated in Figure 24.8, show how the action of many types of hormones causes changes that end up stopping their production in their original endocrine gland once a target metabolic effect level has been reached. Compared to the model in Figure 24.8, the exact workings of the endocrine system have added complexity, allowing for a greater degree of control of homeostasis.

An example showing more details of homeostasis control involves the action of the **hypothalamus** on **endocrine glands** via the **pituitary gland** (Figure 24.9). The central nervous system sends a signal to the hypothalamus. The hypothalamus secretes a hormone-releasing factor, which in turn stimulates the release of a **tropic hormone** by the anterior pituitary (Table 24.3). (The action of the hypothalamus on the posterior pituitary is mediated by nerve impulses.) Tropic hormones act on specific endocrine glands, which release the hormones to be transported to target organs. Negative feedback control is exerted at every

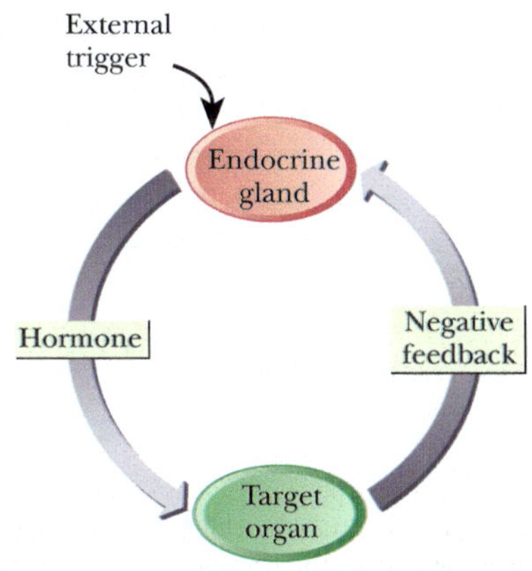

Figure 24.8 A simple negative feedback control system involving an endocrine gland and a target organ.

hypothalamus the portion of the brain that controls, among other things, many of the workings of the endocrine system

endocrine glands specialized organs that produce and secrete hormones directly into the blood to regulate various physiological processes

pituitary gland a central endocrine gland that produces and secretes hormones that regulate other endocrine glands; composed of the anterior pituitary and the posterior pituitary

tropic hormones hormones that are produced by the pituitary gland under the direction of the hypothalamus and that, in turn, cause the release of specific hormones by individual endocrine glands

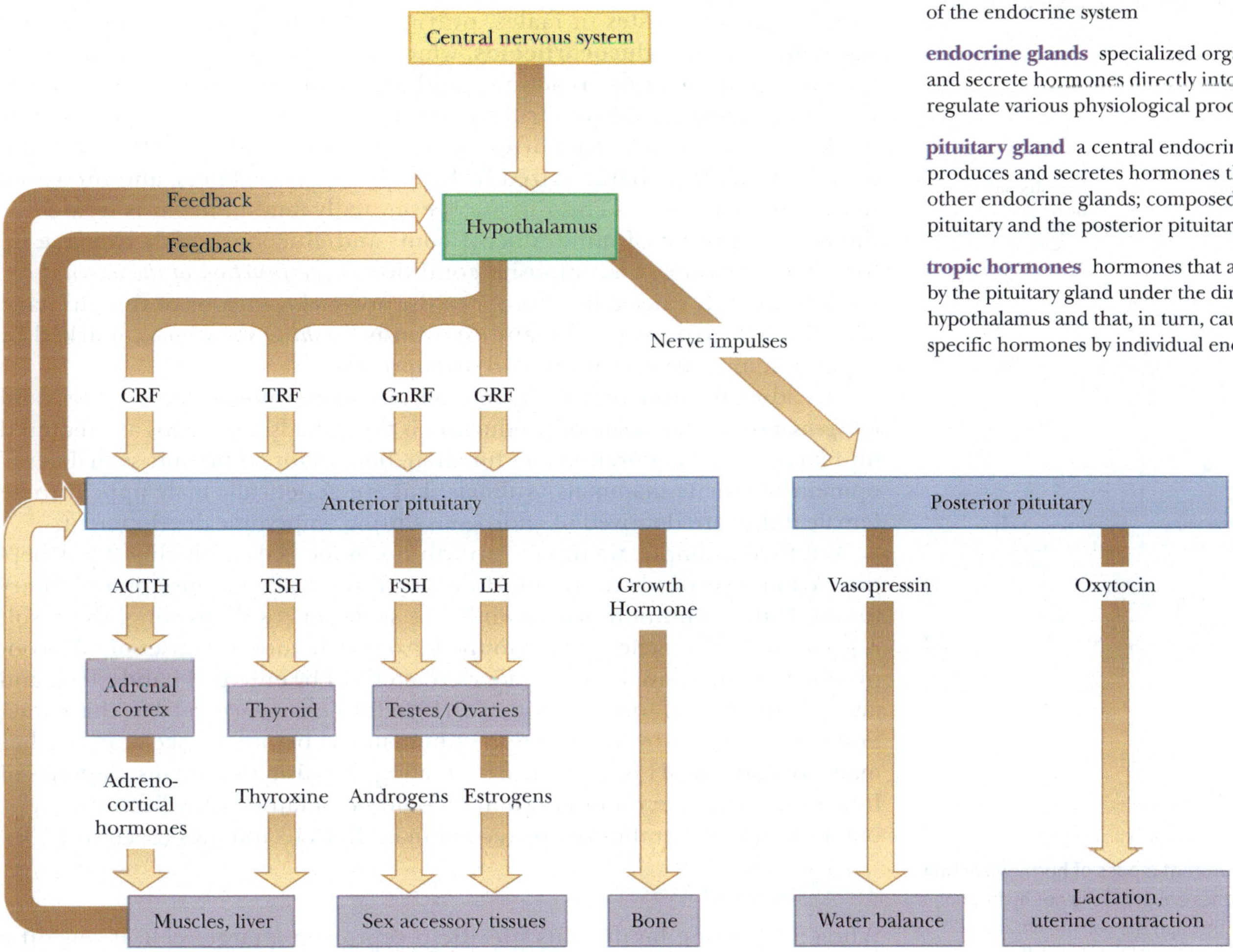

Figure 24.9 Hormonal control pathways. These hormonal control pathways show the role of the hypothalamus on the target endocrine glands via the pituitary gland. See Table 24.3 for the names of the hormones. CRF, TRF, GnRF, and GRF are peptide hormones that stimulate the anterior pituitary to release tropic hormones (ACTH, TSH, FSH, LH, and growth hormone) that act on endocrine glands (adrenal cortex, thyroid, testes, and ovaries) to produce and secrete hormones that regulate physiological processes in target tissues. Negative feedback mechanisms stop the production of these various hormones in the hypothalamus or anterior pituitary when a target effect level has been reached.

stage of the process (Figure 24.9, *left*). Additionally, more fine-tuning is possible with zymogen activation mechanisms (Section 7-4), which exist for many hormones.

The chemical natures of hormones determine how they play their roles in cell signaling (Figure 24.10). Steroid hormones, with their hydrophobic, four-ring cholesterol-derived structure (Section 8-2G), can enter the cell by diffusing through the plasma membrane, or they can bind to plasma-membrane receptors (Figure 24.10, *top left*). Nonsteroid hormones, however, enter the cell exclusively as a result of binding to plasma-membrane receptors (Figure 24.10, *top right*).

The releasing factors and tropic hormones listed in Table 24.3 tend to be polypeptides, but the chemical natures of the hormones released by specific endocrine glands show greater variation. Thyroxine, for example, produced by the thyroid, is an iodinated derivative of the amino acid tyrosine. Abnormally low levels of thyroxine lead to **hypothyroidism**, characterized by lethargy and obesity, whereas increased levels produce the opposite effect (*hyperthyroidism*). Low levels of iodine in the diet often lead to hypothyroidism and an enlarged thyroid gland (*goiter*). This condition has largely been eliminated by the addition of sodium iodide to commercial table salt ("iodized" salt).

hypothyroidism a state in which an individual produces lower than the normal levels of thyroid hormones

adrenocortical hormones steroid hormones secreted by the adrenal cortex that affect inflammation and salt and water balance

glucocorticoids steroid hormones involved in the metabolism of sugars

mineralocorticoids steroid hormones involved in the regulation of levels of inorganic ions or "minerals"

Steroid hormones (Section 8-2G) are produced by the adrenal cortex and the gonads (testes in males, ovaries in females). The **adrenocortical hormones** include **glucocorticoids**, which affect carbohydrate metabolism, modulate inflammatory reactions, and are involved in reactions to stress. The **mineralocorticoids** control the level of excretion of water and salt by the kidneys. If the adrenal cortex does not function adequately, one result is *Addison's disease*, characterized by hypoglycemia, weakness, and increased susceptibility to stress. This disease is eventually fatal unless it is treated by the administration of mineralocorticoids and glucocorticoids to make up for what is missing. The opposite condition, *hyperfunction of the adrenal cortex*, is frequently caused by a tumor of the adrenal cortex or of the pituitary. The characteristic clinical manifestation is *Cushing's syndrome*, marked by hyperglycemia, hypertension, and osteoporosis.

The adrenal cortex produces some steroid sex hormones, the *androgens* and *estrogens*, but the main site of production is the gonads. Estrogens are required for female sexual maturation and function, but not for embryonic sexual development of female mammals. Animals that are genetically male appear to be female if they are deprived of androgens during embryonic development.

As a final example, we discuss growth hormone (GH), which is a polypeptide. When overproduction of GH occurs, it is usually because of a pituitary tumor. If this condition occurs while the skeleton is still growing, the result is *gigantism*. If the skeleton has stopped growing before the onset of GH overproduction, the result is *acromegaly*, characterized by enlarged hands, feet, and facial features. Underproduction of GH leads to *achondroplasia*, but this condition can be treated by the injection of human GH before the skeleton reaches maturity. Animal GH is ineffective in treating dwarfism in humans. Supplies of human GH were very limited when it could be obtained only from cadavers, but it can now be synthesized by recombinant DNA techniques (Section 13-3).

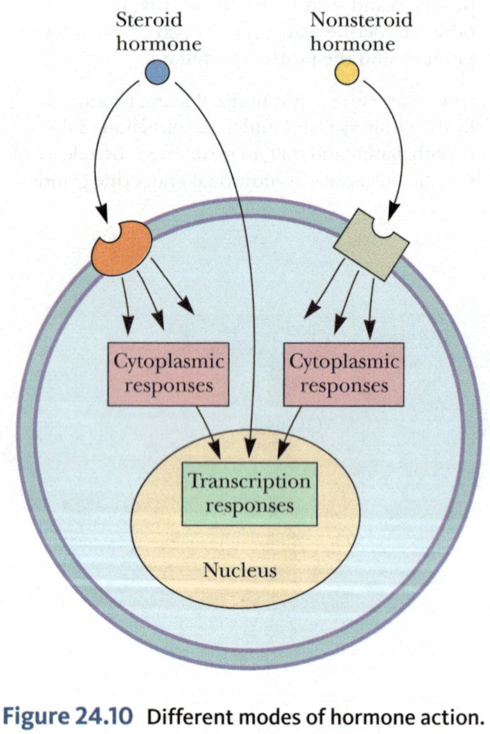

Figure 24.10 Different modes of hormone action. Nonsteroid hormones are generally not hydrophobic enough to diffuse through the cell membrane, so they bind exclusively to plasma-membrane receptors, which mediate the cellular responses to the hormone. Steroid hormones, however, are hydrophobic in nature, so they can exert their effects either by binding to plasma-membrane receptors or by diffusing to the nucleus, where they modulate transcriptional events.

24-3B Second Messengers

When a hormone binds to its specific receptor on a target cell, it sets off a chain of events within the cell. Several kinds of membrane receptors and intracellular receptors are known. The intracellular receptors for steroid hormones that diffuse through the plasma membrane are found in the cytoplasm or the nucleus. These receptor proteins bind the steroid hormones to form activated steroid–receptor complexes that affect the transcription of specific genes. More frequently, the receptor proteins are a part of the plasma membrane, and the binding of hormones to these receptors triggers the production of a **second messenger**. The second messenger brings about the changes within the cell as a result of a series of reactions.

second messenger a substance produced or released by a cell in response to hormone binding to a receptor on the cell surface; it elicits the actual response in the cell

▶ How is the cyclic adenosine monophosphate (cAMP) messenger produced?

Cyclic adenosine monophosphate (cAMP) is one example of a second messenger (Figure 24.11). Recall from Section 21-2 that we saw the hormone-dependent release of fatty acids from triacylglycerols and that cAMP was produced in the process, where it played a role in triggering a cascade of events that led to the release of fatty acids. Figure 24.12 shows more specifics related to cAMP production. The production of cAMP most often starts with the binding of a hormone to a G protein-coupled receptor (GPCR), which triggers the production of cAMP from ATP, catalyzed by *adenylyl cyclase.* This reaction is mediated by a stimulatory G protein, a trimer consisting of α, β, and γ subunits. Binding of the hormone to the receptor activates the G_α protein, and the G_α protein then binds GTP while releasing GDP, hence the name of the protein. The G_α (GTP) complex dissociates from the other two subunits ($G_{\beta\gamma}$) and binds to adenylyl cyclase, stimulating the synthesis of cAMP. The active protein has GTPase activity and slowly hydrolyzes GTP, returning the G protein to the inactive state. GDP remains bound to the α subunit and must be exchanged for GTP when the protein is activated the next time. The newly formed cAMP is released into the interior of the cell as a second messenger that stimulates protein kinase A, which phosphorylates a host of enzymes and transcription factors that regulate specific metabolic processes depending on the identity of the hormone.

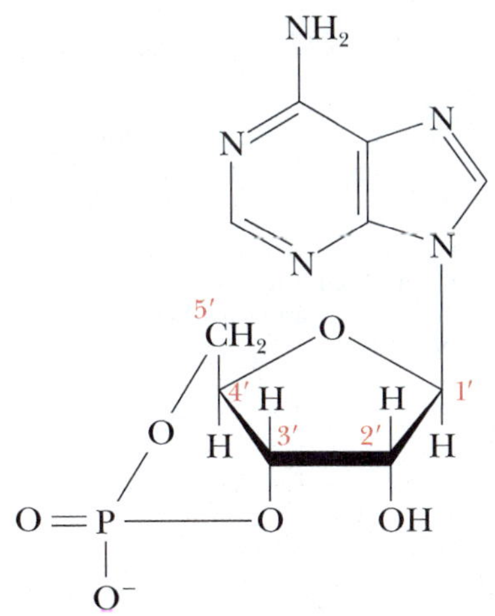

Figure 24.11 Cyclic adenosine monophosphate (cAMP). The phosphate forms a ring by connecting to the 3′ and 5′ carbons of the ribose moiety of cAMP.

To specify different modes of hormone action further, we must distinguish different types of GPCRs and different types of G_α proteins. The production of cAMP in response to stimulatory hormone binding occurs with specific GPCRs called β-adrenergic receptors, and it involves a stimulatory G_α protein ($G_{s\alpha}$) (Figure 24.13, *left*). However, when an inhibitory hormone binds to another type of GPCR called an α_2-adrenergic receptor, it leads to inhibition rather than stimulation of adenylyl cyclase (Figure 24.13, *right*). In these cases, a G protein with a different kind of α subunit (an inhibitory G_α protein, $G_{i\alpha}$) mediates the process.

Figure 24.14 shows more details of the interaction between cAMP and protein kinase A (PKA). PKA is a tetramer consisting of two regulatory subunits and two catalytic subunits. When cAMP binds to the dimer of regulatory subunits, the two active catalytic subunits are released. The active kinase catalyzes the phosphorylation of a target enzyme or transcription factor (Figure 24.14, *bottom right*). In the scheme shown in Figure 24.14, phosphorylation activates the enzyme, but cases are also known in which phosphorylation inactivates a target enzyme (e.g., glycogen synthase; Section 24-4). The usual site of phosphorylation is the hydroxyl group of a serine or a threonine. ATP is the source of the phosphate group that is transferred to the enzyme. The target enzyme then elicits the cellular response, or the target enzyme sets off a cascade that leads to a cellular response.

G proteins are very important signaling molecules in eukaryotes. They can be activated by combinations of hormones. For example, both epinephrine and glucagon act via a stimulatory G protein in liver cells. The effect can be cumulative, so that if both glucagon and epinephrine have been released, the cellular effect is greater. Besides their effect on cAMP, G proteins are involved in activating many other cellular processes, including stimulating phospholipase C and opening or closing membrane ion channels. They are also involved in vision and smell. There are currently more than 100 known GPCRs and more than 20 known G proteins. For more information on GPCRS, refer to the Hot Topic at the end of this chapter.

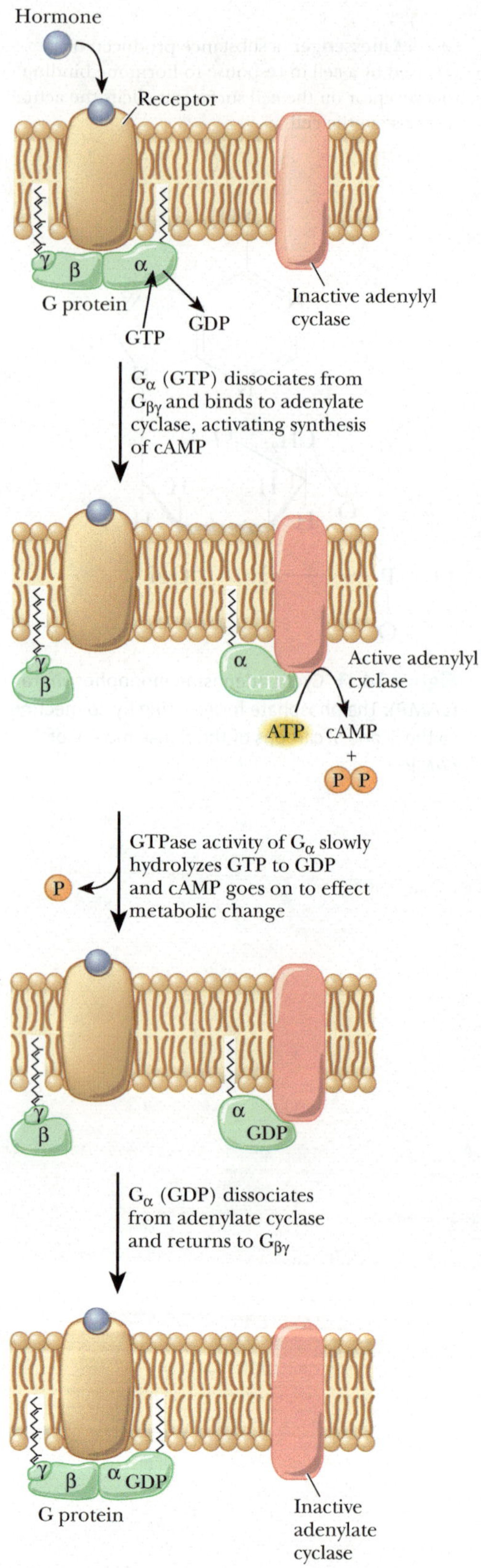

Figure 24.12 The production of the cyclic adenosine monophosphate (cAMP) second messenger after activation of adenylyl cyclase by heterotrimeric G proteins. Binding of a hormone from an endocrine gland to a G protein-coupled receptor (GPCR) causes a conformational change that induces the receptor to catalyze the replacement of GDP by GTP on the G_α subunit of a heterotrimeric G protein. The G_α (GTP) complex dissociates from the other two subunits ($G_{\beta\gamma}$) and binds to adenylyl cyclase, stimulating the synthesis of cAMP, which goes on to cause downstream effects that regulate metabolic processes in the cell. Bound GTP is slowly hydrolyzed to GDP by the intrinsic GTPase activity of G_α. G_α (GDP) dissociates from adenylyl cyclase and reassociates with $G_{\beta\gamma}$. G_α and G_γ are lipid-anchored proteins. Adenylyl cyclase is an integral membrane protein consisting of 12 transmembrane α-helical segments.

protein kinase C (PKC) a family of protein kinase enzymes involved in the control of metabolism, which are themselves affected by levels of other compounds, such as calcium ions and diacylglycerol

▶ How does calcium ion work as a second messenger?

Calcium ion (Ca^{2+}) is involved in another ubiquitous second-messenger scheme called the IP_3-DAG-Ca^{2+} second-messenger pathway, where IP_3 refers to inositol 1,4,5-triphosphate and DAG refers to diacylglycerol (Figure 24.15). Many of the calcium-mediated response pathways depend on the release of Ca^{2+} from intracellular reservoirs, and a component of the inner layer of the phospholipid bilayer, *phosphatidylinositol 4,5-bisphosphate* (PIP_2), is used in the IP_3-DAG-Ca^{2+} pathway (Figure 24.15, *top*).

When the external trigger binds to its receptor on the cell membrane, it activates *phospholipase C* (Section 21-2), which hydrolyzes PIP_2 to *inositol 1,4,5-triphosphate* (IP_3) and *diacylglycerol* (DAG), in a process mediated by a different member of the family of G proteins (Figure 24.15, *middle*). The IP_3 and DAG molecules are second messengers that go on to effect different changes. IP_3 diffuses through the cytosol to the endoplasmic reticulum (ER), where it stimulates the release of Ca^{2+}, the third second messenger in the pathway (Figure 24.15, *bottom right*). A complex is formed between the calcium-binding protein calmodulin and Ca^{2+}. This calcium–calmodulin complex activates a cytosolic protein kinase, which phosphorylates target enzymes in the same fashion as in the cAMP second-messenger scheme. The DAG second messenger diffuses through the plasma membrane to bind to and activate **protein kinase C (PKC)**. PKC also phosphorylates target enzymes, including channel proteins that control the flow of Ca^{2+} into and out of the cell. By controlling the flow of Ca^{2+}, this second-messenger system can produce sustained responses even when the supply of Ca^{2+} in the intracellular reservoirs becomes exhausted.

▶ How do receptor tyrosine kinases work in second-messenger pathways?

Another important type of second-messenger system involves receptors called **receptor tyrosine kinases (RTKs)**. RTKs span the membrane of the cell and have a hormone receptor on the outside and a tyrosine kinase portion on the inside. There are around 20 classes of RTKs, and three of them are shown in Figure 24.16. The best known of these is class II, which includes the insulin receptor (which we look at in more detail in Section 24-5).

RTKs are allosteric enzymes. When the hormone binds to the binding region on the outside of the cell, it induces a conformational change in the tyrosine kinase domain that activates the kinase activity. The activated tyrosine kinases phosphorylate tyrosines on a variety of target proteins, causing alterations in the membrane transport of ions and amino acids and the transcription of certain genes. Phospholipase C is one of the targets of activated RTKs, leading to the production of IP_3, DAG, and Ca^{2+} second messengers (Figure 24.15). Another is an insulin-sensitive protein kinase, which phosphorylates and activates protein phosphatase 1.

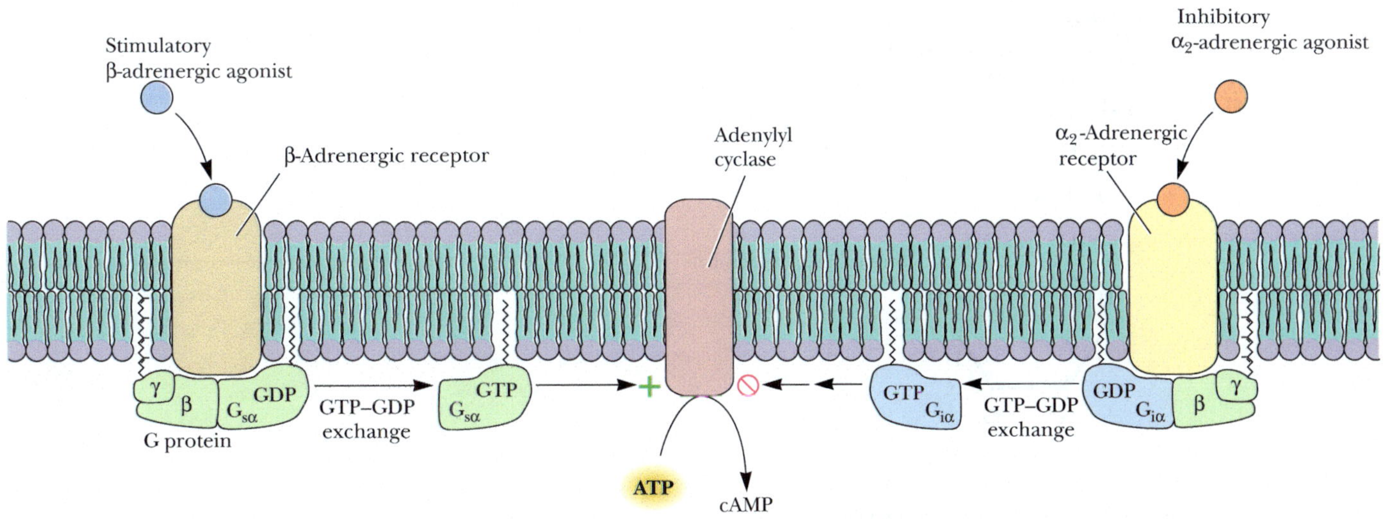

Figure 24.13 Control of adenylyl cyclase by stimulatory or inhibitory hormones. Adenylyl cyclase activity is modulated by the interplay of stimulatory (G_s) and inhibitory (G_i) G proteins. Binding of stimulatory hormones (β-adrenergic agonists) to β-adrenergic receptors activates adenylyl cyclase via $G_{s\alpha}$, whereas inhibitory hormone (α_2-adrenergic agonist) binding to α_2-adrenergic receptors leads to the inhibition of adenylyl cyclase via $G_{i\alpha}$.

24-4 Hormones and the Control of Carbohydrate Metabolism

PKC a family of protein kinase enzymes involved in the control of metabolism, which are themselves affected by levels of other compounds, such as calcium ions and diacylglycerol

receptor tyrosine kinases (RTKs) cell receptors that phosphorylate their own tyrosine residues or tyrosine residues on other proteins when they bind a signaling molecule

Learning Objectives

After reading this section, students will be able to:

LO 24.4.1 Describe how the epinephrine hormone affects carbohydrate metabolism.

LO 24.4.2 Explain how the glucagon hormone affects carbohydrate metabolism.

Now that we know something about the effects of hormones in triggering responses within the cell, we can return to and expand on some earlier points about metabolic control. In Sections 17-6, 18-3, and 19-5, we discussed some

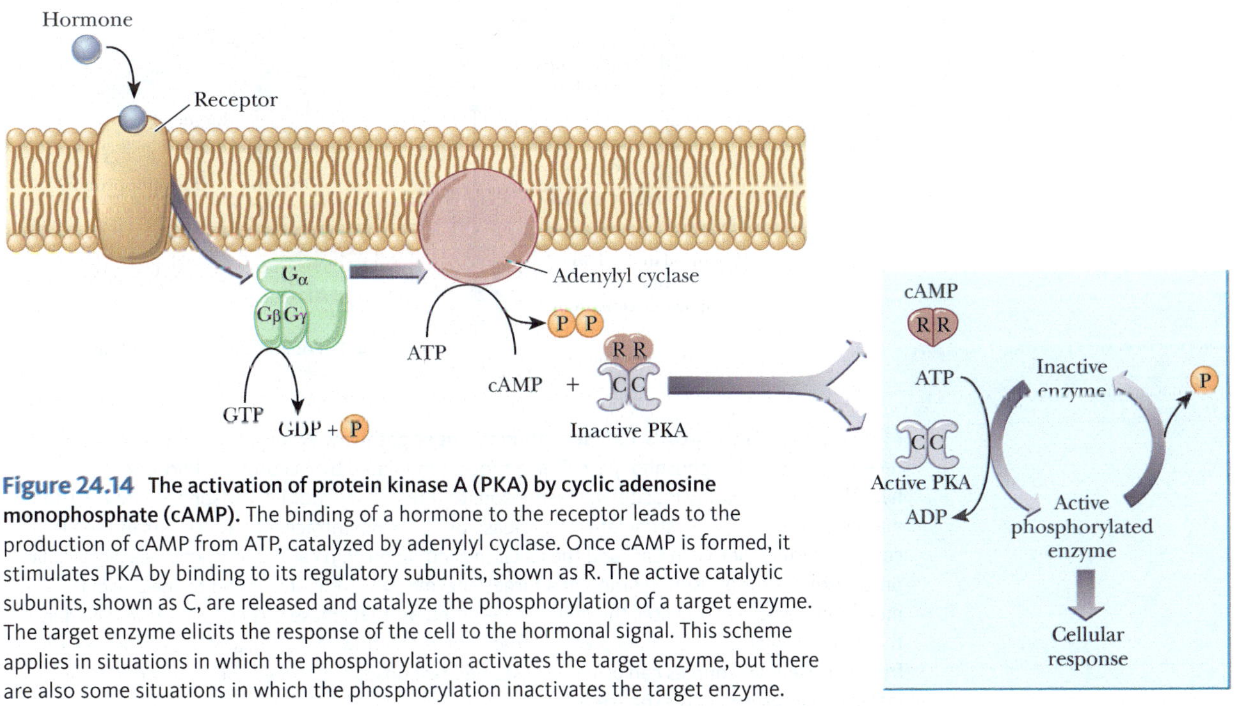

Figure 24.14 The activation of protein kinase A (PKA) by cyclic adenosine monophosphate (cAMP). The binding of a hormone to the receptor leads to the production of cAMP from ATP, catalyzed by adenylyl cyclase. Once cAMP is formed, it stimulates PKA by binding to its regulatory subunits, shown as R. The active catalytic subunits, shown as C, are released and catalyze the phosphorylation of a target enzyme. The target enzyme elicits the response of the cell to the hormonal signal. This scheme applies in situations in which the phosphorylation activates the target enzyme, but there are also some situations in which the phosphorylation inactivates the target enzyme.

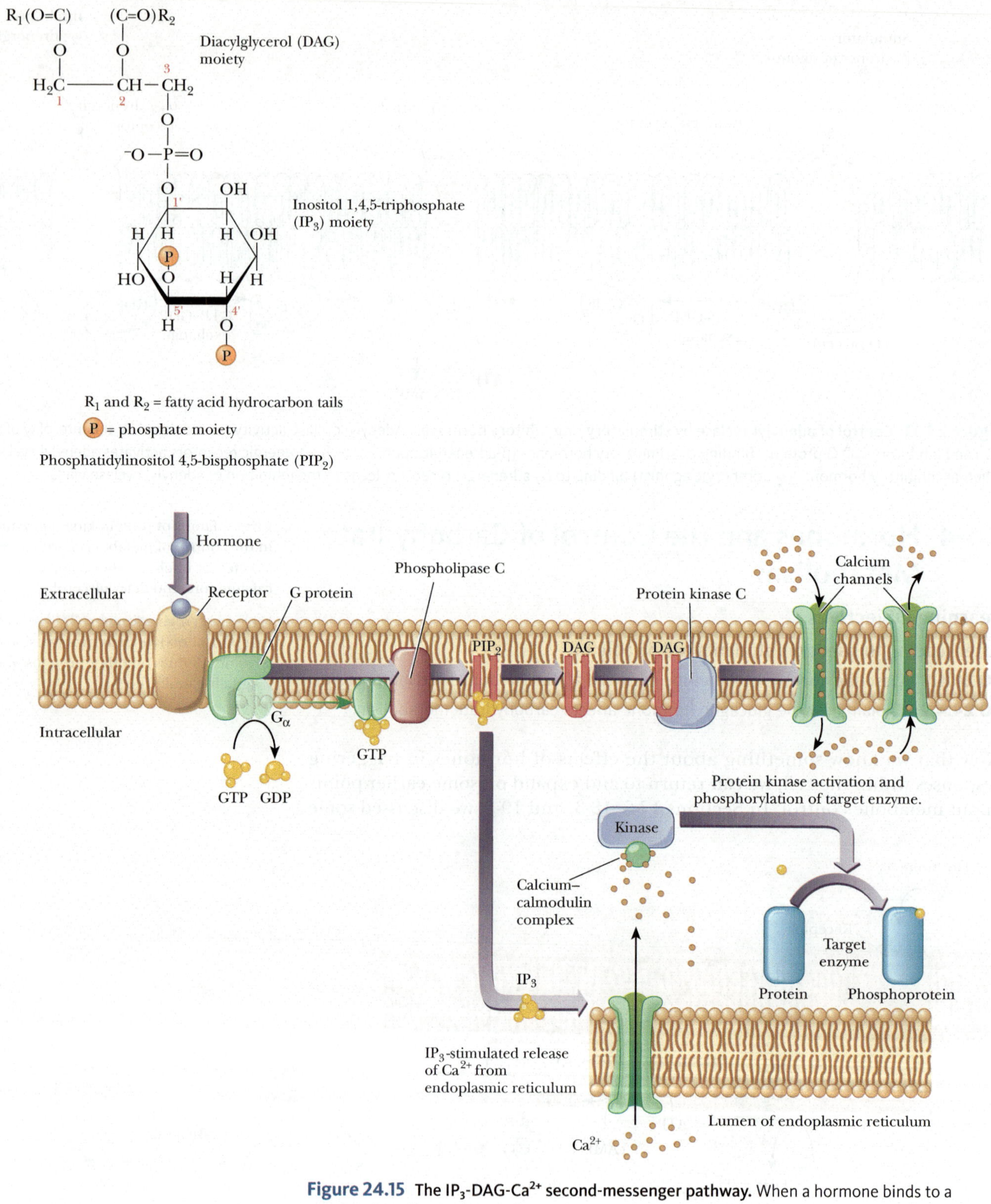

Figure 24.15 The IP_3-DAG-Ca^{2+} second-messenger pathway. When a hormone binds to a receptor, it activates phospholipase C, in a process mediated by a G protein. Phospholipase C hydrolyzes phosphatidylinositol 4,5-bisphosphate (PIP_2) to inositol 1,4,5-trisphosphate (IP_3) and diacylglycerol (DAG). IP_3 stimulates the release of Ca^{2+} from intracellular reservoirs in the ER. A complex formed between Ca^{2+} and the calcium-binding protein calmodulin activates a cytosolic protein kinase for phosphorylation of a target enzyme. DAG remains bound to the plasma membrane, where it activates the membrane-bound protein kinase C (PKC). PKC is involved in the phosphorylation of channel proteins that control the flow of Ca^{2+} in and out of the cell. Ca^{2+} from extracellular sources can produce sustained responses even when the supply of Ca^{2+} in intracellular reservoirs is exhausted.

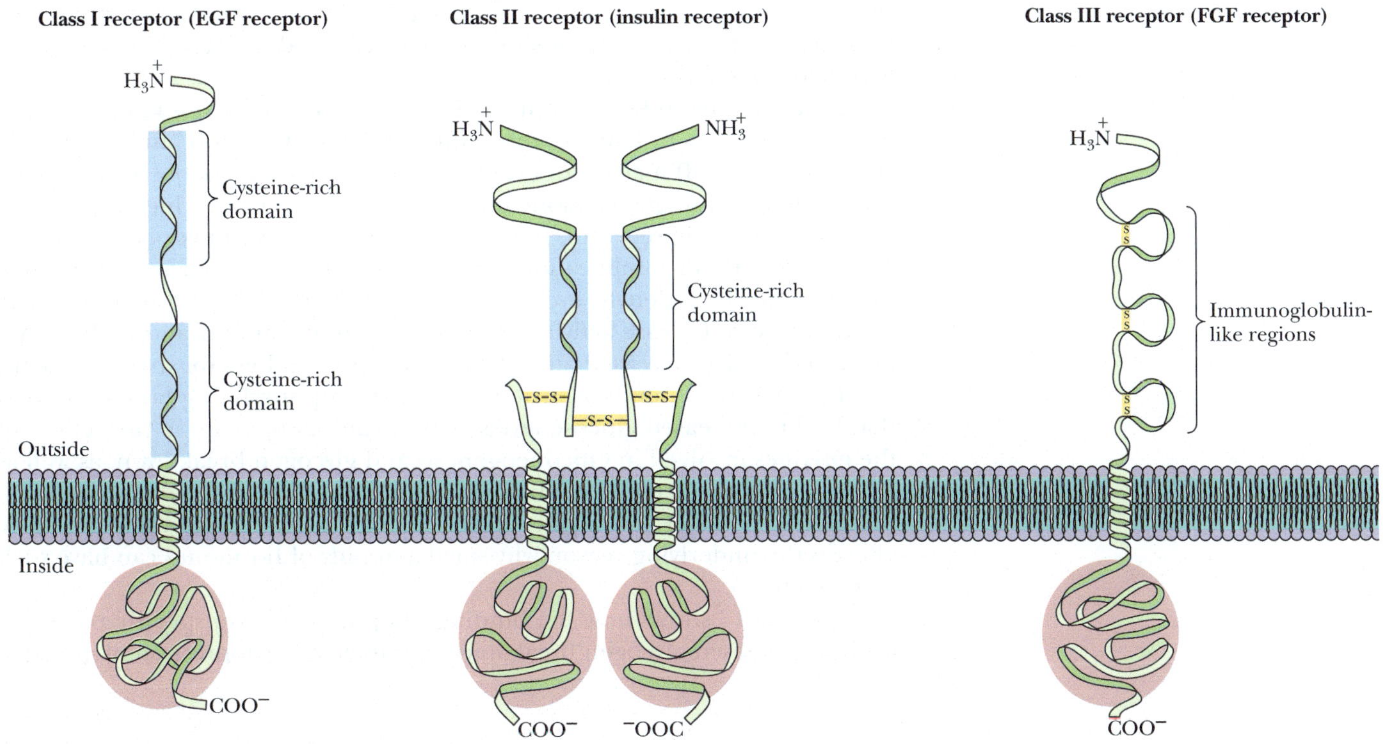

Figure 24.16 The three classes of receptor tyrosine kinases (RTKs). The tyrosine kinase domains (pink) of each class of RTK are where phosphorylation activity occurs upon the binding of a signaling molecule. Class I RTKs are monomeric and contain a pair of Cys-rich repeat sequences. The insulin receptor, a typical class II RTK, is a glycoprotein composed of two kinds of subunits in an $\alpha_2\beta_2$-tetramer. The α and β subunits are synthesized as a single peptide chain together with an N-terminal signal sequence. Subsequent proteolytic processing yields the separate α and β subunits. The β subunits, consisting of 620 residues each, are integral transmembrane proteins, with only a single transmembrane α-helix and with the amino terminus outside the cell and the carboxyl terminus inside. The α subunits, consisting of 735 residues each, are extracellular proteins that are linked to the β subunits and to each other by disulfide bonds. The insulin-binding domain is located in a cysteine-rich region on the α subunits. Class III RTKs contain multiple immunoglobulin-like domains. Shown here is the fibroblast growth factor (FGF) receptor, which has three immunoglobulin-like domains. *(Adapted from Ullrich, A., and Schlessinger, J. (1990). Signal transduction by receptors with tyrosine kinase activity.* Cell *61, 203–212.)*

points about control mechanisms in carbohydrate metabolism. We saw how glycolysis and gluconeogenesis can be regulated and how glycogen synthesis and breakdown can respond to the body's needs. Phosphorylation and dephosphorylation of the appropriate enzymes played a large role there, and that whole scheme is subject to hormonal action.

▶ What hormones control carbohydrate metabolism?

Three of the hormones that play a part in the regulation of carbohydrate metabolism are epinephrine, glucagon, and insulin. Epinephrine acts on muscle tissue to raise the level of glucose on demand, while glucagon acts on the liver, also to increase the availability of glucose. Feedback control (Figure 24.10) plays a role in these processes and ensures that the amount of glucose made available does not reach an excessive level. The role of insulin is to trigger the feedback response that achieves this further control.

Epinephrine (also called adrenaline) is structurally related to the amino acid tyrosine (Figure 24.17). Epinephrine is released from the adrenal glands in response to stress (the "fight-or-flight" response). When it binds to specific

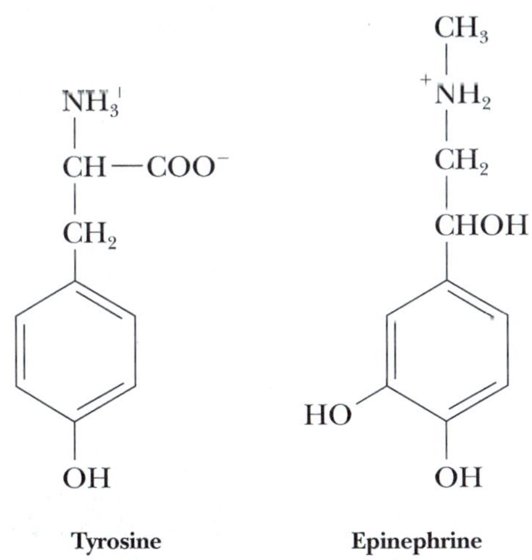

Figure 24.17 Structures of tyrosine and epinephrine. The hormone epinephrine is metabolically derived from the amino acid tyrosine.

receptors, it sets off a chain of events that lead to increased levels of glucose in the blood, increased glycolysis in muscle cells, and increased breakdown of fatty acids for energy.

Glucagon (a peptide that contains 29 amino acid residues) is released by the α-cells of the islets of Langerhans in the pancreas, and it too binds to specific receptors to set off a chain of events to make glucose available to the organism. Each time a single hormone molecule, whether epinephrine or glucagon, binds to its specific receptor, it activates a number of stimulatory G proteins. This effect starts an amplification of the hormonal signal. Each active G protein in turn stimulates adenylyl cyclase several times before the G protein is inactivated by its own GTPase activity, leading to still more amplification. The cAMP produced by the increased adenylyl cyclase activity allows for increased activity of the cAMP-dependent protein kinase, phosphorylating target enzymes that lead to increased glucose levels. This means an increase in the activity of the enzymes involved in gluconeogenesis and glycogen breakdown, as well as a decrease in the activity of enzymes involved in glycolysis and glycogen synthesis. The series of amplifying steps is called a cascade, and the cumulative effect is the underlying reason why small amounts of hormones can have such marked effects.

Figure 24.18 shows how the binding of epinephrine to specific receptors leads to increased glycogen breakdown in muscle and suppression of glycogen

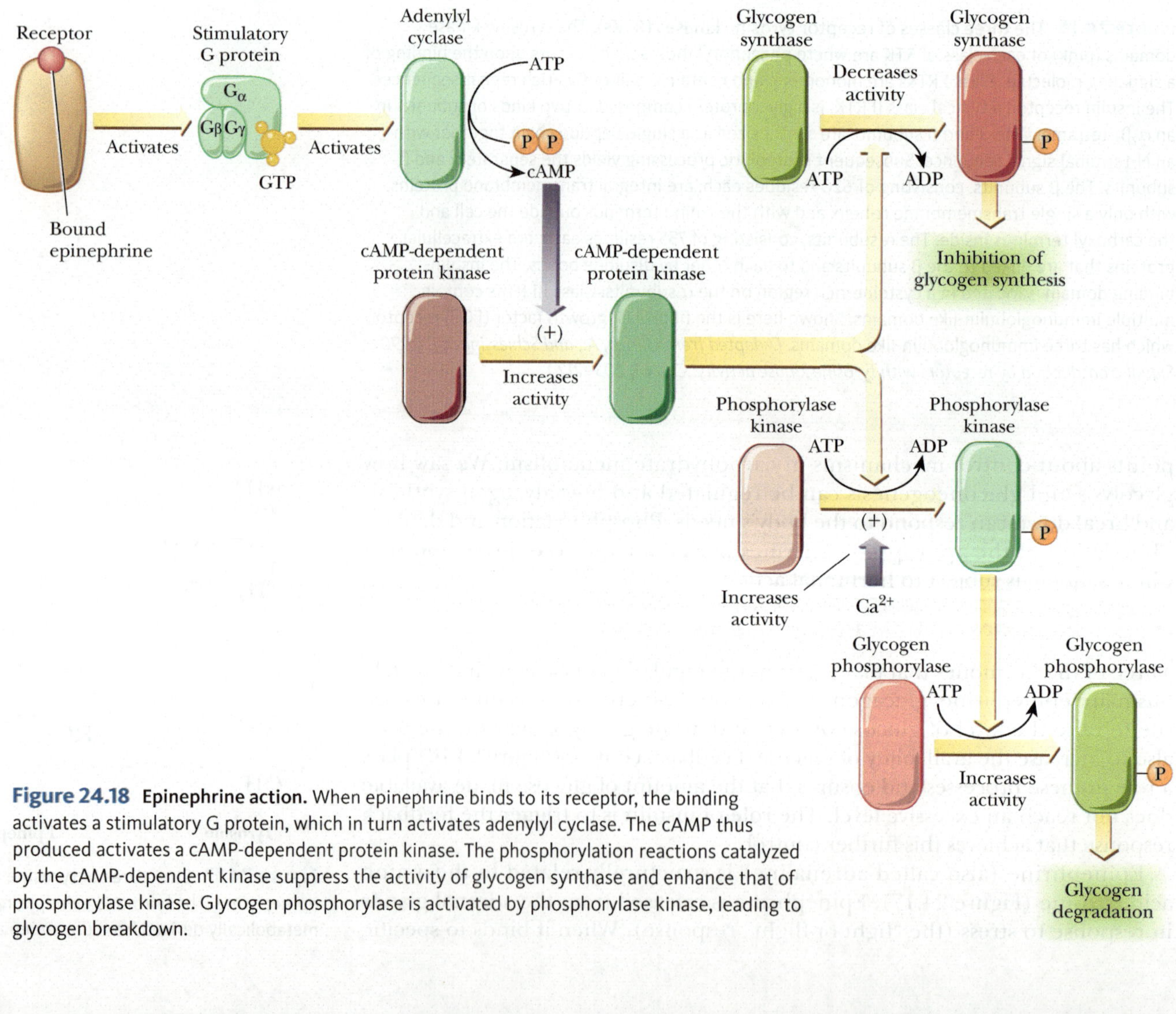

Figure 24.18 **Epinephrine action.** When epinephrine binds to its receptor, the binding activates a stimulatory G protein, which in turn activates adenylyl cyclase. The cAMP thus produced activates a cAMP-dependent protein kinase. The phosphorylation reactions catalyzed by the cAMP-dependent kinase suppress the activity of glycogen synthase and enhance that of phosphorylase kinase. Glycogen phosphorylase is activated by phosphorylase kinase, leading to glycogen breakdown.

synthesis. The hormonal stimulation leads to activation of adenylyl cyclase, which in turn activates the cAMP-dependent protein kinase responsible for activating glycogen phosphorylase and inactivating glycogen synthase.

Figure 24.19 shows the effect of glucagon binding to receptors in stimulating gluconeogenesis in the liver and suppressing glycolysis. The effect depends on changes in the concentration of the key allosteric effector, fructose-2,6-bisphosphate (F2,6BP). Recall from Section 18-3 that F2,6BP is an important allosteric activator of phosphofructokinase, the key enzyme of glycolysis; it is also an inhibitor of fructose-bisphosphate phosphatase, which plays a role in gluconeogenesis. Therefore, a high concentration of F2,6BP stimulates glycolysis, whereas a low concentration stimulates gluconeogenesis. The concentration of F2,6BP in a cell depends on the balance between its synthesis [catalyzed by phosphofructokinase-2 (PFK-2)] and its breakdown [catalyzed

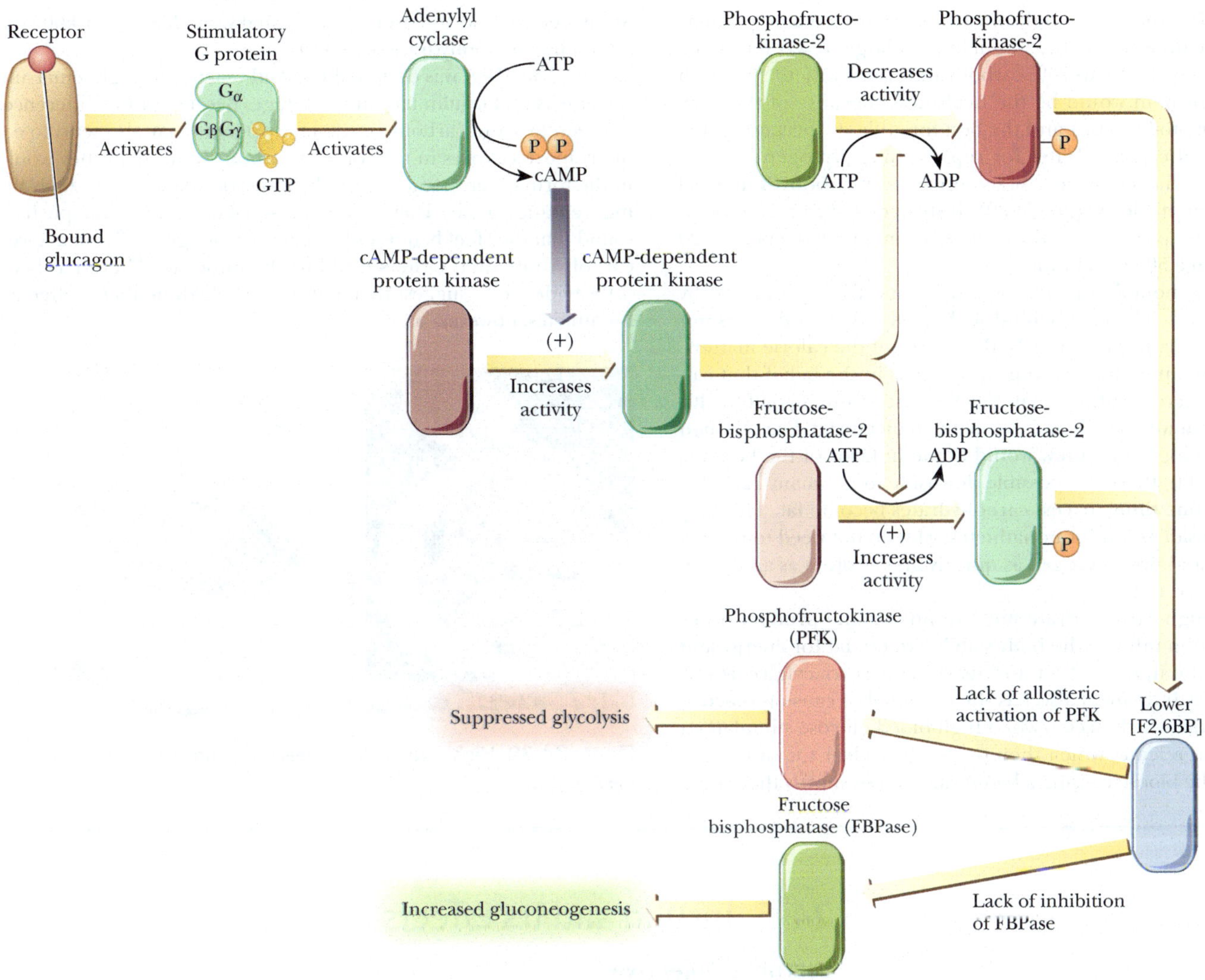

Figure 24.19 Glucagon action. Binding of glucagon to its receptor sets off the chain of events that leads to the activation of a cAMP-dependent protein kinase. The enzymes phosphorylated in this case are phosphofructokinase-2, which is inactivated, and fructose-bisphosphatase-2, which is activated. The combined result of phosphorylating these two enzymes is to lower the concentration of fructose-2,6-bisphosphate (F2,6BP). A lower concentration of F2,6BP leads to less inhibition of the enzyme fructose-bisphosphatase by F2,6BP, thus enhancing gluconeogenesis. At the same time, the lower concentration of F2,6BP implies that phosphofructokinase is lacking a potent allosteric activator, with the result that glycolysis is suppressed.

by fructose-bisphosphatase-2 (FBPase-2)]. Glucagon binding to a receptor causes a cascade that leads to the activation of protein kinase A (PKA), a cAMP-dependent protein kinase (Figure 24.19, *middle*). Activated PKA then causes phosphorylations that lead to a lower F2,6BP concentration (Figure 24.19, *right*) and increased gluconeogenesis and suppressed glycolysis (Figure 24.19, *bottom*). Biochemical Connections 24C discusses the role of insulin in overall metabolism and some current dieting trends.

reactive hypoglycemia a lowering of blood sugar due to the ingestion of a high-carbohydrate meal, which then raises insulin levels

24C Biochemical Connections • Nutrition

Insulin and Low-Carbohydrate Diets

In the 1970s, diets very high in carbohydrates became popular with both athletes and the population at large. It was felt that a diet consisting of 60% to 70% carbohydrates and 15% to 20% each of fat and protein would be the healthiest (because of the high carbohydrate-to-fat ratio) and the best for athletes (because of the high levels of carbohydrates for replenishing glycogen). In the 1990s, newer diets that were based on a lower-carbohydrate level became fashionable (Figure 24.20). Instead of a 70/15/15 ratio of carbohydrate/protein/fat, these diets recommended a 60/20/20 ratio or even a 50/25/25 ratio.

The most notable of these was the Zone Diet, promoted by Dr. Barry Sears. The idea behind such diets is that a calorie is not always a calorie. In other words, the source of the calorie matters. People maintained high-carbohydrate diets in the belief that carbohydrates were healthier than fats because of the myriad health problems that were attributed to too much dietary fat. Even though this seemed logical and few would argue in favor of the benefits of excessive fat, there is a possible downside to too many carbohydrates. For one thing, excess carbohydrates become fat. This may be a big consideration for nonathletes, who do not need to replenish muscle and liver glycogen as quickly and as often as an endurance athlete would.

Also, a high-carbohydrate meal stimulates the production of insulin. Insulin inhibits the body's ability to use fat for energy and stimulates the uptake of fat and its storage as triacylglycerol. A high-carbohydrate meal also has the potential of causing **reactive hypoglycemia**, which occurs when high blood glucose stimulates a large insulin release, which then proceeds to clear too much glucose from the blood, causing a blood sugar crash shortly thereafter. Many people find themselves weak, shaky, or sleepy by 10:00 a.m. after a high-carbohydrate breakfast.

The Zone Diet was designed to avoid reactive hypoglycemia and the effects that insulin has on fat storage. Because of the differences between fats and carbohydrates, many people may also tend to eat many more calories in the form of carbohydrates than they would in the form of fats because carbohydrates do not give the same "filling" sensation as fats. Diet is a very personal thing. Many people have found that they feel better and actually lose weight while on a lower-carbohydrate diet. Others find just the opposite. There has been little evidence to suggest that a lower-carbohydrate diet is effective for athletes, however.

Lightworks/AlamyStock Photo

Jupiter Images/Comstock/Alamy Stock Photo

Figure 24.20 **High-carbohydrate meal (left) and low-carbohydrate meal (right).**

24-5 Insulin and Its Effects

Learning Objectives

After reading this section, students will be able to:

LO 24.5.1 Explain how the insulin hormone affects carbohydrate metabolism.

LO 24.5.2 Describe the relationship between insulin and diabetes.

Insulin is best known as the hormone that is deficient in people with diabetes, and it was this relationship that spurred the study of this fascinating hormone. Research has shown that insulin is involved in many cellular processes and in many different ways from what was previously thought.

24-5A Insulin and Its Receptors

Insulin is a peptide hormone secreted from the pancreas. Its active form is a 51-amino-acid peptide with two different chains, the A chain and the B chain, held together by disulfide bonds. As described in Section 13-4A, insulin was one of the first proteins to be cloned and expressed for human use. Insulin is expressed as an 86-residue precursor called *proinsulin*. Residues 31 to 65 of proinsulin are then removed proteolytically to give the active form. The sequence of human insulin is shown in Figure 24.21.

As we saw in Section 24-3, the insulin receptor is a member of class II receptor tyrosine kinases. When insulin binds to receptor sites on the extracellular side of a cell membrane, it triggers the β subunit to autophosphorylate a tyrosine residue on its interior portion, as shown in Figure 24.22. Once the tyrosines on the receptor are phosphorylated, the receptor then phosphorylates tyrosines on target proteins, called **insulin receptor substrate proteins (IRSs)**, which then act as *adapter proteins* to produce a wide variety of cellular effects.

insulin receptor substrate proteins (IRSs) proteins that are phosphorylated by tyrosine receptor kinases and then cause a wide variety of cellular effects revolving around insulin

24-5B Insulin in Glucose Uptake and Enzyme Interactions

▶ What does insulin do?

The body cannot tolerate great changes in the level of blood glucose. Insulin's primary job is to clear glucose out of the blood, and it does so by increasing the transport of glucose from the blood to muscle cells and adipocytes.

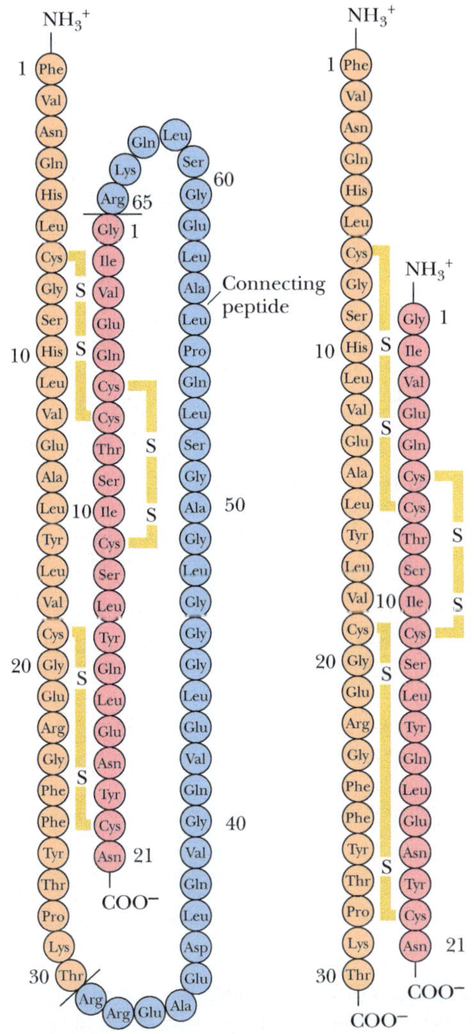

Figure 24.21 Insulin sequence. Proinsulin is an 86-residue precursor to insulin (the sequence shown here is human proinsulin). Proteolytic removal of residues 31 through 65 yields insulin. Residues 1 through 30 (the B chain) remain linked to residues 66 through 86 by a pair of interchain disulfide bridges.

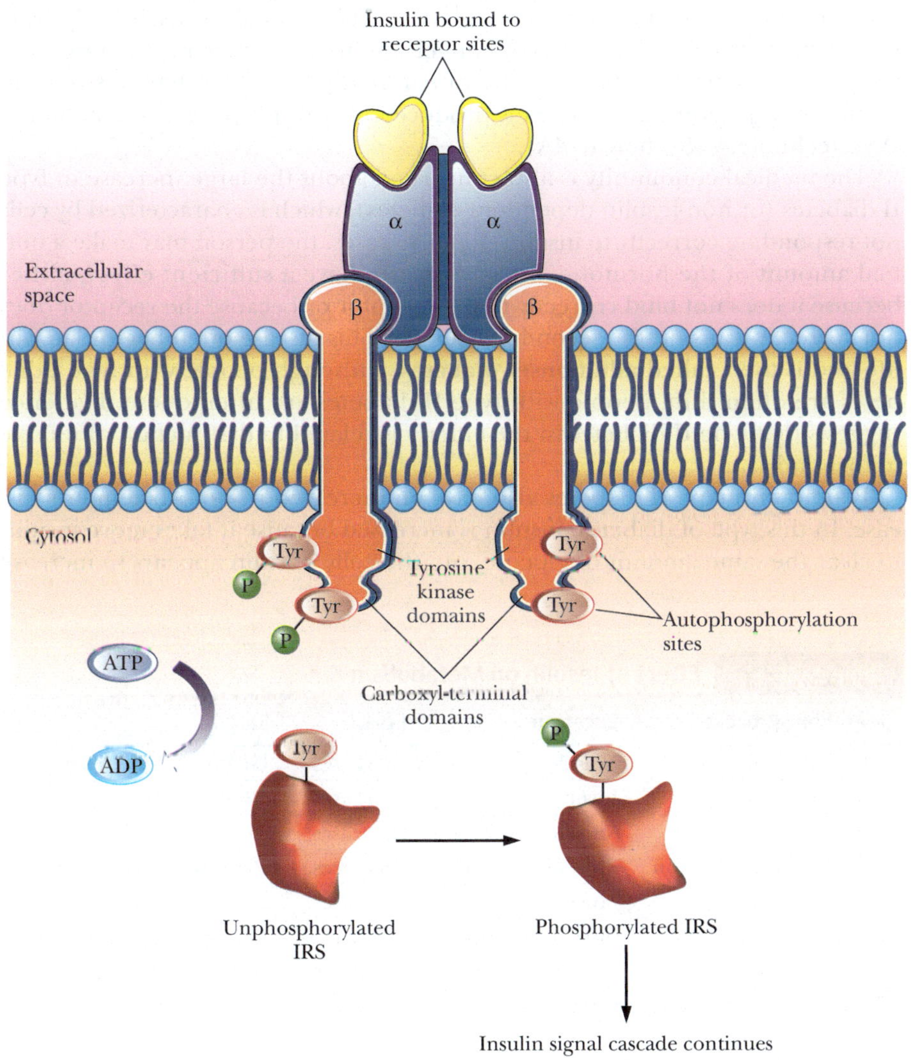

Figure 24.22 The insulin receptor has two types of subunits, *α* and *β*. The α subunit is on the extracellular side of the membrane, and it binds to insulin. The β subunit spans the membrane. When insulin binds to the α subunit, the β subunits autophosphorylate on tyrosine residues. These then phosphorylate target proteins called insulin receptor substrate proteins (IRSs). Phosphorylated IRSs act as signal adapter proteins that cause the production of second-messenger molecules, leading to effects on carbohydrate metabolism. *(From* Principles of Biochemistry, *by Albert L. Lehninger, David L. Nelson and Michael M. Cox. Copyright © 1993 by W. H. Freeman & Co. Used with permission.)*

GLUT4 a specific glucose transporter whose cellular positioning is affected by insulin levels

Using mechanisms still being studied, insulin signaling leads to the movement of a glucose-transporter protein called **GLUT4** from intracellular vesicles to the cell membrane. Once in the cell membrane, the GLUT4 protein allows more glucose to enter the cell, lowering the blood glucose level. Insulin is best known for this effect. Failure of glucose transport is the main characteristic of, and the acute risk associated with, diabetes.

Insulin affects the activity of many enzymes, most of which are involved in getting rid of glucose. However, fat metabolism is also affected. Glucokinase is a liver enzyme that phosphorylates glucose to glucose-6-phosphate (see **Biochemical Connections 6C**). It is induced by insulin, so that when insulin is present, glucose in the liver is sent toward catabolic pathways, such as the pentose phosphate pathway or glycolysis. Insulin also activates liver glycogen synthase and deactivates glycogen phosphorylase, causing glucose to be put into a polymeric form. In addition, insulin stimulates glycolysis through the activation of phosphofructokinase and pyruvate dehydrogenase. Insulin also has a large effect on fatty acid metabolism. It increases fatty acid synthesis by stimulating acetyl-CoA carboxylase (ACC), and it increases triacylglycerol synthesis in the liver by activating lipoprotein lipase. It also increases cholesterol synthesis by activating hydroxymethylglutaryl-CoA reductase (Section 21-8). Table 24.4 summarizes the effects of insulin on metabolism.

24-5C Insulin and Diabetes

Much has been learned about insulin because of its relationship with diabetes. In classical type I diabetes (or insulin-dependent diabetes), the individual does not make insulin or at least not enough of it. This is usually caused by destruction of the beta cells of the islets of Langerhans in the pancreas, resulting from a type of autoimmune disease. The only remedy for type I diabetes is regular insulin injections, and insulin is produced for this purpose by recombinant DNA technology (Section 13-4A).

The medical community is also concerned about the large increase in type II diabetes (or non-insulin-dependent diabetes), which is characterized by cells not responding correctly to insulin. In these cases, the person may make a normal amount of the hormone, but it does not have a sufficient effect, either because it does not bind correctly to the receptor or because the receptor does not correctly transmit the second messenger. This disease often begins later in life and is then called **adult-onset diabetes**. Although people with type I diabetes are often thin, people with type II diabetes are often obese. Evidence suggests that type II diabetes in the elderly is related to dysfunction of muscle mitochondria.

adult-onset diabetes another term for type II diabetes; a diabetes that starts later in life and is based on the body not responding correctly to normal insulin levels

People with type II diabetes also have an increased risk of Alzheimer's disease. In this type of diabetes, insulin is increased because it takes more insulin to clear the same amount of glucose into the cells. Insulin appears to increase

Table 24.4 Effect of Insulin on Metabolism

Metabolic Process	Location	Effect	Target
Glucose uptake	Muscle	Increases	GLUT4 transporter
Glucose breakdown	Liver	Increases	Glucokinase
Glycolysis	Muscle and liver	Increases	PFK-1
Acetyl-CoA production	Muscle and liver	Increases	Pyruvate dehydrogenase
Glycogen synthesis	Muscle and liver	Increases	Glycogen synthase
Glycogen breakdown	Muscle and liver	Decreases	Glycogen phosphorylase
Fatty acid synthesis	Liver and muscle	Increases	Acetyl-CoA carboxylase
Triacylglycerol synthesis	Adipocytes	Increases	Lipoprotein lipase

levels of the β-amyloid protein that forms plaques in the brain. A brain protein called insulin-degrading enzyme (IDE) is involved in binding to and degrading insulin. This enzyme also binds to β-amyloid protein and clears it from the brain. When insulin levels are high, IDE spends more time tied up with insulin and less time clearing the β-amyloid protein. Because insulin is produced in high quantities when a person eats a high-carbohydrate meal, the increasing number of people getting Alzheimer's disease could be tied to our fast-food, high-sugar diet and lifestyle. **Biochemical Connections 24D** describes a link between working out and insulin.

24D Biochemical Connections • Allied Health

A Workout a Day Keeps Diabetes Away?

There seems to be a link between obesity and type II diabetes, although it is not clear whether diabetes leads to obesity or vice versa. The GLUT4 transporter is one of many glucose transporters, and it is the one most affected by insulin levels. It is also a protein whose levels can be affected by physical training. Studies have shown that one of the major changes associated with physical activity is an increase in the amount of GLUT4 in the muscle. In the trained state, a person transports more glucose into the cell than when untrained. Studies have shown that only one week of moderate exercise (one to two hours a day at 70% of maximal oxygen uptake) can double the GLUT4 protein content of the muscles of sedentary people.

By definition, loss of function of glucose transport is type II diabetes. The training effect is such that it takes only a few days without training for the activity of GLUT4 to decrease to only half its normal level (see Figure 24.23). Fortunately, the intensity of the training has less to do with the effect, at least in young to middle-aged people. With the apparent link between type II diabetes and obesity, one method of maintaining proper glucose transport appears to be staying light and fit. ◗

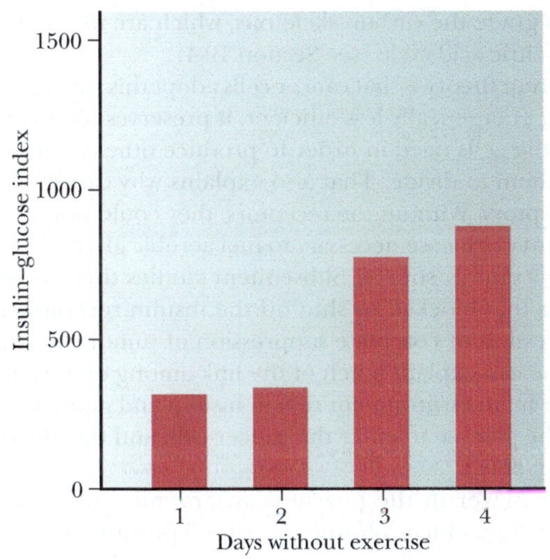

Figure 24.23 Insulin–glucose index versus detraining time. Moderately trained middle-aged males were tested for the effect of detraining on their muscles' ability to clear glucose out of the blood (measured as the insulin–glucose index—the amount of insulin it takes to clear glucose from the blood). This shows that on the third day without training, there is a pronounced increase in the amount of insulin required to clear glucose. *(Adapted from Hardman, A. (1998). Metabolic basis of the health benefits of exercise.* The Biochemist, *20(3), 18–22.)*

24-5D Insulin and Sports

Athletes must be able to control their diets for maximum performance. Insulin plays a large role in the choice of a pre-race breakfast for aerobic athletes. If an athlete eats a large, carbohydrate-filled breakfast in the morning, there is a rise in blood glucose, followed by a rise in insulin. This often leads to a fall in blood glucose below the baseline level. It may take hours for the blood glucose levels to come back up. In addition, the high insulin level would cause the activation of fat and glycogen synthesis and would inhibit glycogen breakdown. Thus, for a period of time after the high-carbohydrate breakfast, the athlete would be essentially running on empty. For this reason, many runners do not eat before a morning event, or, if they do, they eat little and avoid large amounts of high-glycemic-index foods. Athletes often drink coffee or tea in the morning. Besides the general stimulation of the central nervous system, caffeine also inhibits insulin production and stimulates fat mobilization.

In this chapter, and indeed in this entire book, we have explored many topics, most of which are interrelated. In **Biochemical Connections 24E**, we put many of them together to discuss evidence that links insulin, diabetes, and cancer.

24E Biochemical Connections • Allied Health

Insulin, Diabetes, and Cancer

A large body of evidence is piling up that suggests that insulin and a related hormone, called insulin-like growth factor (IGF), play a role in tumor growth, as well as linking obesity, diabetes, and cancer. For many years, people who studied breast tumor cells in the lab have used a cocktail that contains large quantities of glucose, insulin, and a growth factor called EGF. The tumor cells thrive on that cocktail, but when deprived of insulin, the tumor cells die. That insulin is important is interesting, as few cell types, other than adipocytes, liver cells, and muscle cells, respond to insulin or even have insulin receptors. Without the receptors, most cells are insensitive to insulin levels because they cannot translate insulin levels outside the cell into a reaction inside the cell. Normal cells thrive without insulin, yet these tumor cells require it.

Studying this phenomenon led researchers to look closely at the signaling pathway that involves inositol 1,4,5-triphosphate (IP_3) (see Section 24-3 and Figure 24.15), which also happens to be one of the most mutated pathways in cancer cells. In fact, researchers are now tying several observations together to help explain a long-known fact: Individuals with obesity and diabetes have a much higher risk of getting cancer than lean, healthy people, and when they do get cancer, their risk of dying from it is greater. And with the rates of obesity and type II diabetes skyrocketing worldwide, and even more so in the United States, this is a hotbed of activity. The message from a mounting pool of research is clear. Excess body fat accounts for one quarter to one half of the occurrences of many types of cancer, such as breast cancer and colorectal cancer, and the list is growing.

Besides the link between obesity and cancer, there is also a link between diabetes and cancer, and this link seems to center around the circulating levels of insulin and IGF. People with type II diabetes who receive insulin treatment or receive drugs that stimulate their own insulin secretion are much more likely to get cancer. Patients who receive a drug called metformin, which lowers circulating insulin levels, are much less likely to get cancer. It appears that cancer cells "burn in a flame of insulin."

It has been known since 1942 that there is a link between longevity in animals and calorie restriction, after work done by Albert Tannenbaum. Mice fed a diet just sufficient to keep them alive live much longer, mostly because they don't get tumors, which ultimately affect older mice. Researchers now think that the underlying difference in tumor progression has to do with the way glucose is metabolized. When bathed in large amounts of glucose and with significant insulin to allow the glucose to enter the cells, cancer cells adopt a strategy of running most of the glucose through aerobic glycolysis (or the Warburg effect), as shown in Figure 24.24. This type of glycolysis is very inefficient compared to aerobic respiration, and it yields much less ATP. However, unlike aerobic respiration, it also does not degrade the carbon skeletons, which are ultimately lost as CO_2 in the citric acid cycle (see Section 19-4).

The current theory is that cancer cells adopt this strategy because, while being energetically less efficient, it preserves the carbon skeletons that the cells need in order to produce other molecules necessary for them to divide. That also explains why cancer cells have insulin receptors. Without the receptors, they could not take up the large amount of glucose necessary to fuel aerobic glycolysis to lactate as their only energy source. Subsequent studies then showed that using drugs that blocked or shut off the insulin receptors led to a high suppression or complete suppression of tumor growth. These results alone can explain much of the link among obesity, diabetes, and cancer. In an environment rich in insulin and glucose, the insulin allows the glucose to enter the cancer cell, and the glucose fuels the Warburg effect.

Another player in the puzzle is an enzyme called PI_3 kinase (PI3K), which produces phosphoinositol 3-phosphate as a second messenger to regulate the cell's sensitivity to insulin. When PI3K is active, the insulin receptors are more effective at transporting glucose. The PI3K enzyme is linked to a tumor suppressor called PTEN, which is a phosphatase (see Chapter 14). The gene for PTEN is often found to have been deleted in a wide variety of cancers. PTEN's role in tumor suppression appears to occur via counteracting the action of PI_3 kinase. The sum of these experiments was to demonstrate two different methods of arriving at the same place—the stimulation of the PI_3 kinase pathway. In one mechanism, the mutation or deletion of PTEN leads to increased activity of the enzyme. In another, abnormal levels of insulin or IGF lead to the same result. As we now know, people with obesity or those with type II diabetes have high levels of insulin. ◗

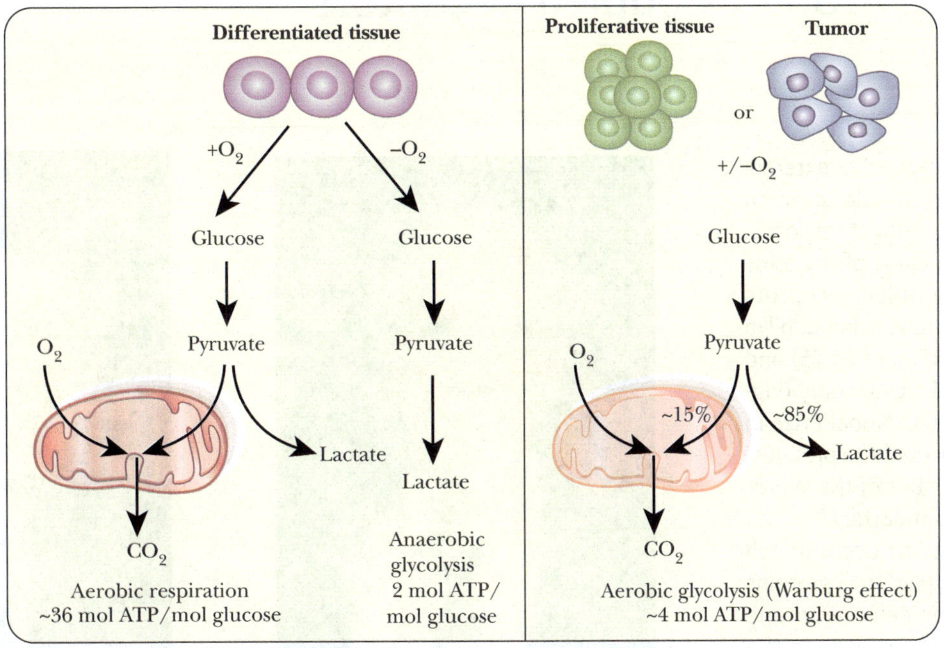

Figure 24.24 Cancer cells preferentially adopt aerobic glycolysis (the Warburg effect) in the presence of excess glucose and insulin. In differentiated tissue, where cell proliferation does not normally happen, cells use different glucose metabolism pathways depending on the presence or absence of oxygen. In the presence of oxygen, they use aerobic respiration and oxidative phosphorylation for ATP production. However, cancer cells, even in the presence of oxygen, preferentially choose aerobic glycolysis to lactate in a phenomenon called the Warburg effect, leading to less ATP produced per glucose molecule.

Hot Topic

G Protein-Coupled Receptors

When our bodies respond to external stimuli such as light or odor, or when we experience the fight-or-flight response, the widely distributed class of proteins known as G protein-coupled receptors (GPCRs) plays a central role. Brian Kobilka of Stanford University (Figure 24.25) and Robert Lefkowitz of Duke University (Figure 24.26) shared the 2012 Nobel Prize in Chemistry for their work on these proteins. What are the characteristics of these proteins that make them so important?

These receptor proteins occur in cell membranes and are coupled to G proteins on the inner side of the cell membrane (see Section 24-3). All GPCRs span the membrane and consist of seven α-helical segments. They are also called 7-transmembrane segment proteins (7-TMS). Rhodopsin, the key protein in vision, and the α- and β-adrenergic receptors, both of which bind the hormone epinephrine, are examples of GPCRs. As an example, Figure 24.27 shows the structure of a β-adrenergic receptor.

When a signaling molecule binds to a GPCR, as shown in Figure 24.28, it sets off a series of events that lead to the transmission and amplification of the signal within the cell. The binding event activates the G protein—thus, the name G protein-coupled receptor. The result can be inhibitory or stimulatory, depending on the specific kind of receptor. Whether the binding is inhibitory or stimulatory, it brings about dissociation of the heterotrimeric subunit structure of the G protein. The G protein subunits are designated α, β, and γ. The β and γ subunits remain bound to each other. The α subunit goes on to signal the next step in the cell, such as inhibition or activation of a specific enzyme. In the process, an exchange of GDP bound to the α subunit for GTP takes place. The bound GTP is slowly hydrolyzed by the α subunit. This GTPase activity gives rise to the name G protein. Several signaling reactions are described in detail in Sections 24-3, 24-4, and 24-5. The wide range of effects depends on the fact that a number

JONATHAN NACKSTRAND/AFP/Getty Images

Figure 24.25 Brian Kobilka of Stanford University. Dr. Kobilka was a co-recipient of the 2012 Nobel Prize in Chemistry for his work on G protein-coupled receptors.

AFP/Getty Images

Figure 24.26 Robert Lefkowitz of Duke University. Dr. Lefkowitz was a co-recipient of the 2012 Nobel Prize in Chemistry for his work on G protein-coupled receptors.

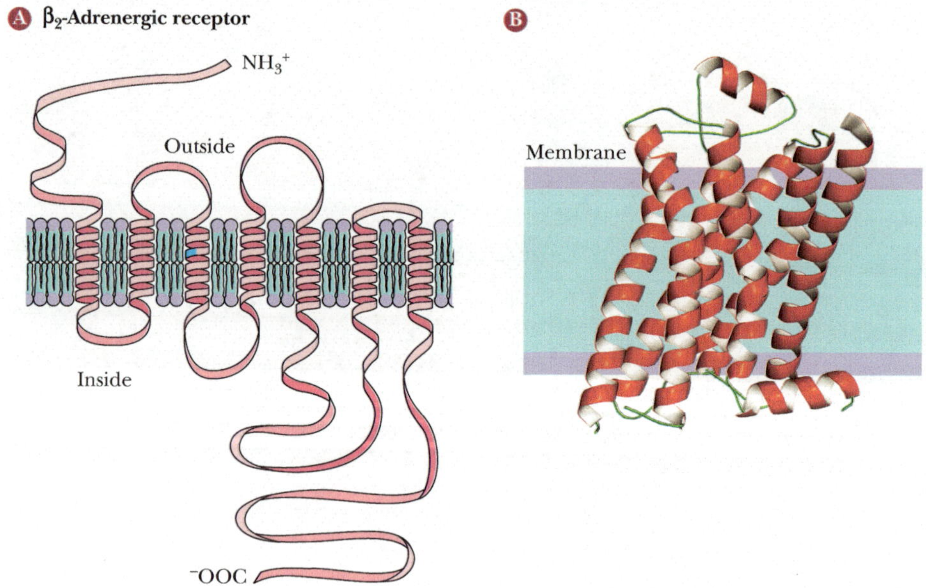

Figure 24.27 Structure of a β-adrenergic receptor. β-Adrenergic receptors are members of the G protein-coupled receptor family, also called 7-transmembrane segment (7-TMS) proteins. (A) Schematic diagram of the seven transmembrane helical segments and their connecting loops along with the extracellular and intracellular segments of the receptor. (B) The main features of the structure of a β-adrenergic receptor. *(Adapted from GARRETT/GRISHAM,* Biochemistry, *4E. 2009 Cengage Learning.)*

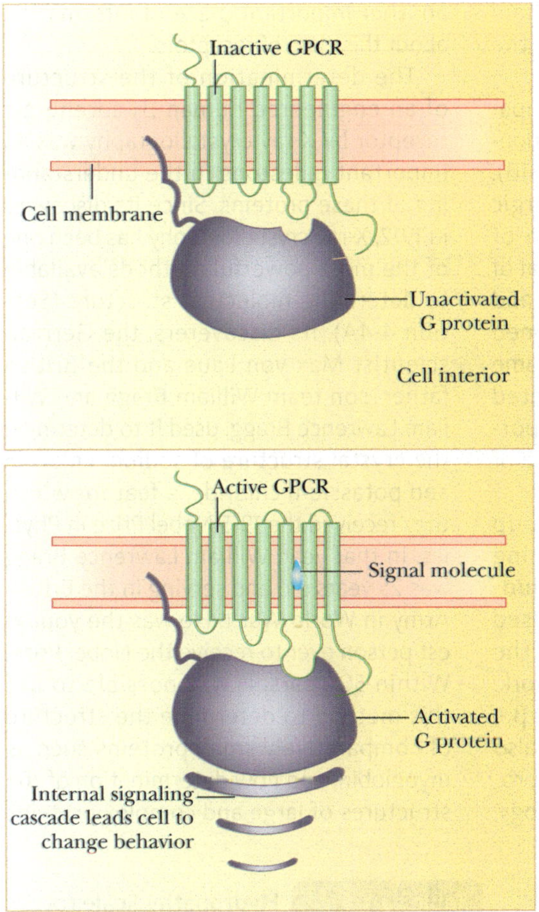

Figure 24.28 The binding of a signal molecule to a G protein-coupled receptor (GPCR). This leads to a conformational change in the GPCR that creates a binding surface for a G protein to bind and become activated, which leads, in turn, to an effect within the cell, usually by a cascade mechanism.

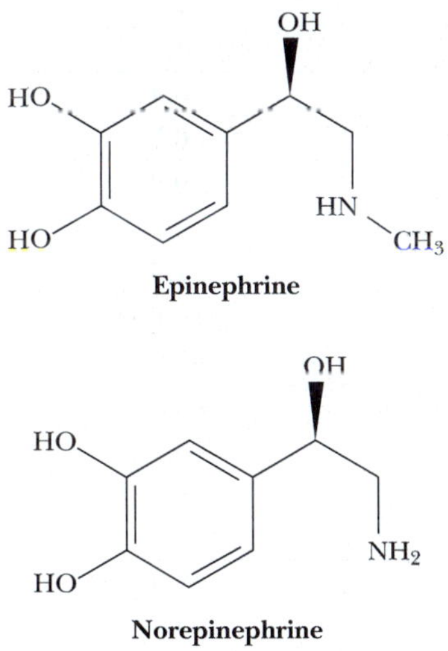

Figure 24.29 Catecholamine structures.

of different receptors exist, all with several subtypes. A number of types of G proteins exist as well, with most of the variation in the α subunit. The various combinations of receptors and G proteins can give rise to a number of responses. The discovery and characterization of G proteins led to the award of the 1994 Nobel Prize in Physiology or Medicine to Alfred Gilman of the University of Texas Southwestern Medical Center and Martin Rodbell of the National Institutes of Health.

The number of GPCRs runs into the hundreds, with many roles in the body. They are grouped into classes depending on the similarity of the amino acid sequence and functional specificity. Of this enormous group, we concentrate on the α- and β-adrenergic receptors—the two main kinds of receptors for the neurotransmitter hormone epinephrine. Catecholamines bind to these receptors—namely, epinephrine and its analog norepinephrine, the hormones of the fight-or-flight response (Figure 24.29).

The α- and β-adrenergic receptors have a number of subgroups (Figure 24.30). The members of these subgroups interact with different kinds of G proteins and have different effects. The α_1 receptor increases intracellular calcium ion concentration and stimulates smooth muscle contraction. The α_2 receptor group plays a number of roles, many of them connected with the gastrointestinal tract. The β receptors are divided into three main types: β_1, β_2, and β_3, and the results of their operation range from heart muscle contraction to smooth muscle relaxation.

A good way to understand GPCRs, their structure, and their many functions is to look at the methods that were used to determine their properties. What genes encode these proteins? What is the amino acid sequence specified by these genes? What is the three-dimensional structure of GPCRs? How does the structure of these receptors determine their functions?

The cloning, sequencing, and expression of the gene coding for the human platelet α_2-adrenergic receptor were an important early step in understanding

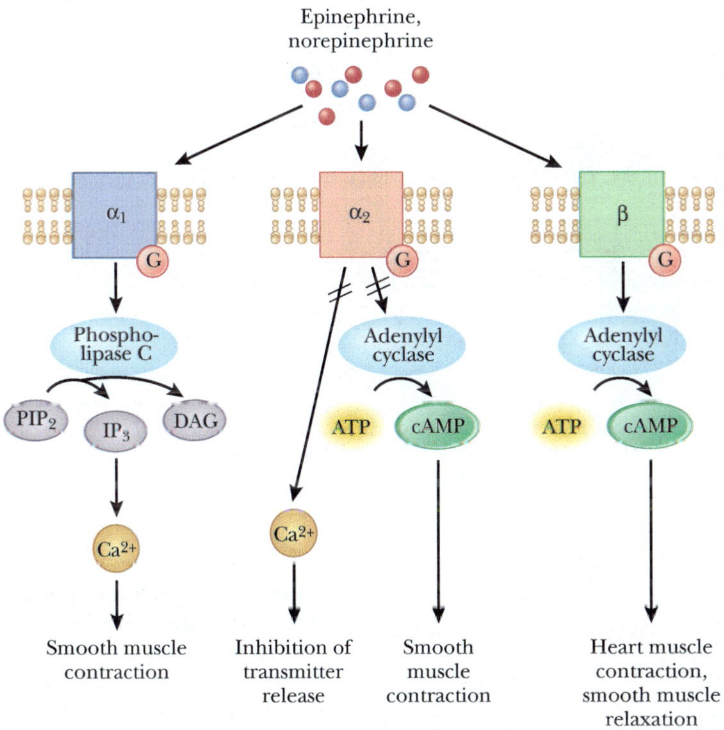

Figure 24.30 A summary of the effects of binding of catecholamines (epinephrine or norepinephrine) to adrenergic receptors. Binding of catecholamines to α_1- or α_2-adrenergic receptors activates phospholipase C or inhibits adenylyl cyclase, respectively. Binding of catecholamines to each type of β-adrenergic receptor (β_1, β_2, or β_3) activates adenylyl cyclase. See Section 24-3 for a description of the reactions of phospholipase C and adenylyl cyclase.

relationships among these receptors. Using the amino acid sequence of the purified protein as a guide, Kobilka, Lefkowitz, and their colleagues synthesized corresponding oligonucleotide probes. The probes were used to screen a human genomic DNA library, using techniques described in detail in Section 13-5. The gene that was recovered codes for a 450-amino-acid protein. The next step was to use hydrophobicity analysis of the amino acid sequence to predict regions that were likely to represent transmembrane helical segments (Figure 24.31A). This analysis depends on the tendency of amino acids with hydrophobic side chains to be in locations removed from the aqueous medium and the corresponding tendency of amino acids with polar side chains to be in contact with the aqueous medium. Table 24.5 lists the hydrophobicity index for the standard amino acids. A positive number indicates a side chain that is likely to be in the hydrophobic portion of a protein or in the hydrophobic environment of a membrane, whereas a negative value indicates a side chain that will be found in contact with the aqueous medium.

Results were available from a comparable analysis of the vision protein rhodopsin, which is also a GPCR (Figure 24.31B). Sequence analysis of the α_2-adrenergic receptor showed a similar pattern of seven helical regions, with a good deal of sequence homology with rhodopsin and other GPCRs. Expression of the cloned gene gave a protein that showed the same ligand behavior as the protein isolated from nature. This work was an important step in understanding the nature of GPCRs.

At about the same time, a group of researchers in Belgium was cloning the genes for several GPCRs, including β-adrenergic receptors. They used techniques similar to those used for the α-adrenergic receptor in the other work. They were able to clone genes for the β_1-, β_2-, and α_2-adrenergic receptors. They also discovered four new GPCRs. All these proteins had significant sequence homology, another important piece of information about this class of receptors.

The determination of the structure of an engineered human β_2-adrenergic receptor by X-ray crystallography was an important milestone in the understanding of these proteins. Since its discovery in 1912, X-ray crystallography has been one of the most powerful methods available to determine molecular structure (Section 4-4A). Its discoverers, the German scientist Max von Laue and the British father–son team William Bragg and William Lawrence Bragg, used it to determine the crystal structure of sodium chloride and potassium chloride, a feat for which they received the 1915 Nobel Prize in Physics. In that year, William Lawrence Bragg was 25 years old and serving in the British Army in World War I. He was the youngest person ever to receive the Nobel Prize. Within 50 years, it was possible to use this method to determine the structure of comparatively small proteins such as myoglobin, and now determination of the structures of large and complex proteins

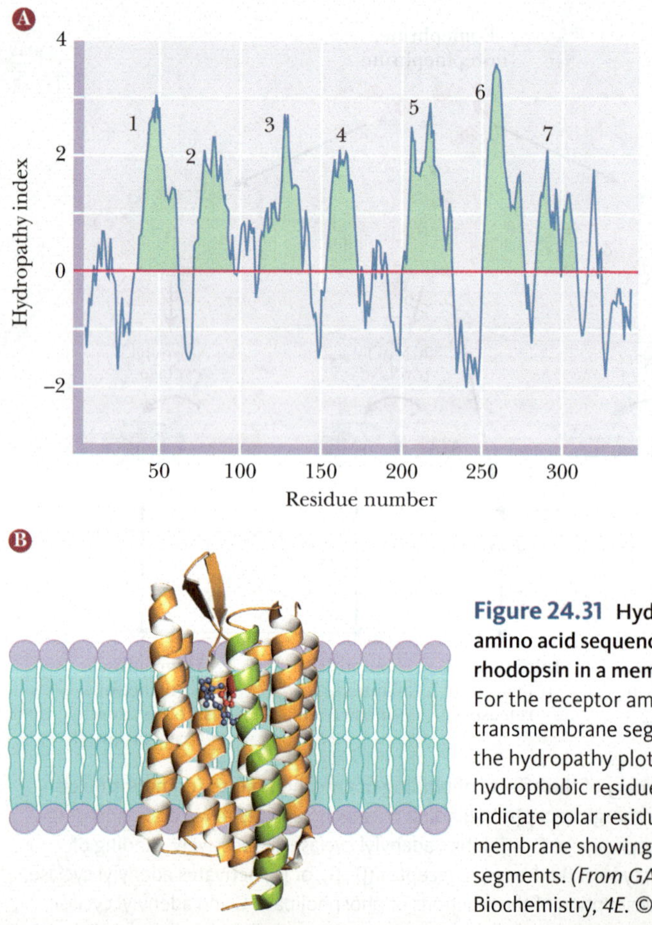

Figure 24.31 Hydropathy plot of a receptor amino acid sequence and structure of rhodopsin in a membrane environment. (A) For the receptor amino acid sequence, seven transmembrane segments are predicted by the hydropathy plot (positive values indicate hydrophobic residues, whereas negative values indicate polar residues). (B) Rhodopsin in a membrane showing seven membrane-spanning segments. *(From GARRETT/GRISHAM,* Biochemistry, *4E. © 2009 Cengage Learning.)*

Table 24.5 Hydropathy Scale for Amino Acid Side Chains in Proteins*

Side Chain	Hydropathy Index
Isoleucine	4.5
Valine	4.2
Leucine	3.8
Phenylalanine	2.8
Cysteine	2.5
Methionine	1.9
Alanine	1.8
Glycine	−0.4
Threonine	−0.7
Serine	−0.8
Tryptophan	−0.9
Tyrosine	−1.3
Proline	−1.6
Histidine	−3.2
Glutamic acid	−3.5
Glutamine	−3.5
Aspartic acid	−3.5
Asparagine	−3.5
Lysine	−3.9
Arginine	−4.5

*From Kyte, J., and Doolittle, R., 1982. A simple method for displaying the hydropathic character of a protein. *Journal of Molecular Biology* 157:105–132.

has become a routine, but far from trivial, operation. In the case of the β_2-adrenergic receptor, detailed information about the three-dimensional structure gives insight into the details of ligand binding. A clear picture of binding can, in turn, be useful in drug design.

An important question for researchers is how to use the properties of GPCRs in medicine. Several possibilities come to mind. One makes use of the fact that they are allosteric proteins, and the general mode of action of allosteric modulators compared to standard drugs is shown in Figure 24.32.

The main point is that the allosteric effector reduces binding of the signal molecule by changing the conformation of the GPCR, not by physically blocking the binding site. An important example of this effect is one connected to the treatment of AIDS. The AIDS virus binds to helper T cells of the immune system by interacting with the cell surface protein CD4, but a GPCR called CCR5 also plays a role in the binding. CCR5 has a binding site for gp120, the notorious viral coat protein that plays a key role in HIV infection. Allosteric modulators that bring about a conformational change in CCR5 to keep it from binding gp120 could be an important step in treating AIDS. Drugs based on this concept are being developed and have reached clinical trials. (People who have a genetic mutation that encodes a nonfunctional form of CCR5 tend to be highly resistant to HIV infection.) The connection between AIDS research and work on GPCRs is an example of the way in which all aspects of biochemistry are related to one another. The divisions among topics are simply a way of making it convenient for us to study this vast subject.

GPCRs play so many roles that it is possible to envision many kinds of medications that bind to or affect this class of receptor in some way. One approach is to devise agents that will induce the cell to absorb the receptor. Some CCR5-targeting drugs trigger absorption of the CCR5 protein. The AIDS virus, no matter how it mutates, cannot bind to a receptor that is no longer on the surface of the cell. Another line of research addresses the situation of a constitutive receptor, one that is active all the time, even in the absence of a signal molecule. A receptor such as this triggers uncontrolled signaling by the G protein with which it interacts. An antagonist, which inhibits the formation of the active conformation of the receptor, would be ineffective when the receptor is locked in the active conformation. The goal in this case is to find an inverse agonist, a compound that will lock the receptor in the inactive conformation.

One last example for our discussion is opioid receptors, which are members of the GPCR family. Through investigation of opioid receptors, researchers have contrived a model called biased agonism or functional selectivity. GPCRs can be influenced by allosteric influences, but it has now been demonstrated that agonists targeting the same receptor site on opioid GPCRs can selectively activate certain signaling pathways rather than activating all possible ones equally. There is much to be done for these promising findings to lead to new drugs, but the implications considering the negative side effects of widely used opioids like morphine may be groundbreaking. The goal is to develop opioid GPCR agonists that provide analgesic effects without addiction, nausea, dizziness, or other commonly encountered issues. Kise and Inoue review research on functional selectivity in opioid GPCRs. Approximately 30–40% of all approved medications act on GPCRs.

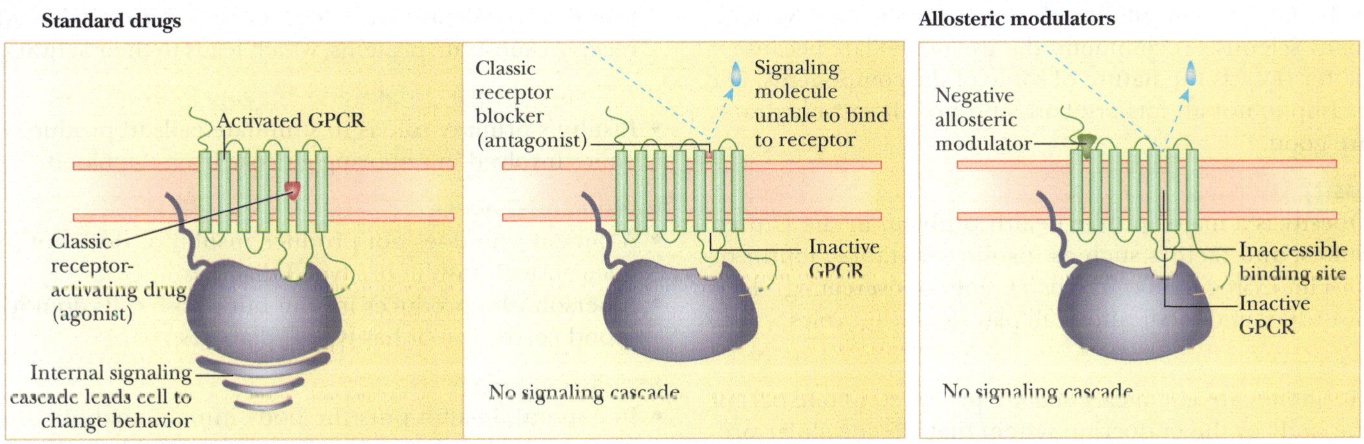

Figure 24.32 Comparison of the mode of action of standard drugs (binding antagonists) with drugs that operate as allosteric modulators. A negative effect is shown for the allosteric modulator, but some allosteric modulators have positive effects that make it easier for the signaling molecule to bind to the receptor.

Summary

Connections Between Metabolic Pathways

- All metabolism is ultimately linked to photosynthesis and energy from the Sun.

Required Nutrients

- Required nutrients are substances that we must include in the diet and that are important to keep track of.
- Macronutrients are nutrients that the body requires in large amounts to provide energy and support essential physiological functions. The three main macronutrients are carbohydrates, fats, and proteins.
- Micronutrients are vitamins and minerals that are needed in small amounts.

Dietary Vitamins

- Vitamins are small substances required for metabolism that our bodies cannot synthesize.
- These include the fat-soluble vitamins A, D, and E, and the water-soluble C and B vitamins.
- The B vitamins are precursors to important coenzymes such as NAD^+, FAD, and TPP.

Dietary Minerals

- In a nutritional sense, minerals refer to inorganic substances such as sodium, potassium, chloride, magnesium, phosphorus, and calcium.
- Minimum daily values for minerals and other required nutrients have been established, and food labels give the quantities of these substances.

Food Guides

- Although you can still find the 1992 Food Guide Pyramid, many scientists recommend the use of MyPlate because it better reflects the nature of some of the components. For example, not all fats are bad, and not all carbohydrates are good.

Obesity

- Obesity is a major public health problem in the United States, and factors such as insulin resistance, leptin or ghrelin dysregulation, the psychology of overeating, sleep disorders, and medications all play significant roles.

Hormones

- Hormones are chemicals that are produced in one part of the body by the endocrine system that affect cellular processes in other parts of the body.
- Endocrine glands produce hormones that circulate in the blood until they find cells that have the correct receptors.
- Once the hormone binds to the cell receptor, it effects a change in the metabolism of the cell.
- Chemically, hormones can be peptides, steroids, or amino acid derivatives.

Second Messengers

- A second messenger is a molecule that is produced or released inside the cell, usually in response to extracellular ligand binding to a receptor.
- Common second messengers include cyclic adenosine monophosphate (cAMP) and Ca^{2+}.
- When cAMP is the second messenger, a hormone binds to a receptor and activates a G protein. The G protein stimulates the enzyme adenylyl cyclase, which produces cAMP. The cAMP then affects target enzymes that lead to the metabolic effect.

Hormones and the Control of Carbohydrate Metabolism

- The principal hormones that control carbohydrate metabolism are epinephrine, glucagon, and insulin.
- Epinephrine promotes the usage of glucose for energy, stimulating the catabolic metabolism of carbohydrates in muscle.
- Glucagon is triggered when the body needs blood sugar. It stimulates gluconeogenesis in the liver.
- Insulin is a peptide hormone produced in the pancreas.
- In its active form, insulin is a 51-amino acid peptide with one A chain and one B chain held together by disulfide bonds.
- The insulin receptor is a member of a class of receptor tyrosine kinases.
- Insulin binding to the insulin receptor stimulates interaction of the receptor with adapter proteins called insulin receptor substrate proteins, which leads to their activation.

Insulin in Glucose Uptake and Enzyme Interactions

- Insulin's primary role is to stimulate cells to produce enzymes involved in removing glucose from the blood.

Insulin and Diabetes

- A person who does not produce insulin or does not produce enough insulin has type I diabetes.
- A person who produces insulin but whose cells do not respond correctly to it has type II diabetes.

Insulin and Sports

- In general, insulin puts the body into an anabolic state, stimulating the production of glycogen and fat and inhibiting the use of glycogen and the breakdown of fat. Pairing the right diet with training can maximize insulin's anabolic and fuel management roles, boosting athletic performance and recovery.

Review Exercises

24-1 Connections Between Metabolic Pathways

1. **Recall** What are the two primary molecules that link anabolic and catabolic reactions? Explain.
2. **Recall** Name some of the key metabolic intermediates that are seen in more than one pathway.
3. **Recall** Many components of the glycolytic pathway and the citric acid cycle are direct exit or entry points to metabolic pathways of other substances. Indicate pathway(s) other than glycolysis or the citric acid cycle that the following compounds are involved in:
   (a) Fructose-6-phosphate
   (b) Oxaloacetate
   (c) Glucose-6-phosphate
   (d) Acetyl-CoA
   (e) Glyceraldehyde-3-phosphate
   (f) α-Ketoglutarate
   (g) Dihydroxyacetone phosphate
   (h) Succinyl-CoA
   (i) 3-Phosphoglycerate
   (j) Fumarate
   (k) Phosphoenolpyruvate
   (l) Citrate
   (m) Pyruvate
4. **Reflect and Apply** People who begin to lose weight often have a rapid weight loss in the first few days. Why might this be due to a loss of water from the body?
5. **Reflect and Apply** The functioning of a particular pathway often depends not only on control enzymes in that pathway but also on control enzymes of other pathways. What happens in the following pathways under the indicated conditions? Suggest what other pathway or pathways might be influenced.
   (a) High ATP or NADH concentration and the citric acid cycle
   (b) High ATP concentration and glycolysis
   (c) High NADPH concentration and the pentose phosphate pathway
   (d) High fructose-2,6-bisphosphate concentration and gluconeogenesis
6. **Reflect and Apply** Why is it important to consider the interconnections between biochemical pathways rather than considering them separately from each other?
7. **Reflect and Apply** To what extent can metabolic pathways be considered reversible? Why?
8. **Reflect and Apply** In eukaryotic cells, metabolic pathways occur in specific locations, such as the mitochondrion or the cytosol. What sort of transport processes are required as a result?
9. **Reflect and Apply** Why is it advantageous for a metabolic pathway to have many steps?

24-2 Biochemistry and Nutrition

10. **Recall** What are the differences between the 1992 Food Guide Pyramid and MyPlate?
11. **Recall** What do we mean when we say that there is no storage form for protein? How is this different from fats and carbohydrates?
12. **Recall** What is the relationship between saturated fatty acids and LDL?
13. **Recall** What is leptin, and how does it work?
14. **Recall** Why might some scientists classify vitamin D as a hormone rather than a vitamin?
15. **Biochemical Connections** People who experience alcoholism and who are exposed to halogen compounds often die of liver failure. Why might this be a logical ultimate result?
16. **Reflect and Apply** What is a possible reason for placing limits on the dose in vitamin A supplements sold in stores?
17. **Recall** In the early twentieth century, goiter was relatively common in the Midwest. Why was this so? How has it been eliminated?
18. **Reflect and Apply** A cat named Lucullus is so spoiled that he will eat nothing but freshly opened canned tuna. Another cat, Griselda, is given only dry cat food by her far less indulgent owner. Canned tuna is essentially all protein, whereas dry cat food can be considered 70% carbohydrate and 30% protein. Assuming that these animals have no other sources of food, what can you say about the differences and similarities in their catabolic activities? (The pun is intended.)
19. **Reflect and Apply** Immature rats are fed all the essential amino acids but one. Six hours later, they are fed the missing amino acid. The rats fail to grow. Explain this observation.
20. **Reflect and Apply** Kwashiorkor is a protein-deficiency disease that occurs most commonly in small children, who characteristically have thin arms and legs and bloated, distended abdomens due to fluid imbalance. When such children are placed on adequate diets, they tend to lose weight at first. Explain this observation.
21. **Reflect and Apply** Why are amino acids such as arginine and histidine required in relatively large amounts by children but in smaller amounts by adults?
22. **Biochemical Connections** During colonial times, iron-deficiency anemia was almost unknown in America. Why?
23. **Reflect and Apply** People on high-fiber diets often have less cancer (especially of the colon) and lower blood cholesterol levels than people on low-fiber diets, even though fiber is not digestible. Suggest reasons for the benefits of fiber.
24. **Reflect and Apply** Most calcium supplements have calcium carbonate as the main ingredient. Other supplements that have calcium citrate as the main ingredient are advertised as being more easily absorbed. Do you consider this a valid claim? Why?
25. **Biochemical Connections** People experiencing alcoholism tend to be malnourished, with thiamine deficiency being a particularly severe problem. Suggest a reason why this is so.
26. **Reflect and Apply** Biologically and nutritionally important trace elements tend to be metals. What is their likely biochemical function?
27. **Reflect and Apply** Over a period of several decades, an adult human consumes tons of nutrients and more than 20,000 L of water without significant weight gain. How is this possible? Is it an example of chemical equilibrium?

24-3 Hormones and Second Messengers

28. **Recall** Are all hormones closely related in their chemical structure?
29. **Recall** How is hormone production affected by damage to the pituitary gland? To the adrenal cortex?
30. **Recall** How do the actions of the hypothalamus and the pituitary gland affect the workings of endocrine glands?
31. **Recall** The hormone thyroxine is given as an oral dose, but insulin needs to be injected into the body. Why?
32. **Recall** What is the difference between a G protein and a receptor tyrosine kinase? Give an example of a hormone that uses each.
33. **Recall** Give three examples of second messengers.

24-4 Hormones and the Control of Carbohydrate Metabolism

34. **Recall** List two hormones that work through the cAMP second messenger.
35. **Recall** How does glucagon affect the following enzymes?
    (a) Glycogen phosphorylase
    (b) Glycogen synthase
    (c) Phosphofructokinase-1
36. **Recall** How does epinephrine affect the enzymes listed in Question 35?
37. **Recall** What stops a hormone response when a G protein is involved?
38. **Reflect and Apply** When PIP_2 is hydrolyzed, why does IP_3 diffuse into the cytosol while DAG remains in the membrane?
39. **Reflect and Apply** Briefly describe the series of events that takes place when cAMP acts as a second messenger.
40. **Reflect and Apply** For each of the three hormones discussed in this chapter, give its source and chemical nature; also discuss the mode of action of each hormone.
41. **Reflect and Apply** Is it likely that any metabolic pathway can exist without control mechanisms?
42. **Reflect and Apply** Cholera harms the body by its effect on a second messenger. Describe how this takes place.
43. **Reflect and Apply** The epinephrine-mediated "amplification cascade" of Figure 24.18 has six steps, all of which are catalytic with one exception. This cascade leads to the activation of glycogen phosphorylase. This enzyme acts in turn on glycogen to yield glucose-1-phosphate (G-1-P).
    (a) Which step is not catalytic?
    (b) If each catalytic step had a turnover number (molecules of substrate acted on per molecule of enzyme) of 10, how many molecules of G-1-P would result from one molecule of epinephrine?
    (c) What is the biochemical advantage of such a cascade?
44. **Reflect and Apply** How is the amplification cascade of Question 43 reversed?
45. **Biochemical Connections** Explain what insulin and low-carbohydrate diets have to do with one another.

24-5 Insulin and Its Effects

46. **Recall** What is the primary function of insulin?
47. **Recall** What is the second messenger for the insulin response?
48. **Recall** What is the link between insulin binding to the receptor and the eventual second messenger?
49. **Recall** What is insulin's effect on the following?
    (a) Glycogen breakdown
    (b) Glycogen synthesis
    (c) Glycolysis
    (d) Fatty acid synthesis
    (e) Fatty acid storage
50. **Reflect and Apply** How is it possible that both insulin and epinephrine stimulate muscle glycolysis?
51. **Reflect and Apply** Why would a runner, who has a morning 5-km race to run, be concerned about insulin?
52. **Biochemical Connections** Why do some people call GLUT4 the "training" glucose transporter?
53. **Reflect and Apply** How are insulin, GLUT4, obesity, and type II diabetes related?
54. **Biochemical Connections** What types of cells typically have insulin receptors?
55. **Biochemical Connections** What is the relationship between obesity, diabetes, and cancer?
56. **Biochemical Connections** What is the Warburg effect? Why would cancer cells favor such inefficient metabolism?
57. **Biochemical Connections** What is PTEN and what is its relationship to cancer?
58. **Biochemical Connections** What is the natural function of PTEN?
59. **Conjectural Question** If you had your choice of doing research on any topic in this book, which would you choose? What are some important considerations?

Further Reading

Bock, A., R. Irannejad, and J. D. Scott. cAMP Signaling: A Remarkably Regional Affair. *Tr. Bioch. Sci.* **49**, 305–317 (2024). This review highlights developments in our understanding of cyclic AMP (cAMP) generation and its spatial utilization within cells.

Bose, A., A. Guilherme, S. I. Robida, S. M. C. Nicoloro, Q. L. Zhou, Z. Y. Jiang, D. P. Pomerleau, and M. P. Czech. Glucose Transporter Recycling in Response to Insulin Is Facilitated by Myosin Myo1c. *Nature* **420**, 821–824 (2002). An article about the transcription factors that have been shown to affect glucose transport.

Cohen, P., M. Miyazaki, N. D. Socci, A. Hagge-Greenberg, W. Liedtke, A. A. Soukas, R. Sharma, L. C. Hudgins, J. M. Ntambi, and J. M. Friedman. Role for Stearoyl-CoA Desaturase-1 in Leptin-Mediated Weight Loss. *Science* **297**, 240–243 (2002). An article about one mechanism by which leptin effects weight loss.

Cowley, M. A., J. L. Smart, M. Rubinstein, M. G. Cerdan, S. Diano, T. L. Horvath, R. D. Cone, and M. J. Low. Leptin Activates Anorexigenic POMC Neurons through a Neural Network in the Arcuate Nucleus. *Nature* **411**, 480–484 (2001). An article about leptin and obesity.

Friedman, J. Fat in All the Wrong Places. *Nature* **415**, 268–269 (2002). A review article about obesity.

Gura, T. Obesity Sheds Its Secrets. *Science* **275**, 751–753 (1997). A Research News article on the role of the protein hormone leptin in obesity research.

Kaeberlein, M., and P. Kapahi. Aging is RSKy Business. *Science* **326**, 55–56 (2009). An article about how signaling through the RSK (ribosomal S6 kinase) pathway influences the molecular mechanisms of aging and lifespan regulation.

Le, T. K. C., X. D. Dao, D. V. Nguyen, D. H. Luu, T. M. H. Bui, T. H. Le, H. T. Nguyen, T. N. Le, T. Hosaka, and T. T. T. Nguyen. Insulin signaling and its application. *Front. Endocrinol.* **14**, Article ID: 1226655 (2023). This mini-review provides an overview of the insulin signaling pathway. It synthesizes updated mechanism discoveries and informs the development of novel applications in diagnosing and treating diseases related to insulin resistance.

Minokoshi, Y., Y. B. Kim, D. P. Odile, L. G. D. Fryer, C. Muller, D. Carling, and B. B. Kahn. Leptin Stimulates Fatty-Acid Oxidation by Activating AMP-Activated Protein Kinase. *Nature* **415**, 339–343 (2002). An article about another mechanism for leptin-associated weight loss.

Rasmussen, H. The Cycling of Calcium as an Intracellular Messenger. *Sci. Amer.* **261** (4), 66–73 (1989). An article on the role of calcium as a second messenger.

Saunders, L. R., and E. Verdin Stress Response and Aging. *Science* **323**, 1021–1022 (2009). This article explores the relationship between cellular stress responses and the aging process.

Schwartz, M. W., and G. J. Morton. Keeping Hunger at Bay. *Nature* **418**, 595–597 (2002). An article about hunger, hormones, and weight loss.

Selman, C., J. M. A. Tullet, D. Wieser, E. Irvine, S. J. Lingard, A. I. Choudhury, M. Claret, H. Al-Qassab, D. Carmignac, F. Ramadani, A. Woods, I. C. A. Robinson, E. Schuster, R. L Batterham, S. C. Kozma, G. Thomas, D. Carling, K. Okkenhaug, J. M. Thornton, L. Partridge, D. Gems, and D. J. Withers. Ribosomal Protein S6 Kinase 1 Signaling Regulates Mammalian Life Span. *Science* **326**, 140–144 (2009). An article that investigates the role of the S6K1 protein kinase in mammalian aging signaling pathways.

Sinclair, D. A., and L. Guarente. Unlocking the Secrets of Longevity Genes. *Sci. Amer.* **294** (3), 48–57 (2006). An article about a family of genes that confer longevity on species from yeast to mammals.

Taubes, G. Insulin Insults May Spur Alzheimer's Disease. *Science* **301**, 40–41 (2003). An article linking Alzheimer's disease and high insulin levels.

White, M. F. Insulin Signaling in Health and Disease. *Science* **302**, 1710–1711 (2003). A review of insulin and its effects on health.

Willett, W. Diet and Health: What Should We Eat? *Science* **264**, 532–537 (1994). A summary of many aspects of a complex topic.

Willett, W. C., and M. J. Stampfer. Rebuilding the Food Pyramid. *Sci. Amer.* **288** (1), 64–71 (2003). A thorough explanation of the changes in nutrition recommendations after the 1992 Food Guide Pyramid was published.

Hot Topic

Cherezov, V., D. Rosenbaum, M. Hanson, S. Rasmussen, F. Thian, T. Kobilka, H. Shoi, P. Kuhn, W. Weis, B. Kobilka, and R. Stevens High-Resolution Crystal Structure of an Engineered Human β_2-Adrenergic G Protein–Coupled Receptor. *Science* **318**, 1258–1265 (2007). The detailed structure of the receptor as determined by X-ray crystallography.

Kenakin, T. New Bull's-Eyes for Drugs. *Sci. Amer.* **93** (4), 51–57 (2005). A review of GPCR operation.

Kise, R. and A. Inoue GPCR Signaling Bias: An Emerging Framework for Opioid Drug Development. *J. Biochem.* **175**, 367–376 (2024). This review provides an overview of biased signaling in opioid receptors and discusses methods to evaluate signaling bias in the GPCR field. It also explores future directions for rational drug development through the integration of diverse signal datasets.

Kobilka, B., H. Matsui, T. Kobilka, T. Yang-Feng, U. Francke, M. Caron, R. Lefkowitz, and J. Regan". Cloning, Sequencing, and Expression of the Gene Coding for the Human Platelet α_2-Adrenergic Receptor. *Science* **238**, 650–687 (1987). A research report on groundbreaking work on the structure and function of the receptor.

Libert, F., M. Permentier, A. Lefort, C. Dinsart, J. Van Sande, C. Maenhaut, M. Simons, J. Dumont, and G. Vassart Selective Cloning and Amplification of Four New Members of the G Protein–Coupled Receptor Family. *Science* **244**, 569–572 (1989). A study of similarities among GPCRs.

Service, R. Receptor Scientists to Receive Chemistry Nobel. *Science* **338**, 313–314 (2012). A news article about Nobel Prize–winning work on GPCRs.

Thompson, G. L., E. Kelly, A. Christopoulos, and M. Canals. Novel GPCR paradigms at the μ-opioid receptor. *British Journal of Pharmacology* **172**, 287–296 (2015). A review article describing pathways to functional selectivity in opioid GPCRs.

Answers to Review Exercises

Chapter 1

1-1 Basic Themes

1. Water, methane, carbon dioxide, ammonia, nitrogen, and hydrogen.
2. See Figure 1.1.

1-2 Chemical Foundations of Biochemistry

3. (b) contain carbon and hydrogen.
4. Amino acids have a central carbon atom that is bonded to a carboxyl group, an amino group, a hydrogen, and a variable group called the side chain. Carbohydrates are compounds made up of carbon, hydrogen, and oxygen, with a general formula of $(CH_2O)_n$, where n is at least 3. Nucleotides are composed of a five-carbon sugar, a nitrogen-containing ring, and one or more phosphate groups. Lipids are composed of long chains of hydrocarbons, although they do have small polar portions as well.
5. (c) A specific arrangement of atoms within a molecule that determines its chemical properties.
6. The correct match of functional groups and the compounds containing those functional groups is as follows:

Amino group	$CH_3CH_2NH_2$
Carbonyl group (ketone)	CH_3COCH_3
Hydroxyl group	CH_3OH
Carboxyl group	CH_3COOH
Carbonyl group (aldehyde)	CH_3CH_2CHO
Thiol group	CH_3SH
Ester linkage	$CH_3COOCH_2CH_3$
Double bond	$CH_3CH{=}CHCH_3$
Amide linkage	$CH_3CON(CH_3)_2$
Ether	$CH_3CH_2OCH_2CH_3$

7. The functional groups in the compounds are as follows:

Glucose

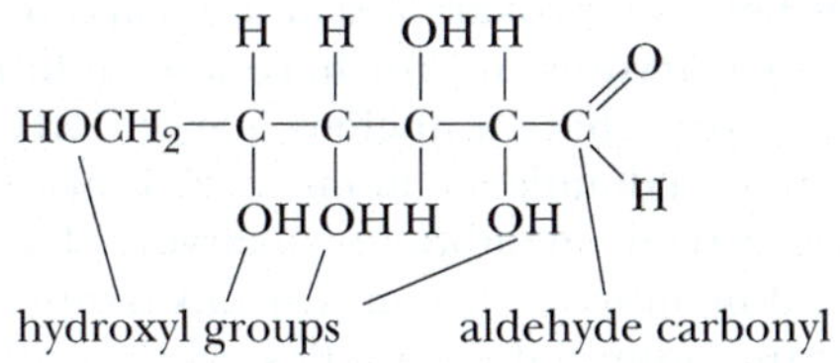

A triglyceride

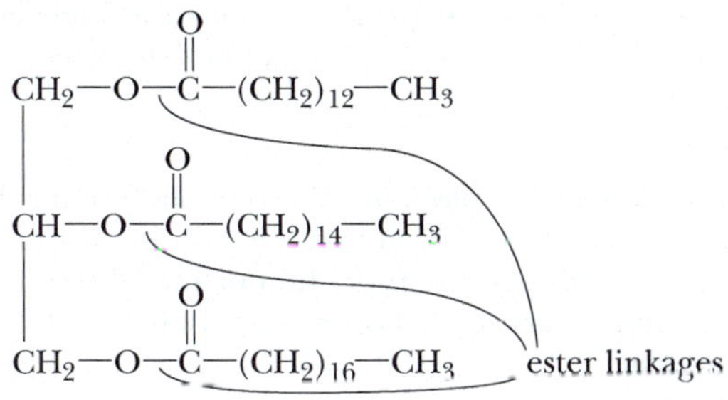

A peptide

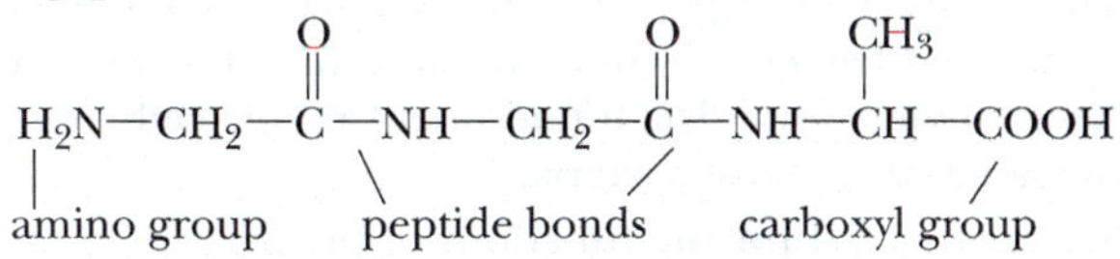

Vitamin A

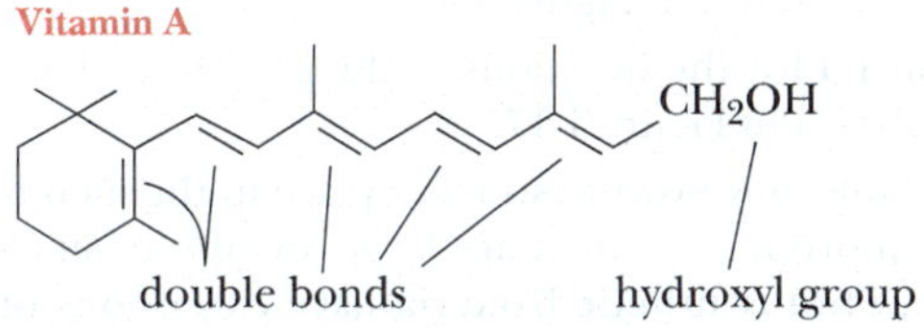

8. Serine, Ribose, and Cholesterol

1-3 The Beginnings of Biology

9. A polymer is a very large molecule formed by linking smaller units (monomers) together. A protein is a polymer formed by linking amino acids together. A nucleic acid is a polymer formed by linking nucleotides together. Catalysis is the process that increases the rate of chemical reactions compared with the rate of the uncatalyzed reaction. Biological catalysts are proteins in almost all cases; the only exceptions are a few types of RNA, which can catalyze some of the reactions of their own metabolism. The genetic code is the means by which the information for the structure and function of all living things is passed from one generation to the next. The sequence of purines and pyrimidines in DNA carries the genetic code. (RNA is the coding material in some viruses.)
10. See Figure 1.10.
11. The number is 4^{40}, or 1.2×10^{24}, which is twice Avogadro's number.
12. RNA is capable of both coding and catalysis.

13. It is unlikely that cells could have arisen as bare cytoplasm without a plasma membrane. The presence of the membrane protects cellular components from the environment and prevents them from diffusing away from each other. The molecules within a cell can react more easily if they are closer to each other.

1-4 The Biggest Biological Distinction—Prokaryotes and Eukaryotes

14. Five differences between prokaryotes and eukaryotes are as follows: (1) Prokaryotes do not have a well-defined nucleus, but eukaryotes have a nucleus marked off from the rest of the cell by a double membrane. (2) Prokaryotes have only a plasma (cell) membrane; eukaryotes have an extensive internal membrane system. (3) Eukaryotic cells contain membrane-bounded organelles, while prokaryotic cells do not. (4) Eukaryotic cells are normally larger than those of prokaryotes. (5) Prokaryotes are single-celled organisms, while eukaryotes can be either single-celled or multicellular.
15. (a) Ribosomes
16. Protein synthesis takes place on ribosomes both in prokaryotes and in eukaryotes. In eukaryotes, ribosomes may be bound to the endoplasmic reticulum or found free in the cytoplasm; in prokaryotes, ribosomes are only found free in the cytoplasm.
17. It is unlikely that mitochondria would be found in bacteria. These eukaryotic organelles are enclosed by a double membrane, and bacteria do not have an internal membrane system. The mitochondria found in eukaryotic cells are about the same size as most bacteria.
18. See Section 1-4 for the functions of the parts of an animal cell, which are shown in Figure 1.17.
19. See Section 1-4 for the functions of the parts of a plant cell, which are shown in Figure 1.17.
20. In green plants, photosynthesis takes place in the membrane system of chloroplasts, which are large membrane-enclosed organelles. Photosynthetic bacteria have extensions of the plasma membrane into the interior of the cell called chromatophores, which are the sites of photosynthesis.
21. Nuclei, mitochondria, and chloroplasts are all enclosed by a double membrane.
22. Nuclei, mitochondria, and chloroplasts all contain DNA. The DNA found in mitochondria and in chloroplasts differs from that found in the nucleus.
23. Mitochondria carry out a high percentage of the oxidation (energy-releasing) reactions of the cell. They are the primary sites of ATP synthesis.
24. The Golgi apparatus is involved in binding carbohydrates to proteins and in exporting substances from the cell. Lysosomes contain hydrolytic enzymes, peroxisomes contain catalase (needed for the metabolism of peroxides), and glyoxysomes contain enzymes needed by plants for the glyoxylate cycle. All of these organelles have the appearance of flattened sacs, and each is enclosed by a single membrane.
25. (b) Genes are smaller than genomes.

1-5 How We Classify Eukaryotes and Prokaryotes

26. Monera includes bacteria (e.g., *E. coli*) and cyanobacteria. Protista includes such organisms as *Euglena, Volvox, Amoeba,* and *Paramecium.* Fungi includes molds and mushrooms. Plantae includes club mosses and oak trees. Animalia includes spiders, earthworms, salmon, rattlesnakes, robins, and dogs.
27. The kingdom Monera consists of prokaryotes. Each of the other four kingdoms consists of eukaryotes.
28. The major advantage of being eukaryotic is that of having compartments (organelles) with specialized functions (and thus division of labor). Another advantage is that cells can be much larger without surface area–to–volume considerations being critical because of compartmentalization.
29. See the discussion of the endosymbiotic theory in Section 1-5.
30. See Question 27. The division of labor in cells gives rise to greater efficiency and a larger number of individuals. This, in turn, allows more opportunity for evolution and speciation.

1-6 Biochemical Energetics

31. See the margin definitions in Section 1-6.
32. Processes that release energy are favored.
33. The term *spontaneous* means energetically favored. It does not necessarily mean fast.
34. In all cases, there is an increase in entropy, and the final state has more possible random arrangements than the initial state.
35. Because the equation involves multiplication of ΔS by T, the value of ΔG is temperature dependent.
36. If entropy is a measure of the dispersion of energy, then molecules at higher temperatures would have more possible arrangements due to increased molecular motion.
37. Assuming the value of ΔS is positive, an increase in temperature increases the $-\Delta G$ contribution of the entropy component to the overall energy change.
38. The heat exchange, getting colder, reflects only the enthalpy or ΔH component of the total energy change. The entropy change must be high enough to offset the enthalpy component and to add up to an overall $-\Delta G$.
39. Entropy would increase. Two molecules, ADP and P_i, can be randomized in more ways than a single molecule, ATP, can.
40. The lowering of entropy needed to give rise to organelles leads to higher entropy in the surroundings, thus increasing the entropy of the universe as a whole.
41. DNA would have higher entropy with the strands separated. There are two single strands instead of one double strand, and the single strands have more conformational mobility.

Chapter 2

2-1 Water and Polarity

1. van der Waals forces are weak intermolecular forces that do not involve a full ionic interaction such as dipole–dipole and dipole–induced dipole interactions.

2. Dipole–dipole interactions, hydrogen bonds, dipole–induced dipole interactions, induced dipole–induced dipole interactions (London dispersion force).
3. An induced dipole is a momentary dipole created by the transient distortion of an electron cloud of one atom due to the proximity of another atom.
4. A salt bridge refers to electrostatic attractions between oppositely charged groups where the two groups are also close enough to form a hydrogen bond.
5. If the dipole is canceled out by another of equal and opposite orientation. The classic example is CO_2. Each carbon–oxygen bond is a dipole, but the two cancel each other out, and the molecule is nonpolar. Another example is CCl_4. Each carbon–chlorine bond is a dipole, but the four cancel each other out due to the tetrahedral geometry of the molecule.
6. The sodium ion is a fully charged anion. Therefore, its interaction with the partial negative charge on the oxygen of ethanol would be stronger than the interaction of the same oxygen with a partially positive charge from the hydrogen of another ethanol molecule.
7. Dipole–dipole > dipole–induced dipole > induced dipole–induced dipole.
8. Ionic bond (1-strongest), salt bridge (2), ion–dipole interactions (3), hydrogen bond (4), dipole–dipole interactions (5), dipole–induced dipole interactions (6), induced dipole–induced dipole interactions (London dispersion force) (7-weakest)
9. Cholesterol (No), Sodium lauryl sulfate (Yes), Lecithin (a phosphatidyl choline) (Yes), Glucose (No), Sodium palmitate (Yes), Urea (No)

2-2 Hydrogen Bonds

10. Dipole–dipole bond
11. Proteins and nucleic acids have hydrogen bonds as an important part of their structures.
12. The C—H bond is insufficiently polar for a greatly unequal distribution of electrons at its two ends. Also, there are no unshared pairs of electrons to serve as hydrogen bond acceptors.
13. Three possible examples are H_2O, CH_3OH, or NH_3.
14. For a bond to be called a hydrogen bond, it must have a hydrogen covalently bonded to O, N, or F. This hydrogen must then form a hydrogen bond with another O, N, or F.
15. In a hydrogen-bonded dimer of acetic acid, the —OH portion of the carboxyl group on molecule 1 is hydrogen-bonded to the —C═O portion of the carboxyl group on molecule 2 and vice versa.

2-3 Acids, Bases, and pH

16. (a) $(CH_3)_3NH^+$ (conjugate acid), $(CH_3)_3N$ (conjugate base)
(b) $^+H_3N—CH_2—COOH$ (conjugate acid), $^+H_3N—CH_2—COO^-$ (conjugate base)
(c) $^+H_3N—CH_2—COO^-$ (conjugate acid), $H_2N—CH_2—COO^-$ (conjugate base)
(d) $^-OOC—CH_2—COOH$ (conjugate acid), $^-OOC—CH_2—COO^-$ (conjugate base)
(e) $^-OOC—CH_2—COOH$ (conjugate base), $HOOC—CH_2—COOH$ (conjugate acid)
17. $(HOCH_2)_3CNH_3^+$ **Conjugate acid** $(HOCH_2)_3CNH_2$ **Conjugate base**

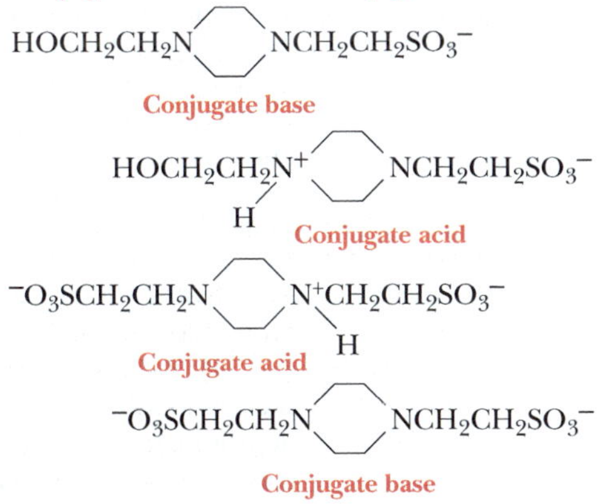

18. The definition of pH is $-\log[H^+]$. By definition of the log function, a change in concentration of 10 leads to a change in pH of 1. The log of 10 is 1, the log of 100 is 2, etc.

19.	Blood plasma, pH 7.4	$[H^+] = 4.0 \times 10^{-8}\ M$
	Orange juice, pH 3.5	$[H^+] = 3.2 \times 10^{-4}\ M$
	Human urine, pH 6.2	$[H^+] = 6.3 \times 10^{-7}\ M$
	Household ammonia, pH 11.5	$[H^+] = 3.2 \times 10^{-12}\ M$
	Gastric juice, pH 1.8	$[H^+] = 1.6 \times 10^{-2}\ M$
20.	Saliva, pH 6.5	$[H^+] = 3.2 \times 10^{-7}\ M$
	Intracellular fluid (liver), pH 6.9	$[H^+] = 1.6 \times 10^{-7}\ M$
	Tomato juice, pH 4.3	$[H^+] = 5.0 \times 10^{-5}\ M$
	Grapefruit juice, pH 3.2	$[H^+] = 6.3 \times 10^{-4}\ M$
21.	Saliva, pH 6.5	$[OH^-] = 3.2 \times 10^{-8}\ M$
	Intracellular fluid (liver), pH 6.9	$[OH^-] = 7.9 \times 10^{-8}\ M$
	Tomato juice, pH 4.3	$[OH^-] = 2.0 \times 10^{-10}\ M$
	Grapefruit juice, pH 3.2	$[OH^-] = 1.6 \times 10^{-11}\ M$

22. pH = 2.28

2-4 Titration Curves

23. (a) The amount of acid or base that can be added to a buffer before experiencing a sharp pH change. (b) The point in a titration curve at which the added acid or base equals the amount of buffer originally present. (c) An experiment in which acid or base is added stepwise to a solution of a compound and the pH is measured as a function of the added substance.
24.

Options	Correct answer	Solution/Feedback
Benzoic acid	ionizable	The carboxylic acid group has a pK_a of approximately 4. It would be primarily deprotonated at pH > 4 and would have a negative charge.
Indole	nonionizable	The lone pair of electrons on the nitrogen in indole is part of the aromatic pi system and is *not* available to accept a proton.

Options	Correct answer	Solution/Feedback
Imidazole	ionizable	The lone pair of electrons on the nitrogen that is *not* part of the aromatic pi system is available to accept a proton. Once protonated, it would contribute a positive charge to the molecule.
Benzyl alcohol	nonionizable	The alcohol functional group has a pK_a value around 16. This means that the group will not become deprotonated and will not contribute a negative charge in the pH 2–12 range.
Phenol	ionizable	Phenols have pK_a values around 10 and would be deprotonated at pH > 10. This would contribute a negative charge.
Benzamide	nonionizable	The lone pair on the nitrogen is *not* available to accept a proton due to resonance with the carbonyl group.

25. Aspirin is electrically neutral at the pH of the stomach and can pass through the membrane more easily than in the small intestine.

26.

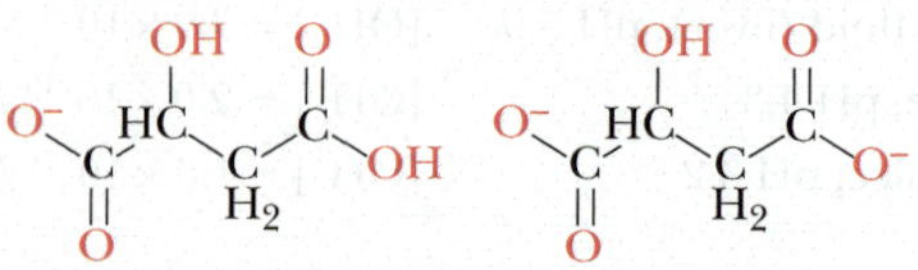

27. (a) Yes.

(b) If x = 60.0 mL, you would be at point D.

2-5 Buffers

28. The pK_a of the buffer should be close to the desired buffer pH, and the substance chosen should not interfere with the reaction being studied.

29. The useful pH range of a buffer is one pH unit above and below its pK_a.

30. Use the Henderson–Hasselbalch equation:

$$\text{pH} = \text{p}K_a + \log\left(\frac{[\text{CH}_3\text{COO}^-]}{[\text{CH}_3\text{COOH}]}\right)$$

$$5.00 = 4.76 + \log\left(\frac{[\text{CH}_3\text{COO}^-]}{[\text{CH}_3\text{COOH}]}\right)$$

$$0.24 = \log\left(\frac{[\text{CH}_3\text{COO}^-]}{[\text{CH}_3\text{COOH}]}\right)$$

$$\frac{[\text{CH}_3\text{COO}^-]}{[\text{CH}_3\text{COOH}]} = \text{inverse log of } 0.24 = \frac{1.7}{1}$$

31. Use the Henderson–Hasselbalch equation:

$$\text{pH} = \text{p}K_a + \log\left(\frac{[\text{CH}_3\text{COO}^-]}{[\text{CH}_3\text{COOH}]}\right)$$

$$4.00 = 4.76 + \log\left(\frac{[\text{CH}_3\text{COO}^-]}{[\text{CH}_3\text{COOH}]}\right)$$

$$-.076 = \log\left(\frac{[\text{CH}_3\text{COO}^-]}{[\text{CH}_3\text{COOH}]}\right)$$

$$\frac{[\text{CH}_3\text{COO}^-]}{[\text{CH}_3\text{COOH}]} = \text{inverse log of } -0.76 = \frac{0.17}{1}$$

32. Use the Henderson–Hasselbalch equation:

$$\text{pH} = \text{p}K_a + \log\left(\frac{[\text{TRIS}]}{[\text{TRIS-H}^+]}\right)$$

$$8.7 = 8.3 + \log\left(\frac{[\text{TRIS}]}{[\text{TRIS-H}^+]}\right)$$

$$0.4 = \log\left(\frac{[\text{TRIS}]}{[\text{TRIS-H}^+]}\right)$$

$$\left(\frac{[\text{TRIS}]}{[\text{TRIS-H}^+]}\right) = \text{inverse log of } 0.4 = \frac{2.5}{1}$$

33. Use the Henderson–Hasselbalch equation:

$$\text{pH} = \text{p}K_a + \log\left(\frac{[\text{HEPES}]}{[\text{HEPES-H}^+]}\right)$$

$$7.9 = 7.55 + \log\left(\frac{[\text{HEPES}]}{[\text{HEPES-H}^+]}\right)$$

$$0.35 = \log\left(\frac{[\text{HEPES}]}{[\text{HEPES-H}^+]}\right)$$

$$\left(\frac{[\text{HEPES}]}{[\text{HEPES-H}^+]}\right) = \text{inverse log of } 0.35 = \frac{2.2}{1}$$

34. At pH 7.5, the ratio of $[HPO_4^{2-}]/[H_2PO_4^-]$ is $\frac{2}{1}$(pK_a of $H_2PO_4^-$ = 7.2), as calculated using the Henderson–Hasselbalch equation. K_2HPO_4 is a source of the base form, and HCl must be added to convert one-third of it to the acid form, according to the $\frac{2}{1}$ base/acid ratio. Weigh out 8.7 g of K_2HPO_4 (0.05 mol, based on a formula weight of 174 g/mol), dissolve it in a small quantity of distilled water, add 16.7 mL of 1 M HCl (this gives $\frac{1}{3}$ of 0.05 mol of hydrogen ion, which converts $\frac{1}{3}$ of the 0.05 mol of HPO_4^{2-} to $H_2PO_4^-$), and dilute the resulting mixture to 1 L.

35. A $\frac{2}{1}$ ratio of the base form to the acid form is still needed because the pH of the buffer is the same in both problems. NaH_2PO_4 is a source of the acid form, and NaOH must be added to convert two-thirds of it to the base form. Weigh out 6.0 g of NaH_2PO_4 (0.05 mol, based on a formula weight of 120 g/mol), dissolve it in a small quantity of distilled water, add 33.3 mL of 1 M NaOH (this gives $\frac{2}{3}$ of 0.05 mol of hydroxide ion, which converts $\frac{2}{3}$ of the 0.05 mol of $H_2PO_4^-$ to HPO_4^{2-}), and dilute the resulting mixture to 1 L.

36. After mixing, the buffer solution (100 mL) contains 0.75 M lactic acid and 0.25 M sodium lactate. The pK_a of lactic acid is 3.86. Use the Henderson–Hasselbalch equation:

$$pH = pK_a + \log\left(\frac{[CH_3CHOHCOO^-]}{[CH_3CHOHCOOH]}\right)$$

$$pH = 3.86 + \log\left(\frac{[CH_3CHOHCOO^-]}{[CH_3CHOHCOOH]}\right)$$

$$pH = 3.86 + \log(0.25\ M/0.75\ M)$$
$$pH = 3.86 + (-0.48)$$
$$pH = 3.38$$

37. After mixing, the buffer solution (100 mL) contains 0.25 *M* lactic acid and 0.75 *M* sodium lactate. The pK_a of lactic acid is 3.86. Use the Henderson–Hasselbalch equation:

$$pH = pK_a + \log\left(\frac{[CH_3CHOHCOO^-]}{[CH_3CHOHCOOH]}\right)$$

$$pH = 3.86 + \log\left(\frac{[CH_3CHOHCOO^-]}{[CH_3CHOHCOOH]}\right)$$

$$pH = 3.86 + \log(0.75\ M/0.25\ M)$$
$$pH = 3.86 + (0.48)$$
$$pH = 4.34$$

38. Use the Henderson–Hasselbalch equation:

$$pH = pK_a + \log\left(\frac{[CH_3COO^-]}{[CH_3COOH]}\right)$$

$$pH = 4.76 + \log\left(\frac{[CH_3COO^-]}{[CH_3COOH]}\right)$$

$$pH = 4.76 + \log(0.25\ M/0.10\ M)$$
$$pH = 4.76 + 0.40$$
$$pH = 5.16$$

39. The solution is a buffer because it contains equal concentrations of TRIS in its acid and free amine forms. When the two solutions are mixed, the concentrations of the resulting solution (in the absence of reaction) are 0.05 *M* HCl and 0.1 *M* TRIS because of dilution. The HCl reacts with half of the TRIS present, giving 0.05 *M* TRIS (protonated form) and 0.05 *M* TRIS (free amine form).

40. Any buffer that has equal concentrations of the acid and basic forms has a pH equal to its pK_a. Therefore, the buffer from Question 36 has a pH of 8.3.

41. First calculate the moles of buffer that you have: 100 mL = 0.1 L, and 0.1 L of 0.1 *M* TRIS buffer is 0.01 mol. Because the buffer is at its pK_a, there are equal concentrations of the acid and basic forms, so the amount of TRIS is 0.005 mol, and the amount of TRIS-H^+ is 0.005 mol. If you then add 3 mL of 1 *M* HCl, you will be adding 0.003 mol of H^+. This reacts as shown:

$$TRIS + H^+ \rightarrow TRIS\text{-}H^+$$

until you run out of something, which will be the H^+, because it is the limiting reagent. The new amounts can be calculated as shown:

$$TRIS\text{-}H^+ = 0.005\ mol + 0.003\ mol = 0.008\ mol$$
$$TRIS = 0.005\ mol - 0.003\ mol = 0.002\ mol$$

Now plug these values into the Henderson–Hasselbalch equation:

$$pH = 8.3 + \log([TRIS]/[TRIS\text{-}H^+]) = 8.3 + \log(0.002/0.008)$$
$$pH = 7.70$$

42. First calculate the moles of buffer that you have (we are going to do some rounding off): 100 mL = 0.1 L, and 0.1 L of 0.1 *M* TRIS buffer is 0.01 mol. Because the buffer is at pH 7.70, we saw in Question 38 that the amount of TRIS is 0.002 mol, and the amount of TRIS-H^+ is 0.008 mol. If you then add 1 mL of 1 *M* HCl, you will be adding 0.001 mol of H^+. This reacts as shown:

$$TRIS + H^+ \rightarrow TRIS\text{-}H^+$$

until you run out of something, which will be the TRIS, because it is the limiting reagent. All the TRIS is converted to TRIS-H^+:

$$TRIS\text{-}H^+ = 0.01\ mol$$
$$TRIS = {\sim}0\ mol$$

We have used up the buffer capacity of the TRIS. We now have 0.001 mol of H^+ in approximately 0.1 L of solution. This is approximately 0.01 *M* H^+.

$$pH = -\log 0.01$$
$$pH = 2.0$$

43. Use the Henderson–Hasselbalch equation:

[Acetate ion]/[acetic acid] = 2.3/1

44. A substance with a pK_a of 3.9 has a buffer range of 2.9 to 4.9. It does not buffer effectively at pH 7.5.

45. Use the Henderson–Hasselbalch equation. The ratio of $[A^-]/[HA]$ would be 3981 to 1.

46. In all cases, the suitable buffer range covers a pH range of $pK_a \pm 1$ pH unit.
   (a) Lactic acid ($pK_a = 3.86$) and its sodium salt: pH 2.86–4.86.
   (b) Acetic acid ($pK_a = 4.76$) and its sodium salt: pH 3.76–5.76.
   (c) TRIS (see Table 2.8, $pK_a = 8.3$) in its protonated form and its free amine form: pH 7.3–9.3.
   (d) HEPES (see Table 2.8, $pK_a = 7.55$) in its zwitterionic form and its anionic form: pH 6.55–8.55.

47. Several of the buffers would be suitable—namely TES, HEPES, MOPS, and PIPES—but the best buffer would be MOPS because its pK_a of 7.2 is closest to the desired pH of 7.3.

48. Buffering capacity is based on the amounts of the acid and base forms present in the buffer solution. A solution with a high buffering capacity can react with a large amount of added acid or base without drastic changes in pH. A solution with a low buffering capacity can react with only comparatively small amounts of acid or base before showing changes in pH. The more concentrated the buffer, the higher its buffering capacity. Buffer (a) has one-tenth the buffering capacity of buffer (b), which in turn has one-tenth the buffering capacity of buffer (c). All three buffers have the same pH because they all have the same relative amounts of the acid and base forms.

49. It would be more effective to start with the HEPES base. You want a buffer at a pH above the pK_a, which means that the base form predominates when you have finished preparing it. It is easier to convert some of the base form to the acid form than to convert most of the acid form to the base form.

50. Hypoventilation decreases the pH of blood.

Chapter 3

3-1 Amino Acids Are Three Dimensional

1. D- and L-amino acids have different stereochemistry around the α-carbon. Peptides that contain D-amino acids are found in bacterial cell walls and in some antibiotics.

2. 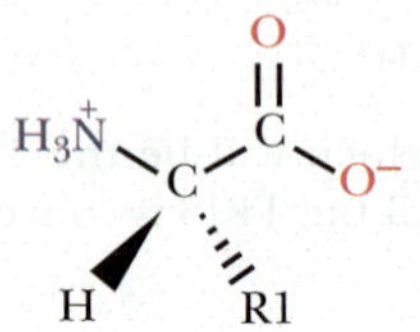

3. 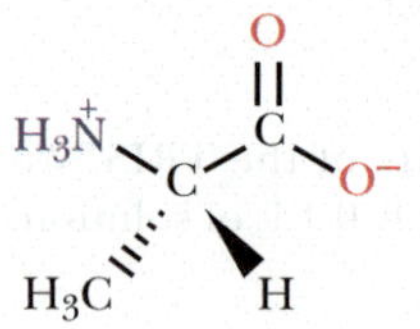

4. 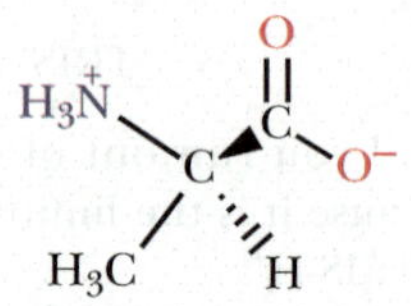

5. 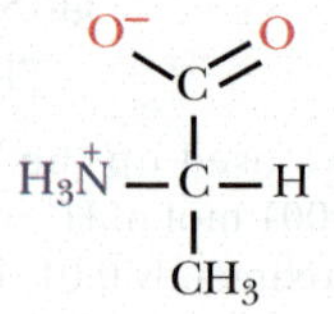

6. 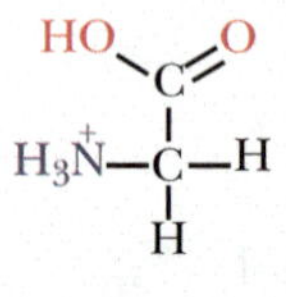 achiral

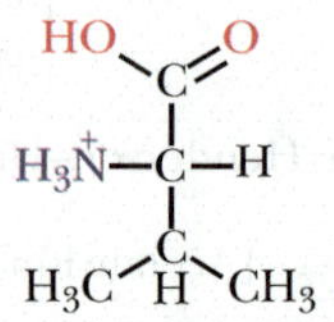 chiral

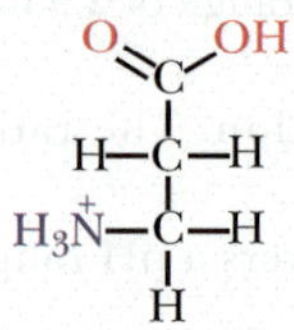 achiral

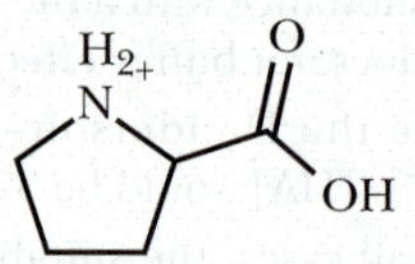 chiral

7. 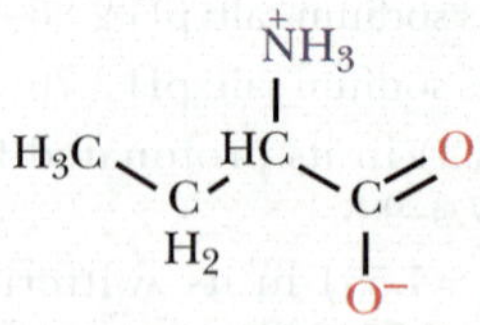

8. 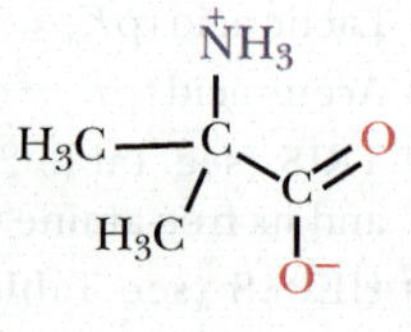

3-2 Structures and Properties of Amino Acids

9. Listed here are amino acids in which the R group contains the following: a hydroxyl group (serine, threonine, or tyrosine); a sulfur atom (cysteine or methionine); a second chiral carbon atom (isoleucine or threonine); an amino group (lysine); an amide group (asparagine or glutamine); an acid group (aspartate or glutamate); an aromatic ring (phenylalanine, tyrosine, or tryptophan); and a branched side chain (leucine or valine).

10. Glycine: Gly, G. For the other three- and one-letter codes, see Table 3.1.

11. In the peptide Val—Met—Ser—Ile—Phe—Arg—Cys—Tyr—Leu, the polar amino acids are Ser, Arg, Cys, and Tyr; the aromatic amino acids are Phe and Tyr; and the sulfur-containing amino acids are Met and Cys.

12. In the peptide Glu—Thr—Val—Asp—Ile—Ser—Ala, the nonpolar amino acids are Val, Ile, and Ala; the acidic amino acids are Glu and Asp.

13. (d) The nature of the side chain of the amino acid.

3-3 Amino Acids Can Act as Both Acids and Bases

14. The ionized forms of each of the following amino acids at pH 7—glutamic acid, leucine, threonine, histidine, and arginine—are as follows:

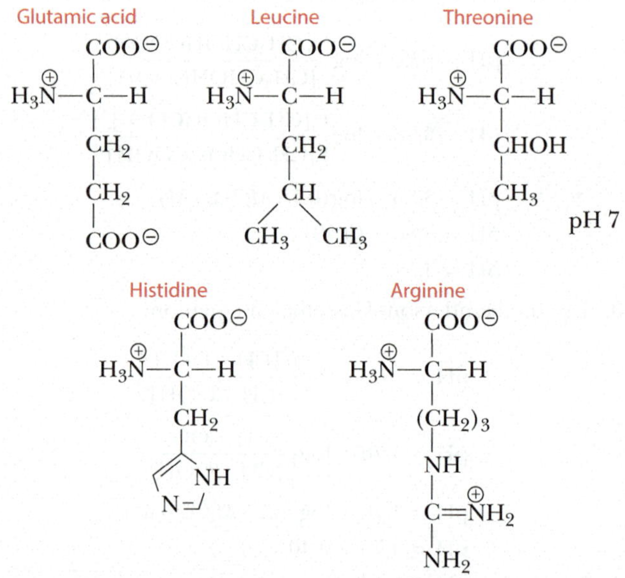

15. 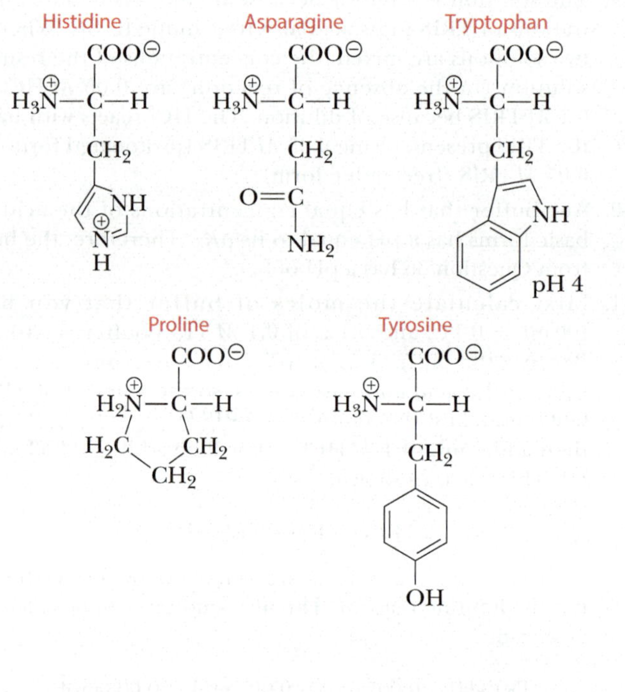

16. Here are the structural details for these amino acids at pH 10 compared to their structures above at pH 4. Histidine: imidazole is deprotonated; the α-amino group is predominantly deprotonated. Asparagine: α-amino group is deprotonated. Tryptophan: α-amino group is predominantly deprotonated. Proline: α-amino group is partially deprotonated. Tyrosine: the α-amino group is predominantly deprotonated; the phenolic hydroxyl is approximately a 50–50 mixture of protonated and deprotonated forms.

17. **Glutamic Acid**

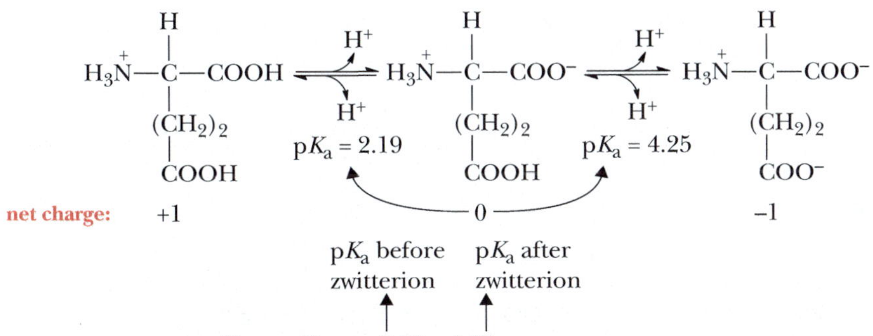

$$pI = \frac{pK_{a1} + pK_{a2}}{2} = \frac{2.19 + 4.25}{2} = 3.22$$

Serine Because serine only has two pK_a values, just plug those into the pI equation.

$$pI = \frac{2.21 + 9.15}{2} = 5.68$$

Tyrosine

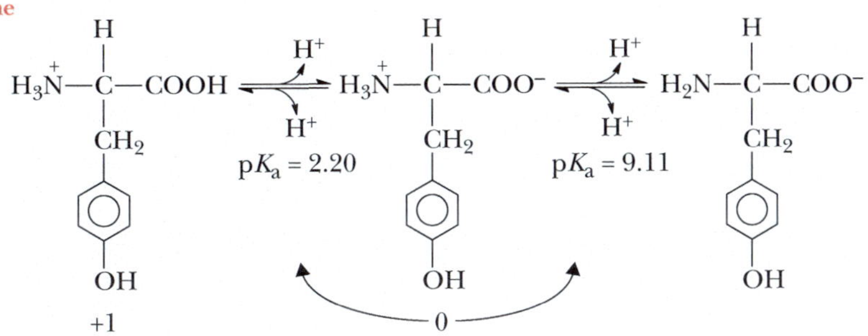

$$pI = \frac{2.20 + 9.11}{2} = 5.66$$

Arginine

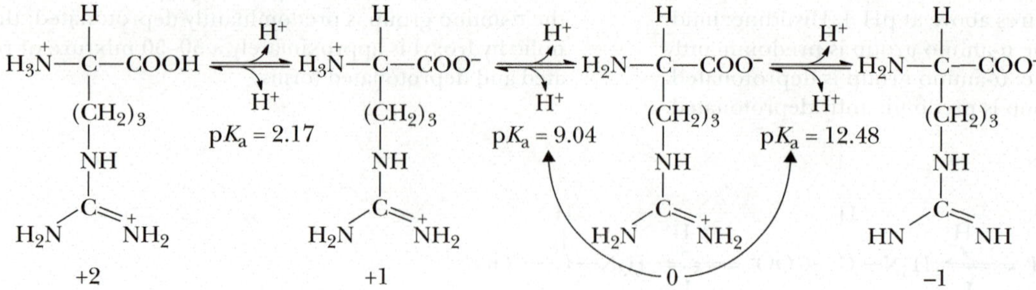

$$pI = \frac{9.04 + 12.48}{2} = 10.76$$

18. Step 1: Write the ionization chemistry for cysteine.

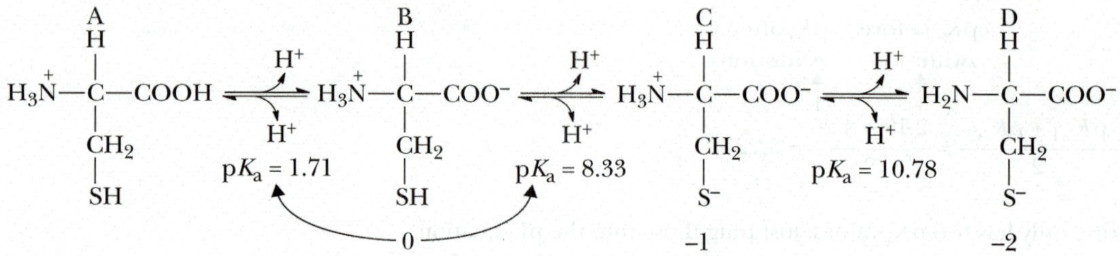

Step 2: Calculate pI.

$$pI = \frac{1.71 + 8.33}{2} = 5.02 \longrightarrow \text{pH at which cysteine has no net charge}$$

Step 3: Draw the titration curve.

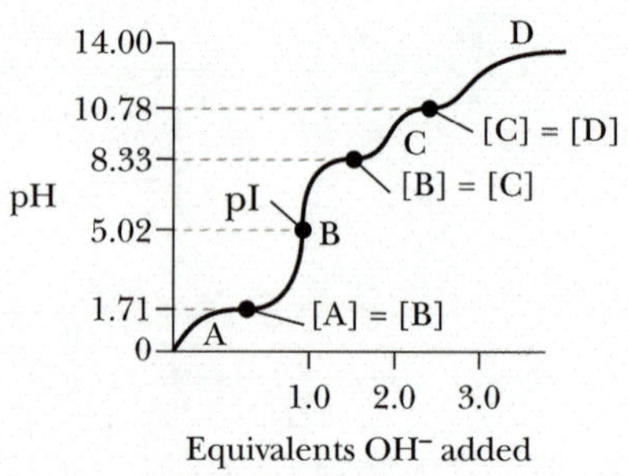

19. The conjugate acid–base pair acts as a buffer in the pH range 1.09–3.09.

20. They have a net charge at pH extremes, and the molecules tend to repel each other. When the molecular charge is zero, the amino acids can aggregate more easily.

21. The ionic dissociation reactions of the amino acids aspartic acid, valine, histidine, serine, and lysine are as follows:

Aspartic acid

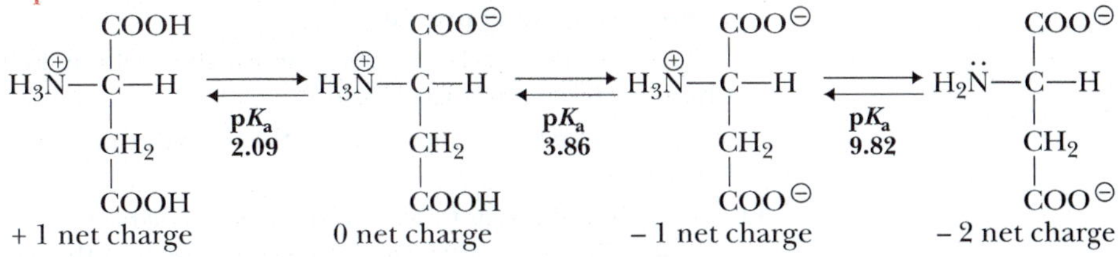

Valine

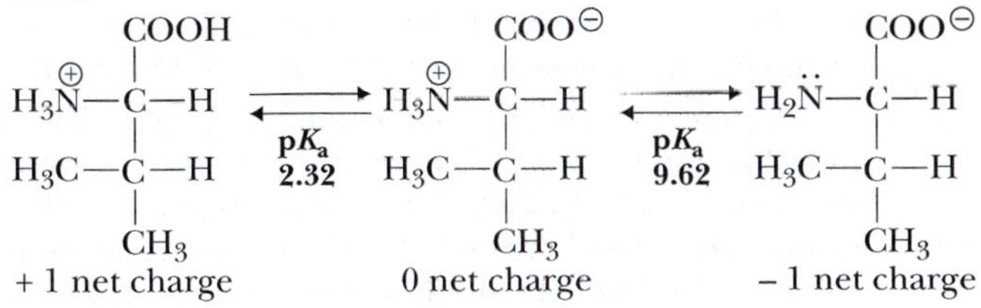

Histidine

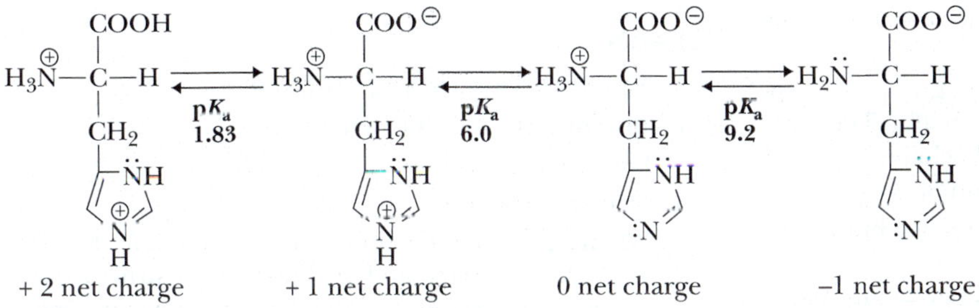

Serine

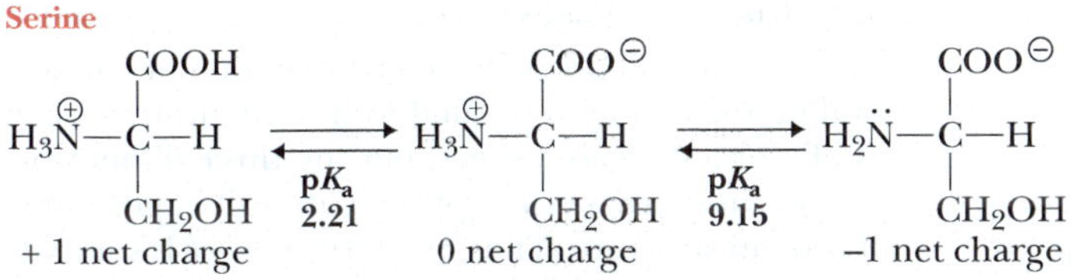

Lysine

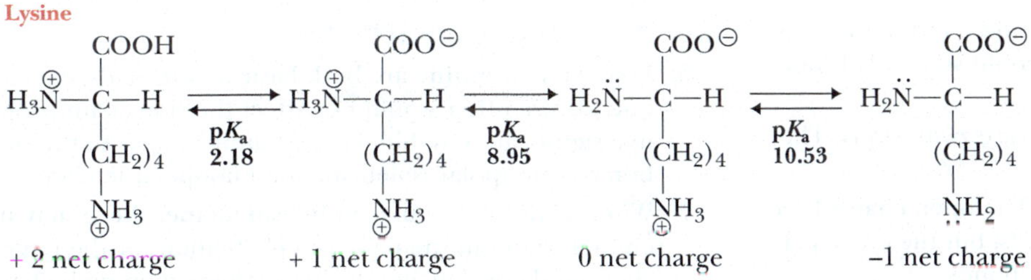

22. The pK_a for the ionization of the thiol group of cysteine is 8.33, so this amino acid could serve as a buffer in the —SH and S^{2-} forms over the pH range 7.33–9.33. The α-amino groups of asparagine and lysine have pK_a values of 8.80 and 8.95, respectively; these are also possible buffers, but they are both near the end of their buffer ranges.

23. At pH 1, the charged groups are the N-terminal NH_3^+ on valine and the protonated guanidino group on arginine, giving a net charge of +2. The charged groups at pH 7 are the same as those at pH 1, with the addition of the carboxylate group on the C-terminal leucine, giving a net charge of +1.

24. Both peptides, Phe—Glu—Ser—Met and Val—Trp—Cys—Leu, have a charge of +1 at pH 1 because of the protonated N-terminal amino group. At pH 7, the peptide on the right has no net charge because of the protonated N-terminal amino group and the ionized C-terminal carboxylate negative charge. The peptide on the left has a net charge of –1 at pH 7 because of the side-chain carboxylate group on the glutamate in addition to the charges on the N-terminal and C-terminal groups.

25. (a) Lysine because of the side-chain amino group.
(b) Serine because of the lack of a side-chain carboxyl.

26. The useful buffer ranges for glycine are pH 1.3 to 3.3 (carboxyl group) and pH 8.6–10.6 (amino group).
27. Part A:

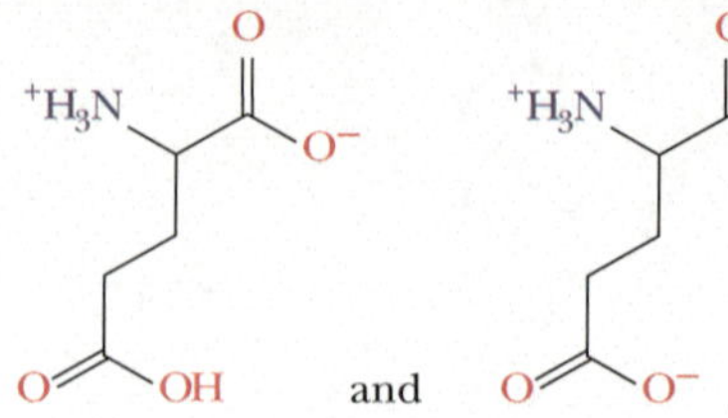

Part B: The average charge would be –0.5.
28. Part A: 4.25

Part B: The second inflection point on the titration curve.

3-4 The Peptide Bond

29. See Figure 3.12.
30. The resonance structures contribute to the planar arrangement by giving the CON bond partial double-bond character.
31. The peptide group would still be planar because the atoms that form the bond are the same.
32. The free carboxyl group and the free amino group of a peptide are both titratable and could serve as a buffer. The same is true for any titratable groups on the side chains. It would not necessarily be a particularly effective buffer.
33. The two peptides differ in amino acid sequence but not in composition.
34. Asp—Leu—Phe; Leu—Asp—Phe; Phe—Asp—Leu; Asp—Phe—Leu; Leu—Phe—Asp; Phe—Leu—Asp.
35. DLF; LDF; FDL; DFL; LFD; FLD.
36. There are two possible products: alanylglycine (free amino group on alanine, free carboxyl on glycine) and glycylalanine (free amino group on glycine, free carboxyl group on alanine).
37. Side chains are not directly bonded to any of the atoms of the peptide bond.
38. Side chains do not enter into the peptide bond themselves, but large groups with chiral centers could sterically hinder bond formation.
39. They are relatively stable because they are zwitterions. They typically have high melting points.
40. These two peptides differ chemically. The open chains have charged C-terminal and N-terminal ends, but the cyclic peptide has no charged ends, only peptide bonds.
41. Carbohydrates are *not* a source of the nitrogen needed for the biosynthesis of amino acids.
42. Suggest that your friend show the carboxyl group as a charged carboxylate (—COO^-) and the amino group in its charged form (—NH_3^+).
43. Very few side chains have functional groups to form cross-links.
44. Many more conformations would be possible because of free rotation around the peptide bond.

3-5 Small Peptides with Physiological Activity

45. Oxytocin has an isoleucine at position 3 and a leucine at position 8. Vasopressin has a phenylalanine at position 3 and an arginine at position 8.
46. Oxytocin stimulates smooth muscle contraction in the uterus during labor and in the mammary glands during lactation. Vasopressin stimulates resorption of water by the kidneys, thus raising blood pressure.
47. The disulfide bond is responsible for the cyclic structure of both oxytocin and vasopressin.
48. Peptide bonds double back on themselves to form a cyclic structure.
49. Oxytocin is thought to be involved in social behaviors. It affects personality, trust, altruism, social bonding, and the ability to recognize and understand the facial expressions and feelings of others. It influences female social and sexual behaviors, including intercourse, maternal attachment, and pair bonding.
50. In nonhumans, studies have included infusing the hormones into the brains of animals to see the alteration of behaviors. Receptors can be blocked with chemicals to gauge the effect of the loss of the hormone. With humans, nasal inhalation of hormones has been used, followed by studies and tests to judge behavior modifications.

Chapter 4

4-1 Protein Structure and Function

1. (a) (iii); (b) (i); (c) (iv); (d) (ii).
2. Quaternary structure
3. Primary structure
4. When a protein is denatured, the interactions that determine secondary, tertiary, and any quaternary structures are overcome by the presence of the denaturing agent. Only the primary structure remains intact.
5. The random coil portions of a protein do not contain structural motifs that are repeated within the protein, such as an α-helix or a β-pleated sheet, but the three-dimensional features of these parts of the protein are still important to overall protein structure. Thus, the term *random* is somewhat of a misnomer.

4-2 Primary Structure of Proteins

6. Lysine and arginine are both basic amino acids; exchanging one for the other would not affect the side chain properties in a significant way. Similar reasoning applies to the substitution of a nonpolar isoleucine for a nonpolar leucine.
7. When alanine is replaced by isoleucine, there may not be enough room in the native conformation for the larger side chain of the isoleucine. If there is a great enough change in the conformation of the protein, then it could lose its activity. When glycine is substituted in turn for isoleucine, the presence of the smaller side chain could restore the active conformation.

4-3 Secondary Structure of Proteins

8. Shape, solubility, and type of biological function (static, structural versus dynamic, catalytic).
9. Hydrogen bonding interactions between atoms of the peptide backbone
10. The angles of the amide planes as they rotate about the α-carbon. The angles are both defined as zero when the two planes would be overlapping such that the carbonyl group of one contacts the N—H of the other.

11. A β-bulge is a common nonrepetitive irregularity found in antiparallel β-sheets. A misalignment occurs between strands of the β-sheet, causing one side to bow outward.
12. A reverse turn is a region of a polypeptide where the direction changes by about 180°. There are two kinds—those that contain proline and those that do not. See Figure 4.7 for examples.
13. The α-helix is not fully extended, and its hydrogen bonds are parallel to the helix. The β-sheet structure is almost fully extended, and its hydrogen bonds are perpendicular to the protein chains in the sheet.
14. The αα unit, the βαβ unit, the β-meander, the Greek key, and the β-barrel.
15. The geometry of the proline residue is such that it does not fit into the α-helices of these proteins, but it does fit exactly in a reverse turn. See Figure 4.7.
16. Glycine is the only residue small enough to fit at crucial points in the collagen triple helix.
17. Wool, which consists largely of the protein keratin, shrinks because of its α-helical conformation. It can stretch and then shrink. Silk consists largely of the protein fibroin, which has the fully extended β-sheet conformation, with far less tendency to stretch or shrink.

4-4 Tertiary Structure of Proteins

18. See Figure 4.2 for a hydrogen bond that is part of an α-helix (secondary structure). See Figure 4.13 for a hydrogen bond that is part of tertiary structure (side-chain hydrogen bonding).
19. See Figure 4.13 for electrostatic interactions such as might be seen between the side chains of lysine and aspartate.
20. See Figure 4.13 for an example of a disulfide bond.
21. See Figure 4.13 for an example of hydrophobic bonds.
22. Fibrous proteins
23. Collagen
24. (c) They are usually involved in structural functions.
25. Primary structure, secondary structure, and tertiary structure
26. (c) The E7 histidine residue
27. Technically, collagen has quaternary structure because it has multiple polypeptide chains. However, most discussions of quaternary structure involve subunits of globular proteins, not fibrous ones like collagen. Many scientists consider the collagen triple helix to be an example of a secondary structure.

4-5 Quaternary Structure of Proteins

28. *Similarities*: Both contain a heme group; both are oxygen binding; secondary structure is primarily α-helix. *Differences*: Hemoglobin is a tetramer, while myoglobin is a monomer; oxygen binding to hemoglobin is cooperative but noncooperative in myoglobin.
29. Quaternary structure.
30. The crucial residues are histidines in both proteins.
31. Myoglobin's highest level of organization is tertiary. Hemoglobin's is quaternary.
32. The function of hemoglobin is oxygen transport; its sigmoidal binding curve reflects the fact that it can bind easily to oxygen at comparatively high pressures and release oxygen at lower pressures. The function of myoglobin is oxygen storage; as a result, it is easily saturated with oxygen at low pressures, as shown by its hyperbolic binding curve.
33. In the presence of H^+ and CO_2, both of which bind to hemoglobin, the oxygen-binding capacity of hemoglobin decreases.
34. In the absence of 2,3-bisphosphoglycerate, the binding of oxygen by hemoglobin resembles that of myoglobin, characterized by a lack of cooperativity. 2,3-Bisphosphoglycerate binds at the center of the hemoglobin molecule, increases cooperativity, stabilizes the deoxy conformation of hemoglobin, and modulates the binding of oxygen so that it can be easily released in the capillaries.
35. Fetal hemoglobin binds oxygen more strongly than adult hemoglobin. See Figure 4.29.
36. Histidine 143 in the beta chain is replaced by a serine in the gamma chain.
37. Deoxygenated hemoglobin is a weaker acid (has a higher pK_a) than oxygenated hemoglobin. In other words, deoxygenated hemoglobin binds more strongly to H^+ than does oxygenated hemoglobin. The binding of H^+ (and of CO_2) to hemoglobin favors the change in quaternary structure to the deoxygenated form of hemoglobin.
38. The change of a histidine to a serine in the gamma chains of Hb F removes a positively charged amino acid that could have interacted with BPG. Thus, fetal hemoglobin does *not* bind BPG as strongly and can therefore bind oxygen more strongly, leading to oxygen transfer from maternal hemoglobin.
39. In fetal hemoglobin, the subunit composition is $\alpha_2\gamma_2$ with replacement of the beta chains by the gamma chains. The sickle-cell mutation affects the beta chain, so a fetus homozygous for Hb S has normal fetal hemoglobin.
40. Because people with sickle-cell disease are chronically anemic, some cells with fetal Hb are produced to help overcome the impaired oxygen delivery system.
41. On the beta chain of Hb S, there is a nonpolar valine at position 6, where there is a polar glutamic acid in the normal form.
42. Valine is a hydrophobic amino acid. The valine is on the outside of the globular protein, where the amino acids have to interact with water. The presence of the hydrophobic amino acid at that position causes hemoglobins to clump together via hydrophobic interactions. This clumping causes the cell deformation.
43. In the heterozygous genotype, it appears that there is an advantage in areas prone to malaria.
44. Hydroxyurea stimulates the bone marrow to produce fetal hemoglobin. Fetal hemoglobin has no β-chain, so the effect of the interactions of those chains is reduced in the presence of increased Hb F.

4-6 Protein Folding Dynamics

45. Protein folding is driven by many processes. The intuitive ones are the direct interactions of functional groups through covalent bonds, electrostatic attractions, and hydrogen bonding. These explain why parts of the protein are attracted to each other and why a protein would tend to adopt a shape that makes these interactions possible. However, much of the protein-folding process is also driven by an entropy effect. We refer to hydrophobic interactions as an explanation of why nonpolar

regions of the protein tend to cluster together, usually in the interior of the protein. However, in reality, it is not the interaction of nonpolar amino acids that drives this process. It is actually the increase in entropy of the solvent, water. When the hydrophobic regions of the protein are isolated to the interior, the water molecules surrounding the protein are freer to rotate and move in less restricted ways. Thus, what drives much of protein folding is not a ΔH change with the bonding of specific amino acids but rather a ΔS increase of the solvent.

46. The formation of ice-like hydrate "cages" around nonpolar solute molecules, leading to a loss of entropy.
47. A molecular chaperone is a protein that aids another protein in folding correctly and keeps it from associating with other proteins before it has reached its final, mature form.
48. A prion is a potentially infectious protein found in multiple forms in mammals, often concentrated in nervous tissue. It is an abnormal form of a normal cellular protein. It tends to form plaques that destroy nervous tissue. Prions have been found to be transmissible across species.
49. The normal form of the prion protein has a higher α-helix content compared to the β-sheet content. The abnormal one has an increased β-sheet content.
50. Alzheimer's, Parkinson's, and Huntington's diseases are caused by accumulation of protein deposits from aggregates caused by misfolded proteins. Chapter 4 also looked at prion diseases. When prions are misfolded, they can cause spongiform encephalopathies, such as mad cow disease, and the human form, Creutzfeldt-Jakob disease.
51. Protein aggregates form when there are exposed areas on a protein surface that are nonpolar. Proteins then stick together via these nonpolar regions, causing the aggregates. An example is prion disease, in which an area of the normal molecule that should be an α-helix adopts a β-sheet conformation instead.
52. The spongiform encephalopathies that we know of have characteristics of both inherited diseases and transmissible diseases. On the one hand, animals can be infected by consuming meat or other tissues that are themselves carrying mutant prion proteins. As the example of the New Zealand sheep showed, even those that are very susceptible to a prion disease can remain disease-free if they are never exposed. However, the predisposition to acquire a prion disease has a hereditary component as well. The prion protein has many known mutations, some of which render the individual very susceptible to the disease. These mutations can be tracked, and they are passed along family lines.
53. Although there may be a strong genetic predisposition to acquire scrapie, that alone will not cause the disease. The disease must be started by ingesting a prion that already has the altered conformation, PrPsc.

Chapter 5

5-1 Extracting Pure Proteins from Cells

1. Using a blender, a Potter–Elvejhem homogenizer, or a sonicator.
2. If you needed to maintain the structural integrity of the subcellular organelles, a Potter–Elvejhem homogenizer would be better because it is gentler. The tissue, such as liver, must be soft enough to use with this device.
3. Salting out is a process whereby a high concentration of salt is used to reduce the solubility of a protein until it comes out of solution and can be centrifuged. The salt forms ion–dipole bonds with the water in the solution, which leaves less water available to hydrate the protein. Nonpolar side chains begin to interact between protein molecules, and they become insoluble.
4. Their amino acid content and arrangements make some proteins more soluble than others. A protein with more highly polar amino acids on the surface is more soluble than one with more hydrophobic ones on the surface.
5. First, homogenize the liver cells using a Potter–Elvejhem homogenizer. Then spin the homogenate at 500 × *g* to sediment the unbroken cells and nuclei. Centrifuge the supernatant at 15,000 × *g* and collect the pellet, which contains the mitochondria.
6. No, peroxisomes and mitochondria have overlapping sedimentation characteristics. Other techniques, such as sucrose-gradient centrifugation, would have to be used to separate the two organelles.
7. If the protein were cytosolic, then once the cells were broken open, you could centrifuge at 100,000 × *g*, and all the organelles would be in the pellet. Your enzyme would be in the supernatant, along with all the other cytosolic ones.
8. Isolate the mitochondria via differential or sucrose-gradient centrifugation. Use another homogenization technique, combined with a strong detergent, to release the enzyme from the membrane.
9. (a)

1. Purification Step	2. Total Activity of Enzyme (micromoles product/min)	3. Total Protein Concentration of Sample (mg/mL)	4. Specific Activity (micromole product/(min mg protein))	5. Fold Purification
Crude homogenate	150,000	56	2679	NA
Salt precipitation	76,000	25	3040	1.1
Ion exchange chromatography	58,000	8.4	6905	2.3
Gel filtration chromatography	37,000	3.2	11563	1.7
Affinity chromatography	29,000	0.4	72500	6.3

(b) Affinity Chromatography

(c) Salt precipitation

10. 57%

5-2 Column Chromatography

11. (a) Size

(b) Specific ligand-binding ability

(c) Net charge

(d) Polarity or hydrophobicity

12. The largest proteins elute first; the smallest elute last. Larger proteins are excluded from the interior of the gel bead, so they have less available column volume to travel. Essentially, they travel a shorter distance through a smaller volume and elute first.

13. A compound can be eluted by raising the salt concentration or by adding a mobile ligand that has a higher affinity for the bound protein than the stationary resin ligand does. Salt is cheaper but less specific. A specific ligand may be more specific, but it is likely to be expensive.
14. A compound can be eluted by raising the salt concentration or by changing the pH. Salt is cheap, but it might not be as specific for a particular protein. Changing the pH may be more specific for a tight pI range, but extremes of pH may also denature the protein.
15. The basis of most resins is agarose, cellulose, dextran, or polyacrylamide.
16. See Figure 5.7.
17. Within the fractionation range of a gel-filtration column, molecules elute with a linear relationship of log MW versus their elution volumes. A series of standards can be run to standardize the column, and then an unknown can be determined by measuring its elution volume and comparing it to a standard curve.
18. Both proteins would elute in the void volume together and would not be separated.
19. Yes, the β-amylase would come out in the void volume, but the bovine serum albumin would enter the pores in the column beads and would elute more slowly.
20. In most chromatography systems, the ligands and solvents are polar. In reverse-phase HPLC, a solution of nonpolar compounds is put through a column that has a nonpolar liquid immobilized on an inert matrix. A more polar liquid serves as the mobile phase and is passed over the matrix. The solute molecules are eluted in proportion to their solubility in the more polar liquid.
21. Ion-exchange chromatography is a specific type of separation based on the net charge of the molecules being separated. The term HPLC refers to chromatography procedures carried out under high pressure, but the basis of the separation could be ion-exchange, gel-filtration, reverse-phase, or affinity chromatography.
22. Set up an anion-exchange column, such as Q-sepharose (quaternary amine). Run the column at pH 8.5, a pH at which the protein X has a net negative charge (proteins are positively charged at pH values below their pI and negatively charged at pH values above their pI). Adjust the pH of the homogenate containing protein X to 8.5, and apply it to the column. Protein X will bind to the column. Then elute by running a salt gradient.
23. Use a cation-exchange column, such as CM-sepharose, and run it at pH 6. Protein X will have a positive charge and will stick to the column.
24. Glutamic acid will be have a net negative charge and will elute first because it will be repelled by the cation-exchange column, which is negatively charged. Leucine will elute next because it will have a net charge of 0 at pH 3.5, and it will not be attracted or repelled by the column to the extent of the other amino acids in the sample. Lysine will be positively charged and will stick to the column. To elute lysine, raise the pH to around 11.
25. The polar glutamic acid will move fastest and elute first because the stationary phase is nonpolar in reverse-phase HPLC. Glycine and phenylalanine are both nonpolar amino acids. However, glycine is smaller and contains less nonpolar surface, so it will elute next, followed by phenylalanine.
26. A protein solution from an ammonium sulfate preparation is passed over a gel-filtration column, where the proteins of interest will elute in the void volume. The salt, being very small, will move through the column slowly. In this way, the proteins will leave the salt behind and exit the column without it.
27.

Pair	Answer	Feedback
A and B	Concanavalin A-agarose: Protein A would be retained the longest.	Concanavalin A-agarose is an affinity resin that would retain protein A (the glycoprotein) on the column while protein B is eluted.
A and C	Concanavalin A-agarose: Protein A would be retained the longest. Sephadex G-75: Protein C would be retained the longest.	Because the two proteins differ significantly in their size as well as their surface appearance (glycoprotein), either Concanavalin A-agarose or Sephadex G-75 should work to separate them. In the case of Concanavalin A-agarose, protein A would be retained on the column. In the case of Sephadex G-75, the smaller protein would be retained the longest.
B and D	Sephadex G-75: Protein D would be retained the longest. Carboxymethyl (CM) cellulose (pH 8.0): Protein D would be retained the longest. Diethylaminoethyl (DEAE) cellulose (pH 8.0): Protein B would be retained the longest.	Because the two proteins differ significantly in their size and pI value, either gel filtration or ion-exchange chromatography would work to separate them. In the case of Sephadex G-75, the smaller protein would be retained the longest, and in the case of CM-Cellulose (cation-exchange resin) at pH 8.0, Protein D would have a net positive charge and be retained the longest. Using diethylaminoethyl (DEAE) cellulose (anion-exchange resin) at pH 8.0, protein B would have a net negative charge and would be retained the longest.

28. (a) Iminodiacetate-agarose
    (b) Increase the imidazole concentration of the mobile phase

5-3 Electrophoresis

29. Size, shape, and charge.
30. Agarose and polyacrylamide.
31. DNA is the molecule most often separated on agarose electrophoresis, although proteins can also be separated.
32. Those with the highest charge/mass ratio would move the fastest. There are three variables to consider, and most electrophoreses are done in a way to eliminate two of the variables so that the separation is by size or by charge, but not by both.

33. Sodium dodecyl sulfate polyacrylamide gel electrophoresis (SDS–PAGE). With SDS–PAGE, the charge and shape differences of proteins are eliminated so that the only parameter determining the migration is the size of the protein.
34. SDS binds to the protein in a constant ratio of 1.4 g SDS per gram of protein. It coats the protein with negative charges and puts it into a random coil shape. Thus, charge and shape differences among proteins in the sample are eliminated.
35. (a) different pI values
36. In a polyacrylamide gel used for gel-filtration chromatography, the larger proteins can travel around the beads, thereby having a shorter path to travel and therefore eluting faster. With electrophoresis, the proteins are forced to go through the matrix, so the larger ones travel more slowly because there is more friction.
37. The MW is 33,000 Da.
38. (b) A sample of DNA
39. (e) The 84 kDa protein is a heterotrimer.

5-4 Determining the Primary Structure of a Protein

40.

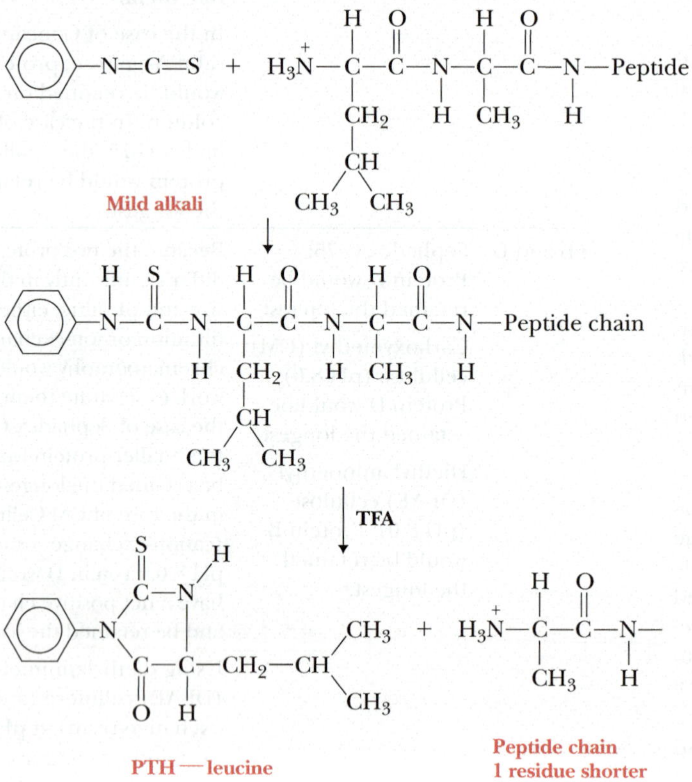

41. Val—Leu—Gly—Met—Ser—Arg—Asn—Thr—Trp—Met—Ile—Lys—Gly—Tyr—Met—Gln—Phe.
42. Met—Val—Ser—Thr—Lys—Leu—Phe—Asn—Glu—Ser—Arg—Val—Ile—Trp—Thr—Leu—Met—Ile.
43. It is possible that your protein is not pure and needs additional purification steps to arrive at a single polypeptide. It is also possible that the protein has subunits, so multiple polypeptide chains could be yielding the contradictory results.
44. There are two fragments that have C-termini that are not lysine or arginine, which is what trypsin is specific for. Normally, there would be only one fragment ending with an amino acid that was not Arg or Lys, and we would immediately know that it was the C-terminus. Histidine is a basic amino acid, although it is usually neutral and therefore does not react with trypsin. It is possible that, in the pH environment of the reaction, the histidine was positively charged and was recognized by trypsin.
45. It would tell you a relative concentration of the various amino acids. This is important because it would help you plan your sequencing experiment better. For example, if you had a protein whose composition showed no aromatic amino acids, it would be a waste of time to use a chymotrypsin digestion.
46. Cyanogen bromide would be useless because there is no methionine. Trypsin would be little better because the protein is 35% basic residues. Trypsin would shred the protein into more than 30 pieces, which would be very hard to analyze.
47. Chymotrypsin would be a good choice. There are more than four residues of aromatic amino acids. The protein, containing 100 amino acids, would be cut four times, possibly yielding nice fragments roughly 20–30 amino acids long, which can be sequenced effectively by the Edman degradation.

5-5 Proteomics

48. Electrospray ionization mass spectrometry (ESI-MS) and matrix-assisted laser desorption ionization-time of flight mass spectrometry (MALDI-TOF MS).
49. MALDI-TOF MS is very sensitive and very accurate. Attomole (10^{-18}) quantities of a molecule can be detected.
50. ELISA is based on antibody–protein interactions. Specific antibodies, called primary antibodies, are put into microtiter plates to localize target proteins. A secondary antibody carrying some type of tag to render it visible is also added. If the target protein is there, then the combination of the protein–primary antibody–secondary antibody will be visible.
51. A primary antibody is specific for a target protein that a researcher is looking for. A secondary antibody will react with the primary antibody. The secondary antibody carries the tag that makes the complex visible.
52. The protein–antibody complexes can be seen based on the nature of the tag carried by the secondary antibody. This tag may be an enzyme that produces a visible color when provided with its substrates, a fluorescent marker, or a radioactive compound.
53. The first step in a Western blot is the separation of proteins via electrophoresis. The next step takes the gel from the electrophoresis and transfers the proteins from the gel onto a thin membrane of nitrocellulose or another absorbing compound. Once the proteins are transferred, they are incubated with the primary antibody. Then they are incubated with the secondary antibody. Lastly, the bands are made visible by reacting with the substrates for the secondary antibody enzyme tag, or they are visualized with a fluorometer or X-ray paper.
54. Western blot got its name as a humorous derivation from the original blotting technique called Southern blotting. The original blotting technique was for DNA and was developed by a researcher named Southern, so they called the technique the Southern blot. The next type of molecule to be blotted was RNA, so to distinguish it from a Southern blot, it

was called a northern blot. After that, the technique for blotting protein was developed and called a western blot.

55. Proteins are transferred to nitrocellulose because all of the protein ends up layered on top of a very thin membrane. This means that small volumes of the antibody solutions can be used to bind to the proteins. These antibodies are very expensive, so the less used, the better. Also, because the original proteins were embedded in a gel, if the gel were reacted directly with the antibodies, the antibodies would not have easy access to the gel because they would not percolate easily through the pores of the gel without the benefit of an electric current to push them.
56. There are thousands of primary antibodies that can be purchased commercially. Other research requires that a new primary antibody be created. The process of attaching an enzyme, a fluorescent marker, or a radioactive compound is a long and difficult task. If every primary antibody had to be tagged, it would be a daunting exercise for the company trying to do it. Instead, a company can specialize in tagging a secondary antibody directed against goat, rabbit, or mouse antibodies. These secondary antibodies can then be used for any experiment in which the primary antibody came from one of these animals. Also, there is a signal multiplication effect of using a primary antibody because several secondary antibodies will attach to a single molecule of the primary antibody.
57. Proteomics is the systematic analysis of an organism's complete complement of proteins, or its proteome. The technology now available has allowed scientists to describe all the DNA of an organism as its genome, all of the RNA as its transcriptome, and all of the proteins produced as its proteome. To understand the flux of proteins in a cell is to understand its metabolism.

Chapter 6

6-1 Enzyme Kinetics vs Thermodynamics

1. Enzymes are many orders of magnitude more effective as catalysts than are nonenzymatic catalysts.
2. The reaction of glucose with oxygen is thermodynamically favored, as shown by the negative Gibbs energy change. The fact that glucose can be maintained in an oxygen atmosphere reflects the slow kinetics of the reaction in the absence of catalysts, requiring the overcoming of a large activation-energy barrier.
3. Heating a protein or RNA molecule denatures it. Enzymatic activity depends on the correct three-dimensional structure of the protein or RNA, which is determined by noncovalent bonds that will be broken by heat.
4. The standard Gibbs energy change is determined by the stability of the products and reactants, which remains the same on the energy profiles in the presence and absence of a catalyst. As the catalyst only stabilizes the transition state, the Gibbs energy of the products and reactants will be unchanged by the presence of a catalyst.
5. The presence of a catalyst lowers the activation energy of a reaction in both directions. This is because the catalyst stabilizes the transition state, which will decrease the difference between the energy of that transition state and the reactants in either direction.
6. The amount of product obtained in a reaction depends on the equilibrium constant. A catalyst does not affect that. A catalyst does decrease the amount of time needed to reach equilibrium to produce the product but not the total amount of product produced.

6-2 Rate of Enzyme-Catalyzed Reactions

7. The reaction is first order with respect to A, first order with respect to B, and second order overall. Recall that to determine the overall order of the reaction, you sum the exponents for [A] and [B], which are both 1 in this hypothetical reaction.
8. "Zero" or "0" or "zero order."
9. The easiest way to follow the rate of this reaction is to monitor the decrease in absorbance at 340 nm, reflecting the disappearance of NADH. The average rate equation for the disappearance of the reactant, NADH, will be equal to the average rate of the appearance of the undetected product.
10. As the lactate dehydrogenase reaction uses hydrogen ions, the pH would increase as the reaction proceeds, which could affect the structure of the enzyme and reduce enzyme efficiency. This could lead to a decrease in the reaction rate measured compared with the rates in a buffered solution.

6-3 Enzyme–Substrate Binding

11. In the lock-and-key model, the enzyme has a rigid active site that is shaped to bind the substrate even in the absence of substrate binding. In the induced-fit model, the enzyme active site and substrate undergo conformational changes when they bind.
12. A perfect fit of the ES complex would stabilize the enzyme and the substrate, further decreasing its energy on the energy diagram, which would increase the activation energy required to reach the transition state.

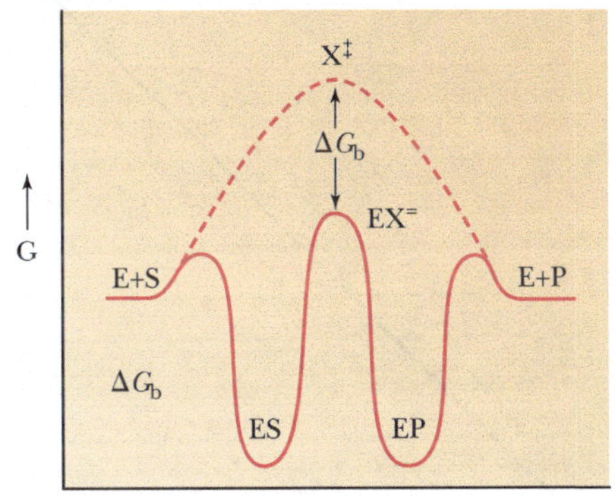

No destabilization,
thus no catalysis

13. Blank 1: Transition state; Blank 2: EX complex
14. The residues are likely all in the active site because they are important for catalysis. This is possible because amino acids that are far apart in the primary structure can be close to each other in the three-dimensional tertiary structure because of protein folding. Considering the induced-fit model, the residues could be in the active site or could be critical for the protein conformational changes needed for catalysis.

6-4 The Michaelis–Menten Approach to Enzyme Kinetics

15. A Lineweaver–Burk plot is useful because it gives a straight line for the kinetics data. See Section 6-4B for more reasoning.
16. Increasing enzyme concentration will increase the observed V_{max}, while K_M will remain unchanged by enzyme

concentration. This is because more enzyme means higher concentrations of active sites producing product under conditions of saturating substrate concentrations.

17. Turnover number $= V_{max}/[E]_T$.
18. See graph: $V_{max} = 0.681\,\mathrm{m}M\,\mathrm{min}^{-1}$, $K_M = 0.421\,M$.

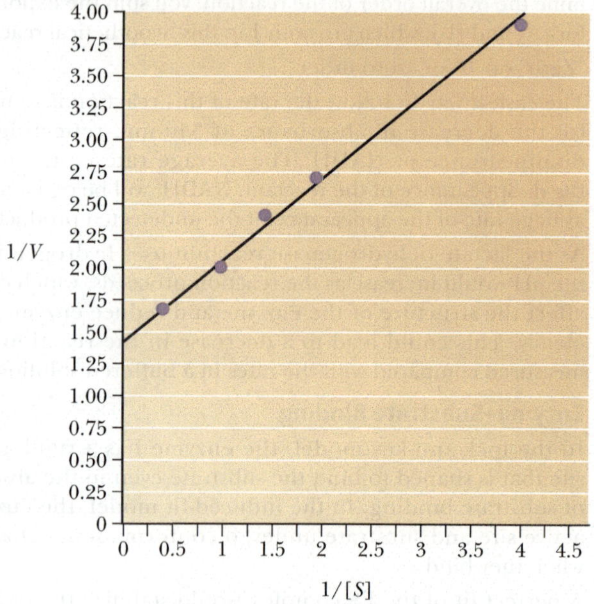

19. See graph: $V_{max} = 2.5 \times 10^{-4}\,M\,\mathrm{s}^{-1}$, $K_M = 10\,\mathrm{m}M$.

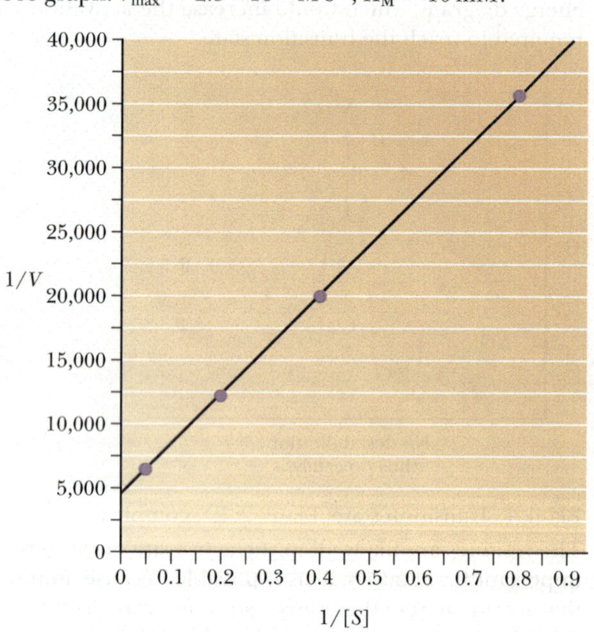

20. The turnover number is $2.5\,\mathrm{s}^{-1}$.
21. The enzyme concentration is $1.56\,\mu M$. The turnover number is $64{,}102\,\mathrm{min}^{-1}$.
22. The low K_M for the aromatic amino acids indicates that they will be oxidized preferentially.
23. Z would work best at high [substrate] because, at high substrate concentrations, V_0 is indistinguishable from V_{max}.
24. Under conditions of nonsaturating substrate concentration.
25. Ordered, random, and ping-pong.
26. With a ping-pong mechanism, one product is released prior to the binding of the second substrate. With ordered and random mechanisms, both substrates are bound before any product is released.
27. Whenever there are multiple substrates, the trick to determining the K_M of one of them is to run the reaction with saturating concentrations of the other.

6-5 Examples of Enzyme-Catalyzed Reactions

28. See Figures 6.11 and 6.12.
29. Not all enzymes follow Michaelis–Menten kinetics. The kinetic behavior of allosteric enzymes does not obey the Michaelis–Menten equation.
30. The graph of initial velocity against substrate concentration is sigmoidal for an allosteric enzyme but hyperbolic for an enzyme that obeys the Michaelis–Menten equation.
31. Recall the behavior of hemoglobin binding to oxygen; you can think of enzymes similarly. Enzymes that exhibit cooperativity have multiple subunits that can influence each other. Most enzymes that are cooperative exhibit positive cooperativity, which means that the binding of substrate to one subunit makes it easier to bind the substrate to another subunit.
32. The enzyme is behaving allosterically, so it likely has more than one subunit because it must contain more than one substrate-binding site.

6-6 Enzyme Inhibition

33. K_M increases in the presence of a competitive inhibitor, while the value of K_M remains unchanged in noncompetitive inhibition.
34. A competitive inhibitor blocks binding, not catalysis. V_{max} is determined at infinite [S], and at these substrate concentrations, the substrate will outcompete the competitive inhibitor.
35. A noncompetitive inhibitor does not change the affinity of the enzyme for its substrate. The noncompetitive inhibitor binds to an allosteric site, so substrate binding is unchanged.
36. A competitive inhibitor binds to the active site of an enzyme, preventing the binding of the substrate. A noncompetitive inhibitor binds at a site different from the active site and instead affects the catalytic efficiency of the enzyme.
37. Irreversible inhibitors bind to the enzyme covalently and remain attached perpetually. Competitive inhibitors can be reversed by increasing substrate concentration or decreasing the concentration of the inhibitor.
38. In a Lineweaver–Burk plot for competitive inhibition, the lines intersect at the y-axis intercept, which is equal to $1/V_{max}$. In a Lineweaver–Burk plot for noncompetitive inhibition, the lines intersect at the x-axis intercept, which is equal to $-1/K_M$.
39. With pure noncompetitive inhibition, the binding of the inhibitor does not change the affinity of the enzyme for the substrate at all and vice versa; thus, the K_M does not change. With mixed inhibition, the substrate and inhibitor do affect the binding of each other such that the K_M for the substrate is different in the presence of the inhibitor.
40. Because the inhibitor can bind to E or to ES equally well, any time there is inhibitor present, some of the enzyme will be tied up in the EIS form, which does not lead to catalysis. For this reason, it would appear that less enzyme is present.

41. The inhibitor now described would be an uncompetitive inhibitor, and the plot would resemble Figure 6.18.
42. The binding of the inhibitor to the ES complex to form EIS removes some of the ES. By Le Châtelier's principle, this will tend to force the reaction to the right, forming more ES. By stimulating the binding of E and S in this manner, the graph will show that the K_M is reduced.
43. This is a noncompetitive inhibitor because K_M is unchanged and V_{max} decreases.
44. Hydroxymethylaspartate is a competitive inhibitor because V_{max} is the same with and without the inhibitor within the limits of error. This means the mechanism is to bind to the aspartase active site and inhibit substrate binding.
45. Both the slope and the intercepts will change. The lines will intersect above the *x*-axis at negative values of $1/[S]$. This is mixed noncompetitive inhibition.
46. An irreversible inhibitor is bound by covalent bonds. Noncovalent interactions are relatively weak and easily broken.
47. A noncompetitive inhibitor does not bind to the active site of an enzyme. Its structure needs to bear no relation to that of the substrate.
48. If A, the noncompetitive inhibitor, binds to an allosteric site, then its ability to inhibit will *not* be affected by high concentrations of substrate. The competitive inhibitor (B), on the other hand, will be outcompeted by high substrate concentrations for the shared binding residues at the active site.
49. "Competitive inhibitor" or "competitive."

Chapter 7

7-1 Behavior of Allosteric Enzymes

1. Allosteric enzymes display sigmoidal kinetics when initial reaction velocity is plotted versus substrate concentration. Michaelis–Menten enzymes exhibit hyperbolic kinetics. Allosteric enzymes usually have multiple subunits, and the binding of substrates or effector molecules to one subunit changes the binding behavior of the other subunits.
2. It is an enzyme used in the early stages of cytidine nucleotide synthesis.
3. ATP acts as a positive effector of ATCase, whereas CTP acts as an inhibitor.
4. The term K_M should be used for enzymes that display Michaelis–Menten kinetics. Thus, it is not used with allosteric enzymes. Technically, competitive and noncompetitive inhibition are also terms that are restricted to Michaelis–Menten enzymes, although the concepts are applicable to any enzyme. An inhibitor that binds to an allosteric enzyme at the same site as the substrate is similar to a classical competitive inhibitor. One that binds at a different site is similar to a noncompetitive inhibitor, but the equations and the graphs characteristic of competitive and noncompetitive inhibition do not work the same way with an allosteric enzyme.
5. A K system is an allosteric enzyme in which the binding of an inhibitor alters the apparent substrate concentration needed to reach one-half V_{max}, $K_{0.5}$.
6. A V system is an allosteric enzyme in which the binding of an inhibitor changes the V_{max} of the enzyme but not the $K_{0.5}$.
7. Homotropic effects are allosteric interactions that occur when several identical molecules are bound to a protein. The binding of substrate molecules to different sites on an enzyme, such as the binding of aspartate to ATCase, is an example of a homotropic effect. Heterotropic effects are allosteric interactions that occur when different substances (such as an inhibitor and a substrate) are bound to the protein. In the ATCase reaction, inhibition by CTP and activation by ATP are both heterotropic effects.
8. ATCase is made up of two different types of subunits. One of them is the catalytic subunit, and there are six of them organized into two trimers. The other is the regulatory subunit, which consists of six protein subunits organized into three dimers.
9. Enzymes that exhibit cooperativity do not show hyperbolic curves of rate versus substrate concentration. Their curves are sigmoidal. The level of cooperativity can be seen by the shape of the sigmoidal curve.
10. CTP makes the shape of the curve more sigmoidal and thus more cooperative.
11. ATP makes the shape of the curve less sigmoidal and thus less cooperative.
12. $K_{0.5}$ is the substrate concentration that leads to half of the maximal velocity. This term is used with allosteric enzymes, where the term K_M is inappropriate.
13. K_M–$K_{0.5}$, V_{max}–V_{max}, Competitive inhibition–K system, Noncompetitive inhibition–V system
14. V system; Homotropic effect; Allosteric inhibition

7-2 The Concerted and Sequential Models for Allosteric Enzymes

15. In the concerted model, all the subunits in an allosteric enzyme are found in the same form—either the T form or the R form. They are in equilibrium, with each enzyme having a characteristic ratio of T/R. In the sequential model, the subunits change individually from T to R.
16. The sequential model can explain negative cooperativity because a substrate binding to the T form could induce other subunits to switch to the T form, thereby reducing binding affinity.
17. Greater cooperativity is favored by having a higher ratio of the T/R form. It is also favored by having a higher dissociation constant for the substrate binding to the T form.
18. The L value is the equilibrium ratio of the T/R form. The *c* value is the ratio of the substrate dissociation constants for the two forms of the enzyme, such that $c = K_R/K_T$.
19. The sequential model is more likely, given that only the first subunit is in the high-affinity state and that the second subunit changes conformation upon substrate binding to the first subunit.

7-3 Control of Enzyme Activity by Phosphorylation

20. A protein kinase is an enzyme that phosphorylates a protein using a high-energy phosphate, such as ATP, as the phosphate donor.
21. Serine, threonine, and tyrosine are the three most often phosphorylated amino acids in proteins that are acted upon by kinases. Aspartate is another one that can be phosphorylated.

22. As part of the mechanism, the sodium–potassium pump has an aspartate residue that becomes phosphorylated. This phosphorylation alters the conformation of the enzyme and causes it to close on one side of the membrane and open on the other, moving ions in the process.
23. K system; the ATP analog will interact with the ATP-binding site in the active site.

7-4 Zymogens

24. Zymogens are often seen with digestive enzymes that are produced in one tissue and used in another. If the enzyme were active immediately upon production, it would digest other cellular proteins, where it would cause great damage. By having it produced as a zymogen, it can be safely made and then transported to the digestive tissue, such as the stomach or small intestine, where it can then be activated.
25. Zymogens are all activated by protein cleavage. Chymotrypsinogen is activated when trypsin cleaves it to form chymotrypsin.

7-5 The Nature of the Active Site

26. Ser 195 and His 57 are the two most critical amino acids in the active site of chymotrypsin.
27. The three-dimensional structure of the enzyme can be determined by X-ray crystallography so that you can visualize the arrangement and proximity of active-site residues. You can also identify active-site residues because they will be more reactive and thus more readily modified by chemical agents that can fit into the active site. If the critical active-site residues are chemically modified, then the enzyme will be inactive.
28. See Figure 7.21.
29. In the first step of the reaction, the serine hydroxyl is the nucleophile that attacks the substrate peptide bond. In the second step, water is the nucleophile that attacks the acyl-enzyme intermediate.
30. Histidine 57 performs a series of steps involving general base catalysis followed by general acid catalysis. In the first phase, it takes a hydrogen from serine 195, acting as a general base. This is followed immediately by an acid catalysis step, in which it gives the hydrogen to the amide group of the peptide bond that is breaking. A similar scheme takes place in the second phase of the reaction.
31. A pK_a near physiological pH ensures the residue can behave as both an acid and a base in different steps of the reaction. For example, histidine 57 is used in both the protonated and deprotonated forms during the chymotrypsin reaction.

7-6 Chemical Reactions Involved in Enzyme Mechanisms

32. They act as Lewis acids (electron-pair acceptors) and can take part in enzyme catalysis mechanisms. In carboxypeptidase A, the zinc ion accepts electrons from the penultimate carbonyl group of a protein substrate, thereby increasing the polarity of the carbonyl group and making it more susceptible to attack by a water molecule.
33. The carbon of a carbonyl group is often attacked by a nucleophile.
34. (1) Brønsted–Lowry acid, (2) Lewis acid, (3) Lewis base, (4) Brønsted–Lowry base.
35. General acid catalysis is the part of an enzyme mechanism in which an amino acid or other molecule donates a hydrogen ion to another molecule.
36. S_N1 stands for unimolecular nucleophilic substitution. The unimolecular part means that it obeys first-order kinetics. If the reaction is R:X + Z: → R:Z + X: and it takes place via an S_N1 reaction, then the rate depends on the speed with which the X breaks away from the R. The Z group comes in later and quickly, compared with the breakdown of R:X. S_N2 stands for bimolecular nucleophilic substitution. This happens with the same reaction scheme if the Z attacks the R:X molecule before it breaks down. Thus, the concentration of both R:X and Z: is important, and the rate displays second-order kinetics.
37. The S_N1 reaction leads to loss of stereospecificity, because the X group leaves before the entering nucleophile. This means that the nucleophile can enter from different angles, leading to different isomers.

7-7 The Active Site and Transition States

38. An effective transition-state analog would have to have a tetrahedral atom where the amide carbonyl group was originally found because the transition state involves a momentary tetrahedral carbon. It would also have to have oxygens on the tetrahedral atom so that there would be sufficient specificity for the active site.
39. A transition-state analog will bind with high affinity to the active site and thus serve as a K system inhibitor. The analog should contain a tetrahedral atom attached to three partially negative atoms (such as oxygen) to mimic the peptide substrate after attack by a water molecule.
40. The induced-fit model assumes that the enzyme and substrate must both move and change to conform to each other perfectly. An enzyme should then have a better fit and thus higher affinity for the transition-state analog than it does for the imperfectly fit substrate.

7-8 Coenzymes

41. See Table 7.1 for coenzyme examples. The inorganic cofactor would be a metal ion such as zinc, magnesium, or iron.
42. Coenzymes; cofactors; cofactors; coenzymes.
43. Enzymes that require coenzymes would be most affected. This is because most coenzymes are derivatives of compounds we call vitamins. For example, nicotinamide adenine dinucleotide is produced from the B vitamin niacin. Flavin adenine dinucleotide comes from riboflavin.

Chapter 8

8-1 The Definition of a Lipid

1. Solubility properties (insoluble in aqueous or polar solvents, soluble in nonpolar solvents). Some lipids are not at all structurally related except for having a prevalence of nonpolar groups present.
2. Carbon and hydrogen are found in all lipids and make up the majority of the hydrophobic portion that characterizes lipids. Several other elements can be commonly found in some lipids, depending on the type; oxygen, phosphorus, nitrogen, and sulfur are examples we see in Chapter 8.

8-2 The Chemical Natures of the Lipid Types

3. In both types of lipids, glycerol is esterified to fatty acids, with three such ester linkages formed in triacylglycerols and two in phosphatidylethanolamine (PE). The structural difference comes in the nature of the third ester linkage to glycerol. In PE, the third hydroxyl group of glycerol is esterified not to a carboxylic acid of a fatty acid but to phosphoric acid. The phosphoric acid moiety is esterified, in turn, to ethanolamine. (See Figures 8.2 and 8.5.)

4.

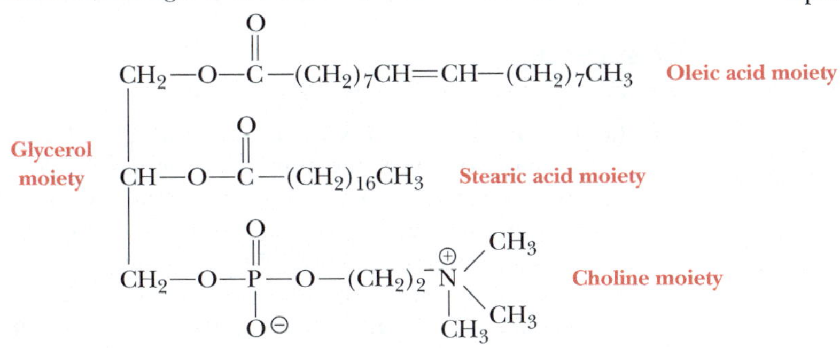

5. Both sphingomyelins and phosphatidylcholine (PC) contain phosphoric acid esterified to an amino alcohol, which must be choline in the case of PC and may be choline in the case of a sphingomyelin. They differ in the second alcohol to which phosphoric acid is esterified. In PC, the second alcohol is glycerol, which has also formed ester bonds to two fatty acids. In sphingomyelins, the second alcohol is another amino alcohol, sphingosine, which has formed an amide bond to a fatty acid. (See Figure 8.7.)
6. Both contain long-chain fatty acids and carbohydrates as building blocks. However, phosphatidylinositol contains a glycerol backbone and phosphate, whereas gangliosides contain a sphingosine backbone and more than one carbohydrate.
7. This lipid is a ceramide, which is one kind of sphingolipid.
8. Sphingolipids contain amide bonds, as do proteins. Both can have hydrophobic and hydrophilic parts, and both can occur in cell membranes.
9. A glycerol backbone and any combination of three fatty acids (in terms of length and extent of unsaturation) are possible. The chemical formula consists of six oxygens and varying amounts of carbon and hydrogen, depending on the lengths and saturation of the fatty acids.
10. Steroids contain a characteristic fused-ring structure, which other lipids do not (see Figure 8.10).
11. Waxes are esters of long-chain carboxylic acids and long-chain alcohols. They tend to be found as protective coatings.
12. A glycoprotein is formed by covalent bonding between a carbohydrate and a protein, whereas a glycolipid is formed by covalent bonding between a carbohydrate and a lipid.
13. Phospholipids are more hydrophilic than cholesterol. The phosphate group is charged, and the attached alcohol is charged or polar. These groups interact readily with water. Cholesterol has only a single, uncharged polar group, an —OH.
14. See Figure 8.3.
15. The waxy surface coating is a barrier that prevents loss of water due to the hydrophobicity of the lipid.

8-3 Biological Membranes

16. Triacylglycerols are not found in animal membranes.
17. Statements (c) and (d) are consistent with what is known about membranes. Covalent bonding between lipids and proteins [statement (e)] occurs in some anchoring motifs but is not widespread otherwise. Proteins span across the lipid bilayers rather than being sandwiched between them [statement (a)]. Bulkier molecules tend to be found in the outer lipid layer [statement (b)].
18. Cholesterol decreases membrane fluidity at typical animal growth temperatures (1) by containing ring structures, which themselves have little flexibility for bond rotation, and (2) by stabilizing the extended conformation of saturated fatty acids to allow for closer packing of the surrounding phospholipids.
19. The transition temperature is lower in a lipid bilayer with mostly unsaturated fatty acids compared with one with a high percentage of saturated fatty acids. The bilayer with the unsaturated fatty acids is already more disordered (has more disrupted packing) than the one with a high percentage of saturated fatty acids.
20. At the lower temperature, the membrane would tend to be less fluid. The presence of more unsaturated fatty acids would maintain the same fluidity of the membrane seen in bacteria grown at higher temperatures with more saturated fatty acids.
21. Hydrophobic interactions among the hydrocarbon tails are the main energetic driving force in the formation of lipid bilayers.

8-4 Membrane Proteins

22. Lipids can be "tagged" with a fluorescent group to observe their motion in membranes. Proteins often have intrinsic fluorescence and can be monitored directly.
23. Not all integral membrane proteins span the membrane; some are partially embedded in one of the two bilayers (making them integral membrane proteins but not transmembrane proteins). Peripheral membrane proteins associate with the membrane by noncovalent interactions with its polar surfaces.
24. Lipid-linked proteins are covalently bonded to a lipid via thioester linkages, whereas transmembrane proteins bond with lipids through hydrophobic interactions (i.e., van der Waals interactions).
25. Perilipin is a peripheral membrane protein because if the residues critical for binding are charged and polar, they must interact with the phospholipid head groups on the membrane. Integral membrane proteins would use hydrophobic residues to interact with the interior of the membrane or lipid droplet.
26. A 100-g sample of membrane contains 50 g of protein and 50 g of phosphoglycerides.

$$50 \text{ g lipid} \times \frac{1 \text{ mol lipid}}{800 \text{ g lipid}} = 0.0625 \text{ mol lipid}$$

$$50 \text{ g protein} \times \frac{1 \text{ mol protein}}{50{,}000 \text{ g protein}} = 0.001 \text{ mol protein}$$

The molar ratio of lipid to protein is 0.0625/0.001 or 62.5/1. Alternatively, you could have recognized that the two components have equal percentages by weight and simply divided 50,000 g by 800 g to determine that, for every 1 protein, there must be 62.5 lipids.

27. In an integral membrane protein that spans a membrane, the nonpolar residues line the exterior surface of the protein; they interact with the hydrocarbon tails of the cell membrane lipids. The polar residues are on the interior surface of the integral protein, lining the channel through which the ions enter and leave the cell.

28. The membranes that contain phosphatidylinositol are more fluid than the membranes that contain sphingolipid.

29. Statements (c) and (d) are correct. Transverse diffusion is only rarely observed [statement (b)], and the term *mosaic* refers to the pattern of distribution of proteins in the lipid bilayer [statement (e)]. Peripheral proteins are also considered part of the membrane but do not interact with the membrane interior [statement (a)].

8-5 The Functions of Membranes

30. Ions are charged and thus tend to be excluded from the nonpolar environment of the interior of the membrane.

31. Statements (a) and (c) are correct; statement (b) is incorrect because passive transport does *not* require energy to be expended.

32. Binding of the hormone causes a movement of the dimeric transmembrane protein that is transmitted to the intracellular side. This causes a movement of the kinase and catalytic domains, activating the kinase activity of the receptor.

33. JAK2 stands for Janus kinase 2. The hormone binding to the outside of the receptor causes an activation of the kinase on the inside of the cell, which is critical for transducing the signal of hGH. Activation of this and many kinases is the key signaling event in many pathways. Such kinases produce a second messenger that begins a cascade of molecular changes that affect the overall changes seen when hormones are released.

8-6 Lipid-Soluble Vitamins and Their Functions

34. Cholesterol is a precursor of vitamin D_3; the conversion reaction involves ring opening.

35. Vitamin E is an antioxidant, which means that it is a reducing agent that removes potentially damaging oxidizing agents.

36. Isoprene units are five-carbon moieties that play a role in the structure of a number of natural products, including fat-soluble vitamins.

37. See Table 8.5.

38. The cis–trans isomerization of retinal in rhodopsin triggers the transmission of an impulse to the optic nerve and is the primary photochemical event in vision.

39. Because vitamins A, D, E, and K are lipids and uncharged, they can pass through the hydrophobic portion of the membrane without requiring carrier proteins.

40. Vitamin K plays a role in the blood-clotting process. Blocking its mode of action can have an anticoagulant effect.

8-7 Arachidonic Acid Derivatives

41. An omega-3 fatty acid has a double bond at the third carbon from the methyl end.

42. Leukotrienes contain a carboxylic acid and three conjugated double bonds in their structure.

43. Prostaglandins contain a carboxylic acid and a five-membered ring in their structure.

44. Prostaglandins, leukotrienes, and thromboxanes are derived from arachidonic acid. They play a role in inflammation and in allergy and asthma attacks.

45. Prostaglandins in blood platelets can inhibit their aggregation, which helps prevent unwanted blood clots from forming.

Chapter 9

9-1 Levels of Structure in Nucleic Acids

1. (a) Double-stranded DNA is usually thought of as having primary structure (the nucleotide sequence), secondary structure (its base pairing), tertiary structure (its double-helix structure and its supercoiling), and quaternary structure (the two strands folded together).

(b) tRNA has primary structure (the nucleotide sequence), secondary structure (intrachain base pairing and loops), and tertiary structure, with many folds and twists in three dimensions.

(c) mRNA is usually considered only a primary structure (the nucleotide sequence), although some mRNA adopts regions of higher-order structure.

2. False. While we often think of RNA as "single stranded," there are many times that RNA complexes with other nucleotides to form quaternary structures. A couple of critical examples are RNA complexed with DNA during transcription and rRNA complexed with other rRNA and proteins to form the ribosome.

9-2 The Covalent Structure of Polynucleotides

3. Thymine has a methyl group attached to carbon 5; uracil does not.

4. In adenine, carbon 6 has an amino group attached; in hypoxanthine, carbon 6 is a carbonyl group.

5.

A	Adenine	Adenosine or deoxyadenosine	Adenosine-5′-triphosphate or deoxyadenosine-5′-triphosphate
G	Guanine	Guanosine or deoxyguanosine	Guanosine-5′-triphosphate or deoxyguanosine-5′-triphosphate
C	Cytosine	Cytidine or deoxycytidine	Cytidine-5′-triphosphate or deoxycytidine-5′-triphosphate
T	Thymine	Deoxythymidine	Deoxythymidine-5′-triphosphate
U	Uracil	Uridine	Uridine-5′-triphosphate

6. ATP is made from adenine, ribose, and three phosphates linked to the 5′-hydroxyl of the ribose. dATP is the same, except that the sugar is deoxyribose.

7. See Figure 9.5, except that the 3′ end will occur after G, which in your drawing will have a hydroxyl on its 3′ carbon instead of being attached to the phosphate of uracil.

8. (a) Definitely yes! If there is anything that you don't want falling apart, it's your storehouse of genetic instructions. (Compare the effectiveness of a computer if all the *.exe files were deleted.)

(b) Yes. The mRNA is the transmitter of information for protein synthesis, but it is needed only as long as a particular protein is needed. If it were long-lived, the protein would continue to be synthesized even when not needed; this would waste energy and could cause more direct detrimental effects. Thus, most mRNAs are short-lived (minutes); if more protein is needed, more mRNA is made.

9-3 The Structure of DNA

9.

Structure	Kind of Nucleic Acid
A-form helices	Double-stranded RNA
B-form helices	DNA
Z-form helices	DNA with repeating CGCGCG sequences
Nucleosomes	Eukaryotic chromosomes
Circular DNA	Bacterial, mitochondrial, and plasmid DNA

10. See Figure 9.9.
11. The sequence on the opposite strand for each of the following (all read 5′ → 3′) is (a) ATACGT, (b) AGATCT, and (c) TACCAG.
12. They are DNA sequences because of the presence of thymine rather than uracil.
13. Statements (c) and (d) are true; statements (a) and (b) are not but would accurately describe eukaryotes.
14. The major groove in B-DNA is much wider than the minor groove; the grooves in A-DNA are much closer in width.
15. Only statement (c) is true. Statements (a) and (b) are false. Statement (d) is true for the B form of DNA but not for the A and Z forms.
16. Supercoiling refers to twists in DNA over and above those of the double helix. Positive supercoiling refers to an extra twist in DNA caused by overwinding of the helix before sealing the ends to produce circular DNA. A topoisomerase is an enzyme that induces a single-strand break in supercoiled DNA, relaxes the supercoiling, and reseals the break. Negative supercoiling refers to the unwinding of the double helix before sealing the ends to produce circular DNA.
17. Propeller twist is a movement of the two bases in a base pair away from being in the same plane.
18. An AG/CT step is a small section of double-stranded DNA where one strand is 5′-AG-3′, and the other is 5′-CT-3′. The exact nature of such steps greatly influences the overall shape of a double helix.
19. Propeller twist reduces the strength of the hydrogen bond but moves the hydrophobic region of the base out of the aqueous environment, thus being more entropically favorable.
20. B-DNA is a right-handed helix with specified dimensions (10 base pairs per turn, significant differences between the major and minor grooves, etc.). Z-DNA is a left-handed double helix with different dimensions (12 base pairs per turn, similar major and minor grooves, etc.).
21. Positive supercoils in circular DNA will be left-handed.
22. Chromatin is the complex consisting of DNA and basic proteins found in eukaryotic nuclei (see Figure 9.17).
23. See Figure 9.9. If T and C are swapped, then partial negative charges from carbonyl groups on both bases will be in close proximity, as will partial positive charges on amides. This results in repulsion between these pairs of like charges instead of hydrogen bond formation!
24. Negative supercoiling, nucleosome winding, Z-form DNA.
25. It binds to the DNA, forming loops around itself. It then cuts both strands of DNA on one part of the loop, passes the ends across another loop, and reseals.
26. Histones are very basic proteins with many arginine and lysine residues. These residues have positively charged side chains under physiological pH. This is a source of attraction between the DNA and histones because the DNA has negatively charged phosphates. For example, histone-NH_3^+ groups attract $^-$O—P—O—DNA backbone groups.

    When the histones become acetylated, they lose their positive charge: histone—NH—$COCH_3$. They therefore have no attraction to the phosphates on the DNA. The situation is even less favorable if the histones get phosphorylated because then both the histones and the DNA carry negative charges and would repel each other.
27. Adenine–guanine base pairs occupy more space (two double-ring nucleobases) than is available in the interior of the double helix, whereas cytosine–thymine base pairs are too small (two single-ring nucleobases) to span the distance between the sites to which complementary bases are bonded. You would not normally expect to find such base pairs in DNA. The normal A–T and G–C base pairs each have a single-ring nucleobase paired with a double-ring nucleobase.
28. The phosphate groups in DNA are negatively charged at physiological pH. If they were grouped together closely, as in the center of a long fiber, the result would be considerable electrostatic repulsion. Such a structure would be unstable.
29. The percentage of cytosine equals that of guanine, 22%. Thus, this DNA has a 44% G–C content, implying a 56% A–T content. The percentage of adenine equals that of thymine, so adenine and thymine are 28% each.
30. If the DNA were not double-stranded, the requirements that G = C and A = T would no longer exist.

9-4 Denaturation of DNA

31. Blank 1: heat or temperature; Blank 2: absorbance; Blank 3: increase.
32. A–T base pairs have two hydrogen bonds, whereas G–C base pairs have three. It takes more energy and a higher temperature to disrupt the structure of DNA rich in G–C base pairs.
33. Polyethylene glycol decreases T_m because less heat is required to break the weakened van der Waals forces that promote base stacking.

9-5 The Principal Kinds of RNA and Their Structures

34. Messenger RNA (mRNA), transfer RNA (tRNA), and ribosomal RNA (rRNA).
35. The base sequence of all RNA types is a reflection of the corresponding DNA sequence that leads to the production of the RNA.
36. You would add siRNA, which falls into the broader category of RNAi. While miRNA works similarly, it is coded for by the cell and is not inserted to test gene function.
37. Transcription.
38. tRNA is bound to an amino acid and brings the correct amino acid to the growing peptide during the process of translation (protein synthesis).

39. The base sequence of the mRNA.
40. Translation.
41. The mRNA of eukaryotes is extensively processed compared with that of prokaryotes. The mRNA in eukaryotes is initially transcribed with numerous introns that later must be spliced out. Also, the transcripts have protections at both ends in the form of a 5′ cap and a 3′ poly-A tail. Neither of these occurs in prokaryotic mRNA.
42. These are RNAs found in the nucleus of eukaryotic cells that are 100 to 200 base pairs long and aid in the processing of mRNA.
43. These are small RNAs of 20 to 22 base pairs that primarily bind to mRNA and alter translation. They are important in the development and regulation of many processes.

9-6 Roles for RNA

44. See Figures 9.23 and 9.28.
45. Four different kinds of bases—adenine, cytosine, guanine, and uracil—make up the preponderant majority of the bases found in RNA, but they are not the only ones. Modified bases occur to some extent, principally in tRNA.
46. Small nuclear RNA (snRNA) is found in the eukaryotic nucleus and is involved in the splicing reactions of other RNA types. An snRNP is a small nuclear ribonucleoprotein particle, and it is a complex of small nuclear RNA and protein that catalyzes the splicing of RNA.
47. Ribosomal RNA (rRNA) is the largest. siRNA and miRNA are the smallest at 20 to 30 nucleotides.
48. Messenger RNA (mRNA) has the least amount of secondary structure (base pairing).
49. The enzyme adds the RNA sequence CCA to the 3′ end of the tRNA molecule once it is transcribed.
50. RNA interference is the process by which small RNAs prevent or regulate the expression of genes.
51. More extensive hydrogen bonding occurs in tRNA than in mRNA. The folded structure of tRNA, which determines its binding to ribosomes in the course of protein synthesis, depends on its hydrogen-bonded arrangement of atoms. Unlike tRNA, where only the anticodon needs to be accessible for hydrogen bonding, the coding sequences of mRNA must all be accessible to hydrogen bond with the anticodon of the tRNAs that deliver the amino acids in the protein being synthesized.
52. They prevent intramolecular hydrogen bonding (which occurs in tRNA via the usual A–U and C–G associations), thus permitting loops that are critical for function, the most important being the anticodon loop.
53. Turnover of mRNA should be rapid to ensure that the cell can respond quickly when specific proteins are needed. Ribosomal subunits, including their rRNA component, can be recycled for many rounds of protein synthesis. mRNA is also less structured than rRNA. As a result, mRNA is degraded more rapidly than rRNA.
54. The mistake in the DNA would be more harmful because every cell division would propagate the mistake. A mistake in transcription would lead to one wrong RNA molecule that can be replaced with a correct version in the next transcription.
55. Eukaryotic mRNA is initially formed in the nucleus by transcription of DNA. The mRNA transcript is then spliced to remove introns, a poly-A tail is added at the 3′ end, and a 5′ cap is put on. This is the final mRNA, which is then transported, in most cases, out of the nucleus for translation by the ribosomes.
56. The numbers 50S, 30S, etc., refer to a relative rate of sedimentation in an ultracentrifuge and cannot be added directly. Many things besides molecular weight influence the sedimentation characteristics, such as shape and density.

9-7 RNA and Medical Applications

57. Using CRISPR/Cas, researchers can create designer mRNA guides that lead the CRISPR complex to specific DNA sequences. These can be cut up and degraded or even changed to different sequences. Studies are currently looking at sickle-cell disease and cystic fibrosis as potentially treatable by this process.
58. mRNA vaccines use a copy of an mRNA molecule that produces a viral protein or a cancer cell protein. When the vaccine is delivered, the patient's immune system can use the viral or cancer cell protein produced from the vaccine to learn how to identify and destroy virus particles or cancer cells.
59. Unlike proteins with secondary, tertiary, and quaternary structure, mRNA is largely unstructured and thus more susceptible to degradation.

Chapter 10

10-1 The Flow of Genetic Information in the Cell

1. Replication is the production of new DNA from a DNA template. Transcription is the production of RNA from a DNA template. Translation is the synthesis of proteins directed by mRNA, which reflects the base sequence of DNA.
2. This theory is called the central dogma of molecular biology.
3. Yes, the first generation of cells would incorporate the ^{31}P phosphorus into nucleotides, and the alpha phosphate remains in the backbone of the newly formed DNA to generate hybrid complexes with the new light strand and the parental heavy strand.
4. False. In retroviruses, the flow of information is RNA → DNA.
5. DNA represents the permanent copy of genetic information, whereas RNA is transient. The cell could survive mistakes in the production of RNA molecules because they are constantly degraded and synthesized.

10-2 Replication of DNA

6. The semiconservative replication of DNA means that a newly formed DNA molecule has one new strand and one strand from the original DNA. The experimental evidence for semiconservative replication comes from density-gradient centrifugation (Figure 10.3). If replication were a conservative process, the original DNA would have two heavy strands, and all newly formed DNA would have light strands.
7. A replication fork is the site of formation of new DNA. The two strands of the original DNA separate, and a new strand is formed on each original strand.
8. An origin of replication consists of a bubble in the DNA. There are two places at opposite ends where new polynucleotide chains are formed (Figure 10.4).

10-3 DNA Polymerase

9. Most DNA polymerase enzymes also have exonuclease activity.
10. DNA polymerase I is primarily a repair enzyme. DNA polymerase III is mainly responsible for the synthesis of new DNA. See Table 10.1.
11. The processivity of a DNA polymerase is the number of nucleotides incorporated before the enzyme dissociates from the template. The higher this number, the more efficient the replication process.
12. The reactants are deoxyribonucleotide triphosphates. They provide not only the moiety to be inserted (the deoxyribonucleotide) but also the energy to drive the reaction (dNTP → inserted dNMP + PP_i).
13. One strand of newly formed DNA uses the 3′-to-5′ strand as a template. The problem arises with the 5′-to-3′ strand. Nature deals with this issue by using short stretches of this strand for a number of chunks of newly formed DNA. They are then linked by DNA ligase (Figure 10.6).
14. The hydroxyl group at the 3′ end is needed for nucleophilic attack on the next nucleotide to be added to the growing DNA chain. A number of antiviral drugs remove the 3′ hydroxyl group in some way.
15. DNA polymerases replicating the lagging strand are closer to the 5′ ends of the Okazaki fragments, which each contain a primer, while the 5′ end of the leading strand may be far from the DNA polymerase replicating it because replication is continuous from a single primer on the leading strand. See Figure 10.10.

10-4 Proteins Required for DNA Replication

16. All four deoxyribonucleoside triphosphates (dATP, dCTP, dTTP, and dGTP), template DNA, DNA polymerase, all four ribonucleoside triphosphates (ATP, CTP, UTP, and GTP), primase, helicase, single-strand binding protein, DNA gyrase, and DNA ligase.
17. See Figure 10.6.
18. DNA gyrase releases the torsional strain caused by unwinding of the DNA double helix during replication. Primase synthesizes the RNA primer. DNA ligase links the Okazaki fragments.
19. In the replication process, the single-stranded portions of DNA are protected from nuclease digestion by single-strand binding proteins (SSBs).
20. The primer in DNA replication is a short sequence of RNA to which the growing DNA chain is bonded.
21. A primer is needed in replication because the main replicative DNA polymerases can only add nucleotides to an existing strand of nucleotides.
22. DNA polymerases have a very common structure that is often compared to a right hand, with domains referred to as the fingers, palm, and thumb. The active site where the polymerase reaction is catalyzed lies in the crevice within the palm domain. The fingers domain acts in deoxynucleotide recognition and binding, and the thumb is responsible for DNA binding.
23. The most significant differences are in the active site located in the crevice within the palm domain. The finger and thumb domains responsible for binding work as expected.
24. A clamp loader is necessary because the sliding clamp of DNA polymerase is a closed circle. It would not be able to get around the DNA without an enzyme to open it up.

10-5 Proofreading and Repair

25. When an incorrect nucleotide is introduced into a growing DNA chain as a result of mismatched base pairing, DNA polymerase acts as a 3′-exonuclease, removing the incorrect nucleotide. The same enzyme then incorporates the correct nucleotide.
26. An exonuclease nicks the DNA near the site of the thymine dimers. Polymerase I acts as a nuclease, excises the incorrect nucleotides, and then acts as a polymerase to incorporate the correct ones. DNA ligase seals the nick.
27. In DNA, cytosine spontaneously deaminates to uracil. The presence of the extra methyl group is a clear indication that a thymine really belongs in that position, not a cytosine that has been deaminated.
28. Prokaryotes methylate their DNA soon after replication. This aids the process of mismatch repair. The enzymes that carry out the process can recognize the correct parental strand by its methyl groups. The newly formed daughter strand, which contains the incorrect base, does not have methyl groups initially.
29. Prokaryotes have a last-resort mechanism for dealing with drastic DNA damage. This mechanism is called the SOS response (see Biochemical Connections 10B).
30. Nonhomologous DNA end joining (NHEJ) or recombination.
31. Ku70/80, DNA ligase IV, and several others.
32. It is error-prone because the repair proceeds without a template.
33. Both base excision repair and nucleotide excision repair require an intact template on the opposite strand that DNA polymerase can use to replace the damaged base(s). Because both strands contain damaged bases, these repair mechanisms cannot be used.
34. It binds the broken ends of the DNA so that replication can continue.

10-6 DNA Recombination

35. They used two different phages to infect bacteria. One of the phages had light DNA, and one had heavy DNA. Without recombination, the light DNA would always package into light virus particles, and the heavy would package into heavy particles. This would lead to only two populations of phages after infection. Their results showed, however, that there were intermediate combinations that had DNA of different weights. This demonstrated that the phage DNA was recombining.
36. Similar to the experiment described in Problem 35, experiments using heavy isotopes demonstrated the semiconservative nature of replication. Intermediate-weight products of replication demonstrated that progeny DNA contains one parental strand and one new strand.
37. Crossover occurs at hot spots during meiosis to generate genetically distinct gametes, so more hot spots will lead to more genetic diversity within the offspring and thus within a species as a whole.

38. Recombination occurs by the breakage and reunion of DNA strands so that physical exchange of DNA parts takes place (see Figure 10.21). The mechanism was deduced in 1964 by Robin Holliday and is referred to as the Holliday model.

10-7 Eukaryotic DNA Replication

39. Eukaryotes usually have several origins of replication, whereas prokaryotes have only one.
40. The general features of DNA replication are similar in prokaryotes and eukaryotes. The main differences are that eukaryotic DNA polymerases do not have exonuclease activity. After synthesis, eukaryotic DNA is complexed with proteins; prokaryotic DNA is not.
41. Histones are protein complexed to eukaryotic DNA, and they can be modified in such a way that they can release the portions of DNA that are being replicated. Their synthesis must take place at the same rate as DNA synthesis. The proteins and DNA must then assemble in proper fashion.
42. Eukaryotes have more DNA polymerases, which tend to be larger molecules. Eukaryotic DNA polymerases tend not to have exonuclease activity. There are more origins of replication in eukaryotes and shorter Okazaki fragments. See Table 10.5.
43. Mechanisms exist to ensure that DNA synthesis takes place only once in the eukaryotic cell cycle, during the S phase. Preparation for DNA synthesis can and does take place in the G1 phase, but the timing of actual synthesis is strictly controlled.
44. If the telomerase enzyme were inactivated, DNA synthesis would involve significant deletions at the ends of the chromosomes. This enzyme maintains the telomere repeats at the 3′ template end strand so that it does not undergo gene-sequence degradation with each round of DNA synthesis (see Biochemical Connections 10C).
45. If histone synthesis took place faster than DNA synthesis, it would be highly disadvantageous to invest the energy required for protein synthesis. The histones would have no DNA with which to bind.
46. Circular DNA does not have ends. This removes the necessity for maintaining the 3′ template end on removal of the RNA primer. Telomeres and telomerase are not needed with circular DNA.
47. Homologous recombination requires an intact homologous double strand of DNA to recombine with, preferably with the same sequence as the damaged strand. This is only possible after the DNA has been duplicated. In other words, DNA repair by homologous recombination during G_1 would lead to a copy of a homologous, but not identical, region of DNA and thus loss of genetic information.
48. The presence of a DNA polymerase that operates only in mitochondria is consistent with the theory that these organelles are derived from energy-producing bacteria incorporated by endosymbiosis.
49. The hypothesis states that RNA was the original molecule of heredity and was the first molecule that took simple compounds and turned them into larger molecules with a function.

Chapter 11

11-2 Transcription in Prokaryotes

1. Whereas a primer is required for DNA replication, no primer is required for the transcription of DNA into RNA.
2. RNA polymerase from *E. coli* has a molecular weight of about 500,000 and four different kinds of subunits. It uses one strand of the DNA template to direct RNA synthesis. It catalyzes polymerization from the 5′ end to the 3′ end.
3. The subunit composition of the holoenzyme is $\alpha_2\beta\beta'\sigma$.
4. The core enzyme lacks the σ subunit; the holoenzyme has it.
5. The DNA strands unwind near the −10 region (the Pribnow box) to form the open complex.
6. The strand that the RNA polymerase uses as a template for its RNA synthesis is called the template strand, the noncoding strand, the antisense strand, and the (−) strand. The other strand, whose sequence matches the RNA produced except for the T–U change, is called the nontemplate strand, the coding strand, the sense strand, and the (+) strand.
7. The promoter region is the portion of DNA to which RNA polymerase binds at the start of transcription. This region lies upstream (nearer the 3′ end of the template DNA) of the actual gene for the RNA. The promoter regions of DNA from many organisms have sequences in common (consensus sequences). The consensus sequences frequently lie 10 base pairs and 35 base pairs upstream of the start of transcription.
8. Moving from 5′ to 3′ on the coding strand, the order is as follows: Fis site, UP element, −35 region, Pribnow box, TSS.
9. Intrinsic termination of transcription involves the formation of a hairpin loop in the RNA being formed, which stalls the RNA polymerase over a region rich in A–U base pairs. This causes termination of transcription and release of the transcript. Rho-dependent termination often involves a similar hairpin loop, but, in addition, a Rho protein binds to the RNA and moves along it toward the transcription bubble. When the Rho protein reaches the transcription bubble, it causes termination.
10. See Figure 11.1. The top DNA strand is the nontemplate strand because it is not used to create the RNA. It is called the coding strand because it has the same sequence as the RNA produced, except for the change of T for U. It is called the sense strand because its sequence would give the correct amino acid sequence of the protein product. It is called the (+) strand because it has the correct sequence. The bottom strand is called the template strand because it is the one used to make the RNA. It is also the noncoding strand because its sequence does not match the RNA produced. It is the antisense and the (−) strand for the same reason.

11-3 Transcription Regulation in Prokaryotes

11. An inducer is a substance that leads to transcription of the structural genes in an operon. A repressor is a substance that prevents transcription of the structural genes in an operon.
12. Allolactose binds to the *lac* repressor, causing it to release from the operator, which allows transcription to occur. When glucose is low, cAMP levels rise, and cAMP forms a complex with the CAP protein. The CAP–cAMP complex binds to an enhancer region and enhances transcription. Therefore, transcription of the *lac* operon is most active when allolactose is present and glucose is absent.
13. The σ factor is a subunit of prokaryotic RNA polymerase. It directs the polymerase to specific promoters and is one of the ways that gene expression is controlled in prokaryotes.

14. σ^{70} is the normal σ subunit for RNA polymerase in *E. coli*. It directs RNA polymerase to most of the genes that are transcribed under normal circumstances. σ^{32} is an alternate subunit that is produced when the cells are grown at higher temperatures. It directs the RNA polymerase to other genes that need to be expressed during heat shock conditions.
15. The catabolite activator protein is a transcription factor in *E. coli* that stimulates transcription of the *lac* operon structural genes. It responds to cAMP levels such that the *lac* operon is transcribed only when the cells must use lactose as a fuel source.
16. Transcription attenuation is the process found in prokaryotes in which transcription can continue or be prematurely aborted based on the concurrent translation of the mRNA produced. This is often seen in genes whose protein products lead to amino acid synthesis.
17. An operon consists of an operator gene, a promoter gene, and structural genes. When a repressor is bound to the operator, RNA polymerase cannot bind to the promoter to start transcription of the structural genes. When an inducer is present, it binds to the repressor, rendering it inactive. The inactive repressor can no longer bind to the operator. As a result, RNA polymerase can bind to the promoter, leading to transcription and eventual translation (protein synthesis) of the structural genes of the operon.
18. See Figure 11.6.
19. With phage SPO1, which infects the bacterium *B. subtilis*, the virus has a set of genes called the early genes that are transcribed by the host's RNA polymerase, using its regular σ subunit. One of the viral early genes codes for a protein called gp28. This protein is another σ subunit, which directs the RNA polymerase to preferentially transcribe more of the viral genes during the middle phase. Products of the middle-phase transcription are gp33 and gp34, which together make up another σ factor that directs the transcription of the late genes.
20. See Figure 11.17. When the level of tryptophan is low, the Trp-$tRNA^{Trp}$ becomes limiting. This stalls the ribosome over the tryptophan codons on the mRNA. By stalling the ribosome there, the antitermination loop can form, transcription is not aborted, and the full mRNA is produced. If the ribosome does not stall there, the termination loop forms, and the leader mRNA dissociates.

11-4 Transcription in Eukaryotes

21. It is the sensing domain of a riboswitch found at the 5′ end.
22. It is mRNA that has two functions: sensing and decision-making.
23. Translation can be prevented when a hairpin loop forms that blocks the Shine–Dalgarno sequence (the translation initiation site). Another process that halts translation occurs when a terminator hairpin is created, similar to the one that forms during transcription attenuation. Also, in the presence of a certain metabolite, the mRNA can be susceptible to faster degradation.
24. Researchers are hoping to find molecules that can act like a competitive inhibitor and make the riboswitch act as though the natural substrate were present. If the riboswitch controlled a vital process, then shutting off the riboswitch would kill the pathogen.
25. Exons are the portions of DNA that are expressed, which means that they are reflected in the base sequence of the final mRNA product. Introns are the intervening sequences that do not appear in the final product but are removed during the splicing of mRNA.
26. There are three RNA polymerases in eukaryotes, compared with one in prokaryotes. There are many more transcription factors in eukaryotes, including complexes of them necessary for polymerase recruitment. RNA is extensively processed after transcription in eukaryotes, and, in most cases, the mRNA must leave the nucleus to be translated, whereas translation and transcription can occur at the same time in prokaryotes.
27. RNA polymerase I produces most of the rRNA. RNA polymerase II produces mRNA, and RNA polymerase III produces tRNA, the 5S ribosomal subunit, and snRNA.
28. The first component includes a variety of upstream elements, which act as enhancers and silencers. Two common ones are close to the core promoter and are the GC box (–40), which has a consensus sequence of GGGCGG, and the CAAT box (extending to –110), which has a consensus sequence of GGCCAATCT. The second component, found at position –25, is the TATA box, which has a consensus sequence of TATAA(T/A). The third component includes the transcription start site at position +1 and is surrounded by a sequence called the initiator element (*Inr*). The final component is a possible downstream regulator.
29. TFIIA, TFIIB, TFIID, TFIIE, TFIIF, and TFIIH are the general transcription factors. TFIID is also the TATA-box binding protein and is associated with TAFs (TBP-associated factors).
30. Its primary function is as a general transcription factor involved in the formation of the open complex for transcription initiation. It binds to the basal unit and is involved in DNA melting through helicase activity, as well as promoter clearance via phosphorylation of the CTD of RNA polymerase. In addition, it has a cyclin-dependent kinase activity. Thus, TFIIH is involved in tying transcription and cell division together. It is also involved in DNA repair mechanisms.

11-5 Transcription Regulation in Eukaryotes

31. Histone acetylation on lysine residues is generally associated with increased transcriptional activity. It reduces the positive charge on histone tails, weakening their interaction with negatively charged DNA. The weakened interaction between histones and DNA opens chromatin structure and allows transcription factors and RNA polymerase access to the DNA in order to transcribe genes.
32. The heat-shock response element responds to increased temperature. The metal-response element responds to the presence of heavy metals such as cadmium, and the cyclic-AMP-response element controls a wide variety of genes based on cAMP levels in the cell.
33. CREB is a transcription factor that binds to the cAMP-response element. It is involved in the transcription of hundreds of genes based on the cAMP levels of the cell. When there is cAMP, CREB is phosphorylated, which allows it to bind the CREB-binding protein, which connects the CRE to the basal transcription machinery, stimulating transcription.

34. Regulation in eukaryotes is much more complicated. Prokaryotic regulation is controlled by the choice of σ subunit, the nature of the promoters, and the use of repressors/inducers. In eukaryotes, there are many more promoter elements, transcription factors, and coactivators. In addition, the DNA must be released from histone proteins, so transcription of DNA is linked to histone modifications.
35. Assuming that there is a basal transcription rate for a particular gene, an enhancer would bind to a transcription factor and lead to a greater level of transcription, while a silencer would bind to a transcription factor and reduce the level of transcription below the basal rate.
36. A response element is an enhancer element that binds to a specific transcription factor and increases the level of transcription of target genes. In the case of response elements, however, this is in response to a more general cell signal such as the presence of cAMP, glucocorticoids, or heavy metals. Response elements may control a large set of genes, and a given gene may be under the control of more than one response element.
37. See Figure 11.27. The CREB protein binds to the CRE DNA sequence. When phosphorylated, CREB also binds to CBP and bridges to the basal transcription complex.
38. TFIID is one of the general transcription factors for RNA polymerase II. Part of it is a protein that binds to the TATA box in eukaryotic promoters. Associated in complex with the TATA box and the TBP are many proteins called TAFs, for TBP-associated factors.
39. Transcription elongation in eukaryotes is controlled in several ways. There are pause sites at which RNA polymerase tends to hesitate. There is also antitermination, at which RNA polymerase can transcribe past a normal termination point. The general transcription factor TFIIF stimulates elongation as well as initiation by helping RNA polymerase II read through pause sites. A separate elongation factor, TFIIS, is called an arrest release factor because it stimulates RNA polymerase to resume transcription once it has hesitated at a pause site. Separate proteins also exist, called P-TEF and NELF, that act to positively or negatively affect elongation, respectively.
40. CREB is a ubiquitous transcription factor that has been found to be involved in many genes. It is phosphorylated when cAMP levels are high, which triggers the activation of the genes. CREB-mediated transcription has been implicated in cell proliferation, cell differentiation, spermatogenesis, release of somatostatin, development of mature T cells, protection of nerve cells under hypoxic conditions, circadian rhythms, adaptation to exercise, regulation of gluconeogenesis, transcription regulation of phosphoenolpyruvate carboxykinase and lactate dehydrogenase, and learning and storage in long-term memory.
41. Acidic domains, glutamine-rich domains, and proline-rich domains.
42. Mediator is a giant complex with a mass of over 1 million daltons, comprising over 20 distinct subunits in yeast and more than 30 subunits in humans. Mediator bridges the promoter, RNA polymerase, and the general transcription machinery with specific remote enhancers and silencers.
43. Mediator bridges the promoter region with the enhancer region to activate transcription, or, in the opposite case, it binds to the silencer element but does not recruit RNA Pol II to the promoter in this case.
44. Eukaryotic DNA is wound tightly in the nucleosomes. Before the DNA can be transcribed, it has to be available to the transcription machinery, so opening up of the nucleosomes is a necessary first step.
45. The first is the release of DNA from chromatin structure. The second is the recruitment and interaction of RNA polymerase with the promoter and transcription machinery.
46. Chromatin remodeling complexes and histone-modifying enzymes.
47. Chromatin remodeling complexes are huge assemblies containing ATP-dependent enzymes that loosen the DNA–protein interactions in nucleosomes by a variety of mechanisms involving sliding, ejecting, inserting, and otherwise restructuring the core octamers.
48. Transcription activation via modification of chromatin involves the covalent modification of histone proteins. In the transcriptionally inactive state, the negative charges on the phosphates of the DNA backbone are tightly bound to the positive charges on the basic proteins of the histones. To activate transcription, this tight binding must be relaxed.
49. The most important modification of the histones is the acetylation of the ε-amino groups of lysine on the histone tails. Acetylating the lysine removes the positive charge and loosens the binding to the DNA. Other modifications also play a role in transcription regulation through histones, including phosphorylation of serine residues and methylation of lysine and arginine residues.
50. Histones are acetylated by histone acetyltransferases (HATs). They are deacetylated by histone deacetylases (HDACs).

11-6 Noncoding RNAs

51. MicroRNAs (miRNAs) are about 22 nucleotides long, and they are typically endogenously produced. They repress the transcription of multiple genes by partial complementarity to multiple mRNAs.
52. Small interfering RNAs (siRNAs) are about 22 nucleotides long, and they are typically exogenously produced. siRNAs repress the transcription of one gene by perfect or near-perfect complementarity to one mRNA.
53. Noncoding RNAs (ncRNAs) have been linked to many processes, including regular transcription, gene silencing, replication, RNA processing, RNA modification, translation, protein stability, and protein translocation.
54. RNA interference occurs when small RNA molecules bind to mRNA, either causing its destruction so it cannot be translated or blocking its translation directly.
55. Loss of miRNA-101 leads to the overexpression of a particular histone methyltransferase that helps the progression of several types of cancer.
56. In normal mice, sciatic nerve injury results in loss of nerve function in the muscle and leads to an increase in miRNA-206. Using a line of mice that had ALS and inactivated miRNA-206 led to a shortening of the time from onset of the disease, indicating this miRNA had a protective effect on nerves in the muscle.

57. Work with ALS mice supports a growing body of evidence of the importance of miRNA to neurological function and susceptibility to disease. MicroRNA networks have been implicated in Parkinson's disease, Huntington's disease, and Alzheimer's disease.
58. Dicer and RISC.
59. The guide strand is the antisense strand of a double-stranded RNA molecule. The passenger strand is the sense strand.
60. The binding of the siRNA is a perfect match to the mRNA, whereas the binding of miRNA is not perfect, forming a loop.
61. When siRNA binds to the mRNA, the mRNA is degraded by the Slicer enzyme. When miRNA is bound to the mRNA, there is no translation of the message, but the mRNA is usually not degraded by Slicer.

11-7 Structural Motifs in DNA-Binding Proteins

62. Helix–turn–helix motifs, zinc fingers, and basic-region leucine zippers.
63. See Figures 11.31–11.37, as well as Table 11.5. The major DNA binding protein motifs are helix–turn–helix, zinc fingers, and basic-region leucine zippers. The helix–turn–helix motifs are organized so that the two helices of the protein fit into the major groove of the DNA. Zinc fingers are formed by combinations of cysteine and/or histidine complexed with zinc ions. A loop of protein forms around this complex, and these loops fit into the major groove of the DNA. Several such loops can be found spiraling around the DNA in the major groove. The basic-region leucine zipper has two domains. One is an area of leucines spaced out every seven amino acids. This puts them on the same side of an α-helix, which allows them to dimerize with another such protein. The basic region is high in lysine and arginine, which bind tightly to the DNA backbone via electrostatic attraction.

11-8 Posttranscriptional RNA Modification

64. In eukaryotic mRNA processing, introns are spliced out, a poly-A tail is put on the 3′ end of the mRNA, and a 5′ cap is put on the mRNA. mRNA base modifications sometimes happen, but this is not as common as in rRNA or tRNA.
65. Addition of a CCA sequence at the 3′ end, which is where the amino acid gets attached to form an amino acyl-tRNA molecule.
66. Trimming is necessary to obtain RNA transcripts of the proper size. Frequently, several tRNAs are transcribed in one long RNA molecule and must be trimmed to obtain active tRNAs.
67. The snRNPs are small nuclear ribonucleoprotein particles. They convert heterogeneous nuclear RNA (hnRNA) to mature mRNA through splicing.
68. Besides its traditional role in mRNA, tRNA, and rRNA, RNA serves other functions such as splicing reactions, trimming reactions, and the peptide synthesis reaction of peptidyl transferase. It has also been shown that some small RNAs are produced; they act as gene silencers by binding to specific DNA sequences and blocking their transcription.
69. See Figure 11.42.
70. The Human Genome Project concluded that humans had far fewer genes than previously thought, yet we seem to be more biologically and biochemically complex. One possibility suggested to explain how so few genes could lead to so many proteins is that more proteins may be produced via differential splicing of mRNA. Thus, the same amount of DNA could lead to more gene products.

11-9 Ribozymes

71. A ribozyme is RNA that has catalytic activity without the intervention of protein at the active site. The catalytic portion of RNase P is a ribozyme. The self-splicing rRNA of *Tetrahymena* is the classic example, and it has been shown that the peptidyl transferase activity of the ribosome is actually a ribozyme.
72. Two mechanisms for RNA self-splicing are known. In Group I ribozymes, an external guanosine is covalently bonded at the splice site, releasing one end of the intron. The free end of the exon thus produced attacks the end of the other exon to splice the two. The intron cyclizes in the process. (See Figure 11.43.) Group II ribozymes display a lariat mechanism. The 2′-OH of an internal adenosine attacks the splice site. (See Figure 11.42.)
73. Proteins are more efficient catalysts than RNA because they have wider variations in structure and thus can tailor the active site for maximum efficiency for a given reaction.
74. Epigenetics roughly translates to heritable changes in DNA that do not involve a change in the primary structure or sequence of the DNA.
75. Epimutations are similar to DNA mutations but affect the DNA scaffolding or modifications without affecting the sequence.
76. Over 30 molecules associated with cancer have been found to be chromatin remodelers.

Chapter 12

12-1 Translating the Genetic Message

1. See Figure 12.1.

12-2 The Genetic Code

2. A code in which two bases code for a single amino acid allows for only 16 (4 × 4) possible codons, which is inadequate to code for 20 amino acids.
3. A degenerate code is one in which more than one triplet can specify a given amino acid.
4. *Nonoverlapping* means that the genetic code is read in sets of three nucleotides (codons) without reusing any bases between codons. For example, the sequence AUGGUC would be read as AUG followed by GUC, never AUG followed by GGU, where the first G is overlapping. This is important because it ensures that the ribosome translates mRNA in a consistent reading frame, producing the correct sequence of amino acids. If the code were overlapping, a single mutation could affect multiple amino acids, increasing the chance of harmful errors.
5. In the binding assay technique, various tRNA molecules, one of which is radioactively labeled with ^{14}C, are mixed with ribosomes and synthetic trinucleotides bound to a filter. If the radioactive label is detected on the filter, then it is known that the particular tRNA is bound to that triplet. The binding experiments can be repeated until all the triplets are assigned.

6. The wobble bases that can break Watson–Crick rules can be uracil, guanine, or hypoxanthine. Adenine and cytosine do not break Watson–Crick rules at the wobble position.
7. The codons UAA, UAG, and UGA are the stop signals. These codons are not recognized by any tRNAs, but they are recognized by proteins called release factors. A release factor not only blocks the binding of a new aminoacyl-tRNA but also affects the activity of the peptidyl transferase, so the bond between the carboxyl end of the peptide and the tRNA is hydrolyzed.
8. The sequence in the codon of mRNA is reversed because mRNA synthesis is antiparallel.
   (a) Position 1 on the template strand will be position 3 (the wobble position) on the mRNA. Therefore, there is a high probability of getting the same amino acid even if position 1 on the template strand is mutated, and the protein thus has a very good chance of functioning.
   (b) Position 2 is the most informational: a different amino acid results from any change. In this case, however, the chances are high (75%) that the mutation would be a conservative one, with one hydrophobic amino acid replacing another one, so the protein would still have a good chance of functioning. A change involving serine or threonine (25% chance) would alter the polarity but would not introduce a charge on the side chain; the protein might still function.
   (c) There is a high probability of a change in the type of amino acid, including differences in charge; the probability of the resulting protein having proper function is considerably lower.
   (d) Position 3 on the template strand will be position 1 on the mRNA, and it has an intermediate effect. For purine changes, a different amino acid results in all cases. The changes tend to be conservative, with only four of the 16 possible changes leading to hydrophobic–hydrophilic differences. For our purposes, glycine is considered neither hydrophobic nor hydrophilic. The resulting protein would have a better chance of functioning than in a second-base change but a lesser probability than in a third-base change.
9. Hypoxanthine is the most versatile of the wobble anticodon bases; it can base pair with adenine, cytosine, or uracil.
10. It is quite reasonable. When codons for a given amino acid have one or two nucleotides in common, a mutation is less likely to give rise to a nonfunctional protein. The survival value of such a feature guarantees its selection in evolution.
11. An ambiguous code would allow for variation in the amino acid sequence of proteins. Consequently, there would be variation in function, including a number of nonfunctional proteins.
12. Variations in the genetic code in mitochondria support the idea of their existence as free-living bacteria early in evolutionary history.
13. A virus that is too lethal will kill its host before having time to spread to other hosts. With modern science, such a virus would be isolated and destroyed quickly. By being less lethal, a virus can pass from host to host and make it difficult to isolate.
14. The IAV protease mRNA, when translated, yields 191 amino acids from the normal reading frame of the mRNA. However, when the ribosome encounters a CGU codon, which is a rare codon for arginine (R), the ribosome stalls on the mRNA due to the relative lack of tRNA-Arg that matches the rare codon. This gives the ribosome time to reset to a new frame, called the +1 frame, where it continues translating, yielding the next 61 amino acids, which are different compared to the normal protease because they are read out of frame.

12-3 Aminoacyl-tRNA Formation

15. In the first aminoacyl-tRNA synthetase step, the amino acid is activated by ATP, forming an aminoacyl-adenylate intermediate (aminoacyl-AMP). In the second step, the enzyme transfers the activated amino acid to the 3′ end of the tRNA, attaching it to the 2′ or 3′ hydroxyl group of the terminal adenosine, charging the tRNA to form an aminoacyl-tRNA.
16. The hydrolysis of ATP to AMP and PP_i provides the energy to drive the activation step.
17. Proofreading in aminoacyl-tRNA formation takes place in two stages. The first requires a hydrolytic site on the aminoacyl-tRNA synthetase; *incorrect amino acids* that have become esterified to the tRNA are removed here. The second stage of proofreading requires the recognition site on the aminoacyl-tRNA synthetase for the tRNA itself. *Incorrect tRNA molecules* do not bind tightly to the enzyme, so they dissociate before becoming charged with an amino acid.
18. The following factors ensure fidelity in protein synthesis. Aminoacyl-tRNA formation includes a high degree of enzyme specificity to connect the right amino acid to the right tRNA, proofreading in the formation of some aminoacyl-adenylates, and kinetic proofreading or "energy overkill." Other factors include proper hydrogen bonding of mRNA to the ribosome and between codon and anticodon. (The latter is a relatively slow association, allowing time for mismatches to dissociate before the peptide bond is formed.) The fidelity of protein synthesis is low compared with DNA synthesis, which has proofreading procedures in addition to kinetic proofreading and proper base pairing. The fidelity of protein synthesis is relatively high compared with RNA synthesis, which has only kinetic proofreading and proper base pairing. This makes sense because RNA molecules are temporary.
19. No, a separate synthetase exists for each amino acid, and this synthetase functions for all of the different tRNA molecules for that amino acid.
20. The linkage of amino acids to tRNA is as an aminoacyl ester.
21. The overall process of aminoacyl-tRNA formation is energetically favored because of the energy contributed by the hydrolysis of two phosphate bonds after the amino acid activation step. Without that input of energy, it would not be favorable.

12-4 Prokaryotic Translation

22. The prokaryotic ribosome has a 30S small subunit and a 50S large subunit. The small subunit contains a 16S rRNA and 21 proteins, while the large subunit contains a 5S rRNA, a 23S rRNA, and up to 34 proteins. The total size of the prokaryotic ribosome is 70S.
23. Peptidyl transferase catalyzes the formation of a new peptide bond in protein synthesis. The elongation factors, EF-Tu and

EF-Ts, are required for the binding of aminoacyl tRNA to the A site. The third elongation factor, EF-G, is needed for the translocation step in which the mRNA moves with respect to the ribosome, exposing the codon for the next amino acid. EF-P is thought to help facilitate peptide bond formation.

24. The prokaryotic translation initiation complex in *E. coli* requires mRNA, the 30S ribosomal subunit, fMet-tRNA^{fmet}, GTP, and three protein-initiation factors, called IF-1, IF-2, and IF-3. The IF-3 protein is needed for the binding of mRNA to the ribosomal subunit. The other two protein factors are required for the binding of fMet-tRNA^{fmet} to the mRNA-30S complex.
25. Attachment of the 50S ribosomal subunit to the 30S subunit in the initiation complex is needed for protein synthesis to proceed to the elongation phase.
26. The A site and the P site on the ribosome are both binding sites for charged tRNAs taking part in protein synthesis. The P (peptidyl) site binds a tRNA to which the growing polypeptide chain is bonded. The A (aminoacyl) site binds to an aminoacyl tRNA. The amino acid moiety is the next one added to the nascent protein. The E (exit) site binds the uncharged tRNA until it is released from the ribosome.
27. The stop codons bind to release factors, proteins that block the binding of aminoacyl tRNAs to the ribosome and release the newly formed protein.
28. The Shine–Dalgarno sequence is a purine-rich leader segment of prokaryotic mRNA. It binds to a pyrimidine-rich sequence on the 16S rRNA part of the 30S ribosomal subunit and aligns it for proper translation, beginning with the AUG start codon.
29. Your friend is mistaken. The hydrogen-bonded regions contribute to the overall shape of the tRNA. Hydrogen-bonded regions are also important in the recognition of tRNAs by aminoacyl-tRNA synthetases.
30. Methionine bound to tRNA^{fmet} can be formylated, but methionine bonded to tRNA^{met} cannot be.
31. Different tRNAs and different factors are involved. Initiation requires IF-2, which recognizes fMet-tRNA^{fmet} but not Met-tRNA^{fmet}. Conversely, in elongation, EF-Tu recognizes Met-tRNA^{met} but not fMet-tRNA^{fmet}.
32. The methionine anticodon (UAC) on the tRNA base pairs with the methionine codon AUG in the mRNA sequence that signals the start of protein synthesis.
33. (a) Amino acid activation cycles needed for a protein with 150 AA: 150.
    (b) Initiation cycles needed for a protein with 150 AA: 1.
    (c) Elongation cycles needed for a protein with 150 AA: 149.
    (d) Termination cycles needed for a protein with 150 AA: 1.
34. Four high-energy phosphate bonds per amino acid: two in aminoacyl-tRNA formation, one in elongation with EF-Tu, and one in translocation from the A to the P site, involving EF-G. Forming a peptide bond requires about 5 kcal/mol. This is an expenditure of about 30 kcal/mol of peptide bonds. This is the price of decreasing entropy by connecting amino acids in an ordered, linear chain; it is also the price of high fidelity.
35. Not very precisely. Ignoring any editing or proofreading costs, a maximum value can be calculated in terms of high-energy phosphate bonds. We will designate each phosphate bond as ~P. Four are needed per amino acid, and two are needed per ribonucleotide or deoxyribonucleotide. Therefore, four ~P per amino acid × six ~P per codon × six ~P per DNA triplet = 144 ~P per amino acid (approximately 1050 kcal per mole of amino acid). However, the actual value would be much less because of several factors. A single mRNA molecule can be involved in the synthesis of several to many protein molecules before it is degraded. One gene can be involved in the synthesis of many mRNA molecules, with replication taking place only once per cell generation. In addition, rRNA and tRNA are relatively long-lived and available for repeated protein syntheses.
36. The fact that peptidyl transferase is one of the most conserved sequences in all of biology may indicate that it evolved very early in evolution and that it is so critical for all living organisms that it cannot be modified.
37. Electron microscopy can give information about ribosomal structure and function, but X-ray crystallography has given far more detailed information.
38. Because the tRNAs are bound in proximity to each other on the ribosome, the growing polypeptide chain and the amino acid to be added are also close to each other. This facilitates the formation of the next peptide bond.
39. A virus takes over the protein-synthesizing machinery of the cell. It uses its own nucleic acids and the cell's ribosomes.

12-5 Eukaryotic Translation

40. The other amino acids found in proteins are created by modifying one of the 20 standard amino acids after the protein is made. Selenocysteine is formed while the amino acid is bound to tRNA. Thus, this amino acid is inserted into the protein during translation just like the standard 20 amino acids are.
41. It is unique chemically because it has a selenium ion in it, which takes the place of the sulfur in cysteine. It is also unique in that it appears to be coded for in the DNA sequence, even though the codon would normally be a stop codon.
42. Pro-X-Thr is conserved in RF-1, and Ser-Pro-Phe is conserved in RF-2.
43. The sequence Gly-Gly-Gln (or the GGQ motif).
44. Similarities between protein synthesis in bacteria and protein synthesis in eukaryotes, same start and stop codons; same genetic code; same chemical mechanisms of synthesis; interchangeable tRNAs. Major differences: in prokaryotes, the Shine–Dalgarno sequence and no introns; in eukaryotes, the 5′ cap and 3′ tail on mRNA, and introns have been spliced out.
45. Protein synthesis in prokaryotes takes place as a coupled process, with simultaneous transcription of mRNA and translation of the message in protein synthesis. This is possible because of the lack of compartmentalization in prokaryotic cells. In eukaryotes, mRNA is transcribed and processed in the nucleus and only then exported to the cytoplasm to direct protein synthesis.
46. Some mutations can introduce stop codons. It is useful to a cell to have some mechanism to suppress the formation of incomplete proteins.
47. New protein synthesis is involved in long-term memories.

48. A critical role is played by the transcription factor CREB in turning short-term memories into long-term ones.
49. Animals given drugs that block protein synthesis cannot form new long-term memories, yet their ability to make short-term memories is preserved.
50. The strength, whether due to a single strong impulse or repeated impulses, depolarizes the cell membrane of the nerve. Only ones that are strong enough will be passed along to a receiving neuron. This controls whether a memory can become long term.

12-6 Posttranslational Modification of Proteins

51. Insulin is initially synthesized as preproinsulin, which has a signal peptide that is *cleaved* and removed to produce proinsulin. Three *disulfide bonds* also form, and further *cleavage* removes a central segment to produce mature, active insulin.
52. Hydroxyproline is formed from proline, an amino acid for which there are four codons, by posttranslational modification of the collagen precursor.
53. Chaperones are proteins that bind to proteins while they are being synthesized and aid in their correct folding and keep them from associating with other proteins they should not.
54. Protein synthesis would be less efficient, and many proteins would be misfolded and useless.
55. By studying heat shock proteins in bacteria.
56. No, it was discovered that the ribosomes themselves aid in protein folding.

12-7 Protein Degradation

57. Ubiquitin is a small polypeptide (76 amino acids) that is highly conserved in eukaryotes. When ubiquitin is linked to a protein, it marks that protein for degradation in a proteasome.
58. If proteins to be degraded did not have some signal marking them, the process would take place more randomly and thus be less efficient.
59. Proteasomes degrade proteins in the cytosol and nucleus that have been tagged with ubiquitin; the process is ATP dependent. Lysosomes are organelles that degrade proteins in a nonspecific manner that does not involve ubiquitin tagging; the process involves acidic hydrolases in the acidic environment inside the organelle.
60. If protein degradation took place at any location in a cell, indiscriminate breakdown of functional proteins could take place, so this is an unlikely occurrence. It is much more useful to the cell to have a mechanism for tagging proteins to be degraded and to do so at a specific location in the cell.
61. They are specific sequences that tell the splicing machinery where to splice out introns. A mutation in an ESE could lead to incorrect removal of introns and an entire exon being left out of the final mRNA.
62. A silent mutation is a change in the DNA sequence of a codon that should lead to the same amino acid being inserted. It is a misnomer because we now know that sometimes differences in codons for the same amino acid do affect the overall protein product.
63. If the silent mutation is in an exonic splicing enhancer, then the splicing out of introns could be incorrect, and the correct exon could be skipped.
64. Silent mutations in the mRNA for the enzyme control secondary structures of the mRNA, which controls how fast and how often the mRNA is translated, leading to different levels of the enzyme related to pain tolerance.
65. Marfan syndrome, androgen-insensitivity syndrome, cholesteryl ester storage disease, McArdle disease, and phenylketonuria.
66. Hypoxia-inducible factors (HIFs).
67. When oxygen is low, a prolyl hydroxylase cannot function to hydroxylate a specific proline residue on HIFα. When hydroxylated, HIFα is targeted for protein degradation. Thus, in low oxygen, HIFα survives long enough to do its job, which is to bind to HIFβ and activate transcription of the genes that produce more red blood cells.

Chapter 13

13-1 Purification and Detection of Nucleic Acids

1. The DNA run on electrophoresis gels is usually cleaved with restriction enzymes to give linear pieces; thus, the shape is uniform for the DNA. The charge is a constant for DNA in that every nucleotide has the same charge due to the phosphate groups; thus, DNA has a uniform shape and a uniform charge-to-mass ratio, so it separates solely on size, with the shorter fragments traveling fastest through the gel.

13-2 Restriction Endonucleases

2. Restriction endonucleases typically recognize and cleave specific palindromic DNA sequences. In biochemistry, they are used to cut DNA in a predictable way for cloning, mapping, or analysis.
3. The use of restriction endonucleases with different specificities gives overlapping sequences that can be combined to give an overall sequence.
4. Restriction endonucleases do not hydrolyze a methylated restriction site.
5. The restriction site of the DNA of the organism that produces a restriction endonuclease is modified, usually by methylation.
6. An endonuclease is an enzyme that cuts nucleic acid chains in the middle, as opposed to cleaving from the ends inward. The term *restriction* came from the restricted growth seen in host cells that are infected by bacteriophages when the bacteria have restriction enzymes that can cleave the viral DNA.
7. Three of the many examples of DNA palindromes are GGATCC, GAATTC, and AAGCTT. (Remember: These are listed 5′ to 3′, so you must read the complementary strand 5′ to 3′ to see that the sequence is the same.)
8. *Hae*III cuts at a sequence of four bases, cuts in the middle of the sequence, and leaves blunt ends. *Bam*HI cuts at a sequence of six bases, cuts on the second base from the 5′ end, and leaves sticky ends.
9. Sticky ends are short regions of single-stranded DNA extending from the ends of double-stranded DNA molecules. These are produced by some restriction enzymes or can be added chemically to blunt-ended double-stranded DNA. They are important because they provide a means for DNA from different sources (e.g., insert DNA and plasmid, both containing sticky ends) to find each other by hydrogen bonding

between complementary bases. A ligase is then used to covalently link the two molecules.

10. An advantage of using *Hae*III is that it yields blunt ends. Thus, one could combine DNA cut with this enzyme with any other DNA that also had blunt ends. Enzymes exist that quickly remove the sticky overhangs from other restriction enzymes. The disadvantage is that *Hae*III is specific for a four-base sequence that is likely to occur many times in a genome, so the target DNA may also be cleaved somewhere in the middle. Also, the blunt ends make it more difficult to get specific ligation of two DNA types.

13-3 Cloning

11. To produce recombinant DNA, a restriction enzyme is used to cut both the plasmid and the insert DNA at specific sequences. The fragments are then joined using DNA ligase, forming a recombinant plasmid. The recombinant plasmid can be introduced into bacterial cells using a process called transformation.
12. A portion of exogenous DNA is introduced into a suitable vector, frequently a bacterial plasmid, and many copies of the DNA are produced when the bacteria grow.
13. The plasmid to be used as a vector needs markers both for the uptake of the target DNA sequence into the plasmid and for the insertion of the plasmid into host cells. Typically, a plasmid has a gene for ampicillin resistance. Only cells that have taken up a plasmid can grow on ampicillin plates. The foreign DNA is usually inserted into a second marker to select for those plasmids that took up the target DNA. This second marker may be another gene, such as the β-galactosidase gene, which allows for blue/white screening (see Figure 13.11).
14. See Figure 13.11. The key feature of a plasmid capable of blue/white screening is the gene for the α-subunit of the enzyme β-galactosidase. These plasmids are used with a strain of *E. coli* that is deficient in the α-subunit of this enzyme. β-Galactosidase can convert a colorless sugar derivative, called X-gal, to a blue one. The site for cleavage of the plasmid by a restriction endonuclease lies within the β-galactosidase gene. Cells that have acquired a plasmid can grow on ampicillin. If the plasmid reclosed on itself without the target DNA, the colonies that took up that plasmid would grow blue. Cells that have acquired the DNA insert cannot produce a blue color because the DNA insert has broken up the β-galactosidase gene. Therefore, the β-galactosidase enzyme cannot be made, and X-gal cannot be converted to its blue derivative.
15. Restriction enzymes to cut DNA, DNA ligase to rejoin DNA, a suitable vector to carry the insert DNA, a cell line to accept the vector, and a way of selecting for the correct transformants.

13-4 Genetic Engineering

16. To increase disease resistance, resistance to pests, shelf life, level of nitrogen fixation (protein content), and resistance to temperature extremes.
17. The corn being grown in the field has been genetically engineered. The gene that was introduced came from the bacterium *Bacillus thuringiensis.*
18. Each of the subunits is coded for by a separate gene, so in order to clone LDH 3, one would have to clone the gene for the M subunit and also separately clone the gene for the H subunit. Each gene would be cloned into an expression cell line, and the proteins would be expressed and purified. The individual subunits could then be combined in a buffer solution, and they would form tetramers, some of which would be LDH 3. This could be verified by native gel electrophoresis.
19. An expression vector, such as the pET 5 plasmid, has the components of any normal cloning vector (e.g., an origin of replication, a selectable marker, and a multiple cloning site), but it also has the ability to have the inserted DNA transcribed by having a promoter for RNA polymerase, such as T7 polymerase, and a termination sequence. These border the multiple cloning site. These vectors are used with a cell line that makes T7 RNA polymerase when induced.
20. A distinguishing feature of a protein expression vector is a strong, inducible promoter, such as a T7 RNA polymerase promoter. This is important because it allows for a high level of transcription of the gene of interest, and it can be regulated to control the timing and amount of protein expression.
21. The DNA sequence to be inserted in the bacterial plasmid to direct the production of α-globin should be cDNA, which is a sequence complementary to the mRNA for α-globin. The α-globin cDNA would not have introns because it would be produced from an mRNA template in a reaction catalyzed by reverse transcriptase.
22. CRISPR-Cas9 uses a guide RNA to direct the Cas9 enzyme to a specific DNA sequence, where the enzyme creates a double-strand break. The cell then repairs the double-strand break in ways that can disrupt, introduce, or modify genes for gene therapy purposes.

13-5 DNA Libraries

23. A DNA library is a collection of cells that carry cloned pieces of the entire DNA genome of an organism. A cDNA library is made by taking the mRNA from an organism, converting it to cDNA using the reverse transcriptase enzyme, and cloning only the cDNA for the library. In this way, active DNA sequences that lead to expressed proteins are stored in cDNA libraries.
24. The amount of work involved in constructing a DNA library makes it desirable to have such libraries available to the entire scientific community, thus avoiding duplication of effort.
25. A fusion protein is a combination of a protein coded for by an expression vector and the target gene. A common example is a histidine tag and enterokinase, which are linked to the target protein when transcribed and translated. They are used to help with the eventual purification of the target protein. The overexpressed target protein can be quickly separated from the rest of the host's proteins by purifying the fusion protein, which has characteristics that make it easy to purify.

13-6 The Polymerase Chain Reaction

26. The polymerase chain reaction depends on repeated cycles of separation of DNA strands, followed by annealing of

primers and extension of the primers by DNA polymerase. Each step requires significantly different temperatures, giving rise to the requirement for strict temperature control.

27. Part of the procedure of the polymerase chain reaction requires the use of high temperatures. When a temperature-stable DNA polymerase is used, there is no need to add fresh batches of enzyme for each round of amplification. This would be necessary, however, if the DNA polymerase could not withstand the high temperatures.

28. Good primers have similar G–C contents so as to have similar annealing temperatures. They have minimal secondary structure possibilities with each other or with themselves and are long enough to give sufficient specificity for the gene to be duplicated without costing too much.

29. The contaminating DNA as well as the desired DNA are amplified at each stage of the polymerase chain reaction, giving rise to an impure product.

30. (a) The primers have very different G–C contents.
    (b) The forward primer will have significant secondary structure with itself (hairpin loop due to inverted Gs and Cs on the ends).
    (c) The forward and reverse primers will bind to each other.

31. It is a technique that allows the PCR reaction to generate time-point data that can be used to determine how much DNA was in the original sample.

32. Regular PCR is designed to create large quantities of DNA, so the reaction is allowed to go to completion. With qPCR, the reaction does not go to completion because it is the time-point data that are needed in order to determine the amount of the starting material.

13-7 DNA Fingerprinting

33. The polymerase chain reaction can increase the amount of a desired DNA sample by a considerable factor, making possible the definite identification of DNA samples that were too small to be characterized by other means. It can be used on hair and blood samples found at the scene of a crime to establish the presence of a suspect. This method can also be used to identify the remains of possible murder victims.

34. It is easier to show that two DNA samples do not match than to prove that they are identical.

13-8 Sequencing DNA

35. 5′GATGCCTACG3′.

36. Two factors are involved here. First, large polymers must be cleaved into smaller, manageable fragments for sequencing. Enzymes (endoproteases) that cleave proteins, while showing some specificity, are far from absolutely specific, and messy mixtures result. On the other hand, restriction endonucleases are absolutely specific for palindromic base sequences in DNA, and "clean" cuts result, allowing for easier purification. (If the gene for a protein isn't available but the mRNA is, then the resulting cDNA can be made using reverse transcriptase.) A second factor is that only relatively short fragments of protein can be sequenced without additional internal cleavage. For example, the Edman degradation is limited to peptides of about 50 amino acids or fewer. With DNA, the dideoxy method, coupled with polyacrylamide-gel separation, can handle DNA fragments 10 to 20 times longer.

37. DNA often has introns in the gene, so knowing the DNA sequence may give the wrong answer for the final protein sequence. Also, proteins are modified posttranslationally, so there may be modifications to the protein sequence not reflected in the DNA.

13-9 Genomics and Other "-omics"

38. The genome is the total DNA of a cell, containing all the genes of that organism. The proteome is the total complement of proteins.

39. Yes, a proteomic analysis has been done on the fruit fly, *Drosophila melanogaster*.

13-10 DNA and Protein Microarrays

40. Using robotic technology, a slide or "chip" is loaded with thousands of specific single-stranded DNA sequences. RNA is collected from samples to be tested and converted to cDNA carrying a fluorescent tag. The sample is placed over the chip, and the cDNA is allowed to bind. A fluorometer measures the fluorescence from the chip and indicates which DNA sequences were bound with their corresponding cDNA. This tells researchers which genes were active because only the active genes would produce RNA.

41. Yeast could be grown under the two conditions, and the mRNA collected. The mRNA could then be converted to cDNA, and each population could be labeled with a different color fluorescent marker. These samples could then be overlaid on a gene chip containing the yeast genome. The color of the fluorescence on the gene chip would then indicate which genes were active under the two conditions.

42. Cancerous cells have altered metabolism at the genetic level. The gene expression patterns seen in patients with known types of cancer act like a fingerprint of that type of cancer. Tissue samples from patients to be diagnosed can be used to collect the RNA and convert it to cDNA. These cDNA samples are then overlaid on a gene chip of the human genome, and the binding pattern is analyzed through fluorescence. The pattern seen can then be compared to the patterns seen in the known cancers to aid in the diagnosis.

43. DNA microarrays have thousands of bound single-stranded DNA spots. They are used to test for the presence of the corresponding mRNA in a biological sample via cDNA produced from the mRNA. Protein arrays, on the other hand, have applied samples of very specific and pure antibodies. Biological tissue samples are placed on the protein chip. If the antigens for the specific antibodies are present, they bind to the antibodies. Another set of antibodies with fluorescent labels is then added, and the chip is analyzed with a fluorometer. The patterns seen show which antigens the tissue sample contains, which can be used to diagnose the patient.

Chapter 14

14-1 Viruses

1. Some viruses have DNA, and some have RNA. In some cases, a viral genome is single-stranded, and in others, it is double-stranded.

2. (a) The virion is the entire virus particle.
   (b) The capsid is the protein coat that surrounds the viral nucleic acid.

(c) The nucleocapsid is the combination of the nucleic acid and the capsid.

(d) A protein spike is a membrane-bound protein that is used to help the virus attach to its host.

3. Three factors determining the family of a virus are whether its genome is DNA or RNA, whether it has a membrane envelope, and whether the nucleic acid is single- or double-stranded.
4. The virus attaches to a specific protein on the host cell's membrane and injects its nucleic acid into the cell.
5. In the lytic pathway, the viral nucleic acid is replicated in the host cell and packaged into new virus particles that lyse the host cell. In the lysogenic pathway, the viral DNA is incorporated into the host DNA, and the host cell is not destroyed.
6. One good choice would be a drug that attacks one of the specific protein spikes on the virus. This may be an antibody that attacks it or a drug that blocks its ability to attach to the host cell. Another choice would be a drug that inhibits a key viral enzyme, such as the reverse transcriptase of a retrovirus, or the enzymes involved in repackaging the viruses.
7. Viruses can switch from one pathway to another, based on the condition of the host cells. If the host is healthy, there is sufficient material to allow the virus to replicate and to produce new virions. If the host cell is starved or unhealthy, there may be insufficient energy and material to do so. In this case, lysogeny allows the DNA to incorporate into the host cell, where it can wait until the cell's health improves.
8. One example would be someone who had helper T cells lacking a CD4 receptor. The HIV virus must bind to the CD4 receptor as part of its attachment process.

14-2 Retroviruses

9. A retrovirus has an RNA genome that must pass through a stage in which it is reverse-transcribed to DNA, and this DNA must recombine with the host's DNA.
10. Reverse transcriptase.
11. One reason is that retroviruses have been linked to cancer. A second reason is that the human immunodeficiency virus (HIV) is a retrovirus. A third reason is that retroviruses can be used in gene therapy.
12. Gene therapy is the introduction, removal, or modification of genetic material in a patient's cells to treat a genetic disorder.
13. Two types are the Moloney murine leukemia virus (MMLV) and adenovirus. Both must be manipulated so that the critical genes for replication are removed and replaced with an expression cassette containing the therapeutic gene.
14. When retroviruses, such as MMLV, are used, there is the danger that the therapeutic gene will incorporate in a place that will disrupt another gene. In more cases than would be predicted by random chance, this seems to occur in a place that disrupts a tumor-suppressor gene, causing cancer. There is also the danger that the patient will have a strong reaction to the virus used to introduce the therapeutic gene. In at least one case, this has had fatal consequences.

14-3 The Immune System

15. Innate immunity refers to a variety of protective processes, including skin, mucus, and tears as a first line of defense, and dendritic cells, phagocytes, macrophages, and natural killer cells as a second line of defense. These are always present, and the innate-immunity cells are always circulating in the body. Acquired immunity refers to the processes involving B cells and T cells, in which specific sets of them are activated in response to an antigen challenge, and these subsets then multiply.
16. One part includes physical barriers such as skin, mucus, and tears. The cells of the innate immune system are dendritic cells, macrophages, and natural killer (NK) cells.
17. B cells, which make antibodies; killer T cells, which attack infected cells; and helper T cells, which help activate B cells.
18. T cells mature in the thymus, and B cells mature in the bone marrow. Both types of cells originally come from stem cells located in the bone marrow.
19. MHCs are receptors on antigen-presenting cells. They bind to fragments of antigens that have been degraded by the infected cell and display them on their surface. T cells then bind to the infected cells.
20. Clonal selection refers to the process in which a particular T cell or B cell is stimulated to divide. Only the one bearing the correct receptor for the antigens being presented is selected.
21. The cells of the innate system initially attack a pathogen such as a virus, bacterium, or even a cancerous cell. They then present antigens from the pathogen on their surfaces via their MHC proteins. The acquired immunity cells then recognize the MHC/antigen complex, bind to it, and begin the involvement of the acquired immunity system.
22. Interferon is a cytokine produced in very small quantities that stimulates natural killer cells, which attack cancerous cells. One of the first treatments for cancer was to give the patient interferon to stimulate NK cells. Having a large supply of cloned interferon is helpful, therefore, in fighting cancer.
23. When T cells and B cells are developing, they are, in a sense, "trained." If they contain receptors that recognize self-antigens, they are eliminated when they are still young. If they don't ever see any antigens they recognize, then they die by neglect. This leaves a set of precursors to T cells and B cells with receptors that recognize foreign antigens but not self-antigens.
24. Macrophages, part of the innate immune system, are a "double-edged sword." Their presence is important to attack cancer cells, and if they do a thorough job, then the cancer cells are all destroyed. However, they also cause inflammation, which has recently been shown to indirectly lead to the progression of the cancer cells that survive.
25. An antibody is made up of four polypeptide chains: two identical heavy chains and two identical light chains. These chains are bonded through disulfide bonds to form a Y-shaped structure.
26. The small noncoding RNA (ncRNA) of the herpes virus has been linked with its ability to evade the immune system.
27. The herpes virus produces an ncRNA that stabilizes the respiratory chain of the mitochondria of the host cell. This prevents the early destruction of the infected cell by the host's immune system. At the same time, a microRNA (miRNA) produced by the virus inhibits production of a protein on the surface of the cell that would otherwise attract NK cells.

14-4 Cancer

28. Cancer cells continue to grow and divide in situations in which normal cells do not, such as when they are not receiving growth signals from surrounding cells. They also continue to grow even if surrounding tissues are sending out "stop growth" signals. Cancer cells can co-opt the body's vascular system, causing the growth of new blood vessels to supply the cancerous cells with nutrients. Cancer cells are essentially immortal. They can continue to grow and to divide indefinitely. Cancer cells can break loose, travel to other parts of the body, and create new cancerous areas, a process known as metastasis.
29. A tumor suppressor is a molecule that restricts the ability of a cell to grow and to divide. An oncogene is a gene whose product stimulates a cell to grow and to divide.
30. The protein called p53 is a tumor suppressor. Mutations of p53 have been found in more than half of all human cancers. Ras is involved in cell division, and mutations in this protein are involved in 30% of human tumors.
31. Viruses have been implicated in many cancers. Retroviruses are particularly dangerous because they insert their DNA into the host's DNA. When this happens in a tumor-suppressor gene, the tumor suppressor is inactivated, causing cancer. Also, the homology between proto-oncogenes and oncogenes makes it likely that the infection cycle of viruses may be responsible for some proto-oncogenes becoming oncogenic.
32. Virotherapy is the process of using a virus to attempt to treat cancer. There are two strategies for virotherapy. One is to use the virus to attack and kill cancer cells directly. In this case, the virus has a protein on its surface that is specific for a cancer cell. Once inside, it kills the cancer cell. The second is to have the virus ferry a gene into the cancer cell that makes the cell more susceptible to a chemotherapeutic agent.
33. A tumor suppressor is a protein that helps control cell growth and division. It is like the brakes on a car, trying to slow down a process. Many cancers are related to mutations and inactivation of tumor suppressors. An oncogene produces something that stimulates growth and division. This is like the accelerator of the car. Many other cancers are caused ultimately by overactivation of an oncogene.
34. When a growth factor binds to a receptor tyrosine kinase, the receptor dimerizes and autophosphorylates. The phosphorylated receptor is then recognized by GRB2, which binds to the Ras exchanger SOS. SOS is activated to exchange GDP for GTP on Ras, activating it. Ras moves Raf to the cell membrane, where it becomes active. Raf phosphorylates and activates mitogen-activated protein kinase kinase (MAPKK) and the rest of the pathway (see Figure 14.29).
35. *Ras*, *Jun*, and *Fos* are all considered oncogenes. In the process of cell division, the Ras protein is a necessary component, but it is usually active only when the cell should be dividing. Oncogenic forms of Ras are overactive and lead to too much cell division. Ras is an early step in the process. Jun and Fos are transcription factors that together make up AP-1, which is involved in the transcription activation pathway involving CBP.
36. Because p53 is a tumor suppressor that is inactivated in many human cancers, p53 reactivation is a cancer-fighting strategy aimed at restoring the tumor-suppressing activity of p53.
37. p53 can be restored in several ways. One way would be through gene therapy to give the patient functioning copies of the p53 gene if they lack it. Another way is to neutralize molecules that naturally inhibit p53. Another is to give the patient drugs that stimulate the production of p53 by stimulating the transcription of the p53 gene. Finally, one could use drugs that would inhibit the transcription of molecules that act as inhibitors of p53.
38. The innate immune system is instrumental in fighting cancer cells. Cells that turn cancerous display antigens on their surfaces that act as a help signal. Cells of the innate immune system, such as macrophages and natural killer cells, attack cells that display these cancer-linked antigens on their surfaces. Often, they destroy the cancerous cell, ending the threat. However, if they do not, the presence of the innate immune cell can lead to inflammation. More and more research is showing that inflammation is the switch that takes a precancerous cell and turns it into a full-fledged cancer cell. Thus, innate immune cells that attack a cancer cell but fail to kill it may just make it stronger.
39. They found more than 30,000 mutations in the melanoma genome and more than 23,000 in lung cancer. This knowledge is helping researchers understand cancer development and improve early detection methods.
40. Most of the mutations associated with melanoma arise from too much exposure to the sun. Likewise, smoking causes most of the DNA errors in lung cancer.

14-5 AIDS

41. Helper T cells.
42. The virus binds to the CD4 receptor on the helper T cell via a viral protein called gp120.
43. HIV is difficult to kill because it is difficult to find. The reverse transcriptase of HIV is very inaccurate in replication. The result is rapid mutation of HIV. There is also a conformational change of the gp120 protein when it binds to the CD4 receptor on a T cell, making it difficult to create antibodies that will mount a useful attack. HIV is also adept at evading the innate immune system. Lastly, HIV hides from the immune system by cloaking its outer membrane in sugars that are very similar to the natural sugars found on most of its host's cells, making it harder for the immune system to recognize it.
44. The two main approaches are creating vaccines to the virus and using antiretroviral therapy.
45. No, the virus does not kill its host directly. It weakens the immune system via its attack on helper T cells. When the immune system is sufficiently weak, the host dies from opportunistic infections.
46. The gp120 protein on the HIV virus is the molecule that docks with the CD4 receptor on the helper T cell, so it is very important for the action of HIV. The problem is that the gp120 protein changes conformation when it binds, making an attempt to use purified gp120 to make antibodies ineffective. There is also a high mutation rate of the gp120 protein.
47. ART stands for antiretroviral therapy. It is based on using a combination of drugs to attack multiple parts of the HIV life cycle.

Chapter 15

15-1 Standard States for Gibbs Energy Changes

1. There is a connection, and it is one of the most important points in this chapter. It can be expressed by the equation $\Delta G^{\circ\prime} = -RT \ln K'_{eq}$.
2. Reaction (a) is an endergonic reaction ($\Delta G > 0$), and it would take place only if it is coupled to an exergonic reaction. Reaction (b) would proceed as written.
3. The information given here deals with the thermodynamics of the reaction, not the kinetics. It is not possible to predict the rate of the reaction.
4. The Gibbs energy change can be used to predict the spontaneity of a reaction under conditions of constant temperature and pressure.
5. Many endergonic reactions are necessary for life processes. They need a source for the energy they require, and that source is the energy released by exergonic reactions.

15-2 A Modified Standard State for Biochemical Applications

6. In biochemistry, the standard conditions used to define the standard Gibbs energy change ($\Delta G^{\circ\prime}$) are 25°C, 1 atm, 1 *M* concentrations of all solutes, and pH 7.
7. The usual thermodynamic standard state refers to pH = 0. This is not very useful in biochemistry because most cells have a pH value near 7.
8. Statement (a) is true, but statement (b) is not. The standard state of solutes is normally defined as unit activity (1 *M* for all but the most careful work). In biological systems, the pH is frequently in the neutral range (i.e., H^+ is close to 10^{-7} *M*); the modification is a matter of convenience.
9. The designation $\Delta G^{\circ\prime}$ indicates a biological standard state. If the prime is omitted, then it is for chemical standard states.
10. No, there is no relationship between the thermodynamic quantity $\Delta G^{\circ\prime}$ and the speed. The $\Delta G^{\circ\prime}$ reflects the thermodynamic possibility under biological standard state. Speed is a kinetic quantity that is based on the ability of an enzyme to catalyze the reaction and the real substrate concentrations in the cell.
11. See Apply Your Knowledge 15.1 for sample work. Assuming one significant figure, the $\Delta G^{\circ\prime}$ values here are −20 kJ/mol, 0 kJ/mol, and 30 kJ/mol, respectively.
12. Adding the $\Delta G^{\circ\prime}$ values for the reactions gives an overall $\Delta G^{\circ\prime} = -260.4$ kJ/mol or −62.3 kcal/mol. The conversion of citrate to α-ketoglutarate is exergonic because it has a large, negative $\Delta G^{\circ\prime}$.
13. Reaction (a) will not proceed as written; $\Delta G^{\circ\prime} = +12.6$ kJ. Reaction (b) will proceed as written; $\Delta G^{\circ\prime} = -20.8$ kJ. Reaction (c) will not proceed as written; $\Delta G^{\circ\prime} = +31.4$ kJ. Reaction (d) will proceed as written; $\Delta G^{\circ\prime} = -18.0$ kJ.
14. Two considerations are involved here. (a) Very rarely, if ever, are *in vivo* concentrations standard 1 *M* concentrations; actual ΔG (not ΔG°) values are very dependent on local concentrations, especially if the number of reactant molecules and product molecules is not the same. (b) Values of ΔG° rigorously apply only to *closed* systems that can reach equilibrium. Biochemical systems, however, are *open* systems and do not reach equilibrium.

15-3 The Nature of Metabolism

15. Metabolism is the sum of all catabolic and anabolic reactions in a cell. Catabolism involves breaking down molecules to release energy, whereas anabolism involves building complex molecules from reactions that use energy.
16. Group 1: catabolism, oxidative, energy-yielding. Group 2: anabolism, reductive, energy-requiring.
17. A closed system can exchange energy but not matter with its surroundings, while an open system can exchange both energy and matter with its surroundings.
18. The local decrease in entropy associated with living organisms is balanced by the increase in the entropy of the surroundings caused by their presence. Coupling reactions lead to an overall dispersal of energy in the Universe.
19. The synthesis of sugars by plants in photosynthesis is endergonic and requires light energy from the Sun.
20. The biosynthesis of proteins is endergonic and is accompanied by a large decrease in entropy.
21. The ATP constantly generated by living organisms is used as a source of chemical energy for endergonic processes. There is a good deal of turnover of molecules but no net change.

15-4 The Role of Oxidation and Reduction in Metabolism

22. An alkane carbon is first oxidized to an alkene carbon and then hydrated to a hydroxyl carbon. Finally, the hydroxyl carbon is oxidized to a carbonyl carbon.
23. (a) NADH is oxidized, $H^+ + NADH \rightarrow NAD^+ + 2e^- + 2H^+$. The aldehyde is reduced, $CH_3CH_2CHO + 2e^- + 2H^+ \rightarrow CH_3CH_2CH_2OH$.
    (b) Fe^{2+} is oxidized, $Fe^{2+} \rightarrow Fe^{3+} + e^-$. Cu^{2+} is reduced, $Cu^{2+} + e^- \rightarrow Cu^+$.
24. (a) The aldehyde is the oxidizing agent; NADH is the reducing agent.
    (b) Cu^{2+} is the oxidizing agent; Fe^{2+} is the reducing agent.

15-5 Coenzymes in Biologically Important Oxidation–Reduction Reactions

25. See Figure 15.6. NAD^+ acts as an oxidizing agent by accepting the electrons of a hydride ion (H^-) at its nicotinamide ring, forming NADH, and thereby allowing a substrate to lose electrons and be oxidized.
26. NAD^+, $NADP^+$, and FAD all contain an adenosine diphosphate (ADP) moiety.
27. In NADPH, the 2′ hydroxyl of the ribose attached to the adenine has a phosphate attached.
28. There is no effect on how the structural difference affects redox chemistry, but the presence of the phosphate in NADPH can be distinguished by enzymes in different pathways. For example, NADH is mainly used as a redox coenzyme in catabolic pathways, whereas NADPH is mainly used as a redox coenzyme in anabolic pathways.
29. NAD^+ is the oxidizing agent in reactions in which a nutrient is the substance to be oxidized, so it is a reactant in such reactions. Like all oxidizing agents, NAD^+ is reduced, producing NADH.

30. None of these statements is true. Some coenzymes are involved in group-transfer reactions such as one-carbon transfer reactions (see Section 7-8). Many coenzymes contain phosphate groups, and coenzyme A contains sulfur. ATP does not represent stored energy but is generated on demand.

31. The second half reaction (the one involving NADH) is that of oxidation; the first half reaction (the one involving O_2) is that of reduction. The overall reaction is $\frac{1}{2}O_2 + NADH + H^+ \rightarrow H_2O + NAD^+$. O_2 is the oxidizing agent, and NADH is the reducing reagent.

32. See Figures 15.6 and 15.7.

33. Glucose-6-phosphate is oxidized, and $NADP^+$ is reduced. $NADP^+$ is the oxidizing agent, and glucose-6-phosphate is the reducing agent.

34. FAD is reduced, and succinate is oxidized. FAD is the oxidizing agent, and succinate is the reducing agent.

35. It is important to have two different pools of redox coenzymes. In the cytosol, the NAD^+/NADH ratio is high, and the NADPH/$NADP^+$ ratio is also high. This means that anabolic reactions using NADPH can take place in the cytosol, while catabolic reactions using NAD^+, such as glycolysis, can also take place in the cytosol. If there were not two different pools of these coenzymes, no single cell location could have both catabolism and anabolism. Having two different, but structurally related, reducing agents helps keep anabolic and catabolic reactions distinct from each other.

15-6 Coupling of Energy Production and Energy Use

36. Only a portion of the energy released by exergonic reactions drives endergonic reactions. An example seen in Section 15-6 is the exergonic oxidation of glucose to two lactate ions, which releases 184.5 kJ for each mole of glucose. This reaction is coupled to the phosphorylation of two ADP to two ATP, an endergonic reaction that requires 61 kJ for each mole of glucose. The two amounts of energy are not the same.

37. Without the release of chemical energy in exergonic reactions, the endergonic reactions of metabolism, especially those that synthesize DNA and proteins, could not take place.

38.

Creatine phosphate + ADP → Creatine + ATP; $\Delta G^{\circ\prime} = -12.6$ kJ

ATP + Glycerol → ADP + Glycerol – 3 – phosphate; $\Delta G^{\circ\prime} = -20.8$ kJ

Overall:
Creatine phosphate + Glycerol → Creatine + Glycerol-3-phosphate;

$\Delta G^{\circ\prime}$ overall $= -33.4$ kJ

39. Glucose-1-phosphate → Glucose + P_i; $\Delta G^{\circ\prime} = -20.9$ kJ/mol

Glucose + P_i → Glucose-6-phosphate; $\Delta G^{\circ\prime} = +12.5$ kJ/mol

Glucose-1-phosphate → Glucose 6-phosphate; $\Delta G^{\circ\prime} = -8.4$ kJ/mol

40. In both pathways, the overall reaction is $ATP + 2\,H_2O \rightarrow AMP + 2\,P_i$. Thermodynamic parameters, such as energy, are additive. The overall energy is the same because the overall pathway is the same.

41. Phosphoarginine + ADP → Arginine + ATP; $\Delta G^{\circ\prime} = -1.7$ kJ

ATP + H_2O → ADP + P_i; $\Delta G^{\circ\prime} = -30.5$ kJ

Phosphoarginine + H_2O → Arginine + P_i; $\Delta G^{\circ\prime} = -32.2$ kJ

42. ATP is less stable than ADP and P_i because of the charge distribution and loss of the resonance stabilization in the phosphate ion. There is stabilization (dispersal of energy) when ATP is hydrolyzed, leading to a negative Gibbs energy change.

$$ADP + P_i + H^+ \longrightarrow ATP + H_2O \qquad \Delta G^{\circ\prime} = 30.5\ \text{kJ mol}^{-1} = 7.3\ \text{kcal mol}^{-1}$$

or in structural form,

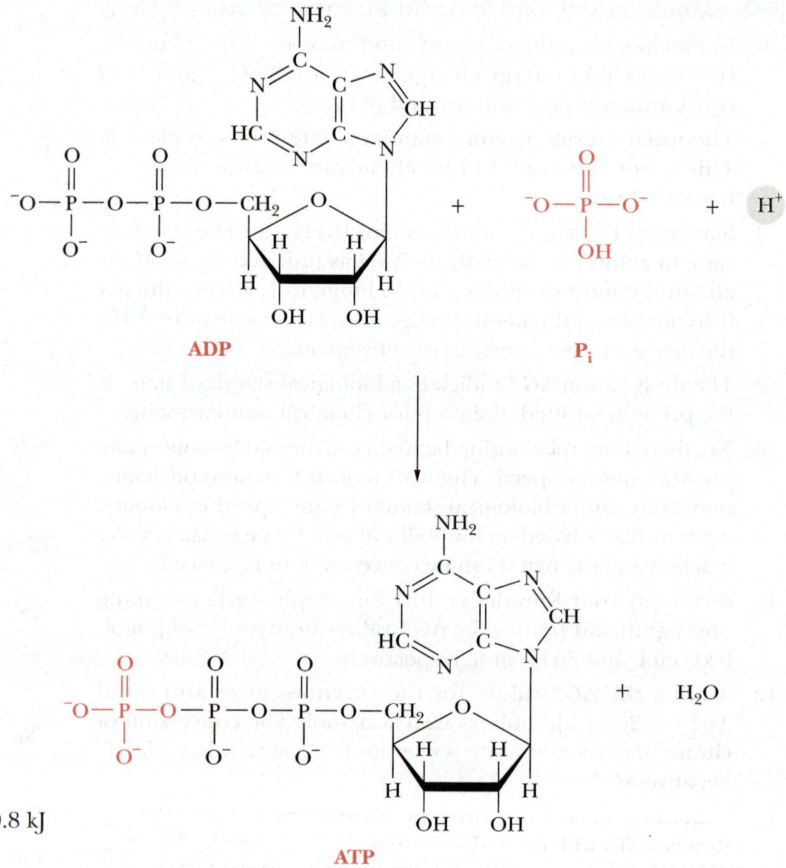

43. It is intermediate; thus, ATP is ideally positioned to serve as a phosphate donor or (as ADP) as a phosphate acceptor, depending on local concentrations.

44. There is a large increase in entropy accompanying the hydrolysis of one molecule to five separate molecules.

45. PEP is a high-energy compound because energy is released upon its hydrolysis, owing to the resonance stabilization of the inorganic phosphate released and the possible keto–enol tautomerization of its product, pyruvate. See Figure 15.12.

15-7 Metabolic Activation

46. An activation step leads to an exergonic next step in a pathway. It is similar to the way in which organic chemists want to attach a good leaving group for the next step in a series of reactions.
47. Small energy changes generally involve mild conditions. Also, such reactions are more sensitive to relatively small changes in concentration and thus are easier to control.
48. Thioesters are high-energy compounds because they easily dissociate into stable reaction products. In general, the more stable the products of a reaction, the less energy is required for the reaction to proceed, and the more energy is released during the reaction.
49. Coenzyme A serves several purposes. It forms high-energy thioester bonds with acetyl or acyl groups, activating molecules for further reactions. It is also used as a tag to "earmark" a molecule for a particular pathway. It is large and cannot cross membranes, contributing to the compartmentalization of pathways.

Chapter 16

16-1 Sugars: Their Structures and Stereochemistry

1. A **polysaccharide** is a polymer of more than 10 simple sugars bonded together. A **furanose** is a cyclic sugar that contains a five-membered ring similar to that of furan. A **pyranose** is a cyclic sugar that contains a six-membered ring similar to that of pyran. An **aldose** is a sugar that contains an aldehyde group, and a **ketose** is a sugar that contains a ketone group. A **glycosidic bond** is the acetal linkage that joins two sugars. An **oligosaccharide** is a compound formed by the linking of 2–10 simple sugars (monosaccharides) through glycosidic bonds. A **glycoprotein** is formed by the covalent bonding of sugars to a protein.
2. All monosaccharides contain an aldehyde or ketone group and a hydroxyl group on each of their remaining carbon atoms.

Furan

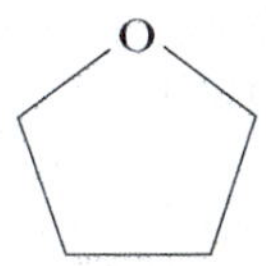
Furanose

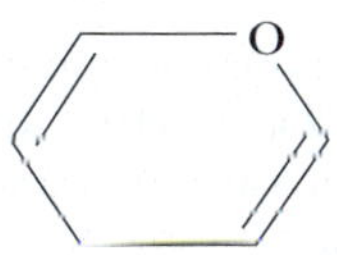
Pyran

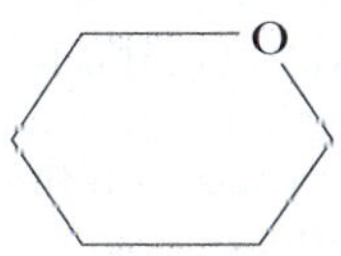
Pyranose

3. D-Mannose and D-galactose are both epimers of D-glucose, with inversion of configuration around carbon atoms 2 and 4, respectively. D-ribose has only five carbons, whereas the rest of the sugars named in this question have six.
4. **Enantiomers** are stereoisomers that are nonsuperimposable mirror images, differing at all chiral centers. **Diastereomers** are stereoisomers that are *not* mirror images and differ at one or more chiral centers. **Epimers** are a specific type of diastereomer that differ in configuration at only one chiral center.
5. Four epimers of D-glucose exist, each with inversion of configuration at a single carbon. The carbons at which this is possible are those numbered two through five.
6. Furanoses and pyranoses have five-membered and six-membered rings, respectively. It is known from organic chemistry that rings of this size are the most stable and the most readily formed.
7. There are four chiral centers in the open-chain form of glucose (carbons two through five). Cyclization introduces another chiral center at the carbon involved in hemiacetal formation, giving a total of five chiral centers in the cyclic form.
8. Enantiomers: (A) and (F), (B) and (D). Epimers: (A) and (C), (A) and (D), (A) and (E), (B) and (F).

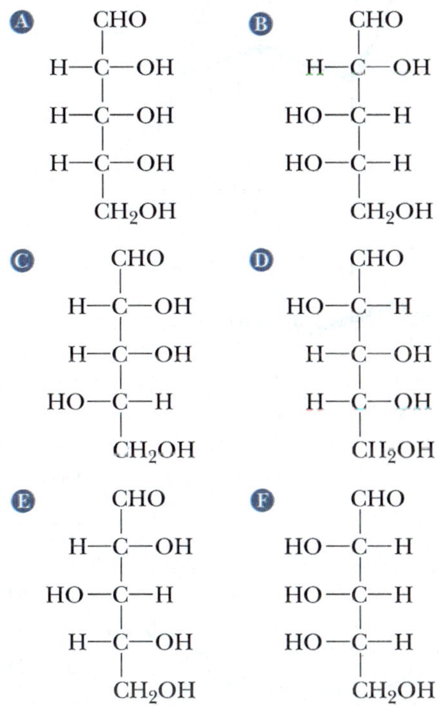

9. Arabinose is an epimer of ribose. Nucleosides in which arabinose is substituted for ribose act as inhibitors in reactions of ribonucleosides.

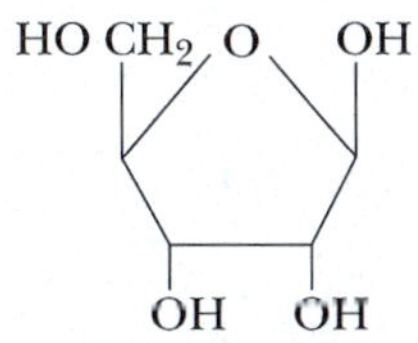
Ribose

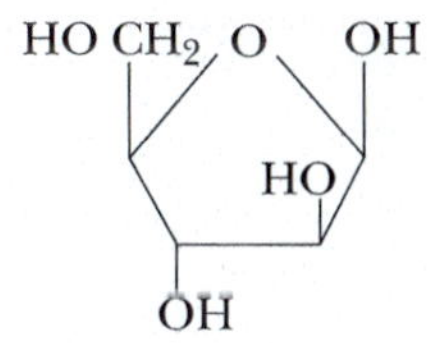
Arabinose

10. Converting a sugar to an epimer requires inversion of configuration at a chiral center. This can be done only by breaking and reforming covalent bonds.
11. Two different orientations with respect to the sugar ring are possible for the hydroxyl group at the anomeric carbon. These two possibilities give rise to a new chiral center.
12. 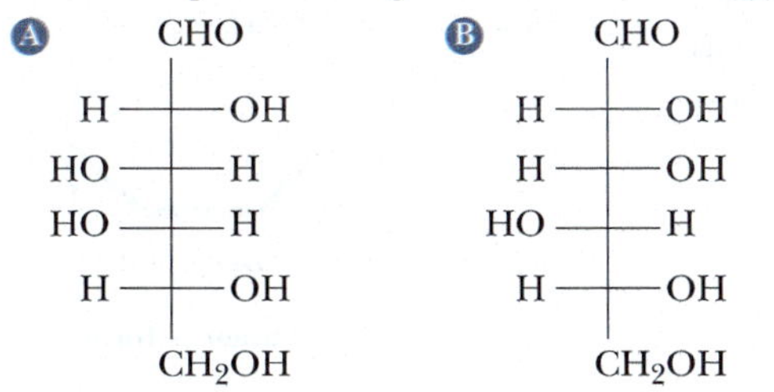

D-Galactose D-Gulose

13. **A** 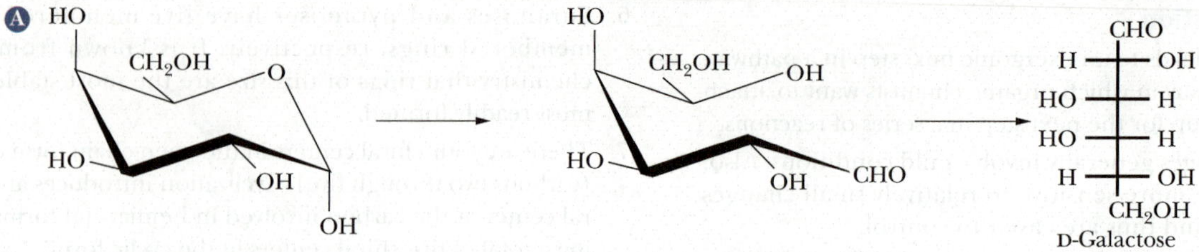

D-Galactose

B 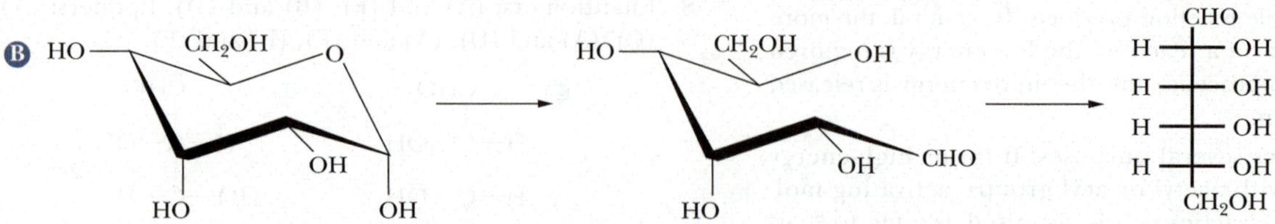

D-Allose

14. 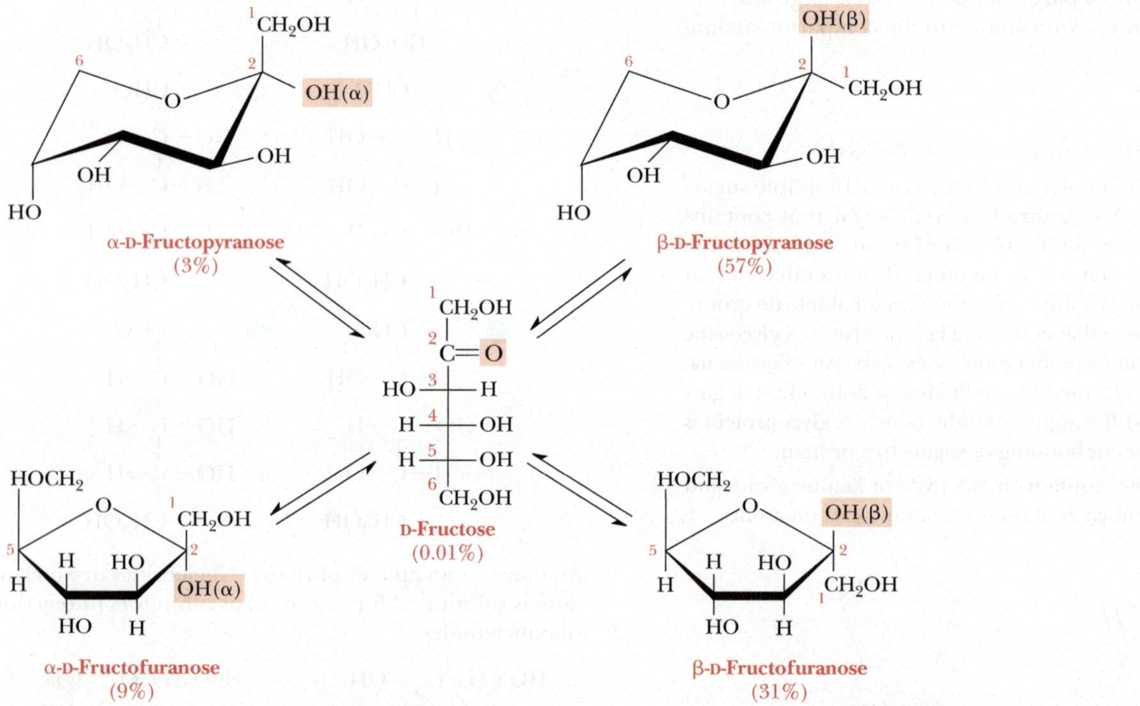

15. 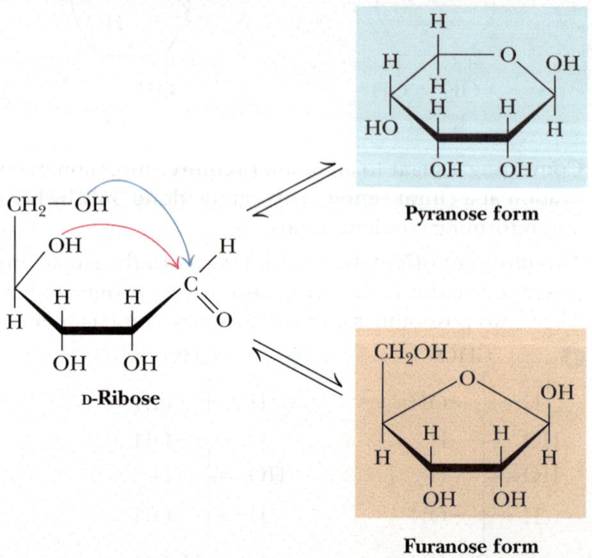

16-2 Reactions of Monosaccharides

16. A glycosidic bond is formed, and the reaction is a condensation reaction in which a hydroxyl group from one monosaccharide reacts with the anomeric carbon of another monosaccharide, releasing a molecule of water. See Figure 16.17 for a close example.
17. GlcNAc differs from glucose by having an *N*-acetyl group replace the hydroxyl group on carbon 2. MurNAc contains the same modification at carbon 2 and also has a lactic acid group attached to carbon 3.
18. A reducing sugar is one that has a free hemiacetal group (or a free hemiketal group). The hemiacetal or hemiketal group is easily oxidized, thus reducing the oxidizing agent.
19. Vitamin C is a lactone (a cyclic ester) with a double bond between two of the ring carbons, making it an unsaturated lactone. Sugars have saturated rings. The presence of the double bond makes it susceptible to air oxidation.

16-3 Some Important Oligosaccharides

20. Differences: sucrose contains fructose, whereas lactose contains galactose. Sucrose has an α,β(1 → 2) glycosidic linkage, whereas lactose has a β(1 → 4) glycosidic linkage. Similarities: sucrose and lactose are both disaccharides, and both contain glucose.

21. Structure of gentibiose

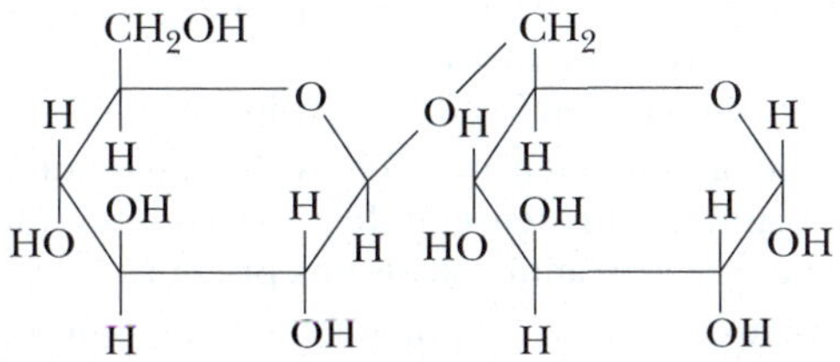

β(1→6)

22. In some cases, the enzyme that degrades lactose (milk sugar) to its components—glucose and galactose—is missing. In other cases, the enzyme isomerizes galactose to glucose for further metabolic breakdown.

23. (A)

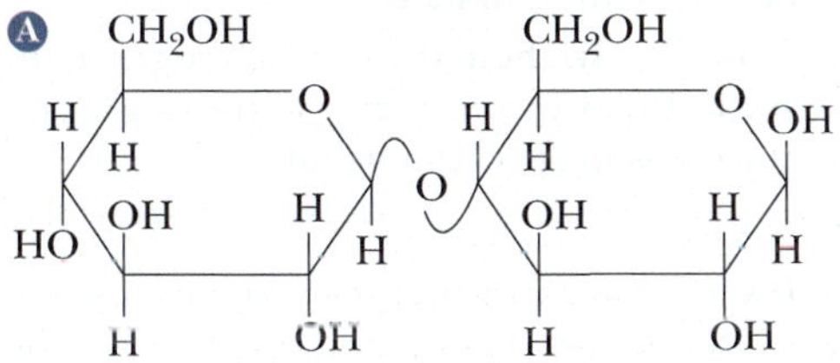

β(1→4)

(B)

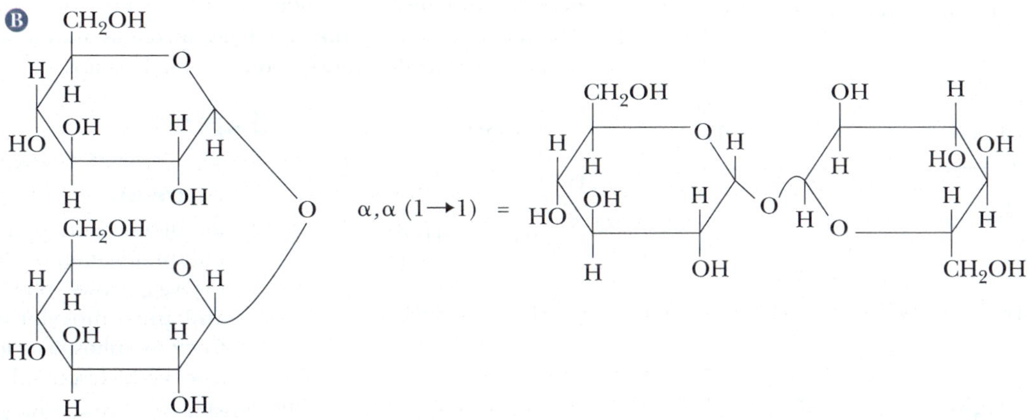

(C)

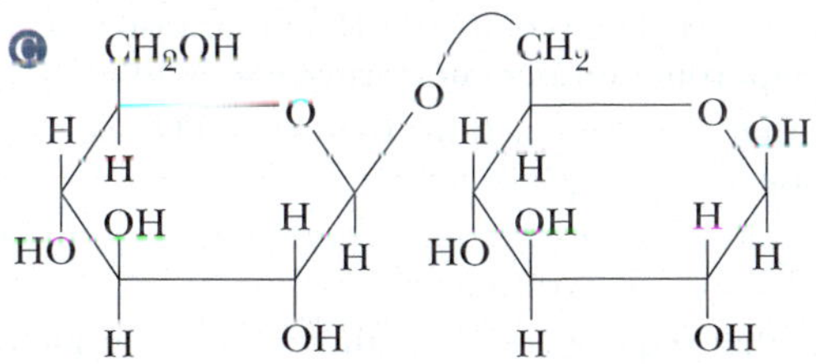

β(1→6)

16-4 Structures and Functions of Polysaccharides

24. The cell walls of plants consist mainly of cellulose (a polysaccharide), whereas those of bacteria consist mainly of polysaccharides with peptide crosslinks.

25. Chitin is a polymer of *N*-acetylglucosamine, whereas cellulose is a polymer of D-glucose. Both polymers play a structural role, but chitin occurs in the exoskeletons of invertebrates and cellulose primarily in plants.

26. Glycogen and starch differ mainly in the degree of chain branching (glycogen is more branched). Starch has both linear (amylose) and branched (amylopectin) forms. Both polymers serve as vehicles for energy storage—glycogen in animals and starch in plants.

27. Cellulose is composed of glucose monomers linked by β(1→4) glycosidic bonds, resulting in a straight, rigid structure. In contrast, starch contains glucose monomers connected mainly by α(1→4) glycosidic bonds, leading to a more helical, flexible shape.

28. A two-residue segment of pectin:

Galacturonic acid (α form)

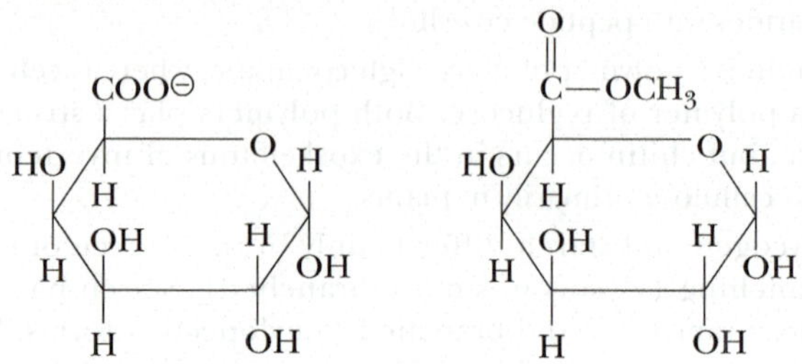

unmethylated methylated

Repeating disaccharide

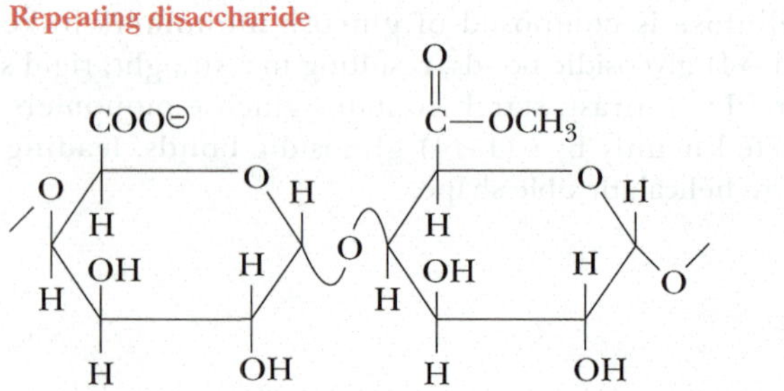

α(1→4) α-anomeric end

29. The concentration of reducing groups is too small to detect because the reducing groups are found at the ends of large polysaccharide molecules.

30. To 2500, one place (0.02%). To 1000, four places (0.08%). To 200, 24 places (0.48%).

31.

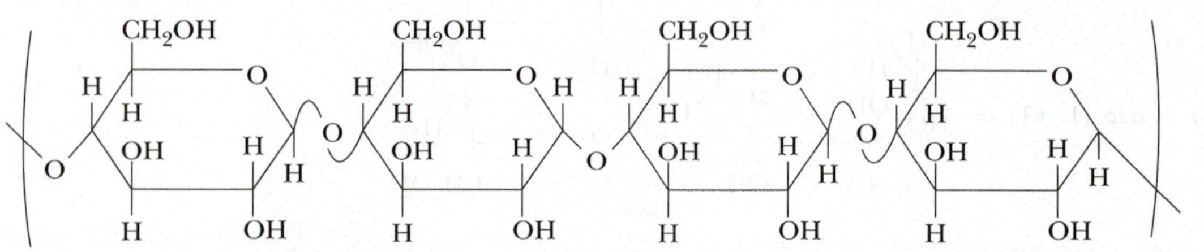

α(1→4) β(1→4) α(1→4) β(1→4) α(1→4)

32. Because of the branching, the glycogen molecule gives rise to a number of available glucose molecules (found at the ends of the branches) at a time when it is being hydrolyzed to provide energy. A linear molecule could produce only one available glucose at a time (the one at the nonreducing end of the molecule).

33. The digestive tract of these animals contains bacteria that have the enzyme to hydrolyze cellulose.

34. Humans can digest glucose polymers containing alpha bonds but not beta bonds. For this reason, glycogen and starch are digestible, whereas cellulose is not. In addition, cellulose is fibrous rather than branched, forming hydrogen bonds between the polymer chains, which makes cellulose insoluble in water and inaccessible to digestion by humans.

35. The enzyme β-amylase is an exoglycosidase, degrading polysaccharides from the ends. The enzyme α-amylase is an endoglycosidase, cleaving internal glycosidic bonds.

36. Fiber binds many toxic substances in the gut and decreases the transit time of ingested food in the digestive tract, so that harmful compounds such as carcinogens are removed from the body more quickly than would be the case with a low-fiber diet.

37. Compare the structures seen in Figures 16.26 and 16.27. A cellulase (an enzyme that degrades cellulose) needs an active site that can recognize glucose residues joined in a β-glycosidic linkage and hydrolyze that linkage. An enzyme that degrades starch has the same requirements with regard to glucose residues joined in an α-glycosidic linkage.

38. The sequence of monomers in a polysaccharide is not genetically coded, and, in this sense, it does not contain information, as is the case with amino acids in a protein.

39. Peptidoglycans are polymers of *N*-acetylglucosamine and *N*-acetylmuramic acid cross-linked by short peptides, and they are found in bacterial cell walls.

40. Bacterial cell walls are not likely to consist largely of protein. The polysaccharides that comprise the cell walls are required to provide sufficient mechanical strength necessary for the structural integrity of the cell walls.

41. Athletes try to increase their stores of glycogen before an event. The most direct way to increase the amount of this polymer of glucose is to eat carbohydrates.

16-5 Glycoproteins

42. Glycoproteins are ones in which carbohydrates are covalently bonded to proteins. They play a role in eukaryotic cell membranes, frequently as recognition sites for external molecules. Antibodies (immunoglobulins) are glycoproteins.

43. The sugar portions of the blood glycoproteins are the source of the antigenic difference between blood groups.

Chapter 17

17-1 The Overall Pathway in Glycolysis

1. Reactions that require ATP: phosphorylation of glucose to give glucose-6-phosphate and phosphorylation of fructose-6-phosphate to give fructose-1,6-bisphosphate.

   Reactions that produce ATP: transfer of a phosphate group from 1,3-bisphosphoglycerate to ADP and transfer of a phosphate group from phosphoenolpyruvate to ADP.

   Enzymes that catalyze reactions requiring ATP: hexokinase, glucokinase, and phosphofructokinase-1.

   Enzymes that catalyze reactions producing ATP: phosphoglycerate kinase and pyruvate kinase.

2. Reactions that require NADH: reduction of pyruvate to lactate and reduction of acetaldehyde to ethanol.

   Reactions that require NAD^+: oxidation of glyceraldehyde-3-phosphate to give 1,3-bisphosphoglycerate.

   Enzymes that catalyze reactions requiring NADH: lactate dehydrogenase and alcohol dehydrogenase.

   Enzymes that catalyze reactions requiring NAD^+: glyceraldehyde-3-phosphate dehydrogenase.

3. Pyruvate can be converted to lactate, ethanol, or acetyl-CoA.

17-2 Conversion of Six-Carbon Glucose to Three-Carbon Glyceraldehyde-3-Phosphate

4. Glucose gets phosphorylated by hexokinase to produce glucose-6-phosphate, which gets isomerized by glucose-6-phosphate isomerase to produce fructose-6-phosphate. This compound is then phosphorylated by phosphofructokinase-1 to produce fructose-1,6-bisphosphate.
5. Aldolase catalyzes the reverse of an aldol condensation. The name "aldol" comes from combining "aldehyde" and "alcohol." In glycolysis, aldolase breaks fructose-1,6-bisphosphate into glyceraldehyde-3-phosphate (an aldehyde) and dihydroxyacetone phosphate (which has a hydroxyl group, making it an alcohol).
6. Isozymes are multimeric enzymes that have slightly different amino acid compositions in different organs. Lactate dehydrogenase is an example, as is phosphofructokinase.
7. Isozymes allow for subtle control of the enzyme to respond to different cellular needs. In the liver, for example, lactate dehydrogenase is most often used to convert lactate to pyruvate, but the reaction is often reversed in the muscle. Having a different isozyme in the muscle and liver allows for those reactions to be optimized.
8. Fructose-1,6-bisphosphate can only undergo the reactions of glycolysis. The components of the pathway up to this point can have other metabolic fates.
9. Add the $\Delta G^{\circ\prime}$/mol values seen in Table 17.1 for the reactions from glucose to glyceraldehyde-3-phosphate. The result is 2.2 kJ/mol = 0.5 kcal/mol.
10. The two enzymes can have different tissue locations and kinetic parameters. Glucokinase has a higher K_M for glucose than hexokinase. Thus, under conditions of low glucose, the liver does not convert glucose to glucose-6-phosphate, which would use substrate that is needed elsewhere. When the glucose concentration is much higher, however, glucokinase helps phosphorylate glucose so that it can be stored as glycogen.
11. Individuals who lack the gene that directs the synthesis of the M form of phosphofructokinase can carry on glycolysis in their livers but experience muscle weakness because they lack the enzyme in muscle.
12. The hexokinase molecule changes shape on binding to substrate (see Figure 17.6), consistent with the induced-fit theory of an enzyme adapting itself to its substrate.
13. ATP inhibits phosphofructokinase, consistent with the fact that ATP is produced by later reactions of glycolysis.

17-3 Glyceraldehyde-3-Phosphate Is Converted to Pyruvate

14. From the point at which triose phosphate isomerase converts dihydroxyacetone phosphate to glyceraldehyde-3-phosphate, all reactions of the glycolysis pathway are considered doubled (only the path from one glyceraldehyde-3-phosphate is usually shown).
15. In the first ATP-forming reaction of glycolysis, 1,3-bisphosphoglycerate is converted to 3-phosphoglycerate by the enzyme phosphoglycerate kinase to produce two molecules of ATP for each molecule of glucose.
16. The NAD-linked dehydrogenases discussed in this chapter are glyceraldehyde-3-phosphate dehydrogenase, lactate dehydrogenase, and alcohol dehydrogenase.
17. The Gibbs energy of hydrolysis of a high-energy intermediate provides the energetic driving force in substrate-level phosphorylation. An example is 1,3-bisphosphoglycerate donating a phosphate group to ADP to form ATP and 3-phosphoglycerate in the reaction catalyzed by phosphoglycerate kinase.
18. The control points in glycolysis are the reactions catalyzed by hexokinase, phosphofructokinase, and pyruvate kinase.
19. Hexokinase is inhibited by glucose-6-phosphate. Phosphofructokinase is inhibited by ATP and citrate. Pyruvate kinase is inhibited by ATP, acetyl-CoA, and alanine. Phosphofructokinase is stimulated by AMP and fructose-2,6-bisphosphate. Pyruvate kinase is stimulated by AMP and fructose-1,6-bisphosphate.
20. The part of the active site that binds to NAD would be the part that is most conserved because many dehydrogenases use that coenzyme.
21. (a) Use of a high-energy phosphate from ATP to phosphorylate a substrate.
    (b) Changing the bond arrangement of a molecule without changing its molecular formula (i.e., replacing one isomer with another).
    (c) Changing the oxidation state of a substrate while simultaneously changing the oxidation state of a coenzyme (NAD, FAD, etc.).
22. An *isomerase* is a general term for an enzyme that changes the bond arrangement of a substrate without changing its molecular formula. A *mutase* is an enzyme of the isomerase class that moves a reactive functional group, such as a phosphate, to a new location in a substrate molecule.
23. Dehydrogenases catalyze redox reactions involving a metabolite paired with a coenzyme. Therefore, pyruvate dehydrogenase could catalyze the oxidation of pyruvate with concomitant reduction of a coenzyme such as NAD.
24. The reaction of 2-phosphoglycerate to phosphoenolpyruvate is a dehydration (loss of water) rather than a redox reaction.
25. Carbon-1 of glyceraldehyde is the aldehyde group. It changes oxidation state to a carboxylic acid, which is phosphorylated simultaneously.
26. ATP is an inhibitor of several steps of glycolysis as well as other catabolic pathways. The purpose of catabolic pathways is to produce energy, and high levels of ATP mean the cell already has sufficient energy. Glucose-6-phosphate inhibits hexokinase and is an example of product inhibition. If the glucose-6-phosphate level is high, it may indicate that sufficient glucose is available from glycogen breakdown or that the subsequent enzymatic steps of glycolysis are going slowly. Either way, there is no reason to produce more glucose-6-phosphate. Phosphofructokinase is inhibited by a special effector molecule, fructose-2,6-bisphosphate, whose levels are controlled by hormones. It is also inhibited by citrate, which indicates that there is sufficient energy from the citric acid cycle, probably from fat and amino acid degradation. Pyruvate kinase is also inhibited by acetyl-CoA, the presence of which indicates that fatty acids are being used to generate energy for the citric acid cycle. The main function of glycolysis is to feed carbon units to the citric acid cycle. When these carbon skeletons can come from other sources, glycolysis is inhibited to spare glucose for other purposes.

27. There would be 15 possible isozymes of LDH, combining three different subunits into combinations of four. Besides the five isozymes containing only M and H, there would also be C_4, CH_3, C_2H_2, C_3H, CH_2M, C_2HM, C_3M, CHM_2, C_2M_2, and CM_3.

28. Glutamic acid has an acidic side chain with a pK_a of 4.25. Therefore, it would be negatively charged at pH 8.6, and the H subunit would move more toward the anode (+) than the M subunit would in native gel electrophoresis. Thus, LDH 1, which is H_4, would move the farthest. LDH 5, which is M_4, would move the least, with the other isozymes migrating between those two extremes proportional to their H content.

29. With few exceptions, a biochemical reaction typically results in only one chemical modification of the substrate. Accordingly, several to many steps are needed to reach the ultimate goal.

17-4 Anaerobic Metabolism of Pyruvate

30. The purpose of the step that produces lactate is to reduce pyruvate so that NADH can be oxidized to NAD^+, which is needed to sustain glycolysis in the absence of oxygen. NAD^+ is used in the glycolysis step catalyzed by glyceraldehyde-3-phosphate dehydrogenase.

31.

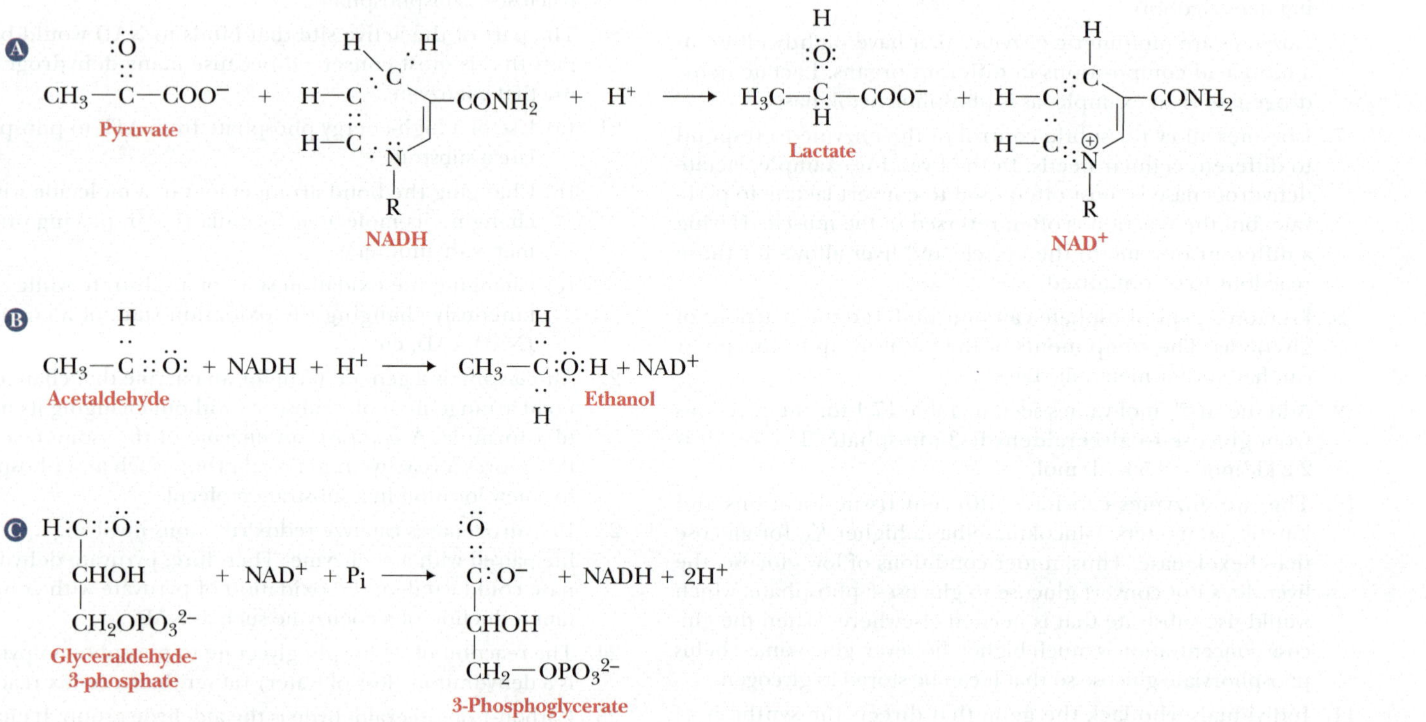

32. Thiamine pyrophosphate (TPP) is a coenzyme that stabilizes carbanion intermediates during the transfer or removal of carbon fragments. It is required for the decarboxylation of pyruvate catalyzed by pyruvate decarboxylase in alcoholic fermentation.

33. The important part of TPP is the five-membered ring, in which a carbon is found between nitrogen and sulfur. This carbon forms a carbanion that is extremely reactive, making it able to perform a nucleophilic attack on carbonyl groups, leading to decarboxylation of several compounds in different pathways.

34. Animals that have been run to death have accumulated large amounts of lactic acid in their muscle tissue, accounting for the sour taste of the meat.

35. This is possible, and it is done. These drugs also affect other tissues, including skin, hair, cells of the intestinal lining, and especially the immune system and red blood cells. People on these treatments are usually more susceptible to infectious diseases than healthy people and are often somewhat anemic.

36. The Warburg effect is the high level of glycolysis seen in many cancer cells, giving rise to pyruvate, followed by lactic acid fermentation. This aerobic glycolysis is observed even at high levels of oxygen, where further oxidation to carbon dioxide and water is expected. This is one of the many aspects of glycolysis, which is the main subject of this chapter.

17-5 Energy Production in Glycolysis

37. There is a net gain of two ATP molecules per glucose molecule consumed in glycolysis.

38. Cells produce much less ATP per glucose during anaerobic glycolysis because, without oxygen, glucose is converted to lactate and not completely broken down to carbon dioxide and water. Therefore, anaerobic glycolysis uses glucose less efficiently, yielding only 2 ATP per glucose compared to 30–32 ATP in aerobic respiration. The advantage of anaerobic glycolysis is that ATP can be generated quickly during strenuous activity or in environments where oxygen is limited.

39. Coupled reactions in biochemistry occur when an energetically unfavorable reaction is driven forward by being linked to

an energetically favorable reaction. In glycolysis, the coupled reactions are those catalyzed by hexokinase, phosphofructokinase-1, phosphoglycerate kinase, and pyruvate kinase. The hexokinase and phosphofructokinase-1 reactions use ATP energy to add phosphate groups to sugars, while the phosphoglycerate kinase and pyruvate kinase reactions make ATP using energy released from breaking high-energy phosphate bonds.

40. The steps catalyzed by hexokinase, phosphofructokinase, and pyruvate kinase are the irreversible steps in glycolysis. These steps have large negative Gibbs energy changes, making the reverse reactions very unfavorable thermodynamically.

41. Phosphoenolpyruvate $\rightarrow$ pyruvate + P_i

$\Delta G^{\circ\prime} = -61.9$ kJ/mol $= -14.8$ kcal/mol

ADP + P_i $\rightarrow$ ATP

$\Delta G^{\circ\prime} = 30.5$ kJ/mol $= 7.3$ kcal/mol

Phosphoenolpyruvate + ADP $\rightarrow$ Pyruvate + ATP

$\Delta G^{\circ\prime} = -31.4$ kJ/mol $= -7.5$ kcal/mol

42. The net yield of ATP from glycolysis is the same, two ATP, when any of the three substrates is used. The energetics of the conversion of hexoses to pyruvate are the same, regardless of hexose type.

43. Starting with glucose-1-phosphate, the net yield is three ATP because one of the priming reactions is no longer used. Thus, glycogen is a more efficient fuel for glycolysis than free glucose.

44. Phosphoenolpyruvate + ADP $\rightarrow$ Pyruvate + ATP

	$\Delta G^{\circ\prime} = -31.4$
ADP + P_i $\rightarrow$ ATP	$\Delta G^{\circ\prime} = 30.5$ kJ/mol
Sum	$\Delta G^{\circ\prime} = -0.9$ kJ/mol

Thus, the reaction is thermodynamically possible under standard conditions.

45. A positive $\Delta G^{\circ\prime}$ does not necessarily mean that the reaction has a positive ΔG. Substrate concentrations can make a negative ΔG out of a positive $\Delta G^{\circ\prime}$.

46. The entire pathway can be looked at as a large, coupled reaction. Thus, if the overall pathway has a negative ΔG, an individual step may be able to have a positive ΔG, and the pathway can operate.

17-6 Control of Glycolysis

47. When G6P levels are high, the enzyme hexokinase is inhibited. If glycolysis reactions begin to increase, then G6P would be consumed and G6P levels would drop. This would reduce the inhibition of hexokinase, allowing more glucose to enter the glycolysis pathway.

48. Hexokinase, phosphofructokinase-1 (PFK-1), and pyruvate kinase are the glycolysis enzymes that are regulated. PFK-1 is the most important control point because it catalyzes the step in which the sugar is committed to glycolysis. Regulation at the hexokinase step controls entry into glycolysis and other pathways like glycogen production (Section 18-1C) and the pentose phosphate pathway (Section 18-4). Regulation at the pyruvate kinase step helps conserve metabolites for pathways like gluconeogenesis (Section 18-2).

49. Hormones exercise another level of control over metabolism, beyond that of allosteric and feedback control. Glycolysis can certainly be affected in this manner. The effect of insulin on carbohydrate metabolism is well established.

50. Regulating a metabolic pathway at the beginning helps the cell avoid wasting energy and resources on unnecessary reactions.

Chapter 18

18-1 How Glycogen Is Degraded and Produced

1. When blood glucose levels are low, glycogen in the liver is broken down into free glucose in a process called glycogenolysis. The enzyme glycogen phosphorylase removes glucose residues from the nonreducing ends of glycogen, releasing glucose-1-phosphate as the product, which then gets converted to glucose-6-phosphate by the enzyme phosphoglucomutase. Debranching enzymes are also needed for cleaving the $\alpha(1 \rightarrow 6)$ glycosidic bonds of glycogen. Finally, free glucose can enter the bloodstream after the enzyme glucose-6-phosphatase removes the phosphate from glucose-6-phosphate.

2. These two pathways occur in the same cellular compartment, and, if both are on at the same time, a futile ATP hydrolysis cycle results. Using the same mechanism to turn them on and off or off and on is highly efficient.

3. In phosphorolysis, a bond is cleaved by adding the elements of phosphoric acid across that bond, whereas in hydrolysis, the cleavage takes place by adding the elements of water across the bond.

4. Glucose-6-phosphate is already phosphorylated and able to enter glycolysis after the hexokinase step. This "saves" one ATP equivalent in the early stages of glycolysis. There is nuance to this "saving" because of the energy that is used to store glucose residues in glycogen chains in the first place.

5. UDPG plays a role as the activated form of glucose that can be used to extend glycogen chains.

6. Glycogen synthase is subject to covalent modification and to allosteric control. The enzyme is active in its dephosphorylated form and inactive when phosphorylated. Glucose-6-phosphate is an allosteric activator of glycogen synthase.

7. When blood glucose is low, glucagon is released, activating a series of events that leads to phosphorylation and activation of glycogen phosphorylase. The activated glycogen phosphorylase then breaks down glycogen to glucose-1-phosphate, which eventually gets converted to free glucose in the blood.

8. Eating high-carbohydrate foods for several days before strenuous activity is intended to build up glycogen stores in the body. Glycogen will be available to supply the required energy.

9. The disaccharide sucrose can be hydrolyzed to glucose and fructose, which can both be readily converted to glucose-1-phosphate, the immediate precursor of glycogen. However, this is not the usual form of "glycogen loading," because complex carbohydrates are digested more slowly and stored more effectively.

10. Probably not, because the sugar spike initially results in a rapid increase in insulin levels, which results in lowering blood glucose levels and increased glycogen storage in the liver.

11. The energy to incorporate glucose into glycogen comes from UTP. Glucose-1-phosphate reacts with UTP to form uridine diphosphate glucose (UDPG) and PP_i. UDPG is an activated

form of glucose that can be added to a glycogen chain with the help of hydrolysis of the PP_i to two inorganic phosphates.

12. The enzyme that catalyzes the addition of glucose residues to a growing glycogen chain cannot form a bond between isolated glucose residues; thus the need for a primer.
13. The glycogen synthase reaction is exergonic overall because it is coupled to phosphate ester hydrolysis.
14. High ATP levels signal energy abundance. Therefore, this promotes gluconeogenesis, which requires energy. High ATP levels also promote glycogen synthesis because glycolysis (glucose breakdown) would be inhibited and available glucose can be stored.

18-2 Gluconeogenesis Produces Glucose from Pyruvate

15. In the first reaction of gluconeogenesis, pyruvate gets carboxylated by the enzyme pyruvate carboxylase to produce oxaloacetate. The enzyme requires the biotin coenzyme. In the last step of gluconeogenesis, glucose-6-phosphate is dephosphorylated by glucose-6-phosphatase, producing glucose.
16. Reactions that require acetyl-CoA: none. Reactions that require biotin: carboxylation of pyruvate to oxaloacetate.
17. Three reactions of glycolysis are irreversible under physiological conditions. They are the production of pyruvate and ATP from phosphoenolpyruvate, the production of fructose-1,6-bisphosphate from fructose-6-phosphate, and the production of glucose-6-phosphate from glucose. The reactions of gluconeogenesis differ from those of glycolysis at these points and are catalyzed by different enzymes.
18. In gluconeogenesis, biotin is a coenzyme for the pyruvate carboxylase enzyme. Biotin carries carbon dioxide for transfer to pyruvate, producing oxaloacetate, which then undergoes further reactions of gluconeogenesis.
19. In gluconeogenesis, glucose-6-phosphate is dephosphorylated to glucose (the last step of the pathway); in glycolysis, it isomerizes to fructose-6-phosphate (an early step in the pathway).
20. Of the three processes—glycogen formation, gluconeogenesis, and the pentose phosphate pathway—only one, gluconeogenesis, involves an enzyme (pyruvate carboxylase) that requires biotin. Therefore, gluconeogenesis would be the only process that could possibly be inhibited if avidin levels are high.
21. The fructose-1,6-bisphosphatase enzyme of gluconeogenesis catalyzes the conversion of fructose-1,6-bisphosphate to fructose-6-phosphate, bypassing the PFK-1 step of glycolysis.

18-3 Control of Carbohydrate Metabolism

22. Reactions that require ATP: formation of UDP-glucose from glucose-1-phosphate and UTP (indirect requirement because ATP is needed to regenerate UTP), regeneration of UTP, and carboxylation of pyruvate to oxaloacetate.

    Reactions that produce ATP: none.

    Enzymes that catalyze ATP-requiring reactions: UDP-glucose pyrophosphorylase (indirect requirement), nucleoside diphosphate kinase, and pyruvate carboxylase.

    Enzymes that catalyze ATP-producing reactions: none.
23. Fructose-2,6-bisphosphate is an allosteric activator of phosphofructokinase (a glycolytic enzyme) and an allosteric inhibitor of fructose-1,6-bisphosphatase (an enzyme in the pathway of gluconeogenesis).
24. The pyruvate carboxylase enzyme, found in the gluconeogenesis pathway, is activated by high acetyl-CoA levels. This makes sense because when acetyl-CoA levels are high, the citric acid cycle is not using up acetyl-CoA. Moreover, ATP levels are high, allowing the liver to produce glucose in the gluconeogenesis pathway and support blood glucose needs.
25. The Cori cycle is a pathway in which there is cycling of glucose between glycolysis in muscle and gluconeogenesis in the liver. The blood transports lactate from muscle glycolysis to the liver and glucose from liver gluconeogenesis to muscle for glycolysis and continued energy production.
26. The sprint is essentially anaerobic and produces lactate from glucose by glycolysis. Lactate is then recycled to glucose by gluconeogenesis (the Cori cycle).
27. Having two control mechanisms allows for fine-tuning of control and for the possibility of amplification. Both mechanisms are capable of rapid response to conditions—milliseconds in the case of allosteric control and seconds to minutes in the case of covalent modification.
28. Glycogen metabolism control mechanisms affect agents (enzymes) that are catalysts. Therefore, an enhancement by a control mechanism on a catalyst of several powers of 10 can add to a catalyst's effect that already increases the rate of a reaction by several powers of 10.
29. Enzymes, like all catalysts, speed up the forward and reverse reactions to the same extent, so use of the same enzymes over the entirety of two opposing pathways would not allow independent regulation. Most steps in opposing pathways use the same enzymes, but having different catalysts at a few key irreversible steps is the only way to ensure independent control over the rates of the forward and reverse processes, preventing futile cycling.
30. Muscle tissue uses large quantities of glucose, producing lactate in the process. The liver is an important site of gluconeogenesis to recycle the lactate to glucose that can be transported back to muscle tissue.
31. Fructose-2,6-bisphosphate is an allosteric activator of phosphofructokinase (a glycolytic enzyme) and an allosteric inhibitor of fructose-1,6-bisphosphatase (an enzyme in the pathway of gluconeogenesis). It thus plays a role in two pathways that are not exactly the reverse of each other.
32. The concentration of fructose-2,6-bisphosphate in a cell depends on the balance between its synthesis (catalyzed by phosphofructokinase-2) and its breakdown (catalyzed by fructose-1,6-bisphosphatase). The separate enzymes that control the formation and breakdown of fructose-2,6-bisphosphate are themselves controlled by a phosphorylation/dephosphorylation mechanism.
33. Glycogen is more extensively branched than starch. Therefore, it is a more useful storage form of glucose for animals because glucose can be mobilized more easily when there is a need for energy.
34. In the glucose-6-phosphatase reaction, the concentration of substrate is the main determinant of reaction velocity. In the fructose-1,6-bisphosphatase reaction, allosteric effects are the main determinant of reaction velocity.

35. AMP and fructose-2,6-bisphosphate are allosteric activators of phosphofructokinase, stimulating glycolysis when energy is low. In contrast, they are allosteric inhibitors of fructose-1,6-bisphosphatase, suppressing gluconeogenesis under the same conditions.
36. Allosteric effects and covalent modification are two important forms of control of enzymatic action. Covalent modification plays a more important role than allosteric effects in glycogen breakdown.
37. Insulin triggers the series of events that lead to glycogen synthesis.
38. Glucagon and epinephrine start the chain of events leading to glycogen breakdown.
39. Glucagon is a peptide composed of 29 amino acids, whereas epinephrine is a single amino acid derivative.

18-4 Glucose Is Sometimes Diverted Through the Pentose Phosphate Pathway

40. In the first reaction of the pentose phosphate pathway, glucose-6-phosphate is oxidized by the enzyme glucose-6-phosphate dehydrogenase to produce 6-phosphogluconate.

    (b) By the nonoxidative transaldolase and transketolase reactions of the PPP in reverse to convert glycolytic intermediates into pentose phosphates without producing NADPH.

45. Transketolase catalyzes the transfer of a two-carbon unit (a ketol group), whereas transaldolase catalyzes the transfer of a three-carbon unit (an aldol group).
46. In red blood cells, the presence of the reduced form of glutathione is necessary for the maintenance of the sulfhydryl groups of hemoglobin and other proteins in their reduced forms, as well as for keeping the Fe(II) of hemoglobin in its reduced form. Glutathione also maintains the integrity of red blood cells by reacting with peroxides that would otherwise degrade fatty acid side chains in the cell membrane.
47. Thiamine pyrophosphate is a cofactor necessary for the function of transketolase, an enzyme that catalyzes one of the reactions in the nonoxidative part of the pentose phosphate pathway.
48.

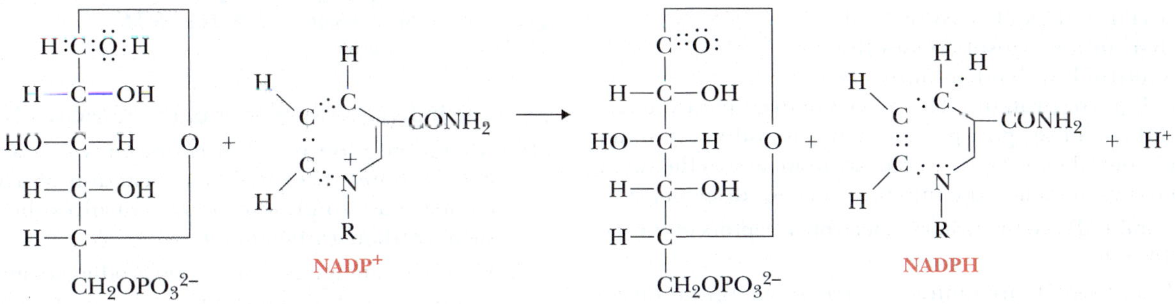

NADP⁺ — NADPH

Glucose-6-phosphate — 6-Phosphoglucono-δ-actone

The enzyme requires the NADP (or $NADP^+$) coenzyme. In the pentose phosphate pathway, two important products are NADPH for reductive anabolic pathways and ribose-5-phosphate for nucleic acid biosynthesis.

41. NADPH has one more phosphate group than NADH (at the 2′ position of the ribose ring of the adenine nucleotide portion of the molecule). NADH is used in redox reactions that give rise to ATP. NADPH is used as a reducing agent in biosynthesis. The enzymes that use NADH as a coenzyme are different from those that require NADPH.
42. Glucose-6-phosphate can be converted to glucose (gluconeogenesis), glycogen (glycogen synthesis), pentose phosphates (pentose phosphate pathway), or pyruvate (glycolysis).
43. Hemolytic anemia is caused by defective functioning of the pentose phosphate pathway. There is a deficiency of NADPH, which indirectly contributes to the integrity of red blood cells. The pentose phosphate pathway is the only source of NADPH in red blood cells.
44. (a) By using glucose-6-phosphate and the oxidative reactions of the PPP to generate NADPH and ribulose-5-phosphate. Excess pentose phosphates are converted back to glucose-6-phosphate by transaldolase and transketolase reactions of the PPP and gluconeogenesis.

49. Having different reducing agents for anabolic and catabolic pathways keeps the pathways metabolically separate. Thus, they are subject to independent control and do not waste energy.
50. If a cell needs NADPH, the oxidative reactions of the pentose phosphate pathway take place. If a cell needs ribose-5-phosphate, the oxidative portion of the pathway can be bypassed; only the nonoxidative reshuffling reactions of glycolytic intermediates take place to produce ribose-5-phosphate. The pentose phosphate pathway does not have a significant effect on the cell's supply of ATP.
51. The reshuffling reactions of the pentose phosphate pathway involve both an epimerase and an isomerase. Without an isomerase, all the sugars involved are keto sugars, which are not substrates for transaldolase, one of the key enzymes in the reshuffling process.

Chapter 19

19-1 The Central Role of the Citric Acid Cycle in Metabolism

1. Anaerobic glycolysis, which produces lactate from glucose, is the principal pathway for the anaerobic metabolism of glucose. The pentose phosphate pathway can also be considered. Glycolysis, the citric acid cycle, the electron transport

chain, and oxidative phosphorylation are responsible for the aerobic metabolism of glucose.

2. Anaerobically, two ATPs can be produced from one glucose molecule. Aerobically, this figure is 30 to 32, depending on which tissue it is occurring.

3. The citric acid cycle is also called the Krebs cycle, after Sir Hans Krebs, who first elucidated the sequence of reactions. It is also called the tricarboxylic acid (TCA) cycle because the structures of citric acid and a few other metabolites in the cycle contain three carboxyl groups.

4. Amphibolic means that the pathway is involved in both catabolism and anabolism.

19-2 The Overall Pathway of the Citric Acid Cycle

5. The citric acid cycle takes place in the mitochondrial matrix because the NADH and $FADH_2$ produced by citric acid cycle enzymes can deliver electrons to the nearby electron transport chain embedded in the inner mitochondrial membrane. Glycolysis takes place in the cytosol.

6. The pyruvate dehydrogenase complex (PDC) is located in the mitochondrial matrix. Pyruvate from glycolysis in the cytosol crosses the outer mitochondrial membrane through porin proteins. The mitochondrial pyruvate carrier (MPC), a transport protein that is found in the inner mitochondrial membrane carries pyruvate across the inner mitochondrial membrane into the mitochondrial matrix.

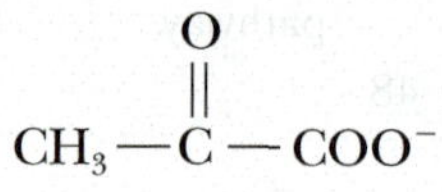

Pyruvate

7. NAD^+ and FAD are the primary electron acceptors of the citric acid cycle.

8. NADH and $FADH_2$ are indirect sources of energy produced in the TCA cycle. They carry electrons to the electron transport chain, which drives oxidative phosphorylation to produce ATP. GTP produced in the cycle is a direct source of energy.

19-3 How Pyruvate Is Converted to Acetyl-CoA

9. In the reaction that occurs right before the citric acid cycle, pyruvate is oxidatively decarboxylated by the pyruvate dehydrogenase complex to produce acetyl-CoA and CO_2. The enzyme requires coenzyme A as well as the NAD^+ coenzyme, which is reduced to NADH. This reaction is important because it connects glycolysis to the citric acid cycle, produces acetyl-CoA for the citric acid cycle and other pathways, and produces NADH, which contributes to ATP production.

10. Five enzymes are involved in the pyruvate dehydrogenase complex of mammals. Pyruvate dehydrogenase (PDH) transfers a two-carbon unit to TPP and releases CO_2 from pyruvate. Dihydrolipoyl transacetylase transfers the two-carbon acetyl unit to lipoamide and then to coenzyme A, producing acetyl-CoA. Dihydrolipoyl dehydrogenase reoxidizes lipoamide and reduces NAD^+ to NADH. Pyruvate dehydrogenase kinase phosphorylates and inactivates PDH. PDH phosphatase removes the phosphate, activating PDH.

11. Lipoamide plays a role both in redox and in acetyl-transfer reactions.

12. All five enzymes are in proximity for efficient shuttling of intermediates between active sites without releasing them into the surroundings and for efficient control of the complex by covalent modification.

13. Thiamine pyrophosphate comes from vitamin B_1 (thiamine). Lipoamide (lipoic acid attached to the dihydrolipoyl transacetylase enzyme of the complex) is a vitamin-like cofactor. Coenzyme A (CoA) comes from vitamin B_5 (pantothenic acid). NAD^+ comes from vitamin B_3 (niacin). FAD comes from vitamin B_2 (riboflavin).

14.

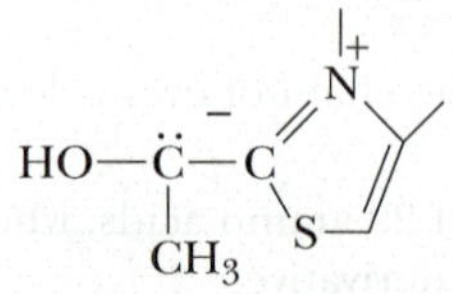

15. See Figure 19.5 for the individual reactions that give rise to the overall reaction below.

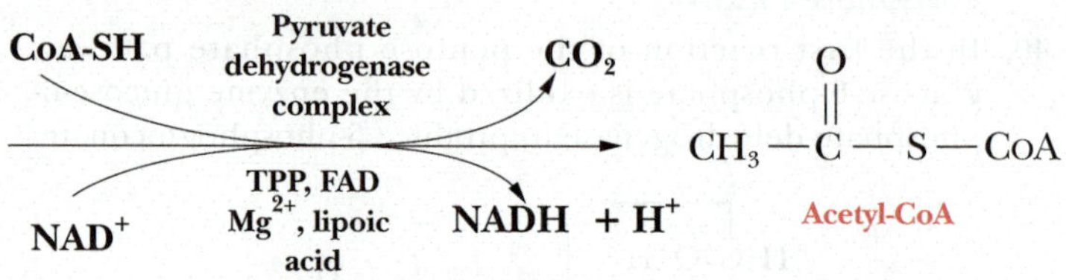

19-4 The Individual Reactions of the Citric Acid Cycle

16. A condensation reaction is one in which a new carbon–carbon bond is formed. The reaction of acetyl-CoA and oxaloacetate to produce citrate involves the formation of such a carbon–carbon bond.

17. Fluoroacetate is a poison that is produced naturally in some plants and is also used as a poison against undesirable pests. It is poisonous because it is used by citrate synthase to make fluorocitrate, which is an inhibitor of the citric acid cycle.

18. In the second reaction of the citric acid cycle, citrate is isomerized to isocitrate by the enzyme aconitase. The conversion of reactant to product involves the movement of a hydroxyl group from the central carbon to a neighboring carbon, enabling the oxidative decarboxylation that occurs in the next step of the cycle.

19. The reaction involves an achiral molecule (citrate) being converted into a chiral one (isocitrate).

20. CO_2 is produced in the oxidative decarboxylation of pyruvate to acetyl-CoA catalyzed by pyruvate dehydrogenase, in the oxidative decarboxylation of isocitrate to α-ketoglutarate catalyzed by isocitrate dehydrogenase, and in the oxidative decarboxylation of α-ketoglutarate to succinyl-CoA catalyzed by α-ketoglutarate dehydrogenase.

21. Reduced electron carriers are produced in the oxidative decarboxylation of pyruvate to acetyl-CoA (producing NADH), in the oxidative decarboxylation of isocitrate to α-ketoglutarate (producing NADH), in the oxidative decarboxylation of α-ketoglutarate to succinyl-CoA (producing NADH), in the oxidation of succinate to fumarate (producing $FADH_2$), and in the oxidation of malate to oxaloacetate (producing NADH).

22. These enzymes catalyze oxidative decarboxylations that release CO_2 and produce the reduced NADH electron carrier.

23. Both are large multienzyme complexes that catalyze oxidative decarboxylations of α-keto acids and use the same cofactors. The differences are the substrates, which are pyruvate or α-ketoglutarate, and the products, which are acetyl-CoA or succinyl-CoA. During the course of the reaction, pyruvate dehydrogenase shuttles a two-carbon acetyl unit, while α-ketoglutarate dehydrogenase shuttles a four-carbon succinyl unit. The pyruvate dehydrogenase complex links glycolysis to the citric acid cycle and is tightly regulated, while the α-ketoglutarate dehydrogenase complex is part of the citric acid cycle and is less tightly regulated.

24. A synthase-catalyzed reaction is where the enzyme forms new chemical bonds in joining two molecules and does not require a direct input of energy from a high-energy phosphate. Thus, citrate synthase catalyzes the synthesis of citrate without using ATP. A synthetase also forms new chemical bonds in joining two molecules but uses a high-energy phosphate from a nucleoside triphosphate like ATP or GTP in the process.

25. In substrate-level phosphorylation, the energy of hydrolysis of a high-energy intermediate provides sufficient energy to allow the endergonic phosphorylation of ADP to ATP. In oxidative phosphorylation linked to the electron transport chain, redox reactions generate a proton-gradient energy to allow the conversion of ADP to ATP.

26. An example from glycolysis is the transfer of a phosphate from 1,3-bisphosphoglycerate to ADP, producing 3-phosphoglycerate and ATP.

27. GTP is equivalent to ATP because an enzyme, nucleoside diphosphate kinase, can interconvert GTP and ATP.

28. The enzymes that reduce NAD^+ are all soluble, matrix enzymes, while succinate dehydrogenase is membrane-bound. The NAD^+-linked dehydrogenases all catalyze oxidations that involve carbons and oxygens such as an alcohol group being oxidized to an aldehyde or an aldehyde to a carboxylic acid. The FAD-linked dehydrogenase oxidizes a carbon–carbon single bond to a double bond.

29. There is an adenine nucleotide portion in the structure of NADH, with a specific binding site on NAD-linked dehydrogenases for this portion of NADH. Therefore, the ATP, ADP, and AMP adenine nucleotides are able to bind this site and inhibit NAD-linked dehydrogenases.

30. No, the conversion of fumarate to malate is a hydration reaction, not a redox reaction.

31.

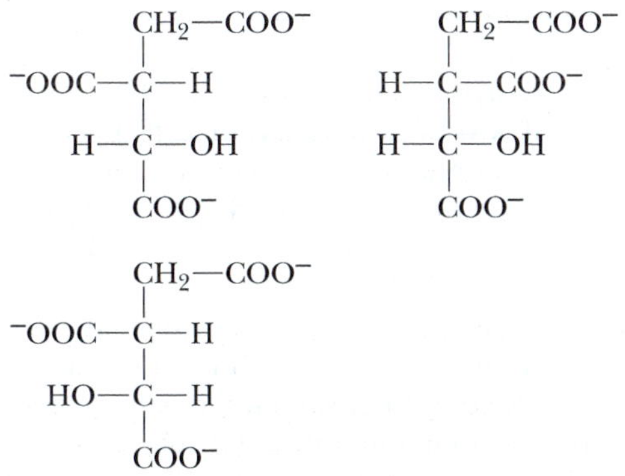

32.

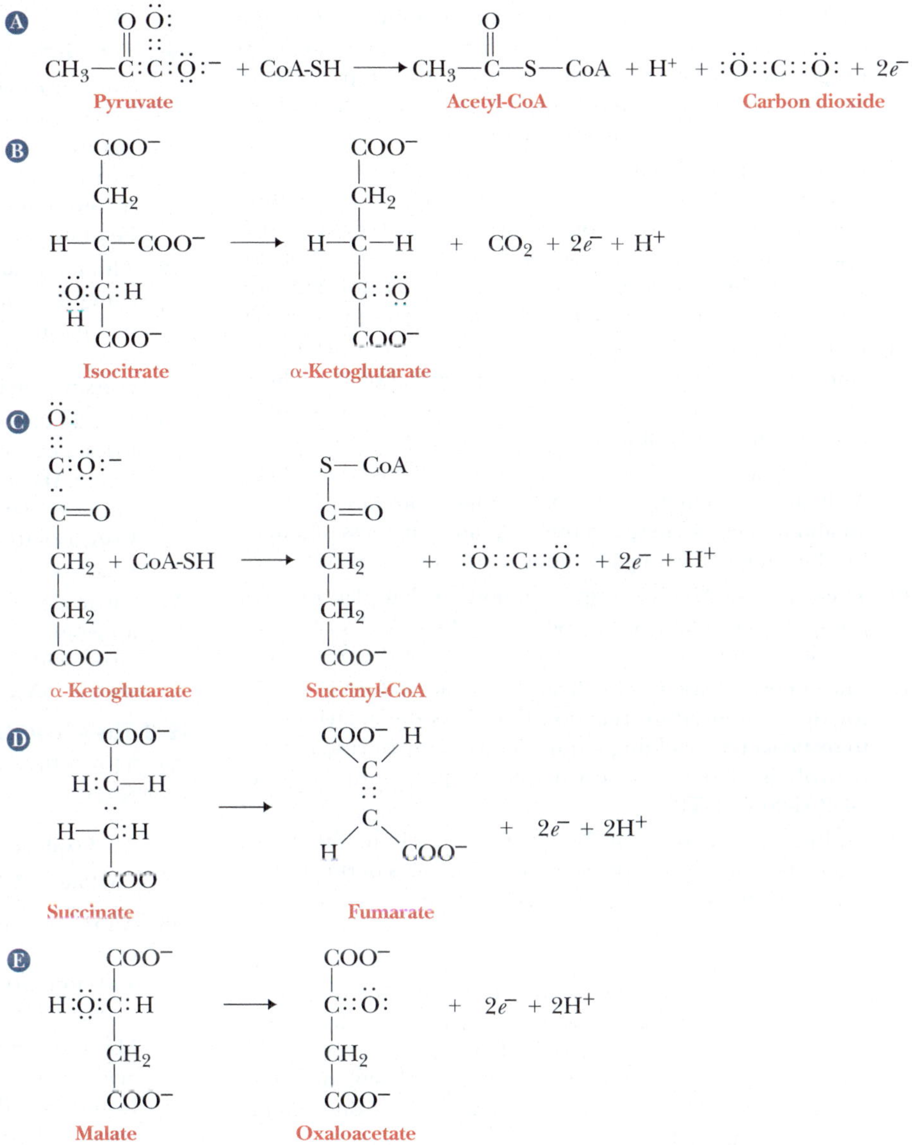

19-5 Energetics and Control of the Citric Acid Cycle

33. The pyruvate through the citric acid cycle control points are the reactions catalyzed by pyruvate dehydrogenase, citrate

synthase, isocitrate dehydrogenase, and α-ketoglutarate dehydrogenase.

34. The pyruvate dehydrogenase complex (PDC) is inhibited by ATP and NADH in prokaryotic cells because both compounds are abundant when a cell has energy readily available, so the use of the citric acid cycle is unnecessary. Conversely, the PDC is activated by ADP and AMP, which are abundant when a cell needs energy.

35. Pyruvate dehydrogenase (PDH) is controlled allosterically in prokaryotes and eukaryotes, being inhibited by ATP, acetyl-CoA, and NADH. In eukaryotes, it is subject to additional control by covalent modification (phosphorylation). When PDH kinase phosphorylates PDH, it becomes inactive. Removal of the phosphate by PDH phosphatase reactivates it.

36. ATP and NADH are the two most common inhibitors of the citric acid cycle and the reaction catalyzed by pyruvate dehydrogenase. This makes sense because both compounds are abundant when a cell has energy readily available, and the use of the citric acid cycle is unnecessary.

37. If the amount of ADP in a cell increases relative to the amount of ATP, the cell needs energy (ATP). This situation not only favors the reactions of the citric acid cycle, which release energy and activate isocitrate dehydrogenase, but also stimulates the formation of NADH and $FADH_2$ for ATP production by electron transport and oxidative phosphorylation.

38. If the amount of NADH in a cell increases relative to the amount of NAD^+, the cell has energy readily available. Therefore, there is less need for the citric acid cycle to be active; as a result, the activity of pyruvate dehydrogenase decreases.

39. The citric acid cycle is less active when a cell has a high ATP/ADP ratio and a high NADH/NAD^+ ratio. Both ratios indicate an abundance of energy in the cell, indicating less of a need for the energy-releasing reactions of the citric acid cycle.

40. Thioesters are "high-energy" compounds that play a role in group-transfer reactions; consequently, their $\Delta G^{\circ\prime}$ of hydrolysis is large and negative to provide energy for the reaction.

41. The energy released by the hydrolysis of acetyl-CoA is needed for the condensation reaction that links the acetyl moiety to oxaloacetate, yielding citrate. The energy released by the hydrolysis of succinyl-CoA drives the phosphorylation of GDP, yielding GTP.

42. Adding the reactions of the citric acid cycle in Table 19.2 (Steps 1–8) shows that the sum of the energies of the individual reactions is −44.3 kJ (−10.6 kcal) for each mole of acetyl-CoA that enters the cycle.

43. Lactose is a disaccharide of glucose and galactose. There is no energy cost in the hydrolysis of the bond between the two monosaccharides, so essentially there are two hexoses to consider. Because the processing of any of the hexoses yields the same amount of energy, the aerobic processing of lactose would lead to 60 to 64 ATPs, depending on the tissue and the shuttle system used.

19-6 The Glyoxylate Cycle: Producing Glucose from Fatty Acid Catabolism

44. The glyoxylate cycle allows organisms such as plants, fungi, and some bacteria to convert fatty acids into carbohydrates by bypassing the CO_2-producing steps of the citric acid cycle.

45. The citric acid cycle enzymes that are bypassed in the glyoxylate cycle are isocitrate dehydrogenase, α-ketoglutarate dehydrogenase, succinyl-CoA synthetase, succinate dehydrogenase, and fumarase. The main reason for bypassing these enzymes is to prevent carbon loss in the decarboxylation steps of the citric acid cycle and facilitate the conversion of fatty acids to glucose.

46. The unique reactions of the glyoxylate cycle are the conversion of isocitrate to succinate and glyoxylate catalyzed by isocitrate lyase and the conversion of glyoxylate and acetyl-CoA to malate catalyzed by malate synthase.

47. Bacteria that have a glyoxylate cycle can convert the acetic acid to amino acids, carbohydrates, and lipids, but humans can use acetic acid only as an energy source or to make lipids.

19-7 The Citric Acid Cycle in Catabolism

48. The citric acid cycle is the central metabolic pathway and an indirect producer of energy. It receives fuels from the other pathways at many points and generates reduced electron carriers that go into the electron transport chain. It is also involved in anabolism because many of its intermediates can be drawn off to synthesize other compounds.

49. The citric acid cycle occurs in the mitochondrial matrix, which is more selective in its permeability than the plasma membrane. Therefore, adding a negative charge with a phosphate group to prevent citric acid cycle metabolites from crossing the inner mitochondrial membrane is unnecessary.

50. In oxidative decarboxylation, the molecule that is oxidized loses a carboxyl group as carbon dioxide, and a coenzyme like NAD^+ or FAD gets reduced. Examples of oxidative decarboxylation include the conversion of pyruvate to acetyl-CoA, isocitrate to α-ketoglutarate, and α-ketoglutarate to succinyl-CoA.

51. Citric acid does not provide significant nutritional value as an essential nutrient. Although it is an intermediate in the citric acid cycle, cells generate citric acid on their own from acetyl-CoA and oxaloacetate during aerobic respiration.

19-8 The Citric Acid Cycle in Anabolism

52. The following series of reactions exchanges NADH for NADPH:

$$\text{Oxaloacetate} + \text{NADH} + \text{H}^+ \rightarrow \text{Malate} + \text{NAD}^+$$

$$\text{Malate} + \text{NADP}^+ \rightarrow \text{Pyruvate} + \text{CO}_2 + \text{NADPH} + \text{H}^+$$

53. A variety of reactions in which amino acids are converted to citric acid cycle intermediates are considered anaplerotic. In addition, pyruvate + CO_2 can form oxaloacetate via pyruvate carboxylase.

54. Many compounds, such as fats, carbohydrates, and amino acids, can be broken down to acetyl-CoA. Acetyl-CoA can also form fats and ketone bodies, as well as feed directly into the citric acid cycle.

19-9 The Link to Oxygen

55. The NADH and $FADH_2$ produced by the citric acid cycle are the electron donors in the electron transport chain linked to oxygen. Because of this connection, the citric acid cycle is considered part of aerobic metabolism.

Chapter 20

20-1 The Role of Electron Transport in Metabolism

1. Electron transport reactions pump protons from the mitochondrial matrix to the intermembrane space (see Figure 20.1), creating the proton gradient. The gradient is important because protons are then driven back into the matrix through ATP synthase, powering the synthesis of ATP from ADP and inorganic phosphate.
2. Electrons are passed from NADH to a flavin-containing protein and then to coenzyme Q. From coenzyme Q, the electrons pass to cytochrome *b*, then to cytochrome *c*, via the Q cycle, followed by cytochromes *a* and a_3. From the cytochrome aa_3 complex, the electrons are finally passed to oxygen.
3. Electron transport and oxidative phosphorylation are different processes. Electron transport requires the respiratory complexes of the inner mitochondrial membrane, whereas oxidative phosphorylation requires ATP synthase, also located on the inner mitochondrial membrane. Electron transport can take place in the absence of oxidative phosphorylation.
4. Mitochondrial structure confines the reduced electron carriers produced by the citric acid cycle to the matrix. There, they are close to the respiratory complexes of the electron transport chain that will pass the electrons from the carriers produced by the citric acid cycle to oxygen, the ultimate recipient of electrons and hydrogens.

20-2 Reduction Potentials in the Electron Transport Chain

5. The electron transport chain translocates charged particles by chemical means. Interconversion of chemical and electrical energy is exactly what a battery does.
6. The reactions are all written in the same direction for purposes of comparison. By convention, they are written as reduction, rather than oxidation, reactions.
7. In the mitochondrial electron transport chain, electrons flow from carriers with more negative reduction potentials to carriers with more positive reduction potentials.
8. We fundamentally add the half reactions in Table 20.1.

	$E°'$ (V)
$NAD^+ + 2H^+ + 2e^- \rightarrow NADH + H^+$	−0.320

This is the wrong direction, so we reverse the equation and the sign of the potential difference.

$NADH + H^+ \rightarrow NAD^+ + 2H^+ + 2e^-$	0.320
$\frac{1}{2}O_2 + 2H^+ + 2e^- \rightarrow H_2O$	0.186
$NADH + H^+ + \frac{1}{2}O_2 \rightarrow NAD^+ + H_2O$	**1.136**

9. We fundamentally add the half reactions in Table 20.1.

	$E°'$ (V)
$NAD^+ + 2H^+ + 2e^- \rightarrow NADH + H^+$	−0.320

This is the wrong direction, so we reverse the equation and the sign of the potential difference.

$NADH + H^+ \rightarrow NAD^+ + 2H^+ + 2e^-$	0.320
$Pyruvate + 2H_+ + 2e^- \rightarrow Lactate$	−0.185
$NADH + H^+ + Pyruvate \rightarrow NAD^+ + Lactate$	**0.135**

10. We fundamentally add the half reactions in Table 20.1.

	$E°'$ (V)
$Fumarate^+ + 2H^+ + 2e^- \rightarrow Succinate$	0.031

This is the wrong direction, so we reverse the equation and the sign of the potential difference.

$Succinate \rightarrow Fumarate + 2H^+ + 2e^-$	−0.031
$\frac{1}{2}O_2 + 2H^+ + 2e– \rightarrow H_2O$	0.186
$Succinate + \frac{1}{2}O_2 \rightarrow Fumarate + H_2O$	**0.785**

11. The cytochrome is the electron donor, and the flavin moiety is the electron acceptor. We add the half reactions in Table 20.1.

	$E°'$ (V)
$Cytochrome\ c(Fe^{3+}) + e^- \rightarrow Cytochrome\ c(Fe^{2+})$	0.254

This is the wrong direction, so we reverse the equation and the sign of the potential difference.

$2\ Cytochrome\ c(Fe^{2+}) + 2e^- \rightarrow 2\ Cytochrome\ c(Fe^{3+})$	−0.254
$[FAD] + 2H^+ + 2e^- \rightarrow [FADH_2]$	0.091
$[FAD] + 2\ Cyt\ c(Fe^{2+}) + 2H^+ \rightarrow [FADH_2] + 2\ Cyt\ c(Fe^{3+})$	**−0.163**

This was the maximum value for a bound flavin. The negative sign indicates that this reaction will not take place as written because it is not energetically favorable.

12. Here is an illustration based on standard reduction potentials.

	$E°'$ (V)
$Fumarate + 2H^+ + 2e^- \rightarrow Succinate$	0.031

This is the wrong direction, so we reverse the equation and the sign of the potential difference.

$Succinate \rightarrow Fumarate + 2H^+ + 2e^-$	−0.031
$FAD + 2H^+ + 2e^- \rightarrow FADH_2$	−0.219
$Succinate + FAD \rightarrow Fumarate + FADH_2$	**−0.250**

The other possibility can be calculated in the same way.

$Succinate \rightarrow Fumarate + 2H^+ + 2e^-$	−0.031
$NAD^+ + 2H^+ + 2e^- \rightarrow NADH + H^+$	−0.320
$Succinate + NAD^+ \rightarrow Fumarate + NADH$	**−0.351**

Both reduction potentials indicate a reaction that is not energetically favorable, but less so with FAD than with NAD^+. Other factors enter consideration, however, in a living cell. The first is that the reactions do not take place under standard conditions, altering the values of reduction potentials. The second is that the reduced electron carriers (NADH and $FADH_2$) are reoxidized. Coupling the reactions we have looked at here with others also makes them less unfavorable.

20-3 The Organization of Electron Transport Complexes

13. In the first electron transport reaction of Complex I, also called NADH-CoQ oxidoreductase, electrons are transferred from NADH to FMN, forming NAD^+ (oxidized) and $FMNH_2$ (reduced). In the final electron transport reaction of Complex I, electrons are transferred from iron–sulfur clusters to CoQ, forming oxidized iron–sulfur clusters and $CoQH_2$.

14. In all reactions, electrons are passed from the reduced form of one reactant to the oxidized form of the next reactant in the chain. The notation [Fe—S] refers to any one of a number of iron–sulfur proteins.

Reactions of Complex I

$NADH + E\text{-}FMN \rightarrow NAD^+ + E\text{-}FMNH_2$	**Liberation of**
$E\text{-}FMNH_2 + 2[Fe—S]_{ox} \rightarrow E\text{-}FMN + 2[Fe—S]_{red}$	**enough energy to produce ATP**

Transfer to Coenzyme Q

$2[Fe—S]_{red} + CoQ \rightarrow 2[Fe—S]_{ox} + CoQH_2$

Reactions of Complex III

Q cycle reactions

$[Fe—S]_{red} + \text{cyt } c_{1ox} \rightarrow [Fe—S]_{ox} + \text{cyt} c_{2red}$	**Liberation of enough energy to produce ATP**

Transfer to Cytochrome *c*

$\text{cyt } c_{2red} + \text{cyt } c_{ox} \rightarrow \text{cyt } c_{1ox} + \text{cyt } c_{red}$

Reactions of Complex IV

$\text{cyt } c_{red} + \text{cyt } a\text{-}a_{3ox} \rightarrow \text{cyt } c_{ox} + \text{cyt } a\text{-}a_{3red}$	**Liberation of**
$\text{cyt } a\text{-}a_{1red} + \frac{1}{2}O_2 \rightarrow \text{cyt } a\text{-}a_{1ox}H_2O$	**enough energy to produce ATP**

15. When $FADH_2$ is the starting point for electron transport, electrons are passed from $FADH_2$ to coenzyme Q in a reaction carried out by Complex II that bypasses Complex I.

$$FADH_2 + 2[Fe—S]_{ox} \rightarrow FAD + 2[Fe—S]_{red}$$
$$2[Fe—S]_{red} + CoQ \rightarrow 2[Fe—S]_{ox} + CoQH_2$$

16. They all contain the heme group, with minor differences in the heme side chains in most cytochromes.

17. Cytochromes are proteins of electron transport; the heme ion alternates between the Fe(II) and Fe(III) states. The function of hemoglobin and myoglobin is oxygen transport and storage, respectively. The heme iron in hemoglobin and myoglobin remains in the Fe(II) state.

18. Coenzyme Q is not bound to any of the respiratory complexes. It moves freely in the inner mitochondrial membrane.

19. A part of Complex II catalyzes the conversion of succinate to fumarate in the citric acid cycle.

20. Three of the four respiratory complexes generate enough energy to phosphorylate ADP to ATP. Complex II is the sole exception.

21. Cytochrome *c* is not tightly bound to the mitochondrial membrane and can easily be lost in the course of cell fractionation. This protein is so similar in most aerobic organisms that cytochrome *c* from one source can easily be substituted for that from another source.

22. The components are in the proper orientation for the electrons to be transferred rapidly from one component to the next; if the components were in solution, speed would be limited to the rate of diffusion. A second advantage, which is a necessity, is that the components are properly positioned to facilitate the transport of protons from the matrix to the intermembrane space.

23. The key point here is not the active site, which has a low tolerance for mutations, but the molecules with which the proteins in question are associated. Cytochromes are membrane-bound and must associate with other members of the electron transport chain; most mutations are likely to interfere with the close fit, and thus, they are not preserved (because they are lethal). Globins, although soluble, still form some associations, so more mutations can be tolerated, within some limits. Hydrolytic enzymes are soluble and not likely to associate with other polypeptides except substrates. They can tolerate a higher proportion of mutations.

24. Having mobile electron carriers in addition to membrane-bound respiratory complexes allows electron transport to use the most readily available complex rather than to use the same one all the time.

25. The Q cycle enables the transfer of electrons from two-electron carriers like $CoQH_2$ to one-electron carriers like cytochrome *c* by splitting the electrons into two separate pathways. This allows Complex III to efficiently couple electron transfer with proton pumping, effectively doubling the number of protons pumped per electron pair transferred. Increased proton pumping leads to higher amounts of ATP synthesis.

26. The protein environment of the iron differs in each of the cytochromes, causing differences in the reduction potential.

27. The heme groups differ slightly in the various kinds of cytochromes. This is the main difference, with some additional modification due to the different protein environments.

28. Like most proteins bound to membranes, the components of respiratory complexes will denature upon removal from a membrane environment. Additionally, preserving the natural interactions of enzymes from the complexes with cofactors and partner proteins is difficult.

29. Adding the $\Delta G^{\circ\prime}$ values for the reactions in rows 4, 5, and 6 of Table 20.2 gives $\Delta G^{\circ\prime} = -48.2$ kJ/mol.

20-4 The Connection Between Electron Transport and the Phosphorylation of ADP to ATP

30. The F_1 portion of the mitochondrial ATP synthase, which projects into the matrix, is the site of ATP synthesis.

31. Yes, because the F_0 portion of mitochondrial ATP synthase lies within the inner mitochondrial membrane. The F_1 portion projects into the matrix, but the entire complex is an integral membrane protein complex.

32. The P/O ratio gives the number of moles of P_i consumed in the reaction $ADP + P_i \rightarrow ATP$ for each mole of oxygen atoms consumed in the reaction $\frac{1}{2}O_2 + 2H^+ \rightarrow 2H_2O$. It is a measure of the coupling of ATP production to electron transport.

33. The F_1 part of mitochondrial ATP synthase has a stationary domain (the $\alpha_3\beta_3\delta$ domain) and a domain that rotates (the $\gamma\varepsilon$ domain). Motion of the rotor domain powers ATP synthesis.

34. A P/O ratio of 1.5 can be expected because oxidation of succinate passes electrons to coenzyme Q via a flavoprotein intermediate, bypassing the first respiratory complex.

35. Exact values for P/O ratios are difficult to determine because of the complexity of the systems that pump protons and phosphorylate ADP. The number of ADP molecules

phosphorylated is directly related to the number of protons pumped across the membrane. This figure has been a matter of some controversy. The stoichiometry was uncertain for decades, but research continues to progress toward a clearer consensus.

36. The difficulties in determining the number of protons pumped across the inner mitochondrial membrane by respiratory complexes are those inherent in working with large assemblies of proteins that must be bound in a membrane environment to be active. As experimental methods improve, the task becomes less difficult.

20-5 The Mechanism of Coupling in Oxidative Phosphorylation

37. The chemiosmotic coupling mechanism is based on the difference in hydrogen ion concentration between the intermembrane space and the matrix of actively respiring mitochondria. The hydrogen ion gradient is created by the proton pumping that accompanies the transfer of electrons. The flow of hydrogen ions back into the matrix through a channel in the ATP synthase is directly coupled to the phosphorylation of ADP.
38. An intact mitochondrial membrane is necessary for compartmentalization, which in turn is necessary for proton pumping and the establishment of a proton gradient.
39. Uncouplers overcome the proton gradient on which oxidative phosphorylation depends.
40. In the production of ATP from electron transport and oxidative phosphorylation, protons flow from the intermembrane space through the ATP synthase enzyme into the mitochondrial matrix. The proton flow powers the synthesis of ATP from ADP and inorganic phosphate.
41. No, uncouplers do not directly inhibit electron transport between components of the electron transport chain. Uncouplers lead to increased permeability of the inner mitochondrial membrane, dissipating the proton gradient and making oxidative phosphorylation less effective.
42. Electron micrographs show different conformations in active and resting mitochondria.
43. Experiments with model systems in artificial systems have shown that electron transport and proton pumping can take place separately, but they cannot take place separately under normal physiological conditions in mitochondria.

20.6 Shuttle Mechanisms for Transferring Electrons from Cytosolic NADH to the Mitochondria

44. The purpose of the glycerol-3-phosphate shuttle is that it is one way the cell transfers electrons from cytosolic NADH into the mitochondrial matrix for the electron transport chain. In this pathway, cytosolic NADH electrons are transferred to dihydroxyacetone phosphate, forming glycerol-3-phosphate, which makes its way into the mitochondrion, where it donates electrons to FAD, producing $FADH_2$. $FADH_2$ then donates its electrons to CoQ via Complex II of the electron transport chain.
45. The complete oxidation of glucose produces 30 molecules of ATP in the muscle and brain and 32 molecules of ATP in the liver, heart, and kidneys. The underlying reason is the different shuttle mechanisms for the transfer of electrons to mitochondria from the NADH produced in the cytosol by glycolysis.
46. The transport "product" (in the matrix) of the malate–aspartate shuttle is NADH, whereas that of the glycerol–phosphate shuttle is $FADH_2$. The latter shuttle can thus go *against* a transmembrane NADH concentration gradient, whereas the former cannot.

20.7 The ATP Yield from the Complete Oxidation of Glucose

47. (a) 34
    (b) 32
    (c) 13.5
    (d) 17
    (e) 2.5
    (f) 12.5

Chapter 21

21-1 Lipids Are Involved in the Generation and Storage of Energy

1. (a) For mobile organisms—such as a migrating hummingbird—weight can be a critical factor, and packing the most energy into the least weight is decidedly advantageous. A 2.5-g hummingbird needs to add about 2 g of fat for migration energy, which would increase its body weight by 80%. The equivalent amount of energy stored as glycogen would be about 5 g, which would increase its body weight by 200% and make it much harder for the bird to get off the ground!
   (b) For immobile plants, weight is not a critical factor, and it takes more energy to make fat or oil than it does to make starch. In the case of plant *seeds,* "compact" energy is beneficial because the seed must be self-sufficient until enough growth has occurred to permit photosynthesis.

21-2 Catabolism of Lipids

2. Phospholipase A_1 hydrolyzes the ester bond to carbon 1 of the glycerol backbone; phospholipase A_2 hydrolyzes the ester bond to carbon 2 of the glycerol backbone.
3. Phospholipase A_1 or A_2 acts on a phosphoacylglycerol to produce one fatty acid and a lysophospholipid (a phosphoacylglycerol that has only one fatty acyl chain instead of the usual two). In contrast, a lipase acts on a triacylglycerol to produce one fatty acid and a diacylglycerol.
4. A hormone signal activates adenylate cyclase, which makes cAMP. This activates protein kinase A, which helps activate the lipases.
5. Acyl-CoA molecules are high-energy compounds that have sufficient energy to initiate the β-oxidation process. Coenzyme A also serves as a tag indicating that the molecule is destined for oxidation.
6. See Figure 21.5. The fatty acyl group is transferred from CoA to carnitine in a reaction catalyzed by carnitine palmitoyltransferase I, which is located on the outer mitochondrial membrane. The fatty acyl-carnitine molecule diffuses across the outer mitochondrial membrane and gets shuttled across the inner mitochondrial membrane by a membrane protein called carnitine-acylcarnitine translocase. Inside the matrix, the fatty acyl group is transferred from carnitine back to CoA

in a reaction catalyzed by carnitine palmitoyltransferase II. That makes a total of three enzymes involved, and the fatty acyl-CoA molecule then undergoes β-oxidation in the matrix.

7. In the first reaction of the β-oxidation pathway, the single bond between the alpha and beta carbons of a fatty acyl-CoA molecule gets oxidized to a double bond, resulting in a *trans*-Δ^2-enoyl-CoA molecule. The reaction is catalyzed by acyl-CoA dehydrogenase, and FAD is a coenzyme that gets reduced to $FADH_2$.
8. Acyl-CoA dehydrogenase removes hydrogens from adjacent carbons, creating a double bond and using FAD as a coenzyme. β-Hydroxy-CoA dehydrogenase oxidizes an alcohol group to a ketone group and uses NAD^+ as a coenzyme.
9. $$CH_3CH_2\mathbf{C}H_2\mathbf{C}H_2CH_2{-}\underset{\displaystyle O}{\underset{\|}{C}}{-}S{-}CoA$$

   The two carbons shown in boldface type are the ones that will have the double bond between them. The orientation will be trans.
10. The eighth two-carbon unit will remain esterified to CoA after the seventh cycle of β-oxidation of palmitic acid (see Figure 21.7).
11. Under conditions that activate hormone-sensitive lipases (low insulin, high glucagon/epinephrine), glycogen breakdown and gluconeogenesis are activated in the liver, while glycogen breakdown and glycolysis are activated in the muscle (Section 18-1D).

21-3 The Energy Yield from the Oxidation of Fatty Acids

12. One obtains 6.7 ATP per carbon and 0.42 ATP per gram for stearic acid versus 5 ATP per carbon and 0.17 ATP per gram for glucose. More energy is available from stearic acid than from glucose.
13. The processing of acetyl-CoA through the citric acid cycle and the electron transport chain (ETC) produces more energy than the processing of the NADH and $FADH_2$ produced during β-oxidation because each acetyl-CoA produces more NADH and $FADH_2$ molecules in the citric acid cycle, increasing the ETC ATP yield compared with the single NADH and $FADH_2$ generated per β-oxidation cycle.
14. From seven cycles of β-oxidation: 8 acetyl-CoA, 7 $FADH_2$, and 7 NADH. From the processing of 8 acetyl-CoA in the citric acid cycle: 8 $FADH_2$, 24 NADH, and 8 GTP. From the reoxidation of all $FADH_2$ and NADH: 22.5 ATP from 15 $FADH_2$ and 77.5 ATP from 31 NADH. From 8 GTP: 8 ATP. Subtotal: 108 ATP. A 2-ATP equivalent was used in the activation step. Grand total: 106 ATP. The grand total for stearic acid was 120 ATP.
15. The humps of camels contain lipids that can be degraded as a source of metabolic water, rather than water as such.

21-4 Catabolism of Unsaturated Fatty Acids and Odd-Carbon Fatty Acids

16. For an odd-chain fatty acid, β-oxidation proceeds normally until the last round. When five carbons are left, that round of β-oxidation releases one acetyl-CoA and one propionyl-CoA. Propionyl-CoA cannot be further metabolized by β-oxidation; however, a separate set of enzymes converts propionyl-CoA into succinyl-CoA, which can then enter the citric acid cycle.
17. The complete oxidation of a fatty acid with 17 carbons produces *seven* molecules of *acetyl-CoA* that can enter the citric acid cycle and *one* molecule of propionyl-*CoA* that is converted in a series of reactions to *succinyl-CoA*, which can enter the citric acid cycle.
18. False. The oxidation of unsaturated fatty acids to acetyl-CoA requires a cis–trans isomerization, a reduction, and occasionally an epimerization, reactions catalyzed by enzymes that are not found in the oxidation of saturated fatty acids.
19. For a monounsaturated fatty acid, an additional enzyme is needed, the enoyl-CoA isomerase.
20. For a polyunsaturated fatty acid, two additional enzymes are needed, the enoyl-CoA isomerase and 2,4-dienoyl-CoA reductase.
21. From seven cycles of β-oxidation: 7 acetyl-CoA, 1 propionyl-CoA, 7 $FADH_2$, and 7 NADH. From the processing of 7 acetyl-CoA in the citric acid cycle: 7 $FADH_2$, 21 NADH, and 7 GTP. From the processing of the propionyl-CoA: −1 ATP for conversion to succinyl-CoA, −1 GTP from the citric acid cycle, and 1 NADH and 1 $FADH_2$ from the citric acid cycle. From the reoxidation of all $FADH_2$ and NADH: 22.5 ATP from 15 $FADH_2$ and 72.5 ATP from 29 NADH. From 8 GTP: 8 ATP. Subtotal: 103 ATP. Subtract a 2-ATP equivalent used in the activation step and a 1-ATP equivalent used in the conversion to succinyl-CoA for a grand total of 100 ATP.
22. An 18-carbon saturated fatty acid yields 120 ATP. For a monounsaturated fatty acid, the double bond eliminates the step that produces $FADH_2$, so there would be 1.5 ATP fewer for oleic acid, or 118.5 ATP total.
23. An 18-carbon saturated fatty acid yields 120 ATP. For a diunsaturated fatty acid with the bonds in the Δ^9 and Δ^{12} positions, the first double bond eliminates an $FADH_2$. The second double bond eliminates another $FADH_2$ and uses an NADPH, which we are assuming is the same cost as using an NADH. Thus, a total of 5.5 ATP are lost, compared with a saturated fatty acid, so the total is 114.5 ATP.
24. It would take seven cycles of β-oxidation to release 14 carbons as acetyl-CoA, with the last three being released as propionyl-CoA.
25. Fats cannot produce a net yield of glucose because they must enter the citric acid cycle as the two-carbon unit acetyl-CoA. In the first few steps, two carbons are released as CO_2. However, an odd-chain fatty acid can be considered partially glucogenic because the final three carbons become succinyl-CoA and enter the citric acid cycle after the decarboxylation steps. Thus, if an extra succinyl-CoA is added, it can then be drawn off later as malate and used for gluconeogenesis without removing the steady-state level of citric acid cycle intermediates.

21-5 Ketone Bodies

26. Ketone bodies are substances related to acetone. They are produced in the liver when an excess of acetyl-CoA arises from the β-oxidation pathway. This condition occurs when an organism has a high intake of lipids and a low intake of carbohydrates.
27. Two acetyl-CoA molecules combine to form acetoacetyl-CoA. This can then release coenzyme A to yield acetoacetate, which can be converted either to β-hydroxybutyrate or to acetone.

28. If the reason for passing out is uncontrolled diabetes, the doctor expects to smell acetone on the breath because the otherwise unused sugars are being converted to fats and ketone bodies.
29. Chronic ethanol metabolism increases NADH in the liver due to alcohol dehydrogenase and aldehyde dehydrogenase activity (Section 17-4B). High [NADH] inhibits the citric acid cycle, diverting acetyl-CoA into fatty acid and triacylglycerol synthesis. This leads to a "fatty liver."
30. The metallic taste may be due to acetone, which is produced when carbohydrate intake is low and fatty acid oxidation increases (leading to excess acetyl-CoA and the production of ketone bodies).

21-6 Fatty Acid Biosynthesis

31. In the first step of fatty acid biosynthesis, acetyl-CoA gets carboxylated by the acetyl-CoA carboxylase enzyme complex to produce malonyl-CoA. The enzyme complex requires the biotin coenzyme.
32. The two pathways have in common the involvement of acetyl-CoA and thioesters, and each round of breakdown or synthesis involves two-carbon units. The differences are many: malonyl-CoA is involved in biosynthesis, not in breakdown; thioesters involve CoA in breakdown and involve acyl carrier proteins (in addition to CoA) in biosynthesis; biosynthesis occurs in the cytosol, but breakdown occurs in the mitochondrial matrix; breakdown is an oxidative process that requires NAD^+ and FAD and produces ATP by electron transport and oxidative phosphorylation, whereas biosynthesis is a reductive process that requires NADPH and ATP.
33. Step 1: Biotin is carboxylated using the bicarbonate ion (HCO_3^-) as the source of the carboxyl group. Step 2: The carboxylated biotin is brought into proximity with enzyme-bound acetyl-CoA by a biotin carrier protein. Step 3: The carboxyl group is transferred to acetyl-CoA, forming malonyl-CoA.
34. Malonyl-CoA is a molecule that commits itself to fatty acid synthesis. It is also a potent inhibitor of carnitine acyltransferase I, thereby shutting down β-oxidation.
35. In β-oxidation, FAD is the coenzyme for the first redox reaction, while NAD^+ is the coenzyme for the second. In fatty acid synthesis, NADPH is the coenzyme for both redox reactions. The β-hydroxy-acyl group in β-oxidation has the L configuration, while it has the D-configuration in fatty acid synthesis.
36. Both have a phosphopantetheine group at the active end. In coenzyme A, this group is attached to 2′-phospho-AMP; in ACP, it is attached to a serine residue of a protein.
37. ACP is a molecule that earmarks acyl groups for fatty acid synthesis. It can be managed separately from acyl-CoA groups. Also, the ACP attaches to the acyl groups like a "swinging arm" that tethers it to the fatty acid synthase complex.
38. Linoleate and linolenate cannot be synthesized by the body and must therefore be obtained from dietary sources. Mammals cannot produce a double bond beyond carbon atom 9 of fatty acids, preventing production of the double bonds at carbons 12 and 15 that are found in these polyunsaturated fatty acids.
39. No, acetyl-CoA intermediates are required as building blocks in the conversion of fatty acids to other lipids.
40. Acetyl groups condense with oxaloacetate to form citrate, which can cross the mitochondrial membrane. Acetyl groups are regenerated in the cytosol by the reverse reaction.
41. If acetyl-carnitine forms in the matrix of the mitochondrion, it can be translocated to the cytosol via the carnitine translocase. Thus, this could represent another way of shuttling acetyl units out of the mitochondria for synthesis.
42. Energy is needed to condense an acetyl group to the growing fatty acid. In theory, this could be done with acetyl-CoA, using ATP. In practice, ATP is used to convert acetyl-CoA to malonyl-CoA; the condensation of the acetyl moiety of malonyl-CoA is driven in part by the accompanying decarboxylation and requires no additional energy. A possible reason for this is to avoid a metabolic confusion of pathways, perhaps particularly important in (uncompartmented) prokaryotes; one could envision an acetyl-CoA from degradation being used immediately for synthesis. Malonyl-CoA says "synthesis"; acetyl-CoA says "degradation."
43. (a) The lipoamide "swinging arm" of the pyruvate dehydrogenase complex in Section 19-3.
    (b) The "arm" or ACP carries the group to be acted on from one enzyme to another (avoiding a diffusion-limited process and positioning key groups correctly). In the case of ACP, the group to be acted on (the β-carbon) is always the same distance from the ACP, regardless of the length of the growing fatty acid, and thus the critical group is always in proximity to the active sites of the several pertinent enzymes.

21-7 Biosynthesis of Acylglycerols and Compound Lipids

44. The glycerol comes from the degradation of other acylglycerols or from glycerol-3-phosphate derived from glycolysis.
45. The activating group found on the acylglycerol is cytidine diphosphate.
46. In prokaryotes, CTP reacts with phosphatidic acid to give a CDP-diacylglycerol. This reacts with serine to give phosphatidylserine, which decarboxylates to phosphatidylethanolamine. In eukaryotes, phosphatidylserine is decarboxylated to give phosphatidylethanolamine in mitochondria or CDP-ethanolamine reacts with a diacylglycerol to give phosphatidylethanolamine.

21-8 Steroid Biosynthesis

47. In steroid biosynthesis, three acetyl-CoA molecules condense to form the six-carbon mevalonate, which then gives rise to a five-carbon isoprenoid unit. A second and then a third isoprenoid units condense, giving rise to a 10-carbon and then a 15-carbon unit. Two of the 15-carbon units condense, forming a 30-carbon precursor of cholesterol.
48.

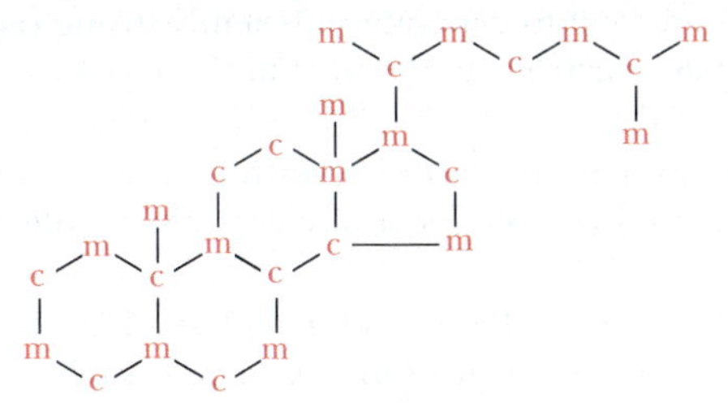

49. Bile acids, steroid hormones, and vitamin D.
50. All steroids have a characteristic fused-ring structure, implying a common biosynthetic origin.
51. One oxygen atom from O_2 is needed to form the epoxide. NADPH is needed to reduce the other oxygen atom to water.
52. Cholesterol is nonpolar and cannot dissolve in blood, which is an aqueous medium.
53. Bile salts are made from cholesterol, and cholesterol is taken from the body into the intestine via bile salts in bile fluid for excretion.

Chapter 22

22-1 Chloroplasts Are the Site of Photosynthesis

1. See Figure 22.4A. Chlorophyll absorbs light most strongly in the red and blue regions of the visible spectrum but reflects light in the green region, which is why it appears green.
2. Iron and manganese are in chloroplasts; iron and copper are in mitochondria. These are transition metals, which can easily undergo redox reactions.
3. Both chloroplasts and mitochondria have an inner and an outer membrane. Both have their own DNA and ribosomes. Chloroplasts, however, have a third membrane, the thylakoid membrane.
4. Chlorophyll has a cyclopentanone ring fused to the tetrapyrrole ring, a feature that does not exist in heme. Chlorophyll contains magnesium, whereas heme contains iron. Chlorophyll has a long side chain based on isoprenoid units, which is not found in heme.
5. Only a relatively small portion of the visible spectrum is absorbed by chlorophylls. The accessory pigments absorb light at additional wavelengths. As a result, most of the visible spectrum can be harnessed in light-dependent reactions.
6. Recall from Section 12-4B that *N*-formylmethionine is the first amino acid in prokaryotic protein synthesis. The fact that this also occurs in chloroplasts is consistent with the endosymbiotic theory, which proposes that chloroplasts evolved from independent bacterial organisms.

22-2 Photosystems I and II and the Light-Dependent Reactions of Photosynthesis

7. The net electron flow in chloroplasts is the reverse of that in the mitochondria because water is oxidized to reduce $NADP^+$ to NADPH, whereas mitochondria oxidize NADH to reduce oxygen. However, this is *not* simply the reverse process because the carriers differ, and chloroplasts require light energy to drive the redox reactions.
8. See Figure 22.11A and its related text.
9. In photosystem I and photosystem II, light energy is needed to raise the reaction-center chlorophylls to a higher energy level. Energy is needed to generate sufficiently strong reducing agents to pass electrons to the next in the series of components in the pathway.
10. No. Most chlorophylls are light-harvesting molecules that transfer energy to the special pair of chlorophyll molecules in each photosystem.
11. The electron transport chain in chloroplasts, like that in mitochondria, consists of protein complexes, such as the photosystems and the cytochrome b_6f complex. It also contains mobile electron carriers, such as pheophytin and plastoquinone (equivalent to coenzyme Q), which is also true of the mitochondrial electron transport chain.
12. Probably the electron transport chain in chloroplasts. Chloroplasts generate molecular oxygen; mitochondria use it. The early atmosphere almost certainly lacked molecular oxygen. Only when early photosynthetic organisms introduced oxygen into the atmosphere could it be used in mitochondria.
13. Electron transport and ATP production are coupled to each other by the same mechanism in mitochondria and chloroplasts. In both cases, the coupling depends on the generation of a proton gradient across the inner mitochondrial membrane or across the thylakoid membrane.
14. In mitochondria, both a proton gradient (chemical) and an electrochemical gradient (based on charge) are formed, both contributing to the total potential energy. In chloroplasts, only a proton gradient is formed because ions move across the thylakoid membrane and neutralize charge. The proton gradient alone is considerably less efficient.
15. With very few exceptions, life directly or indirectly depends on photosynthesis. The electric current is the flow of electrons from water to $NADP^+$, a light-requiring process. The "current" continues in the light-independent reactions, with electrons flowing from NADPH to bisphosphoglycerate, which ultimately yields glucose.
16. Photosystem II requires more energy than photosystem I. The shorter wavelength of light means a higher frequency. Frequency, in turn, is directly proportional to energy.
17. A photosynthetic reaction center is analogous to a battery because its reactions produce a charge separation. The charge separation is comparable to the stored energy of a battery.
18. The electron transport chains of mitochondria and chloroplasts are similar. In mitochondria, antimycin A inhibits electron transfer from cytochrome *b* to coenzyme Q in the Q cycle. By analogy, it can be argued that antimycin A inhibits electron flow from plastoquinone to cytochrome b_6f. A Q cycle may also operate in chloroplasts.
19. Oxygen produced in photosynthesis comes from water. The oxygen-evolving complex is part of the series of electron-transfer reactions from water to NADPH. Carbon dioxide is involved in the dark reactions, which are different reactions that take place in another part of the chloroplast.
20. It is well established that the path of electrons in photosynthesis goes from photosystem II to photosystem I. The reason for the nomenclature is that photosystem I is easier to isolate than photosystem II and was studied more extensively at an earlier date.
21. It would take much work to establish the number of protons pumped across the thylakoid membrane. This is partly the result of experience with mitochondria and partly a prediction based on the greater complexity of structure in the chloroplast (three membranes in chloroplasts compared to two membranes in mitochondria).
22. The oxygen-evolving complex of photosystem II passes through a series of five oxidation states (designated as S_0 through S_4) in the transfer of four electrons in the process of evolving oxygen (Figure 22.8). One electron is passed

from water to photosystem II for each quantum of light. In the process, the components of the reaction center go successively through oxidation states S_1 through S_4. The S_4 state decays spontaneously to the S_0 state and, in the process, oxidizes two water molecules to one oxygen molecule. Four protons are released simultaneously.

23. The loosely bound cytochrome acts as a mobile electron donor, essential for continuous cyclic photophosphorylation in these bacteria.

22-3 Photosynthesis and ATP Production

24. In cyclic photophosphorylation, the excited chlorophyll of photosystem I passes electrons directly to the electron transport chain that normally links photosystem II to photosystem I. This electron transport chain is coupled to ATP production (see Figure 22.10).
25. Both depend on a proton gradient resulting from the flow of electrons. In chloroplasts, protons come from the splitting of water to produce oxygen. In mitochondria, protons come from the oxidation of NADH and ultimately consume oxygen and produce water.
26. The proton gradient is created by the operation of the electron transport chain that links the two photosystems.
27. ATP can be produced by chloroplasts in the absence of light if some way exists to form a proton gradient.
28. Cyclic photophosphorylation can take place when the plant needs ATP but does not have a great need for NADPH. Noncyclic photophosphorylation can take place when the plant needs both.

22-4 Evolutionary Implications of Photosynthesis with and Without Oxygen

29. Many electron donors other than water are possible in photosynthesis. This is especially the case in bacteria, whose photosystems do not have oxidizing agents strong enough to oxidize water. Some of the alternative electron donors are H_2S and organic compounds.
30. A prokaryotic organism that contains both chlorophyll *a* and chlorophyll *b* could be a relic of an evolutionary way station in the development of chloroplasts.

22-5 The Light-Independent (Calvin Cycle) Reactions of Photosynthesis

31. Rubisco is the principal protein in the chloroplasts of all green plants. This wide distribution makes it likely to be the most abundant protein in nature.
32. The amino acid sequence of the catalytic subunits of rubisco is encoded by chloroplast genes, whereas that of the regulatory subunits is encoded by nuclear genes.
33. Gluconeogenesis and the pentose phosphate pathway have a number of reactions similar to those of the Calvin cycle reactions of photosynthesis.
34. From the standpoint of thermodynamics, the production of sugars in photosynthesis is the reverse of the complete oxidation of a sugar such as glucose to CO_2 and water. The complete oxidation reaction produces six moles of CO_2 for each mole of glucose oxidized. To get the energy change for the fixation of one mole of CO_2, change the sign of the energy for the complete oxidation of glucose and divide by 6.
35. After six Calvin cycles of growth in the presence of $^{14}CO_2$, the glucose molecules synthesized by photosynthesis will not be fully labeled with radioactive carbon. This is because each cycle incorporates only one molecule of CO_2 into each molecule of ribulose-1,5-bisphosphate, which then goes on to give rise to sugars. It would take at least 10–20 cycles for most glucose molecules to be fully labeled.
36. Their DNA is circular. Their ribosomes are more like those of bacteria than those of eukaryotes. Their aminoacyl-tRNA synthetases use bacterial tRNAs but not eukaryotic tRNAs. In general, they do not have introns in their genomes. Their mRNA uses a Shine–Dalgarno sequence.
37. The regeneration pathway borrows heavily from the nonoxidative branch of the pentose phosphate pathway and from gluconeogenesis. These pathways yield sugars as well as NADPH for reductive biosynthesis. Thus, only a few new enzymes would need to evolve through mutations to enable the complete Calvin cycle to function.
38. After photosynthetic organisms appeared, atmospheric oxygen increased. The oxygenase activity that produces 2-phosphoglycolate and 3-phosphoglycerate from ribulose-1,5-bisphosphate became a problem because it wastes carbon and energy. Therefore, rubisco is thought to have evolved lower oxygen sensitivity so that its carbon-fixing activity can be favored.
39. The condensation of ribulose-1,5-bisphosphate with carbon dioxide to form two molecules of 3-phosphoglycerate is the actual carbon dioxide fixation. The rest of the Calvin cycle regenerates ribulose-1,5-bisphosphate.
40. Six molecules of carbon dioxide fixed in the Calvin cycle do not end up in the same glucose molecule. However, labeling experiments show that six carbon atoms are incorporated into sugars for every six carbon dioxide molecules that enter the Calvin cycle, but the carbons are distributed among different glucose molecules.

22-6 CO_2 Fixation into Sugars in Tropical Plants

41. In tropical plants (C_4 plants), the C_4 pathway is used, in which CO_2 enters the leaf through pores in the outer mesophyll cells. The CO_2 reacts first with phosphoenolpyruvate to produce the four-carbon oxaloacetate compound and P_i. Oxaloacetate is then reduced to malate, with the concomitant oxidation of NADPH. Malate is then transported to the bundle-sheath cells (the next layer), where CO_2 is released for the Calvin cycle. These reactions do not take place in non-tropical (C_3) plants, where CO_2 is directly fixed into the Calvin cycle.
42. Photorespiration is when rubisco acts as an oxygenase, fixing O_2 instead of CO_2. Photorespiration is inefficient because it releases previously fixed CO_2, consumes ATP and NADPH, and reduces the level of sugars produced compared to the Calvin cycle.
43. High temperatures and strong sunlight increase the oxygenase activity of rubisco, leading to photorespiration, which wastes carbon and energy compared to the Calvin cycle. The C_4 pathway in tropical plants increases CO_2 concentration in the bundle sheath cells, favoring the Calvin cycle instead of photorespiration.
44. Most plants would be more productive in the absence of photorespiration. There is another side to this picture, however. The oxygenase activity appears to be an unavoidable,

wasteful activity of rubisco. Photorespiration is a salvage pathway that saves some of the carbon that would be lost due to the oxygenase activity of rubisco. Photorespiration is essential to plants, even though the plant pays the price in the loss of ATP and reducing power; mutations that affect this pathway can be lethal.

Chapter 23

23-1 Nitrogen Metabolism: An Overview

1. Nitrogen fixation is the process by which inorganic molecular nitrogen from the atmosphere (N_2), which is unusable by organisms, is incorporated first into the ammonium ion (NH_4^+) (Figure 23.1, *right*) and then into organic compounds that organisms can use.
2. Nitrogen-fixing bacteria (symbiotic organisms that form nodules on the roots of leguminous plants such as beans and alfalfa) and some free-living microbes and cyanobacteria can fix nitrogen. Plants and animals cannot.

23-2 Nitrogen Fixation

3. Nitrogen is fixed by the nitrogenase reaction, in which N_2 is converted to NH_4^+. Very few organisms have this enzyme, which can catalyze the breaking of the triple bond in molecular nitrogen. The glutamate dehydrogenase reaction and the glutamine synthase reaction assimilate nitrogen:

$$NH_4^+ + \alpha\text{-Ketoglutarate} \rightleftharpoons \text{Glutamate} + \text{Water(NADPH required)}$$
$$NH_4^+ + \text{Glutamate} \rightleftharpoons \text{Glutamine (ATP required)}$$

4. The chemical synthesis of the ammonium ion from H_2 and N_2.
5. $N_2 + 8e^- + 16ATP + 10H^+ \rightarrow 2NH_4^+ + 16ADP + 16P_i + H_2$ is the half-reaction for reduction via nitrogenase. The oxidation reaction varies with species.
6. The nitrogenase complex is made up of two protein components: dinitrogenase reductase and nitrogenase. A soluble electron carrier, called ferredoxin, delivers electrons to the dinitrogenase reductase component, which is an iron–sulfur protein. The electrons are then transferred (with ATP hydrolysis) to the nitrogenase component, which is an iron–molybdenum protein that houses the unique metal centers that reduce N_2 to NH_4^+.

23-3 Feedback Inhibition in Nitrogen Metabolism

7. Pathways that use nitrogen to make amino acids, purines, and pyrimidines are controlled by feedback inhibition because nitrogen assimilation is energetically expensive. Feedback inhibition ensures a balanced supply of nitrogenous compounds without wasting energy and nitrogen.
8. Because all the components of a cycle are regenerated, only small amounts ("catalytic quantities") are needed. This is important from an energy standpoint and, perhaps for some compounds, because of insolubility problems.

23-4 Amino Acid Biosynthesis

9. They are all interrelated. α-Ketoglutarate can be changed to glutamate via transamination or glutamate dehydrogenase. Glutamine synthetase makes glutamine out of glutamate.

10.

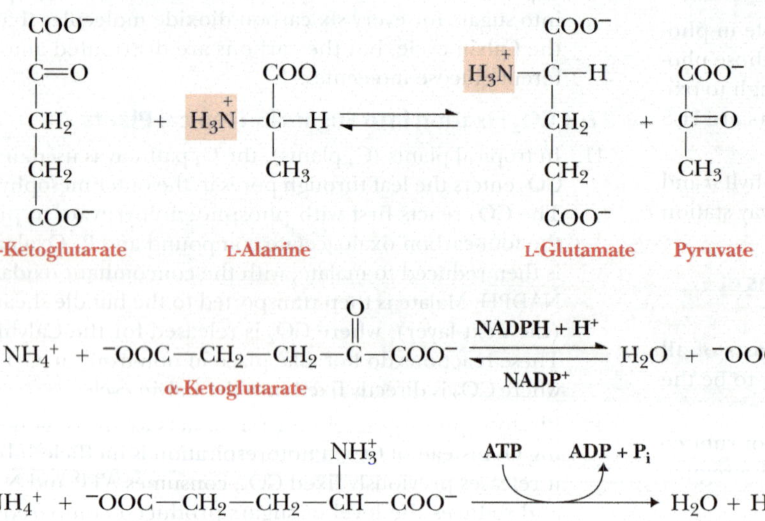

α-Ketoglutarate L-Alanine L-Glutamate Pyruvate

11.

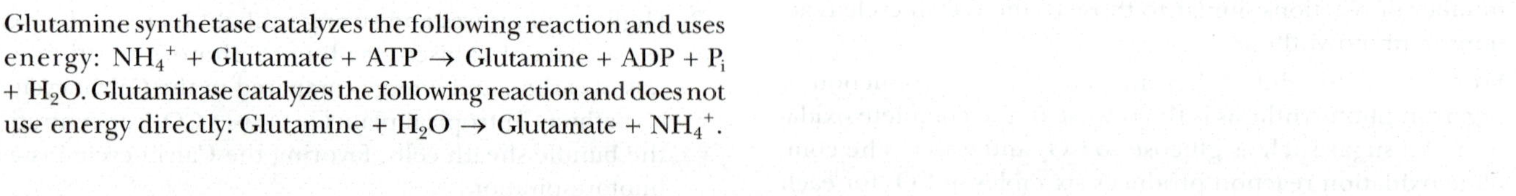

12. Glutamine synthetase catalyzes the following reaction and uses energy: NH_4^+ + Glutamate + ATP → Glutamine + ADP + P_i + H_2O. Glutaminase catalyzes the following reaction and does not use energy directly: Glutamine + H_2O → Glutamate + NH_4^+.

13. See Figure 23.9.
14. The principal ones are tetrahydrofolate and *S*-adenosylmethionine.
15. (a) See the top of Figure 23.13A; (b) see the middle of Figure 23.13A; (c) see the bottom of Figure 23.13A for six ways that THF can serve as a carrier of one-carbon groups.
16. The conversion of homocysteine to methionine using *S*-adenosylmethionine as the methyl donor gives no net gain; one methionine is needed to produce another methionine.
17. Glutamate + α-Keto acid → α-Ketoglutarate + Amino acid
18.

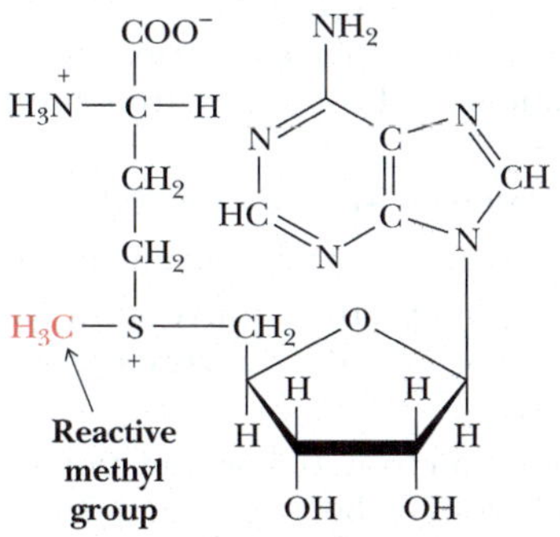

S-Adenosylmethionine

19. Sulfanilamide inhibits folic acid biosynthesis.
20. Methionine can play a dual role. In addition to providing a hydrophobic group for protein structure, methionine (in the form of *S*-adenosylmethionine) can act as a methyl group donor.

23-5 Essential Amino Acids

21. Essential amino acids are those that the human body cannot synthesize in sufficient amounts, so they must be obtained from the diet for protein synthesis and other metabolic functions.
22. The essential amino acids are those with branched chains, aromatic rings, or basic side chains.
23. The additional dietary requirement for individuals with phenylketonuria is tyrosine because they are unable to efficiently convert phenylalanine to tyrosine.

23-6 Amino Acid Catabolism and Nitrogen Excretion

24. An α-amino acid has an amino group attached to an α-carbon (recall that an α-carbon is a carbon that is adjacent to the carbonyl carbon of a larger functional group; in the case of α-amino acids, the carbonyl group is part of a carboxylic acid group). Five α-amino acids are involved directly in the urea cycle (ornithine, citrulline, aspartate, argininosuccinate, and arginine). Of those, only aspartate and arginine are also found in proteins.
25. $H^+ + HCO_3^- + 2NH_3 + 3ATP \rightarrow NH_2CONH_2 + 2ADP + 2P_i + AMP + PP_i + 2H_2O$. The urea cycle is linked to the citric acid cycle by fumarate and by aspartate, which can be converted to malate by transamination (see Figure 23.21).
26. Ornithine is similar to lysine, but it has one fewer methylene group in the side chain. Citrulline is a keto version of arginine, with a side chain $C{=}NH_2^+$ replaced by $C{=}O$.
27. Aspartate and argininosuccinate are the amino acids that link the two pathways. Aspartate is made by transamination of OAA. The aspartate then combines with citrulline to form argininosuccinate, which then releases a fumarate to go back to the TCA cycle.
28. Each round of the urea cycle costs 3 ATP—two to make carbamoyl phosphate and effectively one (ATP → AMP) to make argininosuccinate.
29. It is controlled by a special effector molecule, *N*-acetylglutamate, which is itself controlled by the levels of arginine.
30. When arginine levels build up, it means that the urea cycle is going too slowly because not enough carbamoyl phosphate is available to react with ornithine. Carbamoyl phosphate levels are increased when excess arginine activates carbamoyl phosphate synthetase I.
31. Glutamate brings ammonium ions to the matrix of the mitochondria when it gets oxidatively deaminated by glutamate dehydrogenase. The ammonium ions are then used by carbamoyl phosphate synthetase I to make carbamoyl phosphate for the urea cycle. Therefore, high levels of glutamate stimulate the urea cycle.
32. Glucogenic amino acids are degraded to pyruvate or one of the citric acid cycle intermediates found after the decarboxylation steps such as succinate or malate. Ketogenic amino acids are degraded to acetyl-CoA or acetoacetyl-CoA.
33. (a) Glucogenic
    (b) Glucogenic
    (c) Glucogenic
    (d) Ketogenic
    (e) Glucogenic
    (f) Ketogenic
34. Fish and amphibians in water-rich environments excrete excess nitrogen as the ammonium ion. Birds and reptiles excrete it as uric acid. Mammals and amphibians (in their terrestrial stages) excrete it as urea.
35. Because ostriches don't fly, one could argue that they would excrete their excess nitrogen as urea. On the other hand, they are birds and as such probably have the same metabolism as their lighter counterparts, which is an argument for predicting that they would excrete it as uric acid (which they do).
36. The amounts of arginine necessary in the urea cycle are only catalytic. If arginine from the cycle is used for protein synthesis, the cycle becomes depleted.
37. A high-protein diet leads to increased production of urea. Drinking more water increases the volume of urine, ensuring elimination of the urea from the body with less strain on the kidneys than if urea were at a higher concentration.
38. The metabolism of amino acids encourages urine formation and actually causes greater thirst and a need for water.

23-7 Purine Biosynthesis and Catabolism

39. Because folic acid is critical to the formation of purines, antagonists of folic acid metabolism are used as chemotherapy drugs to inhibit nucleic acid synthesis and cell growth. Rapidly dividing cells, such as those found in cancer and tumors, are more susceptible to these antagonists.
40. All four nitrogen atoms of the purine ring are derived from amino acids: two from glutamine, one from aspartate, and one from glycine. Two of the five carbon atoms (adjacent

to the glycine nitrogen) also come from glycine, two more come from tetrahydrofolate derivatives, and the fifth comes from CO_2.

41. In inosine, carbon-6 of the ring is a ketone group; in adenosine, carbon-6 is bound to an amino group.
42. Tetrahydrofolate is a carrier of carbon groups. Two of the carbons in the purine ring are donated by tetrahydrofolate.
43. The conversion of IMP to GMP produces one NADH and uses the equivalent of 2 ATP because an ATP is converted to AMP. If the NADH goes into the electron transport chain, then it would give rise to 2.5 ATP, and we could say that the conversion of IMP to GMP results in a net production of 0.5 ATP.
44. There is a complicated system of feedback inhibition for the production of purine-containing nucleotides. The final products, ATP and GTP, feed back to inhibit the first steps starting from ribose-5-phosphate. In addition, each intermediate, such as AMP or ADP, can also inhibit the first step. Also, each of the three forms of each nucleotide inhibits the committed reaction from IMP that eventually decides which purine nucleotide is made.
45. The purine salvage reaction that produces GMP requires hydrolysis of three high-energy phosphate bonds ($ATP \rightarrow PRPP + AMP$, $PRPP \rightarrow GMP + PP_i$, and $PP_i \rightarrow 2P_i$; see Figure 23.27 for an analogous example). The pathway to IMP and then to GMP requires the hydrolysis of seven high-energy bonds.
46. In most mammals, uric acid is converted to allantoic acid, which is much more water soluble than uric acid.

23-8 Pyrimidine Biosynthesis and Catabolism

47. In purine nucleotide biosynthesis, the growing purine ring is covalently bonded to ribose; the ribose is added after the ring is synthesized in pyrimidine nucleotide biosynthesis.
48. Purines break down to various products, depending on the species. These products are then excreted, representing a major means of nitrogen excretion for many organisms. Pyrimidine catabolism yields, in addition to NH_4^+ and CO_2, the salvageable product β-alanine, which is a breakdown product of both cytosine and uracil.

23-9 Conversion of Ribonucleotides to Deoxyribonucleotides

49. See Figure 23.34. Thioredoxin donates electrons to ribonucleotide reductase, which catalyzes the reduction of ribonucleotides to deoxyribonucleotides for DNA synthesis. Thioredoxin reductase uses NADPH to regenerate reduced thioredoxin.
50. Fluorouracil substitutes for thymine in DNA synthesis. In rapidly dividing cells, such as cancer cells, the result is the production of defective DNA, causing cell death.
51. The DNA of fast-growing cells, such as those of the hair follicles, is damaged by chemotherapeutic agents.

Chapter 24

24-1 Connections Between Metabolic Pathways

1. ATP and NADPH are the two molecules that link the most pathways. Chemical energy in the form of ATP is generated in catabolism and consumed to drive anabolic reactions.
2. Acetyl-CoA, pyruvate, PEP, α-ketoglutarate, succinyl-CoA, oxaloacetate, and several sugar phosphates such as glucose-6-phosphate and fructose-6-phosphate.
3. Fructose-6-phosphate—the pentose phosphate pathway (PPP).
   (a) Oxaloacetate—gluconeogenesis, conversion to and from aspartate, and conversion from the glyoxylate cycle via citrate.
   (b) Glucose-6-phosphate—the PPP, conversion to and from glycogen in animals, conversion to starch in plants.
   (c) Acetyl-CoA—conversion to and from fatty acids, conversion to steroids (and isoprenoids), conversion from some amino acid degradations, and conversion from the glyoxylate cycle via citrate.
   (d) Glyceraldehyde-3-phosphate—the reverse PPP.
   (e) α-Ketoglutarate—conversion to and from glutamate.
   (f) Dihydroxyacetone phosphate—conversion to and from the glycerol moiety of triacylglycerols and phosphoacylglycerols.
   (g) Succinyl-CoA—conversion from the degradation of fatty acids with odd numbers of carbon atoms and from some amino acid degradations.
   (h) 3-Phosphoglycerate—the Calvin cycle.
   (i) Fumarate—the Urea cycle.
   (j) Phosphoenolpyruvate—conversion from oxaloacetate in gluconeogenesis.
   (k) Citrate—the glyoxylate cycle and transport across the mitochondrial membrane for fatty acid and steroid synthesis.
   (l) Pyruvate—fermentation, gluconeogenesis, and conversion to and from alanine.
4. When the body breaks down proteins to supply material for gluconeogenesis, the increased urea output results in greater urine production, which uses water stored in the body. Another factor in the initial water loss is glycogen depletion. Each gram of glycogen is stored with about three to four grams of water, so when the body breaks down its glycogen stores to provide glucose, a lot of water is also released and excreted.
5. High ATP or NADH concentration and the citric acid cycle: isocitrate dehydrogenase (and the citric acid cycle) would be inhibited. The resulting pileup of acetyl-CoA (or citrate) would stimulate fatty acid and steroid synthesis, gluconeogenesis, and (in plants and some microorganisms) the glyoxylate cycle.
   (a) High ATP concentration and glycolysis: phosphofructokinase-1 (and glycolysis) would be inhibited. Glucose-6-phosphate would pile up, stimulating glycogen (or starch) synthesis, the oxidative pentose phosphate pathway, or gluconeogenesis.
   (b) High NADPH concentration and the pentose phosphate pathway: the oxidative branch of the pentose phosphate pathway would be inhibited, thus making glucose-6-phosphate available for other purposes. These include glycolysis, glycogen synthesis, glucose synthesis, and the "reverse" pentose phosphate pathway (yielding only pentose phosphate).

(c) High fructose-2,6-bisphosphate concentration and gluconeogenesis: fructose-2,6-bisphosphate inhibits fructose-1,6-bisphosphatase and activates phosphofructokinase-1. Gluconeogenesis would thus be inhibited, and glycolysis would be stimulated, as would the reverse pentose phosphate pathway and the production of glycerol phosphate for lipids.

6. Many compounds, such as oxaloacetate, pyruvate, and acetyl-CoA, play a role in a number of reactions in different pathways. Additionally, the end products of some pathways are the starting points of others. Each pathway is one aspect of an interconnected metabolic scheme that cells use to maintain homeostasis.
7. The *effects* of biochemical pathways can be reversed. Examples include glycolysis and gluconeogenesis, glycogen formation and synthesis, and the pentose phosphate pathway. The details are not completely reversible. An irreversible step in one pathway tends to be replaced with another reaction, catalyzed by another enzyme.
8. Transport processes are especially important for substances, such as oxaloacetate, which cannot cross the mitochondrial membrane. The same is true for electrons. Shuttle mechanisms must exist to transport electrons as the reduced forms of important compounds. Compounds that cannot cross the membrane must be converted to ones that can, and then must, be converted back to their original form on the other side of the membrane.
9. When a pathway has many steps, it is possible for energy changes to take place in steps of manageable size. It also allows for control of a pathway to be exercised at more points than would be the case if there were only a few steps.

24-2 Biochemistry and Nutrition

10. The 1992 pyramid treated most carbohydrates as desirable and recommended limiting fats, without distinguishing between healthy and unhealthy types of each. MyPlate emphasizes healthy proportions of each food group at each meal, focusing on balanced meals instead of a strict hierarchy.
11. Fats and carbohydrates can be stored when they are consumed in excess. Fats are stored as triacylglycerols, and carbohydrates are stored as glycogen. However, proteins consumed in excess are not stored. The extra protein is broken down. The amino groups are released as urea, and the carbon skeletons are stored as carbohydrate or fat.
12. Saturated fatty acids have been correlated with increased levels of LDL, which have been shown to be an indicator of a high risk for heart disease.
13. Leptin is a hormone that affects metabolism. It affects the brain to suppress appetite, and it affects metabolism directly by stimulating fatty-acid oxidation and inhibiting fatty-acid synthesis.
14. Unlike typical vitamins, which must be obtained from the diet, vitamin D is made in the body, and many of its functions are hormone-like in nature. It circulates in the blood and binds to receptors to regulate gene expression.
15. The liver is the primary organ for alcohol metabolism and for disposing of drugs (legal, illegal, and accidental) and halocarbon compounds. When the liver spends its time dealing with these other tasks, it may not be able to carry out its other normal functions; in essence, prolonged exposure to any such "toxin" overworks the liver.
16. Vitamin A is a lipid-soluble vitamin, which can accumulate in the body. Overdoses of vitamin A can be toxic.
17. Low levels of iodine in the diet often lead to hypothyroidism and an enlarged thyroid gland (goiter). This condition has largely been eliminated by the addition of sodium iodide to commercial table salt.
18. Lucullus breaks down the protein in the tuna to amino acids, which in turn undergo the urea cycle and the breakdown of the carbon skeleton, eventually leading to the citric acid cycle and electron transport. In addition to protein catabolism, Griselda breaks down the carbohydrates into sugars, which then undergo glycolysis and enter the citric acid cycle. (Gratuitous information: Lucullus was a notorious Roman gourmand. In medieval literature, Griselda was the name usually given to a forbearing, long-suffering woman.)
19. All amino acids must be present at the same time for protein synthesis to occur. Newly synthesized proteins are necessary for growth in immature rats.
20. The weight loss is due to the correction of the bloating caused by retention of liquids.
21. Children need relatively large amounts of arginine and histidine because their rapid growth increases the demand for protein synthesis. After a person is fully grown, many amino acids are scavenged and recycled by the body.
22. The early colonists always cooked in iron pots; enough iron is leached out to supply the required amounts, as long as the body is able to absorb it. (Glass cookware did not become available until after World War I, and aluminum cookware was not available until after World War II.)
23. Diets high in fiber are usually lower in fats, especially saturated fats; fiber adsorbs many potentially toxic substances, such as cholesterol and halocarbons, preventing their absorption into the body; fiber decreases transit time through the intestine, so any toxic materials in food remain in the body for less time and have a smaller chance of being absorbed or otherwise causing problems.
24. This claim has a chemical basis. Calcium carbonate requires stomach acid to dissolve and release calcium ions, so absorption can be reduced in people with low stomach acid. Calcium citrate is more soluble and does not depend on stomach acid, making it easier to absorb under those conditions.
25. Chronic alcoholism promotes thiamine deficiency because ethanol represents empty calories replacing food, and it also impairs thiamine absorption, activation, and storage in the intestine and liver. Thiamine (as TPP) is required for key oxidative enzymes such as pyruvate dehydrogenase, α-ketoglutarate dehydrogenase, and transketolase. Therefore, TPP deficiency disproportionately disrupts energy metabolism in the brain, which relies heavily on glucose oxidation.
26. Metal ions play a role in the structure and function of proteins and some coenzymes. They tend to do so because they operate as Lewis acids (i.e., substances that accept a pair of electrons in chemical reactions).
27. Nutrients and water turn over in the body, sometimes very frequently. This implies that an organism is an open system.

Equilibrium requires a closed system. Consequently, an organism maintains homeostasis, which is a dynamic steady state, but never equilibrium.

24-3 Hormones and Second Messengers

28. No, hormones can have several different kinds of chemical structures, including steroids, polypeptides, and amino acid derivatives.
29. Pituitary gland damage would reduce or eliminate the release of tropic hormones that normally stimulate other endocrine glands; the workings of the adrenal cortex, the thyroid, and the gonads can all be affected as a result. Adrenal cortex damage would reduce production of adrenocortical hormones, including glucocorticoids (involved in carbohydrate metabolism, inflammatory reactions, and reaction to stress) and mineralocorticoids, which control the level of excretion of water and salt by the kidney. If the adrenal cortex does not function adequately, one result is Addison's disease, characterized by hypoglycemia, weakness, and increased susceptibility to stress. The opposite condition, hyperadrenocorticism, is Cushing's syndrome.
30. The hypothalamus secretes hormone-releasing factors. Under the influence of these factors, the pituitary secretes tropic hormones, which act on specific endocrine glands. Individual hormones are then released by the specific endocrine glands.
31. Thyroxine is an amino acid derivative and is absorbed directly from the gut into the bloodstream. If insulin were taken orally, it would be hydrolyzed to amino acids in the stomach and intestine.
32. G proteins get their name because they bind GTP as part of their effect. An example is the G protein that is linked to the epinephrine receptor and leads to the production of cAMP as a second messenger. Receptor tyrosine kinases have a different mode of action. When they bind their hormone, they phosphorylate tyrosine residues on themselves and other target proteins, which then act as a second messenger. Insulin is an example of a hormone that binds to a receptor tyrosine kinase.
33. cAMP, Ca^{2+}, and diacylglycerol (DAG).

24-4 Hormones and the Control of Carbohydrate Metabolism

34. Epinephrine and glucagon are the two that were discussed most in this book.
35. Glucagon causes the activation of glycogen phosphorylase, inhibition of glycogen synthase, and inhibition of phosphofructokinase-1.
36. Epinephrine causes the activation of glycogen phosphorylase and inhibition of glycogen synthase. It causes activation of phosphofructokinase-1 (PFK-1) in muscle but inhibition of PFK-1 in the liver.
37. The G protein is bound to GTP. Eventually, the GTP is hydrolyzed to GDP, which causes it to dissociate from adenylate cyclase. This stops the hormone response until the hormone dissociates from the receptor, the G protein trimers are rejoined, and the process starts over again.
38. IP_3 is a polar compound and can dissolve in the aqueous environment of the cytosol; DAG is nonpolar and interacts with the side chains of the membrane phospholipids.
39. When a stimulatory hormone binds to its receptor on a cell surface, it stimulates the action of adenylate cyclase, mediated by the G protein. The cAMP that is produced elicits the desired effect on the cell by stimulating a kinase that phosphorylates a target enzyme.
40. See Table 24.3.
41. It is most unlikely that a metabolic pathway could exist without control mechanisms. Many pathways require energy, so it is advantageous for an organism to shut down a pathway when its products are not needed. Even if a pathway does not require large amounts of energy, the many connections among pathways make it likely that control is established over the levels of important metabolites.
42. In cholera, adenylate cyclase is permanently "turned on." This, in turn, stimulates active transport of Na^+ and water from epithelial cells, leading to diarrhea.
43. (a) The binding of epinephrine to its receptor is not catalytic.
    (b) Five catalytic steps, including the reaction catalyzed by glycogen phosphorylase, with 10 molecules acted on in each step, would result in 10^5 (i.e., 100,000) G-1-P molecules for each epinephrine.
    (c) Signal cascades allow signal amplification to occur, where a single hormone triggers a large response. They also allow rapid, regulated, and reversible control of biochemical pathways.
44. A phosphatase dephosphorylates glycogen phosphorylase and glycogen synthetase, inactivating and activating them, respectively. The phosphatase becomes active in response to high concentrations of glucose.
45. Low-carbohydrate diets can prevent the high blood sugar levels that arise when large quantities of carbohydrates are consumed. High blood sugar leads to a rapid rise in insulin. Insulin is known to stimulate fat synthesis and to inhibit fatty acid oxidation.

24-5 Insulin and Its Effects

46. Insulin's primary function is to stimulate the transport of glucose out of the blood and into the cell.
47. The second messenger is phosphatidylinositol 3,4,5-trisphosphate (PIP_3).
48. When insulin binds to its receptor, the β subunit of the receptor tyrosine kinase autophosphorylates. When this happens, the receptor kinase is able to phosphorylate tyrosines on the insulin receptor substrates (IRS proteins). Phosphorylated IRS proteins activate phosphoinositide 3-kinase (PI3K), which catalyzes the formation of the second messenger, phosphatidylinositol 3,4,5-trisphosphate (PIP_3).
49. Insulin causes the following effects:
    (a) Glycogen breakdown is decreased.
    (b) Glycogen synthesis is increased.
    (c) Glycolysis is increased.
    (d) Fatty acid synthesis is increased.
    (e) Fatty acid storage is increased.
50. Insulin and epinephrine normally have opposite effects, but they both stimulate muscle glycolysis. Epinephrine is the hormone that signals the need for quick energy, which means the muscle cells must be able to use glucose via glycolysis.

Insulin stimulates pathways that use up glucose so that blood glucose levels decrease, so it makes sense for it to stimulate glycolysis as well. Epinephrine stimulates muscle glycolysis by activating adenylate cyclase, which makes cAMP; cAMP then activates protein kinase A, which phosphorylates phosphofructokinase-2 and fructose bisphosphatase-2. In the muscle, phosphorylation of phosphofructokinase-2 activates it, producing more fructose-2,6-bisphosphate, which activates phosphofructokinase-1 and glycolysis. In muscle, insulin stimulates glycolysis by activating phosphofructokinase and pyruvate dehydrogenase.

51. Prerace diet can be critical to a runner. If the runner eats a typical American breakfast of cereal, toast, or pancakes, they will have a high blood-sugar level within half an hour, which will lead to a high insulin level shortly thereafter. If the race starts under these conditions, the runner's metabolism will be dedicated to fat and glycogen synthesis instead of burning fat or carbohydrates. The runner will be like a car with a full tank of gas and a clogged fuel line.

52. It has been shown that the GLUT4 transporter responds to physical activity. When a person is active, the transporter is active and responds well to insulin. After a few days of detraining, this transporter shows only half the activity it did before.

53. GLUT4 is one of the glucose transporters in muscle cells. It responds to insulin by moving glucose out of the blood and into the cells. In type II diabetes, insulin is present, but it does not have the same effect. It takes more insulin to accomplish the same movement out of the blood and into the cell. People with type II diabetes often show classical signs of obesity, and there is a correlation between diminishing GLUT4 activity, obesity, and diabetes.

54. Adipocytes, liver cells, and muscle cells.

55. Obesity often leads to insulin resistance and chronically high insulin levels, which can contribute to type II diabetes. Obesity and diabetes create inflammation, which can increase the risk of certain cancers.

56. The Warburg effect describes a situation where, when bathed in large amounts of glucose, and with significant insulin to allow the glucose to enter the cells, cancer cells adopt a strategy of running most of the glucose through aerobic glycolysis, where lactate is produced from glucose even when oxygen is abundant. While energetically inefficient, aerobic glycolysis in cancer cells provides metabolic intermediates that are needed for rapid cell growth and division.

57. PTEN is a tumor suppressor gene often found to have been mutated in many cancers.

58. PTEN's role in tumor suppression appears to be via counteracting the action of PI_3 kinase.

59. Thankfully, the possibilities for biochemical research are limitless! Choose a topic that genuinely interests you, and think about why it matters biologically or medically. Consider how the topic connects to other pathways, processes, or scientific disciplines to reveal integrated insights. Carefully read the research literature on the topic so that you can use the knowledge to guide your research and build upon previous discoveries. You may even discover something new.

Index

Page numbers in **boldface** indicate a major discussion of the entry, page numbers followed by an *f* indicate a figure, page numbers followed by a *t* indicate a table, page numbers followed by a *ph* indicate a photograph, HT preceding a page number refers to the Hot Topics section of the book. Positional and configurational designations in chemical names (e.g. 3-, *N*, a-) are ignored in alphabetization.

C

H

I

P

S

T

U

V

W

X

Y

Z

The Standard Genetic Code

First Position (5' End)	Second Position U	Second Position C	Second Position A	Second Position G	Third Position (3' End)
	U	C	A	G	
U	UUU Phe	UCU Ser	UAU Tyr	UGU Cys	U
	UUC Phe	UCC Ser	UAC Tyr	UGC Cys	C
	UUA Leu	UCA Ser	UAA Stop	UGA Stop	A
	UUG Leu	UCG Ser	UAG Stop	UGG Trp	G
C	CUU Leu	CCU Pro	CAU His	CGU Arg	U
	CUC Leu	CCC Pro	CAC His	CGC Arg	C
	CUA Leu	CCA Pro	CAA Gln	CGA Arg	A
	CUG Leu	CCG Pro	CAG Gln	CGG Arg	G
A	AUU Ile	ACU Thr	AAU Asn	AGU Ser	U
	AUC Ile	ACC Thr	AAC Asn	AGC Ser	C
	AUA Ile	ACA Thr	AAA Lys	AGA Arg	A
	AUG Met*	ACG Thr	AAG Lys	AGG Arg	G
G	GUU Val	GCU Ala	GAU Asp	GGU Gly	U
	GUC Val	GCC Ala	GAC Asp	GGC Gly	C
	GUA Val	GCA Ala	GAA Glu	GGA Gly	A
	GUG Val	GCG Ala	GAG Glu	GGG Gly	G

* AUG forms part of the initiation signal as well as coding for internal methionine residues.